Cutnell & Johnson Physics

Twelfth Edition

Cutnell & Johnson Physics

Twelfth Edition

DAVID YOUNG

SHANE STADLER

Louisiana State University

WILEY

VICE PRESIDENT AND GENERAL MANAGER	Aurora Martinez
EXECUTIVE EDITOR	John LaVacca
SENIOR EDITOR	Jennifer Yee
ASSOCIATE EDITOR	Georgia Larsen
ASSISTANT EDITOR	Samantha Hart
SENIOR MANAGING EDITOR	Mary Donovan
MARKETING MANAGER	Sean Willey
SENIOR MANAGER, COURSE DEVELOPMENT AND PRODUCTION	Svetlana Barskaya
SENIOR COURSE PRODUCTION OPERATIONS SPECIALIST	Patricia Gutierrez
COURSE CONTENT DEVELOPER	Corrina Santos
DESIGNER	Wendi Lai
COVER IMAGES CREDIT	Digital Composite Image Of Iris © Erwin Matuschat/ EyeEm/Getty Images, DNA structure © KTSDESIGN/ SCIENCE PHOTO LIBRARY/Getty Images

This book was set in 9.5/12 STIX Two Text by Lumina Datamatics®, Inc.

Founded in 1807, John Wiley & Sons, Inc. has been a valued source of knowledge and understanding for more than 200 years, helping people around the world meet their needs and fulfill their aspirations. Our company is built on a foundation of principles that include responsibility to the communities we serve and where we live and work. In 2008, we launched a Corporate Citizenship Initiative, a global effort to address the environmental, social, economic, and ethical challenges we face in our business. Among the issues we are addressing are carbon impact, paper specifications and procurement, ethical conduct within our business and among our vendors, and community and charitable support. For more information, please visit our Web site: www.wiley.com/go/citizenship.

Evaluation copies are provided to qualified academics and professionals for review purposes only, for use in their courses during the next academic year. These copies are licensed and may not be sold or transferred to a third party. Upon completion of the review period, please return the evaluation copy to Wiley. Return instructions and a free-of-charge return shipping label are available at: www.wiley.com/go/returnlabel. If you have chosen to adopt this textbook for use in your course, please accept this book as your complimentary desk copy. Outside of the United States, please contact your local representative.

Main Text binder Edition: 9781119773610
(Volume 1): 9781119803690
(Volume 2): 9781119803775
ePDF: 9781119788591
epub: 9781119773535

The inside back cover will contain printing identification and country of origin if omitted from this page. In addition, if the ISBN on the cover differs from the ISBN on this page, the one on the cover is correct.

Printed in the United States of America

SKY10029036_081221

Courtesy David Young

DAVID YOUNG received his Ph.D. in experimental condensed matter physics from Florida State University in 1998. He then held a post-doc position in the Department of Chemistry and the Princeton Materials Institute at Princeton University before joining the faculty in the Department of Physics and Astronomy at Louisiana State University in 2000. His research focuses on the synthesis and characterization of high-quality single crystals of novel electronic and magnetic materials. The goal of his research group is to understand the physics of electrons in materials under extreme conditions, i.e., at temperatures close to absolute zero, in high magnetic fields, and under high pressure. He is the coauthor of over 250 research publications that have appeared in peer-reviewed journals, such as *Physical Review B, Physical Review Letters*, and *Nature*. Professor Young has taught introductory physics with the Cutnell & Johnson text since he was a senior undergraduate almost 30 years ago. He routinely lectures to large sections, often in excess of 300 students. To engage such a large number of students, he uses *WileyPLUS*, electronic response systems, tutorial-style recitation sessions, and in-class demonstrations. Professor Young has received multiple awards for outstanding teaching of undergraduates. David enjoys spending his free time with his family, playing basketball, and working on his house.

I would like to thank my family for their continuous love and support.   —David Young

Courtesy Shane Stadler

SHANE STADLER Shane Stadler earned a Ph.D. in experimental condensed matter physics from Tulane University in 1998. Afterwards, he accepted a National Research Council Postdoctoral Fellowship with the Naval Research Laboratory in Washington, DC, where he conducted research on artificially structured magnetic materials. Three years later, he joined the faculty in the Department of Physics at Southern Illinois University (the home institution of John Cutnell and Ken Johnson, the original authors of this textbook), before joining the Department of Physics and Astronomy at Louisiana State University in 2008. His research group studies novel magnetic materials for applications in the areas of spintronics and magnetic cooling.

Over the past twenty years, Professor Stadler has taught the full spectrum of physics courses, from physics for students outside the sciences, to graduate-level physics courses, such as classical electrodynamics. He teaches classes that range from fewer than ten students to those with enrollments of over 300. His educational interests are focused on developing teaching tools and methods that apply to both small and large classes, and which are applicable to emerging teaching strategies, such as "flipping the classroom."

In his spare time, Shane writes science fiction/thriller novels.

I would like to thank my parents, George and Elissa, for their constant support and encouragement.
                                                        —Shane Stadler

Dear Students and Instructors:

Welcome to college physics! To the students: We know there is a negative stigma associated with physics, and you yourself may harbor some trepidation as you begin this course. But fear not! We are here to help. Whether you are worried about your math proficiency, understanding the concepts, or developing your problem-solving skills, the resources available to you are designed to address all of these areas and more. When we were students and had to take introductory physics, we had a printed textbook, a pencil, and some paper. That was it! Can you learn physics this way? You bet! We did! But research has shown that learning styles vary greatly among students. Maybe some of you have a more visual preference, or auditory preference, or some other preferred learning modality. In any case, the resources available to you in this course will satisfy all of these preferences and improve your chance of success. Take a moment to explore below what the textbook and online course have to offer. We suspect that, as you continue to improve throughout the course, some of that initial trepidation will be replaced with confidence and excitement.

To start, 12e will continue to offer a new learning medium unique to this book in the form of a comprehensive set of lecture videos — one for every section (259 in all). These animated lectures (created and narrated by the authors) are 2–10 minutes in length, and explain the basic concepts and learning objectives of each section. They are assignable within WileyPLUS and can be paired with follow-up questions that are gradable. In addition to supplementing traditional lecturing, the videos can be used in a variety of ways, including, flipping the classroom, lectures for online courses, and reviewing for exams.

Next, we created a new end-of-chapter section of problems called *Physics in Biology, Medicine, and Sports*. The text has always offered real-world examples of physics applications in the life sciences, not only to generate interest for the many students in one of the premedical fields, but also to underscore the fact that the human body, like any biological system, is governed by the laws of physics. Overall, in this edition, we have more than doubled the number of bio-inspired examples and problems to more than 300 total for the text, which reflects our commitment to showing students how relevant physics is in their lives. Although they are of general instructional value, many of the end-of-chapter problems are also similar to what premed students will encounter in the Chemical and Physical Foundations of Biological Systems Passages section of the MCAT.

Building on the success of the *Team Problems* introduced in the previous edition, we have doubled their number in each chapter. These are context-rich end-of-chapter problems of variable difficulty designed for group cooperation, but they may also be tackled by the individual student. We have also added a "worksheet" activity to each new Team Problem, which helps to guide the team of students along the way through the solution. These problems offer a great way to promote peer-to-peer instruction in the classroom, as well as serve as the focal point of a group activity during a breakout session as part of an online course.

One of the great strengths of this text is the synergistic relationship it develops between problem solving and conceptual understanding. For instance, available in WileyPLUS are animated *Chalkboard Videos*, which consist of short (2–3 min) videos narrated by the authors that demonstrate step-by-step practical solutions to typical homework problems. We have more than doubled the number of *Chalkboard Videos* in the new edition, so that every section of the book has at least one, which provides comprehensive coverage of the entire course material in this format.

One of the most important techniques developed in the text for solving problems involving multiple forces is the *free-body diagram (FBD)*. Many problems in the force-intensive chapters, such as Chapters 4 and 18, take advantage of the new FBD drawing tools now available online in WileyPLUS, where students can construct the FBDs for a select number of problems and be graded on them. Also available are numerous *Guided Online (GO) Tutorials* that implement a step-by-step pedagogical approach, which provides students a low-stakes environment for refining their problem-solving skills. Finally, new to this edition will be *Adaptive Assignments*. Seamlessly integrated into WileyPLUS, *Adaptive Assignments,* which are powered by Knewton's world-class recommendation engine, offer the most targeted, efficient way for instructors to tailor practice and preparation. This new assignment type pinpoints learner knowledge gaps and offers personalized, just-in-time support, targeted instruction, and detailed answer explanations.

Over the last 20 years nothing has had a more significant impact on the way students learn than the World Wide Web. Students essentially have 24/7 access to countless sources of digital multimedia. They complete homework assignments online with their PC's, tablets, and smart phones. Online homework systems are no longer "in the future," but are now the norm. Physics is no exception. Unfortunately, having all of this information readily available comes at a price. Students have fundamentally changed the way in which they approach their homework assignments. Instead of struggling through the entire solution to a problem from scratch, where much of the learning process takes place, they default to online resources, where they can pay for access to written solutions to the end-of-chapter problems. Alternatively, many students find solutions by simply searching the questions on Google, for example. As a result, a student's online-homework grade has become a rather poor measure of their knowledge of the course material. What's even worse is the false sense of security the students feel as a result of their inflated homework grades. They feel confident and prepared for exams, because they have 95%–100% on their homework. Unfortunately, a poor performance on the first exam is often the initial indicator of their level of misunderstanding. The content outlined below, and the functionality of WileyPLUS and the *Adaptive Assignments,* will provide students with all the resources they need to be successful in the course:

- The *Lecture Videos* created by the authors for each section include questions with intelligent feedback when a student enters the wrong answer.
- The multistep *GO Tutorial* problems created in WileyPLUS are designed to provide targeted, intelligent feedback.
- The *Free-body Diagram* vector drawing tools provide students an easy way to enter answers requiring vector drawing, and also provide enhanced feedback.
- *Chalkboard Video Solutions* take the students step-by-step through the solution and the thought process of the authors. Problem-solving strategies are discussed, and common misconceptions and potential pitfalls are addressed. The students can then apply these techniques to solve similar, but different, problems.

All of these features are designed to encourage students to remain within the WileyPLUS environment, as opposed to pursuing the "pay-for solutions" websites that short circuit the learning process.

To the students—We strongly recommend that you take this honest approach to the course. Take full advantage of the many features and learning resources that accompany the text and the online content. Be engaged with the material and push yourself to work through the exercises. Physics may not be the easiest subject to understand initially, but with the Wiley resources at your disposal, and your hard work, you CAN be successful.

We are immensely grateful to all of you who have provided feedback as we've worked on this new edition, and to our students who have taught us how to teach. Thank you for your guidance, and keep the feedback coming. Best wishes for success in this course and wherever your major may take you!

Brief Contents

Contents

12 Temperature and Heat 356

13 The Transfer of Heat 394

14 The Ideal Gas Law and Kinetic Theory 416

15 Thermodynamics 441

16 Waves and Sound 478

17 The Principle of Linear Superposition and Interference Phenomena   513

18 Electric Forces and Electric Fields   541

19 Electric Potential Energy and the Electric Potential   578

20 Electric Circuits   609

21 Magnetic Forces and Magnetic Fields   653

32 Ionizing Radiation, Nuclear Energy, and Elementary Particles 1007

Appendixes A-1

Note: Chapter sections marked with an asterisk (*) can be omitted with little impact to the overall development of the material.

Our Vision and WileyPLUS with Adaptive Assignments

Our Vision

Our goal is to provide students with the skills they need to succeed in this course, and instructors with the tools they need to develop those skills.

Skills Development

One of the great strengths of this text is the synergistic relationship between conceptual understanding, problem solving, and establishing relevance. We identify here some of the core features of the text that support these synergies.

Conceptual Understanding Students often regard physics as a collection of equations that can be used blindly to solve problems. However, a good problem-solving technique does not begin with equations. It starts with a firm grasp of physics concepts and how they fit together to provide a coherent description of natural phenomena. Helping students develop a conceptual understanding of physics principles is a primary goal of this text. The features in the text that work toward this goal are:

- *Lecture Videos (one for each section of the text)*
- *Conceptual Examples*
- *Concepts & Calculations* problems (now with video solutions)
- *Focus on Concepts* homework material
- *Check Your Understanding* questions
- *Concept Simulations* (an online feature)

Problem Solving The ability to reason in an organized and mathematically correct manner is essential to solving problems, and helping students to improve their reasoning skills is also one of our primary goals. To this end, we have included the following features:

- *Math Skills boxes* for just-in-time delivery of math support
- *Explicit reasoning steps* in all examples
- *Reasoning Strategies* for solving certain classes of problems
- *Analyzing Multiple-Concept Problems*
- *Video Support and Tutorials* (in *WileyPLUS*)
    Chalkboard Video Solutions
    Physics Demonstration Videos
    Video Help
    Concept Simulations
- *Problem Solving Insights*

Relevance Since it is always easier to learn something new if it can be related to day-to-day living, we want to show students that physics principles come into play over and over again in their lives. To emphasize this goal, we have included a wide range of applications of physics principles. Many of these applications are biomedical in nature (for example, wireless capsule endoscopy). Others deal with modern technology (for example, 3-D movies). Still others focus on things that we take for granted in our lives (for example, household plumbing). To call attention to the applications we have used the label The Physics of.

WileyPLUS with Adaptive Assignments

WileyPLUS is an innovative, research-based online environment for effective teaching and learning. The hallmark of *WileyPLUS* with Adaptive Assignments for this text is that the media- and text-based resources are all created by the authors of the project, providing a seamless presentation of content.

WileyPLUS builds students' confidence because it takes the guesswork out of studying by providing students with a clear roadmap: **what to do, how to do it, if they did it right**.

With *WileyPLUS*, our efficacy research shows that students improve their outcomes by as much as one letter grade. *WileyPLUS* helps students take more initiative, so you'll have greater impact on their achievement in the classroom and beyond.

With WileyPLUS, instructors receive:

- **Breadth and Depth of Assessment:** *WileyPLUS* contains a wealth of online questions and problems for creating online homework and assessment including:
    - ALL end-of-chapter questions, plus favorites from past editions not found in the printed text, coded algorithmically, each with at least one form of instructor-controlled question assistance (GO tutorials, hints, link to text, video help)
    - Simulation, animation, and video-based questions
    - Free body and vector drawing questions
    - Test bank questions

- **Gradebook:** *WileyPLUS* provides instant access to reports on trends in class performance, student use of course materials, and progress toward learning objectives, thereby helping instructors' decisions and driving classroom discussion.

With **WileyPLUS**, students receive:

- The complete digital textbook, including the Lecture Videos, saving students up to 60% off the cost of a printed text
- Question assistance, including links to relevant sections in the online digital textbook
- Immediate feedback and proof of progress, 24/7
- Integrated, multimedia resources—including animations, simulations, video demonstrations, and much more—that provide multiple study paths and encourage more active learning
- GO Tutorials
- Chalkboard Videos
- Free Body Diagram/Vector Drawing Questions

New to WileyPLUS for the Twelfth Edition

Team Problems   In each chapter, the end-of-chapter problems contain four *Team Problems* that are designed for group problem-solving exercises. These are context-rich problems of medium difficulty designed for group cooperation, but may also be tackled by the individual student. Accompanying some of these team problems are "worksheets" that serve to guide the students through the solution in a tutorial fashion. Many of these problems read like parts of an adventure story, where the student (or their team) is the main character. The motivation for each problem is clear and personal—the pronoun "you" is used throughout, and the problem statements often start with "You and your team need to ...". Pictures and diagrams are not given with these problems except in rare cases. Students must visualize the problems and discuss strategies with their team members to solve them. The problems require two or more steps/multiple concepts (hence the "medium" difficulty level) and may require basic principles learned earlier. Sometimes, there is no specific target variable given, but rather questions like *Will it work?* or *Is it safe?* Suggested solutions are given in the Instructor Solutions Manual.

"The Physics of . . ." Examples   The text now contains over 300 real-world application examples that reflect our commitment to showing students how relevant physics is in their lives. Each application is identified in the text with the label The Physics of. A subset of these examples focuses on biomedical applications, and we have more than doubled their number in the new edition. In fact, we have created a new end-of-chapter problem section called the *Physics in Biology, Medicine, and Sports*. Students majoring in biomedical and life sciences will find new examples in every chapter covering topics such as cooling the human brain, abdominal aortic aneurysms, the mechanical properties of bone, and many more! The application of physics principles to biomedical problems in these examples is similar to what premed students will encounter in the *Chemical and Physical Foundations of Biological Systems Passages* section of the MCAT. All biomedical examples and end-of-chapter problems will be marked with the **BIO** icon.

EXAMPLE 8 | **BIO** The Physics of Hearing Loss—Standing Waves in the Ear

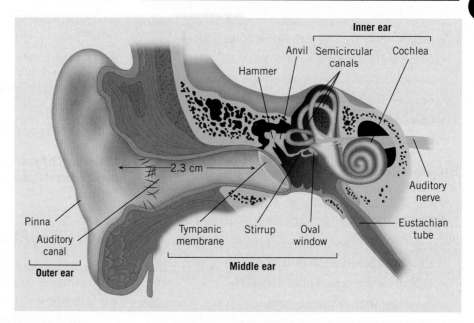

Interactive Graphics   The online reading experience within WileyPLUS has been enhanced with the addition of "Interactive Graphics." Several static figures in each chapter have been transformed to include interactive elements. These graphics drive students to be more engaged with the extensive art program and allow them to more easily absorb complex and/or long multi-part figures.

Use the checkboxes to select the correct vectors as instructed above. Move the vectors to the correct starting points and orient them in the correct direction as instructed.

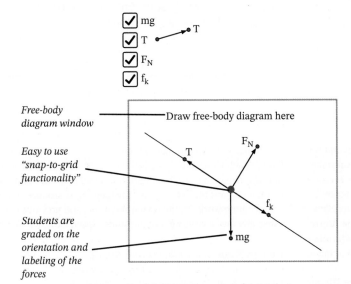

Key to Force Labels
mg = Gravitational force
T = Tension force
F_N = Normal force
f_k = Kinetic friction force

Free-body diagram window

Easy to use "snap-to-grid functionality"

Students are graded on the orientation and labeling of the forces

Draw free-body diagram here

Also Available in WileyPLUS

Free-Body Diagram (FBD) Tools   For many problems involving multiple forces, an interactive free-body diagram tool in *WileyPLUS* is used to construct the diagram. It is essential for students to practice drawing FBDs, as that is the critical first step in solving many equilibrium and non-equilibrium problems with Newton's second law.

GO Tutorial Problems   Some of the homework problems found in the collection at the end of each chapter are marked with a special GO icon. All of these problems (550 of them) are available for assignment online via *WileyPLUS*. Each of these problems in *WileyPLUS* includes a guided tutorial option that instructors can make available for student access with or without penalty.

The GO tutorial

Multiple-choice questions in the GO tutorial include extensive feedback for both correct and incorrect answers

Answer input, including direction and units

Multiple-choice questions guide students to the proper conceptual basis for the problem. The GO tutorial also includes calculational steps

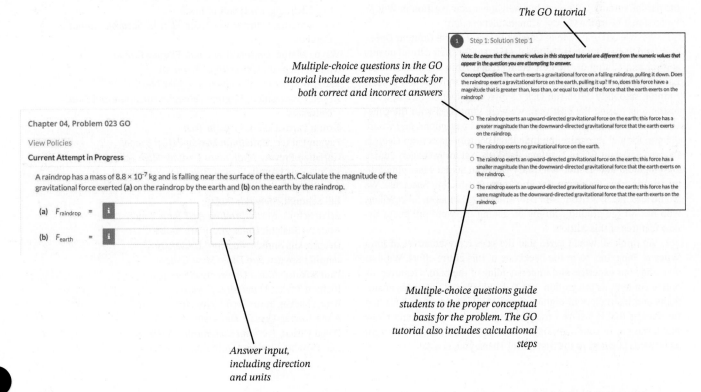

Adaptive Assignments

Seamlessly integrated into WileyPLUS and based on learning science, Adaptive Assignments are the most targeted, efficient way for instructors to tailor practice and preparation. This new assignment type pinpoints learner knowledge gaps and offers personalized, just-in-time support, targeted instruction and detailed answer explanations.

Acknowledgments

The effective revision of a textbook in this age goes well beyond the efforts of the authors and requires the expertise of a great number of people. Their abilities must range from brainstorming revision concepts to editing written and digital content to creating marketing strategies. They have to learn how the needs of our readers have changed and to anticipate how they might transform in the future. As the authors, we are extremely grateful to the people who have provided direction, talent, and effort as a part of the team that made this revision possible. Well done and thank you!

We would especially like to acknowledge John LaVacca III, our Senior Editor. His knowledge of the textbook landscape has offered crucial guidance in developing this revision in order to meet the needs of students and instructors alike. He has provided us ample encouragement and flexibility to pursue the creative aspects of this work within the constraints of a well-planned strategy.

It was a great pleasure working with Jennifer Yee, Senior Editor, on this edition, and we would like to thank her for keeping this edition on track and coordinating our efforts with the other facets of this production. Keeping us informed of the timelines for all aspects of the project made the process smooth and devoid of stress (at least on the part of authors!). We wish her well at her new position in Wiley's Professional Learning Group! Congratulations Jen!

We owe our gratitude to Corrina Santos, Course Content Developer, and Samantha Hart, Editorial Assistant for their careful quality control and editing of the new content, and Anju Joshi and Wendi Lai for layout and design, as well as keeping the production process running smoothly. With the lack of direct interaction due to travel restrictions during the past year, we were impressed with the cohesive team effort, despite the lack of in-person interactions, and we are grateful for and recognize the individual contributions from Georgia Larson, Assistant Editor, Svetlana Barskaya, Senior Manager, Course Development and Production, Mary Donovan,Senior Managing Editor, and Patricia Gutierrez, Senior Course Production Specialist. We would also like to welcome our new Marketing Manager, Sean Willey, who has hit the ground running on getting the word out about the new features of this edition.

We think all would agree that the sales representatives of John Wiley & Sons, Inc. form the backbone of this entire effort. Without you, and your expertise and understanding of the book's features, its online content, its integration into WileyPLUS, and the needs of students and instructors in higher education, it would never have had the success that is has over the past 30 years! And you should know that it was one of your own, Dayna Leaman, our good friend, who got us into this business in the first place! Thank you, Dayna!

Finally, we would like to thank our physics colleagues and their students who have kindly shared their ideas about effective pedagogy, and improved the text by identifying errors. We are grateful for all of their suggestions as they have helped us to produce a more clear and accurate work. To the reviewers of this and previous editions, we especially owe a large debt of gratitude. Specifically, we thank:

Lai Cao, *Baton Rouge Magnet High School*
Candee Chambers-Colbeck, *Maryville University*
Diana Driscoll, *Case Western Reserve University*
Costas Efthimiou, *University of Central Florida*
Robert Egler, *North Carolina State University*
Sambandamurthy Ganapathy, *The State University of New York at Buffalo*
Joseph Ganem, *Loyola University, Maryland*
Jasper Halekas, *University of Iowa*
Lilit Haroyan, *East Los Angeles College*
Richard Holland, *Rend Lake College and SIUC*
Klaus Honscheid, *Ohio State University*
Shyang Huang, *Missouri State University*
Craig Kleitzing, *University of Iowa*
Wayne Mathe, *University of South Florida, Sarasota-Manatee Campus*
Wayne Mathé, *University of South Florida Tampa*
Samuel Mensah, *University of Memphis*
Mark Morgan-Tracy, *University of New Mexico*
Kriton Papavasiliou, *Virginia Polytechnic Institute and State University*
Kriton Papavasiliou, *Virginia Tech*
Payton Parker, *Midlothian Heritage High School*
Christian Prewitt, *Midlothian Heritage High School*
Joshua Ravenscraft, *Vernon Hills High School and College of Lake County*
Bill Schmidt, *Meredith College*
Brian Schuft, *North Carolina A&T State University*
Andreas Shalchi, *University of Manitoba*
Deepshikha Shukla, *Rockford University*
Jennifer Snyder, *San Diego Mesa College*
Paul Sokol, *Indiana University Bloomington*
Richard Taylor, *University of Oregon*
Beth Thacker, *Texas Tech University*
Anne Topper, *Queen's University*
David Ulrich, *Portland Community College*
Luc T. Wille, *Florida Atlantic University*

About the cover: One of the great successes of physics as a science is its ability to accurately predict the behavior of the natural world across all length scales, from the atomically small to the astronomically large. The cover image represents this achievement with an artist's rendition of a human eye drifting in the cosmos. At the center of the eye's pupil is the famous double helix of DNA. The motion of the planets and stars is largely determined by the gravitational force. The formation of an image by our eyes is governed by the laws of optics, and the amino acid base pairs in DNA are held together by electrostatic forces. In this course, you will study all of these topics and learn how to apply them to solve a variety of problems.

In spite of our best efforts to produce an error-free book, errors no doubt remain. They are solely our responsibility, and we would appreciate hearing of any that you find. We hope that this text makes learning and teaching physics easier and more enjoyable, and we look forward to hearing about your experiences with it. Please feel free to write us care of Physics Editor, Global Education, John Wiley & Sons, Inc., 111 River Street, Hoboken, NJ 07030, or contact the authors at **dyoun14@gmail.com** or **sstadler23@gmail.com**.

The animation techniques and special effects used in the film *Avengers: Infinity War* rely on computers and mathematical concepts such as trigonometry and vectors. Such mathematical concepts will be useful throughout this book in our discussion of physics.

Lifestyle pictures/Alamy Stock Photo

Introduction and Mathematical Concepts

LEARNING OBJECTIVES

After reading this module, you should be able to...

1.1 Describe the fundamental nature of physics.

1.2 Describe different systems of units.

1.3 Solve unit conversion problems.

1.4 Solve trigonometry problems.

1.5 Distinguish between vectors and scalars.

1.6 Solve vector addition and subtraction problems by graphical methods.

1.7 Calculate vector components.

1.8 Solve vector addition and subtraction problems using components.

1.1 The Nature of Physics

Physics is the most basic of the sciences, and it is at the very root of subjects like chemistry, engineering, astronomy, and even biology. The discipline of physics has developed over many centuries, and it continues to evolve. It is a mature science, and its laws encompass a wide scope of phenomena that range from the formation of galaxies to the interactions of particles in the nuclei of atoms. Perhaps the most visible evidence of physics in everyday life is the eruption of new applications that have improved our quality of life, such as new medical devices, and advances in computers and high-tech communications.

The exciting feature of physics is its capacity for predicting how nature will behave in one situation on the basis of experimental data obtained in another situation. Such predictions place physics at the heart of modern technology and, therefore, can have a tremendous impact on our lives. Rocketry and the development of space travel have their roots firmly planted in the physical laws of Galileo Galilei (1564–1642) and Isaac Newton (1642–1727). The transportation industry relies heavily on physics in the development of engines and the design of aerodynamic vehicles. Entire electronics and computer industries owe their existence to the invention of the transistor, which grew directly out of the laws of physics that describe the electrical behaviors of solids. The telecommunications industry depends extensively on electromagnetic waves, whose existence was predicted by James Clerk Maxwell (1831–1879) in his theory of electricity and magnetism. The medical profession uses X-ray, ultrasonic, and magnetic resonance methods for obtaining images of the interior of the human body, and physics lies at the core of all of these. Perhaps the most widespread impact in modern technology is that due to the laser. Fields ranging from space exploration to medicine benefit from this incredible device, which is a direct application of the principles of atomic physics.

Because physics is so fundamental, it is a required course for students in a wide range of major areas. We welcome you to the study of this fascinating topic. You will learn how to see the world through the "eyes" of physics and to reason as a physicist does. In the process, you will learn how to apply physics principles to a wide range of problems. We hope that you will come to recognize that physics has important things to say about your environment.

1.2 Units

Physics experiments involve the measurement of a variety of quantities, and a great deal of effort goes into making these measurements as accurate and reproducible as possible. The first step toward ensuring accuracy and reproducibility is defining the units in which the measurements are made.

In this text, we emphasize the system of units known as **SI units,** which stands for the French phrase "Le **S**ystème **I**nternational d'Unités." By international agreement, this system employs the **meter** (m) as the unit of length, the **kilogram** (kg) as the unit of mass, and the **second** (s) as the unit of time. Two other systems of units are also in use, however. The CGS system utilizes the centimeter (cm), the gram (g), and the second for length, mass, and time, respectively, and the BE or British Engineering system (the gravitational version) uses the foot (ft), the slug (sl), and the second. **Table 1.1** summarizes the units used for length, mass, and time in the three systems.

Originally, the meter was defined in terms of the distance measured along the earth's surface between the north pole and the equator. Eventually, a more accurate measurement standard was needed, and by international agreement the meter became the distance between two marks on a bar of platinum–iridium alloy (see **Figure 1.1**) kept at a temperature of 0 °C. Today, to meet further demands for increased accuracy, the meter is defined as the distance that light travels in a vacuum in a time of 1/299 792 458 second. This definition arises because the speed of light is a universal constant that is defined to be 299 792 458 m/s.

Courtesy NIST Archives

FIGURE 1.1 The standard platinum–iridium meter bar.

TABLE 1.1 Units of Measurement

	System		
	SI	CGS	BE
Length	Meter (m)	Centimeter (cm)	Foot (ft)
Mass	Kilogram (kg)	Gram (g)	Slug (sl)
Time	Second (s)	Second (s)	Second (s)

The definition of a kilogram as a unit of mass has also undergone changes over the years. As Chapter 4 discusses, the mass of an object indicates the tendency of the object to continue in motion with a constant velocity. Originally, the kilogram was expressed in terms of a specific amount of water. Since 1879, and until very recently, one kilogram was defined to be the mass of a standard cylinder of platinum–iridium alloy, like the one in **Figure 1.2**. Currently, one kilogram is defined in terms of two other **base** units (see below)—the meter and the second, which are based on fixed fundamental constants of nature. The kilogram is defined by setting Planck's constant h exactly equal to $6.62607015 \times 10^{-34}$ J · s. We will see in Section 29.2 that Planck's constant is important in the field of quantum mechanics. This new definition of the kilogram took effect on May 20, 2019.

As with the units for length and mass, the present definition of the second as a unit of time is different from the original definition. Originally, the second was defined according to the average time for the earth to rotate once about its axis, one day being set equal to 86 400 seconds. The earth's rotational motion was chosen because it is naturally repetitive, occurring over and over again. Today, we still use a naturally occurring repetitive phenomenon to define the second, but of a very different kind. We use the electromagnetic waves emitted by cesium-133 atoms in an atomic clock like that in **Figure 1.3**. One second is defined as the time needed for 9 192 631 770 wave cycles to occur.*

The units for length, mass, and time, along with a few other units that will arise later, are regarded as **base** SI units. The word "base" refers to the fact that these units are used along with various laws to define additional units for other important physical quantities, such as force and energy. The units for such other physical quantities are referred to as **derived** units, since they are combinations of the base units. Derived units will be introduced from time to time, as they arise naturally along with the related physical laws.

The value of a quantity in terms of base or derived units is sometimes a very large or very small number. In such cases, it is convenient to introduce larger or smaller units that are related to the normal units by multiples of ten. **Table 1.2** summarizes the prefixes that are used to denote multiples of ten. For example, 1000 or 10^3 meters are referred to as 1 kilometer (km), and 0.001 or 10^{-3} meter is called 1 millimeter (mm). Similarly, 1000 grams and 0.001 gram are referred to as 1 kilogram (kg) and 1 milligram (mg), respectively. Appendix A contains a discussion of scientific notation and powers of ten, such as 10^3 and 10^{-3}.

Science Source

FIGURE 1.2 The standard platinum–iridium kilogram is kept at the International Bureau of Weights and Measures in Sèvres, France. This copy of it was assigned to the United States in 1889 and is housed at the National Institute of Standards and Technology.

Geoffrey Wheeler

FIGURE 1.3 This atomic clock, the NIST-F1, keeps time with an uncertainty of about one second in sixty million years.

TABLE 1.2	Standard Prefixes Used to Denote Multiples of Ten	
Prefix	**Symbol**	**Factor**[a]
tera	T	10^{12}
giga	G	10^{9}
mega	M	10^{6}
kilo	k	10^{3}
hecto	h	10^{2}
deka	da	10^{1}
deci	d	10^{-1}
centi	c	10^{-2}
milli	m	10^{-3}
micro	μ	10^{-6}
nano	n	10^{-9}
pico	p	10^{-12}
femto	f	10^{-15}

1.3 The Role of Units in Problem Solving

The Conversion of Units

Since any quantity, such as length, can be measured in several different units, it is important to know how to convert from one unit to another. For instance, the foot can be used to express the distance between the two marks on the standard platinum–iridium meter bar. There are 3.281 feet in one meter, and this number can be used to convert from meters to feet, as the following example demonstrates.

*See Chapter 16 for a discussion of waves in general and Chapter 24 for a discussion of electromagnetic waves in particular.

[a]Appendix A contains a discussion of powers of ten and scientific notation.

EXAMPLE 1 | The World's Highest Waterfall

The highest waterfall in the world is Angel Falls in Venezuela, with a total drop of 979.0 m (see **Figure 1.4**). Express this drop in feet.

Reasoning   When converting between units, we write down the units explicitly in the calculations and treat them like any algebraic quantity. In particular, we will take advantage of the following algebraic fact: Multiplying or dividing an equation by a factor of 1 does not alter an equation.

Solution   Since 3.281 feet = 1 meter, it follows that (3.281 feet)/(1 meter) = 1. Using this factor of 1 to multiply the equation "Length = 979.0 meters," we find that

$$\text{Length} = (979.0 \text{ m})(1) = (979.0 \text{ meters})\left(\frac{3.281 \text{ feet}}{1 \text{ meter}}\right) = \boxed{3212 \text{ feet}}$$

The colored lines emphasize that the units of meters behave like any algebraic quantity and cancel when the multiplication is performed, leaving only the desired unit of feet to describe the answer. In this regard, note that 3.281 feet = 1 meter also implies that (1 meter)/(3.281 feet) = 1. However, we chose not to multiply by a factor of 1 in this form, because the units of meters would not have canceled.

   A calculator gives the answer as 3212.099 feet. Standard procedures for significant figures, however, indicate that the answer should be rounded off to four significant figures, since the value of 979.0 meters is accurate to only four significant figures. In this regard, the "1 meter" in the denominator does not limit the significant figures of the answer, because this number is precisely one meter by definition of the conversion factor. Appendix B contains a review of significant figures.

Andoni Canela/Age Fotostock

FIGURE 1.4   Angel Falls in Venezuela is the highest waterfall in the world.

Problem-Solving Insight   In any conversion, if the units do not combine algebraically to give the desired result, the conversion has not been carried out properly.

With this in mind, the next example stresses the importance of writing down the units and illustrates a typical situation in which several conversions are required.

EXAMPLE 2 | BIO The World's Fastest Land Animal

The cheetah can accelerate for a short time to a top speed of 75 miles/hour, making it the fastest animal on land (see **Figure 1.5**). Express 75 miles/hour in meters/second.

Reasoning   As in Example 1, it is important to write down the units explicitly in the calculations and treat them like any

DrZoltan/Pixabay

FIGURE 1.5   The cheetah is the world's fastest land animal.

algebraic quantity. Here, we take advantage of two well-known relationships—namely, 5280 feet = 1 mile and 3600 seconds = 1 hour. As a result, (5280 feet)/(1 mile) = 1 and (3600 seconds)/(1 hour) = 1. In our solution we will use the fact that multiplying and dividing by these factors of unity does not alter an equation.

Solution   Multiplying and dividing by factors of unity, we find the cheetah's speed in feet per second as shown here:

$$\text{Speed} = \left(75 \frac{\text{miles}}{\text{hour}}\right)(1)(1) =$$

$$\left(75 \frac{\text{miles}}{\text{hour}}\right)\left(\frac{5280 \text{ feet}}{1 \text{ mile}}\right)\left(\frac{1 \text{ hour}}{3600 \text{ seconds}}\right) = 110 \frac{\text{feet}}{\text{second}}$$

To convert feet into meters, we use the fact that (1 meter)/(3.281 feet) = 1:

$$\text{Speed} = \left(110 \frac{\text{feet}}{\text{second}}\right)(1) =$$

$$\left(110 \frac{\text{feet}}{\text{second}}\right)\left(\frac{1 \text{ meter}}{3.281 \text{ feet}}\right) = \boxed{34 \frac{\text{meters}}{\text{second}}}$$

In addition to their role in guiding the use of conversion factors, units serve a useful purpose in solving problems. They can provide an internal check to eliminate errors, if they are carried along during each step of a calculation and treated like any algebraic factor.

> **Problem-Solving Insight**   In particular, remember that only quantities with the same units can be added or subtracted.

Thus, at one point in a calculation, if you find yourself adding 12 miles to 32 kilometers, stop and reconsider. Either miles must be converted into kilometers or kilometers must be converted into miles before the addition can be carried out. A collection of useful conversion factors is given on the page facing the inside of the front cover.

The reasoning strategy that we have followed in Examples 1 and 2 for converting between units is outlined as follows:

> **REASONING STRATEGY   Converting Between Units**
>
> 1. In all calculations, write down the units explicitly.
> 2. Treat all units as algebraic quantities. In particular, when identical units are divided, they are eliminated algebraically.
> 3. Use the conversion factors located on the page facing the inside of the front cover of the textbook. Be guided by the fact that multiplying or dividing an equation by a factor of 1 does not alter the equation. For instance, the conversion factor of 3.281 feet = 1 meter might be applied in the form (3.281 feet)/(1 meter) = 1. This factor of 1 would be used to multiply an equation such as "Length = 5.00 meters" in order to convert meters to feet.
> 4. Check to see that your calculations are correct by verifying that the units combine algebraically to give the desired unit for the answer. Only quantities with the same units can be added or subtracted.

Sometimes an equation is expressed in a way that requires specific units to be used for the variables in the equation. In such cases it is important to understand why only certain units can be used in the equation, as the following example illustrates.

EXAMPLE 3 | BIO The Physics of the Body Mass Index

The body mass index (BMI) takes into account your mass in kilograms (kg) and your height in meters (m) and is defined as follows:

$$\text{BMI} = \frac{\text{Mass in kg}}{(\text{Height in m})^2}$$

However, the BMI is often computed using the weight* of a person in pounds (lb) and his or her height in inches (in.). Thus, the expression for the BMI incorporates these quantities, rather than the mass in kilograms and the height in meters. Starting with the definition above, determine the expression for the BMI that uses pounds and inches.

Reasoning   We will begin with the BMI definition and work separately with the numerator and the denominator. We will determine the mass in kilograms that appears in the numerator from the weight in pounds by using the fact that 1 kg corresponds to 2.205 lb. Then, we will determine the height in meters that appears in the denominator from the height in inches with the aid of the facts that 1 m = 3.281 ft and 1 ft = 12 in. These conversion factors are located on the page facing the inside of the front cover of the text.

Solution   Since 1 kg corresponds to 2.205 lb, the mass in kilograms can be determined from the weight in pounds in the following way:

$$\text{Mass in kg} = (\text{Weight in lb})\left(\frac{1 \text{ kg}}{2.205 \text{ lb}}\right)$$

Since 1 ft = 12 in. and 1 m = 3.281 ft, we have

$$\text{Height in m} = (\text{Height in in.})\left(\frac{1 \text{ ft}}{12 \text{ in.}}\right)\left(\frac{1 \text{ m}}{3.281 \text{ ft}}\right)$$

Substituting these results into the numerator and denominator of the BMI definition gives

$$\text{BMI} = \frac{\text{Mass in kg}}{(\text{Height in m})^2} = \frac{(\text{Weight in lb})\left(\dfrac{1 \text{ kg}}{2.205 \text{ lb}}\right)}{(\text{Height in in.})^2 \left(\dfrac{1 \text{ ft}}{12 \text{ in.}}\right)^2 \left(\dfrac{1 \text{ m}}{3.281 \text{ ft}}\right)^2}$$

$$= \left(\frac{1 \text{ kg}}{2.205 \text{ lb}}\right)\left(\frac{12 \text{ in.}}{1 \text{ ft}}\right)^2 \left(\frac{3.281 \text{ ft}}{1 \text{ m}}\right)^2 \frac{(\text{Weight in lb})}{(\text{Height in in.})^2}$$

$$\boxed{\text{BMI} = \left(703.0 \ \frac{\text{kg} \cdot \text{in.}^2}{\text{lb} \cdot \text{m}^2}\right) \frac{(\text{Weight in lb})}{(\text{Height in in.})^2}}$$

For example, if your weight and height are 180 lb and 71 in., your body mass index is 25 kg/m². The BMI can be used to assess approximately whether your weight is normal for your height (see Table 1.3).

TABLE 1.3   The Body Mass Index

BMI (kg/m²)	Evaluation
Below 18.5	Underweight
18.5–24.9	Normal
25.0–29.9	Overweight
30.0–39.9	Obese
40 and above	Morbidly obese

---

*Weight and mass are different concepts, and the relationship between them will be discussed in Section 4.7.

Dimensional Analysis

We have seen that many quantities are denoted by specifying both a number and a unit. For example, the distance to the nearest telephone may be 8 meters, or the speed of a car might be 25 meters/second. Each quantity, according to its physical nature, requires a certain *type* of unit. Distance must be measured in a length unit such as meters, feet, or miles, and a time unit will not do. Likewise, the speed of an object must be specified as a length unit divided by a time unit. In physics, the term **dimension** is used to refer to the physical nature of a quantity and the type of unit used to specify it. Distance has the dimension of length, which is symbolized as [L], while speed has the dimensions of length [L] divided by time [T], or [L/T]. Many physical quantities can be expressed in terms of a combination of fundamental dimensions such as length [L], time [T], and mass [M]. Later on, we will encounter certain other quantities, such as temperature, which are also fundamental. A fundamental quantity like temperature cannot be expressed as a combination of the dimensions of length, time, mass, or any other fundamental dimension.

Dimensional analysis is used to check mathematical relations for the consistency of their dimensions. As an illustration, consider a car that starts from rest and accelerates to a speed v in a time t. Suppose we wish to calculate the distance x traveled by the car but are not sure whether the correct relation is $x = \frac{1}{2}vt^2$ or $x = \frac{1}{2}vt$. We can decide by checking the quantities on both sides of the equals sign to see whether they have the same dimensions. If the dimensions are not the same, the relation is incorrect. For $x = \frac{1}{2}vt^2$, we use the dimensions for distance [L], time [T], and speed [L/T] in the following way:

$$x = \tfrac{1}{2}vt^2$$

Dimensions $\qquad\qquad\qquad [\text{L}] \overset{?}{=} \left[\frac{\text{L}}{\text{T}}\right][\text{T}]^2 = [\text{L}][\text{T}]$

Dimensions cancel just like algebraic quantities, and pure numerical factors like $\frac{1}{2}$ have no dimensions, so they can be ignored. The dimension on the left of the equals sign does not match those on the right, so the relation $x = \frac{1}{2}vt^2$ cannot be correct. On the other hand, applying dimensional analysis to $x = \frac{1}{2}vt$, we find that

$$x = \tfrac{1}{2}vt$$

Dimensions $\qquad\qquad\qquad [\text{L}] \overset{?}{=} \left[\frac{\text{L}}{\text{T}}\right][\text{T}] = [\text{L}]$

> **Problem-Solving Insight**   You can check for errors that may have arisen during algebraic manipulations by performing a dimensional analysis on the final expression.

The dimension on the left of the equals sign matches that on the right, so this relation is dimensionally correct. If we know that one of our two choices is the right one, then $x = \frac{1}{2}vt$ is it. In the absence of such knowledge, however, dimensional analysis cannot identify the correct relation. It can only identify which choices *may be* correct, since it does not account for numerical factors like $\frac{1}{2}$ or for the manner in which an equation was derived from physics principles.

Check Your Understanding

(The answers are given at the end of the book.)

1. **(a)** Is it possible for two quantities to have the same dimensions but different units?

    (b) Is it possible for two quantities to have the same units but different dimensions?

2. You can always add two numbers that have the same units (such as 6 meters + 3 meters). Can you always add two numbers that have the same dimensions, such as two numbers that have the dimensions of length [L]?

3. The following table lists four variables, along with their units:

Variable	Units
x	Meters (m)
v	Meters per second (m/s)
t	Seconds (s)
a	Meters per second squared (m/s^2)

These variables appear in the following equations, along with a few numbers that have no units. In which of the equations are the units on the left side of the equals sign consistent with the units on the right side?

(a) $x = vt$

(b) $x = vt + \frac{1}{2}at^2$

(c) $v = at$

(d) $v = at + \frac{1}{2}at^3$

(e) $v^3 = 2ax^2$

(f) $t = \sqrt{\frac{2x}{a}}$

4. In the equation $y = c^n at^2$ you wish to determine the integer value (1, 2, etc.) of the exponent n. The dimensions of y, a, and t are known. It is also known that c has no dimensions. Can dimensional analysis be used to determine n?

1.4 | Trigonometry

Scientists use mathematics to help them describe how the physical universe works, and trigonometry is an important branch of mathematics. Three trigonometric functions are utilized throughout this text. They are the sine, the cosine, and the tangent of the angle θ (Greek theta), abbreviated as sin θ, cos θ, and tan θ, respectively. These functions are defined below in terms of the symbols given along with the right triangle in **Interactive Figure 1.6**.

h = hypotenuse

h_o = length of side opposite the angle θ

h_a = length of side adjacent to the angle θ

INTERACTIVE FIGURE 1.6 A right triangle.

DEFINITION OF SIN θ, COS θ, AND TAN θ

$$\sin \theta = \frac{h_o}{h} \tag{1.1}$$

$$\cos \theta = \frac{h_a}{h} \tag{1.2}$$

$$\tan \theta = \frac{h_o}{h_a} \tag{1.3}$$

h = **length of the hypotenuse of a right triangle**
h_o = **length of the side opposite the angle θ**
h_a = **length of the side adjacent to the angle θ**

The sine, cosine, and tangent of an angle are numbers without units, because each is the ratio of the lengths of two sides of a right triangle. Example 4 illustrates a typical application of Equation 1.3.

EXAMPLE 4 | Using Trigonometric Functions

On a sunny day, a tall building casts a shadow that is 67.2 m long. The angle between the sun's rays and the ground is $\theta = 50.0°$, as **Figure 1.7** shows. Determine the height of the building.

Reasoning   We want to find the height of the building. Therefore, we begin with the colored right triangle in **Figure 1.7** and identify the height as the length h_o of the side opposite the angle θ. The length of the shadow is the length h_a of the side that is adjacent to the angle θ. The ratio of the length of the opposite side to the length of the adjacent side is the tangent of the angle θ, which can be used to find the height of the building.

Solution   We use the tangent function in the following way, with $\theta = 50.0°$ and $h_a = 67.2$ m:

$$\tan \theta = \frac{h_o}{h_a} \tag{1.3}$$

$$h_o = h_a \tan \theta = (67.2 \text{ m})(\tan 50.0°) = (67.2 \text{ m})(1.19) = \boxed{80.0 \text{ m}}$$

The value of tan 50.0° is found by using a calculator.

FIGURE 1.7   From a value for the angle θ and the length h_a of the shadow, the height h_o of the building can be found using trigonometry.

The sine, cosine, or tangent may be used in calculations such as that in Example 4, depending on which side of the triangle has a known value and which side is asked for.

> **Problem-Solving Insight**   However, the choice of which side of the triangle to label h_o (opposite) and which to label h_a (adjacent) can be made only after the angle θ is identified.

Often the values for two sides of the right triangle in **Interactive Figure 1.6** are available, and the value of the angle θ is unknown. The concept of **inverse trigonometric functions** plays an important role in such situations. Equations 1.4–1.6 give the inverse sine, inverse cosine, and inverse tangent in terms of the symbols used in the drawing. For instance, Equation 1.4 is read as "θ equals the angle whose sine is h_o/h."

$$\theta = \sin^{-1}\left(\frac{h_o}{h}\right) \tag{1.4}$$

$$\theta = \cos^{-1}\left(\frac{h_a}{h}\right) \tag{1.5}$$

$$\theta = \tan^{-1}\left(\frac{h_o}{h_a}\right) \tag{1.6}$$

The use of −1 as an exponent in Equations 1.4–1.6 *does not mean* "take the reciprocal." For instance, $\tan^{-1}(h_o/h_a)$ does not equal $1/\tan(h_o/h_a)$. Another way to express the inverse trigonometric functions is to use arc sin, arc cos, and arc tan instead of $\sin^{-1}$, $\cos^{-1}$, and $\tan^{-1}$. Example 5 illustrates the use of an inverse trigonometric function.

EXAMPLE 5 | Using Inverse Trigonometric Functions

A lakefront drops off gradually at an angle θ, as **Figure 1.8** indicates. For safety reasons, it is necessary to know how deep the lake is at various distances from the shore. To provide some information about the depth, a lifeguard rows straight out from the shore a distance of 14.0 m and drops a weighted fishing line. By measuring the length of the line, the lifeguard determines the depth to be 2.25 m. **(a)** What is the value of θ? **(b)** What would be the depth d of the lake at a distance of 22.0 m from the shore?

Reasoning   Near the shore, the lengths of the opposite and adjacent sides of the right triangle in **Figure 1.8** are $h_o = 2.25$ m and $h_a = 14.0$ m, relative to the angle θ. Having made this identification, we can use the inverse tangent to find the angle in part (a). For part (b) the opposite and adjacent sides farther from the shore become $h_o = d$ and $h_a = 22.0$ m. With the value for θ obtained in part (a), the tangent function can be used to find the unknown depth. Considering the way in which the lake bottom drops off in **Figure 1.8**, we expect the unknown depth to be greater than 2.25 m.

Solution **(a)** Using the inverse tangent given in Equation 1.6, we find that

$$\theta = \tan^{-1}\left(\frac{h_o}{h_a}\right) = \tan^{-1}\left(\frac{2.25 \text{ m}}{14.0 \text{ m}}\right) = \boxed{9.13°}$$

(b) With $\theta = 9.13°$, the tangent function given in Equation 1.3 can be used to find the unknown depth farther from the shore, where $h_o = d$ and $h_a = 22.0$ m. Since $\tan \theta = h_o/h_a$, it follows that

$$h_o = h_a \tan \theta$$

$$d = (22.0 \text{ m})(\tan 9.13°) = \boxed{3.54 \text{ m}}$$

which is greater than 2.25 m, as expected.

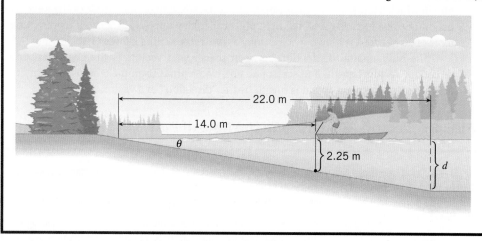

FIGURE 1.8 If the distance from the shore and the depth of the water at any one point are known, the angle θ can be found with the aid of trigonometry. Knowing the value of θ is useful, because then the depth d at another point can be determined.

The right triangle in **Interactive Figure 1.6** provides the basis for defining the various trigonometric functions according to Equations 1.1–1.3. These functions always involve an angle and two sides of the triangle. There is also a relationship among the lengths of the three sides of a right triangle. This relationship is known as the **Pythagorean theorem** and is used often in this text.

PYTHAGOREAN THEOREM

The square of the length of the hypotenuse of a right triangle is equal to the sum of the squares of the lengths of the other two sides:

$$h^2 = h_o^2 + h_a^2 \qquad (1.7)$$

1.5 | Scalars and Vectors

The volume of water in a swimming pool might be 50 cubic meters, or the winning time of a race could be 11.3 seconds. In cases like these, only the size of the numbers matters. In other words, *how much* volume or time is there? The 50 specifies the amount of water in units of cubic meters, while the 11.3 specifies the amount of time in seconds. Volume and time are examples of scalar quantities. A **scalar quantity** is one that can be described with a single number (including any units) giving its size or magnitude. Some other common scalars are temperature (e.g., 20 °C) and mass (e.g., 85 kg).

While many quantities in physics are scalars, there are also many that are not, and for these quantities the magnitude tells only part of the story. Consider **Figure 1.9**, which depicts a car that has moved 2 km along a straight line from start to finish. When describing the motion, it is incomplete to say that "the car moved a distance of 2 km." This statement would indicate only that the car ends up somewhere on a circle whose center is at the starting point and whose radius is 2 km. A complete description must include the direction along with the distance, as in the statement "the car moved a distance of 2 km in a direction 30° north of east." A quantity that deals inherently with *both magnitude and direction* is called a **vector quantity**. Because direction is an important characteristic of vectors, arrows are used to represent them; *the direction of the arrow gives the direction of the vector*. The colored arrow in **Figure 1.9**, for example, is called the *displacement vector*, because it shows how the car is displaced from its starting point. Chapter 2 discusses this particular vector.

FIGURE 1.9 A vector quantity has a magnitude and a direction. The colored arrow in this drawing represents a displacement vector.

The length of the arrow in **Figure 1.9** represents the magnitude of the displacement vector. If the car had moved 4 km instead of 2 km from the starting point, the arrow would have been drawn twice as long. ***By convention, the length of a vector arrow is proportional to the magnitude of the vector.***

In physics there are many important kinds of vectors, and the practice of using the length of an arrow to represent the magnitude of a vector applies to each of them. All forces, for instance, are vectors. In common usage a force is a push or a pull, and the direction in which a force acts is just as important as the strength or magnitude of the force. The magnitude of a force is measured in SI units called newtons (N). An arrow representing a force of 20 newtons is drawn twice as long as one representing a force of 10 newtons.

The fundamental distinction between scalars and vectors is the characteristic of direction. Vectors have it, and scalars do not. Conceptual Example 6 helps to clarify this distinction and explains what is meant by the "direction" of a vector.

CONCEPTUAL EXAMPLE 6 | Vectors, Scalars, and the Role of Plus and Minus Signs

There are places where the temperature is +20 °C at one time of the year and −20 °C at another time. Do the plus and minus signs that signify positive and negative temperatures imply that temperature is a vector quantity? **(a)** Yes **(b)** No

Reasoning A hallmark of a vector is that there is both a magnitude and a physical direction associated with it, such as 20 meters due east or 20 meters due west.

Answer (a) is incorrect. The plus and minus signs associated with +20 °C and −20 °C do not convey a physical direction, such as due east or due west. Therefore, temperature cannot be a vector quantity.

Answer (b) is correct. On a thermometer, the algebraic signs simply mean that the temperature is a number less than or greater than zero on the temperature scale being used and have nothing to do with east, west, or any other physical direction. Temperature, then, is not a vector. It is a scalar, and scalars can sometimes be negative.

Often, for the sake of convenience, quantities such as volume, time, displacement, velocity, and force are represented in physics by symbols. In this text, we write vectors in boldface symbols (**this is boldface**) with arrows above them* and write scalars in italic symbols (*this is italic*). Thus, a displacement vector is written as "$\vec{\mathbf{A}}$ = 750 m, due east," where the $\vec{\mathbf{A}}$ is a boldface symbol. By itself, however, separated from the direction, the magnitude of this vector is a scalar quantity. Therefore, the magnitude is written as "A = 750 m," where the A is an italic symbol without an arrow.

Check Your Understanding

(*The answer is given at the end of the book.*)

5. Which of the following statements, if any, involves a vector? **(a)** I walked 2 miles along the beach. **(b)** I walked 2 miles due north along the beach. **(c)** I jumped off a cliff and hit the water traveling at 17 miles per hour. **(d)** I jumped off a cliff and hit the water traveling straight down at a speed of 17 miles per hour. **(e)** My bank account shows a negative balance of −25 dollars.

1.6 | Vector Addition and Subtraction

Addition

Often it is necessary to add one vector to another, and the process of addition must take into account both the magnitude and the direction of the vectors. The simplest situation occurs when the vectors point along the same direction—that is, when they are colinear,

*Vectors are also sometimes written in other texts as boldface symbols without arrows above them.

as in **Figure 1.10**. Here, a car first moves along a straight line, with a displacement vector $\vec{A}$ of 275 m, due east. Then the car moves again in the same direction, with a displacement vector $\vec{B}$ of 125 m, due east. These two vectors add to give the total displacement vector $\vec{R}$, which would apply if the car had moved from start to finish in one step. The symbol $\vec{R}$ is used because the total vector is often called the **resultant vector**. With the tail of the second arrow located at the head of the first arrow, the two lengths simply add to give the length of the total displacement. This kind of vector addition is identical to the familiar addition of two scalar numbers (2 + 3 = 5) *and can be carried out here only because the vectors point along the same direction.* In such cases we add the individual magnitudes to get the magnitude of the total, knowing in advance what the direction must be. Formally, the addition is written as follows:

$$\vec{R} = \vec{A} + \vec{B}$$

$$\vec{R} = 275 \text{ m, due east} + 125 \text{ m, due east} = 400 \text{ m, due east}$$

Perpendicular vectors are frequently encountered, and **Figure 1.11** indicates how they can be added. This figure applies to a car that first travels with a displacement vector $\vec{A}$ of 275 m, due east, and then with a displacement vector $\vec{B}$ of 125 m, due north. The two vectors add to give a resultant displacement vector $\vec{R}$. Once again, the vectors to be added are arranged in a tail-to-head fashion, and the resultant vector points from the tail of the first to the head of the last vector added. The resultant displacement is given by the vector equation

$$\vec{R} = \vec{A} + \vec{B}$$

The addition in this equation cannot be carried out by writing $R = 275 \text{ m} + 125 \text{ m}$, because the vectors have different directions. Instead, we take advantage of the fact that the triangle in **Figure 1.11** is a right triangle and use the Pythagorean theorem (Equation 1.7). According to this theorem, the magnitude of $\vec{R}$ is

$$R = \sqrt{(275 \text{ m})^2 + (125 \text{ m})^2} = 302 \text{ m}$$

The angle θ in **Figure 1.11** gives the direction of the resultant vector. Since the lengths of all three sides of the right triangle are now known, $\sin \theta$, $\cos \theta$, or $\tan \theta$ can be used to determine θ. Noting that $\tan \theta = B/A$ and using the inverse trigonometric function, we find that:

$$\theta = \tan^{-1}\left(\frac{B}{A}\right) = \tan^{-1}\left(\frac{125 \text{ m}}{275 \text{ m}}\right) = 24.4°$$

Thus, the resultant displacement of the car has a magnitude of 302 m and points north of east at an angle of 24.4°. This displacement would bring the car from the start to the finish in **Figure 1.11** in a single straight-line step.

When two vectors to be added are not perpendicular, the tail-to-head arrangement does not lead to a right triangle, and the Pythagorean theorem cannot be used. **Figure 1.12a** illustrates such a case for a car that moves with a displacement $\vec{A}$ of 275 m, due east, and then with a displacement $\vec{B}$ of 125 m, in a direction 55.0° north of west. As usual, the resultant displacement vector $\vec{R}$ is directed from the tail of the first to the head of the last vector added. The vector addition is still given according to

$$\vec{R} = \vec{A} + \vec{B}$$

However, the magnitude of $\vec{R}$ is not $R = A + B$, because the vectors $\vec{A}$ and $\vec{B}$ do not have the same direction, and neither is it $R = \sqrt{A^2 + B^2}$, because the vectors are not perpendicular, so the Pythagorean theorem does not apply. Some other means must be used to find the magnitude and direction of the resultant vector.

One approach uses a graphical technique. In this method, a diagram is constructed in which the arrows are drawn tail to head. The lengths of the vector arrows are drawn to scale, and the angles are drawn accurately (with a protractor, perhaps). Then the length of the arrow representing the resultant vector is measured with a ruler. This length is converted to the magnitude of the resultant vector by using the scale factor with which

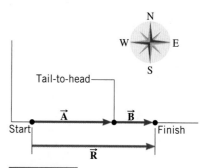

FIGURE 1.10 Two colinear displacement vectors $\vec{A}$ and $\vec{B}$ add to give the resultant displacement vector $\vec{R}$.

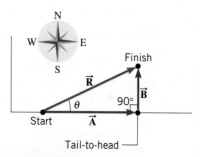

FIGURE 1.11 The addition of two perpendicular displacement vectors $\vec{A}$ and $\vec{B}$ gives the resultant vector $\vec{R}$.

(a)

(b)

FIGURE 1.12 (a) The two displacement vectors $\vec{A}$ and $\vec{B}$ are neither colinear nor perpendicular, but even so they add to give the resultant vector $\vec{R}$. (b) In one method for adding them together, a graphical technique is used.

the drawing is constructed. In **Figure 1.12b**, for example, a scale of one centimeter of arrow length for each 10.0 m of displacement is used, and it can be seen that the length of the arrow representing $\vec{R}$ is 22.8 cm. Since each centimeter corresponds to 10.0 m of displacement, the magnitude of $\vec{R}$ is 228 m. The angle θ, which gives the direction of $\vec{R}$, can be measured with a protractor to be $\theta = 26.7°$ north of east.

Subtraction

The subtraction of one vector from another is carried out in a way that depends on the following fact. *When a vector is multiplied by −1, the magnitude of the vector remains the same, but the direction of the vector is reversed.* Conceptual Example 7 illustrates the meaning of this statement.

CONCEPTUAL EXAMPLE 7 | Multiplying a Vector by −1

Consider two vectors described as follows:

1. A woman climbs 1.2 m up a ladder, so that her displacement vector $\vec{D}$ is 1.2 m, upward along the ladder, as in **Figure 1.13a**.

2. A man is pushing with 450 N of force on his stalled car, trying to move it eastward. The force vector $\vec{F}$ that he applies to the car is 450 N, due east, as in **Figure 1.14a**.

What are the physical meanings of the vectors $-\vec{D}$ and $-\vec{F}$?
(a) $-\vec{D}$ points upward along the ladder and has a magnitude of −1.2 m; $-\vec{F}$ points due east and has a magnitude of −450 N.
(b) $-\vec{D}$ points downward along the ladder and has a magnitude of −1.2 m; $-\vec{F}$ points due west and has a magnitude of −450 N.

(c) $-\vec{D}$ points downward along the ladder and has a magnitude of 1.2 m; $-\vec{F}$ points due west and has a magnitude of 450 N.

Reasoning A displacement vector of $-\vec{D}$ is $(-1)\,\vec{D}$. The presence of the (-1) factor reverses the direction of the vector, but does not change its magnitude. Similarly, a force vector of $-\vec{F}$ has the same magnitude as the vector $\vec{F}$ but has the opposite direction.

Answer (a) and (b) are incorrect. While scalars can sometimes be negative, magnitudes of vectors are never negative.

Answer (c) is correct. The vectors $-\vec{D}$ and $-\vec{F}$ have the same magnitudes as $\vec{D}$ and $\vec{F}$, but point in the opposite direction, as indicated in **Figures 1.13b** and **1.14b**.

(a)

(b)

FIGURE 1.14 *(a)* The force vector for a man pushing on a car with 450 N of force in a direction due east is $\vec{F}$. *(b)* The force vector for a man pushing on a car with 450 N of force in a direction due west is $-\vec{F}$.

(a)   *(b)*

FIGURE 1.13 *(a)* The displacement vector for a woman climbing 1.2 m up a ladder is $\vec{D}$. *(b)* The displacement vector for a woman climbing 1.2 m down a ladder is $-\vec{D}$.

In practice, vector subtraction is carried out exactly like vector addition, except that one of the vectors added is multiplied by a scalar factor of −1. To see why, look at the two vectors $\vec{A}$ and $\vec{B}$ in **Figure 1.15a**. These vectors add together to give a third vector $\vec{C}$, according to $\vec{C} = \vec{A} + \vec{B}$. Therefore, we can calculate vector $\vec{A}$ as $\vec{A} = \vec{C} - \vec{B}$, which is an example of vector subtraction. However, we can also write this result as $\vec{A} = \vec{C} + (-\vec{B})$ and treat it as vector addition. **Figure 1.15b** shows how to calculate vector $\vec{A}$ by adding the vectors $\vec{C}$ and $-\vec{B}$. Notice that vectors $\vec{C}$ and $-\vec{B}$ are arranged tail to head and that any suitable method of vector addition can be employed to determine $\vec{A}$.

Check Your Understanding

(The answers are given at the end of the book.)

6. Two vectors $\vec{A}$ and $\vec{B}$ are added together to give a resultant vector $\vec{R}$: $\vec{R} = \vec{A} + \vec{B}$. The magnitudes of $\vec{A}$ and $\vec{B}$ are 3 m and 8 m, respectively, but the vectors can have any orientation. What are **(a)** the maximum possible value and **(b)** the minimum possible value for the magnitude of $\vec{R}$?

7. Can two nonzero perpendicular vectors be added together so their sum is zero?

8. Can three or more vectors with unequal magnitudes be added together so their sum is zero?

9. In preparation for this question, review Conceptual Example 7. Vectors $\vec{A}$ and $\vec{B}$ satisfy the vector equation $\vec{A} + \vec{B} = 0$. **(a)** How does the magnitude of $\vec{B}$ compare with the magnitude of $\vec{A}$? **(b)** How does the direction of $\vec{B}$ compare with the direction of $\vec{A}$?

10. Vectors $\vec{A}$, $\vec{B}$, and $\vec{C}$ satisfy the vector equation $\vec{A} + \vec{B} = \vec{C}$, and their magnitudes are related by the scalar equation $A^2 + B^2 = C^2$. How is vector $\vec{A}$ oriented with respect to vector $\vec{B}$?

11. Vectors $\vec{A}$, $\vec{B}$, and $\vec{C}$ satisfy the vector equation $\vec{A} + \vec{B} = \vec{C}$, and their magnitudes are related by the scalar equation $A + B = C$. How is vector $\vec{A}$ oriented with respect to vector $\vec{B}$?

1.7 | The Components of a Vector

Vector Components

Suppose a car moves along a straight line from start to finish, as in **Figure 1.16**, the corresponding displacement vector being $\vec{r}$. The magnitude and direction of the vector $\vec{r}$ give the distance and direction traveled along the straight line. However, the car could also arrive at the finish point by first moving due east, turning through 90°, and then moving due north. This alternative path is shown in the drawing and is associated with the two displacement vectors $\vec{x}$ and $\vec{y}$. The vectors $\vec{x}$ and $\vec{y}$ are called the *x* vector component and the *y* vector component of $\vec{r}$.

Vector components are very important in physics and have two basic features that are apparent in **Figure 1.16**. One is that the components add together to equal the original vector:

$$\vec{r} = \vec{x} + \vec{y}$$

The components $\vec{x}$ and $\vec{y}$, when added vectorially, convey exactly the same meaning as does the original vector $\vec{r}$: they indicate how the finish point is displaced relative to the starting point. The other feature of vector components that is apparent in **Figure 1.16** is that $\vec{x}$ and $\vec{y}$ are not just any two vectors that add together to give the original vector $\vec{r}$: they are perpendicular vectors. This perpendicularity is a valuable characteristic, as we will soon see.

Any type of vector may be expressed in terms of its components, in a way similar to that illustrated for the displacement vector in **Figure 1.16**. **Interactive Figure 1.17** shows an arbitrary vector $\vec{A}$ and its vector components $\vec{A}_x$ and $\vec{A}_y$. The components are drawn parallel to convenient *x* and *y* axes and are perpendicular. They add vectorially to equal the original vector $\vec{A}$:

$$\vec{A} = \vec{A}_x + \vec{A}_y$$

There are times when a drawing such as **Interactive Figure 1.17** is not the most convenient way to represent vector components, and **Figure 1.18** presents an alternative method. The disadvantage of this alternative is that the tail-to-head arrangement of $\vec{A}_x$ and $\vec{A}_y$ is missing, an arrangement that is a nice reminder that $\vec{A}_x$ and $\vec{A}_y$ add together to equal $\vec{A}$.

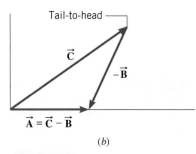

FIGURE 1.15 (*a*) Vector addition according to $\vec{C} = \vec{A} + \vec{B}$. (*b*) Vector subtraction according to $\vec{A} = \vec{C} - \vec{B} = \vec{C} + (-\vec{B})$.

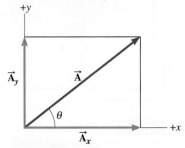

FIGURE 1.16 The displacement vector $\vec{r}$ and its vector components $\vec{x}$ and $\vec{y}$.

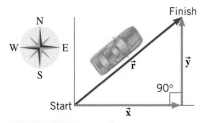

INTERACTIVE FIGURE 1.17 An arbitrary vector $\vec{A}$ and its vector components $\vec{A}_x$ and $\vec{A}_y$.

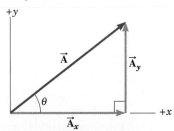

FIGURE 1.18 This alternative way of drawing the vector $\vec{A}$ and its vector components is completely equivalent to that shown in **Interactive Figure 1.17**.

FIGURE 1.19 The vector components of the vector depend on the orientation of the axes used as a reference.

The definition that follows summarizes the meaning of vector components:

> **DEFINITION OF VECTOR COMPONENTS**
>
> In two dimensions, the vector components of a vector $\vec{A}$ are two perpendicular vectors $\vec{A}_x$ and $\vec{A}_y$ that are parallel to the x and y axes, respectively, and add together vectorially according to $\vec{A} = \vec{A}_x + \vec{A}_y$.

> **Problem-Solving Insight** In general, the components of any vector can be used in place of the vector itself in any calculation where it is convenient to do so.

The values calculated for vector components depend on the orientation of the vector relative to the axes used as a reference. **Figure 1.19** illustrates this fact for a vector $\vec{A}$ by showing two sets of axes, one set being rotated clockwise relative to the other. With respect to the black axes, vector $\vec{A}$ has perpendicular vector components $\vec{A}_x$ and $\vec{A}_y$; with respect to the colored rotated axes, vector $\vec{A}$ has different vector components $\vec{A}'_x$ and $\vec{A}'_y$. The choice of which set of components to use is purely a matter of convenience.

Scalar Components

It is often easier to work with the **scalar components**, A_x and A_y (note the italic symbols), rather than the vector components $\vec{A}_x$ and $\vec{A}_y$. Scalar components are positive or negative numbers (with units) that are defined as follows: The scalar component A_x has a magnitude equal to that of $\vec{A}_x$ and is given a positive sign if $\vec{A}_x$ points along the $+x$ axis and a negative sign if $\vec{A}_x$ points along the $-x$ axis. The scalar component A_y is defined in a similar manner. The following table shows an example of vector and scalar components:

Vector Components	Scalar Components	Unit Vectors
$\vec{A}_x$ = 8 meters, directed along the $+x$ axis	$A_x = +8$ meters	$\vec{A}_x = (+8 \text{ meters})\hat{x}$
$\vec{A}_y$ = 10 meters, directed along the $-y$ axis	$A_y = -10$ meters	$\vec{A}_y = (-10 \text{ meters})\hat{y}$

In this text, when we use the term "component," we will be referring to a scalar component, unless otherwise indicated.

Another method of expressing vector components is to use unit vectors. A **unit vector** is a vector that has a magnitude of 1, but no dimensions. We will use a caret (^) to distinguish it from other vectors. Thus,

$\hat{x}$ is a dimensionless unit vector of length 1 that points in the positive x direction, and

$\hat{y}$ is a dimensionless unit vector of length 1 that points in the positive y direction.

These unit vectors are illustrated in **Figure 1.20**. With the aid of unit vectors, the vector components of an arbitrary vector $\vec{A}$ can be written as $\vec{A}_x = A_x\hat{x}$ and $\vec{A}_y = A_y\hat{y}$, where A_x and A_y are its scalar components (see the drawing and the third column of the table above). The vector $\vec{A}$ is then written as $\vec{A} = A_x\hat{x} + A_y\hat{y}$.

FIGURE 1.20 The dimensionless unit vectors $\hat{x}$ and $\hat{y}$ have magnitudes equal to 1, and they point in the $+x$ and $+y$ directions, respectively. Expressed in terms of unit vectors, the vector components of the vector $\vec{A}$ are $A_x\hat{x}$ and $A_y\hat{y}$.

Resolving a Vector into Its Components

If the magnitude and direction of a vector are known, it is possible to find the components of the vector. The process of finding the components is called "resolving the vector into its components." As Example 8 illustrates, this process can be carried out with the aid of trigonometry, because the two perpendicular vector components and the original vector form a right triangle.

EXAMPLE 8 | Finding the Components of a Vector

A displacement vector $\vec{r}$ has a magnitude of $r = 175$ m and points at an angle of 50.0° relative to the x axis in **Figure 1.21**. Find the x and y components of this vector.

Reasoning We will base our solution on the fact that the triangle formed in **Figure 1.21** by the vector $\vec{r}$ and its components $\vec{x}$ and $\vec{y}$ is a right triangle. This fact enables us to use the trigonometric sine and cosine functions, as defined in Equations 1.1 and 1.2.

> **Problem-Solving Insight** You can check to see whether the components of a vector are correct by substituting them into the Pythagorean theorem in order to calculate the magnitude of the original vector.

FIGURE 1.21 The x and y components of the displacement vector $\vec{r}$ can be found using trigonometry.

Solution The y component can be obtained using the 50.0° angle and Equation 1.1, $\sin\theta = y/r$:

$$y = r\sin\theta = (175\text{ m})(\sin 50.0°) = \boxed{134\text{ m}}$$

In a similar fashion, the x component can be obtained using the 50.0° angle and Equation 1.2, $\cos\theta = x/r$:

$$x = r\cos\theta = (175\text{ m})(\cos 50.0°) = \boxed{112\text{ m}}$$

Math Skills Either acute angle of a right triangle can be used to determine the components of a vector. The choice of angle is a matter of convenience. For instance, instead of the 50.0° angle, it is also possible to use the angle α in **Figure 1.21**. Since $\alpha + 50.0° = 90.0°$, it follows that $\alpha = 40.0°$. The solution using α yields the same answers as the solution using the 50.0° angle:

$$\cos\alpha = \frac{y}{r}$$

$$y = r\cos\alpha = (175\text{ m})(\cos 40.0°) = \boxed{134\text{ m}}$$

$$\sin\alpha = \frac{x}{r}$$

$$x = r\sin\alpha = (175\text{ m})(\sin 40.0°) = \boxed{112\text{ m}}$$

Since the vector components and the original vector form a right triangle, the Pythagorean theorem can be applied to check the validity of calculations such as those in Example 8. Thus, with the components obtained in Example 8, the theorem can be used to verify that the magnitude of the original vector is indeed 175 m, as given initially:

$$r = \sqrt{(112\text{ m})^2 + (134\text{ m})^2} = 175\text{ m}$$

It is possible for one of the components of a vector to be zero. This does not mean that the vector itself is zero, however.

> **Problem-Solving Insight** For a vector to be zero, every vector component must individually be zero.

Thus, in two dimensions, saying that $\vec{A} = 0$ is equivalent to saying that $\vec{A}_x = 0$ and $\vec{A}_y = 0$. Or, stated in terms of scalar components, if $\vec{A} = 0$, then $A_x = 0$ and $A_y = 0$.

> **Problem-Solving Insight** Two vectors are equal if, and only if, they have the same magnitude and direction.

Thus, if one displacement vector points east and another points north, they are *not* equal, even if each has the same magnitude of 480 m. In terms of vector components, two vectors $\vec{A}$ and $\vec{B}$ are equal if, and only if, each vector component of one is equal to the corresponding vector component of the other. In two dimensions, if $\vec{A} = \vec{B}$, then $\vec{A}_x = \vec{B}_x$ and $\vec{A}_y = \vec{A}_y$. Alternatively, using scalar components, we write that $A_x = B_x$ and $A_y = B_y$.

Check Your Understanding

(The answers are given at the end of the book.)

12. Which of the following displacement vectors (if any) are equal?

Variable	Magnitude	Direction
$\vec{A}$	100 m	30° north of east
$\vec{B}$	100 m	30° south of west
$\vec{C}$	50 m	30° south of west
$\vec{D}$	100 m	60° east of north

13. Two vectors, $\vec{A}$ and $\vec{B}$, are shown in **CYU Figure 1.1**. **(a)** What are the signs (+ or −) of the scalar components, A_x and A_y, of the vector $\vec{A}$? **(b)** What are the signs of the scalar components, B_x and B_y, of the vector $\vec{B}$? **(c)** What are the signs of the scalar components, R_x and R_y, of the vector $\vec{R}$, where $\vec{R} = \vec{A} + \vec{B}$?

14. Are two vectors with the same magnitude necessarily equal?

15. The magnitude of a vector has doubled, its direction remaining the same. Can you conclude that the magnitude of each component of the vector has doubled?

16. The tail of a vector is fixed to the origin of an x, y axis system. Originally the vector points along the $+x$ axis and has a magnitude of 12 units. As time passes, the vector rotates counterclockwise. What are the sizes of the x and y components of the vector for the following rotational angles? **(a)** 90° **(b)** 180° **(c)** 270° **(d)** 360°

17. A vector has a component of zero along the x axis of a certain axis system. Does this vector necessarily have a component of zero along the x axis of another (rotated) axis system?

CYU FIGURE 1.1

1.8 Addition of Vectors by Means of Components

The components of a vector provide the most convenient and accurate way of adding (or subtracting) any number of vectors. For example, suppose that vector $\vec{A}$ is added to vector $\vec{B}$. The resultant vector is $\vec{C}$, where $\vec{C} = \vec{A} + \vec{B}$. **Interactive Figure 1.22a** illustrates this vector addition, along with the x and y vector components of $\vec{A}$ and $\vec{B}$. In part b of the drawing, the vectors $\vec{A}$ and $\vec{B}$ have been removed, because we can use the vector components of these vectors in place of them. The vector component $\vec{B}_x$ has been shifted downward and arranged tail to head with vector component $\vec{A}_x$. Similarly, the vector component $\vec{A}_y$ has been shifted to the right and arranged tail to head with the vector component $\vec{B}_y$. The x components are colinear and add together to give the x component of the resultant vector $\vec{C}$. In like fashion, the y components are colinear and add together to give the y component of $\vec{C}$. In terms of scalar components, we can write

$$C_x = A_x + B_x \quad \text{and} \quad C_y = A_y + B_y$$

The vector components $\vec{C}_x$ and $\vec{C}_y$ of the resultant vector form the sides of the right triangle shown in **Interactive Figure 1.22c**. Thus, we can find the magnitude of $\vec{C}$ by using the Pythagorean theorem:

$$C = \sqrt{C_x^2 + C_y^2}$$

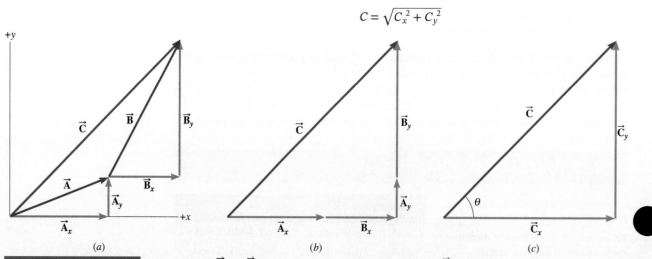

(a) (b) (c)

INTERACTIVE FIGURE 1.22 (a) The vectors $\vec{A}$ and $\vec{B}$ add together to give the resultant vector $\vec{C}$. The x and y components of $\vec{A}$ and $\vec{B}$ are also shown. (b) The drawing illustrates that $\vec{C}_x = \vec{A}_x + \vec{B}_x$ and $\vec{C}_y = \vec{A}_y + \vec{B}_y$. (c) Vector $\vec{C}$ and its components form a right triangle.

The angle θ that $\vec{\mathbf{C}}$ makes with the x axis is given by $\theta = \tan^{-1}(C_y/C_x)$. Example 9 illustrates how to add several vectors using the component method.

Analyzing Multiple-Concept Problems

EXAMPLE 9 | The Component Method of Vector Addition

A jogger runs 145 m in a direction 20.0° east of north (displacement vector $\vec{\mathbf{A}}$) and then 105 m in a direction 35.0° south of east (displacement vector $\vec{\mathbf{B}}$). Using components, determine the magnitude and direction of the resultant vector $\vec{\mathbf{C}}$ for these two displacements.

Reasoning **Figure 1.23** shows the vectors $\vec{\mathbf{A}}$ and $\vec{\mathbf{B}}$, assuming that the y axis corresponds to the direction due north. The vectors are arranged in a tail-to-head fashion, with the resultant vector $\vec{\mathbf{C}}$ drawn from the tail of $\vec{\mathbf{A}}$ to the head of $\vec{\mathbf{B}}$. The components of the vectors are also shown in the figure. Since $\vec{\mathbf{C}}$ and its components form a right triangle (red in the drawing), we will use the Pythagorean theorem and trigonometry to express the magnitude and directional angle θ for $\vec{\mathbf{C}}$ in terms of its components. The components of $\vec{\mathbf{C}}$ will then be obtained from the components of $\vec{\mathbf{A}}$ and $\vec{\mathbf{B}}$ and the data given for these two vectors.

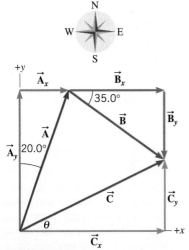

FIGURE 1.23 The vectors $\vec{\mathbf{A}}$ and $\vec{\mathbf{B}}$ add together to give the resultant vector $\vec{\mathbf{C}}$. The vector components of $\vec{\mathbf{A}}$ and $\vec{\mathbf{B}}$ are also shown. The resultant vector $\vec{\mathbf{C}}$ can be obtained once its components have been found.

Knowns and Unknowns   The data for this problem are listed in the table that follows:

Description	Symbol	Value	Comment
Magnitude of vector $\vec{\mathbf{A}}$		145 m	
Direction of vector $\vec{\mathbf{A}}$		20.0° east of north	See **Figure 1.23**.
Magnitude of vector $\vec{\mathbf{B}}$		105 m	
Direction of vector $\vec{\mathbf{B}}$		35.0° south of east	See **Figure 1.23**.
Unknown Variable			
Magnitude of resultant vector	C	?	
Direction of resultant vector	θ	?	

Modeling the Problem

STEP 1 Magnitude and Direction of $\vec{\mathbf{C}}$ In **Figure 1.23** the vector $\vec{\mathbf{C}}$ and its components $\vec{\mathbf{C}}_x$ and $\vec{\mathbf{C}}_y$ form a right triangle, as the red arrows show. Applying the Pythagorean theorem to this right triangle shows that the magnitude of $\vec{\mathbf{C}}$ is given by Equation 1a at the right. From the red triangle it also follows that the directional angle θ for the vector $\vec{\mathbf{C}}$ is given by Equation 1b at the right.

$$C = \sqrt{C_x^2 + C_y^2} \quad \text{(1a)}$$

$$\theta = \tan^{-1}\left(\frac{C_y}{C_x}\right) \quad \text{(1b)}$$

STEP 2 Components of $\vec{\mathbf{C}}$ Since vector $\vec{\mathbf{C}}$ is the resultant of vectors $\vec{\mathbf{A}}$ and $\vec{\mathbf{B}}$, we have $\vec{\mathbf{C}} = \vec{\mathbf{A}} + \vec{\mathbf{B}}$ and can write the scalar components of $\vec{\mathbf{C}}$ as the sum of the scalar components of $\vec{\mathbf{A}}$ and $\vec{\mathbf{B}}$:

$$\boxed{C_x = A_x + B_x} \quad \text{and} \quad \boxed{C_y = A_y + B_y}$$

These expressions can be substituted into Equations 1a and 1b for the magnitude and direction of $\vec{\mathbf{C}}$, as shown at the right.

$$C = \sqrt{C_x^2 + C_y^2} \quad \text{(1a)}$$

$$\boxed{C_x = A_x + B_x} \quad \boxed{C_y = A_y + B_y} \quad \text{(2)}$$

$$\theta = \tan^{-1}\left(\frac{C_y}{C_x}\right) \quad \text{(1b)}$$

$$\boxed{C_x = A_x + B_x} \quad \boxed{C_y = A_y + B_y} \quad \text{(2)}$$

Solution   Algebraically combining the results of each step, we find that

$$\overset{\text{STEP 1}}{\downarrow} \quad \overset{\text{STEP 2}}{\downarrow}$$

$$C = \sqrt{C_x^2 + C_y^2} = \sqrt{(A_x + B_x)^2 + (A_y + B_y)^2}$$

$$\theta = \tan^{-1}\left(\frac{C_y}{C_x}\right) = \tan^{-1}\left(\frac{A_y + B_y}{A_x + B_x}\right)$$

$$\underset{\text{STEP 1}}{\uparrow} \quad \underset{\text{STEP 2}}{\uparrow}$$

To use these results we need values for the individual components of $\vec{A}$ and $\vec{B}$. Referring to **Figure 1.23**, we find these values to be

$A_x = (145 \text{ m}) \sin 20.0° = 49.6 \text{ m}$ and
$A_y = (145 \text{ m}) \cos 20.0° = 136 \text{ m}$

$B_x = (105 \text{ m}) \cos 35.0° = 86.0 \text{ m}$ and
$B_y = -(105 \text{ m}) \sin 35.0° = -60.2 \text{ m}$

Note that the component B_y is negative, because $\vec{B}_y$ points downward, in the negative y direction in the drawing. Substituting these values into the results for C and θ gives

$$C = \sqrt{(A_x + B_x)^2 + (A_y + B_y)^2}$$

$$= \sqrt{(49.6 \text{ m} + 86.0 \text{ m})^2 + (136 \text{ m} - 60.2 \text{ m})^2} = \boxed{155 \text{ m}}$$

$$\theta = \tan^{-1}\left(\frac{A_y + B_y}{A_x + B_x}\right)$$

$$= \tan^{-1}\left(\frac{136 \text{ m} - 60.2 \text{ m}}{49.6 \text{ m} + 86.0 \text{ m}}\right) = \boxed{29°}$$

Math Skills According to the definitions given in Equations 1.1 and 1.2, the sine and cosine functions are $\sin \phi = \dfrac{h_o}{h}$ and $\cos \phi = \dfrac{h_a}{h}$, where h_o is the length of the side of a right triangle that is opposite the angle ϕ, h_a is the length of the side adjacent to the angle ϕ, and h is the length of the hypotenuse (see **Figure 1.24a**). Applications of the sine and cosine functions to determine the scalar components of a vector occur frequently. In such applications we begin by identifying the angle ϕ. **Figure 1.24b** shows the relevant portion of **Figure 1.23** and indicates that $\phi = 20.0°$ for the vector $\vec{A}$. In this case we have $h_o = A_x$, $h_a = A_y$, and $h = A = 145$ m; it follows that

$$\sin 20.0° = \frac{h_o}{h} = \frac{A_x}{A} \text{ or } A_x = A \sin 20.0° = (145 \text{ m}) \sin 20.0° = 49.6 \text{ m}$$

$$\cos 20.0° = \frac{h_a}{h} = \frac{A_y}{A} \text{ or } A_y = A \cos 20.0° = (145 \text{ m}) \cos 20.0° = 136 \text{ m}$$

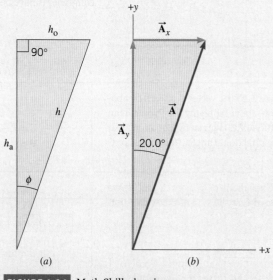

(a)   (b)

FIGURE 1.24 Math Skills drawing.

Related Homework: Problems 42, 44, 47, 50

Check Your Understanding

(The answer is given at the end of the book.)

18. Two vectors, $\vec{A}$ and $\vec{B}$, have vector components that are shown (to the same scale) in CYU Figure 1.2. The resultant vector is labeled $\vec{R}$. Which drawing shows the correct vector sum of $\vec{A} + \vec{B}$? **(a)** 1, **(b)** 2, **(c)** 3, **(d)** 4

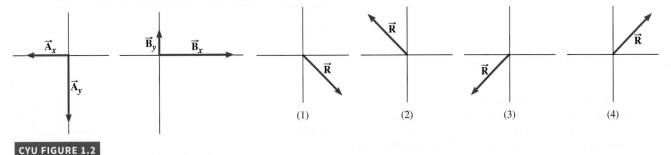

(1)   (2)   (3)   (4)

CYU FIGURE 1.2

In later chapters we will often use the component method for vector addition. For future reference, the main features of the reasoning strategy used in this technique are summarized below.

> **REASONING STRATEGY** **The Component Method of Vector Addition**
>
> 1. **For each vector to be added, determine the x and y components relative to a conveniently chosen x, y coordinate system. Be sure to take into account the directions of the components by using plus and minus signs to denote whether the components point along the positive or negative axes.**
>
> 2. **Find the algebraic sum of the x components, which is the x component of the resultant vector. Similarly, find the algebraic sum of the y components, which is the y component of the resultant vector.**
>
> 3. **Use the x and y components of the resultant vector and the Pythagorean theorem to determine the magnitude of the resultant vector.**
>
> 4. **Use the inverse sine, inverse cosine, or inverse tangent function to find the angle that specifies the direction of the resultant vector.**

EXAMPLE 10 | BIO Multi-joint Movements

Figure 1.25 shows an example of a multi-joint movement involving the shoulder and elbow joints. The view from above shows a person holding a ball in a position that involves both shoulder flexion and elbow extension. Vector $\vec{A}$ represents the position of the elbow joint relative to the shoulder joint, and vector $\vec{B}$ represents the position of the ball relative to the elbow joint. Use the component method of vector addition and the angles given in the figure to find the magnitude and direction (θ) of vector $\vec{C}$, which represents the position of the ball relative to the shoulder joint. The magnitude of vector $\vec{A}$ is 35.6 cm, and the magnitude of vector $\vec{B}$ is 31.2 cm. The angle θ is measured relative to a vertical anatomical plane known as the frontal or coronal plane.

Reasoning Similar to Example 9, the vectors $\vec{A}$ and $\vec{B}$ in Figure 1.25 are drawn tail-to-head. Thus, the resultant vector $\vec{C} = \vec{A} + \vec{B}$. The components of vector $\vec{C}$ will be obtained from the components of vectors $\vec{A}$ and $\vec{B}$. Once C_x and C_y are known, we can calculate the magnitude of vector $\vec{C}$ using the Pythagorean theorem. The directional angle θ will be determined from the components of vector $\vec{C}$.

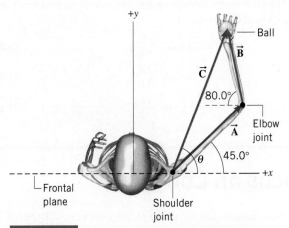

FIGURE 1.25 Top view of a multi-joint movement involving the shoulder and elbow joints. The vectors $\vec{A}$ and $\vec{B}$ add together to give the resultant vector $\vec{C}$, which represents the position vector of the ball with respect to the shoulder joint. The resultant vector $\vec{C}$ can be obtained, once its components are determined.

Solution Applying the component method of vector addition, we know that $C_x = A_x + B_x$, and $C_y = A_y + B_y$. From **Figure 1.25**, we see that $A_x = (35.6 \text{ cm}) \cos 45° = 25.2 \text{ cm}$, $B_x = -(31.2) \cos 80° = -5.42 \text{ cm}$, $A_y = (35.6 \text{ cm}) \sin 45° = 25.2 \text{ cm}$, and $B_y = (31.2 \text{ cm}) \sin 80° = 30.7 \text{ cm}$. Therefore, $C_x = 25.2 \text{ cm} + (-5.42 \text{ cm}) = 19.8 \text{ cm}$, and $C_y = 25.2 \text{ cm} + 30.7 \text{ cm} = 55.9 \text{ cm}$. The magnitude of vector $\vec{\mathbf{C}}$ is now found by the Pythagorean theorem:

$C = \sqrt{C_x^2 + C_y^2} = \sqrt{(19.8 \text{ cm})^2 + (55.9 \text{ cm})^2} = \boxed{59.3 \text{ cm}}$ The directional angle θ is found by using the tangent function:

$$\theta = \tan^{-1}\left(\frac{C_y}{C_x}\right) = \tan^{-1}\left(\frac{55.9 \text{ cm}}{19.8 \text{ cm}}\right) = \boxed{70.5°}.$$

Concept Summary

1.2 Units The SI system of units includes the meter (m), the kilogram (kg), and the second (s) as the base units for length, mass, and time, respectively. One meter is the distance that light travels in a vacuum in a time of 1/299 792 458 second. One kilogram is the mass of a standard cylinder of platinum–iridium alloy kept at the International Bureau of Weights and Measures. One second is the time for a certain type of electromagnetic wave emitted by cesium-133 atoms to undergo 9 192 631 770 wave cycles.

1.3 The Role of Units in Problem Solving To convert a number from one unit to another, multiply the number by the ratio of the two units. For instance, to convert 979 meters to feet, multiply 979 meters by the factor (3.281 foot/1 meter).

The dimension of a quantity represents its physical nature and the type of unit used to specify it. Three such dimensions are length [L], mass [M], time [T]. Dimensional analysis is a method for checking mathematical relations for the consistency of their dimensions.

1.4 Trigonometry The sine, cosine, and tangent functions of an angle θ are defined in terms of a right triangle that contains θ, as in Equations 1.1–1.3, where h_o and h_a are, respectively, the lengths of the sides opposite and adjacent to the angle θ, and h is the length of the hypotenuse.

$$\sin \theta = \frac{h_o}{h} \tag{1.1}$$

$$\cos \theta = \frac{h_a}{h} \tag{1.2}$$

$$\tan \theta = \frac{h_o}{h_a} \tag{1.3}$$

The inverse sine, inverse cosine, and inverse tangent functions are given in Equations 1.4–1.6.

$$\theta = \sin^{-1}\left(\frac{h_o}{h}\right) \tag{1.4}$$

$$\theta = \cos^{-1}\left(\frac{h_a}{h}\right) \tag{1.5}$$

$$\theta = \tan^{-1}\left(\frac{h_o}{h_a}\right) \tag{1.6}$$

The Pythagorean theorem states that the square of the length of the hypotenuse of a right triangle is equal to the sum of the squares of the lengths of the other two sides, according to Equation 1.7.

$$h^2 = h_o^2 + h_a^2 \tag{1.7}$$

1.5 Scalars and Vectors A scalar quantity is described by its size, which is also called its magnitude. A vector quantity has both a magnitude and a direction. Vectors are often represented by arrows, the length of the arrow being proportional to the magnitude of the vector and the direction of the arrow indicating the direction of the vector.

1.6 Vector Addition and Subtraction One procedure for adding vectors utilizes a graphical technique, in which the vectors to be added are arranged in a tail-to-head fashion. The resultant vector is drawn from the tail of the first vector to the head of the last vector. The subtraction of a vector is treated as the addition of a vector that has been multiplied by a scalar factor of -1. Multiplying a vector by -1 reverses the direction of the vector.

1.7 The Components of a Vector In two dimensions, the vector components of a vector $\vec{\mathbf{A}}$ are two perpendicular vectors $\vec{\mathbf{A}}_x$ and $\vec{\mathbf{A}}_y$ that are parallel to the x and y axes, respectively, and that add together vectorially so that $\vec{\mathbf{A}} = \vec{\mathbf{A}}_x + \vec{\mathbf{A}}_y$. The scalar component A_x has a magnitude that is equal to that of $\vec{\mathbf{A}}_x$ and is given a positive sign if $\vec{\mathbf{A}}_x$ points along the $+x$ axis and a negative sign if $\vec{\mathbf{A}}_x$ points along the $-x$ axis. The scalar component A_y is defined in a similar manner.

Two vectors are equal if, and only if, they have the same magnitude and direction. Alternatively, two vectors are equal in two dimensions if the x vector components of each are equal and the y vector components of each are equal. A vector is zero if, and only if, each of its vector components is zero.

1.8 Addition of Vectors by Means of Components If two vectors $\vec{\mathbf{A}}$ and $\vec{\mathbf{B}}$ are added to give a resultant $\vec{\mathbf{C}}$ such that $\vec{\mathbf{C}} = \vec{\mathbf{A}} + \vec{\mathbf{B}}$, then $C_x = A_x + B_x$ and $C_y = A_y + B_y$, where C_x, A_x, and B_x are the scalar components of the vectors along the x direction, and C_y, A_y, and B_y are the scalar components of the vectors along the y direction.

Focus on Concepts

Online

Additional questions are available for assignment in WileyPLUS.

Section 1.6 Vector Addition and Subtraction

1. During a relay race, runner A runs a certain distance due north and then hands off the baton to runner B, who runs for the same distance in a direction south of east. The two displacement vectors $\vec{\mathbf{A}}$ and $\vec{\mathbf{B}}$ can be added together to give a resultant vector $\vec{\mathbf{R}}$. Which drawing correctly shows the resultant vector? **(a)** 1 **(b)** 2 **(c)** 3 **(d)** 4

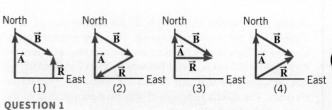

QUESTION 1

2. How is the magnitude R of the resultant vector $\vec{\mathbf{R}}$ in the drawing related to the magnitudes A and B of the vectors $\vec{\mathbf{A}}$ and $\vec{\mathbf{B}}$?

(a) The magnitude of the resultant vector $\vec{R}$ is equal to the sum of the magnitudes of $\vec{A}$ and $\vec{B}$, or $R = A + B$.
(b) The magnitude of the resultant vector $\vec{R}$ is greater than the sum of the magnitudes of $\vec{A}$ and $\vec{B}$, or $R > A + B$. (c) The magnitude of the resultant vector $\vec{R}$ is less than the sum of the magnitudes of $\vec{A}$ and $\vec{B}$, or $R < A + B$.

QUESTION 2

3. The first drawing shows three displacement vectors, $\vec{A}$, $\vec{B}$, and $\vec{C}$, which are added in a tail-to-head fashion. The resultant vector is labeled $\vec{R}$. Which of the following drawings shows the correct resultant vector for $\vec{A} + \vec{B} - \vec{C}$? (a) 1 (b) 2 (c) 3

QUESTION 3

4. The first drawing shows the sum of three displacement vectors, $\vec{A}$, $\vec{B}$, and $\vec{C}$. The resultant vector is labeled $\vec{R}$. Which of the following drawings shows the correct resultant vector for $\vec{A} - \vec{B} - \vec{C}$? (a) 1 (b) 2 (c) 3

QUESTION 4

Section 1.7   The Components of a Vector

5. A person is jogging along a straight line, and her displacement is denoted by the vector $\vec{A}$ in the drawings. Which drawing represents the correct vector components, $\vec{A}_x$ and $\vec{A}_y$, for the vector $\vec{A}$? (a) 1 (b) 2 (c) 3 (d) 4

 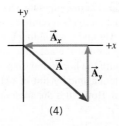

QUESTION 5

6. A person drives a car for a distance of 450.0 m. The displacement $\vec{A}$ of the car is illustrated in the drawing. What are the *scalar* components of this displacement vector?

(a) $A_x = 0$ m and $A_y = +450.0$ m

(b) $A_x = 0$ m and $A_y = -450.0$ m

(c) $A_x = +450.0$ m and $A_y = +450.0$ m

(d) $A_x = -450.0$ m and $A_y = 0$ m

(e) $A_x = -450.0$ m and $A_y = +450.0$ m

QUESTION 6

7. Drawing *a* shows a displacement vector $\vec{A}$ (450.0 m along the $-y$ axis). In this x, y coordinate system the scalar components are $A_x = 0$ m and $A_y = -450.0$ m. Suppose that the coordinate system is rotated counterclockwise by 35.0°, but the magnitude (450.0 m) and direction of vector $\vec{A}$ remain unchanged, as in drawing

b. What are the scalar components, $A_{x'}$ and $A_{y'}$, of the vector $\vec{A}$ in the rotated x', y' coordinate system?

(a) $A_{x'} = -369$ m and $A_{y'} = -258$ m

(b) $A_{x'} = +369$ m and $A_{y'} = -258$ m

(c) $A_{x'} = +258$ m and $A_{y'} = +369$ m

(d) $A_{x'} = +258$ m and $A_{y'} = -369$ m

(e) $A_{x'} = -258$ m and $A_{y'} = -369$ m

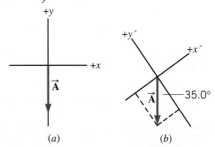

QUESTION 7   (*a*)   (*b*)

8. Suppose the vectors $\vec{A}$ and $\vec{B}$ in the drawing have magnitudes of 6.0 m and are directed as shown. What are A_x and B_x, the scalar components of $\vec{A}$ and $\vec{B}$ along the x axis?

	A_x	B_x
(a)	$+(6.0$ m$)\cos 35° = +4.9$ m	$-(6.0$ m$)\cos 35° = -4.9$ m
(b)	$+(6.0$ m$)\sin 35° = +3.4$ m	$-(6.0$ m$)\cos 35° = -4.9$ m
(c)	$-(6.0$ m$)\cos 35° = -4.9$ m	$+(6.0$ m$)\sin 35° = +3.4$ m
(d)	$-(6.0$ m$)\cos 35° = -4.9$ m	$+(6.0$ m$)\cos 35° = +4.9$ m
(e)	$-(6.0$ m$)\sin 35° = -3.4$ m	$+(6.0$ m$)\sin 35° = +3.4$ m

QUESTION 8

Section 1.8   Addition of Vectors by Means of Components

9. Drawing *a* shows two vectors $\vec{A}$ and $\vec{B}$, and drawing *b* shows their components. The scalar components of these vectors are as follows:

$$A_x = -4.9 \text{ m} \qquad A_y = +3.4 \text{ m}$$
$$B_x = +4.9 \text{ m} \qquad B_y = +3.4 \text{ m}$$

When the vectors $\vec{A}$ and $\vec{B}$ are added, the resultant vector is $\vec{R}$, so that $\vec{R} = \vec{A} + \vec{B}$. What are the values of R_x and R_y, the x and y components of $\vec{R}$?

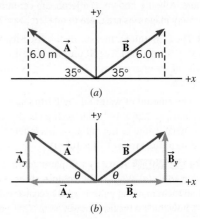

QUESTION 9   (*b*)

10. The displacement vectors $\vec{A}$ and $\vec{B}$, when added together, give the resultant vector $\vec{R}$, so that $\vec{R} = \vec{A} + \vec{B}$. Use the data in the drawing to find the magnitude R of the resultant vector and the angle θ that it makes with the +x axis.

QUESTION 10

Problems

Online

Additional questions are available for assignment in WileyPLUS.

SSM Student Solutions Manual

BIO Biomedical application

MMH Problem-solving help

E Easy

GO Guided Online Tutorial

M Medium

V-HINT Video Hints

H Hard

CHALK Chalkboard Videos

WS Worksheet

T Team Problem

Section 1.2 Units

Section 1.3 The Role of Units in Problem Solving

1. **E** **GO** A student sees a newspaper ad for an apartment that has 1330 square feet (ft^2) of floor space. How many square meters of area are there?

2. **E** Bicyclists in the Tour de France reach speeds of 34.0 miles per hour (mi/h) on flat sections of the road. What is this speed in **(a)** kilometers per hour (km/h) and **(b)** meters per second (m/s)?

3. **E** **SSM** Vesna Vulovic survived the longest fall on record without a parachute when her plane exploded and she fell 6 miles, 551 yards. What is this distance in meters?

4. **E** Suppose a man's scalp hair grows at a rate of 0.35 mm per day. What is this growth rate in feet per century?

5. **E** Given the quantities $a = 9.7$ m, $b = 4.2$ s, $c = 69$ m/s, what is the value of the quantity $d = a^3/(cb^2)$?

6. **E** Consider the equation $v = \frac{1}{3}zxt^2$. The dimensions of the variables v, x, and t are [L]/[T], [L], and [T], respectively. The numerical factor 3 is dimensionless. What must be the dimensions of the variable z, such that both sides of the equation have the same dimensions? Show how you determined your answer.

7. **E** **SSM** A bottle of wine known as a magnum contains a volume of 1.5 liters. A bottle known as a jeroboam contains 0.792 U.S. gallons. How many magnums are there in one jeroboam?

8. **E** The CGS unit for measuring the viscosity of a liquid is the poise (P): 1 P = 1 g/(s · cm). The SI unit for viscosity is the kg/(s · m). The viscosity of water at 0 °C is 1.78×10^{-3} kg/(s · m). Express this viscosity in poise.

9. **E** The amount of water on the Earth's surface is 326 million trillion gallons, or 3.26×10^{20} gallons. If all of this water was placed in a sphere, what would be the radius of this sphere in m? Compare this to the radius of the Moon (1.74×10^6 m).

10. **M** **GO** **CHALK** A partly full paint can has 0.67 U.S. gallons of paint left in it. **(a)** What is the volume of the paint in cubic meters? **(b)** If all the remaining paint is used to coat a wall evenly (wall area = 13 m^2), how thick is the layer of wet paint? Give your answer in meters.

11. **M** **SSM** **CHALK** A spring is hanging down from the ceiling, and an object of mass m is attached to the free end. The object is pulled down, thereby stretching the spring, and then released. The object oscillates up and down, and the time T required for one complete up-and-down oscillation is given by the equation $T = 2\pi\sqrt{m/k}$, where k is known as the spring constant. What must be the dimension of k for this equation to be dimensionally correct?

Section 1.4 Trigonometry

12. **E** You are driving into St. Louis, Missouri, and in the distance you see the famous Gateway to the West arch. This monument rises to a height of 192 m. You estimate your line of sight with the top of the arch to be 2.0° above the horizontal. Approximately how far (in kilometers) are you from the base of the arch?

13. **E** **SSM** A highway is to be built between two towns, one of which lies 35.0 km south and 72.0 km west of the other. What is the shortest length of highway that can be built between the two towns, and at what angle would this highway be directed with respect to due west?

14. **E** **GO** A hill that has a 12.0% grade is one that rises 12.0 m vertically for every 100.0 m of distance in the horizontal direction. At what angle is such a hill inclined above the horizontal?

15. **E** **GO** The corners of a square lie on a circle of diameter $D = 0.35$ m. Each side of the square has a length L. Find L.

16. **E** **GO** The drawing shows a person looking at a building on top of which an antenna is mounted. The horizontal distance between the person's eyes and the building is 85.0 m. In part a the person is looking at the base of the antenna, and his line of sight makes an angle of 35.0° with the horizontal. In part b the person is looking at the top of the antenna, and his line of sight makes an angle of 38.0° with the horizontal. How tall is the antenna?

(a)   (b)

PROBLEM 16

17. **E** **CHALK** The two hot-air balloons in the drawing are 48.2 and 61.0 m above the ground. A person in the left balloon observes that the right balloon is 13.3° above the horizontal. What is the horizontal distance x between the two balloons?

PROBLEM 17

18. **M** **MMH** The drawing shows sodium and chloride ions positioned at the corners of a cube that is part of the crystal structure of sodium chloride (common table salt). The edges of the cube are each 0.281 nm (1 nm = 1 nanometer = 10^{-9} m) in length. What is the value of the angle θ in the drawing?

PROBLEM 18

19. **M** **GO** A person is standing at the edge of the water and looking out at the ocean (see the drawing). The height of the person's eyes above the water is $h = 1.6$ m, and the radius of the earth is $R = 6.38 \times 10^6$ m. **(a)** How far is it to the horizon? In other words, what is the distance d from the person's eyes to the horizon? *(Note: At the horizon the angle between the line of sight and the radius of the earth is 90°.)* **(b)** Express this distance in miles.

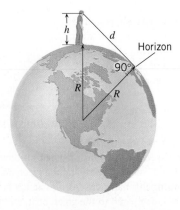

PROBLEM 19

20. **M** **SSM** Three deer, A, B, and C, are grazing in a field. Deer B is located 62 m from deer A at an angle of 51° north of west. Deer C is located 77° north of east relative to deer A. The distance between deer B and C is 95 m. What is the distance between deer A and C? *(Hint: Consider the law of cosines given in Appendix E.)*

21. **H** An aerialist on a high platform holds on to a trapeze attached to a support by an 8.0-m cord. (See the drawing.) Just before he jumps off the platform, the cord makes an angle of 41° with the vertical. He jumps, swings down, then back up, releasing the trapeze at the instant it is 0.75 m below its initial height. Calculate the angle θ that the trapeze cord makes with the vertical at this instant.

PROBLEM 21

Section 1.6   Vector Addition and Subtraction

22. **E** **SSM** **(a)** Two workers are trying to move a heavy crate. One pushes on the crate with a force $\vec{\mathbf{A}}$, which has a magnitude of 445 newtons and is directed due west. The other pushes with a force $\vec{\mathbf{B}}$, which has a magnitude of 325 newtons and is directed due north. What are the magnitude and direction of the resultant force $\vec{\mathbf{A}} + \vec{\mathbf{B}}$ applied to the crate? **(b)** Suppose that the second worker applies a force $-\vec{\mathbf{B}}$ instead of $\vec{\mathbf{B}}$. What then are the magnitude and direction of the resultant force $\vec{\mathbf{A}} - \vec{\mathbf{B}}$ applied to the crate? In both cases express the direction relative to due west.

23. **E** A force vector $\vec{\mathbf{F}}_1$ points due east and has a magnitude of 200 newtons. A second force $\vec{\mathbf{F}}_2$ is added to $\vec{\mathbf{F}}_1$. The resultant of the two vectors has a magnitude of 400 newtons and points along the east/west line. Find the magnitude and direction of $\vec{\mathbf{F}}_2$. Note that there are two answers.

24. **E** **SSM** Consider the following four force vectors:

$$\vec{\mathbf{F}}_1 = 50.0 \text{ newtons, due east}$$
$$\vec{\mathbf{F}}_2 = 10.0 \text{ newtons, due east}$$
$$\vec{\mathbf{F}}_3 = 40.0 \text{ newtons, due west}$$
$$\vec{\mathbf{F}}_4 = 30.0 \text{ newtons, due west}$$

Which two vectors add together to give a resultant with the smallest magnitude, and which two vectors add to give a resultant with the largest magnitude? In each case specify the magnitude and direction of the resultant.

25. **E** **GO** Vector $\vec{\mathbf{A}}$ has a magnitude of 63 units and points due west, while vector $\vec{\mathbf{B}}$ has the same magnitude and points due south. Find the magnitude and direction of **(a)** $\vec{\mathbf{A}} + \vec{\mathbf{B}}$ and **(b)** $\vec{\mathbf{A}} - \vec{\mathbf{B}}$. Specify the directions relative to due west.

26. **E** Two bicyclists, starting at the same place, are riding toward the same campground by two different routes. One cyclist rides 1080 m due east and then turns due north and travels another 1430 m before reaching the campground. The second cyclist starts out by heading due north for 1950 m and then turns and heads directly toward the campground. **(a)** At the turning point, how far is the second cyclist from the campground? **(b)** In what direction (measured relative to due east) must the second cyclist head during the last part of the trip?

27. **E** **GO** The drawing shows a triple jump on a checkerboard, starting at the center of square A and ending on the center of square B. Each side of a square measures 4.0 cm. What is the magnitude of the displacement of the colored checker during the triple jump?

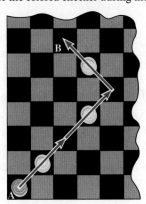

PROBLEM 27

28. **E** Given the vectors $\vec{\mathbf{P}}$ and $\vec{\mathbf{Q}}$ shown on the grid, sketch and calculate the magnitudes of the vectors **(a)** $\vec{\mathbf{M}} = \vec{\mathbf{P}} + \vec{\mathbf{Q}}$ and **(b)** $\vec{\mathbf{K}} = 2\vec{\mathbf{P}} - \vec{\mathbf{Q}}$. Use the tail-to-head method and express the magnitudes in centimeters with the aid of the grid scale shown in the drawing.

PROBLEM 28

29. **M** **MMH** **CHALK** Vector $\vec{A}$ has a magnitude of 12.3 units and points due west. Vector $\vec{B}$ points due north. **(a)** What is the magnitude of $\vec{B}$ if $\vec{A} + \vec{B}$ has a magnitude of 15.0 units? **(b)** What is the direction of $\vec{A} + \vec{B}$ relative to due west? **(c)** What is the magnitude of $\vec{B}$ if $\vec{A} - \vec{B}$ has a magnitude of 15.0 units? **(d)** What is the direction of $\vec{A} - \vec{B}$ relative to due west?

30. **M** **SSM** A car is being pulled out of the mud by two forces that are applied by the two ropes shown in the drawing. The dashed line in the drawing bisects the 30.0° angle. The magnitude of the force applied by each rope is 2900 newtons. Arrange the force vectors tail to head and use the graphical technique to answer the following questions. **(a)** How much force would a single rope need to apply to accomplish the same effect as the two forces added together? **(b)** How would the single rope be directed relative to the dashed line?

2900 newtons

30.0°

2900 newtons

PROBLEM 30

31. **M** **GO** A jogger travels a route that has two parts. The first is a displacement $\vec{A}$ of 2.50 km due south, and the second involves a displacement $\vec{B}$ that points due east. **(a)** The resultant displacement $\vec{A} + \vec{B}$ has a magnitude of 3.75 km. What is the magnitude of $\vec{B}$, and what is the direction of $\vec{A} + \vec{B}$ relative to due south? **(b)** Suppose that $\vec{A} - \vec{B}$ had a magnitude of 3.75 km. What then would be the magnitude of $\vec{B}$, and what is the direction of $\vec{A} - \vec{B}$ relative to due south?

32. **M** At a picnic, there is a contest in which hoses are used to shoot water at a beach ball from three directions. As a result, three forces act on the ball, $\vec{F}_1$, $\vec{F}_2$, and $\vec{F}_3$ (see the drawing). The magnitudes of $\vec{F}_1$ and $\vec{F}_2$ are $F_1 = 50.0$ newtons and $F_2 = 90.0$ newtons. Using a scale drawing and the graphical technique, determine **(a)** the magnitude of $\vec{F}_3$ and **(b)** the angle θ such that the resultant force acting on the ball is zero.

PROBLEM 32

Section 1.7 The Components of a Vector

33. **E** **GO** A force vector has a magnitude of 575 newtons and points at an angle of 36.0° below the positive x axis. What are **(a)** the x scalar component and **(b)** the y scalar component of the vector?

34. **E** **SSM** Vector $\vec{A}$ points along the $+y$ axis and has a magnitude of 100.0 units. Vector $\vec{B}$ points at an angle of 60.0° above the $+x$ axis and has a magnitude of 200.0 units. Vector $\vec{C}$ points along the $+x$ axis and has a magnitude of 150.0 units. Which vector has **(a)** the largest x component and **(b)** the largest y component?

35. **E** Soccer player #1 is 8.6 m from the goal (see the drawing). If she kicks the ball directly into the net, the ball has a displacement labeled $\vec{A}$. If, on the other hand, she first kicks it to player #2, who then kicks it into the net, the ball undergoes two successive displacements, $\vec{A}_y$ and $\vec{A}_x$. What are the magnitudes and directions of $\vec{A}_x$ and $\vec{A}_y$?

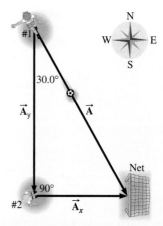

PROBLEM 35

36. **E** **GO** The components of vector $\vec{A}$ are A_x and A_y (both positive), and the angle that it makes with respect to the positive x axis is θ. Find the angle θ if the components of the displacement vector $\vec{A}$ are **(a)** $A_x = 12$ m and $A_y = 12$ m, **(b)** $A_x = 17$ m and $A_y = 12$ m, and **(c)** $A_x = 12$ m and $A_y = 17$ m.

37. **E** During takeoff, an airplane climbs with a speed of 180 m/s at an angle of 34° above the horizontal. The speed and direction of the airplane constitute a vector quantity known as the velocity. The sun is shining directly overhead. How fast is the shadow of the plane moving along the ground? (That is, what is the magnitude of the horizontal component of the plane's velocity?)

38. **E** **SSM** The x vector component of a displacement vector $\vec{r}$ has a magnitude of 125 m and points along the negative x axis. The y vector component has a magnitude of 184 m and points along the negative y axis. Find the magnitude and direction of $\vec{r}$. Specify the direction with respect to the negative x axis.

39. **E** Your friend has slipped and fallen. To help her up, you pull with a force $\vec{F}$, as the drawing shows. The vertical component of this force is 130 newtons, and the horizontal component is 150 newtons. Find **(a)** the magnitude of $\vec{F}$ and **(b)** the angle θ.

PROBLEM 39

40. **M** **GO** Two racing boats set out from the same dock and speed away at the same constant speed of 101 km/h for half an hour (0.500 h), the blue boat headed 25.0° south of west, and the green boat headed 37.0° south of west. During this half hour **(a)** how much farther west does the blue boat travel, compared to the green boat, and **(b)** how much farther south does the green boat travel, compared to the blue boat? Express your answers in km.

41. **M** **SSM** **MMH** **CHALK** The magnitude of the force vector $\vec{F}$ is 82.3 newtons. The x component of this vector is directed along the $+x$ axis and has a magnitude of 74.6 newtons. The y component points along the $+y$ axis. **(a)** Find the direction of $\vec{F}$ relative to the $+x$ axis. **(b)** Find the component of $\vec{F}$ along the $+y$ axis.

Section 1.8 Addition of Vectors by Means of Components

42. **E** **SSM** Consult Multiple-Concept Example 9 in preparation for this problem. A golfer, putting on a green, requires three strokes to "hole the ball." During the first putt, the ball rolls 5.0 m due east. For the second putt, the ball travels 2.1 m at an angle of 20.0° north of east. The third putt is 0.50 m due north. What displacement (magnitude and direction relative to due east) would have been needed to "hole the ball" on the very first putt?

43. **E** **CHALK** The three displacement vectors in the drawing have magnitudes of $A = 5.00$ m, $B = 5.00$ m, and $C = 4.00$ m. Find the resultant (magnitude and directional angle) of the three vectors by means of the component method. Express the directional angle as an angle above the positive or negative x axis.

PROBLEM 43

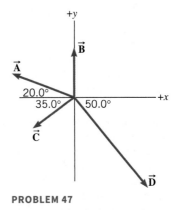

PROBLEM 47

44. **E** **MMH** Multiple-Concept Example 9 reviews the concepts that play a role in this problem. Two forces are applied to a tree stump to pull it out of the ground. Force $\vec{F}_A$ has a magnitude of 2240 newtons and points 34.0° south of east, while force $\vec{F}_B$ has a magnitude of 3160 newtons and points due south. Using the component method, find the magnitude and direction of the resultant force $\vec{F}_A + \vec{F}_B$ that is applied to the stump. Specify the direction with respect to due east.

45. **E** **GO** A baby elephant is stuck in a mud hole. To help pull it out, game keepers use a rope to apply a force $\vec{F}_A$, as part a of the drawing shows. By itself, however, force $\vec{F}_A$ is insufficient. Therefore, two additional forces $\vec{F}_B$ and $\vec{F}_C$ are applied, as in part b of the drawing. Each of these additional forces has the same magnitude F. The magnitude of the resultant force acting on the elephant in part b of the drawing is k times larger than that in part a. Find the ratio F/F_A when $k = 2.00$.

(a)   (b)

PROBLEM 45

46. **E** Displacement vector $\vec{A}$ points due east and has a magnitude of 2.00 km. Displacement vector $\vec{B}$ points due north and has a magnitude of 3.75 km. Displacement vector $\vec{C}$ points due west and has a magnitude of 2.50 km. Displacement vector $\vec{D}$ points due south and has a magnitude of 3.00 km. Find the magnitude and direction (relative to due west) of the resultant vector $\vec{A} + \vec{B} + \vec{C} + \vec{D}$.

47. **E** Multiple-Concept Example 9 provides background pertinent to this problem. The magnitudes of the four displacement vectors shown in the drawing are $A = 16.0$ m, $B = 11.0$ m, $C = 12.0$ m, and $D = 26.0$ m. Determine the magnitude and directional angle for the resultant that occurs when these vectors are added together.

48. **M** **GO** Two geological field teams are working in a remote area. A global positioning system (GPS) tracker at their base camp shows the location of the first team as 38 km away, 19° north of west, and the second team as 29 km away, 35° east of north. When the first team uses its GPS to check the position of the second team, what does the GPS give for the second team's **(a)** distance from them and **(b)** direction, measured from due east?

49. **M** **SSM** A sailboat race course consists of four legs, defined by the displacement vectors $\vec{A}$, $\vec{B}$, $\vec{C}$, and $\vec{D}$, as the drawing indicates. The magnitudes of the first three vectors are $A = 3.20$ km, $B = 5.10$ km, and $C = 4.80$ km. The finish line of the course coincides with the starting line. Using the data in the drawing, find the distance of the fourth leg and the angle θ.

PROBLEM 49

50. **M** Multiple-Concept Example 9 deals with the concepts that are important in this problem. A grasshopper makes four jumps. The displacement vectors are (1) 27.0 cm, due west; (2) 23.0 cm, 35.0° south of west; (3) 28.0 cm, 55.0° south of east; and (4) 35.0 cm, 63.0° north of east. Find the magnitude and direction of the resultant displacement. Express the direction with respect to due west.

51. **M** The route followed by a hiker consists of three displacement vectors $\vec{A}$, $\vec{B}$, and $\vec{C}$. Vector $\vec{A}$ is along a measured trail and is 1550 m in a direction 25.0° north of east. Vector $\vec{B}$ is not along a measured trail, but the hiker uses a compass and knows that the direction is 41.0° east of south. Similarly, the direction of vector $\vec{C}$ is 35.0° north of west. The hiker ends up back where she started. Therefore, it follows that the resultant displacement is zero, or $\vec{A} + \vec{B} + \vec{C} = 0$. Find the magnitudes of **(a)** vector $\vec{B}$ and **(b)** vector $\vec{C}$.

Additional Problems

Online

52. **E** A monkey is chained to a stake in the ground. The stake is 3.00 m from a vertical pole, and the chain is 3.40 m long. How high can the monkey climb up the pole?

53. **M** A car drives from its initial position for 150 km in a direction that is 20° north of east and stops. The car then drives due north for 125 km and stops again. After the car comes to rest the second time, what is the displacement **(a)** magnitude and **(b)** direction of the car from its initial position? State your direction relative to due East.

54. **E** An ocean liner leaves New York City and travels 18.0° north of east for 155 km. How far east and how far north has it gone? In other words, what are the magnitudes of the components of the ship's displacement vector in the directions **(a)** due east and **(b)** due north?

55. **E** **GO** A pilot flies her route in two straight-line segments. The displacement vector $\vec{A}$ for the first segment has a magnitude of 244 km and a direction 30.0° north of east. The displacement vector $\vec{B}$ for the second segment has a magnitude of 175 km and a direction due west. The resultant displacement vector is $\vec{R} = \vec{A} + \vec{B}$ and makes an angle θ with the direction due east. Using the component method, find the magnitude of $\vec{R}$ and the directional angle θ.

56. Ⓜ SSM Vector $\vec{A}$ has a magnitude of 6.00 units and points due east. Vector $\vec{B}$ points due north. **(a)** What is the magnitude of $\vec{B}$, if the vector $\vec{A} + \vec{B}$ points 60.0° north of east? **(b)** Find the magnitude of $\vec{A} + \vec{B}$.

57. Ⓜ GO Three forces act on an object, as indicated in the drawing. Force $\vec{F_1}$ has a magnitude of 21.0 newtons (21.0 N) and is directed 30.0° to the left of the +y axis. Force $\vec{F_2}$ has a magnitude of 15.0 N and points along the +x axis. What must be the magnitude and direction (specified by the angle θ in the drawing) of the third force $\vec{F_3}$ such that the vector sum of the three forces is 0 N?

58. Ⓜ CHALK You live in the building on the left in the drawing, and a

PROBLEM 57

friend lives in the other building. The two of you are having a discussion about the heights of the buildings, and your friend claims that the height of his building is more than 1.50 times the height of yours. To resolve the issue you climb to the roof of your building and estimate that your line of sight to the top edge of the other building makes an angle of 21° above the horizontal, whereas your line of sight to the base of the other building makes an angle of 52° below the horizontal. Determine the ratio of the height of the taller building to the height of the shorter building. State whether your friend is right or wrong.

PROBLEM 58

Physics in Biology, Medicine, and Sports

59. Ⓔ BIO **1.3** Azelastine hydrochloride is an antihistamine nasal spray. A standard size container holds one fluid ounce (oz) of the liquid. You are searching for this medication in a European drugstore and are asked how many milliliters (mL) there are in one fluid ounce. Using the following conversion factors, determine the number of milliliters in a volume of one fluid ounce: 1 gallon (gal) = 128 oz, 3.785 × 10^{-3} cubic meters (m^3) = 1 gal, and 1 mL = 10^{-6} m^3.

60. Ⓔ BIO **1.3** The volume of liquid flowing per second is called the volume flow rate Q and has the dimensions of $[L]^3/[T]$. The flow rate of a liquid through a hypodermic needle during an injection can be estimated with the following equation:

$$Q = \frac{\pi R^n (P_2 - P_1)}{8\eta L}$$

The length and radius of the needle are L and R, respectively, both of which have the dimension [L]. The pressures at opposite ends of the needle are P_2 and P_1, both of which have the dimensions of $[M]/\{[L][T]^2\}$. The symbol η represents the viscosity of the liquid and has the dimensions of $[M]/\{[L][T]\}$. The symbol π stands for pi and, like the number 8 and the exponent n, has no dimensions. Using dimensional analysis, determine the value of n in the expression for Q.

61. Ⓔ BIO **1.3** During their lifetime, a person will consume approximately 1.2 × 10^4 gallons of water. An Olympic-size swimming pool, which has a volume of 2.5 million liters (L), can provide a lifetime of water for how many people? Recall that 1 gallon (gal) = 3.785 × 10^{-3} m^3, and 1 mL = 10^{-6} m^3.

62. Ⓔ BIO **1.3** Deoxyribonucleic acid, or DNA, is a molecule composed of two chains that wrap around each other to form a double helix, as the figure shows. DNA carries the genetic information for the growth and development of all known organisms. If you were to uncoil all of the DNA chains in a human body and lay them end-to-end, they would stretch for a distance of 1.6 × 10^{13} m. What is this

distance in miles? For comparison, the average distance from Earth to Pluto is about 3.7 billion miles.

PROBLEM 62   2.2–2.6 nm

63. Ⓜ BIO **1.3** An owl's keen eyesight is not the only ability that assists them in locating prey. They also have exceptional hearing. Take the barn owl, for example. Their entire head is designed to focus in on low frequency sounds. They have a dish-shaped face that collects and amplifies sound in much the same way a dish antenna collects electromagnetic signals from orbiting satellites (see images below). Their ears are hidden behind feathers on each side of their head behind their eyes. The position of their ear openings is not symmetrical, as their left ear is slightly higher than their right one. This asymmetry allows them to determine if a source of sound is above or below their line of sight. Furthermore, they use a technique known as *triangulation* to pinpoint the location of prey in front of them. Consider the distance between the owl's ears to correspond to the base of a triangle, with the long sides extending from each ear to a mouse located some distance away (see figure). If the mouse is directly in front of the owl, then the time it takes the sound to travel from the mouse to its left and right ear will be the same. However, if the mouse is not directly in front, such as in the figure, then the sound will reach the owl's left and right ears at different times. The owl's auditory system is extremely sensitive to this time difference. In fact, it can detect a time difference between its left and right ear of 30 millionths of a second! The owl turns its head until the sound difference is eliminated, which indicates its prey is directly in front.

The figure shows an example of the mouse being located to the left of the owl's forward direction (dashed line). In this case, the sound created by the mouse would reach the owl's left ear a fraction of a second sooner than it reaches the right ear. Based on the information shown in the figure (not to scale) calculate the distance between the owl's ears (x).

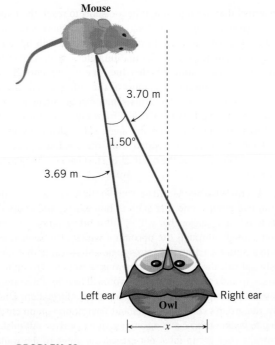

PROBLEM 63

64. **E** **BIO** **1.3** The total surface area of the United States is 9.6×10^6 km². Seventy-four percent of this land can support spiders. Given that there are approximately 5.0×10^4 spiders per acre, how many spiders are there in the United States? Recall that 1 acre = 0.004 km².

65. **M** **BIO** **1.4** Scoliosis is a common disease of the spine, where the spine is not straight, but forms an "S" or "C" shape while viewing a patient from the front. One of the most widely used indicators of scoliosis is a measurement of the Cobb angle (a), as it is a quantitative measure of the spinal curvature. The Cobb angle is calculated directly on X-ray radiographs of the spine (see photo). To determine the Cobb angle, the health professional must decide which two vertebrae are at each end of the deformity. They correspond to the two vertebrae that are rotated the most from a vertical position. In the following sample X-ray, these are shown with red arrows. Lines are then drawn (solid blue) along the endplates of these vertebrae until they intersect as shown, forming the Cobb angle a. In this example, an isosceles triangle is formed, where the two equivalent sides have a length h, and the third side has a length of $0.65h$. If the angle θ in the figure has a value of 38°, what is the value of the Cobb angle, a?

Case courtesy of A.Prof Frank Gaillard, Radiopaedia.org. rID: 49374

PROBLEM 65

Concepts and Calculations Problems   **Online**

This chapter has presented an introduction to the mathematics of trigonometry and vectors, which will be used throughout the text. In this section we apply some of the important features of this mathematics, and review some concepts that can help in anticipating some of the characteristics of the numerical answers.

66. **M** **CHALK** The figure shows two displacement vectors $\vec{A}$ and $\vec{B}$. Vector $\vec{A}$ points at an angle of 22° above the x axis and has an unknown magnitude. Vector $\vec{B}$ has an x component $B_x = 35.0$ m and has an unknown y component B_y. These two vectors are equal. *Concepts:* (i) What does the condition that vector $\vec{A}$ equals $\vec{B}$ imply about the magnitudes and directions of the vectors? (ii) What does the condition that vector $\vec{A}$ equals $\vec{B}$ imply about the x and y components of the vectors? *Calculations:* Find the magnitude of $\vec{A}$ and the value of B_y.

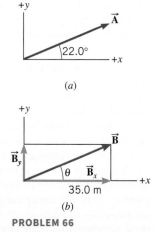

PROBLEM 66

67. **M** **CHALK** **SSM** The figure shows three displacement vectors $\vec{A}$, $\vec{B}$, and $\vec{C}$. These vectors are arranged in tail-to-head fashion, and they add together to give a resultant displacement $\vec{R}$, which lies along the x axis. Note that the vector $\vec{B}$ is parallel to the x axis. *Concepts:* (i) How is the magnitude of $\vec{A}$ related to its scalar components A_x and A_y? (ii) Do any of the vectors in the figure have a zero value for either their x or y components? (If so, which ones?) (iii) What does the fact that $\vec{A}$, $\vec{B}$, and $\vec{C}$ add together to give $\vec{R}$ tell you about the components of these vectors? *Calculations:* What is the magnitude of the vector $\vec{A}$ and its directional angle θ?

PROBLEM 67

Team Problems

Online

68. **E** **BIO** **WS** **T** **Recovering a Downed Eagle.** You are on a road that runs east and west, and another member of your team is on the same road, 1.20 km east of you. Someone reported seeing an eagle strike a wire and eventually go down somewhere in a dense forest north of the road. Your objective is to rescue the downed eagle. The bird is fitted with a transmitter, and you have a detector that indicates that it is located at a distance of 6.00×10^2 m from you. The team member located 1.20 km east of you has a similar detector that indicates that the eagle is 9.00×10^2 m away from them. Note that neither detector gives directional information. **(a)** Assuming a straight-line path, determine the direction you must go in order to find the eagle. **(b)** Assuming you want to minimize your walking distance in the thick forest, how should you proceed? Find the minimum distance traveled in the forest.

69. **E** **BIO** **WS** **T** **Focusing a Gamma Knife.** A gamma knife is a medical device used to deliver high intensity radiation to a tumor inside the human body while minimizing damage to the surrounding tissue. The device directs multiple radiation beams through the same point (e.g., at the position of a tumor) so that the intensity in the volume where the beams intersect is extremely high. Suppose the position of a tumor relative to the origin of an xy coordinate system is represented by a vector of magnitude 32.5 cm that makes an angle of $\theta_0 = +76.4°$ with respect to the $+x$ axis. If two radiation sources are located on the x axis at positions $x_1 = +10.7$ cm and $x_2 = -14.3$ cm, at what angles relative to the $+x$ axis must the radiation beams from these two sources be directed so that they intersect at the tumor?

70. **M** **T** **The Waterfall.** You and your team are exploring a river in South America when you come to the bottom of a tall waterfall. You estimate the cliff over which the water flows to be about 100 feet tall. You have to choose between climbing the cliff or backtracking and taking another route, but climbing the cliff would cut two hours off of your trip. There is only one experienced climber in the group: she would climb the cliff alone and drop a rope over the edge to lift supplies and allow the others to climb without packs. The climber estimates it will take her 45 minutes to get to the top. However, you are concerned that the rope might be too short to reach the bottom of the cliff (it is exactly 30.0 m long). If it is too short, she'll have to climb back down (another 45 minutes) and you will be too far behind schedule to get to your destination before dark. As you contemplate how to determine whether the rope is long enough, you notice that the late afternoon shadow of the cliff grows as the sun descends over its edge. You suddenly remember your trigonometry. You measure the length of the shadow from the base of the cliff to the shadow's edge (144 ft), and the angle subtended between the base and top of the cliff measured from the shadow's edge. The angle is 38.1°. Do you send the climber, or start backtracking to take another route?

71. **M** **T** **The Weather Monitor.** Your South American expedition splits into two groups: one that stays at home base, and yours that goes off to set up a sensor that will monitor precipitation, temperature, and sunlight through the upcoming winter. The sensor must link up to a central communications system at base camp that simultaneously uploads the data from numerous sensors to a satellite. In order to set up and calibrate the sensor, you will have to communicate with base camp to give them specific location information. Unfortunately, the group's communication and navigation equipment has dwindled to walkie-talkies and a compass due to a river-raft mishap, which means your group must not exceed the range of the walkie-talkies (3.0 miles). However, you do have a laser rangefinder to help you measure distances as you navigate with the compass. After a few hours of hiking, you find the perfect plateau on which to mount the sensor. You have carefully mapped your path from base camp around lakes and other obstacles: 550.0 m West (W), 275 m S, 750.0 m W, 900.0 m NE, 800.0 m W, and 400.0 m 30.0° W of S. The final leg is due south, 2.20 km up a constant slope and ending at a plateau that is 320.0 m above the level of base camp. **(a)** How far are you from base camp? Will you be able to communicate with home base using the walkie-talkies? **(b)** What is the geographical direction from base camp to the sensor (expressed in the form $\theta°$ south of west, etc.)? **(c)** What is the angle of inclination from base camp to the detector?

The pilots in the United States Navy's Blue Angels can perform high-speed maneuvers in perfect unison. They do so by controlling the displacement, velocity, and acceleration of their jet aircraft. These three concepts and the relationships among them are the focus of this chapter.   erniedecker/iStockphoto

Kinematics in One Dimension

2.1 Displacement

There are two aspects to any motion. In a purely descriptive sense, there is the movement itself. Is it rapid or slow, for instance? Then, there is the issue of what causes the motion or what changes it, which requires that forces be considered. **Kinematics** deals with the concepts that are needed to describe motion, without any reference to forces. The present chapter discusses these concepts as they apply to motion in one dimension, and the next chapter treats two-dimensional motion. **Dynamics** deals with the effect that forces have on motion, a topic that is considered in Chapter 4. Together, kinematics and dynamics form the branch of physics known as **mechanics**. We turn now to the first of the kinematics concepts to be discussed, which is displacement.

To describe the motion of an object, we must be able to specify the location of the object at all times, and **Figure 2.1** shows how to do this for one-dimensional motion. In this drawing, the initial position of a car is indicated by the vector labeled $\vec{x}_0$. The length of $\vec{x}_0$ is the distance of the car from an arbitrarily chosen origin. At a later time the car has moved

LEARNING OBJECTIVES

After reading this module, you should be able to...

2.1 Define one-dimensional displacement.

2.2 Discriminate between speed and velocity.

2.3 Define one-dimensional acceleration.

2.4 Use one-dimensional kinematic equations to predict future or past values of variables.

2.5 Solve one-dimensional kinematic problems.

2.6 Solve one-dimensional free-fall problems.

2.7 Predict kinematic quantities using graphical analysis.

Origin

FIGURE 2.1 The displacement $\Delta\vec{x}$ is a vector that points from the initial position $\vec{x}_0$ to the final position $\vec{x}$.

to a new position, which is indicated by the vector $\vec{x}$. The **displacement** of the car $\Delta\vec{x}$ (read as "delta x" or "the change in x") is a vector drawn from the initial position to the final position. Displacement is a vector quantity in the sense discussed in Section 1.5, for it conveys both a magnitude (the distance between the initial and final positions) and a direction. The displacement can be related to $\vec{x}_0$ and $\vec{x}$ by noting from the drawing that

$$\vec{x}_0 + \Delta\vec{x} = \vec{x} \qquad \text{or} \qquad \Delta\vec{x} = \vec{x} - \vec{x}_0$$

Thus, the displacement $\Delta\vec{x}$ is the difference between $\vec{x}$ and $\vec{x}_0$, and the Greek letter delta (Δ) is used to signify this difference. It is important to note that the change in any variable is always the final value minus the initial value.

> **DEFINITION OF DISPLACEMENT**
>
> **The displacement is a vector that points from an object's initial position to its final position and has a magnitude that equals the shortest distance between the two positions.**
>
> **SI Unit of Displacement: meter (m)**

The SI unit for displacement is the meter (m), but there are other units as well, such as the centimeter and the inch. When converting between centimeters (cm) and inches (in.), remember that 2.54 cm = 1 in.

Often, we will deal with motion along a straight line. In such a case, a displacement in one direction along the line is assigned a positive value, and a displacement in the opposite direction is assigned a negative value. For instance, assume that a car is moving along an east/west direction and that a positive (+) sign is used to denote a direction due east. Then, $\Delta\vec{x} = +500$ m represents a displacement that points to the east and has a magnitude of 500 meters. Conversely, $\Delta\vec{x} = -500$ m is a displacement that has the same magnitude but points in the opposite direction, due west.

The magnitude of the displacement vector is the shortest distance between the initial and final positions of the object. However, this does not mean that displacement and distance are the same physical quantities. In **Figure 2.1**, for example, the car could reach its final position after going forward and backing up several times. In that case, the total distance traveled by the car would be greater than the magnitude of the displacement vector.

Check Your Understanding

(The answer is given at the end of the book.)

1. A honeybee leaves the hive and travels a total distance of 2 km before returning to the hive. What is the magnitude of the displacement vector of the bee?

2.2 | Speed and Velocity

Average Speed

One of the most obvious features of an object in motion is how fast it is moving. If a car travels 200 meters in 10 seconds, we say its average speed is 20 meters per second, the *average speed* being the distance traveled divided by the time required to cover the distance:

$$\text{Average speed} = \frac{\text{Distance}}{\text{Elapsed time}} \qquad (2.1)$$

Equation 2.1 indicates that the unit for average speed is the unit for distance divided by the unit for time, or meters per second (m/s) in SI units. Example 1 illustrates how the idea of average speed is used.

EXAMPLE 1 | **Distance Run by a Jogger**

How far does a jogger run in 1.5 hours (5400 s) if his average speed is 2.22 m/s?

Reasoning The average speed of the jogger is the average distance per second that he travels. Thus, the distance covered by the jogger is equal to the average distance per second (his average speed) multiplied by the number of seconds (the elapsed time) that he runs.

Solution To find the distance run, we rewrite Equation 2.1 as

Distance = (Average speed)(Elapsed time) = (2.22 m/s)(5400 s)

$$= \boxed{12\ 000\ \text{m}}$$

Speed is a useful idea, because it indicates how fast an object is moving. However, speed does not reveal anything about the direction of the motion. To describe both how fast an object moves and the direction of its motion, we need the vector concept of velocity.

Average Velocity

To define the velocity of an object, we will use two concepts that we have already encountered, displacement and time. The building of new concepts from more basic ones is a common theme in physics. In fact the great strength of physics as a science is that it builds a coherent understanding of nature through the development of interrelated concepts.

Suppose that the initial position of the car in **Figure 2.1** is $\vec{x}_0$ when the time is t_0. A little later that car arrives at the final position $\vec{x}$ at the time t. The difference between these times is the time required for the car to travel between the two positions. We denote this difference by the shorthand notation Δt (read as "delta t"), where Δt represents the final time t minus the initial time t_0:

$$\Delta t = \underbrace{t - t_0}_{\text{Elapsed time}}$$

Note that Δt is defined in a manner analogous to $\Delta \vec{x}$, which is the final position minus the initial position $(\Delta \vec{x} = \vec{x} - \vec{x}_0)$. Dividing the displacement $\Delta \vec{x}$ of the car by the elapsed time Δt gives the **average velocity** of the car. It is customary to denote the average value of a quantity by placing a horizontal bar above the symbol representing the quantity. The average velocity, then, is written as $\overline{\vec{v}}$, as specified in Equation 2.2:

DEFINITION OF AVERAGE VELOCITY

$$\text{Average velocity} = \frac{\text{Displacement}}{\text{Elapsed time}}$$

$$\overline{\vec{v}} = \frac{\vec{x} - \vec{x}_0}{t - t_0} = \frac{\Delta \vec{x}}{\Delta t} \qquad (2.2)$$

SI Unit of Average Velocity: meter per second (m/s)

Equation 2.2 indicates that the unit for average velocity is the unit for length divided by the unit for time, or meters per second (m/s) in SI units. Velocity can also be expressed in other units, such as kilometers per hour (km/h) or miles per hour (mi/h).

Average velocity is a vector that points in the same direction as the displacement in Equation 2.2. **Figure 2.2** illustrates that the velocity of an object confined to move along a line can point either in one direction or in the opposite direction. As with displacement, we will use plus and minus signs to indicate the two possible directions. If the displacement points in the positive direction, the average velocity is positive. Conversely, if the displacement points in the negative direction, the average velocity is negative. Example 2 illustrates these features of average velocity.

Michael Dietrich/Age Fotostock

FIGURE 2.2 The boats in this photograph are traveling in opposite directions; in other words, the velocity of one boat points in a direction that is opposite to the velocity of the other boat.

EXAMPLE 2 | The World's Fastest Jet-Engine Car

Andy Green in the car *ThrustSSC* set a world record of 341.1 m/s (763 mi/h) in 1997. The car was powered by two jet engines, and it was the first one officially to exceed the speed of sound. To establish such a record, the driver makes two runs through the course, one in each direction, to nullify wind effects. **Figure 2.3a** shows that the car first travels from left to right and covers a distance of 1609 m (1 mile) in a time of 4.740 s. **Figure 2.3b** shows that in the

(a)

(b)

FIGURE 2.3 The arrows in the box at the top of the drawing indicate the positive and negative directions for the displacements of the car, as explained in Example 2.

reverse direction, the car covers the same distance in 4.695 s. From these data, determine the average velocity for each run.

Reasoning Average velocity is defined as the displacement divided by the elapsed time. In using this definition we recognize that the displacement is not the same as the distance traveled. Displacement takes the direction of the motion into account, and distance does not. During both runs, the car covers the same distance of 1609 m. However, for the first run the displacement is $\Delta\vec{x} = +1609$ m, while for the second it is $\Delta\vec{x} = -1609$ m. The plus and minus signs are essential, because the first run is to the right, which is the positive direction, and the second run is in the opposite or negative direction.

Solution According to Equation 2.2, the average velocities are

Run 1
$$\bar{\vec{v}} = \frac{\Delta\vec{x}}{\Delta t} = \frac{+1609 \text{ m}}{4.740 \text{ s}} = \boxed{+339.5 \text{ m/s}}$$

Run 2
$$\bar{\vec{v}} = \frac{\Delta\vec{x}}{\Delta t} = \frac{-1609 \text{ m}}{4.695 \text{ s}} = \boxed{-342.7 \text{ m/s}}$$

In these answers the algebraic signs convey the directions of the velocity vectors. In particular, for run 2 the minus sign indicates that the average velocity, like the displacement, points to the left in **Figure 2.3b**. The magnitudes of the velocities are 339.5 and 342.7 m/s. The average of these numbers is 341.1 m/s, and this is recorded in the record book.

Instantaneous Velocity

Suppose the magnitude of your average velocity for a long trip was 20 m/s. This value, being an average, does not convey any information about how fast you were moving or the direction of the motion at any instant during the trip. Both can change from one instant to another. Surely there were times when your car traveled faster than 20 m/s and times when it traveled more slowly. The **instantaneous velocity** $\vec{v}$ of the car indicates how fast the car moves and the direction of the motion at each instant of time. The magnitude of the instantaneous velocity is called the **instantaneous speed**, and it is the number (with units) indicated by the speedometer.

The instantaneous velocity at any point during a trip can be obtained by measuring the time interval Δt for the car to travel a *very small* displacement $\Delta\vec{x}$. We can then compute the average velocity over this interval. If the time Δt is small enough, the instantaneous velocity does not change much during the measurement. Then, the instantaneous velocity $\vec{v}$ at the point of interest is approximately equal to ($\approx$) the average velocity $\bar{\vec{v}}$ computed over the interval, or $\vec{v} \approx \bar{\vec{v}} = \Delta\vec{x}/\Delta t$ (for sufficiently small Δt). In fact, in the limit that Δt becomes infinitesimally small, the instantaneous velocity and the average velocity become equal, so that

$$\vec{v} = \lim_{\Delta t \to 0} \frac{\Delta\vec{x}}{\Delta t} \tag{2.3}$$

The notation $\lim_{\Delta t \to 0} \frac{\Delta\vec{x}}{\Delta t}$ means that the ratio $\Delta\vec{x}/\Delta t$ is defined by a limiting process in which smaller and smaller values of Δt are used, so small that they approach zero. As smaller values of Δt are used, $\Delta\vec{x}$ also becomes smaller. However, the ratio $\Delta\vec{x}/\Delta t$ does *not* become zero but, rather, approaches the value of the instantaneous velocity. For brevity, we will use the word *velocity* to mean "instantaneous velocity" and *speed* to mean "instantaneous speed."

Check Your Understanding

(The answers are given at the end of the book.)

2. Is the average speed of a vehicle a vector or a scalar quantity?

3. Two buses depart from Chicago, one going to New York and one to San Francisco. Each bus travels at a speed of 30 m/s. Do they have equal velocities?

4. One of the following statements is incorrect. **(a)** The car traveled around the circular track at a constant velocity. **(b)** The car traveled around the circular track at a constant speed. Which statement is incorrect?

5. A straight track is 1600 m in length. A runner begins at the starting line, runs due east for the full length of the track, turns around and runs halfway back. The time for this run is five minutes. What is the runner's average velocity, and what is his average speed?

6. The average velocity for a trip has a positive value. Is it possible for the instantaneous velocity at a point during the trip to have a negative value?

2.3 | Acceleration

In a wide range of motions, the velocity changes from moment to moment, such as in the case of the sprinter in **Photo 2.1**. To describe the manner in which it changes, the concept of acceleration is needed. The velocity of a moving object may change in a number of ways. For example, it may increase, as it does when the driver of a car steps on the gas pedal to pass the car ahead. Or it may decrease, as it does when the driver applies the brakes to stop at a red light. In either case, the change in velocity may occur over a short or a long time interval. To describe how the velocity of an object changes during a given time interval, we now introduce the new idea of acceleration. This idea depends on two concepts that we have previously encountered, velocity and time. Specifically, the notion of acceleration emerges when the *change* in the velocity is combined with the time during which the change occurs.

The meaning of **average acceleration** can be illustrated by considering a plane during takeoff. **Figure 2.4** focuses attention on how the plane's velocity changes along the runway. During an elapsed time interval $\Delta t = t - t_0$, the velocity changes from an initial value of $\vec{v}_0$ to a final velocity of $\vec{v}$. The change $\Delta \vec{v}$ in the plane's velocity is its final velocity minus its initial velocity, so that $\Delta \vec{v} = \vec{v} - \vec{v}_0$. The average acceleration $\vec{a}$ is defined in the following manner, to provide a measure of how much the velocity changes per unit of elapsed time.

DEFINITION OF AVERAGE ACCELERATION

$$\text{Average acceleration} = \frac{\text{Change in velocity}}{\text{Elapsed time}}$$

$$\vec{a} = \frac{\vec{v} - \vec{v}_0}{t - t_0} = \frac{\Delta \vec{v}}{\Delta t} \tag{2.4}$$

SI Unit of Average Acceleration: meter per second squared (m/s²)

The average acceleration $\vec{a}$ is a vector that points in the same direction as $\Delta \vec{v}$, the change in the velocity. Following the usual custom, plus and minus signs indicate the two possible directions for the acceleration vector when the motion is along a straight line.

Kaliva/Shutterstock.com

PHOTO 2.1 As this sprinter explodes out of the starting block, his velocity is changing, which means that he is accelerating.

FIGURE 2.4 During takeoff, the plane accelerates from an initial velocity $\vec{v}_0$ to a final velocity $\vec{v}$ during the time interval $\Delta t = t - t_0$.

We are often interested in an object's acceleration at a particular instant of time. The **instantaneous acceleration $\vec{a}$** can be defined by analogy with the procedure used in Section 2.2 for instantaneous velocity:

$$\vec{a} = \lim_{\Delta t \to 0} \frac{\Delta \vec{v}}{\Delta t} \qquad (2.5)$$

Equation 2.5 indicates that the instantaneous acceleration is a limiting case of the average acceleration. When the time interval Δt for measuring the acceleration becomes extremely small (approaching zero in the limit), the average acceleration and the instantaneous acceleration become equal. Moreover, in many situations the acceleration is constant, so the acceleration has the same value at any instant of time. In the future, we will use the word *acceleration* to mean "instantaneous acceleration." Example 3 deals with the acceleration of a plane during takeoff.

EXAMPLE 3 | Acceleration and Increasing Velocity

Suppose the plane in **Figure 2.4** starts from rest ($\vec{v}_0 = 0$ m/s) when $t_0 = 0$ s. The plane accelerates down the runway and at $t = 29$ s attains a velocity of $\vec{v} = +260$ km/h, where the plus sign indicates that the velocity points to the right. Determine the average acceleration of the plane.

Reasoning The average acceleration of the plane is defined as the change in its velocity divided by the elapsed time. The change in the plane's velocity is its final velocity $\vec{v}$ minus its initial velocity $\vec{v}_0$, or $\vec{v} - \vec{v}_0$. The elapsed time is the final time t minus the initial time t_0, or $t - t_0$.

Problem-Solving Insight The change in any variable is the final value minus the initial value: for example, the change in velocity is $\Delta \vec{v} = \vec{v} - \vec{v}_0$, and the change in time is $\Delta t = t - t_0$.

Solution The average acceleration is expressed by Equation 2.4 as

$$\overline{\vec{a}} = \frac{\vec{v} - \vec{v}_0}{t - t_0} = \frac{260 \text{ km/h} - 0 \text{ km/h}}{29 \text{ s} - 0 \text{ s}} = \boxed{+9.0 \frac{\text{km/h}}{\text{s}}}$$

The average acceleration calculated in Example 3 is read as "nine kilometers per hour per second." Assuming the acceleration of the plane is constant, a value of $+9.0 \frac{\text{km/h}}{\text{s}}$ means the velocity changes by $+9.0$ km/h during each second of the motion. During the first second, the velocity increases from 0 to 9.0 km/h; during the next second, the velocity increases by another 9.0 km/h to 18 km/h, and so on. **Figure 2.5** illustrates how the velocity changes during the first two seconds. By the end of the 29th second, the velocity is 260 km/h.

$$\overline{\vec{a}} = \frac{+9.0 \text{ km/h}}{\text{s}}$$

$\Delta t = 0$ s
$\vec{v}_0 = 0$ m/s

$\Delta t = 1.0$ s
$\vec{v} = +9.0$ km/h

$\Delta t = 2.0$ s
$\vec{v} = +18$ km/h

FIGURE 2.5 An acceleration of $+9.0 \frac{\text{km/h}}{\text{s}}$ means that the velocity of the plane changes by $+9.0$ km/h during each second of the motion. The "+" direction for $\overline{\vec{a}}$ and $\vec{v}$ is to the right.

It is customary to express the units for acceleration solely in terms of SI units. One way to obtain SI units for the acceleration in Example 3 is to convert the velocity units from km/h to m/s:

$$\left(260 \frac{\text{km}}{\text{h}}\right)\left(\frac{1000 \text{ m}}{1 \text{ km}}\right)\left(\frac{1 \text{ h}}{3600 \text{ s}}\right) = 72 \frac{\text{m}}{\text{s}}$$

The average acceleration then becomes

$$\overline{\vec{a}} = \frac{72 \text{ m/s} - 0 \text{ m/s}}{29 \text{ s} - 0 \text{ s}} = +2.5 \text{ m/s}^2$$

where we have used $2.5 \frac{m/s}{s} = 2.5 \frac{m}{s \cdot s} = 2.5 \frac{m}{s^2}$. An acceleration of $2.5 \frac{m}{s^2}$ is read as "2.5 meters per second per second" (or "2.5 meters per second squared") and means that the velocity changes by 2.5 m/s during each second of the motion.

Example 4 deals with a case where the motion becomes slower as time passes.

EXAMPLE 4 | Acceleration and Decreasing Velocity

A drag racer crosses the finish line, and the driver deploys a parachute and applies the brakes to slow down, as **Figure 2.6** illustrates. The driver begins slowing down when $t_0 = 9.0$ s and the car's velocity is $\vec{v}_0 = +28$ m/s. When $t = 12.0$ s, the velocity has been reduced to $\vec{v} = +13$ m/s. What is the average acceleration of the dragster?

Reasoning The average acceleration of an object is always specified as its change in velocity, $\vec{v} - \vec{v}_0$, divided by the elapsed time, $t - t_0$. This is true whether the final velocity is less than the initial velocity or greater than the initial velocity.

Solution The average acceleration is, according to Equation 2.4,

$$\bar{\vec{a}} = \frac{\vec{v} - \vec{v}_0}{t - t_0} = \frac{13 \text{ m/s} - 28 \text{ m/s}}{12.0 \text{ s} - 9.0 \text{ s}} = \boxed{-5.0 \text{ m/s}^2}$$

$t_0 = 9.0$ s $t = 12$ s

$\bar{\vec{a}} = -5.0$ m/s^2

$\vec{v}_0 = +28$ m/s $\vec{v} = +13$ m/s

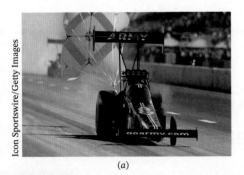

(a)   (b)

FIGURE 2.6 (a) To slow down, a drag racer deploys a parachute and applies the brakes. (b) The velocity of the car is decreasing, giving rise to an average acceleration $\bar{\vec{a}}$ that points opposite to the velocity.

Figure 2.7 shows how the velocity of the dragster in Example 4 changes during the braking, assuming that the acceleration is constant throughout the motion. The acceleration calculated in Example 4 is negative, indicating that the acceleration points to the left in the drawing. As a result, the acceleration and the velocity point in *opposite* directions.

| **Problem-Solving Insight**   Whenever the acceleration and velocity vectors have opposite directions, the object slows down and is said to be "decelerating."

In contrast, the acceleration and velocity vectors in **Figure 2.5** point in the *same* direction, and the object speeds up.

$\bar{\vec{a}} = -5.0$ m/s^2

$\Delta t = 0$ s $\vec{v}_0 = +28$ m/s

$\Delta t = 1.0$ s $\vec{v} = +23$ m/s

$\Delta t = 2.0$ s $\vec{v} = +18$ m/s

FIGURE 2.7 Here, an acceleration of -5.0 m/s^2 means the velocity decreases by 5.0 m/s during each second of elapsed time.

Check Your Understanding

(The answers are given at the end of the book.)

7. At one instant of time, a car and a truck are traveling side by side in adjacent lanes of a highway. The car has a greater velocity than the truck has. Does the car necessarily have the greater acceleration?

8. Two cars are moving in the same direction (the positive direction) on a straight road. The acceleration of each car also points in the positive direction. Car 1 has a greater acceleration than car 2 has. Which one of the following statements is true? **(a)** The velocity of car 1 is always greater than the velocity of car 2. **(b)** The velocity of car 2 is always greater than the velocity of car 1. **(c)** In the same time interval, the velocity of car 1 changes by a greater amount than the velocity of car 2 does. **(d)** In the same time interval, the velocity of car 2 changes by a greater amount than the velocity of car 1 does.

9. An object moving with a constant acceleration slows down if the acceleration points in the direction opposite to the direction of the velocity. But can an object ever come to a permanent halt if its acceleration truly remains constant?

10. A runner runs half the remaining distance to the finish line every ten seconds. She runs in a straight line and does not ever reverse her direction. Does her acceleration have a constant magnitude?

2.4 Equations of Kinematics for Constant Acceleration

It is now possible to describe the motion of an object traveling with a constant acceleration along a straight line. To do so, we will use a set of equations known as the equations of kinematics for constant acceleration. These equations entail no new concepts, because they will be obtained by combining the familiar ideas of displacement, velocity, and acceleration. However, they will provide a very convenient way to determine certain aspects of the motion, such as the final position and velocity of a moving object.

In discussing the equations of kinematics, it will be convenient to assume that the object is located at the origin $\vec{x}_0 = 0$ m when $t_0 = 0$ s. With this assumption, the displacement $\Delta \vec{x} = \vec{x} - \vec{x}_0$ becomes $\Delta \vec{x} = \vec{x}$. Furthermore, in the equations that follow, as is customary, we dispense with the use of boldface symbols overdrawn with small arrows for the displacement, velocity, and acceleration vectors. We will, however, continue to convey the directions of these vectors with plus or minus signs.

Consider an object that has an initial velocity of v_0 at time $t_0 = 0$ s and moves for a time t with a constant acceleration a. For a complete description of the motion, it is also necessary to know the final velocity and displacement at time t. The final velocity v can be obtained directly from Equation 2.4:

$$\bar{a} = a = \frac{v - v_0}{t} \quad \text{or} \quad v = v_0 + at \qquad \text{(constant acceleration)} \qquad (2.4)$$

The displacement x at time t can be obtained from Equation 2.2, if a value for the average velocity $\bar{v}$ can be obtained. Considering the assumption that $x_0 = 0$ m at $t_0 = 0$ s, we have

$$\bar{v} = \frac{x - x_0}{t - t_0} = \frac{x}{t} \quad \text{or} \quad x = \bar{v}t \qquad (2.2)$$

Because the acceleration is constant, the velocity increases at a constant rate. Thus, the average velocity $\bar{v}$ is midway between the initial and final velocities:

$$\bar{v} = \tfrac{1}{2}(v_0 + v) \qquad \text{(constant acceleration)} \qquad (2.6)$$

Equation 2.6, like Equation 2.4, applies only if the acceleration is constant and cannot be used when the acceleration is changing. The displacement at time t can now be determined as

$$x = \bar{v}t = \tfrac{1}{2}(v_0 + v)t \qquad \text{(constant acceleration)} \qquad \textbf{(2.7)}$$

Notice in Equations 2.4 ($v = v_0 + at$) and 2.7 [$x = \tfrac{1}{2}(v_0 + v)t$] that there are five kinematic variables:

1. $x =$ displacement
2. $a = \bar{a} =$ acceleration (constant)
3. $v =$ final velocity at time t
4. $v_0 =$ initial velocity at time $t_0 = 0$ s
5. $t =$ time elapsed since $t_0 = 0$ s

Each of the two equations contains four of these variables, so if three of them are known, the fourth variable can always be found. Example 5 illustrates how Equations 2.4 and 2.7 are used to describe the motion of an object.

Analyzing Multiple-Concept Problems

EXAMPLE 5 | The Displacement of a Speedboat

The speedboat in **Figure 2.8** has a constant acceleration of $+2.0 \text{ m/s}^2$. If the initial velocity of the boat is $+6.0 \text{ m/s}$, find the boat's displacement after 8.0 seconds.

Reasoning As the speedboat accelerates, its velocity is changing. The displacement of the speedboat during a given time interval is equal to the product of its average velocity during that interval and the time. The average velocity, on the other hand, is just one-half the sum of the boat's initial and final velocities. To obtain the final velocity, we will use the definition of constant acceleration, as given in Equation 2.4.

Knowns and Unknowns The numerical values for the three known variables are listed in the table:

Description	Symbol	Value	Comment
Acceleration	a	$+2.0 \text{ m/s}^2$	Positive, because the boat is accelerating to the right, which is the positive direction. See Figure 2.8b.
Initial velocity	v_0	$+6.0 \text{ m/s}$	Positive, because the boat is moving to the right, which is the positive direction. See Figure 2.8b.
Time interval	t	8.0 s	
Unknown Variable			
Displacement of boat	x	?	

(a)　　(b)

FIGURE 2.8 (a) An accelerating speedboat. (b) The boat's displacement x can be determined if the boat's acceleration, initial velocity, and time of travel are known.

Modeling the Problem

STEP 1 Displacement Since the acceleration is constant, the displacement x of the boat is given by Equation 2.7, in which v_0 and v are the initial and final velocities, respectively. In this equation, two of the variables, v_0 and t, are known (see the Knowns and Unknowns table), but the final velocity v is not. However, the final velocity can be determined by employing the definition of acceleration, as Step 2 shows.	$x = \frac{1}{2}(v_0 + v)t$ **(2.7)** ↑ ?
STEP 2 Acceleration According to Equation 2.4, which is just the definition of constant acceleration, the final velocity v of the boat is $\boxed{v = v_0 + at}$ **(2.4)** All the variables on the right side of the equals sign are known, and we can substitute this relation for v into Equation 2.7, as shown at the right.	$x = \frac{1}{2}(v_0 + v)t$ **(2.7)** ↑ $\boxed{v = v_0 + at}$ **(2.4)**

Solution   Algebraically combining the results of Steps 1 and 2, we find that

$$\overset{\text{STEP 1}}{\downarrow} \quad \overset{\text{STEP 2}}{\downarrow}$$

$$x = \frac{1}{2}(v_0 + v)t = \frac{1}{2}[v_0 + (v_0 + at)]t = v_0 t + \frac{1}{2}at^2$$

The displacement of the boat after 8.0 s is

$$x = v_0 t + \frac{1}{2}at^2 = (+6.0 \text{ m/s})(8.0 \text{ s}) + \frac{1}{2}(+2.0 \text{ m/s}^2)(8.0 \text{ s})^2 = \boxed{+112 \text{ m}}$$

Related Homework: Problems 36, 50

In Example 5, we combined two equations $[x = \frac{1}{2}(v_0 + v)t$ and $v = v_0 + at]$ into a single equation by algebraically eliminating the final velocity v of the speedboat (which was not known). The result was the following expression for the displacement x of the speedboat:

$$x = v_0 t + \frac{1}{2}at^2 \quad \text{(constant acceleration)} \tag{2.8}$$

The first term ($v_0 t$) on the right side of this equation represents the displacement that would result if the acceleration were zero and the velocity remained constant at its initial value of v_0. The second term ($\frac{1}{2}at^2$) gives the additional displacement that arises because the velocity changes (a is not zero) to values that are different from its initial value.

Equation 2.8 allows us to calculate the displacement of an object undergoing constant acceleration, given its acceleration (a), the initial velocity (v_0), and the time (t). However, even if the time interval is not given, we can still calculate the displacement, provided we also know the final velocity (v) of the object. We can solve Equation 2.4 for the time t, and then make this substitution into Equation 2.8. The result is $x = (v^2 - v_0^2)/2a$. Solving for v^2 gives

$$v^2 = v_0^2 + 2ax \quad \text{(constant acceleration)} \tag{2.9}$$

Equation 2.9 is often used when the time involved in the motion is unknown.

Table 2.1 presents a summary of the equations that we have been considering. These equations are called the **equations of kinematics**. Each equation contains four of the five kinematic variables, as indicated by the check marks (✓) in the table. Example 6 illustrates how the kinematic equations can be used to analyze the motion of the

TABLE 2.1   Equations of Kinematics for Constant Acceleration

Equation Number	Equation	\multicolumn{5}{c}{Variables}
		x	a	v	v_0	t
(2.4)	$v = v_0 + at$	—	✓	✓	✓	✓
(2.7)	$x = \frac{1}{2}(v + v_0)t$	✓	—	✓	✓	✓
(2.8)	$x = v_0 t + \frac{1}{2}at^2$	✓	✓	—	✓	✓
(2.9)	$v^2 = v_0^2 + 2ax$	✓	✓	✓	✓	—

fastest animal in the world, the peregrine falcon. The next section shows how to apply the equations of kinematics.

EXAMPLE 6 | BIO The Peregrine Falcon

The peregrine falcon (see **Figure 2.9**) is the fastest animal on the planet. Although it flies at speeds less than 60.0 mph during normal flight, its wing shape and aerodynamic features allow it to attain typical speeds of up to 225 mph during dives. The highest velocity recorded for a stooping (diving) peregrine falcon is 242 mph! Other birds cannot even breathe at much lesser speeds. Suppose a peregrine falcon reaches 230.0 mph while performing a vertical dive to intercept a starling that is in straight and level flight. The starling sees the falcon when it is 22.0 m above, and tries to make a move to avoid it. **(a)** How much time (in seconds) does the starling have to avoid a direct hit? **(b)** If the falcon is capable of decelerating with an average magnitude of 15.0g, where $g = 9.80$ m/s^2 is the acceleration due to gravity, what is its stopping distance in meters?

Reasoning   (a) The speed of the falcon and its distance from the starling are given. We can therefore apply a kinematic equation that relates the time, speed, and distance, that is, Equation 2.7. **(b)** We will define the $+y$ direction to be the same as that of the initial velocity, which is downward. Therefore the sign of the acceleration will be

negative since the falcon is decelerating. Assuming a constant deceleration of 15.0g, given the initial speed (230.0 mph) and the final speed (0 mph), we can use Equation 2.9 to find the stopping distance.

Solution   (a) Since we want to know the time in seconds, we want to express the speed in m/s: 230.0 mph = 102.8 m/s. Knowing the falcon's initial speed in the y direction and its distance to the starling, we use Equation 2.7 (with y in place of x):

$$y = \tfrac{1}{2}(v_0 + v)t$$

Since the speed of the falcon is constant, $v_0 = v$, and the equation becomes

$$y = \tfrac{1}{2}(v_0 + v_0)t = v_0 t$$

Solving for t, and substituting the known values, we obtain

$$t = \frac{y}{v_0} = \frac{22.0 \text{ m}}{102.8 \text{ m/s}} = \boxed{0.214 \text{ s}}$$

(b) We can use Equation 2.9 to find the stopping distance, since the acceleration is given:

$$v^2 = v_0^2 + 2ay$$

Solving for y, and setting the final velocity $v = 0$ m/s and $a = -15.0g$, we get

$$y = -\frac{v_0^2}{a} = -\frac{v_0^2}{(-15.0g)} = \frac{(102.8 \text{ m/s})^2}{(15.0)(9.80 \text{ m/s}^2)} = \boxed{71.9 \text{ m}}$$

The acceleration a is negative since it points in a direction opposite to that of the velocity vector. By this calculation, if the falcon cannot maneuver to intercept the starling as it tries to escape, the falcon will overshoot its prey's position by nearly 50 meters. It turns out, however, that peregrine falcons can accelerate *laterally* with magnitudes that exceed 15.0g during their dives, and have the innate ability to tweak their intercept angles to outmaneuver their prey.

Jerry Ting/Flickr

FIGURE 2.9   The peregrine falcon is the fastest animal on the planet.

Check Your Understanding

(*The answers are given at the end of the book.*)

11. The muzzle velocity of a gun is the velocity of the bullet when it leaves the barrel. The muzzle velocity of one rifle with a short barrel is greater than the muzzle velocity of another rifle that has a longer barrel. In which rifle is the acceleration of the bullet larger?

12. A motorcycle starts from rest and has a constant acceleration. In a time interval t, it undergoes a displacement x and attains a final velocity v. Then t is increased so that the displacement is $3x$. In this same increased time interval, what final velocity does the motorcycle attain?

2.5 Applications of the Equations of Kinematics

The equations of kinematics can be applied to any moving object, as long as the acceleration of the object is constant. However, remember that each equation contains four variables. Therefore, numerical values for three of the four must be available if an equation is to be used to calculate the value of the remaining variable. To avoid errors when using these equations, it helps to follow a few sensible guidelines and to be alert for a few situations that can arise during your calculations.

> **Problem-Solving Insight** Decide at the start which directions are to be called positive (+) and negative (−) relative to a conveniently chosen coordinate origin.

This decision is arbitrary, but important because displacement, velocity, and acceleration are vectors, and their directions must always be taken into account. In the examples that follow, the positive and negative directions will be shown in the drawings that accompany the problems. It does not matter which direction is chosen to be positive. However, once the choice is made, it should not be changed during the course of the calculation.

> **Problem-Solving Insight** As you reason through a problem before attempting to solve it, be sure to interpret the terms "decelerating" or "deceleration" correctly, should they occur in the problem statement.

These terms are the source of frequent confusion, and Conceptual Example 7 offers help in understanding them.

CONCEPTUAL EXAMPLE 7 | Deceleration Versus Negative Acceleration

A car is traveling along a straight road and is decelerating. Which one of the following statements correctly describes the car's acceleration? **(a)** It must be positive. **(b)** It must be negative. **(c)** It could be positive or negative.

Reasoning The term "decelerating" means that the acceleration vector points opposite to the velocity vector and indicates that the car is slowing down. One possibility is that the velocity vector of the car points to the right, in the positive direction, as **Figure 2.10a** shows. The term "decelerating" implies, then, that the acceleration vector of the car points to the left, which is the negative direction. Another possibility is that the car is traveling to the left, as in **Figure 2.10b**. Now, since the velocity vector points to the left, the acceleration vector would point opposite, or to the right, which is the positive direction.

Answers (a) and (b) are incorrect. The term "decelerating" means only that the acceleration vector points opposite to the velocity vector. It is not specified whether the velocity vector of the car points in the positive or negative direction. Therefore, it is not possible to know whether the acceleration is positive or negative.

Answer (c) is correct. As shown in **Figure 2.10**, the acceleration vector of the car could point in the positive or the negative

(a)

(b)

FIGURE 2.10 When a car decelerates along a straight road, the acceleration vector points opposite to the velocity vector, as Conceptual Example 7 discusses.

direction, so that the acceleration could be either positive or negative, depending on the direction in which the car is moving.

Related Homework: Problem 13

> **Problem-Solving Insight** Sometimes there are two possible answers to a kinematics problem, each answer corresponding to a different situation.

Example 8 discusses one such case.

EXAMPLE 8 | The Physics of Spacecraft Retrorockets

The spacecraft shown in **Figure 2.11a** is traveling with a velocity of +3250 m/s. Suddenly the retrorockets are fired, and the spacecraft begins to slow down with an acceleration whose magnitude is 10.0 m/s². What is the velocity of the spacecraft when the displacement of the craft is +215 km, relative to the point where the retrorockets began firing?

Spacecraft Data				
x	a	v	v_0	t
+215 000 m	−10.0 m/s²	?	+3250 m/s	

Reasoning Since the spacecraft is slowing down, the acceleration must be opposite to the velocity. The velocity points to the right in the drawing, so the acceleration points to the left, in the negative direction; thus, $a = -10.0$ m/s². The three known variables are listed as follows:

The final velocity v of the spacecraft can be calculated using Equation 2.9, since it contains the four pertinent variables.

Solution From Equation 2.9 ($v^2 = v_0^2 + 2ax$), we find that

$$v = \pm\sqrt{v_0^2 + 2ax}$$

$$= \pm\sqrt{(3250 \text{ m/s})^2 + 2(-10.0 \text{ m/s}^2)(215\,000 \text{ m})}$$

$$= \boxed{+2500 \text{ m/s}} \text{ and } \boxed{-2500 \text{ m/s}}$$

Both of these answers correspond to the *same* displacement ($x = +215$ km), but each arises in a different part of the motion. The answer $v = +2500$ m/s corresponds to the situation in **Figure 2.11a**, where the spacecraft has slowed to a speed of 2500 m/s, but is still traveling to the right. The answer $v = -2500$ m/s arises because the retrorockets eventually bring the spacecraft to a momentary halt and cause it to reverse its direction. Then it moves to the left, and its speed increases due to the continually firing rockets. After a time, the velocity of the craft becomes $v = -2500$ m/s, giving rise to the situation in **Figure 2.11b**. In both parts of the drawing the spacecraft has the same displacement, but a greater travel time is required in part b compared to part a.

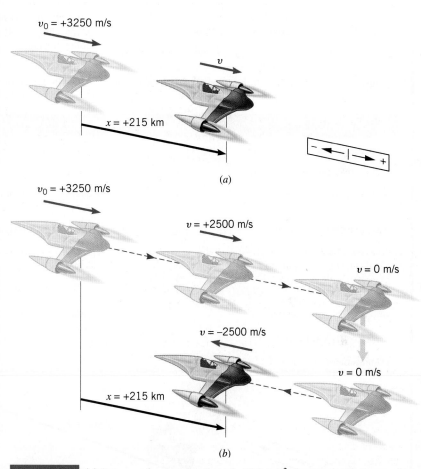

(a)

(b)

FIGURE 2.11 (a) Because of an acceleration of −10.0 m/s², the spacecraft changes its velocity from v_0 to v. (b) Continued firing of the retrorockets changes the direction of the craft's motion.

Problem-Solving Insight The motion of two objects may be interrelated, so that they share a common variable. The fact that the motions are interrelated is an important piece of information. In such cases, data for only two variables need be specified for each object.

Problem-Solving Insight Often the motion of an object is divided into segments, each with a different acceleration. When solving such problems, it is important to realize that the final velocity for one segment is the initial velocity for the next segment, as Example 9 illustrates.

Analyzing Multiple-Concept Problems

EXAMPLE 9 | A Motorcycle Ride

A motorcycle ride consists of two segments, as shown in **Interactive Figure 2.12a**. During segment 1, the motorcycle starts from rest, has an acceleration of $+2.6$ m/s^2, and has a displacement of $+120$ m. Immediately after segment 1, the motorcycle enters segment 2 and begins slowing down with an acceleration of -1.5 m/s^2 until its velocity is $+12$ m/s. What is the displacement of the motorcycle during segment 2?

Reasoning We can use an equation of kinematics from **Table 2.1** to find the displacement x_2 for segment 2. To do this, it will be necessary to have values for three of the

variables that appear in the equation. Values for the acceleration ($a_2 = -1.5$ m/s^2) and final velocity ($v_2 = +12$ m/s) are given. A value for a third variable, the initial velocity v_{02}, can be obtained by noting that it is also the final velocity of segment 1. The final velocity of segment 1 can be found by using an appropriate equation of kinematics, since three variables (x_1, a_1, and v_{01}) are known for this part of the motion, as the following table reveals.

Knowns and Unknowns The data for this problem are listed in the table:

Description	Symbol	Value	Comment
Explicit Data			
Displacement for segment 1	x_1	$+120$ m	
Acceleration for segment 1	a_1	$+2.6$ m/s^2	Positive, because the motorcycle moves in the $+x$ direction and speeds up.
Acceleration for segment 2	a_2	-1.5 m/s^2	Negative, because the motorcycle moves in the $+x$ direction and slows down.
Final velocity for segment 2	v_2	$+12$ m/s	
Implicit Data			
Initial velocity for segment 1	v_{01}	0 m/s	The motorcycle starts from rest.
Unknown Variable			
Displacement for segment 2	x_2	?	

(a)

(b) Segment 2

(c) Segment 1

INTERACTIVE FIGURE 2.12 (a) This motorcycle ride consists of two segments, each with a different acceleration. (b) The variables for segment 2. (c) The variables for segment 1.

Modeling the Problem

STEP 1 Displacement During Segment 2   Interactive Figure 2.12*b* shows the situation during segment 2. Two of the variables—the final velocity v_2 and the acceleration a_2—are known, and for convenience we choose Equation 2.9 to find the displacement x_2 of the motorcycle:

$$v_2^2 = v_{02}^2 + 2a_2x_2 \qquad (2.9)$$

Solving this relation for x_2 yields Equation 1 at the right. Although the initial velocity v_{02} of segment 2 is not known, we will be able to determine its value from a knowledge of the motion during segment 1, as outlined in Steps 2 and 3.

$$x_2 = \frac{v_2^2 - v_{02}^2}{2a_2} \qquad (1)$$

$\boxed{?}$

STEP 2 Initial Velocity of Segment 2   Since the motorcycle enters segment 2 immediately after leaving segment 1, the initial velocity v_{02} of segment 2 is equal to the final velocity v_1 of segment 1, or $v_{02} = v_1$. Squaring both sides of this equation gives

$$\boxed{v_{02}^2 = v_1^2}$$

This result can be substituted into Equation 1 as shown at the right. In the next step v_1 will be determined.

$$x_2 = \frac{v_2^2 - v_{02}^2}{2a_2} \qquad (1)$$

$$\boxed{v_{02}^2 = v_1^2} \qquad (2)$$

$\boxed{?}$

STEP 3 Final Velocity of Segment 1   Interactive Figure 2.12*c* shows the motorcycle during segment 1. Since we know the initial velocity v_{01}, the acceleration a_1, and the displacement x_1, we can employ Equation 2.9 to find the final velocity v_1 at the end of segment 1:

$$\boxed{v_1^2 = v_{01}^2 + 2a_1x_1}$$

This relation for v_1^2 can be substituted into Equation 2, as shown at the right.

$$x_2 = \frac{v_2^2 - v_{02}^2}{2a_2} \qquad (1)$$

$$\boxed{v_{02}^2 = v_1^2} \qquad (2)$$

$$\boxed{v_1^2 = v_{01}^2 + 2a_1x_1} \qquad (3)$$

Solution   Algebraically combining the results of each step, we find that

$$\underset{\text{STEP 1}}{\downarrow} \quad \underset{\text{STEP 2}}{\downarrow} \quad \underset{\text{STEP 3}}{\downarrow}$$
$$x_2 = \frac{v_2^2 - v_{02}^2}{2a_2} = \frac{v_2^2 - v_1^2}{2a_2} = \frac{v_2^2 - (v_{01}^2 + 2a_1x_1)}{2a_2}$$

The displacement x_2 of the motorcycle during segment 2 is

$$x_2 = \frac{v_2^2 - (v_{01}^2 + 2a_1x_1)}{2a_2}$$

$$= \frac{(+12 \text{ m/s})^2 - [(0 \text{ m/s})^2 + 2(+2.6 \text{ m/s}^2)(+120 \text{ m})]}{2(-1.5 \text{ m/s}^2)} = \boxed{+160 \text{ m}}$$

Math Skills   Interactive Figure 2.12*a* shows which direction has been chosen as the positive direction, and the choice leads to a value of $x_2 = +160$ m for the displacement of the motorcycle during segment 2. This answer means that displacement is 160 m in the positive direction, which is upward and to the right in the drawing. The choice for the positive direction is completely arbitrary, however. The meaning of the answer to the problem will be the same no matter what choice is made. Suppose that the choice for the positive direction were opposite to that shown in Interactive Figure 2.12*a*. Then, all of the data listed in the Knowns and Unknowns table would appear with algebraic signs opposite to those specified, and the calculation of the displacement x_2 would appear as follows:

$$x_2 = \frac{(-12 \text{ m/s})^2 - [(0 \text{ m/s})^2 + 2(-2.6 \text{ m/s}^2)(-120 \text{ m})]}{2(+1.5 \text{ m/s}^2)} = \boxed{-160 \text{ m}}$$

The value for x_2 has the same magnitude of 160 m as before, but it is now negative. The negative sign means that the displacement of the motorcycle during segment 2 is 160 m *in the negative direction*. But the negative direction is now upward and to the right in Interactive Figure 2.12*a*, so the *meaning* of the result is exactly the same as it was before.

Now that we have seen how the equations of kinematics are applied to various situations, it's a good idea to summarize the reasoning strategy that has been used. This strategy, which is outlined below, will also be used when we consider freely falling bodies in Section 2.6 and two-dimensional motion in Chapter 3.

REASONING STRATEGY   Applying the Equations of Kinematics

1. **Make a drawing to represent the situation being studied. A drawing helps us to see what's happening.**

2. **Decide which directions are to be called positive (+) and negative (−) relative to a conveniently chosen coordinate origin. Do not change your decision during the course of a calculation.**

3. **In an organized way, write down the values (with appropriate plus and minus signs) that are given for any of the five kinematic variables (x, a, v, v_0, and t). Be on the alert for implicit data, such as the phrase "starts from rest," which means that the value of the initial velocity is $v_0 = 0$ m/s. The data summary tables used in the examples in the text are a good way to keep track of this information. In addition, identify the variables that you are being asked to determine.**

4. **Before attempting to solve a problem, verify that the given information contains values for at least three of the five kinematic variables. Once the three known variables are identified along with the desired unknown variable, the appropriate relation from Table 2.1 can be selected. Remember that the motions of two objects may be interrelated, so they may share a common variable. The fact that the motions are interrelated is an important piece of information. In such cases, data for only two variables need be specified for each object.**

5. **When the motion of an object is divided into segments, as in Example 9, remember that the final velocity of one segment is the initial velocity for the next segment.**

6. **Keep in mind that there may be two possible answers to a kinematics problem as, for instance, in Example 8. Try to visualize the different physical situations to which the answers correspond.**

2.6 | Freely Falling Bodies

Everyone has observed the effect of gravity as it causes objects to fall downward. In the absence of air resistance, it is found that all bodies at the same location above the earth fall vertically with the same acceleration. Furthermore, if the distance of the fall is small compared to the radius of the earth, the acceleration remains essentially constant throughout the descent. This idealized motion, in which air resistance is neglected and the acceleration is nearly constant, is known as **free-fall**. Since the acceleration is constant in free-fall, the equations of kinematics can be used.

The acceleration of a freely falling body is called the **acceleration due to gravity**, and its magnitude (without any algebraic sign) is denoted by the symbol g. The acceleration due to gravity is directed downward, toward the center of the earth. Near the earth's surface, g is approximately

$$g = 9.80 \text{ m/s}^2 \quad \text{or} \quad 32.2 \text{ ft/s}^2$$

Unless circumstances warrant otherwise, we will use either of these values for g in subsequent calculations. In reality, however, g decreases with increasing altitude and varies slightly with latitude.

Figure 2.13a shows the well-known phenomenon of a rock falling faster than a sheet of paper. The effect of air resistance is responsible for the slower fall of the paper, for when air is removed from the tube, as in **Figure 2.13b**, the rock and the paper have exactly the same acceleration due to gravity. In the absence of air, the rock and the paper both exhibit free-fall motion. Free-fall is closely approximated for objects falling near the surface of the moon, where there is no air to retard the motion. A nice demonstration of free-fall was performed on the moon by astronaut David Scott, who dropped a hammer and a feather simultaneously from the same height. Both experienced the same

Air-filled
tube
(a)

Evacuated
tube
(b)

FIGURE 2.13 (a) In the presence of air resistance, the acceleration of the rock is greater than that of the paper. (b) In the absence of air resistance, both the rock and the paper have the same acceleration.

acceleration due to lunar gravity and consequently hit the ground at the same time. The acceleration due to gravity near the surface of the moon is approximately one-sixth as large as that on the earth.

When the equations of kinematics are applied to free-fall motion, it is natural to use the symbol y for the displacement, since the motion occurs in the vertical or y direction. Thus, when using the equations in **Table 2.1** for free-fall motion, we will simply replace x with y. There is no significance to this change. The equations have the same algebraic form for either the horizontal or vertical direction, provided that the acceleration remains constant during the motion. We now turn our attention to several examples that illustrate how the equations of kinematics are applied to freely falling bodies.

EXAMPLE 10 | A Falling Stone

A stone is dropped from rest from the top of a tall building, as **Animated Figure 2.14** indicates. After 3.00 s of free-fall, what is the displacement y of the stone?

Reasoning The upward direction is chosen as the positive direction. The three known variables are shown in the table below. The initial velocity v_0 of the stone is zero, because the stone is dropped from rest. The acceleration due to gravity is negative, since it points downward in the negative direction.

Stone Data				
y	a	v	v_0	t
?	-9.80 m/s^2		0 m/s	3.00 s

Equation 2.8 contains the appropriate variables and offers a direct solution to the problem. Since the stone moves downward, and upward is the positive direction, we expect the displacement y to have a negative value.

> **Problem-Solving Insight** It is only when values are available for at least three of the five kinematic variables (y, a, v, v_0, and t) that the equations in Table 2.1 can be used to determine the fourth and fifth variables.

Solution Using Equation 2.8, we find that

$$y = v_0 t + \tfrac{1}{2} a t^2 = (0 \text{ m/s})(3.00 \text{ s}) + \tfrac{1}{2}(-9.80 \text{ m/s}^2)(3.00 \text{ s})^2$$

$$= \boxed{-44.1 \text{ m}}$$

The answer for y is negative, as expected.

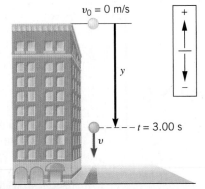

ANIMATED FIGURE 2.14 The stone, starting with zero velocity at the top of the building, is accelerated downward by gravity.

EXAMPLE 11 | The Velocity of a Falling Stone

After 3.00 s of free-fall, what is the velocity v of the stone in **Animated Figure 2.14**?

Reasoning Because of the acceleration due to gravity, the magnitude of the stone's downward velocity increases by 9.80 m/s during each second of free-fall. The data for the stone are the same as in Example 10, and Equation 2.4 offers a direct solution for the final velocity. Since the stone is moving downward in the negative direction, the value determined for v should be negative.

Solution Using Equation 2.4, we obtain

$$v = v_0 + at = 0 \text{ m/s} + (-9.80 \text{ m/s}^2)(3.00 \text{ s}) = \boxed{-29.4 \text{ m/s}}$$

The velocity is negative, as expected.

The acceleration due to gravity is always a downward-pointing vector. It describes how the speed increases for an object that is falling freely downward. This same acceleration also describes how the speed decreases for an object moving upward under the influence of gravity alone, in which case the object eventually comes to a momentary halt and then falls back to earth. Examples 12 and 13 show how the equations of kinematics are applied to an object that is moving upward under the influence of gravity.

EXAMPLE 12 | How High Does It Go?

A football game customarily begins with a coin toss to determine who kicks off. The referee tosses the coin up with an initial speed of 5.00 m/s. In the absence of air resistance, how high does the coin go above its point of release?

Reasoning The coin is given an upward initial velocity, as in **Figure 2.15**. But the acceleration due to gravity points downward. Since the velocity and acceleration point in opposite directions, the coin slows down as it moves upward. Eventually, the velocity of the coin becomes $v = 0$ m/s at the highest point. Assuming that the upward direction is positive, the data can be summarized as shown below:

Coin Data				
y	a	v	v_0	t
?	-9.80 m/s^2	0 m/s	$+5.00$ m/s	

With these data, we can use Equation 2.9 ($v^2 = v_0^2 + 2ay$) to find the maximum height y.

$v = 0$ m/s

y

$v_0 = +5.00$ m/s

FIGURE 2.15 At the start of a football game, a referee tosses a coin upward with an initial velocity of $v_0 = +5.00$ m/s. The velocity of the coin is momentarily zero when the coin reaches its maximum height.

Problem-Solving Insight Implicit data are important. In Example 12, for instance, the phrase "how high does the coin go" refers to the maximum height, which occurs when the final velocity v in the vertical direction is $v = 0$ m/s.

Solution Rearranging Equation 2.9, we find that the maximum height of the coin above its release point is

$$y = \frac{v^2 - v_0^2}{2a} = \frac{(0 \text{ m/s})^2 - (5.00 \text{ m/s})^2}{2(-9.80 \text{ m/s}^2)} = \boxed{1.28 \text{ m}}$$

Math Skills The rearrangement of algebraic equations is a problem-solving step that occurs often. The guiding rule in such a procedure is that whatever you do to one side of an equation, you must also do to the other side. Here in Example 12, for instance, we need to rearrange $v^2 = v_0^2 + 2ay$ (Equation 2.9) in order to determine y. The part of the equation that contains y is $2ay$, and we begin by isolating that part on one side of the equals sign. To do this we subtract v_0^2 from each side of the equation:

$$v^2 - v_0^2 = \cancel{v_0^2} + 2ay - \cancel{v_0^2} = 2ay$$

Thus, we see that $2ay = v^2 - v_0^2$. Then, we eliminate the term $2a$ from the left side of the equals sign by dividing both sides of the equation by $2a$.

$$\frac{2\cancel{a}y}{2\cancel{a}} = \frac{v^2 - v_0^2}{2a} \quad \text{or} \quad y = \frac{v^2 - v_0^2}{2a}$$

The term $2a$ occurs both in the numerator and the denominator on the left side of this result and can be eliminated algebraically, leaving the desired expression for y.

EXAMPLE 13 | How Long Is It in the Air?

In **Figure 2.15**, what is the total time the coin is in the air before returning to its release point?

Reasoning During the time the coin travels upward, gravity causes its speed to decrease to zero. On the way down, however, gravity causes the coin to regain the lost speed. Thus, the time for the coin to go up is equal to the time for it to come down. In other words, the total travel time is twice the time for the upward motion. The data for the coin during the upward trip are the same as in Example 12. With these data, we can use Equation 2.4 ($v = v_0 + at$) to find the upward travel time.

Solution Rearranging Equation 2.4, we find that

$$t = \frac{v - v_0}{a} = \frac{0 \text{ m/s} - 5.00 \text{ m/s}}{-9.80 \text{ m/s}^2} = 0.510 \text{ s}$$

The total up-and-down time is twice this value, or $\boxed{1.02 \text{ s}}$.

It is possible to determine the total time by another method. When the coin is tossed upward and returns to its release point, the displacement for the *entire trip* is $y = 0$ m. With this value for the displacement, Equation 2.8 ($y = v_0 t + \frac{1}{2}at^2$) can be used to find the time for the entire trip directly.

Examples 12 and 13 illustrate that the expression "freely falling" does not necessarily mean an object is falling down. A freely falling object is any object moving either upward or downward under the influence of gravity alone. In either case, the object always experiences the same *downward acceleration* due to gravity, a fact that is the focus of the next example.

CONCEPTUAL EXAMPLE 14 | Acceleration Versus Velocity

There are three parts to the motion of the coin in **Figure 2.15**, in which air resistance is being ignored. On the way up, the coin has an upward-pointing velocity vector with a decreasing magnitude. At the top of its path, the velocity vector of the coin is momentarily zero. On the way down, the coin has a downward-pointing velocity vector with an increasing magnitude. Compare the acceleration vector of the coin with the velocity vector. **(a)** Do the direction and magnitude of the acceleration vector behave in the same fashion as the direction and magnitude of the velocity vector or **(b)** does the acceleration vector have a constant direction and a constant magnitude throughout the motion?

Reasoning   Since air resistance is being ignored, the coin is in free-fall motion. This means that the acceleration vector of the coin is the acceleration due to gravity. Acceleration is the rate at

which velocity *changes* and is not the same concept as velocity itself.

Answer (a) is incorrect.   During the upward and downward parts of the motion, and also at the top of the path, the acceleration due to gravity has a constant downward direction and a constant magnitude of 9.80 m/s². In other words, the acceleration vector of the coin does not behave as the velocity vector does. In particular, the acceleration vector is not zero at the top of the motional path just because the velocity vector is zero there. Acceleration is the rate at which the velocity is changing, and the velocity is changing at the top even though at one instant it is zero.

Answer (b) is correct.   The acceleration due to gravity has a constant downward direction and a constant magnitude of 9.80 m/s² at all times during the motion.

The motion of an object that is thrown upward and eventually returns to earth has a symmetry that is useful to keep in mind from the point of view of problem solving. The calculations just completed indicate that a time symmetry exists in free-fall motion, in the sense that the time required for the object to reach maximum height equals the time for it to return to its starting point.

A type of symmetry involving the speed also exists. **Figure 2.16** shows the coin considered in Examples 12 and 13. At any displacement *y* above the point of release, the coin's speed during the upward trip equals the speed at the same point during the downward trip. For instance, when $y = +1.04$ m, Equation 2.9 gives two possible values for the final velocity v, assuming that the initial velocity is $v_0 = +5.00$ m/s:

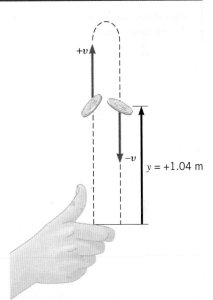

$$v^2 = v_0^2 + 2ay = (5.00 \text{ m/s})^2 + 2(-9.80 \text{ m/s}^2)(1.04 \text{ m}) = 4.62 \text{ m}^2/\text{s}^2$$

$$v = \pm 2.15 \text{ m/s}$$

The value $v = +2.15$ m/s is the velocity of the coin on the upward trip, and $v = -2.15$ m/s is the velocity on the downward trip. The speed in both cases is identical and equals 2.15 m/s. Likewise, the speed just as the coin returns to its point of release is 5.00 m/s, which equals the initial speed. This symmetry involving the speed arises because the coin loses 9.80 m/s in speed each second on the way up and gains back the same amount each second on the way down. In Conceptual Example 15, we use just this kind of symmetry to guide our reasoning as we analyze the motion of a pellet shot from a gun.

FIGURE 2.16   For a given displacement along the motional path, the upward speed of the coin is equal to its downward speed, but the two velocities point in opposite directions.

CONCEPTUAL EXAMPLE 15 | Taking Advantage of Symmetry

Interactive Figure 2.17a shows a pellet that has been fired straight upward from a gun at the edge of a cliff. The initial speed of the pellet is 30 m/s. It goes up and then falls back down, eventually hitting the ground beneath the cliff. In **Interactive Figure 2.17b** the pellet has been fired straight downward at the same initial speed. In the absence of air resistance, would the

INTERACTIVE FIGURE 2.17
(*a*) From the edge of a cliff, a pellet is fired straight upward from a gun. The pellet's initial speed is 30 m/s. (*b*) The pellet is fired straight downward with an initial speed of 30 m/s. (*c*) In Conceptual Example 15 this drawing plays the central role in reasoning that is based on symmetry.

(*a*)   (*b*)   (*c*)

pellet in **Interactive Figure 2.17b** strike the ground with **(a)** a smaller speed than, **(b)** the same speed as, or **(c)** a greater speed than the pellet in **Interactive Figure 2.17a**?

Reasoning In the absence of air resistance, the motion is that of free-fall, and the symmetry inherent in free-fall motion offers an immediate answer.

Answers (a) and (c) are incorrect. These answers are incorrect, because they are inconsistent with the symmetry that is discussed next in connection with the correct answer.

Answer (b) is correct. **Interactive Figure 2.17c** shows the pellet after it has been fired upward and has fallen back down to its starting point. Symmetry indicates that the speed in **Interactive Figure 2.17c** is the same as in **Interactive Figure 2.17a**—namely, 30 m/s, as is also the case when the pellet has been actually fired downward. Consequently, whether the pellet is fired as in **Interactive Figure 2.17a** or *b*, it begins to move downward from the cliff edge at a speed of 30 m/s. In either case, there is the same acceleration due to gravity and the same displacement from the cliff edge to the ground below. Given these conditions, when the pellet reaches the ground, it has the same speed in both **Interactive Figures 2.17a** and *b*.

Related Homework: Problems 42, 49

Check Your Understanding

(The answers are given at the end of the book.)

13. An experimental vehicle slows down and comes to a halt with an acceleration whose magnitude is 9.80 m/s². After reversing direction in a negligible amount of time, the vehicle speeds up with an acceleration of 9.80 m/s². Except for being horizontal, is this motion **(a)** the same as or **(b)** different from the motion of a ball that is thrown straight upward, comes to a halt, and falls back to earth? Ignore air resistance.

14. A ball is thrown straight upward with a velocity $\vec{v}_0$ and in a time t reaches the top of its flight path, which is a displacement $\vec{y}$ above the launch point. With a launch velocity of $2\vec{v}_0$, what would be the time required to reach the top of its flight path and what would be the displacement of the top point above the launch point? **(a)** $4t$ and $2\vec{y}$ **(b)** $2t$ and $4\vec{y}$ **(c)** $2t$ and $2\vec{y}$ **(d)** $4t$ and $4\vec{y}$ **(e)** t and $2\vec{y}$.

15. Two objects are thrown vertically upward, first one, and then, a bit later, the other. Is it **(a)** possible or **(b)** impossible that both objects reach the same maximum height at the same instant of time?

16. A ball is dropped from rest from the top of a building and strikes the ground with a speed v_f. From ground level, a second ball is thrown straight upward at the same instant that the first ball is dropped. The initial speed of the second ball is $v_0 = v_f$, the same speed with which the first ball eventually strikes the ground. Ignoring air resistance, decide whether the balls cross paths **(a)** at half the height of the building, **(b)** above the halfway point, or **(c)** below the halfway point.

2.7 Graphical Analysis of Velocity and Acceleration

Graphical techniques are helpful in understanding the concepts of velocity and acceleration. Suppose a bicyclist is riding with a constant velocity of $v = +4$ m/s. The position x of the bicycle can be plotted along the vertical axis of a graph, while the time t is plotted along the horizontal axis. Since the position of the bike increases by 4 m every second, the graph of x versus t is a straight line. Furthermore, if the bike is assumed to be at $x = 0$ m when $t = 0$ s, the straight line passes through the origin, as **Figure 2.18** shows. Each point on this line gives the position of the bike at a particular time. For instance, at $t = 1$ s the position is 4 m, while at $t = 3$ s the position is 12 m.

In constructing the graph in **Figure 2.18**, we used the fact that the velocity was $+4$ m/s. Suppose, however, that we were given this graph, but did not have prior knowledge of the velocity. The velocity could be determined by considering what happens to the bike between the times of 1 and 3 s, for instance. The change in time is $\Delta t = 2$ s. During this time interval, the position of the bike changes from $+4$ to $+12$ m, and the change in position is $\Delta x = +8$ m. The ratio $\Delta x / \Delta t$ is called the **slope** of the straight line.

$$\text{Slope} = \frac{\Delta x}{\Delta t} = \frac{+8 \text{ m}}{2 \text{ s}} = +4 \text{ m/s}$$

Notice that the slope is equal to the velocity of the bike. This result is no accident, because $\Delta x / \Delta t$ is the definition of average velocity (see Equation 2.2). ***Thus, for an object moving***

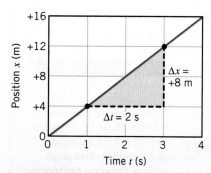

FIGURE 2.18 Position–time graph for an object moving with a constant velocity of $v = \Delta x / \Delta t = +4$ m/s.

with a constant velocity, the slope of the straight line in a position–time graph gives the velocity. Since the position–time graph is a straight line, any time interval Δt can be chosen to calculate the velocity. Choosing a different Δt will yield a different Δx, but the velocity $\Delta x/\Delta t$ will not change. In the real world, objects rarely move with a constant velocity at all times, as the next example illustrates.

EXAMPLE 16 | A Bicycle Trip

A bicyclist maintains a constant velocity on the outgoing leg of a trip, zero velocity while stopped, and another constant velocity on the way back. **Interactive Figure 2.19** shows the corresponding position–time graph. Using the time and position intervals indicated in the drawing, obtain the velocities for each segment of the trip.

Reasoning The average velocity $\bar{v}$ is equal to the displacement Δx divided by the elapsed time Δt, $\bar{v} = \Delta x/\Delta t$. The displacement is the final position minus the initial position, which is a positive number for segment 1 and a negative number for segment 3. Note for segment 2 that $\Delta x = 0$ m, since the bicycle is at rest. The drawing shows values for Δt and Δx for each of the three segments.

Solution The average velocities for the three segments are

Segment 1 $\bar{v} = \dfrac{\Delta x}{\Delta t} = \dfrac{800 \text{ m} - 400 \text{ m}}{400 \text{ s} - 200 \text{ s}} = \dfrac{+400 \text{ m}}{200 \text{ s}} = \boxed{+2 \text{ m/s}}$

Segment 2 $\bar{v} = \dfrac{\Delta x}{\Delta t} = \dfrac{1200 \text{ m} - 1200 \text{ m}}{1000 \text{ s} - 600 \text{ s}} = \dfrac{0 \text{ m}}{400 \text{ s}} = \boxed{0 \text{ m/s}}$

Segment 3 $\bar{v} = \dfrac{\Delta x}{\Delta t} = \dfrac{400 \text{ m} - 800 \text{ m}}{1800 \text{ s} - 1400 \text{ s}} = \dfrac{-400 \text{ m}}{400 \text{ s}} = \boxed{-1 \text{ m/s}}$

In the second segment of the journey the velocity is zero, reflecting the fact that the bike is stationary. Since the position of the bike does not change, segment 2 is a horizontal line that has a zero slope. In the third part of the motion the velocity is negative, because the position of the bike decreases from $x = +800$ m to $x = +400$ m during the 400-s interval shown in the graph. As a result, segment 3 has a negative slope, and the velocity is negative.

INTERACTIVE FIGURE 2.19 This position–time graph consists of three straight-line segments, each corresponding to a different constant velocity.

If the object is accelerating, its velocity is changing. When the velocity is changing, the position–time graph is not a straight line, but is a curve, perhaps like that in **Figure 2.20**. This curve was drawn using Equation 2.8 ($x = v_0 t + \frac{1}{2}at^2$), assuming an acceleration of $a = 0.26$ m/s^2 and an initial velocity of $v_0 = 0$ m/s. The velocity at any instant of time can be determined by measuring the slope of the curve at that instant. The slope at any point along the curve is defined to be the slope of the tangent line drawn to the curve at that point. For instance, in **Figure 2.20** a tangent line is drawn at $t = 20.0$ s. To determine the slope of the tangent line, a triangle is constructed using an arbitrarily chosen time interval of $\Delta t = 5.0$ s. The change in x associated with this time interval can be read from the tangent line as $\Delta x = +26$ m. Therefore,

$$\text{Slope of tangent line} = \frac{\Delta x}{\Delta t} = \frac{+26 \text{ m}}{5.0 \text{ s}} = +5.2 \text{ m/s}$$

The slope of the tangent line is the instantaneous velocity, which in this case is $v = +5.2$ m/s. This graphical result can be verified by using Equation 2.4 with $v_0 = 0$ m/s: $v = at = (+0.26 \text{ m/s}^2)(20.0 \text{ s}) = +5.2$ m/s.

FIGURE 2.20 When the velocity is changing, the position–time graph is a curved line. The slope $\Delta x/\Delta t$ of the tangent line drawn to the curve at a given time is the instantaneous velocity at that time.

FIGURE 2.21 A velocity–time graph that applies to an object with an acceleration of $\Delta v / \Delta t = +6 \ \text{m/s}^2$. The initial velocity is $v_0 = +5 \ \text{m/s}$ when $t = 0 \ \text{s}$.

Insight into the meaning of acceleration can also be gained with the aid of a graphical representation. Consider an object moving with a constant acceleration of $a = +6 \ \text{m/s}^2$. If the object has an initial velocity of $v_0 = +5 \ \text{m/s}$, its velocity at any time is represented by Equation 2.4 as

$$v = v_0 + at = 5 \ \text{m/s} + (6 \ \text{m/s}^2)t$$

This relation is plotted as the velocity–time graph in **Figure 2.21**. The graph of v versus t is a straight line that intercepts the vertical axis at $v_0 = 5 \ \text{m/s}$. The slope of this straight line can be calculated from the data shown in the drawing:

$$\text{Slope} = \frac{\Delta v}{\Delta t} = \frac{+12 \ \text{m/s}}{2 \ \text{s}} = +6 \ \text{m/s}^2$$

The ratio $\Delta v / \Delta t$ is, by definition, equal to the average acceleration (Equation 2.4), so *the slope of the straight line in a velocity–time graph is the average acceleration.*

EXAMPLE 17 | BIO The Average Acceleration of a Sprinter

Figure 2.22 shows the velocity-time graph for a sprinter competing in a 100-m race. At the beginning of the race, usually initiated by the report of a starter's pistol, the sprinter accelerates from rest for a period of 4.00 s. After accelerating, the sprinter runs at constant speed, before slowing (decelerating) near the end of the race. Use the velocity and time intervals shown in the graph to determine the sprinter's average acceleration during the first 4.00 s of the race.

Reasoning The slope of the line during the first 4-second interval of the race will be given by $\frac{\Delta v}{\Delta t}$, which is equivalent to the sprinter's average acceleration, $\bar{a} = \frac{\Delta v}{\Delta t}$.

Solution Using the fact that the average acceleration is given by $\bar{a} = \frac{\Delta v}{\Delta t}$, we just need to calculate the slope of the velocity-time graph over the period from 0 to 4.00 s.

$$\bar{a} = \frac{\Delta v}{\Delta t} = \frac{+12.5 \ \text{m/s} - 0}{4.00 \ \text{s} - 0} = \boxed{+3.13 \ \text{m/s}^2}$$

FIGURE 2.22 The velocity-time graph for a track and field athlete running a 100-m sprint. At the start of the race, the sprinter accelerates from rest for 4.00 s and then maintains a constant speed of 12.5 m/s for approximately four more seconds, before decelerating near the end of the race.

Concept Summary

2.1 Displacement Displacement is a vector that points from an object's initial position to its final position. The magnitude of the displacement is the shortest distance between the two positions.

2.2 Speed and Velocity The average speed of an object is the distance traveled by the object divided by the time required to cover the distance, as shown in Equation 2.1.

The average velocity $\vec{v}$ of an object is the object's displacement $\Delta \vec{x}$ divided by the elapsed time Δt, as shown in Equation 2.2. Average velocity is a vector that has the same direction as the displacement. When the elapsed time becomes infinitesimally small, the average velocity becomes equal to the instantaneous velocity $\vec{v}$, the velocity at an instant of time, as indicated in Equation 2.3.

$$\text{Average speed} = \frac{\text{Distance}}{\text{Elapsed time}} \quad (2.1)$$

$$\vec{v} = \frac{\Delta \vec{x}}{\Delta t} \quad (2.2)$$

$$\vec{v} = \lim_{\Delta t \to 0} \frac{\Delta \vec{x}}{\Delta t} \quad (2.3)$$

2.3 Acceleration The average acceleration $\vec{a}$ is a vector. It equals the change $\Delta \vec{v}$ in the velocity divided by the elapsed time Δt, the change in the velocity being the final minus the initial velocity; see Equation 2.4. When Δt becomes infinitesimally small, the average acceleration becomes equal to the instantaneous acceleration $\vec{a}$, as indicated in Equation 2.5. Acceleration is the rate at which the velocity is changing.

$$\vec{a} = \frac{\Delta\vec{v}}{\Delta t} \qquad (2.4)$$

$$\vec{a} = \lim_{\Delta t \to 0} \frac{\Delta\vec{v}}{\Delta t} \qquad (2.5)$$

2.4 Equations of Kinematics for Constant Acceleration/ 2.5 Applications of the Equations of Kinematics The equations of kinematics apply when an object moves with a constant acceleration along a straight line. These equations relate the displacement $x - x_0$, the acceleration a, the final velocity v, the initial velocity v_0, and the elapsed time $t - t_0$. Assuming that $x_0 = 0$ m at $t_0 = 0$ s, the equations of kinematics are as shown in Equations 2.4 and 2.7–2.9.

$$v = v_0 + at \qquad (2.4)$$

$$x = \tfrac{1}{2}(v_0 + v)t \qquad (2.7)$$

$$x = v_0 t + \tfrac{1}{2}at^2 \qquad (2.8)$$

$$v^2 = v_0^2 + 2ax \qquad (2.9)$$

2.6 Freely Falling Bodies In free-fall motion, an object experiences negligible air resistance and a constant acceleration due to gravity. All objects at the same location above the earth have the same acceleration due to gravity. The acceleration due to gravity is directed toward the center of the earth and has a magnitude of approximately 9.80 m/s² near the earth's surface.

2.7 Graphical Analysis of Velocity and Acceleration The slope of a plot of position versus time for a moving object gives the object's velocity. The slope of a plot of velocity versus time gives the object's acceleration.

Focus on Concepts

Online

Additional questions are available for assignment in WileyPLUS.

Section 2.1 Displacement

1. What is the difference between distance and displacement? **(a)** Distance is a vector, while displacement is not a vector. **(b)** Displacement is a vector, while distance is not a vector. **(c)** There is no difference between the two concepts; they may be used interchangeably.

Section 2.2 Speed and Velocity

2. A jogger runs along a straight and level road for a distance of 8.0 km and then runs back to her starting point. The time for this round-trip is 2.0 h. Which one of the following statements is true? **(a)** Her average speed is 8.0 km/h, but there is not enough information to determine her average velocity. **(b)** Her average speed is 8.0 km/h, and her average velocity is 8.0 km/h. **(c)** Her average speed is 8.0 km/h, and her average velocity is 0 km/h.

Section 2.3 Acceleration

3. The velocity of a train is 80.0 km/h, due west. One and a half hours later its velocity is 65.0 km/h, due west. What is the train's average acceleration? **(a)** 10.0 km/h², due west **(b)** 43.3 km/h², due west **(c)** 10.0 km/h², due east **(d)** 43.3 km/h², due east **(e)** 53.3 km/h², due east.

Section 2.4 Equations of Kinematics for Constant Acceleration

4. In which one of the following situations can the equations of kinematics *not* be used? **(a)** When the velocity changes from moment to moment, **(b)** when the velocity remains constant, **(c)** when the acceleration changes from moment to moment, **(d)** when the acceleration remains constant.

5. In a race two horses, Silver Bullet and Shotgun, start from rest and each maintains a constant acceleration. In the same elapsed time Silver Bullet runs 1.20 times farther than Shotgun. According to the equations of kinematics, which one of the following is true concerning the accelerations of the horses? **(a)** $a_{\text{Silver Bullet}} = 1.44\, a_{\text{Shotgun}}$ **(b)** $a_{\text{Silver Bullet}} = a_{\text{Shotgun}}$ **(c)** $a_{\text{Silver Bullet}} = 2.40\, a_{\text{Shotgun}}$ **(d)** $a_{\text{Silver Bullet}} = 1.20\, a_{\text{Shotgun}}$ **(e)** $a_{\text{Silver Bullet}} = 0.72\, a_{\text{Shotgun}}$

Section 2.6 Freely Falling Bodies

6. A rocket is sitting on the launch pad. The engines ignite, and the rocket begins to rise straight upward, picking up speed as it goes. At about 1000 m above the ground the engines shut down, but the rocket continues straight upward, losing speed as it goes. It reaches the top of its flight path and then falls back to earth. Ignoring air resistance, decide which one of the following statements is true. **(a)** All of the rocket's motion, from the moment the engines ignite until just before the rocket lands, is free-fall. **(b)** Only part of the rocket's motion, from just after the engines shut down until just before it lands, is free-fall. **(c)** Only the rocket's motion while the engines are firing is free-fall. **(d)** Only the rocket's motion from the top of its flight path until just before landing is free-fall. **(e)** Only part of the rocket's motion, from just after the engines shut down until it reaches the top of its flight path, is free-fall.

7. The top of a cliff is located a distance H above the ground. At a distance $H/2$ there is a branch that juts out from the side of the cliff, and on this branch a bird's nest is located. Two children throw stones at the nest with the same initial speed, one stone straight downward from the top of the cliff and the other stone straight upward from the ground. In the absence of air resistance, which stone hits the nest in the least amount of time? **(a)** There is insufficient information for an answer. **(b)** Both stones hit the nest in the same amount of time. **(c)** The stone thrown from the ground. **(d)** The stone thrown from the top of the cliff.

Section 2.7 Graphical Analysis of Velocity and Acceleration

8. The graph accompanying this problem shows a three-part motion. For each of the three parts, A, B, and C, identify the direction of the motion. A positive velocity denotes motion to the right. **(a)** A right, B left, C right **(b)** A right, B right, C left **(c)** A right, B left, C left **(d)** A left, B right, C left **(e)** A left, B right, C right

QUESTION 8

Problems

Online

Additional questions are available for assignment in WileyPLUS.

SSM	Student Solutions Manual	**BIO**	Biomedical application
MMH	Problem-solving help	**E**	Easy
GO	Guided Online Tutorial	**M**	Medium
V-HINT	Video Hints	**H**	Hard
CHALK	Chalkboard Videos	**WS**	Worksheet
		T	Team Problem

Section 2.1 Displacement,

Section 2.2 Speed and Velocity

1. **E** **SSM** The space shuttle travels at a speed of about 7.6×10^3 m/s. The blink of an astronaut's eye lasts about 110 ms. How many football fields (length = 91.4 m) does the shuttle cover in the blink of an eye?

2. **E** **GO** For each of the three pairs of positions listed in the following table, determine the magnitude and direction (positive or negative) of the displacement.

	Initial position x_0	Final position x
(a)	+2.0 m	+6.0 m
(b)	+6.0 m	+2.0 m
(c)	−3.0 m	+7.0 m

3. **E** **SSM** Due to continental drift, the North American and European continents are drifting apart at an average speed of about 3 cm per year. At this speed, how long (in years) will it take for them to drift apart by another 1500 m (a little less than a mile)?

4. **E** **GO** The data in the following table describe the initial and final positions of a moving car. The elapsed time for each of the three pairs of positions listed in the table is 0.50 s. Review the concept of average velocity in Section 2.2 and then determine the average velocity (magnitude and direction) for each of the three pairs. Note that the algebraic sign of your answers will convey the direction.

	Initial position x_0	Final position x
(a)	+2.0 m	+6.0 m
(b)	+6.0 m	+2.0 m
(c)	−3.0 m	+7.0 m

5. **E** **CHALK** One afternoon, a couple walks three-fourths of the way around a circular lake, the radius of which is 1.50 km. They start at the west side of the lake and head due south to begin with. **(a)** What is the distance they travel? **(b)** What are the magnitude and direction (relative to due east) of the couple's displacement?

6. **E** The three-toed sloth is the slowest-moving land mammal. On the ground, the sloth moves at an average speed of 0.037 m/s, considerably slower than the giant tortoise, which walks at 0.076 m/s. After 12 minutes of walking, how much further would the tortoise have gone relative to the sloth?

7. **E** **GO** An 18-year-old runner can complete a 10.0-km course with an average speed of 4.39 m/s. A 50-year-old runner can cover the same distance with an average speed of 4.27 m/s. How much later (in seconds) should the younger runner start in order to finish the course *at the same time* as the older runner?

8. **E** A tourist being chased by an angry bear is running in a straight line toward his car at a speed of 4.0 m/s. The car is a distance d away. The bear is 26 m behind the tourist and running at 6.0 m/s. The tourist reaches the car safely. What is the maximum possible value for d?

9. **M** **V-HINT** In reaching her destination, a backpacker walks with an average velocity of 1.34 m/s, due west. This average velocity results because she hikes for 6.44 km with an average velocity of 2.68 m/s, due west, turns around, and hikes with an average velocity of 0.447 m/s, due east. How far east did she walk?

10. **M** **SSM** **CHALK** A bicyclist makes a trip that consists of three parts, each in the same direction (due north) along a straight road. During the first part, she rides for 22 minutes at an average speed of 7.2 m/s. During the second part, she rides for 36 minutes at an average speed of 5.1 m/s. Finally, during the third part, she rides for 8.0 minutes at an average speed of 13 m/s. **(a)** How far has the bicyclist traveled during the entire trip? **(b)** What is her average velocity for the trip?

11. **M** **V-HINT** A car makes a trip due north for three-fourths of the time and due south one-fourth of the time. The average northward velocity has a magnitude of 27 m/s, and the average southward velocity has a magnitude of 17 m/s. What is the average velocity (magnitude and direction) for the entire trip?

12. **H** You are on a train that is traveling at 3.0 m/s along a level straight track. Very near and parallel to the track is a wall that slopes upward at a 12° angle with the horizontal. As you face the window (0.90 m high, 2.0 m wide) in your compartment, the train is moving to the left, as the drawing indicates. The top edge of the wall first appears at window corner A and eventually disappears at window corner B. How much time passes between appearance and disappearance of the upper edge of the wall?

PROBLEM 12

Section 2.3 Acceleration

13. **E** Review Conceptual Example 7 as background for this problem. A car is traveling to the left, which is the negative direction. The direction of travel remains the same throughout this problem. The car's initial speed is 27.0 m/s, and during a 5.0-s interval, it changes to a final speed of **(a)** 29.0 m/s and **(b)** 23.0 m/s. In each case, find the acceleration (magnitude and algebraic sign) and state whether or not the car is decelerating.

14. **E** **GO** **(a)** Suppose that a NASCAR race car is moving to the right with a constant velocity of +82 m/s. What is the average acceleration of the car? **(b)** Twelve seconds later, the car is halfway around the track and traveling in the opposite direction with the same speed. What is the average acceleration of the car?

15. **E** Over a time interval of 2.16 years, the velocity of a planet orbiting a distant star reverses direction, changing from +20.9 km/s to −18.5 km/s. Find **(a)** the total change in the planet's velocity (in m/s) and **(b)** its average acceleration (in m/s²) during this interval. Include the correct algebraic sign with your answers to convey the directions of the velocity and the acceleration.

16. **E** **SSM** A motorcycle has a constant acceleration of 2.5 m/s². Both the velocity and acceleration of the motorcycle point in the same direction. How much time is required for the motorcycle to change its speed from **(a)** 21 to 31 m/s, and **(b)** 51 to 61 m/s?

17. E A sprinter explodes out of the starting block with an acceleration of +2.3 m/s^2, which she sustains for 1.2 s. Then, her acceleration drops to zero for the rest of the race. What is her velocity **(a)** at $t = 1.2$ s and **(b)** at the end of the race?

18. E GO The initial velocity and acceleration of four moving objects at a given instant in time are given in the following table. Determine the final *speed* of each of the objects, assuming that the time elapsed since $t = 0$ s is 2.0 s.

	Initial velocity v_0	Acceleration a
(a)	+12 m/s	+3.0 m/s^2
(b)	+12 m/s	−3.0 m/s^2
(c)	−12 m/s	+3.0 m/s^2
(d)	−12 m/s	−3.0 m/s^2

19. E GO An Australian emu is running due north in a straight line at a speed of 13.0 m/s and slows down to a speed of 10.6 m/s in 4.0 s. **(a)** What is the direction of the bird's acceleration? **(b)** Assuming that the acceleration remains the same, what is the bird's velocity after an additional 2.0 s has elapsed?

20. E SSM CHALK For a standard production car, the highest road-tested acceleration ever reported occurred in 1993, when a Ford RS200 Evolution went from zero to 26.8 m/s (60 mi/h) in 3.275 s. Find the magnitude of the car's acceleration.

21. M MMH A car is traveling along a straight road at a velocity of +36.0 m/s when its engine cuts out. For the next twelve seconds the car slows down, and its average acceleration is $\bar{a}_1$. For the next six seconds the car slows down further, and its average acceleration is $\bar{a}_2$. The velocity of the car at the end of the eighteen-second period is +28.0 m/s. The ratio of the average acceleration values is $\bar{a}_1/\bar{a}_2 = 1.50$. Find the velocity of the car at the end of the initial twelve-second interval.

22. H Two motorcycles are traveling due east with different velocities. However, four seconds later, they have the same velocity. During this four-second interval, cycle A has an average acceleration of 2.0 m/s^2 due east, while cycle B has an average acceleration of 4.0 m/s^2 due east. By how much did the speeds *differ* at the beginning of the four-second interval, and which motorcycle was moving faster?

Section 2.4 Equations of Kinematics for Constant Acceleration, Section 2.5 Applications of the Equations of Kinematics

23. E In getting ready to slam-dunk the ball, a basketball player starts from rest and sprints to a speed of 6.0 m/s in 1.5 s. Assuming that the player accelerates uniformly, determine the distance he runs.

24. E SSM A jogger accelerates from rest to 3.0 m/s in 2.0 s. A car accelerates from 38.0 to 41.0 m/s also in 2.0 s. **(a)** Find the acceleration (magnitude only) of the jogger. **(b)** Determine the acceleration (magnitude only) of the car. **(c)** Does the car travel farther than the jogger during the 2.0 s? If so, how much farther?

25. E GO A VW Beetle goes from 0 to 60.0 mi/h with an acceleration of +2.35 m/s^2. **(a)** How much time does it take for the Beetle to reach this speed? **(b)** A top-fuel dragster can go from 0 to 60.0 mi/h in 0.600 s. Find the acceleration (in m/s^2) of the dragster.

26. E **(a)** What is the magnitude of the average acceleration of a skier who, starting from rest, reaches a speed of 8.0 m/s when going down a slope for 5.0 s? **(b)** How far does the skier travel in this time?

27. E SSM A jetliner, traveling northward, is landing with a speed of 69 m/s. Once the jet touches down, it has 750 m of runway in which to

reduce its speed to 6.1 m/s. Compute the average acceleration (magnitude and direction) of the plane during landing.

28. E V-HINT MMH The Kentucky Derby is held at the Churchill Downs track in Louisville, Kentucky. The track is one and one-quarter miles in length. One of the most famous horses to win this event was Secretariat. In 1973 he set a Derby record that would be hard to beat. His average acceleration during the last four quarter-miles of the race was +0.0105 m/s^2. His velocity at the start of the final mile ($x = +1609$ m) was about +16.58 m/s. The acceleration, although small, was very important to his victory. To assess its effect, determine the difference between the time he would have taken to run the final mile at a constant velocity of +16.58 m/s and the time he actually took. Although the track is oval in shape, assume it is straight for the purpose of this problem.

29. E SSM MMH A cart is driven by a large propeller or fan, which can accelerate or decelerate the cart. The cart starts out at the position $x = 0$ m, with an initial velocity of +5.0 m/s and a constant acceleration due to the fan. The direction to the right is positive. The cart reaches a maximum position of $x = +12.5$ m, where it begins to travel in the negative direction. Find the acceleration of the cart.

30. E MMH Two rockets are flying in the same direction and are side by side at the instant their retrorockets fire. Rocket A has an initial velocity of +5800 m/s, while rocket B has an initial velocity of +8600 m/s. After a time t both rockets are again side by side, the displacement of each being zero. The acceleration of rocket A is −15 m/s^2. What is the acceleration of rocket B?

31. M CHALK MMH A car is traveling at 20.0 m/s, and the driver sees a traffic light turn red. After 0.530 s (the reaction time), the driver applies the brakes, and the car decelerates at 7.00 m/s^2. What is the stopping distance of the car, as measured from the point where the driver first sees the red light?

32. M GO MMH A race driver has made a pit stop to refuel. After refueling, he starts from rest and leaves the pit area with an acceleration whose magnitude is 6.0 m/s^2; after 4.0 s he enters the main speedway. At the same instant, another car on the speedway and traveling at a constant velocity of 70.0 m/s overtakes and passes the entering car. The entering car maintains its acceleration. How much time is required for the entering car to catch the other car?

33. M V-HINT In a historical movie, two knights on horseback start from rest 88.0 m apart and ride directly toward each other to do battle. Sir George's acceleration has a magnitude of 0.300 m/s^2, while Sir Alfred's has a magnitude of 0.200 m/s^2. Relative to Sir George's starting point, where do the knights collide?

34. M GO Two soccer players start from rest, 48 m apart. They run directly toward each other, both players accelerating. The first player's acceleration has a magnitude of 0.50 m/s^2. The second player's acceleration has a magnitude of 0.30 m/s^2. **(a)** How much time passes before the players collide? **(b)** At the instant they collide, how far has the first player run?

35. M MMH CHALK A car is traveling at a constant speed of 33 m/s on a highway. At the instant this car passes an entrance ramp, a second car enters the highway from the ramp. The second car starts from rest and has a constant acceleration. What acceleration must it maintain, so that the two cars meet for the first time at the next exit, which is 2.5 km away?

36. M GO Refer to Multiple-Concept Example 5 to review a method by which this problem can be solved. You are driving your car, and the traffic light ahead turns red. You apply the brakes for 3.00 s, and the velocity of the car decreases to +4.50 m/s. The car's deceleration has a magnitude of 2.70 m/s^2 during this time. What is the car's displacement?

37. H A Boeing 747 "Jumbo Jet" has a length of 59.7 m. The runway on which the plane lands intersects another runway. The width of the intersection is 25.0 m. The plane decelerates through the intersection at a rate of 5.70 m/s^2 and clears it with a final speed of 45.0 m/s. How much time is needed for the plane to clear the intersection?

38. **H** **SSM** A locomotive is accelerating at 1.6 m/s². It passes through a 20.0-m-wide crossing in a time of 2.4 s. After the locomotive leaves the crossing, how much time is required until its speed reaches 32 m/s?

Section 2.6   Freely Falling Bodies

39. **E** **SSM** The greatest height reported for a jump into an airbag is 99.4 m by stuntman Dan Koko. In 1948 he jumped from rest from the top of the Vegas World Hotel and Casino. He struck the airbag at a speed of 39 m/s (88 mi/h). To assess the effects of air resistance, determine how fast he would have been traveling on impact had air resistance been absent.

40. **E** A dynamite blast at a quarry launches a chunk of rock straight upward, and 2.0 s later it is rising at a speed of 15 m/s. Assuming air resistance has no effect on the rock, calculate its speed **(a)** at launch and **(b)** 5.0 s after launch.

41. **E** **GO** **MMH** A ball is thrown vertically upward, which is the positive direction. A little later it returns to its point of release. The ball is in the air for a total time of 8.0 s. What is its initial velocity? Neglect air resistance.

42. **E** Review Conceptual Example 15 before attempting this problem. Two identical pellet guns are fired simultaneously from the edge of a cliff. These guns impart an initial speed of 30.0 m/s to each pellet. Gun A is fired straight upward, with the pellet going up and then falling back down, eventually hitting the ground beneath the cliff. Gun B is fired straight downward. In the absence of air resistance, how long after pellet B hits the ground does pellet A hit the ground?

43. **E** **MMH** An astronaut on a distant planet wants to determine its acceleration due to gravity. The astronaut throws a rock straight up with a velocity of +15 m/s and measures a time of 20.0 s before the rock returns to his hand. What is the acceleration (magnitude and direction) due to gravity on this planet?

44. **E** **SSM** A hot-air balloon is rising upward with a constant speed of 2.50 m/s. When the balloon is 3.00 m above the ground, the balloonist accidentally drops a compass over the side of the balloon. How much time elapses before the compass hits the ground?

45. **E** **GO** A ball is thrown straight upward and rises to a maximum height of 16 m above its launch point. At what height above its launch point has the speed of the ball decreased to one-half of its initial value?

46. **E** A diver springs upward with an initial speed of 1.8 m/s from a 3.0-m board. **(a)** Find the velocity with which he strikes the water. *[Hint: When the diver reaches the water, his displacement is y = −3.0 m (measured from the board), assuming that the downward direction is chosen as the negative direction.]* **(b)** What is the highest point he reaches above the water?

47. **E** **GO** A ball is thrown straight upward. At 4.00 m above its launch point, the ball's speed is one-half its launch speed. What maximum height above its launch point does the ball attain?

48. **E** **SSM** From her bedroom window a girl drops a water-filled balloon to the ground, 6.0 m below. If the balloon is released from rest, how long is it in the air?

49. **E** Before working this problem, review Conceptual Example 15. A pellet gun is fired straight downward from the edge of a cliff that is 15 m above the ground. The pellet strikes the ground with a speed of 27 m/s. How far above the cliff edge would the pellet have gone had the gun been fired straight upward?

50. **E** **V-HINT** Consult Multiple-Concept Example 5 in preparation for this problem. The velocity of a diver just before hitting the water is −10.1 m/s, where the minus sign indicates that her motion is directly downward. What is her displacement during the last 1.20 s of the dive?

51. **M** **V-HINT** A golf ball is dropped from rest from a height of 9.50 m. It hits the pavement, then bounces back up, rising just 5.70 m before falling back down again. A boy then catches the ball on the way down when it is 1.20 m above the pavement. Ignoring air resistance, calculate the total amount of time that the ball is in the air, from drop to catch.

52. **M** **MMH** A woman on a bridge 75.0 m high sees a raft floating at a constant speed on the river below. Trying to hit the raft, she drops a stone from rest when the raft has 7.00 m more to travel before passing under the bridge. The stone hits the water 4.00 m in front of the raft. Find the speed of the raft.

53. **M** **GO** **CHALK** Two stones are thrown simultaneously, one straight upward from the base of a cliff and the other straight downward from the top of the cliff. The height of the cliff is 6.00 m. The stones are thrown with the same speed of 9.00 m/s. Find the location (above the base of the cliff) of the point where the stones cross paths.

54. **M** **GO** Two arrows are shot vertically upward. The second arrow is shot after the first one, but while the first is still on its way up. The initial speeds are such that both arrows reach their maximum heights at the same instant, although these heights are different. Suppose that the initial speed of the first arrow is 25.0 m/s and that the second arrow is fired 1.20 s after the first. Determine the initial speed of the second arrow.

55. **M** **CHALK** **SSM** A cement block accidentally falls from rest from the ledge of a 53.0-m-high building. When the block is 14.0 m above the ground, a man, 2.00 m tall, looks up and notices that the block is directly above him. How much time, at most, does the man have to get out of the way?

56. **M** **V-HINT** A model rocket blasts off from the ground, rising straight upward with a constant acceleration that has a magnitude of 86.0 m/s² for 1.70 seconds, at which point its fuel abruptly runs out. Air resistance has no effect on its flight. What maximum altitude (above the ground) will the rocket reach?

57. **H** While standing on a bridge 15.0 m above the ground, you drop a stone from rest. When the stone has fallen 3.20 m, you throw a second stone straight down. What initial velocity must you give the second stone if they are both to reach the ground at the same instant? Take the downward direction to be the negative direction.

Section 2.7   Graphical Analysis of Velocity and Acceleration

58. **E** **SSM** A person who walks for exercise produces the position–time graph given with this problem. **(a)** Without doing any calculations, decide which segments of the graph (A, B, C, or D) indicate positive, negative, and zero average velocities. **(b)** Calculate the average velocity for each segment to verify your answers to part (a).

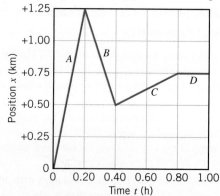

PROBLEM 58

59. **E** Starting at x = −16 m at time t = 0 s, an object takes 18 s to travel 48 m in the +x direction at a constant velocity. Make a position–time graph of the object's motion and calculate its velocity.

60. **E** **MMH** A snowmobile moves according to the velocity–time graph shown in the drawing. What is the snowmobile's average acceleration during each of the segments A, B, and C?

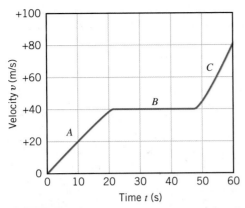

PROBLEM 60

61. **E** A bus makes a trip according to the position–time graph shown in the drawing. What is the average velocity (magnitude and direction) of the bus during each of the segments A, B, and C? Express your answers in km/h.

PROBLEM 61

62. **M** **GO** **CHALK** A bus makes a trip according to the position–time graph shown in the illustration. What is the average acceleration (in km/h²) of the bus for the entire 3.5-h period shown in the graph?

PROBLEM 62

63. **M** A runner is at the position $x = 0$ m when time $t = 0$ s. One hundred meters away is the finish line. Every ten seconds, this runner runs half the remaining distance to the finish line. During each ten-second segment, the runner has a constant velocity. For the first forty seconds of the motion, construct (a) the position–time graph and (b) the velocity-time graph.

Additional Problems

Online

64. **E** **GO** The data in the following table represent the initial and final velocities for a boat traveling along the x axis. The elapsed time for each of the four pairs of velocities in the table is 2.0 s. Review the concept of average acceleration in Section 2.3 and then determine the average acceleration (magnitude and direction) for each of the four pairs. Note that the algebraic sign of your answers will convey the direction.

	Initial velocity v_0	Final velocity v
(a)	+2.0 m/s	+5.0 m/s
(b)	+5.0 m/s	+2.0 m/s
(c)	−6.0 m/s	−3.0 m/s
(d)	+4.0 m/s	−4.0 m/s

65. **E** In 1954 the English runner Roger Bannister broke the four-minute barrier for the mile with a time of 3:59.4 s (3 min and 59.4 s). In 1999 the Moroccan runner Hicham el-Guerrouj set a record of 3:43.13 s for the mile. If these two runners had run in the same race, each running the entire race at the average speed that earned him a place in the record books, el-Guerrouj would have won. By how many meters?

66. **E** **MMH** At the beginning of a basketball game, a referee tosses the ball straight up with a speed of 4.6 m/s. A player cannot touch the ball until after it reaches its maximum height and begins to fall down. What is the minimum time that a player must wait before touching the ball?

67. **E** Electrons move through a certain electric circuit at an average speed of 1.1×10^{-2} m/s. How long (in minutes) does it take an electron to traverse 1.5 m of wire in the filament of a light bulb?

68. **E** **V-HINT** **MMH** A cheetah is hunting. Its prey runs for 3.0 s at a constant velocity of +9.0 m/s. Starting from rest, what constant acceleration must the cheetah maintain in order to run the same distance as its prey runs in the same time?

69. **M** **GO** The leader of a bicycle race is traveling with a constant velocity of +11.10 m/s and is 10.0 m ahead of the second-place cyclist. The second-place cyclist has a velocity of +9.50 m/s and an acceleration of +1.20 m/s². How much time elapses before he catches the leader?

70. **M** **V-HINT** A golfer rides in a golf cart at an average speed of 3.10 m/s for 28.0 s. She then gets out of the cart and starts walking at an average speed of 1.30 m/s. For how long (in seconds) must she walk if her average speed for the entire trip, riding and walking, is 1.80 m/s?

71. **M** **GO** Two cars cover the same distance in a straight line. Car A covers the distance at a constant velocity. Car B starts from rest and maintains a constant acceleration. Both cars cover a distance of 460 m in 210 s. Assume that they are moving in the +x direction. Determine (a) the constant velocity of car A, (b) the final velocity of car B, and (c) the acceleration of car B.

72. **M** **V-HINT** **MMH** A police car is traveling at a velocity of 18.0 m/s due north, when a car zooms by at a constant velocity of 42.0 m/s due north. After a reaction time of 0.800 s the policeman begins to pursue the speeder with an acceleration of 5.00 m/s². Including the reaction time, how long does it take for the police car to catch up with the speeder?

73. **M** **GO** **SSM** A jet is taking off from the deck of an aircraft carrier, as shown in the image. Starting from rest, the jet is catapulted with a constant acceleration of $+31$ m/s^2 along a straight line and reaches a velocity of $+62$ m/s. Find the displacement of the jet.

(*a*) A jet is being launched from an aircraft carrier. (*b*) During the launch, a catapult accelerates the jet down the flight deck.

PROBLEM 73

Physics in Biology, Medicine, and Sports

74. **E** **BIO** **2.6** The drawing shows a device that you can make with a piece of cardboard, which can be used to measure a person's reaction time. Hold the card at the top and suddenly drop it. Ask a friend to try to catch the card between his or her thumb and index finger. Initially, your friend's fingers must be level with the asterisks at the bottom. By noting where your friend catches the card, you can determine his or her reaction time in milliseconds (ms). Calculate the distances d_1, d_2, and d_3.

PROBLEM 74

75. **E** **BIO** **2.2** You step onto a hot beach with your bare feet. A nerve impulse, generated in your foot, travels through your nervous system at an average speed of 110 m/s. How much time does it take for the impulse, which travels a distance of 1.8 m, to reach your brain?

76. **E** **BIO** **2.4** The left ventricle of the heart accelerates blood from rest to a velocity of $+26$ cm/s. **(a)** If the displacement of the blood during the acceleration is $+2.0$ cm, determine its acceleration (in cm/s^2). **(b)** How much time does the blood take to reach its final velocity?

77. **M** **BIO** **2.5** During the late 1940s, Colonel John Paul Stapp was a pioneer in studying the effects of acceleration and deceleration on the human body. He made multiple runs strapped to a rocket sled that quickly accelerated him to high speeds along a straight track (see the images below). His research led to improvements in restraining harnesses and seatbelts for pilots and automobile occupants. During his final run, he reached a maximum speed of 632 mph. When the sled's braking system brought it to rest, Colonel Stapp experienced a deceleration of magnitude 46.2*g*, or 46.2 times the acceleration due to gravity at the Earth's surface. Although he survived, he did sustain injuries, such as a fractured wrist, broken ribs, and bleeding in his eyes. Calculate how long it took to bring the rocket sled to rest. Assume the deceleration was constant during the braking period.

PROBLEM 77

78. **E** **BIO** **2.3** What is the acceleration of a thoroughbred racehorse? Starting from rest out of the gate, thoroughbreds can reach a maximum speed of 19 m/s (~43 mph.) over a distance of 35 m.

PROBLEM 78

79. **E** **BIO** **2.3** What is the acceleration of the fastest human sprinters? Olympic sprinter Usain Bolt reaches his maximum acceleration during the first 20.0 m of the sprint, at which time he is moving with a speed of 6.94 m/s. In reality, Bolt continues to accelerate for the first 60.0 m into the race, reaching a top speed near 28 mph.

PROBLEM 79

80. Ⓜ BIO **2.5**   Almost 80% of all traffic accidents involve distracted driving. Looking at a cell phone text message for even a couple of seconds is very dangerous. A typical human reaction time is 190 milliseconds. Assume a car is traveling at 71 mph. If the maximum deceleration of the car is a constant 8.0 m/s^2, what will be the stopping distance of the car? Take into account the reaction time.

81. Ⓜ BIO **2.6**   In 2012, Austrian skydiver Felix Baumgartner set a record for the highest exit altitude of 127 852 feet. He jumped from a helium balloon located in the earth's stratosphere (see the photo). At that altitude the atmosphere is too thin to breathe, so Baumgartner wore a special pressurized suit. During his free-fall toward the earth, he reached a maximum speed of 843.6 mph., which is greater than the speed of sound. This set a record for greatest speed achieved during free-fall and also set off a small sonic boom heard by his friends and family on the ground. Assume the effect of air resistance can be neglected and the acceleration due to gravity remains constant during his fall with a value of 9.807 m/s^2, and calculate the distance he fell before reaching 843.6 mph.

PROBLEM 81

Concepts and Calculations Problems

Online

In this chapter we have studied the displacement, velocity, and acceleration vectors. The concept questions are designed to review some of the important concepts that can help in anticipating some of the characteristics of the numerical answers.

82. Ⓜ CHALK   A skydiver is falling straight down, along the negative y direction. **(a)** During the initial part of the fall, her speed increases from 16 to 28 m/s in 1.5 s, as in part a of the figure. **(b)** Later, her parachute opens, and her speed decreases from 48 to 26 m/s in 11 s, as in part b of the drawing. *Concepts:* (i) Is her average acceleration positive or negative when her speed is increasing in part a of the figure? (ii) Is her average acceleration positive or negative when her speed is decreasing in part b of the figure? *Calculations:* In both instances (parts a and b of the figure) determine the magnitude and direction of her average acceleration.

83. Ⓜ CHALK SSM   A top-fuel dragster starts from rest and has a constant acceleration of 40.0 m/s^2. *Concepts:* (i) At time t the dragster has a certain velocity. Keeping in mind that the dragster starts from rest, when the time doubles to $2t$, does the velocity also double? (ii) When the time doubles to $2t$, does the displacement of the dragster also double? *Calculations:* What are the **(a)** final velocities and **(b)** displacements of the dragster at the end of 2.0 s and at the end of twice this time, or 4.0 s?

(a)

PROBLEM 82

(b)

Team Problems

84. **E** **T** **WS** **BIO** **A Fall on the Moon.** You and your team are stationed at an earth-based mission control facility and have been tasked with assisting astronauts via radio communication as they exit a spacecraft that has recently landed on the moon. The astronauts intended to explore the lunar surface but, unfortunately, they hit a snag even before they opened the hatch. Their craft has landed right on the edge of a 15-ft cliff, so that there is no way off the lander except to drop that distance from the bottom rung of a ladder on the side of the ship. You look up the structural parameters of the space suits, the gear they are wearing, and the physical limitations of the crew, and find that, on earth, the maximum drop they can safely endure is 4.0 ft. Is it safe for them to make the 15-ft drop on the moon? Explain your reasoning using quantitative information. Note that the acceleration due to gravity on the moon is one-sixth as large as that on earth.

85. **E** **T** **WS** **BIO** **Helmets to Prevent Concussions** In contact sports such as American football, hockey, and rugby, players routinely experience collisions in which considerable contact is made with the head, and therefore the brain can be subjected to significant forces. Repetitive collisions in which the brain experiences accelerations of 10 times the acceleration due to gravity (i.e., 10g) or greater can result in accumulative permanent brain damage. With current helmet design, the effective stopping distance for a football player undergoing a head-to-head collision with an initial speed of 8.00 mph. is 2.25 inches. **(a)** What is the average magnitude of the acceleration during the collision (in terms of g)? Is it a safe value? **(b)** The effective stopping distance can be increased by adding better cushioning features. You and your team are assigned to determine the increase in the effective stopping distance required for a new helmet design that will result in an effective average acceleration of less than 10.0g (with an initial speed of 8.00 mph.), thereby reducing the accumulative effects of head contact.

86. **M** **T** **A Deadly Virus.** On your flight out of South America your cargo plane emergency-lands on a desert island somewhere in the South Pacific. Because it was a beach landing, the pilot skids the plane on its belly rather than lowering the landing gear. The landing is rough: an initial vertical drop followed by an abrupt horizontal deceleration. When the dust settles, the pilot and your scientist colleagues are all okay, but you know there might be a big problem. Your team is transporting samples of a deadly virus from a recent breakout in a small village. The samples were packed in a cryogenically cooled container that has a special seal that may leak if it exceeds an acceleration greater than 10.0g. The container has a mechanical accelerometer gauge that measures the maximum vertical deceleration in case the container is dropped. It reads a value of $a = 7.30g$. However, you are still worried: there was another component of the acceleration. You get out of the plane and measure the length of the skid the plane made in the sand as it landed: 154 ft. The pilot claims the plane's horizontal speed on impact was 160.0 mph. Assuming the worst case scenario (i.e., that the vertical and horizontal accelerations occurred simultaneously), and that the horizontal deceleration was uniform, is it likely that the seal on the virus container is compromised? Explain.

87. **M** **T** **The Lost Drone.** You and your team are exploring the edge of an Antarctic mountain range and you send a drone ahead to help navigate. After takeoff you lose sight of the drone and, a few seconds later, the controls malfunction and the drone stops sending visual images and navigational information except for speed and directional data. Changing speeds erratically, the drone heads west until it makes a drastic turn at the 5-minute mark to 35.0° east of south. After nearly ten minutes, the speed drops to zero and the drone stops sending data. It has crashed. Using the speed/directional data, the team draws up the graph shown in the drawing. How far is the drone from you, and in what direction must you go to retrieve it? Express your result as a geographical direction (i.e., in the form 30° north of east, etc.).

PROBLEM 87

The beautiful water fountain at the World War II Memorial in Washington, D.C., is illuminated at night. The arching streams of water follow parabolic paths whose sizes depend on the launch velocity of the water and the acceleration due to gravity, assuming that the effects of air resistance are negligible.

threespeedjones/iStockphoto

Kinematics in Two Dimensions

3.1 Displacement, Velocity, and Acceleration

In Chapter 2 the concepts of displacement, velocity, and acceleration are used to describe an object moving in one dimension. There are also situations in which the motion is along a curved path that lies in a plane. Such two-dimensional motion can be described using the same concepts. In Grand Prix racing, for example, the course follows a curved road, and **Figure 3.1** shows a race car at two different positions along it. These positions are identified by the vectors $\vec{r}$ and $\vec{r}_0$, which are drawn from an arbitrary coordinate origin. The **displacement** $\Delta\vec{r}$ of the car is the vector drawn from the initial position $\vec{r}_0$ at time t_0 to the final position $\vec{r}$ at time t. The magnitude of $\Delta\vec{r}$ is the shortest distance between the two positions. In the drawing, the vectors $\vec{r}_0$ and $\Delta\vec{r}$ are drawn tail to head, so it is evident that $\vec{r}$ is the vector sum of $\vec{r}_0$ and $\Delta\vec{r}$. (See Sections 1.5 and 1.6 for a review of vectors and vector addition.) This means that $\vec{r} = \vec{r}_0 + \Delta\vec{r}$, or

$$\text{Displacement} = \Delta\vec{r} = \vec{r} - \vec{r}_0$$

LEARNING OBJECTIVES

After reading this module, you should be able to...

3.1 Define two-dimensional kinematic variables.

3.2 Use two-dimensional kinematic equations to predict future or past values of variables.

3.3 Analyze projectile motion to predict future or past values of variables.

3.4 Apply relative velocity equations.

FIGURE 3.1 The displacement $\Delta\vec{r}$ of the car is a vector that points from the initial position of the car at time t_0 to the final position at time t. The magnitude of $\Delta\vec{r}$ is the shortest distance between the two positions.

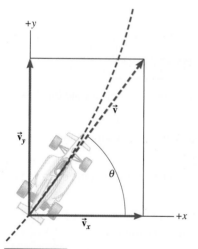

FIGURE 3.2 The instantaneous velocity $\vec{v}$ and its two vector components $\vec{v}_x$ and $\vec{v}_y$.

The displacement here is defined as it is in Chapter 2. Now, however, the displacement vector may lie anywhere in a plane, rather than just along a straight line.

The **average velocity** $\overline{\vec{v}}$ of the car in **Figure 3.1** between the two positions is defined in a manner similar to that in Equation 2.2, as the displacement $\Delta\vec{r} = \vec{r} - \vec{r}_0$ divided by the elapsed time $\Delta t = t - t_0$:

$$\overline{\vec{v}} = \frac{\vec{r} - \vec{r}_0}{t - t_0} = \frac{\Delta\vec{r}}{\Delta t} \tag{3.1}$$

Since both sides of Equation 3.1 must agree in direction, the average velocity vector has the same direction as the displacement $\Delta\vec{r}$. The velocity of the car at an instant of time is its **instantaneous velocity** $\vec{v}$. The average velocity becomes equal to the instantaneous velocity $\vec{v}$ in the limit that Δt becomes infinitesimally small ($\Delta t \to 0$ s):

$$\vec{v} = \lim_{\Delta t \to 0} \frac{\Delta\vec{r}}{\Delta t}$$

Figure 3.2 illustrates that the instantaneous velocity $\vec{v}$ is tangent to the path of the car. The drawing also shows the vector components $\vec{v}_x$ and $\vec{v}_y$ of the velocity, which are parallel to the x and y axes, respectively.

The **average acceleration** $\overline{\vec{a}}$ is defined just as it is for one-dimensional motion—namely, as the change in velocity, $\Delta\vec{v} = \vec{v} - \vec{v}_0$, divided by the elapsed time Δt:

$$\overline{\vec{a}} = \frac{\vec{v} - \vec{v}_0}{t - t_0} = \frac{\Delta\vec{v}}{\Delta t} \tag{3.2}$$

The average acceleration has the same direction as the change in velocity $\Delta\vec{v}$. In the limit that the elapsed time becomes infinitesimally small, the average acceleration becomes equal to the **instantaneous acceleration** $\vec{a}$:

$$\vec{a} = \lim_{\Delta t \to 0} \frac{\vec{v}}{\Delta t}$$

The acceleration has a vector component $\vec{a}_x$ along the x direction and a vector component $\vec{a}_y$ along the y direction.

Check Your Understanding

(*The answer is given at the end of the book.*)

1. Suppose you are driving due east, traveling a distance of 1500 m in 2 minutes. You then turn due north and travel the same distance in the same time. What can be said about the average speeds and the average velocities for the two segments of the trip? **(a)** The average speeds are the same, and the average velocities are the same. **(b)** The average speeds are the same, but the average velocities are different. **(c)** The average speeds are different, but the average velocities are the same.

3.2 | Equations of Kinematics in Two Dimensions

To understand how displacement, velocity, and acceleration are applied to two-dimensional motion, consider a spacecraft equipped with two engines that are mounted perpendicular to each other. These engines produce the only forces that the craft experiences, and the spacecraft is assumed to be at the coordinate origin when $t_0 = 0$ s, so that $\vec{r}_0 = 0$ m. At a later time t, the spacecraft's displacement is $\Delta\vec{r} = \vec{r} - \vec{r}_0 = \vec{r}$. Relative to the x and y axes, the displacement $\vec{r}$ has vector components of $\vec{x}$ and $\vec{y}$, respectively.

FIGURE 3.3 The spacecraft is moving with a constant acceleration a_x parallel to the x axis. There is no motion in the y direction, and the y engine is turned off.

In **Figure 3.3** only the engine oriented along the x direction is firing, and the vehicle accelerates along this direction. It is assumed that the velocity in the y direction is zero, and it remains zero, since the y engine is turned off. The motion of the spacecraft along the x direction is described by the five kinematic variables x, a_x, v_x, v_{0x}, and t. Here the symbol "x" reminds us that we are dealing with the x components of the displacement, velocity, and acceleration vectors. (See Sections 1.7 and 1.8 for a review of vector components.) The variables x, a_x, v_x, and v_{0x} are scalar components (or "components," for short). As Section 1.7 discusses, these components are positive or negative numbers (with units), depending on whether the associated vector components point in the $+x$ or the $-x$ direction. If the spacecraft has a constant acceleration along the x direction, the motion is exactly like that described in Chapter 2, and the equations of kinematics can be used. For convenience, these equations are written in the first four rows of **Table 3.1**, where each equation contains four of the five kinematic variables.

TABLE 3.1	Equations of Kinematics for Constant Acceleration in Two-Dimensional Motion					

| Equation Number | Equation | Variables | | | | |
		x	a_x	v_x	v_{0x}	t
(3.3a)	$v_x = v_{0x} + a_x t$	—	✓	✓	✓	✓
(3.4a)	$x = \frac{1}{2}(v_{0x} + v_x)t$	✓	—	✓	✓	✓
(3.5a)	$x = v_{0x}t + \frac{1}{2}a_x t^2$	✓	✓	—	✓	✓
(3.6a)	$v_x^2 = v_{0x}^2 + 2a_x x$	✓	✓	✓	✓	—

| Equation Number | Equation | Variables | | | | |
		y	a_y	v_y	v_{0y}	t
(3.3b)	$v_y = v_{0y} + a_y t$	—	✓	✓	✓	✓
(3.4b)	$y = \frac{1}{2}(v_{0y} + v_y)t$	✓	—	✓	✓	✓
(3.5b)	$y = v_{0y}t + \frac{1}{2}a_y t^2$	✓	✓	—	✓	✓
(3.6b)	$v_y^2 = v_{0y}^2 + 2a_y y$	✓	✓	✓	✓	—

Figure 3.4 is analogous to **Figure 3.3**, except that now only the y engine is firing, and the spacecraft accelerates along the y direction. Such a motion can be described in terms of the kinematic variables y, a_y, v_y, v_{0y}, and t. And if the acceleration along the y direction is constant, these variables are related by the equations of kinematics, as written in the last four rows of **Table 3.1**. Like their counterparts in the x direction, the scalar components, y, a_y, v_y, and v_{0y}, may be positive $(+)$ or negative $(-)$ numbers (with units).

If both engines of the spacecraft are firing *at the same time,* the resulting motion takes place in part along the x axis and in part along the y axis, as **Interactive Figure 3.5** illustrates. The thrust of each engine gives the vehicle a corresponding acceleration component. The x engine accelerates the craft in the x direction and causes a change in the x component of the velocity. Likewise, the y engine causes a change in the y component of the velocity.

FIGURE 3.4 The spacecraft is moving with a constant acceleration a_y parallel to the y axis. There is no motion in the x direction, and the x engine is turned off.

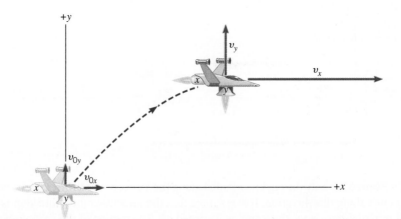

The two-dimensional motion of the spacecraft can be viewed as the combination of the separate x and y motions.

> **Problem-Solving Insight** It is important to realize that the x part of the motion occurs exactly as it would if the y part did not occur at all. Similarly, the y part of the motion occurs exactly as it would if the x part of the motion did not exist.

In other words, the x and y motions are independent of each other.

The independence of the x and y motions lies at the heart of two-dimensional kinematics. It allows us to treat two-dimensional motion as two distinct one-dimensional motions, one for the x direction and one for the y direction. Everything that we have learned in Chapter 2 about kinematics in one dimension will now be applied separately to each of the two directions. In so doing, we will be able to describe the x and y variables separately and then bring these descriptions together to understand the two-dimensional picture. Examples 1 and 2 take this approach in dealing with a moving spacecraft.

EXAMPLE 1 | The Displacement of a Spacecraft

In **Interactive Figure 3.5**, the directions to the right and upward are the positive directions. In the x direction, the spacecraft has an initial velocity component of $v_{0x} = +22$ m/s and an acceleration component of $a_x = +24$ m/s². In the y direction, the analogous quantities are $v_{0y} = +14$ m/s and $a_y = +12$ m/s². At a time of $t = 7.0$ s, find the x and y components of the spacecraft's displacement.

Reasoning The motion in the x direction and the motion in the y direction can be treated separately, each as a one-dimensional motion subject to the equations of kinematics for constant acceleration (see **Table 3.1**). By following this procedure we will be able to determine x and y, which specify the spacecraft's location after an elapsed time of 7.0 s.

> **Problem-Solving Insight** When the motion is two-dimensional, the time variable t has the same value for both the x and y directions.

Solution The data for the motion in the x direction are listed in the following table:

x-Direction Data				
x	a_x	v_x	v_{0x}	t
?	+24 m/s²		+22 m/s	7.0 s

The x component of the craft's displacement can be found by using Equation 3.5a:

$$x = v_{0x}t + \tfrac{1}{2}a_x t^2 = (22 \text{ m/s})(7.0 \text{ s}) + \tfrac{1}{2}(24 \text{ m/s}^2)(7.0 \text{ s})^2 = \boxed{+740 \text{ m}}$$

The data for the motion in the y direction are listed in the following table:

y-Direction Data				
y	a_y	v_y	v_{0y}	t
?	+12 m/s²		+14 m/s	7.0 s

The y component of the craft's displacement can be found by using Equation 3.5b:

$$y = v_{0y}t + \tfrac{1}{2}a_y t^2 = (14 \text{ m/s})(7.0 \text{ s}) + \tfrac{1}{2}(12 \text{ m/s}^2)(7.0 \text{ s})^2 = \boxed{+390 \text{ m}}$$

After 7.0 s, the spacecraft is 740 m to the right and 390 m above the origin.

Analyzing Multiple-Concept Problems

EXAMPLE 2 | The Velocity of a Spacecraft

This example also deals with the spacecraft in **Interactive Figure 3.5**. As in Example 1, the x components of the craft's initial velocity and acceleration are $v_{0x} = +22$ m/s and $a_x = +24$ m/s^2, respectively. The corresponding y components are $v_{0y} = +14$ m/s and $a_y = +12$ m/s^2. At a time of $t = 7.0$ s, find the spacecraft's final velocity (magnitude and direction).

Reasoning **Figure 3.6** shows the final velocity vector, which has components v_x and v_y and a magnitude v. The final velocity is directed at an angle θ above the $+x$ axis. The vector and its components form a right triangle, the hypotenuse being the magnitude of the velocity and the components being the other two sides. Thus, we can use the Pythagorean theorem to determine the magnitude v from values for the components v_x and v_y. We can also use trigonometry to determine the directional angle θ.

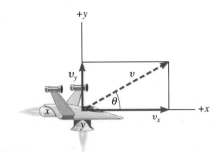

FIGURE 3.6 The velocity vector has components v_x and v_y and a magnitude v. The magnitude gives the speed of the spacecraft, and the angle θ gives the direction of travel relative to the positive x direction.

Knowns and Unknowns The data for this problem are listed in the table that follows:

Description	Symbol	Value	Comment
x component of acceleration	a_x	+24 m/s^2	
x component of initial velocity	v_{0x}	+22 m/s	
y component of acceleration	a_y	+12 m/s^2	
y component of initial velocity	v_{0y}	+14 m/s	
Time	t	7.0 s	Same time for x and y directions
Unknown Variables			
Magnitude of final velocity	v	?	
Direction of final velocity	θ	?	

Modeling the Problem

STEP 1 Final Velocity In **Figure 3.6** the final velocity vector and its components v_x and v_y form a right triangle. Applying the Pythagorean theorem to this right triangle shows that the magnitude v of the final velocity is given in terms of the components by Equation 1a at the right. From the right triangle in **Figure 3.6** it also follows that the directional angle θ is given by Equation 1b at the right.

$$v = \sqrt{v_x^2 + v_y^2} \qquad \text{(1a)}$$

$$\theta = \tan^{-1}\left(\frac{v_y}{v_x}\right) \qquad \text{(1b)}$$

STEP 2 The Components of the Final Velocity Values are given for the kinematic variables a_x, v_{0x}, and t in the x direction and for the corresponding variables in the y direction (see the table of knowns and unknowns). For each direction, then, these values allow us to calculate the final velocity components v_x and v_y by using Equations 3.3a and 3.3b from the equations of kinematics.

$$v_x = v_{0x} + a_x t \qquad \text{(3.3a)}$$

$$v_y = v_{0y} + a_y t \qquad \text{(3.3b)}$$

These expressions can be substituted into Equations 1a and 1b for the magnitude and direction of the final velocity, as shown at the right.

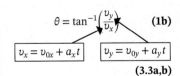

$$v = \sqrt{v_x^2 + v_y^2} \qquad \text{(1a)}$$

$$\boxed{v_x = v_{0x} + a_x t} \qquad \boxed{v_y = v_{0y} + a_y t}$$
$$\text{(3.3a,b)}$$

$$\theta = \tan^{-1}\left(\frac{v_y}{v_x}\right) \qquad \text{(1b)}$$

$$\boxed{v_x = v_{0x} + a_x t} \qquad \boxed{v_y = v_{0y} + a_y t}$$
$$\text{(3.3a,b)}$$

Solution Algebraically combining the results of each step, we find that

STEP 1   STEP 2

$$v = \sqrt{v_x^2 + v_y^2} = \sqrt{(v_{0x} + a_x t)^2 + (v_{0y} + a_y t)^2}$$

$$\theta = \tan^{-1}\left(\frac{v_y}{v_x}\right) = \tan^{-1}\left(\frac{v_{0y} + a_y t}{v_{0x} + a_x t}\right)$$

STEP 1   STEP 2

With the data given for the kinematic variables in the x and y directions, we find that the magnitude and direction of the final velocity of the spacecraft are

$$v = \sqrt{(v_{0x} + a_x t)^2 + (v_{0y} + a_y t)^2}$$

$$= \sqrt{[(22 \text{ m/s}) + (24 \text{ m/s}^2)(7.0 \text{ s})]^2 + [(14 \text{ m/s}) + (12 \text{ m/s}^2)(7.0 \text{ s})]^2} = \boxed{210 \text{ m/s}}$$

$$\theta = \tan^{-1}\left(\frac{v_{0y} + a_y t}{v_{0x} + a_x t}\right) = \tan^{-1}\left[\frac{(14 \text{ m/s}) + (12 \text{ m/s}^2)(7.0 \text{ s})}{(22 \text{ m/s}) + (24 \text{ m/s}^2)(7.0 \text{ s})}\right] = \boxed{27°}$$

After 7.0 s, the spacecraft, at the position determined in Example 1, has a velocity of 210 m/s in a direction of 27° above the positive x axis.

Related Homework: Problem 59

The following Reasoning Strategy gives an overview of how the equations of kinematics are applied to describe motion in two dimensions, as in Examples 1 and 2.

REASONING STRATEGY   **Applying the Equations of Kinematics in Two Dimensions**

1. **Make a drawing to represent the situation being studied.**
2. **Decide which directions are to be called positive (+) and negative (−) relative to a conveniently chosen coordinate origin. Do not change your decision during the course of a calculation.**
3. **Remember that the time variable t has the same value for the part of the motion along the x axis and the part along the y axis.**
4. **In an organized way, write down the values (with appropriate + and − signs) that are given for any of the five kinematic variables associated with the x direction and the y direction. Be on the alert for implied data, such as the phrase "starts from rest," which means that the values of the initial velocity components are zero: $v_{0x} = 0$ m/s and $v_{0y} = 0$ m/s. The data summary tables that are used in the examples are a good way of keeping track of this information. In addition, identify the variables that you are being asked to determine.**
5. **Before attempting to solve a problem, verify that the given information contains values for at least three of the kinematic variables. Do this for the x and the y direction of the motion. Once the three known variables are identified along with the desired unknown variable, the appropriate relations from Table 3.1 can be selected.**
6. **When the motion is divided into segments, remember that the final velocity for one segment is the initial velocity for the next segment.**
7. **Keep in mind that a kinematics problem may have two possible answers. Try to visualize the different physical situations to which the answers correspond.**

Check Your Understanding

(*The answer is given at the end of the book.*)

2. A power boat, starting from rest, maintains a constant acceleration. After a certain time t, its displacement and velocity are $\vec{r}$ and $\vec{v}$. At time $2t$, what would be its displacement and velocity, assuming the acceleration remains the same? **(a)** $2\vec{r}$ and $2\vec{v}$ **(b)** $2\vec{r}$ and $4\vec{v}$ **(c)** $4\vec{r}$ and $2\vec{v}$ **(d)** $4\vec{r}$ and $4\vec{v}$

3.3 Projectile Motion

The biggest thrill in baseball is a home run. The motion of the ball on its curving path into the stands is a common type of two-dimensional motion called "projectile motion." A good description of such motion can often be obtained with the assumption that air resistance is absent.

Using the equations in **Table 3.1**, we consider the horizontal and vertical parts of the motion separately. In the horizontal or x direction, the moving object (the projectile) does not slow down in the absence of air resistance. Thus, the x component of the velocity remains constant at its initial value or $v_x = v_{0x}$, and the x component of the acceleration is $a_x = 0$ m/s². In the vertical or y direction, however, the projectile experiences the effect of gravity. As a result, the y component of the velocity v_y is not constant and changes. The y component of the acceleration a_y is the downward acceleration due to gravity. If the path or trajectory of the projectile is near the earth's surface, a_y has a magnitude of 9.80 m/s². In this text, then, the phrase "projectile motion" means that $a_x = 0$ m/s² and a_y equals the acceleration due to gravity. Example 3 and other examples in this section illustrate how the equations of kinematics are applied to projectile motion.

EXAMPLE 3 | A Falling Care Package

Figure 3.7 shows an airplane moving horizontally with a constant velocity of +115 m/s at an altitude of 1050 m. The directions to the right and upward have been chosen as the positive directions. The plane releases a "care package" that falls to the ground along a curved trajectory. Ignoring air resistance, determine the time required for the package to hit the ground.

Reasoning The time required for the package to hit the ground is the time it takes for the package to fall through a vertical distance of 1050 m. In falling, it moves to the right, as well as downward, but these two parts of the motion occur independently. Therefore, we can focus solely on the vertical part. We note that the package is moving initially in the horizontal or x direction, not in the y direction, so that $v_{0y} = 0$ m/s. Furthermore, when the package hits the ground, the y component of its displacement is $y = -1050$ m, as the drawing shows. The acceleration is that due to gravity, so $a_y = -9.80$ m/s². These data are summarized as follows:

y-Direction Data				
y	a_y	v_y	v_{0y}	t
−1050 m	−9.80 m/s²		0 m/s	?

With these data, Equation 3.5b ($y = v_{0y}t + \frac{1}{2}a_y t^2$) can be used to find the fall time.

> **Problem-Solving Insight** The variables y, a_y, v_y, and v_{0y} are scalar components. Therefore, an algebraic sign (+ or −) must be included with each one to denote direction.

Solution Since $v_{0y} = 0$ m/s, it follows from Equation 3.5b that $y = \frac{1}{2}a_y t^2$ and

$$t = \sqrt{\frac{2y}{a_y}} = \sqrt{\frac{2(-1050 \text{ m})}{-9.80 \text{ m/s}^2}} = \boxed{14.6 \text{ s}}$$

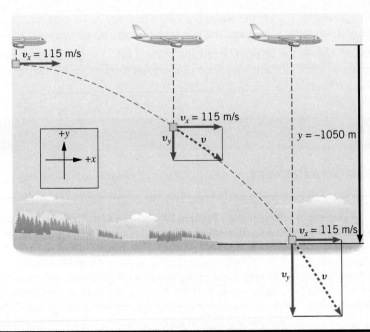

$v_x = 115$ m/s

$v_x = 115$ m/s

$y = -1050$ m

$v_x = 115$ m/s

FIGURE 3.7 The package falling from the plane is an example of projectile motion, as Examples 3 and 4 discuss.

FIGURE 3.8 Package A and package B are released simultaneously at the same height and strike the ground at the same time because their y variables (y, a_y, and v_{0y}) are the same.

The freely falling package in Example 3 picks up vertical speed on the way downward. The horizontal component of the velocity, however, retains its initial value of $v_{0x} = +115$ m/s throughout the entire descent. Since the plane also travels at a constant horizontal velocity of $+115$ m/s, it remains directly above the falling package. The pilot always sees the package directly beneath the plane, as the dashed vertical lines in **Figure 3.7** show. This result is a direct consequence of the fact that the package has no acceleration in the horizontal direction. In reality, air resistance would slow down the package, and it would not remain directly beneath the plane during the descent.

Figure 3.8 illustrates what happens to two packages that are released simultaneously from the same height, in order to emphasize that the vertical and horizontal parts of the motion in Example 3 occur independently. Package A is dropped from a stationary balloon and falls straight downward toward the ground, since it has no horizontal velocity component ($v_{0x} = 0$ m/s). Package B, on the other hand, is given an initial velocity component of $v_{0x} = +115$ m/s in the horizontal direction, as in Example 3, and follows the path shown in the figure. Both packages hit the ground at the same time. Not only do the packages in **Figure 3.8** reach the ground at the same time, but the y components of their velocities are also equal at all points on the way down. However, package B does hit the ground with a greater speed than does package A. Remember, speed is the magnitude of the velocity vector, and the velocity of B has an x component, whereas the velocity of A does not. The magnitude and direction of the velocity vector for package B at the instant just before the package hits the ground is computed in Example 4.

Analyzing Multiple-Concept Problems

EXAMPLE 4 | The Velocity of the Care Package

Figure 3.7 shows a care package falling from a plane, and **Figure 3.8** shows this package as package B. As in Example 3, the directions to the right and upward are chosen as the positive directions, and the plane is moving horizontally with a constant velocity of $+115$ m/s at an altitude of 1050 m. Ignoring air resistance, find the magnitude v and the directional angle θ of the final velocity vector that the package has just before it strikes the ground.

Reasoning Figures 3.7 and 3.8 show the final velocity vector, which has components v_x and v_y and a magnitude v. The vector is directed at an angle θ below the horizontal or x direction. We note the right triangle formed by the vector and its components. The hypotenuse of the triangle is the magnitude of the velocity, and the components are the other two sides. As in Example 2, we can use the Pythagorean theorem to express the magnitude or speed v in terms of the components v_x and v_y, and we can use trigonometry to determine the directional angle θ.

Knowns and Unknowns   The data for this problem are listed in the table that follows:

Description	Symbol	Value	Comment
Explicit Data, x Direction			
x component of initial velocity	v_{0x}	+115 m/s	Package has plane's horizontal velocity at instant of release
Implicit Data, x Direction			
x component of acceleration	a_x	0 m/s²	No horizontal acceleration, since air resistance is ignored
Explicit Data, y Direction			
y component of displacement	y	−1050 m	Negative, since upward is positive and package falls downward
Implicit Data, y Direction			
y component of initial velocity	v_{0y}	0 m/s	Package traveling horizontally in x direction at instant of release, not in y direction
y component of acceleration	a_y	−9.80 m/s²	Acceleration vector for gravity points downward in the negative direction
Unknown Variables			
Magnitude of final velocity	v	?	
Direction of final velocity	θ	?	

Modeling the Problem

STEP 1 Final Velocity   Using the Pythagorean theorem to express the speed v in terms of the components v_x and v_y (see **Figure 3.7** or **3.8**), we obtain Equation 1a at the right. Furthermore, in a right triangle, the cosine of an angle is the side adjacent to the angle divided by the hypotenuse. With this in mind, we see in **Figure 3.7** or **Figure 3.8** that the directional angle θ is given by Equation 1b at the right.

$$v = \sqrt{v_x^{\,2} + v_y^{\,2}} \qquad \text{(1a)}$$

$$\theta = \cos^{-1}\left(\frac{v_x}{v}\right) = \cos^{-1}\left(\frac{v_x}{\sqrt{v_x^{\,2} + v_y^{\,2}}}\right) \qquad \text{(1b)}$$

STEP 2 The Components of the Final Velocity   Reference to the table of knowns and unknowns shows that, in the x direction, values are available for the kinematic variables v_{0x} and a_x. Since the acceleration a_x is zero, the final velocity component v_x remains unchanged from its initial value of v_{0x}, so we have

$$\boxed{v_x = v_{0x}}$$

In the y direction, values are available for y, v_{0y}, and a_y, so that we can determine the final velocity component v_y using Equation 3.6b from the equations of kinematics:

$$\boxed{v_y^{\,2} = v_{0y}^{\,2} + 2a_y y} \qquad \text{(3.6b)}$$

These results for v_x and v_y can be substituted into Equations 1a and 1b, as shown at the right.

$$v = \sqrt{v_x^{\,2} + v_y^{\,2}} \qquad \text{(1a)}$$

$$\boxed{v_x = v_{0x}} \quad \boxed{v_y^{\,2} = v_{0y}^{\,2} + 2a_y y}$$
$$\text{(3.6b)}$$

$$\theta = \cos^{-1}\left(\frac{v_x}{\sqrt{v_x^{\,2} + v_y^{\,2}}}\right) \qquad \text{(1b)}$$

$$\boxed{v_x = v_{0x}} \quad \boxed{v_y^{\,2} = v_{0y}^{\,2} + 2a_y y} \quad \text{(3.6b)}$$

Solution   Algebraically combining the results of each step, we find that

$$\underset{\text{STEP 1}}{\qquad}\underset{\text{STEP 2}}{\qquad}$$
$$v = \sqrt{v_x^{\,2} + v_y^{\,2}} = \sqrt{v_{0x}^{\,2} + v_{0y}^{\,2} + 2a_y y}$$

$$\theta = \cos^{-1}\left(\frac{v_x}{\sqrt{v_x^{\,2} + v_y^{\,2}}}\right) = \cos^{-1}\left(\frac{v_{0x}}{\sqrt{v_{0x}^{\,2} + v_{0y}^{\,2} + 2a_y y}}\right)$$
$$\underset{\text{STEP 1}}{\qquad}\underset{\text{STEP 2}}{\qquad}$$

With the data given for the kinematic variables in the x and y directions, we find that the magnitude and direction of the final velocity of the package are

$$v = \sqrt{v_{0x}^{\,2} + v_{0y}^{\,2} + 2a_y y} = \sqrt{(115 \text{ m/s})^2 + (0 \text{ m/s})^2 + 2(-9.80 \text{ m/s}^2)(-1050 \text{ m})} = \boxed{184 \text{ m/s}}$$

$$\theta = \cos^{-1}\left(\frac{v_{0x}}{\sqrt{v_{0x}^{\,2} + v_{0y}^{\,2} + 2a_y y}}\right) = \cos^{-1}\left[\frac{115 \text{ m/s}}{\sqrt{(115 \text{ m/s})^2 + (0 \text{ m/s})^2 + 2(-9.80 \text{ m/s}^2)(-1050 \text{ m})}}\right] = \boxed{51.3°}$$

Problem-Solving Insight
The speed of a projectile at any location along its path is the magnitude v of its velocity at that location: $v = \sqrt{v_x^{\,2} + v_y^{\,2}}$. Both the horizontal and vertical velocity components contribute to the speed.

Related Homework: Problems 32, 37, 64

An important feature of projectile motion is that there is no acceleration in the horizontal, or x, direction. Conceptual Example 5 discusses an interesting implication of this feature.

CONCEPTUAL EXAMPLE 5 | Shot a Bullet Into the Air . . .

Suppose you are driving in a convertible with the top down. The car is moving to the right at a constant velocity. As **Figure 3.9** illustrates, you point a rifle straight upward and fire it. In the absence of air resistance, would the bullet land **(a)** behind you, **(b)** ahead of you, or **(c)** in the barrel of the rifle?

FIGURE 3.9 The car is moving with a constant velocity to the right, and the rifle is pointed straight up. In the absence of air resistance, a bullet fired from the rifle has no acceleration in the horizontal direction. Example 5 discusses what happens to the bullet.

Reasoning Because there is no air resistance to slow it down, the bullet experiences no horizontal acceleration. Thus, the bullet's horizontal velocity component does not change, and it stays the same as that of the rifle and the car.

Answers (a) and (b) are incorrect. If air resistance were present, it would slow down the bullet and cause it to land behind you, toward the rear of the car. However, air resistance is absent. If the bullet were to land ahead of you, its horizontal velocity component would have to be greater than that of the rifle and the car. This cannot be, since the bullet's horizontal velocity component never changes.

Answer (c) is correct. Since the bullet's horizontal velocity component does not change, it retains its initial value, and remains matched to that of the rifle and the car. As a result, the bullet remains directly above the rifle at all times and would fall directly back into the barrel of the rifle. This situation is analogous to that in **Figure 3.7**, where the care package, as it falls, remains directly below the plane.

Related Homework: Problem 38

Often projectiles, such as footballs and baseballs, are sent into the air at an angle with respect to the ground. From a knowledge of the projectile's initial velocity, a wealth of information can be obtained about the motion. For instance, Example 6 demonstrates how to calculate the maximum height reached by the projectile.

EXAMPLE 6 | The Height of a Kickoff

A placekicker kicks a football at an angle of $\theta = 40.0°$ above the horizontal axis, as **Animated Figure 3.10** shows. The initial speed of the ball is $v_0 = 22$ m/s. Ignore air resistance, and find the maximum height H that the ball attains.

Reasoning The maximum height is a characteristic of the vertical part of the motion, which can be treated separately from the horizontal part. In preparation for making use of this fact, we calculate the vertical component of the initial velocity:

$$v_{0y} = v_0 \sin \theta = +(22 \text{ m/s}) \sin 40.0° = +14 \text{ m/s}$$

The vertical component of the velocity, v_y, decreases as the ball moves upward. Eventually, $v_y = 0$ m/s at the maximum height H. The data below can be used in Equation 3.6b ($v_y^2 = v_{0y}^2 + 2a_y y$) to find the maximum height:

y-Direction Data				
y	a_y	v_y	v_{0y}	t
$H = ?$	-9.80 m/s^2	0 m/s	$+14$ m/s	

ANIMATED FIGURE 3.10 A football is kicked with an initial speed of v_0 at an angle of θ above the ground. The ball attains a maximum height H and a range R.

Problem-Solving Insight   When a projectile reaches maximum height, the vertical component of its velocity is momentarily zero ($v_y = 0$ m/s). However, the horizontal component of its velocity is not zero.

The height H depends only on the y variables; the same height would have been reached had the ball been thrown *straight up* with an initial velocity of $v_{0y} = +14$ m/s.

Solution   From Equation 3.6b, we find that

$$y = H = \frac{v_y^2 - v_{0y}^2}{2a_y} = \frac{(0 \text{ m/s})^2 - (14 \text{ m/s})^2}{2(-9.80 \text{ m/s}^2)} = \boxed{+10 \text{ m}}$$

It is also possible to find the total time or "hang time" during which the football in **Animated Figure 3.10** is in the air. Example 7 shows how to determine this time.

EXAMPLE 7 | The Physics of the "Hang Time" of a Football

For the motion illustrated in **Animated Figure 3.10**, ignore air resistance and use the data from Example 6 to determine the time of flight between kickoff and landing.

Reasoning   Given the initial velocity, it is the acceleration due to gravity that determines how long the ball stays in the air. Thus, to find the time of flight we deal with the vertical part of the motion. Since the ball starts at and returns to ground level, the displacement in the y direction is zero. The initial velocity component in the y direction is the same as that in Example 6; that is, $v_{0y} = +14$ m/s. Therefore, we have

		y-Direction Data		
y	a_y	v_y	v_{0y}	t
0 m	-9.80 m/s^2		$+14$ m/s	?

The time of flight can be determined from Equation 3.5b ($y = v_{0y}t + \frac{1}{2}a_y t^2$).

Solution   Using Equation 3.5b and the fact that $y = 0$ m, we find that

$$y = 0 = v_{0y}t + \tfrac{1}{2}a_y t^2 = \left(v_{0y} + \tfrac{1}{2}a_y t\right)t$$

There are two solutions to this equation. One is given by $v_{0y} + \frac{1}{2}a_y t = 0$, with the result that

$$t = \frac{-2v_{0y}}{a_y} = \frac{-2(14 \text{ m/s})}{-9.80 \text{ m/s}^2} = 2.9 \text{ s}$$

The other is given by $t = 0$ s. The solution we seek is $\boxed{t = 2.9 \text{ s}}$, because $t = 0$ s corresponds to the initial kickoff.

Another important feature of projectile motion is called the "range." The range, as **Animated Figure 3.10** shows, is the horizontal distance traveled between launching and landing, assuming the projectile returns to the *same vertical level* at which it was fired. Example 8 shows how to obtain the range.

EXAMPLE 8 | The Range of a Kickoff

For the motion shown in **Animated Figure 3.10** and discussed in Examples 6 and 7, ignore air resistance and calculate the range R of the projectile.

Reasoning   The range is a characteristic of the horizontal part of the motion. Thus, our starting point is to determine the horizontal component of the initial velocity:

$$v_{0x} = v_0 \cos \theta = +(22 \text{ m/s}) \cos 40.0° = +17 \text{ m/s}$$

Recall from Example 7 that the time of flight is $t = 2.9$ s. Since there is no acceleration in the x direction, v_x remains constant, and the range is simply the product of $v_x = v_{0x}$ and the time.

Solution   The range is

$$x = R = v_{0x}t = +(17 \text{ m/s})(2.9 \text{ s}) = \boxed{+49 \text{ m}}$$

The range in the previous example depends on the angle θ at which the projectile is fired above the horizontal. When air resistance is absent, the maximum range results when $\theta = 45°$.

Math Skills To show that a projectile launched from and returning to ground level has its maximum range when $\theta = 45°$, we begin with the expression for the range R from Example 8:

$$R = v_{0x}t = (v_0 \cos \theta)t$$

Example 7 shows that the time of flight of the projectile is

$$t = \frac{-2v_{0y}}{a_y} = \frac{-2v_{0y}}{-9.80 \text{ m/s}^2} = \frac{2v_{0y}}{9.80 \text{ m/s}^2}$$

According to Example 6, the velocity component v_{0y} in this result is $v_{0y} = v_0 \sin \theta$, so that

$$t = \frac{2v_0 \sin \theta}{9.80 \text{ m/s}^2}$$

Substituting this expression for t into the range expression gives

$$R = (v_0 \cos \theta)t = (v_0 \cos \theta)\frac{2v_0 \sin \theta}{9.80 \text{ m/s}^2} = \frac{v_0^2(2 \sin \theta \cos \theta)}{9.80 \text{ m/s}^2}$$

Equation 6 (Other Trigonometric Identities) in Appendix E.2 shows that $2 \sin \theta \cos \theta = \sin 2\theta$, so the range expression becomes

$$R = \frac{v_0^2(2 \sin \theta \cos \theta)}{9.80 \text{ m/s}^2} = \frac{v_0^2 \sin 2\theta}{9.80 \text{ m/s}^2}$$

In this result R has its maximum value when $\sin 2\theta$ has its maximum value of 1. This occurs when $2\theta = 90°$, or when $\theta = 45°$.

In projectile motion, the magnitude of the acceleration due to gravity affects the trajectory in a significant way. For example, a baseball or a golf ball would travel much farther and higher on the moon than on the earth, when launched with the same initial velocity. The reason is that the moon's gravity is only about one-sixth as strong as the earth's.

Section 2.6 points out that certain types of symmetry with respect to time and speed are present for freely falling bodies. These symmetries are also found in projectile motion, since projectiles are falling freely in the vertical direction. In particular, the time required for a projectile to reach its maximum height H is equal to the time spent returning to the ground. In addition, **Figure 3.11** shows that the speed v of the object at any height above the ground on the upward part of the trajectory is equal to the speed v at the same height on the downward part. Although the two speeds are the same, the velocities are different, because they point in different directions. Conceptual Example 9 shows how to use this type of symmetry in your reasoning.

FIGURE 3.11 The speed v of a projectile at a given height above the ground is the same on the upward and downward parts of the trajectory. The velocities are different, however, since they point in different directions.

CONCEPTUAL EXAMPLE 9 | Two Ways to Throw a Stone

From the top of a cliff overlooking a lake, a person throws two stones. The stones have identical initial speeds v_0, but stone 1 is thrown downward at an angle θ below the horizontal, while stone 2 is thrown upward at the same angle above the horizontal, as

Figure 3.12 shows. Neglect air resistance and decide which stone, if either, strikes the water with the greater velocity: **(a)** both stones strike the water with the same velocity, **(b)** stone 1 strikes with the greater velocity, **(c)** stone 2 strikes with the greater velocity.

Reasoning   Note point P in the drawing, where stone 2 returns to its initial height; here the speed of stone 2 is v_0 (the same as its initial speed), but its velocity is directed at an angle θ below the horizontal. This is exactly the type of projectile symmetry illustrated in **Figure 3.11**, and this symmetry will lead us to the correct answer.

Answers (b) and (c) are incorrect.   You might guess that stone 1, being hurled downward, would strike the water with the greater velocity. Or, you might think that stone 2, having reached a greater height than stone 1, would hit the water with the greater velocity. To understand why neither of these answers is correct, see the response for answer (a) below.

Answer (a) is correct.   Let's follow the path of stone 2 as it rises to its maximum height and falls back to earth. When it reaches point P in the drawing, stone 2 has a velocity that is identical to the velocity with which stone 1 is thrown downward from the top of the cliff (see the drawing). From this point on, the velocity of stone 2 changes in exactly the same way as that for stone 1, so both stones strike the water with the same velocity.

FIGURE 3.12   Two stones are thrown off the cliff with identical initial speeds v_0, but at equal angles θ that are below and above the horizontal. Conceptual Example 9 compares the velocities with which the stones hit the water below.

Related Homework: Problems 23, 67

In all the examples in this section, the projectiles follow a curved trajectory. In general, if the only acceleration is that due to gravity, the shape of the path can be shown to be a *parabola*.

Check Your Understanding

(The answers are given at the end of the book.)

3. A projectile is fired into the air, and it follows the parabolic path shown in **CYU Figure 3.1**, landing on the right. There is no air resistance. At any instant, the projectile has a velocity $\vec{v}$ and an acceleration $\vec{a}$. Which one or more of the drawings could *not* represent the directions for $\vec{v}$ and $\vec{a}$ at any point on the trajectory?

CYU FIGURE 3.1

4. An object is thrown upward at an angle θ above the ground, eventually returning to earth. **(a)** Is there any place along the trajectory where the velocity and acceleration are perpendicular? If so, where? **(b)** Is there any place where the velocity and acceleration are parallel? If so, where?

5. Is the acceleration of a projectile equal to zero when the projectile reaches the top of its trajectory?

6. In baseball, the pitcher's mound is raised to compensate for the fact that the ball falls downward as it travels from the pitcher toward the batter. If baseball were played on the moon, would the pitcher's mound have to be **(a)** higher than, **(b)** lower than, or **(c)** the same height as it is on earth?

7. A tennis ball is hit upward into the air and moves along an arc. Neglecting air resistance, where along the arc is the speed of the ball **(a)** a minimum and **(b)** a maximum?

8. A wrench is accidentally dropped from the top of the mast on a sailboat. Air resistance is negligible. Will the wrench hit at the same place on the deck whether the sailboat is at rest or moving with a constant velocity?

9. A rifle, at a height H above the ground, fires a bullet parallel to the ground. At the same instant and at the same height, a second bullet is dropped from rest. In the absence of air resistance, which bullet, if either, strikes the ground first?

10. A stone is thrown horizontally from the top of a cliff and eventually hits the ground below. A second stone is dropped from rest from the same cliff, falls through the same height, and also hits the ground below. Ignore air resistance. Is each of the following quantities different or the same in the two cases? **(a)** Displacement **(b)** Speed just before impact with the ground **(c)** Time of flight

11. A leopard springs upward at a 45° angle and then falls back to the ground. Air resistance is negligible. Does the leopard, at any point on its trajectory, ever have a speed that is one-half its initial value?

12. Two balls are launched upward from the same spot at different angles with respect to the ground. Both balls rise to the same maximum height. Ball A, however, follows a trajectory that has a greater range than that of ball B. Ignoring air resistance, decide which ball, if either, has the greater launch speed.

3.4 | Relative Velocity

To someone hitchhiking along a highway, two cars speeding by in adjacent lanes seem like a blur. But if the cars have the same velocity, each driver sees the other remaining in place, one lane away. The hitchhiker observes a velocity of perhaps 30 m/s, but each driver observes the other's velocity to be zero. Clearly, the velocity of an object is relative to the observer who is making the measurement.

Figure 3.13 illustrates the concept of relative velocity by showing a passenger walking toward the front of a moving train. The people sitting on the train see the passenger walking with a velocity of +2.0 m/s, where the plus sign denotes a direction to the right. Suppose the train is moving with a velocity of +9.0 m/s relative to an observer standing on the ground. Then the ground-based observer would see the passenger moving with a velocity of +11 m/s, due in part to the walking motion and in part to the train's motion. As an aid in describing relative velocity, let us define the following symbols:

$$\vec{v}_{PT} = \text{velocity of the } \boxed{\text{Passenger}} \text{ relative to the } \boxed{\text{Train}} = +2.0 \text{ m/s}$$
$$\vec{v}_{TG} = \text{velocity of the } \boxed{\text{Train}} \text{ relative to the } \boxed{\text{Ground}} = +9.0 \text{ m/s}$$
$$\vec{v}_{PG} = \text{velocity of the } \boxed{\text{Passenger}} \text{ relative to the } \boxed{\text{Ground}} = +11 \text{ m/s}$$

In terms of these symbols, the situation in Figure 3.13 can be summarized as follows:

$$\vec{v}_{PG} = \vec{v}_{PT} + \vec{v}_{TG} \tag{3.7}$$

or

$$\vec{v}_{PG} = (2.0 \text{ m/s}) + (9.0 \text{ m/s}) = +11 \text{ m/s}$$

According to Equation 3.7*, $\vec{v}_{PG}$ is the vector sum of $\vec{v}_{PT}$ and $\vec{v}_{TG}$, and this sum is shown in the drawing. Had the passenger been walking toward the rear of the train, rather than

FIGURE 3.13 The velocity of the passenger relative to the ground-based observer is $\vec{v}_{PG}$. It is the vector sum of the velocity $\vec{v}_{PT}$ of the passenger relative to the train and the velocity $\vec{v}_{TG}$ of the train relative to the ground: $\vec{v}_{PG} = \vec{v}_{PT} + \vec{v}_{TG}$.

---

*This equation assumes that the train and the ground move in a straight line relative to one another.

toward the front, the velocity relative to the ground-based observer would have been $\vec{\mathbf{v}}_{PG} = (-2.0 \text{ m/s}) + (9.0 \text{ m/s}) = +7.0 \text{ m/s}$.

Each velocity symbol in Equation 3.7 contains a two-letter subscript. The first letter in the subscript refers to the body that is moving, while the second letter indicates the object relative to which the velocity is measured. For example, $\vec{\mathbf{v}}_{TG}$ and $\vec{\mathbf{v}}_{PG}$ are the velocities of the **T**rain and **P**assenger measured relative to the **G**round. Similarly, $\vec{\mathbf{v}}_{PT}$ is the velocity of the **P**assenger measured by an observer sitting on the **T**rain.

The ordering of the subscript symbols in Equation 3.7 follows a definite pattern. The first subscript (P) on the left side of the equation is also the first subscript on the right side of the equation. Likewise, the last subscript (G) on the left side is also the last subscript on the right side. The third subscript (T) appears only on the right side of the equation as the two "inner" subscripts. The colored boxes below emphasize the pattern of the symbols in the subscripts:

$$\vec{\mathbf{v}}_{\boxed{PG}} = \vec{\mathbf{v}}_{\boxed{P}T} + \vec{\mathbf{v}}_{T\boxed{G}}$$

TAL COHEN/AFP/Getty Images

PHOTO 3.1 In landing on a moving aircraft carrier, the pilot of the helicopter must match the helicopter's horizontal velocity to the carrier's velocity, so that the relative velocity of the helicopter and the carrier is zero.

In other situations such as in the case of a helicopter landing on an aircraft carrier (**Photo 3.1**), the subscripts will not necessarily be P, G, and T, but will be compatible with the names of the objects involved in the motion.

Equation 3.7 has been presented in connection with one-dimensional motion, but the result is also valid for two-dimensional motion. **Figure 3.14** depicts a common situation that deals with relative velocity in two dimensions. Part *a* of the drawing shows a boat being carried downstream by a river; the engine of the boat is turned off. In part *b*, the engine is turned on, and now the boat moves across the river in a diagonal fashion because of the combined motion produced by the current and the engine. The list below gives the velocities for this type of motion and the objects relative to which they are measured:

FIGURE 3.14 (*a*) A boat with its engine turned off is carried along by the current. (*b*) With the engine turned on, the boat moves across the river in a diagonal fashion.

$\vec{\mathbf{v}}_{\boxed{BW}}$ = velocity of the $\boxed{\text{Boat}}$ relative to the $\boxed{\text{Water}}$

$\vec{\mathbf{v}}_{\boxed{WS}}$ = velocity of the $\boxed{\text{Water}}$ relative to the $\boxed{\text{Shore}}$

$\vec{\mathbf{v}}_{\boxed{BS}}$ = velocity of the $\boxed{\text{Boat}}$ relative to the $\boxed{\text{Shore}}$

The velocity $\vec{\mathbf{v}}_{BW}$ of the boat relative to the water is the velocity measured by an observer who, for instance, is floating on an inner tube and drifting downstream with the current. When the engine is turned off, the boat also drifts downstream with the current, and $\vec{\mathbf{v}}_{BW}$ is zero. When the engine is turned on, however, the boat can move relative to the water, and $\vec{\mathbf{v}}_{BW}$ is no longer zero. The velocity $\vec{\mathbf{v}}_{WS}$ of the water relative to the shore is the velocity of the current measured by an observer on the shore. The velocity $\vec{\mathbf{v}}_{BS}$ of the boat relative to the shore is due to the combined motion of the boat relative to the water and the motion of the water relative to the shore. In symbols,

$$\vec{\mathbf{v}}_{\boxed{BS}} = \vec{\mathbf{v}}_{\boxed{B}W} + \vec{\mathbf{v}}_{W\boxed{S}}$$

The ordering of the subscripts in this equation is identical to that in Equation 3.7, although the letters have been changed to reflect a different physical situation. Example 10 illustrates the concept of relative velocity in two dimensions.

EXAMPLE 10 | Crossing a River

The engine of a boat drives it across a river that is 1800 m wide. The velocity $\vec{\mathbf{v}}_{BW}$ of the boat relative to the water is 4.0 m/s, directed perpendicular to the current, as in **Interactive Figure 3.15**. The velocity $\vec{\mathbf{v}}_{WS}$ of the water relative to the shore is 2.0 m/s. **(a)** What is the velocity $\vec{\mathbf{v}}_{BS}$ of the boat relative to the shore? **(b)** How long does it take for the boat to cross the river?

Reasoning **(a)** The velocity of the boat relative to the shore is $\vec{\mathbf{v}}_{BS}$. It is the vector sum of the velocity $\vec{\mathbf{v}}_{BW}$ of the boat relative to the water and the velocity $\vec{\mathbf{v}}_{WS}$ of the water relative to the shore: $\vec{\mathbf{v}}_{BS} = \vec{\mathbf{v}}_{BW} + \vec{\mathbf{v}}_{WS}$. Since $\vec{\mathbf{v}}_{BW}$ and $\vec{\mathbf{v}}_{WS}$ are both known, we can use this relation among the velocities, with the aid of trigonometry, to find the magnitude and directional angle of $\vec{\mathbf{v}}_{BS}$. **(b)** The component of $\vec{v}_{BS}$ that is parallel to the width of the river (see **Interactive Figure 3.15**) determines how fast the boat crosses the river; this parallel component is $v_{BS} \sin \theta = v_{BW} = 4.0$ m/s. The time for the boat to cross the river is equal to the width of the river divided by the magnitude of this velocity component.

Solution **(a)** Since the vectors $\vec{v}_{BW}$ and $\vec{v}_{WS}$ are perpendicular (see **Interactive Figure 3.15**), the magnitude of $\vec{v}_{BS}$ can be determined by using the Pythagorean theorem:

$$v_{BS} = \sqrt{(v_{BW})^2 + (v_{WS})^2} = \sqrt{(4.0 \text{ m/s})^2 + (2.0 \text{ m/s})^2} = \boxed{4.5 \text{ m/s}}$$

Thus, the boat moves at a speed of 4.5 m/s with respect to an observer on shore. The direction of the boat relative to the shore is given by the angle θ in the drawing:

$$\tan \theta = \frac{v_{BW}}{v_{WS}} \quad \text{or} \quad \theta = \tan^{-1}\left(\frac{v_{BW}}{v_{WS}}\right) = \tan^{-1}\left(\frac{4.0 \text{ m/s}}{2.0 \text{ m/s}}\right) = \boxed{63°}$$

(b) The time t for the boat to cross the river is

$$t = \frac{\text{Width}}{v_{BS} \sin \theta} = \frac{1800 \text{ m}}{4.0 \text{ m/s}} = \boxed{450 \text{ s}}$$

INTERACTIVE FIGURE 3.15 The velocity of the boat relative to the shore is $\vec{v}_{BS}$. It is the vector sum of the velocity $\vec{v}_{BW}$ of the boat relative to the water and the velocity $\vec{v}_{WS}$ of the water relative to the shore: $\vec{v}_{BS} = \vec{v}_{BW} + \vec{v}_{WS}$.

Sometimes, situations arise when two vehicles are in relative motion, and it is useful to know the relative velocity of one with respect to the other. Example 11 considers this type of relative motion.

EXAMPLE 11 | Approaching an Intersection

Figure 3.16a shows two cars approaching an intersection along perpendicular roads. The cars have the following velocities:

$\vec{v}_{AG}$ = velocity of ⌐car A⌐ relative to the ⌐Ground⌐ = 25.0 m/s, eastward

$\vec{v}_{BG}$ = velocity of ⌐car B⌐ relative to the ⌐Ground⌐ = 15.8 m/s, northward

Find the magnitude and direction of $\vec{v}_{AB}$, where

$\vec{v}_{AB}$ = velocity of ⌐car A⌐ as measured by a passenger in ⌐car B⌐

Reasoning To find $\vec{v}_{AB}$, we use an equation whose subscripts follow the order outlined earlier. Thus,

$$\vec{v}_{\boxed{AB}} = \vec{v}_{\boxed{A}G} + \vec{v}_{G\boxed{B}}$$

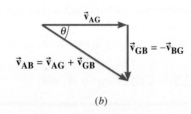

(b)

(a)

FIGURE 3.16 Two cars are approaching an intersection along perpendicular roads.

In this equation, the term $\vec{\mathbf{v}}_{GB}$ is the velocity of the ground relative to a passenger in car B, rather than $\vec{\mathbf{v}}_{BG}$, which is given as 15.8 m/s, northward. In other words, the subscripts are reversed. However, $\vec{\mathbf{v}}_{GB}$ is related to $\vec{\mathbf{v}}_{BG}$ according to

$$\vec{\mathbf{v}}_{GB} = -\vec{\mathbf{v}}_{BG}$$

This relationship reflects the fact that a passenger in car B, moving northward relative to the ground, looks out the car window and sees the ground moving southward, in the opposite direction. Therefore, the equation $\vec{\mathbf{v}}_{AB} = \vec{\mathbf{v}}_{AG} + \vec{\mathbf{v}}_{GB}$ may be used to find $\vec{\mathbf{v}}_{AB}$, provided we recognize $\vec{\mathbf{v}}_{GB}$ as a vector that points opposite to the given velocity $\vec{\mathbf{v}}_{BG}$. With this in mind, **Figure 3.16b** illustrates how $\vec{\mathbf{v}}_{AG}$ and $\vec{\mathbf{v}}_{GB}$ are added vectorially to give $\vec{\mathbf{v}}_{AB}$.

Problem-Solving Insight In general, the velocity of object R relative to object S is always the negative of the velocity of object S relative to R: $\vec{\mathbf{v}}_{RS} = -\vec{\mathbf{v}}_{SR}$.

Solution From the vector triangle in **Figure 3.16b**, the magnitude and direction of $\vec{\mathbf{v}}_{AB}$ can be calculated as

$$v_{AB} = \sqrt{(v_{AG})^2 + (v_{GB})^2} = \sqrt{(25.0 \text{ m/s})^2 + (-15.8 \text{ m/s})^2}$$
$$= \boxed{29.6 \text{ m/s}}$$

and

$$\cos\theta = \frac{v_{AG}}{v_{AB}} \quad \text{or} \quad \theta = \cos^{-1}\left(\frac{v_{AG}}{v_{AB}}\right) = \cos^{-1}\left(\frac{25.0 \text{ m/s}}{29.6 \text{ m/s}}\right) = \boxed{32.4°}$$

THE PHYSICS OF . . . raindrops falling on car windows. While driving a car, have you ever noticed that the rear window sometimes remains dry, even though rain is falling? This phenomenon is a consequence of relative velocity, as **Figure 3.17** helps to explain. Part *a* shows a car traveling horizontally with a velocity of $\vec{\mathbf{v}}_{CG}$ and a raindrop falling vertically with a velocity of $\vec{\mathbf{v}}_{RG}$. Both velocities are measured relative to the ground. To determine whether the raindrop hits the window, however, we need to consider the velocity of the raindrop relative to the car, not to the ground. This velocity is $\vec{\mathbf{v}}_{RC}$, and we know that

$$\vec{\mathbf{v}}_{RC} = \vec{\mathbf{v}}_{RG} + \vec{\mathbf{v}}_{GC} = \vec{\mathbf{v}}_{RG} - \vec{\mathbf{v}}_{CG}$$

Here, we have used the fact that $\vec{\mathbf{v}}_{GC} = -\vec{\mathbf{v}}_{CG}$. Part *b* of the drawing shows the tail-to-head arrangement corresponding to this vector subtraction and indicates that the direction of $\vec{\mathbf{v}}_{RC}$ is given by the angle θ_R. In comparison, the rear window is inclined at an angle θ_W with respect to the vertical (see the blowup in part *a*). When θ_R is greater than θ_W, the raindrop will miss the window. However, θ_R is determined by the speed v_{RG} of the raindrop and the speed v_{CG} of the car, according to $\theta_R = \tan^{-1}(v_{CG}/v_{RG})$. At higher car speeds, the angle θ_R becomes too large for the drop to hit the window. At a high enough speed, then, the car simply drives out from under each falling drop!

(a)   (b)

FIGURE 3.17 (*a*) With respect to the ground, a car is traveling at a velocity of $\vec{\mathbf{v}}_{CG}$ and a raindrop is falling at a velocity of $\vec{\mathbf{v}}_{RG}$. The rear window of the car is inclined at an angle θ_W with respect to the vertical. (*b*) This tail-to-head arrangement of vectors corresponds to the equation $\vec{\mathbf{v}}_{RC} = \vec{\mathbf{v}}_{RG} - \vec{\mathbf{v}}_{CG}$.

Check Your Understanding

(*The answers are given at the end of the book.*)

13. Three cars, A, B, and C, are moving along a straight section of a highway. The velocity of A relative to B is $\vec{\mathbf{v}}_{AB}$, the velocity of A relative to C is $\vec{\mathbf{v}}_{AC}$, and the velocity of C relative to B is $\vec{\mathbf{v}}_{CB}$. Fill in the missing velocities in the table.

	$\vec{\mathbf{v}}_{AB}$	$\vec{\mathbf{v}}_{AC}$	$\vec{\mathbf{v}}_{CB}$
(a)	?	+40 m/s	+30 m/s
(b)	?	+50 m/s	−20 m/s
(c)	+60 m/s	+20 m/s	?
(d)	−50 m/s	?	+10 m/s

14. On a riverboat cruise, a plastic bottle is accidentally dropped overboard. A passenger on the boat estimates that the boat pulls ahead of the bottle by 5 meters each second. Is it possible to conclude that the magnitude of the velocity of the boat with respect to the shore is 5 m/s?

15. A plane takes off at St. Louis, flies straight to Denver, and then returns the same way. The plane flies at the same speed with respect to the ground during the entire flight, and there are no head winds or tail winds. Since the earth revolves around its axis once a day, you might expect that the times for the outbound trip and the return trip differ, depending on whether the plane flies against the earth's rotation or with it. Is this true, or are the two times the same?

16. A child is playing on the floor of a recreational vehicle (RV) as it moves along the highway at a constant velocity. He has a toy cannon, which shoots a marble at a fixed angle and speed with respect to the floor. The cannon can be aimed toward the front or the rear of the RV. Is the range toward the front the same as, less than, or greater than the range toward the rear? Answer this question **(a)** from the child's point of view and **(b)** from the point of view of an observer standing still on the ground.

17. Three swimmers can swim equally fast relative to the water. They have a race to see who can swim across a river in the least time. Swimmer A swims perpendicular to the current and lands on the far shore downstream, because the current has swept him in that direction. Swimmer B swims upstream at an angle to the current, choosing the angle so that he lands on the far shore directly opposite the starting point. Swimmer C swims downstream at an angle to the current in an attempt to take advantage of the current. Who crosses the river in the least time?

EXAMPLE 12 | BIO The Physics of Blood Splatter

The shape of the splatter made by a falling drop of blood on a surface can provide information on the velocity of the drop at impact. By measuring the physical dimensions of the pattern, forensic scientists can determine the drop's angle of impact with the surface. If a drop of blood falls straight downward and strikes a horizontal surface, the pattern it leaves is fairly circular (**Figure 3.18a**). However, if the drop strikes the surface at some angle less than 90°, then the pattern will be elongated with a tail (**Figure 3.18a**). Consider the following situation where a chef at a restaurant accidentally cuts her finger while slicing vegetables. She quickly gets in her car to drive to the closest emergency room for stitches, but she keeps her hand outside the window, so as to not drip blood inside the car. Based on the elongation of the drops of blood on the roadway, the impact angle was determined to be 15°. Assume the drops of blood behave like a projectile released from a height of 1.1 m and determine the speed of the car.

Reasoning This problem is similar to the physics of falling raindrops. We can analyze the blood splatter by using relative velocity. Here, we assume the chef's car is traveling horizontally with a velocity relative to ground of $\vec{v}_{CG}$, and the blood drop is falling vertically and impacts the ground with a velocity relative to the car of $\vec{v}_{BC}$ (**Figure 3.18b**). The velocity of the blood drop relative to the ground, which determines its elongation on impact, will be given by:

$$\vec{v}_{BG} = \vec{v}_{BC} + \vec{v}_{CG}$$

Notice how the subscripts follow the order outlined earlier. This equation can be solved for the velocity of the car relative to ground ($\vec{v}_{CG}$). The velocity of the blood drop relative to the car ($\vec{v}_{BC}$) at impact can be determined using the kinematic equations for freely falling bodies.

Solution From the vector triangle in **Figure 3.18c**, $\tan 15° = \dfrac{v_{BC}}{v_{CG}}$, therefore, $v_{CG} = \dfrac{v_{BC}}{\tan 15°}$. To calculate this we have to find $\vec{v}_{BC}$. This represents the speed of the blood drop relative to the car at impact. The speed of the blood drop at impact will be equal to the speed of an object that falls from rest through a distance of

(a)

(b)   (c)

FIGURE 3.18 (a) The shape of a drop of blood splatter on a horizontal surface depends on the angle of impact (θ) of the drop. (b) The velocity of the car relative to the ground is $\vec{v}_{CG}$, and the velocity of the blood drop relative to the car is $\vec{v}_{BC}$. (c) This figure illustrates how $\vec{v}_{BC}$ and $\vec{v}_{CG}$ add vectorially to give $\vec{v}_{BG}$.

1.1 m. Using Equation 2.9, we have $v_{BC}^2 = v_{BC_0}^2 + 2ay$ or $v_{BC} = \sqrt{v_{BC_0}^2 + 2ay}$. Therefore,

$$v_{BC} = \sqrt{(0 \text{ m/s})^2 + 2(-9.8 \text{ m/s}^2)(-1.1 \text{ m})} = 4.6 \text{ m/s}$$

Finally, $v_{CG} = \dfrac{v_{BC}}{\tan 15°} = \dfrac{4.6 \text{ m/s}}{\tan 15°} = \boxed{17 \text{ m/s}}$

Concept Summary

3.1 Displacement, Velocity, and Acceleration The position of an object is located with a vector $\vec{r}$ drawn from the coordinate origin to the object. The displacement $\Delta \vec{r}$ of the object is defined as $\Delta \vec{r} = \vec{r} - \vec{r}_0$, where $\vec{r}$ and $\vec{r}_0$ specify its final and initial positions, respectively.

The average velocity $\overline{\vec{v}}$ of an object moving between two positions is defined as its displacement $\Delta \vec{r} = \vec{r} - \vec{r}_0$ divided by the elapsed time $\Delta t = t - t_0$, as in Equation 3.1.

$$\overline{\vec{v}} = \frac{\vec{r} - \vec{r}_0}{t - t_0} = \frac{\Delta \vec{r}}{\Delta t} \tag{3.1}$$

The instantaneous velocity $\vec{v}$ is the velocity at an instant of time. The average velocity becomes equal to the instantaneous velocity in the limit that Δt becomes infinitesimally small ($\Delta t \to 0$ s), as shown in Equation 1.

$$\vec{v} = \lim_{\Delta t \to 0} \frac{\Delta \vec{r}}{\Delta t} \tag{1}$$

The average acceleration $\overline{\vec{a}}$ of an object is the change in its velocity, $\Delta \vec{v} = \vec{v} - \vec{v}_0$ divided by the elapsed time $\Delta t = t - t_0$, as in Equation 3.2.

$$\overline{\vec{a}} = \frac{\vec{v} - \vec{v}_0}{t - t_0} = \frac{\Delta \vec{v}}{\Delta t} \tag{3.2}$$

The instantaneous acceleration $\vec{a}$ is the acceleration at an instant of time. The average acceleration becomes equal to the instantaneous acceleration in the limit that the elapsed time Δt becomes infinitesimally small, as shown in Equation 2.

$$\vec{a} = \lim_{\Delta t \to 0} \frac{\Delta \vec{v}}{\Delta t} \tag{2}$$

3.2 Equations of Kinematics in Two Dimensions Motion in two dimensions can be described in terms of the time t and the x and y components of four vectors: the displacement, the acceleration, and the initial and final velocities. The x part of the motion occurs exactly as it would if the y part did not occur at all. Similarly, the y part of the motion occurs exactly as it would if the x part of the motion did not exist. The motion can be analyzed by treating the x and y components of the four vectors separately and realizing that the time t is the same for each component.

When the acceleration is constant, the x components of the displacement, the acceleration, and the initial and final velocities are related by the equations of kinematics, and so are the y components:

x Component		y Component	
$v_x = v_{0x} + a_x t$	(3.3a)	$v_y = v_{0y} + a_y t$	(3.3b)
$x = \frac{1}{2}(v_{0x} + v_x)t$	(3.4a)	$y = \frac{1}{2}(v_{0y} + v_y)t$	(3.4b)
$x = v_{0x}t + \frac{1}{2}a_x t^2$	(3.5a)	$y = v_{0y}t + \frac{1}{2}a_y t^2$	(3.5b)
$v_x^2 = v_{0x}^2 + 2a_x x$	(3.6a)	$v_y^2 = v_{0y}^2 + 2a_y y$	(3.6b)

The directions of the components of the displacement, the acceleration, and the initial and final velocities are conveyed by assigning a plus (+) or minus (−) sign to each one.

3.3 Projectile Motion Projectile motion is an idealized kind of motion that occurs when a moving object (the projectile) experiences only the acceleration due to gravity, which acts vertically downward. If the trajectory of the projectile is near the earth's surface, the vertical component a_y of the acceleration has a magnitude of 9.80 m/s^2. The acceleration has no horizontal component ($a_x = 0 \text{ m/s}^2$), the effects of air resistance being negligible.

There are several symmetries in projectile motion: (1) The time to reach maximum height from any vertical level is equal to the time spent returning from the maximum height to that level. (2) The speed of a projectile depends only on its height above its launch point, and not on whether it is moving upward or downward.

3.4 Relative Velocity The velocity of object A relative to object B is $\vec{v}_{AB}$, and the velocity of object B relative to object C is $\vec{v}_{BC}$. The velocity of A relative to C is shown in Equation 3 (note the ordering of the subscripts). While the velocity of object A relative to object B is $\vec{v}_{AB}$, the velocity of B relative to A is $\vec{v}_{BA} = -\vec{v}_{AB}$.

$$\vec{v}_{AC} = \vec{v}_{AB} + \vec{v}_{BC} \tag{3}$$

Focus on Concepts

Online

Additional questions are available for assignment in WileyPLUS.

Section 3.3   Projectile Motion

1. The drawing shows projectile motion at three points along the trajectory. The speeds at the points are v_1, v_2, and v_3. Assume there is no air resistance and rank the speeds, largest to smallest. (Note that the symbol > means "greater than.") (a) $v_1 > v_3 > v_2$ (b) $v_1 > v_2 > v_3$ (c) $v_2 > v_3 > v_1$ (d) $v_2 > v_1 > v_3$ (e) $v_3 > v_2 > v_1$

QUESTION 1

2. Two balls are thrown from the top of a building, as in the drawing. Ball 1 is thrown straight down, and ball 2 is thrown with the same speed, but upward at an angle θ with respect to the horizontal. Consider the motion of the balls after they are released. Which one of the following statements is true? (a) The acceleration of ball 1 becomes larger and larger as it falls, because the ball is going faster and faster. (b) The acceleration of ball 2 decreases as it rises, becomes zero at the top of the trajectory, and then increases as the ball begins to fall toward the ground. (c) Both balls have the same acceleration at all times. (d) Ball 2 has an acceleration in both the horizontal and vertical directions, but ball 1 has an acceleration only in the vertical direction.

QUESTION 2

3. Each drawing shows three points along the path of a projectile, one on its way up, one at the top, and one on its way down. The launch point is on the left in each drawing. Which drawing correctly represents the acceleration $\vec{a}$ of the projectile at these three points? **(a)** 1 **(b)** 2 **(c)** 3 **(d)** 4 **(e)** 5

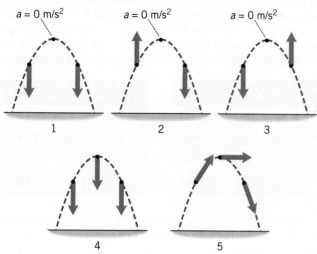

QUESTION 3

4. Ball 1 is thrown into the air and it follows the trajectory for projectile motion shown in the drawing. At the instant that ball 1 is at the top of its trajectory, ball 2 is dropped from rest from the same height. Which ball reaches the ground first? **(a)** Ball 1 reaches the ground first, since it

QUESTION 4

is moving at the top of the trajectory, while ball 2 is dropped from rest. **(b)** Ball 2 reaches the ground first, because it has the shorter distance to travel. **(c)** Both balls reach the ground at the same time. **(d)** There is not enough information to tell which ball reaches the ground first.

5. Two objects are fired into the air, and the drawing shows the projectile motions. Projectile 1 reaches the greater height, but projectile 2 has the greater range. Which one is in the air for the greater amount of time? **(a)** Projectile 1, because it travels higher than projectile 2. **(b)** Projectile 2, because it has the greater range.

QUESTION 5

(c) Both projectiles spend the same amount of time in the air. **(d)** Projectile 2, because it has the smaller initial speed and, therefore, travels more slowly than projectile 1.

Section 3.4 Relative Velocity

6. A slower-moving car is traveling behind a faster-moving bus. The velocities of the two vehicles are as follows:

$$\vec{v}_{CG} = \text{velocity of the Car relative to the Ground} = +12 \text{ m/s}$$
$$\vec{v}_{BG} = \text{velocity of the Bus relative to the Ground} = +16 \text{ m/s}$$

A passenger on the bus gets up and walks toward the front of the bus with a velocity of $\vec{v}_{PB}$, where $\vec{v}_{PB}$ = velocity of the **P**assenger relative to the **B**us = +2 m/s. What is $\vec{v}_{PC}$, the velocity of the **P**assenger relative to the **C**ar?

    (a) +2 m/s + 16 m/s + 12 m/s = +30 m/s
    (b) −2 m/s + 16 m/s + 12 m/s = +26 m/s
    (c) +2 m/s + 16 m/s − 12 m/s = +6 m/s
    (d) −2 m/s + 16 m/s − 12 m/s = +2 m/s

7. Your car is traveling behind a jeep. Both are moving at the same speed, so the velocity of the jeep relative to you is zero. A spare tire is strapped to the back of the jeep. Suddenly the strap breaks, and the tire falls off the jeep. Will your car hit the spare tire before the tire hits the road? Assume that air resistance is absent. **(a)** Yes. As long as the car doesn't slow down, it will hit the tire. **(b)** No. The car will not hit the tire before the tire hits the ground, no matter how close you are to the jeep. **(c)** If the tire falls from a great enough height, the car will hit the tire. **(d)** If the car is far enough behind the jeep, the car will not hit the tire.

8. The drawing shows two cars traveling in different directions with different speeds. Their velocities are:

$$\vec{v}_{AG} = \text{velocity of car A relative to the Ground} = 27.0 \text{ m/s, due east}$$
$$\vec{v}_{BG} = \text{velocity of car B relative to the Ground} = 21.0 \text{ m/s, due north}$$

The passenger of car B looks out the window and sees car A. What is the velocity (magnitude and direction) of car A as observed by the passenger of car B? In other words, what is the velocity $\vec{v}_{AB}$ of car A relative to car B? Give the directional angle of $\vec{v}_{AB}$ with respect to due east.

QUESTION 8

Problems

Online

Additional questions are available for assignment in WileyPLUS.

  SSM Student Solutions Manual

  MMH Problem-solving help

  GO Guided Online Tutorial

  V-HINT Video Hints

  CHALK Chalkboard Videos

  BIO Biomedical application

  E Easy

  M Medium

  H Hard

  WS Worksheet

  T Team Problem

Section 3.1 Displacement, Velocity, and Acceleration

1. **E** **SSM** Two trees have perfectly straight trunks and are both growing perpendicular to the flat horizontal ground beneath them. The sides of the trunks that face each other are separated by 1.3 m. A frisky squirrel makes three jumps in rapid succession. First, he leaps from the foot of one tree to a spot that is 1.0 m above the ground on the other tree. Then, he jumps back to the first tree, landing on it at a spot that is 1.7 m above the ground. Finally, he leaps back to the other tree, now landing at a spot that is 2.5 m above the ground. What is the magnitude of the squirrel's displacement?

2. **E** A meteoroid is traveling east through the atmosphere at 18.3 km/s while descending at a rate of 11.5 km/s. What is its speed, in km/s?

3. **E** **GO** In a football game a kicker attempts a field goal. The ball remains in contact with the kicker's foot for 0.050 s, during which time it experiences an acceleration of 340 m/s². The ball is launched at an angle of 51° above the ground. Determine the horizontal and vertical components of the launch velocity.

4. **E** A baseball player hits a triple and ends up on third base. A baseball "diamond" is a square, each side of length 27.4 m, with home plate and the three bases on the four corners. What is the magnitude of the player's displacement?

5. **E** **SSM** In diving to a depth of 750 m, an elephant seal also moves 460 m due east of his starting point. What is the magnitude of the seal's displacement?

6. **E** A mountain-climbing expedition establishes two intermediate camps, labeled A and B in the drawing, above the base camp. What is the magnitude Δr of the displacement between camp A and camp B?

PROBLEM 6

7. **E** **SSM** A radar antenna is tracking a satellite orbiting the earth. At a certain time, the radar screen shows the satellite to be 162 km away. The radar antenna is pointing upward at an angle of 62.3° from the ground. Find the x and y components (in km) of the position vector of the satellite, relative to the antenna.

8. **E** **GO** In a mall, a shopper rides up an escalator between floors. At the top of the escalator, the shopper turns right and walks 9.00 m to a store. The magnitude of the shopper's displacement from the bottom of the escalator to the store is 16.0 m. The vertical distance between the floors is 6.00 m. At what angle is the escalator inclined above the horizontal?

9. **E** **SSM** A skateboarder, starting from rest, rolls down a 12.0-m ramp. When she arrives at the bottom of the ramp her speed is 7.70 m/s. **(a)** Determine the magnitude of her acceleration, assumed to be constant. **(b)** If the ramp is inclined at 25.0° with respect to the ground, what is the component of her acceleration that is parallel to the ground?

10. **M** **V-HINT** **MMH** A bird watcher meanders through the woods, walking 0.50 km due east, 0.75 km due south, and 2.15 km in a direction 35.0° north of west. The time required for this trip is 2.50 h. Determine the magnitude and direction (relative to due west) of the bird watcher's **(a)** displacement and **(b)** average velocity. Use kilometers and hours for distance and time, respectively.

11. **M** **MMH** **CHALK** The earth moves around the sun in a nearly circular orbit of radius 1.50×10^{11} m. During the three summer months (an elapsed time of 7.89×10^6 s), the earth moves one-fourth of the distance around the sun. **(a)** What is the average

speed of the earth? **(b)** What is the magnitude of the average velocity of the earth during this period?

Section 3.2 Equations of Kinematics in Two Dimensions/Section 3.3 Projectile Motion

12. **E** A spacecraft is traveling with a velocity of $v_{0x} = 5480$ m/s along the $+x$ direction. Two engines are turned on for a time of 842 s. One engine gives the spacecraft an acceleration in the $+x$ direction of $a_x = 1.20$ m/s², while the other gives it an acceleration in the $+y$ direction of $a_y = 8.40$ m/s². At the end of the firing, find **(a)** v_x and **(b)** v_y.

13. **E** **SSM** A volleyball is spiked so that it has an initial velocity of 15 m/s directed downward at an angle of 55° below the horizontal. What is the horizontal component of the ball's velocity when the opposing player fields the ball?

14. **E** **CHALK** As a tennis ball is struck, it departs from the racket horizontally with a speed of 28.0 m/s. The ball hits the court at a horizontal distance of 19.6 m from the racket. How far above the court is the tennis ball when it leaves the racket?

15. **E** A skateboarder shoots off a ramp with a velocity of 6.6 m/s, directed at an angle of 58° above the horizontal. The end of the ramp is 1.2 m above the ground. Let the x axis be parallel to the ground, the $+y$ direction be vertically upward, and take as the origin the point on the ground directly below the top of the ramp. **(a)** How high above the ground is the highest point that the skateboarder reaches? **(b)** When the skateboarder reaches the highest point, how far is this point horizontally from the end of the ramp?

16. **E** **GO** A puck is moving on an air hockey table. Relative to an x, y coordinate system at time $t = 0$ s, the x components of the puck's initial velocity and acceleration are $v_{0x} = +1.0$ m/s and $a_x = +2.0$ m/s². The y components of the puck's initial velocity and acceleration are $v_{0y} = +2.0$ m/s and $a_y = -2.0$ m/s². Find the magnitude and direction of the puck's velocity at a time of $t = 0.50$ s. Specify the direction relative to the $+x$ axis.

17. **E** **SSM** A spider crawling across a table leaps onto a magazine blocking its path. The initial velocity of the spider is 0.870 m/s at an angle of 35.0° above the table, and it lands on the magazine 0.0770 s after leaving the table. Ignore air resistance. How thick is the magazine? Express your answer in millimeters.

18. **E** A horizontal rifle is fired at a bull's-eye. The muzzle speed of the bullet is 670 m/s. The gun is pointed directly at the center of the bull's-eye, but the bullet strikes the target 0.025 m below the center. What is the horizontal distance between the end of the rifle and the bull's-eye?

19. **E** **MMH** A golfer imparts a speed of 30.3 m/s to a ball, and it travels the maximum possible distance before landing on the green. The tee and the green are at the same elevation. **(a)** How much time does the ball spend in the air? **(b)** What is the longest hole in one that the golfer can make, if the ball does not roll when it hits the green?

20. **E** **GO** A golfer hits a shot to a green that is elevated 3.0 m above the point where the ball is struck. The ball leaves the club at a speed of 14.0 m/s at an angle of 40.0° above the horizontal. It rises to its maximum height and then falls down to the green. Ignoring air resistance, find the speed of the ball just before it lands.

21. **E** **SSM** In the aerials competition in skiing, the competitors speed down a ramp that slopes sharply upward at the end. The sharp upward slope launches them into the air, where they perform acrobatic maneuvers. The end of a launch ramp is directed 63° above the horizontal. With this launch angle, a skier attains a height of 13 m above the end of the ramp. What is the skier's launch speed?

22. **E** **GO** A space vehicle is coasting at a constant velocity of 21.0 m/s in the +*y* direction relative to a space station. The pilot of the vehicle fires a RCS (reaction control system) thruster, which causes it to accelerate at 0.320 m/s² in the +*x* direction. After 45.0 s, the pilot shuts off the RCS thruster. After the RCS thruster is turned off, find **(a)** the magnitude and **(b)** the direction of the vehicle's velocity relative to the space station. Express the direction as an angle measured from the +*y* direction.

23. **E** **SSM** As preparation for this problem, review Conceptual Example 9. The drawing shows two planes each about to drop an empty fuel tank. At the moment of release each plane has the same speed of 135 m/s, and each tank is at the same height of 2.00 km above the ground. Although the speeds are the same, the velocities are different at the instant of release, because one plane is flying at an angle of 15.0° above the horizontal and the other is flying at an angle of 15.0° below the horizontal. Find the magnitude and direction of the

Fuel tank
Plane A

Plane B

PROBLEM 23

velocity with which the fuel tank hits the ground if it is from **(a)** plane A and **(b)** plane B. In each part, give the directional angles with respect to the horizontal.

24. **E** **GO** A criminal is escaping across a rooftop and runs off the roof horizontally at a speed of 5.3 m/s, hoping to land on the roof of an adjacent building. Air resistance is negligible. The horizontal distance between the two buildings is *D*, and the roof of the adjacent building is 2.0 m below the jumping-off point. Find the maximum value for *D*.

25. **E** On a spacecraft, two engines are turned on for 684 s at a moment when the velocity of the craft has *x* and *y* components of $v_{0x} = 4370$ m/s and $v_{0y} = 6280$ m/s. While the engines are firing, the craft undergoes a displacement that has components of $x = 4.11 \times 10^6$ m and $y = 6.07 \times 10^6$ m. Find the *x* and *y* components of the craft's acceleration.

26. **E** **GO** In the absence of air resistance, a projectile is launched from and returns to ground level. It follows a trajectory similar to that shown in **Animated Figure 3.10** and has a range of 23 m. Suppose the launch speed is doubled, and the projectile is fired at the same angle above the ground. What is the new range?

27. **E** **SSM** A fire hose ejects a stream of water at an angle of 35.0° above the horizontal. The water leaves the nozzle with a speed of 25.0 m/s. Assuming that the water behaves like a projectile, how far from a building should the fire hose be located to hit the highest possible fire?

28. **E** A major-league pitcher can throw a baseball in excess of 41.0 m/s. If a ball is thrown horizontally at this speed, how much will it drop by the time it reaches a catcher who is 17.0 m away from the point of release?

29. **E** A quarterback claims that he can throw the football a horizontal distance of 183 m (200 yd). Furthermore, he claims that he can do this by launching the ball at the relatively low angle of 30.0° above the horizontal. To evaluate this claim, determine the speed with which this quarterback must throw the ball. Assume that the ball is launched and caught at the same vertical level and that air resistance can be ignored. For comparison, a baseball pitcher who can accurately throw a fastball at 45 m/s (100 mph) would be considered exceptional.

30. **E** **SSM** An eagle is flying horizontally at 6.0 m/s with a fish in its claws. It accidentally drops the fish. **(a)** How much time passes

before the fish's speed doubles? **(b)** How much additional time would be required for the fish's speed to double again?

31. **E** **GO** The highest barrier that a projectile can clear is 13.5 m, when the projectile is launched at an angle of 15.0° above the horizontal. What is the projectile's launch speed?

32. **E** Consult Multiple-Concept Example 4 for background before beginning this problem. Suppose the water at the top of Niagara Falls has a horizontal speed of 2.7 m/s just before it cascades over the edge of the falls. At what vertical distance below the edge does the velocity vector of the water point downward at a 75° angle below the horizontal?

33. **E** **MMH** On a distant planet, golf is just as popular as it is on earth. A golfer tees off and drives the ball 3.5 times as far as he would have on earth, given the same initial velocities on both planets. The ball is launched at a speed of 45 m/s at an angle of 29° above the horizontal. When the ball lands, it is at the same level as the tee. On the distant planet, what are **(a)** the maximum height and **(b)** the range of the ball?

34. **E** **CHALK** A rocket is fired at a speed of 75.0 m/s from ground level, at an angle of 60.0° above the horizontal. The rocket is fired toward an 11.0-m-high wall, which is located 27.0 m away. The rocket attains its launch speed in a negligibly short period of time, after which its engines shut down and the rocket coasts. By how much does the rocket clear the top of the wall?

35. **E** A rifle is used to shoot twice at a target, using identical cartridges. The first time, the rifle is aimed parallel to the ground and directly at the center of the bull's-eye. The bullet strikes the target at a distance of H_A below the center, however. The second time, the rifle is similarly aimed, but from twice the distance from the target. This time the bullet strikes the target at a distance of H_B below the center. Find the ratio H_B/H_A.

36. **M** **SSM** An airplane with a speed of 97.5 m/s is climbing upward at an angle of 50.0° with respect to the horizontal. When the plane's altitude is 732 m, the pilot releases a package. **(a)** Calculate the distance along the ground, measured from a point directly beneath the point of release, to where the package hits the earth. **(b)** Relative to the ground, determine the angle of the velocity vector of the package just before impact.

37. **M** **V-HINT** Multiple-Concept Example 4 deals with a situation similar to that presented here. A marble is thrown horizontally with a speed of 15 m/s from the top of a building. When it strikes the ground, the marble has a velocity that makes an angle of 65° with the horizontal. From what height above the ground was the marble thrown?

38. **M** **V-HINT** **MMH** Review Conceptual Example 5 before beginning this problem. You are traveling in a convertible with the top down. The car is moving at a constant velocity of 25 m/s, due east along flat ground. You throw a tomato straight upward at a speed of 11 m/s. How far has the car moved when you get a chance to catch the tomato?

39. **M** **V-HINT** A diver springs upward from a diving board. At the instant she contacts the water, her speed is 8.90 m/s, and her body is extended at an angle of 75.0° with respect to the horizontal surface of the water. At this instant her vertical displacement is −3.00 m, where downward is the negative direction. Determine her initial velocity, both magnitude and direction.

40. **M** **GO** **MMH** A soccer player kicks the ball toward a goal that is 16.8 m in front of him. The ball leaves his foot at a speed of 16.0 m/s and an angle of 28.0° above the ground. Find the speed of the ball when the goalie catches it in front of the net.

41. **M** **V-HINT** In the javelin throw at a track-and-field event, the javelin is launched at a speed of 29 m/s at an angle of 36° above the horizontal. As the javelin travels upward, its velocity points above the horizontal at an angle that decreases as time passes. How much time is required for the angle to be reduced from 36° at launch to 18°?

42. **M** **SSM** An airplane is flying with a velocity of 240 m/s at an angle of 30.0° with the horizontal, as the drawing shows. When the altitude of the plane is 2.4 km, a flare is released from the plane. The flare hits the target on the ground. What is the angle θ?

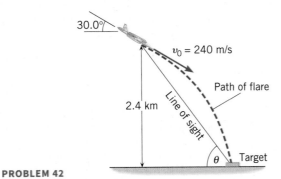

PROBLEM 42

43. **M** **V-HINT** A child operating a radio-controlled model car on a dock accidentally steers it off the edge. The car's displacement 1.1 s after leaving the dock has a magnitude of 7.0 m. What is the car's speed at the instant it drives off the edge of the dock?

44. **M** **MMH** After leaving the end of a ski ramp, a ski jumper lands downhill at a point that is displaced 51.0 m horizontally from the end of the ramp. His velocity, just before landing, is 23.0 m/s and points in a direction 43.0° below the horizontal. Neglecting air resistance and any lift he experiences while airborne, find his initial velocity (magnitude and direction) when he left the end of the ramp. Express the direction as an angle relative to the horizontal.

45. **M** **GO** Stones are thrown horizontally with the same velocity from the tops of two different buildings. One stone lands twice as far from the base of the building from which it was thrown as does the other stone. Find the ratio of the height of the taller building to the height of the shorter building.

46. **H** **SSM** The drawing shows an exaggerated view of a rifle that has been "sighted in" for a 91.4-meter target. If the muzzle speed of the bullet is $v_0 = 427$ m/s, what are the two possible angles θ_1 and θ_2 between the rifle barrel and the horizontal such that the bullet will hit the target? One of these angles is so large that it is never used in target shooting. (*Hint: The following trigonometric identity may be useful:* $2 \sin \theta \cos \theta = \sin 2\theta$.)

PROBLEM 46

47. **H** **MMH** A projectile is launched from ground level at an angle of 12.0° above the horizontal. It returns to ground level. To what value should the launch angle be adjusted, without changing the launch speed, so that the range doubles?

48. **H** **SSM** From the top of a tall building, a gun is fired. The bullet leaves the gun at a speed of 340 m/s, parallel to the ground. As the drawing shows, the bullet puts a hole in a window of another building and hits the wall that faces the window. Using the data in the drawing,

determine the distances D and H, which locate the point where the gun was fired. Assume that the bullet does not slow down as it passes through the window.

PROBLEM 48

Section 3.4   Relative Velocity

49. **E** In a marathon race Chad is out in front, running due north at a speed of 4.00 m/s. John is 95 m behind him, running due north at a speed of 4.50 m/s. How long does it take for John to pass Chad?

50. **E** **SSM** A swimmer, capable of swimming at a speed of 1.4 m/s in still water (i.e., the swimmer can swim with a speed of 1.4 m/s relative to the water), starts to swim directly across a 2.8-km-wide river. However, the current is 0.91 m/s, and it carries the swimmer downstream. **(a)** How long does it take the swimmer to cross the river? **(b)** How far downstream will the swimmer be upon reaching the other side of the river?

51. **E** **GO** Two friends, Barbara and Neil, are out rollerblading. With respect to the ground, Barbara is skating due south at a speed of 4.0 m/s. Neil is in front of her. With respect to the ground, Neil is skating due west at a speed of 3.2 m/s. Find Neil's velocity (magnitude and direction relative to due west), as seen by Barbara.

52. **E** At some airports there are speed ramps to help passengers get from one place to another. A speed ramp is a moving conveyor belt on which you can either stand or walk. Suppose a speed ramp has a length of 105 m and is moving at a speed of 2.0 m/s relative to the ground. In addition, suppose you can cover this distance in 75 s when walking on the ground. If you walk at the same rate with respect to the speed ramp that you walk on the ground, how long does it take for you to travel the 105 m using the speed ramp?

53. **E** You are in a hot-air balloon that, relative to the ground, has a velocity of 6.0 m/s in a direction due east. You see a hawk moving directly away from the balloon in a direction due north. The speed of the hawk relative to you is 2.0 m/s. What are the magnitude and direction of the hawk's velocity relative to the ground? Express the directional angle relative to due east.

54. **E** **GO** On a pleasure cruise a boat is traveling relative to the water at a speed of 5.0 m/s due south. Relative to the boat, a passenger walks toward the back of the boat at a speed of 1.5 m/s. **(a)** What are the magnitude and direction of the passenger's velocity relative to the water? **(b)** How long does it take for the passenger to walk a distance of 27 m on the boat? **(c)** How long does it take for the passenger to cover a distance of 27 m on the water?

55. **E** **GO** The captain of a plane wishes to proceed due west. The cruising speed of the plane is 245 m/s relative to the air. A weather report indicates that a 38.0-m/s wind is blowing from the south to the north. In what direction, measured with respect to due west, should the pilot head the plane?

56. **M** **V-HINT** A person looking out the window of a stationary train notices that raindrops are falling vertically down at a speed of 5.0 m/s relative to the ground. When the train moves at a constant velocity, the raindrops make an angle of 25° when they move past the window, as the drawing shows. How fast is the train moving?

PROBLEM 56

57. **M** **SSM** **CHALK** Mario, a hockey player, is skating due south at a speed of 7.0 m/s relative to the ice. A teammate passes the puck to him. The puck has a speed of 11.0 m/s and is moving in a direction of 22° west of south, relative to the ice. What are the magnitude and direction (relative to due south) of the puck's velocity, as observed by Mario?

58. **H** A jetliner can fly 6.00 hours on a full load of fuel. Without any wind it flies at a speed of 2.40×10^2 m/s. The plane is to make a round-trip by heading due west for a certain distance, turning around, and then heading due east for the return trip. During the entire flight, however, the plane encounters a 57.8-m/s wind from the jet stream, which blows from west to east. What is the maximum distance that the plane can travel due west and just be able to return home?

Additional Problems

59. **E** Useful background for this problem can be found in Multiple-Concept Example 2. On a spacecraft two engines fire for a time of 565 s. One gives the craft an acceleration in the x direction of $a_x = 5.10$ m/s², while the other produces an acceleration in the y direction of $a_y = 7.30$ m/s². At the end of the firing period, the craft has velocity components of $v_x = 3775$ m/s and $v_y = 4816$ m/s. Find the magnitude and direction of the initial velocity. Express the direction as an angle with respect to the $+x$ axis.

60. **E** A dolphin leaps out of the water at an angle of 35° above the horizontal. The horizontal component of the dolphin's velocity is 7.7 m/s. Find the magnitude of the vertical component of the velocity.

61. **E** A hot-air balloon is rising straight up with a speed of 3.0 m/s. A ballast bag is released from rest relative to the balloon at 9.5 m above the ground. How much time elapses before the ballast bag hits the ground?

62. **E** **CHALK** A golf ball rolls off a horizontal cliff with an initial speed of 11.4 m/s. The ball falls a vertical distance of 15.5 m into a lake below. **(a)** How much time does the ball spend in the air? **(b)** What is the speed v of the ball just before it strikes the water?

63. **E** **GO** When chasing a hare along a flat stretch of ground, a greyhound leaps into the air at a speed of 10.0 m/s, at an angle of 31.0° above the horizontal. **(a)** What is the range of his leap and **(b)** for how much time is he in the air?

64. **E** **SSM** Multiple-Concept Example 4 provides useful background for this problem. A diver runs horizontally with a speed of 1.20 m/s off a platform that is 10.0 m above the water. What is his speed just before striking the water?

65. **E** **GO** A ball is thrown upward at a speed v_0 at an angle of 52° above the horizontal. It reaches a maximum height of 7.5 m. How high would this ball go if it were thrown straight upward at speed v_0?

66. **M** **MMH** A golfer, standing on a fairway, hits a shot to a green that is elevated 5.50 m above the point where she is standing. If the ball leaves her club with a velocity of 46.0 m/s at an angle of 35.0° above the ground, find the time that the ball is in the air before it hits the green.

67. **M** **SSM** As preparation for this problem, review Conceptual Example 9. The two stones described there have identical initial speeds of $v_0 = 13.0$ m/s and are thrown at an angle $\theta = 30.0°$, one below the horizontal and one above the horizontal. What is the distance *between* the points where the stones strike the ground?

68. **M** **GO** Relative to the ground, a car has a velocity of 16.0 m/s, directed due north. Relative to this car, a truck has a velocity of 24.0 m/s, directed 52.0° north of east. What is the magnitude of the truck's velocity relative to the ground?

69. **M** **V-HINT** The lob in tennis is an effective tactic when your opponent is near the net. It consists of lofting the ball over his head, forcing him to move quickly away from the net (see the drawing). Suppose that you lob the ball with an initial speed of 15.0 m/s, at an angle of 50.0° above the horizontal. At this instant your opponent is 10.0 m away from the ball. He begins moving away from you 0.30 s later, hoping to reach the ball and hit it back at the moment that it is 2.10 m above its launch point. With what minimum average speed must he move? (Ignore the fact that he can stretch, so that his racket can reach the ball before he does.)

15.0 m/s

50.0°

2.10 m

—10.0 m—

PROBLEM 69

70. **M** **GO** A baseball player hits a home run, and the ball lands in the left-field seats, 7.5 m above the point at which it was hit. It lands with a velocity of 36 m/s at an angle of 28° below the horizontal (see the drawing). The positive directions are upward and to the right in the drawing. Ignoring air resistance, find the magnitude and direction of the initial velocity with which the ball leaves the bat.

$+y$

$+x$

v_x

28°

$y = 7.5$ m

v_y

$v = 36$ m/s

v_{0y}

v_0

θ

v_{0x}

PROBLEM 70

Physics in Biology, Medicine, and Sports

71. Ⓜ **BIO** **3.3**   The Pacific white-sided dolphin (see photo) can jump straight out of the water to incredible heights near 9.10 m (~30 feet). If a dolphin jumps out of the water at a launch angle of 45.0°, **(a)** what will be its maximum height, and **(b)** how far will the dolphin jump horizontally? Assume the dolphin leaves the water at the same speed as when it jumps straight upward.

PROBLEM 71

72. Ⓜ **BIO** **3.3**   A spear-throwing lever, called an atlatl, is a device used in creating greater velocity of a thrown spear or dart (see the figure) compared to when thrown with just the arm. Used primarily in hunting, as well as a weapon, the atlatl can produce spear speeds near 90 mph. Historical evidence suggests that the atlatl was used by *Homo sapiens* dating back to 30 000 to 42 000 years ago. Suppose an atlatl is used to launch a spear at 40.0 m/s with a launch angle of 10.0° above the horizontal. What is the speed of the spear 1.00 s after it is thrown? Assume air resistance is negligible.

PROBLEM 72

73. Ⓜ **BIO** **3.4**   One of the world's greatest spectacles of animal migration is the annual circular trek of the wildebeest through the Serengeti National Park in Tanzania, Africa. Each year the availability of grazing land drives nearly 1.7 million wildebeest into river crossings, such as that shown in the photo. The wildebeest begin by walking across the river in the shallow waters close to the bank but end up swimming across in the middle of the fast-moving river, where the water is deeper. Eventually they reach the shallow area near the other bank, where they can walk out on the other side. This produces the characteristic zigzag pattern seen in the photo. The drawing shows a view of the river crossing from above and the path traced out by the wildebeest. Assume that the wildebeest walk at 1.00 m/s and can swim at 2.00 m/s with respect to still water. Use the distances in the drawing to calculate **(a)** how long it takes them to cross the river and **(b)** the speed of the river current.

PROBLEM 73

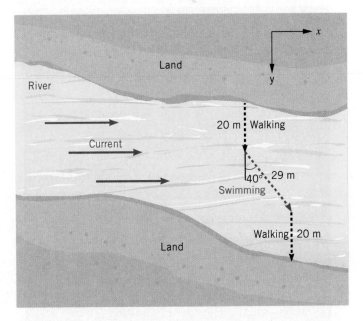

PROBLEM 73

74. **M** **BIO** **3.3** On average out of every 1000 newborn babies, one to two will suffer from a condition known as *pyloric stenosis*. This is characterized by a narrowing of the opening between the stomach and small intestine. One of the symptoms of this condition can be extremely forceful vomiting after feeding, known as projectile vomiting (see the drawing). The condition can usually be treated by opening the gastric outlet through laparoscopic surgery, and once corrected, it rarely

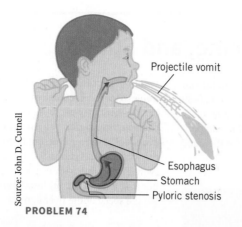

Source: John D. Cutnell

PROBLEM 74

PROBLEM 74

causes any long-term side effects. Imagine an infant is sitting upright, so that his mouth is 41 cm above the floor. He suddenly vomits horizontally, with it landing 1.2 m on the floor in front of him, as shown in the drawing. As the name suggests, treat the vomit as a projectile and calculate the velocity at which it leaves his mouth.

75. **M** **BIO** **3.3** Archerfish are one of the few fish that prey on land insects and other small animals. They do this by shooting a jet of water from their specialized mouths at an insect resting on a leaf or plant stem above the water (see the photo). This often stuns the insect, causing it to fall into the water, where it can be eaten. In the drawing, an archerfish is shooting water toward an unsuspecting horsefly at rest on a blade of marsh grass above the surface of a pond. Use the information in the drawing to calculate how high (h) the horsefly is above the water. The initial velocity of the water (v_0) is 3.2 m/s, and assume that the projected water reaches its maximum height above the pond's surface when it strikes the horsefly.

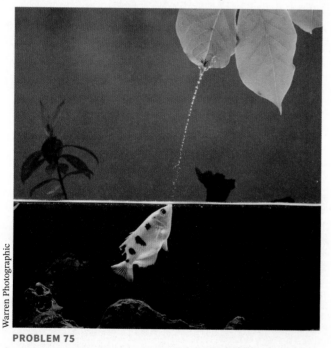

Warren Photographic

PROBLEM 75

PROBLEM 75

Concepts and Calculations Problems

Online

A primary focus of this chapter has been projectile motion. These additional problems serve as a review of the basic features of this type of motion.

76. **M** **CHALK** **SSM** Three balls are launched simultaneously, as in the figure. The red ball is launched at an initial velocity $\vec{v}_0$ directed at an angle θ above the horizontal. The blue ball is launched vertically with a speed of 10.0 m/s, and the yellow ball is launched horizontally with a speed of 4.6 m/s. The blue and red balls reach identical maximum heights and return to the ground at the same time, and the red ball returns to the ground to meet the yellow ball at the same place along the horizontal. *Concepts:* (i) Since the blue and red balls reach the same maximum height, what can be said about the y components of their velocities? (ii) Since the red and yellow balls travel equal distances along the x direction in the same amount of time, what can be said about the x components of their initial velocities? *Calculations:* Find the launch speed v_0 and launch angle θ for the red ball.

PROBLEM 76

77. **M** **CHALK** A projectile is launched from and returns to ground level, as the figure shows. There is no air resistance. The horizontal range of the projectile is measured to be $R = 175$ m, and the horizontal component of the launch velocity is $v_{0x} = +25$ m/s. *Concepts:* (i) What is the final value of the horizontal component v_x of the projectile's velocity? (ii) Can the time be determined for the horizontal part of the motion? (iii) Is the time for the horizontal part of the motion the same as that for the vertical part? (iv) For the vertical part of the motion, what is the displacement of the projectile? *Calculations:* Find the vertical component v_{0y} of the projectile.

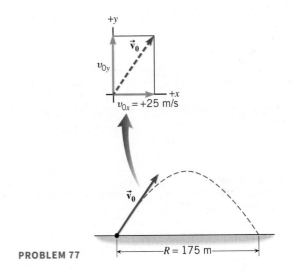

PROBLEM 77

Team Problems

Online

78. **E** **T** **WS** **A Starship Getaway.** You and your crew are on a spaceship that is being pursued by an enemy craft. Your ship is currently hiding behind an enormous space station, looking to make a run for the gate of a wormhole that will take you to safety. As the enemy ship passes to the other side of the space station, you will have exactly 5.00 s to burn your x and y thrusters, but must then turn them off so that you are not detected by the enemy craft as it comes around the other side of the space station. The gate is located at coordinates $x = 455$ m, $y = 633$ m, and your ship is located at the origin. The other requirement is that you enter the gate with a speed of exactly 220.0 m/s. Assuming that you keep your x and y thrusters on for the full 5.00 s (they turn on and off simultaneously), and that the acceleration is constant during this time, determine **(a)** the angle that your acceleration vector must make with the x axis, **(b)** the magnitude of the acceleration, and **(c)** the x and y components of the acceleration so that you know how much acceleration that each thruster should provide. **(d)** How far are you from the gate when the thrusters are turned off?

79. **E** **T** **WS** **A Water Slide.** You and your team are designing a water slide that launches a rider at a maximum speed of 30.0 mph. **(a)** To be safe, the maximum vertical component of the velocity of impact must be less than or equal to 20.0 mph. What launch angle results in a maximum vertical impact velocity of 20.0 mph? **(b)** Next, you have to design the pool in which the rider lands. For safety, the rider should land in the center of the pool when launched at maximum speed. Assuming one edge of the pool is at the launch position of the slide, how long should you make the pool (in meters)?

(c) What is the maximum height the rider can reach with this design (in meters)? Assume that the launch height is at water level.

80. **M** **T** **A Water Balloon Battle.** You are launching water balloons at a rival team using a large slingshot. The other team is set up on the opposite side of a flat-topped building that is 30.0 ft tall and 50.0 ft wide. Your reconnaissance team has reported that the opposition is set up 10.0 m from the wall of the building. Your balloon launcher is calibrated for launch speeds that can reach as high as 100 mph at angles between 0 and 80.0° from the horizontal. Since a direct shot is not possible (the opposing team is on the opposite side of the building), you plan to splash the other team by making a balloon explode on the ground near them. If your launcher is located 55.0 m from the building (opposite side as the opposing team), what should your launch velocity be (magnitude and direction) to land a balloon 5.0 meters beyond the opposing team with maximum impact (i.e., maximum vertical speed)?

81. **M** **T** **A Motorcycle Jump.** You are planning to make a jump with your motorcycle by driving over a ramp that will launch you at an angle of 30.0° with respect to the horizontal. The front edge of the ramp on which you are supposed to land, however, is 25.0 ft lower than the edge of the launch ramp (i.e., your launch height). **(a)** Assuming a launch speed of 65.0 mph, at what horizontal distance from your launch point should the landing ramp be placed? **(b)** In order to land smoothly, the angle of the landing ramp should match the direction of your velocity vector when you touch down. What should be the angle of the landing ramp?

CHAPTER 4

In order to successfully land the planetary rover, *Curiosity*, on the surface of Mars, NASA scientists and engineers had to take into account many forces, such as the thrust from the "sky crane" rocket engines, the tension in its towing cables, air resistance, and the weight of the rover due to the gravity on Mars, just to name a few. This chapter will discuss how forces influence the motion of objects.

NASA/JPL-Caltec

LEARNING OBJECTIVES

After reading this module, you should be able to...

4.1 Discuss the concepts of force and mass.

4.2 Define Newton's first law of motion.

4.3 Define Newton's second law of motion.

4.4 Apply Newton's second law of motion in two dimensions.

4.5 Apply Newton's third law of motion.

4.6 Identify types of forces.

4.7 Define Newton's law of universal gravitation.

4.8 Solve problems using the normal force.

4.9 Solve problems involving friction.

4.10 Solve problems involving tension.

4.11 Apply Newton's first law to equilibrium problems.

4.12 Apply Newton's second law to nonequilibrium problems.

Forces and Newton's Laws of Motion

4.1 The Concepts of Force and Mass

In common usage, a **force** is a push or a pull, as the examples in **Figure 4.1** illustrate. In football, an offensive lineman pushes against his opponent. The tow bar attached to a speeding boat pulls a water skier. Forces such as those that push against the football player or pull the skier are called **contact forces**, because they arise from the physical contact between two objects. There are circumstances, however, in which two objects exert forces on one another even though they are not touching. Such forces are referred to as **noncontact forces** or **action-at-a-distance forces**. One example of such a noncontact force occurs when a diver is pulled toward the earth because of the force of gravity. The earth exerts this force even when it is not in direct contact with the diver. In **Figure 4.1**, arrows are used to represent the forces. It is appropriate to use arrows, because a force is a vector quantity and has both a magnitude and a direction. The direction of the arrow gives the direction of the force, and the length is proportional to its strength or magnitude.

The word **mass** is just as familiar as the word *force*. A massive supertanker, for instance, is one that contains an enormous amount of mass. As we will see in the next section, it is difficult to set such a massive object into motion and difficult to bring it to a halt once it is moving. In comparison, a penny does not contain much mass. The emphasis here is on the amount of mass, and the idea of direction is of no concern. Therefore, mass is a scalar quantity.

During the seventeenth century, Isaac Newton, building on the work of Galileo, developed three important laws that deal with force and mass. Collectively they are called "Newton's laws of motion" and provide the basis for understanding the effect that forces have on an object. Because of the importance of these laws, a separate section will be devoted to each one.

4.2 | Newton's First Law of Motion

The First Law

To gain some insight into Newton's first law, think about the game of ice hockey (**Figure 4.2**). If a player does not hit a stationary puck, it will remain at rest on the ice. After the puck is struck, however, it coasts on its own across the ice, slowing down only slightly because of friction. Since ice is very slippery, there is only a relatively small amount of friction to slow down the puck. In fact, if it were possible to remove all friction and wind resistance, and if the rink were infinitely large, the puck would coast forever in a straight line at a constant speed. Left on its own, the puck would lose none of the velocity imparted to it at the time it was struck. This is the essence of Newton's first law of motion:

> **NEWTON'S FIRST LAW OF MOTION**
>
> **An object continues in a state of rest or in a state of motion at a constant velocity (constant speed in a constant direction), unless compelled to change that state by a net force.**

In the first law the phrase "net force" is crucial. Often, several forces act simultaneously on a body, and *the net force is the vector sum of all of them*. Individual forces matter only to the extent that they contribute to the total. For instance, if friction and other opposing forces were absent, a car could travel forever at 30 m/s in a straight line, without using any gas after it had come up to speed. In reality gas is needed, but only so that the engine can produce the necessary force to cancel opposing forces such as friction. This cancellation ensures that there is no net force to change the state of motion of the car.

(a)

David Dermer/Diamond Images/Getty Images

(b)

age fotostock

(c)

Ikon Images/SuperStock

FIGURE 4.1 The arrow labeled $\vec{\mathbf{F}}$ represents the force that acts on (*a*) the football player on the left, (*b*) the water skier, and (*c*) the cliff diver.

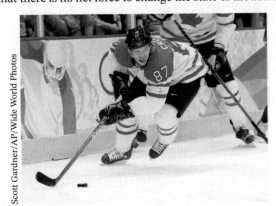

Scott Gardner/AP/Wide World Photos

FIGURE 4.2 The game of ice hockey can give some insight into Newton's laws of motion.

When an object moves at a constant speed in a constant direction, its velocity is constant. Newton's first law indicates that a state of rest (zero velocity) and a state of constant velocity are completely equivalent, in the sense that neither one requires the application of a net force to sustain it. ***The purpose served when a net force acts on an object is not to sustain the object's velocity, but, rather, to change it.***

Inertia and Mass

A greater net force is required to change the velocity of some objects than of others. For instance, a net force that is just enough to cause a bicycle to pick up speed will cause only an imperceptible change in the motion of a freight train. In comparison to the bicycle, the train has a much greater tendency to remain at rest. Accordingly, we say that the train has more **inertia** than the bicycle. Quantitatively, the inertia of an object is measured by its **mass**. The following definition of inertia and mass indicates why Newton's first law is sometimes called the law of inertia:

> **DEFINITION OF INERTIA AND MASS**
>
> **Inertia is the natural tendency of an object to remain at rest or in motion at a constant velocity. The mass of an object is a quantitative measure of inertia.**
>
> **SI Unit of Inertia and Mass: kilogram (kg)**

The SI unit for mass is the kilogram (kg), whereas the units in the CGS system and the BE system are the gram (g) and the slug (sl), respectively. Conversion factors between these units are given on the page facing the inside of the front cover. **Interactive Figure 4.3** gives the typical masses of various objects, ranging from a penny to a supertanker. The larger the mass, the greater is the inertia. Often the words "mass" and "weight" are used interchangeably, but this is incorrect. Mass and weight are different concepts, and Section 4.7 will discuss the distinction between them.

THE PHYSICS OF . . . seat belts. Figure 4.4 shows a useful application of inertia. Automobile seat belts unwind freely when pulled gently, so they can be buckled. But in an accident, they hold you safely in place. One seat-belt mechanism consists of a ratchet wheel, a locking bar, and a pendulum. The belt is wound around a spool mounted on the ratchet wheel. While the car is at rest or moving at a constant velocity, the pendulum hangs straight down, and the locking bar rests horizontally, as the gray part of the drawing shows. Consequently, nothing prevents the ratchet wheel from turning, and the seat belt can be pulled out easily. When the car suddenly slows down in an accident, however, the relatively massive lower part of the pendulum keeps moving forward because of its inertia. The pendulum swings on its pivot into the position shown in color and causes the locking bar to block the rotation of the ratchet wheel, thus preventing the seat belt from unwinding.

An Inertial Reference Frame

Newton's first law (and also the second law) can appear to be invalid to certain observers. Suppose, for instance, that you are a passenger riding in a friend's car. While the car moves at a constant speed along a straight line, you do not feel the seat pushing against your back to any unusual extent. This experience is consistent with the first law, which indicates that in the absence of a net force you should move with a constant velocity. Suddenly the driver floors the gas pedal. Immediately you feel the seat pressing against your back as the car accelerates. Therefore, you sense that a force is being applied to you. The first law leads you to believe that your motion should change, and, relative to the ground outside, your motion does change. But *relative to the car,* you can see that your motion does *not* change, because you remain stationary with respect to the car. Clearly, Newton's first law does not hold for observers who use the accelerating car as a frame of reference. As a result, such a reference frame is said to be *noninertial*. All accelerating reference frames are noninertial. In contrast, observers for whom the law of inertia is

Penny
(0.003 kg)

Book
(2 kg)

Bicycle
(15 kg)

Car
(2000 kg)

Jetliner
$(1.2 \times 10^5$ kg)

Supertanker
$(1.5 \times 10^8$ kg)

INTERACTIVE FIGURE 4.3 Typical masses of various objects.

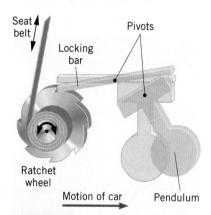

Seat belt

Pivots

Locking bar

Ratchet wheel

Motion of car

Pendulum

FIGURE 4.4 Inertia plays a central role in one seat-belt mechanism. The gray part of the drawing applies when the car is at rest or moving at a constant velocity. The colored parts show what happens when the car suddenly slows down, as in an accident.

valid are said to be using **inertial reference frames** for their observations, as defined below:

> **DEFINITION OF AN INERTIAL REFERENCE FRAME**
>
> **An inertial reference frame is one in which Newton's law of inertia is valid.**

The acceleration of an inertial reference frame is zero, so it moves with a constant velocity. All of Newton's laws of motion are valid in inertial reference frames, and when we apply these laws, we will be assuming such a reference frame. In particular, the earth itself is a good approximation of an inertial reference frame.

Check Your Understanding

(*The answer is given at the end of the book.*)

1. Which of the following statements can be explained by Newton's first law? (A): When your car suddenly comes to a halt, you lunge forward. (B): When your car rapidly accelerates, you are pressed backward against the seat. **(a)** Neither A nor B **(b)** Both A and B **(c)** A but not B **(d)** B but not A

4.3 | Newton's Second Law of Motion

Newton's first law indicates that if no net force acts on an object, then the velocity of the object remains unchanged. The second law deals with what happens when a net force does act. Consider a hockey puck once again. When a player strikes a stationary puck, he causes the velocity of the puck to change. In other words, he makes the puck accelerate. The cause of the acceleration is the force that the hockey stick applies. As long as this force acts, the velocity increases, and the puck accelerates. Now, suppose another player strikes the puck and applies twice as much force as the first player does. The greater force produces a greater acceleration. In fact, if the friction between the puck and the ice is negligible, and if there is no wind resistance, the acceleration of the puck is directly proportional to the force. Twice the force produces twice the acceleration. Moreover, the acceleration is a vector quantity, just as the force is, and points in the same direction as the force.

Often, several forces act on an object simultaneously. Friction and wind resistance, for instance, do have some effect on a hockey puck. In such cases, it is the net force, or the vector sum of all the forces acting, that is important. Mathematically, the net force is written as $\Sigma\vec{F}$, where the Greek capital letter Σ (sigma) denotes the vector sum. Newton's second law states that the acceleration is proportional to the net force acting on the object.

In Newton's second law, the net force is only one of two factors that determine the acceleration. The other is the inertia or mass of the object. After all, the same net force that imparts an appreciable acceleration to a hockey puck (small mass) will impart very little acceleration to a semitrailer truck (large mass). Newton's second law states that for a given net force, the magnitude of the acceleration is inversely proportional to the mass. Twice the mass means one-half the acceleration, if the same net force acts on both objects. Thus, the second law shows how the acceleration depends on both the net force and the mass, as given in Equation 4.1.

> **NEWTON'S SECOND LAW OF MOTION**
>
> When a net external force $\Sigma\vec{F}$ acts on an object of mass m, the acceleration $\vec{a}$ that results is directly proportional to the net force and has a magnitude that is inversely proportional to the mass. The direction of the acceleration is the same as the direction of the net force.
>
> $$\vec{a} = \frac{\Sigma\vec{F}}{m} \quad \text{or} \quad \Sigma\vec{F} = m\vec{a} \qquad (4.1)$$
>
> **SI Unit of Force:** $\text{kg} \cdot \text{m}/\text{s}^2 = \text{newton (N)}$

TABLE 4.1 Units for Mass, Acceleration, and Force

System	Mass	Acceleration	Force
SI	kilogram (kg)	meter/second^2 (m/s^2)	newton (N)
CGS	gram (g)	centimeter/second^2 (cm/s^2)	dyne (dyn)
BE	slug (sl)	foot/second^2 (ft/s^2)	pound (lb)

Note that the net force in Equation 4.1 includes only the forces that the environment exerts on the object of interest. Such forces are called **external forces**. In contrast, **internal forces** are forces that one part of an object exerts on another part of the object and are not included in Equation 4.1.

According to Equation 4.1, the SI unit for force is the unit for mass (kg) times the unit for acceleration (m/s^2), or

$$\text{SI unit for force} = (\text{kg})\left(\frac{m}{s^2}\right) = \frac{\text{kg} \cdot \text{m}}{s^2}$$

The combination of kg · m/s^2 is called a *newton* (N) and is a derived SI unit, not a base unit; 1 newton = 1 N = 1 kg · m/s^2.

In the CGS system, the procedure for establishing the unit of force is the same as with SI units, except that mass is expressed in grams (g) and acceleration in cm/s^2. The resulting unit for force is the *dyne*; 1 dyne = 1 g · cm/s^2.

In the BE system, the unit for force is defined to be the pound (lb),* and the unit for acceleration is ft/s^2. With this procedure, Newton's second law can then be used to obtain the BE unit for mass:

$$\text{BE unit for force} = \text{lb} = (\text{unit for mass})\left(\frac{\text{ft}}{s^2}\right)$$

$$\text{Unit for mass} = \frac{\text{lb} \cdot s^2}{\text{ft}}$$

The combination of lb · s^2/ft is the unit for mass in the BE system and is called the *slug* (sl); 1 slug = 1 sl = 1 lb · s^2/ft.

Table 4.1 summarizes the various units for mass, acceleration, and force. Conversion factors between force units from different systems are provided on the page facing the inside of the front cover.

When using the second law to calculate the acceleration, it is necessary to determine the net force that acts on the object. In this determination a **free-body diagram** helps enormously. A free-body diagram is a diagram that represents the object and the forces that act on it. Only the forces that *act on the object* appear in a free-body diagram. Forces that the object exerts on its environment are not included. Example 1 illustrates the use of a free-body diagram.

EXAMPLE 1 | Pushing a Stalled Car

Two people are pushing a stalled car, as **Animated Figure 4.5a** indicates. The mass of the car is 1850 kg. One person applies a force of 275 N to the car, while the other applies a force of 395 N. Both forces act in the same direction. A third force of 560 N also acts on the car, but in a direction opposite to that in which the people are pushing. This force arises because of friction and the extent to which the pavement opposes the motion of the tires. Find the acceleration of the car.

Reasoning According to Newton's second law, the acceleration is the net force divided by the mass of the car. To determine the net force, we use the free-body diagram in **Animated Figure 4.5b**. In this diagram, the car is represented as a dot, and its motion is along the +x axis. The diagram makes it clear that the forces all act along one direction. Therefore, they can be added as colinear vectors to obtain the net force.

---

*We refer here to the gravitational version of the BE system, in which a force of one pound is defined to be the pull of the earth on a certain standard body at a location where the acceleration due to gravity is 32.174 ft/s^2.

(a)

(b) Free-body diagram of the car

ANIMATED FIGURE 4.5 (a) Two people push a stalled car. A force created by friction and the pavement opposes their efforts. (b) A free-body diagram that shows the horizontal forces acting on the car.

Problem-Solving Insight The direction of the acceleration is always the same as the direction of the net force.

Solution From Equation 4.1, the acceleration is $a = (\Sigma F)/m$. The net force is

$$\Sigma F = +275\ \text{N} + 395\ \text{N} - 560\ \text{N} = +110\ \text{N}$$

The acceleration can now be obtained:

$$a = \frac{\Sigma F}{m} = \frac{+110\ \text{N}}{1850\ \text{kg}} = \boxed{+0.059\ \text{m/s}^2} \qquad \textbf{(4.1)}$$

The plus sign indicates that the acceleration points along the $+x$ axis, in the same direction as the net force.

Check Your Understanding

(The answers are given at the end of the book.)

2. The net external force acting on an object is zero. Which one of the following statements is true? **(a)** The object can only be stationary. **(b)** The object can only be traveling with a constant velocity. **(c)** The object can be either stationary or traveling with a constant velocity. **(d)** The object can only be traveling with a velocity that is changing.

3. In Case A an object is moving straight downward with a constant speed of 9.80 m/s, while in Case B an object is moving straight downward with a constant acceleration of magnitude 9.80 m/s². Which one of the following is true? **(a)** A nonzero net external force acts on the object in both cases. **(b)** A nonzero net external force acts on the object in neither case. **(c)** A nonzero net external force acts on the object in Case A only. **(d)** A nonzero net external force acts on the object in Case B only.

4.4 The Vector Nature of Newton's Second Law of Motion

When a football player throws a pass, the direction of the force he applies to the ball is important. Both the force and the resulting acceleration of the ball are vector quantities, as are all forces and accelerations. The directions of these vectors can be taken into account in two dimensions by using x and y components. The net force $\Sigma \vec{F}$ in Newton's second law has components ΣF_x and ΣF_y, while the acceleration $\vec{a}$ has components a_x and a_y. Consequently, Newton's second law, as expressed in Equation 4.1, can be written in an equivalent form as two equations, one for the x components and one for the y components:

$$\Sigma F_x = ma_x \qquad \textbf{(4.2a)}$$

$$\Sigma F_y = ma_y \qquad \textbf{(4.2b)}$$

This procedure is similar to that employed in Chapter 3 for the equations of two-dimensional kinematics (see **Table 3.1**). The components in Equations 4.2a and 4.2b are scalar components and will be either positive or negative numbers, depending on whether they point along the positive or negative x or y axis. The remainder of this section deals with examples that show how these equations are used.

EXAMPLE 2 | Applying Newton's Second Law Using Components

A man is stranded on a raft (mass of man and raft = 1300 kg), as shown in **Figure 4.6a**. By paddling, he causes an average force $\vec{P}$ of 17 N to be applied to the raft in a direction due east (the $+x$ direction). The wind also exerts a force $\vec{A}$ on the raft. This force has a magnitude of 15 N and points 67° north of east. Ignoring any resistance from the water, find the x and y components of the raft's acceleration.

Reasoning Since the mass of the man and the raft is known, Newton's second law can be used to determine the acceleration components from the given forces. According to the form of the second law in Equations 4.2a and 4.2b, the acceleration component in a given direction is the component of the net force in that direction divided by the mass. As an aid in determining the components ΣF_x and ΣF_y of the net force, we use the free-body diagram in **Figure 4.6b**. In this diagram, the directions due east and due north are the $+x$ and $+y$ directions, respectively.

Solution **Figure 4.6b** shows the force components:

Force	x Component	y Component
$\vec{P}$	+17 N	0 N
$\vec{A}$	+(15 N) cos 67° = +6 N	+(15 N) sin 67° = +14 N
	ΣF_x = +17 N + 6 N = +23 N	ΣF_y = +14 N

The plus signs indicate that ΣF_x points in the direction of the $+x$ axis and ΣF_y points in the direction of the $+y$ axis. The x and y components of the acceleration point in the directions of ΣF_x and ΣF_y, respectively, and can now be calculated:

$$a_x = \frac{\Sigma F_x}{m} = \frac{+23 \text{ N}}{1300 \text{ kg}} = \boxed{+0.018 \text{ m/s}^2} \tag{4.2a}$$

$$a_y = \frac{\Sigma F_y}{m} = \frac{+14 \text{ N}}{1300 \text{ kg}} = \boxed{+0.011 \text{ m/s}^2} \tag{4.2b}$$

These acceleration components are shown in **Figure 4.6c**.

(a)   (b)   (c)   (d)

FIGURE 4.6 (a) A man is paddling a raft, as in Examples 2 and 3. (b) The free-body diagram shows the forces $\vec{P}$ and $\vec{A}$ that act on the raft. Forces acting on the raft in a direction perpendicular to the surface of the water play no role in the examples and are omitted for clarity. (c) The raft's acceleration components a_x and a_y. (d) In 65 s, the components of the raft's displacement are $x = 48$ m and $y = 23$ m.

EXAMPLE 3 | The Displacement of a Raft

At the moment that the forces $\vec{P}$ and $\vec{A}$ begin acting on the raft in Example 2, the velocity of the raft is 0.15 m/s, in a direction due east (the $+x$ direction). Assuming that the forces are maintained for 65 s, find the x and y components of the raft's displacement during this time interval.

Reasoning Once the net force acting on an object and the object's mass have been used in Newton's second law to determine the acceleration, it becomes possible to use the equations of kinematics to describe the resulting motion. We know from Example 2 that the acceleration components are $a_x = +0.018$ m/s² and $a_y = +0.011$ m/s², and it is given here that the initial velocity

components are $v_{0x} = +0.15$ m/s and $v_{0y} = 0$ m/s. Thus, Equation 3.5a ($x = v_{0x}t + \frac{1}{2}a_xt^2$) and Equation 3.5b ($y = v_{0y}t + \frac{1}{2}a_yt^2$) can be used with $t = 65$ s to determine the x and y components of the raft's displacement.

Solution According to Equations 3.5a and 3.5b, the x and y components of the displacement are

$$x = v_{0x}t + \frac{1}{2}a_xt^2 = (0.15 \text{ m/s})(65 \text{ s}) + \frac{1}{2}(0.018 \text{ m/s}^2)(65 \text{ s})^2 = \boxed{48 \text{ m}}$$

$$y = v_{0y}t + \frac{1}{2}a_yt^2 = (0 \text{ m/s})(65 \text{ s}) + \frac{1}{2}(0.011 \text{ m/s}^2)(65 \text{ s})^2 = \boxed{23 \text{ m}}$$

Figure 4.6d shows the final location of the raft.

Check Your Understanding

(The answers are given at the end of the book.)

4. Newton's second law indicates that when a net force acts on an object, it must accelerate. Does this mean that when two or more forces are applied to an object simultaneously, it must accelerate?

5. All of the following, except one, cause the acceleration of an object to double. Which one is the exception? **(a)** All forces acting on the object double. **(b)** The net force acting on the object doubles. **(c)** Both the net force acting on the object and the mass of the object double. **(d)** The net force acting on the object remains the same, while the mass of the object is reduced by a factor of two.

4.5 | Newton's Third Law of Motion

Imagine you are in a football game. You line up facing your opponent, the ball is snapped, and the two of you crash together. No doubt, you feel a force. But think about your opponent. He too feels something, for while he is applying a force to you, you are applying a force to him. In other words, there isn't just one force on the line of scrimmage; there is a pair of forces. Newton was the first to realize that all forces occur in pairs and there is no such thing as an isolated force, existing all by itself. His third law of motion deals with this fundamental characteristic of forces.

> **NEWTON'S THIRD LAW OF MOTION**
> **Whenever one object exerts a force on a second object, the second object exerts an oppositely directed force of equal magnitude on the first object.**

The third law is often called the "action–reaction" law, because it is sometimes quoted as follows: "For every action (force) there is an equal, but opposite, reaction."

Figure 4.7 illustrates how the third law applies to an astronaut who is drifting just outside a spacecraft and who pushes on the spacecraft with a force $\vec{P}$. According to the third law, the spacecraft pushes back on the astronaut with a force $-\vec{P}$ that is equal in magnitude but opposite in direction. In Example 4, we examine the accelerations produced by each of these forces.

FIGURE 4.7 The astronaut pushes on the spacecraft with a force $+\vec{P}$. According to Newton's third law, the spacecraft simultaneously pushes back on the astronaut with a force $-\vec{P}$.

EXAMPLE 4 | The Accelerations Produced by Action and Reaction Forces

Suppose that the mass of the spacecraft in **Figure 4.7** is $m_S =$ 11 000 kg and that the mass of the astronaut is $m_A = 92$ kg. In addition, assume that the astronaut pushes with a force of $\vec{P} =$ +36 N on the spacecraft. Find the accelerations of the spacecraft and the astronaut.

Reasoning   Consistent with Newton's third law, when the astronaut applies the force $\vec{P} = +36$ N to the spacecraft, the spacecraft applies a reaction force $-\vec{P} = -36$ N to the astronaut. As a result, the spacecraft and the astronaut accelerate in opposite directions. Although the action and reaction forces have the same magnitude, they do not create accelerations of the same magnitude, because the spacecraft and the astronaut have different masses. According to Newton's second law, the astronaut, having a much smaller mass, will experience a much larger acceleration. In applying the second law, we note that the net force acting on the spacecraft is $\Sigma\vec{F} = \vec{P}$, while the net force acting on the astronaut is $\Sigma\vec{F} = -\vec{P}$.

Problem-Solving Insight Even though the magnitudes of the action and reaction forces are always equal, these forces do not necessarily produce accelerations that have equal magnitudes, since each force acts on a different object that may have a different mass.

Solution Using the second law, we find that the acceleration of the spacecraft is

$$\vec{a}_S = \frac{\vec{P}}{m_S} = \frac{+36\text{ N}}{11\,000\text{ kg}} = \boxed{+0.0033\text{ m/s}^2}$$

The acceleration of the astronaut is

$$\vec{a}_A = \frac{-\vec{P}}{m_A} = \frac{-36\text{ N}}{92\text{ kg}} = \boxed{-0.39\text{ m/s}^2}$$

THE PHYSICS OF . . . automatic trailer brakes. There is a clever application of Newton's third law in some rental trailers. As **Figure 4.8** illustrates, the tow bar connecting the trailer to the rear bumper of a car contains a mechanism that can automatically actuate brakes on the trailer wheels. This mechanism works without the need for electrical connections between the car and the trailer. When the driver applies the car brakes, the car slows down. Because of inertia, however, the trailer continues to roll forward and begins pushing against the bumper. In reaction, the bumper pushes back on the tow bar. The reaction force is used by the mechanism in the tow bar to "push a brake pedal" for the trailer.

Mechanism for actuating
trailer brakes

FIGURE 4.8 Some rental trailers include an automatic brake-actuating mechanism.

Check Your Understanding

(*The answer is given at the end of the book.*)

6. A father and his seven-year-old daughter are facing each other on ice skates. With their hands, they push off against one another. Which one or more of the following statements is (are) true? **(a)** Each experiences an acceleration that has a different magnitude. **(b)** Each experiences an acceleration of the same magnitude. **(c)** Each experiences a pushing force that has a different magnitude. **(d)** Each experiences a pushing force that has the same magnitude.

4.6 Types of Forces: An Overview

Newton's three laws of motion make it clear that forces play a central role in determining the motion of an object. In the next four sections some common forces will be discussed: the gravitational force (Section 4.7), the normal force (Section 4.8), frictional forces (Section 4.9), and the tension force (Section 4.10). In later chapters, we will encounter still others, such as electric and magnetic forces. It is important to realize that Newton's second law is always valid, regardless of which of these forces may act on an object. One does not have a different law for every type of common force. Thus, we need only to determine what forces are acting on an object, add them together to form the net force, and then use Newton's second law to determine the object's acceleration.

In nature there are two general types of forces, fundamental and nonfundamental. Fundamental forces are the ones that are truly unique, in the sense that all other forces can be explained in terms of them. Only three fundamental forces have been discovered:

1. Gravitational force
2. Strong nuclear force
3. Electroweak force

The gravitational force is discussed in the next section. The strong nuclear force plays a primary role in the stability of the nucleus of the atom (see Section 31.2). The electroweak force is a single force that manifests itself in two ways (see Section 32.6). One manifestation is the electromagnetic force that electrically charged particles exert on one another (see Sections 18.5, 21.2, and 21.8). The other manifestation is the so-called weak nuclear force that plays a role in the radioactive disintegration of certain nuclei (see Section 31.5).

Except for the gravitational force, all of the forces discussed in this chapter are nonfundamental, because they are related to the electromagnetic force. They arise from the interactions between the electrically charged particles that comprise atoms and molecules. Our understanding of which forces are fundamental, however, is continually evolving. For instance, in the 1860s and 1870s James Clerk Maxwell showed that the electric force and the magnetic force could be explained as manifestations of a single electromagnetic force. Then, in the 1970s, Sheldon Glashow (1932–), Abdus Salam (1926–1996), and Steven Weinberg (1933–) presented the theory that explains how the electromagnetic force and the weak nuclear force are related to the electroweak force. They received a Nobel Prize in 1979 for their achievement. Today, efforts continue that have the goal of further reducing the number of fundamental forces.

4.7 | The Gravitational Force

Newton's Law of Universal Gravitation

Objects fall downward because of gravity, and Chapters 2 and 3 discuss how to describe the effects of gravity by using a value of $g = 9.80$ m/s^2 for the downward acceleration it causes. However, nothing has been said about why g is 9.80 m/s^2. The reason is fascinating, as we will see later in this section.

The acceleration due to gravity is like any other acceleration, and Newton's second law indicates that it must be caused by a net force. In addition to his famous three laws of motion, Newton also provided a coherent understanding of the **gravitational force**. His "law of universal gravitation" is stated as follows:

NEWTON'S LAW OF UNIVERSAL GRAVITATION

Every particle in the universe exerts an attractive force on every other particle. A particle is a piece of matter, small enough in size to be regarded as a mathematical point. For two particles that have masses m_1 and m_2 and are separated by a distance r, the force that each exerts on the other is directed along the line joining the particles (see Figure 4.9) and has a magnitude given by

$$F = G\frac{m_1 m_2}{r^2} \qquad (4.3)$$

The symbol G denotes the universal gravitational constant, whose value is found experimentally to be

$$G = 6.674 \times 10^{-11}\,\text{N} \cdot \text{m}^2/\text{kg}^2$$

FIGURE 4.9 The two particles, whose masses are m_1 and m_2, are attracted by gravitational forces $+\vec{F}$ and $-\vec{F}$.

The constant G that appears in Equation 4.3 is called the **universal gravitational constant**, because it has the same value for all pairs of particles anywhere in the universe, no matter what their separation. The value for G was first measured in an experiment by the English scientist Henry Cavendish (1731–1810), more than a century after Newton proposed his law of universal gravitation.

To see the main features of Newton's law of universal gravitation, look at the two particles in **Figure 4.9**. They have masses m_1 and m_2 and are separated by a distance r. In the picture, it is assumed that a force pointing to the right is positive. The gravitational forces point along the line joining the particles and are

$+\vec{F}$, the gravitational force exerted on particle 1 by particle 2

$-\vec{F}$, the gravitational force exerted on particle 2 by particle 1

These two forces have equal magnitudes and opposite directions. They act on different bodies, causing them to be mutually attracted. In fact, these forces are an action–reaction pair, as required by Newton's third law. Example 5 shows that the magnitude of the gravitational force is extremely small for ordinary values of the masses and the distance between them.

EXAMPLE 5 | Gravitational Attraction

What is the magnitude of the gravitational force that acts on each particle in **Figure 4.9**, assuming $m_1 = 12$ kg (approximately the mass of a bicycle), $m_2 = 25$ kg, and $r = 1.2$ m?

Reasoning and Solution   The magnitude of the gravitational force can be found using Equation 4.3:

$$F = G\frac{m_1 m_2}{r^2} = (6.67 \times 10^{-11}\,\text{N} \cdot \text{m}^2/\text{kg}^2)\frac{(12\,\text{kg})(25\,\text{kg})}{(1.2\,\text{m})^2}$$
$$= \boxed{1.4 \times 10^{-8}\,\text{N}}$$

For comparison, you exert a force of about 1 N when pushing a doorbell, so that the gravitational force is exceedingly small in circumstances such as those here. This result is due to the fact that G itself is very small. However, if one of the bodies has a large mass, like that of the earth (5.98×10^{24} kg), the gravitational force can be large.

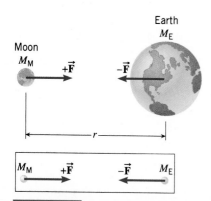

FIGURE 4.10   The gravitational force that each uniform sphere of matter exerts on the other is the same as if each sphere were a particle with its mass concentrated at its center. The earth (mass M_E) and the moon (mass M_M) approximate such uniform spheres.

As expressed by Equation 4.3, Newton's law of gravitation applies only to particles. However, most familiar objects are too large to be considered particles. Nevertheless, the law of universal gravitation can be applied to such objects with the aid of calculus. Newton was able to prove that an object of finite size can be considered to be a particle for purposes of using the gravitation law, provided the mass of the object is distributed with spherical symmetry about its center. Thus, Equation 4.3 can be applied when each object is a sphere whose mass is spread uniformly over its entire volume. **Figure 4.10** shows this kind of application, assuming that the earth and the moon are such uniform spheres of matter. In this case, r is the distance *between the centers of the spheres* and not the distance between the outer surfaces. The gravitational forces that the spheres exert on each other are the same as if the entire mass of each were concentrated at its center. Even if the objects are not uniform spheres, Equation 4.3 can be used to a good degree of approximation if the sizes of the objects are small relative to the distance of separation r.

Weight

The weight of an object exists because of the gravitational pull of the earth, according to the following definition:

DEFINITION OF WEIGHT

The weight of an object on or above the earth is the gravitational force that the earth exerts on the object. The weight always acts downward, toward the center of the earth. On or above another astronomical body, the weight is the gravitational force exerted on the object by that body.

SI Unit of Weight: newton (N)

Using W for the magnitude of the weight,* m for the mass of the object, and M_E for the mass of the earth, it follows from Equation 4.3 that

$$W = G\frac{M_E m}{r^2} \qquad (4.4)$$

Equation 4.4 and **Figure 4.11** both emphasize that an object has weight whether or not it is resting on the earth's surface, because the gravitational force is acting even when the distance r is not equal to the radius R_E of the earth. However, the gravitational force becomes weaker as r increases, since r is in the denominator of Equation 4.4. **Figure 4.12**, for example, shows how the weight of the Hubble Space Telescope becomes smaller as the distance r from the center of the earth increases. In Example 6 the telescope's weight is determined when it is on earth and in orbit.

Object of mass m

Mass of earth $= M_E$

FIGURE 4.11 On or above the earth, the weight $\vec{W}$ of an object is the gravitational force exerted on the object by the earth.

EXAMPLE 6 | The Hubble Space Telescope

The mass of the Hubble Space Telescope is 11 600 kg. Determine the weight of the telescope **(a)** when it was resting on the earth and **(b)** as it is in its orbit 598 km above the earth's surface.

Reasoning The weight of the Hubble Space Telescope is the gravitational force exerted on it by the earth. According to Equation 4.4, the weight varies inversely as the square of the radial distance r. Thus, we expect the telescope's weight on the earth's surface (r smaller) to be greater than its weight in orbit (r larger).

> **Problem-Solving Insight** When applying Newton's gravitation law to uniform spheres of matter, remember that the distance r is between the centers of the spheres, not between the surfaces.

Solution **(a)** On the earth's surface, the weight is given by Equation 4.4 with $r = 6.38 \times 10^6$ m (the earth's radius):

$$W = G\frac{M_E m}{r^2} = \frac{(6.67 \times 10^{-11}\,\text{N} \cdot \text{m}^2/\text{kg}^2)(5.98 \times 10^{24}\,\text{kg})(11\,600\,\text{kg})}{(6.38 \times 10^6\,\text{m})^2}$$

$$\boxed{W = 1.14 \times 10^5\,\text{N}}$$

(b) When the telescope is 598 km above the surface, its distance from the center of the earth is

$$r = 6.38 \times 10^6\,\text{m} + 598 \times 10^3\,\text{m} = 6.98 \times 10^6\,\text{m}$$

The weight now can be calculated as in part (a), except that the new value of r must be used: $\boxed{W = 0.950 \times 10^5\,\text{N}}$. As expected, the weight is less in orbit.

The space age has forced us to broaden our ideas about weight. For instance, an astronaut weighs only about one-sixth as much on the moon as on the earth. To obtain his weight on the moon from Equation 4.4, it is only necessary to replace M_E by M_M (the mass of the moon) and let $r = R_M$ (the radius of the moon).

Relation Between Mass and Weight

Although massive objects weigh a lot on the earth, mass and weight are not the same quantity. As Section 4.2 discusses, mass is a quantitative measure of inertia. As such, mass is an intrinsic property of matter and does not change as an object is moved from one location to another. Weight, in contrast, is the gravitational force acting on the object and can vary, depending on how far the object is above the earth's surface or whether it is located near another body such as the moon.

The relation between weight W and mass m can be written in two ways:

$$W = \boxed{G\frac{M_E}{r^2}}\,m \qquad (4.4)$$

$$W = m\,\boxed{g} \qquad (4.5)$$

FIGURE 4.12 The weight of the Hubble Space Telescope decreases as the telescope gets farther from the earth. The distance from the center of the earth to the telescope is r.

$R_E = 6.38 \times 10^6$ m

---

*Often, the word "weight" and the phrase "magnitude of the weight" are used interchangeably, even though weight is a vector. Generally, the context makes it clear when the direction of the weight vector must be taken into account.

Equation 4.4 is Newton's law of universal gravitation, and Equation 4.5 is Newton's second law (net force equals mass times acceleration) incorporating the acceleration g due to gravity. These expressions make the distinction between mass and weight stand out. The weight of an object whose mass is m depends on the values for the universal gravitational constant G, the mass M_E of the earth, and the distance r. These three parameters together determine the acceleration g due to gravity. The specific value of $g = 9.80$ m/s^2 applies only when r equals the radius R_E of the earth. For larger values of r, as would be the case on top of a mountain, the effective value of g is less than 9.80 m/s^2. The fact that g decreases as the distance r increases means that the weight likewise decreases. The mass of the object, however, does not depend on these effects and does not change. Conceptual Example 7 further explores the difference between mass and weight.

CONCEPTUAL EXAMPLE 7 | Mass Versus Weight

A vehicle designed for exploring the moon's surface (see **Photo 4.1**) is being tested on earth, where it weighs roughly six times more than it will on the moon. The acceleration of the vehicle along the ground is measured. To achieve the same acceleration on the moon, will the required net force be **(a)** the same as, **(b)** greater than, or **(c)** less than that on earth?

Reasoning Do not be misled by the fact that the vehicle weighs more on earth. The greater weight occurs only because the mass and radius of the earth are different from the mass and radius of the moon. In any event, in Newton's second law the net force is proportional to the vehicle's mass, not its weight.

Answers (b) and (c) are incorrect. According to Newton's second law, for a given acceleration, the net force depends only on the mass. If the required net force were greater or smaller on the moon than it is on the earth, the implication would be that the vehicle's mass is different on the moon than it is on earth, which is contrary to fact.

Answer (a) is correct. The net force $\Sigma\vec{F}$ required to accelerate the vehicle is specified by Newton's second law as $\Sigma\vec{F} = m\vec{a}$, where m is the vehicle's mass and $\vec{a}$ is the acceleration. For a given acceleration, the net force depends only on the mass, which is the same on the moon as it is on the earth. Therefore, the required net force is the same on the moon as it is on the earth.

Related Homework: Problems 24, 84

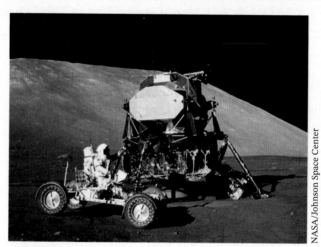

NASA/Johnson Space Center

PHOTO 4.1 On the moon the Lunar Roving Vehicle that astronaut Eugene Cernan is driving and the Lunar Excursion Module (behind the Roving Vehicle) have the same masses that they have on the earth. However, their weights are different on the moon than on the earth, as Conceptual Example 7 discusses.

Check Your Understanding

(The answers are given at the end of the book.)

7. When a body is moved from sea level to the top of a mountain, what changes: **(a)** the body's mass, **(b)** its weight, or **(c)** both its mass and its weight?

8. Object A weighs twice as much as object B at the same spot on the earth. Would the same be true at a given spot on Mars?

9. Three particles have identical masses. Each particle experiences only the gravitational forces due to the other two particles. How should the particles be arranged so each one experiences a net gravitational force that has the same magnitude? **(a)** On the corners of an equilateral triangle **(b)** On three of the four corners of a square **(c)** On the corners of a right triangle

10. Two objects with masses m_1 and m_2 are separated by a distance $2d$ (see **CYU Figure 4.1**). Mass m_2 is greater than mass m_1. A third object has a mass m_3. All three objects are located on the same straight line. The net gravitational force acting on the third object is zero. Which one of the drawings correctly represents the locations of the objects?

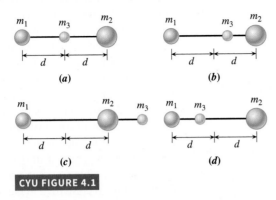

CYU FIGURE 4.1

| 4.8 | **The Normal Force** |

The Definition and Interpretation of the Normal Force

In many situations, an object is in contact with a surface, such as a tabletop. Because of the contact, there is a force acting on the object. The present section discusses only one component of this force, the component that acts perpendicular to the surface. The next section discusses the component that acts parallel to the surface. The perpendicular component is called the **normal force**.

> **DEFINITION OF THE NORMAL FORCE**
>
> The normal force $\vec{\mathbf{F}}_N$ is one component of the force that a surface exerts on an object with which it is in contact—namely, the component that is perpendicular to the surface.

Figure 4.13 shows a block resting on a horizontal table and identifies the two forces that act on the block, the weight $\vec{\mathbf{W}}$ and the normal force $\vec{\mathbf{F}}_N$. To understand how an inanimate object, such as a tabletop, can exert a normal force, think about what happens when you sit on a mattress. Your weight causes the springs in the mattress to compress. As a result, the compressed springs exert an upward force (the normal force) on you. In a similar manner, the weight of the block causes invisible "atomic springs" in the surface of the table to compress, thus producing a normal force on the block.

Newton's third law plays an important role in connection with the normal force. In **Figure 4.13**, for instance, the block exerts a force on the table by pressing down on it. Consistent with the third law, the table exerts an oppositely directed force of equal magnitude on the block. This reaction force is the normal force. The magnitude of the normal force indicates how hard the two objects press against each other.

If an object is resting on a horizontal surface and there are no vertically acting forces except the object's weight and the normal force, the magnitudes of these two forces are equal; that is, $F_N = W$. This is the situation in **Figure 4.13**. The weight must be balanced by the normal force for the object to remain at rest on the table. If the magnitudes of these forces were not equal, there would be a net force acting on the block, and the block would accelerate either upward or downward, in accord with Newton's second law.

If other forces in addition to $\vec{\mathbf{W}}$ and $\vec{\mathbf{F}}_N$ act in the vertical direction, the magnitudes of the normal force and the weight are no longer equal. In **Figure 4.14a**, for instance, a box whose weight is 15 N is being pushed downward against a table. The pushing force has a magnitude of 11 N. Thus, the total downward force exerted on the box is 26 N, and this must be balanced by the upward-acting normal force if the box is to remain at rest. In this situation, then, the normal force is 26 N, which is considerably larger than the weight of the box.

FIGURE 4.13 Two forces act on the block, its weight $\vec{\mathbf{W}}$ and the normal force $\vec{\mathbf{F}}_N$ exerted by the surface of the table.

FIGURE 4.14 (a) The normal force $\vec{F}_N$ is greater than the weight of the box, because the box is being pressed downward with an 11-N force. (b) The normal force is smaller than the weight, because the rope supplies an upward force of 11 N that partially supports the box.

Figure 4.14b illustrates a different situation. Here, the box is being pulled upward by a rope that applies a force of 11 N. The net force acting on the box due to its weight and the rope is only 4 N, downward. To balance this force, the normal force needs to be only 4 N. It is not hard to imagine what would happen if the force applied by the rope were increased to 15 N—exactly equal to the weight of the box. In this situation, the normal force would become zero. In fact, the table could be removed, since the block would be supported entirely by the rope. The situations in **Figure 4.14** are consistent with the idea that the magnitude of the normal force indicates how hard two objects press against each other. Clearly, the box and the table press against each other harder in part *a* of the picture than in part *b*.

Like the box and the table in **Figure 4.14**, various parts of the human body press against one another and exert normal forces. Example 8 illustrates the remarkable ability of the human skeleton to withstand a wide range of normal forces.

EXAMPLE 8 | BIO The Physics of the Human Skeleton

In a circus balancing act, a woman performs a headstand on top of a standing performer's head, as **Figure 4.15a** illustrates. The woman weighs 490 N, and the standing performer's head and neck weigh 50 N. It is primarily the seventh cervical vertebra in the spine that supports all the weight above the shoulders. What is the normal force that this vertebra exerts on the neck and head of the standing performer **(a)** before the act and **(b)** during the act?

Reasoning To begin, we draw a free-body diagram for the neck and head of the standing performer. Before the act, there are only two forces, the weight of the standing performer's head and neck, and the normal force. During the act, an additional force is present due to the woman's weight. In both cases, the upward and downward

forces must balance for the head and neck to remain at rest. This condition of balance will lead us to values for the normal force.

Solution **(a)** **Figure 4.15b** shows the free-body diagram for the standing performer's head and neck before the act. The only forces acting are the normal force $\vec{F}_N$ and the 50-N weight. These two forces must balance for the standing performer's head and neck to remain at rest. Therefore, the seventh cervical vertebra exerts a normal force of $\boxed{F_N = 50 \text{ N}}$.

(b) **Figure 4.15c** shows the free-body diagram that applies during the act. Now, the total downward force exerted on the standing performer's head and neck is 50 N + 490 N = 540 N, which must be balanced by the upward normal force, so that $\boxed{F_N = 540 \text{ N}}$.

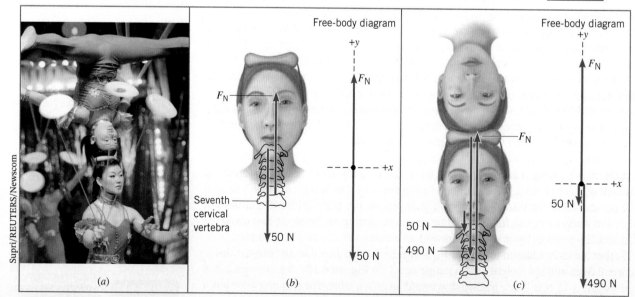

FIGURE 4.15 (a) A young woman keeps her balance during a performance by China's Sichuan Acrobatic group. A free-body diagram is shown for the standing performer's body above the shoulders (b) before the act and (c) during the act. For convenience, the scales used for the vectors in parts b and c are different.

In summary, the normal force does not necessarily have the same magnitude as the weight of the object. The value of the normal force depends on what other forces are present. It also depends on whether the objects in contact are accelerating. In one situation that involves accelerating objects, the magnitude of the normal force can be regarded as a kind of "apparent weight," as we will now see.

Apparent Weight

Usually, the weight of an object can be determined with the aid of a scale. However, even if a scale is working properly, there are situations in which it does not give the correct weight. In such situations, the reading on the scale gives only the "apparent" weight, rather than the gravitational force or "true" weight. The apparent weight is the force that the object exerts on the scale with which it is in contact.

(a) No acceleration ($\vec{v}$ = constant)   (b) Upward acceleration   (c) Downward acceleration   (d) Free-fall

INTERACTIVE FIGURE 4.16 (a) When the elevator is not accelerating, the scale registers the true weight ($W = 700$ N) of the person. (b) When the elevator accelerates upward, the apparent weight (1000 N) exceeds the true weight. (c) When the elevator accelerates downward, the apparent weight (400 N) is less than the true weight. (d) The apparent weight is zero if the elevator falls freely—that is, if it falls with the acceleration due to gravity.

To see the discrepancies that can arise between true weight and apparent weight, consider the scale in the elevator in **Interactive Figure 4.16**. The reasons for the discrepancies will be explained shortly. A person whose true weight is 700 N steps on the scale. If the elevator is at rest or moving with a constant velocity (either upward or downward), the scale registers the true weight, as **Interactive Figure 4.16a** illustrates.

If the elevator is accelerating, the apparent weight and the true weight are not equal. When the elevator accelerates upward, the apparent weight is greater than the true weight, as **Interactive Figure 4.16b** shows. Conversely, if the elevator accelerates downward, as in part c, the apparent weight is less than the true weight. In fact, if the elevator falls freely, so its acceleration is equal to the acceleration due to gravity, the apparent weight becomes zero, as part d indicates. In a situation such as this, where the apparent weight is zero, the person is said to be "weightless." The apparent weight, then, does not equal the true weight if the scale and the person on it are accelerating.

The discrepancies between true weight and apparent weight can be understood with the aid of Newton's second law. **Figure 4.17** shows a free-body diagram of the person in the elevator. The two forces that act on him are the true weight $\vec{W} = m\vec{g}$ and the normal force $\vec{F}_N$ exerted by the platform of the scale. Applying Newton's second law in the vertical direction gives

$$\Sigma F_y = +F_N - mg = ma$$

where a is the acceleration of the elevator and person. In this result, the symbol g stands for the magnitude of the acceleration due to gravity and can never be a negative quantity. However, the acceleration a may be either positive or negative, depending on whether

FIGURE 4.17 A free-body diagram showing the forces acting on the person riding in the elevator of **Interactive Figure 4.16**. $\vec{W}$ is the true weight, and $\vec{F}_N$ is the normal force exerted on the person by the platform of the scale.

the elevator is accelerating upward (+) or downward (−). Solving for the normal force F_N shows that

$$\underbrace{F_N}_{\substack{\text{Apparent} \\ \text{weight}}} = \underbrace{mg}_{\substack{\text{True} \\ \text{weight}}} + ma \qquad \text{(4.6)}$$

In Equation 4.6, F_N is the magnitude of the normal force exerted on the person by the scale. But in accord with Newton's third law, F_N is also the magnitude of the downward force that the person exerts on the scale—namely, the apparent weight.

Equation 4.6 contains all the features shown in **Interactive Figure 4.16**. If the elevator is not accelerating, $a = 0$ m/s^2, and the apparent weight equals the true weight. If the elevator accelerates upward, a is positive, and the equation shows that the apparent weight is greater than the true weight. If the elevator accelerates downward, a is negative, and the apparent weight is less than the true weight. If the elevator falls freely, $a = -g$, and the apparent weight is zero. The apparent weight is zero because when both the person and the scale fall freely, they cannot push against one another. In this text, when the weight is given, it is assumed to be the true weight, unless stated otherwise.

Check Your Understanding

(The answers are given at the end of the book.)

11. A stack of books whose true weight is 165 N is placed on a scale in an elevator. The scale reads 165 N. From this information alone, can you tell whether the elevator is moving with a constant velocity of 2 m/s upward, is moving with a constant velocity of 2 m/s downward, or is at rest?

12. A 10-kg suitcase is placed on a scale that is in an elevator. In which direction is the elevator accelerating when the scale reads 75 N and when it reads 120 N? **(a)** Downward when it reads 75 N and upward when it reads 120 N **(b)** Upward when it reads 75 N and downward when it reads 120 N **(c)** Downward in both cases **(d)** Upward in both cases

13. You are standing on a scale in an elevator that is moving upward with a constant velocity. The scale reads 600 N. The following table shows five options for what the scale reads when the elevator slows down as it comes to a stop, when it is stopped, and when it picks up speed on its way back down. Which one of the five options correctly describes the scale's readings? Note that the symbol < means "less than" and > means "greater than."

Option	Elevator slows down as it comes to a halt	Elevator is stopped	Elevator picks up speed on its way back down
(a)	> 600 N	> 600 N	> 600 N
(b)	< 600 N	600 N	< 600 N
(c)	> 600 N	600 N	< 600 N
(d)	< 600 N	< 600 N	< 600 N
(e)	< 600 N	600 N	> 600 N

Stocksnapper/123RF

FIGURE 4.18 This photo, shot parallel to the road surface, shows a tire rolling under wet conditions. The channels in the tire collect and divert water away from the regions where the tire contacts the surface, thus providing better traction.

4.9 | Static and Kinetic Frictional Forces

When an object is in contact with a surface, there is a force acting on the object. The previous section discusses the component of this force that is perpendicular to the surface, which is called the normal force. When the object moves or attempts to move along the surface, there is also a component of the force that is parallel to the surface. This parallel force component is called the **frictional force**, or simply **friction**.

In many situations considerable engineering effort is expended trying to reduce friction. For example, oil is used to reduce the friction that causes wear and tear in the pistons and cylinder walls of an automobile engine. Sometimes, however, friction is absolutely necessary. Without friction, car tires could not provide the traction needed to move the car. In fact, the raised tread on a tire is designed to maintain friction. On a wet road, the spaces in the tread pattern (see **Figure 4.18**) provide channels for the water to collect and be diverted away. Thus, these channels largely prevent the water from coming between the tire surface and the road surface, where it would reduce friction and allow the tire to skid.

Surfaces that appear to be highly polished can actually look quite rough when examined under a microscope. Such an examination reveals that two surfaces in contact touch only at relatively few spots, as **Figure 4.19** illustrates. The microscopic area of contact for these spots is substantially less than the apparent macroscopic area of contact between the surfaces—perhaps thousands of times less. At these contact points the molecules of the different bodies are close enough together to exert strong attractive intermolecular forces on one another, leading to what are known as "cold welds." Frictional forces are associated with these welded spots, but the exact details of how frictional forces arise are not well understood. However, some empirical relations have been developed that make it possible to account for the effects of friction.

Interactive Figure 4.20 helps to explain the main features of the type of friction known as **static friction**. The block in this drawing is initially at rest on a table, and as long as there is no attempt to move the block, there is no static frictional force. Then, a horizontal force $\vec{F}$ is applied to the block by means of a rope. If $\vec{F}$ is small, as in part a, experience tells us that the block still does not move. Why? It does not move because the static frictional force $\vec{f}_s$ exactly cancels the effect of the applied force. The direction of $\vec{f}_s$ is opposite to that of $\vec{F}$ and the magnitude of $\vec{f}_s$ equals the magnitude of the applied force, $f_s = F$. Increasing the applied force in **Interactive Figure 4.20** by a small amount still does not cause the block to move. There is no movement because the static frictional force also increases by an amount that cancels out the increase in the applied force (see part b of the drawing). If the applied force continues to increase, however, there comes a point when the block finally "breaks away" and begins to slide. The force just before breakaway represents the *maximum static frictional force* $\vec{f}_s^{MAX}$ that the table can exert on the block (see part c of the drawing). Any applied force that is greater than $\vec{f}_s^{MAX}$ cannot be balanced by static friction, and the resulting net force accelerates the block to the right.

Experimental evidence shows that, to a good degree of approximation, the maximum static frictional force between a pair of dry, unlubricated surfaces has two main characteristics. It is independent of the apparent macroscopic area of contact between the objects, provided that the surfaces are hard or nondeformable. For instance, in **Figure 4.21** the maximum static frictional force that the surface of the table can exert on a block is the same, whether the block is resting on its largest or its smallest side. The other main characteristic of $\vec{f}_s^{MAX}$ is that its magnitude is proportional to the magnitude of the normal force $\vec{F}_N$. As Section 4.8 points out, the magnitude of the normal force indicates how hard two surfaces are being pressed together. The harder they are pressed, the larger is f_s^{MAX}, presumably because the number of "cold-welded," microscopic contact points is increased. Equation 4.7 expresses the proportionality between f_s^{MAX} and F_N with the aid of a proportionality constant μ_s, which is called the **coefficient of static friction**.

STATIC FRICTIONAL FORCE

The magnitude f_s of the static frictional force can have any value from zero up to a maximum value of f_s^{MAX}, depending on the applied force. In other words, $f_s \leq f_s^{MAX}$, where the symbol $\leq$ is read as "less than or equal to." The equality holds only when f_s attains its maximum value, which is

$$f_s^{MAX} = \mu_s F_N \qquad (4.7)$$

In Equation 4.7, μ_s is the coefficient of static friction, and F_N is the magnitude of the normal force.

It should be emphasized that Equation 4.7 relates only the *magnitudes* of $\vec{f}_s^{MAX}$ and $\vec{F}_N$, *not the vectors themselves*. This equation does not imply that the directions of the vectors are the same. In fact, $\vec{f}_s^{MAX}$ is parallel to the surface, while $\vec{F}_N$ is perpendicular to it.

The coefficient of static friction, being the ratio of the magnitudes of two forces ($\mu_s = f_s^{MAX}/F_N$), has no units. Also, it depends on the type of material from which each surface is made (steel on wood, rubber on concrete, etc.), the condition of the

FIGURE 4.19 Even when two highly polished surfaces are in contact, they touch only at relatively few points.

No movement
(a)

No movement
(b)

When movement just begins
(c)

INTERACTIVE FIGURE 4.20 (a) and (b) Applying a small force $\vec{F}$ to the block produces no movement, because the static frictional force $\vec{f}_s$ exactly balances the applied force. (c) The block just begins to move when the applied force is slightly greater than the maximum static frictional force $\vec{f}_s^{MAX}$.

FIGURE 4.21 The maximum static frictional force $\vec{f}_s^{MAX}$ would be the same, no matter which side of the block is in contact with the table.

TABLE 4.2   **Approximate Values of the Coefficients of Friction for Various Surfaces[a]**

Materials	Coefficient of Static Friction, μ_s	Coefficient of Kinetic Friction, μ_k
Glass on glass (dry)	0.94	0.4
Ice on ice (clean, 0 °C)	0.1	0.02
Rubber on dry concrete	1.0	0.8
Rubber on wet concrete	0.7	0.5
Steel on ice	0.1	0.05
Steel on steel (dry hard steel)	0.78	0.42
Teflon on Teflon	0.04	0.04
Wood on wood	0.35	0.3

[a]The last column gives the coefficients of kinetic friction, a concept that will be discussed shortly.

surfaces (polished, rough, etc.), and other variables such as temperature. **Table 4.2** gives some typical values of μ_s for various surfaces. Example 9 illustrates the use of Equation 4.7 for determining the maximum static frictional force.

Analyzing Multiple-Concept Problems

EXAMPLE 9 | The Force Needed to Start a Skier Moving

A skier is standing motionless on a horizontal patch of snow. She is holding onto a horizontal tow rope, which is about to pull her forward (see **Figure 4.22a**). The skier's mass is 59 kg, and the coefficient of static friction between the skis and snow is 0.14. What is the magnitude of the maximum force that the tow rope can apply to the skier without causing her to move?

Reasoning   When the rope applies a relatively small force, the skier does not accelerate. The reason is that the static frictional force opposes the applied force and the two forces have the same magnitude. We can apply Newton's second law in the horizontal direction to this situation. In order for the rope to pull the skier forward, it must exert a force large enough to overcome the *maximum* static frictional force acting on the skis. The magnitude of the maximum static frictional force depends on the coefficient of static friction (which is known) and on the magnitude of the normal force. We can determine the magnitude of the normal force by using Newton's second law, along with the fact that the skier does not accelerate in the vertical direction.

Knowns and Unknowns   The data for this problem are as follows:

Description	Symbol	Value
Mass of skier	m	59 kg
Coefficient of static friction	μ_s	0.14
Unknown Variable		
Magnitude of maximum horizontal force that tow rope can apply	F	?

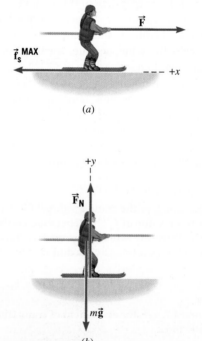

FIGURE 4.22   (a) Two forces act on the skier in the horizontal direction just before she begins to move. (b) Two vertical forces act on the skier.

Modeling the Problem

STEP 1 Newton's Second Law (Horizontal Direction)   **Figure 4.22a** shows the two horizontal forces that act on the skier just before she begins to move: the force $\vec{F}$ applied by the tow rope and the maximum static frictional force $\vec{f}_s^{\,\text{MAX}}$. Since the skier is standing motionless, she is not accelerating in the horizontal or x direction, so $a_x = 0 \text{ m/s}^2$. Applying Newton's second law (Equation 4.2a) to this situation, we have

$$\Sigma F_x = ma_x = 0$$

Since the net force ΣF_x in the x direction is $\Sigma F_x = +F - f_s^{\,\text{MAX}}$, Newton's second law can be written as $+F - f_s^{\,\text{MAX}} = 0$. Thus,

$$F = f_s^{\,\text{MAX}}$$

We do not know $f_s^{\,\text{MAX}}$, but its value will be determined in Steps 2 and 3.

$$F = f_s^{\,\text{MAX}} \qquad \text{(1)}$$

$$\boxed{?}$$

STEP 2 The Maximum Static Frictional Force   The magnitude $f_s^{\,\text{MAX}}$ of the maximum static frictional force is related to the coefficient of static friction μ_s and the magnitude F_N of the normal force by Equation 4.7:

$$\boxed{f_s^{\,\text{MAX}} = \mu_s F_N} \qquad \text{(4.7)}$$

We now substitute this result into Equation 1, as indicated in the right column. The coefficient of static friction is known, but F_N is not. An expression for F_N will be obtained in the next step.

$$F = f_s^{\,\text{MAX}} \qquad \text{(1)}$$

$$\boxed{f_s^{\,\text{MAX}} = \mu_s F_N} \qquad \text{(4.7)}$$

$$\boxed{?}$$

STEP 3 Newton's Second Law (Vertical Direction)   We can find the magnitude F_N of the normal force by noting that the skier does not accelerate in the vertical or y direction, so $a_y = 0 \text{ m/s}^2$. **Figure 4.22b** shows the two vertical forces that act on the skier: the normal force $\vec{F}_N$ and her weight $m\vec{g}$. Applying Newton's second law (Equation 4.2b) in the vertical direction gives

$$\Sigma F_y = ma_y = 0$$

The net force in the y direction is $\Sigma F_y = +F_N - mg$, so Newton's second law becomes $+F_N - mg = 0$. Thus,

$$\boxed{F_N = mg}$$

We now substitute this result into Equation 4.7, as shown at the right.

$$F = f_s^{\,\text{MAX}} \qquad \text{(1)}$$

$$\boxed{f_s^{\,\text{MAX}} = \mu_s F_N} \qquad \text{(4.7)}$$

$$\boxed{F_N = mg}$$

Solution   Algebraically combining the results of the three steps, we have

$$\underset{\text{STEP 1}}{\Big\downarrow} \qquad \underset{\text{STEP 2}}{\Big\downarrow} \qquad \underset{\text{STEP 3}}{\Big\downarrow}$$

$$F = f_s^{\,\text{MAX}} = \mu_s F_N = \mu_s mg$$

The magnitude F of the maximum force is

$$F = \mu_s mg = (0.14)(59 \text{ kg})(9.80 \text{ m/s}^2) = \boxed{81 \text{ N}}$$

If the force exerted by the tow rope exceeds this value, the skier will begin to accelerate forward.

Related Homework: Problem 39

BIO **THE PHYSICS OF . . . rock climbing.**   Static friction is often essential, as it is to the rock climber in **Figure 4.23**, for instance. She presses outward against the walls of the rock formation with her hands and feet to create sufficiently large normal forces, so that the static frictional forces help support her weight.

Once two surfaces begin sliding over one another, the static frictional force is no longer of any concern. Instead, a type of friction known as **kinetic* friction** comes

---

*The word "kinetic" is derived from the Greek word *kinetikos*, meaning "pertaining to motion."

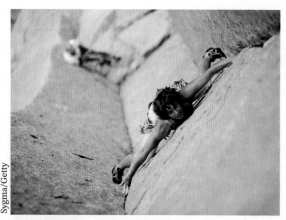

FIGURE 4.23 In maneuvering her way up Devil's Tower at Devil's Tower National Monument in Wyoming, this rock climber uses the static frictional forces between her hands and feet and the vertical rock walls to help support her weight.

into play. The kinetic frictional force opposes the relative sliding motion. If you have ever pushed an object across a floor, you may have noticed that it takes less force to keep the object sliding than it takes to get it going in the first place. In other words, the kinetic frictional force is usually less than the static frictional force.

Experimental evidence indicates that the kinetic frictional force $\vec{f}_k$ has three main characteristics, to a good degree of approximation. It is independent of the apparent area of contact between the surfaces (see **Figure 4.21**). It is independent of the speed of the sliding motion, if the speed is small. And last, the magnitude of the kinetic frictional force is proportional to the magnitude of the normal force. Equation 4.8 expresses this proportionality with the aid of a proportionality constant μ_k, which is called the **coefficient of kinetic friction**.

KINETIC FRICTIONAL FORCE

The magnitude f_k of the kinetic frictional force is given by

$$f_k = \mu_k F_N \tag{4.8}$$

In Equation 4.8, μ_k is the coefficient of kinetic friction, and F_N is the magnitude of the normal force.

Equation 4.8, like Equation 4.7, is a relationship between only the magnitudes of the frictional and normal forces. The directions of these forces are perpendicular to each other. Moreover, like the coefficient of static friction, the coefficient of kinetic friction is a number without units and depends on the type and condition of the two surfaces that are in contact. As indicated in **Table 4.2**, values for μ_k are typically less than those for μ_s, reflecting the fact that kinetic friction is generally less than static friction. The next example illustrates the effect of kinetic friction.

Analyzing Multiple-Concept Problems

EXAMPLE 10 | Sled Riding

A sled and its rider are moving at a speed of 4.0 m/s along a horizontal stretch of snow, as **Figure 4.24a** illustrates. The snow exerts a kinetic frictional force on the runners of the sled, so the sled slows down and eventually comes to a stop. The coefficient of kinetic friction is 0.050. What is the displacement x of the sled?

Reasoning As the sled slows down, its velocity is decreasing. As our discussions in Chapters 2 and 3 indicate, the changing velocity is described by an acceleration (which in this case is a deceleration since the sled is slowing down). Assuming that the acceleration is constant, we can use one of the equations of kinematics from Chapter 3 to relate the displacement to the initial and final velocities and to the acceleration. The acceleration of the sled is not given directly. However, we can determine it by using Newton's second law of motion, which

(a)

(b) Free-body diagram for the sled and rider

FIGURE 4.24 (a) The moving sled decelerates because of the kinetic frictional force. (b) Three forces act on the moving sled: the weight $\vec{W}$ of the sled and its rider, the normal force $\vec{F}_N$, and the kinetic frictional force $\vec{f}_k$. The free-body diagram for the sled and rider shows these forces.

relates the acceleration to the net force (which is the kinetic frictional force in this case) acting on the sled and to the mass of the sled (plus rider).

Knowns and Unknowns   The data for this problem are listed in the table:

Description	Symbol	Value	Comment
Explicit Data			
Initial velocity	v_{0x}	+4.0 m/s	Positive, because the velocity points in the $+x$ direction. See the drawing.
Coefficient of kinetic friction	μ_k	0.050	
Implicit Data			
Final velocity	v_x	0 m/s	The sled comes to a stop.
Unknown Variable			
Displacement	x	?	

Modeling the Problem

STEP 1 Displacement   To obtain the displacement x of the sled we will use Equation 3.6a from the equations of kinematics:

$$v_x^2 = v_{0x}^2 + 2a_x x$$

Solving for the displacement x gives the result shown at the right. This equation is useful because two of the variables, v_{0x} and v_x, are known and the acceleration a_x can be found by applying Newton's second law to the accelerating sled (see Step 2).

$$x = \frac{v_x^2 - v_{0x}^2}{2a_x} \tag{1}$$

$\boxed{?}$

STEP 2 Newton's Second Law   Newton's second law, as given in Equation 4.2a, states that the acceleration a_x is equal to the net force ΣF_x divided by the mass m:

$$a_x = \frac{\Sigma F_x}{m}$$

The free-body diagram in **Figure 4.24b** shows that the only force acting on the sled in the horizontal or x direction is the kinetic frictional force $\vec{\mathbf{f}}_k$. We can write this force as $-f_k$, where f_k is the magnitude of the force and the minus sign indicates that it points in the $-x$ direction. Since the net force is $\Sigma F_x = -f_k$, Equation 4.2a becomes

$$\boxed{a_x = \frac{-f_k}{m}}$$

This result can now be substituted into Equation 1, as shown at the right.

$$x = \frac{v_x^2 - v_{0x}^2}{2a_x} \tag{1}$$

$$\boxed{a_x = \frac{-f_k}{m}} \tag{2}$$

$\boxed{?}$

STEP 3 Kinetic Frictional Force   We do not know the magnitude f_k of the kinetic frictional force, but we do know the coefficient of kinetic friction μ_k. According to Equation 4.8, the two are related by

$$\boxed{f_k = \mu_k F_N} \tag{4.8}$$

where F_N is the magnitude of the normal force. This relation can be substituted into Equation 2, as shown at the right. An expression for F_N will be obtained in the next step.

$$x = \frac{v_x^2 - v_{0x}^2}{2a_x} \tag{1}$$

$$a_x = \frac{-f_k}{m} \tag{2}$$

$$\boxed{f_k = \mu_k F_N} \tag{4.8}$$

$\boxed{?}$

STEP 4 Normal Force   The magnitude F_N of the normal force can be found by noting that the sled does not accelerate in the vertical or y direction ($a_y = 0$ m/s²). Thus, Newton's second law, as given in Equation 4.2b, becomes

$$\Sigma F_y = ma_y = 0$$

There are two forces acting on the sled in the y direction: the normal force $\vec{\mathbf{F}}_N$ and its weight $\vec{\mathbf{W}}$ (see **Figure 4.24b**). Therefore, the net force in the y direction is

$$\Sigma F_y = +F_N - W$$

where $W = mg$ (Equation 4.5). Thus, Newton's second law becomes

$$+F_N - mg = 0 \quad \text{or} \quad \boxed{F_N = mg}$$

This result for F_N can be substituted into Equation 4.8, as shown at the right.

$$x = \frac{v_x^2 - v_{0x}^2}{2a_x} \tag{1}$$

$$a_x = \frac{-f_k}{m} \tag{2}$$

$$f_k = \mu_k F_N \tag{4.8}$$

$$\boxed{F_N = mg}$$

Solution Algebraically combining the results of each step, we find that

$$x = \underset{\text{STEP 1}}{\frac{v_x^2 - v_{0x}^2}{2a_x}} = \underset{\text{STEP 2}}{\frac{v_x^2 - v_{0x}^2}{2\left(\frac{-f_k}{m}\right)}} = \underset{\text{STEP 3}}{\frac{v_x^2 - v_{0x}^2}{2\left(\frac{-\mu_k F_N}{m}\right)}} = \underset{\text{STEP 4}}{\frac{v_x^2 - v_{0x}^2}{2\left(\frac{-\mu_k mg}{m}\right)}} = \frac{v_x^2 - v_{0x}^2}{2(-\mu_k g)}$$

Note that the mass m of the sled and rider is algebraically eliminated from the final result. Thus, the displacement of the sled is

$$x = \frac{v_x^2 - v_{0x}^2}{2(-\mu_k g)} = \frac{(0 \text{ m/s})^2 - (+4.0 \text{ m/s})^2}{2[-(0.050)(9.80 \text{ m/s}^2)]} = \boxed{+16 \text{ m}}$$

Related Homework: Problems 42, 87

Static friction opposes the impending relative motion between two objects, while kinetic friction opposes the relative sliding motion that actually does occur. In either case, *relative motion* is opposed. However, this opposition to relative motion does not mean that friction prevents or works against the motion of *all* objects. For instance, consider what happens when you walk.

BIO **THE PHYSICS OF . . . walking.** Your foot exerts a force on the earth, and the earth exerts a reaction force on your foot. This reaction force is a static frictional force, and it opposes the impending backward motion of your foot, propelling you forward in the process. Kinetic friction can also cause an object to move, all the while opposing relative motion, as it does in Example 10. In this example the kinetic frictional force acts on the sled and opposes the relative motion of the sled and the earth. Newton's third law indicates, however, that since the earth exerts the kinetic frictional force on the sled, the sled must exert a reaction force on the earth. In response, the earth accelerates, but because of the earth's huge mass, the motion is too slight to be noticed.

Example 11 discusses the low-friction conditions on the inside of an important structural component of the human body.

EXAMPLE 11 | BIO The Amazing Synovial Joint

A synovial joint is a joint between bones that rub against each other, such as in elbows, hips, and knees. In a healthy joint, the inner contact surfaces are covered with cartilage and immersed in a lubricant called synovial fluid, which provides an extremely low coefficient of kinetic friction estimated to be $\mu_k \approx 2.00 \times 10^{-3}$, a value that has been difficult to match in the development of artificial joints. To get some intuition as to how "slippery" this is, consider a hockey puck coated with a material such that the coefficient of kinetic friction between it and the surface on which it slides is equal to 2.00×10^{-3}. Neglecting air resistance and assuming the surface on which the puck slides is horizontal and level, **(a)** how far will the puck travel and **(b)** how long will it take before it comes to rest if its initial speed is 15.0 mi/h (6.71 m/s)?

REASONING Knowing the coefficient of kinetic friction, we can determine the force of kinetic friction and then the acceleration (which will be negative). Since the mass of the puck was not given, we should expect it to be eliminated from the problem, and our solution should therefore be independent of the puck's mass. Once we know the (constant) acceleration, we can apply the kinematic equations to get both the distance it travels and the time it takes to come to rest.

SOLUTION **(a)** Here the force of kinetic friction acts opposite to the direction of motion and has a magnitude given by Equation 4.8:

$$f_k = \mu_k F_N$$

Since, in this situation, the weight and the normal force are equal and opposite in magnitude, $F_N = mg$, and the frictional force becomes

$$f_k = \mu_k mg$$

From Newton's second law (Equation 4.2a), the net force is equal to the mass times the acceleration, and we have

$$\Sigma F_x = \mu_k mg = ma_x$$

And therefore, since the masses cancel, the magnitude of the acceleration is

$$a_x = \mu_k g = (2.00 \times 10^{-3})(9.80 \text{ m/s}^2) = 0.0196 \text{ m/s}^2$$

This acceleration will be opposite to the direction of motion, which we will define to be in the $+x$ direction, and thus, the acceleration will be negative, that is, $a_x = -0.0196 \text{ m/s}^2$.

Now that the acceleration is known, as well as the initial velocity and final velocity (0 m/s), we can apply the kinematic equations to find the distance traveled. In this case, we use Equation 3.6a:

$$v_x^2 = v_{0x}^2 + 2a_x x$$

Setting the final velocity equal to zero and solving for x gives

$$x = -\frac{v_{0x}^2}{2a_x} = -\frac{(6.71 \text{ m/s})^2}{2(-0.0196 \text{ m/s}^2)} = \boxed{1150 \text{ m}}$$

(b) The time it takes for the puck to come to rest can be found from Equation 3.3a:

$$v_x = v_{0x} + a_x t$$

Setting the final velocity equal to zero and solving for the time gives

$$t = \frac{-v_{0x}}{a_x} = \frac{-6.71 \text{ m/s}}{-0.0196 \text{ m/s}^2} = \boxed{342 \text{ s}}$$

So the puck will travel a distance of over a kilometer and will remain in motion for 5 minutes and 42 seconds!

Although low-friction materials have been developed that perform better than the synovial joint, materials that go into the human body must satisfy strict criteria in regard to toxicity and compatibility with the human immune system. In addition, the durability of artificial joints must be exceptional, as they are subjected to large forces and repetitive motion.

Check Your Understanding

(The answers are given at the end of the book.)

14. Suppose that the coefficients of static and kinetic friction have values such that $\mu_s = 1.4\mu_k$ for a crate in contact with a cement floor. Which one of the following statements is true? **(a)** The magnitude of the static frictional force is always 1.4 times the magnitude of the kinetic frictional force. **(b)** The magnitude of the kinetic frictional force is always 1.4 times the magnitude of the static frictional force. **(c)** The magnitude of the maximum static frictional force is 1.4 times the magnitude of the kinetic frictional force.

15. A person has a choice of either pushing or pulling a sled at a constant velocity, as **CYU Figure 4.2** illustrates. Friction is present. If the angle θ is the same in both cases, does it require less force to push or to pull the sled?

CYU FIGURE 4.2

16. A box has a weight of 150 N and is being pulled across a horizontal floor by a force that has a magnitude of 110 N. The pulling force can point horizontally, or it can point above the horizontal at an angle θ. When the pulling force points horizontally, the kinetic frictional force acting on the box is twice as large as when the pulling force points at the angle θ. Find θ.

17. A box rests on the floor of an elevator. Because of static friction, a force is required to start the box sliding across the floor when the elevator is **(a)** stationary, **(b)** accelerating upward, and **(c)** accelerating downward. Rank the forces required in these three situations in ascending order—that is, smallest first.

(a)

(b)

4.10 The Tension Force

Forces are often applied by means of cables or ropes that are used to pull an object. For instance, **Figure 4.25a** shows a force $\vec{T}$ being applied to the right end of a rope attached to a box. Each particle in the rope in turn applies a force to its neighbor. As a result, the force is applied to the box, as part *b* of the drawing shows.

In situations such as that in **Figure 4.25**, we say that the force $\vec{T}$ is applied to the box because of the tension in the rope, meaning that the tension and the force applied to the box have the same magnitude. However, the word "tension" is commonly used to mean the tendency of the rope to be pulled apart. To see the relationship between these two uses of the word "tension," consider the left end of the rope, which applies the force $\vec{T}$ to the box. In accordance with Newton's third law, the box applies a reaction force to the rope. The reaction force has the same magnitude as $\vec{T}$ but is oppositely directed. In other words, a force $-\vec{T}$ acts on the left end of the rope. Thus, forces of equal magnitude act on opposite ends of the rope, as in **Figure 4.25c**, and tend to pull it apart.

FIGURE 4.25 (*a*) A force $\vec{T}$ is being applied to the right end of a rope. (*b*) The force is transmitted to the box. (*c*) Forces are applied to both ends of the rope. These forces have equal magnitudes and opposite directions.

FIGURE 4.26 The force $\vec{T}$ applied at one end of a massless rope is transmitted undiminished to the other end, even when the rope bends around a pulley, provided the pulley is also massless and friction is absent.

In the previous discussion, we have used the concept of a "massless" rope ($m = 0$ kg) without saying so. In reality, a massless rope does not exist, but it is useful as an idealization when applying Newton's second law. According to the second law, a net force is required to accelerate an object that has mass. In contrast, no net force is needed to accelerate a massless rope, since $\Sigma\vec{F} = m\vec{a}$ and $m = 0$ kg. Thus, when a force $\vec{T}$ is applied to one end of a massless rope, none of the force is needed to accelerate the rope. As a result, the force $\vec{T}$ is also applied undiminished to the object attached at the other end, as we assumed in **Figure 4.25**.* If the rope had mass, however, some of the force $\vec{T}$ would have to be used to accelerate the rope. The force applied to the box would then be less than $\vec{T}$, and the tension would be different at different locations along the rope. In this text we will assume that a rope connecting one object to another is massless, unless stated otherwise. The ability of a massless rope to transmit tension undiminished from one end to the other is not affected when the rope passes around objects such as the pulley in **Figure 4.26** (provided the pulley itself is massless and frictionless).

Check Your Understanding

(The answer is given at the end of the book.)

18. A rope is used in a tug-of-war between two teams of five people each. Both teams are equally strong, so neither team wins. An identical rope is tied to a tree, and the same ten people pull just as hard on the loose end as they did in the contest. In both cases, the people pull steadily with no jerking. Which rope sustains the greater tension, **(a)** the rope tied to the tree or **(b)** the rope in the tug-of-war, or **(c)** do the ropes sustain the same tension?

4.11 Equilibrium Applications of Newton's Laws of Motion

Have you ever been so upset that it took days to recover your "equilibrium"? In this context, the word "equilibrium" refers to a balanced state of mind, one that is not changing wildly. In physics, the word "equilibrium" also refers to a lack of change, but in the sense that the velocity of an object isn't changing. If its velocity doesn't change, an object is not accelerating. Our definition of equilibrium, then, is as follows:

> **DEFINITION OF EQUILIBRIUM†**
>
> **An object is in equilibrium when it has zero acceleration.**

Newton's laws of motion apply whether or not an object is in equilibrium. For an object in equilibrium the acceleration is zero ($\vec{a} = 0$ m/s^2) in Newton's second law, and the present section presents several examples of this type. In the nonequilibrium case the acceleration of the object is not zero ($\vec{a} \neq 0$ m/s^2) in the second law, and Section 4.12 deals with these kinds of situations.

Since the acceleration is zero for an object in equilibrium, all of the acceleration components are also zero. In two dimensions, this means that $a_x = 0$ m/s^2 and $a_y = 0$ m/s^2. Substituting these values into the second law ($\Sigma F_x = ma_x$ and $\Sigma F_y = ma_y$) shows that the x component and the y component of the net force must each be zero. In other words,

---

*If a rope is not accelerating, $\vec{a}$ is zero in the second law, and $\Sigma\vec{F} = m\vec{a} = 0$, regardless of the mass of the rope. Then, the rope can be ignored, no matter what mass it has.

†In this discussion of equilibrium we ignore rotational motion, which is discussed in Chapters 8 and 9. In Section 9.2 a more complete treatment of the equilibrium of a rigid object is presented and takes into account the concept of torque and the fact that objects can rotate.

the forces acting on an object in equilibrium must balance. Thus, in two dimensions, the equilibrium condition is expressed by two equations:

$$\Sigma F_x = 0 \qquad\qquad \text{(4.9a)}$$

$$\Sigma F_y = 0 \qquad\qquad \text{(4.9b)}$$

In using Equations 4.9a and 4.9b to solve equilibrium problems, we will use the following five-step reasoning strategy:

REASONING STRATEGY **Analyzing Equilibrium Situations**

1. Select the object (often called the "system") to which Equations 4.9a and 4.9b are to be applied. It may be that two or more objects are connected by means of a rope or a cable. If so, it may be necessary to treat each object separately according to the following steps.

2. Draw a free-body diagram for each object chosen above. Be sure to include only forces that act on the object. *Do not include forces that the object exerts on its environment.*

3. Choose a set of x, y axes for each object and resolve all forces in the free-body diagram into components that point along these axes. Select the axes so that as many forces as possible point along one or the other of the two axes. Such a choice minimizes the calculations needed to determine the force components.

4. Apply Equations 4.9a and 4.9b by setting the sum of the x components and the sum of the y components of the forces each equal to zero.

5. Solve the two equations obtained in Step 4 for the desired unknown quantities, remembering that two equations can yield answers for only two unknowns at most.

Example 12 illustrates how these steps are followed. It deals with a traction device in which three forces act together to bring about the equilibrium.

EXAMPLE 12 | BIO The Physics of Traction for the Foot

Figure 4.27a shows a traction device used with a foot injury. The weight of the 2.2-kg object creates a tension in the rope that passes around the pulleys. Therefore, tension forces $\vec{T}_1$ and $\vec{T}_2$ are applied to the pulley on the foot. (It may seem surprising that the rope applies a force to either side of the foot pulley. A similar effect occurs when you place a finger inside a rubber band and push downward. You can feel each side of the rubber band pulling upward on the finger.) The foot pulley is kept in equilibrium because the foot also applies a force $\vec{F}$ to it. This force arises in reaction (Newton's third law) to the pulling effect of the forces $\vec{T}_1$ and $\vec{T}_2$. Ignoring the weight of the foot, find the magnitude of $\vec{F}$.

(a)

(b) Free-body diagram for the foot pulley

FIGURE 4.27 (a) A traction device for the foot. (b) The free-body diagram for the pulley on the foot.

Reasoning   The forces $\vec{T}_1$, $\vec{T}_2$, and $\vec{F}$ keep the pulley on the foot at rest. The pulley, therefore, has no acceleration and is in equilibrium. As a result, the sum of the x components and the sum of the y components of the three forces must each be zero. **Figure 4.27b** shows the free-body diagram of the pulley on the foot. The x axis is chosen to be along the direction of force $\vec{F}$, and the components of the forces $\vec{T}_1$ and $\vec{T}_2$ are indicated in the drawing. (See Section 1.7 for a review of vector components.)

Problem-Solving Insight   Choose the orientation of the x, y axes for convenience. In Example 12, the axes have been rotated so the force $\vec{F}$ points along the x axis. Since $\vec{F}$ does not have a component along the y axis, the analysis is simplified.

Solution   Since the sum of the y components of the forces is zero, it follows that

$$\Sigma F_y = +T_1 \sin 35° - T_2 \sin 35° = 0 \qquad \text{(4.9b)}$$

or $T_1 = T_2$. In other words, the magnitudes of the tension forces are equal. In addition, the sum of the x components of the forces is zero, so we have that

$$\Sigma F_x = +T_1 \cos 35° + T_2 \cos 35° - F = 0 \qquad \text{(4.9a)}$$

Solving for F and letting $T_1 = T_2 = T$, we find that $F = 2T \cos 35°$. However, the tension T in the rope is determined by the weight of the 2.2-kg object: $T = mg$, where m is its mass and g is the acceleration due to gravity. Therefore, the magnitude of $\vec{F}$ is

$$F = 2T \cos 35° = 2mg \cos 35° = 2(2.2 \text{ kg})(9.80 \text{ m/s}^2) \cos 35° = \boxed{35 \text{ N}}$$

Example 13 presents another situation in which three forces are responsible for the equilibrium of an object. However, in this example all the forces have different magnitudes.

EXAMPLE 13 | Replacing an Engine

An automobile engine has a weight $\vec{W}$, whose magnitude is $W = 3150$ N. This engine is being positioned above an engine compartment, as **Figure 4.28a** illustrates. To position the engine, a worker is using a rope. Find the tension $\vec{T}_1$ in the supporting cable and the tension $\vec{T}_2$ in the positioning rope.

Reasoning   Under the influence of the forces $\vec{W}$, $\vec{T}_1$, and $\vec{T}_2$ the ring in **Figure 4.28a** is at rest and, therefore, in equilibrium. Consequently, the sum of the x components and the sum of the y components of these forces must each be zero; $\Sigma F_x = 0$ and $\Sigma F_y = 0$. By using these relations, we can find T_1 and T_2. **Figure 4.28b** shows the free-body diagram of the ring and the force components for a suitable x, y axis system.

Problem-Solving Insight   When an object is in equilibrium, as here in Example 13, the net force is zero, $\Sigma \vec{F} = 0$. This does not mean that each individual force is zero. It means that the vector sum of all the forces is zero.

Solution   The free-body diagram shows the components for each of the three forces, and the components are listed in the following table:

Force	x Component	y Component
$\vec{T}_1$	$-T_1 \sin 10.0°$	$+T_1 \cos 10.0°$
$\vec{T}_2$	$+T_2 \sin 80.0°$	$-T_2 \cos 80.0°$
$\vec{W}$	0	$-W$

The plus signs in the table denote components that point along the positive axes, and the minus signs denote components that point along the negative axes. Setting the sum of the x components and the sum of the y components equal to zero leads to the following equations:

$$\Sigma F_x = -T_1 \sin 10.0° + T_2 \sin 80.0° = 0 \qquad \text{(4.9a)}$$

$$\Sigma F_y = +T_1 \cos 10.0° - T_2 \cos 80.0° - W = 0 \qquad \text{(4.9b)}$$

(a)

(b) Free-body diagram for the ring

FIGURE 4.28   (a) The ring is in equilibrium because of the three forces $\vec{T}_1$ (the tension force in the supporting cable), $\vec{T}_2$ (the tension force in the positioning rope), and $\vec{W}$ (the weight of the engine). (b) The free-body diagram for the ring.

Solving the first of these equations for T_1 shows that

$$T_1 = \left(\frac{\sin 80.0°}{\sin 10.0°}\right)T_2$$

Substituting this expression for T_1 into the second equation gives

$$\left(\frac{\sin 80.0°}{\sin 10.0°}\right)T_2 \cos 10.0° - T_2 \cos 80.0° - W = 0$$

$$T_2 = \frac{W}{\left(\dfrac{\sin 80.0°}{\sin 10.0°}\right)\cos 10.0° - \cos 80.0°}$$

Setting $W = 3150$ N in this result yields $\boxed{T_2 = 582 \text{ N}}$.

Since $T_1 = \left(\dfrac{\sin 80.0°}{\sin 10.0°}\right)T_2$ and $T_2 = 582$ N, it follows that

$$\boxed{T_1 = 3.30 \times 10^3 \text{ N}}$$

An object can be moving and still be in equilibrium, provided there is no acceleration. Example 14 illustrates such a case, and the solution is again obtained using the five-step reasoning strategy summarized at the beginning of the section.

EXAMPLE 14 | Equilibrium at Constant Velocity

A jet plane is flying with a constant speed along a straight line, at an angle of 30.0° above the horizontal, as **Figure 4.29a** indicates. The plane has a weight $\vec{W}$ whose magnitude is $W = 86\,500$ N, and its engines provide a forward thrust $\vec{T}$ of magnitude $T = 103\,000$ N. In addition, the lift force $\vec{L}$ (directed perpendicular to the wings) and the force $\vec{R}$ of air resistance (directed opposite to the motion) act on the plane. Find $\vec{L}$ and $\vec{R}$.

Reasoning **Figure 4.29b** shows the free-body diagram of the plane, including the forces $\vec{W}$, $\vec{L}$, $\vec{T}$, and $\vec{R}$. Since the plane is flying with a constant speed along a straight line, it is not accelerating, it is in equilibrium, and the sum of the x components and the sum of the y components of these forces must be zero. The weight $\vec{W}$ and the thrust $\vec{T}$ are known, so the lift force $\vec{L}$ and the force $\vec{R}$ of air resistance can be obtained from these equilibrium conditions. To calculate the components, we have chosen axes in the free-body diagram that are rotated by 30.0° from their usual horizontal–vertical positions. This has been done purely for convenience, since the weight $\vec{W}$ is then the only force that does not lie along either axis.

> **Problem-Solving Insight** A moving object is in equilibrium if it moves with a constant velocity; then its acceleration is zero. A zero acceleration is the fundamental characteristic of an object in equilibrium.

Solution When determining the components of the weight, it is necessary to realize that the angle β in **Figure 4.29a** is 30.0°. Part c of the drawing focuses attention on the geometry that is responsible for this fact. There it can be seen that $\alpha + \beta = 90°$ and $\alpha + 30.0° = 90°$, with the result that $\beta = 30.0°$. The following table lists the components of the forces acting on the jet.

Force	x Component	y Component
$\vec{W}$	$-W \sin 30.0°$	$-W \cos 30.0°$
$\vec{L}$	0	$+L$
$\vec{T}$	$+T$	0
$\vec{R}$	$-R$	0

Setting the sum of the x component of the forces to zero gives

$$\Sigma F_x = -W \sin 30.0° + T - R = 0 \qquad \textbf{(4.9a)}$$

$$R = T - W \sin 30.0° = 103\,000 \text{ N} - (86\,500 \text{ N}) \sin 30.0° = \boxed{59\,800 \text{ N}}$$

Setting the sum of the y component of the forces to zero gives

$$\Sigma F_y = -W \cos 30.0° + L = 0 \qquad \textbf{(4.9b)}$$

$$L = W \cos 30.0° = (86\,500 \text{ N}) \cos 30.0° = \boxed{74\,900 \text{ N}}$$

(a) *(b) Free-body diagram* *(c)*

FIGURE 4.29 (a) A plane moves with a constant velocity at an angle of 30.0° above the horizontal due to the action of four forces, the weight $\vec{W}$, the lift $\vec{L}$, the engine thrust $\vec{T}$, and the air resistance $\vec{R}$. (b) The free-body diagram for the plane. (c) This geometry occurs often in physics.

Check Your Understanding

(The answers are given at the end of the book.)

19. In which one of the following situations could an object possibly be in equilibrium? **(a)** Three forces act on the object; the forces all point along the same line but may have different directions. **(b)** Two perpendicular forces act on the object. **(c)** A single force acts on the object. **(d)** In none of the situations described in (a), (b), and (c) could the object possibly be in equilibrium.

20. A stone is thrown from the top of a cliff. Air resistance is negligible. As the stone falls, is it **(a)** in equilibrium or **(b)** not in equilibrium?

21. During the final stages of descent, a sky diver with an open parachute approaches the ground with a constant velocity. There is no wind to blow him from side to side. Which one of the following statements is true? **(a)** The sky diver is not in equilibrium. **(b)** The force of gravity is the only force acting on the sky diver, so that he is in equilibrium. **(c)** The sky diver is in equilibrium because no forces are acting on him. **(d)** The sky diver is in equilibrium because two forces act on him, the downward-acting force of gravity and the upward-acting force of the parachute.

22. A crate hangs from a ring at the middle of a rope, as **CYU Figure 4.3** illustrates. A person is pulling on the right end of the rope to keep the crate in equilibrium. Can the rope ever be made to be perfectly horizontal?

CYU FIGURE 4.3

4.12 Nonequilibrium Applications of Newton's Laws of Motion

When an object is accelerating, it is not in equilibrium. The forces acting on it are not balanced, so the net force is not zero in Newton's second law. However, with one exception, the reasoning strategy followed in solving nonequilibrium problems is identical to that used in equilibrium situations. The exception occurs in Step 4 of the five steps outlined at the beginning of the previous section. Since the object is now accelerating, the representation of Newton's second law in Equations 4.2a and 4.2b applies instead of Equations 4.9a and 4.9b:

$$\Sigma F_x = ma_x \quad \textbf{(4.2a)} \qquad \text{and} \qquad \Sigma F_y = ma_y \quad \textbf{(4.2b)}$$

Example 15 uses these equations in a situation where the forces are applied in directions similar to those in Example 12, except that now an acceleration is present.

EXAMPLE 15 | Towing a Supertanker

A supertanker of mass $m = 1.50 \times 10^8$ kg is being towed by two tugboats, as in **Figure 4.30a**. The tensions in the towing cables apply the forces $\vec{T}_1$ and $\vec{T}_2$ at equal angles of 30.0° with respect to the tanker's axis. In addition, the tanker's engines produce a forward drive force $\vec{D}$, whose magnitude is $D = 75.0 \times 10^3$ N. Moreover, the water applies an opposing force $\vec{R}$, whose magnitude is $R = 40.0 \times 10^3$ N. The tanker moves forward with an acceleration that points along the tanker's axis and has a magnitude of 2.00×10^{-3} m/s². Find the magnitudes of the tensions $\vec{T}_1$ and $\vec{T}_2$.

Reasoning The unknown forces $\vec{T}_1$ and $\vec{T}_2$ contribute to the net force that accelerates the tanker. To determine $\vec{T}_1$ and $\vec{T}_2$, therefore, we analyze the net force, which we will do using components. The various force components can be found by referring to the free-body diagram for the tanker in **Figure 4.30b**, where the ship's axis is chosen as the x axis. We will then use Newton's second law in its component form, $\Sigma F_x = ma_x$ and $\Sigma F_y = ma_y$, to obtain the magnitudes of $\vec{T}_1$ and $\vec{T}_2$.

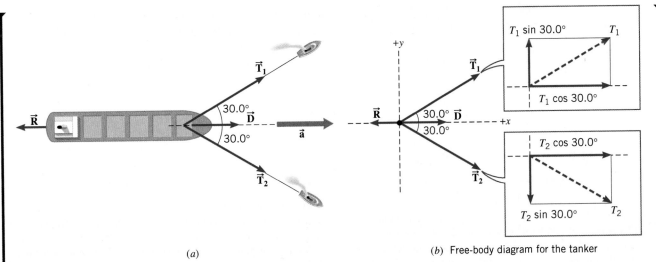

(a)

(b) Free-body diagram for the tanker

FIGURE 4.30 (a) Four forces act on a supertanker: $\vec{T}_1$ and $\vec{T}_2$ are the tension forces due to the towing cables, $\vec{D}$ is the forward drive force produced by the tanker's engines, and $\vec{R}$ is the force with which the water opposes the tanker's motion. (b) The free-body diagram for the tanker.

Solution The individual force components are summarized as follows:

Force	x Component	y Component
$\vec{T}_1$	$+T_1 \cos 30.0°$	$+T_1 \sin 30.0°$
$\vec{T}_2$	$+T_2 \cos 30.0°$	$-T_2 \sin 30.0°$
$\vec{D}$	$+D$	0
$\vec{R}$	$-R$	0

Since the acceleration points along the x axis, there is no y component of the acceleration ($a_y = 0$ m/s^2). Consequently, the sum of the y components of the forces must be zero:

$$\Sigma F_y = +T_1 \sin 30.0° - T_2 \sin 30.0° = 0$$

This result shows that the magnitudes of the tensions in the cables are equal, $T_1 = T_2$. Since the ship accelerates along the x direction, the sum of the x components of the forces is not zero. The second law indicates that

$$\Sigma F_x = T_1 \cos 30.0° + T_2 \cos 30.0° + D - R = ma_x$$

Since $T_1 = T_2$, we can replace the two separate tension symbols by a single symbol T, the magnitude of the tension. Solving for T gives

$$T = \frac{ma_x + R - D}{2 \cos 30.0°}$$

$$= \frac{(1.50 \times 10^8 \text{ kg})(2.00 \times 10^{-3} \text{ m/s}^2) + 40.0 \times 10^3 \text{ N} - 75.0 \times 10^3 \text{ N}}{2 \cos 30.0°}$$

$$= \boxed{1.53 \times 10^5 \text{ N}}$$

Math Skills The sine and cosine functions are defined in Equations 1.1 and 1.2 as $\sin \theta = \dfrac{h_o}{h}$ and $\cos \theta = \dfrac{h_a}{h}$, where h_o is the length of the side of a right triangle that is opposite the angle θ, h_a is the length of the side adjacent to the angle θ, and h is the length of the hypotenuse (see **Figure 4.31a**). When using the sine and cosine functions to determine the scalar components of a vector, we begin by identifying the angle θ. **Figure 4.31b** indicates that $\theta = 30.0°$ for the vector $\vec{T}_1$. The components of $\vec{T}_1$ are T_{1x} and T_{1y}. Comparing the shaded triangles in **Figure 4.31**, we can see that $h_o = T_{1y}$, $h_a = T_{1x}$, and $h = T_1$. Therefore, we have

$$\cos 30.0° = \frac{h_a}{h} = \frac{T_{1x}}{T_1} \quad \text{or} \quad T_{1x} = T_1 \cos 30.0°$$

$$\sin 30.0° = \frac{h_o}{h} = \frac{T_{1y}}{T_1} \quad \text{or} \quad T_{1y} = T_1 \sin 30.0°$$

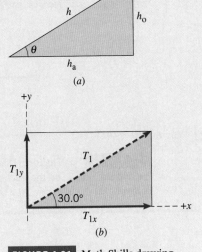

FIGURE 4.31 Math Skills drawing.

It often happens that two objects are connected somehow, perhaps by a drawbar like that used when a truck pulls a trailer. If the tension in the connecting device is of no interest, the objects can be treated as a single composite object when applying Newton's second law. However, if it is necessary to find the tension, as in the next example, then the second law must be applied separately to at least one of the objects.

EXAMPLE 16 | Hauling a Trailer

A truck is hauling a trailer along a level road, as **Figure 4.32a** illustrates. The mass of the truck is $m_1 = 8500$ kg and that of the trailer is $m_2 = 27\,000$ kg. The two move along the x axis with an acceleration of $a_x = 0.78$ m/s^2. Ignoring the retarding forces of friction and air resistance, determine **(a)** the tension $\vec{T}$ in the horizontal drawbar between the trailer and the truck and **(b)** the force $\vec{D}$ that propels the truck forward.

Reasoning   Since the truck and the trailer accelerate along the horizontal direction and friction is being ignored, only forces that have components in the horizontal direction are of interest. Therefore, **Figure 4.32** omits the weight and the normal force, which act vertically. To determine the tension force $\vec{T}$ in the drawbar, we draw the free-body diagram for the trailer and apply Newton's second law, $\Sigma F_x = ma_x$. Similarly, we can determine the propulsion force $\vec{D}$ by drawing the free-body diagram for the truck and applying Newton's second law.

> **Problem-Solving Insight**   A free-body diagram is very helpful when applying Newton's second law. Always start a problem by drawing the free-body diagram.

Solution   **(a)** The free-body diagram for the trailer is shown in **Figure 4.32b**. There is only one horizontal force acting on the trailer, the tension force $\vec{T}$ due to the drawbar. Therefore, it is straightforward to obtain the tension from $\Sigma F_x = m_2 a_x$, since the mass of the trailer and the acceleration are known:

$$\Sigma F_x = T = m_2 a_x = (27\,000 \text{ kg})(0.78 \text{ m/s}^2) = \boxed{21\,000 \text{ N}}$$

(b) Two horizontal forces act on the truck, as the free-body diagram in **Figure 4.32b** shows. One is the desired force $\vec{D}$. The other is the force $\vec{T}'$. According to Newton's third law, $\vec{T}'$ is the force with which the trailer pulls back on the truck, in reaction to the truck pulling forward. If the drawbar has negligible mass, the magnitude of $\vec{T}'$ is equal to the magnitude of $\vec{T}$—namely, 21 000 N. Since the magnitude of $\vec{T}'$, the mass of the truck, and the acceleration are known, $\Sigma F_x = m_1 a_x$ can be used to determine the drive force:

$$\Sigma F_x = +D - T' = m_1 a_x$$

$$D = m_1 a_x + T' = (8500 \text{ kg})(0.78 \text{ m/s}^2) + 21\,000 \text{ N} = \boxed{28\,000 \text{ N}}$$

(a)

(b) Free-body diagrams

FIGURE 4.32   (a) The force $\vec{D}$ acts on the truck and propels it forward. The drawbar exerts the tension force $\vec{T}'$ on the truck and the tension force $\vec{T}$ on the trailer. (b) The free-body diagrams for the trailer and the truck, ignoring the vertical forces.

In Section 4.11 we examined situations where the net force acting on an object is zero, and in the present section we have considered two examples where the net force is not zero. Conceptual Example 17 illustrates a common situation where the net force is zero at certain times but is not zero at other times.

CONCEPTUAL EXAMPLE 17 | The Motion of a Water Skier

Figure 4.33 shows a water skier at four different moments:

a. The skier is floating motionless in the water.

b. The skier is being pulled out of the water and up onto the skis.

c. The skier is moving at a constant speed along a straight line.

d. The skier has let go of the tow rope and is slowing down.

For each moment, explain whether the net force acting on the skier is zero.

Reasoning and Solution According to Newton's second law, if an object has zero acceleration, the net force acting on it is zero. In such a case, the object is in equilibrium. In contrast, if the object has an acceleration, the net force acting on it is not zero. Such an object is not in equilibrium. We will consider the acceleration in each of the four phases of the motion to decide whether the net force is zero.

a. The skier is floating motionless in the water, so her velocity and acceleration are both zero. Therefore, the net force acting on her is zero, and she is in equilibrium.

b. As the skier is being pulled up and out of the water, her velocity is increasing. Thus, she is accelerating, and the net force acting on her is not zero. The skier is not in equilibrium. The direction of the net force is shown in **Figure 4.33b**.

c. The skier is now moving at a constant speed along a straight line (**Figure 4.33c**), so her velocity is constant. Since her velocity is constant, her acceleration is zero. Thus, the net force acting on her is zero, and she is again in equilibrium, even though she is moving.

d. After the skier lets go of the tow rope, her speed decreases, so she is decelerating. Thus, the net force acting on her is not zero, and she is not in equilibrium. The direction of the net force is shown in **Figure 4.33d**.

Related Homework: Problem 60

(a)   *(b)*   *(c)*   *(d)*

FIGURE 4.33 A water skier (*a*) floating in the water, (*b*) being pulled up by the boat, (*c*) moving at a constant velocity, and (*d*) slowing down.

The force of gravity is often present among the forces that affect the acceleration of an object. Examples 18 and 19 deal with typical situations.

EXAMPLE 18 | Accelerating Blocks

Block 1 (mass $m_1 = 8.00$ kg) is moving on a frictionless 30.0° incline. This block is connected to block 2 (mass $m_2 = 22.0$ kg) by a massless cord that passes over a massless and frictionless pulley (see **Figure 4.34a**). Find the acceleration of each block and the tension in the cord.

Reasoning Since both blocks accelerate, there must be a net force acting on each one. The key to this problem is to realize that Newton's second law can be used separately for each block to relate the net force and the acceleration. Note also that both blocks have accelerations of the same magnitude *a*, since they move as a

Block 1   Block 2

(b) Free-body diagrams

(a)

FIGURE 4.34 (*a*) Three forces act on block 1: its weight $\vec{W}_1$, the normal force $\vec{F}_N$, and the force $\vec{T}$ due to the tension in the cord. Two forces act on block 2: its weight $\vec{W}_2$ and the force $\vec{T}'$ due to the tension. The acceleration is labeled according to its magnitude *a*. (*b*) Free-body diagrams for the two blocks.

unit. We assume that block 1 accelerates up the incline and choose this direction to be the +x axis. If block 1 in reality accelerates down the incline, then the value obtained for the acceleration will be a negative number.

> **Problem-Solving Insight** Mass and weight are different quantities. They cannot be interchanged when solving problems.

Solution   Three forces act on block 1: (1) $\vec{W}_1$ is its weight [$W_1 = m_1 g = (8.00\ \text{kg})(9.80\ \text{m/s}^2) = 78.4\ \text{N}$], (2) $\vec{T}$ is the force applied because of the tension in the cord, and (3) $\vec{F}_N$ is the normal force the incline exerts. **Figure 4.34b** shows the free-body diagram for block 1. The weight is the only force that does not point along the x, y axes, and its x and y components are given in the diagram. Applying Newton's second law ($\Sigma F_x = m_1 a_x$) to block 1 shows that

$$\Sigma F_x = -W_1 \sin 30.0° + T = m_1 a$$

where we have set $a_x = a$. This equation cannot be solved as it stands, since both T and a are unknown quantities. To complete the solution, we next consider block 2.

Two forces act on block 2, as the free-body diagram in **Figure 4.34b** indicates: (1) $\vec{W}_2$ is its weight [$W_2 = m_2 g = (22.0\ \text{kg})(9.80\ \text{m/s}^2) = 216\ \text{N}$] and (2) $\vec{T}'$ is exerted as a result of block 1 pulling back on the connecting cord. Since the cord and the frictionless pulley are massless, the magnitudes of $\vec{T}'$ and $\vec{T}$ are the same: $T' = T$. Applying Newton's second law ($\Sigma F_y = m_2 a_y$) to block 2 reveals that

$$\Sigma F_y = T - W_2 = m_2(-a)$$

The acceleration a_y has been set equal to $-a$ since block 2 moves downward along the $-y$ axis in the free-body diagram, consistent with the assumption that block 1 moves up the incline. Now

there are two equations in two unknowns, and they may be solved simultaneously (see Appendix C) to give T and a:

$$\boxed{T = 86.3\ \text{N}} \quad \text{and} \quad \boxed{a = 5.89\ \text{m/s}^2}$$

Math Skills   The two equations containing the unknown quantities T and a are

$$-W_1 \sin 30.0° + T = m_1 a \qquad (1)$$
and
$$T - W_2 = m_2(-a) \qquad (2)$$

Neither equation by itself can yield numerical values for T and a. However, the two equations can be solved simultaneously in the following manner. To begin with, we rearrange Equation (1) to give $T = m_1 a + W_1 \sin 30.0°$. Next, we substitute this result into Equation (2) and obtain a result containing only the unknown acceleration a:

$$\underbrace{m_1 a + W_1 \sin 30.0°}_{T} - W_2 = m_2(-a)$$

Rearranging this equation so that the terms $m_1 a$ and $m_2 a$ stand alone on the left of the equals sign, we have

$$m_1 a + m_2 a = W_2 - W_1 \sin 30.0° \quad \text{or} \quad a = \frac{W_2 - W_1 \sin 30.0°}{m_1 + m_2}$$

This result yields the value of $\boxed{a = 5.89\ \text{m/s}^2}$. We can also substitute the expression for a into either one of the two starting equations and obtain a result containing only the unknown tension T. We choose Equation (1) and find that

$$-W_1 \sin 30.0° + T = m_1 \underbrace{\left(\frac{W_2 - W_1 \sin 30.0°}{m_1 + m_2}\right)}_{a}$$

Solving for T gives

$$T = W_1 \sin 30.0° + m_1\left(\frac{W_2 - W_1 \sin 30.0°}{m_1 + m_2}\right)$$

This expression yields the value of $\boxed{T = 86.3\ \text{N}}$.

EXAMPLE 19 | Hoisting a Scaffold

A window washer on a scaffold is hoisting the scaffold up the side of a building by pulling downward on a rope, as in **Figure 4.35a**. The magnitude of the pulling force is 540 N, and the combined mass of the worker and the scaffold is 155 kg. Find the upward acceleration of the unit.

Reasoning   The worker and the scaffold form a single unit, on which the rope exerts a force in three places. The left end of the rope exerts an upward force $\vec{T}$ on the worker's hands. This force arises because he pulls downward with a 540-N force, and the rope exerts an oppositely directed force of equal magnitude on him, in accord with Newton's third law. Thus, the magnitude T of the upward force is $T = 540\ \text{N}$ and is the magnitude of the tension in the rope. If the masses of the rope and each pulley are negligible and if the pulleys are friction-free, the tension is transmitted undiminished along the rope. Then, a 540-N tension force $\vec{T}$ acts

upward on the left side of the scaffold pulley (see part a of the drawing). A tension force is also applied to the point P, where the rope attaches to the roof. The roof pulls back on the rope in accord with the third law, and this pull leads to the 540-N tension force $\vec{T}$ that acts on the right side of the scaffold pulley. In addition to the three upward forces, the weight of the unit must be taken into account [$W = mg = (155\ \text{kg})(9.80\ \text{m/s}^2) = 1520\ \text{N}$]. Part b of the drawing shows the free-body diagram.

Solution   Newton's second law ($\Sigma F_y = ma_y$) can be applied to calculate the acceleration a_y:

$$\Sigma F_y = +T + T + T - W = ma_y$$

$$a_y = \frac{3T - W}{m} = \frac{3(540\ \text{N}) - 1520\ \text{N}}{155\ \text{kg}} = \boxed{0.65\ \text{m/s}^2}$$

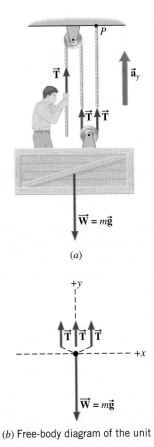

$$\vec{T}$$ $$\vec{a}_y$$

$$\vec{T}\ \vec{T}$$

$$\vec{W} = m\vec{g}$$

(a)

+y

$$\boxed{T}\ \boxed{T}\ \boxed{T}$$

------------•------------ +x

$$\vec{W} = m\vec{g}$$

(b) Free-body diagram of the unit

FIGURE 4.35 (a) A window washer pulls down on the rope to hoist the scaffold up the side of a building. The force $\vec{T}$ results from the effort of the window washer and acts on him and the scaffold in three places, as discussed in Example 19. (b) The free-body diagram of the unit comprising the man and the scaffold.

EXAMPLE 20 | BIO The Importance of Seatbelts

Police departments around the country use a device, called *The Convincer*, to educate the public on the safety of seatbelts (see **Figure 4.36a**). The device consists of a sled that is free to move on rails that are inclined at 6.5° above the horizontal. A passenger is placed in the sled and secured with a seatbelt. The sled is released from rest and travels a distance of 2.3 m along the rails before slamming into a stop at the bottom, thereby simulating a car collision. Assume the rails are frictionless and calculate the speed of the sled at impact.

Reasoning The free-body diagram for the sled is shown in **Figure 4.36b**. The x-component of the sled's weight is the only force acting on the sled in the x-direction. Once released, the sled will accelerate down the rails. Thus, this is a nonequilibrium

problem, and the sum of the forces acting on the sled in the x-direction will be equal to its mass times its acceleration in that direction. Since the sled's acceleration down the rails will be constant, we can use our relationships from kinematics to calculate its speed when it strikes the stop.

Solution Applying Newton's second law to the sled in the x-direction, we have $\Sigma F_x = W \sin 6.5° = ma_x$. We can solve this expression for the acceleration of the sled in the x-direction: $a_x = \dfrac{W \sin 6.5°}{m} = \dfrac{mg \sin 6.5°}{m} = g \sin 6.5° = (9.8 \text{ m/s}^2)(\sin 6.5°) = 1.1 \text{ m/s}^2$. Now that we have the acceleration of the sled, we can use Equation 3.6a to calculate its speed when it strikes the stop: $v_x^2 = v_{0x}^2 + 2a_x x$. Solving for v_x, and using the fact that the sled begins

Passenger

Sled

Rails

2.3 m

$$\vec{a}$$

Stop

6.5°

(a)

+y

$$\vec{F}_N$$

W sin 6.5°

+x

W cos 6.5°

6.5°

$$\vec{W}$$

(b) Free-body diagram

FIGURE 4.36 (a) The passenger in the sled accelerates down the rails and collides with a stop in order to simulate a car collision. (b) Free-body diagram showing the forces acting on the sled. The only force providing its acceleration down the rails is the x-component of its weight.

from rest ($v_{0x} = 0$), we get the following result for its speed at impact: $v_x = \sqrt{2a_x x} = \sqrt{2(1.1 \text{ m/s}^2)(2.3 \text{ m})} = \boxed{2.2 \text{ m/s}}$. Consider the magnitude of this speed. This is approximately equal to 5 mi/h, which is a relatively low speed for a car. Even so, the passenger still receives quite a jolt when the sled strikes the stop. Without

the seatbelt to decelerate the passenger upon impact, she would continue to move in a straight line and strike the front of the sled (Newton's first law). Since many car collisions occur at speeds much larger than 5 mi/h, *The Convincer* demonstrates that seatbelt use is essential for passenger safety.

Check Your Understanding

(The answers are given at the end of the book.)

23. A circus performer hangs stationary from a rope. She then begins to climb upward by pulling herself up, hand over hand. When she starts climbing, is the tension in the rope **(a)** less than, **(b)** equal to, or **(c)** greater than it is when she hangs stationary?

24. A freight train is accelerating on a level track. Other things being equal, would the tension in the coupling between the engine and the first car change if some of the cargo in the last car were transferred to any one of the other cars?

25. Two boxes have masses m_1 and m_2, and m_2 is greater than m_1. The boxes are being pushed across a frictionless horizontal surface (see **CYU Figure 4.4**). As the drawing shows, there are two possible arrangements, and the pushing force is the same in each. In which arrangement, **(a)** or **(b)**, does the force that the left box applies to the right box have a greater magnitude, or **(c)** is the magnitude the same in both cases?

(a)   *(b)*

CYU FIGURE 4.4

Concept Summary

4.1 The Concepts of Force and Mass A force is a push or a pull and is a vector quantity. Contact forces arise from the physical contact between two objects. Noncontact forces are also called action-at-a-distance forces, because they arise without physical contact between two objects.

Mass is a property of matter that determines how difficult it is to accelerate or decelerate an object. Mass is a scalar quantity.

4.2 Newton's First Law of Motion Newton's first law of motion, sometimes called the law of inertia, states that an object continues in a state of rest or in a state of motion at a constant velocity unless compelled to change that state by a net force.

Inertia is the natural tendency of an object to remain at rest or in motion at a constant velocity. The mass of a body is a quantitative measure of inertia and is measured in an SI unit called the kilogram (kg). An inertial reference frame is one in which Newton's law of inertia is valid.

4.3 Newton's Second Law of Motion/4.4 The Vector Nature of Newton's Second Law of Motion Newton's second law of motion states that when a net force $\Sigma\vec{F}$ acts on an object of mass m, the acceleration $\vec{a}$ of the object can be obtained from Equation 4.1. This is a

vector equation and, for motion in two dimensions, is equivalent to Equations 4.2a and 4.2b. In these equations the x and y subscripts refer to the scalar components of the force and acceleration vectors. The SI unit of force is the newton (N).

$$\Sigma\vec{F} = m\vec{a} \qquad (4.1)$$
$$\Sigma F_x = ma_x \qquad (4.2a)$$
$$\Sigma F_y = ma_y \qquad (4.2b)$$

When determining the net force, a free-body diagram is helpful. A free-body diagram is a diagram that represents the object and the forces acting on it.

4.5 Newton's Third Law of Motion Newton's third law of motion, often called the action–reaction law, states that whenever one object exerts a force on a second object, the second object exerts an oppositely directed force of equal magnitude on the first object.

4.6 Types of Forces: An Overview Only three fundamental forces have been discovered: the gravitational force, the strong nuclear force, and the electroweak force. The electroweak force manifests itself as either the electromagnetic force or the weak nuclear force.

4.7 The Gravitational Force Newton's law of universal gravitation states that every particle in the universe exerts an attractive force on every other particle. For two particles that are separated by a distance r and have masses m_1 and m_2, the law states that the magnitude of this attractive force is as given in Equation 4.3. The direction of this force lies along the line between the particles. The constant G has a value of $G = 6.674 \times 10^{-11} \ \mathrm{N \cdot m^2/kg^2}$ and is called the universal gravitational constant.

$$F = G \frac{m_1 m_2}{r^2} \qquad (4.3)$$

The weight W of an object on or above the earth is the gravitational force that the earth exerts on the object and can be calculated from the mass m of the object and the magnitude g of the acceleration due to the earth's gravity, according to Equation 4.5.

$$W = mg \qquad (4.5)$$

4.8 The Normal Force The normal force $\vec{\mathbf{F}}_N$ is one component of the force that a surface exerts on an object with which it is in contact—namely, the component that is perpendicular to the surface.

The apparent weight is the force that an object exerts on the platform of a scale and may be larger or smaller than the true weight mg if the object and the scale have an acceleration a (+ if upward, − if downward). The apparent weight is given by Equation 4.6.

$$\text{Apparent weight} = mg + ma \qquad (4.6)$$

4.9 Static and Kinetic Frictional Forces A surface exerts a force on an object with which it is in contact. The component of the force perpendicular to the surface is called the normal force. The component parallel to the surface is called friction.

The force of static friction between two surfaces opposes any impending relative motion of the surfaces. The magnitude of the static frictional force depends on the magnitude of the applied force and can assume any value up to the maximum specified in Equation 4.7,

where μ_s is the coefficient of static friction and F_N is the magnitude of the normal force.

$$f_s^{\text{MAX}} = \mu_s F_N \qquad (4.7)$$

The force of kinetic friction between two surfaces sliding against one another opposes the relative motion of the surfaces. This force has a magnitude given by Equation 4.8, where μ_k is the coefficient of kinetic friction.

$$f_k = \mu_k F_N \qquad (4.8)$$

4.10 The Tension Force The word "tension" is commonly used to mean the tendency of a rope to be pulled apart due to forces that are applied at each end. Because of tension, a rope transmits a force from one end to the other. When a rope is accelerating, the force is transmitted undiminished only if the rope is massless.

4.11 Equilibrium Applications of Newton's Laws of Motion An object is in equilibrium when the object has zero acceleration, or, in other words, when it moves at a constant velocity. The constant velocity may be zero, in which case the object is stationary. The sum of the forces that act on an object in equilibrium is zero. Under equilibrium conditions in two dimensions, the separate sums of the force components in the x direction and in the y direction must each be zero, as in Equations 4.9a and 4.9b.

$$\Sigma F_x = 0 \qquad (4.9a)$$
$$\Sigma F_y = 0 \qquad (4.9b)$$

4.12 Nonequilibrium Applications of Newton's Laws of Motion If an object is not in equilibrium, then Newton's second law, as expressed in Equations 4.2a and 4.2b, must be used to account for the acceleration.

$$\Sigma F_x = ma_x \qquad (4.2a)$$
$$\Sigma F_y = ma_y \qquad (4.2b)$$

Focus on Concepts

Online

Addtional questions are available for assignment in WileyPLUS

Section 4.2   Newton's First Law of Motion

1. An object is moving at a constant velocity. All but one of the following statements could be true. Which one cannot be true? **(a)** No forces act on the object. **(b)** A single force acts on the object. **(c)** Two forces act simultaneously on the object. **(d)** Three forces act simultaneously on the object.

2. A cup of coffee is sitting on a table in a recreational vehicle (RV). The cup slides toward the rear of the RV. According to Newton's first law, which one or more of the following statements could describe the motion of the RV? (A) The RV is at rest, and the driver suddenly accelerates. (B) The RV is moving forward, and the driver suddenly accelerates. (C) The RV is moving backward, and the driver suddenly hits the brakes. **(a)** A **(b)** B **(c)** C **(d)** A and B **(e)** A, B, and C

Section 4.4   The Vector Nature of Newton's Second Law of Motion

3. Two forces act on a moving object that has a mass of 27 kg. One force has a magnitude of 12 N and points due south, while the other

force has a magnitude of 17 N and points due west. What is the acceleration of the object? **(a)** 0.63 m/s² directed 55° south of west **(b)** 0.44 m/s² directed 24° south of west **(c)** 0.77 m/s² directed 35° south of west **(d)** 0.77 m/s² directed 55° south of west **(e)** 1.1 m/s² directed 35° south of west

Section 4.5   Newton's Third Law of Motion

4. Which one of the following is true, according to Newton's laws of motion? Ignore friction. **(a)** A sports utility vehicle (SUV) hits a stationary motorcycle. Since it is stationary, the motorcycle sustains a greater force than the SUV does. **(b)** A semitrailer truck crashes all the way through a wall. Since the wall collapses, the wall sustains a greater force than the truck does. **(c)** Sam (18 years old) and his sister (9 years old) go ice skating. They push off against each other and fly apart. Sam flies off with the greater acceleration. **(d)** Two astronauts on a space walk are throwing a ball back and forth between each other. In this game of catch the distance between them remains constant. **(e)** None of the above is true, according to the third law.

5. Two ice skaters, Paul and Tom, are each holding on to opposite ends of the same rope. Each pulls the other toward him. The magnitude of

Paul's acceleration is 1.25 times greater than the magnitude of Tom's acceleration. What is the ratio of Paul's mass to Tom's mass? **(a)** 0.67 **(b)** 0.80 **(c)** 0.25 **(d)** 1.25 **(e)** 0.50

Section 4.7 The Gravitational Force

6. In another solar system a planet has twice the earth's mass and three times the earth's radius. Your weight on this planet is _____ times your earth-weight. Assume that the masses of the earth and of the other planet are uniformly distributed. **(a)** 0.667 **(b)** 2.000 **(c)** 0.111 **(d)** 0.444 **(e)** 0.222

7. What is the mass on Mercury of an object that weighs 784 N on the earth's surface? **(a)** 80.0 kg **(b)** 48.0 kg **(c)** 118 kg **(d)** 26.0 kg **(e)** There is not enough information to calculate the mass.

Section 4.8 The Normal Force

8. The apparent weight of a passenger in an elevator is greater than his true weight. Which one of the following is true? **(a)** The elevator is either moving upward with an increasing speed or moving upward with a decreasing speed. **(b)** The elevator is either moving upward with an increasing speed or moving downward with an increasing speed. **(c)** The elevator is either moving upward with a decreasing speed or moving downward with a decreasing speed. **(d)** The elevator is either moving upward with an increasing speed or moving downward with a decreasing speed. **(e)** The elevator is either moving upward with a decreasing speed or moving downward with an increasing speed.

9. The drawings show three examples of the force with which someone pushes against a vertical wall. In each case the magnitude of the pushing force is the same. Rank the normal forces that the wall applies to the pusher in ascending order (smallest first). **(a)** C, B, A **(b)** B, A, C **(c)** A, C, B **(d)** B, C, A **(e)** C, A, B

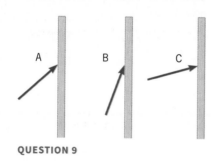

QUESTION 9

Section 4.9 Static and Kinetic Frictional Forces

10. The drawing shows three blocks, each with the same mass, stacked one upon the other. The bottom block rests on a frictionless horizontal surface and is being pulled by a force $\vec{F}$ that is parallel to this surface. The surfaces where the blocks touch each other have identical coefficients of static friction. Which one of the following correctly describes the magnitude of the *net* force of static friction f_s that acts on each block?

(a) $f_{s, A} = f_{s, B} = f_{s, C}$

(b) $f_{s, A} = f_{s, B} = \frac{1}{2}f_{s, C}$

(c) $f_{s, A} = 0$ and $f_{s, B} = \frac{1}{2}f_{s, C}$

(d) $f_{s, C} = 0$ and $f_{s, A} = \frac{1}{2}f_{s, B}$

(e) $f_{s, A} = f_{s, C} = \frac{1}{2}f_{s, B}$

QUESTION 10

11. Three identical blocks are being pulled or pushed across a horizontal surface by a force $\vec{F}$, as shown in the drawings. The force $\vec{F}$ in each case has the same magnitude. Rank the kinetic frictional forces that act on the blocks in ascending order (smallest first).

(a) B, C, A **(b)** C, A, B **(c)** B, A, C **(d)** C, B, A **(e)** A, C, B

QUESTION 11

Section 4.10 The Tension Force

12. A heavy block is suspended from a ceiling using pulleys in three different ways, as shown in the drawings. Rank the tension in the rope that passes over the pulleys in ascending order (smallest first).

(a) B, A, C **(b)** C, B, A **(c)** A, B, C **(d)** C, A, B **(e)** B, C, A

QUESTION 12

Section 4.11 Equilibrium Applications of Newton's Laws of Motion

13. A certain object is in equilibrium. Which one of the following statements is *not* true? **(a)** The object must be at rest. **(b)** The object has a constant velocity. **(c)** The object has no acceleration. **(d)** No net force acts on the object.

14. Two identical boxes are being pulled across a horizontal floor at a constant velocity by a horizontal pulling force of 176 N that is applied to one of the boxes, as the drawing shows. There is kinetic friction between each box and the floor. Find the tension in the rope between the boxes. **(a)** 176 N **(b)** 88.0 N **(c)** 132 N **(d)** 44.0 N **(e)** There is not enough information to calculate the tension.

QUESTION 14

Section 4.12 Nonequilibrium Applications of Newton's Laws of Motion

15. A man is standing on a platform that is connected to a pulley arrangement, as the drawing shows. By pulling upward on the rope with a force $\vec{P}$ the man can raise the platform and himself. The total mass of the man plus the platform is 94.0 kg. What pulling force should the man apply to create an upward acceleration of 1.20 m/s²?

QUESTION 15

Problems

Note to Instructors: *Most of the homework problems in this chapter are available for assignment via WileyPLUS. See the Preface for additional details. Additional problems are available for assignment in WileyPLUS*

SSM Student Solutions Manual

MMH Problem-solving help

GO Guided Online Tutorial

V-HINT Video Hints

CHALK Chalkboard Videos

BIO Biomedical application

E Easy

M Medium

H Hard

WS Worksheet

T Team Problem

Section 4.3   Newton's Second Law of Motion

1. **E** An airplane has a mass of 3.1×10^4 kg and takes off under the influence of a constant net force of 3.7×10^4 N. What is the net force that acts on the plane's 78-kg pilot?

2. **E** **MMH** A boat has a mass of 6800 kg. Its engines generate a drive force of 4100 N due west, while the wind exerts a force of 800 N due east and the water exerts a resistive force of 1200 N due east. What are the magnitude and direction of the boat's acceleration?

3. **E** **GO** Two horizontal forces, $\vec{F}_1$ and $\vec{F}_2$, are acting on a box, but only $\vec{F}_1$ is shown in the drawing. $\vec{F}_2$ can point either to the right or to the left. The box moves only along the x axis. There is no friction between the box and the surface. Suppose that $\vec{F}_1 = +9.0$ N and the mass of the box is 3.0 kg. Find the magnitude and direction of $\vec{F}_2$ when the acceleration of the box is (a) $+5.0$ m/s^2, (b) -5.0 m/s^2, and (c) 0 m/s^2.

$\vec{F}_1$ ---+x

PROBLEM 3

4. **E** In the amusement park ride known as Magic Mountain Superman, powerful magnets accelerate a car and its riders from rest to 45 m/s (about 100 mi/h) in a time of 7.0 s. The combined mass of the car and riders is 5.5×10^3 kg. Find the average net force exerted on the car and riders by the magnets.

5. **E** **SSM** A person in a kayak starts paddling, and it accelerates from 0 to 0.60 m/s in a distance of 0.41 m. If the combined mass of the person and the kayak is 73 kg, what is the magnitude of the net force acting on the kayak?

6. **E** Scientists are experimenting with a kind of gun that may eventually be used to fire payloads directly into orbit. In one test, this gun accelerates a 5.0-kg projectile from rest to a speed of 4.0×10^3 m/s. The net force accelerating the projectile is 4.9×10^5 N. How much time is required for the projectile to come up to speed?

7. **E** **SSM** **MMH** A 1580-kg car is traveling with a speed of 15.0 m/s. What is the magnitude of the horizontal net force that is required to bring the car to a halt in a distance of 50.0 m?

8. **E** **GO** The space probe *Deep Space 1* was launched on October 24, 1998. Its mass was 474 kg. The goal of the mission was to test a new kind of engine called an ion propulsion drive. This engine generated only a weak thrust, but it could do so over long periods of time with the consumption of only small amounts of fuel. The mission was spectacularly successful. At a thrust of 56 mN how many days were required for the probe to attain a velocity of 805 m/s (1800 mi/h), assuming that the probe started from rest and that the mass remained nearly constant?

9. **M** **SSM** Two forces $\vec{F}_A$ and $\vec{F}_B$ are applied to an object whose mass is 8.0 kg. The larger force is $\vec{F}_A$. When both forces point due east, the object's acceleration has a magnitude of 0.50 m/s^2. However, when $\vec{F}_A$ points due east and $\vec{F}_B$ points due west, the acceleration is 0.40 m/s^2, due east. Find (a) the magnitude of $\vec{F}_A$ and (b) the magnitude of $\vec{F}_B$.

10. **M** **V-HINT** An electron is a subatomic particle ($m = 9.11 \times 10^{-31}$ kg) that is subject to electric forces. An electron moving in the $+x$ direction accelerates from an initial velocity of $+5.40 \times 10^5$ m/s to a final velocity of $+2.10 \times 10^6$ m/s while traveling a distance of 0.038 m. The electron's acceleration is due to two electric forces parallel to the x axis: $\vec{F}_1 = +7.50 \times 10^{-17}$ N, and $\vec{F}_2$, which points in the $-x$ direction. Find the magnitudes of (a) the net force acting on the electron and (b) the electric force $\vec{F}_2$.

Section 4.4   The Vector Nature of Newton's Second Law of Motion, Section 4.5   Newton's Third Law of Motion

11. **E** Only two forces act on an object (mass = 3.00 kg), as in the drawing. Find the magnitude and direction (relative to the x axis) of the acceleration of the object.

PROBLEM 11

12. **E** **SSM** **MMH** A rocket of mass 4.50×10^5 kg is in flight. Its thrust is directed at an angle of 55.0° above the horizontal and has a magnitude of 7.50×10^6 N. Find the magnitude and direction of the rocket's acceleration. Give the direction as an angle above the horizontal.

13. **E** When a parachute opens, the air exerts a large drag force on it. This upward force is initially greater than the weight of the sky diver and, thus, slows him down. Suppose the weight of the sky diver is 915 N and the drag force has a magnitude of 1027 N. The mass of the sky diver is 93.4 kg. What are the magnitude and direction of his acceleration?

14. **E** **GO** **CHALK** Two skaters, a man and a woman, are standing on ice. Neglect any friction between the skate blades and the ice. The mass of the man is 82 kg, and the mass of the woman is 48 kg. The woman pushes on the man with a force of 45 N due east. Determine the acceleration (magnitude and direction) of (a) the man and (b) the woman.

15. **M** **V-HINT** **CHALK** A space probe has two engines. Each generates the same amount of force when fired, and the directions of these forces can be independently adjusted. When the engines are fired simultaneously and each applies its force in the same direction, the probe, starting from rest, takes 28 s to travel a certain distance. How long does it take to travel the same distance, again starting from rest, if the engines are fired simultaneously and the forces that they apply to the probe are perpendicular?

16. **H** At a time when mining asteroids has become feasible, astronauts have connected a line between their 3500-kg space tug and a 6200-kg asteroid. Using their tug's engine, they pull on the asteroid with a force of 490 N. Initially the tug and the asteroid are at rest, 450 m apart. How much time does it take for the tug and the asteroid to meet?

17. **H** **SSM** A 325-kg boat is sailing 15.0° north of east at a speed of 2.00 m/s. Thirty seconds later, it is sailing 35.0° north of east at a speed of 4.00 m/s. During this time, three forces act on the boat: a 31.0-N force directed 15.0° north of east (due to an auxiliary engine), a

23.0-N force directed 15.0° south of west (resistance due to the water), and $\vec{F}_W$ (due to the wind). Find the magnitude and direction of the force $\vec{F}_W$. Express the direction as an angle with respect to due east.

Section 4.7 The Gravitational Force

18. **E** **GO** A 5.0-kg rock and a 3.0×10^{-4}-kg pebble are held near the surface of the earth. **(a)** Determine the magnitude of the gravitational force exerted on each by the earth. **(b)** Calculate the magnitude of the acceleration of each object when released.

19. **E** Mars has a mass of 6.46×10^{23} kg and a radius of 3.39×10^6 m. **(a)** What is the acceleration due to gravity on Mars? **(b)** How much would a 65-kg person weigh on this planet?

20. **E** On earth, two parts of a space probe weigh 11 000 N and 3400 N. These parts are separated by a center-to-center distance of 12 m and may be treated as uniform spherical objects. Find the magnitude of the gravitational force that each part exerts on the other out in space, far from any other objects.

21. **E** **GO** A raindrop has a mass of 5.2×10^{-7} kg and is falling near the surface of the earth. Calculate the magnitude of the gravitational force exerted **(a)** on the raindrop by the earth and **(b)** on the earth by the raindrop.

22. **E** The weight of an object is the same on two different planets. The mass of planet A is only sixty percent that of planet B. Find the ratio r_A/r_B of the radii of the planets.

23. **E** **SSM** A bowling ball (mass = 7.2 kg, radius = 0.11 m) and a billiard ball (mass = 0.38 kg, radius = 0.028 m) may each be treated as uniform spheres. What is the magnitude of the maximum gravitational force that each can exert on the other?

24. **E** Review Conceptual Example 7 in preparation for this problem. In tests on earth a lunar surface exploration vehicle (mass = 5.90×10^3 kg) achieves a forward acceleration of 0.220 m/s². To achieve this same acceleration on the moon, the vehicle's engines must produce a drive force of 1.43×10^3 N. What is the magnitude of the frictional force that acts on the vehicle on the moon?

25. **E** **SSM** Synchronous communications satellites are placed in a circular orbit that is 3.59×10^7 m above the surface of the earth. What is the magnitude of the acceleration due to gravity at this distance?

26. **E** The drawing (not to scale) shows one alignment of the sun, earth, and moon. The gravitational force $\vec{F}_{SM}$ that the sun exerts on the moon is perpendicular to the force $\vec{F}_{EM}$ that the earth exerts on the moon. The masses are: mass of sun = 1.99×10^{30} kg, mass of earth = 5.98×10^{24} kg, mass of moon = 7.35×10^{22} kg. The distances shown in the drawing are $r_{SM} = 1.50 \times 10^{11}$ m and $r_{EM} = 3.85 \times 10^8$ m. Determine the magnitude of the net gravitational force on the moon.

Moon

$\vec{F}_{SM}$

Sun

$\vec{F}_{EM}$

r_{EM}

Earth

PROBLEM 26

27. **E** The drawing shows three particles far away from any other objects and located on a straight line. The masses of these particles are $m_A = 363$ kg, $m_B = 517$ kg, and $m_C = 154$ kg. Find the magnitude and direction of the net gravitational force acting on **(a)** particle A, **(b)** particle B, and **(c)** particle C.

PROBLEM 27

28. **E** A space traveler weighs 540.0 N on earth. What will the traveler weigh on another planet whose radius is twice that of earth and whose mass is three times that of earth?

29. **M** **CHALK** A spacecraft is on a journey to the moon. At what point, as measured from the center of the earth, does the gravitational force exerted on the spacecraft by the earth balance that exerted by the moon? This point lies on a line between the centers of the earth and the moon. The distance between the earth and the moon is 3.85×10^8 m, and the mass of the earth is 81.4 times as great as that of the moon.

30. **M** **GO** A neutron star has a mass of 2.0×10^{30} kg (about the mass of our sun) and a radius of 5.0×10^3 m (about the height of a good-sized mountain). Suppose an object falls from rest near the surface of such a star. How fast would this object be moving after it had fallen a distance of 0.010 m? (Assume that the gravitational force is constant over the distance of the fall and that the star is not rotating.)

31. **M** **SSM** The sun is more massive than the moon, but the sun is farther from the earth. Which one exerts a greater gravitational force on a person standing on the earth? Give your answer by determining the ratio F_{sun}/F_{moon} of the magnitudes of the gravitational forces. Use the data on the inside of the front cover.

32. **H** Two particles are located on the x axis. Particle 1 has a mass m and is at the origin. Particle 2 has a mass $2m$ and is at $x = +L$. A third particle is placed between particles 1 and 2. Where on the x axis should the third particle be located so that the magnitude of the gravitational force on *both* particle 1 and particle 2 doubles? Express your answer in terms of L.

Section 4.8 The Normal Force, Section 4.9 Static and Kinetic Frictional Forces

33. **E** A 35-kg crate rests on a horizontal floor, and a 65-kg person is standing on the crate. Determine the magnitude of the normal force that **(a)** the floor exerts on the crate and **(b)** the crate exerts on the person.

34. **E** **SSM** A 60.0-kg crate rests on a level floor at a shipping dock. The coefficients of static and kinetic friction are 0.760 and 0.410, respectively. What horizontal pushing force is required to **(a)** just start the crate moving and **(b)** slide the crate across the dock at a constant speed?

35. **E** **CHALK** A rocket blasts off from rest and attains a speed of 45 m/s in 15 s. An astronaut has a mass of 57 kg. What is the astronaut's apparent weight during takeoff?

36. **E** **GO** A car is traveling up a hill that is inclined at an angle θ above the horizontal. Determine the ratio of the magnitude of the normal force to the weight of the car when **(a)** $\theta = 15°$ and **(b)** $\theta = 35°$.

37. **E** A woman stands on a scale in a moving elevator. Her mass is 60.0 kg, and the combined mass of the elevator and scale is an additional 815 kg. Starting from rest, the elevator accelerates upward. During the acceleration, the hoisting cable applies a force of 9410 N. What does the scale read during the acceleration?

38. **E** A Mercedes-Benz 300SL ($m = 1700$ kg) is parked on a road that rises 15° above the horizontal. What are the magnitudes of **(a)** the normal force and **(b)** the static frictional force that the ground exerts on the tires?

39. **E** **GO** Consult Multiple-Concept Example 9 to explore a model for solving this problem. A person pushes on a 57-kg refrigerator with a horizontal force of −267 N; the minus sign indicates that the force points in the −x direction. The coefficient of static friction is 0.65. **(a)** If the refrigerator does not move, what are the magnitude and direction of the static frictional force that the floor exerts on the

refrigerator? **(b)** What is the magnitude of the largest pushing force that can be applied to the refrigerator before it just begins to move?

40. **E** **SSM** A 6.00-kg box is sliding across the horizontal floor of an elevator. The coefficient of kinetic friction between the box and the floor is 0.360. Determine the kinetic frictional force that acts on the box when the elevator is **(a)** stationary, **(b)** accelerating upward with an acceleration whose magnitude is 1.20 m/s², and **(c)** accelerating downward with an acceleration whose magnitude is 1.20 m/s².

41. **E** **GO** A cup of coffee is on a table in an airplane flying at a constant altitude and a constant velocity. The coefficient of static friction between the cup and the table is 0.30. Suddenly, the plane accelerates forward, its altitude remaining constant. What is the maximum acceleration that the plane can have without the cup sliding backward on the table?

42. **M** **V-HINT** Consult Multiple-Concept Example 10 in preparation for this problem. Traveling at a speed of 16.1 m/s, the driver of an automobile suddenly locks the wheels by slamming on the brakes. The coefficient of kinetic friction between the tires and the road is 0.720. What is the speed of the automobile after 1.30 s have elapsed? Ignore the effects of air resistance.

43. **M** **SSM** **CHALK** A person is trying to judge whether a picture (mass = 1.10 kg) is properly positioned by temporarily pressing it against a wall. The pressing force is perpendicular to the wall. The coefficient of static friction between the picture and the wall is 0.660. What is the minimum amount of pressing force that must be used?

44. **M** **V-HINT** Air rushing over the wings of high-performance race cars generates unwanted horizontal air resistance but also causes a vertical *downforce,* which helps the cars hug the track more securely. The coefficient of static friction between the track and the tires of a 690-kg race car is 0.87. What is the magnitude of the maximum acceleration at which the car can speed up without its tires slipping when a 4060-N downforce and an 1190-N horizontal-air-resistance force act on it?

45. **H** While moving in, a new homeowner is pushing a box across the floor at a constant velocity. The coefficient of kinetic friction between the box and the floor is 0.41. The pushing force is directed downward at an angle θ below the horizontal. When θ is greater than a certain value, it is not possible to move the box, no matter how large the pushing force is. Find that value of θ.

Section 4.10   The Tension Force,
Section 4.11   Equilibrium Applications
of Newton's Laws of Motion

46. **E** **MMH** The helicopter in the drawing is moving horizontally to the right at a constant velocity $\vec{V}$. The weight of the helicopter is $W =$ 53 800 N. The lift force $\vec{L}$ generated by the rotating blade makes an angle of 21.0° with respect to the vertical. **(a)** What is the magnitude of the lift force? **(b)** Determine the magnitude of the air resistance $\vec{R}$ that opposes the motion.

PROBLEM 46

47. **E** **SSM** Three forces act on a moving object. One force has a magnitude of 80.0 N and is directed due north. Another has a magnitude of 60.0 N and is directed due west. What must be the magnitude and direction of the third force, such that the object continues to move with a constant velocity?

48. **E** **SSM** A worker stands still on a roof sloped at an angle of 36° above the horizontal. He is prevented from slipping by a static frictional force of 390 N. Find the mass of the worker.

49. **E** A stuntman is being pulled along a rough road at a constant velocity by a cable attached to a moving truck. The cable is parallel to the ground. The mass of the stuntman is 109 kg, and the coefficient of kinetic friction between the road and him is 0.870. Find the tension in the cable.

50. **E** **GO** The drawing shows a circus clown who weighs 890 N. The coefficient of static friction between the clown's feet and the ground is 0.53. He pulls vertically downward on a rope that passes around three pulleys and is tied around his feet. What is the minimum pulling force that the clown must exert to yank his feet out from under himself?

PROBLEM 50

51. **M** **V-HINT** **MMH** **CHALK** During a storm, a tree limb breaks off and comes to rest across a barbed wire fence at a point that is not in the middle between two fence posts. The limb exerts a downward force of 151 N on the wire. The left section of the wire makes an angle of 14.0° relative to the horizontal and sustains a tension of 447 N. Find the magnitude and direction of the tension that the right section of the wire sustains.

52. **M** **GO** **MMH** **CHALK** A toboggan slides down a hill and has a constant velocity. The angle of the hill is 8.00° with respect to the horizontal. What is the coefficient of kinetic friction between the surface of the hill and the toboggan?

53. **H** **SSM** A bicyclist is coasting straight down a hill at a constant speed. The combined mass of the rider and bicycle is 80.0 kg, and the hill is inclined at 15.0° with respect to the horizontal. Air resistance opposes the motion of the cyclist. Later, the bicyclist climbs the same hill at the same constant speed. How much force (directed parallel to the hill) must be applied to the bicycle in order for the bicyclist to climb the hill?

54. **H** **MMH** A kite is hovering over the ground at the end of a straight 43-m line. The tension in the line has a magnitude of 16 N. Wind blowing on the kite exerts a force of 19 N, directed 56° above the horizontal. Note that the line attached to the kite is *not* oriented at an angle of 56° above the horizontal. Find the height of the kite, relative to the person holding the line.

Section 4.12   Nonequilibrium Applications
of Newton's Laws of Motion

55. **E** A 1450-kg submarine rises straight up toward the surface. Seawater exerts both an upward buoyant force of 16 140 N on the submarine and a downward resistive force of 1030 N. What is the submarine's acceleration?

56. **E** **SSM** A 15-g bullet is fired from a rifle. It takes 2.50×10^{-3} s for the bullet to travel the length of the barrel, and it exits the barrel with a speed of 715 m/s. Assuming that the acceleration of the bullet is constant, find the average net force exerted on the bullet.

57. **E** A fisherman is fishing from a bridge and is using a "45-N test line." In other words, the line will sustain a maximum force of 45 N without breaking. What is the weight of the heaviest fish that can be pulled up vertically when the line is reeled in **(a)** at a constant speed and **(b)** with an acceleration whose magnitude is 2.0 m/s²?

58. **E** **SSM** Only two forces act on an object (mass = 4.00 kg), as in the drawing. Find the magnitude and direction (relative to the *x* axis) of the acceleration of the object.

PROBLEM 58

59. **E** A helicopter flies over the arctic ice pack at a constant altitude, towing an airborne 129-kg laser sensor that measures the thickness of the ice (see the drawing). The helicopter and the sensor both move only in the horizontal direction and have a horizontal acceleration of magnitude 2.84 m/s². Ignoring air resistance, find the tension in the cable towing the sensor.

PROBLEM 59

60. **E** Review Conceptual Example 17 as background for this problem. The water skier there has a mass of 73 kg. Find the magnitude of the net force acting on the skier when **(a)** she is accelerated from rest to a speed of 11 m/s in 8.0 s and **(b)** she lets go of the tow rope and glides to a halt in 21 s.

61. **E** A rescue helicopter is lifting a man (weight = 822 N) from a capsized boat by means of a cable and harness. **(a)** What is the tension in the cable when the man is given an initial upward acceleration of 1.10 m/s²? **(b)** What is the tension during the remainder of the rescue when he is pulled upward at a constant velocity?

62. **E** A car is towing a boat on a trailer. The driver starts from rest and accelerates to a velocity of +11 m/s in a time of 28 s. The combined mass of the boat and trailer is 410 kg. The frictional force acting on the trailer can be ignored. What is the tension in the hitch that connects the trailer to the car?

63. **E** **GO** A 292-kg motorcycle is accelerating up along a ramp that is inclined 30.0° above the horizontal. The propulsion force pushing the motorcycle up the ramp is 3150 N, and air resistance produces a force of 250 N that opposes the motion. Find the magnitude of the motorcycle's acceleration.

64. **E** **SSM** A student is skateboarding down a ramp that is 6.0 m long and inclined at 18° with respect to the horizontal. The initial speed of the skateboarder at the top of the ramp is 2.6 m/s. Neglect friction and find the speed at the bottom of the ramp.

65. **E** A man seeking to set a world record wants to tow a 109 000-kg airplane along a runway by pulling horizontally on a cable attached to the airplane. The mass of the man is 85 kg, and the coefficient of static friction between his shoes and the runway is 0.77. What is the greatest acceleration the man can give the airplane? Assume that the airplane is on wheels that turn without any frictional resistance.

66. **M** **V-HINT** A 205-kg log is pulled up a ramp by means of a rope that is parallel to the surface of the ramp. The ramp is inclined at 30.0° with respect to the horizontal. The coefficient of kinetic friction between the log and the ramp is 0.900, and the log has an acceleration of magnitude 0.800 m/s². Find the tension in the rope.

67. **M** **GO** To hoist himself into a tree, a 72.0-kg man ties one end of a nylon rope around his waist and throws the other end over a branch of the tree. He then pulls downward on the free end of the rope with a force of 358 N. Neglect any friction between the rope and the branch, and determine the man's upward acceleration.

68. **M** **CHALK** **SSM** Two objects (45.0 and 21.0 kg) are connected by a massless string that passes over a massless, frictionless pulley. The pulley hangs from the ceiling. Find **(a)** the acceleration of the objects and **(b)** the tension in the string.

69. **M** **GO** A train consists of 50 cars, each of which has a mass of 6.8×10^3 kg. The train has an acceleration of $+8.0 \times 10^{-2}$ m/s². Ignore friction and determine the tension in the coupling **(a)** between the 30th and 31st cars and **(b)** between the 49th and 50th cars.

70. **M** In Problem 65, an 85-kg man plans to tow a 109 000-kg airplane along a runway by pulling horizontally on a cable attached to it. Suppose that he instead attempts the feat by pulling the cable at an angle of 9.0° above the horizontal. The coefficient of static friction between his shoes and the runway is 0.77. What is the greatest acceleration the man can give the airplane? Assume that the airplane is on wheels that turn without any frictional resistance.

71. **M** **MMH** The drawing shows a large cube (mass = 25 kg) being accelerated across a horizontal frictionless surface by a horizontal force $\vec{P}$. A small cube (mass = 4.0 kg) is in contact with the front surface of the large cube and will slide downward unless $\vec{P}$ is sufficiently large. The coefficient of static friction between the cubes is 0.71. What is the smallest magnitude that $\vec{P}$ can have in order to keep the small cube from sliding downward?

PROBLEM 71

72. **M** **V-HINT** The alarm at a fire station rings and an 86-kg fireman, starting from rest, slides down a pole to the floor below (a distance of 4.0 m). Just before landing, his speed is 1.4 m/s. What is the magnitude of the kinetic frictional force exerted on the fireman as he slides down the pole?

73. **M** **GO** Two blocks are sliding to the right across a horizontal surface, as the drawing shows. In Case A the mass of each block is 3.0 kg. In Case B the mass of block 1 (the block behind) is 6.0 kg, and the mass of block 2 is 3.0 kg. No frictional force acts on block 1 in either Case A or Case B. However, a kinetic frictional force of 5.8 N does act on block 2 in both cases and opposes the motion. For both Case A and Case B determine **(a)** the magnitude of the forces with which the blocks push against each other and **(b)** the magnitude of the acceleration of the blocks.

PROBLEM 73

74. **M** **SSM** A person whose weight is 5.20×10^2 N is being pulled up vertically by a rope from the bottom of a cave that is 35.1 m deep. The maximum tension that the rope can withstand without breaking is 569 N. What is the shortest time, starting from rest, in which the person can be brought out of the cave?

75. **M** **CHALK** A girl is sledding down a slope that is inclined at 30.0° with respect to the horizontal. The wind is aiding the motion by providing a steady force of 105 N that is parallel to the motion of the sled. The combined mass of the girl and the sled is 65.0 kg, and the coefficient of kinetic friction between the snow and the runners of the sled is 0.150. How much time is required for the sled to travel down a 175-m slope, starting from rest?

76. **H** A small sphere is hung by a string from the ceiling of a van. When the van is stationary, the sphere hangs vertically. However, when the van accelerates, the sphere swings backward so that the string makes an angle of θ with respect to the vertical. **(a)** Derive an expression for the magnitude a of the acceleration of the van in terms of the angle θ and the magnitude g of the acceleration due to gravity. **(b)** Find the acceleration of the van when $\theta = 10.0°$. **(c)** What is the angle θ when the van moves with a constant velocity?

77. **H** **SSM** A penguin slides at a constant velocity of 1.4 m/s down an icy incline. The incline slopes above the horizontal at an angle of 6.9°. At the bottom of the incline, the penguin slides onto a horizontal patch of ice. The coefficient of kinetic friction between the penguin and the ice is the same for the incline as for the horizontal patch. How much time is required for the penguin to slide to a halt after entering the horizontal patch of ice?

Additional Problems

78. **E** **GO** A person with a black belt in karate has a fist that has a mass of 0.70 kg. Starting from rest, this fist attains a velocity of 8.0 m/s in 0.15 s. What is the magnitude of the average net force applied to the fist to achieve this level of performance?

79. **E** **SSM** A 95.0-kg person stands on a scale in an elevator. What is the apparent weight when the elevator is **(a)** accelerating upward with an acceleration of 1.80 m/s², **(b)** moving upward at a constant speed, and **(c)** accelerating downward with an acceleration of 1.30 m/s²?

80. **E** **MMH** Two forces, $\vec{F}_1$ and $\vec{F}_2$, act on the 7.00-kg block shown in the drawing. The magnitudes of the forces are $F_1 = 59.0$ N and $F_2 = 33.0$ N. What is the horizontal acceleration (magnitude and direction) of the block?

PROBLEM 80

81. **E** **SSM** A student presses a book between his hands, as the drawing indicates. The forces that he exerts on the front and back covers of the book are perpendicular to the book and are horizontal. The book weighs 31 N. The coefficient of static friction between his hands and the book is 0.40. To keep the book from falling, what is the magnitude of the minimum pressing force that each hand must exert?

PROBLEM 81

82. **E** In a European country a bathroom scale displays its reading in kilograms. When a man stands on this scale, it reads 92.6 kg. When he pulls down on a chin-up bar installed over the scale, the reading decreases to 75.1 kg. What is the magnitude of the force he exerts on the chin-up bar?

83. **E** **SSM** A 1380-kg car is moving due east with an initial speed of 27.0 m/s. After 8.00 s the car has slowed down to 17.0 m/s. Find the magnitude and direction of the net force that produces the deceleration.

84. **E** In preparation for this problem, review Conceptual Example 7. A space traveler whose mass is 115 kg leaves earth. What are his weight and mass **(a)** on earth and **(b)** in interplanetary space where there are no nearby planetary objects?

85. **E** **(a)** Calculate the magnitude of the gravitational force exerted on a 425-kg satellite that is a distance of two earth radii from the center of the earth. **(b)** What is the magnitude of the gravitational force exerted on the earth by the satellite? **(c)** Determine the magnitude of the satellite's acceleration. **(d)** What is the magnitude of the earth's acceleration?

86. **M** **SSM** **MMH** The drawing shows Robin Hood (mass = 77.0 kg) about to escape from a dangerous situation. With one hand, he is gripping the rope that holds up a chandelier (mass = 195 kg). When he cuts the rope where it is tied to the floor, the chandelier will fall, and he will be pulled up toward a balcony above. Ignore the friction between the rope and the beams over which it slides, and find **(a)** the acceleration with which Robin is pulled upward and **(b)** the tension in the rope while Robin escapes.

87. **M** **V-HINT** Consult Multiple-Concept Example 10 for insight into solving this type of problem. A box is sliding up an incline that makes an angle of 15.0° with respect to the horizontal. The coefficient of kinetic friction between the box and the surface of the incline is 0.180. The initial speed of the box at the bottom of the incline is 1.50 m/s. How far does the box travel along the incline before coming to rest?

PROBLEM 86

88. **H** As part *a* of the drawing shows, two blocks are connected by a rope that passes over a set of pulleys. One block has a weight of 412 N, and the other has a weight of 908 N. The rope and the pulleys are massless and there is no friction. **(a)** What is the acceleration of the lighter block? **(b)** Suppose that the heavier block is removed, and a downward force of 908 N is provided by someone pulling on the rope, as part *b* of the drawing shows. Find the acceleration of the remaining block. **(c)** Explain why the answers in (a) and (b) are different.

(a) *(b)*

PROBLEM 88

89. **H** **SSM** The three objects in the drawing are connected by strings that pass over massless and friction-free pulleys. The objects move, and the coefficient of kinetic friction between the middle object and the surface of the table is 0.100. **(a)** What is the acceleration of the three objects? **(b)** Find the tension in each of the two strings.

PROBLEM 89

90. Ⓜ GO A flatbed truck is carrying a crate up a hill of angle of inclination $\theta = 10.0°$, as the figure illustrates. The coefficient of static friction between the truck bed and the crate is $\mu_s = 0.350$. Find the maximum acceleration that the truck can attain before the crate begins to slip backward relative to the truck.

$\theta = 10.0°$ $\vec{W} = m\vec{g}$ $\theta = 10.0°$

(a)

$+y$

$\vec{F}_N$

$mg \sin \theta$ $\vec{f}_s^{MAX}$ $+x$

$mg \cos \theta$

$\vec{W} = m\vec{g}$ $\theta = 10.0°$

(b) Free-body diagram of the crate

PROBLEM 90

Physics in Biology, Medicine, and Sports

91. Ⓜ BIO **4.3** To this point, we have neglected the effects of air resistance (or drag) on falling bodies, which acts in the opposite direction of the object's weight. However, drag can significantly affect the motion of a falling object. The drag force (F_d) due to air resistance can be represented by the following relationship: $F_d = CAv^n$, where C is the coefficient of air friction, A is the cross-sectional area of the falling object, and v is the object's speed relative to the air. The value of n will vary depending on the object's speed. It typically has a value near 1 at low speeds and 2 at high speeds. The important point is that, as the speed of the object increases, the drag force increases, and the acceleration decreases. The figure shows the free-body diagram for a skydiver, where drag has been included. Eventually, the drag force will be equal to the weight, and the sum of the forces acting on the skydiver will be zero. Thus, the acceleration will be zero, and the skydiver will fall at constant speed. This speed is referred to as the *terminal velocity*. What is the terminal velocity of a 95-kg skydiver? Assume the cross-sectional area of the skydiver is 0.20 m², the coefficient of air friction is 0.90 kg/m³, and $n = 2$.

F_d $+y$

W

PROBLEM 91

92. Ⓔ BIO **4.8** Walking up an inclined surface can have tremendous health benefits. Your leg muscles and hip flexors work harder to propel you up the incline than when you walk across level terrain. Likewise, your abdominal muscles must contract to keep your body upright and stable, since it is not perpendicular to the inclined surface, which helps to strengthen your core. Imagine a person in leather boots standing on an inclined oak plank, as the figure shows. If the coefficient of static friction between her boots and the plank is 0.60, what is the angle of the incline, assuming she is at the maximum angle before she begins to slide?

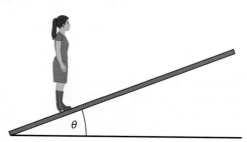

θ

PROBLEM 92

93. Ⓜ BIO **4.10** The longest bone in the human body is the femur. It is the only bone located in the upper leg. Fractures, or breaking of the femur bone, often involve high-speed accidents, such as from skiing or in an automobile collision. An alternative way to apply skeletal traction to the injured leg, as compared to Example 12, is shown in the figure. **(a)** What is the tension in the rope, and **(b)** what is the traction force (F_T) that acts on the leg?

F_T 60°

3.0 kg

PROBLEM 93

94. Ⓜ BIO **4.3** The sport of American football often involves athletes running at high speeds and smashing into each other. While they wear protective pads and helmets, their bodies still experience large decelerations and forces. In fact, some of the collisions between players, or between a player and the ground, are comparable to what a person might endure in an automobile accident. In 2017, a large study found that 110 out of 111 deceased NFL players had *chronic traumatic encephalopathy* (CTE), a degenerative brain disorder associated with repetitive head trauma (see photo). Other studies have linked CTE to higher rates of early onset dementia and memory loss. Consider an NFL player running at 3.92 m/s. During a collision and tackle, his head is brought to a stop in a distance of 0.125 m. (a) Calculate the magnitude of the deceleration of the player's head (assume the deceleration to be constant and treat this as a one-dimensional problem). How does it compare to the acceleration of gravity near the surface of Earth? (b) Assume the mass of the player's brain is 1.35 kg and calculate the force experienced by the brain.

Normal Brain   Advanced CTE

PROBLEM 94

95. **M** **BIO** **4.8** Doctor Doom's Fearfall ride at Universal Studios in Orlando, Florida launches thrill seekers straight upward with accelerations greater than what astronauts experienced during the launch of the Space Shuttle. Assume a 75.0-kg rider undergoes a constant acceleration and reaches a maximum speed of 40.0 mph during the first 5.10 m of the trip. What is the apparent weight of the rider during this time?

PROBLEM 95

96. **E** **BIO** **4.4** At an instant when a soccer ball is in contact with the foot of a player kicking it, the horizontal or x component of the ball's acceleration is 810 m/s^2 and the vertical or y component of its acceleration is 1100 m/s^2. The ball's mass is 0.43 kg. What is the magnitude of the net force acting on the soccer ball at this instant?

97. **E** **GO** **BIO** **4.4** A billiard ball strikes and rebounds from the cushion of a pool table perpendicularly. The mass of the ball is 0.38 kg. The ball approaches the cushion with a velocity of +2.1 m/s and rebounds with a velocity of −2.0 m/s. The ball remains in contact with the cushion for a time of 3.3×10^{-3} s. What is the average net force (magnitude and direction) exerted on the ball by the cushion?

98. **E** **MMH** **BIO** **4.9** An 81-kg baseball player slides into second base. The coefficient of kinetic friction between the player and the ground is 0.49. (a) What is the magnitude of the frictional force? (b) If the player comes to rest after 1.6 s, what was his initial velocity?

99. **E** **MMH** **BIO** **4.9** The speed of a bobsled is increasing because it has an acceleration of 2.4 m/s^2. At a given instant in time, the forces resisting the motion, including kinetic friction and air resistance, total 450 N. The combined mass of the bobsled and its riders is 270 kg. (a) What is the magnitude of the force propelling the bobsled forward? (b) What is the magnitude of the net force that acts on the bobsled?

100. **E** **BIO** **4.3** When a 58-g tennis ball is served, it accelerates from rest to a speed of 45 m/s. The impact with the racket gives the ball a constant acceleration over a distance of 44 cm. What is the magnitude of the net force acting on the ball?

101. **M** **GO** **BIO** **4.9** A skater with an initial speed of 7.60 m/s stops propelling himself and begins to coast across the ice, eventually coming to rest. Air resistance is negligible. (a) The coefficient of kinetic friction between the ice and the skate blades is 0.100. Find the deceleration caused by kinetic friction. (b) How far will the skater travel before coming to rest?

102. **M** **GO** **BIO** **4.11** A mountain climber, in the process of crossing between two cliffs by a rope, pauses to rest. She weighs 535 N. As the drawing shows, she is closer to the left cliff than to the right cliff, with the result that the tensions in the left and right sides of the rope are not the same. Find the tensions in the rope to the left and to the right of the mountain climber.

65.0°   80.0°

PROBLEM 102

103. **M** **BIO** **V-HINT** **4.11** The person in the drawing is standing on crutches. Assume that the force exerted on each crutch by the ground is directed along the crutch, as the force vectors in the drawing indicate. If the coefficient of static friction between a crutch and the ground is 0.90, determine the largest angle θ^{MAX} that the crutch can have just before it begins to slip on the floor.

θ

$\vec{F}$

PROBLEM 103

Concepts and Calculations Problems

Online

Newton's three laws of motion provide the basis for understanding the effect of forces on the motion of an object. The second law is especially important, because it provides the quantitative relationship between net force and acceleration. The following problems serve as a review of the essential features of this relationship.

104. **M** **CHALK** **SSM** The figure shows two forces, $\vec{F}_1 = +3000$ N and $\vec{F}_2 = +5000$ N, acting on a spacecraft; the plus signs indicate that the forces are directed along the +x axis. A third force $\vec{F}_3$ also acts on the spacecraft but is not shown in the drawing. *Concepts:* (i) Suppose the spacecraft were stationary. What would be the

direction of $\vec{F}_3$? (ii) When the spacecraft is moving at a constant velocity of +850 m/s, what is the direction of $\vec{F}_3$? *Calculations:* Find the direction and magnitude of $\vec{F}_3$.

PROBLEM 104

105. [M] [CHALK] On earth a block has a weight of 88 N. This block is sliding on a horizontal surface on the moon, where the acceleration due to gravity is 1.60 m/s². As the figure shows, the block is being pulled by a horizontal rope in which the tension is $T = 24$ N. The coefficient of kinetic friction between the block and the surface is $\mu_k = 0.20$. *Concepts:* (i) Which of Newton's laws of motion provides the best way to determine the acceleration of the block? (ii) Does the net force in the x direction equal the tension T? *Calculations:* Determine the acceleration of the block.

(a)

(b) Free-body diagram for the block

PROBLEM 105

Team Problems

Online

106. [E] [T] [WS] **A Water Balloon Launcher.** Two metal poles sticking vertically out of the ground are separated by 1.50 m. A giant slingshot is formed by attaching two pieces of elastic surgical tubing to the poles (one each) with their opposite ends attached to a launch basket that holds a water balloon. The two elastic tubes are the same length, but are different (one is harder to stretch than the other). The launching mechanism requires that the basket be pulled along a line that bisects, and is perpendicular to, the line connecting the two poles. You and your team are trying to determine how to aim this balloon launcher taking into consideration that the two sides of the slingshot provide forces of different magnitudes. When the launch basket is pulled 2.00 meters straight back (parallel to the ground) and connected to the triggering mechanism, you find that the magnitude of the pulling force exerted on the basket by the tubing on the left pole is 1.33 times larger than that exerted by the tubing connected to the right pole. How will the direction of the initial acceleration (the instant after the trigger is pulled) of the launch basket deviate from the straight-line path, that is, from the line that bisects the poles? Express your answer in terms of the angle of deviation to the left or right of the bisecting line.

107. [M] [T] [WS] **Extracting a Truck Engine.** You and your team removed a 175-kg engine from the chassis of a truck and hung it from a chain of length $L = 2.80$ m from the 4.00-m high ceiling of a garage. You now want to move the truck out of the garage and move in another vehicle into which you intend to place the engine. To make the exchange, you must pull the hanging engine to the side so that the chain makes an angle of $\theta = 30.0°$ with respect to the vertical. Located on the wall directly to the right of the engine (i.e., in the direction you want to pull it) is a pulley mounted 1.20 m below the ceiling. You measure a horizontal distance of 2.35 m between the wall and the point where the chain is connected to the engine while it is hanging vertically. You get the idea to thread a rope through the pulley on the wall, connect it to the engine at the same place where the chain is connected, and then pull the rope to move the engine to the side. However, you need a rope that can safely do the job. What is the minimum tension the rope should be rated to handle? Give your answer in both newtons and pounds.

108. [E] [T] **Towing a Detector.** You are in a helicopter towing a 129-kg laser detector that is mapping out the thickness of the Brunt Ice Shelf along the coast of Antarctica. The original cable used to suspend the detector was damaged and replaced by a lighter one with a maximum tension rating of 310.0 pounds, not much more than the weight of the detector. The replacement cable would work without question in the case that the detector and helicopter were not accelerating. However, some acceleration of the helicopter is inevitable. In order to monitor the tension force on the cable to make sure the maximum is not exceeded (and therefore to not lose the very expensive detector) you calculate the maximum angle the cable can make with the vertical without the cable exceeding the tension limit. (a) Assuming straight and level flight of the helicopter, what is that maximum angle? (b) What is the corresponding acceleration? (c) Your colleague wants to add a 10.0-kg infrared camera to the detector. What is the maximum allowable angle now?

109. [M] [T] **Stuck on a Ledge.** You and your friends pull your car off a mountain road onto an overlook to take pictures of the frozen lake below. You lose your footing on the packed snow, fall over the cliff, and land in a soft snow bank on a ledge, 10 feet below the edge. The cliff's wall is too smooth to climb upwards and too steep and dangerous to climb the remaining 300 feet to the bottom. While trying to think your way out of your predicament, you realize one reason you had slipped was that you had misjudged how the cliff sloped towards the edge (a decline of about 5–10 degrees). There is nothing near the edge to which your friends can tie a rope, and they fear the footing is too slippery for them to pull you to the top. However, they can get the car close enough to the edge to tie the rope to the bumper and hang it over the edge. You are worried, however, that the car might slip over the edge as well. On your phone (which has survived the fall) you look up the coefficient of static friction of snow tires on packed snow and find that it ranges from 0.22 to 0.26, and that the model of your friends' car weighs 3856 lb. Your own weight fluctuates between 120–130 lb. (a) What is the worst-case scenario in terms of the parameters given? (b) Assuming the worst-case scenario, should your friends bring the car close to the edge? (c) Next, should you risk tying the rope to the trailer hitch and trying to climb out? You assume the hitch is low enough to the ground so that the tension force is parallel to the slope near the edge of the cliff and that you climb up the rope at a constant velocity.

Crew members are refueling a V-22 Osprey helicopter. The green lights near the tips of its rotating blades experience a net force and acceleration that points toward the center of the circle. We will now see how and why this net force and acceleration arise.

Chief Petty Officer Joe Kane (U.S. Navy), Image taken from https://commons.wikimedia.org/wiki/File:20080406165033%21V-22_Osprey_refueling_edit1.jpg

● Dynamics of Uniform Circular Motion

LEARNING OBJECTIVES

After reading this module, you should be able to...

5.1 Define uniform circular motion.

5.2 Solve uniform circular motion kinematic problems.

5.3 Solve uniform circular motion dynamic problems.

5.4 Solve problems involving banked curves.

5.5 Analyze circular gravitational orbits.

5.6 Solve application problems involving gravity and uniform circular motion.

5.7 Analyze vertical circular motion.

5.1 Uniform Circular Motion

There are many examples of motion on a circular path. Of the many possibilities, we single out those that satisfy the following definition:

> **DEFINITION OF UNIFORM CIRCULAR MOTION**
> **Uniform circular motion is the motion of an object traveling at a constant (uniform) speed on a circular path.**

As an example of uniform circular motion, **Figure 5.1** shows a model airplane on a guideline. The speed of the plane is the magnitude of the velocity vector $\vec{v}$, and since the speed is constant, the vectors in the drawing have the same magnitude at all points on the circle.

Sometimes it is more convenient to describe uniform circular motion by specifying the period of the motion, rather than the speed. The **period T** is the time required to travel once around the circle—that is, to make one complete revolution. There is a relationship between

FIGURE 5.1 The motion of a model airplane flying at a constant speed on a horizontal circular path is an example of uniform circular motion.

period and speed, since speed v is the distance traveled (here, the circumference of the circle $= 2\pi r$) divided by the time T:

$$v = \frac{2\pi r}{T} \tag{5.1}$$

If the radius is known, as in Example 1, the speed can be calculated from the period, or vice versa.

EXAMPLE 1 | A Tire-Balancing Machine

The wheel of a car has a radius of $r = 0.29$ m and is being rotated at 830 revolutions per minute (rpm) on a tire-balancing machine. Determine the speed (in m/s) at which the outer edge of the wheel is moving.

Reasoning The speed v can be obtained directly from $v = 2\pi r/T$, but first the period T is needed. The period is the time for one revolution, and it must be expressed in seconds, because the problem asks for the speed in meters per second.

Solution Since the tire makes 830 revolutions in one minute, the number of minutes required for a single revolution is

$$\frac{1}{830 \text{ revolutions/min}} = 1.2 \times 10^{-3} \text{ min/revolution}$$

Therefore, the period is $T = 1.2 \times 10^{-3}$ min, which corresponds to 0.072 s. Equation 5.1 can now be used to find the speed:

$$v = \frac{2\pi r}{T} = \frac{2\pi (0.29 \text{ m})}{0.072 \text{ s}} = \boxed{25 \text{ m/s}}$$

(a)

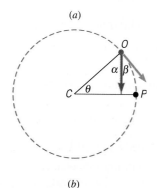

(b)

FIGURE 5.2 (a) For an object (•) in uniform circular motion, the velocity $\vec{v}$ has different directions at different places on the circle. (b) The velocity vector has been removed from point P, shifted parallel to itself, and redrawn with its tail at point O.

The definition of uniform circular motion emphasizes that the speed, or the magnitude of the velocity vector, is constant. It is equally significant that the direction of the vector is *not constant*. In **Figure 5.1**, for instance, the velocity vector changes direction as the plane moves around the circle. Any change in the velocity vector, even if it is only a change in direction, means that an acceleration is occurring. This particular acceleration is called "centripetal acceleration," because it points toward the center of the circle, as the next section explains.

5.2 Centripetal Acceleration

In this section we determine how the magnitude a_c of the centripetal acceleration depends on the speed v of the object and the radius r of the circular path. We will see that $a_c = v^2/r$.

In **Figure 5.2a** an object (symbolized by a dot •) is in uniform circular motion. At a time t_0 the velocity is tangent to the circle at point O, and at a later time t the velocity is tangent at point P. As the object moves from O to P, the radius traces out the angle θ, and the velocity vector changes direction. To emphasize the change, part b of the picture shows the velocity vector removed from point P, shifted parallel to itself, and redrawn with its tail at point O. The angle β between the two vectors indicates the change in direction. Since the radii CO and CP are perpendicular to the tangents at points O and P, respectively, it follows that $\alpha + \beta = 90°$ and $\alpha + \theta = 90°$. Therefore, angle β and angle θ are equal.

As always, acceleration is the change $\Delta\vec{v}$ in velocity divided by the elapsed time Δt, or $\vec{a} = \Delta\vec{v}/\Delta t$. **Figure 5.3a** shows the two velocity vectors oriented at the angle θ with respect to one another, together with the vector $\Delta\vec{v}$ that represents the change

(a)

(b)

FIGURE 5.3 (a) The directions of the velocity vector at times t and t_0 differ by the angle θ. (b) When the object moves along the circle from O to P, the radius r traces out the same angle θ. Here, the sector COP has been rotated clockwise by 90° relative to its orientation in **Figure 5.2**.

in velocity. The change $\Delta\vec{v}$ is the increment that must be added to the velocity at time t_0, so that the resultant velocity has the new direction after an elapsed time $\Delta t = t - t_0$. **Figure 5.3b** shows the sector of the circle COP. In the limit that Δt is very small, the arc length OP is approximately a straight line whose length is the distance $v\,\Delta t$ traveled by the object. In this limit, COP is an isosceles triangle, as is the triangle in part a of the drawing. Note that both triangles have equal angles θ. This means that they are similar triangles, so that

$$\frac{\Delta v}{v} = \frac{v\,\Delta t}{r}$$

This equation can be solved for $\Delta v/\Delta t$, to show that the magnitude a_c of the centripetal acceleration is given by $a_c = v^2/r$.

Centripetal acceleration is a vector quantity and, therefore, has a direction as well as a magnitude. The direction is toward the center of the circle, and Conceptual Example 2 helps us to set the stage for explaining this important fact.

CONCEPTUAL EXAMPLE 2 | Which Way Will the Object Go?

In **Figure 5.4** an object, such as a model airplane on a guideline, is in uniform circular motion. The object is symbolized by a dot ($\cdot$), and at point O it is released suddenly from its circular path. For instance, suppose that the guideline for a model plane is cut suddenly. Does the object move **(a)** along the straight tangent line between points O and A or **(b)** along the circular arc between points O and P?

Reasoning Newton's first law of motion (see Section 4.2) guides our reasoning. This law states that an object continues in a state of rest or in a state of motion at a constant velocity (i.e., at a constant speed along a straight line) unless compelled to change that state by a net force. When an object is suddenly released from its circular path, there is no longer a net force being applied to the object. In the case of the model airplane, the guideline cannot apply a force, since it is cut. Gravity certainly acts on the plane, but the wings provide a lift force that balances the weight of the plane.

Answer (b) is incorrect. An object such as a model airplane will remain on a circular path only if a net force keeps it there. Since there is no net force, it cannot travel on the circular arc.

Answer (a) is correct. In the absence of a net force, the plane or any object would continue to move at a constant speed along a straight line in the direction it had at the time of release, consistent with Newton's first law. This speed and direction are given in **Figure 5.4** by the velocity vector $\vec{v}$.

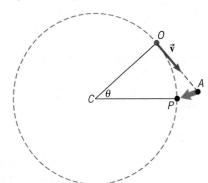

FIGURE 5.4 If an object ($\cdot$) moving on a circular path were released from its path at point O, it would move along the straight tangent line OA in the absence of a net force.

As Example 2 discusses, the object in **Figure 5.4** would travel on a tangent line if it were released from its circular path suddenly at point O. It would move in a straight line to point A in the time it would have taken to travel on the circle to point P. It is as if, in the process of remaining on the circle, the object drops through the distance AP, and AP is directed toward the center of the circle in the limit that the angle θ is small. Thus, the object in uniform circular motion accelerates toward the center of the circle at every moment. Since the word "centripetal" means "moving toward a center," the acceleration is called **centripetal acceleration**.

CENTRIPETAL ACCELERATION

Magnitude: **The centripetal acceleration of an object moving with a speed v on a circular path of radius r has a magnitude a_c given by**

$$a_c = \frac{v^2}{r} \qquad (5.2)$$

Direction: **The centripetal acceleration vector always points toward the center of the circle and continually changes direction as the object moves.**

The following example illustrates the effect of the radius r on the centripetal acceleration.

EXAMPLE 3 | The Physics of a Bobsled Track

The bobsled track at the 1994 Olympics in Lillehammer, Norway, contained turns with radii of 33 m and 24 m, as **Interactive Figure 5.5** illustrates. Find the centripetal acceleration at each turn for a speed of 34 m/s, a speed that was achieved in the two-man event. Express the answers as multiples of $g = 9.8$ m/s^2.

Reasoning   In each case, the magnitude of the centripetal acceleration can be obtained from $a_c = v^2/r$. Since the radius r is in the denominator on the right side of this expression, we expect the acceleration to be smaller when r is larger.

Solution   From $a_c = v^2/r$ it follows that

Radius = 33 m $a_c = \dfrac{(34 \text{ m/s})^2}{33 \text{ m}} = 35 \text{ m/s}^2 = \boxed{3.6\ g}$

Radius = 24 m $a_c = \dfrac{(34 \text{ m/s})^2}{24 \text{ m}} = 48 \text{ m/s}^2 = \boxed{4.9\ g}$

The centripetal acceleration is indeed smaller when the radius is larger. In fact, with r in the denominator on the right of $a_c = v^2/r$, the acceleration approaches zero when the radius becomes very large. Uniform circular motion along the arc of an infinitely large circle entails no acceleration, because it is just like motion at a constant speed along a straight line.

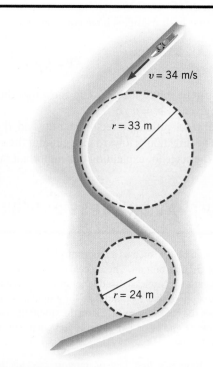

INTERACTIVE FIGURE 5.5   This bobsled travels at the same speed around two curves with different radii. For the turn with the larger radius, the sled has a smaller centripetal acceleration.

In Section 4.11 we learned that an object is in equilibrium when it has zero acceleration. Conceptual Example 4 discusses whether an object undergoing uniform circular motion can ever be at equilibrium.

CONCEPTUAL EXAMPLE 4 | Uniform Circular Motion and Equilibrium

A car moves at a constant speed along a straight line as it approaches a circular turn. In which of the following parts of the motion is the car in equilibrium? **(a)** As it moves along the straight line toward the circular turn, **(b)** as it is going around the turn, **(c)** as it moves away from the turn along a straight line.

Reasoning   An object is in equilibrium when it has no acceleration, according to the definition given in Section 4.11. If the object's velocity remains constant, both in magnitude and direction, its acceleration is zero.

Answer (b) is incorrect.   As the car goes around the turn, the direction of travel changes, so the car has a centripetal acceleration that is characteristic of uniform circular motion. Because of this acceleration, the car is not in equilibrium during the turn.

Answers (a) and (c) are correct.   As the car either approaches the turn or moves away from the turn it is traveling along a straight line, and both the speed and direction of the motion are constant. Thus, the velocity vector does not change, and there is no acceleration. Consequently, for these parts of the motion, the car is in equilibrium.

Related Homework: Problem 10

We have seen that going around tight turns (smaller r) and gentle turns (larger r) at the same speed entails different centripetal accelerations. And most drivers know that such turns "feel" different. This feeling is associated with the force that is present in uniform circular motion, and we turn to this topic in the next section.

Check Your Understanding

(*The answers are given at the end of the book.*)

1. The car in **CYU Figure 5.1** is moving clockwise around a circular section of road at a constant speed. What are the directions of its velocity and acceleration at **(a)** position 1 and **(b)** position 2? Specify your responses as north, east, south, or west.

2. The speedometer of your car shows that you are traveling at a constant speed of 35 m/s. Is it possible that your car is accelerating?

3. Consider two people, one on the earth's surface at the equator and the other at the north pole. Which has the larger centripetal acceleration?

4. Which of the following statements about centripetal acceleration is true? **(a)** An object moving at a constant velocity cannot have a centripetal acceleration. **(b)** An object moving at a constant speed may have a centripetal acceleration.

5. A car is traveling at a constant speed along the road *ABCDE* shown in **CYU Figure 5.2**. Sections *AB* and *DE* are straight. Rank the accelerations in each of the four sections according to magnitude, listing the smallest first.

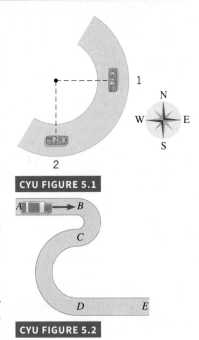

CYU FIGURE 5.1

CYU FIGURE 5.2

5.3 Centripetal Force

Newton's second law indicates that whenever an object accelerates, there must be a net force to create the acceleration. Thus, in uniform circular motion there must be a net force to produce the centripetal acceleration. The second law gives this net force as the product of the object's mass m and its acceleration v^2/r. The net force causing the centripetal acceleration is called the **centripetal force $\vec{F}_c$** and points in the same direction as the acceleration—that is, toward the center of the circle.

> **CENTRIPETAL FORCE**
>
> *Magnitude:* **The centripetal force is the name given to the net force required to keep an object of mass m, moving at a speed v, on a circular path of radius r, and it has a magnitude of**
>
> $$F_c = \frac{mv^2}{r} \tag{5.3}$$
>
> *Direction:* **The centripetal force always points toward the center of the circle and continually changes direction as the object moves.**

The phrase "centripetal force" does not denote a new and separate force created by nature. The phrase merely labels the net force pointing toward the center of the circular path, and this net force is the vector sum of all the force components that point along the radial direction. Sometimes the centripetal force consists of a single force such as tension (see Example 5), friction (see Example 7), the normal force or a component thereof (see Examples 9 and 14), or the gravitational force (see Examples 10–12). However, there are circumstances when a number of different forces contribute simultaneously to the centripetal force (see Section 5.7).

In some cases, it is easy to identify the source of the centripetal force, as when a model airplane on a guideline flies in a horizontal circle. The only force pulling the plane inward is the tension in the line, so this force alone (or a component of it) is the centripetal force. Example 5 illustrates the fact that higher speeds require greater tensions.

EXAMPLE 5 | The Effect of Speed on Centripetal Force

The model airplane in **Figure 5.6** has a mass of 0.90 kg and moves at a constant speed on a circle that is parallel to the ground. The path of the airplane and its guideline lie in the same horizontal plane, because the weight of the plane is balanced by the lift generated by its wings. Find the tension in the guideline (length = 17 m) for speeds of 19 and 38 m/s.

Reasoning Since the plane flies on a circular path, it experiences a centripetal acceleration that is directed toward the center of the circle. According to Newton's second law of motion, this acceleration is produced by a net force that acts on the plane, and this net force is called the centripetal force. The centripetal force is also directed toward the center of the circle. Since the tension in the guideline is the only force pulling the plane inward, it must be the centripetal force.

Solution Equation 5.3 gives the tension directly: $F_c = T = mv^2/r$.

$$\text{Speed} = 19 \text{ m/s} \quad T = \frac{(0.90 \text{ kg})(19 \text{ m/s})^2}{17 \text{ m}} = \boxed{19 \text{ N}}$$

$$\text{Speed} = 38 \text{ m/s} \quad T = \frac{(0.90 \text{ kg})(38 \text{ m/s})^2}{17 \text{ m}} = \boxed{76 \text{ N}}$$

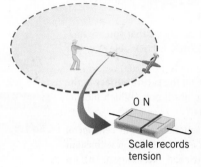

FIGURE 5.6 The scale records the tension in the guideline. See Example 5.

Scale records tension

Conceptual Example 6 deals with another case where it is easy to identify the source of the centripetal force.

CONCEPTUAL EXAMPLE 6 | The Physics of a Trapeze Act

In a circus, a man hangs upside down from a trapeze, legs bent over the bar and arms downward, holding his partner (see **Figure 5.7**). Is it harder for the man to hold his partner **(a)** when the partner hangs straight down and is stationary or **(b)** when the partner is swinging through the straight-down position?

Reasoning Whenever an object moves on a circular path, it experiences a centripetal acceleration that is directed toward the center of the path. A net force, known as the centripetal force, is required to produce this acceleration.

Answer (a) is incorrect. When the man and his partner are stationary, the man's arms must support only his partner's weight. When the two are swinging, however, the man's arms must provide the additional force required to produce the partner's centripetal acceleration. Thus, it is easier, not harder, for the man to hold his partner when the partner hangs straight down and is stationary.

Answer (b) is correct. When the two are swinging, the partner is moving on a circular arc and, therefore, has a centripetal acceleration. The man's arms must support the partner's weight and must

simultaneously exert an additional pull to provide the centripetal force that produces this acceleration. Because of the additional pull, it is harder for the man to hold his partner while swinging.

Related Homework: Problems 14, 16

INTERCHROME/SuperStock

FIGURE 5.7 See Example 6 for a discussion of the role of centripetal force in this trapeze act.

FIGURE 5.8 When the car moves without skidding around a curve, static friction between the road and the tires provides the centripetal force to keep the car on the road.

Static frictional force

When a car moves at a steady speed around an unbanked curve, the centripetal force keeping the car on the curve comes from the static friction between the road and the tires, as **Figure 5.8** indicates. It is static, rather than kinetic friction, because the tires are not slipping with respect to the radial direction. If the static frictional force is insufficient, given the speed and the radius of the turn, the car will skid off the road. Example 7 shows how an icy road can limit safe driving.

Analyzing Multiple-Concept Problems

EXAMPLE 7 | Centripetal Force and Safe Driving

At what maximum speed can a car safely negotiate a horizontal unbanked turn (radius = 51 m) in dry weather (coefficient of static friction = 0.95) and icy weather (coefficient of static friction = 0.10)?

Reasoning   The speed v at which the car of mass m can negotiate a turn of radius r is related to the centripetal force F_c that is available, according to $F_c = mv^2/r$ (Equation 5.3). Static friction provides the centripetal force and can provide a maximum force f_s^{MAX} given by $f_s^{MAX} = \mu_s F_N$ (Equation 4.7), in which μ_s is the coefficient of static friction and F_N is the normal force. Thus, we will use the fact that $F_c = f_s^{MAX}$ to obtain the maximum speed from Equation 5.3. It will also be necessary to evaluate the normal force, which we will do by considering that it must balance the weight of the car. Experience indicates that the maximum speed should be greater for the dry road than for the icy road.

Knowns and Unknowns   The data for this problem are as follows:

Description	Symbol	Value	Comment
Radius of turn	r	51 m	
Coefficient of static friction	μ_s	0.95	Dry conditions
Coefficient of static friction	μ_s	0.10	Icy conditions
Unknown Variable			
Speed of car	v	?	

Modeling the Problem

STEP 1 Speed and Centripetal Force   According to Equation 5.3, we have

$$F_c = \frac{mv^2}{r}$$

Solving for the speed v gives Equation 1 at the right. To use this result, it is necessary to consider the centripetal force F_c, which we do in Step 2.

$$v = \sqrt{\frac{rF_c}{m}} \qquad (1)$$

$\boxed{?}$

STEP 2 Static Friction   The force of static friction is the centripetal force, and the greater its value, the greater is the speed at which the car can negotiate the turn. The maximum force f_s^{MAX} of static friction is given by $f_s^{MAX} = \mu_s F_N$ (Equation 4.7). Thus, the maximum available centripetal force is

$$\boxed{F_c = f_s^{MAX} = \mu_s F_N}$$

which we can substitute into Equation 1, as shown at the right. The next step considers the normal force F_N.

$$v = \sqrt{\frac{rF_c}{m}} \qquad (1)$$

$$\boxed{F_c = \mu_s F_N} \qquad (2)$$

$\boxed{?}$

STEP 3 The Normal Force   The fact that the car does not accelerate in the vertical direction tells us that the normal force balances the car's weight mg, or

$$\boxed{F_N = mg}$$

This result for the normal force can now be substituted into Equation 2, as indicated at the right.

$$v = \sqrt{\frac{rF_c}{m}} \qquad (1)$$

$$\boxed{F_c = \mu_s F_N} \qquad (2)$$

$$\boxed{F_N = mg}$$

Solution   Algebraically combining the results of each step, we find that

$$\overset{\text{STEP 1}}{\downarrow} \quad \overset{\text{STEP 2}}{\downarrow} \quad \overset{\text{STEP 3}}{\downarrow}$$

$$v = \sqrt{\frac{rF_c}{m}} = \sqrt{\frac{r\mu_s F_N}{m}} = \sqrt{\frac{r\mu_s mg}{m}}$$

The mass m of the car is eliminated algebraically from this result, and we find that the maximum speeds are

Dry road ($\mu_s = 0.95$) $v = \sqrt{r\mu_s g} = \sqrt{(51\text{ m})(0.95)(9.8\text{ m/s}^2)} = \boxed{22\text{ m/s}}$

Icy road ($\mu_s = 0.10$) $v = \sqrt{r\mu_s g} = \sqrt{(51\text{ m})(0.10)(9.8\text{ m/s}^2)} = \boxed{7.1\text{ m/s}}$

As expected, the dry road allows a greater maximum speed.

Problem-Solving Insight
When using an equation to obtain a numerical answer, algebraically solve for the unknown variable in terms of the known variables. Then substitute in the numbers for the known variables, as this example shows.

Related Homework: Problems 12, 17, 22

A passenger in **Figure 5.8** must also experience a centripetal force to remain on the circular path. However, if the upholstery is very slippery, there may not be enough static friction to keep him in place as the car makes a tight turn at high speed. Then, when viewed from inside the car, he appears to be thrown toward the outside of the curve. What really happens is that the passenger slides off on a tangent to the circle, until he encounters a source of centripetal force to keep him in place while the car turns. This occurs when the passenger bumps into the side of the car, which pushes on him with the necessary force. While rounding a curve in a car can be exciting, moving on a circular path at high speed can have significant physiological effects, as the next example illustrates.

THE PHYSICS OF . . . flying an airplane in a banked turn.   Sometimes the source of the centripetal force is not obvious. A pilot making a turn, for instance, banks or tilts the plane at an angle to create the centripetal force. As a plane flies, the air pushes upward on the wing surfaces with a net lifting force $\vec{L}$ that is perpendicular to the wing surfaces, as **Animated Figure 5.9a** shows. Part *b* of the drawing illustrates that when the plane is banked at an angle θ, a component $L \sin \theta$ of the lifting force is directed toward the center of the turn. It is this component that provides the centripetal force. Greater speeds and/or tighter turns require greater centripetal forces. In such situations, the pilot must bank the plane at a larger angle, so that a larger component of the lift points toward the center of the turn. The technique of banking into a turn also has an application in the construction of high-speed roadways, where the road itself is banked to achieve a similar effect, as the next section discusses.

ANIMATED FIGURE 5.9 (*a*) The air exerts an upward lifting force $\frac{1}{2}\vec{L}$ on each wing. (*b*) When a plane executes a circular turn, the plane banks at an angle θ. The lift component $L \sin \theta$ is directed toward the center of the circle and provides the centripetal force.

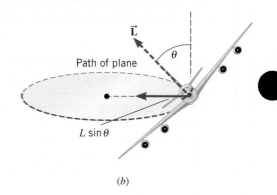

(a)

(b)

EXAMPLE 8 | BIO The Physics of *g*-LOC

The National Aeronautics and Space Administration (NASA) studies the physiological effects of large accelerations on astronauts. Some of these studies use a machine known as a centrifuge. This machine consists of a long arm, rotating about an axis through its center. At the two ends of the arm, and other distances from the center, are chambers in which the astronauts sit. The astronauts move on a circular path, much like a model airplane flying in a circle on a guideline (**Figure 5.10**). One setup inside the chamber is a seat that is free to pivot, so that when the centrifuge is rotating, the astronaut's head will point toward the center of the circular motion. The normal force from the astronaut's chair provides the centripetal force. The greater the astronaut's speed, the greater the normal force needs to be. The normal force is responsible for the astronaut's apparent weight. As the speed increases, the astronaut will feel heavier. This is referred to as "pulling g's", where *g* represents the acceleration due to gravity near earth's surface. If the centripetal acceleration is 1*g*, the astronaut will feel her normal weight. If it is 2*g*, she will feel twice her normal weight, and so on. One side effect of pulling a high number of *g*'s is the potential loss of consciousness known as *g*-LOC. This occurs because the fluids, primarily the blood, in an astronaut's body can move around. Just like a spinning washing machine drains the water from wet clothes, the astronaut's blood drains away from her head, which causes the blood to collect in her lower extremities, such as the legs and feet. The heart has to work against this motion, and at some point, is unable to maintain sufficient blood flow, and thus oxygen to the brain, which results in the astronaut blacking out or losing consciousness. The centrifuge helps astronauts and fighter jet pilots to train in techniques to prevent *g*-LOC. Other techniques to combat *g*-LOC include the use of special suits called anti-*g* suits (see Example 15 and Problem 52). If an astronaut in the centrifuge is rotating at 22.5 rpm, how heavy does she feel, compared to her real weight? The radius of the centrifuge is 8.84 m.

Reasoning   How heavy the astronaut feels will be determined by the magnitude of the normal force from the chair. The normal force also provides the centripetal force in this problem, so we can apply Equation 5.3. This requires us to know the speed of the astronaut, which we can determine from the period of the circular motion and Equation 5.1.

(a)

(b)

FIGURE 5.10 (a) The 20-g centrifuge at the NASA Ames Research Center. The arrows in the picture show the locations of chambers, called cabs, where the astronaut can sit. (b) Orientation of the astronaut while the centrifuge is rotating. Notice that the normal force from the chair that acts on the astronaut points inward, toward the center of the circular motion.

Solution We begin by converting the rotation rate to the period:

$$22.5 \text{ rpm} = \frac{22.5 \text{ revolutions}}{\text{min}} \times \frac{1 \text{ min}}{60 \text{ s}} = 0.375 \text{ rev/s}$$

The period (T) is simply the reciprocal of this value: $T = \frac{1}{0.375 \text{ rev/s}} = 2.67$ s. **Figure 5.10b** shows the normal force acting on the astronaut is equal to the centripetal force: $F_N = F_C = \frac{mv^2}{r}$, by applying Equation 5.3. We can write the speed (v) in terms of the period by using Equation 5.1: $v = \frac{2\pi r}{T}$. Substituting this into the expression for F_N, we find

$$F_N = \frac{4m\pi^2 r^2}{r T^2} = \frac{4m\pi^2 r}{T^2} = m\left(\frac{4\pi^2 r}{T^2}\right).$$

Since F_N determines the astronaut's apparent weight, the term in parentheses represents "g effective." Comparing this to $g = 9.80$ m/s², we have

$$\text{"g effective"} = \frac{4\pi^2 r}{T^2} = \frac{4\pi^2 (8.84 \text{ m})}{(2.67 \text{ s})^2} = 48.9 \text{ m/s}^2 = \boxed{5g}$$

Thus, the astronaut will feel as if she weighs five times her true weight. We would say the astronaut is pulling 5 g's.

Related Homework: Problems 51, 52, and 53

Check Your Understanding

(The answers are given at the end of the book.)

6. A car is traveling in uniform circular motion on a section of road whose radius is *r* (see **CYU Figure 5.3**). The road is slippery, and the car is just on the verge of sliding. **(a)** If the car's speed were doubled, what would be the smallest radius at which the car does not slide? Express your answer in terms of *r*. **(b)** What would be your answer to part (a) if the car were replaced by one that weighed twice as much, the car's speed still being doubled?

CYU FIGURE 5.3

7. Other things being equal, would it be easier to drive at high speed around an unbanked horizontal curve on the moon than to drive around the same curve on the earth?

8. What is the chance of a light car safely rounding an unbanked curve on an icy road as compared to that of a heavy car: worse, the same, or better? Assume that both cars have the same speed and are equipped with identical tires.

9. A penny is placed on a rotating turntable. Where on the turntable does the penny require the largest centripetal force to remain in place: at the center of the turntable or at the edge of the turntable?

5.4 | Banked Curves

When a car travels without skidding around an unbanked curve, the static frictional force between the tires and the road provides the centripetal force. The reliance on friction can be eliminated completely for a given speed, however, if the curve is banked at an angle relative to the horizontal, much in the same way that a plane is banked while making a turn.

Interactive Figure 5.11a shows a car going around a friction-free banked curve. The radius of the curve is r, where r is measured parallel to the horizontal and not to the slanted surface. Part b shows the normal force $\vec{F}_N$ that the road applies to the car, the normal force being perpendicular to the road. Because the roadbed makes an angle θ with respect to the horizontal, the normal force has a component $F_N \sin \theta$ that points toward the center C of the circle and provides the centripetal force:

$$F_c = F_N \sin \theta = \frac{mv^2}{r}$$

The vertical component of the normal force is $F_N \cos \theta$ and, since the car does not accelerate in the vertical direction, this component must balance the weight mg of the car. Therefore, $F_N \cos \theta = mg$. Dividing this equation into the previous one shows that

$$\frac{F_N \sin \theta}{F_N \cos \theta} = \frac{mv^2/r}{mg}$$

$$\tan \theta = \frac{v^2}{rg} \qquad (5.4)$$

Equation 5.4 indicates that, for a given speed v, the centripetal force needed for a turn of radius r can be obtained from the normal force by banking the turn at an angle θ, independent of the mass of the vehicle. Greater speeds and smaller radii require more steeply banked curves—that is, larger values of θ. At a speed that is too small for a given θ, a car would slide down a frictionless banked curve; at a speed that is too large, a car would slide off the top. The next example deals with a famous banked curve.

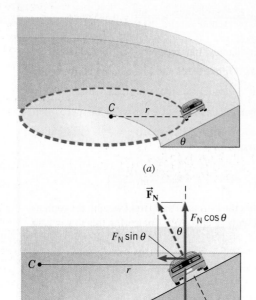

(a)

(b)

INTERACTIVE FIGURE 5.11 (a) A car travels on a circle of radius r on a frictionless banked road. The banking angle is θ, and the center of the circle is at C. (b) The forces acting on the car are its weight $m\vec{g}$ and the normal force $\vec{F}_N$. A component $F_N \sin \theta$ of the normal force provides the centripetal force.

EXAMPLE 9 | The Physics of the Daytona International Speedway

The Daytona 500 is the major event of the NASCAR (National Association for Stock Car Auto Racing) season. It is held at the Daytona International Speedway in Daytona, Florida. The turns in this oval track have a maximum radius (at the top) of $r = 316$ m and are banked steeply, with $\theta = 31°$ (see **Interactive Figure 5.11**). Suppose these maximum-radius turns were frictionless. At what speed would the cars have to travel around them?

Reasoning In the absence of friction, the horizontal component of the normal force that the track exerts on the car must provide the centripetal force. Therefore, the speed of the car is given by Equation 5.4.

Solution From Equation 5.4, it follows that

$$v = \sqrt{rg \tan \theta} = \sqrt{(316 \text{ m})(9.80 \text{ m/s}^2) \tan 31°}$$

$$= \boxed{43 \text{ m/s } (96 \text{ mph})}$$

Drivers actually negotiate the turns at speeds up to 195 mph, however, which requires a greater centripetal force than that implied by Equation 5.4 for frictionless turns. Static friction provides the additional force.

Check Your Understanding

(The answer is given at the end of the book.)

10. Go to **Concept Simulation 5.2** at **www.wiley.com/college/cutnell** to review the concepts involved in this question. Two cars are identical, except for the type of tread design on their tires. The cars are driven at the same speed and enter the same unbanked horizontal turn. Car A cannot negotiate the turn, but car B can. Which tread design, the one on car A or the one on car B, yields a larger coefficient of static friction between the tires and the road?

5.5 | Satellites in Circular Orbits

Today there are many satellites in orbit about the earth. The ones in circular orbits are examples of uniform circular motion. Like a model airplane on a guideline, each satellite is kept on its circular path by a centripetal force. The gravitational pull of the earth provides the centripetal force and acts like an invisible guideline for the satellite.

There is only one speed that a satellite can have if the satellite is to remain in an orbit with a fixed radius. To see how this fundamental characteristic arises, consider the gravitational force acting on the satellite of mass m in **Figure 5.12**. Since the gravitational force is the only force acting on the satellite in the radial direction, it alone provides the centripetal force. Therefore, using Newton's law of gravitation (Equation 4.3), we have

$$F_c = G\frac{mM_E}{r^2} = \frac{mv^2}{r}$$

where G is the universal gravitational constant, M_E is the mass of the earth, and r is the distance from the center of the earth to the satellite. Solving for the speed v of the satellite gives

$$v = \sqrt{\frac{GM_E}{r}} \tag{5.5}$$

If the satellite is to remain in an orbit of radius r, the speed must have precisely this value. Note that the radius r of the orbit is in the denominator in Equation 5.5. This means that the closer the satellite is to the earth, the smaller is the value for r and the greater the orbital speed must be.

The mass m of the satellite does not appear in Equation 5.5, having been eliminated algebraically. *Consequently, for a given orbit, a satellite with a large mass has exactly the same orbital speed as a satellite with a small mass.* However, more effort is certainly required to lift the larger-mass satellite into orbit. The orbital speed of one famous artificial satellite is determined in the following example.

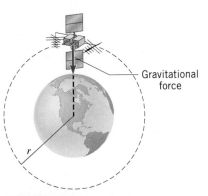

FIGURE 5.12 For a satellite in circular orbit around the earth, the gravitational force provides the centripetal force.

EXAMPLE 10 | The Physics of the Hubble Space Telescope

Determine the speed of the Hubble Space Telescope (see **Figure 5.13**) orbiting at a height of 598 km above the earth's surface.

Reasoning Before Equation 5.5 can be applied, the orbital radius r must be determined *relative to the center of the earth*. Since the radius of the earth is approximately 6.38×10^6 m, and the height of the telescope above the earth's surface is 0.598×10^6 m, the orbital radius is $r = 6.98 \times 10^6$ m.

> **Problem-Solving Insight** The orbital radius r that appears in the relation $v = \sqrt{GM_E/r}$ is the distance from the satellite to the center of the earth (not to the surface of the earth).

Solution The orbital speed is

$$v = \sqrt{\frac{GM_E}{r}} = \sqrt{\frac{(6.67 \times 10^{-11}\ \text{N} \cdot \text{m}^2/\text{kg}^2)(5.98 \times 10^{24}\ \text{kg})}{6.98 \times 10^6\ \text{m}}}$$

$$v = \boxed{7.56 \times 10^3\ \text{m/s}\ (16\,900\ \text{mi/h})}$$

FIGURE 5.13 The Hubble Space Telescope orbiting the earth.

Frank Whitney/The Image Bank/Getty Images

FIGURE 5.14 The Global Positioning System (GPS) of satellites can be used with a GPS receiver to locate an object, such as a car, on the earth. (*a*) One satellite identifies the car as being somewhere on a circle. (*b*) A second places it on another circle, which identifies two possibilities for the exact spot. (*c*) A third provides the means for deciding where the car is.

THE PHYSICS OF . . . the Global Positioning System. Many applications of satellite technology affect our lives. An increasingly important application is the network of 24 satellites called the Global Positioning System (GPS), which can be used to determine the position of an object to within 5 m or less. **Figure 5.14** illustrates how the system works. Each GPS satellite carries a highly accurate atomic clock, whose time is transmitted to the ground continually by means of radio waves. In the drawing, a car carries a computerized GPS receiver that can detect the waves and is synchronized to the satellite clock. The receiver can therefore determine the distance between the car and a satellite from a knowledge of the travel time of the waves and the speed at which they move. This speed, as we will see in Chapter 24, is the speed of light and is known with great precision. A measurement using a single satellite locates the car somewhere on a circle, as **Figure 5.14a** shows, while a measurement using a second satellite locates the car on another circle. The intersection of the circles reveals two possible positions for the car, as in **Figure 5.14b**. With the aid of a third satellite, a third circle can be established, which intersects the other two and identifies the car's exact position, as in **Figure 5.14c**. The use of ground-based radio beacons to provide additional reference points leads to a system called Differential GPS, which can locate objects even more accurately than the satellite-based system alone. Navigational systems for automobiles and portable systems that tell hikers and people with visual impairments where they are located are two of the many uses for the GPS technique. GPS applications are so numerous that they have developed into a multibillion dollar industry.

Equation 5.5 applies to human-made satellites or to natural satellites like the moon. It also applies to circular orbits about any astronomical object, provided M_E is replaced by the mass of the object on which the orbit is centered. Example 11, for instance, shows how scientists have applied this equation to conclude that a supermassive black hole is probably located at the center of the galaxy known as M87. This galaxy is located at a distance of about 50 million light-years away from the earth. (One light-year is the distance that light travels in a year, or 9.5×10^{15} m.)

EXAMPLE 11 | The Physics of Locating a Black Hole

The Hubble Telescope has detected the light being emitted from different regions of galaxy M87, which is shown in **Figure 5.15**. The black circle identifies the center of the galaxy. From the characteristics of this light, astronomers have determined that the orbiting speed is 7.5×10^5 m/s for matter located at a distance of 5.7×10^{17} m from the center. Find the mass M of the object located at the galactic center.

Reasoning and Solution   Replacing M_E in Equation 5.5 with M gives $v = \sqrt{GM/r}$, which can be solved to show that

$$M = \frac{v^2 r}{G} = \frac{(7.5 \times 10^5 \text{ m/s})^2 (5.7 \times 10^{17} \text{ m})}{6.67 \times 10^{-11} \text{ N} \cdot \text{m}^2/\text{kg}^2}$$

$$= \boxed{4.8 \times 10^{39} \text{ kg}}$$

Since the mass of our sun is 2.0×10^{30} kg, matter equivalent to 2.4 billion suns is located at the center of galaxy M87. The volume of space in which this matter is located contains relatively few visible stars, so researchers believe that the data provide strong evidence for the existence of a supermassive black hole. The term "black hole" is used because the tremendous mass prevents even light from escaping. The light that forms the image in **Figure 5.15** comes not from the black hole itself, but from matter that surrounds it.*

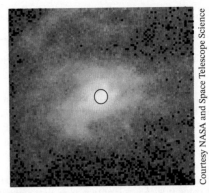

FIGURE 5.15   This image of the ionized gas (yellow) at the heart of galaxy M87 was obtained by the Hubble Space Telescope. The circle identifies the center of the galaxy, at which a black hole is thought to exist.

Courtesy NASA and Space Telescope Science Institute

The period T of a satellite is the time required for one orbital revolution. As in any uniform circular motion, the period is related to the speed of the motion by $v = 2\pi r/T$. Substituting v from Equation 5.5 shows that

$$\sqrt{\frac{GM_E}{r}} = \frac{2\pi r}{T}$$

Solving this expression for the period T gives

$$T = \frac{2\pi r^{3/2}}{\sqrt{GM_E}} \qquad\qquad (5.6)$$

Although derived for earth orbits, Equation 5.6 can also be used for calculating the periods of those planets in nearly circular orbits about the sun, if M_E is replaced by the mass M_S of the sun and r is interpreted as the distance between the center of the planet and the center of the sun. The fact that the period is proportional to the three-halves power of the orbital radius is known as Kepler's third law, and it is one of the laws discovered by Johannes Kepler (1571–1630) during his studies of planetary motion. Kepler's third law also holds for elliptical orbits, which will be discussed in Chapter 9.

An important application of Equation 5.6 occurs in the field of communications, where "synchronous satellites" are put into a circular orbit that is in the plane of the equator, as **Figure 5.16** shows. The orbital period is chosen to be one day, which is also the time it takes for the earth to turn once about its axis. Therefore, these satellites move around their orbits in a way that is synchronized with the rotation of the earth. For earth-based observers, synchronous satellites have the useful characteristic of appearing in fixed positions in the sky and can serve as "stationary" relay stations for communication signals sent up from the earth's surface.

THE PHYSICS OF . . . digital satellite system TV.   This is exactly what is done in the digital satellite systems that are a popular alternative to cable TV. As the blowup in **Figure 5.16** indicates, a small "dish" antenna on your house picks up the digital TV signals relayed back to earth by the satellite. After being decoded, these signals are delivered to your TV set. All synchronous satellites are in orbit at the same height above the earth's surface, as Example 12 shows.

Small dish antenna

TV signals from synchronous satellite

Equator

FIGURE 5.16   A synchronous satellite orbits the earth once per day on a circular path in the plane of the equator. Digital satellite system television uses such satellites as relay stations. TV signals are sent up from the earth's surface and then rebroadcast down to your own small dish antenna.

---

*More complete analysis shows that the mass of the black hole is equivalent to 6.6 billion suns (K. Gebhardt et al., *The Astrophysical Journal*, Vol. 729, 2011, p. 119).

EXAMPLE 12 | The Orbital Radius for Synchronous Satellites

What is the height H above the earth's surface at which all synchronous satellites (regardless of mass) must be placed in orbit?

Reasoning The period T of a synchronous satellite is one day, so we can use Equation 5.6 to find the distance r from the satellite to the center of the earth. To find the height H of the satellite above the earth's surface we will have to take into account the fact that the earth itself has a radius of 6.38×10^6 m.

Solution A period of one day* corresponds to $T = 8.64 \times 10^4$ s. In using this value it is convenient to rearrange the equation $T = 2\pi r^{3/2}/\sqrt{GM_E}$ as follows:

$$r^{3/2} = \frac{T\sqrt{GM_E}}{2\pi}$$

$$= \frac{(8.64 \times 10^4 \text{ s})\sqrt{(6.67 \times 10^{-11} \text{ N} \cdot \text{m}^2/\text{kg}^2)(5.98 \times 10^{24} \text{ kg})}}{2\pi}$$

By squaring and then taking the cube root, we find that $r = 4.23 \times 10^7$ m. Since the radius of the earth is approximately 6.38×10^6 m, the height of the satellite above the earth's surface is

$$H = 4.23 \times 10^7 \text{ m} - 0.64 \times 10^7 \text{ m} = \boxed{3.59 \times 10^7 \text{ m (22 300 mi)}}$$

Math Skills In solving $T = \dfrac{2\pi r^{3/2}}{\sqrt{GM_E}}$ (Equation 5.6) for r, we remember that whatever is done to one side of the equals sign must also be done to the other side. To isolate the term $r^{3/2}$ on one side we multiply both sides by $\sqrt{GM_E}$ and divide both sides by 2π:

$$T\frac{\sqrt{GM_E}}{2\pi} = \frac{2\pi r^{3/2}}{\sqrt{GM_E}}\frac{\sqrt{GM_E}}{2\pi} \quad \text{or} \quad r^{3/2} = \frac{T\sqrt{GM_E}}{2\pi}$$

Next, we square both sides of the result for $r^{3/2}$:

$$(r^{3/2})^2 = \left(\frac{T\sqrt{GM_E}}{2\pi}\right)^2 \quad \text{or} \quad r^3 = \frac{T^2 GM_E}{4\pi^2}$$

Finally, taking the cube root of the result for r^3 gives

$$\sqrt[3]{r^3} = \sqrt[3]{\frac{T^2 GM_E}{4\pi^2}} \quad \text{or} \quad (r^3)^{1/3} = \left(\frac{T^2 GM_E}{4\pi^2}\right)^{1/3} \quad \text{or}$$

$$r = \frac{T^{2/3} G^{1/3} M_E^{1/3}}{4^{1/3}\pi^{2/3}}$$

Using values for T, G, and M_E in the result for r shows that $r = 4.23 \times 10^7$ m.

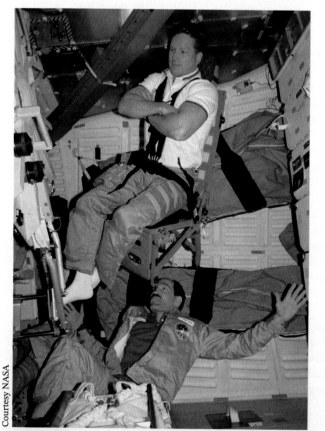

Courtesy NASA

FIGURE 5.17 In a state of apparent weightlessness in orbit, astronaut and pilot Guy S. Gardner and mission specialist William M. Shepherd (in chair) float, with Gardner appearing to "balance" the chair on his nose.

Check Your Understanding

(The answer is given at the end of the book.)

11. Two satellites are placed in orbit, one about Mars and the other about Jupiter, such that the orbital speeds are the same. Mars has the smaller mass. Is the radius of the satellite in orbit about Mars less than, greater than, or equal to the radius of the satellite orbiting Jupiter?

5.6 | Apparent Weightlessness and Artificial Gravity

The idea of life on board an orbiting satellite conjures up visions of astronauts floating around in a state of "weightlessness," as in **Figure 5.17**. Actually, this state should be called "apparent weightlessness," because it is similar to the condition of zero apparent weight that occurs in an elevator during free-fall. Conceptual Example 13 explores this similarity.

*Successive appearances of the sun define the solar day of 24 h or 8.64×10^4 s. The sun moves against the background of the stars, however, and the time required for the earth to turn once on its axis relative to the fixed stars is 23 h 56 min, which is called the sidereal day. The sidereal day should be used in Example 12, but the neglect of this effect introduces an error of less than 0.4% in the answer.

CONCEPTUAL EXAMPLE 13 | The Physics of Apparent Weightlessness

Figure 5.18 shows a person on a scale in a freely falling elevator and in a satellite in a circular orbit. Assume that when the person is standing stationary on the earth, his weight is 800 N (180 lb). In each case, what apparent weight is recorded by the scale? **(a)** The scale in the elevator records 800 N while that in the satellite records 0 N. **(b)** The scale in the elevator records 0 N while that in the satellite records 800 N. **(c)** Both scales record 0 N.

Reasoning As Section 4.8 discusses, apparent weight is the force that an object exerts on the platform of a scale. This force depends on whether or not the object and the platform are accelerating together.

Answer (a) is incorrect. The scale and the person in the free-falling elevator are accelerating toward the earth at the same rate. Therefore, they cannot push against one another, and so the scale does not record an apparent weight of 800 N.

Answer (b) is incorrect. The scale and the person in the satellite are accelerating toward the center of the earth at the same rate (they have the same centripetal acceleration). Therefore, they cannot push against one another, and so the scale does not record an apparent weight of 800 N.

Answer (c) is correct. The scale and the person in the elevator fall together and, therefore, they cannot push against one another. Therefore, the scale records an apparent weight of 0 N. In the orbiting satellite in **Figure 5.18b**, both the person and the scale are in uniform circular motion. Objects in uniform circular motion continually accelerate or "fall" toward the center of the circle in order to remain on the circular path. Consequently, both the person and the scale "fall" with the same acceleration toward the center of the earth and cannot push against one another. Thus, the apparent weight in the satellite is zero, just as it is in the freely falling elevator.

Orbit

Free-fall
(a)

Earth

Free-fall
(b)

FIGURE 5.18 (a) During free-fall, the elevator accelerates downward with the acceleration due to gravity, and the apparent weight of the person is zero. (b) The orbiting space station is also in free-fall toward the center of the earth.

THE PHYSICS OF . . . artificial gravity. The physiological effects of prolonged apparent weightlessness are only partially known. To minimize such effects, it is likely that artificial gravity will be provided in large space stations of the future. To help explain artificial gravity, **Figure 5.19** shows a space station rotating about an axis. Because of the rotational motion, any object located at a point P on the interior surface of the station experiences a centripetal force directed toward the axis. The surface of the station provides this force by pushing on the feet of an astronaut, for instance. The centripetal force can be adjusted to match the astronaut's earth-weight by properly selecting the rotational speed of the space station, as Example 14 illustrates.

FIGURE 5.19 The surface of the rotating space station pushes on an object with which it is in contact and thereby provides the centripetal force that keeps the object moving on a circular path.

Analyzing Multiple-Concept Problems

EXAMPLE 14 | Artificial Gravity

At what speed must the interior surface of the space station ($r = 1700$ m) move in Figure 5.19, so that the astronaut located at point P experiences a push on his feet that equals his weight on earth?

Reasoning The floor of the rotating space station exerts a normal force on the feet of the astronaut. This is the centripetal force ($F_c = mv^2/r$) that keeps the astronaut moving on a circular path. Since the magnitude of the normal force equals the astronaut's weight on earth, we can determine the speed v of the space station's floor.

Knowns and Unknowns The data for this problem are given in the following table:

Description	Symbol	Value
Radius of space station	r	1700 m
Unknown Variable		
Speed of space station's floor	v	?

Modeling the Problem

STEP 1 Speed and Centripetal Force The centripetal force acting on the astronaut is given by Equation 5.3 as

$$F_c = \frac{mv^2}{r}$$

Solving for the speed v gives Equation 1 at the right. Step 2 considers the centripetal force F_c that appears in this result.

$$v = \sqrt{\frac{rF_c}{m}} \qquad (1)$$

$$\boxed{?}$$

STEP 2 Magnitude of the Centripetal Force The normal force applied to the astronaut's feet by the floor is the centripetal force and has a magnitude equal to the astronaut's earth-weight. This earth-weight is given by Equation 4.5 as the astronaut's mass m times the magnitude g of the acceleration due to the earth's gravity. Thus, we have for the centripetal force that

$$\boxed{F_c = mg}$$

which can be substituted into Equation 1, as indicated at the right.

$$v = \sqrt{\frac{rF_c}{m}} \qquad (1)$$

$$\boxed{F_c = mg}$$

Solution Algebraically combining the results of each step, we find that

$$\overset{\text{STEP 1}}{\underset{\downarrow}{}} \quad \overset{\text{STEP 2}}{\underset{\downarrow}{}}$$
$$v = \sqrt{\frac{rF_c}{m}} = \sqrt{\frac{rmg}{m}}$$

The astronaut's mass m is eliminated algebraically from this result, so the speed of the space station floor is

$$v = \sqrt{rg} = \sqrt{(1700 \text{ m})(9.80 \text{ m/s}^2)} = \boxed{130 \text{ m/s}}$$

Check Your Understanding

(The answer is given at the end of the book.)

12. The acceleration due to gravity on the moon is one-sixth that on earth. **(a)** Is the true weight of a person on the moon less than, greater than, or equal to the true weight of the same person on earth? **(b)** Is the apparent weight of a person in orbit about the moon less than, greater than, or equal to the apparent weight of the same person in orbit about the earth?

5.7 *Vertical Circular Motion

Motorcycle stunt drivers perform a feat in which they drive their cycles around a vertical circular track, as in **Interactive Figure 5.20a**. Usually, the speed varies in this stunt. When the speed of travel on a circular path changes from moment to moment, the motion is said to be nonuniform. Nonetheless, we can use the concepts that apply to uniform circular motion to gain considerable insight into the motion that occurs on a vertical circle.

There are four points on a vertical circle where the centripetal force can be identified easily, as **Interactive Figure 5.20b** indicates. As you look at **Interactive Figure 5.20b**, keep in mind that the centripetal force is not a new and separate force of nature. Instead, at each point the centripetal force is the net sum of all the force components oriented along the radial direction, and it points toward the center of the circle. The drawing shows only the weight of the cycle plus rider (magnitude = mg) and the normal force pushing on the cycle (magnitude = F_N). The propulsion and braking forces are omitted for simplicity, because they do not act in the radial direction. The magnitude of the centripetal force at each of the four points is given as follows in terms of mg and F_N:

(1) $\underbrace{F_{N1} - mg}_{= F_{c1}} = \dfrac{mv_1^2}{r}$

(3) $\underbrace{F_{N3} + mg}_{= F_{c3}} = \dfrac{mv_3^2}{r}$

(2) $\underbrace{F_{N2}}_{= F_{c2}} = \dfrac{mv_2^2}{r}$

(4) $\underbrace{F_{N4}}_{= F_{c4}} = \dfrac{mv_4^2}{r}$

> **Problem-Solving Insight** Centripetal force is the name given to the net force that points toward the center of a circular path. As shown here, there may be more than one force that contributes to this net force.

THE PHYSICS OF . . . the loop-the-loop motorcycle stunt. As the cycle goes around, the magnitude of the normal force changes. It changes because the speed changes and because the weight does not have the same effect at every point. At the bottom, the normal force and the weight oppose one another, giving a centripetal force of magnitude $F_{N1} - mg$. At the top, the normal force and the weight reinforce each other to provide a centripetal force whose magnitude is $F_{N3} + mg$. At points 2 and 4 on either side, only F_{N2} and F_{N4} provide the centripetal force. The weight is tangent to the circle at points 2 and 4 and has no component pointing toward the center. If the speed at each of the four places is known, along with the mass and the radius, the normal forces can be determined.

Riders who perform the loop-the-loop trick must have at least a minimum speed at the top of the circle to remain on the track. This speed can be determined by considering the centripetal force at point 3. The speed v_3 in the equation $F_{N3} + mg = mv_3^2/r$ is a minimum when F_{N3} is zero. Then, the speed is given by $v_3 = \sqrt{rg}$. At this speed, the track does not exert a normal force to keep the cycle on the circle, because the weight mg provides all the centripetal force. Under these conditions, the rider experiences an apparent weightlessness like that discussed in Section 5.6, because for an instant the rider and the cycle are falling freely toward the center of the circle.

(a)

(b)

INTERACTIVE FIGURE 5.20 (a) A vertical loop-the-loop motorcycle stunt. (b) The normal force $\vec{F}_N$ and the weight $m\vec{g}$ of the cycle and the rider are shown here at four locations.

Check Your Understanding

(*The answers are given at the end of the book.*)

13. Would a change in the earth's mass affect **(a)** the banking of airplanes as they turn, **(b)** the banking of roadbeds, **(c)** the speeds with which satellites are put into circular orbits, and **(d)** the performance of the loop-the-loop motorcycle stunt?

14. A stone is tied to a string and whirled around in a circle at a constant speed. Is the string more likely to break when the circle is horizontal or when it is vertical? Assume that the constant speed is the same in each case.

EXAMPLE 15 | BIO The Physics of *g*-Suits

Jet fighter pilots and astronauts often experience large accelerations from centripetal forces during flight. This is commonly referred to as "pulling *g*'s", as mentioned in Example 8. For example, pilots flying along the path of a vertical circle are susceptible to the loss of consciousness (blackout) due to blood flowing from the top of their head and pooling in their lower body. To help prevent this potentially dangerous situation, the pilot wears a device called a *g*-suit. The suit contains air-filled bladders that expand and squeeze on a pilot's legs and abdomen. This helps prevent the blood from draining from the upper part of the body, thereby allowing the pilot to remain conscious during higher levels of acceleration. Consider the following situation shown in **Figure 5.21**, where a fighter pilot is flying upside down along a vertical circle whose radius is 1.2 km. What must be the speed of the plane, so that the normal force acting on the pilot is equal to his weight?

Reasoning The forces acting on the pilot are shown in **Figure 5.21**. Since the jet is undergoing circular motion, we should be able to identify the centripetal force. We can then apply Equation 5.3 to calculate the speed of the plane.

Solution Since both forces on the pilot in **Figure 5.21** point in toward the center of the circular motion, the centripetal force will be equal to the sum of the normal force and the weight: $F_C = F_N + mg$. Equation 5.3 tells us that the centripetal force is equal to mv^2/r. Therefore, we find that $\frac{mv^2}{r} = F_N + mg$. The problem tells us we want the normal force on the pilot to be equal to his weight. Thus, we make that substitution and find the following: $\frac{mv^2}{r} = mg + mg = 2mg$. Notice that the mass of the pilot drops out, and we can solve for the speed of the plane:

$$v = \sqrt{2rg} = \sqrt{2(1200 \text{ m})(9.8 \text{ m/s}^2)} = \boxed{150 \text{ m/s}}$$

In addition to the importance of *g*-suits in this type of maneuver, this particular combination of forces can be very alarming for young pilots. Since the normal force on the pilot is equal to his weight, he feels as if he is sitting right side up, when in fact, he is upside down. This can lead to spatial disorientation.

FIGURE 5.21 The normal force and weight of a pilot both point down toward the center of the circular motion while flying upside down at the top of a vertical circle. The speed of the plane will change the magnitude of the normal force (i.e., how much the seat pushes against the pilot).

Concept Summary

5.1 Uniform Circular Motion Uniform circular motion is the motion of an object traveling at a constant (uniform) speed on a circular path. The period T is the time required for the object to travel once around the circle. The speed v of the object is related to the period and the radius r of the circle by Equation 5.1.

$$v = \frac{2\pi r}{T} \qquad (5.1)$$

5.2 Centripetal Acceleration An object in uniform circular motion experiences an acceleration, known as centripetal acceleration. The magnitude a_c of the centripetal acceleration is given by Equation 5.2, where v is the speed of the object and r is the radius of the circle. The direction of the centripetal acceleration vector always points toward the center of the circle and continually changes as the object moves.

$$a_c = \frac{v^2}{r} \qquad (5.2)$$

5.3 Centripetal Force To produce a centripetal acceleration, a net force pointing toward the center of the circle is required. This net force is called the centripetal force, and its magnitude F_c is given by Equation 5.3, where m and v are the mass and speed of the object, and r is the radius of the circle. The direction of the centripetal force vector, like that of the centripetal acceleration vector, always points toward the center of the circle.

$$F_c = \frac{mv^2}{r} \qquad (5.3)$$

5.4 Banked Curves A vehicle can negotiate a circular turn without relying on static friction to provide the centripetal force, provided the turn is banked at an angle relative to the horizontal. The angle θ at which a friction-free curve must be banked is related to the speed v of the vehicle, the radius r of the curve, and the magnitude g of the acceleration due to gravity by Equation 5.4.

$$\tan \theta = \frac{v^2}{rg} \qquad (5.4)$$

5.5 Satellites in Circular Orbits When a satellite orbits the earth, the gravitational force provides the centripetal force that keeps the satellite moving in a circular orbit. The speed v and period T of a satellite depend on the mass M_E of the earth and the radius r of the orbit according to Equations 5.5 and 5.6, where G is the universal gravitational constant.

$$v = \sqrt{\frac{GM_E}{r}} \qquad (5.5)$$

$$T = \frac{2\pi r^{3/2}}{\sqrt{GM_E}} \qquad (5.6)$$

5.6 Apparent Weightlessness and Artificial Gravity The apparent weight of an object is the force that it exerts on a scale with which it is in contact. All objects, including people, on board an orbiting satellite are in free-fall, since they experience negligible air resistance and they have an acceleration that is equal to the acceleration due to gravity. When a person is in free-fall, his or her apparent weight is zero, because both the person and the scale fall freely and cannot push against one another.

5.7 Vertical Circular Motion Vertical circular motion occurs when an object, such as a motorcycle, moves on a vertical circular path. The speed of the object often varies from moment to moment, and so do the magnitudes of the centripetal acceleration and centripetal force.

Focus on Concepts

Online

Additional questions are available for assignment in WileyPLUS.

Section 5.2   Centripetal Acceleration

1. Two cars are traveling at the same constant speed v. As the drawing indicates, car A is moving along a straight section of the road, while car B is rounding a circular turn. Which statement is true about the accelerations of the cars? (a) The acceleration of both cars is zero, since they are traveling at a constant speed. (b) Car A is accelerating, but car B is not accelerating. (c) Car A is not accelerating, but car B is accelerating. (d) Both cars are accelerating.

QUESTION 2

Section 5.3   Centripetal Force

3. The drawing shows two identical stones attached to cords that are being whirled on a tabletop at the same speed. The radius of the larger circle is twice that of the smaller circle. How is the tension T_1 in the longer cord related to the tension T_2 in the shorter cord? (a) $T_1 = T_2$ (b) $T_1 = 2T_2$ (c) $T_1 = 4T_2$ (d) $T_1 = \frac{1}{2}T_2$ (e) $T_1 = \frac{1}{4}T_2$

QUESTION 1   A

QUESTION 3   Path 1

2. Two cars are driving at the same constant speed v around a racetrack. However, they are traveling through turns that have different radii, as shown in the drawing. Which statement is true about the magnitude of the centripetal acceleration of each car? (a) The magnitude of the centripetal acceleration of each car is the same, since the cars are moving at the same speed. (b) The magnitude of the centripetal acceleration of the car at A is greater than that of the car at B, since the radius of the circular track is smaller at A. (c) The magnitude of the centripetal acceleration of the car at A is greater than that of the car at B, since the radius of the circular track is greater at A. (d) The magnitude of the centripetal acceleration of the car at A is less than that of the car at B, since the radius of the circular track is smaller at A.

4. Three particles have the following masses (in multiples of m_0) and move on three different circles with the following speeds (in multiples of v_0) and radii (in multiples of r_0):

Particle	Mass	Speed	Radius
1	m_0	$2v_0$	r_0
2	m_0	$3v_0$	$3r_0$
3	$2m_0$	$2v_0$	$4r_0$

Rank the particles according to the magnitude of the centripetal force that acts on them, largest first. **(a)** 1, 2, 3 **(b)** 1, 3, 2 **(c)** 2, 1, 3 **(d)** 2, 3, 1 **(e)** 3, 2, 1

Section 5.4   Banked Curves

5. Two identical cars, one on the moon and one on the earth, have the same speed and are rounding banked turns that have the same radius *r*. There are two forces acting on each car, its weight *mg* and the normal force F_N exerted by the road. Recall that the weight of an object on the moon is about one-sixth of its weight on earth. How does the centripetal force on the moon compare with that on the earth? **(a)** The centripetal forces are the same. **(b)** The centripetal force on the moon is less than that on the earth. **(c)** The centripetal force on the moon is greater than that on the earth.

Section 5.5   Satellites in Circular Orbits

6. Two identical satellites are in orbit about the earth. As the drawing shows, one orbit has a radius *r* and the other 2*r*. The centripetal force on the satellite in the larger orbit is _____ as that on the satellite in the smaller orbit. **(a)** the same **(b)** twice as great **(c)** four times as great **(d)** half as great **(e)** one-fourth as great

QUESTION 6

Section 5.7   Vertical Circular Motion

7. The drawing shows an extreme skier at the bottom of a ski jump. At this point the track is circular with a radius *r*. Two forces act on the skier, her weight *mg* and the normal force F_N. Which relation describes how the net force acting on her is related to her mass *m* and speed *v* and to the radius *r*? Assume that "up" is the positive direction.

(a) $F_N + mg = \frac{mv^2}{r}$ **(b)** $F_N - mg = \frac{mv^2}{r}$

(c) $F_N = \frac{mv^2}{r}$ **(d)** $-mg = \frac{mv^2}{r}$

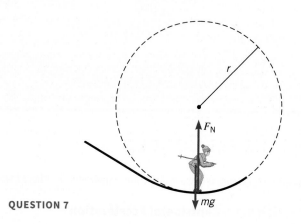

QUESTION 7

Problems

Online

Additional questions are available for assignment in WileyPLUS.

SSM Student Solutions Manual	**BIO** Biomedical application
MMH Problem-solving help	**E** Easy
GO Guided Online Tutorial	**M** Medium
V-HINT Video Hints	**H** Hard
CHALK Chalkboard Videos	**WS** Worksheet
	T Team Problem

	Radius	Speed
Example 1	0.50 m	12 m/s
Example 2	Infinitely large	35 m/s
Example 3	1.8 m	2.3 m/s

Section 5.1   Uniform Circular Motion,

Section 5.2   Centripetal Acceleration

1. **E** **SSM** A car travels at a constant speed around a circular track whose radius is 2.6 km. The car goes once around the track in 360 s. What is the magnitude of the centripetal acceleration of the car?

2. **E** **GO** The following table lists data for the speed and radius of three examples of uniform circular motion. Find the magnitude of the centripetal acceleration for each example.

3. **E** **GO** Speedboat A negotiates a curve whose radius is 120 m. Speedboat B negotiates a curve whose radius is 240 m. Each boat experiences the same centripetal acceleration. What is the ratio v_A/v_B of the speeds of the boats?

4. **E** **SSM** How long does it take a plane, traveling at a constant speed of 110 m/s, to fly once around a circle whose radius is 2850 m?

5. **E** **MMH** The blade of a windshield wiper moves through an angle of 90.0° in 0.40 s. The tip of the blade moves on the arc of a circle that has a radius of 0.45 m. What is the magnitude of the centripetal acceleration of the tip of the blade?

6. **E** **V-HINT** There is a clever kitchen gadget for drying lettuce leaves after you wash them. It consists of a cylindrical container mounted so that it can be rotated about its axis by turning a hand crank. The outer wall of the cylinder is perforated with small holes. You put the wet leaves in the container and turn the crank to spin off the water. The radius of the container is 12 cm. When the cylinder is rotating at 2.0 revolutions per second, what is the magnitude of the centripetal acceleration at the outer wall?

7. **E** **SSM** Computer-controlled display screens provide drivers in the Indianapolis 500 with a variety of information about how their cars are performing. For instance, as a car is going through a turn, a speed of 221 mi/h (98.8 m/s) and centripetal acceleration of 3.00 g (three times the acceleration due to gravity) are displayed. Determine the radius of the turn (in meters).

8. **M** **V-HINT** **CHALK** A computer is reading data from a rotating CD-ROM (an antiquated disc-shaped data storage device). At a point that is 0.030 m from the center of the disc, the centripetal acceleration is 120 m/s². What is the centripetal acceleration at a point that is 0.050 m from the center of the disc?

9. **M** **CHALK** The earth rotates once per day about an axis passing through the north and south poles, an axis that is perpendicular to the plane of the equator. Assuming the earth is a sphere with a radius of 6.38×10^6 m, determine the speed and centripetal acceleration of a person situated **(a)** at the equator and **(b)** at a latitude of 30.0° north of the equator.

Section 5.3 Centripetal Force

10. **E** **SSM** Review Example 3, which deals with the bobsled in **Interactive Figure 5.5**. Also review Conceptual Example 4. The mass of the sled and its two riders in **Interactive Figure 5.5** is 350 kg. Find the magnitude of the centripetal force that acts on the sled during the turn with a radius of **(a)** 33 m and **(b)** 24 m.

11. **E** At an amusement park there is a ride in which cylindrically shaped chambers spin around a central axis. People sit in seats facing the axis, their backs against the outer wall. At one instant the outer wall moves at a speed of 3.2 m/s, and an 83-kg person feels a 560-N force pressing against his back. What is the radius of the chamber?

12. **E** **MMH** Multiple-Concept Example 7 reviews the concepts that play a role in this problem. Car A uses tires for which the coefficient of static friction is 1.1 on a particular unbanked curve. The maximum speed at which the car can negotiate this curve is 25 m/s. Car B uses tires for which the coefficient of static friction is 0.85 on the same curve. What is the maximum speed at which car B can negotiate the curve?

13. **E** A speed skater goes around a turn that has a radius of 31 m. The skater has a speed of 14 m/s and experiences a centripetal force of 460 N. What is the mass of the skater?

14. **E** For background pertinent to this problem, review Conceptual Example 6. In **Figure 5.7** the man hanging upside down is holding a partner who weighs 475 N. Assume that the partner moves on a circle that has a radius of 6.50 m. At a swinging speed of 4.00 m/s, what force must the man apply to his partner in the straight-down position?

15. **E** **MMH** A car is safely negotiating an unbanked circular turn at a speed of 21 m/s. The road is dry, and the maximum static frictional force acts on the tires. Suddenly a long wet patch in the road decreases the maximum static frictional force to one-third of its dry-road value. If the car is to continue safely around the curve, to what speed must the driver slow the car?

16. **E** See Conceptual Example 6 to review the concepts involved in this problem. A 9.5-kg monkey is hanging by one arm from a branch and is swinging on a vertical circle. As an approximation, assume a radial distance of 85 cm between the branch and the point where the monkey's mass is located. As the monkey swings through the lowest point on the circle, it has a speed of 2.8 m/s. Find **(a)** the magnitude of the centripetal force acting on the monkey and **(b)** the magnitude of the tension in the monkey's arm.

17. **E** **GO** Multiple-Concept Example 7 deals with the concepts that are important in this problem. A penny is placed at the outer edge of a disk (radius = 0.150 m) that rotates about an axis perpendicular to the plane of the disk at its center. The period of the rotation is 1.80 s. Find the minimum coefficient of friction necessary to allow the penny to rotate along with the disk.

18. **M** **CHALK** **SSM** The hammer throw is a track-and-field event in which a 7.3-kg ball (the "hammer") is whirled around in a circle several times and released. It then moves upward on the familiar curving path of projectile motion and eventually returns to earth some distance away. The world record for this distance is 86.75 m, achieved in 1986 by Yuriy Sedykh. Ignore air resistance and the fact that the ball is released above the ground rather than at ground level. Furthermore, assume that the ball is whirled on a circle that has a radius of 1.8 m and that its velocity at the instant of release is directed 41° above the horizontal. Find the magnitude of the centripetal force acting on the ball just prior to the moment of release.

19. **M** **V-HINT** An 830-kg race car can drive around an unbanked turn at a maximum speed of 58 m/s without slipping. The turn has a radius of curvature of 160 m. Air flowing over the car's wing exerts a downward-pointing force (called the *downforce*) of 11 000 N on the car. **(a)** What is the coefficient of static friction between the track and the car's tires? **(b)** What would be the maximum speed if no downforce acted on the car?

20. **M** **CHALK** **MMH** A "swing" ride at a carnival consists of chairs that are swung in a circle by 15.0-m cables attached to a vertical rotating pole, as the drawing shows. Suppose the total mass of a chair and its occupant is 179 kg. **(a)** Determine the tension in the cable attached to the chair. **(b)** Find the speed of the chair.

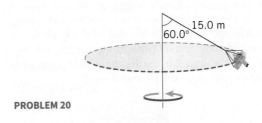

PROBLEM 20

Section 5.4 Banked Curves

21. **E** On a banked race track, the smallest circular path on which cars can move has a radius of 112 m, while the largest has a radius of 165 m, as the drawing illustrates. The height of the outer wall is 18 m. Find **(a)** the smallest and **(b)** the largest speed at which cars can move on this track without relying on friction.

165 m

18 m

112 m

PROBLEM 21

22. **E** **V-HINT** Before attempting this problem, review Examples 7 and 9. Two curves on a highway have the same radii. However, one is unbanked and the other is banked at an angle θ. A car can safely travel along the unbanked curve at a maximum speed v_0 under conditions when the coefficient of static friction between the tires and the road is $\mu_s = 0.81$. The banked curve is frictionless, and the car can negotiate it at the same maximum speed v_0. Find the angle θ of the banked curve.

23. **E** **GO** A woman is riding a Jet Ski at a speed of 26 m/s and notices a seawall straight ahead. The farthest she can lean the craft in order to make a turn is 22°. This situation is like that of a car on a curve that is banked at an angle of 22°. If she tries to make the turn without slowing down, what is the minimum distance from the seawall that she can begin making her turn and still avoid a crash?

24. **E** Two banked curves have the same radius. Curve A is banked at an angle of 13°, and curve B is banked at an angle of 19°. A car can travel around curve A without relying on friction at a speed of 18 m/s. At what speed can this car travel around curve B without relying on friction?

25. **M** **MMH** **CHALK** A racetrack has the shape of an inverted cone, as the drawing shows. On this surface the cars race in circles that are parallel to the ground. For a speed of 34.0 m/s, at what value of the distance d should a driver locate his car if he wishes to stay on a circular path without depending on friction?

PROBLEM 25   40.0°   40.0°

26. **M** **GO** A jet flying at 123 m/s banks to make a horizontal circular turn. The radius of the turn is 3810 m, and the mass of the jet is 2.00×10^5 kg. Calculate the magnitude of the necessary lifting force.

27. **H** The drawing shows a baggage carousel at an airport. Your suitcase has not slid all the way down the slope and is going around at a constant speed on a circle ($r = 11.0$ m) as the carousel turns. The coefficient of static friction between the suitcase and the carousel is 0.760, and the angle θ in the drawing is 36.0°. How much time is required for your suitcase to go around once?

PROBLEM 27

Section 5.5   Satellites in Circular Orbits,
Section 5.6   Apparent Weightlessness and Artificial Gravity

28. **E** **GO** Two satellites are in circular orbits around the earth. The orbit for satellite A is at a height of 360 km above the earth's surface, while that for satellite B is at a height of 720 km. Find the orbital speed for each satellite.

29. **E** A rocket is used to place a synchronous satellite in orbit about the earth. What is the speed of the satellite in orbit?

30. **E** **SSM** A satellite is in a circular orbit around an unknown planet. The satellite has a speed of 1.70×10^4 m/s, and the radius of the orbit is 5.25×10^6 m. A second satellite also has a circular orbit around this same planet. The orbit of this second satellite has a radius of 8.60×10^6 m. What is the orbital speed of the second satellite?

31. **E** **GO** A satellite is in a circular orbit about the earth ($M_E = 5.98 \times 10^{24}$ kg). The period of the satellite is 1.20×10^4 s. What is the speed at which the satellite travels?

32. **E** A satellite circles the earth in an orbit whose radius is twice the earth's radius. The earth's mass is 5.98×10^{24} kg, and its radius is 6.38×10^6 m. What is the period of the satellite?

33. **M** **GO** **CHALK** A satellite has a mass of 5850 kg and is in a circular orbit 4.10×10^5 m above the surface of a planet. The period of the orbit is 2.00 hours. The radius of the planet is 4.15×10^6 m. What would be the true weight of the satellite if it were at rest on the planet's surface?

34. **M** **V-HINT** Two newly discovered planets follow circular orbits around a star in a distant part of the galaxy. The orbital speeds of the planets are determined to be 43.3 km/s and 58.6 km/s. The slower planet's orbital period is 7.60 years. **(a)** What is the mass of the star? **(b)** What is the orbital period of the faster planet, in years?

Section 5.7   Vertical Circular Motion

35. **E** **SSM** A motorcycle has a constant speed of 25.0 m/s as it passes over the top of a hill whose radius of curvature is 126 m. The mass of the motorcycle and driver is 342 kg. Find the magnitudes of **(a)** the centripetal force and **(b)** the normal force that acts on the cycle.

36. **E** **SSM** For the normal force in **Interactive Figure 5.20** to have the same magnitude at all points on the vertical track, the stunt driver must adjust the speed to be different at different points. Suppose, for example, that the track has a radius of 3.0 m and that the driver goes past point 1 at the bottom with a speed of 15 m/s. What speed must she have at point 3, so that the normal force at the top has the same magnitude as it did at the bottom?

37. **E** **GO** A special electronic sensor is embedded in the seat of a car that takes riders around a circular loop-the-loop ride at an amusement park. The sensor measures the magnitude of the normal force that the seat exerts on a rider. The loop-the-loop ride is in the vertical plane and its radius is 21 m. Sitting on the seat before the ride starts, a rider is level and stationary, and the electronic sensor reads 770 N. At the top of the loop, the rider is upside down and moving, and the sensor reads 350 N. What is the speed of the rider at the top of the loop?

38. **E** **GO** A 0.20-kg ball on a stick is whirled on a vertical circle at a constant speed. When the ball is at the three o'clock position, the stick tension is 16 N. Find the tensions in the stick when the ball is at the twelve o'clock and at the six o'clock positions.

39. **M** **GO** **MMH** A stone is tied to a string (length = 1.10 m) and whirled in a circle at the same constant speed in two different ways. First, the circle is horizontal and the string is nearly parallel to the ground. Next, the circle is vertical. In the vertical case the maximum tension in the string is 15.0% larger than the tension that exists when the circle is horizontal. Determine the speed of the stone.

40. **M** **GO** **CHALK** A motorcycle is traveling up one side of a hill and down the other side. The crest of the hill is a circular arc with a radius of 45.0 m. Determine the maximum speed that the cycle can have while moving over the crest without losing contact with the road.

Additional Problems

Online

41. **E** **SSM** In a skating stunt known as crack-the-whip, a number of skaters hold hands and form a straight line. They try to skate so that the line rotates about the skater at one end, who acts as the pivot. The skater farthest out has a mass of 80.0 kg and is 6.10 m from the pivot. He is skating at a speed of 6.80 m/s. Determine the magnitude of the centripetal force that acts on him.

42. **E** **GO** **MMH** The second hand and the minute hand on one type of clock are the same length. Find the ratio $(a_{c, \text{second}}/a_{c, \text{minute}})$ of the centripetal accelerations of the tips of the second hand and the minute hand.

43. **E** **MMH** A child is twirling a 0.0120-kg plastic ball on a string in a horizontal circle whose radius is 0.100 m. The ball travels once around the circle in 0.500 s. **(a)** Determine the centripetal force acting on the ball. **(b)** If the speed is doubled, does the centripetal force double? If not, by what factor does the centripetal force increase?

44. **E** **GO** **MMH** Two cars are traveling at the same speed of 27 m/s on a curve that has a radius of 120 m. Car A has a mass of 1100 kg, and car B has a mass of 1600 kg. Find the magnitude of the centripetal acceleration and the magnitude of the centripetal force for each car.

45. **E** **SSM** A roller coaster at an amusement park has a dip that bottoms out in a vertical circle of radius r. A passenger feels the seat of the car pushing upward on her with a force equal to twice her weight as she goes through the dip. If $r = 20.0$ m, how fast is the roller coaster traveling at the bottom of the dip?

46. **M** **V-HINT** The large blade of a helicopter is rotating in a horizontal circle. The length of the blade is 6.7 m, measured from its tip to the center of the circle. Find the ratio of the centripetal acceleration at the end of the blade to that which exists at a point located 3.0 m from the center of the circle.

47. **M** A block is hung by a string from the inside roof of a van. When the van goes straight ahead at a speed of 28 m/s, the block hangs vertically down. But when the van maintains this same speed around an unbanked curve (radius = 150 m), the block swings toward the outside of the curve. Then the string makes an angle θ with the vertical. Find θ.

48. **M** **GO** A rigid massless rod is rotated about one end in a horizontal circle. There is a particle of mass m_1 attached to the center of the rod and a particle of mass m_2 attached to the outer end of the rod. The inner section of the rod sustains a tension that is three times as great as the tension that the outer section sustains. Find the ratio m_1/m_2.

49. **M** **GO** A space laboratory is rotating to create artificial gravity, as the figure indicates. Its period of rotation is chosen so that the outer ring ($r_O = 2150$ m) simulates the acceleration due to gravity on earth (9.80 m/s^2). What should be the radius r_1 of the inner ring, so it simulates the acceleration due to gravity on the surface of Mars (3.72 m/s^2)?

PROBLEM 49

Physics in Biology, Medicine, and Sports

50. **E** **BIO** **5.2** The aorta is a major artery, rising upward from the left ventricle of the heart and curving down to carry blood to the abdomen and lower half of the body. The curved artery can be approximated as a semicircular arch whose diameter is 5.0 cm. If blood flows through the aortic arch at a speed of 0.32 m/s, what is the magnitude (in m/s^2) of the blood's centripetal acceleration?

51. **M** **BIO** **5.2** A centrifuge is a device in which a small container of material is rotated at a high speed on a circular path. Such a device is used in medical laboratories, for instance, to cause the more dense red blood cells to settle through the less dense blood serum and collect at the bottom of the container. Suppose the centripetal acceleration of the sample is 6.25×10^3 times as large as the acceleration due to gravity. How many revolutions per minute is the sample making, if it is located at a radius of 5.00 cm from the axis of rotation?

52. **E** **BIO** **V-HINT** **5.7** Pilots of high-performance fighter planes can be subjected to large centripetal accelerations during high-speed turns (see Example 8). Because of these accelerations, the pilots are subjected to forces that can be much greater than their body weight, leading to an accumulation of blood in the abdomen and legs. As a result, the brain becomes starved for blood, and the pilot can lose consciousness ("black out"). The pilots wear "anti-g suits" to help keep the blood from draining out of the brain. To appreciate the forces that a fighter pilot must endure, consider the magnitude F_N of the normal force that the pilot's seat exerts on him at the bottom of a dive. The magnitude of the pilot's weight is W. The plane is traveling at 230 m/s on a vertical circle of radius 690 m. Determine the ratio F_N/W. For comparison, note that blackout can occur for values of F_N/W as small as 2 if the pilot is not wearing an anti-G suit.

53. **E** **BIO** **5.3** Consider again the "human centrifuge" in Example 8. If the chamber is located 15 m from the center of the circle. At what speed must the chamber move so that an astronaut is subjected to 7.5 times the acceleration due to gravity?

54. **M** **BIO** **5.6** Astronauts in the International Space Station (ISS) are in free-fall as they orbit the earth, so their apparent weight is zero. While floating around weightless may appear exciting and fun, there can be significant negative health effects on astronauts who are in space for long periods of time. Muscles atrophy (including the heart), bone density decreases, and flattening of the eyeballs occurs along with changes to the retina. On earth, gravity pulls downward on the blood and other fluids in the body. However, in orbit, the fluids redistribute themselves, which leads to more fluids in the upper half of the body and an increase in intercranial pressure. Perhaps one of the most surprising effects of long-term weightlessness is a change at the genetic level. In 2015 and 2016, former NASA astronaut Scott Kelly participated in a year-long "twins experiment" with his twin brother Mark. Scott spent a year onboard the ISS, while Mark remained on the ground. After returning to earth, medical examination of the brothers showed that certain genes turn on and off in space. Telomeres, which form the end of chromosomes and slow down their deterioration, temporarily got longer in Scott while he was in space. If the human race is to survive and we are to travel from earth to other planets, we will need to live and work in space. One possible solution to the negative physiological effects of space travel is to simulate gravity, as discussed in Example 13. In the 2015 movie *The Martian* the Hermes spaceship that took the crew to Mars and back had a rotating section to create artificial gravity of 1g (see photo, highlighted section). If the speed of the outer wall of the rotating section of the Hermes is 14 m/s, what is the diameter (d) of the section?

PROBLEM 54

55. **M** **BIO** **5.3** Adhesive capsulitis, also known as a frozen shoulder, is a condition that affects the motion of the shoulder joint. The articular shoulder capsule becomes inflamed, stiff, and restricts a person's mobility. Physical therapy provides one course of treatment for a frozen shoulder. One of the exercises often performed is the pendulum. Here, the patient bends forward and lets the injured arm hang downward and swing freely like a pendulum (see figure). The patient slowly swings their arm in a circular motion, for say 10 cycles, and then switches direction and completes 10 more. The patient may hold a light weight dumbbell while performing this motion, or just use an empty hand. Imagine the situation shown in the figure. A patient is moving a 7.0-lb (3.2-kg) dumbbell at constant speed and completes one circular motion in 1.2 s. What is the radius (r) of the circular motion?

PROBLEM 55

56. **E** **BIO** **5.2** A dental drill operates at incredibly high speeds. It uses compressed air to do this, and it can accommodate multiple drilling heads. The head shown in the figure is called a dental burr, and it has a diameter of 0.30 mm. If the drill operates at 1.8×10^5 rpm, what is the centripetal acceleration of a point on the outside edge of the burr?

0.3 mm

PROBLEM 56

57. **M** **BIO** **5.3** Does your apparent weight depend on where you are located on the earth's surface? Yes! Your apparent weight on a scale standing at the equator will be less than your apparent weight while standing on a scale at the North Pole (see figure). This difference occurs due to the earth's rotation. How much extra mass would a 65.0-kg person have to hold while weighing himself at the equator, so that his apparent weight there is equal to his apparent weight at the North Pole?

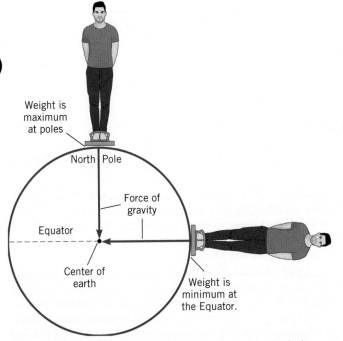

Weight is maximum at poles

North Pole

Force of gravity

Equator

Center of earth

Weight is minimum at the Equator.

Source: Md. Khaja Shareef, Measure the Mass of the Earth, Science Inspiration. http://scienceinspiration.blogspot.com/2012/05/measure-mass-of-earth.html?

PROBLEM 57

58. **M** **BIO** **5.3** A variation of the amusement park ride discussed in Problem 11 is the Gravitron, which consists of a rotating cylinder, where riders stand with their backs against the inner surface of the cylinder (see image). As the cylinder rotates faster, riders get stuck to the wall. The effect is great enough that on some rides, the floor even drops away. Let's model the Gravitron as a rotating cylinder with vertical walls. The radius of the cylinder is 5.5 m. At one point during the ride, the speed of the Gravitron is 24 rpm, at which point a rider slides down the wall at constant speed. What is the coefficient of kinetic friction between the rider and the wall?

PROBLEM 58

Concepts and Calculations Problems

Online

In uniform circular motion the concepts of acceleration and force play central roles. Problem 59 deals with acceleration, particularly in connection with the equations of kinematics. Problem 60 deals with force, and it stresses that centripetal force is the net force along the radial direction. This means that all forces along the radial direction must be taken into account when identifying the centripetal force.

59. **M** **CHALK** **SSM** At time $t = 0$ s, automobile A is traveling at a speed of 18 m/s along a straight road and is picking up speed with an acceleration that has a magnitude of 3.5 m/s² as in part *a* of the figure. At $t = 0$ s, automobile B is traveling at a speed of 18 m/s in uniform circular motion as it negotiates a turn (part *b* of the figure). It has a centripetal acceleration whose magnitude is also 3.5 m/s². *Concepts:* (i) Which automobile has a constant acceleration? (ii) For which automobile do the equations of kinematics apply? *Calculations:* Determine the speed of each automobile when $t = 2$ s.

PROBLEM 59   (*b*) Automobile B

60. **M** **CHALK** Ball A is attached to one end of a rigid massless rod, while an identical ball B is attached to the center of the rod, as shown in the figure. Each ball has a mass of $m = 0.50$ kg, and the length of each half of the rod is $L = 0.40$ m. This arrangement is held by the empty end and is whirled around in a horizontal circle at a constant rate, so that each ball is in uniform circular motion. Ball A travels at a constant speed of $v_A = 5.0$ m/s. *Concepts:* (i) How many tension forces contribute to the centripetal force that acts on ball A? (ii) How many tension forces contribute to the centripetal force that acts on ball B? (iii) Is the speed of ball A the same as that of ball B? *Calculations:* Find the tension in each half of the rod.

PROBLEM 60

Team Problems   Online

61. **E** **T** **WS** **A Ball Bearing Launch System.** A steel ball bearing of mass $m = 0.100$ kg has a tiny loop on its surface through which is connected a hook that protrudes from the end of a light cable. On the opposite end of the cable (length $L = 0.850$ m) is a trigger that extracts the hook from the loop, and thereby releases the ball. Built into the cable is an in-line digital scale that reports the tension in the cable when the trigger is pulled. You and your team take turns launching the ball bearing by twirling it in a horizontal circle above your head and pulling the release trigger. **(a)** If the largest value recorded by the scale is 28.5 lb (the units given by the digital in-line scale), what is the speed of the ball bearing at launch? Assume that the mass of the cable is negligible. **(b)** What is the maximum horizontal range with this launch speed? For simplicity, you can assume that the ball bearing is launched from ground level, and you can ignore air resistance.

62. **E** **T** **WS** **Not Breaking an Egg.** The House of Fabergé in Russia created the first Fabergé egg in the late 1800s. It is an intricate golden egg, laden with jewels, the first of which was crafted for Tsar Alexander III as a gift for his wife. You and your team are transporting an expensive Fabergé egg that is so fragile that it may be damaged if it experiences an apparent weight of magnitude greater than 2.50 times that of its actual weight on the surface of the earth. The egg is to be

shipped in a military transport aircraft that travels at 370.0 mi/h. Calculate the minimum allowable radius (in meters) of a horizontal turn of the plane. Assume that the plane banks so that the normal force acting on the egg from the internal surface of the plane is the source of the centripetal force required for the turn.

63. **M** **T** **Multilevel Artificial Gravity.** You and your team have the task of designing a space station that will mimic the force due to gravity on the earth, the moon, and Mars. The station's purpose is to physically prepare astronauts for the transition to the different gravitational forces they will experience on Mars and the moon, but also to acclimate them to earth's gravity when they return. The space station should be constructed of concentric cylinders, with a maximum allowed diameter of 300 m. **(a)** Draw a sketch of the space station and label the levels (i.e., the cylinders) according to the celestial body whose gravitational force each cylinder simulates. **(b)** Determine the required radii of each cylinder. Assume the outermost cylinder has the maximum allowable diameter. **(c)** What is the period of rotation of the station? ($g_{earth} = 9.80$ m/s², $g_{Mars} = 3.76$ m/s², and $g_{moon} = 1.62$ m/s²)

64. **M** **T** **A Dangerous Ride.** You and your exploration team are stuck on a steep slope in the Andes Mountains in Argentina. A deadly winter storm is approaching and you must get down the mountain before the storm hits. Your path leads you around an extremely slippery, horizontal curve with a diameter of 90.0 m and banked at an angle of 40.0° relative to the horizontal. You get the idea to unpack the toboggan that you have been using to haul supplies, load your team upon it, and ride it down the mountain to get enough speed to get around the banked curve. You must be extremely careful, however, not to slide down the bank: At the bottom of the curve is a steep cliff. **(a)** Neglecting friction and air resistance, what must be the speed of your toboggan in order to get around the curve without sliding up or down its bank? Express your answer in m/s and mph.

(b) You will need to climb up the mountain and ride the toboggan down in order to attain the speed you need to safely navigate the curve (from part (a)). The mountain slope leading into the curve is at an angle of 30.0° relative to the horizontal, and the coefficient of kinetic friction between the toboggan and the surface of the slope is ($\mu_{\text{mountain}} = 0.150$). How far up the mountain (distance along the slope, not elevation) from the curve should you start your ride? Note: the path down the mountain levels off at the bottom so that the toboggan enters the curve moving in the horizontal plane (i.e., in the same plane as the curve).

CHAPTER 6

On the popular Behemoth roller coaster at Canada's Wonderland amusement park, riders cresting the top of the hill will pick up speed on the way down, as they convert gravitational potential energy into kinetic energy. This conversion is governed by the principle of conservation of energy, a central topic in this chapter.

Oleksiy Maksymenko Photography/Alamy Images

LEARNING OBJECTIVES

After reading this module, you should be able to...

6.1 Determine the work done by a constant force.

6.2 Relate work and kinetic energy.

6.3 Use gravitational potential energy.

6.4 Identify conservative and nonconservative forces.

6.5 Use the law of conservation of mechanical energy.

6.6 Solve problems involving nonconservative forces.

6.7 Solve problems involving power.

6.8 Employ the law of conservation of energy.

6.9 Determine the work done by a variable force.

Work and Energy

6.1 Work Done by a Constant Force

Work is a familiar concept. For example, it takes work to push a stalled car. In fact, more work is done when the pushing force is greater or when the displacement of the car is greater. Force and displacement are, in fact, the two essential elements of work, as **Figure 6.1** illustrates. The drawing shows a constant pushing force $\vec{\mathbf{F}}$ that points in the same direction as the resulting displacement $\vec{\mathbf{s}}$.* In such a case, the work W is defined

FIGURE 6.1 Work is done when a force $\vec{\mathbf{F}}$ pushes a car through a displacement $\vec{\mathbf{s}}$.

*When discussing work, it is customary to use the symbol $\vec{\mathbf{s}}$ for the displacement, rather than $\vec{\mathbf{x}}$ or $\vec{\mathbf{y}}$.

as the magnitude F of the force times the magnitude s of the displacement: $W = Fs$. The work done to push a car is the same whether the car is moved north to south or east to west, provided that the amount of force used and the distance moved are the same. Since work does not convey directional information, it is a scalar quantity.

The equation $W = Fs$ indicates that the unit of work is the unit of force times the unit of distance, or the newton · meter in SI units. One newton · meter is referred to as a *joule* (J) (rhymes with "cool"), in honor of James Joule (1818–1889) and his research into the nature of work, energy, and heat. **Table 6.1** summarizes the units for work in several systems of measurement.

TABLE 6.1   Units of Measurement for Work

System	Force	×	Distance	=	Work
SI	newton (N)		meter (m)		joule (J)
CGS	dyne (dyn)		centimeter (cm)		erg
BE	pound (lb)		foot (ft)		foot · pound (ft · 1b)

The definition of work as $W = Fs$ does have one surprising feature: If the distance s is zero, the work is zero, even if a force is applied. Pushing on an immovable object, such as a brick wall, may tire your muscles, but there is no work done of the type we are discussing. In physics, the idea of work is intimately tied up with the idea of motion. If the object does not move, the force acting on the object does no work.

Often, the force and displacement do not point in the same direction. For instance, **Figure 6.2a** shows a suitcase-on-wheels being pulled to the right by a force that is applied along the handle. The force is directed at an angle θ relative to the displacement. In such a case, only the component of the force along the displacement is used in defining work. As **Figure 6.2b** shows, this component is $F \cos \theta$, and it appears in the general definition below:

DEFINITION OF WORK DONE BY A CONSTANT* FORCE

The work done on an object by a constant force $\vec{F}$ is

$$W = (F \cos \theta)s \qquad (6.1)$$

where F is the magnitude of the force, s is the magnitude of the displacement, and θ is the angle between the force and the displacement.

SI Unit of Work: newton · meter = joule (J)

When the force points in the same direction as the displacement, then $\theta = 0°$, and Equation 6.1 reduces to $W = Fs$. Example 1 shows how Equation 6.1 is used to calculate work.

EXAMPLE 1 | Pulling a Suitcase-on-Wheels

Find the work done by a 45.0-N force in pulling the suitcase in **Figure 6.2a** at an angle $\theta = 50.0°$ for a distance $s = 75.0$ m.

Reasoning The pulling force causes the suitcase to move a distance of 75.0 m and does work. However, the force makes an angle of 50.0° with the displacement, and we must take this angle into account by using the definition of work given by Equation 6.1.

Solution The work done by the 45.0-N force is

$$W = (F \cos \theta)s = [(45.0 \text{ N}) \cos 50.0°](75.0 \text{ m}) = \boxed{2170 \text{ J}}$$

The answer is expressed in newton · meters or joules (J).

(a)

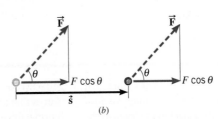

(b)

FIGURE 6.2 (*a*) Work can be done by a force $\vec{F}$ that points at an angle θ relative to the displacement $\vec{s}$. (*b*) The force component that points along the displacement is $F \cos \theta$.

---

*Section 6.9 considers the work done by a variable force.

The definition of work in Equation 6.1 takes into account only the component of the force in the direction of the displacement. The force component perpendicular to the displacement does no work. To do work, there must be a force *and* a displacement, and since there is no displacement in the perpendicular direction, there is no work done by the perpendicular component of the force. If the entire force is perpendicular to the displacement, the angle θ in Equation 6.1 is 90°, and the force does no work at all.

Work can be either positive or negative, depending on whether a component of the force points in the same direction as the displacement or in the opposite direction. Example 2 illustrates how positive and negative work arise.

EXAMPLE 2 | BIO The Physics of Weight Lifting

The weight lifter in **Interactive Figure 6.3a** is bench-pressing a barbell whose weight is 710 N. In part *b* of the figure, he raises the barbell a distance of 0.65 m above his chest, and in part *c* he lowers it the same distance. The weight is raised and lowered at a constant velocity. Determine the work done on the barbell by the weight lifter during **(a)** the lifting phase and **(b)** the lowering phase.

Reasoning To calculate the work, it is necessary to know the force exerted by the weight lifter. The barbell is raised and lowered at a constant velocity and, therefore, is in equilibrium. Consequently, the force $\vec{F}$ exerted by the weight lifter must balance the weight of the barbell, so $F = 710$ N. During the lifting phase, the force $\vec{F}$ and displacement $\vec{s}$ are in the same direction, as **Interactive Figure 6.3b** shows. The angle between them is $\theta = 0°$. When the barbell is lowered, however, the force and displacement are in opposite directions, as in **Interactive Figure 6.3c**. The angle

between the force and the displacement is now $\theta = 180°$. With these observations, we can find the work.

Solution **(a)** During the lifting phase, the work done by the force $\vec{F}$ is given by Equation 6.1 as

$$W = (F \cos \theta)s = [(710 \text{ N}) \cos 0°](0.65 \text{ m}) = \boxed{460 \text{ J}}$$

(b) The work done during the lowering phase is

$$W = (F \cos \theta)s$$

$$= [(710 \text{ N}) \cos 180°](0.65 \text{ m}) = \boxed{-460 \text{ J}}$$

since $\cos 180° = -1$. The work is negative, because the force is opposite to the displacement. Weight lifters call each complete up-and-down movement of the barbell a repetition, or "rep." The lifting of the weight is referred to as the positive part of the rep, and the lowering is known as the negative part.

(a)   (b)   (c)

INTERACTIVE FIGURE 6.3 (a) In the bench press, work is done during both the lifting and lowering phases. (b) During the lifting phase, the force $\vec{F}$ does positive work. (c) During the lowering phase, the force does negative work.

Dan Galic/Alamy Images

Example 3 deals with the work done by a static frictional force when it acts on a crate that is resting on the bed of an accelerating truck.

EXAMPLE 3 | Accelerating a Crate

Figure 6.4a shows a 120-kg crate on the flatbed of a truck that is moving with an acceleration of $a = +1.5$ m/s² along the positive *x* axis. The crate does not slip with respect to the truck as the truck undergoes a displacement whose magnitude is $s = 65$ m. What is the total work done on the crate by all of the forces acting on it?

Reasoning The free-body diagram in **Figure 6.4b** shows the forces that act on the crate: (1) the weight $\vec{W}$ of the crate, (2) the normal force $\vec{F_N}$ exerted by the flatbed, and (3) the static frictional force $\vec{f_s}$, which is exerted by the flatbed in the forward direction and keeps the crate from slipping backward. The weight and the normal force are perpendicular to the displacement, so they do no work. Only the static frictional force does work, since it acts in the x direction. To determine the frictional force, we note that the crate does not slip and, therefore, must have the same acceleration of $a = +1.5$ m/s^2 as does the truck. The force creating this acceleration is the static frictional force, and, knowing the mass of the crate and its acceleration, we can use Newton's second law to obtain its magnitude. Then, knowing the frictional force and the displacement, we can determine the total work done on the crate.

Solution From Newton's second law, we find that the magnitude f_s of the static frictional force is

$$f_s = ma = (120 \text{ kg})(1.5 \text{ m/s}^2) = 180 \text{ N}$$

The total work is that done by the static frictional force and is

$$W = (f_s \cos \theta)s = (180 \text{ N})(\cos 0°)(65 \text{ m}) = \boxed{1.2 \times 10^4 \text{ J}} \qquad \textbf{(6.1)}$$

The work is positive, because the frictional force is in the same direction as the displacement ($\theta = 0°$).

(a)

(b) Free-body diagram for the crate

FIGURE 6.4 (a) The truck and crate are accelerating to the right for a distance of $s = 65$ m. (b) The free-body diagram for the crate.

Check Your Understanding

(The answers are given at the end of the book.)

1. Two forces $\vec{F}_1$ and $\vec{F}_2$ are acting on the box shown in **CYU Figure 6.1**, causing the box to move across the floor. The two force vectors are drawn to scale. Which one of the following statements is correct? **(a)** $\vec{F}_2$ does more work than $\vec{F}_1$ does. **(b)** $\vec{F}_1$ does more work than $\vec{F}_2$ does. **(c)** Both forces do the same amount of work. **(d)** Neither force does any work.

CYU FIGURE 6.1

2. A box is being moved with a velocity $\vec{v}$ by a force $\vec{P}$ (in the same direction as $\vec{v}$) along a level horizontal floor. The normal force is $\vec{F}_N$, the kinetic frictional force is $\vec{f}_k$, and the weight is $m\vec{g}$. Which one of the following statements is correct? **(a)** $\vec{P}$ does positive work, $\vec{F}_N$ and $\vec{f}_k$ do zero work, and $m\vec{g}$ does negative work. **(b)** $\vec{F}_N$ does positive work, $\vec{P}$ and $\vec{f}_k$ do zero work, and $m\vec{g}$ does negative work. **(c)** $\vec{f}_k$ does positive work, $\vec{F}_N$ and $m\vec{g}$ do zero work, and $\vec{P}$ does negative work. **(d)** $\vec{P}$ does positive work, $\vec{F}_N$ and $m\vec{g}$ do zero work, and $\vec{f}_k$ does negative work.

3. A force does positive work on a particle that has a displacement pointing in the $+x$ direction. This same force does negative work on a particle that has a displacement pointing in the $+y$ direction. In which quadrant of the x, y coordinate system does the force lie? **(a)** First **(b)** Second **(c)** Third **(d)** Fourth

4. A suitcase is hanging straight down from your hand as you ride an escalator. Your hand exerts a force on the suitcase, and this force does work. This work is **(a)** positive when you ride up and negative when you ride down, **(b)** negative when you ride up and positive when you ride down, **(c)** positive when you ride up or down, **(d)** negative when you ride up or down.

6.2 | The Work–Energy Theorem and Kinetic Energy

Most people expect that if you do work, you get something as a result. In physics, when a net force performs work on an object, there is always a result from the effort. The result is a change in the **kinetic energy** of the object. As we will now see, the relationship that relates work to the change in kinetic energy is known as the **work–energy theorem**. This theorem is obtained by bringing together three basic concepts that we've already learned

INTERACTIVE FIGURE 6.5 A constant net external force $\Sigma\vec{F}$ acts over a displacement $\vec{s}$ and does work on the plane. As a result of the work done, the plane's kinetic energy changes.

about. First, we'll apply Newton's second law of motion, $\Sigma F = ma$, which relates the net force ΣF to the acceleration a of an object. Then, we'll determine the work done by the net force when the object moves through a certain distance. Finally, we'll use Equation 2.9, one of the equations of kinematics, to relate the distance and acceleration to the initial and final speeds of the object. The result of this approach will be the work–energy theorem.

To gain some insight into the idea of kinetic energy and the work–energy theorem, look at **Interactive Figure 6.5**, where a constant net external force $\Sigma\vec{F}$ acts on an airplane of mass m. This net force is the vector sum of all the external forces acting on the plane, and, for simplicity, it is assumed to have the same direction as the displacement $\vec{s}$. According to Newton's second law, the net force produces an acceleration a, given by $a = \Sigma F/m$. Consequently, the speed of the plane changes from an initial value of v_0 to a final value of v_f.* Multiplying both sides of $\Sigma F = ma$ by the distance s gives

$$\underbrace{(\Sigma F)s}_{\substack{\text{Work done by} \\ \text{net ext. force}}} = mas$$

The left side of this equation is the work done by the net external force. The term as on the right side can be related to v_0 and v_f by using $v_f^2 = v_0^2 + 2as$ (Equation 2.9) from the equations of kinematics. Solving this equation gives

$$as = \tfrac{1}{2}\left(v_f^2 - v_0^2\right)$$

Substituting this result into $(\Sigma F)s = mas$ shows that

$$\underbrace{(\Sigma F)s}_{\substack{\text{Work done by} \\ \text{net ext. force}}} = \underbrace{\tfrac{1}{2}mv_f^2}_{\substack{\text{Final} \\ \text{KE}}} - \underbrace{\tfrac{1}{2}mv_0^2}_{\substack{\text{Initial} \\ \text{KE}}}$$

This expression is the work–energy theorem. Its left side is the work W done by the net external force, whereas its right side involves the difference between two terms, each of which has the form $\tfrac{1}{2}(\text{mass})(\text{speed})^2$. The quantity $\tfrac{1}{2}(\text{mass})(\text{speed})^2$ is called kinetic energy (KE) and plays a significant role in physics, as we will see in this chapter and later on in other chapters as well.

DEFINITION OF KINETIC ENERGY

The kinetic energy KE of an object with mass m and speed v is given by

$$\text{KE} = \tfrac{1}{2}mv^2 \tag{6.2}$$

SI Unit of Kinetic Energy: joule (J)

The SI unit of kinetic energy is the same as the unit for work, the joule. Kinetic energy, like work, is a scalar quantity. These are not surprising observations, because work and kinetic energy are closely related, as is clear from the following statement of the work–energy theorem.

THE WORK–ENERGY THEOREM

When a net external force does work W on an object, the kinetic energy of the object changes from its initial value of KE_0 to a final value of KE_f, the difference between the two values being equal to the work:

$$W = \text{KE}_f - \text{KE}_0 = \tfrac{1}{2}mv_f^2 - \tfrac{1}{2}mv_0^2 \tag{6.3}$$

*For extra emphasis, the final speed is now represented by the symbol v_f, rather than v.

The work–energy theorem may be derived for any direction of the force relative to the displacement, not just the situation in **Interactive Figure 6.5**. In fact, the force may even vary from point to point along a path that is curved rather than straight, and the theorem remains valid. According to the work–energy theorem, a moving object has kinetic energy, because work was done to accelerate the object from rest to a speed v_f.[†] Conversely, an object with kinetic energy can perform work, if it is allowed to push or pull on another object. Example 4 illustrates the work–energy theorem and considers a single force that does work to change the kinetic energy of a space probe.

Analyzing Multiple-Concept Problems

EXAMPLE 4 | The Physics of an Ion Propulsion Drive

The space probe *Deep Space 1* was launched October 24, 1998, and it used a type of engine called an ion propulsion drive. An ion propulsion drive generates only a weak force (or thrust), but can do so for long periods of time using only small amounts of fuel. Suppose the probe, which has a mass of 474 kg, is traveling at an initial speed of 275 m/s. No forces act on it except the 5.60×10^{-2}-N thrust of its engine. This external force $\vec{F}$ is directed parallel to the displacement $\vec{s}$, which has a magnitude of 2.42×10^9 m (see **Figure 6.6**). Determine the final speed of the probe, assuming that its mass remains nearly constant.

FIGURE 6.6 An ion propulsion drive generates a single force $\vec{F}$ that points in the same direction as the displacement $\vec{s}$. The force performs positive work, causing the space probe to gain kinetic energy.

Reasoning If we can determine the final kinetic energy of the space probe, we can determine its final speed, since kinetic energy is related to mass and speed according to Equation 6.2 and the mass of the probe is known. We will use the work–energy theorem ($W = KE_f - KE_0$), along with the definition of work, to find the final kinetic energy.

Knowns and Unknowns The following list summarizes the data for this problem:

Description	Symbol	Value	Comment
Explicit Data			
Mass	m	474 kg	
Initial speed	v_0	275 m/s	
Magnitude of force	F	5.60×10^{-2} N	
Magnitude of displacement	s	2.42×10^9 m	
Implicit Data			
Angle between force $\vec{F}$ and displacement $\vec{s}$	θ	0°	The force is parallel to the displacement.
Unknown Variable			
Final speed	v_f	?	

[†]Strictly speaking, the work–energy theorem, as given by Equation 6.3, applies only to a single particle, which occupies a mathematical point in space. A macroscopic object, however, is a collection or system of particles and is spread out over a region of space. Therefore, when a force is applied to a macroscopic object, the point of application of the force may be anywhere on the object. To take into account this and other factors, a discussion of work and energy is required that is beyond the scope of this text. The interested reader may refer to A. B. Arons, *The Physics Teacher*, October 1989, p. 506.

Modeling the Problem

STEP 1 Kinetic Energy   An object of mass m and speed v has a kinetic energy KE given by Equation 6.2 as $KE = \frac{1}{2}mv^2$. Using the subscript f to denote the final kinetic energy and the final speed of the probe, we have that

$$KE_f = \frac{1}{2}mv_f^2$$

Solving for v_f gives Equation 1 at the right. The mass m is known but the final kinetic energy KE_f is not, so we will turn to Step 2 to evaluate it.

$$v_f = \sqrt{\frac{2(KE_f)}{m}} \qquad (1)$$

$$\boxed{?}$$

STEP 2 The Work–Energy Theorem   The work–energy theorem relates the final kinetic energy KE_f of the probe to its initial kinetic energy KE_0 and the work W done by the force of the engine. According to Equation 6.3, this theorem is $W = KE_f - KE_0$. Solving for KE_f shows that

$$KE_f = KE_0 + W$$

The initial kinetic energy can be expressed as $KE_0 = \frac{1}{2}mv_0^2$, so the expression for the final kinetic energy becomes

$$\boxed{KE_f = \frac{1}{2}mv_0^2 + W}$$

This result can be substituted into Equation 1, as indicated at the right. Note from the data table that we know the mass m and the initial speed v_0. The work W is not known and will be evaluated in Step 3.

$$v_f = \sqrt{\frac{2(KE_f)}{m}} \qquad (1)$$

$$\boxed{KE_f = \frac{1}{2}mv_0^2 + W} \qquad (2)$$

$$\boxed{?}$$

STEP 3 Work   The work W is that done by the net external force acting on the space probe. Since there is only the one force $\vec{F}$ acting on the probe, it is the net force. The work done by this force is given by Equation 6.1 as

$$\boxed{W = (F \cos \theta)s}$$

where F is the magnitude of the force, θ is the angle between the force and the displacement, and s is the magnitude of the displacement. All the variables on the right side of this equation are known, so we can substitute it into Equation 2, as shown in the right column.

$$v_f = \sqrt{\frac{2(KE_f)}{m}} \qquad (1)$$

$$\boxed{KE_f = \frac{1}{2}mv_0^2 + W} \qquad (2)$$

$$\boxed{W = (F \cos \theta)s}$$

Solution   Algebraically combining the results of the three steps, we have

$$\overset{\textbf{STEP 1}}{\big\downarrow} \qquad \overset{\textbf{STEP 2}}{\big\downarrow} \qquad \overset{\textbf{STEP 3}}{\big\downarrow}$$

$$v_f = \sqrt{\frac{2(KE_f)}{m}} = \sqrt{\frac{2\left(\frac{1}{2}mv_0^2 + W\right)}{m}} = \sqrt{\frac{2\left[\frac{1}{2}mv_0^2 + (F \cos \theta)s\right]}{m}}$$

The final speed of the space probe is

$$v_f = \sqrt{\frac{2\left[\frac{1}{2}mv_0^2 + (F \cos \theta)s\right]}{m}}$$

$$= \sqrt{\frac{2\left[\frac{1}{2}(474 \text{ kg})(275 \text{ m/s})^2 + (5.60 \times 10^{-2}\text{N})(\cos 0°)(2.42 \times 10^9 \text{ m})\right]}{474 \text{ kg}}} = \boxed{805 \text{ m/s}}$$

Related Homework: Problem 22

In Example 4 only the force of the engine does work. If several external forces act on an object, they must be added together vectorially to give the net force. The work done by the net force can then be related to the change in the object's kinetic energy by using the work–energy theorem, as in the next example.

Analyzing Multiple-Concept Problems

EXAMPLE 5 | Downhill Skiing

A 58-kg skier is coasting down a 25° slope, as **Figure 6.7a** shows. Near the top of the slope, her speed is 3.6 m/s. She accelerates down the slope because of the gravitational force, even though a kinetic frictional force of magnitude 71 N opposes her motion. Ignoring air resistance, determine the speed at a point that is displaced 57 m downhill.

Reasoning The skier's speed at a point 57 m downhill (her final speed) depends on her final kinetic energy. According to the work–energy theorem, her final kinetic energy is related to her initial kinetic energy (which we can calculate directly) and the work done by the net external force that acts on her. The work can be evaluated directly from its definition.

(a)

(b) Free-body diagram for the skier

FIGURE 6.7 (a) A skier coasting downhill. (b) The free-body diagram for the skier.

Knowns and Unknowns   The data for this problem are listed below:

Description	Symbol	Value	Comment
Explicit Data			
Mass	m	58 kg	
Initial speed	v_0	3.6 m/s	
Magnitude of kinetic frictional force	f_k	71 N	
Magnitude of skier's displacement	s	57 m	
Angle of slope above horizontal		25°	
Implicit Data			
Angle between net force acting on skier and her displacement	θ	0°	The skier is accelerating down the slope, so the direction of the net force is parallel to her displacement.
Unknown Variable			
Final speed	v_f	?	

Modeling the Problem

STEP 1 Kinetic Energy   The final speed v_f of the skier is related to her final kinetic energy KE_f and mass m by Equation 6.2:

$$KE_f = \tfrac{1}{2}mv_f^2$$

Solving for v_f gives Equation 1 at the right. Her mass is known, but her final kinetic energy is not, so we turn to Step 2 to evaluate it.

$$v_f = \sqrt{\frac{2(KE_f)}{m}} \qquad (1)$$

?

STEP 2 The Work–Energy Theorem   The final kinetic energy KE_f of the skier is related to her initial kinetic energy KE_0 and the work W done by the net external force acting on her. This is according to the work–energy theorem (Equation 6.3): $KE_f = KE_0 + W$.

The initial kinetic energy can be written as $KE_0 = \frac{1}{2}mv_0^2$, so the expression for the final kinetic energy becomes

$$KE_f = \tfrac{1}{2}mv_0^2 + W$$

This result can be substituted into Equation 1, as indicated at the right. The mass m and the initial speed v_0 are known. The work W is not known, and will be evaluated in Steps 3 and 4.

$$v_f = \sqrt{\frac{2(KE_f)}{m}} \quad \text{(1)}$$

$$KE_f = \tfrac{1}{2}mv_0^2 + W \quad \text{(2)}$$

$$\boxed{?}$$

STEP 3 Work The work W done by the net external force acting on the skier is given by Equation 6.1 s

$$W = (\Sigma F \cos\theta)s$$

where ΣF is the magnitude of the net force, θ is the angle between the net force and the displacement, and s is the magnitude of the displacement. This result can be substituted into Equation 2, as shown at the right. The variables θ and s are known (see the table of knowns and unknowns), and the net external force will be determined in Step 4.

$$v_f = \sqrt{\frac{2(KE_f)}{m}} \quad \text{(1)}$$

$$KE_f = \tfrac{1}{2}mv_0^2 + W \quad \text{(2)}$$

$$W = (\Sigma F \cos\theta)s \quad \text{(3)}$$

$$\boxed{?}$$

STEP 4 The Net External Force Figure 6.7b is a free-body diagram for the skier and shows the three external forces acting on her: the gravitational force $m\vec{g}$, the kinetic frictional force $\vec{f}_k$ and the normal force $\vec{F}_N$. The net external force along the y axis is zero, because there is no acceleration in that direction (the normal force balances the component $mg\cos 25°$ of the weight perpendicular to the slope). Using the data from the table of knowns and unknowns, we find that the net external force along the x axis is

$$\Sigma F = mg\sin 25° - f_k$$
$$= (58\text{ kg})(9.80\text{ m/s}^2)\sin 25° - 71\text{ N} = 170\text{ N}$$

All the variables on the right side of this equation are known, so we substitute this expression into Equation 3 for the net external force (see the right column).

$$v_f = \sqrt{\frac{2(KE_f)}{m}} \quad \text{(1)}$$

$$KE_f = \tfrac{1}{2}mv_0^2 + W \quad \text{(2)}$$

$$W = (\Sigma F \cos\theta)s \quad \text{(3)}$$

$$\Sigma F = mg\sin 25° - f_k = 170\text{ N}$$

Math Skills In Figure 6.7b the angle between the gravitational force and the $-y$ axis equals the 25° angle of the slope. Figure 6.8 uses β for this angle and helps to explain why $\beta = 25°$. It shows the vertical directional line along which the gravitational force acts, as well as the slope and the x, y axes. Since this directional line is perpendicular to the ground, the triangle ABC is a right triangle. It follows, then, that $\alpha + 25° = 90°$ or $\alpha = 65°$. Since the y axis is perpendicular to the slope, it is also true that $\alpha + \beta = 90°$. Solving this equation for the angle β and substituting $\alpha = 65°$ gives

$$\beta = 90° - \alpha = 90° - 65° = 25°$$

FIGURE 6.8 Math Skills drawing.

Solution   Algebraically combining the four steps, we arrive at the following expression for the final speed of the skier:

STEP 1   STEP 2   STEP 3 and 4

$$v_f = \sqrt{\frac{2(KE_f)}{m}} = \sqrt{\frac{2\left(\frac{1}{2}mv_0^2 + W\right)}{m}} = \sqrt{\frac{2\left[\frac{1}{2}mv_0^2 + (170\ \text{N})(\cos\theta)s\right]}{m}}$$

Referring to the table of knowns and unknowns for the values of the symbols in this result, we find that the final speed of the skier is

$$v_f = \sqrt{\frac{2\left[\frac{1}{2}mv_0^2 + (170\ \text{N})(\cos\theta)s\right]}{m}}$$

$$= \sqrt{\frac{2\left[\frac{1}{2}(58\ \text{kg})(3.6\ \text{m/s})^2 + (170\ \text{N})(\cos 0°)(57\ \text{m})\right]}{58\ \text{kg}}} = \boxed{19\ \text{m/s}}$$

Related Homework: Problem 22

Problem-Solving Insight   Example 5 emphasizes that the *work–energy theorem deals with the work done by the net external force. The work–energy theorem does not apply to the work done by an individual force*, unless that force happens to be the only one present, in which case it is the net force.

If the work done by the net force is *positive,* as in Example 5, the kinetic energy of the object *increases.* If the work done is *negative,* the kinetic energy *decreases.* If the work is zero, the kinetic energy remains the same. Conceptual Example 6 explores these ideas further in a situation where the component of the net force along the direction of the displacement changes during the motion.

CONCEPTUAL EXAMPLE 6 | Work and Kinetic Energy

Figure 6.9 illustrates a satellite moving about the earth in a circular orbit and in an elliptical orbit. The only external force that acts on the satellite is the gravitational force. In which orbit does the kinetic energy of the satellite change, **(a)** the circular orbit or **(b)** the elliptical orbit?

Reasoning   The gravitational force is the only force acting on the satellite, so it is the net force. With this fact in mind, we will apply the work–energy theorem, which states that the work done by the net force equals the change in the kinetic energy.

Answer (a) is incorrect.   For the circular orbit in **Figure 6.9a**, the gravitational force $\vec{\mathbf{F}}$ does no work on the satellite since the force is perpendicular to the instantaneous displacement $\vec{\mathbf{s}}$ at all times. Thus, the work done by the net force is zero, and according to the work–energy theorem (Equation 6.3), the kinetic energy of the satellite (and, hence, its speed) remains the same everywhere on the orbit.

Answer (b) is correct.   For the elliptical orbit in **Figure 6.9b**, the gravitational force $\vec{\mathbf{F}}$ does do work. For example, as the satellite moves toward the earth in the top part of **Figure 6.9b**, there is a component of $\vec{\mathbf{F}}$ that points in the same direction as the

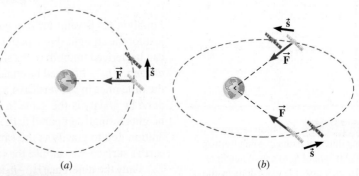

(a)   *(b)*

FIGURE 6.9   (a) In a circular orbit, the gravitational force $\vec{\mathbf{F}}$ is always perpendicular to the displacement $\vec{\mathbf{s}}$ of the satellite and does no work. (b) In an elliptical orbit, there can be a component of the force along the displacement, and, consequently, work can be done.

displacement. Consequently, $\vec{\mathbf{F}}$ does positive work during this part of the orbit, and the kinetic energy of the satellite increases. When the satellite moves away from the earth, as in the lower part of **Figure 6.9b**, $\vec{\mathbf{F}}$ has a component that points opposite to the displacement and, therefore, does negative work. As a result, the kinetic energy of the satellite decreases.

Related Homework: Problem 18

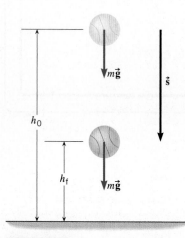

FIGURE 6.10 Gravity exerts a force $m\vec{g}$ on the basketball. Work is done by the gravitational force as the basketball falls from a height of h_0 to a height of h_f.

Check Your Understanding

(The answers are given at the end of the book.)

5. A sailboat is moving at a constant velocity. Is work being done by a net external force acting on the boat?

6. A ball has a speed of 15 m/s. Only one external force acts on the ball. After this force acts, the speed of the ball is 7 m/s. Has the force done **(a)** positive, **(b)** zero, or **(c)** negative work on the ball?

7. A rocket is at rest on the launch pad. When the rocket is launched, its kinetic energy increases. Consider all of the forces acting on the rocket during the launch, and decide whether the following statement is true or false: The amount by which the kinetic energy of the rocket increases during the launch is equal to the work done by the force generated by the rocket's engine.

8. A net external force acts on a particle that is moving along a straight line. This net force is not zero. Which one of the following statements is correct? **(a)** The velocity, but not the kinetic energy, of the particle is changing. **(b)** The kinetic energy, but not the velocity, of the particle is changing. **(c)** Both the velocity and the kinetic energy of the particle are changing.

6.3 | Gravitational Potential Energy

Work Done by the Force of Gravity

The gravitational force is a well-known force that can do positive or negative work, and **Figure 6.10** helps to show how the work can be determined. This drawing depicts a basketball of mass m moving vertically downward, the force of gravity $m\vec{g}$ being the only force acting on the ball. The initial height of the ball is h_0, and the final height is h_f, both distances measured from the earth's surface. The displacement $\vec{s}$ is downward and has a magnitude of $s = h_0 - h_f$. To calculate the work $W_{gravity}$ done on the ball by the force of gravity, we use $W = (F \cos \theta)s$ with $F = mg$ and $\theta = 0°$, since the force and displacement are in the same direction:

$$W_{gravity} = (mg \cos 0°)(h_0 - h_f) = mg(h_0 - h_f) \qquad (6.4)$$

Equation 6.4 is valid for *any path* taken between the initial and final heights, and not just for the straight-down path shown in **Figure 6.10**. For example, the same expression can be derived for both paths shown in **Figure 6.11**. Thus, only the *difference in vertical distances* $(h_0 - h_f)$ need be considered when calculating the work done by gravity. Since the difference in the vertical distances is the same for each path in the drawing, the work done by gravity is the same in each case. We are assuming here that the difference in heights is small compared to the radius of the earth, so that the magnitude g of the acceleration due to gravity is the same at every height. Moreover, for positions close to the earth's surface, we can use the value of $g = 9.80 \text{ m/s}^2$.

Only the difference $h_0 - h_f$ appears in Equation 6.4, so h_0 and h_f themselves need not be measured from the earth. For instance, they could be measured relative to a level that is one meter above the ground, and $h_0 - h_f$ would still have the same value. Example 7 illustrates how the work done by gravity is used with the work–energy theorem.

FIGURE 6.11 An object can move along different paths in going from an initial height of h_0 to a final height of h_f. In each case, the work done by the gravitational force is the same [$W_{gravity} = mg(h_0 - h_f)$], since the change in vertical distance $(h_0 - h_f)$ is the same.

EXAMPLE 7 | A Gymnast on a Trampoline

A gymnast springs vertically upward from a trampoline as in **Figure 6.12a**. The gymnast leaves the trampoline at a height of 1.20 m and reaches a maximum height of 4.80 m before falling back down. All heights are measured with respect to the ground. Ignoring air resistance, determine the initial speed v_0 with which the gymnast leaves the trampoline.

Reasoning We can find the initial speed of the gymnast (mass = m) by using the work–energy theorem, provided the work done by the net external force can be determined. Since only the gravitational force acts on the gymnast in the air, it is the net force, and we can evaluate the work by using the relation $W_{gravity} = mg(h_0 - h_f)$.

Solution **Figure 6.12b** shows the gymnast moving upward. The initial and final heights are $h_0 = 1.20$ m and $h_f = 4.80$ m, respectively. The initial speed is v_0 and the final speed is $v_f = 0$ m/s, since the gymnast comes to a momentary halt at the highest point. Since $v_f = 0$ m/s, the final kinetic energy is $KE_f = 0$ J, and the work–energy theorem becomes $W = KE_f - KE_0 = -KE_0$. The work W is that due to gravity, so this theorem reduces to

$$W_{gravity} = mg(h_0 - h_f) = -\tfrac{1}{2}mv_0^2$$

Solving for v_0 gives

$$v_0 = \sqrt{-2g(h_0 - h_f)} = \sqrt{-2(9.80 \text{ m/s}^2)(1.20 \text{ m} - 4.80 \text{ m})}$$

$$= \boxed{8.40 \text{ m/s}}$$

$v_f = 0$ m/s

h_f

h_0

v_0

(a)   (b)

FIGURE 6.12 (a) A gymnast bounces on a trampoline. (b) The gymnast moves upward with an initial speed v_0 and reaches maximum height with a final speed of zero.

Gravitational Potential Energy

An object in motion has kinetic energy. There are also other types of energy. For example, an object may possess energy by virtue of its position relative to the earth and is said to have gravitational potential energy. A pile driver, for instance, is used to pound "piles," or structural support beams, into the ground. The device contains a massive hammer that is raised to a height h and dropped (see **Figure 6.13**), so the hammer has the potential to do the work of driving the pile into the ground. The greater the height h, the greater is the potential for doing work, and the greater is the gravitational potential energy.

Now, let's obtain an expression for the gravitational potential energy. Our starting point is Equation 6.4 for the work done by the gravitational force as an object moves from an initial height h_0 to a final height h_f:

$$W_{gravity} = \underbrace{mgh_0}_{\substack{\text{Gravitational} \\ \text{potential energy} \\ PE_0 \text{ (initial)}}} - \underbrace{mgh_f}_{\substack{\text{Gravitational} \\ \text{potential energy} \\ PE_f \text{ (final)}}} \qquad \textbf{(6.4)}$$

This shows that the work done by the gravitational force is equal to the difference between the initial and final values of the quantity mgh. The value of mgh is larger when the height is larger and smaller when the height is smaller. We identify the quantity mgh as the **gravitational potential energy**. The concept of potential energy is associated only with a type of force known as a "conservative" force, as we will discuss in Section 6.4.

Hammer

$m\vec{g}$

h

Pile

FIGURE 6.13 In a pile driver, the gravitational potential energy of the hammer relative to the ground is $PE = mgh$.

DEFINITION OF GRAVITATIONAL POTENTIAL ENERGY

The gravitational potential energy PE is the energy that an object of mass m has by virtue of its position relative to the surface of the earth. That position is measured by the height h of the object relative to an arbitrary zero level:

$$PE = mgh \qquad \textbf{(6.5)}$$

SI Unit of Gravitational Potential Energy: joule (J)

Gravitational potential energy, like work and kinetic energy, is a scalar quantity and has the same SI unit as they do—the joule. It is the *difference* between two potential energies that is related by Equation 6.4 to the work done by the force of gravity. Therefore, the zero level for the heights can be taken anywhere, as long as both h_0 and h_f are measured relative to the same zero level. The gravitational potential energy depends on both the object and the earth (m and g, respectively), as well as the height h. Therefore, the gravitational potential energy belongs to the object and the earth as a system, although one often speaks of the object alone as possessing the gravitational potential energy.

Check Your Understanding

(The answer is given at the end of the book.)

9. In a simulation on earth, an astronaut in his space suit climbs up a vertical ladder. On the moon, the same astronaut makes the exact same climb. Which one of the following statements correctly describes how the gravitational potential energy of the astronaut changes during the climb? **(a)** It changes by a greater amount on the earth. **(b)** It changes by a greater amount on the moon. **(c)** The change is the same in both cases.

6.4 Conservative Versus Nonconservative Forces

The gravitational force has an interesting property that when an object is moved from one place to another, the work done by the gravitational force does not depend on the choice of path. In **Figure 6.11**, for instance, an object moves from an initial height h_0 to a final height h_f along two different paths. As Section 6.3 discusses, the work done by gravity depends only on the initial and final heights, and not on the path between these heights. For this reason, the gravitational force is called a **conservative force**, according to version 1 of the following definition:

> **DEFINITION OF A CONSERVATIVE FORCE**
>
> **Version 1** A force is conservative when the work it does on a moving object is independent of the path between the object's initial and final positions.
>
> **Version 2** A force is conservative when it does no net work on an object moving around a closed path, starting and finishing at the same point.

Figure 6.14 helps us to illustrate version 2 of the definition of a conservative force. The picture shows a roller coaster car racing through dips and double dips, ultimately

Start

FIGURE 6.14 A roller coaster track is an example of a closed path.

returning to its starting point. This kind of path, which begins and ends at the same place, is called a *closed* path. Gravity provides the only force that does work on the car, assuming that there is no friction or air resistance. Of course, the track exerts a normal force, but this force is always directed perpendicular to the motion and does no work. On the downward parts of the trip, the gravitational force does positive work, increasing the car's kinetic energy. Conversely, on the upward parts of the motion, the gravitational force does negative work, decreasing the car's kinetic energy. Over the entire trip, the gravitational force does as much positive work as negative work, so the net work is zero, and the car returns to its starting point with the same kinetic energy it had at the start. Therefore, consistent with version 2 of the definition of a conservative force, $W_{\text{gravity}} = 0$ J for a closed path.

The gravitational force is our first example of a conservative force. Later, we will encounter others, such as the elastic force of a spring and the electrical force of electrically charged particles. With each conservative force we will associate a potential energy, as we have already done in the gravitational case (see Equation 6.5). For other conservative forces, however, the algebraic form of the potential energy will differ from that in Equation 6.5.

Not all forces are conservative. A force is nonconservative if the work it does on an object moving between two points depends on the path of the motion between the points. The kinetic frictional force is one example of a nonconservative force. When an object slides over a surface and kinetic friction is present, the frictional force points opposite to the sliding motion and does negative work. Between any two points, greater amounts of work are done over longer paths between the points, so that the work depends on the choice of path. Thus, the kinetic frictional force is nonconservative. Air resistance is another nonconservative force. ***The concept of potential energy is not defined for a nonconservative force.***

For a closed path, the total work done by a nonconservative force is not zero as it is for a conservative force. In **Figure 6.14**, for instance, a frictional force would oppose the motion and slow down the car. Unlike gravity, friction would do negative work on the car throughout the entire trip, on *both* the up and down parts of the motion. Assuming that the car makes it back to the starting point, the car would have *less* kinetic energy than it had originally. **Table 6.2** gives some examples of conservative and nonconservative forces.

In normal situations, conservative forces (such as gravity) and nonconservative forces (such as friction and air resistance) act simultaneously on an object. Therefore, we write the work W done by the net external force as $W = W_{\text{c}} + W_{\text{nc}}$, where W_{c} is the work done by the conservative forces and W_{nc} is the work done by the nonconservative forces. According to the work–energy theorem, the work done by the net external force is equal to the change in the object's kinetic energy, or $W_{\text{c}} + W_{\text{nc}} = \frac{1}{2}mv_{\text{f}}^2 - \frac{1}{2}mv_0^2$. If the only conservative force acting is the gravitational force, then $W_{\text{c}} = W_{\text{gravity}} = mg(h_0 - h_{\text{f}})$, and the work–energy theorem becomes

$$mg(h_0 - h_{\text{f}}) + W_{\text{nc}} = \tfrac{1}{2}mv_{\text{f}}^2 - \tfrac{1}{2}mv_0^2$$

The work done by the gravitational force can be moved to the right side of this equation, with the result that

$$W_{\text{nc}} = \left(\tfrac{1}{2}mv_{\text{f}}^2 - \tfrac{1}{2}mv_0^2\right) + (mgh_{\text{f}} - mgh_0) \tag{6.6}$$

In terms of kinetic and potential energies, we find that

$$\underbrace{W_{\text{nc}}}_{\substack{\text{Net work done by} \\ \text{nonconservative} \\ \text{forces}}} = \underbrace{(\text{KE}_{\text{f}} - \text{KE}_0)}_{\substack{\text{Change in kinetic} \\ \text{energy}}} + \underbrace{(\text{PE}_{\text{f}} - \text{PE}_0)}_{\substack{\text{Change in} \\ \text{gravitational} \\ \text{potential energy}}} \tag{6.7a}$$

Equation 6.7a states that the net work W_{nc} done by all the external nonconservative forces equals the change in the object's kinetic energy plus the change in its gravitational potential energy. It is customary to use the delta symbol (Δ) to denote such changes;

TABLE 6.2	Some Conservative and Nonconservative Forces

Conservative Forces

Gravitational force (Ch. 4)

Elastic spring force (Ch. 10)

Electric force (Ch. 18, 19)

Nonconservative Forces

Static and kinetic frictional forces

Air resistance

Tension

Normal force

Propulsion force of a rocket

thus, $\Delta \text{KE} = (\text{KE}_f - \text{KE}_0)$ and $\Delta \text{PE} = (\text{PE}_f - \text{PE}_0)$. With the delta notation, the work–energy theorem takes the form

$$W_{nc} = \Delta \text{KE} + \Delta \text{PE} \qquad \text{(6.7b)}$$

In the next two sections, we will show why the form of the work–energy theorem expressed by Equations 6.7a and 6.7b is useful.

Math Skills   The change in both the kinetic and the potential energy is *always* the final value minus the initial value: $\Delta \text{KE} = \text{KE}_f - \text{KE}_0$ and $\Delta \text{PE} = \text{PE}_f - \text{PE}_0$. For example, consider a skier starting from rest at the top of a steep ski jumping ramp, gathering speed on the way down, and leaving the end of the ramp at a high speed. The initial kinetic energy is $\text{KE}_0 = 0$ J, since the skier is at rest at the top. The final kinetic energy, however, is $\text{KE}_f = \frac{1}{2}mv_f^2$, since the skier leaves the ramp at a high speed v_f. The kinetic energy increases from zero to a nonzero value, so the *change* in kinetic energy has a positive value of

$$\Delta \text{KE} = \text{KE}_f - \text{KE}_0 = \tfrac{1}{2}mv_f^2 - 0 \text{ J} = \tfrac{1}{2}mv_f^2$$

If ΔKE were written incorrectly as the initial value minus the final value or $\Delta \text{KE} = \text{KE}_0 - \text{KE}_f$, the skier's kinetic energy would appear to change by the following amount:

Incorrect Equation

$$\Delta \text{KE} = \text{KE}_0 - \text{KE}_f = 0 \text{ J} - \tfrac{1}{2}mv_f^2 - \tfrac{1}{2}mv_f^2$$

This change is negative. It would imply that the skier, although starting at the top with zero kinetic energy, nevertheless loses kinetic energy on the way down! This is nonsense and would lead to serious errors in problem solving.

6.5 The Conservation of Mechanical Energy

The concept of work and the work–energy theorem have led us to the conclusion that an object can possess two kinds of energy: kinetic energy, KE, and gravitational potential energy, PE. The sum of these two energies is called the **total mechanical energy E**, so that $E = \text{KE} + \text{PE}$. Later on we will update this definition to include other types of potential energy in addition to the gravitational form. The concept of total mechanical energy will be extremely useful in describing the motion of objects.

By rearranging the terms on the right side of Equation 6.7a, the work–energy theorem can be expressed in terms of the total mechanical energy:

$$W_{nc} = (\text{KE}_f - \text{KE}_0) + (\text{PE}_f - \text{PE}_0) \qquad \text{(6.7a)}$$
$$= \underbrace{(\text{KE}_f + \text{PE}_f)}_{E_f} - \underbrace{(\text{KE}_0 + \text{PE}_0)}_{E_0}$$

or

$$W_{nc} = E_f - E_0 \qquad \text{(6.8)}$$

Remember: Equation 6.8 is just another form of the work–energy theorem. It states that W_{nc}, the net work done by external nonconservative forces, changes the total mechanical energy from an initial value of E_0 to a final value of E_f.

The conciseness of the work–energy theorem in the form $W_{nc} = E_f - E_0$ allows an important basic principle of physics to stand out. This principle is known as the conservation of mechanical energy. Suppose that the net work W_{nc} done by external nonconservative forces is zero, so $W_{nc} = 0$ J. Then, Equation 6.8 reduces to

$$E_f = E_0 \qquad \text{(6.9a)}$$
$$\underbrace{\tfrac{1}{2}mv_f^2 + mgh_f}_{E_f} = \underbrace{\tfrac{1}{2}mv_0^2 + mgh_0}_{E_0} \qquad \text{(6.9b)}$$

KE	PE	E = KE + PE	
			$\vec{v}_0 = 0$ m/s
0 J	600 000 J	600 000 J	
200 000 J	400 000 J	600 000 J	
400 000 J	200 000 J	600 000 J	
600 000 J	0 J	600 000 J	

INTERACTIVE FIGURE 6.15 If friction and wind resistance are ignored, a bobsled run illustrates how kinetic and potential energy can be interconverted, while the total mechanical energy remains constant. The total mechanical energy is 600 000 J, being all potential energy at the top and all kinetic energy at the bottom.

Equation 6.9a indicates that the final mechanical energy is equal to the initial mechanical energy. Consequently, the total mechanical energy *remains constant all along the path between the initial and final points,* never varying from the initial value of E_0. A quantity that remains constant throughout the motion is said to be "conserved." The fact that the total mechanical energy is conserved when $W_{nc} = 0$ J is called the **principle of conservation of mechanical energy.**

> **THE PRINCIPLE OF CONSERVATION OF MECHANICAL ENERGY**
>
> **The total mechanical energy (E = KE + PE) of an object remains constant as the object moves, provided that the net work done by external nonconservative forces is zero, $W_{nc} = 0$ J.**

The principle of conservation of mechanical energy offers keen insight into the way in which the physical universe operates. While the sum of the kinetic and potential energies at any point is conserved, the two forms may be interconverted or transformed into one another. Kinetic energy of motion is converted into potential energy of position, for instance, when a moving object coasts up a hill. Conversely, potential energy is converted into kinetic energy when an object is allowed to fall. **Interactive Figure 6.15** shows such transformations of energy for a bobsled run, assuming that nonconservative forces, such as friction and wind resistance, can be ignored. The normal force, being directed perpendicular to the path, does no work. Only the force of gravity does work, so the total mechanical energy E remains constant at all points along the run. The conservation principle is well known for the ease with which it can be applied, as in the following example.

EXAMPLE 8 | A Daredevil Motorcyclist

A motorcyclist is trying to leap across the canyon shown in **Figure 6.16** by driving horizontally off the cliff at a speed of 38.0 m/s. Ignoring air resistance, find the speed with which the cycle strikes the ground on the other side.

Reasoning Once the cycle leaves the cliff, no forces other than gravity act on the cycle, since air resistance is being ignored. Thus, the work done by external nonconservative forces is zero, $W_{nc} = 0$ J. Accordingly, the principle of conservation of mechanical energy

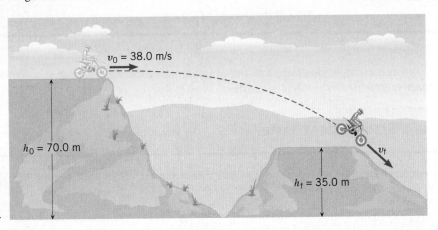

FIGURE 6.16 A daredevil jumping a canyon.

holds, so the total mechanical energy is the same at the final and initial positions of the motorcycle. We will use this important observation to determine the final speed of the cyclist.

> **Problem-Solving Insight** Be on the alert for factors, such as the mass m here in Example 8, that sometimes can be eliminated algebraically when using the conservation of mechanical energy.

Solution The principle of conservation of mechanical energy is written as

$$\underbrace{\tfrac{1}{2}mv_f^2 + mgh_f}_{E_f} = \underbrace{\tfrac{1}{2}mv_0^2 + mgh_0}_{E_0} \tag{6.9b}$$

The mass m of the rider and cycle can be eliminated algebraically from this equation, since m appears as a factor in every term. Solving for v_f gives

$$v_f = \sqrt{v_0^2 + 2g(h_0 - h_f)}$$

$$v_f = \sqrt{(38.0\ \text{m/s})^2 + 2(9.80\ \text{m/s}^2)(70.0\ \text{m} - 35.0\ \text{m})} = \boxed{46.2\ \text{m/s}}$$

Examples 9 and 10 emphasize that the principle of conservation of mechanical energy can be applied even when forces act perpendicular to the path of a moving object.

CONCEPTUAL EXAMPLE 9 | The Favorite Swimming Hole

A rope is tied to a tree limb. It is used by a swimmer who, starting from rest, swings down toward the water below, as in **Figure 6.17**. Only two forces act on him during his descent, the nonconservative force $\vec{T}$, which is due to the tension in the rope, and his weight, which is due to the conservative gravitational force. There is no air resistance. His initial height h_0 and final height h_f are known. Can we use the principle of conservation of mechanical energy to find his speed v_f at the point where he lets go of the rope, even though a nonconservative external force is present? **(a)** Yes. **(b)** No.

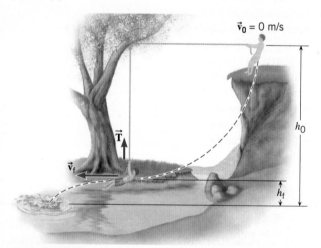

$\vec{v}_0 = 0$ m/s

h_0

$\vec{T}$

$\vec{v}_f$

h_f

FIGURE 6.17 During the downward swing, the tension $\vec{T}$ in the rope acts perpendicular to the circular arc and, hence, does no work on the person.

Reasoning The principle of conservation of mechanical energy can be used even in the presence of nonconservative forces, provided that the net work W_{nc} done by the nonconservative forces is zero. $W_{nc} = 0$ J.

Answer (b) is incorrect. The mere presence of nonconservative forces does not prevent us from applying the principle of conservation of mechanical energy. The deciding factor is whether the net work done by the nonconservative forces is zero. In this case, the tension force $\vec{T}$ is always perpendicular to the circle (see **Figure 6.17**). Thus, the angle θ between $\vec{T}$ and the instantaneous displacement of the person is always 90°. According to Equation 6.1, the work is proportional to the cosine of this angle, or cos 90°, which is zero, with the result that the work is also zero. As a result, the tension here does not prevent us from using the conservation principle.

> **Problem-Solving Insight** When nonconservative forces are perpendicular to the motion, we can still use the principle of conservation of mechanical energy, because such "perpendicular" forces do no work.

Answer (a) is correct. As the person swings downward, he follows a circular path, as shown in **Figure 6.17**. Since the tension force is the only nonconservative force acting and is always perpendicular to this path, the corresponding net work is $W_{nc} = 0$ J, and we can apply the principle of conservation of mechanical energy to find the speed v_f.

Related Homework: Problem 38

The next example illustrates how the conservation of mechanical energy is applied to the breathtaking drop of a roller coaster.

EXAMPLE 10 | The Physics of a Giant Roller Coaster

The Kingda Ka is a giant among roller coasters (see **Figure 6.18**). Located in Jackson Township, New Jersey, the ride includes a vertical drop of 127 m. Suppose that the coaster has a speed of 6.00 m/s at the top of the drop. Neglect friction and air resistance and find the speed of the riders at the bottom.

Reasoning Since we are neglecting friction and air resistance, we may set the work done by these forces equal to zero. A normal force from the seat acts on each rider, but this force is perpendicular to the motion, so it does not do any work. Thus, the work done by external nonconservative forces is zero, $W_{nc} = 0$ J, and we may use the principle of conservation of mechanical energy to find the speed of the riders at the bottom.

Solution The principle of conservation of mechanical energy states that

$$\underbrace{\tfrac{1}{2}mv_f^2 + mgh_f}_{E_f} = \underbrace{\tfrac{1}{2}mv_0^2 + mgh_0}_{E_0} \qquad \textbf{(6.9b)}$$

The mass m of the rider appears as a factor in every term in this equation and can be eliminated algebraically. Solving for the final speed gives

$$v_f = \sqrt{v_0^2 + 2g(h_0 - h_f)}$$

$$v_f = \sqrt{(6.00 \text{ m/s})^2 + 2(9.80 \text{ m/s}^2)(127 \text{ m})}$$

$$= \boxed{50.3 \text{ m/s } (112 \text{ mi/h})}$$

where the vertical drop is $h_0 - h_f = 127$ m.

FIGURE 6.18 The Kingda Ka roller coaster in Six Flags Great Adventure, located in Jackson Township, New Jersey, is a giant. It includes a vertical drop of 127 m.

Six Flags/Splash News/Newscom

When applying the principle of conservation of mechanical energy in solving problems, we have been using the following reasoning strategy:

> **REASONING STRATEGY   Applying the Principle of Conservation of Mechanical Energy**
>
> 1. Identify the external conservative and nonconservative forces that act on the object. For this principle to apply, the total work done by nonconservative forces must be zero, $W_{nc} = 0$ J. A nonconservative force that is perpendicular to the displacement of the object does no work, for example.
>
> 2. Choose the location where the gravitational potential energy is taken to be zero. This location is arbitrary but must not be changed during the course of solving a problem.
>
> 3. Set the final total mechanical energy of the object equal to the initial total mechanical energy, as in Equations 6.9a and 6.9b. The total mechanical energy is the sum of the kinetic and potential energies.

Check Your Understanding

(The answers are given at the end of the book.)

10. Suppose the total mechanical energy of an object is conserved. Which one or more of the following statements is/are true? **(a)** If the kinetic energy decreases, the gravitational potential energy increases. **(b)** If the gravitational potential energy decreases, the kinetic energy increases. **(c)** If the kinetic energy does not change, the gravitational potential energy also does not change.

11. In Example 10 the Kingda Ka roller coaster starts with a speed of 6.0 m/s at the top of the drop and attains a speed of 50.3 m/s when it reaches the bottom. If the roller coaster were then to start up an identical hill, what speed would it attain when it reached the top? Assume that friction and air resistance are absent. **(a)** Greater than 6.0 m/s **(b)** Exactly 6.0 m/s **(c)** Between 0 m/s and 6.0 m/s **(d)** 0 m/s

12. CYU Figure 6.2 shows an empty fuel tank about to be released by three different jet planes. At the moment of release, each plane has the same speed, and each tank is at the same height above the ground. However, the direc-

Fuel tank

A   B   C

CYU FIGURE 6.2

tions of the velocities of the planes are different. Which tank has the largest speed upon hitting the ground? Ignore friction and air resistance. **(a)** A **(b)** B **(c)** C **(d)** Each tank hits the ground with the same speed.

13. In which one or more of the following situations is the principle of conservation of mechanical energy obeyed? **(a)** An object moves uphill with an increasing speed. **(b)** An object moves uphill with a decreasing speed. **(c)** An object moves uphill with a constant speed. **(d)** An object moves downhill with an increasing speed. **(e)** An object moves downhill with a decreasing speed. **(f)** An object moves downhill with a constant speed.

6.6 | Nonconservative Forces and the Work–Energy Theorem

Most moving objects experience nonconservative forces, such as friction, air resistance, and propulsive forces, and the work W_{nc} done by the net external nonconservative force is not zero. In these situations, the difference between the final and initial total mechanical energies is equal to W_{nc} according to $W_{nc} = E_f - E_0$ (Equation 6.8). Consequently, the total mechanical energy is not conserved. The next two examples illustrate how Equation 6.8 is used when nonconservative forces are present and do work.

EXAMPLE 11 | The Kingda Ka Revisited

In Example 10, we ignored nonconservative forces, such as friction. In reality, however, such forces are present when the roller coaster descends. The actual speed of the riders at the bottom is 45.0 m/s, which is less than that determined in Example 10. Assuming again that the coaster has a speed of 6.00 m/s at the top, find the work done by nonconservative forces on a 55.0-kg rider during the descent from a height h_0 to a height h_f, where $h_0 - h_f = 127$ m.

Reasoning   Since the speed at the top, the final speed, and the vertical drop are given, we can determine the initial and final total mechanical energies of the rider. The work–energy theorem, $W_{nc} = E_f - E_0$, can then be used to determine the work W_{nc} done by the nonconservative forces.

> **Problem-Solving Insight**   As illustrated here and in Example 3, a nonconservative force such as friction can do negative or positive work. It does

negative work when it has a component opposite to the displacement and slows down the object. It does positive work when it has a component in the direction of the displacement and speeds up the object.

Solution   The work–energy theorem is

$$W_{nc} = \underbrace{\left(\tfrac{1}{2}mv_f^2 + mgh_f\right)}_{E_f} - \underbrace{\left(\tfrac{1}{2}mv_0^2 + mgh_0\right)}_{E_0} \quad (6.8)$$

Rearranging the terms on the right side of this equation gives

$$W_{nc} = \tfrac{1}{2}m(v_f^2 - v_0^2) - mg(h_0 - h_f)$$

$$W_{nc} = \tfrac{1}{2}(55.0 \text{ kg})[(45.0 \text{ m/s})^2 - (6.00 \text{ m/s})^2] -$$

$$(55.0 \text{ kg})(9.80 \text{ m/s}^2)(127 \text{ m}) = \boxed{-1.38 \times 10^4 \text{ J}}$$

EXAMPLE 12 | Fireworks

A 0.20-kg rocket in a fireworks display is launched from rest and follows an erratic flight path to reach the point P, as **Figure 6.19** shows. Point P is 29 m above the starting point. In the process, 425 J of work is done on the rocket by

the nonconservative force generated by the burning propellant. Ignoring air resistance and the mass lost due to the burning propellant, find the speed v_f of the rocket at the point P.

Reasoning The only nonconservative force acting on the rocket is the force generated by the burning propellant, and the work done by this force is $W_{nc} = 425$ J. Because work is done by a nonconservative force, we use the work–energy theorem in the form $W_{nc} = E_f - E_0$ to find the final speed v_f of the rocket.

Solution From the work–energy theorem we have

$$W_{nc} = \left(\tfrac{1}{2}mv_f^2 + mgh_f\right) - \left(\tfrac{1}{2}mv_0^2 + mgh_0\right) \tag{6.8}$$

Solving this expression for the final speed of the rocket and noting that the initial speed of the rocket at rest is $v_0 = 0$ m/s, we get

$$v_f = \sqrt{\frac{2[W_{nc} + \tfrac{1}{2}mv_0^2 - mg(h_f - h_0)]}{m}}$$

$$v_f = \sqrt{\frac{2[425\ \text{J} + \tfrac{1}{2}(0.20\ \text{kg})(0\ \text{m/s})^2 - (0.20\ \text{kg})(9.80\ \text{m/s}^2)(29\ \text{m})]}{0.20\ \text{kg}}}$$

$$= \boxed{61\ \text{m/s}}$$

FIGURE 6.19 A fireworks rocket, moving along an erratic flight path, reaches a point P that is 29 m above the launch point.

Check Your Understanding

(The answers are given at the end of the book.)

14. A net external nonconservative force does positive work on a particle. Based solely on this information, you are justified in reaching only one of the following conclusions. Which one is it? **(a)** The kinetic and potential energies of the particle both decrease. **(b)** The kinetic and potential energies of the particle both increase. **(c)** Neither the kinetic nor the potential energy of the particle changes. **(d)** The total mechanical energy of the particle decreases. **(e)** The total mechanical energy of the particle increases.

15. In one case, a sports car, its engine running, is driven up a hill at a constant speed. In another case, a truck approaches a hill, and its driver turns off the engine at the bottom of the hill. The truck then coasts up the hill. Which vehicle is obeying the principle of conservation of mechanical energy? Ignore friction and air resistance. **(a)** Both the sports car and the truck **(b)** Only the sports car **(c)** Only the truck

6.7 | Power

In many situations, the time it takes to do work is just as important as the amount of work that is done. Consider two automobiles that are identical in all respects (e.g., both have the same mass), except that one has a "souped-up" engine. The car with the souped-up engine can go from 0 to 27 m/s (60 mph) in 4 seconds, while the other car requires 8 seconds to achieve the same speed. Each engine does work in accelerating its car, but one does it more quickly. Where cars are concerned, we associate the quicker performance with an engine that has a larger horsepower rating. A large horsepower rating means that the engine can do a large amount of work in a short time. In physics, the horsepower rating is just one way to measure an engine's ability to generate power. The idea of **power** incorporates both the concepts of work and time, for power is work done per unit time.

> **DEFINITION OF AVERAGE POWER**
>
> Average power $\overline{P}$ is the average rate at which work W is done, and it is obtained by dividing W by the time t required to perform the work:
>
> $$\overline{P} = \frac{\text{Work}}{\text{Time}} = \frac{W}{t} \tag{6.10a}$$
>
> SI Unit of Power: joule/s = watt (W)

The definition of average power presented in Equation 6.10a involves work. However, the work–energy theorem relates the work done by a net external force to the change in the energy of the object (see, for example, Equations 6.3 and 6.8). Therefore, we can also define average power as the rate at which the energy is changing, or as the change in energy divided by the time during which the change occurs:

$$\bar{P} = \frac{\text{Change in energy}}{\text{Time}} \tag{6.10b}$$

Since work, energy, and time are scalars, power is also a scalar. The unit in which power is expressed is that of work divided by time, or a joule per second in SI units. One joule per second is called a watt (W), in honor of James Watt (1736–1819), developer of the steam engine. The unit of power in the BE system is the foot · pound per second (ft · lb/s), although the familiar horsepower (hp) unit is used frequently for specifying the power generated by electric motors and internal combustion engines:

$$1 \text{ horsepower} = 550 \text{ foot} \cdot \text{pounds/second} = 745.7 \text{ watts}$$

Table 6.3 summarizes the units for power in the various systems of measurement.

TABLE 6.3 Units of Measurement for Power

System	Work	÷	Time	=	Power
SI	joule (J)		second (s)		watt (W)
CGS	erg		second (s)		erg per second (erg/s)
BE	foot · pound (ft · lb)		second (s)		foot · pound per second (ft · lb/s)

BIO THE PHYSICS OF . . . human metabolism. Equation 6.10b provides the basis for understanding the production of power in the human body. In this context the "Change in energy" on the right-hand side of the equation refers to the energy produced by metabolic processes, which, in turn, is derived from the food we eat. Table 6.4 gives typical metabolic rates of energy production needed to sustain various activities. Running at 15 km/h (9.3 mi/h), for example, requires metabolic power sufficient to operate eighteen 75-watt light bulbs, and the metabolic power used in sleeping would operate a single 75-watt bulb.

An alternative expression for power can be obtained from Equation 6.1, which indicates that the work W done when a constant net force of magnitude F points in the same direction as the displacement is $W = (F \cos 0°)s = Fs$. Dividing both sides of this equation by the time t it takes for the force to move the object through the distance s, we obtain

$$\frac{W}{t} = \frac{Fs}{t}$$

But W/t is the average power $\bar{P}$, and s/t is the average speed $\bar{v}$, so that

$$\bar{P} = F\bar{v} \tag{6.11}$$

Example 13 applies the concepts of work and power to the performance of the human heart.

TABLE 6.4 Human Metabolic Rates[a]

Activity	Rate (watts)
Running (15 km/h)	1340 W
Skiing	1050 W
Biking	530 W
Walking (5 km/h)	280 W
Sleeping	77 W

[a]For a young 70-kg male.

EXAMPLE 13 | BIO Power Output of the Human Heart

The total average work done during one contraction and relaxation (i.e., during one beat) of the heart of a person at rest is about 1.33 J. This is split between the work done by the right ventricle (0.22 J) and the left ventricle (1.11 J). **(a)** If the efficiency of the heart is 10.0% (i.e., its work output is 1/10th of its energy input), what is the total energy expended in one day by the heart of someone who has an average heart rate of 70.0 beats/minute? **(b)** What is the average output power of the heart during this time?

Reasoning (a) Knowing the total work done during a single beat of the heart and the heart rate, the total work done by the heart in one day can be determined. Since we know the efficiency, we can

then determine the total energy expended. **(b)** The average power output is equal to the total work done divided by the total time.

Solution (a) The total work done in one day is the total number of heartbeats (N) in one day multiplied by the work done during one heartbeat (W_B):

$$W_T = NW_B$$

The total number of heartbeats is given by the number of beats per minute multiplied by the number of minutes in one day:

$$N = \left(\frac{70 \text{ beats}}{\text{min}}\right)\left(\frac{60 \text{ min}}{\text{hr}}\right)\left(\frac{24 \text{ hr}}{\text{day}}\right) = 100\ 804 \frac{\text{beats}}{\text{day}}$$

The total work done by the heart in one day is then given by

$$W_T = NW_B = (100\ 804\ \text{beats})\left(1.33\ \frac{\text{J}}{\text{beats}}\right) = 1.34 \times 10^5\ \text{J}$$

Since the energy input is 10 times the work output, the total energy (E_T) expended by the heart in one day is given by

$$E_T = (10)\,W_T = \boxed{1.34 \times 10^6\ \text{J}}$$

The energy expended by the heart ultimately comes from food, the energy of which is measured in units of kcal (or the Calorie). Since 1 kcal = 4186 J, the heart burns 320.0 kcal per day in this example.

(b) The average power is equal to the work done divided by the time (Equation 6.10a). In order to determine the output power in watts (W), we divide the total work done by the heart in one day by the number of seconds in one day (86 400 s) to obtain

$$P = \frac{W_T}{t} = \frac{1.34 \times 10^5\ \text{J}}{8.64 \times 10^4\ \text{s}} = \boxed{1.55\ \text{W}}$$

This is the output power. Considering that the energy efficiency of the heart is about 10%, the required input power delivered to the heart by the body is 15.5 W. The total metabolic rate for a resting person is about 100 W, that is, about the same value as that of the electrical power delivered to a bright light bulb. The heart therefore accounts for about 15% of the body's total power output.

Check Your Understanding

(The answers are given at the end of the book.)

16. Engine A has a greater power rating than engine B. Which one of the following statements correctly describes the abilities of these engines to do work? **(a)** Engines A and B can do the same amount of work, but engine A can do it more quickly. **(b)** Engines A and B can do the same amount of work in the same amount of time. **(c)** In the same amount of time, engine B can do more work than engine A.

17. Is it correct to conclude that one engine is doing twice the work that another is doing just because it is generating twice the power? **(a)** Yes **(b)** No

6.8 | Other Forms of Energy and the Conservation of Energy

Up to now, we have considered only two types of energy, kinetic energy and gravitational potential energy. There are many other types, however. Electrical energy is used to run electrical appliances. Energy in the form of heat is utilized in cooking food. Moreover, the work done by the kinetic frictional force often appears as heat, as you can experience by rubbing your hands back and forth. When gasoline is burned, some of the stored chemical energy is released and does the work of moving cars, airplanes, and boats. The chemical energy stored in food provides the energy needed for metabolic processes.

The research of many scientists, most notably Albert Einstein, led to the discovery that mass itself is one manifestation of energy. Einstein's famous equation, $E_0 = mc^2$, describes how mass m and energy E_0 are related, where c is the speed of light in a vacuum and has a value of 3.00×10^8 m/s. Because the speed of light is so large, this equation implies that very small masses are equivalent to large amounts of energy. The relationship between mass and energy will be discussed further in Chapter 28.

We have seen that kinetic energy can be converted into gravitational potential energy and vice versa. In general, energy of all types can be converted from one form to another.

BIO **THE PHYSICS OF . . . transforming chemical energy in food into mechanical energy.** Part of the chemical energy stored in food, for example, is transformed into gravitational potential energy when a hiker climbs a mountain. Suppose a 65-kg hiker eats a 250-Calorie* snack, which contains 1.0×10^6 J of chemical energy. If this were 100% converted into potential energy $mg(h_f - h_0)$, the change in height would be

$$h_f - h_0 = \frac{1.0 \times 10^6\ \text{J}}{(65\ \text{kg})(9.8\ \text{m/s}^2)} = 1600\ \text{m}$$

*Energy content in food is typically given in units called Calories, which we will discuss in Section 12.7.

At a more realistic conversion efficiency of 50%, the change in height would be 800 m. Similarly, in a moving car the chemical energy of gasoline is converted into kinetic energy, as well as into electrical energy and heat.

Whenever energy is transformed from one form to another, it is found that no energy is gained or lost in the process; the total of all the energies before the process is equal to the total of the energies after the process. This observation leads to the following important principle:

> **THE PRINCIPLE OF CONSERVATION OF ENERGY**
>
> **Energy can neither be created nor destroyed, but can only be converted from one form to another.**

Learning how to convert energy from one form to another more efficiently is one of the main goals of modern science and technology.

6.9 | Work Done by a Variable Force

The work W done by a constant force (constant in both magnitude and direction) is given by Equation 6.1 as $W = (F \cos \theta)s$. Quite often, situations arise in which the force is not constant but changes with the displacement of the object. For instance, **Animated Figure 6.20a** shows an archer using a high-tech compound bow. This type of bow consists of a series of pulleys and strings that produce a force–displacement graph like that in **Animated Figure 6.20b**. One of the key features of the compound bow is that the force rises to a maximum as the string is drawn back, and then falls to 60% of this maximum value when the string is fully drawn. The reduced force at $s = 0.500$ m makes it much easier for the archer to hold the fully drawn bow while aiming the arrow.

ANIMATED FIGURE 6.20 (a) A compound bow. (b) A plot of $F \cos \theta$ versus s as the bowstring is drawn back.

When the force varies with the displacement, as in **Animated Figure 6.20b**, we cannot use the relation $W = (F \cos \theta)s$ to find the work, because this equation is valid only when the force is constant. However, we can use a graphical method. In this method we divide the total displacement into very small segments, Δs_1, Δs_2, and so on (see **Figure 6.21a**). For each segment, the *average value* of the force component is indicated by a short horizontal line. For example, the short horizontal line for segment Δs_1 is labeled $(F \cos \theta)_1$ in **Figure 6.21a**. We can then use this average value as the constant-force component in Equation 6.1 and determine an approximate value for the work ΔW_1 done during the first segment: $\Delta W_1 = (F \cos \theta)_1 \Delta s_1$. But this work is just the area of the colored rectangle in the drawing. The word "area" here refers to the area of a rectangle that has a width of Δs_1 and a height of $(F \cos \theta)_1$; it does not mean an area in square meters, such as the area of a parcel of land. In a like manner, we can calculate an approximate

FIGURE 6.21 (a) The work done by the average-force component $(F \cos \theta)_1$ during the small displacement Δs_1 is $(F \cos \theta)_1 \Delta s_1$, which is the area of the colored rectangle. (b) The work done by a variable force is equal to the colored area under the graph of $F \cos \theta$ versus s.

value for the work for each segment. Then we add the results for the segments to get, approximately, the work W done by the variable force:

$$W \approx (F \cos \theta)_1 \, \Delta s_1 + (F \cos \theta)_2 \, \Delta s_2 + \cdots$$

The symbol $\approx$ means "approximately equal to." The right side of this equation is the sum of all the rectangular areas in **Figure 6.21a** and is an approximate value for the area shaded in color under the graph in **Figure 6.21b**. If the rectangles are made narrower and narrower by decreasing each Δs, the right side of this equation eventually becomes equal to the area under the graph.

> **Problem-Solving Insight** The work done by a variable force in moving an object is equal to the area under the graph of $F \cos \theta$ versus s.

Example 14 illustrates how to use this graphical method to determine the approximate work done when a high-tech compound bow is drawn.

EXAMPLE 14 | The Physics of a Compound Bow

Find the work that the archer must do in drawing back the string of the compound bow in **Animated Figure 6.20a** from 0 to 0.500 m.

Reasoning The work is equal to the colored area under the curved line in **Animated Figure 6.20b**. For convenience, this area is divided into a number of small squares, each having an area of $(9.00 \text{ N})(2.78 \times 10^{-2} \text{ m}) = 0.250$ J. The area can be found by counting the number of squares under the curve and multiplying by the area per square.

Solution We estimate that there are 242 colored squares in the drawing. Since each square represents 0.250 J of work, the total work done is

$$W = (242 \text{ squares})\left(0.250 \, \frac{\text{J}}{\text{square}}\right) = \boxed{60.5 \text{ J}}$$

When the arrow is fired, part of this work is imparted to it as kinetic energy.

EXAMPLE 15 | BIO The Physics of Seatbelts

Seatbelts are the most important safety feature for automobile drivers and passengers. Roughly 50% of passengers killed in auto accidents were not wearing seatbelts. In modern cars, seatbelts, along with sensors and airbags, function as an advanced safety restraining system. The graph below shows the force applied to the chest wall as a function of displacement during a frontend collision. Modern seatbelts use a pretensioner effect when the collision is high-speed, and therefore particularly dangerous. This reduces the maximum force experienced by the chest wall from the seatbelt during the impact. Using the data in the graph, calculate the work done on the chest wall from the variable force applied by the seatbelt. Assume the force from the seatbelt is applied parallel to the chest wall displacement.

Reasoning The work is equal to the area under the solid curve. We see from the graph that this will consist of the combined area of two rectangles (Areas I and III), and one triangle (Area II).

Solution The total area under the curve will be equal to Area I + Area II + Area III. From the data in **Figure 6.22**, we see that Area I $= (5000 \text{ N} - 0 \text{ N})(0.20 \text{ m} - 0.00 \text{ m}) = 1000$ J. Area II $= \frac{1}{2}(0.20 \text{ m} - $

FIGURE 6.22 The variable force applied by a seatbelt to the chest wall as a function of the wall's displacement. The work done by the seatbelt will be equal to the total area under the solid curve.

$0.00 \text{ m})(13\,000 \text{ N} - 5000 \text{ N}) = 800$ J. And finally, Area III $= (13\,000 \text{ N} - 0 \text{ N})(0.45 \text{ m} - 0.20 \text{ m}) = 3300$ J. The work done is therefore equal to $1000 \text{ J} + 800 \text{ J} + 3300 \text{ J} = \boxed{5100 \text{ J}}$

Concept Summary

6.1 Work Done by a Constant Force The work W done by a constant force acting on an object is given by Equation 6.1, where F is the magnitude of the force, s is the magnitude of the displacement, and θ is the angle between the force and the displacement vectors. Work is a scalar quantity and can be positive or negative, depending on whether the force has a component that points, respectively, in the same direction as the displacement or in the opposite direction. The work is zero if the force is perpendicular ($\theta = 90°$) to the displacement.

$$W = (F \cos \theta)s \qquad \textbf{(6.1)}$$

6.2 The Work–Energy Theorem and Kinetic Energy The kinetic energy KE of an object of mass m and speed v is given by Equation 6.2. The work–energy theorem states that the work W done by the net external force acting on an object equals the difference between the object's final kinetic energy KE_f and initial kinetic energy KE_0, according to Equation 6.3. The kinetic energy increases when the net force does positive work and decreases when the net force does negative work.

$$KE = \tfrac{1}{2}mv^2 \qquad \textbf{(6.2)}$$

$$W = KE_f - KE_0 \qquad \textbf{(6.3)}$$

6.3 Gravitational Potential Energy Work done by the force of gravity on an object of mass m is given by Equation 6.4, where h_0 and h_f are the initial and final heights of the object, respectively.

$$W_{gravity} = mg(h_0 - h_f) \qquad \textbf{(6.4)}$$

$$PE = mgh \qquad \textbf{(6.5)}$$

Gravitational potential energy PE is the energy that an object has by virtue of its position. For an object near the surface of the earth, the gravitational potential energy is given by Equation 6.5, where h is the height of the object relative to an arbitrary zero level.

6.4 Conservative Versus Nonconservative Forces A conservative force is one that does the same work in moving an object between two points, independent of the path taken between the points. Alternatively, a force is conservative if the work it does in moving an object

around any closed path is zero. A force is nonconservative if the work it does on an object moving between two points depends on the path of the motion between the points.

6.5 The Conservation of Mechanical Energy The total mechanical energy E is the sum of the kinetic energy and potential energy: $E = KE + PE$. The work–energy theorem can be expressed in an alternate form as shown in Equation 6.8, where W_{nc} is the net work done by the external nonconservative forces, and E_f and E_0 are the final and initial total mechanical energies, respectively.

$$W_{nc} = E_f - E_0 \qquad \textbf{(6.8)}$$

The principle of conservation of mechanical energy states that the total mechanical energy E remains constant along the path of an object, provided that the net work done by external nonconservative forces is zero. Whereas E is constant, KE and PE may be transformed into one another.

6.6 Nonconservative Forces and the Work–Energy Theorem/

6.7 Power Average power $\overline{P}$ is the work done per unit time or the rate at which work is done, as shown in Equation 6.10a. It is also the rate at which energy changes, as shown in Equation 6.10b. When a force of magnitude F acts on an object moving with an average speed $\overline{v}$, the average power is given by Equation 6.11.

$$\overline{P} = \frac{\text{Work}}{\text{Time}} \qquad \textbf{(6.10a)}$$

$$\overline{P} = \frac{\text{Change in energy}}{\text{Time}} \qquad \textbf{(6.10b)}$$

$$\overline{P} = F\overline{v} \qquad \textbf{(6.11)}$$

6.8 Other Forms of Energy and the Conservation of Energy The principle of conservation of energy states that energy can neither be created nor destroyed, but can only be transformed from one form to another.

6.9 Work Done by a Variable Force The work done by a variable force of magnitude F in moving an object through a displacement of magnitude s is equal to the area under the graph of $F \cos \theta$ versus s. The angle θ is the angle between the force and displacement vectors.

Focus on Concepts

Online

Additional questions are available for assignment in WileyPLUS.

Section 6.1 Work Done by a Constant Force

1. The same force $\vec{\mathbf{F}}$ pushes in three different ways on a box moving with a velocity $\vec{\mathbf{v}}$, as the drawings show. Rank the work done by the force $\vec{\mathbf{F}}$ in ascending order (smallest first): **(a)** A, B, C **(b)** A, C, B **(c)** B, A, C **(d)** C, B, A **(e)** C, A, B

A   B   C

QUESTION 1

2. Consider the force $\vec{\mathbf{F}}$ shown in the drawing. This force acts on an object that can move along the positive or negative x axis, or along the positive or negative y axis. The work done by this force is positive when the displacement of the object is along the _____ axis or along the _____ axis: **(a)** $-x$, $-y$ **(b)** $-x$, $+y$ **(c)** $+x$, $+y$ **(d)** $+x$, $-y$

QUESTION 2

Section 6.2 The Work-Energy Theorem and Kinetic Energy

3. Two forces $\vec{\mathbf{F}}_1$ and $\vec{\mathbf{F}}_2$ act on a particle. As a result the speed of the particle increases. Which one of the following is NOT possible?

(a) The work done by $\vec{F}_1$ is positive, and the work done by $\vec{F}_2$ is zero. (b) The work done by $\vec{F}_1$ is zero, and the work done by $\vec{F}_2$ is positive. (c) The work done by each force is positive. (d) The work done by each force is negative. (e) The work done by $\vec{F}_1$ is positive, and the work done by $\vec{F}_2$ is negative.

4. Force $\vec{F}_1$ acts on a particle and does work W_1. Force $\vec{F}_2$ acts simultaneously on the particle and does work W_2. The speed of the particle does not change. Which one of the following must be true? (a) W_1 is zero and W_2 is positive (b) $W_1 = -W_2$ (c) W_1 is positive and W_2 is zero (d) W_1 is positive and W_2 is positive.

Section 6.4   Conservative Versus Nonconservative Forces

5. A person is riding on a Ferris wheel. When the wheel makes one complete turn, the net work done on the person by the gravitational force _____. (a) is positive (b) is negative (c) is zero (d) depends on how fast the wheel moves (e) depends on the diameter of the wheel

Section 6.5   The Conservation of Mechanical Energy

6. In which one of the following circumstances could mechanical energy not possibly be conserved, even if friction and air resistance are absent? (a) A car moves up a hill, its velocity continually decreasing along the way. (b) A car moves down a hill, its velocity continually increasing along the way. (c) A car moves along level ground at a constant velocity. (d) A car moves up a hill at a constant velocity.

7. A ball is fixed to the end of a string, which is attached to the ceiling at point P. As the drawing shows, the ball is projected downward at A with the launch speed v_0. Traveling on a circular path, the ball comes to a halt at point B. What enables

QUESTION 7

the ball to reach point B, which is above point A? Ignore friction and air resistance. (a) The work done by the tension in the string (b) The ball's initial gravitational potential energy (c) The ball's initial kinetic energy (d) The work done by the gravitational force

Section 6.6   Nonconservative Forces and the Work–Energy Theorem

8. In which one of the following circumstances does the principle of conservation of mechanical energy apply, even though a nonconservative force acts on the moving object? (a) The nonconservative force points in the same direction as the displacement of the object. (b) The nonconservative force is perpendicular to the displacement of the object. (c) The nonconservative force has a direction that is opposite to the displacement of the object. (d) The nonconservative force has a component that points in the same direction as the displacement of the object. (e) The nonconservative force has a component that points opposite to the displacement of the object.

9. A 92.0-kg skydiver with an open parachute falls straight downward through a vertical height of 325 m. The skydiver's velocity remains constant. What is the work done by the nonconservative force of air resistance, which is the only nonconservative force acting? (a) -2.93×10^5 J (b) 0 J (c) $+2.93 \times 10^5$ J (d) The answer is not obtainable, because insufficient information about the skydiver's speed is given.

Section 6.7   Power

10. The power needed to accelerate a projectile from rest to its launch speed v in a time t is 43.0 W. How much power is needed to accelerate the same projectile from rest to a launch speed of $2v$ in a time of $\frac{1}{2}t$?

Problems

Online

Additional questions are available for assignment in WileyPLUS.

SSM Student Solutions Manual

MMH Problem-solving help

GO Guided Online Tutorial

V-HINT Video Hints

CHALK Chalkboard Videos

BIO Biomedical application

E Easy

M Medium

H Hard

WS Worksheet

T Team Problem

Section 6.1   Work Done by a Constant Force

1. **E SSM** During a tug-of-war, team A pulls on team B by applying a force of 1100 N to the rope between them. The rope remains parallel to the ground. How much work does team A do if they pull team B toward them a distance of 2.0 m?

2. **E GO** You are moving into an apartment and take the elevator to the 6th floor. Suppose your weight is 685 N and that of your belongings is 915 N. (a) Determine the work done by the elevator in lifting you and your belongings up to the 6th floor (15.2 m) at a constant velocity. (b) How much work does the elevator do on you alone (without belongings) on the downward trip, which is also made at a constant velocity?

3. **E** The brakes of a truck cause it to slow down by applying a retarding force of 3.0×10^3 N to the truck over a distance of 850 m. What is the work done by this force on the truck? Is the work positive or negative? Why?

4. **E** A 75.0-kg man is riding an escalator in a shopping mall. The escalator moves the man at a constant velocity from ground level to the floor above, a vertical height of 4.60 m. What is the work done on the man by (a) the gravitational force and (b) the escalator?

5. **E SSM** Suppose in **Figure 6.2** that $+1.10 \times 10^3$ J of work is done by the force $\vec{F}$ (magnitude = 30.0 N) in moving the suitcase a distance of 50.0 m. At what angle θ is the force oriented with respect to the ground?

6. **E** A person pushes a 16.0-kg shopping cart at a constant velocity for a distance of 22.0 m. She pushes in a direction 29.0° below the horizontal. A 48.0-N frictional force opposes the motion of the cart. (a) What is the magnitude of the force that the shopper exerts? Determine the work done by (b) the pushing force, (c) the frictional force, and (d) the gravitational force.

7. **E SSM MMH** The drawing shows a plane diving toward the ground and then climbing back upward. During each of these motions, the lift force $\vec{L}$ acts perpendicular to the displacement $\vec{s}$, which has the

same magnitude, 1.7×10^3 m, in each case. The engines of the plane exert a thrust $\vec{\mathbf{T}}$, which points in the direction of the displacement and has the same magnitude during the dive and the climb. The weight $\vec{\mathbf{W}}$ of the plane has a magnitude of 5.9×10^4 N. In both motions, net work is performed due to the combined action of the forces $\vec{\mathbf{L}}$, $\vec{\mathbf{T}}$, and $\vec{\mathbf{W}}$. **(a)** Is more net work done during the dive or the climb? Explain. **(b)** Find the difference between the net work done during the dive and the climb.

(a) Dive   (b) Climb

PROBLEM 7

8. **E** A person pulls a toboggan for a distance of 35.0 m along the snow with a rope directed 25.0° above the snow. The tension in the rope is 94.0 N. **(a)** How much work is done on the toboggan by the tension force? **(b)** How much work is done if the same tension is directed parallel to the snow?

9. **M** **V-HINT** As a sailboat sails 52 m due north, a breeze exerts a constant force $\vec{\mathbf{F}}_1$ on the boat's sails. This force is directed at an angle west of due north. A force $\vec{\mathbf{F}}_2$ of the same magnitude directed due north would do the same amount of work on the sailboat over a distance of just 47 m. What is the angle between the direction of the force $\vec{\mathbf{F}}_1$ and due north?

10. **M** **GO** A 55-kg box is being pushed a distance of 7.0 m across the floor by a force $\vec{\mathbf{P}}$ whose magnitude is 160 N. The force $\vec{\mathbf{P}}$ is parallel to the displacement of the box. The coefficient of kinetic friction is 0.25. Determine the work done on the box by each of the *four* forces that act on the box. Be sure to include the proper plus or minus sign for the work done by each force.

11. **M** **V-HINT** A 1.00×10^2-kg crate is being pushed across a horizontal floor by a force $\vec{\mathbf{P}}$ that makes an angle of 30.0° below the horizontal. The coefficient of kinetic friction is 0.200. What should be the magnitude of $\vec{\mathbf{P}}$, so that the net work done by it and the kinetic frictional force is zero?

12. **H** A 1200-kg car is being driven up a 5.0° hill. The frictional force is directed opposite to the motion of the car and has a magnitude of $f = 524$ N. A force $\vec{\mathbf{F}}$ is applied to the car by the road and propels the car forward. In addition to these two forces, two other forces act on the car: its weight $\vec{\mathbf{W}}$ and the normal force $\vec{\mathbf{F}}_N$ directed perpendicular to the road surface. The length of the road up the hill is 290 m. What should be the magnitude of $\vec{\mathbf{F}}$, so that the net work done by all the forces acting on the car is +150 kJ?

Section 6.2 The Work–Energy Theorem and Kinetic Energy

13. **E** A fighter jet is launched from an aircraft carrier with the aid of its own engines and a steam-powered catapult. The thrust of its engines is 2.3×10^5 N. In being launched from rest it moves through a distance of 87 m and has a kinetic energy of 4.5×10^7 J at lift-off. What is the work done on the jet by the catapult?

14. **E** **MMH** A golf club strikes a 0.045-kg golf ball in order to launch it from the tee. For simplicity, assume that the average net force applied to the ball acts parallel to the ball's motion, has a magnitude of 6800 N, and is in contact with the ball for a distance of 0.010 m. With what speed does the ball leave the club?

15. **E** **SSM** It takes 185 kJ of work to accelerate a car from 23.0 m/s to 28.0 m/s. What is the car's mass?

16. **E** Starting from rest, a 1.9×10^{-4}-kg flea springs straight upward. While the flea is pushing off from the ground, the ground exerts an average upward force of 0.38 N on it. This force does $+2.4 \times 10^{-4}$ J of work on the flea. **(a)** What is the flea's speed when it leaves the ground? **(b)** How far upward does the flea move while it is pushing off? Ignore both air resistance and the flea's weight.

17. **E** **GO** A water-skier is being pulled by a tow rope attached to a boat. As the driver pushes the throttle forward, the skier accelerates. A 70.3-kg water-skier has an initial speed of 6.10 m/s. Later, the speed increases to 11.3 m/s. Determine the work done by the net external force acting on the skier.

18. **E** As background for this problem, review Conceptual Example 6. A 7420-kg satellite has an elliptical orbit, as in **Figure 6.9b**. The point on the orbit that is farthest from the earth is called the *apogee* and is at the far right side of the drawing. The point on the orbit that is closest to the earth is called the *perigee* and is at the left side of the drawing. Suppose that the speed of the satellite is 2820 m/s at the apogee and 8450 m/s at the perigee. Find the work done by the gravitational force when the satellite moves from **(a)** the apogee to the perigee and **(b)** the perigee to the apogee.

19. **E** **GO** A 16-kg sled is being pulled along the horizontal snow-covered ground by a horizontal force of 24 N. Starting from rest, the sled attains a speed of 2.0 m/s in 8.0 m. Find the coefficient of kinetic friction between the runners of the sled and the snow.

20. **E** **GO** An asteroid is moving along a straight line. A force acts along the displacement of the asteroid and slows it down. The asteroid has a mass of 4.5×10^4 kg, and the force causes its speed to change from 7100 to 5500 m/s. **(a)** What is the work done by the force? **(b)** If the asteroid slows down over a distance of 1.8×10^6 m, determine the magnitude of the force.

21. **M** The concepts in this problem are similar to those in Multiple-Concept Example 4, except that the force doing the work in this problem is the tension in the cable. A rescue helicopter lifts a 79-kg person straight up by means of a cable. The person has an upward acceleration of 0.70 m/s² and is lifted from rest through a distance of 11 m. **(a)** What is the tension in the cable? How much work is done by **(b)** the tension in the cable and **(c)** the person's weight? **(d)** Use the work–energy theorem and find the final speed of the person.

22. **M** **V-HINT** Consult Multiple-Concept Example 5 for insight into solving this problem. A skier slides horizontally along the snow for a distance of 21 m before coming to rest. The coefficient of kinetic friction between the skier and the snow is $\mu_k = 0.050$. Initially, how fast was the skier going?

23. **M** **GO** **CHALK** Under the influence of its drive force, a snowmobile is moving at a constant velocity along a horizontal patch of snow. When the drive force is shut off, the snowmobile coasts to a halt. The snowmobile and its rider have a mass of 136 kg. Under the influence of a drive force of 205 N, it is moving at a constant velocity whose magnitude is 5.50 m/s. The drive force is then shut off. Find **(a)** the distance in which the snowmobile coasts to a halt and **(b)** the time required to do so.

24. **H** The model airplane in Figure 5.6 is flying at a speed of 22 m/s on a horizontal circle of radius 16 m. The mass of the plane is 0.90 kg. The person holding the guideline pulls it in until the radius becomes 14 m. The plane speeds up, and the tension in the guideline becomes four times greater. What is the net work done on the plane?

Section 6.3 Gravitational Potential Energy,
Section 6.4 Conservative Versus Nonconservative Forces

25. **E** **CHALK** A 75.0-kg skier rides a 2830-m-long lift to the top of a mountain. The lift makes an angle of 14.6° with the horizontal. What is the change in the skier's gravitational potential energy?

26. **E** Juggles and Bangles are clowns. Juggles stands on one end of a teeter-totter at rest on the ground. Bangles jumps off a platform 2.5 m above the ground and lands on the other end of the teeter-totter, launching Juggles into the air. Juggles rises to a height of 3.3 m above the ground, at which point he has the same amount of gravitational potential energy as Bangles had before he jumped, assuming both potential energies are measured using the ground as the reference level. Bangles' mass is 86 kg. What is Juggles' mass?

27. **E** A 0.60-kg basketball is dropped out of a window that is 6.1 m above the ground. The ball is caught by a person whose hands are 1.5 m above the ground. **(a)** How much work is done on the ball by its weight? What is the gravitational potential energy of the basketball, relative to the ground, when it is **(b)** released and **(c)** caught? **(d)** How is the change ($PE_f - PE_0$) in the ball's gravitational potential energy related to the work done by its weight?

28. **E** A pole-vaulter just clears the bar at 5.80 m and falls back to the ground. The change in the vaulter's potential energy during the fall is -3.70×10^3 J. What is his weight?

29. **E** **SSM** A bicyclist rides 5.0 km due east, while the resistive force from the air has a magnitude of 3.0 N and points due west. The rider then turns around and rides 5.0 km due west, back to her starting point. The resistive force from the air on the return trip has a magnitude of 3.0 N and points due east. **(a)** Find the work done by the resistive force during the round trip. **(b)** Based on your answer to part (a), is the resistive force a conservative force? Explain.

30. **E** **GO** **CHALK** "Rocket Man" has a propulsion unit strapped to his back. He starts from rest on the ground, fires the unit, and accelerates straight upward. At a height of 16 m, his speed is 5.0 m/s. His mass, including the propulsion unit, has the approximately constant value of 136 kg. Find the work done by the force generated by the propulsion unit.

31. **E** **SSM** A 55.0-kg skateboarder starts out with a speed of 1.80 m/s. He does +80.0 J of work on himself by pushing with his feet against the ground. In addition, friction does −265 J of work on him. In both cases, the forces doing the work are nonconservative. The final speed of the skateboarder is 6.00 m/s. **(a)** Calculate the change ($\Delta PE = PE_f - PE_0$) in the gravitational potential energy. **(b)** How much has the vertical height of the skater changed, and is the skater above or below the starting point?

Section 6.5 The Conservation of Mechanical Energy

32. **E** A 35-kg girl is bouncing on a trampoline. During a certain interval after she leaves the surface of the trampoline, her kinetic energy decreases to 210 J from 440 J. How high does she rise during this interval? Neglect air resistance.

33. **E** **SSM** **MMH** A gymnast is swinging on a high bar. The distance between his waist and the bar is 1.1 m, as the drawing shows. At the top of the swing his speed is momentarily zero. Ignoring friction and treating the gymnast as if all of his mass is located at his waist, find his speed at the bottom of the swing.

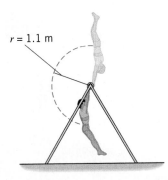

$r = 1.1$ m

PROBLEM 33

34. **E** **CHALK** **GO** The skateboarder in the drawing starts down the left side of the ramp with an initial speed of 5.4 m/s. Neglect nonconservative forces, such as friction and air resistance, and find the height h of the highest point reached by the skateboarder on the right side of the ramp.

h

PROBLEM 34

35. **E** **MMH** A slingshot fires a pebble from the top of a building at a speed of 14.0 m/s. The building is 31.0 m tall. Ignoring air resistance, find the speed with which the pebble strikes the ground when the pebble is fired **(a)** horizontally, **(b)** vertically straight up, and **(c)** vertically straight down.

36. **E** **GO** The drawing shows two boxes resting on frictionless ramps. One box is relatively light and sits on a steep ramp. The other box is heavier and rests on a ramp that is less steep. The boxes are released from rest at A and allowed to slide down the ramps. The two boxes have masses of 11 and 44 kg. If A and B are 4.5 and 1.5 m, respectively, above the ground, determine the speed of **(a)** the lighter box and **(b)** the heavier box when each reaches B. **(c)** What is the ratio of the kinetic energy of the heavier box to that of the lighter box at B?

$v_A = 0$ m/s $v_A = 0$ m/s

A

$h_A = 4.5$ m

B

$h_B = 1.5$ m

PROBLEM 36

37. **E** A 47.0-g golf ball is driven from the tee with an initial speed of 52.0 m/s and rises to a height of 24.6 m. **(a)** Neglect air resistance and determine the kinetic energy of the ball at its highest point. **(b)** What is its speed when it is 8.0 m below its highest point?

38. **E** **MMH** Consult Conceptual Example 9 in preparation for this problem. The drawing shows a person who, starting from rest at the top of a cliff, swings down at the end of a rope, releases it, and falls into the water below. There are two paths by which the person can enter the water. Suppose he enters the water at a speed of 13.0 m/s via path 1. How fast is he moving on path 2 when he releases the rope at a height of 5.20 m above the water? Ignore the effects of air resistance.

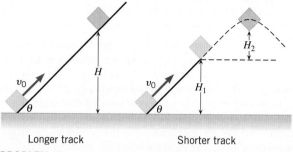

all kinetic energy. The initial speed of each block is $v_0 = 7.00$ m/s, and each incline slopes upward at an angle of $\theta = 50.0°$. The block on the shorter track leaves the track at a height of $H_1 = 1.25$ m above the ground. Find **(a)** the height H for the block on the longer track and **(b)** the total height $H_1 + H_2$ for the block on the shorter track.

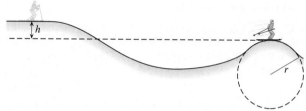

Longer track   Shorter track
PROBLEM 43

PROBLEM 38

39. **M** **GO** **SSM** The drawing shows a skateboarder moving at 5.4 m/s along a horizontal section of a track that is slanted upward by 48° above the horizontal at its end, which is 0.40 m above the ground. When she leaves the track, she follows the characteristic path of projectile motion. Ignoring friction and air resistance, find the maximum height H to which she rises above the end of the track.

44. **H** A skier starts from rest at the top of a hill. The skier coasts down the hill and up a second hill, as the drawing illustrates. The crest of the second hill is circular, with a radius of $r = 36$ m. Neglect friction and air resistance. What must be the height h of the first hill so that the skier just loses contact with the snow at the crest of the second hill?

PROBLEM 39

40. **M** **CHALK** **GO** A small lead ball, attached to a 0.75-m rope, is being whirled in a circle that lies in the vertical plane. The ball is whirled at a constant rate of three revolutions per second and is released on the upward part of the circular motion when it is 1.5 m above the ground. The ball travels straight upward. In the absence of air resistance, to what maximum height above the ground does the ball rise?

41. **M** **CHALK** A pendulum consists of a small object hanging from the ceiling at the end of a string of negligible mass. The string has a length of 0.75 m. With the string hanging vertically, the object is given an initial velocity of 2.0 m/s parallel to the ground and swings upward on a circular arc. Eventually, the object comes to a momentary halt at a point where the string makes an angle θ with its initial vertical orientation and then swings back downward. Find the angle θ.

42. **M** **V-HINT** A semitrailer is coasting downhill along a mountain highway when its brakes fail. The driver pulls onto a runaway-truck ramp that is inclined at an angle of 14.0° above the horizontal. The semitrailer coasts to a stop after traveling 154 m along the ramp. What was the truck's initial speed? Neglect air resistance and friction.

43. **M** **GO** The drawing shows two frictionless inclines that begin at ground level ($h = 0$ m) and slope upward at the same angle θ. One track is longer than the other, however. Identical blocks are projected up each track with the same initial speed v_0. On the longer track the block slides upward until it reaches a maximum height H above the ground. On the shorter track the block slides upward, flies off the end of the track at a height H_1 above the ground, and then follows the familiar parabolic trajectory of projectile motion. At the highest point of this trajectory, the block is a height H_2 above the end of the track. The initial total mechanical energy of each block is the same and is

PROBLEM 44

45. **H** A person starts from rest at the top of a large frictionless spherical surface, and slides into the water below (see the drawing). At what angle θ does the person leave the surface? (*Hint: When the person leaves the surface, the normal force is zero.*)

PROBLEM 45

Section 6.6 Nonconservative Forces and the Work–Energy Theorem

46. **E** **SSM** A projectile of mass 0.750 kg is shot straight up with an initial speed of 18.0 m/s. **(a)** How high would it go if there were no air resistance? **(b)** If the projectile rises to a maximum height of only 11.8 m, determine the magnitude of the average force due to air resistance.

47. **E** A basketball player makes a jump shot. The 0.600-kg ball is released at a height of 2.00 m above the floor with a speed of 7.20 m/s. The ball goes through the net 3.10 m above the floor at a speed of 4.20 m/s. What is the work done on the ball by air resistance, a nonconservative force?

48. **E** Starting from rest, a 93-kg firefighter slides down a fire pole. The average frictional force exerted on him by the pole has a magnitude of 810 N, and his speed at the bottom of the pole is 3.4 m/s. How far did he slide down the pole?

49. **E** **GO** A student, starting from rest, slides down a water slide. On the way down, a kinetic frictional force (a nonconservative force) acts on her. The student has a mass of 83.0 kg, and the height of the water slide is 11.8 m. If the kinetic frictional force does -6.50×10^3 J of work, how fast is the student going at the bottom of the slide?

50. **E** **SSM** The (nonconservative) force propelling a 1.50×10^3-kg car up a mountain road does 4.70×10^6 J of work on the car. The car starts from rest at sea level and has a speed of 27.0 m/s at an altitude of 2.00×10^2 m above sea level. Obtain the work done on the car by the combined forces of friction and air resistance, both of which are nonconservative forces.

51. **M** **V-HINT** In attempting to pass the puck to a teammate, a hockey player gives it an initial speed of 1.7 m/s. However, this speed is inadequate to compensate for the kinetic friction between the puck and the ice. As a result, the puck travels only one-half the distance between the players before sliding to a halt. What minimum initial speed should the puck have been given so that it reached the teammate, assuming that the same force of kinetic friction acted on the puck everywhere between the two players?

52. **M** **MMH** A 67.0-kg person jumps from rest off a 3.00-m-high tower straight down into the water. Neglect air resistance. She comes to rest 1.10 m under the surface of the water. Determine the magnitude of the average force that the water exerts on the diver. This force is nonconservative.

53. **M** At a carnival, you can try to ring a bell by striking a target with a 9.00-kg hammer. In response, a 0.400-kg metal piece is sent upward toward the bell, which is 5.00 m above. Suppose that 25.0% of the hammer's kinetic energy is used to do the work of sending the metal piece upward. How fast must the hammer be moving when it strikes the target so that the bell just barely rings?

Section 6.7 Power

54. **E** A person is making homemade ice cream. She exerts a force of magnitude 22 N on the free end of the crank handle on the ice-cream maker, and this end moves on a circular path of radius 0.28 m. The force is always applied parallel to the motion of the handle. If the handle is turned once every 1.3 s, what is the average power being expended?

55. **E** **MMH** A car accelerates uniformly from rest to 20.0 m/s in 5.6 s along a level stretch of road. Ignoring friction, determine the average power required to accelerate the car if **(a)** the weight of the car is 9.0×10^3 N and **(b)** the weight of the car is 1.4×10^4 N.

56. **E** **GO** A helicopter, starting from rest, accelerates straight up from the roof of a hospital. The lifting force does work in raising the helicopter. An 810-kg helicopter rises from rest to a speed of 7.0 m/s in a time of 3.5 s. During this time it climbs to a height of 8.2 m. What is the average power generated by the lifting force?

57. **M** **V-HINT** The cheetah is one of the fastest-accelerating animals, because it can go from rest to 27 m/s (about 60 mi/h) in 4.0 s. If its mass is 110 kg, determine the average power developed by the cheetah during the acceleration phase of its motion. Express your answer in **(a)** watts and **(b)** horsepower.

58. **M** **V-HINT** **CHALK** In 2.0 minutes, a ski lift raises four skiers at constant speed to a height of 140 m. The average mass of each skier is 65 kg. What is the average power provided by the tension in the cable pulling the lift?

Section 6.9 Work Done by a Variable Force

59. **E** The graph shows how the force component $F \cos \theta$ along the displacement varies with the magnitude s of the displacement. Find the work done by the force. (*Hint: Recall how the area of a triangle is related to the triangle's base and height.*)

PROBLEM 59

60. **E** **CHALK** The force component along the displacement varies with the magnitude of the displacement, as shown in the graph. Find the work done by the force in the interval from **(a)** 0 to 1.0 m, **(b)** 1.0 to 2.0 m, and **(c)** 2.0 to 4.0 m. (*Note: In the last interval the force component is negative, so the work is negative.*)

PROBLEM 60

61. **E** **GO** A net external force is applied to a 6.00-kg object that is initially at rest. The net force component along the displacement of the object varies with the magnitude of the displacement as shown in the drawing. What is the speed of the object at $s = 20.0$ m?

PROBLEM 61

Additional Problems

Online

62. **E** A cable lifts a 1200-kg elevator at a constant velocity for a distance of 35 m. What is the work done by **(a)** the tension in the cable and **(b)** the elevator's weight?

63. **E** **SSM** A 2.00-kg rock is released from rest at a height of 20.0 m. Ignore air resistance and determine the kinetic energy, gravitational potential energy, and total mechanical energy at each of the following heights: 20.0, 10.0, and 0 m.

64. **E GO** When an 81.0-kg adult uses a spiral staircase to climb to the second floor of his house, his gravitational potential energy increases by 2.00×10^3 J. By how much does the potential energy of an 18.0-kg child increase when the child climbs a normal staircase to the second floor?

65. **M GO** A husband and wife take turns pulling their child in a wagon along a horizontal sidewalk. Each exerts a constant force and pulls the wagon through the same displacement. They do the same amount of work, but the husband's pulling force is directed 58° above the horizontal, and the wife's pulling force is directed 38° above the horizontal. The husband pulls with a force whose magnitude is 67 N. What is the magnitude of the pulling force exerted by his wife?

66. **M V-HINT** Some gliders are launched from the ground by means of a winch, which rapidly reels in a towing cable attached to the glider. What average power must the winch supply in order to accelerate a 184-kg ultralight glider from rest to 26.0 m/s over a horizontal distance of 48.0 m? Assume that friction and air resistance are negligible, and that the tension in the winch cable is constant.

67. **M V-HINT** A 63-kg skier coasts up a snow-covered hill that makes an angle of 25° with the horizontal. The initial speed of the skier is 6.6 m/s. After coasting 1.9 m up the slope, the skier has a speed of 4.4 m/s. **(a)** Find the work done by the kinetic frictional force that acts on the skis. **(b)** What is the magnitude of the kinetic frictional force?

68. **M V-HINT** A water slide is constructed so that swimmers, starting from rest at the top of the slide, leave the end of the slide traveling horizontally. As the drawing shows, one person hits the water 5.00 m from the end of the slide in a time of 0.500 s after leaving the slide. Ignoring friction and air resistance, find the height H in the drawing.

Water slide

H

← 5.00 m →

PROBLEM 68

69. **M GO** A car, starting from rest, accelerates in the $+x$ direction as in the figure. It has a mass of 1.10×10^3 kg and maintains an acceleration of $+4.60$ m/s^2 for 5.00 s. Assume that a single horizontal force (not shown) accelerates the vehicle. Determine the average power generated by this force.

$\vec{a}$ $\vec{v}$

At rest $+x$

PROBLEM 69

Physics in Biology, Medicine, and Sports

70. **E SSM BIO 6.2** The hammer throw is a track-and-field event in which a 7.3 kg ball (the "hammer"), starting from rest, is whirled around in a circle several times and released. It then moves upward on the familiar curving path of projectile motion. In one throw, the hammer is given a speed of 29 m/s. For comparison, a .22 caliber bullet has a mass of 2.6 g and, starting from rest, exits the barrel of a gun at a speed of 410 m/s. Determine the work done to launch the motion of **(a)** the hammer and **(b)** the bullet.

71. **E GO BIO 6.6** In the sport of skeleton a participant jumps onto a sled (known as a skeleton) and proceeds to slide down an icy track, belly down and head first. In the 2010 Winter Olympics, the track had sixteen turns and dropped 126 m in elevation from top to bottom. **(a)** In the absence of nonconservative forces, such as friction and air resistance, what would be the speed of a rider at the bottom of the track? Assume that the speed at the beginning of the run is relatively small and can be ignored. **(b)** In reality, the gold-medal winner (Canadian Jon Montgomery) reached the bottom in one heat with a speed of 40.5 m/s (about 91 mi/h). How much work was done on him and his sled (assuming a total mass of 118 kg) by nonconservative forces during this heat?

PROBLEM 71

Richard Heathcote / Staff / Getty Images

72. **E BIO SSM 6.7** Bicyclists in the Tour de France do enormous amounts of work during a race. For example, the average power per kilogram generated by seven-time-winner Lance Armstrong ($m = 75.0$ kg) is 6.50 W per kilogram of his body mass. **(a)** How much work does he do during a 135-km race in which his average speed is 12.0 m/s? **(b)** Often, the work done is expressed in nutritional Calories rather than in joules. Express the work done in part (a) in terms of nutritional Calories, noting that 1 joule = 2.389×10^{-4} nutritional Calories.

73. **E** **BIO** **GO** **6.7** You are working out on a rowing machine. Each time you pull the rowing bar (which simulates the oars) toward you, it moves a distance of 1.2 m in a time of 1.5 s. The readout on the display indicates that the average power you are producing is 82 W. What is the magnitude of the force that you exert on the handle?

74. **M** **BIO** **6.3** A 65.0-kg mountain climber just ate a Johnny Rocket's bacon-cheddar-double-burger meal along with two peanut butter milkshakes, which have 5810 Calories (5810 kilocalories). Assume she can convert these calories into energy with 100% efficiency and use the fact that 1 kcal = 4186 J. Calculate how many times she could climb up Mt. Everest—the world's tallest mountain. The height of Mt. Everest is 8840 m.

75. **M** **BIO** **6.2** Which has more kinetic energy: the world's fastest man or a 9-mm bullet fired from a handgun? Usain Bolt holds the world record in the 100-m run. His top speed was recorded at 27.4 mph (12.3 m/s), and his mass is 94.0 kg. In comparison, a 9-mm bullet is fired at a speed of 791 mph (354 m/s), and it has a mass of 7.45×10^{-3} kg. Calculate **(a)** the kinetic energy of Usain Bolt, and **(b)** the kinetic energy of the 9-mm bullet.

76. **M** **BIO** **6.7** A typical human brain has a mass of 1.4 kg, which makes up about 2% of a person's mass. However, the brain requires 20% of the body's power consumption. In essence, the brain requires 10 times as much power as any other organ in the body. The average power output of the human brain is about 2.0×10^1 W. Assume that the brain's power output could be converted into mechanical work and calculate how high we could use this power to lift an apple ($m = 0.20$ kg) in one minute.

77. **M** **BIO** **6.1** The barbell curl (see the image) is an exercise that can strengthen and enlarge the biceps muscle in the upper arm. During a single contraction, the biceps muscle in each arm can exert a force of 850 N, as the muscle contracts through a distance of 7.2 cm. How much work can the biceps muscle perform in one contraction?

PROBLEM 77

Anush Gorak/Pexels

78. **M** **BIO** **6.7** In the motion picture *Iron Man*, Robert Downey, Jr.'s character, Tony Stark, creates the original arc reactor to power his Iron Man suit (see the image). Tony calculated the energy output of the arc reactor to be 3.00 GJ/s (gigajoules per second), and his math is always right. In comparison, calculate how much energy is produced by a human heart over an 80.0-year lifetime. The average output power of the heart is 1.33 W.

PROBLEM 78

Concepts and Calculations Problems

Online

Problem 79 reviews the important concept of work, and illustrates how forces can give rise to positive, negative, and zero work. Problem 80 examines the all-important conservation of mechanical energy.

79. **M** **CHALK** **SSM** The skateboarder in the figure is coasting down a ramp, and there are three forces acting on her: her weight $\vec{W}$ (magnitude = 675 N), a frictional force $\vec{f}$ (magnitude = 125 N) that opposes her motion, and a normal force $\vec{F}_N$ (magnitude = 612 N) *Concepts:* Part b of the figure shows each force, along with the displacement $\vec{s}$ of the skateboarder. By examining these diagrams and without doing any numerical calculations, determine whether the work done by each force is positive, negative, or zero. Provide a reason for each answer. *Calculations:* Determine the net work done by the three forces when she coasts for a distance of 9.2 m.

80. **M** **CHALK** The figure shows a 0.41-kg block sliding from A to B along a frictionless surface. When the block reaches B, it continues to slide along the horizontal surface BC where the kinetic frictional force acts. As a result, the block slows down, coming to rest at C. The kinetic energy of

PROBLEM 79

the block at A is 37 J, and the heights of A and B are 12.0 and 7.0 m above the ground, respectively. *Concepts:* (i) Is the total mechanical energy of the block conserved as the block goes from A to B? Why or why not? (ii) When the block reaches point B, has its kinetic energy increased, decreased, or remained the same relative to what it was at A? Give a reason for your answer. (iii) Is the total mechanical energy of the block conserved as the block goes from B to C? Justify your answer. *Calculations:* (a) What is the value of the kinetic energy of the block when it reaches B? (b) How much work does the kinetic frictional force do during the BC segment of the trip?

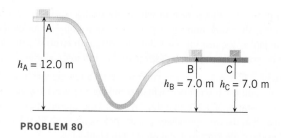

$h_A = 12.0$ m

$h_B = 7.0$ m $h_C = 7.0$ m

PROBLEM 80

Team Problems

Online

81. **E** **T** **WS** **A Runaway Sled.** You and your team are on the top of an icy slope that has a steep grade (i.e., angle of ascent) of 24.0°, and need to pull up a heavy sled loaded with supplies. You fasten a pulley into the ice at the top of the slope through which you thread a 150.0-ft rope (which is just long enough to reach the sled) and tie it to the sled. The rope has a strict tension rating of 450.0 lb, and a warning sticker on one end claims that it will break if this limit is exceeded by just 10.0 lb. Lined up like a tug-o-war team, you and six others in your group steadily pull on the rope with a nearly constant force, resulting in a constant rate of ascent. However, after exactly 45.0 s of pulling, the rope suddenly snaps, and the sled and its crucial supplies plummet down the hill and disappear into a snow bank. You measure the length of rope pulled through the pulley before it failed to be 85.0 ft. (a) Neglecting friction, how fast was the sled going when it reached the bottom of the slope? (b) For the next attempt to pull the sled up the slope, you get a stronger rope and want to use a motor-driven winch (it was tiring work by hand). What should be the minimum power rating of the motor (in horsepower) in order that the winch pulls the sled at the same rate as your team had been pulling it by hand? (*Note: 1 horsepower = 745.7 watts.*)

82. **E** **T** **WS** **A Tire Swing.** As background for this problem, review Conceptual Example 9. You and your team are designing a tire swing that hangs from a sturdy branch that is located 22.0 feet above the base of a tree. You tie one end of a rope to the branch and the other to a 35.0-lb tire, and adjust the length of the swing (i.e., the distance from the branch to the center of the tire) so that the center of the tire is 1.20 m above the ground at its lowest point. Near the tree is a grassy hill, the crest of which is 12.0 feet above the base of the tree, which makes for a perfect takeoff point. (a) If starting from rest at the top of the hill (i.e., the center of the swing is 12.0 feet above the base of the tree), how fast would a rider be moving when they reached the bottom? (b) If the maximum allowable weight of a rider is 160.0 lb, what should be the maximum tension rating for the rope?

83. **M** **T** **A Makeshift Elevator.** While exploring an elaborate tunnel system, you and your team get lost and find yourselves at the bottom of a 450-m vertical shaft. Suspended from a thick rope (near the floor) is a large rectangular bucket that looks like it had been used

to transport tools and debris up and down the tunnel. Mounted on the floor near one of the walls is a gasoline engine (3.5 hp) that turns a pulley and rope, and a sign that reads "Emergency Lift." It is clear that the engine is used to drive the bucket up the shaft. On the wall next to the engine is a sign indicating that a full tank of gas will last exactly 15 minutes when the engine is running at full power. You open the engine's gas tank and estimate that it is ¼ full, and there are no other sources of gasoline. (a) Assuming zero friction, if you send your team's lightest member (who weighs 125 lb), and the bucket weighs 150 lb when empty, how far up the shaft will the engine take her (and the bucket)? Will it get her out of the mine? (b) Assuming an effective collective friction (from the pulleys, etc.) of $\mu_{eff} = 0.10$ (so that $F_f = \mu_{eff}Mg$, where M is the total mass of the bucket plus team member), will the engine (with a ¼-full tank of gas) lift her to the top of the shaft?

84. **M** **T** **A Sledding Contest.** You are in a sledding contest where you start at a height of 40.0 m above the bottom of a valley and slide down a hill that makes an angle of 25.0° with respect to the horizontal. When you reach the valley, you immediately slide up a second hill that makes an angle of 15.0° with respect to the horizontal. The winner of the contest will be the contestant who travels the greatest distance up the second hill. You must now choose between using your flat-bottomed plastic sled, or your "Blade Runner," which glides on two steel rails. The hill you will ride down is covered with loose snow. However, the hill you will climb on the other side is a popular sledding hill, and is packed hard and is slick. The two sleds perform very differently on the two surfaces, the plastic one performing better on loose snow, and the Blade Runner doing better on hard-packed snow or ice. The performances of each sled can be quantified in terms of their respective coefficients of kinetic friction on the two surfaces. For the plastic sled: $\mu = 0.17$ on loose snow, and $\mu = 0.15$ on packed snow or ice. For the Blade Runner, $\mu = 0.19$ on loose snow, and $\mu = 0.070$ on packed snow or ice. Assuming the two hills are shaped like inclined planes, and neglecting air resistance, (a) how far does each sled make it up the second hill before stopping? (b) Assuming the total mass of the sled plus rider is 55.0 kg in both cases, how much work is done by nonconservative forces (over the total trip) in each case?

Traffic accidents often involve the collision between two or more motor vehicles. Although increased seat belt use, air bags, and other technologies have greatly improved the safety of modern automobiles, almost 90 Americans lose their lives everyday in traffic accidents. Much of the kinetic energy in the collision goes into permanently damaging the vehicles. In physics such a collision is classified as being inelastic. Inelastic collisions are one of the two basic types that this chapter introduces.

Impulse and Momentum

7.1 The Impulse–Momentum Theorem

There are many situations in which the force acting on an object is not constant, but varies with time. For instance, **Figure 7.1a** shows a baseball being hit, and part *b* of the figure illustrates approximately how the force applied to the ball by the bat changes during the time of contact. The magnitude of the force is zero at the instant t_0 just before the bat touches the ball. During contact, the force rises to a maximum and then returns to zero at the time t_f when the ball leaves the bat. The time interval $\Delta t = t_f - t_0$ during which the bat and ball are in contact is quite short, being only a few thousandths of a second, although the maximum force can be very large, often exceeding thousands of newtons. For comparison, the graph also shows the magnitude $\overline{F}$ of the average force exerted on the ball during the time of contact. **Figure 7.2** depicts other situations in which a time-varying force is applied to a ball.

LEARNING OBJECTIVES

After reading this module, you should be able to...

7.1 Apply the impulse–momentum theorem.

7.2 Apply the law of conservation of linear momentum.

7.3 Analyze one-dimensional collisions.

7.4 Analyze two-dimensional collisions.

7.5 Determine the location and the velocity of the center of mass.

Chuck Savage/Corbis/Getty Images

(a)

Mike Powell/Getty Images

Frank Gunn/AP Images

FIGURE 7.2 In each of these situations, the force applied to the ball varies with time. The time of contact is small, but the maximum force can be large.

(b)

FIGURE 7.1 (a) The collision time between a bat and a ball is very short, often less than a millisecond, but the force can be quite large. (b) When the bat strikes the ball, the magnitude of the force exerted on the ball rises to a maximum and then returns to zero when the ball leaves the bat. The time interval during which the force acts is Δt, and the magnitude of the average force is $\overline{F}$.

FIGURE 7.3 When a bat hits a ball, an average force $\overline{\mathbf{F}}$ is applied to the ball by the bat. As a result, the ball's velocity changes from an initial value of $\vec{\mathbf{v}}_0$ (top drawing) to a final value of $\vec{\mathbf{v}}_f$ (bottom drawing).

To describe how a time-varying force affects the motion of an object, we will introduce two new ideas: the impulse of a force and the linear momentum of an object. These ideas will be used with Newton's second law of motion to produce an important result known as the impulse–momentum theorem. This theorem plays a central role in describing collisions, such as that between a ball and a bat. Later on, we will see also that the theorem leads in a natural way to one of the most fundamental laws in physics, the conservation of linear momentum.

If a baseball is to be hit well, both the magnitude of the force and the time of contact are important. When a large average force acts on the ball for a long enough time, the ball is hit solidly. To describe such situations, we bring together the average force and the time of contact, calling the product of the two the **impulse** of the force.

DEFINITION OF IMPULSE

The impulse $\vec{\mathbf{J}}$ of a force is the product of the average force $\overline{\mathbf{F}}$ and the time interval Δt during which the force acts:

$$\vec{\mathbf{J}} = \overline{\mathbf{F}} \, \Delta t \qquad (7.1)$$

Impulse is a vector quantity and has the same direction as the average force.

SI Unit of Impulse: newton · second (N · s)

When a ball is hit, it responds to the value of the impulse. A large impulse produces a large response; that is, the ball departs from the bat with a large velocity. However, we know from experience that the more massive the ball, the less velocity it has after leaving the bat. Both mass and velocity play a role in how an object responds to a given impulse, and the effect of each of them is included in the concept of **linear momentum**, which is defined as follows:

DEFINITION OF LINEAR MOMENTUM

The linear momentum $\vec{\mathbf{p}}$ of an object is the product of the object's mass m and velocity $\vec{\mathbf{v}}$:

$$\vec{\mathbf{p}} = m\vec{\mathbf{v}} \qquad (7.2)$$

Linear momentum is a vector quantity that points in the same direction as the velocity.

SI Unit of Linear Momentum: kilogram · meter/second (kg · m/s)

Newton's second law of motion can now be used to reveal a relationship between impulse and momentum. **Figure 7.3** shows a ball with an initial velocity $\vec{\mathbf{v}}_0$ approaching a bat, being struck by the bat, and then departing with a final velocity $\vec{\mathbf{v}}_f$. When the velocity of an object changes from $\vec{\mathbf{v}}_0$ to $\vec{\mathbf{v}}_f$ during a time interval Δt, the average acceleration $\vec{\mathbf{a}}$ is given by Equation 2.4 as

$$\vec{\mathbf{a}} = \frac{\vec{\mathbf{v}}_f - \vec{\mathbf{v}}_0}{\Delta t}$$

According to Newton's second law, $\Sigma\vec{\overline{F}} = m\vec{\overline{a}}$, the average acceleration is produced by the net average force $\Sigma\vec{\overline{F}}$. Here $\Sigma\vec{\overline{F}}$ represents the vector sum of all the average forces that act on the object. Thus,

$$\Sigma\vec{\overline{F}} = m\left(\frac{\vec{v}_f - \vec{v}_0}{\Delta t}\right) = \frac{m\vec{v}_f - m\vec{v}_0}{\Delta t} \qquad (7.3)$$

In this result, the numerator on the far right is the final momentum minus the initial momentum, which is the change in momentum. Thus, the net average force is given by the change in momentum per unit of time.* Multiplying both sides of Equation 7.3 by Δt yields Equation 7.4, which is known as the **impulse–momentum theorem**.

IMPULSE–MOMENTUM THEOREM

When a net average force $\Sigma\vec{\overline{F}}$ acts on an object during a time interval Δt the impulse of this force is equal to the change in momentum of the object:

$$\underbrace{(\Sigma\vec{\overline{F}})\Delta t}_{\text{Impulse}} = \underbrace{m\vec{v}_f}_{\substack{\text{Final} \\ \text{momentum}}} - \underbrace{m\vec{v}_0}_{\substack{\text{Initial} \\ \text{momentum}}} \qquad (7.4)$$

Impulse = Change in momentum

During a collision, it is often difficult to measure the net average force $\Sigma\vec{\overline{F}}$, so it is not easy to determine the impulse $(\Sigma\vec{\overline{F}})\,\Delta t$ directly. On the other hand, it is usually straightforward to measure the mass and velocity of an object, so that its momentum $m\vec{v}_f$ just after the collision and $m\vec{v}_0$ just before the collision can be found. Thus, the impulse–momentum theorem allows us to gain information about the impulse indirectly by measuring the change in momentum that the impulse causes (see **Photo 7.1**). Then, armed with a knowledge of the contact time Δt, we can evaluate the net average force. Examples 1 and 2 illustrate how the theorem is used in this way.

PHOTO 7.1 During the launch of SpaceX's *Falcon Heavy*, the engines fire and create a force that applies an impulse to the rocket. This impulse causes the momentum of the rocket to increase in the upward direction.

John Raoux/AP Images

EXAMPLE 1 | BIO The Physics of Sneezing—A Violent Respiratory Event

Contagious respiratory diseases, such as SARS (severe acute respiratory syndrome), H5N1 and H1N1 influenza, and more recently coronavirus disease 2019 (COVID-19), have motivated researchers to study the physics of human sneezing, as water droplets ejected into the air from a person's nose and mouth are a common mode of viral transmission. One of the factors that determines social distancing guidelines during a pandemic is how far the droplets travel in the air after sneezing. Studies have shown that this distance can vary greatly (from less than 1 m to over 8 m), and it depends on many variables, such as the force on the droplets during the sneeze, sneeze velocity, droplet size, wind conditions, and so on (see **Figure 7.4**). A typical sneeze can contain as many as 40 000 droplets of various sizes, with diameters ranging from <10 μm to >1000 μm. Calculate the magnitude of the average force applied to a single droplet during a sneeze that takes place over a time duration of 0.200 s. Assume the diameter of an average droplet is 15.0 μm (mass of 1.77×10^{-12} kg), and it is ejected with a speed of 100 mph (45.0 m/s).

Reasoning We can use the impulse—momentum theorem (Equation 7.4) to calculate the average net force on the droplet. During the sneeze, the forces acting on the droplet include its weight, the propelling force (due to a difference in air pressure, $\vec{F}_P$), and possibly forces between droplets and the walls of the nasal cavity. During the sneeze, the propelling force is much greater than these other forces, so we neglect them here. Therefore, the net force $\Sigma\vec{\overline{F}} = \vec{\overline{F}}_P$ will be equal to the propelling force. The change in momentum of the droplet can be calculated from its mass and change in speed.

Solution From Equation 7.4, $\vec{\overline{F}}\Delta t = m\vec{v}_f - m\vec{v}_0$. Here, the initial velocity of the droplet is zero, and the final velocity is given in the problem. Since we want to calculate the magnitude of the average force, we drop the vector notation. Solving for $\overline{F}$ we find:

$$\overline{F}_P = \overline{F} = \frac{mv_f}{\Delta t} = \frac{(1.77 \times 10^{-12}\ \text{kg})(45.0\ \text{m/s})}{(0.200\ \text{s})} = \boxed{3.98 \times 10^{-10}\ \text{N}}$$

Even though the average force applied to the droplet is quite small, its acceleration will be large ($>200\ \text{m/s}^2$) due to its very small mass.

---

*The equality between the net force and the change in momentum per unit time is the version of the second law of motion presented originally by Newton.

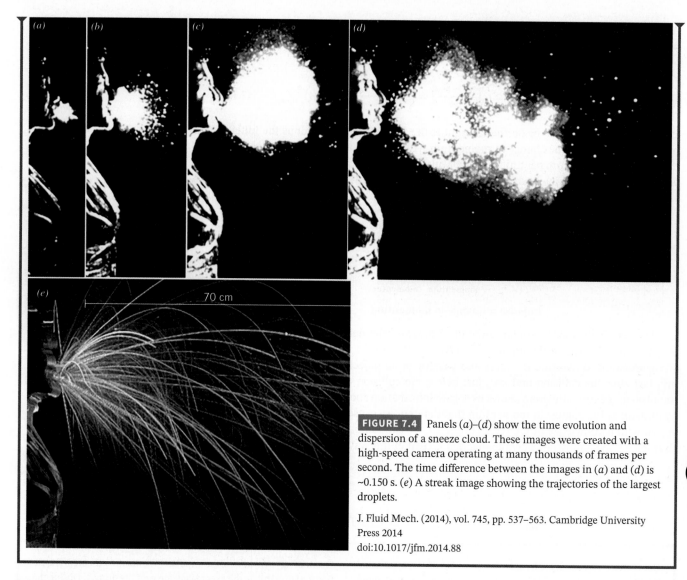

(a) (b) (c) (d)

(e)

70 cm

FIGURE 7.4 Panels (a)–(d) show the time evolution and dispersion of a sneeze cloud. These images were created with a high-speed camera operating at many thousands of frames per second. The time difference between the images in (a) and (d) is ~0.150 s. (e) A streak image showing the trajectories of the largest droplets.

J. Fluid Mech. (2014), vol. 745, pp. 537–563. Cambridge University Press 2014
doi:10.1017/jfm.2014.88

EXAMPLE 2 | A Well-Hit Ball

A baseball ($m = 0.14$ kg) has an initial velocity of $\vec{\mathbf{v}}_0 = -38$ m/s as it approaches a bat. We have chosen the direction of approach as the negative direction. The bat applies an average force $\overline{\vec{\mathbf{F}}}$ that is much larger than the weight of the ball, and the ball departs from the bat with a final velocity of $\vec{\mathbf{v}}_f = +58$ m/s. **(a)** Determine the impulse applied to the ball by the bat. **(b)** Assuming that the time of contact is $\Delta t = 1.6 \times 10^{-3}$ s, find the average force exerted on the ball by the bat.

Reasoning   Two forces act on the ball during impact, and together they constitute the net average force: the average force $\overline{\vec{\mathbf{F}}}$ exerted by the bat, and the weight of the ball. As in Example 1, since $\overline{\vec{\mathbf{F}}}$ is much greater than the weight of the ball, we neglect the weight. Thus, the net average force is equal to $\overline{\vec{\mathbf{F}}}$, or $\Sigma\overline{\vec{\mathbf{F}}} = \overline{\vec{\mathbf{F}}}$. In hitting the ball, the bat imparts an impulse to it. We cannot use Equation 7.1 ($\vec{\mathbf{J}} = \overline{\vec{\mathbf{F}}}\,\Delta t$) to determine the impulse $\overline{\vec{\mathbf{F}}}$ directly, since $\overline{\vec{\mathbf{F}}}$ is not known. We can find the impulse indirectly, however, by turning to the impulse–momentum theorem, which states that the impulse is equal to the ball's final momentum minus its initial momentum. With values for the impulse and the time of contact, Equation 7.1 can be used to determine the average force applied by the bat to the ball.

> **Problem-Solving Insight**   Momentum is a vector and, as such, has a magnitude and a direction. For

motion in one dimension, be sure to indicate the direction by assigning a plus or minus sign to it, as in this example.

Solution   **(a)** According to the impulse–momentum theorem, the impulse $\vec{\mathbf{J}}$ applied to the ball is

$$\vec{\mathbf{J}} = m\vec{\mathbf{v}}_f - m\vec{\mathbf{v}}_0$$

$$= \underbrace{(0.14\text{ kg})(+58\text{ m/s})}_{\text{Final momentum}} - \underbrace{(0.14\text{ kg})(-38\text{ m/s})}_{\text{Initial momentum}}$$

$$= \boxed{+13.4\text{ kg}\cdot\text{m/s}}$$

(b) Now that the impulse is known, the contact time can be used in Equation 7.1 to find the average force $\overline{\vec{\mathbf{F}}}$ exerted by the bat on the ball:

$$\overline{\vec{\mathbf{F}}} = \frac{\vec{\mathbf{J}}}{\Delta t} = \frac{+13.4\text{ kg}\cdot\text{m/s}}{1.6\times 10^{-3}\text{ s}}$$

$$= \boxed{+8400\text{ N}}$$

The force is positive, indicating that it points opposite to the velocity of the approaching ball. A force of 8400 N corresponds to 1900 lb, such a large value being necessary to change the ball's momentum during the brief contact time.

EXAMPLE 3 | A Rainstorm

During a storm, rain comes straight down with a velocity of $\vec{v}_0 = -15$ m/s and hits the roof of a car perpendicularly (see Figure 7.5). The mass of rain per second that strikes the car roof is 0.060 kg/s. Assuming that the rain comes to rest upon striking the car ($\vec{v}_f = 0$ m/s), find the average force exerted by the rain on the roof.

Reasoning This example differs from Example 2 in an important way. Example 2 gives information about the ball and asks for the force applied to the ball. In contrast, the present example gives information about the rain but asks for the force acting on the roof. However, the force exerted on the roof by the rain and the force exerted on the rain by the roof have equal magnitudes and opposite directions, according to Newton's law of action and reaction (see Section 4.5). Thus, we will find the force exerted on the rain and then apply the law of action and reaction to obtain the force on the roof. Two forces act on the rain while it impacts with the roof: the average force $\vec{F}$ exerted by the roof and the weight of the rain. These two forces constitute the net average force. By comparison, however, the force $\vec{F}$ is much greater than the weight, so we may neglect the weight. Thus, the net average force becomes equal to $\vec{F}$, or $\Sigma\vec{F} = \vec{F}$. The value of $\vec{F}$ can be obtained by applying the impulse–momentum theorem to the rain.

Solution The average force $\vec{F}$ needed to reduce the rain's velocity from $\vec{v}_0 = -15$ m/s to $\vec{v}_f = 0$ m/s is given by Equation 7.4 as

$$\vec{F} = \frac{m\vec{v}_f - m\vec{v}_0}{\Delta t} = -\left(\frac{m}{\Delta t}\right)\vec{v}_0$$

FIGURE 7.5 A raindrop falling on a car roof has an initial velocity of $\vec{v}_0$ just before striking the roof. The final velocity of the raindrop is $\vec{v}_f = 0$ m/s, because it comes to rest on the roof.

The term $m/\Delta t$ is the mass of rain per second that strikes the roof, so that $m/\Delta t = 0.060$ kg/s. Thus, the average force exerted on the rain by the roof is

$$\vec{F} = -(0.060 \text{ kg/s})(-15 \text{ m/s}) = +0.90 \text{ N}$$

This force is in the positive or upward direction, which is reasonable since the roof exerts an upward force on each falling drop in order to bring it to rest. According to the action–reaction law, the force exerted on the roof by the rain also has a magnitude of 0.90 N but points downward: Force on roof = $\boxed{-0.90 \text{ N}}$.

As you reason through problems such as those in Examples 1, 2, and 3, take advantage of the impulse–momentum theorem. It is a powerful statement that can lead to significant insights. The following Conceptual Example further illustrates its use.

CONCEPTUAL EXAMPLE 4 | Hailstones Versus Raindrops

In Example 3 rain is falling on the roof of a car and exerts a force on it. Instead of rain, suppose hail is falling. The hail comes straight down at a mass rate of $m/\Delta t = 0.060$ kg/s and an initial velocity of $\vec{v}_0 = -15$ m/s and strikes the roof perpendicularly, just as the rain does in Example 3. However, unlike rain, hail usually does not come to rest after striking a surface. Instead, the hailstones bounce off the roof of the car. If hail fell instead of rain, would the force on the roof be **(a)** smaller than, **(b)** equal to, or **(c)** greater than that calculated in Example 3?

Reasoning The raindrops and the hailstones fall in the same manner. That is, both fall with the same initial velocity and mass rate, and both strike the roof perpendicularly. However, there is an important difference: the raindrops come to rest (see **Figure 7.5**) after striking the roof, whereas the hailstones bounce upward (see **Figure 7.6**). According to the impulse–momentum theorem (Equation 7.4), the impulse that acts on an object is equal to the change in the object's momentum. This change is $m\vec{v}_f - m\vec{v}_0 = m \Delta\vec{v}$ and is proportional to the change $\Delta\vec{v}$ in the velocity. For a raindrop, the

Hailstone

FIGURE 7.6 Hailstones have a downward velocity of $\vec{v}_0$ just before striking this car roof. They rebound with an upward velocity of $\vec{v}_f$.

change in velocity is from $\vec{v}_0$ (downward) to zero. For a hailstone, the change is from $\vec{v}_0$ (downward) to $\vec{v}_f$ (upward). Thus, a raindrop and a hailstone experience different changes in velocity, and, hence, different changes in momentum and different impulses.

Answers (a) and (b) are incorrect. The change $\Delta\vec{v}$ in the velocity of a raindrop is smaller than that of a hailstone, since a raindrop does not rebound after striking the roof. According to the impulse–momentum theorem, a smaller impulse acts on a raindrop. But impulse is the product of the average force and the time interval Δt. Since the same amount of mass falls in the same time interval in either case, Δt is the same for both a raindrop and a hailstone. The smaller impulse acting on a raindrop, then, means that the car roof exerts a smaller force on it. By Newton's third law, this implies that a raindrop, not a hailstone, exerts a smaller force on the roof.

Answer (c) is correct. A hailstone experiences a larger change in momentum than does a raindrop, since a hailstone rebounds after striking the roof. Therefore, according to the impulse–momentum theorem, a greater impulse acts on a hailstone. An impulse is the product of the average force and the time interval Δt. Since the same amount of mass falls in the same time interval in either case, Δt is the same for a hailstone as for a raindrop. The greater impulse acting on a hailstone means that the car roof exerts a greater force on a hailstone than on a raindrop. According to Newton's action–reaction law (see Section 4.5) then, the car roof experiences a greater force from the hail than from the rain.

Related Homework: Problems 3, 7

Check Your Understanding

(The answers are given at the end of the book.)

1. Two identical automobiles have the same speed, one traveling east and one traveling west. Do these cars have the same momentum?

2. In Times Square in New York City, people celebrate on New Year's Eve. Some just stand around, but many move about randomly. Consider a group consisting of all of these people. Approximately, what is the total linear momentum of this group at any given instant?

3. Two objects have the same momentum. Do the velocities of these objects necessarily have **(a)** the same directions and **(b)** the same magnitudes?

4. (a) Can a single object have a kinetic energy but no momentum? **(b)** Can a group of two or more objects have a total kinetic energy that is not zero but a total momentum that is zero?

5. Suppose you are standing on the edge of a dock and jump straight down. If you land on sand your stopping time is much shorter than if you land on water. Using the impulse–momentum theorem as a guide, determine which one of the following statements is correct. **(a)** In bringing you to a halt, the sand exerts a greater impulse on you than does the water. **(b)** In bringing you to a halt, the sand and the water exert the same impulse on you, but the sand exerts a greater average force. **(c)** In bringing you to a halt, the sand and the water exert the same impulse on you, but the sand exerts a smaller average force.

6. An airplane is flying horizontally with a constant momentum during a time interval Δt. **(a)** Is there a *net* impulse acting on the plane during this time? Use the impulse–momentum theorem to guide your thinking. **(b)** In the horizontal direction, both the thrust generated by the engines and air resistance act on the plane. Considering your answer to part (a), how is the impulse of the thrust related (in magnitude and direction) to the impulse of the force due to the air resistance?

(a) Before collision

(b) During collision

(c) After collision

INTERACTIVE FIGURE 7.7 (a) The velocities of the two objects before the collision are $\vec{v}_{01}$ and $\vec{v}_{02}$. (b) During the collision, each object exerts a force on the other. These forces are $\vec{F}_{12}$ and $\vec{F}_{21}$. (c) The velocities after the collision are $\vec{v}_{f1}$ and $\vec{v}_{f2}$.

7.2 | The Principle of Conservation of Linear Momentum

It is worthwhile to compare the impulse–momentum theorem to the work–energy theorem discussed in Chapter 6. The impulse–momentum theorem states that the impulse produced by a net force is equal to the change in the object's momentum, whereas the work–energy theorem states that the work done by a net force is equal to the change in the object's kinetic energy. The work–energy theorem leads directly to the principle of conservation of mechanical energy (see Section 6.5), and, as we will shortly see, the impulse–momentum theorem also leads to a conservation principle, known as the conservation of linear momentum.

We begin by applying the impulse–momentum theorem to a midair collision between two objects. The two objects (masses m_1 and m_2) are approaching each other with initial velocities $\vec{v}_{01}$ and $\vec{v}_{02}$, as **Interactive Figure 7.7a** shows. The collection of objects being studied is referred to as the "system." In this case, the system contains only the two objects. They interact during the collision in part *b* and then depart with the final velocities $\vec{v}_{f1}$ and $\vec{v}_{f2}$ shown in part *c*. Because of the collision, the initial and final velocities are not the same.

Two types of forces act on the system:

1. *Internal forces:* Forces that the objects within the system exert on each other.

2. *External forces:* Forces exerted on the objects by agents external to the system.

During the collision in **Interactive Figure 7.7b**, $\vec{F}_{12}$ is the force exerted on object 1 by object 2, while $\vec{F}_{21}$ is the force exerted on object 2 by object 1. These forces are action–reaction forces that are equal in magnitude but opposite in direction, so $\vec{F}_{12} = -\vec{F}_{21}$. They are internal forces, since they are forces that the two objects within the system exert on each other. The force of gravity also acts on the objects, their weights being $\vec{\mathbf{W}}_1$ and $\vec{\mathbf{W}}_2$. These weights, however, are external forces, because they are applied by the earth, which is outside the system. Friction and air resistance would also be considered external forces, although these forces are ignored here for the sake of simplicity. The impulse–momentum theorem, as applied to each object, gives the following results:

Object 1
$$(\underbrace{\vec{\mathbf{W}}_1}_{\substack{\text{External} \\ \text{force}}} + \underbrace{\vec{\mathbf{F}}_{12}}_{\substack{\text{Internal} \\ \text{force}}})\,\Delta t = m_1\vec{\mathbf{v}}_{f1} - m_1\vec{\mathbf{v}}_{01}$$

Object 2
$$(\underbrace{\vec{\mathbf{W}}_2}_{\substack{\text{External} \\ \text{force}}} + \underbrace{\vec{\mathbf{F}}_{21}}_{\substack{\text{Internal} \\ \text{force}}})\,\Delta t = m_2\vec{\mathbf{v}}_{f2} - m_2\vec{\mathbf{v}}_{02}$$

Adding these equations produces a single result for the system as a whole:

$$(\underbrace{\vec{\mathbf{W}}_1 + \vec{\mathbf{W}}_2}_{\substack{\text{External} \\ \text{forces}}} + \underbrace{\vec{\mathbf{F}}_{12} + \vec{\mathbf{F}}_{21}}_{\substack{\text{Internal} \\ \text{forces}}})\,\Delta t = \underbrace{(m_1\vec{\mathbf{v}}_{f1} + m_2\vec{\mathbf{v}}_{f2})}_{\substack{\text{Total final} \\ \text{momentum } \vec{\mathbf{P}}_f}} - \underbrace{(m_1\vec{\mathbf{v}}_{01} + m_2\vec{\mathbf{v}}_{02})}_{\substack{\text{Total initial} \\ \text{momentum } \vec{\mathbf{P}}_0}}$$

On the right side of this equation, the quantity $m_1\vec{\mathbf{v}}_{f1} + m_2\vec{\mathbf{v}}_{f2}$ is the vector sum of the final momenta for each object, or the total final momentum $\vec{\mathbf{P}}_f$ of the system. Likewise, the quantity $m_1\vec{\mathbf{v}}_{01} + m_2\vec{\mathbf{v}}_{02}$ is the total initial momentum $\vec{\mathbf{P}}_0$. Therefore, the result above becomes

$$\left(\begin{matrix}\textbf{Sum of average} \\ \textbf{external forces}\end{matrix} + \begin{matrix}\textbf{Sum of average} \\ \textbf{internal forces}\end{matrix}\right)\Delta t = \vec{\mathbf{P}}_f - \vec{\mathbf{P}}_0 \qquad (7.5)$$

The advantage of the internal/external force classification is that the internal forces always add together to give zero, as a consequence of Newton's law of action–reaction; $\vec{\mathbf{F}}_{12} = -\vec{\mathbf{F}}_{21}$ so that $\vec{\mathbf{F}}_{12} + \vec{\mathbf{F}}_{21} = 0$. Cancellation of the internal forces occurs no matter how many parts there are to the system and allows us to ignore the internal forces, as Equation 7.6 indicates:

$$\textbf{(Sum of average external forces)}\,\Delta t = \vec{\mathbf{P}}_f - \vec{\mathbf{P}}_0 \qquad (7.6)$$

We developed this result with gravity as the only external force. But, in general, the sum of the external forces on the left includes *all* external forces.

With the aid of Equation 7.6, it is possible to see how the conservation of linear momentum arises. Suppose that the sum of the external forces is zero. A system for which this is true is called an **isolated system**. Then Equation 7.6 indicates that

$$0 = \vec{\mathbf{P}}_f - \vec{\mathbf{P}}_0 \quad \text{or} \quad \vec{\mathbf{P}}_f = \vec{\mathbf{P}}_0 \qquad (7.7a)$$

Thus, the final total momentum of the isolated system after the objects in **Interactive Figure 7.7** collide is the same as the initial total momentum.* Explicitly writing out the final and initial momenta for the two-body collision, we obtain from Equation 7.7a that

$$\underbrace{m_1\vec{\mathbf{v}}_{f1} + m_2\vec{\mathbf{v}}_{f2}}_{\vec{\mathbf{P}}_f} = \underbrace{m_1\vec{\mathbf{v}}_{01} + m_2\vec{\mathbf{v}}_{02}}_{\vec{\mathbf{P}}_0} \qquad (7.7b)$$

This result is an example of a general principle known as the **principle of conservation of linear momentum**.

PRINCIPLE OF CONSERVATION OF LINEAR MOMENTUM

The total linear momentum of an isolated system remains constant (is conserved). An isolated system is one for which the vector sum of the average external forces acting on the system is zero.

---

*Technically, the initial and final momenta are equal when the impulse of the sum of the external forces is zero—that is, when the left-hand side of Equation 7.6 is zero. Sometimes, however, the initial and final momenta are very nearly equal even when the sum of the external forces is not zero. This occurs when the time Δt during which the forces act is so short that it is effectively zero. Then, the left-hand side of Equation 7.6 is approximately zero.

The conservation-of-momentum principle applies to a system containing any number of objects, regardless of the internal forces, provided the system is isolated. Whether the system is isolated depends on whether the vector sum of the external forces is zero. Judging whether a force is internal or external depends on which objects are included in the system, as Conceptual Example 5 illustrates.

CONCEPTUAL EXAMPLE 5 | Is the Total Momentum Conserved?

Imagine two balls colliding on a billiard table that is friction-free. Using the momentum-conservation principle as a guide, decide which statement is correct: **(a)** The total momentum of the system that contains only one of the two balls is the same before and after the collision. **(b)** The total momentum of the system that contains both of the two balls is the same before and after the collision.

Reasoning The total momentum of an isolated system is the same before and after the collision; in such a situation the total momentum is said to be conserved. An isolated system is one for which the vector sum of the average external forces acting on the system is zero. To decide whether statement (a) or (b) is correct, we need to examine the one-ball and two-ball systems and see if they are, in fact, isolated.

Answer (a) is incorrect. In **Figure 7.8a** only one ball is included in the system, as indicated by the rectangular dashed box. The forces acting on this system are all external and include the weight $\vec{W}_1$ of the ball and the normal force $\vec{F}_{N1}$ due to the table. Since the ball does not accelerate in the vertical direction, the normal force must balance the weight, so the vector sum of these two vertical forces is zero. However, there is a third external force to consider. Ball 2 is outside the system, so the force $\vec{F}_{12}$ that it applies to the system (ball 1) during the collision is an external force. As a result, the vector sum of the three external forces is not zero. Thus, the one-ball system is not an isolated system, and its momentum is not the same before and after the collision.

Answer (b) is correct. The rectangular dashed box in Figure 7.8b shows that both balls are included in the system. The collision forces are not shown, because they are internal forces and cannot cause the total momentum of the two-ball system to change. The external forces include the weights $\vec{W}_1$ and $\vec{W}_2$

(a)

(b)

FIGURE 7.8 Two billiard balls collide on a pool table. (a) The rectangular dashed box emphasizes that only one of the balls (ball 1) is included in the system; $\vec{W}_1$ is its weight, $\vec{F}_{N1}$ is the normal force exerted on ball 1 by the pool table, and $\vec{F}_{12}$ is the force exerted on ball 1 by ball 2. (b) Now both balls are included in the system; $\vec{W}_2$ is the weight of ball 2, and $\vec{F}_{N2}$ is the normal force acting on it.

of the balls and the upward-pointing normal forces $\vec{F}_{N1}$ and $\vec{F}_{N2}$. Since the balls do not accelerate in the vertical direction, $\vec{F}_{N1}$ balances $\vec{W}_1$, and $\vec{F}_{N2}$ balances $\vec{W}_2$. Furthermore, the table is friction-free. Thus, there is no net external force to change the total momentum of the two-ball system, and, as a result, the total momentum is the same before and after the collision.

Next, we apply the principle of conservation of linear momentum to the problem of assembling a freight train.

EXAMPLE 6 | Assembling a Freight Train

A freight train is being assembled in a switching yard, and **Animated Figure 7.9** shows two boxcars. Car 1 has a mass of $m_1 = 65 \times 10^3$ kg and moves at a velocity of $v_{01} = +0.80$ m/s. Car 2, with a mass of $m_2 = 92 \times 10^3$ kg and a velocity of $v_{02} = +1.3$ m/s,

overtakes car 1 and couples to it. Neglecting friction, find the common velocity v_f of the cars after they become coupled.

Reasoning The two boxcars constitute the system. The sum of the external forces acting on the system is zero, because

(a) Before coupling

(b) After coupling

ANIMATED FIGURE 7.9 (a) The boxcar on the left eventually catches up with the other boxcar and (b) couples to it. The coupled cars move together with a common velocity after the collision.

the weight of each car is balanced by a corresponding normal force, and friction is being neglected. Thus, the system is isolated, and the principle of conservation of linear momentum applies. The coupling forces that each car exerts on the other are internal forces and do not affect the applicability of this principle.

> **Problem-Solving Insight** The conservation of linear momentum is applicable only when the net external force acting on the system is zero. Therefore, the first step in applying momentum conservation is to be sure that the net external force is zero.

Solution Momentum conservation indicates that

$$\underbrace{(m_1 + m_2)v_f}_{\substack{\text{Total momentum} \\ \text{after collision}}} = \underbrace{m_1 v_{01} + m_2 v_{02}}_{\substack{\text{Total momentum} \\ \text{before collision}}}$$

This equation can be solved for v_f, the common velocity of the two cars after the collision:

$$v_f = \frac{m_1 v_{01} + m_2 v_{02}}{m_1 + m_2}$$

$$= \frac{(65 \times 10^3 \text{ kg})(0.80 \text{ m/s}) + (92 \times 10^3 \text{ kg})(1.3 \text{ m/s})}{(65 \times 10^3 \text{ kg} + 92 \times 10^3 \text{ kg})}$$

$$= \boxed{+1.1 \text{ m/s}}$$

In the previous example it can be seen that the velocity of car 1 increases, while the velocity of car 2 decreases as a result of the collision. The acceleration and deceleration arise at the moment the cars become coupled, because the cars exert internal forces on each other. The powerful feature of the momentum-conservation principle is that it allows us to determine the changes in velocity without knowing what the internal forces are. Example 7 further illustrates this feature.

> **Problem-Solving Insight** It is important to realize that the total linear momentum may be conserved even when the kinetic energies of the individual parts of a system change.

EXAMPLE 7 | Ice Skaters

Starting from rest, two skaters push off against each other on smooth level ice, where friction is negligible. As **Interactive Figure 7.10a** shows, one is a woman ($m_1 = 54$ kg), and one is a man ($m_2 = 88$ kg). Part b of the drawing shows that the woman moves away with a velocity of $v_{f1} = +2.5$ m/s. Find the "recoil" velocity v_{f2} of the man.

Reasoning For a system consisting of the two skaters on level ice, the sum of the external forces is zero. This is because the weight of each skater is balanced by a corresponding normal force and friction is negligible. The skaters, then, constitute an isolated system, and the principle of conservation of linear momentum applies. We expect the man to have a smaller recoil speed for the following reason: The internal forces that the man and woman exert on each other during pushoff have equal magnitudes but opposite directions, according to Newton's action–reaction law. The man, having the larger mass, experiences a smaller acceleration according to Newton's second law. Hence, he acquires a smaller recoil speed.

Solution The total momentum of the skaters before they push on each other is zero, since they are at rest. Momentum conservation requires that the total momentum remains zero after the skaters have separated, as in **Interactive Figure 7.10b**:

$$\underbrace{m_1 v_{f1} + m_2 v_{f2}}_{\substack{\text{Total momentum} \\ \text{after pushing}}} = \underbrace{0}_{\substack{\text{Total momentum} \\ \text{before pushing}}}$$

Solving for the recoil velocity of the man gives

$$v_{f2} = \frac{-m_1 v_{f1}}{m_2} = \frac{-(54 \text{ kg})(+2.5 \text{ m/s})}{88 \text{ kg}} = \boxed{-1.5 \text{ m/s}}$$

The minus sign indicates that the man moves to the left in the drawing. After the skaters separate, the total momentum of the system remains zero, because momentum is a vector quantity, and the momenta of the man and the woman have equal magnitudes but opposite directions.

Math Skills When you use the momentum-conservation principle, you must choose which direction to call positive. The choice is arbitrary, but it is important. *The results of using the principle can only be interpreted with respect to your choice of the positive direction.* In Example 7, the direction to the right has been selected as the positive direction (see **Interactive Figure 7.10**), with the result that the velocity v_{f1} of the woman is given as $v_{f1} = +2.5$ m/s, and the answer for the velocity v_{f2} of the man is $v_{f2} = -1.5$ m/s. This answer means that the man moves off in the negative direction, which is to the left in the drawing. Suppose, however, that the positive direction in **Interactive Figure 7.10** had been chosen to be to the left. Then, the velocity v_{f1} of the woman would have been given as $v_{f1} = -2.5$ m/s, and the calculation for the velocity of the man would have revealed that the answer for v_{f2} is

$$v_{f2} = \frac{-m_1 v_{f1}}{m_2} = \frac{-(54 \text{ kg})(-2.5 \text{ m/s})}{88 \text{ kg}} = +1.5 \text{ m/s}$$

This answer means that the man moves off in the positive direction, which is now the direction to the left. Thus, the momentum-conservation principle leads to the same conclusion for the man. He moves off to the left, irrespective of the arbitrary choice of the positive direction.

(a) Before pushoff   (b) After pushoff

INTERACTIVE FIGURE 7.10 (a) In the absence of friction, two skaters pushing on each other constitute an isolated system. (b) As the skaters move away, the total linear momentum of the system remains zero, which is what it was initially.

In Example 7, for instance, the initial kinetic energy is zero, since the skaters are stationary. But after they push off, the skaters are moving, so each has kinetic energy. The kinetic energy changes because work is done by the internal force that each skater exerts on the other. However, internal forces cannot change the total linear momentum of a system, since the total linear momentum of an isolated system is conserved in the presence of such forces.

When applying the principle of conservation of linear momentum, we have been following a definite reasoning strategy that is summarized as follows:

> **REASONING STRATEGY   Applying the Principle of Conservation of Linear Momentum**
>
> 1. Decide which objects are included in the system.
> 2. Relative to the system that you have chosen, identify the internal forces and the external forces.
> 3. Verify that the system is isolated. In other words, verify that the sum of the external forces applied to the system is zero. Only if this sum is zero can the conservation principle be applied. If the sum of the average external forces is not zero, consider a different system for analysis.
> 4. Set the total final momentum of the isolated system equal to the total initial momentum. Remember that linear momentum is a vector. If necessary, apply the conservation principle separately to the various vector components.

Check Your Understanding

(*The answers are given at the end of the book.*)

7. An object slides along the surface of the earth and slows down because of kinetic friction. If the object alone is considered as the system, the kinetic frictional force must be identified as an external force that, according to Equation 7.4, decreases the momentum of the system. **(a)** If *both* the object and the earth are considered to be the system, is the force of kinetic friction still an external force? **(b)** Can the frictional force change the total linear momentum of the two-body system?

8. A satellite explodes in outer space, far from any other body, sending thousands of pieces in all directions. Is the linear momentum of the satellite before the explosion less than, equal to, or greater than the total linear momentum of all the pieces after the explosion?

9. On a distant asteroid, a large catapult is used to throw chunks of stone into space. Could such a device be used as a propulsion system to move the asteroid closer to the earth?

10. A canoe with two people aboard is coasting with an initial momentum of +110 kg · m/s. Then, one of the people (person 1) dives off the back of the canoe. During this time, the net average external force acting on the system (the canoe and the two people) is zero. The table lists four possibilities for the final momentum of person 1 and the final momentum of person 2 plus the canoe, immediately after person 1 dives off. Only one possibility could be correct. Which one is it?

	Final Momenta	
	Person 1	**Person 2 and Canoe**
(a)	$-60 \text{ kg} \cdot \text{m/s}$	$+170 \text{ kg} \cdot \text{m/s}$
(b)	$-30 \text{ kg} \cdot \text{m/s}$	$+110 \text{ kg} \cdot \text{m/s}$
(c)	$-40 \text{ kg} \cdot \text{m/s}$	$-70 \text{ kg} \cdot \text{m/s}$
(d)	$+80 \text{ kg} \cdot \text{m/s}$	$-30 \text{ kg} \cdot \text{m/s}$

11. You are a passenger on a jetliner that is flying at a constant velocity. You get up from your seat and walk toward the front of the plane. Because of this action, your forward momentum increases. Does the forward momentum of the plane itself decrease, remain the same, or increase?

12. An ice boat is coasting on a frozen lake. Friction between the ice and the boat is negligible, and so is air resistance. Nothing is propelling the boat. From a bridge someone jumps straight down into the boat, which continues to coast straight ahead. **(a)** Does the total horizontal momentum of the boat plus the jumper change? **(b)** Does the speed of the boat itself increase, decrease, or remain the same?

13. Concept Simulation 7.1 at **www.wiley.com/college/cutnell** reviews the concepts that are pertinent in this question. In movies, Superman hovers in midair, grabs a villain by the neck, and throws him forward. Superman, however, remains stationary. This is not possible, because it violates which one or more of the following: **(a)** The law of conservation of energy **(b)** Newton's second law **(c)** Newton's third law **(d)** The principle of conservation of linear momentum

14. The energy released by the exploding gunpowder in a cannon propels the cannonball forward. Simultaneously, the cannon recoils. The mass of the cannonball is less than that of the cannon. Which has the greater kinetic energy, the launched cannonball or the recoiling cannon? Assume that momentum conservation applies.

7.3 | Collisions in One Dimension

As discussed in the previous section, the total linear momentum is conserved when two objects collide, provided they constitute an isolated system. When the objects are atoms or subatomic particles, the total kinetic energy of the system is often conserved also. In other words, the total kinetic energy of the particles before the collision equals the total kinetic energy of the particles after the collision, so that kinetic energy gained by one particle is lost by another.

In contrast, when two macroscopic objects collide, such as two cars, the total kinetic energy after the collision is generally less than that before the collision. Kinetic energy is lost mainly in two ways: (1) It can be converted into heat because of friction, and (2) it is spent in creating permanent distortion or damage, as in an automobile collision. With very hard objects, such as a solid steel ball and a marble floor, the permanent distortion suffered upon collision is much smaller than with softer objects and, consequently, less kinetic energy is lost.

Collisions are often classified according to whether the total kinetic energy changes during the collision:

1. *Elastic collision:* One in which the total kinetic energy of the system after the collision is equal to the total kinetic energy before the collision.

2. *Inelastic collision:* One in which the total kinetic energy of the system is *not* the same before and after the collision; if the objects stick together after colliding, the collision is said to be *completely inelastic*.

> **Problem-Solving Insight**   As long as the net external force is zero, the conservation of linear momentum applies to any type of collision. This is true whether the collision is elastic or inelastic.

The boxcars coupling together in **Animated Figure 7.9** provide an example of a completely inelastic collision. When a collision is completely inelastic, the greatest amount of kinetic energy is lost. Example 8 shows how one particular elastic collision is described using the conservation of linear momentum and the fact that no kinetic energy is lost.

Analyzing Multiple-Concept Problems

EXAMPLE 8 | A Head-On Collision

Figure 7.11 illustrates an elastic head-on collision between two balls. One ball has a mass of $m_1 = 0.250$ kg and an initial velocity of $+5.00$ m/s. The other has a mass of $m_2 = 0.800$ kg and is initially at rest. No external forces act on the balls. What are the velocities of the balls after the collision?

Reasoning   Three facts will guide our solution. The first is that the collision is elastic, so that the kinetic energy of the two-ball system is the same before and after the balls collide. The second fact is that the collision occurs head-on. This means that the velocities before and after the balls collide all point along the same line. In other words, the collision occurs in one dimension. Last, no external forces act on the balls, with the result that the two-ball system is isolated and its total linear momentum is conserved. We expect that ball 1, having the smaller mass, will rebound to the left after striking ball 2, which is more massive. Ball 2 will be driven to the right in the process.

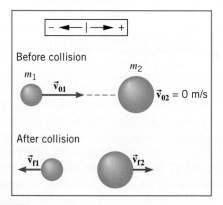

FIGURE 7.11   A 0.250-kg ball, traveling with an initial velocity of $v_{01} = +5.00$ m/s, undergoes an elastic collision with a 0.800-kg ball that is initially at rest.

Knowns and Unknowns   The data for this problem are given in the following table:

Description	Symbol	Value	Comment
Explicit Data			
Mass of ball 1	m_1	0.250 kg	
Initial velocity of ball 1	v_{01}	+5.00 m/s	Before collision occurs.
Mass of ball 2	m_2	0.800 kg	
Implicit Data			
Initial velocity of ball 2	v_{02}	0 m/s	Before collision occurs, ball is initially at rest.
Unknown Variables			
Final velocity of ball 1	v_{f1}	?	After collision occurs.
Final velocity of ball 2	v_{f2}	?	After collision occurs.

Modeling the Problem

STEP 1 Elastic Collision   Since the collision is elastic, the kinetic energy of the two-ball system is the same before and after the balls collide:

$$\underbrace{\tfrac{1}{2}m_1 v_{f1}^2 + \tfrac{1}{2}m_2 v_{f2}^2}_{\substack{\text{Total kinetic energy} \\ \text{after collision}}} = \underbrace{\tfrac{1}{2}m_1 v_{01}^2 + 0}_{\substack{\text{Total kinetic energy} \\ \text{before collision}}}$$

Here we have utilized the fact that ball 2 is at rest before the collision. Thus, its initial velocity v_{02} is zero and so is its initial kinetic energy. Solving this equation for v_{f1}^2, the square of the velocity of ball 1 after the collision, gives Equation 1 at the right. To use this result, we need a value for v_{f2}, which we will obtain in Step 2.

$$v_{f1}^2 = v_{01}^2 - \frac{m_2}{m_1}v_{f2}^2 \qquad (1)$$

$\boxed{?}$

STEP 2 Conservation of Linear Momentum   Total linear momentum of the two-ball system is conserved since no external forces act on the system. This fact does not depend on whether the collision is elastic. Conservation of linear momentum indicates that

$$\underbrace{m_1 v_{f1} + m_2 v_{f2}}_{\substack{\text{Total momentum} \\ \text{after collision}}} = \underbrace{m_1 v_{01} + 0}_{\substack{\text{Total momentum} \\ \text{before collision}}}$$

We have again utilized the fact that ball 2 is at rest before the collision; since its initial velocity v_{02} is zero, so is its initial momentum. Solving the expression above for v_{f2} gives

$$v_{f2} = \frac{m_1}{m_2}(v_{01} - v_{f1})$$

which can be substituted into Equation 1, as shown at the right.

$$v_{f1}^2 = v_{01}^2 - \frac{m_2}{m_1}v_{f2}^2 \qquad \textbf{(1)}$$

$$v_{f2} = \frac{m_1}{m_2}(v_{01} - v_{f1}) \qquad \textbf{(2)}$$

Solution   Algebraically combining the results of each step, we find that

$$\overset{\textbf{STEP 1}}{\underset{\downarrow}{}} \qquad \overset{\textbf{STEP 2}}{\underset{\downarrow}{}}$$

$$v_{f1}^2 = v_{01}^2 - \frac{m_2}{m_1}v_{f2}^2 = v_{01}^2 - \frac{m_2}{m_1}\left[\frac{m_1}{m_2}(v_{01} - v_{f1})\right]^2$$

Solving for v_{f1} shows that

$$v_{f1} = \left(\frac{m_1 - m_2}{m_1 + m_2}\right)v_{01} = \left(\frac{0.250 \text{ kg} - 0.800 \text{ kg}}{0.250 \text{ kg} + 0.800 \text{ kg}}\right)(+5.00 \text{ m/s}) = \boxed{-2.62 \text{ m/s}} \qquad \textbf{(7.8a)}$$

Math Skills   The equation $v_{f1}^2 = v_{01}^2 - \frac{m_2}{m_1}\left[\frac{m_1}{m_2}(v_{01} - v_{f1})\right]^2$ is a quadratic equation, because it contains the square of the unknown variable v_{f1}. Such equations can always be solved using the quadratic formula (see Appendix C.4, Equation C-2), although it is often awkward and time consuming to do so. A more sensible approach in this case is to begin by rearranging the equation as follows:

$$v_{f1}^2 = v_{01}^2 - \frac{m_2}{m_1}\frac{m_1}{m_2}\frac{m_1}{m_2}(v_{01} - v_{f1})^2 \quad \text{or} \quad v_{f1}^2 = v_{01}^2 - \frac{m_1}{m_2}(v_{01} - v_{f1})^2$$

Further rearrangement and use of the fact that $v_{01}^2 - v_{f1}^2 = (v_{01} - v_{f1})(v_{01} + v_{f1})$ produces

$$v_{01}^2 - v_{f1}^2 = \frac{m_1}{m_2}(v_{01} - v_{f1})^2 \quad \text{or} \quad (\cancel{v_{01} - v_{f1}})(v_{01} + v_{f1}) = \frac{m_1}{m_2}(\cancel{v_{01} - v_{f1}})(v_{01} - v_{f1})$$

This result is equivalent to

$$v_{01} + v_{f1} = \frac{m_1}{m_2}v_{01} - \frac{m_1}{m_2}v_{f1} \quad \text{or} \quad \left(\frac{m_1}{m_2} + 1\right)v_{f1} = \left(\frac{m_1}{m_2} - 1\right)v_{01}$$

Thus, we find that

$$v_{f1} = \frac{\left(\frac{m_1}{m_2} - 1\right)}{\left(\frac{m_1}{m_2} + 1\right)}v_{01} \quad \text{or} \quad v_{f1} = \left(\frac{m_1 - m_2}{m_1 + m_2}\right)v_{01}$$

Substituting the expression for v_{f1} into Equation 2 gives

$$v_{f2} = \left(\frac{2m_1}{m_1 + m_2}\right)v_{01} = \left[\frac{2(0.250 \text{ kg})}{0.250 \text{ kg} + 0.800 \text{ kg}}\right](+5.00 \text{ m/s}) = \boxed{+2.38 \text{ m/s}} \qquad \textbf{(7.8b)}$$

The negative value for v_{f1} indicates that ball 1 rebounds to the left after the collision in **Figure 7.11**, while the positive value for v_{f2} indicates that ball 2 moves to the right, as expected.

Related Homework: Problem 35

We can get a feel for an elastic collision by dropping a steel ball onto a hard surface, such as a marble floor. If the collision were elastic, the ball would rebound to its original height, as **Figure 7.12a** illustrates. In contrast, a partially deflated basketball exhibits little rebound from a relatively soft asphalt surface, as in part *b*, indicating that a fraction of the ball's kinetic energy is dissipated during the inelastic collision. The very deflated basketball in part *c* has no bounce at all, and a maximum amount of kinetic energy is lost during the completely inelastic collision.

The next example illustrates a completely inelastic collision in a device called a "ballistic pendulum." This device can be used to measure the speed of a bullet.

(a) Elastic collision

(b) Inelastic collision

(c) Completely inelastic collision

FIGURE 7.12 (a) A hard steel ball would rebound to its original height after striking a hard marble surface if the collision were elastic. (b) A partially deflated basketball has little bounce on a soft asphalt surface. (c) A very deflated basketball has no bounce at all.

Analyzing Multiple-Concept Problems

EXAMPLE 9 | The Physics of Measuring the Speed of a Bullet

A ballistic pendulum can be used to measure the speed of a projectile, such as a bullet. The ballistic pendulum shown in **Figure 7.13a** consists of a stationary 2.50-kg block of wood suspended by a wire of negligible mass. A 0.0100-kg bullet is fired into the block, and the block (with the bullet in it) swings to a maximum height of 0.650 m above the initial position (see part b of the drawing). Find the speed with which the bullet is fired, assuming that air resistance is negligible.

Reasoning The physics of the ballistic pendulum can be divided into two parts. The first is the completely inelastic collision between the bullet and the block. The total linear momentum of the system (block plus bullet) is conserved during the collision, because the suspension wire supports the system's weight, which means that the sum of the external forces acting on the system is nearly zero. The second part of the physics is the resulting motion of the block and bullet as they swing upward after the collision. As the system swings upward, the principle of conservation of mechanical energy applies, since nonconservative forces do no work (see Section 6.5). The tension force in the wire does no work because it acts perpendicular to the motion. Since air resistance is negligible, we can ignore the work it does. The conservation principles for linear momentum and mechanical energy provide the basis for our solution.

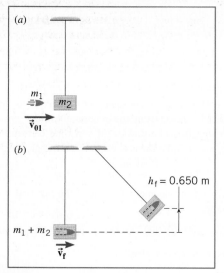

FIGURE 7.13 (a) A bullet approaches a ballistic pendulum. (b) The block and bullet swing upward after the collision.

Knowns and Unknowns The data for this problem are as follows:

Description	Symbol	Value	Comment
Explicit Data			
Mass of bullet	m_1	0.0100 kg	
Mass of block	m_2	2.50 kg	
Height to which block plus bullet swings	h_f	0.650 m	Maximum height of swing.
Implicit Data			
Initial velocity of block	v_{02}	0 m/s	Before collision, block is stationary.
Unknown Variable			
Speed with which bullet is fired	v_{01}	?	Before collision with block.

Modeling the Problem

STEP 1 Completely Inelastic Collision Just after the bullet collides with it, the block (with the bullet in it) has a speed v_f. Since linear momentum is conserved, the total momentum of the block–bullet system after the collision is the same as it is before the collision:

$$\underbrace{(m_1 + m_2)v_f}_{\substack{\text{Total momentum} \\ \text{after collision}}} = \underbrace{m_1 v_{01} + 0}_{\substack{\text{Total momentum} \\ \text{before collision}}}$$

$$v_{01} = \left(\frac{m_1 + m_2}{m_1}\right)v_f \qquad \text{(1)}$$

?

Note that the block is at rest before the collision ($v_{02} = 0$ m/s), so its initial momentum is zero. Solving this equation for v_{01} gives Equation 1 at the right. To find a value for the speed v_f in this equation, we turn to Step 2.

STEP 2 Conservation of Mechanical Energy The speed v_f immediately after the collision can be obtained from the maximum height h_f to which the system swings, by using the principle of conservation of mechanical energy:

$$\underbrace{(m_1 + m_2)gh_f}_{\substack{\text{Total mechanical energy} \\ \text{at top of swing, all} \\ \text{potential}}} = \underbrace{\tfrac{1}{2}(m_1 + m_2)v_f^2}_{\substack{\text{Total mechanical energy} \\ \text{at bottom of swing, all} \\ \text{kinetic}}}$$

This result can be solved for v_f to show that

$$\boxed{v_f = \sqrt{2gh_f}}$$

which can be substituted into Equation 1, as shown at the right. In applying the energy-conservation principle, it is tempting to say that the total potential energy at the top of the swing is equal to the total kinetic energy of the bullet before it strikes the block $[(m_1 + m_2)gh_f = \tfrac{1}{2}m_1v_{01}^2]$ and solve directly for v_{01}. This is incorrect, however, because the collision between the bullet and the block is inelastic, so that some of the bullet's initial kinetic energy is dissipated during the collision (due to friction and structural damage to the block and bullet).

$$v_{01} = \left(\frac{m_1 + m_2}{m_1}\right)v_f \qquad (1)$$

$$\boxed{v_f = \sqrt{2gh_f}}$$

Solution Algebraically combining the results of each step, we find that

$$\overset{\textbf{STEP 1}}{\underset{\downarrow}{}} \qquad \overset{\textbf{STEP 2}}{\underset{\downarrow}{}}$$

$$v_{01} = \left(\frac{m_1 + m_2}{m_1}\right)v_f = \left(\frac{m_1 + m_2}{m_1}\right)\sqrt{2gh_f}$$

$$v_{01} = \left(\frac{0.0100 \text{ kg} + 2.50 \text{ kg}}{0.0100 \text{ kg}}\right)\sqrt{2(9.80 \text{ m/s}^2)(0.650 \text{ m})} = \boxed{+896 \text{ m/s}}$$

Related Homework: Problem 27

Check Your Understanding

(The answers are given at the end of the book.)

15. Two balls collide in a one-dimensional elastic collision. The two balls constitute a system, and the net external force acting on them is zero. The table shows four possible sets of values for the initial and final momenta of the two balls, as well as their initial and final kinetic energies. Only one set of values could be correct. Which set is it?

		Initial (Before Collision)		Final (After Collision)	
		Momentum	**Kinetic Energy**	**Momentum**	**Kinetic Energy**
(a)	Ball 1:	+4 kg · m/s	12 J	−5 kg · m/s	10 J
	Ball 2:	−3 kg · m/s	5 J	−1 kg · m/s	7 J
(b)	Ball 1:	+7 kg · m/s	22 J	+5 kg · m/s	18 J
	Ball 2:	+2 kg · m/s	8 J	+4 kg · m/s	15 J
(c)	Ball 1:	−5 kg · m/s	12 J	−6 kg · m/s	15 J
	Ball 2:	−8 kg · m/s	31 J	−9 kg · m/s	25 J
(d)	Ball 1:	+9 kg · m/s	25 J	+6 kg · m/s	18 J
	Ball 2:	+4 kg · m/s	15 J	+7 kg · m/s	22 J

16. In an elastic collision, is the kinetic energy of *each* object the same before and after the collision?

17. Concept Simulation 7.2 at **www.wiley.com/college/cutnell** illustrates the concepts that are involved in this question. Also review Multiple-Concept Example 8. Suppose two objects collide head on, as in Example 8, where initially object 1 (mass $= m_1$) is moving and object 2 (mass $= m_2$) is stationary. Now assume that they have the same mass, so $m_1 = m_2$. Which one of the following statements is true? **(a)** Both objects have the same velocity (magnitude and direction) after the collision. **(b)** Object 1 rebounds with one-half its initial speed, while object 2 moves to the right, as in **Figure 7.11**, with one-half the speed that object 1 had before the collision. **(c)** Object 1 stops completely, while object 2 acquires the same velocity (magnitude and direction) that object 1 had before the collision.

FIGURE 7.14 (a) Top view of two balls colliding on a horizontal frictionless table. (b) This part of the drawing shows the x and y components of the velocity of ball 1 after the collision.

7.4 | Collisions in Two Dimensions

The collisions discussed so far have been head-on, or one-dimensional, because the velocities of the objects all point along a single line before and after contact. Collisions often occur, however, in two or three dimensions. **Figure 7.14** shows a two-dimensional case in which two balls collide on a horizontal frictionless table.

For the system consisting of the two balls, the external forces include the weights of the balls and the corresponding normal forces produced by the table. Since each weight is balanced by a normal force, the sum of the external forces is zero, and the total momentum of the system is conserved, as Equation 7.7b indicates. Momentum is a vector quantity, however, and in two dimensions the x and y components of the total momentum are conserved separately. In other words, Equation 7.7b is equivalent to the following two equations:

x Component

$$\underbrace{m_1 v_{f1x} + m_2 v_{f2x}}_{P_{fx}} = \underbrace{m_1 v_{01x} + m_2 v_{02x}}_{P_{0x}} \qquad (7.9a)$$

y Component

$$\underbrace{m_1 v_{f1y} + m_2 v_{f2y}}_{P_{fy}} = \underbrace{m_1 v_{01y} + m_2 v_{02y}}_{P_{0y}} \qquad (7.9b)$$

These equations are written for a system that contains two objects. If a system contains more than two objects, a mass-times-velocity term must be included for each additional object on either side of Equations 7.9a and 7.9b.

7.5 | Center of Mass

In previous sections, we have encountered situations in which objects interact with one another, such as the two skaters pushing off in Example 7. In these situations, the mass of the system is located in several places, and the various objects move relative to each other before, after, and even during the interaction. It is possible, however, to speak of a kind of average location for the total mass by introducing a concept known as the **center of mass** (abbreviated as "cm"). With the aid of this concept, we will be able to gain additional insight into the principle of conservation of linear momentum.

The center of mass is a point that represents the average location for the total mass of a system. **Figure 7.15**, for example, shows two particles of mass m_1 and m_2 that are located on the x axis at the positions x_1 and x_2, respectively. The position x_{cm} of the center-of-mass point from the origin is defined to be

FIGURE 7.15 The center of mass cm of the two particles is located on a line between them and lies closer to the more massive particle.

Center of mass

$$x_{cm} = \frac{m_1 x_1 + m_2 x_2}{m_1 + m_2} \qquad (7.10)$$

Each term in the numerator of this equation is the product of a particle's mass and position, while the denominator is the total mass of the system. If the two masses are equal, we expect the average location of the total mass to be midway between the particles. With $m_1 = m_2 = m$, Equation 7.10 becomes $x_{cm} = (mx_1 + mx_2)/(m + m) = \frac{1}{2}(x_1 + x_2)$, which indeed corresponds to the point midway between the particles. Alternatively, suppose that $m_1 = 5.0$ kg and $x_1 = 2.0$ m, while $m_2 = 12$ kg and $x_2 = 6.0$ m. Then we expect the average location of the total mass to be located closer to particle 2, since it is more massive. Equation 7.10 is also consistent with this expectation, for it gives

$$x_{cm} = \frac{(5.0 \text{ kg})(2.0 \text{ m}) + (12 \text{ kg})(6.0 \text{ m})}{5.0 \text{ kg} + 12 \text{ kg}} = 4.8 \text{ m}$$

If a system contains more than two particles, the center-of-mass point can be determined by generalizing Equation 7.10. For three particles, for instance, the numerator would contain a third term m_3x_3, and the total mass in the denominator would be $m_1 + m_2 + m_3$. For a macroscopic object, which contains many, many particles, the center-of-mass point is located at the geometric center of the object, provided that the mass is distributed symmetrically about the center. Such would be the case for a billiard ball. For objects such as a golf club, the mass is not distributed symmetrically, and the center-of-mass point is not located at the geometric center of the club. The driver used to launch a golf ball from the tee, for instance, has more mass in the club head than in the handle, so the center-of-mass point is closer to the head than to the handle.

Equation 7.10 (and its generalization to more than two particles) deals with particles that lie along a straight line. A system of particles, however, may include particles that do not all lie along a single straight line. For particles lying in a plane, an equation like Equation 7.10 uses the x coordinates of each particle to give the x coordinate of the center of mass. A similar equation uses the y coordinates of each particle to give the y coordinate of the center of mass. If a system contains rigid objects, each of them may be treated as if it were a particle with all of the mass located at the object's own center of mass. In this way, a collection of rigid objects becomes a collection of particles, and the x and y coordinates of the center of mass of the collection can be determined as described earlier in this paragraph.

To see how the center-of-mass concept is related to momentum conservation, suppose that the two particles in a system are moving, as they would be during a collision. With the aid of Equation 7.10, we can determine the velocity v_{cm} of the center-of-mass point. During a time interval Δt, the particles experience displacements of Δx_1 and Δx_2, as **Interactive Figure 7.16** shows. They have different displacements during this time because they have different velocities. Equation 7.10 can be used to find the displacement Δx_{cm} of the center of mass by replacing x_{cm} by Δx_{cm}, x_1 by Δx_1, and x_2 by Δx_2:

$$\Delta x_{cm} = \frac{m_1 \Delta x_1 + m_2 \Delta x_2}{m_1 + m_2}$$

Now we divide both sides of this equation by the time interval Δt. In the limit as Δt becomes infinitesimally small, the ratio $\Delta x_{cm}/\Delta t$ becomes equal to the instantaneous velocity v_{cm} of the center of mass. (See Section 2.2 for a review of instantaneous velocity.) Likewise, the ratios $\Delta x_1/\Delta t$ and $\Delta x_2/\Delta t$ become equal to the instantaneous velocities v_1 and v_2, respectively. Thus, we have

Velocity of center of mass $\qquad v_{cm} = \dfrac{m_1 v_1 + m_2 v_2}{m_1 + m_2}$ $\qquad\qquad$ (7.11)

The numerator $(m_1 v_1 + m_2 v_2)$ on the right-hand side in Equation 7.11 is the momentum of particle 1 $(m_1 v_1)$ plus the momentum of particle 2 $(m_2 v_2)$, which is the total linear momentum of the system. In an isolated system, the total linear momentum does not change because of an interaction such as a collision. Therefore, Equation 7.11 indicates that the velocity v_{cm} of the center of mass does not change either. To emphasize this important point, consider the collision discussed in Example 8. With the data from that example, we can apply Equation 7.11 to determine the velocity of the center of mass before and after the collision:

Before collision

$$v_{cm} = \frac{(0.250 \text{ kg})(+5.00 \text{ m/s}) + (0.800 \text{ kg})(0 \text{ m/s})}{0.250 \text{ kg} + 0.800 \text{ kg}} = +1.19 \text{ m/s}$$

INTERACTIVE FIGURE 7.16 During a time interval Δt, the displacements of the particles are Δx_1 and Δx_2, while the displacement of the center of mass is Δx_{cm}.

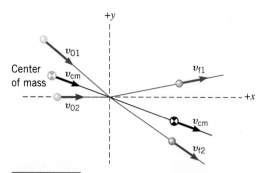

FIGURE 7.17 This drawing shows the two balls in Figure 7.14a and the path followed by the center-of-mass point as the balls approach the collision point and then move away from it. Because of momentum conservation, the velocity of the center of mass of the balls is the same before and after the collision (see the vectors labeled v_{cm}). As a result, the center of mass moves along the same straight-line path before and after the collision.

After collision

$$v_{cm} = \frac{(0.250 \text{ kg})(-2.62 \text{ m/s}) + (0.800 \text{ kg})(+2.38 \text{ m/s})}{0.250 \text{ kg} + 0.800 \text{ kg}} = +1.19 \text{ m/s}$$

Thus, the velocity of the center of mass is the same before and after the objects interact during a collision in which the total linear momentum is conserved.

In a two-dimensional collision, the velocity of the center of mass is also the same before and after the collision, provided that the total linear momentum is conserved. **Figure 7.17** illustrates this fact. The drawing reproduces **Figure 7.14a**, except that the initial and final center-of-mass locations are indicated, along with the velocity vector for the center of mass at these two places (see the vectors labeled v_{cm}). As the drawing shows, this velocity vector is the same before and after the balls collide. Thus, the center of mass moves along a straight line as the two balls approach the collision point and continues along the same straight line following the collision.

Check Your Understanding

(The answers are given at the end of the book.)

18. Would you expect the center of mass of a baseball bat to be located halfway between the ends of the bat, nearer the lighter end, or nearer the heavier end?

19. A sunbather is lying on a floating raft that is stationary. She then gets up and walks to one end of the raft. Consider the sunbather and raft as an isolated system. **(a)** What is the velocity of the center of mass of this system while she is walking? **(b)** Does the raft itself move while she is walking? If so, what is the direction of the raft's velocity relative to that of the sunbather?

20. Water, dripping at a constant rate from a faucet, falls to the ground. At any instant there are many drops in the air between the faucet and the ground. Where does the center of mass of the drops lie relative to the halfway point between the faucet and the ground? **(a)** Above it **(b)** Below it **(c)** Exactly at the halfway point

---

EXAMPLE 10 | BIO The Physics of High Jumpers—The "Fosbury Flop"

Track athletes participating in the high jump often utilize a technique known as the Fosbury Flop, which is named for Dick Fosbury, who first brought attention to the technique in the 1968 Summer Olympics. Here, the athlete positions the body in the air so as to actually travel "backwards" over the bar, as shown in **Figure 7.18a**. We can see why this is such an effective technique by calculating the position of the athlete's center of mass, as the body maintains this curved shape while passing over the bar. In **Figure 7.18b** we approximate the athlete's body as consisting of three different linear mass segments. Segment A represents the mass of the head, neck, trunk, and arms. Segment B represents the mass from the upper and lower legs, and Segment C represents the mass from the feet. The masses and positions of the centers of mass of each segment are indicated in **Figure 7.18b**. Use this information to calculate the position of the overall center of mass of the three segments.

Reasoning This is a 2-dimensional center of mass problem, where we can find the x and y coordinates of the position of the overall center of mass by applying Equation 7.10 twice, once for the x-direction and once for the y-direction.

Solution We have three mass segments, so we will have three terms in the numerator and denominator of Equation 7.10. Let's begin with the x-direction:

$$x_{cm} = \frac{m_A x_A + m_B x_B + m_C x_C}{m_A + m_B + m_C}$$

$$= \frac{(39 \text{ kg})(99 \text{ cm}) + (20 \text{ kg})(37 \text{ cm}) + (1.9 \text{ kg})(0 \text{ cm})}{39 \text{ kg} + 20 \text{ kg} + 1.9 \text{ kg}} = \boxed{76 \text{ cm}}$$

And now for the y-direction:

$$y_{cm} = \frac{m_A y_A + m_B y_B + m_C y_C}{m_A + m_B + m_C}$$

$$= \frac{(39 \text{ kg})(36 \text{ cm}) + (20 \text{ kg})(37 \text{ cm}) + (1.9 \text{ kg})(0 \text{ cm})}{39 \text{ kg} + 20 \text{ kg} + 1.9 \text{ kg}} = \boxed{35 \text{ cm}}$$

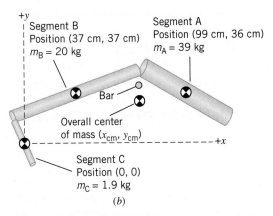

FIGURE 7.18 (a) A high jumper clearing the bar by performing the Fosbury Flop. This technique allows her to keep her center of mass closer to the ground. (b) The approximate distribution of the high jumper's mass as she passes over the bar. The masses and locations of different parts of her body are represented by the three segments in the figure.

Therefore, the position of the overall center of mass is given by (x_{cm} = 76 cm, y_{cm} = 35 cm). What makes this technique so effective is that it keeps the high jumper's center of mass closer to the ground. In fact, her center of mass actually passes below the bar. Keeping her center of mass closer to the ground reduces the total energy she needs to clear the bar.

Concept Summary

7.1 The Impulse–Momentum Theorem The impulse $\vec{J}$ of a force is the product of the average force $\vec{F}$ and the time interval Δt during which the force acts, according to Equation 7.1. Impulse is a vector that points in the same direction as the average force.

The linear momentum $\vec{p}$ of an object is the product of the object's mass m and velocity $\vec{v}$, according to Equation 7.2. Linear momentum is a vector that points in the same direction as the velocity. The total linear momentum of a system of objects is the vector sum of the momenta of the individual objects.

The impulse–momentum theorem states that when a net average force $\Sigma\vec{F}$ acts on an object, the impulse of this force is equal to the change in momentum of the object, as in Equation 7.4.

$$\vec{J} = \vec{F}\Delta t \tag{7.1}$$

$$\vec{p} = m\vec{v} \tag{7.2}$$

$$(\Sigma\vec{F})\Delta t = m\vec{v}_f - m\vec{v}_0 \tag{7.4}$$

7.2 The Principle of Conservation of Linear Momentum External forces are those forces that agents external to the system exert on objects within the system. An isolated system is one for which the vector sum of the external forces acting on the system is zero.

The principle of conservation of linear momentum states that the total linear momentum of an isolated system remains constant. For a two-body system, the conservation of linear momentum can be written as in Equation 7.7b, where m_1 and m_2 are the masses, $\vec{v}_{f1}$ and $\vec{v}_{f2}$ are the final velocities, and $\vec{v}_{01}$ and $\vec{v}_{02}$ are the initial velocities of the objects.

$$\underbrace{m_1\vec{v}_{f1} + m_2\vec{v}_{f2}}_{\text{Final total linear momentum}} = \underbrace{m_1\vec{v}_{01} + m_2\vec{v}_{02}}_{\text{Initial total linear momentum}} \tag{7.7b}$$

7.3 Collisions in One Dimension An elastic collision is one in which the total kinetic energy of the system after the collision is equal to the total kinetic energy of the system before the collision. An inelastic collision is one in which the total kinetic energy of the system is not the same before and after the collision. If the objects stick together after the collision, the collision is said to be completely inelastic.

7.4 Collisions in Two Dimensions When the total linear momentum is conserved in a two-dimensional collision, the x and y components of the total linear momentum are conserved separately. For a collision between two objects, the conservation of total linear momentum can be written as in Equations 7.9a and 7.9b.

$$\underbrace{m_1 v_{f1x} + m_2 v_{f2x}}_{\substack{x\text{ component of final}\\ \text{total linear momentum}}} = \underbrace{m_1 v_{01x} + m_2 v_{02x}}_{\substack{x\text{ component of initial}\\ \text{total linear momentum}}} \tag{7.9a}$$

$$\underbrace{m_1 v_{f1y} + m_2 v_{f2y}}_{\substack{y\text{ component of final}\\ \text{total linear momentum}}} = \underbrace{m_1 v_{01y} + m_2 v_{02y}}_{\substack{y\text{ component of initial}\\ \text{total linear momentum}}} \tag{7.9b}$$

7.5 Center of Mass The location of the center of mass of two particles lying on the x axis is given by Equation 7.10, where m_1 and m_2 are the masses of the particles and x_1 and x_2 are their positions relative to the coordinate origin. If the particles move with velocities v_1 and v_2, the velocity v_{cm} of the center of mass is given by Equation 7.11. If the total linear momentum of a system of particles remains constant during an interaction such as a collision, the velocity of the center of mass also remains constant.

$$x_{cm} = \frac{m_1 x_1 + m_2 x_2}{m_1 + m_2} \tag{7.10}$$

$$v_{cm} = \frac{m_1 v_1 + m_2 v_2}{m_1 + m_2} \tag{7.11}$$

Focus on Concepts

Additional questions are available for assignment in WileyPLUS.

Section 7.1 The Impulse–Momentum Theorem

1. Two identical cars are traveling at the same speed. One is heading due east and the other due north, as the drawing shows. Which statement is true regarding the kinetic energies and momenta of the cars? **(a)** They both have the same kinetic energies and the same momenta. **(b)** They have the same kinetic energies, but different momenta. **(c)** They have different kinetic energies, but the same momenta. **(d)** They have different kinetic energies and different momenta.

QUESTION 1

2. Six runners have the mass (in multiples of m_0), speed (in multiples of v_0), and direction of travel that are indicated in the table. Which two runners have identical momenta? **(a)** B and C **(b)** A and C **(c)** C and D **(d)** A and E **(e)** D and F

Runner	Mass	Speed	Direction of Travel
A	$\frac{1}{2}m_0$	v_0	Due north
B	m_0	v_0	Due east
C	m_0	$2v_0$	Due south
D	$2m_0$	v_0	Due west
E	m_0	$\frac{1}{2}v_0$	Due north
F	$2m_0$	$2v_0$	Due west

3. A particle is moving along the $+x$ axis, and the graph shows its momentum p as a function of time t. Use the impulse–momentum theorem to rank (largest to smallest) the three regions according to the *magnitude* of the impulse applied to the particle. **(a)** A, B, C **(b)** A, C, B **(c)** A and C (a tie), B **(d)** C, A, B **(e)** B, A, C

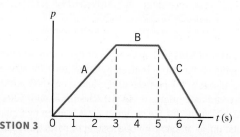

QUESTION 3

4. A particle moves along the $+x$ axis, and the graph shows its momentum p as a function of time t. In each of the four regions a force, which may or may not be nearly zero, is applied to the particle. In which region is the *magnitude* of the force largest and in which region is it smallest? **(a)** B largest, D smallest **(b)** C largest, B smallest

(c) A largest, D smallest **(d)** C largest, A smallest **(e)** A largest, C smallest

QUESTION 4

Section 7.2 The Principle of Conservation of Linear Momentum

5. As the text discusses, the conservation of linear momentum is applicable only when the system of objects is an isolated system. Which of the systems listed below are isolated systems?

  1. A ball is dropped from the top of a building. The system is the ball.
  2. A ball is dropped from the top of a building. The system is the ball and the earth.
  3. A billiard ball collides with a stationary billiard ball on a frictionless pool table. The system is the moving ball.
  4. A car slides to a halt in an emergency. The system is the car.
  5. A space probe is moving in deep space where gravitational and other forces are negligible. The system is the space probe.

(a) Only 2 and 5 are isolated systems. **(b)** Only 1 and 3 are isolated systems. **(c)** Only 3 and 5 are isolated systems. **(d)** Only 4 and 5 are isolated systems. **(e)** Only 5 is an isolated system.

Section 7.3 Collisions in One Dimension

6. Two objects are involved in a completely inelastic one-dimensional collision. The net external force acting on them is zero. The table lists four possible sets of the initial and final momenta and kinetic energies of the two objects. Which is the only set that could occur?

		Initial (Before Collision)		Final (After Collision)	
		Momentum	Kinetic Energy	Momentum	Kinetic Energy
a.	Object 1:	$+6$ kg · m/s	15 J	$+8$ kg · m/s	9 J
	Object 2:	0 kg · m/s	0 J		
b.	Object 1:	$+8$ kg · m/s	5 J	$+6$ kg · m/s	12 J
	Object 2:	-2 kg · m/s	7 J		
c.	Object 1:	-3 kg · m/s	1 J	$+1$ kg · m/s	4 J
	Object 2:	$+4$ kg · m/s	6 J		
d.	Object 1:	0 kg · m/s	3 J	-8 kg · m/s	11 J
	Object 2:	-8 kg · m/s	8 J		

Section 7.4   Collisions in Two Dimensions

7. Object 1 is moving along the *x* axis with an initial momentum of +16 kg · m/s, where the + sign indicates that it is moving to the right. As the drawing shows, object 1 collides with a second object that is initially at rest. The collision is not head-on, so the objects move off in different directions after the collision. The net external force acting on the two-object system is zero. After the collision, object 1 has a momentum whose *y* component is −5 kg · m/s. What is the *y* component of the momentum of object 2 after the collision? **(a)** 0 kg · m/s **(b)** +16 kg · m/s **(c)** +5 kg · m/s **(d)** −16 kg · m/s **(e)** The *y* component of the momentum of object 2 cannot be determined.

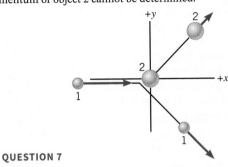

QUESTION 7

Section 7.5   Center of Mass

8. The drawing shows three particles that are moving with different velocities. Two of the particles have mass *m*, and the third has a mass 2*m*. At the instant shown, the center of mass (cm) of the three particles is at the coordinate origin. What is the velocity v_{cm} (magnitude and direction) of the center of mass?

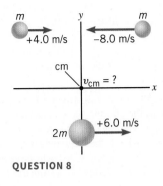

QUESTION 8

Problems

Section 7.1   The Impulse–Momentum Theorem

1. E SSM A 46-kg skater is standing still in front of a wall. By pushing against the wall she propels herself backward with a velocity of −1.2 m/s. Her hands are in contact with the wall for 0.80 s. Ignore friction and wind resistance. Find the magnitude and direction of the average force she exerts on the wall (which has the same magnitude as, but opposite direction to, the force that the wall applies to her).

2. E A model rocket is constructed with a motor that can provide a total impulse of 29.0 N · s. The mass of the rocket is 0.175 kg. What is the speed that this rocket achieves when launched from rest? Neglect the effects of gravity and air resistance.

3. E Before starting this problem, review Conceptual Example 4. Suppose that the hail described there bounces off the roof of the car with a velocity of +15 m/s. Ignoring the weight of the hailstones, calculate the force exerted by the hail on the roof. Compare your answer to that obtained in Example 2 for the rain, and verify that your answer is consistent with the conclusion reached in Conceptual Example 4.

4. E GO In a performance test, each of two cars takes 9.0 s to accelerate from rest to 27 m/s. Car A has a mass of 1400 kg, and car B has a mass of 1900 kg. Find the net average force that acts on each car during the test.

Online

5. E SSM A volleyball is spiked so that its incoming velocity of +4.0 m/s is changed to an outgoing velocity of −21 m/s. The mass of the volleyball is 0.35 kg. What impulse does the player apply to the ball?

6. E Two arrows are fired horizontally with the same speed of 30.0 m/s. Each arrow has a mass of 0.100 kg. One is fired due east and the other due south. Find the magnitude and direction of the total momentum of this two-arrow system. Specify the direction with respect to due east.

7. E Refer to Conceptual Example 4 as an aid in understanding this problem. A hockey goalie is standing on ice. Another player fires a puck (*m* = 0.17 kg) at the goalie with a velocity of +65 m/s. **(a)** If the goalie catches the puck with his glove in a time of 5.0×10^{-3} s, what is the average force (magnitude and direction) exerted on the goalie by the puck? **(b)** Instead of catching the puck, the goalie slaps it with his stick and returns the puck straight back to the player with a velocity of −65 m/s. The puck and stick are in contact for a time of 5.0×10^{-3} s. Now what is the average force exerted on the goalie by the puck? Verify that your answers to parts (a) and (b) are consistent with the conclusion of Conceptual Example 4.

8. E A space probe is traveling in outer space with a momentum that has a magnitude of 7.5×10^7 kg · m/s. A retrorocket is fired to slow down the probe. It applies a force to the probe that has a magnitude of 2.0×10^6 N and a direction opposite to the probe's motion. It fires for a period of 12 s. Determine the momentum of the probe after the retrorocket ceases to fire.

9. M MMH A stream of water strikes a stationary turbine blade horizontally, as the drawing illustrates. The incident water stream has a velocity of +16.0 m/s, while the exiting water stream has a velocity of −16.0 m/s. The mass of water per second that strikes the blade is 30.0 kg/s. Find the magnitude of the average force exerted on the water by the blade.

Stationary turbine blade

$v_0 = +16.0$ m/s

$v_f = -16.0$ m/s

PROBLEM 9

10. **M** **GO** A student ($m = 63$ kg) falls freely from rest and strikes the ground. During the collision with the ground, he comes to rest in a time of 0.040 s. The average force exerted on him by the ground is +18 000 N, where the upward direction is taken to be the positive direction. From what height did the student fall? Assume that the only force acting on him during the collision is that due to the ground.

11. **M** **V-HINT** **CHALK** A golf ball strikes a hard, smooth floor at an angle of 30.0° and, as the drawing shows, rebounds at the same angle. The mass of the ball is 0.047 kg, and its speed is 45 m/s just before and after striking the floor. What is the magnitude of the impulse applied to the golf ball by the floor? (*Hint: Note that only the vertical component of the ball's momentum changes during impact with the floor, and ignore the weight of the ball.*)

45 m/s

30.0° 30.0°

45 m/s

PROBLEM 11

12. **M** **V-HINT** An 85-kg jogger is heading due east at a speed of 2.0 m/s. A 55-kg jogger is heading 32° north of east at a speed of 3.0 m/s. Find the magnitude and direction of the sum of the momenta of the two joggers.

13. **M** **GO** A basketball ($m = 0.60$ kg) is dropped from rest. Just before striking the floor, the ball has a momentum whose magnitude is 3.1 kg · m/s. At what height was the basketball dropped?

14. **H** **SSM** A dump truck is being filled with sand. The sand falls straight downward from rest from a height of 2.00 m above the truck bed, and the mass of sand that hits the truck per second is 55.0 kg/s. The truck is parked on the platform of a weight scale. By how much does the scale reading exceed the weight of the truck and sand?

Section 7.2 The Principle of Conservation of Linear Momentum

15. **E** **MMH** In a science fiction novel two enemies, Bonzo and Ender, are fighting in outer space. From stationary positions they push against each other. Bonzo flies off with a velocity of +1.5 m/s, while Ender recoils with a velocity of −2.5 m/s. **(a)** Without doing any calculations, decide which person has the greater mass. Give your reasoning. **(b)** Determine the ratio $m_{\text{Bonzo}}/m_{\text{Ender}}$ of the masses of these two enemies.

16. **E** A 2.3-kg cart is rolling across a frictionless, horizontal track toward a 1.5-kg cart that is held initially at rest. The carts are loaded with strong magnets that cause them to attract one another. Thus, the speed of each cart increases. At a certain instant before the carts collide, the first cart's velocity is +4.5 m/s, and the second cart's velocity is −1.9 m/s. **(a)** What is the total momentum of the system of the two carts at this instant? **(b)** What was the velocity of the first cart when the second cart was still at rest?

17. **E** **GO** As the drawing illustrates, two disks with masses m_1 and m_2 are moving horizontally to the right at a speed of v_0. They are on an air-hockey table, which supports them with an essentially frictionless cushion of air. They move as a unit, with a compressed spring between them, which has a negligible mass. Then the spring is released and allowed to push the disks outward. Consider the situation where disk 1 comes to a momentary halt shortly after the spring is released. Assuming that $m_1 = 1.2$ kg, $m_2 = 2.4$ kg, and $v_0 = +5.0$ m/s, find the velocity of disk 2 at that moment.

m_1 m_2 v_0

PROBLEM 17

18. **E** **SSM** A lumberjack (mass = 98 kg) is standing at rest on one end of a floating log (mass = 230 kg) that is also at rest. The lumberjack runs to the other end of the log, attaining a velocity of +3.6 m/s relative to the shore, and then hops onto an identical floating log that is initially at rest. Neglect any friction and resistance between the logs and the water. **(a)** What is the velocity of the first log just before the lumberjack jumps off? **(b)** Determine the velocity of the second log if the lumberjack comes to rest on it.

19. **E** **GO** **MMH** An astronaut in his space suit and with a propulsion unit (empty of its gas propellant) strapped to his back has a mass of 146 kg. The astronaut begins a space walk at rest, with a completely filled propulsion unit. During the space walk, the unit ejects some gas with a velocity of +32 m/s. As a result, the astronaut recoils with a velocity of −0.39 m/s. After the gas is ejected, the mass of the astronaut (now wearing a partially empty propulsion unit) is 165 kg. What percentage of the gas was ejected from the completely filled propulsion unit?

20. **E** **SSM** A two-stage rocket moves in space at a constant velocity of 4900 m/s. The two stages are then separated by a small explosive charge placed between them. Immediately after the explosion the velocity of the 1200-kg upper stage is 5700 m/s in the same direction as before the explosion. What is the velocity (magnitude and direction) of the 2400-kg lower stage after the explosion?

21. **M** **V-HINT** A 40.0-kg boy, riding a 2.50-kg skateboard at a velocity of +5.30 m/s across a level sidewalk, jumps forward to leap over a wall. Just after leaving contact with the board, the boy's velocity relative to the sidewalk is 6.00 m/s, 9.50° above the horizontal. Ignore any friction between the skateboard and the sidewalk. What is the skateboard's velocity relative to the sidewalk at this instant? Be sure to include the correct algebraic sign with your answer.

22. **M** **GO** The lead female character in the movie *Diamonds Are Forever* is standing at the edge of an offshore oil rig. As she fires a gun, she is driven back over the edge and into the sea. Suppose the mass of a bullet is 0.010 kg and its velocity is +720 m/s. Her mass (including the gun) is 51 kg. **(a)** What recoil velocity does she acquire in response to a single shot from a stationary position, assuming that no external force keeps her in place? **(b)** Under the same assumption, what would be her recoil velocity if, instead, she shoots a blank cartridge that ejects a mass of 5.0×10^{-4} kg at a velocity of +720 m/s?

23. **M** **MMH** A 0.015-kg bullet is fired straight up at a falling wooden block that has a mass of 1.8 kg. The bullet has a speed of 810 m/s when it strikes the block. The block originally was dropped from rest from the top of a building and has been falling for a time t when the collision with the bullet occurs. As a result of the collision, the block (with the bullet in it) reverses direction, rises, and comes to a momentary halt at the top of the building. Find the time t.

24. **M** **SSM** By accident, a large plate is dropped and breaks into three pieces. The pieces fly apart parallel to the floor. As the plate falls, its momentum has only a vertical component and no component parallel to the floor. After the collision, the component of the total momentum parallel to the floor must remain zero, since the net external force acting on the plate has no component parallel to the floor. Using the data shown in the drawing, find the masses of pieces 1 and 2.

3.00 m/s

25.0°

m_1

1.79 m/s

m_2

45.0°

$m_3 = 1.30$ kg

3.07 m/s

PROBLEM 24

25. **H** Adolf and Ed are wearing harnesses and are hanging at rest from the ceiling by means of ropes attached to them. Face to face, they push off against one another. Adolf has a mass of 120 kg, and Ed has a mass of 78 kg. Following the push, Adolf swings upward to a height of 0.65 m above his starting point. To what height above his own starting point does Ed rise?

Section 7.3 Collisions in One Dimension,

Section 7.4 Collisions in Two Dimensions

26. **E** After sliding down a snow-covered hill on an inner tube, Ashley is coasting across a level snowfield at a constant velocity of +2.7 m/s. Miranda runs after her at a velocity of +4.5 m/s and hops on the inner tube. How fast do the two slide across the snow together on the inner tube? Ashley's mass is 71 kg and Miranda's is 58 kg. Ignore the mass of the inner tube and any friction between the inner tube and the snow.

27. **E** Consult Multiple-Concept Example 9 for background pertinent to this problem. A 2.50-g bullet, traveling at a speed of 425 m/s, strikes the wooden block of a ballistic pendulum, such as that in **Figure 7.13**. The block has a mass of 215 g. **(a)** Find the speed of the bullet–block combination immediately after the collision. **(b)** How high does the combination rise above its initial position?

28. **E** **GO** One object is at rest, and another is moving. The two collide in a one-dimensional, completely inelastic collision. In other words, they stick together after the collision and move off with a common velocity. Momentum is conserved. The speed of the object that is moving initially is 25 m/s. The masses of the two objects are 3.0 and 8.0 kg. Determine the final speed of the two-object system after the collision for the case when the large-mass object is the one moving initially and the case when the small-mass object is the one moving initially.

29. **E** **SSM** Batman (mass = 91 kg) jumps straight down from a bridge into a boat (mass = 510 kg) in which a criminal is fleeing. The velocity of the boat is initially +11 m/s. What is the velocity of the boat after Batman lands in it?

30. **E** A car (mass = 1100 kg) is traveling at 32 m/s when it collides head-on with a sport utility vehicle (mass = 2500 kg) traveling in the opposite direction. In the collision, the two vehicles come to a halt. At what speed was the sport utility vehicle traveling?

31. **E** **SSM** A 5.00-kg ball, moving to the right at a velocity of +2.00 m/s on a frictionless table, collides head-on with a stationary 7.50-kg ball. Find the final velocities of the balls if the collision is **(a)** elastic and **(b)** completely inelastic.

32. **E** **CHALK** The drawing shows a collision between two pucks on an air-hockey table. Puck A has a mass of 0.025 kg and is moving along the *x* axis with a velocity of +5.5 m/s. It makes a collision with puck B, which has a mass of 0.050 kg and is initially at rest. The collision is not head-on. After the collision, the two pucks fly apart with

the angles shown in the drawing. Find the final speeds of **(a)** puck A and **(b)** puck B.

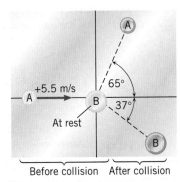

+5.5 m/s

A

At rest

B

65°

37°

A

B

Before collision After collision

PROBLEM 32

33. **E** **SSM** A projectile (mass = 0.20 kg) is fired at and embeds itself in a stationary target (mass = 2.50 kg). With what percentage of the projectile's incident kinetic energy does the target (with the projectile in it) fly off after being struck?

34. **E** **GO** Object A is moving due east, while object B is moving due north. They collide and stick together in a completely inelastic collision. Momentum is conserved. Object A has a mass of $m_A = 17.0$ kg and an initial velocity of $\vec{v}_{0A} = 8.00$ m/s, due east. Object B, however, has a mass of $m_B = 29.0$ kg and an initial velocity of $\vec{v}_{0B} = 5.00$ m/s, due north. Find the magnitude and direction of the total momentum of the two-object system after the collision.

35. **E** Multiple-Concept Example 8 deals with some of the concepts that are used to solve this problem. A cue ball (mass = 0.165 kg) is at rest on a frictionless pool table. The ball is hit dead center by a pool stick, which applies an impulse of +1.50 N · s to the ball. The ball then slides along the table and makes an elastic head-on collision with a second ball of equal mass that is initially at rest. Find the velocity of the second ball just after it is struck.

36. **M** **CHALK** **GO** A ball is attached to one end of a wire, the other end being fastened to the ceiling. The wire is held horizontal, and the ball is released from rest (see the drawing). It swings downward and strikes a block initially at rest on a horizontal frictionless surface. Air resistance is negligible, and the collision is elastic. The masses of the ball and block are, respectively, 1.60 kg and 2.40 kg, and the length of the wire is 1.20 m. Find the velocity (magnitude and direction) of the ball **(a)** just before the collision, and **(b)** just after the collision.

+y

+x

PROBLEM 36

37. **M** **V-HINT** A girl is skipping stones across a lake. One of the stones accidentally ricochets off a toy boat that is initially at rest in the water (see the drawing). The 0.072-kg stone strikes the boat at a velocity of 13 m/s, 15° below due east, and ricochets off at a velocity of 11 m/s, 12° above due east. After being struck by the stone, the boat's velocity is 2.1 m/s, due east. What is the mass of the boat? Assume the water offers no resistance to the boat's motion.

East

15°

13 m/s

11 m/s

East

12°

PROBLEM 37

38. **M** **GO** A mine car (mass = 440 kg) rolls at a speed of 0.50 m/s on a horizontal track, as the drawing shows. A 150-kg chunk of coal has a speed of 0.80 m/s when it leaves the chute. Determine the speed of the car–coal system after the coal has come to rest in the car.

PROBLEM 38

39. **M** **SSM** A 50.0-kg skater is traveling due east at a speed of 3.00 m/s. A 70.0-kg skater is moving due south at a speed of 7.00 m/s. They collide and hold on to each other after the collision, managing to move off at an angle θ south of east, with a speed of v_f. Find (a) the angle θ and (b) the speed v_f, assuming that friction can be ignored.

40. **M** **CHALK** **GO** A 4.00-g bullet is moving horizontally with a velocity of +355 m/s, where the +sign indicates that it is moving to the right (see part a of the drawing). The bullet is approaching two blocks resting on a horizontal frictionless surface. Air resistance is negligible. The bullet passes completely through the first block (an inelastic collision) and embeds itself in the second one, as indicated in part b. Note that both blocks are moving after the collision with the bullet. The mass of the first block is 1150 g, and its velocity is +0.550 m/s after the bullet passes through it. The mass of the second block is 1530 g. (a) What is the velocity of the second block after the bullet embeds itself? (b) Find the ratio of the total kinetic energy after the collisions to that before the collisions.

(a) Before collisions

(b) After collisions

$m_{\text{block 1}} = 1150$ g $m_{\text{block 2}} = 1530$ g
                    $m_{\text{bullet}} = 4.00$ g

PROBLEM 40

41. **M** **V-HINT** An electron collides elastically with a stationary hydrogen atom. The mass of the hydrogen atom is 1837 times that of the electron. Assume that all motion, before and after the collision, occurs along the same straight line. What is the ratio of the kinetic energy of the hydrogen atom after the collision to that of the electron before the collision?

42. **M** **GO** A 60.0-kg person, running horizontally with a velocity of +3.80 m/s, jumps onto a 12.0-kg sled that is initially at rest. (a) Ignoring the effects of friction during the collision, find the velocity of the

sled and person as they move away. (b) The sled and person coast 30.0 m on level snow before coming to rest. What is the coefficient of kinetic friction between the sled and the snow?

43. **H** **SSM** Starting with an initial speed of 5.00 m/s at a height of 0.300 m, a 1.50-kg ball swings downward and strikes a 4.60-kg ball that is at rest, as the drawing shows. (a) Using the principle of conservation of mechanical energy, find the speed of the 1.50-kg ball just before impact. (b) Assuming that the collision is elastic, find the velocities (magnitude and direction) of both balls just after the collision. (c) How high does each ball swing after the collision, ignoring air resistance?

PROBLEM 43

Section 7.5 Center of Mass

44. **E** Two particles are moving along the x axis. Particle 1 has a mass m_1 and a velocity $v_1 = +4.6$ m/s. Particle 2 has a mass m_2 and a velocity $v_2 = -6.1$ m/s. The velocity of the center of mass of these two particles is zero. In other words, the center of mass of the particles remains stationary, even though each particle is moving. Find the ratio m_1/m_2 of the masses of the particles.

45. **E** **GO** John's mass is 86 kg, and Barbara's is 55 kg. He is standing on the x axis at $x_J = +9.0$ m, while she is standing on the x axis at $x_B = 12.0$ m. They switch positions. How far and in which direction does their center of mass move as a result of the switch?

46. **E** **GO** Two stars in a binary system orbit around their center of mass. The centers of the two stars are 7.17×10^{11} m apart. The larger of the two stars has a mass of 3.70×10^{30} kg, and its center is 2.08×10^{11} m from the system's center of mass. What is the mass of the smaller star?

47. **M** **GO** **CHALK** The drawing shows a sulfur dioxide molecule. It consists of two oxygen atoms and a sulfur atom. A sulfur atom is twice as massive as an oxygen atom. Using this information and the data provided in the drawing, find (a) the x coordinate and (b) the y coordinate of the center of mass of the sulfur dioxide molecule. Express your answers in nanometers (1 nm = 10^{-9} m).

PROBLEM 47

Additional Problems

Online

48. **E** **GO** A golf ball bounces down a flight of steel stairs, striking several steps on the way down, but never hitting the edge of a step. The ball starts at the top step with a vertical velocity component of zero. If all the collisions with the stairs are elastic, and if the vertical height of the staircase is 3.00 m, determine the bounce height when the ball reaches the bottom of the stairs. Neglect air resistance.

49. **E** **MMH** A baseball ($m = 149$ g) approaches a bat horizontally at a speed of 40.2 m/s (90 mi/h) and is hit straight back at a speed of 45.6 m/s (102 mi/h). If the ball is in contact with the bat for a time of 1.10 ms, what is the average force exerted on the ball by the bat? Neglect the weight of the ball, since it is so much less than the force of the bat. Choose the direction of the incoming ball as the positive direction.

50. **E** **GO** A wagon is rolling forward on level ground. Friction is negligible. The person sitting in the wagon is holding a rock. The total mass of the wagon, rider, and rock is 95.0 kg. The mass of the rock is 0.300 kg. Initially the wagon is rolling forward at a speed of 0.500 m/s. Then the person throws the rock with a speed of 16.0 m/s. Both speeds are relative to the ground. Find the speed of the wagon after the rock is thrown directly forward in one case and directly backward in another.

51. **M** **V-HINT** Two ice skaters have masses m_1 and m_2 and are initially stationary. Their skates are identical. They push against one another, as in **Interactive Figure 7.10**, and move in opposite directions with different speeds. While they are pushing against each other, any kinetic frictional forces acting on their skates can be ignored. However, once the skaters separate, kinetic frictional forces eventually bring them to a halt. As they glide to a halt, the magnitudes of their accelerations are equal, and skater 1 glides twice as far as skater 2. What is the ratio m_1/m_2 of their masses?

52. **M** **GO** For the situation depicted in the figure, use momentum conservation to determine the magnitude and direction of the final velocity of ball 1 after the collision.

(a)

(b)

(a) Top view of two balls colliding on a horizontal surface. (b) This part of the drawing shows the x and y components of the velocity of ball 1 after the collision.

PROBLEM 52

Physics in Biology, Medicine, and Sports

53. **E** **BIO** **7.1** When jumping straight down, you can be seriously injured if you land stiff-legged. One way to avoid injury is to bend your knees upon landing to reduce the force of the impact. A 75-kg man just before contact with the ground has a speed of 6.4 m/s. **(a)** In a stiff-legged landing he comes to a halt in 2.0 ms. Find the average net force that acts on him during this time. **(b)** When he bends his knees, he comes to a halt in 0.10 s. Find the average net force now. **(c)** During the landing, the force of the ground on the man points upward, while the force due to gravity points downward. The average net force acting on the man includes both of these forces. Taking into account the directions of the forces, find the force of the ground on the man in parts (a) and (b).

54. **E** **BIO** **MMH** **7.2** For tests using a *ballistocardiograph*, a patient lies on a horizontal platform that is supported on jets of air. Because of the air jets, the friction impeding the horizontal motion of the platform is negligible. Each time the heart beats, blood is pushed out of the heart in a direction that is nearly parallel to the platform. Since momentum must be conserved, the body and the platform recoil, and this recoil can be detected to provide information about the heart. For each beat, suppose that 0.050 kg of blood is pushed out of the heart with a velocity of +0.25 m/s and that the mass of the patient and platform is 85 kg. Assuming that the patient does not slip with respect to the platform, and that the patient and platform start from rest, determine the recoil velocity.

55. **M** **BIO** **V-HINT** **7.5** The drawing shows a human figure in a sitting position. For purposes of this problem, there are three parts to the figure, and the center of mass of each one is shown in the drawing. These parts are: (1) the torso, neck, and head (total mass = 41 kg) with a center of mass located on the y axis at a point 0.39 m above the origin, (2) the upper legs (mass = 17 kg) with a center of mass located on the x axis at a point 0.17 m to the right of the origin, and (3) the lower legs and feet (total mass = 9.9 kg) with a center of mass located 0.43 m to the right of and 0.26 m below the origin. Find the x and y coordinates of the center of mass of the human figure. Note that the mass of the arms and hands (approximately 12% of the whole-body mass) has been ignored to simplify the drawing.

PROBLEM 55

56. **M** **BIO** **7.1** As of 2008, 42 people had survived a fall from a height above 10 000 feet without a parachute. Without air resistance, a person would hit the ground with a speed of 240 m/s (540 mph!). However, by placing your body in a skydiving position (see photo), you can maximize your cross-sectional area and take advantage of the drag force to limit your speed. In this position, your terminal speed will only be 54 m/s (120 mph). This is still a high rate of speed. Just imagine being in an automobile accident at 120 mph! Nevertheless, there is a higher probability of survival, depending upon what you land on. After all, it is not the fall that will kill you, but the impact. If you strike concrete at 54 m/s, we will say nice things about you at your funeral. You can take comfort in the fact that you won't even feel it. Contrary to popular belief, hitting water at this speed is not much better. Your body will have to displace and compress the water as you enter it. But water, and most liquids, are difficult to compress. The surface of the liquid does not "give" nearly enough to cushion your fall. It is no surprise that the survivors fell into deep snow, trees and bushes, and marsh mud. **(a)** Calculate the average force exerted on a 75-kg person who strikes concrete at 54 m/s and comes to rest in 0.0052 s. **(b)** In comparison, calculate the average force exerted on the same person who strikes deep snow at 54 m/s and comes to rest 5.7 m deep in the snow. Assume the person's deceleration is constant and the only force that acts on him when he is brought to rest is the snow.

skeeze/Pixabay

PROBLEM 56

57. **M** **BIO** **7.3** Cephalopods, such as octopi and squids, can use a form of jet propulsion to move quickly through the water. This technique of locomotion is useful in providing short bursts of speed to catch prey and to avoid predators, and it gives them the title of fastest marine invertebrates. Oxygenated water is taken in by the cephalopod and then expelled by muscular contraction through an opening called the hyponome. Assume a 4.2-kg squid draws in 0.32 kg of water and then quickly (in 0.12 s) expels this water at a speed of 11 m/s. **(a)** What is the speed of the squid just after the water is expelled, and **(b)** what is the average force exerted on the expelled water?

joakant/Pixabay

PROBLEM 57

58. **M** **BIO** **7.3** Frank "Cannonball" Richards was an American entertainer and vaudeville performer. He was famous for having incredibly strong abdominal muscles. In feats of strength, he would often let people punch him in the stomach, jump on him, and even hit him with sledgehammers. However, his greatest claim to fame was being shot in the stomach with a 47-kg (104 lb) cannonball (see the photo). While performing, Richards limited this demonstration to just twice a day, since it was quite painful. He would live a long life and pass away at age 81 in 1969. If the speed of the cannonball was 11 m/s right before it collided

with Richards and the speed of the cannonball was 0.75 m/s right after the collision, what was Richards's speed immediately after being struck by the cannonball? Assume Richards's mass was 86 kg.

Unknown stock footage film-maker, Image taken from https://en.wikipedia.org/wiki/File:Frank_%-22Cannonball%22_Richards_shot_in_stomach_with_cannonball.jpg

PROBLEM 58

59. **M** **BIO** **7.1** A professional golfer hits a golf ball off the tee a horizontal distance of 310 m (see the photo). The ball leaves the tee at an angle of 45° with respect to the horizontal. If the head of the driver is in contact with the 45-g ball for 1.0 ms, calculate the average force exerted on the ball by the driver.

Leonard Kamsler/Popperfoto/Getty Images

PROBLEM 59

60. **M** **BIO** **7.3** One common inaccuracy in action movies occurs when someone gets shot. Oftentimes, when struck by the bullet, the person goes flying backward. Let's consider an average person ($m = 75$ kg) standing at rest on frictionless ice who is shot at point-blank range with a high-energy handgun—a .357 Magnum. The mass of the bullet is 8.0 g and its muzzle velocity is approximately 1450 ft/s (440 m/s). **(a)** What is the velocity of the person after he is shot? Assume the bullet remains lodged in his body. You should find that your answer to part (a) is quite small—far too small to send the body flying backward, as is often portrayed in the movies. Does this mean getting shot is not that dangerous? No! The bullet transfers a tremendous amount of kinetic energy to a very small area of the body. This energy can do damaging work on the body's internal organs and structures. **(b)** Calculate the kinetic energy of the bullet just before it strikes the body. Now assume the person gets "shot" with a baseball ($m = 140$ g) that has the same kinetic energy as the bullet. **(c)** What would be the speed of the baseball in mph?

Concepts and Calculations Problems

Online

Momentum and energy are two of the most fundamental concepts in physics. As we have seen in this chapter, momentum is a vector and, like all vectors, has a magnitude and a direction. In contrast, energy is a scalar quantity, as Chapter 6 discusses, and does not have a direction associated with it. Problem 61 provides the opportunity to review how the vector nature of momentum and the scalar nature of kinetic energy influence calculations using these quantities. Problem 62 explores some further differences between momentum and kinetic energy.

61. **M** **CHALK** **SSM** Two joggers, Jim and Tom, are both running at a speed of 4.00 m/s (see the figure). Jim has a mass of 90.0 kg, and Tom has a mass of 55.0 kg. *Concepts:* (i) Does the total kinetic energy of the two-jogger system have a smaller value in part *a* or *b* of the figure, or is it the same in both cases? (ii) Does their total momentum have a smaller magnitude in part *a* or *b* of the figure, or is it the same for each? *Calculations:* Find the kinetic energy and momentum of the two-jogger system when **(a)** Jim and Tom are both running due north as in part *a* of the figure, and **(b)** Jim is running due north and Tom is running due south, as in part *b* of the figure.

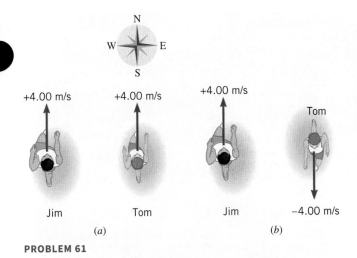

PROBLEM 61

62. **M** **CHALK** The table gives mass and speed data for the two objects in the figure. *Concepts:* (i) Is it possible for two objects to have different speeds when their momenta have the same amplitude? Explain your answer. (ii) If two objects have the same momenta do they necessarily have the same kinetic energy? Explain. *Calculations:* Find the magnitude of the momentum and the kinetic energy for each object.

	Mass	Speed
Object A	2.0 kg	6.0 m/s
Object B	6.0 kg	2.0 m/s

PROBLEM 62

Team Problems

Online

63. **E** **T** **WS** **A Balancing Act.** A heavy object that is part of a gigantic particle detector that will be used in a new particle accelerator is composed of four identically shaped disks and an aluminum plate. Each disk is composed of a single pure element: one each of lead (Pb), copper (Cu), tungsten (W), and titanium (Ti). The disks are 60.0 cm in diameter and 25.0 cm thick, and are bolted through their centers to the corners of a square piece of aluminum (Al) plate that is 12.0 cm thick. When bolted in place, the curved edges of the disks touch those of their neighbors mounted on adjacent corners. If a typical right-handed xy coordinate system is defined such that its origin is at the center of the square plate, the tungsten disk is located in the first quadrant (i.e., the $+x$, $+y$ quadrant), titanium in the second ($-x$, $+y$), copper in the third ($-x$, $-y$), and, finally, the lead disk is in the fourth quadrant ($+x$, $-y$). It is crucial, in order to place the object where it belongs on the detector, that the object be lifted with the aluminum plate parallel to the ground using a single chain. Your team is tasked with locating the position at which a hook should be placed in order to connect the chain that will be used to lift the awkward object with a crane. Determine the x and y coordinates of the hook position in centimeters. The densities of the materials in g/cm^3 are as follows: lead (11.3), copper (8.96), tungsten (19.3), titanium (4.51), and aluminum (2.70). (*Hint: you can treat each object as a point mass with its total mass concentrated at its geometric center.*)

64. **E** **T** **WS** **Which one is heavier?** Two smooth disks about the size of drink coasters are identical in size and overall appearance, except that one is composed of radioactive cobalt (Co) and the other of iron (Fe), which is not radioactive. Since the two disks are exactly the same size, their weights are slightly different because they are made from different materials. However, no scale is available to measure the weight and you are unable to discern which one is heavier by handling them (something you must avoid since one is radioactive). However, there is an air hockey table available (i.e., an air table on which the disks can move with nearly zero friction). You and your team have been given the task of determining which of the two disks is radioactive (a radiation detector is not available). Describe a one-dimensional collision experiment that does not require measuring the masses or speeds of the disks and that uses the air hockey table to reveal which disk is heavier. Explain in detail (with mathematical reasoning) how it will provide the information that you seek. The densities of the materials in g/cm^3 are: Fe (7.86) and Co (8.90).

65. **M** **T** **Docking a Spaceship.** You and your crew must dock your 2.50×10^4 kg spaceship at Spaceport Alpha, which is orbiting Mars. In the process, Alpha's control tower has requested that you ram another vessel, a freight ship of mass 1.65×10^4 kg, latch onto it, and use your combined momentum to bring it into dock. The freight ship is not moving with respect to the colossal Spaceport Alpha, which has a mass of 1.85×10^7 kg. Alpha's automated system that guides incoming spacecraft into dock requires that the incoming speed is less than 2.00 m/s. **(a)** Assuming a perfectly linear alignment of your ship's velocity vector with the freight ship (which is stationary with respect to Alpha) and Alpha's docking port, what must be your ship's speed (before colliding with the freight ship) so that the combination of the freight ship and your ship arrives at Alpha's docking port with a speed of 1.50 m/s? **(b)** How does the velocity of Spaceport Alpha change when the combination of your vessel and the freight ship successfully docks with it? **(c)** Suppose you made a mistake while maneuvering your vessel in an attempt to ram the freight ship and, rather than latching on to it and making a perfectly inelastic collision, you strike it and knock it in the direction of the spaceport with a perfectly elastic collision. What is the speed of the freight ship in that case (assuming your ship had the same initial velocity as that calculated in part (a))?

66. **M** **T** **Spaceship Malfunction.** You and your team get a distress call from another ship indicating that there has been an explosion and their propulsion and navigation systems are now off line. A large piece of their ship has been jettisoned into space, and their life support systems are failing. They need help as soon as possible. You try to track them, but are only able to pick up the piece of the craft that had been projected away from them in the explosion. You look up the make and model of their ship, a Galaxy Explorer 5000 (GE 5000), which has a total mass of 5.50×10^4 kg. The piece heading in your direction is the ship's storage module, and has a mass of 7.50×10^3 kg. The GE 5000 was at rest with respect to your ship at the time of the explosion. **(a)** Assuming you received the distress call 20.0 minutes before the jettisoned storage module passed by your ship at a speed of 750.0 m/s, what is the speed of the damaged craft? **(b)** How far away is the Galaxy Explorer 5000 when its storage module passes by? **(c)** In which direction, and how fast must you go, in order to catch up with the GE 5000 before its life support systems fail. (The crew had 80.0 minutes of life support left when they made the distress call, so you have 60.0 minutes to get to them.)

CHAPTER 8

The figure shows the front view of a turbine jet engine on a commercial aircraft. The rotating fan blades collect air into the engine before it is compressed, mixed with fuel, and ignited to produce thrust. The rotational motion of the blades can be described using the concepts of angular displacement, angular velocity, and angular acceleration within the framework of rotational kinematics.

Mikhail Starodubov/Shutterstock.com

Rotational Kinematics

LEARNING OBJECTIVES

After reading this module, you should be able to...

8.1 Define angular displacement.

8.2 Define angular velocity and angular acceleration.

8.3 Solve rotational kinematics problems.

8.4 Relate angular and tangential variables.

8.5 Distinguish between centripetal and tangential accelerations.

8.6 Analyze rolling motion.

8.7 Use the right-hand rule to determine the direction of angular vectors.

8.1 Rotational Motion and Angular Displacement

In the simplest kind of rotation, points on a rigid object move on circular paths. In **Figure 8.1**, for example, we see the circular paths for points A, B, and C on a spinning skater. The centers of all such circular paths define a line, called the **axis of rotation**.

The angle through which a rigid object rotates about a fixed axis is called the **angular displacement**. **Figure 8.2** shows how the angular displacement is measured for a thin rotating disc. Here, the axis of rotation passes through the center of the disc and is perpendicular to its surface. On the surface of the disc we draw a radial line, which is a line that intersects the axis of rotation perpendicularly. As the disc turns, we observe the angle through which this line moves relative to a convenient reference line that does not rotate. The radial line moves from its initial orientation at angle θ_0 to a final orientation at angle θ (Greek letter theta). In the process, the line sweeps out the angle $\theta - \theta_0$. As with other differences that we have encountered ($\Delta x = x - x_0$, $\Delta v = v - v_0$, $\Delta t = t - t_0$), it is customary to denote the difference between the final and initial angles by the notation $\Delta\theta$ (read as "delta theta"): $\Delta\theta = \theta - \theta_0$. The angle $\Delta\theta$ is the angular displacement. A rotating object may rotate either counterclockwise or clockwise, and standard convention calls a counterclockwise displacement positive and a clockwise displacement negative.

DEFINITION OF ANGULAR DISPLACEMENT

When a rigid body rotates about a fixed axis, the angular displacement is the angle $\Delta\theta$ swept out by a line passing through any point on the body and intersecting the axis of rotation perpendicularly. By convention, the angular displacement is positive if it is counterclockwise and negative if it is clockwise.

SI Unit of Angular Displacement: radian (rad)*

Angular displacement is often expressed in one of three units. The first is the familiar **degree**, and it is well known that there are 360 degrees in a circle. The second unit is the **revolution (rev)**, one revolution representing one complete turn of 360°. The most useful unit from a scientific viewpoint, however, is the SI unit called the **radian (rad)**. **Figure 8.3** shows how the radian is defined, again using a disc as an example. The picture focuses attention on a point P on the disc. This point starts out on the stationary reference line, so that $\theta_0 = 0$ rad, and the angular displacement is $\Delta\theta = \theta - \theta_0 = \theta$. As the disc rotates, the point traces out an arc of length s, which is measured along a circle of radius r. Equation 8.1 defines the angle θ in radians:

$$\theta \text{(in radians)} = \frac{\text{Arc length}}{\text{Radius}} = \frac{s}{r} \tag{8.1}$$

According to this definition, an angle in radians is the ratio of two lengths; for example, meters/meters. In calculations, therefore, the radian is treated as a number without units and has no effect on other units that it multiplies or divides.

FIGURE 8.1 When a rigid object rotates, points on the object, such as A, B, or C, move on circular paths. The centers of the circles form a line that is the axis of rotation.

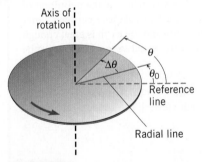

FIGURE 8.2 The angular displacement of a rotating disc is the angle $\Delta\theta$ swept out by a radial line as the disc turns about its axis of rotation.

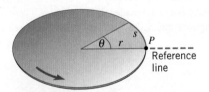

FIGURE 8.3 In radian measure, the angle θ is defined to be the arc length s divided by the radius r.

To convert between degrees and radians, it is only necessary to remember that the arc length of an entire circle of radius r is the circumference $2\pi r$. Therefore, according to Equation 8.1, *the number of radians that corresponds to 360°, or one revolution, is*

$$\theta = \frac{2\pi r}{r} = 2\pi \text{ rad}$$

Since 2π rad corresponds to 360°, the number of degrees in one radian is

$$1 \text{ rad} = \frac{360°}{2\pi} = 57.3°$$

It is useful to express an angle θ in radians, because then the arc length s subtended at any radius r can be calculated by multiplying θ by r. Example 1 illustrates this point and also shows how to convert between degrees and radians.

---

*The radian is neither a base SI unit nor a derived one. It is regarded as a supplementary SI unit.

EXAMPLE 1 | The Physics of Synchronous Communications Satellites

Synchronous or "stationary" communications satellites are put into an orbit whose radius is $r = 4.23 \times 10^7$ m. The orbit is in the plane of the equator, and two adjacent satellites have an angular separation of $\theta = 2.00°$, as **Figure 8.4** illustrates. Find the arc length s (see the drawing) that separates the satellites.

Reasoning Since the radius r and the angle θ are known, we may find the arc length s by using the relation θ (in radians) $= s/r$. But first, the angle must be converted to radians from degrees.

Solution To convert 2.00° into radians, we use the fact that 2π radians is equivalent to 360°:

$$2.00° = (2.00 \text{ degrees})\left(\frac{2\pi \text{ radians}}{360 \text{ degrees}}\right) = 0.0349 \text{ radians}$$

From Equation 8.1, it follows that the arc length between the satellites is

$$s = r\theta = (4.23 \times 10^7 \text{ m})(0.0349 \text{ rad}) = \boxed{1.48 \times 10^6 \text{ m (920 miles)}}$$

The radian, being a unitless quantity, is dropped from the final result, leaving the answer expressed in meters.

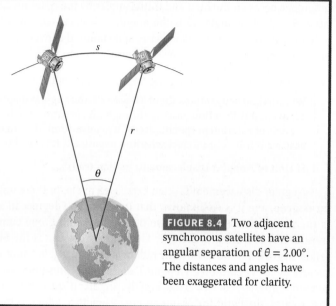

FIGURE 8.4 Two adjacent synchronous satellites have an angular separation of $\theta = 2.00°$. The distances and angles have been exaggerated for clarity.

Conceptual Example 2 takes advantage of the radian as a unit for measuring angles and explains the spectacular phenomenon of a total solar eclipse.

CONCEPTUAL EXAMPLE 2 | The Physics of a Total Solar Eclipse

The diameter of the sun is about 400 times greater than that of the moon. By coincidence, the sun is also about 400 times farther from the earth than is the moon. For an observer on earth, compare the angles subtended by the sun and the moon. **(a)** The angle subtended by the sun is much greater than that subtended by the moon. **(b)** The angle subtended by the sun is much smaller than that subtended by the moon. **(c)** The angles subtended by the sun and the moon are approximately equal.

Reasoning Equation 8.1 ($\theta = s/r$), which is the definition of an angle in radians, will guide us. The distance r between either the sun and the earth or the moon and the earth is great enough that the arc length s is very nearly equal to the diameter of the sun or the moon.

Answers (a) and (b) are incorrect. **Interactive Figure 8.5a** shows a person on earth viewing the sun and the moon.

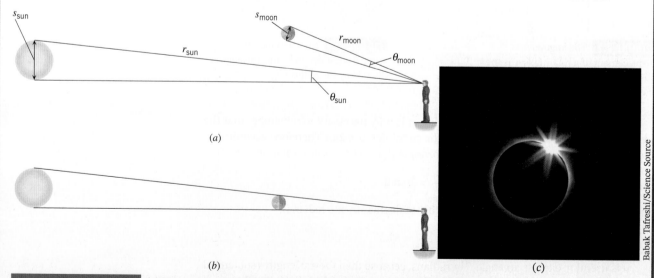

(a)

(b)

(c)

INTERACTIVE FIGURE 8.5 (a) The angles subtended by the moon and sun at the eyes of the observer are θ_{moon} and θ_{sun}. (The distances and angles are exaggerated for the sake of clarity.) (b) Since the moon and sun subtend approximately the same angle, the moon blocks nearly all the sun's light from reaching the observer's eyes. (c) The result is a total solar eclipse.

Babak Tafreshi/Science Source

For the diameters s of the sun and the moon, we know that $s_{\text{sun}} \approx 400 s_{\text{moon}}$ (where the symbol $\approx$ means "approximately equal to"). For the distances r between the sun and the earth or the moon and the earth, we know that $r_{\text{sun}} \approx 400 r_{\text{moon}}$. Because of these facts, the ratio s/r is approximately the same for the sun and for the moon. Therefore, the subtended angle ($\theta = s/r$) is approximately the same for the sun and the moon.

Answer (c) is correct. Applying Equation 8.1 to the case of the sun and the moon gives $\theta_{\text{sun}} = s_{\text{sun}}/r_{\text{sun}}$ and $\theta_{\text{moon}} = s_{\text{moon}}/r_{\text{moon}}$.

We know, however, that $s_{\text{sun}} \approx 400 s_{\text{moon}}$ and $r_{\text{sun}} \approx 400 r_{\text{moon}}$. Substitution into the expression for θ_{sun} gives

$$\theta_{\text{sun}} = \frac{s_{\text{sun}}}{r_{\text{sun}}} \approx \frac{400\, s_{\text{moon}}}{400\, r_{\text{moon}}} = \theta_{\text{moon}}$$

Interactive Figure 8.5b shows what happens when the moon comes between the sun and the earth. Since the angles subtended by the sun and the moon are nearly equal, the moon blocks most of the sun's light from reaching the observer's eyes, and a total solar eclipse like that in **Interactive Figure 8.5c** occurs.

Related Homework: Problem 7, 10

Check Your Understanding

(The answers are given at the end of the book.)

1. In **CYU Figure 8.1**, the flat triangular sheet ABC is lying in the plane of the paper. This sheet is going to rotate about first one axis and then another axis. Both of these axes lie in the plane of the paper and pass through point A. For each of the axes the points B and C move on separate circular paths that have the same radii. Identify these two axes.

CYU FIGURE 8.1

2. Three objects are visible in the night sky. They have the following diameters (in multiples of d) and subtend the following angles (in multiples of θ_0) at the eye of the observer. Object A has a diameter of $4d$ and subtends an angle of $2\theta_0$. Object B has a diameter of $3d$ and subtends an angle of $\theta_0/2$. Object C has a diameter of $d/2$ and subtends an angle of $\theta_0/8$. Rank them in descending order (greatest first) according to their distance from the observer.

8.2 | Angular Velocity and Angular Acceleration

Angular Velocity

In Section 2.2 we introduced the idea of linear velocity to describe how fast an object moves and the direction of its motion. The average linear velocity $\vec{\mathbf{v}}$ was defined as the linear displacement $\Delta \vec{\mathbf{x}}$ of the object divided by the time Δt required for the displacement to occur, or $\vec{\mathbf{v}} = \Delta \vec{\mathbf{x}}/\Delta t$ (see Equation 2.2). We now introduce the analogous idea of angular velocity to describe the motion of a rigid object rotating about a fixed axis. The **average angular velocity** $\overline{\omega}$ (Greek letter omega) is defined as the angular displacement $\Delta\theta = \theta - \theta_0$ divided by the elapsed time Δt during which the displacement occurs.

DEFINITION OF AVERAGE ANGULAR VELOCITY

$$\text{Average angular velocity} = \frac{\text{Angular displacement}}{\text{Elapsed time}}$$

$$\overline{\omega} = \frac{\theta - \theta_0}{t - t_0} = \frac{\Delta\theta}{\Delta t} \tag{8.2}$$

SI Unit of Angular Velocity: radian per second (rad/s)

The SI unit for angular velocity is the radian per second (rad/s), although other units such as revolutions per minute (rev/min or rpm) are also used. In agreement with the sign convention adopted for angular displacement, angular velocity is positive when the rotation is counterclockwise and negative when it is clockwise. Example 3 shows how the concept of average angular velocity is applied to a gymnast.

EXAMPLE 3 | Gymnast on a High Bar

A gymnast on a high bar swings through two revolutions in a time of 1.90 s, as **Figure 8.6** suggests. Find the average angular velocity (in rad/s) of the gymnast.

Reasoning The average angular velocity of the gymnast in rad/s is the angular displacement in radians divided by the elapsed time. However, the angular displacement is given as two revolutions, so we begin by converting this value into radian measure.

Solution The angular displacement (in radians) of the gymnast is

$$\Delta\theta = -2.00 \text{ revolutions} \left(\frac{2\pi \text{ radians}}{1 \text{ revolution}} \right) = -12.6 \text{ radians}$$

where the minus sign denotes that the gymnast rotates clockwise (see the drawing). The average angular velocity is

$$\overline{\omega} = \frac{\Delta\theta}{\Delta t} = \frac{-12.6 \text{ rad}}{1.90 \text{ s}} = \boxed{-6.63 \text{ rad/s}} \qquad (8.2)$$

FIGURE 8.6 Swinging on a high bar.

The **instantaneous angular velocity** ω is the angular velocity that exists at any given instant. To measure it, we follow the same procedure used in Chapter 2 for the instantaneous linear velocity. In this procedure, a small angular displacement $\Delta\theta$ occurs during a small time interval Δt. The time interval is so small that it approaches zero ($\Delta t \to 0$), and in this limit, the measured average angular velocity, $\overline{\omega} = \Delta\theta/\Delta t$, becomes the instantaneous angular velocity ω:

$$\omega = \lim_{\Delta t \to 0} \overline{\omega} = \lim_{\Delta t \to 0} \frac{\Delta\theta}{\Delta t} \qquad (8.3)$$

The magnitude of the instantaneous angular velocity, without reference to whether it is a positive or negative quantity, is called the **instantaneous angular speed**. If a rotating object has a constant angular velocity, the instantaneous value and the average value are the same.

Angular Acceleration

In linear motion, a changing velocity means that an acceleration is occurring. Such is also the case in rotational motion; a changing angular velocity means that an **angular acceleration** is occurring. There are many examples of angular acceleration. For instance, when the push buttons of an electric blender are changed from a lower setting to a higher setting, the angular velocity of the blades increases. We will define the average angular acceleration in a fashion analogous to that used for the average linear acceleration. Recall that the average linear acceleration $\vec{\mathbf{a}}$ is equal to the change $\Delta\vec{\mathbf{v}}$ in the linear velocity of an object divided by the elapsed time Δt: $\vec{\mathbf{a}} = \Delta\vec{\mathbf{v}}/\Delta t$ (Equation 2.4). When the angular velocity changes from an initial value of ω_0 at time t_0 to a final value of ω at time t, the average angular acceleration $\overline{\alpha}$ (Greek letter alpha) is defined similarly:

DEFINITION OF AVERAGE ANGULAR ACCELERATION

$$\text{Average angular acceleration} = \frac{\text{Change in angular velocity}}{\text{Elapsed time}}$$

$$\overline{\alpha} = \frac{\omega - \omega_0}{t - t_0} = \frac{\Delta\omega}{\Delta t} \qquad (8.4)$$

SI Unit of Average Angular Acceleration: radian per second squared (rad/s²)

The SI unit for average angular acceleration is the unit for angular velocity divided by the unit for time, or $(\text{rad/s})/\text{s} = \text{rad/s}^2$. An angular acceleration of $+5 \text{ rad/s}^2$, for example, means that the angular velocity of the rotating object increases by $+5$ radians per second during each second of acceleration.

The **instantaneous angular acceleration** α is the angular acceleration at a given instant. In discussing linear motion, we assumed a condition of constant acceleration, so that the average and instantaneous accelerations were identical ($\vec{\bar{a}} = \vec{a}$). Similarly, we assume that the angular acceleration is constant, so that the instantaneous angular acceleration α and the average angular acceleration $\bar{\alpha}$ are the same ($\bar{\alpha} = \alpha$). The next example illustrates the concept of angular acceleration.

EXAMPLE 4 | A Jet Revving Its Engines

A jet awaiting clearance for takeoff is momentarily stopped on the runway. As seen from the front of one engine, the fan blades are rotating with an angular velocity of -110 rad/s, where the negative sign indicates a clockwise rotation (see **Figure 8.7**). As the plane takes off, the angular velocity of the blades reaches -330 rad/s in a time of 14 s. Find the angular acceleration, assuming it to be constant.

Reasoning Since the angular acceleration is constant, it is equal to the average angular acceleration. The average acceleration is the change in the angular velocity, $\omega - \omega_0$, divided by the elapsed time, $t - t_0$.

Solution Applying the definition of average angular acceleration given in Equation 8.4, we find that

$$\bar{\alpha} = \frac{\omega - \omega_0}{t - t_0} = \frac{(-330 \text{ rad/s}) - (-110 \text{ rad/s})}{14 \text{ s}} = \boxed{-16 \text{ rad/s}^2}$$

FIGURE 8.7 The fan blades of a jet engine have an angular velocity ω in a clockwise direction.

Thus, the magnitude of the angular velocity increases by 16 rad/s during each second that the blades are accelerating. The negative sign in the answer indicates that the direction of the angular acceleration is also in the clockwise direction.

Check Your Understanding

(The answers are given at the end of the book.)

3. A pair of scissors is being used to cut a piece of paper in half. Does each blade of the scissors have the same angular velocity (both magnitude and direction) at a given instant?

4. An electric clock is hanging on a wall. As you are watching the second hand rotate, the clock's battery stops functioning, and the second hand comes to a halt over a brief period of time. Which one of the following statements correctly describes the angular velocity ω and angular acceleration α of the second hand as it slows down? **(a)** ω and α are both negative. **(b)** ω is positive and α is negative. **(c)** ω is negative and α is positive. **(d)** ω and α are both positive.

8.3 | The Equations of Rotational Kinematics

In Chapters 2 and 3 the concepts of displacement, velocity, and acceleration were introduced. We then combined these concepts and developed a set of equations called the equations of kinematics for constant acceleration (see Tables 2.1 and 3.1). These equations are a great aid in solving problems involving linear motion in one and two dimensions.

We now take a similar approach for rotational motion by bringing together the ideas of angular displacement, angular velocity, and angular acceleration to produce a set of equations called the equations of rotational kinematics for constant angular acceleration. These equations, like those developed in Chapters 2 and 3 for linear motion, will prove very useful in solving problems that involve rotational motion.

A complete description of rotational motion requires values for the angular displacement $\Delta\theta$, the angular acceleration α, the final angular velocity ω, the initial angular velocity ω_0, and the elapsed time Δt. In Example 4, for instance, only the angular displacement of the fan blades during the 14-s interval is missing. Such missing information can be calculated, however. For convenience in the calculations, we assume that the orientation of the rotating object is given by $\theta_0 = 0$ rad at time $t_0 = 0$ s. Then, the angular displacement becomes $\Delta\theta = \theta - \theta_0 = \theta$, and the time interval becomes $\Delta t = t - t_0 = t$.

In Example 4, the angular velocity of the fan blades changes at a constant rate from an initial value of $\omega_0 = -110$ rad/s to a final value of $\omega = -330$ rad/s. Therefore, the average angular velocity is midway between the initial and final values:

$$\overline{\omega} = \tfrac{1}{2}[(-110 \text{ rad/s}) + (-330 \text{ rad/s})] = -220 \text{ rad/s}$$

In other words, when the angular acceleration is constant, the average angular velocity is given by

$$\overline{\omega} = \tfrac{1}{2}(\omega_0 + \omega) \tag{8.5}$$

With a value for the average angular velocity, Equation 8.2 can be used to obtain the angular displacement of the fan blades:

$$\theta = \overline{\omega}t = (-220 \text{ rad/s})(14 \text{ s}) = -3100 \text{ rad}$$

In general, when the angular acceleration is constant, the angular displacement can be obtained from

$$\theta = \overline{\omega}t = \tfrac{1}{2}(\omega_0 + \omega)t \tag{8.6}$$

This equation and Equation 8.4 provide a complete description of rotational motion under the condition of constant angular acceleration. Equation 8.4 (with $t_0 = 0$ s) and Equation 8.6 are compared with the analogous results for linear motion in the first two rows of **Table 8.1**. The purpose of this comparison is to emphasize that the mathematical forms of Equations 8.4 and 2.4 are identical, as are the forms of Equations 8.6 and 2.7. Of course, the symbols used for the rotational variables are different from those used for the linear variables, as **Table 8.2** indicates.

TABLE 8.1 The Equations of Kinematics for Rotational and Linear Motion

Rotational Motion (α = constant)		Linear Motion (a = constant)	
$\omega = \omega_0 + \alpha t$	(8.4)	$v = v_0 + at$	(2.4)
$\theta = \tfrac{1}{2}(\omega_0 + \omega)t$	(8.6)	$x = \tfrac{1}{2}(v_0 + v)t$	(2.7)
$\theta = \omega_0 t + \tfrac{1}{2}\alpha t^2$	(8.7)	$x = v_0 t + \tfrac{1}{2}at^2$	(2.8)
$\omega^2 = \omega_0^2 + 2\alpha\theta$	(8.8)	$v^2 = v_0^2 + 2ax$	(2.9)

TABLE 8.2 Symbols Used in Rotational and Linear Kinematics

Rotational Motion	Quantity	Linear Motion
θ	Displacement	x
ω_0	Initial velocity	v_0
ω	Final velocity	v
α	Acceleration	a
t	Time	t

In Chapter 2, Equations 2.4 and 2.7 are used to derive the remaining two equations of kinematics (Equations 2.8 and 2.9). These additional equations convey no new information but are convenient to have when solving problems. Similar derivations can be carried out here. The results are listed as Equations 8.7 and 8.8 below and in **Table 8.1**; they can be inferred directly from their counterparts in linear motion by making the substitution of symbols indicated in **Table 8.2**:

$$\theta = \omega_0 t + \tfrac{1}{2}\alpha t^2 \tag{8.7}$$
$$\omega^2 = \omega_0^2 + 2\alpha\theta \tag{8.8}$$

The four equations in the left column of **Table 8.1** are called the **equations of rotational kinematics for constant angular acceleration**. The following example illustrates that they are used in the same fashion as the equations of linear kinematics.

EXAMPLE 5 | Blending with a Blender

The blades of an electric blender are whirling with an angular velocity of +375 rad/s while the "puree" button is pushed in, as Figure 8.8 shows. When the "blend" button is pressed, the blades accelerate and reach a greater angular velocity after the blades have rotated through an angular displacement of +44.0 rad. The angular acceleration has a constant value of +1740 rad/s². Find the final angular velocity of the blades.

Reasoning   The three known variables are listed in the following table, along with a question mark indicating that a value for the final angular velocity ω is being sought.

θ	α	ω	ω_0	t
+44.0 rad	+1740 rad/s²	?	+375 rad/s	

We can use Equation 8.8, because it relates the angular variables θ, α, ω, and ω_0.

> **Problem-Solving Insight**   Each equation of rotational kinematics contains four of the five kinematic variables, θ, α, ω, ω_0, and t. Therefore, it is necessary to have values for three of these variables if one of the equations is to be used to determine a value for an unknown variable.

> Axis of rotation

FIGURE 8.8   The angular velocity of the blades in an electric blender changes each time a different push button is chosen.

Solution   From Equation 8.8 ($\omega^2 = \omega_0^2 + 2\alpha\theta$) it follows that

$$\omega = +\sqrt{\omega_0^2 + 2\alpha\theta} = +\sqrt{(375 \text{ rad/s})^2 + 2(1740 \text{ rad/s}^2)(44.0 \text{ rad})}$$

$$= \boxed{+542 \text{ rad/s}}$$

The negative root is disregarded, since the blades do not reverse their direction of rotation.

> **Problem-Solving Insight**   *The equations of rotational kinematics can be used with any self-consistent set of units for θ, α, ω, ω_0, and t.* Radians are used in Example 5 only because data are given in terms of radians. Had the data for θ, α, and ω_0 been provided in rev, rev/s², and rev/s, respectively, then Equation 8.8 could have been used to determine the answer for ω directly in rev/s. In any case, the reasoning strategy for applying the kinematics equations is summarized as follows.

REASONING STRATEGY   Applying the Equations of Rotational Kinematics

1. Make a drawing to represent the situation being studied, showing the direction of rotation.

2. Decide which direction of rotation is to be called positive (+) and which direction is to be called negative (−). In this text we choose the counterclockwise direction to be positive and the clockwise direction to be negative, but this is arbitrary. However, do not change your decision during the course of a calculation.

3. In an organized way, write down the values (with appropriate + and − signs) that are given for any of the five rotational kinematic variables (θ, α, ω, ω_0, and t). Be on the alert for implied data, such as the phrase "starts from rest," which means that the value of the initial angular velocity is $\omega_0 = 0$ rad/s. The data table in Example 5 is a good way to keep track of this information. In addition, identify the variable(s) that you are being asked to determine.

4. Before attempting to solve a problem, verify that the given information contains values for at least three of the five kinematic variables. Once the three variables are identified for which values are known, the appropriate relation from Table 8.1 can be selected.

5. When the rotational motion is divided into segments, the final angular velocity of one segment is the initial angular velocity for the next segment.

6. Keep in mind that there may be two possible answers to a kinematics problem. Try to visualize the different physical situations to which the answers correspond.

Check Your Understanding

(The answers are given at the end of the book.)

5. The blades of a ceiling fan start from rest and, after two revolutions, have an angular speed of 0.50 rev/s. The angular acceleration of the blades is constant. What is the angular speed after eight revolutions?

6. Equation 8.7 ($\theta = \omega_0 t + \frac{1}{2}\alpha t^2$) is being used to solve a problem in rotational kinematics. Which one of the following sets of values for the variables ω_0, α, and t *cannot* be substituted directly into this equation to calculate a value for θ? **(a)** $\omega_0 = 1.0$ rad/s, $\alpha = 1.8$ rad/s^2, and $t = 3.8$ s **(b)** $\omega_0 = 0.16$ rev/s, $\alpha = 1.8$ rad/s^2, and $t = 3.8$ s **(c)** $\omega_0 = 0.16$ rev/s, $\alpha = 0.29$ rev/s^2, and $t = 3.8$ s

8.4 Angular Variables and Tangential Variables

In the familiar ice-skating stunt known as crack-the-whip, a number of skaters attempt to maintain a straight line as they skate around the one person (the pivot) who remains in place. **Figure 8.9** shows each skater moving on a circular arc and includes the corresponding velocity vector at the instant portrayed in the picture. For every individual skater, the vector is drawn tangent to the appropriate circle and, therefore, is called the **tangential velocity** $\vec{v}_T$. The magnitude of the tangential velocity is referred to as the **tangential speed**.

THE PHYSICS OF . . . "crack-the-whip." Of all the skaters involved in the stunt, the one farthest from the pivot has the hardest job. Why? Because, in keeping the line straight, this skater covers more distance than anyone else. To accomplish this, he must skate faster than anyone else and, thus, must have the largest tangential speed. In fact, the line remains straight only if each person skates with the correct tangential speed. The skaters closer to the pivot must move with smaller tangential speeds than those farther out, as indicated by the magnitudes of the tangential velocity vectors in **Figure 8.9**.

With the aid of **Figure 8.10**, it is possible to show that the tangential speed of any skater is directly proportional to the skater's distance r from the pivot, assuming a given angular speed for the rotating line. When the line rotates as a rigid unit for a time t, it sweeps out the angle θ shown in the drawing. The distance s through which a skater moves along a circular arc can be calculated from $s = r\theta$ (Equation 8.1), provided that θ is measured in radians. Dividing both sides of this equation by t, we find that $s/t = r(\theta/t)$. The term s/t is the tangential speed v_T (e.g., in meters/second) of the skater, while θ/t is the angular speed ω (in radians/second) of the line:

$$v_T = r\omega \qquad (\omega \text{ in rad/s}) \tag{8.9}$$

In this expression, the terms v_T and ω refer to the magnitudes of the tangential and angular velocities, respectively, and are numbers without algebraic signs.

It is important to emphasize that the angular speed ω in Equation 8.9 must be expressed in radian measure (e.g., in rad/s); no other units, such as revolutions per second, are acceptable. This restriction arises because the equation was derived by using the definition of radian measure, $s = r\theta$.

The real challenge for the crack-the-whip skaters is to keep the line straight while making it pick up angular speed—that is, while giving it an angular acceleration. To make the angular speed of the line increase, each skater must increase his tangential speed, since the two speeds are related according to $v_T = r\omega$. Of course, the fact that a skater must skate faster and faster means that he must accelerate, and his tangential

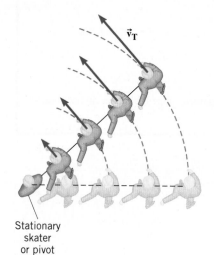

FIGURE 8.9 When doing the stunt known as crack-the-whip, each skater along the radial line moves on a circular arc. The tangential velocity $\vec{v}_T$ of each skater is represented by an arrow that is tangent to each arc.

FIGURE 8.10 During a time t, the line of skaters sweeps through an angle θ. An individual skater, located at a distance r from the stationary skater, moves through a distance s on a circular arc.

acceleration a_T can be related to the angular acceleration α of the line. If the time is measured relative to $t_0 = 0$ s, the definition of linear acceleration is given by Equation 2.4 as $a_T = (v_T - v_{T0})/t$, where v_T and v_{T0} are the final and initial tangential speeds, respectively. Substituting $v_T = r\omega$ for the tangential speed shows that

$$a_T = \frac{v_T - v_{T0}}{t} = \frac{(r\omega) - (r\omega_0)}{t} = r\left(\frac{\omega - \omega_0}{t}\right)$$

Since $\alpha = (\omega - \omega_0)/t$ according to Equation 8.4, it follows that

$$a_T = r\alpha \qquad (\alpha \text{ in rad/s}^2) \tag{8.10}$$

This result shows that, for a given value of α, the tangential acceleration a_T is proportional to the radius r, so the skater farthest from the pivot must have the largest tangential acceleration. In this expression, the terms a_T and α refer to the magnitudes of the numbers involved, without reference to any algebraic sign. Moreover, as is the case for ω in $v_T = r\omega$, only radian measure can be used for α in Equation 8.10.

There is an advantage to using the angular velocity ω and the angular acceleration α to describe the rotational motion of a rigid object. The advantage is that these angular quantities describe the motion of the *entire object*. In contrast, the tangential quantities v_T and a_T describe only the motion of a single point on the object, and Equations 8.9 and 8.10 indicate that different points located at different distances r have different tangential velocities and accelerations. Example 6 stresses this advantage.

EXAMPLE 6 | A Helicopter Blade

A helicopter blade has an angular speed of $\omega = 6.50$ rev/s and has an angular acceleration of $\alpha = 1.30$ rev/s^2. For points 1 and 2 on the blade in **Figure 8.11**, find the magnitudes of **(a)** the tangential speeds and **(b)** the tangential accelerations.

Reasoning Since the radius r for each point and the angular speed ω of the helicopter blade are known, we can find the tangential speed v_T for each point by using the relation $v_T = r\omega$. However, since this equation can be used only with radian measure, the angular speed ω must be converted to rad/s from rev/s. In a similar manner, the tangential acceleration a_T for points 1 and 2 can be found using $a_T = r\alpha$, provided the angular acceleration α is expressed in rad/s^2 rather than in rev/s^2.

> **Problem-Solving Insight When using Equations 8.9 and 8.10 to relate tangential and angular quantities, remember that the angular quantity is always expressed in radian measure. These equations are not valid if angles are expressed in degrees or revolutions.**

FIGURE 8.11 Points 1 and 2 on the rotating blade of the helicopter have the same angular speed and acceleration, but they have *different* tangential speeds and accelerations.

The rad unit, being dimensionless, does not appear in the final answers.

(b) Converting the angular acceleration α to rad/s^2 from rev/s^2, we find

$$\alpha = \left(1.30\,\frac{\text{rev}}{\text{s}^2}\right)\left(\frac{2\pi\,\text{rad}}{1\,\text{rev}}\right) = 8.17\,\frac{\text{rad}}{\text{s}^2}$$

The tangential accelerations can now be determined:

Point 1 $a_T = r\alpha = (3.00\text{ m})(8.17\text{ rad/s}^2) = \boxed{24.5\text{ m/s}^2}$ **(8.10)**

Point 2 $a_T = r\alpha = (6.70\text{ m})(8.17\text{ rad/s}^2) = \boxed{54.7\text{ m/s}^2}$ **(8.10)**

Solution **(a)** Converting the angular speed ω to rad/s from rev/s, we obtain

$$\omega = \left(6.50\,\frac{\text{rev}}{\text{s}}\right)\left(\frac{2\pi\,\text{rad}}{1\,\text{rev}}\right) = 40.8\,\frac{\text{rad}}{\text{s}}$$

The tangential speed for each point is

Point 1 $v_T = r\omega = (3.00\text{ m})(40.8\text{ rad/s})$
$= \boxed{122\text{ m/s (273 mph)}}$ **(8.9)**

Point 2 $v_T = r\omega = (6.70\text{ m})(40.8\text{ rad/s})$
$= \boxed{273\text{ m/s (611 mph)}}$ **(8.9)**

EXAMPLE 7 | BIO Wing Speed of the Rufous Hummingbird

Rufous hummingbirds have a top linear speed of about 50 mi/h, and fly over 2000 miles during their migratory transits. However, what stands out compared to other birds is their ability to dart and hover. This is due in part to their small body mass (3.4 g) combined with their high wingbeat frequency (41 beat cycles per second). If the length of a single hummingbird wing is 47 mm, and the angular displacement of the wing from the top to the bottom of the stroke (i.e., half of a full wingbeat) is 105°, determine **(a)** the average angular speed (in rad/s) of the wing and **(b)** the average linear speed (in m/s) of the wingtip.

Reasoning **(a)** Since we know the wingbeat frequency in cycle/s and the angular displacement from the top to the bottom of the stroke, we can calculate the average angular speed using Equation 8.2. **(b)** Once we have the average angular speed, then we can find the average linear speed (i.e., the tangential speed) of the wingtip using Equation 8.9.

Solution **(a)** The total angular displacement during a full cycle is equal to $2 \times 105° = 210°$. Since the wingbeat frequency is 41 cycles/s, the magnitude of the total angular displacement in one second is given by

$$\Delta\theta = (41 \text{ cycles})(210°) = 8610°$$

Next, we can write this in radians by multiplying by the proper conversion factor as follows:

$$\Delta\theta = (8610 \text{ deg})\left(\frac{2\pi \text{ rad}}{360 \text{ deg}}\right) = 150 \text{ rad}$$

The average angular speed is then given by Equation 8.2 as

$$\bar{\omega} = \frac{\Delta\theta}{\Delta t} = \frac{150 \text{ rad}}{1 \text{ s}} = \boxed{150 \text{ rad/s}}$$

This is the average angular speed, which means that much higher values are achieved during a complete wingbeat cycle. Peak angular velocities of hummingbird wings can exceed 200 rad/s! Compare this to the angular speed of the helicopter blades (40.8 rad/s) in Example 6!

(b) The average tangential speed of the wingtips can be calculated from the average angular speed and the wing length (r) using Equation 8.9:

$$\bar{v}_T = r\bar{\omega} = (0.047 \text{ m})(150 \text{ rad/s}) = \boxed{7.1 \text{ m/s}},$$

where we have modified the equation by placing bars over the variables to specify that they are average values. This is equal to 16 mi/h. Again, this is an average value, so the maximum speed is much larger. Compare this to the speed of the tips of the helicopter blades (611 mi/h) in Example 6. Although the angular speed of the hummingbird wings is much greater than that of the helicopter blades, the helicopter blades are much longer, and this is what accounts for the much greater tangential speed of the helicopter blade tips.

Check Your Understanding

(*The answers are given at the end of the book.*)

7. A thin rod rotates at a constant angular speed. In case A the axis of rotation is perpendicular to the rod at its center. In case B the axis is perpendicular to the rod at one end. In which case, if either, are there points on the rod that have the same tangential speeds?

8. It is possible to build a clock in which the tips of the hour hand and the second hand move with the same tangential speed. This is normally never done, however. Why? **(a)** The length of the hour hand would be 3600 times greater than the length of the second hand. **(b)** The hour hand and the second hand would have the same length. **(c)** The length of the hour hand would be 3600 times smaller than the length of the second hand.

9. The earth rotates once per day about its axis, which is perpendicular to the plane of the equator and passes through the north geographic pole. Where on the earth's surface should you stand in order to have the smallest possible tangential speed?

10. A building is located on the earth's equator. As the earth rotates about its axis, which floor of the building has the greatest tangential speed? **(a)** The first floor **(b)** The tenth floor **(c)** The twentieth floor

11. The blade of a lawn mower is rotating at an angular speed of 17 rev/s. The tangential speed of the outer edge of the blade is 32 m/s. What is the radius of the blade?

8.5 | Centripetal Acceleration and Tangential Acceleration

When an object picks up speed as it moves around a circle, it has a tangential acceleration, as discussed in the previous section. In addition, the object also has a centripetal acceleration, as emphasized in Chapter 5. That chapter deals with **uniform circular motion**, in which a particle moves at a constant tangential speed on a circular path. The tangential speed v_T is the magnitude of the tangential velocity vector. Even when the magnitude of the tangential velocity is constant, an acceleration is present, since the direction of the velocity changes continually. Because the resulting acceleration points toward the center of the circle, it is called the centripetal acceleration. **Interactive Figure 8.12a** shows the centripetal acceleration $\vec{a}_c$ for a model airplane flying in uniform circular motion on a guide wire. The magnitude of $\vec{a}_c$ is

$$a_c = \frac{v_T^2}{r} \qquad (5.2)$$

The subscript "T" has now been included in this equation as a reminder that it is the tangential speed that appears in the numerator.

The centripetal acceleration can be expressed in terms of the angular speed ω by using $v_T = r\omega$ (Equation 8.9):

$$a_c = \frac{v_T^2}{r} = \frac{(r\omega)^2}{r} = r\omega^2 \qquad (\omega \text{ in rad/s}) \qquad (8.11)$$

Only radian measure (rad/s) can be used for ω in this result, since the relation $v_T = r\omega$ presumes radian measure.

While considering uniform circular motion in Chapter 5, we ignored the details of how the motion is established in the first place. In **Interactive Figure 8.12b**, for instance, the engine of the plane produces a thrust in the tangential direction, and this force leads to a tangential acceleration. In response, the tangential speed of the plane increases from moment to moment, until the situation shown in the drawing results. While the tangential speed is changing, the motion is called **nonuniform circular motion**.

Interactive Figure 8.12b illustrates an important feature of nonuniform circular motion. Since the direction and the magnitude of the tangential velocity are both changing, the airplane experiences two acceleration components simultaneously. The changing direction means that there is a centripetal acceleration $\vec{a}_c$. The magnitude of $\vec{a}_c$ at any moment can be calculated using the value of the instantaneous angular speed and the radius: $a_c = r\omega^2$. The fact that the magnitude of the tangential velocity is changing means that there is also a tangential acceleration $\vec{a}_T$. The magnitude of $\vec{a}_T$ can be determined from the angular acceleration α according to $a_T = r\alpha$, as the previous section explains. If the magnitude F_T of the net tangential force and the mass m are known, a_T also can be calculated using Newton's second law, $F_T = ma_T$. **Interactive Figure 8.12b** shows the two acceleration components. The total acceleration is given by the vector sum of $\vec{a}_c$ and $\vec{a}_T$. Since $\vec{a}_c$ and $\vec{a}_T$ are perpendicular, the magnitude of the total acceleration $\vec{a}$ can be obtained from the Pythagorean theorem as $a = \sqrt{a_c^2 + a_T^2}$, while the angle ϕ in the drawing can be determined from $\tan \phi = a_T/a_c$. The next example applies these concepts to a discus thrower.

(a) Uniform circular motion

(b) Nonuniform circular motion

INTERACTIVE FIGURE 8.12 (a) If a model airplane flying on a guide wire has a constant tangential speed, the motion is uniform circular motion, and the plane experiences only a centripetal acceleration $\vec{a}_c$. (b) Nonuniform circular motion occurs when the tangential speed changes. Then there is a tangential acceleration $\vec{a}_T$ in addition to the centripetal acceleration.

Analyzing Multiple-Concept Problems

EXAMPLE 8 | A Discus Thrower

Discus throwers often warm up by throwing the discus with a twisting motion of their bodies. **Figure 8.13a** illustrates a top view of such a warm-up throw. Starting from rest, the thrower accelerates the discus to a final angular velocity of +15.0 rad/s in a time of 0.270 s before releasing it. During the acceleration, the discus moves on a circular arc of radius 0.810 m. Find the magnitude a of the total acceleration of the discus just before the discus is released.

Reasoning Since the tangential speed of the discus increases as the thrower turns, the discus simultaneously experiences a tangential acceleration $\vec{a}_T$ and a centripetal acceleration $\vec{a}_c$ that are oriented at right angles to each other (see the drawing). The magnitude a of the total acceleration is $a = \sqrt{a_c^2 + a_T^2}$, where a_c and a_T are the magnitudes of the centripetal and tangential accelerations. The magnitude of the centripetal acceleration can be evaluated from Equation 8.11 ($a_c = r\omega^2$),

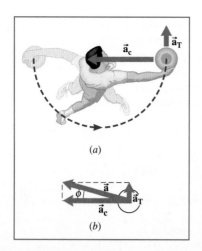

(a)

(b)

and the magnitude of the tangential acceleration follows from Equation 8.10 ($a_T = r\alpha$). The angular acceleration α can be found from its definition in Equation 8.4.

FIGURE 8.13 (a) A discus thrower and the centripetal acceleration $\vec{a}_c$ and tangential acceleration $\vec{a}_T$ of the discus. (b) The total acceleration $\vec{a}$ of the discus just before the discus is released is the vector sum of $\vec{a}_c$ and $\vec{a}_T$.

Knowns and Unknowns   The data for this problem are:

Description	Symbol	Value	Comment
Explicit Data			
Final angular velocity	ω	+15.0 rad/s	Positive, because discus moves counterclockwise (see drawing).
Time	t	0.270 s	
Radius of circular arc	r	0.810 m	
Implicit Data			
Initial angular velocity	ω_0	0 rad/s	Discus starts from rest.
Unknown Variable			
Magnitude of total acceleration	a	?	

Modeling the Problem

STEP 1 Total Acceleration   **Figure 8.13b** shows the two perpendicular components of the acceleration of the discus at the moment of release. The centripetal acceleration $\vec{a}_c$ arises because the discus is traveling on a circular path; this acceleration always points toward the center of the circle (see Section 5.2). The tangential acceleration $\vec{a}_T$ arises because the tangential velocity of the discus is increasing; this acceleration is tangent to the circle (see Section 8.4). Since $\vec{a}_c$ and $\vec{a}_T$ are perpendicular, we can use the Pythagorean theorem to find the magnitude a of the total acceleration, as given by Equation 1 at the right. Values for a_c and a_T will be obtained in the next two steps.

$$a = \sqrt{a_c^2 + a_T^2} \qquad (1)$$

$\boxed{?} \quad \boxed{?}$

STEP 2 Centripetal Acceleration   The magnitude a_c of the centripetal acceleration can be evaluated from Equation 8.11 as

$$\boxed{a_c = r\omega^2}$$

where r is the radius and ω is the angular speed of the discus. Both r and ω are known, so we can substitute this expression for a_c into Equation 1, as indicated at the right. In Step 3, we will evaluate the magnitude a_T of the tangential acceleration.

$$a = \sqrt{a_c^2 + a_T^2} \qquad (1)$$

$\boxed{a_c = r\omega^2} \quad \boxed{?}$

STEP 3 Tangential Acceleration   According to Equation 8.10, the magnitude a_T of the tangential acceleration is $a_T = r\alpha$, where r is the radius of the path and α is the magnitude of the angular acceleration. The angular acceleration is defined (see Equation 8.4) as the change $\omega - \omega_0$ in the angular velocity divided by the time t, or $\alpha = (\omega - \omega_0)/t$. Thus, the tangential acceleration can be written as

$$\boxed{a_T = r\alpha = r\left(\frac{\omega - \omega_0}{t}\right)}$$

All the variables on the right side of this equation are known, so we can substitute this expression for a_T into Equation 1 (see the right column).

$$a = \sqrt{a_c^2 + a_T^2} \qquad (1)$$

$\boxed{a_c = r\omega^2} \quad \boxed{a_T = r\left(\frac{\omega - \omega_0}{t}\right)}$

Solution   Algebraically combining the results of the three steps, we have:

STEP 1   STEP 2   STEP 3

$$a = \sqrt{a_c^2 + a_T^2} = \sqrt{r^2\omega^4 + a_T^2} = \sqrt{r^2\omega^4 + r^2\left(\frac{\omega - \omega_0}{t}\right)^2}$$

The magnitude of the total acceleration of the discus is

$$a = \sqrt{r^2\omega^4 + r^2\left(\frac{\omega - \omega_0}{t}\right)^2}$$

$$= \sqrt{(0.810 \text{ m})^2 (15.0 \text{ rad/s})^4 + (0.810 \text{ m})^2 \left(\frac{15.0 \text{ rad/s} - 0 \text{ rad/s}}{0.270 \text{ s}}\right)^2} = \boxed{188 \text{ m/s}^2}$$

Note that we can also determine the angle ϕ between the total acceleration of the discus and its centripetal acceleration (see **Figure 8.13b**). From trigonometry, we have that $\tan\phi = a_T/a_c$, so

$$\phi = \tan^{-1}\left(\frac{a_T}{a_c}\right) = \tan^{-1}\left[\frac{r\left(\frac{\omega - \omega_0}{t}\right)}{r\omega^2}\right] = \tan^{-1}\left[\frac{\left(\frac{15.0 \text{ rad/s} - 0 \text{ rad/s}}{0.270 \text{ s}}\right)}{(15.0 \text{ rad/s})^2}\right] = 13.9°$$

Related Homework: Problems 41, 43, 56

Math Skills   The tangent function ($\tan\phi$) is defined as $\tan\phi = \dfrac{h_o}{h_a}$ (Equation 1.3). As shown in **Figure 8.14a**, h_o is the length of the side of a right triangle opposite the angle ϕ, and h_a is the length of the side adjacent to the angle ϕ. As always, the first step in applying such a trigonometric function is to identify this angle and its associated right triangle. **Figure 8.13b** establishes the angle ϕ and is reproduced in **Figure 8.14b**. A comparison of the shaded right triangles in the drawing reveals that $h_o = a_T$ and $h_a = a_c$. Thus, it follows that $\tan\phi = \dfrac{h_o}{h_a} = \dfrac{a_T}{a_c}$, and ϕ is the angle whose tangent is $\dfrac{a_T}{a_c}$. This result can be expressed by using the inverse tangent function ($\tan^{-1}$):

$$\phi = \tan^{-1}\left(\frac{a_T}{a_c}\right)$$

(a)

(b)

FIGURE 8.14   Math Skills drawing.

Check Your Understanding

(The answers are given at the end of the book.)

12. A car is up on a hydraulic lift at a garage. The wheels are free to rotate, and the drive wheels are rotating with a constant angular velocity. Which one of the following statements is true? **(a)** A point on the rim has no tangential and no centripetal acceleration. **(b)** A point on the rim has both a nonzero tangential acceleration and a nonzero centripetal acceleration. **(c)** A point on the rim has a nonzero tangential acceleration but no centripetal acceleration. **(d)** A point on the rim has no tangential acceleration but does have a nonzero centripetal acceleration.

13. Section 5.6 discusses how the uniform circular motion of a space station can be used to create artificial gravity. This can be done by adjusting the angular speed of the space station, so the centripetal acceleration at an astronaut's feet equals g, the magnitude of the acceleration due to the earth's gravity (see **Figure 5.19**). If such an adjustment is made, will the acceleration at the astronaut's head due to the artificial gravity be **(a)** greater than, **(b)** equal to, or **(c)** less than g?

14. A bicycle is turned upside down, the front wheel is spinning (see **CYU Figure 8.2**), and there is an angular acceleration. At the instant shown, there are six points on the wheel that have arrows associated with them. Which one of the following quantities could the arrows *not* represent? **(a)** Tangential velocity **(b)** Centripetal acceleration **(c)** Tangential acceleration

CYU FIGURE 8.2

15. A rotating object starts from rest and has a constant angular acceleration. Three seconds later the centripetal acceleration of a point on the object has a magnitude of 2.0 m/s². What is the magnitude of the centripetal acceleration of this point six seconds after the motion begins?

8.6 Rolling Motion

(a)

(b)

FIGURE 8.15 (*a*) An automobile moves with a linear speed v. (*b*) If the tires roll and do not slip, the distance d through which an axle moves equals the circular arc length s along the outer edge of a tire.

Rolling motion is a familiar situation that involves rotation, as **Figure 8.15** illustrates for the case of an automobile tire. The essence of rolling motion is that there is *no slipping* at the point of contact where the tire touches the ground. To a good approximation, the tires on a normally moving automobile roll and do not slip. In contrast, the squealing tires that accompany the start of a drag race are rotating, but they are not rolling while they rapidly spin and slip against the ground.

When the tires in **Figure 8.15** roll, there is a relationship between the angular speed at which the tires rotate and the linear speed (assumed constant) at which the car moves forward. In part *b* of the drawing, consider the points labeled *A* and *B* on the left tire. Between these points we apply a coat of red paint to the tread of the tire; the length of this circular arc of paint is *s*. The tire then rolls to the right until point *B* comes in contact with the ground. As the tire rolls, all the paint comes off the tire and sticks to the ground, leaving behind the horizontal red line shown in the drawing. The axle of the wheel moves through a linear distance *d*, which is equal to the length of the horizontal strip of paint. Since the tire does not slip, the distance *d* must be equal to the circular arc length *s*, measured along the outer edge of the tire: $d = s$. Dividing both sides of this equation by the elapsed time *t* shows that $d/t = s/t$. The term d/t is the speed at which the axle moves parallel to the ground—namely, the linear speed v of the car. The term s/t is the tangential speed v_T at which a point on the outer edge of the tire moves relative to the axle. In addition, v_T is related to the angular speed ω about the axle according to $v_T = r\omega$ (Equation 8.9). Therefore, it follows that

$$\underbrace{v}_{\substack{\text{Linear} \\ \text{speed}}} = \underbrace{v_T = r\omega}_{\substack{\text{Tangential} \\ \text{speed}}} \quad (\omega \text{ in rad/s}) \tag{8.12}$$

If the car in **Figure 8.15** has a linear acceleration $\vec{a}$ parallel to the ground, a point on the tire's outer edge experiences a tangential acceleration $\vec{a}_T$ relative to the axle. The same kind of reasoning as that used in the previous paragraph reveals that the magnitudes of these accelerations are the same and that they are related to the angular acceleration α of the wheel relative to the axle:

$$\underbrace{a}_{\substack{\text{Linear} \\ \text{acceleration}}} = \underbrace{a_T = r\alpha}_{\substack{\text{Tangential} \\ \text{acceleration}}} \quad (\alpha \text{ in rad/s}^2) \tag{8.13}$$

Equations 8.12 and 8.13 may be applied to any rolling motion, because the object does not slip against the surface on which it is rolling.

(*The answers are given at the end of the book.*)

16. The speedometer of a truck is set to read the linear speed of the truck, but uses a device that actually measures the angular speed of the rolling tires that came with the truck. However, the owner replaces the tires with larger-diameter versions. Does the reading on the speedometer after the replacement give a speed that is **(a)** less than, **(b)** equal to, or **(c)** greater than the true linear speed of the truck?

17. Rolling motion is an example that involves rotation about an axis that is not fixed. Give three other examples of rotational motion about an axis that is not fixed.

8.7 | *The Vector Nature of Angular Variables

We have presented angular velocity and angular acceleration by taking advantage of the analogy between angular variables and linear variables. Like the linear velocity and the linear acceleration, the angular quantities are also vectors and have a direction as well as a magnitude. As yet, however, we have not discussed the directions of these vectors.

When a rigid object rotates about a fixed axis, it is the axis that identifies the motion, and the angular velocity vector points along this axis. **Interactive Figure 8.16** shows how to determine the direction using a *right-hand rule:*

> ***Right-Hand Rule*** Grasp the axis of rotation with your right hand, so that your fingers circle the axis in the same sense as the rotation. Your extended thumb points along the axis in the direction of the angular velocity vector.

No part of the rotating object moves in the direction of the angular velocity vector.

Angular acceleration arises when the angular velocity changes, and the acceleration vector also points along the axis of rotation. The acceleration vector has the same direction as the *change* in the angular velocity. That is, when the magnitude of the angular velocity (which is the angular speed) is increasing, the angular acceleration vector points in the same direction as the angular velocity. Conversely, when the magnitude of the angular velocity is decreasing, the angular acceleration vector points in the direction opposite to the angular velocity.

Right hand

Right hand

INTERACTIVE FIGURE 8.16 The angular velocity vector $\vec{\omega}$ of a rotating object points along the axis of rotation. The direction along the axis depends on the sense of the rotation and can be determined with the aid of a right-hand rule (see the text).

EXAMPLE 9 | BIO The Physics of a Centrifuge

One popular piece of equipment in research laboratories is a centrifuge (**Figure 8.17a**). Typical applications of a centrifuge include separating immiscible liquids or suspended solids like blood, for example. Samples are placed in tubes (**Figure 8.17b**) and then rotated at high speeds around a fixed axis. The centripetal acceleration causes the denser substances and particles to separate and settle at the bottom of the tubes. Suppose a centrifuge starts from rest and reaches its maximum angular speed of 4510 rpm in 12.2 s. Through how many revolutions does the centrifuge spin during this time?

Reasoning This is a standard rotational kinematics problem, with the following known variables:

scotth23/Pixabay

(a)   (b)

FIGURE 8.17 (*a*) Common laboratory centrifuge. (*b*) Drawing of a typical sample tube used in the centrifuge. The tubes are rotated at high speed. The centripetal acceleration separates the materials in the tube by density, with the more dense particles settling at the bottom of the tube.

θ	α	ω	ω_0	t
?		+4510 rpm	0	12.2 s

Given the information in the table, we can simply use Equation 8.6 to calculate the angular displacement, θ.

Solution In order to apply Equation 8.6, we need the value of the average angular speed, $\overline{\omega}$. We can calculate $\overline{\omega}$ using Equation 8.5:

$$\overline{\omega} = \tfrac{1}{2}(\omega_0 + \omega) = \tfrac{1}{2}(0 + 4510 \text{ rpm}) = 2260 \text{ rpm}.$$

Now we can solve for the angular displacement. However, before we can apply Equation 8.6, we need to have consistent units. We need to convert the time given from seconds to minutes:

$$t = (12.2 \text{ s})\left(\frac{1 \text{ min}}{60 \text{ s}}\right) = 0.203 \text{ min}.$$

Now we can calculate the angular displacement:

$$\theta = \overline{\omega}t = (2260 \text{ rpm})(0.203 \text{ min}) = \boxed{459 \text{ revolutions}}$$

Concept Summary

8.1 Rotational Motion and Angular Displacement When a rigid body rotates about a fixed axis, the angular displacement is the angle swept out by a line passing through any point on the body and intersecting the axis of rotation perpendicularly. By convention, the angular displacement is positive if it is counterclockwise and negative if it is clockwise.

The radian (rad) is the SI unit of angular displacement. The angle θ in radians is defined in Equation 8.1 as the circular arc of length s traveled by a point on the rotating body divided by the radial distance r of the point from the axis.

$$(\theta \text{ in radians}) = \tfrac{s}{r} \qquad \textbf{(8.1)}$$

8.2 Angular Velocity and Angular Acceleration The average angular velocity $\overline{\omega}$ is the angular displacement $\Delta\theta$ divided by the elapsed time Δt, according to Equation 8.2. As Δt approaches zero, the average angular velocity becomes equal to the instantaneous angular velocity ω. The magnitude of the instantaneous angular velocity is called the instantaneous angular speed.

The average angular acceleration $\overline{\alpha}$ is the change $\Delta\omega$ in the angular velocity divided by the elapsed time Δt, according to Equation 8.4. As Δt approaches zero, the average angular acceleration becomes equal to the instantaneous angular acceleration α.

$$\overline{\omega} = \frac{\Delta\theta}{\Delta t} \qquad \textbf{(8.2)}$$

$$\overline{\alpha} = \frac{\Delta\omega}{\Delta t} \qquad \textbf{(8.4)}$$

8.3 The Equations of Rotational Kinematics The equations of rotational kinematics apply when a rigid body rotates with a constant angular acceleration about a fixed axis. These equations relate the angular displacement $\theta - \theta_0$, the angular acceleration α, the final angular velocity ω, the initial angular velocity ω_0, and the elapsed time $t - t_0$. Assuming that $\theta_0 = 0$ rad at $t_0 = 0$ s, the equations of rotational kinematics are written as in Equations 8.4 and 8.6–8.8. These equations may be used with any self-consistent set of units and are not restricted to radian measure.

$$\omega = \omega_0 + \alpha t \qquad \textbf{(8.4)}$$

$$\theta = \tfrac{1}{2}(\omega + \omega_0)t \qquad \textbf{(8.6)}$$

$$\theta = \omega_0 t + \tfrac{1}{2}\alpha t^2 \qquad \textbf{(8.7)}$$

$$\omega^2 = \omega_0^2 + 2\alpha\theta \qquad \textbf{(8.8)}$$

8.4 Angular Variables and Tangential Variables When a rigid body rotates through an angle θ about a fixed axis, any point on the

body moves on a circular arc of length s and radius r. Such a point has a tangential velocity (magnitude $= v_T$) and, possibly, a tangential acceleration (magnitude $= a_T$). The angular and tangential variables are related by Equations 8.1, 8.9, and 8.10. These equations refer to the magnitudes of the variables involved, without reference to positive or negative signs, and only radian measure can be used when applying them.

$$s = r\theta \qquad (\theta \text{ in rad}) \qquad \textbf{(8.1)}$$

$$v_T = r\omega \qquad (\omega \text{ in rad/s}) \qquad \textbf{(8.9)}$$

$$a_T = r\alpha \qquad (\alpha \text{ in rad/s}^2) \qquad \textbf{(8.10)}$$

8.5 Centripetal Acceleration and Tangential Acceleration The magnitude a_c of the centripetal acceleration of a point on an object rotating with uniform or nonuniform circular motion can be expressed in terms of the radial distance r of the point from the axis and the angular speed ω, as shown in Equation 8.11. This point experiences a total acceleration $\vec{a}$ that is the vector sum of two perpendicular acceleration components, the centripetal acceleration $\vec{a}_c$ and the tangential acceleration $\vec{a}_T$; $\vec{a} = \vec{a}_c + \vec{a}_T$.

$$a_c = r\omega^2 \qquad (\omega \text{ in rad/s}) \qquad \textbf{(8.11)}$$

8.6 Rolling Motion The essence of rolling motion is that there is no slipping at the point where the object touches the surface upon which it is rolling. As a result, the tangential speed v_T of a point on the outer edge of a rolling object, measured relative to the axis through the center of the object, is equal to the linear speed v with which the object moves parallel to the surface. In other words, we have Equation 8.12.

The magnitudes of the tangential acceleration a_T and the linear acceleration a of a rolling object are similarly related, as in Equation 8.13.

$$v = v_T = r\omega \qquad (\omega \text{ in rad/s}) \qquad \textbf{(8.12)}$$

$$a = a_T = r\alpha \qquad (\alpha \text{ in rad/s}^2) \qquad \textbf{(8.13)}$$

8.7 The Vector Nature of Angular Variables The direction of the angular velocity vector is given by a right-hand rule. Grasp the axis of rotation with your right hand, so that your fingers circle the axis in the same sense as the rotation. Your extended thumb points along the axis in the direction of the angular velocity vector. The angular acceleration vector has the same direction as the change in the angular velocity.

Focus on Concepts

Additional questions are available for assignment in WileyPLUS.

Section 8.1   Rotational Motion and Angular Displacement

1. The moon is 3.85×10^8 m from the earth and has a diameter of 3.48×10^6 m. You have a pea (diameter = 0.50 cm) and a dime (diameter = 1.8 cm). You close one eye and hold each object at arm's length (71 cm) between your open eye and the moon. Which objects, if any, completely cover your view of the moon? Assume that the moon and both objects are sufficiently far from your eye that the given diameters are equal to arc lengths when calculating angles. **(a)** Both **(b)** Neither **(c)** Pea **(d)** Dime

Section 8.2   Angular Velocity and Angular Acceleration

2. The radius of the circle traced out by the second hand on a clock is 6.00 cm. In a time t the tip of the second hand moves through an arc length of 24.0 cm. Determine the value of t in seconds.

3. A rotating object has an angular acceleration of $\alpha = 0$ rad/s^2. Which one or more of the following three statements is consistent with a zero angular acceleration? A. The angular velocity is $\omega = 0$ rad/s at all times. B. The angular velocity is $\omega = 10$ rad/s at all times. C. The angular displacement θ has the same value at all times. **(a)** A, B, and C **(b)** A and B, but not C **(c)** A only **(d)** B only **(e)** C only

Section 8.3   The Equations of Rotational Kinematics

4. A rotating wheel has a constant angular acceleration. It has an angular velocity of 5.0 rad/s at time $t = 0$ s, and 3.0 s later has an angular velocity of 9.0 rad/s. What is the angular displacement of the wheel during the 3.0-s interval? **(a)** 15 rad **(b)** 21 rad **(c)** 27 rad **(d)** There is not enough information given to determine the angular displacement.

5. A rotating object starts from rest at $t = 0$ s and has a constant angular acceleration. At a time of $t = 7.0$ s the object has an angular velocity of $\omega = 16$ rad/s. What is its angular velocity at a time of $t = 14$ s?

Section 8.4   Angular Variables and Tangential Variables

6. A merry-go-round at a playground is a circular platform that is mounted parallel to the ground and can rotate about an axis that is perpendicular to the platform at its center. The angular speed of the merry-go-round is constant, and a child at a distance of 1.4 m from the axis has a tangential speed of 2.2 m/s. What is the tangential speed of another child, who is located at a distance of 2.1 m from the axis? **(a)** 1.5 m/s **(b)** 3.3 m/s **(c)** 2.2 m/s **(d)** 5.0 m/s **(e)** 0.98 m/s

7. A small fan has blades that have a radius of 0.0600 m. When the fan is turned on, the tips of the blades have a tangential acceleration of 22.0 m/s^2 as the fan comes up to speed. What is the angular acceleration α of the blades?

Section 8.5   Centripetal Acceleration and Tangential Acceleration

8. A wheel rotates with a constant angular speed ω. Which one of the following is true concerning the angular acceleration α of the wheel, the tangential acceleration a_T of a point on the rim of the wheel, and the centripetal acceleration a_c of a point on the rim?

   (a) $\alpha = 0$ rad/s^2, $a_T = 0$ m/s^2, and $a_c = 0$ m/s^2

   (b) $\alpha = 0$ rad/s^2, $a_T \neq 0$ m/s^2, and $a_c = 0$ m/s^2

   (c) $\alpha \neq 0$ rad/s^2, $a_T = 0$ m/s^2, and $a_c = 0$ m/s^2

   (d) $\alpha = 0$ rad/s^2, $a_T = 0$ m/s^2, and $a_c \neq 0$ m/s^2

   (e) $\alpha \neq 0$ rad/s^2, $a_T \neq 0$ m/s^2, and $a_c \neq 0$ m/s^2

9. A platform is rotating with an angular speed of 3.00 rad/s and an angular acceleration of 11.0 rad/s^2. At a point on the platform that is 1.25 m from the axis of rotation, what is the magnitude of the *total* acceleration a?

Section 8.6   Rolling Motion

10. The radius of each wheel on a bicycle is 0.400 m. The bicycle travels a distance of 3.0 km. Assuming that the wheels do not slip, how many revolutions does each wheel make?

   (a) 1.2×10^3 revolutions

   (b) 2.4×10^2 revolutions

   (c) 6.0×10^3 revolutions

   (d) 8.4×10^{-4} revolutions

   (e) Since the time of travel is not given, there is not enough information for a solution.

Problems

Additional questions are available for assignment in WileyPLUS.

SSM Student Solutions Manual	**BIO** Biomedical application
MMH Problem-solving help	**E** Easy
GO Guided Online Tutorial	**M** Medium
V-HINT Video Hints	**H** Hard
CHALK Chalkboard Videos	**WS** Worksheet
	T Team Problem

Section 8.1   Rotational Motion and Angular Displacement,

Section 8.2   Angular Velocity and Angular Acceleration

1. **E SSM CHALK** A pitcher throws a curveball that reaches the catcher in 0.60 s. The ball curves because it is spinning at an average angular velocity of 330 rev/min (assumed constant) on its way to

the catcher's mitt. What is the angular displacement of the baseball (in radians) as it travels from the pitcher to the catcher?

2. **E** **GO** The table that follows lists four pairs of initial and final angles of a wheel on a moving car. The elapsed time for each pair of angles is 2.0 s. For each of the four pairs, determine the average angular velocity (magnitude and direction as given by the algebraic sign of your answer).

	Initial angle θ_0	Final angle θ
(a)	0.45 rad	0.75 rad
(b)	0.94 rad	0.54 rad
(c)	5.4 rad	4.2 rad
(d)	3.0 rad	3.8 rad

3. **E** The earth spins on its axis once a day and orbits the sun once a year ($365\frac{1}{4}$ days). Determine the average angular velocity (in rad/s) of the earth as it **(a)** spins on its axis and **(b)** orbits the sun. In each case, take the positive direction for the angular displacement to be the direction of the earth's motion.

4. **E** Our sun rotates in a circular orbit about the center of the Milky Way galaxy. The radius of the orbit is 2.2×10^{20} m, and the angular speed of the sun is 1.1×10^{-15} rad/s. How long (in years) does it take for the sun to make one revolution around the center?

5. **E** **GO** The initial angular velocity and the angular acceleration of four rotating objects at the same instant in time are listed in the table that follows. For each of the objects (a), (b), (c), and (d), determine the final angular *speed* after an elapsed time of 2.0 s.

	Initial angular velocity ω_0	Angular acceleration α
(a)	+12 rad/s	+3.0 rad/s^2
(b)	+12 rad/s	−3.0 rad/s^2
(c)	−12 rad/s	+3.0 rad/s^2
(d)	−12 rad/s	−3.0 rad/s^2

6. **E** **GO** The table that follows lists four pairs of initial and final angular velocities for a rotating fan blade. The elapsed time for each of the four pairs of angular velocities is 4.0 s. For each of the four pairs, find the average angular acceleration (magnitude and direction as given by the algebraic sign of your answer).

	Initial angular velocity ω_0	Final angular velocity ω
(a)	+2.0 rad/s	+5.0 rad/s
(b)	+5.0 rad/s	+2.0 rad/s
(c)	−7.0 rad/s	−3.0 rad/s
(d)	+4.0 rad/s	−4.0 rad/s

7. **E** Conceptual Example 2 provides some relevant background for this problem. A jet is circling an airport control tower at a distance of 18.0 km. An observer in the tower watches the jet cross in front of the moon. As seen from the tower, the moon subtends an angle of 9.04×10^{-3} radians. Find the distance traveled (in meters) by the jet as the observer watches the nose of the jet cross from one side of the moon to the other.

8. **E** **SSM** A Ferris wheel rotates at an angular velocity of 0.24 rad/s. Starting from rest, it reaches its operating speed with an average

angular acceleration of 0.030 rad/s^2. How long does it take the wheel to come up to operating speed?

9. **E** **GO** A floor polisher has a rotating disk that has a 15-cm radius. The disk rotates at a constant angular velocity of 1.4 rev/s and is covered with a soft material that does the polishing. An operator holds the polisher in one place for 45 s, in order to buff an especially scuffed area of the floor. How far (in meters) does a spot on the outer edge of the disk move during this time?

10. **M** The sun appears to move across the sky, because the earth spins on its axis. To a person standing on the earth, the sun subtends an angle of $\theta_{\text{sun}} = 9.28 \times 10^{-3}$ rad (see Conceptual Example 2). How much time (in seconds) does it take for the sun to move a distance equal to its own diameter?

11. **M** **GO** A propeller is rotating about an axis perpendicular to its center, as the drawing shows. The axis is parallel to the ground. An arrow is fired at the propeller, travels parallel to the axis, and passes through one of the open spaces between the propeller blades. The angular open spaces between the three propeller blades are each $\pi/3$ rad (60.0°). The vertical drop of the arrow may be ignored. There is a maximum value ω for the angular speed of the propeller, beyond which the arrow cannot pass through an open space without being struck by one of the blades. Find this maximum value when the arrow has the lengths L and speeds v shown in the following table.

	L	v
(a)	0.71 m	75.0 m/s
(b)	0.71 m	91.0 m/s
(c)	0.81 m	91.0 m/s

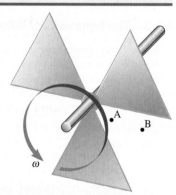

PROBLEM 11

12. **M** **CHALK** Two people start at the same place and walk around a circular lake in opposite directions. One walks with an angular speed of 1.7×10^{-3} rad/s, while the other has an angular speed of 3.4×10^{-3} rad/s. How long will it be before they meet?

13. **M** A space station consists of two donut-shaped living chambers, A and B, that have the radii shown in the drawing. As the station rotates, an astronaut in chamber A is moved 2.40×10^2 m along a circular arc. How far along a circular arc is an astronaut in chamber B moved during the same time?

$r_A = 3.20 \times 10^2$ m

PROBLEM 13 $r_B = 1.10 \times 10^3$ m

14. **M** **V-HINT** The drawing shows a device that can be used to measure the speed of a bullet. The device consists of two rotating disks, separated by a distance of $d = 0.850$ m, and rotating with an angular speed of 95.0 rad/s. The bullet first passes through the left disk and then through the right disk. It is found that the angular displacement between the two bullet holes is $\theta = 0.240$ rad. From these data, determine the speed of the bullet.

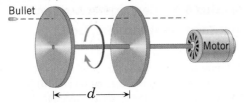

Bullet

Motor

PROBLEM 14

$\longleftarrow d \longrightarrow$

15. **M** **GO** An automatic dryer spins wet clothes at an angular speed of 5.2 rad/s. Starting from rest, the dryer reaches its operating speed with an average angular acceleration of 4.0 rad/s². How long does it take the dryer to come up to speed?

16. **M** **SSM** A stroboscope is a light that flashes on and off at a constant rate. It can be used to illuminate a rotating object, and if the flashing rate is adjusted properly, the object can be made to appear stationary. **(a)** What is the shortest time between flashes of light that will make a three-bladed propeller appear stationary when it is rotating with an angular speed of 16.7 rev/s? **(b)** What is the next shortest time?

17. **H** The drawing shows a golf ball passing through a windmill at a miniature golf course. The windmill has 8 blades and rotates at an angular speed of 1.25 rad/s. The opening between successive blades is equal to the width of a blade. A golf ball (diameter 4.50×10^{-2} m) has just reached the edge of one of the rotating blades (see the drawing). Ignoring the thickness of the blades, find the *minimum* linear speed with which the ball moves along the ground, such that the ball will not be hit by the next blade.

Golf ball

PROBLEM 17

Section 8.3 The Equations of Rotational Kinematics

18. **E** A figure skater is spinning with an angular velocity of +15 rad/s. She then comes to a stop over a brief period of time. During this time, her angular displacement is +5.1 rad. Determine **(a)** her average angular acceleration and **(b)** the time during which she comes to rest.

19. **E** **SSM** A gymnast is performing a floor routine. In a tumbling run she spins through the air, increasing her angular velocity from 3.00 to 5.00 rev/s while rotating through one-half of a revolution. How much time does this maneuver take?

20. **E** The angular speed of the rotor in a centrifuge increases from 420 to 1420 rad/s in a time of 5.00 s. **(a)** Obtain the angle through which the rotor turns. **(b)** What is the magnitude of the angular acceleration?

21. **E** A wind turbine is initially spinning at a constant angular speed. As the wind's strength gradually increases, the turbine experiences a constant angular acceleration of 0.140 rad/s². After making 2870 revolutions, its angular speed is 137 rad/s. **(a)** What is the initial angular velocity of the turbine? **(b)** How much time elapses while the turbine is speeding up?

22. **E** **GO** A car is traveling along a road, and its engine is turning over with an angular velocity of +220 rad/s. The driver steps on the accelerator, and in a time of 10.0 s the angular velocity increases to +280 rad/s. **(a)** What would have been the angular displacement of the engine if its angular velocity had remained constant at the initial value of +220 rad/s during the entire 10.0-s interval? **(b)** What would have been the angular displacement if the angular velocity had been equal to its final value of +280 rad/s during the entire 10.0-s interval? **(c)** Determine the actual value of the angular displacement during the 10.0-s interval.

23. **E** **SSM** The wheels of a bicycle have an angular velocity of +20.0 rad/s. Then, the brakes are applied. In coming to rest, each wheel makes an angular displacement of +15.92 revolutions. **(a)** How much time does it take for the bike to come to rest? **(b)** What is the angular acceleration (in rad/s²) of each wheel?

24. **M** **SSM** **MMH** A motorcyclist is traveling along a road and accelerates for 4.50 s to pass another cyclist. The angular acceleration of each wheel is +6.70 rad/s², and, just after passing, the angular velocity of each wheel is +74.5 rad/s, where the plus signs indicate counterclockwise directions. What is the angular displacement of each wheel during this time?

25. **M** **V-HINT** A top is a toy that is made to spin on its pointed end by pulling on a string wrapped around the body of the top. The string has a length of 64 cm and is wound around the top at a spot where its radius is 2.0 cm. The thickness of the string is negligible. The top is initially at rest. Someone pulls the free end of the string, thereby unwinding it and giving the top an angular acceleration of +12 rad/s². What is the final angular velocity of the top when the string is completely unwound?

26. **M** **MMH** The drive propeller of a ship starts from rest and accelerates at 2.90×10^{-3} rad/s² for 2.10×10^{3} s. For the next 1.40×10^{3} s the propeller rotates at a constant angular speed. Then it decelerates at 2.30×10^{-3} rad/s² until it slows (without reversing direction) to an angular speed of 4.00 rad/s. Find the total angular displacement of the propeller.

27. **M** **GO** The drawing shows a graph of the angular velocity of a rotating wheel as a function of time. Although not shown in the graph, the angular velocity continues to increase at the same rate until $t = 8.0$ s. What is the angular displacement of the wheel from 0 to 8.0 s?

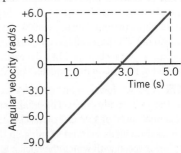

PROBLEM 27

28. **M** **V-HINT** At the local swimming hole, a favorite trick is to run horizontally off a cliff that is 8.3 m above the water. One diver runs off the edge of the cliff, tucks into a "ball," and rotates on the way down with an average angular speed of 1.6 rev/s. Ignore air resistance and determine the number of revolutions she makes while on the way down.

29. **M** **V-HINT** A spinning wheel on a fireworks display is initially rotating in a counterclockwise direction. The wheel has an angular acceleration of −4.00 rad/s². Because of this acceleration, the angular velocity of the wheel changes from its initial value to a final value of −25.0 rad/s. While this change occurs, the angular displacement of the wheel is zero. (Note the similarity to that of a ball being thrown vertically upward, coming to a momentary halt, and then falling downward to its initial position.) Find the time required for the change in the angular velocity to occur.

Section 8.4　Angular Variables and Tangential Variables

30. **E** **GO** A fan blade is rotating with a constant angular acceleration of $+12.0 \, \text{rad/s}^2$. At what point on the blade, as measured from the axis of rotation, does the magnitude of the tangential acceleration equal that of the acceleration due to gravity?

31. **E** An auto race takes place on a circular track. A car completes one lap in a time of 18.9 s, with an average tangential speed of 42.6 m/s. Find **(a)** the average angular speed and **(b)** the radius of the track.

32. **E** **SSM** A string trimmer is a tool for cutting grass and weeds; it utilizes a length of nylon "string" that rotates about an axis perpendicular to one end of the string. The string rotates at an angular speed of 47 rev/s, and its tip has a tangential speed of 54 m/s. What is the length of the rotating string?

33. **E** **V-HINT** In 9.5 s a fisherman winds 2.6 m of fishing line onto a reel whose radius is 3.0 cm (assumed to be constant as an approximation). The line is being reeled in at a constant speed. Determine the angular speed of the reel.

34. **E** The earth has a radius of 6.38×10^6 m and turns on its axis once every 23.9 h. **(a)** What is the tangential speed (in m/s) of a person living in Ecuador, a country that lies on the equator? **(b)** At what latitude (i.e., the angle θ in the drawing) is the tangential speed one-third that of a person living in Ecuador?

PROBLEM 34

35. **M** **SSM** A baseball pitcher throws a baseball horizontally at a linear speed of 42.5 m/s (about 95 mi/h). Before being caught, the baseball travels a horizontal distance of 16.5 m and rotates through an angle of 49.0 rad. The baseball has a radius of 3.67 cm and is rotating about an axis as it travels, much like the earth does. What is the tangential speed of a point on the "equator" of the baseball?

36. **M** **GO** A person lowers a bucket into a well by turning the hand crank, as the drawing illustrates. The crank handle moves with a constant tangential speed of 1.20 m/s on its circular path. The rope holding the bucket unwinds without slipping on the barrel of the crank. Find the linear speed with which the bucket moves down the well.

PROBLEM 36

37. **M** A thin rod (length = 1.50 m) is oriented vertically, with its bottom end attached to the floor by means of a frictionless hinge. The mass of the rod may be ignored, compared to the mass of an object

fixed to the top of the rod. The rod, starting from rest, tips over and rotates downward. **(a)** What is the angular speed of the rod just before it strikes the floor? (*Hint: Consider using the principle of conservation of mechanical energy.*) **(b)** What is the magnitude of the angular acceleration of the rod just before it strikes the floor?

Section 8.5　Centripetal Acceleration and Tangential Acceleration

38. **E** **SSM** A racing car travels with a constant tangential speed of 75.0 m/s around a circular track of radius 625 m. Find **(a)** the magnitude of the car's total acceleration and **(b)** the direction of its total acceleration relative to the radial direction.

39. **E** **GO** Two Formula One racing cars are negotiating a circular turn, and they have the same centripetal acceleration. However, the path of car A has a radius of 48 m, while that of car B is 36 m. Determine the ratio of the angular speed of car A to the angular speed of car B.

40. **E** **SSM** The earth orbits the sun once a year (3.16×10^7 s) in a nearly circular orbit of radius 1.50×10^{11} m. With respect to the sun, determine **(a)** the angular speed of the earth, **(b)** the tangential speed of the earth, and **(c)** the magnitude and direction of the earth's centripetal acceleration.

41. **E** **MMH** Review Multiple-Concept Example 8 in this chapter as an aid in solving this problem. In a fast-pitch softball game the pitcher is impressive to watch, as she delivers a pitch by rapidly whirling her arm around so that the ball in her hand moves on a circle. In one instance, the radius of the circle is 0.670 m. At one point on this circle, the ball has an angular acceleration of $64.0 \, \text{rad/s}^2$ and an angular speed of 16.0 rad/s. **(a)** Find the magnitude of the total acceleration (centripetal plus tangential) of the ball. **(b)** Determine the angle of the total acceleration relative to the radial direction.

42. **M** **GO** A rectangular plate is rotating with a constant angular speed about an axis that passes perpendicularly through one corner, as the drawing shows. The centripetal acceleration measured at corner A is n times as great as that measured at corner B. What is the ratio L_1/L_2 of the lengths of the sides of the rectangle when $n = 2.00$?

PROBLEM 42

43. **M** **V-HINT** Multiple-Concept Example 8 explores the approach taken in problems such as this one. The blades of a ceiling fan have a radius of 0.380 m and are rotating about a fixed axis with an angular velocity of $+1.50$ rad/s. When the switch on the fan is turned to a higher speed, the blades acquire an angular acceleration of $+2.00 \, \text{rad/s}^2$. After 0.500 s has elapsed since the switch was reset, what is **(a)** the total acceleration (in m/s²) of a point on the tip of a blade and **(b)** the angle ϕ between the total acceleration $\vec{a}$ and the centripetal acceleration $\vec{a}_c$? (See **Interactive Figure 8.12b**.)

44. **M** **CHALK** **SSM** The sun has a mass of 1.99×10^{30} kg and is moving in a circular orbit about the center of our galaxy, the Milky Way. The radius of the orbit is 2.3×10^4 light-years (1 light-year = 9.5×10^{15} m), and the angular speed of the sun is 1.1×10^{-15} rad/s. **(a)** Determine the tangential speed of the sun. **(b)** What is the magnitude of the net force that acts on the sun to keep it moving around the center of the Milky Way?

Section 8.6 Rolling Motion

Note: All problems in this section assume that there is no slipping of the surfaces in contact during the rolling motion.

45. **E** **SSM** A motorcycle accelerates uniformly from rest and reaches a linear speed of 22.0 m/s in a time of 9.00 s. The radius of each tire is 0.280 m. What is the magnitude of the angular acceleration of each tire?

46. **E** An automobile tire has a radius of 0.330 m, and its center moves forward with a linear speed of $v = 15.0$ m/s. **(a)** Determine the angular speed of the wheel. **(b)** Relative to the axle, what is the tangential speed of a point located 0.175 m from the axle?

47. **E** **MMH** A car is traveling with a speed of 20.0 m/s along a straight horizontal road. The wheels have a radius of 0.300 m. If the car speeds up with a linear acceleration of 1.50 m/s² for 8.00 s, find the angular displacement of each wheel during this period.

48. **E** Suppose you are riding a stationary exercise bicycle, and the electronic meter indicates that the wheel is rotating at 9.1 rad/s. The wheel has a radius of 0.45 m. If you ride the bike for 35 min, how far would you have gone if the bike could move?

49. **M** **V-HINT** A motorcycle, which has an initial linear speed of 6.6 m/s, decelerates to a speed of 2.1 m/s in 5.0 s. Each wheel has a radius of 0.65 m and is rotating in a counterclockwise (positive) direction. What are **(a)** the constant angular acceleration (in rad/s²) and **(b)** the angular displacement (in rad) of each wheel?

50. **M** **GO** A dragster starts from rest and accelerates down a track. Each tire has a radius of 0.320 m and rolls without slipping. At a distance of 384 m, the angular speed of the wheels is 288 rad/s. Determine **(a)** the linear speed of the dragster and **(b)** the magnitude of the angular acceleration of its wheels.

51. **M** **GO** A bicycle is rolling down a circular portion of a path; this portion of the path has a radius of 9.00 m. As the drawing illustrates, the angular displacement of the bicycle is 0.960 rad. What is the angle (in radians) through which each bicycle wheel (radius = 0.400 m) rotates?

PROBLEM 51

52. **M** **SSM** The penny-farthing is a bicycle that was popular between 1870 and 1890. As the drawing shows, this type of bicycle has a large front wheel and a small rear wheel. During a ride, the front wheel (radius = 1.20 m) makes 276 revolutions. How many revolutions does the rear wheel (radius = 0.340 m) make?

PROBLEM 52

53. **M** **CHALK** **GO** A ball of radius 0.200 m rolls with a constant linear speed of 3.60 m/s along a horizontal table. The ball rolls off the edge and falls a vertical distance of 2.10 m before hitting the floor. What is the angular displacement of the ball while the ball is in the air?

Additional Problems

Online

54. **E** A flywheel has a constant angular deceleration of 2.0 rad/s². **(a)** Find the angle through which the flywheel turns as it comes to rest from an angular speed of 220 rad/s. **(b)** Find the time for the flywheel to come to rest.

55. **E** **SSM** An electric fan is running on HIGH. After the LOW button is pressed, the angular speed of the fan decreases to 83.8 rad/s in 1.75 s. The deceleration is 42.0 rad/s². Determine the initial angular speed of the fan.

56. **E** Refer to Multiple-Concept Example 8 for insight into this problem. During a tennis serve, a racket is given an angular acceleration of magnitude 160 rad/s². At the top of the serve, the racket has an angular speed of 14 rad/s. If the distance between the top of the racket and the shoulder is 1.5 m, find the magnitude of the total acceleration of the top of the racket.

57. **M** **SSM** The compact disc (or CD) is an outdated storage medium that was used prominently in the 1990s to store digital data, including music, on a spiral track. Music is put onto a CD with the assumption that, while the disc spins during playback, the music will be detected at a *constant tangential speed* at any point. Since $v_T = r\omega$, a CD rotates at a smaller angular speed for music near the outer edge and a larger angular speed for music near the inner part of the disc. For music at the outer edge ($r = 0.0568$ m), the angular speed is 3.50 rev/s. Find **(a)** the constant tangential speed at which music is detected and **(b)** the angular speed (in rev/s) for music at a distance of 0.0249 m from the center of a CD.

58. **M** **V-HINT** After 10.0 s, a spinning roulette wheel at a casino has slowed down to an angular velocity of +1.88 rad/s. During this time, the wheel has an angular acceleration of −5.04 rad/s². Determine the angular displacement of the wheel.

59. **M** **GO** At a county fair there is a betting game that involves a spinning wheel. As the drawing shows, the wheel is set into rotational motion with the beginning of the angular section labeled "1" at the marker at the top of the wheel. The wheel then decelerates and eventually comes to a halt on one of the numbered sections. The wheel in the drawing is divided into twelve sections, each of which is an angle of 30.0°. Determine the numbered section on which the wheel comes to a halt when the deceleration of the wheel has a magnitude of 0.200 rev/s² and the initial angular velocity is **(a)** +1.20 rev/s and **(b)** +1.47 rev/s.

PROBLEM 59

60. **M** **V-HINT** **MMH** A racing car, starting from rest, travels around a circular turn of radius 23.5 m. At a certain instant, the car is still accelerating, and its angular speed is 0.571 rad/s. At this time, the total acceleration (centripetal plus tangential) makes an angle of 35.0° with respect to the radius. (The situation is similar to that in **Interactive Figure 8.12b**.) What is the magnitude of the total acceleration?

61. **M** **GO** **SSM** An automobile, starting from rest, has a linear acceleration to the right whose magnitude is 0.800 m/s² (see the figure). During the next 20.0 s, the tires roll and do not slip. The radius of each wheel is 0.330 m. At the end of this time, what is the angle through which each wheel has rotated?

At rest
$t_0 = 0$ s

$t = 20.0$ s

θ

PROBLEM 61

Physics in Biology, Medicine, and Sports

62. **M** **BIO** 8.3 A dentist causes the bit of a high-speed drill to accelerate from an angular speed of 1.05×10^4 rad/s to an angular speed of 3.14×10^4 rad/s. In the process, the bit turns through 1.88×10^4 rad. Assuming a constant angular acceleration, how long would it take the bit to reach its maximum speed of 7.85×10^4 rad/s, starting from rest?

63. **E** **BIO** 8.4 Some bacteria are propelled by biological motors that spin hairlike flagella. A typical bacterial motor turning at a constant angular velocity has a radius of 1.5×10^{-8} m, and a tangential speed at the rim of 2.3×10^{-5} m/s. **(a)** What is the angular speed (the magnitude of the angular velocity) of this bacterial motor? **(b)** How long does it take the motor to make one revolution?

64. **M** **BIO** **GO** 8.5 In a large centrifuge used for training pilots and astronauts, a small chamber is fixed at the end of a rigid arm that rotates in a horizontal circle. A trainee riding in the chamber of a centrifuge rotating with a constant angular speed of 2.5 rad/s experiences a centripetal acceleration of 3.2 times the acceleration due to gravity. In a second training exercise, the centrifuge speeds up from rest with a constant angular acceleration. When the centrifuge reaches an angular speed of 2.5 rad/s, the trainee experiences a total acceleration equal to 4.8 times the acceleration due to gravity. **(a)** How long is the arm of the centrifuge? **(b)** What is the angular acceleration of the centrifuge in the second training exercise?

65. **M** **BIO** 8.2 On January 19, 2015, at the National Stadium in Warsaw, Poland, figure skater Olivia Rybicka-Oliver, at only 11 years old, shattered the Guinness World Record for the fastest spin on ice at 340 rpm (see the photo). **(a)** What was her angular speed in rad/s? **(b)** Assume she comes to rest from this maximum angular speed in 3.2 s and calculate the number of rotations she completes during this time.

Olivia Oliver

PROBLEM 65

66. **M** **BIO** 8.4 Evan Nagao (see the photo) is an incredible talent who can make a yo-yo perform amazing tricks. He is a multiple former U.S. national champion who has even launched his own yo-yo brand. One basic yo-yo trick is to make it "sleep." One throws the yo-yo straight downward by quickly snapping the elbow and wrist. The yo-yo then spins at the bottom of the string very quickly. Assume the yo-yo begins from rest and is thrown downward in 5.00 ms, at which point it rotates at a constant 4650 rpm while hanging at the bottom of the string. **(a)** What was the average angular acceleration of the yo-yo in rad/s² during the throw? **(b)** If the diameter of the yo-yo is 7.00 cm, calculate the tangential speed of a point on its outer edge while it is sleeping.

PROBLEM 66

Ken McKay/ITV/Shutterstock.com

67. **M** **BIO** 8.3 *Helicobacter pylori* (*H. pylori*) is a helically shaped bacterium that is usually found in the stomach. It burrows through the gastric mucous lining to establish an infection in the stomach's epithelial cells (see the photo). Approximately 90% of the people infected with *H. pylori* will never experience symptoms. Others may develop peptic ulcers and show symptoms of chronic gastritis. The method of motility of *H. pylori* is a prokaryotic flagellum attached to the back of the bacterium that rigidly rotates like a propeller on a ship. The flagellum is composed of proteins, is approximately 40.0 nm in diameter, and can reach rotation speeds as high as 1.50×10^3 rpm. If the speed of the bacterium is 10.0 μm/s, how far has it moved in the time it takes the flagellum to rotate through an angular displacement of 5.00×10^2 rad?

H. PYLORI CROSSING MUCUS LAYER OF STOMACH

H.pylori   Gastric   Epithelial   H.pylori raises pH,
mucin   cells   mucin de-gels
gel

PROBLEM 67

68. **M** **BIO** **8.3**   A stationary bicycle is a great option to improve one's fitness level. The rider's pedaling action rotates a weighted flywheel that simulates the resistance a person would feel while riding outside. The Keiser M3 stationary bike uses one of the industry's lightest flywheels at just 8 lb (see the photo). This allows the rider to reach maximum speed in less time. Assume a rider begins from rest and accelerates the flywheel at a constant rate to 110 rpm in 3.7 s. **(a)** What is the angular acceleration of the flywheel in rad/s^2, and **(b)** how many rotations does the wheel complete during this time?

PROBLEM 68

69. **M** **BIO** **8.3**   Wind dispersal, also called *anemochory*, is one of the most primitive ways that a plant spreads its seeds. Maple trees do this, in which they have winged seeds called *samaras* that spin like a helicopter as they drop toward the ground (see the photo). This causes the seeds to drop slowly, thereby increasing the likelihood that the wind carries them farther away from their parent tree. As the seed falls, it quickly reaches terminal velocity due to air resistance, and it rotates at a constant angular speed of 25 rad/s. If a seed falls from a height of 9.2 m and falls with a terminal speed of 1.1 m/s, how many rotations does the seed complete during its fall?

PROBLEM 69

Concepts and Calculations Problems   **Online**

In this chapter we have studied the concepts of angular displacement, angular velocity, and angular acceleration. We now review some important aspects of these quantities. Problem 70 illustrates that the angular acceleration and the angular velocity can have the same or opposite direction, depending on whether the angular speed is increasing or decreasing. Problem 71 reviews the two different types of acceleration, centripetal and tangential, that a car can have when it travels on a circular road.

70. **M** **CHALK**   A rider on a mountain bike is traveling to the left in the figure. Each wheel has an angular velocity of +21.7 rad/s, where, as usual, the plus sign indicates that the wheel is rotating in the counterclockwise direction. **(a)** To pass another cyclist, the rider pumps harder, and the angular velocity of the wheels increases from +21.7 to +28.5 rad/s in a time of 3.50 s. **(b)** After passing the cyclist, the rider begins to coast, and the angular velocity of the wheels decreases from +28.5 to +15.3 rad/s in a time of 10.7 s. *Concepts:* (i) Is the angular acceleration positive or negative when the rider is passing the cyclist

and the angular speed of the wheels is increasing? (ii) Is the angular acceleration positive or negative when the rider is coasting and the angular speed of the wheels is decreasing? *Calculations:* In both instances, **(a)** and **(b)**, determine the magnitude and direction of the angular acceleration (assumed constant) of the wheels.

(a) Angular speed   (b) Angular speed
increasing   decreasing

PROBLEM 70

71. M CHALK SSM Suppose you are driving a car in a counter-clockwise direction on a circular road whose radius is $r = 390$ m (see the figure). You look at the speedometer and it reads a steady 32 m/s (about 72 mi/h). *Concepts:* (i) Does an object traveling at a constant tangential speed (for example, $v_T = 32$ m/s) along a circular path have an acceleration? (ii) Is there a tangential acceleration $\vec{a}_T$ when the angular speed of an object changes (e.g., when the car's angular speed decreases to 4.9×10^{-2} rad/s)? *Calculations:* **(a)** What is the angular speed of the car? **(b)** Determine the acceleration (magnitude and direction) of the car. **(c)** To avoid a rear-end collision with the vehicle ahead, you apply the brakes and reduce your angular speed to 4.9×10^{-2} rad/s in a time of 4.0 s. What is the tangential acceleration (magnitude and direction) of the car?

(*a*) Constant angular speed   (*b*) Decreasing angular speed

PROBLEM 71

Team Problems   Online

72. E T WS **A Rotating Scientific Satellite.** You and your team are designing a device that is to be placed into orbit to measure the earth's magnetic field. It is constructed from two identical spheres, each of diameter 0.550 m and mass 270.0 kg, connected to each other via a cable so that the distance between the sphere centers is 22.0 m. The composite object is then set into rotational motion, resembling a rotating dumbbell, where the cable both holds the spheres together and provides the centripetal force that keeps them on a circular path of diameter 22.0 m. Your task is to determine the maximum angular velocity of the device if the tension in the cable is not to exceed 15 500 N. Express your answer in both rad/s and rpm.

73. E T WS **Soft Braking.** A circular space station with a diameter of 3.20 km is spinning at a rate such that its crew on the outer rim experiences artificial gravity equal to that on the surface of the earth. But something has gone wrong, and you and your team have been assigned the task of reducing the station's rotation to zero so that crucial repairs can be made. The problem is that the crew cannot leave the outer rim while this happens. In order to bring the station to a halt in the least amount of time that will also be safe to the people inside, the linear acceleration at the rim where they are located cannot exceed $g = 9.80$ m/s². **(a)** What constant value of the angular acceleration (which will be negative, since the rate of rotation must decrease) will accomplish this? **(b)** How long will it take the space station to come to rest?

74. M T **Energy of a Bullet Dissipated by Plywood.** As part of a criminal investigation, you need to determine how much of a bullet's

energy is dissipated by a 0.500-inch piece of plywood. You construct a device that consists of three disks that are separated by a distance $d = 0.950$ m and rotate on a common axis. The bullet is fired through the first disk (a few inches above its center), which is composed of a light plastic that has a negligible effect on the speed of the bullet. The bullet then passes through the second disk, which is composed of 0.500-inch plywood. Finally, the bullet strikes the third disk, where it becomes embedded. The disks rotate with an angular velocity of $\omega = 92.0$ rad/s. The angular displacement between holes in the first and second disks is $\Delta\theta_{12} = 0.255$ rad, and the angular displacement between the holes in the second and third disks is $\Delta\theta_{23} = 0.273$ rad. If the mass of the bullet is 15.0 g, find **(a)** the initial speed of the bullet and **(b)** the energy dissipated by the 0.500-inch plywood.

75. M T **Three Wheels.** Three rubber wheels are mounted on axles so that their outer edges make tight contact with each other and their centers are on a line. The wheel on the far left axle is connected to a motor that rotates it at 25 rpm, and drives the wheel in contact with it on its right, which, in turn, drives the wheel on its right. The left wheel (Wheel 1) has a diameter of $d_1 = 0.20$ m, the middle wheel (Wheel 2) has $d_2 = 0.30$ m, and the far right wheel (Wheel 3) has $d_3 = 0.45$ m. **(a)** If Wheel 1 rotates clockwise, in which direction does Wheel 3 rotate? **(b)** What is the angular speed of Wheel 3, and what is the tangential speed on its outer edge? **(c)** What arrangement of the wheels gives the largest tangential speed on the outer edge of the wheel in the far right position (assuming the wheel in the far left position is driven at 25 rpm)?

A hurricane, or typhoon, is composed of large rotating masses of air with a very low pressure eye at its center. When the rotating air near the outer edge of the storm is forced inward due to the lower pressure, its moment of inertia decreases. As a consequence of this, its angular speed increases to conserve angular momentum. This effect produces the very high and dangerous winds in the eye wall of the storm (see photograph). Moment of inertia and conservation of angular momentum are two important topics in rotational dynamics, which we will study in this chapter.

skeeze/Pixabay

Rotational Dynamics

After reading this module, you should be able to...

9.1 Distinguish between torque and force.

9.2 Analyze rigid objects in equilibrium.

9.3 Determine the center of gravity of rigid objects.

9.4 Analyze rotational dynamics using moments of inertia.

9.5 Apply the relation between rotational work and energy.

9.6 Solve problems using the conservation of angular momentum.

9.1 The Action of Forces and Torques on Rigid Objects

The mass of most rigid objects, such as a propeller or a wheel, is spread out and not concentrated at a single point. These objects can move in a number of ways. **Figure 9.1a** illustrates one possibility called translational

(*a*) Translation   (*b*) Combined translation and rotation

FIGURE 9.1 Examples of (*a*) translational motion and (*b*) combined translational and rotational motions.

INTERACTIVE FIGURE 9.2 With a force of a given magnitude, a door is easier to open by (*a*) pushing at the outer edge than by (*b*) pushing closer to the axis of rotation (the hinge). (*c*) Pushing into the hinge makes it difficult to open the door.

motion, in which all points on the body travel on parallel paths (not necessarily straight lines). In pure translation there is no rotation of any line in the body. Because translational motion can occur along a curved line, it is often called curvilinear motion or linear motion. Another possibility is rotational motion, which may occur in combination with translational motion, as is the case for the somersaulting gymnast in **Figure 9.1***b*.

We have seen many examples of how a net force affects linear motion by causing an object to accelerate. We now need to take into account the possibility that a rigid object can also have an angular acceleration. A net external force causes linear motion to change, but what causes rotational motion to change? For example, something causes the rotational velocity of a speedboat's propeller to change when the boat accelerates. Is it simply the net force? As it turns out, it is not the net external force, but rather the net external torque that causes the rotational velocity to change. Just as greater net forces cause greater linear accelerations, greater net torques cause greater rotational or angular accelerations.

Interactive Figure 9.2 helps to explain the idea of torque. When you push on a door with a force $\vec{F}$, as in part *a*, the door opens more quickly when the force is larger. Other things being equal, a larger force generates a larger torque. However, the door does not open as quickly if you apply the same force at a point closer to the hinge, as in part *b*, because the force now produces less torque. Furthermore, if your push is directed nearly at the hinge, as in part *c*, you will have a hard time opening the door at all, because the torque is nearly zero. In summary, the torque depends on the magnitude of the force, on the point where the force is applied relative to the axis of rotation (the hinge in **Interactive Figure 9.2**), and on the direction of the force.

For simplicity, we deal with situations in which the force lies in a plane that is perpendicular to the axis of rotation. In **Figure 9.3**, for instance, the axis is perpendicular to the page and the force lies in the plane of the paper. The drawing shows the line of action and the lever arm of the force, two concepts that are important in the definition of torque. The **line of action** is an extended line drawn colinear with the force. The **lever arm** is the distance ℓ between the line of action and the axis of rotation, measured on a line that is perpendicular to both. The torque is represented by the symbol τ (Greek letter *tau*), and its magnitude is defined as the magnitude of the force times the lever arm:

DEFINITION OF TORQUE

Magnitude of torque = (Magnitude of the force) × (Lever arm) = $F\ell$ (9.1)

Direction: The torque τ **is positive when the force tends to produce a counterclockwise rotation about the axis, and negative when the force tends to produce a clockwise rotation.**

SI Unit of Torque: newton · meter (N · m)

Equation 9.1 indicates that forces of the same magnitude can produce *different* torques, depending on the value of the lever arm, and Example 1 illustrates this important feature.

EXAMPLE 1 | Different Lever Arms, Different Torques

In **Figure 9.3** a force (magnitude = 55 N) is applied to a door. However, the lever arms are different in the three parts of the drawing: **(a)** $\ell = 0.80$ m, **(b)** $\ell = 0.60$ m, and **(c)** $\ell = 0$ m. Find the torque in each case.

Reasoning In each case the lever arm is the perpendicular distance between the axis of rotation and the line of action of the force. In part *a* this perpendicular distance is equal to the width of the door. In parts *b* and *c*, however, the lever arm is less than the width. Because the lever arm is different in each case, the torque is different, even though the magnitude of the applied force is the same.

Solution Using Equation 9.1, we find the following values for the torques:

(a) $\tau = +F\ell = +(55 \text{ N})(0.80 \text{ m}) = \boxed{+44 \text{ N} \cdot \text{m}}$

(b) $\tau = +F\ell = +(55 \text{ N})(0.60 \text{ m}) = \boxed{+33 \text{ N} \cdot \text{m}}$

(c) $\tau = +F\ell = +(55 \text{ N})(0 \text{ m}) = \boxed{0 \text{ N} \cdot \text{m}}$

In parts *a* and *b* the torques are positive, since the forces tend to produce a counterclockwise rotation of the door. In part *c* the line of action of $\vec{F}$ passes through the axis of rotation (the hinge). Hence, the lever arm is zero, and the torque is zero.

(a) (b) (c)

FIGURE 9.3 In this top view, the hinges of a door appear as a black dot (•) and define the axis of rotation. The line of action and lever arm ℓ are illustrated for a force applied to the door (a) perpendicularly and (b) at an angle. (c) The lever arm is zero because the line of action passes through the axis of rotation.

In our bodies, muscles and tendons produce torques about various joints. Example 2 illustrates how the Achilles tendon produces a torque about the ankle joint.

EXAMPLE 2 | BIO The Physics of the Achilles Tendon

Figure 9.4a shows the ankle joint and the Achilles tendon attached to the heel at point P. The tendon exerts a force $\vec{\mathbf{F}}$ (magnitude = 720 N), as **Figure 9.4b** indicates. Determine the torque (magnitude and direction) of this force about the ankle joint, which is located 3.6×10^{-2} m away from point P.

Reasoning To calculate the magnitude of the torque, it is necessary to have a value for the lever arm ℓ. However, the lever arm is not the given distance of 3.6×10^{-2} m. Instead, the lever arm is the perpendicular distance between the axis of rotation at the

ankle joint and the line of action of the force $\vec{\mathbf{F}}$. In **Figure 9.4b** this distance is indicated by the dashed red line.

Solution From the drawing, it can be seen that the lever arm is $\ell = (3.6 \times 10^{-2}$ m) cos 55°. The magnitude of the torque is

$$F\ell = (720 \text{ N})(3.6 \times 10^{-2} \text{ m}) \cos 55° = 15 \text{ N} \cdot \text{m} \qquad (9.1)$$

The force $\vec{\mathbf{F}}$ tends to produce a clockwise rotation about the ankle joint, so the torque is negative: $\boxed{\tau = -15 \text{ N} \cdot \text{m}}$.

(a)

(b)

FIGURE 9.4 The force $\vec{\mathbf{F}}$ generated by the Achilles tendon produces a clockwise (negative) torque about the ankle joint.

Check Your Understanding

(The answers are given at the end of the book.)

1. **CYU Figure 9.1** shows an overhead view of a horizontal bar that is free to rotate about an axis perpendicular to the page. Two forces act on the bar, and they have the same magnitude. However, one force is perpendicular to the bar, and the other makes an angle ϕ with respect to it. The angle ϕ can be 90°, 45°, or 0°. Rank the values of ϕ according to the magnitude of the net torque (the sum of the torques) that the two forces produce, largest net torque first.

CYU FIGURE 9.1

2. Sometimes, even with a wrench, one cannot loosen a nut that is frozen tightly to a bolt. It is often possible to loosen the nut by slipping one end of a long pipe over the wrench handle and pushing at the other end of the pipe. With the aid of the pipe, does the applied force produce a smaller torque, a greater torque, or the same torque on the nut?

3. Is it possible **(a)** for a large force to produce a small, or even zero, torque and **(b)** for a small force to produce a large torque?

4. **CYU Figure 9.2** shows a woman struggling to keep a stack of boxes balanced on a dolly. The woman's left foot is on the axle of the dolly. Assuming that the boxes are identical, which one creates the greatest torque with respect to the axle?

Howard Berman/Stone/Getty Images

CYU FIGURE 9.2

9.2 | Rigid Objects in Equilibrium

If a rigid body is in equilibrium, neither its linear motion nor its rotational motion changes. This lack of change leads to certain equations that apply for rigid-body equilibrium. For instance, an object whose linear motion is not changing has no acceleration $\vec{a}$. Therefore, the net force $\Sigma\vec{F}$ applied to the object must be zero, since $\Sigma\vec{F} = m\vec{a}$ and $\vec{a} = 0$. For two-dimensional motion the x and y components of the net force are separately zero: $\Sigma F_x = 0$ and $\Sigma F_y = 0$. In calculating the net force, we include only forces from external agents, or *external forces.** In addition to linear motion, we must consider rotational motion, which also does not change under equilibrium conditions. This means that the net external torque acting on the object must be zero, because torque is what causes rotational motion to change. Using the symbol $\Sigma\tau$ to represent the net external torque (the sum of all positive and negative torques), we have

$$\Sigma\tau = 0 \qquad (9.2)$$

We define rigid-body equilibrium, then, in the following way.

> **EQUILIBRIUM OF A RIGID BODY**
>
> **A rigid body is in equilibrium if it has zero translational acceleration and zero angular acceleration. In equilibrium, the sum of the externally applied forces is zero, and the sum of the externally applied torques is zero:**
>
> $$\Sigma F_x = 0 \quad \text{and} \quad \Sigma F_y = 0 \qquad \text{(4.9a and 4.9b)}$$
>
> $$\Sigma\tau = 0 \qquad (9.2)$$

*We ignore internal forces that one part of an object exerts on another part, because they occur in action–reaction pairs, each of which consists of oppositely directed forces of equal magnitude. The effect of one force cancels the effect of the other, as far as the acceleration of the entire object is concerned.

The reasoning strategy for analyzing the forces and torques acting on a body in equilibrium is given below. The first four steps of the strategy are essentially the same as those outlined in Section 4.11, where only forces are considered. Steps 5 and 6 have been added to account for any external torques that may be present. Example 3 illustrates how this reasoning strategy is applied to a diving board.

REASONING STRATEGY Applying the Conditions of Equilibrium to a Rigid Body

1. **Select the object to which the equations for equilibrium are to be applied.**

2. **Draw a free-body diagram that shows all the external forces acting on the object.**

3. **Choose a convenient set of x, y axes and resolve all forces into components that lie along these axes.**

4. **Apply the equations that specify the balance of forces at equilibrium: $\Sigma F_x = 0$ and $\Sigma F_y = 0$.**

5. **Select a convenient axis of rotation. The choice is arbitrary. Identify the point where each external force acts on the object, and calculate the torque produced by each force about the chosen axis. Set the sum of the torques equal to zero: $\Sigma \tau = 0$.**

6. **Solve the equations in Steps 4 and 5 for the desired unknown quantities.**

EXAMPLE 3 | A Diving Board

A woman whose weight is 530 N is poised at the right end of a diving board with a length of 3.90 m. The board has negligible weight and is bolted down at the left end, while being supported 1.40 m away by a fulcrum, as **Figure 9.5a** shows. Find the forces $\vec{F}_1$ and $\vec{F}_2$ that the bolt and the fulcrum, respectively, exert on the board.

Reasoning   Part b of the figure shows the free-body diagram of the diving board. Three forces act on the board: $\vec{F}_1$, $\vec{F}_2$, and the force due to the diver's weight $\vec{W}$. In choosing the directions of $\vec{F}_1$ and $\vec{F}_2$, we have used our intuition: $\vec{F}_1$ points downward because the bolt must pull in that direction to counteract the tendency of the board to rotate clockwise about the fulcrum; $\vec{F}_2$ points upward, because the board pushes downward against the fulcrum, which, in reaction, pushes upward on the board. Since the board is stationary, it is in equilibrium.

Solution   Since the board is in equilibrium, the sum of the vertical forces must be zero:

$$\Sigma F_y = -F_1 + F_2 - W = 0 \qquad \textbf{(4.9b)}$$

Similarly, the sum of the torques must be zero, $\Sigma \tau = 0$. For calculating torques, we select an axis that passes through the left end of the board and is perpendicular to the page. (We will see shortly that this choice is arbitrary.) The force $\vec{F}_1$ produces no torque since it passes through the axis and, therefore, has a zero lever arm, while $\vec{F}_2$ creates a counterclockwise (positive) torque, and $\vec{W}$ produces a clockwise (negative) torque. The free-body diagram shows the lever arms for the torques:

$$\Sigma \tau = +F_2 \ell_2 - W \ell_w = 0 \qquad \textbf{(9.2)}$$

Solving this equation for F_2 yields

$$F_2 = \frac{W \ell_w}{\ell_2} = \frac{(530 \text{ N})(3.90 \text{ m})}{1.40 \text{ m}} = \boxed{1480 \text{ N}}$$

This value for F_2, along with $W = 530$ N, can be substituted into Equation 4.9b to show that $\boxed{F_1 = 950 \text{ N}}$.

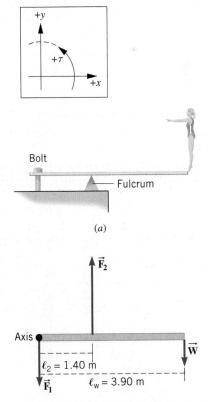

(a)

(b) Free-body diagram of the diving board

FIGURE 9.5 (a) A diver stands at the end of a diving board. (b) The free-body diagram for the diving board. The box at the top left shows the positive x and y directions for the forces, as well as the positive direction (counterclockwise) for the torques.

In Example 3 the sum of the external torques is calculated using an axis that passes through the left end of the diving board.

> **Problem-Solving Insight** The choice of the axis is completely arbitrary, because if an object is in equilibrium, it is in equilibrium with respect to any axis whatsoever.

Thus, the sum of the external torques is zero, no matter where the axis is placed. One usually chooses the location so that the lines of action of one or more of the unknown forces pass through the axis. Such a choice simplifies the torque equation, because the torques produced by these forces are zero. For instance, in Example 3 the torque due to the force $\vec{F}_1$ does not appear in Equation 9.2, because the lever arm of this force is zero.

In a calculation of torque, the lever arm of the force must be determined relative to the axis of rotation. In Example 3 the lever arms are obvious, but sometimes a little care is needed in determining them, as in the next example.

EXAMPLE 4 | Fighting a Fire

In **Animated Figure 9.6a** an 8.00-m ladder of weight $W_L = 355$ N leans against a smooth vertical wall. The term "smooth" means that the wall can exert only a normal force directed perpendicular to the wall and cannot exert a frictional force parallel to it. A firefighter, whose weight is $W_F = 875$ N, stands 6.30 m up from the bottom of the ladder. Assume that the ladder's weight acts at the ladder's center, and neglect the hose's weight. Find the forces that the wall and the ground exert on the ladder.

Reasoning Animated Figure 9.6b shows the free-body diagram of the ladder. The following forces act on the ladder:

1. Its weight $\vec{W}_L$
2. A force due to the weight $\vec{W}_F$ of the firefighter
3. The force $\vec{P}$ applied to the top of the ladder by the wall and directed perpendicular to the wall
4. The forces $\vec{G}_x$ and $\vec{G}_y$, which are the horizontal and vertical components of the force exerted by the ground on the bottom of the ladder

The ground, unlike the wall, is not smooth, so that the force $\vec{G}_x$ is produced by static friction and prevents the ladder from slipping. The force $\vec{G}_y$ is the normal force applied to the ladder by the ground. The ladder is in equilibrium, so the sum of these forces and the sum of the torques produced by them must be zero.

Solution Since the net force acting on the ladder is zero, we have

$$\Sigma F_x = G_x - P = 0 \tag{4.9a}$$

$$\Sigma F_y = G_y - W_L - W_F = 0 \tag{4.9b}$$

Solving Equation 4.9b gives

$$G_y = W_L + W_F = 355 \text{ N} + 875 \text{ N} = \boxed{1230 \text{ N}}$$

Equation 4.9a cannot be solved as it stands, because it contains two unknown variables. However, we know that the net torque acting on an object in equilibrium is zero. In calculating torques, we use an axis at the left end of the ladder, directed perpendicular to the page, as **Animated Figure 9.6c** indicates. This axis is convenient, because $\vec{G}_x$ and $\vec{G}_y$ produce no torques about it, their

(a)   (b) Free-body diagram of the ladder   (c)

ANIMATED FIGURE 9.6 (a) A ladder leaning against a smooth wall. (b) The free-body diagram for the ladder. (c) Three of the forces that act on the ladder and their lever arms. The axis of rotation is at the lower end of the ladder and is perpendicular to the page.

lever arms being zero. Consequently, these forces will not appear in the equation representing the balance of torques. The lever arms for the remaining forces are shown in **Animated Figure 9.6c** as red dashed lines of length ℓ. The following list summarizes these forces, their lever arms, and the torques:

Force	Lever Arm	Torque
$W_L = 355$ N	$\ell_L = (4.00$ m$)\cos 50.0°$	$-W_L\ell_L$
P	$\ell_P = (8.00$ m$)\sin 50.0°$	$+P\ell_P$
$W_F = 875$ N	$\ell_F = (6.30$ m$)\cos 50.0°$	$-W_F\ell_F$

Setting the sum of the torques equal to zero gives

$$\Sigma\tau = -W_L\ell_L - W_F\ell_F + P\ell_P = 0 \qquad (9.2)$$

Solving this equation for P gives

$$P = \frac{W_L\ell_L + W_F\ell_F}{\ell_P}$$

$$= \frac{(355 \text{ N})(4.00 \text{ m})\cos 50.0° + (875 \text{ N})(6.30 \text{ m})\cos 50.0°}{(8.00 \text{ m})\sin 50.0°} = \boxed{727 \text{ N}}$$

Substituting $P = 727$ N into Equation 4.9a reveals that

$$G_x = P = \boxed{727 \text{ N}}$$

Math Skills Sometimes it is necessary to use trigonometry to determine the lever arms from the distances given in a problem. As examples, consider the lever arms for the forces $\vec{W}_L$ and $\vec{P}$ (see **Animated Figure 9.6c**). To determine the lever arm ℓ_L for the force $\vec{W}_L$, we will use the cosine function. Equation 1.2 defines the cosine of the angle θ as $\cos\theta = \dfrac{h_a}{h}$, where h_a is the side of a right triangle adjacent

to the angle θ and h is the hypotenuse of the right triangle, as shown by the smaller triangle in **Figure 9.7a**. By comparing this triangle with the smaller triangle in **Figure 9.7b** it can be seen that $\theta = 50.0°$ and that $h_a = \ell_L$ and $h = 4.00$ m. As a result we have

$$\cos 50.0° = \frac{h_a}{h} = \frac{\ell_L}{4.00 \text{ m}} \quad \text{or} \quad \ell_L = (4.00 \text{ m})\cos 50.0°$$

To determine the lever arm ℓ_P for the force $\vec{P}$, we will use the sine function. Equation 1.1 defines the sine of the angle θ as $\sin\theta = \dfrac{h_o}{h}$, where h_o is the side of a right triangle opposite the angle θ and h is the hypotenuse of the right triangle, as shown by the larger triangle in **Figure 9.7a**. By comparing this triangle with the larger triangle in **Figure 9.7b** it can be seen that $\theta = 50.0°$ and that $h_o = \ell_P$ and $h = 8.00$ m. As a result we have

$$\sin 50.0° = \frac{h_o}{h} = \frac{\ell_P}{8.00 \text{ m}} \quad \text{or} \quad \ell_P = (8.00 \text{ m})\sin 50.0°$$

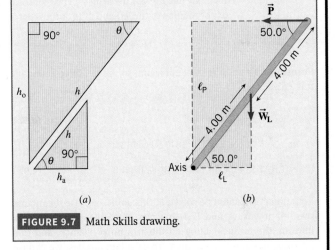

FIGURE 9.7 Math Skills drawing.

To a large extent the directions of the forces acting on an object in equilibrium can be deduced using intuition. Sometimes, however, the direction of an unknown force is not obvious, and it is inadvertently drawn reversed in the free-body diagram. This kind of mistake causes no difficulty.

Problem-Solving Insight Choosing the direction of an unknown force backward in the free-body diagram simply means that the value determined for the force will be a negative number.

EXAMPLE 5 | BIO The Physics of Bodybuilding

A bodybuilder holds a dumbbell of weight $\vec{W}_d$ as in **Figure 9.8a**. His arm is horizontal and weighs $W_a = 31.0$ N. The deltoid muscle is assumed to be the only muscle acting and is attached to the arm as shown. The maximum force $\vec{M}$ that the deltoid muscle can supply has a magnitude of 1840 N. **Figure 9.8b** shows the distances that locate where the various forces act on the arm. What is the weight of the heaviest dumbbell that can be held, and what are the horizontal and vertical force components, $\vec{S}_x$ and $\vec{S}_y$, that the shoulder joint applies to the left end of the arm?

Reasoning Figure 9.8b is the free-body diagram for the arm. Note that $\vec{S}_x$ is directed to the right, because the deltoid

muscle pulls the arm in toward the shoulder joint, and the joint pushes back in accordance with Newton's third law. The direction of the force $\vec{S}_y$, however, is less obvious, and we are alert for the possibility that the direction chosen in the free-body diagram is backward. If so, the value obtained for $\vec{S}_y$ will be negative.

Problem-Solving Insight When a force is negative, such as the vertical force $S_y = -297$ N in this example, it means that the direction of the force is opposite to that chosen originally.

FIGURE 9.8 (a) The fully extended, horizontal arm of a bodybuilder supports a dumbbell. (b) The free-body diagram for the arm. (c) Three of the forces that act on the arm and their lever arms. The axis of rotation at the left end of the arm is perpendicular to the page. Force vectors are not to scale.

Solution The arm is in equilibrium, so the net force acting on it is zero:

$$\Sigma F_x = S_x - M \cos 13.0° = 0 \qquad \textbf{(4.9a)}$$

$$\text{or} \quad S_x = M \cos 13.0° = (1840 \text{ N}) \cos 13.0° = \boxed{1790 \text{ N}}$$

$$\Sigma F_y = S_y + M \sin 13.0° - W_a - W_d = 0 \qquad \textbf{(4.9b)}$$

Equation 4.9b cannot be solved at this point, because it contains two unknowns, S_y and W_d. However, since the arm is in equilibrium, the torques acting on the arm must balance, and this fact provides another equation. To calculate torques, we choose an axis through the left end of the arm and perpendicular to the page. With this axis, the torques due to $\vec{S}_x$ and $\vec{S}_y$ are zero, because the line of action of each force passes through the axis and the lever arm of each force is zero. The list below summarizes the remaining forces, their lever arms (see **Figure 9.8c**), and the torques.

Force	Lever Arm	Torque
$W_a = 31.0$ N	$\ell_a = 0.280$ m	$-W_a\ell_a$
W_d	$\ell_d = 0.620$ m	$-W_d\ell_d$
$M = 1840$ N	$\ell_M = (0.150 \text{ m}) \sin 13.0°$	$+M\ell_M$

The condition specifying a zero net torque is

$$\Sigma \tau = -W_a\ell_a - W_d\ell_d + M\ell_M = 0 \qquad \textbf{(9.2)}$$

Solving this equation for W_d yields

$$W_d = \frac{-W_a\ell_a + M\ell_M}{\ell_d}$$

$$= \frac{-(31.0 \text{ N})(0.280 \text{ m}) + (1840 \text{ N})(0.150 \text{ m}) \sin 13.0°}{0.620 \text{ m}} = \boxed{86.1 \text{ N}}$$

Substituting this value for W_d into Equation 4.9b and solving for S_y gives $\boxed{S_y = -297 \text{ N}}$. The minus sign indicates that the choice of direction for S_y in the free-body diagram is wrong. In reality, S_y has a magnitude of 297 N but is directed downward, not upward.

Check Your Understanding

(The answers are given at the end of the book.)

5. Three forces (magnitudes either F or $2F$) in **CYU Figure 9.3** act on each of the thin, square sheets shown in the drawing. In parts A and B of the drawing, the force labeled $2\vec{F}$ acts at the center of the sheet. When considering angular acceleration, use an axis of rotation that is perpendicular to the plane of a sheet at its center. Determine in which drawing **(a)** the translational acceleration is equal to zero, but the angular acceleration is not equal to zero; **(b)** the translational acceleration is not equal to zero, but the angular acceleration is equal to zero; and **(c)** both the translational and angular accelerations are zero.

6. The free-body diagram in **CYU Figure 9.4** shows the forces that act on a thin rod. The three forces are drawn to scale and lie in the plane of the paper. Are these forces sufficient to keep the rod in equilibrium, or are additional forces necessary?

CYU FIGURE 9.3

CYU FIGURE 9.4

Center of gravity

$\frac{1}{2}L$

Hinge (axis)

$\vec{W}$

FIGURE 9.9 A thin, uniform, horizontal rod of length L is attached to a vertical wall by a hinge. The center of gravity of the rod is at its geometrical center.

9.3 Center of Gravity

Often, it is important to know the torque produced by the weight of an *extended* body. In Examples 4 and 5, for instance, it is necessary to determine the torques caused by the weight of the ladder and the arm, respectively. In both cases the weight is considered to act at a definite point for the purpose of calculating the torque. This point is called the **center of gravity** (abbreviated "cg").

> **DEFINITION OF CENTER OF GRAVITY**
>
> **The center of gravity of a rigid body is the point at which its weight can be considered to act when the torque due to the weight is being calculated.**

When an object has a symmetrical shape and its weight is distributed uniformly, the center of gravity lies at its geometrical center. For instance, **Figure 9.9** shows a thin, uniform, horizontal rod of length L attached to a vertical wall by a hinge. The center of gravity of the rod is located at the geometrical center. The lever arm for the weight $\vec{W}$ is $L/2$, and the magnitude of the torque is $W(L/2)$. In a similar fashion, the center of gravity of any symmetrically shaped and uniform object, such as a sphere, disk, cube, or cylinder, is located at its geometrical center. However, this does not mean that the center of gravity must lie within the object itself. The center of gravity of a compact disc recording, for instance, lies at the center of the hole in the disc and is, therefore, "outside" the object.

Suppose we have a group of objects, with known weights and centers of gravity, and it is necessary to know the center of gravity for the group as a whole. As an example, **Figure 9.10a** shows a group composed of two parts: a horizontal uniform board (weight $\vec{W}_1$) and a uniform box (weight $\vec{W}_2$) near the left end of the board. The center of gravity can be determined by calculating the net torque created by the board and box about an axis that is picked arbitrarily to be at the right end of the board. Part *a* of the figure shows the weights $\vec{W}_1$ and $\vec{W}_2$ and their corresponding lever arms x_1 and x_2. The net torque is $\Sigma\tau = W_1x_1 + W_2x_2$. It is also possible to calculate the net torque by treating the total weight $\vec{W}_1 + \vec{W}_2$ as if it were located at the center of gravity and had the lever arm x_{cg}, as part *b* of the drawing indicates: $\Sigma\tau = (W_1 + W_2)x_{cg}$. The two values for the net torque must be the same, so that

$$W_1x_1 + W_2x_2 = (W_1 + W_2)x_{cg}$$

This expression can be solved for x_{cg}, which locates the center of gravity relative to the axis:

Center of gravity

$$x_{cg} = \frac{W_1x_1 + W_2x_2 + \cdots}{W_1 + W_2 + \cdots} \tag{9.3}$$

The notation "$+ \cdots$" indicates that Equation 9.3 can be extended to account for any number of weights distributed along a horizontal line. **Figure 9.10c** illustrates that the group can be balanced by a single external force (due to the index finger), if the line of action of the force passes through the center of gravity, and if the force is equal in magnitude, but opposite in direction, to the weight of the group. Example 6 demonstrates how to calculate the center of gravity for the human arm.

x_2

x_1

Axis

$\vec{W}_2$

$\vec{W}_1$

(a)

x_{cg}

cg

(b)

$\vec{W}_1 + \vec{W}_2$

cg

(c)

FIGURE 9.10 (*a*) A box rests near the left end of a horizontal board. (*b*) The total weight ($\vec{W}_1 + \vec{W}_2$) acts at the center of gravity of the group. (*c*) The group can be balanced by applying an external force (due to the index finger) at the center of gravity.

EXAMPLE 6 | BIO The Center of Gravity of an Arm

The horizontal arm illustrated in **Figure 9.11** is composed of three parts: the upper arm (weight $W_1 = 17$ N), the lower arm ($W_2 = 11$ N), and the hand ($W_3 = 4.2$ N). The drawing shows the center of gravity of each part, measured with respect to the shoulder joint. Find the center of gravity of the entire arm, relative to the shoulder joint.

Reasoning and Solution The coordinate x_{cg} of the center of gravity is given by

$$x_{cg} = \frac{W_1 x_1 + W_2 x_2 + W_3 x_3}{W_1 + W_2 + W_3} \qquad (9.3)$$

$$= \frac{(17 \text{ N})(0.13 \text{ m}) + (11 \text{ N})(0.38 \text{ m}) + (4.2 \text{ N})(0.61 \text{ m})}{17 \text{ N} + 11 \text{ N} + 4.2 \text{ N}} = \boxed{0.28 \text{ m}}$$

FIGURE 9.11 The three parts of a human arm, and the weight and center of gravity for each.

The center of gravity plays an important role in determining whether a group of objects remains in equilibrium as the weight distribution within the group changes. A change in the weight distribution causes a change in the position of the center of gravity, and if the change is too great, the group will not remain in equilibrium. Conceptual Example 7 discusses a shift in the center of gravity that led to an embarrassing result.

CONCEPTUAL EXAMPLE 7 | Overloading a Cargo Plane

Figure 9.12a shows a stationary cargo plane with its front landing gear 9 meters off the ground. This accident occurred because the plane was overloaded toward the rear. How did a shift in the center of gravity of the loaded plane cause the accident?

Reasoning and Solution **Figure 9.12b** shows a drawing of a correctly loaded plane, with the center of gravity located between the front and the rear landing gears. The weight $\vec{W}$ of the plane and cargo acts downward at the center of gravity, and the normal

(a)

Kevork Djansazian/AP Images

(b)   (c)

FIGURE 9.12 (a) This stationary cargo plane is sitting on its tail at Los Angeles International Airport, after being overloaded toward the rear. (b) In a correctly loaded plane, the center of gravity is between the front and the rear landing gears. (c) When the plane is overloaded toward the rear, the center of gravity shifts behind the rear landing gear, and the accident in part a occurs.

forces $\vec{F}_{N1}$ and $\vec{F}_{N2}$ act upward at the front and at the rear landing gear, respectively. With respect to an axis at the rear landing gear, the counterclockwise torque due to $\vec{F}_{N1}$ balances the clockwise torque due to $\vec{W}$, and the plane remains in equilibrium. **Figure 9.12c** shows the plane with too much cargo loaded toward the rear, just after the plane has begun to rotate counterclockwise. Because of the overloading, the center of gravity has shifted behind the rear landing gear. The torque due to $\vec{W}$ is now counterclockwise and is

not balanced by any clockwise torque. Due to the unbalanced counterclockwise torque, the plane rotates until its tail hits the ground, which applies an upward force to the tail. The clockwise torque due to this upward force balances the counterclockwise torque due to $\vec{W}$, and the plane comes again into an equilibrium state, this time with the front landing gear 9 meters off the ground.

Related Homework: Problems 12, 15, 55

As we have seen in Example 7, the center of gravity plays an important role in the equilibrium orientation of airplanes. It also plays a similar role in the design of vehicles that can be safely driven with minimal risk to the passengers. Example 8 discusses this role in the context of sport utility vehicles.

Analyzing Multiple-Concept Problems

EXAMPLE 8 | The Physics of the Static Stability Factor and Rollover

Figure 9.13a shows a sport utility vehicle (SUV) that is moving away from you and negotiating a horizontal turn. The radius of the turn is 16 m, and its center is on the right in the drawing. The center of gravity of the vehicle is 0.94 m above the ground and, as

an approximation, is assumed to be located midway between the wheels on the left and right sides. The separation between these wheels is the track width and is 1.7 m. What is the greatest speed at which the SUV can negotiate the turn without rolling over?

FIGURE 9.13 (a) A sport utility vehicle (SUV) is shown moving away from you, following a road curving to the right. The radius of the turn is r. The SUV's center of gravity ($\odot$) is located at a height h above the ground, and its wheels are separated by a distance d. (b) This free-body diagram shows the SUV at the instant just before rollover toward the outside of the turn begins. The forces acting on it are its weight $m\vec{g}$, the total normal force $\vec{F}_N$ (acting only on the left-side tires), and the total force of static friction $\vec{f}_s$ (also acting only on the left-side tires). (c) The lever arms ℓ_N (for $\vec{F}_N$) and ℓ_s (for $\vec{f}_s$) are for an axis passing through the center of gravity and perpendicular to the page.

Reasoning The free-body diagram in **Figure 9.13b** shows the SUV at the instant just before it begins to roll over toward the outside of the turn, which is on the left side of the drawing. At this moment the right-side wheels have just lost contact with the ground, so no forces are acting on them. The forces acting on the SUV are its weight $m\vec{g}$, the total normal force $\vec{F}_N$ (acting

only on the left-side tires), and the total force of static friction $\vec{f}_s$ (also acting only on the left-side tires). Since the SUV is moving around the turn, it has a centripetal acceleration and must be experiencing a centripetal force. The static frictional force $\vec{f}_s$ alone provides the centripetal force. The speed v at which the SUV (mass m) negotiates the turn of radius r is related to the magnitude F_c of the centripetal force by $F_c = mv^2/r$ (Equation 5.3). After applying this relation, we will consider rollover by analyzing the torques acting on the SUV with respect to an axis through the center of gravity and perpendicular to the drawing. The normal

force $\vec{F}_N$ will play a role in this analysis, and it will be evaluated by using the fact that it must balance the weight of the vehicle.

Knowns and Unknowns   We are given the following data:

Description	Symbol	Value	Comment
Radius of turn	r	16 m	
Location of center of gravity	h	0.94 m	Height above ground.
Track width	d	1.7 m	Distance between left- and right-side wheels.
Unknown Variable			
Speed of SUV	v	?	

Modeling the Problem

STEP 1 Speed and Centripetal Force   According to Equation 5.3, the magnitude of the centripetal force (provided solely by the static frictional force of magnitude f_s) is related to the speed v of the car by

$$F_c = f_s = \frac{mv^2}{r}$$

Solving for the speed gives Equation 1 at the right. To use this result, we must have a value for f_s, which we obtain in Step 2.

$$v = \sqrt{\frac{rf_s}{m}} \quad (1)$$

$\boxed{?}$

STEP 2 Torques   Figure 9.13b shows the free-body diagram for the SUV at the instant just before rollover begins, when the total normal force $\vec{F}_N$ and the total static frictional force $\vec{f}_s$ are acting only on the left-side wheels, the right-side wheels having just lost contact with the ground. At this moment, the sum of the torques due to these two forces is zero for an axis through the center of gravity and perpendicular to the page. Each torque has a magnitude given by Equation 9.1 as the magnitude of the force times the lever arm for the force, and part c of the drawing shows the lever arms. No lever arm is shown for the weight, because it passes through the axis and, therefore, contributes no torque. The force $\vec{f}_s$ produces a positive torque, since it causes a counterclockwise rotation about the chosen axis, while the force $\vec{F}_N$ produces a negative torque, since it causes a clockwise rotation. Thus, we have

$$f_s \ell_s - F_N \ell_N = 0 \quad \text{or} \quad \boxed{f_s = \frac{F_N \ell_N}{\ell_s} = \frac{F_N d}{2h}}$$

This result for f_s can be substituted into Equation 1, as illustrated at the right. We now proceed to Step 3, to obtain a value for F_N.

$$v = \sqrt{\frac{rf_s}{m}} \quad (1)$$

$$f_s = \frac{F_N d}{2h} \quad (2)$$

$\boxed{?}$

STEP 3 The Normal Force   The SUV does not accelerate in the vertical direction, so the normal force must balance the car's weight mg, or

$$\boxed{F_N = mg}$$

The substitution of this result into Equation 2 is shown at the right.

$$v = \sqrt{\frac{rf_s}{m}} \quad (1)$$

$$f_s = \frac{F_N d}{2h} \quad (2)$$

$$\boxed{F_N = mg}$$

Solution   The results of each step can be combined algebraically to show that

STEP 1 STEP 2 STEP 3

$$v = \sqrt{\frac{rf_s}{m}} = \sqrt{\frac{rF_N d}{2hm}} = \sqrt{\frac{rmgd}{2hm}}$$

The final expression is independent of the mass of the SUV, since m has been eliminated algebraically. The greatest speed at which the SUV can negotiate the turn without rolling over is

$$v = \sqrt{rg\left(\underbrace{\frac{d}{2h}}_{\text{SSF}}\right)} = \sqrt{(16 \text{ m})(9.80 \text{ m/s}^2)\left[\frac{1.7 \text{ m}}{2(0.94 \text{ m})}\right]} = \boxed{12 \text{ m/s}}$$

At this speed, the SUV will negotiate the turn just on the verge of rolling over. In the final result and beneath the term $\frac{d}{2h}$, we have included the label SSF, which stands for *static stability factor*. The SSF provides one measure of how susceptible a vehicle is to rollover. Higher values of the SSF are better, because they lead to larger values for v, the greatest speed at which the vehicle can negotiate a turn without rolling over. Note that greater values for the track width d (more widely separated wheels) and smaller values for the height h (center of gravity closer to the ground) lead to higher values of the SSF. A Formula One racing car, for example, with its low center of gravity and large track width, is much less prone to rolling over than is an SUV, when both are driven at the same speed around the same curve.

The center of gravity of an object with an irregular shape and a nonuniform weight distribution can be found by suspending the object from two different points P_1 and P_2, one at a time. **Figure 9.14a** shows the object at the moment of release, when its weight $\vec{W}$, acting at the center of gravity, has a nonzero lever arm ℓ relative to the axis shown in the drawing. At this instant the weight produces a torque about the axis. The tension force $\vec{T}$ applied to the object by the suspension cord produces no torque because its line of action passes through the axis. Hence, in part a there is a net torque applied to the object, and the object begins to rotate. Friction eventually brings the object to rest as in part b, where the center of gravity lies directly below the point of suspension. In such an orientation, the line of action of the weight passes through the axis, so there is no longer any net torque. In the absence of a net torque the object remains at rest. By suspending the object from a second point P_2 (see **Figure 9.14c**), a second line through the object can be established, along which the center of gravity must also lie. The center of gravity, then, must be at the intersection of the two lines.

The center of gravity is closely related to the center-of-mass concept discussed in Section 7.5. To see why they are related, let's replace each occurrence of the weight in Equation 9.3 by $W = mg$, where m is the mass of a given object and g is the magnitude of the acceleration due to gravity at the location of the object. Suppose that g has the same value everywhere the objects are located. Then it can be algebraically eliminated from each term on the right side of Equation 9.3. The resulting equation, which contains only masses and distances, is the same as Equation 7.10, which defines the location of the center of mass. Thus, the two points are identical. For ordinary-sized objects, like cars and boats, the center of gravity coincides with the center of mass.

The center of gravity of the human body will vary depending on the body's shape. Therefore, it figures prominently in athletic events, such as that discussed within the context of center of mass in Example 10 of Chapter 7. The center of gravity of the body also impacts everyday activities, such as lifting a heavy object, as the following example illustrates.

FIGURE 9.14 The center of gravity (cg) of an object can be located by suspending the object from two different points, P_1 and P_2, one at a time.

CONCEPTUAL EXAMPLE 9 | BIO The Importance of Proper Lifting Technique

Over one million people injure their back in the workplace annually, and an estimated 80% of the adult population will injure their back during their lifetime. Americans spend more than $50 billion a year on treating back pain. Many of these injuries could be avoided by using proper lifting techniques. What is the best technique to use to reduce the risk of a back injury?

Reasoning and Solution **Figure 9.15a** shows proper body position when picking up a heavy object. The lifter keeps their spine straight, bends at the knees, and keeps the object close to their body. The weight of the person's upper body ($\vec{W}_U$) and the weight of the box ($\vec{W}_B$) produce clockwise torques about the axis at the hip joint, with their forces acting at their respective centers of gravity. To maintain this position, the lower back muscles must apply a force to the spine ($\vec{F}_B$). This force produces the counterclockwise torque necessary to balance the two torques from the weights. **Figure 9.15b** shows a very poor technique for lifting a heavy object. The back is no longer straight, but rounded. The person is leaning more forward and holding the box farther from their body. This increases the lever arm, and thus the torques, for both weight forces. In order to compensate and maintain equilibrium about the hip joint, the force from the lower back muscles must increase, placing them at a greater risk of injury.

(a)   (b)

FIGURE 9.15 (a) Proper technique in lifting a heavy box. The person keeps their back straight, bends their knees, and holds the box close to their body. The lift is made with the legs, with very little rotation about the hip joint. (b) Poor lifting technique. The person is leaning too far forward, so that the back is more rounded and less vertical, as well as holding the box further from their body. This greatly increases the force required by the muscles in the lower back. Notice too how the person in (b) has to slide their left foot forward, in order to keep a point of support below their center of gravity.

Check Your Understanding

(The answers are given at the end of the book.)

7. Starting in the spring, fruit begins to grow on the outer end of a branch on a pear tree. As the fruit grows, does the center of gravity of the pear-growing branch **(a)** move toward the pears at the end of the branch, **(b)** move away from the pears, or **(c)** not move at all?

8. CYU Figure 9.5 shows a wine rack for a single bottle of wine that seems to defy common sense as it balances on a tabletop. Where is the center of gravity of the combined wine rack and bottle of wine located?

Wine rack

Rack is not attached to the tabletop.

CYU FIGURE 9.5

(a) At the neck of the bottle where it passes through the wine rack **(b)** Directly above the point where the wine rack touches the tabletop **(c)** At a location to the right of where the wine rack touches the tabletop

9. Bob and Bill have the same weight and wear identical shoes. When they both keep their feet flat on the floor and their bodies straight, Bob can lean forward farther than Bill can before falling. Other things being equal, whose center of gravity is closer to the ground when both are standing erect?

$\vec{F}_T$

r 90°

FIGURE 9.16 A model airplane on a guideline has a mass m and is flying on a circle of radius r (top view). A net tangential force $\vec{F}_T$ acts on the plane.

9.4 | Newton's Second Law for Rotational Motion About a Fixed Axis

The goal of this section is to put Newton's second law into a form suitable for describing the rotational motion of a rigid object about a fixed axis. We begin by considering a particle moving on a circular path. **Figure 9.16** presents a good approximation of this situation by using a small model plane on a guideline of negligible mass. The plane's engine produces a net external tangential force F_T that gives the plane a tangential acceleration a_T. In accord with Newton's second law, it follows that $F_T = ma_T$. The torque τ produced by this force is $\tau = F_T r$, where the radius r of the circular path is also the lever arm. As a result, the torque is $\tau = ma_T r$. However, the tangential acceleration is related to the angular acceleration α according to $a_T = r\alpha$ (Equation 8.10), where α must be expressed in rad/s². With this substitution for a_T, the torque becomes

$$\tau = \underbrace{(mr^2)}_{\substack{\text{Moment of} \\ \text{inertia } I}}\alpha \tag{9.4}$$

Equation 9.4 is the form of Newton's second law we have been seeking. It indicates that the net external torque τ is directly proportional to the angular acceleration α. The constant of proportionality is $I = mr^2$, which is called the **moment of inertia of the particle**. The SI unit for moment of inertia is kg · m².

If all objects were single particles, it would be just as convenient to use the second law in the form $F_T = ma_T$ as in the form $\tau = I\alpha$. The advantage in using $\tau = I\alpha$ is that it can be applied to any rigid body rotating about a fixed axis, and not just to a particle. To illustrate how this advantage arises, **Figure 9.17a** shows a flat sheet of material that rotates about an axis perpendicular to the sheet. The sheet is composed of a number of mass particles, $m_1, m_2, \ldots, m_N$, where N is very large. Only four particles are shown for the sake of clarity. Each particle behaves in the same way as the model airplane in **Figure 9.16** and obeys the relation $\tau = (mr^2)\alpha$:

$$\tau_1 = (m_1 r_1^2)\alpha$$
$$\tau_2 = (m_2 r_2^2)\alpha$$
$$\cdot$$
$$\cdot$$
$$\cdot$$
$$\tau_N = (m_N r_N^2)\alpha$$

In these equations each particle has the same angular acceleration α, since the rotating object is assumed to be rigid. Adding together the N equations and factoring out the common value of α, we find that

$$\underbrace{\Sigma\tau}_{\substack{\text{Net} \\ \text{external torque}}} = \underbrace{(\Sigma mr^2)}_{\substack{\text{Moment} \\ \text{of inertia}}}\alpha \tag{9.5}$$

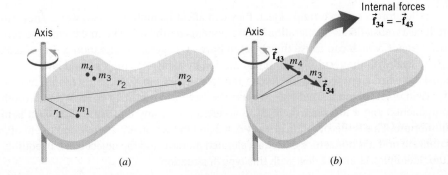

(a)   (b)

FIGURE 9.17 (a) A rigid body consists of a large number of particles, four of which are shown. (b) The internal forces that particles 3 and 4 exert on each other obey Newton's law of action and reaction.

where the expression $\Sigma\tau = \tau_1 + \tau_2 + \cdots + \tau_N$ is the sum of the external torques, and $\Sigma mr^2 = m_1r_1^2 + m_2r_2^2 + \cdots + m_Nr_N^2$ represents the sum of the individual moments of inertia. The latter quantity is the **moment of inertia I of the body**:

Moment of inertia of a body	$I = m_1r_1^2 + m_2r_2^2 + \cdots + m_Nr_N^2 = \Sigma mr^2$ (9.6)

In this equation, r is the perpendicular radial distance of each particle from the axis of rotation. Combining Equation 9.6 with Equation 9.5 gives the following result:

ROTATIONAL ANALOG OF NEWTON'S SECOND LAW FOR A RIGID BODY ROTATING ABOUT A FIXED AXIS

$$\text{Net external torque} = \begin{pmatrix} \text{Moment of} \\ \text{inertia} \end{pmatrix} \times \begin{pmatrix} \text{Angular} \\ \text{acceleration} \end{pmatrix}$$

$$\Sigma\tau = I\alpha \qquad (9.7)$$

Requirement: α must be expressed in rad/s^2.

The version of Newton's second law given in Equation 9.7 applies only for rigid bodies. The word "rigid" means that the distances r_1, r_2, r_3, etc. that locate each particle m_1, m_2, m_3, etc. (see **Figure 9.17a**) do not change during the rotational motion. In other words, a rigid body is one that does not change its shape while undergoing an angular acceleration in response to an applied net external torque.

The form of the second law for rotational motion, $\Sigma\tau = I\alpha$, is similar to the equation for translational (linear) motion, $\Sigma F = ma$, and is valid only in an inertial frame. The moment of inertia I plays the same role for rotational motion that the mass m does for translational motion. Thus, I is a measure of the rotational inertia of a body. When using Equation 9.7, α must be expressed in rad/s^2, because the relation $a_T = r\alpha$ (which requires radian measure) was used in the derivation.

When calculating the sum of torques in Equation 9.7, it is necessary to include only the *external torques*, those applied by agents outside the body. The torques produced by internal forces need not be considered, because they always combine to produce a net torque of zero. Internal forces are those that one particle within the body exerts on another particle. They always occur in pairs of oppositely directed forces of equal magnitude, in accord with Newton's third law (see m_3 and m_4 in **Figure 9.17b**). The forces in such a pair have the same line of action, so they have identical lever arms and produce torques of equal magnitudes. One torque is counterclockwise, while the other is clockwise, the net torque from the pair being zero.

It can be seen from Equation 9.6 that the moment of inertia depends on both the mass of each particle and its distance from the axis of rotation. The farther a particle is from the axis, the greater is its contribution to the moment of inertia. Therefore, although a rigid object possesses a unique total mass, it does not have a unique moment of inertia, as indicated by Example 10. This example shows how the moment of inertia can change when the axis of rotation changes. The procedure illustrated in Example 10 can be extended using integral calculus to evaluate the moment of inertia of a rigid object with a continuous mass distribution, and **Table 9.1** gives some typical results. These results depend on the total mass of the object, its shape, and the location and orientation of the axis.

When forces act on a rigid object, they can affect its motion in two ways. They can produce a translational acceleration a (components a_x and a_y). The forces can also produce torques, which can cause the object to have an angular acceleration α. In general, we can deal with the resulting combined motion by using Newton's second law. For the translational motion, we use the law in the form $\Sigma F = ma$. For the rotational motion of a rigid object about a fixed axis, we use the law in the form $\Sigma\tau = I\alpha$. When a (both components) and α are zero, there is no acceleration of any kind, and the object is in equilibrium. This is the situation already discussed in Section 9.2. If any component of a is nonzero or if α is nonzero, we have accelerated motion, and the object is not in equilibrium. Examples 11 and 12 deal with this type of situation.

TABLE 9.1	**Moments of Inertia I for Various Rigid Objects of Mass M**

Thin-walled hollow cylinder or hoop

$I = MR^2$

Solid cylinder or disk

$I = \frac{1}{2}MR^2$

Thin rod, axis perpendicular to rod and passing through center

$I = \frac{1}{12}ML^2$

Thin rod, axis perpendicular to rod and passing through one end

$I = \frac{1}{3}ML^2$

Solid sphere, axis through center

$I = \frac{2}{5}MR^2$

Solid sphere, axis tangent to surface

$I = \frac{7}{5}MR^2$

Thin-walled spherical shell, axis through center

$I = \frac{2}{3}MR^2$

Thin rectangular sheet, axis parallel to one edge and passing through center of other edge

$I = \frac{1}{12}ML^2$

Thin rectangular sheet, axis along one edge

$I = \frac{1}{3}ML^2$

EXAMPLE 10 | The Moment of Inertia Depends on Where the Axis Is

Two particles each have a mass M and are fixed to the ends of a thin rigid rod, whose mass can be ignored. The length of the rod is L. Find the moment of inertia when this object rotates relative to an axis that is perpendicular to the rod at **(a)** one end and **(b)** the center. (See **Interactive Figure 9.18**.)

Reasoning   When the axis of rotation changes, the distance r between the axis and each particle changes. In determining the moment of inertia using $I = \Sigma mr^2$, we must be careful to use the distances that apply for each axis.

> **Problem-Solving Insight**   The moment of inertia depends on the location and orientation of the axis relative to the particles that make up the object.

Solution   **(a)** Particle 1 lies on the axis, as part a of the drawing shows, and has a zero radial distance: $r_1 = 0$. In contrast, particle 2 moves on a circle whose radius is $r_2 = L$. Noting that $m_1 = m_2 = M$, we find that the moment of inertia is

$$I = \Sigma mr^2 = m_1 r_1^2 + m_2 r_2^2 = M(0)^2 + M(L)^2 = \boxed{ML^2} \quad (9.6)$$

(b) Part b of the drawing shows that particle 1 no longer lies on the axis but now moves on a circle of radius $r_1 = L/2$. Particle 2 moves on a circle with the same radius, $r_2 = L/2$. Therefore,

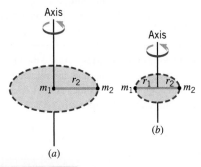

INTERACTIVE FIGURE 9.18 Two particles, masses m_1 and m_2, are attached to the ends of a massless rigid rod. The moment of inertia of this object is different, depending on whether the rod rotates about an axis through (a) the end or (b) the center of the rod.

$$I = \Sigma mr^2 = m_1 r_1^2 + m_2 r_2^2 = M(L/2)^2 + M(L/2)^2 = \boxed{\tfrac{1}{2}ML^2}$$

This value differs from the value in part (a) because the axis of rotation is different, and the distances of the particles from the axis are different.

Analyzing Multiple-Concept Problems

EXAMPLE 11 | The Torque of an Electric Saw Motor

The motor in an electric saw brings the circular blade from rest up to the rated angular velocity of 80.0 rev/s in 240.0 rev. One type of blade has a moment of inertia of 1.41×10^{-3} kg · m². What net torque (assumed constant) must the motor apply to the blade?

Reasoning   Newton's second law for rotational motion, $\Sigma\tau = I\alpha$ (Equation 9.7), can be used to find the net torque $\Sigma\tau$. However,

when using the second law, we will need a value for the angular acceleration α, which can be obtained by using one of the equations of rotational kinematics. In addition, we must remember that the value for α must be expressed in rad/s², not rev/s², because Equation 9.7 requires radian measure.

Knowns and Unknowns   The given data are summarized in the following table:

Description	Symbol	Value	Comment
Explicit Data			
Final angular velocity	ω	80.0 rev/s	Must be converted to rad/s.
Angular displacement	θ	240.0 rev	Must be converted to rad.
Moment of inertia	I	1.41×10^{-3} kg · m²	
Implicit Data			
Initial angular velocity	ω_0	0 rad/s	Blade starts from rest.
Unknown Variable			
Net torque applied to blade	$\Sigma\tau$	?	

Modeling the Problem

STEP 1 Newton's Second Law for Rotation Newton's second law for rotation (Equation 9.7) specifies the net torque $\Sigma\tau$ applied to the blade in terms of the blade's moment of inertia I and angular acceleration α. In Step 2, we will obtain the value for α that is needed in Equation 9.7.	$$\Sigma\tau = I\alpha \qquad (9.7)$$ $\uparrow$ $\boxed{?}$
STEP 2 Rotational Kinematics As the data table indicates, we have data for the angular displacement θ, the final angular velocity ω, and the initial angular velocity ω_0. With these data, Equation 8.8 from the equations of rotational kinematics can be used to determine the angular acceleration α: $$\omega^2 = \omega_0^2 + 2\alpha\theta \qquad (8.8)$$ Solving for α gives $$\boxed{\alpha = \frac{\omega^2 - \omega_0^2}{2\theta}}$$ which can be substituted into Equation 9.7, as shown at the right.	$$\Sigma\tau = I\alpha \qquad (9.7)$$ $\uparrow$ $$\boxed{\alpha = \frac{\omega^2 - \omega_0^2}{2\theta}}$$

Solution The results of each step can be combined algebraically to show that

$$\begin{array}{ccccc} & \textbf{STEP 1} & & \textbf{STEP 2} & \\ & \downarrow & & \downarrow & \\ \Sigma\tau & = & I\alpha & = & I\left(\dfrac{\omega^2 - \omega_0^2}{2\theta}\right) \end{array}$$

In this result for $\Sigma\tau$, we must use radian measure for the variables ω, ω_0, and θ. To convert from revolutions (rev) to radians (rad), we will use the fact that 1 rev = 2π rad. Thus, the net torque applied by the motor to the blade is

$$\Sigma\tau = I\left(\frac{\omega^2 - \omega_0^2}{2\theta}\right)$$

$$= (1.41 \times 10^{-3} \text{ kg} \cdot \text{m}^2)\left\{\frac{\left[\left(80.0 \,\frac{\text{rev}}{\text{s}}\right)\left(\frac{2\pi \text{ rad}}{1 \text{ rev}}\right)\right]^2 - (0 \text{ rad/s})^2}{2(240.0 \text{ rev})\left(\frac{2\pi \text{ rad}}{1 \text{ rev}}\right)}\right\} = \boxed{0.118 \text{ N} \cdot \text{m}}$$

Related Homework: Problems 23, 27, 29, 67

BIO **THE PHYSICS OF . . . wheelchairs.** To accelerate a wheelchair, the rider applies a force to a handrail on each wheel. The magnitude of the torque generated by the force is the product of the force-magnitude and the lever arm. As **Figure 9.19** illustrates, the lever arm is the radius of the circular rail, which is designed to be as large as possible. Thus, a relatively large torque can be generated for a given force, allowing for rapid acceleration.

JACKIE JOHNSTON/AP Images

FIGURE 9.19 A rider applies a force $\vec{\mathbf{F}}$ to the circular handrail. The magnitude of the torque produced by this force is the product of the force-magnitude and the lever arm ℓ about the axis of rotation.

Example 11 shows how Newton's second law for rotational motion is used when design considerations demand an adequately large angular acceleration. There are also situations in which it is desirable to have as little angular acceleration as possible, and Conceptual Example 12 deals with one of them.

CONCEPTUAL EXAMPLE 12 | The Physics of Archery and Bow Stabilizers

Archers can shoot with amazing accuracy, especially using modern bows such as the one in **Figure 9.20**. Notice the bow stabilizer, a long, thin rod that extends from the front of the bow and has a relatively massive cylinder at the tip. Advertisements claim that the stabilizer helps to steady the archer's aim. Which of the following explains why this is true? The addition of the stabilizer **(a)** decreases the bow's moment of inertia, making it easier for the archer to hold the bow steady; **(b)** has nothing to do with the bow's moment of inertia; **(c)** increases the bow's moment of inertia, making it easier for the archer to hold the bow steady.

Reasoning An axis of rotation (the black dot) has been added to **Figure 9.20**. This axis passes through the archer's left shoulder and is perpendicular to the plane of the paper. Any angular acceleration of the archer's body about this axis will lead to a rotation of the bow and, thus, will degrade the archer's aim. The angular acceleration will depend on any unbalanced torques that occur while the archer's tensed muscles try to hold the drawn bow, as well as the bow's moment of inertia.

Answers (a) and (b) are incorrect. Adding a stabilizer to the bow increases its mass. According to the definition of the moment of inertia of a body (Equation 9.6), the increase in mass leads to an increase in the bow's moment of inertia.

Answer (c) is correct. Newton's second law for rotational motion states that the angular acceleration α of the bow is given by $\alpha = (\Sigma\tau)/I$ (Equation 9.7), where $\Sigma\tau$ is the net torque acting on the bow and I is its moment of inertia. The stabilizer increases I, especially the relatively massive cylinder at the tip, since it is so far from the axis. Note that the moment of inertia is in the denominator on the right side of this equation. Therefore, to the extent that I is larger, a given net torque $\Sigma\tau$ will create a smaller angular acceleration and, hence, less disturbance of the aim.

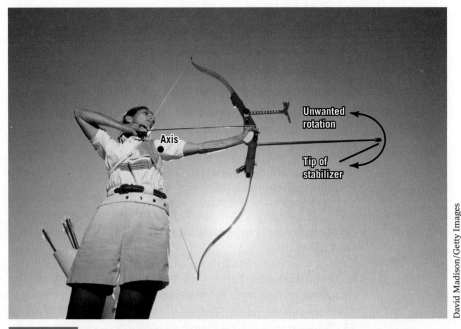

FIGURE 9.20 The long, thin rod extending from the front of the bow is a stabilizer. The stabilizer helps to steady the archer's aim, as Conceptual Example 12 discusses.

We have seen that Newton's second law for rotational motion, $\Sigma\tau = I\alpha$, has the same form as the law for translational motion, $\Sigma F = ma$, so each rotational variable has a translational analog: torque τ and force F are analogous quantities, as are moment of inertia I and mass m, and angular acceleration α and linear acceleration a. The other physical concepts developed for studying translational motion, such as kinetic energy and momentum, also have rotational analogs. For future reference, **Table 9.2** itemizes these concepts and their rotational analogs.

TABLE 9.2   Analogies Between Rotational and Translational Concepts

Physical Concept	Rotational	Translational
Displacement	θ	s
Velocity	ω	v
Acceleration	α	a
The cause of acceleration	Torque τ	Force F
Inertia	Moment of inertia I	Mass m
Newton's second law	$\Sigma\tau = I\alpha$	$\Sigma F = ma$
Work	$\tau\theta$	Fs
Kinetic energy	$\frac{1}{2}I\omega^2$	$\frac{1}{2}mv^2$
Momentum	$L = I\omega$	$p = mv$

Check Your Understanding

(The answers are given at the end of the book.)

10. Three massless rods (A, B, and C) are free to rotate about an axis at their left end (see **CYU Figure 9.6**). The same force $\vec{F}$ is applied to the right end of each rod. Objects with different masses are attached to the rods, but the total mass ($3m$) of the objects is the same for each rod. Rank the angular acceleration of the rods, largest to smallest.

11. A flat triangular sheet of uniform material is shown in **CYU Figure 9.7**. There are three possible axes of rotation, each perpendicular to the sheet and passing through one corner, A, B, or C. For which axis is the greatest net external torque required to bring the triangle up to an angular speed of 10.0 rad/s in 10.0 s, starting from rest? Assume that the net torque is kept constant while it is being applied.

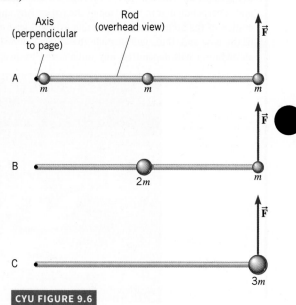

CYU FIGURE 9.6

12. At a given instant an object has an angular velocity. It also has an angular acceleration due to torques that are present. Therefore, the angular velocity is changing. Does the angular velocity at this instant increase, decrease, or remain the same **(a)** if additional torques are applied so as to make the net torque suddenly equal to zero and **(b)** if all the torques are suddenly removed?

13. The space probe in **CYU Figure 9.8** is initially moving with a constant translational velocity and zero angular velocity. **(a)** When the two engines are fired, each generating a thrust of magnitude T, does the translational velocity increase, decrease, or remain the same? **(b)** Does the angular velocity increase, decrease, or remain the same?

CYU FIGURE 9.7

CYU FIGURE 9.8

9.5 # Rotational Work and Energy

Work and energy are among the most fundamental and useful concepts in physics. Chapter 6 discusses their application to translational motion. These concepts are equally useful for rotational motion, provided they are expressed in terms of angular variables.

The work W done by a constant force that points in the same direction as the displacement is $W = Fs$ (Equation 6.1), where F and s are the magnitudes of the force and the displacement, respectively. To see how this expression can be rewritten using angular variables, consider **Figure 9.21**. Here a rope is wrapped around a wheel and is under a constant tension F. If the rope is pulled out a distance s, the wheel rotates through an angle $\theta = s/r$ (Equation 8.1), where r is the radius of the wheel and θ is in radians. Thus, $s = r\theta$, and the work done by the tension force in turning the wheel is $W = Fs = Fr\theta$. However, Fr is the torque τ applied to the wheel by the tension, so the rotational work can be written as follows:

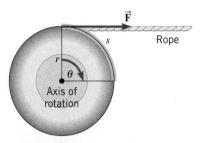

FIGURE 9.21 The force $\vec{F}$ does work in rotating the wheel through the angle θ.

DEFINITION OF ROTATIONAL WORK

The rotational work W_{R} done by a constant torque τ in turning an object through an angle θ is

$$W_{R} = \tau\theta \qquad (9.8)$$

Requirement: θ must be expressed in radians.

SI Unit of Rotational Work: joule (J)

Section 6.2 discusses the work–energy theorem and kinetic energy. There we saw that the work done on an object by a net external force causes the translational kinetic energy ($\frac{1}{2}mv^2$) of the object to change. In an analogous manner, the rotational work done by a net external torque causes the rotational kinetic energy to change. A rotating body possesses kinetic energy, because its constituent particles are moving. If the body is rotating with an angular speed ω, the tangential speed v_{T} of a particle at a distance r from the axis is $v_{T} = r\omega$ (Equation 8.9). **Figure 9.22** shows two such particles. If a particle's mass is m, its kinetic energy is $\frac{1}{2}mv_{T}^2 = \frac{1}{2}mr^2\omega^2$. The kinetic energy of the entire rotating body, then, is the sum of the kinetic energies of the particles:

FIGURE 9.22 The rotating wheel is composed of many particles, two of which are shown.

$$\text{Rotational KE} = \Sigma\left(\tfrac{1}{2}mr^2\omega^2\right) = \tfrac{1}{2}\underbrace{(\Sigma mr^2)}_{\substack{\text{Moment of} \\ \text{inertia, } I}}\omega^2$$

In this result, the angular speed ω is the same for all particles in a rigid body and, therefore, has been factored outside the summation. According to Equation 9.6, the term in parentheses is the moment of inertia, $I = \Sigma mr^2$, so the rotational kinetic energy takes the following form:

DEFINITION OF ROTATIONAL KINETIC ENERGY

The rotational kinetic energy KE_{R} of a rigid object rotating with an angular speed ω about a fixed axis and having a moment of inertia I is

$$\text{KE}_{R} = \tfrac{1}{2}I\omega^2 \qquad (9.9)$$

Requirement: ω must be expressed in rad/s.

SI Unit of Rotational Kinetic Energy: joule (J)

Kinetic energy is one part of an object's total mechanical energy. The total mechanical energy is the sum of the kinetic and potential energies and obeys the principle of conservation of mechanical energy (see Section 6.5). Specifically, we need to remember that translational and rotational motion can occur simultaneously. When a bicycle coasts down a hill, for instance, its tires are both translating and rotating. An object such as a rolling bicycle tire has both translational and rotational kinetic energies, so that the total mechanical energy is

$$\underbrace{E}_{\substack{\text{Total mechanical} \\ \text{energy}}} = \underbrace{\tfrac{1}{2}mv^2}_{\substack{\text{Translational} \\ \text{kinetic energy}}} + \underbrace{\tfrac{1}{2}I\omega^2}_{\substack{\text{Rotational} \\ \text{kinetic energy}}} + \underbrace{mgh}_{\substack{\text{Gravitational} \\ \text{potential energy}}}$$

Here m is the mass of the object, v is the translational speed of its center of mass, I is its moment of inertia about an axis through the center of mass, ω is its angular speed, and h is the height of the object's center of mass relative to an arbitrary zero level. Mechanical energy is conserved if W_{nc}, the net work done by external nonconservative forces and external torques, is zero. If the total mechanical energy is conserved as an object moves, its final total mechanical energy E_f equals its initial total mechanical energy E_0: $E_f = E_0$.

Example 13 illustrates the effect of combined translational and rotational motion in the context of how the total mechanical energy of a cylinder is conserved as it rolls down an incline.

EXAMPLE 13 | Rolling Cylinders

A thin-walled hollow cylinder (mass $= m_h$, radius $= r_h$) and a solid cylinder (mass $= m_s$, radius $= r_s$) start from rest at the top of an incline (**Figure 9.23**). Both cylinders start at the same vertical height h_0 and roll down the incline without slipping. All heights are measured relative to an arbitrarily chosen zero level that passes through the center of mass of a cylinder when it is at the bottom of the incline (see the drawing). Ignoring energy losses due to retarding forces, determine which cylinder has the greatest translational speed on reaching the bottom.

Reasoning Only the conservative force of gravity does work on the cylinders, so the total mechanical energy is conserved as they roll down. The total mechanical energy E at any height h above the zero level is the sum of the translational kinetic energy ($\frac{1}{2}mv^2$), the rotational kinetic energy ($\frac{1}{2}I\omega^2$), and the gravitational potential energy (mgh):

$$E = \tfrac{1}{2}mv^2 + \tfrac{1}{2}I\omega^2 + mgh$$

As the cylinders roll down, potential energy is converted into kinetic energy, but the kinetic energy is shared between the translational form ($\frac{1}{2}mv^2$) and the rotational form ($\frac{1}{2}I\omega^2$). The object with more of its kinetic energy in the translational form will have the greater translational speed at the bottom of the incline. We expect the solid cylinder to have the greater translational speed, because more of its mass is located near the rotational axis and, thus, possesses less rotational kinetic energy.

Solution The total mechanical energy E_f at the bottom ($h_f = 0$ m) is the same as the total mechanical energy E_0 at the top ($h = h_0$, $v_0 = 0$ m/s, $\omega_0 = 0$ rad/s):

$$\tfrac{1}{2}mv_f^2 + \tfrac{1}{2}I\omega_f^2 + mgh_f = \tfrac{1}{2}mv_0^2 + \tfrac{1}{2}I\omega_0^2 + mgh_0$$

$$\tfrac{1}{2}mv_f^2 + \tfrac{1}{2}I\omega_f^2 = mgh_0$$

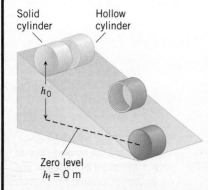

Solid cylinder   Hollow cylinder

h_0

Zero level
$h_f = 0$ m

FIGURE 9.23 A hollow cylinder and a solid cylinder start from rest and roll down the incline plane. The conservation of mechanical energy can be used to show that the solid cylinder, having the greater translational speed, reaches the bottom first.

Since each cylinder rolls without slipping, the final rotational speed ω_f and the final translational speed v_f of its center of mass are related by Equation 8.12, $\omega_f = v_f/r$, where r is the radius of the cylinder. Substituting this expression for ω_f into the energy-conservation equation and solving for v_f yields

$$v_f = \sqrt{\frac{2mgh_0}{m + \dfrac{I}{r^2}}}$$

Setting $m = m_h$, $r = r_h$, and $I = m r_h^2$ for the hollow cylinder and then setting $m = m_s$, $r = r_s$, and $I = \frac{1}{2}m r_s^2$ for the solid cylinder (see **Table 9.1**), we find that the two cylinders have the following translational speeds at the bottom of the incline:

Hollow cylinder $v_f = \sqrt{gh_0}$

Solid cylinder $v_f = \sqrt{\dfrac{4gh_0}{3}} = 1.15\sqrt{gh_0}$

The solid cylinder has the greater translational speed at the bottom and, thus, arrives there first.

Math Skills To obtain the desired expression for the translational speed v_f, we proceed as follows. The energy-conservation equation is $\frac{1}{2}mv_f^2 + \frac{1}{2}I\omega_f^2 = mgh_0$. Substituting $\omega_f = \dfrac{v_f}{r}$ (Equation 8.12) into this equation gives

$$\tfrac{1}{2}mv_f^2 + \tfrac{1}{2}I\left(\frac{v_f}{r}\right)^2 = mgh_0 \quad \text{or} \quad \tfrac{1}{2}\left(m + \frac{I}{r^2}\right)v_f^2 = mgh_0$$

Multiplying both sides of the right-hand equation by 2 and dividing both sides by $m + \dfrac{I}{r^2}$ shows that

$$\frac{2\tfrac{1}{2}\left(m + \frac{I}{r^2}\right)v_f^2}{m + \frac{I}{r^2}} = \frac{2mgh_0}{m + \frac{I}{r^2}} \quad \text{or} \quad v_f^2 = \frac{2mgh_0}{m + \frac{I}{r^2}}$$

Taking the square root of both sides of the right-hand equation shows that

$$v_f = \sqrt{\frac{2mgh_0}{m + \dfrac{I}{r^2}}}$$

Related Homework: Problems 42, 77

Check Your Understanding

(*The answers are given at the end of the book.*)

14. Two uniform solid balls are placed side by side at the top of an incline plane and, starting from rest, are allowed to roll down the incline. Which ball, if either, has the greater translational speed at the bottom if **(a)** they have the same radii, but one is more massive than the other and **(b)** they have the same mass, but one has a larger radius?

15. A thin sheet of plastic is uniform and has the shape of an equilateral triangle. Consider two axes for rotation. Both are perpendicular to the plane of the triangle, axis A passing through the center of the triangle and axis B passing through one corner. If the angular speed ω about each axis is the same, for which axis does the triangle have the greater rotational kinetic energy?

16. A hoop, a solid cylinder, a spherical shell, and a solid sphere are placed at rest at the top of an incline. All the objects have the same radius. They are then released at the same time. What is the order in which they reach the bottom (fastest first)?

9.6 | Angular Momentum

In Chapter 7 the linear momentum p of an object is defined as the product of its mass m and linear velocity v; that is, $p = mv$. For rotational motion the analogous concept is called the **angular momentum** L. The mathematical form of angular momentum is analogous to that of linear momentum, with the mass m and the linear velocity v being replaced with their rotational counterparts, the moment of inertia I and the angular velocity ω.

DEFINITION OF ANGULAR MOMENTUM

The angular momentum L of a body rotating about a fixed axis is the product of the body's moment of inertia I and its angular velocity ω with respect to that axis:

$$L = I\omega \tag{9.10}$$

Requirement: ω must be expressed in rad/s.

SI Unit of Angular Momentum: $\text{kg} \cdot \text{m}^2/\text{s}$

Linear momentum is an important concept in physics because the total linear momentum of a system is conserved when the sum of the average external forces acting on the system is zero. Then, the final total linear momentum P_f and the initial total linear momentum P_0 are the same: $P_f = P_0$. In the case of angular momentum, a similar line of reasoning indicates that when the sum of the average external torques is zero, the final and initial angular momenta are the same: $L_f = L_0$, which is the **principle of conservation of angular momentum**.

PRINCIPLE OF CONSERVATION OF ANGULAR MOMENTUM

The total angular momentum of a system remains constant (is conserved) if the net average external torque acting on the system is zero.

Example 14 illustrates an interesting consequence of the conservation of angular momentum.

CONCEPTUAL EXAMPLE 14 | BIO The Physics of a Spinning Skater

In **Figure 9.24a** an ice skater is spinning with both arms and a leg outstretched. In **Figure 9.24b** she pulls her arms and leg inward. As a result of this maneuver, her angular velocity ω increases dramatically. Why? Neglect any air resistance and assume that friction between her skates and the ice is negligible. **(a)** A net external torque acts on the skater, causing ω to increase. **(b)** No net external torque acts on her; she is simply obeying the conservation of angular momentum. **(c)** Due to the movements of her arms and legs, a net internal torque acts on the skater, causing her angular momentum and ω to increase.

Reasoning Considering the skater as the system, we will use the conservation of angular momentum as a guide; it indicates that only a net external torque can cause the total angular momentum of a system to change.

Answer (a) is incorrect. There is no net external torque acting on the skater, because air resistance and the friction between her skates and the ice are negligible.

Answer (c) is incorrect. The movements of her arms and legs do produce internal torques. However, only *external* torques, not internal torques, can change the angular momentum of a system.

Answer (b) is correct. Since any air resistance and friction are negligible, the net external torque acting on the skater is zero, and the skater's angular momentum is conserved as she pulls her arms and leg inward. However, angular momentum is the product of the moment of inertia I and the angular velocity ω (see Equation 9.10). By moving the mass of her arms and leg inward, the skater decreases the distance r of the mass from the axis of rotation and, consequently, decreases her moment of inertia

(a)   (b)

FIGURE 9.24 (a) A skater spins slowly on one skate, with both arms and one leg outstretched. (b) As she pulls her arms and leg in toward the rotational axis, her angular velocity ω increases.

I ($I = \Sigma mr^2$). Since the product of I and ω is constant, then ω must increase as I decreases. Thus, as she pulls her arms and leg inward, she spins with a larger angular velocity.

Related Homework: Problem 48

The next example involves a satellite and illustrates another application of the principle of conservation of angular momentum.

EXAMPLE 15 | The Physics of a Satellite in an Elliptical Orbit

An artificial satellite is placed into an elliptical orbit about the earth, as illustrated in **Figure 9.25**. Telemetry data indicate that its point of closest approach (called the *perigee*) is $r_P = 8.37 \times 10^6$ m from the center of the earth, and its point of greatest distance (called the *apogee*) is $r_A = 25.1 \times 10^6$ m from the center of the earth. The speed of the satellite at the perigee is $v_P = 8450$ m/s. Find its speed v_A at the apogee.

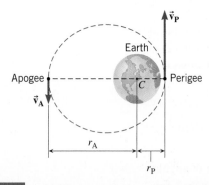

FIGURE 9.25 A satellite is moving in an elliptical orbit about the earth. The gravitational force exerts no torque on the satellite, so the angular momentum of the satellite is conserved.

Reasoning The only force of any significance that acts on the satellite is the gravitational force of the earth. However, at any instant, this force is directed toward the center of the earth and passes through the axis about which the satellite instantaneously rotates. Therefore, the gravitational force exerts *no torque* on the satellite (the lever arm is zero). Consequently, the net average external torque acting on the satellite is zero, and the angular momentum of the satellite remains constant at all times.

Solution Since the angular momentum is the same at the apogee (A) and the perigee (P), it follows that $I_A\omega_A = I_P\omega_P$. Furthermore, the orbiting satellite can be considered a point mass, so its moment of inertia is $I = mr^2$ (see Equation 9.4). In addition, the angular speed ω of the satellite is related to its tangential speed v_T by $\omega = v_T/r$ (Equation 8.9). If Equation 9.4 and Equation 8.9 are used at the apogee and perigee, the conservation of angular momentum gives the following result:

$$I_A\omega_A = I_P\omega_P \quad \text{or} \quad (mr_A^2)\left(\frac{v_A}{r_A}\right) = (mr_P^2)\left(\frac{v_P}{r_P}\right)$$

$$v_A = \frac{r_P v_P}{r_A} = \frac{(8.37 \times 10^6 \text{ m})(8450 \text{ m/s})}{25.1 \times 10^6 \text{ m}} = \boxed{2820 \text{ m/s}}$$

The answer is independent of the mass of the satellite. The satellite behaves just like the skater in **Figure 9.24**, because its speed is smaller at the apogee, where the moment of inertia is greater, and greater at the perigee, where the moment of inertia is smaller.

The result in Example 15 indicates that a satellite does not have a constant speed in an elliptical orbit. The speed changes from a maximum at the perigee to a minimum at the apogee; the closer the satellite comes to the earth, the faster it travels. Planets moving around the sun in elliptical orbits exhibit the same kind of behavior, and Johannes

Kepler (1571–1630) formulated his famous second law based on observations of such characteristics of planetary motion. Kepler's second law states that, in a given amount of time, a line joining any planet to the sun sweeps out the same amount of area no matter where the planet is on its elliptical orbit, as **Figure 9.26** illustrates. The conservation of angular momentum can be used to show why the law is valid, by means of a calculation similar to that in Example 15.

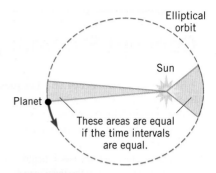

FIGURE 9.26 Kepler's second law of planetary motion states that a line joining a planet to the sun sweeps out equal areas in equal time intervals.

Check Your Understanding

(*The answers are given at the end of the book.*)

17. A woman is sitting on the spinning seat of a piano stool with her arms folded. Ignore any friction in the spinning stool. What happens to her **(a)** angular velocity and **(b)** angular momentum when she extends her arms outward?

18. Review Conceptual Example 14 as an aid in answering this question. Suppose the ice cap at the South Pole were to melt and the water were distributed uniformly over the earth's oceans. Would the earth's angular velocity increase, decrease, or remain the same?

19. Conceptual Example 14 provides background for this question. A cloud of interstellar gas is rotating. Because the gravitational force pulls the gas particles together, the cloud shrinks, and, under the right conditions, a star may ultimately be formed. Would the angular velocity of the star be less than, equal to, or greater than the angular velocity of the rotating gas?

20. A person is hanging motionless from a vertical rope over a swimming pool. She lets go of the rope and drops straight down. After letting go, is it possible for her to curl into a ball and start spinning?

EXAMPLE 16 | BIO The Physics of the Rotational Kinetic Energy of a Figure Skater

In Conceptual Example 14 we saw how a spinning figure skater can change her angular velocity by changing her moment of inertia. This occurs because, in the absence of external torques, the skater's angular momentum is conserved. But what about her rotational kinetic energy? Is that also conserved? Assume the skater in **Figure 9.24a** has a moment of inertia I_0 and is rotating with an angular velocity ω_0. When she brings her arms in (**Figure 9.24b**) assume her rotational velocity doubles. Calculate the ratio of her final rotational kinetic energy to her initial rotational kinetic energy (KE_{Rf}/KE_{R0}) and answer the question of whether or not her rotational kinetic energy is conserved.

Reasoning We can calculate the skater's rotational kinetic energy, both before and after she brings her arms in, by using Equation 9.9. To simplify the ratio of her rotational kinetic energies, we can apply conservation of angular momentum.

Solution Applying Equation 9.9, the ratio of her rotational kinetic energy will be given by the following:

$$\frac{KE_{R_f}}{KE_{R_0}} = \frac{\frac{1}{2}I_f\omega_f^2}{\frac{1}{2}I_0\omega_0^2}.$$

We can simplify this expression and rewrite it in the following way:

$$\frac{KE_{R_f}}{KE_{R_0}} = \frac{\frac{1}{2}I_f\omega_f^2}{\frac{1}{2}I_0\omega_0^2} = \frac{(I_f\omega_f)\omega_f}{(I_0\omega_0)\omega_0}.$$

The product of $I\omega$ in the numerator and denominator represents the angular momentum, L, after and before the skater brings her arms in, respectively. We know these two quantities are equal,

(a)   (b)

FIGURE 9.24 (REPEATED) (a) A skater spins slowly on one skate, with both arms and one leg outstretched. (b) As she pulls her arms and leg in toward the rotational axis, her angular velocity ω increases.

since the angular momentum in this problem is conserved. Therefore, the ratio of her rotational kinetic energies reduces to the ratio of her final and initial angular velocities:

$$\frac{KE_{R_f}}{KE_{R_0}} = \frac{\omega_f}{\omega_0}.$$

Here, we are told that her angular velocity doubles. Thus, $\omega_f = 2\omega_0$, and $\frac{KE_{R_f}}{KE_{R_0}} = \frac{2\omega_0}{\omega_0} = \boxed{2}$. While her angular momentum during this maneuver is conserved, her rotational kinetic energy is not. Where does the extra rotational kinetic energy come from? It comes from the work that she does internally to move her arms closer to the axis of rotation.

Concept Summary

9.1 The Action of Forces and Torques on Rigid Objects The line of action of a force is an extended line that is drawn colinear with the force. The lever arm ℓ is the distance between the line of action and the axis of rotation, measured on a line that is perpendicular to both.

The torque of a force has a magnitude that is given by the magnitude F of the force times the lever arm ℓ. The magnitude of the torque τ is given by Equation 9.1, and τ is positive when the force tends to produce a counterclockwise rotation about the axis, and negative when the force tends to produce a clockwise rotation.

$$\text{Magnitude of torque} = F\ell \qquad (9.1)$$

9.2 Rigid Objects in Equilibrium A rigid body is in equilibrium if it has zero translational acceleration and zero angular acceleration. In equilibrium, the net external force and the net external torque acting on the body are zero, according to Equations 4.9a, 4.9b, and 9.2.

$$\Sigma F_x = 0 \quad \text{and} \quad \Sigma F_y = 0 \qquad \textbf{(4.9a and 4.9b)}$$

$$\Sigma \tau = 0 \qquad (9.2)$$

9.3 Center of Gravity The center of gravity of a rigid object is the point where its entire weight can be considered to act when calculating the torque due to the weight. For a symmetrical body with uniformly distributed weight, the center of gravity is at the geometrical center of the body. When a number of objects whose weights are $W_1, W_2, \ldots$ are distributed along the x axis at locations $x_1, x_2, \ldots$, the center of gravity x_{cg} is given by Equation 9.3. The center of gravity is identical to the center of mass, provided the acceleration due to gravity does not vary over the physical extent of the objects.

$$x_{\text{cg}} = \frac{W_1 x_1 + W_2 x_2 + \cdots}{W_1 + W_2 + \cdots} \qquad (9.3)$$

9.4 Newton's Second Law for Rotational Motion About a Fixed Axis The moment of inertia I of a body composed of N particles is given by Equation 9.6, where m is the mass of a particle and r is the perpendicular distance of the particle from the axis of rotation.

$$I = m_1 r_1^2 + m_2 r_2^2 + \cdots + m_N r_N^2 = \Sigma m r^2 \qquad (9.6)$$

For a rigid body rotating about a fixed axis, Newton's second law for rotational motion is stated as in Equation 9.7, where $\Sigma \tau$ is the net external torque applied to the body, I is the moment of inertia of the body, and α is its angular acceleration.

$$\Sigma \tau = I\alpha \quad (\alpha \text{ in rad/s}^2) \qquad (9.7)$$

9.5 Rotational Work and Energy The rotational work W_R done by a constant torque τ in turning a rigid body through an angle θ is specified by Equation 9.8.

The rotational kinetic energy KE_R of a rigid object rotating with an angular speed ω about a fixed axis and having a moment of inertia I is specified by Equation 9.9.

The total mechanical energy E of a rigid body is the sum of its translational kinetic energy $(\frac{1}{2}mv^2)$, its rotational kinetic energy $(\frac{1}{2}I\omega^2)$, and its gravitational potential energy (mgh), according to Equation 1, where m is the mass of the object, v is the translational speed of its center of mass, I is its moment of inertia about an axis through the center of mass, ω is its angular speed, and h is the height of the object's center of mass relative to an arbitrary zero level.

The total mechanical energy is conserved if the net work done by external nonconservative forces and external torques is zero. When the total mechanical energy is conserved, the final total mechanical energy E_f equals the initial total mechanical energy E_0: $E_f = E_0$.

$$W_R = \tau\theta \quad (\theta \text{ in radians}) \qquad (9.8)$$

$$\text{KE}_R = \tfrac{1}{2}I\omega^2 \quad (\omega \text{ in rad/s}) \qquad (9.9)$$

$$E = \tfrac{1}{2}mv^2 + \tfrac{1}{2}I\omega^2 + mgh \qquad (1)$$

9.6 Angular Momentum The angular momentum of a rigid body rotating with an angular velocity ω about a fixed axis and having a moment of inertia I with respect to that axis is given by Equation 9.10.

$$L = I\omega \quad (\omega \text{ in rad/s}) \qquad (9.10)$$

The principle of conservation of angular momentum states that the total angular momentum of a system remains constant (is conserved) if the net average external torque acting on the system is zero. When the total angular momentum is conserved, the final angular momentum L_f equals the initial angular momentum L_0: $L_f = L_0$.

Focus on Concepts

Online

Additional questions are available for assignment in WileyPLUS.

Section 9.1 The Action of Forces and Torques on Rigid Objects

1. The wheels on a moving bicycle have both translational (or linear) and rotational motions. What is meant by the phrase "a rigid body, such as a bicycle wheel, is in equilibrium"? (a) The body cannot have translational or rotational motion of any kind. (b) The body can have translational motion, but it cannot have rotational motion. (c) The body cannot have translational motion, but it can have rotational motion. (d) The body can have translational and rotational motions, as long as its translational acceleration and angular acceleration are zero.

Section 9.2 Rigid Objects in Equilibrium

2. The drawing illustrates an overhead view of a door and its axis of rotation. The axis is perpendicular to the page. There are four forces acting on the door, and they have the same magnitude. Rank the torque τ that each force produces, largest to smallest.

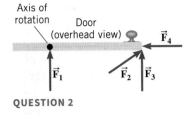

QUESTION 2

(a) $\tau_4, \tau_3, \tau_2, \tau_1$ (b) τ_3, τ_2, τ_1 and τ_4 (a two-way tie) (c) $\tau_2, \tau_4, \tau_3, \tau_1$ (d) τ_1, τ_4, τ_3, τ_2 (e) τ_2, τ_3 and τ_4 (a two-way tie), τ_1

3. Five hockey pucks are sliding across frictionless ice. The drawing shows a top view of the pucks and the three forces that act on each one. As shown, the forces have different magnitudes (F, $2F$, or $3F$), and are applied at different points on the pucks. Only one of the five pucks can be in equilibrium. Which puck is it? (a) 1 (b) 2 (c) 3 (d) 4 (e) 5

QUESTION 3

4. The drawing shows a top view of a square box lying on a frictionless floor. Three forces, which are drawn to scale, act on the box. Consider an angular acceleration with respect to an axis through the center of the box (perpendicular to the page). Which one of the following statements is correct? (a) The box will have a translational acceleration but not an angular acceleration. (b) The box will have both a translational and an angular acceleration. (c) The box will have an angular acceleration but not a translational acceleration. (d) The box will have neither a translational nor an angular acceleration. (e) It is not possible to determine whether the box will have a translational or an angular acceleration.

QUESTION 4

Section 9.4 Newton's Second Law for Rotational Motion About a Fixed Axis

5. The drawing shows three objects rotating about a vertical axis. The mass of each object is given in terms of m_0, and its perpendicular distance from the axis is specified in terms of r_0. Rank the three objects according to their moments of inertia, largest to smallest. (a) A, B, C (b) A, C, B (c) B, A, C (d) B, C, A (e) C, A, B

QUESTION 5

6. Two blocks are placed at the ends of a horizontal massless board, as in the drawing. The board is kept from rotating and rests on a support that serves as an axis of rotation. The moment of inertia of this system relative to the axis is $12 \text{ kg} \cdot \text{m}^2$. Determine the magnitude of the angular acceleration when the system is allowed to rotate.

QUESTION 6

7. The same force $\vec{F}$ is applied to the edge of two hoops (see the drawing). The hoops have the same mass, whereas the radius of the larger hoop is twice the radius of the smaller one. The entire mass of each hoop is concentrated at its rim, so the moment of inertia is $I = Mr^2$, where M is the mass and r is the radius. Which hoop has the greater angular acceleration, and how many times as great is it compared to the angular acceleration of the other hoop? (a) The smaller hoop; two times as great (b) The smaller hoop; four times as great (c) The larger hoop; two times as great (d) The larger hoop; four times as great (e) Both have the same angular acceleration.

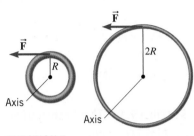

QUESTION 7

Section 9.5 Rotational Work and Energy

8. Two hoops, starting from rest, roll down identical inclined planes. The work done by nonconservative forces, such as air resistance, is zero ($W_{nc} = 0 \text{ J}$). Both have the same mass M, but, as the drawing shows, one hoop has twice the radius of the other. The moment of inertia for each hoop is $I = Mr^2$, where r is its radius. Which hoop, if either, has the greater total kinetic energy (translational plus rotational) at the bottom of the incline? (a) The larger hoop (b) The smaller hoop (c) Both have the same total kinetic energy.

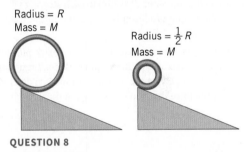

QUESTION 8

Section 9.6 Angular Momentum

9. Under what condition(s) is the angular momentum of a rotating body, such as a spinning ice skater, conserved? (a) Each external force acting on the body must be zero. (b) Each external force and each external torque acting on the body must be zero. (c) Each external force may be nonzero, but the sum of the forces must be zero. (d) Each external torque may be nonzero, but the sum of the torques must be zero.

10. An ice skater is spinning on frictionless ice with her arms extended outward. She then pulls her arms in toward her body, reducing her moment of inertia. Her angular momentum is conserved, so as she reduces her moment of inertia, her angular velocity increases and she spins faster. Compared to her initial rotational kinetic energy, her final rotational kinetic energy is (a) the same (b) larger, because her angular speed is larger (c) smaller, because her moment of inertia is smaller.

Problems

Online

Additional questions are available for assignment in WileyPLUS.

SSM Student Solutions Manual	**BIO** Biomedical application
MMH Problem-solving help	**E** Easy
GO Guided Online Tutorial	**M** Medium
V-HINT Video Hints	**H** Hard
CHALK Chalkboard Videos	**WS** Worksheet
	T Team Problem

Section 9.1 The Action of Forces and Torques on Rigid Objects

1. **E** **SSM** **MMH** The wheel of a car has a radius of 0.350 m. The engine of the car applies a torque of 295 N · m to this wheel, which does not slip against the road surface. Since the wheel does not slip, the road must be applying a force of static friction to the wheel that produces a countertorque. Moreover, the car has a constant velocity, so this countertorque balances the applied torque. What is the magnitude of the static frictional force?

2. **E** The steering wheel of a car has a radius of 0.19 m, and the steering wheel of a truck has a radius of 0.25 m. The same force is applied in the same direction to each steering wheel. What is the ratio of the torque produced by this force in the truck to the torque produced in the car?

3. **E** **SSM** You are installing a new spark plug in your car, and the manual specifies that it be tightened to a torque that has a magnitude of 45 N · m. Using the data in the drawing, determine the magnitude F of the force that you must exert on the wrench.

PROBLEM 3

4. **E** **GO** Two children hang by their hands from the same tree branch. The branch is straight, and grows out from the tree trunk at an angle of 27.0° above the horizontal. One child, with a mass of 44.0 kg, is hanging 1.30 m along the branch from the tree trunk. The other child, with a mass of 35.0 kg, is hanging 2.10 m from the tree trunk. What is the magnitude of the net torque exerted on the branch by the children? Assume that the axis is located where the branch joins the tree trunk and is perpendicular to the plane formed by the branch and the trunk.

5. **E** **SSM** The drawing shows a jet engine suspended beneath the wing of an airplane. The weight $\vec{W}$ of the engine is 10 200 N and acts as shown in the drawing. In flight the engine produces a thrust $\vec{T}$ of 62 300 N that is parallel to the ground. The rotational axis in the drawing is perpendicular to the plane of the paper. With respect to this axis, find the magnitude of the torque due to (a) the weight and (b) the thrust.

PROBLEM 5

6. **E** A square, 0.40 m on a side, is mounted so that it can rotate about an axis that passes through the center of the square. The axis is perpendicular to the plane of the square. A force of 15 N lies in this plane and is applied to the square. What is the magnitude of the maximum torque that such a force could produce?

7. **M** **SSM** A pair of forces with equal magnitudes, opposite directions, and different lines of action is called a "couple." When a couple acts on a rigid object, the couple produces a torque that does *not* depend on the location of the axis. The drawing shows a couple acting on a tire wrench, each force being perpendicular to the wrench. Determine an expression for the torque produced by the couple when the axis is perpendicular to the tire and passes through (a) point A, (b) point B, and (c) point C. Express your answers in terms of the magnitude F of the force and the length L of the wrench.

PROBLEM 7

8. **M** **GO** One end of a meter stick is pinned to a table, so the stick can rotate freely in a plane parallel to the tabletop. Two forces, both parallel to the tabletop, are applied to the stick in such a way that the net torque is zero. The first force has a magnitude of 2.00 N and is applied perpendicular to the length of the stick at the free end. The second force has a magnitude of 6.00 N and acts at a 30.0° angle with respect to the length of the stick. Where along the stick is the 6.00-N force applied? Express this distance with respect to the end of the stick that is pinned.

9. **M** **V-HINT** **MMH** A rod is lying on the top of a table. One end of the rod is hinged to the table so that the rod can rotate freely on the tabletop. Two forces, both parallel to the tabletop, act on the rod at the same place. One force is directed perpendicular to the rod and has a magnitude of 38.0 N. The second force has a magnitude of 55.0 N and is directed at an angle θ with respect to the rod. If the sum of the torques due to the two forces is zero, what must be the angle θ?

10. **H** A rotational axis is directed perpendicular to the plane of a square and is located as shown in the drawing. Two forces, $\vec{F}_1$ and $\vec{F}_2$, are applied to diagonally opposite corners, and act along the sides of the square, first as shown in part *a* and then as shown in part *b* of the drawing. In each case the net torque produced by the forces is zero. The square is one meter on a side, and the magnitude of $\vec{F}_2$ is three times that of $\vec{F}_1$. Find the distances *a* and *b* that locate the axis.

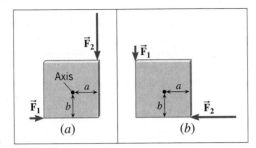

PROBLEM 10

Section 9.2 Rigid Objects in Equilibrium,

Section 9.3 Center of Gravity

11. **E SSM MMH** A hiker, who weighs 985 N, is strolling through the woods and crosses a small horizontal bridge. The bridge is uniform, weighs 3610 N, and rests on two concrete supports, one at each end. He stops one-fifth of the way along the bridge. What is the magnitude of the force that a concrete support exerts on the bridge **(a)** at the near end and **(b)** at the far end?

12. **E** Conceptual Example 7 provides useful background for this problem. Workers have loaded a delivery truck in such a way that its center of gravity is only slightly forward of the rear axle, as shown in the drawing. The mass of the truck and its contents is 7460 kg. Find the magnitudes of the forces exerted by the ground on **(a)** the front wheels and **(b)** the rear wheels of the truck.

PROBLEM 12

13. **E GO** The drawing shows a rectangular piece of wood. The forces applied to corners B and D have the same magnitude of 12 N and are directed parallel to the long and short sides of the rectangle. The long side of the rectangle is twice as long as the short side. An axis of rotation is shown perpendicular to the plane of the rectangle at its center. A third force (not shown in the drawing) is applied to corner A, directed along the short side of the rectangle (either toward B or away from B), such that the piece of wood is at equilibrium. Find the magnitude and direction of the force applied to corner A.

PROBLEM 13

14. **E GO** The wheels, axle, and handles of a wheelbarrow weigh 60.0 N. The load chamber and its contents weigh 525 N. The drawing shows these two forces in two different wheelbarrow designs. To support the wheelbarrow in equilibrium, the man's hands apply a force $\vec{F}$ to the handles that is directed vertically upward. Consider a rotational axis at the point where the tire contacts the ground, directed perpendicular to the plane of the paper. Find the magnitude of the man's force for both designs.

PROBLEM 14

15. **E MMH** Review Conceptual Example 7 as background material for this problem. A jet transport has a weight of 1.00×10^6 N and is at rest on the runway. The two rear wheels are 15.0 m behind the front wheel, and the plane's center of gravity is 12.6 m behind the front wheel. Determine the normal force exerted by the ground on **(a)** the front wheel and on **(b)** *each* of the two rear wheels.

16. **E GO** The drawing shows a uniform horizontal beam attached to a vertical wall by a frictionless hinge and supported from below at an angle $\theta = 39°$ by a brace that is attached to a pin. The beam has a weight of 340 N. Three additional forces keep the beam in equilibrium. The brace applies a force $\vec{P}$ to the right end of the beam that is directed upward at the angle θ with respect to the horizontal. The hinge applies a force to the left end of the beam that has a horizontal component $\vec{H}$ and a vertical component $\vec{V}$. Find the magnitudes of these three forces.

PROBLEM 16

17. **M SSM CHALK** A uniform board is leaning against a smooth vertical wall. The board is at an angle θ above the horizontal ground. The coefficient of static friction between the ground and the lower end of the board is 0.650. Find the smallest value for the angle θ, such that the lower end of the board does not slide along the ground.

18. **M V-HINT** The drawing shows a bicycle wheel resting against a small step whose height is $h = 0.120$ m. The weight and radius of the wheel are $W = 25.0$ N and $r = 0.340$ m, respectively. A horizontal force $\vec{F}$ is applied to the axle of the wheel. As the magnitude of $\vec{F}$ increases, there comes a time when the wheel just begins to rise up and loses contact with the ground. What is the magnitude of the force when this happens?

PROBLEM 18

19. **M SSM CHALK** A 1220-N uniform beam is attached to a vertical wall at one end and is supported by a cable at the other end. A 1960-N crate hangs from the far end of the beam. Using the data shown in the drawing, find **(a)** the magnitude of the tension in the wire and **(b)** the

magnitudes of the horizontal and vertical components of the force that the wall exerts on the left end of the beam.

PROBLEM 19

20. **M** **V-HINT** A wrecking ball (weight = 4800 N) is supported by a boom, which may be assumed to be uniform and has a weight of 3600 N. As the drawing shows, a support cable runs from the top of the boom to the tractor. The angle between the support cable and the horizontal is 32°, and the angle between the boom and the horizontal is 48°. Find **(a)** the tension in the support cable and **(b)** the magnitude of the force exerted on the lower end of the boom by the hinge at point *P*.

PROBLEM 20

21. **H** A man drags a 72-kg crate across the floor at a constant velocity by pulling on a strap attached to the bottom of the crate. The crate is tilted 25° above the horizontal, and the strap is inclined 61° above the horizontal. The center of gravity of the crate coincides with its geometrical center, as indicated in the drawing. Find the magnitude of the tension in the strap.

PROBLEM 21

22. **H** The drawing shows an A-shaped stepladder. Both sides of the ladder are equal in length. This ladder is standing on a frictionless horizontal surface, and only the crossbar (which has a negligible mass) of the "A" keeps the ladder from collapsing. The ladder is uniform and has a mass of 20.0 kg. Determine the tension in the crossbar of the ladder.

PROBLEM 22

Section 9.4 Newton's Second Law for Rotational Motion About a Fixed Axis

23. **E** Consult Multiple-Concept Example 11 to review an approach to problems such as this. A CD has a mass of 17 g and a radius of 6.0 cm. When inserted into a player, the CD starts from rest and accelerates to an angular velocity of 21 rad/s in 0.80 s. Assuming the CD is a uniform solid disk, determine the net torque acting on it.

24. **E** A clay vase on a potter's wheel experiences an angular acceleration of 8.00 rad/s² due to the application of a 10.0-N · m net torque. Find the total moment of inertia of the vase and potter's wheel.

25. **E** A solid circular disk has a mass of 1.2 kg and a radius of 0.16 m. Each of three identical thin rods has a mass of 0.15 kg. The rods are attached perpendicularly to the plane of the disk at its outer edge to form a three-legged stool (see the drawing). Find the moment of inertia of the stool with respect to an axis that is perpendicular to the plane of the disk at its center. (*Hint: When considering the moment of inertia of each rod, note that all of the mass of each rod is located at the same perpendicular distance from the axis.*)

PROBLEM 25

26. **E** A ceiling fan is turned on and a net torque of 1.8 N · m is applied to the blades. The blades have a total moment of inertia of 0.22 kg · m². What is the angular acceleration of the blades?

27. **E** **GO** Multiple-Concept Example 11 provides one model for solving this type of problem. Two wheels have the same mass and radius of 4.0 kg and 0.35 m, respectively. One has the shape of a hoop and the other the shape of a solid disk. The wheels start from rest and have a constant angular acceleration with respect to a rotational axis that is perpendicular to the plane of the wheel at its center. Each turns through an angle of 13 rad in 8.0 s. Find the net external torque that acts on each wheel.

28. **E** A 9.75-m ladder with a mass of 23.2 kg lies flat on the ground. A painter grabs the top end of the ladder and pulls straight upward with a force of 245 N. At the instant the top of the ladder leaves the ground, the ladder experiences an angular acceleration of 1.80 rad/s² about an axis passing through the bottom end of the ladder. The ladder's center of gravity lies halfway between the top and bottom ends. **(a)** What is the net torque acting on the ladder? **(b)** What is the ladder's moment of inertia?

29. **E** **SSM** **MMH** Multiple-Concept Example 11 offers useful background for problems like this. A cylinder is rotating about an axis that passes through the center of each circular end piece. The cylinder has a radius of 0.0830 m, an angular speed of 76.0 rad/s, and a moment of inertia of 0.615 kg · m². A brake shoe presses against the surface of the cylinder and applies a tangential frictional force to it. The frictional force reduces the angular speed of the cylinder by a factor of two during a time of 6.40 s. **(a)** Find the magnitude of the angular deceleration of the cylinder. **(b)** Find the magnitude of the force of friction applied by the brake shoe.

30. **M** **V-HINT** A 15.0-m length of hose is wound around a reel, which is initially at rest. The moment of inertia of the reel is 0.44 kg · m², and its radius is 0.160 m. When the reel is turning, friction at the axle exerts a torque of magnitude 3.40 N · m on the reel. If the hose is pulled so that the tension in it remains a constant 25.0 N, how long does it take to completely unwind the hose from the reel? Neglect the mass and thickness of the hose on the reel, and assume that the hose unwinds without slipping.

31. **E** **GO** The drawing shows two identical systems of objects; each consists of the same three small balls connected by massless rods. In both systems the axis is perpendicular to the page, but it is located at a different place, as shown. The same force of magnitude F is applied to the same ball in each system (see the drawing). The masses of the balls are $m_1 = 9.00$ kg, $m_2 = 6.00$ kg, and $m_3 = 7.00$ kg. The magnitude of the force is $F = 424$ N. **(a)** For each of the two systems, determine the moment of inertia about the given axis of rotation. **(b)** Calculate the torque (magnitude and direction) acting on each system. **(c)** Both systems start from rest, and the direction of the force moves with the system and always points along the 4.00-m rod. What is the angular velocity of each system after 5.00 s?

PROBLEM 31

32. M GO The drawing shows the top view of two doors. The doors are uniform and identical. Door A rotates about an axis through its left edge, and door B rotates about an axis through its center. The same force $\vec{F}$ is applied perpendicular to each door at its right edge, and the force remains perpendicular as the door turns. No other force affects the rotation of either door. Starting from rest, door A rotates through a certain angle in 3.00 s. How long does it take door B (also starting from rest) to rotate through the same angle?

PROBLEM 32

33. M SSM A stationary bicycle is raised off the ground, and its front wheel ($m = 1.3$ kg) is rotating at an angular velocity of 13.1 rad/s (see the drawing). The front brake is then applied for 3.0 s, and the wheel slows down to 3.7 rad/s. Assume that all the mass of the wheel is concentrated in the rim, the radius of which is 0.33 m. The coefficient of kinetic friction between each brake pad and the rim is $\mu_k = 0.85$. What is the magnitude of the normal force that *each* brake pad applies to the rim?

PROBLEM 33

34. M The **parallel axis theorem** provides a useful way to calculate the moment of inertia I about an arbitrary axis. The theorem states that $I = I_{cm} + Mh^2$, where I_{cm} is the moment of inertia of the object relative to an axis that passes through the center of mass and is parallel to the axis of interest, M is the total mass of the object, and h is the perpendicular distance between the two axes. Use this theorem and information to determine an expression for the moment of inertia of a solid cylinder of radius R relative to an axis that lies on the surface of the cylinder and is perpendicular to the circular ends.

35. H The crane shown in the drawing is lifting a 180-kg crate upward with an acceleration of 1.2 m/s². The cable from the crate passes over

a solid cylindrical pulley at the top of the boom. The pulley has a mass of 130 kg. The cable is then wound onto a hollow cylindrical drum that is mounted on the deck of the crane. The mass of the drum is 150 kg, and its radius is 0.76 m. The engine applies a counterclockwise torque to the drum in order to wind up the cable. What is the magnitude of this torque? Ignore the mass of the cable.

PROBLEM 35

Section 9.5 Rotational Work and Energy

36. H Calculate the kinetic energy that the earth has because of **(a)** its rotation about its own axis and **(b)** its motion around the sun. Assume that the earth is a uniform sphere and that its path around the sun is circular. For comparison, the total energy used in the United States in one year is about 1.1×10^{20} J.

37. E SSM Three objects lie in the x, y plane. Each rotates about the z axis with an angular speed of 6.00 rad/s. The mass m of each object and its perpendicular distance r from the z axis are as follows: (1) $m_1 = 6.00$ kg and $r_1 = 2.00$ m, (2) $m_2 = 4.00$ kg and $r_2 = 1.50$ m, (3) $m_3 = 3.00$ kg and $r_3 = 3.00$ m. **(a)** Find the tangential speed of each object. **(b)** Determine the total kinetic energy of this system using the expression $KE = \frac{1}{2}m_1v_1^2 + \frac{1}{2}m_2v_2^2 + \frac{1}{2}m_3v_3^2$. **(c)** Obtain the moment of inertia of the system. **(d)** Find the rotational kinetic energy of the system using the relation $KE_R = \frac{1}{2}I\omega^2$ to verify that the answer is the same as the answer to (b).

38. E GO Two thin rods of length L are rotating with the same angular speed ω (in rad/s) about axes that pass perpendicularly through one end. Rod A is massless but has a particle of mass 0.66 kg attached to its free end. Rod B has a mass of 0.66 kg, which is distributed uniformly along its length. The length of each rod is 0.75 m, and the angular speed is 4.2 rad/s. Find the kinetic energies of rod A with its attached particle and of rod B.

39. E SSM A flywheel is a solid disk that rotates about an axis that is perpendicular to the disk at its center. Rotating flywheels provide a means for storing energy in the form of rotational kinetic energy and are being considered as a possible alternative to batteries in electric cars. The gasoline burned in a 300-mile trip in a typical midsize car produces about 1.2×10^9 J of energy. How fast would a 13-kg flywheel with a radius of 0.30 m have to rotate to store this much energy? Give your answer in rev/min.

40. E A helicopter has two blades (see Figure 8.11); each blade has a mass of 240 kg and can be approximated as a thin rod of length 6.7 m. The blades are rotating at an angular speed of 44 rad/s. **(a)** What is the total moment of inertia of the two blades about the axis of rotation? **(b)** Determine the rotational kinetic energy of the spinning blades.

41. M MMH A solid sphere is rolling on a surface. What fraction of its total kinetic energy is in the form of rotational kinetic energy about the center of mass?

42. M GO Review Example 13 before attempting this problem. A marble and a cube are placed at the top of a ramp. Starting from rest at the same height, the marble rolls without slipping and the cube slides (no kinetic friction) down the ramp. Determine the ratio of the center-of-mass speed of the cube to the center-of-mass speed of the marble at the bottom of the ramp.

43. M V-HINT Starting from rest, a basketball rolls from the top of a hill to the bottom, reaching a translational speed of 6.6 m/s. Ignore frictional losses. **(a)** What is the height of the hill? **(b)** Released from rest at the same height, a can of frozen juice rolls to the bottom of the same hill. What is the translational speed of the frozen juice can when it reaches the bottom?

44. M CHALK SSM A bowling ball encounters a 0.760-m vertical rise on the way back to the ball rack, as the drawing illustrates. Ignore frictional losses and assume that the mass of the ball is distributed uniformly. The translational speed of the ball is 3.50 m/s at the bottom of the rise. Find the translational speed at the top.

0.760 m

PROBLEM 44

45. H A tennis ball, starting from rest, rolls down the hill in the drawing. At the end of the hill the ball becomes airborne, leaving at an angle of 35° with respect to the ground. Treat the ball as a thin-walled spherical shell, and determine the range x.

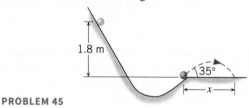

1.8 m

35°

x

PROBLEM 45

Section 9.6 Angular Momentum

46. E SSM Two disks are rotating about the same axis. Disk A has a moment of inertia of 3.4 kg · m² and an angular velocity of +7.2 rad/s. Disk B is rotating with an angular velocity of −9.8 rad/s. The two disks are then linked together without the aid of any external torques, so that they rotate as a single unit with an angular velocity of −2.4 rad/s. The axis of rotation for this unit is the same as that for the separate disks. What is the moment of inertia of disk B?

47. E When some stars use up their fuel, they undergo a catastrophic explosion called a *supernova*. This explosion blows much or all of the star's mass outward, in the form of a rapidly expanding spherical shell. As a simple model of the supernova process, assume that the star is a solid sphere of radius R that is initially rotating at 2.0 revolutions per day. After the star explodes, find the angular velocity, in revolutions per day, of the expanding supernova shell when its radius is $4.0R$. Assume that all of the star's original mass is contained in the shell.

48. E GO Conceptual Example 14 provides useful background for this problem. A playground carousel is free to rotate about its center

on frictionless bearings, and air resistance is negligible. The carousel itself (without riders) has a moment of inertia of 125 kg · m². When one person is standing on the carousel at a distance of 1.50 m from the center, the carousel has an angular velocity of 0.600 rad/s. However, as this person moves inward to a point located 0.750 m from the center, the angular velocity increases to 0.800 rad/s. What is the person's mass?

49. E MMH A thin rod has a length of 0.25 m and rotates in a circle on a frictionless tabletop. The axis is perpendicular to the length of the rod at one of its ends. The rod has an angular velocity of 0.32 rad/s and a moment of inertia of 1.1×10^{-3} kg · m². A bug standing on the axis decides to crawl out to the other end of the rod. When the bug (mass = 4.2×10^{-3} kg) gets where it's going, what is the angular velocity of the rod?

50. E GO As seen from above, a playground carousel is rotating counterclockwise about its center on frictionless bearings. A person standing still on the ground grabs onto one of the bars on the carousel very close to its outer edge and climbs aboard. Thus, this person begins with an angular speed of zero and ends up with a nonzero angular speed, which means that he underwent a counterclockwise angular acceleration. The carousel has a radius of 1.50 m, an initial angular speed of 3.14 rad/s, and a moment of inertia of 125 kg · m². The mass of the person is 40.0 kg. Find the final angular speed of the carousel after the person climbs aboard.

51. M GO A thin, uniform rod is hinged at its midpoint. To begin with, one-half of the rod is bent upward and is perpendicular to the other half. This bent object is rotating at an angular velocity of 9.0 rad/s about an axis that is perpendicular to the left end of the rod and parallel to the rod's upward half (see the drawing). Without the aid of external torques, the rod suddenly assumes its straight shape. What is the angular velocity of the straight rod?

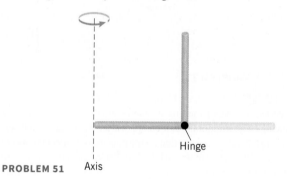

Hinge

PROBLEM 51 Axis

52. H A small 0.500-kg object moves on a frictionless horizontal table in a circular path of radius 1.00 m. The angular speed is 6.28 rad/s. The object is attached to a string of negligible mass that passes through a small hole in the table at the center of the circle. Someone under the table begins to pull the string downward to make the circle smaller. If the string will tolerate a tension of no more than 105 N, what is the radius of the smallest possible circle on which the object can move?

Additional Problems

Online

53. E A solid disk rotates in the horizontal plane at an angular velocity of 0.067 rad/s with respect to an axis perpendicular to the disk at its center. The moment of inertia of the disk is 0.10 kg · m². From above, sand is dropped straight down onto this rotating disk, so that a thin uniform ring of sand is formed at a distance of 0.40 m from the axis. The sand in the ring has a mass of 0.50 kg. After all the sand is in place, what is the angular velocity of the disk?

54. E A solid cylindrical disk has a radius of 0.15 m. It is mounted to an axle that is perpendicular to the circular end of the disk at its

center. When a 45-N force is applied tangentially to the disk, perpendicular to the radius, the disk acquires an angular acceleration of 120 rad/s². What is the mass of the disk?

55. E GO Review Conceptual Example 7 before starting this problem. A uniform plank of length 5.0 m and weight 225 N rests horizontally on two supports, with 1.1 m of the plank hanging over the right support (see the drawing). To what distance x can a person who weighs 450 N walk on the overhanging part of the plank before it just begins to tip?

PROBLEM 55

56. E SSM A rotating door is made from four rectangular sections, as indicated in the drawing. The mass of each section is 85 kg. A person pushes on the outer edge of one section with a force of $F = 68$ N that is directed perpendicular to the section. Determine the magnitude of the door's angular acceleration.

PROBLEM 56

57. M V-HINT A block (mass = 2.0 kg) is hanging from a massless cord that is wrapped around a pulley (moment of inertia = 1.1×10^{-3} kg · m²), as the drawing shows. Initially the pulley is prevented from rotating and the block is stationary. Then, the pulley is allowed to rotate as the block falls. The cord does not slip relative to the pulley as the block falls. Assume that the radius of the cord around the pulley remains constant at a value of 0.040 m during the block's descent. Find the angular acceleration of the pulley and the tension in the cord.

PROBLEM 57

58. M GO A thin, rigid, uniform rod has a mass of 2.00 kg and a length of 2.00 m. **(a)** Find the moment of inertia of the rod relative to an axis that is perpendicular to the rod at one end. **(b)** Suppose all the mass of the rod were located at a single point. Determine the perpendicular distance of this point from the axis in part (a), such that this point particle has the same moment of inertia as the rod does. This distance is called the **radius of gyration** of the rod.

59. M GO Two identical wheels are moving on horizontal surfaces. The center of mass of each has the same linear speed. However, one wheel is rolling, while the other is sliding on a frictionless surface without rolling. Each wheel then encounters an incline plane. One continues to roll up the incline, while the other continues to slide up. Eventually they come to a momentary halt, because the gravitational force slows them down. Each wheel is a disk of mass 2.0 kg. On the horizontal surfaces the center of mass of each wheel moves with a linear speed of 6.0 m/s. **(a)** What is the total kinetic energy of each wheel? **(b)** Determine the maximum height reached by each wheel as it moves up the incline.

60. H MMH By means of a rope whose mass is negligible, two blocks are suspended over a pulley, as the drawing shows. The pulley can be treated as a uniform solid cylindrical disk. The downward acceleration of the 44.0-kg block is observed to be exactly one-half the acceleration due to gravity. Noting that the tension in the rope is not the same on each side of the pulley, find the mass of the pulley.

11.0 kg
44.0 kg

PROBLEM 60

61. M GO SSM A crate of mass 451 kg is being lifted by the mechanism shown in part *a* of the figure. The two cables are wrapped around their respective pulleys, which have radii of 0.600 and 0.200 m. The pulleys are fastened together to form a dual pulley and turn as a single unit about the center axle, relative to which the combined moment of inertia is 46.0 kg · m². The cables roll on the dual pulley without slipping. A tension of magnitude 2150 N is maintained in the cable attached to the motor. Find the angular acceleration of the dual pulley and the tension in the cable attached to the crate.

PROBLEM 61   *(a)*

(b) Free-body diagram of pulley   *(c)* Free-body diagram of crate

Physics in Biology, Medicine, and Sports

62. **E** **BIO** **9.3** A person is standing on a level floor. His head, upper torso, arms, and hands together weigh 438 N and have a center of gravity that is 1.28 m above the floor. His upper legs weigh 144 N and have a center of gravity that is 0.760 m above the floor. Finally, his lower legs and feet together weigh 87.0 N and have a center of gravity that is 0.250 m above the floor. Relative to the floor, find the location of the center of gravity for his entire body.

63. **E** **BIO** **9.3** The drawing shows a person (weight, $W = 584$ N) doing push-ups. Find the normal force exerted by the floor on *each* hand and *each* foot, assuming that the person holds this position.

PROBLEM 63

64. **E** **BIO** **MMH** **9.3** A person exerts a horizontal force of 190 N in the test apparatus shown in the drawing. Find the horizontal force $\overrightarrow{\mathbf{M}}$ (magnitude and direction) that his flexor muscle exerts on his forearm.

PROBLEM 64

65. **E** **BIO** **CHALK** **9.3** A man holds a 178-N ball in his hand, with the forearm horizontal (see the drawing). He can support the ball in this position because of the flexor muscle force $\overrightarrow{\mathbf{M}}$, which is applied perpendicular to the forearm. The forearm weighs 22.0 N and has a center of gravity as indicated. Find **(a)** the magnitude of $\overrightarrow{\mathbf{M}}$ and **(b)** the magnitude and direction of the force applied by the upper arm bone to the forearm at the elbow joint.

PROBLEM 65

66. **M** **BIO** **GO** **9.3** A person is sitting with one leg outstretched and stationary, so that it makes an angle of 30.0° with the horizontal, as the drawing indicates. The weight of the leg below the knee is 44.5 N, with the center of gravity located below the knee joint. The leg is being held in this position because of the force $\overrightarrow{\mathbf{M}}$ applied by the quadriceps

muscle, which is attached 0.100 m below the knee joint (see the drawing). Obtain the magnitude of $\overrightarrow{\mathbf{M}}$.

PROBLEM 66

67. **E** **BIO** **CHALK** **9.4** Multiple-Concept Example 11 reviews the approach and some of the concepts that are pertinent to this problem. The drawing shows a model for the motion of the human forearm in throwing a dart. Because of the force $\overrightarrow{\mathbf{M}}$ applied by the triceps muscle, the forearm can rotate about an axis at the elbow joint. Assume that the forearm has the dimensions shown in the drawing and a moment of inertia of 0.065 kg · m² (including the effect of the dart) relative to the axis at the elbow. Assume also that the force $\overrightarrow{\mathbf{M}}$ acts perpendicular to the forearm. Ignoring the effect of gravity and any frictional forces, determine the magnitude of the force $\overrightarrow{\mathbf{M}}$ needed to give the dart a tangential speed of 5.0 m/s in 0.10 s, starting from rest.

PROBLEM 67

68. **E** **BIO** **9.1** The drawing shows a lower leg being exercised. It has a 49-N weight attached to the foot and is extended at an angle θ with respect to the vertical. Consider a rotational axis at the knee. **(a)** When $\theta = 90.0°$, find the magnitude of the torque that the weight creates. **(b)** At what angle θ does the magnitude of the torque equal 15 N · m?

PROBLEM 68

69. M BIO V-HINT **9.2** The drawing shows an outstretched arm (0.61 m in length) that is parallel to the floor. The arm is pulling downward against the ring attached to the pulley system, in order to hold the 98-N weight stationary. To pull the arm downward, the latissimus dorsi muscle applies the force $\overrightarrow{\mathbf{M}}$ in the drawing, at a point that is 0.069 m from the shoulder joint and oriented at an angle of 29°. The arm has a weight of 47 N and a center of gravity (cg) that is located 0.28 m from the shoulder joint. Find the magnitude of $\overrightarrow{\mathbf{M}}$.

PROBLEM 69

70. M BIO **9.2** The figure shows the finishing position of the arm of a bodybuilder who is performing a standing lateral raise with a 5.00-kg dumbbell. This is an excellent exercise that isolates the lateral deltoid muscle. The weight of the arm itself is 42.5 N, and it acts at point A in the figure. Given the information shown, and the fact that $\theta = 35.0°$, **(a)** find the magnitude of the tension force in the deltoid ($\overrightarrow{\mathbf{F}}_{d}$) and **(b)** the normal force from the shoulder joint ($\overrightarrow{\mathbf{F}}_{s}$) acting on the humerus bone that holds the arm in this equilibrium position.

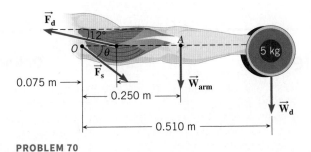

PROBLEM 70

71. E BIO **9.2** Contraction of the quadriceps muscle on the front of the upper leg results in lower-leg extension. This action creates tension in the quadriceps tendon, which is transmitted to the patellar tendon through the patella, or kneecap bone. Tension in the patellar tendon applies a force to the tibia bone in the lower leg, producing a rotation about the knee joint. Imagine an athlete is performing an exercise on a leg extension machine. At the moment shown in the drawing, she is holding the weight at rest. Use the information in the figure, and the fact that the magnitude of $\overrightarrow{\mathbf{F}}$ is 490 N, to calculate the magnitude of the tension in the patellar tendon ($\overrightarrow{\mathbf{T}}_{p}$) at this position. Neglect the weight of the lower leg.

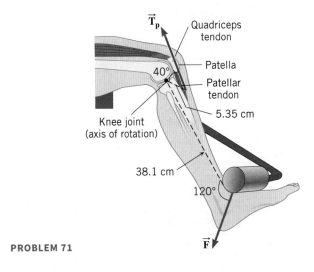

PROBLEM 71

72. M BIO **9.4** Raise your arm straight away from your side until it is in a horizontal position, as shown in the figure. Now relax your shoulder muscles and let your arm "free-fall" as it rotates back to your side. Calculate the tangential acceleration of your fingertips right as your arm begins to fall. Write your answer in terms of g, the acceleration of falling objects near the surface of the earth. Treat your arm as a uniform rod that rotates about an axis at one end (hinged at your shoulder). How does your answer compare with g?

PROBLEM 72

73. E BIO **9.5** The title of the *fastest spinning animal on earth* may go to the *flattie spider* (see the figure). They can anchor one of their legs and pivot around it at incredible speeds to attack prey in any direction. Some of these spiders can spin with angular speeds of 3000°/s. **(a)** How many rotations does a flattie spider complete in the blink of an eye (300 ms)? **(b)** Assume the spider is rotating at a constant angular speed of 3000°/s. What is the spider's rotational kinetic energy? Assume the moment of inertia of the spider while pivoting about one leg is 9.4×10^{-5} kg·m².

magnitude of the tension in the elephant's trunk $\vec{\mathbf{T}}$ ($m = 4540$ kg, $r_{FN} = r_{fs} = 133$ cm, $r_{cg} = 76.0$ cm, $r_T = 229$ cm).

Yu Zeng

PROBLEM 73

Michael Poliza/Caters News Agency

PROBLEM 74

74. H BIO **9.2** African elephants are the largest land animals. They consume approximately 10% of their body weight in food each day, which for an adult male, can be 1000 lb of vegetation! Their diet consists mostly of grasses, bamboo, tree bark, and fruit. They also like to dine on tree leaves. To reach these leaves, they often stand on their hind legs and extend their trunks (see the figure). The elephant in the figure is in equilibrium. The location of the elephant's center of gravity is shown, and the axis of rotation has been chosen to correspond to the hip joint. The forces in the elephant's free-body diagram are shown, and there is a static friction force between the elephant's back feet and the ground. Use the following information and calculate the

Concepts and Calculations Problems

Online

In this chapter we have seen that a rotational or angular acceleration results when a net external torque acts on an object. In contrast, when a net external force acts on an object, it leads to translational or linear acceleration, as discussed in Chapter 4. Torque and force are, then, fundamentally different concepts, and Problem 75 focuses on this fact. For rotational motion, the moment of inertia and the net external torque determine the angular acceleration of a rotating object, according to Newton's second law. Newton's second law together with the equations of rotational kinematics are useful in accounting for a wide variety of rotational motion. Problem 76 utilizes these concepts in a situation where a rotating object is slowing down.

75. M CHALK The figure shows a uniform crate resting on a horizontal surface. The crate has a square cross section and a weight of $W = 580$ N, which is uniformly distributed. At the bottom right edge of the surface is a small obstruction that prevents the crate from sliding when a horizontal pushing force $\vec{\mathbf{P}}$ is applied to the left side. However, if this force is great enough, the crate will begin to tip and rotate over the obstruction. *Concepts:* (i) What causes the tipping—the force $\vec{\mathbf{P}}$ or the torque that it creates? (ii) Where should $\vec{\mathbf{P}}$ be applied so that a minimum force will be necessary to provide the necessary torque? In

(*a*) A horizontal pushing force $\vec{\mathbf{P}}$ is applied to a uniform crate, which has a square cross section and a weight $\vec{\mathbf{W}}$. The crate rests on the ground, up against a small obstruction. (*b*) Some of the forces acting on the crate and their

PROBLEM 75 lever arms.

other words, should the lever arm be a minimum or a maximum? *Calculations:* Determine the minimum pushing force that leads to tipping.

76. **M** **CHALK** **SSM** Two spheres are each rotating at an angular speed of 24 rad/s about axes that pass through their centers. Each has a radius of 0.20 m and a mass of 1.5 kg. However, as the figure shows, one is solid and the other is a thin-walled spherical shell. Suddenly, a net external torque due to friction (magnitude = 0.12 N · m) begins to act on each sphere and slows the motion down. *Concepts:* (i) Which sphere has the greater moment of inertia and why? (ii) Which sphere has the angular acceleration (a deceleration) with the smaller magnitude? (iii) Which sphere takes a longer time to come to a halt? *Calculations:* How long does it take each sphere to come to a halt?

Solid sphere   Thin-walled
                    spherical shell

PROBLEM 76

Team Problems

Online

77. **M** **T** **WS** **A Car Race.** You and your team are designing a small wooden car for a race where the cars are released from rest at the top of an inclined plane, and the first to get to the bottom wins. The rules for the design of the cars are as follows: (1) the body is made of a single piece of carved wood that weighs 0.75 kg, and (2) the combined mass of the wheels, regardless of shape, must be 0.12 kg, making the total mass of the car 0.87 kg. The four wheels, however, must be identical to each other. The car must also be exactly 18.0 cm in length, with no other specifications on the shape. You are free to use any type of wood that you like for the body, and any material, including metals, for the wheels, as long as the constituents conform to the mass requirements. You recognize that you should choose the most dense wood available (such as hickory) so as to minimize the cross-sectional area of your car body and reduce air resistance. Your choice for the wheels, however, needs some thought. You do not have time to construct your own wheels, but you have some available that all meet the mass requirements. Your choices are as follows: (1) solid wood disks of radius $r = 2.50$ cm, (2) solid aluminum (Al) disks ($r = 1.25$ cm), (3) solid lead (Pb) disks ($r = 1.25$ cm, but thinner than the Al disks, since Pb is denser than Al), (4) Al hollow cylinders ($r = 1.25$ cm), (5) Pb hollow cylinders ($r = 1.25$ cm), or (6) Pb solid disks ($r = 1.00$ cm). Which set of wheels do you choose, and why? Make a mathematical argument to justify your choice. Consult Example 13 for useful insight into this problem.

78. **E** **T** **WS** **A Spinning Space Station.** You and your team are on a space station in deep space that is shaped like a ring and is spinning about its central axis. The crew resides on the inner surface of the outer edge and experiences artificial gravity with an acceleration of 9.80 m/s². The station has a diameter of 750.0 m, and a mass of 3.70×10^7 kg. However, there is a plan to attach thousands of tungsten radiation shields by setting them moving toward the ring from above the plane of rotation. When they come into contact with the ring, they will adhere to it via precisely positioned magnets located on the shields and on the ring. With no external torques acting on the system, the shields will attach simultaneously, be uniformly distributed, and have a combined total mass of 9.98×10^6 kg. It can be assumed that the station is a thin ring, and has the same radius before and after the shields are added. You and your team must decide whether the addition of the shields will affect the station's

artificial gravity and, if so, what must be done to correct it. Find the new value of the centripetal acceleration (i.e., the new artificial gravity) felt by the crew in the ring before any corrections are made.

79. **M** **T** **Smugglers.** Rumor has it that a company has been smuggling gold out of the country using sealed, cylindrical barrels with hollow walls. They pour molten gold into the hollows, and then fill the remainder of the barrel's internal volume with packing peanuts. The total mass of the gold-walled barrel was designed so that it exactly matches those used to transport a volatile chemical that cannot be exposed to air (and therefore the barrel cannot be opened and checked). The X-ray machine usually used to screen containers is suspiciously damaged and not available. **(a)** There are 20 barrels total, and they are all identical: mass $m = 50.0$ kg, height $h = 1.20$ m, and diameter $D = 0.250$ m. How do you determine which ones have walls filled with gold (and are essentially hollow on the interior except for packing peanuts) and those completely filled with the volatile chemical (a tightly-packed powder) where the mass is uniformly distributed? Hint: apply the concepts of moment of inertia. Assume that, in the case of the gold-filled barrels, the entire mass is concentrated at the outer wall of the barrel and, in the case of the barrels filled with the chemical, the mass is distributed evenly throughout the volume of the cylinder. You can neglect the circular bottoms and the lids of the barrels, and assume there is no slipping. **(b)** What is the acceleration of the center of mass of each of the barrels as they roll down a 30.0° inclined plane? **(c)** How much time does it take each barrel to roll 10.0 m down the 30.0° plane?

80. **M** **T** **A Ride Inside a Tractor Tire.** You and your friends plan to roll down a hill on the inside of 600.0-pound tractor tire (diameter $D = 1.80$ m). The hill is inclined at an angle of 25.0° and you initially plan to start from a distance $L = 100.0$ m up the hill, but decide to first check whether it will be safe. **(a)** Assuming the masses of the tire and your 105-pound body are concentrated at the outer rim of a thin-walled cylinder/hoop, what is the effective acceleration your body experiences at the bottom of the hill where your angular speed is greatest, that is, how many "g's" will you experience? Assuming the human body can withstand a g-force of 8.00 g's (1 g = 9.80 m/s²), is it safe to make the ride from 100.0 m up the hill? **(b)** What is the maximum starting distance (L_{max}) up the hill that is safe?

CHAPTER **10**

Don Bartletti/Los Angeles Times/Getty Images

Sarah Reinertsen is a professional athlete who holds numerous world records in her disability division. Her athletic performance is made possible by a high-tech prosthetic leg made of carbon fiber, which flexes and stores elastic potential energy, like a spring does. The elastic potential energy stored by a spring is one of the topics in this chapter.

Simple Harmonic Motion and Elasticity

10.1 The Ideal Spring and Simple Harmonic Motion

Springs are familiar objects that have many applications, ranging from push-button switches on electronic components, to automobile suspension systems, to mattresses. In use, they can be stretched or compressed. For example, the top drawing in **Figure 10.1** shows a spring being stretched. Here a hand applies a pulling force F_x^{Applied} to the spring. The subscript x reminds us that F_x^{Applied} lies along the x axis (not shown in the drawing), which is parallel to the length of the spring. In response, the spring stretches and undergoes a displacement of x from its original, or "unstrained," length. The bottom drawing in **Figure 10.1** illustrates the spring being compressed. Now the hand applies a pushing force to the spring, and it again undergoes a displacement from its unstrained length.

Experiment reveals that for relatively small displacements, the force F_x^{Applied} required to stretch or compress a spring is directly proportional

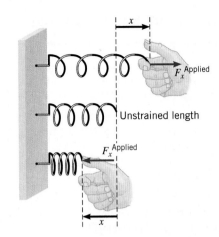

FIGURE 10.1 An ideal spring obeys the equation $F_x^{\text{Applied}} = kx$, where F_x^{Applied} is the force applied to the spring, x is the displacement of the spring from its unstrained length, and k is the spring constant.

to the displacement x, or $F_x^{\text{Applied}} \propto x$. As is customary, this proportionality may be converted into an equation by introducing a proportionality constant k:

$$F_x^{\text{Applied}} = kx \qquad\qquad (10.1)$$

The constant k is called the **spring constant**, and Equation 10.1 shows that it has the dimensions of force per unit length (N/m). A spring that behaves according to $F_x^{\text{Applied}} = kx$ is said to be an **ideal spring**. Example 1 illustrates one application of such a spring.

EXAMPLE 1 | The Physics of a Tire Pressure Gauge

When a tire pressure gauge is pressed against a tire valve, as in **Figure 10.2**, the air in the tire pushes against a plunger attached to a spring. Suppose the spring constant of the spring is $k = 320$ N/m and the bar indicator of the gauge extends 2.0 cm when the gauge is pressed against the tire valve. What force does the air in the tire apply to the spring?

Reasoning We assume that the spring is an ideal spring, so that the relation $F_x^{\text{Applied}} = kx$ is obeyed. The spring constant k is known, as is the displacement x. Therefore, we can determine the force applied to the spring.

Solution The force needed to compress the spring is given by Equation 10.1 as

$$F_x^{\text{Applied}} = kx = (320 \text{ N/m})(0.020 \text{ m}) = \boxed{6.4 \text{ N}}$$

Thus, the exposed length of the bar indicator indicates the force that the air pressure in the tire exerts on the spring. We will see later that pressure is force per unit area, so force is pressure times area. Since the area of the plunger surface is fixed, the bar indicator can be marked in units of pressure.

FIGURE 10.2 In a tire pressure gauge, the pressurized air from the tire exerts a force F_x^{Applied} that compresses a spring.

Sometimes the spring constant k is referred to as the **stiffness** of the spring, because a large value for k means the spring is "stiff," in the sense that a large force is required to stretch or compress it. Conceptual Example 2 examines what happens to the stiffness of a spring when the spring is cut into two shorter pieces.

CONCEPTUAL EXAMPLE 2 | Are Shorter Springs Stiffer Springs?

Figure 10.3a shows a 10-coil spring that has a spring constant k. When this spring is cut in half, so there are two 5-coil springs, is the spring constant of each of the shorter springs **(a)** $\frac{1}{2}k$ or **(b)** $2k$?

Reasoning When the length of a spring is increased or decreased, the change in length is distributed over the entire spring. Greater forces are required to cause changes that are a greater fraction of the spring's initial length, since such changes distort the atomic structure of the spring material to a greater extent.

Answer (a) is incorrect. When a force F_x^{Applied} is applied to the 10-coil spring, as in **Figure 10.3a**, the displacement of the spring from its unstrained length is x. If this same force were applied to a 5-coil spring that had a spring constant of $\frac{1}{2}k$, the displacement would be $2x$ because the spring is only half as stiff. Since this displacement would be a larger fraction of the length of the 5-coil spring, it would require a force greater than F_x^{Applied}, not equal to F_x^{Applied}. Therefore, the spring constant cannot be $\frac{1}{2}k$.

Answer (b) is correct. As indicated in **Figure 10.3a**, the displacement of the spring from its unstrained length is x when a force F_x^{Applied} is applied to the 10-coil spring. **Figure 10.3b** shows the spring divided in half between the fifth and sixth coils (counting from the right). The spring is in equilibrium, so the net force acting on the right half (coils 1–5) must be zero. Thus, as part b shows, a force of $-F_x^{\text{Applied}}$ must act on coil 5 in order to balance the force F_x^{Applied} that acts on coil 1. It is the adjacent coil 6 that exerts the force $-F_x^{\text{Applied}}$, and Newton's action–reaction law now comes into play. It tells us that coil 5, in response, exerts an oppositely directed force of equal magnitude on coil 6. In other words,

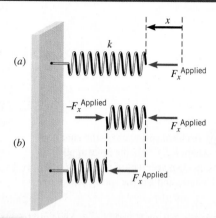

(a)

(b)

FIGURE 10.3 (a) The 10-coil spring has a spring constant k. The applied force is F_x^{Applied}, and the displacement of the spring from its unstrained length is x. (b) The spring in part a is divided in half, so that the forces acting on the two 5-coil springs can be analyzed.

the force F_x^{Applied} is also exerted on the left half of the spring, as part b also indicates. As a result, the left half compresses by an amount that is one-half the displacement x experienced by the 10-coil spring. We conclude, then, that the 5-coil spring must be twice as stiff as the 10-coil spring. In general, the spring constant is inversely proportional to the number of coils in the spring, so shorter springs are stiffer springs, other things being equal.

Related Homework: Problems 10, 18

To stretch or compress a spring, a force must be applied to it. In accord with Newton's third law, the spring exerts an oppositely directed force of equal magnitude. This reaction force is applied by the spring to the agent that does the pulling or pushing. In other words, the reaction force is applied to the object attached to the spring. The reaction force is also called a "restoring force," for a reason that will be clarified shortly. The restoring force of an ideal spring is obtained from the relation $F_x^{\text{Applied}} = kx$ by including the minus sign required by Newton's action–reaction law, as indicated in Equation 10.2.

> **HOOKE'S LAW* RESTORING FORCE OF AN IDEAL SPRING**
>
> **The restoring force of an ideal spring is**
>
> $$F_x = -kx \qquad (10.2)$$
>
> **where k is the spring constant and x is the displacement of the spring from its unstrained length. The minus sign indicates that the restoring force always points in a direction opposite to the displacement of the spring from its unstrained length.**

In Chapter 4 we encountered four types of forces: the gravitational force, the normal force, frictional forces, and the tension force. These forces can contribute to the net external force, which Newton's second law relates to the mass and acceleration of an object. The restoring force of a spring can also contribute to the net external force. Once again, we see the unifying theme of Newton's second law, in that individual forces contribute to the net force, which, in turn, is responsible for the acceleration. Newton's second law plays a central role in describing the motion of objects attached to springs.

---

*As we will see in Section 10.8, Equation 10.2 is similar to a relationship first discovered by Robert Hooke (1635–1703).

Interactive Figure 10.4 helps to explain why the phrase "restoring force" is used. In the picture, an object of mass m is attached to a spring on a frictionless table. In part A, the spring has been stretched to the right, so it exerts the leftward-pointing force F_x. When the object is released, this force pulls it to the left, restoring it toward its equilibrium position. However, consistent with Newton's first law, the moving object has inertia and coasts beyond the equilibrium position, compressing the spring as in part B. The force exerted by the spring now points to the right and, after bringing the object to a momentary halt, acts to restore the object to its equilibrium position. But the object's inertia again carries it beyond the equilibrium position, this time stretching the spring and leading to the restoring force F_x shown in part C. The back-and-forth motion illustrated in the drawing then repeats itself, continuing forever, since no friction acts on the object or the spring.

When the restoring force has the mathematical form given by $F_x = -kx$, *the type of friction-free motion illustrated in* **Interactive Figure 10.4** *is designated as "simple harmonic motion."*

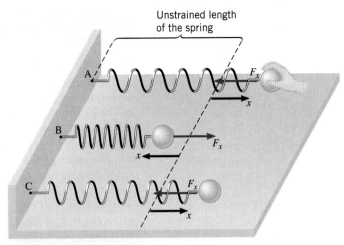

INTERACTIVE FIGURE 10.4 The restoring force F_x (see blue arrows) produced by an ideal spring always points opposite to the displacement x (see black arrows) of the spring and leads to a back-and-forth motion of the object.

By attaching a pen to the object and moving a strip of paper past it at a steady rate, we can record the position of the vibrating object as time passes. **Figure 10.5** illustrates the resulting graphical record of simple harmonic motion. The maximum excursion from equilibrium is the **amplitude** A of the motion. The shape of this graph is characteristic of simple harmonic motion and is called "sinusoidal," because it has the shape of a trigonometric sine or cosine function.

The restoring force also leads to simple harmonic motion when the object is attached to a vertical spring, just as it does when the spring is horizontal. When the spring is vertical, however, the weight of the object causes the spring to stretch, and the motion occurs with respect to the equilibrium position of the object on the stretched spring, as **Animated Figure 10.6** indicates. The amount of initial stretching d_0 due to the weight can be calculated by equating the weight to the magnitude of the restoring force that supports it; thus, $mg = kd_0$, which gives $d_0 = mg/k$.

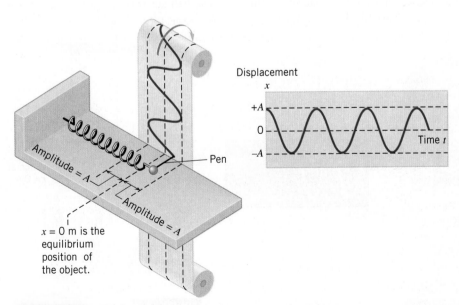

FIGURE 10.5 When an object moves in simple harmonic motion, a graph of its position as a function of time has a sinusoidal shape with an amplitude A. A pen attached to the object records the graph.

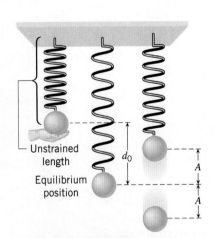

ANIMATED FIGURE 10.6 The weight of an object on a vertical spring stretches the spring by an amount d_0. Simple harmonic motion of amplitude A occurs with respect to the equilibrium position of the object on the stretched spring.

Check Your Understanding

(The answers are given at the end of the book.)

1. A steel ball is dropped onto a very hard floor. Over and over again, the ball rebounds to its original height (assuming that no energy is lost during the collision with the floor). Is the motion of the ball simple harmonic motion?

2. CYU Figure 10.1 shows identical springs that are attached to a box in two different ways. Initially, the springs are unstrained. The box is then pulled to the right and released. In each case the initial displacement of the box is the same. At the moment of release, which box, if either, experiences the greater net force due to the springs?

CYU FIGURE 10.1

3. A 0.42-kg block is attached to the end of a horizontal ideal spring and rests on a frictionless surface. The block is pulled so that the spring stretches for 2.1 cm relative to its unstrained length. When the block is released, it moves with an acceleration of 9.0 m/s². What is the spring constant of the spring?

4. Two people pull on a horizontal spring that is attached to an immovable wall. Then, they detach it from the wall and pull on opposite ends of the horizontal spring. They pull just as hard in each case. In which situation, if either, does the spring stretch more?

10.2 Simple Harmonic Motion and the Reference Circle

Simple harmonic motion can be described in terms of displacement, velocity, and acceleration, and the model in **Figure 10.7** is helpful in explaining these characteristics. The model consists of a small ball attached to the top of a rotating turntable. The ball is moving in uniform circular motion (see Section 5.1) on a path known as the **reference circle**. As the ball moves, its shadow falls on a strip of film, which is moving upward at a steady rate and recording where the shadow is. A comparison of the film with the paper in **Figure 10.5** reveals the same kind of patterns, suggesting that the model is useful.

Displacement

Figure 10.8 shows the reference circle (radius = A) and indicates how to determine the displacement of the shadow on the film. The ball starts on the x axis at $x = +A$

FIGURE 10.7 The ball mounted on the turntable moves in uniform circular motion, and its shadow, projected on a moving strip of film, executes simple harmonic motion.

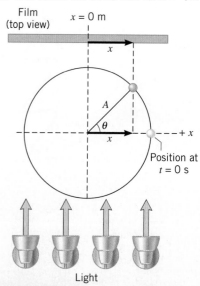

FIGURE 10.8 A top view of a ball on a turntable. The ball's shadow on the film has a displacement x that depends on the angle θ through which the ball has moved on the reference circle.

and moves through the angle θ in a time t. The circular motion is uniform, so the ball moves with a constant angular speed ω (in rad/s), and the angle has a value (in rad) of $\theta = \omega t$. The displacement x of the shadow is just the projection of the radius A onto the x axis:

$$x = A \cos \theta = A \cos \omega t \qquad (10.3)$$

Figure 10.9 shows a graph of this equation. As time passes, the shadow of the ball oscillates between the values of $x = +A$ and $x = -A$, corresponding to the limiting values of $+1$ and -1 for the cosine of an angle. The radius A of the reference circle, then, is the amplitude of the simple harmonic motion.

As the ball moves one revolution or cycle around the reference circle, its shadow executes one cycle of back-and-forth motion. For any object in simple harmonic motion, the time required to complete one cycle is the **period** T, as **Figure 10.9** indicates. The value of T depends on the angular speed ω of the ball because the greater the angular speed, the shorter the time it takes to complete one revolution. We can obtain the relationship between ω and T by recalling that $\omega = \Delta\theta/\Delta t$ (Equation 8.2), where $\Delta\theta$ is the angular displacement of the ball and Δt is the time interval. For one cycle, $\Delta\theta = 2\pi$ rad and $\Delta t = T$, so that

$$\omega = \frac{2\pi}{T} \qquad (\omega \text{ in rad/s}) \qquad (10.4)$$

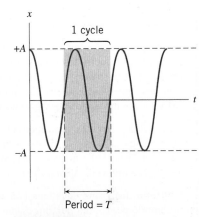

FIGURE 10.9 For simple harmonic motion, the graph of displacement x versus time t is a sinusoidal curve. The period T is the time required for one complete motional cycle.

Instead of the period, we often speak of the **frequency** f of the motion, the frequency being the number of cycles of the motion per second. For example, if an object on a spring completes 10 cycles in one second, the frequency is $f = 10$ cycles/s. The period T, or the time for one cycle, would be $\frac{1}{10}$ s. Thus, frequency and period are related according to

$$f = \frac{1}{T} \qquad (10.5)$$

Usually, one cycle per second is referred to as one hertz (Hz), the unit being named after Heinrich Hertz (1857–1894). One thousand cycles per second is called one kilohertz (kHz). Thus, five thousand cycles per second, for instance, can be written as 5 kHz.

Using the relationships $\omega = 2\pi/T$ and $f = 1/T$, we can relate the angular speed ω (in rad/s) to the frequency f (in cycles/s or Hz):

$$\omega = \frac{2\pi}{T} = 2\pi f \qquad (\omega \text{ in rad/s}) \qquad (10.6)$$

Because ω is directly proportional to the frequency f, ω is often called the **angular frequency**.

Velocity

The reference circle model can also be used to determine the velocity of an object in simple harmonic motion. **Figure 10.10** shows the tangential velocity $\vec{v}_T$ of the ball on the reference circle. The drawing indicates that the velocity $\vec{v}_x$ of the shadow is just the x component of the vector $\vec{v}_T$; that is, $v_x = -v_T \sin\theta$, where $\theta = \omega t$. The minus sign is necessary because $\vec{v}_x$ points to the left, in the direction of the negative x axis. Since the tangential speed v_T is related to the angular speed ω by $v_T = r\omega$ (Equation 8.9) and since $r = A$, it follows that $v_T = A\omega$. Therefore, the velocity in simple harmonic motion is given by

$$v_x = -v_T \sin\theta = -A\omega \sin\omega t \qquad (\omega \text{ in rad/s}) \qquad (10.7)$$

This velocity is *not* constant, but varies between maximum and minimum values as time passes. When the shadow changes direction at either end of the oscillatory motion, the velocity

FIGURE 10.10 The velocity $\vec{v}_x$ of the ball's shadow is the x component of the tangential velocity $\vec{v}_T$ of the ball on the reference circle.

is momentarily zero. When the shadow passes through the $x = 0$ m position, the velocity has a maximum magnitude of $A\omega$, since the sine of an angle is between $+1$ and -1:

$$v_{max} = A\omega \qquad (\omega \text{ in rad/s}) \qquad (10.8)$$

Both the amplitude A and the angular frequency ω determine the maximum velocity, as Example 3 emphasizes.

EXAMPLE 3 | The Physics of a Loudspeaker Diaphragm

The diaphragm of a loudspeaker moves back and forth in simple harmonic motion to create sound, as in **Figure 10.11**. The frequency of the motion is $f = 1.0$ kHz and the amplitude is $A = 0.20$ mm. **(a)** What is the maximum speed of the diaphragm? **(b)** Where in the motion does this maximum speed occur?

Reasoning The maximum speed of an object vibrating in simple harmonic motion is $v_{max} = A\omega$ (ω in rad/s), according to Equation 10.8. The angular frequency ω is related to the frequency f by $\omega = 2\pi f$, according to Equation 10.6.

Solution **(a)** Using Equations 10.8 and 10.6, we find that the maximum speed of the vibrating diaphragm is

$$v_{max} = A\omega = A(2\pi f) = (0.20 \times 10^{-3}\,\text{m})(2\pi)(1.0 \times 10^3\,\text{Hz})$$
$$= \boxed{1.3\,\text{m/s}}$$

(b) The speed of the diaphragm is zero when the diaphragm momentarily comes to rest at either end of its motion: $x = +A$ and

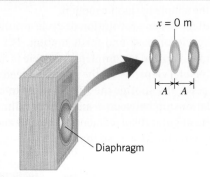

$x = -A$. Its maximum speed occurs midway between these two positions, or at $x = 0$ m.

FIGURE 10.11 The diaphragm of a loudspeaker generates a sound by moving back and forth in simple harmonic motion.

Simple harmonic motion is not just any kind of vibratory motion. It is a very specific kind and, among other things, must have the velocity given by Equation 10.7. For instance, advertising signs often use a "moving light" display to grab your attention. Conceptual Example 4 examines the back-and-forth motion in one such display, to see whether it is simple harmonic motion.

CONCEPTUAL EXAMPLE 4 | Moving Lights

Over the entrance to a restaurant is mounted a strip of equally spaced light bulbs, as **Figure 10.12a** illustrates. Starting at the left end, each bulb turns on in sequence for one-half second. Thus, a lighted bulb appears to move from left to right. After the last bulb on the right turns on, the apparent motion reverses. The lighted bulb then appears to move to the left, as part b of the drawing indicates. As a result, the lighted bulb appears to oscillate back and forth. Is the apparent motion simple harmonic motion? **(a)** No **(b)** Yes

Reasoning In simple harmonic motion the velocity of the moving object must have the velocity specified by $v = -A\omega \sin \omega t$ (Equation 10.7). This velocity is not constant as time passes.

Answer (b) is incorrect. Since the bulbs are equally spaced and each one remains lit for the same amount of time, the apparent motion in **Figure 10.12a** (or in **Figure 10.12b**) occurs at a constant velocity and, therefore, cannot be simple harmonic motion.

Answer (a) is correct. The apparent motion is not simple harmonic motion. If it were, the speed would be zero at each end of

(a)

(b)

FIGURE 10.12 The motion of a lighted bulb is from (a) left to right and then from (b) right to left.

the sign and would increase to a maximum speed at the center, consistent with Equation 10.7. However, the speed is constant because the bulbs are equally spaced and each remains on for the same amount of time.

Acceleration

In simple harmonic motion, the velocity is not constant; consequently, there must be an acceleration. This acceleration can also be determined with the aid of the reference-circle model. As **Figure 10.13** shows, the ball on the reference circle moves in uniform circular motion, and, therefore, has a centripetal acceleration $\vec{a}_c$ that points toward the center of the circle. The acceleration $\vec{a}_x$ of the shadow is the x component of the centripetal acceleration; $a_x = -a_c \cos \theta$. The minus sign is needed because the acceleration of the shadow points to the left. Recalling that the centripetal acceleration is related to the angular speed ω by $a_c = r\omega^2$ (Equation 8.11) and using $r = A$, we find that $a_c = A\omega^2$. With this substitution and the fact that $\theta = \omega t$, the acceleration in simple harmonic motion becomes

$$a_x = -a_c \cos \theta = -A\omega^2 \cos \omega t \quad (\omega \text{ in rad/s}) \tag{10.9}$$

The acceleration, like the velocity, does *not* have a constant value as time passes. The maximum magnitude of the acceleration is

$$a_{max} = A\omega^2 \quad (\omega \text{ in rad/s}) \tag{10.10}$$

Although both the amplitude A and the angular frequency ω determine the maximum value, the frequency has a particularly strong effect, because it is squared. Example 5 shows that the acceleration can be remarkably large in a practical situation.

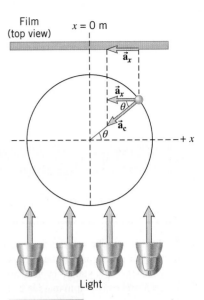

FIGURE 10.13 The acceleration $\vec{a}_x$ of the ball's shadow is the x component of the centripetal acceleration $\vec{a}_c$ of the ball on the reference circle.

EXAMPLE 5 | The Loudspeaker Revisited—The Maximum Acceleration

The loudspeaker diaphragm in **Figure 10.11** is vibrating at a frequency of $f = 1.0$ kHz, and the amplitude of the motion is $A = 0.20$ mm.
(a) What is the maximum acceleration of the diaphragm, and
(b) where does this maximum acceleration occur?

Reasoning The maximum acceleration of an object vibrating in simple harmonic motion is $a_{max} = A\omega^2$ (ω in rad/s), according to Equation 10.10. Equation 10.6 shows that the angular frequency ω is related to the frequency f by $\omega = 2\pi f$.

Solution **(a)** Using Equations 10.10 and 10.6, we find that the maximum acceleration of the vibrating diaphragm is

$$a_{max} = A\omega^2 = A(2\pi f)^2 = (0.20 \times 10^{-3} \text{m})[2\pi(1.0 \times 10^3 \text{Hz})]^2$$

$$= \boxed{7.9 \times 10^3 \text{m/s}^2}$$

This is an incredible acceleration, being more than 800 times the acceleration due to gravity, and the diaphragm must be built to withstand it.

(b) The maximum acceleration occurs when the force acting on the diaphragm is a maximum. The maximum force arises when the diaphragm is at the ends of its path, where the displacement is greatest. Thus, the maximum acceleration occurs at $x = +A$ and $x = -A$ in **Figure 10.11**.

Math Skills The angular frequency ω and the vibration frequency f are not the same thing, so you will need to be careful to distinguish between them when solving problems. The two quantities are proportional, but they differ by a factor of 2π. The relation between them is

$$\omega = 2\pi f \tag{10.6}$$

You can always tell which quantity is being referred to in a problem by looking at the given data. A value for ω is given in radians per second, whereas a value for f is given in hertz (cycles per second).

Frequency of Vibration

With the aid of Newton's second law ($\Sigma F_x = ma_x$), it is possible to determine the frequency at which an object of mass m vibrates on a spring. We assume that the mass of the spring itself is negligible and that the only force acting on the object in the horizontal direction is due to the spring—that is, the Hooke's law restoring force. Thus, the net force is $\Sigma F_x = -kx$, and Newton's second law becomes $-kx = ma_x$, where a_x is the acceleration of the object. The displacement and acceleration of an oscillating spring are, respectively, $x = A \cos \omega t$ (Equation 10.3) and $a_x = -A\omega^2 \cos \omega t$ (Equation 10.9). Substituting these expressions for x and a_x into the relation $-kx = ma_x$, we find that

$$-k(A \cos \omega t) = m(-A\omega^2 \cos \omega t)$$

which yields

$$\omega = \sqrt{\frac{k}{m}} \qquad (\omega \text{ in rad/s}) \qquad \text{(10.11)}$$

In this expression, the angular frequency ω must be in radians per second. Larger spring constants k and smaller masses m result in larger frequencies. Example 6 illustrates an application of Equation 10.11.

Analyzing Multiple-Concept Problems

EXAMPLE 6 | BIO The Physics of a Body-Mass Measurement Device

Astronauts who spend a long time in orbit measure their body masses as part of their health-maintenance programs. On earth, it is simple to measure body weight W with a scale and convert it to mass m using the magnitude g of the acceleration due to gravity, since $W = mg$. However, this procedure does not work in orbit, because both the scale and the astronaut are in free fall and cannot press against each other (see Conceptual Example 13 in Chapter 5). Instead, astronauts use a body-mass measurement device, as **Figure 10.14** illustrates. The device consists of a spring-mounted chair in which the astronaut sits. The chair is then started oscillating in simple harmonic motion. The period of the motion is measured electronically and is automatically converted into a value of the astronaut's mass, after the mass of the chair is taken into account. The spring used in one such device has a spring constant of 606 N/m, and the mass of the chair is 12.0 kg. The measured oscillation period is 2.41 s. Find the mass of the astronaut.

Reasoning Since the astronaut and chair are oscillating in simple harmonic motion, the total mass m_{total} of

the two is related to the angular frequency ω of the motion by $\omega = \sqrt{k/m_{\text{total}}}$, or $m_{\text{total}} = k/\omega^2$. The angular frequency, in turn, is related to the period of the motion by $\omega = 2\pi/T$. These two relations will enable us to find the total mass of the astronaut and chair. From this result, we will subtract the mass of the chair (m_{chair}) to obtain the mass of the astronaut (m_{astro}).

Knowns and Unknowns The data for this problem are:

Description	Symbol	Value
Spring constant	k	606 N/m
Mass of chair	m_{chair}	12.0 kg
Period of oscillation	T	2.41 s
Unknown Variable		
Mass of astronaut	m_{astro}	?

Courtesy NASA

FIGURE 10.14 Astronaut Millie Hughes-Fulford trains to use a body-mass measurement device developed for determining mass in orbit.

Modeling the Problem

STEP 1 Angular Frequency of Vibration The mass m_{astro} of the astronaut is equal to the total mass m_{total} of the astronaut and chair minus the mass m_{chair} of the chair:

$$m_{\text{astro}} = m_{\text{total}} - m_{\text{chair}} \qquad \text{(1)}$$

The mass of the chair is known. Since the astronaut and chair oscillate in simple harmonic motion, the angular frequency ω of the oscillation is related to the spring constant k of the spring and the total mass m_{total} by Equation 10.11:

$$\omega = \sqrt{\frac{k}{m_{\text{total}}}} \quad \text{or} \quad m_{\text{total}} = \frac{k}{\omega^2}$$

Substituting this result for m_{total} into Equation 1 gives Equation 2 at the right. The spring constant and mass of the chair are known, but the angular frequency ω is not; we will evaluate it in Step 2.

$$m_{\text{astro}} = \frac{k}{\omega^2} - m_{\text{chair}} \qquad \text{(2)}$$

?

STEP 2 Angular Frequency and Period   The angular frequency ω of the oscillating motion is inversely related to its period T by

$$\boxed{\omega = \frac{2\pi}{T}}$$ (10.4)

All the variables on the right side of this equation are known, and we substitute it into Equation 2, as indicated in the right column.

$$m_{astro} = \frac{k}{\omega^2} - m_{chair}$$ (2)

$$\boxed{\omega = \frac{2\pi}{T}}$$ (10.4)

Solution   Algebraically combining the results of the steps above, we have:

$$\underset{\text{STEP 1}}{\downarrow} \qquad \underset{\text{STEP 2}}{\downarrow}$$

$$m_{astro} = \frac{k}{\omega^2} - m_{chair} = \frac{k}{\left(\frac{2\pi}{T}\right)^2} - m_{chair} = \frac{kT^2}{4\pi^2} - m_{chair}$$

$$= \frac{(606 \text{ N/m})(2.41 \text{ s})^2}{4\pi^2} - 12.0 \text{ kg} = \boxed{77.2 \text{ kg}}$$

Related Homework: Problem 68

THE PHYSICS OF . . . detecting and measuring small amounts of chemicals.   Example 6 indicates that the mass of the vibrating object influences the frequency of simple harmonic motion. Electronic sensors are being developed that take advantage of this effect in detecting and measuring small amounts of chemicals. These sensors utilize tiny quartz crystals that vibrate when an electric current passes through them. If a crystal is coated with a substance that absorbs a particular chemical, then the mass of the coated crystal increases as the chemical is absorbed and, according to the relation $f = \frac{1}{2\pi}\sqrt{k/m}$ (Equations 10.6 and 10.11), the frequency of the simple harmonic motion decreases. The change in frequency is detected electronically, and the sensor is calibrated to give the mass of the absorbed chemical.

Check Your Understanding

(The answers are given at the end of the book.)

5. CYU Figure 10.2 shows plots of the displacement x versus the time t for three objects undergoing simple harmonic motion. Which object—I, II, or III—has the greatest maximum velocity?

6. In **Figure 10.13** the shadow moves in simple harmonic motion. Where on the path of the shadow is the acceleration equal to zero?

7. A particle is oscillating in simple harmonic motion. The time required for the particle to travel through one complete cycle is equal to the period of the motion, no matter what the amplitude is. But how can this be, since larger amplitudes mean that the particle travels farther?

CYU FIGURE 10.2

FIGURE 10.15 A door-closing unit. The elastic potential energy stored in the compressed spring is used to close the door.

FIGURE 10.16 When the object is released, its displacement changes from an initial value of x_0 to a final value of x_f.

10.3 | Energy and Simple Harmonic Motion

We saw in Chapter 6 that an object above the surface of the earth has gravitational potential energy. Therefore, when the object is allowed to fall, like the hammer of the pile driver in Figure 6.13, it can do work. A spring also has potential energy when the spring is stretched or compressed, which we refer to as **elastic potential energy**. Because of elastic potential energy, a stretched or compressed spring can do work on an object that is attached to the spring.

THE PHYSICS OF . . . a door-closing unit. For instance, **Figure 10.15** shows a door-closing unit that is often found on screen doors. When the door is opened, a spring inside the unit is compressed and has elastic potential energy. When the door is released, the compressed spring expands and does the work of closing the door.

To find an expression for the elastic potential energy, we will determine the work done by the spring force on an object. **Figure 10.16** shows an object attached to one end of a stretched spring. When the object is released, the spring contracts and pulls the object from its initial position x_0 to its final position x_f. The work W done by a constant force is given by Equation 6.1 as $W = (F \cos \theta)s$, where F is the magnitude of the force, s is the magnitude of the displacement ($s = x_0 - x_f$), and θ is the angle between the force and the displacement. The magnitude of the spring force is not constant, however. Equation 10.2 gives the spring force as $F_x = -kx$, and as the spring contracts, the magnitude of this force changes from kx_0 to kx_f. In using Equation 6.1 to determine the work, we can account for the changing magnitude by using an average magnitude $\overline{F}_x$ in place of the constant magnitude F_x. Because the dependence of the spring force on x is linear, the magnitude of the average force is just one-half the sum of the initial and final values, or $\overline{F}_x = \frac{1}{2}(kx_0 + kx_f)$. The work W_{elastic} done by the average spring force is, then,

$$W_{\text{elastic}} = (\overline{F}_x \cos \theta)s = \tfrac{1}{2}(kx_0 + kx_f)(\cos 0°)(x_0 - x_f)$$

$$W_{\text{elastic}} = \underbrace{\tfrac{1}{2}kx_0^2}_{\substack{\text{Initial elastic} \\ \text{potential energy}}} - \underbrace{\tfrac{1}{2}kx_f^2}_{\substack{\text{Final elastic} \\ \text{potential energy}}} \tag{10.12}$$

In the calculation above, θ is 0°, since the spring force has the same direction as the displacement. Equation 10.12 indicates that the work done by the spring force is equal to the difference between the initial and final values of the quantity $\frac{1}{2}kx^2$. The quantity $\frac{1}{2}kx^2$ is analogous to the quantity mgh, which we identified in Section 6.3 as the gravitational potential energy. Here, the quantity $\frac{1}{2}kx^2$ is the elastic potential energy. Equation 10.13 indicates that the elastic potential energy is a maximum for a fully stretched or compressed spring and zero for a spring that is neither stretched nor compressed ($x = 0$ m).

DEFINITION OF ELASTIC POTENTIAL ENERGY

The elastic potential energy PE_{elastic} is the energy that a spring has by virtue of being stretched or compressed. For an ideal spring that has a spring constant k and is stretched or compressed by an amount x relative to its unstrained length, the elastic potential energy is

$$PE_{\text{elastic}} = \tfrac{1}{2}kx^2 \tag{10.13}$$

SI Unit of Elastic Potential Energy: joule (J)

The total mechanical energy E is a familiar idea that we originally defined to be the sum of the translational kinetic energy and the gravitational potential energy (see Section 6.5). Then, we included the rotational kinetic energy (see Section 9.5). We now expand the total mechanical energy to include the elastic potential energy, as shown in Equation 10.14.

$$\underbrace{E}_{\substack{\text{Total} \\ \text{mechanical} \\ \text{energy}}} = \underbrace{\tfrac{1}{2}mv^2}_{\substack{\text{Translational} \\ \text{kinetic} \\ \text{energy}}} + \underbrace{\tfrac{1}{2}I\omega^2}_{\substack{\text{Rotational} \\ \text{kinetic} \\ \text{energy}}} + \underbrace{mgh}_{\substack{\text{Gravitational} \\ \text{potential} \\ \text{energy}}} + \underbrace{\tfrac{1}{2}kx^2}_{\substack{\text{Elastic} \\ \text{potential} \\ \text{energy}}} \tag{10.14}$$

As Section 6.5 discusses, the total mechanical energy is conserved when external nonconservative forces (such as friction) do no net work; that is, when $W_{nc} = 0$ J. Then, the final and initial values of E are the same: $E_f = E_0$. The principle of conservation of total mechanical energy is the subject of the next example.

EXAMPLE 7 | An Object on a Horizontal Spring

Interactive Figure 10.17 shows an object of mass $m = 0.200$ kg that is vibrating on a horizontal frictionless table. The spring has a spring constant of $k = 545$ N/m. The spring is stretched initially to $x_0 = 4.50$ cm and is then released from rest (see part A of the drawing). Determine the final translational speed v_f of the object when the final displacement of the spring is **(a)** $x_f = 2.25$ cm and **(b)** $x_f = 0$ cm.

Reasoning The conservation of mechanical energy indicates that, in the absence of friction (a nonconservative force), the final and initial total mechanical energies are the same:

$$E_f = E_0$$

$$\tfrac{1}{2}mv_f^2 + \tfrac{1}{2}I\omega_f^2 + mgh_f + \tfrac{1}{2}kx_f^2 = \tfrac{1}{2}mv_0^2 + \tfrac{1}{2}I\omega_0^2 + mgh_0 + \tfrac{1}{2}kx_0^2$$

Since the object is moving on a horizontal table, the final and initial heights are the same: $h_f = h_0$. The object is not rotating, so its angular speed is zero: $\omega_f = \omega_0 = 0$ rad/s. Also, as the problem states, the initial translational speed of the object is zero, $v_0 = 0$ m/s. With these substitutions, the conservation-of-energy equation becomes

$$\tfrac{1}{2}mv_f^2 + \tfrac{1}{2}kx_f^2 = \tfrac{1}{2}kx_0^2$$

from which we can obtain v_f:

$$v_f = \sqrt{\frac{k}{m}(x_0^2 - x_f^2)}$$

Solution **(a)** Since $x_0 = 0.0450$ m and $x_f = 0.0225$ m, the final translational speed is

$$v_f = \sqrt{\frac{545 \text{ N/m}}{0.200 \text{ kg}}[(0.0450 \text{ m})^2 - (0.0225 \text{ m})^2]} = \boxed{2.03 \text{ m/s}}$$

The total mechanical energy at this point is composed partly of translational kinetic energy ($\tfrac{1}{2}mv_f^2 = 0.414$ J) and partly of elastic potential energy ($\tfrac{1}{2}kx_f^2 = 0.138$ J), as indicated in part B of

Interactive Figure 10.17. The total mechanical energy E is the sum of these two energies: $E = 0.414$ J $+ 0.138$ J $= 0.552$ J. Because the total mechanical energy remains constant during the motion, this value equals the initial total mechanical energy when the object is stationary and the energy is entirely elastic potential energy ($E_0 = \tfrac{1}{2}kx_0^2 = 0.552$ J).

(b) When $x_0 = 0.0450$ m and $x_f = 0$ m, we have

$$v_f = \sqrt{\frac{k}{m}(x_0^2 - x_f^2)} = \sqrt{\frac{545 \text{ N/m}}{0.200 \text{ kg}}[(0.0450 \text{ m})^2 - (0 \text{ m})^2]}$$

$$= \boxed{2.35 \text{ m/s}}$$

Now the total mechanical energy is due entirely to the translational kinetic energy ($\tfrac{1}{2}mv_f^2 = 0.552$ J), since the elastic potential energy is zero (see part C of **Interactive Figure 10.17**). Note that the total mechanical energy is the same as it is in Solution part (a). In the absence of friction, the simple harmonic motion of a spring converts the different types of energy between one form and another, the total always remaining the same.

INTERACTIVE FIGURE 10.17 The total mechanical energy of this system is entirely elastic potential energy (A), partly elastic potential energy and partly kinetic energy (B), and entirely kinetic energy (C).

Conceptual Example 8 takes advantage of energy conservation to illustrate what happens to the maximum speed, amplitude, and angular frequency of a simple harmonic oscillator when its mass is changed suddenly at a certain point in the motion.

CONCEPTUAL EXAMPLE 8 | Changing the Mass of a Simple Harmonic Oscillator

Figure 10.18a shows a box of mass m attached to a spring that has a force constant k. The box rests on a horizontal, frictionless surface. The spring is initially stretched to $x = A$ and then released from rest. The box executes simple harmonic motion that is characterized by a maximum speed v_x^{max}, an amplitude A, and an angular frequency ω. When the box is passing through the point where the spring is unstrained ($x = 0$ m), a second box of

the same mass m and speed v_x^{max}, is attached to it, as in part b of the drawing. Discuss what happens to **(a)** the maximum speed, **(b)** the amplitude, and **(c)** the angular frequency of the subsequent simple harmonic motion.

Reasoning and Solution **(a)** The maximum speed of an object in simple harmonic motion occurs when the object is passing through the point where the spring is unstrained ($x = 0$ m), as in **Figure 10.18b**. Since the second box is attached at this point with the same speed, ***the maximum speed of the two-box system remains the same as that of the one-box system.***

(b) At the same speed, the maximum kinetic energy of the two boxes is twice that of a single box, since the mass is twice as much. Subsequently, when the two boxes move to the left and compress the spring, their kinetic energy is converted into elastic potential energy. Since the two boxes have twice as much kinetic energy as one box alone, the two will have twice as much elastic potential energy when they come to a halt at the extreme left. Here, we are using the principle of conservation of mechanical energy, which applies since friction is absent. But the elastic potential energy is proportional to the amplitude squared (A^2) of the motion, ***so the amplitude of the two-box system is $\sqrt{2}$ times as great as that of the one-box system.***

(c) The angular frequency ω of a simple harmonic oscillator is $\omega = \sqrt{k/m}$ (Equation 10.11). Since the mass of the two-box system is twice the mass of the one-box system, ***the angular frequency of the two-box system is $\sqrt{2}$ times as small as that of the one-box system.***

(a)

(b)

FIGURE 10.18 (a) A box of mass m, starting from rest at $x = A$, undergoes simple harmonic motion about $x = 0$ m. (b) When $x = 0$ m, a second box, with the same mass and speed, is attached.

In the previous two examples, gravitational potential energy plays no role because the spring is horizontal. The next example illustrates that gravitational potential energy must be taken into account when a spring is oriented vertically.

EXAMPLE 9 | A Falling Ball on a Vertical Spring

A 0.20-kg ball is attached to a vertical spring, as in **Figure 10.19**. The spring constant of the spring is 28 N/m. The ball, supported initially so that the spring is neither stretched nor compressed, is released from rest. In the absence of air resistance, how far does the ball fall before being brought to a momentary stop by the spring?

Reasoning Since air resistance is absent, only the conservative forces of gravity and the spring act on the ball. Therefore, the principle of conservation of mechanical energy applies:

$$E_f = E_0$$

$$\tfrac{1}{2}mv_f^2 + \tfrac{1}{2}I\omega_f^2 + mgh_f + \tfrac{1}{2}ky_f^2 = \tfrac{1}{2}mv_0^2 + \tfrac{1}{2}I\omega_0^2 + mgh_0 + \tfrac{1}{2}ky_0^2$$

Note that we have replaced x with y in the elastic-potential-energy terms ($\tfrac{1}{2}ky_f^2$ and $\tfrac{1}{2}ky_0^2$), in recognition of the fact that the spring is moving in the vertical direction. The problem states that the final and initial translational speeds of the ball are zero: $v_f = v_0 = 0$ m/s. The ball and spring do not rotate; therefore, the final and initial angular speeds are also zero: $\omega_f = \omega_0 = 0$ rad/s. As **Figure 10.19** indicates, the initial height of the ball is h_0, and the final height is $h_f = 0$ m. In addition, the spring is unstrained ($y_0 = 0$ m) to begin with, and so it has no elastic potential energy initially. With these substitutions, the conservation-of-mechanical-energy equation reduces to

$$\tfrac{1}{2}ky_f^2 = mgh_0$$

This result shows that the initial gravitational potential energy (mgh_0) is converted into elastic potential energy ($\tfrac{1}{2}ky_f^2$). When the ball falls to its lowest point, its displacement is $y_f = -h_0$, where the minus sign indicates that the displacement is downward.

FIGURE 10.19 The ball is supported initially so that the spring is unstrained. After being released from rest, the ball falls through the distance h_0 before being momentarily stopped by the spring.

Substituting this result into the equation above and solving for h_0 yields $h_0 = 2mg/k$.

> **Problem-Solving Insight** When evaluating the total mechanical energy E, always include a potential energy term for every conservative force acting on the system. In Example 9 there are two such terms, gravitational and elastic.

Solution The distance that the ball falls before coming to a momentary halt is

$$h_0 = \frac{2mg}{k} = \frac{2(0.20 \text{ kg})(9.8 \text{ m/s}^2)}{28 \text{ N/m}} = \boxed{0.14 \text{ m}}$$

Check Your Understanding

(The answers are given at the end of the book.)

8. Is more elastic potential energy stored in a spring when the spring is compressed by one centimeter than when it is stretched by the same amount?

9. A block is attached to the end of a horizontal ideal spring and rests on a frictionless surface. The block is pulled so that the spring stretches relative to its unstrained length. In each of the following three cases, the spring is stretched initially by the same amount. Rank the amplitudes of the resulting simple harmonic motion in decreasing order (largest first). **(a)** The block is released from rest. **(b)** The block is given an initial speed v_0. **(c)** The block is given an initial speed $\frac{1}{2}v_0$.

10. A block is attached to a horizontal spring and slides back and forth on a frictionless horizontal surface. A second identical block is suddenly attached to the first block. The attachment is accomplished by joining the blocks at one extreme end of the oscillation cycle. The velocities of the blocks are exactly matched at the instant of joining. How do the **(a)** amplitude, **(b)** frequency, and **(c)** maximum speed of the simple harmonic motion change?

10.4 | The Pendulum

A **simple pendulum** consists of a particle of mass m, attached to a frictionless pivot P by a cable of length L and negligible mass. When the particle is pulled away from its equilibrium position by an angle θ and released, it swings back and forth as **Figure 10.20** shows. By attaching a pen to the bottom of the swinging particle and moving a strip of paper beneath it at a steady rate, we can record the position of the particle as time passes. The graphical record reveals a pattern that is similar (but not identical) to the sinusoidal pattern for simple harmonic motion.

Gravity causes the back-and-forth rotation about the axis at P. The rotation speeds up as the particle approaches the lowest point and slows down on the upward part of the swing. Eventually the angular speed is reduced to zero, and the particle swings back. As Section 9.4 discusses, a net torque is required to change the angular speed. The gravitational force $m\vec{g}$ produces this torque. (The tension $\vec{T}$ in the cable creates no torque, because it points directly at the pivot P and, therefore, has a zero lever arm.) According to Equation 9.1, the magnitude of the torque τ is the product of the magnitude mg of the gravitational force and the lever arm ℓ, so that $\tau = -(mg)\ell$. The minus sign is included since the torque is a restoring torque; it acts to reduce the angle θ [the angle θ is positive (counterclockwise), while the torque is negative (clockwise)]. The lever arm ℓ is the perpendicular distance between the line of action of $m\vec{g}$ and the pivot P. In **Figure 10.20**, ℓ is very nearly equal to the arc length s of the circular path when the angle θ is small (about 10° or less). Furthermore, if θ is expressed in radians, the arc length and the radius L of the circular path are related, according to $s = L\theta$ (Equation 8.1). It follows, then, that $\ell \approx s = L\theta$, and the gravitational torque is

$$\tau \approx \underbrace{-mgL}_{k'}\theta$$

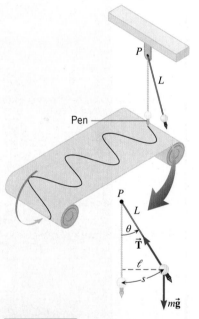

FIGURE 10.20 A simple pendulum swinging back and forth about the pivot P. If the angle θ is small (about 10° or less), the swinging is approximately simple harmonic motion.

In the equation above, the term mgL has a constant value k', independent of θ. *For small angles,* then, the torque that restores the pendulum to its vertical equilibrium position is proportional to the angular displacement θ. The expression $\tau = -k'\theta$ has the same form as the Hooke's law restoring force for an ideal spring, $F = -kx$. Therefore, we expect the frequency of the back-and-forth movement of the pendulum to be given by an equation analogous to Equation 10.11 ($\omega = 2\pi f = \sqrt{k/m}$). In place of the spring constant k, the constant $k' = mgL$ will appear, and, as usual in rotational motion, in place of the mass m, the moment of inertia I will appear:

$$\omega = 2\pi f = \sqrt{\frac{mgL}{I}} \qquad \text{(small angles only)} \qquad \textbf{(10.15)}$$

The moment of inertia of a particle of mass m, rotating at a radius $r = L$ about an axis, is given by $I = mL^2$ (Equation 9.6). Substituting this expression for I into Equation 10.15 reveals that for a simple pendulum

Simple pendulum $\omega = 2\pi f = \sqrt{\dfrac{g}{L}}$ (small angles only)   **(10.16)**

The mass of the particle has been eliminated algebraically from this expression, so only the length L and the magnitude g of the acceleration due to gravity determine the frequency of a simple pendulum. If the angle of oscillation is large, the pendulum does not exhibit simple harmonic motion, and Equation 10.16 does not apply. Equation 10.16 provides the basis for using a pendulum to keep time, as Example 10 demonstrates.

EXAMPLE 10 | Keeping Time

Figure 10.21 shows a clock that uses a pendulum to keep time. Determine the length of a simple pendulum that will swing back and forth in simple harmonic motion with a period of 1.00 s.

Reasoning When a simple pendulum is swinging back and forth in simple harmonic motion, its frequency f is given by Equation 10.16 as $f = \frac{1}{2\pi}\sqrt{g/L}$, where g is the magnitude of the acceleration due to gravity and L is the length of the pendulum. We also

know from Equation 10.5 that the frequency is the reciprocal of the period T, so $f = 1/T$. Thus, the equation above becomes $1/T = \frac{1}{2\pi}\sqrt{g/L}$. We can solve this equation for the length L.

Solution The length of the pendulum is

$$L = \frac{T^2 g}{4\pi^2} = \frac{(1.00\text{ s})^2(9.80\text{ m/s}^2)}{4\pi^2} = \boxed{0.248\text{ m}}$$

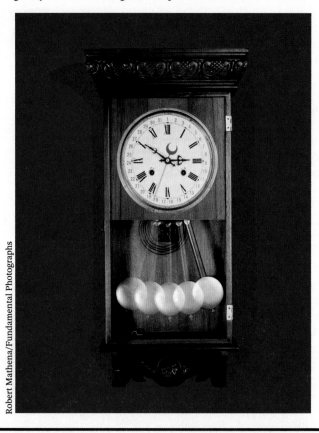

Robert Mathena/Fundamental Photographs

FIGURE 10.21 This pendulum clock keeps time as the pendulum swings back and forth.

It is not necessary that the object in **Figure 10.20** be a particle at the end of a cable. It may be a rigid extended object, in which case the pendulum is called a **physical pendulum**. For small oscillations, Equation 10.15 still applies, but the moment of inertia I is no longer mL^2. The proper value for the rigid object must be used. (See Section 9.4 for a discussion of moment of inertia.) In addition, the length L for a physical pendulum is the distance between the axis at P and the center of gravity of the object. The next example deals with an important type of physical pendulum.

Analyzing Multiple-Concept Problems

EXAMPLE 11 | BIO The Physics of Walking

When we walk, our legs alternately swing forward about the hip joint as a pivot. In this motion the leg is acting approximately as a physical pendulum. Treating the leg as a thin uniform rod of length 0.80 m, find the time it takes for the leg to swing forward.

Reasoning The time it takes for the leg to swing forward is one-half of the period T of the pendulum motion, which is related to the frequency f of the motion by $T = 1/f$ (Equation 10.5). The frequency of a physical pendulum is given by $f = \frac{1}{2\pi}\sqrt{mgL/I}$ (Equation 10.15), where m and I are, respectively, its mass and moment of inertia, and L is the distance

between the pivot at the hip joint and the center of gravity of the leg. By combining these two relations we will be able to find the time it takes for the leg to swing forward.

Knowns and Unknowns The data for this problem are:

Description	Symbol	Value
Length of leg	L_{leg}	0.80 m
Unknown Variable		
Time for leg to swing forward	t	?

Modeling the Problem

STEP 1 Period and Frequency The time t it takes for the leg to swing forward is one-half the period T, or

$$t = \tfrac{1}{2}T$$

The period, in turn, is related to the frequency f by $T = 1/f$ (see Equation 10.5). Substituting this expression for T into the equation above for t gives Equation 1 at the right. The frequency of oscillation will be evaluated in Step 2.

$$t = \frac{1}{2}\left(\frac{1}{f}\right) \qquad (1)$$

$$\boxed{?}$$

STEP 2 The Frequency of a Physical Pendulum According to Equation 10.15, the frequency of a physical pendulum is given by

$$f = \frac{1}{2\pi}\sqrt{\frac{mgL}{I}} \qquad (10.15)$$

Since the leg is being approximated as a thin uniform rod, the distance L between the pivot and the center of gravity is one-half the length L_{leg} of the leg, so $L = \frac{1}{2}L_{\text{leg}}$. Using this relation in Equation 10.15 gives Equation 2 at the right, which we then substitute into Equation 1, as shown. The moment of inertia I will be evaluated in Step 3.

$$t = \frac{1}{2}\left(\frac{1}{f}\right) \qquad (1)$$

$$\boxed{f = \frac{1}{2\pi}\sqrt{\frac{mg\left(\frac{1}{2}L_{\text{leg}}\right)}{I}}} \qquad (2)$$

$$\boxed{?}$$

STEP 3 Moment of Inertia The leg is being approximated as a thin uniform rod of length L_{leg} that rotates about an axis that is perpendicular to one end. The moment of inertia I for such an object is given in Table 9.1 as

$$\boxed{I = \tfrac{1}{3}mL_{\text{leg}}^2}$$

We substitute this expression for I into Equation 2, as shown in the right column.

$$t = \frac{1}{2}\left(\frac{1}{f}\right) \qquad (1)$$

$$f = \frac{1}{2\pi}\sqrt{\frac{mg\left(\frac{1}{2}L_{\text{leg}}\right)}{I}} \qquad (2)$$

$$\boxed{I = \tfrac{1}{3}mL_{\text{leg}}^2}$$

Solution Algebraically combining the results of the steps above, we have:

$$\overset{\text{STEP 1}}{\underset{\downarrow}{}}\ \overset{\text{STEP 2}}{\underset{\downarrow}{}}\qquad\qquad \overset{\text{STEP 3}}{\underset{\downarrow}{}}$$

$$t = \frac{1}{2}\left(\frac{1}{f}\right) = \frac{1}{2}\left[\frac{1}{\frac{1}{2\pi}\sqrt{\frac{mg\left(\frac{1}{2}L_{\text{leg}}\right)}{I}}}\right] = \frac{1}{2}\left[\frac{1}{\frac{1}{2\pi}\sqrt{\frac{mg\left(\frac{1}{2}L_{\text{leg}}\right)}{\frac{1}{3}mL_{\text{leg}}^2}}}\right]$$

$$= \pi\sqrt{\frac{2L_{\text{leg}}}{3g}} = \pi\sqrt{\frac{2(0.80\text{ m})}{3(9.80\text{ m/s}^2)}} = \boxed{0.73\text{ s}}$$

Note that the mass m is algebraically eliminated from the final result.

Check Your Understanding

(The answers are given at the end of the book.)

11. Suppose that a grandfather clock (a simple pendulum) is running slowly; that is, the time it takes to complete each cycle is greater than it should be. Should you **(a)** shorten or **(b)** lengthen the pendulum to make the clock keep correct time?

12. Consult **Concept Simulation 10.2** to review the concept that is important here. In principle, the motions of a simple pendulum and an object on an ideal spring can both be used to provide the basic time interval or period used in a clock. Which of the two kinds of clocks becomes more inaccurate when carried to the top of a high mountain?

13. Concept Simulation 10.2 deals with the concept on which this question is based. Suppose you were kidnapped and held prisoner by space invaders in a completely isolated room, with nothing but a watch and a pair of shoes (including shoe laces of known length). How could you determine whether you were on earth or on the moon?

14. Two people are sitting on identical playground swings. One is pulled back 4° from the vertical and the other is pulled back 8°. They are both released at the same instant. Will they both come back to their starting points at the same time? Assume simple-pendulum motion.

Shock absorber

(a) Suspension spring

Oil

Hole

Piston

(b)

FIGURE 10.22 (a) A shock absorber mounted in the suspension system of an automobile and (b) a simplified, cutaway view of the shock absorber.

10.5 Damped Harmonic Motion

In simple harmonic motion, an object oscillates with a constant amplitude, because there is no mechanism for dissipating energy. In reality, however, friction or some other energy-dissipating mechanism is always present. In the presence of energy dissipation, the oscillation amplitude decreases as time passes, and the motion is no longer simple harmonic motion. Instead, it is referred to as **damped harmonic motion**, the decrease in amplitude being called "damping."

THE PHYSICS OF . . . a shock absorber. One widely used application of damped harmonic motion is in the suspension system of an automobile. **Figure 10.22a** shows a shock absorber attached to a main suspension spring of a car. A shock absorber is designed to introduce damping forces, which reduce the vibrations associated with a bumpy ride. As part *b* of the drawing shows, a shock absorber consists of a piston in a reservoir of oil. When the piston moves in response to a bump in the road, holes in the piston head permit the piston to pass through the oil. Viscous forces that arise during this movement cause the damping.

Figure 10.23 illustrates the different degrees of damping that can exist. As applied to the example of a car's suspension system, these graphs show the vertical position of the chassis after it has been pulled upward by an amount A_0 at time $t_0 = 0$ s and then released. Part *a* of the figure compares undamped or simple harmonic motion in curve 1 (red) to slightly damped motion in curve 2 (green). In damped harmonic motion, the chassis oscillates with decreasing amplitude and eventually comes to rest. As the degree of damping is increased from curve 2 to curve 3 (gold), the car makes fewer oscillations before coming to a halt.

Figure 10.23b shows that as the degree of damping is increased further, there comes a point when the car does not oscillate at all after it is released but, rather, settles directly back to its equilibrium position, as in curve 4 (blue). The smallest degree of damping that completely eliminates the oscillations is termed "critical damping," and the motion is said to be *critically damped*. **Figure 10.23b** also shows that the car takes the longest time to return to its equilibrium position in curve 5 (purple), where the degree of damping is above the value for critical damping. When the damping exceeds the critical value, the motion is said to be *overdamped*. In contrast, when the damping

(a)

(b)

FIGURE 10.23 Damped harmonic motion. The degree of damping increases from curve 1 to curve 5. Curve 1 (red) represents undamped or simple harmonic motion. Curves 2 (green) and 3 (gold) show underdamped motion. Curve 4 (blue) represents critically damped harmonic motion. Curve 5 (purple) shows overdamped motion.

is less than the critical level, the motion is said to be *underdamped* (curves 2 and 3). Typical automobile shock absorbers are designed to produce underdamped motion somewhat like that in curve 3.

Check Your Understanding

(*The answer is given at the end of the book.*)

15. The shock absorbers on a car are badly in need of replacement and introduce very little damping. Does the number of occupants in the car affect the vibration frequency of the car's suspension system?

Driven Harmonic Motion and Resonance

In damped harmonic motion, a mechanism such as friction dissipates or reduces the energy of an oscillating system, with the result that the amplitude of the motion decreases in time. This section discusses the opposite effect—namely, the increase in amplitude that results when energy is continually added to an oscillating system.

To set an object on an ideal spring into simple harmonic motion, some agent must apply a force that stretches or compresses the spring initially. Suppose that this force is applied at all times, not just for a brief initial moment. The force could be provided, for example, by a person who simply pushes and pulls the object back and forth. The resulting motion is known as **driven harmonic motion**, because the additional force drives or controls the behavior of the object to a large extent. The additional force is identified as the **driving force**.

Figure 10.24 illustrates one particularly important example of driven harmonic motion. Here, the driving force has the same frequency as the spring system and always points in the direction of the object's velocity. The frequency of the spring system is $f = \frac{1}{2\pi} \sqrt{k/m}$ and is called a **natural frequency** because it is the frequency at which the spring system naturally oscillates. Since the driving force and the velocity always have the same direction, positive work is done on the object at all times, and the total mechanical energy of the system increases. As a result, the amplitude of the vibration becomes larger and will increase without limit if there is no damping force to dissipate the energy being added by the driving force. The situation depicted in **Figure 10.24** is known as **resonance**.

FIGURE 10.24 Resonance occurs when the frequency of the driving force (blue arrows) matches a frequency at which the object naturally vibrates. The red arrows represent the velocity of the object.

> RESONANCE
>
> **Resonance is the condition in which a time-dependent force can transmit large amounts of energy to an oscillating object, leading to a large-amplitude motion. In the absence of damping, resonance occurs when the frequency of the force matches a natural frequency at which the object will oscillate.**

The role played by the frequency of a driving force is a critical one. The matching of this frequency with a natural frequency of vibration allows even a relatively weak force to produce a large-amplitude vibration, because the effect of each push–pull cycle is cumulative.

THE PHYSICS OF . . . high tides at the Bay of Fundy.   Resonance can occur with any object that can oscillate, and springs need not be involved. The greatest tides in the world occur in the Bay of Fundy, which lies between the Canadian provinces of New Brunswick and Nova Scotia. **Figure 10.25** shows the enormous difference between the water level at high and low tides, a difference that in some locations averages about 15 m. This phenomenon is partly due to resonance. The time, or period, that it takes for the tide to flow into and ebb out of a bay depends on the size of the bay, the topology of the bottom, and the configuration of the shoreline. The ebb and flow of the water in the Bay of Fundy has a period of 12.5 hours, which is very close to the lunar tidal period of 12.42 hours. The tide then "drives" water into and out of the Bay of Fundy at a frequency (once per 12.42 hours) that nearly matches the natural frequency of the bay (once per 12.5 hours). The result is the extraordinary high tide in the bay. (You can create a similar effect in a bathtub full of water by moving back and forth in synchronism with the waves you're causing.)

FIGURE 10.25   The Bay of Fundy, Canada, at (*left*) high tide and (*right*) low tide. In some places the water level changes by almost 15 m.

Check Your Understanding

(*The answer is given at the end of the book.*)

16. A car travels at a constant speed over a road that contains a series of equally spaced bumps. The spacing between bumps is d. The mass of the car is m, and the spring constant of the car's suspension springs is k. Because of resonance, a particularly jarring ride results. Ignoring the effect of the car's shock absorbers, derive an expression for the car's speed v in terms of d, m, and k, as well as some numerical constants.

10.7 Elastic Deformation

Stretching, Compression, and Young's Modulus

We have seen that a spring returns to its original shape when the force compressing or stretching it is removed. In fact, all materials become distorted in some way when they are squeezed or stretched, and many of them, such as rubber, return to their original shape when the squeezing or stretching is removed. Such materials are said to be "elastic." From an atomic viewpoint, elastic behavior has its origin in the forces that atoms exert on each other, and **Figure 10.26** symbolizes these forces as springs. It is because of these atomic-level "springs" that a material tends to return to its initial shape once the forces that cause the deformation are removed.

The interatomic forces that hold the atoms of a solid together are particularly strong, so considerable force must be applied to stretch or compress a solid object. Experiments have shown that the magnitude of the force can be expressed by the following relation, provided that the amount of stretch or compression is small compared to the original length of the object:

$$F = Y\left(\frac{\Delta L}{L_0}\right)A \qquad (10.17)$$

As **Figure 10.27** shows, F denotes the magnitude of the stretching force applied perpendicularly to the surface at the end, A is the cross-sectional area of the rod, ΔL is the increase in length, and L_0 is the original length. An analogous picture applies in the case of a force that causes compression. The term Y is a proportionality constant called *Young's modulus*, after Thomas Young (1773–1829). Solving Equation 10.17 for Y shows that Young's modulus has units of force per unit area (N/m^2).

> **Problem-Solving Insight** It should be noted that the magnitude of the force in Equation 10.17 is proportional to the fractional increase (or decrease) in length $\Delta L/L_0$, rather than the absolute change in length ΔL.

The magnitude of the force is also proportional to the cross-sectional area A, which need not be circular, but can have any shape (e.g., rectangular).

BIO THE PHYSICS OF . . . surgical implants. Table 10.1 reveals that the value of Young's modulus depends on the nature of the material. The values for metals are much larger than those for bone, for example. Equation 10.17 indicates that, for a given force, the material with the greater value of Y undergoes the smaller change in length. This difference between the changes in length is the reason why surgical implants (e.g., artificial hip joints), which are often made from stainless steel or titanium alloys, can lead to chronic deterioration of the bone that is in contact with the implanted prosthesis.

BIO THE PHYSICS OF . . . bone structure. Forces that are applied as in **Figure 10.27** and cause stretching are called "tensile" forces, because they create a tension in the material, much like the tension in a rope. Equation 10.17 also applies when the force compresses the material along its length. In this situation, the force is applied in a direction opposite to that shown in **Figure 10.27**, and ΔL stands for the amount by which the original length L_0 decreases. **Table 10.1** indicates, for example, that bone has different values of Young's modulus for compression and tension, the value for tension being greater. Such differences are related to the structure of the material. The solid part of bone consists of collagen fibers (a protein material) distributed throughout hydroxyapatite (a mineral). The collagen acts like the steel rods in reinforced concrete and increases the value of Y for tension relative to the value of Y for compression.

Most solids have Young's moduli that are rather large, reflecting the fact that a large force is needed to change the length of a solid object by even a small amount, as Example 12 illustrates.

FIGURE 10.26 The forces between atoms act like springs. The atoms are represented by red spheres, and the springs between some atoms have been omitted for clarity.

FIGURE 10.27 In this diagram, $\vec{F}$ denotes the stretching force, A the cross-sectional area, L_0 the original length of the rod, and ΔL the amount of stretch.

Material	Young's Modulus Y (N/m^2)
Aluminum	6.9×10^{10}
Bone	
Compression	9.4×10^9
Tension	1.6×10^{10}
Brass	9.0×10^{10}
Brick	1.4×10^{10}
Copper	1.1×10^{11}
Mohair	2.9×10^9
Nylon	3.7×10^9
Pyrex glass	6.2×10^{10}
Steel	2.0×10^{11}
Teflon	3.7×10^8
Titanium	1.2×10^{11}
Tungsten	3.6×10^{11}

TABLE 10.1 Values for the Young's Modulus of Solid Materials

EXAMPLE 12 | BIO The Physics of Bone Compression

A circus performer supports the combined weight (1080 N) of a number of colleagues (see **Figure 10.28**). Each thighbone (femur) of this performer has a length of 0.55 m and an effective cross-sectional area of 7.7×10^{-4} m². Determine the amount by which each thighbone compresses under the extra weight.

Reasoning The additional weight supported by each thighbone is $F = \frac{1}{2}(1080 \text{ N}) = 540$ N, and **Table 10.1** indicates that Young's modulus for bone compression is 9.4×10^9 N/m². Since the length and cross-sectional area of the thighbone are also known, we can use Equation 10.17 to find the amount by which the additional weight compresses the thighbone.

Solution The amount of compression ΔL of each thighbone is

$$\Delta L = \frac{FL_0}{YA} = \frac{(540 \text{ N})(0.55 \text{ m})}{(9.4 \times 10^9 \text{ N/m}^2)(7.7 \times 10^{-4} \text{ m}^2)}$$

$$= \boxed{4.1 \times 10^{-5} \text{ m}}$$

This is a very small change, the fractional decrease being $\Delta L / L_0 = 0.000\ 075$.

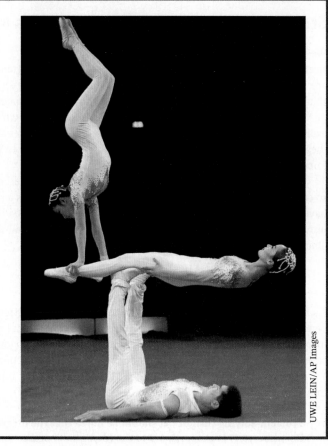

FIGURE 10.28 The entire weight of the balanced group is supported by the legs of the performer who is lying on his back.

UWE LEIN/AP Images

Just as Example 12 showed us that the fractional change in the length of the human femur changes very little in response to a large compressive force, Example 13 now illustrates that human ligaments both are strong and can endure relatively large elastic deformations.

EXAMPLE 13 | BIO The Iliofemoral Ligament

Also known as the "Y-ligament," the iliofemoral ligament in the hip joins the ilium to the femur (see **Figure 10.29**). It is the strongest ligament in the human body, and can withstand stretching forces exceeding 770.0 lb! It has a Young's modulus of 5.00×10^7 N/m² and an effective cross-sectional area of about 5.50 cm² in an average male adult. Assuming a uniform cross-sectional area and an unstrained length of 9.40 cm, determine **(a)** how far the ligament stretches beyond its unstrained length when subjected to a 770.0-lb stretching force, and **(b)** the percent change in its length under this load.

Reasoning **(a)** We are given the force F, the unstrained length L_0, the effective cross-sectional area A, and Young's modulus Y of the ligament material. Assuming that the effective cross-sectional area is uniform, and that the force is a tensile force (i.e., a "stretching" force), we can calculate the change in length (ΔL) using Equation 10.17. **(b)** To find the percent change in length, we need to divide the value of ΔL calculated in part (a) by the unstrained length L_0, and multiply the result by 100.

Solution **(a)** This is an elastic deformation problem where the deformation results from a stretching force. We can therefore apply Equation 10.17, but we first have to convert the force F into units of N, L_0 into m, and the cross-sectional area A into m²:

$$F = (770.0 \text{ lb})\left(4.45 \frac{\text{N}}{\text{lb}}\right) = 3427 \text{ N}$$

$$L_0 = (9.40 \text{ cm})\left(\frac{\text{m}}{100 \text{ cm}}\right) = 9.40 \times 10^{-2} \text{ m}$$

and

$$A = (5.50 \text{ cm}^2)\left(\frac{1 \text{ m}}{100 \text{ cm}}\right)\left(\frac{1 \text{ m}}{100 \text{ cm}}\right) = 5.50 \times 10^{-4} \text{ m}^2$$

Solving Equation 10.17 for ΔL and plugging in the known values gives

$$\Delta L = \frac{FL_0}{YA} = \frac{(3427 \text{ N})(0.0940 \text{ m})}{(5.00 \times 10^7 \text{ N/m}^2)(5.50 \times 10^{-4} \text{ m}^2)} = \boxed{0.0117 \text{ m}}$$

or about 1.17 cm.

(b) The percent change is given by the change in length divided by the unstrained length, multiplied by 100:

$$\% \text{ change in length} = \left(\frac{\Delta L}{L_0}\right) \times 100 = \left(\frac{0.0117 \text{ m}}{0.0940 \text{ m}}\right) \times 100 = \boxed{12.4\%}$$

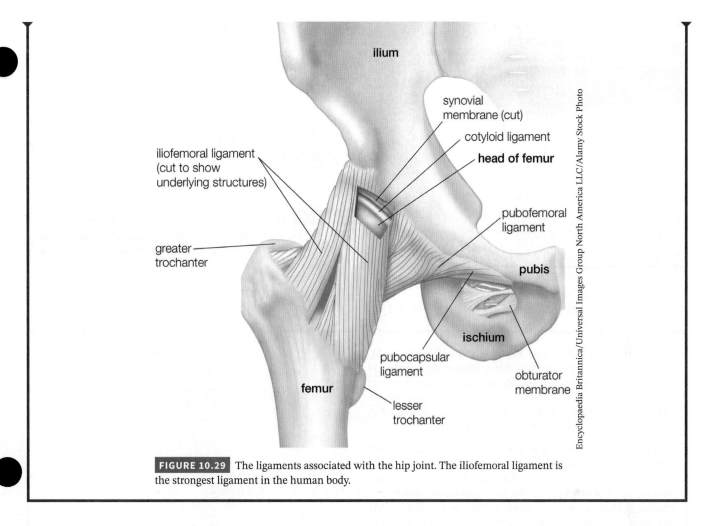

ilium

synovial
membrane (cut)

cotyloid ligament

head of femur

iliofemoral ligament
(cut to show
underlying structures)

pubofemoral
ligament

pubis

greater
trochanter

ischium

pubocapsular
ligament

obturator
membrane

femur

lesser
trochanter

Encyclopaedia Britannica/Universal Images Group North America LLC/Alamy Stock Photo

FIGURE 10.29 The ligaments associated with the hip joint. The iliofemoral ligament is the strongest ligament in the human body.

Shear Deformation and the Shear Modulus

It is possible to deform a solid object in a way other than by stretching or compressing it. For instance, place a book on a rough table and push on the top cover, as in **Figure 10.30a**. Notice that the top cover, and the pages below it, become shifted relative to the stationary bottom cover. The resulting deformation is called a **shear deformation** and occurs because of the combined effect of the force $\vec{F}$ applied (by the hand) to the top of the book and the force $-\vec{F}$ applied (by the table) to the bottom of the book. In general, shearing forces cause a solid object to change its shape. In **Figure 10.30** the directions of the forces are parallel to the covers of the book, each of which has an area A, as illustrated in part b of the drawing. These two forces have equal magnitudes, but opposite directions, so the book remains in equilibrium. Equation 10.18 gives the magnitude F of the force needed to produce an amount of shear ΔX for an object with thickness L_0:

$$F = S\left(\frac{\Delta X}{L_0}\right)A \qquad (10.18)$$

This equation is very similar to Equation 10.17. The constant of proportionality S is called the **shear modulus** and, like Young's modulus, has units of force per unit area (N/m^2). The value of S depends on the nature of the material, and **Table 10.2** gives some representative values. Example 14 illustrates how to determine the shear modulus of a familiar dessert.

TABLE 10.2	Values for the Shear Modulus of Solid Materials	
Material	**Shear Modulus S (N/m^2)**
Aluminum	2.4×10^{10}
Bone	1.2×10^{10}
Brass	3.5×10^{10}
Copper	4.2×10^{10}
Lead	5.4×10^{9}
Nickel	7.3×10^{10}
Steel	8.1×10^{10}
Tungsten	1.5×10^{11}

(a)　　　　　(b)

FIGURE 10.30 (a) An example of a shear deformation. The shearing forces $\vec{F}$ and $-\vec{F}$ are applied parallel to the top and bottom covers of the book. (b) The shear deformation is ΔX. The area of each cover is A, and the thickness of the book is L_0.

EXAMPLE 14 | J-E-L-L-O

A block of Jell-O is resting on a plate. **Figure 10.31a** gives the dimensions of the block. You are bored, impatiently waiting for dinner, and push tangentially across the top surface with a force of $F = 0.45$ N, as in part b of the drawing. The top surface moves a distance $\Delta X = 6.0 \times 10^{-3}$ m relative to the bottom surface. Use this idle gesture to measure the shear modulus of Jell-O.

Reasoning The finger applies a force that is parallel to the top surface of the Jell-O block. The shape of the block changes, because the top surface moves a distance ΔX relative to the bottom surface. The magnitude of the force required to produce this change in shape is given by Equation 10.18 as $F = S(\Delta X/L_0)A$. We know the values for all the variables in this relation except S, which, therefore, can be determined.

Solution Solving Equation 10.18 for the shear modulus S, we find that $S = FL_0/(A\,\Delta X)$, where $A = (0.070$ m$)(0.070$ m$)$ is the area of the top surface, and $L_0 = 0.030$ m is the thickness of the block:

$$S = \frac{FL_0}{A\,\Delta X} = \frac{(0.45\text{ N})(0.030\text{ m})}{(0.070\text{ m})(0.070\text{ m})(6.0 \times 10^{-3}\text{m})} = \boxed{460\text{ N/m}^2}$$

(a)

(b)

FIGURE 10.31 (a) A block of Jell-O and (b) a shearing force applied to it.

Jell-O can be deformed easily, so its shear modulus is significantly less than that of a more rigid material like steel (see **Table 10.2**).

Although Equations 10.17 and 10.18 are algebraically similar, they refer to different kinds of deformations. The tensile force in **Figure 10.27** is perpendicular to the surface whose area is A, whereas the shearing force in **Figure 10.30** is parallel to that surface. Furthermore, the ratio $\Delta L/L_0$ in Equation 10.17 is different from the ratio $\Delta X/L_0$ in Equation 10.18. The distances ΔL and L_0 are parallel, whereas ΔX and L_0 are perpendicular. Young's modulus refers to a *change in length* of one dimension of a solid object as a result of tensile or compressive forces. The shear modulus refers to a *change in shape* of a solid object as a result of shearing forces.

Volume Deformation and the Bulk Modulus

When a compressive force is applied along one dimension of a solid, the length of that dimension decreases. It is also possible to apply compressive forces so that the size of every dimension (length, width, and depth) decreases, leading to a decrease in volume, as **Figure 10.32** illustrates. This kind of overall compression occurs, for example, when an object is submerged in a liquid, and the liquid presses inward everywhere on the object. The forces acting in such situations are applied perpendicular to every surface, and it is more convenient to speak of the perpendicular force per unit area, rather than the amount of any one force in particular. The magnitude of the perpendicular force per unit area is called the **pressure P**.

DEFINITION OF PRESSURE

The pressure P is the magnitude F of the force acting perpendicular to a surface divided by the area A over which the force acts:

$$P = \frac{F}{A} \qquad (10.19)$$

Pressure is a scalar, not a vector, quantity.

SI Unit of Pressure: $N/m^2 = $ pascal (Pa)

Equation 10.19 indicates that the SI unit for pressure is the unit of force divided by the unit of area, or newton/meter^2 (N/m^2). This unit of pressure is often referred to as a *pascal* (Pa), named after the French scientist Blaise Pascal (1623–1662).

Suppose we change the pressure on an object by an amount ΔP, where ΔP represents the final pressure P minus the initial pressure P_0: $\Delta P = P - P_0$. Because of this change in pressure, the volume of the object changes by an amount $\Delta V = V - V_0$, where V and V_0 are the final and initial volumes, respectively. Such a pressure change occurs, for example, when a swimmer dives deeper into the water. Experiment reveals that the change ΔP in pressure needed to change the volume by an amount ΔV is directly proportional to the fractional change $\Delta V/V_0$ in the volume:

$$\Delta P = -B\left(\frac{\Delta V}{V_0}\right) \qquad (10.20)$$

This relation is analogous to Equations 10.17 and 10.18, except that the area A in those equations does not appear here explicitly; the area is already taken into account by the concept of pressure (magnitude of the force per unit area). The proportionality constant B is known as the **bulk modulus**. The minus sign occurs because an increase in pressure (ΔP positive) always creates a decrease in volume (ΔV negative), and B is given as a positive quantity. Like Young's modulus and the shear modulus, the bulk modulus has units of force per unit area (N/m^2), and its value depends on the nature of the material. **Table 10.3** gives representative values of the bulk modulus.

FIGURE 10.32 The arrows denote the forces that push perpendicularly on every surface of an object immersed in a liquid. The magnitude of the force per unit area is the pressure. When the pressure increases, the volume of the object decreases.

TABLE 10.3 Values for the Bulk Modulus of Solid and Liquid Materials

Material	Bulk Modulus B [N/m^2 (=Pa)]
Solids	
Aluminum	7.1×10^{10}
Brass	6.7×10^{10}
Copper	1.3×10^{11}
Diamond	4.43×10^{11}
Lead	4.2×10^{10}
Nylon	6.1×10^{9}
Osmium	4.62×10^{11}
Pyrex glass	2.6×10^{10}
Steel	1.4×10^{11}
Liquids	
Ethanol	8.9×10^{8}
Oil	1.7×10^{9}
Water	2.2×10^{9}

Check Your Understanding

(The answers are given at the end of the book.)

17. Young's modulus for steel is greater than that for a particular unknown material. What does this mean about how these materials compress when used in construction? **(a)** Steel compresses much more easily than the unknown material does. **(b)** The unknown material compresses more easily than steel does. **(c)** Young's modulus has nothing to do with compression, so not enough information is given for an answer.

18. Two rods are made from the same material. One has a circular cross section, and the other has a square cross section. The circle just fits within the square. When the same force is applied to stretch these rods, they each stretch by the same amount. Which rod, if either, is longer?

19. A trash compactor crushes empty aluminum cans, thereby reducing the total volume, so that $\Delta V/V_0 = -0.75$ in Equation 10.20. Can the value given in **Table 10.3** for the bulk modulus of aluminum be used to calculate the change ΔP in pressure generated in the trash compactor?

20. Both sides of the relation $F = S(\Delta X/L_0)A$ (Equation 10.18) can be divided by the area A to give F/A on the left side. Can this F/A term be called a pressure, such as the pressure that appears in $\Delta P = -B(\Delta V/V_0)$ (Equation 10.20)?

10.8 | Stress, Strain, and Hooke's Law

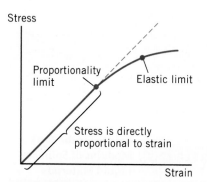

TABLE 10.4	**Stress and Strain Relations for Elastic Behavior**		
$\dfrac{F}{A}$	$= Y$	$\left(\dfrac{\Delta L}{L_0}\right)$	(10.17)
$\dfrac{F}{A}$	$= S$	$\left(\dfrac{\Delta X}{L_0}\right)$	(10.18)
ΔP	$= B$	$\left(\dfrac{-\Delta V}{V_0}\right)$	(10.20)

Stress is proportional to Strain

Equations 10.17, 10.18, and 10.20 specify the amount of force needed for a given amount of elastic deformation, and they are repeated in **Table 10.4** to emphasize their common features. The left side of each equation is the magnitude of the force per unit area required to cause an elastic deformation. In general, the ratio of the magnitude of the force to the area is called the **stress**. The right side of each equation involves the change in a quantity (ΔL, ΔX, or ΔV) divided by a quantity (L_0 or V_0) relative to which the change is compared. The terms $\Delta L/L_0$, $\Delta X/L_0$, and $\Delta V/V_0$ are unitless ratios, and each is referred to as the **strain** that results from the applied stress. In the case of stretch and compression, the strain is the fractional change in length, whereas in volume deformation it is the fractional change in volume. In shear deformation the strain refers to a change in shape of the object. Experiments show that these three equations, with constant values for Young's modulus, the shear modulus, and the bulk modulus, apply to a wide range of materials. Therefore, stress and strain are directly proportional to one another, a relationship first discovered by Robert Hooke (1635–1703) and now referred to as **Hooke's law**.

> **HOOKE'S LAW FOR STRESS AND STRAIN**
>
> **Stress is directly proportional to strain.**
>
> **SI Unit of Stress: newton per square meter (N/m²) = pascal (Pa)**
>
> **SI Unit of Strain: Strain is a unitless quantity.**

In reality, materials obey Hooke's law only up to a certain limit, as **Figure 10.33** shows. As long as stress remains proportional to strain, a plot of stress versus strain is a straight line. The point on the graph where the material begins to deviate from straight-line behavior is called the "proportionality limit." Beyond the proportionality limit stress and strain are no longer directly proportional. However, if the stress does not exceed the "elastic limit" of the material, the object will return to its original size and shape once the stress is removed. The "elastic limit" is the point beyond which the object no longer returns to its original size and shape when the stress is removed; the object remains permanently deformed.

FIGURE 10.33 Hooke's law (stress is directly proportional to strain) is valid only up to the proportionality limit of a material. Beyond this limit, Hooke's law no longer applies. Beyond the elastic limit, the material remains deformed even when the stress is removed.

Check Your Understanding

(The answer is given at the end of the book.)

21. The block in **CYU Figure 10.3** rests on the ground. Which face—A, B, or C—experiences the largest stress and which face experiences the smallest stress when the block is resting on it?

CYU FIGURE 10.3

EXAMPLE 15 | BIO The Physics of the Mechanical Properties of Bone

Human bones also follow Hooke's Law for Stress and Strain, as indicated in **Figure 10.34**. This is essentially **Figure 10.33**, where we plot the stress versus strain (under compression) specifically for a human femur—the largest bone in the body. In Section 6.9, we plotted the average force versus displacement for a *variable* force. The area under the curve was equal to the work done by the force. Now let's apply a similar analysis to **Figure 10.34**. Calculate the area under the curve where Hooke's law is valid (shaded region), and use unit analysis to describe what this area represents.

Reasoning   As we did in Section 6.9, we need to calculate the area of the shaded region. Since we are to consider only the range where Hooke's law is valid, the shaded region will have the shape of a triangle whose area is equal to $\frac{1}{2}$(base × height).

Solution   Using the data in **Figure 10.34** we calculate the area of the shaded region:

Area = $\frac{1}{2}$(base × height) = $\frac{1}{2}$(0.010)(200 × 10^6 Pa) = $\boxed{1.00 \times 10^6 \text{ Pa}}$

FIGURE 10.34   Stress versus strain for a human femur bone. Hooke's Law is obeyed up to the proportionality limit.

Since stress has the units of pressure, and strain is unitless, our answer also has units of pressure, or Pa. However, we can gain a greater understanding of the significance of the area under the curve if we write pressure as the stress $P = \frac{F}{A}$ and the strain as $\frac{\Delta L}{L_0}$. Now when we calculate the area under the curve in general, we have:

$$\text{Area} = \frac{1}{2}\left(\frac{F}{A}\right)\left(\frac{\Delta L}{L_0}\right)$$

Here, the force is given by Hooke's law:

$$F = kx = k\,\Delta L$$

If we substitute this into our equation for the area above, we get the following:

$$\text{Area} = \frac{\frac{1}{2}k(\Delta L)^2}{AL_0}$$

The numerator is equivalent to Equation 10.13 and represents the elastic potential energy. The denominator is equal to an area times a length, which has the units of volume. Therefore, in general, the area under the curve is equivalent to energy per unit volume, or energy density:

$$\text{Area} = \frac{\text{PE [J]}}{\text{Vol [m}^3]}$$

For our simple example above, the result indicates that one million J/m^3 of elastic potential energy is stored in the bone under compression. Of course, the stored energy in a human femur is much smaller than one million joules, as the change in volume under compression is much smaller than 1 m^3. However, if the energy delivered to the bone is greater than the energy it can absorb, the bone will fail, and a fracture will occur.

Concept Summary

10.1 The Ideal Spring and Simple Harmonic Motion   The force that must be applied to stretch or compress an ideal spring is given by Equation 10.1, where k is the spring constant and x is the displacement of the spring from its unstrained length.

$$F_x^{\text{Applied}} = kx \tag{10.1}$$

A spring exerts a restoring force on an object attached to the spring. The restoring force F_x produced by an ideal spring is given by Equation 10.2, where the minus sign indicates that the restoring force points opposite to the displacement of the spring.

$$F_x = -kx \tag{10.2}$$

Simple harmonic motion is the oscillatory motion that occurs when a restoring force of the form $F_x = -kx$ acts on an object. A graphical record of position versus time for an object in simple harmonic motion is sinusoidal. The amplitude A of the motion is the

maximum distance that the object moves away from its equilibrium position.

10.2 Simple Harmonic Motion and the Reference Circle   The period T of simple harmonic motion is the time required to complete one cycle of the motion, and the frequency f is the number of cycles per second that occurs. Frequency and period are related, according to Equation 10.5. The frequency f (in Hz) is related to the angular frequency ω (in rad/s), according to Equation 10.6.

$$f = \frac{1}{T} \tag{10.5}$$

$$\omega = 2\pi f \quad (\omega \text{ in rad/s}) \tag{10.6}$$

The maximum speed v_{max} of an object moving in simple harmonic motion is given by Equation 10.8, where A is the amplitude of the motion.

$$v_{\text{max}} = A\omega \quad (\omega \text{ in rad/s}) \tag{10.8}$$

The maximum acceleration a_{max} of an object moving in simple harmonic motion is given by Equation 10.10.

$$a_{max} = A\omega^2 \qquad (\omega \text{ in rad/s}) \qquad \textbf{(10.10)}$$

The angular frequency of simple harmonic motion is given by Equation 10.11.

$$\omega = \sqrt{\frac{k}{m}} \qquad (\omega \text{ in rad/s}) \qquad \textbf{(10.11)}$$

10.3 Energy and Simple Harmonic Motion The elastic potential energy of an object attached to an ideal spring is given by Equation 10.13. The total mechanical energy E of such a system is the sum of its translational and rotational kinetic energies, gravitational potential energy, and elastic potential energy, according to Equation 10.14. If external nonconservative forces like friction do no net work, the total mechanical energy of the system is conserved, as indicated by Equation 1.

$$PE_{elastic} = \tfrac{1}{2}kx^2 \qquad \textbf{(10.13)}$$

$$E = \tfrac{1}{2}mv^2 + \tfrac{1}{2}I\omega^2 + mgh + \tfrac{1}{2}kx^2 \qquad \textbf{(10.14)}$$

$$E_f = E_0 \qquad \textbf{(1)}$$

10.4 The Pendulum A simple pendulum is a particle of mass m attached to a frictionless pivot by a cable whose length is L and whose mass is negligible. The small-angle ($\leq 10°$) back-and-forth swinging of a simple pendulum is simple harmonic motion, but large-angle movement is not. The frequency f of the small-angle motion is given by Equation 10.16.

$$2\pi f = \sqrt{\frac{g}{L}} \qquad (\text{small angles only}) \qquad \textbf{(10.16)}$$

A physical pendulum consists of a rigid object, with moment of inertia I and mass m, suspended from a frictionless pivot. For small-angle displacements, the frequency f of simple harmonic motion for a physical pendulum is given by Equation 10.15, where L is the distance between the axis of rotation and the center of gravity of the rigid object.

$$2\pi f = \sqrt{\frac{mgL}{I}} \qquad (\text{small angles only}) \qquad \textbf{(10.15)}$$

10.5 Damped Harmonic Motion Damped harmonic motion is motion in which the amplitude of oscillation decreases as time passes. Critical damping is the minimum degree of damping that eliminates any oscillations in the motion as the object returns to its equilibrium position.

10.6 Driven Harmonic Motion and Resonance Driven harmonic motion occurs when a driving force acts on an object along with the restoring force. Resonance is the condition under which the driving force can transmit large amounts of energy to an oscillating object, leading to large-amplitude motion. In the absence of damping, resonance occurs when the frequency of the driving force matches a natural frequency at which the object oscillates.

10.7 Elastic Deformation One type of elastic deformation is stretch and compression. The magnitude F of the force required to stretch or compress an object of length L_0 and cross-sectional area A by an amount ΔL (see **Figure 10.27**) is given by Equation 10.17, where Y is a constant called Young's modulus.

$$F = Y\left(\frac{\Delta L}{L_0}\right)A \qquad \textbf{(10.17)}$$

Another type of elastic deformation is shear. The magnitude F of the shearing force required to create an amount of shear ΔX for an object of thickness L_0 and cross-sectional area A is (see **Figure 10.30**) given by Equation 10.18, where S is a constant called the shear modulus.

$$F = S\left(\frac{\Delta X}{L_0}\right)A \qquad \textbf{(10.18)}$$

A third type of elastic deformation is volume deformation, which has to do with pressure. The pressure P is the magnitude F of the force acting perpendicular to a surface divided by the area A over which the force acts, according to Equation 10.19. The SI unit for pressure is N/m², a unit known as a pascal (Pa): 1 Pa = 1 N/m². The change ΔP in pressure needed to change the volume V_0 of an object by an amount ΔV (see **Figure 10.32**) is given by Equation 10.20, where B is a constant known as the bulk modulus.

$$P = \frac{F}{A} \qquad \textbf{(10.19)}$$

$$\Delta P = -B\left(\frac{\Delta V}{V_0}\right) \qquad \textbf{(10.20)}$$

10.8 Stress, Strain, and Hooke's Law Stress is the magnitude of the force per unit area applied to an object and causes strain. For stretch/compression, the strain is the fractional change $\Delta L/L_0$ in length. For shear, the strain reflects the change in shape of the object and is given by $\Delta X/L_0$ (see **Figure 10.30**). For volume deformation, the strain is the fractional change in volume $\Delta V/V_0$. Hooke's law states that stress is directly proportional to strain.

Focus on Concepts

Online

Additional questions are available for assignment in WileyPLUS.

Section 10.1 The Ideal Spring and Simple Harmonic Motion

1. Which one of the following graphs correctly represents the restoring force F of an ideal spring as a function of the displacement x of the spring from its unstrained length?

QUESTION 1

Section 10.2   Simple Harmonic Motion and the Reference Circle

2. You have two springs. One has a greater spring constant than the other. You also have two objects, one with a greater mass than the other. Which object should be attached to which spring, so that the resulting spring–object system has the greatest possible period of oscillation? **(a)** The object with the greater mass should be attached to the spring with the greater spring constant. **(b)** The object with the greater mass should be attached to the spring with the smaller spring constant. **(c)** The object with the smaller mass should be attached to the spring with the smaller spring constant. **(d)** The object with the smaller mass should be attached to the spring with the greater spring constant.

3. An object is oscillating in simple harmonic motion with an amplitude A and an angular frequency ω. What should you do to increase the maximum speed of the motion? **(a)** Reduce both A and ω by 10%. **(b)** Increase A by 10% and reduce ω by 10%. **(c)** Reduce A by 10% and increase ω by 10%. **(d)** Increase both A and ω by 10%.

Section 10.3   Energy and Simple Harmonic Motion

4. The kinetic energy of an object attached to a horizontal ideal spring is denoted by KE and the elastic potential energy by PE. For the simple harmonic motion of this object the maximum kinetic energy and the maximum elastic potential energy during an oscillation cycle are KE_{max} and PE_{max}, respectively. In the absence of friction, air resistance, and any other nonconservative forces, which of the following equations applies to the object–spring system?

A. KE + PE = constant
B. $KE_{max} = PE_{max}$
(a) A, but not B **(b)** B, but not A **(c)** A and B **(d)** Neither A nor B

5. A block is attached to a horizontal spring. On top of this block rests another block. The two-block system slides back and forth in simple harmonic motion on a frictionless horizontal surface. At one extreme end of the oscillation cycle, where the blocks come to a momentary halt before reversing the direction of their motion, the top block is suddenly lifted vertically upward, without changing the zero velocity of the bottom block. The simple harmonic motion then continues. What happens to the amplitude and the angular frequency of the ensuing motion? **(a)** The amplitude remains the same, and the angular frequency increases. **(b)** The amplitude increases, and the angular frequency remains the same. **(c)** Both the amplitude and the angular frequency remain the same. **(d)** Both the amplitude and the angular frequency decrease. **(e)** Both the amplitude and the angular frequency increase.

Section 10.4   The Pendulum

6. Five simple pendulums are shown in the drawings. The lengths of the pendulums are drawn to scale, and the masses are either m or $2m$, as shown. Which pendulum has the smallest angular frequency of oscillation?   **(a)** A **(b)** B **(c)** C **(d)** D **(e)** E

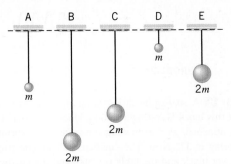

QUESTION 6

Section 10.5   Damped Harmonic Motion

7. An object on a spring is oscillating in simple harmonic motion. Suddenly, friction appears and causes the energy of the system to be dissipated. The system now exhibits _____. **(a)** driven harmonic motion **(b)** Hooke's-law type of motion **(c)** damped harmonic motion

Section 10.6   Driven Harmonic Motion and Resonance

8. An external force (in addition to the spring force) is continually applied to an object of mass m attached to a spring that has a spring constant k. The frequency of this external force is such that resonance occurs. Then the frequency of this external force is doubled, and the force is applied to one of the spring systems shown in the drawing. With which system would resonance occur?   **(a)** A   **(b)** B   **(c)** C **(d)** D   **(e)** E

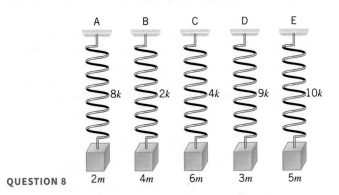

QUESTION 8

Section 10.7   Elastic Deformation

9. Drawings A and B show two cylinders that are identical in all respects, except that one is hollow. Identical forces are applied to each cylinder in order to stretch them. Which cylinder, if either, stretches more? **(a)** A and B both stretch by the same amount. **(b)** A stretches more than B. **(c)** B stretches more than A. **(d)** Insufficient information is given for an answer.

QUESTION 9

Section 10.8   Stress, Strain, and Hooke's Law

10. A material has a shear modulus of $5.0 \times 10^9 \, N/m^2$. A shear stress of $8.5 \times 10^6 \, N/m^2$ is applied to a piece of the material. What is the resulting shear strain?

Problems

Online

Additional questions are available for assignment in WileyPLUS.

SSM Student Solutions Manual	**BIO** Biomedical application
MMH Problem-solving help	**E** Easy
GO Guided Online Tutorial	**M** Medium
V-HINT Video Hints	**H** Hard
CHALK Chalkboard Videos	**WS** Worksheet
	T Team Problem

Section 10.1 The Ideal Spring and Simple Harmonic Motion

1. **E** **SSM** A hand exerciser utilizes a coiled spring. A force of 89.0 N is required to compress the spring by 0.0191 m. Determine the force needed to compress the spring by 0.0508 m.

2. **E** The drawing shows three identical springs hanging from the ceiling. Nothing is attached to the first spring, whereas a 4.50-N block hangs from the second spring. A block of unknown weight hangs from the third spring. From the drawing, determine **(a)** the spring constant (in N/m) and **(b)** the weight of the block hanging from the third spring.

PROBLEM 2

3. **E** In a room that is 2.44 m high, a spring (unstrained length = 0.30 m) hangs from the ceiling. A board whose length is 1.98 m is attached to the free end of the spring. The board hangs straight down, so that its 1.98-m length is perpendicular to the floor. The weight of the board (104 N) stretches the spring so that the lower end of the board just extends to, but does not touch, the floor. What is the spring constant of the spring?

4. **E** **GO** A spring lies on a horizontal table, and the left end of the spring is attached to a wall. The other end is connected to a box. The box is pulled to the right, stretching the spring. Static friction exists between the box and the table, so when the spring is stretched only by a small amount and the box is released, the box does not move. The mass of the box is 0.80 kg, and the spring has a spring constant of 59 N/m. The coefficient of static friction between the box and the table on which it rests is $\mu_s = 0.74$. How far can the spring be stretched from its unstrained position without the box moving when it is released?

5. **E** **SSM** A person who weighs 670 N steps onto a spring scale in the bathroom, and the spring compresses by 0.79 cm. **(a)** What is the spring constant? **(b)** What is the weight of another person who compresses the spring by 0.34 cm?

6. **E** A spring ($k = 830$ N/m) is hanging from the ceiling of an elevator, and a 5.0-kg object is attached to the lower end. By how much does the spring stretch (relative to its unstrained length) when the elevator is accelerating upward at $a = 0.60$ m/s^2?

7. **E** **CHALK** A 0.70-kg block is hung from and stretches a spring that is attached to the ceiling. A second block is attached to the first one, and the amount that the spring stretches from its unstrained length triples. What is the mass of the second block?

8. **M** **V-HINT** A uniform 1.4-kg rod that is 0.75 m long is suspended at rest from the ceiling by two springs, one at each end of the rod. Both springs hang straight down from the ceiling. The springs have identical lengths when they are unstretched. Their spring constants are 59 N/m and 33 N/m. Find the angle that the rod makes with the horizontal.

9. **M** **SSM** In 0.750 s, a 7.00-kg block is pulled through a distance of 4.00 m on a frictionless horizontal surface, starting from rest. The block has a constant acceleration and is pulled by means of a horizontal spring that is attached to the block. The spring constant of the spring is 415 N/m. By how much does the spring stretch?

10. **M** **GO** Review Conceptual Example 2 as an aid in solving this problem. An object is attached to the lower end of a 100-coil spring that is hanging from the ceiling. The spring stretches by 0.160 m. The spring is then cut into two identical springs of 50 coils each. As the drawing shows, each spring is attached between the ceiling and the object. By how much does each spring stretch?

50-coil spring

PROBLEM 10

11. **M** **CHALK** **SSM** A small ball is attached to one end of a spring that has an unstrained length of 0.200 m. The spring is held by the other end, and the ball is whirled around in a horizontal circle at a speed of 3.00 m/s. The spring remains nearly parallel to the ground during the motion and is observed to stretch by 0.010 m. By how much would the spring stretch if it were attached to the ceiling and the ball allowed to hang straight down, motionless?

12. **M** **MMH** To measure the static friction coefficient between a 1.6-kg block and a vertical wall, the setup shown in the drawing is used. A spring (spring constant = 510 N/m) is attached to the block. Someone pushes on the end of the spring in a direction perpendicular to the wall until the block does not slip downward. The spring is compressed by 0.039 m. What is the coefficient of static friction?

PROBLEM 12

13. **H** A 30.0-kg block is resting on a flat horizontal table. On top of this block is resting a 15.0-kg block, to which a horizontal spring is attached, as the drawing illustrates. The spring constant of the spring is 325 N/m. The coefficient of kinetic friction between the lower block and the table is 0.600, and the coefficient of static friction between the two blocks is 0.900. A horizontal force $\vec{F}$ is applied to the lower block as shown. This force is increasing in such a way as to keep the blocks moving at a *constant speed*. At the point where the upper block begins to slip on the lower block, determine (a) the amount by which the spring is compressed and (b) the magnitude of the force $\vec{F}$.

PROBLEM 13

14. Ⓗ A 15.0-kg block rests on a horizontal table and is attached to one end of a massless, horizontal spring. By pulling horizontally on the other end of the spring, someone causes the block to accelerate uniformly and reach a speed of 5.00 m/s in 0.500 s. In the process, the spring is stretched by 0.200 m. The block is then pulled at a *constant speed* of 5.00 m/s, during which time the spring is stretched by only 0.0500 m. Find (a) the spring constant of the spring and (b) the coefficient of kinetic friction between the block and the table.

Section 10.2   Simple Harmonic Motion and the Reference Circle

15. Ⓔ The fan blades on a jet engine make one thousand revolutions in a time of 50.0 ms. Determine (a) the period (in seconds) and (b) the frequency (in Hz) of the rotational motion. (c) What is the angular frequency of the blades?

16. Ⓔ MMH A block of mass $m = 0.750$ kg is fastened to an unstrained horizontal spring whose spring constant is $k = 82.0$ N/m. The block is given a displacement of +0.120 m, where the + sign indicates that the displacement is along the +x axis, and then released from rest. (a) What is the force (magnitude and direction) that the spring exerts on the block just before the block is released? (b) Find the angular frequency ω of the resulting oscillatory motion. (c) What is the maximum speed of the block? (d) Determine the magnitude of the maximum acceleration of the block.

17. Ⓔ MMH An 0.80-kg object is attached to one end of a spring, as in Figure 10.5, and the system is set into simple harmonic motion. The displacement x of the object as a function of time is shown in the drawing. With the aid of these data, determine (a) the amplitude A of the motion, (b) the angular frequency ω, (c) the spring constant k, (d) the speed of the object at $t = 1.0$ s, and (e) the magnitude of the object's acceleration at $t = 1.0$ s.

PROBLEM 17

18. Ⓔ Refer to Conceptual Example 2 as an aid in solving this problem. A 100-coil spring has a spring constant of 420 N/m. It is cut into four shorter springs, each of which has 25 coils. One end of a 25-coil spring is attached to a wall. An object of mass 46 kg is attached to the other end of the spring, and the system is set into horizontal oscillation. What is the angular frequency of the motion?

19. Ⓜ GO Objects of equal mass are oscillating up and down in simple harmonic motion on two different vertical springs. The spring constant of spring 1 is 174 N/m. The motion of the object on spring 1 has twice the amplitude as the motion of the object on spring 2. The magnitude of the maximum velocity is the same in each case. Find the spring constant of spring 2.

20. Ⓜ SSM MMH CHALK A spring stretches by 0.018 m when a 2.8-kg object is suspended from its end. How much mass should be attached to this spring so that its frequency of vibration is $f = 3.0$ Hz?

21. Ⓜ V-HINT An object attached to a horizontal spring is oscillating back and forth along a frictionless surface. The maximum speed of the object is 1.25 m/s, and its maximum acceleration is 6.89 m/s². How much time elapses between an instant when the object's speed is at a maximum and the next instant when its acceleration is at a maximum?

22. Ⓜ GO MMH A vertical spring (spring constant = 112 N/m) is mounted on the floor. A 0.400-kg block is placed on top of the spring and pushed down to start it oscillating in simple harmonic motion. The block is not attached to the spring. (a) Obtain the frequency (in Hz) of the motion. (b) Determine the amplitude at which the block will lose contact with the spring.

23. Ⓗ A tray is moved horizontally back and forth in simple harmonic motion at a frequency of $f = 2.00$ Hz. On this tray is an empty cup. Obtain the coefficient of static friction between the tray and the cup, given that the cup begins slipping when the amplitude of the motion is 5.00×10^{-2} m.

Section 10.3   Energy and Simple Harmonic Motion

24. Ⓔ A pen contains a spring with a spring constant of 250 N/m. When the tip of the pen is in its retracted position, the spring is compressed 5.0 mm from its unstrained length. In order to push the tip out and lock it into its writing position, the spring must be compressed an additional 6.0 mm. How much work is done by the spring force to ready the pen for writing? Be sure to include the proper algebraic sign with your answer.

25. Ⓔ GO The drawing shows three situations in which a block is attached to a spring. The position labeled "0 m" represents the unstrained position of the spring. The block is moved from an initial position x_0 to a final position x_f, the magnitude of the displacement being denoted by the symbol s. Suppose the spring has a spring constant of $k = 46.0$ N/m. Using the data provided in the drawing, determine the total work done by the restoring force of the spring for each situation.

PROBLEM 25

26. Ⓔ A spring is hung from the ceiling. A 0.450-kg block is then attached to the free end of the spring. When released from rest, the block drops 0.150 m before momentarily coming to rest, after which it moves back upward. (a) What is the spring constant of the spring? (b) Find the angular frequency of the block's vibrations.

27. Ⓔ A 3.2-kg block is hanging stationary from the end of a vertical spring that is attached to the ceiling. The elastic potential energy of

this spring-block system is 1.8 J. What is the elastic potential energy of the system when the 3.2-kg block is replaced by a 5.0-kg block?

28. **E** **SSM** A vertical spring with a spring constant of 450 N/m is mounted on the floor. From directly above the spring, which is unstrained, a 0.30-kg block is dropped from rest. It collides with and sticks to the spring, which is compressed by 2.5 cm in bringing the block to a momentary halt. Assuming air resistance is negligible, from what height (in cm) above the compressed spring was the block dropped?

29. **E** In preparation for shooting a ball in a pinball machine, a spring ($k = 675$ N/m) is compressed by 0.0650 m relative to its unstrained length. The ball ($m = 0.0585$ kg) is at rest against the spring at point A. When the spring is released, the ball slides (without rolling). It leaves the spring and arrives at point B, which is 0.300 m higher than point A. Ignore friction, and find the ball's speed at point B.

30. **E** **MMH** A heavy-duty stapling gun uses a 0.140-kg metal rod that rams against the staple to eject it. The rod is attached to and pushed by a stiff spring called a "ram spring" ($k = 32\,000$ N/m). The mass of this spring may be ignored. The ram spring is compressed by 3.0×10^{-2} m from its unstrained length and then released from rest. Assuming that the ram spring is oriented vertically and is still compressed by 0.8×10^{-2} m when the downward-moving ram hits the staple, find the speed of the ram at the instant of contact.

31. **E** A rifle fires a 2.10×10^{-2}-kg pellet straight upward, because the pellet rests on a compressed spring that is released when the trigger is pulled. The spring has a negligible mass and is compressed by 9.10×10^{-2} m from its unstrained length. The pellet rises to a maximum height of 6.10 m above its position on the compressed spring. Ignoring air resistance, determine the spring constant.

32. **E** **SSM** A 1.00×10^{-2}-kg block is resting on a horizontal frictionless surface and is attached to a horizontal spring whose spring constant is 124 N/m. The block is shoved parallel to the spring axis and is given an initial speed of 8.00 m/s, while the spring is initially unstrained. What is the amplitude of the resulting simple harmonic motion?

33. **M** **V-HINT** An 86.0-kg climber is scaling the vertical wall of a mountain. His safety rope is made of nylon that, when stretched, behaves like a spring with a spring constant of 1.20×10^3 N/m. He accidentally slips and falls freely for 0.750 m before the rope runs out of slack. How much is the rope stretched when it breaks his fall and momentarily brings him to rest?

34. **M** **V-HINT** A horizontal spring is lying on a frictionless surface. One end of the spring is attached to a wall, and the other end is connected to a movable object. The spring and object are compressed by 0.065 m, released from rest, and subsequently oscillate back and forth with an angular frequency of 11.3 rad/s. What is the speed of the object at the instant when the spring is *stretched* by 0.048 m relative to its unstrained length?

35. **M** **GO** A spring is resting vertically on a table. A small box is dropped onto the top of the spring and compresses it. Suppose the spring has a spring constant of 450 N/m and the box has a mass of 1.5 kg. The speed of the box just before it makes contact with the spring is 0.49 m/s. (a) Determine the magnitude of the spring's displacement at an instant when the acceleration of the box is zero. (b) What is the magnitude of the spring's displacement when the spring is fully compressed?

36. **M** **CHALK** **SSM** **MMH** A spring is compressed by 0.0620 m and is used to launch an object horizontally with a speed of 1.50 m/s. If the object were attached to the spring, at what angular frequency (in rad/s) would it oscillate?

37. **M** **GO** A 0.60-kg metal sphere oscillates at the end of a vertical spring. As the spring stretches from 0.12 to 0.23 m (relative to its unstrained length), the speed of the sphere decreases from 5.70 to 4.80 m/s. What is the spring constant of the spring?

38. **M** A 1.1-kg object is suspended from a vertical spring whose spring constant is 120 N/m. (a) Find the amount by which the spring is stretched from its unstrained length. (b) The object is pulled straight down by an additional distance of 0.20 m and released from rest. Find the speed with which the object passes through its original position on the way up.

39. **H** **SSM** A 70.0-kg circus performer is fired from a cannon that is elevated at an angle of 40.0° above the horizontal. The cannon uses strong elastic bands to propel the performer, much in the same way that a slingshot fires a stone. Setting up for this stunt involves stretching the bands by 3.00 m from their unstrained length. At the point where the performer flies free of the bands, his height above the floor is the same as the height of the net into which he is shot. He takes 2.14 s to travel the horizontal distance of 26.8 m between this point and the net. Ignore friction and air resistance and determine the effective spring constant of the firing mechanism.

40. **H** A 1.00×10^{-2}-kg bullet is fired horizontally into a 2.50-kg wooden block attached to one end of a massless horizontal spring ($k = 845$ N/m). The other end of the spring is fixed in place, and the spring is unstrained initially. The block rests on a horizontal, frictionless surface. The bullet strikes the block perpendicularly and quickly comes to a halt within it. As a result of this completely inelastic collision, the spring is compressed along its axis and causes the block/bullet to oscillate with an amplitude of 0.200 m. What is the speed of the bullet?

Section 10.4   The Pendulum

41. **E** A simple pendulum is made from a 0.65-m-long string and a small ball attached to its free end. The ball is pulled to one side through a small angle and then released from rest. After the ball is released, how much time elapses before it attains its greatest speed?

42. **E** **GO** **MMH** Astronauts on a distant planet set up a simple pendulum of length 1.2 m. The pendulum executes simple harmonic motion and makes 100 complete vibrations in 280 s. What is the magnitude of the acceleration due to gravity on this planet?

43. **E** **GO** The length of a simple pendulum is 0.79 m and the mass of the particle (the "bob") at the end of the cable is 0.24 kg. The pendulum is pulled away from its equilibrium position by an angle of 8.50° and released from rest. Assume that friction can be neglected and that the resulting oscillatory motion is simple harmonic motion. (a) What is the angular frequency of the motion? (b) Using the position of the bob at its lowest point as the reference level, determine the total mechanical energy of the pendulum as it swings back and forth. (c) What is the bob's speed as it passes through the lowest point of the swing?

44. **E** **CHALK** A spiral staircase winds up to the top of a tower in an old castle. To measure the height of the tower, a rope is attached to the top of the tower and hung down the center of the staircase. However, nothing is available with which to measure the length of the rope. Therefore, at the bottom of the rope a small object is attached so as to form a simple pendulum that just clears the floor. The period of the pendulum is measured to be 9.2 s. What is the height of the tower?

45. **E** **GO** Two physical pendulums (not simple pendulums) are made from meter sticks that are suspended from the ceiling at one end. The sticks are uniform and are identical in all respects, except that one is made of wood (mass = 0.17 kg) and the other of metal (mass = 0.85 kg). They are set into oscillation and execute simple harmonic motion. Determine the period of (a) the wood pendulum and (b) the metal pendulum.

46. **H** A small object oscillates back and forth at the bottom of a frictionless hemispherical bowl, as the drawing illustrates. The radius of the bowl is R, and the angle θ is small enough that the object oscillates in simple harmonic motion. Derive an expression for the angular frequency ω of the motion. Express your answer in terms of R and g, the magnitude of the acceleration due to gravity.

PROBLEM 46

Section 10.7   Elastic Deformation,

Section 10.8   Stress, Strain, and Hooke's Law

47. **E** **SSM** A tow truck is pulling a car out of a ditch by means of a steel cable that is 9.1 m long and has a radius of 0.50 cm. When the car just begins to move, the tension in the cable is 890 N. How much has the cable stretched?

48. **E** Two stretched cables both experience the same stress. The first cable has a radius of 3.5×10^{-3} m and is subject to a stretching force of 270 N. The radius of the second cable is 5.1×10^{-3} m. Determine the stretching force acting on the second cable.

49. **E** **SSM** **CHALK** The pressure increases by 1.0×10^4 N/m² for every meter of depth beneath the surface of the ocean. At what depth does the volume of a Pyrex glass cube, 1.0×10^{-2} m on an edge at the ocean's surface, decrease by 1.0×10^{-10} m³?

50. **E** A 59-kg water skier is being pulled by a nylon tow rope that is attached to a boat. The unstretched length of the rope is 12 m, and its cross-sectional area is 2.0×10^{-5} m². As the skier moves, a resistive force (due to the water) of magnitude 130 N acts on her; this force is directed opposite to her motion. What is the change in the length of the rope when the skier has an acceleration whose magnitude is 0.85 m/s²?

51. **E** A solid steel cylinder is standing (on one of its ends) vertically on the floor. The length of the cylinder is 3.6 m and its radius is 65 cm. When an object is placed on top of the cylinder, the cylinder compresses by an amount of 5.7×10^{-7} m. What is the weight of the object?

52. **E** The drawing shows a 160-kg crate hanging from the end of a steel bar. The length of the bar is 0.10 m, and its cross-sectional area is 3.2×10^{-4} m². Neglect the weight of the bar itself and determine **(a)** the shear stress on the bar and **(b)** the vertical deflection ΔY of the right end of the bar.

PROBLEM 52

53. **E** A copper cube, 0.30 m on a side, is subjected to two shearing forces, each of which has a magnitude $F = 6.0 \times 10^6$ N (see the drawing). Find the angle θ (in degrees), which is one measure of how the shape of the block has been altered by shear deformation.

PROBLEM 53

54. **E** **SSM** Two metal beams are joined together by four rivets, as the drawing indicates. Each rivet has a radius of 5.0×10^{-3} m and is to

be exposed to a shearing stress of no more than 5.0×10^8 Pa. What is the maximum tension $\vec{T}$ that can be applied to each beam, assuming that each rivet carries one-fourth of the total load?

PROBLEM 54

55. **E** A copper cylinder and a brass cylinder are stacked end to end, as in the drawing. Each cylinder has a radius of 0.25 cm. A compressive force of $F = 6500$ N is applied to the right end of the brass cylinder. Find the amount by which the length of the stack decreases.

PROBLEM 55

56. **E** A piece of aluminum is surrounded by air at a pressure of 1.01×10^5 Pa. The aluminum is placed in a vacuum chamber where the pressure is reduced to zero. Determine the fractional change $\Delta V/V_0$ in the volume of the aluminum.

57. **E** **GO** One end of a piano wire is wrapped around a cylindrical tuning peg and the other end is fixed in place. The tuning peg is turned so as to stretch the wire. The piano wire is made from steel ($Y = 2.0 \times 10^{11}$ N/m²). It has a radius of 0.80 mm and an unstrained length of 0.76 m. The radius of the tuning peg is 1.8 mm. Initially, there is no tension in the wire, but when the tuning peg is turned, tension develops. Find the tension in the wire when the tuning peg is turned through two revolutions. Ignore the radius of the wire compared to the radius of the tuning peg.

58. **M** **GO** A piece of mohair (see Table 10.1) taken from an Angora goat has a radius of 31×10^{-6} m. What is the least number of identical pieces of mohair needed to suspend a 75-kg person, so the strain experienced by each piece is less than 0.010? Assume that the tension is the same in all the pieces.

59. **M** **GO** **CHALK** A square plate is 1.0×10^{-2} m thick, measures 3.0×10^{-2} m on a side, and has a mass of 7.2×10^{-2} kg. The shear modulus of the material is 2.0×10^{10} N/m². One of the square faces rests on a flat horizontal surface, and the coefficient of static friction between the plate and the surface is 0.91. A force is applied to the top of the plate, as in **Figure 10.30a**. Determine **(a)** the maximum possible amount of shear stress, **(b)** the maximum possible amount of shear strain, and **(c)** the maximum possible amount of shear deformation ΔX (see **Figure 10.30b**) that can be created by the applied force just before the plate begins to move.

60. **H** A 6.8-kg bowling ball is attached to the end of a nylon cord with a cross-sectional area of 3.4×10^{-5} m². The other end of the cord is fixed to the ceiling. When the bowling ball is pulled to one side and released from rest, it swings downward in a circular arc. At the instant it reaches its lowest point, the bowling ball is 1.4 m lower than the point from which it was released, and the cord is stretched 2.7×10^{-3} m from its unstrained length. What is the unstrained length of the cord? (*Hint: When calculating any quantity other than the strain, ignore the increase in the length of the cord.*)

61. **H** **SSM** A solid brass sphere is subjected to a pressure of 1.0×10^5 Pa due to the earth's atmosphere. On Venus the pressure due to the atmosphere is 9.0×10^6 Pa. By what fraction $\Delta r/r_0$ (including the algebraic sign) does the radius of the sphere change when it is exposed to the Venusian atmosphere? Assume that the change in radius is very small relative to the initial radius.

Additional Problems

Online

62. **E** A loudspeaker diaphragm is producing a sound for 2.5 s by moving back and forth in simple harmonic motion. The angular frequency of the motion is 7.54×10^4 rad/s. How many times does the diaphragm move back and forth?

63. **E** A person bounces up and down on a trampoline, while always staying in contact with it. The motion is simple harmonic motion, and it takes 1.90 s to complete one cycle. The height of each bounce above the equilibrium position is 45.0 cm. Determine **(a)** the amplitude and **(b)** the angular frequency of the motion. **(c)** What is the maximum speed attained by the person?

64. **E** A simple pendulum is swinging back and forth through a small angle, its motion repeating every 1.25 s. How much longer should the pendulum be made in order to increase its period by 0.20 s?

65. **E** An archer, about to shoot an arrow, is applying a force of $+240$ N to a drawn bowstring. The bow behaves like an ideal spring whose spring constant is 480 N/m. What is the displacement of the bowstring?

66. **M** **V-HINT** A block rests on a frictionless horizontal surface and is attached to a spring. When set into simple harmonic motion, the block oscillates back and forth with an angular frequency of 7.0 rad/s. The drawing indicates the position of the block when the spring is unstrained. This position is labeled "$x = 0$ m." The drawing also shows a small bottle located 0.080 m to the right of this position. The block is pulled to the right, stretching the spring by 0.050 m, and is then thrown to the left. In order for the block to knock over the bottle, it must be thrown with a speed exceeding v_0. Ignoring the width of the block, find v_0.

PROBLEM 66

67. **M** **GO** A vertical ideal spring is mounted on the floor and has a spring constant of 170 N/m. A 0.64-kg block is placed on the spring in two different ways. **(a)** In one case, the block is placed on the spring and not released until it rests stationary on the spring in its equilibrium position. Determine the amount (magnitude only) by which the spring is compressed. **(b)** In a second situation, the block is released from rest immediately after being placed on the spring and falls downward until it comes to a momentary halt. Determine the amount (magnitude only) by which the spring is now compressed.

68. **M** Multiple-Concept Example 6 reviews the principles that play roles in this problem. A bungee jumper, whose mass is 82 kg, jumps from a tall platform. After reaching his lowest point, he continues to oscillate up and down, reaching the low point two more times in 9.6 s. Ignoring air resistance and assuming that the bungee cord is an ideal spring, determine its spring constant.

69. **M** **GO** When an object of mass m_1 is hung on a vertical spring and set into vertical simple harmonic motion, it oscillates at a frequency of 12.0 Hz. When another object of mass m_2 is hung on the spring along with the first object, the frequency of the motion is 4.00 Hz. Find the ratio m_2/m_1 of the masses.

70. **H** **SSM** The drawing shows a top view of a frictionless horizontal surface, where there are two springs with particles of mass m_1 and m_2 attached to them. Each spring has a spring constant of 120 N/m. The particles are pulled to the right and then released from the positions shown in the drawing. How much time passes before the particles are side by side for the first time at $x = 0$ m if **(a)** $m_1 = m_2 = 3.0$ kg and **(b)** $m_1 = 3.0$ kg and $m_2 = 27$ kg?

Position of
unstrained
spring
PROBLEM 70 $(x = 0$ m$)$

71. **M** **GO** A helicopter is using a steel cable to lift a 2100-kg jeep (see the figure). The unstretched length of the cable is 16 m, and its radius is 5.0×10^{-3} m. By what amount does the cable stretch when the jeep is hoisted straight upward with an acceleration of $+1.5$ m/s^2?

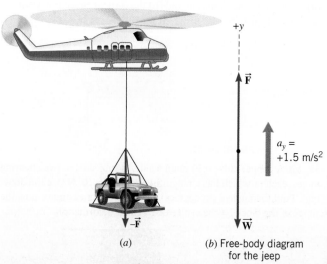

(a) (b) Free-body diagram
for the jeep

(a) The jeep applies a force $-\vec{F}$ to the lower end of the cable, thereby stretching it. (b) The free-body diagram for the jeep, showing the two forces acting on it.

PROBLEM 71

Physics in Biology, Medicine, and Sports

72. **E** **BIO** **SSM** **10.2**   When responding to sound, the human eardrum vibrates about its equilibrium position. Suppose an eardrum is vibrating with an amplitude of 6.3×10^{-7} m and a maximum speed of 2.9×10^{-3} m/s. **(a)** What is the frequency (in Hz) of the eardrum's vibration? **(b)** What is the maximum acceleration of the eardrum?

73. **E** **BIO** **V-HINT** **10.7**   When subjected to a force of compression, the length of a bone decreases by 2.7×10^{-5} m. When this same bone is subjected to a tensile force of the same magnitude, by how much does it stretch?

74. **M** **BIO** **10.8**   A gymnast does a one-arm handstand. The humerus, which is the upper arm bone (between the elbow and the shoulder joint), may be approximated as a 0.30-m-long cylinder with an outer radius of 1.00×10^{-2} m and a hollow inner core with a radius of 4.0×10^{-3} m. Excluding the arm, the mass of the gymnast is 63 kg. **(a)** What is the compressional strain of the humerus? **(b)** By how much is the humerus compressed?

75. **M** **BIO** **GO** **10.8**   Depending on how you fall, you can break a bone easily. The severity of the break depends on how much energy the bone absorbs in the accident, and to evaluate this let us treat the bone as an ideal spring. The maximum applied force of compression that one man's thighbone can endure without breaking is 7.0×10^4 N. The minimum effective cross-sectional area of the bone is 4.0×10^{-4} m², its length is 0.55 m, and Young's modulus is $Y = 9.4 \times 10^9$ N/m². The mass of the man is 65 kg. He falls straight down without rotating, strikes the ground stiff-legged on one foot, and comes to a halt without rotating. To see that it is easy to break a thighbone when falling in this fashion, find the maximum distance through which his center of gravity can fall without his breaking a bone.

76. **E** **BIO** **10.8**   The femur is a bone in the leg whose minimum cross-sectional area is about 4.0×10^{-4} m². A compressional force in excess of 6.8×10^4 N will fracture this bone. **(a)** Find the maximum stress that this bone can withstand. **(b)** What is the strain that exists under a maximum-stress condition?

77. **E** **BIO** **10.7**   Between each pair of vertebrae in the spinal column is a cylindrical disc of cartilage. Typically, this disc has a radius of about 3.0×10^{-2} m and a thickness of about 7.0×10^{-3} m. The shear modulus of cartilage is 1.2×10^7 N/m². Suppose that a shearing force of magnitude 11 N is applied parallel to the top surface of the disc while the bottom surface remains fixed in place. How far does the top surface move relative to the bottom surface?

78. **H** **BIO** **10.7**   A cylindrically shaped piece of collagen (a substance found in the body in connective tissue) is being stretched by a force that increases from 0 to 3.0×10^{-2} N. The length and radius of the collagen are, respectively, 2.5 and 0.091 cm, and Young's modulus is 3.1×10^6 N/m². **(a)** If the stretching obeys Hooke's law, what is the spring constant k for collagen? **(b)** How much work is done by the variable force that stretches the collagen? (See Section 6.9 for a discussion of the work done by a variable force.)

79. **E** **BIO** **10.7**   Elastic therapeutic tape, also called kineso tape or KT-tape, consists of an elastic cotton weave with an adhesive backing. It is placed directly on the skin to provide support to injured or painful joints, pulled muscles, and other ailments (see the photo). While there is no definitive scientific evidence that KT-tape offers significant health benefits, its popularity surged after many athletes wore it in the 2008 Olympic Games. Standard KT-tape is 5.08 cm wide and 0.180 mm thick. One end of a 10.0-cm length of KT-tape is hung from the ceiling, and a 1.00-kg mass is attached to the other end, causing the tape to stretch by 2.14 mm. What is the Young's modulus of the tape?

PROBLEM 79

80. **H** **BIO** **10.7**   The silk of Darwin's bark spider (*Caerostris darwini*), named after the naturalist Charles Darwin, is the strongest biological material ever discovered (see the photo). In fact, many of the mechanical properties of this silk supersede those of steel. While the Young's modulus for steel (2.00×10^{11} N/m²) is larger than that for silk (1.15×10^{10} N/m²), the density of the silk is much less, such that the strength-to-density ratio of the silk is about five times that of steel. Consider a 10.0-cm unstrained length of spider silk with a circular cross section and a radius of 2.50×10^{-6} m. When this segment supports the weight of a dragonfly ($m = 3.00 \times 10^{-3}$ kg), **(a)** what is the percent change in the length of the silk? **(b)** What force would have to be applied to a steel filament of the same dimensions in order to produce the same percent change in length?

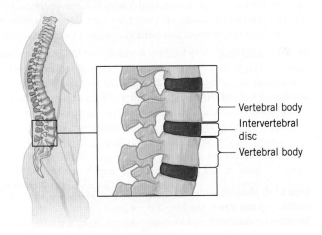

PROBLEM 80

81. **E** **BIO** **10.8**   The intervertebral discs in the human spine (see the figure) are multilayered elastic structures composed of collagen that act like shock absorbers between the vertebrae. While lifting heavy loads improperly, such as lifting a heavy box with your back arched instead of using your legs, the compressive forces on the discs can be quite large. In one case, the compressive force on a disc is 3100 N. If the disc is compressed a distance of 1.5 mm, what is the effective spring constant of the disc?

—— Vertebral body

—— Intervertebral disc

—— Vertebral body

PROBLEM 81

82. **M** **BIO** **10.4** Most injuries to the shoulder will require rehab and physical therapy. Pendulum exercises are a great way to increase strength as well as to improve flexibility and range of motion in an injured shoulder. Here, the patient bends forward until their back is horizontal. The injured arm is hung vertically downward and is allowed to swing freely back and forth like a pendulum (see the figure). Treat the arm (length of 69 cm) in the figure as a physical pendulum—a uniform rod rotating about an axis through one of its ends—and calculate the period of the arm's small oscillations.

69 cm

PROBLEM 82

83. **M** **BIO** **10.3** Skeletal muscles are attached to bones by tendons. Tendons are composed of elastic tissues that, for small stretching displacements, behave like a spring and follow Hooke's law. The largest and strongest tendon in the human body is the Achilles tendon (see the figure), which attaches the muscles of the lower leg to the heel bone (calcaneus). While the Achilles tendon is incredibly strong, it is often under high tension and has a low blood supply. This makes the tendon prone to injuries, such as tears and ruptures. In fact, 1 out of every 10 000 people rupture their Achilles tendon each year. One unfortunate soul had his Achilles tendon rupture when the tension in the tendon was 28 500 N. If the effective spring constant of the tendon is 4.60×10^6 N/m, **(a)** how far did the tendon stretch before it ruptured? **(b)** How much elastic potential energy was stored in the tendon when it ruptured?

Anatomy of the Lower Leg

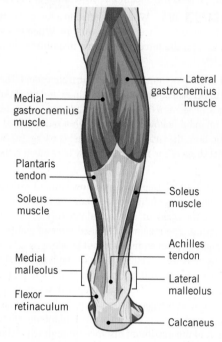

Medial gastrocnemius muscle

Lateral gastrocnemius muscle

Plantaris tendon

Soleus muscle

Soleus muscle

Medial malleolus

Achilles tendon

Flexor retinaculum

Lateral malleolus

Calcaneus

PROBLEM 83

Concepts and Calculations Problems

Online

This chapter has examined an important kind of vibratory motion known as simple harmonic motion. Specifically, it has discussed how the motion's displacement, velocity, and acceleration vary with time, and explained what determines the frequency of the motion. In addition, we saw that the elastic force is conservative, so that the total mechanical energy is conserved if nonconservative forces, such as friction and air resistance, are absent. We conclude now with some problems that review important features of simple harmonic motion.

84. **M** **CHALK** **SSM** A 75-kg diver is standing at the end of a diving board while it is vibrating up and down in simple harmonic motion, as indicated in the figure. The diving board has an effective spring constant of $k = 4100$ N/m, and the vertical distance between the highest and lowest points in the motion is 0.30 m. *Concepts:* (i) How is the amplitude A related to the vertical distance between the highest and lowest points of the diver's motion? (ii) Starting from the top, where is the diver located one-quarter of a period later, and what can be said about his speed at this point? (iii) If the amplitude were to double, would the period also double? Explain. *Calculations:* **(a)** What is the amplitude of the motion? **(b)** Starting when the diver is at the highest point, what is his speed one-quarter of a period later? **(c)** If the vertical distance between his highest and lowest points were changed to 0.10 m, what would be the time required for the diver to make one complete motional cycle?

0.30 m

PROBLEM 84

85. **M** **CHALK** A 68.0-kg bungee jumper is standing on a tall platform ($h_0 = 46.0$ m), as indicated in the figure. The bungee cord has a natural length of $L_0 = 9.00$ m and, when stretched, behaves like an ideal spring with a spring constant of $k = 66.0$ N/m. The jumper falls from rest, and it is assumed that the only forces acting on him are his weight and, for the latter part of the descent, the elastic force of the bungee cord. *Concepts:* (i) Can we use the conservation of mechanical energy to find his speed at any point along the descent? Explain your answer. (ii) What type of energy does he have when he is standing on the platform? (iii) What types of energy does he have at point A? (iv) What types of energy does he have at point B? *Calculations:* What is his speed when he is at the following heights above the water: (a) $h_A = 37.0$ m, and (b) $h_B = 15.0$ m?

PROBLEM 85

Team Problems

Online

86. 🄴🅃 🆆🆂 **Moving a Fragile Particle Detector.** You and your team are assigned to design a structure that protects a heavy, fragile detector as it is moved from one lab to another. The idea is to suspend it by springs inside a large cubical box. You decide to use four identical springs, one attached to each of the four upper corners of the box. The opposite end of each spring is connected to a single hook on the top of the instrument that will hang from the springs, suspending it inside the box. The box has sides of length $L = 1.00$ m, and the springs have an unstrained length of 70.0 cm. Assuming that you want the hook on the top of the instrument to which all of the springs are connected to be located at the geometric center of the box when the 78.0-kg instrument is suspended, what should you choose for the spring constant of the springs?

87. 🄴🅃 🆆🆂 **The Spring and the Pendulum.** A steel ball of mass $m = 6.25$ kg hangs from the end of a vertical spring of spring constant $k = 112$ N/m. You and your team have been given the task of designing a simple pendulum that utilizes the same mass m and has the same frequency of oscillation f as the spring–mass system. (a) What is the frequency of oscillation of the spring–mass system? (b) Find the required length L and mass m of the pendulum so that it has the same frequency of oscillation as that found in part (a). (c) What is the relationship between the length of the pendulum and the equilibrium displacement of the spring from its unstrained length when the 6.25-kg mass hangs from it? *Hint: Assume that the frequency of oscillation of a vertical spring system is the same as that of a horizontal spring system.*

88. 🄼🅃 **Bungee Cord Escape.** You are running from pirates on a tropical island somewhere in the Caribbean. You have somehow become separated from the rest of your group and now find yourself on the edge of a cliff with your pursuers less than 10 minutes behind you. According to a sign posted on the guardrail at the cliff's edge, the drop to the beach below is $h = 140.0$ feet. Your team members (waiting for you on the beach, near your boat) have a rope, but there is no time for anyone to climb the cliff to save you. You break into a deserted cabin nearby, and rummage around for a rope. Instead, you find a brand new, still-in-package, bungee cord that must have been intended for tourists jumping from a nearby bridge. You figure you might be able to attach it to the guardrail and jump to the beach, letting go at the bottom before it reverses your motion. You read the bungee cord specifications on the package: the total length of the cord is $L_0 = 100.0$ m, the maximum elastic deformation is 200% (i.e., it can safely triple its length), and the elastic constant is $k = 75.0$ N/m. (a) If you weigh 170 lb, how far is the bungee cord designed to let you fall before it stops you and reverses your direction? Will this afford you a safe landing? (b) You realize that you don't have to hang from the very end of the bungee, but rather from some point in the middle. How far from the attached end should you grasp the unstretched bungee cord so that you land softly on the beach? Will you be able to perform the jump and stay under the elastic deformation limit?

89. 🄼🅃 **A Light-Beam Metronome.** You are given the task of opening an antiquated "light lock," which is unlocked by shining red light pulses of a certain frequency into a light sensor on the lock. You are given a red laser pointer, a spring of unstretched length $L = 15.0$ cm and spring constant $k = 7.20$ N/m, a sheet of steel ($\rho = 7.60$ g/cm³) that is 0.125 inches thick, and some tools. You come up with the idea to take a piece of the steel sheet (of mass m), cut a slot in it, and hang it from the spring. If you shine the laser through the slot and onto the sensor, and then stretch the spring and let it go, the steel plate will oscillate and cause the beam to pass through the slot periodically. (a) Assuming the beam is passing through the slot (and onto the lock's sensor) when the spring-mass system is in equilibrium, how is the frequency at which the light pulses hit the sensor related to the frequency of the spring/mass (i.e., steel plate) system. (b) Based on your answer for (a), what should be the frequency of the spring/mass system if the unlocking frequency is 3.50 Hz? (c) What should be the mass m of the steel plate? (d) Calculate some reasonable dimensions for the steel plate (i.e., they should be consistent with the mass that is required for the spring-mass system). You may assume the material removed to make the slot in the steel plate is of negligible mass.

CHAPTER 11

The air is a fluid, and this chapter examines the forces and pressures that fluids exert when they are at rest and when they are in motion. In a tornado the air is moving very rapidly and, as we will see, moving air has a lower pressure than that of stationary air. This difference in air pressure is one of the reasons that tornadoes, such as the one in this photograph, are so destructive.

Design Pics Inc/Alamy Stock Photo

LEARNING OBJECTIVES

After reading this module, you should be able to...

11.1 Define mass density.

11.2 Define pressure.

11.3 Relate pressure and depth in a static fluid.

11.4 Discriminate between gauge, atmospheric, and absolute pressures.

11.5 Apply Pascal's principle.

11.6 Solve statics problems using the buoyant force.

11.7 Identify types of fluid flow.

11.8 Apply the equation of continuity.

11.9 Use Bernoulli's equation to relate speed, pressure, and elevation.

11.10 Apply Bernoulli's equation for moving fluids.

11.11 Analyze viscous flow in moving liquids.

Fluids

11.1 Mass Density

Fluids are materials that can flow, and they include both gases and liquids. Air is the most common gas, and flows from place to place as wind. Water is the most familiar liquid, and flowing water has many uses, from generating hydroelectric power to white-water rafting. The **mass density** of a liquid or gas is one of the important factors that determine its behavior as a fluid. As indicated below, the mass density is the mass per unit volume and is denoted by the Greek letter rho (ρ).

DEFINITION OF MASS DENSITY

The mass density ρ is the mass m of a substance divided by its volume V:

$$\rho = \frac{m}{V} \qquad (11.1)$$

SI Unit of Mass Density: kg/m³

Equal volumes of different substances generally have different masses, so the density depends on the nature of the material, as **Table 11.1** indicates. Gases have the smallest densities because gas molecules are relatively far apart and a gas contains a large fraction of empty space. In contrast, the molecules are much more tightly packed in liquids and solids, and the tighter packing leads to larger densities.

TABLE 11.1 **Mass Densities^a of Common Substances**

Substance	Mass Density ρ (kg/m³)	Substance	Mass Density ρ (kg/m³)
Solids		**Liquids**	
Aluminum	2700	Blood (whole, 37 °C)	1060
Brass	8470	Ethyl alcohol	806
Concrete	2200	Mercury	13 600
Copper	8890	Oil (hydraulic)	800
Diamond	3520	Water (4 °C)	1.000×10^3
Gold	19 300	**Gases**	
Ice	917	Air	1.29
Iron (steel)	7860	Carbon dioxide	1.98
Lead	11 300	Helium	0.179
Quartz	2660	Hydrogen	0.0899
Silver	10 500	Nitrogen	1.25
Wood (yellow pine)	550	Oxygen	1.43

^{a}Unless otherwise noted, densities are given at 0 °C and 1 atm pressure.

The densities of gases are very sensitive to changes in temperature and pressure. However, for the range of temperatures and pressures encountered in this text, the densities of liquids and solids do not differ much from the values in **Table 11.1**.

It is the mass of a substance, not its weight, that enters into the definition of density. In situations where weight is needed, it can be calculated from the mass density, the volume, and the acceleration due to gravity, as Example 1 illustrates.

EXAMPLE 1 | Blood as a Fraction of Body Weight

The body of a man whose weight is 690 N contains about 5.2×10^{-3} m³ (5.5 qt) of blood. **(a)** Find the blood's weight and **(b)** express it as a percentage of the body weight.

Reasoning To find the weight W of the blood, we need the mass m, since $W = mg$ (Equation 4.5), where g is the magnitude of the acceleration due to gravity. According to **Table 11.1**, the density of blood is 1060 kg/m³, so the mass of the blood can be found by using the given volume of 5.2×10^{-3} m³ in Equation 11.1.

Solution **(a)** According to Equation 4.5, the blood's weight is $W = mg$. Equation 11.1 can be solved for m to show that the mass is $m = \rho V$. Substituting this result into Equation 4.5 gives

$$W = mg = (\rho V)g = (1060 \text{ kg/m}^3)(5.2 \times 10^{-3} \text{ m}^3)(9.80 \text{ m/s}^2)$$
$$= \boxed{54 \text{ N}}$$

(b) The percentage of body weight contributed by the blood is

$$\text{Percentage} = \frac{54 \text{ N}}{690 \text{ N}} \times 100 = \boxed{7.8\%}$$

A convenient way to compare densities is to use the concept of **specific gravity**. The specific gravity of a substance is its density divided by the density of a standard reference material, usually chosen to be water at 4 °C.

$$\text{Specific gravity} = \frac{\text{Density of substance}}{\text{Density of water at 4 °C}} = \frac{\text{Density of substance}}{1.000 \times 10^3 \text{ kg/m}^3} \quad (11.2)$$

Being the ratio of two densities, specific gravity has no units. For example, **Table 11.1** reveals that diamond has a specific gravity of 3.52, since the density of diamond is 3.52 times the density of water at 4 °C.

The next two sections deal with the important concept of pressure. We will see that the density of a fluid is one factor determining the pressure that a fluid exerts.

11.2 Pressure

People who have fixed a flat tire know something about pressure. The final step in the job is to reinflate the tire to the proper pressure. The underinflated tire is soft because it contains an insufficient number of air molecules to push outward against the rubber and give the tire that solid feel. When air is added from a pump, the number of molecules and the collective force they exert are increased. The air molecules within a tire are free to wander throughout its entire volume, and in the course of their wandering they collide with one another and the inner walls of the tire. The collisions with the walls allow the air to exert a force against every part of the wall surface, as **Figure 11.1** shows. The pressure *P* exerted by a fluid is defined in Section 10.7 (Equation 10.19) as the magnitude *F* of the force acting perpendicular to a surface divided by the area *A* over which the force acts:

$$P = \frac{F}{A} \tag{11.3}$$

FIGURE 11.1 In colliding with the inner walls of the tire, the air molecules (blue dots) exert a force on every part of the wall surface. If a small cube were inserted inside the tire, the cube would experience forces (blue arrows) acting perpendicular to each of its six faces.

The SI unit for pressure is a newton/meter^2 (N/m^2), a combination that is referred to as a *pascal* (Pa). A pressure of 1 Pa is a very small amount. Many common situations involve pressures of approximately 10^5 Pa, an amount referred to as one *bar* of pressure. Alternatively, force can be measured in pounds and area in square inches, so another unit for pressure is pounds per square inch (lb/in.^{2}), often abbreviated as "psi."

Because of its pressure, the air in a tire applies a force to any surface with which the air is in contact. Suppose, for instance, that a small cube is inserted inside the tire. As **Figure 11.1** shows, the air pressure causes a force to act perpendicularly on each face of the cube. In a similar fashion, a liquid such as water also exerts pressure. A swimmer, for example, feels the water pushing perpendicularly inward everywhere on her body, as **Figure 11.2** illustrates. In general, a static fluid cannot produce a force parallel to a surface, for if it did, the surface would apply a reaction force to the fluid, consistent with Newton's action–reaction law. In response, the fluid would flow and would not then be static.

While fluid pressure can generate a force, which is a vector quantity, ***pressure itself is not a vector***. In the definition of pressure, *P* = *F*/*A*, the symbol *F* refers only to the magnitude of the force, so that pressure has no directional characteristic. The force generated by the pressure of a static fluid is always perpendicular to the surface that the fluid contacts, as Example 2 illustrates.

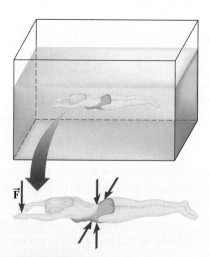

FIGURE 11.2 Water applies a force perpendicular to each surface within the water, including the walls and bottom of the swimming pool, and all parts of the swimmer's body.

EXAMPLE 2 | The Force on a Swimmer

Suppose that the pressure acting on the back of a swimmer's hand is 1.2×10^5 Pa, a realistic value near the bottom of the diving end of a pool. The surface area of the back of the hand is 8.4×10^{-3} m². **(a)** Determine the magnitude of the force that acts on it. **(b)** Discuss the direction of the force.

Reasoning From the definition of pressure in Equation 11.3, we can see that the magnitude of the force is the pressure times the area. The direction of the force is always perpendicular to the surface that the water contacts.

| **Problem-Solving Insight** Force is a vector, but pressure is not.

Solution **(a)** A pressure of 1.2×10^5 Pa is 1.2×10^5 N/m². From Equation 11.3, we find

$$F = PA = (1.2 \times 10^5 \, \text{N/m}^2)(8.4 \times 10^{-3} \, \text{m}^2) = \boxed{1.0 \times 10^3 \, \text{N}}$$

This is a rather large force, about 230 lb.

(b) In **Figure 11.2**, the hand (palm downward) is oriented parallel to the bottom of the pool. Since the water pushes perpendicularly against the back of the hand, the force $\vec{F}$ is directed downward in the drawing. This downward-acting force is balanced by an upward-acting force on the palm, so that the hand is in equilibrium. If the hand were rotated by 90°, the directions of these forces would also be rotated by 90°, always being perpendicular to the hand.

A person need not be under water to experience the effects of pressure. Walking about on land, we are at the bottom of the earth's atmosphere, which is a fluid and pushes inward on our bodies just like the water in a swimming pool. As **Figure 11.3** indicates, there is enough air above the surface of the earth to create the following pressure at sea level:

Atmospheric pressure at sea level 1.013×10^5 Pa = 1 atmosphere

This pressure corresponds to 14.70 lb/in.² and is referred to as one *atmosphere* (atm), a significant amount of pressure. Look, for instance, in **Figure 11.3** at the results of pumping out the air from within a gasoline can. With no internal air to push outward, the inward push of the external air is unbalanced and is strong enough to crumple the can.

THE PHYSICS OF . . . lynx paws.   In contrast to the situation in **Figure 11.3**, reducing the pressure is sometimes beneficial. Lynx, for example, are well suited for hunting on snow because of their oversize paws (see **Figure 11.4**). The large paws act as snowshoes and distribute the weight over a large area. Thus, they reduce the weight per unit area, or the pressure that the cat applies to the surface, which helps to keep it from sinking into the snow.

FIGURE 11.3 Atmospheric pressure at sea level is 1.013×10^5 Pa, which is sufficient to crumple a can if the inside air is pumped out.

Area of base = 1.000 m²
Crumpled can   Pump
Force = 1.013×10^5 N

Michael Lane/123RF

FIGURE 11.4 Lynx have large paws that act as natural snowshoes.

Check Your Understanding

(The answers are given at the end of the book.)

1. **BIO** As you climb a mountain, your ears "pop" because of the changes in atmospheric pressure. In which direction, outward or inward, does your eardrum move **(a)** as you climb up and **(b)** as you climb down?

2. A bottle of juice is sealed under partial vacuum, with a lid on which a red dot or "button" is painted. Around the button the following phrase is printed: "Button pops up when seal is broken." Why does the button remain pushed in when the seal is intact? **(a)** The pressure inside the bottle is greater than the pressure outside the bottle. **(b)** The pressure inside the bottle is less than the pressure outside the bottle. **(c)** There is a greater force acting on the interior surface of the seal than acts on the exterior surface.

3. A method for resealing a partially full bottle of wine under a vacuum uses a specially designed rubber stopper to close the bottle. A simple pump is attached to the stopper, and to remove air from the bottle, the plunger of the pump is pulled up and then released. After about 15 pull-and-release cycles the wine is under a partial vacuum. On the fifteenth pull-and-release cycle, does it require **(a)** more force, **(b)** less force, or **(c)** the same force to pull the plunger up than it did on the first cycle?

11.3 | Pressure and Depth in a Static Fluid

The deeper an underwater swimmer goes, the more strongly the water pushes on his body and the greater is the pressure that he experiences. To determine the relation between pressure and depth, we turn to Newton's second law ($\Sigma\vec{F} = m\vec{a}$). In using the second law, we will focus on two external forces that act on the fluid. One is the gravitational force—that is, the weight of the fluid. The other is the collisional force that is responsible for fluid pressure, as Section 11.2 discusses. Since the fluid is at rest, its acceleration is zero ($\vec{a} = 0$ m/s^2), and it is in equilibrium. By applying the second law in the form $\Sigma\vec{F} = 0$, we will derive a relation between pressure and depth. This relation is especially important because it leads to Pascal's principle (Section 11.5) and Archimedes' principle (Section 11.6), both of which are essential in describing the properties of static fluids.

Interactive Figure 11.5 shows a container of fluid and focuses attention on one column of the fluid. The free-body diagram in the figure shows all the vertical forces acting on the column. On the top face (area $= A$), the fluid pressure P_1 generates a downward force whose magnitude is P_1A. Similarly, on the bottom face, the pressure P_2 generates an upward force of magnitude P_2A. The pressure P_2 is greater than the pressure P_1 because the bottom face supports the weight of more fluid than the upper one does. In fact, the excess weight supported by the bottom face is exactly the weight of the fluid within the column. As the free-body diagram indicates, this weight is mg, where m is the mass of the fluid and g is the magnitude of the acceleration due to gravity. Since the column is in equilibrium, we can set the sum of the vertical forces equal to zero and find that

$$\Sigma F_y = P_2A - P_1A - mg = 0 \quad \text{or} \quad P_2A = P_1A + mg$$

The mass m is related to the density ρ and the volume V of the column by $m = \rho V$ (Equation 11.1). Since the volume is the cross-sectional area A times the vertical dimension h, we have $m = \rho A h$. With this substitution, the condition for equilibrium becomes $P_2A = P_1A + \rho A h g$. The area A can be eliminated algebraically from this expression, with the result that

$$P_2 = P_1 + \rho g h \tag{11.4}$$

Equation 11.4 indicates that if the pressure P_1 is known at a higher level, the larger pressure P_2 at a deeper level can be calculated by adding the increment $\rho g h$. In determining the pressure increment $\rho g h$, we assumed that the density ρ is the same at any vertical distance h or, in other words, the fluid is incompressible. The assumption is reasonable for liquids, since the bottom layers can support the upper layers with little compression. In a gas, however, the lower layers are compressed markedly by the weight of the upper layers, with the result that the density varies with vertical distance. For example, the density of our atmosphere is larger near the earth's surface than it is at higher altitudes.

Pressure = P_1
Area = A

$+y$ axis

P_1A

mg

P_2A

Pressure = P_2
Area = A

(a)

(b) Free-body diagram of the column

INTERACTIVE FIGURE 11.5
(a) A container of fluid in which one column of the fluid is outlined. The fluid is at rest. (b) The free-body diagram, showing the vertical forces acting on the column.

When applied to gases, the relation $P_2 = P_1 + \rho gh$ can be used only when h is small enough that any variation in ρ is negligible.

A significant feature of Equation 11.4 is that the pressure increment ρgh is affected by the vertical distance h, but not by any horizontal distance within the fluid. Conceptual Example 3 helps to clarify this feature.

CONCEPTUAL EXAMPLE 3 | The Hoover Dam

Lake Mead is the largest wholly artificial reservoir in the United States and was formed after the completion of the Hoover Dam in 1936. As **Figure 11.6a** suggests, the water in the reservoir backs up behind the dam for a considerable distance (about 200 km or 120 miles). Suppose that all the water were removed, except for a relatively narrow vertical column in contact with the dam. **Figure 11.6b** shows a side view of this hypothetical situation, in which the water against the dam has the same depth as in **Figure 11.6a**. How would the dam needed to contain the water in this hypothetical reservoir compare with the Hoover Dam? Would it need to be **(a)** less massive or **(b)** equally massive?

Reasoning Imagine a small square in the inner face of the dam, located beneath the water. The magnitude of the force on this square is the product of its area and the pressure of the water, according to Equation 11.3. However, the relation $P_2 = P_1 + \rho gh$ (Equation 11.4) indicates that the pressure at a given point depends on the vertical distance h that the small square is below the water. Thus, the force exerted by the water at any given location depends on the depth at that location.

Answer (a) is incorrect. Since our hypothetical reservoir contains much less water than Lake Mead, it is tempting to say that a less massive structure would be required. Note from the Reasoning section that the force exerted on a given section of the dam depends on how far below the surface the section is located. The horizontal distance of the water backed up behind the dam does not appear in Equation 11.4, and, therefore, has no effect on the pressure and, hence, on the force.

Answer (b) is correct. The force exerted on a given section of the dam depends only on how far that section is located vertically below the surface (see the Reasoning section). Certainly, as one goes deeper and deeper, the water pressure and force become greater. But no matter how deep one goes, the force that the water applies on a given section of the dam does not depend on the amount of water backed up behind the dam. Thus, the dam for our imaginary reservoir would sustain the same forces that the Hoover Dam sustains and, therefore, would need to be equally massive.

(a)   (b)

FIGURE 11.6 (a) The Hoover Dam in Nevada and Lake Mead behind it. (b) A hypothetical reservoir formed by removing most of the water from Lake Mead. Conceptual Example 3 compares the dam needed for this hypothetical reservoir with the Hoover Dam.

Adam G. Sylvester/Science Source

The next example deals further with the relationship between pressure and depth given by Equation 11.4.

EXAMPLE 4 | The Swimming Hole

Figure 11.7 shows the cross section of a swimming hole. Points *A* and *B* are both located at a distance of *h* = 5.50 m below the surface of the water. Find the pressure at each of these two points.

Reasoning The pressure at point *B* is the same as that at point *A*, since both are located at the *same vertical distance* beneath the surface and only the vertical distance *h* affects the pressure increment $\rho g h$ in Equation 11.4. To understand this important feature more clearly, consider the path *AA'B'B* in **Figure 11.7**. The pressure decreases on the way up along the vertical segment *AA'* and increases by the same amount on the way back down along segment *B'B*. Since no change in pressure occurs along the horizontal segment *A'B'*, the pressure is the same at *A* and *B*.

> **Problem-Solving Insight** The pressure at any point in a fluid depends on the vertical distance *h* of the point beneath the surface. However, for a given vertical distance, the pressure is the same, no matter where the point is located horizontally in the fluid.

Solution The pressure acting on the surface of the water is the atmospheric pressure of 1.01×10^5 Pa. Using this value as P_1 in Equation 11.4, we can determine a value for the pressure P_2 at

FIGURE 11.7 The pressures at points *A* and *B* are the same, since both points are located at the same vertical distance of 5.50 m beneath the surface of the water.

either point *A* or *B*, both of which are located 5.50 m under the water. **Table 11.1** gives the density of water as 1.000×10^3 kg/m³.

$$P_2 = P_1 + \rho g h$$
$$P_2 = 1.01 \times 10^5 \text{ Pa} + (1.000 \times 10^3 \text{ kg/m}^3)(9.80 \text{ m/s}^2)(5.50 \text{ m})$$
$$= \boxed{1.55 \times 10^5 \text{ Pa}}$$

Figure 11.8 shows an irregularly shaped container of liquid. Reasoning similar to that used in Example 4 leads to the conclusion that the pressure is the same at points *A*, *B*, *C*, and *D*, since each is at the same vertical distance *h* beneath the surface. In effect, the arteries in our bodies are also an irregularly shaped "container" for the blood. The next example examines the blood pressure at different places in this "container."

Richard Megna/Fundamental Photographs

FIGURE 11.8 Since points *A*, *B*, *C*, and *D* are at the same distance *h* beneath the liquid surface, the pressure at each of them is the same.

EXAMPLE 5 | BIO   The Physics of Blood Pressure

Blood in the arteries is flowing, but as a first approximation, the effects of this flow can be ignored and the blood treated as a static fluid. Estimate the amount by which the blood pressure P_2 in the anterior tibial artery at the foot exceeds the blood pressure P_1 in the aorta at the heart when a person is **(a)** reclining horizontally as in **Figure 11.9a** and **(b)** standing as in **Figure 11.9b**.

Reasoning and Solution   **(a)** When the body is horizontal, there is little or no vertical separation between the feet and the heart. Since $h = 0$ m,

$$P_2 - P_1 = \rho gh = \boxed{0 \text{ Pa}} \qquad \text{(11.4)}$$

(b) When an adult is standing up, the vertical separation between the feet and the heart is about 1.35 m, as **Figure 11.9b** indicates. **Table 11.1** gives the density of blood as 1060 kg/m³, so that

$$P_2 - P_1 = \rho gh = (1060 \text{ kg/m}^3)(9.80 \text{ m/s}^2)(1.35 \text{ m})$$

$$= \boxed{1.40 \times 10^4 \text{ Pa}}$$

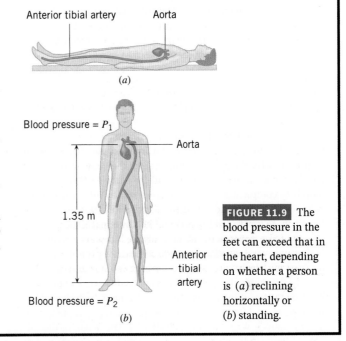

(a)

Blood pressure = P_1

1.35 m

Blood pressure = P_2

(b)

FIGURE 11.9   The blood pressure in the feet can exceed that in the heart, depending on whether a person is (a) reclining horizontally or (b) standing.

Sometimes fluid pressure places limits on how a job can be done. Conceptual Example 6 and Example 7 illustrate how fluid pressure restricts the height to which water can be pumped, as well as placing a limit on the length of a straw.

CONCEPTUAL EXAMPLE 6 | The Physics of Pumping Water

Figure 11.10 shows two methods for pumping water from a well. In one method, the pump is submerged in the water at the bottom of the well, while in the other, it is located at ground level. If the well is shallow, either technique can be used. However, if the well is very deep, only one of the methods works. Which pumping method works, **(a)** the submerged pump or **(b)** the pump located at ground level?

Reasoning   To answer this question, we need to examine the nature of the job done by the pump in each place. The pump at the bottom of the well pushes water up the pipe, while the pump at ground level does not push water at all. Instead, the ground-level pump removes air from the pipe, creating a partial vacuum within it. (It's acting just like you do when drinking through a straw. You draw some of the air out of the straw, and the external air pressure pushes the liquid up into it, as discussed in Example 7.)

Answer (b) is incorrect.   As the pump at ground level removes air from the pipe, the pressure above the water within the pipe is reduced (see point *A* in the drawing). The greater air pressure outside the pipe (see point *B*) pushes water up the pipe. However, even the strongest pump can only remove *all* of the air. Once the air is completely removed, an increase in pump strength does not increase the height to which the water is pushed by the external air pressure. Thus, the ground-level pump can only cause water to rise to a certain maximum height and cannot be used for very deep wells.

Answer (a) is correct.   For a very deep well, the column of water becomes very tall, and the pressure at the bottom of the pipe becomes large, due to the pressure increment ρgh in the relation

FIGURE 11.10   A water pump can be placed at the bottom of a well or at ground level. Conceptual Example 6 discusses the two placements.

$P_2 = P_1 + \rho gh$ (Equation 11.4). However, as long as the pump can push with sufficient strength to overcome the large pressure, it can shove the next increment of water into the pipe, so the method can be used for very deep wells.

Related Homework: Problem 20

EXAMPLE 7 | BIO The Physics of Drinking Straws—How Long Can They Be?

Consider for a moment how your favorite cold beverage ends up in your mouth, when you drink it with a straw. Many people would say that you "suck" the liquid up the straw with your mouth. Based on the fact that the liquid moves up the straw due to a difference in pressure, what is the maximum length of a vertical drinking straw?

Reasoning and Solution Figure 11.11*a* shows that when an open-ended drinking straw is placed in a liquid, the liquid quickly fills the straw to the same height as the top of the liquid in the glass. One atmosphere of pressure is pushing down on the top of the liquid in the glass, as well as the liquid in the open straw. In Figure 11.11*b*, the person drinking from the straw forms a seal around the straw with their lips and then inhales through their mouth by contracting their diaphragm muscle. This increases the volume of the lungs and lowers the air pressure inside them. Since the air in the straw and lungs share the same space, the air pressure in the straw above the liquid is also reduced. As a result, the higher atmospheric pressure outside of the straw pushes the liquid up the straw. The height that the column of liquid in the straw rises above the liquid in the glass depends on the difference in pressure between the top of the column (P_2) and the liquid in the glass (P_1). If the pressure difference is great enough, the liquid is pushed up into the mouth, where it can be swallowed. The height of the column of liquid in the straw can be determined from Equation 11.4:

$$P_2 = P_1 + \rho g h \Rightarrow h = \frac{P_2 - P_1}{\rho g}$$

The maximum value of h will be determined by the maximum value of the pressure difference, $P_2 - P_1$. The value of P_2 is 1 atmosphere, or 1.103×10^5 Pa, and the smallest pressure for P_1 is, in principle, 0 Pa, which assumes that all of the air above the liquid in the straw is removed. Thus, under the best of circumstances, the greatest pressure difference can only be 1 atmosphere. Using this information, we can calculate the maximum value for h.

$$h = \frac{1.01 \times 10^5 \text{ Pa} - 0 \text{ Pa}}{(1.000 \times 10^3 \text{ kg/m}^3)(9.80 \text{ m/s}^2)} = \boxed{10.3 \text{ m}}$$

Here, we have assumed that the density of the liquid in the glass is equal to that of fresh water. This is a fairly long straw at over 33 feet! Since the act of breathing in only lowers the pressure in the straw, but does not create the condition of zero pressure, we would expect the maximum length of the straw to be somewhat less than 10.3 m. Simple experiments, which you could recreate for yourself, show that liquid can be drawn up to a height of at least 6 m. This result also holds for the situation described in Example 6 for the case in which the pump is located at ground level (see Figure 11.10). The maximum depth from which the ground-level can pump fresh water is 10.3 m.

(a)   (b)

FIGURE 11.11 (*a*) An open-ended straw is placed in a glass of liquid. The liquid rises in the straw to match the height of the top of the liquid in the glass. The pressure at the top of the liquid is 1 atmosphere both inside and outside of the straw. (*b*) When the drinker inhales, the air pressure above the liquid inside the straw is reduced below atmospheric pressure. The higher atmospheric pressure outside pushes the liquid up the straw. The height of the liquid column in the straw depends on the pressure difference between the top of the liquid in the glass and the top of the liquid in the straw.

Related Homework: Problem 20

Check Your Understanding

(The answers are given at the end of the book.)

4. A scuba diver is swimming under water, and **CYU Figure 11.1** shows a plot of the water pressure acting on the diver as a function of time. In each of the three regions, **(a)** A → B, **(b)** B → C, and **(c)** C → D, does the depth of the diver increase, decrease, or remain constant?

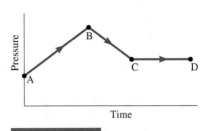

CYU FIGURE 11.1

5. A 15-meter-high tank is closed and completely filled with water. A valve is then opened at the bottom of the tank and water begins to flow out. When the water stops flowing, will the tank be completely empty, or will there still be a noticeable amount of water in it?

6. Could you use a straw to sip a drink on the moon? **(a)** Yes. It would be no different than drinking with a straw on earth. **(b)** No, because there is no air on the moon and, therefore, no air pressure to push the liquid up the straw. **(c)** Yes, and it is easier on the moon because the acceleration due to gravity on the moon is only $\frac{1}{6}$ of that on the earth.

7. A scuba diver is below the surface of the water when a storm approaches, dropping the air pressure above the water. Would a sufficiently sensitive pressure gauge attached to his wrist register this drop in air pressure? Assume that the diver's wrist does not move as the storm approaches.

11.4 | Pressure Gauges

One of the simplest pressure gauges is the mercury barometer used for measuring atmospheric pressure. This device is a tube sealed at one end, filled completely with mercury, and then inverted, so that the open end is under the surface of a pool of mercury (see **Figure 11.12**). Except for a negligible amount of mercury vapor, the space above the mercury in the tube is empty, and the pressure P_1 is nearly zero there. The pressure P_2 at point A at the bottom of the mercury column is the same as the pressure at point B—namely, atmospheric pressure—because these two points are at the same level. With $P_1 = 0$ Pa and $P_2 = P_{atm}$, it follows from Equation 11.4 that $P_{atm} = 0$ Pa $+ \rho g h$. Thus, the atmospheric pressure can be determined from the height h of the mercury in the tube, the density ρ of mercury, and the acceleration due to gravity. Usually, weather forecasters report the pressure in terms of the height h, expressing it in millimeters or inches of mercury. For instance, using $P_{atm} = 1.013 \times 10^5$ Pa and $\rho = 13.6 \times 10^3$ kg/m³ for the density of mercury, we find that $h = P_{atm}/(\rho g) = 760$ mm (29.9 inches).* Slight variations from this value occur, depending on weather conditions and altitude.

Figure 11.13 shows another kind of pressure gauge, the open-tube manometer. The phrase "open-tube" refers to the fact that one side of the U-tube is open to atmospheric pressure. The tube contains a liquid, often mercury, and its other side is connected to the container whose pressure P_2 is to be measured. When the pressure in the container is equal to the atmospheric pressure, the liquid levels in both sides of the U-tube are the same. When the pressure in the container is greater than atmospheric pressure, as in **Figure 11.13**, the liquid in the tube is pushed downward on the left side and upward on the right side. The relation $P_2 = P_1 + \rho g h$ can be used to determine the container pressure. Atmospheric pressure exists at the top of the right column, so that $P_1 = P_{atm}$. The pressure P_2 is the same at points A and B, so we find that $P_2 = P_{atm} + \rho g h$, or

$$P_2 - P_{atm} = \rho g h$$

The height h is proportional to $P_2 - P_{atm}$, which is called the **gauge pressure**. The gauge pressure is the amount by which the container pressure differs from atmospheric pressure. The actual value for P_2 is called the **absolute pressure**.

FIGURE 11.12 A mercury barometer.

FIGURE 11.13 The U-shaped tube is called an open-tube manometer and can be used to measure the pressure P_2 in a container.

---

*A pressure of one millimeter of mercury is sometimes referred to as one *torr*, to honor the inventor of the barometer, Evangelista Torricelli (1608–1647). Thus, one atmosphere of pressure is 760 torr.

Cuff

Squeeze — Release
bulb — valve

Stethoscope

Open-tube
manometer

FIGURE 11.14 A sphygmomanometer
is used to measure blood pressure.

(a)

(b)

FIGURE 11.15 (a) An external
force $\vec{F}_1$ is applied to the piston on the
left. As a result, a force $\vec{F}_2$ is exerted on
the cap on the chamber on the right.
(b) The familiar hydraulic car lift.

Problem-Solving Insight   When solving problems that deal with pressure, be
sure to note the distinction between gauge pressure and absolute pressure.

BIO **THE PHYSICS OF . . . a sphygmomanometer.**   The sphygmomanometer
is a familiar device for measuring blood pressure. As **Figure 11.14** illustrates, a squeeze
bulb can be used to inflate the cuff with air, which cuts off the flow of blood through the
artery below the cuff. When the release valve is opened, the cuff pressure drops. Blood
begins to flow again when the pressure created by the heart at the peak of its beating cycle
exceeds the cuff pressure. Using a stethoscope to listen for the initial flow, the operator can
measure the corresponding cuff gauge pressure with, for example, an open-tube manometer. This cuff gauge pressure is called the *systolic* pressure. Eventually, there comes a point
when even the pressure created by the heart at the low point of its beating cycle is sufficient
to cause blood to flow. Identifying this point with the stethoscope, the operator can measure the corresponding cuff gauge pressure, which is referred to as the *diastolic* pressure.
The systolic and diastolic pressures are reported in millimeters of mercury, and values of
less than 120 and 80, respectively, are typical of a young, healthy heart.

11.5 Pascal's Principle

As we have seen, the pressure in a fluid increases with depth, due to the weight of the
fluid above the point of interest. A completely enclosed fluid may be subjected to an additional pressure by the application of an external force. For example, **Figure 11.15a** shows
two interconnected cylindrical chambers. The chambers have different diameters and,
together with the connecting tube, are completely filled with a liquid. The larger chamber is sealed at the top with a cap, while the smaller one is fitted with a movable piston.
Consider the pressure P_1 at a point immediately beneath the piston. According to the
definition of pressure, it is the magnitude F_1 of the external force divided by piston area
A_1, or $P_1 = F_1/A_1$. If it is necessary to know the pressure P_2 *at any deeper place in the liquid,*
we just add to the value of P_1 the increment ρgh, which takes into account the depth h
below the piston: $P_2 = P_1 + \rho gh$. The important feature here is this: The pressure P_1 adds
to the pressure ρgh due to the depth of the liquid at any point, whether that point is in
the smaller chamber, the connecting tube, or the larger chamber. Therefore, if the applied
pressure P_1 is increased or decreased, the pressure at any other point within the confined
liquid changes correspondingly. This behavior is described by **Pascal's principle**.

PASCAL'S PRINCIPLE

**Any change in the pressure applied to a completely enclosed fluid is transmitted
undiminished to all parts of the fluid and the enclosing walls.**

The usefulness of the arrangement in **Figure 11.15a** becomes apparent when we
calculate the force F_2 applied by the liquid to the cap on the right side. The area of the
cap is A_2 and the pressure there is P_2. As long as the tops of the left and right chambers
are at the same level, the pressure increment ρgh is zero, so that the relation $P_2 = P_1 + \rho gh$
becomes $P_2 = P_1$. Consequently, $F_2/A_2 = F_1/A_1$, and

$$F_2 = F_1\left(\frac{A_2}{A_1}\right) \tag{11.5}$$

If area A_2 is larger than area A_1, a large force $\vec{F}_2$ can be applied to the cap on the right chamber, starting with a smaller force $\vec{F}_1$ on the left. Depending on the ratio of the areas A_2/A_1, the
force $\vec{F}_2$ can be large indeed, as in the familiar hydraulic car lift shown in **Figure 11.15b**. In
this device the force $\vec{F}_2$ is not applied to a cap that seals the larger chamber, but, rather, to a
movable plunger that lifts a car. Examples 8 and 9 deal with a hydraulic car lift.

EXAMPLE 8 | The Physics of a Hydraulic Car Lift

In the hydraulic car lift shown in **Figure 11.15b**, the input piston
on the left has a radius of $r_1 = 0.0120$ m and a negligible weight.
The output plunger on the right has a radius of $r_2 = 0.150$ m. The
combined weight of the car and the plunger is 20 500 N. Since the

output force has a magnitude of $F_2 = 20\ 500$ N, it supports the car.
Suppose that the bottom surfaces of the piston and plunger are at
the same level, so that $h = 0$ m in **Figure 11.15b**. What is the magnitude F_1 of the input force needed so that $F_2 = 20\ 500$ N?

Reasoning   When the bottom surfaces of the piston and plunger are at the same level, as in **Figure 11.15a**, Equation 11.5 applies, and we can use it to determine F_1.

> **Problem-Solving Insight**   Note that the relation $F_1 = F_2(A_1/A_2)$, which results from Pascal's principle, applies only when the points 1 and 2 lie at the same depth ($h = 0$ m) in the fluid.

Solution   According to Equation 11.5, we have

$$F_2 = F_1\left(\frac{A_2}{A_1}\right) \quad \text{or} \quad F_1 = F_2\left(\frac{A_1}{A_2}\right)$$

Using $A = \pi r^2$ for the circular areas of the piston and plunger, we find that

$$F_1 = F_2\left(\frac{A_1}{A_2}\right) = F_2\left(\frac{\pi r_1^2}{\pi r_2^2}\right) = (20\ 500\ \text{N})\frac{(0.0120\ \text{m})^2}{(0.150\ \text{m})^2} = \boxed{131\ \text{N}}$$

(a)

(b)

FIGURE 11.15 (REPEATED)
(a) An external force $\vec{F}_1$ is applied to the piston on the left. As a result, a force $\vec{F}_2$ is exerted on the cap on the chamber on the right. (b) The familiar hydraulic car lift.

Analyzing Multiple-Concept Problems

EXAMPLE 9 | A Car Lift Revisited

The data are the same as in Example 8. Suppose now, however, that the bottom surfaces of the piston and plunger are at *different* levels, such that $h = 1.10$ m in **Figure 11.15b**. The car lift uses hydraulic oil that has a density of 8.00×10^2 kg/m³. What is the magnitude F_1 of the input force that is now needed to produce an output force having a magnitude of $F_2 = 20\ 500$ N?

Reasoning   Since the bottom surfaces of the piston and plunger are at different levels, Equation 11.5 no longer

applies. Our approach will be based on the definition of pressure (Equation 11.3) and the relation between pressure and depth given in Equation 11.4. It is via Equation 11.4 that we will take into account the fact that the bottom of the output plunger is $h = 1.10$ m below the bottom of the input piston.

Knowns and Unknowns   We have the following data:

Description	Symbol	Value	Comment
Radius of input piston	r_1	0.0120 m	
Radius of output piston	r_2	0.150 m	
Magnitude of output force	F_2	20 500 N	
Density of hydraulic oil	ρ	8.00×10^2 kg/m³	
Level difference between input piston and output plunger	h	1.10 m	See **Figure 11.15b**.
Unknown Variables			
Magnitude of input force	F_1	?	

Modeling the Problem

STEP 1 Force and Pressure   The input force F_1 is determined by the pressure P_1 acting on the bottom surface of the input piston, which has an area A_1. According to Equation 11.3, the input force is $F_1 = P_1 A_1$, where $A_1 = \pi r_1^2$ because the piston has a circular cross section with a radius r_1. With this expression for the area A_1, the input force can be written as in Equation 1 at the right. We have a value for r_1, but P_1 is unknown, so we turn to Step 2 in order to obtain it.

$$F_1 = P_1 \pi r_1^2 \qquad (1)$$

$$\boxed{?}$$

STEP 2 Pressure and Depth in a Static Fluid In **Figure 11.15b**, the bottom surface of the plunger at point B is at the same level as point A, which is at a depth h beneath the input piston. Equation 11.4 applies, so that

$$P_2 = P_1 + \rho g h \quad \text{or} \quad \boxed{P_1 = P_2 - \rho g h}$$

This result for P_1 can be substituted into Equation 1, as shown at the right. Now it is necessary to have a value for P_2, the pressure at the bottom surface of the output plunger, which we obtain in Step 3.

$$F_1 = P_1 \pi r_1^2 \qquad (1)$$
$$\boxed{P_1 = P_2 - \rho g h} \qquad (2)$$
$$\boxed{?}$$

STEP 3 Pressure and Force By using Equation 11.3, we can relate the unknown pressure P_2 to the given output force F_2 and the area A_2 of the bottom surface of the output plunger: $P_2 = F_2/A_2$. The area A_2 is a circle with a radius r_2, so $A_2 = \pi r_2^2$. Thus, the pressure P_2 is

$$\boxed{P_2 = \frac{F_2}{\pi r_2^2}}$$

At the right, this result is substituted into Equation 2.

$$F_1 = P_1 \pi r_1^2 \qquad (1)$$
$$P_1 = P_2 - \rho g h \qquad (2)$$
$$\boxed{P_2 = \frac{F_2}{\pi r_2^2}}$$

Solution Combining the results of each step algebraically, we find that

$$\underset{\text{STEP 1}}{\downarrow} \quad \underset{\text{STEP 2}}{\downarrow} \qquad \underset{\text{STEP 3}}{\downarrow}$$
$$F_1 = P_1 \pi r_1^2 = (P_2 - \rho g h)\pi r_1^2 = \left(\frac{F_2}{\pi r_2^2} - \rho g h\right)\pi r_1^2$$

Thus, we find that the necessary input force is

$$F_1 = \frac{F_2 r_1^2}{r_2^2} - \rho g h \pi r_1^2$$

$$= \frac{(20\,500\text{ N})(0.0120\text{ m})^2}{(0.150\text{ m})^2} - (8.00 \times 10^2\text{ kg/m}^3)(9.80\text{ m/s}^2)(1.10\text{ m})\pi(0.0120\text{ m})^2$$

$$= \boxed{127\text{ N}}$$

The answer here is less than the answer in Example 8 because the weight of the 1.10-m column of hydraulic oil provides some of the input force to support the car and plunger.

Related Homework: Problem 31

In a device such as a hydraulic car lift, the same amount of work is done by both the input and output forces in the absence of friction. The larger output force $\vec{F}_2$ moves through a smaller distance, while the smaller input force $\vec{F}_1$ moves through a larger distance. The work, being the product of the magnitude of the force and the distance, is the same in either case since mechanical energy is conserved (assuming that friction is negligible).

An enormous variety of clever devices use hydraulic fluids, just as the car lift does. In **Figure 11.16**, for instance, a hydraulic fluid multiplies a small input force into the large output force required to operate the digging scoop of the excavator.

FIGURE 11.16 Excavators such as this one are a familiar sight at construction jobs. They use hydraulic fluids to generate the large output forces necessary for digging. The shiny cylinder attached to the digging scoop is a sure sign that a hydraulic fluid is at work.

Heinrich van den Berg/Gallo Images ROOTS RF collection/Getty Images

11.6 Archimedes' Principle

Anyone who has tried to push a beach ball under the water has felt how the water pushes back with a strong upward force. This upward force is called the **buoyant force**, and all fluids apply such a force to objects that are immersed in them. The buoyant force exists because fluid pressure is larger at greater depths.

In **Interactive Figure 11.17** a cylinder of height h is being held under the surface of a liquid. The pressure P_1 on the top face generates the downward force P_1A, where A is the area of the face. Similarly, the pressure P_2 on the bottom face generates the upward force P_2A. Since the pressure is greater at greater depths, the upward force exceeds the downward force. Consequently, the liquid applies to the cylinder a net upward force, or buoyant force, whose magnitude F_B is

$$F_B = P_2A - P_1A = (P_2 - P_1)A = \rho g h A$$

We have substituted $P_2 - P_1 = \rho g h$ from Equation 11.4 into this result. In so doing, we find that the buoyant force equals $\rho g h A$. The quantity hA is the volume of liquid that the cylinder moves aside or displaces in being submerged, and ρ denotes the density of the liquid, not the density of the material from which the cylinder is made. Therefore, $\rho h A$ gives the mass m of the displaced fluid, so that the buoyant force equals mg, the weight of the displaced fluid. The phrase "weight of the displaced fluid" refers to the weight of the fluid that would spill out if the container were filled to the brim before the cylinder is inserted into the liquid. The buoyant force is not a new type of force. It is just the name given to the net upward force exerted by the fluid on the object.

The shape of the object in **Interactive Figure 11.17** is not important. No matter what its shape, the buoyant force pushes it upward in accord with **Archimedes' principle**. It was an impressive accomplishment that the Greek scientist Archimedes (ca. 287–212 BC) discovered the essence of this principle so long ago.

ARCHIMEDES' PRINCIPLE

Any fluid applies a buoyant force to an object that is partially or completely immersed in it; the magnitude of the buoyant force equals the weight of the fluid that the object displaces:

$$\underbrace{F_B}_{\substack{\text{Magnitude of} \\ \text{buoyant force}}} = \underbrace{W_{\text{fluid}}}_{\substack{\text{Weight of} \\ \text{displaced fluid}}} \qquad (11.6)$$

The effect that the buoyant force has depends on its strength compared with the strengths of the other forces that are acting. For example, if the buoyant force is strong enough to balance the force of gravity, an object will float in a fluid. **Figure 11.18** explores this possibility. In part a, a block that weighs 100 N displaces some liquid, and the liquid applies a buoyant force F_B to the block, according to Archimedes' principle. Nevertheless, if the block were released, it would fall further into the liquid because the buoyant force is not sufficiently strong to balance the weight of the block. In part b, however, enough of the block is submerged to provide a buoyant force that can balance the 100-N weight, so the block is in equilibrium and floats when released. If the buoyant force were not large enough to balance the weight, even with the block completely submerged, the block would sink. Even if an object sinks, there is still a buoyant force acting on it; it's just that the buoyant force is not large enough to balance the weight. Example 9 provides additional insight into what determines whether an object floats or sinks in a fluid.

INTERACTIVE FIGURE 11.17 The fluid applies a downward force P_1A to the top face of the submerged cylinder and an upward force P_2A to the bottom face.

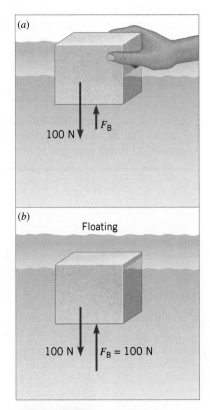

FIGURE 11.18 (*a*) An object of weight 100 N is being immersed in a liquid. The deeper the object is, the more liquid it displaces, and the stronger the buoyant force is. (*b*) The buoyant force matches the 100-N weight, so the object floats.

EXAMPLE 10 | A Swimming Raft

A solid, square pinewood raft measures 4.0 m on a side and is 0.30 m thick. **(a)** Determine whether the raft floats in water, and **(b)** if so, how much of the raft is beneath the surface (see the distance h in **Figure 11.19**).

Reasoning To determine whether the raft floats, we will compare the weight of the raft to the maximum possible buoyant force and see whether there could be enough buoyant force to balance the weight. If so, then the value of the distance h can be obtained by utilizing the fact that the floating raft is in equilibrium, with the magnitude of the buoyant force equaling the raft's weight.

Solution **(a)** The weight of the raft is $W_{\text{raft}} = m_{\text{pine}}g$ (Equation 4.5), where m_{pine} is the mass of the raft and can be calculated as $m_{\text{pine}} = \rho_{\text{pine}}V_{\text{pine}}$ (Equation 11.1). The pinewood's density is $\rho_{\text{pine}} = 550 \text{ kg/m}^3$ (**Table 11.1**), and its volume is $V_{\text{pine}} = 4.0 \text{ m} \times 4.0 \text{ m} \times 0.30 \text{ m} = 4.8 \text{ m}^3$. Thus, we find the weight of the raft to be

$$W_{\text{raft}} = m_{\text{pine}}g = (\rho_{\text{pine}}V_{\text{pine}})g$$
$$= (550 \text{ kg/m}^3)(4.8 \text{ m}^3)(9.80 \text{ m/s}^2) = 26\,000 \text{ N}$$

The maximum possible buoyant force occurs when the entire raft is under the surface, displacing a volume of water that is $V_{\text{water}} = V_{\text{pine}} = 4.8 \text{ m}^3$. According to Archimedes' principle, the weight of this volume of water is the maximum buoyant force F_B^{MAX}. It can be obtained using the density of water:

$$F_B^{\text{MAX}} = \rho_{\text{water}}V_{\text{water}}g$$
$$= (1.000 \times 10^3 \text{ kg/m}^3)(4.8 \text{ m}^3)(9.80 \text{ m/s}^2) = 47\,000 \text{ N}$$

Since the maximum possible buoyant force exceeds the 26 000-N weight of the raft, the raft will float only partially submerged at a distance h beneath the water.

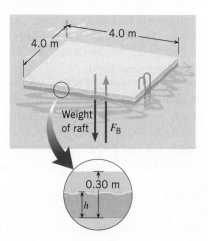

FIGURE 11.19 A raft floating with a distance h beneath the water.

(b) We now find the value of h. The buoyant force balances the raft's weight, so $F_B = 26\,000 \text{ N}$. However, according to Equation 11.6, the magnitude of the buoyant force is also the weight of the displaced water, so $F_B = 26\,000 \text{ N} = W_{\text{water}}$. Using the density of water, we can also express the weight of the displaced water as $W_{\text{water}} = \rho_{\text{water}}V_{\text{water}}g$, where the water volume is $V_{\text{water}} = 4.0 \text{ m} \times 4.0 \text{ m} \times h$. As a result,

$$26\,000 \text{ N} = W_{\text{water}} = \rho_{\text{water}}(4.0 \text{ m} \times 4.0 \text{ m} \times h)g$$

$$h = \frac{26\,000 \text{ N}}{\rho_{\text{water}}(4.0 \text{ m} \times 4.0 \text{ m})g}$$

$$= \frac{26\,000 \text{ N}}{(1.000 \times 10^3 \text{ kg/m}^3)(4.0 \text{ m} \times 4.0 \text{ m})(9.80 \text{ m/s}^2)}$$

$$= \boxed{0.17 \text{ m}}$$

In order to decide whether the raft will float in part (a) of Example 10, we compared the weight of the raft $(\rho_{\text{pine}}V_{\text{pine}})g$ to the maximum possible buoyant force $(\rho_{\text{water}}V_{\text{water}})g = (\rho_{\text{water}}V_{\text{pine}})g$. The comparison depends only on the densities ρ_{pine} and ρ_{water}. The take-home message is this: **Any object that is *solid throughout* will float in the liquid if its density is less than the density of the liquid, and will sink in the liquid if its density is greater than the density of the liquid.** For instance, at 0 °C ice has a density of 917 kg/m³, whereas water has a density of 1000 kg/m³. Therefore, ice floats in water. In the case that the object and liquid have equal densities, the weight of the object and the buoyant force have equal magnitudes, and the object will suspend motionless in the medium, that is, it will neither sink nor float. In such a case, the object is said to have **neutral buoyancy** in the liquid.

Although a solid piece of a high-density material like steel will sink in water, such materials can, nonetheless, be used to make floating objects. A supertanker, for example, floats because it is *not* solid metal. It contains enormous amounts of empty space and, because of its shape, displaces enough water to balance its own large weight. Conceptual Example 11 focuses on an interesting aspect of a floating ship.

CONCEPTUAL EXAMPLE 11 | How Much Water Is Needed to Float a Ship?

A ship floating in the ocean is a familiar sight. But is all that water really necessary? Can an ocean vessel float in the amount of water that a swimming pool contains, for instance?

Reasoning and Solution In principle, a ship can float in much less than the amount of water in a swimming pool. To see why,

look at **Figure 11.20**. Part *a* shows the ship floating in the ocean because it contains empty space within its hull and displaces enough water to balance its own weight. Part *b* shows the water that the ship displaces, which, according to Archimedes' principle, has a weight that equals the ship's weight. Although this wedge-shaped portion of water represents the water *displaced* by the ship, it is not the amount that must be present to float the ship, as part *c* illustrates. This part of the drawing shows a canal, the cross section of which matches the shape in part *b*. **All that is needed, in** *principle, is a thin section of water that separates the hull of the floating ship from the sides of the canal.* This thin section of water could have a very small volume indeed.

(a)   (b)   (c)

FIGURE 11.20 (*a*) A ship floating in the ocean. (*b*) This is the water that the ship displaces. (*c*) The ship floats here in a canal that has a cross section similar in shape to that in part *b*.

THE PHYSICS OF . . . a state-of-charge battery indicator. Archimedes' principle is used in some car batteries to alert the owner that recharging is necessary, via a state-of-charge indicator, such as the one illustrated in **Figure 11.21**. The battery includes a viewing port that looks down through a plastic rod, which extends into the battery acid. Attached to the end of this rod is a "cage" containing a green ball. The cage has holes in it that allow the acid to enter. When the battery is charged, the density of the acid is great enough that its buoyant force makes the ball rise to the top of the cage, to just beneath the plastic rod. The viewing port shows a green dot. As the battery discharges, the density of the acid decreases. Since the buoyant force is the weight of the acid displaced by the ball, the buoyant force also decreases. As a result, the ball sinks into one of the two chambers oriented at an angle beneath the plastic rod (see **Figure 11.21**). With the ball no longer visible, the viewing port shows a dark or black dot, warning that the battery charge is low.

Port for viewing state-of-charge indicator

Green dot   Black dot

Plastic rod
Battery acid
Cage

Charged   Discharged

FIGURE 11.21 A state-of-charge indicator for a car battery.

Archimedes' principle has allowed us to determine how an object can float in a liquid. This principle also applies to gases, as the next example illustrates.

Analyzing Multiple-Concept Problems

EXAMPLE 12 | The Physics of a Goodyear Airship

Normally, a Goodyear airship, such as the one in **Figure 11.22**, contains about 5.40×10^3 m^3 of helium (He), whose density is 0.179 kg/m^3. Find the weight W_L of the load that the airship can carry in equilibrium at an altitude where the density of air is 1.20 kg/m^3.

Reasoning The airship and its load are in equilibrium. Thus, the buoyant force F_B applied to the airship by the surrounding air balances the weight W_{He} of the helium and the weight W_L of the load, including the solid parts of the airship. The free-body diagram in **Figure 11.22b** shows these forces. The buoyant force is the weight of the air that the ship displaces, according to Archimedes' principle. The weight of air or helium is given as $W = mg$ (Equation 4.5). Using the density ρ and volume V, we can express the mass m as $m = \rho V$ (Equation 11.1).

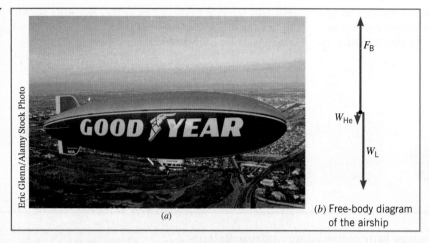

(a)

F_B

W_{He}

W_L

(b) Free-body diagram of the airship

Eric Glenn/Alamy Stock Photo

FIGURE 11.22 (*a*) A helium-filled Goodyear airship. (*b*) The free-body diagram of the airship, including the load weight W_L, the weight W_{He} of the helium, and the buoyant force F_B.

Knowns and Unknowns The following table summarizes the data:

Description	Symbol	Value
Volume of helium in airship	V_{He}	$5.40 \times 10^3 \ m^3$
Density of helium	ρ_{He}	$0.179 \ kg/m^3$
Density of air	ρ_{air}	$1.20 \ kg/m^3$
Unknown Variables		
Weight of load	W_L	?

Modeling the Problem

STEP 1 Equilibrium Because the forces in the free-body diagram (see **Figure 11.22b**) balance at equilibrium, we have

$$W_{He} + W_L = F_B$$

Rearranging this result gives Equation 1 at the right. Neither the buoyant force F_B nor the weight W_{He} of the helium is known. We will determine them in Steps 2 and 3.

$$W_L = F_B - W_{He} \qquad (1)$$

STEP 2 Weight and Density According to Equation 4.5, the weight is given by $W = mg$. On the other hand, the density ρ is defined as the mass m divided by the volume V (see Equation 11.1), so we know that $m = \rho V$. Thus, the weight of the helium in the airship is

$$\boxed{W_{He} = m_{He}g = \rho_{He}V_{He}g}$$

The result can be substituted into Equation 1, as shown at the right. We turn now to Step 3 to evaluate the buoyant force.

$$W_L = F_B - W_{He} \qquad (1)$$
$$W_{He} = \rho_{He}V_{He}g$$

STEP 3 Archimedes' Principle The buoyant force is given by Archimedes' principle as the weight of the air displaced by the airship. Thus, following the approach in Step 2, we can write the buoyant force as follows:

$$F_B = W_{air} = \rho_{air}V_{air}g$$

In this result, V_{air} is very nearly the same as V_{He}, since the volume occupied by the materials of the ship's outer structure is negligible compared to V_{He}. Assuming that $V_{air} = V_{He}$, we see that the expression for the buoyant force becomes

$$\boxed{F_B = \rho_{air}V_{He}g}$$

Substitution of this value for the buoyant force into Equation 1 is shown at the right.

$$W_L = F_B - W_{He} \qquad (1)$$
$$W_{He} = \rho_{He}V_{He}g$$
$$F_B = \rho_{air}V_{He}g$$

Solution   Combining the results of each step algebraically, we find that

STEP 1   STEP 2   STEP 3

$$W_L = F_B - W_{He} = F_B - \rho_{He}V_{He}g = \rho_{air}V_{He}\,g - \rho_{He}V_{He}g$$

The weight of the load that the airship can carry at an altitude where $\rho_{air} = 1.20 \text{ kg/m}^3$ is, then,

$$W_L = (\rho_{air} - \rho_{He})V_{He}g = (1.20 \text{ kg/m}^3 - 0.179 \text{ kg/m}^3)(5.40 \times 10^3 \text{ m}^3)(9.80\text{m/s}^2)$$

$$= \boxed{5.40 \times 10^4 \text{ N}}$$

Related Homework: Problem 43

Check Your Understanding

(The answers are given at the end of the book.)

8. A glass is filled to the brim with water and has an ice cube floating in it. When the ice cube melts, what happens? **(a)** Water spills out of the glass. **(b)** The water level in the glass drops. **(c)** The water level in the glass does not change.

9. A steel beam is suspended completely under water by a cable that is attached to one end of the beam, so it hangs vertically. Another identical beam is also suspended completely under water, but by a cable that is attached to the center of the beam, so it hangs horizontally. Which beam, if either, experiences the greater buoyant force? Neglect any change in water density with depth.

10. A glass beaker, filled to the brim with water, is resting on a scale. A solid block is then placed in the water, causing some of it to spill over. The water that spills is wiped away, and the beaker is still filled to the brim. How do the initial and final readings on the scale compare if the block is made from **(a)** wood (whose density is less than that of water) and **(b)** iron (whose density is greater than that of water)?

11. On a distant planet the acceleration due to gravity is less than it is on earth. Would you float more easily in water on this planet than on earth?

12. As a person dives toward the bottom of a swimming pool, the pressure increases noticeably. Does the buoyant force acting on her also increase? Neglect any change in water density with depth.

13. A pot is partially filled with water, in which a plastic cup is floating. Inside the floating cup is a small block of lead. When the lead block is removed from the cup and placed into the water, it sinks to the bottom. When this happens, does the water level in the pot **(a)** rise, **(b)** fall, or **(c)** remain the same?

FIGURE 11.23 Two fluid particles in a stream. At different locations in the stream the particle velocities may be different, as indicated by $\vec{v}_1$ and $\vec{v}_2$.

11.7 | Fluids in Motion

Fluids can move or flow in many ways. Water may flow smoothly and slowly in a quiet stream or violently over a waterfall. The air may form a gentle breeze or a raging tornado. To deal with such diversity, it helps to identify some of the basic types of fluid flow.

Fluid flow can be steady or unsteady. In **steady flow** the velocity of the fluid particles at any point is constant as time passes. For instance, in **Figure 11.23** a fluid particle flows with a velocity of $\vec{v}_1 = +2$ m/s past point 1. In steady flow every particle passing through this point has this same velocity. At another location the velocity may be different, as in a river, which usually flows fastest near its center and slowest near its banks. Thus, at point 2 in the figure, the fluid velocity is $\vec{v}_2 = +0.5$ m/s, and if the flow is steady, all particles passing through this point have a velocity of +0.5 m/s. **Unsteady flow** exists whenever the velocity at a point in the fluid changes as time passes. **Turbulent flow** is an extreme kind of unsteady flow and occurs when there are sharp obstacles or bends in the path of a fast-moving fluid, as in the rapids in **Figure 11.24**. In turbulent flow, the velocity at a point changes erratically from moment to moment, both in magnitude and in direction.

FIGURE 11.24 The flow of water in white-water rapids is an example of turbulent flow.

Fluid flow can be compressible or incompressible. Most liquids are nearly incompressible; that is, the density of a liquid remains almost constant as the pressure changes. To a good approximation, then, liquids flow in an incompressible manner. In contrast, gases are highly compressible. However, there are situations in which the density of a flowing gas remains constant enough that the flow can be considered incompressible.

Fluid flow can be viscous or nonviscous. A viscous fluid, such as honey, does not flow readily and is said to have a large viscosity.* In contrast, water is less viscous and flows more readily; water has a smaller viscosity than honey. The flow of a viscous fluid is an energy-dissipating process. The viscosity hinders neighboring layers of fluid from sliding freely past one another. A fluid with zero viscosity flows in an unhindered manner with no dissipation of energy. Although no real fluid has zero viscosity at normal temperatures, some fluids have negligibly small viscosities. An incompressible, nonviscous fluid is called an **ideal fluid**.

When the flow is steady, **streamlines** are often used to represent the trajectories of the fluid particles. A streamline is a line drawn in the fluid such that a tangent to the streamline at any point is parallel to the fluid velocity at that point. **Figure 11.25** shows the velocity vectors at three points along a streamline. The fluid velocity can vary (in both magnitude and direction) from point to point along a streamline, but at any given point, the velocity is constant in time, as required by the condition of steady flow. In fact, steady flow is often called **streamline flow**.

Figure 11.26a illustrates a method for making streamlines visible by using small tubes to release a colored dye into the moving liquid. The dye does not immediately mix with the liquid and is carried along a streamline. In the case of a flowing gas, such as that in a wind tunnel, streamlines are often revealed by smoke streamers, as part *b* of the figure shows.

In steady flow, the pattern of streamlines is steady in time, and, as **Figure 11.26a** indicates, no two streamlines cross one another. If they did cross, every particle arriving at the crossing point could go one way or the other. This would mean that the velocity at the crossing point would change from moment to moment, a condition that does not exist in steady flow.

FIGURE 11.25 At any point along a streamline, the velocity vector of the fluid particle at that point is tangent to the streamline.

(a)   (b)

FIGURE 11.26 (a) In the steady flow of a liquid, a colored dye reveals the streamlines. (b) A smoke streamer reveals a streamline pattern for the air flowing around this skier, as he tests his position for air resistance in a wind tunnel.

Check Your Understanding

(The answer is given at the end of the book.)

14. In steady flow, the velocity $\vec{v}$ of a fluid particle at any point is constant in time. On the other hand, the fluid in a pipe accelerates when it moves from a larger-diameter section of the pipe into a smaller-diameter section, so the velocity is increasing during the transition. Does the condition of steady flow rule out such an acceleration?

*See Section 11.11 for a discussion of viscosity.

11.8 The Equation of Continuity

Have you ever used your thumb to control the water flowing from the end of a hose, as in **Figure 11.27**? If so, you have seen that the water velocity increases when your thumb reduces the cross-sectional area of the hose opening. This kind of fluid behavior is described by the **equation of continuity**. This equation expresses the following simple idea: If a fluid enters one end of a pipe at a certain rate (e.g., 5 kilograms per second), then fluid must also leave at the same rate, assuming that there are no places between the entry and exit points to add or remove fluid. The mass of fluid per second (e.g., 5 kg/s) that flows through a tube is called the **mass flow rate**.

Animated **Figure 11.28** shows a small mass of fluid or fluid element (dark blue) moving along a tube. Upstream at position 2, where the tube has a cross-sectional area A_2, the fluid has a speed v_2 and a density ρ_2. Downstream at location 1, the corresponding quantities are A_1, v_1, and ρ_1. During a small time interval Δt, the fluid at point 2 moves a distance of $v_2\Delta t$, as the drawing shows. The volume of fluid that has flowed past this point is the cross-sectional area times this distance, or $A_2v_2\Delta t$. The mass Δm_2 of this fluid element is the product of the density and volume: $\Delta m_2 = \rho_2 A_2 v_2 \Delta t$. Dividing Δm_2 by Δt gives the mass flow rate (the mass per second):

$$\text{Mass flow rate at position 2} = \frac{\Delta m_2}{\Delta t} = \rho_2 A_2 v_2 \qquad \textbf{(11.7a)}$$

Similar reasoning leads to the mass flow rate at position 1:

$$\text{Mass flow rate at position 1} = \frac{\Delta m_1}{\Delta t} = \rho_1 A_1 v_1 \qquad \textbf{(11.7b)}$$

Since no fluid can cross the sidewalls of the tube, the mass flow rates at positions 1 and 2 must be equal. However, these positions were selected arbitrarily, so the mass flow rate has the same value everywhere in the tube, an important result known as the **equation of continuity**. The equation of continuity is an expression of the fact that mass is conserved (i.e., neither created nor destroyed) as the fluid flows.

> **EQUATION OF CONTINUITY**
>
> **The mass flow rate ($\rho A v$) has the same value at every position along a tube that has a single entry and a single exit point for fluid flow. For two positions along such a tube**
>
> $$\rho_1 A_1 v_1 = \rho_2 A_2 v_2 \qquad \textbf{(11.8)}$$
>
> **where** ρ = **fluid density (kg/m³)**
> A = **cross-sectional area of tube (m²)**
> v = **fluid speed (m/s)**
>
> **SI Unit of Mass Flow Rate: kg/s**

The density of an incompressible fluid does not change during flow, so that $\rho_1 = \rho_2$, and the equation of continuity reduces to

Incompressible fluid $$A_1 v_1 = A_2 v_2 \qquad \textbf{(11.9)}$$

The quantity Av represents the volume of fluid per second (measured in m³/s, for instance) that passes through the tube and is referred to as the **volume flow rate Q**:

$$Q = \text{Volume flow rate} = Av \qquad \textbf{(11.10)}$$

FIGURE 11.27 When the end of a hose is partially closed off, thus reducing its cross-sectional area, the fluid velocity increases.

ANIMATED FIGURE 11.28 In general, a fluid flowing in a tube that has different cross-sectional areas A_1 and A_2 at positions 1 and 2 also has different velocities $\vec{v}_1$ and $\vec{v}_2$ at these positions.

Equation 11.9 shows that where the tube's cross-sectional area is large, the fluid speed is small, and, conversely, where the tube's cross-sectional area is small, the speed is large. Example 13 explores this behavior in more detail for the hose in **Figure 11.27**.

EXAMPLE 13 | A Garden Hose

A garden hose has an unobstructed opening with a cross-sectional area of 2.85×10^{-4} m², from which water fills a bucket in 30.0 s. The volume of the bucket is 8.00×10^{-3} m³ (about two gallons). Find the speed of the water that leaves the hose through **(a)** the unobstructed opening and **(b)** an obstructed opening with half as much area.

Reasoning If we can determine the volume flow rate Q, we can obtain the speed of the water from Equation 11.10 as $v = Q/A$, since the area A is given. We can find the volume flow rate from the volume of the bucket and its fill time.

Solution **(a)** The volume flow rate Q is equal to the volume of the bucket divided by the fill time. Therefore, the speed of the water is

$$v = \frac{Q}{A} = \frac{(8.00 \times 10^{-3}\,\text{m}^3)/(30.0\,\text{s})}{2.85 \times 10^{-4}\,\text{m}^2} = \boxed{0.936\,\text{m/s}} \quad \textbf{(11.10)}$$

(b) Water can be considered incompressible, so the equation of continuity can be applied in the form $A_1 v_1 = A_2 v_2$. When the opening of the hose is unobstructed, its area is A_1, and the speed of the water is that determined in part (a)—namely, $v_1 = 0.936$ m/s. Since $A_2 = \frac{1}{2}A_1$, we find that

$$v_2 = \left(\frac{A_1}{A_2}\right)v_1 = \left(\frac{A_1}{\frac{1}{2}A_1}\right)(0.936\,\text{m/s}) = \boxed{1.87\,\text{m/s}} \quad \textbf{(11.9)}$$

The next example applies the equation of continuity to the flow of blood.

EXAMPLE 14 | BIO The Physics of a Clogged Artery

In the condition known as atherosclerosis, a deposit, or atheroma, forms on the arterial wall and reduces the opening through which blood can flow. In the carotid artery in the neck, blood flows three times faster through a partially blocked region than it does through an unobstructed region. Determine the ratio of the effective radii of the artery at the two places.

Reasoning Blood, like most liquids, is incompressible, and the equation of continuity in the form of $A_1 v_1 = A_2 v_2$ (Equation 11.9) can be applied. In applying this equation, we use the fact that the area of a circle is πr^2.

> **Problem-Solving Insight** The equation of continuity in the form $A_1 v_1 = A_2 v_2$ applies only when the density of the fluid is constant. If the density is not constant, the equation of continuity is $\rho_1 A_1 v_1 = \rho_2 A_2 v_2$.

Solution From Equation 11.9, it follows that

$$\underbrace{(\pi r_U^2) v_U}_{\substack{\text{Unobstructed} \\ \text{volume flow rate}}} = \underbrace{(\pi r_O^2) v_O}_{\substack{\text{obstructed} \\ \text{volume flow rate}}}$$

The ratio of the radii is

$$\frac{r_U}{r_O} = \sqrt{\frac{v_O}{v_U}} = \sqrt{3} = \boxed{1.7}$$

Check Your Understanding

(The answers are given at the end of the book.)

15. Water flows from left to right through the five sections (A, B, C, D, E) of the pipe shown in CYU Figure 11.2. In which section(s) does the water speed increase, decrease, and remain constant? Treat the water as an incompressible fluid.

	Speed Increases	**Speed Decreases**	**Speed Is Constant**
(a)	A, B	D, E	C
(b)	D	B	A, C, E
(c)	D, E	A, B	C
(d)	B	D	A, C, E
(e)	A, B	C, D	E

CYU FIGURE 11.2

16. In case A, water falls downward from a faucet. In case B, water shoots upward, as in a fountain. In each case, as the water moves downward or upward, it has a cross-sectional area that **(a)** does not change, **(b)** becomes larger in A and smaller in B, **(c)** becomes smaller in A and larger in B, **(d)** becomes larger in both cases, **(e)** becomes smaller in both cases.

11.9 Bernoulli's Equation

For *steady flow,* the speed, pressure, and elevation of an *incompressible and nonviscous* fluid are related by an equation discovered by Daniel Bernoulli (1700–1782). To derive **Bernoulli's equation,** we will use the work–energy theorem. This theorem, which is introduced in Chapter 6, states that the net work W_{nc} done on an object by external non-conservative forces is equal to the change in the total mechanical energy of the object (see Equation 6.8). As mentioned earlier, the pressure within a fluid is caused by colli-sional forces, which are nonconservative. Therefore, when a fluid is accelerated because of a difference in pressures, work is being done by nonconservative forces ($W_{nc} \neq 0$ J), and this work changes the total mechanical energy of the fluid from an initial value of E_0 to a final value of E_f. The total mechanical energy is not conserved. We will now see how the work–energy theorem leads directly to Bernoulli's equation.

To begin with, let us make two observations about a moving fluid. First, whenever a fluid is flowing in a horizontal pipe and encounters a region of reduced cross-sectional area, the pressure of the fluid drops, as the pressure gauges in **Figure 11.29a** indicate. The reason for this follows from Newton's second law. When moving from the wider region 2 to the narrower region 1, the fluid speeds up or accelerates, consistent with the conservation of mass (as expressed by the equation of continuity). According to the second law, the accelerating fluid must be subjected to an unbalanced force. However, there can be an unbalanced force only if the pressure in region 2 exceeds the pressure in region 1. We will see that the difference in pressures is given by Bernoulli's equation. The second observation is that if the fluid moves to a higher elevation, the pressure at the lower level is greater than the pressure at the higher level, as in **Figure 11.29b**. The basis for this observation is our previous study of static fluids, and Bernoulli's equation will confirm it, provided that the cross-sectional area of the pipe does not change.

(a)

(b)

FIGURE 11.29 (*a*) In this horizontal pipe, the pressure in region 2 is greater than that in region 1. The difference in pressures leads to the net force that accelerates the fluid to the right. (*b*) When the fluid changes elevation, the pressure at the bottom is greater than the pressure at the top, assuming that the cross-sectional area of the pipe is constant.

To derive Bernoulli's equation, consider **Figure 11.30a**. This drawing shows a fluid element of mass m, upstream in region 2 of a pipe. Both the cross-sectional area and the elevation are different at different places along the pipe. The speed, pressure, and eleva-tion in this region are v_2, P_2, and y_2, respectively. Downstream in region 1 these variables have the values v_1, P_1, and y_1. As Chapter 6 discusses, an object moving under the influ-ence of gravity has a total mechanical energy E that is the sum of the kinetic energy KE and the gravitational potential energy PE: $E = KE + PE = \frac{1}{2}mv^2 + mgy$. When work W_{nc}

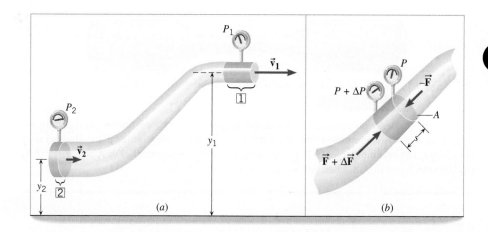

FIGURE 11.30 (a) A fluid element (dark blue) moving through a pipe whose cross-sectional area and elevation change. (b) The fluid element experiences a force $-\vec{\mathbf{F}}$ on its top surface due to the fluid above it, and a force $\vec{\mathbf{F}} + \Delta\vec{\mathbf{F}}$ on its bottom surface due to the fluid below it.

is done on the fluid element by external nonconservative forces, the total mechanical energy changes. According to the work–energy theorem, the work equals the change in the total mechanical energy:

$$W_{nc} = E_1 - E_2 = \underbrace{\left(\tfrac{1}{2}mv_1^2 + mgy_1\right)}_{\substack{\text{Total mechanical}\\\text{energy in region 1}}} - \underbrace{\left(\tfrac{1}{2}mv_2^2 + mgy_2\right)}_{\substack{\text{Total mechanical}\\\text{energy in region 2}}} \qquad \textbf{(6.8)}$$

Figure 11.30b helps us understand how the work W_{nc} arises. On the top surface of the fluid element, the surrounding fluid exerts a pressure P. This pressure gives rise to a force of magnitude $F = PA$, where A is the cross-sectional area. On the bottom surface, the surrounding fluid exerts a slightly greater pressure, $P + \Delta P$, where ΔP is the pressure difference between the ends of the element. As a result, the force on the bottom surface has a magnitude of $F + \Delta F = (P + \Delta P)A$. The magnitude of the *net* force pushing the fluid element up the pipe is $\Delta F = (\Delta P)A$. When the fluid element moves through its own length s, the work done is the product of the magnitude of the net force and the distance: Work $= (\Delta F)s = (\Delta P)As$. The quantity As is the volume V of the element, so the work is $(\Delta P)V$. The total work done on the fluid element in moving it from region 2 to region 1 is the sum of the small increments of work $(\Delta P)V$ done as the element moves along the pipe. This sum amounts to $W_{nc} = (P_2 - P_1)V$, where $P_2 - P_1$ is the pressure difference between the two regions. With this expression for W_{nc}, the work–energy theorem becomes

$$W_{ne} = (P_2 - P_1)V = \left(\tfrac{1}{2}mv_1^2 + mgy_1\right) - \left(\tfrac{1}{2}mv_2^2 + mgy_2\right)$$

By dividing both sides of this result by the volume V, recognizing that m/V is the density ρ of the fluid, and rearranging terms, we obtain Bernoulli's equation.

> **BERNOULLI'S EQUATION**
>
> **In the steady flow of a nonviscous, incompressible fluid of density ρ, the pressure P, the fluid speed v, and the elevation y at any two points (1 and 2) are related by**
>
> $$P_1 + \tfrac{1}{2}\rho v_1^2 + \rho g y_1 = P_2 + \tfrac{1}{2}\rho v_2^2 + \rho g y_2 \qquad \textbf{(11.11)}$$

Since the points 1 and 2 were selected arbitrarily, the term $P + \tfrac{1}{2}\rho v^2 + \rho g y$ has a constant value at all positions in the flow. For this reason, Bernoulli's equation is sometimes expressed as $P + \tfrac{1}{2}\rho v^2 + \rho g y = \text{constant}$.

Equation 11.11 can be regarded as an extension of the earlier result that specifies how the pressure varies with depth in a static fluid ($P_2 = P_1 + \rho g h$), the terms $\tfrac{1}{2}\rho v_1^2$ and $\tfrac{1}{2}\rho v_2^2$ accounting for the effects of fluid speed. Bernoulli's equation reduces to the result for static fluids when the speed of the fluid is the same everywhere ($v_1 = v_2$), as it is when the cross-sectional area remains constant. Under such conditions, Bernoulli's equation is $P_1 + \rho g y_1 = P_2 + \rho g y_2$. After rearrangement, this result becomes

$$P_2 = P_1 + \rho g(y_1 - y_2) = P_1 + \rho g h$$

which is the result (Equation 11.4) for static fluids.

11.10 Applications of Bernoulli's Equation

When a moving fluid is contained in a horizontal pipe, all parts of it have the same elevation ($y_1 = y_2$), and Bernoulli's equation simplifies to

$$P_1 + \tfrac{1}{2}\rho v_1^2 = P_2 + \tfrac{1}{2}\rho v_2^2 \qquad (11.12)$$

Thus, the quantity $P + \tfrac{1}{2}\rho v^2$ remains constant throughout a horizontal pipe; if v increases, P decreases, and vice versa. This is the result that we deduced qualitatively from Newton's second law (Section 11.9). Conceptual Example 15 illustrates it.

CONCEPTUAL EXAMPLE 15 | Tarpaulins and Bernoulli's Equation

A tarpaulin is a piece of canvas that is used to cover a cargo, like that pulled by the truck in **Figure 11.31**. Whenever the truck stops, the tarpaulin lies flat. Why does it bulge outward whenever the truck is speeding down the highway? **(a)** The air rushing over the outside surface of the canvas creates a higher pressure than does the stationary air inside the cargo area. **(b)** The air rushing over the outside surface of the canvas creates a lower pressure than does the stationary air inside the cargo area. **(c)** The air inside the cargo area heats up, thus increasing the pressure on the tarp and pushing it outward.

Reasoning When the truck is stationary, the air outside and inside the cargo area is stationary, so the pressure is the same in both places. This pressure applies the same force to the outer and inner surfaces of the canvas, with the result that the tarpaulin lies flat. When the truck is moving, the outside air rushes over the top surface of the canvas, and the pressure generated by the moving air is different than the pressure of the stationary air.

Answer (a) is incorrect. A higher pressure outside and a lower pressure in the cargo area would cause the tarpaulin to sink inward, not bulge outward.

Answer (c) is incorrect. A heating effect would not disappear every time the truck stops and reappear only when the truck is moving.

FIGURE 11.31 The tarpaulin that covers the cargo is flat when the truck is stationary but bulges outward when the truck is moving.

Answer (b) is correct. In accord with Bernoulli's equation (Equation 11.12), the moving air outside the canvas has a lower pressure than does the stationary air inside the cargo area. The greater inside pressure generates a greater force on the inner surface of the canvas, and the tarpaulin bulges outward.

Related Homework: Problems 54, 64

THE PHYSICS OF . . . household plumbing.   The impact of fluid flow on pressure is widespread. **Figure 11.32**, for instance, illustrates how household plumbing takes into account the implications of Bernoulli's equation. The U-shaped section of pipe beneath the sink is called a "trap," because it traps water, which serves as a barrier to prevent sewer gas

FIGURE 11.32 In a household plumbing system, a vent is necessary to equalize the pressures at points A and B, thus preventing the trap from being emptied. An empty trap allows sewer gas to enter the house.

from leaking into the house. Part *a* of the drawing shows poor plumbing. When water from the clothes washer rushes through the sewer pipe, the high-speed flow causes the pressure at point A to drop. The pressure at point B in the sink, however, remains at the higher atmospheric pressure. As a result of this pressure difference, the water is pushed out of the trap and into the sewer line, leaving no protection against sewer gas. A correctly designed system is vented to the outside of the house, as in **Figure 11.32b**. The vent ensures that the pressure at A remains the same as that at B (atmospheric pressure), even when water from the clothes washer is rushing through the pipe. Thus, the purpose of the vent is to prevent the trap from being emptied, not to provide an escape route for sewer gas.

THE PHYSICS OF . . . airplane wings.   One of the most spectacular examples of how fluid flow affects pressure is the dynamic lift on airplane wings. **Figure 11.33a** shows a wing moving to the right, with the air flowing leftward past the wing. Hence, according to Bernoulli's equation, the pressures on the top of and beneath the wing are both lower than atmospheric pressure.* However, the pressure above the wing is reduced relative to the pressure under the wing. This is due to the wing's shape, which causes the air to travel faster (more reduction in pressure) over the curved top surface and more slowly (less reduction in pressure) over the flatter bottom. Thus, the wing is lifted upward. Part *b* of the figure shows the wing of an airplane.

FIGURE 11.33 (*a*) Air flowing around an airplane wing. The wing is moving to the right. (*b*) The end of this wing has roughly the shape indicated in part *a*.

(a)   *(b)*

THE PHYSICS OF . . . a curveball.   The curveball, one of a baseball pitcher's main weapons, is another illustration of the effects of fluid flow. **Figure 11.34a** shows a baseball moving to the right with no spin. The view is from above, looking down toward the ground. Here, air flows with the same speed around both sides of the ball, and the pressure is reduced on both sides by the same amount relative to atmospheric pressure. No net force exists to make the ball curve to either side. However, when the ball is given a spin, the air close to its surface is dragged around with it, and the situation changes. The speed of the air on one-half of the ball is increased, and the pressure there is even more reduced relative to atmospheric pressure. On the other half of the ball, the speed of the air is decreased,

(*a*) Without spin   (*b*) With spin   (*c*)

FIGURE 11.34 These views of a baseball are from above, looking down toward the ground, with the ball moving to the right. (*a*) Without spin, the ball does not curve to either side. (*b*) A spinning ball curves in the direction of the deflection force. (*c*) The spin in part *b* causes the ball to curve as shown here.

*The application of Bernoulli's equation to an airplane wing is only an approximation, as air is both viscous and compressible.

leading to a lesser reduction in pressure than occurs without spin. Part *b* of the picture shows the effects of a counterclockwise spin. The baseball experiences a net deflection force and curves on its way from the pitcher's mound to the plate, as part *c* shows.*

As a final application of Bernoulli's equation, **Figure 11.35a** shows a large tank from which water is emerging through a small pipe near the bottom. Bernoulli's equation can be used to determine the speed (called the efflux speed) at which the water leaves the pipe, as the next example shows.

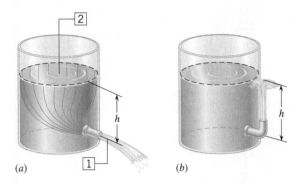

(a)

(b)

FIGURE 11.35 (*a*) Bernoulli's equation can be used to determine the speed of the liquid leaving the small pipe. (*b*) An ideal fluid (no viscosity) will rise to the fluid level in the tank after leaving a vertical outlet nozzle.

EXAMPLE 16 | Efflux Speed

The tank in **Figure 11.35a** is open to the atmosphere at the top. Find an expression for the speed of the liquid leaving the pipe at the bottom.

Reasoning   We assume that the liquid behaves as an ideal fluid. Therefore, we can apply Bernoulli's equation, and in preparation for doing so, we locate two points in the liquid in **Figure 11.35a**. Point 1 is just outside the efflux pipe, and point 2 is at the top surface of the liquid. The pressure at each of these points is equal to the atmospheric pressure, a fact that will be used to simplify Bernoulli's equation.

Solution   Since the pressures at points 1 and 2 are the same, we have $P_1 = P_2$, and Bernoulli's equation becomes $\frac{1}{2}\rho v_1^2 + \rho g y_1 = \frac{1}{2}\rho v_2^2 + \rho g y_2$. The density ρ can be eliminated algebraically from this result, which can then be solved for the square of the efflux speed v_1:

$$v_1^2 = v_2^2 + 2g(y_2 - y_1) = v_2^2 + 2gh$$

We have substituted $h = y_2 - y_1$ for the height of the liquid above the efflux tube. If the tank is very large, the liquid level changes only slowly, and the speed at point 2 can be set equal to zero, so that $\boxed{v_1 = \sqrt{2gh}}$.

Math Skills   To obtain the desired expression for v_1^2, we start with Bernoulli's equation in the following form:

$$\frac{1}{2}\rho v_1^2 + \rho g y_1 = \frac{1}{2}\rho v_2^2 + \rho g y_2$$

To solve this equation for v_1^2, we first isolate the term $\frac{1}{2}\rho v_1^2$ on the left side of the equals sign. To do this, we subtract $\rho g y_1$ from both sides of the equals sign:

$$\frac{1}{2}\rho v_1^2 + \rho g y_1 - \rho g y_1 = \frac{1}{2}\rho v_2^2 + \rho g y_2 - \rho g y_1 \quad \text{or}$$
$$\frac{1}{2}\rho v_1^2 = \frac{1}{2}\rho v_2^2 + \rho g y_2 - \rho g y_1$$

Note that the density ρ appears in each term of the result. Therefore, it can be eliminated algebraically:

$$\frac{1}{2}\rho v_1^2 = \frac{1}{2}\rho v_2^2 + \rho g y_2 - \rho g y_1 \quad \text{or} \quad \frac{1}{2}v_1^2 = \frac{1}{2}v_2^2 + g y_2 - g y_1$$

We can now simplify the resulting equation by factoring out the term g that appears on the right side in the terms $g y_2$ and $g y_1$:

$$\frac{1}{2}v_1^2 = \frac{1}{2}v_2^2 + g(y_2 - y_1)$$

Multiplying both sides of the equals sign by 2 yields

$$2\frac{1}{2}v_1^2 = 2\frac{1}{2}v_2^2 + 2g(y_2 - y_1) \quad \text{or} \quad v_1^2 = v_2^2 + 2g(y_2 - y_1)$$

In Example 16 the liquid is assumed to be an ideal fluid, and the speed with which it leaves the pipe is the same as if the liquid had freely fallen through a height h (see Equation 2.9 with $x = h$ and $a = g$). This result is known as **Torricelli's theorem**. If the outlet pipe were pointed directly upward, as in **Figure 11.35b**, the liquid would rise to a height h equal to the fluid level above the pipe. However, if the liquid is not an ideal fluid, its viscosity cannot be neglected. Then, the efflux speed would be less than that given by Bernoulli's equation, and the liquid would rise to a height less than h. Example 17 illustrates how the Bernoulli equation can be used to analyze blood flow in the case of an abdominal aortic aneurysm.

*In the jargon used in baseball, the pitch shown in **Figure 11.34** parts *b* and *c* is called a "slider."

EXAMPLE 17 | BIO The Physics of an Abdominal Aortic Aneurysm

Example 14 demonstrated how the velocity of blood in an artery changes, as clogging reduces the effective cross-sectional area of the artery. Here we show that an *increase* in the cross-sectional area of an artery can be equally as dangerous. **Figure 11.36** represents a person who has a potentially fatal condition known as an abdominal aortic aneurysm (AAA). The aorta is the largest blood vessel in the body that carries blood away from the heart, and factors such as smoking, high blood pressure, and genetic history can lead to a weakening of the arterial wall that results in the formation of a large bulge (aneurysm) in the artery (**Figure 11.36a**). If the wall weakens too much, the blood pressure can cause it to rupture, leading to a significant loss of blood within the circulatory system. Because the cross-sectional area of the artery increases in the region of the aneurysm, the pressure will be different there. We can model the normal aorta and the aneurysm as two circular tubes with different diameters (**Figure 11.36b**). Assume the diameter of a normal aorta in an adult is $d_1 = 2.0$ cm, and the diameter of the aneurysm is $d_2 = 3.5$ cm. The velocity of the blood in the normal part of the aorta is $v_1 = 0.40$ m/s. Use the equation of continuity and Bernoulli's equation to calculate the difference in pressure ($\Delta P = P_2 - P_1$) between the aneurysm and the normal aorta. Assume the blood is an ideal fluid and the patient is lying down horizontally, so that the normal portion of the aorta and the aneurysm are at the same height.

Reasoning We can first apply the Equation of Continuity (Equation 11.8) in order to find v_2—the speed of the blood in the region of the aneurysm. Once we know v_2, then we can apply the Bernoulli equation (Equation 11.11) to find the pressure difference ΔP.

Solution We begin with Equation 11.8: $\rho_1 A_1 v_1 = \rho_2 A_2 v_2$. Here, the density of the fluid is constant, and each area is circular, or equal to $\pi(d/2)^2$. Therefore, Equation 11.8 reduces to the following: $d_1^2 v_1 = d_2^2 v_2$. Thus, $v_2 = (d_1/d_2)^2 v_1$. Now we apply the Bernoulli equation: $P_1 + \frac{1}{2}\rho v_1^2 + \rho g y_1 = P_2 + \frac{1}{2}\rho v_2^2 + \rho g y_2$. We now solve this for $\Delta P = P_2 - P_1$, where we note that $y_1 = y_2$: $\Delta P = \frac{1}{2}\rho(v_1^2 v_2^2)$. We can substitute in our expression above for v_2 and obtain our final result:

$$\Delta P = \frac{1}{2}\rho v_1^2(1 - (d_1/d_2)^4)$$

$$= \frac{1}{2}(1060 \text{ kg/m}^3)(0.40 \text{ m/s})^2(1 - (2.0 \text{ cm}/3.5 \text{ cm})^4) = \boxed{76\text{Pa}}$$

Heart

Aneurysm

Normal aorta

d_2

v_2

P_2

P_1

v_1

d_1

(b)

(a)

FIGURE 11.36 (a) Diagram of the human circulatory system showing the most common location of an AAA. (b) The normal aorta and aneurysm represented as circular tubes with different cross-sectional areas. The different areas lead to different blood velocities and pressures in each region.

Because the velocity of the blood slows down in the aneurysm due to its larger cross-sectional area, the blood pressure increases. This creates a dangerous cycle: As the aneurysm continues to grow, the pressure keeps increasing. If left untreated, the increasing pressure can rupture the arterial wall.

Related Homework: Problem 77

Check Your Understanding

(The answers are given at the end of the book.)

17. Fluid is flowing from left to right through a pipe (see **CYU Figure 11.3**). Points A and B are at the same elevation, but the cross-sectional areas of the pipe are different. Points B and C are at different elevations, but the cross-sectional areas are the same. Rank the pressures at the three points, highest to lowest: **(a)** A and B (a tie), C **(b)** C, A and B (a tie) **(c)** B, C, A **(d)** C, B, A **(e)** A, B, C

C

B

A

CYU FIGURE 11.3

18. Have you ever had a large truck (traveling in your direction) pass you from behind? You probably noticed that your car was pulled toward the truck as it passed. How does the speed of the air (and, hence, its pressure) between your car and the truck compare to the speed of the air on the opposite side of your car? The speed of the air between the two vehicles is **(a)** less and produces a smaller air pressure **(b)** less and produces a greater air pressure **(c)** greater and produces a smaller air pressure **(d)** greater and produces a greater air pressure.

19. Hold two sheets of paper by adjacent corners, so that they hang downward. The sheets should be parallel and slightly separated, so that you can see the floor through the gap between them. Blow air strongly down through the gap. What happens to the sheets? **(a)** Nothing **(b)** The sheets move further apart. **(c)** The sheets come closer together.

20. Suppose that you are a right-handed batter in a baseball game, so the ball is moving from your left to your right. You are caught unprepared, looking directly at the ball as it passes by for a strike. If the ball curves upward on its way to the plate, which way is it spinning? **(a)** Clockwise **(b)** Counterclockwise

21. You are sitting on a stationary train next to an *open* window, and the pressure of the air in your inner ear is equal to the pressure outside your ear. The train starts up, and as it accelerates to a high speed, your ears "pop." Your eardrums respond to a decrease or increase in the air pressure by "popping" outward or inward, respectively. Assume that the air pressure in your inner ear has not had time to change, so it remains the same as when the train was stationary. Do your eardrums "pop" **(a)** outward or **(b)** inward?

22. Sometimes the weather conditions at an airport give rise to air that has an unusually low density. What effect does such a low-density condition have on a plane's ability to generate the required lift force for takeoff? **(a)** It has no effect. **(b)** It makes it easier. **(c)** It makes it harder.

(a)

$\vec{v}$ = 0 m/s at wall

$\vec{v}$ is a maximum
(b)   at the center

FIGURE 11.37 *(a)* In ideal (nonviscous) fluid flow, all fluid particles across the pipe have the same velocity. *(b)* In viscous flow, the speed of the fluid is zero at the surface of the pipe and increases to a maximum along the center axis.

11.11 | *Viscous Flow

In an ideal fluid there is no viscosity to hinder the fluid layers as they slide past one another. Within a pipe of uniform cross section, every layer of an ideal fluid moves with the same velocity, even the layer next to the wall, as **Figure 11.37a** shows. When viscosity is present, the fluid layers have different velocities, as part *b* of the drawing illustrates. The fluid at the center of the pipe has the greatest velocity. In contrast, the fluid layer next to the wall surface does not move at all, because it is held tightly by intermolecular forces. So strong are these forces that if a solid surface moves, the adjacent fluid layer moves along with it and remains at rest *relative* to the moving surface.

To help introduce viscosity in a quantitative fashion, **Figure 11.38a** shows a viscous fluid between two parallel plates. The top plate is free to move, while the bottom one is stationary. If the top plate is to move with a velocity $\vec{v}$ relative to the bottom plate, a force $\vec{F}$ is required. For a highly viscous fluid, like honey, a large force is needed; for a less viscous fluid, like water, a smaller one will do. As part *b* of the drawing suggests, we may imagine the fluid to be composed of many thin horizontal layers. When the top plate moves, the intermediate fluid layers slide over each other. The velocity of each layer is different, changing uniformly from $\vec{v}$ at the top plate to zero at the bottom plate. The resulting flow is called **laminar flow**, since a thin layer is often referred to as a *lamina*. As each layer moves, it is subjected to viscous forces from its neighbors. The purpose of the force $\vec{F}$ is to compensate for the effect of these forces, so that any layer can move with a constant velocity.

The amount of force required in **Figure 11.38a** depends on several factors. Larger areas A, being in contact with more fluid, require larger forces, so that the force is proportional to the contact area ($F \propto A$). For a given area, greater speeds require larger forces, with the result that the force is proportional to the speed ($F \propto v$). The force is also inversely proportional to the perpendicular distance y between the top and bottom plates ($F \propto 1/y$). The larger the distance y, the smaller is the force required to achieve a given speed with a given contact area. These three proportionalities can be expressed simultaneously in the following way: $F \propto Av/y$. Equation 11.13 expresses this relationship with the aid of a proportionality constant η (Greek letter *eta*), which is called the **coefficient of viscosity** or simply the **viscosity**.

Area = A $\vec{F}$

Viscous fluid

Stationary plate, $\vec{v}$ = 0 m/s
(a)

y $\vec{v}$

$\vec{v}$ = 0 m/s
(b)

FIGURE 11.38 *(a)* A force $\vec{F}$ is applied to the top plate, which is in contact with a viscous fluid. *(b)* Because of the force $\vec{F}$, the top plate and the adjacent layer of fluid move with a constant velocity $\vec{v}$.

FORCE NEEDED TO MOVE A LAYER OF VISCOUS FLUID WITH A CONSTANT VELOCITY

The magnitude of the tangential force $\vec{F}$ required to move a fluid layer at a constant speed v, when the layer has an area A and is located a perpendicular distance y from an immobile surface, is given by

$$F = \frac{\eta A v}{y} \qquad (11.13)$$

where η is the coefficient of viscosity.

SI Unit of Viscosity: Pa · s

Common or CGS Unit of Viscosity: poise (P)

By solving Equation 11.13 for the viscosity, $\eta = Fy/(vA)$, it can be seen that the SI unit for viscosity is N · m/[(m/s) · m²] = Pa · s. Another common unit for viscosity is the *poise* (P), which is used in the CGS system of units and is named after the French physician Jean Poiseuille (1799–1869; pronounced, approximately, as Pwah-zoy′). The following relation exists between the two units:

$$1 \text{ poise (P)} = 0.1 \text{ Pa} \cdot \text{s}$$

Values of viscosity depend on the nature of the fluid. Under ordinary conditions, the viscosities of liquids are significantly *larger* than those of gases. Moreover, the viscosities of either liquids or gases depend markedly on temperature. Usually, the viscosities of liquids decrease as the temperature is increased. Anyone who has heated honey or oil, for example, knows that these fluids flow much more freely at an elevated temperature. In contrast, the viscosities of gases increase as the temperature is raised. An ideal fluid has $\eta = 0$ P.

Viscous flow occurs in a wide variety of situations, such as oil moving through a pipeline or a liquid being forced through the needle of a hypodermic syringe. **Figure 11.39** identifies the factors that determine the volume flow rate Q (in m³/s) of the fluid. First, a difference in pressures $P_2 - P_1$ must be maintained between any two locations along the pipe for the fluid to flow. In fact, Q is proportional to $P_2 - P_1$, a greater pressure difference leading to a larger flow rate. Second, a long pipe offers greater resistance to the flow than a short pipe does, and Q is inversely proportional to the length L. Third, high-viscosity fluids flow less readily than low-viscosity fluids, and Q is inversely proportional to the viscosity η. Finally, the volume flow rate is larger in a pipe of larger radius, other things being equal. The dependence on the radius R is a surprising one, Q being proportional to the fourth power of the radius, or R^4. If, for instance, the pipe radius is reduced to one-half of its original value, the volume flow rate is reduced to one-sixteenth of its original value, assuming the other variables remain constant. The mathematical relation for Q in terms of these parameters was discovered by Poiseuille and is known as **Poiseuille's law.**

FIGURE 11.39 For viscous flow, the difference in pressures $P_2 - P_1$, the radius R and length L of the tube, and the viscosity η of the fluid influence the volume flow rate.

POISEUILLE'S LAW

A fluid whose viscosity is η, flowing through a pipe of radius R and length L, has a volume flow rate Q given by

$$Q = \frac{\pi R^4 (P_2 - P_1)}{8 \eta L} \qquad (11.14)$$

where P_1 and P_2 are the pressures at the ends of the pipe.

THE PHYSICS OF . . . pipeline pumping stations. The fact that Q is inversely proportional to L in Equation 11.14 has important consequences for long pipelines, such as the Alaskan pipeline (see **Figure 11.40**). Solving Equation 11.14 for P_1 shows that

$$P_1 = P_2 - \frac{Q 8 \eta L}{\pi R^4}$$

Thus, the pressure P_1 at the downstream end of a length of pipe is less than the pressure P_2 at the upstream end. Long pipelines must have pumping stations at various places along the line to compensate for the drop in pressure.

Example 18 illustrates the use of Poiseuille's law.

FIGURE 11.40 As oil flows along the Alaskan pipeline, the pressure drops because oil is a viscous fluid. Pumping stations are located along the pipeline to compensate for the drop in pressure.

EXAMPLE 18 | BIO The Physics of a Hypodermic Syringe

A hypodermic syringe is filled with a solution whose viscosity is 1.5×10^{-3} Pa · s. As **Figure 11.41** shows, the plunger area of the syringe is 8.0×10^{-5} m^2, and the length of the needle is 0.025 m. The internal radius of the needle is 4.0×10^{-4} m. The gauge pressure in the vein that is injected is 1900 Pa (14 mm of mercury). What force must be applied to the plunger, so that 1.0×10^{-6} m^3 of solution can be injected in 3.0 s?

Reasoning The necessary force is the pressure applied to the plunger times the area of the plunger. Since viscous flow is occurring, the pressure is different at different points along the syringe. However, the barrel of the syringe is so wide that little pressure difference is required to sustain the flow up to point 2, where the fluid encounters the narrow needle. Consequently, the pressure applied to the plunger is nearly equal to the pressure P_2 at point 2. To find this pressure, we apply Poiseuille's law to the needle. Poiseuille's law indicates that $P_2 - P_1 = 8\eta LQ/(\pi R^4)$. We note that the pressure P_1 is given as a gauge pressure, which, in this case, is the amount of pressure in excess of atmospheric pressure. This causes no difficulty, because we need to find the amount of force in excess of the force applied to the plunger by the atmosphere. The volume flow rate Q can be obtained from the time needed to inject the known volume of solution.

Solution The volume flow rate is $Q = (1.0 \times 10^{-6}$ m$^3)/(3.0$ s$) = 3.3 \times 10^{-7}$ m^3/s. According to Poiseuille's law (Equation 11.14), the required pressure difference is

FIGURE 11.41 The difference in pressure $P_2 - P_1$ required to sustain the fluid flow through a hypodermic needle can be found with the aid of Poiseuille's law.

$$P_2 - P_1 = \frac{8\eta LQ}{\pi R^4} = \frac{8(1.5 \times 10^{-3} \text{ Pa} \cdot \text{s})(0.025 \text{ m})(3.3 \times 10^{-7} \text{ m}^3/\text{s})}{\pi (4.0 \times 10^{-4} \text{ m})^4}$$

$$= 1200 \text{ Pa}$$

Since $P_1 = 1900$ Pa, the pressure P_2 must be $P_2 = 1200$ Pa $+ 1900$ Pa $= 3100$ Pa. The force that must be applied to the plunger is this pressure times the plunger area:

$$F = (3100 \text{ Pa})(8.0 \times 10^{-5} \text{ m}^2) = \boxed{0.25 \text{ N}}$$

Concept Summary

11.1 Mass Density Fluids are materials that can flow, and they include gases and liquids. The mass density ρ of a substance is its mass m divided by its volume V, according to Equation 11.1.

$$\rho = \frac{m}{V} \quad (11.1)$$

The specific gravity of a substance is its mass density divided by the density of water at 4 °C (1.000×10^3 kg/m^3), according to Equation 11.2.

$$\text{Specific gravity} = \frac{\text{Density of substance}}{1.000 \times 10^3 \text{kg/m}^3} \quad (11.2)$$

11.2 Pressure The pressure P exerted by a fluid is the magnitude F of the force acting perpendicular to a surface embedded in the fluid divided by the area A over which the force acts, as shown by Equation 11.3. The SI unit for measuring pressure is the pascal (Pa); 1 Pa = 1 N/m^2. One atmosphere of pressure is 1.013×10^5 Pa or 14.7 lb/in.^{2}

$$P = \frac{F}{A} \quad (11.3)$$

11.3 Pressure and Depth in a Static Fluid In the presence of gravity, the upper layers of a fluid push downward on the layers beneath, with the result that fluid pressure is related to depth. In an incompressible static fluid whose density is ρ, the relation is given by Equation 11.4,

where P_1 is the pressure at one level, P_2 is the pressure at a level that is h meters deeper, and g is the magnitude of the acceleration due to gravity.

$$P_2 = P_1 + \rho gh \quad (11.4)$$

11.4 Pressure Gauges Two basic types of pressure gauges are the mercury barometer and the open-tube manometer.

The gauge pressure is the amount by which a pressure P differs from atmospheric pressure. The absolute pressure is the actual value for P.

11.5 Pascal's Principle Pascal's principle states that any change in the pressure applied to a completely enclosed fluid is transmitted undiminished to all parts of the fluid and the enclosing walls.

11.6 Archimedes' Principle The buoyant force is the upward force that a fluid applies to an object that is partially or completely immersed in it. Archimedes' principle states that the magnitude of the buoyant force equals the weight of the fluid that the partially or completely immersed object displaces, as indicated by Equation 11.6.

$$\underbrace{F_B}_{\substack{\text{Magnitude of} \\ \text{buoyant force}}} = \underbrace{W_{\text{fluid}}}_{\substack{\text{Weight of} \\ \text{displaced fluid}}} \quad (11.6)$$

11.7 Fluids in Motion/11.8 The Equation of Continuity In steady flow, the velocity of the fluid particles at any point is constant as time passes.

An incompressible, nonviscous fluid is known as an ideal fluid.

The mass flow rate of a fluid with a density ρ, flowing with a speed v in a pipe of cross-sectional area A, is the mass per second (kg/s) flowing past a point and is given by Equation 11.7.

$$\text{Mass flow rate} = \rho A v \qquad (11.7)$$

The equation of continuity expresses the fact that mass is conserved: what flows into one end of a pipe flows out the other end, assuming there are no additional entry or exit points in between. Expressed in terms of the mass flow rate, the equation of continuity is given by Equation 11.8, where the subscripts 1 and 2 denote any two points along the pipe.

$$\rho_1 A_1 v_1 = \rho_2 A_2 v_2 \qquad (11.8)$$

If a fluid is incompressible, the density at any two points is the same, $\rho_1 = \rho_2$. For an incompressible fluid, the equation of continuity is expressed as in Equation 11.9. The product Av is known as the volume flow rate Q (in m³/s), according to Equation 11.10.

$$A_1 v_1 = A_2 v_2 \qquad (11.9)$$

$$Q = \text{Volume flow rate} = Av \qquad (11.10)$$

11.9 Bernoulli's Equation/11.10 Applications of Bernoulli's Equation In the steady flow of an ideal fluid whose density is ρ, the pressure P, the fluid speed v, and the elevation y at any two points (1 and 2) in the fluid are related by Bernoulli's equation (see Equation 11.11). When the flow is horizontal ($y_1 = y_2$), Bernoulli's equation indicates that higher fluid speeds are associated with lower fluid pressures.

$$P_1 + \tfrac{1}{2}\rho v_1^2 + \rho g y_1 = P_2 + \tfrac{1}{2}\rho v_2^2 + \rho g y_2 \qquad (11.11)$$

11.11 Viscous Flow The magnitude F of the tangential force required to move a fluid layer at a constant speed v, when the layer has an area A and is located a perpendicular distance y from an immobile surface, is given by Equation 11.13, where η is the coefficient of viscosity.

$$F = \frac{\eta A v}{y} \qquad (11.13)$$

A fluid whose viscosity is η, flowing through a pipe of radius R and length L, has a volume flow rate Q given by Poiseuille's law (see Equation 11.14), where P_1 and P_2 are the pressures at the downstream and upstream ends of the pipe, respectively.

$$Q = \frac{\pi R^4 (P_2 - P_1)}{8 \eta L} \qquad (11.14)$$

Focus on Concepts

Online

Additional questions are available for assignment in WileyPLUS.

Section 11.3 Pressure and Depth in a Static Fluid

1. The drawing shows three containers filled to the same height with the same fluid. In which container, if any, is the pressure at the bottom greatest? **(a)** Container A, because its bottom has the greatest surface area. **(b)** All three containers have the same pressure at the bottom. **(c)** Container A, because it has the greatest volume of fluid. **(d)** Container B, because it has the least volume of fluid. **(e)** Container C, because its bottom has the least surface area.

QUESTION 1   A   B   C

2. Two liquids, 1 and 2, are in equilibrium in a U-tube that is open at both ends, as in the drawing. The liquids do not mix, and liquid 1 rests on top of liquid 2. How is the density ρ_1 of liquid 1 related to the density ρ_2 of liquid 2? **(a)** ρ_1 is equal to ρ_2 because the liquids are in equilibrium. **(b)** ρ_1 is greater than ρ_2. **(c)** ρ_1 is less than ρ_2. **(d)** There is not enough information to tell which liquid has the greater density.

Liquid 1

Liquid 2

QUESTION 2

Section 11.6 Archimedes' Principle

3. A beaker is filled to the brim with water. A solid object of mass 3.00 kg is lowered into the beaker so that the object is fully submerged in the water (see the drawing). During this process, 2.00 kg of water flows over the rim and out of the beaker. What is the buoyant force that acts on the submerged object, and, when released, does the object rise, sink, or remain in place? **(a)** 29.4 N; the object rises. **(b)** 29.4 N; the object sinks. **(c)** 19.6 N; the object rises. **(d)** 19.6 N; the object sinks. **(e)** 19.6 N; the object remains in place.

Object

QUESTION 3

4. Three solid objects are floating in a liquid, as in the drawing. They have different weights and volumes, but have the same thickness (the dimension perpendicular to the page). Rank the objects according to their density, largest first. **(a)** A, B, C **(b)** A, C, B **(c)** B, A, C **(d)** B, C, A **(e)** C, A, B

QUESTION 4

Section 11.8 The Equation of Continuity

5. A hollow pipe is submerged in a stream of water so that the length of the pipe is parallel to the velocity of the water. If the water speed doubles and the cross-sectional area of the pipe triples, what happens to the volume flow rate of the water passing through the pipe? **(a)** The volume flow rate does not change. **(b)** The volume flow rate increases by a factor of 2. **(c)** The volume flow rate increases by a factor of 3. **(d)** The volume flow rate increases by a factor of 4. **(e)** The volume flow rate increases by a factor of 6.

6. In the drawing, water flows from a wide section of a pipe to a narrow section. In which part of the pipe is the volume flow rate the greatest? **(a)** The wide section **(b)** The narrow section **(c)** The volume flow rate is the same in both sections of the pipe.

QUESTION 6

Section 11.9 Bernoulli's Equation

7. Blood flows through a section of a horizontal artery that is partially blocked by a deposit along the artery wall. A hemoglobin molecule moves from the narrow region into the wider region. What happens to the pressure acting on the molecule? **(a)** The pressure increases. **(b)** The pressure decreases. **(c)** There is no change in the pressure.

QUESTION 7

8. Water is flowing down through the pipe shown in the drawing. Point A is at a higher elevation than B and C are. The cross-sectional areas are the same at A and B but are wider at C. Rank the pressures at the three points, largest first. **(a)** P_A, P_B, P_C **(b)** P_C, P_B, P_A **(c)** P_B, P_C, P_A

QUESTION 8

Section 11.11 Viscous Flow

9. A *viscous* fluid is flowing through two horizontal pipes. The pressure difference $P_2 - P_1$ between the ends of each pipe is the same. The pipes have the same radius, although one is twice as long as the other. How does the volume flow rate Q_B in the longer pipe compare with the rate Q_A in the shorter pipe? **(a)** Q_B is the same as Q_A. **(b)** Q_B is twice as large as Q_A. **(c)** Q_B is four times as large as Q_A. **(d)** Q_B is one-half as large as Q_A. **(e)** Q_B is one-fourth as large as Q_A.

QUESTION 9

Problems

Online

Additional questions are available for assignment in WileyPLUS.

SSM Student Solutions Manual

MMH Problem-solving help

GO Guided Online Tutorial

V-HINT Video Hints

CHALK Chalkboard Videos

BIO Biomedical application

E Easy

M Medium

H Hard

WS Worksheet

T Team Problem

Section 11.1 Mass Density

1. **E** **SSM** One of the concrete pillars that support a house is 2.2 m tall and has a radius of 0.50 m. The density of concrete is about 2.2×10^3 kg/m³. Find the weight of this pillar in pounds (1 N = 0.2248 lb).

2. **E** A cylindrical storage tank has a radius of 1.22 m. When filled to a height of 3.71 m, it holds 14 300 kg of a liquid industrial solvent. What is the density of the solvent?

3. **E** **SSM** Accomplished silver workers in India can pound silver into incredibly thin sheets, as thin as 3.00×10^{-7} m (about one-hundredth of the thickness a sheet of paper). Find the area of such a sheet that can be formed from 1.00 kg of silver.

4. **E** Neutron stars consist only of neutrons and have unbelievably high densities. A typical mass and radius for a neutron star might be 2.7×10^{28} kg and 1.2×10^3 m. **(a)** Find the density of such a star. **(b)** If a dime ($V = 2.0 \times 10^{-7}$ m³) were made from this material, how much would it weigh (in pounds)?

5. **E** One end of a wire is attached to a ceiling, and a solid brass ball is tied to the lower end. The tension in the wire is 120 N. What is the radius of the brass ball?

6. **M** **GO** Planners of an experiment are evaluating the design of a sphere of radius R that is to be filled with helium (0 °C, 1 atm pressure). Ultrathin silver foil of thickness T will be used to make the sphere, and the designers claim that the mass of helium in the sphere will equal the mass of silver used. Assuming that T is much less than R, calculate the ratio T/R for such a sphere.

7. **M** **SSM** A bar of gold measures 0.15 m × 0.050 m × 0.050 m. How many gallons of water have the same mass as this bar?

8. **M** A full can of black cherry soda has a mass of 0.416 kg. It contains 3.54×10^{-4} m³ of liquid. Assuming that the soda has the same density as water, find the volume of aluminum used to make the can.

9. **M** **V-HINT** **CHALK** A hypothetical spherical planet consists entirely of iron. What is the period of a satellite that orbits this planet just above its surface? Consult **Table 11.1** as necessary.

10. **H** An antifreeze solution is made by mixing ethylene glycol ($\rho = 1116$ kg/m³) with water. Suppose that the specific gravity of such a solution is 1.0730. Assuming that the total volume of the solution is the sum of its parts, determine the volume percentage of ethylene glycol in the solution.

Section 11.2 Pressure

11. **E** **SSM** An airtight box has a removable lid of area 1.3×10^{-2} m² and negligible weight. The box is taken up a mountain where the air pressure outside the box is 0.85×10^5 Pa. The inside of the box is completely evacuated. What is the magnitude of the force required to pull the lid off the box?

12. E A person who weighs 625 N is riding a 98-N mountain bike. Suppose that the entire weight of the rider and bike is supported equally by the two tires. If the pressure in each tire is 7.60×10^5 Pa, what is the area of contact between each tire and the ground?

13. E MMH A solid concrete block weighs 169 N and is resting on the ground. Its dimensions are 0.400 m × 0.200 m × 0.100 m. A number of identical blocks are stacked on top of this one. What is the smallest number of whole blocks (including the one on the ground) that can be stacked so that their weight creates a pressure of at least two atmospheres on the ground beneath the first block?

14. E United States currency is printed using intaglio presses that generate a printing pressure of 8.0×10^4 lb/in.² A $20 bill is 6.1 in. by 2.6 in. Calculate the magnitude of the force that the printing press applies to one side of the bill.

15. E SSM A glass bottle of soda is sealed with a screw cap. The absolute pressure of the carbon dioxide inside the bottle is 1.80×10^5 Pa. Assuming that the top and bottom surfaces of the cap each have an area of 4.10×10^{-4} m², obtain the magnitude of the force that the screw thread exerts on the cap in order to keep it on the bottle. The air pressure outside the bottle is one atmosphere.

16. M GO A 58-kg skier is going down a slope oriented 35° above the horizontal. The area of each ski in contact with the snow is 0.13 m². Determine the pressure that each ski exerts on the snow.

17. M GO A suitcase (mass $m = 16$ kg) is resting on the floor of an elevator. The part of the suitcase in contact with the floor measures 0.50 m × 0.15 m. The elevator is moving upward with an acceleration of magnitude 1.5 m/s². What pressure (in excess of atmospheric pressure) is applied to the floor beneath the suitcase?

18. M CHALK A cylinder is fitted with a piston, beneath which is a spring, as in the drawing. The cylinder is open to the air at the top. Friction is absent. The spring constant of the spring is 3600 N/m. The piston has a negligible mass and a radius of 0.024 m. **(a)** When the air beneath the piston is completely pumped out, how much does the atmospheric pressure cause the spring to compress? **(b)** How much work does the atmospheric pressure do in compressing the spring?

PROBLEM 18

Section 11.3 Pressure and Depth in a Static Fluid,

Section 11.4 Pressure Gauges

19. E The Mariana trench is located in the floor of the Pacific Ocean at a depth of about 11 000 m below the surface of the water. The density of seawater is 1025 kg/m³. **(a)** If an underwater vehicle were to explore such a depth, what force would the water exert on the vehicle's observation window (radius = 0.10 m)? **(b)** For comparison, determine the weight of a jetliner whose mass is 1.2×10^5 kg.

20. E Review Conceptual Example 6 as an aid in understanding this problem. Consider the pump on the right side of **Figure 11.10**, which acts to reduce the air pressure in the pipe. The air pressure outside the pipe is one atmosphere. Find the maximum depth from which this pump can extract water from the well.

21. E GO A meat baster consists of a squeeze bulb attached to a plastic tube. When the bulb is squeezed and released, with the open end of the tube under the surface of the basting sauce, the sauce rises in the tube to a distance h, as the drawing shows. Using 1.013×10^5 Pa

for the atmospheric pressure and 1200 kg/m³ for the density of the sauce, find the absolute pressure in the bulb when the distance h is **(a)** 0.15 m and **(b)** 0.10 m.

PROBLEM 21

22. E SSM The main water line enters a house on the first floor. The line has a gauge pressure of 1.90×10^5 Pa. **(a)** A faucet on the second floor, 6.50 m above the first floor, is turned off. What is the gauge pressure at this faucet? **(b)** How high could a faucet be before no water would flow from it, even if the faucet were open?

23. E SSM A water tower is a familiar sight in many towns. The purpose of such a tower is to provide storage capacity and to provide sufficient pressure in the pipes that deliver the water to customers. The drawing shows a spherical reservoir that contains 5.25×10^5 kg of water when full. The reservoir is vented to the atmosphere at the top. For a full reservoir, find the gauge pressure that the water has at the faucet in **(a)** house A and **(b)** house B. Ignore the diameter of the delivery pipes.

PROBLEM 23

24. M GO CHALK A mercury barometer reads 747.0 mm on the roof of a building and 760.0 mm on the ground. Assuming a constant value of 1.29 kg/m³ for the density of air, determine the height of the building.

25. M SSM A 1.00-m-tall container is filled to the brim, partway with mercury and the rest of the way with water. The container is open to the atmosphere. What must be the depth of the mercury so that the absolute pressure on the bottom of the container is twice the atmospheric pressure?

26. M CHALK Two identical containers are open at the top and are connected at the bottom via a tube of negligible volume and a valve that is closed. Both containers are filled initially to the same height of 1.00 m, one with water, the other with mercury, as the drawing indicates. The valve is then opened. Water and mercury are immiscible. Determine the fluid level in the left container when equilibrium is reestablished.

PROBLEM 26

27. **M** **V-HINT** The vertical surface of a reservoir dam that is in contact with the water is 120 m wide and 12 m high. The air pressure is one atmosphere. Find the magnitude of the total force acting on this surface in a completely filled reservoir. (*Hint: The pressure varies linearly with depth, so you must use an average pressure.*)

28. **H** A house has a roof with the dimensions shown in the drawing. Determine the magnitude and direction of the net force that the atmosphere applies to the roof when the outside pressure drops suddenly by 75.0 mm of mercury before the air pressure in the attic can adjust. Express your answer in (a) newtons and (b) pounds.

14.5 m   4.21 m

30.0°

30.0°

PROBLEM 28

Section 11.5   Pascal's Principle

29. **E** **SSM** The atmospheric pressure above a swimming pool changes from 755 to 765 mm of mercury. The bottom of the pool is a rectangle (12 m × 24 m). By how much does the force on the bottom of the pool increase?

30. **E** A barber's chair with a person in it weighs 2100 N. The output plunger of a hydraulic system begins to lift the chair when the barber's foot applies a force of 55 N to the input piston. Neglect any height difference between the plunger and the piston. What is the ratio of the radius of the plunger to the radius of the piston?

31. **E** **MMH** Multiple-Concept Example 9 presents an approach to problems of this kind. The hydraulic oil in a car lift has a density of 8.30×10^2 kg/m^3. The weight of the input piston is negligible. The radii of the input piston and output plunger are 7.70×10^{-3} m and 0.125 m, respectively. What input force F is needed to support the 24 500-N combined weight of a car and the output plunger, when (a) the bottom surfaces of the piston and plunger are at the same level, and (b) the bottom surface of the output plunger is 1.30 m *above* that of the input piston?

32. **E** **GO** In the process of changing a flat tire, a motorist uses a hydraulic jack. She begins by applying a force of 45 N to the input piston, which has a radius r_1. As a result, the output plunger, which has a radius r_2, applies a force to the car. The ratio r_2/r_1 has a value of 8.3. Ignore the height difference between the input piston and output plunger and determine the force that the output plunger applies to the car.

33. **M** **V-HINT** A dump truck uses a hydraulic cylinder, as the drawing illustrates. When activated by the operator, a pump injects hydraulic oil into the cylinder at an absolute pressure of 3.54×10^6 Pa and drives the output plunger, which has a radius of 0.150 m. Assuming that the plunger remains perpendicular to the floor of the load bed, find the torque that the plunger creates about the axis identified in the drawing.

3.50 m   Output plunger

Axis

Load bed
Input for
hydraulic oil

PROBLEM 33

34. **M** **GO** **CHALK** The drawing shows a hydraulic chamber with a spring (spring constant = 1600 N/m) attached to the input piston and a rock of mass 40.0 kg resting on the output plunger. The piston and plunger are nearly at the same height, and each has a negligible mass. By how much is the spring compressed from its unstrained position?

PROBLEM 34   Area = 15 cm^2 Area = 65 cm^2

35. **M** **SSM** The drawing shows a hydraulic system used with disc brakes. The force $\vec{F}$ is applied perpendicularly to the brake pedal. The pedal rotates about the axis shown in the drawing and causes a force to be applied perpendicularly to the input piston (radius = 9.50×10^{-3} m) in the master cylinder. The resulting pressure is transmitted by the brake fluid to the output plungers (radii = 1.90×10^{-2} m), which are covered with the brake linings. The linings are pressed against both sides of a disc attached to the rotating wheel. Suppose that the magnitude of $\vec{F}$ is 9.00 N. Assume that the input piston and the output plungers are at the same vertical level, and find the force applied to each side of the rotating disc.

0.100 m
0.0500 m
Brake
fluid
Axis
F
Brake
pedal
Master
cylinder
Plunger
Lining
Edge view
of rotating
disc attached
to wheel

PROBLEM 35

Section 11.6   Archimedes' Principle

36. **E** The density of ice is 917 kg/m^3, and the density of seawater is 1025 kg/m^3. A swimming polar bear climbs onto a piece of floating ice that has a volume of 5.2 m^3. What is the weight of the heaviest bear that the ice can support without sinking completely beneath the water?

37. **E** **SSM** A 0.10-m × 0.20-m × 0.30-m block is suspended from a wire and is completely under water. What buoyant force acts on the block?

38. **E** **GO** A hydrometer is a device used to measure the density of a liquid. It is a cylindrical tube weighted at one end, so that it floats with the heavier end downward. The tube is contained inside a large "medicine dropper," into which the liquid is drawn using the squeeze bulb (see the drawing). For use with your car, marks are put on the tube so that the level at which it floats indicates whether the liquid is battery acid (more dense) or antifreeze (less dense). The hydrometer has a weight of $W = 5.88 \times 10^{-2}$ N and a cross-sectional area of $A = 7.85 \times 10^{-5}$ m^2. How far from the bottom of the tube should the mark be put that denotes (a) battery acid ($\rho = 1280$ kg/m^3) and (b) antifreeze ($\rho = 1073$ kg/m^3)?

Hydrometer

Marks for
two types of
liquids.

PROBLEM 38

39. **E** A duck is floating on a lake with 25% of its volume beneath the water. What is the average density of the duck?

40. **E** A paperweight, when weighed in air, has a weight of $W = 6.9$ N. When completely immersed in water, however, it has a weight of $W_{\text{in water}} = 4.3$ N. Find the volume of the paperweight.

41. **E** **SSM** An 81-kg person puts on a life jacket, jumps into the water, and floats. The jacket has a volume of 3.1×10^{-2} m³ and is completely submerged under the water. The volume of the person's body that is under water is 6.2×10^{-2} m³. What is the density of the life jacket?

42. **E** **V-HINT** A lost shipping container is found resting on the ocean floor and completely submerged. The container is 6.1 m long, 2.4 m wide, and 2.6 m high. Salvage experts attach a spherical balloon to the top of the container and inflate it with air pumped down from the surface. When the balloon's radius is 1.5 m, the shipping container just begins to rise toward the surface. What is the mass of the container? Ignore the mass of the balloon and the air within it. Do *not* neglect the buoyant force exerted on the shipping container by the water. The density of seawater is 1025 kg/m³.

43. **M** **CHALK** **SSM** Refer to Multiple-Concept Example 12 to see a problem similar to this one. What is the smallest number of whole logs ($\rho = 725$ kg/m³, radius = 0.0800 m, length = 3.00 m) that can be used to build a raft that will carry four people, each of whom has a mass of 80.0 kg?

44. **M** **GO** A hot-air balloon is accelerating upward under the influence of two forces, its weight and the buoyant force. For simplicity, consider the weight to be only that of the hot air within the balloon, thus ignoring the balloon fabric and the basket. The hot air inside the balloon has a density of $\rho_{\text{hot air}} = 0.93$ kg/m³, and the density of the cool air outside is $\rho_{\text{cool air}} = 1.29$ kg/m³. What is the acceleration of the rising balloon?

45. **M** A hollow cubical box is 0.30 m on an edge. This box is floating in a lake with one-third of its height beneath the surface. The walls of the box have a negligible thickness. Water from a hose is poured into the open top of the box. What is the depth of the water in the box just at the instant that water from the lake begins to pour into the box from the lake?

46. **M** **GO** To verify her suspicion that a rock specimen is hollow, a geologist weighs the specimen in air and in water. She finds that the specimen weighs twice as much in air as it does in water. The density of the solid part of the specimen is 5.0×10^3 kg/m³. What fraction of the specimen's apparent volume is solid?

47. **H** **SSM** A solid cylinder (radius = 0.150 m, height = 0.120 m) has a mass of 7.00 kg. This cylinder is floating in water. Then oil ($\rho = 725$ kg/m³) is poured on top of the water until the situation shown in the drawing results. How much of the height of the cylinder is in the oil?

PROBLEM 47

48. **H** One kilogram of glass ($\rho = 2.60 \times 10^3$ kg/m³) is shaped into a hollow spherical shell that just barely floats in water. What are the inner and outer radii of the shell? Do not assume that the shell is thin.

Section 11.8 The Equation of Continuity

49. **E** A fuel pump sends gasoline from a car's fuel tank to the engine at a rate of 5.88×10^{-2} kg/s. The density of the gasoline is 735 kg/m³, and the radius of the fuel line is 3.18×10^{-3} m. What is the speed at which the gasoline moves through the fuel line?

50. **E** A room has a volume of 120 m³. An air-conditioning system is to replace the air in this room every twenty minutes, using ducts that have a square cross section. Assuming that air can be treated as an incompressible fluid, find the length of a side of the square if the air speed within the ducts is **(a)** 3.0 m/s and **(b)** 5.0 m/s.

51. **E** **GO** Water flows straight down from an open faucet. The cross-sectional area of the faucet is 1.8×10^{-4} m², and the speed of the water is 0.85 m/s as it leaves the faucet. Ignoring air resistance, find the cross-sectional area of the water stream at a point 0.10 m below the faucet.

52. **M** **GO** **MMH** **CHALK** Three fire hoses are connected to a fire hydrant. Each hose has a radius of 0.020 m. Water enters the hydrant through an underground pipe of radius 0.080 m. In this pipe the water has a speed of 3.0 m/s. **(a)** How many kilograms of water are poured onto a fire in one hour by all three hoses? **(b)** Find the water speed in each hose.

Section 11.9 Bernoulli's Equation,

Section 11.10 Applications of Bernoulli's Equation

53. **E** **SSM** Prairie dogs are burrowing rodents. They do not suffocate in their burrows, because the effect of air speed on pressure creates sufficient air circulation. The animals maintain a difference in the shapes of two entrances to the burrow, and because of this difference, the air ($\rho = 1.29$ kg/m³) blows past the openings at different speeds, as the drawing indicates. Assuming that the openings are at the same vertical level, find the difference in air pressure between the openings and indicate which way the air circulates.

PROBLEM 53

54. **E** Review Conceptual Example 15 before attempting this problem. The truck in that example is traveling at 27 m/s. The density of air is 1.29 kg/m³. By how much does the pressure inside the cargo area beneath the tarpaulin exceed the outside pressure?

55. **E** **SSM** An airplane wing is designed so that the speed of the air across the top of the wing is 251 m/s when the speed of the air below the wing is 225 m/s. The density of the air is 1.29 kg/m³. What is the lifting force on a wing of area 24.0 m²?

56. **E** **GO** Water flowing out of a horizontal pipe emerges through a nozzle. The radius of the pipe is 1.9 cm, and the radius of the nozzle is 0.48 cm. The speed of the water in the pipe is 0.62 m/s. Treat the water as an ideal fluid, and determine the absolute pressure of the water in the pipe.

57. **E** **MMH** A small crack occurs at the base of a 15.0-m-high dam. The effective crack area through which water leaves is 1.30×10^{-3} m². **(a)** Ignoring viscous losses, what is the speed of water flowing through the crack? **(b)** How many cubic meters of water per second leave the dam?

58. **E** Water is circulating through a closed system of pipes in a two-floor apartment. On the first floor, the water has a gauge pressure of 3.4×10^5 Pa and a speed of 2.1 m/s. However, on the second floor, which is 4.0 m higher, the speed of the water is 3.7 m/s. The speeds are different because the pipe diameters are different. What is the gauge pressure of the water on the second floor?

59. **E GO** A ship is floating on a lake. Its hold is the interior space beneath its deck; the hold is empty and is open to the atmosphere. The hull has a hole in it, which is below the water line, so water leaks into the hold. The effective area of the hole is 8.0×10^{-3} m^2 and is located 2.0 m beneath the surface of the lake. What volume of water per second leaks into the ship?

60. **M SSM** A Venturi meter is a device that is used for measuring the speed of a fluid within a pipe. The drawing shows a gas flowing at speed v_2 through a horizontal section of pipe whose cross-sectional area is $A_2 = 0.0700$ m^2. The gas has a density of $\rho = 1.30$ kg/m^3. The Venturi meter has a cross-sectional area of $A_1 = 0.0500$ m^2 and has been substituted for a section of the larger pipe. The pressure difference between the two sections is $P_2 - P_1 = 120$ Pa. Find **(a)** the speed v_2 of the gas in the larger, original pipe and **(b)** the volume flow rate Q of the gas.

PROBLEM 60

61. **M V-HINT CHALK** A hand-pumped water gun is held level at a height of 0.75 m above the ground and fired. The water stream from the gun hits the ground a horizontal distance of 7.3 m from the muzzle. Find the gauge pressure of the water gun's reservoir at the instant when the gun is fired. Assume that the speed of the water in the reservoir is zero and that the water flow is steady. Ignore both air resistance and the height difference between the reservoir and the muzzle.

62. **M CHALK** A liquid is flowing through a horizontal pipe whose radius is 0.0200 m. The pipe bends straight upward through a height of 10.0 m and joins another horizontal pipe whose radius is 0.0400 m. What volume flow rate will keep the pressures in the two horizontal pipes the same?

63. **M V-HINT** An airplane has an effective wing surface area of 16 m^2 that is generating the lift force. In level flight the air speed over the top of the wings is 62.0 m/s, while the air speed beneath the wings is 54.0 m/s. What is the weight of the plane?

64. **M MMH** The construction of a flat rectangular roof (5.0 m × 6.3 m) allows it to withstand a maximum net outward force that is 22 000 N. The density of the air is 1.29 kg/m^3. At what wind speed will this roof blow outward?

65. **M GO** A pump and its horizontal intake pipe are located 12 m beneath the surface of a large reservoir. The speed of the water in the intake pipe causes the pressure there to decrease, in accord with

Bernoulli's principle. Assuming nonviscous flow, what is the maximum speed with which water can flow through the intake pipe?

66. **H SSM** A uniform rectangular plate is hanging vertically downward from a hinge that passes along its left edge. By blowing air at 11.0 m/s *over the top of the plate only*, it is possible to keep the plate in a horizontal position, as illustrated in part *a* of the drawing. To what value should the air speed be reduced so that the plate is kept at a 30.0° angle with respect to the vertical, as in part *b* of the drawing? (*Hint: Apply Bernoulli's equation in the form of Equation 11.12.*)

PROBLEM 66

Section 11.11 Viscous Flow

67. **E** A pipe is horizontal and carries oil that has a viscosity of 0.14 Pa · s. The volume flow rate of the oil is 5.3×10^{-5} m^3/s. The length of the pipe is 37 m, and its radius is 0.60 cm. At the output end of the pipe the pressure is atmospheric pressure. What is the absolute pressure at the input end?

68. **M GO CHALK** Two hoses are connected to the same outlet using a Y-connector, as the drawing shows. The hoses A and B have the same length, but hose B has the larger radius. Each is open to the atmosphere at the end where the water exits. Water flows through both hoses as a viscous fluid, and Poiseuille's law [$Q = \pi R^4 (P_2 - P_1)/(8\eta L)$] applies to each. In this law, P_2 is the pressure upstream, P_1 is the pressure downstream, and Q is the volume flow rate. The ratio of the radius of hose B to the radius of hose A is $R_B/R_A = 1.50$. Find the ratio of the speed of the water in hose B to the speed in hose A.

PROBLEM 68   Hose A   Hose B

Additional Problems

Online

69. **E SSM** Measured along the surface of the water, a rectangular swimming pool has a length of 15 m. Along this length, the flat bottom of the pool slopes downward at an angle of 11° below the horizontal, from one end to the other. By how much does the pressure at the bottom of the deep end exceed the pressure at the bottom of the shallow end?

70. **E** A water bed for sale has dimensions of 1.83 m × 2.13 m × 0.229 m. The floor of the bedroom will tolerate an additional weight of no more than 6660 N. Find the weight of the water in the bed and determine whether the bed should be purchased.

71. **E GO** An underground pump initially forces water through a horizontal pipe at a flow rate of 740 gallons per minute. After several years of

operation, corrosion and mineral deposits have reduced the inner radius of the pipe to 0.19 m from 0.24 m, but the pressure difference between the ends of the pipe is the same as it was initially. Find the final flow rate in the pipe in gallons per minute. Treat water as a viscous fluid.

72. **E** (a) The mass and the radius of the sun are, respectively, 1.99×10^{30} kg and 6.96×10^8 m. What is its density? (b) If a solid object is made from a material that has the same density as the sun, would it sink or float in water? Why? (c) Would a solid object sink or float in water if it were made from a material whose density was the same as that of the planet Saturn (mass = 5.7×10^{26} kg, radius = 6.0×10^7 m)? Provide a reason for your answer.

73. **M** **SSM** A water line with an internal radius of 6.5×10^{-3} m is connected to a shower head that has 12 holes. The speed of the water in the line is 1.2 m/s. (a) What is the volume flow rate in the line?

(b) At what speed does the water leave one of the holes (effective hole radius = 4.6×10^{-4} m) in the head?

74. **M** **V-HINT** A log splitter uses a pump with hydraulic oil to push a piston, which is attached to a chisel. The pump can generate a pressure of 2.0×10^7 Pa in the hydraulic oil, and the piston has a radius of 0.050 m. In a stroke lasting 25 s, the piston moves 0.60 m. What is the power needed to operate the log splitter's pump?

75. **M** An object is solid throughout. When the object is completely submerged in ethyl alcohol, its apparent weight is 15.2 N. When completely submerged in water, its apparent weight is 13.7 N. What is the volume of the object?

76. **M** **GO** A gold prospector finds a solid rock that is composed solely of quartz and gold. The mass and volume of the rock are, respectively, 12.0 kg and 4.00×10^{-3} m³. Find the mass of the gold in the rock.

Physics in Biology, Medicine, and Sports

77. **M** **GO** **SSM** **BIO** **11.10**   Review Example 17 as an aid in understanding this problem. An aneurysm is an abnormal enlargement of a blood vessel such as the aorta. Because of the aneurysm, the normal cross-sectional area A_1 of the aorta increases to a value of $A_2 = 1.7A_1$. The speed of the blood ($\rho = 1060$ kg/m³) through a normal portion of the aorta is $v_1 = 0.40$ m/s. Assuming that the aorta is horizontal (the person is lying down), determine the amount by which the pressure P_2 in the enlarged region exceeds the pressure P_1 in the normal region.

78. **M** **BIO** **V-HINT** **11.2**   As the initially empty urinary bladder fills with urine and expands, its internal pressure increases by 3300 Pa, which triggers the micturition reflex (the feeling of the need to urinate). The drawing shows a horizontal, square section of the bladder wall with an edge length of 0.010 m. Because the bladder is stretched, four tension forces of equal magnitude T act on the square section, one at each edge, and each force is directed at an angle θ below the horizontal. What is the magnitude T of the tension force acting on one edge of the section when the internal bladder pressure is 3300 Pa and each of the four tension forces is directed 5.0° below the horizontal?

PROBLEM 78

79. **E** **BIO** **11.4**   The drawing shows an intravenous feeding. With the distance shown, nutrient solution ($\rho = 1030$ kg/m³) can just barely enter the blood in the vein. What is the gauge pressure of the venous blood? Express your answer in millimeters of mercury.

Atmospheric pressure

0.610 m

PROBLEM 79

80. **E** **BIO** **11.3**   The human lungs can function satisfactorily up to a limit where the pressure difference between the outside and inside of the lungs is one-twentieth of an atmosphere. If a diver uses a snorkel for breathing, how far below the water can she swim? Assume the diver is in salt water whose density is 1025 kg/m³.

81. **E** **BIO** **11.3**   At a given instant, the blood pressure in the heart is 1.6×10^4 Pa. If an artery in the brain is 0.45 m above the heart, what is the pressure in the artery? Ignore any pressure changes due to blood flow.

82. **E** **BIO** **SSM** **11.8**   A patient recovering from surgery is being given fluid intravenously. The fluid has a density of 1030 kg/m³, and 9.5×10^{-4} m³ of it flows into the patient every six hours. Find the mass flow rate in kg/s.

83. **E** **BIO** **MMH** **11.8**   (a) The volume flow rate in an artery supplying the brain is 3.6×10^{-6} m³/s. If the radius of the artery is 5.2 mm, determine the average blood speed. (b) Find the average blood speed at a constriction in the artery if the constriction reduces the radius by a factor of 3. Assume that the volume flow rate is the same as that in part (a).

84. **M** **BIO** **V-HINT** **11.8**   The aorta carries blood away from the heart at a speed of about 40 cm/s and has a radius of approximately 1.1 cm. The aorta branches eventually into a large number of tiny capillaries that distribute the blood to the various body organs. In a capillary, the blood speed is approximately 0.07 cm/s, and the radius is about 6×10^{-4} cm. Treat the blood as an incompressible fluid, and use these data to determine the approximate number of capillaries in the human body.

85. **E** **BIO** **11.10**   The blood speed in a normal segment of a horizontal artery is 0.11 m/s. An abnormal segment of the artery is narrowed down by an arteriosclerotic plaque to one-fourth the normal cross-sectional area. What is the difference in blood pressures between the normal and constricted segments of the artery?

86. **E** **BIO** **11.11**   In the human body, blood vessels can dilate, or increase their radii, in response to various stimuli, so that the volume flow rate of the blood increases. Assume that the pressure at either end of a blood vessel, the length of the vessel, and the viscosity of the blood remain the same, and determine the factor $R_{\text{dilated}}/R_{\text{normal}}$ by which the radius of a vessel must change in order to double the volume flow rate of the blood through the vessel.

87. **E** **BIO** **GO** **11.11**   A blood transfusion is being set up in an emergency room for an accident victim. Blood has a density of 1060 kg/m³ and a viscosity of 4.0×10^{-3} Pa · s. The needle being used has a

length of 3.0 cm and an inner radius of 0.25 mm. The doctor wishes to use a volume flow rate through the needle of 4.5×10^{-8} m³/s. What is the distance h above the victim's arm where the level of the blood in the transfusion bottle should be located? As an approximation, assume that the level of the blood in the transfusion bottle and the point where the needle enters the vein in the arm have the same pressure of one atmosphere. (In reality, the pressure in the vein is slightly above atmospheric pressure.)

88. **E** **BIO** **11.9**   One way to administer an inoculation is with a "gun" that shoots the vaccine through a narrow opening. No needle is necessary, for the vaccine emerges with sufficient speed to pass directly into the tissue beneath the skin. The speed is high, because the vaccine ($\rho = 1100$ kg/m³) is held in a reservoir where a high pressure pushes it out. The pressure on the surface of the vaccine in one gun is 4.1×10^6 Pa above the atmospheric pressure outside the narrow opening. The dosage is small enough that the vaccine's surface in the reservoir is nearly stationary during an inoculation. The vertical height between the vaccine's surface in the reservoir and the opening can be ignored. Find the speed at which the vaccine emerges.

89. **E** **BIO** **11.3**   If a scuba diver descends too quickly into the sea, the internal pressure on each eardrum remains at atmospheric pressure, while the external pressure increases due to the increased water depth. At sufficient depths, the difference between the external and internal pressures can rupture an eardrum. Eardrums can rupture when the pressure difference is as little as 35 kPa. What is the depth at which this pressure difference could occur? The density of seawater is 1025 kg/m³.

90. **E** **BIO** **11.3**   The Mariana snailfish (see the photo) holds the record for the world's deepest living fish. The snailfish has been found in the Mariana Trench at a depth of 26 500 feet below the water's surface. **(a)** Relative to atmospheric pressure at the surface of the ocean, what is the water pressure at the depth where the snailfish lives? Use the density of sea water (1030 kg/m³) in your calculation. **(b)** How many total pounds of force would be applied to a human head at this depth? Treat the head as a sphere with a radius of 9.80 cm.

Geringer M. E., Linley T. D., Jamieson A. J., Goetze E. Drazen J. C., Image taken from https://en.wikipedia.org/wiki/Pseudoliparis_swirei#/media/File:Pseudoliparis_swirei.png. Licensed under CC by 3.0

PROBLEM 90

91. **M** **BIO** **11.1**   Astronauts living and working on the International Space Station in orbit above the earth will experience many negative physiological effects from microgravity. One of the more serious ones is a decrease in bone density. Our bones are composed of living tissue that is constantly being broken down and replaced with new bone. On the earth's surface, these two processes occur at roughly the same rate, so our bone density remains fairly constant for long periods of time. In space, the muscles and bones are not subjected to the constant weights, forces, and stresses that come with living in gravity. This lack of constant stress leads to a slower rate of new bone growth and a decrease in overall bone density (see the figure). The longest and strongest bone in the human body—the femur—has a volume of 118 cm³ and a density of 1.94 g/cm³. If an astronaut on the space station experiences a net loss of 20.0% of the bone mass in their femur after 5 months, what is the bone density at that time?

Normal Bone   Decrease in Bone Density

Tefi/Shutterstock.com

PROBLEM 91

92. **M** **BIO** **11.3**   The tympanic membrane, or eardrum, is a structure that separates the external and middle parts of the ear (see the figure). It is sensitive to and vibrates in response to changes in air pressure and transmits these vibrations to other structures in the inner ear that lead to the sensation of hearing. Under normal conditions, the pressures on the inside and outside of the tympanic membrane are kept approximately equal. The auditory tube, also called the Eustachian tube, is responsible for this equilibration. However, rapid changes in external pressure can cause large pressure differentials on the tympanic membrane, causing it to rupture. A differential force across the eardrum membrane as little as 5.0 N can cause a rupture. **(a)** If the cross-sectional area of the membrane is 1.0 cm², what is the maximum tolerable pressure difference between the external and inner ear? **(b)** Based on your answer in part (a), to what maximum depth could a person dive in fresh water before rupturing an eardrum? Experienced divers can actually go much deeper than the answer to part (b) by using techniques to equilibrate the pressure on the inside and outside of the membrane as they dive.

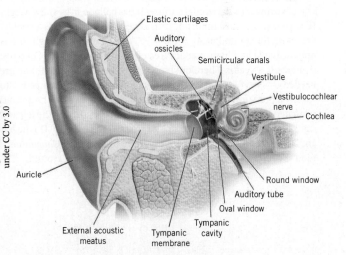

Elastic cartilages
Auditory ossicles
Semicircular canals
Vestibule
Vestibulocochlear nerve
Cochlea
Round window
Auditory tube
Oval window
Tympanic cavity
Tympanic membrane
External acoustic meatus
Auricle

The Anatomy of the Ear

PROBLEM 92

93. **M** **BIO** **11.6**   The blue whale is the largest animal known to have ever existed (see the photo). It can be almost 30 m long with a mass of 1.73×10^5 kg. Imagine a blue whale is completely immersed and at rest under the surface of salt water. What volume of salt water is displaced by the whale? Calculate your answer in gallons. Take the density of salt water to be 1030 kg/m³.

PROBLEM 93

94. M BIO **11.11** A patient is receiving medication through an intravenous infusion, as the figure shows. The medication has a viscosity of 2.2×10^{-3} N $\cdot$ s/m^2 and a density of 1050 kg/m^3. The needle being used has a length of 0.035 m, has an inner radius of 5.0×10^{-3} m, and has been inserted into a vein that has a gauge pressure of 2.0×10^3 Pa. The IV bag hangs 0.45 m above the location of the needle. What volume of medication enters the vein in 10.0 minutes?

Give your answer in cm^3. Assume the height of the column of medication in the IV bag and connecting tube remains constant during the 10.0 minutes.

PROBLEM 94

Concepts and Calculations Problems

Online

Pressure plays an important role in the behavior of fluids. As we have seen in this chapter, pressure is the magnitude of the force acting perpendicular to a surface divided by the area of the surface. Pressure should not be confused, however, with the force itself. Problem 95 serves to emphasize that pressure and force are different concepts. Problem 96 focuses on the essence of Archimedes' principle and its application to the situation of a buoyant force acting on a submerged object.

95. M CHALK The figure shows a rear view of a loaded two-wheeled wheelbarrow on a horizontal surface. It has balloon tires and a weight $W = 684$ N, which is uniformly distributed. The left tire has a contact area with the ground of $A_L = 6.6 \times 10^{-4}$ m^2, whereas the right tire is underinflated and has a contact area of $A_R = 9.9 \times 10^{-4}$ m^2. *Concepts:* (i) Force is a vector. Therefore both a direction and a magnitude are needed to specify it. Are both a direction and magnitude needed to specify a pressure? (ii) How is the force each tire applies to the ground related to the force the ground applies to each tire? (iii) Do the left and right tires apply the same force to the ground? Explain. (iv) Do the left and right tires apply the same pressure to the ground? *Calculations:* Find the force and pressure that each tire applies to the ground.

(a)   (b) Free-body diagram for the wheelbarrow

(a) Rear view of a wheelbarrow with balloon tires. (b) Free-body diagram of the wheelbarrow, showing its weight $\vec{W}$ and the forces $\vec{F}_L$ and $\vec{F}_R$ that the ground applies, respectively, to the left and right tires.

PROBLEM 95

96. M CHALK SSM A father (weight $W = 830$ N) and his daughter (weight $W = 340$ N) are spending the day at the lake. They are each sitting on a beach ball that is just submerged beneath the water (see the figure). *Concepts:* (i) Each ball is in equilibrium, being stationary and having no acceleration. Thus, the net force acting on each ball is zero. What balances the downward-acting weight in each case? (ii) In which case is the buoyant force greater? (iii) In this situation, what determines the magnitude of the buoyant force? (iv) Which beach ball has the larger radius? *Calculations:* Ignoring the weight of the air in each ball, and the volumes of their legs that are under the water, find the radius of each ball.

PROBLEM 96

Team Problems

97. [M] [T] [WS] **A Fire Hose.** You and your team need to put out a fire on a rooftop that is currently too high for the water stream emerging from a fire hose. With no nozzle attached, and the 4.00-in.-diameter hose aimed at a 60.0° angle above the horizontal, the peak of the stream's trajectory barely makes it to the bottom of the third floor, which is 20.0 feet above the ground. Assuming the flow rate does not change, what diameter nozzle (in cm) would be sufficient to get water to the roof of the five-story building (i.e., 60.0 feet above the ground) when the hose is directed 60.0° above the horizontal? Treat the water trajectory like projectile motion, and neglect air resistance.

98. [E] [T] [WS] **A Malfunctioning Submarine.** You and your team are inside a small research submarine in the Weddell Sea off the coast of western Antarctica, searching for undersea tunnels leading inland, when the sub suffers serious malfunctions. First, power fails, and you are left with only emergency batteries that will provide dim light and run the air purifying system for no more than 2 hours. Next, the valves to the ballast tanks (i.e., the mechanism used to control the buoyancy of the craft) inadvertently opened and the tanks completely filled with seawater, making the sub sink to the seabed, 70 feet below the surface. All safety devices, such as inflatable buoys that could be sent to warn those on the surface of the emergency, are nonfunctional. You find the submarine's operation manual and learn that there is a manual pump that can be used to expel water from the ballast tanks at a rate of about 1.0 gallon per minute (gpm). In the manual, you find that the mass of the submarine, with crew and supplies, and with empty ballast tanks, is 9250 kg. The shape of the sub is roughly cylindrical, with a length of 5.0 m and a diameter of 1.6 m. The ballast tanks are contained inside the submarine's body, and account for 15% of its total volume. Using the manual pump, how long will it take to remove enough water to float the submarine? The density of seawater is 1040 kg/m^3.

99. [M] [T] **Crossing a River.** You and your team come to a slow flowing river that you need to cross. The nearest bridge is 20 miles to the north, too dangerous and too far to trek with your group. You explore the area down river and discover an abandoned shed with a stash of 55-gallon drums (all empty and weighing 35.0 lb each) and a stack of 10-foot planks: You will build a raft. The six members of your team have a combined weight of 925 pounds (assuming everyone was truthful). You also have a four-wheeler (all-terrain vehicle), which weighs 450.0 lb, and other gear, which adds another 315 lb. Your simple raft design is as follows: a platform of planks with barrels strapped to the bottom. You estimate that the planks weigh about 45.0 lb each, and you will need 20 of them to make a platform that can accommodate everything for the one-way trip. You measure the dimensions of the cylindrical drums and find they have diameter $D = 22.5$ inches and height $h = 33.5$ inches (note: not exactly "55 gallons"). **(a)** What is the minimum number of barrels that you will need so that the raft will float when fully loaded? **(b)** What is the minimum number of barrels you will need if you want the platform to be at least 6 inches above the water when fully loaded?

100. [M] [T] **A Spring Gun.** A hydraulic press is used to compress a spring that will then be used to project a 25.0-kg steel ball. The system is similar, but smaller in scale, to the car jack illustrated in **Figure 11.15**. In this case, the smaller cylinder has a diameter of $d_1 = 0.750$ cm and has a manually operated plunger. The larger cylinder has a diameter of $d_2 = 10.0$ cm, and its piston compresses the spring. The idea is that the gun operator pulls a lever that pushes the plunger on the small cylinder, which transmits a pressure to the larger piston that, in turn, exerts a force on the spring and compresses it. Once the spring is compressed, the steel ball is loaded and the spring is released, ejecting the ball. **(a)** If a force of 950.0 N is exerted on the primary (smaller) piston to compress the spring 1.25 m from its equilibrium (uncompressed) position, what is the spring constant k of the spring? **(b)** What is the velocity of the steel ball just after it is ejected? **(c)** Neglecting air resistance, what is the maximum range of this "spring gun"?

As this artist's rendition shows, the underside of the Space Shuttle consists of special ceramic tiles that shield the Shuttle's interior and flight crew from the high temperatures and heat produced upon reentering the Earth's atmosphere after a mission in space. In 2003 the Space Shuttle *Columbia's* seven-member crew was killed when the shuttle disintegrated on reentry after suffering damage to the heat shield during launch.

NASA/Science Source

LEARNING OBJECTIVES

After reading this module, you should be able to...

12.1 Convert temperature scales.

12.2 Define the Kelvin scale and absolute zero.

12.3 Identify types of thermometers.

12.4 Analyze linear thermal expansion.

12.5 Analyze volume thermal expansion.

12.6 Define heat and internal energy.

12.7 Solve specific heat problems.

12.8 Solve heat problems with phase changes.

12.9 Analyze phase equilibrium.

12.10 Solve humidity problems.

Temperature and Heat

12.1 Common Temperature Scales

To measure temperature we use a thermometer. Many thermometers make use of the fact that materials usually expand with increasing temperature. For example, **Figure 12.1** shows the common mercury-in-glass thermometer, which consists of a mercury-filled glass bulb connected to a capillary tube. When the mercury is heated, it expands into the capillary tube, the amount of expansion being proportional to the change in temperature. The outside of the glass is marked with an appropriate scale for reading the temperature.

A number of different temperature scales have been devised, two popular choices being the **Celsius** (formerly, centigrade) and **Fahrenheit scales. Figure 12.1** illustrates these scales. Historically,* both scales were defined by assigning two temperature points on the scale and then dividing the distance between them into a number of equally spaced intervals. One

---

*Today, the Celsius and Fahrenheit scales are defined in terms of the Kelvin temperature scale; Section 12.2 discusses the Kelvin scale.

point was chosen to be the temperature at which ice melts under one atmosphere of pressure (the "ice point"), and the other was the temperature at which water boils under one atmosphere of pressure (the "steam point"). On the Celsius scale, an ice point of 0 °C (0 degrees Celsius) and a steam point of 100 °C were selected. On the Fahrenheit scale, an ice point of 32 °F (32 degrees Fahrenheit) and a steam point of 212 °F were chosen. The Celsius scale is used worldwide, while the Fahrenheit scale is used mostly in the United States, often in home medical thermometers.

There is a subtle difference in the way the temperature of an object is reported, as compared to a *change* in its temperature. For example, the temperature of the human body is about 37 °C, where the symbol °C stands for "degrees Celsius." However, the *change* between two temperatures is specified in "Celsius degrees" (C°)—not in "degrees Celsius." Thus, if the body temperature rises to 39 °C, the change in temperature is 2 Celsius degrees or 2 C°, not 2 °C.

As **Figure 12.1** indicates, the separation between the ice and steam points on the Celsius scale is divided into 100 Celsius degrees, while on the Fahrenheit scale the separation is divided into 180 Fahrenheit degrees. Therefore, the size of the Celsius degree is larger than that of the Fahrenheit degree by a factor of $\frac{180}{100}$, or $\frac{9}{5}$. Examples 1 and 2 illustrate how to convert between the Celsius and Fahrenheit scales using this factor.

The reasoning strategy used in Examples 1 and 2 for converting between different temperature scales is summarized below.

REASONING STRATEGY   **Converting Between Different Temperature Scales**

1. **Determine the magnitude of the difference between the stated temperature and the ice point on the initial scale.**

2. **Convert this number of degrees from one scale to the other scale by using the appropriate conversion factor. For conversion between the Celsius and Fahrenheit scales, the factor is based on the fact that 1 C° = $\frac{9}{5}$ F°**

3. **Add or subtract the number of degrees on the new scale to or from the ice point on the new scale.**

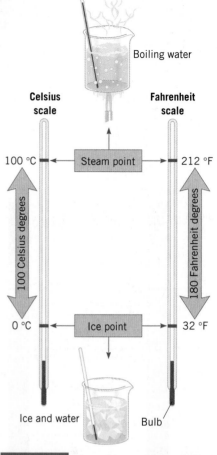

FIGURE 12.1 The Celsius and Fahrenheit temperature scales.

EXAMPLE 1 | Converting from a Fahrenheit to a Celsius Temperature

A healthy person has an oral temperature of 98.6 °F. What would this reading be on the Celsius scale?

Reasoning and Solution A temperature of 98.6 °F is 66.6 Fahrenheit degrees above the ice point of 32.0 °F. Since 1 C° = $\frac{9}{5}$ F°, the difference of 66.6 F° is equivalent to

$$(66.6 \text{ F°})\left(\frac{1 \text{ C}}{\frac{9}{5} \text{ F°}}\right) = 37.0 \text{ C°}$$

Thus, the person's temperature is 37.0 Celsius degrees above the ice point. Adding 37.0 Celsius degrees to the ice point of 0 °C on the Celsius scale gives a Celsius temperature of $\boxed{37.0 \text{ °C}}$.

EXAMPLE 2 | Converting from a Celsius to a Fahrenheit Temperature

A time and temperature sign on a bank indicates that the outdoor temperature is −20.0 °C. Find the corresponding temperature on the Fahrenheit scale.

Reasoning and Solution The temperature of −20.0 °C is 20.0 Celsius degrees *below* the ice point of 0 °C. This number of Celsius degrees corresponds to

$$(20.0 \text{ C°})\left(\frac{\frac{9}{5} \text{ F°}}{1 \text{ C°}}\right) = 36.0 \text{ F°}$$

The temperature, then, is 36.0 Fahrenheit degrees below the ice point. Subtracting 36.0 Fahrenheit degrees from the ice point of 32.0 °F on the Fahrenheit scale gives a Fahrenheit temperature of $\boxed{-4.0 \text{ °F}}$.

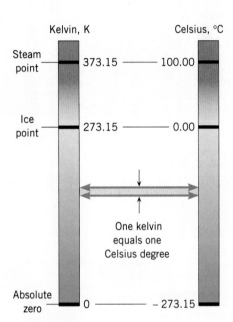

FIGURE 12.2 A comparison of the Kelvin and Celsius temperature scales.

12.2 The Kelvin Temperature Scale

Although the Celsius and Fahrenheit scales are widely used, the **Kelvin temperature scale** has greater scientific significance. It was introduced by the Scottish physicist William Thompson (Lord Kelvin, 1824–1907), and in his honor each degree on the scale is called a kelvin (K). By international agreement, the symbol K is not written with a degree sign (°), nor is the word "degrees" used when quoting temperatures. For example, a temperature of 300 K (not 300 °K) is read as "three hundred kelvins," not "three hundred degrees kelvin." The kelvin is the SI base unit for temperature.

Figure 12.2 compares the Kelvin and Celsius scales. The size of one kelvin is identical to the size of one Celsius degree because there are one hundred divisions between the ice and steam points on both scales. As we will discuss shortly, experiments have shown that there exists a lowest possible temperature, below which no substance can be cooled. This lowest temperature is defined to be the zero point on the Kelvin scale and is referred to as *absolute zero*.

The ice point (0 °C) occurs at 273.15 K on the Kelvin scale. Thus, the Kelvin temperature T and the Celsius temperature T_c are related by

$$T = T_c + 273.15 \qquad (12.1)$$

The number 273.15 in Equation 12.1 is an experimental result, obtained in studies that utilize a gas-based thermometer.

When a gas confined to a fixed volume is heated, its pressure increases. Conversely, when the gas is cooled, its pressure decreases. For example, the air pressure in automobile tires can rise by as much as 20% after the car has been driven and the tires have become warm. The change in gas pressure with temperature is the basis for the **constant-volume gas thermometer**.

A constant-volume gas thermometer consists of a gas-filled bulb to which a pressure gauge is attached, as in **Figure 12.3**. The gas is often hydrogen or helium at a low density, and the pressure gauge can be a U-tube manometer filled with mercury. The bulb is placed in thermal contact with the substance whose temperature is being measured. The volume of the gas is held constant by raising or lowering the *right* column of the U-tube manometer in order to keep the mercury level in the *left* column at the same reference level. The absolute pressure of the gas is proportional to the height h of the mercury on the right. As the temperature changes, the pressure changes and can be used to indicate the temperature, once the constant-volume gas thermometer has been calibrated.

Suppose that the absolute pressure of the gas in **Figure 12.3** is measured at different temperatures. If the results are plotted on a pressure-versus-temperature graph, a straight line is obtained, as in **Figure 12.4**. If the straight line is extended or extrapolated to lower and lower temperatures, the line crosses the temperature axis at −273.15 °C. In reality, no gas can be cooled to this temperature, because all gases liquify before reaching it. However, helium and hydrogen liquify at such low temperatures that they are often used in the thermometer. This kind of graph can be obtained for different amounts and types of low-density gases. In all cases, a straight line is found that extrapolates to

FIGURE 12.3 A constant-volume gas thermometer.

−273.15 °C on the temperature axis, which suggests that the value of −273.15 °C has fundamental significance. The significance of this number is that it is the **absolute zero point** for temperature measurement. The phrase "absolute zero" means that temperatures lower than −273.15 °C cannot be reached by continually cooling a gas or any other substance. If lower temperatures could be reached, then further extrapolation of the straight line in **Figure 12.4** would suggest that negative absolute gas pressures could exist. Such a situation would be impossible, because a negative absolute gas pressure has no meaning. Thus, the Kelvin scale is chosen so that its zero temperature point is the lowest temperature attainable.

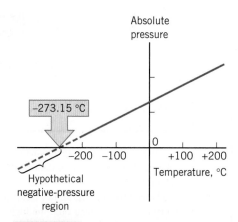

FIGURE 12.4 A plot of absolute pressure versus temperature for a low-density gas at constant volume. The graph is a straight line and, when extrapolated (dashed line), crosses the temperature axis at −273.15 °C.

12.3 Thermometers

All thermometers make use of the change in some physical property with temperature. A property that changes with temperature is called a **thermometric property**. For example, the thermometric property of the mercury thermometer is the length of the mercury column, while in the constant-volume gas thermometer it is the pressure of the gas. Several other important thermometers and their thermometric properties will now be discussed.

The *thermocouple* is a thermometer used extensively in scientific laboratories. It consists of thin wires of different metals, welded together at the ends to form two junctions, as **Figure 12.5** illustrates. Often the metals are copper and constantan (a copper–nickel alloy). One of the junctions, called the "hot" junction, is placed in thermal contact with the object whose temperature is being measured. The other junction, termed the "reference" junction, is kept at a known constant temperature (usually an ice–water mixture at 0 °C). The thermocouple generates a voltage that depends on the *difference in temperature* between the two junctions. This voltage is the thermometric property and is measured by a voltmeter, as the drawing indicates. With the aid of calibration tables, the temperature of the hot junction can be obtained from the voltage. Thermocouples are used to measure temperatures as high as 2300 °C or as low as −270 °C.

Most substances offer resistance to the flow of electricity, and this resistance changes with temperature. As a result, electrical resistance provides another thermometric property. *Electrical resistance thermometers* are often made from platinum wire, because platinum has excellent mechanical and electrical properties in the temperature range from −270 °C to +700 °C. The resistance of platinum wire is known as a function of temperature. Thus,

(a)

(b)

Courtesy of David Young and Shane Stadler

FIGURE 12.5 (a) A thermocouple is made from two different types of wires, copper and constantan in this case. (b) A thermocouple junction between two different wires.

the temperature of a substance can be determined by placing the resistance thermometer in thermal contact with the substance and measuring the resistance of the wire.

Almost anything that has a predictable response to temperature can be used as a thermometer. As Example 3 shows, this even applies to the chirp of the snowy tree cricket.

EXAMPLE 3 | BIO A Cricket Thermometer

Snowy tree crickets are green, are about 1.5 cm long, and have an orange mark and long antennae on their heads. They can be found in most of North America, and have been extensively studied. One interesting characteristic of the snowy tree cricket is the temperature dependence of their chirp rate: the higher the ambient temperature, the higher the chirp rate. This temperature dependence can serve as a crude audible thermometer. On a particularly cold night ($T = 13\ °C$), suppose the total number of chirps recorded in your back yard in one minute was 119. The next day, a warm front comes through, the nighttime temperature increases to 26 °C, and the number of chirps recorded in one minute is 258. Assuming a linear relationship between the chirp rate and the temperature, and that the number of crickets does not change, what is the temperature on the next night when 204 chirps are counted in one minute?

Reasoning Since we assume a linear relationship between the number of chirps per minute N and the temperature T, we can apply the point-slope formula to obtain the slope, and then a linear equation that relates T and N.

Solution The point-slope formula for a line passing through points (x_1, y_1) and (x_2, y_2) is given by $y_2 - y_1 = m(x_2 - x_1)$, where m is the slope, y is the dependent variable N, and x is the independent variable T. Writing this as $N_2 - N_1 = m(T_2 - T_1)$ and solving for the slope, we have

$$m = \frac{N_2 - N_1}{T_2 - T_1} = \frac{258\ \text{chirps/min} - 119\ \text{chirps/min}}{26\ °C - 13\ °C} = 10.7\ \frac{\text{chirps}}{\text{min} \cdot °C}$$

The equation of a line is given by $y = mx + b$, which we write as $N = mT + b$, where b is the N-axis intercept. Since we know the slope (m), we can solve this equation for b, and plug in one of the two sets of known values (say $N = 119$ and $T = 13\ °C$) to obtain the following:

$$b = N - mT = 119\ \text{chirps/min} - [10.7\ \text{chirps/(min} \cdot °C)](13\ °C)$$

$$= -20.1\ \text{chirps/min}$$

Note that this negative number of chirps has no physical meaning, but is just the y-intercept of the line. Solving $N = mT + b$ for T, and substituting the values for m and b, we have

$$T = \frac{N - b}{m} = \frac{N}{m} - \frac{b}{m} = \frac{N}{[10.7\ \text{chirps/(min} \cdot °C)]}$$

$$- \frac{-20.1\ \text{chirps/min}}{[10.7\ \text{chirps/(min} \cdot °C)]}$$

or

$$T = \left(0.0935\ \frac{\text{min} \cdot °C}{\text{chirps}}\right)N + 1.88\ °C$$

Substituting $N = 204$ chirps/min into this equation, and taking 2 significant figures, we have

$$T = \left(0.0935\ \frac{\text{min} \cdot °C}{\text{chirps}}\right)\left(204\ \frac{\text{chirps}}{\text{min}}\right) + 1.88\ °C = \boxed{21\ °C}$$

This result is only an approximation, and does not work, for instance, at temperatures below about 10 °C or above 38 °C, where snowy tree crickets stop chirping altogether.

SPL/Science Source

FIGURE 12.6 The presence of breast cancer can be detected in a thermogram via the elevated temperatures associated with the malignant tissue. The healthy breast on the left side of the thermogram registers predominantly blue (see the temperature coding bar to the right of the image), indicating temperatures that are not unusually elevated. The breast on the right side of the thermogram, however, registers the colors yellow, orange, and red, which indicate the elevated temperatures due to breast cancer.

BIO THE PHYSICS OF . . . thermography. Radiation emitted by an object can also be used to indicate temperature. At low to moderate temperatures, the predominant radiation emitted is infrared. As the temperature is raised, the intensity of the radiation increases substantially. In one interesting application, an infrared camera registers the intensity of the infrared radiation produced at different locations on the human body. The camera is connected to a color monitor that displays the different infrared intensities as different colors. This "thermal painting" is called a *thermograph* or *thermogram*. Thermography is an important diagnostic tool in medicine. For example, breast cancer is indicated in the thermogram in **Figure 12.6** by the elevated temperatures associated with malignant tissue. In another application, **Figure 12.7** shows thermographic images of a smoker's forearms before (left) and 5 minutes after (right) he has smoked a cigarette. After smoking, the forearms are cooler due to the effect of nicotine, which causes vasoconstriction (narrowing of the blood vessels) and reduces blood flow, a result that can lead to a higher risk from blood clotting. Temperatures in these images range from over 34 °C to about 28 °C and are indicated in decreasing order by the colors white, red, yellow, green, and blue.

Oceanographers and meteorologists also use thermograms extensively, to map the temperature distribution on the surface of the earth. For example, **Figure 12.8** shows a satellite image of the sea-surface temperature of the Pacific Ocean. The region depicted in red is the 1997/98 El Niño, a large area of

FIGURE 12.7 Thermogram showing a smoker's forearms before (*left*) and 5 minutes after (*right*) he has smoked a cigarette. Temperatures decrease from white (over 34 °C) to yellow, to red, to green, to blue (about 28 °C).

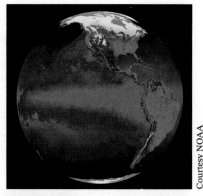

FIGURE 12.8 A thermogram of the 1997/98 El Niño (red), a large region of abnormally high temperatures in the Pacific Ocean.

the ocean, approximately twice the width of the United States, where temperatures reached abnormally high values. This El Niño caused major weather changes in certain regions of the earth.

12.4 | Linear Thermal Expansion

Normal Solids

Have you ever found the metal lid on a glass jar too tight to open? One solution is to run hot water over the lid, which loosens it because the metal expands more than the glass does. To varying extents, most materials expand when heated and contract when cooled. The increase in any one dimension of a solid is called **linear expansion**, linear in the sense that the expansion occurs along a line. **Figure 12.9** illustrates the linear expansion of a rod whose length is L_0 when the temperature is T_0. When the temperature of the rod increases to $T_0 + \Delta T$, the length becomes $L_0 + \Delta L$, where ΔT and ΔL are the changes in temperature and length, respectively. Conversely, when the temperature decreases to $T_0 - \Delta T$, the length decreases to $L_0 - \Delta L$.

For modest temperature changes, experiments show that the change in length is directly proportional to the change in temperature ($\Delta L \propto \Delta T$). In addition, the change in length is proportional to the initial length of the rod, a fact that can be understood with the aid of **Figure 12.10**. Part *a* of the drawing shows two identical rods. Each rod has a length L_0 and expands by ΔL when the temperature increases by ΔT. Part *b* shows the two heated rods combined into a single rod, for which the total expansion is the sum of the expansions of each part—namely, $\Delta L + \Delta L = 2\Delta L$. Clearly, the amount of expansion doubles if the rod is twice as long to begin with. In other words, the change in length is directly proportional to the original length ($\Delta L \propto L_0$). Equation 12.2 expresses the fact that ΔL is proportional to both L_0 and ΔT ($\Delta L \propto L_0 \Delta T$) by using a proportionality constant α, which is called the **coefficient of linear expansion**.

Temperature = T_0

Temperature = $T_0 + \Delta T$

L_0 ΔL

FIGURE 12.9 When the temperature of a rod is raised by ΔT, the length of the rod increases by ΔL.

L_0

ΔL

L_0 (a)

ΔL

is equivalent to

L_0

ΔL

$2L_0$

(b)

$2\,\Delta L$

FIGURE 12.10 (*a*) Each of two identical rods expands by ΔL when heated. (*b*) When the rods are combined into a single rod of length $2L_0$, the "combined" rod expands by $2\,\Delta L$.

LINEAR THERMAL EXPANSION OF A SOLID

The length L_0 of an object changes by an amount ΔL when its temperature changes by an amount ΔT:

$$\Delta L = \alpha L_0\, \Delta T \qquad (12.2)$$

where α is the coefficient of linear expansion.

Common Unit for the Coefficient of Linear Expansion: $\frac{1}{\text{C}°} = (\text{C}°)^{-1}$

TABLE 12.1 **Coefficients of Thermal Expansion for Solids and Liquids[a]**

| Substance | Coefficient of Thermal Expansion (C°)$^{-1}$ | |
	Linear (α)	Volume (β)
Solids		**Liquids**
Aluminum	23×10^{-6}	69×10^{-6}
Brass	19×10^{-6}	57×10^{-6}
Concrete	12×10^{-6}	36×10^{-6}
Copper	17×10^{-6}	51×10^{-6}
Glass (common)	8.5×10^{-6}	26×10^{-6}
Glass (Pyrex)	3.3×10^{-6}	9.9×10^{-6}
Gold	14×10^{-6}	42×10^{-6}
Iron or steel	12×10^{-6}	36×10^{-6}
Lead	29×10^{-6}	87×10^{-6}
Nickel	13×10^{-6}	39×10^{-6}
Quartz (fused)	0.50×10^{-6}	1.5×10^{-6}
Silver	19×10^{-6}	57×10^{-6}
Liquids[b]		
Benzene	—	1240×10^{-6}
Carbon tetrachloride	—	1240×10^{-6}
Ethyl alcohol	—	1120×10^{-6}
Gasoline	—	950×10^{-6}
Mercury	—	182×10^{-6}
Methyl alcohol	—	1200×10^{-6}
Water	—	207×10^{-6}

[a]The values for α and β pertain to a temperature near 20°C.

[b]Since liquids do not have fixed shapes, the coefficient of linear expansion is not defined for them.

Solving Equation 12.2 for α shows that $\alpha = \Delta L/(L_0 \Delta T)$. Since the length units of ΔL and L_0 algebraically cancel, the coefficient of linear expansion α has the unit of $(\text{C}°)^{-1}$ when the temperature difference ΔT is expressed in Celsius degrees (C°). Different materials with the same initial length expand and contract by different amounts as the temperature changes, so the value of α depends on the nature of the material. Table 12.1 shows some typical values. Coefficients of linear expansion also vary somewhat depending on the range of temperatures involved, but the values in Table 12.1 are adequate approximations. Example 4 deals with a situation in which a dramatic effect due to thermal expansion can be observed, even though the change in temperature is small.

EXAMPLE 4 | Buckling of a Sidewalk

A concrete sidewalk is constructed between two buildings on a day when the temperature is 25 °C. The sidewalk consists of two slabs, each three meters in length and of negligible thickness (Figure 12.11a). As the temperature rises to 38 °C, the slabs expand, but no space is provided for thermal expansion. The buildings do not move, so the slabs buckle upward. Determine the vertical distance y in part b of the drawing.

Reasoning The expanded length of each slab is equal to its original length plus the change in length ΔL due to the rise in temperature. We know the original length, and Equation 12.2 can be used to find the change in length. Once the expanded length has been determined, the Pythagorean theorem can be employed to find the vertical distance y in **Figure 12.11***b*.

Solution The change in temperature is $\Delta T = 38\ °C - 25\ °C = 13\ C°$, and the coefficient of linear expansion for concrete is given in **Table 12.1**. The change in length of each slab associated with this temperature change is

$$\Delta L = \alpha L_0\, \Delta T = [12 \times 10^{-6}(C°)^{-1}](3.0\ \text{m})(13\ C°)$$

$$= 0.000\ 47\ \text{m} \tag{12.2}$$

The expanded length of each slab is, thus, 3.000 47 m. We can now determine the vertical distance y by applying the Pythagorean theorem to the right triangle in **Figure 12.11***b*:

$$y = \sqrt{(3.000\ 47\ \text{m})^2 - (3.000\ 00\ \text{m})^2} = \boxed{0.053\ \text{m}}$$

Math Skills The calculation of the answer for y involves two numbers, each of which has six significant figures. Yet the answer has only two significant figures. This reduction in significant figures occurs because the calculation includes a subtraction step. *Be on the lookout for subtractions, because they may reduce the number of significant figures in your answers.* For instance, in the present case we have $y = \sqrt{(3.000\ 47\ \text{m})^2 - (3.000\ 00\ \text{m})^2}$, which has the form of $y = \sqrt{a^2 - b^2} = \sqrt{(a+b)(a-b)}$. Therefore, we can write the expression for y in the following way:

$$y = \sqrt{(3.000\ 47\ \text{m})^2 - (3.000\ 00\ \text{m})^2}$$

$$= \sqrt{(3.000\ 47\ \text{m} + 3.000\ 00\ \text{m})(3.000\ 47\ \text{m} - 3.000\ 00\ \text{m})}$$

In this result, the subtraction reduces the number of significant figures from six to two:

$$3.000\ 47\ \text{m} - 3.000\ 00\ \text{m} = 0.000\ 47\ \text{m}$$

As a result, our expression for y becomes

$$y = \sqrt{(6.000\ 47\ \text{m})(0.000\ 47\ \text{m})} = \boxed{0.053\ \text{m}}$$

which limits the answer to two significant figures.

FIGURE 12.11 (*a*) Two concrete slabs completely fill the space between the buildings. (*b*) When the temperature increases, each slab expands, causing the sidewalk to buckle.

The buckling of a sidewalk is one consequence of not providing sufficient room for thermal expansion. To eliminate such problems, engineers incorporate expansion joints or spaces at intervals along bridge roadbeds, as **Figure 12.12** shows.

BIO THE PHYSICS OF . . . an antiscalding device. Although Example 4 shows how thermal expansion can cause problems, there are also times when it can be useful. For instance, each year thousands of children are taken to emergency rooms suffering from burns caused by scalding tap water. Such accidents can be reduced with the aid of the antiscalding device shown in **Figure 12.13**. This device screws onto the end of a faucet and quickly shuts off the flow of water when it becomes too hot. As the water temperature rises, the actuator spring expands and pushes the plunger forward, shutting off the flow. When the water cools, the spring contracts and the water flow resumes.

Thermal Stress

If the concrete slabs in **Figure 12.11** had not buckled upward, they would have been subjected to immense forces from the buildings. The forces needed to prevent a solid object from expanding must be strong enough to counteract any change in length that

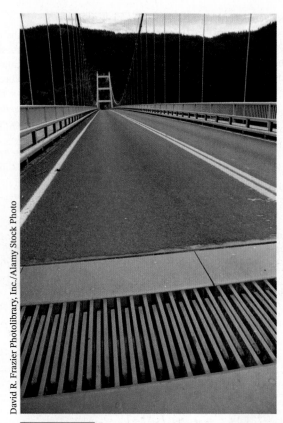

David R. Frazier Photolibrary, Inc./Alamy Stock Photo

FIGURE 12.12   An expansion joint in a bridge.

Movable plunger   Actuator spring

Water flow

FIGURE 12.13   An antiscalding device.

would occur due to a change in temperature. Although the change in temperature may be small, the forces—and hence the stresses—can be enormous. They can, in fact, lead to serious structural damage. Example 5 illustrates just how large the stresses can be.

Analyzing Multiple-Concept Problems

EXAMPLE 5 | The Physics of Thermal Stress

A steel beam is used in the roadbed of a bridge. The beam is mounted between two concrete supports when the temperature is 23 °C, with no room provided for thermal expansion (see **Figure 12.14**). What compressional stress must the concrete supports apply to each end of the beam, if they are to keep the beam from expanding when the temperature rises to 42 °C? Assume that the distance between the concrete supports does not change as the temperature rises.

Reasoning   When the temperature rises by an amount ΔT, the natural tendency of the beam is to expand. If the beam were free to expand, it would lengthen by an amount $\Delta L = \alpha L_0 \, \Delta T$ (Equation 12.2). However, the concrete supports prevent this expansion from occurring by exerting a compressional force on each end of the beam. The magnitude F of this force depends on ΔL through the relation $F = YA(\Delta L/L_0)$ (Equation 10.17), where Y is Young's modulus for steel and A

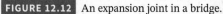

Beam

Concrete support   Concrete support

FIGURE 12.14   A steel beam is mounted between concrete supports with no room provided for thermal expansion.

is the cross-sectional area of the beam. According to the discussion in Section 10.8, the compressional stress is equal to the magnitude of the compressional force divided by the cross-sectional area, or Stress = F/A.

Knowns and Unknowns   The data for this problem are listed in the table:

Description	Symbol	Value
Initial temperature	T_0	23 °C
Final temperature	T	42 °C
Unknown Variables		
Stress	—	?

Modeling the Problem

STEP 1 Stress and Force   The compressional stress is defined as the magnitude F of the compressional force divided by the cross-sectional area A of the beam (see Section 10.8), or

$$\text{Stress} = \frac{F}{A}$$

According to Equation 10.17, the magnitude of the compressional force that the concrete supports exert on each end of the steel beam is given by

$$F = YA\left(\frac{\Delta L}{L_0}\right)$$

where Y is Young's modulus, ΔL is the change in length, and L_0 is the original length of the beam. Substituting this expression for F into the definition of stress gives Equation 1 in the right column. Young's modulus Y for steel is known (see Table 10.1), but we do not know either ΔL or L_0. However, ΔL is proportional to L_0, so we will focus on ΔL in Step 2.

$$\text{Stress} = Y\frac{\Delta L}{L_0} \qquad (1)$$

$$\boxed{?}$$

STEP 2 Linear Thermal Expansion   If it were free to do so, the beam would expand by an amount $\Delta L = \alpha L_0 \, \Delta T$ (Equation 12.2), where α is the coefficient of linear expansion. The change in temperature is the final temperature T minus the initial temperature T_0, or $\Delta T = T - T_0$. Thus, the beam would have expanded by an amount

$$\boxed{\Delta L = \alpha L_0 (T - T_0)}$$

In this expression the variables T and T_0 are known, and α is available in **Table 12.1**. We substitute this relation for ΔL into Equation 1, as indicated in the right column.

$$\text{Stress} = Y\frac{\Delta L}{L_0} \qquad (1)$$

$$\boxed{\Delta L = \alpha L_0 (T - T_0)}$$

Solution   Algebraically combining the results of each step, we have

$$\overset{\text{STEP 1}}{\downarrow} \quad \overset{\text{STEP 2}}{\downarrow}$$

$$\text{Stress} = Y\frac{\Delta L}{L_0} = Y\frac{\alpha \cancel{L_0}(T - T_0)}{\cancel{L_0}} = Y\alpha(T - T_0)$$

Note that the original length L_0 of the beam is eliminated algebraically from this result. Taking the value of $Y = 2.0 \times 10^{11}$ N/m² from Table 10.1 and $\alpha = 12 \times 10^{-6}$ (C°)⁻¹ from **Table 12.1**, we find that

$$\text{Stress} = Y\alpha(T - T_0)$$

$$= (2.0 \times 10^{11} \text{ N/m}^2)[12 \times 10^{-6} \text{ (C°)}^{-1}](42 \text{ °C} - 23 \text{ °C}) = \boxed{4.6 \times 10^7 \text{ N/m}^2}$$

This is a large stress, equivalent to nearly one million pounds per square foot.

Related Homework: Problems 17, 20

The Bimetallic Strip

A **bimetallic strip** is made from two thin strips of metal that have *different* coefficients of linear expansion, as **Animated Figure 12.15a** shows. Often brass [$\alpha = 19 \times 10^{-6}$ (C°)⁻¹] and steel [$\alpha = 12 \times 10^{-6}$ (C°)⁻¹] are selected. The two pieces are welded or riveted together. When the bimetallic strip is heated, the brass, having the larger value of α, expands more than the steel. Since the two metals are bonded together,

(a)   (b) Heated   (c) Cooled

ANIMATED FIGURE 12.15

(a) A bimetallic strip and how it behaves when (b) heated and (c) cooled.

the bimetallic strip bends into an arc as in part b, with the longer brass piece having a larger radius than the steel piece. When the strip is cooled, the bimetallic strip bends in the opposite direction, as in part c.

THE PHYSICS OF . . . an automatic coffee maker.   Bimetallic strips are frequently used as adjustable automatic switches in electrical appliances. **Figure 12.16** shows an automatic coffee maker that turns off when the coffee is brewed to the selected strength. In part a, while the brewing cycle is on, electricity passes through the heating coil that heats the water. The electricity can flow because the contact mounted on the bimetallic strip touches the contact mounted on the "strength" adjustment knob, thus providing a continuous path for the electricity. When the bimetallic strip gets hot enough to bend away, as in part b of the drawing, the contacts separate. The electricity stops because it no longer has a continuous path along which to flow, and the brewing cycle is shut off. Turning the "strength" knob adjusts the brewing time by adjusting the distance through which the bimetallic strip must bend for the contact points to separate.

FIGURE 12.16   A bimetallic strip controls whether this coffee pot is (a) "on" (strip cold, straight) or (b) "off" (strip hot, bent).

The Expansion of Holes

An interesting example of linear expansion occurs when there is a hole in a piece of solid material. We know that the material itself expands when heated, but what about the hole? Does it expand, contract, or remain the same? Conceptual Example 6 provides some insight into the answer to this question.

CONCEPTUAL EXAMPLE 6 | Do Holes Expand or Contract When the Temperature Increases?

Interactive Figure 12.17a shows eight square tiles that are attached together and arranged to form a square pattern with a hole in the center. If the tiles are heated, does the size of the hole **(a)** decrease or **(b)** increase?

Reasoning   We can analyze this problem by disassembling the pattern into separate tiles, heating them, and then reassembling the pattern. What happens to each of the individual tiles can be explained using what we know about linear expansion.

Answer (a) is incorrect.   When a tile is heated both its length and width expand. It is tempting to think, therefore, that the

hole in the pattern decreases as the surrounding tiles expand into it. However, this is not correct, because any one tile is prevented from expanding into the hole by the expansion of the tiles next to it.

Answer (b) is correct.   Since each tile expands upon heating, the pattern also expands, and the hole along with it, as shown in **Interactive Figure 12.17b**. In fact, if we had a ninth tile that was identical to the others and heated it to the same extent, it would fit exactly into the hole, as **Interactive Figure 12.17c** indicates. Thus, not only does the hole expand, it does so exactly

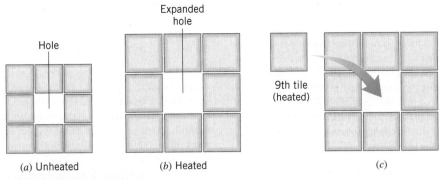

(a) Unheated (b) Heated (c)

INTERACTIVE FIGURE 12.17 (a) The tiles are arranged to form a square pattern with a hole in the center. (b) When the tiles are heated, the pattern, including the hole in the center, expands. (c) The expanded hole is the same size as a heated tile.

as each of the tiles does. Since the ninth tile is made of the same material as the others, we see that the hole expands just as if it were made of the material of the surrounding tiles. The thermal expansion of the hole and the surrounding material is analogous to a photographic enlargement: everything is enlarged, including holes.

Related Homework: Problem 10

Instead of the separate tiles in Example 6, we could have used a square plate with a square hole in the center. The hole in the plate would have expanded just like the hole in the pattern of tiles. Furthermore, the same conclusion applies to a hole of any shape. Thus, it follows that

> **Problem-Solving Insight** A hole in a piece of solid material expands when heated and contracts when cooled, just as if it were filled with the material that surrounds it.

The equation $\Delta L = \alpha L_0 \, \Delta T$ can be used to find the change in any linear dimension of the hole, such as its radius or diameter, if the hole is circular. Example 7 illustrates this type of linear expansion.

EXAMPLE 7 | A Heated Engagement Ring

A gold engagement ring has an inner diameter of 1.5×10^{-2} m and a temperature of 27 °C. The ring falls into a sink of hot water whose temperature is 49 °C. What is the change in the diameter of the hole in the ring?

Reasoning The hole expands as if it were filled with gold, so the change in the diameter is given by $\Delta L = \alpha L_0 \Delta T$, where $\alpha = 14 \times 10^{-6} \, (\text{C}°)^{-1}$ is the coefficient of linear expansion for gold

(Table 12.1), L_0 is the original diameter, and ΔT is the change in temperature.

Solution The change in the ring's diameter is

$$\Delta L = \alpha L_0 \, \Delta T = [14 \times 10^{-6} \, (\text{C}°)^{-1}](1.5 \times 10^{-2} \text{ m})(49 \,°\text{C} - 27 \,°\text{C})$$

$$= \boxed{4.6 \times 10^{-6} \text{ m}}$$

The previous two examples illustrate that holes expand like the surrounding material when heated. Therefore, holes in materials with larger coefficients of linear expansion expand more than those in materials with smaller coefficients of linear expansion. Conceptual Example 8 explores this aspect of thermal expansion.

CONCEPTUAL EXAMPLE 8 | Expanding Cylinders

Figure 12.18 shows a cross-sectional view of three cylinders, A, B, and C. One is made from lead, one from brass, and one from steel. All three have the same temperature, and they barely fit inside each other. As the cylinders are heated to the same higher temperature, C falls off, while A becomes tightly wedged against B. Which cylinder is made from which material? **(a)** A is brass, B is lead, C is steel **(b)** A is lead, B is brass, C is steel **(c)** A is lead, B is steel, C is brass **(d)** A is brass, B is steel, C is lead **(e)** A is steel, B is brass, C is lead **(f)** A is steel, B is lead, C is brass

Reasoning We will consider how the outer and inner diameters of each cylinder change as the temperature is raised. In particular, with regard to the inner diameter we note that a hole expands as if it were filled with the surrounding material. According to **Table 12.1**, lead has the greatest coefficient of linear expansion, followed by brass, and then by steel. Thus, the outer and inner diameters of the lead cylinder change the most, while those of the steel cylinder change the least.

Answers (a), (b), (e), and (f) are incorrect. Since the steel cylinder expands the least, it cannot be the outer one, for if it were, the greater expansion of the middle cylinder would prevent the

steel cylinder from falling off, as outer cylinder C actually does. The steel cylinder also cannot be the inner one, because then the greater expansion of the middle cylinder would allow the steel cylinder to fall out, contrary to what is observed for inner cylinder A.

Answers (c) and (d) are correct. Since the steel cylinder cannot be on the outside or on the inside, it must be the middle cylinder B. **Figure 12.18a** shows the lead cylinder as the outer cylinder C. It will fall off as the temperature is raised, since lead expands more than steel. The brass inner cylinder A expands more than the steel cylinder that surrounds it and becomes tightly wedged, as observed. Similar reasoning applies also to **Figure 12.18b**, which shows the brass cylinder as the outer cylinder and the lead cylinder as the inner one, since both brass and lead expand more than steel.

(a)

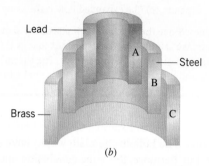

(b)

FIGURE 12.18 Conceptual Example 8 discusses the arrangements of the three cylinders shown in cutaway views in parts *a* and *b*.

Check Your Understanding

(The answers are given at the end of the book.)

2. A rod is hung from an aluminum frame, as **CYU Figure 12.1** shows. The rod and the frame have the same temperature, and there is a small gap between the rod and the floor. The frame and rod are then heated uniformly. Will the rod ever touch the floor if the rod is made from **(a)** aluminum, **(b)** lead, **(c)** brass?

3. A simple pendulum is made using a long, thin metal wire. When the temperature drops, does the period of the pendulum increase, decrease, or remain the same?

CYU FIGURE 12.1

4. For added strength, many highways and buildings are constructed with reinforced concrete (concrete that is reinforced with embedded steel rods). **Table 12.1** shows that the coefficient of linear expansion α for concrete is the same as that for steel. Why is it important that these two coefficients be the same?

5. One type of cooking pot is made from stainless steel and has a copper coating over the outside of the bottom. At room temperature the bottom of this pot is flat, but when heated the bottom is not flat. When the bottom of this pot is heated, is it bowed outward or inward?

6. A metal ball has a diameter that is slightly *greater* than the diameter of a hole that has been cut into a metal plate. The coefficient of linear expansion for the metal from which the ball is made is *greater* than that for the metal of the plate. Which one or more of the following procedures can be used to make the ball pass through the hole? **(a)** Raise the temperatures of the ball and the plate by the same amount. **(b)** Lower the temperatures of the ball and the plate by the same amount. **(c)** Heat the ball and cool the plate. **(d)** Cool the ball and heat the plate.

7. A hole is cut through an aluminum plate. A brass ball has a diameter that is slightly *smaller* than the diameter of the hole. The plate and the ball have the same temperature at all times. Should the plate and ball both be heated or both be cooled to *prevent* the ball from falling through the hole?

12.5 | Volume Thermal Expansion

The volume of a normal material increases as the temperature increases. Most solids and liquids behave in this fashion. By analogy with linear thermal expansion, the change in volume ΔV is proportional to the change in temperature ΔT and to the initial volume V_0, provided the change in temperature is not too large. These two proportionalities can be converted into Equation 12.3 with the aid of a proportionality constant β, known as the **coefficient of volume expansion**. The algebraic form of this equation is similar to that for linear expansion, $\Delta L = \alpha L_0 \, \Delta T$.

VOLUME THERMAL EXPANSION

The volume V_0 of an object changes by an amount ΔV when its temperature changes by an amount ΔT:

$$\Delta V = \beta V_0 \, \Delta T \qquad (12.3)$$

where β is the coefficient of volume expansion.

Common Unit for the Coefficient of Volume Expansion: $(C°)^{-1}$

The unit for β, like that for α, is $(C°)^{-1}$. Values for β depend on the nature of the material, and **Table 12.1** lists some examples measured near 20 °C. The values of β for liquids are substantially larger than those for solids, because liquids typically expand more than solids, given the same initial volumes and temperature changes. **Table 12.1** also shows that, for most solids, the coefficient of volume expansion is three times as much as the coefficient of linear expansion: $\beta = 3\alpha$.

If a cavity exists within a solid object, the volume of the cavity increases when the object expands, just as if the cavity were filled with the surrounding material. The expansion of the cavity is analogous to the expansion of a hole in a sheet of material. Accordingly, the change in volume of a cavity can be found using the relation $\Delta V = \beta V_0 \, \Delta T$, where β is the coefficient of volume expansion of the material that surrounds the cavity. Example 9 illustrates this point.

EXAMPLE 9 | The Physics of the Overflow of an Automobile Radiator

A small plastic container, called the coolant reservoir, catches the radiator fluid that overflows when an automobile engine becomes hot (see **Figure 12.19**). The radiator is made of copper, and the coolant has a coefficient of volume expansion of $\beta = 4.10 \times 10^{-4}\ (C°)^{-1}$. If the radiator is filled to its 15-quart capacity when the engine is cold (6.0 °C), how much overflow will spill into the reservoir when the coolant reaches its operating temperature of 92 °C?

Reasoning When the temperature increases, both the coolant and the radiator expand. If they were to expand by the same amount, there would be no overflow. However, the liquid coolant expands more than the radiator, and the overflow volume is the amount of coolant expansion *minus* the amount of the radiator cavity expansion.

Solution When the temperature increases by 86 C°, the coolant expands by an amount

$$\Delta V = \beta V_0 \, \Delta T = [4.10 \times 10^{-4}\ (C°)^{-1}](15\ \text{quarts})(86\ C°) \quad \textbf{(12.3)}$$

$$= 0.53\ \text{quarts}$$

The radiator cavity expands as if it were filled with copper $[\beta = 51 \times 10^{-6}\ (C°)^{-1}$; see **Table 12.1**]. The expansion of the radiator cavity is

$$\Delta V = \beta V_0 \, \Delta T = [51 \times 10^{-6}\ (C°)^{-1}](15\ \text{quarts})(86\ C°)$$

$$= 0.066\ \text{quarts}$$

Coolant reservoir

Radiator

FIGURE 12.19 An automobile radiator and a coolant reservoir for catching the overflow from the radiator.

The overflow volume is 0.53 quarts − 0.066 quarts = $\boxed{0.46\ \text{quarts}}$.

FIGURE 12.20 The density of water in the temperature range from 0 to 10 °C. At 4 °C water has a maximum density of 999.973 kg/m³. (This value is equivalent to the often-quoted density of 1.000 00 grams per milliliter.)

FIGURE 12.21 As water freezes and expands, enormous pressure is applied to the liquid water between the ice and the closed faucet.

Although most substances expand when heated, a few do not. For instance, if water at 0 °C is heated, its volume *decreases* until the temperature reaches 4 °C. Above 4 °C water behaves normally, and its volume increases as the temperature increases. Because a given mass of water has a minimum volume at 4 °C, the density (mass per unit volume) of water is greatest at 4 °C, as **Figure 12.20** shows.

BIO **THE PHYSICS OF . . . ice formation and the survival of aquatic life.** The fact that water has its greatest density at 4 °C, rather than at 0 °C, has important consequences for the way in which a lake freezes. When the air temperature drops, the surface layer of water is chilled. As the temperature of the surface layer drops toward 4 °C, this layer becomes more dense than the warmer water below. The denser water sinks and pushes up the deeper and warmer water, which in turn is chilled at the surface. This process continues until the temperature of the entire lake reaches 4 °C. Further cooling of the surface water below 4 °C makes it *less dense* than the deeper layers; consequently, the surface layer does not sink but stays on top. Continued cooling of the top layer to 0 °C leads to the formation of ice that floats on the water, because ice has a smaller density than water at any temperature. Below the ice, however, the water temperature remains above 0 °C. The sheet of ice acts as an insulator that reduces the loss of heat from the lake, especially if the ice is covered with a blanket of snow, which is also an insulator. As a result, lakes usually do not freeze solid, even during prolonged cold spells, so fish and other aquatic life can survive.

THE PHYSICS OF . . . bursting water pipes. The fact that the density of ice is smaller than the density of water has an important consequence for homeowners, who have to contend with the possibility of bursting water pipes during severe winters. Water often freezes in a section of pipe exposed to unusually cold temperatures. The ice can form an immovable plug that prevents the subsequent flow of water, as **Figure 12.21** illustrates. When water (larger density) turns to ice (smaller density), its volume expands by 8.3%. Therefore, when more water freezes at the left side of the plug, the expanding ice pushes liquid back into the pipe leading to the street connection, and no damage is done. However, when ice forms on the right side of the plug, the expanding ice pushes liquid to the right—but it has nowhere to go if the faucet is closed. As ice continues to form and expand, the water pressure between the plug and faucet rises. Even a small increase in the amount of ice produces a large increase in the pressure. This situation is analogous to the thermal stress discussed in Multiple-Concept Example 5, where a small change in the length of the steel beam produces a large stress on the concrete supports. The entire section of pipe to the right of the blockage experiences the same elevated pressure, according to Pascal's principle (Section 11.5). Therefore, the pipe can burst at any point where it is structurally weak, even within the heated space of the building. If you should lose heat during the winter, there is a simple way to prevent pipes from bursting. Simply open the faucet so it drips a little. The excessive pressure will be relieved.

Check Your Understanding

(The answers are given at the end of the book.)

8. Suppose that liquid mercury and glass both had the same coefficient of volume expansion. Would a mercury-in-glass thermometer still work?

9. Is the buoyant force provided by warmer water (above 4 °C) greater than, less than, or equal to the buoyant force provided by cooler water (also above 4 °C)?

12.6 | Heat and Internal Energy

An object with a high temperature is said to be hot, and the word "hot" brings to mind the word "heat." **Heat** flows from a hotter object to a cooler object when the two are placed in contact. It is for this reason that a cup of hot coffee feels hot to the touch, while a glass of ice water feels cold. When the person in **Figure 12.22a** touches the coffee cup, heat flows from the hotter cup into the cooler hand. When the person touches the glass in part *b* of the drawing, heat again flows from hot to cold, in this case from the warmer hand into the colder glass. The response of the nerves in the hand to the arrival or departure of heat prompts the brain to identify the coffee cup as being hot and the glass as being cold.

What exactly is heat? As the following definition indicates, heat is a form of energy, energy in transit from hot to cold.

> **DEFINITION OF HEAT**
>
> **Heat is energy that flows from a higher-temperature object to a lower-temperature object because of the difference in temperatures.**
>
> **SI Unit of Heat: joule (J)**

Being a kind of energy, heat is measured in the same units used for work, kinetic energy, and potential energy. Thus, the SI unit for heat is the joule.

The heat that flows from hot to cold in **Figure 12.22** originates in the **internal energy** of the hot substance. The internal energy of a substance is the sum of the molecular kinetic energy (due to random motion of the molecules), the molecular potential energy (due to forces that act between the atoms of a molecule and between molecules), and other kinds of molecular energy. When heat flows in circumstances where no work is done, the internal energy of the hot substance decreases and the internal energy of the cold substance increases. Although heat may originate in the internal energy supply of a substance, *it is not correct to say that a substance contains heat.* The substance has internal energy, not heat. The word "heat" only refers to the energy actually in transit from hot to cold.

The next two sections consider some effects of heat. For instance, when preparing spaghetti, the first thing that a cook does is to heat the water. Heat from the stove causes the internal energy of the water to increase. Associated with this increase is a rise in temperature. After a while, the temperature reaches 100 °C, and the water begins to boil. During boiling, the added heat causes the water to change from a liquid to a vapor phase (steam). The next section investigates how the addition (or removal) of heat causes the temperature of a substance to change. Then, Section 12.8 discusses the relationship between heat and phase change, such as that which occurs when water boils.

FIGURE 12.22 Heat is energy in transit from hot to cold. (*a*) Heat flows from the hotter coffee cup to the colder hand. (*b*) Heat flows from the warmer hand to the colder glass of ice water.

12.7 | Heat and Temperature Change: Specific Heat Capacity

Solids and Liquids

Greater amounts of heat are needed to raise the temperature of solids or liquids to higher values. A greater amount of heat is also required to raise the temperature of a greater mass of material. Similar comments apply when the temperature is lowered, except that heat must be removed. For limited temperature ranges, experiment shows that the heat Q is directly proportional to the change in temperature ΔT and to the mass m. These two proportionalities are expressed below in Equation 12.4, with the help of a proportionality constant c that is referred to as the **specific heat capacity** of the material.

> **HEAT SUPPLIED OR REMOVED IN CHANGING THE TEMPERATURE OF A SUBSTANCE**
>
> **The heat Q that must be supplied or removed to change the temperature of a substance of mass m by ΔT degrees is**
>
> $$Q = cm\,\Delta T \qquad (12.4)$$
>
> **where c is the specific heat capacity of the substance.**
>
> **Common Unit for Specific Heat Capacity: J/(kg · C°)**

Solving Equation 12.4 for the specific heat capacity shows that $c = Q/(m\,\Delta T)$, so the unit for specific heat capacity is J/(kg · C°). **Table 12.2** reveals that the value of the

TABLE 12.2	Specific Heat Capacities[a] of Some Solids and Liquids
Substance	**Specific Heat Capacity, c J/(kg · C°)**
Solids	
Aluminum	9.00×10^2
Copper	387
Glass	840
Human body (37 °C, average)	3500
Ice (−15 °C)	2.00×10^3
Iron or steel	452
Lead	128
Silver	235
Liquids	
Benzene	1740
Ethyl alcohol	2450
Glycerin	2410
Mercury	139
Water (15 °C)	4186

[a]Except as noted, the values are for 25 °C and 1 atm of pressure.

specific heat capacity depends on the nature of the material. Examples 10 and 11 illustrate the use of Equation 12.4.

EXAMPLE 10 | BIO A Hot Jogger

In a half hour, a 65-kg jogger can generate 8.0×10^5 J of heat. This heat is removed from the jogger's body by a variety of means, including the body's own temperature-regulating mechanisms. If the heat were not removed, how much would the jogger's body temperature increase?

Reasoning The increase in body temperature depends on the amount of heat Q generated by the jogger, her mass m, and the specific heat capacity c of the human body. Since numerical values are known for these three variables, we can determine the potential rise in temperature by using Equation 12.4.

Solution Table 12.2 gives the average specific heat capacity of the human body as 3500 J/(kg · C°). With this value, Equation 12.4 shows that

$$\Delta T = \frac{Q}{cm} = \frac{8.0 \times 10^5 \text{ J}}{[3500 \text{ J/(kg} \cdot \text{C°)}](65 \text{ kg})} = \boxed{3.5 \text{ C°}}$$

An increase in body temperature of 3.5 °C could be life-threatening. One way in which the jogger's body prevents it from occurring is to remove excess heat by perspiring. In contrast, cats, such as the

FIGURE 12.23 Cats, such as this lion, often pant to get rid of excess heat.

one in **Figure 12.23**, do not perspire but often pant to remove excess heat.

EXAMPLE 11 | Taking a Hot Shower

Cold water at a temperature of 15 °C enters a heater, and the resulting hot water has a temperature of 61 °C. A person uses 120 kg of hot water in taking a shower. **(a)** Find the energy needed to heat the water. **(b)** Assuming that the utility company charges $0.10 per kilowatt · hour for electrical energy, determine the cost of heating the water.

Reasoning The amount Q of heat needed to raise the water temperature can be found from the relation $Q = cm \, \Delta T$, since the specific heat capacity, mass, and temperature change of the water are known. To determine the cost of this energy, we multiply the cost per unit of energy ($0.10 per kilowatt · hour) by the amount of energy used, expressed in energy units of kilowatt · hours.

Solution **(a)** From equation 12.4, the amount of heat needed to heat the water is

$Q = cm \, \Delta T = [4186 \text{ J/(kg} \cdot \text{C°)}](120 \text{ kg})(61 \text{ °C} - 15 \text{ °C})$
$= \boxed{2.3 \times 10^7 \text{ J}}$

(b) The kilowatt · hour (kWh) is the unit of energy that utility companies use in your electric bill. To calculate the cost, we need to determine the number of joules in one kilowatt · hour. Recall that 1 kilowatt is 1000 watts (1 kW = 1000 W), 1 watt is 1 joule per second (1 W = 1 J/s; see Table 6.3), and 1 hour is equal to 3600 seconds (1 h = 3600 s). Thus,

$$1 \text{ kWh} = (1 \text{ kWh})\left(\frac{1000 \text{ W}}{1 \text{ kW}}\right)\left(\frac{1 \text{ J/s}}{1 \text{ W}}\right)\left(\frac{3600 \text{ s}}{1 \text{ h}}\right) = 3.60 \times 10^6 \text{ J}$$

The number of kilowatt · hours of energy used to heat the water is

$$(2.3 \times 10^7 \text{ J})\left(\frac{1 \text{ kWh}}{3.60 \times 10^6 \text{ J}}\right) = 6.4 \text{ kWh}$$

At a cost of $0.10 per kWh, the bill for the heat is $\boxed{\$0.64}$ or 64 cents.

Gases

As we will see in Section 15.6, the value of the specific heat capacity depends on whether the pressure or volume is held constant while energy in the form of heat is added to or removed from a substance. The distinction between constant pressure and constant volume is usually not important for solids and liquids but is significant for gases. In Section 15.6, we will see that a greater value for the specific heat capacity is obtained for a gas at constant pressure than for a gas at constant volume.

Heat Units Other than the Joule

There are three heat units other than the joule in common use. One kilocalorie (1 kcal) was defined historically as the amount of heat needed to raise the temperature of one kilogram of water by one Celsius degree.* With $Q = 1.00$ kcal, $m = 1.00$ kg, and $\Delta T = 1.00$ C°, the equation $Q = cm\,\Delta T$ shows that such a definition is equivalent to a specific heat capacity for water of $c = 1.00$ kcal/(kg · C°). Similarly, one calorie (1 cal) was defined as the amount of heat needed to raise the temperature of one gram of water by one Celsius degree, which yields a value of $c = 1.00$ cal/(g · C°). (Nutritionists use the word "Calorie," with a capital C, to specify the energy content of foods; this use is unfortunate, since 1 Calorie = 1000 calories = 1 kcal.) The British thermal unit (Btu) is the other commonly used heat unit and was defined historically as the amount of heat needed to raise the temperature of one pound of water by one Fahrenheit degree.

It was not until the time of James Joule (1818–1889) that the relationship between energy in the form of work (in units of joules) and energy in the form of heat (in units of kilocalories) was firmly established. Joule's experiments revealed that the performance of mechanical work, like rubbing your hands together, can make the temperature of a substance rise, just as the absorption of heat can. His experiments and those of later workers have shown that

$$1 \text{ kcal} = 4186 \text{ joules} \qquad \text{or} \qquad 1 \text{ cal} = 4.186 \text{ joules}$$

Because of its historical significance, this conversion factor is known as the **mechanical equivalent of heat**.

Calorimetry

In Section 6.8 we encountered the principle of conservation of energy, which states that energy can be neither created nor destroyed, but can only be converted from one form to another. There we dealt with kinetic and potential energies. In this chapter we have expanded our concept of energy to include heat, which is energy that flows from a higher-temperature object to a lower-temperature object because of the difference in temperature. No matter what its form, whether kinetic energy, potential energy, or heat, energy can be neither created nor destroyed. This fact governs the way objects at different temperatures come to an equilibrium temperature when they are placed in contact. If there is no heat loss to the external surroundings, the amount of heat gained by the cooler objects equals the amount of heat lost by the hotter ones, a process that is consistent with the conservation of energy. Just this kind of process occurs within a thermos. A perfect thermos would prevent any heat from leaking out or in. However, energy in the form of heat can flow *between* materials inside the thermos to the extent that they have different temperatures; for example, between ice cubes and warm tea. The transfer of energy continues until a common temperature is reached at thermal equilibrium.

The kind of heat transfer that occurs within a thermos of iced tea also occurs within a calorimeter, which is the experimental apparatus used in a technique known as **calorimetry**. **Figure 12.24** shows that, like a thermos, a calorimeter is essentially an insulated container. It can be used to determine the specific heat capacity of a substance, as the next example illustrates.

Thermometer

Calorimeter cup

Insulating container

Unknown material

FIGURE 12.24 A calorimeter can be used to measure the specific heat capacity of an unknown material.

---

*From 14.5 to 15.5 °C.

EXAMPLE 12 | Measuring the Specific Heat Capacity

The calorimeter cup in **Figure 12.24** is made from 0.15 kg of aluminum and contains 0.20 kg of water. Initially, the water and the cup have a common temperature of 18.0 °C. A 0.040-kg mass of unknown material is heated to a temperature of 97.0 °C and then added to the water. The temperature of the water, the cup, and the unknown material is 22.0 °C after thermal equilibrium is reestablished. Ignoring the small amount of heat gained by the thermometer, find the specific heat capacity of the unknown material.

Reasoning Since energy is conserved and there is negligible heat flow between the calorimeter and the outside surroundings, the amount of heat gained by the cold water and the aluminum cup as they warm up is equal to the amount of heat lost by the unknown material as it cools down. Each amount of heat can be calculated using the relation $Q = cm\,\Delta T$, if we are careful to write the change in temperature ΔT as the higher temperature minus the lower temperature. The equation "Heat gained = Heat lost" contains a single unknown quantity, the desired specific heat capacity.

> **Problem-Solving Insight** In the equation "Heat gained = Heat lost," both sides must have the same algebraic sign. Therefore, when calculating

heat contributions for use in this equation, write any temperature changes as the higher minus the lower temperature.

Solution

$$\underbrace{(cm\,\Delta T)_{\text{Al}} + (cm\,\Delta T)_{\text{water}}}_{\substack{\text{Heat gained by}\\\text{aluminum and water}}} = \underbrace{(cm\,\Delta T)_{\text{unknown}}}_{\substack{\text{Heat lost by}\\\text{unknown material}}}$$

$$c_{\text{unknown}} = \frac{C_{\text{Al}}\,m_{\text{Al}}\,\Delta T_{\text{Al}} + C_{\text{water}}\,m_{\text{water}}\,\Delta T_{\text{water}}}{m_{\text{unknown}}\,\Delta T_{\text{unknown}}}$$

The changes in temperature of the substances are $\Delta T_{\text{Al}} = \Delta T_{\text{water}} = 22.0\ °\text{C} - 18.0\ °\text{C} = 4.0\ \text{C}°$, and $\Delta T_{\text{unknown}} = 97.0\ °\text{C} - 22.0\ °\text{C} = 75.0\ \text{C}°$. **Table 12.2** contains values for the specific heat capacities of aluminum and water. Substituting these data into the equation above, we find that

$$c_{\text{unknown}} = \frac{[9.00 \times 10^2\ \text{J/(kg} \cdot \text{C}°)](0.15\ \text{kg})(4.0\ \text{C}°)}{(0.040\ \text{kg})(75.0\ \text{C}°)}$$

$$+ \frac{[4186\ \text{J/(kg} \cdot \text{C})](0.20\ \text{kg})(4.0\ \text{C}°)}{(0.040\ \text{kg})(75.0\ \text{C}°)}$$

$$= \boxed{1300\ \text{J/(kg} \cdot \text{C}°)}$$

Check Your Understanding

(The answers are given at the end of the book.)

10. Two different objects are supplied with equal amounts of heat. Which one or more of the following statements explain why their temperature changes would not necessarily be the same? **(a)** The objects have the same mass but are made from materials that have different specific heat capacities. **(b)** The objects are made from the same material but have different masses. **(c)** The objects have the same mass and are made from the same material.

11. Two objects are made from the same material but have different masses. The two are placed in contact, and neither one loses any heat to the environment. Which object experiences the temperature change with the greater magnitude, or does each object experience a temperature change of the same magnitude?

12. Consider an object of mass m that experiences a change ΔT in its temperature. Various possibilities for these variables are listed in **CYU Table 12.1**. Rank these possibilities in descending order (largest first) according to how much heat is needed to bring about the change in temperature.

CYU TABLE 12.1

	m (kg)	ΔT (C°)
(a)	2.0	15
(b)	1.5	40
(c)	3.0	25
(d)	2.5	20

12.8 | Heat and Phase Change: Latent Heat

Surprisingly, there are situations in which the addition or removal of heat does not cause a temperature change. Consider a well-stirred glass of iced tea that has come to thermal equilibrium. Even though heat enters the glass from the warmer room, the temperature of the tea does not rise above 0 °C as long as ice cubes are present.

Apparently the heat is being used for some purpose other than raising the temperature. In fact, the heat is being used to melt the ice, and only when all of it is melted will the temperature of the liquid begin to rise.

An important point illustrated by the iced tea example is that there is more than one type or phase of matter. For instance, some of the water in the glass is in the solid phase (ice) and some is in the liquid phase. The gas or vapor phase is the third familiar phase of matter. In the gas phase, water is referred to as water vapor or steam. All three phases of water are present in the scene depicted in **Figure 12.25**, although the water vapor in the air is not visible in the photograph.

Matter can change from one phase to another, and heat plays a role in the change. **Figure 12.26** summarizes the various possibilities. A solid can **melt** or **fuse** into a liquid if heat is added, while the liquid can **freeze** into a solid if heat is removed. Similarly, a liquid can **evaporate** into a gas if heat is supplied, while the gas can **condense** into a liquid if heat is taken away. Rapid evaporation, with the formation of vapor bubbles within the liquid, is called **boiling**. Finally, a solid can sometimes change directly into a gas if heat is provided. We say that the solid **sublimes** into a gas. Examples of sublimation are (1) solid carbon dioxide, CO_2 (dry ice), turning into gaseous CO_2 and (2) solid naphthalene (moth balls) turning into naphthalene fumes. Conversely, if heat is removed under the right conditions, the gas will condense directly into a solid.

Figure 12.27 displays a graph that indicates what typically happens when heat is added to a material that changes phase. The graph records temperature versus heat added and refers to water at the normal atmospheric pressure of 1.01×10^5 Pa. The water starts off as ice at the subfreezing temperature of $-30\ °C$. As heat is added, the temperature of the ice increases, in accord with the specific heat capacity of ice [2000 J/(kg · C°)]. Not until the temperature reaches the normal melting/freezing point of 0 °C does the water begin to change phase. Then, when heat is added, the solid changes into the liquid, the temperature staying at 0 °C until *all the ice has melted*. Once all the material is in the liquid phase, additional heat causes the temperature to increase again, now in accord with the specific heat capacity of liquid water [4186 J/(kg · C°)]. When the temperature reaches the normal boiling/condensing point of 100 °C, the water begins to change from the liquid to the gas phase and continues to do so as long as heat is added. The temperature remains at 100 °C *until all liquid is gone*. When all of the material is in the gas phase, additional heat once again causes the temperature to rise, this time according to the specific heat capacity of water vapor at constant atmospheric pressure [2020 J/(kg · C°)]. Conceptual Example 13 applies the information in **Figure 12.27** to a familiar situation.

FIGURE 12.25 This antarctic scene shows resting crabeater seals. The three phases of water are present: solid ice is floating in liquid water, and water vapor is present in the air (invisible) and in the clouds in the sky.

Martin Schneiter/123RF

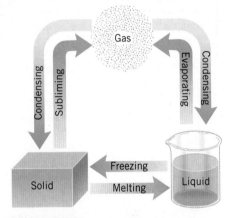

FIGURE 12.26 Three familiar phases of matter—solid, liquid, and gas—and the phase changes that can occur between any two of them.

FIGURE 12.27 The graph shows the way the temperature of water changes as heat is added, starting with ice at $-30\ °C$. The pressure is atmospheric pressure.

CONCEPTUAL EXAMPLE 13 | Saving Energy

Suppose you are cooking spaghetti, and the instructions say "boil the pasta in water for ten minutes." To cook spaghetti in an open pot using the least amount of energy, should you **(a)** turn up the burner to its fullest so the water vigorously boils or **(b)** turn down the burner so the water barely boils?

Reasoning The spaghetti needs to cook at the temperature of boiling water for ten minutes. In an open pot the pressure is atmospheric pressure, and water boils at 100 °C, regardless of whether it is vigorously boiling or just barely boiling. To convert water into steam requires energy in the form of heat from the burner, and the

greater the amount of water converted, the greater the amount of energy needed.

Answer (a) is incorrect. Causing the water to boil vigorously just wastes energy unnecessarily. All it accomplishes is to convert more water into steam.

Answer (b) is correct. Keeping the water just barely boiling uses the least amount of energy to keep the spaghetti at 100 °C, because it minimizes the amount of water converted into steam.

When a substance changes from one phase to another, the amount of heat that must be added or removed depends on the type of material and the nature of the phase change. The heat per kilogram associated with a phase change is referred to as **latent heat**:

HEAT SUPPLIED OR REMOVED IN CHANGING THE PHASE OF A SUBSTANCE

The heat Q that must be supplied or removed to change the phase of a mass m of a substance is

$$Q = mL \qquad (12.5)$$

where L is the latent heat of the substance.

SI Unit of Latent Heat: J/kg

The **latent heat of fusion** L_f refers to the change between solid and liquid phases, the **latent heat of vaporization** L_v applies to the change between liquid and gas phases, and the **latent heat of sublimation** L_s refers to the change between solid and gas phases.

Table 12.3 gives some typical values of latent heats of fusion and vaporization. For instance, the latent heat of fusion for water is $L_f = 3.35 \times 10^5$ J/kg. Thus, 3.35×10^5 J of heat must be supplied to melt one kilogram of ice at 0 °C into liquid water at 0 °C; conversely, this amount of heat must be removed from one kilogram of liquid water at 0 °C to freeze the liquid into ice at 0 °C.

TABLE 12.3   Latent Heats[a] of Fusion and Vaporization

Substance	Melting Point (°C)	Latent Heat of Fusion, L_f (J/kg)	Boiling Point (°C)	Latent Heat of Vaporization, L_v (J/kg)
Ammonia	−77.8	33.2×10^4	−33.4	13.7×10^5
Benzene	5.5	12.6×10^4	80.1	3.94×10^5
Copper	1083	20.7×10^4	2566	47.3×10^5
Ethyl alcohol	−114.4	10.8×10^4	78.3	8.55×10^5
Gold	1063	6.28×10^4	2808	17.2×10^5
Lead	327.3	2.32×10^4	1750	8.59×10^5
Mercury	−38.9	1.14×10^4	356.6	2.96×10^5
Nitrogen	−210.0	2.57×10^4	−195.8	2.00×10^5
Oxygen	−218.8	1.39×10^4	−183.0	2.13×10^5
Water	0.0	33.5×10^4	100.0	22.6×10^5

[a]The values pertain to 1 atm pressure.

BIO **THE PHYSICS OF . . . steam burns.** In comparison to the latent heat of fusion, Table 12.3 indicates that the latent heat of vaporization for water has the much larger value of $L_v = 22.6 \times 10^5$ J/kg. When water boils at 100 °C, 22.6×10^5 J of heat must be supplied for each kilogram of liquid turned into steam. And when steam condenses at 100 °C, this amount of heat is released from each kilogram of steam that changes back into liquid. Liquid water at 100 °C is hot enough by itself to cause a bad burn, and the additional effect of the large latent heat can cause severe tissue damage if condensation occurs on the skin.

THE PHYSICS OF . . . high-tech clothing. By taking advantage of the latent heat of fusion, designers can now engineer clothing that can absorb or release heat to help maintain a comfortable and approximately constant temperature close to your body. As the photograph in **Figure 12.28** shows, the fabric in this type of clothing is coated with microscopic balls of heat-resistant plastic that contain a substance known as a "phase-change material" (PCM). When you are enjoying your favorite winter sport, for example, it is easy to become overheated. The PCM prevents this by melting, absorbing excess body heat in the process. When you are taking a break and cooling down, however, the PCM freezes and releases heat to keep you warm. The temperature range over which the PCM can maintain a comfort zone is related to its melting/freezing temperature, which is determined by its chemical composition.

Examples 14 and 15 illustrate how to take into account the effect of latent heat when using the conservation-of-energy principle.

Courtesy Outlast Technologies LLC,Golden, CO

FIGURE 12.28 This highly magnified image shows a high-tech fabric that can automatically adjust in reaction to your body heat and help maintain a constant temperature next to your skin. See the text discussion.

EXAMPLE 14 | Ice-Cold Lemonade

Ice at 0 °C is placed in a Styrofoam cup containing 0.32 kg of lemonade at 27 °C. The specific heat capacity of lemonade is virtually the same as that of water; that is, $c = 4186$ J/(kg · C°). After the ice and lemonade reach an equilibrium temperature, some ice still remains. The latent heat of fusion for water is $L_f = 3.35 \times 10^5$ J/kg. Assume that the mass of the cup is so small that it absorbs a negligible amount of heat, and ignore any heat lost to the surroundings. Determine the mass of ice that has melted.

Reasoning According to the principle of energy conservation, the amount of heat gained by the melting ice equals the amount of heat lost by the cooling lemonade. According to Equation 12.5, the amount of heat gained by the melting ice is $Q = mL_f$, where m is the mass of the melted ice, and L_f is the latent heat of fusion for water. The amount of heat lost by the lemonade is given by

$Q = cm\,\Delta T$, where ΔT is the higher temperature of 27 °C minus the lower equilibrium temperature. The equilibrium temperature is 0 °C, because there is some ice remaining, and ice is in equilibrium with liquid water when the temperature is 0 °C.

Solution

$$\underbrace{(mL_f)_{ice}}_{\substack{\text{Heat gained} \\ \text{by ice}}} = \underbrace{(cm\,\Delta T)_{lemonade}}_{\substack{\text{Heat lost by} \\ \text{lemonade}}}$$

The mass m_{ice} of ice that has melted is

$$m_{ice} = \frac{(cm\,\Delta T)_{lemonade}}{L_f}$$
$$= \frac{[4186\text{ J/(kg · C°)}](0.32\text{ kg})(27\text{ °C} - 0\text{ °C})}{3.35 \times 10^5\text{ J/kg}} = \boxed{0.11\text{ kg}}$$

EXAMPLE 15 | Getting Ready for a Party

A 7.00-kg glass bowl [$c = 840$ J/(kg·C°)] contains 16.0 kg of punch at 25.0 °C. Two-and-a-half kilograms of ice [$c = 2.00 \times 10^3$ J/(kg·C°)] are added to the punch. The ice has an initial temperature of −20.0 °C, having been kept in a very cold freezer. The punch may be treated as if it were water [$c = 4186$ J/(kg · C°)], and it may be

assumed that there is no heat flow between the punch bowl and the external environment. The latent heat of fusion for water is 3.35×10^5 J/kg. When thermal equilibrium is reached, all the ice has melted, and the final temperature of the mixture is above 0 °C. Determine this temperature.

Reasoning The final temperature can be found by using the conservation of energy principle: the amount of heat gained is equal to the amount of heat lost. Heat is gained (a) by the ice in warming up to the melting point, (b) by the ice in changing phase from a solid to a liquid, and (c) by the liquid that results from the ice warming up to the final temperature; heat is lost (d) by the punch and (e) by the bowl in cooling down. The amount of heat gained or lost by each component in changing temperature can be determined from the relation $Q = cm \, \Delta T$, where ΔT is the higher temperature minus the lower temperature. The amount of heat gained when water changes phase from a solid to a liquid at 0 °C is $Q = mL_f$, where m is the mass of water and L_f is the latent heat of fusion.

Solution The amount of heat gained or lost by each component is listed as follows:

(a) Heat gained when ice warms to 0.0 °C
$= [2.00 \times 10^3 \text{ J/(kg} \cdot \text{C°})](2.50 \text{ kg})$
$[0.0 \text{ °C} - (-20.0 \text{ °C})]$

(b) Heat gained when ice melts at 0.0 °C
$= (2.50 \text{ kg})(3.35 \times 10^5 \text{ J/kg})$

(c) Heat gained when melted ice (liquid) warms to temperature T
$= [4186 \text{ J/(kg} \cdot \text{C°})](2.50 \text{ kg})$
$(T - 0.0 \text{ °C})$

(d) Heat lost when punch cools to temperature T
$= [4186 \text{ J/(kg} \cdot \text{C°})](16.0 \text{ kg})$
$(25.0 \text{ °C} - T)$

(e) Heat lost when bowl cools to temperature T
$= [840 \text{ J/(kg} \cdot \text{C°})](7.00 \text{ kg})$
$(25.0 \text{ °C} - T)$

Setting the amount of heat gained equal to the amount of heat lost gives:

$$\underbrace{\text{(a)} + \text{(b)} + \text{(c)}}_{\text{Heat gained}} = \underbrace{\text{(d)} + \text{(e)}}_{\text{Heat lost}}$$

This equation can be solved to show that $\boxed{T = 11 \text{ °C}}$.

Math Skills Carrying out the multiplications listed for the terms (a), (b), (c), (d), and (e), we find that the heat-gained-equals-heat-lost equation can be written as follows:

$$\underbrace{(1.00 \times 10^5)}_{\text{(a)}} + \underbrace{(8.38 \times 10^5)}_{\text{(b)}} + \underbrace{(1.05 \times 10^4)T}_{\text{(c)}}$$

$$= \underbrace{\boxed{(1.67 \times 10^6)} - (6.70 \times 10^4)\, T}_{\text{(d)}}$$

$$+ \underbrace{\boxed{(1.47 \times 10^5)} - (5.88 \times 10^3)\, T}_{\text{(e)}}$$

Units have been omitted for the sake of clarity. Combining terms shows that

$$(9.38 \times 10^5) + (1.05 \times 10^4)T = (1.82 \times 10^6)$$
$$- (7.29 \times 10^4)T \qquad \textbf{(1)}$$

In combining terms, we have paid particular attention to the fact that not all of the exponential powers of ten are the same. For example, in the boxes on the right-hand side of our starting equation, we have

$$(1.67 \times 10^6) + (1.47 \times 10^5) = 1.82 \times 10^6$$

We do not have

$$(1.67 \times 10^6) + (1.47 \times 10^6) = 3.14 \times 10^6 \qquad \textbf{or}$$
$$(1.67 \times 10^5) + (1.47 \times 10^5) = 3.14 \times 10^5$$

Rearranging and further combining terms in Equation 1 yields

$$(8.34 \times 10^4)\, T = 8.8 \times 10^5 \qquad \text{or} \qquad T = \boxed{11 \text{ °C}}$$

THE PHYSICS OF . . . a dye-sublimation color printer. An interesting application of the phase change between a solid and a gas is found in one kind of color printer used with computers. A dye-sublimation printer uses a thin plastic film coated with separate panels of cyan (blue), yellow, and magenta pigment or dye. A full spectrum of colors is produced by using combinations of tiny spots of these dyes. As **Figure 12.29** shows, the coated film passes in front of a print head that extends across the width of the paper and contains 2400 heating elements. When a heating element is turned on, the dye in front of it absorbs heat and goes from a solid to a gas—it sublimes—with no liquid phase in between. A coating on the paper absorbs the gaseous dye on contact, producing a small spot of color. The intensity of the spot is controlled by the heating element, since each element can produce 256 different temperatures; the hotter the element, the greater the amount of dye transferred to the paper. The paper makes three separate passes across the print head, once for each of the dyes. The final result is an image of near-photographic quality. Some printers also employ a fourth pass, in which a clear plastic coating is deposited over the photograph. This coating makes the print waterproof and also helps to prevent premature fading.

Dye removed from film

Heating element

Print head

Coated paper

Plastic film

Print head

Gaseous dye

FIGURE 12.29 A dye-sublimation printer. As the plastic film passes in front of the print head, the heat from a given heating element causes one of three pigments or dyes on the film to sublime from a solid to a gas. The gaseous dye is absorbed onto the coated paper as a dot of color. The size of the dots on the paper has been exaggerated for clarity.

Check Your Understanding

(The answers are given at the end of the book.)

13. Fruit blossoms are permanently damaged at temperatures of about −4 °C (a hard freeze). Orchard owners sometimes spray a film of water over the blossoms to protect them when a hard freeze is expected. Why does this technique offer protection?

14. When ice cubes are used to cool a drink, both their mass and temperature are important in how effective they are. **CYU Table 12.2** lists several possibilities for the mass and temperature of the ice cubes used to cool one particular drink. Rank the possibilities in descending order (best first) according to their cooling effectiveness. Note that the latent heat of phase change and the specific heat capacity must be considered.

CYU TABLE 12.2

	Mass of ice cubes	Temperature of ice cubes
(a)	m	−6.0 °C
(b)	$\frac{1}{2}m$	−12 °C
(c)	$2m$	−3.0 °C

12.9 *Equilibrium Between Phases of Matter

Under specific conditions of temperature and pressure, a substance can exist at equilibrium in more than one phase at the same time. Consider **Interactive Figure 12.30**, which shows a container kept at a constant temperature by a large reservoir of heated sand. Initially the container is evacuated. Part *a* shows it just after it has been partially filled with a liquid and a few fast-moving molecules are escaping the liquid and forming a vapor phase. These molecules pick up the required energy (the latent heat of vaporization) during collisions with neighboring molecules in the liquid. However, the reservoir of heated sand replenishes the energy carried away, thus maintaining the constant temperature. At first, the movement of molecules is predominantly from liquid to vapor, although some molecules in the vapor phase do reenter the liquid. As the molecules accumulate in the vapor, the number reentering the liquid eventually equals the number entering the vapor, at which point equilibrium is established, as in part *b*. From this point on, the concentration of molecules in the vapor phase does not change, and the vapor pressure remains constant. The pressure of the vapor that coexists in equilibrium with the liquid is called the **equilibrium vapor pressure** of the liquid.

The equilibrium vapor pressure does not depend on the volume of space above the liquid. If more space were provided, more liquid would vaporize, until equilibrium was reestablished at the same vapor pressure, assuming the same temperature is maintained. In fact, the equilibrium vapor pressure depends only on the temperature of the liquid; a higher temperature causes a higher pressure, as the graph in **Figure 12.31** indicates for the specific case of water. Only when the temperature and vapor pressure correspond to a

Constant-temperature heated sand

(a)

Constant-temperature heated sand

(b)

INTERACTIVE FIGURE 12.30
(*a*) Some of the molecules begin entering the vapor phase in the evacuated space above the liquid. (*b*) Equilibrium is reached when the number of molecules entering the vapor phase equals the number returning to the liquid.

FIGURE 12.31 A plot of the equilibrium vapor pressure versus temperature is called the vapor pressure curve or the vaporization curve, the example shown being that for the liquid/vapor equilibrium of water.

point on the curved line, which is called the **vapor pressure curve** or the **vaporization curve**, do liquid and vapor phases coexist at equilibrium.

To illustrate the use of a vaporization curve, let's consider what happens when water boils in a pot that is *open to the air*. Assume that the air pressure acting on the water is 1.01×10^5 Pa (one atmosphere). When boiling occurs, bubbles of water vapor form throughout the liquid, rise to the surface, and break. For these bubbles to form and rise, the pressure of the vapor inside them must at least equal the air pressure acting on the surface of the water. According to **Figure 12.31**, a value of 1.01×10^5 Pa corresponds to a temperature of 100 °C. Consequently, water boils at 100 °C at one atmosphere of pressure. In general, a ***liquid boils at the temperature for which its vapor pressure equals the external pressure***. Water will not boil, then, at sea level if the temperature is only 83 °C, because at this temperature the vapor pressure of water is only 0.53×10^5 Pa (see **Figure 12.31**), a value less than the external pressure of 1.01×10^5 Pa. However, water does boil at 83 °C on a mountain at an altitude of just under five kilometers, because the atmospheric pressure there is 0.53×10^5 Pa.

The fact that water can boil at a temperature less than 100 °C leads to an interesting phenomenon that Conceptual Example 16 discusses.

CONCEPTUAL EXAMPLE 16 | How to Boil Water That Is Cooling Down

Figure 12.32a shows water boiling in an open flask. Shortly after the flask is removed from the burner, the boiling stops. A cork is then placed in the neck of the flask to seal it, and water is poured over the neck of the flask, as in part *b* of the drawing. To restart the boiling, should the water poured over the neck be **(a)** cold or **(b)** hot—but not boiling?

Reasoning   When the open flask is removed from the burner, the water begins to cool and the pressure above its surface is one atmosphere (1.01×10^5 Pa). Boiling quickly stops, because water cannot boil when its temperature is less than 100 °C and the pressure above its surface is one atmosphere. To restart the boiling, it is necessary either to reheat the water to 100 °C or reduce the pressure above the water in the corked flask to something less than one atmosphere so that boiling can occur at a temperature less than 100 °C.

Answer (b) is incorrect.   Certainly, pouring hot water over the corked flask will reheat the water. However, since the water being poured is not boiling, its temperature must be less than 100 °C. Therefore, it cannot reheat the water within the flask to 100 °C and restart the boiling.

Answer (a) is correct.   When cold water is poured over the corked flask, it causes some of the water vapor inside to condense. Consequently, the pressure above the liquid in the flask drops. When it drops to the value of the vapor pressure of the

Cork —

— Water

(a) Water boiling   *(b)* Water boiling again

FIGURE 12.32   (*a*) Water is boiling at a temperature of 100 °C and a pressure of one atmosphere. (*b*) The water boils at a temperature that is less than 100 °C, because the cool water reduces the pressure above the water in the flask.

water in the flask at its current temperature (which is now less than 100 °C), the boiling restarts.

THE PHYSICS OF . . . spray cans.   The operation of spray cans is based on the equilibrium between a liquid and its vapor. **Figure 12.33a** shows that a spray can contains a liquid propellant that is mixed with the product (such as hair spray). Inside the can, propellant vapor forms over the liquid. A propellant is chosen that has an equilibrium vapor pressure that is greater than atmospheric pressure at room temperature. Consequently, when the nozzle of the can is pressed, as in part *b* of the drawing, the vapor pressure forces the liquid propellant and product up the tube in the can and out the nozzle as a spray. When the nozzle is released, the coiled spring reseals the can and the propellant vapor builds up once again to its equilibrium value.

FIGURE 12.33 (*a*) A closed spray can containing liquid and vapor in equilibrium. (*b*) An open spray can.

As is the case for liquid/vapor equilibrium, a solid can be in equilibrium with its liquid phase only at specific conditions of temperature and pressure. For each temperature, there is a single pressure at which the two phases can coexist in equilibrium. A plot of the equilibrium pressure versus equilibrium temperature is referred to as the **fusion curve**, and **Figure 12.34a** shows a typical curve for a normal substance. A normal substance expands on melting (e.g., carbon dioxide and sulfur). Since higher pressures make it more difficult for such materials to expand, a higher melting temperature is needed for a higher pressure, and the fusion curve slopes upward to the right. Part *b* of the picture illustrates the fusion curve for water, one of the few substances that contract when they melt. Higher pressures make it easier for such substances to melt. Consequently, a lower melting temperature is associated with a higher pressure, and the fusion curve slopes downward to the right.

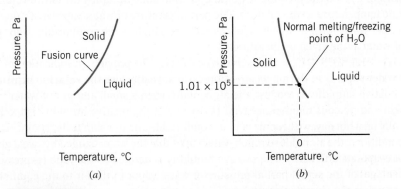

FIGURE 12.34 (*a*) The fusion curve for a normal substance that expands on melting. (*b*) The fusion curve for water, one of the few substances that contract on melting.

It should be noted that just because two phases can coexist in equilibrium does not necessarily mean that they will. Other factors may prevent it. For example, water in an *open* bowl may never come into equilibrium with water vapor if air currents are present. Under such conditions the liquid, perhaps at a temperature of 25 °C, attempts to establish the corresponding equilibrium vapor pressure of 3.2×10^3 Pa. If air currents continually blow the water vapor away, however, equilibrium will never be established, and eventually the water will evaporate completely. Each kilogram of water that goes into the vapor phase takes along the latent heat of vaporization. Because of this heat loss, the remaining liquid would become cooler, except for the fact that the surroundings replenish the loss.

BIO THE PHYSICS OF . . . evaporative cooling of the human body.   In the case of the human body, water is exuded by the sweat glands and evaporates from a much larger area than the surface of a typical bowl of water. The removal of heat along with the water vapor is called evaporative cooling and is one mechanism that the body uses to maintain its constant temperature.

Check Your Understanding

(The answers are given at the end of the book.)

15. A camping stove is used to boil water high on a mountain, where the atmospheric pressure is lower than it is at sea level. Does it necessarily follow that the same stove can boil water at sea level?

16. **BIO** Medical instruments are sterilized at a high temperature in an autoclave, which is essentially a pressure cooker that heats the instruments in water under a pressure greater than one atmosphere. Why is the water in an autoclave able to reach a very high temperature, while water in an open pot can only be heated to 100 °C?

17. A jar is half filled with boiling water. The lid is then screwed on the jar. After the jar has cooled to room temperature, the lid is difficult to remove. Why? Ignore the thermal expansion and contraction of the jar and the lid.

18. A bottle of carbonated soda (sealed and under a pressure greater than one atmosphere) is left outside in subfreezing temperatures, although the soda remains liquid. When the soda is brought inside and opened immediately, it suddenly freezes. Why?

19. When a bowl of water is placed in a closed container and the water vapor is pumped away rapidly enough, why does the remaining liquid turn into ice?

12.10 *Humidity

Air is a mixture of gases, including nitrogen, oxygen, and water vapor. The total pressure of the mixture is the sum of the partial pressures of the component gases. The **partial pressure** of a gas is the pressure it would exert if it alone occupied the entire volume at the same temperature as the mixture. The partial pressure of water vapor in air depends on weather conditions. It can be as low as zero or as high as the equilibrium vapor pressure of water at the given temperature.

THE PHYSICS OF . . . relative humidity.   To provide an indication of how much water vapor is in the air, weather forecasters usually give the **relative humidity**. If the relative humidity is too low, the air contains such a small amount of water vapor that skin and mucous membranes tend to dry out. If the relative humidity is too high, especially on a hot day, we become very uncomfortable and our skin feels "sticky." Under such conditions, the air holds so much water vapor that the water exuded by sweat glands cannot evaporate efficiently. The relative humidity is defined as the ratio (expressed as a percentage) of the actual partial pressure of water vapor in the air to the equilibrium vapor pressure at a given temperature.

$$\text{Percent relative humidity} = \frac{\text{Partial pressure of water vapor}}{\text{Equilibrium vapor pressure of water at the existing temperature}} \times 100 \tag{12.6}$$

The term in the denominator on the right of Equation 12.6 is given by the vaporization curve of water and is the pressure of the water vapor in equilibrium with the liquid. At a given temperature, the partial pressure of the water vapor in the air cannot exceed this value. If it did, the vapor would not be in equilibrium with the liquid and would condense as dew or rain to reestablish equilibrium.

When the partial pressure of the water vapor equals the equilibrium vapor pressure of water at a given temperature, the relative humidity is 100%. In such a situation, the vapor is said to be *saturated* because it is present in the maximum amount, as it would be above a pool of liquid at equilibrium in a closed container. If the relative humidity is less than 100%, the water vapor is said to be *unsaturated*. Example 17 demonstrates how to find the relative humidity.

EXAMPLE 17 | Relative Humidities

One day, the partial pressure of water vapor in the air is 2.0×10^3 Pa. Using the vaporization curve for water in **Figure 12.35**, determine the relative humidity if the temperature is **(a)** 32 °C and **(b)** 21 °C.

Reasoning and Solution **(a)** According to **Figure 12.35**, the equilibrium vapor pressure of water at 32 °C is 4.8×10^3 Pa. Equation 12.6 reveals that the relative humidity is

$$\text{Relative humidity at } 32\,°C = \frac{2.0 \times 10^3 \text{ Pa}}{4.8 \times 10^3 \text{ Pa}} \times 100 = \boxed{42\%}$$

(b) A similar calculation shows that

$$\text{Relative humidity at } 21\,°C = \frac{2.0 \times 10^3 \text{ Pa}}{2.5 \times 10^3 \text{ Pa}} \times 100 = \boxed{80\%}$$

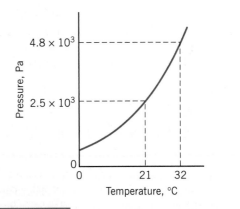

FIGURE 12.35 The vaporization curve of water.

THE PHYSICS OF . . . fog formation.   When air containing a given amount of water vapor is cooled, a temperature is reached in which the partial pressure of the vapor equals the equilibrium vapor pressure. This temperature is known as the **dew point**. For instance, **Figure 12.36** shows that if the partial pressure of water vapor is 3.2×10^3 Pa, the dew point is 25 °C. This partial pressure would correspond to a relative humidity of 100%, if the ambient temperature were equal to the dew-point temperature. Hence, the dew point is the temperature below which water vapor in the air condenses in the form of liquid drops (dew or fog). The closer the actual temperature is to the dew point, the closer the relative humidity is to 100%. Thus, for fog to form, the air temperature must drop below the dew point. Similarly, water condenses on the outside of a cold glass when the temperature of the air next to the glass falls below the dew point.

THE PHYSICS OF . . . a home dehumidifier.   The cold coils in a home dehumidifier function very much in the same way that the cold glass does. The coils are kept cold by a circulating refrigerant (see **Figure 12.37**). When the air blown across them by the fan cools below the dew point, water vapor condenses in the form of droplets, which collect in a receptacle.

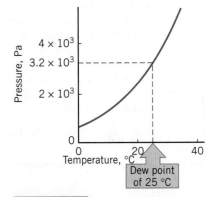

FIGURE 12.36 On the vaporization curve of water, the dew point is the temperature that corresponds to the actual partial pressure of water vapor in the air.

FIGURE 12.37 The cold coils of a dehumidifier cool the air blowing across them to below the dew point, and water vapor condenses out of the air.

Check Your Understanding

(The answers are given at the end of the book.)

20. A bowl of water is covered tightly and allowed to sit at a constant temperature of 23 °C for a long time. What is the relative humidity in the space between the surface of the water and the cover?

21. Is it possible for dew to form on Tuesday night and not on Monday night, even though Monday night is the cooler night?

22. Two rooms in a mansion have the same temperature. One of these rooms contains an indoor swimming pool. On a cold day the windows of one of the two rooms are "steamed up." Which room is it likely to be? Explain.

EXAMPLE 18 | BIO Heating Up Your Brain

The human brain is a remarkable organ. It only requires approximately 20 W of power to function normally. For comparison, a man-made computer processor with the same computational power as your brain would consume 10–20 MW of power! With that being said, the brain is the most energy-hungry organ in the human body. It only accounts for 1/50 of the body's total weight, but it requires 1/5 (20%) of the body's total energy. This power output from the brain produces heat, which raises the temperature of the brain and the surrounding bone and tissues. The brain contains approximately 1.2 kg of water, which accounts for 77% of its total mass. If a human brain was operating at a constant power output of 20 W for one hour, what would be the increase in its temperature? Assume it is composed of only the 1.2 kg of water, and no heat transfer occurs between the brain and its surroundings.

Reasoning We can calculate the change in temperature (ΔT) of the water inside the brain by knowing the total heat (Q) produced

in 1 hour. Equation 12.4 then provides the relationship between Q and ΔT.

Solution We can relate the power output of the brain to the total heat produced, since $P = Q/t$. Therefore, $Q = Pt$. Now we apply Equation 12.4: $Q = cm\,\Delta T$. Substituting in for Q and solving for ΔT, we get the following:

$$\Delta T = Pt/mc = (20\ \text{W})(60\ \text{s})/(1.2\ \text{kg})(4186\ \text{J/(kg}\cdot\text{C}°)) = \boxed{0.24\ °\text{C}}$$

Although this is a small temperature change, we have done the calculation assuming the entire brain has a high specific heat capacity equal to that of water. The actual rise in temperature will be higher if we consider the rest of the brain that is not water. Studies have shown that after behavioral stimuli, it is not unusual to observe a 2–3 °C temperature increase in the brain. This is a significant change and must be controlled by the body's temperature regulation system that consists of the circulatory system and the surface of the skin.

Concept Summary

12.1 Common Temperature Scales On the Celsius temperature scale, there are 100 equal divisions between the ice point (0 °C) and the steam point (100 °C). On the Fahrenheit temperature scale, there are 180 equal divisions between the ice point (32 °F) and the steam point (212 °F).

12.2 The Kelvin Temperature Scale For scientific work, the Kelvin temperature scale is the scale of choice. One kelvin (K) is equal in size to one Celsius degree. However, the temperature T on the Kelvin scale differs from the temperature T_c on the Celsius scale by an additive constant of 273.15, as indicated by Equation 12.1. The lower limit of temperature is called absolute zero and is designated as 0 K on the Kelvin scale.

$$T = T_c + 273.15 \qquad (12.1)$$

12.3 Thermometers The operation of any thermometer is based on the change in some physical property with temperature; this physical property is called a thermometric property. Examples of thermometric properties are the length of a column of mercury, electrical voltage, and electrical resistance.

12.4 Linear Thermal Expansion Most substances expand when heated. For linear expansion, an object of length L_0 experiences a change ΔL in length when the temperature changes by ΔT, as shown in Equation 12.2, where α is the coefficient of linear expansion.

$$\Delta L = \alpha L_0\,\Delta T \qquad (12.2)$$

For an object held rigidly in place, a thermal stress can occur when the object attempts to expand or contract. The stress can be large, even for small temperature changes.

When the temperature changes, a hole in a plate of solid material expands or contracts as if the hole were filled with the surrounding material.

12.5 Volume Thermal Expansion For volume expansion, the change ΔV in the volume of an object of volume V_0 is given by Equation 12.3, where β is the coefficient of volume expansion. When the temperature changes, a cavity in a piece of solid material expands or contracts as if the cavity were filled with the surrounding material.

$$\Delta V = \beta V_0\,\Delta T \qquad (12.3)$$

12.6 Heat and Internal Energy The internal energy of a substance is the sum of the kinetic, potential, and other kinds of energy that the molecules of the substance have. Heat is energy that flows from a higher-temperature object to a lower-temperature object because of the difference in temperatures. The SI unit for heat is the joule (J).

12.7 Heat and Temperature Change: Specific Heat Capacity The heat Q that must be supplied or removed to change the temperature of a substance of mass m by an amount ΔT is given by Equation 12.4, where c is a constant known as the specific heat capacity.

$$Q = cm\,\Delta T \qquad (12.4)$$

When materials are placed in thermal contact within a perfectly insulated container, the principle of energy conservation requires that the amount of heat lost by warmer materials equals the amount of heat gained by cooler materials.

Heat is sometimes measured with a unit called the kilocalorie (kcal). The conversion factor between kilocalories and joules is known as the mechanical equivalent of heat: 1 kcal = 4186 joules.

12.8 Heat and Phase Change: Latent Heat Heat must be supplied or removed to make a material change from one phase to another. The heat Q that must be supplied or removed to change the phase of a mass m of a substance is given by Equation 12.5, where L is the latent heat of the substance and has SI units of J/kg. The latent heats of fusion, vaporization, and sublimation refer, respectively, to the solid/liquid, the liquid/vapor, and the solid/vapor phase changes.

$$Q = mL \qquad (12.5)$$

12.9 Equilibrium Between Phases of Matter The equilibrium vapor pressure of a substance is the pressure of the vapor phase that is in equilibrium with the liquid phase. For a given substance, vapor pressure depends only on temperature. For a liquid, a plot of the equilibrium vapor pressure versus temperature is called the vapor pressure curve or vaporization curve.

The fusion curve gives the combinations of temperature and pressure for equilibrium between solid and liquid phases.

12.10 Humidity The relative humidity is defined as in Equation 12.6.

$$\text{Percent relative humidity} = \frac{\text{Partial pressure of water vapor}}{\text{Equilibrium vapor pressure of water at the existing temperature}} \times 100 \qquad (12.6)$$

The dew point is the temperature below which the water vapor in the air condenses. On the vaporization curve of water, the dew point is the temperature that corresponds to the actual pressure of water vapor in the air.

Focus on Concepts   Online

Additional questions are available for assignment in WileyPLUS.

Section 12.2   The Kelvin Temperature Scale

1. Which one of the following statements correctly describes the Celsius and the Kelvin temperature scales? (a) The size of the degree on the Celsius scale is larger than that on the Kelvin scale by a factor of 9/5. (b) Both scales assign the same temperature to the ice point, but they assign different temperatures to the steam point. (c) Both scales assign the same temperature to the steam point, but they assign different temperatures to the ice point. (d) The Celsius scale assigns the same values to the ice and the steam points that the Kelvin scale assigns. (e) The size of the degree on each scale is the same.

Section 12.4   Linear Thermal Expansion

2. The drawing shows two thin rods, one made from aluminum [$\alpha = 23 \times 10^{-6}\,(\text{C}°)^{-1}$] and the other from steel [$\alpha = 12 \times 10^{-6}\,(\text{C}°)^{-1}$]. Each rod has the same length and the same initial temperature and is attached

QUESTION 2

at one end to an immovable wall, as shown. The temperatures of the rods are increased, both by the same amount, until the gap between the rods is closed. Where do the rods meet when the gap is closed? (a) The rods meet exactly at the midpoint. (b) The rods meet to the right of the midpoint. (c) The rods meet to the left of the midpoint.

3. A ball is slightly too large to fit through a hole in a flat plate. The drawing shows two arrangements of this situation. In Arrangement I the ball is made from metal A and the plate from metal B. When both the ball and the plate are cooled by the same number of Celsius degrees, the ball passes through the hole. In Arrangement II the ball is also made from metal A,

QUESTION 3

but the plate is made from metal C. Here, the ball passes through the hole when both the ball and the plate are heated by the same number of Celsius degrees. Rank the coefficients of linear thermal expansion of metals A, B, and C in descending order (largest first): (a) α_B, α_A, α_C (b) α_B, α_C, α_A (c) α_C, α_B, α_A (d) α_C, α_A, α_B (e) α_A, α_B, α_C

Section 12.5   Volume Thermal Expansion

4. A solid sphere and a solid cube are made from the same material. The sphere would just fit within the cube, if it could. Both begin at the same temperature, and both are heated to the same temperature. Which object, if either, has the greater change in volume? (a) The sphere. (b) The cube. (c) Both have the same change in volume. (d) Insufficient information is given for an answer.

5. A container can be made from steel [$\beta = 36 \times 10^{-6}\,(\text{C}°)^{-1}$] or lead [$\beta = 87 \times 10^{-6}\,(\text{C}°)^{-1}$]. A liquid is poured into the container, filling it to the brim. The liquid is either water [$\beta = 207 \times 10^{-6}\,(\text{C}°)^{-1}$] or ethyl alcohol [$\beta = 1120 \times 10^{-6}\,(\text{C}°)^{-1}$]. When the full container is heated, some liquid spills out. To keep the overflow to a minimum, from what material should the container be made and what should the liquid be? (a) Lead, water (b) Steel, water (c) Lead, ethyl alcohol (d) Steel, ethyl alcohol

Section 12.7 Heat and Temperature Change: Specific Heat Capacity

6. Which of the following cases (if any) requires the greatest amount of heat? In each case the material is the same. **(a)** 1.5 kg of the material is to be heated by 7.0 C°. **(b)** 3.0 kg of the material is to be heated by 3.5 C°. **(c)** 0.50 kg of the material is to be heated by 21 C°. **(d)** 0.75 kg of the material is to be heated by 14 C°. **(e)** The amount of heat required is the same in each of the four previous cases.

7. The following three hot samples have the same temperature. The same amount of heat is removed from each sample. Which one experiences the smallest drop in temperature, and which one experiences the largest drop?

> Sample A. 4.0 kg of water $[c = 4186 \text{ J/(kg} \cdot \text{C°)}]$
>
> Sample B. 2.0 kg of oil $[c = 2700 \text{ J/(kg} \cdot \text{C°)}]$
>
> Sample C. 9.0 kg of dirt $[c = 1050 \text{ J/(kg} \cdot \text{C°)}]$

(a) C smallest and A largest **(b)** B smallest and C largest **(c)** A smallest and B largest **(d)** C smallest and B largest **(e)** B smallest and A largest

Section 12.8 Heat and Phase Change: Latent Heat

8. The latent heat of fusion for water is 33.5×10^4 J/kg, while the latent heat of vaporization is 22.6×10^5 J/kg. What mass m of water must be frozen in order to release the amount of heat that 1.00 kg of steam releases when it condenses?

Section 12.9 Equilibrium Between Phases of Matter

9. Which one or more of the following techniques can be used to freeze water?

> **A.** Cooling the water below its normal freezing point of 0 °C at the normal atmospheric pressure of 1.01×10^5 Pa
>
> **B.** Cooling the water below its freezing point of −1 °C at a pressure greater than 1.01×10^5 Pa
>
> **C.** Rapidly pumping away the water vapor above the liquid in an insulated container (The insulation prevents heat flowing from the surroundings into the remaining liquid.)

(a) Only A **(b)** Only B **(c)** Only A and B **(d)** A, B, and C **(e)** Only C

Section 12.10 Humidity

10. Which of the following three statements concerning relative humidity values of 30% and 40% are true? Note that when the relative humidity is 30%, the air temperature may be different than it is when the relative humidity is 40%.

> **A.** It is possible that at a relative humidity of 30% there is a smaller partial pressure of water vapor in the air than there is at a relative humidity of 40%.
>
> **B.** It is possible that there is the same partial pressure of water vapor in the air at 30% and at 40% relative humidity.
>
> **C.** It is possible that at a relative humidity of 30% there is a greater partial pressure of water vapor in the air than there is at a relative humidity of 40%.

(a) A, B, and C **(b)** Only A and B **(c)** Only A and C **(d)** Only B and C **(e)** Only A

Problems

Online

Additional questions are available for assignment in WileyPLUS.

Note: *For problems in this set, use the values of α and β in **Table 12.1**, and the values of c, L_f, and L_v in **Tables 12.2** and **12.3**, unless stated otherwise.*

SSM	Student Solutions Manual	**BIO**	Biomedical application
MMH	Problem-solving help	**E**	Easy
GO	Guided Online Tutorial	**M**	Medium
V-HINT	Video Hints	**H**	Hard
CHALK	Chalkboard Videos	**WS**	Worksheet
		T	Team Problem

Section 12.1 Common Temperature Scales,

Section 12.2 The Kelvin Temperature Scale,

Section 12.3 Thermometers

1. **E** **SSM** Suppose you are hiking down the Grand Canyon. At the top, the temperature early in the morning is a cool 3 °C. By late afternoon, the temperature at the bottom of the canyon has warmed to a sweltering 34 °C. What is the *difference* between the higher and lower temperatures in **(a)** Fahrenheit degrees and **(b)** kelvins?

2. **E** You are sick, and your temperature is 312.0 kelvins. Convert this temperature to the Fahrenheit scale.

3. **E** **CHALK** On the moon the surface temperature ranges from 375 K during the day to 1.00×10^2 K at night. What are these temperatures on the **(a)** Celsius and **(b)** Fahrenheit scales?

4. **E** **GO** **CHALK** The drawing shows two thermometers, A and B, whose temperatures are measured in °A and °B. The ice and boiling points of water are also indicated. **(a)** Using the data in the drawing, determine the number of B degrees on the B scale that correspond to 1 A° on the A scale. **(b)** If the temperature of a substance reads +40.0 °A on the A scale, what would that temperature read on the B scale?

PROBLEM 4

5. **M GO** A copper–constantan thermocouple generates a voltage of 4.75×10^{-3} volts when the temperature of the hot junction is 110.0 °C and the reference junction is kept at a temperature of 0.0 °C. If the voltage is proportional to the difference in temperature between the junctions, what is the temperature of the hot junction when the voltage is 1.90×10^{-3} volts?

6. **M V-HINT** If a nonhuman civilization were to develop on Saturn's largest moon, Titan, its scientists might well devise a temperature scale based on the properties of methane, which is much more abundant on the surface than water is. Methane freezes at −182.6 °C on Titan, and boils at −155.2 °C. Taking the boiling point of methane as 100.0 °M (degrees Methane) and its freezing point as 0 °M, what temperature on the Methane scale corresponds to the absolute zero point of the Kelvin scale?

Section 12.4   Linear Thermal Expansion

7. **E** A steel section of the Alaskan pipeline had a length of 65 m and a temperature of 18 °C when it was installed. What is its change in length when the temperature drops to a frigid −45 °C?

8. **E SSM** A steel aircraft carrier is 370 m long when moving through the icy North Atlantic at a temperature of 2.0 °C. By how much does the carrier lengthen when it is traveling in the warm Mediterranean Sea at a temperature of 21 °C?

9. **E** The Eiffel Tower is a steel structure whose height increases by 19.4 cm when the temperature changes from −9 to +41 °C. What is the approximate height (in meters) at the lower temperature?

10. **E** Conceptual Example 6 provides background for this problem. A hole is drilled through a copper plate whose temperature is 11 °C. (a) When the temperature of the plate is increased, will the radius of the hole be larger or smaller than the radius at 11 °C? Why? (b) When the plate is heated to 110 °C, by what fraction $\Delta r/r_0$ will the radius of the hole change?

11. **E** A commonly used method of fastening one part to another part is called "shrink fitting." A steel rod has a diameter of 2.0026 cm, and a flat plate contains a hole whose diameter is 2.0000 cm. The rod is cooled so that it just fits into the hole. When the rod warms up, the enormous thermal stress exerted by the plate holds the rod securely to the plate. By how many Celsius degrees should the rod be cooled?

12. **E SSM** When the temperature of a coin is raised by 75 C°, the coin's diameter increases by 2.3×10^{-5} m. If the original diameter of the coin is 1.8×10^{-2} m, find the coefficient of linear expansion.

13. **E** One January morning in 1943, a warm chinook wind rapidly raised the temperature in Spearfish, South Dakota, from below freezing to +12.0 °C. As the chinook died away, the temperature fell to −20.0 °C in 27.0 minutes. Suppose that a 19-m aluminum flagpole were subjected to this temperature change. Find the average speed at which its height would decrease, assuming the flagpole responded instantaneously to the changing temperature.

14. **E GO** One rod is made from lead and another from quartz. The rods are heated and experience the same change in temperature. The change in length of each rod is the same. If the initial length of the lead rod is 0.10 m, what is the initial length of the quartz rod?

15. **M GO** A thin rod consists of two parts joined together. One-third of it is silver and two-thirds is gold. The temperature decreases by 26 C°. Determine the fractional decrease $\dfrac{\Delta L}{L_{0,\,\text{Silver}} + L_{0,\,\text{Gold}}}$ in the rod's length, where $L_{0,\,\text{Silver}}$ and $L_{0,\,\text{Gold}}$ are the initial lengths of the silver and gold rods.

16. **M CHALK** The brass bar and the aluminum bar in the drawing are each attached to an immovable wall. At 28 °C the air gap between the rods is 1.3×10^{-3} m. At what temperature will the gap be closed?

PROBLEM 16

17. **M V-HINT** Multiple-Concept Example 5 reviews the concepts that are involved in this problem. A ruler is accurate when the temperature is 25 °C. When the temperature drops to −14 °C, the ruler shrinks and no longer measures distances accurately. However, the ruler can be made to read correctly if a force of magnitude 1.2×10^3 N is applied to each end so as to stretch it back to its original length. The ruler has a cross-sectional area of 1.6×10^{-5} m², and it is made from a material whose coefficient of linear expansion is 2.5×10^{-5} (C°)⁻¹. What is Young's modulus for the material from which the ruler is made?

18. **M SSM** A simple pendulum consists of a ball connected to one end of a thin brass wire. The period of the pendulum is 2.0000 s. The temperature rises by 140 C°, and the length of the wire increases. Determine the period of the heated pendulum.

19. **M V-HINT** As the drawing shows, two thin strips of metal are bolted together at one end; both have the same temperature. One is steel, and the other is aluminum. The steel strip is 0.10% longer than the aluminum strip. By how much should the temperature of the strips be increased, so that the strips have the same length?

PROBLEM 19

20. **M GO** Consult Multiple-Concept Example 5 for insight into solving this problem. A copper rod is fastened securely at both ends to immovable supports. When this rod is stretched, it will rupture when a tensile stress of 2.3×10^7 N/m² is applied at each end. The rod just fits between the supports, so initially there is no stress applied to the rod. The rod is then cooled. What is the magnitude $|\Delta T|$ of the change in temperature of the rod when it ruptures?

21. **M GO** A ball and a thin plate are made from different materials and have the same initial temperature. The ball does not fit through a hole in the plate, because the diameter of the ball is slightly larger than the diameter of the hole. However, the ball will pass through the hole when the ball and the plate are both heated to a common higher temperature. In each of the arrangements in the drawing the diameter of the ball is 1.0×10^{-5} m larger than the diameter of the hole in the thin plate, which has a diameter of 0.10 m. The initial temperature of each arrangement is 25.0 °C. At what temperature will the ball fall through the hole in each arrangement?

PROBLEM 21

22. **H** An 85.0-N backpack is hung from the middle of an aluminum wire, as the drawing shows. The temperature of the wire then drops by 20.0 C°. Find the tension in the wire at the lower temperature. Assume that the distance between the supports does not change, and ignore any thermal stress.

PROBLEM 22

Section 12.5 Volume Thermal Expansion

23. **E** A flask is filled with 1.500 L (L = liter) of a liquid at 97.1 °C. When the liquid is cooled to 15.0 °C, its volume is only 1.383 L, however. Neglect the contraction of the flask and use **Table 12.1** to identify the liquid.

24. **E** **MMH** A thin spherical shell of silver has an inner radius of 2.0×10^{-2} m when the temperature is 18 °C. The shell is heated to 147 °C. Find the change in the interior volume of the shell.

25. **E** A test tube contains 2.54×10^{-4} m^3 of liquid carbon tetrachloride at a temperature of 75.0 °C. The test tube and the carbon tetrachloride are cooled to a temperature of −13.0 °C, which is above the freezing point of carbon tetrachloride. Find the volume of carbon tetrachloride in the test tube at −13.0 °C.

26. **E** **SSM** A copper kettle contains water at 24 °C. When the water is heated to its boiling point of 100.0 °C, the volume of the kettle expands by 1.2×10^{-5} m^3. Determine the volume of the kettle at 24 °C.

27. **E** Suppose you are selling apple cider for two dollars a gallon when the temperature is 4.0 °C. The coefficient of volume expansion of the cider is 280×10^{-6} (C°)$^{-1}$. How much more money (in pennies) would you make per gallon by refilling the container on a day when the temperature is 26 °C? Ignore the expansion of the container.

28. **E** **MMH** During an all-night cram session, a student heats up a one-half liter (0.50×10^{-3} m^3) glass (Pyrex) beaker of cold coffee. Initially, the temperature is 18 °C, and the beaker is filled to the brim. A short time later when the student returns, the temperature has risen to 92 °C. The coefficient of volume expansion of coffee is the same as that of water. How much coffee (in cubic meters) has spilled out of the beaker?

29. **E** **MMH** Many hot-water heating systems have a reservoir tank connected directly to the pipeline, to allow for expansion when the water becomes hot. The heating system of a house has 76 m of copper pipe whose inside radius is 9.5×10^{-3} m. When the water and pipe are heated from 24 to 78 °C, what must be the minimum volume of the reservoir tank to hold the overflow of water?

30. **E** **SSM** Suppose that the steel gas tank in your car is completely filled when the temperature is 17 °C. How many gallons will spill out of the twenty-gallon tank when the temperature rises to 35 °C?

31. **E** **GO** An aluminum can is filled to the brim with a liquid. The can and the liquid are heated so their temperatures change by the same amount. The can's initial volume at 5 °C is 3.5×10^{-4} m^3. The coefficient of volume expansion for aluminum is 69×10^{-6} (C°)$^{-1}$. When the can and the liquid are heated to 78 °C, 3.6×10^{-6} m^3 of liquid spills over. What is the coefficient of volume expansion of the liquid?

32. **M** **V-HINT** A spherical brass shell has an interior volume of 1.60×10^{-3} m^3. Within this interior volume is a solid steel ball that has a volume of 0.70×10^{-3} m^3. The space between the steel ball and the inner surface of the brass shell is filled completely with mercury. A small hole is drilled through the brass, and the temperature of the arrangement is increased by 12 C°. What is the volume of the mercury that spills out of the hole?

33. **M** **GO** At the bottom of an old mercury-in-glass thermometer is a 45-mm^3 reservoir filled with mercury. When the thermometer was placed under your tongue, the warmed mercury would expand into a very narrow cylindrical channel, called a capillary, whose radius was 1.7×10^{-2} mm. Marks were placed along the capillary that indicated the temperature. Ignore the thermal expansion of the glass and determine how far (in mm) the mercury would expand into the capillary when the temperature changed by 1.0 C°.

34. **M** **SSM** The bulk modulus of water is $B = 2.2 \times 10^9$ N/m^2. What change in pressure ΔP (in atmospheres) is required to keep water from expanding when it is heated from 15 to 25 °C?

35. **M** **GO** **CHALK** The density of mercury is 13 600 kg/m^3 at 0 °C. What would its density be at 166 °C?

36. **H** **SSM** Two identical thermometers made of Pyrex glass contain, respectively, identical volumes of mercury and methyl alcohol. If the expansion of the glass is taken into account, how many times greater is the distance between the degree marks on the methyl alcohol thermometer than the distance on the mercury thermometer?

37. **H** The column of mercury in a barometer (see Figure 11.12) has a height of 0.760 m when the pressure is one atmosphere and the temperature is 0.0 °C. Ignoring any change in the glass containing the mercury, what will be the height of the mercury column for the same one atmosphere of pressure when the temperature rises to 38.0 °C on a hot day? (*Hint: The pressure in the barometer is given by Pressure = $\rho g h$, and the density ρ of the mercury changes when the temperature changes.*)

Section 12.6 Heat and Internal Energy,

Section 12.7 Heat and Temperature Change: Specific Heat Capacity

38. **E** **SSM** Ideally, when a thermometer is used to measure the temperature of an object, the temperature of the object itself should not change. However, if a significant amount of heat flows from the object to the thermometer, the temperature will change. A thermometer has a mass of 31.0 g, a specific heat capacity of $c = 815$ J/(kg · C°), and a temperature of 12.0 °C. It is immersed in 119 g of water, and the final temperature of the water and thermometer is 41.5 °C. What was the temperature of the water before the insertion of the thermometer?

39. **E** **SSM** Blood can carry excess energy from the interior to the surface of the body, where the energy is dispersed in a number of ways. While a person is exercising, 0.6 kg of blood flows to the body's surface and releases 2000 J of energy. The blood arriving at the surface has the temperature of the body's interior, 37.0 °C. Assuming that blood has the same specific heat capacity as water, determine the temperature of the blood that leaves the surface and returns to the interior.

40. **E** An ice chest at a beach party contains 12 cans of soda at 5.0 °C. Each can of soda has a mass of 0.35 kg and a specific heat capacity of 3800 J/(kg · C°). Someone adds a 6.5-kg watermelon at 27 °C to the chest. The specific heat capacity of watermelon is nearly the same as that of water. Ignore the specific heat capacity of the chest and determine the final temperature T of the soda and watermelon.

41. **E** A piece of glass has a temperature of 83.0 °C. Liquid that has a temperature of 43.0 °C is poured over the glass, completely covering it, and the temperature at equilibrium is 53.0 °C. The mass of the glass and the liquid is the same. Ignoring the container that holds the glass and liquid and assuming that the heat lost to or gained from the surroundings is negligible, determine the specific heat capacity of the liquid.

42. **E** **GO** Two bars of identical mass are at 25 °C. One is made from glass and the other from another substance. The specific heat capacity of glass is 840 J/(kg · C°). When identical amounts of heat are supplied to each, the glass bar reaches a temperature of 88 °C, while the other bar reaches 250.0 °C. What is the specific heat capacity of the other substance?

43. **E** **SSM** At a fabrication plant, a hot metal forging has a mass of 75 kg and a specific heat capacity of 430 J/(kg · C°). To harden it, the forging is immersed in 710 kg of oil that has a temperature of 32 °C and a specific heat capacity of 2700 J/(kg · C°). The final temperature of the oil and forging at thermal equilibrium is 47 °C. Assuming that heat flows only between the forging and the oil, determine the initial temperature of the forging.

44. **M** **MMH** **CHALK** A 0.35-kg coffee mug is made from a material that has a specific heat capacity of 920 J/(kg · C°) and contains 0.25 kg of water. The cup and water are at 15 °C. To make a cup of coffee, a small electric heater is immersed in the water and brings it to a boil in three minutes. Assume that the cup and water always have the same temperature and determine the minimum power rating of this heater.

45. **M** **GO** Three portions of the same liquid are mixed in a container that prevents the exchange of heat with the environment. Portion A has a mass m and a temperature of 94.0 °C, portion B also has a mass m but a temperature of 78.0 °C, and portion C has a mass m_C and a temperature of 34.0 °C. What must be the mass of portion C so that the final temperature T_f of the three-portion mixture is $T_f = 50.0$ °C? Express your answer in terms of m; for example, $m_C = 2.20\ m$.

46. **M** **V-HINT** The heating element of a water heater in an apartment building has a maximum power output of 28 kW. Four residents of the building take showers at the same time, and each receives heated water at a volume flow rate of 14×10^{-5} m³/s. If the water going into the heater has a temperature of 11 °C, what is the maximum possible temperature of the hot water that each showering resident receives?

47. **M** **SSM** A rock of mass 0.20 kg falls from rest from a height of 15 m into a pail containing 0.35 kg of water. The rock and water have the same initial temperature. The specific heat capacity of the rock is 1840 J/(kg · C°). Ignore the heat absorbed by the pail itself, and determine the rise in the temperature of the rock and water.

48. **H** A steel rod ($\rho = 7860$ kg/m³) has a length of 2.0 m. It is bolted at both ends between immobile supports. Initially there is no tension in the rod, because the rod just fits between the supports. Find the tension that develops when the rod loses 3300 J of heat.

Section 12.8 Heat and Phase Change: Latent Heat

49. **E** **SSM** How much heat must be added to 0.45 kg of aluminum to change it from a solid at 130 °C to a liquid at 660 °C (its melting point)? The latent heat of fusion for aluminum is 4.0×10^5 J/kg.

50. **E** Suppose that the amount of heat removed when 3.0 kg of water freezes at 0.0 °C were removed from ethyl alcohol at its freezing/melting point of −114.4 °C. How many kilograms of ethyl alcohol would freeze?

51. **E** To help prevent frost damage, fruit growers sometimes protect their crop by spraying it with water when overnight temperatures are expected to go below freezing. When the water turns to ice during the night, heat is released into the plants, thereby giving a measure of protection against the cold. Suppose a grower sprays 7.2 kg of water at 0 °C onto a fruit tree. **(a)** How much heat is released by the water when it freezes? **(b)** How much would the temperature of a 180-kg tree rise if it absorbed the heat released in part (a)? Assume that the specific heat capacity of the tree is 2.5×10^3 J/(kg · C°) and that no phase change occurs within the tree itself.

52. **E** **GO** **(a)** Objects A and B have the same mass of 3.0 kg. They melt when 3.0×10^4 J of heat is added to object A and when 9.0×10^4 J

is added to object B. Determine the latent heat of fusion for the substance from which each object is made. **(b)** Find the heat required to melt object A when its mass is 6.0 kg.

53. **E** **SSM** Find the mass of water that vaporizes when 2.10 kg of mercury at 205 °C is added to 0.110 kg of water at 80.0 °C.

54. **E** A mass $m = 0.054$ kg of benzene vapor at its boiling point of 80.1 °C is to be condensed by mixing the vapor with water at 41 °C. What is the minimum mass of water required to condense all of the benzene vapor? Assume that the mixing and condensation take place in a perfectly insulating container.

55. **E** A certain quantity of steam has a temperature of 100.0 °C. To convert this steam into ice at 0.0 °C, energy in the form of heat must be removed from the steam. If this amount of energy were used to accelerate the ice from rest, what would be the linear speed of the ice? For comparison, bullet speeds of about 700 m/s are common.

56. **E** A thermos contains 150 cm³ of coffee at 85 °C. To cool the coffee, you drop two 11-g ice cubes into the thermos. The ice cubes are initially at 0 °C and melt completely. What is the final temperature of the coffee? Treat the coffee as if it were water.

57. **M** **V-HINT** A snow maker at a resort pumps 130 kg of lake water per minute and sprays it into the air above a ski run. The water droplets freeze in the air and fall to the ground, forming a layer of snow. If all the water pumped into the air turns to snow, and the snow cools to the ambient air temperature of −7.0 °C, how much heat does the snow-making process release each minute? Assume that the temperature of the lake water is 12.0 °C, and use 2.00×10^3 J/(kg · C°) for the specific heat capacity of snow.

58. **M** **SSM** **CHALK** **MMH** A 42-kg block of ice at 0 °C is sliding on a horizontal surface. The initial speed of the ice is 7.3 m/s and the final speed is 3.5 m/s. Assume that the part of the block that melts has a very small mass and that all the heat generated by kinetic friction goes into the block of ice. Determine the mass of ice that melts into water at 0 °C.

59. **M** **GO** Water at 23.0 °C is sprayed onto 0.180 kg of molten gold at 1063 °C (its melting point). The water boils away, forming steam at 100.0 °C and leaving solid gold at 1063 °C. What is the minimum mass of water that must be used?

60. **M** **SSM** An unknown material has a normal melting/freezing point of −25.0 °C, and the liquid phase has a specific heat capacity of 160 J/(kg · C°). One-tenth of a kilogram of the solid at −25.0 °C is put into a 0.150-kg aluminum calorimeter cup that contains 0.100 kg of glycerin. The temperature of the cup and the glycerin is initially 27.0 °C. All the unknown material melts, and the final temperature at equilibrium is 20.0 °C. The calorimeter neither loses energy to nor gains energy from the external environment. What is the latent heat of fusion of the unknown material?

61. **M** **GO** When it rains, water vapor in the air condenses into liquid water, and energy is released. **(a)** How much energy is released when 0.0254 m (one inch) of rain falls over an area of 2.59×10^6 m² (one square mile)? **(b)** If the average energy needed to heat one home for a year is 1.50×10^{11} J, how many homes could be heated for a year with the energy determined in part (a)?

62. **M** **SSM** It is claimed that if a lead bullet goes fast enough, it can melt completely when it comes to a halt suddenly, and all its kinetic energy is converted into heat via friction. Find the minimum speed of a lead bullet (initial temperature = 30.0 °C) for such an event to happen.

63. **M** **V-HINT** Equal masses of two different liquids have the same temperature of 25.0 °C. Liquid A has a freezing point of −68.0 °C and a specific heat capacity of 1850 J/(kg · C°). Liquid B has a freezing point of −96.0 °C and a specific heat capacity of 2670 J/(kg · C°). The same amount of heat must be removed from each liquid in order to freeze it into a solid at its respective freezing point. Determine the difference $L_{f,A} - L_{f,B}$ between the latent heats of fusion for these liquids.

64. **M** **CHALK** Occasionally, huge icebergs are found floating on the ocean's currents. Suppose one such iceberg is 120 km long, 35 km wide, and 230 m thick. **(a)** How much heat would be required to melt this iceberg (assumed to be at 0 °C) into liquid water at 0 °C? The density of ice is 917 kg/m³. **(b)** The annual energy consumption by the United States is about 1.1×10^{20} J. If this energy were delivered to the iceberg every year, how many years would it take before the ice melted?

Section 12.9 Equilibrium Between Phases of Matter,

Section 12.10 Humidity

65. **E** **CHALK** What atmospheric pressure would be required for carbon dioxide to boil at a temperature of 20 °C? See the vapor pressure curve for carbon dioxide in the drawing.

PROBLEM 65

66. **E** **SSM** **CHALK** At a temperature of 10 °C the percent relative humidity is R_{10}, and at 40 °C it is R_{40}. At each of these temperatures the partial pressure of water vapor in the air is the same. Using the

vapor pressure curve for water that accompanies this problem, determine the ratio R_{10}/R_{40} of the two humidity values.

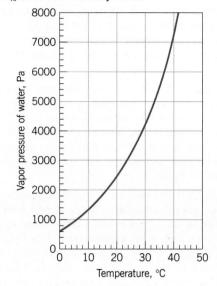

PROBLEM 66

67. **E** What is the relative humidity on a day when the temperature is 30 °C and the dew point is 10 °C? Use the vapor pressure curve that accompanies Problem 66.

68. **E** **GO** The vapor pressure of water at 10 °C is 1300 Pa. **(a)** What percentage of atmospheric pressure is this? Take atmospheric pressure to be 1.013×10^5 Pa. **(b)** What percentage of the total air pressure at 10 °C is due to water vapor when the relative humidity is 100%? **(c)** The vapor pressure of water at 35 °C is 5500 Pa. What is the relative humidity at this temperature if the partial pressure of water in the air has not changed from what it was at 10 °C when the relative humidity was 100%?

69. **E** **GO** The temperature of 2.0 kg of water is 100.0 °C, but the water is not boiling, because the external pressure acting on the water surface is 3.0×10^5 Pa. Using the vapor pressure curve for water given in Figure 12.31, determine the amount of heat that must be added to the water to bring it to the point where it just begins to boil.

Additional Problems

Online

70. **E** An aluminum baseball bat has a length of 0.86 m at a temperature of 17 °C. When the temperature of the bat is raised, the bat lengthens by 0.000 16 m. Determine the final temperature of the bat.

71. **E** A 0.200-kg piece of aluminum that has a temperature of −155 °C is added to 1.5 kg of water that has a temperature of 3.0 °C. At equilibrium the temperature is 0.0 °C. Ignoring the container and assuming that the heat exchanged with the surroundings is negligible, determine the mass of water that has been frozen into ice.

72. **E** A thick, vertical iron pipe has an inner diameter of 0.065 m. A thin aluminum disk, heated to a temperature of 85 °C, has a diameter that is 3.9×10^{-5} m greater than the pipe's inner diameter. The disk is laid on top of the open upper end of the pipe, perfectly centered on it, and allowed to cool. What is the temperature of the aluminum disk when the disk falls into the pipe? Ignore the temperature change of the pipe.

73. **E** **SSM** A lead object and a quartz object each have the same initial volume. The volume of each increases by the same amount, because the temperature increases. If the temperature of the lead object increases by 4.0 C°, by how much does the temperature of the quartz object increase?

74. **E** **GO** If the price of electrical energy is $0.10 per kilowatt · hour, what is the cost of using electrical energy to heat the water in a swimming pool (12.0 m × 9.00 m × 1.5 m) from 15 to 27 °C?

75. **M** Concrete sidewalks are always laid in sections, with gaps between each section. For example, the drawing shows three identical 2.4-m sections, the outer two of which are against immovable walls. The two identical gaps between the sections are provided so that thermal expansion will not create the thermal stress that could lead to cracks. What is the minimum gap width necessary to account for an increase in temperature of 32 C°?

PROBLEM 75

76. Ⓜ Ⓖⓞ A constant-volume gas thermometer (see Figures 12.3 and 12.4) has a pressure of 5.00×10^3 Pa when the gas temperature is 0.00 °C. What is the temperature (in °C) when the pressure is 2.00×10^3 Pa?

77. Ⓜ When 4200 J of heat are added to a 0.15-m-long silver bar, its length increases by 4.3×10^{-3} m. What is the mass of the bar?

78. Ⓗ A wire is made by attaching two segments together, end to end. One segment is made of aluminum and the other is steel. The effective coefficient of linear expansion of the two-segment wire is 19×10^{-6} (C°)$^{-1}$. What fraction of the length is aluminum?

79. Ⓗ An insulated container is partly filled with oil. The lid of the container is removed, 0.125 kg of water heated to 90.0 °C is poured in, and the lid is replaced. As the water and the oil reach equilibrium, the volume of the oil increases by 1.20×10^{-5} m³. The density of the oil is 924 kg/m³, its specific heat capacity is 1970 J/(kg · C°), and its coefficient of volume expansion is 721×10^{-6} (C°)$^{-1}$. What is the temperature when the oil and the water reach equilibrium?

80. Ⓗ A steel bicycle wheel (without the rubber tire) is rotating freely with an angular speed of 18.00 rad/s. The temperature of the wheel changes from −100.0 to +300.0 °C. No net external torque acts on the wheel, and the mass of the spokes is negligible. **(a)** Does the angular speed increase or decrease as the wheel heats up? Why? **(b)** What is the angular speed at the higher temperature?

81. Ⓜ Ⓖⓞ Ⓢⓢⓜ The figure shows a swimming pool on a sunny day. If the water absorbs 2.00×10^9 J of heat from the sun, what is the change in volume of the water?

PROBLEM 81

Physics in Biology, Medicine, and Sports

82. Ⓔ Ⓑⓘⓞ Ⓢⓢⓜ **12.2** Dermatologists often remove small precancerous skin lesions by freezing them quickly with liquid nitrogen, which has a temperature of 77 K. What is this temperature on the **(a)** Celsius and **(b)** Fahrenheit scales?

83. Ⓔ Ⓑⓘⓞ Ⓜⓜⓗ **12.7** When resting, a person has a metabolic rate of about 3.0×10^5 joules per hour. The person is submerged neck-deep into a tub containing 1.2×10^3 kg of water at 21.00 °C. If the heat from the person goes only into the water, find the water temperature after half an hour.

84. Ⓔ Ⓑⓘⓞ Ⓖⓞ **12.6** When you drink cold water, your body must expend metabolic energy in order to maintain normal body temperature (37 °C) by warming up the water in your stomach. Could drinking ice water, then, substitute for exercise as a way to "burn calories?" Suppose you expend 430 kilocalories during a brisk hour-long walk. How many liters of ice water (0 °C) would you have to drink in order to use up 430 kilocalories of metabolic energy? For comparison, the stomach can hold about 1 liter.

85. Ⓜ Ⓑⓘⓞ Ⓜⓜⓗ **12.6** One ounce of a well-known breakfast cereal contains 110 Calories (1 food Calorie = 4186 J). If 2.0% of this energy could be converted by a weight lifter's body into work done in lifting a barbell, what is the heaviest barbell that could be lifted a distance of 2.1 m?

86. Ⓔ Ⓑⓘⓞ **12.8** The latent heat of vaporization of H_2O at body temperature (37.0 °C) is 2.42×10^6 J/kg. To cool the body of a 75-kg jogger [average specific heat capacity = 3500 J/(kg · C°)] by 1.5 C°, how many kilograms of water in the form of sweat have to be evaporated?

87. Ⓔ Ⓑⓘⓞ **12.10** Suppose that air in the human lungs has a temperature of 37 °C, and the partial pressure of water vapor has a value of 5.5×10^3 Pa. What is the relative humidity in the lungs? Consult the vapor pressure curve for water that accompanies Problem 75.

88. Ⓔ Ⓑⓘⓞ **12.7** A person eats a container of strawberry yogurt. The Nutritional Facts label states that it contains 240 Calories (1 Calorie = 4186 J). What mass of perspiration would one have to lose to get rid of this energy? At body temperature, the latent heat of vaporization of water is 2.42×10^6 J/kg.

89. Ⓜ Ⓑⓘⓞ **12.8** The human brain accounts for 1/5 of the body's total power output, with an average power of 20.0 W. How long, in hours, would it take the heat produced by the brain to vaporize a cup of water with an initial temperature of 25.0 °C into steam at 1.00×10^2 °C? Assume the mass of the water in the cup is 237 g and all of the heat produced by the brain is transferred to the water.

90. Ⓔ Ⓑⓘⓞ **12.2** Since 1871, the normal internal body temperature for human beings has been 98.6 °F. This average was based on over a million measurements analyzed by a German physician, Carl Reinhold August Wunderlich. However, in 2020 researchers at Stanford University published a report claiming normal body temperature has dropped since the Industrial Revolution. Analyzing almost 700 000 temperature readings from 1862 to 2017, the researchers found the average body temperature has decreased by 0.03 °C per decade. The new average body temperature was found to be 97.9 °F. What is this temperature in **(a)** Celsius, and **(b)** Kelvin?

91. Ⓜ Ⓑⓘⓞ **12.6** The Peloton spin cycle is an upscale stationary bike that you can purchase to use in your home. It features a 22-inch sweat-proof screen, where the rider interacts with a live instructor, effectively recreating a studio spin class at home. The Peloton is great for improving cardiovascular fitness. If the Peloton instructor maintains an average power output of 435 W for 30.0 minutes, how many Calories does the instructor burn during this time?

92. Ⓜ Ⓑⓘⓞ **12.7** Imagine a physics lecture hall with 100 students who are settling in for a 1-hour lecture. At the start of the lecture, the temperature of the air in the room is a comfortable 21.0 °C. Unfortunately, the room's air conditioner breaks right as the lecture begins. Each student has an average power output of about 60.0 W at room temperature. Imagine the energy released by each student goes into heating just the air in the room, which has a volume of 9.50×10^2 m³ and a density of 1.20 kg/m³. Assume the volume of the air remains constant and the specific heat capacity of the air is 718 J/(kg · °C). Calculate the room's temperature at the end of the lecture. In reality, a significant portion of the heat produced would be absorbed by the walls, ceiling, floors, chairs, desks, and so on, which we are neglecting.

93. **M** **BIO** **12.7** Fever, or pyrexia, is the body's natural response to infection, which leads to an increase in the body's setpoint temperature. This was one of the most common symptoms experienced by people infected with the COVID-19 virus during the pandemic in 2020. Normal body temperature was recently established to be 97.9 °F (36.6 °C), and during a high fever, the body temperature may rise to 103 °F (39.4 °C) or more. This heat is produced by the body ramping up its metabolic rate and burning more Calories. How many Calories would a 75.0-kg person have to consume in order to raise their temperature from 36.6 °C to 39.4 °C ? Take the average specific heat capacity of the human body to be 3470 J/(kg · C°).

Concepts and Calculations Problems

Online

Problem 94 provides insight on the variables involved when the length and volume of an object change due to a temperature change. Problem 95 discusses how different factors affect the temperature change of an object to which heat is being added.

94. **M** **CHALK** The figure shows three rectangular blocks made from the same material. The initial dimensions of each are expressed as multiples of D, where $D = 2.00$ cm. The blocks are heated and their temperatures increase by 35.0 C°. The coefficients of linear and volume expansion are $\alpha = 1.50 \times 10^{-5}$ (C°)$^{-1}$ and $\beta = 4.50 \times 10^{-5}$ (C°)$^{-1}$, respectively. *Concepts:* (i) Does the change in the vertical height of a block depend only on its height, or does it also depend on its width and depth? Without doing any calculations, rank the blocks according to their change in height, largest first. (ii) Does the change in the volume of a block depend only on its height, or does it also depend on its width and depth? Without doing any calculations, rank the blocks according to their change in volume, largest first. *Calculations:* Determine the change in the (a) vertical heights and (b) volumes of the blocks.

PROBLEM 94

95. **M** **CHALK** **SSM** Objects A and B in the figure are made from copper, but the mass of object B is three times that of object A. Object C is made from glass and has the same mass as object B. The *same amount* of heat Q is supplied to each one: $Q = 14$ J. *Concepts:* (i) Which object, A or B, experiences a greater rise in temperature, and why? (ii) Which object, B or C, experiences a greater rise in temperature, and why? *Calculations:* Determine the rise in temperature of each block.

PROBLEM 95

Team Problems

Online

96. **M** **T** **WS** **Rule of Mixtures.** Two barrels each contain a liquid that is a mixture of water and acetone, a clear reagent that mixes well with water. One barrel contains 10.0% water and 90.0% acetone by volume, and the other 10.0% acetone and 90.0% water by volume. The problem is that you do not know which is which and, although acetone has a distinct odor, you cannot tell them apart by smelling them. Although the density of acetone ($\rho_A = 784$ kg/m³) is less than that of water ($\rho_W = 1.00 \times 10^3$ kg/m³), you do not have a scale with which to measure their weights. However, you have a 1000-mL beaker marked with lines every 100 mL, a 1400-mL calorimeter vessel, a small freezer, a set of brass masses from an old balance scale, and a thermometer. You look up the specific heat capacities and find that they are very different for water ($c_{p,w} = 4128$ J/kg · C°) and acetone ($c_{p,A} = 2160$ J/kg · C°), and devise an idea to take advantage of that difference in order to identify the mixtures. You and your team plan to use the "rule of mixtures," which states that the effective specific heat capacity of a mixture is the weighted average of the specific heats of the components, that is,

$$c_{p,T} = \left(\frac{m_1}{m_T}\right)c_{p1} + \left(\frac{m_2}{m_T}\right)c_{p2}$$

where $c_{p,T}$ is the effective specific heat of the total mixture, c_{p1} and c_{p2} are the specific heats of the two liquid components, m_1 and m_2 are the respective masses of the components, and m_T is the mass of the entire mixture (i.e., $m_T = m_1 + m_2$). You put a 1.00×10^3-g brass mass ($c_{p,B} = 920.0$ J/kg · C°) in the freezer ($T = -11.0$ °C) and wait for it to equilibrate. Next, using the beaker, you measure out 1.00×10^3 mL of one of the mixtures and pour it into the calorimeter. The mixtures both start at room temperature (28.2 °C). Finally, you take the brass weight out of the freezer, submerge it in the liquid in the calorimeter, close the lid, and insert the thermometer into the liquid through a hole in the lid. You wait for the temperature to equilibrate, record the temperature, and then repeat the process for the other mixture. **(a)** Which mixture will equilibrate at a lower temperature? Explain your answer. **(b)** What are the respective masses of acetone and water in 1.00×10^3 mL of each mixture, and what are the total masses of each mixture? **(c)** What are the effective specific heats of each mixture? **(d)** Calculate the expected final temperature of each mixture using this procedure. The calorimeter vessel is made of Styrofoam, and you can neglect any heat transfer to or from the vessel.

97. **E** **T** **WS** **A Giant Frisbee.** A solid copper disk of radius 10.50 m is launched toward the sun like a gigantic Frisbee with a constant linear velocity of 350 m/s and a rotation rate of exactly 438.7 rpm about its central axis. At the time of launch, the disk was very cold (-191.1 °C). As it approaches the sun, however, it will get very hot. You and your team have been brought in to determine whether the increased temperature will affect the motion of the disk. **(a)** Will the rate of rotation be affected when the disk becomes hot? If so, how and why? **(b)** What will be the rate of rotation (in rad/s) when the copper disk reaches a temperature of 785.9 °C? There are no external torques acting on the disk.

98. **M** **T** **A Crude Thermometer.** You and your team are given the task of constructing a crude thermometer that covers a temperature range from 0 °C to 60.0 °C. You have at your disposal an aluminum rod of length $L_0 = 2.00$ m (at $T = 0$ °C) and diameter $D = 0.500$ cm, and a broken clock that is missing its hour hand, and its longer minute hand is hanging loose and pointing in the six o'clock direction. The long hand of the clock is 8.20 inches long and pivots loosely at the clock's center. You get the idea to mount the rod horizontally so that one end butts against a wall and the other end pushes against the dangling minute hand of the clock. A temperature-induced change in the length of the rod will then be reflected in a change in the angle that the minute hand makes with the vertical (i.e., relative to the six o'clock position). **(a)** If you want the full temperature range to span

the angle between the 6 and 7 markings (uniformly spaced) on the clock, how far from the central pivot should the end of the rod make contact with the minute hand? **(b)** The current temperature in the room is 65.0 °F. At what angle relative to the vertical (the six o'clock position) should the minute hand point at this temperature if it is to point directly at 6 when $T = 0$ °C? **(c)** What would be the angular range of your "clock-thermometer" if the rod were made of steel, rather than aluminum? Assume that it is placed at the same position on the minute hand as determined in (a).

99. **M** **T** **A Submerged Machine.** You and your team are testing a device that is to be submerged in the cold waters of Antarctica. It is designed to rotate a small wheel at a very precise rate. One component of the device consists of a steel wheel (diameter $D_{steel} = 3.0000$ cm at $T = 78.400$ °F) and a large aluminum wheel that drives it (diameter $D_{Al} = 150.00$ cm at $T = 78.400$ °F). **(a)** If the aluminum wheel rotates at 35.000 RPM, at what rate does the smaller, steel wheel turn while both wheels are at $T = 78.400$ °F? **(b)** You submerge the device in a large vat of water held at 32.800 °C, simulating its working environment in the Antarctic seas. Assuming the larger, aluminum driving wheel still rotates at 35.000 RPM, at what rate does the smaller, steel wheel rotate when the device is submerged in the cold water? **(c)** How should you adjust the rate of the aluminum driving wheel so that the steel wheel rotates at the same rate as it had at $T = 78.400$ °F? Note: $\alpha_{Al} = 23 \times 10^{-6}$ and $\alpha_{steel} = 12 \times 10^{-6}$.

33.4°C

32

30

28

26

24

22

21.6°C

Wikimedia Commons

All objects with a temperature above absolute zero (−273.15 °C) emit infrared or thermal radiation. The image above, known as a thermal image, or thermogram, is made with a special camera that is sensitive to this type of radiation. A colored scale represents the relative temperature of the objects in the image. Warmer objects, such as people, most animals, and other sources of heat, stand out in contrast to the cooler background and colder objects. The image above shows a snake wrapped around a human hand and wrist. The snake, being a coldblooded animal, matches its temperature to its environment and appears much cooler than the warm hand that is holding it. In this chapter, we will study radiation, as well as the other processes of heat transfer.

LEARNING OBJECTIVES

After reading this module, you should be able to...

13.1 Define convection.

13.2 Solve conduction problems.

13.3 Solve radiation problems.

13.4 Analyze heat transfer applications.

The Transfer of Heat

13.1 Convection

When heat is transferred to or from a substance, the internal energy of the substance can change, as we saw in Chapter 12. This change in internal energy is accompanied by a change in temperature or a change in phase. The transfer of heat affects us in many ways. For instance, within our homes, furnaces distribute heat on cold days, and air conditioners remove it on hot days. Our bodies constantly transfer heat in one direction or another, to prevent the adverse effects of hypo- and hyperthermia. And virtually all our energy originates in the sun and is transferred to us over a distance of 150 million kilometers through the void of space. Today's sunlight provides the energy to drive photosynthesis in the plants that provide our food and, hence, metabolic energy. Ancient sunlight nurtured the organic matter that became the fossil fuels of oil, natural gas, and coal. This chapter examines the three processes by which heat is transferred: convection, conduction, and radiation.

When part of a fluid is warmed, such as the air above a fire, the volume of that part of the fluid expands, and the density decreases.

FIGURE 13.1 A plume of smoke originates from an oil burn near the site of the Deepwater Horizon oil spill disaster in the Gulf of Mexico in 2010. The plume rises hundreds of meters into the air because of convection.

FIGURE 13.2 Convection currents are set up when a pot of water is heated.

Charlie Neibergall/AP Images

According to Archimedes' principle (see Section 11.6), the surrounding cooler and denser fluid exerts a buoyant force on the warmer fluid and pushes it upward. As warmer fluid rises, the surrounding cooler fluid replaces it. This cooler fluid, in turn, is warmed and pushed upward. Thus, a continuous flow is established, which carries along heat. Whenever heat is transferred by the bulk movement of a gas or a liquid, the heat is said to be transferred by **convection**. The fluid flow itself is called a **convection current**.

> **CONVECTION**
>
> **Convection is the process in which heat is carried from place to place by the bulk movement of a fluid.**

The smoke rising from a fire, like the one in **Figure 13.1**, is one visible result of convection. **Figure 13.2** shows the less visible example of convection currents in a pot of water being heated on a gas burner. The currents distribute the heat from the burning gas to all parts of the water. Conceptual Example 1 deals with some of the important roles that convection plays in the home.

CONCEPTUAL EXAMPLE 1 | The Physics of Heating and Cooling by Convection

Hot water baseboard heating units are frequently used in homes, and a cooling coil is a major component of a refrigerator. The locations of these heating and cooling devices are different because each is designed to maximize the production of convection currents. Where should the heating unit and the cooling coil be located? **(a)** Heating unit near the floor of the room and cooling coil near the top of the refrigerator **(b)** Heating unit near the ceiling of the room and cooling coil near the bottom of the refrigerator

Reasoning An important goal for the heating system is to distribute heat throughout a room. The analogous goal for the cooling coil is to remove heat from all of the space within a refrigerator. In each case, the heating or cooling device must be positioned so that convection makes the goal achievable.

Answer (b) is incorrect. If the heating unit were placed near the ceiling of the room, warm air from the unit would remain there, because warm air does not fall (it rises). Thus, there would be very little natural movement (or convection) of air to distribute the heat throughout the room. If the cooling coil were located near the bottom of the refrigerator, the cool air would remain there, because cool air does not rise (it sinks). There would be very little convection to carry the heat from other parts of the refrigerator to the coil for removal.

Answer (a) is correct. The air above the baseboard unit is heated, like the air above a fire. Buoyant forces from the surrounding cooler air push the warm air upward. Cooler air near the ceiling is displaced downward and then warmed by the baseboard heating unit, causing the convection current illustrated in **Animated Figure 13.3a**. Within the refrigerator, air in contact with the top-mounted coil is cooled, its volume decreases, and its density increases. The surrounding warmer and less dense air cannot provide sufficient buoyant force to support the cooler air, which sinks downward. In the process, warmer air near the bottom of the refrigerator is displaced upward and is then cooled by the coil, establishing the convection current shown in **Animated Figure 13.3b**.

ANIMATED FIGURE 13.3 (*a*) Air warmed by the baseboard heating unit is pushed to the top of the room by the cooler and denser air. (*b*) Air cooled by the cooling coil sinks to the bottom of the refrigerator. In both (*a*) and (*b*) a convection current is established.

FIGURE 13.4 Updrafts, or thermals, are caused by the convective movement of air that the ground has warmed.

THE PHYSICS OF . . . thermals. Another example of convection occurs when the ground, heated by the sun's rays, warms the neighboring air. Surrounding cooler and denser air pushes the heated air upward. The resulting updraft or "thermal" can be quite strong, depending on the amount of heat that the ground can supply. As **Figure 13.4** illustrates, these thermals can be used by glider pilots to gain considerable altitude. Birds such as eagles utilize thermals in a similar fashion.

THE PHYSICS OF . . . an inversion layer. Sometimes meteorological conditions cause a layer to form in the atmosphere where the temperature increases with increasing altitude. Such a layer is called an **inversion layer** because its temperature profile is inverted compared to the usual situation, wherein the air temperature decreases with increasing altitude. In the usual situation, upward convection currents occur and are important for dispersing pollutants from industrial sources and automobile exhaust systems. An inversion layer, in contrast, arrests the normal upward convection currents, causing a stagnant-air condition in which the concentration of pollutants increases substantially. This condition leads to a smog layer that can often be seen hovering over large cities.

We have been discussing **natural convection**, in which a temperature difference causes the density at one place in a fluid to be different from the density at another. Sometimes, natural convection is inadequate to transfer sufficient amounts of heat. In such cases **forced convection** is often used, and an external device such as a pump or a fan mixes the warmer and cooler portions of the fluid.

BIO **THE PHYSICS OF . . . rapid thermal exchange.** **Figure 13.5** shows an application of forced convection that is revolutionizing the way in which the effects of overheating are being treated. Athletes, for example, are especially prone to overheating, and the device illustrated in **Figure 13.5** is appearing more and more frequently at athletic events. The technique is known as rapid thermal exchange and takes advantage of specialized blood vessels called arteriovenous anastomoses (AVAs) that are found in the palms of the hands (and soles of the feet). These blood vessels are used to help dissipate unwanted heat from the body. The device in the drawing consists of a small chamber containing a curved metal plate, through which cool water is circulated from a refrigerated supply. The overheated athlete inserts his hand into the chamber and places his palm on the plate. The chamber seals around the wrist and is evacuated slightly to reduce the air pressure and thereby promote circulation of blood through the hand. Forced convection plays two roles in this treatment. It causes the water to circulate through the metal plate and remove heat from the blood in the AVAs. Also, the cooled blood returns through veins to the heart, which pumps it throughout the body, thus lowering the body temperature and relieving the effects of overheating.

FIGURE 13.5 An overheated athlete uses a rapid-thermal-exchange device to cool down. He places the palm of his hand on a curved metal plate within a slightly evacuated chamber. Forced convection circulates cool water through the plate, which cools the blood flowing through the hand. The cooled blood returns through veins to the heart, which circulates it throughout the body.

FIGURE 13.6 The forced convection generated by a pump circulates radiator fluid through an automobile engine to remove excess heat.

Figure 13.6 shows the application of forced convection in an automobile engine. As in the previous application, forced convection occurs in two ways. First, a pump circulates radiator fluid (water and antifreeze) through the engine to remove excess heat from the combustion process. Second, a radiator fan draws air through the radiator. Heat is transferred from the hotter radiator fluid to the cooler air, thereby cooling the fluid.

BIO THE PHYSICS OF . . . the windchill factor. Forced convection also plays the principal role in the windchill factor that is often mentioned in weather reports. The wind mixes the cold ambient air with the warm layer of air that immediately surrounds the exposed portions of your body. The forced convection removes heat from exposed body surfaces, thereby making you feel colder than you would if there were no wind.

Check Your Understanding

(*The answer is given at the end of the book.*)

1. The transfer of heat by convection is *smallest* in (a) solids, (b) liquids, (c) gases.

13.2 Conduction

Anyone who has fried a hamburger in an all-metal skillet knows that the metal handle becomes hot. Somehow, heat is transferred from the burner to the handle. Clearly, heat is not being transferred by the bulk movement of the metal or the surrounding air, so convection can be ruled out. Instead, heat is transferred directly through the metal by a process called **conduction**.

CONDUCTION
Conduction is the process whereby heat is transferred directly through a material, with any bulk motion of the material playing no role in the transfer.

One mechanism for conduction occurs when the atoms or molecules in a hotter part of the material vibrate or move with greater energy than those in a cooler part. By means of collisions, the more energetic molecules pass on some of their energy to their less energetic neighbors. For example, imagine a gas filling the space between two walls that face each other and are maintained at different temperatures. Molecules strike the hotter wall, absorb energy from it, and rebound with a greater kinetic energy than when they arrived. As these more energetic molecules collide with their less energetic neighbors, they transfer some of their energy to them. Eventually, this energy is passed on until it reaches the molecules next to the cooler wall. These molecules, in turn, collide with the wall, giving up some of their energy to it in the process. Through such molecular collisions, heat is conducted from the hotter to the cooler wall.

A similar mechanism for the conduction of heat occurs in metals. Metals are different from most substances in having a pool of electrons that are more or less free to wander throughout the metal. These free electrons can transport energy and allow metals to transfer heat very well. The free electrons are also responsible for the excellent electrical conductivity that metals have.

Those materials that conduct heat well are called **thermal conductors**, and those that conduct heat poorly are known as **thermal insulators**. Most metals are excellent thermal conductors; wood, glass, and most plastics are common thermal insulators. Thermal insulators have many important applications. Virtually all new housing construction incorporates thermal insulation in attics and walls to reduce heating and cooling costs. And the wooden or plastic handles on many pots and pans reduce the flow of heat to the cook's hand.

INTERACTIVE FIGURE 13.7 Heat is conducted through the bar when the ends of the bar are maintained at different temperatures. The heat flows from the warmer to the cooler end.

To illustrate the factors that influence the conduction of heat, **Interactive Figure 13.7** displays a rectangular bar. The ends of the bar are in thermal contact with two bodies, one of which is kept at a constant higher temperature, while the other is kept at a constant lower temperature. Although not shown for the sake of clarity, the sides of the bar are insulated, so the heat lost through them is negligible. The amount of heat Q conducted through the bar from the warmer end to the cooler end depends on a number of factors:

1. Q is proportional to the time t during which conduction takes place ($Q \propto t$). More heat flows in longer time periods.

2. Q is proportional to the temperature difference ΔT between the ends of the bar ($Q \propto \Delta T$). A larger difference causes more heat to flow. No heat flows when both ends have the same temperature and $\Delta T = 0$ C°.

3. Q is proportional to the cross-sectional area A of the bar ($Q \propto A$). **Interactive Figure 13.8** helps to explain this fact by showing two identical bars (insulated sides not shown) placed between the warmer and cooler bodies. Clearly, twice as much heat flows through two bars as through one, because the cross-sectional area has been doubled.

4. Q is inversely proportional to the length L of the bar ($Q \propto 1/L$). Greater lengths of material conduct less heat. To experience this effect, put two insulated mittens (the

INTERACTIVE FIGURE 13.8 Twice as much heat flows through two identical bars as through one.

pot holders that cooks keep near the stove) on the *same hand*. Then, touch a hot pot and notice that it feels cooler than when you wear only one mitten, signifying that less heat passes through the greater thickness ("length") of material.

These proportionalities can be stated together as $Q \propto (A\,\Delta T)t/L$. Equation 13.1 expresses this result with the aid of a proportionality constant k, which is called the **thermal conductivity**.

> **CONDUCTION OF HEAT THROUGH A MATERIAL**
> The heat Q conducted during a time t through a bar of length L and cross-sectional area A is
>
> $$Q = \frac{(kA\,\Delta T)t}{L} \qquad (13.1)$$
>
> where ΔT is the temperature difference between the ends of the bar (the higher temperature minus the lower temperature) and k is the thermal conductivity of the material.
>
> **SI Unit of Thermal Conductivity:** $J/(s \cdot m \cdot C°)$

Since $k = QL/(tA\,\Delta T)$, the SI unit for thermal conductivity is $J \cdot m/(s \cdot m^2 \cdot C°)$ or $J/(s \cdot m \cdot C°)$. The SI unit of power is the joule per second (J/s), or watt (W), so the thermal conductivity is also given in units of $W/(m \cdot C°)$.

Different materials have different thermal conductivities, and **Table 13.1** gives some representative values. Because metals are such good thermal conductors, they have large thermal conductivities. In comparison, liquids and gases generally have small thermal conductivities. In fact, in most fluids the heat transferred by conduction is negligible compared to that transferred by convection when there are strong convection currents. Air, for instance, with its small thermal conductivity, is an excellent thermal insulator when confined to small spaces where no appreciable convection currents can be established. Goose down, Styrofoam, and wool derive their fine insulating properties in part from the small dead-air spaces within them, as **Figure 13.9** illustrates.

THE PHYSICS OF . . . dressing warmly. We also take advantage of dead-air spaces when we dress "in layers" during very cold weather and put on several layers of relatively thin clothing rather than one thick layer. The air trapped between the layers acts as an excellent insulator.

Example 2 deals with the role that conduction through body fat plays in regulating body temperature.

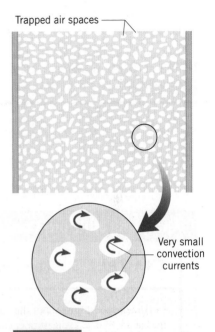

FIGURE 13.9 Styrofoam is an excellent thermal insulator because it contains many small, dead-air spaces. These small spaces inhibit heat transfer by convection currents, and air itself has a very low thermal conductivity.

TABLE 13.1 Thermal Conductivities[a] of Selected Materials

Substance	Thermal Conductivity, k [$J/(s \cdot m \cdot C°)$]	Substance	Thermal Conductivity, k [$J/(s \cdot m \cdot C°)$]
Metals		**Other Materials**	
Aluminum	240	Asbestos	0.090
Brass	110	Body fat	0.20
Copper	390	Concrete	1.1
Iron	79	Diamond	2450
Lead	35	Glass	0.80
Silver	420	Goose down	0.025
Steel (stainless)	14	Ice (0 °C)	2.2
Gases		Styrofoam	0.010
Air	0.0256	Water	0.60
Hydrogen (H_2)	0.180	Wood (oak)	0.15
Nitrogen (N_2)	0.0258	Wool	0.040
Oxygen (O_2)	0.0265		

[a]Except as noted, the values pertain to temperatures near 20 °C.

EXAMPLE 2 | BIO The Physics of Heat Transfer in the Human Body

When excessive heat is produced within the body, it must be transferred to the skin and dispersed if the temperature at the body interior is to be maintained at the normal value of 37.0 °C. One possible mechanism for transfer is conduction through body fat. Suppose that heat travels through 0.030 m of fat in reaching the skin, which has a total surface area of 1.7 m² and a temperature of 34.0 °C. Find the amount of heat that reaches the skin in half an hour (1800 s).

Reasoning and Solution In **Table 13.1** the thermal conductivity of body fat is given as $k = 0.20$ J/(s · m · C°). According to Equation 13.1,

$$Q = \frac{(kA\,\Delta T)t}{L}$$

$$Q = \frac{[0.20\ \text{J/(s} \cdot \text{m} \cdot \text{C°})](1.7\,\text{m}^2)(37.0\,°C - 34.0\,°C)(1800\ \text{s})}{0.030\ \text{m}}$$

$$= \boxed{6.1 \times 10^4\ \text{J}}.$$

For comparison, a jogger can generate over ten times this amount of heat in a half hour. Thus, conduction through body fat is not a particularly effective way of removing excess heat. Heat transfer via blood flow to the skin is more effective and has the added advantage that the body can vary the blood flow as needed (see Problem 36).

Example 3 uses Equation 13.1 to determine what the temperature is at a point between the warmer and cooler ends of the bar in **Interactive Figure 13.7**.

Analyzing Multiple-Concept Problems

EXAMPLE 3 | The Temperature at a Point Between the Ends of a Bar

In **Interactive Figure 13.7** the temperatures at the ends of the bar are 85.0 °C at the warmer end and 27.0 °C at the cooler end. The bar has a length of 0.680 m. What is the temperature at a point that is 0.220 m from the cooler end of the bar?

Reasoning The point in question is closer to the cooler end than to the warmer end of the bar. It might be expected,

therefore, that the temperature at this point is less than halfway between 27.0 °C and 85.0 °C. We will demonstrate that this is, in fact, the case, by applying Equation 13.1. This expression applies because no heat escapes through the insulated sides of the bar, and we will use it twice to determine the desired temperature.

Knowns and Unknowns The available data are as follows:

Description	Symbol	Value
Temperature at warmer end	T_W	85.0 °C
Temperature at cooler end	T_C	27.0 °C
Length of bar	L	0.680 m
Distance from cooler end	D	0.220 m
Unknown Variable		
Temperature at distance D from cooler end	T	?

Modeling the Problem

STEP 1 The Conduction of Heat The heat Q conducted in a time t past the point in question (which is a distance D from the cooler end of the bar) is given by Equation 13.1 as

$$Q = \frac{kA(T - T_C)t}{D}$$

where k is the thermal conductivity of the material from which the bar is made, A is the bar's cross-sectional area, and T and T_C are, respectively, the temperature at the point in question and at the cooler end of the bar. Solving for T gives Equation 1 at the right. The variables Q, k, A, and t are unknown, so we proceed to Step 2 to deal with them.

$$T = T_C + \frac{QD}{kAt} \qquad (1)$$

STEP 2 The Conduction of Heat Revisited The heat Q that is conducted from the point in question to the cooler end of the bar originates at the warmer end of the bar. Thus, since no heat is lost through the sides, we may apply Equation 13.1 a second time to obtain an expression for Q:

$$Q = \frac{kA(T_W - T_C)t}{L}$$

where T_W and T_C are, respectively, the temperatures at the warmer and cooler ends of the bar, which has a length L. This expression for Q can be substituted into Equation 1, as indicated at the right. The terms k, A, and t remain to be dealt with. Fortunately, however, values for them are unnecessary, because they can be eliminated algebraically from the final calculation.

$$T = T_C + \frac{QD}{kAt} \quad (1)$$

$$Q = \frac{kA(T_W - T_C)t}{L}$$

Solution Combining the results of each step algebraically, we find that

$$\overset{\text{STEP 1}}{\downarrow} \quad \overset{\text{STEP 2}}{\downarrow}$$
$$T = T_C + \frac{QD}{kAt} = T_C + \frac{\left[\frac{kA(T_W - T_C)t}{L}\right]D}{kAt}$$

Simplifying this result gives

$$T = T_C + \frac{\frac{\cancel{kA}(T_W - T_C)\cancel{t}}{L}D}{\cancel{kAt}} = T_C + \frac{(T_W - T_C)D}{L}$$

$$= 27.0\,°\text{C} + \frac{(85.0\,°\text{C} - 27.0\,°\text{C})(0.220\,\text{m})}{0.680\,\text{m}} = \boxed{45.8\,°\text{C}}$$

As expected, this temperature is less than halfway between 27.0 °C and 85.0 °C.

Related Homework: Problem 32

Virtually all homes contain insulation in the walls to reduce heat loss. Example 4 illustrates how to determine this loss with and without insulation.

EXAMPLE 4 | The Physics of Layered Insulation

One wall of a house consists of 0.019-m-thick plywood backed by 0.076-m-thick insulation, as **Figure 13.10** shows. The temperature at the inside surface is 25.0 °C, while the temperature at the outside surface is 4.0 °C, both being constant. The thermal conductivities of the insulation and the plywood are, respectively, 0.030 and 0.080 J/(s · m · C°), and the area of the wall is 35 m². Find the heat conducted through the wall in one hour **(a)** with the insulation and **(b)** without the insulation.

Reasoning The temperature T at the insulation–plywood interface (see **Figure 13.10**) must be determined before the heat conducted through the wall can be obtained. In calculating this temperature, we use the fact that no heat is accumulating in the wall because the inner and outer temperatures are constant. Therefore, the heat conducted through the insulation must equal the heat conducted through the plywood during the same time; that is, $Q_{\text{insulation}} = Q_{\text{plywood}}$. Each of the Q values can be expressed as $Q = (kA\,\Delta T)t/L$, according to Equation 13.1, leading to an expression that can be solved for the interface temperature. Once a value for T is available, Equation 13.1 can be used to obtain the heat conducted through the wall.

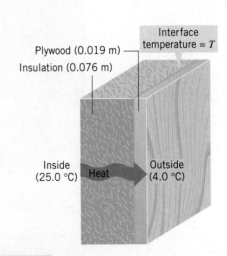

FIGURE 13.10 Heat flows through the insulation and plywood from the warmer inside to the cooler outside.

Problem-Solving Insight When heat is conducted through a multi-layered material (such as the plywood/insulation in this example) and the high and low temperatures are constant, the heat conducted through each layer is the same.

Solution **(a)** Using Equation 13.1 and the fact that $Q_{insulation} = Q_{plywood}$, we find that

$$\left[\frac{(kA\,\Delta T)t}{L}\right]_{insulation} = \left[\frac{(kA\,\Delta T)t}{L}\right]_{plywood}$$

$$\frac{[0.030\ \text{J/(s}\cdot\text{m}\cdot\text{C°})]A(25.0\ \text{°C}-T)t}{0.076\ \text{m}} = \frac{[0.080\ \text{J/(s}\cdot\text{m}\cdot\text{C°})]A(T-4.0\ \text{°C})t}{0.019\ \text{m}}$$

Note that on each side of the equals sign we have written ΔT as the higher temperature minus the lower temperature. Eliminating the area A and time t algebraically and solving this equation for T reveals that the temperature at the insulation–plywood interface is $T = 5.8\ \text{°C}$.

The heat conducted through the wall is either $Q_{insulation}$ or $Q_{plywood}$, since the two quantities are equal. Choosing $Q_{insulation}$ and using $T = 5.8\ \text{°C}$ in Equation 13.1, we find that

$$Q_{insulation} = \frac{[0.030\ \text{J/(s}\cdot\text{m}\cdot\text{C°})](35\ \text{m}^2)(25.0\ \text{°C}-5.8\ \text{°C})(3600\ \text{s})}{0.076\ \text{m}}$$

$$= \boxed{9.5\times 10^5\ \text{J}}$$

(b) It is straightforward to use Equation 13.1 to calculate the amount of heat that would flow through the plywood in one hour if the insulation were absent:

$$Q_{plywood} = \frac{[0.080\ \text{J/(s}\cdot\text{m}\cdot\text{C°})](35\ \text{m}^2)(25.0\ \text{°C}-4.0\ \text{°C})(3600\ \text{s})}{0.019\ \text{m}}$$

$$= \boxed{110\times 10^5\ \text{J}}$$

Without insulation, the heat loss is increased by a factor of about 12.

Math Skills Without units (omitted for the sake of clarity), the equation that needs to be solved for the temperature T is

$$\frac{0.030A(25.0-T)t}{0.076} = \frac{0.080A(T-4.0)t}{0.019} \quad \text{or}$$

$$\frac{0.030(25.0-T)}{0.076} = \frac{0.080(T-4.0)}{0.019}$$

The terms A and t appear on both sides of the equation as factors in the numerator and, therefore, have been eliminated algebraically. Rearranging the result slightly and carrying out the indicated divisions, we obtain

$$\frac{0.030}{0.076}(25.0-T) = \frac{0.080}{0.019}(T-4.0) \quad \text{or}$$

$$0.394(25.0-T) = 4.21(T-4.0)$$

Expanding the terms on each side of the equals sign and rearranging the result shows that

$$9.85 - 0.394T = 4.21T - 16.84 \quad \text{or} \quad 26.69 = 4.604T$$

In these results, in order to avoid round-off errors, we have carried along more significant figures than only the two justified by the original data. However, in solving the final equation for T we round off to two significant figures and obtain

$$T = \frac{26.69}{4.604} = 5.8\ \text{°C}.$$

THE PHYSICS OF . . . protecting fruit plants from freezing.   Fruit growers sometimes protect their crops by spraying them with water when overnight temperatures are expected to drop below freezing. Some fruit crops, like the strawberries in Figure 13.11, can withstand temperatures down to freezing (0 °C), but not below freezing. When water is sprayed on the plants, it can freeze and release heat (see Section 12.8),

Pierre Ducharme/Reuters/Newscom

FIGURE 13.11 When the temperature dips below freezing, strawberry growers spray their plants with water to put a coat of ice on them. Because of the heat released when the water freezes and because of the relatively small thermal conductivity of ice, this procedure protects the plants against subfreezing temperatures. The photograph shows berries that survived a temperature of −3 °C with little or no damage.

some of which goes into warming the plant. In addition, water and ice have relatively small thermal conductivities, as **Table 13.1** indicates. Thus, they also protect the crop by acting as thermal insulators that reduce heat loss from the plants.

Although a layer of ice may be beneficial to strawberries, it is not so desirable inside a refrigerator, as Conceptual Example 5 discusses.

CONCEPTUAL EXAMPLE 5 | An Iced-up Refrigerator

In a refrigerator, heat is removed by a cold refrigerant fluid that circulates within a tubular space embedded inside a metal plate, as **Figure 13.12** illustrates. A good refrigerator cools food as quickly as possible. Which arrangement works best: **(a)** an aluminum plate coated with ice, **(b)** an aluminum plate without ice, **(c)** a stainless steel plate coated with ice, or **(d)** a stainless steel plate without ice?

Reasoning Figure 13.12 (see the blow-ups) shows the metal cooling plate with and without a layer of ice. Without ice, heat passes by conduction through the metal plate to the refrigerant fluid within. For a given temperature difference across the thickness of the metal, the rate of heat transfer depends on the thermal conductivity of the metal. When the plate becomes coated with ice, any heat that is removed by the refrigerant fluid must first be transferred by conduction through the ice before it encounters the metal plate.

FIGURE 13.12 In a refrigerator, cooling is accomplished by a cold refrigerant fluid that circulates through a tubular space embedded within a metal plate. Sometimes the plate becomes coated with a layer of ice.

Answers (a), (c), and (d) are incorrect. For answers (a) and (c), the relation $Q = \frac{(kA\,\Delta T)t}{L}$ (Equation 13.1) indicates that the heat conducted per unit time (Q/t) is inversely proportional to the thickness L of the ice. As ice builds up, the heat removed per unit time by the cooling plate decreases. Thus, when covered with ice, the cooling plate—regardless of whether it's made from aluminum or stainless steel—does not work as well as a plate that is ice-free. Answer (d)—the stainless steel plate without ice—is incorrect, because heat is transferred more readily through a plate that has a greater thermal conductivity, and stainless steel has a smaller thermal conductivity than does aluminum (see **Table 13.1**).

Answer (b) is correct. The relation $Q = \frac{(kA\,\Delta T)t}{L}$ (Equation 13.1) shows that the heat conducted per unit time (Q/t) is directly proportional to the thermal conductivity k of the metal plate. Since the thermal conductivity of aluminum is more than 17 times greater than the thermal conductivity of stainless steel (see **Table 13.1**), aluminum is the preferred plate. The aluminum plate arrangement works best without an ice buildup. When ice builds up, the heat removed per unit time decreases because of the increased thickness of material through which the heat must pass.

Related Homework: Problem 10

Check Your Understanding

(The answers are given at the end of the book.)

2. A poker used in a fireplace is held at one end, while the other end is in the fire. In terms of being cooler to the touch, should a poker be made from **(a)** a high-thermal-conductivity material, **(b)** a low-thermal-conductivity material, or **(c)** can either type be used?

3. Several days after a snowstorm, the outdoor temperature remains below freezing. The roof on one house is uniformly covered with snow. On a neighboring house, however, the snow on the roof has completely melted. Which house is better insulated?

4. Concrete walls often contain steel reinforcement bars. Does the steel **(a)** enhance, **(b)** degrade, or **(c)** have no effect on the insulating value of the concrete wall? (Consult **Table 13.1**.)

5. To keep your hands as warm as possible during skiing, should you wear mittens or gloves? (Mittens, except for the thumb, do not have individual finger compartments.) Assume that the mittens and gloves are the same size and are made of the same material. You should wear: **(a)** gloves,

(Continued)

because the individual finger compartments mean that the gloves have a smaller thermal conductivity; **(b)** gloves, because the individual finger compartments mean that the gloves have a larger thermal conductivity; **(c)** mittens, because they have less surface area exposed to the cold air.

6. A water pipe is buried slightly beneath the ground. The ground is covered with a thick layer of snow, which contains a lot of small dead-air spaces within it. The air temperature suddenly drops to well below freezing. The accumulation of snow **(a)** has no effect on whether the water in the pipe freezes, **(b)** causes the water in the pipe to freeze more quickly than if the snow were not there, **(c)** helps prevent the water in the pipe from freezing.

7. Some animals have hair strands that are hollow, air-filled tubes. Others have hair strands that are solid. Which kind, if either, would be more likely to give an animal an advantage for surviving in very cold climates?

8. Two bars are placed between plates whose temperatures are T_{hot} and T_{cold} (see **CYU Figure 13.1**). The thermal conductivity of bar 1 is six times that of bar 2 ($k_1 = 6k_2$), but bar 1 has only one-third the cross-sectional area ($A_1 = \frac{1}{3}A_2$). Ignore any heat loss through the sides of the bars. What can you conclude about the amounts of heat Q_1 and Q_2, respectively, that bar 1 and bar 2 conduct in a given amount of time? **(a)** $Q_1 = \frac{1}{4}Q_2$ **(b)** $Q_1 = \frac{1}{8}Q_2$ **(c)** $Q_1 = 2Q_2$ **(d)** $Q_1 = 4Q_2$ **(e)** $Q_1 = Q_2$

CYU FIGURE 13.1

9. A piece of Styrofoam and a piece of wood are joined together to form a layered slab. The two pieces have the same thickness and cross-sectional area, but the Styrofoam has the smaller thermal conductivity. The temperature of the exposed Styrofoam surface is greater than the temperature of the exposed wood surface, both temperatures being constant. Is the temperature of the Styrofoam–wood interface **(a)** closer to the higher temperature of the exposed Styrofoam surface, **(b)** closer to the lower temperature of the exposed wood surface, or **(c)** halfway between the two temperatures?

13.3 Radiation

Energy from the sun is brought to earth by large amounts of visible light waves, as well as by substantial amounts of infrared and ultraviolet waves. These waves are known as electromagnetic waves, a class that also includes the microwaves used for cooking and the radio waves used for AM and FM broadcasts. The sunbather in **Figure 13.13** feels hot because her body absorbs energy from the sun's electromagnetic waves. Anyone who has stood by a roaring fire or put a hand near an incandescent light bulb has experienced a similar effect. Thus, fires and light bulbs also emit electromagnetic waves, and when the energy of such waves is absorbed, it can have the same effect as heat.

The process of transferring energy via electromagnetic waves is called **radiation**, and, unlike convection or conduction, it does not require a material medium. Electromagnetic waves from the sun, for example, travel through the void of space during their journey to earth.

FIGURE 13.13 Suntans are produced by ultraviolet rays.

> **RADIATION**
> Radiation is the process in which energy is transferred by means of electromagnetic waves.

All bodies continuously radiate energy in the form of electromagnetic waves. Even an ice cube radiates energy, although so little of it is in the form of visible light that an ice cube cannot be seen in the dark. Likewise, the human body emits insufficient visible light to be seen in the dark. However, as **Figures 12.6** and **12.7** illustrate, the infrared waves radiating from the body can be detected in the dark by electronic cameras. Generally, an object does not emit much visible light until the temperature of the object exceeds about 1000 K. Then a characteristic red glow appears, like that of a heating coil on an electric stove. When its temperature reaches about 1700 K, an object begins to glow white-hot, like the tungsten filament in an incandescent light bulb.

In the transfer of energy by radiation, the absorption of electromagnetic waves is just as important as their emission. The surface of an object plays a significant role in determining how much radiant energy the object will absorb or emit. The two blocks in sunlight in **Interactive Figure 13.14**, for example, are identical, except that one has a rough surface coated with lampblack (a fine black soot), while the other has a highly polished silver surface. As the thermometers indicate, the temperature of the black block rises at a much faster rate than that of the silvery block. This is because lampblack absorbs about 97% of the incident radiant energy, while the silvery surface absorbs only about 10%. The remaining part of the incident energy is reflected in each case. We see the lampblack as black in color because it reflects so little of the light falling on it, while the silvery surface looks like a mirror because it reflects so much light. Since the color black is associated with nearly complete absorption of visible light, the term **perfect blackbody** or, simply, **blackbody** is used when referring to an object that absorbs *all* the electromagnetic waves falling on it.

All objects emit and absorb electromagnetic waves simultaneously. When a body has the same constant temperature as its surroundings, the amount of radiant energy being absorbed must balance the amount being emitted in a given interval of time. The block coated with lampblack absorbs and emits the same amount of radiant energy, and the silvery block does too. In either case, if absorption were greater than emission, the block would experience a net gain in energy. As a result, the temperature of the block would rise and not be constant. Similarly, if emission were greater than absorption, the temperature would fall. Since absorption and emission are balanced, *a material that is a good absorber, like lampblack, is also a good emitter, and a material that is a poor absorber, like polished silver, is also a poor emitter.* A perfect blackbody, being a perfect absorber, is also a perfect emitter.

THE PHYSICS OF . . . summer clothing. The fact that a black surface is both a good absorber and a good emitter is the reason people are uncomfortable wearing dark clothes during the summer. Dark clothes absorb a large fraction of the sun's radiation and then reemit it in all directions. About one-half of the emitted radiation is directed inward toward the body and creates the sensation of warmth. Light-colored clothes, in contrast, are cooler to wear, because they absorb and reemit relatively little of the incident radiation.

THE PHYSICS OF . . . a white sifaka lemur warming up. The use of light colors for comfort also occurs in nature. Most lemurs, for instance, are nocturnal and have dark fur like the lemur shown in **Figure 13.15a**. Since they are active at night, the dark fur poses no disadvantage in absorbing excessive sunlight. **Figure 13.15b** shows a species of lemur called the white sifaka, which lives in semiarid regions where there is little shade. The white color of the fur may help in thermoregulation, by reflecting sunlight, but during the cool mornings, reflection of sunlight would hinder warming up. However, these lemurs have black skin and only sparse fur on their bellies, and to warm up in the morning, they turn their dark bellies toward the sun. The dark color enhances the absorption of sunlight.

The amount of radiant energy Q emitted by a perfect blackbody is proportional to the radiation time interval t ($Q \propto t$). The longer the time, the greater is the amount of energy radiated. Experiment shows that Q is also proportional to the surface area A ($Q \propto A$). An object with a large surface area radiates more energy than one with a small

INTERACTIVE FIGURE 13.14 The temperature of the block coated with lampblack rises faster than the temperature of the block coated with silver because the black surface absorbs radiant energy from the sun at the greater rate.

(a)   (b)

FIGURE 13.15 (a) Most lemurs, like this one, are nocturnal and have dark fur. (b) The species of lemur called the white sifaka, however, is active during the day and has white fur.

surface area, other things being equal. Finally, experiment reveals that Q is proportional to the *fourth power of the Kelvin temperature* T ($Q \propto T^4$), so the emitted energy increases markedly with increasing temperature. If, for example, the Kelvin temperature of an object doubles, the object emits 2^4 or 16 times more energy. Combining these factors into a single proportionality, we see that $Q \propto T^4 At$. This proportionality is converted into an equation by inserting a proportionality constant σ, known as the *Stefan–Boltzmann constant*. It has been found experimentally that $\sigma = 5.67 \times 10^{-8}$ J/(s · m² · K⁴):

$$Q = \sigma T^4 At$$

The relationship above holds only for a perfect emitter. Most objects are not perfect emitters, however. Suppose that an object radiates only about 80% of the visible light energy that a perfect emitter would radiate, so Q (for the object) $= (0.80)\sigma T^4 At$. The factor such as the 0.80 in this equation is called the **emissivity** e and is a dimensionless number between zero and one. The emissivity is the ratio of the energy an object actually radiates to the energy the object would radiate if it were a perfect emitter. For visible light, the value of e for the human body, for instance, varies between about 0.65 and 0.80, the smaller values pertaining to lighter skin colors. For infrared radiation, e is nearly one for all skin colors. For a perfect blackbody emitter, $e = 1$. Including the factor e on the right side of the expression $Q = \sigma T^4 At$ leads to the **Stefan–Boltzmann law of radiation**.

THE STEFAN–BOLTZMANN LAW OF RADIATION

The radiant energy Q, emitted in a time t by an object that has a Kelvin temperature T, a surface area A, and an emissivity e, is given by

$$Q = e\sigma T^4 At \qquad (13.2)$$

where $\sigma = 5.67 \times 10^{-8}$ J/(s · m² · K⁴) is the Stefan–Boltzmann constant.

In Equation 13.2, the Stefan–Boltzmann constant σ is a universal constant in the sense that its value is the same for all bodies, regardless of the nature of their surfaces. The emissivity e, however, depends on the condition of the surface. Example 6 shows how the Stefan–Boltzmann law can be used to determine the size of a star.

EXAMPLE 6 | A Supergiant Star

The supergiant star Betelgeuse has a surface temperature of about 2900 K (about one-half that of our sun) and emits a radiant power (in joules per second, or watts) of approximately 4×10^{30} W (about 10 000 times as great as that of our sun). Assuming that Betelgeuse is a perfect emitter (emissivity $e = 1$) and spherical, find its radius.

Reasoning According to the Stefan–Boltzmann law, the power emitted is $Q/t = e\sigma T^4 A$. A star with a relatively small temperature T can have a relatively large radiant power Q/t only if its surface area A is large. As we will see, Betelgeuse has a very large surface area, so its radius is enormous.

> **Problem-Solving Insight** First solve an equation for the unknown in terms of the known variables. Then substitute numbers for the known variables, as this example shows.

Solution Solving the Stefan–Boltzmann law for the surface area, we find

$$A = \frac{Q/t}{e\sigma T^4}$$

But the surface area of a sphere is $A = 4\pi r^2$, so $r = \sqrt{A/(4\pi)}$. Therefore, we have

$$r = \sqrt{\frac{Q/t}{4\pi e\sigma T^4}} = \sqrt{\frac{4 \times 10^{30}\text{ W}}{4\pi(1)[5.67 \times 10^{-8}\text{ J/(s · m}^2\text{ · K}^4)](2900\text{ K})^4}}$$

$$= \boxed{3 \times 10^{11}\text{ m}}$$

For comparison, Mars orbits the sun at a distance of 2.28×10^{11} m. Betelgeuse is certainly a "supergiant."

The next example explains how to apply the Stefan–Boltzmann law when an object, such as a wood stove, simultaneously emits and absorbs radiant energy.

EXAMPLE 7 | An Unused Wood-Burning Stove

An unused wood-burning stove has a constant temperature of 18 °C (291 K), which is also the temperature of the room in which the stove stands. The stove has an emissivity of 0.900 and a surface area of 3.50 m². What is the *net* radiant power generated by the stove?

Reasoning Power is the change in energy per unit time (Equation 6.10b), or Q/t, which, according to the Stefan–Boltzmann law, is $Q/t = e\sigma T^4 A$ (Equation 13.2). In this problem, however, we need to find the *net* power produced by the stove. The net power is the power the stove emits minus the power the stove absorbs. The power the stove absorbs comes from the walls, ceiling, and floor of the room, all of which emit radiation.

Solution   Remembering that temperature must be expressed in kelvins when using the Stefan–Boltzmann law, we find that

$$\text{Power emitted by unheated stove at 18 °C} = \frac{Q}{t} = e\sigma T^4 A \qquad \textbf{(13.2)}$$

$= (0.900)[5.67 \times 10^{-8}\,\text{J}/(\text{s}\cdot\text{m}^2\cdot\text{K}^4)](291\,\text{K})^4(3.50\,\text{m}^2) = 1280\,\text{W}$

The fact that the unheated stove emits 1280 W of power and yet maintains a constant temperature means that the stove also

absorbs 1280 W of radiant power from its surroundings. Thus, the *net* power generated by the unheated stove is zero:

$$\text{Net power generated by stove at 18 °C} = \underbrace{1280\,\text{W}}_{\substack{\text{Power emitted}\\\text{by stove at}\\\text{18 °C}}} - \underbrace{1280\,\text{W}}_{\substack{\text{Power emitted by}\\\text{room at 18 °C and}\\\text{absorbed by stove}}} = \boxed{0\,\text{W}}$$

Probem-Solving Insight   In applying the Stefan–Boltzmann law (Equation 13.2), the temperature must be expressed in kelvins.

When an object has a higher temperature than its surroundings, the object emits a net radiant power $P_\text{net} = (Q/t)_\text{net}$. The net power is the power the object emits minus the power it absorbs. Applying the Stefan–Boltzmann law leads to the following expression for P_net when the temperature of the object is T and the temperature of the environment is T_0:

$$P_\text{net} = e\sigma A\left(T^4 - T_0^4\right) \qquad \textbf{(13.3)}$$

Check Your Understanding

(The answers are given at the end of the book.)

10. **BIO** One way that heat is transferred from place to place inside the human body is by the flow of blood. Which one of the following heat transfer processes—forced convection, conduction, or radiation—best describes this action of the blood?

11. Two strips of material, A and B, are identical, except that they have emissivities of 0.4 and 0.7, respectively. The strips are heated to the same temperature and have a red glow. A brighter glow signifies that more energy per second is being radiated. Which strip has the brighter glow?

12. One day during the winter the sun has been shining all day. Toward sunset a light snow begins to fall. It collects without melting on a cement playground, but melts instantly on contact with a black asphalt road adjacent to the playground. Why the difference? **(a)** Being black, asphalt has a higher emissivity than cement, so the asphalt absorbs more radiant energy from the sun during the day and, consequently, warms above the freezing point. **(b)** Being black, asphalt has a lower emissivity than cement, so it absorbs more radiant energy from the sun during the day and, consequently, warms above the freezing point.

13. **BIO** If you were stranded in the mountains in cold weather, it would help to minimize energy losses from your body if you curled up into the tightest ball possible. Which factor in the relation $Q = e\sigma T^4 A t$ (Equation 13.2) are you using to the best advantage by curling into a ball? **(a)** e **(b)** σ **(c)** T **(d)** A **(e)** t

14. Two identical cubes have the same temperature. One of them, however, is cut in two and the pieces are separated (see **CYU Figure 13.2**). The radiant energy emitted by the cube cut into two pieces is $Q_\text{two pieces}$ and that emitted by the uncut cube is Q_cube. What is true about the radiant energy emitted in a given time?

(a) $Q_\text{two pieces} = 2Q_\text{cube}$ **(b)** $Q_\text{two pieces} = \frac{4}{3}Q_\text{cube}$

(c) $Q_\text{two pieces} = Q_\text{cube}$ **(d)** $Q_\text{two pieces} = \frac{1}{2}Q_\text{cube}$

(e) $Q_\text{two pieces} = \frac{1}{3}Q_\text{cube}$

Cube cut into two pieces   Uncut cube

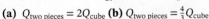
CYU FIGURE 13.2

15. Two objects have the same size and shape. Object A has an emissivity of 0.3, and object B has an emissivity of 0.6. Each radiates the same power. How is the Kelvin temperature T_A of A related to the Kelvin temperature T_B of B? **(a)** $T_A = T_B$ **(b)** $T_A = 2T_B$ **(c)** $T_A = \frac{1}{2}T_B$ **(d)** $T_A = \sqrt{2}\,T_B$ **(e)** $T_A = \sqrt[4]{2}\,T_B$

13.4 Applications

THE PHYSICS OF . . . rating thermal insulation by *R* values. To keep heating and air conditioning bills to a minimum, it pays to use good thermal insulation in your home. Insulation inhibits convection between inner and outer walls and minimizes heat transfer by conduction. With respect to conduction, the logic behind home insulation ratings comes directly from Equation 13.1. According to this equation, the heat per unit time Q/t flowing through a thickness of material is $Q/t = kA\,\Delta T/L$. Keeping the value for Q/t to a minimum means using materials that have small thermal conductivities k and large thicknesses L. Construction engineers, however, prefer to use Equation 13.1 in the slightly different form shown below:

$$\frac{Q}{t} = \frac{A\,\Delta T}{L/k}$$

The term L/k in the denominator is called the *R value*. An *R* value expresses in a single number the combined effects of thermal conductivity and thickness. Larger *R* values reduce the heat per unit time flowing through the material and, therefore, mean better insulation. It is also convenient to use *R* values to describe layered slabs formed by sandwiching together a number of materials with different thermal conductivities and different thicknesses. The *R* values for the individual layers can be added to give a single *R* value for the entire slab. It should be noted, however, that *R* values are expressed using units of feet, hours, F°, and BTU for thickness, time, temperature, and heat, respectively.

THE PHYSICS OF . . . regulating the temperature of an orbiting satellite. When it is in the earth's shadow, an orbiting satellite is shielded from the intense electromagnetic waves emitted by the sun. But when it moves out of the earth's shadow, the satellite experiences the full effect of these waves. As a result, the temperature within a satellite would decrease and increase sharply during an orbital period and sensitive electronic circuitry would suffer, unless precautions are taken. To minimize temperature fluctuations, satellites are often covered with a highly reflecting and, hence, poorly absorbing metal foil, as **Figure 13.16** shows. By reflecting much of the sunlight, the foil minimizes temperature rises. Being a poor absorber, the foil is also a poor emitter and reduces radiant energy losses. Reducing these losses keeps the temperature from falling excessively.

THE PHYSICS OF . . . a thermos bottle. A thermos bottle, sometimes referred to as a Dewar flask, reduces the rate at which hot liquids cool down or cold liquids warm up. A thermos usually consists of a double-walled glass vessel with silvered inner walls (see **Interactive Figure 13.17**) and minimizes heat transfer via convection, conduction, and radiation. The space between the walls is evacuated to reduce energy losses due to conduction and convection. The silvered surfaces reflect most of the radiant energy that would otherwise enter or leave the liquid in the thermos. Finally, little heat is lost through the glass or the rubberlike gaskets and stopper, because these materials have relatively small thermal conductivities.

THE PHYSICS OF . . . a halogen cooktop stove. Halogen cooktops use radiant energy to heat pots and pans. A halogen cooktop uses several quartz–iodine lamps, like the ones in ultra-bright automobile headlights. These lamps are electrically powered and mounted below a ceramic top (see **Figure 13.18**). The electromagnetic energy they radiate

FIGURE 13.16 The highly reflective metal foil covering this satellite (the Hubble Space Telescope) minimizes temperature changes.

INTERACTIVE FIGURE 13.17 A thermos bottle minimizes energy transfer due to convection, conduction, and radiation.

FIGURE 13.18 In a halogen cooktop, quartz–iodine lamps emit a large amount of electromagnetic energy that is absorbed directly by a pot or pan.

passes through the ceramic top and is absorbed directly by the bottom of the pot. Consequently, the pot heats up very quickly, rivaling the time of a pot on an open gas burner.

EXAMPLE 8 | BIO Staying Cool While Exercising

As warm-blooded animals, humans regulate their internal body temperature through the metabolism of food. These baseline metabolic processes occur continuously in the body, even when the body is at rest or sleeping. However, while exercising, the body's muscles convert even more of the chemical energy stored in food into mechanical work. Furthermore, the body's metabolic energy conversion is rather poor, where 60% of the available energy is converted to heat. As an example, let's say a jogger is doing work with a power output of 250 W. She is running on a day when the air temperature is 18 °C (291 K), and her skin temperature is 34 °C (307 K). How much water must she evaporate per hour by sweating if she is to maintain a skin temperature of 34 °C? Assume the surface area of her skin is 1.4 m² with an emissivity of 0.65, and she only loses heat by radiation.

Reasoning Heat is carried away from her skin by radiation, and we can calculate the rate of heat loss by using Equation 13.3. Any excess heat produced by physical activity must be removed by water evaporation. By knowing the quantity of excess heat, we

can calculate the amount of water we need to evaporate by using Equation 12.5 and the latent heat of fusion for water.

Solution We begin by calculating the radiation heat loss from the skin due to the difference in the skin and air temperature. Applying Equation 13.3, we have:

$$P_{net} = e\sigma A(T^4 - T_0^4) = (0.65)[(5.67 \times 10^{-8} \text{ J})/(\text{s} \cdot \text{m}^2 \cdot \text{K}^4)](1.4 \text{ m}^2)$$
$$[(307 \text{ K})^4 - (291 \text{ K})^4] = 88 \text{ W}.$$

Any excess power output above 88 W will produce heat that has to be removed to maintain her current skin temperature. Her total power output that produces heat will be 60% of 250 W, or (0.60) (250 W) = 150 W. Therefore, the additional power output is 150 W − 88 W = 62 W, or 62 J/s. Thus, in one hour, or 3600 s, this is an extra $Q = (62 \text{ J/s})(3600 \text{ s}) = 2.2 \times 10^5$ J of heat that must be removed by evaporation. Finally, we apply Equation 12.5 to calculate the mass of water evaporated:

$$m = Q/L = 2.2 \times 10^5 \text{ J}/22.6 \times 10^5 \text{ J} = \boxed{0.097 \text{ kg}}$$

The process of radiation is important in regulating the temperature of our bodies, as the previous example discussed. The radiation emitted by human beings can also indicate potential health issues, as the following example shows.

EXAMPLE 9 | BIO The Physics of Thermography—Revisited

In Section 12.3, the medical diagnostic technique of thermography was introduced. The intensity of the radiant heat emitted by different parts of the body depends on temperature, in accordance with the Stefan–Boltzmann law. By measuring the radiated power from the body, color-contrast images like Figure 12.6 can be created. The sensitivity of thermography has increased significantly, to where it now can often diagnose medical conditions, such as breast cancer, before more traditional techniques, like mammography. In fact, the minimum resolvable temperature difference of modern thermal cameras is tens of millikelvins. Consider a tumor just below the surface of the skin that creates an increased skin temperature of 41.0 °C over an area of 6.25 cm². What is the difference in the net power radiated by this region and a region of the same area with a normal skin temperature of 36.5 °C? Assume that the emissivity of the skin from both areas is 0.650.

Reasoning The net power radiated from the skin can be calculated with Equation 13.3. This relationship can be used

twice—once for each area of the skin with its respective temperature. The difference in these two values gives us the desired result.

Solution We apply Equation 13.3 to calculate the net power radiated from the hotter area of the skin P_{net-H} and the cooler area P_{net-C}. We calculate the difference in these two quantities, remembering to express the temperatures in kelvins and to convert the area to m²:

$$P_{net-H} - P_{net-C} = e\sigma A(T_H^4 - T_0^4) - e\sigma A(T_C^4 - T_0^4) = e\sigma A(T_H^4 - T_C^4)$$

$$= (0.650)[5.67 \times 10^{-8} \text{ J}/(\text{s} \cdot \text{m}^2 \cdot \text{K}^4)](6.25 \times 10^{-4} \text{ m}^2)[(314 \text{ K})^4$$

$$- (309.5 \text{ K})^4] = \boxed{1.26 \times 10^{-2} \text{ W}}$$

Notice in the final expression that the temperature of the ambient air in the room (T_0) is eliminated, since we are calculating the *difference* in the radiated net power.

Concept Summary

13.1 Convection Convection is the process in which heat is carried from place to place by the bulk movement of a fluid. During natural convection, the warmer, less dense part of a fluid is pushed upward by the buoyant force provided by the surrounding cooler and denser part.

Forced convection occurs when an external device, such as a fan or a pump, causes the fluid to move.

13.2 Conduction Conduction is the process whereby heat is transferred directly through a material, with any bulk motion of the material playing no role in the transfer. Materials that conduct heat well, such as most metals, are known as thermal conductors. Materials that conduct heat poorly, such as wood, glass, and most plastics, are referred to as thermal insulators. The heat Q conducted during a time t through a bar of length L and cross-sectional area A is given by Equation 13.1, where ΔT is the temperature difference between the ends of the bar and k is the thermal conductivity of the material.

$$Q = \frac{(kA\,\Delta T)t}{L} \qquad (13.1)$$

13.3 Radiation Radiation is the process in which energy is transferred by means of electromagnetic waves. All objects, regardless of their temperature, simultaneously absorb and emit electromagnetic waves. Objects that are good absorbers of radiant energy are also good emitters, and objects that are poor absorbers are also poor emitters. An object that absorbs all the radiation incident upon it is called a perfect blackbody. A perfect blackbody, being a perfect absorber, is also a perfect emitter.

The radiant energy Q emitted during a time t by an object whose surface area is A and whose Kelvin temperature is T is given by the Stefan–Boltzmann law of radiation (see Equation 13.2), where $\sigma = 5.67 \times 10^{-8}$ J/(s $\cdot$ m^2 $\cdot$ K^4) is the Stefan–Boltzmann constant and e is the emissivity, a dimensionless number characterizing the surface of the object. The emissivity lies between 0 and 1, being zero for a nonemitting surface and one for a perfect blackbody.

$$Q = e\sigma T^4 At \qquad (13.2)$$

The net radiant power is the power an object emits minus the power it absorbs. The net radiant power P_{net} emitted by an object with a temperature T located in an environment with a temperature T_0 is given by Equation 13.3.

$$P_{\text{net}} = e\sigma A(T^4 - T_0^4) \qquad (13.3)$$

Focus on Concepts

Online

Additional questions are available for assignment in WileyPLUS.

Section 13.2 Conduction

1. The heat conducted through a bar depends on which of the following?

A. The coefficient of linear expansion

B. The thermal conductivity

C. The specific heat capacity

D. The length of the bar

E. The cross-sectional area of the bar

(a) A, B, and D (b) A, C, and D (c) B, C, D, and E (d) B, D, and E (e) C, D, and E

2. Two bars are conducting heat from a region of higher temperature to a region of lower temperature. The bars have identical lengths and cross-sectional areas, but are made from different materials. In the drawing they are placed "in parallel" between the two temperature regions in arrangement A, whereas they are placed end to end in arrangement B. In which arrangement is the heat that is conducted the greatest? (a) The heat conducted is the same in both arrangements. (b) Arrangement A (c) Arrangement B (d) It is not possible to determine which arrangement conducts more heat.

QUESTION 2

3. The drawing shows a composite slab consisting of three materials through which heat is conducted from left to right. The materials have identical thicknesses and cross-sectional areas. Rank the materials according to their thermal conductivities, largest first. (a) k_1, k_2, k_3 (b) k_1, k_3, k_2 (c) k_2, k_1, k_3 (d) k_2, k_3, k_1 (e) k_3, k_2, k_1

QUESTION 3

4. The long single bar on the left in the drawing has a thermal conductivity of 240 J/(s $\cdot$ m $\cdot$ C°). The ends of the bar are at temperatures of 400 and 200 °C, and the temperature of its midpoint is halfway between these two temperatures, or 300 °C. The two bars on the right are half as long as the bar on the left, and the thermal conductivities of these bars are different (see the drawing). All of the bars have the same cross-sectional area. What can be said about the temperature at the point where the two bars on the right are joined together? (a) The temperature at the point where the two bars are joined together is 300 °C. (b) The temperature at the point where the two bars are joined together is greater than 300 °C. (c) The temperature at the point where the two bars are joined together is less than 300 °C.

QUESTION 4

Section 13.3 Radiation

5. Three cubes are made from the same material. As the drawing indicates, they have different sizes and temperatures. Rank the cubes according to the radiant energy they emit per second, largest first. (a) A, B, C (b) A, C, B (c) B, A, C (d) B, C, A (e) C, B, A

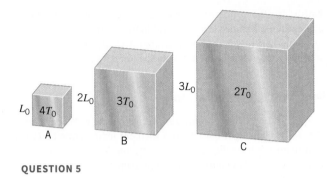

QUESTION 5

6. An astronaut on a space station has two objects that are identical in all respects, except that one is painted black and the other is painted silver. Initially, they are at the same temperature. When taken from inside the space station and placed in outer space, which object, if either, cools down at a faster rate? (a) The object painted black (b) The object painted silver (c) Both objects cool down at the same rate. (d) It is not possible to determine which object cools down at the faster rate.

7. The emissivity e of object B is $\frac{1}{16}$ that of object A, although both objects are identical in size and shape. If the objects radiate the same energy per second, what is the ratio T_B/T_A of their Kelvin temperatures? (a) $\frac{1}{16}$ (b) $\frac{1}{4}$ (c) $\frac{1}{2}$ (d) 2 (e) 4

Problems

Online

Additional questions are available for assignment in WileyPLUS.

Note: *For problems in this set, use the values for thermal conductivities given in* **Table 13.1** *unless stated otherwise.*

SSM Student Solutions Manual	**BIO** Biomedical application
MMH Problem-solving help	**E** Easy
GO Guided Online Tutorial	**M** Medium
V-HINT Video Hints	**H** Hard
CHALK Chalkboard Videos	**WS** Worksheet
	T Team Problem

Section 13.2   Conduction

1. **E** In an electrically heated home, the temperature of the ground in contact with a concrete basement wall is 12.8 °C. The temperature at the inside surface of the wall is 20.0 °C. The wall is 0.10 m thick and has an area of 9.0 m². Assume that one kilowatt · hour of electrical energy costs $0.10. How many hours are required for one dollar's worth of energy to be conducted through the wall?

2. **E** **SSM** **MMH** A person's body is covered with 1.6 m² of wool clothing. The thickness of the wool is 2.0×10^{-3} m. The temperature at the outside surface of the wool is 11 °C, and the skin temperature is 36 °C. How much heat per second does the person lose due to conduction?

3. **E** **GO** Two objects are maintained at constant temperatures, one hot and one cold. Two identical bars can be attached end to end, as in part a of the drawing, or one on top of the other, as in part b. When either of these arrangements is placed between the hot and the cold objects for the same amount of time, heat Q flows from left to right. Find the ratio Q_a/Q_b.

      (a)   (b)

PROBLEM 3

4. **E** **SSM** One end of an iron poker is placed in a fire where the temperature is 502 °C, and the other end is kept at a temperature of 26 °C. The poker is 1.2 m long and has a radius of 5.0×10^{-3} m. Ignoring the heat lost along the length of the poker, find the amount of heat conducted from one end of the poker to the other in 5.0 s.

5. **E** **GO** The block in the drawing has dimensions $L_0 \times 2L_0 \times 3L_0$, where $L_0 = 0.30$ m. The block has a thermal conductivity of 250 J/ (s · m · C°). In drawings A, B, and C, heat is conducted through the block in three different directions; in each case the temperature of the warmer surface is 35 °C and that of the cooler surface is 19 °C. Determine the heat that flows in 5.0 s for each case.

PROBLEM 5

6. **E** **GO** A closed box is filled with dry ice at a temperature of −78.5 °C, while the outside temperature is 21.0 °C. The box is cubical, measuring 0.350 m on a side, and the thickness of the walls is 3.00×10^{-2} m. In one day, 3.10×10^6 J of heat is conducted through the six walls. Find the thermal conductivity of the material from which the box is made.

7. **E** **MMH** One end of a brass bar is maintained at 306 °C, while the other end is kept at a constant, but lower, temperature. The cross-sectional area of the bar is 2.6×10^{-4} m². Because of insulation, there is negligible heat loss through the sides of the bar. Heat flows through the bar, however, at the rate of 3.6 J/s. What is the temperature of the bar at a point 0.15 m from the hot end?

8. **E** **GO** A wall in a house contains a single window. The window consists of a single pane of glass whose area is 0.16 m² and whose thickness is 2.0 mm. Treat the wall as a slab of the insulating material Styrofoam whose area and thickness are 18 m² and 0.10 m, respectively. Heat is lost via conduction through the wall and the window. The temperature difference between the inside and outside is the same for the wall and the window. Of the total heat lost by the wall and the window, what is the percentage lost by the window?

9. **E** **V-HINT** **MMH** A composite rod is made from stainless steel and iron and has a length of 0.50 m. The cross section of this composite rod is shown in the drawing and consists of a square within a circle. The square cross section of the steel is 1.0 cm on a side. The temperature at one end of the rod is 78 °C, while it is 18 °C at the other end. Assuming that no heat exits through the cylindrical outer surface, find the total amount of heat conducted through the rod in two minutes.

Iron

Stainless
steel

PROBLEM 9

10. **M** **CHALK** Review Conceptual Example 5 before attempting this problem. To illustrate the effect of ice on the aluminum cooling plate, consider the drawing shown here and the data that it contains. Ignore any limitations due to significant figures. **(a)** Calculate the heat per second per square meter that is conducted through the ice–aluminum combination.

Aluminum

Ice

−10.0 °C   −25.0 °C

|← 0.0050 m →|←

0.0015 m

PROBLEM 10

(b) Calculate the heat per second per square meter that would be conducted through the aluminum if the ice were not present. Notice how much larger the answer is in (b) as compared to (a).

11. **M** A cubical piece of heat-shield tile from the space shuttle measures 0.10 m on a side and has a thermal conductivity of 0.065 J/ (s · m · C°). The outer surface of the tile is heated to a temperature of 1150 °C, while the inner surface is maintained at a temperature of 20.0 °C. **(a)** How much heat flows from the outer to the inner surface of the tile in five minutes? **(b)** If this amount of heat were transferred to two liters (2.0 kg) of liquid water, by how many Celsius degrees would the temperature of the water rise?

12. **M** **GO** Two pots are identical except that the flat bottom of one is aluminum, whereas that of the other is copper. Water in these pots is boiling away at 100.0 °C at the same rate. The temperature of the heating element on which the aluminum bottom is sitting is 155.0 °C. Assume that heat enters the water only through the bottoms of the pots and find the temperature of the heating element on which the copper bottom rests.

13. **M** **GO** A pot of water is boiling under one atmosphere of pressure. Assume that heat enters the pot only through its bottom, which is copper and rests on a heating element. In two minutes, the mass of water boiled away is $m = 0.45$ kg. The radius of the pot bottom is $R = 6.5$ cm, and the thickness is $L = 2.0$ mm. What is the temperature T_E of the heating element in contact with the pot?

14. **M** **V-HINT** In a house the temperature at the surface of a window is 25 °C. The temperature outside at the window surface is 5.0 °C. Heat is lost through the window via conduction, and the heat lost per second has a certain value. The temperature outside begins to fall, while the conditions inside the house remain the same. As a result, the heat lost per second increases. What is the temperature at the outside window surface when the heat lost per second doubles?

15. **H** The drawing shows a solid cylindrical rod made from a center cylinder of lead and an outer concentric jacket of copper. Except for its ends, the rod is insulated (not shown), so

Copper   Lead

r_1
r_2

PROBLEM 15

that the loss of heat from the curved surface is negligible. When a temperature difference is maintained between its ends, this rod conducts one-half the amount of heat that it would conduct if it were solid copper. Determine the ratio of the radii r_1/r_2.

Section 13.3 Radiation

16. **E** **GO** Light bulb 1 operates with a filament temperature of 2700 K, whereas light bulb 2 has a filament temperature of 2100 K. Both filaments have the same emissivity, and both bulbs radiate the same power. Find the ratio A_1/A_2 of the filament areas of the bulbs.

17. **E** **SSM** The amount of radiant power produced by the sun is approximately 3.9×10^{26} W. Assuming the sun to be a perfect blackbody sphere with a radius of 6.96×10^8 m, find its surface temperature (in kelvins).

18. **E** In an old house, the heating system uses radiators, which are hollow metal devices through which hot water or steam circulates. In one room the radiator has a dark color (emissivity = 0.75). It has a temperature of 62 °C. The new owner of the house paints the radiator a lighter color (emissivity = 0.50). Assuming that it emits the same radiant power as it did before being painted, what is the temperature (in degrees Celsius) of the newly painted radiator?

19. **E** **SSM** A person is standing outdoors in the shade where the temperature is 28 °C. **(a)** What is the radiant energy absorbed per second by his head when it is covered with hair? The surface area of the hair (assumed to be flat) is 160 cm² and its emissivity is 0.85. **(b)** What would be the radiant energy absorbed per second by the same person if he were bald and the emissivity of his head were 0.65?

20. **E** The Kelvin temperature of an object is T_1, and the object radiates a certain amount of energy per second. The Kelvin temperature of the object is then increased to T_2, and the object radiates twice the energy per second that it radiated at the lower temperature. What is the ratio T_2/T_1?

21. **E** A baking dish is removed from a hot oven and placed on a cooling rack. As the dish cools down to 35 °C from 175 °C, its net radiant power decreases to 12.0 W. What was the net radiant power of the baking dish when it was first removed from the oven? Assume that the temperature in the kitchen remains at 22 °C as the dish cools down.

22. **E** **GO** Sirius B is a white star that has a surface temperature (in kelvins) that is four times that of our sun. Sirius B radiates only 0.040 times the power radiated by the sun. Our sun has a radius of 6.96×10^8 m. Assuming that Sirius B has the same emissivity as the sun, find the radius of Sirius B.

23. **E** **GO** A solar collector is placed in direct sunlight where it absorbs energy at the rate of 880 J/s for each square meter of its surface. The emissivity of the solar collector is $e = 0.75$. What equilibrium temperature does the collector reach? Assume that the only energy loss is due to the emission of radiation.

24. **M** **GO** Liquid helium is stored at its boiling-point temperature of 4.2 K in a spherical container ($r = 0.30$ m). The container is a perfect blackbody radiator. The container is surrounded by a spherical shield whose temperature is 77 K. A vacuum exists in the space between the container and the shield. The latent heat of vaporization for helium is 2.1×10^4 J/kg. What mass of liquid helium boils away through a venting valve in one hour?

25. **M** **CHALK** A solid sphere has a temperature of 773 K. The sphere is melted down and recast into a cube that has the same emissivity and emits the same radiant power as the sphere. What is the cube's temperature?

Additional Problems

26. **E** **SSM** Due to a temperature difference ΔT, heat is conducted through an aluminum plate that is 0.035 m thick. The plate is then replaced by a stainless steel plate that has the same temperature difference and cross-sectional area. How thick should the steel plate be so that the same amount of heat per second is conducted through it?

27. **E** **GO** A copper pipe with an outer radius of 0.013 m runs from an outdoor wall faucet into the interior of a house. The temperature of the faucet is 4.0 °C, and the temperature of the pipe, at 3.0 m from the faucet, is 25 °C. In fifteen minutes, the pipe conducts a total of 270 J of heat to the outdoor faucet from the house interior. Find the inner radius of the pipe. Ignore any water inside the pipe.

28. **E** How many days does it take for a perfect blackbody cube (0.0100 m on a side, 30.0 °C) to radiate the same amount of energy that a one-hundred-watt light bulb uses in one hour?

29. **E** **GO** An object is inside a room that has a constant temperature of 293 K. Via radiation, the object emits three times as much power as it absorbs from the room. What is the temperature (in kelvins) of the object? Assume that the temperature of the object remains constant.

30. **E** The concrete wall of a building is 0.10 m thick. The temperature inside the building is 20.0 °C, while the temperature outside is 0.0 °C. Heat is conducted through the wall. When the building is unheated, the inside temperature falls to 0.0 °C, and heat conduction ceases. However, the wall does emit radiant energy when its temperature is 0.0 °C. The radiant energy emitted per second per square meter at 0.0 °C is the same as the heat lost per second per square meter due to conduction when the temperature inside the building is 20.0 °C. What is the emissivity of the wall?

31. **M** **GO** Part *a* of the drawing shows a rectangular bar whose dimensions are $L_0 \times 2L_0 \times 3L_0$. The bar is at the same constant temperature as the room (not shown) in which it is located. The bar is then cut, lengthwise, into two identical pieces, as shown in part *b* of the drawing. The temperature of each piece is the same as that of the original bar. **(a)** What is the ratio of the power absorbed by the two bars in part *b* of the drawing to the single bar in part *a*? **(b)** Suppose that the temperature of the single bar in part *a* is 450.0 K. What would the temperature (in kelvins) of the room and the two bars in part *b* have to be so that the two bars absorb the same power as the single bar in part *a*?

PROBLEM 31   (a)   (b)

32. **M** **V-HINT** Multiple-Concept Example 3 discusses an approach to problems such as this. The ends of a thin bar are maintained at different temperatures. The temperature of the cooler end is 11 °C, while the temperature at a point 0.13 m from the cooler end is 23 °C and the temperature of the warmer end is 48 °C. Assuming that heat flows only along the length of the bar (the sides are insulated), find the length of the bar.

33. **M** **CHALK** **GO** A copper rod has a length of 1.5 m and a cross-sectional area of 4.0×10^{-4} m^2. One end of the rod is in contact with boiling water and the other with a mixture of ice and water. What is the mass of ice per second that melts? Assume that no heat is lost through the side surface of the rod.

34. **M** **GO** **SSM** A wood-burning stove (emissivity = 0.900 and surface area = 3.50 m^2) is being used to heat a room. The fire keeps the stove surface at a constant 198 °C (471 K) and the room at a constant 29 °C (302 K). Determine the *net* radiant power generated by the stove.

Physics in Biology, Medicine, and Sports

35. **E** **BIO** **13.2** The amount of heat per second conducted from the blood capillaries beneath the skin to the surface is 240 J/s. The energy is transferred a distance of 2.0×10^{-3} m through a body whose surface area is 1.6 m^2. Assuming that the thermal conductivity is that of body fat, determine the temperature difference between the capillaries and the surface of the skin.

36. **E** **BIO** **SSM** **13.2** In the conduction equation $Q = (kA\,\Delta T)t/L$, the combination of terms kA/L is called the *conductance*. The human body has the ability to vary the conductance of the tissue beneath the skin by means of vasoconstriction and vasodilation, in which the flow of blood to the veins and capillaries is decreased and increased, respectively. The conductance can be adjusted over a range such that the tissue beneath the skin is equivalent to a thickness of 0.080 mm of Styrofoam or 3.5 mm of air. By what factor (high/low) can the body adjust the conductance?

37. **E** **BIO** **SSM** **13.3** A person eats a dessert that contains 260 Calories. (This "Calorie" unit, with a capital C, is the one used by nutritionists; 1 Calorie = 4186 J. See Section 12.7.) The skin temperature of this individual is 36 °C and that of her environment is 21 °C. The emissivity of her skin is 0.75 and its surface area is 1.3 m^2. How much time would it take for her to emit a *net* radiant energy from her body that is equal to the energy contained in this dessert?

38. **E** **BIO** **13.3** A person's body is producing energy internally due to metabolic processes. If the body loses more energy than metabolic processes are generating, its temperature will drop. If the drop is severe, it can be life-threatening. Suppose that a person is unclothed and energy is being lost via radiation from a body surface area of 1.40 m^2, which has a temperature of 34 °C and an emissivity of 0.700. Also suppose that metabolic processes are producing energy at a rate of 115 J/s. What is the temperature of the coldest room in which this person could stand and not experience a drop in body temperature?

39. **E** **BIO** **MMH** **13.3** Suppose the skin temperature of a naked person is 34 °C when the person is standing inside a room whose temperature is 25 °C. The skin area of the individual is 1.5 m^2. **(a)** Assuming the emissivity is 0.80, find the net loss of radiant power from the body. **(b)** Determine the number of food Calories of energy (1 food Calorie = 4186 J) that are lost in one hour due to the net loss rate obtained in part (a). Metabolic conversion of food into energy replaces this loss.

40. **BIO** **M** **13.3** A person infected with the SARS-CoV-2 virus is running a fever with an internal body temperature of 103 °F (39.4 °C).

This raises the person's skin temperature to 36.5 °C. The person is standing unclothed in a room with an ambient temperature of 21.0 °C. If the temperature of the person's skin is maintained at 36.5 °C solely by the energy obtained from the metabolic conversion of food, how many Calories would this person have to consume per hour? Assume heat is lost only by radiation, and the surface area and emissivity of the person's skin are 1.50 m² and 0.800, respectively.

41. **BIO** **M** **13.3** Hypothermia is a dangerous condition in which the core temperature of the body falls to an unsafe level. In humans, this occurs around 35 °C. Typically, the cause of hypothermia is exposure to cold. Extremely cold conditions, like being outside without protection in below-freezing temperatures or falling into cold water, are certainly dangerous. However, even less extreme conditions can be serious. In fact, any situation that leads to the body's heat loss can result in hypothermia. Furthermore, the outside temperature may not be the most significant factor that contributes to hypothermia. Consider a very cold day, where the ambient air temperature is 4.20 °C, and an unclothed person is outside with a skin temperature of 39.5 °C. Find the ratio of the rate of heat loss of the person completely submerged in water to the rate of heat loss of the same person standing in the open air. Assume the water has the same temperature as the ambient air, there are no wind or water currents, and the heat is lost through the same thickness of water and air. The thermal conductivities of air and water at 4.20 °C are 2.42×10^{-2} W/(m · C°) and 5.74×10^{-1} W/(m · C°), respectively.

42. **BIO** **M** **13.3** Shape memory polymers are a class of materials that can remember their original shape and transition between shapes when stretched, twisted, or compressed. They have recently found uses in orthopedics and as other surgical and dental implants. One of the most promising new materials is PEEK (polyether ether ketone). It is a very good electrical and thermal insulator. As a dental implant material, it provides the patient a comfortable implant that does not conduct heat or alter taste, which can be a constant bother with traditional materials (see the photos). Imagine a PEEK implant that is 1.50 cm in length with a circular diameter of 0.25 cm. An ice cube is in contact with the end of the implant. How much heat is transferred along the length of the implant in 35 seconds, if the change in

temperature across the implant is 27 °C? The thermal conductivity of PEEK is 0.25 W/(m · C°).

Extraction of a tooth followed by the insertion of a PEEK implant. The image of the X-ray shows the implant inserted into a tapered implant body that is secured in the jaw bone.

PROBLEM 42

43. **BIO** **M** **13.3** What is the rate of heat transfer in J/s through the body's skin and the fat layer just beneath its surface? Treat the skin and fat as a single layer that is 1.20 cm thick. Assume the temperature of the inner surface of the layer corresponds to internal body temperature, or 36.7 °C, and the temperature of the outer surface of the layer corresponds to skin temperature, or 34.0 °C. Take the total surface area of the layer to be 1.50 m² and use a value of 0.243 W/(m · C°) for the average thermal conductivity of the layer.

44. **BIO** **M** **13.3** Many surgeries require the use of tools that cut through or remove parts of skeletal bones. These procedures can locally generate a lot of heat. For example, thermal bone necrosis will begin if the bone is kept at 47.0 °C for at least one minute. One of the most important factors that determines how hot the bone gets is its thermal conductivity. Imagine a bone saw cutting through a 2.50-cm length of cylindrical bone with a radius of 1.20 cm. How much heat is conducted along the bone in 1.00 minute of cutting, if the temperature in the bone at the location of the cut is 47.0 °C, and the temperature at the other end of the length of bone is 37 °C? The thermal conductivity of the bone is 0.625 W/(m · C°). Assume all of the heat created by the saw is transferred to the bone by conduction.

Concepts and Calculations Problems

Online

Heat conduction is governed by Equation 13.1, as we have seen. Problem 45 illustrates a familiar application of this relation in the kitchen. It also gives us the opportunity to review the idea of latent heat of vaporization. Problem 46 deals with a case in which heat loss by radiation leads to freezing of water. It stresses the importance of the area from which the radiation occurs, and reviews the concept of the latent heat of fusion.

45. **M** **CHALK** Two pots are identical, except that in one case the flat bottom is aluminum and in the other it is copper. Each pot contains the same amount of boiling water and sits on a heating element that has a temperature of 155 °C. In the aluminum pot, the water boils away completely in 360 s. *Concepts:* (i) Is the heat needed to boil away the water in the aluminum-bottom pot less than, greater than, or the same as the heat needed in the copper-bottom pot? (ii) One of the factors in Equation 13.1 that influences the amount of heat conducted through the bottom of each pot is the temperature difference ΔT between the upper and lower surfaces of the pot's bottom. Is this temperature difference for the aluminum-bottom pot less than, greater than, or the same as that for the copper-bottom pot? (iii) Is

the time required to boil away the water completely in the copper-bottom pot less than, greater than, or the same as that required for the aluminum-bottom pot? *Calculations:* Find the time it takes the water in the copper-bottom pot to boil away completely.

46. **M** **CHALK** **SSM** One half of a kilogram of liquid water at 273 K (0 °C) is placed outside on a day when the temperature is 261 K (−12 °C). Assume that the heat is lost from the water only by means of radiation and that the emissivity of the radiating surface is 0.60. Consider two cases: when the surface area of the water is **(a)** 0.035 m² (as it might be in a cup) and **(b)** 1.5 m² (as it could be if the water were spilled out to form a thin sheet). *Concepts:* (i) In case (a) is the heat that must be removed to freeze the water less than, greater than, or the same as in case (b)? (ii) The loss of heat by radiation depends on the temperature of the radiating object. Does the temperature of the water change as it freezes? (iii) The water both loses and gains heat by radiation. How, then, can heat transfer by radiation lead to freezing of the water? (iv) Will it take longer for the water to freeze in case (a) when the area is smaller or in case (b) when the area is larger? *Calculations:* For each case, (a) and (b), determine the time it takes the water to freeze into ice at 0 °C.

Team Problems

47. **E** **T** **WS** **Flying Close to the Sun.** A space probe you designed is approaching the sun, and you and your team need to determine whether it needs to deviate from its current course. The probe has a large square detector 10.0 meters on a side (area $= 1.00 \times 10^2$ m^2) that consists of a thin sheet composed of strong carbon fibers. The carbon sheet can withstand high temperatures with a melting point of 3600 °C, but aluminum brackets that support the carbon sheet have a relatively lower melting temperature of 660.3 °C. Based on the size and temperature of the star, at the closest distance of approach the sheet will absorb energy at a rate of 48 250 J/s for each square meter of its surface. Assuming that the aluminum brackets are in thermal contact with the sheet and will reach the same temperature, and that the emissivity of the carbon-fiber sheet is $e = 0.880$, what equilibrium temperature will the sheet and aluminum brackets reach? Should you deviate from the current course? You can assume that the only energy loss is due to the emission of radiation, and you can neglect any effects of the small aluminum brackets.

48. **M** **T** **WS** **The Cold Night Sky.** You and your team are on an expedition on a high-altitude mesa and need ice to cool a small vial that contains a perishable biological sample that you have collected. Even though you only need a few grams of ice in order to keep the small, highly insulated sample container cold enough for the trip home the next day, you do not have a freezer with which to make ice. In addition, even though you are at high altitude, the nighttime temperature will reach a low of 4.00 °C, well above the freezing point of water (0 °C). The weather report not only calls for that low temperature (4.00 °C) to be sustained for 6.0 hours during the night, but also that there will be a perfectly clear sky with no moon visible. You then recall something you once read about how ancient Persians made ice in the desert by pouring shallow puddles of water on flat stones in the evening and waking up to find them frozen, even though the ambient temperature never dropped below freezing. You know that at your elevation, and with perfectly clear conditions, the night sky acts like a blackbody with a temperature of −31.0 °C. You take a thick slab of Styrofoam and carve a square cavity into its top surface that is 10.0 cm on a side and 2.0 cm deep. You set the Styrofoam, which acts as an excellent thermal insulator, on flat ground. Next, you pour in 25.0 mL (25.0 g) of water, which covers the bottom of the cavity. **(a)** Starting at a temperature of 4.00 °C, how much heat must be removed from the water to make it freeze? **(b)** If the emissivity of water is $e = 0.950$, what is the rate at which heat is being removed from the water through radiation? **(c)** How long (in hours) will it take the water to freeze with the starting temperature equal to 4.00 °C?

Will the water freeze before the ambient temperature starts to increase, that is, in less than 6.0 hours? (*Hint: You will need to apply Equation 13.3. As an approximation, you can take the starting temperature of the water to be T = 0 °C, since this is the temperature of the water at which most of the heat (i.e., the latent heat of the phase change) is removed.*)

49. **M** **T** **The Diamond Ring Solution.** The processing chip on the computer that controls the navigation equipment on your spacecraft is overheating. Unless you fix the problem, the chip will be damaged and the navigation system will shut down. You open the panel and find that the small copper disk that was supposed to bridge the gap between the smooth top of the chip and the cooling plate is missing, leaving a 2.0 mm gap between them. In this configuration, the heat cannot escape the chip at the required rate. You notice by the thin smudge of thermal grease (a highly thermally conductive material used to promote good thermal contact between surfaces) that the missing copper disk was 2.0 mm thick and had a diameter of 1.0 cm. You know that the chip is designed to run below 75 °C, and the copper cooling plate is held at a constant 5.0 °C. **(a)** What was the rate of heat flow from the chip to the copper plate when the original copper disk was in place and the chip was at its maximum operating temperature? **(b)** The only material that you have available on board to bridge the gap between the chip and copper plate is lead. If the cross-sectional area of the lead piece you plan to wedge into the gap is 1.0 cm^2, what is the rate of heat flow from the chip to the copper plate? Does it match the value calculated in part (a)? **(c)** While brainstorming for other possible solutions to your problem, you happen to glance down at the engagement ring on your finger: a large, glittering diamond. The top and bottom surfaces are flat and nearly rectangular ($L = 0.75$ cm and $W = 0.50$ cm), and the thickness looks to be about 2.0 mm, just right to bridge the gap. You pry the diamond out of its holder and press it into the gap. What is the rate of heat flow now? Good enough?

50. **M** **T** **Indirect Cooling With Liquid Nitrogen.** You are designing a system to cool an insulated silver plate of dimensions 2.00 cm × 2.00 cm × 0.500 cm. One end of a thermally insulated copper wire (diameter $D = 2.50$ mm and length $L = 15.0$ cm) is dipped into a vat of liquid nitrogen ($T = 77.2$ K), and the other end is attached to the bottom of the silver plate. **(a)** If the silver plate starts at room temperature (72.0 °F), what is the initial rate of heat flow between the plate and the liquid nitrogen reservoir? **(b)** Assuming the rate of heat flow calculated in part (a), estimate the temperature of the silver plate after 30.0 seconds.

CHAPTER 14

A scuba diver carries his air supply in the tank on his back. To the extent that the air in a scuba tank behaves like an ideal gas, its pressure, volume, and temperature are related by the ideal gas law. We will see that it is possible to use this law to estimate how long a diver, using a tank of a given size, can stay under the water at a given depth.

Greg Amptman/Shutterstock.com

LEARNING OBJECTIVES

After reading this module, you should be able to...

14.1 Express the amount of a substance in moles.

14.2 Apply the ideal gas law.

14.3 Apply the kinetic theory of gases to ideal gases.

14.4 Solve diffusion problems.

The Ideal Gas Law and Kinetic Theory

14.1 Molecular Mass, the Mole, and Avogadro's Number

Often, we wish to compare the mass of one atom with another. To facilitate the comparison, a mass scale known as the **atomic mass scale** has been established. To set up this scale, a reference value (along with a unit) is chosen for one of the elements. The unit is called the **atomic mass unit** (symbol: u). By international agreement, the reference element is chosen to be the most abundant type or isotope* of carbon, which is called carbon-12. Its atomic mass† is defined to be exactly twelve atomic mass units, or 12 u. The relationship between the atomic mass unit and the kilogram is

$$1\ \mathrm{u} = 1.6605 \times 10^{-27}\ \mathrm{kg}$$

---

*Isotopes are discussed in Section 31.1.
†In chemistry the expression "atomic weight" is frequently used in place of "atomic mass."

The atomic masses of all the elements are listed in the periodic table, part of which is shown in **Interactive Figure 14.1**. The complete periodic table is given on the inside of the back cover of the textbook. In general, the masses listed are average values and take into account the various isotopes of an element that exist naturally. For brevity, the unit "u" is often omitted from the table. For example, a magnesium atom (Mg) has an average atomic mass of 24.305 u, whereas a lithium atom (Li) has an average atomic mass of 6.941 u. Thus, atomic magnesium is more massive than atomic lithium by a factor of (24.305 u)/(6.941 u) = 3.502. In the periodic table, the atomic mass of carbon (C) is given as 12.011 u, rather than exactly 12 u. This is because a small amount (about 1%) of the naturally occurring material is an isotope called carbon-13. The value of 12.011 u is an average that reflects the small contribution of carbon-13.

The molecular mass of a molecule is the sum of the atomic masses of its atoms. For instance, the elements hydrogen and oxygen have atomic masses of 1.007 94 u and 15.9994 u, respectively, so the molecular mass of a water molecule (H_2O) is, therefore, 2(1.007 94 u) + 15.9994 u = 18.0153 u.

INTERACTIVE FIGURE 14.1 A portion of the periodic table showing the atomic number and atomic mass of each element. In the periodic table it is customary to omit the symbol "u" denoting the atomic mass unit.

Macroscopic amounts of materials contain large numbers of atoms or molecules. Even in a small volume of gas, 1 cm^3, for example, the number is enormous. It is convenient to express such large numbers in terms of a single unit, the **gram-mole**, or simply the **mole** (symbol: mol). *One gram-mole of a substance contains as many particles (atoms or molecules) as there are atoms in 12 grams of the isotope carbon-12.* Experiment shows that 12 grams of carbon-12 contain 6.022×10^{23} atoms. The number of atoms per mole is known as **Avogadro's number N_A**, after the Italian scientist Amedeo Avogadro (1776–1856):

$$N_A = 6.022 \times 10^{23}\,\text{mol}^{-1}$$

Thus, the number of moles n contained in any sample is the number of particles N in the sample divided by the number of particles per mole N_A (Avogadro's number):

$$n = \frac{N}{N_A}$$

Although defined in terms of carbon atoms, the concept of a mole can be applied to any collection of objects by noting that one mole contains Avogadro's number of objects. Thus, one mole of atomic sulfur contains 6.022×10^{23} sulfur atoms, one mole of water contains 6.022×10^{23} H_2O molecules, and one mole of golf balls contains 6.022×10^{23} golf balls. The mole is the SI base unit for expressing "the amount of a substance."

The number n of moles contained in a sample can also be found from its mass. To see how, multiply and divide the right-hand side of the previous equation by the mass $m_{particle}$ of a single particle, expressed in grams:

$$n = \frac{m_{particle}\,N}{m_{particle}\,N_A} = \frac{m}{\text{Mass per mole}}$$

The numerator $m_{particle}N$ is the mass of a particle times the number of particles in the sample, which is the mass m of the sample expressed in grams. The denominator $m_{particle}N_A$ is the mass of a particle times the number of particles per mole, which is the mass per mole, expressed in grams per mole.

> **Problem-Solving Insight** The mass per mole (in g/mol) of any substance has the same numerical value as the atomic or molecular mass of the substance (in atomic mass units).

To understand this fact, consider the carbon-12 and sodium atoms as examples. The mass per mole of carbon-12 is 12 g/mol, since, by definition, 12 grams of carbon-12 contain one mole of atoms. On the other hand, the mass per mole of sodium (Na) is 22.9898 g/mol for the following reason: as indicated in **Interactive Figure 14.1**, a sodium atom is more massive than a carbon-12 atom by the ratio of their atomic masses, (22.9898 u)/(12 u) = 1.915 82. Therefore, the mass per mole of sodium is 1.915 82 times as great as that of carbon-12, which means equivalently that (1.915 82)(12 g/mol) = 22.9898 g/mol. Thus, the numerical value of the mass per mole of sodium (22.9898) is the same as the numerical value of its atomic mass.

Since one gram-mole of a substance contains Avogadro's number of particles (atoms or molecules), the mass $m_{particle}$ of a particle (in grams) can be obtained by dividing the mass per mole (in g/mol) by Avogadro's number:

$$m_{particle} = \frac{\text{Mass per mole}}{N_A}$$

Example 1 illustrates how to use the concepts of the mole, atomic mass, and Avogadro's number to determine the number of atoms and molecules present in two famous gemstones.

EXAMPLE 1 | The Physics of Gemstones

Figure 14.2a shows the Hope diamond (44.5 carats), which is almost pure carbon. **Figure 14.2b** shows the Rosser Reeves ruby (138 carats), which is primarily aluminum oxide (Al_2O_3). One carat is equivalent to a mass of 0.200 g. Determine **(a)** the number of carbon atoms in the diamond and **(b)** the number of Al_2O_3 molecules in the ruby.

Reasoning The number N of atoms (or molecules) in a sample is the number of moles n times the number of atoms per mole N_A (Avogadro's number); $N = nN_A$. We can determine the number of moles by dividing the mass of the sample m by the mass per mole of the substance.

Solution **(a)** The Hope diamond's mass is m = (44.5 carats) [(0.200 g)/(1 carat)] = 8.90 g. Since the average atomic mass of naturally occurring carbon is 12.011 u (see the periodic table on the inside of the back cover), the mass per mole of this substance is 12.011 g/mol. The number of moles of carbon in the Hope diamond is

$$n = \frac{m}{\text{Mass per mole}} = \frac{8.90\ \text{g}}{12.011\ \text{g/mol}} = 0.741\ \text{mol}$$

The number of carbon atoms in the Hope diamond is

$$N = nN_A = (0.741\ \text{mol})(6.022 \times 10^{23}\ \text{atoms/mol})$$
$$= \boxed{4.46 \times 10^{23}\ \text{atoms}}$$

(b) The mass of the Rosser Reeves ruby is m = (138 carats)[(0.200 g)/(1 carat)] = 27.6 g. The molecular mass of an aluminum oxide molecule (Al_2O_3) is the sum of the atomic masses of its atoms, which are 26.9815 u for aluminum and 15.9994 u for oxygen (see the periodic table on the inside of the back cover):

$$\text{Molecular mass} = \underbrace{2(26.9815\ \text{u})}_{\substack{\text{Mass of 2} \\ \text{aluminum atoms}}} + \underbrace{3(15.9994\ \text{u})}_{\substack{\text{Mass of 3} \\ \text{oxygen atoms}}} = 101.9612\ \text{u}$$

Thus, the mass per mole of Al_2O_3 is 101.9612 g/mol. Calculations like those in part (a) reveal that the Rosser Reeves ruby contains 0.271 mol or $\boxed{1.63 \times 10^{23}\ \text{molecules of } Al_2O_3}$.

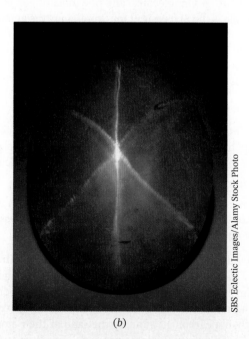

Chip Clark/Smithsonian Institute

SBS Eclectic Images/Alamy Stock Photo

(a)   (b)

FIGURE 14.2 (a) The Hope diamond surrounded by 16 smaller diamonds. (b) The Rosser Reeves ruby. Both gems are on display at the Smithsonian Institution in Washington, D.C.

Check Your Understanding

(*The answers are given at the end of the book.*)

1. Consider one mole of hydrogen (H_2) and one mole of oxygen (O_2). Which, if either, has the greater number of molecules and which, if either, has the greater mass?

2. The molecules of substances A and B are composed of different atoms. However, the two substances have the same mass densities. Consider the possibilities for the molecular masses of the two types of molecules and decide whether 1 m^3 of substance A contains the same number of molecules as 1 m^3 of substance B.

3. A gas mixture contains equal masses of the monatomic gases argon (atomic mass = 39.948 u) and neon (atomic mass = 20.179 u). These two are the only gases present. Of the total number of atoms in the mixture, what percentage is neon?

14.2 | The Ideal Gas Law

An **ideal gas** is an idealized model for real gases that have sufficiently low densities. The condition of low density means that the molecules of the gas are so far apart that they do not interact (except during collisions that are effectively elastic). The ideal gas law expresses the relationship between the absolute pressure, the Kelvin temperature, the volume, and the number of moles of the gas.

In discussing the constant-volume gas thermometer, Section 12.2 has already explained the relationship between the absolute pressure and Kelvin temperature of a low-density gas. This thermometer utilizes a small amount of gas (e.g., hydrogen or helium) placed inside a bulb and kept at a constant volume. Since the density is low, the gas behaves as an ideal gas. Experiment reveals that a plot of gas pressure versus temperature is a straight line, as in Figure 12.4. This plot is redrawn in **Animated Figure 14.3**, with the change that the temperature axis is now labeled in kelvins rather than in degrees Celsius. The graph indicates that the absolute pressure P is directly proportional to the Kelvin temperature T ($P \propto T$), for a fixed volume and a fixed number of molecules.

The relation between absolute pressure and the number of molecules of an ideal gas is simple. Experience indicates that it is possible to increase the pressure of a gas by adding more molecules; this is exactly what happens when a tire is pumped up. When the volume and temperature of a low-density gas are kept constant, doubling the number of molecules doubles the pressure. Thus, the absolute pressure of an ideal gas at constant temperature and constant volume is proportional to the number of molecules or, equivalently, to the number of moles n of the gas ($P \propto n$).

To see how the absolute pressure of a gas depends on the volume of the gas, look at the partially filled balloon in **Figure 14.4a**. This balloon is "soft," because the pressure of the air is low. However, if all the air in the balloon is squeezed into a smaller "bubble," as in part b of the figure, the "bubble" has a tighter feel. This tightness indicates that the pressure in the smaller volume is high enough to stretch the rubber substantially. Thus, it is possible to increase the pressure of a gas by reducing its volume, and if the number of molecules and the temperature are kept constant, the absolute pressure of an ideal gas is inversely proportional to its volume V ($P \propto 1/V$).

The three relations just discussed for the absolute pressure of an ideal gas can be expressed as a single proportionality, $P \propto nT/V$. This proportionality can be written as an equation by inserting a proportionality constant R, called the **universal gas constant**. Experiments have shown that $R = 8.31$ J/(mol · K) for any real gas with a density sufficiently low to ensure ideal gas behavior. The resulting equation is called the **ideal gas law**.

ANIMATED FIGURE 14.3 The pressure inside a constant-volume gas thermometer is directly proportional to the Kelvin temperature, a proportionality that is characteristic of an ideal gas.

FIGURE 14.4 (*a*) A partially filled balloon. (*b*) The air pressure in the partially filled balloon can be increased by decreasing the volume of the balloon as shown.

(*a*)   (*b*)

IDEAL GAS LAW

The absolute pressure *P* of an ideal gas is directly proportional to the Kelvin temperature *T* and the number of moles *n* of the gas and is inversely proportional to the volume *V* of the gas: $P = R(nT/V)$. In other words,

$$PV = nRT \qquad\qquad (14.1)$$

where *R* is the universal gas constant and has the value of 8.31 J/(mol · K).

Sometimes, it is convenient to express the ideal gas law in terms of the total number of particles *N*, instead of the number of moles *n*. To obtain such an expression, we multiply and divide the right side of Equation 14.1 by Avogadro's number $N_A = 6.022 \times 10^{23}$ particles/mol* and recognize that the product nN_A is equal to the total number *N* of particles:

$$PV = nRT = nN_A\left(\frac{R}{N_A}\right)T = N\left(\frac{R}{N_A}\right)T$$

The constant term R/N_A is referred to as **Boltzmann's constant**, in honor of the Austrian physicist Ludwig Boltzmann (1844–1906), and is represented by the symbol *k*:

$$k = \frac{R}{N_A} = \frac{8.31 \text{ J/(mol} \cdot \text{K)}}{6.022 \times 10^{23} \text{ mol}^{-1}} = 1.38 \times 10^{-23} \text{ J/K}$$

With this substitution, the ideal gas law becomes

$$PV = NkT \qquad\qquad (14.2)$$

Example 2 presents an application of the ideal gas law.

EXAMPLE 2 | BIO The Physics of Oxygen in the Lungs

In the lungs, a thin respiratory membrane separates tiny sacs of air (absolute pressure = 1.00×10^5 Pa) from the blood in the capillaries. These sacs are called alveoli, and it is from them that oxygen enters the blood. The average radius of the alveoli is 0.125 mm, and the air inside contains 14% oxygen. Assuming that the air behaves as an ideal gas at body temperature (310 K), find the number of oxygen molecules in one of the sacs.

Reasoning The pressure and temperature of the air inside an alveolus are known, and its volume can be determined since we know the radius. Thus, the ideal gas law in the form $PV = NkT$ can be used directly to find the number *N* of air particles inside one of the sacs. The number of oxygen molecules is 14% of the number of air particles.

Problem-Solving Insight In the ideal gas law, the temperature *T* must be expressed on the Kelvin scale. The Celsius and Fahrenheit scales cannot be used.

Solution The volume of a spherical sac is $V = \frac{4}{3}\pi r^3$, where *r* is the radius. Solving Equation 14.2 for the number of air particles, we have

$$N = \frac{PV}{kT} = \frac{(1.00 \times 10^5 \text{ Pa})\left[\frac{4}{3}\pi(0.125 \times 10^{-3}\text{ m})^3\right]}{(1.38 \times 10^{-23} \text{ J/K})(310 \text{ K})} = 1.9 \times 10^{14}$$

The number of oxygen molecules is 14% of this value, or $0.14N = \boxed{2.7 \times 10^{13}}$.

---

*"Particles" is not an SI unit and is often omitted. Then, particles/mol = 1/mol = mol⁻¹.

With the aid of the ideal gas law, it can be shown that one mole of an ideal gas occupies a volume of 22.4 liters at a temperature of 273 K (0 °C) and a pressure of one atmosphere (1.013×10^5 Pa). These conditions of temperature and pressure are known as **standard temperature and pressure (STP)**. Conceptual Example 3 discusses the physics of beer bubbles, and Example 4 applies the ideal gas law to hyperbaric medicine.

CONCEPTUAL EXAMPLE 3 | The Physics of Rising Beer Bubbles

If you look carefully at the bubbles rising in a glass of beer (see **Figure 14.5**), you'll see them grow in size as they move upward, often doubling in volume by the time they reach the surface. Beer bubbles contain mostly carbon dioxide (CO_2), a gas that is dissolved in the beer because of the fermentation process. Which variable describing the gas is responsible for the growth of the rising bubbles? **(a)** The Kelvin temperature T **(b)** The absolute pressure P **(c)** The number of moles n

Reasoning   The variables T, P, and n are related to the volume V of a bubble by the ideal gas law ($V = nRT/P$). We assume that this law applies and use it to guide our thinking. According to this law, an increase in temperature, a decrease in pressure, or an increase in the number of moles could account for the growth in size of the upward-moving bubbles.

Answers (a) and (b) are incorrect.   Temperature can be eliminated immediately, since it is constant throughout the beer. Pressure cannot be dismissed so easily. As a bubble rises, its depth decreases, and so does the fluid pressure that a bubble experiences. Since volume is inversely proportional to pressure according to the ideal gas law, at least part of the bubble growth is due to the decreasing pressure of the surrounding beer. However, some bubbles *double in volume* on the way up. To account for the doubling, there would need to be two atmospheres of pressure at the bottom of the glass, compared to the one atmosphere at the top. The pressure increment due to depth is ρgh (see Equation 11.4), so an extra pressure of one atmosphere at the bottom would mean 1.01×10^5 Pa $= \rho gh$. Solving for h with ρ equal to the density of water reveals that $h = 10.3$ m. Since most beer glasses are only about 0.2 m tall, we can rule out a change in pressure as the major cause of the change in volume.

Answer (c) is correct.   The process of elimination brings us to the conclusion that the number of moles of CO_2 in a bubble must somehow be increasing on the way up. This is, in fact, the case. Each bubble acts as a nucleation site for CO_2 molecules dissolved in the surrounding beer, so as a bubble moves upward, it accumulates carbon dioxide and grows larger.

Courtesy Richard Zare, Stanford University

FIGURE 14.5   The bubbles in a glass of beer grow larger as they move upward.

Related Homework: Problem 22

EXAMPLE 4 | BIO Hyperbaric Medicine

Hyperbaric Medicine   A hyperbaric chamber like that shown in **Figure 14.6** provides a high-pressure, high-oxygen environment to treat medical conditions such as carbon monoxide poisoning, cancer, burns, and wounds that are difficult to heal. One of the more prominent uses, however, is to treat decompression sickness that sometimes occurs in diving accidents when a person is brought to the surface too quickly. The process involves placing a patient in a chamber and increasing the pressure, while supplying them with pure oxygen, often through a tightly fitted mask. The pressures can reach as high as six times the pressure at sea level, that is, 6 atmospheres! Suppose a cylindrical hyperbaric chamber has a radius $r = 1.10$ m and a length $L = 2.70$ m and is to be filled with pure nitrogen (N_2) gas to a pressure $P = 6.00$ atm. The temperature of the interior of the chamber is set to 22.0 °C. Assuming that nitrogen gas (N_2) behaves like an ideal gas with a molar mass of 28.0 g/mol, determine the mass of nitrogen inside the chamber in kg.

FIGURE 14.6 A hyperbaric chamber for the treatment of ailments such as carbon monoxide poisoning and decompression sickness.

Reasoning The pressure and temperature are known, and the volume of the chamber can be calculated since its dimensions are given. To find the mass m, we can apply the ideal gas law in the form $PV = nRT$, and use the definition of the number of moles $n = m/m_{mol}$, where m_{mol} is the molar mass of nitrogen (N_2).

Solution Solving Equation 14.1 (the ideal gas law) for n, and substituting for the volume of the chamber ($V = \pi r^2 L$), we have

$$n = \frac{PV}{RT} = \frac{\pi r^2 LP}{RT}$$

The radius of the chamber is $r = D/2 = 0.550$ m. Note that we must also convert the pressure to Pa (6.00 atm $= 6.08 \times 10^5$ Pa) and the temperature to Kelvin (22.0 °C $= 295$ K). Since $n = m/m_{mol}$, and the molar mass of nitrogen is $m_{mol} = 28.0$ g/mol, we have

$$m = \left(\frac{\pi r^2 LP}{RT}\right) m_{mol}$$

$$= \left(\frac{\pi(0.550 \text{ m})^2(2.70 \text{ m})(6.08 \times 10^5 \text{ Pa})}{(8.31 \text{ J} \cdot \text{mol}^{-1} \cdot \text{K}^{-1})(295 \text{ K})}\right)(28.0 \text{ g/mol})$$

$$= 17.8 \times 10^3 \text{ g} = \boxed{17.8 \text{ kg}}$$

FIGURE 14.7 A pressure-versus-volume plot for a gas at a constant temperature is called an isotherm. For an ideal gas, each isotherm is a plot of the equation $P = nRT/V = $ constant/V.

Historically, the work of several investigators led to the formulation of the ideal gas law. The Irish scientist Robert Boyle (1627–1691) discovered that at a constant temperature, the absolute pressure of a fixed mass (fixed number of moles) of a low-density gas is inversely proportional to its volume ($P \propto 1/V$). This fact is often called Boyle's law and can be derived from the ideal gas law by noting that $P = nRT/V = $ constant/V when n and T are constants. Alternatively, if an ideal gas changes from an initial pressure and volume (P_i, V_i) to a final pressure and volume (P_f, V_f), it is possible to write $P_i V_i = nRT$ and $P_f V_f = nRT$. Since the right sides of these equations are equal, we may equate the left sides to give the following concise way of expressing **Boyle's law:**

Constant T, constant n $P_i V_i = P_f V_f$ **(14.3)**

Figure 14.7 illustrates how pressure and volume change according to Boyle's law for a fixed number of moles of an ideal gas at a constant temperature of 100 K. The gas begins with an initial pressure and volume of P_i and V_i and is compressed. The pressure increases as the volume decreases, according to $P = nRT/V$, until the final pressure and volume of P_f and V_f are reached. The curve that passes through the initial and final points is called an **isotherm**, meaning "same temperature." If the temperature had been 300 K, rather than 100 K, the compression would have occurred along the 300-K isotherm. Different isotherms do not intersect. Example 5 deals with an application of Boyle's law to scuba diving.

Analyzing Multiple-Concept Problems

EXAMPLE 5 | The Physics of Scuba* Diving

When a scuba diver descends to greater depths, the water pressure increases. The air pressure inside the body cavities (e.g., lungs, sinuses) must be maintained at the same pressure as that of the surrounding water; otherwise the cavities would collapse. A special valve automatically adjusts the pressure of the air coming from the scuba tank to ensure that the air pressure equals the water pressure at all times. The scuba gear in **Figure 14.8a** consists of a 0.0150-m³ tank filled with compressed air at an absolute pressure of 2.02×10^7 Pa. Assume that the diver consumes air at the rate of 0.0300 m³ per minute and that the temperature of the air does not change as the diver goes deeper into the water. How long (in minutes) can a diver stay under water at a depth of 10.0 m? Take the density of seawater to be 1025 kg/m³.

Reasoning The time (in minutes) that a scuba diver can remain under water is equal to the volume of air that is

*The word is an acronym for *self-contained underwater breathing apparatus.*

FIGURE 14.8 (*a*) The air pressure inside the body cavities of a scuba diver must be maintained at the same value as the pressure of the surrounding water. (*b*) The pressure P_2 at a depth h is greater than the pressure P_1 at the surface.

available divided by the volume per minute consumed by the diver. The volume of air available to the diver depends on the volume and pressure of the air in the scuba tank, as well as the pressure of the air inhaled by the diver, according to

Boyle's law. The pressure of the air inhaled equals the water pressure that acts on the diver. This pressure can be found from a knowledge of the diver's depth beneath the surface of the water.

Knowns and Unknowns   The data for this problem are listed below:

Description	Symbol	Value	Comment
Explicit Data			
Volume of air in tank	V_i	0.0150 m^3	
Pressure of air in tank	P_i	$2.02 \times 10^7 \text{ Pa}$	
Rate of air consumption	C	$0.0300 \text{ m}^3/\text{min}$	
Mass density of seawater	ρ	1025 kg/m^3	
Depth of diver	h	10.0 m	
Implicit Data			
Air pressure at surface of water	P_1	$1.01 \times 10^5 \text{ Pa}$	Atmospheric pressure at sea level (see Section 11.2).
Unknown Variable			
Time that diver can remain at 10.0-m depth	t	?	

Modeling the Problem

STEP 1 Duration of the Dive   The air inside the scuba tank has an initial pressure of P_i and a volume of V_i (the volume of the tank). A scuba diver does *not* breathe the air *directly* from the tank, because the tank pressure of 2.02×10^7 Pa is nearly 200 times atmospheric pressure and would cause his lungs to explode. Instead, a valve on the tank adjusts the pressure of the air being sent to the diver so it equals the surrounding water pressure P_f. The time t (in minutes) that the diver can remain under water is equal to the total volume of air consumed by the diver divided by the rate C (in cubic meters per minute) at which the air is consumed:

$$t = \frac{\text{Total volume of air consumed}}{C}$$

The total volume of air consumed is the volume V_f available at the breathing pressure P_f minus the volume V_i of the scuba tank, because this amount of air always remains behind in the tank. Thus, we have Equation 1 at the right. The volume V_i and the rate C are known, but the final volume V_f is not, so we turn to Step 2 to evaluate it.

$$t = \frac{V_f - V_i}{\boxed{?} \, C} \tag{1}$$

STEP 2 Boyle's Law Since the temperature of the air remains constant, the air volume V_f available to the diver at the pressure P_f is related to the initial pressure P_i and volume V_i of air in the tank by Boyle's law $P_iV_i = P_fV_f$ (Equation 14.3). Solving for V_f yields

$$\boxed{V_f = \frac{P_iV_i}{P_f}}$$

This expression for V_f can be substituted into Equation 1, as indicated at the right. The initial pressure P_i and volume V_i are given. However, we still need to determine the pressure P_f of the air inhaled by the diver, and we will evaluate it in the next step.

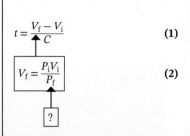

$$t = \frac{V_f - V_i}{C} \qquad (1)$$

$$V_f = \frac{P_iV_i}{P_f} \qquad (2)$$

$$\boxed{?}$$

STEP 3 Pressure and Depth in a Static Fluid Figure 14.8b shows the diver at a depth h below the surface of the water. The absolute pressure P_2 at this depth is related to the pressure P_1 at the surface of the water by Equation 11.4: $P_2 = P_1 + \rho g h$, where ρ is the mass density of seawater and g is the magnitude of the acceleration due to gravity. Since P_1 is the air pressure at the surface of the water, it is atmospheric pressure. Recall that the valve on the scuba tank adjusts the pressure P_f of the air inhaled by the diver to be equal to the pressure P_2 of the surrounding water. Thus, $P_2 = P_f$, and Equation 11.4 becomes

$$\boxed{P_f = P_1 + \rho g h}$$

We now substitute this expression for P_f into Equation 2, as indicated in the right column.

$$t = \frac{V_f - V_i}{C} \qquad (1)$$

$$V_f = \frac{P_iV_i}{P_f} \qquad (2)$$

$$P_f = P_1 + \rho g h$$

Problem-Solving Insight When using the ideal gas law, either directly or in the form of Boyle's law, remember that the pressure P must be the absolute pressure, not the gauge pressure.

Solution Algebraically combining the results of the three steps, we have

$$\overset{\text{STEP 1}}{\underset{t=}{\downarrow}} \overset{\text{STEP 2}}{\underset{\frac{V_f - V_i}{C}}{\downarrow}} = \overset{\text{STEP 3}}{\underset{\frac{\frac{P_iV_i}{P_f} - V_i}{C}}{\downarrow}} = \frac{\frac{P_iV_i}{P_1 + \rho g h} - V_i}{C}$$

The time that the diver can remain at a depth of 10.0 m is

$$t = \frac{\dfrac{P_iV_i}{P_1 + \rho g h} - V_i}{C}$$

$$= \frac{\dfrac{(2.02 \times 10^7\,\text{Pa})(0.0150\,\text{m}^3)}{1.01 \times 10^5\,\text{Pa} + (1025\,\text{kg/m}^3)(9.80\,\text{m/s}^2)(10.0\,\text{m})} - 0.0150\,\text{m}^3}{0.0300\,\text{m}^3/\text{min}} = \boxed{49.6\,\text{min}}$$

Note that at a fixed consumption rate C, greater values for h lead to smaller values for t. In other words, a deeper dive must have a shorter duration.

Related Homework: Problems 18, 22

Another investigator whose work contributed to the formulation of the ideal gas law was the Frenchman Jacques Charles (1746–1823). He discovered that at a constant pressure, the volume of a fixed mass (fixed number of moles) of a low-density gas is directly proportional to the Kelvin temperature ($V \propto T$). This relationship is known as Charles' law and can be obtained from the ideal gas law by noting that $V = nRT/P = (\text{constant})T$, if n and P are constant. Equivalently, when an ideal gas changes from an initial volume and temperature (V_i, T_i) to a final volume and temperature (V_f, T_f), it is possible to write $V_i/T_i = nR/P$ and $V_f/T_f = nR/P$. Thus, one way of stating **Charles' law** is

Constant P, constant n $$\frac{V_i}{T_i} = \frac{V_f}{T_f} \qquad (14.4)$$

Check Your Understanding

(The answers are given at the end of the book.)

4. A tightly sealed house has a large ceiling fan that blows air out of the house and into the attic. The owners turn the fan on and forget to open any windows or doors. What happens to the air pressure in the house after the fan has been on for a while, and does it become easier or harder for the fan to do its job?

5. Above the liquid in a can of hair spray is a gas at a relatively high pressure. The label on the can includes the warning "DO NOT STORE AT HIGH TEMPERATURES." Why is the warning given?

6. What happens to the pressure in a tightly sealed house when the electric furnace turns on and runs for a while?

7. BIO When you climb a mountain, your eardrums "pop" outward as the air pressure decreases. When you come down, they pop inward as the pressure increases. At the sea coast, you swim through a completely submerged passage and emerge into a pocket of air trapped within a cave. As the tide comes in, the water level in the cave rises, and your eardrums pop. Is this popping analogous to what happens as you climb up or climb down a mountain?

8. Atmospheric pressure decreases with increasing altitude. Given this fact, explain why helium-filled weather balloons are underinflated when they are launched from the ground. Assume that the temperature does not change much as the balloon rises.

9. A slippery cork is being pressed into an almost full (but not 100% full) bottle of wine. When released, the cork slowly slides back out. However, if half the wine is removed from the bottle before the cork is inserted, the cork does not slide out. Explain.

10. Consider equal masses of three monatomic gases: argon (atomic mass = 39.948 u), krypton (atomic mass = 83.80 u), and xenon (atomic mass = 131.29 u). The pressure and volume of each gas is the same. Which gas has the greatest and which the smallest temperature?

14.3 | Kinetic Theory of Gases

As useful as it is, the ideal gas law provides no insight as to how pressure and temperature are related to properties of the molecules themselves, such as their masses and speeds. To show how such microscopic properties are related to the pressure and temperature of an ideal gas, this section examines the dynamics of molecular motion. The pressure that a gas exerts on the walls of a container is due to the force exerted by the gas molecules when they collide with the walls. Therefore, we will begin by combining the notion of collisional forces exerted by a fluid (Section 11.2) with Newton's second and third laws of motion (Sections 4.3 and 4.5). These concepts will allow us to obtain an expression for the pressure in terms of microscopic properties. We will then combine this with the ideal gas law to show that the average translational kinetic energy $\overline{KE}$ of a particle in an ideal gas is $\overline{KE} = \frac{3}{2}kT$ where k is Boltzmann's constant and T is the Kelvin temperature. In the process, we will also see that the internal energy U of a monatomic ideal gas is $U = \frac{3}{2}nRT$ where n is the number of moles and R is the universal gas constant.

The Distribution of Molecular Speeds

A macroscopic container filled with a gas at standard temperature and pressure contains a large number of particles (atoms or molecules). These particles are in constant, random motion, colliding with each other and with the walls of the container. In the course of one second, a particle undergoes many collisions, and each one changes the particle's speed and direction of motion. As a result, the atoms or molecules have different speeds. It is possible, however, to speak about an average particle speed. At any given instant, some particles have speeds less than, some near, and some greater than the average. For conditions of low gas density, the distribution of speeds within a large collection of molecules at a constant temperature was calculated by the Scottish physicist James Clerk

FIGURE 14.9 The Maxwell distribution curves for molecular speeds in oxygen gas at temperatures of 300 and 1200 K.

Maxwell (1831–1879). **Figure 14.9** displays the Maxwell speed distribution curves for O_2 gas at two different temperatures. When the temperature is 300 K, the maximum in the curve indicates that the most probable speed is about 400 m/s. At a temperature of 1200 K, the distribution curve shifts to the right, and the most probable speed increases to about 800 m/s. One particularly useful type of average speed, known as the rms speed and written as v_{rms}, is also shown in the drawing. When the temperature of the oxygen gas is 300 K the rms speed is 484 m/s, and it increases to 967 m/s when the temperature rises to 1200 K. The meaning of the rms speed and the reason why it is so important will be discussed shortly.

Kinetic Theory

FIGURE 14.10 The pressure that a gas exerts is caused by the collisions of its molecules with the walls of the container.

If a ball is thrown against a wall, it exerts a force on the wall. As **Figure 14.10** suggests, gas particles do the same thing, except that their masses are smaller and their speeds are greater. The number of particles is so great and they strike the wall so often that the effect of their individual impacts appears as a continuous force. Dividing the magnitude of this force by the area of the wall gives the pressure exerted by the gas.

To calculate the force, consider an ideal gas composed of N identical particles in a cubical container whose sides have length L. Except for elastic* collisions, these particles do not interact. **Figure 14.11** focuses attention on one particle of mass m as it strikes the right wall perpendicularly and rebounds elastically. While approaching the wall, the particle has a velocity $+v$ and linear momentum $+mv$ (see Section 7.1 for a review of linear momentum). The particle rebounds with a velocity $-v$ and momentum $-mv$, travels to the left wall, rebounds again, and heads back toward the right. The time t between collisions with the right wall is the round-trip distance $2L$ divided by the speed of the particle; that is, $t = 2L/v$. According to Newton's second law of motion, in the form of the impulse-momentum, the average force exerted on the particle by the wall is given by the change in the particle's momentum per unit time:

FIGURE 14.11 A gas particle is shown colliding elastically with the right wall of the container and rebounding from it.

$$\text{Average force} = \frac{\text{Final momentum} - \text{Initial momentum}}{\text{Time between successive collisions}} \tag{7.4}$$

$$= \frac{(-mv) - (+mv)}{2L/v} = \frac{-mv^2}{L}$$

According to Newton's law of action–reaction, the force applied to the wall by the particle is equal in magnitude to this value, but oppositely directed (i.e., $+mv^2/L$).

*The term "elastic" is used here to mean that *on the average*, in a large number of particles, there is no gain or loss of translational kinetic energy because of collisions.

Math Skills   It is important to note that the average of the squared speed $\overline{v^2}$ is **not the same thing** as the square of the average speed, which is $\overline{v}^2$. To see why these two quantities are not equal, consider two particles with the following speeds: v_A is the speed of particle A and v_B is the speed of particle B. To compute the average of the squared speed and the square of the average speed, we proceed in the usual fashion when computing averages. We calculate each average by adding the two values (one for particle A and one for particle B) and then dividing the result by 2.

$$\overline{v^2} = \frac{v_A^2 + v_B^2}{2}$$

$$\overline{v}^2 = \left(\frac{v_A + v_B}{2}\right)^2 = \frac{v_A^2 + 2v_A v_B + v_B^2}{4}$$

Clearly, $\overline{v^2}$ is not equal to $\overline{v}^2$.

The magnitude F of the *total* force exerted on the right wall is equal to the number of particles that collide with the wall during the time t multiplied by the average force exerted by each particle. Since the N particles move randomly in three dimensions, one-third of them on the average strike the right wall during the time t. Therefore, the total force is

$$F = \left(\frac{N}{3}\right)\left(\frac{m\overline{v^2}}{L}\right)$$

In this result v^2 has been replaced by $\overline{v^2}$, *the average value* of the squared speed. The collection of particles possesses a Maxwell distribution of speeds, so an average value for v^2 must be used, rather than a value for any individual particle. The square root of the quantity $\overline{v^2}$ is called the **root-mean-square speed**, or, for short, the **rms speed**; $v_{rms} = \sqrt{\overline{v^2}}$. With this substitution, the total force becomes

$$F = \left(\frac{N}{3}\right)\left(\frac{mv_{rms}^2}{L}\right)$$

Pressure is force per unit area, so the pressure P acting on a wall of area L^2 is

$$P = \frac{F}{L^2} = \left(\frac{N}{3}\right)\left(\frac{mv_{rms}^2}{L^3}\right) = \left(\frac{N}{3}\right)\left(\frac{mv_{rms}^2}{V}\right)$$

where $V = L^3$ is the volume of the box. This result can be written as

$$PV = \tfrac{2}{3}N\left(\tfrac{1}{2}mv_{rms}^2\right) \tag{14.5}$$

Equation 14.5 relates the macroscopic properties of the gas—its pressure and volume—to the microscopic properties of the constituent particles—their mass and speed. Since the term $\tfrac{1}{2}mv_{rms}^2$ is the average translational kinetic energy $\overline{KE}$ of an individual particle, it follows that

$$PV = \tfrac{2}{3}N\,(\overline{KE})$$

This result is similar to the ideal gas law, $PV = NkT$ (Equation 14.2). Both equations have identical terms on the left, so the terms on the right must be equal: $\tfrac{2}{3}N(\overline{KE}) = NkT$. Therefore,

$$\overline{KE} = \tfrac{1}{2}mv_{rms}^2 = \tfrac{3}{2}kT \tag{14.6}$$

Equation 14.6 is significant, because it allows us to interpret temperature in terms of the motion of gas particles. This equation indicates that the Kelvin temperature is directly proportional to the average translational kinetic energy per particle in an ideal gas, no matter what the pressure and volume are. On the average, the particles have greater kinetic energies when the gas is hotter than when it is cooler. Conceptual Example 6 discusses a common misconception about the relation between kinetic energy and temperature.

CONCEPTUAL EXAMPLE 6 | Does a Single Particle Have a Temperature?

Each particle in a gas has kinetic energy. Furthermore, the equation $\frac{1}{2}mv_{rms}^2 = \frac{3}{2}kT$ establishes the relationship between the average kinetic energy per particle and the temperature of an ideal gas. Is it valid, then, to conclude that a single particle has a temperature?

Reasoning and Solution We know that a gas contains an enormous number of particles that are traveling with a distribution of speeds, such as those indicated by the graphs in Figure 14.9. Therefore, the particles do not all have the same kinetic energy, but possess a distribution of kinetic energies ranging from very nearly zero to extremely large values. If each particle had a temperature that was associated with its kinetic energy, there would

be a whole range of different temperatures within the gas. This is not so, for a gas at thermal equilibrium has only one temperature (see Section 15.2), a temperature that would be registered by a thermometer placed in the gas. Thus, temperature is a property that characterizes the gas as a whole, a fact that is inherent in the relation $\frac{1}{2}mv_{rms}^2 = \frac{3}{2}kT$. The term v_{rms} is a kind of *average particle speed*. Therefore, $\frac{1}{2}mv_{rms}^2$ is the *average kinetic energy* per particle and is characteristic of the gas as a whole. Since the Kelvin temperature is proportional to $\frac{1}{2}mv_{rms}^2$ it is also a characteristic of the gas as a whole and cannot be ascribed to each gas particle individually. Thus, *a single gas particle does not have a temperature*.

If two ideal gases have the same temperature, the relation $\frac{1}{2}mv_{rms}^2 = \frac{3}{2}kT$ indicates that the average kinetic energy of each kind of gas particle is the same. In general, however, the rms speeds of the different particles are not the same, because the masses may be different. Air, for example, is primarily a mixture of nitrogen N_2 (molecular mass = 28.0 u) and oxygen O_2 (molecular mass = 32.0 u). If we assume that each behaves like an ideal gas and apply Equation 14.6, we find the rms speeds of nitrogen and oxygen at room temperature (293 K) are 511 m/s and 478 m/s, respectively. As a comparison, the speed of sound in air at the same temperature is 343 m/s.

The equation $\overline{KE} = \frac{3}{2}kT$ has also been applied to particles much larger than atoms or molecules. The English botanist Robert Brown (1773–1858) observed through a microscope that pollen grains suspended in water move on very irregular, zigzag paths. This Brownian motion can also be observed with other particle suspensions, such as fine smoke particles in air. In 1905, Albert Einstein (1879–1955) showed that Brownian motion could be explained as a response of the large suspended particles to impacts from the moving molecules of the fluid medium (e.g., water or air). As a result of the impacts, the suspended particles have the same average translational kinetic energy as the fluid molecules—namely, $\overline{KE} = \frac{3}{2}kT$. Unlike the molecules, however, the particles are large enough to be seen through a microscope and, because of their relatively large mass, have a comparatively small average speed.

The Internal Energy of a Monatomic Ideal Gas

Chapter 15 deals with the science of thermodynamics, in which the concept of internal energy plays an important role. Using the results just developed for the average translational kinetic energy, we conclude this section by expressing the internal energy of a monatomic ideal gas in a form that is suitable for use later on.

The internal energy of a substance is the sum of the various kinds of energy that the atoms or molecules of the substance possess. A monatomic ideal gas is composed of single atoms. These atoms are assumed to be so small that the mass is concentrated at a point, with the result that the moment of inertia I about the center of mass is negligible. Thus, the rotational kinetic energy $\frac{1}{2}I\omega^2$ is also negligible. Vibrational kinetic and potential energies are absent, because the atoms are not connected by chemical bonds and, except for elastic collisions, do not interact. As a result, the internal energy U is the total translational kinetic energy of the N atoms that constitute the gas: $U = N(\frac{1}{2}mv_{rms}^2)$.

Since $\frac{1}{2}mv_{rms}^2 = \frac{3}{2}kT$ according to Equation 14.6, the internal energy can be written in terms of the Kelvin temperature as

$$U = N\left(\tfrac{3}{2}kT\right)$$

Usually, U is expressed in terms of the number of moles n, rather than the number of atoms N. Using the fact that Boltzmann's constant is $k = R/N_A$, where R is the universal gas constant and N_A is Avogadro's number, and realizing that $N/N_A = n$, we find that

Monatomic ideal gas $\qquad\qquad\qquad U = \tfrac{3}{2}nRT \qquad\qquad$ **(14.7)**

Thus, the internal energy depends on the number of moles and the Kelvin temperature of the gas. In fact, it can be shown that the internal energy is proportional to the Kelvin temperature for *any type* of ideal gas (e.g., monatomic, diatomic, etc.). For example, when hot-air balloonists turn on the burner, they increase the temperature, and hence the internal energy per mole, of the air inside the balloon (see **Figure 14.12**).

Example 7 illustrates how Equation 14.7 can be used to calculate how much internal energy is contained in the air inside human lungs.

MUSTAFA QURAISHI/AP Images

FIGURE 14.12 When the burner is turned on to heat the air within a hot-air balloon, the temperature of the air rises. Air behaves approximately as an ideal gas, for which the internal energy per mole is proportional to the Kelvin temperature.

EXAMPLE 7 | BIO The Energy in Your Lungs

How much internal energy is contained in the air within your lungs? The result might surprise you. The volume of human adult lungs is approximately 6 liters ($6.0 \times 10^{-3}\,\text{m}^3$), and the pressure in your lungs is approximately 1 atm ($1.0 \times 10^5\,\text{Pa}$). Assume the temperature of the air is close to internal body temperature or 37 °C (310 K). To simplify the calculation, we will assume your lungs are filled with helium gas, which is something many of you might have done to make your voice sound funny!

Reasoning Since we are using helium gas in our calculation, which is monatomic, we can apply Equation 14.7 to calculate its internal energy. The internal energy depends on n, the number of moles of the gas. This quantity can be determined by applying the ideal gas law (Equation 14.1) with the information given in the problem.

Solution We begin by calculating the number of moles of helium gas in the lungs using Equation 14.1:

$$n = \frac{PV}{RT} = \frac{(1.0 \times 10^5\,\text{Pa})(6.0 \times 10^{-3}\,\text{m}^3)}{[8.31\,\text{J/(mol} \cdot \text{K)}](310\,\text{K})} = 0.23\,\text{mol}.$$

Using this value for n, we can apply Equation 14.7 to calculate the internal energy of the gas:

$$U = \tfrac{3}{2}nRT = \tfrac{3}{2}(0.23\,\text{mol})[8.31\,\text{J/(mol} \cdot \text{K)}](310\,\text{K}) = \boxed{890\,\text{J}}$$

Is this a large energy? It is! This value is equivalent to the kinetic energy of a baseball moving at 110 m/s, or almost 250 mph!

Check Your Understanding

(The answers are given at the end of the book.)

11. The kinetic theory of gases assumes that, for a given collision time, a gas molecule rebounds with the same speed after colliding with the wall of a container. If the speed after the collision were less than the speed before the collision, the duration of the collision remaining the same, would the pressure of the gas be greater than, equal to, or less than the pressure predicted by kinetic theory?

12. If the temperature of an ideal gas were doubled from 50 to 100 °C, would the average translational kinetic energy of the gas particles also double?

13. The pressure of a monatomic ideal gas doubles, while the volume decreases to one-half its initial value. Does the internal energy of the gas increase, decrease, or remain unchanged?

14. The atoms in a container of helium (He) have the same translational rms speed as the atoms in a container of argon (Ar). Treating each gas as an ideal gas, decide which, if either, has the greater temperature.

15. The pressure of a monatomic ideal gas is doubled, while its volume is reduced by a factor of four. What is the ratio of the new rms speed of the atoms to the initial rms speed?

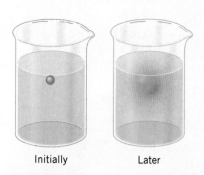

FIGURE 14.13 A drop of ink placed in water eventually becomes completely dispersed because of diffusion.

Initially   Later

14.4 *Diffusion

You can smell the fragrance of a perfume at some distance from an open bottle because perfume molecules leave the space above the liquid in the bottle, where they are relatively concentrated, and spread out into the air, where they are less concentrated. During their journey, they collide with other molecules, so their paths resemble the zigzag paths characteristic of Brownian motion. The process in which molecules move from a region of higher concentration to one of lower concentration is called **diffusion**. Diffusion also occurs in liquids and solids, and **Figure 14.13** illustrates ink diffusing through water. However, compared to the rate of diffusion in gases, the rate is generally smaller in liquids and even smaller in solids. The host medium, such as the air or water in the examples above, is referred to as the **solvent**, while the diffusing substance, like the perfume molecules or the ink in **Figure 14.13**, is known as the **solute**. Relatively speaking, diffusion is a slow process, even in a gas. Conceptual Example 8 illustrates why.

CONCEPTUAL EXAMPLE 8 | Why Diffusion Is Relatively Slow

The fragrance from an open bottle of perfume takes several seconds or sometimes even minutes to reach the other side of a room by the process of diffusion. Which of the following accounts for the fact that diffusion is relatively slow? **(a)** The nature of Brownian motion **(b)** The relatively slow translational rms speeds that characterize gas molecules at room temperature

Reasoning The important characteristic of the paths followed by objects in Brownian motion is their zigzag shapes.

Answer (b) is incorrect. A gas molecule near room temperature has a translational rms speed of hundreds of meters per second. Such speeds are not slow. It would take a molecule traveling at such a speed just a fraction of a second to cross an ordinary room.

Answer (a) is correct. When a perfume molecule diffuses through air, it makes millions of collisions each second with air molecules. As **Figure 14.14** illustrates, the velocity of the molecule changes abruptly because of each collision, but between collisions, it moves in a straight line. Although it does move very fast between collisions, a perfume molecule wanders only slowly away from the bottle because of the zigzag path. It would take a

long time indeed to diffuse in this manner across a room. Usually, however, convection currents are present and carry the fragrance to the other side of the room in a matter of seconds or minutes.

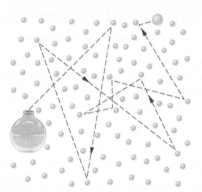

FIGURE 14.14 A perfume molecule collides with millions of air molecules during its journey, so the path has a zigzag shape. Although the air molecules are shown as stationary, they are also moving.

BIO **THE PHYSICS OF . . . drug delivery systems.** Diffusion is the basis for drug delivery systems that bypass the need to administer medication orally or via injections. **Interactive Figure 14.15** shows one such system, the transdermal patch. The word "transdermal" means "across the skin." Such patches, for example, are used to deliver nicotine in programs designed to help you stop smoking. The patch is attached to the skin using an adhesive, and the backing of the patch contains the drug within a reservoir. The concentration of the drug in the reservoir is relatively high, just like the concentration of perfume molecules above the liquid in a bottle. The drug diffuses slowly through a control membrane and directly into the skin, where its concentration is relatively low. Diffusion carries it into the blood vessels present in the skin. The purpose of the control membrane is to limit the rate of diffusion, which can also be adjusted in the reservoir by dissolving the drug in a neutral material to lower its initial concentration. Another diffusion-controlled drug delivery system utilizes capsules that are inserted surgically beneath the skin. Some contraceptives are administered in this fashion. The drug in the capsule diffuses slowly into the bloodstream over extended periods that can be as long as a year.

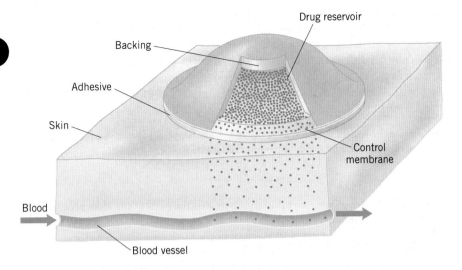

Using diffusion, a transdermal patch delivers a drug directly into the skin, where it enters blood vessels. The backing contains the drug within the reservoir, and the control membrane limits the rate of diffusion into the skin.

The diffusion process can be described in terms of the arrangement in **Interactive Figure 14.16a**. A hollow channel of length L and cross-sectional area A is filled with a fluid. The left end of the channel is connected to a container in which the solute concentration C_2 is relatively high, while the right end is connected to a container in which the solute concentration C_1 is lower. These concentrations are defined as the total mass of the solute molecules divided by the volume of the solution (e.g., 0.1 kg/m³). Because of the difference in concentration between the ends of the channel, $\Delta C = C_2 - C_1$, there is a net diffusion of the solute from the left end to the right end.

Interactive Figure 14.16a is similar to **Interactive Figure 13.7** for the conduction of heat along a bar, which, for convenience, is reproduced in **Interactive Figure 14.16b**. When the ends of the bar are maintained at different temperatures, T_2 and T_1, the heat Q conducted along the bar in a time t is

$$Q = \frac{(kA\,\Delta T)t}{L} \qquad (13.1)$$

where $\Delta T = T_2 - T_1$, and k is the thermal conductivity. Whereas conduction is the flow of heat from a region of higher temperature to a region of lower temperature, diffusion

(a)

(b)

(a) Solute diffuses through the channel from the region of higher concentration to the region of lower concentration. (b) Heat is conducted along a bar whose ends are maintained at different temperatures.

is the mass flow of solute from a region of higher concentration to a region of lower concentration. By analogy with Equation 13.1, it is possible to write an equation for diffusion: (1) replace Q by the mass m of solute that is diffusing through the channel, (2) replace $\Delta T = T_2 - T_1$ by the difference in concentrations $\Delta C = C_2 - C_1$, and (3) replace k by a constant known as the diffusion constant D. The resulting equation, first formulated by the German physiologist Adolf Fick (1829–1901), is referred to as **Fick's law of diffusion**.

FICK'S LAW OF DIFFUSION

The mass m of solute that diffuses in a time t through a solvent contained in a channel of length L and cross-sectional area A is*

$$m = \frac{(DA\,\Delta C)t}{L} \tag{14.8}$$

where ΔC is the concentration difference between the ends of the channel and D is the diffusion constant.

SI Unit for the Diffusion Constant: m^2/s

It can be verified from Equation 14.8 that the diffusion constant has units of m^2/s, the exact value depending on the nature of the solute and the solvent. For example, the diffusion constant for ink in water is different from that for ink in benzene. Example 9 illustrates an important application of Fick's law.

EXAMPLE 9 | BIO The Physics of Water Loss from Plant Leaves

Large amounts of water can be given off by plants. It has been estimated, for instance, that a single sunflower plant can lose up to a pint of water a day during the growing season. **Figure 14.17** shows a cross-sectional view of a leaf. Inside the leaf, water passes from the liquid phase to the vapor phase at the walls of the mesophyll cells. The water vapor then diffuses through the intercellular air spaces and eventually exits the leaf through small openings, called stomatal pores. The diffusion constant for water vapor in air is $D = 2.4 \times 10^{-5}\ m^2/s$. A stomatal pore has a cross-sectional area of about $A = 8.0 \times 10^{-11}\ m^2$ and a length of about $L = 2.5 \times 10^{-5}\ m$. The concentration of water vapor on the interior side of a pore is roughly $C_2 = 0.022\ kg/m^3$, whereas the concentration on the outside is approximately $C_1 = 0.011\ kg/m^3$. Determine the mass of water vapor that passes through a stomatal pore in one hour.

FIGURE 14.17 A cross-sectional view of a leaf. Water vapor diffuses out of the leaf through a stomatal pore.

Reasoning and Solution   Fick's law of diffusion shows that

$$m = \frac{(DA\,\Delta C)t}{L} \tag{14.8}$$

$$= \frac{(2.4 \times 10^{-5}\,m^2/s)(8.0 \times 10^{-11}\,m^2)(0.022\ kg/m^3 - 0.011\ kg/m^3)(3600\ s)}{2.5 \times 10^{-5}\,m}$$

$$= \boxed{3.0 \times 10^{-9}\,kg}$$

This amount of water may not seem significant. However, a single leaf may have as many as a million stomatal pores, so the water lost by an entire plant can be substantial.

---

*Fick's law assumes that the temperature of the solvent is constant throughout the channel. Experiments indicate that the diffusion constant depends strongly on the temperature.

Check Your Understanding

(The answers are given at the end of the book.)

16. BIO In the lungs, oxygen in very small sacs called alveoli diffuses into the blood. The diffusion occurs directly through the walls of the sacs, which have a thickness L. The total effective area A across which diffusion occurs is the sum of the individual areas (each quite small) of the various sac walls. Considering the fact that the mass m of oxygen that enters the blood per second needs to be large and referring to Fick's law of diffusion, what can you deduce about L and about the total number of sacs present in the lungs?

17. The same solute is diffusing through the same solvent in each of three cases. For each case, **CYU Table 14.1** gives the length and cross-sectional area of the diffusion channel. The concentration difference between the ends of the diffusion channel is the same in each case. Rank the diffusion rates (in kg/s) in descending order (largest first).

CYU TABLE 14.1

Case	Length	Cross-Sectional Area
(a)	$\frac{1}{2}L$	A
(b)	L	$\frac{1}{2}A$
(c)	$\frac{1}{3}L$	$2A$

Concept Summary

14.1 Molecular Mass, the Mole, and Avogadro's Number   Each element in the periodic table is assigned an atomic mass. One atomic mass unit (u) is exactly one-twelfth the mass of an atom of carbon-12. The molecular mass of a molecule is the sum of the atomic masses of its atoms.

The number of moles n contained in a sample is equal to the number of particles N (atoms or molecules) in the sample divided by the number of particles per mole N_A, as shown in Equation 1, where N_A is called Avogadro's number and has a value of $N_A = 6.022 \times 10^{23}$ particles per mole.

$$n = \frac{N}{N_A} \tag{1}$$

The number of moles n is also equal to the mass m of the sample (expressed in grams) divided by the mass per mole (expressed in grams per mole), as shown in Equation 2. The mass per mole (in g/mol) of a substance has the same numerical value as the atomic or molecular mass of one of its particles (in atomic mass units).

$$n = \frac{m}{\text{Mass per mole}} \tag{2}$$

The mass m_{particle} of a particle (in grams) can be obtained by dividing the mass per mole (in g/mol) by Avogadro's number, according to Equation 3.

$$m_{\text{particle}} = \frac{\text{Mass per mole}}{N_A} \tag{3}$$

14.2 The Ideal Gas Law   The ideal gas law relates the absolute pressure P, the volume V, the number of moles n, and the Kelvin temperature T of an ideal gas, according to Equation 14.1, where $R = 8.31$ J/(mol · K) is the universal gas constant. An alternative form of the ideal gas law is given by Equation 14.2, where N is the number of particles and $k = \dfrac{R}{N_A}$ is Boltzmann's constant. A real gas behaves as an ideal gas when its density is low enough that its particles do not interact, except via elastic collisions.

$$PV = nRT \tag{14.1}$$

$$PV = NkT \tag{14.2}$$

A form of the ideal gas law that applies when the number of moles and the temperature are constant is known as Boyle's law. Using the subscripts "i" and "f" to denote, respectively, initial and final conditions, we can write Boyle's law as in Equation 14.3. A form of the ideal gas law that applies when the number of moles and the pressure are constant is called Charles' law and is given by Equation 14.4.

$$P_i V_i = P_f V_f \tag{14.3}$$

$$\frac{V_i}{T_i} = \frac{V_f}{T_f} \tag{14.4}$$

14.3 Kinetic Theory of Gases   The distribution of particle speeds in an ideal gas at constant temperature is the Maxwell speed distribution (see **Figure 14.9**). The kinetic theory of gases indicates that the Kelvin temperature T of an ideal gas is related to the average translational kinetic energy $\overline{\text{KE}}$ of a particle, according to Equation 14.6, where v_{rms} is the root-mean-square speed of the particles.

$$\overline{\text{KE}} = \tfrac{1}{2} m v_{\text{rms}}^2 = \tfrac{3}{2} kT \tag{14.6}$$

The internal energy U of n moles of a monatomic ideal gas is given by Equation 14.7. The internal energy of any type of ideal gas (e.g., monatomic, diatomic) is proportional to its Kelvin temperature.

$$U = \tfrac{3}{2} nRT \tag{14.7}$$

14.4 Diffusion   Diffusion is the process whereby solute molecules move through a solvent from a region of higher solute concentration to a region of lower solute concentration. Fick's law of diffusion states that the mass m of solute that diffuses in a time t through the solvent in a channel of length L and cross-sectional area A is given by Equation 14.8, where ΔC is the solute concentration difference between the ends of the channel and D is the diffusion constant.

$$m = \frac{(DA\,\Delta C)t}{L} \tag{14.8}$$

Focus on Concepts

Additional questions are available for assignment in WileyPLUS.

Section 14.1 Molecular Mass, the Mole, and Avogadro's Number

1. All but one of the following statements are true. Which one is not true? **(a)** A mass (in grams) equal to the molecular mass (in atomic mass units) of a pure substance contains the same number of molecules, no matter what the substance is. **(b)** One mole of any pure substance contains the same number of molecules. **(c)** Ten grams of a pure substance contains twice as many molecules as five grams of the substance. **(d)** Ten grams of a pure substance contains the same number of molecules, no matter what the substance is. **(e)** Avogadro's number of molecules of a pure substance and one mole of the substance have the same mass.

2. A mixture of ethyl alcohol (molecular mass = 46.1 u) and water (molecular mass = 18.0 u) contains one mole of molecules. The mixture contains 20.0 g of ethyl alcohol. What mass m of water does it contain?

Section 14.2 The Ideal Gas Law

3. For an ideal gas, each of the following unquestionably leads to an increase in the pressure of the gas, except one. Which one is it? **(a)** Increasing the temperature and decreasing the volume, while keeping the number of moles of the gas constant **(b)** Increasing the temperature, the volume, and the number of moles of the gas **(c)** Increasing the temperature, while keeping the volume and the number of moles of the gas constant **(d)** Increasing the number of moles of the gas, while keeping the temperature and the volume constant **(e)** Decreasing the volume, while keeping the temperature and the number of moles of the gas constant.

4. The cylinder in the drawing contains 3.00 mol of an ideal gas. By moving the piston, the volume of the gas is reduced to one-fourth its initial value, while the temperature is held constant. How many moles Δn of the gas must be allowed to escape through the valve, so that the pressure of the gas does not change?

Valve

5. Carbon monoxide is a gas at 0 °C and a pressure of 1.01×10^5 Pa. It is a diatomic gas, each of its molecules consisting of one carbon atom (atomic mass = 12.0 u) and one oxygen atom (atomic mass = 16.0 u). Assuming that carbon monoxide is an ideal gas, calculate its density ρ.

Section 14.3 Kinetic Theory of Gases

6. If the speed of every atom in a monatomic ideal gas were doubled, by what factor would the Kelvin temperature of the gas be multiplied? **(a)** 4 **(b)** 2 **(c)** 1 **(d)** $\frac{1}{2}$ **(e)** $\frac{1}{4}$

7. The atomic mass of a nitrogen atom (N) is 14.0 u, while that of an oxygen atom (O) is 16.0 u. Three diatomic gases have the same temperature: nitrogen (N_2), oxygen (O_2), and nitric oxide (NO). Rank these gases in ascending order (smallest first), according to the values of their translational rms speeds: **(a)** O_2, N_2, NO **(b)** NO, N_2, O_2 **(c)** N_2, NO, O_2 **(d)** O_2, NO, N_2 **(e)** N_2, O_2, NO

8. The pressure of a monatomic ideal gas is doubled, while the volume is cut in half. By what factor is the internal energy of the gas multiplied? **(a)** $\frac{1}{4}$ **(b)** $\frac{1}{2}$ **(c)** 1 **(d)** 2 **(e)** 4

Section 14.4 Diffusion

9. The following statements concern how to increase the rate of diffusion (in kg/s). All but one statement are always true. Which one is not necessarily true? **(a)** Increase the cross-sectional area of the diffusion channel, keeping constant its length and the difference in solute concentrations between its ends. **(b)** Increase the difference in solute concentrations between the ends of the diffusion channel, keeping constant its cross-sectional area and its length. **(c)** Decrease the length of the diffusion channel, keeping constant its cross-sectional area and the difference in solute concentrations between its ends. **(d)** Increase the cross-sectional area of the diffusion channel, and decrease its length, keeping constant the difference in solute concentrations between its ends. **(e)** Increase the cross-sectional area of the diffusion channel, increase the difference in solute concentrations between its ends, and increase its length.

10. The diffusion rate for a solute is 4.0×10^{-11} kg/s in a solvent-filled channel that has a cross-sectional area of 0.50 cm^2 and a length of 0.25 cm. What would be the diffusion rate m/t in a channel with a cross-sectional area of 0.30 cm^2 and a length of 0.10 cm?

Problems

Additional questions are available for assignment in WileyPLUS.

Note: *The pressures referred to in these problems are absolute pressures, unless indicated otherwise.*

SSM Student Solutions Manual	**MMH** Problem-solving help

GO Guided Online Tutorial	**M** Medium
V-HINT Video Hints	**H** Hard
CHALK Chalkboard Videos	**WS** Worksheet
BIO Biomedical application	**T** Team Problem
E Easy	

Section 14.1 Molecular Mass, the Mole, and Avogadro's Number

1. **E** **SSM** Hemoglobin has a molecular mass of 64 500 u. Find the mass (in kg) of one molecule of hemoglobin.

2. **E** A mass of 135 g of a certain element is known to contain 30.1 × 10^{23} atoms. What is the element?

3. **E** **GO** A certain element has a mass per mole of 196.967 g/mol. What is the mass of a single atom in (a) atomic mass units and (b) kilograms? (c) How many moles of atoms are in a 285-g sample?

4. **E** **GO** The chlorophyll-*a* molecule ($C_{55}H_{72}MgN_4O_5$) is important in photosynthesis. (a) Determine its molecular mass (in atomic mass units). (b) What is the mass (in grams) of 3.00 moles of chlorophyll-*a* molecules?

5. **M** **V-HINT** A runner weighs 580 N (about 130 lb), and 71% of this weight is water. (a) How many moles of water are in the runner's body? (b) How many water molecules (H_2O) are there?

6. **M** **GO** Consider a mixture of three different gases: 1.20 g of argon (molecular mass = 39.948 g/mol), 2.60 g of neon (molecular mass = 20.180 g/mol), and 3.20 g of helium (molecular mass = 4.0026 g/mol). For this mixture, determine the percentage of the total number of atoms that corresponds to each of the components.

7. **M** **V-HINT** **CHALK** A cylindrical glass of water (H_2O) has a radius of 4.50 cm and a height of 12.0 cm. The density of water is 1.00 g/cm^3. How many moles of water molecules are contained in the glass?

Section 14.2 The Ideal Gas Law

8. **E** **SSM** It takes 0.16 g of helium (He) to fill a balloon. How many grams of nitrogen (N_2) would be required to fill the balloon to the same pressure, volume, and temperature?

9. **E** A 0.030-m^3 container is initially evacuated. Then, 4.0 g of water is placed in the container, and, after some time, all the water evaporates. If the temperature of the water vapor is 388 K, what is its pressure?

10. **E** An ideal gas at 15.5 °C and a pressure of 1.72 × 10^5 Pa occupies a volume of 2.81 m^3. (a) How many moles of gas are present? (b) If the volume is raised to 4.16 m^3 and the temperature raised to 28.2 °C, what will be the pressure of the gas?

11. **E** **GO** Four closed tanks, A, B, C, and D, each contain an ideal gas. The table gives the absolute pressure and volume of the gas in each tank. In each case, there is 0.10 mol of gas. Using this number and the data in the table, compute the temperature of the gas in each tank.

	A	B	C	D
Absolute pressure (Pa)	25.0	30.0	20.0	2.0
Volume (m^3)	4.0	5.0	5.0	75

12. **E** A Goodyear blimp typically contains 5400 m^3 of helium (He) at an absolute pressure of 1.1 × 10^5 Pa. The temperature of the helium is 280 K. What is the mass (in kg) of the helium in the blimp?

13. **E** A clown at a birthday party has brought along a helium cylinder, with which he intends to fill balloons. When full, each balloon contains 0.034 m^3 of helium at an absolute pressure of 1.2 × 10^5 Pa. The cylinder contains helium at an absolute pressure of 1.6 × 10^7 Pa and has a volume of 0.0031 m^3. The temperature of the helium in the tank and in the balloons is the same and remains constant. What is the maximum number of balloons that can be filled?

14. **E** **GO** The volume of an ideal gas is held constant. Determine the ratio P_2/P_1 of the final pressure to the initial pressure when the temperature of the gas rises (a) from 35.0 to 70.0 K and (b) from 35.0 to 70.0 °C.

15. **E** Two ideal gases have the same mass density and the same absolute pressure. One of the gases is helium (He), and its temperature is 175 K. The other gas is neon (Ne). What is the temperature of the neon?

16. **E** On the sunlit surface of Venus, the atmospheric pressure is 9.0 × 10^6 Pa, and the temperature is 740 K. On the earth's surface the atmospheric pressure is 1.0 × 10^5 Pa, while the surface temperature can reach 320 K. These data imply that Venus has a "thicker" atmosphere at its surface than does the earth, which means that the number of molecules per unit volume (N/V) is greater on the surface of Venus than on the earth. Find the ratio $(N/V)_{Venus}/(N/V)_{Earth}$.

17. **E** **V-HINT** A tank contains 0.85 mol of molecular nitrogen (N_2). Determine the mass (in grams) of nitrogen that must be *removed* from the tank in order to lower the pressure from 38 to 25 atm. Assume that the volume and temperature of the nitrogen in the tank do not change.

18. **M** **V-HINT** Multiple-Concept Example 5 reviews the principles that play roles in this problem. A primitive diving bell consists of a cylindrical tank with one end open and one end closed. The tank is lowered into a freshwater lake, open end downward. Water rises into the tank, compressing the trapped air, whose temperature remains constant during the descent. The tank is brought to a halt when the distance between the surface of the water in the tank and the surface of the lake is 40.0 m. Atmospheric pressure at the surface of the lake is 1.01 × 10^5 Pa. Find the fraction of the tank's volume that is filled with air.

19. **M** **GO** A tank contains 11.0 g of chlorine gas (Cl_2) at a temperature of 82 °C and an absolute pressure of 5.60 × 10^5 Pa. The mass per mole of Cl_2 is 70.9 g/mol. (a) Determine the volume of the tank. (b) Later, the temperature of the tank has dropped to 31 °C and, due to a leak, the pressure has dropped to 3.80 × 10^5 Pa. How many grams of chlorine gas have leaked out of the tank?

20. **M** **CHALK** **MMH** The dimensions of a room are 2.5 m × 4.0 m × 5.0 m. Assume that the air in the room is composed of 79% nitrogen (N_2) and 21% oxygen (O_2). At a temperature of 22 °C and a pressure of 1.01 × 10^5 Pa, what is the mass (in grams) of the air?

21. **M** **GO** The drawing shows an ideal gas confined to a cylinder by a massless piston that is attached to an ideal spring. Outside the cylinder is a vacuum. The cross-sectional area of the piston is $A = 2.50 \times 10^{-3}$ m^2. The initial pressure, volume, and temperature of the gas are, respectively, P_0, $V_0 = 6.00 \times 10^{-4}$ m^3, and $T_0 = 273$ K, and the spring is initially stretched by an amount $x_0 = 0.0800$ m with respect to its unstrained length. The gas is heated, so that its final pressure, volume, and temperature are P_f, V_f, and T_f, and the spring is stretched by an amount $x_f = 0.1000$ m with respect to its unstrained length. What is the final temperature of the gas?

PROBLEM 21

22. **M** **SSM** Multiple-Concept Example 5 and Conceptual Example 3 are pertinent to this problem. A bubble, located 0.200 m beneath the surface in a glass of beer, rises to the top. The air pressure at the top is 1.01 × 10^5 Pa. Assume that the density of beer is the same as that of fresh water. If the temperature and number of moles of CO_2 in the bubble remain constant as the bubble rises, find the ratio of the bubble's volume at the top to its volume at the bottom.

23. M V-HINT The relative humidity is 55% on a day when the temperature is 30.0 °C. Using the graph that accompanies Problem 66 in Chapter 12, determine the number of moles of water vapor per cubic meter of air.

24. M MMH One assumption of the ideal gas law is that the atoms or molecules themselves occupy a negligible volume. Verify that this assumption is reasonable by considering gaseous xenon (Xe). Xenon has an atomic radius of 2.0×10^{-10} m. For STP conditions, calculate the percentage of the total volume occupied by the atoms.

25. H A spherical balloon is made from an amount of material whose mass is 3.00 kg. The thickness of the material is negligible compared to the 1.50-m radius of the balloon. The balloon is filled with helium (He) at a temperature of 305 K and just floats in air, neither rising nor falling. The density of the surrounding air is 1.19 kg/m³. Find the absolute pressure of the helium gas.

26. H The mass of a hot-air balloon and its occupants is 320 kg (excluding the hot air inside the balloon). The air outside the balloon has a pressure of 1.01×10^5 Pa and a density of 1.29 kg/m³. To lift off, the air inside the balloon is heated. The volume of the heated balloon is 650 m³. The pressure of the heated air remains the same as the pressure of the outside air. To what temperature (in kelvins) must the air be heated so that the balloon just lifts off? The molecular mass of air is 29 u.

Section 14.3 Kinetic Theory of Gases

27. E SSM MMH Very fine smoke particles are suspended in air. The translational rms speed of a smoke particle is 2.8×10^{-3} m/s, and the temperature is 301 K. Find the mass of a particle.

28. E GO Four tanks A, B, C, and D are filled with monatomic ideal gases. For each tank, the mass of an individual atom and the rms speed of the atoms are expressed in terms of m and v_{rms}, respectively (see the table). Suppose that $m = 3.32 \times 10^{-26}$ kg, and $v_{rms} = 1223$ m/s. Find the temperature of the gas in each tank.

	A	B	C	D
Mass	m	m	$2m$	$2m$
Rms speed	v_{rms}	$2v_{rms}$	v_{rms}	$2v_{rms}$

29. E GO Suppose a tank contains 680 m³ of neon (Ne) at an absolute pressure of 1.01×10^5 Pa. The temperature is changed from 293.2 to 294.3 K. What is the increase in the internal energy of the neon?

30. E Two moles of an ideal gas are placed in a container whose volume is 8.5×10^{-3} m³. The absolute pressure of the gas is 4.5×10^5 Pa. What is the average translational kinetic energy of a molecule of the gas?

31. E Two gas cylinders are identical. One contains the monatomic gas argon (Ar), and the other contains an equal mass of the monatomic gas krypton (Kr). The pressures in the cylinders are the same, but the temperatures are different. Determine the ratio $\dfrac{\overline{KE}_{Krypton}}{\overline{KE}_{Argon}}$ of the average kinetic energy of a krypton atom to the average kinetic energy of an argon atom.

32. E A container holds 2.0 mol of gas. The total average kinetic energy of the gas molecules in the container is equal to the kinetic energy of an 8.0×10^{-3}-kg bullet with a speed of 770 m/s. What is the Kelvin temperature of the gas?

33. M CHALK MMH The temperature near the surface of the earth is 291 K. A xenon atom (atomic mass = 131.29 u) has a kinetic energy equal to the average translational kinetic energy and is moving straight up. If the atom does not collide with any other atoms or molecules, how high up will it go before coming to rest? Assume that the acceleration due to gravity is constant throughout the ascent.

34. M GO Compressed air can be pumped underground into huge caverns as a form of energy storage. The volume of a cavern is 5.6×10^5 m³, and the pressure of the air in it is 7.7×10^6 Pa. Assume that air is a diatomic ideal gas whose internal energy U is given by $U = \frac{5}{2} nRT$. If one home uses 30.0 kW · h of energy per day, how many homes could this internal energy serve for one day?

35. H A cubical box with each side of length 0.300 m contains 1.000 moles of neon gas at room temperature (293 K). What is the average rate (in atoms/s) at which neon atoms collide with one side of the container? The mass of a single neon atom is 3.35×10^{-26} kg.

Section 14.4 Diffusion

36. E A tube has a length of 0.015 m and a cross-sectional area of 7.0×10^{-4} m². The tube is filled with a solution of sucrose in water. The diffusion constant of sucrose in water is 5.0×10^{-10} m²/s. A difference in concentration of 3.0×10^{-3} kg/m³ is maintained between the ends of the tube. How much time is required for 8.0×10^{-13} kg of sucrose to be transported through the tube?

37. E GO The diffusion constant for the amino acid glycine in water has a value of 1.06×10^{-9} m²/s. In a 2.0-cm-long tube with a cross-sectional area of 1.5×10^{-4} m², the mass rate of diffusion is $m/t = 4.2 \times 10^{-14}$ kg/s, because the glycine concentration is maintained at a value of 8.3×10^{-3} kg/m³ at one end of the tube and at a lower value at the other end. What is the lower concentration?

38. E SSM A large tank is filled with methane gas at a concentration of 0.650 kg/m³. The valve of a 1.50-m pipe connecting the tank to the atmosphere is inadvertently left open for twelve hours. During this time, 9.00×10^{-4} kg of methane diffuses out of the tank, leaving the concentration of methane in the tank essentially unchanged. The diffusion constant for methane in air is 2.10×10^{-5} m²/s. What is the cross-sectional area of the pipe? Assume that the concentration of methane in the atmosphere is zero.

39. M V-HINT CHALK Carbon tetrachloride (CCl_4) is diffusing through benzene (C_6H_6), as the drawing illustrates. The concentration of CCl_4 at the left end of the tube is maintained at 1.00×10^{-2} kg/m³, and the diffusion constant is 20.0×10^{-10} m²/s. The CCl_4 enters the tube at a mass rate of 5.00×10^{-13} kg/s. Using these data and those shown in the drawing, find (a) the mass of CCl_4 per second that passes point A and (b) the concentration of CCl_4 at point A.

PROBLEM 39

Additional Problems

40. ⓔ At the start of a trip, a driver adjusts the absolute pressure in her tires to be 2.81×10^5 Pa when the outdoor temperature is 284 K. At the end of the trip she measures the pressure to be 3.01×10^5 Pa. Ignoring the expansion of the tires, find the air temperature inside the tires at the end of the trip.

41. ⓔ ⓖⓞ When you push down on the handle of a bicycle pump, a piston in the pump cylinder compresses the air inside the cylinder. When the pressure in the cylinder is greater than the pressure inside the inner tube to which the pump is attached, air begins to flow from the pump to the inner tube. As a biker slowly begins to push down the handle of a bicycle pump, the pressure inside the cylinder is 1.0×10^5 Pa, and the piston in the pump is 0.55 m above the bottom of the cylinder. The pressure inside the inner tube is 2.4×10^5 Pa. How far down must the biker push the handle before air begins to flow from the pump to the inner tube? Ignore the air in the hose connecting the pump to the inner tube, and assume that the temperature of the air in the pump cylinder does not change.

42. ⓔ ⓢⓢⓜ In a diesel engine, the piston compresses air at 305 K to a volume that is one-sixteenth of the original volume and a pressure that is 48.5 times the original pressure. What is the temperature of the air after the compression?

43. ⓔ ⓖⓞ When a gas is diffusing through air in a diffusion channel, the diffusion rate is the number of gas atoms per second diffusing from one end of the channel to the other end. The faster the atoms move, the greater is the diffusion rate, so the diffusion rate is proportional to the rms speed of the atoms. The atomic mass of ideal gas A is 1.0 u, and that of ideal gas B is 2.0 u. For diffusion through the same channel under the same conditions, find the ratio of the diffusion rate of gas A to the diffusion rate of gas B.

44. ⓔ Near the surface of Venus, the rms speed of carbon dioxide molecules (CO_2) is 650 m/s. What is the temperature (in kelvins) of the atmosphere at that point?

45. Ⓜ ⓖⓞ At the normal boiling point of a material, the liquid phase has a density of 958 kg/m^3, and the vapor phase has a density of 0.598 kg/m^3. The average distance between neighboring molecules in the vapor phase is d_{vapor}. The average distance between neighboring molecules in the liquid phase is d_{liquid}. Determine the ratio d_{vapor}/d_{liquid}. (*Hint: Assume that the volume of each phase is filled with many cubes, with one molecule at the center of each cube.*)

46. Ⓜ ⓖⓞ Helium (He), a monatomic gas, fills a 0.010-m^3 container. The pressure of the gas is 6.2×10^5 Pa. How long would a 0.25-hp engine have to run (1 hp = 746 W) to produce an amount of energy equal to the internal energy of this gas?

47. Ⓗ A gas fills the right portion of a horizontal cylinder whose radius is 5.00 cm. The initial pressure of the gas is 1.01×10^5 Pa. A frictionless movable piston separates the gas from the left portion of the cylinder, which is evacu-

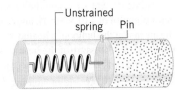

PROBLEM 47

ated and contains an ideal spring, as the drawing shows. The piston is initially held in place by a pin. The spring is initially unstrained, and the length of the gas-filled portion is 20.0 cm. When the pin is removed and the gas is allowed to expand, the length of the gas-filled chamber doubles. The initial and final temperatures are equal. Determine the spring constant of the spring.

48. Ⓜ ⓖⓞ ⓢⓢⓜ Air is primarily a mixture of nitrogen N_2 (molecular mass = 28.0 u) and oxygen O_2 (molecular mass = 32.0 u). Assume that each behaves like an ideal gas and determine the rms speed of the nitrogen and oxygen molecules when the air temperature is 293 K.

Physics in Biology, Medicine, and Sports

49. ⓔ ⓑⓘⓞ **14.1** Manufacturers of headache remedies routinely claim that their own brands are more potent pain relievers than the competing brands. Their way of making the comparison is to compare the number of molecules in the standard dosage. Tylenol uses 325 mg of acetaminophen ($C_8H_9NO_2$) as the standard dose, whereas Advil uses 2.00×10^2 mg of ibuprofen ($C_{13}H_{18}O_2$). Find the number of molecules of pain reliever in the standard doses of **(a)** Tylenol and **(b)** Advil.

50. ⓔ ⓑⓘⓞ ⓢⓢⓜ **14.1** The active ingredient in the allergy medication Claritin contains carbon (C), hydrogen (H), chlorine (Cl), nitrogen (N), and oxygen (O). Its molecular formula is $C_{22}H_{23}ClN_2O_2$. The standard adult dosage utilizes 1.572×10^{19} molecules of this species. Determine the mass (in grams) of the active ingredient in the standard dosage.

51. Ⓜ ⓑⓘⓞ ⓢⓢⓜ **14.1** The preparation of homeopathic "remedies" involves the repeated dilution of solutions containing an active ingredient such as arsenic trioxide (As_2O_3). Suppose one begins with 18.0 g of arsenic trioxide dissolved in water, and repeatedly dilutes the

solution with pure water, each dilution reducing the amount of arsenic trioxide remaining in the solution by a factor of 100. Assuming perfect mixing at each dilution, what is the maximum number of dilutions one may perform so that at least one molecule of arsenic trioxide remains in the diluted solution? For comparison, homeopathic "remedies" are commonly diluted 15 or even 30 times.

52. ⓔ ⓑⓘⓞ ⓒⓗⓐⓛⓚ ⓢⓢⓜ **14.2** A young male adult takes in about 5.0×10^{-4} m^3 of fresh air during a normal breath. Fresh air contains approximately 21% oxygen. Assuming that the pressure in the lungs is 1.0×10^5 Pa and that air is an ideal gas at a temperature of 310 K, find the number of oxygen molecules in a normal breath.

53. ⓔ ⓑⓘⓞ **14.4** Insects do not have lungs as we do, nor do they breathe through their mouths. Instead, they have a system of tiny tubes, called tracheae, through which oxygen diffuses into their bodies. The tracheae begin at the surface of an insect's body and penetrate into the interior. Suppose that a trachea is 1.9 mm long with a cross-sectional area of 2.1×10^{-9} m². The concentration of oxygen in the air outside the insect is 0.28 kg/m³, and the diffusion constant is 1.1×10^{-5} m²/s. If the mass per second of oxygen diffusing through a trachea is 1.7×10^{-12} kg/s, find the oxygen concentration at the interior end of the tube.

54. E BIO SSM **14.2** Oxygen for hospital patients is kept in special tanks, where the oxygen has a pressure of 65.0 atmospheres and a temperature of 288 K. The tanks are stored in a separate room, and the oxygen is pumped to the patient's room, where it is administered at a pressure of 1.00 atmosphere and a temperature of 297 K. What volume does 1.00 m³ of oxygen in the tanks occupy at the conditions in the patient's room?

55. M BIO **14.1** *Remdesivir* is a broad-spectrum antiviral medication. It has been used to treat certain RNA viruses, such as SARS and MERS. As of early 2020, the drug was being tested as a specific treatment for the SARS-CoV-2 virus. The chemical formula of the drug is $C_{27}H_{35}N_6O_8P$. One common delivery method of the drug is by intravenous solution. The mass of the drug in the dose is 1.00×10^2 mg. (a) How many moles of Remdesivir are there in a standard dose? (b) How many molecules of Remdesivir are contained in this dose?

56. E BIO **14.2** The lungs of a blue whale are enormous. Their total volume is capable of holding 13 000 gallons (49.2 m³) of air. Nevertheless, relative to body size, human lungs are actually larger. Human lungs occupy 7% of the body's internal cavity, while the blue whale's lungs occupy only 3%. However, a whale's lungs are highly efficient. With every breath, they are able to pass 90% of the inhaled oxygen into their blood, as compared to just 10% to 15% for humans. Furthermore, a blue whale's blood is 60% hemoglobin, compared to only 30% in humans. These and other adaptations allow the whales to store twice as much oxygen per volume of blood than humans. These abilities allow the whale to remain under water for extended periods of time—some as long as 90 minutes! Assume with each breath, the blue whale takes in 10.3 m³ of oxygen. Treat the oxygen in the air as an ideal gas at a temperature of 20.0 °C. The partial pressure of oxygen in the atmosphere at sea level is 21.3 kPa. Calculate what mass of oxygen is transferred to the whale's blood with every breath.

57. M BIO **14.3** The air we breathe into our lungs is a mixture of gases. While the atmosphere is mostly composed of nitrogen and oxygen, there are other gases too. One of these is a monatomic gas that has an rms speed of 428 m/s in your lungs at 20.0 °C. What gas is this?

58. M BIO **14.3** The earth's atmosphere is composed of a mixture of gases. The oxygen (O_2) we need to sustain our lives makes up only 21.0% of the air we breathe. On a hot summer day, when the temperature outside is 39.0 °C, with what rms speed do the oxygen molecules strike the surface of your skin?

59. H BIO **14.4** While the majority of the symptoms experienced by people infected with the SARS-CoV-2 (COVID-19) virus are mild, some develop the very serious condition of ARDS (acute respiratory distress syndrome), which results in a rapid and widespread inflammation of the lungs. The major function of the lungs—the process of breathing—facilitates the vital exchange of gases between the environment and the body. The rate of exchange of these gases is largely governed by diffusion, which takes place in the extensive network of tiny sacs in the lungs called alveoli—hundreds of millions of them! In fact, the surface area of the alveoli in the human lungs is so large that if every alveoli were laid flat, they would cover a surface area as large as 70 to 80 m². The diffusion of oxygen and carbon dioxide takes place over this large surface. Conditions like ARDS, and other diseases in the lungs, can reduce this surface area, resulting in an insufficient amount of oxygen being transferred to the blood during respiration, a condition known as hypoxemia (see the photo), which is why patients are given pure oxygen to breathe. ARDS results in considerable fluid accumulation in the lungs, which floods the alveoli and inhibits respiration. ARDS patients often have rapid breathing and are gasping for breath. This requires considerable physical exertion, and many patients quickly become exhausted. At this point, patients may be moved into an intensive care unit and placed on a mechanical ventilator to help them breathe. This is a very serious condition, and the overall prognosis for someone with ARDS is poor, with a mortality rate near 40%. For those who survive, they often experience a lower quality of life, due to irreversible damage.

Consider a small section (5.2 μm²) of the respiratory membrane that covers each alveolus. It is composed of a layer of squamous epithelial cells with a thickness of 1.0 μm. If the concentration of oxygen on one side of the membrane is 4.5×10^{-17} mol/μm³ and 7.8×10^{-18} mol/μm³ on the other side, what mass of oxygen (in micrograms) is transported across the membrane each second? The diffusion constant for oxygen across the membrane is 1.7×10^{-9} m²/s.

A common indicator of hypoxemia (or hypoxia) is a condition called cyanosis, which results in blue/purple coloring of the skin, often at peripheral areas, such as fingers, toes, and lips. The coloring is a result of an increase in the concentration of deoxygenated hemoglobin in the blood.

PROBLEM 59

Concepts and Calculations Problems

Online

This chapter introduces the ideal gas law, which is a relation between the pressure, volume, temperature, and number of moles of an ideal gas. In Section 10.1, we examined how the compression of a spring depends on the forced applied to it. Problem 60 reviews how a gas produces a force and why an ideal gas at different temperatures causes a spring to compress by different amounts. The kinetic theory of gases is important because it allows us to understand the relation between the macroscopic properties of a gas, such as pressure and temperature, and the microscopic properties of molecules, such as speed and mass. Problem 61 reveals the essential features of this theory.

60. M CHALK The figure shows three identical chambers containing a piston and a spring whose spring constant is $k = 5.8 \times 10^4$ N/m. The chamber in part a is completely evacuated, and the piston just touches its left end. In this position, the spring is unstrained. In part b of the drawing, 0.75 mol of ideal gas 1 is introduced into the space between the left end of the chamber and the piston, and the spring compresses by $x_1 = 15$ cm. In part c, 0.75 mol of ideal gas 2 is introduced into the space between the left end of the chamber and the piston, and the spring compresses by x_2 = 24 cm. *Concepts*: (i) Which gas exerts a greater force on the piston? (ii) How is the force required to compress a spring related to the displacement of the spring from the unstrained position? (iii) Which gas exerts the greater pressure on the piston? (iv) Which gas has the greater temperature? *Calculations*: Find the temperature of each gas.

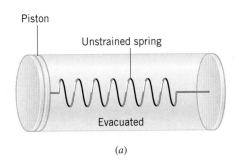

(a)

PROBLEM 60

61. M CHALK SSM In outer space the density of matter is extremely low, about one atom per cm³. The matter is mainly hydrogen atoms ($m = 1.67 \times 10^{-27}$ kg) whose rms speed is 260 m/s. A cubical box, 2.0 m on a side, is placed in outer space, and the hydrogen atoms are allowed to enter. *Concepts*: (i) Why do hydrogen atoms exert a force on the wall of the box? (ii) Do the atoms generate a pressure on the walls of the box? (iii) Do hydrogen atoms in outer space have a temperature? If so, how is the temperature related to the microscopic properties of the atoms? *Calculations*: (a) What is the magnitude of the force that the atoms exert on one wall of the box? (b) Determine the pressure that the atoms exert. (c) Does outer space have a temperature and, if so, what is it?

Gas 1   (b)

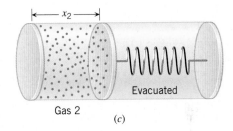

Gas 2   (c)

Team Problems

Online

62. E T WS **Danger on a Spacecraft.** You and your team are on a spacecraft and discover that an external argon gas storage tank is about to fail, and you need to reduce the pressure by releasing some of the gas into space. The pressure gauge for this tank is no longer operable, but you know that the last pressure reading for the tank was 27.2 atm, and it should still be at that pressure since it has not leaked, and nothing else has changed. You and your team must reduce the pressure to 18.0 atm using only a mass flow meter that tells you how much mass of the gas flows through an exit valve. You also know that the volume of the tank is 4.35 m³ and that the temperature is 140.5 K, which has not changed since the last pressure measurement. (a) What is the mass of argon (in kg) in the tank before you open the valve? (b) What mass of argon gas (in kg) should you release from the tank in order to reduce the pressure to 18.0 atm? Assume that the gas behaves like an ideal gas and that the volume and temperature of the tank do not change.

63. M T WS **Dalton's Law of Partial Pressures.** You and your team are on a space station that has been damaged. The air in one entire section of the station has leaked through a crack, and the

volume is now at vacuum (i.e., $P = 0$ atm). The leak has since been repaired, and you and your team have been given the task of making this part of the station livable again by refilling it with one atmosphere of breathable air: a mixture of nitrogen (75.0%) and oxygen (25.0%), where the percentages are by weight (wt%). The problem is that you have separate sources for nitrogen and oxygen, and have to determine how much of each must be introduced into the void in order to have the proper mix of gases at a pressure of one atmosphere (1.013×10^5 Pa). One of your team members suggests that you use *Dalton's Rule of Partial Pressures*, which states that, for ideal gases, the total pressure of a gas mixture is the sum of the partial pressures that each amount of the constituent gases would have in the same volume and at the same temperature. For instance, in this case, the total pressure $P_T = P_N(75.0 \text{ wt\%}) + P_O(25.0 \text{ wt\%}) = 1.00$ atm, where P_N would be the pressure with only the given amount of nitrogen in the mixture, and P_O would be the pressure with only the given amount of oxygen in the mixture. If the volume of the section that needs to be filled is 98.0 m³, and is to be held at a comfortable temperature of 23.0 °C, determine the masses (in kg) of nitrogen and oxygen that must be introduced into the evacuated chamber. Assume that N_2 and O_2 behave like ideal gases with molar masses of 28.0 g/mol (N_2) and 32.0 g/mol (O_2).

64. Ⓜ Ⓣ **A Gas Dosing Device.** You and your team are given the task of designing a crude system to portion out a specific amount (moles) of krypton (Kr) gas. You brainstorm and come up with the idea of a device that consists of a cylinder ($D = 11.0$ cm) with a spring-loaded piston. The side of the piston with the spring is at atmospheric pressure. Krypton gas is introduced into the cylinder on the other side of the piston (without the spring), and the piston moves and compresses from its equilibrium position. When the spring is in its equilibrium position, the piston is at the very end of the cylinder, and the gas volume is zero. **(a)** If the spring ($k = 8.20 \times 10^4$ N/m) compresses 12.5 cm when the Kr gas is introduced into the cylinder, how many moles of Kr are contained in the cylinder? Treat it as an ideal, monatomic gas, and assume that the temperature of the gas has equilibrated with the outside environment ($T = 72.0$ °F) when you measure the compressed distance. **(b)** How far should the spring compress for 1.50 moles of Kr gas (at $T = 72.0$ °F)?

65. Ⓜ Ⓣ **An Underwater Gas Piston.** You and your team have been given the task of preparing a gas piston device that is to be deployed to the bottom of the Southern Sea, beneath the Brunt Ice Shelf off the coast of Antarctica. The inner volume of the cylinder has a diameter of 9.00 cm and a length of 25.0 cm, and is fitted with a movable piston that compresses the argon (Ar) gas that it contains. Before deployment, the piston is located and held at the very top of the cylinder, providing the maximum volume for the gas. When the device is submerged, however, the piston will move inward due to the water pressure created by the depth, and the change in temperature of the gas. The device is to be taken to a depth of 350.0 m in salt water (density $\rho = 1.03$ g/cm³) at a temperature of 33.0 °F. If the maximum allowable compression distance of the piston is 13.0 cm, what is the minimum gas pressure in the cylinder that must be attained when it is filled at the surface (at atmospheric pressure and $T = 59.0$ °F) to prevent the maximum displacement of the piston from exceeding its maximum (13.0 cm) when submerged to a depth of 350.0 m?

NASA/SDO

The sun fuses 600 million tons of hydrogen into helium every second, and about 4 million tons of this matter is converted into energy. This is an irreversible process that transfers heat from a hot source, the sun, to a cold source, space. Some of this energy will reach the earth and other planets in the solar system, allowing the formation of highly ordered and complex biological systems, like plants and animals. However, during this process, the total disorder, or entropy, of the universe increases. In this chapter, we will study the laws that govern heat transfer and its relationship to work and energy in physical systems.

Thermodynamics

15.1 Thermodynamic Systems and Their Surroundings

We have studied heat (Chapter 12) and work (Chapter 6) as separate topics. Often, however, they occur simultaneously. In an automobile engine, for instance, fuel is burned at a relatively high temperature, some of its internal energy is used for doing the work of driving the pistons up and down, and the excess heat is removed by the cooling system to prevent overheating. **Thermodynamics** is the branch of physics that is built upon the fundamental laws that heat and work obey.

In thermodynamics the collection of objects on which attention is being focused is called the **system**, while everything else in the environment is called the **surroundings**. For example, the system in an automobile engine could be the burning gasoline, whereas the surroundings would then include the pistons, the exhaust system, the radiator, and the outside air. The system and its surroundings are separated by walls of some kind. Walls that permit heat to flow through

LEARNING OBJECTIVES

After reading this module, you should be able to...

15.1 Identify thermodynamic systems and their surroundings.

15.2 Define the zeroth law of thermodynamics.

15.3 Define the first law of thermodynamics.

15.4 Analyze thermal processes.

15.5 Analyze thermal processes in an ideal gas.

15.6 Distinguish between different forms of heat capacity.

15.7 Define the second law of thermodynamics.

15.8 Analyze heat engines.

15.9 Analyze Carnot engines.

15.10 Analyze the operation of refrigerators, air conditioners, and heat pumps.

15.11 Calculate the change in entropy in a thermodynamic process.

15.12 Define the third law of thermodynamics.

FIGURE 15.1 The air in one of these hot-air balloons is one example of a thermodynamic system.

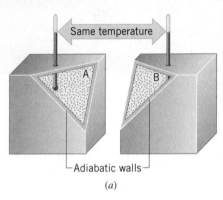

Same temperature

A   B

Adiabatic walls

(a)

A   B

Diathermal walls
(silver)   (b)

FIGURE 15.2 (a) Systems A and B are surrounded by adiabatic walls and register the same temperature on a thermometer. (b) When system A is put into thermal contact with system B through diathermal walls, no net flow of heat occurs between the systems.

them, such as those of the engine block, are called **diathermal walls**. Perfectly insulating walls that do not permit heat to flow between the system and its surroundings are known as **adiabatic walls**.

To understand what the laws of thermodynamics have to say about the relationship between heat and work, it is necessary to describe the physical condition or **state of a system**. We might be interested, for instance, in the hot air within one of the balloons in **Figure 15.1**. The hot air itself would be the system, and the skin of the balloon provides the walls that separate this system from the surrounding cooler air. The state of the system would be specified by giving values for the pressure, volume, temperature, and mass of the hot air.

As this chapter discusses, there are four laws of thermodynamics. We begin with the one known as the zeroth law and then consider the remaining three.

15.2 The Zeroth Law of Thermodynamics

The zeroth law of thermodynamics deals with the concept of **thermal equilibrium**. Two systems are said to be in thermal equilibrium if there is no net flow of heat between them when they are brought into thermal contact. For instance, you are definitely *not* in thermal equilibrium with the water in Lake Michigan in January. Just dive into it, and you will find out how quickly your body loses heat to the frigid water. To help explain the central idea of the zeroth law of thermodynamics, **Figure 15.2a** shows two systems labeled A and B. Each is within a container whose adiabatic walls are made from insulation that prevents the flow of heat, and each has the same temperature, as indicated by a thermometer. In part *b*, one wall of each container is replaced by a thin silver sheet, and the two sheets are touched together. Silver has a large thermal conductivity, so heat flows through it readily and the silver sheets behave as diathermal walls. Even though the diathermal walls would permit it, no net flow of heat occurs in part *b*, indicating that the two systems are in thermal equilibrium. There is no net flow of heat because the two systems have the same temperature. We see, then, that *temperature is the indicator of thermal equilibrium in the sense that there is no net flow of heat between two systems in thermal contact that have the same temperature.*

In **Figure 15.2** the thermometer plays an important role. System A is in equilibrium with the thermometer, and so is system B. In each case, the thermometer registers the same temperature, thereby indicating that the two systems are equally hot. Consequently, systems A and B are found to be in thermal equilibrium with each other. In effect, the thermometer is a third system. The fact that system A and system B are each in thermal equilibrium with this third system at the same temperature means that they are in thermal equilibrium with each other. This finding is an example of the **zeroth law of thermodynamics**.

> **THE ZEROTH LAW OF THERMODYNAMICS**
>
> **Two systems individually in thermal equilibrium with a third system* are in thermal equilibrium with each other.**

The zeroth law establishes temperature as the indicator of thermal equilibrium and implies that all parts of a system must be in thermal equilibrium if the system is to have a definable single temperature. In other words, there can be no flow of heat within a system that is in thermal equilibrium.

15.3 The First Law of Thermodynamics

The atoms and molecules of a substance have kinetic and potential energy. These and other kinds of molecular energy constitute the internal energy of a substance. When a substance participates in a process involving energy in the form of work and heat, the internal

---

*The state of the third system is the same when it is in thermal equilibrium with either of the two systems. In **Figure 15.2**, for example, the mercury level is the same in the thermometer in either system.

energy of the substance can change. The relationship between work, heat, and changes in the internal energy is known as the first law of thermodynamics. We will now see that the first law of thermodynamics is an expression of the conservation of energy.

Suppose that a system gains heat Q and that this is the only effect occurring. Consistent with the law of conservation of energy, the internal energy of the system increases from an initial value of U_i to a final value of U_f, the change being $\Delta U = U_f - U_i = Q$. In writing this equation, we use the following convention.

> **Problem-Solving Insight**   Heat Q is positive when the system gains heat and negative when the system loses heat.

The internal energy of a system can also change because of work. If a system does work W on its surroundings and there is no heat flow, energy conservation indicates that the internal energy of the system decreases from U_i to U_f, the change now being $\Delta U = U_f - U_i = -W$. The minus sign is included with the work because we employ the following convention.

> **Problem-Solving Insight**   Work is positive when it is done by the system and negative when it is done on the system.

A system can gain or lose energy simultaneously in the form of heat Q and work W. The change in internal energy due to both factors is given by Equation 15.1. Thus, the **first law of thermodynamics** is just the conservation-of-energy principle applied to heat, work, and the change in the internal energy.

> **THE FIRST LAW OF THERMODYNAMICS**
>
> **The internal energy of a system changes from an initial value U_i to a final value of U_f due to heat Q and work W:**
>
> $$\Delta U = U_f - U_i = Q - W \qquad (15.1)$$
>
> Q **is positive when the system gains heat and negative when it loses heat. W is positive when work is done by the system and negative when work is done on the system.**

Examples 1 and 2 illustrate the use of Equation 15.1 and the sign conventions for Q and W.

EXAMPLE 1 | Positive and Negative Work

Interactive Figure 15.3 illustrates a system and its surroundings. In part a, the system gains 1500 J of heat from its surroundings, and 2200 J of work is done *by* the system on the surroundings. In part b, the system also gains 1500 J of heat, but 2200 J of work is done *on* the system by the surroundings. In each case, determine the change in the internal energy of the system.

Reasoning   In **Interactive Figure 15.3a** the system loses more energy in doing work than it gains in the form of heat, so the internal energy of the system decreases. Thus, we expect the change in the internal energy, $\Delta U = U_f - U_i$, to be negative. In part b of the drawing, the system gains energy in the form of both heat and work. The internal energy of the system increases, and we expect ΔU to be positive.

> **Problem-Solving Insight**   When using the first law of thermodynamics, as expressed by Equation 15.1, be careful to follow the proper sign conventions for the heat Q and the work W.

(a)   *(b)*

INTERACTIVE FIGURE 15.3   (*a*) The system gains energy in the form of heat but loses energy because work is done by the system. (*b*) The system gains energy in the form of heat and also gains energy because work is done on the system.

Solution   (a) The heat is positive, $Q = +1500$ J, since it is gained by the system. The work is positive, $W = +2200$ J, since it is done *by* the system. According to the first law of thermodynamics

$$\Delta U = Q - W = (+1500 \text{ J}) - (+2200 \text{ J}) = \boxed{-700 \text{ J}} \qquad (15.1)$$

The minus sign for ΔU indicates that the internal energy has decreased, as expected.

(b) The heat is positive, $Q = +1500$ J, since it is gained by the system. But the work is negative, $W = -2200$ J, since it is done *on* the system. Thus,

$$\Delta U = Q - W = (+1500 \text{ J}) - (-2200 \text{ J}) = \boxed{+3700 \text{ J}} \qquad (15.1)$$

The plus sign for ΔU indicates that the internal energy has increased, as expected.

EXAMPLE 2 | BIO Refueling Your Lost Internal Energy

Suppose you spend 1 hour jogging and do work with a power output of 170 W. During this time, you also give off 4.2×10^5 J of heat to your surroundings. How many bananas would you have to eat to replace the internal energy you lost by jogging? Assume one banana contains 100 Calories.

Reasoning We can use the first law of thermodynamics (Equation 15.1) to calculate the change in the jogger's internal energy. This depends on the heat (Q) lost or gained by the system and the work (W) done on or by the system. Once we know the jogger's change in internal energy, then we can use the mechanical equivalent of heat to calculate the Calories (bananas) she must eat to replace this lost energy.

Solution We begin by using the first law of thermodynamics (Equation 15.1): $\Delta U = Q - W$. The heat from the system (the jogger) is given in the problem, and it will be negative, since it is lost

by the system. The work (W) will be positive, since it is done by the system. We can calculate the work using the power output and the time she runs:

$$W = Pt = (170 \text{ W})(3600 \text{ s}) = 6.1 \times 10^5 \text{ J}$$

Calculating the change in the internal energy, we get:

$$\Delta U = -4.2 \times 10^5 \text{ J} - 6.1 \times 10^5 \text{ J} = -10.3 \times 10^5 \text{ J}$$

As expected, her change in internal energy is negative, and it's this quantity that must be replaced by the food she eats. Each banana contains 100 Calories, which is equal to 100 kcal. This is equivalent to $(100 \text{ kcal}) \times (4186 \text{ J/kcal}) = 4.186 \times 10^5$ J per banana. The number of bananas required to replace her lost internal energy will be:

$$(10.3 \times 10^5 \text{ J})/(4.186 \times 10^5 \text{ J/banana}) = \boxed{2.5 \text{ bananas}}$$

There is a lot of energy content in food!

In the first law of thermodynamics, the internal energy U, heat Q, and work W are energy quantities, and each is expressed in energy units such as joules. However, there is a fundamental difference between U, on the one hand, and Q and W on the other. The next example sets the stage for explaining this difference.

EXAMPLE 3 | An Ideal Gas

The temperature of three moles of a monatomic ideal gas is reduced from $T_i = 540$ K to $T_f = 350$ K by two different methods. In the first method 5500 J of heat flows into the gas, whereas in the second, 1500 J of heat flows into it. In each case find (a) the change in the internal energy and (b) the work done by the gas.

Reasoning Since the internal energy of a monatomic ideal gas is $U = \frac{3}{2}nRT$ (Equation 14.7) and since the number of moles n is fixed, only a change in temperature T can alter the internal energy. Because the change in T is the same in both methods, the change in U is also the same. From the given temperatures, the change ΔU in internal energy can be determined. Then, the first law of thermodynamics can be used with ΔU and the given heat values to calculate the work for each of the methods.

Solution (a) Using Equation 14.7 for the internal energy of a monatomic ideal gas, we find for each method of adding heat that

$$\Delta U = \frac{3}{2}nR(T_f - T_i) = \frac{3}{2}(3.0 \text{ mol})[8.31 \text{ J/(mol} \cdot \text{K)}](350 \text{ K} - 540 \text{ K})$$
$$= \boxed{-7100 \text{ J}}$$

(b) Since ΔU is now known and the heat is given in each method, we can use Equation 15.1 ($\Delta U = Q - W$) to determine the work:

1st method $W = Q - \Delta U = 5500 \text{ J} - (-7100 \text{ J}) = \boxed{12\,600 \text{ J}}$

2nd method $W = Q - \Delta U = 1500 \text{ J} - (-7100 \text{ J}) = \boxed{8600 \text{ J}}$

In each method the gas does work, but it does more in the first method.

To understand the difference between U and either Q or W, consider the value for ΔU in Example 3. In both methods ΔU is the same. Its value is determined once the initial and final temperatures are specified because the internal energy of an ideal gas depends only on the temperature. Temperature is one of the variables (along with pressure and volume) that define the state of a system.

> **Problem-Solving Insight** The internal energy depends only on the state of a system, not on the method by which the system arrives at a given state.

In recognition of this characteristic, internal energy is referred to as a ***function of state***.* In contrast, heat and work are not functions of state because they have different values for each different method used to make the system change from one state to another, as in Example 3.

*The fact that an ideal gas is used in Example 3 does not restrict our conclusion. Had a real (nonideal) gas or other material been used, the only difference would have been that the expression for the internal energy would have been more complicated. It might have involved the volume V, as well as the temperature T, for instance.

Check Your Understanding

(The answer is given at the end of the book.)

1. A gas is enclosed within a chamber that is fitted with a frictionless piston. The piston is then pushed in, thereby compressing the gas. Which statement below regarding this process is consistent with the first law of thermodynamics? **(a)** The internal energy of the gas will increase. **(b)** The internal energy of the gas will decrease. **(c)** The internal energy of the gas will not change. **(d)** The internal energy of the gas may increase, decrease, or remain the same, depending on the amount of heat that the gas gains or loses.

15.4 | Thermal Processes

A system can interact with its surroundings in many ways, and the heat and work that come into play always obey the first law of thermodynamics. This section introduces four common thermal processes. In each case, the process is assumed to be **quasi-static**, which means that it occurs slowly enough that a uniform pressure and temperature exist throughout all regions of the system at all times.

An isobaric process is one that occurs at constant pressure. For instance, **Figure 15.4** shows a substance (solid, liquid, or gas) contained in a chamber fitted with a frictionless piston. The pressure P experienced by the substance is always the same, because it is determined by the external atmosphere and the weight of the piston and the block resting on it. Heating the substance makes it expand and do work W in lifting the piston and block through the displacement $\vec{s}$. The work can be calculated from $W = Fs$ (Equation 6.1), where F is the magnitude of the force and s is the magnitude of the displacement. The force is generated by the pressure P acting on the bottom surface of the piston (area $= A$), according to $F = PA$ (Equation 10.19). With this substitution for F, the work becomes $W = (PA)s$. But the product $A \cdot s$ is the change in volume of the material, $\Delta V = V_f - V_i$, where V_f and V_i are the final and initial volumes, respectively. Thus, the relation is

Isobaric process
$$W = P\,\Delta V = P(V_f - V_i) \qquad (15.2)$$

Consistent with our sign convention, this result predicts a positive value for the work done *by a system* when it expands isobarically (V_f exceeds V_i). Equation 15.2 also applies to an isobaric compression (V_f less than V_i). Then, the work is negative, since work must be done *on the system* to compress it. Example 4 emphasizes that $W = P\,\Delta V$ applies to any system, solid, liquid, or gas, as long as the pressure remains constant while the volume changes.

FIGURE 15.4 The substance in the chamber is expanding isobarically because the pressure is held constant by the external atmosphere and the weight of the piston and the block.

Analyzing Multiple-Concept Problems

EXAMPLE 4 | Isobaric Expansion of Water

One gram of water is placed in the cylinder in **Figure 15.4**, and the pressure is maintained at 2.0×10^5 Pa. The temperature of the water is raised by 31 C°. In one case, the water is in the liquid phase, expands by the small amount of 1.0×10^{-8} m³, and has a specific heat capacity of 4186 J/(kg · C°). In another case, the water is in the gas phase, expands by the much greater amount of 7.1×10^{-5} m³, and has a specific heat capacity of 2020 J/(kg · C°). Determine the change in the internal energy of the water in each case.

Reasoning The change ΔU in the internal energy is given by the first law of thermodynamics as $\Delta U = Q - W$ (Equation 15.1). The heat Q may be evaluated as $Q = cm\,\Delta T$ (Equation 12.4). Finally, since the process occurs at a constant pressure (isobaric), the work W may be found using $W = P\,\Delta V$ (Equation 15.2).

Knowns and Unknowns The following table summarizes the given data:

Description	Symbol	Value	Comment
Mass of water	m	1.0 g	0.0010 kg
Pressure on water	P	2.0×10^5 Pa	Pressure is constant.
Increase in temperature	ΔT	31 C°	
Increase in volume of liquid	ΔV_{liquid}	1.0×10^{-8} m³	Expansion occurs.
Specific heat capacity of liquid	c_{liquid}	4186 J/(kg · C°)	
Increase in volume of gas	ΔV_{gas}	7.1×10^{-5} m³	Expansion occurs.
Specific heat capacity of gas	c_{gas}	2020 J/(kg · C°)	
Unknown Variables			
Change in internal energy of liquid	ΔU_{liquid}	?	
Change in internal energy of gas	ΔU_{gas}	?	

Modeling the Problem

STEP 1 The First Law of Thermodynamics   The change ΔU in the internal energy is given by the first law of thermodynamics, as shown at the right. In Equation 15.1, neither the heat Q nor the work W is known, so we turn to Steps 2 and 3 to evaluate them.

$$\Delta U = Q - W \qquad (15.1)$$
$$\boxed{?} \quad \boxed{?}$$

STEP 2 Heat and Specific Heat Capacity   According to Equation 12.4, the heat Q needed to raise the temperature of a mass m of material by an amount ΔT is

$$\boxed{Q = cm\,\Delta T}$$

where c is the material's specific heat capacity. Data are available for all of the terms on the right side of this expression, which can be substituted into Equation 15.1, as shown at the right. The remaining unknown variable in Equation 15.1 is the work W, and we evaluate it in Step 3.

$$\Delta U = Q - W \qquad (15.1)$$
$$\boxed{Q = cm\,\Delta T} \qquad (12.4)$$
$$\boxed{?}$$

STEP 3 Work Done at Constant Pressure   Under constant-pressure, or isobaric, conditions, the work W done is given by Equation 15.2 as

$$\boxed{W = P\,\Delta V}$$

where P is the pressure acting on the material and ΔV is the change in the volume of the material. Substitution of this expression into Equation 15.1 is shown at the right.

$$\Delta U = Q - W \qquad (15.1)$$
$$\boxed{Q = cm\,\Delta T} \qquad (12.4)$$
$$\boxed{W = P\,\Delta V} \qquad (15.1)$$

Solution   Combining the results of each step algebraically, we find that

$$\overset{\text{STEP 1}}{\downarrow} \quad \overset{\text{STEP 2}}{\downarrow} \quad \overset{\text{STEP 3}}{\downarrow}$$
$$\Delta U = Q - W = cm\,\Delta T - W = cm\,\Delta T - P\,\Delta V$$

Applying this result to the liquid and to the gaseous water gives

$$\Delta U_{liquid} = c_{liquid}\, m\, \Delta T - P\, \Delta V_{liquid}$$

$$= [4186 \text{ J/(kg · C°)}](0.0010 \text{ kg})(31 \text{ C°}) - (2.0 \times 10^5 \text{ Pa})(1.0 \times 10^{-8} \text{ m}^3)$$

$$= 130 \text{ J} - 0.0020 \text{ J} = \boxed{130 \text{ J}}$$

$$\Delta U_{gas} = c_{gas}\, m\, \Delta T - P\, \Delta V_{gas}$$

$$= [2020 \text{ J/(kg · C°)}](0.0010 \text{ kg})(31 \text{ C°}) - (2.0 \times 10^5 \text{ Pa})(7.1 \times 10^{-5} \text{ m}^3)$$

$$= 63 \text{ J} - 14 \text{ J} = \boxed{49 \text{ J}}$$

For the liquid, virtually all the 130 J of heat serves to change the internal energy, since the volume change and the corresponding work of expansion are so small. In contrast, a significant fraction of the 63 J of heat added to the gas causes work of expansion to be done, so that only 49 J is left to change the internal energy.

Related Homework: Problem 15

It is often convenient to display thermal processes graphically. For instance, **Interactive Figure 15.5** shows a plot of pressure versus volume for an isobaric expansion. Since the pressure is constant, the graph is a horizontal straight line, beginning at the initial volume V_i and ending at the final volume V_f. In terms of such a plot, the work $W = P(V_f - V_i)$ is the area under the graph, which is the shaded rectangle of height P and width $V_f - V_i$.

Another common thermal process is an ***isochoric process, one that occurs at constant volume***. **Figure 15.6a** illustrates an isochoric process in which a substance (solid, liquid, or gas) is heated. The substance would expand if it could, but the rigid container keeps the volume constant, so the pressure–volume plot shown in **Figure 15.6b** is a vertical straight line. Because the volume is constant, the pressure inside rises, and the substance exerts more and more force on the walls. Although enormous forces can be generated in the closed container, no work is done ($W = 0$ J), since the walls do not move. Consistent with zero work being done, the area under the vertical straight line in **Figure 15.6b** is zero. Since no work is done, the first law of thermodynamics indicates that the heat in an isochoric process serves only to change the internal energy: $\Delta U = Q - W = Q$.

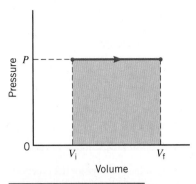

INTERACTIVE FIGURE 15.5 For an isobaric process, a pressure–volume plot is a horizontal straight line. The work done $[W = P(V_f - V_i)]$ is the colored rectangular area under the graph.

FIGURE 15.6 (a) The substance in the chamber is being heated isochorically because the rigid chamber keeps the volume constant. (b) The pressure–volume plot for an isochoric process is a vertical straight line. The area under the graph is zero, indicating that no work is done.

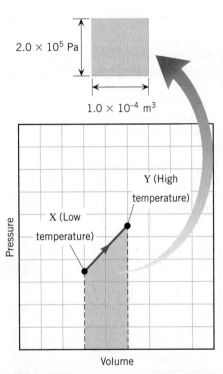

FIGURE 15.7 The colored area gives the work done by the gas for the process from X to Y.

A third important thermal process is an ***isothermal process, one that takes place at constant temperature***. The next section illustrates the important features of an isothermal process when the system is an ideal gas.

Last, there is the ***adiabatic process, one that occurs without the transfer of heat***. Since there is no heat transfer, Q equals zero, and the first law indicates that $\Delta U = Q - W = -W$. Thus, when work is done by a system adiabatically, W is positive and the internal energy decreases by exactly the amount of the work done. When work is done on a system adiabatically, W is negative and the internal energy increases correspondingly. The next section discusses an adiabatic process for an ideal gas.

A process may be complex enough that it is not recognizable as one of the four just discussed. For instance, **Figure 15.7** shows a process for a gas in which the pressure, volume, and temperature are changed along the straight line from X to Y. With the aid of integral calculus, the following can be proved.

| **Problem-Solving Insight** The area under a pressure–volume graph is the work for any kind of process. |

Thus, the area representing the work has been colored in **Figure 15.7**. The volume increases, so that work is done by the gas. This work is positive by convention, as is the area. In contrast, if a process reduces the volume, work is done on the gas, and this work is negative by convention. Correspondingly, the area under the pressure–volume graph would be assigned a negative value. In Example 5, we determine the work for the case shown in **Figure 15.7**.

EXAMPLE 5 | Work and the Area Under a Pressure–Volume Graph

Determine the work for the process in which the pressure, volume, and temperature of a gas are changed along the straight line from X to Y in **Figure 15.7**.

Reasoning The work is given by the area (in color) under the straight line between X and Y. Since the volume increases, work is done by the gas on the surroundings, so the work is positive. The

area can be found by counting squares in **Figure 15.7** and multiplying by the area per square.

Solution We estimate that there are 8.9 colored squares in the drawing. The area of one square is $(2.0 \times 10^5 \text{ Pa})(1.0 \times 10^{-4} \text{ m}^3) = 2.0 \times 10^1$ J, so the work is

$$W = +(8.9 \text{ squares})(2.0 \times 10^1 \text{ J/square}) = \boxed{+180 \text{ J}}$$

Check Your Understanding

(The answers are given at the end of the book.)

2. Is it possible for the temperature of a substance to rise without heat flowing into the substance? **(a)** Yes, provided that the volume of the substance does not change. **(b)** Yes, provided that the substance expands and does positive work. **(c)** Yes, provided that work is done on the substance and it contracts.

3. CYU **Figure 15.1** shows a pressure-versus-volume plot for a three-step process: A → B, B → C, and C → A. For each step, the work can be positive, negative, or zero. Which answer in **CYU Table 15.1** correctly describes the work for the three steps?

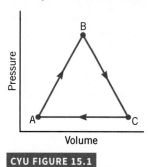

CYU FIGURE 15.1

CYU TABLE 15.1	Work Done by the System		
	A → B	B → C	C → A
(a)	Positive	Negative	Negative
(b)	Positive	Positive	Negative
(c)	Negative	Negative	Positive
(d)	Positive	Negative	Zero
(e)	Negative	Positive	Zero

4. CYU **Figure 15.2** shows a pressure–volume graph in which a gas expands at constant pressure from *A* to *B*, and then goes from *B* to *C* at constant volume. Complete **CYU Table 15.2** by deciding whether each of the four unspecified quantities is positive (+), negative (−), or zero (0).

CYU FIGURE 15.2

CYU TABLE 15.2			
	ΔU	Q	W
$A \to B$	+	?	?
$B \to C$	?	+	?

5. When a solid melts at constant pressure, the volume of the resulting liquid does not differ much from the volume of the solid. According to the first law of thermodynamics, how does the internal energy of the liquid compare to the internal energy of the solid? The internal energy of the liquid is **(a)** greater than, **(b)** the same as, **(c)** less than the internal energy of the solid.

15.5 Thermal Processes Using an Ideal Gas

Isothermal Expansion or Compression

When a system performs work isothermally, the temperature remains constant. In **Animated Figure 15.8a**, for instance, a metal cylinder contains n moles of an ideal gas, and the large mass of hot water maintains the cylinder and gas at a constant Kelvin temperature T. The piston is held in place initially so the volume of the gas is V_i. As the external force applied to the piston is reduced quasi-statically, the pressure decreases as the gas expands to the final volume V_f. **Animated Figure 15.8b** gives a plot of pressure ($P = nRT/V$) versus volume for the process. The solid red line in the graph is called an isotherm (meaning "constant temperature") and represents the relation between pressure and volume when the temperature is held constant. The work W done by the gas is *not* given by $W = P\,\Delta V = P(V_f - V_i)$ because the pressure is not constant. Nevertheless, the work is equal to the area under the graph. The techniques of integral calculus lead to the following result* for W:

Isothermal expansion or compression of an ideal gas

$$W = nRT \ln\left(\frac{V_f}{V_i}\right) \qquad (15.3)$$

Where does the energy for this work originate? Since the internal energy of any ideal gas is proportional to the Kelvin temperature ($U = \frac{3}{2}nRT$ for a monatomic ideal gas, for example), the internal energy remains constant throughout an isothermal process, and the change in internal energy is zero. As a result, the first law of thermodynamics becomes $\Delta U = 0 = Q - W$. In other words, $Q = W$, and the energy for the work originates in the hot water. Heat flows into the gas from the water, as **Animated Figure 15.8a** illustrates. If the gas is compressed isothermally, Equation 15.3 still applies, and heat flows out of the gas into the water. Example 6 deals with the isothermal expansion of an ideal gas.

Metal cylinder

Hot water at temperature T

(a)

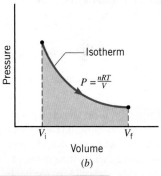

Pressure

Isotherm

$P = \frac{nRT}{V}$

V_i V_f

Volume

(b)

ANIMATED FIGURE 15.8 (*a*) The ideal gas in the cylinder is expanding isothermally at temperature T. The force holding the piston in place is reduced slowly, so the expansion occurs quasi-statically. (*b*) The work done by the gas is given by the colored area.

EXAMPLE 6 | Isothermal Expansion of an Ideal Gas

Two moles of the monatomic gas argon expand isothermally at 298 K, from an initial volume of $V_i = 0.025$ m³ to a final volume of $V_f = 0.050$ m³. Assuming that argon behaves as an ideal gas, find **(a)** the work done by the gas, **(b)** the change in the internal energy of the gas, and **(c)** the heat supplied to the gas.

Reasoning and Solution **(a)** The work done by the gas can be found from Equation 15.3:

$$W = nRT \ln\left(\frac{V_f}{V_i}\right) = (2.0 \text{ mol})[8.31 \text{ J/(mol} \cdot \text{K)}](298 \text{ K}) \ln\left(\frac{0.050 \text{ m}^3}{0.025 \text{ m}^3}\right)$$

$$= \boxed{+3400 \text{ J}}$$

(b) The internal energy of a monatomic ideal gas is $U = \frac{3}{2}nRT$ (Equation 14.7) and does not change when the temperature is constant. Therefore, $\boxed{\Delta U = 0 \text{ J}}$.

(c) The heat Q supplied can be determined from the first law of thermodynamics:

$$Q = \Delta U + W = 0 \text{ J} + 3400 \text{ J} = \boxed{+3400 \text{ J}} \qquad (15.1)$$

Adiabatic Expansion or Compression

When a system performs work adiabatically, no heat flows into or out of the system. **Figure 15.9a** shows an arrangement in which n moles of an ideal gas do work under adiabatic conditions, expanding quasi-statically from an initial volume V_i to a final volume

---

*In this result, "ln" denotes the natural logarithm to the base $e = 2.71828$. The natural logarithm is related to the common logarithm to the base ten by $\ln(V_f/V_i) = 2.303 \log(V_f/V_i)$.

Metal cylinder

Insulating material

(a)

(b)

FIGURE 15.9 (a) The ideal gas in the cylinder is expanding adiabatically. The force holding the piston in place is reduced slowly, so the expansion occurs quasi-statically. (b) A plot of pressure versus volume yields the adiabatic curve shown in red, which intersects the isotherms (blue) at the initial temperature T_i and the final temperature T_f. The work done by the gas is given by the colored area.

V_f. The arrangement is similar to that in **Animated Figure 15.8** for isothermal expansion. However, a different amount of work is done here, because the cylinder is now surrounded by insulating material that prevents the flow of heat from occurring, so $Q = 0$ J. According to the first law of thermodynamics, the change in internal energy is $\Delta U = Q - W = -W$. Since the internal energy of an ideal monatomic gas is $U = \frac{3}{2}nRT$ (Equation 14.7), it follows directly that $\Delta U = U_f - U_i = \frac{3}{2}nR(T_f - T_i)$, where T_i and T_f are the initial and final Kelvin temperatures. With this substitution, the relation $\Delta U = -W$ becomes

Adiabatic expansion or compression of a monatomic ideal gas

$$W = \frac{3}{2}nR(T_i - T_f) \tag{15.4}$$

When an ideal gas expands adiabatically, it does positive work, so W is positive in Equation 15.4. Therefore, the term $T_i - T_f$ is also positive, so the final temperature of the gas must be less than the initial temperature. The internal energy of the gas is reduced to provide the necessary energy to do the work, and because the internal energy is proportional to the Kelvin temperature, the temperature decreases. **Figure 15.9b** shows a plot of pressure versus volume for an adiabatic process. The adiabatic curve (red) intersects the isotherms (blue) at the higher initial temperature [$T_i = P_iV_i/(nR)$] and also at the lower final temperature [$T_f = P_fV_f/(nR)$]. The colored area under the adiabatic curve represents the work done.

The reverse of an adiabatic expansion is an adiabatic compression (W is negative), and Equation 15.4 indicates that the final temperature of the ideal gas exceeds the initial temperature. The energy provided by the agent doing the work increases the internal energy of the gas. As a result, the gas becomes hotter.

The equation that gives the adiabatic curve (red) between the initial pressure and volume (P_i, V_i) and the final pressure and volume (P_f, V_f) in **Figure 15.9b** can be derived using integral calculus. The result is

Adiabatic expansion or compression of an ideal gas

$$P_iV_i^\gamma = P_fV_f^\gamma \tag{15.5}$$

where the exponent γ (Greek gamma) is the ratio of the specific heat capacities at constant pressure and constant volume, $\gamma = c_P/c_V$. Equation 15.5 applies in conjunction with the ideal gas law, because *each point* on the adiabatic curve satisfies the relation $PV = nRT$.

Math Skills Suppose it is necessary to solve Equation 15.5 for the variable V_i. If the value for γ were $\gamma = 2$ (which it never is for an ideal gas), we would begin by taking the square root of both sides of the equals sign in Equation 15.5 and would deal with the square root of V_i^2 (or V_f^2) as follows:

$$\sqrt{V_i^2} = (V_i^2)^{1/2} = V_i^{2(1/2)} = V_i^1 = V_i$$

We will see in the next section that $\gamma = \frac{5}{3}$ for a monatomic ideal gas, and we deal with such a case in a similar way. We begin by raising both sides of Equation 15.5 to the power $1/\gamma$:

$$P_iV_i^\gamma = P_fV_f^\gamma \quad \text{or} \quad (P_iV_i^\gamma)^{1/\gamma} = (P_fV_f^\gamma)^{1/\gamma}$$

Simplifying the terms in the right-hand equation, we have

$$(P_iV_i^\gamma)^{1/\gamma} = (P_fV_f^\gamma)^{1/\gamma} \quad \text{or} \quad P_i^{1/\gamma}V_i^{\gamma(1/\gamma)} = P_f^{1/\gamma}V_f^{\gamma(1/\gamma)} \quad \text{or} \quad P_i^{1/\gamma}V_i = P_f^{1/\gamma}V_f$$

Dividing both sides of the far-right equation by $P_i^{1/\gamma}$ gives

$$\frac{P_i^{1/\gamma}V_i}{P_i^{1/\gamma}} = \frac{P_f^{1/\gamma}V_f}{P_i^{1/\gamma}} \quad \text{or} \quad V_i = \frac{P_f^{1/\gamma}V_f}{P_i^{1/\gamma}} \quad \text{or} \quad V_i = \left(\frac{P_f}{P_i}\right)^{1/\gamma}V_f$$

For $\gamma = \frac{5}{3}$, the exponent of $1/\gamma$ in the final expression for V_i is $\frac{1}{\gamma} = \frac{3}{5}$.

Table 15.1 summarizes the work done in the four types of thermal processes that we have been considering. For each process it also shows how the first law of thermodynamics depends on the work and other variables.

TABLE 15.1   Summary of Thermal Processes

Type of Thermal Process	Work Done	First Law of Thermodynamics $(\Delta U = Q - W)$
Isobaric (constant pressure)	$W = P(V_f - V_i)$	$\Delta U = Q - \underbrace{P(V_f - V_i)}_{W}$
Isochoric (constant volume)	$W = 0\ \text{J}$	$\Delta U = Q - \underbrace{0\ \text{J}}_{Q}$
Isothermal (constant temperature)	$W = nRT \ln\left(\frac{V_f}{V_i}\right)$ (for an ideal gas)	$\underbrace{0\ \text{J}}_{\substack{\Delta U \text{ for an} \\ \text{ideal gas}}} = Q - \underbrace{nRT \ln\left(\frac{V_f}{V_i}\right)}_{W}$
Adiabatic (no heat flow)	$W = \frac{3}{2}nR(T_i - T_f)$ (for a monatomic ideal gas)	$\Delta U = \underbrace{0\ \text{J}}_{Q} - \underbrace{\frac{3}{2}nR(T_i - T_f)}_{W}$

Check Your Understanding

(The answers are given at the end of the book.)

6. One hundred joules of heat is added to a gas, and the gas expands at constant pressure. Is it possible that the internal energy increases by 100 J? **(a)** Yes **(b)** No; the increase in the internal energy is less than 100 J, since work is done by the gas. **(c)** No; the increase in the internal energy is greater than 100 J, since work is done by the gas.

7. A gas is compressed isothermally, and its internal energy increases. Is the gas an ideal gas? **(a)** No, because if the temperature of an ideal gas remains constant, its internal energy must also remain constant. **(b)** No, because if the temperature of an ideal gas remains constant, its internal energy must decrease. **(c)** Yes, because if the temperature of an ideal gas remains constant, its internal energy must increase.

8. A material undergoes an isochoric process that is also adiabatic. Is the internal energy of the material at the end of the process **(a)** greater than, **(b)** less than, or **(c)** the same as it was at the start?

9. CYU Figure 15.3 shows an arrangement for an adiabatic free expansion or "throttling" process. The process is adiabatic because the entire arrangement is contained within perfectly insulating walls. The gas in chamber A rushes suddenly into chamber B through a hole in the partition. Chamber B is initially evacuated, so the gas expands there under zero external pressure and the work ($W = P\,\Delta V$) it does is zero. Assume that the gas is an ideal gas. How does the final temperature of the gas after expansion compare to its initial temperature? The final temperature is **(a)** greater than, **(b)** less than, **(c)** the same as the initial temperature.

CYU FIGURE 15.3

15.6 | Specific Heat Capacities

In this section the first law of thermodynamics is used to gain an understanding of the factors that determine the specific heat capacity of a material. Remember, when the temperature of a substance changes as a result of heat flow, the change in temperature ΔT and the amount of heat Q are related according to $Q = cm\,\Delta T$ (Equation 12.4). In this expression c denotes the specific heat capacity in units of J/(kg · C°), and m is the mass in kilograms. Now, however, it is more convenient to express the amount of material as the number of moles n, rather than the number of kilograms. Therefore, we replace the expression $Q = cm\,\Delta T$ with the following analogous expression:

$$Q = Cn\,\Delta T \tag{15.6}$$

where the capital letter C (as opposed to the lowercase c) refers to the **molar specific heat capacity** in units of J/(mol · K). In addition, the unit for measuring the temperature change ΔT is the kelvin (K) rather than the Celsius degree (C°), and $\Delta T = T_f - T_i$, where T_f and T_i are the final and initial temperatures. For gases it is necessary to distinguish between the molar specific heat capacities C_P and C_V, which apply, respectively, to conditions of constant pressure and constant volume. With the help of the first law of thermodynamics and an ideal gas as an example, it is possible to see why C_P and C_V differ.

To determine the molar specific heat capacities, we must first calculate the heat Q needed to raise the temperature of an ideal gas from T_i to T_f. According to the first law, $Q = \Delta U + W$. We also know that the internal energy of a monatomic ideal gas is $U = \frac{3}{2}nRT$ (Equation 14.7). As a result, $\Delta U = U_f - U_i = \frac{3}{2}nR(T_f - T_i)$. When the heating process occurs at constant pressure, the work done is given by Equation 15.2: $W = P\,\Delta V = P(V_f - V_i)$. For an ideal gas, $PV = nRT$, so the work becomes $W = nR(T_f - T_i)$. On the other hand, when the volume is constant, $\Delta V = 0\ \text{m}^3$, and the work done is zero. The calculation of the heat is summarized below:

$$Q = \Delta U + W$$
$$Q_{\text{constant pressure}} = \tfrac{3}{2}nR(T_f - T_i) + nR(T_f - T_i) = \tfrac{5}{2}nR(T_f - T_i)$$
$$Q_{\text{constant volume}} = \tfrac{3}{2}nR(T_f - T_i) + 0$$

The molar specific heat capacities can now be determined, since Equation 15.6 indicates that $C = Q/[n(T_f - T_i)]$:

Constant pressure for a monatomic ideal gas
$$C_P = \frac{Q_{\text{constant pressure}}}{n(T_f - T_i)} = \tfrac{5}{2}R \tag{15.7}$$

Constant volume for a monatomic ideal gas
$$C_V = \frac{Q_{\text{constant volume}}}{n(T_f - T_i)} = \tfrac{3}{2}R \tag{15.8}$$

The ratio γ of the specific heats is

Monatomic ideal gas
$$\gamma = \frac{C_P}{C_V} = \frac{\tfrac{5}{2}R}{\tfrac{3}{2}R} = \frac{5}{3} \tag{15.9}$$

For real monatomic gases near room temperature, experimental values of C_P and C_V give ratios very close to the theoretical value of $\gamma = \tfrac{5}{3}$.

Many gases are not monatomic. Instead, they consist of molecules formed from more than one atom. The oxygen in our atmosphere, for example, is a diatomic gas, because it consists of molecules formed from two oxygen atoms. Similarly, atmospheric nitrogen is a diatomic gas consisting of molecules formed from two nitrogen atoms. Whereas the individual atoms in a monatomic ideal gas can exhibit only translational motion, the molecules in a diatomic ideal gas can exhibit translational and rotational motion, as well as vibrational motion at sufficiently high temperatures. The result of such additional motions is that Equations 15.7–15.9 do not apply to a diatomic ideal gas. Instead, if the temperature is sufficiently moderate that the diatomic molecules do not vibrate, the molar specific heat capacities of a diatomic ideal gas are $C_P = \tfrac{7}{2}R$ and $C_V = \tfrac{5}{2}R$, with the result that $\gamma = \frac{C_P}{C_V} = \tfrac{7}{5}$.

The difference between C_P and C_V arises because work is done when the gas expands in response to the addition of heat under conditions of constant pressure, whereas no work is done under conditions of constant volume. For a monatomic ideal gas, C_P exceeds C_V by an amount equal to R, the ideal gas constant:

$$C_p - C_V = R \tag{15.10}$$

In fact, it can be shown that Equation 15.10 applies to any kind of ideal gas—monatomic, diatomic, etc.

Check Your Understanding

(The answers are given at the end of the book.)

10. Suppose that a material contracts when it is heated. Following the same line of reasoning used in the text to reach Equations 15.7 and 15.8, deduce the relationship between the specific heat capacity at constant pressure (C_P) and the specific heat capacity at constant volume (C_V). Which of the following describes the relationship? **(a)** $C_P = C_V$ **(b)** C_P is greater than C_V **(c)** C_P is less than C_V.

11. You want to heat a gas so that its temperature will be as high as possible. Should you heat the gas under conditions of **(a)** constant pressure or **(b)** constant volume? **(c)** It does not matter what the conditions are.

15.7 | The Second Law of Thermodynamics

Ice cream melts when left out on a warm day. A cold can of soda warms up on a hot day at a picnic. Ice cream and soda never become colder when left in a hot environment, for heat always flows spontaneously from hot to cold, and never from cold to hot. The spontaneous flow of heat is the focus of one of the most profound laws in all of science, the **second law of thermodynamics**.

> **THE SECOND LAW OF THERMODYNAMICS: THE HEAT FLOW STATEMENT**
>
> **Heat flows spontaneously from a substance at a higher temperature to a substance at a lower temperature and does not flow spontaneously in the reverse direction.**

It is important to realize that the second law of thermodynamics deals with a different aspect of nature than does the first law of thermodynamics. The second law is a statement about the natural tendency of heat to flow from hot to cold, whereas the first law deals with energy conservation and focuses on both heat and work. A number of important devices depend on heat and work in their operation, and to understand such devices both laws are needed. For instance, an automobile engine is a type of heat engine because it uses heat to produce work. In discussing heat engines, Sections 15.8 and 15.9 will bring together the first and second laws to analyze engine efficiency. Then, in Section 15.10 we will see that refrigerators, air conditioners, and heat pumps also utilize heat and work and are closely related to heat engines. The way in which these three appliances operate also depends on both the first and second laws of thermodynamics. Before moving on to our study of heat engines, we discuss an important thermodynamic system, where heat from a hotter source, the Sun, is transferred to a colder object, planet earth.

THE PHYSICS OF . . . global climate change. Earth, the Sun, and the surrounding space constitute a thermodynamic system, and our planet is warming. The surface temperature of Earth is determined primarily by the radiation energy absorbed from the Sun. Energy produced by the decay of radioactive elements (Chapter 31) in Earth's interior also heat the surface, but this contributes only a small fraction, about 1/1000, of the heating provided by the Sun. Since Earth's surface is in thermal equilibrium with its surroundings, it must be emitting radiant energy into space at the same rate at which it is absorbing it, as described in Section 13.3. The rate of the emitted radiation depends on the surface temperature (Equation 13.2), and if the absorption and emission of radiation were the only two factors in determining the surface temperature of Earth, its average temperature would be −18 °C, which is far too cold to support life as we know it. However, since the average temperature of Earth's surface is closer to 15 °C, there must be another mechanism that contributes to warming. This is where Earth's atmosphere comes into play. Gases in Earth's atmosphere, such as carbon dioxide (CO_2), water vapor

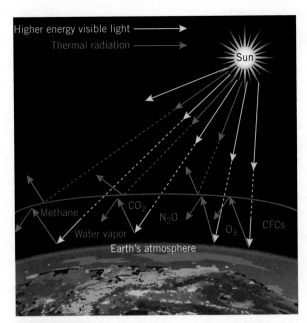

FIGURE 15.10 An atmosphere populated with greenhouse gases absorbs more of the thermal radiation emitted from the oceans and land masses and redirects a portion of it back to the earth, thereby warming its surface.

(H_2O), chlorofluorocarbons (CF_2Cl_2), methane (CH_4), nitrous oxide (N_2O), and ozone (O_3)-the so-called **greenhouse gases**-allow the Sun's shorter wavelength visible light to pass through and reach Earth, while the longer wavelength thermal radiation is absorbed and reemitted, much of it back into space (see **Figure 15.10**). The sunlight that does reach Earth warms its surface, and then the surface reemits a portion of this energy as thermal radiation. Just as the greenhouse gases in the atmosphere prevented the thermal radiation from the Sun from reaching Earth's surface, they now prevent the thermal radiation emitted by Earth's surface from leaving the atmosphere. This effectively acts like a blanket over Earth, trapping additional heat and raising the surface temperature. This process, called the **greenhouse effect**, is what produces the moderate temperature on Earth's surface.

Venus, a planet similar in size and density to Earth, is nearly 30% closer to the Sun than Earth is, and yet its average surface temperature, based solely on the absorption and emission of the Sun's radiant energy (neglecting atmospheric effects), should be −41 °C. This is even colder than the value predicted for Earth, and it results from the fact that Venus reflects a greater fraction of the Sun's energy than Earth does. Nevertheless, the average surface temperature on Venus is 460 °C, which makes it the hottest planet in the solar system! The atmosphere of Venus is composed of greater than 96% CO_2 (compared to just 0.04% on Earth), and this high concentration of greenhouse gases effectively traps the heat from the Sun's radiant energy, resulting in the planet's high surface temperature. The conditions on Venus correspond to a **runaway greenhouse effect**, which occurs when the concentration of greenhouse gases in a planet's atmosphere are so high that it blocks the thermal radiation from the planet and prevents the planet from cooling.

Since the late 1880's, the average surface temperature of Earth has increased by over 1.8 °F (**Figure 15.11**). While pollution, overpopulation, and deforestation have all contributed to the effects of this global warming, the most significant cause of the increasing temperature is the greenhouse effect. And while it is true that the average surface temperature of Earth and the level of carbon dioxide in the atmosphere have naturally fluctuated over time, since 1750 the level of CO_2 has increased by over 45%. Ice-core data over the last 800 000 years show the average level of CO_2 in the atmosphere was about 225 ppm (parts per million), while never spiking above 300 ppm. Just in the last 270 years, the level has increased from 280 ppm to over 400 ppm (**Figure 15.12**). In other words, human beings as a species have never lived on Earth during a time when the CO_2 levels in the atmosphere were this high. Human activities after the Industrial Revolution, primarily in the form of the extraction and burning of fossil fuels, such as oil, coal, and natural gas, are responsible for the vast majority of the increase in CO_2. To date, there are no alternative explanations, natural or otherwise, that can reasonably account for the rapid change in climate. The Intergovernmental Panel on Climate Change (IPCC) concluded that "human influence on climate has been the dominant cause of observed global warming since the mid-20th century".[1]

An increase of 1.8 °F does not seem like much, so why should we be concerned? The last decade (2000–2019) was the hottest decade on record. In fact, the past five years have been the hottest five years on record for the second year running. With this rise in surface temperature, vast amounts of the polar ice caps, sea ice, glaciers, and permafrost are melting, which is leading to a rise in sea levels. One conservative estimate suggests a 4.2-foot rise in sea level for every 1 °F increase in average surface

FIGURE 15.11 The annual average change in the surface temperature of the earth relative to 1880. Reprinted from USGRCP and Wuebbles *et al.* (eds.), *Climate Science Special Report: Fourth National Climate Assessment,* Vol. 1 (Washington, DC: USGRCP), 2017.)

[1]Allen *et al.*, "Chapter 1: Framing and Context," IPCC SR15, 2018, pp. 49–91.

FIGURE 15.12 The level of carbon dioxide gas in the atmosphere over time. The average concentration was about 225 ppm dating back 800 000 years. However, after the year 1750, the level increased, and it is presently over 400 ppm- a level never before reached in the history of human beings. Adapted from NOAA Climate.gov based on data from Lüthi, *et al.*, 2008, via NOAA NCEI Paleoclimatology Program.

temperature.[2] Widespread coastal flooding is expected in the future, with 150 million people being displaced during high tide by the year 2050.[3] A warmer climate also results in more extreme weather, with heavy rain and flooding, more intense hurricanes, tornados, and snow storms, as well as longer and more widespread periods of drought. These changes in weather will ultimately displace our coastal populations and affect our food supply, both in agriculture and livestock production, as well as have negative impacts on our health.

The effects of global warming cannot be stopped completely, but the impacts of climate change can be mitigated by reducing our emission of greenhouse gases. For example, in order to keep the mean surface temperature of Earth from rising more than 2.7 °F (1.5 °C), we need to bring our CO_2 emissions to zero by 2050. Climate experts agree that an increase in temperature above 2.7 °F would result in catastrophic changes in climate. This is a huge challenge for humankind, and the fate of countless other animal species and entire ecosystems will depend on our ability to act. We will need to reduce our dependence on fossil fuels and develop alternative sources of renewable energy, such as wind and solar. Nuclear energy offers certain advantages over fossil fuels, but it also has significant disadvantages, such as the creation and storage of nuclear waste. However, more than just using different sources of energy, combating global climate change will require us to change our behavior, in terms of how we live, work, travel, and play. Fortunately, we still have time, but the window for action is quickly closing. This will not be solved by one country alone, but will require a concerted effort of all nations. The Paris Agreement, which was created under the United Nations Framework Convention on Climate Change, is a multinational effort aimed at limiting the negative impacts of global warming. It proposed both energy and climate policy with the goals of 1) reducing CO_2 emissions by 20%, 2) increasing the renewable energy market by 20%, and 3) improving the efficiency of existing energy technologies by 20%. The agreement was signed in April 2016 by 196 nations. The energy carried by sunlight and its impact on the greenhouse effect is discussed more in Section 24.4.

15.8 | Heat Engines

THE PHYSICS OF . . . a heat engine.   A **heat engine** is any device that uses heat to perform work. It has three essential features:

1. Heat is supplied to the engine at a relatively high input temperature from a place called the *hot reservoir.*

2. Part of the input heat is used to perform work by the *working substance* of the engine, which is the material within the engine that actually does the work (e.g., the gasoline–air mixture in an automobile engine).

3. The remainder of the input heat is rejected to a place called the *cold reservoir,* which has a temperature lower than the input temperature.

[2]U.S. Global Change Research Program (USGCRP), "Climate Science Special Report: Chapter 12. Sea Level Rise," retrieved December 27, 2018, from science2017.globalchange.gov.
[3]Kulp, S. A., and Strauss, B. H., "New Elevation Data Triple Estimates of Global Vulnerability to Seal-Level Rise and Coastal Flooding," *Nature Communications*, Vol. 10, no. 1, p. 4844 (2019).

Hot reservoir

$|Q_H|$

Engine $|W|$

$|Q_C|$

Cold reservoir

INTERACTIVE FIGURE 15.13 This schematic representation of a heat engine shows the input heat (magnitude = $|Q_H|$) that originates from the hot reservoir, the work (magnitude = $|W|$) that the engine does, and the heat (magnitude = $|Q_C|$) that the engine rejects to the cold reservoir.

Interactive Figure 15.13 illustrates these features. The symbol Q_H refers to the input heat, and the subscript H indicates the hot reservoir. Similarly, the symbol Q_C stands for the rejected heat, and the subscript C denotes the cold reservoir. The symbol W refers to the work done. The vertical bars enclosing each of these three symbols in the drawing are included to emphasize that we are concerned here with the absolute values, or magnitudes, of the symbols. Thus, $|Q_H|$ indicates the magnitude of the input heat, $|Q_C|$ denotes the magnitude of the rejected heat, and $|W|$ stands for the magnitude of the work done.

To be highly efficient, a heat engine must produce a relatively large amount of work from as little input heat as possible. Therefore, the **efficiency** e of a heat engine is defined as the ratio of the magnitude of the work $|W|$ done by the engine to the magnitude of the input heat $|Q_H|$:

$$e = \frac{|W|}{|Q_H|} \tag{15.11}$$

If the input heat were converted entirely into work, the engine would have an efficiency of 1.00, since $|W| = |Q_H|$; such an engine would be 100% efficient. *Efficiencies are often quoted as percentages obtained by multiplying the ratio $|W|/|Q_H|$ by a factor of 100.* Thus, an efficiency of 68% would mean that a value of 0.68 is used for the efficiency in Equation 15.11.

An engine, like any device, must obey the principle of conservation of energy. Some of the engine's input heat $|Q_H|$ is converted into work $|W|$, and the remainder $|Q_C|$ is rejected to the cold reservoir. If there are no other losses in the engine, the principle of energy conservation requires that

$$|Q_H| = |W| + |Q_C| \tag{15.12}$$

Solving this equation for $|W|$ and substituting the result into Equation 15.11 leads to the following alternative expression for the efficiency e of a heat engine:

$$e = \frac{|Q_H| - |Q_C|}{|Q_H|} = 1 - \frac{|Q_C|}{|Q_H|} \tag{15.13}$$

Example 7 illustrates how the concepts of efficiency and energy conservation are applied to a heat engine.

Math Skills The symbols $|Q_H|$, $|Q_C|$, and $|W|$ refer to absolute values or magnitudes only. It is essential to remember that these symbols never have negative values assigned to them when they appear in equations. For example, the value assigned to $|W|$ is the same if $W = -1830$ J or $W = +1830$ J.
This follows because

$$|W| = |-1830 \text{ J}| = 1830 \text{ J}$$

and

$$|W| = |+1830 \text{ J}| = 1830 \text{ J}$$

Analyzing Multiple-Concept Problems

EXAMPLE 7 | An Automobile Engine

An automobile engine has an efficiency of 22.0% and produces 2510 J of work. How much heat is rejected by the engine?

Reasoning Energy conservation indicates that the amount of heat rejected to the cold reservoir is the part of the input heat that is *not* converted into work. The work is given, and the input heat can be obtained since the efficiency of the engine is also given.

Knowns and Unknowns The following data are available:

Description	Symbol	Value
Efficiency of engine	e	22.0% (0.220)
Magnitude of work	$|W|$	2510 J
Unknown Variable		
Magnitude of rejected heat	$|Q_C|$	?

Modeling the Problem

STEP 1 The Conservation of Energy According to the energy-conservation principle, the magnitudes of the input heat $|Q_H|$, the work done $|W|$, and the rejected heat $|Q_C|$ are related according to $|Q_H| = |W| + |Q_C|$ (Equation 15.12). Solving for $|Q_C|$ gives Equation 1 at the right. In this result, $|W|$ is known, but $|Q_H|$ is not, although it will be evaluated in Step 2.

$$|Q_C| = |Q_H| - |W| \qquad (1)$$

$$\uparrow$$

$$\boxed{?}$$

STEP 2 Engine Efficiency Equation 15.11 gives the engine efficiency as $e = |W|/|Q_H|$. Solving for $|Q_H|$, we find that

$$\boxed{|Q_H| = \frac{|W|}{e}}$$

which can be substituted into Equation 1 as shown in the right column.

$$|Q_C| = |Q_H| - |W| \qquad (1)$$

$$\uparrow$$

$$\boxed{|Q_H| = \frac{|W|}{e}}$$

> **Problem-Solving Insight** When efficiency is stated as a percentage (e.g., 22.0%), it must be converted to a decimal fraction (e.g., 0.220) before being used in an equation.

Solution Combining the results of each step algebraically, we find that

$$\overset{\text{STEP 1}}{\downarrow} \qquad \overset{\text{STEP 2}}{\downarrow}$$

$$|Q_C| = |Q_H| - |W| = \left|\frac{|W|}{e}\right| - |W|$$

The magnitude of the rejected heat, then, is

$$|Q_C| = |W|\left(\frac{1}{e} - 1\right) = (2510 \text{ J})\left(\frac{1}{0.220} - 1\right) = \boxed{8900 \text{ J}}$$

Related Homework: Problem 41

In Example 7, less than one-quarter of the input heat is converted into work because the efficiency of the automobile engine is only 22.0%. If the engine were 100% efficient, all the input heat would be converted into work. Unfortunately, nature does not permit 100%-efficient heat engines to exist, as the next section discusses.

15.9 Carnot's Principle and the Carnot Engine

What is it that allows a heat engine to operate with maximum efficiency? The French engineer Sadi Carnot (1796–1832) proposed that a heat engine has maximum efficiency when the processes within the engine are reversible. *A reversible process is one in which both the system and its environment can be returned to exactly the states they were in before the process occurred.*

In a reversible process, *both* the system and its environment can be returned to their initial states. Therefore, a process that involves an energy-dissipating mechanism, such as friction, cannot be reversible because the energy wasted due to friction would alter the system or the environment or both. There are also reasons other than friction why a process may not be reversible. For instance, the spontaneous flow of heat from a hot substance to a cold substance is irreversible, even though friction is not present. For heat to flow in the reverse direction, work must be done, as we will see in Section 15.10. The agent doing such work must be located in the environment of the hot and cold substances, and, therefore, the environment must change while the heat is moved back from cold to hot. Since the system and the environment cannot *both* be returned to their initial states, the process of spontaneous heat flow is irreversible. In fact, all spontaneous processes are irreversible, such as the explosion of an unstable chemical or the bursting of a bubble. When the word "reversible" is used in connection with engines, it does not just mean a gear that allows the engine to operate a device in reverse. All cars have a reverse gear, for instance, but no automobile engine is thermodynamically reversible, since friction exists no matter which way the car moves.

Today, the idea that the efficiency of a heat engine is a maximum when the engine operates reversibly is referred to as **Carnot's principle**.

CARNOT'S PRINCIPLE: AN ALTERNATIVE STATEMENT OF THE SECOND LAW OF THERMODYNAMICS

No irreversible engine operating between two reservoirs at constant temperatures can have a greater efficiency than a reversible engine operating between the same temperatures. Furthermore, all reversible engines operating between the same temperatures have the same efficiency.

Carnot's principle is quite remarkable, for no mention is made of the working substance of the engine. It does not matter whether the working substance is a gas, a liquid, or a solid. As long as the process is reversible, the efficiency of the engine is a maximum. However, Carnot's principle does *not* state, or even imply, that a reversible engine has an efficiency of 100%.

It can be shown that if Carnot's principle were not valid, it would be possible for heat to flow spontaneously from a cold substance to a hot substance, in violation of the second law of thermodynamics. In effect, then, Carnot's principle is another way of expressing the second law.

No real engine operates reversibly. Nonetheless, the idea of a reversible engine provides a useful standard for judging the performance of real engines. **Figure 15.14** shows a reversible engine, called a **Carnot engine**, that is particularly useful as an idealized model. An important feature of a Carnot engine is that all input heat (magnitude = $|Q_H|$) originates from a hot reservoir *at a single temperature* T_H and all rejected heat (magnitude = $|Q_C|$) goes into a cold reservoir *at a single temperature* T_C.

Carnot's principle implies that the efficiency of a reversible engine is independent of the working substance of the engine, and therefore can depend only on the temperatures of the hot and cold reservoirs. Since efficiency is $e = 1 - |Q_C|/|Q_H|$ according to Equation 15.13, the ratio $|Q_C|/|Q_H|$ can depend only on the reservoir temperatures. This observation led Lord Kelvin to propose a **thermodynamic temperature scale**. He proposed that the thermodynamic temperatures of the cold and hot reservoirs be defined such that their ratio is equal to $|Q_C|/|Q_H|$. Thus, the thermodynamic temperature scale is related to the heats absorbed and rejected by a Carnot engine, and is independent of the working substance. If a reference temperature is properly chosen, it can be shown that the thermodynamic temperature scale is identical to the Kelvin scale introduced in Section 12.2 and used in the ideal gas law. As a result, the ratio of the magnitude of the rejected heat $|Q_C|$ to the magnitude of the input heat $|Q_H|$ is

Temperature = T_H

Hot reservoir

$|Q_H|$

Engine $|W|$

$|Q_C|$

Cold reservoir

Temperature = T_C

FIGURE 15.14 A Carnot engine is a reversible engine in which all input heat (magnitude = $|Q_H|$) originates from a hot reservoir at a single temperature T_H, and all rejected heat (magnitude = $|Q_C|$) goes into a cold reservoir at a single temperature T_C.

$$\frac{|Q_C|}{|Q_H|} = \frac{T_C}{T_H} \qquad (15.14)$$

where the temperatures T_C and T_H *must be expressed in kelvins.*

Math Skills When determining the efficiency of a Carnot engine with Equation 15.15, be sure the temperatures T_C and T_H of the cold and hot reservoirs are expressed in kelvins; degrees Celsius or Fahrenheit will lead to errors. Suppose, for example, that a Carnot engine has reservoir temperatures of $T_C = 254$ K (-19 °C) and $T_H = 294$ K (21 °C). According to Equation 15.15, the efficiency of this engine is

$$e_{Carnot} = 1 - \frac{T_C}{T_H} = 1 - \frac{254 \text{ K}}{294 \text{ K}} = 0.136 \text{ or } 13.6\%$$

For comparison, the incorrect calculation using Celsius degrees is shown below:

Incorrect Calculation $\qquad e_{Carnot} = 1 - \frac{T_C}{T_H} = 1 - \frac{-19 \text{ °C}}{21 \text{ °C}} = 1.90 \text{ or } 190\%$

This result makes no sense, since it implies an impossible efficiency of greater than 100%. The trouble with the Celsius scale in the calculation is that it allows values that are below zero, or negative. The Kelvin scale, in contrast, is chosen so that its zero temperature point is the lowest temperature attainable. Therefore, it cannot give efficiencies greater than 100% when used in Equation 15.15.

The efficiency e_{Carnot} of a Carnot engine can be written in a particularly useful way by substituting Equation 15.14 into Equation 15.13 for the efficiency, $e = 1 - |Q_C|/|Q_H|$:

Efficiency of a Carnot engine $\qquad\qquad e_{Carnot} = 1 - \frac{T_C}{T_H} \qquad\qquad$ **(15.15)**

This relation gives the *maximum possible efficiency* for a heat engine operating between two Kelvin temperatures T_C and T_H, and the next example illustrates its application.

EXAMPLE 8 | The Physics of Extracting Work from a Warm Ocean

Water near the surface of a tropical ocean has a temperature of 298.2 K (25.0 °C), whereas water 700 m beneath the surface has a temperature of 280.2 K (7.0 °C). It has been proposed that the warm water be used as the hot reservoir and the cool water as the cold reservoir of a heat engine. Find the maximum possible efficiency for such an engine.

Reasoning The maximum possible efficiency is the efficiency that a Carnot engine would have (Equation 15.15) operating between temperatures of $T_H = 298.2$ K and $T_C = 280.2$ K.

Solution Using $T_H = 298.2$ K and $T_C = 280.2$ K in Equation 15.15, we find that

$$e_{Carnot} = 1 - \frac{T_C}{T_H} = 1 - \frac{280.2 \text{ K}}{298.2 \text{ K}} = \boxed{0.060 \ (6.0\%)}$$

In Example 8 the maximum possible efficiency is only 6.0%. The small efficiency arises because the Kelvin temperatures of the hot and cold reservoirs are so close. A greater efficiency is possible only when there is a greater difference between the reservoir temperatures. However, there are limits on how large the efficiency of a heat engine can be, as Conceptual Example 9 discusses.

CONCEPTUAL EXAMPLE 9 | Natural Limits on the Efficiency of a Heat Engine

Consider a hypothetical engine that receives 1000 J of heat as input from a hot reservoir and delivers 1000 J of work, rejecting no heat to a cold reservoir whose temperature is above 0 K. Which law of thermodynamics does this engine violate? **(a)** The first law **(b)** The second law **(c)** Both the first and second laws

Reasoning The first law of thermodynamics is an expression of energy conservation. The second law states that no irreversible engine operating between two reservoirs at constant temperatures can have a greater efficiency than a reversible engine operating between the same two reservoirs. The efficiency of such a reversible engine is e_{Carnot}, the efficiency of a Carnot engine.

Answers (a) and (c) are incorrect. From the point of view of energy conservation, nothing is wrong with an engine that converts 1000 J of heat into 1000 J of work. Energy has been neither created nor destroyed; it has only been transformed from one form (heat) into another form (work). Therefore, this engine does not violate the first law of thermodynamics.

Answer (b) is correct. Since all of the input heat is converted into work, the efficiency of the engine is 1, or 100%. But Equation

15.15, which is based on the second law of thermodynamics, indicates that the maximum possible efficiency is $e_{Carnot} = 1 - T_C/T_H$, where T_C and T_H are the temperatures of the cold and hot reservoirs, respectively. Since we are told that T_C is above 0 K, it is clear that the ratio T_C/T_H is greater than zero, so the maximum possible efficiency is less than 1 (or less than 100%). The hypothetical engine, therefore, violates the second law of thermodynamics, which limits the efficiencies of heat engines to values less than 100%.

THE PHYSICS OF . . . thermal pollution. Example 9 has emphasized that *even a perfect heat engine has an efficiency that is less than 1.0 or 100%*. In this regard, we note that the maximum possible efficiency, as given by Equation 15.15, approaches 1.0 when T_C approaches absolute zero (0 K). However, experiments have shown that it is not possible to cool a substance to absolute zero (see Section 15.12), so nature does not allow a 100%-efficient heat engine to exist. As a result, there will always be heat rejected to a cold reservoir whenever a heat engine is used to do work, even if friction and other irreversible processes are eliminated completely. This rejected heat is a form of thermal pollution. The second law of thermodynamics requires that at least some thermal pollution be generated whenever heat engines are used to perform work. This kind of thermal pollution can be reduced only if society reduces its dependence on heat engines to do work.

Check Your Understanding

(The answers are given at the end of the book.)

12. The second law of thermodynamics, in the form of Carnot's principle, indicates that the most efficient heat engine operating between two temperatures is a reversible one. Does this mean that a reversible engine operating between the temperatures of 600 and 400 K must be more efficient than an *irreversible* engine operating between 700 and 300 K?

13. Three reversible engines, A, B, and C, use the same cold reservoir for their exhaust heats. However, they use different hot reservoirs that have the following temperatures: (A) 1000 K, (B) 1100 K, and (C) 900 K. Rank these engines in order of increasing efficiency (smallest efficiency first). (a) A, C, B (b) C, B, A (c) B, A, C (d) C, A, B

14. Suppose that you wish to improve the efficiency of a Carnot engine. Which answer describes the best way? (a) Lower the Kelvin temperature of the cold reservoir by a factor of four. (b) Raise the Kelvin temperature of the hot reservoir by a factor of four. (c) Cut the Kelvin temperature of the cold reservoir in half and double the Kelvin temperature of the hot reservoir. (d) All three choices give the same improvement in efficiency.

15. Consider a hypothetical device that takes 10 000 J of heat from a hot reservoir and 5000 J of heat from a cold reservoir (whose temperature is greater than 0 K) and produces 15 000 J of work. What can be said about this device? (a) It violates the first law of thermodynamics but not the second law. (b) It violates the second law of thermodynamics but not the first law. (c) It violates both the first and second laws of thermodynamics. (d) It does not violate either the first or the second law of thermodynamics.

15.10 Refrigerators, Air Conditioners, and Heat Pumps

The natural tendency of heat is to flow from hot to cold, as indicated by the second law of thermodynamics. However, if work is used, heat can be *made* to flow from cold to hot, against its natural tendency. Refrigerators, air conditioners, and heat pumps are, in fact, devices that do just that. As **Figure 15.15** illustrates, these devices use work (magnitude = $|W|$) to extract heat (magnitude = $|Q_C|$) from the cold reservoir and deposit heat (magnitude = $|Q_H|$) into the hot reservoir. Generally speaking, such a process is called a **refrigeration process**. A comparison of the left and right sides of this drawing shows that the directions of the arrows symbolizing heat and work in a refrigeration process are opposite to those in an engine process. Nonetheless, energy is conserved during a refrigeration process, just as it is in an engine process, so $|Q_H| = |W| + |Q_C|$. Moreover, if the process occurs reversibly, we have ideal devices that are called Carnot refrigerators, Carnot air conditioners, and Carnot heat pumps. For these ideal devices, the relation $|Q_C|/|Q_H| = T_C/T_H$ (Equation 15.14) applies, just as it does for a Carnot engine.

THE PHYSICS OF . . . refrigerators. In a **refrigerator**, the interior of the unit is the cold reservoir, while the warmer exterior is the hot reservoir. As **Figure 15.16** illustrates, the refrigerator takes heat from the food inside and deposits it into the kitchen, along with the energy needed to do the work of making the heat flow from cold to hot. For this reason, the outside surfaces (usually the sides and back) of most refrigerators are warm to the touch while the units operate.

THE PHYSICS OF . . . air conditioners. An **air conditioner** is like a refrigerator, except that the room itself is the cold reservoir and the outdoors is the hot reservoir. **Figure 15.17** shows a window unit, which cools a room by removing heat and depositing it outside, along with the work used to make the heat flow from cold to hot. Conceptual Example 10 considers a common misconception about refrigerators and air conditioners.

Refrigeration Process   **Engine Process**

FIGURE 15.15 In the refrigeration process on the left, work (magnitude = $|W|$) is used to remove heat (magnitude = $|Q_C|$) from the cold reservoir and deposit heat (magnitude = $|Q_H|$) into the hot reservoir. Compare this with the engine process on the right.

FIGURE 15.16 A refrigerator.

FIGURE 15.17 A window air conditioner removes heat from a room, which is the cold reservoir, and deposits heat outdoors, which is the hot reservoir.

CONCEPTUAL EXAMPLE 10 | You Can't Beat the Second Law of Thermodynamics

Is it possible (A) to cool your kitchen by leaving the refrigerator door open or (B) to cool your bedroom by putting a window air conditioner on the floor by the bed? **(a)** Only A is possible. **(b)** Only B is possible. **(c)** Both are possible. **(d)** Neither is possible.

Reasoning During a refrigeration process (be it in a refrigerator or in an air conditioner), heat (magnitude = $|Q_C|$) is removed from a cold reservoir and heat (magnitude = $|Q_H|$) is deposited into a hot reservoir. Moreover, according to the second law of thermodynamics, work (magnitude = $|W|$) is required to move this heat from the cold reservoir to the hot reservoir. The principle of conservation of energy states that $|Q_H| = |W| + |Q_C|$ (Equation 15.12), and we will use this as a guide in assessing the possibilities.

Answers (a), (b), and (c) are incorrect. If you wanted to cool your kitchen by leaving the refrigerator door open, the refrigerator would have to take heat from directly in front of the open door and pump *less* heat out the back of the unit and into the kitchen (since the refrigerator is supposed to be cooling the entire kitchen). Likewise, if you tried to cool your entire bedroom by placing the air conditioner on the floor by the bed, the air conditioner would have

to take heat (magnitude = $|Q_C|$) from directly in front of the unit and deposit *less* heat (magnitude = $|Q_H|$) out the back. According to the second law of thermodynamics this cannot happen, since $|Q_H| = |W| + |Q_C|$; that is, $|Q_H|$ is greater than (not less than) $|Q_C|$ because $|W|$ is greater than zero.

Answer (d) is correct. The heat (magnitude = $|Q_C|$) removed from the air directly in front of the open refrigerator is deposited back into the kitchen at the rear of the unit. Moreover, according to the second law of thermodynamics, work (magnitude = $|W|$) is needed to move that heat from cold to hot, and the energy from this work is also deposited into the kitchen as additional heat. Thus, the open refrigerator puts into the kitchen an amount of heat $|Q_H| = |W| + |Q_C|$, which is *more* than it removes from in front of the open refrigerator. Thus, rather than cooling the kitchen, the open refrigerator warms it up. Putting an air conditioner on the floor to cool your bedroom is similarly a no-win game. The heat pumped out the back of the air conditioner and into the bedroom is greater than the heat pulled into the front of the unit. Consequently, the air conditioner actually warms the bedroom.

$|W|$ = Magnitude of work done by electrical energy

$|Q_H| = |W| + |Q_C|$ (Warm house)

Heat pump

$|Q_C|$ (Cold outdoors)

FIGURE 15.18 In a heat pump the cold reservoir is the wintry outdoors, and the hot reservoir is the inside of the house.

The quality of a refrigerator or air conditioner is rated according to its coefficient of performance. Such appliances perform well when they remove a relatively large amount of heat (magnitude = $|Q_C|$) from a cold reservoir by using as small an amount of work (magnitude = $|W|$) as possible. Therefore, the coefficient of performance is defined as the ratio of $|Q_C|$ to $|W|$, and the greater this ratio is, the better the performance is:

$$\text{\textbf{\textit{Refrigerator or air conditioner}}} \quad \frac{\text{Coefficient of}}{\text{performance}} = \frac{|Q_C|}{|W|} \qquad (15.16)$$

Commercially available refrigerators and air conditioners have coefficients of performance in the range 2 to 6, depending on the temperatures involved. The coefficients of performance for these real devices are less than those for ideal, or Carnot, refrigerators and air conditioners.

In a sense, refrigerators and air conditioners operate like pumps. They pump heat "uphill" from a lower temperature to a higher temperature, just as a water pump forces water uphill from a lower elevation to a higher elevation. It would be appropriate to call them heat pumps. However, the name "heat pump" is reserved for the device illustrated in **Figure 15.18**, which is a home heating appliance.

THE PHYSICS OF . . . heat pumps. The **heat pump** uses work (magnitude = $|W|$) to make heat (magnitude = $|Q_C|$) from the wintry outdoors (the cold reservoir) flow up the temperature "hill" into a warm house (the hot reservoir). According to the conservation of energy, the heat pump deposits inside the house an amount of heat $|Q_H| = |W| + |Q_C|$. The air conditioner and the heat pump do closely related jobs. The air conditioner refrigerates the inside of the house and heats up the outdoors, whereas the heat pump refrigerates the outdoors and heats up the inside. These jobs are so closely related that most heat pump systems serve in a dual capacity, being equipped with a switch that converts them from heaters in the winter into air conditioners in the summer.

Heat pumps are popular for home heating in today's energy-conscious world, and it is easy to understand why. Suppose that 1000 J of energy is available for home heating. **Figure 15.19** shows that a conventional electric heating system uses this 1000 J to heat a coil of wire, just as in a toaster. A fan blows air across the hot coil, and forced convection carries the 1000 J of heat into the house. In contrast, the heat pump in **Figure 15.18** does not use the 1000 J directly as heat. Instead, it uses the 1000 J to do the work (magnitude = $|W|$) of pumping heat (magnitude = $|Q_C|$) from the cooler outdoors into the warmer house and, in so doing, delivers an amount of energy $|Q_H| = |W| + |Q_C|$. With $|W| = 1000$ J, this becomes $|Q_H| = 1000$ J $+ |Q_C|$, so that the heat pump delivers more than 1000 J of heat, whereas the conventional electric heating system delivers only 1000 J.

It is also possible to specify a coefficient of performance for heat pumps. However, unlike refrigerators and air conditioners, the job of a heat pump is to heat, not to cool. As a result, the coefficient of performance of a heat pump is the ratio of the magnitude of the heat $|Q_H|$ delivered into the house to the magnitude of the work $|W|$ required to deliver it:

Heat pump $\dfrac{\text{Coefficient of}}{\text{performance}} = \dfrac{|Q_H|}{|W|}$ (15.17)

The coefficient of performance depends on the indoor and outdoor temperatures. Commercial units have coefficients of about 3 to 4 under favorable conditions.

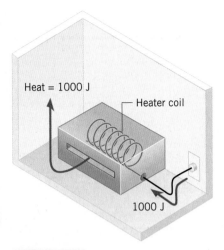

FIGURE 15.19 This conventional electric heating system is delivering 1000 J of heat to the living room.

Check Your Understanding

(*The answers are given at the end of the book.*)

16. Each drawing in **CYU Figure 15.4** represents a hypothetical heat engine or a hypothetical heat pump and shows the corresponding heats and work. Only one of these hypothetical situations is allowed in nature. Which is it?

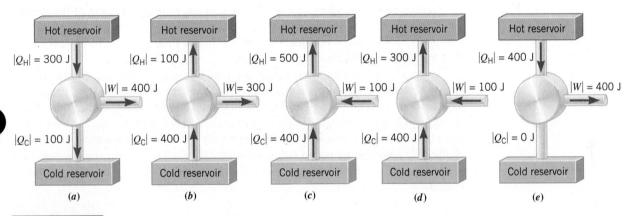

CYU FIGURE 15.4

17. A refrigerator is kept in a garage that is not heated in the cold winter or air-conditioned in the hot summer. Does it cost more for this refrigerator to make a kilogram of ice cubes in the winter or in the summer? **(a)** In the summer **(b)** In the winter **(c)** It costs the same in both seasons.

18. The coefficient of performance of a heat pump that is removing heat from the cold outdoors **(a)** must always be less than one, **(b)** can be either less than or greater than one, **(c)** must always be greater than one.

19. A kitchen air conditioner and a refrigerator both remove heat from a cold reservoir and deposit it in a hot reservoir. The air conditioner _____ the kitchen, whereas the refrigerator _____ the kitchen. **(a)** cools, cools **(b)** cools, warms **(c)** warms, warms **(d)** warms, cools

20. On a summer day a window air conditioner cycles on and off, according to how the temperature within the room changes. When are you more likely to be able to fry an egg on the outside part of the unit? **(a)** When the unit is on **(b)** When the unit is off **(c)** It does not matter whether the unit is on or off.

15.11 | Entropy

A Carnot engine has the maximum possible efficiency for its operating conditions because the processes occurring within it are reversible. Irreversible processes, such as friction, cause real engines to operate at less than maximum efficiency, for they reduce our ability to use heat to perform work. As an extreme example, imagine that a hot object is placed in thermal contact with a cold object, so heat flows spontaneously, and hence irreversibly, from hot to cold. Eventually both objects reach the same temperature, and $T_C = T_H$. A Carnot engine using these two objects as heat reservoirs is unable to do work, because the efficiency of the engine is zero [$e_{Carnot} = 1 - (T_C/T_H) = 0$]. In general, irreversible processes cause us to lose some, but not necessarily all, of the ability to perform work. This partial loss can be expressed in terms of a concept called **entropy**.

To introduce the idea of entropy we recall the relation $|Q_C|/|Q_H| = T_C/T_H$ (Equation 15.14) that applies to a Carnot engine. It is possible to rearrange this equation as $|Q_C|/T_C = |Q_H|/T_H$, which focuses attention on the heat Q divided by the Kelvin temperature T. The quantity Q/T is called the change in the entropy ΔS:

$$\Delta S = \left(\frac{Q}{T}\right)_R \qquad (15.18)$$

In this expression the temperature T must be in kelvins, and the subscript R refers to the word "reversible." It can be shown that Equation 15.18 applies to any process in which heat enters (Q is positive) or leaves (Q is negative) a system reversibly at a constant temperature. Such is the case for the heat that flows into and out of the reservoirs of a Carnot engine. Equation 15.18 indicates that the SI unit for entropy is a joule per kelvin (J/K).

Entropy, like internal energy, is a function of the state or condition of the system. Only the state of a system determines the entropy S that a system has. Therefore, the change in entropy ΔS is equal to the entropy of the final state of the system minus the entropy of the initial state.

We can now describe what happens to the entropy of a Carnot engine. As the engine operates, the entropy of the hot reservoir decreases, since heat of magnitude $|Q_H|$ departs reversibly at a Kelvin temperature T_H. The corresponding change in the entropy is $\Delta S_H = -|Q_H|/T_H$, where the minus sign is needed to indicate a decrease, since the symbol $|Q_H|$ denotes only the magnitude of the heat. In contrast, the entropy of the cold reservoir increases by an amount $\Delta S_C = +|Q_C|/T_C$, for the rejected heat reversibly enters the cold reservoir at a Kelvin temperature T_C. The total change in entropy is

$$\Delta S_C + \Delta S_H = +\frac{|Q_C|}{T_C} - \frac{|Q_H|}{T_H} = 0$$

because $|Q_C|/T_C = |Q_H|/T_H$ according to Equation 15.14.

The fact that the total change in entropy is zero for a Carnot engine is a specific illustration of a general result. It can be proved that when *any* reversible process occurs, the change in the entropy of the universe is zero; $\Delta S_{universe} = 0$ J/K for a reversible process. The word "universe" means that $\Delta S_{universe}$ takes into account the entropy changes of all parts of the system and all parts of the environment. ***Reversible processes do not alter the total entropy of the universe.*** To be sure, the entropy of one part of the universe may change because of a reversible process, but if so, the entropy of another part changes in the opposite way by the same amount.

What happens to the entropy of the universe when an *irreversible* process occurs is more complex, because the expression $\Delta S = (Q/T)_R$ does not apply directly. However, if a system changes irreversibly from an initial state to a final state, this expression can be used to calculate ΔS indirectly, as **Figure 15.20** indicates. We imagine a hypothetical reversible process that causes the system to change between *the same initial and final states* and then find ΔS for this reversible process. The value obtained for ΔS also applies to the irreversible process that actually occurs, since only the nature of the initial and final states, and not the path between them, determines ΔS. Example 11 illustrates this indirect method and shows that spontaneous (irreversible) processes increase the entropy of the universe.

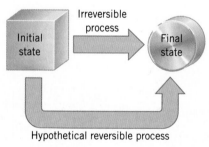

ΔS for irreversible = ΔS for hypothetical
process   reversible process

FIGURE 15.20 Although the relation $\Delta S = (Q/T)_R$ applies to reversible processes, it can be used as part of an indirect procedure to find the entropy change for an irreversible process. This drawing illustrates the procedure discussed in the text.

EXAMPLE 11 | The Entropy of the Universe Increases

Figure 15.21 shows 1200 J of heat flowing spontaneously through a copper rod from a hot reservoir at 650 K to a cold reservoir at 350 K. Determine the amount by which this irreversible process changes the entropy of the universe, assuming that no other changes occur.

Hot reservoir
$T_H = 650$ K

Copper rod
1200 J

Cold reservoir
$T_C = 350$ K

FIGURE 15.21 Heat flows spontaneously from a hot reservoir to a cold reservoir.

Reasoning The hot-to-cold heat flow is irreversible, so the relation $\Delta S = (Q/T)_R$ is applied to a hypothetical process whereby the 1200 J of heat is taken reversibly from the hot reservoir and added reversibly to the cold reservoir.

Solution The total entropy change of the universe is the algebraic sum of the entropy changes for each reservoir:

$$\Delta S_{universe} = -\underbrace{\frac{1200\ J}{650\ K}}_{\substack{\text{Entropy lost by} \\ \text{hot reservoir}}} + \underbrace{\frac{1200\ J}{350\ K}}_{\substack{\text{Entropy gained} \\ \text{by cold reservoir}}} = \boxed{+1.6\ J/K}$$

The irreversible process causes the entropy of the universe to increase by 1.6 J/K.

Example 11 is a specific illustration of a general result: ***Any irreversible process increases the entropy of the universe.*** In other words, $\Delta S_{universe} > 0$ J/K for an irreversible process. Reversible processes do not alter the entropy of the universe, whereas irreversible processes cause the entropy to increase. Therefore, the entropy of the universe continually increases, like time itself, and entropy is sometimes called "time's arrow." It can be shown that this behavior of the entropy of the universe provides a completely general statement of the second law of thermodynamics, which applies not only to heat flow but also to all kinds of other processes.

> **THE SECOND LAW OF THERMODYNAMICS STATED IN TERMS OF ENTROPY**
>
> The total entropy of the universe does not change when a reversible process occurs ($\Delta S_{universe} = 0$ J/K) and increases when an irreversible process occurs ($\Delta S_{universe} > 0$ J/K).

When an irreversible process occurs and the entropy of the universe increases, the energy available for doing work decreases, as the next example illustrates.

EXAMPLE 12 | Energy Unavailable for Doing Work

Suppose that 1200 J of heat is used as input for an engine under two different conditions. In **Figure 15.22a** the heat is supplied by a hot reservoir whose temperature is 650 K. In part *b* of the drawing, the heat flows irreversibly through a copper rod into a second reservoir whose temperature is 350 K and then enters the engine. In either case, a 150-K reservoir is used as the cold reservoir. For each case, determine the maximum amount of work that can be obtained from the 1200 J of heat.

Reasoning According to Equation 15.11, the work (magnitude = $|W|$) obtained from the engine is the product of its efficiency e and the input heat (magnitude = $|Q_H|$), or $|W| = e|Q_H|$. For a given input heat, the maximum amount of work is obtained when the efficiency is a maximum; that is, when the engine is a Carnot engine. The efficiency of a Carnot engine is given by Equation 15.15 as $e_{Carnot} = 1 - T_C/T_H$. Therefore, the efficiency

may be determined from the Kelvin temperatures of the hot and cold reservoirs.

Solution

Before irreversible heat flow
$$e_{Carnot} = 1 - \frac{T_C}{T_H} = 1 - \frac{150\ K}{650\ K} = 0.77$$
$$|W| = (e_{Carnot})(1200\ J) = (0.77)(1200\ J) = \boxed{920\ J}$$

After irreversible heat flow
$$e_{Carnot} = 1 - \frac{T_C}{T_H} = 1 - \frac{150\ K}{350\ K} = 0.57$$
$$|W| = (e_{Carnot})(1200\ J) = (0.57)(1200\ J) = \boxed{680\ J}$$

When the 1200 J of input heat is taken from the 350-K reservoir instead of the 650-K reservoir, the efficiency of the Carnot engine is smaller. As a result, less work (680 J versus 920 J) can be extracted from the input heat.

Hot reservoir
$T_H = 650$ K

$|Q_H| = 1200$ J

Carnot engine

$|W| = 920$ J

$|Q_C| = 280$ J

Cold reservoir
$T_C = 150$ K

(a)

650 K

Copper rod
1200 J

Hot reservoir
$T_H = 350$ K

$|Q_H| = 1200$ J

Carnot engine

$|W| = 680$ J

$|Q_C| = 520$ J

Cold reservoir
$T_C = 150$ K

(b)

FIGURE 15.22 Heat in the amount of $|Q_H| = 1200$ J is used as input for an engine under two different conditions in parts a and b.

Example 12 shows that 240 J less work (920 J − 680 J) can be performed when the input heat is obtained from the hot reservoir with the lower temperature. In other words, the irreversible process of heat flow through the copper rod causes energy to become unavailable for doing work in the amount of $W_{unavailable} = 240$ J. Example 11 shows that this irreversible process simultaneously causes the entropy of the universe to increase by an amount $\Delta S_{universe} = +1.6$ J/K. These values for $W_{unavailable}$ and $\Delta S_{universe}$ are in fact related. If you multiply $\Delta S_{universe}$ by 150 K, which is the lowest Kelvin temperature in Example 12, you obtain $W_{unavailable} = (150 \text{ K}) \times (1.6 \text{ J/K}) = 240$ J. This illustrates the following general result:

$$W_{unavailable} = T_0 \, \Delta S_{universe} \tag{15.19}$$

where T_0 is the Kelvin temperature of the coldest heat reservoir. Since irreversible processes cause the entropy of the universe to increase, they cause energy to be degraded, in the sense that part of the energy becomes unavailable for the performance of work. In contrast, there is no penalty when reversible processes occur, because for them $\Delta S_{universe} = 0$ J/K, and there is no loss of work.

Entropy can also be interpreted in terms of order and disorder. As an example, consider a block of ice (**Figure 15.23**) with each of its H_2O molecules fixed rigidly in place in a highly structured and ordered arrangement. In comparison, the puddle of water into which the ice melts is disordered and unorganized, because the molecules in a liquid are free to move from place to place. Heat is required to melt the ice and produce the disorder. Moreover, heat flow into a system increases the entropy of the system, according to $\Delta S = (Q/T)_R$. We associate an increase in entropy, then, with an increase in disorder. Conversely, we associate a decrease

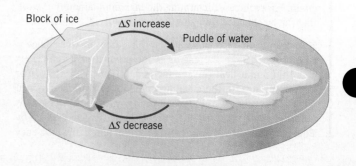

Block of ice

ΔS increase

Puddle of water

ΔS decrease

FIGURE 15.23 A block of ice is an example of an ordered system relative to a puddle of water.

in entropy with a decrease in disorder or a greater degree of order. Example 13 illustrates an order-to-disorder change and the increase of entropy that accompanies it.

EXAMPLE 13 | Order to Disorder

Find the change in entropy that results when a 2.3-kg block of ice melts slowly (reversibly) at 273 K (0 °C).

Reasoning Since the phase change occurs reversibly at a constant temperature, the change in entropy can be found by using Equation 15.18, $\Delta S = (Q/T)_R$, where Q is the heat absorbed by the melting ice. This heat can be determined by using the relation $Q = mL_f$ (Equation 12.5), where m is the

mass and $L_f = 3.35 \times 10^5$ J/kg is the latent heat of fusion of water (see Table 12.3).

Solution Using Equation 15.18 and Equation 12.5, we find that the change in entropy is

$$\Delta S = \left(\frac{Q}{T}\right)_R = \frac{mL_f}{T} = \frac{(2.3 \text{ kg})(3.35 \times 10^5 \text{ J/kg})}{273 \text{ K}} = \boxed{+2.8 \times 10^3 \text{ J/K}}$$

a result that is positive, since the ice absorbs heat as it melts.

Figure 15.24 shows another order-to-disorder change that can be described in terms of entropy.

Martin Hunter/Getty Images

FIGURE 15.24 With the aid of explosives, demolition experts caused this building to go from the ordered state (lower entropy), top photograph, to the disordered state (higher entropy), bottom photograph.

Check Your Understanding

(The answers are given at the end of the book.)

21. Two equal amounts of water are mixed together in an insulated container, and no work is done in the process. The initial temperatures of the water are different, but the mixture reaches a uniform temperature. Do the internal energy and entropy of the water increase, decrease, or remain constant as a result of the mixing process (refer to **CYU Table 15.3**)?

CYU TABLE 15.3

	Internal Energy of the Water	Entropy of the Water
(a)	Increases	Increases
(b)	Decreases	Decreases
(c)	Remains constant	Decreases
(d)	Remains constant	Increases
(e)	Remains constant	Remains constant

22. An event happens somewhere in the universe and, as a result, the entropy of an object changes by -5 J/K. Consistent with the second law of thermodynamics, which one (or more) of the following is a possible value for the entropy change for the rest of the universe? **(a)** -5 J/K **(b)** 0 J/K **(c)** $+5$ J/K **(d)** $+10$ J/K

23. In each of the following cases, which has the greater entropy, a handful of popcorn kernels or the popcorn that results from them; a salad before or after it has been tossed; and a messy apartment or a neat apartment?

24. A glass of water contains a teaspoon of dissolved sugar. After a while, the water evaporates, leaving behind sugar crystals. The entropy of the sugar crystals is less than the entropy of the dissolved sugar because the sugar crystals are in a more ordered state. Why doesn't this process violate the second law of thermodynamics? **(a)** Because, considering what happens to the water, the total entropy of the universe also decreases. **(b)** Because, considering what happens to the water, the total entropy of the universe increases. **(c)** Because the second law does not apply to this situation.

25. A builder uses lumber to construct a building, which is unfortunately destroyed in a fire. Thus, the lumber existed at one time or another in three different states: (A) as unused building material, (B) as a building, and (C) as a burned-out shell of a building. Rank these three states in order of decreasing entropy (largest first). **(a)** C, B, A **(b)** A, B, C **(c)** C, A, B **(d)** A, C, B **(e)** B, A, C

15.12 The Third Law of Thermodynamics

To the zeroth, first, and second laws of thermodynamics we add the third (and last) law. The **third law of thermodynamics** indicates that it is impossible to reach a temperature of absolute zero.

> **THE THIRD LAW OF THERMODYNAMICS**
>
> It is not possible to lower the temperature of any system to absolute zero ($T = 0$ K) in a finite number of steps.

This law, like the second law, can be expressed in a number of ways, but a discussion of them is beyond the scope of this text. The third law is needed to explain a number of experimental observations that cannot be explained by the other laws of thermodynamics.

Concept Summary

15.1 Thermodynamic Systems and Their Surroundings A thermodynamic system is the collection of objects on which attention is being focused, and the surroundings are everything else in the environment. The state of a system is the physical condition of the system, as described by values for physical parameters, often pressure, volume, and temperature.

15.2 The Zeroth Law of Thermodynamics Two systems are in thermal equilibrium if there is no net flow of heat between them when they are brought into thermal contact. Temperature is the indicator of thermal equilibrium in the sense that there is no net flow of heat between two systems in thermal contact that have the same temperature. The zeroth law of thermodynamics states that two systems individually in thermal equilibrium with a third system are in thermal equilibrium with each other.

15.3 The First Law of Thermodynamics The first law of thermodynamics states that due to heat Q and work W, the internal energy of a system changes from its initial value of U_i to a final value of U_f according to Equation 15.1. In this equation Q is positive when the system gains heat and negative when it loses heat. W is positive when work is done by the system and negative when work is done on the system. The first law of thermodynamics is the conservation-of-energy principle applied to heat, work, and the change in the internal energy.

$$\Delta U = U_f - U_i = Q - W \qquad (15.1)$$

The internal energy is called a function of state because it depends only on the state of the system and not on the method by which the system came to be in a given state.

15.4 Thermal Processes A thermal process is quasi-static when it occurs slowly enough that a uniform pressure and temperature exist throughout the system at all times. An isobaric process is one that occurs at constant pressure. The work W done when a system changes at a constant pressure P from an initial volume V_i to a final volume V_f is given by Equation 15.2. An isochoric process is one that takes place at constant volume, and no work is done in such a process. An isothermal process is one that takes place at constant temperature. An adiabatic process is one that takes place without the transfer of heat. The work done in any kind of quasi-static process is given by the area under the corresponding pressure–volume graph.

$$W = P\,\Delta V = P(V_f - V_i) \qquad (15.2)$$

15.5 Thermal Processes Using an Ideal Gas When n moles of an ideal gas change quasi-statically from an initial volume V_i to a final volume V_f at a constant Kelvin temperature T, the work done is given by Equation 15.3, and the process is said to be isothermal.

$$W = nRT \ln\left(\frac{V_f}{V_i}\right) \qquad (15.3)$$

When n moles of a monatomic ideal gas change quasi-statically and adiabatically from an initial temperature T_i to a final temperature T_f, the work done is given by Equation 15.4. During an adiabatic process, and in addition to the ideal gas law, an ideal gas obeys Equation 15.5, where $\gamma = c_P/c_V$ is the ratio of the specific heat capacities at constant pressure and constant volume.

$$W = \tfrac{3}{2}nR(T_i - T_f) \qquad (15.4)$$

$$P_i V_i^{\gamma} = P_f V_f^{\gamma} \qquad (15.5)$$

15.6 Specific Heat Capacities The molar specific heat capacity C of a substance determines how much heat Q is added or removed when the temperature of n moles of the substance changes by an amount ΔT, according to Equation 15.6. For a monatomic ideal gas,

the molar specific heat capacities at constant pressure and constant volume are given by Equations 15.7 and 15.8, respectively, where R is the ideal gas constant. For a diatomic ideal gas at moderate temperatures that do not allow vibration to occur, these values are $C_P = \tfrac{7}{2}R$ and $C_V = \tfrac{5}{2}R$. For any type of ideal gas, the difference between C_P and C_V is given by Equation 15.10.

$$Q = Cn\,\Delta T \qquad (15.6)$$

$$C_P = \tfrac{5}{2}R \qquad (15.7)$$

$$C_V = \tfrac{3}{2}R \qquad (15.8)$$

$$C_P - C_V = R \qquad (15.10)$$

15.7 The Second Law of Thermodynamics The second law of thermodynamics can be stated in a number of equivalent forms. In terms of heat flow, the second law declares that heat flows spontaneously from a substance at a higher temperature to a substance at a lower temperature and does not flow spontaneously in the reverse direction.

15.8 Heat Engines A heat engine produces work (magnitude = $|W|$) from input heat (magnitude = $|Q_H|$) that is extracted from a heat reservoir at a relatively high temperature. The engine rejects heat (magnitude = $|Q_C|$) into a reservoir at a relatively low temperature. The efficiency e of a heat engine is given by Equation 15.11.

$$e = \frac{\text{Work done}}{\text{Input heat}} = \frac{|W|}{|Q_H|} \qquad (15.11)$$

The conservation of energy requires that $|Q_H|$ must be equal to $|W|$ plus $|Q_C|$, as in Equation 15.12. By combining Equation 15.12 with Equation 15.11, the efficiency of a heat engine can also be written as shown in Equation 15.13.

$$|Q_H| = |W| + |Q_C| \qquad (15.12)$$

$$e = 1 - \frac{|Q_C|}{|Q_H|} \qquad (15.13)$$

15.9 Carnot's Principle and the Carnot Engine A reversible process is one in which *both* the system and its environment can be returned to exactly the states they were in before the process occurred.

Carnot's principle is an alternative statement of the second law of thermodynamics. It states that no irreversible engine operating between two reservoirs at constant temperatures can have a greater efficiency than a reversible engine operating between the same temperatures. Furthermore, all reversible engines operating between the same temperatures have the same efficiency.

A Carnot engine is a reversible engine in which all input heat (magnitude = $|Q_H|$) originates from a hot reservoir at a single Kelvin temperature T_H and all rejected heat (magnitude = $|Q_C|$) goes into a cold reservoir at a single Kelvin temperature T_C. For a Carnot engine, Equation 15.14 applies. The efficiency e_{Carnot} of a Carnot engine is the maximum efficiency that an engine operating between two fixed temperatures can have and is given by Equation 15.15.

$$\frac{|Q_C|}{|Q_H|} = \frac{T_C}{T_H} \qquad (15.14)$$

$$e_{\text{Carnot}} = 1 - \frac{T_C}{T_H} \qquad (15.15)$$

15.10 Refrigerators, Air Conditioners, and Heat Pumps Refrigerators, air conditioners, and heat pumps are devices that utilize work (magnitude = $|W|$) to make heat (magnitude = $|Q_C|$) flow from a lower Kelvin temperature T_C to a higher Kelvin temperature T_H. In the process (the refrigeration process) they deposit heat (magnitude = $|Q_H|$)

at the higher temperature. The principle of the conservation of energy requires that $|Q_H| = |W| + |Q_C|$.

$$\text{Coefficient of performance of a refrigerator} = \frac{|Q_C|}{|W|} \qquad (15.16)$$

If the refrigeration process is ideal, in the sense that it occurs reversibly, the devices are called Carnot devices and the relation $|Q_C|/|Q_H| = T_C/T_H$ (Equation 15.14) holds.

The coefficient of performance of a refrigerator or an air conditioner is given by Equation 15.16. The coefficient of performance of a heat pump, however, is given by Equation 15.17.

$$\text{Coefficient of performance of a heat pump} = \frac{|Q_H|}{|W|} \qquad (15.17)$$

15.11 Entropy The change in entropy ΔS for a process in which heat Q enters or leaves a system reversibly at a constant Kelvin temperature T is given by Equation 15.18, where the subscript R stands for "reversible."

$$\Delta S = \left(\frac{Q}{T}\right)_R \qquad (15.18)$$

The second law of thermodynamics can be stated in a number of equivalent forms. In terms of entropy, the second law states that the total entropy of the universe does not change when a reversible process occurs ($\Delta S_{universe} = 0$ J/K) and increases when an irreversible process occurs ($\Delta S_{universe} > 0$ J/K).

Irreversible processes cause energy to be degraded in the sense that part of the energy becomes unavailable for the performance of work. The energy $W_{unavailable}$ that is unavailable for doing work because of an irreversible process is shown in Equation 15.19, where $\Delta S_{universe}$ is the total entropy change of the universe and T_0 is the Kelvin temperature of the coldest reservoir into which heat can be rejected.

$$W_{unavailable} = T_0\,\Delta S_{universe} \qquad (15.19)$$

Increased entropy is associated with a greater degree of disorder and decreased entropy with a lesser degree of disorder (more order).

15.12 The Third Law of Thermodynamics The third law of thermodynamics states that it is not possible to lower the temperature of any system to absolute zero ($T = 0$ K) in a finite number of steps.

Focus on Concepts

Online

Additional questions are available for assignment in WileyPLUS.

Section 15.3   The First Law of Thermodynamics

1. The first law of thermodynamics states that the change ΔU in the internal energy of a system is given by $\Delta U = Q - W$, where Q is the heat and W is the work. Both Q and W can be positive or negative numbers. Q is a positive number if _____, and W is a positive number if _____. (a) the system *loses* heat; work is done *by* the system (b) the system *loses* heat; work is done *on* the system (c) the system *gains* heat; work is done *by* the system (d) the system *gains* heat; work is done *on* the system

Section 15.4   Thermal Processes

2. The drawing shows the expansion of three ideal gases. Rank the gases according to the work they do, largest to smallest. (a) A, B, C (b) A and B (a tie), C (c) B and C (a tie), A (d) B, C, A (e) C, A, B

QUESTION 2

3. The pressure–volume graph shows three paths in which a gas expands from an initial state A to a final state B. The change $\Delta U_{A \to B}$ in internal energy is the same for each of the paths. Rank the paths according to the heat Q added to the gas, largest to smallest.
(a) 1, 2, 3 (b) 1, 3, 2 (c) 2, 1, 3 (d) 3, 1, 2 (e) 3, 2, 1

QUESTION 3

Section 15.5   Thermal Processes Using an Ideal Gas

4. An ideal monatomic gas expands isothermally from A to B, as the graph shows. What can be said about this process? (a) The gas does no work. (b) No heat enters or leaves the gas. (c) The first law of thermodynamics does not apply to an isothermal process. (d) The ideal gas law is not valid during an isothermal process. (e) There is no change in the internal energy of the gas.

QUESTION 4

5. A monatomic ideal gas is thermally insulated, so no heat can flow between it and its surroundings. Is it possible for the temperature of the gas to rise? (a) Yes. The temperature can rise if work is done *by* the gas. (b) No. The only way that the temperature can rise is if heat is added to the gas. (c) Yes. The temperature can rise if work is done *on* the gas.

Section 15.8   Heat Engines

6. A heat engine takes heat Q_H from a hot reservoir and uses part of this energy to perform work W. Assuming that Q_H cannot be changed, how can the efficiency of the engine be improved? (a) Increase the work W; the heat Q_C rejected to the cold reservoir increases as a result. (b) Increase the work W; the heat Q_C rejected to the cold reservoir remains unchanged. (c) Increase the work W; the heat Q_C rejected to the cold reservoir decreases as a result. (d) Decrease the work W; the heat Q_C rejected to the cold reservoir remains unchanged. (e) Decrease the work W; the heat Q_C rejected to the cold reservoir decreases as a result.

Section 15.9   Carnot's Principle and the Carnot Engine

7. The three Carnot engines shown in the drawing operate with hot and cold reservoirs whose temperature differences are 100 K. Rank the efficiencies of the engines, largest to smallest. **(a)** All engines have the same efficiency. **(b)** A, B, C **(c)** B, A, C **(d)** C, B, A **(e)** C, A, B

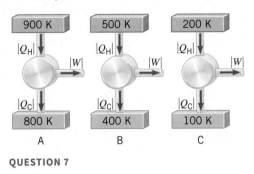

QUESTION 7

Section 15.10   Refrigerators, Air Conditioners, and Heat Pumps

8. A refrigerator operates for a certain time, and the work done by the electrical energy during this time is $W = 1000$ J. What can be said about the heat delivered to the room containing the refrigerator? **(a)** The heat delivered to the room is less than 1000 J. **(b)** The heat delivered to the room is equal to 1000 J. **(c)** The heat delivered to the room is greater than 1000 J.

Section 15.11   Entropy

9. Heat is transferred from the sun to the earth via electromagnetic waves (see Chapter 24). Because of this transfer, the entropy of the sun _____, the entropy of the earth _____, and the entropy of the sun–earth system _____. **(a)** increases, decreases, decreases **(b)** decreases, increases, increases **(c)** increases, increases, increases **(d)** increases, decreases, increases **(e)** decreases, increases, decreases.

Problems   Online

Additional questions are available for assignment in WileyPLUS.

SSM Student Solutions Manual	**BIO** Biomedical application
MMH Problem-solving help	**E** Easy
GO Guided Online Tutorial	**M** Medium
V-HINT Video Hints	**H** Hard
CHALK Chalkboard Videos	**WS** Worksheet
	T Team Problem

Section 15.3   The First Law of Thermodynamics

1. **E** In moving out of a dormitory at the end of the semester, a student does 1.6×10^4 J of work. In the process, his internal energy decreases by 4.2×10^4 J. Determine each of the following quantities (including the algebraic sign): **(a)** W **(b)** ΔU **(c)** Q

2. **E** The internal energy of a system changes because the system gains 165 J of heat and performs 312 J of work. In returning to its initial state, the system loses 114 J of heat. During this return process, **(a)** what work is involved, and **(b)** is the work done by the system or on the system?

3. **E** **SSM** A system does 164 J of work on its environment and gains 77 J of heat in the process. Find the change in the internal energy of **(a)** the system and **(b)** the environment.

4. **E** **GO** A system does 4.8×10^4 J of work, and 7.6×10^4 J of heat flows into the system during the process. Find the change in the internal energy of the system.

5. **E** **V-HINT** Three moles of an ideal monatomic gas are at a temperature of 345 K. Then, 2438 J of heat is added to the gas, and 962 J of work is done on it. What is the final temperature of the gas?

Section 15.4   Thermal Processes

6. **E** A system undergoes a two-step process. In the first step, the internal energy of the system increases by 228 J when 166 J of work is done on the system. In the second step, the internal energy of the

system increases by 115 J when 177 J of work is done on the system. For the overall process, find the heat. What type of process is the overall process? Explain.

7. **E** **SSM** When a .22-caliber rifle is fired, the expanding gas from the burning gunpowder creates a pressure behind the bullet. This pressure causes the force that pushes the bullet through the barrel. The barrel has a length of 0.61 m and an opening whose radius is 2.8×10^{-3} m. A bullet (mass = 2.6×10^{-3} kg) has a speed of 370 m/s after passing through this barrel. Ignore friction and determine the average pressure of the expanding gas.

8. **E** **GO** A system gains 2780 J of heat at a constant pressure of 1.26×10^5 Pa, and its internal energy increases by 3990 J. What is the change in the volume of the system, and is it an increase or a decrease?

9. **E** **V-HINT** A system gains 1500 J of heat, while the internal energy of the system increases by 4500 J and the volume decreases by 0.010 m³. Assume that the pressure is constant and find its value.

PROBLEM 10

10. **E** The volume of a gas is changed along the curved line between A and B in the drawing. Do not assume that the curved line is an isotherm or that the gas is ideal. **(a)** Find the magnitude of the work for the process, and **(b)** determine whether the work is positive or negative.

11. **E** **SSM** **(a)** Using the data presented in the accompanying pressure–volume graph, estimate the magnitude of the

PROBLEM 11

work done when the system changes from *A* to *B* to *C* along the path shown. **(b)** Determine whether the work is done by the system or on the system and, hence, whether the work is positive or negative.

12. **E** Sections 14.2 and 14.3 provide useful information for this problem. When a monatomic ideal gas expands at a constant pressure of 2.6×10^5 Pa, the volume of the gas increases by 6.2×10^{-3} m³. **(a)** Determine the heat that flows into or out of the gas. **(b)** Specify the direction of the flow.

13. **E** **CHALK** A gas is contained in a chamber such as that in Figure 15.4. Suppose that the region outside the chamber is evacuated and the total mass of the block and the movable piston is 135 kg. When 2050 J of heat flows into the gas, the internal energy of the gas increases by 1730 J. What is the distance *s* through which the piston rises?

14. **M** A piece of aluminum has a volume of 1.4×10^{-3} m³. The coefficient of volume expansion for aluminum is $\beta = 69 \times 10^{-6}\,(\text{C}°)^{-1}$. The temperature of this object is raised from 20 to 320 °C. How much work is done by the expanding aluminum if the air pressure is 1.01×10^5 Pa?

15. **M** **SSM** Refer to Multiple-Concept Example 4 to see how the concepts pertinent to this problem are used. The pressure of a gas remains constant while the temperature, volume, and internal energy of the gas increase by 53.0 C°, 1.40×10^{-3} m³, and 939 J, respectively. The mass of the gas is 24.0 g, and its specific heat capacity is 1080 J/(kg · C°). Determine the pressure.

16. **M** **GO** Refer to the drawing that accompanies Problem 11. When a system changes from *A* to *B* along the path shown on the pressure-versus-volume graph, it gains 2700 J of heat. What is the change in the internal energy of the system?

17. **H** Water is heated in an open pan where the air pressure is one atmosphere. The water remains a liquid, which expands by a small amount as it is heated. Determine the ratio of the work done by the water to the heat absorbed by the water.

Section 15.5 Thermal Processes Using an Ideal Gas

18. **E** Six grams of helium (molecular mass = 4.0 u) expand isothermally at 370 K and do 9600 J of work. Assuming that helium is an ideal gas, determine the ratio of the final volume of the gas to the initial volume.

19. **E** **SSM** Five moles of a monatomic ideal gas expand adiabatically, and its temperature decreases from 370 to 290 K. Determine **(a)** the work done (including the algebraic sign) by the gas, and **(b)** the change in its internal energy.

20. **E** **GO** Three moles of neon expand isothermally to 0.250 from 0.100 m³. Into the gas flows 4.75×10^3 J of heat. Assuming that neon is an ideal gas, find its temperature.

21. **E** The temperature of a monatomic ideal gas remains constant during a process in which 4700 J of heat flows out of the gas. How much work (including the proper + or − sign) is done?

22. **E** **GO** One-half mole of a monatomic ideal gas expands adiabatically and does 610 J of work. By how many kelvins does its temperature change? Specify whether the change is an increase or a decrease.

23. **E** **SSM** A monatomic ideal gas has an initial temperature of 405 K. This gas expands and does the same amount of work whether the expansion is adiabatic or isothermal. When the expansion is adiabatic, the final temperature of the gas is 245 K. What is the ratio of the final to the initial volume when the expansion is isothermal?

24. **E** **V-HINT** Heat is added isothermally to 2.5 mol of a monatomic ideal gas. The temperature of the gas is 430 K. How much heat must be added to make the volume of the gas double?

25. **M** **MMH** **CHALK** A diesel engine does not use spark plugs to ignite the fuel and air in the cylinders. Instead, the temperature required to ignite the fuel occurs because the pistons compress the air in the cylinders. Suppose that air at an initial temperature of 21 °C is compressed adiabatically to a temperature of 688 °C. Assume the air to be an ideal gas for which $\gamma = \frac{7}{5}$. Find the compression ratio, which is the ratio of the initial volume to the final volume.

26. **M** A monatomic ideal gas expands from point *A* to point *B* along the path shown in the drawing. **(a)** Determine the work done by the gas. **(b)** The temperature of the gas at point *A* is 185 K. What is its temperature at point *B*? **(c)** How much heat has been added to or removed from the gas during the process?

PROBLEM 26

27. **M** **SSM** The drawing refers to one mole of a monatomic ideal gas and shows a process that has four steps, two isobaric (*A* to *B*, *C* to *D*) and two isochoric (*B* to *C*, *D* to *A*). Complete the following table by calculating ΔU, *W*, and *Q* (including the algebraic signs) for each of the four steps.

PROBLEM 27

	ΔU	*W*	*Q*
A to *B*			
B to *C*			
C to *D*			
D to *A*			

28. **M** **GO** A monatomic ideal gas ($\gamma = \frac{5}{3}$) is contained within a perfectly insulated cylinder that is fitted with a movable piston. The initial pressure of the gas is 1.50×10^5 Pa. The piston is pushed so as to compress the gas, with the result that the Kelvin temperature doubles. What is the final pressure of the gas?

29. **M** The pressure and volume of an ideal monatomic gas change from *A* to *B* to *C*, as the drawing shows. The curved line between *A* and *C* is an isotherm. **(a)** Determine the total heat for the process and **(b)** state whether the flow of heat is into or out of the gas.

PROBLEM 29

30. **H** The work done by one mole of a monatomic ideal gas ($\gamma = \frac{5}{3}$) in expanding adiabatically is 825 J. The initial temperature and volume of the gas are 393 K and 0.100 m^3. Obtain **(a)** the final temperature and **(b)** the final volume of the gas.

31. **H** **SSM** The drawing shows an adiabatically isolated cylinder that is divided initially into two identical parts by an adiabatic partition. Both sides contain one mole of a monatomic ideal gas ($\gamma = \frac{5}{3}$), with the initial temperature being 525 K on the left and 275 K on the right. The partition is then allowed to move slowly (i.e., quasi-statically) to the right, until the pressures on each side of the partition are the same. Find the final temperatures on the **(a)** left and **(b)** right.

525 K — — 275 K

Partition

PROBLEM 31

Section 15.6   Specific Heat Capacities

32. **E** Argon is a monatomic gas whose atomic mass is 39.9 u. The temperature of eight grams of argon is raised by 75 K under conditions of constant pressure. Assuming that argon behaves as an ideal gas, how much heat is required?

33. **E** **SSM** The temperature of 2.5 mol of a monatomic ideal gas is 350 K. The internal energy of this gas is doubled by the addition of heat. How much heat is needed when it is added at **(a)** constant volume and **(b)** constant pressure?

34. **E** **GO** Under constant-volume conditions, 3500 J of heat is added to 1.6 moles of an ideal gas. As a result, the temperature of the gas increases by 75 K. How much heat would be required to cause the same temperature change under constant-pressure conditions? Do not assume anything about whether the gas is monatomic, diatomic, etc.

35. **E** **SSM** Heat is added to two identical samples of a monatomic ideal gas. In the first sample the heat is added while the volume of the gas is kept constant, and the heat causes the temperature to rise by 75 K. In the second sample, an identical amount of heat is added while the pressure (but not the volume) of the gas is kept constant. By how much does the temperature of this sample increase?

36. **E** **GO** A monatomic ideal gas in a rigid container is heated from 217 K to 279 K by adding 8500 J of heat. How many moles of gas are there in the container?

37. **E** Three moles of a monatomic ideal gas are heated at a constant volume of 1.50 m^3. The amount of heat added is 5.24×10^3 J. **(a)** What is the change in the temperature of the gas? **(b)** Find the change in its internal energy. **(c)** Determine the change in pressure.

38. **M** **V-HINT** A monatomic ideal gas expands at constant pressure. **(a)** What percentage of the heat being supplied to the gas is used to increase the internal energy of the gas? **(b)** What percentage is used for doing the work of expansion?

39. **M** **CHALK** Suppose a monatomic ideal gas is contained within a vertical cylinder that is fitted with a movable piston. The piston is frictionless and has a negligible mass. The area of the piston is 3.14×10^{-2} m^2, and the pressure outside the cylinder is 1.01×10^5 Pa. Heat (2093 J) is removed from the gas. Through what distance does the piston drop?

40. **M** **V-HINT** A monatomic ideal gas is heated while at a constant volume of 1.00×10^{-3} m^3, using a ten-watt heater. The pressure of the gas increases by 5.0×10^4 Pa. How long was the heater on?

Section 15.8   Heat Engines

41. **E** Multiple-Concept Example 7 provides a review of the concepts that play roles here. An engine has an efficiency of 64% and produces 5500 J of work. Determine **(a)** the input heat and **(b)** the rejected heat.

42. **E** **GO** Engine 1 has an efficiency of 0.18 and requires 5500 J of input heat to perform a certain amount of work. Engine 2 has an efficiency of 0.26 and performs the same amount of work. How much input heat does the second engine require?

43. **E** Due to a tune-up, the efficiency of an automobile engine increases by 5.0%. For an input heat of 1300 J, how much more work does the engine produce after the tune-up than before?

44. **M** **SSM** **CHALK** Due to design changes, the efficiency of an engine increases from 0.23 to 0.42. For the same input heat $|Q_H|$, these changes increase the work done by the more efficient engine and reduce the amount of heat rejected to the cold reservoir. Find the ratio of the heat rejected to the cold reservoir for the improved engine to that for the original engine.

45. **H** Engine A receives three times more input heat, produces five times more work, and rejects two times more heat than engine B. Find the efficiency of **(a)** engine A and **(b)** engine B.

Section 15.9   Carnot's Principle and the Carnot Engine

46. **E** A Carnot engine operates with an efficiency of 27.0% when the temperature of its cold reservoir is 275 K. Assuming that the temperature of the hot reservoir remains the same, what must be the temperature of the cold reservoir in order to increase the efficiency to 32.0%?

47. **E** An engine has a hot-reservoir temperature of 950 K and a cold-reservoir temperature of 620 K. The engine operates at three-fifths maximum efficiency. What is the efficiency of the engine?

48. **E** **SSM** A Carnot engine has an efficiency of 0.700, and the temperature of its cold reservoir is 378 K. **(a)** Determine the temperature of its hot reservoir. **(b)** If 5230 J of heat is rejected to the cold reservoir, what amount of heat is put into the engine?

49. **E** An engine does 18 500 J of work and rejects 6550 J of heat into a cold reservoir whose temperature is 285 K. What would be the smallest possible temperature of the hot reservoir?

50. **E** **GO** A Carnot engine has an efficiency of 0.40. The Kelvin temperature of its hot reservoir is quadrupled, and the Kelvin temperature of its cold reservoir is doubled. What is the efficiency that results from these changes?

51. **E** **MMH** A Carnot engine operates between temperatures of 650 and 350 K. To improve the efficiency of the engine, it is decided either to raise the temperature of the hot reservoir by 40 K or to lower the temperature of the cold reservoir by 40 K. Which change gives the greater improvement? Justify your answer by calculating the efficiency in each case.

52. **M** The hot reservoir for a Carnot engine has a temperature of 890 K, while the cold reservoir has a temperature of 670 K. The heat input for this engine is 4800 J. The 670-K reservoir also serves as the hot reservoir for a second Carnot engine. This second engine uses the rejected heat of the first engine as input and extracts additional work from it. The rejected heat from the second engine goes into a reservoir that has a temperature of 420 K. Find the total work delivered by the two engines.

53. **M** **SSM** Suppose that the gasoline in a car engine burns at 631 °C, while the exhaust temperature (the temperature of the cold reservoir) is 139 °C and the outdoor temperature is 27 °C. Assume that the engine can be treated as a Carnot engine (a gross oversimplification). In an attempt to increase mileage performance, an inventor builds a second engine that functions between the exhaust and outdoor temperatures and uses the exhaust heat to produce additional work. Assume that the inventor's engine can also be treated as a Carnot engine. Determine the ratio of the total work produced by both engines to that produced by the first engine alone.

54. **M** **V-HINT** A power plant taps steam superheated by geothermal energy to 505 K (the temperature of the hot reservoir) and uses the steam to do work in turning the turbine of an electric generator. The steam is then converted back into water in a condenser at 323 K (the temperature of the cold reservoir), after which the water is pumped back down into the earth where it is heated again. The output power (work per unit time) of the plant is 84 000 kilowatts. Determine **(a)** the maximum efficiency at which this plant can operate and **(b)** the minimum amount of rejected heat that must be removed from the condenser every twenty-four hours.

Section 15.10 Refrigerators, Air Conditioners, and Heat Pumps

55. **E** **SSM** A Carnot air conditioner maintains the temperature in a house at 297 K on a day when the temperature outside is 311 K. What is the coefficient of performance of the air conditioner?

56. **E** **GO** The inside of a Carnot refrigerator is maintained at a temperature of 277 K, while the temperature in the kitchen is 299 K. Using 2500 J of work, how much heat can this refrigerator remove from its inside compartment?

57. **E** A refrigerator operates between temperatures of 296 and 275 K. What would be its maximum coefficient of performance?

58. **E** **GO** Two Carnot air conditioners, A and B, are removing heat from different rooms. The outside temperature is the same for both rooms, 309.0 K. The room serviced by unit A is kept at a temperature of 294.0 K, while the room serviced by unit B is kept at 301.0 K. The heat removed from either room is 4330 J. For both units, find the magnitude of the work required and the magnitude of the heat deposited outside.

59. **E** The water in a deep underground well is used as the cold reservoir of a Carnot heat pump that maintains the temperature of a house at 301 K. To deposit 14 200 J of heat in the house, the heat pump requires 800 J of work. Determine the temperature of the well water.

60. **E** **GO** A Carnot engine has an efficiency of 0.55. If this engine were run backward as a heat pump, what would be the coefficient of performance?

61. **E** A Carnot refrigerator is used in a kitchen in which the temperature is kept at 301 K. This refrigerator uses 241 J of work to remove 2561 J of heat from the food inside. What is the temperature inside the refrigerator?

62. **M** **V-HINT** The wattage of a commercial ice maker is 225 W and is the rate at which it does work. The ice maker operates just like a refrigerator or an air conditioner and has a coefficient of performance of 3.60. The water going into the unit has a temperature of 15.0 °C, and the ice maker produces ice cubes at 0.0 °C. Ignoring the work needed to keep stored ice from melting, find the maximum amount (in kg) of ice that the unit can produce in one day of continuous operation.

63. **M** **SSM** **CHALK** A Carnot refrigerator transfers heat from its inside (6.0 °C) to the room air outside (20.0 °C). **(a)** Find the coefficient of performance of the refrigerator. **(b)** Determine the magnitude of the minimum work needed to cool 5.00 kg of water from 20.0 to 6.0 °C when it is placed in the refrigerator.

Section 15.11 Entropy

64. **E** Consider three engines that each use 1650 J of heat from a hot reservoir (temperature = 550 K). These three engines reject heat to a cold reservoir (temperature = 330 K). Engine I rejects 1120 J of heat. Engine II rejects 990 J of heat. Engine III rejects 660 J of heat. One of the engines operates reversibly, and two operate irreversibly. However, of the two irreversible engines, one violates the second law of thermodynamics and could not exist. For each of the engines determine the total entropy change of the universe, which is the sum of the entropy changes of the hot and cold reservoirs. On the basis of your calculations, identify which engine operates reversibly, which operates irreversibly and could exist, and which operates irreversibly and could not exist.

65. **E** Heat Q flows spontaneously from a reservoir at 394 K into a reservoir at 298 K. Because of the spontaneous flow, 2800 J of energy is rendered unavailable for work when a Carnot engine operates between the reservoir at 298 K and a reservoir at 248 K. Find Q.

66. **E** **SSM** Find the change in entropy of the H_2O molecules when **(a)** three kilograms of ice melts into water at 273 K and **(b)** three kilograms of water changes into steam at 373 K. **(c)** On the basis of the answers to parts (a) and (b), discuss which change creates more disorder in the collection of H_2O molecules.

67. **E** **GO** On a cold day, 24 500 J of heat leaks out of a house. The inside temperature is 21 °C, and the outside temperature is −15 °C. What is the increase in the entropy of the universe that this heat loss produces?

68. **M** **MMH** **(a)** After 6.00 kg of water at 85.0 °C is mixed in a perfect thermos with 3.00 kg of ice at 0.0 °C, the mixture is allowed to reach equilibrium. When heat is added to or removed from a solid or liquid of mass m and specific heat capacity c, the change in entropy can be shown to be $\Delta S = mc \ln(T_f/T_i)$, where T_i and T_f are the initial and final Kelvin temperatures. Using this expression and the change in entropy for melting, find the change in entropy that occurs. **(b)** Should the entropy of the universe increase or decrease as a result of the mixing process? Give your reasoning and state whether your answer in part (a) is consistent with your answer here.

69. **M** **V-HINT** **CHALK** The sun is a sphere with a radius of 6.96×10^8 m and an average surface temperature of 5800 K. Determine the amount by which the sun's thermal radiation increases the entropy of the entire universe each second. Assume that the sun is a perfect blackbody, and that the average temperature of the rest of the universe is 2.73 K. Do not consider the thermal radiation absorbed by the sun from the rest of the universe.

70. **M** **GO** An irreversible engine operates between temperatures of 852 and 314 K. It absorbs 1285 J of heat from the hot reservoir and does 264 J of work. **(a)** What is the change $\Delta S_{universe}$ in the entropy of the universe associated with the operation of this engine? **(b)** If the engine were reversible, what would be the magnitude $|W|$ of the work it would have done, assuming that it operated between the same temperatures and absorbed the same heat as the irreversible engine? **(c)** Using the results of parts (a) and (b), find the difference between the work produced by the reversible and irreversible engines.

Additional Problems

71. **E** The pressure of a monatomic ideal gas ($\gamma = \frac{5}{3}$) doubles during an adiabatic compression. What is the ratio of the final volume to the initial volume?

72. **E** **SSM** One-half mole of a monatomic ideal gas absorbs 1200 J of heat while 2500 J of work is done by the gas. **(a)** What is the temperature change of the gas? **(b)** Is the change an increase or a decrease?

73. **E** A gas, while expanding under isobaric conditions, does 480 J of work. The pressure of the gas is 1.6×10^5 Pa, and its initial volume is 1.5×10^{-3} m^3. What is the final volume of the gas?

74. **E** **V-HINT** A lawnmower engine with an efficiency of 0.22 rejects 9900 J of heat every second. What is the magnitude of the work that the engine does in one second?

75. **E** **SSM** A process occurs in which the entropy of a system increases by 125 J/K. During the process, the energy that becomes unavailable for doing work is zero. **(a)** Is this process reversible or irreversible? Give your reasoning. **(b)** Determine the change in the entropy of the surroundings.

76. **E** A Carnot heat pump operates between an outdoor temperature of 265 K and an indoor temperature of 298 K. Find its coefficient of performance.

77. **E** The temperatures indoors and outdoors are 299 and 312 K, respectively. A Carnot air conditioner deposits 6.12×10^5 J of heat outdoors. How much heat is removed from the house?

78. **E** **GO** Carnot engine A has an efficiency of 0.60, and Carnot engine B has an efficiency of 0.80. Both engines utilize the same hot reservoir, which has a temperature of 650 K and delivers 1200 J of heat to each engine. Find the magnitude of the work produced by each engine and the temperatures of the cold reservoirs that they use.

79. **E** **SSM** The pressure and volume of a gas are changed along the path *ABCA*. Using the data shown in the graph, determine the work done (including the algebraic sign) in each segment of the path: **(a)** *A* to *B*, **(b)** *B* to *C*, and **(c)** *C* to *A*.

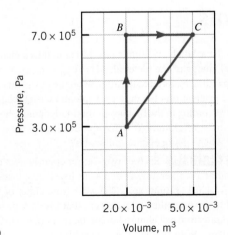

PROBLEM 79

80. **M** **GO** An ideal gas is taken through the three processes (A → B, B → C, and C → A) shown in the drawing. In general, for each process the internal energy *U* of the gas can change because heat *Q* can be added to or removed from the gas and work *W* can be done by the gas or on the gas. For the three processes shown in the drawing, fill in the five missing entries in the following table.

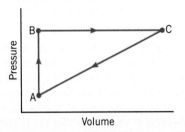

PROBLEM 80

Process	ΔU	Q	W
A → B	**(b)**	+561 J	**(a)**
B → C	+4303 J	**(c)**	+3740 J
C → A	**(d)**	**(e)**	−2867 J

81. **H** **SSM** An engine has an efficiency e_1. The engine takes input heat of magnitude $|Q_H|$ from a hot reservoir and delivers work of magnitude $|W_1|$. The heat rejected by this engine is used as input heat for a second engine, which has an efficiency e_2 and delivers work of magnitude $|W_2|$. The overall efficiency of this two-engine device is the magnitude of the total work delivered ($|W_1| + |W_2|$) divided by the magnitude $|Q_H|$ of the input heat. Find an expression for the overall efficiency e in terms of e_1 and e_2.

82. **M** **GO** **SSM** An ideal, or Carnot, heat pump is used to heat a house to a temperature of 294 K (21 °C). How much work must the pump do to deliver 3350 J of the heat into the house on a day when the outdoor temperature is 273 K (0 °C), and on another day when the temperature is 252 K (−21 °C)?

Physics in Biology, Medicine, and Sports

83. E BIO GO 15.3 In a game of football outdoors on a cold day, a player will begin to feel exhausted after using approximately 8.0×10^5 J of internal energy. **(a)** One player, dressed too lightly for the weather, has to leave the game after losing 6.8×10^5 J of heat. How much work has he done? **(b)** Another player, wearing clothes that offer better protection against heat loss, is able to remain in the game long enough to do 2.1×10^5 J of work. What is the magnitude of the heat that he has lost?

84. M BIO CHALK SSM 15.3 In exercising, a weight lifter loses 0.150 kg of water through evaporation, the heat required to evaporate the water coming from the weight lifter's body. The work done in lifting weights is 1.40×10^5 J. **(a)** Assuming that the latent heat of vaporization of perspiration is 2.42×10^6 J/kg, find the change in the internal energy of the weight lifter. **(b)** Determine the minimum number of nutritional Calories of food (1 nutritional Calorie = 4186 J) that must be consumed to replace the loss of internal energy.

85. E BIO SSM 15.8 Heat engines take input energy in the form of heat, use some of that energy to do work, and exhaust the remainder. Similarly, a person can be viewed as a heat engine that takes an input of internal energy, uses some of it to do work, and gives off the rest as heat. Suppose that a trained athlete can function as a heat engine with an efficiency of 0.11. **(a)** What is the magnitude of the internal energy that the athlete uses in order to do 5.1×10^4 J of work? **(b)** Determine the magnitude of the heat the athlete gives off.

86. M BIO GO 15.8 A 52-kg mountain climber, starting from rest, climbs a vertical distance of 730 m. At the top, she is again at rest. In the process, her body generates 4.1×10^6 J of energy via metabolic processes. In fact, her body acts like a heat engine, the efficiency of which is given by Equation 15.11 as $e = |W|/|Q_H|$, where $|W|$ is the magnitude of the work she does and $|Q_H|$ is the magnitude of the input heat. Find her efficiency as a heat engine.

87. M BIO 15.3 A weightlifter drinks a protein shake that contains 2.00×10^2 Calories. She then performs multiple repetitions on the bench press and does 2.75×10^5 J of work. After her workout, her net change in internal energy is $+1.50 \times 10^5$ J. During her workout, she loses heat to the environment, which results in the vaporization of perspiration from the surface of her skin. What mass of water did she lose due to perspiration? Assume the latent heat of vaporization of the perspiration is 2.42×10^6 J/kg.

88. M BIO 15.3 A long-distance runner performs work in the form of exercise with an average power output of 650 W. In 35 minutes of running, his body loses 5.2×10^5 J of heat. **(a)** What was the decrease in the internal energy of the runner? **(b)** How many calories would the runner have to ingest to replace the energy lost in part (a)?

89. H BIO 15.8 Imagine a person lifts 50 bags of landscaping rock, each with a mass of 35.0 kg, onto a table that is 1.00 m high. During this activity, 6.20×10^4 J of heat is exhausted by the body through perspiration. **(a)** Treating the body as a heat engine, what is its efficiency? **(b)** How much energy was extracted from the body's internal heat reservoir, which is created by the metabolism of food?

90. M BIO 15.3 During an intense 30-minute fat-burning cardio workout class, the instructor maintains an average power output of 2.0×10^2 W, and she loses heat to the environment at an average rate of 2.5×10^2 W. Ignoring heat loss due to perspiration, how much fat will the instructor lose during the workout? The energy content of fat is 9.3 kcal/g.

Concepts and Calculations Problems

Online

The first law of thermodynamics is basically a restatement of the conservation-of-energy principle in terms of heat and work. Problem 91 emphasizes this fact by showing that the conservation principle and the first law provide the same approach to the problem. It also provides a review of the concept of the latent heat of sublimation and the ideal gas law. Problem 92 reviews some of the central features of heat engines, as well as kinetic energy and the work–energy theorem.

91. M CHALK The sublimation of zinc (mass per mole = 0.0654 kg/mol) takes place at a temperature of 6.00×10^2 K, and the latent heat of sublimation is 1.99×10^6 J/kg. The pressure remains constant during the sublimation. Assume that the zinc vapor can be treated as a monatomic ideal gas and that the volume of solid zinc is negligible compared to the corresponding vapor. *Concepts:* (i) What is sublimation, and what is the latent heat of sublimation? (ii) When a solid phase changes to a gas phase, does the volume of the material increase or decrease, and by how much? (iii) As the material changes from a solid to a gas, does it do work on the environment, or does the environment do work on it? How much work is involved? (iv) In this problem we begin with heat Q and realize that it is used for two purposes: First, it makes the solid change into a gas, which entails a change ΔU in the internal energy of the material, $\Delta U = U_{gas} - U_{solid}$. Second, it allows the expanding material to do work W on the environment. According to the conservation-of-energy principle, how is Q related to ΔU and W? (v) According to the first law of thermodynamics, how is Q related to ΔU and W? *Calculations:* What is the change in the internal energy of zinc when 1.50 kg of zinc sublimates?

92. M CHALK SSM Each of two Carnot engines uses the same cold reservoir at a temperature of 275 K for its exhaust heat. Each engine receives 1450 J of input heat. The work from either of these engines is used to drive a pulley arrangement that uses a rope to accelerate a 125-kg crate from rest along a horizontal frictionless surface, as shown in the figure. With engine 1 the crate attains a speed of 2.00 m/s, while with engine 2 it attains a speed of 3.00 m/s. *Concepts:* (i) With which engine is the change in the crate's energy greater? (ii) Which engine does more work? Explain your answer. (iii) For which engine is the temperature of the hot reservoir greater? *Calculations:* Find the temperature of the hot reservoir for each engine.

Temperature = T_H

Hot reservoir

$|Q_H|$

Engine $|W|$

$|Q_C|$

Cold reservoir

Temperature = T_C

PROBLEM 92

Team Problems

93. E T WS **Relieve the Pressure.** A pressure relief valve is designed to prevent a gas containment vessel (a metal cylinder) from exploding due to overpressurization. The idea is that when the internal pressure exceeds a certain level, the valve automatically opens to let out some of the gas, and then closes again when the pressure is reduced to a safe level. You and your team have been assigned to test a pressure relief valve on an insulated neon gas storage tank to make sure that it is operable. The tank has a volume of 2.75 m³ and is currently at a pressure of 5.20×10^5 Pa, and the pressure relief valve is set to open at a pressure of 8.50×10^5 Pa. To increase the pressure, you turn on a 250.0-watt heater located inside the tank. How long after you turn on the heater should the relief valve open? Neon is a monatomic gas.

94. E T WS **A Water Chiller.** You and your team have been assigned to set up a special medical X-ray device that requires chilled water to be circulated through it in order that it not overheat. The problem is that a water chiller was not included with the device and you need to determine the type of chiller that is needed, and then obtain one. You read in the operation manual of the device that the temperature of the water going into the X-ray system must be no greater than 18.0 °C, and must circulate at a rate of no less than 8.00 liters per minute. The temperature of the water entering the chiller is 29.0 °C. If the coefficient of performance (COP) of the water chiller is 3.65, what is the minimum wattage required for a chiller that will do the job, where the wattage is the rate at which the chiller can do work? You can treat the water chiller like a typical refrigeration system.

95. T M **A High-Performance Engine.** You and your team are tasked with evaluating the parameters of a high-performance engine. The engine compresses the fuel-air mixture in the cylinder to make it ignite rather than igniting it with a spark plug (this is how diesel engines operate). You are given the *compression ratio* for the cylinders in the engine, which is the ratio of the initial volume of the cylinder (before the piston compresses the gas) to the final volume of the cylinder (after the gas is compressed): $V_i:V_f$. Assume that the fuel-gas mixture enters the cylinder at a temperature of 22.0 °C, and that the gas behaves like an ideal gas with $\gamma = 7/5$. **(a)** If the compression ratio is 15.4:1, what is the final temperature of the gas if the compression is adiabatic? **(b)** With everything else the same as in (a), what is the final temperature of the gas if the compression ratio is increased to 17.0:1? **(c)** Everything else being the same, for which compression ratio do you think the engine runs more efficiently? Give a qualitative argument for your answer.

96. T M **A Gas Lift.** A monatomic gas is contained in a long, vertical cylinder of inner radius $r = 14.0$ cm. A movable piston of negligible mass is inserted in the cylinder at a height $h = 1.20$ m. Initially, the gas inside the piston is at ambient temperature (23.5 °C) and standard pressure (1 atm). You place a 250.0-kg mass on the top of the piston and allow it to isothermally compress the gas below it. **(a)** What is the final pressure of the gas? **(b)** How far is the piston displaced from the equilibrium position? **(c)** Assuming an isobaric expansion, to what temperature should the gas inside the cylinder be heated in order to lift the mass 15.0 cm above the original position of the cylinder (i.e., to a height of 1.35 m)?

CHAPTER 16

NASA; https://upload.wikimedia.org/wikipedia/commons/thumb/1/13/Ares1-X%2828OCT2009%29.jpg/180px-Ares1-X%2828OCT2009%29.jpg

In this chapter we will see that sound is a longitudinal wave of pressure fluctuations and travels through air at a speed of 343 m/s when the temperature is 20 °C. Sound is produced by a vibrating object, such as the surfaces of a rocket in flight. When a rocket approaches and then exceeds the speed of sound, it is said to break through the sound barrier. The so-called barrier is formed due to sound waves previously emitted by the rocket at speeds less than the speed of sound. Here, an Aries 1-X rocket emerges from a cloud called a vapor cone or vapor egg during launch. These often occur several seconds after the rocket breaks the sound barrier and passes through what flight control refers to as "Max Q," which is the time during the launch that the vehicle experiences maximum aerodynamic pressure. After this time, the pressure around the rocket decreases, which is accompanied by a drop in temperature. Under the right atmospheric conditions, water vapor will condense on the surface of the shock wave.

LEARNING OBJECTIVES

After reading this module, you should be able to...

16.1 Distinguish between transverse and longitudinal waves.

16.2 Calculate speed, frequency, and wavelength for a wave.

16.3 Calculate the speed of a wave on a string.

16.4 Use wave functions to describe a wave.

16.5 Examine sound waves.

16.6 Calculate the speed of a sound wave.

16.7 Analyze sound intensity.

16.8 Calculate sound intensity level.

16.9 Solve Doppler effect problems.

16.10 Describe applications of sound in medicine.

16.11 Examine the sensitivity of the human ear using Fletcher–Munson curves.

Waves and Sound

16.1 The Nature of Waves

There are two features common to all waves:

1. A wave is a traveling disturbance.
2. A wave carries energy from place to place.

Consider a water wave, for instance. In **Figure 16.1** the wave created by the motorboat travels across the lake and disturbs the fisherman. However, *there is no bulk flow of water* outward from the motorboat. The wave is not a bulk movement of water such as a river, but, rather, a disturbance traveling on the surface of the lake. Part of the wave's energy in **Figure 16.1** is transferred to the fisherman and his boat.

We will consider two basic types of waves, transverse and longitudinal. **Interactive Figure 16.2** illustrates how a transverse wave can

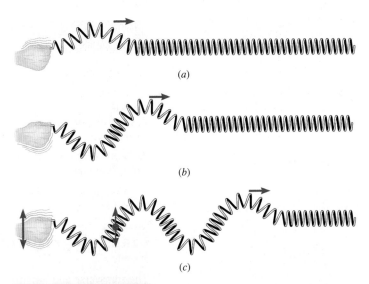

(*a*) An upward pulse moves to the right, followed by (*b*) a downward pulse. (*c*) When the end of the Slinky is moved up and down continuously, a transverse wave is produced.

FIGURE 16.1 The wave created by the motorboat travels across the lake and disturbs the fisherman.

be generated using a Slinky, a remarkable toy in the form of a long, loosely coiled spring. If one end of the Slinky is jerked up and down, as in part *a*, an upward pulse is sent traveling toward the right. If the end is then jerked down and up, as in part *b*, a downward pulse is generated and also moves to the right. If the end is continually moved up and down in simple harmonic motion, an entire wave is produced. As part *c* illustrates, the wave consists of a series of alternating upward and downward sections that propagate to the right, disturbing the vertical position of the Slinky in the process. To focus attention on the disturbance, a colored dot is attached to the Slinky in part *c* of the drawing. As the wave advances, the dot is displaced up and down in simple harmonic motion. The motion of the dot occurs perpendicular, or transverse, to the direction in which the wave travels. Thus, *a transverse wave is one in which the disturbance occurs perpendicular to the direction of travel of the wave*. Radio waves, light waves, and microwaves are transverse waves. Transverse waves also travel on the strings of instruments such as guitars and banjos.

A longitudinal wave can also be generated with a Slinky, and **Interactive Figure 16.3** demonstrates how. When one end is pushed forward along its length (i.e., longitudinally) and then pulled back to its starting point, as in part *a*, a region where the coils are compressed together is sent traveling to the right. If the end is pulled backward and then pushed forward to its starting point, as in part *b*, a region where the coils are stretched apart is formed and also moves to the right. If the end is continually moved back and forth in simple harmonic motion, an entire wave is created. As part *c* shows, the wave consists of a series of alternating compressed and stretched regions that travel to the right and disturb the coils. A colored dot is once again attached to the Slinky to emphasize the vibratory nature of the disturbance. In response to the wave, the dot moves back and forth in simple harmonic motion along the line of travel of the wave. Thus, *a longitudinal wave is one in which the disturbance occurs parallel to the line of travel of the wave*. A sound wave is a longitudinal wave.

A water wave is neither transverse nor longitudinal, since the motion of the water particles is not strictly perpendicular or strictly parallel to the line along which the wave travels. Instead, the motion includes both transverse and longitudinal components, since the water particles at the surface move on nearly circular paths, as **Figure 16.4** indicates.

(*a*) A compressed region moves to the right, followed by (*b*) a stretched region. (*c*) When the end of the Slinky is moved back and forth continuously, a longitudinal wave is produced.

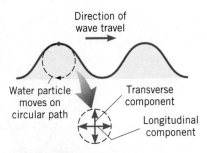

FIGURE 16.4 A water wave is neither transverse nor longitudinal, since water particles at the surface move clockwise on nearly circular paths as the wave moves from left to right.

Check Your Understanding

(The answers are given at the end of the book.)

1. Considering the nature of a water wave (see **Figure 16.4**), which of the following statements correctly describes how a fishing float moves on the surface of a lake when a wave passes beneath it? **(a)** It bobs up and down vertically. **(b)** It moves back and forth horizontally. **(c)** It moves in a vertical plane, exhibiting both motions described in (a) and (b) simultaneously.

2. Suppose that the longitudinal wave in **Interactive Figure 16.3c** moves to the right at a speed of 1 m/s. Does one coil of the Slinky move a distance of 1 mm to the right in a time of 1 ms?

16.2 Periodic Waves

The transverse and longitudinal waves that we have been discussing are called **periodic waves** because they consist of cycles or patterns that are produced over and over again by the source. In **Figures 16.2** and **16.3** the repetitive patterns occur as a result of the simple harmonic motion of the left end of the Slinky, so that every segment of the Slinky vibrates in simple harmonic motion. Sections 10.1 and 10.2 discuss the simple harmonic motion of an object on a spring and introduce the concepts of cycle, amplitude, period, and frequency. This same terminology is used to describe periodic waves, such as the sound waves we hear (discussed later in this chapter) and the light waves we see (discussed in Chapter 24).

    Figure 16.5 uses a graphical representation of a transverse wave on a Slinky to review the terminology. One **cycle** of a wave is shaded in color in both parts of the drawing. A wave is a series of many cycles. In part *a* the vertical position of the Slinky is plotted on the vertical axis, and the corresponding distance along the length of the Slinky is plotted on the horizontal axis. Such a graph is equivalent to a photograph of the wave taken at one instant in time and shows the disturbance that exists at each point along the Slinky's length. As marked on this graph, the **amplitude** *A* is the maximum excursion of a particle of the medium (i.e., the Slinky) in which the wave exists from the particle's undisturbed position. The amplitude is the distance between a crest, or highest point on the wave pattern, and the undisturbed position; it is also the distance between a trough, or lowest point on the wave pattern, and the undisturbed position. The **wavelength** λ is the horizontal length of one cycle of the wave, as shown in **Figure 16.5a**. The wavelength is also the horizontal distance between two successive crests, two successive troughs, or any two successive equivalent points on the wave.

    Part *b* of **Figure 16.5** shows a graph in which time, rather than distance, is plotted on the horizontal axis. This graph is obtained by observing a single point on the Slinky. As the wave passes, the point under observation oscillates up and down in simple harmonic motion. As indicated on the graph, the **period** *T* is the time required for one complete up/down cycle, just as it is for an object vibrating on a spring. The period *T* is related to the **frequency** *f*, just as it is for any example of simple harmonic motion:

$$f = \frac{1}{T} \tag{10.5}$$

    (a) At a particular time   (b) At a particular location

FIGURE 16.5 One cycle of the wave is shaded in color, and the amplitude of the wave is denoted as *A*.

The period is commonly measured in seconds, and frequency is measured in cycles per second, or hertz (Hz). If, for instance, one cycle of a wave takes one-tenth of a second to pass an observer, then ten cycles pass the observer per second, as Equation 10.5 indicates [$f = 1/(0.1\ s) = 10\ cycles/s = 10\ Hz$].

A simple relation exists between the period, the wavelength, and the speed of any periodic wave, a relation that **Figure 16.6** helps to introduce. Imagine waiting at a railroad crossing, while a freight train moves by at a constant speed v. The train consists of a long line of identical boxcars, each of which has a length λ and requires a time T to pass, so the speed is $v = \lambda/T$. This same equation applies for a wave and relates the speed of the wave to the wavelength λ and the period T. Since the frequency of a wave is $f = 1/T$, the expression for the speed is

$$v = \frac{\lambda}{T} = f\lambda \qquad (16.1)$$

The terminology just discussed and the fundamental relations $f = 1/T$ and $v = f\lambda$ apply to longitudinal as well as to transverse waves. Example 1 shows how the wavelength of a wave is determined by the wave speed and the frequency established by the source.

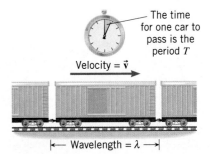

FIGURE 16.6 A train moving at a constant speed serves as an analogy for a traveling wave.

EXAMPLE 1 | The Wavelengths of Radio Waves

AM and FM radio waves are transverse waves consisting of electric and magnetic disturbances traveling at a speed of 3.00×10^8 m/s. A station broadcasts an AM radio wave whose frequency is 1230×10^3 Hz (1230 kHz on the dial) and an FM radio wave whose frequency is 91.9×10^6 Hz (91.9 MHz on the dial). Find the distance between adjacent crests in each wave.

Reasoning   The distance between adjacent crests is the wavelength λ. Since the speed of each wave is $v = 3.00 \times 10^8$ m/s and the frequencies are known, the relation $v = f\lambda$ can be used to determine the wavelengths.

> **Problem-Solving Insight**   The equation $v = f\lambda$ applies to any kind of periodic wave.

Solution

AM
$$\lambda = \frac{v}{f} = \frac{3.00 \times 10^8\,\text{m/s}}{1230 \times 10^3\,\text{Hz}} = \boxed{244\ \text{m}}$$

FM
$$\lambda = \frac{v}{f} = \frac{3.00 \times 10^8\,\text{m/s}}{91.9 \times 10^6\,\text{Hz}} = \boxed{3.26\ \text{m}}$$

Notice that the wavelength of an AM radio wave is longer than two and one-half football fields!

Check Your Understanding

(The answer is given at the end of the book.)

3. A sound wave (a periodic longitudinal wave) from a loudspeaker travels from air into water. The frequency of the wave does not change, because the loudspeaker producing the sound determines the frequency. The speed of sound in air is 343 m/s, whereas the speed in fresh water is 1482 m/s. When the sound wave enters the water, does its wavelength increase, decrease, or remain the same?

16.3 | The Speed of a Wave on a String

The properties of the material* or medium through which a wave travels determine the speed of the wave. For example, **Figure 16.7** shows a transverse wave on a string and draws attention to four string particles that have been drawn as colored dots. As the wave moves to the right, each particle is displaced, one after the other, from its undisturbed

---

*Electromagnetic waves (discussed in Chapter 24) can move through a vacuum, as well as through materials such as glass and water.

FIGURE 16.7 As a transverse wave moves to the right with speed v, each string particle is displaced, one after the other, from its undisturbed position.

position. In the drawing, particles 1 and 2 have already been displaced upward, while particles 3 and 4 are not yet affected by the wave. Particle 3 will be next to move because the section of string immediately to its left (i.e., particle 2) will pull it upward.

Figure 16.7 leads us to conclude that the speed with which the wave moves to the right depends on how quickly one particle of the string is accelerated upward in response to the net pulling force exerted by its adjacent neighbors. In accord with Newton's second law, a stronger net force results in a greater acceleration, and, thus, a faster-moving wave. The ability of one particle to pull on its neighbors depends on how tightly the string is stretched—that is, on the tension (see Section 4.10 for a review of tension). The greater the tension, the greater the pulling force the particles exert on each other and the faster the wave travels, other things being equal. Along with the tension, a second factor influences the wave speed. According to Newton's second law, the inertia or mass of particle 3 in **Figure 16.7** also affects how quickly it responds to the upward pull of particle 2. For a given net pulling force, a smaller mass has a greater acceleration than a larger mass. Therefore, other things being equal, a wave travels faster on a string whose particles have a small mass, or, as it turns out, on a string that has a small mass per unit length. The mass per unit length is called the *linear density* of the string. It is the mass m of the string divided by its length L, or m/L. Effects of the tension F and the mass per unit length are evident in the following expression for the speed v of a small-amplitude wave on a string:

$$v = \sqrt{\frac{F}{m/L}} \tag{16.2}$$

The motion of transverse waves along a string is important in the operation of musical instruments, such as the guitar, the violin, and the piano. In these instruments, the strings are either plucked, bowed, or struck to produce transverse waves. Example 2 discusses the speed of the waves on the strings of a guitar.

EXAMPLE 2 | The Physics of Waves on Guitar Strings

Transverse waves travel on each string of an electric guitar after the string is plucked (see **Figure 16.8**). The length of each string between its two fixed ends is 0.628 m, and the mass is 0.208 g for the highest pitched E string and 3.32 g for the lowest pitched E string. Each string is under a tension of 226 N. Find the speeds of the waves on the two strings.

Reasoning The speed of a wave on a guitar string, as expressed by Equation 16.2, depends on the tension F in the string and its linear density m/L. Since the tension is the same for both strings, and smaller linear densities give rise to greater speeds, we expect the wave speed to be greatest on the string with the smallest linear density.

Solution The speeds of the waves are given by Equation 16.2 as

High-pitched E $\quad v = \sqrt{\dfrac{F}{m/L}} = \sqrt{\dfrac{226 \text{ N}}{(0.208 \times 10^{-3} \text{ kg})/(0.628 \text{ m})}} = \boxed{826 \text{ m/s}}$

Low-pitched E $\quad v = \sqrt{\dfrac{F}{m/L}} = \sqrt{\dfrac{226 \text{ N}}{(3.32 \times 10^{-3} \text{ kg})/(0.628 \text{ m})}} = \boxed{207 \text{ m/s}}$

Notice how fast the waves move: the speeds correspond to 1850 and 463 mi/h.

FIGURE 16.8 Plucking a guitar string generates transverse waves.

Conceptual Example 3 offers additional insight into the nature of a wave as a traveling disturbance.

CONCEPTUAL EXAMPLE 3 | Wave Speed Versus Particle Speed

As indicated in **Figure 16.9**, the speed of a transverse wave on a string is v_{wave}, and the speed at which a string particle moves is $v_{particle}$. Which of the following statements is correct? **(a)** The speeds v_{wave} and $v_{particle}$ are identical. **(b)** The speeds v_{wave} and $v_{particle}$ are different.

Reasoning A wave moves on a string at a speed v_{wave} that is determined by the properties of the string and has a constant value everywhere on the string at all times, assuming that these properties are the same everywhere on the string. Each particle on the string, however, moves in simple harmonic motion, assuming that the source generating the wave (e.g., the hand in **Interactive Figure 16.2c**) moves in simple harmonic motion. Each particle has a speed $v_{particle}$ that is characteristic of simple harmonic motion.

Answer (a) is incorrect. The speed v_{wave} has a constant value at all times. In contrast, $v_{particle}$ is not constant at all times, because it is the speed that characterizes simple harmonic motion and that speed varies as time passes. Thus, the two speeds are not identical.

Answer (b) is correct. The speed v_{wave} is determined by the tension F and the mass per unit length m/L of the string, according to $v_{wave} = \sqrt{\dfrac{F}{m/L}}$ (see Equation 16.2). The speed $v_{particle}$ is characteristic of simple harmonic motion, according to $v_{particle} = A\omega \sin \omega t$ (Equation 10.7 without the minus sign, since we deal

FIGURE 16.9 A transverse wave on a string is moving to the right with a constant speed v_{wave}. A string particle moves up and down in simple harmonic motion about the undisturbed position of the string. A string particle moves with a speed $v_{particle}$.

here only with speed, which is the magnitude of the velocity). The particle speed depends on the amplitude A and the angular frequency ω of the simple harmonic motion, as well as the time t; the speed is greatest when the particle is passing through the undisturbed position of the string, and it is zero when the particle has its maximum displacement. Thus, the two speeds are different, because v_{wave} depends on the properties of the string and $v_{particle}$ depends on the properties of the source creating the wave.

Related Homework: Problem 19

Check Your Understanding

(The answers are given at the end of the book.)

4. One end of each of two identical strings is attached to a wall. Each string is being pulled equally tight by someone at the other end. A transverse pulse is sent traveling along string A. A bit later an identical pulse is sent traveling along string B. What, if anything, can be done to make the pulse on string B catch up with and pass the pulse on string A?

5. In Section 4.10 the concept of a *massless* rope is discussed. Considering Equation 16.2, would it take any time for a transverse wave to travel the length of a truly massless rope?

6. A wire is strung tightly between two immovable posts. Review Section 12.4 and decide whether the speed of a transverse wave on this wire would increase, decrease, or remain the same when the temperature increases. Ignore any change in the mass per unit length of the wire.

7. Examine Conceptual Example 3 before addressing this question. A wave moves on a string with a constant velocity. Does this mean that the particles of the string always have zero acceleration?

8. A rope of mass m is hanging down from the ceiling. Nothing is attached to the loose end of the rope. As a transverse wave travels upward on the rope, does the speed of the wave increase, decrease, or remain the same?

9. String I and string II have the same length. However, the mass of string I is twice the mass of string II, and the tension in string I is eight times the tension in string II. A wave of the same amplitude and frequency travels on each of these strings. Which of the drawings in **CYU Figure 16.1** correctly shows the waves: **(a)** A **(b)** B **(c)** C?

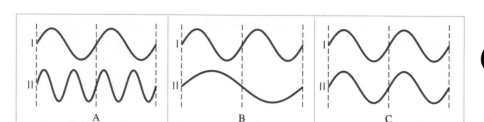

CYU FIGURE 16.1

16.4 | *The Mathematical Description of a Wave

When a wave travels through a medium, it displaces the particles of the medium from their undisturbed positions. Suppose that a particle is located at a distance x from a coordinate origin. We would like to know the displacement y of this particle from its undisturbed position at any time t as the wave passes. For periodic waves that result from simple harmonic motion of the source, the expression for the displacement involves a sine or cosine, a fact that is not surprising. After all, in Chapter 10 simple harmonic motion is described using sinusoidal equations, and the graphs for a wave in **Figure 16.5** look like a plot of displacement versus time for an object oscillating on a spring (see **Figure 10.5**).

Our approach will be to present the expression for the displacement and then show graphically that it gives a correct description. Equation 16.3 represents the displacement of a particle caused by a wave traveling in the $+x$ direction (to the right), with an amplitude A, frequency f, and wavelength λ. Equation 16.4 applies to a wave moving in the $-x$ direction (to the left).

Wave motion toward +x $\qquad y = A \sin\left(2\pi ft - \dfrac{2\pi x}{\lambda}\right)$ (16.3)

Wave motion toward −x $\qquad y = A \sin\left(2\pi ft + \dfrac{2\pi x}{\lambda}\right)$ (16.4)

These equations apply to transverse or longitudinal waves and assume that $y = 0$ m when $x = 0$ m and $t = 0$ s. The term $(2\pi ft - 2\pi x/\lambda)$ in Equation 16.3 or the term $(2\pi ft + 2\pi x/\lambda)$ in Equation 16.4 is called the **phase angle** of the wave. In either case the phase angle is measured in *radians*, not in degrees.

Consider a transverse wave moving in the $+x$ direction along a string. A string particle located at the origin ($x = 0$ m) exhibits simple harmonic motion with a phase angle of $2\pi ft$; that is, its displacement as a function of time is $y = A \sin(2\pi ft)$. A particle located at a distance x also exhibits simple harmonic motion, but its phase angle in Equation 16.3 is

$$2\pi ft - \frac{2\pi x}{\lambda} = 2\pi f\left(t - \frac{x}{f\lambda}\right) = 2\pi f\left(t - \frac{x}{v}\right)$$

The quantity x/v is the time needed for the wave to travel the distance x. In other words, the simple harmonic motion that occurs at x is delayed by the time interval x/v compared to the motion at the origin.

Figure 16.10 shows the displacement y plotted as a function of position x along the string at a series of time intervals separated by one-fourth of the period T ($t = 0$ s, $\frac{1}{4}T$, $\frac{2}{4}T$, $\frac{3}{4}T$, T). These graphs are constructed by substituting the corresponding value for t into Equation 16.3, remembering that $f = 1/T$, and then calculating y at a series of values for x. The graphs are like photographs taken at various times as the wave moves to the right. For reference, the colored square on each graph marks the place on the wave that

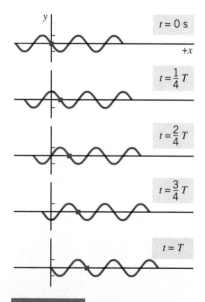

FIGURE 16.10 Equation 16.3 is plotted here at a series of times separated by one-fourth of the period T. The colored square in each graph marks the place on the wave that is located at $x = 0$ m when $t = 0$ s. As time passes, the wave moves to the right.

is located at $x = 0$ m when $t = 0$ s. As time passes, the colored square moves to the right, along with the wave. In a similar manner, it can be shown that Equation 16.4 represents a wave moving in the $-x$ direction.

16.5 The Nature of Sound

Longitudinal Sound Waves

Sound is a longitudinal wave that is created by a vibrating object, such as a guitar string, the human vocal cords, or the diaphragm of a loudspeaker. Moreover, sound can be created or transmitted only in a medium, such as a gas, liquid, or solid. As we will see, the particles of the medium must be present for the disturbance of the wave to move from place to place. Sound cannot exist in a vacuum.

THE PHYSICS OF . . . a loudspeaker diaphragm. To see how sound waves are produced and why they are longitudinal, consider the vibrating diaphragm of a loudspeaker. When the diaphragm moves outward, it compresses the air directly in front of it, as in **Interactive Figure 16.11a**. This compression causes the air pressure to rise slightly. The region of increased pressure is called a **condensation**, and it travels away from the speaker at the speed of sound. The condensation is analogous to the compressed region of coils in a longitudinal wave on a Slinky, which is included in **Interactive Figure 16.11a** for comparison. After producing a condensation, the diaphragm reverses its motion and moves inward, as in part *b* of the drawing. The inward motion produces a region known as a **rarefaction**, where the air pressure is slightly less than normal. The rarefaction is similar to the stretched region of coils in a longitudinal Slinky wave. Following immediately behind the condensation, the rarefaction also travels away from the speaker at the speed of sound. **Figure 16.12** further emphasizes the similarity between a sound wave and a longitudinal Slinky wave. As the wave passes, the colored dots attached both to the Slinky and to an air molecule execute simple harmonic motion about their undisturbed positions. The colored arrows on either side of the dots indicate that the simple harmonic motion occurs parallel to the line of travel. The drawing also shows that the wavelength λ is the distance between the centers of two successive condensations; λ is also the distance between the centers of two successive rarefactions.

Figure 16.13 illustrates a sound wave spreading out in space after being produced by a loudspeaker. When the condensations and rarefactions arrive at the ear, they force the eardrum to vibrate at the same frequency as the speaker diaphragm. The vibratory motion of the eardrum is interpreted by the brain as sound. It should be emphasized that sound is not a mass movement of air, like the wind. As the condensations and rarefactions of the sound wave travel outward from the vibrating diaphragm in **Figure 16.13**, the individual air molecules are not carried along with the wave. Rather, each molecule executes simple harmonic motion about a fixed location. In doing so, one molecule collides with its neighbor and passes the condensations and rarefactions forward. The neighbor, in turn, repeats the process.

The Frequency of a Sound Wave

Each cycle of a sound wave includes one condensation and one rarefaction, and the **frequency** is the number of cycles per second that passes by a given location. For example, if the diaphragm of a speaker vibrates back and forth in simple harmonic motion at a frequency of 1000 Hz, then 1000 condensations, each followed by a rarefaction, are generated every second, thus forming a sound wave whose frequency is also 1000 Hz. A sound with a single frequency is called a **pure tone**. Experiments have shown that a healthy young person hears all sound frequencies from approximately 20 to 20 000 Hz (20 kHz). The ability to hear the high frequencies decreases with age, however, and a normal middle-aged adult hears frequencies only up to 12–14 kHz.

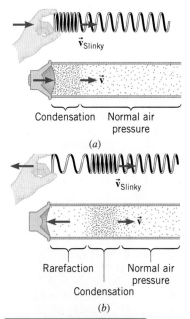

INTERACTIVE FIGURE 16.11 (*a*)When the speaker diaphragm moves outward, it creates a condensation. (*b*) When the diaphragm moves inward, it creates a rarefaction. The condensation and rarefaction on the Slinky are included for comparison. In reality, the velocity of the wave on the Slinky $\vec{v}_{\text{Slinky}}$ is much smaller than the velocity of sound in air $\vec{v}$. For simplicity, the two waves are shown here to have the same velocity.

FIGURE 16.12 Both the wave on the Slinky and the sound wave are longitudinal. The colored dots attached to the Slinky and to an air molecule vibrate back and forth parallel to the line of travel of the wave.

FIGURE 16.13 Condensations and rarefactions travel from the speaker to the listener, but the individual air molecules do not move with the wave. A given molecule vibrates back and forth about a fixed location.

FIGURE 16.14 A touch-tone telephone and a schematic showing the two pure tones produced when each button is pressed.

THE PHYSICS OF . . . a touch-tone telephone. Pure tones are used in touch-tone telephones, such as the one shown in **Figure 16.14**. These phones simultaneously produce two pure tones when each button is pressed, a different pair of tones for each different button. The tones are transmitted electronically to the central telephone office, where they activate switching circuits that complete the call. For example, the drawing indicates that pressing the "5" button produces pure tones of 770 and 1336 Hz simultaneously. These frequencies are characteristic of the second row and second column of buttons, respectively. Similarly, the "9" button generates tones of 852 and 1477 Hz.

Sound can be generated whose frequency lies below 20 Hz or above 20 kHz, although humans normally do not hear it. Sound waves with frequencies below 20 Hz are said to be **infrasonic**, while those with frequencies above 20 kHz are referred to as **ultrasonic**. Some species of bats known as microbats use ultrasonic frequencies up to 120 kHz for locating prey and for navigating (**Figure 16.15**), while rhinoceroses use infrasonic frequencies as low as 5 Hz to call one another (**Figure 16.16**).

Frequency is an objective property of a sound wave because frequency can be measured with an electronic frequency counter. A listener's perception of frequency, however, is subjective. The brain interprets the frequency detected by the ear primarily in terms of the subjective quality called **pitch**. A pure tone with a large (high) frequency is interpreted as a high-pitched sound, while a pure tone with a small (low) frequency is interpreted as a low-pitched sound. For instance, a piccolo produces high-pitched sounds, and a tuba produces low-pitched sounds.

FIGURE 16.15 Some bats use ultrasonic sound waves for locating prey and for navigating. This bat has captured a katydid.

FIGURE 16.16 Rhinoceroses call to one another using infrasonic sound waves.

The Pressure Amplitude of a Sound Wave

Figure 16.17 illustrates a pure-tone sound wave traveling in a tube. Attached to the tube is a series of gauges that indicate the pressure variations along the wave. The graph shows that the air pressure varies sinusoidally along the length of the tube. Although this graph has the appearance of a transverse wave, remember that the sound itself is a longitudinal wave. The graph also shows the **pressure amplitude** of the wave, which is the magnitude of the maximum change in pressure, measured relative to the undisturbed or atmospheric pressure. The pressure fluctuations in a sound wave are normally very small. For instance, in a typical conversation between two people the pressure amplitude is about 3×10^{-2} Pa, certainly a small amount compared with the atmospheric pressure of $1.01 \times 10^{+5}$ Pa. The ear is remarkable in being able to detect such small changes.

Loudness is an attribute of sound that depends primarily on the amplitude of the wave: the larger the amplitude, the louder the sound. The pressure amplitude is an objective property of a sound wave, since it can be measured. Loudness, on the other hand, is subjective. Each individual determines what is loud, depending on the acuteness of his or her hearing.

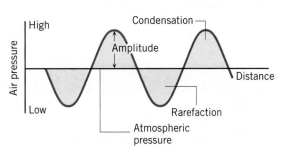

FIGURE 16.17 A sound wave is a series of alternating condensations and rarefactions. The graph shows that the condensations are regions of higher than normal air pressure, and the rarefactions are regions of lower than normal air pressure.

Check Your Understanding

(*The answer is given at the end of the book.*)

10. In a traveling sound wave, are there any particles that are *always* at rest as the wave passes by?

16.6 | The Speed of Sound

Gases

Sound travels through gases, liquids, and solids at considerably different speeds, as Table 16.1 reveals. Near room temperature, the speed of sound in air is 343 m/s (767 mi/h) and is markedly greater in liquids and solids. For example, sound travels more than four times faster in water and more than seventeen times faster in steel than it does in air. In general, sound travels slowest in gases, faster in liquids, and fastest in solids.

Like the speed of a wave on a guitar string, the speed of sound depends on the properties of the medium. In a gas, it is only when molecules collide that the condensations and rarefactions of a sound wave can move from place to place. It is reasonable, then, to expect the speed of sound in a gas to have the same order of magnitude as the average molecular speed between collisions. For an ideal gas this average speed is the translational rms speed given by Equation 14.6: $v_{rms} = \sqrt{3kT/m}$, where T is the Kelvin temperature, m is the mass of a molecule, and k is Boltzmann's constant. Although the expression for v_{rms} overestimates the speed of sound, it does give the correct dependence on Kelvin temperature and particle mass. Careful analysis shows that the speed of sound in an ideal gas is given by

Ideal gas
$$v = \sqrt{\frac{\gamma kT}{m}}$$
(16.5)

where $\gamma = c_P/c_V$ is the ratio of the specific heat capacity at constant pressure c_P to the specific heat capacity at constant volume c_V.

The factor γ is introduced in Section 15.5, where the adiabatic compression and expansion of an ideal gas are discussed. In Section 15.6 it is shown that γ has the value of $\gamma = \frac{5}{3}$ for ideal monatomic gases and a value of $\gamma = \frac{7}{5}$ for ideal diatomic gases. The value of γ appears in Equation 16.5 because the condensations and rarefactions of a sound wave are formed by adiabatic compressions and expansions of the gas. The regions that

TABLE 16.1	Speed of Sound in Gases, Liquids, and Solids
Substance	**Speed (m/s)**
Gases	
Air (0 °C)	331
Air (20 °C)	343
Carbon dioxide (0 °C)	259
Oxygen (0 °C)	316
Helium (0 °C)	965
Liquids	
Chloroform (20 °C)	1004
Ethyl alcohol (20 °C)	1162
Mercury (20 °C)	1450
Fresh water (20 °C)	1482
Seawater (20 °C)	1522
Solids	
Copper	5010
Glass (Pyrex)	5640
Lead	1960
Steel	5960

are compressed (the condensations) become slightly warmed, and the regions that are expanded (the rarefactions) become slightly cooled. However, no appreciable heat flows from a condensation to an adjacent rarefaction because the distance between the two (half a wavelength) is relatively large for most audible sound waves and a gas is a poor thermal conductor. Thus, the compression and expansion process is adiabatic. Example 4 illustrates the use of Equation 16.5.

Analyzing Multiple-Concept Problems

EXAMPLE 4 | The Physics of an Ultrasonic Ruler

Figure 16.18 shows an ultrasonic ruler that is used to measure the distance to a target, such as a wall. To initiate the measurement, the ruler generates a pulse of ultrasonic sound that travels to the wall and, much like an echo, reflects from it. The reflected pulse returns to the ruler, which measures the time it takes for the round-trip. Using a preset value for the speed of sound, the unit determines the distance to the wall and displays it on a digital readout. Suppose that the round-trip travel time is 20.0 ms on a day when the air temperature is 32 °C. Assuming that air is an ideal diatomic gas

$(\gamma = \frac{7}{5})$ and that the average molecular mass of air is 28.9 u, find the distance between the ultrasonic ruler and the wall.

Reasoning   The distance between the ruler and the wall is equal to the speed of sound multiplied by the time it takes for the sound pulse to reach the wall. The speed v of sound can be determined from a knowledge of the air temperature T and the average mass m of an air molecule by using the relation $v = \sqrt{\gamma k T/m}$ (Equation 16.5). The time can be deduced from the given data.

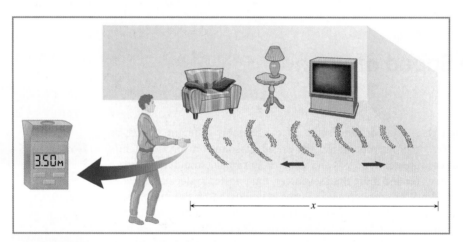

FIGURE 16.18   An ultrasonic ruler uses sound with a frequency greater than 20 kHz to measure the distance x to the wall. The blue arcs and blue arrow denote the outgoing sound wave, and the red arcs and red arrow denote the wave reflected from the wall.

Knowns and Unknowns   The data for this problem are listed below:

Description	Symbol	Value	Comment
Round-trip time of sound	t_{RT}	20.0 ms	20.0 ms = 20.0×10^{-3} s
Air temperature	T_c	32 °C	
Ratio of specific heats for air	γ	$\frac{7}{5}$	
Average molecular mass of air	m	28.9 u	Must convert "u" to kilograms.
Unknown Variable			
Distance between ruler and wall	x	?	

Modeling the Problem

STEP 1 Kinematics Since sound moves at a constant speed, the distance x it travels is the product of its speed v and the time t, or $x = vt$. The time for the sound to reach the wall is one-half the round-trip time t_{RT}, so $t = \frac{1}{2}t_{RT}$. Thus, the distance to the wall is

$$x = v\left(\tfrac{1}{2}t_{RT}\right)$$

The round-trip time t_{RT} is known, but the speed of sound in air at 32 °C is not. We will find an expression for this speed in Step 2.

STEP 2 Speed of Sound Since the air is assumed to be an ideal gas, the speed v of sound is related to the Kelvin temperature T and the average mass m of an air molecule by

$$v = \sqrt{\frac{\gamma kT}{m}} \qquad \text{(16.5)}$$

where γ is the ratio of the specific heat capacity of air at constant pressure to that at constant volume (see Section 15.5), and k is Boltzmann's constant. The temperature in this expression must be the Kelvin temperature of the air, which is related to its Celsius temperature T_c by $T = T_c + 273.15$ (Equation 12.1). Thus, the speed of sound in air is

$$v = \sqrt{\frac{\gamma k(T_c + 273.15)}{m}}$$

This expression for v can be substituted into Equation 1, as shown on the right.

> **Problem-Solving Insight** When using the equation $v = \sqrt{\gamma kT/m}$ to calculate the speed of sound in an ideal gas, be sure to express the temperature T in kelvins and not in degrees Celsius or Fahrenheit.

Solution Combining the results of the modeling steps, we have

$$\overset{\text{STEP 1}}{\downarrow} \quad \overset{\text{STEP 2}}{\downarrow}$$
$$x = v\left(\tfrac{1}{2}t_{RT}\right) = \sqrt{\frac{\gamma k(T_c + 273.15)}{m}}\left(\tfrac{1}{2}t_{RT}\right)$$

Since the average mass of an air molecule is given in atomic mass units (28.9 u), we must convert it to kilograms by using the conversion factor $1\ \mathrm{u} = 1.6605 \times 10^{-27}\ \mathrm{kg}$ (see Section 14.1). Thus,

$$m = (28.9\ \mathrm{u})\left(\frac{1.6605 \times 10^{-27}\ \mathrm{kg}}{1\ \mathrm{u}}\right) = 4.80 \times 10^{-26}\ \mathrm{kg}$$

The distance from the ultrasonic ruler to the wall is

$$x = \sqrt{\frac{\gamma k(T_c + 273.15)}{m}}\left(\tfrac{1}{2}t_{RT}\right)$$

$$= \sqrt{\frac{\frac{7}{5}(1.38 \times 10^{-23}\ \mathrm{J/K})(32\ ^\circ\mathrm{C} + 273.15)}{4.80 \times 10^{-2}\ \mathrm{kg}}}\left[\tfrac{1}{2}(20.0 \times 10^{-3}\ \mathrm{s})\right] = \boxed{3.50\ \mathrm{m}}$$

Related Homework: Problem 46

THE PHYSICS OF . . . sonar. Sonar (**so**und **na**vigation **r**anging) is a technique for determining water depth and locating underwater objects, such as reefs, submarines, and schools of fish. The core of a sonar unit consists of an ultrasonic transmitter and receiver mounted on the bottom of a ship. The transmitter emits a short pulse of ultrasonic sound, and at a later time the reflected pulse returns and is detected by the receiver. The distance to the object is determined from the electronically measured round-trip time of the pulse and a knowledge of the speed of sound in water; the distance registers automatically on an appropriate meter. Such a measurement is similar to the distance measurement discussed for the ultrasonic ruler in Example 4.

Conceptual Example 5 illustrates how the speed of sound in air can be used to estimate the distance to a thunderstorm, using a handy rule of thumb.

CONCEPTUAL EXAMPLE 5 | Lightning, Thunder, and a Rule of Thumb

In a thunderstorm, lightning and thunder occur nearly simultaneously. The light waves from the lightning travel at a speed of $v_{light} = 3.0 \times 10^8$ m/s, whereas the sound waves from the thunder travel at $v_{sound} = 343$ m/s. There is a rule of thumb for estimating how far away a storm is. After you see a lightning flash, count the seconds until you hear the thunder; divide the number of seconds by five to get the approximate distance (in miles) to the storm. In this rule, which of the two speeds plays a role? **(a)** Both v_{sound} and v_{light} **(b)** Only v_{sound} **(c)** Only v_{light}

Reasoning   At a distance of one mile from a storm, the observer in **Figure 16.19** detects either type of wave only after a time that is equal to the distance divided by the speed at which the wave travels. This fact will guide our analysis.

Answers (b) and (c) are incorrect.   The rule involves the time that passes *between* seeing the lightning flash and hearing the thunder, not just the time at which either type of wave is detected. Therefore, both the speeds v_{light} and v_{sound} must play a role in the rule.

Answer (a) is correct.   Light from the flash travels so rapidly that it reaches the observer almost instantaneously; its travel time for one mile (1.6×10^3 m) is only

$$t_{light} = \frac{1.6 \times 10^3\,\text{m}}{v_{light}} = \frac{1.6 \times 10^3\,\text{m}}{3.0 \times 10^8\,\text{m/s}} = 5 \times 10^{-6}\,\text{s}$$

FIGURE 16.19   A lightning bolt from a thunderstorm generates a flash of light and sound (thunder). The speed of light is much greater than the speed of sound. Therefore, the light reaches the person first, followed later by the sound.

In comparison, the sound of the thunder travels very slowly; its travel time for one mile is

$$t_{sound} = \frac{1.6 \times 10^3\,\text{m}}{v_{sound}} = \frac{1.6 \times 10^3\,\text{m}}{343\,\text{m/s}} = 5\,\text{s}$$

Since t_{light} is negligible compared to t_{sound}, the time between seeing the lightning flash and hearing the thunder is about 5 s for every mile of distance from the storm.

Liquids

In a liquid, the speed of sound depends on the density ρ and the *adiabatic* bulk modulus B_{ad} of the liquid:

Liquid $$v = \sqrt{\frac{B_{ad}}{\rho}} \qquad\qquad (16.6)$$

The bulk modulus is introduced in Section 10.7 in a discussion of the volume deformation of liquids and solids. There it is tacitly assumed that the temperature remains constant while the volume of the material changes; that is, the compression or expansion is isothermal. However, the condensations and rarefactions in a sound wave occur under *adiabatic* rather than isothermal conditions. Thus, the adiabatic bulk modulus B_{ad} must be used when calculating the speed of sound in liquids. Values of B_{ad} will be provided as needed in this text.

Table 16.1 gives some data for the speed of sound in liquids. In seawater, for instance, the speed is 1522 m/s, which is more than four times as great as the speed in air. The speed of sound is an important parameter in the measurement of distance, as discussed for the ultrasonic ruler in Example 4.

BIO THE PHYSICS OF . . . cataract surgery.   Accurate distance measurements using ultrasonic sound also play an important role in medicine, where the sound often travels through liquid-like materials in the body. A routine preoperative procedure in cataract surgery, for example, uses an ultrasonic probe called an A-scan to measure the length of the eyeball in front of the lens, the thickness of the lens, and the length of the eyeball between the lens and the retina (see **Figure 16.20**). The measurement is similar to that discussed in Example 4 and relies on the fact that the speed of sound in the material in front of and behind the lens of the eye is 1532 m/s, whereas that within the lens is 1641 m/s. In cataract surgery, the cataractous lens is removed and often replaced with an implanted artificial lens. Data provided by the A-scan facilitate the design of the lens implant (its size and the optical correction that it introduces).

FIGURE 16.20 A cross-sectional view of the human eye.

Solid Bars

When sound travels through a long, slender, solid bar, the speed of the sound depends on the properties of the medium according to

Long, slender, solid bar $$v = \sqrt{\frac{Y}{\rho}}$$ (16.7)

where Y is Young's modulus (defined in Section 10.7) and ρ is the density.

Check Your Understanding

(The answers are given at the end of the book.)

11. Do you expect an echo to return to you more quickly on a hot day or a cold day, other things being equal?

12. Carbon monoxide (CO), hydrogen (H_2), and nitrogen (N_2) may be treated as ideal gases. Each has the same temperature and nearly the same value for the ratio of the specific heat capacities at constant pressure and constant volume. In which two of the three gases is the speed of sound approximately the same?

13. Jell-O starts out as a liquid and then sets to a gel. As the Jell-O sets and becomes more solid, does the speed of sound in this material increase, decrease, or remain the same?

16.7 Sound Intensity

Sound waves carry energy that can be used to do work, like forcing the eardrum to vibrate. In an extreme case such as a sonic boom, the energy can be sufficient to cause damage to windows and buildings. The amount of energy transported per second by a sound wave is called the **power** of the wave and is measured in SI units of joules per second (J/s) or watts (W).

When a sound wave leaves a source, such as the loudspeaker in **Figure 16.21**, the power spreads out and passes through imaginary surfaces that have increasingly larger areas. For instance, the same sound power passes through the surfaces labeled 1 and 2 in the drawing. However, the power is spread out over a greater area in surface 2. We will bring together the ideas of sound power and the area through which the power passes and, in the process, will formulate the concept of sound intensity. The idea of wave intensity is not confined to sound waves. It will recur, for example, in Chapter 24 when we discuss another important type of waves, electromagnetic waves.

The **sound intensity** I is defined as the sound power P that passes perpendicularly through a surface divided by the area A of that surface:

$$I = \frac{P}{A}$$ (16.8)

FIGURE 16.21 The power carried by a sound wave spreads out after leaving a source, such as a loudspeaker. Thus, the power passes perpendicularly through surface 1 and then through surface 2, which has the larger area.

The unit of sound intensity is power per unit area, or W/m^2. The next example illustrates how the sound intensity changes as the distance from a loudspeaker changes.

EXAMPLE 6 | Sound Intensities

In **Figure 16.21**, 12×10^{-5} W of sound power passes perpendicularly through the surfaces labeled 1 and 2. These surfaces have areas of $A_1 = 4.0$ m^2 and $A_2 = 12$ m^2. Determine the sound intensity at each surface and discuss why listener 2 hears a quieter sound than listener 1.

Reasoning The sound intensity I is the sound power P passing perpendicularly through a surface divided by the area A of that surface. Since the same sound power passes through both surfaces and surface 2 has the greater area, the sound intensity is less at surface 2.

> **Problem-Solving Insight** Sound intensity I and sound power P are different concepts. They are related, however, since intensity equals power per unit area.

Solution The sound intensity at each surface follows from Equation 16.8:

Surface 1
$$I_1 = \frac{P}{A_1} = \frac{12 \times 10^{-5}\,\text{W}}{4.0\,\text{m}^2} = \boxed{3.0 \times 10^{-5}\,\text{W/m}^2}$$

Surface 2
$$I_2 = \frac{P}{A_2} = \frac{12 \times 10^{-5}\,\text{W}}{12\,\text{m}^2} = \boxed{1.0 \times 10^{-5}\,\text{W/m}^2}$$

The sound intensity is less at the more distant surface, where the same power passes through a threefold greater area. The ear of a listener, with its fixed area, intercepts less power where the intensity, or power per unit area, is smaller. Thus, listener 2 intercepts less of the sound power than listener 1. With less power striking the ear, the sound is quieter.

Sound source at center of sphere

FIGURE 16.22 The sound source at the center of the sphere emits sound uniformly in all directions. In this drawing, only a hemisphere is shown for clarity.

For a 1000-Hz tone, the smallest sound intensity that the human ear can detect is about 1×10^{-12} W/m^2; this intensity is called the **threshold of hearing**. On the other extreme, continuous exposure to intensities greater than 1 W/m^2 can be painful and can result in permanent hearing damage. The human ear is remarkable for the wide range of intensities to which it is sensitive.

If a source emits sound *uniformly in all directions,* the intensity depends on distance in a simple way. **Figure 16.22** shows such a source at the center of an imaginary sphere (for clarity only a hemisphere is shown). The radius of the sphere is r. Since all the radiated power P passes through the spherical surface of area $A = 4\pi r^2$, the intensity at a distance r is

Spherically uniform radiation
$$I = \frac{P}{4\pi r^2} \qquad (16.9)$$

From this we see that the intensity of a source that radiates sound uniformly in all directions varies as $1/r^2$. For example, if the distance increases by a factor of two, the sound intensity decreases by a factor of $2^2 = 4$. Example 7 illustrates the effect of the $1/r^2$ dependence of intensity on distance.

EXAMPLE 7 | Fireworks

During a fireworks display, a rocket explodes high in the air above the observers. Assume that the sound spreads out uniformly in all directions and that reflections from the ground can be ignored. When the sound reaches listener 2 in **Figure 16.23**, who is $r_2 = 640$ m away from the explosion, the sound has an intensity of $I_2 = 0.10$ W/m^2. What is the sound intensity detected by listener 1, who is $r_1 = 160$ m away from the explosion?

Reasoning Listener 1 is four times closer to the explosion than listener 2. Therefore, the sound intensity detected by listener 1 is $4^2 = 16$ times greater than the sound intensity detected by listener 2.

> **Problem-Solving Insight** Equation 16.9 can be used only when the sound spreads out uniformly in all directions and there are no reflections of the sound waves.

FIGURE 16.23 If an explosion in a fireworks display radiates sound uniformly in all directions, the intensity at any distance r is $I = P/(4\pi r^2)$, where P is the sound power of the explosion.

Solution The ratio of the sound intensities can be found using Equation 16.9:

$$\frac{I_1}{I_2} = \frac{\dfrac{P}{4\pi r_1^2}}{\dfrac{P}{4\pi r_2^2}} = \frac{r_2^2}{r_1^2} = \frac{(640\text{ m})^2}{(160\text{ m})^2} = 16$$

As a result, $I_1 = (16)I_2 = (16)(0.10\text{ W/m}^2) = \boxed{1.6\text{ W/m}^2}$.

Equation 16.9 is valid only when no walls, ceilings, floors, etc. are present to reflect the sound and cause it to pass through the same surface more than once. Conceptual Example 8 demonstrates why this is so. Example 9 illustrates the application of Equation 16.9 to analyze the limitations of echolocation used by bats to locate prey.

CONCEPTUAL EXAMPLE 8 | Reflected Sound and Sound Intensity

Suppose that the person singing in the shower in **Figure 16.24** produces a sound power P. Sound reflects from the surrounding shower stall. At a distance r in front of the person, does the expression $I = P/(4\pi r^2)$ (Equation 16.9) **(a)** overestimate, **(b)** underestimate, or **(c)** give the correct total sound intensity?

Reasoning In arriving at Equation 16.9, it was assumed that the sound spreads out uniformly from the source and passes only once through the imaginary surface that surrounds it (see **Figure 16.22**). In **Figure 16.24**, only part of this imaginary surface (colored blue) is shown, but nonetheless, if Equation 16.9 is to apply, the same assumption must hold.

Answers (a) and (c) are incorrect. Equation 16.9 cannot overestimate the sound intensity, because it assumes that the sound passes through the imaginary surface only once and, hence, does not take into account the reflected sound within the shower stall. For the same reason, neither can Equation 16.9 give the correct sound intensity.

Answer (b) is correct. Figure 16.24 illustrates three paths by which the sound passes through the imaginary surface. The "direct" sound travels along a path from its source directly to the surface. It is the intensity of this sound that is given by $I = P/(4\pi r^2)$. The remaining paths are two of the many that characterize the sound reflected from the shower stall. The *total* sound

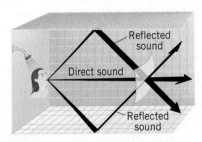

FIGURE 16.24 When someone sings in the shower, the sound power passing through part of an imaginary spherical surface (shown in blue) is the sum of the direct sound power and the reflected sound power.

power that passes through the surface is the sum of the direct and reflected powers. Thus the total sound intensity at a distance r from the source is greater than the intensity of the direct sound alone, so Equation 16.9 underestimates the sound intensity from the singing. People like to sing in the shower because their voices sound so much louder due to the enhanced intensity caused by the reflected sound.

Related Homework: Problem 61

EXAMPLE 9 | BIO Ultrasonic Waves and the Echolocation of Bats

A bat like the one shown in **Figure 16.15** emits ultrasonic pulses through its larynx at frequencies up to 120 kHz. The emitted sound waves scatter from objects and then "echo" back to the bat, whose brain then analyzes the time of travel, frequency, and intensity of the echo in order to echolocate obstacles and prey. The properties of the emitted sound, and the sensitivity of the bat's ears, set the size and distance limits of what the bat can detect. For instance, only objects of linear physical dimensions greater than or equal to the wavelength of the sound can be detected. Also, if the cross-sectional area of the target object is too small, the intensity of the echo will be undetectable by the bat's ears. Consider a stationary bat that emits a 65.0-kHz sound wave of power $P_0 = 1.90 \times 10^{-5}$ W, measured at a point just in front of the bat's mouth. At this frequency, the lowest intensity the bat can detect is 1.20×10^{-12} W/m². At a distance $r = 1.50$ m, what is the smallest diameter D of a spherical object that the bat can detect based on **(a)** the frequency f and **(b)** the intensity I_0 of the sound waves the bat emits? **(c)** Is it the wavelength or the intensity of the sound the bat emits that limits the size of an object than can be detected at a distance of 1.50 m? Assume that the speed of sound in air is 343 m/s, and that all of the power striking the object is reflected from it, that is, no energy is absorbed.

Reasoning **(a)** Knowing the speed of sound in air, and the frequency of the waves, the wavelength can be determined using Equation 16.1, $v = f\lambda$. The wavelength sets the lower size limit of what can be detected, and the wavelength will be the minimum diameter of a spherical object that can be detected based on the frequency of the waves. **(b)** Assuming that the pulse emitted by the bat is a spherical wave, we can calculate the intensity of the emitted pulse at the location of the object using Equation 16.9. We can then calculate the reflected power in terms of the unknown cross-sectional area of the object, and then the intensity of this echo at the bat's location, again using Equation 16.9. Finally, we can determine the cross-sectional area of the target that delivers the minimum detectable power to the bat and, finally, solve for the diameter of that circular area. **(c)** Comparing the results of (a) and (b) will reveal whether it is the emitted frequency or emitted intensity that limits the minimum size of an object that the bat can detect.

Solution **(a)** The wavelength λ of the emitted sound waves will be equal to the minimum diameter (D) of the spherical object that the bat can detect. Solving Equation 16.1 for the wavelength, we have

$$\lambda = \frac{v}{f} = \frac{343 \text{ m/s}}{65.0 \times 10^3 \text{ 1/s}} = \boxed{5.28 \times 10^{-3} \text{ m}}$$

Therefore, for the given frequency, D can be no smaller than 5.28×10^{-3} m.

FIGURE 16.15 (REPEATED) Some bats use ultrasonic sound waves for locating prey and for navigating. This bat has captured a katydid.

MerlinTuttle.org/Science Source

(b) Assuming that the sound waves emitted by the bat behave like spherical waves, following Equation 16.9, the intensity I_r at the location of the target object a distance r from the bat is given by

$$I_r = \frac{P_0}{4\pi r^2}$$

where P_0 is the power emitted by the bat. From this equation, it can be seen that the intensity of the wave decreases as the distance r between the source (the bat) and the object increases, as expected. The power reflected by the object will be the incident intensity I_r impinging upon the object multiplied by the cross-sectional area A of the object, or

$$P_r = I_r A = \left(\frac{P_0}{4\pi r^2}\right) A \qquad (1)$$

Assuming that the sound waves reflected from the object propagate as spherical waves, the intensity of the reflected waves will also diminish as they travel back to the bat. Therefore, Equation 16.9 must be applied again, this time treating the reflection as the source. The final intensity I_F detected by the bat is then given by

$$I_F = \frac{P_r}{4\pi r^2}$$

Substituting Equation 1 for P_r, we have

$$I_F = \frac{P_r}{4\pi r^2} = \frac{P_0 A}{(4\pi r^2)^2} = \frac{P_0 \pi \left(\frac{D}{2}\right)^2}{(4\pi r^2)^2}$$

where D is the diameter of the spherical object. Solving for D, we have

$$D = \left(\frac{64\pi r^4 I_F}{P_0}\right)^{\frac{1}{2}} = \left[\frac{64\pi (1.50 \text{ m})^4 (1.20 \times 10^{-12} \text{ W/m}^2)}{1.90 \times 10^{-5} \text{ W}}\right]^{\frac{1}{2}}$$

$$= \boxed{8.02 \times 10^{-3} \text{ m}}$$

(c) Since the minimum size of an object that can be detected by the bat at a distance of 1.50 m is limited to $D = 5.28 \times 10^{-3}$ m due to the wavelength of the sound, and to $D = 8.02 \times 10^{-3}$ m due to the emitted intensity, it is the **emitted intensity** that limits the minimum size of an object that the bat can detect.

Check Your Understanding

(The answers are given at the end of the book.)

14. BIO Some animals rely on an acute sense of hearing for survival, and the visible parts of the ears on such animals are often relatively large. How does this anatomical feature help to increase the sensitivity of the animal's hearing for low-intensity sounds?

15. A source is emitting sound uniformly in all directions. There are no reflections anywhere. A *flat* surface faces the source. Is the sound intensity the same at all points on the surface?

16.8 | Decibels

The **decibel** (dB) is a measurement unit used when comparing two sound intensities. The simplest method of comparison would be to compute the ratio of the intensities. For instance, we could compare $I = 8 \times 10^{-12}$ W/m² to $I_0 = 1 \times 10^{-12}$ W/m² by computing $I/I_0 = 8$ and stating that I is eight times as great as I_0. However, because of the way in which the human hearing mechanism responds to intensity, it is more appropriate to use a logarithmic scale for the comparison. For this purpose, the **intensity level** β (expressed in decibels) is defined as follows:

$$\beta = (10 \text{ dB}) \log\left(\frac{I}{I_0}\right) \qquad (16.10)$$

where "log" denotes the common logarithm to the base ten. I_0 is the intensity of the reference level to which I is being compared and is sometimes the threshold of hearing; that is, $I_0 = 1.00 \times 10^{-12}$ W/m². With the aid of a calculator, the intensity level can be evaluated for the values of I and I_0 given above:

$$\beta = (10 \text{ dB}) \log\left(\frac{8 \times 10^{-12} \text{ W/m}^2}{1 \times 10^{-12} \text{ W/m}^2}\right) = (10 \text{ dB}) \log 8 = (10 \text{ dB})(0.9) = 9 \text{ dB}$$

This result indicates that I is 9 decibels greater than I_0. Although β is called the "intensity level," it is *not* an intensity and does *not* have intensity units of W/m². In fact, the decibel, like the radian, is dimensionless.

Notice that if both I and I_0 are at the threshold of hearing, then $I = I_0$, and the intensity level is 0 dB according to Equation 16.10:

$$\beta = (10 \text{ dB}) \log\left(\frac{I_0}{I_0}\right) = (10 \text{ dB}) \log 1 = 0 \text{ dB}$$

since log 1 = 0. Thus,

> **Problem-Solving Insight**   An intensity level of zero decibels does not mean that the sound intensity I is zero; it means that $I = I_0$.

Intensity levels can be measured with a sound-level meter, such as the one in **Figure 16.25**. The intensity level β is displayed on its scale, assuming that the threshold of hearing is 0 dB. **Table 16.2** lists the intensities I and the associated intensity levels β for some common sounds, using the threshold of hearing as the reference level.

When a sound wave reaches a listener's ear, the sound is interpreted by the brain as loud or soft, depending on the intensity of the wave. Greater intensities give rise to

> **Math Skills**   In Equation 16.10, $\log\left(\frac{I}{I_0}\right)$ refers to the common logarithm and **not** the natural logarithm, which is $\ln\left(\frac{I}{I_0}\right)$. Since most calculators provide single-button access to both types of logarithm, **be careful to hit the "log" button and not the "ln" button when you evaluate the logarithm.** For example, if $\frac{I}{I_0} = 10.00$, it follows that log (10.00) = 1.000. But if you mistakenly hit the "ln" button, your calculator will display 2.303, and the value that you determine for the intensity level β in Equation 16.10 will be incorrect. When using Equation 16.10, also note that the ratio $\frac{I}{I_0}$ cannot be a negative number, since the logarithm of a negative number is not defined.

FIGURE 16.25   A sound-level meter and a close-up view of its decibel scale.

TABLE 16.2	Typical Sound Intensities and Intensity Levels Relative to the Threshold of Hearing	
	Intensity I (W/m²)	**Intensity Level β (dB)**
Threshold of hearing	1.0×10^{-12}	0
Rustling leaves	1.0×10^{-11}	10
Whisper	1.0×10^{-10}	20
Normal conversation (1 meter)	3.2×10^{-6}	65
Inside car in city traffic	1.0×10^{-4}	80
Car without muffler	1.0×10^{-2}	100
Live rock concert	1.0	120
Threshold of pain	10	130

louder sounds. However, the relation between intensity and loudness is not a simple proportionality, because doubling the intensity does *not* double the loudness, as we will now see.

Suppose you are sitting in front of a stereo system that is producing an intensity level of 90 dB. If the volume control on the amplifier is turned up slightly to produce a 91-dB level, you would just barely notice the change in loudness. ***Hearing tests have revealed that a one-decibel (1-dB) change in the intensity level corresponds to approximately the smallest change in loudness that an average listener with normal hearing can detect.*** Since 1 dB is the smallest perceivable increment in loudness, a change of 3 dB—say, from 90 to 93 dB—is still a rather small change in loudness. Example 10 determines the factor by which the sound intensity must be increased to achieve such a change.

EXAMPLE 10 | Comparing Sound Intensities

Audio system 1 produces an intensity level of $\beta_1 = 90.0$ dB, and system 2 produces an intensity level of $\beta_2 = 93.0$ dB. The corresponding intensities (in W/m^2) are I_1 and I_2. Determine the ratio I_2/I_1.

Reasoning Intensity levels are related to intensities by logarithms (see Equation 16.10), and it is a property of logarithms (see Appendix D) that $\log A - \log B = \log (A/B)$. Subtracting the two intensity levels and using this property, we find that

$$\beta_2 - \beta_1 = (10\ \text{dB}) \log \left(\frac{I_2}{I_0}\right) - (10\ \text{dB}) \log \left(\frac{I_1}{I_0}\right) = (10\ \text{dB}) \log \left(\frac{I_2/I_0}{I_1/I_0}\right)$$

$$= (10\ \text{dB}) \log \left(\frac{I_2}{I_1}\right)$$

Solution Using the result just obtained, we find

$$93.0\ \text{dB} - 90.0\ \text{dB} = (10\ \text{dB}) \log \left(\frac{I_2}{I_1}\right)$$

$$0.30 = \log \left(\frac{I_2}{I_1}\right) \quad \text{or} \quad \frac{I_2}{I_1} = 10^{0.30} = \boxed{2.0}$$

Doubling the intensity changes the loudness by only a small amount (3 dB) and does not double it, so there is no simple proportionality between intensity and loudness.

FIGURE 16.26 In spite of its tenfold greater power, the 200-watt audio system has only about double the loudness of the 20-watt system, when both are set for maximum volume.

To double the loudness of a sound, the intensity must be increased by more than a factor of two. ***Experiment shows that if the intensity level increases by 10 dB, the new sound seems approximately twice as loud as the original sound.*** For instance, a 70-dB intensity level sounds about twice as loud as a 60-dB level, and an 80-dB intensity level sounds about twice as loud as a 70-dB level. The factor by which the sound intensity must be increased to double the loudness can be determined as in Example 10:

$$\beta_2 - \beta_1 = 10.0\ \text{dB} = (10\ \text{dB}) \left[\log \left(\frac{I_2}{I_0}\right) - \log \left(\frac{I_1}{I_0}\right)\right]$$

Solving this equation as in Example 10 reveals that $I_2/I_1 = 10.0$. Thus, increasing the sound intensity by a factor of ten will double the perceived loudness. Consequently, with both audio systems in **Figure 16.26** set at maximum volume, the 200-watt system will sound only twice as loud as the much cheaper 20-watt system.

Check Your Understanding

(The answers are given at the end of the book.)

16. If two people talk simultaneously and each creates an intensity level of 65 dB at a certain point, does the total intensity level at this point equal 130 dB?

17. Two observation points are located at distances r_1 and r_2 from a source of sound. The sound spreads out uniformly from the source, and there are no reflecting surfaces in the environment. The sound intensity level at distance r_2 is 6 dB less than the level at distance r_1. **(a)** What is the ratio I_2/I_1 of the sound intensities at the two distances? **(b)** What is the ratio r_2/r_1 of the distances?

16.9 The Doppler Effect

Have you ever heard an approaching fire truck and noticed the distinct change in the sound of the siren as the truck passes? The effect is similar to what you get when you put together the two syllables "eee" and "yow" to produce "eee-yow." While the truck approaches, the pitch of the siren is relatively high ("eee"), but as the truck passes and moves away, the pitch suddenly drops ("yow"). Something similar, but less familiar, occurs when an observer moves toward or away from a stationary source of sound. Such phenomena were first identified in 1842 by the Austrian physicist Christian Doppler (1803–1853) and are collectively referred to as the Doppler effect.

To explain why the Doppler effect occurs, we will bring together concepts that we have discussed previously—namely, the velocity of an object and the wavelength and frequency of a sound wave (Section 16.5). We will combine the effects of the velocities of the source and observer of the sound with the definitions of wavelength and frequency. In so doing, we will see that the **Doppler effect** is the change in frequency or pitch of the sound detected by an observer because the sound source and the observer have different velocities with respect to the medium of sound propagation.

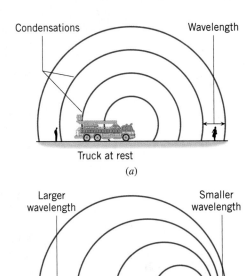

ANIMATED FIGURE 16.27 (a) When the truck is stationary, the wavelength of the sound is the same in front of and behind the truck. (b) When the truck is moving, the wavelength in front of the truck becomes smaller, while the wavelength behind the truck becomes larger.

Moving Source

To see how the Doppler effect arises, consider the sound emitted by a siren on the stationary fire truck in **Animated Figure 16.27a**. Like the truck, the air is assumed to be stationary with respect to the earth. Each solid blue arc in the drawing represents a condensation of the sound wave. Since the sound pattern is symmetrical, listeners standing in front of or behind the truck detect the same number of condensations per second and, consequently, hear the same frequency. Once the truck begins to move, the situation changes, as part b of the picture illustrates. Ahead of the truck, the condensations are now closer together, resulting in a decrease in the wavelength of the sound. This "bunching-up" occurs because the moving truck "gains ground" on a previously emitted condensation before emitting the next one. Since the condensations are closer together, the observer standing in front of the truck senses more of them arriving per second than she does when the truck is stationary. The increased rate of arrival corresponds to a greater sound frequency, which the observer hears as a higher pitch. Behind the moving truck, the condensations are farther apart than they are when the truck is stationary. This increase in the wavelength occurs because the truck pulls away from condensations emitted toward the rear. Consequently, fewer condensations per second arrive at the ear of an observer behind the truck, corresponding to a smaller sound frequency or lower pitch.

If the stationary siren in **Animated Figure 16.27a** emits a condensation at the time $t = 0$ s, it will emit the next one at time T, where T is the period of the wave. The distance between these two condensations is the wavelength λ of the sound produced by the stationary source, as **Figure 16.28a** indicates. When the truck is moving with a speed v_s (the subscript "s" stands for the "source" of sound) toward a stationary observer, the siren also emits condensations at $t = 0$ s and at time T. However, prior to emitting the second condensation, the truck moves closer to the observer by a distance $v_s T$, as **Figure 16.28b** shows. As a result, the distance between successive condensations is no longer the wavelength λ created by the stationary siren, but, rather, a wavelength λ' that is shortened by the amount $v_s T$:

$$\lambda' = \lambda - v_s T$$

Let's denote the frequency perceived by the stationary observer as f_o, where the subscript "o" stands for "observer." According to Equation 16.1, f_o is equal to the speed of sound v divided by the shortened wavelength λ':

$$f_o = \frac{v}{\lambda'} = \frac{v}{\lambda - v_s T}$$

But for the stationary siren, we have $\lambda = v/f_s$ and $T = 1/f_s$, where f_s is the frequency of the sound emitted by the source (not the frequency f_o perceived by the observer). With

FIGURE 16.28 (a) When the fire truck is stationary, the distance between successive condensations is one wavelength λ. (b) When the truck moves with a speed v_s, the wavelength of the sound in front of the truck is shortened to λ'.

the aid of these substitutions for λ and T, the expression for f_o can be arranged to give the following result:

Source moving toward
stationary observer
$$f_o = f_s\left(\frac{1}{1 - \frac{v_s}{v}}\right)$$
(16.11)

Since the term $1 - v_s/v$ is in the denominator in Equation 16.11 and is less than one, the frequency f_o heard by the observer is *greater* than the frequency f_s emitted by the source. The difference between these two frequencies, $f_o - f_s$, is called the **Doppler shift**, and its magnitude depends on the ratio of the speed of the source v_s to the speed of sound v.

When the siren moves away from, rather than toward, the observer, the wavelength λ' becomes *greater* than λ according to

$$\lambda' = \lambda + v_s T$$

Notice the presence of the "+" sign in this equation, in contrast to the "−" sign that appeared earlier. The same reasoning that led to Equation 16.11 can be used to obtain an expression for the observed frequency f_o:

Source moving away from
stationary observer
$$f_o = f_s\left(\frac{1}{1 + \frac{v_s}{v}}\right)$$
(16.12)

The denominator $1 + v_s/v$ in Equation 16.12 is greater than one, so the frequency f_o heard by the observer is *less* than the frequency f_s emitted by the source. The next example illustrates how large the Doppler shift is in a familiar situation.

EXAMPLE 11 | The Sound of a Passing Train

A high-speed train is traveling at a speed of 44.7 m/s (100 mi/h) when the engineer sounds the 415-Hz warning horn. The speed of sound is 343 m/s. What are the frequency and wavelength of the sound, as perceived by a person standing at a crossing, when the train is **(a)** approaching and **(b)** leaving the crossing?

Reasoning When the train approaches, the person at the crossing hears a sound whose frequency is greater than 415 Hz because of the Doppler effect. As the train moves away, the person hears a frequency that is less than 415 Hz. We may use Equations 16.11 and 16.12, respectively, to determine these frequencies. In either case, the observed wavelength can be obtained according to Equation 16.1 as the speed of sound divided by the observed frequency.

Solution **(a)** When the train approaches, the observed frequency is

$$f_o = f_s\left(\frac{1}{1 - \frac{v_s}{v}}\right)$$

$$= (415 \text{ Hz})\left(\frac{1}{1 - \frac{44.7 \text{ m/s}}{343 \text{ m/s}}}\right) = \boxed{477 \text{ Hz}}$$
(16.11)

The observed wavelength is

$$\lambda' = \frac{v}{f_o} = \frac{343 \text{ m/s}}{477 \text{ Hz}} = \boxed{0.719 \text{ m}}$$
(16.1)

(b) When the train leaves the crossing, the observed frequency is

$$f_o = f_s\left(\frac{1}{1 + \frac{v_s}{v}}\right)$$

$$= (415 \text{ Hz})\left(\frac{1}{1 + \frac{44.7 \text{ m/s}}{343 \text{ m/s}}}\right) = \boxed{367 \text{ Hz}}$$
(16.12)

In this case, the observed wavelength is

$$\lambda' = \frac{v}{f_o} = \frac{343 \text{ m/s}}{367 \text{ Hz}} = \boxed{0.935 \text{ m}}$$
(16.1)

Moving Observer

Figure 16.29 shows how the Doppler effect arises when the sound source is stationary and the observer moves. As in the case of the moving source, we assume that the air is stationary. The observer moves with a speed v_o ("o" stands for "observer") toward the stationary source and covers a distance $v_o t$ in a time t. During this time, the moving observer encounters all the condensations that he would if he were stationary, *plus an additional number*. The additional number of condensations encountered is the distance $v_o t$ divided by the distance λ between successive condensations, or $v_o t/\lambda$. Thus, the additional number of condensations encountered per second by the moving observer is v_0/λ. Since a stationary observer would hear a frequency f_s emitted by the source, the moving observer hears a greater frequency f_o given by

$$f_o = f_s + \frac{v_0}{\lambda} = f_s\left(1 + \frac{v_0}{f_s\lambda}\right)$$

Using the fact that $v = f_s\lambda$, where v is the speed of sound, we find that

Observer moving toward $$f_o = f_s\left(1 + \frac{v_0}{v}\right) \qquad\qquad \textbf{(16.13)}$$
stationary source

An observer moving *away from* a stationary source moves in the same direction as the sound wave and, as a result, intercepts *fewer* condensations per second than a stationary observer does. In this case, the moving observer hears a smaller frequency f_o that is given by

Observer moving away $$f_o = f_s\left(1 - \frac{v_0}{v}\right) \qquad\qquad \textbf{(16.14)}$$
from stationary source

It should be noted that the physical mechanism producing the Doppler effect is different when the source moves and the observer is stationary from when the observer moves and the source is stationary. When the source moves, as in **Figure 16.28b**, the wavelength of the sound perceived by the observer changes from λ to λ'. When the wavelength changes, the stationary observer hears a different frequency f_o than the frequency produced by the source. On the other hand, when the observer moves and the source is stationary, the *wavelength λ does not change* (see, for example, **Figure 16.29**). Instead, the moving observer intercepts a different number of wave condensations per second than does a stationary observer and therefore detects a different frequency f_o.

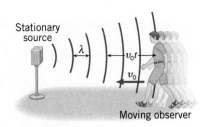

Stationary source

Moving observer

FIGURE 16.29 An observer moving with a speed v_0 toward the stationary source intercepts more wave condensations per unit of time than does a stationary observer.

General Case

It is possible for *both* the sound source and the observer to move with respect to the medium of sound propagation. If the medium is stationary, Equations 16.11–16.14 may be combined to give the observed frequency f_o as

Source and observer $$f_o = f_s\left(\frac{1 \pm \frac{v_0}{v}}{1 \mp \frac{v_s}{v}}\right) \qquad\qquad \textbf{(16.15)}$$
both moving

In the numerator, the plus sign applies when the observer moves toward the source, and the minus sign applies when the observer moves away from the source. In the denominator, the minus sign is used when the source moves toward the observer, and the plus sign is used when the source moves away from the observer. The symbols v_0, v_s, and v denote numbers without an algebraic sign because the direction of travel has been taken into account by the plus and minus signs that appear directly in this equation.

NEXRAD

THE PHYSICS OF . . . NEXRAD. NEXRAD stands for **Nex**t Generation Weather **Rad**ar and is a nationwide system used by the National Weather Service to provide dramatically improved early warning of severe storms, such as the tornado in **Figure 16.30**. The system is based on radar waves, which are a type of electromagnetic wave (see Chapter 24) and, like sound waves, can exhibit the Doppler effect. The Doppler effect is at the heart of NEXRAD. As the drawing illustrates, a tornado is a swirling mass of air and water droplets. Radar pulses are sent out by a NEXRAD unit, whose protective covering is shaped like a soccer ball. The waves reflect from the water droplets and return to the unit, where the frequency is observed and compared to the outgoing frequency. For instance, droplets at point A in the drawing are moving toward the unit, and the radar waves reflected from them have their frequency Doppler-shifted to higher values. Droplets at point B, however, are moving away from the unit, and the frequency of the waves reflected from these droplets is Doppler-shifted to lower values. Computer processing of the Doppler frequency shifts leads to color-enhanced views on display screens (see

Wayne Hanna / Design Pics/Getty Images

(a)

(b)

FIGURE 16.30 (a) A tornado is one of nature's most dangerous storms. (b) The National Weather Service uses the NEXRAD system, which is based on Doppler-shifted radar, to identify the storms that are likely to spawn tornadoes.

Courtesy Kurt Hondl, National Severe Storms Laboratory, Norman, OK

FIGURE 16.31 This color-enhanced view of a tornado shows circulating winds detected by a NEXRAD station, which is located below and to the right of the figure. The white dot and arrow indicate the storm center and the direction of wind circulation.

Figure 16.31). These views reveal the direction and magnitude of the wind velocity and can identify, from distances up to 140 mi, the swirling air masses that are likely to spawn tornadoes. The equations that specify the Doppler frequency shifts of radar waves are different from those given for sound waves by Equations 16.11–16.15. The reason for the difference is that radar waves propagate from one place to another by a different mechanism than that of sound waves (see Section 24.5).

Check Your Understanding

(The answers are given at the end of the book.)

18. At a swimming pool, a music fan up on a diving platform is listening to a radio. As the radio is playing a tone that has a constant frequency f_s, it is accidentally knocked off the platform. Describe the Doppler effect heard by **(a)** the person on the platform and **(b)** a person down below in the water. In each case, state whether the observed frequency f_o is greater or smaller than f_s and describe how f_o changes (if it changes) as the radio falls.

19. When a car is at rest, its horn emits a frequency of 600 Hz. A person standing in the middle of the street with this car behind him hears the horn with a frequency of 580 Hz. Does he need to jump out of the way?

20. A source of sound produces the same frequency under water as it does in air. This source has the same velocity in air as it does under water. Consider the ratio f_o/f_s of the observed frequency f_o to the source frequency f_s. Is this ratio greater in air or under water when the source **(a)** approaches and **(b)** moves away from the observer?

21. Two cars, one behind the other, are traveling in the same direction at the same speed. Does either driver hear the other's horn at a frequency that is different from the frequency heard when both cars are at rest?

22. When a truck is stationary, its horn produces a frequency of 500 Hz. You are driving your car, and this truck is following behind. You hear its horn at a frequency of 520 Hz. **(a)** Refer to Equation 16.15 and decide which algebraic sign should be used in the numerator and which in the denominator. **(b)** Which driver, if either, is driving faster?

16.10 Applications of Sound in Medicine

BIO **THE PHYSICS OF . . . ultrasonic imaging.** When ultrasonic waves are used in medicine for diagnostic purposes, high-frequency sound pulses are produced by a transmitter and directed into the body. As in sonar, reflections occur. They occur each time a pulse encounters a boundary between two tissues that have different densities or a boundary between a tissue and the adjacent fluid. By scanning ultrasonic waves across the body and detecting the echoes generated from various internal locations, it is possible to obtain an image, or sonogram, of the inner anatomy. Ultrasonic imaging is employed extensively in obstetrics to examine the developing fetus (**Figure 16.32**). The fetus, surrounded by the amniotic sac, can be distinguished from other anatomical features so that fetal size, position, and possible abnormalities can be detected. Ultrasound is also used in other medically related areas. For instance, tumors in the liver, kidney, brain, and pancreas can be detected with ultrasound. Yet another application involves monitoring the real-time movement of pulsating structures, such as heart valves ("echocardiography") and large blood vessels.

FIGURE 16.32 An ultrasonic scanner can be used to produce an image of the fetus as it develops in the uterus. Conventional scanning produces two-dimensional images. Three-dimensional scanning uses computer technology to produce images that are more detailed than the 2-D variety. The image on the right, for example, was obtained using 3-D scanning in the second trimester of pregnancy and illustrates the kind of detail that is achievable.

When ultrasound is used to form images of internal anatomical features or foreign objects in the body, the wavelength of the sound wave must be about the same size as, or smaller than, the object to be located. Therefore, high frequencies in the range from 1 to 15 MHz (1 MHz = 1 megahertz = 1×10^6 Hz) are the norm. For instance, the wavelength of 5-MHz ultrasound is $\lambda = v/f = 0.3$ mm, if a value of 1540 m/s is used for the speed of sound through tissue. A sound wave with a frequency higher than 5 MHz and a correspondingly shorter wavelength is required for locating objects smaller than 0.3 mm.

BIO THE PHYSICS OF . . . the cavitron ultrasonic surgical aspirator.   Neuro-surgeons use a device called a **c**avitron **u**ltrasonic **s**urgical **a**spirator (CUSA) to remove brain tumors once thought to be inoperable. Ultrasonic sound waves cause the slender tip of the CUSA probe (see **Figure 16.33**) to vibrate at approximately 23 kHz. The probe shatters any section of the tumor that it touches, and the fragments are flushed out of the brain with a saline solution. Because the tip of the probe is small, the surgeon can selectively remove small bits of malignant tissue without damaging the surrounding healthy tissue.

BIO THE PHYSICS OF . . . bloodless surgery with HIFU.   Another application of ultrasound is in a new type of bloodless surgery, which can eliminate abnormal cells, such as those in benign hyperplasia of the prostate gland. This technique is known as HIFU (**h**igh-**i**ntensity **f**ocused **u**ltrasound). It is analogous to focusing the sun's electromagnetic waves by using a magnifying glass and producing a small region where the energy carried by the waves can cause localized heating. Ultrasonic waves can be used in a similar fashion. The waves enter directly through the skin and come into focus inside the body over a region that is sufficiently well defined to be surgically useful. Within this region the energy of the waves causes localized heating, leading to a temperature of about 56 °C (normal body temperature is 37 °C), which is sufficient to kill abnormal cells. The killed cells are eventually removed by the body's natural processes.

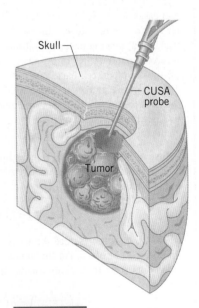

FIGURE 16.33 Neurosurgeons use a cavitron ultrasonic surgical aspirator (CUSA) to "cut out" brain tumors without adversely affecting the surrounding healthy tissue.

Transmitter Receiver

Skin

Incident
sound

Reflected
sound

v_s

Red blood cell

FIGURE 16.34 A Doppler flow meter measures the speed of red blood cells.

BIO **THE PHYSICS OF . . . the Doppler flow meter.** The Doppler flow meter is a particularly interesting medical application of the Doppler effect. This device measures the speed of blood flow, using transmitting and receiving elements that are placed directly on the skin, as in **Figure 16.34**. The transmitter emits a continuous sound whose frequency is typically about 5 MHz. When the sound is reflected from the red blood cells, its frequency is changed in a kind of Doppler effect because the cells are moving. The receiving element detects the reflected sound, and an electronic counter measures its frequency. From the change in frequency the speed of the blood flow can be determined. Typically, the change in frequency is around 600 Hz for flow speeds of about 0.1 m/s. The Doppler flow meter can be used to locate regions where blood vessels have narrowed, since greater flow speeds occur in the narrowed regions, according to the equation of continuity (see Section 11.8). In addition, the Doppler flow meter can be used to detect the motion of a fetal heart as early as 8–10 weeks after conception.

EXAMPLE 12 | **BIO** The Physics of Color Doppler Ultrasound

As discussed in Section 16.10, the technique of medical ultrasound can be used to image many different internal structures of the body. When combined with the Doppler effect, the technique of Color Doppler Ultrasound can provide information on the velocity of fluids, such as blood, moving in these structures. The Doppler information is converted by a computer into a false-color image that is overlaid on the ultrasound image of the structure, such as a blood vessel. The colors indicate the speed and direction of the blood flow (**Figure 16.35**). As an example, consider an ultrasound image created with sound waves at a frequency of 5.00 MHz. The sound waves travel at a speed of 1540 m/s through the body and reflect off of red blood cells flowing toward the transmitter. A sound receiver, also located in the transmitter, detects the reflected sound waves with a frequency that is 980 Hz higher than the sound sourced by the transmitter. Calculate the speed of the blood moving toward the transmitter.

Reasoning This is a straightforward application of the Doppler effect equations for sound, where we have to take into account two cases. First, as the sound wave is sent from the transmitter toward the red blood cells, we have the case of the observer (the blood) moving toward a stationary source (the transmitter/receiver), or Equation 16.13. Next, the sound wave is reflected off the red blood cell back toward the receiver. Thus, the blood cell acts as the source, and we have the case of a source moving toward a stationary observer, or Equation 16.11. Therefore, we will need to use both Equation 16.13 and Equation 16.11 to calculate the speed of the blood (v_B).

Solution The sound will be Doppler shifted on the way to the blood cells and also Doppler shifted upon reflection. For the sound traveling from the transmitter to the red blood cells, we use Equation 16.13:

$$f_o = f_s\left(1 + \frac{v_o}{v}\right) = f_s\left(1 + \frac{v_B}{v}\right)$$

where we have represented the speed of the observer (v_o) as v_B, the speed of the blood. For the sound traveling back from the red blood cells to the receiver, we use Equation 16.11:

$$f_o = f_s\left(\frac{1}{1 - \frac{v_s}{v}}\right) = f_s\left(\frac{1}{1 - \frac{v_B}{v}}\right)$$

FIGURE 16.35 Color Doppler ultrasound image showing the red and blue color overlay that provides information on the speed and direction of blood flow within internal body structures. Blue indicates blood flow toward the ultrasonic transducer, and red indicates flow away. This particular image shows increased vascularity in the right fallopian tube of an expecting mother, which, unfortunately, is evidence of an ectopic pregnancy.

Hong Pham, Eugene C. Lin

where now, since the source of the sound is the reflection off of the red blood cells, we replace v_S with v_B. We can combine these two expressions by recognizing that the observed frequency in Equation 16.13 becomes the source frequency in Equation 16.11. Making this substitution, we have:

$$f_o = f_s\left(\frac{1 + \frac{v_B}{v}}{1 - \frac{v_B}{v}}\right)$$

We can now rearrange this equation and solve for v_B:

$$\frac{f_o}{f_s}\left(1 - \frac{v_B}{v}\right) = \left(1 + \frac{v_B}{v}\right) \Rightarrow v_B = \frac{v(f_o - f_s)}{(f_o + f_s)} = \frac{(1540 \text{ m/s})(980 \text{ Hz})}{(10\,000\,980 \text{ Hz})}$$

$$= \boxed{0.15 \text{ m/s}}$$

16.11 *The Sensitivity of the Human Ear

BIO **THE PHYSICS OF . . . hearing.** Although the ear is capable of detecting sound intensities as small as 1×10^{-12} W/m², it is *not* equally sensitive to all frequencies, as **Figure 16.36** shows. This figure displays a series of graphs known as the **Fletcher–Munson curves**, after H. Fletcher and M. Munson, who first determined them in 1933. In these graphs the audible sound frequencies are plotted on the horizontal axis, and the sound intensity levels (in decibels) are plotted on the vertical axis. Each curve is a *constant loudness* curve because it shows the sound intensity level needed at each frequency to make the sound appear to have the same loudness. For example, the lowest (red) curve represents the threshold of hearing. It shows the intensity levels at which sounds of different frequencies just become audible. The graph indicates that the intensity level of a 100-Hz sound must be about 37 dB greater than the intensity level of a 1000-Hz sound to be at the threshold of hearing. Therefore, the ear is *less sensitive* to a 100-Hz sound than it is to a 1000-Hz sound. In general, **Figure 16.36** reveals that the ear is most sensitive in the range of about 1–5 kHz, and becomes progressively less sensitive at higher and lower frequencies.

Each curve in **Figure 16.36** represents a different loudness, and each is labeled according to its intensity level at 1000 Hz. For instance, the curve labeled "60" represents all sounds that have the same loudness as a 1000-Hz sound whose intensity level is 60 dB. These constant-loudness curves become flatter as the loudness increases, the relative flatness indicating that the ear is nearly equally sensitive to all frequencies when the sound is loud. Thus, when you listen to loud sounds, you hear the low frequencies, the middle frequencies, and the high frequencies about equally well. However, when you listen to quiet sounds, the high and low frequencies seem to be absent, because the ear is relatively insensitive to these frequencies under such conditions.

FIGURE 16.36 Each curve represents the intensity levels at which sounds of various frequencies have the same loudness. The curves are labeled by their intensity levels at 1000 Hz and are known as the Fletcher–Munson curves.

Concept Summary

16.1 The Nature of Waves A wave is a traveling disturbance and carries energy from place to place. In a transverse wave, the disturbance occurs perpendicular to the direction of travel of the wave. In a longitudinal wave, the disturbance occurs parallel to the line along which the wave travels.

16.2 Periodic Waves A periodic wave consists of cycles or patterns that are produced over and over again by the source of the wave. The amplitude of the wave is the maximum excursion of a particle of the medium from the particle's undisturbed position. The wavelength λ is the distance along the length of the wave between two successive equivalent points, such as two crests or two troughs. The period T is the time required for the wave to travel a distance of one wavelength. The frequency f (in hertz) is the number of wave cycles per second that passes an observer and is the reciprocal of the period (in seconds), as shown in Equation 10.5. The speed v of a wave is related to its wavelength and frequency according to Equation 16.1.

$$f = \frac{1}{T} \tag{10.5}$$

$$v = f\lambda \tag{16.1}$$

16.3 The Speed of a Wave on a String The speed of a wave depends on the properties of the medium in which the wave travels. For a transverse wave on a string that has a tension F and a mass per unit length m/L, the wave speed is given by Equation 16.2. The mass per unit length is also called the linear density.

$$v = \sqrt{\frac{F}{m/L}} \tag{16.2}$$

16.4 The Mathematical Description of a Wave When a wave of amplitude A, frequency f, and wavelength λ moves in the $+x$ direction through a medium, the wave causes a displacement y of a particle at position x according to Equation 16.3. For a wave moving in the $-x$ direction, the displacement y is given by Equation 16.4.

$$y = A \sin\left(2\pi ft - \frac{2\pi x}{\lambda}\right) \qquad (16.3)$$

$$y = A \sin\left(2\pi ft + \frac{2\pi x}{\lambda}\right) \qquad (16.4)$$

16.5 The Nature of Sound Sound is a longitudinal wave that can be created only in a medium; it cannot exist in a vacuum. Each cycle of a sound wave includes one condensation (a region of greater than normal pressure) and one rarefaction (a region of less than normal pressure).

A sound wave with a single frequency is called a pure tone. Frequencies less than 20 Hz are called infrasonic. Frequencies greater than 20 kHz are called ultrasonic. The brain interprets the frequency detected by the ear primarily in terms of the subjective quality known as pitch. A high-pitched sound is one with a large frequency (e.g., piccolo). A low-pitched sound is one with a small frequency (e.g., tuba).

The pressure amplitude of a sound wave is the magnitude of the maximum change in pressure, measured relative to the undisturbed pressure. The pressure amplitude is associated with the subjective quality of loudness. The larger the pressure amplitude, the louder the sound.

16.6 The Speed of Sound The speed of sound v depends on the properties of the medium. In an ideal gas, the speed of sound is given by Equation 16.5, where $\gamma = c_P/c_V$ is the ratio of the specific heat capacities at constant pressure and constant volume, k is Boltzmann's constant, T is the Kelvin temperature, and m is the mass of a molecule of the gas. In a liquid, the speed of sound is given by Equation 16.6, where B_{ad} is the adiabatic bulk modulus and ρ is the mass density. In a solid that has a Young's modulus of Y and the shape of a long slender bar, the speed of sound is given by Equation 16.7.

$$v = \sqrt{\frac{\gamma k T}{m}} \qquad (16.5)$$

$$v = \sqrt{\frac{B_{ad}}{\rho}} \qquad (16.6)$$

$$v = \sqrt{\frac{Y}{\rho}} \qquad (16.7)$$

16.7 Sound Intensity The intensity I of a sound wave is the power P that passes perpendicularly through a surface divided by the area A of the surface, as shown in Equation 16.8. The SI unit for intensity is watts per square meter (W/m²). The smallest sound intensity that the human ear can detect is known as the threshold of hearing and is about 1×10^{-12} W/m² for a 1-kHz sound. When a source radiates sound uniformly in all directions and no reflections are present, the intensity of the sound is inversely proportional to the square of the distance from the source, according to Equation 16.9.

$$I = \frac{P}{A} \qquad (16.8)$$

$$I = \frac{P}{4\pi r^2} \qquad (16.9)$$

16.8 Decibels The intensity level β (in decibels) is used to compare a sound intensity I to the sound intensity I_0 of a reference level, as indicated by Equation 16.10. The decibel, like the radian, is dimensionless. An intensity level of zero decibels means that $I = I_0$. One decibel is approximately the smallest change in loudness that an average listener with healthy hearing can detect. An increase of ten decibels in the intensity level corresponds approximately to a doubling of the loudness of the sound.

$$\beta = (10 \text{ dB}) \log\left(\frac{I}{I_0}\right) \qquad (16.10)$$

16.9 The Doppler Effect The Doppler effect is the change in frequency detected by an observer because the sound source and the observer have different velocities with respect to the medium of sound propagation. If the observer and source move with speeds v_o and v_s, respectively, and if the medium is stationary, the frequency f_o detected by the observer is given by Equation 16.15, where f_s is the frequency of the sound emitted by the source and v is the speed of sound. In the numerator, the plus sign applies when the observer moves toward the source, and the minus sign applies when the observer moves away from the source. In the denominator, the minus sign is used when the source moves toward the observer, and the plus sign is used when the source moves away from the observer.

$$f_o = f_s \left(\frac{1 \pm \frac{v_o}{v}}{1 \mp \frac{v_s}{v}}\right) \qquad (16.15)$$

Focus on Concepts

Additional questions are available for assignment in WileyPLUS.

Section 16.1 The Nature of Waves

1. Domino toppling is an event in which a large number of dominoes are lined up close together and then allowed to topple, one after the other. The disturbance that propagates along the line of dominoes is _____. (a) partly transverse and partly longitudinal (b) transverse (c) longitudinal

Section 16.2 Periodic Waves

2. A transverse wave on a string has an amplitude A. A tiny spot on the string is colored red. As one cycle of the wave passes by, what is the total distance traveled by the red spot? (a) A (b) $2A$ (c) $\frac{1}{2}A$ (d) $4A$ (e) $\frac{1}{4}A$

Section 16.3 The Speed of a Wave on a String

3. As a wave moves through a medium at a speed v, the particles of the medium move in simple harmonic motion about their undisturbed

positions. The maximum speed of the simple harmonic motion is v_{max}. When the amplitude of the wave doubles, _____. **(a)** v doubles, but v_{max} remains the same **(b)** v remains unchanged, but v_{max} doubles **(c)** both v and v_{max} remain unchanged **(d)** both v and v_{max} double

4. A rope is attached to a hook in the ceiling and is hanging straight down. The rope has a mass m, and nothing is attached to the free end of the rope. As a transverse wave travels down the rope from the top, _____. **(a)** the speed of the wave does not change **(b)** the speed of the wave increases **(c)** the speed of the wave decreases

Section 16.4   The Mathematical Description of a Wave

5. The equation that describes a transverse wave on a string is

$$y = (0.0120 \text{ m}) \sin[(483 \text{ rad/s})t - (3.00 \text{ rad/m})x]$$

where y is the displacement of a string particle and x is the position of the particle on the string. The wave is traveling in the $+x$ direction. What is the speed v of the wave?

Section 16.5   The Nature of Sound

6. As the amplitude of a sound wave in air decreases to zero, _____. **(a)** nothing happens to the condensations and rarefactions of the wave **(b)** the condensations and rarefactions of the wave occupy more and more distance along the direction in which the wave is traveling **(c)** the condensations of the wave disappear, but nothing happens to the rarefactions **(d)** nothing happens to the condensations of the wave, but the rarefactions disappear **(e)** both the condensations and the rarefactions of the wave disappear

Section 16.6   The Speed of Sound

7. An echo is sound that returns to you after being reflected from a distant surface (e.g., the side of a cliff). Assuming that the distances involved are the same, an echo under water and an echo in air return to you _____. **(a)** at different times, the echo under water returning more slowly **(b)** at different times, the echo under water returning more quickly **(c)** at the same time

8. A horn on a boat sounds a warning, and the sound penetrates the water. How does the frequency of the sound in the air compare to its frequency in the water? How does the wavelength in the air compare to the wavelength in the water? **(a)** The frequency in the air is smaller than the frequency in the water, and the wavelength in the air is greater than the wavelength in the water. **(b)** The frequency in the air is greater than the frequency in the water, and the wavelength in the air is smaller than the wavelength in the water. **(c)** The frequency in the air is the same as the frequency in the water, and the wavelength in the air is the same as the wavelength in the water. **(d)** The frequency in the air is the same as the frequency in the water, and the wavelength in the air is smaller than the wavelength in the water. **(e)** The frequency in the air is the same as the frequency in the water, and the wavelength in the air is greater than the wavelength in the water.

Section 16.7   Sound Intensity

9. A source emits sound uniformly in all directions. There are no reflections of the sound. At a distance of 12 m from the source, the intensity of the sound is 5.0×10^{-3} W/m^2. What is the total sound power P emitted by the source?

Section 16.8   Decibels

10. A source emits sound uniformly in all directions. There are no reflections of the sound. At a distance r_1 from the source, the sound is 7.0 dB louder than it is at a distance r_2 from the source. What is the ratio r_1/r_2?

Section 16.9   The Doppler Effect

11. A red car and a blue car can move along the same straight one-lane road. Both cars can move only at one speed when they move (e.g., 60 mph). The driver of the red car sounds his horn. In which one of the following situations does the driver of the blue car hear the highest horn frequency? **(a)** Both cars are moving at the same speed, and they are moving apart. **(b)** Both cars are moving in the same direction at the same speed. **(c)** Both cars are moving at the same speed, and they are moving toward each other. **(d)** The red car is moving toward the blue car, which is stationary. **(e)** The blue car is moving toward the red car, which is stationary.

12. What happens to the Doppler effect in air (i.e., the shift in frequency of a sound wave) as the temperature increases? **(a)** It is greater at higher temperatures, but only in the case of a moving source and a stationary observer. **(b)** It is greater at higher temperatures, but only in the case of a moving observer and a stationary source. **(c)** It is greater at higher temperatures than at lower temperatures. **(d)** It is less at higher temperatures than at lower temperatures. **(e)** The Doppler effect does not change as the temperature increases.

Problems

Online

Additional questions are available for assignment in WileyPLUS.

SSM Student Solutions Manual
MMH Problem-solving help
GO Guided Online Tutorial
V-HINT Video Hints
CHALK Chalkboard Videos

BIO Biomedical application
E Easy
M Medium
H Hard
WS Worksheet
T Team Problem

Section 16.1   The Nature of Waves,

Section 16.2   Periodic Waves

1. **E** **SSM** Light is an electromagnetic wave and travels at a speed of 3.00×10^8 m/s. The human eye is most sensitive to yellow-green light, which has a wavelength of 5.45×10^{-7} m. What is the frequency of this light?

2. **E** Consider the freight train in **Figure 16.6**. Suppose that 15 boxcars pass by in a time of 12.0 s and each has a length of 14.0 m.

(a) What is the frequency at which each boxcar passes? **(b)** What is the speed of the train?

3. **E** A woman is standing in the ocean, and she notices that after a wave crest passes, five more crests pass in a time of 50.0 s. The distance between two successive crests is 32 m. Determine, if possible, the wave's **(a)** period, **(b)** frequency, **(c)** wavelength, **(d)** speed, and **(e)** amplitude. If it is not possible to determine any of these quantities, then so state.

4. **E** Tsunamis are fast-moving waves often generated by underwater earthquakes. In the deep ocean their amplitude is barely noticeable, but upon reaching shore, they can rise up to the astonishing height of a six-story building. One tsunami, generated off the Aleutian islands in Alaska, had a wavelength of 750 km and traveled a distance of 3700 km in 5.3 h. **(a)** What was the speed (in m/s) of the wave? For reference, the speed of a 747 jetliner is about 250 m/s. Find the wave's **(b)** frequency and **(c)** period.

5. **E** **SSM** In **Interactive Figure 16.2c** the hand moves the end of the Slinky up and down through two complete cycles in one second. The wave moves along the Slinky at a speed of 0.50 m/s. Find the distance between two adjacent crests on the wave.

6. **E** **GO** **CHALK** A person fishing from a pier observes that four wave crests pass by in 7.0 s and estimates the distance between two successive crests to be 4.0 m. The timing starts with the first crest and ends with the fourth. What is the speed of the wave?

7. **E** Using the data in the graphs that accompany this problem, determine the speed of the wave.

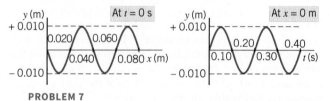

PROBLEM 7

8. **M** A 3.49-rad/s ($33\frac{1}{3}$ rpm) record has a 5.00-kHz tone cut in the groove. If the groove is located 0.100 m from the center of the record (see the drawing), what is the wavelength in the groove?

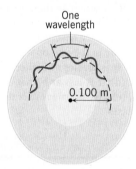

PROBLEM 8

9. **M** **V-HINT** **CHALK** The speed of a transverse wave on a string is 450 m/s, and the wavelength is 0.18 m. The amplitude of the wave is 2.0 mm. How much time is required for a particle of the string to move through a total distance of 1.0 km?

10. **M** **GO** A jetskier is moving at 8.4 m/s in the direction in which the waves on a lake are moving. Each time he passes over a crest, he feels a bump. The bumping frequency is 1.2 Hz, and the crests are separated by 5.8 m. What is the wave speed?

Section 16.3 The Speed of a Wave on a String

11. **E** The mass of a string is 5.0×10^{-3} kg, and it is stretched so that the tension in it is 180 N. A transverse wave traveling on this string

has a frequency of 260 Hz and a wavelength of 0.60 m. What is the length of the string?

12. **E** **SSM** The middle C string on a piano is under a tension of 944 N. The period and wavelength of a wave on this string are 3.82 ms and 1.26 m, respectively. Find the linear density of the string.

13. **E** A wire is stretched between two posts. Another wire is stretched between two posts that are twice as far apart. The tension in the wires is the same, and they have the same mass. A transverse wave travels on the shorter wire with a speed of 240 m/s. What would be the speed of the wave on the longer wire?

14. **E** **MMH** To measure the acceleration due to gravity on a distant planet, an astronaut hangs a 0.055-kg ball from the end of a wire. The wire has a length of 0.95 m and a linear density of 1.2×10^{-4} kg/m. Using electronic equipment, the astronaut measures the time for a transverse pulse to travel the length of the wire and obtains a value of 0.016 s. The mass of the wire is negligible compared to the mass of the ball. Determine the acceleration due to gravity.

15. **E** Two wires are parallel, and one is directly above the other. Each has a length of 50.0 m and a mass per unit length of 0.020 kg/m. However, the tension in wire A is 6.00×10^2 N, and the tension in wire B is 3.00×10^2 N. Transverse wave pulses are generated simultaneously, one at the left end of wire A and one at the right end of wire B. The pulses travel toward each other. How much time does it take until the pulses pass each other?

16. **E** **SSM** The drawing shows two transverse waves traveling on strings. The linear density of each string is 0.065 kg/m. The tension is provided by a 26-N block that is hanging from the string. Find the speed of the wave in part **(a)** and in part **(b)** of the drawing.

PROBLEM 16

17. **M** **GO** A steel cable has a cross-sectional area 2.83×10^{-3} m² and is kept under a tension of 1.00×10^4 N. The density of steel is 7860 kg/m³. *Note that this value is not the linear density of the cable.* At what speed does a transverse wave move along the cable?

18. **M** **GO** The drawing shows a graph of two waves traveling to the right at the same speed. **(a)** Using the data in the drawing, determine the wavelength of each wave. **(b)** The speed of the waves is 12 m/s; calculate the frequency of each one. **(c)** What is the maximum speed for a particle attached to each wave?

PROBLEM 18

19. **M** **V-HINT** Review Conceptual Example 3 before starting this problem. The amplitude of a transverse wave on a string is 4.5 cm. The ratio of the maximum particle speed to the speed of the wave is 3.1. What is the wavelength (in cm) of the wave?

20. **M** **CHALK** The drawing shows a frictionless incline and pulley. The two blocks are connected by a wire (mass per unit length = 0.0250 kg/m) and remain stationary. A transverse wave on the wire has a speed of 75.0 m/s. Neglecting the weight of the wire relative to the tension in the wire, find the masses m_1 and m_2 of the blocks.

PROBLEM 20   30.0°

21. **H** The drawing shows a 15.0-kg ball being whirled in a circular path on the end of a string. The motion occurs on a frictionless, horizontal table. The angular speed of the ball is $\omega = 12.0$ rad/s. The string has a mass of 0.0230 kg. How much time does it take for a wave on the string to travel from the center of the circle to the ball?

PROBLEM 21

Section 16.4   The Mathematical Description of a Wave

(Note: The phase angles $(2\pi ft - 2\pi x/\lambda)$ and $(2\pi ft + 2\pi x/\lambda)$ are measured in radians, not degrees.)

22. **E GO** A wave traveling along the x axis is described mathematically by the equation $y = 0.17 \sin (8.2\pi t + 0.54\pi x)$, where y is the displacement (in meters), t is in seconds, and x is in meters. What is the speed of the wave?

23. **E V-HINT** A wave has the following properties: amplitude = 0.37 m, period = 0.77 s, wave speed = 12 m/s. The wave is traveling in the $-x$ direction. What is the mathematical expression (similar to Equation 16.3 or 16.4) for the wave?

24. **E** The drawing shows a graph that represents a transverse wave on a string. The wave is moving in the $+x$ direction with a speed of 0.15 m/s. Using the information contained in the graph, write the mathematical expression (similar to Equation 16.3 or 16.4) for the wave.

PROBLEM 24

25. **E SSM** A wave traveling in the $+x$ direction has an amplitude of 0.35 m, a speed of 5.2 m/s, and a frequency of 14 Hz. Write the equation of the wave in the form given by either Equation 16.3 or 16.4.

26. **M GO CHALK** A transverse wave is traveling on a string. The displacement y of a particle from its equilibrium position is given by $y = (0.021 \text{ m}) \sin (25t - 2.0 x)$. Note that the phase angle $25t - 2.0x$ is in radians, t is in seconds, and x is in meters. The linear density of the string is 1.6×10^{-2} kg/m. What is the tension in the string?

27. **M SSM** The tension in a string is 15 N, and its linear density is 0.85 kg/m. A wave on the string travels toward the $-x$ direction; it has an amplitude of 3.6 cm and a frequency of 12 Hz. What are the (a) speed and (b) wavelength of the wave? (c) Write down a mathematical expression (like Equation 16.3 or 16.4) for the wave, substituting numbers for the variables A, f, and λ.

28. **H** A transverse wave on a string has an amplitude of 0.20 m and a frequency of 175 Hz. Consider the particle of the string at $x = 0$ m. It begins with a displacement of $y = 0$ m when $t = 0$ s, according to Equation 16.3 or 16.4. How much time passes between the first two instants when this particle has a displacement of $y = 0.10$ m?

Section 16.5   The Nature of Sound,

Section 16.6   The Speed of Sound

29. **E SSM** For research purposes a sonic buoy is tethered to the ocean floor and emits an infrasonic pulse of sound (speed = 1522 m/s). The period of this sound is 71 ms. Determine the wavelength of the sound.

30. **E CHALK** To navigate, a porpoise emits a sound wave that has a wavelength of 1.5 cm. The speed at which the wave travels in seawater is 1522 m/s. Find the period of the wave.

31. **E MMH** At what temperature is the speed of sound in helium (ideal gas, $\gamma = 1.67$, atomic mass = 4.003 u) the same as its speed in oxygen at 0 °C?

32. **E** Have you ever listened for an approaching train by kneeling next to a railroad track and putting your ear to the rail? Young's modulus for steel is $Y = 2.0 \times 10^{11}$ N/m², and the density of steel is $\rho = 7860$ kg/m³. On a day when the temperature is 20 °C, how many times greater is the speed of sound in the rail than in the air?

33. **E SSM** The speed of a sound in a container of hydrogen at 201 K is 1220 m/s. What would be the speed of sound if the temperature were raised to 405 K? Assume that hydrogen behaves like an ideal gas.

34. **E GO** Suppose you are part of a team that is trying to break the sound barrier with a jet-powered car, which means that it must travel faster than the speed of sound in air. In the morning, the air temperature is 0 °C, and the speed of sound is 331 m/s. What speed must your car exceed if it is to break the sound barrier when the temperature has risen to 43 °C in the afternoon? Assume that air behaves like an ideal gas.

35. **E** As the drawing illustrates, a siren can be made by blowing a jet of air through 20 equally spaced holes in a rotating disk. The time it takes for successive holes to move past the air jet is the period of the sound. The siren is to produce a 2200-Hz tone. What must be the angular speed ω (in rad/s) of the disk?

PROBLEM 35

36. **E** At a height of ten meters above the surface of a freshwater lake, a sound pulse is generated. The echo from the bottom of the lake returns to the point of origin 0.110 s later. The air and water temperature are 20 °C. How deep is the lake?

37. **E** An observer stands 25 m behind a marksman practicing at a rifle range. The marksman fires the rifle horizontally, the speed of the bullets is 840 m/s, and the air temperature is 20 °C. How far does each bullet travel before the observer hears the report of the rifle? Assume that the bullets encounter no obstacles during this interval, and ignore both air resistance and the vertical component of the bullets' motion.

38. **E GO** An ultrasonic ruler, such as the one discussed in Example 4 in Section 16.6, displays the distance between the ruler and an object, such as a wall. The ruler sends out a pulse of ultrasonic sound and measures the time it takes for the pulse to reflect from the object and return. The ruler uses this time, along with a preset value for the speed of sound in air, to determine the distance. Suppose that you use this ruler under water, rather than in air. The actual distance from the ultrasonic ruler to an object is 25.0 m. The adiabatic bulk modulus and density of seawater are $B_{ad} = 2.37 \times 10^9$ Pa and $\rho = 1025$ kg/m^3, respectively. Assume that the ruler uses a preset value of 343 m/s for the speed of sound in air. Determine the distance reading that the ruler displays.

39. **E** An explosion occurs at the end of a pier. The sound reaches the other end of the pier by traveling through three media: air, fresh water, and a slender metal handrail. The speeds of sound in air, water, and the handrail are 343, 1482, and 5040 m/s, respectively. The sound travels a distance of 125 m in each medium. (a) Through which medium does the sound arrive first, second, and third? (b) After the first sound arrives, how much later do the second and third sounds arrive?

40. **M** **V-HINT** A sound wave travels twice as far in neon (Ne) as it does in krypton (Kr) in the same time interval. Both neon and krypton can be treated as monatomic ideal gases. The atomic mass of neon is 20.2 u, and the atomic mass of krypton is 83.8 u. The temperature of the krypton is 293 K. What is the temperature of the neon?

41. **M** **SSM** A hunter is standing on flat ground between two vertical cliffs that are directly opposite one another. He is closer to one cliff than to the other. He fires a gun and, after a while, hears three echoes. The second echo arrives 1.6 s after the first, and the third echo arrives 1.1 s after the second. Assuming that the speed of sound is 343 m/s and that there are no reflections of sound from the ground, find the distance between the cliffs.

42. **M** **GO** A monatomic ideal gas ($\gamma = 1.67$) is contained within a box whose volume is 2.5 m^3. The pressure of the gas is 3.5×10^5 Pa. The total mass of the gas is 2.3 kg. Find the speed of sound in the gas.

43. **M** **V-HINT** A long slender bar is made from an unknown material. The length of the bar is 0.83 m, its cross-sectional area is 1.3×10^{-4} m^2, and its mass is 2.1 kg. A sound wave travels from one end of the bar to the other end in 1.9×10^{-4} s. From which one of the materials listed in Table 10.1 is the bar most likely to be made?

44. **M** **CHALK** As the drawing shows, one microphone is located at the origin, and a second microphone is located on the +y axis. The microphones are separated by a distance of $D = 1.50$ m. A source of sound is located on the +x axis, its distances from microphones 1 and 2 being L_1 and L_2, respectively. The speed of sound is 343 m/s. The sound reaches microphone 1 first, and then, 1.46 ms later, it reaches microphone 2. Find the distances L_1 and L_2.

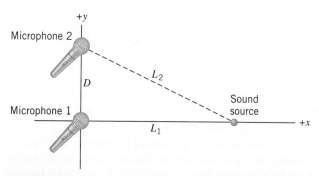

PROBLEM 44

45. **M** **SSM** When an earthquake occurs, two types of sound waves are generated and travel through the earth. The primary, or P, wave has a speed of about 8.0 km/s and the secondary, or S, wave has a speed of about 4.5 km/s. A seismograph, located some distance away, records the arrival of the P wave and then, 78 s later, records the

arrival of the S wave. Assuming that the waves travel in a straight line, how far is the seismograph from the earthquake?

46. **M** **V-HINT** Consult Multiple-Concept Example 4 in order to review a model for solving this type of problem. Suppose that you are standing by the side of a road in the Sahara desert where the temperature has reached a hot 56 °C (130 °F). A truck, traveling at a constant speed, passes by. After 4.00 s have elapsed, you use the ultrasonic ruler discussed in Example 4 to measure the distance to the truck. A sound pulse leaves the ultrasonic ruler and returns 0.120 s later. Assume that the average molecular mass of air is 28.9 u, air is an ideal diatomic gas ($\gamma = \frac{7}{5}$), and the truck moves a negligible distance in the time it takes for the sound pulse to reach it. Determine how fast the truck is moving.

Section 16.7 Sound Intensity

47. **E** **SSM** At a distance of 3.8 m from a siren, the sound intensity is 3.6×10^{-2} W/m^2. Assuming that the siren radiates sound uniformly in all directions, find the total power radiated.

48. **E** **GO** A source of sound is located at the center of two concentric spheres, parts of which are shown in the drawing. The source emits sound uniformly in all directions. On the spheres are drawn three small patches that may or may not have equal areas. However, the same sound power passes through each patch. The source produces 2.3 W of sound power, and the radii of the concentric spheres are $r_A = 0.60$ m and $r_B = 0.80$ m. (a) Determine the sound intensity at each of the three patches. (b) The sound power that passes through each of the patches is 1.8×10^{-3} W. Find the area of each patch.

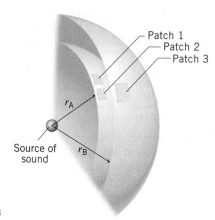

PROBLEM 48

49. **E** Suppose that a public address system emits sound uniformly in all directions and that there are no reflections. The intensity at a location 22 m away from the sound source is 3.0×10^{-4} W/m^2. What is the intensity at a spot that is 78 m away?

50. **E** **SSM** A loudspeaker has a circular opening with a radius of 0.0950 m. The electrical power needed to operate the speaker is 25.0 W. The average sound intensity at the opening is 17.5 W/m^2. What percentage of the electrical power is converted by the speaker into sound power?

51. **E** **GO** A man stands at the midpoint between two speakers that are broadcasting an amplified static hiss uniformly in all directions. The speakers are 30.0 m apart and the total power of the sound coming from each speaker is 0.500 W. Find the total sound intensity that the man hears (a) when he is at his initial position halfway between the speakers, and (b) after he has walked 4.0 m directly toward one of the speakers.

52. **M** **SSM** **MMH** A dish of lasagna is being heated in a microwave oven. The effective area of the lasagna that is exposed to the

microwaves is 2.2×10^{-2} m^2. The mass of the lasagna is 0.35 kg, and its specific heat capacity is 3200 J/(kg · C°). The temperature rises by 72 °C in 8.0 minutes. What is the intensity of the microwaves in the oven?

53. M GO CHALK Two sources of sound are located on the x axis, and each emits power uniformly in all directions. There are no reflections. One source is positioned at the origin and the other at $x = +123$ m. The source at the origin emits four times as much power as the other source. Where on the x axis are the two sounds equal in intensity? Note that there are two answers.

Section 16.8 Decibels

54. E A woman stands a distance d from a loud motor that emits sound uniformly in all directions. The sound intensity at her position is an uncomfortable 3.2×10^{-3} W/m^2. There are no reflections. At a position twice as far from the motor, what are **(a)** the sound intensity and **(b)** the sound intensity level relative to the threshold of hearing?

55. E The volume control on a surround-sound amplifier is adjusted so the sound intensity level at the listening position increases from 23 to 61 dB. What is the ratio of the final sound intensity to the original sound intensity?

56. E GO Using an intensity of 1×10^{-12} W/m^2 as a reference, the threshold of hearing for an average young person is 0 dB. Person 1 and person 2, who are not average, have thresholds of hearing that are $\beta_1 = -8.00$ dB and $\beta_2 = +12.0$ dB. What is the ratio I_1/I_2 of the sound intensity I_1 when person 1 hears the sound at his own threshold of hearing compared to the sound intensity I_2 when person 2 hears the sound at his own threshold of hearing?

57. E A listener doubles his distance from a source that emits sound uniformly in all directions. There are no reflections. By how many decibels does the sound intensity level change?

58. E Sound is passing perpendicularly through an open window whose dimensions are 1.1 m × 0.75 m. The sound intensity level is 95 dB above the threshold of hearing. How much sound *energy* comes through the window in one hour?

59. E The bellow of a territorial bull hippopotamus has been measured at 115 dB above the threshold of hearing. What is the sound intensity?

60. M CHALK SSM When one person shouts at a football game, the sound intensity level at the center of the field is 60.0 dB. When all the people shout together, the intensity level increases to 109 dB. Assuming that each person generates the same sound intensity at the center of the field, how many people are at the game?

61. M GO Review Conceptual Example 8 as background for this problem. A loudspeaker is generating sound in a room. At a certain point, the sound waves coming directly from the speaker (without reflecting from the walls) create an intensity level of 75.0 dB. The waves reflected from the walls create, by themselves, an intensity level of 72.0 dB at the same point. What is the total intensity level? (*Hint: The answer is not 147.0 dB.*)

62. M V-HINT A portable radio is sitting at the edge of a balcony 5.1 m above the ground. The unit is emitting sound uniformly in all directions. By accident, it falls from rest off the balcony and continues to play on the way down. A gardener is working in a flower bed directly below the falling unit. From the instant the unit begins to fall, how much time is required for the sound intensity level heard by the gardener to increase by 10.0 dB?

Section 16.9 The Doppler Effect

63. E A bird is flying directly toward a stationary bird-watcher and emits a frequency of 1250 Hz. The bird-watcher, however, hears a frequency of 1290 Hz. What is the speed of the bird, expressed as a percentage of the speed of sound?

64. E SSM From a vantage point very close to the track at a stock car race, you hear the sound emitted by a moving car. You detect a frequency that is 0.86 times as small as the frequency emitted by the car when it is stationary. The speed of sound is 343 m/s. What is the speed of the car?

65. E Dolphins emit clicks of sound for communication and echolocation. A marine biologist is monitoring a dolphin swimming in seawater where the speed of sound is 1522 m/s. When the dolphin is swimming directly away at 8.0 m/s, the marine biologist measures the number of clicks occurring per second to be at a frequency of 2500 Hz. What is the difference (in Hz) between this frequency and the number of clicks per second actually emitted by the dolphin?

66. E MMH A convertible moves toward you and then passes you; all the while, its loudspeakers are producing a sound. The speed of the car is a constant 9.00 m/s, and the speed of sound is 343 m/s. What is the ratio of the frequency you hear while the car is approaching to the frequency you hear while the car is moving away?

67. E The security alarm on a parked car goes off and produces a frequency of 960 Hz. The speed of sound is 343 m/s. As you drive toward this parked car, pass it, and drive away, you observe the frequency to change by 95 Hz. At what speed are you driving?

68. M V-HINT A car is parked 20.0 m directly south of a railroad crossing. A train is approaching the crossing from the west, headed directly east at a speed of 55.0 m/s. The train sounds a short blast of its 289-Hz horn when it reaches a point 20.0 m west of the crossing. What frequency does the car's driver hear when the horn blast reaches the car? The speed of sound in air is 343 m/s. (*Hint: Assume that only the component of the train's velocity that is directed toward the car affects the frequency heard by the driver.*)

69. M GO A loudspeaker in a parked car is producing sound whose frequency is 20 510 Hz. A healthy young person with normal hearing is standing nearby on the sidewalk but cannot hear the sound because the frequency is too high. When the car is moving, however, this person can hear the sound. **(a)** Is the car moving toward or away from the person? Why? **(b)** If the speed of sound is 343 m/s, what is the minimum speed of the moving car?

70. M SSM MMH Two trucks travel at the same speed. They are far apart on adjacent lanes and approach each other essentially head-on. One driver hears the horn of the other truck at a frequency that is 1.14 times the frequency he hears when the trucks are stationary. The speed of sound is 343 m/s. At what speed is each truck moving?

71. M GO A wireless transmitting microphone is mounted on a small platform that can roll down an incline, directly away from a loudspeaker that is mounted at the top of the incline. The loudspeaker broadcasts a tone that has a fixed frequency of 1.000×10^4 Hz, and the speed of sound is 343 m/s. At a time of 1.5 s following the release of the platform, the microphone detects a frequency of 9939 Hz. At a time of 3.5 s following the release of the platform, the microphone detects a frequency of 9857 Hz. What is the acceleration (assumed constant) of the platform?

72. M V-HINT A car is accelerating while its horn is sounding. Just after the car passes a stationary person, the person hears a frequency of 966.0 Hz. Fourteen seconds later, the frequency heard by the person has decreased to 912.0 Hz. When the car is stationary, its horn emits a sound whose frequency is 1.00×10^3 Hz. The speed of sound is 343 m/s. What is the acceleration of the car?

73. **M** **GO** The siren on an ambulance is emitting a sound whose frequency is 2450 Hz. The speed of sound is 343 m/s. **(a)** If the ambulance is stationary and you (the "observer") are sitting in a parked car, what are the wavelength and the frequency of the sound you hear? **(b)** Suppose that the ambulance is moving toward you at a speed of 26.8 m/s. Determine the wavelength and the frequency of the sound

you hear. **(c)** If the ambulance is moving toward you at a speed of 26.8 m/s and you are moving toward it at a speed of 14.0 m/s, find the wavelength and frequency of the sound you hear.

74. **M** **SSM** Two submarines are under water and approaching each other head-on. Sub A has a speed of 12 m/s and sub B has a speed of 8 m/s. Sub A sends out a 1550-Hz sonar wave that travels at a speed of 1522 m/s. **(a)** What is the frequency detected by sub B? **(b)** Part of the sonar wave is reflected from sub B and returns to sub A. What frequency does sub A detect for this reflected wave?

Additional Problems

Online

75. **E** A recording engineer works in a soundproofed room that is 44.0 dB quieter than the outside. If the sound intensity that leaks into the room is 1.20×10^{-10} W/m², what is the intensity outside?

76. **E** You are flying in an ultralight aircraft at a speed of 39 m/s. An eagle, whose speed is 18 m/s, is flying directly toward you. Each of the given speeds is relative to the ground. The eagle emits a shrill cry whose frequency is 3400 Hz. The speed of sound is 330 m/s. What frequency do you hear?

77. **E** **SSM** Suppose that the linear density of the A string on a violin is 7.8×10^{-4} kg/m. A wave on the string has a frequency of 440 Hz and a wavelength of 65 cm. What is the tension in the string?

78. **E** A car driving along a highway at a speed of 23 m/s strays onto the shoulder. Evenly spaced parallel grooves called "rumble strips" are carved into the pavement of the shoulder. Rolling over the rumble strips causes the car's wheels to oscillate up and down at a frequency of 82 Hz. How far apart are the centers of adjacent rumble-strip grooves?

79. **E** **SSM** When Gloria wears her hearing aid, the sound intensity level increases by 30.0 dB. By what factor does the sound intensity increase?

80. **E** The average sound intensity inside a busy neighborhood restaurant is 3.2×10^{-5} W/m². How much energy goes into each ear (area = 2.1×10^{-3} m²) during a one-hour meal?

81. **E** **SSM** Suppose that the amplitude and frequency of the transverse wave in **Interactive Figure 16.2c** are, respectively, 1.3 cm and 5.0 Hz. Find the *total vertical distance* (in cm) through which the colored dot moves in 3.0 s.

82. **E** A bat emits a sound whose frequency is 91 kHz. The speed of sound in air at 20.0 °C is 343 m/s. However, the air temperature is 35 °C, so the speed of sound is not 343 m/s. Assume that air behaves like an ideal gas, and find the wavelength of the sound.

83. **E** **SSM** You are riding your bicycle directly away from a stationary source of sound and hear a frequency that is 1.0% lower than the emitted frequency. The speed of sound is 343 m/s. What is your speed?

84. **E** **GO** Argon (molecular mass = 39.9 u) is a monatomic gas. Assuming that it behaves like an ideal gas at 298 K ($\gamma = 1.67$), find **(a)** the rms speed of argon atoms and **(b)** the speed of sound in argon.

85. **E** **SSM** The sound intensity level at a rock concert is 115 dB, while that at a jazz fest is 95 dB. Determine the ratio of the sound intensity at the rock concert to the sound intensity at the jazz fest.

86. **E** An amplified guitar has a sound intensity level that is 14 dB greater than the same unamplified sound. What is the ratio of the amplified intensity to the unamplified intensity?

87. **E** In a discussion person A is talking 1.5 dB louder than person B, and person C is talking 2.7 dB louder than person A. What is the ratio of the sound intensity of person C to the sound intensity of person B?

88. **E** In **Interactive Figure 16.3c** the colored dot exhibits simple harmonic motion as the longitudinal wave passes. The wave has an amplitude of 5.4×10^{-3} m and a frequency of 4.0 Hz. Find the maximum acceleration of the dot.

89. **M** **GO** **(a)** A uniform rope of mass m and length L is hanging straight down from the ceiling. A small-amplitude transverse wave is sent up the rope from the bottom end. Derive an expression that gives the speed v of the wave on the rope in terms of the distance y above the bottom end of the rope and the magnitude g of the acceleration due to gravity. **(b)** Use the expression that you have derived to calculate the speeds at distances of 0.50 m and 2.0 m above the bottom end of the rope.

90. **M** **V-HINT** A spider hangs from a strand of silk whose radius is 4.0×10^{-6} m. The density of the silk is 1300 kg/m³. When the spider moves, waves travel along the strand of silk at a speed of 280 m/s. Ignore the mass of the silk strand, and determine the mass of the spider.

91. **M** **V-HINT** A member of an aircraft maintenance crew wears protective earplugs that reduce the sound intensity by a factor of 350. When a jet aircraft is taking off, the sound intensity level experienced by the crew member is 88 dB. What sound intensity level would the crew member experience if he removed the protective earplugs?

92. **M** **GO** **SSM** A speedboat, starting from rest, moves along a straight line away from a dock. The boat has a constant acceleration of +3.00 m/s² (see the figure). Attached to the dock is a siren that is producing a 755-Hz tone. If the air temperature is 20 °C, what is the frequency of the sound heard by a person on the boat when the boat's displacement from the dock is +45.0 m?

PROBLEM 92

Physics in Biology, Medicine, and Sports

93. **E** **BIO** **16.7** A typical adult ear has a surface area of 2.1×10^{-3} m². The sound intensity during a normal conversation is about 3.2×10^{-6} W/m² at the listener's ear. Assume that the sound strikes the surface of the ear perpendicularly. How much power is intercepted by the ear?

94. **M** **BIO** **V-HINT** **16.7** Deep ultrasonic heating is used to promote the healing of torn tendons. It is produced by applying ultrasonic sound over the affected area of the body. The sound transducer (generator) is circular with a radius of 1.8 cm, and it produces a sound intensity of 5.9×10^3 W/m². How much time is required for the transducer to emit 4800 J of sound energy?

95. **E** **BIO** **SSM** **16.8** A middle-aged man typically has poorer hearing than a middle-aged woman. In one case a woman can just begin to hear a musical tone, while a man can just begin to hear the tone only when its intensity level is increased by 7.8 dB relative to the just-audible intensity level for the woman. What is the ratio of the sound intensity just detected by the man to the sound intensity just detected by the woman?

96. **E** **BIO** **V-HINT** **16.8** Hearing damage may occur when a person is exposed to a sound intensity level of 90.0 dB (relative to the threshold of hearing) for a period of 9.0 hours. One particular eardrum has an area of 2.0×10^{-4} m². How much sound energy is incident on this eardrum during this time?

97. **M** **BIO** **16.8** Tortoises do have ears, but they are very different than human ears. They have no outer ear. Their eardrums are located along the side of their head behind their eyes and are covered with a flap of scales (see the photo). The physiology of their ears makes them sensitive to lower frequency sounds that travel well through water and also along the ground as vibrations. Consider a tortoise that hears a 72-dB sound from a boat propeller under water with a frequency of 1.0×10^2 Hz.* If the eardrum of the tortoise is approximately circular with a radius of 3.2 mm, how much power is delivered to the eardrum?

PROBLEM 97

Shelly/totallytortoise.com

*Sound intensity levels underwater are calculated slightly differently than in air, which we won't take into account for this calculation. In fact, the threshold of hearing for animals underwater can be many times higher than in air.

98. **E** **BIO** **16.6** Frequency ranges for hearing vary dramatically across the animal kingdom. For example, turtles are sensitive to lower frequencies (20 to 1000 Hz), humans have a greater range (20 to 20 000 Hz), but the champion of frequency range is the porpoise (75 to 150 000 Hz). What is the wavelength of a 150 000-Hz sound wave in seawater at a temperature of 20 °C?

99. **M** **BIO** **16.10** According to Section 16.10, the smallest detail that can be resolved by medical ultrasound is essentially equal to one wavelength of the sound. (a) If the speed of a 2.20-MHz ultrasonic wave in human tissue is 1540 m/s, what is the size of the smallest structure that can be observed? (b) How far below the surface of the skin can the sound waves penetrate? Assume their maximum penetration depth is 2.00×10^2 wavelengths. (c) How long would it take the sound waves to leave the transmitter, reach their maximum depth, and return to the receiver at the skin's surface?

100. **E** **BIO** **16.8** Review Example 9 as background material for this problem. Dolphins, like bats, use echolocation to detect objects in their respective environments, including prey. This allows them to scan the physical world using sound. While bats emit short pulses of high frequency sound waves with a frequency of about 60 kHz, dolphins use high frequency ultrasonic waves with a frequency of about 10^6 Hz. They produce a series of whistles or clicks at rates of up to 2000 per second. The clicks echo off of objects and prey and travel back to the dolphin, where they are absorbed in the lower jawbone, which transmits the vibrations to the inner ear. While the nature of sound waves in both air and water is similar, there are significant differences in how the intensity level of the sound is determined. The sound intensity of a wave in W/m² can be directly compared between a sound wave in air and one in water. However, the sound intensity level in dB is not the same. The dB is a relative unit of measure, unlike W/m², which is absolute. The reference used for air (typically 1.00×10^{-12} W/m²) is different for water. This fact, along with the speed of the wave in water (about 5 times faster than in air), and the greater density of the water results in an intensity level in water being 61.5 dB greater than in air. In other words, $\beta_{\text{air}} = \beta_{\text{water}} - 61.5$ dB.

If a dolphin detects a reflected sound wave in water with an intensity level of 80.0 dB, what would be the (a) intensity level and (b) intensity of this wave in air?

101. **M** **BIO** **16.8** With the advent of tiny, powerful rare earth magnets, and improvements in rechargeable battery and wireless technologies, wireless earbuds, like Apple's AirPods (see the photo), enhance the enjoyment of listening to music, movies, and podcasts, and allow handsfree conversation. However, since the earbuds focus sound waves directly into the auditory canal at close distances, care must be taken to protect the eardrum from long-term exposure to high volumes (intensity levels). Hearing experts recommend listening at a maximum intensity level of 85 dB for no more than 8.0 hours a day. Assuming a constant intensity level of 85 dB, how much energy does the eardrum absorb during the 8.0 hours? Take the area of the eardrum to be 0.65 cm².

Courtesy of David Young

PROBLEM 101

Concepts and Calculations Problems

Online ●

One of the most important concepts we encountered in this chapter is the transverse wave. For instance, transverse waves travel along a guitar string when it is plucked or along a violin string when it is bowed. Problem 102 reviews how the travel speed depends on the properties of the string and on the tension in it. Problem 103 illustrates how the Doppler effect arises when an observer is moving away from or toward a stationary source of sound. In fact, we will see that it's possible for both situations to occur at the same time.

102. M CHALK The figure shows waves traveling on two strings. Each string is attached to a wall at one end and to a box that has a weight of 28.0 N at the other end. String 1 has a mass of 8.00 g and a length of 4.00 cm, and string 2 has a mass of 12.0 g and a length of 8.00 cm. *Concepts:* (i) Is the tension the same in each string? (ii) Is the speed of each wave the same? (iii) String 1 has a smaller mass and, hence, less inertia than string 2. Does this mean that the speed of the wave on string 1 is greater than the speed on string 2? *Calculations:* Determine the speed of the wave on each string.

103. M CHALK SSM A siren, mounted on a tower, emits a sound whose frequency is 2140 Hz. A person is driving a car away from the tower at a speed of 27.0 m/s. As the figure illustrates, the sound reaches the person by two paths: the sound reflected from the building in front of the car, and the sound coming directly from the siren. The speed of sound is 343 m/s. *Concepts:* (i) One way that the Doppler effect can arise is that the wavelength of the sound changes. For either the direct or reflected sound, does the wavelength change? (ii) Why does the driver hear a frequency for the reflected sound that is different from 2140 Hz, and is it greater or smaller than 2140 Hz? (iii) Why does the driver hear a frequency for the direct sound that is different from 2140 Hz, and is it greater or smaller than 2140 Hz? *Calculations:* What frequency does the person hear for the (a) reflected and (b) direct sound?

PROBLEM 103

PROBLEM 102

Team Problems

Online ●

104. E T WS A Pneumatic Warning Siren. A pneumatic siren can be made by blowing a narrow jet of air through equally spaced holes on a rotating disk. The time it takes for successive holes to pass thought the jet stream is equal to the period of the resulting sound wave. You and your team are designing such a device to be used as a community warning alarm to be activated in the case of severe weather, or other dangerous situations. The siren should produce a tone as close as possible to 710 Hz, and people with average hearing should be able to detect it above normal conversation at a distance of at least 2.0 miles. The circular pattern of holes will be located 24 cm from the center of a thin aluminum disk, and the motor that rotates the disk runs at 1500 rpm. (a) What *even* integer number of equally spaced holes will result in a sound frequency closest to the desired 710 Hz? (b) What should be the minimum sound output power of the siren in watts (W) so that people with average hearing can detect it above normal conversation at a distance of 2.0 miles from the source? You can assume that the output is isotropic, that is, it emanates like a spherical wave, and that the sound intensity of normal conversion is 3.20×10^{-6} W/m^2 (from Table 16.2).

105. E T WS Tuning the Waves. You and your team are designing a system that allows you to control the speed of a transverse wave on a thin tungsten wire by adjusting its tension with a spring. One end of the wire is connected to an inner wall of a large wooden crate and the other end to one side of a spring. The opposite side of the spring is then connected to a screw that is mounted on the opposite wall of the crate. The screw can then be adjusted to stretch the spring, and therefore adjust the tension in the wire. The uniform tungsten wire is 75.6 cm long and has a mass of 3.15×10^{-3} kg. The spring has an unstretched length of 11.0 cm and a spring constant $k = 280.0$ N/m. In the lowest possible tension setting of the screw, that is, that which stretches the spring the least, the spring is stretched by 0.500 cm from

its equilibrium position. The screw can be turned to stretch it an additional 2.75 cm. What are the highest and lowest attainable speeds of transverse waves that can travel on the wire with this setup? You can neglect any stretching of wire.

106. M WS Setting Safety Parameters. You and your team are asked to determine the hearing safety parameters for a loud experiment. Since the sound will persist for only a short time, the largest allowed intensity level that will not result in permanent hearing damage is 135 dB. The experiment consists of a single source with a power output of 5.50×10^2 W that projects sound isotropically (i.e., uniformly in all directions). (a) At what distance should you set the safety perimeter (i.e., the radius of a sphere in meters) centered at the source? (b) How far from the source should you set the perimeter if the power of the source doubles?

107. M WS A Mysterious Underwater Object. You and your team are on a reconnaissance mission in a submarine exploring a mysterious object in the cold waters of the Weddell Sea, off the coast of Antarctica. The sonar indicates that the object, which had otherwise been moving erratically, has changed course and is now on a direct collision course with your sub. The captain issues an "all stop" order, bringing the sub to a halt relative to the water. The sonar operator "pings" the object, which amounts to sending a short blast of sound in the direction of the object. The emitted sound wave has a frequency of 1.550 kHz and a speed of 1552 m/s (the speed of sound in seawater). The sound reflects from the object and returns 2.582 s after it was emitted from your sub, and its frequency has shifted to 1.598 kHz. (a) How far from the sub was the object when the sound reflected from it? (b) What is the object's speed? (c) How long after you receive the return signal will it take the object to reach your submarine?

Each raindrop that strikes the water's surface creates waves that propagate outward in a circular pattern. When two or more of these sets of waves meet on the surface, they overlap, or interfere with each other, in a process called linear superposition. Where the waves meet crest-to-crest or trough-to-trough, the amplitude of the resultant wave increases, and where the waves meet trough-to-crest, the amplitude of the resultant wave decreases. These phenomena are known as constructive and destructive interference, respectively, and they are two important topics in this chapter.

Sandid/1642 images/Pixabay

The Principle of Linear Superposition and Interference Phenomena

17.1 The Principle of Linear Superposition

Often, two or more sound waves are present at the same place at the same time, as is the case with sound waves when everyone is talking at a party or when music plays from the speakers of a stereo system. To illustrate what happens when several waves pass simultaneously through the same region,

LEARNING OBJECTIVES

After reading this module, you should be able to...

17.1 Express the principle of linear superposition.

17.2 Solve spatial interference problems for sound waves.

17.3 Apply wave interference concepts to the diffraction of sound waves.

17.4 Explain beats as a wave interference phenomenon.

17.5 Analyze transverse standing waves.

17.6 Analyze longitudinal standing waves.

17.7 Define the harmonic content of complex sound waves.

(a) Overlap begins

(b) Total overlap; the Slinky has twice the height of either pulse

(c) The receding pulses

ANIMATED FIGURE 17.1 Two transverse "up" pulses passing through each other.

let's consider **Animated Figures 17.1** and **17.2**, which show two transverse pulses of equal heights moving toward each other along a Slinky. In **Animated Figure 17.1** both pulses are "up," whereas in **Animated Figure 17.2** one is "up" and the other is "down." Part *a* of each figure shows the two pulses beginning to overlap. The pulses merge, and the Slinky assumes a shape that is *the sum of the shapes of the individual pulses.* Thus, when the two "up" pulses overlap completely, as in **Animated Figure 17.1b**, the Slinky has a pulse height that is twice the height of an individual pulse. Likewise, when the "up" pulse and the "down" pulse overlap exactly, as in **Animated Figure 17.2b**, they momentarily cancel, and the Slinky becomes straight. In either case, the two pulses move apart after overlapping, and the Slinky once again conforms to the shapes of the individual pulses, as in part *c* of both figures.

The adding together of individual pulses to form a resultant pulse is an example of a more general concept called the **principle of linear superposition**.

THE PRINCIPLE OF LINEAR SUPERPOSITION

When two or more waves are present simultaneously at the same place, the resultant disturbance is the sum of the disturbances from the individual waves.

This principle can be applied to all types of waves, including sound waves, water waves, and electromagnetic waves such as light, radio waves, and microwaves. It embodies one of the most important concepts in physics, and the remainder of this chapter deals with examples related to it.

Check Your Understanding

(The answer is given at the end of the book.)

1. CYU Figure 17.1 shows a graph of two pulses traveling toward each other at $t = 0$ s. Each pulse has a constant speed of 1 cm/s. When $t = 2$ s, what is the height of the resultant pulse at **(a)** $x = 3$ cm and **(b)** $x = 4$ cm?

CYU FIGURE 17.1

(a) Overlap begins

(b) Total overlap

(c) The receding pulses

ANIMATED FIGURE 17.2 Two transverse pulses, one "up" and one "down," passing through each other.

17.2 | Constructive and Destructive Interference of Sound Waves

Suppose that the sounds from two speakers overlap in the middle of a listening area, as in **Interactive Figure 17.3**, and that each speaker produces a sound wave of the same amplitude and frequency. For convenience, the wavelength of the sound is chosen to be $\lambda = 1$ m. In addition, assume that the diaphragms of the speakers vibrate in phase; that is, they move outward together and inward together. If the distance of each speaker from the overlap point is the same (3 m in the drawing), the condensations (C) of one wave always meet the condensations of the other when the waves come together; similarly,

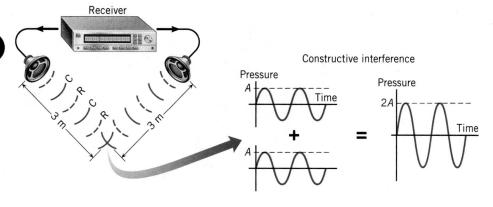

rarefactions (R) always meet rarefactions. According to the principle of linear superposition, the combined pattern is the sum of the individual patterns. As a result, the pressure fluctuations at the overlap point have twice the amplitude A that the individual waves have, and a listener at this spot hears a louder sound than the sound coming from either speaker alone. When two waves always meet condensation-to-condensation and rarefaction-to-rarefaction (or crest-to-crest and trough-to-trough), they are said to be **exactly in phase** and to exhibit **constructive interference**.

Now consider what happens if one of the speakers is moved. The result is surprising. In **Interactive Figure 17.4**, the left speaker is moved away* from the overlap point by a distance equal to one-half of the wavelength, or 0.5 m. Therefore, at the overlap point, a condensation arriving from the left meets a rarefaction arriving from the right. Likewise, a rarefaction arriving from the left meets a condensation arriving from the right. According to the principle of linear superposition, the net effect is a mutual cancellation of the two waves. The condensations from one wave offset the rarefactions from the other, leaving only a *constant air pressure*. A constant air pressure, devoid of condensations and rarefactions, means that a listener detects no sound. When two waves always meet condensation-to-rarefaction (or crest-to-trough), they are said to be **exactly out of phase** and to exhibit **destructive interference**.

INTERACTIVE FIGURE 17.4 The speakers in this drawing vibrate in phase. However, the left speaker is one-half of a wavelength ($\frac{1}{2}$ m) farther from the overlap point than the right speaker. Because of destructive interference, no sound is heard at the overlap point (C, condensation; R, rarefaction).

When two waves meet, they interfere constructively if they always meet exactly in phase and destructively if they always meet exactly out of phase. In either case, this means that the wave patterns do not shift relative to one another as time passes. Sources that produce waves in this fashion are called **coherent sources**.

THE PHYSICS OF . . . noise-canceling headphones. Destructive interference is the basis of a useful technique for reducing the loudness of undesirable sounds. For instance, **Figure 17.5** shows a pair of noise-canceling headphones. Small microphones are mounted inside the headphones and detect noise such as the engine noise

---

*When the left speaker is moved back, its sound intensity and, hence, its pressure amplitude decrease at the overlap point. In this chapter assume that the power delivered to the left speaker by the receiver is increased slightly to keep the amplitudes equal at the overlap point.

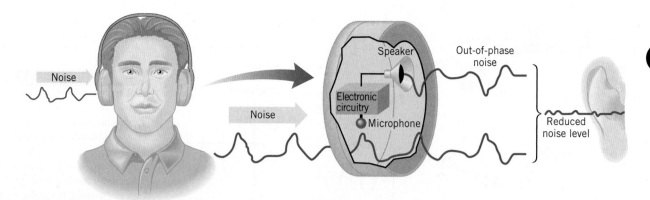

FIGURE 17.5 Noise-canceling headphones utilize destructive interference.

that an airplane pilot would hear. The headphones also contain circuitry to process the electronic signals from the microphones and reproduce the noise in a form that is exactly out of phase compared to the original. This out-of-phase version is played back through the headphone speakers and, because of destructive interference, combines with the original noise to produce a quieter background.

If the left speaker in **Interactive Figure 17.4** were moved away from the overlap point by *another* one-half wavelength $\left(3\frac{1}{2}\text{ m} + \frac{1}{2}\text{ m} = 4\text{ m}\right)$, the two waves would again be in phase, and constructive interference would occur. The listener would hear a loud sound because the left wave travels one whole wavelength ($\lambda = 1$ m) farther than the right wave and, at the overlap point, condensation meets condensation and rarefaction meets rarefaction. In general, the important issue is the *difference* in the path lengths traveled by each wave in reaching the overlap point:

> **Problem-Solving Insight** For two wave sources vibrating in phase, a difference in path lengths that is zero or an integer number (1, 2, 3, . . .) of wavelengths leads to constructive interference; a difference in path lengths that is a half-integer number $\left(\frac{1}{2}, 1\frac{1}{2}, 2\frac{1}{2}, \ldots\right)$ of wavelengths leads to destructive interference.

> **Problem-Solving Insight** For two wave sources vibrating out of phase, a difference in path lengths that is a half-integer number $\left(\frac{1}{2}, 1\frac{1}{2}, 2\frac{1}{2}, \ldots\right)$ of wavelengths leads to constructive interference; a difference in path lengths that is zero or an integer number (1, 2, 3, . . .) of wavelengths leads to destructive interference.

Interference effects can also be detected if the two speakers are fixed in position and the listener moves about the room. Consider **Figure 17.6**, where the sound waves spread outward from each of two in-phase speakers, as indicated by the concentric circular arcs. Each solid arc represents the middle of a condensation, and each dashed arc represents the middle of a rarefaction. Where the two waves overlap, there are places of constructive interference and places of destructive interference. Constructive interference occurs wherever two condensations or two rarefactions intersect, and the drawing shows four such places as solid dots. A listener stationed at any one of these locations hears a loud sound. On the other hand, destructive interference occurs wherever a condensation and a rarefaction intersect, such as the two open dots in the picture. A listener situated at a point of destructive interference hears no sound. At locations where neither constructive nor destructive interference occurs, the two waves partially reinforce or partially cancel, depending on the position relative to the speakers. Thus, it is possible for a listener to walk about the overlap region and hear marked variations in loudness.

The individual sound waves from the speakers in **Figure 17.6** carry energy, and the energy delivered to the overlap region is the sum of the energies of the individual waves. This fact is consistent with the principle of conservation of energy, which we first encountered in Section 6.8. This principle states that energy can neither be created nor destroyed, but can only be converted from one form to another. One of the interesting consequences of interference is that the energy is redistributed, so there are places within the overlap region where the sound is loud and other places where there is no sound at all. Interference, so to speak, "robs Peter to pay Paul," but energy is always conserved in the process. Example 1 illustrates how to decide what a listener hears.

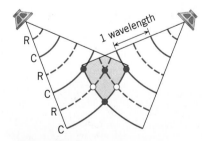

FIGURE 17.6 Two sound waves overlap in the shaded region. The solid lines denote the middle of the condensations (C), and the dashed lines denote the middle of the rarefactions (R). Constructive interference occurs at each solid dot (●) and destructive interference at each open dot (○).

EXAMPLE 1 | What Does a Listener Hear?

In **Figure 17.7** two in-phase loudspeakers, A and B, are separated by 3.20 m. A listener is stationed at point C, which is 2.40 m in front of speaker B. The triangle ABC is a right triangle. Both speakers are playing identical 214-Hz tones, and the speed of sound is 343 m/s. Does the listener hear a loud sound or no sound?

Reasoning The listener will hear either a loud sound or no sound, depending on whether the interference occurring at point C is constructive or destructive. To determine which it is, we need to find the difference in the distances traveled by the two sound waves that reach point C and see whether the difference is an integer or half-integer number of wavelengths. In either event, the wavelength can be found from the relation $\lambda = v/f$ (Equation 16.1).

> **Problem-Solving Insight** To decide whether two sources of sound produce constructive or destructive interference at a point, determine the difference in path lengths between each source and that point and compare it to the wavelength of the sound.

Solution Since the triangle ABC is a right triangle, the distance AC is given by the Pythagorean theorem as $\sqrt{(3.20\,\text{m})^2 + (2.40\,\text{m})^2} = 4.00$ m. The distance BC is given as 2.40 m. Thus, the difference in the travel distances for the waves is 4.00 m − 2.40 m = 1.60 m. The wavelength of the sound is

$$\lambda = \frac{v}{f} = \frac{343 \text{ m/s}}{214 \text{ Hz}} = 1.60 \text{ m} \qquad \textbf{(16.1)}$$

FIGURE 17.7 Example 1 discusses whether this setup leads to constructive or destructive interference at point C for 214-Hz sound waves.

Since the difference in the distances is an integer number ($n = 0, 1, 2 \ldots$) of wavelengths (here $n = 1$), constructive interference occurs at point C, and the *listener hears a loud sound.*

It should be noted that, if the difference in the distances had not been equal to an integer (constructive interference) or an odd half-integer (destructive interference) number of wavelengths, the listener would hear a sound with a loudness somewhere between zero and the maximum which occurs as a result of constructive interference.

Up to this point, we have been assuming that the speaker diaphragms vibrate synchronously, or in phase; that is, they move outward together and inward together. This may not be the case, however, and Conceptual Example 2 considers what happens then.

CONCEPTUAL EXAMPLE 2 | Out-of-Phase Speakers

To make a speaker operate, two wires (one red and one black, for instance) must be connected between the speaker and the receiver (amplifier), as in **Figure 17.8**. Consider one of the speakers in **Interactive Figure 17.4**, where the red wire connects the red terminal of the speaker to the red terminal of the receiver. Similarly, the black wire connects the black terminal of the speaker to the black terminal of the receiver. For the other speaker, however, the wires connect a terminal of one color on the speaker to a terminal of a different color on the receiver. Since the two speakers are not connected to the receiver in exactly the same way, the two diaphragms will vibrate out of phase, one moving outward every time the other moves inward, and vice versa. A listener at the overlap point in **Interactive Figure 17.4** would now hear **(a)** no sound because destructive interference occurs **(b)** a loud sound because constructive interference occurs.

Reasoning Since the speaker diaphragms in **Interactive Figure 17.4** are now vibrating out of phase, one of them is moving exactly opposite to the way it was moving originally; let us assume that it is the left speaker. The effect of this change is that every condensation originating from the left speaker becomes a rarefaction, and every rarefaction becomes a condensation.

FIGURE 17.8 A loudspeaker is connected to a receiver (amplifier) by two wires.

Andy Washnik

Answer (a) is incorrect. When the two speakers in **Interactive Figure 17.4** are wired *in phase*, a condensation from one speaker always meets a rarefaction from the other at the overlap point, and destructive interference occurs. However, if one of the speakers were wired out of phase relative to the other, a condensation from one speaker would meet a condensation from the other, and destructive interference would *not* occur.

Answer (b) is correct. If the left speaker in **Interactive Figure 17.4** were connected out of phase with respect to the right speaker, a condensation from the right speaker would meet a condensation (not a rarefaction) from the left speaker at the overlap point. Similarly, a rarefaction from the right speaker would meet a rarefaction from the left speaker. The result is *constructive* interference, and a loud sound would be heard.

THE PHYSICS OF . . . wiring the speakers in an audio system. Instructions for connecting stereo or surround-sound systems specifically warn owners to avoid out-of-phase vibration of speaker diaphragms. If the wires and the terminals of the speakers and the receiver are not color-coded, you can check for problems in the following way. Play music with a lot of low-frequency bass tones. Set the receiver to its monaural mode, so the same sound comes out of both speakers being tested. If the diaphragms are in phase, the bass sound will either remain the same or become slightly louder as you slide the speakers together. If the diaphragms are out of phase, the bass sound will fade noticeably (due to destructive interference) when the speakers are right next to each other. In this event, simply interchange the wires to the terminals on one (not both) of the speakers.

The phenomena of constructive and destructive interference are exhibited by all types of waves, not just sound waves. We will encounter interference effects again in Chapter 27, in connection with light waves.

Check Your Understanding

(*The answers are given at the end of the book.*)

2. Does the principle of linear superposition imply that two sound waves, passing through the same place at the same time, always create a louder sound than is created by either wave alone?

3. Suppose that you are sitting at the overlap point between the two speakers in **Interactive Figure 17.4**. Because of destructive interference, you hear no sound, even though both speakers are emitting identical sound waves. One of the speakers is suddenly shut off. Will you now hear a sound? **(a)** No. **(b)** Yes. **(c)** Yes, but only if you move a distance of one wavelength closer to the speaker that is still producing sound.

4. Starting at the overlap point in **Interactive Figure 17.3**, you walk along a straight path that is *perpendicular* to the line between the speakers and passes through the midpoint of that line. As you walk, the loudness of the sound **(a)** changes from loud to faint to loud **(b)** changes from faint to loud to faint **(c)** does not change.

5. Starting at the overlap point in **Interactive Figure 17.3**, you walk along a path that is *parallel* to the line between the speakers. As you walk, the loudness of the sound **(a)** changes from loud to faint to loud **(b)** changes from faint to loud to faint **(c)** does not change.

17.3 Diffraction

Section 16.5 discusses the fact that sound is a pressure wave created by a vibrating object, such as a loudspeaker. The previous two sections of this chapter have examined what happens when two sound waves are present simultaneously at the same place; according to the principle of linear superposition, a resultant disturbance is formed from the sum of the individual waves. This principle reveals that overlapping sound waves exhibit interference effects, whereby the sound energy is redistributed within the overlap region. We will now use the principle of linear superposition to explore another interference effect, that of diffraction.

When a wave encounters an obstacle or the edges of an opening, it bends around them. For instance, a sound wave produced by a stereo system bends around the edges of an open doorway, as **Figure 17.9a** illustrates. If such bending did not occur, sound could be heard outside the room only at locations directly in front of the doorway, as part *b* of the drawing suggests. (It is assumed that no sound is transmitted directly through the walls.) The bending of a wave around an obstacle or the edges of an opening is called **diffraction**. All kinds of waves exhibit diffraction.

To demonstrate how the bending of waves arises, **Figure 17.10** shows an expanded view of **Figure 17.9a**. When the sound wave reaches the doorway, the air in the doorway is set into longitudinal vibration. In effect, each molecule of the air in the doorway becomes a source of a sound wave in its own right, and, for purposes of illustration, the drawing shows two of the molecules. Each produces a sound wave that expands outward in three dimensions, much like a water wave does in two dimensions when a stone is dropped into a pond. The sound waves generated by all the molecules in the doorway must be added together to obtain the total sound wave at any location outside the room, in accord with the principle of linear superposition. However, even considering only the waves from the two molecules in the picture, it is clear that the expanding wave patterns reach locations off to either side of the doorway. The net effect is a "bending," or diffraction, of the sound around the edges of the opening. Further insight into the origin of diffraction can be obtained with the aid of Huygens' principle (see Section 27.5).

When the sound waves generated by every molecule in the doorway are added together, it is found that there are places where the intensity is a maximum and places where it is zero, in a fashion similar to that discussed in the previous section. Analysis shows that at a great distance from the doorway the intensity is a maximum directly opposite the center of the opening. As the distance to either side of the center increases, the intensity decreases and reaches zero, then rises again to a maximum, falls again to zero, rises back to a maximum, and so on. Only the maximum at the center is a strong one. The other maxima are weak and become progressively weaker at greater distances from the center. In **Figure 17.10** the angle θ defines the location of the first minimum intensity point on either side of the center. Equation 17.1 gives θ in terms of the wavelength λ and the width D of the doorway and assumes that the doorway can be treated like a slit whose height is very large compared to its width:

Single slit—first minimum $\sin\theta = \dfrac{\lambda}{D}$ (17.1)

Waves also bend around the edges of openings other than single slits. Particularly important is the diffraction of sound by a circular opening, such as that in a loudspeaker. In this case, the angle θ is related to the wavelength λ and the diameter D of the opening by

Circular opening—first minimum $\sin\theta = 1.22\dfrac{\lambda}{D}$ (17.2)

An important point to remember about Equations 17.1 and 17.2 is that the extent of the diffraction depends on the ratio of the wavelength to the size of the opening. If the ratio λ/D is small, then θ is small and little diffraction occurs. The waves are beamed in the forward direction as they leave an opening, much like the light from a flashlight. Such sound waves are said to have "narrow dispersion." Since high-frequency sound has a relatively small wavelength, it tends to have a narrow dispersion. On the other hand, for larger values of the ratio λ/D, the angle θ is larger. The waves spread out over a larger region and are said to have a "wide dispersion." Low-frequency sound, with its relatively large wavelength, typically has a wide dispersion.

In a stereo loudspeaker, a wide dispersion of the sound is desirable. Example 3 illustrates, however, that there are limitations to the dispersion that can be achieved, depending on the loudspeaker design.

(a) With diffraction

(b) Without diffraction

FIGURE 17.9 (*a*) The bending of a sound wave around the edges of the doorway is an example of diffraction. The source of the sound within the room is not shown. (*b*) If diffraction did not occur, the sound wave would not bend as it passed through the doorway.

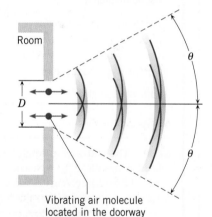

Vibrating air molecule located in the doorway

FIGURE 17.10 Each vibrating molecule of the air in the doorway generates a sound wave that expands outward and bends, or diffracts, around the edges of the doorway. Because of interference effects among the sound waves produced by all the molecules, the sound intensity is mostly confined to the region defined by the angle θ on either side of the doorway.

Analyzing Multiple-Concept Problems

EXAMPLE 3 | The Physics of Loudspeakers

A 1500-Hz sound and a 8500-Hz sound emerge from a loud-speaker through a circular opening that has a diameter of 0.30 m (see **Figure 17.11**). Assuming that the speed of sound in air is 343 m/s, find the diffraction angle θ for each sound.

Reasoning The diffraction angle θ depends on the ratio of the wavelength λ of the sound to the diameter D of the opening, according to $\sin\theta = 1.22(\lambda/D)$ (Equation 17.2). The wavelength for each sound can be obtained from the given frequencies and the speed of sound.

The high frequencies are beamed forward inside this cone

9.4° 68°

This person hears primarily the low frequencies

This person hears the high and low frequencies

FIGURE 17.11 Because the dispersion of high frequencies is less than the dispersion of low frequencies, you should be directly in front of the speaker to hear both the high and low frequencies equally well.

Knowns and Unknowns The following data are available:

Description	Symbol	Value
Sound frequency	f	1500 Hz or 8500 Hz
Diameter of speaker opening	D	0.30 m
Speed of sound	v	343 m/s
Unknown Variables		
Diffraction angle	θ	?

Modeling the Problem

STEP 1 The Diffraction Angle Equation 17.2 indicates that the diffraction angle is related to the ratio of the wavelength λ of the sound to the diameter D of the opening by $\sin\theta = 1.22(\lambda/D)$. Solving this expression for θ gives Equation 1 at the right. A value for D is given. To determine the value for λ, we turn to Step 2.

$$\theta = \sin^{-1}\left(1.22\frac{\lambda}{D}\right) \qquad (1)$$

$$\boxed{?}$$

STEP 2 Wavelength The wavelength λ is related to the frequency f and the speed v of the sound according to $v = f\lambda$ (Equation 16.1). Solving for λ gives

$$\boxed{\lambda = \frac{v}{f}}$$

which can be substituted into Equation 1 as shown at the right.

$$\theta = \sin^{-1}\left(1.22\frac{\lambda}{D}\right) \qquad (1)$$

$$\boxed{\lambda = \frac{v}{f}}$$

> **Problem-Solving Insight** When a wave passes through an opening, the extent of diffraction is greater when the ratio λ/D is greater, where λ is the wavelength of the wave and D is the width or diameter of the opening.

Solution Combining the results of each step algebraically, we find that

$$\overset{\text{STEP 1}}{\downarrow} \quad \overset{\text{STEP 2}}{\downarrow}$$

$$\theta = \sin^{-1}\left(1.22\frac{\lambda}{D}\right) = \sin^{-1}\left(1.22\frac{v/f}{D}\right)$$

Applying the above result to each of the sound frequencies shows that

1500-Hz *sound* $\theta = \sin^{-1}\left(1.22\dfrac{v}{fD}\right) = \sin^{-1}\left[1.22\dfrac{(343 \text{ m/s})}{(1500 \text{ Hz})(0.30 \text{ m})}\right] = \boxed{68°}$

8500-Hz *sound* $\theta = \sin^{-1}\left(1.22\dfrac{v}{fD}\right) = \sin^{-1}\left[1.22\dfrac{(343 \text{ m/s})}{(8500 \text{ Hz})(0.30 \text{ m})}\right] = \boxed{9.4°}$

Figure 17.11 illustrates these results. With a 0.30-m opening, the dispersion of the higher-frequency sound is limited to only 9.4°. To increase the dispersion, a smaller opening is needed. It is for this reason that loudspeaker designers use a small-diameter speaker called a *tweeter* to generate the high-frequency sound (see **Figure 17.12**).

Related Homework: Problems 12, 14

Tweeter

FIGURE 17.12 Small-diameter speakers (called tweeters) are used to produce high-frequency sound. The small diameter helps to promote a wider dispersion of the sound.

JPagetRFphotos/Alamy Stock Photo

As we have seen, diffraction is an interference effect, one in which some of the wave's energy is directed into regions that would otherwise not be accessible. Energy, of course, is conserved during this process, because energy is only redistributed during diffraction; no energy is created or destroyed.

Check Your Understanding

(The answers are given at the end of the book.)

6. At an open-air rock concert you are standing directly in front of a speaker. You hear the high-frequency sounds of a female vocalist as well as the low-frequency sounds of the rhythmic bass. As you walk to one side of the speaker, the sounds of the vocalist _____, and those of the rhythmic bass _____. **(a)** drop off noticeably; also drop off noticeably **(b)** drop off only slightly; drop off noticeably **(c)** drop off only slightly; also drop off only slightly **(d)** drop off noticeably; drop off only slightly

7. Refer to Example 1 in Section 16.2. Which type of radio wave, AM or FM, diffracts more readily around a given obstacle? **(a)** AM, because it has a greater wavelength **(b)** FM, because it has a greater wavelength **(c)** AM, because it has a greater frequency **(d)** FM, because it has a greater frequency

17.4 Beats

In situations where waves with the *same frequency* overlap, we have seen how the principle of linear superposition leads to constructive and destructive interference and how it explains diffraction. We will see in this section that two overlapping waves with *slightly different frequencies* give rise to the phenomenon of beats. However, the principle of linear superposition again provides an explanation of what happens when the waves overlap.

A tuning fork has the property of producing a single-frequency sound wave when struck with a sharp blow. **Figure 17.13** shows sound waves coming from two tuning forks placed side by side. The tuning forks in the drawing are identical, and each is designed to produce a 440-Hz tone. However, a small piece of putty has been attached

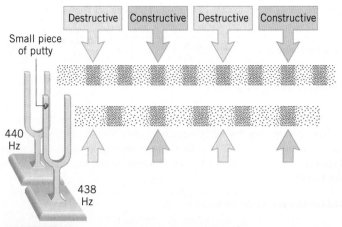

FIGURE 17.13 Two tuning forks have slightly different frequencies of 440 and 438 Hz. The phenomenon of beats occurs when the forks are sounded simultaneously. The sound waves are not drawn to scale.

to one fork, whose frequency is lowered to 438 Hz because of the added mass. When the forks are sounded simultaneously, the loudness of the resulting sound rises and falls periodically—faint, then loud, then faint, then loud, and so on. The periodic variations in loudness are called *beats* and result from the interference between two sound waves with slightly different frequencies.

For clarity, **Figure 17.13** shows the condensations and rarefactions of the sound waves separately. In reality, however, the waves spread out and overlap. In accord with the principle of linear superposition, the ear detects the combined total of the two. Notice that there are places where the waves interfere constructively and places where they interfere destructively. When a region of constructive interference reaches the ear, a loud sound is heard. When a region of destructive interference arrives, the sound intensity drops to zero (assuming each of the waves has the same amplitude). The number of times per second that the loudness rises and falls is the **beat frequency** and is the **difference** between the two sound frequencies. Thus, in the situation illustrated in Figure 17.13, an observer hears the sound loudness rise and fall at the rate of 2 times per second (440 Hz – 438 Hz).

Figure 17.14 helps to explain why the beat frequency is the difference between the two frequencies. The drawing displays graphical representations of the pressure patterns of a 10-Hz wave and a 12-Hz wave, along with the pressure pattern that results when the two overlap. These frequencies have been chosen for convenience, even though they lie below the audio range and are inaudible. Audible sound waves behave in exactly the same way. The top two drawings, in blue, show the pressure variations in a one-second interval of each wave. The third drawing, in red, shows the result of adding together the blue patterns according to the principle of linear superposition. Notice that the amplitude in the red drawing is not constant, as it is in the individual waves. Instead, the amplitude changes from a minimum to a maximum, back to a minimum, and so on. When such pressure variations reach the ear and occur in the audible frequency range, they produce a loud sound when the amplitude is a maximum and a faint sound when the amplitude is a minimum. Two loud–faint cycles, or beats, occur in the one-second interval shown in the drawing, corresponding to a beat frequency of 2 Hz. Thus, the beat frequency is the difference between the frequencies of the individual waves, or 12 Hz – 10 Hz = 2 Hz. Beats will occur whenever two frequencies are slightly different, as demonstrated in the following example.

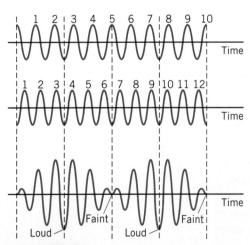

FIGURE 17.14 A 10-Hz and a 12-Hz sound wave, when added together, produce a wave with a beat frequency of 2 Hz. The drawings show the pressure patterns (in blue) of the individual waves and the pressure pattern (in red) that results when the two overlap. The time interval shown is one second.

EXAMPLE 4 | [BIO] The Physics of Detecting Small Differences in Frequency

The ability of our ears to distinguish between two pure tones of slightly different frequencies is known as **frequency discrimination**. Research has shown that for frequencies below 1000-Hz, we are able to distinguish a second tone with a change in frequency of only 1 to 3 Hz, assuming the second tone is emitted right after the first one and at nearly the same intensity. In general, our ability to discriminate two frequencies steadily decreases at higher frequencies, and studies involving people with peripheral hearing loss have shown that auditory pathology can actually improve frequency discrimination, as the brain undergoes neural reorganization after cochlear damage. With this being said, we are able to distinguish two slightly different frequencies *simultaneously*. The rise and fall in intensity of the beat frequency is very noticeable, especially when the difference in frequencies is small. For example, imagine the situation where you are at rest in your car, and you are honking the horn, which has a frequency of 525 Hz. Your friend is driving toward you in the same-model car and honking his horn with the same 525-Hz frequency. You hear a beat frequency of 2.00 Hz. How fast is your friend driving toward you? Assume the speed of sound in air is 343 m/s.

Reasoning Since the two horns produce sounds of the same frequency, the beats indicate a shift in one of the observed frequencies.

Since your friend's car is moving and yours is stationary, the shift is due to the motion of his car and can be attributed to the Doppler effect. Equation 16.11 can be used to find the speed of the moving car from the Doppler shift. Here, we have the case of a source moving toward a stationary observer.

Solution The Doppler effect will result in a higher observed frequency, since the sound source is moving toward the stationary observer. Since the beat frequency is equal to the Doppler shift we have the following:

$$f_o - f_s = 2 \text{ Hz} \Rightarrow f_o = f_s + 2 \text{ Hz} = 525 \text{ Hz} + 2 \text{ Hz} = 527 \text{ Hz}.$$

We can now apply Equation 16.11 and solve for the speed of the source (moving car):

$$f_o = f_s \left(\frac{1}{1 - \frac{v_s}{v}} \right) \Rightarrow 1 - \frac{v_s}{v} = \frac{f_s}{f_o} \Rightarrow v_s = v \left(1 - \frac{f_s}{f_o} \right)$$

$$= (343 \text{ m/s}) \left(1 - \frac{525 \text{ Hz}}{527 \text{ Hz}} \right) = \boxed{1.30 \text{ m/s}}$$

As a final note, car horns are often arranged in pairs—one with a frequency of 500 Hz and the second with a frequency between 405 – 420 Hz. The beat frequency produced by the two horns is more perceptible than a single tone, making it easier to detect, especially in an environment with a high level of ambient noise.

THE PHYSICS OF . . . tuning a musical instrument. Musicians often tune their instruments by listening to a beat frequency. For instance, a guitar player plucks an out-of-tune string along with a tone that has the correct frequency. He then adjusts the string tension until the beats vanish, ensuring that the string is vibrating at the correct frequency.

Check Your Understanding

(The answers are given at the end of the book.)

8. Tuning fork A (frequency unknown) and tuning fork B (frequency = 384 Hz) together produce 6 beats in 2 seconds. When a small piece of putty is attached to tuning fork A, as in **Figure 17.13**, the beat frequency decreases. What is the frequency of tuning fork A before the putty is attached? **(a)** 378 Hz **(b)** 381 Hz **(c)** 387 Hz **(d)** 390 Hz

9. A tuning fork has a frequency of 440 Hz. The string of a violin and this tuning fork, when sounded together, produce a beat frequency of 1 Hz. From these two pieces of information alone, is it possible to determine the exact frequency of the violin string? **(a)** Yes; the frequency of the violin string is 441 Hz. **(b)** No, because the frequency of the violin string could be either 439 or 441 Hz. **(c)** Yes; the frequency of the violin string is 439 Hz.

10. When the regions of constructive and destructive interference in **Figure 17.13** move past a listener's ear, a beat frequency of 2 Hz is heard. Suppose that the tuning forks in the drawing are sounded under water and that the listener is also under water. The forks vibrate at 438 and 440 Hz, just as they do in air. However, sound travels four times faster in water than in air. The beat frequency heard by the underwater listener is **(a)** 16 Hz **(b)** 8 Hz **(c)** 4 Hz **(d)** 2 Hz.

17.5 | Transverse Standing Waves

A standing wave is another interference effect that can occur when two waves overlap. Standing waves can arise with transverse waves, such as those on a guitar string, and also with longitudinal sound waves, such as those in a flute. In any case, the principle

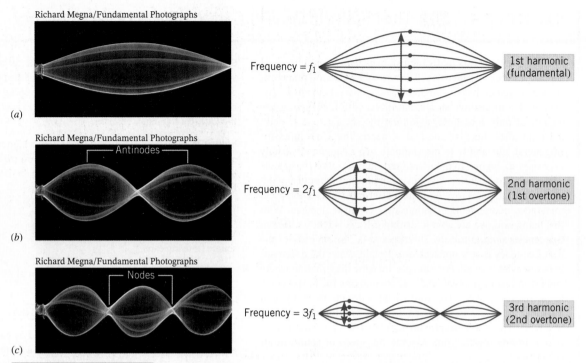

Richard Megna/Fundamental Photographs

(a) Frequency = f_1 — 1st harmonic (fundamental)

Richard Megna/Fundamental Photographs — Antinodes

(b) Frequency = $2f_1$ — 2nd harmonic (1st overtone)

Richard Megna/Fundamental Photographs — Nodes

(c) Frequency = $3f_1$ — 3rd harmonic (2nd overtone)

INTERACTIVE FIGURE 17.15 Vibrating a string at certain unique frequencies sets up transverse standing wave patterns, such as the three shown in the photographs on the left. Each drawing on the right shows the various shapes that the string assumes at various times as it vibrates. The red dots attached to the strings focus attention on the maximum vibration that occurs at an antinode. In each of the drawings, one-half of a wave cycle is outlined in red. (Richard Megna/Fundamental Photographs)

of linear superposition provides an explanation of the effect, just as it does for diffraction and beats.

Interactive Figure 17.15 shows some of the essential features of transverse standing waves. In this figure the left end of each string is vibrated back and forth, while the right end is attached to a wall. Regions of the string move so fast that they appear only as a blur in the photographs. Each of the patterns shown is called a **transverse standing wave pattern**. Notice that the patterns include special places called nodes and antinodes. The **nodes** are places that do not vibrate at all, and the **antinodes** are places where maximum vibration occurs. To the right of each photograph is a drawing that helps us to visualize the motion of the string as it vibrates in a standing wave pattern. These drawings freeze the shape of the string at various times and emphasize the maximum vibration that occurs at an antinode with the aid of a red dot attached to the string.

Each standing wave pattern is produced at a unique frequency of vibration. These frequencies form a series, the smallest frequency f_1 corresponding to the one-loop pattern and the larger frequencies being integer multiples of f_1, as **Interactive Figure 17.15** indicates. Thus, if f_1 is 10 Hz, the frequency needed to establish the 2-loop pattern is $2f_1$ or 20 Hz, whereas the frequency needed to create the 3-loop pattern is $3f_1$ or 30 Hz, and so on. The frequencies in this series (f_1, $2f_1$, $3f_1$, etc.) are called **harmonics**. The lowest frequency f_1 is called the first harmonic, and the higher frequencies are designated as the second harmonic ($2f_1$), the third harmonic ($3f_1$), and so forth. The harmonic number (1st, 2nd, 3rd, etc.) corresponds to the number of loops in the standing wave pattern. The frequencies in this series are also referred to as the fundamental frequency, the first overtone, the second overtone, and so on. Thus, frequencies above the fundamental are **overtones** (see **Interactive Figure 17.15**).

Standing waves arise because identical waves travel on the string in *opposite directions* and combine in accord with the principle of linear superposition. A standing wave is said to be *standing* because it does not travel in one direction or the other, as do the individual waves that produce it. **Figure 17.16** shows why there are waves traveling in both directions on the string. At the top of the picture, one-half of a wave cycle (the remainder of the wave is omitted for clarity) is moving toward the wall on the right. When the half-cycle reaches the wall, it causes the string to pull upward on the wall. Consistent with Newton's action-reaction law, the wall pulls downward on the string, and a downward-pointing half-cycle is sent back toward the left. Thus, the wave reflects from the wall. Upon arriving back at the point of origin, the wave reflects again, this time from the hand vibrating the string. For small vibration amplitudes, the hand is essentially fixed and behaves as the wall does in causing reflections. Repeated reflections at both ends of the string create a multitude of wave cycles traveling in both directions.

FIGURE 17.16 In reflecting from the wall, a forward-traveling half-cycle becomes a backward-traveling half-cycle that is inverted.

As each new cycle is formed by the vibrating hand, previous cycles that have reflected from the wall arrive and reflect again from the hand. Unless the timing is right, however, the new and the reflected cycles tend to offset one another, and a standing wave is not formed. Think about pushing someone on a swing and timing your pushes so that they reinforce one another. Such reinforcement in the case of the wave cycles leads to a large-amplitude standing wave. Suppose that the string has a length L and its left end is being vibrated at a frequency f_1. The time required to create a new wave cycle is the period T of the wave, where $T = 1/f_1$ (Equation 10.5). On the other hand, the time needed for a cycle to travel from the hand to the wall and back, a distance of $2L$, is $2L/v$, where v is the wave speed. Reinforcement between new and reflected cycles occurs if these two times are equal; that is, if $1/f_1 = 2L/v$. Thus, a standing wave is formed when the string is vibrated with a frequency of $f_1 = v/(2L)$.

Repeated reinforcement between newly created and reflected cycles causes a large-amplitude standing wave to develop on the string, *even when the hand itself vibrates with only a small amplitude.* Thus, the motion of the string is a resonance effect, analogous to that discussed in Section 10.6 for an object attached to a spring. The frequency f_1 at which resonance occurs is sometimes called a **natural frequency** of the string, similar to the frequency at which an object oscillates on a spring.

There is a difference between the resonance of the string and the resonance of a spring system, however. An object on a spring has only a single natural frequency, whereas the string has a *series* of natural frequencies. The series arises because a reflected wave cycle need not return to its point of origin in time to reinforce *every* newly created cycle. Reinforcement can occur, for instance, on *every other* new cycle, as it does if the string is vibrated at twice the frequency f_1, or $f_2 = 2f_1$. Likewise, if the vibration frequency is $f_3 = 3f_1$, reinforcement occurs on *every third* new cycle. Similar arguments apply for any frequency $f_n = nf_1$, where n is an integer. As a result, the series of natural frequencies that lead to standing waves on a string fixed at both ends is

String fixed at both ends $$f_n = n\left(\frac{v}{2L}\right) \qquad n = 1, 2, 3, 4, \ldots \qquad \textbf{(17.3)}$$

It is also possible to obtain Equation 17.3 in another way. In **Interactive Figure 17.15**, one-half of a wave cycle is outlined in red for each of the harmonics, to show that each loop in a standing wave pattern corresponds to one-half a wavelength. Since the two fixed ends of the string are nodes, the length L of the string must contain an integer number n of half-wavelengths: $L = n\left(\frac{1}{2}\lambda_n\right)$ or $\lambda_n = 2L/n$. Using this result for the wavelength in the relation $f_n\lambda_n = v$ shows that $f_n(2L/n) = v$, which can be rearranged to give Equation 17.3.

> **Problem-Solving Insight** The distance between two successive nodes (or between two successive antinodes) of a standing wave is equal to one-half of a wavelength.

Standing waves on a string are important in the way many musical instruments produce sound. For instance, a guitar string is stretched between two supports and, when plucked, vibrates according to the series of natural frequencies given by Equation 17.3. The next two examples show how this series of frequencies governs the design of a guitar.

Analyzing Multiple-Concept Problems

EXAMPLE 5 | Playing a Guitar

The heaviest string on an electric guitar has a linear density of 5.28×10^{-3} kg/m and is stretched with a tension of 226 N. This string produces the musical note E when vibrating along its entire length in a standing wave at the fundamental frequency of 164.8 Hz. **(a)** Find the length L of the string between its two fixed ends (see **Figure 17.17a**). **(b)** A guitar player wants the string to vibrate at a fundamental frequency of 2×164.8 Hz $= 329.6$ Hz, as it must if the musical note E is to be sounded one octave higher in pitch. To accomplish this, he presses the string against the proper fret before plucking the string. Find the distance L between the fret and the bridge of the guitar (see **Figure 17.17b**).

Reasoning   The series of natural frequencies (including the fundamental) for a string fixed at both ends is given by $f_n = nv/(2L)$ (Equation 17.3), where $n = 1, 2, 3$, etc. This equation can be solved for the length L. The speed v at which waves travel can be obtained from the tension and the linear density of the string. The fundamental frequencies that are given correspond to $n = 1$.

(a)   (b)

FIGURE 17.17   These drawings show the standing waves (in blue) that exist on a guitar string under different playing conditions.

Knowns and Unknowns   The given data are summarized as follows:

Description	Symbol	Value	Comment
Explicit Data			
Linear density of string	m/L	5.28×10^{-3} kg/m	
Tension in string	F	226 N	
Natural frequency at which string vibrates	f_n	164.8 Hz or 329.6 Hz	These are fundamental frequencies.
Implicit Data			
Integer variable in series of natural frequencies	n	1	Fundamental frequencies are given.
Unknown Variable			
Length	L	?	

Modeling the Problem

STEP 1 Natural Frequencies   According to Equation 17.3, the natural frequencies for a string fixed at both ends are given by $f_n = nv/(2L)$, where n takes on the integer values 1, 2, 3, etc., v is the speed of the waves on the string, and L is the length between the two fixed ends. Solving this expression for L gives Equation 1 at the right. In this equation, only the speed v is unknown. We will obtain a value for it in Step 2.

$$L = \frac{nv}{2f_n} \qquad (1)$$

$$\boxed{?}$$

STEP 2 Speed of the Waves on the String   The speed v of the waves traveling on the string is given by Equation 16.2 as

$$v = \sqrt{\frac{F}{m/L}}$$

where F is the tension and m/L is the linear density, both of which are given. The substitution of this expression for the speed into Equation 1 is shown at the right.

$$L = \frac{nv}{2f_n} \qquad (1)$$

$$v = \sqrt{\frac{F}{m/L}}$$

Solution   Combining the results of each step algebraically, we find that

STEP 1 STEP 2

$$L \;\underset{\downarrow}{=}\; \frac{nv}{2f_n} \;\underset{\downarrow}{=}\; \frac{n}{2f_n}\sqrt{\frac{F}{m/L}}$$

It is given that the two natural frequencies at which the string vibrates are fundamental frequencies. Thus, $n = 1$ and $f_n = f_1$. We now determine the desired lengths in parts (a) and (b).

(a) $\qquad L = \dfrac{n}{2f_n}\sqrt{\dfrac{F}{m/L}} = \dfrac{1}{2(164.8\ \text{Hz})}\sqrt{\dfrac{226\ \text{N}}{5.28 \times 10^{-3}\ \text{kg/m}}} = \boxed{0.628\ \text{m}}$

(b) $\qquad L = \dfrac{n}{2f_n}\sqrt{\dfrac{F}{m/L}} = \dfrac{1}{2(329.6\ \text{Hz})}\sqrt{\dfrac{226\ \text{N}}{5.28 \times 10^{-3}\ \text{kg/m}}} = \boxed{0.314\ \text{m}}$

The length in part (b) is one-half the length in part (a) because the fundamental frequency in part (b) is twice the fundamental frequency in part (a).

Related Homework: Problems 28, 31, 33

CONCEPTUAL EXAMPLE 6 | The Physics of the Frets on a Guitar

Figure 17.18 shows the frets on the neck of a guitar. They allow the player to produce a complete sequence of musical notes using a single string. Starting with the fret at the top of the neck, each successive fret shows where the player should press to get the next note in the sequence. Musicians call the sequence the chromatic scale, and every thirteenth note in it corresponds to one octave, or a doubling of the sound frequency. Which describes the spacing between the frets? It is **(a)** the same everywhere along the neck **(b)** greatest at the top of the neck and decreases with each additional fret further on down toward the bridge **(c)** smallest at the top of the neck and increases with each additional fret further on down toward the bridge.

Reasoning   Our reasoning is based on the relation $f_1 = v/(2L)$ (Equation 17.3, with $n = 1$). The value of n is 1 because a string vibrates mainly at its fundamental frequency when plucked, as mentioned in Example 5. This equation shows that L, which is the length between a given fret and the bridge of the guitar, is inversely proportional to the fundamental frequency f_1, or $L = v/(2f_1)$. In Example 5 we found that the E string has a length of $L = 0.628$ m, corresponding to a fundamental frequency of $f_1 = 164.8$ Hz. We also found that the length between the bridge and the fret that must be pressed to double this frequency

to 2×164.8 Hz $= 329.6$ Hz is one-half of 0.628 m, or $L = 0.314$ m. To understand the spacing between frets as one moves down the neck, consider the fret that must be pressed to double the frequency again, from 329.6 Hz to 659.2 Hz. The length between the bridge and this fret would be one-half of 0.314 m, or $L = 0.157$ m. Thus, the distances of the three frets that we have been discussing are 0.628 m, 0.314 m, and 0.157 m, as indicated in **Figure 17.18**. The distance D_1 between the first two of these frets is $D_1 = 0.628$ m $- 0.314$ m $= 0.314$ m. Similarly, the distance between the second and third of these frets is $D_2 = 0.314$ m $- 0.157$ m $= 0.157$ m.

Answers (a) and (c) are incorrect.   The distances between the frets are shown in **Figure 17.18**. Clearly, the distances D_1 and D_2 are not equal, nor are they smaller at the top of the neck and greater further on down.

Answer (b) is correct.   Figure 17.18 shows that D_1 is greater than D_2. Thus, the spacing between the frets is greatest at the top of the neck and decreases with each additional fret further on down.

Related Homework: Problem 38

FIGURE 17.18   The spacing between the frets on the neck of a guitar changes going down the neck toward the bridge.

Check Your Understanding

(The answers are given at the end of the book.)

11. A standing wave that corresponds to the fourth harmonic is set up on a string that is fixed at both ends. **(a)** How many loops are in this standing wave? **(b)** How many nodes (excluding the nodes at the ends of the string) does this standing wave have? **(c)** Is there a node or an antinode at the midpoint of the string? **(d)** If the frequency of this standing wave is 440 Hz, what is the frequency of the lowest-frequency standing wave that could be set up on this string?

12. The tension in a guitar string is doubled. By what factor does the frequency of the vibrating string change? **(a)** It increases by a factor of 2. **(b)** It increases by a factor of $\sqrt{2}$. **(c)** It decreases by a factor of 2. **(d)** It decreases by a factor of $\sqrt{2}$.

13. A string is vibrating back and forth as in **Interactive Figure 17.15a**. The tension in the string is decreased by a factor of four, with the frequency and length of the string remaining the same. A new standing wave pattern develops on the string. How many loops are in this new pattern? **(a)** 5 **(b)** 4 **(c)** 3 **(d)** 2

14. A rope is hanging vertically straight down. The top end is being vibrated back and forth, and a standing wave with many loops develops on the rope, analogous (but not identical) to a standing wave on a horizontal rope. The rope has mass. The separation between successive nodes is **(a)** everywhere the same along the rope **(b)** greater near the top of the rope than near the bottom **(c)** greater near the bottom of the rope than near the top.

17.6 | Longitudinal Standing Waves

Standing wave patterns can also be formed from longitudinal waves. For example, when sound reflects from a wall, the forward- and backward-going waves can produce a standing wave. **Figure 17.19** illustrates the vibrational motion in a longitudinal standing wave on a Slinky. As in a transverse standing wave, there are nodes and antinodes. At the nodes the Slinky coils do not vibrate at all; that is, they have no displacement. At the antinodes the coils vibrate with maximum amplitude and, thus, have a maximum displacement. The red dots in **Figure 17.19** indicate the lack of vibration at a node and the maximum vibration at an antinode. The vibration occurs along the line of travel of the individual waves, as is to be expected for longitudinal waves. In a standing wave of sound, at the nodes and antinodes, the molecules or atoms of the medium behave as the red dots do.

FIGURE 17.19 A longitudinal standing wave on a Slinky showing the displacement nodes (N) and antinodes (A).

Musical instruments in the wind family depend on longitudinal standing waves in producing sound. Since wind instruments (trumpet, flute, clarinet, pipe organ, etc.) are modified tubes or columns of air, it is useful to examine the standing waves that can be set up in such tubes. **Interactive Figure 17.20** shows two cylindrical columns of air that are open at both ends. Sound waves, originating from a tuning fork, travel up and down within each tube, since they reflect from the ends of the tubes, even though the ends are open. If the frequency f of the tuning fork matches one of the natural frequencies of the air column, the downward- and upward-traveling waves combine to form a standing wave, and the sound of the tuning fork becomes markedly louder. To emphasize the longitudinal nature of the standing wave patterns, the left side of each pair of drawings in **Interactive Figure 17.20** replaces the air in the tubes with Slinkies, on which the nodes and antinodes are indicated with red dots. As an additional aid in visualizing the standing waves, the right side of each pair of drawings shows blurred blue patterns

INTERACTIVE FIGURE 17.20 A pictorial representation of longitudinal standing waves on a Slinky (left side of each pair) and in a tube of air (right side of each pair) that is open at both ends (A, antinode; N, node).

within each tube. These patterns symbolize the amplitude of the vibrating air molecules at various locations. Wherever the pattern is widest, the amplitude of vibration is greatest (a displacement antinode), and wherever the pattern is narrowest there is no vibration (a displacement node).

To determine the natural frequencies of the air columns in **Interactive Figure 17.20**, notice that there is a displacement antinode at each end of the open tube because the air molecules there are free to move.* As in a transverse standing wave, the distance between two successive antinodes is one-half of a wavelength, so the length L of the tube must be an integer number n of half-wavelengths: $L = n\left(\frac{1}{2}\lambda_n\right)$ or $\lambda_n = 2L/n$. Using this wavelength in the relation $f_n = v/\lambda_n$ shows that the natural frequencies f_n of the tube are

Tube open at both ends $\quad f_n = n\left(\dfrac{v}{2L}\right) \qquad n = 1, 2, 3, 4, \ldots$ $\qquad$ **(17.4)**

At these frequencies, large-amplitude standing waves develop within the tube due to resonance. Example 7 illustrates how Equation 17.4 is involved when a flute is played.

EXAMPLE 7 | The Physics of a Flute

When all the holes are closed on one type of flute, the lowest note it can sound is a middle C (fundamental frequency = 261.6 Hz). The air temperature is 293 K, and the speed of sound is 343 m/s. Assuming the flute is a cylindrical tube open at both ends, determine the distance L in **Figure 17.21**—that is, the distance from the mouthpiece to the end of the tube. (This distance is approximate, since the antinode does not occur exactly at the mouthpiece.)

Reasoning For a tube open at both ends, the series of natural frequencies (including the fundamental) is given by $f_n = nv/(2L)$ (Equation 17.4), where n takes on the integer values 1, 2, 3, etc. To obtain a value for L, we can solve this equation, since the given fundamental frequency corresponds to $n = 1$ and the speed v of sound is known.

Solution Solving Equation 17.4 for the length L, we obtain

$$L = \frac{nv}{2f_n} = \frac{(1)(343 \text{ m/s})}{2(261.6 \text{ Hz})} = 0.656 \text{ m}$$

FIGURE 17.21 The length L of a flute between the mouthpiece and the end of the instrument determines the fundamental frequency of the lowest playable note.

Standing waves can also exist in a tube with only one end open, as the patterns in **Figure 17.22** indicate. Note the difference between these patterns and those in **Interactive Figure 17.20**. Here the standing waves have a displacement antinode at the open end and a displacement node at the closed end, where the air molecules are not free to move. Since the distance between a node and an adjacent antinode is one-fourth of a wavelength, the length L of the tube must be an odd number of

---

*In reality, the antinode does not occur exactly at the open end. However, if the tube's diameter is small compared to its length, little error is made in assuming that the antinode is located right at the end.

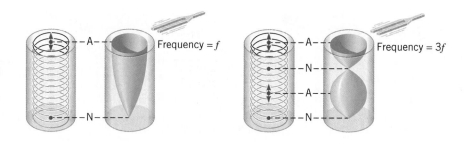

FIGURE 17.22 A pictorial representation of the longitudinal standing waves on a Slinky (left side of each pair) and in a tube of air (right side of each pair) that is open only at one end (A, antinode; N, node).

quarter-wavelengths: $L = 1\left(\frac{1}{4}\lambda\right)$ and $L = 3\left(\frac{1}{4}\lambda\right)$ for the two standing wave patterns in **Figure 17.22**. In general, then, $L = n\left(\frac{1}{4}\lambda_n\right)$, where n is any odd integer ($n = 1, 3, 5, \ldots$). From this result it follows that $\lambda_n = 4L/n$, and the natural frequencies f_n can be obtained from the relation $f_n = v/\lambda_n$:

Tube open at only one end $\qquad f_n = n\left(\dfrac{v}{4L}\right) \qquad n = 1, 3, 5, \ldots \qquad$ **(17.5)**

A tube open at only one end can develop standing waves only at the odd harmonic frequencies f_1, f_3, f_5, etc. In contrast, a tube open at both ends can develop standing waves at all harmonic frequencies f_1, f_2, f_3, etc. Moreover, the fundamental frequency f_1 of a tube open at only one end (Equation 17.5) is one-half that of a tube open at both ends (Equation 17.4). In other words, a tube open only at one end needs to be only one-half as long as a tube open at both ends in order to produce the *same* fundamental frequency.

Energy is also conserved when a standing wave is produced, either on a string or in a tube of air. The energy of the standing wave is the sum of the energies of the individual waves that comprise the standing wave. Once again, interference redistributes the energy of the individual waves to create locations of greatest energy (displacement antinodes) and locations of no energy (displacement nodes). Example 8 shows how the energy contained in standing waves set up in the human ear canal can lead to hearing loss.

EXAMPLE 8 | BIO The Physics of Hearing Loss—Standing Waves in the Ear

The human ear canal (**Figure 17.23**) essentially acts like a tube closed at one end. If the ear canal has a length of 2.3 cm, what are the fundamental wavelength and frequency for standing waves in the ear? Take the speed of sound to be 343 m/s. What is the significance of this result? Consult the graph in **Figure 16.36**.

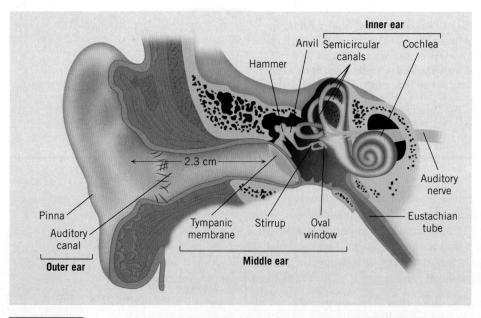

FIGURE 17.23 Diagram of the human ear showing the auditory canal as a tube closed at one end.

Reasoning We treat the auditory canal as a tube closed at one end. We can use Equation 17.5 to calculate the standing wave frequencies.

Solution The fundamental frequency is calculated for $n = 1$. Using this value in Equation 17.5, we get the following:

$$f_1 = (1)\left(\frac{343 \text{ m/s}}{(4)(0.023 \text{ m})}\right) = \boxed{3700 \text{ Hz}}$$

The fundamental wavelength is related to the fundamental frequency by

$$\lambda = \frac{v}{f} = \frac{343 \text{ m/s}}{3700 \text{ Hz}} = 9.3 \times 10^{-2} \text{ m} = \boxed{9.3 \text{ cm}}$$

Looking at the graph in **Figure 16.36**, we see that the intensity curves dip in the range of 2000 – 5000 Hz, with the ear having the greatest sensitivity at the fundamental frequency just below 4000 Hz. In fact, the ear can detect sound intensities below the threshold of hearing at 3700 Hz. Due to the resonance, sounds at this frequency are amplified. Damage to the ear can occur if exposed to high-intensity sounds for prolonged periods of time. People working in high-noise environments without ear protection often show signs of 3700-Hz hearing loss.

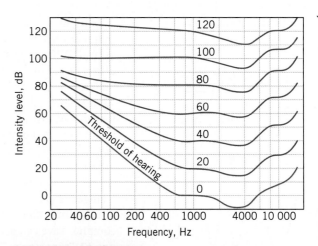

FIGURE 16.36 (REPEATED) Each curve represents the intensity levels at which sounds of various frequencies have the same loudness. The curves are labeled by their intensity levels at 1000 Hz and are known as Fletcher-Munson curves.

Check Your Understanding

(*The answers are given at the end of the book.*)

15. A cylindrical bottle, partially filled with water, is open at the top. When you blow across the top of the bottle a standing wave is set up inside it. Is there a node or an antinode **(a)** at the top of the bottle and **(b)** at the surface of the water? **(c)** If the standing wave is vibrating at its fundamental frequency, what is the distance between the top of the bottle and the surface of the water? Express your answer in terms of the wavelength λ of the standing wave. **(d)** If you take a sip from the bottle, is the fundamental frequency of the standing wave raised, lowered, or does it remain the same?

16. In **Interactive Figure 17.20** both tubes are filled with air, in which the speed of sound is v_{air}. Suppose, instead, that the tube near the tuning fork labeled "Frequency = 2f" is filled not with air, but with another gas in which the speed of sound is v_{gas}. The frequency of each tuning fork remains unchanged. How should v_{gas} compare with v_{air} in order that the standing wave pattern in each tube has the same appearance? **(a)** $v_{gas} = \frac{1}{2}v_{air}$ **(b)** $v_{gas} = 2\,v_{air}$ **(c)** $v_{gas} = \frac{1}{4}v_{air}$ **(d)** $v_{gas} = 4\,v_{air}$

17. Standing waves can ruin the acoustics of a concert hall if there is excessive reflection of the sound waves that the performers generate. For example, suppose that a performer generates a 2093-Hz tone. If a large-amplitude standing wave is present, it is possible for a listener to move a distance of only 4.1 cm and hear the loudness of the tone change from loud to faint. What does the distance of 4.1 cm represent? **(a)** One wavelength of the sound **(b)** One-half the wavelength of the sound **(c)** One-fourth the wavelength of the sound

18. A wind instrument is brought into a warm house from the cold outdoors. What happens to the natural frequencies of the instrument? Neglect any change in the length of the instrument. **(a)** They increase. **(b)** They decrease. **(c)** They remain the same.

17.7 | *Complex Sound Waves

Musical instruments produce sound in a way that depends on standing waves. Examples 5 and 6 illustrate the role of transverse standing waves on the string of an electric guitar, while Example 7 stresses the role of longitudinal standing waves in the air column within a flute. In each example, sound is produced at the fundamental frequency of the instrument.

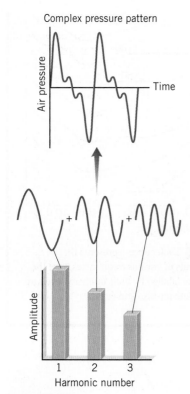

Complex pressure pattern

FIGURE 17.24   The topmost graph shows the pattern of pressure fluctuations such as a singer might produce. The pattern is the sum of the first three harmonics. The relative amplitudes of the harmonics correspond to the heights of the vertical bars in the bar graph.

In general, however, a musical instrument does not produce just the fundamental frequency when it plays a note, but simultaneously generates a number of harmonics as well. Different instruments, such as a violin and a trumpet, generate harmonics to different extents, and the harmonics give the instruments their characteristic sound qualities or timbres. Suppose, for instance, that a violinist and a trumpet player both sound concert A, a note whose fundamental frequency is 440 Hz. Even though both instruments are playing the same note, most people can distinguish the sound of the violin from that of the trumpet. The instruments sound different because the relative amplitudes of the harmonics (880 Hz, 1320 Hz, etc.) that the instruments create are different.

The sound wave corresponding to a note produced by a musical instrument or a singer is called a **complex sound wave** because it consists of a mixture of the fundamental and harmonic frequencies. The pattern of pressure fluctuations in a complex wave can be obtained by using the principle of linear superposition, as **Figure 17.24** indicates. This drawing shows a bar graph in which the heights of the bars give the relative amplitudes of the harmonics contained in a note such as a singer might produce. When the individual pressure patterns for each of the three harmonics are added together, they yield the complex pressure pattern shown at the top of the picture.*

THE PHYSICS OF . . . a spectrum analyzer.   In practice, a bar graph such as that in **Figure 17.24** is determined with the aid of an electronic instrument known as a spectrum analyzer. When the note is produced, the complex sound wave is detected by a microphone that converts the wave into an electrical signal. The electrical signal, in turn, is fed into the spectrum analyzer, as **Figure 17.25** illustrates. The spectrum analyzer then determines the amplitude and frequency of each harmonic present in the complex wave and displays the results on its screen.

FIGURE 17.25   A microphone detects a complex sound wave produced by a singer's voice, and a spectrum analyzer determines the amplitude and frequency of each harmonic present in the wave.

Concept Summary

17.1 The Principle of Linear Superposition   The principle of linear superposition states that when two or more waves are present simultaneously at the same place, the resultant disturbance is the sum of the disturbances from the individual waves.

17.2 Constructive and Destructive Interference of Sound Waves   Constructive interference occurs at a point when two waves meet there crest-to-crest and trough-to-trough, thus reinforcing each

*In carrying out the addition, we assume that each individual pattern begins at zero at the origin when the time equals zero.

other. Destructive interference occurs when the waves meet crest-to-trough and cancel each other.

When waves meet crest-to-crest and trough-to-trough, they are exactly in phase. When they meet crest-to-trough, they are exactly out of phase.

For two wave sources vibrating in phase, a difference in path lengths that is zero or an integer number $(1, 2, 3, \ldots)$ of wavelengths leads to constructive interference; a difference in path lengths that is a half-integer number $\left(\frac{1}{2}, 1\frac{1}{2}, 2\frac{1}{2}, \ldots\right)$ of wavelengths leads to destructive interference.

For two wave sources vibrating out of phase, a difference in path lengths that is a half-integer number $\left(\frac{1}{2}, 1\frac{1}{2}, 2\frac{1}{2}, \ldots\right)$ of wavelengths leads to constructive interference; a difference in path lengths that is zero or an integer number $(1, 2, 3, \ldots)$ of wavelengths leads to destructive interference.

17.3 Diffraction Diffraction is the bending of a wave around an obstacle or the edges of an opening. The angle through which the wave bends depends on the ratio of the wavelength λ of the wave to the width D of the opening; the greater the ratio λ/D, the greater the angle.

When a sound wave of wavelength λ passes through an opening, the first place where the intensity of the sound is a minimum relative to that at the center of the opening is specified by the angle θ. If the opening is a rectangular slit of width D, such as a doorway, the angle is given by Equation 17.1. If the opening is a circular opening of diameter D, such as that in a loudspeaker, the angle is given by Equation 17.2.

$$\sin \theta = \frac{\lambda}{D} \qquad (17.1)$$

$$\sin \theta = 1.22\,\frac{\lambda}{D} \qquad (17.2)$$

17.4 Beats Beats are the periodic variations in amplitude that arise from the linear superposition of two waves that have slightly different frequencies. When the waves are sound waves, the variations in amplitude cause the loudness to vary at the beat frequency, which is the difference between the frequencies of the waves.

17.5 Transverse Standing Waves A standing wave is the pattern of disturbance that results when oppositely traveling waves of the same frequency and amplitude pass through each other. A standing wave has places of minimum and maximum vibration called, respectively, nodes and antinodes. Under resonance conditions, standing waves can be established only at certain natural frequencies. The frequencies in this series $(f_1, 2f_1, 3f_1,$ etc.) are called harmonics. The lowest frequency f_1 is called the first harmonic, the next frequency $2f_1$ is the second harmonic, and so on. For a string that is fixed at both ends and has a length L, the natural frequencies are specified by Equation 17.3, where v is the speed of the wave on the string and n is a positive integer.

$$f_n = n\left(\frac{v}{2L}\right) \quad n = 1, 2, 3, 4, \ldots \qquad (17.3)$$

17.6 Longitudinal Standing Waves For a gas in a cylindrical tube open at both ends, the natural frequencies of vibration are specified by Equation 17.4, where v is the speed of sound in the gas and L is the length of the tube. For a gas in a cylindrical tube open at only one end, the natural frequencies of vibration are given by Equation 17.5.

$$f_n = n\left(\frac{v}{2L}\right) \quad n = 1, 2, 3, 4, \ldots \qquad (17.4)$$

$$f_n = n\left(\frac{v}{4L}\right) \quad n = 1, 3, 5, 7, \ldots \qquad (17.5)$$

17.7 Complex Sound Waves A complex sound wave consists of a mixture of a fundamental frequency and overtone frequencies.

Focus on Concepts

Online

Additional questions are available for assignment in WileyPLUS.

Section 17.1 The Principle of Linear Superposition

1. The drawing shows four moving pulses. Although shown as separated, the four pulses exactly overlap each other at the instant shown. Which combination of these pulses would produce a resultant pulse with the highest peak *and* the deepest valley at this instant? **(a)** 1 and 2 **(b)** 2, 3, and 4 **(c)** 2 and 3 **(d)** 1, 2, and 3 **(e)** 1 and 3

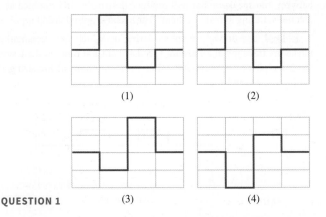

(1)   (2)

QUESTION 1   (3)   (4)

Section 17.2 Constructive and Destructive Interference of Sound Waves

2. Two cellists, one seated directly behind the other in an orchestra, play the same note for the conductor, who is directly in front of them. Because of the separation between the cellists, destructive interference occurs at the conductor. This separation is the smallest that produces destructive interference. Would this separation increase, decrease, or remain the same if the cellists produced a note with a higher frequency? **(a)** The separation between the cellists would remain the same. **(b)** The separation would decrease because the wavelength of the sound is greater. **(c)** The separation would decrease because the wavelength of the sound is smaller. **(d)** The separation would increase because the wavelength of the sound is greater. **(e)** The separation would increase because the wavelength of the sound is smaller.

Section 17.3 Diffraction

3. A loudspeaker is producing sound of a certain wavelength. Which combination of the wavelength λ (expressed as a multiple of λ_0) and the speaker's diameter D (expressed as a multiple of D_0) would exhibit the greatest amount of diffraction when the sound leaves the speaker and enters the room? **(a)** $\lambda = \lambda_0, D = D_0$ **(b)** $\lambda = 2\lambda_0, D = D_0$ **(c)** $\lambda = \lambda_0, D = 2D_0$ **(d)** $\lambda = 2\lambda_0, D = 2D_0$ **(e)** $\lambda = 3\lambda_0, D = 2D_0$

4. Sound of a given frequency leaves a loudspeaker and spreads out due to diffraction. The speaker is placed in a room that contains either air or helium. The speed of sound in helium is about three times as great as the speed of sound in air. In which room, if either, does the sound exhibit the greater diffraction when leaving the speaker? (a) The greater diffraction occurs in the air-filled room, because the wavelength of the sound is smaller in that room. (b) The greater diffraction occurs in the air-filled room, because the wavelength of the sound is greater in that room. (c) The diffraction is the same in both rooms. (d) The greater diffraction occurs in the helium-filled room, because the wavelength of the sound is smaller in that room. (e) The greater diffraction occurs in the helium-filled room, because the wavelength of the sound is greater in that room.

Section 17.4 Beats

5. Two musicians are comparing their trombones. The first produces a tone that is known to be 438 Hz. When the two trombones play together they produce 6 beats every 2 seconds. Which statement is true about the second trombone? (a) It is producing either a 432-Hz sound or a 444-Hz sound. (b) It is producing either a 436-Hz sound or a 440-Hz sound. (c) It is producing a 444-Hz sound, and could be producing no other sound frequency. (d) It is producing either a 435-Hz sound or a 441-Hz sound. (e) It is producing a 441-Hz sound and could be producing no other sound frequency.

Section 17.5 Transverse Standing Waves

6. Two transverse standing waves are shown in the drawing. The strings have the same tension and length, but the bottom string is more massive. Which standing wave, if either, is vibrating at the higher frequency? (a) The top standing wave has the higher frequency, because the traveling waves have a smaller speed due to the smaller mass of the string. (b) The top standing wave has the higher frequency, because the traveling waves have a larger speed due to the

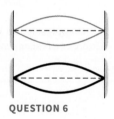

QUESTION 6

smaller mass of the string. (c) Both standing waves have the same frequency, because the frequency of vibration does not depend on the mass of the string. (d) The bottom standing wave has the higher frequency, because the traveling waves have a smaller speed due to the larger mass of the string. (e) The bottom standing wave has the higher frequency, because the traveling waves have a larger speed due to the larger mass of the string.

7. A standing wave on a string fixed at both ends is vibrating at its fourth harmonic. If the length, tension, and linear density are kept constant, what can be said about the wavelength and frequency of the fifth harmonic relative to the fourth harmonic? (a) The wavelength of the fifth harmonic is longer, and its frequency is higher. (b) The wavelength of the fifth harmonic is longer, and its frequency is lower. (c) The wavelength of the fifth harmonic is shorter, and its frequency is higher. (d) The wavelength of the fifth harmonic is shorter, and its frequency is lower.

Section 17.6 Longitudinal Standing Waves

8. A longitudinal standing wave is established in a tube that is open at both ends (see the drawing). The length of the tube is 0.80 m. What is the wavelength of the waves that make up the standing wave? (a) 0.20 m (b) 0.40 m (c) 0.80 m (d) 1.20 m (e) 1.60 m

QUESTION 8

9. A longitudinal standing wave is established in a tube open at only one end (see the drawing). The frequency of the standing wave is 660 Hz, and the speed of sound in air is 343 m/s. What is the length of the tube? (a) 0.13 m (b) 0.26 m (c) 0.39 m (d) 0.52 m (e) 0.65 m

QUESTION 9

Problems

Online

Additional questions are available for assignment in WileyPLUS.

Section 17.1 The Principle of Linear Superposition,

Section 17.2 Constructive and Destructive Interference of Sound Waves

1. **E** In **Figure 17.7**, suppose that the separation between speakers A and B is 5.00 m and the speakers are vibrating in phase. They are playing identical 125-Hz tones, and the speed of sound is 343 m/s.

What is the largest possible distance between speaker B and the observer at C, such that he observes destructive interference?

2. **E** Two speakers, one directly behind the other, are each generating a 245-Hz sound wave. What is the smallest separation distance between the speakers that will produce destructive interference at a listener standing in front of them? The speed of sound is 343 m/s.

3. **E SSM** The drawing graphs a string on which two rectangular pulses are traveling at a constant speed of 1 cm/s at time $t = 0$ s. Using the principle of linear superposition, draw the shape of the string at $t = 1$ s, 2 s, 3 s, and 4 s.

PROBLEM 3

4. **E** Loudspeakers A and B are vibrating in phase and are playing the same tone, which has a frequency of 250 Hz. They are set up as in **Figure 17.7**, and point C is located as shown there. However, the distance between the speakers and the distance between speaker B and point C have the same value *d*. The speed of sound is 343 m/s. What is the smallest value of *d*, such that constructive interference occurs at point C?

5. **E** **SSM** Two waves are traveling in opposite directions on the same string. The displacements caused by the individual waves are given by $y_1 = (24.0 \text{ mm})\sin(9.00\pi t - 1.25\pi x)$ and $y_2 = (35.0 \text{ mm})\sin(2.88\pi t + 0.400\pi x)$. Note that the phase angles $(9.00\pi t - 1.25\pi x)$ and $(2.88\pi t + 0.400\pi x)$ are in radians, *t* is in seconds, and *x* is in meters. At $t = 4.00$ s, what is the net displacement (in mm) of the string at **(a)** $x = 2.16$ m and **(b)** $x = 2.56$ m? Be sure to include the algebraic sign (+ or –) with your answers.

6. **E** **CHALK** **GO** Both drawings show the same square, each of which has a side of length $L = 0.75$ m. An observer O is stationed at one corner of each square. Two loudspeakers are located at corners of the square, as in either drawing 1 or drawing 2. The speakers produce the same single-frequency tone in either drawing and are in phase. The speed of sound is 343 m/s. Find the single smallest frequency that will produce both constructive interference in drawing 1 and destructive interference in drawing 2.

PROBLEM 6   Drawing 1   Drawing 2

7. **E** **SSM** The drawing shows a loudspeaker A and point C, where a listener is positioned. A second loudspeaker B is located somewhere to the right of loudspeaker A. Both speakers vibrate in phase and are playing a 68.6-Hz tone. The speed of sound is 343 m/s. What is the closest to speaker A that speaker B can be located, so that the listener hears no sound?

PROBLEM 7   A   B

8. **E** **GO** Suppose that the two speakers in **Figure 17.7** are separated by 2.50 m and are vibrating exactly *out of phase* at a frequency of 429 Hz. The speed of sound is 343 m/s. Does the observer at C observe constructive or destructive interference when his distance from speaker B is **(a)** 1.15 m and **(b)** 2.00 m?

9. **M** **MMH** Two loudspeakers on a concert stage are vibrating in phase. A listener is 50.5 m from the left speaker and 26.0 m from the right one. The listener can respond to all frequencies from 20 to 20 000 Hz, and the speed of sound is 343 m/s. What are the two lowest frequencies that can be heard loudly due to constructive interference?

10. **M** **GO** **CHALK** A listener is standing in front of two speakers that are producing sound of the same frequency and amplitude, except that they are vibrating out of phase. Initially, the distance between the listener and each speaker is the same (see the drawing). As the listener moves sideways, the sound intensity gradually changes. When the distance *x* in the drawing is 0.92 m, the change reaches the maximum amount (either loud to soft, or soft to loud). Using the data shown in the drawing and 343 m/s for the speed of sound, determine the frequency of the sound coming from the speakers.

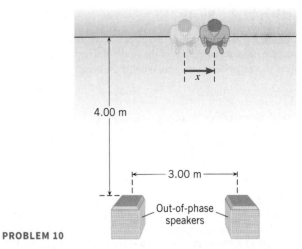

4.00 m

3.00 m

Out-of-phase speakers

PROBLEM 10

11. **H** Speakers A and B are vibrating in phase. They are directly facing each other, are 7.80 m apart, and are each playing a 73.0-Hz tone. The speed of sound is 343 m/s. On the line between the speakers there are three points where constructive interference occurs. What are the distances of these three points from speaker A?

Section 17.3   Diffraction

12. **E** Consult Multiple-Concept Example 3 for background pertinent to this problem. A speaker has a diameter of 0.30 m. **(a)** Assuming that the speed of sound is 343 m/s, find the diffraction angle θ for a 2.0-kHz tone. **(b)** What speaker diameter *D* should be used to generate a 6.0-kHz tone whose diffraction angle is as wide as that for the 2.0-kHz tone in part (a)?

13. **E** **SSM** Sound exits a diffraction horn loudspeaker through a rectangular opening like a small doorway. Such a loudspeaker is mounted outside on a pole. In winter, when the temperature is 273 K, the diffraction angle θ has a value of 15.0°. What is the diffraction angle for the same sound on a summer day when the temperature is 311 K?

14. **E** **GO** For one approach to problems such as this, see Multiple-Concept Example 3. Sound emerges through a doorway, as in **Figure 17.10**. The width of the doorway is 77 cm, and the speed of sound is 343 m/s. Find the diffraction angle θ when the frequency of the sound is **(a)** 5.0 kHz and **(b)** 5.0×10^2 Hz.

15. **E** **GO** The following two lists give the diameters and sound frequencies for three loudspeakers. Pair each diameter with a frequency, so that the diffraction angle is the same for each of the speakers, and then find the common diffraction angle. Take the speed of sound to be 343 m/s.

Diameter, *D*	Frequency, *f*
0.050 m	6.0 kHz
0.10 m	4.0 kHz
0.15 m	12.0 kHz

16. **M** A 3.00-kHz tone is being produced by a speaker with a diameter of 0.175 m. The air temperature changes from 0 to 29 °C. Assuming air to be an ideal gas, find the *change* in the diffraction angle θ.

17. **M** **GO** Sound (speed = 343 m/s) exits a diffraction horn loudspeaker through a rectangular opening like a small doorway. A person is sitting at an angle α off to the side of a diffraction horn that has a width *D* of 0.060 m. This individual does not hear a sound wave that has a frequency of 8100 Hz. When she is sitting at an angle $\alpha/2$, the frequency that she does not hear is different. What is this frequency?

Section 17.4 Beats

18. **E** **SSM** Two pure tones are sounded together. The drawing shows the pressure variations of the two sound waves, measured with respect to atmospheric pressure. What is the beat frequency?

PROBLEM 18

19. **E** **GO** Two pianos each sound the same note simultaneously, but they are both out of tune. On a day when the speed of sound is 343 m/s, piano A produces a wavelength of 0.769 m, while piano B produces a wavelength of 0.776 m. How much time separates successive beats?

20. **E** **V-HINT** **MMH** A 440.0-Hz tuning fork is sounded together with an out-of-tune guitar string, and a beat frequency of 3 Hz is heard. When the string is tightened, the frequency at which it vibrates increases, and the beat frequency is heard to decrease. What was the original frequency of the guitar string?

21. **E** Two ultrasonic sound waves combine and form a beat frequency that is in the range of human hearing for a healthy young person. The frequency of one of the ultrasonic waves is 70 kHz. What are **(a)** the smallest possible and **(b)** the largest possible value for the frequency of the other ultrasonic wave?

22. **E** **SSM** Two out-of-tune flutes play the same note. One produces a tone that has a frequency of 262 Hz, while the other produces 266 Hz. When a tuning fork is sounded together with the 262-Hz tone, a beat frequency of 1 Hz is produced. When the same tuning fork is sounded together with the 266-Hz tone, a beat frequency of 3 Hz is produced. What is the frequency of the tuning fork?

23. **M** **GO** **CHALK** Two cars have identical horns, each emitting a frequency of $f_s = 395$ Hz. One of the cars is moving with a speed of 12.0 m/s toward a bystander waiting at a corner, and the other car is parked. The speed of sound is 343 m/s. What is the beat frequency heard by the bystander?

24. **M** **GO** A sound wave is traveling in seawater, where the adiabatic bulk modulus and density are 2.31×10^9 Pa and 1025 kg/m³, respectively. The wavelength of the sound is 3.35 m. A tuning fork is struck under water and vibrates at 440.0 Hz. What would be the beat frequency heard by an underwater swimmer?

25. **H** Two loudspeakers are mounted on a merry-go-round whose radius is 9.01 m. When stationary, the speakers both play a tone whose frequency is 100.0 Hz. As the drawing illustrates, they are situated at opposite ends of a diameter. The speed of sound is 343.00 m/s, and the merry-go-round revolves once every 20.0 s. What is the beat frequency that is detected by the listener when the merry-go-round is near the position shown?

Merry-go-round
(top view)

PROBLEM 25

Listener

Section 17.5 Transverse Standing Waves

26. **E** The fundamental frequency of a string fixed at both ends is 256 Hz. How long does it take for a wave to travel the length of this string?

27. **E** A string that is fixed at both ends has a length of 2.50 m. When the string vibrates at a frequency of 85.0 Hz, a standing wave with five loops is formed. **(a)** What is the wavelength of the waves that travel on the string? **(b)** What is the speed of the waves? **(c)** What is the fundamental frequency of the string?

28. **E** **SSM** The approach to solving this problem is similar to that taken in Multiple-Concept Example 5. On a cello, the string with the largest linear density (1.56×10^{-2} kg/m) is the C string. This string produces a fundamental frequency of 65.4 Hz and has a length of 0.800 m between the two fixed ends. Find the tension in the string.

29. **E** **GO** Two wires, each of length 1.2 m, are stretched between two fixed supports. On wire A there is a second-harmonic standing wave whose frequency is 660 Hz. However, the same frequency of 660 Hz is the third harmonic on wire B. Find the speed at which the individual waves travel on each wire.

30. **E** **SSM** Suppose that the strings on a violin are stretched with the same tension and each has the same length between its two fixed ends. The musical notes and corresponding fundamental frequencies of two of these strings are G (196.0 Hz) and E (659.3 Hz). The linear density of the E string is 3.47×10^{-4} kg/m. What is the linear density of the G string?

31. **E** To review the concepts that play roles in this problem, consult Multiple-Concept Example 5. Sometimes, when the wind blows across a long wire, a low-frequency "moaning" sound is produced. This sound arises because a standing wave is set up on the wire, like a standing wave on a guitar string. Assume that a wire (linear density = 0.0140 kg/m) sustains a tension of 323 N because the wire is stretched between two poles that are 7.60 m apart. The lowest frequency that an average, healthy human ear can detect is 20.0 Hz. What is the lowest harmonic number n that could be responsible for the "moaning" sound?

32. **E** A string has a linear density of 8.5×10^{-3} kg/m and is under a tension of 280 N. The string is 1.8 m long, is fixed at both ends, and is vibrating in the standing wave pattern shown in the drawing. Determine the **(a)** speed, **(b)** wavelength, and **(c)** frequency of the traveling waves that make up the standing wave.

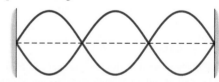

PROBLEM 32

33. **E** **GO** Multiple-Concept Example 5 deals with the same concepts as this problem. A 41-cm length of wire has a mass of 6.0 g. It is stretched between two fixed supports and is under a tension of 160 N. What is the fundamental frequency of this wire?

34. **M** **CHALK** **GO** A copper block is suspended from a wire, as in part 1 of the drawing. A container of mercury is then raised up around the block, as in part 2, so that 50.0% of the block's volume is submerged in the mercury. The density of copper is 8890 kg/m³, and that of mercury is 13 600 kg/m³. Find the ratio of the fundamental frequency of the wire in part 2 to the fundamental frequency of the wire in part 1.

Copper block

Part 1   Part 2 **PROBLEM 34**

35. **M** **V-HINT** The drawing shows two strings that have the same length and linear density. The left end of each string is attached to a wall, while the right end passes over a pulley and is connected to objects of different weights (W_A and W_B). Different standing waves are set up on each string, but their frequencies are the same. If $W_A = 44$ N, what is W_B?

PROBLEM 35

36. **M** **SSM** The E string on an electric bass guitar has a length of 0.628 m and, when producing the note E, vibrates at a fundamental frequency of 41.2 Hz. Players sometimes add to their instruments a device called a "D-tuner." This device allows the E string to be used to produce the note D, which has a fundamental frequency of 36.7 Hz. The D-tuner works by extending the length of the string, keeping all other factors the same. By how much does a D-tuner extend the length of the E string?

37. **M** **GO** Standing waves are set up on two strings fixed at each end, as shown in the drawing. The two strings have the same tension and mass per unit length, but they differ in length by 0.57 cm. The waves on the shorter string propagate with a speed of 41.8 m/s, and the fundamental frequency of the shorter string is 225 Hz. Determine the beat frequency produced by the two standing waves.

PROBLEM 37

38. **H** Review Conceptual Example 6 before attempting this problem. As the drawing shows, the length of a guitar string is 0.628 m. The frets are numbered for convenience. A performer can play a musical scale on a single string because the spacing *between the frets* is designed according to the following rule: When the string is pushed against any fret j, the fundamental frequency of the shortened string is larger by a factor of the twelfth root of two $\left(\sqrt[12]{2}\right)$ than it is when the string is pushed against the fret $j - 1$. Assuming that the tension in the string is the same for any note, find the spacing **(a)** between fret 1 and fret 0 and **(b)** between fret 7 and fret 6.

← 0.628 m →

7 6 5 4 3 2 1 0

PROBLEM 38

Section 17.6 Longitudinal Standing Waves,

Section 17.7 Complex Sound Waves

39. **E** **GO** A tube with a cap on one end, but open at the other end, has a fundamental frequency of 130.8 Hz. The speed of sound is 343 m/s **(a)** If the cap is removed, what is the new fundamental frequency of the tube? **(b)** How long is the tube?

40. **E** An organ pipe is open at both ends. It is producing sound at its third harmonic, the frequency of which is 262 Hz. The speed of sound is 343 m/s. What is the length of the pipe?

41. **E** The range of human hearing is roughly from twenty hertz to twenty kilohertz. Based on these limits and a value of 343 m/s for the speed of sound, what are the lengths of the longest and shortest pipes (open at both ends and producing sound at their fundamental frequencies) that you expect to find in a pipe organ?

42. **E** **MMH** The fundamental frequencies of two air columns are the same. Column A is open at both ends, while column B is open at only one end. The length of column A is 0.70 m. What is the length of column B?

43. **E** **SSM** A tube is open only at one end. A certain harmonic produced by the tube has a frequency of 450 Hz. The next higher harmonic has a frequency of 750 Hz. The speed of sound in air is 343 m/s. **(a)** What is the integer n that describes the harmonic whose frequency is 450 Hz? **(b)** What is the length of the tube?

44. **M** **V-HINT** A thin 1.2-m aluminum rod sustains a longitudinal standing wave with vibration antinodes at each end of the rod. There are no other antinodes. The density and Young's modulus of aluminum are, respectively, 2700 kg/m³ and 6.9×10^{10} N/m². What is the frequency of the rod's vibration?

45. **M** **CHALK** **SSM** A person hums into the top of a well and finds that standing waves are established at frequencies of 42, 70.0, and 98 Hz. The frequency of 42 Hz is not necessarily the fundamental frequency. The speed of sound is 343 m/s. How deep is the well?

46. **M** **GO** A vertical tube is closed at one end and open to air at the other end. The air pressure is 1.01×10^5 Pa. The tube has a length of 0.75 m. Mercury (mass density = 13 600 kg/m³) is poured into it to shorten the effective length for standing waves. What is the absolute pressure at the bottom of the mercury column, when the fundamental frequency of the shortened, air-filled tube is equal to the third harmonic of the original tube?

Additional Problems

Online

47. **E** A string is fixed at both ends and is vibrating at 130 Hz, which is its third harmonic frequency. The linear density of the string is 5.6×10^{-3} kg/m, and it is under a tension of 3.3 N. Determine the length of the string.

48. **E** **GO** One method for measuring the speed of sound uses standing waves. A cylindrical tube is open at both ends, and one end admits sound

from a tuning fork. A movable plunger is inserted into the other end at a distance L from the end of the tube where the tuning fork is. For a fixed frequency, the plunger is moved until the smallest value of L is measured that allows a standing wave to be formed. Suppose that the tuning fork produces a 485-Hz tone, and that the smallest value observed for L is 0.264 m. What is the speed of sound in the gas in the tube?

49. **E** **MMH** The drawing graphs a string on which two pulses (half up and half down) are traveling at a constant speed of 1 cm/s at $t = 0$ s. Using the principle of linear superposition, draw the shape of the string at $t = 1$ s, 2 s, 3 s, and 4 s.

1 cm/s   1 cm/s

0   2   4   6   8   10   12

PROBLEM 49   Distance, cm

50. **E** **SSM** The A string on a string bass vibrates at a fundamental frequency of 55.0 Hz. If the string's tension were increased by a factor of four, what would be the new fundamental frequency?

51. **M** **V-HINT** The two speakers in the drawing are vibrating in phase, and a listener is standing at point P. Does constructive or destructive interference occur at P when the speakers produce sound waves whose frequency is **(a)** 1466 Hz and **(b)** 977 Hz? Justify your answers with appropriate calculations. Take the speed of sound to be 343 m/s.

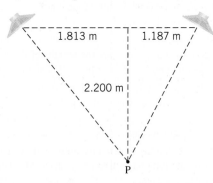

1.813 m   1.187 m

2.200 m

PROBLEM 51   P

52. **H** **SSM** The arrangement in the drawing shows a block (mass = 15.0 kg) that is held in position on a frictionless incline by a cord (length = 0.600 m). The mass per unit length of the cord is 1.20×10^{-2} kg/m, so the mass of the cord is negligible compared to the mass of the block. The cord is being vibrated at a frequency of 165 Hz (vibration source not shown in the drawing). What are the values of the angle θ between 15.0° and 90.0° at which a standing wave exists on the cord?

PROBLEM 52 θ

53. **M** **GO** A flautist is playing a flute as discussed in Example 7, but now the temperature is 305 K instead of 293 K. As a result, the speed of sound is no longer 343 m/s. Therefore, with the length calculated in Example 7, the note middle C does not have the proper fundamental frequency of 261.6 Hz. In other words, the flute is out of tune. To adjust the tuning, the flautist can alter the flute's length by changing the extent to which the head joint (see **Figure 17.21**) is inserted into the main stem of the instrument. To what length must the flute be adjusted to play middle C at its proper frequency?

Physics in Biology, Medicine, and Sports

54. **E** **BIO** **SSM** **17.6** Sound enters the ear, travels through the auditory canal, and reaches the eardrum. The auditory canal is approximately a tube open at only one end. The other end is closed by the eardrum. A typical length for the auditory canal in an adult is about 2.9 cm. The speed of sound is 343 m/s. What is the fundamental frequency of the canal? (Interestingly, the fundamental frequency is in the frequency range where human hearing is most sensitive.)

55. **E** **BIO** **17.6** Divers working in underwater chambers at great depths must deal with the danger of nitrogen narcosis (the "bends"), in which nitrogen dissolves into the blood at toxic levels. One way to avoid this danger is for divers to breathe a mixture containing only helium and oxygen. Helium, however, has the effect of giving the voice a high-pitched quality, like that of Donald Duck's voice. To see why this occurs, assume for simplicity that the voice is generated by the vocal cords vibrating above a gas-filled cylindrical tube that is open only at one end. The quality of the voice depends on the harmonic frequencies generated by the tube; larger frequencies lead to higher-pitched voices. Consider two such tubes at 20 °C. One is filled with air, in which the speed of sound is 343 m/s. The other is filled with helium, in which the speed of sound is 1.00×10^3 m/s. To see the effect of helium on voice quality, calculate the ratio of the nth natural frequency of the helium-filled tube to the nth natural frequency of the air-filled tube.

56. **E** **BIO** **17.6** Human adults are most sensitive to sound waves that satisfy a standing wave condition for certain frequencies. Example 8 showed the fundamental frequency for humans is 3700 Hz.

How does this frequency compare to the fundamental frequency of an elephant's ear, whose auditory canal is 18 cm in length? Again, treat the auditory canal as a tube open only at one end. Use 340 m/s for the speed of sound. Compared to humans, you should find that an elephant's ears are sensitive to much lower frequency sounds.

57. **E** **BIO** **17.4** Binaural beat therapy is a form of sound therapy used to treat anxiety, stress, and sleep disorders. A patient wears headphones that apply slightly different frequencies of sound to the left and right ears. The frequencies are chosen such that the auditory part of the brain only hears a single tone, while the frequencies of the impulses traveling along the left and right auditory nerves are slightly different. There are multiple types of brain waves: alpha, beta, gamma, delta, and theta. Each of these operate over a slightly different frequency range. Synchronizing the beat frequency to those of the different types of brain waves may offer therapeutic benefits, although clinical trials have proved inconclusive. A person undergoing binaural beat therapy is wearing headphones that deliver a single tone with a frequency of 1.200×10^3 Hz into the left ear and a single tone with a frequency of 1.208×10^3 Hz into the right ear. Her therapy is targeting her brain's alpha waves (7 to 13 Hz) in an effort to promote relaxation. How many hours should she listen in order to complete 1.000×10^3 beat cycles?

58. **E** **BIO** **17.4** Sound passing through the vocal cords is modified by a set of resonating cavities (see the figure). These include the pharnyx, oral cavity (mouth), nasal cavity, and paranasal sinuses. The so-called nasal sounds of which examples in English would be "m,"

"n," and "ng," are created by having the air expelled through the vocal cords leave almost entirely through the nose and not the mouth. The pharynx and nasal cavity combine to form a resonating chamber that is open at one end (see the blue outline in the figure). **(a)** If the overall length of this chamber is 22.0 cm, what is the chamber's fundamental frequency? **(b)** What is the frequency of the chamber's third harmonic? The air being expelled through the vocal cords has been warmed by the body's temperature, so assume the speed of sound in air for these conditions is 353 m/s.

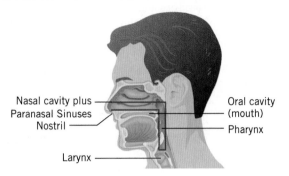

Nasal cavity plus
Paranasal Sinuses
Nostril

Oral cavity
(mouth)

Pharynx

Larynx

PROBLEM 58

59. **E** **BIO** **17.6**   When a male American alligator is attempting to attract a female, he engages in a unique courtship ritual known as *water dancing*. With his head and tail raised up above the water, he lowers his body until his upper back is just below the water's surface. With his mouth closed, he then exhales while emitting a low frequency tone. When the frequency of the sound he emits matches the standing wave condition for his chest cavity, resonance occurs, and the water just above his upper back vibrates violently, producing quite the auditory and visual display (see the photo). Assume the alligator emits the fundamental frequency for resonance (58.3 Hz) and that his chest cavity can be approximated as a tube open only at one end. Use this information to estimate the length of his chest cavity, and take the speed of sound to be 343 m/s.

PROBLEM 59

60. **M** **BIO** **17.3**   Scientists have recently learned that elephants emit infrasonic sound waves with frequencies in the range of 15 to 35 Hz, as a means of communicating with each other. This technique was discovered after researchers observed large herds of African elephants of the savannah communicating over great distances, sometimes as far as 4 km. Infrasonic sounds were detected right before the occurrence of synchronized movements of the entire herd. Elephants living in the Central African rainforests also displayed similar behavior, even though the dense forest vegetation made it impossible for the elephants at the rear of the herd to see the leaders at the front. Diffraction allows the low frequency sound waves to effectively bend around the trees and vegetation, spanning the size of the herd (see the photo). While the range in the forest is reduced compared to the savannah, communications up to 3 km were observed. Imagine an elephant emitting a low frequency sound wave at 27.0 Hz. The wave passes through an opening in a grove of trees and foliage that is 25.0 m in diameter and approximately circular. How far has the wave spread out after traveling 1.00 km from the trees? Assume the speed of sound in air is 343 m/s.

This graphic shows how lower frequency sounds are able to travel farther than higher frequency sounds, largely due to greater diffractive bending. Source: Elephant Listening Project, Cornell Labs. © Cornell University.

PROBLEM 60

Concepts and Calculations Problems

Online

Diffraction is the bending of a traveling wave around an obstacle or around the edges of an opening, and is one of the consequences of superposition. Problem 61 compares diffraction in two different media and reviews some of the fundamental properties of sound

waves. Problem 62 deals with standing waves of sound in a gas. One of the factors that affects the formation of standing waves is the speed at which the individual waves travel. Problem 62 reviews how the speed of sound depends on the properties of the gas.

61. **M** **CHALK** **SSM** A sound wave with a frequency of 15 kHz emerges through a circular opening that has a diameter of 0.20 m. *Concepts:* (i) The diffraction angle for a wave emerging through a circular opening is given by sin $\theta = 1.22 \lambda/D$, where λ is the wavelength of the sound and D is the diameter of the opening. What is meant by the diffraction angle? (ii) How is the wavelength related to the frequency of the sound? (iii) Is the wavelength of the sound in air greater than, smaller than, or equal to the wavelength in water? Why? (Note: The speed of sound in air is 343 m/s and the speed of sound in water is 1482 m/s.) (iv) Is the diffraction angle of the sound in air greater than, smaller than, or equal to the diffraction angle in water? Explain. *Calculations:* Find the diffraction angle θ when the sound travels **(a)** in air and **(b)** in water.

62. **M** **CHALK** Two tubes of gas are identical and are open only at one end. One tube contains neon (Ne) and the other krypton (Kr).

Both are monatomic gases, have the same temperature, and may be assumed to be ideal gases. The fundamental frequency of the tube containing neon is 481 Hz. *Concepts:* (i) For a gas-filled tube open only at one end, the fundamental frequency ($n = 1$) is $f_1 = v/(4L)$, where v is the speed of sound and L is the length of the tube. How is the speed related to the properties of the gas? (ii) All of the factors that affect the speed of sound in this problem are the same except for the atomic masses, which are given by 20.180 u for neon, and 83.80 u for krypton. Is the speed of sound in krypton greater than, smaller than, or equal to the speed of sound in neon? Why? (iii) Is the fundamental frequency of the tube containing krypton greater than, less than, or equal to the fundamental frequency of the tube containing neon? Explain. *Calculations:* What is the fundamental frequency of the tube containing krypton?

Team Problems

Online

63. **M** **T** **WS** **Finding the Dead Frequencies.** A football field is a rectangle of width 53.3 yards (48.8 m) and length 100 yards (91.4 m). The goal lines are on the extremes (100 yards apart) with the 50-yard line at the center. Yard lines are drawn every 5 yards and labeled every 10 yards, counting up from each goal line to the 50-yard line. To the left and right of the midfield line (that is, the 50-yard line) are the 40-yard lines, then the 30-yard lines, and so on. A new announcer box was constructed near the football field, centered on the 50-yard line, and 20.0 yards away from the sideline. Two loudspeakers are mounted on the upper left and right corners of the box, which is 12.0 yards wide. You and your team have been assigned to identify all frequencies between 0 Hz and 1.25 kHz for which dead spots occur due to destructive interference at the 35.0-yard markers on the opposite sideline. Assume that the speakers are in phase, and that the speed of sound in air is 343 m/s. You can neglect the height of the speakers above the ground.

64. **E** **T** **WS** **Finding a Faulty Valve.** You and your team are on a cargo ship and have encountered a problem. A bundle of 15 vertical pipes leads from the deck to the engine room. Each pipe is 15.0 cm in diameter, is 10.5 m long, and has a valve on its lower end, in a difficult to reach place. Unfortunately, one of the valves will not open. You try to look down the pipes using a flashlight, but cannot tell which pipe has the faulty valve. After some thinking, you get the idea to use a function generator with a speaker, a microphone, and some audio software to identify the pipe with the closed valve. The function generator produces sound at a single frequency (tone) that can be swept from 50.0 Hz to 20.0 kHz, and the speaker can be placed near the exposed open ends to send sound waves down the tubes. The microphone can then be placed near the open end to sense the intensity of the sound, which is the sum of the source sound plus the response of the tube, that is, it will be able to detect the resonances, or natural frequencies, of the tubes. The audio software will then reveal the precise frequencies at which these resonances occur. Assuming a pipe with an open valve behaves like a cylindrical tube with two open ends, and a pipe with a closed valve behaves like a cylindrical tube with only one open end, what are the first three resonant frequencies above 50.0 Hz that you should detect for a pipe with **(a)** an open valve, and **(b)** a closed valve? Assume that the speed of sound in air is 343 m/s.

65. **M** **T** **An Acoustic Remote Control.** You and your team are exploring an antiquated research facility in the mountains of southern Argentina that had been abandoned in the 1960s. You come to a giant locked door that has no visible handles or actuators, but you find a handheld device nearby that has two buttons, labeled "Open" and "Close." It looks like some kind of crude remote control, but when you push the buttons they just click and nothing happens. You open the device's top cover and inspect it. The internal mechanism resembles an old acoustic remote control called the "Space Command 600" that your parents had for their ancient TV. Pushing a button on the remote actuated a small hammer on the inside that struck the end of an aluminum rod, about 2 or 3 cm in length. The rod vibrated and emitted an ultrasonic sound wave that actuated an electrical circuit in the TV that was sensitive to that frequency. The TV remote had three buttons, and therefore three rods that vibrated at different frequencies. The first frequency turned the TV on and off, the second made the tuning dial click to the next station, and the third made the dial turn in the opposite direction. You pull off the cover of the device and find that it has places for two ¼-inch diameter rods, but both are missing. However, written on the inside of the cover of the device is the following: "Open = 95.50 kHz" and "Close = 102.5 kHz." Your team members search and eventually find a long piece of ¼-inch diameter aluminum rod. **(a)** To what lengths must you cut the rod in order to get the remote to work properly (i.e., so that the fundamental frequencies of the rods match those utilized by the remote)? **(b)** Suppose you had instead found a titanium rod. What lengths would be required in that case? (Young's moduli are $Y_{Al} = 6.9 \times 10^{10}$ N/m^2 and $Y_{Ti} = 1.2 \times 10^{11}$ N/m^2; the mass densities are $\rho_{Al} = 2700$ kg/m^3 and $\rho_{Ti} = 4500$ kg/m^3.)

66. **M** **T** **A Tunable Inclined Plane.** You and your team are designing a tunable, single-tone, acoustic emitter. One end of a string is connected to a post on the top of a frictionless inclined plane. The string supports a mass at rest below, on the surface of the plane. The angle θ of the incline relative to the horizontal is adjustable. The length of the string is $L = 35.0$ cm, and the mass per unit length of the string is $\lambda = 5.28 \times 10^{-3}$ kg/m. **(a)** When $\theta = 17.0°$, the plucked string is supposed to emit a fundamental frequency of $f = 212$ Hz. What value of the mass (M) is needed under these conditions? **(b)** What is the full frequency range of the device (i.e., by changing the angle)? **(c)** With the mass calculated in part (a), to what value should you set θ so that the fundamental frequency is 330 Hz?

We have all experienced static electricity in our hair and on our clothes, and have been zapped on occasion when touching a doorknob after walking on carpet. These phenomena occur when electric charges, one of the fundamental building blocks of atoms, separate, and one type (either positive or negative) becomes more abundant than the other. As we will see in this chapter, like charges repel, which is why the toddler's hair is standing on end.

Rachel Hopper/Dreamstime.com

Electric Forces and Electric Fields

18.1 The Origin of Electricity

The electrical nature of matter is inherent in atomic structure. An atom consists of a small, relatively massive nucleus that contains particles called protons and neutrons. A proton has a mass of 1.673×10^{-27} kg, and a neutron has a slightly greater mass of 1.675×10^{-27} kg. Surrounding the nucleus is a diffuse cloud of orbiting particles called electrons, as **Figure 18.1** suggests. An electron has a mass of 9.11×10^{-31} kg. Like mass, **electric charge** is an intrinsic property of protons and electrons, and only two types of charge have been discovered, positive and negative. A proton has a positive charge, and an electron has a negative charge. A neutron has no net electric charge.

Experiment reveals that the magnitude of the charge on the proton *exactly equals* the magnitude of the charge on the electron; the proton carries a charge $+e$, and the electron carries a charge $-e$. The SI unit for measuring the magnitude of an electric charge is the **coulomb*** (C), and e has been determined experimentally to have the value

$$e = 1.60 \times 10^{-19} \text{ C}$$

LEARNING OBJECTIVES

After reading this module, you should be able to...

18.1 Define electric charge.

18.2 Describe the electric force between charged particles.

18.3 Distinguish between conductors and insulators.

18.4 Explain charging by contact and charging by induction.

18.5 Use Coulomb's law to calculate the force on a point charge due to other point charges.

18.6 Calculate the net electric field due to a configuration of point charges.

18.7 Draw electric field lines.

18.8 Describe the electric field inside a conductor.

18.9 Use Gauss' law to obtain the value of the electric field due to charge distributions.

*The definition of the coulomb depends on electric currents and magnetic fields, concepts that will be discussed later. Therefore, we postpone its definition until Section 21.7.

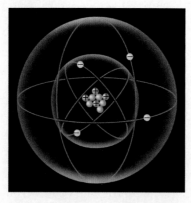

⊖ electron
⊕ proton
⊙ neutron

FIGURE 18.1 An atom contains a small, positively charged nucleus, about which the negatively charged electrons move. The closed-loop paths shown here are symbolic only. In reality, the electrons do not follow discrete paths, as Section 30.5 discusses.

The symbol *e* represents only the magnitude of the charge on a proton or an electron and does not include the algebraic sign that indicates whether the charge is positive or negative.

In nature, atoms are normally found with equal numbers of protons and electrons. Usually, then, an atom carries no net charge because the algebraic sum of the positive charge of the nucleus and the negative charge of the electrons is zero. When an atom, or any object, carries no net charge, the object is said to be *electrically neutral*. The neutrons in the nucleus are electrically neutral particles.

Charges of larger magnitude than the charge on an electron or on a proton are built up on an object by adding or removing electrons. Thus, any charge of magnitude *q* is an integer multiple of *e*; that is, $q = Ne$, where *N* is an integer. Because any electric charge *q* occurs in integer multiples of elementary, indivisible charges of magnitude *e*, electric charge is said to be *quantized*. In Section 19.7, we will discuss the experiment that determined the value of *e*. Example 1 emphasizes the quantized nature of electric charge.

EXAMPLE 1 | A Lot of Electrons

How many electrons are there in one coulomb of negative charge?

Reasoning The negative charge is due to the presence of excess electrons, since they carry negative charge. Because each electron has a charge whose magnitude is $e = 1.60 \times 10^{-19}$ C, the number of electrons is equal to the charge magnitude of one coulomb (1.00 C) divided by *e*.

Solution The number *N* of electrons is

$$N = \frac{1.00 \text{ C}}{e} = \frac{1.00 \text{ C}}{1.60 \times 10^{-19} \text{ C}} = \boxed{6.25 \times 10^{18}}$$

18.2 Charged Objects and the Electric Force

Electricity has many useful applications that have come about because it is possible to transfer electric charge from one object to another. Usually electrons are transferred, and the body that gains electrons acquires an excess of negative charge. The body that loses electrons has an excess of positive charge. Such separation of charge occurs often when two unlike materials are rubbed together. For example, when an ebonite (hard, black rubber) rod is rubbed against animal fur, some of the electrons from atoms of the fur are transferred to the rod. The ebonite becomes negatively charged, and the fur becomes positively charged, as **Figure 18.2** indicates. Similarly, if a glass rod is rubbed with a silk cloth, some of the electrons are removed from the atoms of the glass and deposited on the silk, leaving the silk negatively charged and the glass positively charged. There are many familiar examples of charge separation, as when you walk across a nylon rug or run a comb through dry hair. In each case, objects become "electrified" as surfaces rub against one another.

Animal fur

Ebonite rod

FIGURE 18.2 When an ebonite rod is rubbed against animal fur, electrons from atoms of the fur are transferred to the rod. This transfer gives the rod a negative charge (−) and leaves a positive charge (+) on the fur.

When an ebonite rod is rubbed with animal fur, the rubbing process serves only to separate electrons and protons already present in the materials. No electrons or protons are created or destroyed. Whenever an electron is transferred to the rod, a proton is left behind on the fur. Since the charges on the electron and proton have identical magnitudes but opposite signs, the algebraic sum of the two charges is zero, and the transfer does not change the net charge of the fur/rod system. If each material contains an equal number of protons and electrons to begin with, the net charge of the system is zero initially and remains zero at all times during the rubbing process.

Electric charges play a role in many situations other than rubbing two surfaces together. They are involved, for instance, in chemical reactions, electric circuits, and radioactive decay. A great number of experiments have verified that in any situation, the **law of conservation of electric charge** is obeyed.

> **LAW OF CONSERVATION OF ELECTRIC CHARGE**
>
> **During any process, the net electric charge of an isolated system remains constant (is conserved).**

It is easy to demonstrate that two electrically charged objects exert a force on one another. Consider **Interactive Figure 18.3a**, which shows two small balls that have been *oppositely charged* and are light and free to move. The balls attract each other. On the other hand, balls with the *same* type of charge, either both positive or both negative, repel each other, as parts *b* and *c* of the drawing indicate. The behavior depicted in **Interactive Figure 18.3** illustrates the following fundamental characteristic of electric charges:

Like charges repel and unlike charges attract each other.

Like other forces that we have encountered, the **electric force** (also sometimes called the **electrostatic force**) can alter the motion of an object. It can do so by contributing to the net external force $\sum \vec{F}$ that acts on the object. Newton's second law, $\sum \vec{F} = m\vec{a}$, specifies the acceleration $\vec{a}$ that arises because of the net external force. Any external electric force that acts on an object must be included when determining the net external force to be used in the second law.

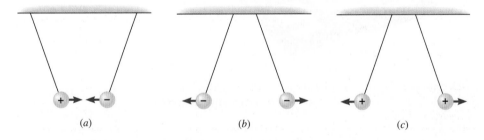

(a)   (b)   (c)

INTERACTIVE FIGURE 18.3
(*a*) A positive charge (–) and a negative charge (+) attract each other. (*b*) Two negative charges repel each other. (*c*) Two positive charges repel each other.

THE PHYSICS OF . . . electronic ink.   The development of this technology began in the late 1990's and, since 2015, has been used in many applications. In fact, electronic ink, which utilizes electric forces, has revolutionized the way books and other printed matter are made, including popular devices used to read electronic books. It allows letters and graphics on a page to be changed instantly, much like the symbols displayed on a computer monitor. **Figure 18.4a** illustrates the essential features of electronic ink. It consists of millions of clear microcapsules, each having the diameter of a human hair and filled with a dark, inky liquid. Inside each microcapsule are several dozen extremely tiny white beads that carry a slightly negative charge. The microcapsules are sandwiched between two sheets, an opaque base layer and a transparent top layer, at which the reader looks. When a positive charge is applied to a small region of the base layer, as shown in part *b* of the drawing, the negatively charged white beads are drawn to it, leaving dark ink at the top layer. Thus, a viewer sees only the dark liquid. When a negative charge is applied to a region of the base layer, the negatively charged white beads are repelled from it and are forced to the top of the microcapsules; now a viewer sees a white area due to the beads. Thus, electronic ink is based on the principle that like charges repel and unlike charges attract each other; a positive charge causes one color to appear, and a negative charge causes another color to appear. Each small region, whether dark or light, is known as a *pixel* (short for "picture element"). Computer chips provide the

See your sales person for the right fit

Base layer

Microcapsules

Beads

(a)

Dark liquid

Base layer

Dark pixel

Light pixel

(b)

FIGURE 18.4 (a) Electronic ink consists of microcapsules filled with a dark, inky liquid and dozens of white beads. (b) Dark and light pixels are formed when positive and negative charges are placed in the base layer by electronic circuitry.

instructions to produce the negative and positive charges on the base layer of each pixel. Letters and graphics are produced by the patterns generated with the two colors.

Check Your Understanding

(*The answers are given at the end of the book.*)

1. An electrically neutral object acquires a net electric charge. Which one of the following statements concerning the mass of the object is true? **(a)** The mass does not change. **(b)** The mass increases if the charge is positive and decreases if it is negative. **(c)** The mass increases if the charge is negative and decreases if it is positive.

2. Object A and object B are each electrically neutral. Two million electrons are removed from object A and placed on object B. Expressed in coulombs, what is the resulting charge (algebraic sign and magnitude) on object A and on object B?

3. Object A has a charge of -1.6×10^{-13} C, and object B is electrically neutral. Two million electrons are removed from object A and placed on object B. Expressed in coulombs, what is the resulting charge (algebraic sign and magnitude) on object A and on object B?

18.3 Conductors and Insulators

Electric charge can not only exist *on an object,* but it can also move *through an object.* However, materials differ vastly in their abilities to allow electric charge to move or be conducted through them. To help illustrate such differences in conductivity, **Figure 18.5a** recalls the conduction of heat through a bar of material whose ends are maintained at different temperatures. As Section 13.2 discusses, metals conduct heat readily and, therefore, are known as thermal conductors. On the other hand, substances that conduct heat poorly are referred to as thermal insulators.

A situation analogous to the conduction of heat arises when a metal bar is placed between two charged objects, as in **Figure 18.5b**. Electrons are conducted through the bar from the negatively charged object toward the positively charged object. Substances that readily conduct electric charge are called **electrical conductors**. Although there are exceptions, good thermal conductors are generally good electrical conductors. Metals such as copper, aluminum, silver, and gold are excellent electrical conductors and, therefore, are used in electrical wiring. Materials that conduct electric charge *poorly* are known as **electrical insulators**. In many cases, thermal insulators are also electrical insulators. Common electrical insulators are rubber, many plastics, and wood. Insulators, such as the rubber or plastic that coats electrical wiring, prevent electric charge from going where it is not wanted.

(a)   (b)

FIGURE 18.5 (a) Heat is conducted from the hotter end of the metal bar to the cooler end. (b) Electrons are conducted from the negatively charged end of the metal bar to the positively charged end.

The difference between electrical conductors and insulators is related to atomic structure. As electrons orbit the nucleus, those in the outer orbits experience a weaker force of attraction to the nucleus than do those in the inner orbits. Consequently, the outermost electrons (also called the valence electrons) can be dislodged more easily than the inner ones. In a good conductor, some valence electrons become detached from a parent atom and wander more or less freely throughout the material, belonging to no one atom in particular. The exact number of electrons detached from each atom depends on the nature of the material, but is usually between one and three. When one end of a conducting bar is placed in contact with a negatively charged object and the other end in contact with a positively charged object, as in **Figure 18.5b**, the "free" electrons are able to move readily away from the negative end and toward the positive end. The ready movement of electrons is the hallmark of a good conductor. In an insulator the situation is different, for there are very few electrons free to move throughout the material. Virtually every electron remains bound to its parent atom. Without the "free" electrons, there is very little flow of charge when the material is placed between two oppositely charged bodies, so the material is an electrical insulator.

THE PHYSICS OF . . . photoconductivity. The conducting properties of some materials can change depending on their environments or external stimuli. For instance, there are materials that change from insulators to conductors depending on their temperature. An example of such a material that exhibits this behavior is vanadium dioxide (VO_2), which is a conductor above 341 K and an insulator below that temperature. Some materials, however, such as selenium, transform from poor conductors to good conductors when exposed to light. Such a material is said to be a photoconductor. At the atomic level, the light provides energy to the least tightly bound electrons so that they become more mobile, and can therefore more readily conduct through the material. The electrical conductivity of selenium can increase by a factor of a thousand when exposed to light!

Some of the earliest and most influential technological applications of the photoconducting properties of selenium are seen in the photocopying process of xerography and in the early development of television. The word "xerography" derives from the Greek *xeros* and *graphos*, meaning "dry writing." At the heart of a xerographic copier (an old technology that has since been replaced by modern printers) is a xerographic drum, an aluminum cylinder coated with a layer of selenium. An image of the document to be copied is focused on the selenium-coated cylinder that initially carries a uniform positive charge density. The portions exposed to the light become conducting and are then electrically neutralized by conducting electrons. This process creates an "image" in positive charge density on the surface of the cylinder that then attracts a negatively charged toner powder, which can then be transferred to positively charged paper. The paper with its toner image is then sent through heated rollers to form the final copy.

The origin of early television can be traced back to the development of selenium photoconducting cells that could be used to convert light into electrical signals. These electrical signals could then be transmitted using radio waves and then detected, processed, and displayed by a receiver.

18.4 Charging by Contact and by Induction

ANIMATED FIGURE 18.6 (*a*) Electrons are transferred by rubbing the negatively charged rod on the metal sphere. (*b*) When the rod is removed, the electrons distribute themselves over the surface of the sphere.

When a negatively charged ebonite rod is rubbed on a metal object, such as the sphere in **Animated Figure 18.6a**, some of the excess electrons from the rod are transferred to the object. Once the electrons are on the metal sphere (where they can move readily) and the rod is removed, they repel one another and spread out over the sphere's surface. The insulated stand prevents them from flowing to the earth, where they could spread out even more. As shown in part *b* of the picture, the sphere is left with a negative charge distributed over its surface. In a similar manner, the sphere would be left with a positive charge after being rubbed with a positively charged rod. In this case, electrons from the sphere would be transferred to the rod. The process of giving one object a net electric charge by placing it in contact with another object that is already charged is known as **charging by contact**.

It is also possible to charge a conductor in a way that does not involve contact. In **Animated Figure 18.7**, a negatively charged rod is brought close to, *but does not touch,* a metal sphere. In the sphere, the free electrons closest to the rod move to the other side, as part *a* of the drawing indicates. As a result, the part of the sphere nearest the rod becomes positively charged and the part farthest away becomes negatively charged. These positively and negatively charged regions have been "induced" or "persuaded" to form because of the repulsive force between the negative rod and the free electrons in the sphere. If the rod were removed, the free electrons would return to their original places, and the charged regions would disappear.

Under most conditions the earth is a good electrical conductor. So when a metal wire is attached between the sphere and the ground, as in **Animated Figure 18.7b**, some of the free electrons leave the sphere and distribute themselves over the much larger earth. If the grounding wire is then removed, followed by the ebonite rod, the sphere is left with

ANIMATED FIGURE 18.7 (*a*) When a charged rod is brought near the metal sphere without touching it, some of the positive and negative charges in the sphere are separated. (*b*) Some of the electrons leave the sphere through the grounding wire, with the result (*c*) that the sphere acquires a positive net charge.

a positive net charge, as part *c* of the picture shows. The process of giving one object a net electric charge *without touching* the object to a second charged object is called **charging by induction**. The process could also be used to give the sphere a negative net charge, if a positively charged rod were used. Then, electrons would be drawn up from the ground through the grounding wire and onto the sphere.

If the sphere in **Animated Figure 18.7** were made from an insulating material like plastic, instead of metal, the method of producing a net charge by induction would not work, because very little charge would flow through the insulating material and down the grounding wire. However, the electric force of the charged rod would have some effect on the insulating material. The electric force would cause the positive and negative charges in the molecules of the material to separate slightly, with the negative charges being "pushed" away from the negative rod, as **Figure 18.8** illustrates. Although no net charge is created, the surface of the plastic does acquire a slight induced positive charge and is attracted to the negative rod. It is attracted in spite of the repulsive force between the negative rod and the negative charges in the plastic. This is because the negative charges in the plastic are further away from the rod than the positive charges are. For a similar reason, one piece of cloth can stick to another in the phenomenon known as "static cling," which occurs when an article of clothing has acquired an electric charge while being tumbled about in a clothes dryer.

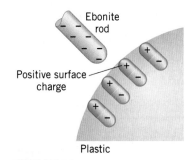

FIGURE 18.8 The negatively charged rod induces a slight positive surface charge on the plastic.

Check Your Understanding

(The answers are given at the end of the book.)

4. Two metal spheres are identical. They are electrically neutral and are touching. An electrically charged ebonite rod is then brought near the spheres without touching them, as **CYU Figure 18.1** shows. After a while, with the rod held in place, the spheres are separated, and the rod is

CYU FIGURE 18.1

A   B

then removed. The following statements refer to the masses m_A and m_B of the spheres after they are separated and the rod is removed. Which one or more of the statements is true? **(a)** $m_A = m_B$ **(b)** $m_A > m_B$ if the rod is positive **(c)** $m_A < m_B$ if the rod is positive **(d)** $m_A > m_B$ if the rod is negative **(e)** $m_A < m_B$ if the rod is negative

5. Blow up a balloon, tie it shut, and rub it against your shirt a number of times, so that the balloon acquires a net electric charge. Now touch the balloon to the ceiling. When released, will the balloon remain stuck to the ceiling?

6. A rod made from insulating material carries a net charge (which may be positive or negative), whereas a copper sphere is electrically neutral. The rod is held close to the sphere but does not touch it. Which one of the following statements concerning the forces that the rod and sphere exert on each other is true? **(a)** The forces are always attractive. **(b)** The forces are always repulsive. **(c)** The forces are attractive when the rod is negative and repulsive when it is positive. **(d)** The forces are repulsive when the rod is negative and attractive when it is positive. **(e)** There are no forces.

18.5 Coulomb's Law

The Force That Point Charges Exert on Each Other

The electrostatic force that stationary charged objects exert on each other depends on the amount of charge on the objects and the distance between them. Experiments reveal that the greater the charge and the closer together they are, the greater is the force. To set the stage for explaining these features in more detail, **Figure 18.9** shows two charged bodies. These objects are so small, compared to the distance r between them, that they

FIGURE 18.9 Each point charge exerts a force on the other. Regardless of whether the forces are (a) attractive or (b) repulsive, they are directed along the line between the charges and have equal magnitudes.

can be regarded as mathematical points. The "point charges" have magnitudes* $|q_1|$ and $|q_2|$. If the charges have *unlike* signs, as in part *a* of the picture, each object is *attracted* to the other by a force that is directed along the line between them; $+\vec{\mathbf{F}}$ is the electric force exerted on object 1 by object 2 and $-\vec{\mathbf{F}}$ is the electric force exerted on object 2 by object 1. If, as in part *b*, the charges have the *same* sign (both positive or both negative), each object is *repelled* from the other. The repulsive forces, like the attractive forces, act along the line between the charges. Whether attractive or repulsive, the two forces are equal in magnitude but opposite in direction. These forces always exist as a pair, each one acting on a different object, in accord with Newton's action–reaction law.

The French physicist Charles Augustin de Coulomb (1736–1806) carried out a number of experiments to determine how the electric force that one point charge applies to another depends on the amount of each charge and the separation between them. His result, now known as **Coulomb's law**, is stated as follows:

COULOMB'S LAW

The magnitude F of the electrostatic force exerted by one point charge q_1 on another point charge q_2 is directly proportional to the magnitudes $|q_1|$ and $|q_2|$ of the charges and inversely proportional to the square of the distance r between them:

$$F = k\frac{|q_1|\,|q_2|}{r^2} \qquad (18.1)$$

where k is a proportionality constant: $k = 8.99 \times 10^9\ \mathrm{N \cdot m^2/C^2}$ in SI units. Equation 18.1 gives only the magnitude of the electrostatic force that each point charge exerts on the other; it does *not* give the direction. The electrostatic force is directed along the line joining the charges, and it is attractive if the charges have unlike signs and repulsive if the charges have like signs.

It is common practice to express k in terms of another constant ε_0, by writing $k = 1/(4\pi\varepsilon_0)$; ε_0 is called the **permittivity of free space** and has a value that is given according to $\varepsilon_0 = 1/(4\pi k) = 8.85 \times 10^{-12}\ \mathrm{C^2/(N \cdot m^2)}$. Example 2 illustrates the use of Coulomb's law.

Math Skills When using Equation 18.1, substitute only the charge magnitudes (without algebraic signs) for $|q_1|$ and $|q_2|$. Do not substitute negative numbers for these symbols. This is because the equation gives only the magnitude of the electrostatic force, and the magnitude of a force cannot be negative. For example, suppose that $q_1 = -5.0 \times 10^{-6}$ C and $q_2 = +7.0 \times 10^{-6}$ C. Then, for use in Equation 18.1, we would have the following values:

$$|q_1| = |-5.0 \times 10^{-6}\ \mathrm{C}| = 5.0 \times 10^{-6}\ \mathrm{C} \quad (not - 5.0 \times 10^{-6}\ \mathrm{C})$$

$$|q_2| = |+7.0 \times 10^{-6}\ \mathrm{C}| = 7.0 \times 10^{-6}\ \mathrm{C}$$

EXAMPLE 2 | A Large Attractive Force

Two objects, whose charges are +1.0 and −1.0 C, are separated by 1.0 km. Compared to 1.0 km, the sizes of the objects are small. Find the magnitude of the attractive force that either charge exerts on the other.

Reasoning Considering that the sizes of the objects are small compared to the separation distance, we can treat the charges as point charges. Coulomb's law may then be used to find the magnitude of

the attractive force, provided that only the *magnitudes of the charges* are used for the symbols $|q_1|$ and $|q_2|$ that appear in the law.

Solution The magnitude of the force is

$$F = k\frac{|q_1|\,|q_2|}{r^2} = \frac{(8.99 \times 10^9\ \mathrm{N \cdot m^2/C^2})(1.0\ \mathrm{C})(1.0\ \mathrm{C})}{(1.0 \times 10^3\ \mathrm{m})^2} \qquad (18.1)$$

$$= \boxed{9.0 \times 10^3\ \mathrm{N}}$$

---

*The magnitude of a variable is sometimes called the absolute value and is symbolized by a vertical bar to the left and to the right of the variable. Thus, $|q|$ denotes the magnitude or absolute value of the variable q, which is the value of q without its algebraic plus or minus sign. For example, if $q = -2.0$ C, then $|q| = 2.0$ C.

The force calculated in Example 2 corresponds to about 2000 pounds and is so large because charges of ± 1.0 C are enormous. Such large charges are rare and are encountered only in the most severe conditions, as in a lightning bolt, where as much as 25 C can be transferred between the cloud and the ground. The typical charges produced in the laboratory are much smaller and are measured conveniently in microcoulombs (1 microcoulomb = 1 μC = 10^{-6} C).

Coulomb's law has a form that is remarkably similar to Newton's law of gravitation ($F = Gm_1m_2/r^2$). The force in both laws depends on the inverse square ($1/r^2$) of the distance between the two objects and is directed along the line between them. In addition, the force is proportional to the product of an intrinsic property of each of the objects, the magnitudes of the charges $|q_1|$ and $|q_2|$ in Coulomb's law and the masses m_1 and m_2 in the gravitation law. However, there is a major difference between the two laws. The electrostatic force can be either repulsive or attractive, depending on whether or not the charges have the same sign; in contrast, the gravitational force is *always* an attractive force.

Section 5.5 discusses how the gravitational attraction between the earth and a satellite provides the centripetal force that keeps a satellite in orbit. Example 3 illustrates that the electrostatic force of attraction plays a similar role in a famous model of the atom created by the Danish physicist Niels Bohr (1885–1962).

Analyzing Multiple-Concept Problems

EXAMPLE 3 | A Model of the Hydrogen Atom

In the Bohr model of the hydrogen atom, the electron (charge = $-e$) is in a circular orbit about the nuclear proton (charge = $+e$) at a radius of 5.29×10^{-11} m, as **Figure 18.10** shows. The mass of the electron is 9.11×10^{-31} kg. Determine the speed of the electron.

Reasoning Recall from Section 5.3 that a net force is required to keep an object such as an electron moving on a circular path. This net force is called the centripetal force and always points toward the center of the circle. The centripetal force has a magnitude given by $F_c = mv^2/r$, where m and v are, respectively, the mass and speed of the electron and r is the radius of the orbit. This equation can be solved for the speed of the electron. Since the mass and orbital radius are known, we can calculate the electron's speed provided that a value for the centripetal force can be found. For the electron in the hydrogen atom, the centripetal force is provided almost exclusively by the electrostatic force that the proton exerts on the electron.

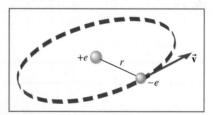

FIGURE 18.10 In the Bohr model of the hydrogen atom, the electron ($-e$) orbits the proton ($+e$) at a distance that is $r = 5.29 \times 10^{-11}$ m. The velocity of the electron is $\vec{v}$.

This attractive force points toward the center of the circle, and its magnitude is given by Coulomb's law. The electron is also pulled toward the proton by the gravitational force. However, the gravitational force is negligible in comparison to the electrostatic force.

Knowns and Unknowns The data for this problem are:

Description	Symbol	Value
Electron charge	$-e$	-1.60×10^{-19} C
Electron mass	m	9.11×10^{-31} kg
Proton charge	$+e$	$+1.60 \times 10^{-19}$ C
Radius of orbit	r	5.29×10^{-11} m
Unknown Variable		
Orbital speed of electron	v	?

Modeling the Problem

STEP 1 Centripetal Force An electron of mass m that moves with a constant speed v on a circular path of radius r experiences a net force, called the centripetal force. The magnitude F_c of this force is given by $F_c = mv^2/r$ (Equation 5.3). By solving this equation for the speed, we obtain Equation 1 at the right. The mass and radius in this expression are known. However, the magnitude of the centripetal force is not known, so we will evaluate it in Step 2.

$$v = \sqrt{\frac{rF_c}{m}} \qquad (1)$$

STEP 2 Coulomb's Law As the electron orbits the proton in the hydrogen atom, it is attracted to the proton by the electrostatic force. The magnitude F of the electrostatic force is given by Coulomb's law as $F = k|q_1| |q_2|/r^2$ (Equation 18.1), where $|q_1|$ and $|q_2|$ are the magnitudes of the charges, r is the orbital radius, and $k = 8.99 \times 10^9 \, \text{N} \cdot \text{m}^2/\text{C}^2$. Since the centripetal force is provided almost entirely by the electrostatic force, it follows that $F_c = F$. Furthermore, $|q_1| = |-e|$ and $|q_2| = |+e|$. With these substitutions, Equation 18.1 becomes

$$F_c = k\frac{|-e| \, |+e|}{r^2}$$

All the variables on the right side of this expression are known, so we substitute it into Equation 1, as indicated in the right column.

$$v = \sqrt{\frac{rF_c}{m}} \qquad (1)$$

$$F_c = k\frac{|-e| \, |+e|}{r^2}$$

Solution Algebraically combining the results of the modeling steps, we have

$$\overset{\textbf{STEP 1}}{\underset{\downarrow}{}} \quad \overset{\textbf{STEP 2}}{\underset{\downarrow}{}}$$

$$v = \sqrt{\frac{rF_c}{m}} = \sqrt{\frac{r\left(k\frac{|-e| \, |+e|}{r^2}\right)}{m}} = \sqrt{\frac{k|-e| \, |+e|}{mr}}$$

The speed of the orbiting electron is

$$v = \sqrt{\frac{k|-e| \, |+e|}{mr}}$$

$$= \sqrt{\frac{(8.99 \times 10^9 \, \text{N} \cdot \text{m}^2/\text{C}^2)|-1.60 \times 10^{-19} \, \text{C}| \, |+1.60 \times 10^{-19} \, \text{C}|}{(9.11 \times 10^{-31} \, \text{kg})(5.29 \times 10^{-11} \, \text{m})}} = \boxed{2.19 \times 10^6 \, \text{m/s}}$$

Related Homework: Problems 19, 23

THE PHYSICS OF . . . adhesion. Since the electrostatic force depends on the inverse square of the distance between the charges, it becomes larger for smaller distances, such as those involved when a strip of adhesive tape is stuck to a smooth surface. Electrons shift over the small distances between the tape and the surface. As a result, the materials become oppositely charged. Since the distance between the charges is relatively small, the electrostatic force of attraction is large enough to contribute to the adhesive bond. **Figure 18.11** shows an image of the sticky surface of a piece of tape after it has been pulled off a metal surface. The image was obtained using an atomic-force microscope and reveals the tiny pits left behind when microscopic portions of the adhesive remain stuck to the metal because of the strong adhesive bonding forces.

The Force on a Point Charge Due to Two or More Other Point Charges

Up to now, we have been discussing the electrostatic force on a point charge (magnitude $|q_1|$) due to another point charge (magnitude $|q_2|$). Suppose that a third point charge (magnitude $|q_3|$) is also present. What would be the net force on q_1 due to both q_2 and q_3? It is convenient to deal with such a problem in parts. First, find the magnitude and direction of the force exerted on q_1 by q_2 (ignoring q_3). Then, determine the force exerted on q_1 by q_3 (ignoring q_2). The *net force* on q_1 is the *vector sum* of these forces. Examples 4 and 5 illustrate this approach when the charges lie along a straight line and on a plane, respectively.

EXAMPLE 4 | Three Charges on a Line

Figure 18.12a shows three point charges that lie along the *x* axis in a vacuum. Determine the magnitude and direction of the net electrostatic force on q_1.

Reasoning Part *b* of the drawing shows a free-body diagram of the forces that act on q_1. Since q_1 and q_2 have opposite signs, they attract one another. Thus, the force exerted on q_1 by q_2 is $\vec{F}_{12}$, and it points to the left. Similarly, the force exerted on q_1 by q_3 is $\vec{F}_{13}$ and is also an attractive force. It points to the right in Figure 18.12b. The magnitudes of these forces can be obtained from Coulomb's law. The net force is the vector sum of $\vec{F}_{12}$ and $\vec{F}_{13}$.

Solution In calculating the magnitudes of the individual forces with the aid of Equation 18.1, we use only the magnitudes of the charges (without algebraic signs):

$$F_{12} = k\frac{|q_1|\,|q_2|}{r_{12}^2} = \frac{(8.99 \times 10^9 \text{ N} \cdot \text{m}^2/\text{C}^2)(3.0 \times 10^{-6} \text{ C})(4.0 \times 10^{-6} \text{ C})}{(0.20 \text{ m})^2}$$

$$= 2.7 \text{ N}$$

$$F_{13} = k\frac{|q_1|\,|q_3|}{r_{13}^2} = \frac{(8.99 \times 10^9 \text{ N} \cdot \text{m}^2/\text{C}^2)(3.0 \times 10^{-6} \text{ C})(7.0 \times 10^{-6} \text{ C})}{(0.15 \text{ m})^2}$$

$$= 8.4 \text{ N}$$

FIGURE 18.12 (*a*) Three charges lying along the *x* axis. (*b*) The force exerted on q_1 by q_2 is $\vec{F}_{12}$, while the force exerted on q_1 by q_3 is $\vec{F}_{13}$.

Since $\vec{F}_{12}$ points in the negative *x* direction, and $\vec{F}_{13}$ points in the positive *x* direction, the net force $\vec{F}$ is

$$\vec{F} = \vec{F}_{12} + \vec{F}_{13} = (-2.7 \text{ N}) + (8.4 \text{ N}) = \boxed{+5.7 \text{ N}}$$

The plus sign in the answer indicates that the net force points to the right in the drawing.

EXAMPLE 5 | Three Charges in a Plane

Figure 18.13a shows three point charges that lie in the *x*, *y* plane in a vacuum. Find the magnitude and direction of the net electrostatic force on q_1.

Reasoning The force exerted on q_1 by q_2 is $\vec{F}_{12}$ and is an attractive force because the two charges have opposite signs. It points

along the line between the charges. The force exerted on q_1 by q_3 is $\vec{F}_{13}$ and is also an attractive force. It points along the line between q_1 and q_3. Coulomb's law specifies the magnitudes of these forces. Since the forces point in different directions (see Figure 18.13b), we will use vector components to find the net force.

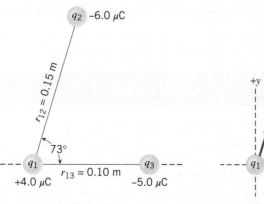

(*a*)

(*b*) Free-body diagram for q_1

FIGURE 18.13 (*a*) Three charges lying in a plane. (*b*) The net force acting on q_1 is $\vec{F} = \vec{F}_{12} + \vec{F}_{13}$. The angle that $\vec{F}$ makes with the +*x* axis is θ.

Solution The magnitudes of the forces are

$$F_{12} = k\frac{|q_1|\,|q_2|}{r_{12}^2} = \frac{(8.99 \times 10^9 \text{ N} \cdot \text{m}^2/\text{C}^2)(4.0 \times 10^{-6} \text{ C})(6.0 \times 10^{-6} \text{ C})}{(0.15 \text{ m})^2}$$

$$= 9.6 \text{ N}$$

$$F_{13} = k\frac{|q_1|\,|q_3|}{r_{13}^2} = \frac{(8.99 \times 10^9 \text{ N} \cdot \text{m}^2/\text{C}^2)(4.0 \times 10^{-6} \text{ C})(5.0 \times 10^{-6} \text{ C})}{(0.10 \text{ m})^2}$$

$$= 18 \text{ N}$$

The net force $\vec{F}$ is the vector sum of $\vec{F}_{12}$ and $\vec{F}_{13}$, as part b of the drawing shows. The components of $\vec{F}$ that lie in the x and y directions are $\vec{F}_x$ and $\vec{F}_y$, respectively. Our approach to finding $\vec{F}$ is the same as that used in Chapters 1 and 4. The forces $\vec{F}_{12}$ and $\vec{F}_{13}$ are resolved into x and y components. Then, the x components are combined to give $\vec{F}_x$, and the y components are combined to give $\vec{F}_y$. Once $\vec{F}_x$ and $\vec{F}_y$ are known, the magnitude and direction of $\vec{F}$ can be determined using trigonometry.

Force	x component	y component
$\vec{F}_{12}$	$+(9.6 \text{ N}) \cos 73° = +2.8 \text{ N}$	$+(9.6 \text{ N}) \sin 73° = +9.2 \text{ N}$
$\vec{F}_{13}$	$+18 \text{ N}$	0 N
$\vec{F}$	$\vec{F}_x = +21 \text{ N}$	$\vec{F}_y = +9.2 \text{ N}$

The magnitude F and the angle θ of the net force are

$$F = \sqrt{F_x^2 + F_y^2} = \sqrt{(21 \text{ N})^2 + (9.2 \text{ N})^2} = \boxed{23 \text{ N}}$$

$$\theta = \tan^{-1}\left(\frac{F_y}{F_x}\right) = \tan^{-1}\left(\frac{9.2 \text{ N}}{21 \text{ N}}\right) = \boxed{24°}$$

Math Skills As discussed in Section 1.7, the vector components of the force $\vec{F}$ (magnitude $= F$) are two perpendicular vectors $\vec{F}_x$ and $\vec{F}_y$ that are parallel to the x and y axes, respectively, and add vectorially so that $\vec{F} = \vec{F}_x + \vec{F}_y$. The scalar component F_x has a magnitude equal to that of the vector component $\vec{F}_x$ and is given a positive sign if $\vec{F}_x$ points along the $+x$ axis and a negative sign if $\vec{F}_x$ points along the $-x$ axis. The scalar component F_y is defined similarly. **Figure 18.14** shows the vector $\vec{F}$ and its components. Since the components are perpendicular, the shaded triangle is a right triangle, and we can use the Pythagorean theorem (Equation 1.7). This theorem states that the square of the length of the hypotenuse of a right triangle is equal to the sum of the squares of the lengths of the other two sides. The hypotenuse of the shaded triangle is F, and the other two sides are F_x and F_y. Thus, we have that

$$F^2 = F_x^2 + F_y^2 \quad \text{or} \quad F = \sqrt{F_x^2 + F_y^2}$$

To determine the angle θ we apply the tangent function ($\tan \theta$). According to Equation 1.3, $\tan \theta$ is the length of the side of the triangle opposite the angle θ divided by the length of the side adjacent to the angle θ. Thus, referring to **Figure 18.14** to identify these lengths, we see that $\tan \theta = \dfrac{F_y}{F_x}$. This equation means that θ is the angle whose tangent is $\dfrac{F_y}{F_x}$, a result that can be expressed by using the inverse tangent function ($\tan^{-1}$):

$$\theta = \tan^{-1}\left(\frac{F_y}{F_x}\right)$$

FIGURE 18.14 Math Skills drawing.

Check Your Understanding

(The answers are given at the end of the book.)

7. Identical point charges are fixed to diagonally opposite corners of a square. Where does a third point charge experience the greater force? **(a)** At the center of the square **(b)** At one of the empty corners **(c)** The question is unanswerable because the polarities of the charges are not given.

8. CYU Figure 18.2 shows three point charges arranged in three different ways. The charges are $+q$, $-q$, and $-q$; each has the same magnitude, with one positive and the other two negative. In each of the arrangements the distance d is the same. Rank the arrangements in descending order (largest first) according to the magnitude of the net electrostatic force that acts on the positive charge.

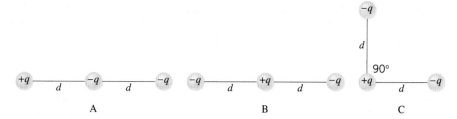

CYU FIGURE 18.2

9. A proton and an electron are held in place on the x axis. The proton is at $x = -d$, while the electron is at $x = +d$. They are released simultaneously, and the only force that affects their motions significantly is the electrostatic force of attraction that each applies to the other. Which particle reaches the origin first?

10. A particle is attached to one end of a horizontal spring, and the other end of the spring is attached to a wall. When the particle is pushed so that the spring is compressed more and more, the particle experiences a greater and greater force from the spring. Similarly, a charged particle experiences a greater and greater force when pushed closer and closer to another particle that is fixed in position and has a charge of the same polarity. Considering this similarity, will the charged particle exhibit simple harmonic motion on being released, as will the particle on the spring?

FIGURE 18.15 A positive charge q_0 experiences an electrostatic force $\vec{F}$ due to the surrounding charges on the ebonite rod and the two spheres.

18.6 | The Electric Field

Definition

As we know, a charge can experience an electrostatic force due to the presence of other charges. For instance, the positive charge q_0 in **Figure 18.15** experiences a force $\vec{F}$, which is the vector sum of the forces exerted by the charges on the rod and the two spheres. It is useful to think of q_0 as a **test charge** for determining the extent to which the surrounding charges generate a force. However, in using a test charge, we must be careful to select one with a very small magnitude, so that it does not alter the locations of the other charges. The next example illustrates how the concept of a test charge is applied.

EXAMPLE 6 | A Test Charge

The positive test charge shown in **Figure 18.15** is $q_0 = +3.0 \times 10^{-8}$ C and experiences a force $F = 6.0 \times 10^{-8}$ N in the direction shown in the drawing. **(a)** Find the *force per coulomb* that the test charge experiences. **(b)** Using the result of part (a), predict the force that a charge of $+12 \times 10^{-8}$ C would experience if it replaced q_0.

Reasoning The charges in the environment apply a force $\vec{F}$ to the test charge q_0. The force per coulomb experienced by the test charge is $\vec{F}/q_0$. If q_0 is replaced by a new charge q, then the force on this new charge is the force per coulomb times q.

Solution **(a)** The force per coulomb of charge is

$$\frac{\vec{F}}{q_0} = \frac{6.0 \times 10^{-8}\,\text{N}}{3.0 \times 10^{-8}\,\text{C}} = \boxed{2.0\,\text{N/C}}$$

The direction of the force per coulomb is the same as the direction of $\vec{F}$ in **Figure 18.15**.

(b) The result from part (a) indicates that the surrounding charges can exert 2.0 newtons of force per coulomb of charge. Thus, a charge of $+12 \times 10^{-8}$ C would experience a force whose magnitude is

$$F = (2.0\,\text{N/C})(12 \times 10^{-8}\,\text{C}) = \boxed{24 \times 10^{-8}\,\text{N}}$$

The direction of this force would be the same as the direction of the force experienced by the test charge, since both have the same positive sign.

The electric force per coulomb, $\vec{F}/q_0$, calculated in Example 6(a) is one illustration of an idea that is very important in the study of electricity. The idea is called the **electric field**. Equation 18.2 presents the definition of the electric field.

DEFINITION OF THE ELECTRIC FIELD

The electric field $\vec{E}$ that exists at a point is the electrostatic force $\vec{F}$ experienced by a small test charge* q_0 placed at that point divided by the charge itself:

$$\vec{E} = \frac{\vec{F}}{q_0} \tag{18.2}$$

The electric field is a vector, and its direction is the same as the direction of the force $\vec{F}$ on a positive test charge.

SI Unit of Electric Field: newton per coulomb (N/C)

Equation 18.2 indicates that the unit for the electric field is that of force divided by charge, which is a newton/coulomb (N/C) in SI units.

It is the surrounding charges that create an electric field at a given point. Any positive or negative charge placed at the point interacts with the field and, as a result, experiences a force, as the next example indicates.

EXAMPLE 7 | An Electric Field Leads to a Force

In **Figure 18.16** the charges on the two metal spheres and the ebonite rod create an electric field $\vec{E}$ at the spot indicated. This field has a magnitude of 2.0 N/C and is directed as in the drawing. Determine the force on a charge placed at that spot, if the charge has a value of **(a)** $q_0 = +18 \times 10^{-8}$ C and **(b)** $q_0 = -24 \times 10^{-8}$ C.

Reasoning The electric field at a given spot can exert a variety of forces, depending on the magnitude and sign of the charge placed there. The charge is assumed to be small enough that it does not alter the locations of the surrounding charges that create the field.

Solution **(a)** The magnitude of the force is the product of the magnitudes of q_0 and $\vec{E}$:

$$F = |q_0| \, E = (18 \times 10^{-8} \text{ C})(2.0 \text{ N/C}) = \boxed{36 \times 10^{-8} \text{ N}} \tag{18.2}$$

Since q_0 is positive, the force points in the same direction as the electric field, as part a of the drawing indicates.

(b) In this case, the magnitude of the force is

$$F = |q_0| \, E = (24 \times 10^{-8} \text{ C})(2.0 \text{ N/C}) = \boxed{48 \times 10^{-8} \text{ N}} \tag{18.2}$$

The force on the negative charge points in the direction *opposite* to the force on the positive charge—that is, opposite to the electric field (see part b of the drawing).

(a)   (b)

FIGURE 18.16 The electric field $\vec{E}$ that exists at a given spot can exert a variety of forces. The force exerted depends on the magnitude and sign of the charge placed at that spot. (*a*) The force on a positive charge points in the same direction as $\vec{E}$, while (*b*) the force on a negative charge points opposite to $\vec{E}$.

---

*As long as the test charge is small enough that it does not disturb the surrounding charges, it may be either positive or negative. Compared to a positive test charge, a negative test charge of the same magnitude experiences a force of the same magnitude that points in the opposite direction. However, the same electric field is given by Equation 18.2, in which $\vec{F}$ is replaced by $-\vec{F}$ and q_0 is replaced by $-q_0$.

At a particular point in space, each of the surrounding charges contributes to the net electric field that exists there. To determine the net field, it is necessary to obtain the various contributions separately and then find the vector sum of them all. Such an approach is an illustration of the principle of linear superposition, as applied to electric fields. (This principle is introduced in Section 17.1, in connection with waves.) Example 8 emphasizes the vector nature of the electric field.

EXAMPLE 8 | Electric Fields Add as Vectors Do

Animated Figure 18.17 shows two charged objects, A and B. Each contributes as follows to the net electric field at point P: $\vec{E}_A = 3.00$ N/C directed to the right, and $\vec{E}_B = 2.00$ N/C directed downward. Thus, $\vec{E}_A$ and $\vec{E}_B$ are perpendicular. What is the net electric field at P?

Reasoning   The net electric field $\vec{E}$ is the vector sum of $\vec{E}_A$ and $\vec{E}_B$: $\vec{E} = \vec{E}_A + \vec{E}_B$. As illustrated in **Animated Figure 18.17**, $\vec{E}_A$ and $\vec{E}_B$ are perpendicular, so $\vec{E}$ is the diagonal of the rectangle shown in the drawing. Thus, we can use the Pythagorean theorem to find the magnitude of $\vec{E}$ and trigonometry to find the directional angle θ.

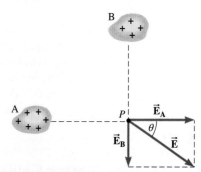

Solution   The magnitude of the net electric field is

$$E = \sqrt{E_A^2 + E_B^2} = \sqrt{(3.00 \text{ N/C})^2 + (2.00 \text{ N/C})^2} = \boxed{3.61 \text{ N/C}}$$

The direction of $\vec{E}$ is given by the angle θ in the drawing:

$$\theta = \tan^{-1}\left(\frac{E_B}{E_A}\right) = \tan^{-1}\left(\frac{2.00 \text{ N/C}}{3.00 \text{ N/C}}\right) = \boxed{33.7°}$$

ANIMATED FIGURE 18.17   The electric field contributions $\vec{E}_A$ and $\vec{E}_B$, which come from the two charge distributions, are added vectorially to obtain the net field $\vec{E}$ at point P.

Point Charges

A more complete understanding of the electric field concept can be gained by considering the field created by a point charge, as in the following example.

EXAMPLE 9 | The Electric Field of a Point Charge

There is an isolated point charge of $q = +15 \, \mu$C in a vacuum at the left in **Figure 18.18a**. Using a test charge of $q_0 = +0.80 \, \mu$C, determine the electric field at point P, which is 0.20 m away.

Reasoning   Following the definition of the electric field, we place the test charge q_0 at point P, determine the force acting on the test charge, and then divide the force by the test charge.

Solution   Coulomb's law (Equation 18.1) gives the magnitude of the force:

$$F = k\frac{|q_0|\,|q|}{r^2} = \frac{(8.99 \times 10^9 \text{ N} \cdot \text{m}^2/\text{C}^2)(0.80 \times 10^{-6} \text{ C})(15 \times 10^{-6} \text{ C})}{(0.20 \text{ m})^2}$$

$$= \boxed{2.7 \text{ N}}$$

Equation 18.2 gives the magnitude of the electric field:

$$E = \frac{F}{|q_0|} = \frac{2.7 \text{ N}}{0.80 \times 10^{-6} \text{ C}} = \boxed{3.4 \times 10^6 \text{ N/C}}$$

The electric field $\vec{E}$ points in the *same direction* as the force $\vec{F}$ on the positive test charge. Since the test charge experiences a force of repulsion directed to the right, the electric field vector also points to the right, as **Figure 18.18b** shows.

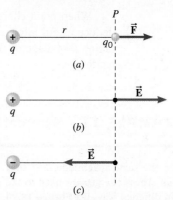

FIGURE 18.18   (a) At location P, a positive test charge q_0 experiences a repulsive force $\vec{F}$ due to the positive point charge q. (b) At P, the electric field $\vec{E}$ is directed to the right. (c) If the charge q were negative rather than positive, the electric field would have the same magnitude as in (b) but would point to the left.

The electric field produced by a point charge q can be obtained in general terms from Coulomb's law. First, note that the magnitude of the force exerted by the charge q on a test charge q_0 is $F = k|q||q_0|/r^2$. Then, divide this value by $|q_0|$ to obtain the magnitude of the field. Since $|q_0|$ is eliminated algebraically from the result, *the electric field does not depend on the test charge:*

Point charge q
$$E = \frac{k|q|}{r^2}$$
(18.3)

> **Problem-Solving Insight**   Equation 18.3 gives only the magnitude of the electric field produced by a point charge. Therefore, do not use negative numbers for the symbol $|q|$ in this equation.

As in Coulomb's law, the symbol $|q|$ denotes the magnitude of q in Equation 18.3, without regard to whether q is positive or negative. If q is positive, then $\vec{E}$ is directed away from q, as in **Figure 18.18b**. On the other hand, if q is negative, then $\vec{E}$ is directed toward q, since a negative charge attracts a positive test charge. For instance, **Figure 18.18c** shows the electric field that would exist at P if there were a charge of $-q$ instead of $+q$ at the left of the drawing. Example 10 reemphasizes the fact that all the surrounding charges make a contribution to the electric field that exists at a given place.

EXAMPLE 10 | The Electric Fields from Separate Charges May Cancel

Two positive point charges, $q_1 = +16\ \mu\text{C}$ and $q_2 = +4.0\ \mu\text{C}$, are separated in a vacuum by a distance of 3.0 m, as **Figure 18.19** illustrates. Find the spot on the line between the charges where the net electric field is zero.

Reasoning   Between the charges the two field contributions have opposite directions, and the net electric field is zero at the place where the magnitude of $\vec{E}_1$ equals the magnitude of $\vec{E}_2$. However, since q_2 is smaller than q_1, this location must be *closer* to q_2, in order that the field of the smaller charge can balance the field of the larger charge. In the drawing, the cancellation spot is labeled P, and its distance from q_1 is d.

Solution   At P, $E_1 = E_2$, and using the expression $E = k|q|/r^2$ (Equation 18.3), we have

$$\frac{k(16 \times 10^{-6}\ \text{C})}{d^2} = \frac{k(4.0 \times 10^{-6}\ \text{C})}{(3.0\ \text{m} - d)^2}$$

Rearranging this expression shows that $4.0(3.0\ \text{m} - d)^2 = d^2$. Taking the square root of each side of this equation reveals that

FIGURE 18.19   The two point charges q_1 and q_2 create electric fields $\vec{E}_1$ and $\vec{E}_2$ that cancel at a location P on the line between the charges.

$$2.0(3.0\ \text{m} - d) = \pm d$$

The plus and minus signs on the right occur because either the positive or negative root can be taken. Therefore, there are two possible values for d: +2.0 m and +6.0 m. The value +6.0 m corresponds to a location off to the right of both charges, where the magnitudes of $\vec{E}_1$ and $\vec{E}_2$ are equal, but where the directions are the same. Thus, $\vec{E}_1$ and $\vec{E}_2$ do not cancel at this spot. The other value for d corresponds to the location shown in the drawing and is the zero-field location: $\boxed{d = +2.0\ \text{m}}$.

When point charges are arranged in a symmetrical fashion, it is often possible to deduce useful information about the magnitude and direction of the electric field by taking advantage of the symmetry. Conceptual Example 11 illustrates the use of this technique.

CONCEPTUAL EXAMPLE 11 | Symmetry and the Electric Field

Four point charges all have the same magnitude, but they do not all have the same sign. These charges are fixed to the corners of a rectangle in two different ways, as **Figure 18.20** shows. Consider the net electric field at the center C of the rectangle in each case. In which case, if either, is the net electric field greater? **(a)** It is greater in **Figure 18.20a**. **(b)** It is greater in **Figure 18.20b**. **(c)** The field has the same magnitude in both cases.

Reasoning The net electric field at C is the vector sum of the individual fields created there by the charges at each corner. Each of the individual fields has the same magnitude, since the charges all have the same magnitude and are equidistant from C. The directions of the individual fields are different, however. The field created by a positive charge points away from the charge, and the field created by a negative charge points toward the charge.

Answers (a) and (c) are incorrect. To see why these answers are incorrect, note that the charges on corners 2 and 4 are identical in both parts of **Figure 18.20**. Moreover, in **Figure 18.20a** the charges at corners 1 and 3 are both $+q$, so they contribute individual fields of the same magnitude at C that have *opposite* directions and, therefore, cancel. However, in **Figure 18.20b** the charges at corners 1 and 3 are $-q$ and $+q$, respectively. They contribute individual fields of the same magnitude at C that have the *same* directions and do not cancel, but combine to produce the field $\vec{E}_{13}$ shown in Figure 18.20b. The fact that this contribution to the net field at C is present in **Figure 18.20b** but not in **Figure 18.20a** means that the net fields in the two cases are different and that the net field in **Figure 18.20a** is less than (not greater than) the net field in **Figure 18.20b**.

Answer (b) is correct. To assess the net field at C, we need to consider the contribution from the charges at corners 2 and 4, which are $-q$ and $+q$, respectively, in both cases. This is just like the arrangement on corners 1 and 3, which was discussed previously. It leads to a contribution to the net field at C that is shown as $\vec{E}_{24}$ in both parts of the figure. In **Figure 18.20a** the net field at C is just $\vec{E}_{24}$, but in **Figure 18.20b** it is the vector sum of $\vec{E}_{13}$ and $\vec{E}_{24}$, which is clearly greater than either of these two values alone.

Related Homework: Problem 37

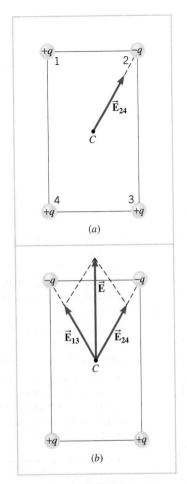

FIGURE 18.20 Charges of identical magnitude are placed at the corners of a rectangle. However, the charges do not all have the same sign and give rise to different electric fields at the center C of the rectangle, depending on the signs that they have.

The Parallel Plate Capacitor

Equation 18.3, which gives the electric field of a point charge, is a very useful result. With the aid of integral calculus, this equation can be applied in a variety of situations where point charges are distributed over one or more surfaces. One such example that has considerable practical importance is the **parallel plate capacitor**. As **Figure 18.21** shows, this device consists of two parallel metal plates, each with area A. A charge $+q$ is spread uniformly over one plate, while a charge $-q$ is spread uniformly over the other plate. In the region between the plates and away from the edges, the electric field points from the positive plate toward the negative plate and is perpendicular to both. It can be shown (see Example 16 in Section 18.9) that this electric field has a magnitude of

Parallel plate capacitor
$$E = \frac{q}{\varepsilon_0 A} = \frac{\sigma}{\varepsilon_0} \tag{18.4}$$

where ε_0 is the permittivity of free space. In this expression the Greek symbol sigma (σ) denotes the charge per unit area ($\sigma = q/A$) and is sometimes called the **charge density**. Except in the region near the edges, the field has the same value at all places between the plates. The field does *not* depend on the distance from the charges, in distinct contrast to the field created by an isolated point charge.

FIGURE 18.21 A parallel plate capacitor.

Check Your Understanding

(The answers are given at the end of the book.)

11. There is an electric field at point P. A very small positive charge is placed at this point and experiences a force. Then the positive charge is replaced by a very small negative charge that has a magnitude different from that of the positive charge. Which one of the following statements is true concerning the forces that these charges experience at *P*? **(a)** They are identical. **(b)** They have the same magnitude but different directions. **(c)** They have different magnitudes but the same direction. **(d)** They have different magnitudes and different directions.

12. Suppose that in **Figure 18.20a** point charges –*q* are fixed in place at corners 1 and 3 and point charges +*q* are fixed in place at corners 2 and 4. What then would be the net electric field at the center *C* of the rectangle?

13. A positive point charge +*q* is fixed in position at the center of a square, as **CYU Figure 18.3** shows. A second point charge is fixed to corner B, C, or D. The net electric field that results at corner A is zero. **(a)** At which corner is the second charge located? **(b)** Is the second charge positive or negative? **(c)** Does the second charge have a greater, a smaller, or the same magnitude as the charge at the center?

CYU FIGURE 18.3

14. A positive point charge is located to the left of a negative point charge. When both charges have the same magnitude, there is no place on the line passing through both charges where the net electric field due to the two charges is zero. Suppose, however, that the negative charge has a greater magnitude than the positive charge. On which part of the line, if any, is a place of zero net electric field now located? **(a)** To the left of the positive charge **(b)** Between the two charges **(c)** To the right of the negative charge **(d)** There is no zero place.

15. Three point charges are fixed to the corners of a square, one to a corner, in such a way that the net electric field at the empty corner is zero. Do these charges all have **(a)** the same sign and **(b)** the same magnitude (but, possibly, different signs)?

16. Consider two identical, thin, and nonconducting rods, A and B. On rod A, positive charge is spread evenly, so that there is the same amount of charge per unit length at every point. On rod B, positive charge is spread evenly over only the left half, and the same amount of negative charge is spread evenly over the right half. For each rod deduce the *direction* of the electric field at a point that is located directly above the midpoint of the rod.

18.7 Electric Field Lines

As we have seen, electric charges create an electric field in the space around them. It is useful to have a kind of "map" that gives the direction and strength of the field at various places. The great English physicist Michael Faraday (1791–1867) proposed an idea that provides such a "map"—the idea of **electric field lines**. Since the electric field is the electric force per unit charge, field lines are also called **lines of force**.

To introduce the electric field line concept, **Figure 18.22a** shows a positive point charge +*q*. At the locations numbered 1–8, a positive test charge would experience a repulsive force, as the arrows in the drawing indicate. Therefore, the electric field created by the charge +*q* is directed radially outward. The electric field lines are lines drawn

FIGURE 18.22 (*a*) At any of the eight marked spots around a positive point charge +*q*, a positive test charge would experience a repulsive force directed radially outward. (*b*) The electric field lines are directed radially outward from a positive point charge +*q*.

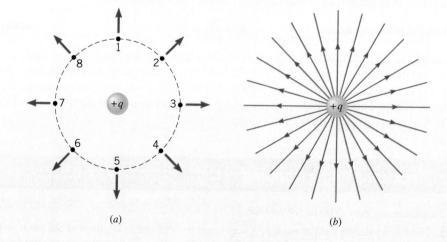

(*a*) (*b*)

to show this direction, as **Figure 18.22b** illustrates. They begin on the charge $+q$ and point radially *outward*. **Figure 18.23** shows the field lines in the vicinity of a negative charge $-q$. In this case they are directed radially *inward* because the force on a positive test charge is one of attraction, indicating that the electric field points inward. In general, ***electric field lines are always directed away from positive charges and toward negative charges.***

The electric field lines in **Figures 18.22** and **18.23** are drawn in only two dimensions, as a matter of convenience. Field lines radiate from the charges in three dimensions, and an infinite number of lines could be drawn. However, for clarity only a small number are ever included in pictures. The number is chosen to be proportional to the magnitude of the charge; thus, five times as many lines would emerge from a $+5q$ charge as from a $+q$ charge.

The pattern of electric field lines also provides information about the magnitude or strength of the field. Notice that in **Figures 18.22** and **18.23**, the lines are closer together near the charges, where the electric field is stronger. At distances far from the charges, where the electric field is weaker, the lines are more spread out. It is true in general that the electric field is stronger in regions where the field lines are closer together. In fact, no matter how many charges are present, the number of lines per unit area passing perpendicularly through a surface is proportional to the magnitude of the electric field.

In regions where the electric field lines are equally spaced, there is the same number of lines per unit area everywhere, and the electric field has the same strength at all points. For example, **Figure 18.24** shows that the field lines between the plates of a parallel plate capacitor are parallel and equally spaced, except near the edges where they bulge outward. The equally spaced, parallel lines indicate that the electric field has the same magnitude and direction at all points in the central region of the capacitor.

Often, electric field lines are curved, as in the case of an **electric dipole**. An electric dipole consists of two separated point charges that have the same magnitude but opposite signs. The electric field of a dipole is proportional to the product of the magnitude of one of the charges and the distance between the charges. This product is called the **dipole moment**. Many molecules, such as H_2O and HCl, have dipole moments. **Figure 18.25** depicts the field lines in the vicinity of a dipole. For a curved field line, the electric field vector at a point is *tangent* to the line at that point (see points 1, 2, and 3 in the drawing). The pattern of the lines for the dipole indicates that the electric field is greatest in the region between and immediately surrounding the two charges, since the lines are closest together there.

Notice in **Figure 18.25** that any given field line starts on the positive charge and ends on the negative charge. This is a fundamental characteristic of electric field lines:

> **Problem-Solving Insight**   **Electric field lines always begin on a positive charge and end on a negative charge and do not start or stop in midspace. Furthermore, the number of lines leaving a positive charge or entering a negative charge is proportional to the magnitude of the charge.**

This means, for example, that if 100 lines are drawn leaving a $+4~\mu C$ charge, then 75 lines would have to end on a $-3~\mu C$ charge and 25 lines on a $-1~\mu C$ charge. Thus, 100 lines leave the charge of $+4~\mu C$ and end on a total charge of $-4~\mu C$, so the lines begin and end on equal amounts of total charge.

The electric field lines are also curved in the vicinity of two identical charges. **Figure 18.26** shows the pattern associated with two positive point charges and reveals that there is an absence of lines in the region between the charges. The absence of lines indicates that the electric field is relatively weak between the charges.

Some of the important properties of electric field lines are considered in Conceptual Example 12, and the role of the electric field in the electrophoresis of DNA is explored in Example 13.

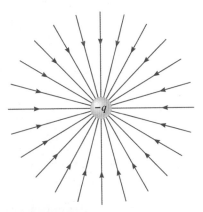

FIGURE 18.23 The electric field lines are directed radially inward toward a negative point charge $-q$.

Edge view

FIGURE 18.24 In the central region of a parallel plate capacitor, the electric field lines are parallel and evenly spaced, indicating that the electric field there has the same magnitude and direction at all points.

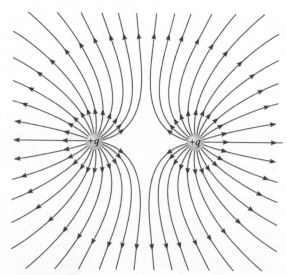

FIGURE 18.25 The electric field lines of an electric dipole are curved and extend from the positive to the negative charge. At any point, such as 1, 2, or 3, the field created by the dipole is tangent to the line through the point.

FIGURE 18.26 The electric field lines for two identical positive point charges. If the charges were both negative, the directions of the lines would be reversed.

CONCEPTUAL EXAMPLE 12 | Drawing Electric Field Lines

Figure 18.27 shows four choices for the electric field lines between three negative point charges ($-q$, $-q$, and $-2q$) and one positive point charge ($+4q$). Which of these choices is the only one of the four that could possibly show a correct representation of the field lines? **(a)** Figure 18.27a **(b)** Figure 18.27b **(c)** Figure 18.27c **(d)** Figure 18.27d

Reasoning Electric field lines begin on positive charges and end on negative charges. The tangent to a field line at a point gives

the direction of the electric field at that point. Equally spaced parallel field lines indicate that the field has a constant value (magnitude and direction) in the corresponding region of space.

Answer (a) is incorrect. Field lines can never cross, as they do at point P in Figure 18.27a. If two field lines were to intersect, there would be two electric fields at the point of intersection, one associated with each line. However, there can only be one value of the electric field at a given point.

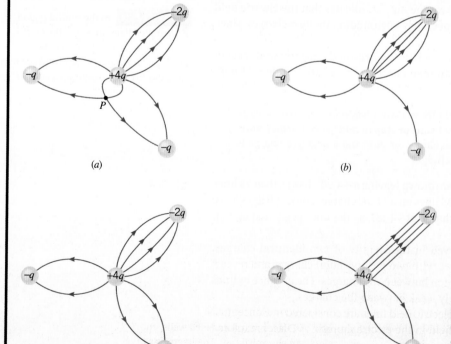

(a)

(b)

(c)

(d)

FIGURE 18.27 Only one of these drawings shows a representation of the electric field lines between the charges that could be correct. (See Conceptual Example 12.)

Answer (b) is incorrect. The number of field lines that leave a positive charge or end on a negative charge is proportional to the magnitude of the charge. Since 8 lines leave the $+4q$ charge, one-half of them (or 4) must end on the $-2q$ charge, and one-fourth of them (or 2) must end on each of the $-q$ charges. **Figure 18.27b** incorrectly shows 5 lines ending on the $-2q$ charge and 1 line ending on one of the $-q$ charges.

Answer (d) is incorrect. **Figure 18.27d** is incorrect because the field lines between the $+4q$ charge and the $-2q$ charge are parallel and evenly spaced, which would indicate that the field everywhere in this region has a constant magnitude and direction. However, the field between the $+4q$ charge and the $-2q$ charge certainly is stronger in places close to either of the charges. The field lines should, therefore, have a curved nature, similar (but not identical) to the field lines that surround a dipole.

Answer (c) is correct. **Figure 18.27c** contains none of the errors discussed previously and, therefore, is the only drawing that could be correct.

Related Homework: Problem 30

EXAMPLE 13 | **BIO** Electrophoresis of DNA

Gel electrophoresis is a method used to electrically separate charged macromolecules such as those in viruses, bacterial cells, proteins, and DNA. When subjected to an electric field, these objects experience a force and migrate. Due to the viscous properties of the medium in which they travel (a gel), smaller objects move with a greater speed than larger ones, assuming they have equal charge densities. This method is used to carry out DNA tests. In this case, restriction enzymes are first used to cut the DNA into pieces related to the specific DNA bases, that is, adenine (A), cytosine (C), guanine (G), and thymine (T). What remains are DNA fragments that can now be sorted according to size (length). The size distribution of these fragments is what is measured in a gel electrophoresis DNA test, and displayed as a series of lines (**Figure 18.28**), with shorter segments on one side and longer segments on the other.

Suppose a DNA fragment has a net negative charge with a magnitude equal to that of one electron ($e = 1.60 \times 10^{-19}$ C) per nanometer of its length. **(a)** If a set of electrodes in a gel sets up a uniform electric field of magnitude $E = 125$ N/C, determine the magnitude of the net electric force acting on a DNA fragment of length 0.220 μm. **(b)** Toward which electrode does the DNA fragment migrate (i.e., toward the negative or positive electrode) if the electric field points from the positive (+) electrode toward the negative (−) electrode?

Reasoning **(a)** From the linear charge density and the DNA fragment length, we can calculate the magnitude of the net charge (q_0)

of a fragment. Since we know the electric field E, the force is given by Equation 18.2 as $F = q_0 E$. **(b)** The net charge of the fragment is negative. Therefore, from the vector nature of the force on a charge due to an electric field expressed by Equation 18.2, the electric force on the fragment is antiparallel to the electric field.

Solution **(a)** The magnitude of the net charge on the DNA fragment is given by

$$|q_0| = (\text{length}) \times (\text{charge per unit length})$$

$$= (0.220 \ \mu m)\left(\frac{e}{1 \ nm}\right) = (0.220 \times 10^{-6} \ m)\left(\frac{1.60 \times 10^{-19} \ C}{1.00 \times 10^{-9} \ m}\right)$$

$$= 3.52 \times 10^{-17} \ C$$

Applying Equation 18.2 to find the magnitude of the total force, we have

$$F = |q_0|E = (3.52 \times 10^{-17} \ C)(125 \ N/C) = \boxed{4.40 \times 10^{-15} \ N}$$

(b) Since the electric field points from positive charges toward negative charges, the electric field points from the positive electrode toward the negative electrode. Since the force on a negatively charged particle points in a direction opposite to that of the field, the DNA fragment migrates toward the positive electrode.

FIGURE 18.28 DNA treated with restriction enzymes is placed into slots called wells, and then subjected to an electric field set up by electrodes in an electrolytic gel. The fragments then migrate and separate according to length. (Modified from *Encyclopedia Britannica* 2007)

Check Your Understanding

(The answers are given at the end of the book.)

17. Drawings A and B in **CYU Figure 18.4** show two examples of electric field lines. Which (one or more) of the following statements are true, and which (one or more) are false? **(a)** In both A and B the electric field is the same everywhere. **(b)** As you move from left to right in each case, the electric field becomes stronger. **(c)** The electric field in A is the same everywhere, but it becomes stronger in B as you move from left to right. **(d)** The electric fields in both A and B could be created by negative charges located somewhere on the left and positive charges somewhere on the right. **(e)** Both A and B arise from a single positive point charge located somewhere on the left.

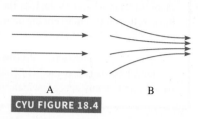

A   B

CYU FIGURE 18.4

18. A positively charged particle is moving horizontally when it enters the region between the plates of a parallel plate capacitor, as **CYU Figure 18.5** illustrates. When the particle is within the capacitor, which of the following vectors, if any, are *parallel* to the electric field lines inside the capacitor? **(a)** The particle's displacement **(b)** Its velocity **(c)** Its linear momentum **(d)** Its acceleration

CYU FIGURE 18.5

18.8 The Electric Field Inside a Conductor: Shielding

In conducting materials such as copper, electric charges move readily in response to the forces that electric fields exert. This property of conducting materials has a major effect on the electric field that can exist within and around them. Suppose that a piece of copper carries a number of excess electrons somewhere within it, as in **Figure 18.29a**. Each electron would experience a force of repulsion because of the electric field of its neighbors. And, since copper is a conductor, the excess electrons move readily in response to that force. In fact, as a consequence of the $1/r^2$ dependence on distance in Coulomb's law, they rush to the surface of the copper. Once static equilibrium is established with all of the excess charge on the surface, no further movement of charge occurs, as part *b* of the drawing indicates. Similarly, excess positive charge also moves to the surface of a conductor. In general, *at equilibrium under electrostatic conditions, any excess charge resides on the surface of a conductor.*

Now consider the interior of the copper in **Figure 18.29b**. The interior is electrically neutral, although there are still free electrons that can move under the influence of an electric field. The absence of a net movement of these free electrons indicates that there is no net electric field present within the conductor. In fact, the excess charges arrange themselves on the conductor surface precisely in the manner needed to make the electric field zero within the material. Thus, *at equilibrium under electrostatic conditions, the electric field is zero at any point within a conducting material.* This fact has some fascinating implications.

Figure 18.30a shows an uncharged, solid, cylindrical conductor at equilibrium in the central region of a parallel plate capacitor. Induced charges on the surface of the cylinder alter the electric field lines of the capacitor. Since an electric field cannot exist within the conductor under these conditions, the electric field lines do not penetrate the cylinder. Instead, they end or begin on the induced charges. Consequently, a test charge placed *inside* the conductor would feel no force due to the presence of the charges on the capacitor. In other words, *the conductor shields any charge within it from electric fields created outside the conductor.* The shielding results from the induced charges on the conductor surface.

(*a*)

(*b*)

FIGURE 18.29 (*a*) Excess charge within a conductor (copper) moves quickly (*b*) to the surface.

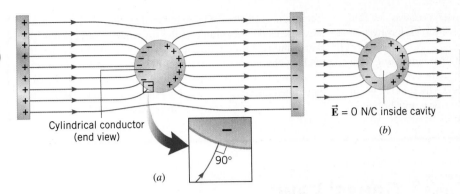

Cylindrical conductor
(end view)

90°

(a)

$\vec{E} = 0$ N/C inside cavity

(b)

FIGURE 18.30 (*a*) A cylindrical conductor (shown as an end view) is placed between the oppositely charged plates of a capacitor. The electric field lines do not penetrate the conductor. The blowup shows that, just outside the conductor, the electric field lines are perpendicular to its surface. (*b*) The electric field is zero in a cavity within the conductor.

THE PHYSICS OF ... shielding electronic circuits. Since the electric field is zero inside the conductor, nothing is disturbed if a cavity is cut from the interior of the material, as in **Figure 18.30b**. Thus, the interior of the cavity is also shielded from external electric fields, a fact that has important applications, particularly for shielding electronic circuits. "Stray" electric fields are produced by various electrical appliances (e.g., hair dryers, blenders, and vacuum cleaners), and these fields can interfere with the operation of sensitive electronic circuits, such as those in stereo amplifiers, televisions, and computers. To eliminate such interference, circuits are often enclosed within metal boxes that provide shielding from external fields.

The blowup in **Figure 18.30a** shows another aspect of how conductors alter the electric field lines created by external charges. The lines are altered because ***the electric field just outside the surface of a conductor is perpendicular to the surface at equilibrium under electrostatic conditions.*** If the field were not perpendicular, there would be a component of the field parallel to the surface. Since the free electrons on the surface of the conductor can move, they would do so under the force exerted by that parallel component. In reality, however, no electron flow occurs at equilibrium. Therefore, there can be no parallel component, and the electric field is perpendicular to the surface.

The preceding discussion deals with features of the electric field within and around a conductor at equilibrium under electrostatic conditions. These features are related to the fact that conductors contain mobile free electrons and *do not apply to insulators,* which contain very few free electrons. Conceptual Example 14 further explores the behavior of a conducting material in the presence of an electric field.

CONCEPTUAL EXAMPLE 14 | A Conductor in an Electric Field

A charge $+q$ is suspended at the center of a hollow, electrically neutral, spherical, metallic conductor, as **Figure 18.31** illustrates. The table shows a number of possibilities for the charges that this suspended charge induces on the interior and exterior surfaces of the conductor. Which one of the possibilities is correct?

	Interior Surface	Exterior Surface
(a)	$-q$	0
(b)	$-\frac{1}{2}q$	$-\frac{1}{2}q$
(c)	$+q$	$-q$
(d)	$-q$	$+q$

FIGURE 18.31
A positive charge $+q$ is suspended at the center of a hollow spherical conductor that is electrically neutral. Induced charges appear on the inner and outer surfaces of the conductor. The electric field within the conductor itself is zero.

Reasoning Three facts will guide our analysis. First: Since the suspended charge does not touch the conductor, the net charge on the conductor must remain zero, because it is electrically neutral to begin with. Second: Electric field lines begin on positive charges and end on negative charges. Third: At equilibrium under electrostatic conditions, there is no electric field and, hence, no field lines inside the solid material of the metallic conductor.

Answers (a) and (b) are incorrect. The net charge on the conductor in each of these answers is $-q$, which cannot be, since the conductor's net charge must remain zero.

Answer (c) is incorrect. Electric field lines emanate from the suspended positive charge $+q$ and must end on the interior surface of the metallic conductor, since they do not penetrate the metal. However, the charge on the interior surface in this answer is positive, and field lines must end on negative charges. Thus, this answer is incorrect.

Answer (d) is correct. Since the field lines emanating from the suspended positive charge $+q$ terminate only on negative charges and do not penetrate the metal, there must be an induced *negative* charge on the interior surface. Furthermore, the lines begin and end on equal amounts of charge, so the

magnitude of the total charge induced on the interior surface is the same as the magnitude of the suspended charge. Thus, the charge induced on the interior surface is $-q$. We know that the total net charge on the metallic conductor must remain at zero. Therefore, if a charge $-q$ is induced on the interior surface, there must also be a charge of $+q$ induced on the exterior surface, because excess charge cannot remain inside of the solid metal at equilibrium.

18.9 Gauss' Law

Section 18.6 discusses how a point charge creates an electric field in the space around the charge. There are also many situations in which an electric field is produced by charges that are spread out over a region, rather than by a single point charge. Such an extended collection of charges is called a *charge distribution*. For example, the electric field within the parallel plate capacitor in **Figure 18.21** is produced by positive charges spread uniformly over one plate and an equal number of negative charges spread over the other plate. As we will see, Gauss' law describes the relationship between a charge distribution and the electric field it produces. This law was formulated by the German mathematician and physicist Carl Friedrich Gauss (1777–1855).

In presenting Gauss' law, it will be necessary to introduce a new idea called **electric flux**. The idea of flux involves both the electric field and the surface through which it passes. By bringing together the electric field and the surface through which it passes, we will be able to define electric flux and then present Gauss' law.

We begin by developing a version of Gauss' law that applies only to a point charge, which we assume to be positive. The electric field lines for a positive point charge radiate outward in all directions from the charge, as **Figure 18.22b** indicates. The magnitude E of the electric field at a distance r from the charge is $E = kq/r^2$, according to Equation 18.3, in which we have replaced the symbol $|q|$ with the symbol q since we are assuming that the charge is positive. As mentioned in Section 18.5, the constant k can be expressed as $k = 1/(4\pi\varepsilon_0)$, where ε_0 is the permittivity of free space. With this substitution, the magnitude of the electric field becomes $E = q/(4\pi\varepsilon_0 r^2)$. We now place this point charge at the center of an imaginary spherical surface of radius r, as **Figure 18.32** shows. Such a hypothetical closed surface is called a **Gaussian surface**, although in general it need not be spherical. The surface area A of a sphere is $A = 4\pi r^2$, and the magnitude of the electric field can be written in terms of this area as $E = q/(A\varepsilon_0)$, or

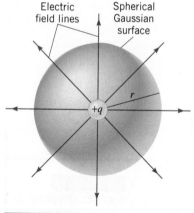

Electric field lines   Spherical Gaussian surface

r

$+q$

FIGURE 18.32 A positive point charge is located at the center of an imaginary spherical surface of radius r. Such a surface is one example of a Gaussian surface. Here the electric field is perpendicular to the surface and has the same magnitude everywhere on it.

Gauss' law for a point charge

$$\underbrace{EA}_{\substack{\text{Electric} \\ \text{flux } \Phi_E}} = \frac{q}{\varepsilon_0} \qquad (18.5)$$

The left side of Equation 18.5 is the product of the magnitude E of the electric field at any point on the Gaussian surface and the area A of the surface. In Gauss' law this product is especially important and is called the **electric flux**, Φ_E: $\Phi_E = EA$. (It will be necessary to modify this definition of flux when we consider the general case of a Gaussian surface with an arbitrary shape.)

Equation 18.5 is the result we have been seeking, for it is the form of Gauss' law that applies to a point charge. This result indicates that, aside from the constant ε_0, the electric flux Φ_E depends only on the charge q within the Gaussian surface and is independent of the radius r of the surface. We will now see how to generalize Equation 18.5 to account for distributions of charges and Gaussian surfaces with arbitrary shapes.

Interactive Figure 18.33 shows a charge distribution whose net charge is labeled Q. The charge distribution is surrounded by a Gaussian surface—that is, an imaginary closed surface. The surface can have *any arbitrary shape* (it need not be spherical), but *it must be closed* (an open surface would be like that of half an eggshell). The direction of the electric field is not necessarily perpendicular to the Gaussian surface. Furthermore, the magnitude of the electric field need not be constant on the surface but can vary from point to point.

To determine the electric flux through such a surface, we divide the surface into many tiny sections with areas ΔA_1, ΔA_2, and so on. Each section is so small that it is essentially flat and the electric field $\vec{E}$ is a constant (both in magnitude and direction) over it. For reference, a dashed line called the "normal" is drawn perpendicular to each section on the outside of the surface. To determine the electric flux for each of the sections, we use only the component of $\vec{E}$ that is perpendicular to the surface—that is, the component of the electric field that passes through the surface. From **Interactive Figure 18.33** it can be seen that this component has a magnitude of $E \cos \phi$, where ϕ is the angle between the electric field and the normal. The electric flux through any one section is then $(E \cos \phi)$ ΔA. The electric flux Φ_E that passes through the entire Gaussian surface is the sum of all of these individual fluxes: $\Phi_E = (E_1 \cos \phi_1)\Delta A_1 + (E_2 \cos \phi_2)\Delta A_2 + \cdots$, or

$$\Phi_E = \Sigma(E \cos \phi)\Delta A \qquad \textbf{(18.6)}$$

where, as usual, the symbol Σ means "the sum of." Gauss' law relates the electric flux Φ_E to the net charge Q enclosed by the arbitrarily shaped Gaussian surface.

GAUSS' LAW

The electric flux Φ_E through a Gaussian surface is equal to the net charge Q enclosed by the surface divided by ε_0, the permittivity of free space:

$$\underbrace{\Sigma(E \cos \phi)\Delta A}_{\text{Electric flux, } \Phi_E} = \frac{Q}{\varepsilon_0} \qquad \textbf{(18.7)}$$

SI Unit of Electric Flux: $\text{N} \cdot \text{m}^2/\text{C}$

Although we arrived at Gauss' law by assuming the net charge Q was positive, Equation 18.7 also applies when Q is negative. In this case, the electric flux Φ_E is also negative. Gauss' law is often used to find the magnitude of the electric field produced by a distribution of charges. The law is most useful when the distribution is uniform and symmetrical. In the next two examples we will see how to apply Gauss' law in such situations.

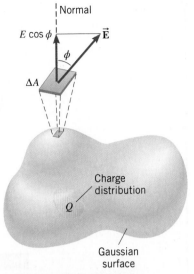

INTERACTIVE FIGURE 18.33 The charge distribution Q is surrounded by an arbitrarily shaped Gaussian surface. The electric flux Φ through any tiny segment of the surface is the product of $E \cos \phi$ and the area ΔA of the segment: $\Phi = (E \cos \phi)\Delta A$. The angle ϕ is the angle between the electric field and the normal to the surface.

EXAMPLE 15 | The Electric Field of a Charged Thin Spherical Shell

Figures 18.34a and *b* show a thin spherical shell of radius R (for clarity, only half of the shell is shown). A positive charge q is spread uniformly over the shell. Find the magnitude of the electric field at any point **(a)** outside the shell and **(b)** inside the shell.

Reasoning Because the charge is distributed uniformly over the spherical shell, the electric field is symmetrical. This means that the electric field is directed radially outward in all directions, and its magnitude is the same at all points that are equidistant from the shell. All such points lie on a sphere, so the symmetry is called *spherical symmetry*. With this symmetry in mind, we will use a spherical Gaussian surface to evaluate the electric flux Φ_E. We will then use Gauss' law to determine the magnitude of the electric field.

Solution **(a)** To find the magnitude of the electric field outside the charged shell, we evaluate the electric flux $\Phi_E = \Sigma(E \cos \phi)\Delta A$ by using a spherical Gaussian surface of radius r $(r > R)$ that is concentric with the shell. See the blue surface labeled S in **Figure 18.34a**. Since the electric field $\vec{E}$ is everywhere perpendicular to the Gaussian surface, $\phi = 0°$ and $\cos \phi = 1$. In addition, E has the same value at all points on the surface, since they are equidistant from the charged shell. Being

constant over the surface, E can be factored outside the summation, with the result that

$$\Phi_E = \Sigma(E \cos 0°)\,\Delta A = E(\Sigma\Delta A) = E\underbrace{(4\pi r^2)}$$

$$\underset{\substack{\text{Area of} \\ \text{Gaussian} \\ \text{surface}}}{} \qquad \underset{\substack{\text{Surface area} \\ \text{of sphere}}}{}$$

The term $\Sigma\Delta A$ is just the sum of the tiny areas that make up the Gaussian surface. Since the area of a spherical surface is $4\pi r^2$, we have $\Sigma\Delta A = 4\pi r^2$. Setting the electric flux equal to Q/ε_0, as specified by Gauss' law, yields $E(4\pi r^2) = Q/\varepsilon_0$. Since the only charge within the Gaussian surface is the charge q on the shell, it follows that the net charge within the Gaussian surface is $Q = q$. Thus, we can solve for E and find that

$$E = \frac{q}{4\pi\varepsilon_0 r^2} \quad \text{(for } r > R\text{)}$$

This is a surprising result, for it is the same as that for a point charge (see Equation 18.3 with $|q| = q$). Thus, the electric field outside a uniformly charged spherical shell is the same as if all the charge q were concentrated as a point charge at the center of the shell.

(b) To find the magnitude of the electric field inside the charged shell, we select a spherical Gaussian surface that lies inside the

FIGURE 18.34 A uniform distribution of positive charge resides on a thin spherical shell of radius R. The spherical Gaussian surfaces S and S_1 are used in Example 15 to evaluate the electric flux (a) outside and (b) inside the shell, respectively. For clarity, only half the shell and the Gaussian surfaces are shown.

shell and is concentric with it. See the blue surface labeled S_1 in **Figure 18.34b**. Inside the charged shell, the electric field (if it exists) must also have spherical symmetry. Therefore, using reasoning like that in part (a), the electric flux through the Gaussian surface is $\Phi_E = \Sigma(E \cos \phi)\Delta A = E(4\pi r_1^2)$. In accord with Gauss' law, the electric flux must be equal to Q/ε_0, where Q is the net charge *inside* the Gaussian surface. But now $Q = 0$ C, since all the

charge lies on the shell that is *outside* the surface S_1. Consequently, we have $E(4\pi r_1^2) = Q/\varepsilon_0 = 0$, or

$$\boxed{E = 0 \text{ N/C} \quad (\text{for } r < R)}$$

Gauss' law allows us to deduce that there is no electric field inside a uniform spherical shell of charge. An electric field exists only on the outside.

EXAMPLE 16 | The Electric Field Inside a Parallel Plate Capacitor

According to Equation 18.4, the electric field inside a parallel plate capacitor, and away from the edges, is constant and has a magnitude of $E = \sigma/\varepsilon_0$, where σ is the charge density (the charge per unit area) on a plate. Use Gauss' law to obtain this result.

Reasoning Figure 18.35a shows the electric field inside a parallel plate capacitor. Because the positive and negative charges are distributed uniformly over the surfaces of the plates, symmetry requires that the electric field be perpendicular to the plates. We will take advantage of this symmetry by choosing our Gaussian surface to be a small cylinder whose axis is perpendicular to the plates (see part b of the figure). With this choice, we will be able to evaluate the electric flux and then, with the aid of Gauss' law, determine E.

Solution Figure 18.35b shows that we have placed our Gaussian cylinder so that its left end is inside the positive metal plate, and the right end is in the space between the plates. To determine the electric flux through this Gaussian surface, we evaluate the flux through each of the three parts—labeled 1, 2, and 3 in the drawing—that make up the total surface of the cylinder and then add up the fluxes.

Surface 1—the flat left end of the cylinder—is embedded inside the positive metal plate. As discussed in Section 18.8, the electric field is zero everywhere inside a conductor that is in equilibrium under electrostatic conditions. Since $E = 0$ N/C, the electric flux through this surface is also zero:

$$\Phi_1 = \Sigma(E \cos \phi)\,\Delta A = \Sigma[(0 \text{ N/C})\cos \phi]\,\Delta A = 0$$

(a)

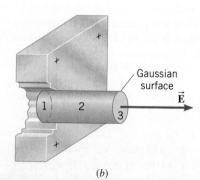

(b)

FIGURE 18.35 (a) A side view of a parallel plate capacitor, showing some of the electric field lines. (b) The Gaussian surface is a cylinder oriented so its axis is perpendicular to the positive plate and its left end is inside the plate.

Surface 2—the curved wall of the cylinder—is everywhere parallel to the electric field between the plates, so that $\cos \phi = \cos 90° = 0$. Therefore, the electric flux through this surface is also zero:

$$\Phi_2 = \Sigma(E \cos \phi) \Delta A = \Sigma(E \cos 90°) \Delta A = 0$$

Surface 3—the flat right end of the cylinder—is perpendicular to the electric field between the plates, so $\cos \phi = \cos 0° = 1$. The electric field is constant over this surface, so E can be taken outside the summation in Equation 18.6. Noting that $\Sigma \Delta A = A$ is the area of surface 3, we find that the electric flux through this surface is

$$\Phi_3 = \Sigma(E \cos \phi) \Delta A = \Sigma(E \cos 0°) \Delta A = E(\Sigma \Delta A) = EA$$

The electric flux through the entire Gaussian cylinder is the sum of the three fluxes determined above:

$$\Phi_E = \Phi_1 + \Phi_2 + \Phi_3 = 0 + 0 + EA = EA$$

According to Gauss' law, we set the electric flux equal to Q/ε_0, where Q is the net charge *inside* the Gaussian cylinder: $EA = Q/\varepsilon_0$. But Q/A is the charge per unit area, σ, on the plate. Therefore, we arrive at the value of the electric field inside a parallel plate capacitor: $\boxed{E = \sigma/\varepsilon_0}$. The distance of the right end of the Gaussian cylinder from the positive plate does not appear in this result, indicating that the electric field is the same everywhere between the plates.

Check Your Understanding

(The answers are given at the end of the book.)

19. A Gaussian surface contains a single charge within it, and as a result an electric flux passes through the surface. Suppose that the charge is then moved to another spot within the Gaussian surface. Does the flux through the surface change?

20. **CYU Figure 18.6** shows an arrangement of three charges. In parts *a* and *b* different Gaussian surfaces (both in blue) are shown. Through which surface, if either, does the greater electric flux pass?

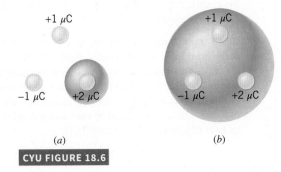

(a) (b)

CYU FIGURE 18.6

21. **CYU Figure 18.7** shows three charges, labeled q_1, q_2, and q_3. A Gaussian surface (in blue) is drawn around q_1 and q_2. **(a)** Which charges determine the electric flux through the Gaussian surface? **(b)** Which charges produce the electric field that exists at the point P?

CYU FIGURE 18.7

Concept Summary

18.1 The Origin of Electricity There are two kinds of electric charge: positive and negative. The SI unit of electric charge is the coulomb (C). The magnitude of the charge on an electron or a proton is $e = 1.60 \times 10^{-19}$ C. Since the symbol e denotes a magnitude, it has no algebraic sign. Thus, the electron carries a charge of $-e$, and the proton carries a charge of $+e$. The charge on any object, whether positive or negative, is quantized, in the sense that the charge consists of an integer number of protons or electrons.

18.2 Charged Objects and the Electric Force The law of conservation of electric charge states that the net electric charge of an isolated system remains constant during any process.

Like charges repel and unlike charges attract each other.

18.3 Conductors and Insulators An electrical conductor is a material, such as copper, that conducts electric charge readily. An electrical insulator is a material, such as rubber, that conducts electric charge poorly.

18.4 Charging by Contact and by Induction Charging by contact is the process of giving one object a net electric charge by placing it in contact with an object that is already charged. Charging by induction is the process of giving an object a net electric charge without touching it to a charged object.

18.5 Coulomb's Law A point charge is a charge that occupies so little space that it can be regarded as a mathematical point. Coulomb's law gives the magnitude F of the electric force that two point charges q_1 and q_2 exert on each other, according to Equation 18.1, where $|q_1|$ and $|q_2|$ are the magnitudes of the charges and have no algebraic sign. The term k is a constant and has the value $k = 8.99 \times 10^9$ N $\cdot$ m^2/C^2. The force specified by Equation 18.1 acts along the line between the two charges.

$$F = k\frac{|q_1|\,|q_2|}{r^2} \qquad (18.1)$$

The permittivity of free space ε_0 is defined by the relation

$$k = \frac{1}{4\pi\varepsilon_0}$$

18.6 The Electric Field The electric field $\vec{E}$ at a given spot is a vector and is the electrostatic force $\vec{F}$ experienced by a very small test charge q_0 placed at that spot divided by the charge itself, as given by Equation 18.2. The direction of the electric field is the same as the direction of the force on a positive test charge. The SI unit for the electric field is the newton per coulomb (N/C). The source of the electric field at any spot is the collection of charged objects surrounding that spot.

$$\vec{E} = \frac{\vec{F}}{q_0} \qquad (18.2)$$

The magnitude of the electric field created by a point charge q is given by Equation 18.3, where $|q|$ is the magnitude of the charge and has no algebraic sign and r is the distance from the charge. The electric field $\vec{E}$ points away from a positive charge and toward a negative charge.

$$E = \frac{k|q|}{r^2} \qquad (18.3)$$

For a parallel plate capacitor that has a charge per unit area of σ on each plate, the magnitude of the electric field between the plates is given by Equation 18.4.

$$E = \frac{\sigma}{\varepsilon_0} \qquad (18.4)$$

18.7 Electric Field Lines Electric field lines are lines that can be thought of as a "map," insofar as the lines provide information about the direction and strength of the electric field. The lines are directed away from positive charges and toward negative charges. The direction of the lines gives the direction of the electric field, since the electric field vector at a point is tangent to the line at that point. The electric field is strongest in regions where the number of lines per unit area passing perpendicularly through a surface is the greatest—that is, where the lines are packed together most tightly.

18.8 The Electric Field Inside a Conductor: Shielding Excess negative or positive charge resides on the surface of a conductor at equilibrium under electrostatic conditions. In such a situation, the electric field at any point within the conducting material is zero, and the electric field just outside the surface of the conductor is perpendicular to the surface.

18.9 Gauss' Law The electric flux Φ_E through a surface is related to the magnitude E of the electric field, the area A of the surface, and the angle ϕ that specifies the direction of the field relative to the normal to the surface, as shown in Equation 18.6. Gauss' law states that the electric flux through a closed surface (a Gaussian surface) is equal to the net charge Q enclosed by the surface divided by ε_0, the permittivity of free space (see Equation 18.7).

$$\Phi_E = \Sigma(E \cos \phi)\Delta A \qquad (18.6)$$

$$\Phi_E = \Sigma(E \cos \phi)\Delta A = \frac{Q}{\varepsilon_0} \qquad (18.7)$$

Focus on Concepts

Online

Additional questions are available for assignment in WileyPLUS.

Section 18.1 The Origin of Electricity

1. An object carries a charge of -8.0 μC, while another carries a charge of -2.0 μC. How many electrons must be transferred from the first object to the second so that both objects have the same charge?

Section 18.2 Charged Objects and the Electric Force

2. Each of three objects carries a charge. As the drawing shows, objects A and B attract each other, and objects C and A also attract each other. Which one of the following statements concerning objects B and C is true? **(a)** They attract each other. **(b)** They repel each other. **(c)** They neither

QUESTION 2

attract nor repel each other. **(d)** This question cannot be answered without additional information.

Section 18.4 Charging by Contact and by Induction

3. Each of two identical objects carries a net charge. The objects are made from conducting material. One object is attracted to a positively charged ebonite rod, and the other is repelled by the rod. After the objects are touched together, it is found that they are each repelled by the rod. What can be concluded about the initial charges on the objects? **(a)** Initially both objects are positive, with both charges having the same magnitude. **(b)** Initially both objects are negative, with both charges having the same magnitude. **(c)** Initially one object is positive and one is negative, with the negative charge having a greater magnitude than the positive charge. **(d)** Initially one object is positive and one is negative, with the positive charge having a greater magnitude than the negative charge.

4. Only one of three balls A, B, and C carries a net charge q. The balls are made from conducting material and are identical. One of the uncharged balls can become charged by touching it to the charged ball and then separating the two. This process of touching one ball to another and then separating the two balls can be repeated over and over again, with the result that the three balls can take on a variety of charges. Which one of the following distributions of charges could not possibly be achieved in this fashion, even if the process were repeated a large number of times?

(a) $q_A = \frac{1}{3}q$, $q_B = \frac{1}{3}q$, $q_C = \frac{1}{3}q$ (c) $q_A = \frac{1}{2}q$, $q_B = \frac{3}{8}q$, $q_C = \frac{1}{4}q$

(b) $q_A = \frac{1}{2}q$, $q_B = \frac{1}{4}q$, $q_C = \frac{1}{4}q$ (d) $q_A = \frac{3}{8}q$, $q_B = \frac{3}{8}q$, $q_C = \frac{1}{4}q$

Section 18.5 Coulomb's Law

5. Three point charges have equal magnitudes and are located on the same line. The separation d between A and B is the same as the separation between B and C. One of the charges is positive and two are negative, as the drawing shows. Consider the net electrostatic force that each charge experiences due to the other two charges. Rank the net forces in descending order (greatest first) according to magnitude. (a) A, B, C (b) B, C, A (c) A, C, B (d) C, A, B (e) B, A, C

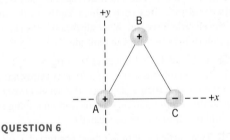

QUESTION 5

6. Three point charges have equal magnitudes and are fixed to the corners of an equilateral triangle. Two of the charges are positive and one is negative, as the drawing shows. At which one of the corners is the net force acting on the charge directed parallel to the x axis? (a) A (b) B (c) C

QUESTION 6

Section 18.6 The Electric Field

7. A positive point charge q_1 creates an electric field of magnitude E_1 at a spot located at a distance r_1 from the charge. The charge is replaced by another positive point charge q_2, which creates a field of magnitude $E_2 = E_1$ at a distance of $r_2 = 2r_1$. How is q_2 related to q_1? (a) $q_2 = \frac{1}{2}q_1$ (b) $q_2 = \frac{1}{2}q_1$ (c) $q_2 = 4q_1$ (d) $q_2 = \frac{1}{4}q_1$ (e) $q_2 = \sqrt{2}q_1$

8. The drawing shows a positive and a negative point charge. The negative charge has the greater magnitude. Where on the line that passes through the charges is the one spot where the total electric field is zero? (a) To the right of the negative charge (b) To the left of the positive charge (c) Between the charges, to the left of the midpoint (d) Between the charges, to the right of the midpoint

QUESTION 8

Section 18.7 Electric Field Lines

9. The drawing shows some electric field lines. For the points indicated, rank the magnitudes of the electric field in descending order (largest first). (a) B, C, A (b) B, A, C (c) A, B, C (d) A, C, B (e) C, A, B

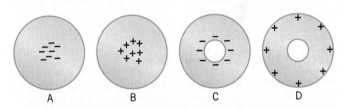

QUESTION 9

Section 18.8 The Electric Field Inside a Conductor: Shielding

10. The drawings show (in cross section) two solid spheres and two spherical shells. Each object is made from copper and has a net charge, as the plus and minus signs indicate. Which drawing correctly shows where the charges reside when they are in equilibrium? (a) A (b) B (c) C (d) D

QUESTION 10

Section 18.9 Gauss' Law

11. A cubical Gaussian surface surrounds two charges, $q_1 = +6.0 \times 10^{-12}$ C and $q_2 = -2.0 \times 10^{-12}$ C, as the drawing shows. What is the electric flux passing through the surface?

Problems

Online

Additional questions are available for assignment in WileyPLUS.

SSM Student Solutions Manual   E Easy

MMH Problem-solving help   M Medium

GO Guided Online Tutorial   H Hard

V-HINT Video Hints   WS Worksheet

CHALK Chalkboard Videos   T Team Problem

BIO Biomedical application

Section 18.1 The Origin of Electricity,

Section 18.2 Charged Objects and the Electric Force,

Section 18.3 Conductors and Insulators,

Section 18.4 Charging by Contact and by Induction

1. E SSM Iron atoms have been detected in the sun's outer atmosphere, some with many of their electrons stripped away. What is the

net electric charge (in coulombs) of an iron atom with 26 protons and 7 electrons? Be sure to include the algebraic sign (+ or –) in your answer.

2. **E** **CHALK** An object has a charge of –2.0 μC. How many electrons must be removed so that the charge becomes +3.0 μC?

3. **E** **CHALK** Four identical metallic objects carry the following charges: +1.6, +6.2, –4.8, and –9.4 μC. The objects are brought simultaneously into contact, so that each touches the others. Then they are separated. **(a)** What is the final charge on each object? **(b)** How many electrons (or protons) make up the final charge on each object?

4. **E** **GO** Four identical metal spheres have charges of q_A = –8.0 μC, q_B = –2.0 μC, q_C = +5.0 μC, and q_D = +12.0 μC. **(a)** Two of the spheres are brought together so they touch, and then they are separated. Which spheres are they, if the final charge on each one is +5.0 μC? **(b)** In a similar manner, which three spheres are brought together and then separated, if the final charge on each of the three is +3.0 μC? **(c)** The final charge on each of the three separated spheres in part (b) is +3.0 μC. How many electrons would have to be added to one of these spheres to make it electrically neutral?

5. **E** **SSM** Consider three identical metal spheres, A, B, and C. Sphere A carries a charge of +5q. Sphere B carries a charge of –q. Sphere C carries no net charge. Spheres A and B are touched together and then separated. Sphere C is then touched to sphere A and separated from it. Last, sphere C is touched to sphere B and separated from it. **(a)** How much charge ends up on sphere C? What is the total charge on the three spheres **(b)** before they are allowed to touch each other and **(c)** after they have touched?

6. **E** **GO** **CHALK** A plate carries a charge of –3.0 μC, while a rod carries a charge of +2.0 μC. How many electrons must be transferred from the plate to the rod, so that both objects have the same charge?

7. **M** **V-HINT** **CHALK** Water has a mass per mole of 18.0 g/mol, and each water molecule (H_2O) has 10 electrons. **(a)** How many electrons are there in one liter (1.00×10^{-3} m^3) of water? **(b)** What is the net charge of all these electrons?

Section 18.5 Coulomb's Law

8. **E** In a vacuum, two particles have charges of q_1 and q_2, where q_1 = +3.5 μC. They are separated by a distance of 0.26 m, and particle 1 experiences an attractive force of 3.4 N. What is q_2 (magnitude and sign)?

9. **E** **SSM** Two spherical objects are separated by a distance that is 1.80×10^{-3} m. The objects are initially electrically neutral and are very small compared to the distance between them. Each object acquires the same negative charge due to the addition of electrons. As a result, each object experiences an electrostatic force that has a magnitude of 4.55×10^{-21} N. How many electrons did it take to produce the charge on one of the objects?

10. **E** Two tiny conducting spheres are identical and carry charges of –20.0 μC and +50.0 μC. They are separated by a distance of 2.50 cm. **(a)** What is the magnitude of the force that each sphere experiences, and is the force attractive or repulsive? **(b)** The spheres are brought into contact and then separated to a distance of 2.50 cm. Determine the magnitude of the force that each sphere now experiences, and state whether the force is attractive or repulsive.

11. **E** **SSM** Two very small spheres are initially neutral and separated by a distance of 0.50 m. Suppose that 3.0×10^{13} electrons are removed from one sphere and placed on the other. **(a)** What is the magnitude of the electrostatic force that acts on each sphere? **(b)** Is the force attractive or repulsive? Why?

QUESTION 11

12. **E** Two charges attract each other with a force of 1.5 N. What will be the force if the distance between them is reduced to one-ninth of its original value?

13. **E** Two point charges are fixed on the y axis: a negative point charge q_1 = –25 μC at y_1 = +0.22 m and a positive point charge q_2 at y_2 = +0.34 m. A third point charge q = +8.4 μC is fixed at the origin. The net electrostatic force exerted on the charge q by the other two charges has a magnitude of 27 N and points in the +y direction. Determine the magnitude of q_2.

14. **E** **GO** The drawings show three charges that have the same magnitude but may have different signs. In all cases the distance d between the charges is the same. The magnitude of the charges is $|q|$ = 8.6 μC, and the distance between them is d = 3.8 mm. Determine the magnitude of the net force on charge 2 for each of the three drawings.

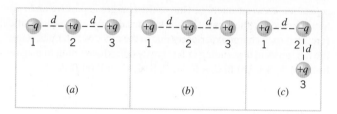

(a)   (b)   (c)

PROBLEM 14

15. **E** **SSM** **MMH** Two tiny spheres have the same mass and carry charges of the same magnitude. The mass of each sphere is 2.0×10^{-6} kg. The gravitational force that each sphere exerts on the other is balanced by the electric force. **(a)** What algebraic signs can the charges have? **(b)** Determine the charge magnitude.

16. **E** **GO** A charge +q is located at the origin, while an identical charge is located on the x axis at x = +0.50 m. A third charge of +2q is located on the x axis at such a place that the net electrostatic force on the charge at the origin doubles, its direction remaining unchanged. Where should the third charge be located?

17. **E** **SSM** Two particles, with identical positive charges and a separation of 2.60×10^{-2} m, are released from rest. Immediately after the release, particle 1 has an acceleration $\vec{a}_1$ whose magnitude is 4.60×10^3 m/s², while particle 2 has an acceleration $\vec{a}_2$ whose magnitude is 8.50×10^3 m/s². Particle 1 has a mass of 6.00×10^{-6} kg. Find **(a)** the charge on each particle and **(b)** the mass of particle 2.

18. **E** A charge of –3.00 μC is fixed at the center of a compass. Two additional charges are fixed on the circle of the compass, which has a radius of 0.100 m. The charges on the circle are –4.00 μC at the position due north and +5.00 μC at the position due east. What are the magnitude and direction of the net electrostatic force acting on the charge at the center? Specify the direction relative to due east.

19. **M** Multiple-Concept Example 3 provides some pertinent background for this problem. Suppose a single electron orbits about a nucleus containing two protons (+2e), as would be the case for a helium atom from which one of the two naturally occurring electrons is removed. The radius of the orbit is 2.65×10^{-11} m. Determine the magnitude of the electron's centripetal acceleration.

20. **M** **V-HINT** The drawing shows an equilateral triangle, each side of which has a length of 2.00 cm. Point charges are fixed to each corner, as shown. The 4.00 μC charge experiences a net force due to the charges q_A and q_B. This net force points vertically downward and has a magnitude of 405 N. Determine the magnitudes and algebraic signs of the charges q_A and q_B.

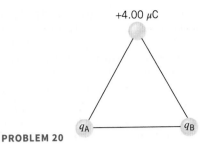

+4.00 μC

q_A q_B

PROBLEM 20

21. **M** **V-HINT** **MMH** The drawing shows three point charges fixed in place. The charge at the coordinate origin has a value of $q_1 = +8.00\ \mu C$; the other two charges have identical magnitudes, but opposite signs: $q_2 = -5.00\ \mu C$ and $q_3 = +5.00\ \mu C$. (a) Determine the net force (magnitude and direction) exerted on q_1 by the other two charges. (b) If q_1 had a mass of 1.50 g and it were free to move, what would be its acceleration?

+y
1.30 m
q_2
23.0°
+x
q_1
23.0°
1.30 m
q_3

PROBLEM 21

22. **M** **GO** An electrically neutral model airplane is flying in a horizontal circle on a 3.0-m guideline, which is nearly parallel to the ground. The line breaks when the kinetic energy of the plane is 50.0 J. Reconsider the same situation, except that now there is a point charge of $+q$ on the plane and a point charge of $-q$ at the other end of the guideline. In this case, the line breaks when the kinetic energy of the plane is 51.8 J. Find the magnitude of the charges.

23. **M** **V-HINT** Multiple-Concept Example 3 illustrates several of the concepts that come into play in this problem. A single electron orbits a lithium nucleus that contains three protons $(+3e)$. The radius of the orbit is 1.76×10^{-11} m. Determine the kinetic energy of the electron.

24. **M** **GO** An unstrained horizontal spring has a length of 0.32 m and a spring constant of 220 N/m. Two small charged objects are attached to this spring, one at each end. The charges on the objects have equal magnitudes. Because of these charges, the spring stretches by 0.020 m relative to its unstrained length. Determine (a) the possible algebraic signs and (b) the magnitude of the charges.

25. **M** **SSM** In the rectangle in the drawing, a charge is to be placed at the empty corner to make the net force on the charge at corner A point along the vertical direction. What charge (magnitude and algebraic sign) must be placed at the empty corner?

+3.0 μC
4d
d
A
PROBLEM 25   +3.0 μC   +3.0 μC

26. **H** **CHALK** **SSM** A small spherical insulator of mass 8.00×10^{-2} kg and charge $+0.600\ \mu C$ is hung by a thread of negligible mass. A charge of $-0.900\ \mu C$ is held 0.150 m away from the sphere and directly to the right of it, so the thread makes an angle θ with the vertical (see the drawing). Find (a) the angle θ and (b) the tension in the thread.

θ
0.150 m
+0.600 μC   -0.900 μC

PROBLEM 26

Section 18.6 The Electric Field,

Section 18.7 Electric Field Lines,

Section 18.8 The Electric Field Inside a Conductor: Shielding

27. **E** **SSM** At a distance r_1 from a point charge, the magnitude of the electric field created by the charge is 248 N/C. At a distance r_2 from the charge, the field has a magnitude of 132 N/C. Find the ratio r_2/r_1.

28. **E** **GO** Suppose you want to determine the electric field in a certain region of space. You have a small object of known charge and an instrument that measures the magnitude and direction of the force exerted on the object by the electric field. (a) The object has a charge of $+20.0\ \mu C$ and the instrument indicates that the electric force exerted on it is 40.0 μN, due east. What are the magnitude and direction of the electric field? (b) What are the magnitude and direction of the electric field if the object has a charge of $-10.0\ \mu C$ and the instrument indicates that the force is 20.0 μN, due west?

29. **E** An electric field of 260 000 N/C points due west at a certain spot. What are the magnitude and direction of the force that acts on a charge of $-7.0\ \mu C$ at this spot?

30. **E** **MMH** Review the important features of electric field lines discussed in Conceptual Example 12. Three point charges $(+q, +2q,$ and $-3q)$ are at the corners of an equilateral triangle. Sketch in six electric field lines between the three charges.

31. **E** Four point charges have the same magnitude of 2.4×10^{-12} C and are fixed to the corners of a square that is 4.0 cm on a side. Three of the charges are positive and one is negative. Determine the magnitude of the net electric field that exists at the center of the square.

32. **E** **GO** The drawing shows two situations in which charges are placed on the x and y axes. They are all located at the same distance of 6.1 cm from the origin O. For each of the situations in the drawing, determine the magnitude of the net electric field at the origin.

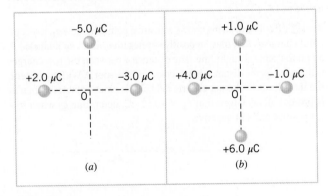

-5.0 μC
+2.0 μC   -3.0 μC
O
(a)

+1.0 μC
+4.0 μC   -1.0 μC
O
+6.0 μC
(b)

PROBLEM 32

33. **E** **MMH** A uniform electric field exists everywhere in the x, y plane. This electric field has a magnitude of 4500 N/C and is directed in the positive x direction. A point charge -8.0×10^{-9} C is placed at

the origin. Determine the magnitude of the net electric field at **(a)** $x = -0.15$ m, **(b)** $x = +0.15$ m, and **(c)** $y = +0.15$ m.

34. **E** **MMH** A 3.0 μC point charge is placed in an external uniform electric field that has a magnitude of 1.6×10^4 N/C. At what distance from the charge is the net electric field zero?

35. **E** **SSM** A tiny ball (mass = 0.012 kg) carries a charge of -18 μC. What electric field (magnitude and direction) is needed to cause the ball to float above the ground?

36. **E** **GO** A proton and an electron are moving due east in a constant electric field that also points due east. The electric field has a magnitude of 8.0×10^4 N/C. Determine the magnitude of the acceleration of the proton and the electron.

37. **E** Review Conceptual Example 11 before attempting to work this problem. The magnitude of each of the charges in **Figure 18.20** is 8.60×10^{-12} C. The lengths of the sides of the rectangles are 3.00 cm and 5.00 cm. Find the magnitude of the electric field at the center of the rectangle in **Figures 18.20a and b**.

38. **E** Two charges are placed between the plates of a parallel plate capacitor. One charge is $+q_1$ and the other is $q_2 = +5.00$ μC. The charge per unit area on each of the plates has a magnitude of $\sigma = 1.30 \times 10^{-4}$ C/m^2. The magnitude of the force on q_1 due to q_2 equals the magnitude of the force on q_1 due to the electric field of the parallel plate capacitor. What is the distance r between the two charges?

39. **M** **MMH** A small object has a mass of 3.0×10^{-3} kg and a charge of -34 μC. It is placed at a certain spot where there is an electric field. When released, the object experiences an acceleration of 2.5×10^3 m/s^2 in the direction of the $+x$ axis. Determine the magnitude and direction of the electric field.

40. **M** **SSM** Two point charges are located along the x axis: $q_1 = +6.0$ μC at $x_1 = +4.0$ cm, and $q_2 = +6.0$ μC at $x_2 = -4.0$ cm. Two other charges are located on the y axis: $q_3 = +3.0$ μC at $y_3 = +5.0$ cm, and $q_4 = -8.0$ μC at $y_4 = +7.0$ cm. Find the net electric field (magnitude and direction) at the origin.

41. **M** **GO** The total electric field $\vec{E}$ consists of the vector sum of two parts. One part has a magnitude of $E_1 = 1200$ N/C and points at an angle $\theta_1 = 35°$ above the $+x$ axis. The other part has a magnitude of $E_2 = 1700$ N/C and points at an angle $\theta_2 = 55°$ above the $+x$ axis. Find the magnitude and direction of the total field. Specify the directional angle relative to the $+x$ axis.

42. **M** **V-HINT** **CHALK** A particle of charge $+12$ μC and mass 3.8×10^{-5} kg is released from rest in a region where there is a constant electric field of $+480$ N/C. What is the displacement of the particle after a time of 1.6×10^{-2} s?

43. **M** **GO** The drawing shows a positive point charge $+q_1$, a second point charge q_2 that may be positive or negative, and a spot labeled P, all on the same straight line. The distance d between the two charges is the same as the distance between q_1 and the spot P. With q_2 present, the magnitude of the net electric field at P is twice what it is when q_1 is present alone. Given that $q_1 = +0.50$ μC, determine q_2 when it is **(a)** positive and **(b)** negative.

PROBLEM 43

44. **M** **CHALK** An electron is released from rest at the negative plate of a parallel plate capacitor. The charge per unit area on each plate is $\sigma = 1.8 \times 10^{-7}$ C/m^2, and the plate separation is 1.5×10^{-2} m. How fast is the electron moving just before it reaches the positive plate?

45. **H** Two particles are in a uniform electric field that points in the $+x$ direction and has a magnitude of 2500 N/C. The mass and charge

of particle 1 are $m_1 = 1.4 \times 10^{-5}$ kg and $q_1 = -7.0$ μC, while the corresponding values for particle 2 are $m_2 = 2.6 \times 10^{-5}$ kg and $q_2 = +18$ μC. Initially the particles are at rest. The particles are both located on the same electric field line but are separated from each other by a distance d. Particle 1 is located to the left of particle 2. When released, they accelerate but always remain at this same distance from each other. Find d.

46. **H** The drawing shows an electron entering the lower left side of a parallel plate capacitor and exiting at the upper right side. The initial speed of the electron is 7.00×10^6 m/s. The capacitor is 2.00 cm long, and its plates are separated by 0.150 cm. Assume that the electric field between the plates is uniform everywhere and find its magnitude.

PROBLEM 46

47. **H** A small plastic ball with a mass of 6.50×10^{-3} kg and with a charge of $+0.150$ μC is suspended from an insulating thread and hangs between the plates of a capacitor (see the drawing). The ball is in equilibrium, with the thread making an angle of $30.0°$ with respect to the vertical. The area of each plate is 0.0150 m^2. What is the magnitude of the charge on each plate?

PROBLEM 47

Section 18.9 Gauss' Law

48. **E** A spherical surface completely surrounds a collection of charges. Find the electric flux through the surface if the collection consists of **(a)** a single $+3.5 \times 10^{-6}$ C charge, **(b)** a single -2.3×10^{-6} C charge, and **(c)** both of the charges in (a) and (b).

49. **E** **SSM** The drawing shows an edge-on view of two planar surfaces that intersect and are mutually perpendicular. Surface 1 has an area of 1.7 m^2, while surface 2 has an area of 3.2 m^2. The electric field $\vec{E}$ in the drawing is uniform and has a magnitude of 250 N/C. Find the magnitude of the electric flux through **(a)** surface 1 and **(b)** surface 2.

PROBLEM 49 Surface 2

50. **E** A surface completely surrounds a $+2.0 \times 10^{-6}$ C charge. Find the electric flux through this surface when the surface is **(a)** a sphere with a radius of 0.50 m, **(b)** a sphere with a radius of 0.25 m, and **(c)** a cube with edges that are 0.25 m long.

51. **E** A circular surface with a radius of 0.057 m is exposed to a uniform external electric field of magnitude 1.44×10^4 N/C. The magnitude of the electric flux through the surface is 78 N · m^2/C. What is the angle (less than 90°) between the direction of the electric field and the normal to the surface?

52. **E** **V-HINT** A charge Q is located inside a rectangular box. The electric flux through each of the six surfaces of the box is: $\Phi_1 = +1500$ N · m²/C, $\Phi_2 = +2200$ N · m²/C, $\Phi_3 = +4600$ N · m²/C, $\Phi_4 = -1800$ N · m²/C, $\Phi_5 = -3500$ N · m²/C, and $\Phi_6 = -5400$ N · m²/C. What is Q?

53. **M** **GO** **MMH** **CHALK** Two spherical shells have a common center. A -1.6×10^{-6} C charge is spread uniformly over the inner shell, which has a radius of 0.050 m. A $+5.1 \times 10^{-6}$ C charge is spread uniformly over the outer shell, which has a radius of 0.15 m. Find the magnitude and direction of the electric field at a distance (measured from the common center) of **(a)** 0.20 m, **(b)** 0.10 m, and **(c)** 0.025 m.

Additional Problems

Online

54. **E** **GO** The masses of the earth and moon are 5.98×10^{24} and 7.35×10^{22} kg, respectively. Identical amounts of charge are placed on each body, such that the net force (gravitational plus electrical) on each is zero. What is the magnitude of the charge placed on each body?

55. **E** A small drop of water is suspended motionless in air by a uniform electric field that is directed upward and has a magnitude of 8480 N/C. The mass of the water drop is 3.50×10^{-9} kg. **(a)** Is the excess charge on the water drop positive or negative? Why? **(b)** How many excess electrons or protons reside on the drop?

56. **E** **SSM** Two charges are placed on the x axis. One of the charges ($q_1 = +8.5$ μC) is at $x_1 = +3.0$ cm and the other ($q_2 = -21$ μC) is at $x_1 = +9.0$ cm. Find the net electric field (magnitude and direction) at **(a)** $x = 0$ cm and **(b)** $x = +6.0$ cm.

57. **E** When point charges $q_1 = +8.4$ μC and $q_2 = +5.6$ μC are brought near each other, each experiences a repulsive force of magnitude 0.66 N. Determine the distance between the charges.

58. **M** **CHALK** **MMH** The drawing shows two positive charges q_1 and q_2 fixed to a circle. At the center of the circle they produce a net electric field that is directed upward along the vertical axis. Determine the ratio $|q_2|/|q_1|$ of the charge magnitudes.

PROBLEM 58

59. **M** **GO** Three point charges have equal magnitudes, two being positive and one negative. These charges are fixed to the corners of an equilateral triangle, as the drawing shows. The magnitude of each of the charges is 5.0 μC, and the lengths of the sides of the triangle are 3.0 cm. Calculate the magnitude of the net force that each charge experiences.

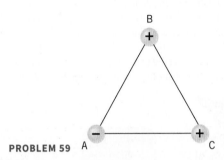

PROBLEM 59 A

60. **M** **GO** Four point charges have equal magnitudes. Three are positive, and one is negative, as the drawing shows. They are fixed in place on the same straight line, and adjacent charges are equally separated by a distance d. Consider the net electrostatic force acting on each charge. Calculate the ratio of the largest to the smallest net force.

PROBLEM 60

61. **H** Two identical small insulating balls are suspended by separate 0.25-m threads that are attached to a common point on the ceiling. Each ball has a mass of 8.0×10^{-4} kg. Initially the balls are uncharged and hang straight down. They are then given identical positive charges and, as a result, spread apart with an angle of 36° between the threads. Determine **(a)** the charge on each ball and **(b)** the tension in the threads.

62. (animated) **M** **GO** In a vacuum, a proton (charge $= +e$, mass $= 1.67 \times 10^{-27}$ kg) is moving parallel to a uniform electric field that is directed along the $+x$ axis (see the figure below). The proton starts with a velocity of $+2.5 \times 10^4$ m/s and accelerates in the same direction as the electric field, which has a value of $+2.3 \times 10^3$ N/C. Find the velocity of the proton when its displacement is $+2.0$ mm from the starting point.

ANIMATED PROBLEM 62

Physics in Biology, Medicine, and Sports

63. **E** **BIO** **GO** **18.6** The membrane surrounding a living cell consists of an inner and an outer wall that are separated by a small space. Assume that the membrane acts like a parallel plate capacitor in which the effective charge density on the inner and outer walls has a magnitude of 7.1×10^{-6} C/m². **(a)** What is the magnitude of the electric field within the cell membrane? **(b)** Find the magnitude of the electric force that would be exerted on a potassium ion (K⁺; charge $= +e$) placed inside the membrane.

64. **M** **BIO** **18.5** The average human body contains 7 billion billion billion atoms, or 7.0×10^{27} atoms. Of these, 65% are hydrogen. Each

hydrogen atom is electrically neutral, as the positive charge on its proton is balanced by the negative charge on its electron. Imagine if just 1.0% of your hydrogen atoms were missing their corresponding electron. This would leave you with a net positive charge. Assume your best friend, who is standing 1.0 cm away from you, also has the same net positive charge. What would be the repulsive force between you? Treat you and your friend as point charges. For comparison, the force required to lift the weight of the entire earth is about 6.0×10^{25} N.

65. **E** **BIO** **18.5** Electrostatic forces play an important role in biological systems, as many macromolecules, like DNA, are heavily charged (see Example 13). In fact, the electrostatic attractive force holds the two strands of DNA together, giving the molecule its strength and the famous structure of its double helix (see the image). The four organic bases of DNA (cytosine [C], guanine [G], adenine [A], and thymine [T]) hold the two strands together by forming pairs of bases that bond through these electrostatic forces. The order of the bases varies along the strand, but the pairing between bases is always the same; C always pairs with G, and A always pairs with T. In order for the pairs of bases to bond, they must be sufficiently close together. A guanine and cytosine, for example (see the image), pair by forming three hydrogen bonds (noncovalent bonds) between oxygen and nitrogen atoms. While a single hydrogen bond is relatively weak, the large number of them between complementary pairs gives the DNA molecule its strength. Calculate the electrostatic force of attraction within one hydrogen bond. Model the bond as a simple electric dipole, where equal and opposite charges of $(+/-e)$ are separated by a distance of 0.30 nm.

PROBLEM 65

66. **E** **BIO** **18.4** Many of a computer's internal components, such as the random access memory (RAM) and central processing unit (CPU), are very sensitive to electrostatic discharge (ESD). You have probably experienced ESDs yourself, if you have ever shuffled your feet across a carpeted floor and then touched a metal doorknob. You experienced a small shock, as the excess electrons you stripped from the carpet are transferred to the door knob. If the relative humidity in the room is low, the buildup of charge on the surface of your body, which acts like an insulator, can be significant—more than enough to damage a computer's components. To prevent ESDs, people assembling computers will wear antistatic equipment, such as that shown in the photo. The wrist strap is said to be grounded, which gives the static charge a safe conducting path away from the body, thereby protecting the computer components. In walking across a carpeted hallway, a computer technician acquires a charge of -4.79×10^{-9} C. Assume he touches one of the contact pins of a RAM module without wearing his antistatic equipment and completely delivers his excess charge through the component. How many electrons enter the RAM module during the discharge? Round your answer to the nearest whole number.

Desco Industries Inc.

PROBLEM 66

67. **M** **BIO** **18.6** During the COVID-19 pandemic in 2020, social distancing and the wearing of protective face masks became a way of life. Normal cloth masks provide protection against the spread of the disease by reducing the transmission of water droplets that contain the virus. However, researchers have developed "smartmasks" that use electric fields to destabilize the coronavirus particles, and while still in the early stages of development, they have shown promise in damaging structures the virus needs to infect host cells. The electroceutical fabric of which the mask is composed consists of a dot matrix of silver and zinc nanoparticles that acts like an array of electric dipoles when exposed to moisture (see the photo and figures below). Consider just one pair of silver and zinc nanoparticles extracted from the mask. Treat each nanoparticle as an equal and opposite point charge. If the value of the electric field midway between them (point P in the figure) is 120 N/C, what is the magnitude of the charge on each nanoparticle? Take the distance r to be 2.0 mm.

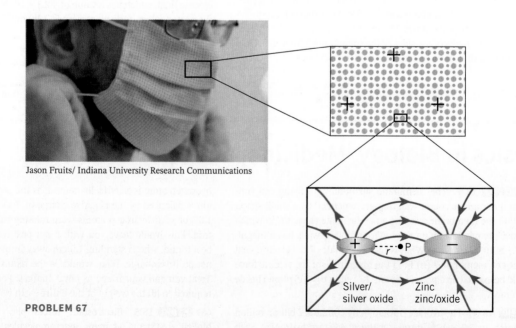

Jason Fruits/ Indiana University Research Communications

PROBLEM 67

68. M BIO **18.6** Natural diffusion of positive sodium ions and negative chlorine ions through the epithelial cells creates an endogenous electric field whenever you cut or wound your skin. Research has shown that electricity has a major impact on the healing of wounds. Certain proteins and genes within our cells respond to the natural electric fields that are produced at the site of a wound, which aids in the healing process (see the figures). In fact, research has shown that patients with slow-healing wounds, like those with diabetes, have smaller endogenous electric fields at the wound site. By exposing the wound to an external electric field and varying the

intensity, the rate of cell migration to the wound can be increased, which promotes healing. One of the most important cells involved in wound healing are basal keratinocytes, which make up 90% of the epidermis. Consider a single keratinocyte with a net charge of $+5e$ and a mass of 1.0×10^2 picograms. Assume it is accelerated from rest through a wound gap by a uniform endogenous electric field of 55 N/C. How long will it take the cell to move through a distance of 1.0 mm, which is half the width of the wound? The cell is moving through a liquid medium of blood plasma and serous fluid, but we will ignore these effects.

PROBLEM 68

69. E BIO **18.6** Electroreception is an animal's ability to detect electrical stimuli (i.e., electric fields in its environment). Certain species of fish, sharks, and dolphins use this ability to locate prey and other objects around them. Studies suggest that sharks are the most electrically sensitive animals known, being able to detect electric fields as low

as 5.0×10^{-7} N/C. If a small fish carrying a charge of -0.1 nC is creating this electric field (like a point charge), what is the maximum distance the shark could be away from the fish and still detect it? Note, we are neglecting the effect of the water, which can significantly reduce the electric field. See Section 19.5.

Concepts and Calculations Problems

Online

In this chapter we have studied electric forces and electric fields. The format of the following problems stresses the role of conceptual understanding in problem solving. The purpose of the conceptual question section is to review the important concepts and to build intuition to help in anticipating some of the characteristics of the numerical answers.

70. M CHALK The charges on three identical metal spheres are $-12\,\mu C$, $+4.0\,\mu C$, and $+2.0\,\mu C$. The spheres are brought together so they simultaneously touch each other. They are then separated and placed on the x and y axes, as shown in the figure. Treat the spheres as if they were particles. *Concepts:* (i) Is the net charge on the system comprising the three spheres the same before and after touching? Why? (ii) After the spheres touch and are separated, how is the charge distributed, and what is the value of the charge on each sphere? (iii) Do q_2 and q_3 exert forces of equal magnitude on q_1? (iv) Is the magnitude of the net force exerted on q_1 equal to $2F$, where F is the magnitude of the force that either q_2 or q_3 exerts on q_1? *Calculations:* What is the net force (magnitude and direction) exerted on the sphere at the origin?

(*a*) Three equal charges lie on the *x* and *y* axes. (*b*) The net force exerted

PROBLEM 70 by the two charges is $\vec{F}$.

71. **M** **CHALK** Two point charges are lying on the y axis as in the figure: $q_1 = -4.00\ \mu C$ and $q_2 = +4.00\ \mu C$. They are equidistant from the point P, which lies on the x axis. *Concepts:* (i) There is no charge at P in part a of the figure. Is there an electric field at P? (ii) What is the direction of the electric field at point P due to charge q_2? (iii) Is the magnitude of the electric field at P equal to $E_1 + E_2$, where E_1 and E_2 are the magnitudes of the electric fields produced by q_1 and q_2? Explain. *Calculations:* (a) What is the net electric field at P? (b) A small object of charge $q_0 = +8.00\ \mu C$ and mass $m = 1.2$ g is placed at P. When it is released, what is its acceleration?

(a) Charges q_1 and q_2 lie on the y axis, and point P lies on the x axis. (b) The net electric field at point P is $\vec{E}$.

PROBLEM 71

Team Problems   Online

72. **M** **T** **WS** **Ion Thrusters I: The Electric Field.** You and your team are designing a new propulsion engine that ejects xenon (Xe) ions in one direction, causing a spacecraft to accelerate in the opposite direction, in accordance with the conservation of linear momentum and the impulse-momentum theorem. This is exactly what happens when a rocket engine expels high velocity particles when burning fuel, except that in the case of the ion drive the ions are accelerated with an electric field. In this process, Xe gas atoms are ionized so that they each acquire a net charge of $+e$. They are then subjected to an electric field, which accelerates and ejects the ions from the engine. The final step is to infuse the ejected ion beam with electrons in order to neutralize the ions so that they no longer electrically interact with the spacecraft, or with each other. In the thruster you are designing, assume that the Xe atoms accelerate from rest in a uniform electric field over a distance of 0.110 m before they are ejected with a velocity of 55.0 km/s relative to the engine. (a) What is the acceleration of the Xe ions? (b) What must be the magnitude of the electric field such that the ejection velocity is 55.0 km/s relative to the engine? (c) If Xe ions are ejected at a rate of 9.05×10^{-6} kg/s, what is the average magnitude of the thrust (i.e., the force) provided by the ion drive? The mass of a Xe ion is 2.18×10^{-25} kg.

73. **E** **T** **WS** **Measuring the Charge.** You and your team are designing an experiment where two spherical insulating beads of diameter 0.45 cm and mass 0.15 g are strung together on a fine thread

that passes through microscopic holes drilled through their centers. When hung vertically, the lower bead rests on a knot at the bottom of the thread and the upper bead rests upon the lower bead. When placed inside a vacuum chamber and exposed to X-rays (highly energetic electromagnetic radiation), the beads acquire equal, uniform distributions of positive charge on their surfaces due to the loss of electrons ejected when struck by the X-rays. During this experiment, the beads are hung vertically, and a gap forms between them as if the top bead were levitating above the lower one, presumably due to the repulsive electrostatic force between them. The gap, that is, the vertical separation between the top of the lower bead and the bottom of the upper bead, is 0.27 cm. (a) How much excess charge is on each bead? You can treat the net excess charge on each bead as a point charge located at its center, and may neglect friction between the beads and the thread. (b) Determine the mass of a replacement upper bead of the same size and charge as those of the original upper bead that would reduce the gap between the beads to 0.15 cm.

74. **M** **T** **An Electron Gun.** You and your team are tasked with designing an "electron gun" that operates in a vacuum chamber, the purpose of which is to direct a beam of electrons toward a tiny metallic plate in order to heat it. The electrons in the beam have a speed of 3.50×10^7 m/s and travel in the positive x direction through the center of a set of deflecting plates (a parallel-plate capacitor) that sets up a uniform electric field in the region between the plates. The target is located 22.0 cm along the x-axis from the trailing edge of plates (i.e., the edge closest to the target), and 11.5 cm above the horizontal (i.e., in the +y direction). The length of the plates (in the x direction) is 2.50 cm. **(a)** In which direction should the electric field between the plates point in order to deflect the electrons toward the target? **(b)** To what magnitude should you set the electric field so that the electron beam hits the target? **(c)** After successfully striking the target using your results from (a) and (b), you realized that the target is not heating up to the required temperature. Since the degree of heating depends on the speed of the electrons, you increase the electron speed to 5.20×10^7 m/s. With the electric field setting from **(b)**, will the electrons still be on target? If not, to what value should you set the electric field?

75. **M** **T** **An Electrostatic Positioner.** You and your team are designing a device that can be used to position a small, plastic object in the region between the plates of a parallel-plate capacitor. A small plastic sphere of mass $m = 1.20 \times 10^{-2}$ kg carries a charge $q = +0.200$ μC and hangs vertically (along the y direction) from a massless, insulating thread (length $l = 10.0$ cm) between two vertical capacitor plates. When there is no electric field, the object resides at the midpoint between the plates (at $x = 0$). However, when there is a field between plates (in the $\pm x$ direction) the object moves to a new equilibrium position. **(a)** To what value should you set the field if you want the object to be located at $x = 2.10$ cm? **(b)** To what value should you set the field if you want the object to be located at $x = -3.30$ cm?

Lightning permeates the sky around the ash plume above the Puyehue-Cordon Caulle volcano in south-central Chile in 2011. Unlike normal lightning associated with rain clouds, where static charges are produced by colliding ice particles, volcanic lightning, or "dirty thunderstorms," can result from frictional charging between colliding ash and dust particles. The natural convective thermal currents in the hot ash cloud aid in the separation of charges. This creates extremely high differences in voltage, or potential, between different parts of the dust cloud or between the cloud and the ground. If the voltage difference is sufficiently large, the insulating properties of the air break down, and it conducts electricity in spectacular fashion. The electric potential, and its relationship to charge, will be one of the topics we study in this chapter.

Carlos Gutierrez/Reuters

LEARNING OBJECTIVES

After reading this module, you should be able to...

19.1 Define electrical potential energy.

19.2 Solve problems involving electric potential and electric potential energy.

19.3 Calculate electric potential created by point charges.

19.4 Relate equipotential surfaces to the electric field.

19.5 Solve problems involving capacitors.

19.6 Describe biomedical applications of electric potential.

19.7 Describe the Millikan oil-drop experiment.

Electric Potential Energy and the Electric Potential

19.1 Potential Energy

In Chapter 18 we discussed the electrostatic force that two point charges exert on each other, the magnitude of which is $F = k|q_1||q_2|/r^2$. The form of this equation is similar to the form for the gravitational force that two particles exert on each other, which is $F = Gm_1m_2/r^2$, according to Newton's law of universal gravitation (see Section 4.7). Both of these forces are conservative and, as Section 6.4 explains, a potential energy can be associated with a conservative force. Thus, an electric potential energy exists that is analogous to the gravitational potential energy. To set the stage for a discussion of the electric potential energy, let's review some of the important aspects of the gravitational counterpart.

Figure 19.1, which is essentially Figure 6.10, shows a basketball of mass m falling from point A to point B. The gravitational force, $m\vec{\mathbf{g}}$, is the only force acting on the ball, where g is the magnitude of the acceleration due to gravity. As Section 6.3 discusses, the work W_{AB} done by the gravitational force when the ball falls from a height of h_A to a height of h_B is

$$W_{AB} = \underbrace{mgh_A}_{\substack{\text{Initial gravitational} \\ \text{potential energy,} \\ \text{GPE}_A}} - \underbrace{mgh_B}_{\substack{\text{Final gravitational} \\ \text{potential energy,} \\ \text{GPE}_B}} = \text{GPE}_A - \text{GPE}_B \qquad \textbf{(6.4)}$$

Recall that the quantity mgh is the gravitational potential energy* of the ball, $\text{GPE} = mgh$ (Equation 6.5), and represents the energy that the ball has by virtue of its position relative to the surface of the earth. Thus, the work done by the gravitational force equals the initial gravitational potential energy minus the final gravitational potential energy.

Interactive Figure 19.2 clarifies the analogy between electric and gravitational potential energies. In this drawing a positive test charge $+q_0$ is situated at point A between two oppositely charged plates. Because of the charges on the plates, an electric field $\vec{\mathbf{E}}$ exists in the region between the plates. Consequently, the test charge experiences an electric force, $\vec{\mathbf{F}} = q_0\vec{\mathbf{E}}$ (Equation 18.2), that is directed downward, toward the lower plate. (The gravitational force is being neglected here.) As the charge moves from A to B, work is done by this force, in a fashion analogous to the work done by the gravitational force in **Figure 19.1**. The work W_{AB} done by the electric force equals the difference between the electric potential energy EPE at A and the electric potential energy at B:

$$W_{AB} = \text{EPE}_A - \text{EPE}_B \qquad \textbf{(19.1)}$$

This expression is similar to Equation 6.4. The path along which the test charge moves from A to B is of no consequence because the electric force, like the gravitational force, is conservative. For such forces, the work W_{AB} is the same for all paths (see Section 6.4).

19.2 The Electric Potential Difference

Since the electric force is $\vec{\mathbf{F}} = q_0\vec{\mathbf{E}}$, the work that it does as the charge moves from A to B in **Interactive Figure 19.2** depends on the charge q_0. It is useful, therefore, to express this work on a per-unit-charge basis, by dividing both sides of Equation 19.1 by the charge:

$$\frac{W_{AB}}{q_0} = \frac{\text{EPE}_A}{q_0} - \frac{\text{EPE}_B}{q_0} \qquad \textbf{(19.2)}$$

Notice that the right-hand side of this equation is the difference between two terms, each of which is an electric potential energy divided by the test charge, EPE/q_0. The quantity EPE/q_0 is the electric potential energy per unit charge and is an important concept in electricity. It is called the **electric potential** or, simply, the **potential** and is referred to with the symbol V, as in Equation 19.3.

DEFINITION OF ELECTRIC POTENTIAL

The electric potential V at a given point is the electric potential energy EPE of a small test charge q_0 situated at that point divided by the charge itself:

$$V = \frac{\text{EPE}}{q_0} \qquad \textbf{(19.3)}$$

SI Unit of Electric Potential: joule/coulomb = volt (V)

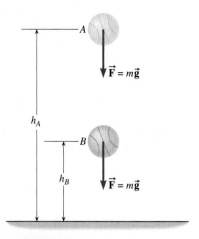

FIGURE 19.1 Gravity exerts a force, $\vec{\mathbf{F}} = m\vec{\mathbf{g}}$, on the basketball of mass m. Work is done by the gravitational force as the ball falls from A to B.

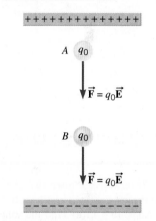

INTERACTIVE FIGURE 19.2 Because of the electric field $\vec{\mathbf{E}}$, an electric force, $\vec{\mathbf{F}} = q_0\vec{\mathbf{E}}$, is exerted on a positive test charge $+q_0$. Work is done by the force as the charge moves from A to B.

---

*The gravitational potential energy is now being denoted by GPE to distinguish it from the electric potential energy EPE.

The SI unit of electric potential is a joule per coulomb, a quantity known as a *volt*. The name honors Alessandro Volta (1745–1827), who invented the voltaic pile, the forerunner of the battery. In spite of the similarity in names, the electric potential energy EPE and the electric potential V are *not* the same. The electric potential energy, as its name implies, is an *energy* and, therefore, is measured in joules. In contrast, the electric potential is an *energy per unit charge* and is measured in joules per coulomb, or volts.

We can now relate the work W_{AB} done by the electric force when a charge q_0 moves from A to B to the potential difference $V_B - V_A$ between the points. Combining Equations 19.2 and 19.3, we have:

$$V_B - V_A = \frac{\text{EPE}_B}{q_0} - \frac{\text{EPE}_A}{q_0} = \frac{-W_{AB}}{q_0} \qquad (19.4)$$

Often, the "delta" notation is used to express the difference (final value minus initial value) in potentials and in potential energies: $\Delta V = V_B - V_A$ and $\Delta(\text{EPE}) = \text{EPE}_B - \text{EPE}_A$. In terms of this notation, Equation 19.4 takes the following more compact form:

$$\Delta V = \frac{\Delta(\text{EPE})}{q_0} = \frac{-W_{AB}}{q_0} \qquad (19.4)$$

Neither the potential V nor the potential energy EPE can be determined in an absolute sense, because only the *differences* ΔV and $\Delta(\text{EPE})$ are measurable in terms of the work W_{AB}. The gravitational potential energy has this same characteristic, since only the value at one height relative to the value at some reference height has any significance. Example 1 emphasizes the relative nature of the electric potential.

EXAMPLE 1 | Work, Electric Potential Energy, and Electric Potential

In **Interactive Figure 19.2**, the work done by the electric force as the test charge ($q_0 = +2.0 \times 10^{-6}$ C) moves from A to B is $W_{AB} = +5.0 \times 10^{-5}$ J. **(a)** Find the value of the difference, $\Delta(\text{EPE}) = \text{EPE}_B - \text{EPE}_A$, in the electric potential energies of the charge between these points. **(b)** Determine the potential difference, $\Delta V = V_B - V_A$, between the points.

Reasoning The work done by the electric force when the charge travels from A to B is $W_{AB} = \text{EPE}_A - \text{EPE}_B$, according to Equation 19.1. Therefore, the difference in the electric potential energies (final value minus initial value) is $\Delta(\text{EPE}) = \text{EPE}_B - \text{EPE}_A = -W_{AB}$. The potential difference, $\Delta V = V_B - V_A$, is the difference in the electric potential energies divided by the charge q_0, according to Equation 19.4.

Solution **(a)** The difference in the electric potential energies of the charge between the points A and B is

$$\underbrace{\text{EPE}_B - \text{EPE}_A}_{= \Delta(\text{EPE})} = -W_{AB} = \boxed{-5.0 \times 10^{-5} \text{ J}} \qquad (19.1)$$

The fact that $\text{EPE}_B - \text{EPE}_A$ is negative means that the charge has a higher electric potential energy at A than at B.

(b) The potential difference ΔV between A and B is

$$\underbrace{V_B - V_A}_{= \Delta V} = \frac{\text{EPE}_B - \text{EPE}_A}{q_0} = \frac{-5.0 \times 10^{-5} \text{ J}}{2.0 \times 10^{-6} \text{ C}} = \boxed{-25 \text{ V}} \qquad (19.4)$$

The fact that $V_B - V_A$ is negative means that the electric potential is higher at A than at B.

The potential difference between two points is measured in volts and, therefore, is often referred to as a "voltage." Everyone has heard of "voltage" because, as we will see in Chapter 20, it is frequently used in connection with everyday devices. For example, your TV requires a "voltage" of 120 V (which is applied between the two prongs of the plug on the power cord when it is inserted into an electrical wall outlet), and your cell phone and laptop computer use batteries that provide, for example, "voltages" of 1.5 V or 9 V (which exist between the two battery terminals).

In **Figure 19.1** the speed of the basketball increases as it falls from A to B. Since point A has a greater gravitational potential energy than point B, we see that an object of mass m accelerates when it moves from a region of higher potential energy toward a region of lower potential energy. Likewise, the positive charge in **Interactive Figure 19.2** accelerates as it moves from A to B because of the electric repulsion from the upper plate and the attraction to the lower plate. Since point A has a higher electric potential than point B, we conclude the following:

> **Problem-Solving Insight**   **A positive charge accelerates from a region of higher electric potential toward a region of lower electric potential.**

On the other hand, a negative charge placed between the plates in **Interactive Figure 19.2** behaves in the opposite fashion, since the electric force acting on the negative charge is directed opposite to the electric force acting on the positive charge.

> **Problem-Solving Insight**   **A negative charge accelerates from a region of lower potential toward a region of higher potential.**

The next example illustrates the way positive and negative charges behave.

CONCEPTUAL EXAMPLE 2 | How Positive and Negative Charges Accelerate

Three points, A, B, and C, are located along a horizontal line, as **Figure 19.3** illustrates. A positive test charge is released from rest at A and accelerates toward B. Upon reaching B, the test charge continues to accelerate toward C. Assuming that only motion along the line is possible, what will a negative test charge do when it is released from rest at B? A negative test charge will **(a)** accelerate toward C, **(b)** remain stationary, **(c)** accelerate toward A.

Reasoning   A positive charge accelerates from a region of higher potential toward a region of lower potential. A negative charge behaves in an opposite manner, because it accelerates from a region of lower potential toward a region of higher potential.

Answers (a) and (b) are incorrect.   The positive charge accelerates from A to B and then from B to C. A negative charge placed at B also accelerates, but in a direction opposite to that of the positive charge. Therefore, a negative charge placed at B will not remain stationary, nor will it accelerate toward C.

Answer (c) is correct.   Since the positive charge accelerates from A to B, the potential at A must exceed the potential at B. And

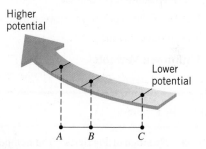

FIGURE 19.3   The electric potentials at points A, B, and C are different. Under the influence of these potentials, positive and negative charges accelerate in opposite directions.

since the positive test charge accelerates from B to C, the potential at B must exceed the potential at C. The potential at point B, then, must lie between the potential at points A and C, as **Figure 19.3** illustrates. When the negative test charge is released from rest at B, it will accelerate toward the region of higher potential, so it will begin moving toward A.

As a familiar application of electric potential energy and electric potential, **Figure 19.4** shows a 12-V automobile battery with a headlight connected between its terminals. The positive terminal, point A, has a potential that is 12 V higher than the potential at the negative terminal, point B; in other words, $V_A - V_B = 12$ V. Positive charges would be repelled from the positive terminal and would travel through the wires and headlight toward the negative terminal.* As the charges pass through the headlight, virtually all their potential energy is converted into heat, which causes the filament to glow "white hot" and emit light. When the charges reach the negative terminal, they no longer have any potential energy. The battery then gives the charges an additional "shot" of potential energy by moving them to the higher-potential positive terminal, and the cycle is repeated. In raising the potential energy of the charges, the battery does work on them and draws from its reserve of chemical energy to do so. Example 3 illustrates the concepts of electric potential energy and electric potential as applied to a battery.

FIGURE 19.4   A headlight connected to a 12-V battery.

---

*Historically, it was believed that positive charges flow in the wires of an electric circuit. Today, it is known that negative charges flow in wires from the negative toward the positive terminal. Here, however, we follow the customary practice of describing the flow of negative charges by specifying the opposite but equivalent flow of positive charges. This hypothetical flow of positive charges is called the "conventional electric current," as we will see in Section 20.1.

Analyzing Multiple-Concept Problems

EXAMPLE 3 | Operating a Headlight

The wattage of the headlight in **Figure 19.4** is 60.0 W. Determine the number of particles, each carrying a charge of 1.60×10^{-19} C (the magnitude of the charge on an electron), that pass between the terminals of the 12-V car battery when the headlight burns for one hour.

Reasoning The number of particles is the total charge that passes between the battery terminals in one hour divided by

the magnitude of the charge on each particle. The total charge is the amount of charge needed to convey the energy used by the headlight in one hour. This energy is related to the wattage of the headlight, which specifies the power or rate at which energy is used, and the time the light is on.

Knowns and Unknowns The following table summarizes the data provided:

Description	Symbol	Value	Comment
Wattage of headlight	P	60.0 W	
Charge magnitude per particle	e	1.60×10^{-19} C	
Electric potential difference between battery terminals	$V_A - V_B$	12 V	See **Figure 19.4**.
Time headlight is on	t	3600 s	One hour.
Unknown Variable			
Number of particles	n	?	

Modeling the Problem

STEP 1 The Number of Particles The number n of particles is the total charge q_0 that passes between the battery terminals in one hour divided by the magnitude e of the charge on each particle, as expressed by Equation 1 at the right. The value of e is given. To evaluate q_0, we proceed to Step 2.

$$n = \frac{q_0}{e} \quad (1)$$

$\boxed{?}$

STEP 2 The Total Charge Provided by the Battery The battery must supply the total energy used by the headlight in one hour. The battery does this by supplying the charge q_0 to convey this energy. The energy is the difference between the electric potential energy EPE_A at terminal A and the electric potential energy EPE_B at terminal B of the battery (see **Figure 19.4**). According to Equation 19.4, this total energy is $\text{EPE}_A - \text{EPE}_B = q_0(V_A - V_B)$, where $V_A - V_B$ is the electric potential difference between the battery terminals. Solving this expression for q_0 gives

$$\boxed{q_0 = \frac{\text{EPE}_A - \text{EPE}_B}{V_A - V_B}}$$

which can be substituted into Equation 1, as shown at the right. As the data table shows, a value is given for $V_A - V_B$. In Step 3, we determine a value for $\text{EPE}_A - \text{EPE}_B$.

$$n = \frac{q_0}{e} \quad (1)$$

$$q_0 = \frac{\text{EPE}_A - \text{EPE}_B}{V_A - V_B} \quad (2)$$

$\boxed{?}$

STEP 3 The Energy Used by the Headlight The rate at which the headlight uses energy is the power P or wattage of the headlight. According to Equation 6.10b, the power is the total energy $\text{EPE}_A - \text{EPE}_B$ divided by the time t, so that $P = (\text{EPE}_A - \text{EPE}_B)/t$. Solving for the total energy gives

$$\boxed{\text{EPE}_A - \text{EPE}_B = Pt}$$

Since P and t are given, we substitute this result into Equation 2, as indicated at the right.

$$n = \frac{q_0}{e} \quad (1)$$

$$q_0 = \frac{\text{EPE}_A - \text{EPE}_B}{V_A - V_B} \quad (2)$$

$$\boxed{\text{EPE}_A - \text{EPE}_B = Pt}$$

Solution Combining the results of each step algebraically, we find that

$$\overset{\text{STEP 1}}{n} = \overset{\text{STEP 2}}{\frac{q_0}{e}} = \overset{}{\frac{(\text{EPE}_A - \text{EPE}_B)/(V_A - V_B)}{e}} = \overset{\text{STEP 2}}{\frac{Pt/(V_A - V_B)}{e}}$$

The number of particles that pass between the battery terminals in one hour is

$$n = \frac{Pt}{e(V_A - V_B)} = \frac{(60.0 \text{ W})(3600 \text{ s})}{(1.60 \times 10^{-19} \text{ C})(12 \text{ V})} = \boxed{1.1 \times 10^{23}}$$

Related Homework: Problem 46

As used in connection with batteries, the volt is a familiar unit for measuring electric potential difference. The word "volt" also appears in another context, as part of a unit that is used to measure energy, particularly the energy of an atomic particle, such as an electron or a proton. This energy unit is called the *electron volt* (eV).

> **Problem-Solving Insight**   One electron volt is the magnitude of the amount by which the potential energy of an electron changes when the electron moves through a potential difference of one volt.

Since the magnitude of the change in potential energy is $|q_0 \Delta V| = |(-1.60 \times 10^{-19} \text{ C}) \times (1.00 \text{ V})| = 1.60 \times 10^{-19}$ J, it follows that

$$1 \text{ eV} = 1.60 \times 10^{-19} \text{ J}$$

One million (10^{+6}) electron volts of energy is referred to as one MeV, and one billion (10^{+9}) electron volts of energy is one GeV, where the "G" stands for the prefix "giga."

In Equation 19.3, we have seen that the electric potential is the electric potential energy per unit charge. In previous chapters, we have seen that the total energy of an object, which is the sum of its kinetic and potential energies, is an important concept. Its significance lies in the fact that the total energy remains the same (is conserved) during the object's motion, provided that nonconservative forces, such as friction, are either absent or do no net work. While the sum of the energies at each instant remains constant, energy may be converted from one form to another; for example, gravitational potential energy is converted into kinetic energy as a ball falls. We now include the electric potential energy EPE as part of the total energy that an object can have:

$$\underbrace{E}_{\substack{\text{Total} \\ \text{energy}}} = \underbrace{\tfrac{1}{2}mv^2}_{\substack{\text{Translational} \\ \text{kinetic energy}}} + \underbrace{\tfrac{1}{2}I\omega^2}_{\substack{\text{Rotational} \\ \text{kinetic energy}}} + \underbrace{mgh}_{\substack{\text{Gravitational} \\ \text{potential} \\ \text{energy}}} + \underbrace{\tfrac{1}{2}kx^2}_{\substack{\text{Elastic} \\ \text{potential} \\ \text{energy}}} + \underbrace{\text{EPE}}_{\substack{\text{Electric} \\ \text{potential} \\ \text{energy}}}$$

If the total energy is conserved as the object moves, then its final energy E_f is equal to its initial energy E_0, or $E_f = E_0$. Example 4 illustrates how the conservation of energy is applied to a charge moving in an electric field.

Analyzing Multiple-Concept Problems

EXAMPLE 4 | The Conservation of Energy

A particle has a mass of 1.8×10^{-5} kg and a charge of $+3.0 \times 10^{-5}$ C. It is released from rest at point A and accelerates until it reaches point B, as **Figure 19.5a** shows. The particle moves on a horizontal straight line and does not rotate. The only forces acting on the particle are the gravitational force and an electrostatic force (neither is shown in the drawing). The electric potential at A is 25 V greater than that at B; in other words, $V_A - V_B = 25$ V. What is the translational speed of the particle at point B?

Reasoning   The translational speed of the particle is related to the particle's translational kinetic energy, which forms one part of the total energy that the particle has. The total energy is conserved, because only the gravitational force and an electrostatic force, both of which are conservative forces, act on the particle (see Section 6.5). Thus, we will determine the speed at point B by utilizing the principle of conservation of energy.

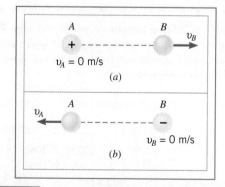

FIGURE 19.5   (a) A positive charge starts from rest at point A and accelerates toward point B. (b) A negative charge starts from rest at B and accelerates toward A.

Knowns and Unknowns   We have the following data:

Description	Symbol	Value	Comment
Explicit Data			
Mass of particle	m	1.8×10^{-5} kg	
Electric charge of particle	q_0	$+3.0 \times 10^{-5}$ C	
Electric potential difference between points A and B	$V_A - V_B$	25 V	See **Figure 19.5a**.
Implicit Data			
Speed at point A	v_A	0 m/s	Particle released from rest.
Vertical height above ground	h	Remains constant	Particle travels horizontally.
Angular speed	ω	0 rad/s	Particle does not rotate during motion.
Elastic force	F_{elastic}	0 N	No elastic force acts on particle.
Unknown Variable			
Speed at point B	v_B	?	

Modeling the Problem

STEP 1 Conservation of Total Energy   The particle's total energy E is

$$E = \underbrace{\tfrac{1}{2}mv^2}_{\substack{\text{Translational}\\\text{kinetic energy}}} + \underbrace{\tfrac{1}{2}I\omega^2}_{\substack{\text{Rotational}\\\text{kinetic energy}}} + \underbrace{mgh}_{\substack{\text{Gravitational}\\\text{potential}\\\text{energy}}} + \underbrace{\tfrac{1}{2}kx^2}_{\substack{\text{Elastic}\\\text{potential}\\\text{energy}}} + \underbrace{\text{EPE}}_{\substack{\text{Electric}\\\text{potential}\\\text{energy}}}$$

Since the particle does not rotate, the angular speed ω is always zero (see the data table) and since there is no elastic force (see the data table), we may omit the terms $\tfrac{1}{2}I\omega^2$ and $\tfrac{1}{2}kx^2$ from this expression. With this in mind, we express the fact that $E_B = E_A$ (energy is conserved) as follows:

$$\tfrac{1}{2}mv_B^2 + mgh_B + \text{EPE}_B = \tfrac{1}{2}mv_A^2 + mgh_A + \text{EPE}_A$$

This equation can be simplified further, since the particle travels horizontally, so that $h_B = h_A$ (see the data table), with the result that

$$\tfrac{1}{2}mv_B^2 + \text{EPE}_B = \tfrac{1}{2}mv_A^2 + \text{EPE}_A$$

Solving for v_B gives Equation 1 at the right. Values for v_A and m are available, and we turn to Step 2 in order to evaluate $\text{EPE}_A - \text{EPE}_B$.

$$v_B = \sqrt{v_A^2 + \frac{2(\text{EPE}_A - \text{EPE}_B)}{m}} \quad (1)$$

$\boxed{?}$

STEP 2 The Electric Potential Difference   According to Equation 19.4, the difference in electric potential energies $\text{EPE}_A - \text{EPE}_B$ is related to the electric potential difference $V_A - V_B$:

$$\boxed{\text{EPE}_A - \text{EPE}_B = q_0(V_A - V_B)}$$

The terms q_0 and $V_A - V_B$ are known, so we substitute this expression into Equation 1, as shown at the right.

$$v_B = \sqrt{v_A^2 + \frac{2(\text{EPE}_A - \text{EPE}_B)}{m}} \quad (1)$$

$$\boxed{\text{EPE}_A - \text{EPE}_B = q_0(V_A - V_B)}$$

> **Problem-Solving Insight**   A positive charge accelerates from a region of higher potential toward a region of lower potential. In contrast, a negative charge accelerates from a region of lower potential toward a region of higher potential.

Solution   Combining the results of each step algebraically, we find that

$$\overset{\textbf{STEP 1}}{\downarrow} \qquad\qquad \overset{\textbf{STEP 2}}{\downarrow}$$
$$v_B = \sqrt{v_A^2 + \frac{2(\text{EPE}_A - \text{EPE}_B)}{m}} = \sqrt{v_A^2 + \frac{2q_0(V_A - V_B)}{m}}$$

The speed of the particle at point B is

$$v_B = \sqrt{v_A^2 + \frac{2q_0(V_A - V_B)}{m}} = \sqrt{(0\ \text{m/s})^2 + \frac{2(+3.0 \times 10^{-5}\ \text{C})(25\ \text{V})}{1.8 \times 10^{-5}\ \text{kg}}} = \boxed{9.1\ \text{m/s}}$$

Note that if the particle had a negative charge of -3.0×10^{-5} C and were released from rest at point B, it would arrive at point A with the same speed of 9.1 m/s (see **Figure 19.5b**). This result can be obtained by returning to Modeling Step 1 and solving for v_A instead of v_B.

Related Homework: Problems 4, 5

Check Your Understanding

(*The answers are given at the end of the book.*)

1. An ion, starting from rest, accelerates from point A to point B due to a potential difference between the two points. Does the electric potential energy of the ion at point B depend on **(a)** the magnitude of its charge and **(b)** its mass? Does the speed of the ion at B depend on **(c)** the magnitude of its charge and **(d)** its mass?

2. **CYU Figure 19.1** shows three possibilities for the potentials at two points, A and B. In each case, the same positive charge is moved from A to B. In which case, if any, is the most work done on the positive charge by the electric force?

A	B	A	B	A	B
•	•	•	•	•	•
150 V	100 V	25 V	−25 V	−10 V	−60 V
Case 1		Case 2		Case 3	

CYU FIGURE 19.1

3. A proton and an electron are released from rest at the midpoint between the plates of a charged parallel plate capacitor (see Section 18.6). Except for these particles, nothing else is between the plates. Ignore the attraction between the proton and the electron, and decide which particle strikes a capacitor plate first.

19.3 The Electric Potential Difference Created by Point Charges

A positive point charge $+q$ creates an electric potential in a way that **Figure 19.6** helps explain. This picture shows two locations A and B, at distances r_A and r_B from the charge. At any position between A and B an electrostatic force of repulsion $\vec{F}$ acts on a positive test charge $+q_0$. The magnitude of the force is given by Coulomb's law as $F = kq_0q/r^2$, where we assume for convenience that q_0 and q are positive, so that $|q_0| = q_0$ and $|q| = q$. When the test charge moves from A to B, work is done by this force. Since r varies between r_A and r_B, the force F also varies, and the work is not the product of the force and the distance between the points. (Recall from Section 6.1 that work is force times distance only if the force is constant.) However, the work W_{AB} can be found with the methods of integral calculus. The result is

$$W_{AB} = \frac{kqq_0}{r_A} - \frac{kqq_0}{r_B}$$

This result is valid whether q is positive or negative, and whether q_0 is positive or negative. The potential difference, $V_B - V_A$, between A and B can now be determined by substituting this expression for W_{AB} into Equation 19.4:

$$V_B - V_A = \frac{-W_{AB}}{q_0} = \frac{kq}{r_B} - \frac{kq}{r_A} \tag{19.5}$$

As point B is located farther and farther from the charge q, r_B becomes larger and larger. In the limit that r_B is infinitely large, the term kq/r_B becomes zero, and it is customary to set V_B equal to zero also. In this limit, Equation 19.5 becomes $V_A = kq/r_A$, and it is standard convention to omit the subscripts and write the potential in the following form:

Potential of a point charge $$V = \frac{kq}{r} \tag{19.6}$$

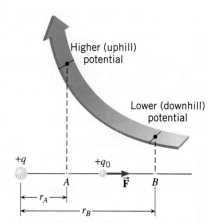

FIGURE 19.6 The positive test charge $+q_0$ experiences a repulsive force $\vec{F}$ due to the positive point charge $+q$. As a result, work is done by this force when the test charge moves from A to B. Consequently, the electric potential is higher (uphill) at A and lower (downhill) at B.

Math Skills *In Equation 19.6 the symbol q denotes the value of the point charge, including both the magnitude of and the algebraic sign of the charge.* Note especially that the symbol q does not have the same meaning as the symbol $|q|$, which denotes only the magnitude of the charge. When you use Equation 19.6 to solve problems dealing with the potential of a point charge, both the magnitude and the algebraic sign of the charge must be taken into account, and the presence of q (rather than $|q|$) in the equation ensures that this will be so. Using only the magnitude of a point charge in Equation 19.6 would ignore the important fact that positive and negative point charges create different potentials, even when the charge magnitudes are the same.

The symbol V in this equation does not refer to the potential in any absolute sense. Rather, $V = kq/r$ stands for the amount by which the potential at a distance r from a point charge differs from the potential at an infinite distance away. In other words, V refers to a potential difference with the arbitrary assumption that the potential at infinity is zero.

With the aid of Equation 19.6, we can describe the effect that a point charge q has on the surrounding space. When q is positive, the value of $V = kq/r$ is also positive, indicating that the positive charge has everywhere raised the potential above the zero reference value. Conversely, when q is negative, the potential V is also negative, indicating that the negative charge has everywhere decreased the potential below the zero reference value. The next example deals with these effects quantitatively.

EXAMPLE 5 | The Potential of a Point Charge

Using a zero reference potential at infinity, determine the amount by which a point charge of 4.0×10^{-8} C alters the electric potential at a spot 1.2 m away when the charge is **(a)** positive and **(b)** negative.

Reasoning A point charge q alters the potential at every location in the surrounding space. In the expression $V = kq/r$, the effect of the charge in increasing or decreasing the potential is conveyed by the algebraic sign for the value of q.

Solution **(a)** Figure 19.7a shows the potential when the charge is positive:

$$V = \frac{kq}{r} = \frac{(8.99 \times 10^9 \ \text{N} \cdot \text{m}^2/\text{C}^2)(+4.0 \times 10^{-8} \ \text{C})}{1.2 \ \text{m}} = \boxed{+300 \ \text{V}} \quad (19.6)$$

(b) Part b of the drawing illustrates the results when the charge is negative. A calculation similar to the one in part (a) shows that the potential is now negative: $\boxed{-300 \ \text{V}}$.

FIGURE 19.7 A point charge of 4.0×10^{-8} C alters the potential at a spot 1.2 m away. The potential is (a) increased by 300 V when the charge is positive and (b) decreased by 300 V when the charge is negative, relative to a zero reference potential at infinity.

A single point charge raises or lowers the potential at a given location, depending on whether the charge is positive or negative.

> **Problem-Solving Insight** When two or more charges are present, the potential due to all the charges is obtained by adding together the individual potentials.

The next two examples will demonstrate this point.

EXAMPLE 6 | The Total Electric Potential

At locations A and B in **Figure 19.8**, find the total electric potential due to the two point charges.

Reasoning At each location, each charge contributes to the total electric potential. We obtain the individual contributions by using $V = kq/r$ and find the total potential by adding the individual contributions algebraically. The two charges have the same magnitude, but different signs. Thus, at A the total potential is positive because this spot is closer to the positive charge, whose effect dominates over that of the more distant negative charge. At B, midway

FIGURE 19.8 Both the positive and negative charges affect the electric potential at locations A and B.

between the charges, the total potential is zero, since the potential of one charge exactly offsets that of the other.

Solution

Location	Contribution from + Charge		Contribution from − Charge	Total Potential
A	$\dfrac{(8.99 \times 10^9 \text{ N} \cdot \text{m}^2/\text{C}^2)(+8.0 \times 10^{-9} \text{ C})}{0.20 \text{ m}}$	+	$\dfrac{(8.99 \times 10^9 \text{ N} \cdot \text{m}^2/\text{C}^2)(-8.0 \times 10^{-9} \text{ C})}{0.60 \text{ m}} =$	$\boxed{+240 \text{ V}}$
B	$\dfrac{(8.99 \times 10^9 \text{ N} \cdot \text{m}^2/\text{C}^2)(+8.0 \times 10^{-9} \text{ C})}{0.40 \text{ m}}$	−	$\dfrac{(8.99 \times 10^9 \text{ N} \cdot \text{m}^2/\text{C}^2)(-8.0 \times 10^{-9} \text{ C})}{0.40 \text{ m}} =$	$\boxed{0 \text{ V}}$

CONCEPTUAL EXAMPLE 7 | Where Is the Potential Zero?

Two point charges are fixed in place, as in **Figure 19.9**. The positive charge is $+2q$ and has twice the magnitude of the negative charge, which is $-q$. On the line that passes through the charges, three spots are identified, P_1, P_2, and P_3. At which of these spots could the potential be equal to zero? **(a)** P_2 and P_3 **(b)** P_1 and P_3 **(c)** P_1 and P_2

FIGURE 19.9 Two point charges, one positive and one negative. The positive charge, $+2q$, has twice the magnitude of the negative charge, $-q$.

Reasoning The total potential is the algebraic sum of the individual potentials created by each charge. It will be zero if the potential due to the positive charge is exactly offset by the potential due to the negative charge. The potential of a point charge is directly proportional to the charge and inversely proportional to the distance from the charge.

Answers (b) and (c) are incorrect. The total potential at P_1 cannot be zero. The positive charge has the larger magnitude and is closer to P_1 than is the negative charge. As a result, the potential of the positive charge at P_1 dominates over the potential of the negative charge, so the total potential cannot be zero.

Answer (a) is correct. Between the charges there is a location at which the individual potentials cancel each other. We saw a similar situation in Example 6, where the canceling occurred at the midpoint between the two charges that had equal magnitudes. Now the charges have unequal magnitudes, so the cancellation point does not occur at the midpoint. Instead, it occurs at the location P_2, which is closer to the charge with the smaller magnitude—namely, the negative charge. At P_2, since the potential of a point charge is inversely proportional to the distance from the charge, the effect of the smaller charge will be able to offset the effect of the more distant larger charge.

To the right of the negative charge there is also a location at which the individual potentials exactly cancel each other. All places on this section of the line are closer to the negative charge than to the positive charge. Therefore, there is a location P_3 in this region at which the potential of the smaller negative charge exactly cancels the potential of the more distant and larger positive charge.

In Example 6 we determined the total potential at a spot due to two point charges. In Example 8 we now extend this technique to find the total potential *energy* of three charges.

EXAMPLE 8 | The Potential Energy of a Group of Charges

Three point charges initially are infinitely far apart. Then, as **Figure 19.10** shows, they are brought together and placed at the corners of an equilateral triangle. Each side of the triangle has a length of 0.50 m. Determine the electric potential energy of the triangular group. In other words, determine the amount by which the electric potential energy of the group differs from that of the three charges in their initial, infinitely separated, locations.

Reasoning We will proceed in steps by adding charges to the triangle one at a time, and then determining the electric potential energy at each step. According to Equation 19.3, EPE $= q_0 V$, the electric potential energy is the product of the charge and the electric potential at the spot where the charge is placed. The total

electric potential energy of the triangular group is the sum of the energies of each step in assembling the group.

Solution The order in which the charges are put on the triangle does not matter; we begin with the charge $q_1 = +5.0 \ \mu\text{C}$. When this charge is placed at a corner of the triangle, it has no electric potential energy, according to $\text{EPE}_1 = q_1(V_2 + V_3) = 0 \text{ J}$ (Equation 19.3). This is because the total potential $V_2 + V_3$ produced by the other two charges is zero at this corner, since they are infinitely far away. Once the charge q_1 is in place, the potential it creates at either empty corner ($r = 0.50 \text{ m}$) is

$$V_1 = \frac{kq_1}{r} = \frac{(8.99 \times 10^9 \text{ N} \cdot \text{m}^2/\text{C}^2)(+5.0 \times 10^{-6} \text{ C})}{0.50 \text{ m}}$$

$$= +9.0 \times 10^4 \text{ V} \tag{19.6}$$

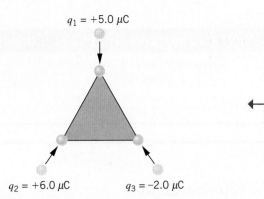

$q_1 = +5.0\ \mu C$

$q_2 = +6.0\ \mu C$ $q_3 = -2.0\ \mu C$

FIGURE 19.10 Three point charges are placed on the corners of an equilateral triangle. Example 8 illustrates how to determine the total electric potential energy of this group of charges.

Therefore, when the charge $q_2 = +6.0\ \mu C$ is placed at the second corner of the triangle, its electric potential energy is

$$\text{EPE}_2 = q_2 V_1 = \frac{q_2 k q_1}{r} = (+6.0 \times 10^{-6}\ \text{C})(+9.0 \times 10^4\ \text{V})$$

$$= +0.54\ \text{J} \qquad\qquad \textbf{(19.3)}$$

The electric potential at the remaining empty corner is the sum of the potentials due to the two charges that are already in place:

$$V_1 + V_2 = \frac{k q_1}{r} + \frac{k q_2}{r}$$

$$= \frac{(8.99 \times 10^9\ \text{N} \cdot \text{m}^2/\text{C}^2)(+5.0 \times 10^{-6}\ \text{C})}{0.50\ \text{m}}$$

$$+ \frac{(8.99 \times 10^9\ \text{N} \cdot \text{m}^2/\text{C}^2)(+6.0 \times 10^{-6}\ \text{C})}{0.50\ \text{m}}$$

$$= +2.0 \times 10^5\ \text{V}$$

When the third charge $q_3 = -2.0\ \mu C$ is placed at the remaining empty corner, its electric potential energy is

$$\text{EPE}_3 = q_3(V_1 + V_2) = \frac{q_3 k q_1}{r} + \frac{q_3 k q_2}{r}$$

$$= (-2.0 \times 10^{-6}\ \text{C})(+2.0 \times 10^5\ \text{V}) = -0.40\ \text{J} \qquad \textbf{(19.3)}$$

> **Problem-Solving Insight** Be careful to distinguish between the concepts of potential V and electric potential energy EPE. Potential is electric potential energy per unit charge: $V = \text{EPE}/q$.

Math Skills To illustrate that the order in which the charges are put on the triangle does not matter, let us obtain the total potential energy of the group when the order begins with q_2, continues with q_3, and concludes with q_1. When q_2 is placed at a corner of the triangle, it has no electric potential energy EPE_2, since it experiences no electric potential due to the other charges, which are infinitely far away. Thus, we have

$$\text{EPE}_2 = 0\ \text{J}$$

Once q_2 is in place, it creates a potential $V_2 = \frac{k q_2}{r}$ (Equation 19.6) at either empty corner located a distance r away. When the charge q_3 is placed at an empty corner, it experiences this potential and has an electric potential energy EPE_3, as specified by Equation 19.3:

$$\text{EPE}_3 = q_3 V_2 = \frac{q_3 k q_2}{r}$$

Each of the charges in place creates a potential at the remaining empty corner. Using Equation 19.6 for each charge, we see that the total potential at the remaining empty corner is $V_2 + V_3 = \frac{k q_2}{r} + \frac{k q_3}{r}$. When charge q_1 is placed on this corner, therefore, it experiences this potential and has an electric potential energy EPE_1, as given by Equation 19.3:

$$\text{EPE}_1 = q_1(V_2 + V_3) = \frac{q_1 k q_2}{r} + \frac{q_1 k q_3}{r}$$

Thus, we obtain the following total electric potential energy of the group:

$$\text{Total potential energy} = 0\ \text{J} + \frac{q_3 k q_2}{r} + \frac{q_1 k q_2}{r} + \frac{q_1 k q_3}{r}$$

This result is exactly the same as that determined in Example 8, where a different order of adding the charges is used: first q_1, second q_2, and third q_3.

The total potential energy of the triangular group differs from that of the widely separated charges by an amount that is the sum of the potential energies calculated previously:

$$\text{Total potential energy} = 0\ \text{J} + \underbrace{\frac{q_2 k q_1}{r}}_{\text{EPE}_1} + \underbrace{\frac{q_3 k q_1}{r}}_{\text{EPE}_2} + \underbrace{\frac{q_3 k q_2}{r}}_{\text{EPE}_3}$$

$$= 0\ \text{J} + 0.54\ \text{J} - 0.40\ \text{J} = \boxed{+0.14\ \text{J}}$$

This energy originates in the work done to bring the charges together.

Check Your Understanding

(The answers are given at the end of the book.)

4. CYU Figure 19.2 shows four arrangements (A–D) of two point charges. In each arrangement consider the total electric potential that the charges produce at location P. Rank the arrangements (largest to smallest) according to the total potential. **(a)** B, C, A and D (a tie) **(b)** D, C, A, B **(c)** A and C (a tie), B, D **(d)** C, D, A, B

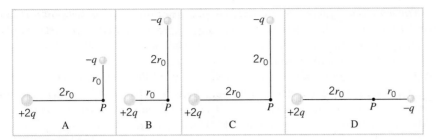

CYU FIGURE 19.2

5. A positive point charge and a negative point charge have equal magnitudes. One charge is fixed to one corner of a square, and the other is fixed to another corner. On which corners should the charges be placed, so that the same potential exists at the empty corners? The charges should be placed at **(a)** adjacent corners, **(b)** diagonally opposite corners.

6. Three point charges have identical magnitudes, but two of the charges are positive and one is negative. These charges are fixed to the corners of a square, one to a corner. No matter how the charges are arranged, the potential at the empty corner is always **(a)** zero, **(b)** negative, **(c)** positive.

7. Consider a spot that is located midway between two identical point charges. Which one of the following statements concerning the electric field and the electric potential at this spot is true? **(a)** The electric field is zero, but the electric potential is not zero. **(b)** The electric field is not zero, but the electric potential is zero. **(c)** Both the electric field and the electric potential are zero. **(d)** Neither the electric field nor the electric potential is zero.

8. Four point charges have the same magnitude (but they may have different signs) and are placed at the corners of a square, as **CYU Figure 19.3** shows. What must be the sign (+ or −) of each charge so that *both* the electric field *and* the electric potential are zero at the center of the square? Assume that the potential has a zero value at infinity.

	q_1	q_2	q_3	q_4
(a)	−	−	−	−
(b)	+	+	−	−
(c)	+	+	+	+
(d)	+	−	+	−

CYU FIGURE 19.3

9. An electric potential energy exists when two protons are separated by a certain distance. Does the electric potential energy increase, decrease, or remain the same **(a)** when both protons are replaced by electrons, and **(b)** when only one of the protons is replaced by an electron?

10. A proton is fixed in place. An electron is released from rest and allowed to collide with the proton. Then the roles of the proton and electron are reversed, and the same experiment is repeated. Which, if either, is traveling faster when the collision occurs, the proton or the electron?

Equipotential surfaces

FIGURE 19.11 The equipotential surfaces that surround the point charge $+q$ are spherical. The electric force does no work as a charge moves on a path that lies on an equipotential surface, such as the path *ABC*. However, work is done by the electric force when a charge moves between two equipotential surfaces, as along the path *AD*.

19.4 | Equipotential Surfaces and Their Relation to the Electric Field

An **equipotential surface** is a surface on which the electric potential is the same everywhere. The easiest equipotential surfaces to visualize are those that surround an isolated point charge. According to Equation 19.6, the potential at a distance r from a point charge q is $V = kq/r$. Thus, wherever r is the same, the potential is the same, and the equipotential surfaces are spherical surfaces centered on the charge. There are an infinite number of such surfaces, one for every value of r, and **Figure 19.11** illustrates two of them. The larger the distance r, the smaller is the potential of the equipotential surface.

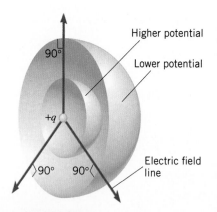

FIGURE 19.12 The radially directed electric field of a point charge is perpendicular to the spherical equipotential surfaces that surround the charge. The electric field points in the direction of *decreasing* potential.

The net electric force does no work as a charge moves on an equipotential surface. This important characteristic arises because when an electric force does work W_{AB} as a charge moves from A to B, the potential changes according to $V_B - V_A = -W_{AB}/q_0$ (Equation 19.4). Since the potential remains the same everywhere on an equipotential surface, $V_A = V_B$, and we see that $W_{AB} = 0$ J. In **Figure 19.11**, for instance, the electric force does no work as a test charge moves along the circular arc ABC, which lies on an equipotential surface. In contrast, the electric force does work when a charge moves *between* equipotential surfaces, as from A to D in the picture.

The spherical equipotential surfaces that surround an isolated point charge illustrate another characteristic of all equipotential surfaces. **Figure 19.12** shows two of the surfaces around a positive point charge, along with some electric field lines. The electric field lines give the direction of the electric field, and for a positive point charge the electric field is directed radially outward. Therefore, at each location on an equipotential sphere the electric field is perpendicular to the surface and points outward in the direction of decreasing potential, as the drawing emphasizes. This perpendicular relation is valid whether or not the equipotential surfaces result from a positive charge or have a spherical shape.

> **Problem-Solving Insight** The electric field created by any charge or group of charges is everywhere perpendicular to the associated equipotential surfaces and points in the direction of decreasing potential.

For example, **Animated Figure 19.13** shows the electric field lines (in red) around an electric dipole, along with some equipotential surfaces (in blue), shown in cross section. Since the field lines are not simply radial, the equipotential surfaces are no longer spherical but, instead, have the shape necessary to be everywhere perpendicular to the field lines.

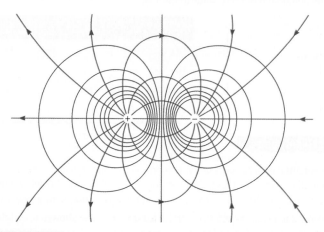

ANIMATED FIGURE 19.13 A cross-sectional view of the equipotential surfaces (in blue) of an electric dipole. The surfaces are drawn to show that at every point they are perpendicular to the electric field lines (in red) of the dipole.

FIGURE 19.14 In this hypothetical situation, the electric field $\vec{E}$ is not perpendicular to the equipotential surface. As a result, there is a component of $\vec{E}$ parallel to the surface.

To see why an equipotential surface must be perpendicular to the electric field, consider **Figure 19.14**, which shows a hypothetical situation in which the perpendicular relation does *not* hold. If $\vec{E}$ were not perpendicular to the equipotential surface, there would be a component of $\vec{E}$ parallel to the surface. This field component would exert an electric force on a test charge placed on the surface. As the charge moved along the surface, work would be done by this component of the electric force. The work, according to Equation 19.4, would cause the potential to change, and, thus, the surface could not be an equipotential surface as assumed. The only way out of the dilemma is for the electric field to be perpendicular to the surface, so there is no component of the field parallel to the surface.

We have already encountered one equipotential surface. In Section 18.8, we found that the direction of the electric field just outside an electrical conductor is perpendicular to the conductor's surface, when the conductor is at equilibrium under electrostatic conditions. Thus, the surface of any conductor is an equipotential surface under such conditions. In fact, since the electric field is zero everywhere inside a conductor whose charges are in equilibrium, the entire conductor can be regarded as an equipotential volume.

There is a quantitative relation between the electric field and the equipotential surfaces. One example that illustrates this relation is the parallel plate capacitor in **Figure 19.15**. As Section 18.6 discusses, the electric field $\vec{E}$ between the metal plates is perpendicular to them and is the same everywhere, ignoring fringe fields at the edges. To be perpendicular to the electric field, the equipotential surfaces must be planes that are parallel to the capacitor plates, which themselves are equipotential surfaces. The potential difference between the plates is given by Equation 19.4 as $\Delta V = V_B - V_A = -W_{AB}/q_0$, where A is a point on the positive plate and B is a point on the negative plate. The work done by the electric force as a positive test charge q_0 moves from A to B is $W_{AB} = F\,\Delta s$, where F refers to the electric force and Δs to the displacement along a line perpendicular to the plates. The force equals the product of the charge and the electric field E ($F = q_0 E$), so the work becomes $W_{AB} = F\,\Delta s = q_0 E\,\Delta s$. Therefore, the potential difference between the capacitor plates can be written in terms of the electric field as $\Delta V = -W_{AB}/q_0 = -q_0 E\,\Delta s/q_0$, or

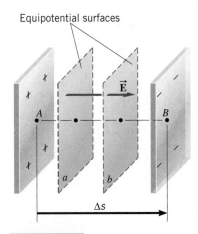

Equipotential surfaces

FIGURE 19.15 The metal plates of a parallel plate capacitor are equipotential surfaces. Two additional equipotential surfaces are shown between the plates. These two equipotential surfaces are parallel to the plates and are perpendicular to the electric field $\vec{E}$ between the plates.

$$E = -\frac{\Delta V}{\Delta s} \qquad\qquad (19.7a)$$

The quantity $\Delta V/\Delta s$ is referred to as the **potential gradient** and has units of volts per meter. In general, the relation $E = -\Delta V/\Delta s$ gives only the component of the electric field along the displacement Δs; it does not give the perpendicular component. The next example deals further with the equipotential surfaces between the plates of a capacitor.

EXAMPLE 9 | The Electric Field and Potential Are Related

The plates of the capacitor in **Figure 19.15** are separated by a distance $\Delta s = (\Delta s)_{\text{plates}} = 0.032$ m, and the potential difference between them is $(\Delta V)_{\text{plates}} = V_B - V_A = -64$ V. Between the two blue equipotential surfaces there is a potential difference of $(\Delta V)_{\text{blue}} = V_b - V_a = -3.0$ V. Find the spacing between the two blue surfaces.

Reasoning To find the spacing between the blue surfaces, we will solve Equation 19.7a for $(\Delta s)_{\text{blue}}$ with $(\Delta V)_{\text{blue}} = -3.0$ V and E equal to the electric field between the plates of the capacitor. The electric field E is the same everywhere between the plates (ignoring the fringe fields). A value for E can be obtained by using Equation 19.7a with the data given for the distance between the plates and the potential difference between them [$(\Delta s)_{\text{plates}} = 0.032$ m and $(\Delta V)_{\text{plates}} = -64$ V].

Solution Solving Equation 19.7a for the spacing between the blue equipotential surfaces gives

$$(\Delta s)_{\text{blue}} = -\frac{(\Delta V)_{\text{blue}}}{E}$$

Using Equation 19.7a to determine the electric field between the plates, we find that

$$E = -\frac{(\Delta V)_{\text{plates}}}{(\Delta s)_{\text{plates}}} = -\frac{-64\text{ V}}{0.032\text{ m}} = 2.0 \times 10^3\text{ V/m}$$

Substituting this result into the equation for $(\Delta s)_{\text{blue}}$ gives

$$(\Delta s)_{\text{blue}} = -\frac{(\Delta V)_{\text{blue}}}{E} = -\frac{-3.0\text{ V}}{2.0 \times 10^3\text{ V/m}} = \boxed{1.5 \times 10^{-3}\text{ m}}$$

Equation 19.7a gives the relationship between the electric field and the electric potential. It gives the component of the electric field along the displacement Δs in a region of space where the electric potential changes from place to place and applies to a wide variety of situations. When applied strictly to a parallel plate capacitor, however, this expression is often used in a slightly different form. In **Figure 19.15**, the metal plates of the capacitor are marked A (higher potential) and B (lower potential). Traditionally, in discussions of such a capacitor, the potential difference between the plates is referred to by using the symbol V to denote the amount by which the higher potential exceeds the lower potential ($V = V_A - V_B$). In this tradition, the symbol V is often referred to as simply the "voltage." For example, if the potential difference between the plates of a capacitor is 5 volts, it is common to say that the "voltage" of the capacitor is 5 volts. In addition, the displacement from plate A to plate B is expressed in terms of the separation d between the plates ($d = s_B - s_A$). With this nomenclature, Equation 19.7a becomes

$$E = -\frac{\Delta V}{\Delta s} = -\frac{V_B - V_A}{s_B - s_A} = \frac{V_A - V_B}{s_B - s_A} = \frac{V}{d} \quad \text{(parallel plate capacitor)} \qquad \textbf{(19.7b)}$$

Check Your Understanding

(The answers are given at the end of the book.)

11. CYU Figure 19.4 shows a cross-sectional view of two spherical equipotential surfaces and two electric field lines that are perpendicular to these surfaces. When an electron moves from point A to point B (against the electric field), the electric force does $+3.2 \times 10^{-19}$ J of work. What are the electric potential differences **(a)** $V_B - V_A$, **(b)** $V_C - V_B$, and **(c)** $V_C - V_A$?

12. The electric potential is constant throughout a given region of space. Is the electric field zero or nonzero in this region?

13. In a region of space where the electric field is constant everywhere, as it is inside a parallel plate capacitor, is the potential constant everywhere? **(a)** Yes. **(b)** No, the potential is greatest at the positive plate. **(c)** No, the potential is greatest at the negative plate.

14. A positive test charge is placed in an electric field. In what direction should the charge be moved relative to the field, so that the charge experiences a constant electric potential? The charge should be moved **(a)** perpendicular to the electric field, **(b)** in the same direction as the electric field, **(c)** opposite to the direction of the electric field.

15. The location marked P in **CYU Figure 19.5** lies midway between the point charges $+q$ and $-q$. The blue lines labeled A, B, and C are edge-on views of three planes. Which of the planes is an equipotential surface? **(a)** A and C **(b)** A, B, and C **(c)** Only B **(d)** None of the planes is an equipotential surface.

Electric field lines

Equipotential surfaces (Cross-sectional view)

CYU FIGURE 19.4

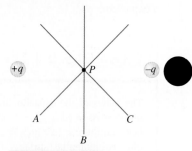

CYU FIGURE 19.5

16. Imagine that you are moving a positive test charge along the line between two identical point charges. With regard to the electric potential, is the midpoint on the line analogous to the top of a mountain or the bottom of a valley when the two point charges are **(a)** positive and **(b)** negative?

19.5 Capacitors and Dielectrics

The Capacitance of a Capacitor

In Section 18.6 we saw that a parallel plate capacitor consists of two parallel metal plates placed near one another but not touching. This type of capacitor is only one among many. In general, a **capacitor** consists of two conductors of any shape placed near one another without touching. For a reason that will become clear later on, it is common practice to fill the region between the conductors or plates with an electrically insulating material called a **dielectric**, as **Figure 19.16** illustrates.

A capacitor stores electric charge. Each capacitor plate carries a charge of the *same magnitude,* one positive and the other negative. Because of the charges, the electric potential of the positive plate exceeds that of the negative plate by an amount *V*, as **Figure 19.16** indicates. Experiment shows that when the magnitude *q* of the charge on each plate is doubled, the magnitude *V* of the electric potential difference is also doubled, so *q* is proportional to *V*: $q \propto V$. Equation 19.8 expresses this proportionality with the aid of a proportionality constant *C*, which is the **capacitance** of the capacitor.

THE RELATION BETWEEN CHARGE AND POTENTIAL DIFFERENCE FOR A CAPACITOR

The magnitude *q* of the charge on each plate of a capacitor is directly proportional to the magnitude *V* of the potential difference between the plates:

$$q = CV \qquad (19.8)$$

where *C* is the capacitance.

SI Unit of Capacitance: coulomb/volt = farad (F)

Equation 19.8 shows that the SI unit of capacitance is the coulomb per volt (C/V). This unit is called the *farad* (F), named after the English scientist Michael Faraday (1791–1867). One farad is an enormous capacitance. Usually smaller amounts, such as a microfarad (1 μF = 10^{-6} F) or a picofarad (1 pF = 10^{-12} F), are used in electric circuits. The capacitance reflects the ability of the capacitor to store charge, in the sense that a larger capacitance *C* allows more charge *q* to be put onto the plates for a given value of the potential difference *V*.

THE PHYSICS OF . . . random-access memory (RAM) chips. The ability of a capacitor to store charge lies at the heart of the random-access memory (RAM) chips used in computers, where information is stored in the form of the "ones" and "zeros" that comprise binary numbers. **Figure 19.17** illustrates the role of a capacitor in a RAM chip. The capacitor is connected to a transistor switch, to which two lines are connected, an address line and a data line. A single RAM chip often contains millions of such transistor–capacitor units. The address line is used by the computer to locate a particular transistor–capacitor combination, and the data line carries the data to be stored. A pulse on the address line turns on the transistor switch. With the switch turned on, a pulse coming in on the data line can cause the capacitor to charge. A charged capacitor means that a "one" has been stored, whereas an uncharged capacitor means that a "zero" has been stored.

The Dielectric Constant

If a dielectric is inserted between the plates of a capacitor, the capacitance can increase markedly because of the way in which the dielectric alters the electric field between the plates. **Interactive Figure 19.18** shows how this effect comes about. In part *a*, the region between the charged plates is empty. The field lines point from the positive toward the negative plate. In part *b*, a dielectric has been inserted between the plates. Since the capacitor is not connected to anything, the charge on the plates remains constant as the dielectric is inserted. In many materials (e.g., water) the molecules possess permanent dipole moments, even though the molecules are electrically neutral. The dipole moment exists because one end of a molecule has a slight excess of negative charge while the other end has a slight excess of positive charge. When such molecules are placed between the charged plates of the capacitor, the negative ends are attracted to the positive plate and the positive ends are attracted to the negative plate. As a result, the dipolar molecules tend to orient themselves end to end, as in part *b*. Whether or not a molecule has a permanent dipole moment, the

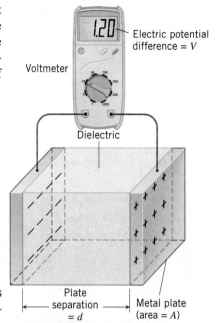

FIGURE 19.16 A parallel plate capacitor consists of two metal plates, one carrying a charge +*q* and the other a charge −*q*. The potential of the positive plate exceeds that of the negative plate by an amount *V*. The region between the plates is filled with a dielectric.

FIGURE 19.17 A transistor–capacitor combination is part of a RAM chip used in computer memories.

(a)

Positive surface charge on dielectric   Negative surface charge on dielectric

Molecule
(b)

(c)

INTERACTIVE FIGURE 19.18

(a) The electric field lines inside an empty capacitor. (b) The electric field produced by the charges on the plates aligns the molecular dipoles within the dielectric end to end. The space between the dielectric and the plates is added for clarity. In reality, the dielectric fills the region between the plates. (c) The surface charges on the dielectric reduce the electric field inside the dielectric.

TABLE 19.1	Dielectric Constants of Some Common Substances[a]	
Substance	**Dielectric Constant, κ**
Vacuum	1
Air	1.000 54
Teflon	2.1
Benzene	2.28
Paper (royal gray)	3.3
Ruby mica	5.4
Neoprene rubber	6.7
Methyl alcohol	33.6
Water	80.4

[a]Near room temperature.

electric field can cause the electrons to shift position within a molecule, making one end slightly negative and the opposite end slightly positive. Because of the end-to-end orientation, the left surface of the dielectric becomes positively charged, and the right surface becomes negatively charged. The surface charges are shown in red in the picture.

Because of the surface charges on the dielectric, not all the electric field lines generated by the charges on the plates pass through the dielectric. As **Interactive Figure 19.18c** shows, some of the field lines end on the negative surface charges and begin again on the positive surface charges. Thus, the electric field inside the dielectric is less strong than the electric field inside the empty capacitor, assuming the charge on the plates remains constant. This reduction in the electric field is described by the **dielectric constant κ**, which is the ratio of the field magnitude E_0 without the dielectric to the field magnitude E inside the dielectric:

$$\kappa = \frac{E_0}{E} \qquad (19.9)$$

Being a ratio of two field strengths, the dielectric constant is a number without units. Moreover, since the field $\vec{E}_0$ without the dielectric is greater than the field $\vec{E}$ inside the dielectric, the dielectric constant is greater than unity. The value of κ depends on the nature of the dielectric material, as **Table 19.1** indicates.

The Capacitance of a Parallel Plate Capacitor

The capacitance of a capacitor is affected by the geometry of the plates and the dielectric constant of the material between them. For example, **Figure 19.16** shows a parallel plate capacitor in which the area of each plate is A and the separation between the plates is d. The magnitude of the electric field inside the dielectric is given by Equation 19.7b as $E = V/d$, where V is the magnitude of the potential difference between the plates. If the charge on each plate is kept fixed, the electric field inside the dielectric is related to the electric field in the absence of the dielectric via Equation 19.9. Therefore,

$$E = \frac{E_0}{\kappa} = \frac{V}{d}$$

Since the electric field within an empty capacitor is $E_0 = q/(\varepsilon_0 A)$ (see Equation 18.4), it follows that $q/(\kappa \varepsilon_0 A) = V/d$, which can be solved for q to give

$$q = \left(\frac{\kappa \varepsilon_0 A}{d}\right) V$$

A comparison of this expression with $q = CV$ (Equation 19.8) reveals that the capacitance C is

Parallel plate capacitor filled with a dielectric
$$C = \frac{\kappa \varepsilon_0 A}{d} \qquad (19.10)$$

Only the geometry of the plates (A and d) and the dielectric constant κ affect the capacitance. With C_0 representing the capacitance of the empty capacitor ($\kappa = 1$), Equation 19.10 shows that $C = \kappa C_0$. In other words, the capacitance with the dielectric present is increased by a factor of κ over the capacitance without the dielectric. It can be shown that the relation $C = \kappa C_0$ applies to any capacitor, not just to a parallel plate capacitor. One reason, then, that capacitors are filled with dielectric materials is to increase the capacitance.

CONCEPTUAL EXAMPLE 10 | The Effect of a Dielectric When a Capacitor Has a Constant Charge

An empty capacitor is connected to a battery and charged up. The capacitor is then disconnected from the battery, and a slab of dielectric material is inserted between the plates. Does the potential difference across the plates (a) increase, (b) remain the same, or (c) decrease?

Reasoning Our reasoning is guided by the following fact: Once the capacitor is disconnected from the battery, the charge on its plates remains constant, for there is no longer any way for charge to be added or removed. According to Equation 19.8, the magnitude q of the charge stored by the capacitor is $q = CV$, where C is its capacitance and V is the magnitude of the potential difference between the plates.

Answers (a) and (b) are incorrect. Placing a dielectric between the plates of a capacitor reduces the electric field in that region (see

Interactive Figure 19.18c). The magnitude V of the potential difference is related to the magnitude E of the electric field by $V = Ed$, where d is the distance between the capacitor plates. Since E decreases when the dielectric is inserted and d is unchanged, the potential difference does not increase or remain the same.

Answer (c) is correct. Inserting the dielectric causes the capacitance C to increase. Since $q = CV$ and q is fixed, the potential difference V across the plates must decrease in order for q to remain unchanged. The amount by which the potential difference decreases from the value initially established by the battery depends on the dielectric constant of the slab.

Related Homework: Problem 44

Capacitors are used often in electronic devices, and Example 11 deals with one familiar application.

EXAMPLE 11 | The Physics of a Computer Keyboard

One kind of computer keyboard is based on the idea of capacitance. Each key is mounted on one end of a plunger, and the other end is attached to a movable metal plate (see **Figure 19.19**). The movable plate is separated from a fixed plate, the two plates forming a capacitor. When the key is pressed, the movable plate is pushed closer to the fixed plate, and the capacitance increases. Electronic circuitry enables the computer to detect the *change* in capacitance, thereby recognizing which key has been pressed. The separation of the plates is normally 5.00×10^{-3} m but decreases to 0.150×10^{-3} m when a key is pressed. The plate area is 9.50×10^{-5} m^2, and the capacitor is filled with a material whose dielectric constant is 3.50. Determine the change in capacitance that is detected by the computer.

Reasoning We can use Equation 19.10 directly to find the capacitance of the key, since the dielectric constant κ, the plate area A, and the plate separation d are known. We will use this relation twice, once to find the capacitance when the key is pressed and once when it is not pressed. The change in capacitance will be the difference between these two values.

Solution When the key is pressed, the capacitance is

$$C = \frac{\kappa \varepsilon_0 A}{d} = \frac{(3.50)[8.85 \times 10^{-12}\ \text{C}^2/(\text{N} \cdot \text{m}^2)](9.50 \times 10^{-5}\ \text{m}^2)}{0.150 \times 10^{-3}\ \text{m}}$$

$$= 19.6 \times 10^{-12}\ \text{F} \quad (19.6\ \text{pF}) \qquad \qquad \textbf{(19.10)}$$

A calculation similar to the one above reveals that when the key is *not* pressed, the capacitance has a value of 0.589×10^{-12} F (0.589 pF). The *change* in capacitance is an increase of

FIGURE 19.19 In one kind of computer keyboard, each key, when pressed, changes the separation between the plates of a capacitor.

$\boxed{19.0 \times 10^{-12}\ \text{F}\,(19.0\ \text{pF})}$. The *change* in the capacitance is greater with the dielectric present, which makes it easier for the circuitry within the computer to detect it.

Energy Storage in a Capacitor

When a capacitor stores charge, it also stores energy. In charging up a capacitor, for example, a battery does work in transferring an increment of charge from one plate of the capacitor to the other plate. The work done is equal to the product of the charge increment and the potential difference between the plates. However, as each increment of charge is moved, the potential difference increases slightly, and a larger amount of work is needed to move the next increment. The total work W done in completely charging the capacitor is the product of the total charge q transferred and the average potential difference $\overline{V}$; $W = q\overline{V}$. Since the average potential difference is one-half the final potential V, or $\overline{V} = \frac{1}{2}V$, the total work done by the battery is $W = \frac{1}{2}qV$. This work does not disappear but is stored as electric potential energy in the capacitor, so that Energy $= \frac{1}{2}qV$. Equation 19.8 indicates that $q = CV$ or, equivalently, that $V = q/C$. We can see, then, that our expression for the energy can be cast into two additional equivalent forms by substituting for q or for V. Equations 19.11a–c summarize these results:

$$\text{Energy} = \tfrac{1}{2}qV \tag{19.11a}$$

$$\text{Energy} = \tfrac{1}{2}(CV)V = \tfrac{1}{2}CV^2 \tag{19.11b}$$

$$\text{Energy} = \tfrac{1}{2}q\left(\frac{q}{C}\right) = \frac{q^2}{2C} \tag{19.11c}$$

It is also possible to regard the energy as being stored in the electric field between the plates. The relation between energy and field strength can be obtained for a parallel plate capacitor by substituting $V = Ed$ (Equation 19.7b) and $C = \kappa\varepsilon_0 A/d$ (Equation 19.10) into Equation 19.11b:

$$\text{Energy} = \tfrac{1}{2}CV^2 = \tfrac{1}{2}\left(\frac{\kappa\varepsilon_0 A}{d}\right)(Ed)^2$$

Since the area A times the separation d is the volume between the plates, the energy per unit volume or **energy density** is

$$\text{Energy density} = \frac{\text{Energy}}{\text{Volume}} = \tfrac{1}{2}\kappa\varepsilon_0 E^2 \tag{19.12}$$

It can be shown that this expression is valid for any electric field strength, not just that between the plates of a capacitor.

THE PHYSICS OF . . . an electronic flash attachment for a camera. The energy-storing capability of a capacitor is often put to good use in electronic circuits. For example, in an electronic flash attachment for a camera, energy from the battery pack is stored in a capacitor. The capacitor is then discharged between the electrodes of the flash tube, which converts the energy into light. Flash duration times range from 1/200 to 1/1 000 000 second or less, with the shortest flashes being used in high-speed photography (see **Figure 19.20**). Some flash attachments automatically control the flash duration by monitoring the light reflected from the photographic subject and quickly stopping or quenching the capacitor discharge when the reflected light reaches a predetermined level.

BIO **THE PHYSICS OF . . . a defibrillator.** During a heart attack, the heart produces a rapid, unregulated pattern of beats, a condition known as cardiac fibrillation. Cardiac fibrillation can often be stopped by sending a very fast discharge of electrical energy through the heart. For this purpose, emergency medical personnel use defibrillators, such as the one being used in **Figure 19.21**. A paddle is connected to each plate

FIGURE 19.20 This time-lapse photo of a figure skater was obtained using a camera with an electronic flash attachment. The energy for each flash of light comes from the electrical energy stored in a capacitor.

FIGURE 19.21 A paramedic is using a portable defibrillator in an attempt to revive a heart attack victim. A defibrillator uses the electrical energy stored in a capacitor to deliver a controlled electric current that can restore normal heart rhythm.

Kyodo News/Getty Images

Adam Davis/Science Source

of a large capacitor, and the paddles are placed on the chest near the heart. The capacitor is charged to a potential difference of about a thousand volts. The capacitor is then discharged in a few thousandths of a second; the discharge current passes through a paddle, the heart, and the other paddle. Within a few seconds, the heart often returns to its normal beating pattern.

Check Your Understanding

(The answers are given at the end of the book.)

17. An empty parallel plate capacitor is connected to a battery that maintains a constant potential difference between the plates. With the battery connected, a dielectric is then inserted between the plates. Do the following quantities decrease, remain the same, or increase when the dielectric is inserted? **(a)** The electric field between the plates **(b)** The capacitance **(c)** The charge on the plates **(d)** The energy stored by the capacitor

18. A parallel plate capacitor is charged up by a battery. The battery is then disconnected, but the charge remains on the plates. The plates are then pulled apart. Do the following quantities decrease, remain the same, or increase as the distance between the plates increases? **(a)** The capacitance of the capacitor **(b)** The potential difference between the plates **(c)** The electric field between the plates **(d)** The electric potential energy stored by the capacitor

19.6 | *Biomedical Applications of Electric Potential Differences

Conduction of Electrical Signals in Neurons

The human nervous system is remarkable for its ability to transmit information in the form of electrical signals. These signals are carried by the nerves, and the concept of electric potential difference plays an important role in the process. For example, sensory information from our eyes and ears is carried to the brain by the optic nerves and auditory nerves, respectively. Other nerves transmit signals from the brain or spinal column to muscles, causing them to contract. Still other nerves carry signals within the brain.

A nerve consists of a bundle of *axons,* and each axon is one part of a nerve cell, or *neuron.* As **Interactive Figure 19.22** illustrates, a typical neuron consists of a cell body with numerous extensions, called *dendrites,* and a single axon. The dendrites convert stimuli, such as pressure or heat, into electrical signals that travel through the neuron. The axon sends the signal to the nerve endings, which transmit the signal across a gap (called a *synapse*) to the next neuron or to a muscle.

The fluid inside a cell, the intracellular fluid, is quite different from that outside the cell, the extracellular fluid. Both fluids contain concentrations of positive and negative ions. However, the extracellular fluid is rich in sodium (Na^+) and chlorine (Cl^-) ions, whereas the intracellular fluid is rich in potassium (K^+) ions and negatively charged proteins. These concentration differences between the fluids are extremely important to the life of the cell. If the cell membrane were freely permeable, the ions would diffuse across it until the concentrations on both sides were equal. (See Section 14.4 for a review of diffusion.) This does not happen, because a living cell has a selectively permeable membrane. Ions can enter or leave the cell only through membrane channels, and the permeability of the channels varies markedly from one ion to another. For example, it is much easier for K^+ ions to diffuse out of the cell than it is for Na^+ to enter the cell. As a result of selective membrane permeability, there is a small buildup of negative charges just on the

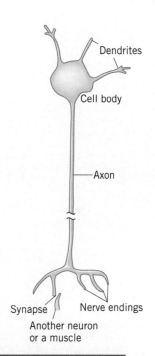

INTERACTIVE FIGURE 19.22 The anatomy of a typical neuron.

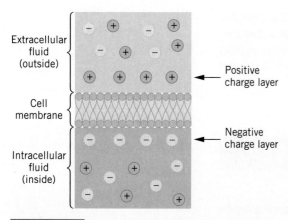

FIGURE 19.23 Positive and negative charge layers form on the outside and inside surfaces of a membrane during its resting state.

inner side of the membrane and an equal amount of positive charges on the outer side (see **Figure 19.23**). The buildup of charge occurs very close to the membrane, so the membrane acts like a capacitor (see Problems 56 and 57). Elsewhere in the intracellular and extracellular fluids, there are equal numbers of positive and negative ions, so the fluids are overall electrically neutral. Such a separation of positive and negative charges gives rise to an electric potential difference across the membrane, called the *resting membrane potential*. In neurons, the resting membrane potential ranges from −40 to −90 mV, with a typical value of −70 mV. The minus sign indicates that the inner side of the membrane is negative relative to the outer side.

BIO THE PHYSICS OF . . . an action potential.   A "resting" neuron is one that is not conducting an electrical signal. The *change* in the resting membrane potential is the key factor in the initiation and conduction of a signal. When a sufficiently strong stimulus is applied to a given point on the neuron, "gates" in the membrane open and sodium ions flood into the cell, as **Figure 19.24** illustrates. The sodium ions are driven into the cell by attraction to the negative ions on the inner side of the membrane as well as by the relatively high concentration of sodium ions outside the cell. The large influx of Na^+ ions first neutralizes the negative ions on the interior of the membrane and then causes it to become positively charged. As a result, the membrane potential in this localized region goes from −70 mV, the resting potential, to about +30 mV in a very short time (see **Figure 19.25**). The sodium gates then close, and the cell membrane quickly returns to its normal resting potential. This change in potential, from −70 mV to +30 mV and back to −70 mV, is known as the *action potential*. The action potential lasts for a few milliseconds, and it is the electrical signal that propagates down the axon, typically at a speed of about 50 m/s, to the next neuron or to a muscle cell.

FIGURE 19.24 When a stimulus is applied to the cell, positive sodium ions (Na^+) rush into the cell, causing the interior surface of the membrane to become momentarily positive.

FIGURE 19.25 The action potential is caused by the rush of positive sodium ions into the cell and a subsequent return of the cell to its resting potential.

Medical Diagnostic Techniques

Several important medical diagnostic techniques depend on the fact that the surface of the human body is *not* an equipotential surface. Between various points on the body there are small potential differences (approximately 30–500 μV), which provide the basis for electrocardiography, electroencephalography, and electroretinography. The potential differences can be traced to the electrical characteristics of muscle cells and nerve cells. In carrying out their biological functions, these cells utilize positively charged sodium and potassium ions and negatively charged chlorine ions that exist within the cells and in the extracellular fluid. As a result of such charged particles, electric fields are generated that extend to the surface of the body and lead to the small potential differences.

FIGURE 19.26 The potential differences generated by heart muscle activity provide the basis for electrocardiography. The normal and abnormal EKG patterns correspond to one heartbeat.

BIO   THE PHYSICS OF . . . electrocardiography.   Figure 19.26 shows electrodes placed on the body to measure potential differences in electrocardiography. The potential difference between two locations changes as the heart beats and forms a repetitive pattern. The recorded pattern of potential difference versus time is called an electrocardiogram (ECG or EKG), and its shape depends on which pair of points in the picture (*A* and *B*, *B* and *C*, etc.) is used to locate the electrodes. The figure also shows some EKGs and indicates the regions (*P*, *Q*, *R*, *S*, and *T*) associated with specific parts of the heart's beating cycle. The differences between the EKGs of normal and abnormal hearts provide physicians with a valuable diagnostic tool.

BIO   THE PHYSICS OF . . . electroencephalography.   In electroencephalography the electrodes are placed at specific locations on the head, as **Figure 19.27** indicates, and they record the potential differences that characterize brain behavior. The graph of potential difference versus time is known as an electroencephalogram (EEG). The various parts of the patterns in an EEG are often referred to as "waves" or "rhythms." The drawing shows an example of the main resting rhythm of the brain, the so-called alpha rhythm, and also illustrates the distinct differences that are found between the EEGs generated by healthy (normal) and diseased (abnormal) tissue.

FIGURE 19.27   In electroencephalography the potential differences created by the electrical activity of the brain are used for diagnosing abnormal behavior.

BIO   THE PHYSICS OF . . . electroretinography.   The electrical characteristics of the retina of the eye lead to the potential differences measured in electroretinography. **Figure 19.28** shows a typical electrode placement used to record the pattern of potential difference versus time that occurs when the eye is stimulated by a flash of light. One electrode is mounted on a contact lens, while the other is often placed on the forehead. The recorded pattern is called an electroretinogram (ERG), and parts of the pattern are referred to as the "*A* wave" and the "*B* wave." As the graphs show, the ERGs of normal and diseased (abnormal) eyes can differ markedly.

FIGURE 19.28   The electrical activity of the retina of the eye generates the potential differences used in electroretinography.

EXAMPLE 12 | BIO Energy Stored in a Defibrillator

A cardiac defibrillator, such as the one shown in **Figure 19.21**, stores energy in a large capacitor connected to its paddles. This energy is delivered to the heart by discharging the capacitor through the chest wall of the patient. If the capacitor connected to the paddles has a value of 22.5 μF, and it is charged to a potential of 1250 V, what is **(a)** the energy delivered by the capacitor, and **(b)** the charge on its plates?

Reasoning **(a)** To find the energy stored in the capacitor, we can simply apply Equation 19.11b. **(b)** The definition of

capacitance (Equation 19.8) can be used to find the charge on each plate.

Solution **(a)** Applying Equation 19.11b with the data given in the problem, we have

$$\text{Energy} = \tfrac{1}{2}CV^2 = \tfrac{1}{2}(22.5 \times 10^{-6}\ \text{F})(1250\ \text{V})^2 = \boxed{17.6\ \text{J}}$$

(b) The charge on the capacitor is given by

$$Q = CV = (22.5 \times 10^{-6}\ \text{F})(1250\ \text{V}) = \boxed{0.0280\ \text{C}}$$

19.7 | *The Millikan Oil-drop Experiment

In Section 18.1 we considered objects with a charge greater in magnitude than the charge on a single electron. Charge builds up on an object by adding or removing electrons, and therefore the total charge is quantized, that is, it consists of an integer multiple of the fundamental electron charge, or -1.602×10^{-19} C. However, we never discussed how this value of the fundamental charge had been determined. An American physicist, Robert A. Millikan (1868–1953), is credited with the first accurate experimental determination of the charge on a single electron.

In 1909, Millikan and his graduate student, Harvey Fletcher, analyzed a series of oil-drop experiments to measure the electron's charge. **Figure 19.29a** shows a schematic of the equipment they used for their measurements. The apparatus consists of a cylindrical container in which an atomizer produces a fine mist of oil droplets, some of which fall between the plates of a parallel plate capacitor. A source of X-rays, which is a high energy form of invisible light that we will study in Chapters 24 and 30, ionizes the air molecules in the space between the plates, which causes the molecules to lose electrons. As the oil droplets fall through this space, some of them capture one or more of these electrons on their surfaces, causing them to become negatively charged. By applying a voltage across the plates as shown, a uniform electric field is created that points downward. This produces an upward force on the negatively charged oil droplets.

Figure 19.29b shows the free-body diagram for one of the oil droplets. The force due to the electric field is upward, and the droplet's weight acts downward. The voltage (V) between the plates, and therefore the electric field, can be varied, such that the upward electrostatic force can be made to balance the weight. At this point, the droplet will appear motionless in the eyepiece. Setting the electrostatic force equal to the weight, we have $F_E = W$. Since $F_E = qE$ (Equation 18.2), and writing the weight as mg, we have $qE = mg$. Solving for the charge on the droplet, we obtain $q = mg/E$. However, in order to calculate the charge, we still need the mass of the droplet.

The mass is related to the volume of the droplet and the density of the oil through Equation 11.1: $m = \rho V$. While the droplet is falling, or held motionless, surface tension makes the shape of the droplet spherical, therefore the volume V of the droplet is that of a sphere ($V_s = \tfrac{4}{3}\pi r^3$) and $m = \rho\left(\tfrac{4}{3}\pi r^3\right)$. The equation for the charge depends on E, which can be written in terms of the voltage across the plates and

FIGURE 19.29 (a) The experimental apparatus used by Robert Millikan to determine the fundamental charge on the electron. Atomized oil droplets fall through the ionized air between the plates of a cylindrical, parallel plate capacitor, collecting electrons as they fall. The downward electric field between the plates can be adjusted so that a particular oil droplet viewed through the microscope can be made stationary in the field of view. At this point, the upward electrostatic force on the droplet balances its weight, as shown by the free-body diagram for the droplet in (b).

the distance (d) between them: $E = \dfrac{V}{d}$. Substituting these results for m and E into the equation for the charge gives the following*:

$$q = \frac{mg}{E} = \frac{\rho\left(\frac{4}{3}\pi r^3\right)gd}{V} = \frac{4\pi\rho g d r^3}{3V} \qquad (19.13)$$

All of the quantities in Equation 19.13 are known or easily measured, except for the radius of the drop. To determine the radius, a second measurement is made. Here, the voltage across the plates is set to zero, so that the electric field, and thus F_E, vanishes. The droplet then falls due to its weight, but quickly reaches terminal velocity due to the upward drag force from the air in the chamber (review Problem 91 in Chapter 4.). The terminal velocity of the droplet can be measured, and this depends on the cross-sectional area of the drop, and thus its radius. Once the radius is determined, the charge on the droplet can be calculated using Equation 19.13.

After measuring many thousands of drops, Millikan and Fletcher determined, to within their experimental uncertainty, that the charge on any drop consisted of a small integer multiple of the same factor: $e = -1.5924 \times 10^{-19}$ C. For example, each drop had a charge of $-e$ (1 electron), $-3e$ (3 electrons), or $-8e$ (8 electrons), etc., but a droplet never had a non-integer multiple of a charge, such as $-1.5e$. Millikan's experiment showed that electric charge is indeed quantized, and his careful measurement produced a value within 1% of the present day accepted value of -1.602×10^{-19} C. Millikan was awarded the Nobel Prize in Physics in 1923 for determining the fundamental charge on the electron, as well as for his work on the photoelectric effect (see Section 29.3).

Concept Summary

19.1 Potential Energy When a positive test charge $+q_0$ moves from point A to point B in the presence of an electric field, work W_{AB} is done by the electric force. The work equals the electric potential energy (EPE) at A minus that at B, as given by Equation 19.1. The electric force is a conservative force, so the path along which the test charge moves from A to B is of no consequence, for the work W_{AB} is the same for all paths.

$$W_{AB} = \text{EPE}_A - \text{EPE}_B \qquad (19.1)$$

19.2 The Electric Potential Difference The electric potential V at a given point is the electric potential energy of a small test charge q_0 situated at that point divided by the charge itself, as shown in Equation 19.3. The SI unit of electric potential is the joule per coulomb (J/C), or volt (V). The electric potential difference between two points A and B is given by Equation 19.4. The potential difference between two points (or between two equipotential surfaces) is often called the "voltage."

$$V = \frac{\text{EPE}}{q_0} \qquad (19.3)$$

$$V_B - V_A = \frac{\text{EPE}_B}{q_0} - \frac{\text{EPE}_A}{q_0} = \frac{-W_{AB}}{q_0} \qquad (19.4)$$

A positive charge accelerates from a region of higher potential toward a region of lower potential. Conversely, a negative charge accelerates from a region of lower potential toward a region of higher potential.

An electron volt (eV) is a unit of energy. The relationship between electron volts and joules is 1 eV = 1.60×10^{-19} J.

The total energy E of a system is the sum of its translational $\left(\frac{1}{2}mv^2\right)$ and rotational $\left(\frac{1}{2}I\omega^2\right)$ kinetic energies, gravitational potential energy (mgh), elastic potential energy $\left(\frac{1}{2}kx^2\right)$, and electric potential energy (EPE), as indicated by Equation 1. If external nonconservative forces like friction do no net work, the total energy of the system is conserved. That is, the final total energy E_f is equal to the initial total energy E_0; $E_f = E_0$.

$$E = \tfrac{1}{2}mv^2 + \tfrac{1}{2}I\omega^2 + mgh + \tfrac{1}{2}kx^2 + \text{EPE} \qquad (1)$$

19.3 The Electric Potential Difference Created by Point Charges The electric potential V at a distance r from a point charge q is given by Equation 19.6, where $k = 8.99 \times 10^9$ N · m²/C². This expression for V assumes that the electric potential is zero at an infinite distance away from the charge. The total electric potential at a given location due to two or more charges is the algebraic sum of the potentials due to each charge.

$$V = \frac{kq}{r} \qquad (19.6)$$

The total potential energy of a group of charges is the amount by which the electric potential energy of the group differs from its initial value when the charges are infinitely far apart and far away. It is also equal to the work required to assemble the group, one charge at a time, starting with the charges infinitely far apart and far away.

19.4 Equipotential Surfaces and Their Relation to the Electric Field An equipotential surface is a surface on which the electric potential is the same everywhere. The electric force does no work as a charge moves on an equipotential surface, because the force is always perpendicular to the displacement of the charge.

The electric field created by any group of charges is everywhere perpendicular to the associated equipotential surfaces and points in

---

*The weight of the droplet that has to be balanced by the electrostatic force is less than the actual weight due to an upward buoyant force on the droplet. The buoyant force, according to Archimedes' principle, will be equal to the weight of the air displaced by the droplet. Thus, the density of the oil (ρ) in Equation 19.13 should be replaced by ($\rho - \rho_{air}$), where ρ_{air} is the density of the air. This is a small correction, as the density of the oil is typically several hundred times larger than the density of air.

the direction of decreasing potential. The electric field is related to two equipotential surfaces by Equation 19.7a, where ΔV is the potential difference between the surfaces and Δs is the displacement. The term $\Delta V/\Delta s$ is called the potential gradient.

$$E = -\frac{\Delta V}{\Delta s} \qquad (19.7a)$$

19.5 Capacitors and Dielectrics A capacitor is a device that stores charge and energy. It consists of two conductors or plates that are near one another, but not touching. The magnitude q of the charge on each plate is given by Equation 19.8, where V is the magnitude of the potential difference between the plates and C is the capacitance. The SI unit for capacitance is the coulomb per volt (C/V), or farad (F).

$$q = CV \qquad (19.8)$$

The insulating material included between the plates of a capacitor is called a dielectric. The dielectric constant κ of the material is defined as shown in Equation 19.9, where E_0 and E are, respectively,

the magnitudes of the electric fields between the plates without and with a dielectric, assuming the charge on the plates is kept fixed.

$$\kappa = \frac{E_0}{E} \qquad (19.9)$$

The capacitance of a parallel plate capacitor filled with a dielectric is given by Equation 19.10, where $\varepsilon_0 = 8.85 \times 10^{-12}$ C^2/(N · m^2) is the permittivity of free space, A is the area of each plate, and d is the distance between the plates.

$$C = \frac{\kappa \varepsilon_0 A}{d} \qquad (19.10)$$

The electric potential energy stored in a capacitor is given by Equations 19.11a–c. The energy density is the energy stored per unit volume and is related to the magnitude E of the electric field, as indicated in Equation 19.12.

$$\text{Energy} = \tfrac{1}{2}qV = \tfrac{1}{2}CV^2 = q^2/(2C) \qquad (19.11a\text{–}c)$$

$$\text{Energy density} = \tfrac{1}{2}\kappa\varepsilon_0 E^2 \qquad (19.12)$$

Focus on Concepts

Online

Additional questions are available for assignment in WileyPLUS.

Section 19.2 The Electric Potential Difference

1. Two different charges, q_1 and q_2, are placed at two different locations, one charge at each location. The locations have the same electric potential V. Do the charges have the same electric potential energy? (a) Yes. If the electric potentials at the two locations are the same, the electric potential energies are also the same, regardless of the type (+ or −) and magnitude of the charges placed at these locations. (b) Yes, because electric potential and electric potential energy are just different names for the same concept. (c) No, because the electric potential V at a given location depends on the charge placed at that location, whereas the electric potential energy EPE does not. (d) No, because the electric potential energy EPE at a given location depends on the charge placed at that location as well as the electric potential V.

2. A proton is released from rest at point A in a constant electric field and accelerates to point B (see part a of the drawing). An electron is released from rest at point B and accelerates to point A (see part b of the drawing). How does the change in the proton's electric potential energy compare with the change in the electron's electric potential energy? (a) The change in the proton's electric potential energy is the same as the change in the electron's electric potential energy. (b) The proton experiences a greater change in electric potential energy, since it has a greater charge magnitude. (c) The proton experiences a smaller change in electric potential energy, since it has a smaller charge magnitude. (d) The proton experiences a smaller change in electric potential energy, since it has a smaller speed at B than the electron has at A. This is due to the larger mass of the proton. (e) One cannot compare the change in electric potential energies because the proton and electron move in opposite directions.

QUESTION 2

Section 19.3 The Electric Potential Difference Created by Point Charges

3. The drawing shows three arrangements of charged particles, all the same distance from the origin. Rank the arrangements, largest to smallest, according to the total electric potential V at the origin.

(a) A, B, C

(b) B, A, C

(c) B, C, A

(d) A, B and C (a tie)

(e) A and C (a tie), B

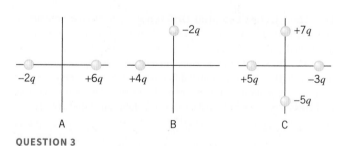

QUESTION 3

4. Four pairs of charged particles with identical separations are shown in the drawing. Rank the pairs according to their electric potential energy EPE, greatest (most positive) first.

(a) A and C (a tie), B and D (a tie)

(b) A, B, C, D

(c) C, B, D, A

(d) B, A, C and D (a tie)

(e) A and B (a tie), C, D

QUESTION 4

Section 19.4   Equipotential Surfaces and Their Relation to the Electric Field

5. The drawing shows edge-on views of three parallel plate capacitors with the same separation between the plates. The potential of each plate is indicated above it. Rank the capacitors as to the *magnitude* of the electric field inside them, largest to smallest.

(a) A, B, C   (b) A, C, B

(c) C, B, A   (d) C, A, B

(e) B, C, A

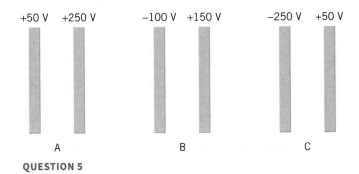

QUESTION 5

6. The drawing shows a plot of the electric potential V versus the displacement s. The plot consists of four segments. Rank the *magnitude* of the electric fields for the four segments, largest to smallest.

(a) D, C, B, A

(b) A and C (a tie), B and D (a tie)

(c) A, B, D, C

(d) B, D, C, A

(e) D, B, A and C (a tie)

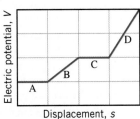

QUESTION 6

Section 19.5   Capacitors and Dielectrics

7. Which two or more of the following actions would increase the energy stored in a parallel plate capacitor when a constant potential difference is applied across the plates?

1. Increasing the area of the plates
2. Decreasing the area of the plates
3. Increasing the separation between the plates
4. Decreasing the separation between the plates
5. Inserting a dielectric between the plates

(a) 2, 4   (b) 2, 3, 5

(c) 1, 4, 5   (d) 1, 3

Problems

Online

Additional questions are available for assignment in WileyPLUS.

SSM Student Solutions Manual	**BIO** Biomedical application
MMH Problem-solving help	**E** Easy
GO Guided Online Tutorial	**M** Medium
V-HINT Video Hints	**H** Hard
CHALK Chalkboard Videos	**WS** Worksheet
	T Team Problem

Section 19.1   Potential Energy

Section 19.2   The Electric Potential Difference

1. **E** During a particular thunderstorm, the electric potential difference between a cloud and the ground is $V_{cloud} - V_{ground} = 1.3 \times 10^8$ V, with the cloud being at the higher potential. What is the change in an electron's electric potential energy when the electron moves from the ground to the cloud?

2. **E** A particle with a charge of -1.5 μC and a mass of 2.5×10^{-6} kg is released from rest at point A and accelerates toward point B, arriving there with a speed of 42 m/s. The only force acting on the particle is the electric force. (a) Which point is at the higher potential? Give your reasoning. (b) What is the potential difference $V_B - V_A$ between A and B?

3. **E GO CHALK** A particle has a charge of $+1.5$ μC and moves from point A to point B, a distance of 0.20 m. The particle experiences a constant electric force, and its motion is along the line of action of the force. The difference between the particle's electric potential energy at A and at B is $EPE_A - EPE_B = +9.0 \times 10^{-4}$ J. (a) Find the magnitude and direction of the electric force that acts on the particle. (b) Find the magnitude and direction of the electric field that the particle experiences.

4. **E GO** Review Multiple-Concept Example 4 to see the concepts that are pertinent here. In a television picture tube, electrons strike the screen after being accelerated from rest through a potential difference of 25 000 V. The speeds of the electrons are quite large, and for accurate calculations of the speeds, the effects of special relativity must be taken into account. Ignoring such effects, find the electron speed just before the electron strikes the screen.

5. **E SSM** Multiple-Concept Example 4 deals with the concepts that are important in this problem. As illustrated in **Figure 19.5b**, a negatively charged particle is released from rest at point B and accelerates until it reaches point A. The mass and charge of the particle are 4.0×10^{-6} kg and -2.0×10^{-5} C, respectively. Only the gravitational force and the electrostatic force act on the particle, which moves on a horizontal straight line without rotating. The electric potential at A is 36 V greater than that at B; in other words, $V_A - V_B = 36$ V. What is the translational speed of the particle at point A?

6. **E GO** An electron and a proton, starting from rest, are accelerated through an electric potential difference of the same magnitude. In the process, the electron acquires a speed v_e, while the proton acquires a speed v_p. Find the ratio v_e/v_p.

7. **M V-HINT CHALK** A moving particle encounters an external electric field that decreases its kinetic energy from 9520 eV to 7060 eV as the particle moves from position A to position B. The electric potential at A is -55.0 V, and the electric potential at B is $+27.0$ V. Determine the charge of the particle. Include the algebraic sign ($+$ or $-$) with your answer.

8. **M GO** During a lightning flash, there exists a potential difference of $V_{cloud} - V_{ground} = 1.2 \times 10^9$ V between a cloud and the ground. As a result, a charge of -25 C is transferred from the ground to the cloud. (a) How much work $W_{ground-cloud}$ is done on the charge by the electric force? (b) If the work done by the electric force were used to accelerate a 1100-kg automobile from rest, what would be its final speed? **(c)** If the work done by the electric force were converted into heat, how many kilograms of water at 0 °C could be heated to 100 °C?

Section 19.3 The Electric Potential Difference Created by Point Charges

9. **E** Two point charges, $+3.40$ μC and -6.10 μC, are separated by 1.20 m. What is the electric potential midway between them?

10. **E** An electron and a proton are initially very far apart (effectively an infinite distance apart). They are then brought together to form a hydrogen atom, in which the electron orbits the proton at an average distance of 5.29×10^{-11} m. What is $EPE_{final} - EPE_{initial}$, which is the change in the electric potential energy?

11. **E** **SSM** Two charges A and B are fixed in place, at different distances from a certain spot. At this spot the potentials due to the two charges are equal. Charge A is 0.18 m from the spot, while charge B is 0.43 m from it. Find the ratio q_B/q_A of the charges.

12. **E** **GO** The drawing shows a square, each side of which has a length of $L = 0.25$ m. On two corners of the square are fixed different positive charges, q_1 and q_2. Find the electric potential energy of a third charge $q_3 = -6.0 \times 10^{-9}$ C placed at corner A and then at corner B.

PROBLEM 12 $q_1 = +1.5 \times 10^{-9}$ C $q_2 = +4.0 \times 10^{-9}$ C

13. **E** **SSM** The drawing shows four point charges. The value of q is 2.0 μC, and the distance d is 0.96 m. Find the total potential at the location P. Assume that the potential of a point charge is zero at infinity.

PROBLEM 13

14. **E** A charge of $+125$ μC is fixed at the center of a square that is 0.64 m on a side. How much work is done by the electric force as a charge of $+7.0$ μC is moved from one corner of the square to any other empty corner? Explain.

15. **E** **CHALK** The drawing shows six point charges arranged in a rectangle. The value of q is 9.0 μC, and the distance d is 0.13 m. Find the total electric potential at location P, which is at the center of the rectangle.

PROBLEM 15

16. **E** Location A is 3.00 m to the right of a point charge q. Location B lies on the same line and is 4.00 m to the right of the charge. The potential difference between the two locations is $V_B - V_A = 45.0$ V. What are the magnitude and sign of the charge?

17. **E** Identical $+1.8$ μC charges are fixed to adjacent corners of a square. What charge (magnitude and algebraic sign) should be fixed to one of the empty corners, so that the total electric potential at the remaining empty corner is 0 V?

18. **E** **GO** Charges of $-q$ and $+2q$ are fixed in place, with a distance of 2.00 m between them. A dashed line is drawn through the negative charge, perpendicular to the line between the charges. On the dashed line, at a distance L from the negative charge, there is at least one spot where the total potential is zero. Find L.

19. **M** **CHALK** **SSM** Determine the electric potential energy for the array of three charges in the drawing, relative to its value when the charges are infinitely far away and infinitely far apart.

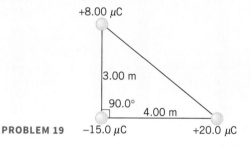

PROBLEM 19 -15.0 μC $+20.0$ μC

20. **M** **V-HINT** Two identical point charges ($q = +7.20 \times 10^{-6}$ C) are fixed at diagonally opposite corners of a square with sides of length 0.480 m. A test charge ($q_0 = -2.40 \times 10^{-8}$ C), with a mass of 6.60×10^{-8} kg, is released from rest at one of the empty corners of the square. Determine the speed of the test charge when it reaches the center of the square.

21. **M** **MMH** Two protons are moving directly toward one another. When they are very far apart, their initial speeds are 3.00×10^6 m/s. What is the distance of closest approach?

22. **M** **V-HINT** Four identical charges ($+2.0$ μC each) are brought from infinity and fixed to a straight line. The charges are located 0.40 m apart. Determine the electric potential energy of this group.

23. **M** **SSM** A charge of -3.00 μC is fixed in place. From a horizontal distance of 0.0450 m, a particle of mass 7.20×10^{-3} kg and charge -8.00 μC is fired with an initial speed of 65.0 m/s directly toward the fixed charge. How far does the particle travel before its speed is zero?

24. **M** **GO** Identical point charges of $+1.7$ μC are fixed to diagonally opposite corners of a square. A third charge is then fixed at the center of the square, such that it causes the potentials at the empty corners to change signs without changing magnitudes. Find the sign and magnitude of the third charge.

25. **H** One particle has a mass of 3.00×10^{-3} kg and a charge of $+8.00$ μC. A second particle has a mass of 6.00×10^{-3} kg and the same charge. The two particles are initially held in place and then released. The particles fly apart, and when the separation between them is 0.100 m, the speed of the 3.00×10^{-3} kg particle is 125 m/s. Find the initial separation between the particles.

Section 19.4 Equipotential Surfaces and Their Relation to the Electric Field

26. **E** Two equipotential surfaces surround a $+1.50 \times 10^{-8}$ C point charge. How far is the 190-V surface from the 75.0-V surface?

27. **E** An equipotential surface that surrounds a point charge q has a potential of 490 V and an area of 1.1 m². Determine q.

28. **E** **GO** A positive point charge ($q = +7.2 \times 10^{-8}$ C) is surrounded by an equipotential surface A, which has a radius of $r_A = 1.8$ m. A positive test charge ($q_0 = +4.5 \times 10^{-11}$ C) moves from surface A to another equipotential surface B, which has a radius r_B. The work done as the test charge moves from surface A to surface B is $W_{AB} = -8.1 \times 10^{-9}$ J. Find r_B.

29. **E** **SSM** A spark plug in an automobile engine consists of two metal conductors that are separated by a distance of 0.75 mm. When an electric spark jumps between them, the magnitude of the electric field is 4.7×10^7 V/m. What is the magnitude of the potential difference ΔV between the conductors?

30. **E** The drawing that accompanies Problem 47 shows a graph of a set of equipotential surfaces in cross section. The grid lines are 2.0 cm apart. Determine the magnitude and direction of the electric field at position D. Specify whether the electric field points toward the top or the bottom of the drawing.

31. **M** **SSM** **MMH** An electric field has a constant value of 4.0×10^3 V/m and is directed downward. The field is the same everywhere. The potential at a point P within this region is 155 V. Find the potential at the following points: **(a)** 6.0×10^{-3} m directly above P, **(b)** 3.0×10^{-3} m directly below P, **(c)** 8.0×10^{-3} m directly to the right of P.

32. **M** **CHALK** **GO** An electron is released from rest at the negative plate of a parallel plate capacitor and accelerates to the positive plate (see the drawing). The plates are separated by a distance of 1.2 cm, and the electric field within the capacitor has a magnitude of 2.1×10^6 V/m. What is the kinetic energy of the electron just as it reaches the positive plate?

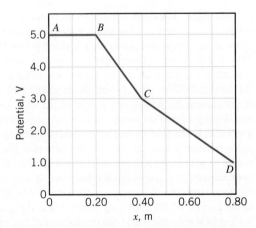

Electric field

Electron

$\vec{F}$

PROBLEM 32

33. **M** **V-HINT** The drawing shows the electric potential as a function of distance along the x axis. Determine the magnitude of the electric field in the region **(a)** A to B, **(b)** B to C, and **(c)** C to D.

PROBLEM 33

34. **M** **GO** At a distance of 1.60 m from a point charge of +2.00 μC, there is an equipotential surface. At greater distances there are additional equipotential surfaces. The potential difference between any two successive surfaces is 1.00×10^3 V. Starting at a distance of 1.60 m and moving radially outward, how many of the additional equipotential surfaces are crossed by the time the electric field has shrunk to one-half of its initial value? Do not include the starting surface.

35. **M** **V-HINT** The drawing shows a uniform electric field that points in the negative y direction; the magnitude of the field is 3600 N/C. Determine the electric potential difference **(a)** $V_B - V_A$ between points A and B, **(b)** $V_C - V_B$ between points B and C, and **(c)** $V_A - V_C$ between points C and A.

PROBLEM 35

Section 19.5   Capacitors and Dielectrics

36. **E** What is the capacitance of a capacitor that stores 4.3 μC of charge on its plates when a voltage of 1.5 V is applied between them?

37. **E** Two identical capacitors store different amounts of energy: capacitor A stores 3.1×10^{-3} J, and capacitor B stores 3.4×10^{-4} J. The voltage across the plates of capacitor B is 12 V. Find the voltage across the plates of capacitor A.

38. **E** **MMH** The electronic flash attachment for a camera contains a capacitor for storing the energy used to produce the flash. In one such unit, the potential difference between the plates of an 850-μF capacitor is 280 V. **(a)** Determine the energy that is used to produce the flash in this unit. **(b)** Assuming that the flash lasts for 3.9×10^{-3} s, find the effective power or "wattage" of the flash.

39. **E** **GO** The same voltage is applied between the plates of two different capacitors. When used with capacitor A, this voltage causes the capacitor to store 11 μC of charge and 5.0×10^{-5} J of energy. When used with capacitor B, which has a capacitance of 6.7 μF, this voltage causes the capacitor to store a charge that has a magnitude of q_B. Determine q_B.

40. **E** **SSM** A parallel plate capacitor has a capacitance of 7.0 μF when filled with a dielectric. The area of each plate is 1.5 m^2 and the separation between the plates is 1.0×10^{-5} m. What is the dielectric constant of the dielectric?

41. **E** **GO** Two capacitors are identical, except that one is empty and the other is filled with a dielectric ($\kappa = 4.50$). The empty capacitor is connected to a 12.0-V battery. What must be the potential difference across the plates of the capacitor filled with a dielectric so that it stores the same amount of electrical energy as the empty capacitor?

42. **M** **GO** **CHALK** Capacitor A and capacitor B both have the same voltage across their plates. However, the energy of capacitor A can melt m kilograms of ice at 0 °C, while the energy of capacitor B can boil away the same amount of water at 100 °C. The capacitance of capacitor A is 9.3 μF. What is the capacitance of capacitor B?

43. **M** **SSM** **MMH** What is the potential difference between the plates of a 3.3-F capacitor that stores sufficient energy to operate a 75-W light bulb for one minute?

44. **M** Review Conceptual Example 10 before attempting this problem. An empty capacitor is connected to a 12.0-V battery and charged up. The capacitor is then disconnected from the battery, and a slab of dielectric material ($\kappa = 2.8$) is inserted between the plates. Find the amount by which the potential difference across the plates changes. Specify whether the change is an increase or a decrease.

45. **M** **V-HINT** An empty parallel plate capacitor is connected between the terminals of a 9.0-V battery and charged up. The capacitor is then disconnected from the battery, and the spacing between the capacitor plates is doubled. As a result of this change, what is the new voltage between the plates of the capacitor?

Additional Problems

Online

46. E GO Refer to Multiple-Concept Example 3 to review the concepts that are needed here. A cordless electric shaver uses energy at a rate of 4.0 W from a rechargeable 1.5-V battery. Each of the charged particles that the battery delivers to the shaver carries a charge that has a magnitude of 1.6×10^{-19} C. A fully charged battery allows the shaver to be used for its maximum operation time, during which 3.0×10^{22} of the charged particles pass between the terminals of the battery as the shaver operates. What is the shaver's maximum operation time?

47. E V-HINT The drawing shows a graph of a set of equipotential surfaces seen in cross section. Each is labeled according to its electric potential. A $+2.8 \times 10^{-7}$ C point charge is placed at position A. Find the work that is done on the point charge by the electric force when it is moved **(a)** from A to B, and **(b)** from A to C.

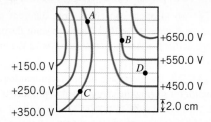

+650.0 V
+550.0 V
+450.0 V

+150.0 V
+250.0 V
+350.0 V

2.0 cm

PROBLEM 47

48. E SSM The work done by an electric force in moving a charge from point A to point B is 2.70×10^{-3} J. The electric potential difference between the two points is $V_A - V_B = 50.0$ V. What is the charge?

49. E GO Two capacitors have the same plate separation, but one has square plates and the other has circular plates. The square plates are a length L on each side, and the diameter of the circular plate is L. The capacitors have the same capacitance because they contain different dielectric materials. The dielectric constant of the material between the square plates has a value of $\kappa_{\text{square}} = 3.00$. What is the dielectric constant κ_{circle} of the material between the circular plates?

50. E Three point charges, -5.8×10^{-9} C, -9.0×10^{-9} C, and $+7.3 \times 10^{-9}$ C, are fixed at different positions on a circle. The total electric potential at the center of the circle is -2100 V. What is the radius of the circle?

51. M GO Equipotential surface A has a potential of 5650 V, while equipotential surface B has a potential of 7850 V. A particle has a mass of 5.00×10^{-2} kg and a charge of $+4.00 \times 10^{-5}$ C. The particle has a speed of 2.00 m/s on surface A. A nonconservative outside force is applied to the particle, and it moves to surface B, arriving there with a speed of 3.00 m/s. How much work is done by the outside force in moving the particle from A to B?

52. M GO SSM The capacitance of an empty capacitor is 1.2 μF. The capacitor is connected to a 12-V battery and charged up. With the capacitor connected to the battery, a slab of dielectric material is inserted between the plates. As a result, 2.6×10^{-5} C of *additional* charge flows from one plate, through the battery, and onto the other plate. What is the dielectric constant of the material?

Physics in Biology, Medicine, and Sports

53. E BIO SSM **19.2** Suppose that the electric potential outside a living cell is higher than that inside the cell by 0.070 V. How much work is done by the electric force when a sodium ion (charge = $+e$) moves from the outside to the inside?

54. E BIO SSM **19.4** The inner and outer surfaces of a cell membrane carry a negative and a positive charge, respectively. Because of these charges, a potential difference of about 0.070 V exists across the membrane. The thickness of the cell membrane is 8.0×10^{-9} m. What is the magnitude of the electric field in the membrane?

55. E BIO SSM **19.5** The electric potential energy stored in the capacitor of a defibrillator is 73 J, and the capacitance is 120 μF. What is the potential difference that exists across the capacitor plates?

56. E BIO **19.5** The membrane that surrounds a certain type of living cell has a surface area of 5.0×10^{-9} m^2 and a thickness of 1.0×10^{-8} m. Assume that the membrane behaves like a parallel plate capacitor and has a dielectric constant of 5.0. **(a)** The potential on the outer surface of the membrane is +60.0 mV greater than that on the inside surface. How much charge resides on the outer surface? **(b)** If the charge in part (a) is due to positive ions (charge $+e$), how many such ions are present on the outer surface?

57. E BIO SSM **19.5** An axon is the relatively long tail-like part of a neuron, or nerve cell. The outer surface of the axon membrane (dielectric constant = 5, thickness = 1×10^{-8} m) is charged positively, and the inner portion is charged negatively. Thus, the membrane is a kind of capacitor. Assuming that the membrane acts like a parallel plate capacitor with a plate area of 5×10^{-6} m^2, what is its capacitance?

58. E BIO **19.4** Probably all of us have experienced small electric shocks after strolling across a carpeted floor and touching something metallic, like a doorknob. Under certain conditions, and depending on the kind of shoes you are wearing, each step you take strips electrons off the carpet and places them uniformly over the surface area of your body. Bringing your finger in close to the doorknob induces a separation of surface charges on the doorknob (see the figure). When the potential difference between the doorknob and the tip of your finger exceeds the breakdown voltage of the air, a spark will occur, transferring some of the excess electrons from your finger to the doorknob. This is essentially a very small form of lightning. Consider that, at a separation distance of 1.0 mm, the potential difference between the tip of your finger and the doorknob is 7500 V. **(a)** If we treat the field between your finger and the doorknob as uniform, what would be the acceleration of a single electron that moves from your finger to the doorknob? **(b)** How long would it take the electron to bridge the gap between your finger and the doorknob? Neglect the potential collisions between the electron and air molecules.

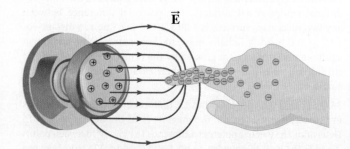

$\vec{E}$

PROBLEM 58

59. **E** **BIO** **19.5** Quite often the best course of treatment for breast cancer is a mastectomy, where the breast and surrounding tissue are surgically removed. This often involves the removal of lymph nodes in the breast area, which can lead to a condition called lymphedema. This is a chronic condition that consists of a buildup of lymphatic fluid that can result in pain, swelling, decreased mobility, and recurring infections. One technique that shows promise in early detection of lymphedema is comparative tissue dielectric constant measurements, or TDC. Here, a probe is placed against the skin and transmits a low power electromagnetic wave (see Chapter 24). The wave reflects back from the skin to a sensor in the probe head that determines information about the average dielectric constant of the skin and tissue beneath the probe. This value strongly depends on how much water is located beneath the skin, which is a strong indicator of lymphedema. As part of the probe's calibration, a parallel plate capacitor is used. The capacitor is connected to a $1.0 \times 10^2 \, \text{V}$ power source, charged, and then disconnected from the source. Different tissue samples with varying water content are then placed between the plates, and then the voltage across the plates is measured. **(a)** When the samples are inserted between the plates, do you expect the voltage across the plates to increase or decrease? **(b)** For one particular tissue sample, the voltage across the plates was measured to be 2.8 V. What is the dielectric constant of this tissue sample?

60. **E** **BIO** **19.5** In previous examples, we have discussed how a cardiac defibrillator uses a capacitor to deliver electrical energy to the heart. However, a battery, or other voltage source, could accomplish the same thing. So why does the defibrillator use a capacitor to deliver the charge, instead of a battery? Restarting a heart often requires a large charge delivered over a short burst, and because many defibrillators need to be portable, they get their power from batteries. However, batteries can only produce a continuous low flow of current. A capacitor solves this problem, as it is able to deliver a large charge very quickly. It is true that the battery will have to charge the capacitor, and this can take some time, but this is necessary to achieve the required charge. Also, by varying the charging time, the energy stored in the capacitor, and subsequently delivered to the heart, can be varied. Defibrillator capacitors come in a variety of values and sizes (see the photo). One of the larger capacitors available is 32 μF. If it is charged until the voltage across its plates is $6.0 \times 10^2 \, \text{V}$, how much energy is stored in the capacitor?

PROBLEM 60   2010-2019 CD Aero, LLC (Formerly Aerovox Corp.)

61. **M** **BIO** **19.2** When lightning strikes a tree, the effects can be explosive (see the photo). A typical flash can transport 30.0 C of charge through a potential difference of 125 MV. **(a)** How much energy would be delivered to a tree by this strike? **(b)** What mass of water in the tree would be converted to steam by this energy? Assume the initial temperature of the water is 20.0 °C.

PROBLEM 61

Photo courtesy of Professor James Klett, Colorado State University

62. **BIO** **19.5** As part of our natural digestive process, our bodies rely on peristalsis—the involuntary muscular contractions that keep food moving through our digestive tract. If these muscular contractions are weak, food may move too slowly, or not at all, which can lead to a condition called *gastroparesis*. Symptoms include nausea, vomiting, abdominal pain, and bloating. Dietary changes and drug therapies are often prescribed as initial treatments. However, if these prove ineffective, implantable gastric stimulation may be used. Here, a small battery-powered electronic device, often referred to as a gastric pacer, is inserted beneath the skin in the lower abdomen (see the figure). Conducting leads from the device provide controlled electrical pulses to the muscles in the lower stomach wall, which cause contractions that aid in peristalsis. The capacitor in one device delivers a short pulse containing 52 μC of charge. **(a)** If the charging voltage across the plates was 5.0 V, what is the value of the device's capacitor? **(b)** How much energy was delivered during the pulse?

PROBLEM 62   © 2020 Henry Ford Health System

Concepts and Calculations Problems

Online

The conservation of energy (Chapter 6) and the conservation of linear momentum (Chapter 7) are two of the most broadly applicable principles in all of science. In this chapter, we have seen that electrically charged particles obey the conservation-of-energy principle, provided that the electric potential is taken into account. The behavior of electrically charged particles, however, must also be consistent with the conservation-of-momentum principle, as

Problem 63 emphasizes. Chapter 18 introduces the electric field, and the present chapter introduces the electric potential. These two concepts are central to the study of electricity, and it is important to distinguish between them, as Problem 64 illustrates.

63. **M** **CHALK** Particle 1 has a mass of $m_1 = 3.6 \times 10^{-6}$ kg, while particle 2 has a mass of $m_2 = 6.2 \times 10^{-6}$ kg. Each has the same electric charge. These particles are initially held at rest, and the two-particle system has an initial electric potential energy of 0.150 J. Suddenly, the particles are released and fly apart because of the repulsive electric force that acts on each one (see the figure). The effects of the gravitational force are negligible, and no other forces act on the particles. *Concepts:* (i) What types of energy does the two-particle system have initially? (ii) What types of energy does the two-particle system have at the instant illustrated in part *b* of the drawing? (iii) Does the principle of conservation of energy apply to this problem? Explain. (iv) Does the conservation of linear momentum apply to the two particles as they fly apart? Explain. *Calculations:* At one instant following the release, the speed of particle 1 is measured to be $v_1 = 170$ m/s. What is the electric potential energy at this instant?

(a) Initial (at rest)

(b) Final

PROBLEM 63

(a) Two particles have different masses, but the same electrical charge q. They are initially at rest. (b) At the instant following the release of the particles, they are flying apart due to the mutual force of electric repulsion.

64. **M** **CHALK** **SSM** Two identical point charges $(+2.4 \times 10^{-9}$ C) are fixed in place, separated by 0.50 m (see the figure). *Concepts:* (i) The electric field is a vector and has a direction. At the midpoint, what are the directions of the individual electric-field contributions from q_A and q_B? (ii) Is the magnitude of the net electric field at the midpoint greater than, less than, or equal to zero? (iii) Is the total electric potential at the midpoint positive, negative, or zero? (iv) Does the electric potential have a direction associated with it? Explain. *Calculations:* Find the electric field and the electric potential at the midpoint of the line between the charges q_A and q_B.

Problem 71 determines the electric field and the electric potential at the midpoint ($r_A = r_B$) between the identical charges ($q_A = q_B$).

PROBLEM 64

Team Problems
Online

65. **E** **T** **WS** **Ion Thrusters II: Electrostatic Potential Energy.** You and your team are designing a new propulsion engine that ejects xenon (Xe) ions in one direction, causing a spacecraft to accelerate in the opposite direction, in accordance with the conservation of linear momentum and the impulse-momentum theorem. This is exactly what happens when a rocket engine expels high velocity particles when burning fuel, except that in the case of the ion drive the ions are accelerated with an electric field. In this process, Xe gas atoms are ionized so that they each acquire a net charge of $+e$. They are then subjected to an electric field, which accelerates and ejects the ions from the engine. In the thruster you are designing, the Xe ions accelerate through a potential difference between an electrode and negatively charged grid through which they are ejected. **(a)** What is the potential difference (V) between the electrode (held at 0 V) and the grid (held at voltage $-V$) in order that the ejection velocity of the ions is 55.0 km/s relative to the grid? **(b)** If Xe ions are ejected at a rate of 9.05×10^{-6} kg/s, what must be the average "wattage" of the power source that accelerates them? The mass of a Xe ion is 2.18×10^{-25} kg.

66. **M** **T** **WS** **A Capacitive Scale.** You and your team are tasked with finding a way to electronically measure weight. You get the idea to measure the capacitance between two circular conducting plates of mass $m = 0.112$ kg and radius $r = 11.5$ cm separated by a nonconducting vertical spring of spring constant $k = 1.35 \times 10^4$ N/m and unstrained length $d_0 = 3.55$ cm. The idea is that the object whose weight is to be determined is placed on the upper plate, thereby compressing the spring and bringing the two plates closer together. This, in turn, changes the capacitance. An electronic instrument that measures capacitance is interfaced with a computer that records the capacitance of the plates and converts it to weight. **(a)** Your task

is to derive an equation for the weight in terms of the capacitance that the computer will use to convert the measured capacitance to weight (in N). You can neglect any electrostatic forces between the plates of the capacitor. **(b)** What is the weight of an object placed on the upper plate that results in a measured capacitance of 16.3 pF?

67. **M** **T** **A Proton Gun.** You and your team are designing a device that creates a beam of high-velocity protons. The device consists of a parallel-plate capacitor in which protons start at rest at the positive plate. The protons accelerate toward the negative plate, which has a hole through which some of the protons pass. The plates are separated by 6.0 cm, and the potential difference between the plates is 5.0 kV. What are **(a)** the kinetic energies and **(b)** the speeds of the protons when they pass through the hole? **(c)** What is the acceleration of the protons between the plates?

68. **M** **T** **Storing Wind Energy.** A certain windmill produces an average of 0.500 MW of electrical power during the late evening hours when energy consumption is at a low (a four-hour span). An industrial capacitor (the "BC125") has a capacitance of $C = 63.0$ F and operates with a maximum voltage of 125 V between its plates. **(a)** You and your team are asked to determine the minimum number of these capacitors that will be needed to store the energy produced during the four off-peak hours in the late evening so that it can be used during the peak hours of the early afternoon. **(b)** The same company that makes the BC125 storage capacitor has discovered a breakthrough dielectric material that they use in their new model of storage capacitor, the SBC240. The new dielectric does two things: (1) Its dielectric constant is 2.5 times that of the dielectric in the BC125, and (2) it can operate at 240 V. What is the minimum number of SBC240 capacitors that will be needed to store the off-peak windmill energy?

The earth at night from space. The earth's major metropolitan areas and modern industrialized regions twinkle brilliantly with electric lights. Without electric circuits, lighting of the night sky would not be possible. Virtually every aspect of life in modern society utilizes or depends upon electric circuits in some way, which is the topic of this chapter.

NASA/NOAA

Electric Circuits

20.1 Electromotive Force and Current

Look around you. Chances are that there is an electrical device nearby—a radio, a hair dryer, a computer—something that uses electrical energy to operate. The energy needed to run an MP3 player, for instance, comes from batteries, as **Figure 20.1** illustrates. The transfer of energy takes place via an electric circuit, in which the energy source (the battery pack) and the energy-consuming device (the MP3 player) are connected by conducting wires, through which electric charges move.

Within a battery, a chemical reaction occurs that transfers electrons from one terminal (leaving it positively charged) to another terminal (leaving it negatively charged). **Figure 20.2** shows the two terminals of a car battery and a flashlight battery. The drawing also illustrates the symbol ($\dashv\vdash$) used to represent a battery in circuit drawings. Because of the positive and negative charges on the battery terminals, an electric potential difference exists between them. The maximum potential difference is called the **electromotive force*** (emf) of the battery, for which the symbol $\mathscr{E}$ is used. In a typical car battery, the chemical reaction maintains the potential of the positive terminal at a maximum of 12 volts (12 joules/coulomb) higher than the potential of the negative terminal,

*The word "force" appears in this context for historical reasons, even though it is incorrect. As we have seen in Section 19.2, electric potential is energy per unit charge, which is not force.

Moving charge

To MP3 player mechanism

Battery pack

Conducting wire

FIGURE 20.1 In an electric circuit, energy is transferred from a source (the battery pack) to a device (the MP3 player) by charges that move through a conducting wire.

Positive (+) terminal

Negative (−) terminal

Positive (+) terminal (raised button)

Negative (−) terminal (metallic bottom surface)

FIGURE 20.2 Typical batteries and the symbol $\left(\dfrac{+}{\vphantom{x}}\big|\dfrac{}{\vphantom{x}}\right)$ used to represent them in electric circuits.

Surface

FIGURE 20.3 The electric current is the amount of charge per unit time that passes through an imaginary surface that is perpendicular to the motion of the charges.

so the emf is $\mathscr{E} = 12$ V. Thus, one coulomb of charge emerging from the battery and entering a circuit has at most 12 joules of energy. In a typical flashlight battery the emf is 1.5 V. In reality, the potential difference between the terminals of a battery is somewhat less than the maximum value indicated by the emf, for reasons that Section 20.9 discusses.

In a circuit such as the one shown in **Figure 20.1**, the battery creates an electric field within and parallel to the wire, directed from the positive toward the negative terminal. The electric field exerts a force on the free electrons in the wire, and they respond by moving. **Figure 20.3** shows charges moving inside a wire and crossing an imaginary surface that is perpendicular to their motion. This flow of charge is known as an **electric current**. The electric current I is defined as the amount of charge per unit time that crosses the imaginary surface, as in **Figure 20.3**, in much the same sense that a river current is the amount of water per unit time that is flowing past a certain point. If the rate is constant, the current is

$$I = \frac{\Delta q}{\Delta t} \tag{20.1}$$

If the rate of flow is not constant, then Equation 20.1 gives the average current. Since the units for charge and time are the coulomb (C) and the second (s), the SI unit for current is a coulomb per second (C/s). One coulomb per second is referred to as an **ampere** (A), after the French mathematician André-Marie Ampère (1775–1836).

If the charges move around a circuit in the same direction at all times, the current is said to be **direct current (dc)**, which is the kind produced by batteries. In contrast, the current is said to be **alternating current (ac)** when the charges move first one way and then the opposite way, changing direction from moment to moment. Many energy sources produce alternating current—for example, generators at power companies and microphones. Example 1 deals with direct current.

EXAMPLE 1 | A Pocket Calculator

The battery pack of a pocket calculator has a voltage* of 3.0 V and delivers a current of 0.17 mA. In one hour of operation, **(a)** how much charge flows in the circuit and **(b)** how much energy does the battery deliver to the calculator circuit?

Reasoning Since current is defined as charge per unit time, the charge that flows in one hour is the product of the current and the time (3600 s). The charge that leaves the 3.0-V battery pack has 3.0 joules of energy per coulomb of charge. Thus, the total energy delivered to the calculator circuit is the charge (in coulombs) times the energy per unit charge (in volts or joules/coulomb).

Solution **(a)** The charge that flows in one hour can be determined from Equation 20.1:

$$\Delta q = I(\Delta t) = (0.17 \times 10^{-3}\text{A})(3600\text{ s}) = \boxed{0.61 \text{ C}}$$

(b) The energy delivered to the calculator circuit is

$$\text{Energy} = \text{Charge} \times \underbrace{\frac{\text{Energy}}{\text{Charge}}}_{\substack{\text{Battery} \\ \text{voltage}}} = (0.61\text{ C})(3.0\text{ V}) = \boxed{1.8 \text{ J}}$$

*The potential difference between two points, such as the terminals of a battery, is commonly called the voltage between the points.

Today, it is known that electrons flow in metal wires. **Figure 20.4** shows the negative electrons emerging from the negative terminal of the battery and moving counterclockwise around the circuit toward the positive terminal. It is customary, however, *not* to use the flow of electrons when discussing circuits. Instead, a so-called **conventional current** is used, for reasons that date to the time when it was believed that positive charges moved through metal wires. Conventional current is the hypothetical flow of positive charges that would have the same effect in the circuit as the movement of negative charges that actually does occur. In **Figure 20.4**, negative electrons leave the negative terminal of the battery, pass through the device, and arrive at the positive terminal. The same effect would have been achieved if an equivalent amount of positive charge had left the positive terminal, passed through the device, and arrived at the negative terminal. Therefore, the drawing shows the conventional current originating from the positive terminal and moving clockwise around the circuit. A conventional current of hypothetical positive charges is consistent with our earlier use of a positive test charge for defining electric fields and potentials. The direction of conventional current is always from a point of higher potential toward a point of lower potential—that is, from the positive toward the negative terminal. In this text, the symbol I stands for conventional current.

FIGURE 20.4 In a circuit, electrons actually flow through the metal wires. However, it is customary to use a conventional current I to describe the flow of charges.

20.2 | Ohm's Law

The current that a battery can push through a wire is analogous to the water flow that a pump can push through a pipe. Greater pump pressures lead to larger water flow rates, and, similarly, greater battery voltages lead to larger electric currents. In the simplest case, the current I is directly proportional to the voltage V; that is, $I \propto V$. Thus, a voltage of 12 V leads to twice as much current as a voltage of 6 V, when each is connected to the same circuit.

In a water pipe, the flow rate is not only determined by the pump pressure but is also affected by the length and diameter of the pipe. Longer and narrower pipes offer higher resistance to the moving water and lead to smaller flow rates for a given pump pressure. A similar situation exists in electric circuits, and to deal with it we introduce the concept of electrical resistance. Electrical resistance is defined in terms of two ideas that have already been discussed—the electric potential difference, or voltage (see Section 19.2), and the electric current (see Section 20.1).

The **resistance** R is defined as the ratio of the voltage V applied across a piece of material to the current I through the material, or $R = V/I$. When only a small current results from a large voltage, there is a high resistance to the moving charge. For many materials (e.g., metals), the ratio V/I is the same for a given piece of material over a wide range of voltages and currents. In such a case, the resistance is a constant. Then, the relation $R = V/I$ is referred to as **Ohm's law**, after the German physicist Georg Simon Ohm (1789–1854), who discovered it.

OHM'S LAW

The ratio V/I is a constant, where V is the voltage applied across a piece of material (such as a wire) and I is the current through the material:

$$\frac{V}{I} = R = \text{constant or } V = IR \qquad (20.2)$$

R **is the resistance of the piece of material.**

SI Unit of Resistance: volt/ampere (V/A) = ohm (Ω)

The SI unit for resistance is a volt per ampere, which is called an *ohm* and is represented by the Greek capital letter omega (Ω). Ohm's law is not a fundamental law of nature like Newton's laws of motion. It is only a statement of the way certain materials behave in electric circuits.

To the extent that a wire or an electrical device offers resistance to the flow of charges, it is called a **resistor**. The resistance can have a wide range of values. The copper wires in a television set, for instance, have a very small resistance. On the other hand, commercial resistors can have resistances up to many kilohms (1 kΩ = 10^3 Ω) or megohms (1 MΩ = 10^6 Ω). Such resistors play an important role in electric circuits, where they are typically used to limit the amount of current and establish the desired voltage levels.

In drawing electric circuits we follow the usual conventions: (1) a zigzag line ($-\!\!\!\bigwedge\!\!\!\!\bigwedge\!\!\!-$) represents a resistor and (2) a straight line ($\!-\!\!\!-\!\!\!-\!$) represents an ideal conducting wire, or one with a negligible resistance. Example 2 illustrates an application of Ohm's law to the circuit in a flashlight.

EXAMPLE 2 | A Flashlight

The filament in a light bulb is a resistor in the form of a thin piece of wire. The wire becomes hot enough to emit light because of the current in it. **Interactive Figure 20.5** shows a flashlight that uses two 1.5-V batteries (effectively a single 3.0-V battery) to provide a current of 0.40 A in the filament. Determine the resistance of the glowing filament.

Reasoning The filament resistance is assumed to be the only resistance in the circuit. The potential difference applied across

the filament is that of the 3.0-V battery. The resistance, given by Equation 20.2, is equal to this potential difference divided by the current.

Solution The resistance of the filament is

$$R = \frac{V}{I} = \frac{3.0\ \text{V}}{0.40\ \text{A}} = \boxed{7.5\ \Omega} \qquad (20.2)$$

INTERACTIVE FIGURE 20.5 The circuit in this flashlight consists of a resistor (the filament of the light bulb) connected to a 3.0-V battery (two 1.5-V batteries).

Check Your Understanding

(The answers are given at the end of the book.)

1. In circuit A the battery that supplies energy has twice as much voltage as the battery in circuit B. However, the current in circuit A is only one-half the current in circuit B. Circuit A presents _____ the resistance to the current that circuit B does. **(a)** twice **(b)** one-half **(c)** the same **(d)** four times **(e)** one-fourth

2. Two circuits present the same resistance to the current. In one circuit the battery causing the flow of current has a voltage of 9.0 V, and the current is 3.0 A. In the other circuit the battery has a voltage of 1.5 V. What is the current in this other circuit?

Resistance and Resistivity

In a water pipe, the length and cross-sectional area of the pipe determine the resistance that the pipe offers to the flow of water. Longer pipes with smaller cross-sectional areas offer greater resistance. Analogous effects are found in the electrical case. For a wide range of materials, the resistance of a piece of material of length L and cross-sectional area A is

$$R = \rho \frac{L}{A} \qquad\qquad (20.3)$$

where ρ is a proportionality constant known as the **resistivity** of the material. It can be seen from Equation 20.3 that the unit for resistivity is the ohm · meter ($\Omega \cdot$ m), and **Table 20.1** lists values for various materials. All the conductors in the table are metals and have small resistivities. Insulators such as rubber have large resistivities. Materials like germanium and silicon have intermediate resistivity values and are, accordingly, called *semiconductors*.

TABLE 20.1   Resistivities[a] of Various Materials

Material	Resistivity ρ ($\Omega \cdot$ m)	Material	Resistivity ρ ($\Omega \cdot$ m)
Conductors		**Semiconductors**	
Aluminum	2.82×10^{-8}	Carbon	3.5×10^{-5}
Copper	1.72×10^{-8}	Germanium	0.5^{b}
Gold	2.44×10^{-8}	Silicon	$20\text{--}2300^{b}$
Iron	9.7×10^{-8}	**Insulators**	
Mercury	95.8×10^{-8}	Mica	$10^{11}\text{--}10^{15}$
Nichrome (alloy)	100×10^{-8}	Rubber (hard)	$10^{13}\text{--}10^{16}$
Silver	1.59×10^{-8}	Teflon	10^{16}
Tungsten	5.6×10^{-8}	Wood (maple)	3×10^{10}

[a]The values pertain to temperatures near 20 °C.
[b]Depending on purity.

Resistivity is an inherent property of a material, inherent in the same sense that density is an inherent property. Resistance, on the other hand, depends on both the resistivity and the geometry of the material. Thus, two wires can be made from copper, which has a resistivity of $1.72 \times 10^{-8}\ \Omega \cdot$ m, but Equation 20.3 indicates that a short wire with a large cross-sectional area has a smaller resistance than does a long, thin wire. Wires that carry large currents, such as main power cables, are thick rather than thin so that the resistance of the wires is kept as small as possible. Similarly, electric tools that are to be used far away from wall sockets require thicker extension cords, as Example 3 illustrates.

EXAMPLE 3 | The Physics of Electrical Extension Cords

The instructions for an electric lawn mower suggest that a 20-gauge extension cord can be used for distances up to 35 m, but a thicker 16-gauge cord should be used for longer distances, to keep the resistance of the wire as small as possible. The cross-sectional area of 20-gauge wire is 5.2×10^{-7} m², while that of 16-gauge wire is 13×10^{-7} m². Determine the resistance of **(a)** 35 m of 20-gauge copper wire and **(b)** 75 m of 16-gauge copper wire.

Reasoning According to Equation 20.3, the resistance of a copper wire depends on the resistivity of copper and the length and cross-sectional area of the wire. The resistivity can be obtained from **Table 20.1**, while the length and cross-sectional area are given in the problem statement.

Solution According to **Table 20.1** the resistivity of copper is $1.72 \times 10^{-8}\ \Omega \cdot$ m. The resistance of the wires can be found using Equation 20.3:

20-gauge wire $\quad R = \dfrac{\rho L}{A} = \dfrac{(1.72 \times 10^{-8}\ \Omega \cdot \text{m})(35\ \text{m})}{5.2 \times 10^{-7}\ \text{m}^2} = \boxed{1.2\ \Omega}$

16-gauge wire $\quad R = \dfrac{\rho L}{A} = \dfrac{(1.72 \times 10^{-8}\ \Omega \cdot \text{m})(75\ \text{m})}{13 \times 10^{-7}\ \text{m}^2} = \boxed{0.99\ \Omega}$

Even though it is more than twice as long, the thicker 16-gauge wire has less resistance than the thinner 20-gauge wire. It is necessary to keep the resistance as low as possible to minimize heating of the wire, thereby reducing the possibility of a fire, as Conceptual Example 7 in Section 20.5 emphasizes.

Ac current source

Pressure cuff inflation system

Electrode

$\vec{L}$

Voltage measurement device

FIGURE 20.6 Using the technique of impedance plethysmography, the electrical resistance of the calf can be measured to diagnose deep venous thrombosis (blood clotting in the veins).

BIO **THE PHYSICS OF . . . impedance plethysmography.** Equation 20.3 provides the basis for an important medical diagnostic technique known as impedance (or resistance) plethysmography. **Figure 20.6** shows how the technique is applied to diagnose blood clotting in the veins (deep venous thrombosis) near the knee. A pressure cuff, like that used in blood pressure measurements, is placed around the midthigh, while electrodes are attached around the calf. The two outer electrodes are connected to a source that supplies a small amount of ac current. The two inner electrodes are separated by a distance L, and the voltage between them is measured. The voltage divided by the current gives the resistance. The key to this technique is the fact that resistance can be related to the volume V_{calf} of the calf between the inner electrodes. The volume is the product of the length L and the calf's cross-sectional area A, or $V_{calf} = LA$. Solving for A and substituting in Equation 20.3 shows that

$$R = \rho \frac{L}{A} = \rho \frac{L}{V_{calf}/L} = \rho \frac{L^2}{V_{calf}}$$

Thus, resistance is inversely proportional to volume, a fact that is exploited in diagnosing deep venous thrombosis. Blood flows from the heart into the calf through arteries in the leg and returns through the system of veins. The pressure cuff in **Figure 20.6** is inflated to the point where it cuts off the venous flow but does not alter the arterial flow. As a result, more blood enters than leaves the calf. Therefore, the volume of the calf increases, and the electrical resistance decreases. When the cuff pressure is removed suddenly, the volume returns to a normal value, and so does the electrical resistance. With healthy (unclotted) veins, there is a rapid return to normal values. A slow return, however, reveals the presence of clotting.

The resistivity of a material depends on temperature. In metals, the resistivity increases with increasing temperature, whereas in semiconductors the reverse is true. For many materials and limited temperature ranges, it is possible to express the temperature dependence of the resistivity as follows:

$$\rho = \rho_0[1 + \alpha(T - T_0)] \tag{20.4}$$

In this expression ρ and ρ_0 are the resistivities at temperatures T and T_0, respectively. The term α has the unit of reciprocal temperature and is the **temperature coefficient of resistivity**. When the resistivity increases with increasing temperature, α is positive, as it is for metals. When the resistivity decreases with increasing temperature, α is negative, as it is for the semiconductors carbon, germanium, and silicon. Since resistance is given by $R = \rho L/A$, both sides of Equation 20.4 can be multiplied by L/A to show that resistance depends on temperature according to

$$R = R_0[1 + \alpha(T - T_0)] \tag{20.5}$$

The next example illustrates the role of the resistivity and its temperature coefficient in determining the electrical resistance of a piece of material.

EXAMPLE 4 | The Physics of a Heating Element on an Electric Stove

Figure 20.7a shows a cherry-red heating element on an electric stove. The element contains a wire (length = 1.1 m, cross-sectional area = 3.1×10^{-6} m^2) through which electric charge flows. As **Figure 20.7b** shows, this wire is embedded within an electrically insulating material that is contained within a metal casing. The wire becomes hot in response to the flowing charge and heats the casing. The material of the wire has a resistivity of $\rho_0 = 6.8 \times 10^{-5}$ $\Omega \cdot$ m at $T_0 = 320$ °C and a temperature coefficient of resistivity of $\alpha = 2.0 \times 10^{-3}$ (C°)$^{-1}$. Determine the resistance of the heater wire at an operating temperature of 420 °C.

Reasoning The expression $R = \rho L/A$ (Equation 20.3) can be used to find the resistance of the wire at 420 °C, once the resistivity ρ is determined at this temperature. Since the resistivity at 320 °C is given, Equation 20.4 can be employed to find the resistivity at 420 °C.

Solution At the operating temperature of 420 °C, the material of the wire has a resistivity of

$$\rho = \rho_0[1 + \alpha(T - T_0)]$$
$$\rho = (6.8 \times 10^{-5} \ \Omega \cdot \text{m})\{1 + [2.0 \times 10^{-3}(\text{C}°)^{-1}](420 \ °\text{C} - 320 \ °\text{C})\}$$
$$= 8.2 \times 10^{-5} \ \Omega \cdot \text{m}$$

This value of the resistivity can be used along with the given length and cross-sectional area to find the resistance of the heater wire:

$$R = \frac{\rho L}{A} = \frac{(8.2 \times 10^{-5}\,\Omega \cdot m)(1.1\,m)}{3.1 \times 10^{-6}\,m^2} = \boxed{29\,\Omega} \quad \text{(20.3)}$$

Heater wire
$(A = 3.1 \times 10^{-6}\,m^2)$

Metal casing

$L = 1.1\,m$

(a)

(b)

FIGURE 20.7 A heating element on an electric stove.

There is an important class of materials whose resistivity suddenly goes to zero below a certain temperature T_c, which is called the *critical temperature* and is commonly a few degrees above absolute zero. Below this temperature, such materials are called *superconductors*. The name derives from the fact that with zero resistivity, these materials offer no resistance to electric current and are, therefore, perfect conductors. Once a current is established in a superconducting ring, the current continues indefinitely without the need for an emf. Currents have persisted in superconductors for many years without measurable decay. In contrast, the current in a nonsuperconducting material drops to zero almost immediately after the emf is removed.

Many metals become superconductors only at very low temperatures, such as aluminum ($T_c = 1.18$ K), tin ($T_c = 3.72$ K), lead ($T_c = 7.20$ K), and niobium ($T_c = 9.25$ K). Materials involving copper oxide complexes have been made that undergo the transition to the superconducting state at 175 K. Superconductors have many technological applications, including magnetic resonance imaging (Section 21.7), magnetic levitation of trains (Section 21.9), cheaper transmission of electric power, powerful (yet small) electric motors, and faster computer chips.

Check Your Understanding

(The answers are given at the end of the book.)

3. Two materials have different resistivities. Two wires of the same length are made, one from each of the materials. Is it possible for each wire to have the same resistance? **(a)** Yes, if the material with the greater resistivity is used for a thinner wire. **(b)** Yes, if the material with the greater resistivity is used for a thicker wire. **(c)** It is not possible.

4. How does the resistance of a copper wire change when both the length and diameter of the wire are doubled? **(a)** It decreases by a factor of two. **(b)** It increases by a factor of two. **(c)** It increases by a factor of four. **(d)** It decreases by a factor of four. **(e)** It does not change.

5. A resistor is connected between the terminals of a battery. This resistor is a wire, and the following five choices give possibilities for its length and radius in multiples of L_0 and r_0, respectively. For which one or more of the possibilities is the current in the resistor a minimum? **(a)** L_0 and r_0 **(b)** $\frac{1}{2}L_0$ and $\frac{1}{2}r_0$ **(c)** $2L_0$ and $2r_0$ **(d)** $2L_0$ and r_0 **(e)** $8L_0$ and $2r_0$

20.4 Electric Power

One of the most important functions of current in an electric circuit is to transfer energy from a source (a battery or a generator) to an electrical device (MP3 player, cell phone, etc.), as **Figure 20.8** illustrates. Note that the positive (+) terminal of the battery is connected by a wire to the terminal labeled A on the device; likewise, the negative terminal (−) of the battery is connected to the B terminal. Thus, the battery maintains a constant potential difference between the terminals A and B, with A being at the higher potential. When an amount of positive charge Δq moves from the higher potential (A) to the lower potential (B), its electric potential energy decreases. In accordance with Equation 19.4, this decrease is $(\Delta q)V$, where V is the amount by which the electric potential at A exceeds that at B or, in other words, the voltage between the two points. Since the change in energy per unit time is the power P (Equation 6.10b), the electric power associated with this change in energy is

$$P = \frac{\text{Change in energy}}{\text{Time interval}} = \frac{(\Delta q)V}{\Delta t} = \underbrace{\frac{\Delta q}{\Delta t}}_{\text{Current } I} V$$

The term $\Delta q/\Delta t$ is the charge per unit time, or the current I in the device, according to Equation 20.1. It follows, then, that the electric power is the product of the current and the voltage.

ELECTRIC POWER
When electric charge flows from point A to point B in a circuit, leading to a current I, and the voltage between the points is V, the electric power associated with this current and voltage is
$$P = IV \qquad\qquad (20.6)$$
SI Unit of Power: watt (W)

Power is measured in watts, and Equation 20.6 indicates that the product of an ampere and a volt is equal to a watt.

FIGURE 20.8 The current I in the circuit delivers energy to the electric device. The voltage between the terminals of the device is V.

When the charge moves through the device in **Figure 20.8**, the charge *loses* electric potential energy. The principle of conservation of energy tells us that the decrease in potential energy must be accompanied by a transfer of energy to some other form (or forms). In a cell phone, for example, the energy transferred appears as light energy (coming from the display screen), sound energy (emanating from the speaker), thermal energy (due to heating of the internal circuitry), and so on.

The charge in a circuit can also gain electrical energy. For example, when it moves through the battery in **Figure 20.8**, the charge goes from a lower to a higher potential, just the opposite of what happens in the electrical device. In this case, the charge *gains* electric potential energy. Consistent with the conservation of energy, this increase in potential energy must come from somewhere; in this case it comes from the chemical energy stored in the battery. Thus, the charge regains the energy it lost to the device, at the expense of the chemical energy of the battery.

Many electrical devices are essentially resistors that become hot when provided with sufficient electric power: toasters, irons, space heaters, heating elements on electric stoves, and incandescent light bulbs, to name a few. In such cases, it is convenient to have additional expressions that are equivalent to the power $P = IV$ but which include the resistance R explicitly. We can obtain two such equations by substituting $V = IR$, or equivalently $I = V/R$, into the relation $P = IV$:

$$P = IV \tag{20.6a}$$

$$P = I(IR) = I^2R \tag{20.6b}$$

$$P = \left(\frac{V}{R}\right)V = \frac{V^2}{R} \tag{20.6c}$$

Example 5 deals with the electric power delivered to the bulb of a flashlight.

EXAMPLE 5 | The Power and Energy Used in a Flashlight

In the flashlight in **Interactive Figure 20.5**, the current is 0.40 A, and the voltage is 3.0 V. Find **(a)** the power delivered to the bulb and **(b)** the electrical energy dissipated in the bulb in 5.5 minutes of operation.

Reasoning   The electric power delivered to the bulb is the product of the current and voltage. Since power is energy per unit time, the energy delivered to the bulb is the product of the power and time.

Solution   **(a)** The power is

$$P = IV = (0.40 \text{ A})(3.0 \text{ V}) = \boxed{1.2 \text{ W}} \tag{20.6a}$$

The "wattage" rating of this bulb would, therefore, be 1.2 W.

(b) The energy consumed in 5.5 minutes (330 s) follows from the definition of power as energy per unit time:

$$\text{Energy} = P \, \Delta t = (1.2 \text{ W})(330 \text{ s}) = \boxed{4.0 \times 10^2 \text{ J}}$$

Monthly electric bills specify the cost for the energy consumed during the month. Energy is the product of power and time, and electric companies compute your energy consumption by expressing power in kilowatts and time in hours. Therefore, a commonly used unit for energy is the *kilowatt · hour* (kWh). For instance, if you used an average power of 1440 watts (1.44 kW) for 30 days (720 h), your energy consumption would be (1.44 kW) (720 h) = 1040 kWh. At a cost of $0.12 per kWh, your monthly bill would be $125. As shown in Example 11 in Chapter 12, 1 kWh = 3.60×10^6 J of energy.

Check Your Understanding

(The answers are given at the end of the book.)

6. A toaster is designed to operate with a voltage of 120 V, and a clothes dryer is designed to operate with a voltage of 240 V. Based solely on this information, which appliance uses more power? **(a)** The toaster **(b)** The dryer **(c)** Insufficient information is given for an answer.

7. When an incandescent light bulb is turned on, a constant voltage is applied across the tungsten filament, which then becomes white hot. The temperature coefficient of resistivity for tungsten is a positive number. What happens to the power delivered to the bulb as the filament heats up? **(a)** It decreases. **(b)** It increases. **(c)** It remains constant.

8. CYU Figure 20.1 shows a circuit that includes a bimetallic strip (made from brass and steel; see Section 12.4) with a resistance heater wire wrapped around it. When the switch is initially closed, a current appears in the circuit because charges flow through the heater wire (which becomes hot), the strip itself, the contact point, and the light bulb. The bulb glows in response. As long as the switch remains closed, does the bulb **(a)** continue to glow, **(b)** eventually turn off permanently, or **(c)** flash on and off?

CYU FIGURE 20.1

20.5 Alternating Current

Many electric circuits use batteries and involve direct current (dc). However, there are considerably more circuits that operate with alternating current (ac), in which the charge flow reverses direction periodically. The common generators that create ac electricity depend on magnetic forces for their operation and are discussed in Chapter 22. In an ac circuit, these generators serve the same purpose as a battery serves in a dc circuit; that is, they give energy to the moving electric charges. This section deals with ac circuits that contain only resistance.

Since the electrical outlets in a house provide alternating current, we all use ac circuits routinely. For example, the heating element of a toaster is essentially a thin wire of resistance R and becomes red hot when electrical energy is dissipated in it. **Figure 20.9** shows the ac circuit that is formed when a toaster is plugged into a wall socket. The circuit schematic in the picture introduces the symbol ⊝ that is used to represent the generator. In this case, the generator is located at the electric power company.

FIGURE 20.9 This circuit consists of a toaster (resistance = R) and an ac generator at the electric power company.

Figure 20.10 shows a graph that records the voltage V produced between the terminals of the ac generator in **Figure 20.9** at each moment of time t. This is the most common type of ac voltage. It fluctuates sinusoidally between positive and negative values:

$$V = V_0 \sin 2\pi f t \qquad (20.7)$$

where V_0 is the maximum or peak value of the voltage, and f is the frequency (in cycles/s or Hz) at which the voltage oscillates. The angle $2\pi f t$ in Equation 20.7 is expressed in radians, so a calculator must be set to its radian mode before the sine of this angle is evaluated. In the United States, the voltage at most home wall outlets has a *peak value* of approximately $V_0 = 170$ volts and oscillates with a frequency of $f = 60$ Hz. Thus, the period of each cycle is $\frac{1}{60}$ s, and the polarity of the generator terminals reverses twice during each cycle, as **Figure 20.10** indicates.

The current in an ac circuit also oscillates. In circuits that contain only resistance, the current reverses direction each time the polarity of the generator terminals reverses. Thus, the current in a circuit like that in **Figure 20.9** would have a frequency of 60 Hz and would change direction twice during each cycle. Substituting $V = V_0 \sin 2\pi ft$ into $V = IR$ shows that the current can be represented as

$$I = \frac{V_0}{R} \sin 2\pi ft = I_0 \sin 2\pi ft \tag{20.8}$$

The peak current is given by $I_0 = V_0/R$, so it can be determined if the peak voltage and the resistance are known.

The power delivered to an ac circuit by the generator is given by $P = IV$, just as it is in a dc circuit. However, since both I and V depend on time, the power fluctuates as time passes. Substituting Equations 20.7 and 20.8 for V and I into $P = IV$ gives

$$P = I_0 V_0 \sin^2 2\pi ft \tag{20.9}$$

This expression is plotted in **Figure 20.11**.

Since the power fluctuates in an ac circuit, it is customary to consider the average power $\bar{P}$, which is one-half the peak power, as **Figure 20.11** indicates:

$$\bar{P} = \tfrac{1}{2} I_0 V_0 \tag{20.10}$$

On the basis of this expression, a kind of average current and average voltage can be introduced that are very useful when discussing ac circuits. A rearrangement of Equation 20.10 reveals that

$$\bar{P} = \left(\frac{I_0}{\sqrt{2}}\right)\left(\frac{V_0}{\sqrt{2}}\right) = I_{rms} V_{rms} \tag{20.11}$$

I_{rms} and V_{rms} are called the **root-mean-square (rms)** current and voltage, respectively, and may be calculated from the peak values by dividing them by $\sqrt{2}$:*

$$I_{rms} = \frac{I_0}{\sqrt{2}} \tag{20.12}$$

$$V_{rms} = \frac{V_0}{\sqrt{2}} \tag{20.13}$$

> **Problem-Solving Insight** The rms values of the ac current and the ac voltage, I_{rms} and V_{rms}, respectively, are not the same as the peak values I_0 and V_0. The rms values are always smaller than the peak values by a factor of $\sqrt{2}$.

Since the normal maximum ac voltage at a home wall socket in the United States is $V_0 = 170$ volts, the corresponding rms voltage is $V_{rms} = (170 \text{ volts})/\sqrt{2} = 120$ volts. Instructions for electrical devices usually specify this rms value. Similarly, when we specify an ac voltage or current in this text, it is an rms value, unless indicated otherwise. When we specify ac power, it is an average power, unless stated otherwise.

Except for dealing with average quantities, the relation $\bar{P} = I_{rms} V_{rms}$ has the same form as Equation 20.6a ($P = IV$). Moreover, Ohm's law can be written conveniently in terms of rms quantities:

$$V_{rms} = I_{rms} R \tag{20.14}$$

Substituting Equation 20.14 into $\bar{P} = I_{rms} V_{rms}$ shows that the average power can be expressed in the following ways:

$$\bar{P} = I_{rms} V_{rms} \tag{20.15a}$$

$$\bar{P} = I_{rms}^2 R \tag{20.15b}$$

$$\bar{P} = \frac{V_{rms}^2}{R} \tag{20.15c}$$

---

*This applies only for sinusoidal current and voltage.

FIGURE 20.10 In the most common case, the ac voltage is a sinusoidal function of time. The relative polarity of the generator terminals during the positive and negative parts of the sine wave is indicated.

FIGURE 20.11 In an ac circuit, the power P delivered to a resistor oscillates between zero and a peak value of $I_0 V_0$, where I_0 and V_0 are the peak current and voltage, respectively.

These expressions are completely analogous to $P = IV = I^2R = V^2/R$ (Equations 20.6a–c) for dc circuits. Example 6 deals with the average power in one familiar ac circuit.

EXAMPLE 6 | Electric Power Sent to a Loudspeaker

A stereo receiver applies a peak ac voltage of 34 V to a speaker. The speaker behaves approximately* as if it had a resistance of 8.0 Ω, as the circuit in **Figure 20.12** indicates. Determine **(a)** the rms voltage, **(b)** the rms current, and **(c)** the average power for this circuit.

Reasoning The rms voltage is, by definition, equal to the peak voltage divided by $\sqrt{2}$. Furthermore, we are assuming that the circuit contains only a resistor. Therefore, we can use Ohm's law (Equation 20.14) to calculate the rms current as the rms voltage divided by the resistance, and we can then determine the average power as the rms current times the rms voltage (Equation 20.15a).

Solution **(a)** The peak value of the voltage is $V_0 = 34$ V, so the corresponding rms value is

$$V_{\text{rms}} = \frac{V_0}{\sqrt{2}} = \frac{34\text{ V}}{\sqrt{2}} = \boxed{24\text{ V}} \qquad \textbf{(20.13)}$$

(b) The rms current can be obtained from Ohm's law:

$$I_{\text{rms}} = \frac{V_{\text{rms}}}{R} = \frac{24\text{ V}}{8.0\text{ }\Omega} = \boxed{3.0\text{ A}} \qquad \textbf{(20.14)}$$

(c) The average power is

$$\overline{P} = I_{\text{rms}}V_{\text{rms}} = (3.0\text{ A})(24\text{ V}) = \boxed{72\text{ W}} \qquad \textbf{(20.15a)}$$

8.0-Ω speaker

Receiver

FIGURE 20.12 A receiver applies an ac voltage (peak value = 34 V) to an 8.0-Ω speaker.

$V_0 = 34$ V $\qquad R = 8.0$ Ω

The electric power dissipated in a resistor causes the resistor to heat up. Excessive power can lead to a potential fire hazard, as Conceptual Example 7 discusses.

CONCEPTUAL EXAMPLE 7 | Extension Cords and a Potential Fire Hazard

During the winter, many people use portable electric space heaters to keep warm. When the heater is located far from a 120-V wall receptacle, an extension cord is necessary (see **Figure 20.13**). To prevent fires, however, manufacturers sometimes caution about using extension cords. To minimize the risk of a fire, should a long extension cord used with a space heater be made from **(a)** larger-gauge or **(b)** smaller-gauge wire?

Reasoning An electric space heater contains a heater element that is a piece of wire of resistance R, which is heated to a high temperature because of the power $I_{\text{rms}}^2 R$ dissipated in it. A typical heater uses a relatively large current I_{rms} of about 12 A. On its way to the heater, this current passes through the wires of the extension cord. Since these additional wires offer resistance to the current, the extension cord can also heat up, just as the heater element does. This unwanted heating depends on the resistance of the wire in the extension cord and could lead to a fire. As Example 3 discusses, the larger-gauge wire is the one that has the smaller cross-sectional area. The cross-sectional area is important because

FIGURE 20.13 When an extension cord is used with a space heater, the cord must have a resistance that is sufficiently small to prevent overheating of the cord.

it is one of the factors that determine the resistance of the wire in the extension cord.

Answer (a) is incorrect. To keep the heating of the extension cord to a safe level, the resistance of the wire must be kept small. Recall from Section 20.3 that the resistance of a wire depends inversely on its cross-sectional area. A larger-gauge wire has a

---

*Other factors besides resistance can affect the current and voltage in ac circuits; they are discussed in Chapter 23.

smaller cross-sectional area and, therefore, has a greater resistance $R_{\text{extension cord}}$. This greater resistance would lead to more (not less) heating of the extension cord, because of the power $I_{\text{rms}}^2 R_{\text{extension cord}}$ dissipated in it.

larger cross-sectional area, has a smaller resistance $R_{\text{extension cord}}$. This smaller resistance would lead to less heating of the extension cord due to the power $I_{\text{rms}}^2 R_{\text{extension cord}}$ dissipated in it, thus minimizing the risk of a fire.

Answer (b) is correct. Since the wire's resistance depends inversely on its cross-sectional area, a smaller-gauge wire, with its

Related Homework: Problem 30

Check Your Understanding

(The answers are given at the end of the book.)

9. CYU Figure 20.2 shows a circuit in which a light bulb is connected to the household ac voltage via two switches, S_1 and S_2. This is the kind of wiring, for example, that allows you to turn on a carport light from either inside the house or out in the carport. Which one or more of the combinations of the switch positions will turn on the light? **(a)** S_1 set to A and S_2 set to B **(b)** S_1 set to B and S_2 set to B **(c)** S_1 set to B and S_2 set to A **(d)** S_1 set to A and S_2 set to A

10. Two light bulbs are designed for use with an ac voltage of 120 V and are rated at 75 W and 150 W. Which bulb, if either, has the greater filament resistance?

CYU FIGURE 20.2

11. An ac circuit contains only a generator and a resistor. Which one of the following changes leads to the greatest average power being delivered to the circuit? **(a)** Double the peak voltage of the generator and double the resistance. **(b)** Double the resistance. **(c)** Double the peak voltage of the generator. **(d)** Double the peak voltage of the generator and reduce the resistance by a factor of two.

20.6 Series Wiring

Thus far, we have dealt with circuits that include only a single device, such as a light bulb or a loudspeaker. There are, however, many circuits in which more than one device is connected to a voltage source. This section introduces one method by which such connections may be made—namely, series wiring. ***Series wiring means that the devices are connected in such a way that there is the same electric current through each device.*** **Figure 20.14** shows a circuit in which two different devices, represented by resistors R_1 and R_2, are connected in series with a battery. Note that if the current in one resistor is interrupted, the current in the other is too. This could occur, for example, if two light bulbs were connected in series and the filament of one bulb broke. Because of the series wiring, the voltage V supplied by the battery is divided between the two resistors. The drawing indicates that the portion of the voltage across R_1 is V_1, while the portion across R_2 is V_2, so $V = V_1 + V_2$. For the individual resistances, the definition of resistance indicates that $R_1 = V_1/I$ and $R_2 = V_2/I$, so that $V_1 = IR_1$ and $V_2 = IR_2$. Therefore, we have

FIGURE 20.14 When two resistors are connected in series, the same current I is in both of them.

$$V = V_1 + V_2 = IR_1 + IR_2 = I(R_1 + R_2) = IR_S$$

where R_S is called the **equivalent resistance** of the series circuit. Thus, two resistors in series are equivalent to a single resistor whose resistance is $R_S = R_1 + R_2$, in the sense that there is the same current through R_S as there is through the series combination of R_1 and R_2. This line of reasoning can be extended to any number of resistors in series if we note the following:

| **Problem-Solving Insight**   The voltage across all the resistors in series is the sum of the individual voltages across each resistor. |

The result for the equivalent resistance is

Series resistors $$R_S = R_1 + R_2 + R_3 + \cdots$$ **(20.16)**

Examples 8 and 9 illustrate the concept of equivalent resistance in series circuits.

EXAMPLE 8 | A Series Circuit

Suppose that the resistances in **Figure 20.14** are $R_1 = 47\,\Omega$ and $R_2 = 86\,\Omega$, and the battery voltage is 24 V. Determine the equivalent resistance of the two resistors and the current in the circuit.

Reasoning The two resistors are wired in series, since there is the same current through each one. The equivalent resistance R_S of a series circuit is the sum of the individual resistances, so $R_S = R_1 + R_2$. The current I can be obtained from Ohm's law as the voltage V divided by the equivalent resistance: $I = V/R_S$.

Solution The equivalent resistance is

$$R_s = R_1 + R_2 = 47\,\Omega + 86\,\Omega = \boxed{133\,\Omega}$$ **(20.16)**

The current in the circuit is

$$I = \frac{V}{R_s} = \frac{24\text{ V}}{133\,\Omega} = \boxed{0.18\text{ A}}$$ **(20.2)**

Analyzing Multiple-Concept Problems

EXAMPLE 9 | Power Delivered to a Series Circuit

A 6.00-Ω resistor and a 3.00-Ω resistor are connected in series with a 12.0-V battery, as **Figure 20.15** indicates. Assuming that the battery contributes no resistance to the circuit, find the power delivered to each of the resistors.

Reasoning The power P delivered to each resistor is the product of the current squared (I^2) and the corresponding

resistance R, or $P = I^2R$. The resistances are known, and Ohm's law can be used to find the current. Ohm's law states that the current in the circuit (which is also the current through each of the resistors) equals the voltage V of the battery divided by the equivalent resistance R_S of the two resistors: $I = V/R_S$. Since the resistors are connected in series, we can obtain the equivalent resistance by adding the two resistances (see **Figure 20.15**).

FIGURE 20.15 A 6.00-Ω and a 3.00-Ω resistor connected in series are equivalent to a single 9.00-Ω resistor.

Knowns and Unknowns The data for this problem are:

Description	Symbol	Value
Resistance of 6.00-Ω resistor	R_1	6.00 Ω
Resistance of 3.00-Ω resistor	R_2	3.00 Ω
Battery voltage	V	12.0 V
Unknown Variables		
Power delivered to 6.00-Ω resistor	P_1	?
Power delivered to 3.00-Ω resistor	P_2	?

Modeling the Problem

STEP 1 Power The power P_1 delivered to the 6.00-Ω resistor is given by $P_1 = I^2R_1$ (Equation 20.6b), where I is the current through the resistor and R_1 is the resistance. In this expression R_1 is a known quantity, and the current I will be determined in Step 2.	$P_1 = I^2R_1$ $\boxed{?}$	**(20.6b)**
STEP 2 Ohm's Law The current I in the circuit depends on the voltage V of the battery and the equivalent resistance R_S of the two resistors in series (see **Figure 20.15**). This dependence is given by Ohm's law (Equation 20.2) as $$\boxed{I = \frac{V}{R_S}}$$ This result for the current can be substituted into Equation 20.6b, as indicated at the right. Note from the data table that the voltage is given. In Step 3 we will evaluate the equivalent resistance from the individual resistances R_1 and R_2.	$P_1 = I^2R_1$ $\boxed{I = \dfrac{V}{R_S}}$ $\boxed{?}$	**(20.6b)** **(1)**
STEP 3 Equivalent Resistance Since the two resistors are wired in series, the equivalent resistance R_S is the sum of the two resistances (Equation 20.16): $$\boxed{R_S = R_1 + R_2}$$ The resistances R_1 and R_2 are known. We substitute this expression for R_S into Equation 1, as shown in the right column.	$P_1 = I^2R_1$ $\boxed{I = \dfrac{V}{R_S}}$ $\boxed{R_S = R_1 + R_2}$	**(20.6b)** **(1)** **(20.16)**

Solution Algebraically combining the results of the three steps, we have

$$\underset{\underset{P_1}{\downarrow}}{\text{STEP 1}} \quad = \quad \underset{\underset{I^2R_1}{\downarrow}}{\text{STEP 2}} \quad = \quad \left(\frac{V}{R_S}\right)^2 R_1 = \underset{\underset{\left(\frac{V}{R_1+R_2}\right)^2 R_1}{\downarrow}}{\text{STEP 2}}$$

The power delivered to the 6.00-Ω resistor is

$$P_1 = \left(\frac{V}{R_1 + R_2}\right)^2 R_1 = \left(\frac{12.0 \text{ V}}{6.00 \text{ }\Omega + 3.00 \text{ }\Omega}\right)^2 (6.00 \text{ }\Omega) = \boxed{10.7 \text{ W}}$$

In a similar fashion, it can be shown that the power delivered to the 3.00-Ω resistor is

$$P_2 = \left(\frac{V}{R_1 + R_2}\right)^2 R_2 = \left(\frac{12.0 \text{ V}}{6.00 \text{ }\Omega + 3.00 \text{ }\Omega}\right)^2 (3.00 \text{ }\Omega) = \boxed{5.3 \text{ W}}$$

Related Homework: Problems 37, 39

In Example 9 the total power sent to the two resistors is $P = 10.7 \text{ W} + 5.3 \text{ W} = 16.0 \text{ W}$. Alternatively, the total power could have been obtained by using the voltage across the two resistors (the battery voltage) and the equivalent resistance R_S:

$$P = \frac{V^2}{R_S} = \frac{(12.0 \text{ V})^2}{6.00 \text{ }\Omega + 3.00 \text{ }\Omega} = 16.0 \text{ W} \qquad \text{(20.6c)}$$

Problem-Solving Insight The total power delivered to any number of resistors in series is equal to the power delivered to the equivalent resistance.

THE PHYSICS OF . . . personal digital assistants. Pressure-sensitive pads form the heart of computer input devices that function as personal digital assistants, or PDAs, and they offer an interesting application of series resistors. These devices are simple to use. You write directly on the pad with a plastic stylus that itself contains no electronics (see **Figure 20.16**). The writing appears as the stylus is moved, and recognition software interprets it as input for the built-in computer. The pad utilizes two transparent conductive layers that are separated by a small distance, except where pressure from the stylus brings them into contact (see point P in the drawing). Current I enters the positive side of the top layer, flows into the bottom layer through point P, and exits that layer through its negative side. Each layer provides resistance to the current, the amounts depending on where the point P is located. As the right side of the drawing indicates, the resistances from the layers are in series, since the same current exists in both of

(a)

FIGURE 20.16 The pressure pad on which the user writes in a personal digital assistant is based on the use of resistances that are connected in series.

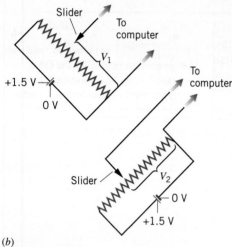

(b)

FIGURE 20.17 (a) A joystick uses two perpendicular movable sliders, and each makes contact with a coil resistor. (b) The sliders allow detection of the voltages V_1 and V_2, which a computer translates into positional data.

them. The voltage across the top-layer resistance is V_T, and the voltage across the bottom-layer resistance is V_B. In each case, the voltage is the current times the resistance. These two voltages are used to locate the point P and to activate (darken) one of the elements or pixels in a liquid crystal display matrix that lies beneath the transparent layers (see Section 24.6 for a discussion of liquid crystal displays). As the stylus is moved, the writing becomes visible as one element after another in the display matrix is activated.

THE PHYSICS OF . . . a joystick. The joystick, found in computer games, also takes advantage of resistors connected in series. A joystick contains two straight coils of resistance wire that are oriented at 90° to each other (see **Figure 20.17a**). When the joystick is moved, it repositions the metallic slider on each of the coils. As part b of the drawing illustrates, each coil is connected across a 1.5-V battery.* Because one end of a coil is at 1.5 V and the other at 0 V, the voltage at the location of the slider is somewhere between these values; the voltage of the left slider in the drawing is labeled V_1 and the voltage of the right slider is V_2. The slider voltages are sent via wires to a computer, which translates them into positional data. In effect, the slider divides each resistance coil into two smaller resistance coils wired in series, and allows the voltage at the point where they are joined together to be detected.

Check Your Understanding

(The answer is given at the end of the book.)

12. The power rating of a 1000-W heater specifies the power that the heater uses when it is connected to an ac voltage of 120 V. What is the total power used by two of these heaters when they are connected in series with a single ac voltage of 120 V? **(a)** 4000 W **(b)** 3000 W **(c)** 2000 W **(d)** 1000 W **(e)** 500 W

20.7 | Parallel Wiring

Parallel wiring is another method of connecting electrical devices. *Parallel wiring means that the devices are connected in such a way that the same voltage is applied across each device.* **Figure 20.18** shows two resistors connected in parallel between the terminals of a battery. Part a of the picture is drawn so as to emphasize that the entire voltage of the battery is applied across each resistor. Actually, parallel connections are

*For clarity, two batteries are shown in **Figure 20.17b**, one associated with each resistance coil. In reality, both coils are connected across a single battery.

rarely drawn in this manner; instead they are drawn as in part *b*, where the dots indicate the points where the wires for the two branches are joined together. **Figures 20.18a and b** are equivalent representations of the same circuit.

Parallel wiring is very common. For example, when an electrical appliance is plugged into a wall socket, the appliance is connected in parallel with other appliances, as in **Figure 20.19**, where the entire voltage of 120 V is applied across each one of the devices: the television, the stereo, and the light bulb (when the switch is turned on). The presence of the unused socket or other devices that are turned off does not affect the operation of those devices that are turned on. Moreover, if the current in one device is interrupted (perhaps by an opened switch or a broken wire), the current in the other devices is not interrupted. In contrast, if household appliances were connected in series, there would be no current through any appliance if the current in the circuit were halted at any point.

When two resistors R_1 and R_2 are connected as in **Figure 20.18**, each receives current from the battery as if the other were not present. Therefore, R_1 and R_2 together draw more current from the battery than does either resistor alone. According to the definition of resistance, $R = V/I$, a larger current implies a smaller resistance. Thus, the two parallel resistors behave as a single equivalent resistance that is *smaller* than either R_1 or R_2. **Figure 20.20** returns to the water-flow analogy to provide additional insight into this important feature of parallel wiring. In part *a*, two sections of pipe that have the same length are connected in parallel with a pump. In part *b* these two sections have been replaced with a single pipe of the same length, whose cross-sectional area equals the combined cross-sectional areas of section 1 and section 2. The pump (analogous to a voltage source) can push more water per second (analogous to current) through the wider pipe in part *b* (analogous to a wider wire) than it can through *either* of the narrower pipes (analogous to narrower wires) in part *a*. In effect, the wider pipe offers less resistance to the flow of water than either of the narrower pipes offers individually.

As in a series circuit, it is possible to replace a parallel combination of resistors with an equivalent resistor that results in the same total current and power for a given voltage as the original combination. To determine the equivalent resistance for the two resistors in **Figure 20.18b**, note that the total current I from the battery is the sum of I_1 and I_2, where I_1 is the current in resistor R_1 and I_2 is the current in resistor R_2: $I = I_1 + I_2$. Since the same voltage V is applied across each resistor, the definition of resistance indicates that $I_1 = V/R_1$ and $I_2 = V/R_2$. Therefore,

$$I = I_1 + I_2 = \frac{V}{R_1} + \frac{V}{R_2} = V\left(\frac{1}{R_1} + \frac{1}{R_2}\right) = V\left(\frac{1}{R_P}\right)$$

where R_P is the equivalent resistance. Hence, when two resistors are connected in parallel, they are equivalent to a single resistor whose resistance R_P can be obtained from $1/R_P = 1/R_1 + 1/R_2$.

| **Problem-Solving Insight** For any number of resistors wired in parallel, the total current from the voltage source is the sum of the currents in the individual resistors. |

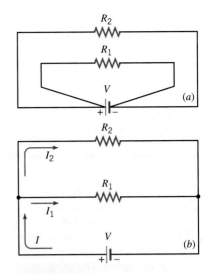

FIGURE 20.18 (*a*) When two resistors are connected in parallel, the same voltage V is applied across each resistor. (*b*) This drawing is equivalent to part *a*. I_1 and I_2 are the currents in R_1 and R_2.

FIGURE 20.19 This drawing shows some of the parallel connections found in a typical home. Each wall socket provides 120 V to the appliance connected to it. In addition, 120 V is applied to the light bulb when the switch is turned on.

Cross-sectional area = A_2

Cross-sectional area = A_1

Pump

(*a*)

Cross-sectional area = $A_1 + A_2$

Pump

(*b*)

FIGURE 20.20 (*a*) Two equally long pipe sections, with cross-sectional areas A_1 and A_2, are connected in parallel to a water pump. (*b*) The two parallel pipe sections in part *a* are equivalent to a single pipe of the same length whose cross-sectional area is $A_1 + A_2$.

Thus, a similar line of reasoning reveals that the equivalent resistance is

Parallel resistors
$$\frac{1}{R_P} = \frac{1}{R_1} + \frac{1}{R_2} + \frac{1}{R_3} + \cdots$$
(20.17)

The next example deals with a parallel combination of resistors that occurs in a stereo system.

EXAMPLE 10 | The Physics of Main and Remote Stereo Speakers

Most receivers allow the user to connect "remote" speakers (to play music in another room, for instance) in addition to the main speakers. Figure 20.21 shows that the remote speaker and the main speaker for the right stereo channel are connected to the receiver in parallel (for clarity, the speakers for the left channel are not shown). At the instant represented in the picture, the ac voltage across the speakers is 6.00 V. The main-speaker resistance is 8.00 Ω, and the remote-speaker resistance is 4.00 Ω.* Determine **(a)** the equivalent resistance of the two speakers, **(b)** the total current supplied by the receiver, **(c)** the current in each speaker, and **(d)** the power dissipated in each speaker.

Reasoning   The total current supplied to the two speakers by the receiver can be calculated as $I_{rms} = V_{rms}/R_P$, where R_P is the equivalent resistance of the two speakers in parallel and can be obtained from $1/R_P = 1/R_1 + 1/R_2$. The current in each speaker is different, however, since the speakers have different resistances. The average power delivered to a given speaker is the product of its current and voltage. In the parallel connection the same voltage is applied to each speaker.

> **Problem-Solving Insight**   The equivalent resistance R_P of a number of resistors in parallel has a reciprocal given by $R_P^{-1} = R_1^{-1} + R_2^{-1} + R_3^{-1} + \cdots$, where R_1, R_2, and R_3 are the individual resistances. After adding together the reciprocals R_1^{-1}, R_2^{-1}, and R_3^{-1}, do not forget to take the reciprocal of the result to find R_P.

Solution   **(a)** According to Equation 20.17, the equivalent resistance of the two speakers is given by

$$\frac{1}{R_P} = \frac{1}{8.00\ \Omega} + \frac{1}{4.00\ \Omega} = \frac{3}{8.00\ \Omega} \quad \text{or} \quad R_P = \frac{8.00\ \Omega}{3} = \boxed{2.67\ \Omega}$$

This result is illustrated in part *b* of the drawing.

(b) Using the equivalent resistance in Ohm's law shows that the total current is

$$I_{rms} = \frac{V_{rms}}{R_P} = \frac{6.00\ \text{V}}{2.67\ \Omega} = \boxed{2.25\ \text{A}} \qquad (20.14)$$

(c) Applying Ohm's law to each speaker gives the individual speaker currents:

8.00-Ω speaker
$$I_{rms} = \frac{V_{rms}}{R} = \frac{6.00\ \text{V}}{8.00\ \Omega} = \boxed{0.750\ \text{A}}$$

4.00-Ω speaker
$$I_{rms} = \frac{V_{rms}}{R} = \frac{6.00\ \text{V}}{4.00\ \Omega} = \boxed{1.50\ \text{A}}$$

The sum of these currents is equal to the total current obtained in part (b).

(d) The average power dissipated in each speaker can be calculated using $\overline{P} = I_{rms}V_{rms}$ with the individual currents obtained in part (c):

8.00-Ω speaker $\overline{P} = (0.750\ \text{A})(6.00\ \text{V}) = \boxed{4.50\ \text{W}}$ (20.15a)

4.00-Ω speaker $\overline{P} = (1.50\ \text{A})(6.00\ \text{V}) = \boxed{9.00\ \text{W}}$ (20.15a)

FIGURE 20.21   (*a*) The main and remote speakers in a stereo system are connected in parallel to the receiver. (*b*) The circuit schematic shows the situation when the ac voltage across the speakers is 6.00 V.

*In reality, frequency-dependent characteristics (see Chapter 23) play a role in the operation of a loudspeaker. We assume here, however, that the frequency of the sound is low enough that the speakers behave as pure resistances.

In Example 10, the total power delivered by the receiver is the sum of the individual values that were found in part (d), $\bar{P} = 4.50\ \text{W} + 9.00\ \text{W} = 13.5\ \text{W}$. Alternatively, the total power can be obtained from the equivalent resistance $R_P = 2.67\ \Omega$ and the total current in part (b):

$$\bar{P} = I_{\text{rms}}^2 R_P = (2.25\ \text{A})^2 (2.67\ \Omega) = 13.5\ \text{W} \qquad \textbf{(20.15b)}$$

| **Problem-Solving Insight** The total power delivered to any number of resistors in parallel is equal to the power delivered to the equivalent resistor. |

In a parallel combination of resistances, it is the *smallest* resistance that has the largest impact in determining the equivalent resistance. In fact, if one resistance approaches zero, then according to Equation 20.17, the equivalent resistance also approaches zero. In such a case, the near-zero resistance is said to *short out* the other resistances by providing a near-zero resistance path for the current to follow as a shortcut around the other resistances.

An interesting application of parallel wiring occurs in a three-way light bulb, as Conceptual Example 11 discusses.

CONCEPTUAL EXAMPLE 11 | The Physics of a Three-Way Light Bulb

Three-way light bulbs are popular because they can provide three levels of illumination (e.g., 50 W, 100 W, and 150 W) using a 120-V socket. The socket, however, must be equipped with a special three-way switch that enables one to select the illumination level. This switch does not select different voltages, because a three-way bulb uses a single voltage of 120 V. Within the bulb are two separate filaments. When the bulb is producing its highest illumination level and one of the filaments burns out (i.e., vaporizes), the bulb shines at one of the other illumination levels (either the lowest or the intermediate one). When the bulb is set to its highest level of illumination, how are the two filaments connected, **(a)** in parallel or **(b)** in series?

Reasoning In a series connection, the filaments would be connected in such a way that there is the same current through each one. The current would enter one filament and then leave that filament and enter into the other one. In a parallel connection, the same voltage would be applied across each filament, but the current through each would, in general, be different, the two currents existing independently of one another.

Answer (b) is incorrect. If the filaments were wired in series and one of them burned out, no current would pass through the bulb and none of the illumination levels would be available, contrary to what is observed. Therefore, the filaments are not wired in series.

Answer (a) is correct. Since the filaments are not in series, they must be in parallel, as **Interactive Figure 20.22** helps to explain. The power dissipated in a resistance R is $\bar{P} = V_{\text{rms}}^2 / R$, according to Equation 20.15c. With a single value of 120 V for the voltage V_{rms}, three different power ratings for the bulb can be obtained only if three different values for the resistance R are available. In a 50-W/100-W/150-W bulb, for example, one resistance R_{50} is provided by the 50-W filament, and the second resistance R_{100} comes from the 100-W filament. The third resistance R_{150} is the parallel combination of the other two and can be obtained from

INTERACTIVE FIGURE 20.22 A three-way light bulb uses two connected filaments. The filaments can be turned on one at a time or both together in parallel.

$1/R_{150} = 1/R_{50} + 1/R_{100}$. **Interactive Figure 20.22** illustrates a simplified version of how the three-way switch operates in such a bulb. The first position of the switch closes contact A and leaves contact B open, energizing only the 50-W filament. The second position closes contact B and leaves contact A open, energizing only the 100-W filament. The third position closes both contacts A and B, so that both filaments light up to give the highest level of illumination.

Related Homework: Problem 44

(The answers are given at the end of the book.)

13. A car has two headlights, and their power is derived from the car's battery. The filament in one burns out, but the other headlight stays on. Are the headlights connected in series or in parallel?

14. Two identical light bulbs are connected to identical batteries in two different ways. In method A the bulbs are connected in parallel, and the parallel combination is connected between the one battery's terminals. In method B the bulbs are connected in series, and the series combination is connected between the other battery's terminals. What is the ratio of the power supplied by the battery in method A to the power supplied in method B? **(a)** $\frac{1}{4}$ **(b)** 4 **(c)** $\frac{1}{2}$ **(d)** 2 **(e)** 1

20.8 | Circuits Wired Partially in Series and Partially in Parallel

Often an electric circuit is wired partially in series and partially in parallel. The key to determining the current, voltage, and power in such a case is to deal with the circuit in parts, with the resistances in each part being either in series or in parallel with each other. Example 12 shows how such an analysis is carried out.

EXAMPLE 12 | A Four-Resistor Circuit

Figure 20.23a shows a circuit composed of a 24-V battery and four resistors, whose resistances are 110, 180, 220, and 250 Ω. Find **(a)** the total current supplied by the battery and **(b)** the voltage between points A and B in the circuit.

Reasoning The total current that is supplied by the battery can be obtained from Ohm's law, $I = V/R$, where R is the equivalent resistance of the four resistors. The equivalent resistance can be calculated by dealing with the circuit in parts. The voltage V_{AB} between the two

FIGURE 20.23 The circuits shown in this picture are equivalent.

points A and B is also given by Ohm's law, $V_{AB} = IR_{AB}$, where I is the current and R_{AB} is the equivalent resistance between the two points.

Solution **(a)** The 220-Ω resistor and the 250-Ω resistor are in series, so they are equivalent to a single resistor whose resistance is 220 Ω + 250 Ω = 470 Ω (see **Figure 20.23a**). The 470-Ω resistor is in parallel with the 180-Ω resistor. Their equivalent resistance can be obtained from Equation 20.17:

$$\frac{1}{R_{AB}} = \frac{1}{470 \ \Omega} + \frac{1}{180 \ \Omega} = 0.0077 \ \Omega^{-1}$$

$$R_{AB} = \frac{1}{0.0077 \ \Omega^{-1}} = 130 \ \Omega$$

The circuit is now equivalent to a circuit containing a 110-Ω resistor in series with a 130-Ω resistor (see **Figure 20.23b**). This combination behaves like a single resistor whose resistance is $R = 110 \ \Omega + 130 \ \Omega = 240 \ \Omega$ (see **Figure 20.23c**). The total current from the battery is, then,

$$I = \frac{V}{R} = \frac{24 \text{ V}}{240 \ \Omega} = \boxed{0.10 \text{ A}}$$

(b) The current $I = 0.10$ A passes through the resistance between points A and B. Therefore, Ohm's law indicates that the voltage across the 130-Ω resistor between points A and B is

$$V_{AB} = IR_{AB} = (0.10 \text{ A})(130 \ \Omega) = \boxed{13 \text{ V}}$$

Check Your Understanding

(The answers are given at the end of the book.)

15. In one of the circuits in **CYU Figure 20.3** none of the resistors is in series or in parallel with any of the other resistors. Which circuit is it?

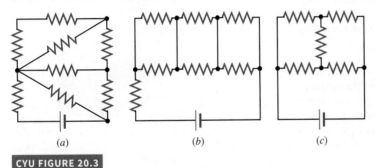

(a)   (b)   (c)

CYU FIGURE 20.3

16. You have three resistors, each of which has a resistance R. By connecting all three together in various ways, which one or more of the following resistance values can you obtain? **(a)** $3R$ **(b)** $\frac{3}{2}R$ **(c)** R **(d)** $\frac{2}{3}R$ **(e)** $\frac{1}{3}R$

17. You have four resistors, each of which has a resistance R. It is possible to connect these four together so that the equivalent resistance of the combination is also R. How many ways can you do it? There is more than one way.

20.9 | Internal Resistance

So far, the circuits we have considered include batteries or generators that contribute only their emfs to a circuit. In reality, however, such devices also add some resistance. This resistance is called the **internal resistance** of the battery or generator because it is located inside the device. In a battery, the internal resistance is due to the chemicals within the battery. In a generator, the internal resistance is the resistance of wires and other components within the generator.

Figure 20.24 presents a schematic representation of the internal resistance r of a battery. The drawing emphasizes that when an external resistance R is connected to the battery, the resistance is connected *in series* with the internal resistance. The internal resistance of a functioning battery is typically small (several thousandths of an ohm for a new car battery). Nevertheless, the effect of the internal resistance may not be negligible. Example 13 illustrates that when current is drawn from a battery, the internal resistance causes the voltage between the terminals to drop below the maximum value specified by the battery's emf. The actual voltage between the terminals of a battery is known as the **terminal voltage**.

FIGURE 20.24 When an external resistance R is connected between the terminals of a battery, the resistance is connected in series with the internal resistance r of the battery.

EXAMPLE 13 | The Physics of Automobile Batteries

Figure 20.25 shows a car battery whose emf is 12.0 V and whose internal resistance is 0.010 Ω. This resistance is relatively large because the battery is old and the terminals are corroded. What is the terminal voltage when the current I drawn from the battery is **(a)** 10.0 A and **(b)** 100.0 A?

Reasoning The voltage between the terminals is not the entire 12.0-V emf, because part of the emf is needed to make the current go through the internal resistance. The amount of voltage needed can be determined from Ohm's law as the current I through the battery times the internal resistance r. For larger currents, a larger amount of voltage is needed, leaving less of the emf between the terminals.

Solution **(a)** The voltage needed to make a current of $I = 10.0$ A go through an internal resistance of $r = 0.010$ Ω is

$$V = Ir = (10.0 \text{ A})(0.010 \text{ Ω}) = 0.10 \text{ V}$$

To find the terminal voltage, remember that the direction of conventional current is always from a higher toward a lower potential. To emphasize this fact in the drawing, plus and minus signs have been included at the right and left ends, respectively, of the resistance r. The terminal voltage can be calculated by starting at the negative terminal and keeping track of how the voltage increases and decreases as we move toward the positive terminal. The voltage rises by 12.0 V due to the battery's emf. However, the voltage drops by 0.10 V because of the potential difference across the internal resistance. Therefore, the terminal voltage is 12.0 V − 0.10 V = $\boxed{11.9 \text{ V}}$.

To car's electrical system
(ignition, lights, radio, etc.)

FIGURE 20.25 A car battery whose emf is 12 V and whose internal resistance is r.

(b) When the current through the battery is 100.0 A, the voltage needed to make the current go through the internal resistance is

$$V = Ir = (100.0 \text{ A})(0.010 \text{ Ω}) = 1.0 \text{ V}$$

The terminal voltage now decreases to 12.0 V − 1.0 V = $\boxed{11.0 \text{ V}}$.

Example 13 indicates that the terminal voltage of a battery is smaller when the current drawn from the battery is larger, an effect that any car owner can demonstrate. Turn the headlights on before starting your car, so that the current through the battery is about 10 A, as in part (a) of Example 13. Then start the car. The starter motor draws a large amount of additional current from the battery, momentarily increasing the total current by an appreciable amount. Consequently, the terminal voltage of the battery decreases, causing the headlights to become dimmer.

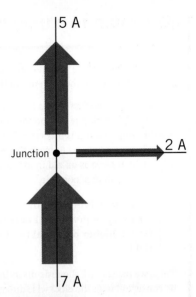

20.10 | Kirchhoff's Rules

Electric circuits that contain a number of resistors can often be analyzed by combining individual groups of resistors in series and parallel, as Section 20.8 discusses. However, there are many circuits in which no two resistors are in series or in parallel. To deal with such circuits it is necessary to employ methods other than the series–parallel method. One alternative is to take advantage of Kirchhoff's rules, named after their developer Gustav Kirchhoff (1824–1887). There are two rules, the **junction rule** and the **loop rule**, and both arise from principles and ideas that we have encountered earlier. The junction rule is an application of the law of conservation of electric charge (see Section 18.2) to the electric current in a circuit. The loop rule is an application of the principle of conservation of energy (see Section 6.8) to the electric potential (see Section 19.2) that exists at various places in a circuit.

Figure 20.26 illustrates in greater detail the basic idea behind Kirchhoff's junction rule. The picture shows a junction where several wires are connected together. As Section 18.2 discusses, electric charge is conserved. Therefore, since there is no accumulation of charges at the junction itself, the total charge per second flowing into the junction must equal the total charge per second flowing out of it. In other words, ***the junction rule states that the total current directed into a junction must equal the total current directed out of the junction,*** or 7 A = 5 A + 2 A for the specific case shown in the picture.

To help explain Kirchhoff's loop rule, **Figure 20.27** shows a circuit in which a 12-V battery is connected to a series combination of a 5-Ω and a 1-Ω resistor. The plus and minus signs associated with each resistor remind us that, outside a battery, conventional current is directed from a higher toward a lower potential. From left to right, there is a potential drop of 10 V across the first resistor and another drop of 2 V across the second resistor. Keeping in mind that potential is the electric potential energy per unit charge, let us follow a positive test charge clockwise* around the circuit. Starting at the negative terminal of the battery, we see that the test charge gains potential energy because of the 12-V rise in potential due to the battery. The test charge then loses potential energy because of the 10-V and 2-V drops in potential across the resistors, ultimately arriving back at the negative terminal. In traversing the closed-circuit loop, the test charge is like a skier gaining gravitational potential energy in going up a hill on a chair lift and then losing it to friction in coming down and stopping. When the skier returns to the starting point, the gain equals the loss, so there is no net change in potential energy. Similarly, when the test charge arrives back at its starting point, there is no net change in electric potential energy, the gains matching the losses. ***The loop rule expresses this example of energy conservation in terms of the electric potential and states that for a closed-circuit loop, the total of all the potential drops is the same as the total of all the potential rises,*** or 10 V + 2 V = 12 V for the specific case in **Figure 20.27**.

Kirchhoff's rules can be applied to any circuit, even when the resistors are not in series or in parallel. The two rules are summarized below, and Examples 14 and 15 illustrate how to use them.

KIRCHHOFF'S RULES

Junction rule. The sum of the magnitudes of the currents directed into a junction equals the sum of the magnitudes of the currents directed out of the junction.

Loop rule. Around any closed-circuit loop, the sum of the potential drops equals the sum of the potential rises.

FIGURE 20.26 A junction is a point in a circuit where a number of wires are connected together. If 7 A of current is directed into the junction, then a total of 7 A (5 A + 2 A) of current must be directed out of the junction.

FIGURE 20.27 Following a positive test charge clockwise around the circuit, we see that the total voltage drop of 10 V + 2 V across the two resistors equals the voltage rise of 12 V due to the battery. The plus and minus signs on the resistors emphasize that, outside a battery, conventional current is directed from a higher potential (+) toward a lower potential (−).

---

*The choice of the clockwise direction is arbitrary.

EXAMPLE 14 | Using Kirchhoff's Loop Rule

Figure 20.28 shows a circuit that contains two batteries and two resistors. Determine the current I in the circuit.

Reasoning The first step is to draw the current, which we have chosen to be clockwise around the circuit. The choice of direction is *arbitrary*, and if it is incorrect, I will turn out to be negative.

The second step is to mark the resistors with plus and minus signs, which serve as an aid in identifying the potential drops and rises for Kirchhoff's loop rule.

> **Problem-Solving Insight** Remember that, outside a battery, conventional current is always directed from a higher potential (+) toward a lower potential (−).

Thus, we *must* mark the resistors as indicated in **Figure 20.28**, to be consistent with the clockwise direction chosen for the current.

We may now apply Kirchhoff's loop rule to the circuit, starting at corner A, proceeding clockwise around the loop, and identifying the potential drops and rises as we go. The potential across each resistor is given by Ohm's law as $V = IR$. The clockwise direction is arbitrary, and the same result is obtained with a counterclockwise path.

Solution Starting at corner A and moving clockwise around the loop, there is

1. A potential drop (+ to −) of $IR = I(12\ \Omega)$ across the 12-Ω resistor
2. A potential drop (+ to −) of 6.0 V across the 6.0-V battery

FIGURE 20.28 A single-loop circuit that contains two batteries and two resistors.

3. A potential drop (+ to −) of $IR = I(8.0\ \Omega)$ across the 8.0-Ω resistor
4. A potential rise (− to +) of 24 V across the 24-V battery

Setting the sum of the potential drops equal to the sum of the potential rises, as required by Kirchhoff's loop rule, gives

$$\underbrace{I(12\ \Omega) + 6.0\ \text{V} + I(8.0\ \Omega)}_{\text{Potential drops}} = \underbrace{24\ \text{V}}_{\text{Potential rises}}$$

Solving this equation for the current yields $\boxed{I = 0.90\ \text{A}}$. The current is a positive number, indicating that our initial choice for the direction (clockwise) of the current was correct.

EXAMPLE 15 | The Physics of an Automobile Electrical System

In a car, the headlights are connected to the battery and would discharge the battery if it were not for the alternator, which is run by the engine. **Interactive Figure 20.29** indicates how the car battery, headlights, and alternator are connected. The circuit includes an internal resistance of 0.0100 Ω for the car battery and its leads and a resistance of 1.20 Ω for the headlights. For the sake of simplicity, the alternator is approximated as an additional 14.00-V battery with an internal resistance of 0.100 Ω. Determine

the currents through the car battery (I_B), the headlights (I_H), and the alternator (I_A).

Reasoning The drawing shows the directions chosen arbitrarily for the currents I_B, I_H, and I_A. If any direction is incorrect, the analysis will reveal a negative value for the corresponding current.

Next, we mark the resistors with the plus and minus signs that serve as an aid in identifying the potential drops and rises for

INTERACTIVE FIGURE 20.29 A circuit showing the headlight(s), battery, and alternator of a car.

the loop rule, recalling that, outside a battery, conventional current is always directed from a higher potential (+) toward a lower potential (−). Thus, given the directions selected for I_B, I_H, and I_A, the plus and minus signs *must* be those indicated in **Interactive Figure 20.29**.

Kirchhoff's junction and loop rules can now be used.

> **Problem-Solving Insight** In problems involving Kirchhoff's rules, it is always helpful to mark the resistors with plus and minus signs to keep track of the potential rises and drops in the circuit.

Solution The junction rule can be applied to junction B or junction E to give the same result:

Junction rule applied at B

$$\underbrace{I_A + I_B}_{\substack{\text{Into} \\ \text{junction}}} = \underbrace{I_H}_{\substack{\text{Out of} \\ \text{junction}}}$$

In applying the loop rule to the lower loop *BEFA,* we start at point B, move clockwise around the loop, and identify the potential drops and rises. There is a potential rise (− to +) of I_B (0.0100 Ω) across the 0.0100-Ω resistor and then a drop (+ to −) of 12.00 V due to the car battery. Continuing around the loop, we find a 14.00-V rise (− to +) across the alternator, followed by a potential drop (+ to −) of I_A (0.100 Ω) across the 0.100-Ω resistor. Setting the sum of the potential drops equal to the sum of the potential rises gives the following result:

Loop rule: BEFA $\underbrace{I_A(0.100\ \Omega) + 12.00\ \text{V}}_{\text{Potential drops}} = \underbrace{I_B(0.0100\ \Omega) + 14.00\ \text{V}}_{\text{Potential rises}}$

Since there are three unknown variables in this problem, I_B, I_H, and I_A, a third equation is needed for a solution. To obtain the third equation, we apply the loop rule to the upper loop *CDEB,* choosing a clockwise path for convenience. The result is

Loop rule: CDEB $\underbrace{I_B(0.0100\ \Omega) + I_H(1.20\ \Omega)}_{\text{Potential drops}} = \underbrace{12.00\ \text{V}}_{\text{Potential rises}}$

These three equations can be solved simultaneously to show that

$$\boxed{I_A = 19.1\ \text{A}, \quad I_B = -9.0\ \text{A}, \quad I_H = 10.1\ \text{A}}$$

The negative answer for I_B indicates that the current through the battery is not directed from right to left, as drawn in **Interactive Figure 20.29**. Instead, the 9.0-A current is directed from left to right, opposite to the way current would be directed if the alternator were not connected. It is the left-to-right current created by the alternator that keeps the battery charged.

Note that we can check our results by applying Kirchhoff's loop rule to the outer loop *ABCDEF.* If our results are correct, then the sum of the potential drops around this loop will be equal to the sum of the potential rises.

Math Skills The three equations that must be solved simultaneously for I_A, I_B, and I_H are

$$I_A + I_B = I_H \tag{1}$$

$$I_A(0.100) + 12.00 = I_B(0.0100) + 14.00 \tag{2}$$

$$I_B(0.0100) + I_H(1.20) = 12.00 \tag{3}$$

For clarity, units have been omitted. Substituting Equation 1 for I_H into Equation 3 gives

$$I_B(0.0100) + \underbrace{(I_A + I_B)}_{I_H}(1.20) = 12.00 \text{ or}$$

$$I_B(1.21) + I_A(1.20) = 12.00 \tag{4}$$

Solving Equation 4 for I_B shows that

$$I_B = \frac{12.00 - I_A(1.20)}{1.21} = 9.92 - I_A(0.992) \tag{5}$$

Substituting Equation 5 for I_B into Equation 2, we find that

$$I_A(0.100) + 12.00 = \underbrace{[9.92 - I_A(0.992)]}_{I_B}(0.0100) + 14.00$$

This result contains only the unknown variable I_A and can be solved to show that $\boxed{I_A = 19.1\ \text{A}}$. Using this value for I_A in Equation 5 reveals that

$$I_B = 9.92 - \underbrace{(19.1)}_{I_A}(0.992) = \boxed{-9.0\ \text{A}}$$

Finally, substituting the values for both I_A and I_B into Equation 1 yields

$$I_H = \underbrace{19.1}_{I_A} + \underbrace{(-9.0)}_{I_B} = \boxed{10.1\ \text{A}}$$

REASONING STRATEGY Applying Kirchhoff's Rules

1. **Draw the current in each branch of the circuit. Choose any direction. If your choice is incorrect, the value obtained for the current will turn out to be a negative number.**

2. **Mark each resistor with a plus sign at one end and a minus sign at the other end, in a way that is consistent with your choice for the current direction in Step 1. Outside a battery, conventional current is always directed from a higher potential (the end marked +) toward a lower potential (the end marked −).**

3. **Apply the junction rule and the loop rule to the circuit, obtaining in the process as many independent equations as there are unknown variables.**

4. **Solve the equations obtained in Step 3 simultaneously for the unknown variables. (See Appendix C.3.)**

Check Your Understanding

(The answer is given at the end of the book.)

18. CYU Figure 20.4 shows a circuit containing three resistors and three batteries. In preparation for applying Kirchhoff's rules, the currents in each resistor have been drawn. For these currents, write down the equations that result from applying Kirchhoff's junction rule and loop rule. Apply the loop rule to loops *ABCD* and *BEFC*.

CYU FIGURE 20.4

FIGURE 20.30 (*a*) A dc galvanometer. The coil of wire and pointer rotate when there is a current in the wire. (*b*) A galvanometer with a coil resistance of R_C is represented in a circuit diagram as shown here.

20.11 | The Measurement of Current and Voltage

Current and voltage can be measured with devices known, respectively, as ammeters and voltmeters. There are two types of such devices: those that use digital electronics and those that do not. The essential feature of nondigital devices is the dc *galvanometer*. As **Figure 20.30a** illustrates, a galvanometer consists of a magnet, a coil of wire, a spring, a pointer, and a calibrated scale. The coil is mounted so that it can rotate, which causes the pointer to move in relation to the scale. The coil rotates in response to the torque applied by the magnet when there is a current in the coil (see Section 21.6). The coil stops rotating when this torque is balanced by the torque of the spring.

Two characteristics of a galvanometer are important when it is used as part of a measurement device. First, the amount of dc current that causes full-scale deflection of the pointer indicates the sensitivity of the galvanometer. For instance, **Figure 20.30a** shows an instrument that deflects full scale when the current in the coil is 0.10 mA. The second important characteristic is the resistance R_C of the wire in the coil. **Figure 20.30b** shows how a galvanometer with a coil resistance of R_C is represented in a circuit diagram.

THE PHYSICS OF ... an ammeter. Since an **ammeter** is an instrument that measures current, it must be inserted in the circuit so the current passes directly through it, as **Figure 20.31** shows. (This is true for both digital and nondigital ammeters; **Figure 20.31** shows a digital instrument.)

A nondigital ammeter includes a galvanometer and one or more *shunt resistors,* which are connected in parallel with the galvanometer and provide a bypass for current in excess of the galvanometer's full-scale limit. The bypass allows the ammeter to be used to measure a current exceeding the full-scale limit. In **Figure 20.32**, for instance, a current of 60.0 mA enters terminal *A* of an ammeter (nondigital), which uses a galvanometer with a full-scale current of 0.100 mA. The shunt resistor *R* can be selected so that 0.100 mA of current enters the galvanometer, while 59.9 mA bypasses it. In such a case, the measurement scale on the ammeter would be labeled 0 to 60.0 mA. To determine the shunt resistance, it is necessary to know the coil resistance R_C (see Problem 76).

When an ammeter is inserted into a circuit, the equivalent resistance of the ammeter adds to the circuit resistance. Any increase in circuit resistance causes a reduction in current, and this is a problem, because an ammeter should only measure the current, not change it. Therefore, an *ideal* ammeter would have zero resistance. In practice, a good ammeter is designed with an equivalent resistance small enough so there is only a negligible reduction of the current in the circuit when the ammeter is inserted.

FIGURE 20.31 An ammeter must be inserted into a circuit so that the current passes directly through it.

Nondigital ammeter

60.0 mA

FIGURE 20.32 If a galvanometer with a full-scale limit of 0.100 mA is to be used to measure a current of 60.0 mA, a shunt resistance R must be used, so the excess current of 59.9 mA can detour around the galvanometer coil.

Digital voltmeter

FIGURE 20.33 To measure the voltage between two points A and B in a circuit, a voltmeter is connected between the points.

THE PHYSICS OF ... a voltmeter. A **voltmeter** is an instrument that measures the voltage between two points, A and B, in a circuit. **Figure 20.33** shows that the voltmeter must be connected between the points and is *not* inserted into the circuit as an ammeter is. (This is true for both digital and nondigital voltmeters; **Figure 20.33** shows a digital instrument.)

A nondigital voltmeter includes a galvanometer whose scale is calibrated in volts. Suppose, for instance, that the galvanometer in **Figure 20.34** has a full-scale current of 0.1 mA and a coil resistance of 50 Ω. Under full-scale conditions, the voltage across the coil would, therefore, be $V = IR_C = (0.1 \times 10^{-3} \text{ A})(50 \text{ Ω}) = 0.005$ V. Thus, this galvanometer could be used to register voltages in the range 0–0.005 V. A nondigital voltmeter, then, is a galvanometer used in this fashion, along with some provision for adjusting the range of voltages to be measured. To adjust the range, an additional resistance R is connected in series with the coil resistance R_C (see Problem 77).

$R_C = 50$ Ω

0.1 mA

0.005 V

FIGURE 20.34 The galvanometer shown has a full-scale deflection of 0.1 mA and a coil resistance of 50 Ω.

Ideally, the voltage registered by a voltmeter should be the same as the voltage that exists when the voltmeter is not connected. However, a voltmeter takes some current from a circuit and, thus, alters the circuit voltage to some extent. An *ideal* voltmeter would have infinite resistance and would draw away only an infinitesimal amount of current. In reality, a good voltmeter is designed with a resistance that is large enough so the unit does not appreciably alter the voltage in the circuit to which it is connected.

Check Your Understanding

(The answers are given at the end of the book.)

19. An ideal ammeter has _____ resistance, whereas an ideal voltmeter has _____ resistance. **(a)** zero, zero **(b)** infinite, infinite **(c)** zero, infinite **(d)** infinite, zero

20. What would happen to the current in a circuit if a voltmeter, inadvertently mistaken for an ammeter, were inserted into the circuit? The current would **(a)** increase markedly **(b)** decrease markedly **(c)** remain the same.

20.12 Capacitors in Series and in Parallel

Figure 20.35 shows two different capacitors connected in parallel to a battery. Since the capacitors are in parallel, they have the same voltage V across their plates. However, the capacitors *contain different amounts of charge*. The charge stored by a capacitor is $q = CV$ (Equation 19.8), so $q_1 = C_1V$ and $q_2 = C_2V$.

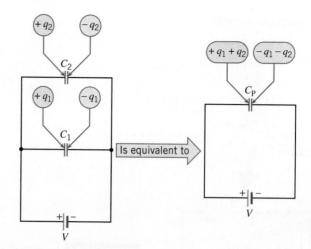

FIGURE 20.35 In a parallel combination of capacitances C_1 and C_2, the voltage V across each capacitor is the same, but the charges q_1 and q_2 on each capacitor are different.

As with resistors, it is always possible to replace a parallel combination of capacitors with an *equivalent capacitor* that stores the same charge and energy for a given voltage as the combination does. To determine the equivalent capacitance C_P, note that the total charge q stored by the two capacitors is

$$q = q_1 + q_2 = C_1 V + C_2 V = (C_1 + C_2)V = C_P V$$

This result indicates that two capacitors in parallel can be replaced by an equivalent capacitor whose capacitance is $C_P = C_1 + C_2$. For any number of capacitors in parallel, the equivalent capacitance is

Parallel capacitors $\qquad\qquad C_P = C_1 + C_2 + C_3 + \cdots$ $\qquad\qquad$ **(20.18)**

Capacitances in parallel simply add together to give an equivalent capacitance. This behavior contrasts with that of resistors in parallel, which combine as reciprocals, according to Equation 20.17. The reason for this difference is that the charge q on a capacitor is directly proportional to the capacitance C ($q = CV$), whereas the current I in a resistor is inversely proportional to the resistance R ($I = V/R$).

The equivalent capacitor not only stores the same amount of charge as the parallel combination of capacitors, but also stores the same amount of energy. For instance, the energy stored in a single capacitor is $\frac{1}{2}CV^2$ (Equation 19.11b), so the total energy stored by two capacitors in parallel is

$$\text{Total energy} = \tfrac{1}{2}C_1 V^2 + \tfrac{1}{2}C_2 V^2 = \tfrac{1}{2}(C_1 + C_2)V^2 = \tfrac{1}{2}C_P V^2$$

which is equal to the energy stored in the equivalent capacitor C_P.

When capacitors are connected in series, the equivalent capacitance is different from when they are in parallel. As an example, **Figure 20.36** shows two capacitors in series and reveals the following important fact.

FIGURE 20.36 In a series combination of capacitances C_1 and C_2, the same amount of charge q is on the plates of each capacitor, but the voltages V_1 and V_2 across each capacitor are different.

Problem-Solving Insight   All capacitors in series, regardless of their capacitances, contain charges of the same magnitude, $+q$ and $-q$, on their plates.

The battery places a charge of $+q$ on plate a of capacitor C_1, and this charge induces a charge of $+q$ to depart from the opposite plate a', leaving behind a charge $-q$. The $+q$ charge that leaves plate a' is deposited on plate b of capacitor C_2 (since these two plates are connected by a wire), where it induces a $+q$ charge to move away from the opposite plate b', leaving behind a charge of $-q$. Thus, all capacitors in series contain charges of the same magnitude on their plates. Note the difference between charging capacitors in parallel and in series. When charging parallel capacitors, the battery moves a charge q that is the sum of the charges moved for each of the capacitors: $q = q_1 + q_2 + q_3 + \cdots$. In contrast, when charging a series combination of n capacitors, the battery only moves a charge q, not nq, because the charge q passes by induction from one capacitor directly to the next one in line.

The equivalent capacitance C_S for the series connection in **Figure 20.36** can be determined by observing that the battery voltage V is shared by the two capacitors. The drawing indicates that the voltages across C_1 and C_2 are V_1 and V_2, so that $V = V_1 + V_2$. Since the voltages across the capacitors are $V_1 = q/C_1$ and $V_2 = q/C_2$, we find that

$$V = V_1 + V_2 = \frac{q}{C_1} + \frac{q}{C_2} = q\left(\frac{1}{C_1} + \frac{1}{C_2}\right) = q\left(\frac{1}{C_S}\right)$$

Thus, two capacitors in series can be replaced by an equivalent capacitor, which has a capacitance C_S that can be obtained from $1/C_S = 1/C_1 + 1/C_2$. For any number of capacitors connected in series the equivalent capacitance is given by

Series capacitors $$\frac{1}{C_S} = \frac{1}{C_1} + \frac{1}{C_2} + \frac{1}{C_3} + \cdots \qquad (20.19)$$

Equation 20.19 indicates that capacitances in series combine as reciprocals and do not simply add together as resistors in series do. It is left as an exercise (Problem 87) to show that the equivalent series capacitance stores the same electrostatic energy as the sum of the energies of the individual capacitors.

It is possible to simplify circuits containing a number of capacitors in the same general fashion as that outlined for resistors in Example 12 and **Figure 20.23**. The capacitors in a parallel grouping can be combined according to Equation 20.18, and those in a series grouping can be combined according to Equation 20.19, as the following example shows.

EXAMPLE 16 | BIO The Physics of Projective Capacitive Touchscreens

Chances are you are the owner of a smartphone, such as an iPhone, and you use your fingers to communicate with the device by touching its front panel. Touchscreens on modern smartphones use capacitive sensing to locate the position of your fingers on the screen. If that position is located on the screen above the icon of your favorite app, then the phone will launch that app. **Figure 20.37a** shows the essential components of the touchscreen, which utilizes a grid of parallel plate capacitors composed of sensing lines and driving lines that are etched into a glass substrate. This is referred to as **projective capacitive touch (PCT) technology**. A voltage is applied across the sensing and driving lines, which places a small charge on their surfaces, effectively creating a grid of capacitors, with positive charges on one grid and negative charges on the other. When a finger, or other conductor, is brought near the surface of the screen, it is conductive enough to disrupt the electric field lines around the grid. Negative charges on your skin are attracted to the positive charges on the driving lines and move toward the tip of your finger. The positive charges on the driving lines are maintained by a constant

current. The negative charges in your finger repel some of the negative charges from the sensing electrode, thereby decreasing the effective capacitance between the electrodes (**Figure 20.37b**).

The electrical conductivity of your skin is essential for the operation of the touchscreen. This is why your smartphone screen doesn't work when you wear gloves, as cloth is a good electrical insulator. However, special touchscreen gloves will work, as they use conductive thread in their fingertips. The drop in capacitance is measured by the phone's sensing electronics, which pinpoint the location of your finger on the screen. By continuously monitoring the mutual capacitance of all the grid crossing points, swiping motions and two-fingered movements, such as zooming in or out, can also be detected.

We can model the finger touch on the screen as a series of capacitors, as shown in **Figure 20.37b**, where C_{E0} is the capacitance between the grid electrodes before the screen is touched, C_{EF} is the capacitance between the grid electrodes after the screen is touched, and C_F is the capacitance between the sensing electrode and the finger. Suppose that, after the screen is touched, the sensing circuit measures the two capacitors in **Figure 20. 38b** in series (C_s) to be 0.19 pF. If the value of $C_{EF} = 0.25$ pF, what is the value of the capacitance between the finger and the sensing electrode (C_F)?

Reasoning The equivalent capacitance for capacitors connected in series is less than the individual capacitance values, as the measurement indicates. We can use Equation 20.19 to calculate the equivalent capacitance in terms of C_{EF} and C_F, which we can then solve for C_F.

Solution Applying Equation 20.19 to the two capacitors in series in **Figure 20. 37b**, we have the following:

$$\frac{1}{C_S} = \frac{1}{C_{EF}} + \frac{1}{C_F} \Rightarrow \frac{1}{C_F} = \frac{1}{C_S} - \frac{1}{C_{EF}} \Rightarrow C_F = \left(\frac{1}{C_S} - \frac{1}{C_{EF}}\right)^{-1}$$

$$\Rightarrow C_F = \left(\frac{1}{0.19 \text{ pF}} - \frac{1}{0.25 \text{ pF}}\right)^{-1} = \boxed{0.79 \text{ pF}}$$

Top view of electrode grid. Blue and red indicate two different layers.

Cover glass
Electrode 1
Bonding layer
Electrode 2
Glass layer
Smartphone body

Liquid crystal display (LCD) layers

(a)

Finger above screen

Finger touching screen

Cover glass

Cover glass

C_F

C_{EO}

C_{EF}

Driving electrode   Sensing electrode

Driving electrode   Sensing electrode

(b)

FIGURE 20.37 (a) The main components of a smartphone touchscreen. The driving and sensing electrodes are composed of transparent conducting layers of indium tin oxide (ITO). (b) Bringing a conducting object, like a bare finger, in contact with the screen changes the capacitance near the finger. The phone's circuitry detects this change, which pinpoints the finger's location.

20.13 RC Circuits

$+q$ $-q$

R

I C

Switch

V_0

(a)

q

$q_0 = CV_0$

t

(b)

FIGURE 20.38 Charging a capacitor.

Many electric circuits contain both resistors and capacitors. **Figure 20.38** illustrates an example of a resistor–capacitor circuit, or *RC* circuit. Part *a* of the drawing shows the circuit at a time *t* after the switch has been closed and the battery has begun to charge up the capacitor plates. The charge on the plates builds up gradually to its equilibrium value of $q_0 = CV_0$, where V_0 is the voltage of the battery. Assuming that the capacitor is uncharged at time $t = 0$ s when the switch is closed, it can be shown that the magnitude q of the charge on the plates at time t is

Capacitor charging $q = q_0\left[1 - e^{-t/(RC)}\right]$ (20.20)

Math Skills If the need arises, you can solve Equation 20.20 for the time *t* with the aid of natural logarithms. The first step is to rearrange the equation so as to isolate the exponential *e* on one side of the equals sign:

$$q = q_0\left[1 - e^{-t/(RC)}\right] \quad \text{or} \quad \frac{q}{q_0} = 1 - e^{-t/(RC)} \quad \text{or} \quad 1 - \frac{q}{q_0} = e^{-t/(RC)}$$

According to Equation D-9 in Appendix D, the natural logarithm of e^Z is $\ln e^Z = Z$. Therefore, we can take the natural logarithm of both sides of the rearranged equation and obtain

$$\ln\left(1 - \frac{q}{q_0}\right) = \ln e^{-t/(RC)} = -\frac{t}{RC} \quad \text{or} \quad t = -RC\left[\ln\left(1 - \frac{q}{q_0}\right)\right]$$

To illustrate this result, suppose that $RC = 5.00$ s and that the capacitor has acquired a charge q that is one-fourth of its final equilibrium value q_0, so that $q = 0.250\, q_0$. The time required for the capacitor to acquire the charge is

$$t = -(5.00 \text{ s})\left[\ln\left(1 - \frac{0.250\, q_0}{q_0}\right)\right]$$

$$= -(5.00 \text{ s}) \ \ln 0.750 = -(5.00 \text{ s})(-0.288) = 1.44 \text{ s}$$

where the exponential e has the value of 2.718. . . . Part b of the drawing shows a graph of this expression, which indicates that the charge is $q = 0$ C when $t = 0$ s and increases gradually toward the equilibrium value of $q_0 = CV_0$. The voltage V across the capacitor at any time can be obtained from Equation 20.20 by dividing the charges q and q_0 by the capacitance C, since $V = q/C$ and $V_0 = q_0/C$.

The term RC in the exponent in Equation 20.20 is called the **time constant** τ of the circuit:

$$\tau = RC \tag{20.21}$$

The time constant is measured in seconds; verification of the fact that an ohm times a farad is equivalent to a second is left as an exercise (see Check Your Understanding Question 21). The time constant is the amount of time required for the capacitor to accumulate 63.2% of its equilibrium charge, as can be seen by substituting $t = \tau = RC$ in Equation 20.20; $q_0(1 - e^{-1}) = q_0(0.632)$. The charge approaches its equilibrium value rapidly when the time constant is small and slowly when the time constant is large.

Figure 20.39a shows a circuit at a time t after a switch is closed to allow a charged capacitor to begin discharging. There is no battery in this circuit, so the charge $+q$ on the left plate of the capacitor can flow counterclockwise through the resistor and neutralize the charge $-q$ on the right plate. Assuming that the capacitor has a charge q_0 at time $t = 0$ s when the switch is closed, it can be shown that

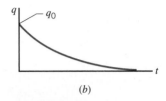

(b)

FIGURE 20.39 Discharging a capacitor.

Capacitor discharging $$q = q_0 e^{-t/(RC)} \tag{20.22}$$

where q is the amount of charge remaining on either plate at time t. The graph of this expression in part b of the drawing shows that the charge begins at q_0 when $t = 0$ s and decreases gradually toward zero. Smaller values of the time constant RC lead to a more rapid discharge. Equation 20.22 indicates that when $t = \tau = RC$, the magnitude of the charge remaining on each plate is $q_0 e^{-1} = q_0(0.368)$. Therefore, the time constant is also the amount of time required for a charged capacitor to *lose* 63.2% of its charge.

EXAMPLE 17 | BIO The Physics of Pacemakers

THE PHYSICS OF . . . heart pacemakers. The charging/discharging of a capacitor has many applications. Heart pacemakers, for instance, incorporate *RC* circuits to control the timing of voltage pulses that are delivered to a malfunctioning heart to regulate its beating cycle. The pulses are delivered by electrodes attached externally to the chest or located internally near the heart when the pacemaker is implanted surgically (see **Figure 20.40**). A voltage pulse is delivered when the capacitor discharges to a preset level, following which the capacitor is recharged rapidly and the cycle repeats. The value of the time constant *RC* controls the pulsing rate, which is about once per second.

One particular pacemaker uses an *RC* circuit that discharges when the charge on the capacitor reaches 75% of its equilibrium charge value. If this occurs every 1.2 seconds, and the capacitance of the *RC* circuit is 110 μF, what is the resistance of the resistor?

Reasoning To calculate the resistance of the *RC* circuit we can use Equation 20.20, which is the charging equation for the capacitor in the circuit.

Solution Beginning with Equation 20.20, we have: $q = q_0[1 - e^{-t/(RC)}]$. We rearrange this equation and solve for the resistance, R:

$$1 - \frac{q}{q_0} = e^{-t/(RC)} \Rightarrow \ln\left(1 - \frac{q}{q_0}\right) = \frac{-t}{RC} \Rightarrow R = \frac{-t}{C\left[\ln\left(1 - \frac{q}{q_0}\right)\right]}$$

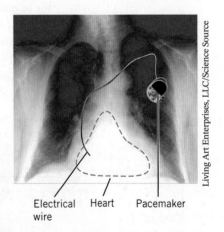

FIGURE 20.40 This color-enhanced frontal X-ray photograph of a human chest shows an enlarged heart and a heart pacemaker that has been implanted surgically.

Using the values given in the problem, we can compute the resistance as follows:

$$R = \frac{-1.2 \text{ s}}{(110 \times 10^{-6} \text{ F})\,[\ln(1 - 0.75)]} = \boxed{7.9 \times 10^3 \ \Omega}$$

Living Art Enterprises, LLC/Science Source

THE PHYSICS OF . . . windshield wipers. The charging/discharging of a capacitor is also used in automobiles that have windshield wipers equipped for intermittent operation during a light drizzle. In this mode of operation, the wipers remain off for a while and then turn on briefly. The timing of the on–off cycle is determined by the time constant of a resistor–capacitor combination.

Check Your Understanding

(The answers are given at the end of the book.)

21. The time constant τ of a series RC circuit is measured in seconds (s) and is given by $\tau = RC$, where the resistance R is measured in ohms (Ω) and the capacitance C is measured in farads (F). Verify that an ohm times a farad is equivalent to a second.

22. CYU Figure 20.5 shows two different resistor–capacitor arrangements. The time constant for arrangement A is 0.20 s. What is the time constant for arrangement B? **(a)** 0.050 s **(b)** 0.10 s **(c)** 0.20 s **(d)** 0.40 s **(e)** 0.80 s

CYU FIGURE 20.5

20.14 Safety and the Physiological Effects of Current

Electric circuits, although very useful, can also be hazardous. To reduce the danger inherent in using circuits, proper **electrical grounding** is necessary. The next two figures help to illustrate what electrical grounding means and how it is achieved.

Figure 20.41a shows a clothes dryer connected to a wall socket via a two-prong plug. The dryer is operating normally; that is, the wires inside are insulated from the metal casing of the dryer, so no charge flows through the casing itself. Notice that one terminal of the ac generator is customarily connected to ground ($\perp$) by the electric power company. Part *b* of the drawing shows the hazardous result that occurs if a wire comes loose and contacts the metal casing. A person touching it receives a shock, since electric charge flows through the casing, the person's body, and the ground on the way back to the generator.

FIGURE 20.41 (*a*) A normally operating clothes dryer that is connected to a wall socket via a two-prong plug. (*b*) An internal wire accidentally touches the metal casing, and a person who touches the casing receives an electrical shock.

THE PHYSICS OF . . . safe electrical grounding. Figure 20.42 shows the same appliance connected to a wall socket via a three-prong plug that provides safe electrical grounding. The third prong connects the metal casing directly to a copper rod driven into the ground or to a copper water pipe that is in the ground. This arrangement protects against electrical shock in the event that a broken wire touches the metal casing. In this event, charge would flow through the casing, through the third prong, and into

FIGURE 20.42 A safely connected dryer. If the dryer malfunctions, a person touching it receives no shock, since electric charge flows through the third prong and into the ground via a copper rod, rather than through the person's body.

the ground, returning eventually to the generator. No charge would flow through the person's body, because the copper rod provides much less electrical resistance than does the body.

BIO **THE PHYSICS OF . . . the physiological effects of current.** Serious and sometimes fatal injuries can result from electrical shock. The severity of the injury depends on the magnitude of the current and the parts of the body through which the moving charges pass. The amount of current that causes a mild tingling sensation is about 0.001 A. Currents on the order of 0.01–0.02 A can lead to muscle spasms, in which a person "can't let go" of the object causing the shock. Currents of approximately 0.2 A are potentially fatal because they can make the heart fibrillate, or beat in an uncontrolled manner. Substantially larger currents stop the heart completely. However, since the heart often begins beating normally again after the current ceases, the larger currents can be less dangerous than the smaller currents that cause fibrillation.

Concept Summary

20.1 Electromotive Force and Current There must be at least one source or generator of electrical energy in an electric circuit. The electromotive force (emf) of a generator, such as a battery, is the maximum potential difference (in volts) that exists between the terminals of the generator.

The rate of flow of charge is called the electric current. If the rate is constant, the current I is given by Equation 20.1, where Δq is the magnitude of the charge crossing a surface in a time Δt, the surface being perpendicular to the motion of the charge. The SI unit for current is the coulomb per second (C/s), which is referred to as an ampere (A). When the charges flow only in one direction around a circuit, the current is called direct current (dc). When the direction of charge flow changes from moment to moment, the current is known as alternating current (ac). Conventional current is the hypothetical flow of positive charges that would have the same effect in a circuit as the movement of negative charges that actually does occur.

$$I = \frac{\Delta q}{\Delta t} \qquad (20.1)$$

20.2 Ohm's Law The definition of electrical resistance R is $R = V/I$, where V (in volts) is the voltage applied across a piece of material and I (in amperes) is the current through the material. Resistance is measured in volts per ampere, a unit called an ohm (Ω). If the ratio of the voltage to the current is constant for all values of voltage and current, the resistance is constant. In this event, the definition of resistance becomes Ohm's law, Equation 20.2.

$$\frac{V}{I} = R = \text{constant} \quad \text{or} \quad V = IR \qquad (20.2)$$

20.3 Resistance and Resistivity The resistance of a piece of material of length L and cross-sectional area A is given by Equation 20.3,

where ρ is the resistivity of the material. The resistivity of a material depends on the temperature. For many materials and limited temperature ranges, the temperature dependence is given by Equation 20.4, where ρ and ρ_0 are the resistivities at the temperatures T and T_0, respectively, and α is the temperature coefficient of resistivity. The temperature dependence of the resistance R is given by Equation 20.5, where R and R_0 are the resistances at the temperatures T and T_0, respectively.

$$R = \rho \frac{L}{A} \qquad (20.3)$$

$$\rho = \rho_0[1 + \alpha(T - T_0)] \qquad (20.4)$$

$$R = R_0[1 + \alpha(T - T_0)] \qquad (20.5)$$

20.4 Electric Power When electric charge flows from point A to point B in a circuit, leading to a current I, and the voltage between the points is V, the electric power associated with this current and voltage is given by Equation 20.6a. For a resistor, Ohm's law applies, and it follows that the power delivered to the resistor is also given by either Equation 20.6b or 20.6c.

$$P = IV \qquad (20.6a)$$

$$P = I^2 R \qquad (20.6b)$$

$$P = \frac{V^2}{R} \qquad (20.6\text{vc})$$

20.5 Alternating Current The alternating voltage between the terminals of an ac generator can be represented by Equation 20.7, where V_0 is the peak value of the voltage, t is the time, and f is the frequency (in hertz) at which the voltage oscillates. Correspondingly, in a circuit containing only resistance, the ac current is given by Equation 20.8, where I_0 is the peak value of the current and is related to the peak voltage via $I_0 = V_0/R$.

$$V = V_0 \sin 2\pi ft \qquad (20.7)$$

$$I = I_0 \sin 2\pi ft \qquad (20.8)$$

For sinusoidal current and voltage, the root mean square (rms) current and voltage are related to the peak values according to Equations 20.12 and 20.13.

$$I_{rms} = \frac{I_0}{\sqrt{2}} \qquad (20.12)$$

$$V_{rms} = \frac{V_0}{\sqrt{2}} \qquad (20.13)$$

The power in an ac circuit is the product of the current and the voltage and oscillates in time. The average power is given by Equation 20.15a. For a resistor, Ohm's law applies, so that $V_{rms} = I_{rms}R$, and the average power delivered to the resistor is also given by Equations 20.15b and 20.15c.

$$\overline{P} = I_{rms}V_{rms} \qquad (20.15a)$$

$$\overline{P} = I_{rms}^2 R \qquad (20.15b)$$

$$\overline{P} = \frac{V_{rms}^2}{R} \qquad (20.15c)$$

20.6 Series Wiring When devices are connected in series, there is the same current through each device. The equivalent resistance R_S of a series combination of resistances (R_1, R_2, R_3, etc.) is given by Equation 20.16. The power delivered to the equivalent resistance is equal to the total power delivered to any number of resistors in series.

$$R_S = R_1 + R_2 + R_3 + \cdots \qquad (20.16)$$

20.7 Parallel Wiring When devices are connected in parallel, the same voltage is applied across each device. In general, devices wired in parallel carry different currents. The reciprocal of the equivalent resistance R_P of a parallel combination of resistances (R_1, R_2, R_3, etc.) is given by Equation 20.17. The power delivered to the equivalent resistance is equal to the total power delivered to any number of resistors in parallel.

$$\frac{1}{R_P} = \frac{1}{R_1} + \frac{1}{R_2} + \frac{1}{R_3} + \cdots \qquad (20.17)$$

20.8 Circuits Wired Partially in Series and Partially in Parallel Sometimes, one section of a circuit is wired in series, while another is wired in parallel. In such cases the circuit can be analyzed in parts, according to the respective series and parallel equivalent resistances of the various sections.

20.9 Internal Resistance The internal resistance of a battery or generator is the resistance within the battery or generator. The terminal voltage is the voltage between the terminals of a battery or generator and is equal to the emf only when there is no current through the device. When there is a current I, the internal resistance r causes the terminal voltage to be less than the emf by an amount Ir.

20.10 Kirchhoff's Rules Kirchhoff's junction rule states that the sum of the magnitudes of the currents directed into a junction equals the sum of the magnitudes of the currents directed out of the junction. Kirchhoff's loop rule states that, around any closed-circuit loop, the sum of the potential drops equals the sum of the potential rises. The Reasoning Strategy given at the end of Section 20.10 explains how these two rules are applied to analyze any circuit.

20.11 The Measurement of Current and Voltage A galvanometer is a device that responds to electric current and is used in nondigital ammeters and voltmeters. An ammeter is an instrument that measures current and must be inserted into a circuit in such a way that the current passes directly through the ammeter. A voltmeter is an instrument for measuring the voltage between two points in a circuit. A voltmeter must be connected between the two points and is not inserted into a circuit as an ammeter is.

20.12 Capacitors in Series and in Parallel The equivalent capacitance C_P for a parallel combination of capacitances (C_1, C_2, C_3, etc.) is given by Equation 20.18. In general, each capacitor in a parallel combination carries a different amount of charge. The equivalent capacitor carries the same total charge and stores the same total energy as the parallel combination.

$$C_P = C_1 + C_2 + C_3 + \cdots \qquad (20.18)$$

The reciprocal of the equivalent capacitance C_S for a series combination (C_1, C_2, C_3, etc.) of capacitances is given by Equation 20.19. The equivalent capacitor carries the same amount of charge as *any one* of the capacitors in the combination and stores the same total energy as the series combination.

$$\frac{1}{C_S} = \frac{1}{C_1} + \frac{1}{C_2} + \frac{1}{C_3} + \cdots \qquad (20.19)$$

20.13 RC Circuits The charging or discharging of a capacitor in a dc series circuit (resistance R, capacitance C) does not occur instantaneously. The charge on a capacitor builds up gradually, as described by Equation 20.20, where q is the charge on the capacitor at time t and q_0 is the equilibrium value of the charge. The time constant τ of the circuit is given by Equation 20.21. The discharging of a capacitor through a resistor is described by Equation 20.22, where q_0 is the charge on the capacitor at time $t = 0$ s.

$$q = q_0[1 - e^{-t/(RC)}] \qquad (20.20)$$

$$\tau = RC \qquad (20.21)$$

$$q = q_0 e^{-t/(RC)} \qquad (20.22)$$

Focus on Concepts

Online

Additional questions are available for assignment in WileyPLUS.

Section 20.1 Electromotive Force and Current

1. In 2.0 s, 1.9×10^{19} electrons pass a certain point in a wire. What is the current I in the wire?

Section 20.2 Ohm's Law

2. Which one of the following graphs correctly represents Ohm's law, where V is the voltage and I is the current? **(a)** A **(b)** B **(c)** C **(d)** D

QUESTION 2

Section 20.3 Resistance and Resistivity

3. Two wires are made from the same material. One wire has a resistance of 0.10 Ω. The other wire is twice as long as the first wire and has a radius that is half as much. What is the resistance of the second wire? **(a)** 0.40 Ω **(b)** 0.20 Ω **(c)** 0.10 Ω **(d)** 0.050 Ω **(e)** 0.80 Ω

Section 20.4 Electric Power

4. A single resistor is connected across the terminals of a battery. Which one or more of the following changes in voltage and current leaves unchanged the electric power dissipated in the resistor?

    (A) Doubling the voltage and reducing the current by a factor of two

    (B) Doubling the voltage and increasing the resistance by a factor of four

    (C) Doubling the current and reducing the resistance by a factor of four

(a) A, B, C **(b)** A, B **(c)** B, C **(d)** A **(e)** B

Section 20.5 Alternating Current

5. The average power dissipated in a 47-Ω resistor is 2.0 W. What is the peak value I_0 of the ac current in the resistor?

Section 20.6 Series Wiring

6. For the circuit shown in the drawing, what is the voltage V_1 across resistance R_1?

(a) $V_1 = \left(\dfrac{R_1}{R_2}\right)V$ **(d)** $V_1 = \left(\dfrac{R_1}{R_1 + R_2}\right)V$

(b) $V_1 = \left(\dfrac{R_2}{R_1}\right)V$ **(e)** $V_1 = \left(\dfrac{R_1 + R_2}{R_1}\right)V$

(c) $V_1 = V$

QUESTION 6

Section 20.7 Parallel Wiring

7. For the circuit shown in the drawing, what is the ratio of the current I_1 in resistance R_1 to the current I_2 in resistance R_2?

(a) $\dfrac{I_1}{I_2} = \dfrac{R_1}{R_2}$ **(d)** $\dfrac{I_1}{I_2} = 1$

(b) $\dfrac{I_1}{I_2} = \dfrac{R_2}{R_1 + R_2}$ **(e)** $\dfrac{I_1}{I_2} = \dfrac{R_2}{R_1}$

(c) $\dfrac{I_1}{I_2} = \dfrac{R_1}{R_1 + R_2}$

QUESTION 7

Section 20.8 Circuits Wired Partially in Series and Partially in Parallel

8. In the following three arrangements each resistor has the same resistance R. Rank the equivalent resistances of the arrangements in descending order (largest first). **(a)** A, B, C **(b)** B, A, C **(c)** B, C, A **(d)** A, C, B **(e)** C, B, A

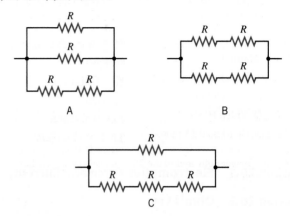

QUESTION 8

Section 20.9 Internal Resistance

9. A battery has an emf of V and an internal resistance of r. What resistance R, when connected across the terminals of this battery, will cause the terminal voltage of the battery to be $\frac{1}{2}V$? **(a)** $R = \frac{1}{2}r$ **(b)** $R = 2r$ **(c)** $R = 4r$ **(d)** $R = r$ **(e)** $R = \frac{1}{4}r$

Section 20.10 Kirchhoff's Rules

10. When applying Kirchhoff's rules, one of the essential steps is to mark each resistor with plus and minus signs to label how the potential changes from one end of the resistor to the other. The circuit in the drawing contains four resistors, each marked with the associated plus and minus signs. However, one resistor is marked incorrectly. Which one is it? **(a)** R_1 **(b)** R_2 **(c)** R_3 **(d)** R_4

QUESTION 10

Section 20.12 Capacitors in Series and in Parallel

11. Three capacitors are identical, each having a capacitance C. Two of them are connected in series. Then, this series combination is connected in parallel with the third capacitor. What is the equivalent capacitance of the entire connection? **(a)** $\frac{1}{2}C$ **(b)** $\frac{1}{3}C$ **(c)** $3C$ **(d)** $\frac{2}{3}C$ **(e)** $\frac{3}{2}C$

Section 20.13 RC Circuits

12. The time constant of an RC circuit is 2.6 s. How much time t is required for the capacitor (uncharged initially) to gain one-half of its full equilibrium charge?

Problems

Online

Additional questions are available for assignment in WileyPLUS.
Note: For problems that involve ac conditions, the current and volt-age are rms values and the power is an average value, unless indi-cated otherwise.

SSM Student Solutions Manual	**BIO** Biomedical application
MMH Problem-solving help	**E** Easy
GO Guided Online Tutorial	**M** Medium
V-HINT Video Hints	**H** Hard
CHALK Chalkboard Videos	**WS** Worksheet
	T Team Problem

Section 20.1 Electromotive Force and Current,

Section 20.2 Ohm's Law

1. **E** An especially violent lightning bolt has an average current of 1.26×10^3 A lasting 0.138 s. How much charge is delivered to the ground by the lightning bolt?

2. **E** **SSM** A battery charger is connected to a dead battery and de-livers a current of 6.0 A for 5.0 hours, keeping the voltage across the battery terminals at 12 V in the process. How much energy is deliv-ered to the battery?

3. **E** **CHALK** A coffee-maker contains a heating element that has a resistance of 14 Ω. This heating element is energized by a 120-V outlet. What is the current in the heating element?

4. **M** A car battery has a rating of 220 ampere · hours (A · h). This rating is one indication of the *total charge* that the battery can provide to a circuit before failing. **(a)** What is the total charge (in coulombs) that this battery can provide? **(b)** Determine the maximum current that the battery can provide for 38 minutes.

5. **M** A resistor is connected across the terminals of a 9.0-V battery, which delivers 1.1×10^5 J of energy to the resistor in six hours. What is the resistance of the resistor?

6. **M** **GO** The resistance of a bagel toaster is 14 Ω. To prepare a bagel, the toaster is operated for one minute from a 120-V outlet. How much energy is delivered to the toaster?

Section 20.3 Resistance and Resistivity

7. **E** **GO** The resistance and the magnitude of the current depend on the path that the current takes. The drawing shows three situations in which the current takes different paths through a piece of material. Each of the rectangular pieces is made from a material whose resis-tivity is $\rho = 1.50 \times 10^{-2}$ Ω · m, and the unit of length in the drawing is $L_0 = 5.00$ cm. Each piece of material is connected to a 3.00-V battery. Find **(a)** the resistance and **(b)** the current in each case.

(a)   (b)   (c)

PROBLEM 7

8. **E** Two wires are identical, except that one is aluminum and one is copper. The aluminum wire has a resistance of 0.20 Ω. What is the resistance of the copper wire?

9. **E** A cylindrical wire has a length of 2.80 m and a radius of 1.03 mm. It carries a current of 1.35 A, when a voltage of 0.0320 V is applied across the ends of the wire. From what material in **Table 20.1** is the wire made?

10. **E** A coil of wire has a resistance of 38.0 Ω at 25 °C and 43.7 Ω at 55 °C. What is the temperature coefficient of resistivity?

11. **E** A large spool in an electrician's workshop has 75 m of insulation-coated wire coiled around it. When the electrician con-nects a battery to the ends of the spooled wire, the resulting current is 2.4 A. Some weeks later, after cutting off various lengths of wire for use in repairs, the electrician finds that the spooled wire carries a 3.1-A current when the same battery is connected to it. What is the length of wire remaining on the spool?

12. **E** High-voltage power lines are a familiar sight throughout the country. The aluminum wire used for some of these lines has a cross-sectional area of 4.9×10^{-4} m². What is the resistance of ten kilometers of this wire?

13. **M** The temperature coefficient of resistivity for the metal gold is 0.0034 (C°)⁻¹, and for tungsten it is 0.0045 (C°)⁻¹. The resistance of a gold wire increases by 7.0% due to an increase in temperature. For the same increase in temperature, what is the percentage increase in the resistance of a tungsten wire?

14. **M** **GO** A tungsten wire has a radius of 0.075 mm and is heated from 20.0 to 1320 °C. The temperature coefficient of resistivity is $\alpha = 4.5 \times 10^{-3}$ (C°)⁻¹. When 120 V is applied across the ends of the hot wire, a current of 1.5 A is produced. How long is the wire? Neglect any effects due to thermal expansion of the wire.

15. **M** **V-HINT** **CHALK** Two cylindrical rods, one copper and the other iron, are identical in lengths and cross-sectional areas. They are joined end to end to form one long rod. A 12-V battery is connected across the free ends of the copper–iron rod. What is the voltage between the ends of the copper rod?

Section 20.4 Electric Power

16. **E** An electric blanket is connected to a 120-V outlet and con-sumes 140 W of power. What is the resistance of the heater wire in the blanket?

17. **E** The heating element in an iron has a resistance of 24 Ω. The iron is plugged into a 120-V outlet. What is the power delivered to the iron?

18. **E** A blow-dryer and a vacuum cleaner each operate with a volt-age of 120 V. The current rating of the blow-dryer is 11 A, and that of the vacuum cleaner is 4.0 A. Determine the power consumed by **(a)** the blow-dryer and **(b)** the vacuum cleaner. **(c)** Determine the ratio of the energy used by the blow-dryer in 15 minutes to the energy used by the vacuum cleaner in one-half hour.

19. **E** There are approximately 110 million households that use TVs in the United States. Each TV uses, on average, 75 W of power and is turned on for 6.0 hours a day. If electrical energy costs $0.12 per kWh, how much money is spent every day in keeping 110 million TVs turned on?

20. **E** An MP3 player operates with a voltage of 3.7 V, and is using 0.095 W of power. Find the current being supplied by the player's battery.

21. **E** **SSM** In doing a load of clothes, a clothes dryer uses 16 A of current at 240 V for 45 min. A personal computer, in contrast, uses 2.7 A of current at 120 V. With the energy used by the clothes dryer, how long (in hours) could you use this computer to "surf" the Internet?

22. **M** **V-HINT** An electric heater used to boil small amounts of water consists of a 15-Ω coil that is immersed directly in the water. It operates from a 120-V socket. How much time is required for this heater to raise the temperature of 0.50 kg of water from 13 °C to the normal boiling point?

23. **M** The rear window of a van is coated with a layer of ice at 0 °C. The density of ice is 917 kg/m^3. The driver of the van turns on the rear-window defroster, which operates at 12 V and 23 A. The defroster directly heats an area of 0.52 m^2 of the rear window. What is the maximum thickness of ice coating this area that the defroster can melt in 3.0 minutes?

24. **M** **GO** **CHALK** A piece of Nichrome wire has a radius of 6.5 × 10^{-4} m. It is used in a laboratory to make a heater that uses 4.00×10^2 W of power when connected to a voltage source of 120 V. Ignoring the effect of temperature on resistance, estimate the necessary length of wire.

25. **M** **SSM** Tungsten has a temperature coefficient of resistivity of 0.0045 (C°)$^{-1}$. A tungsten wire is connected to a source of constant voltage via a switch. At the instant the switch is closed, the temperature of the wire is 28 °C, and the initial power delivered to the wire is P_0. At what wire temperature will the power that is delivered to the wire be decreased to $\frac{1}{2}P_0$?

Section 20.5 Alternating Current

26. **E** According to Equation 20.7, an ac voltage V is given as a function of time t by $V = V_0 \sin 2\pi f t$, where V_0 is the peak voltage and f is the frequency (in hertz). For a frequency of 60.0 Hz, what is the smallest value of the time at which the voltage equals one-half of the peak value?

27. **E** The rms current in a copy machine is 6.50 A, and the resistance of the machine is 18.6 Ω. What are (a) the average power and (b) the peak power delivered to the machine?

28. **E** **GO** The rms current in a 47-Ω resistor is 0.50 A. What is the peak value of the voltage across this resistor?

29. **E** A 550-W space heater is designed for operation in Germany, where household electrical outlets supply 230 V (rms) service. What is the power output of the heater when plugged into a 120-V (rms) electrical outlet in a house in the United States? Ignore the effects of temperature on the heater's resistance.

30. **E** **V-HINT** Review Conceptual Example 7 as an aid in solving this problem. A portable electric heater uses 18 A of current. The manufacturer recommends that an extension cord attached to the heater receive no more than 2.0 W of power per meter of length. What is the smallest radius of copper wire that can be used in the extension cord? (*Note:* An extension cord contains two wires.)

31. **E** **SSM** The average power used by a stereo speaker is 55 W. Assuming that the speaker can be treated as a 4.0-Ω resistance, find the peak value of the ac voltage applied to the speaker.

32. **M** **GO** **CHALK** The *recovery time* of a hot water heater is the time required to heat all the water in the unit to the desired temperature. Suppose that a 52-gal (1.00 gal = 3.79 × 10^{-3} m^3) unit starts with cold water at 11 °C and delivers hot water at 53 °C. The unit is electric and utilizes a resistance heater (120 V ac, 3.0 Ω) to heat the water. Assuming that no heat is lost to the environment, determine the recovery time (in hours) of the unit.

33. **M** **SSM** A light bulb is connected to a 120.0-V wall socket. The current in the bulb depends on the time t according to the relation $I = (0.707 \text{ A}) \sin [(314 \text{ Hz})t]$. (a) What is the frequency f of the alternating current? (b) Determine the resistance of the bulb's filament. (c) What is the average power delivered to the light bulb?

Section 20.6 Series Wiring

34. **E** **SSM** Three resistors, 25, 45, and 75 Ω, are connected in series, and a 0.51-A current passes through them. What are (a) the equivalent resistance and (b) the potential difference across the three resistors?

35. **E** **GO** A 60.0-W lamp is placed in series with a resistor and a 120.0-V source. If the voltage across the lamp is 25 V, what is the resistance R of the resistor?

36. **E** **SSM** The current in a series circuit is 15.0 A. When an additional 8.00-Ω resistor is inserted in series, the current drops to 12.0 A. What is the resistance in the original circuit?

37. **E** **V-HINT** Multiple-Concept Example 9 discusses the physics principles used in this problem. Three resistors, 2.0, 4.0, and 6.0 Ω, are connected in series across a 24-V battery. Find the power delivered to each resistor.

38. **E** The current in a 47-Ω resistor is 0.12 A. This resistor is in series with a 28-Ω resistor, and the series combination is connected across a battery. What is the battery voltage?

39. **M** **V-HINT** Multiple-Concept Example 9 reviews the concepts that are important to this problem. A light bulb is wired in series with a 144-Ω resistor, and they are connected across a 120.0-V source. The power delivered to the light bulb is 23.4 W. What is the resistance of the light bulb? Note that there are two possible answers.

40. **M** **SSM** **MMH** **CHALK** Three resistors are connected in series across a battery. The value of each resistance and its maximum power rating are as follows: 2.0 Ω and 4.0 W, 12.0 Ω and 10.0 W, and 3.0 Ω and 5.0 W. (a) What is the greatest voltage that the battery can have without one of the resistors burning up? (b) How much power does the battery deliver to the circuit in (a)?

41. **M** **V-HINT** One heater uses 340 W of power when connected by itself to a battery. Another heater uses 240 W of power when connected by itself to the same battery. How much total power do the heaters use when they are both connected in series across the battery?

42. **H** Two resistances, R_1 and R_2, are connected in series across a 12-V battery. The current increases by 0.20 A when R_2 is removed, leaving R_1 connected across the battery. However, the current increases by just 0.10 A when R_1 is removed, leaving R_2 connected across the battery. Find (a) R_1 and (b) R_2.

Section 20.7 Parallel Wiring

43. **E** A coffee-maker (14 Ω) and a toaster (19 Ω) are connected in parallel to the same 120-V outlet in a kitchen. How much total power is supplied to the two appliances when both are turned on?

44. **E** For the three-way bulb (50 W, 100 W, 150 W) discussed in Conceptual Example 11, find the resistance of each of the two filaments. Assume that the wattage ratings are not limited by significant figures, and ignore any heating effects on the resistances.

45. **E** **GO** The drawing shows three different resistors in two different circuits. The battery has a voltage of $V = 24.0$ V, and the resistors have values of $R_1 = 50.0 \ \Omega$, $R_2 = 25.0 \ \Omega$, and $R_3 = 10.0 \ \Omega$. (a) For the circuit on the left, determine the current through and the voltage across each resistor. (b) Repeat part (a) for the circuit on the right.

(a)   (b)

PROBLEM 45

46. **E** **SSM** The drawing shows a circuit that contains a battery, two resistors, and a switch. What is the equivalent resistance of the circuit when the switch is (a) open and (b) closed? What is the total power delivered to the resistors when the switch is (c) open and (d) closed?

PROBLEM 46

47. **E** A 16-Ω loudspeaker, an 8.0-Ω loudspeaker, and a 4.0-Ω loudspeaker are connected in parallel across the terminals of an amplifier. Determine the equivalent resistance of the three speakers, assuming that they all behave as resistors.

48. **E** **SSM** Two resistors, 42.0 and 64.0 Ω, are connected in parallel. The current through the 64.0-Ω resistor is 3.00 A. (a) Determine the current in the other resistor. (b) What is the total power supplied to the two resistors?

49. **E** **V-HINT** Two identical resistors are connected in parallel across a 25-V battery, which supplies them with a total power of 9.6 W. While the battery is still connected, one of the resistors is heated so that its resistance doubles. The resistance of the other resistor remains unchanged. Find (a) the initial resistance of each resistor and (b) the total power delivered to the resistors after one resistor has been heated.

50. **E** **MMH** A coffee cup heater and a lamp are connected in parallel to the same 120-V outlet. Together, they use a total of 111 W of power. The resistance of the heater is 4.0×10^2 Ω. Find the resistance of the lamp.

51. **M** **CHALK** Two resistors have resistances R_1 and R_2. When the resistors are connected in series to a 12.0-V battery, the current from the battery is 2.00 A. When the resistors are connected in parallel to the battery, the total current from the battery is 9.00 A. Determine R_1 and R_2.

52. **M** **GO** **SSM** A cylindrical aluminum pipe of length 1.50 m has an inner radius of 2.00×10^{-3} m and an outer radius of 3.00×10^{-3} m. The interior of the pipe is completely filled with copper. What is the resistance of this unit? (*Hint:* Imagine that the pipe is connected between the terminals of a battery and decide whether the aluminum and copper parts of the pipe are in series or in parallel.)

53. **M** **GO** The drawing shows two circuits, and the same battery is used in each. The two resistances R_A in circuit A are the same, and the two resistances R_B in circuit B are the same. Knowing that the same total power is delivered in each circuit, find the ratio R_B/R_A for the circuits.

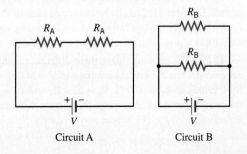

PROBLEM 53   Circuit A   Circuit B

Section 20.8   Circuits Wired Partially in Series and Partially in Parallel

54. **E** A 60.0-Ω resistor is connected in parallel with a 120.0-Ω resistor. This parallel group is connected in series with a 20.0-Ω resistor. The total combination is connected across a 15.0-V battery. Find (a) the current and (b) the power delivered to the 120.0-Ω resistor.

55. **E** **SSM** A 14-Ω coffee maker and a 16-Ω frying pan are connected in series across a 120-V source of voltage. A 23-Ω bread maker is also connected across the 120-V source and is in parallel with the series combination. Find the total current supplied by the source of voltage.

56. **E** **GO** Find the equivalent resistance between points A and B in the drawing.

PROBLEM 56

57. **E** **CHALK** **SSM** Determine the equivalent resistance between the points A and B for the group of resistors in the drawing.

PROBLEM 57

58. **E** **GO** The circuit in the drawing contains three identical resistors. Each resistor has a value of 10.0 Ω. Determine the equivalent resistance between the points a and b, b and c, and a and c.

PROBLEM 58   a   b

59. **E** **GO** Find the equivalent resistance between the points A and B in the drawing.

$R_1 = 16$ Ω $R_2 = 8$ Ω
$R_4 = 26$ Ω   B
A
$R_3 = 48$ Ω

PROBLEM 59

60. **M** **GO** Each resistor in the three circuits in the drawing has the same resistance R, and the batteries have the same voltage V. The values for R and V are 9.0 Ω and 6.0 V, respectively. Determine the total power delivered by the battery in each of the three circuits.

Circuit A   Circuit B

Circuit C

PROBLEM 60

61. [M] Eight different values of resistance can be obtained by connecting together three resistors (1.00, 2.00, and 3.00 Ω) in all possible ways. What are the values?

62. [M] [V-HINT] [MMH] Determine the power supplied to each of the resistors in the drawing.

PROBLEM 62

63. [M] [SSM] The circuit in the drawing contains five identical resistors. The 45-V battery delivers 58 W of power to the circuit. What is the resistance R of each resistor?

PROBLEM 63

Section 20.9 Internal Resistance

64. [E] A 1.40-Ω resistor is connected across a 9.00-V battery. The voltage between the terminals of the battery is observed to be only 8.30 V. Find the internal resistance of the battery.

65. [E] When a light bulb is connected across the terminals of a battery, the battery delivers 24 W of power to the bulb. A voltage of 11.8 V exists between the terminals of the battery, which has an internal resistance of 0.10 Ω. What is the emf of the battery?

66. [E] A battery has an internal resistance of 0.012 Ω and an emf of 9.00 V. What is the maximum current that can be drawn from the battery without the terminal voltage dropping below 8.90 V?

67. [M] [GO] [CHALK] A battery delivering a current of 55.0 A to a circuit has a terminal voltage of 23.4 V. The electric power being dissipated by the internal resistance of the battery is 34.0 W. Find the emf of the battery.

Section 20.10 Kirchhoff's Rules

68. [E] [SSM] Consider the circuit in the drawing. Determine **(a)** the magnitude of the current in the circuit and **(b)** the magnitude of the voltage between the points labeled A and B. **(c)** State which point, A or B, is at the higher potential.

PROBLEM 68

69. [E] [GO] The drawing shows a portion of a larger circuit. Current flows left to right in each resistor. What is the current in the resistor R?

PROBLEM 69

70. [E] [MMH] Find the magnitude and the direction of the current in the 2.0-Ω resistor in the drawing.

PROBLEM 70

71. [E] [GO] Using Kirchhoff's loop rule, find the value of the current I in part c of the drawing, where $R = 5.0$ Ω. (*Note:* Parts a and b of the drawing are used in the online tutorial help that is provided for this problem in the *WileyPLUS* homework management program.)

PROBLEM 71

72. [E] Determine the current (both magnitude and direction) in the 8.0- and 2.0-Ω resistors in the drawing.

PROBLEM 72

73. M V-HINT Determine the voltage across the 5.0-Ω resistor in the drawing. Which end of the resistor is at the higher potential?

PROBLEM 73

74. M CHALK SSM MMH Find the current in the 4.00-Ω resistor in the drawing. Specify the direction of the current.

PROBLEM 74

75. H None of the resistors in the circuit shown in the drawing is connected in series or in parallel with one another. Find (a) the current I_5 and the resistances (b) R_2 and (c) R_3.

PROBLEM 75

Section 20.11 The Measurement of Current and Voltage

76. E SSM The coil of a galvanometer has a resistance of 20.0 Ω, and its meter deflects full scale when a current of 6.20 mA passes through it. To make the galvanometer into a nondigital ammeter, a 24.8-mΩ shunt resistor is added to it. What is the maximum current that this ammeter can read?

77. E The coil of wire in a galvanometer has a resistance of $R_C = 60.0$ Ω. The galvanometer exhibits a full-scale deflection when the current through it is 0.400 mA. A resistor is connected in series with this combination so as to produce a nondigital voltmeter. The voltmeter is to have a full-scale deflection when it measures a potential difference of 10.0 V. What is the resistance of this resistor?

78. E A galvanometer with a coil resistance of 9.00 Ω is used with a shunt resistor to make a nondigital ammeter that has an equivalent resistance of 0.40 Ω. The current in the shunt resistor is 3.00 mA when the galvanometer reads full scale. Find the full-scale current of the galvanometer.

79. M SSM CHALK Two scales on a nondigital voltmeter measure voltages up to 20.0 and 30.0 V, respectively. The resistance connected in series with the galvanometer is 1680 Ω for the 20.0-V scale and 2930 Ω for the 30.0-V scale. Determine the coil resistance and the full-scale current of the galvanometer that is used in the voltmeter.

80. H In measuring a voltage, a voltmeter uses some current from the circuit. Consequently, the voltage measured is only an approximation to the voltage present when the voltmeter is not connected. Consider a circuit consisting of two 1550-Ω resistors connected in series across a 60.0-V battery. (a) Find the voltage across one of the resistors. (b) A nondigital voltmeter has a full-scale voltage of 60.0 V and uses a galvanometer with a full-scale deflection of 5.00 mA. Determine the voltage that this voltmeter registers when it is connected across the resistor used in part (a).

Section 20.12 Capacitors in Series and in Parallel

81. E Two capacitors are connected in parallel across the terminals of a battery. One has a capacitance of 2.0 μF and the other a capacitance of 4.0 μF. These two capacitors together store 5.4×10^{-5} C of charge. What is the voltage of the battery?

82. E Three parallel plate capacitors are connected in series. These capacitors have identical geometries. However, they are filled with three different materials. The dielectric constants of these materials are 3.30, 5.40, and 6.70. It is desired to replace this series combination with a single parallel plate capacitor. Assuming that this single capacitor has the same geometry as each of the other three capacitors, determine the dielectric constant of the material with which it is filled.

83. E SSM Three capacitors are connected in series. The equivalent capacitance of this combination is 3.00 μF. Two of the individual capacitances are 6.00 μF and 9.00 μF. What is the third capacitance (in μF)?

84. E GO Two capacitors are connected to a battery. The battery voltage is $V = 60.0$ V, and the capacitances are $C_1 = 2.00$ μF and $C_2 = 4.00$ μF. Determine the total energy stored by the two capacitors when they are wired (a) in parallel and (b) in series.

85. E Determine the equivalent capacitance between A and B for the group of capacitors in the drawing.

PROBLEM 85

86. E V-HINT A 2.00-μF and a 4.00-μF capacitor are connected to a 60.0-V battery. What is the total charge supplied to the capacitors when they are wired (a) in parallel and (b) in series with each other?

87. E Suppose that two capacitors (C_1 and C_2) are connected in series. Show that the sum of the energies stored in these capacitors is equal to the energy stored in the equivalent capacitor. [Hint: The energy stored in a capacitor can be expressed as $q^2/(2C)$.]

88. M A 3.00-μF and a 5.00-μF capacitor are connected in series across a 30.0-V battery. A 7.00-μF capacitor is then connected in parallel across the 3.00-μF capacitor. Determine the voltage across the 7.00-μF capacitor.

89. M CHALK SSM A 7.0-μF and a 3.0-μF capacitor are connected in series across a 24-V battery. What voltage is required to charge a parallel combination of the two capacitors to the same total energy?

90. H The drawing shows two capacitors that are fully charged ($C_1 = 2.00\ \mu$F, $q_1 = 6.00\ \mu$C; $C_2 = 8.00\ \mu$F, $q_2 = 12.0\ \mu$C). The switch is closed, and charge flows until equilibrium is reestablished (i.e., until both capacitors have the same voltage across their plates). Find the resulting voltage across either capacitor.

PROBLEM 90

Section 20.13   RC Circuits

91. E A circuit contains a resistor in series with a capacitor, the series combination being connected across the terminals of a battery, as in **Figure 20.38a**. The time constant for charging the capacitor is 1.5 s when the resistor has a resistance of $2.0 \times 10^4\ \Omega$. What would the time constant be if the resistance had a value of $5.2 \times 10^4\ \Omega$?

92. E V-HINT The circuit in the drawing contains two resistors and two capacitors that are connected to a battery via a switch. When the switch is closed, the capacitors begin to charge up. What is the time constant for the charging process?

PROBLEM 92

93. M MMH How many time constants must elapse before a capacitor in a series RC circuit is charged to 80.0% of its equilibrium charge?

94. M MMH CHALK Four identical capacitors are connected with a resistor in two different ways. When they are connected as in part a of the drawing, the time constant to charge up this circuit is 0.72 s. What is the time constant when they are connected with the same resistor, as in part b?

PROBLEM 94

Additional Problems   Online

95. E GO Each of the four circuits in the drawing consists of a single resistor whose resistance is either R or $2R$, and a single battery whose voltage is either V or $2V$. The unit of voltage in each circuit is $V = 12.0$ V and the unit of resistance is $R = 6.00\ \Omega$. Determine **(a)** the power supplied to each resistor and **(b)** the current delivered to each resistor.

PROBLEM 95

96. E SSM In the Arctic, electric socks are useful. A pair of socks uses a 9.0-V battery pack for each sock. A current of 0.11 A is drawn from each battery pack by wire woven into the socks. Find the resistance of the wire in one sock.

97. E In Section 12.3 it was mentioned that temperatures are often measured with electrical resistance thermometers made of platinum wire. Suppose that the resistance of a platinum resistance thermometer is 125 Ω when its temperature is 20.0 °C. The wire is then immersed in boiling chlorine, and the resistance drops to 99.6 Ω. The temperature coefficient of resistivity of platinum is $\alpha = 3.72 \times 10^{-3}$ (C°)$^{-1}$. What is the temperature of the boiling chlorine?

98. E The circuit in the drawing shows two resistors, a capacitor, and a battery. When the capacitor is fully charged, what is the magnitude q of the charge on one of its plates?

PROBLEM 98

99. E An 86-Ω resistor and a 67-Ω resistor are connected in series across a battery. The voltage across the 86-Ω resistor is 27 V. What is the voltage across the 67-Ω resistor?

100. M SSM The current in the 8.00-Ω resistor in the drawing is 0.500 A. Find the current in **(a)** the 20.0-Ω resistor and in **(b)** the 9.00-Ω resistor.

101. M SSM An extension cord is used with an electric weed trimmer that has a resistance of 15.0 Ω. The extension cord is made of copper wire that has a cross-sectional area of 1.3×10^{-6} m^2. The combined length of the two wires in the extension cord is 92 m. **(a)** Determine the resistance of the extension cord. **(b)** The extension cord is plugged into a 120-V socket. What voltage is applied to the trimmer itself?

PROBLEM 100

102. [M] [GO] The total current delivered to a number of devices connected in parallel is the sum of the individual currents in each device. Circuit breakers are resettable automatic switches that protect against a dangerously large total current by "opening" to stop the current at a specified safe value. A 1650-W toaster, a 1090-W iron, and a 1250-W microwave oven are turned on in a kitchen. As the drawing shows, they are all connected through a 20-A circuit breaker (which has negligible

PROBLEM 102

resistance) to an ac voltage of 120 V. **(a)** Find the equivalent resistance of the three devices. **(b)** Obtain the total current delivered by the source and determine whether the breaker will "open" to prevent an accident.

103. [M] [V-HINT] The filament in an incandescent light bulb is made from tungsten. The light bulb is plugged into a 120-V outlet and draws a current of 1.24 A. If the radius of the tungsten wire is 0.0030 mm, how long must the wire be?

Physics in Biology, Medicine, and Sports

104. [E] [BIO] **20.2** A defibrillator is used during a heart attack to restore the heart to its normal beating pattern (see Section 19.5). A defibrillator passes 18 A of current through the torso of a person in 2.0 ms. **(a)** How much charge moves during this time? **(b)** How many electrons pass through the wires connected to the patient?

105. [E] [BIO] [GO] **20.2** Suppose that the resistance between the walls of a biological cell is 5.0×10^9 Ω. **(a)** What is the current when the potential difference between the walls is 75 mV? **(b)** If the current is composed of Na^+ ions ($q = +e$), how many such ions flow in 0.50 s?

106. [H] [BIO] **20.3** A digital thermometer employs a thermistor as the temperature-ensing element. A thermistor is a kind of semiconductor and has a large negative temperature coefficient of resistivity α. Suppose that $\alpha = -0.060$ (C°)$^{-1}$ for the thermistor in a digital thermometer used to measure the temperature of a patient. The resistance of the thermistor decreases to 85% of its value at the normal body temperature of 37.0 °C. What is the patient's temperature?

107. [E] [BIO] [SSM] **20.13** In a heart pacemaker, a pulse is delivered to the heart 81 times per minute. The capacitor that controls this pulsing rate discharges through a resistance of 1.8×10^6 Ω. One pulse is delivered every time the fully charged capacitor loses 63.2% of its original charge. What is the capacitance of the capacitor?

108. [M] [BIO] **20.4** The World Meteorological Organization uses a system of satellites to continuously monitor electrical activity in the earth's atmosphere. Very recently, they observed two world-record lightning events. The first record was for the longest (in length) lightning strike, which occurred on October 31, 2018, over Brazil. The length of the flash was an incredible 709 km (441 miles)! This more than doubled the previous record, which was recorded in 2007, where a flash over Oklahoma almost spanned the entire state. The second was for the longest-lasting lightning strike, which occurred in the skies between Cordoba and Buenos Aires in Argentina on March 4, 2019. Typically, lightning flashes last just for a brief moment, but this one lasted for 16.7 s! **(a)** If the current in this strike was 2.00×10^5 A, and the potential difference between its ends was 1.20×10^5 V, what was the power delivered by lightning flash? **(b)** How much energy was released by the flash during the 16.7 s? **(c)** How many years could this lightning flash provide the energy requirements for an average home that uses 11 000 kWh/year?

109. [M] [BIO] **20.12** Electrogenic fish, like electric eels, have the ability to create fairly large voltages that can be used for defense or to stun their prey (see the photo). The electric eel has three organs along the length of its body that allow it to produce electricity. The organs contain special cells, called *electrocytes,* that can generate a voltage of 0.15 V/cell. When aligned, the cells resemble a stack of capacitors connected in series that allow a current of ions to flow through them. Eels are capable of producing a discharge at 860 V with a current of 1.0 A. **(a)** How much electrical power is delivered by the eel during a discharge? Use the information above. **(b)** How many cells would have to be aligned to produce a discharge voltage of 860 V? **(c)** What is the ratio of the capacitance of a single electrocyte to that of the equivalent capacitance of the entire stack of cells?

PROBLEM 109

110. [E] [BIO] **20.2 and 20.14** Human body tissue has a fairly low resistance, as the electrolytic fluids in our tissues are good conductors of electricity. However, the electrical resistance of human skin will largely determine how much current will flow through the body. Skin resistance will vary greatly, depending on the condition of the skin. Thick, dry skin has a much higher resistance than thin skin that is wet from, say, perspiration. Under dry conditions, the resistance of human skin can be 100 kΩ. Under wet conditions, this value can drop below 1 kΩ. What is the minimum skin resistance a person can have while receiving a shock from a 120-V outlet, if the current they receive cannot exceed 2.0×10^2 mA?

111. [M] [BIO] **20.1** In 50.0 ms, a current of 3.70 nA flows through a cell membrane wall as the result of the motion of sodium ions out of the cell. How many moles of sodium pass through the wall during this time?

112. M BIO **20.4** After open-heart surgery, the heart may need to be "kickstarted" or have its normal sinus rhythm restored. This is often done with an internal defibrillator that consists of two conducting metal paddle electrodes that are placed directly on opposite sides of the heart (see the photos). Usually an electric shock with an energy as small as 5.0 J is sufficient to eliminate v-fib, or ventricular fibrillation, in which the ventricular chambers of the heart quiver, as opposed to pumping blood. **(a)** If the capacitor used in the internal defibrillator has a value of 45 μF, what is the charging voltage? **(b)** If the energy stored in the capacitor is sent to the patient's heart in 210 ms, what is the effective electrical resistance of the heart?

Copyright ©2001-2020 DOTmed.com, Inc.

PROBLEM 112

Concepts and Calculations Problems

Online

Series and parallel wiring are two common ways in which devices, such as light bulbs, can be connected in a circuit. Problem 113 reviews the concepts of voltage, resistance, and power in the context of these two types of circuits. Kirchhoff's junction rule and loop rule are important tools for analyzing the currents and voltages in complex circuits. The rules are easy to use, once some of the subtleties are understood. Problem 114 explores these subtleties in a two-loop circuit.

113. M CHALK SSM A circuit contains a 48-V battery and a single light bulb whose resistance is 240 Ω. A second, identical, light bulb can be wired either in series or in parallel with the first one (see the figure). *Concepts:* (i) How is the power P that is delivered to a light bulb related to the bulb's resistance R and the voltage V across it? (ii) When there is only one bulb in the circuit, what is the voltage across it? (iii) The more power delivered to a bulb, the brighter it is. When two bulbs are wired in series, does the brightness of each bulb increase, decrease, or remain the same relative to the brightness of the bulb in the single-bulb circuit? (iv) When two bulbs are wired in parallel, does the brightness of each bulb increase, decrease, or remain the same relative to the brightness of the bulb in the single-bulb circuit? *Calculations:* Determine the power delivered to a single bulb when the circuit contains **(a)** only one bulb, **(b)** two bulbs in series and **(c)** two bulbs in parallel. Assume that the battery has no internal resistance.

114. M CHALK For this problem concerning Kirchhoff's junction rule and loop rule, refer to the figure. *Concepts:* (i) Notice that there are two loops, labeled 1 and 2 in this circuit. Does it matter that there is no battery in loop 1, but only two resistors? Explain. (ii) The currents through the three resistors are labeled as I_1, I_2, and I_3. Does it matter which direction, left-to-right or right-to-left, has been chosen for each circuit? (iii) When we place + and − signs on the ends of each resistor, does it matter which side is + and which is −? (iv) When we evaluate the potential drops and rises around a closed loop, does it matter which direction, clockwise or counterclockwise, is chosen for the evaluation? *Calculations:* Use Kirchhoff's junction and loop rules to determine the currents through the three resistors.

PROBLEM 114

240 Ω

48 V

240 Ω 240 Ω

48 V

Series wiring

240 Ω

240 Ω

48 V

Parallel wiring

PROBLEM 113

Team Problems

115. **E** **T** **WS** **Ion Thrusters III: Current and Power.** You and your team are designing a new propulsion engine that ejects xenon (Xe) ions in one direction, causing a spacecraft to accelerate in the opposite direction, in accordance with the conservation of linear momentum and the impulse-momentum theorem. This is exactly what happens when a rocket engine expels high velocity particles when burning fuel, except that, in the case of the ion drive, the ions are accelerated with an electric field. In this process, Xe gas atoms are ionized so that they each acquire a net charge of $+e$. They are then subjected to an electric field, which accelerates and ejects the ions from the engine. In the thruster you are designing, the Xe atoms accelerate through a potential difference of 2.06 kV between a grounded electrode and a negatively charged grid, through which the ions are ejected at a speed of 55.0 km/s relative to the engine. **(a)** If Xe ions are ejected from the thruster at a rate of 9.05×10^{-6} kg/s, what is the magnitude of the electric current that passes through the grid, that is, ejected from the thruster? **(b)** What is the electrical power in watts supplied by the power source of the ion thruster? The mass of a Xe ion is 2.18×10^{-25} kg.

116. **M** **T** **WS** **Dividing the Voltage.** You and your team are on a science expedition and encounter a problem with one of your instruments. After some troubleshooting, you find that its 5.00 V dc power supply has failed. You rummage around your supplies and find a 10.0-V dc power supply, but that voltage is too large and will damage your instrument. You also discover a bag of fifty 22.0 Ω resistors that have a power rating of 0.250 W, and get the idea to build a circuit to reduce the 10.0 V supply to a 5.00 V supply. The drawing shows a simple circuit called a voltage divider. The idea is that the voltage across the lower resistor in the voltage divider is equal to one-half the voltage (V) of the source, provided that the two resisters in the circuit have equal values. **(a)** Determine the resistance R in the voltage divider shown in the drawing that utilizes the resistors (i.e., R must be an integer number of 22.0 Ω resistors) and the 10.0-V power supply that results in a dc voltage of 5.00 V at the output terminals. The resistor values should be chosen to provide the largest current possible without exceeding one-half the power rating of the resistors (i.e., 0.125 W). **(b)** Draw the circuit and explicitly show the resistors, 10.0-V supply, and output terminals for the 5.00 V supply. **(c)** Determine the total power dissipated by the resistors in your voltage divider when there is nothing connected to the output terminals.

PROBLEM 116

117. **M** **T** **Fixing a Radio.** You and your team are stranded on a tropical island that hosts a deserted military base. In your efforts to get rescued, you find a radio. However, when you power it up some lights turn on, but the radio does not transmit or receive. You open up the back cover and quickly identify the problem: There is a burnt resistor on the circuit board. You have located some basic electronics tools and supplies, but the charred resistor is unidentifiable, so you do not know what to use for its replacement. There is a crude schematic on the inside of the cover of the radio, but it only gives the value of the current through the part of the circuit with the burnt resistor. You sketch the circuit with all the elements you can identify (see the drawing). **(a)** Determine the value of the damaged resistor. **(b)** The next problem is that the resistors that you have found are all 47.0 Ω. The available space will only allow for six or fewer resistors. Find a configuration of six or fewer 47.0-Ω resistors that has an equivalent resistance within 10% of that calculated in part (a).

PROBLEM 117   (Burnt)

118. **M** **T** **A Resistive Heater.** You and your team are designing a small tube heater that consists of a small ceramic tube wrapped with a special heater wire composed of Nichrome. When you run an electrical current through the wire, the wire (and therefore the tube) heats up through resistive heating. Nichrome is an alloy composed of 80% nickel and 20% chromium, and has a resistivity of $\rho = 1.25 \times 10^{-6}\ \Omega \cdot$m. **(a)** What is the resistance per centimeter of 28-gauge Nichrome wire? ("28 gauge" means the wire has a diameter $D = 0.320$ mm.) **(b)** You wrap the tube with 45.0 cm of the Nichrome wire, and you want the power output of your heater to be 120 W. What current is needed? **(c)** What voltage will provide the current calculated in (b)?

Earth

Active regions on the surface of the sun correspond to areas of intense magnetic fields. The looping magnetic field lines above this area are illuminated by the motion of charged particles. The moving charges experience a force in the magnetic field and spiral along them. In this chapter, we will study magnetic fields and the forces they apply to charged particles. The sizes of these magnetic structures can be enormous, as the earth, drawn to scale at the lower right, demonstrates.

Solar Dynamics Observatory, NASA

Magnetic Forces and Magnetic Fields

21.1 | Magnetic Fields

Permanent magnets have long been used in navigational compasses. As **Figure 21.1** illustrates, the compass needle is a permanent magnet supported so it can rotate freely in a plane. When the compass is placed on a horizontal surface, the needle rotates until one end points approximately to the north. The end of the needle that points north is labeled the **north magnetic pole**; the opposite end is the **south magnetic pole**.

Magnets can exert forces on each other. **Figure 21.2** shows that the magnetic forces between north and south poles have the property that

like poles repel each other, and unlike poles attract.

This behavior is similar to that of like and unlike electric charges. However, there is a significant difference between magnetic poles and electric charges. It is possible to separate positive from negative electric charges and produce isolated charges of either kind. In contrast, no one has

LEARNING OBJECTIVES

After reading this module, you should be able to...

21.1 Define magnetic field.

21.2 Calculate the magnetic force on a moving charge in a magnetic field.

21.3 Analyze the motion of a charged particle in a magnetic field.

21.4 Describe how the masses of ions are determined using a mass spectrometer.

21.5 Calculate the magnetic force on a current in a magnetic field.

21.6 Calculate the torque on a current-carrying coil.

21.7 Calculate magnetic fields produced by currents.

21.8 Apply Ampère's law to calculate the magnetic field due to a steady current.

21.9 Describe magnetic materials.

FIGURE 21.1 The needle of a compass is a permanent magnet that has a north magnetic pole (N) at one end and a south magnetic pole (S) at the other.

Like poles repel
(a)

Unlike poles attract
(b)

FIGURE 21.2 Bar magnets have a north magnetic pole at one end and a south magnetic pole at the other end. (a) Like poles repel each other, and (b) unlike poles attract.

Compass

FIGURE 21.3 At any location in the vicinity of a magnet, the north pole (the arrowhead in this drawing) of a small compass needle points in the direction of the magnetic field at that location.

found a magnetic monopole (an isolated north or south pole). Any attempt to separate north and south poles by cutting a bar magnet in half fails, because each piece becomes a smaller magnet with its own north and south poles.

Surrounding a magnet, there is a **magnetic field**. The magnetic field is analogous to the electric field that exists in the space around electric charges. Like the electric field, the magnetic field has both a magnitude and a direction. We postpone a discussion of the magnitude until Section 21.2, concentrating our attention here on the direction. ***The direction of the magnetic field at any point in space is the direction indicated by the north pole of a small compass needle placed at that point.*** In **Figure 21.3** the compass needle is symbolized by an arrow, with the head of the arrow representing the north pole. The drawing shows how compasses can be used to map out the magnetic field in the space around a bar magnet. Since like poles repel and unlike poles attract, the needle of each compass becomes aligned relative to the magnet in the manner shown in the picture. The compass needles provide a visual picture of the magnetic field that the bar magnet creates.

To help visualize the electric field, we introduced electric field lines in Section 18.7. In a similar fashion, it is possible to draw magnetic field lines, and **Figure 21.4a** illustrates some of the lines around a bar magnet. The lines appear to originate from the north pole and end on the south pole; they do not start or stop in midspace. A visual image of the magnetic field lines can be created by sprinkling finely ground iron filings on a piece of paper that covers the magnet. Iron filings in a magnetic field behave like tiny compasses and align themselves along the field lines, as the photograph in **Figure 21.4b** shows.

As is the case with electric field lines, the magnetic field at any point is tangent to the magnetic field line at that point. Furthermore, the strength of the field is proportional to the number of lines per unit area that passes through a surface oriented perpendicular to the lines. Thus, the magnetic field is stronger in regions where the field lines are relatively close together and weaker where they are relatively far apart. For instance, in **Figure 21.4a** the lines are closest together near the north and south poles, reflecting the fact that the strength of the field is greatest in these regions. Away from the poles, the magnetic field becomes weaker. Notice in part *c* of the drawing that the field lines in

Magnetic
field lines

(a)

(b)

Cordelia Molloy/Science Source

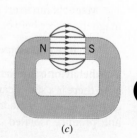

(c)

FIGURE 21.4 (a) The magnetic field lines and (b) the pattern of iron filings (black, curved regions) in the vicinity of a bar magnet. (c) The magnetic field lines in the gap of a horseshoe magnet.

the gap between the poles of the horseshoe magnet are nearly parallel and equally spaced, indicating that the magnetic field there is approximately constant.

Although the north pole of a compass needle points northward, it does not point exactly at the north geographic pole. The north geographic pole is that point where the earth's axis of rotation crosses the surface in the northern hemisphere (see **Figure 21.5**). Measurements of the magnetic field surrounding the earth show that the earth behaves magnetically almost as if it were a bar magnet.* As the drawing illustrates, the orientation of this fictitious bar magnet defines a magnetic axis for the earth. The location at which the magnetic axis crosses the surface in the northern hemisphere is known as the north magnetic pole. It is so named because it is the location toward which the north end of a compass needle points. Since unlike poles attract, the south pole of the earth's fictitious bar magnet lies beneath the north magnetic pole, as **Figure 21.5** indicates.

The north magnetic pole does not coincide with the north geographic pole but, instead, lies at a latitude of nearly 80°, just northwest of Ellef Ringnes Island in extreme northern Canada. It is interesting to note that the position of the north magnetic pole is not fixed, but moves over the years. Pointing as it does at the north magnetic pole, a compass needle deviates from the north geographic pole. The angle that a compass needle deviates is called the *angle of declination*. For New York City, the present angle of declination is about 13° west, meaning that a compass needle points 13° west of geographic north.

Figure 21.5 shows that the earth's magnetic field lines are not parallel to the surface at all points. For instance, near the north magnetic pole the field lines are almost perpendicular to the surface of the earth. The angle that the magnetic field makes with respect to the surface at any point is known as the *angle of dip*.

BIO **THE PHYSICS OF . . . navigation in animals.** Some animals can sense the earth's magnetic field and use it for navigational purposes. Until recently (2004), the only examples of this ability were in vertebrates, or animals that have a backbone, such as migratory birds. Now, however, researchers have found that the spiny lobster (see **Figure 21.6**), which is an invertebrate, can also use the earth's magnetic field to navigate and can determine its geographic location in a way similar to that of a person using the Global Positioning System (see Section 5.5). This ability may be related to the presence in the lobsters of the mineral magnetite, a magnetic material used for compass needles.

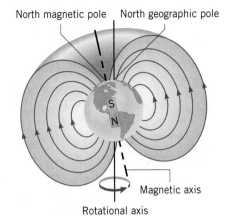

FIGURE 21.5 The earth behaves magnetically almost as if a bar magnet were located near its center. The axis of this fictitious bar magnet does not coincide with the earth's rotational axis; the two axes are currently about 9.4° apart.

FIGURE 21.6 Spiny lobsters use the earth's magnetic field to navigate and determine their geographic position.

21.2 The Force That a Magnetic Field Exerts on a Moving Charge

When a charge is placed in an electric field, it experiences an electric force, as Section 18.6 discusses. When a charge is placed in a magnetic field, it also experiences a force, provided that certain conditions are met, as we will see. The **magnetic force**, like all the forces we have studied (e.g., the gravitational, elastic, and electric forces), may contribute to the net force that causes an object to accelerate. Thus, when present, the magnetic force must be included in Newton's second law.

The following two conditions must be met for a charge to experience a magnetic force when placed in a magnetic field:

1. The charge must be moving, because no magnetic force acts on a stationary charge.

2. The velocity of the moving charge must have a component that is perpendicular to the direction of the magnetic field.

To examine the second condition, consider **Figure 21.7**, which shows a positive test charge $+q_0$ moving with a velocity $\vec{v}$ through a magnetic field $\vec{B}$. The field is produced

*At present it is not known with certainty what causes the earth's magnetic field. The magnetic field seems to originate from electric currents that in turn arise from electric charges circulating within the liquid outer region of the earth's core. Section 21.7 discusses how a current produces a magnetic field.

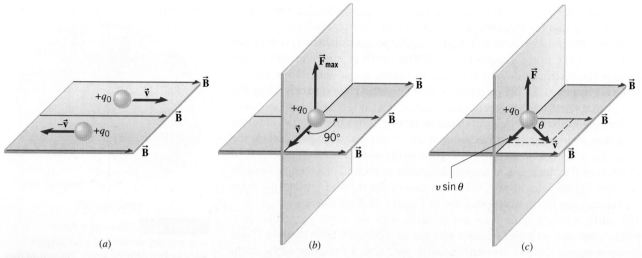

(a)   (b)   (c)

FIGURE 21.7 (a) No magnetic force acts on a charge moving with a velocity $\vec{v}$ that is parallel or antiparallel to a magnetic field $\vec{B}$. (b) The charge experiences a maximum force $\vec{F}_{max}$ when the charge moves perpendicular to the field. (c) If the charge travels at an angle θ with respect to $\vec{B}$, only the velocity component perpendicular to $\vec{B}$ gives rise to a magnetic force $\vec{F}$, which is smaller than $\vec{F}_{max}$. This component is $v \sin \theta$.

by magnets not shown in the drawing and is assumed to be constant in both magnitude and direction. If the charge moves *parallel or antiparallel* to the field, as in **Figure 21.7a**, the charge experiences *no magnetic force*. If, however, the charge moves *perpendicular* to the field, as in **Figure 21.7b**, the charge experiences the *maximum possible force* $\vec{F}_{max}$. In general, if a charge moves at an angle θ^* with respect to the field (see **Figure 21.7c**), only the velocity component $v \sin \theta$, which is perpendicular to the field, gives rise to a magnetic force. This force $\vec{F}$ is smaller than the maximum possible force. The component of the velocity that is parallel to the magnetic field yields no force.

Figure 21.7 shows that the direction of the magnetic force $\vec{F}$ is perpendicular to both the velocity $\vec{v}$ and the magnetic field $\vec{B}$; in other words, $\vec{F}$ is perpendicular to the plane defined by $\vec{v}$ and $\vec{B}$. As an aid in remembering the direction of the force, it is convenient to use *Right-Hand Rule No. 1 (RHR-1)*, as **Animated Figure 21.8** illustrates:

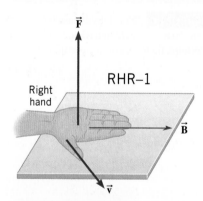

RHR–1

Right hand

> ***Right-Hand Rule No. 1:*** **The RHR to Find the Direction of the Magnetic Force That Acts on Moving Charges and Currents (Section 21.5) in a Magnetic Field.** Extend the right hand so the fingers point along the direction of the magnetic field $\vec{B}$ and the thumb points along the velocity $\vec{v}$ of the charge. The palm of the hand then faces in the direction of the magnetic force $\vec{F}$ that acts on a positive charge.

ANIMATED FIGURE 21.8 Right-Hand Rule No. 1 is illustrated. When the right hand is oriented so the fingers point along the magnetic field $\vec{B}$ and the thumb points along the velocity $\vec{v}$ of a positively charged particle, the palm faces in the direction of the magnetic force $\vec{F}$ applied to the particle.

It is as if the open palm of the right hand pushes on the positive charge in the direction of the magnetic force.

| **Problem-Solving Insight** If the moving charge is negative instead of positive, the direction of the magnetic force is opposite to that predicted by RHR-1. |

Thus, there is an easy method for finding the force on a moving negative charge. First, assume that the charge is positive and use RHR-1 to find the direction of the force. Then, reverse this direction to find the direction of the force acting on the negative charge.

We will now use what we know about the magnetic force to define the magnetic field, in a procedure that is analogous to that used in Section 18.6 to define the electric field. Recall that the electric field at any point in space is the force per unit charge that acts on a test charge q_0 placed at that point. In other words, to determine the electric field $\vec{E}$, we divide the electrostatic force $\vec{F}$ by the charge q_0: $\vec{E} = \vec{F}/q_0$. In the magnetic case, however, the test charge is moving, and the force depends not only on the charge q_0, but also on the velocity component $v \sin \theta$ that is perpendicular to the magnetic field. Therefore, to determine the magnitude of the magnetic field, we divide the magnitude of the magnetic force by the magnitude $|q_0|$ of the charge and also by $v \sin \theta$, according to the following definition:

*The angle θ between the velocity of the charge and the magnetic field is chosen so that it lies in the range $0 \leq \theta \leq 180°$.

DEFINITION OF THE MAGNETIC FIELD

The magnitude B of the magnetic field at any point in space is defined as

$$B = \frac{F}{|q_0|(v \sin \theta)} \qquad (21.1)$$

where F is the magnitude of the magnetic force on a test charge, $|q_0|$ is the magnitude of the test charge, and v is the magnitude of the charge's velocity, which makes an angle θ ($0 \le \theta \le 180°$) with the direction of the magnetic field. The magnetic field $\vec{B}$ is a vector, and its direction can be determined by using a small compass needle.

SI Unit of Magnetic Field: $\dfrac{\text{newton} \cdot \text{second}}{\text{coulomb} \cdot \text{meter}} = 1$ tesla (T)

The unit of magnetic field strength that follows from Equation 21.1 is the N · s/(C · m). This unit is called the *tesla* (T), a tribute to the Croatian-born American engineer Nikola Tesla (1856–1943). Thus, one tesla is the strength of the magnetic field in which a unit test charge, traveling perpendicular to the magnetic field at a speed of one meter per second, experiences a force of one newton. Because a coulomb per second is an ampere (1 C/s = 1 A, see Section 20.1), the tesla is often written as 1 T = 1 N/(A · m).

In many situations the magnetic field has a value that is considerably less than one tesla. For example, the strength of the magnetic field near the earth's surface is approximately 10^{-4} T. In such circumstances, a magnetic field unit called the *gauss* (G) is sometimes used. Although not an SI unit, the gauss is a convenient size for many applications involving magnetic fields. The relation between the gauss and the tesla is

$$1 \text{ gauss} = 10^{-4} \text{ tesla}$$

Example 1 deals with the magnetic force exerted on a moving proton and on a moving electron.

EXAMPLE 1 | Magnetic Forces on Charged Particles

A proton in a particle accelerator has a speed of 5.0×10^6 m/s. The proton encounters a magnetic field whose magnitude is 0.40 T and whose direction makes an angle of $\theta = 30.0°$ with respect to the proton's velocity (see **Figure 21.7c**). Find the magnitude and direction of **(a)** the magnetic force on the proton and **(b)** the acceleration of the proton. **(c)** What would be the force and acceleration if the particle were an electron instead of a proton?

Reasoning For both the proton and the electron, the magnitude of the magnetic force is given by Equation 21.1. The magnetic forces that act on these particles have opposite directions, however, because the charges have opposite signs. In either case, the acceleration is given by Newton's second law, which applies to the magnetic force just as it does to any force. In using the second law, we must take into account the fact that the masses of the proton and the electron are different.

Solution **(a)** The positive charge on a proton is 1.60×10^{-19} C, and according to Equation 21.1, the magnitude of the magnetic force is $F = |q_0|vB \sin \theta$. Therefore,

$$F = (1.60 \times 10^{-19} \text{ C})(5.0 \times 10^6 \text{ m/s})(0.40 \text{ T})(\sin 30.0°)$$

$$= \boxed{1.6 \times 10^{-13} \text{ N}}$$

The direction of the magnetic force is given by RHR-1 and is **upward** in **Figure 21.7c**, with the magnetic field pointing to the right.

(b) The magnitude a of the proton's acceleration follows directly from Newton's second law as the magnitude of the net force divided by the mass m_p of the proton. Since the only force acting on the proton is the magnetic force F, it is the net force. Thus,

$$a = \frac{F}{m_p} = \frac{1.6 \times 10^{-13} \text{ N}}{1.67 \times 10^{-27} \text{ kg}} = \boxed{9.6 \times 10^{13} \text{ m/s}^2} \qquad (4.1)$$

The direction of the acceleration is the same as the direction of the net force (the magnetic force).

(c) The magnitude of the magnetic force on the electron is the same as that on the proton, since both have the same velocity and charge magnitude. However, the direction of the force on the electron is opposite to that on the proton, or **downward** in **Figure 21.7c**, since the electron charge is negative. Furthermore, the electron has a smaller mass m_e and, therefore, experiences a significantly greater acceleration:

$$a = \frac{F}{m_e} = \frac{1.6 \times 10^{-13} \text{ N}}{9.11 \times 10^{-31} \text{ kg}} = \boxed{1.8 \times 10^{17} \text{ m/s}^2}$$

The direction of this acceleration is downward in **Figure 21.7c**.

Check Your Understanding

(The answers are given at the end of the book.)

1. Suppose that you accidentally use your left hand, instead of your right hand, to determine the direction of the magnetic force that acts on a positive charge moving in a magnetic field. Do you get the correct answer? **(a)** Yes, because either hand can be used **(b)** No, because the direction you get will be perpendicular to the correct direction **(c)** No, because the direction you get will be opposite to the correct direction

2. Two particles, having the same charge but different velocities, are moving in a constant magnetic field (see **CYU Figure 21.1**, where the velocity vectors are drawn to scale). Which particle, if either, experiences the greater magnetic force? **(a)** Particle 1 experiences the greater force, because it is moving perpendicular to the magnetic field. **(b)** Particle 2 experiences the greater force, because it has the greater speed. **(c)** Particle 2 experiences the greater force, because a component of its velocity is parallel to the magnetic field. **(d)** Both particles experience the same magnetic force, because the component of each velocity that is perpendicular to the magnetic field is the same.

CYU FIGURE 21.1

3. A charged particle, passing through a certain region of space, has a velocity whose magnitude and direction remain constant. **(a)** If it is known that the external magnetic field is zero everywhere in this region, can you conclude that the external electric field is also zero? **(b)** If it is known that the external electric field is zero everywhere, can you conclude that the external magnetic field is also zero?

FIGURE 21.9 (a) The electric force $\vec{F}$ that acts on a positive charge is parallel to the electric field $\vec{E}$. (b) The magnetic force $\vec{F}$ is perpendicular to both the magnetic field $\vec{B}$ and the velocity $\vec{v}$.

21.3 | The Motion of a Charged Particle in a Magnetic Field

Comparing Particle Motion in Electric and Magnetic Fields

The motion of a charged particle in an electric field is noticeably different from the motion in a magnetic field. For example, **Figure 21.9a** shows a positive charge moving between the plates of a parallel plate capacitor. Initially, the charge is moving perpendicular to the direction of the electric field. Since the direction of the electric force on a positive charge is in the same direction as the electric field, the particle is deflected sideways. Part *b* of the drawing shows the same particle traveling initially at right angles to a magnetic field. An application of RHR-1 shows that when the charge enters the field, the charge is deflected upward (not sideways) by the magnetic force. As the charge moves upward, the direction of the magnetic force changes, always remaining perpendicular to both the magnetic field and the velocity. Conceptual Example 2 focuses on the difference in how electric and magnetic fields apply forces to a moving charge.

CONCEPTUAL EXAMPLE 2 | The Physics of a Velocity Selector

A velocity selector is a device for measuring the velocity of a charged particle. The device operates by applying electric and magnetic forces to the particle in such a way that these forces balance. **Figure 21.10a** shows a particle with a positive charge $+q$ and a velocity $\vec{v}$ that is perpendicular to a constant magnetic field* $\vec{B}$. **Figure 21.10b** illustrates a velocity selector, which is a

---

*In many instances it is convenient to orient the magnetic field $\vec{B}$ so its direction is perpendicular to the page. In these cases it is customary to use a dot to symbolize the magnetic field pointing *out of the page* (toward the reader); this dot symbolizes the tip of the arrow representing the $\vec{B}$ vector. A region in which a magnetic field is directed *into the page* is drawn as a series of crosses that indicate the tail feathers of the arrows representing the $\vec{B}$ vectors. Therefore, regions in which a magnetic field is directed out of the page or into the page are drawn as shown below:

Out of page   Into page

cylindrical tube that is located within the magnetic field. Inside the tube there is a parallel plate capacitor that produces an electric field $\vec{E}$ (not shown) perpendicular to the magnetic field. The charged particle enters the left end of the tube, moving perpendicular to the magnetic field. If the strengths of $\vec{E}$ and $\vec{B}$ are adjusted properly, the electric and magnetic forces acting on the particle will cancel each other. With no net force acting on the particle, the velocity remains unchanged, according to Newton's first law. As a result, the particle moves in a straight line at a constant speed and exits at the right end of the tube. The magnitude of the velocity that is "selected" can be determined from a knowledge of the strengths of the electric and magnetic fields. Particles with velocities different from the one "selected" are deflected and do not exit at the right end of the tube.

How should the electric field $\vec{E}$ be directed so that the force it applies to the particle can balance the magnetic force produced by $\vec{B}$? The electric field should be directed: **(a)** in the same direction as the magnetic field; **(b)** in a direction opposite to that of the magnetic field; **(c)** from the upper plate of the parallel plate capacitor toward the lower plate; **(d)** from the lower plate of the parallel plate capacitor toward the upper plate.

Reasoning If the electric and magnetic forces are to cancel each other, they must have opposite directions. The direction of the magnetic force can be found by applying Right-Hand Rule No. 1 (RHR-1) to the moving charged particle. This rule reveals that the magnetic force acting on the positively charged particle in **Figure 21.10a** is directed upward, toward the top of the page when the particle enters the field region. Since the particle is positively charged, the direction of the electric force is the same as the direction of the electric field produced by the capacitor plates.

Answers (a), (b), and (d) are incorrect. The electric force on the positively charged particle is in the direction of the electric field. Thus, the electric force would point perpendicularly into the page in answer (a), perpendicularly out of the page in answer (b), and upward toward the top of the page in answer (d). Since the magnetic force points upward toward the top of the page when

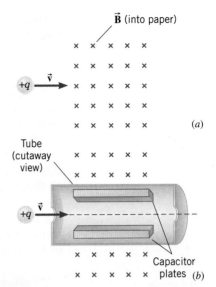

FIGURE 21.10 (*a*) A particle with a positive charge *q* and velocity $\vec{v}$ moves perpendicularly into a magnetic field $\vec{B}$. (*b*) A velocity selector is a tube in which an electric field (not shown) is perpendicular to a magnetic field, and the field magnitudes are adjusted so that the electric and magnetic forces acting on the particle cancel each other.

the particle enters the field region, these electric forces do not have the proper direction to cancel the magnetic force.

Answer (c) is correct. Since the magnetic force is directed upward when the particle enters the field region, the electric force must be directed downward. The force applied to a positive charge by an electric field has the same direction as the field itself, so the electric field must point downward, from the upper plate of the capacitor toward the lower plate. As a result, the upper plate must be positively charged.

Related Homework: Problems 23, 27

We have seen that a charged particle traveling in a magnetic field experiences a magnetic force that is always perpendicular to the field. In contrast, the force applied by an electric field is always parallel (or antiparallel) to the field direction. Because of the difference in the way that electric and magnetic fields exert forces, the work done on a charged particle by each field is different, as we now discuss.

The Work Done on a Charged Particle Moving Through Electric and Magnetic Fields

In **Figure 21.9a** an electric field applies a force to a positively charged particle, and, consequently, the path of the particle bends in the direction of the force. Because there is a component of the particle's displacement in the direction of the electric force, the force does work on the particle, according to Equation 6.1. This work increases the kinetic energy and, hence, the speed of the particle, in accord with the work–energy theorem (see Section 6.2). In contrast, the magnetic force in **Figure 21.9b** always acts in a direction that is perpendicular to the motion of the charge. Consequently, the displacement of the moving charge never has a component in the direction of the magnetic force. As a result, *the magnetic force cannot do work and change the kinetic energy of the charged* *particle* in **Figure 21.9b**. Thus, the speed of the particle *does not* change, although the force does alter the direction of the motion.

The Circular Trajectory

To describe the motion of a charged particle in a constant magnetic field more completely, let's discuss the special case in which the velocity of the particle is perpendicular to a uniform magnetic field. As **Interactive Figure 21.11** illustrates, the magnetic force serves to move the particle in a circular path. To understand why, consider two points on the circumference labeled 1 and 2. When the positively charged particle is at point 1, the magnetic force $\vec{F}$ is perpendicular to the velocity $\vec{v}$ and points directly upward in the drawing. This force causes the trajectory to bend upward. When the particle reaches point 2, the magnetic force still remains perpendicular to the velocity but is now directed to the left in the drawing.

| **Problem-Solving Insight** The magnetic force always remains perpendicular to the velocity and is directed toward the center of the circular path.

To find the radius of the path in **Interactive Figure 21.11**, we use the concept of centripetal force from Section 5.3. The centripetal force is the net force, directed toward the center of the circle, that is needed to keep a particle moving along a circular path. The magnitude F_c of this force depends on the speed v and mass m of the particle, as well as the radius r of the circle:

$$F_c = \frac{mv^2}{r} \tag{5.3}$$

In the present situation, the magnetic force furnishes the centripetal force. Being perpendicular to the velocity, the magnetic force does no work in keeping the charge $+q$ on the circular path. According to Equation 21.1, the magnitude of the magnetic force is given by $|q|vB \sin 90°$, so $|q|vB = mv^2/r$ or

$$r = \frac{mv}{|q|B} \tag{21.2}$$

This result shows that the radius of the circle is inversely proportional to the magnitude of the magnetic field, with stronger fields producing "tighter" circular paths. Example 3 illustrates an application of Equation 21.2.

INTERACTIVE FIGURE 21.11 A positively charged particle is moving perpendicular to a constant magnetic field. The magnetic force $\vec{F}$ causes the particle to move on a circular path (R.H. = right hand).

EXAMPLE 3 | The Motion of a Proton

A proton is released from rest at point *A*, which is located next to the positive plate of a parallel plate capacitor (see **Figure 21.12**). The proton then accelerates toward the negative plate, leaving the capacitor at point *B* through a small hole in the plate. The electric potential of the positive plate is 2100 V greater than that of the negative plate, so $V_A - V_B = 2100$ V. Once outside the capacitor, the proton travels at a constant velocity until it enters a region of constant magnetic field of magnitude 0.10 T. The velocity is perpendicular to the magnetic field, which is directed out of the page in **Figure 21.12**. Find **(a)** the speed v_B of the proton when it leaves the negative plate of the capacitor, and **(b)** the radius r of the circular path on which the proton moves in the magnetic field.

Reasoning The only force that acts on the proton (charge $= +e$) while it is between the capacitor plates is the conservative electric force. Thus, we can use the conservation of energy to find the speed of the proton when it leaves the negative plate. The total energy of the proton is the sum of its kinetic energy, $\frac{1}{2}mv^2$, and its electric potential energy, EPE. Following Example 4 in Chapter 19, we set the total energy at point *B* equal to the total energy at point *A*:

$$\underbrace{\tfrac{1}{2}mv_B^2 + \text{EPE}_B}_{\text{Total energy at } B} = \underbrace{\tfrac{1}{2}mv_A^2 + \text{EPE}_A}_{\text{Total energy at } A}$$

We note that $v_A = 0$ m/s, since the proton starts from rest, and use Equation 19.4 to set $\text{EPE}_A - \text{EPE}_B = e(V_A - V_B)$. Then the conservation

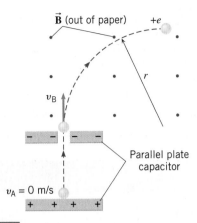

FIGURE 21.12 A proton, starting from rest at the positive plate of the capacitor, accelerates toward the negative plate. After leaving the capacitor, the proton enters a magnetic field, where it moves on a circular path of radius r.

of energy reduces to $\frac{1}{2}mv_B^2 = e(V_A - V_B)$. Solving for v_B gives $v_B = \sqrt{2e(V_A - V_B)/m}$. The proton enters the magnetic field with this speed and moves on a circular path with a radius that is given by Equation 21.2.

Solution **(a)** The speed of the proton is

$$v_B = \sqrt{\frac{2e(V_A - V_B)}{m}} = \sqrt{\frac{2(1.60 \times 10^{-19}\ \text{C})(2100\ \text{V})}{1.67 \times 10^{-27}\ \text{kg}}}$$

$$= \boxed{6.3 \times 10^5\ \text{m/s}}$$

(b) When the proton moves in the magnetic field, the radius of the circular path is

$$r = \frac{mv_B}{eB} = \frac{(1.67 \times 10^{-27}\ \text{kg})(6.3 \times 10^5\ \text{m/s})}{(1.60 \times 10^{-19}\ \text{C})(0.10\ \text{T})}$$

$$= \boxed{6.6 \times 10^{-2}\ \text{m}} \tag{21.2}$$

One of the important and exciting areas in physics today is the study of elementary particles, which are the basic building blocks from which all matter is constructed. Important information about an elementary particle can be obtained from its motion in a magnetic field, with the aid of a device known as a bubble chamber. A bubble chamber contains a superheated liquid such as hydrogen, which will boil and form bubbles readily. When an electrically charged particle passes through the chamber, a thin track of bubbles is left in its wake. This track can be photographed to show how a magnetic field affects the particle motion. Conceptual Example 4 illustrates how physicists deduce information from such photographs.

CONCEPTUAL EXAMPLE 4 | Particle Tracks in a Bubble Chamber

Figure 21.13a shows the bubble-chamber tracks resulting from an event that begins at point A. At this point a gamma ray (emitted by certain radioactive substances), traveling in from the left, spontaneously transforms into two charged particles. There is no track from the gamma ray itself. These particles move away from point A, producing the two spiral tracks. A third charged particle is knocked out of a hydrogen atom and moves forward, producing the long track with the slight upward curvature. Each of the three particles has the same mass and carries a charge of the same magnitude. A uniform magnetic field is directed out of the paper toward you. What is the sign (+ or −) of the charge carried by each particle?

(a)

(b)

FIGURE 21.13 (a) A photograph of tracks in a bubble chamber. A magnetic field is directed perpendicularly out of the paper. At point A, a gamma ray (not visible) spontaneously transforms into two charged particles, and a third charged particle is knocked out of a hydrogen atom in the chamber. (b) In accord with RHR-1, the magnetic field applies a downward force to a positively charged particle that moves to the right.

	Particle 1	Particle 2	Particle 3
(a)	−	−	+
(b)	−	+	−
(c)	+	−	−
(d)	+	−	+

Reasoning Figure 21.13b shows a positively charged particle traveling with a velocity $\vec{v}$ that is perpendicular to a magnetic field. The field is directed out of the paper, just like it is in part a of the drawing. RHR-1 indicates that the magnetic force points downward. This magnetic force provides the centripetal force that causes a particle to move on a circular path (see Section 5.3).

The centripetal force is directed toward the center of the circular path. Thus, in **Figure 21.13a** a positive charge would move on a downward-curving track, and a negative charge would move on an upward-curving track.

Answers (a), (c), and (d) are incorrect. Since particles 1 and 3 move on upward-curving tracks, they are negatively charged, not positively charged.

Answer (b) is correct. A downward-curving track in the photograph indicates a positive charge, while an upward-curving track indicates a negative charge. Thus, particles 1 and 3 carry a negative charge. They are, in fact, electrons (e^-). In contrast, particle 2 carries a positive charge. It is called a positron (e^+), an elementary particle that has the same mass as an electron but an opposite charge.

Related Homework: Check Your Understanding 5, 6, Problem 25

Check Your Understanding

(The answers are given at the end of the book.)

4. Suppose that the positive charge in **Figure 21.9a** were launched from the negative plate toward the positive plate, in a direction *opposite* to the electric field $\vec{E}$. A sufficiently strong electric field would prevent the charge from striking the positive plate. Suppose that the positive charge in **Figure 21.9b** were launched from the south pole toward the north pole, in a direction *opposite* to the magnetic field $\vec{B}$. Would a sufficiently strong magnetic field prevent the charge from reaching the north pole? **(a)** Yes **(b)** No, because a magnetic field cannot exert a force on a charged particle that is moving antiparallel to the field **(c)** No, because the magnetic force would cause the charge to move faster as it moved toward the north pole

5. Review Conceptual Example 4 as background for this question. Three particles move through a constant magnetic field and follow the paths shown in **CYU Figure 21.2**. Determine whether each particle is positively (+) charged, negatively (−) charged, or neutral.

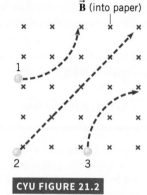

$\vec{B}$ (into paper)

CYU FIGURE 21.2

	Particle 1	Particle 2	Particle 3
(a)	neutral	+	neutral
(b)	−	neutral	+
(c)	−	−	−
(d)	+	neutral	−
(e)	+	+	+

6. Suppose that the three particles in **Figure 21.13a** have identical charge magnitudes and masses. Which particle has the greatest speed? Refer to Conceptual Example 4 as needed.

7. A positive charge moves along a circular path under the influence of a magnetic field. The magnetic field is perpendicular to the plane of the circle, as in **Interactive Figure 21.11**. If the velocity of the particle is reversed at some point along the path, will the particle retrace its path? **(a)** Yes **(b)** No, because it will move around a *different* circle in a counterclockwise direction

8. Refer to **Interactive Figure 21.11**. Assume that the particle in the picture is a proton. If an electron is projected at point 1 with the same velocity $\vec{v}$, it will not follow exactly the same path as the proton, unless the magnetic field is adjusted in the following manner: the magnitude of the magnetic field must be _____, and the direction of the magnetic field must be _____.
(a) the same, reversed **(b)** increased, the same **(c)** reduced, reversed

9. **CYU Figure 21.3** shows a top view of four interconnected chambers. A *negative* charge is fired into chamber 1. By turning on separate magnetic fields in each chamber, the charge can be made to exit from chamber 4, as shown. How should the

CYU FIGURE 21.3

	Chamber 1	Chamber 2	Chamber 3	Chamber 4
(a)	out of	into	out of	into
(b)	into	out of	out of	into
(c)	out of	into	into	out of
(d)	into	out of	into	out of

magnetic field in each chamber be directed: out of the page or into the page?

10. **CYU Figure 21.4** shows a particle carrying a positive charge $+q$ at the coordinate origin, as well as a target located in the third quadrant. A uniform magnetic field is directed perpendicularly into the plane of the paper. The charge can be projected in the plane of the paper only, along the positive or negative x or y axis. There are four possible directions ($+x$, $-x$, $+y$, $-y$) for the initial velocity of the particle. The particle can be made to hit the target for only two of the four directions. Which two directions are they? **(a)** $+y, -y$ **(b)** $-y, +x$ **(c)** $-x, +y$ **(d)** $+x, -x$

CYU FIGURE 21.4

21.4 The Mass Spectrometer

Physicists use mass spectrometers for determining the relative masses and abundances of isotopes.* Chemists use these instruments to help identify unknown molecules produced in chemical reactions. Mass spectrometers are also used during surgery, where they give the anesthesiologist information on the gases, including the anesthetic, in the patient's lungs.

THE PHYSICS OF . . . a mass spectrometer. In the type of mass spectrometer illustrated in **Figure 21.14**, the atoms or molecules are first vaporized and then ionized by the ion source. The ionization process removes one electron from the particle, leaving it with a net positive charge of $+e$. The positive ions are then accelerated through the potential difference V, which is applied between the ion source and the metal plate. With a speed v, the ions pass through a hole in the plate and enter a region of constant magnetic field $\vec{B}$, where they are deflected in semicircular paths. Only those ions following a path with the proper radius r strike the detector, which records the number of ions arriving per second.

The mass m of the detected ions can be expressed in terms of r, B, and v by recalling that the radius of the path followed by a particle of charge $+e$ is $r = mv/(eB)$ (Equation 21.2). In addition, the Reasoning section in Example 3 shows that the ion speed v can be expressed in terms of the potential difference V as $v = \sqrt{2eV/m}$. This expression for the ion speed is the same as that used in Example 3, except that, for convenience, we have replaced the potential difference, $V_A - V_B$, by the symbol V. Eliminating v from these two equations algebraically and solving for the mass gives

$$m = \left(\frac{er^2}{2V}\right)B^2$$

This result shows that the mass of each ion reaching the detector is proportional to B^2. Experimentally changing the value of B and keeping the term in the parentheses constant will allow ions of different masses to enter the detector. A plot of the detector output as a function of B^2 then gives an indication of what masses are present and the abundance of each mass.

Figure 21.15 shows a record obtained by a mass spectrometer for naturally occurring neon gas. The results show that the element neon has three isotopes whose atomic mass numbers are 20, 21, and 22. These isotopes occur because neon atoms exist with different numbers of neutrons in the nucleus. Notice that the isotopes have different abundances, with neon-20 being the most abundant.

FIGURE 21.14 In this mass spectrometer the dashed lines are the paths traveled by ions of different masses. Ions with mass m follow the path of radius r and enter the detector. Ions with the larger mass m_1 follow the outer path and miss the detector.

FIGURE 21.15 The mass spectrum (not to scale) of naturally occurring neon, showing three isotopes whose atomic mass numbers are 20, 21, and 22. The larger the peak, the more abundant the isotope.

*Isotopes are atoms that have the same atomic number but different atomic masses due to the presence of different numbers of neutrons in the nucleus. They are discussed in Section 31.1.

21.5 The Force on a Current in a Magnetic Field

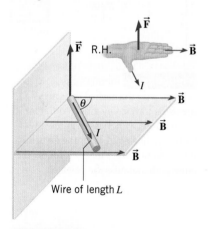

INTERACTIVE FIGURE 21.16 The wire carries a current I, and the bottom segment of the wire is oriented perpendicular to a magnetic field $\vec{B}$. A magnetic force deflects the wire to the right.

As we have seen, a charge moving through a magnetic field can experience a magnetic force. Since an electric current is a collection of moving charges, a current in the presence of a magnetic field can also experience a magnetic force. In **Interactive Figure 21.16**, for instance, a current-carrying wire is placed between the poles of a magnet. When the direction of the current I is as shown, the moving charges experience a magnetic force that pushes the wire to the right in the drawing. The direction of the force is determined in the usual manner by using RHR-1, with the minor modification that the direction of the velocity of a positive charge is replaced by the direction of the conventional current I. If the current in the drawing were reversed by switching the leads to the battery, the direction of the force would be reversed, and the wire would be pushed to the left.

> **Problem-Solving Insight** Whenever the current in a wire reverses direction, the force exerted on the wire by a given magnetic field also reverses direction.

When a charge moves through a magnetic field, the magnitude of the force that acts on the charge is $F = |q|vB \sin \theta$ (Equation 21.1). With the aid of **Figure 21.17**, this expression can be put into a form that is more suitable for use with an electric current. The drawing shows a wire of length L that carries a current I. The wire is oriented at an angle θ with respect to a magnetic field $\vec{B}$. This picture is similar to **Figure 21.7c**, except that now the charges move in a wire. The magnetic force exerted on this length of wire is the net force acting on the total amount of charge moving in the wire. Suppose that an amount of conventional positive charge Δq travels the length of the wire in a time interval Δt. The magnitude of the magnetic force on this amount of charge is given by Equation 21.1 as $F = (\Delta q)vB \sin \theta$. Multiplying and dividing the right side of this equation by Δt, we find that

$$F = \underbrace{\left(\frac{\Delta q}{\Delta t}\right)}_{I} \underbrace{(v\,\Delta t)}_{L} B \sin \theta$$

The term $\Delta q / \Delta t$ is the current I in the wire (see Equation 20.1), and the term $v\,\Delta t$ is the length L of the wire. With these two substitutions, the expression for the magnetic force exerted on a current-carrying wire becomes

Magnetic force on a current-carrying wire of length L

$$F = ILB \sin \theta \qquad (21.3)$$

FIGURE 21.17 The current I in the wire, oriented at an angle θ with respect to a magnetic field $\vec{B}$, is acted upon by a magnetic force $\vec{F}$.

As in the case of a single charge traveling in a magnetic field, the magnetic force on a current-carrying wire is a maximum when the wire is oriented perpendicular to the field ($\theta = 90°$) and vanishes when the current is parallel or antiparallel to the field ($\theta = 0°$ or $180°$). The direction of the magnetic force is given by RHR-1, as **Figure 21.17** indicates.

THE PHYSICS OF . . . a loudspeaker. Most loudspeakers operate on the principle that a magnetic field exerts a force on a current-carrying wire. **Figure 21.18a** shows a speaker design that consists of three basic parts: a cone, a voice coil, and a permanent magnet. The cone is mounted so it can vibrate back and forth. When vibrating, it pushes and pulls on the air in front of it, thereby creating sound waves. Attached to the apex of the cone is the voice coil, which is a hollow cylinder around which coils of wire are wound. The voice coil is slipped over one of the poles of the stationary permanent magnet (the north pole in the drawing) and can move freely. The two ends of the voice-coil wire are connected to the speaker terminals on the back panel of a receiver. The receiver acts as an ac generator, sending an alternating current to the voice coil. The alternating current interacts with the magnetic field to generate an alternating force that pushes and pulls on the voice coil and the attached cone. To see how the

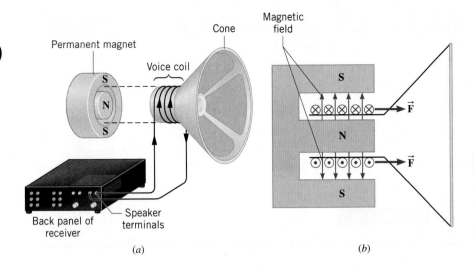

Magnetic field

Cone

Permanent magnet

Voice coil

S
N
S

Back panel of receiver

Speaker terminals

(a)

S

N

$\vec{F}$

$\vec{F}$

S

(b)

FIGURE 21.18 (a) An "exploded" view of one type of speaker design, which shows a cone, a voice coil, and a permanent magnet. (b) Because of the current in the voice coil (shown as ⊗ and ⊙), the magnetic field causes a force $\vec{F}$ to be exerted on the voice coil and cone.

magnetic force arises, consider **Figure 21.18b**, which is a cross-sectional view of the voice coil and the magnet. In the cross-sectional view, the current is directed into the page in the upper half of the voice coil (⊗⊗⊗) and out of the page in the lower half (⊙⊙⊙). In both cases the magnetic field is perpendicular to the current, so the maximum possible force is exerted on the wire. An application of RHR-1 to both the upper and lower halves of the voice coil shows that the magnetic force $\vec{F}$ in the drawing is directed to the right, causing the cone to accelerate in that direction. One-half of a cycle later when the current is reversed, the direction of the magnetic force is also reversed, and the cone accelerates to the left. If, for example, the alternating current from the receiver has a frequency of 1000 Hz, the alternating magnetic force causes the cone to vibrate back and forth at the same frequency, and a 1000-Hz sound wave is produced. Thus, it is the magnetic force on a current-carrying wire that is responsible for converting an electrical signal into a sound wave. In Example 5 a typical force and acceleration in a loudspeaker are determined.

EXAMPLE 5 | The Force and Acceleration in a Loudspeaker

The voice coil of a speaker has a diameter of $d = 0.025$ m, contains 55 turns of wire, and is placed in a 0.10-T magnetic field. The current in the voice coil is 2.0 A. **(a)** Determine the magnetic force that acts on the coil and cone. **(b)** The voice coil and cone have a combined mass of 0.020 kg. Find their acceleration.

Reasoning The magnetic force that acts on the current-carrying voice coil is given by Equation 21.3 as $F = ILB \sin \theta$. The effective length L of the wire in the voice coil is very nearly the number of turns N times the circumference (πd) of one turn: $L = N\pi d$. The acceleration of the voice coil and cone is given by Newton's second law as the magnetic force divided by the combined mass.

Solution **(a)** Since the magnetic field acts perpendicular to all parts of the wire, $\theta = 90°$ and the force on the voice coil is

$$F = ILB \sin \theta = I(N\pi d)B \sin \theta = (2.0 \text{ A})[55\pi(0.025 \text{ m})](0.10 \text{ T}) \sin 90°$$
$$= \boxed{0.86 \text{ N}} \tag{21.3}$$

(b) According to Newton's second law, the acceleration of the voice coil and cone is

$$a = \frac{F}{m} = \frac{0.86 \text{ N}}{0.020 \text{ kg}} = \boxed{43 \text{ m/s}^2} \tag{4.1}$$

This acceleration is more than four times the acceleration due to gravity.

Check Your Understanding

(The answers are given at the end of the book.)

11. Refer to **Interactive Figure 21.16**. **(a)** What happens to the direction of the magnetic force if the current is reversed? **(b)** What happens to the direction of the force if both the current and the magnetic poles are reversed?

12. The same current-carrying wire is placed in the same magnetic field $\vec{\mathbf{B}}$ in four different orientations (see **CYU Figure 21.5**). Rank the orientations according to the magnitude of the magnetic force exerted on the wire, largest to smallest.

CYU FIGURE 21.5

21.6 The Torque on a Current-Carrying Coil

We have seen that a current-carrying wire can experience a force when placed in a magnetic field. If a loop of wire is suspended properly in a magnetic field, the magnetic force produces a torque that tends to rotate the loop. This torque is responsible for the operation of a widely used type of electric motor.

Figure 21.19a shows a rectangular loop of wire attached to a vertical shaft. The shaft is mounted so that it is free to rotate in a uniform magnetic field. When there is a current in the loop, the loop rotates because magnetic forces act on the vertical sides, labeled 1 and 2 in the drawing. Part *b* shows a top view of the loop and the magnetic forces $\vec{\mathbf{F}}$ and $-\vec{\mathbf{F}}$ that act on the two sides. These two forces have the same magnitude, but an application of RHR-1 shows that they point in opposite directions, so the loop experiences no net force. The loop does, however, experience a net torque that tends to rotate it in a clockwise fashion about the vertical shaft. **Figure 21.20a** shows that the torque is maximum when the normal to the plane of the loop is perpendicular to the field. In contrast, part *b* shows that the torque is zero when the normal is parallel to the field. **When a current-carrying loop is placed in a magnetic field, the loop tends to rotate such that its normal becomes aligned with the magnetic field.** In this respect, a current loop behaves like a magnet (e.g., a compass needle) suspended in a magnetic field, since a magnet also rotates to align itself with the magnetic field.

(a) Maximum torque   (b) Zero torque

FIGURE 21.19 (*a*) A current-carrying loop of wire, which can rotate about a vertical shaft, is situated in a magnetic field. (*b*) A top view of the loop. The current in side 1 is directed out of the page (⊙), while the current in side 2 is directed into the page (⊗). The current in side 1 experiences a force $\vec{\mathbf{F}}$ that is opposite to the force $-\vec{\mathbf{F}}$ exerted on side 2. The two forces produce a clockwise torque about the shaft.

FIGURE 21.20 (*a*) Maximum torque occurs when the normal to the plane of the loop is perpendicular to the magnetic field. (*b*) The torque is zero when the normal is parallel to the field.

It is possible to determine the magnitude of the torque on the loop in **Figure 21.19a**. From Equation 21.3 the magnetic force on each vertical side has a magnitude of $F = ILB \sin 90°$, where L is the length of side 1 or side 2, and $\theta = 90°$ because the current I always remains perpendicular to the magnetic field as the loop rotates. As Section 9.1 discusses, the torque produced by a force is the product of the magnitude of the force and the lever arm. In **Figure 21.19b** the lever arm ℓ is the perpendicular distance from the line of action of the force to the shaft. This distance is given by $\ell = (w/2) \sin \phi$, where w is the width of the loop, and ϕ is the angle between the normal to the plane of the loop and the direction of the magnetic field. The net torque is the sum of the torques on the two sides, so

$$\text{Net torque} = \tau = ILB(\tfrac{1}{2}w \sin \phi) + ILB(\tfrac{1}{2}w \sin \phi) = IAB \sin \phi$$

In this result the product Lw has been replaced by the area A of the loop. If the wire is wrapped so as to form a coil containing N loops, each of area A, the force on each side is N times larger, and the torque becomes proportionally greater:

$$\tau = \underbrace{NIA}_{\substack{\text{Magnetic} \\ \text{moment}}}(B \sin \phi) \tag{21.4}$$

Equation 21.4 has been derived for a rectangular coil, but it is valid for any shape of flat coil, such as a circular coil. The torque depends on the geometric properties of the coil and the current in it through the quantity NIA. This quantity is known as the **magnetic moment** of the coil, and its unit is ampere $\cdot$ meter^2 (A $\cdot$ m^2). The greater the magnetic moment of a current-carrying coil, the greater is the torque that the coil experiences when placed in a magnetic field. Example 6 discusses the torque that a magnetic field applies to such a coil.

Math Skills To understand why the lever arm for the torque calculation is $\ell = (w/2) \sin \phi$, it is first necessary to know why the two angles labeled ϕ in **Figure 21.19b** are equal. Referring to **Figure 21.21a** and noting that the normal is perpendicular to the plane of the loop, we can see that $\alpha + \phi = 90°$ or $\alpha = 90° - \phi$. Furthermore, the force $-\vec{F}$ is perpendicular to the magnetic field $\vec{B}$, so the triangle including angles α and β is a right triangle. Therefore, it follows that $\alpha + \beta = 90°$. Solving this equation gives $\beta = 90° - \alpha$, and substituting $\alpha = 90° - \phi$ reveals that

$$\beta = 90° - \alpha = 90° - \underbrace{(90° - \phi)}_{\alpha} = \phi$$

Recognizing that the two angles labeled ϕ in **Figure 21.19b** are equal, we can now understand why the lever arm is $\ell = (w/2)\sin \phi$ for the force $-\vec{F}$. According to the definition given in Equation 1.1, the sine function is $\sin \phi = \dfrac{h_o}{h}$, where h_o is the length of the side of a right triangle that is opposite the angle ϕ and h is the length of the hypotenuse (see **Figure 21.21b**). By comparing **Figure 21.21b** with **Figure 21.21c**, we can see that

$$\sin \phi = \frac{h_o}{h} = \frac{\ell}{w/2} \quad \text{or} \quad \ell = \frac{w}{2} \sin \phi$$

(a)   *(b)*   *(c)*

FIGURE 21.21 Math Skills drawing.

EXAMPLE 6 | The Torque Exerted on a Current-Carrying Coil

A coil of wire has an area of $2.0 \times 10^{-4} \text{ m}^2$, consists of 100 loops or turns, and contains a current of 0.045 A. The coil is placed in a uniform magnetic field of magnitude 0.15 T. **(a)** Determine the magnetic moment of the coil. **(b)** Find the maximum torque that the magnetic field can exert on the coil.

Reasoning and Solution **(a)** The magnetic moment of the coil is

Magnetic moment $= NIA = (100)(0.045 \text{ A})(2.0 \times 10^{-4} \text{ m}^2)$

$$= \boxed{9.0 \times 10^{-4} \text{ A} \cdot \text{m}^2}$$

(b) According to Equation 21.4, the torque is the product of the magnetic moment NIA and $B \sin \phi$. However, the maximum torque occurs when $\phi = 90°$, so

$$\tau = \underbrace{(NIA)}_{\substack{\text{Magnetic} \\ \text{moment}}}(B \sin 90°) = (9.0 \times 10^{-4} \text{ A} \cdot \text{m}^2)(0.15 \text{ T})$$

$$= \boxed{1.4 \times 10^{-4} \text{ N} \cdot \text{m}}$$

CD platter

Armature
(iron core
not shown)

Shaft

Brush

Brush

Half-rings

FIGURE 21.22 The basic components of a dc motor. A CD platter is shown as it might be attached to the motor.

THE PHYSICS OF . . . a direct-current electric motor. The electric motor is found in many devices, such as CD players, automobiles, washing machines, and air conditioners. **Figure 21.22** shows that a direct-current (dc) motor consists of a coil of wire placed in a magnetic field and free to rotate about a vertical shaft. The coil of wire contains many turns and is wrapped around an iron cylinder that rotates with the coil, although these features have been omitted to simplify the drawing. The coil and iron cylinder assembly is known as the armature. Each end of the wire coil is attached to a metallic half-ring. Rubbing against each of the half-rings is a graphite contact called a brush. While the half-rings rotate with the coil, the graphite brushes remain stationary. The two half-rings and the associated brushes are referred to as a split-ring commutator (see below).

The operation of a motor can be understood by considering **Figure 21.23**. In part *a* the current from the battery enters the coil through the left brush and half-ring, goes around the coil, and then leaves through the right half-ring and brush. Consistent with RHR-1, the directions of the magnetic forces $\vec{F}$ and $-\vec{F}$ on the two sides of the coil are as shown in the drawing. These forces produce the torque that turns the coil. Eventually, the coil reaches the position shown in part *b* of the drawing. In this position the half-rings momentarily lose electrical contact with the brushes, so that there is no current in the coil and no applied torque. However, like any moving object, the rotating coil does not stop immediately, for its inertia carries it onward. When the half-rings reestablish contact with the brushes, there again is a current in the coil, and a magnetic torque again rotates the coil in the same direction. The split-ring commutator ensures that the current is always in the proper direction to yield a torque that produces a continuous rotation of the coil.

(a)   (b)

FIGURE 21.23 (*a*) When a current exists in the coil, the coil experiences a torque. (*b*) Because of its inertia, the coil continues to rotate when there is no current.

21.7 Magnetic Fields Produced by Currents

We have seen that a current-carrying wire can experience a magnetic force when placed in a magnetic field that is produced by an external source, such as a permanent magnet. *A current-carrying wire also produces a magnetic field of its own*, as we will see in this section. Hans Christian Oersted (1777–1851) first discovered this effect in 1820 when he observed that a current-carrying wire influences the orientation of a nearby compass needle. The compass needle aligns itself with the net magnetic field produced by the current and the magnetic field of the earth. Oersted's discovery, which linked the motion of electric charges with the creation of a magnetic field, marked the beginning of an important discipline called **electromagnetism**.

A Long, Straight Wire

Figure 21.24a illustrates Oersted's discovery with a very long, straight wire. When a current is present, the compass needles point in a circular pattern about the wire. The pattern indicates that the magnetic field lines produced by the current are circles centered on the wire. If the direction of the current is reversed, the needles also reverse their directions, indicating that the direction of the magnetic field has reversed. The direction of the field can be obtained by using *Right-Hand Rule No. 2 (RHR-2)*, as part *b* of the drawing indicates:

> **Right-Hand Rule No. 2: The RHR for Determining the Direction of the Magnetic Field Near Long, Straight Wires.** Curl the fingers of the right hand into the shape of a half-circle. Point the thumb in the direction of the conventional current *I*, and the tips of the fingers will point in the direction of the magnetic field $\vec{\mathbf{B}}$.

Experimentally, it is found that the magnitude *B* of the magnetic field produced by an infinitely long, straight wire is directly proportional to the current *I* and inversely proportional to the radial distance *r* from the wire: $B \propto I/r$. As usual, this proportionality is converted into an equation by introducing a proportionality constant, which, in this instance, is written as $\mu_0/(2\pi)$. Thus, the magnitude of the magnetic field is

Infinitely long, straight wire
$$B = \frac{\mu_0 I}{2\pi r} \tag{21.5}$$

The constant μ_0 is referred to as the **permeability of free space**, and its value is $\mu_0 = 4\pi \times 10^{-7}$ T · m/A. The magnetic field becomes stronger nearer the wire, where *r* is smaller. Therefore, the field lines near the wire are closer together than those located farther away, where the field is weaker. **Figure 21.25** shows the pattern of field lines.

The magnetic field that surrounds a current-carrying wire can exert a force on a moving charge, as the next example illustrates.

FIGURE 21.24 (*a*) A very long, straight, current-carrying wire produces magnetic field lines that are circular about the wire, as indicated by the compass needles. (*b*) With the thumb of the right hand (R.H.) along the current *I*, the curled fingers point in the direction of the magnetic field, according to RHR-2.

FIGURE 21.25 The magnetic field becomes stronger as the radial distance *r* decreases, so the field lines are closer together near the wire.

Analyzing Multiple-Concept Problems

EXAMPLE 7 | A Current Exerts a Magnetic Force on a Moving Charge

Figure 21.26 shows a very long, straight wire carrying a current of 3.0 A. A particle has a charge of $+6.5 \times 10^{-6}$ C and is moving parallel to the wire at a distance of 0.050 m. The speed of the particle is 280 m/s. Determine the magnitude and direction of the magnetic force exerted on the charged particle by the current in the wire.

Reasoning The current generates a magnetic field in the space around the wire. The charged particle moves in the presence of this field and, therefore, can experience a magnetic force. The magnitude of this force is given by Equation 21.1, and the direction can be determined by applying RHR-1 (see Section 21.2). Note in **Figure 21.26** that the magnetic field $\vec{\mathbf{B}}$ produced by the current lies in a plane that is perpendicular to both the wire and the velocity $\vec{\mathbf{v}}$ of the particle. Thus, the angle between $\vec{\mathbf{B}}$ and $\vec{\mathbf{v}}$ is $\theta = 90.0°$.

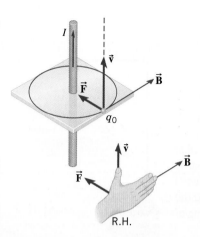

FIGURE 21.26 The positive charge q_0 moves with a velocity $\vec{v}$ and experiences a magnetic force $\vec{F}$ because of the magnetic field $\vec{B}$ produced by the current in the wire.

Knowns and Unknowns   The following list summarizes the data that are given:

Description	Symbol	Value	Comment
Explicit Data			
Current in wire	I	3.0 A	
Electric charge of particle	q_0	$+6.5 \times 10^{-6}$ C	
Distance of particle from wire	r	0.050 m	Particle moves parallel to wire; see **Figure 21.26**.
Speed of particle	v	280 m/s	
Implicit Data			
Directional angle of particle velocity with respect to magnetic field	θ	90.0°	Particle moves parallel to wire; see *Reasoning*.
Unknown Variable			
Magnitude of magnetic force exerted on particle	F	?	

Modeling the Problem

STEP 1 Magnetic Force on the Particle   The magnitude F of the magnetic force acting on the charged particle is given at the right by Equation 21.1, where $|q_0|$ is the magnitude of the charge, v is the particle speed, B is the magnitude of the magnetic field produced by the wire, and θ is the angle between the particle velocity and the magnetic field. Values for $|q_0|$, v, and θ are given. The value of B, however, is unknown, and we determine it in Step 2.

$$F = |q_0|vB \sin \theta \qquad (21.1)$$

$$\boxed{?}$$

STEP 2 Magnetic Field Produced by the Wire   The magnitude B of the magnetic field produced by a current I in an infinitely long, straight wire is given by Equation 21.5:

$$\boxed{B = \frac{\mu_0 I}{2\pi r}} \qquad (21.5)$$

where μ_0 is the permeability of free space and r is the distance from the wire. This expression can be substituted into Equation 21.1, as shown at the right.

$$F = |q_0|vB \sin \theta \qquad (21.1)$$

$$\boxed{B = \frac{\mu_0 I}{2\pi r}} \qquad (21.5)$$

Solution   Combining the results of each step algebraically, we find that

$$\overset{\textstyle \text{STEP 1}}{\downarrow} \quad \overset{\textstyle \text{STEP 2}}{\downarrow}$$

$$F = |q_0|vB \sin \theta = |q_0|v\left(\frac{\mu_0 I}{2\pi r}\right) \sin \theta$$

The magnitude of the magnetic force on the charged particle is

$$F = |q_0| v \left(\frac{\mu_0 I}{2\pi r} \right) \sin \theta$$

$$= (6.5 \times 10^{-6} \text{ C})(280 \text{ m/s}) \frac{(4\pi \times 10^{-7} \text{ T} \cdot \text{m/A})(3.0 \text{ A})}{2\pi(0.050 \text{ m})} \sin 90.0° = \boxed{2.2 \times 10^{-8} \text{ N}}$$

The direction of the magnetic force is predicted by RHR-1 and, as shown in **Figure 21.26**, is radially inward toward the wire.

Related Homework: Problem 66

We have now seen that an electric current can create a magnetic field of its own. Earlier, we have also seen that an electric current can experience a force created by another magnetic field. Therefore, the magnetic field that one current creates can exert a force on another nearby current. Conceptual Example 8 deals with this magnetic interaction between currents.

CONCEPTUAL EXAMPLE 8 | The Net Force That a Current-Carrying Wire Exerts on a Current-Carrying Coil

Figure 21.27 shows a very long, straight wire carrying a current I_1 and a rectangular coil carrying a current I_2. The wire and the coil lie in the same plane, with the wire parallel to the long sides of the rectangle. The coil is **(a)** attracted to the wire, **(b)** repelled from the wire, **(c)** neither attracted to nor repelled from the wire.

Reasoning The current in the straight wire exerts a force on each of the four sides of the coil. The net force acting on the coil is the vector sum of these four forces. To determine whether the net force is attractive, repulsive, or neither, we need to consider the directions and magnitudes of the individual forces. For each side of the rectangular coil we will first use RHR-2 to determine the direction of the magnetic field produced by the long, straight wire. Then, we will employ RHR-1 to find the direction of the magnetic force exerted on each side.

It should be noted that the magnetic forces that act on the two short sides of the rectangular coil play no role. To see why, consider a small segment of each of the short sides, located at the same distance from the straight wire, as indicated by the dashed line in **Figure 21.27**. Each of these segments experiences the same magnetic field from the current I_1. RHR-2 shows that this field is directed downward into the plane of the paper, so that it is perpendicular to the current I_2 in each segment. However, the directions of I_2 in the segments are opposite. As a result, RHR-1 reveals that the magnetic field from the straight wire applies a magnetic force to one segment that is opposite to the magnetic force applied to the other segment. Thus, the forces on the two short sides of the coil cancel.

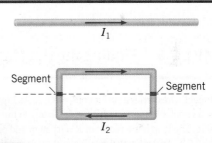

FIGURE 21.27 A very long, straight wire carries a current I_1, and a rectangular coil carries a current I_2. The dashed line is parallel to the wire and locates a small segment on each short side of the coil.

Answers (b) and (c) are incorrect. There is indeed a net magnetic force that acts on the rectangular coil, but it is not a force that repels the coil from the straight wire. To see why, consider the following explanation.

Answer (a) is correct. In the long side of the coil near the wire, the current I_2 has the same direction as the current I_1, and two such currents *attract* each other. In the other long side of the coil the current I_2 has a direction opposite to that of I_1, and they *repel* one another. However, the attractive force is stronger than the repulsive force because the magnetic field produced by the current I_1 is stronger at shorter distances than at greater distances. Consequently, the rectangular coil is attracted to the long, straight wire.

FIGURE 21.28 (*a*) The magnetic field lines in the vicinity of a current-carrying circular loop. (*b*) The direction of the magnetic field at the center of the loop is given by RHR-3.

A Loop of Wire

If a current-carrying wire is bent into a circular loop, the magnetic field lines around the loop have the pattern shown in **Figure 21.28a**. At the *center* of a loop of radius R, the magnetic field is perpendicular to the plane of the loop and has the value $B = \mu_0 I/(2R)$, where I is the current in the loop. Often, the loop consists of N turns of wire that are wound sufficiently close together that they form a flat coil with a single radius. In this case, the magnetic fields of the individual turns add together to give a net field that is N times greater than the field of a single loop. For such a coil the magnetic field at the center is

Center of a circular loop
$$B = N\frac{\mu_0 I}{2R} \tag{21.6}$$

The direction of the field through the loop can be obtained by using *Right-Hand Rule No. 3 (RHR-3)*, as demonstrated in part (b) of the figure:

> **Right-Hand Rule No. 3: The RHR Used to Determine the Direction of the Magnetic Field at the Center of a Loop or Solenoid.** Curl the fingers of the right hand along the direction of the conventional current I around the loop or solenoid, and the thumb will point in the direction of the magnetic field $\vec{\mathbf{B}}$ at the center of the loop or solenoid.

A solenoid, which we will describe later in this section, is just a stack of loops.

We can apply RHR-3 to determine the direction of the magnetic field for the single loop in **Figure 21.28**. If the fingers of the right hand curl along the direction of the current in the loop, as in **Figure 21.28b**, then the thumb indicates the direction the magnetic field points at the center of the loop. Here it points from right to left.

Example 9 shows how the magnetic fields produced by the current in a loop of wire and the current in a long, straight wire combine to form a net magnetic field.

EXAMPLE 9 | Finding the Net Magnetic Field

A long, straight wire carries a current of $I_1 = 8.0$ A. As **Figure 21.29a** illustrates, a circular loop of wire lies immediately to the right of the straight wire. The loop has a radius of $R = 0.030$ m and carries a current of $I_2 = 2.0$ A. Assuming that the thickness of each wire is negligible, find the magnitude and direction of the net magnetic field at the center C of the loop.

Reasoning The net magnetic field at the point C is the sum of two contributions: (1) the field $\vec{\mathbf{B}}_1$ produced by the long, straight wire according to Equation 21.5 $[B_1 = \mu_0 I_1/(2\pi r)]$, and (2) the field $\vec{\mathbf{B}}_2$ produced by the circular loop according to Equation 21.6 with $N = 1$ $[B_2 = \mu_0 I_2/(2R)]$. An application of RHR-2 shows that at point C the field $\vec{\mathbf{B}}_1$ is directed upward, perpendicular to the plane containing the straight wire and the loop (see part *b* of the drawing). Applying RHR-3 to the circular loop carrying a current I_2 shows that the magnetic field $\vec{\mathbf{B}}_2$ is directed downward, opposite to the direction of $\vec{\mathbf{B}}_1$.

> **Problem-Solving Insight** Do not confuse the formula for the magnetic field produced at the center of a circular loop with that of a very long, straight wire. The formulas are similar, differing only by a factor of π in the denominator.

(a)   (b)

FIGURE 21.29 (*a*) A long, straight wire carrying a current I_1 lies next to a circular loop that carries a current I_2. (*b*) The magnetic fields at the center C of the loop produced by the straight wire ($\vec{\mathbf{B}}_1$) and the loop ($\vec{\mathbf{B}}_2$).

$$\underbrace{B = \frac{\mu_0 I_1}{2\pi r}}_{\substack{\text{Long,}\\\text{straight wire}}} - \underbrace{\frac{\mu_0 I_2}{2R}}_{\substack{\text{Center of a}\\\text{circular loop}}} = \frac{\mu_0}{2}\left(\frac{I_1}{\pi r} - \frac{I_2}{R}\right)$$

$$B = \frac{(4\pi \times 10^{-7}\text{ T}\cdot\text{m/A})}{2}\left[\frac{8.0\text{ A}}{\pi(0.030\text{ m})} - \frac{2.0\text{ A}}{0.030\text{ m}}\right] = \boxed{1.1 \times 10^{-5}\text{ T}}$$

Solution If we choose the upward direction in **Figure 21.29b** as positive, the net magnetic field at point C is

The net field is positive, so it is directed upward, perpendicular to the plane.

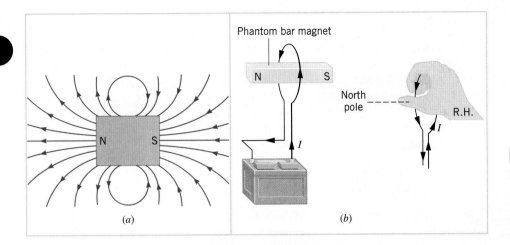

(a)

(b)

FIGURE 21.30 (*a*) The field lines around the bar magnet resemble those around the loop in **Figure 21.28a**. (*b*) The current loop can be imagined to be a "phantom" bar magnet with a north pole and a south pole.

A comparison of the magnetic field lines around the current loop in **Figure 21.28a** with those in the vicinity of the short bar magnet in **Figure 21.30a** shows that the two patterns are similar. Not only are the patterns similar, but the loop itself behaves as a bar magnet with a "north pole" on one side and a "south pole" on the other side. To emphasize that the loop may be imagined to be a bar magnet, **Figure 21.30b** includes a "phantom" bar magnet at the center of the loop. The side of the loop that acts like a north pole can be determined with the aid of RHR-3: curl the fingers of the right hand along the direction of the current in the loop. The thumb not only points in the direction of B at the center of the loop, but it also points toward the north pole of the loop.

Because a current-carrying loop acts like a bar magnet, two adjacent loops can be either attracted to or repelled from each other, depending on the relative directions of the currents. **Figure 21.31** includes a "phantom" magnet for each loop and shows that the loops are attracted to each other when the currents are in the same direction and repelled from each other when the currents are in opposite directions.

A Solenoid

A solenoid is a long coil of wire in the shape of a helix (see **Figure 21.32**). If the wire is wound so the turns are packed close to each other and the solenoid is long compared to its diameter, the magnetic field lines have the appearance shown in the drawing (**Figure 21.32b**). Notice that the field inside the solenoid and away from its ends is nearly constant in magnitude and directed parallel to the axis. The direction of the field inside the solenoid is given by RHR-3 (**Figure 21.32c**), just as it is for a circular current loop. The magnitude of the magnetic field in the interior of a long solenoid is

Interior of a long solenoid

$$B = \mu_0 n I \qquad \text{(21.7)}$$

where n is the number of turns per unit length of the solenoid and I is the current. If, for example, the solenoid contains 100 turns and has a length of 0.05 m, the number of turns per unit length is $n = (100 \text{ turns})/(0.05 \text{ m}) = 2000$ turns/m. The magnetic field outside the solenoid is not constant and is much weaker than the interior field. In fact, the magnetic field outside is nearly zero if the length of the solenoid is much greater than its diameter.

As with a single loop of wire, a solenoid can also be imagined to be a bar magnet, for the solenoid is just an array of connected current loops. And, as with a circular current loop, the location of the north pole can be determined with RHR-3. **Figure 21.32b** shows that the left end of the solenoid acts as a north pole, and the right end behaves as a south pole. Solenoids are often referred to as *electromagnets,* and they have several advantages over permanent magnets. For one thing, the strength of the magnetic field can be altered by changing the current and/or the number of turns per unit length. Furthermore, the north and south poles of an electromagnet can be readily switched by reversing the current.

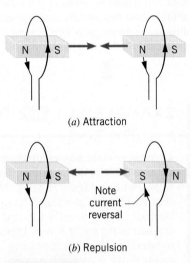

(a) Attraction

(b) Repulsion

Note current reversal

FIGURE 21.31 (*a*) The two current loops attract each other if the directions of the currents are the same and (*b*) repel each other if the directions are opposite. The "phantom" magnets help explain the attraction and repulsion.

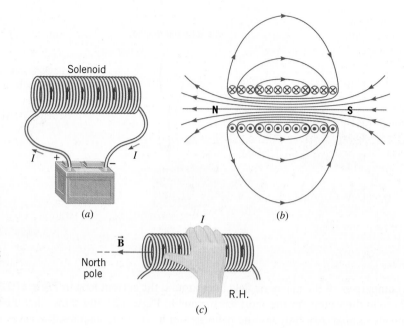

FIGURE 21.32 (a) A solenoid and (b) a cross-sectional view of it, showing the magnetic field lines and north and south poles. (c) Application of RHR-3 to the solenoid. The fingers of the right hand curl around the solenoid in the direction of the current flow, and the thumb points in the direction of the magnetic field everywhere inside the solenoid and toward the north pole.

EXAMPLE 10 | BIO The Physics of · · · Magnetic Resonance Imaging (MRI)

Applications of the magnetic field produced by a current-carrying solenoid are widespread. An important medical application is in magnetic resonance imaging (MRI). With this technique, detailed pictures of the internal parts of the body can be obtained in a noninvasive way that involves none of the risks inherent in the use of X-rays. **Figure 21.33** shows a magnetic resonance imaging machine being used. The circular opening visible in the photograph provides access to the interior of a solenoid, which is typically made from superconducting wire. The superconducting wire facilitates the use of a large current to produce a strong magnetic field. In the presence of this

field, the nuclei of certain atoms can be made to behave as tiny radio transmitters and emit radio waves similar to those used by FM stations. The hydrogen atom, which is so prevalent in the human body, can be made to behave in this fashion. The strength of the magnetic field determines where a given collection of hydrogen atoms will "broadcast" on an imaginary FM dial. With a magnetic field that has a slightly different strength at different places, it is possible to associate the location on this imaginary FM dial with a physical location within the body. Computer processing of these locations produces the magnetic resonance image. When hydrogen atoms are used in this way, the image is essentially a map showing their distribution within the body. Remarkably detailed images can now be obtained, such as the one in **Figure 21.34**. They provide doctors with a powerful diagnostic tool that complements those from X-ray and other techniques. Surgeons can now perform operations

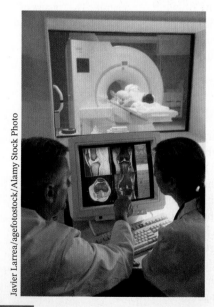

FIGURE 21.33 A magnetic resonance imaging (MRI) machine being used. The circular opening in which the patient is positioned provides access to the interior of a solenoid. MRI scans of the knee are shown on the monitor.

Javier Larrea/agefotostock/Alamy Stock Photo

Living Art Enterprises/Science Source

FIGURE 21.34 Magnetic resonance imaging provides one way to diagnose brain disorders. This color-enhanced MRI scan shows the brain of a patient with a large pineal region meningioma (see the red circular region), which is a usually benign, slow-growing tumor.

more accurately by stepping inside specially designed MRI scanners along with the patient and seeing live images of the area into which they are cutting.

The diameter of an MRI magnet has to be fairly large, so that, if need be, the patient's entire body can fit inside. The machine shown in **Figure 21.33** is an open-bore magnet, meaning the solenoid is open at both ends. Older machines would often use a closed-bore configuration, in which the solenoid was closed at one end. This resulted in many patients feeling claustrophobic. One open-bore MRI magnet with a length of 1.20 m produces a magnetic field of 2.50 T at its center. The coils of the magnet are comprised of a superconducting wire with a total length of 3.75×10^4 m carrying a current of 125 A. **(a)** What is the number of turns in the solenoid? **(b)** What is the diameter of the solenoid?

Reasoning The magnetic field at the center of the solenoid can be determined by Equation 21.7, which we can solve for the number of turns in the solenoid. The total length of wire used to wind the solenoid will be equal to the circumference of each turn

multiplied by the total number of turns. From this, we can calculate the diameter of the solenoid.

Solution **(a)** Equation 21.7 gives the magnetic field at the center of the solenoid: $B = \mu_0 I n$. Here, n is the number of turns per unit length of the solenoid, which we can write as N/ℓ, where ℓ is the solenoid's length. Making this substitution and then solving for N, we have

$$B = \frac{\mu_0 IN}{\ell} \Rightarrow N = \frac{B\ell}{\mu_0 I} = \frac{(2.50 \text{ T})(1.20 \text{ m})}{(4\pi \times 10^{-7} \text{ T} \cdot \text{m/A})(125 \text{ A})}$$

$$= \boxed{1.91 \times 10^4 \text{ turns}}$$

(b) The total length of wire in the solenoid (L_w) will be equal to the circumference of one turn (πd) multiplied by the total number of turns N, we have

$$L_w = \pi d N \Rightarrow d = \frac{L_w}{\pi N} = \frac{3.75 \times 10^4 \text{ m}}{\pi (1.91 \times 10^4 \text{ turns})} = 0.625 \text{ m} = \boxed{62.5 \text{ cm}}$$

THE PHYSICS OF . . . old television screens and computer display monitors. Some older television sets and some computer display monitors use electromagnets (solenoids) to produce images by exerting magnetic forces on moving electrons. An evacuated glass tube, called a cathode-ray tube (CRT), contains an electron gun that sends a narrow beam of high-speed electrons toward the screen of the tube, as **Figure 21.35a** illustrates. The inner surface of the screen is covered with a phosphor coating, and when the electrons strike it, they generate a spot of visible light. This spot is called a pixel (a contraction of "picture element").

To create a black-and-white picture, the electron beam is scanned rapidly from left to right across the screen. As the beam makes each horizontal scan, the number of electrons per second striking the screen is changed by electronics controlling the electron gun, making the scan line brighter in some places and darker in others. When the beam reaches the right side of the screen, it is turned off and returned to the left side slightly below where it started (see **Figure 21.35b**). The beam is then scanned across the next line, and so on. In these older TV sets, a complete picture consists of 525 scan lines (or 625 in Europe) and is formed in $\frac{1}{30}$ of a second. High-definition TV sets that utilized this older technology had about 1100 scan lines, giving a much sharper, more detailed picture.

FIGURE 21.35 (a) A cathode-ray tube contains an electron gun, a magnetic field for deflecting the electron beam, and a phosphor-coated screen. A CRT color TV actually uses three guns, although only one is shown here for clarity. (b) The image is formed by scanning the electron beam across the screen. (c) The red, green, and blue phosphors of a color TV.

The electron beam is deflected by a pair of electromagnets placed around the neck of the tube, between the electron gun and the screen. One electromagnet is responsible for producing the horizontal deflection of the beam and the other for the vertical deflection. For clarity, **Figure 21.35a** shows the net magnetic field generated by the electromagnets at one instant, and not the electromagnets themselves. The electric current in the electromagnets produces a net magnetic field that exerts a force on the moving electrons, causing their trajectories to bend and reach different points on the screen. Changing the current changes the field, so the electrons can be deflected to any point on the screen.

A color TV operates with three electron guns instead of one. Furthermore, the single phosphor of a black-and-white TV is replaced by a large number of three-dot clusters of phosphors that glow red, green, and blue when struck by an electron beam, as indicated in **Figure 21.35c**. Each red, green, and blue color in a cluster is produced when electrons from one of the three guns strike the corresponding phosphor dot. The three dots are so close together that, from a normal viewing distance, they cannot be separately distinguished. Red, green, and blue are primary colors, so virtually all other colors can be created by varying the intensities of the three beams focused on a cluster.

Check Your Understanding

(The answers are given at the end of the book.)

13. **CYU Figure 21.6** shows a conducting wire wound into a helical shape. The helix acts like a spring and expands back toward its original shape after the coils are squeezed together and released. The bottom end of the wire just barely touches the mercury (a good electrical conductor) in the cup. After the switch is closed, current in the circuit causes the light bulb to glow. Does the bulb **(a)** repeatedly turn on and off like a turn signal on a car, **(b)** glow continually, or **(c)** glow briefly and then go out?

CYU FIGURE 21.6

14. For each electromagnet at the left in **CYU Figure 21.7**, will it be attracted to or repelled from the permanent magnet immediately to its right?

15. For each electromagnet at the left in **CYU Figure 21.8**, will it be attracted to or repelled from the electromagnet immediately to its right?

CYU FIGURE 21.7

16. Refer to **Figure 21.5**. If the earth's magnetism is assumed to originate from a large circular loop of current within the earth, then the plane of the current loop must be oriented _____ to the earth's magnetic axis, and the direction of the current around the loop (when looking down on the loop from the north magnetic pole) is _____. **(a)** parallel, clockwise **(b)** parallel, counterclockwise **(c)** perpendicular, clockwise **(d)** perpendicular, counterclockwise

CYU FIGURE 21.8

17. **CYU Figure 21.9** shows an end-on view of three parallel wires that are perpendicular to the plane of the paper. In two of the wires the current is directed into the paper, while in the remaining wire the current is directed out of the paper. The two outermost wires are held rigidly in place. Which way will the middle wire move? **(a)** To the left **(b)** To the right **(c)** It will not move at all.

CYU FIGURE 21.9

18. In **CYU Figure 21.10**, assume that the current I_1 is larger than the current I_2. In parts a and b, decide whether there are places at which the total magnetic field is zero. State whether these places are located to the left of both wires, between the wires, or to the right of both wires.

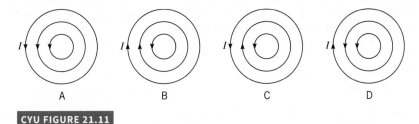

CYU FIGURE 21.10

19. Each of the four drawings **CYU Figure 21.11** shows the same three concentric loops of wire. The currents in the loops have the same magnitude I and have the directions shown. Rank the magnitude of the net magnetic field produced at the center of each of the four drawings, largest to smallest.

CYU FIGURE 21.11

20. There are four wires viewed end-on in **CYU Figure 21.12**. They are long, straight, and perpendicular to the plane of the paper. Their cross sections lie at the corners of a square. The magnitude of the current in each wire is the same. What must be the direction of the current in each wire (into or out of the page), so that when *any* single current is turned off, the total magnetic field at P (the center of the square) is directed toward a corner of the square?

	Wire 1	Wire 2	Wire 3	Wire 4
(a)	out of	out of	into	out of
(b)	into	out of	into	into
(c)	out of	into	out of	out of
(d)	into	into	into	into

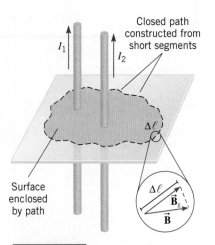

CYU FIGURE 21.12

21.8 Ampère's Law

We have seen that an electric current creates a magnetic field. However, the magnitude and direction of the field at any point in space depends on the specific geometry of the wire carrying the current. For instance, distinctly different magnetic fields surround a long, straight wire, a circular loop of wire, and a solenoid. Although different, each of these fields can be obtained from a general law known as **Ampère's law**, which is valid for a wire of any geometrical shape. Ampère's law specifies the relationship between an electric current and the magnetic field that it creates.

To see how Ampère's law is stated, consider **Figure 21.36**, which shows two wires carrying currents I_1 and I_2. In general, there may be any number of currents. Around the wires we construct an arbitrarily shaped but closed path. This path encloses a surface and is constructed from a large number of short segments, each of length $\Delta\ell$. Ampère's law deals with the product of $\Delta\ell$ and $B_{\parallel}$ for each segment, where $B_{\parallel}$ is the component

FIGURE 21.36 This setup is used in the text to explain Ampère's law.

of the magnetic field that is *parallel* to $\Delta \ell$ (see the blow-up view in the drawing). For magnetic fields that do not change as time passes, the law states that the sum of all the $B_{\parallel} \Delta \ell$ terms is proportional to the net current I passing through the surface enclosed by the path. For the specific example in **Figure 21.36**, we see that $I = I_1 + I_2$. Ampère's law is stated in equation form as follows:

AMPÈRE'S LAW FOR STATIC MAGNETIC FIELDS

For any current geometry that produces a magnetic field that does not change in time,

$$\Sigma B_{\parallel} \Delta \ell = \mu_0 I \qquad (21.8)$$

where $\Delta \ell$ is a small segment of length along a closed path of arbitrary shape around the current, $B_{\parallel}$ is the component of the magnetic field parallel to $\Delta \ell$, I is the net current passing through the surface bounded by the path, and μ_0 is the permeability of free space. The symbol Σ indicates that the sum of all $B_{\parallel} \Delta \ell$ terms must be taken around the closed path.

To illustrate the use of Ampère's law, we apply it in Example 11 to the special case of the current in a long, straight wire and show that it leads to the proper expression for the magnetic field.

EXAMPLE 11 | An Infinitely Long, Straight, Current-Carrying Wire

Use Ampère's law to obtain the magnetic field produced by the current in an infinitely long, straight wire.

Reasoning **Figure 21.24a** shows that compass needles point in a circular pattern around the wire, so we know that the magnetic field lines are circular. Therefore, it is convenient to use a circular path of radius r when applying Ampère's law, as **Figure 21.37** indicates.

Solution Along the circular path in **Figure 21.37**, the magnetic field is everywhere parallel to $\Delta \ell$ and has a constant magnitude, since each point is at the same distance from the wire. Thus, $B_{\parallel} = B$ and, according to Ampère's law, we have

$$\Sigma B_{\parallel} \Delta \ell = B(\Sigma \, \Delta \ell) = \mu_0 I$$

However, $\Sigma \Delta \ell$ is just the circumference $2\pi r$ of the circle, so Ampère's law reduces to

$$B(\Sigma \, \Delta \ell) = B(2\pi r) = \mu_0 I$$

FIGURE 21.37 Example 11 uses Ampère's law to find the magnetic field in the vicinity of this long, straight, current-carrying wire.

Dividing both sides by $2\pi r$ shows that $\boxed{B = \mu_0 I/(2\pi r)}$, as given earlier in Equation 21.5.

21.9 | Magnetic Materials

Ferromagnetism

The similarity between the magnetic field lines in the neighborhood of a bar magnet and those around a current loop suggests that the magnetism in each case arises from a common cause. The field that surrounds the loop is created by the charges moving in the wire. The magnetic field around a bar magnet is also due to the motion of charges, but the motion is not that of a bulk current through the magnetic material. Instead, the motion responsible for the magnetism is that of the electrons within the atoms of the material.

The magnetism produced by electrons within an atom can arise from two motions. First, each electron orbiting the nucleus behaves like an atomic-sized loop of current that generates a small magnetic field, similar to the field created by the current loop in **Figure 21.28a**. Second, each electron possesses a spin that also gives rise to a magnetic field. The net magnetic field created by the electrons within an atom is due to the combined fields created by their orbital and spin motions.

In most substances the magnetism produced at the atomic level tends to cancel out, with the result that the substance is nonmagnetic overall. However, there are some materials, known as **ferromagnetic materials**, in which the cancellation does not occur for groups of approximately 10^{16}–10^{19} neighboring atoms, because they have electron spins that are naturally aligned parallel to each other. This alignment results from a special type of quantum mechanical* interaction between the spins. The result of the interaction is a small but highly magnetized region of about 0.01 to 0.1 mm in size, depending on the nature of the material; this region is called a **magnetic domain**. Each domain behaves as a small magnet with its own north and south poles. Common ferromagnetic materials are iron, nickel, cobalt, chromium dioxide, and alnico (an *a*luminum–*ni*ckel–*co*balt alloy).

Induced Magnetism

Often magnetic domains in ferromagnetic materials are arranged randomly, as **Figure 21.38*a*** illustrates for a piece of iron. In such a situation, the magnetic fields of the domains cancel each other, so the iron displays little, if any, overall magnetism. However, an unmagnetized piece of iron can be magnetized by placing it in an external magnetic field provided by a permanent magnet or an electromagnet. The external magnetic field penetrates the unmagnetized iron and *induces* (or brings about) a state of magnetism in the iron by causing two effects. Those domains whose magnetism is parallel or nearly parallel to the external magnetic field grow in size at the expense of other domains that are not so oriented. In **Figure 21.38*b*** the growing domains are colored gold. In addition, the magnetic alignment of some domains may rotate and become more oriented in the direction of the external field. The resulting preferred alignment of the domains gives the iron an overall magnetism, so the iron behaves like a magnet with associated north and south poles. In some types of ferromagnetic materials, such as the chromium dioxide used in magnetic recording tapes, the domains remain aligned for the most part when the external magnetic field is removed, and the material thus becomes permanently magnetized.

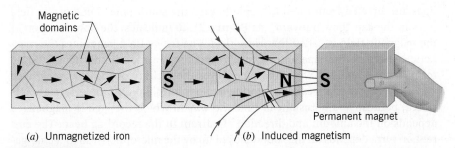

(*a*) Unmagnetized iron (*b*) Induced magnetism

FIGURE 21.38 (*a*) Each magnetic domain is a highly magnetized region that behaves like a small magnet (represented by an arrow whose head indicates a north pole). Unmagnetized iron consists of many domains that are randomly aligned. The size of each domain is exaggerated for clarity. (*b*) The external magnetic field of the permanent magnet causes those domains that are parallel or nearly parallel to the field to grow in size (shown in gold color).

The magnetism induced in a ferromagnetic material can be surprisingly large, even in the presence of a weak external field. For instance, it is not unusual for the induced magnetic field to be a hundred to a thousand times stronger than the external field that causes the alignment. For this reason, high-field electromagnets are constructed by wrapping the current-carrying wire around a solid core made from iron or other ferromagnetic material.

Induced magnetism explains why a permanent magnet sticks to a refrigerator door and why an electromagnet can pick up scrap iron at a junkyard. Notice in **Figure 21.38*b*** that there is a north pole at the end of the iron that is closest to the south pole of the

*The branch of physics called quantum mechanics is mentioned in Section 29.5. A detailed discussion of it is beyond the scope of this book.

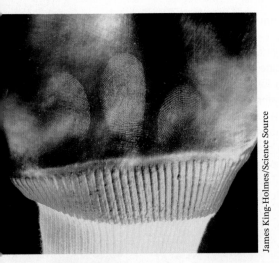

FIGURE 21.39 Magnetic fingerprint powder has been used to reveal fingerprints on this neoprene glove. The powder consists of tiny iron flakes coated with an organic material that enables them to stick to the greasy residue in the print. Because of induced magnetism, a permanent magnet can be used to remove any excess powder that would obscure the evidence. This eliminates the need for brushing and possible damage to the print.

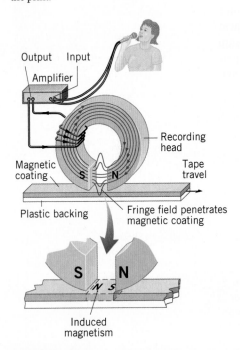

FIGURE 21.40 The magnetic fringe field of the recording head penetrates the magnetic coating on the tape and magnetizes it.

permanent magnet. The net result is that the two opposite poles give rise to an attraction between the iron and the permanent magnet. Conversely, the north pole of the permanent magnet would also attract the piece of iron by inducing a south pole in the nearest side of the iron. In nonferromagnetic materials, such as aluminum and copper, the formation of magnetic domains does not occur, so magnetism cannot be induced into these substances. Consequently, magnets do not stick to aluminum cans.

THE PHYSICS OF . . . detecting fingerprints. The attraction that induced magnetism creates between a permanent magnet and a ferromagnetic material is used in crime-scene investigations, where powder is dusted onto surfaces to make fingerprints visible. Magnetic fingerprint powder allows investigators to recover evidence from surfaces like the neoprene glove in **Figure 21.39**, which are very difficult to examine without damaging the prints in the process. Brushing excess conventional powder away ruins the delicate ridges of the pattern that allow the print to be identified reliably. Magnetic fingerprint powder, however, consists of tiny iron flakes coated with an organic material that allows them to stick to the greasy residue in the print. A permanent magnet eliminates the need for brushing away excess powder by creating induced magnetism in the iron and pulling away the powder not stuck directly to the greasy residue.

Magnetic Tape Recording

THE PHYSICS OF . . . magnetic tape recording. The process of magnetic tape recording uses induced magnetism, as **Figure 21.40** illustrates. The weak electrical signal from a microphone is routed to an amplifier where it is amplified. The current from the output of the amplifier is then sent to the recording head, which is a coil of wire wrapped around an iron core. The iron core has the approximate shape of a horseshoe with a small gap between the two ends. The ferromagnetic iron substantially enhances the magnetic field produced by the current in the coil.

When there is a current in the coil, the recording head becomes an electromagnet with a north pole at one end and a south pole at the other end. The magnetic field lines pass through the iron core and cross the gap. Within the gap, the lines are directed from the north pole toward the south pole. Some of the field lines in the gap "bow outward," as **Figure 21.40** indicates; the bowed region of the magnetic field is called the *fringe field*. The fringe field penetrates the magnetic coating on the tape and induces magnetism in the coating. This induced magnetism is retained when the tape leaves the vicinity of the recording head and, thus, provides a means for storing audio information. Audio information is stored, because at any instant in time the way in which the tape is magnetized depends on the amount and direction of current in the recording head. The current, in turn, depends on the sound picked up by the microphone, so that changes in the sound that occur from moment to moment are preserved as changes in the tape's induced magnetism.

Maglev Trains

THE PHYSICS OF . . . a magnetically levitated train. A magnetically levitated train—maglev, for short—uses forces that arise from induced magnetism to levitate or float above a guideway. Since it rides a few centimeters above the guideway, a maglev does not need wheels. Freed from friction due to the guideway, the train can achieve significantly greater speeds than do conventional trains. For example, the Transrapid maglev in **Figure 21.41** has achieved speeds of 110 m/s (250 mph).

Figure 21.41a shows that the Transrapid maglev achieves levitation with electromagnets mounted on arms that extend around and under the guideway. When a current is sent to an electromagnet, the resulting magnetic field creates induced magnetism in a rail mounted in the guideway. The upward attractive force from the induced magnetism is balanced by the weight of the train, so the train moves without touching the rail or the guideway.

Guideway   Rail

Arm

Levitation
electromagnet

(a)

Guideway

(b)

Bernd Mellmann/Alamy Stock Photo

FIGURE 21.41 (*a*) The Transrapid maglev (a German train) has achieved speeds of 110 m/s (250 mph). The levitation electromagnets are drawn up toward the rail in the guideway, levitating the train. (*b*) The magnetic propulsion system.

Magnetic levitation only lifts the train and does not move it forward. **Figure 21.41*b*** illustrates how magnetic propulsion is achieved. In addition to the levitation electromagnets, propulsion electromagnets are also placed along the guideway. By controlling the direction of the currents in the train and guideway electromagnets, it is possible to create an unlike pole in the guideway just ahead of each electromagnet on the train and a like pole just behind. Each electromagnet on the train is thus both pulled and pushed forward by electromagnets in the guideway. By adjusting the timing of the like and unlike poles in the guideway, the speed of the train can be adjusted. Reversing the poles in the guideway electromagnets serves to brake the train.

Check Your Understanding

(The answers are given at the end of the book.)

21. In a TV commercial that advertises a soda pop, a strong electromagnet picks up a delivery truck carrying cans of the soft drink. The picture switches to the interior of the truck, where cans are seen to fly upward and stick to the roof just beneath the electromagnet. Are these cans made entirely of aluminum?

22. Suppose that you have two bars. Bar 1 is a permanent magnet and bar 2 is not a magnet, but is made from a ferromagnetic material like iron. A third bar (bar 3), which is known to be a permanent magnet, is brought close to one end of bar 1 and then to one end of bar 2. Which one of the following statements is true? **(a)** Bars 1 and 3 will either be attracted to or repelled from each other, while bars 2 and 3 will always be repelled from each other. **(b)** Bars 1 and 3 will either be attracted to or repelled from each other, while bars 2 and 3 will always be attracted to each other. **(c)** Bars 1 and 3 will always be repelled from each other, while bars 2 and 3 will either be attracted to or repelled from each other. **(d)** Bars 1 and 3 will always be attracted to each other, while bars 2 and 3 will either be attracted to or repelled from each other.

EXAMPLE 12 | BIO The Physics of MRI Machines—The Superconducting Magnet

As discussed in Example 10, the medical diagnostic technique of magnetic resonance imaging relies on a large solenoid electromagnet to produce strong magnetic fields. Assume the magnet is wound with a single layer of 2.0-mm diameter wire that is insulated. If the highest magnetic field produced by the magnet is 1.7 T, what is the maximum current in the magnet?

Reasoning Since the magnetic field in the MRI machine is produced by a solenoid, we can apply Equation 21.7 to calculate the current in the magnet. To do this, we will need the value of *n* (the number of turns per unit length). This can be determined from the length of the magnet and the thickness of the wire.

Solution Beginning with Equation 21.7, we have:

$$B = \mu_0 n I = \mu_0 \left(\frac{N}{\ell}\right) I,$$

where $n = \left(\frac{N}{\ell}\right)$ is the number of turns (N) in the solenoid divided by its length (ℓ). N is determined by how many turns of wire in a single layer can fit side-by-side along the length of the solenoid. Therefore, $N = \ell/d$, where we let d represent the diameter of the wire. Making this substitution into Equation 21.7 above, we have the following:

$$B = \mu_0 \left(\frac{\ell/d}{\ell}\right) I = \frac{\mu_0 I}{d}$$

We can now solve this for the current in the solenoid:

$$I = \frac{dB}{\mu_0} = \frac{(0.0020 \text{ m})(1.7 \text{ T})}{4\pi \times 10^{-7} \text{ T} \cdot \text{m/A}} = \boxed{2700 \text{ A}}$$

This is an incredibly large current for a wire that is only 2.0 mm in diameter! As we mentioned previously in the chapter, the niobium-based wire in the MRI magnet can carry such large currents because of its superconducting properties (no loss due to electrical resistance). However, in order to take advantage of the wire's superconducting properties, it must be cooled to temperatures close to absolute zero (0 K) by submersing it in liquid helium.

Concept Summary

21.1 Magnetic Fields A magnet has a north pole and a south pole. The north pole is the end that points toward the north magnetic pole of the earth when the magnet is freely suspended. Like magnetic poles repel each other, and unlike poles attract each other.

A magnetic field exists in the space around a magnet. The magnetic field is a vector whose direction at any point is the direction indicated by the north pole of a small compass needle placed at that point. As an aid in visualizing the magnetic field, magnetic field lines are drawn in the vicinity of a magnet. The lines appear to originate from the north pole and end on the south pole. The magnetic field at any point in space is tangent to the magnetic field line at the point. Furthermore, the strength of the magnetic field is proportional to the number of lines per unit area that passes through a surface oriented perpendicular to the lines.

21.2 The Force That a Magnetic Field Exerts on a Moving Charge The direction of the magnetic force acting on a charge moving with a velocity $\vec{v}$ in a magnetic field $\vec{B}$ is perpendicular to both $\vec{v}$ and $\vec{B}$. For a positive charge the direction can be determined with the aid of Right-Hand Rule No. 1 (see below). The magnetic force on a moving negative charge is opposite to the force on a moving positive charge.

Right-Hand Rule No. 1: Extend the right hand so the fingers point along the direction of the magnetic field $\vec{B}$ and the thumb points along the velocity $\vec{v}$ of the charge. The palm of the hand then faces in the direction of the magnetic force $\vec{F}$ that acts on a positive charge.

The magnitude B of the magnetic field at any point in space is defined according to Equation 21.1, where F is the magnitude of the magnetic force on a test charge, $|q_0|$ is the magnitude of the test charge, and v is the magnitude of the charge's velocity, which makes an angle θ with the direction of the magnetic field. The SI unit for the magnetic field is the tesla (T). Another, smaller unit for the magnetic field is the gauss; 1 gauss $= 10^{-4}$ tesla. The gauss is not an SI unit.

$$B = \frac{F}{|q_0| v \sin \theta} \qquad (21.1)$$

21.3 The Motion of a Charged Particle in a Magnetic Field When a charged particle moves in a region that contains both magnetic and electric fields, the net force on the particle is the vector sum of the magnetic and electric forces.

A magnetic force does no work on a charged particle moving as in Figure 21.9b, because the direction of the force is always perpendicular to the motion of the particle. Being unable to do work, the magnetic force cannot change the kinetic energy, and hence the speed, of the particle; however, the magnetic force does change the direction in which the particle moves.

When a particle of charge q (magnitude $= |q|$) and mass m moves with speed v perpendicular to a uniform magnetic field of magnitude

B, the magnetic force causes the particle to move on a circular path that has a radius given by Equation 21.2.

$$r = \frac{mv}{|q|B} \qquad (21.2)$$

21.4 The Mass Spectrometer The mass spectrometer is an instrument for measuring the abundance of ionized atoms or molecules that have different masses. The atoms or molecules are ionized (charge $= +e$), accelerated to a speed v by a potential difference V, and sent into a uniform magnetic field of magnitude B. The magnetic field causes the particles (each with a mass m) to move on a circular path of radius r. The relation between m and B is given by Equation 1.

$$m = \left(\frac{er^2}{2V}\right) B^2 \qquad (1)$$

21.5 The Force on a Current in a Magnetic Field An electric current, being composed of moving charges, can experience a magnetic force when placed in a magnetic field of magnitude B. For a straight wire that has a length L and carries a current I, the magnetic force has a magnitude that is given by Equation 21.3, where θ is the angle between the directions of the current and the magnetic field. The direction of the force is perpendicular to both the current and the magnetic field and is given by Right-Hand Rule No. 1.

$$F = ILB \sin \theta \qquad (21.3)$$

21.6 The Torque on a Current-Carrying Coil Magnetic forces can exert a torque on a current-carrying loop of wire and thus cause the loop to rotate. When a current I exists in a coil of wire with N turns, each of area A, in the presence of a magnetic field of magnitude B, the coil experiences a net torque that has a magnitude given by Equation 21.4, where ϕ is the angle between the direction of the magnetic field and the normal to the plane of the coil. The quantity NIA is known as the magnetic moment of the coil.

$$\tau = NIAB \sin \phi \qquad (21.4)$$

21.7 Magnetic Fields Produced by Currents An electric current produces a magnetic field, with different current geometries giving rise to different field patterns. For an infinitely long, straight wire, the magnetic field lines are circles centered on the wire, and their direction is given by Right-Hand Rule No. 2 (see below). The magnitude of the magnetic field at a radial distance r from the wire is given by Equation 21.5, where I is the current in the wire and μ_0 is a constant known as the permeability of free space ($\mu_0 = 4\pi \times 10^{-7}$ T $\cdot$ m/A).

$$B = \frac{\mu_0 I}{2\pi r} \qquad (21.5)$$

Right-Hand Rule No. 2: Curl the fingers of the right hand into the shape of a half-circle. Point the thumb in the direction of the conventional current I, and the tips of the fingers will point in the direction of the magnetic field $\vec{\mathbf{B}}$.

The magnitude of the magnetic field at the center of a flat circular loop consisting of N turns, each of radius R and carrying a current I, is given by Equation 21.6.

$$B = N\frac{\mu_0 I}{2R} \qquad (21.6)$$

The loop has associated with it a north pole on one side and a south pole on the other side. The side of the loop that behaves like a north pole can be predicted by using Right-Hand Rule No. 3: Curl the fingers of the right hand along the direction of the conventional current I around the loop or solenoid, and the thumb will point in the direction of the magnetic field at the center of the loop, or everywhere inside the solenoid, and toward their respective north poles.

A solenoid is a coil of wire wound in the shape of a helix. Inside a long solenoid the magnetic field is nearly constant and has a magnitude that is given by Equation 21.7, where n is the number of turns per unit length of the solenoid and I is the current in the wire. One end of the solenoid behaves like a north pole, and the other end like

a south pole. The end that is the north pole can be predicted by using Right-Hand Rule No. 3.

$$B = \mu_0 nI \qquad (21.7)$$

21.8 Ampère's Law Ampère's law specifies the relationship between a current and its associated magnetic field. For any current geometry that produces a magnetic field that does not change in time, Ampère's law is given by Equation 21.8, where $\Delta\ell$ is a small segment of length along a closed path of arbitrary shape around the current, $B_\parallel$ is the component of the magnetic field parallel to $\Delta\ell$, I is the net current passing through the surface bounded by the path, and μ_0 is the permeability of free space. The symbol Σ indicates that the sum of all $B_\parallel \Delta\ell$ terms must be taken around the closed path.

$$\Sigma B_\parallel \Delta\ell = \mu_0 I \qquad (21.8)$$

21.9 Magnetic Materials Ferromagnetic materials, such as iron, are made up of tiny regions called magnetic domains, each of which behaves as a small magnet. In an unmagnetized ferromagnetic material, the domains are randomly aligned. In a permanent magnet, many of the domains are aligned, and a high degree of magnetism results. An unmagnetized ferromagnetic material can be induced into becoming magnetized by placing it in an external magnetic field.

Focus on Concepts   Online

Additional questions are available for assignment in WileyPLUS.

Section 21.2   The Force That a Magnetic Field Exerts on a Moving Charge

1. At a location near the equator, the earth's magnetic field is horizontal and points north. An electron is moving vertically upward from the ground. What is the direction of the magnetic force that acts on the electron? **(a)** North **(b)** East **(c)** South **(d)** West **(e)** The magnetic force is zero.

2. The drawing shows four situations in which a positively charged particle is moving with a velocity $\vec{\mathbf{v}}$ through a magnetic field $\vec{\mathbf{B}}$. In each case, the magnetic field is directed out of the page toward you, and the velocity is directed to the right. In only one of these drawings is the magnetic force $\vec{\mathbf{F}}$ physically reasonable. Which drawing is it? **(a)** 1 **(b)** 2 **(c)** 3 **(d)** 4

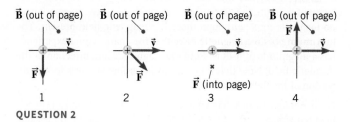

QUESTION 2

Section 21.3   The Motion of a Charged Particle in a Magnetic Field

3. Three particles are moving perpendicular to a uniform magnetic field and travel on circular paths (see the drawing). The particles have the same mass and speed. List the particles in order of their charge magnitude, largest to smallest. **(a)** 3, 2, 1 **(b)** 3, 1, 2 **(c)** 2, 3, 1 **(d)** 1, 3, 2 **(e)** 1, 2, 3

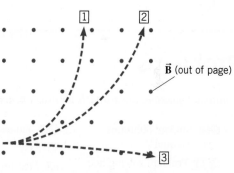

QUESTION 3

4. The drawing shows the circular paths of an electron and a proton. These particles have the same charge magnitudes, but the proton is more massive. They travel at the same speed in a uniform magnetic field $\vec{\mathbf{B}}$, which is directed into the page everywhere. Which particle follows the larger circle, and does it travel clockwise or counterclockwise?

QUESTION 4

	Particle	Direction of Travel
(a)	electron	clockwise
(b)	electron	counterclockwise
(c)	proton	clockwise
(d)	proton	counterclockwise

Section 21.5 The Force on a Current in a Magnetic Field

5. Four views of a horseshoe magnet and a current-carrying wire are shown in the drawing. The wire is perpendicular to the page, and the current is directed out of the page toward you. In which one or more of these situations does the magnetic force on the current point due north? **(a)** 1 and 2 **(b)** 3 and 4 **(c)** 2 **(d)** 3 **(e)** 1

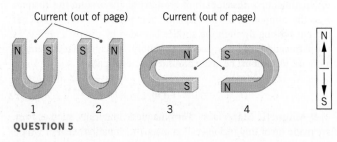

QUESTION 5

Section 21.6 The Torque on a Current-Carrying Coil

6. A square, current-carrying loop is placed in a uniform magnetic field $\vec{B}$ with the plane of the loop parallel to the magnetic field (see the drawing). The dashed line is the axis of rotation. The magnetic field exerts _____. **(a)** a net force and a

Axis through center of loop

QUESTION 6

net torque on the loop **(b)** a net force, but not a net torque, on the loop **(c)** a net torque, but not a net force, on the loop **(d)** neither a net force nor a net torque on the loop

Section 21.7 Magnetic Fields Produced by Currents

7. The drawing shows four situations in which two very long wires are carrying the same current, although the directions of the currents may be different. The point P in the drawings is equidistant from each wire. Which one (or more) of these situations gives rise to a zero net magnetic field at P? **(a)** 2 and 4 **(b)** Only 1 **(c)** Only 2 **(d)** 2 and 3 **(e)** 3 and 4

QUESTION 7

8. Three long, straight wires are carrying currents that have the same magnitude. In C the current is opposite to the current in A and B. The wires are equally spaced. Each wire experiences a net force due to the other two wires. Which wire experiences a net force with the greatest magnitude? **(a)** A **(b)** B **(c)** C **(d)** All three wires experience a net force that has the same magnitude.

QUESTION 8

Problems

Online

Additional questions are available for assignment in WileyPLUS.

SSM	Student Solutions Manual	**BIO**	Biomedical application
MMH	Problem-solving help	**E**	Easy
GO	Guided Online Tutorial	**M**	Medium
V-HINT	Video Hints	**H**	Hard
CHALK	Chalkboard Videos	**WS**	Worksheet
		T	Team Problem

Section 21.1 Magnetic Fields,

Section 21.2 The Force That a Magnetic Field Exerts on a Moving Charge

1. **E** **SSM** In New England, the horizontal component of the earth's magnetic field has a magnitude of 1.6×10^{-5} T. An electron is shot vertically straight up from the ground with a speed of 2.1×10^6 m/s. What is the magnitude of the acceleration caused by the magnetic force? Ignore the gravitational force acting on the electron.

2. **E** **(a)** A proton, traveling with a velocity of 4.5×10^6 m/s due east, experiences a magnetic force that has a maximum magnitude of 8.0×10^{-14} N and a direction of due south. What are the magnitude and direction of the magnetic field causing the force? **(b)** Repeat part (a) assuming the proton is replaced by an electron.

3. **E** **SSM** At a certain location, the horizontal component of the earth's magnetic field is 2.5×10^{-5} T, due north. A proton moves

eastward with just the right speed for the magnetic force on it to balance its weight. Find the speed of the proton.

4. **E** A charge of -8.3 μC is traveling at a speed of 7.4×10^6 m/s in a region of space where there is a magnetic field. The angle between the velocity of the charge and the field is 52°. A force of magnitude 5.4×10^{-3} N acts on the charge. What is the magnitude of the magnetic field?

5. **E** When a charged particle moves at an angle of 25° with respect to a magnetic field, it experiences a magnetic force of magnitude F. At what angle (less than 90°) with respect to this field will this particle, moving at the same speed, experience a magnetic force of magnitude $2F$?

6. **E** **GO** A particle that has an 8.2 μC charge moves with a velocity of magnitude 5.0×10^5 m/s along the $+x$ axis. It experiences no magnetic force, although there is a magnetic field present. The maximum possible magnetic force that the charge could experience has a magnitude of 0.48 N. Find the magnitude and direction of the magnetic field. Note that there are two possible answers for the direction of the field.

7. **E** A magnetic field has a magnitude of 1.2×10^{-3} T, and an electric field has a magnitude of 4.6×10^3 N/C. Both fields point in the same direction. A positive 1.8 μC charge moves at a speed of 3.1×10^6 m/s in a direction that is perpendicular to both fields. Determine the magnitude of the net force that acts on the charge.

8. **E** **GO** Two charged particles move in the same direction with respect to the same magnetic field. Particle 1 travels three times faster than particle 2. However, each particle experiences a magnetic force of the same magnitude. Find the ratio $|q_1|/|q_2|$ of the magnitudes of the charges.

9. **M** **V-HINT** The drawing shows a parallel plate capacitor that is moving with a speed of 32 m/s through a 3.6-T magnetic field. The velocity $\vec{v}$ is perpendicular to the magnetic field. The electric field within the capacitor has a value of 170 N/C, and each plate has an area of 7.5×10^{-4} m². What is the magnetic force (magnitude and direction) exerted on the positive plate of the capacitor?

PROBLEM 9

10. **M** **V-HINT** **CHALK** One component of a magnetic field has a magnitude of 0.048 T and points along the $+x$ axis, while the other component has a magnitude of 0.065 T and points along the $-y$ axis. A particle carrying a charge of $+2.0 \times 10^{-5}$ C is moving along the $+z$ axis at a speed of 4.2×10^{3} m/s. **(a)** Find the magnitude of the net magnetic force that acts on the particle. **(b)** Determine the angle that the net force makes with respect to the $+x$ axis.

11. **M** **SSM** **CHALK** The electrons in the beam of a television tube have a kinetic energy of 2.40×10^{-15} J. Initially, the electrons move horizontally from west to east. The vertical component of the earth's magnetic field points down, toward the surface of the earth, and has a magnitude of 2.00×10^{-5} T. **(a)** In what direction are the electrons deflected by this field component? **(b)** What is the acceleration of an electron in part (a)?

Section 21.3   The Motion of a Charged Particle in a Magnetic Field,

Section 21.4   The Mass Spectrometer

12. **E** An ionized helium atom has a mass of 6.6×10^{-27} kg and a speed of 4.4×10^{5} m/s. It moves perpendicular to a 0.75-T magnetic field on a circular path that has a 0.012-m radius. Determine whether the charge of the ionized atom is $+e$ or $+2e$.

13. **E** **GO** A charged particle with a charge-to-mass ratio of $|q|/m = 5.7 \times 10^{8}$ C/kg travels on a circular path that is perpendicular to a magnetic field whose magnitude is 0.72 T. How much time does it take for the particle to complete one revolution?

14. **E** A charged particle enters a uniform magnetic field and follows the circular path shown in the drawing. **(a)** Is the particle positively or negatively charged? Why? **(b)** The particle's speed is 140 m/s, the magnitude of the magnetic field is 0.48 T, and the radius of the path is 960 m. Determine the mass of the particle, given that its charge has a magnitude of 8.2×10^{-4} C.

PROBLEM 14

15. **E** **GO** A proton is projected perpendicularly into a magnetic field that has a magnitude of 0.50 T. The field is then adjusted so that an electron will follow a circular path of the same radius when it is projected perpendicularly into the field with the same velocity that the proton had. What is the magnitude of the field used for the electron?

16. **E** **SSM** When beryllium-7 ions ($m = 11.65 \times 10^{-27}$ kg) pass through a mass spectrometer, a uniform magnetic field of 0.283 T curves their path directly to the center of the detector (see **Figure 21.14**). For the same accelerating potential difference, what magnetic field should be used to send beryllium-10 ions ($m = 16.63 \times 10^{-27}$ kg) to the same location in the detector? Both types of ions are singly ionized ($q = +e$).

17. **E** **GO** Suppose that an ion source in a mass spectrometer produces *doubly* ionized gold ions (Au^{2+}), each with a mass of 3.27×10^{-25} kg. The ions are accelerated from rest through a potential difference of 1.00 kV. Then, a 0.500-T magnetic field causes the ions to follow a circular path. Determine the radius of the path.

18. **E** An α-particle has a charge of $+2e$ and a mass of 6.64×10^{-27} kg. It is accelerated from rest through a potential difference that has a value of 1.20×10^{6} V and then enters a uniform magnetic field whose magnitude is 2.20 T. The α-particle moves perpendicular to the magnetic field at all times. What is **(a)** the speed of the α-particle, **(b)** the magnitude of the magnetic force on it, and **(c)** the radius of its circular path?

19. **E** **GO** Particle 1 and particle 2 have masses of $m_1 = 2.3 \times 10^{-8}$ kg and $m_2 = 5.9 \times 10^{-8}$ kg, but they carry the same charge q. The two particles accelerate from rest through the same electric potential difference V and enter the same magnetic field, which has a magnitude B. The particles travel perpendicular to the magnetic field on circular paths. The radius of the circular path for particle 1 is $r_1 = 12$ cm. What is the radius (in cm) of the circular path for particle 2?

20. **E** Two of the isotopes of carbon, carbon-12 and carbon-13, have masses of 19.93×10^{-27} kg and 21.59×10^{-27} kg, respectively. These two isotopes are singly ionized ($+e$), each given a speed of 6.667×10^{5} m/s. The ions then enter the bending region of a mass spectrometer where the magnetic field is 0.8500 T. Determine the spatial separation between the two isotopes after they have traveled through a half-circle.

21. **M** **V-HINT** The ion source in a mass spectrometer produces both singly and doubly ionized species, X^{+} and X^{2+}. The difference in mass between these species is too small to be detected. Both species are accelerated through the same electric potential difference, and both experience the same magnetic field, which causes them to move on circular paths. The radius of the path for the species X^{+} is r_1, while the radius for species X^{2+} is r_2. Find the ratio r_1/r_2 of the radii.

22. **M** **SSM** **MMH** **CHALK** A proton with a speed of 3.5×10^{6} m/s is shot into a region between two plates that are separated by a distance of 0.23 m. As the drawing shows, a magnetic field exists between the plates, and it is perpendicular to the velocity of the proton. What must be the magnitude of the magnetic field so the proton just misses colliding with the opposite plate?

PROBLEM 22

23. **M** **V-HINT** Review Conceptual Example 2 as an aid in understanding this problem. A velocity selector has an electric field of magnitude 2470 N/C, directed vertically upward, and a horizontal magnetic field that is directed south. Charged particles, traveling east at a speed of 6.50×10^{3} m/s, enter the velocity selector and are able to pass completely through without being deflected. When a different particle with an electric charge of $+4.00 \times 10^{-12}$ C enters the velocity selector traveling east, the net force (due to the electric and magnetic fields) acting on it is 1.90×10^{-9} N, pointing directly upward. What is the speed of this particle?

24. **M** **SSM** A particle of mass 6.0×10^{-8} kg and charge $+7.2$ μC is traveling due east. It enters perpendicularly a magnetic field whose magnitude is 3.0 T. After entering the field, the particle completes one-half of a circle and exits the field traveling due west. How much time does the particle spend traveling in the magnetic field?

25. **M** **V-HINT** Conceptual Example 4 provides background pertinent to this problem. An electron has a kinetic energy of 2.0×10^{-17} J. It moves on a circular path that is perpendicular to a uniform magnetic field of magnitude 5.3×10^{-5} T. Determine the radius of the path.

26. **M** **CHALK** A positively charged particle of mass 7.2×10^{-8} kg is traveling due east with a speed of 85 m/s and enters a 0.31-T uniform magnetic field. The particle moves through one-quarter of a circle in a time of 2.2×10^{-3} s, at which time it leaves the field heading due south. All during the motion the particle moves perpendicular to the magnetic field. **(a)** What is the magnitude of the magnetic force acting on the particle? **(b)** Determine the magnitude of its charge.

27. **M** Review Conceptual Example 2 as background for this problem. A charged particle moves through a velocity selector at a constant speed in a straight line. The electric field of the velocity selector is 3.80×10^3 N/C, while the magnetic field is 0.360 T. When the electric field is turned off, the charged particle travels on a circular path whose radius is 4.30 cm. Find the charge-to-mass ratio of the particle.

Section 21.5 The Force on a Current in a Magnetic Field

28. **E** At New York City, the earth's magnetic field has a vertical component of 5.2×10^{-5} T that points downward (perpendicular to the ground) and a horizontal component of 1.8×10^{-5} T that points toward geographic north (parallel to the ground). What are the magnitude and direction of the magnetic force on a 6.0-m long, straight wire that carries a current of 28 A perpendicularly into the ground?

29. **E** **SSM** A 45-m length of wire is stretched horizontally between two vertical posts. The wire carries a current of 75 A and experiences a magnetic force of 0.15 N. Find the magnitude of the earth's magnetic field at the location of the wire, assuming the field makes an angle of 60.0° with respect to the wire.

30. **E** A straight wire in a magnetic field experiences a force of 0.030 N when the current in the wire is 2.7 A. The current in the wire is changed, and the wire experiences a force of 0.047 N as a result. What is the new current?

31. **E** A horizontal wire of length 0.53 m, carrying a current of 7.5 A, is placed in a uniform external magnetic field. When the wire is horizontal, it experiences no magnetic force. When the wire is tilted upward at an angle of 19°, it experiences a magnetic force of 4.4×10^{-3} N. Determine the magnitude of the external magnetic field.

32. **E** **GO** The drawing shows a wire composed of three segments, AB, BC, and CD. There is a current of $I = 2.8$ A in the wire. There is also a magnetic field $\vec{B}$ (magnitude = 0.26 T) that is the same everywhere and points in the direction of the +z axis. The lengths of the wire segments are $L_{AB} = 1.1$ m, $L_{BC} = 0.55$ m, and $L_{CD} = 0.55$ m. Find the magnitude of the force that acts on each segment.

PROBLEM 32

33. **E** **SSM** A wire carries a current of 0.66 A. This wire makes an angle of 58° with respect to a magnetic field

of magnitude 4.7×10^{-5} T. The wire experiences a magnetic force of magnitude 7.1×10^{-5} N. What is the length of the wire?

34. **E** A loop of wire has the shape of a right triangle (see the drawing) and carries a current of $I = 4.70$ A. A uniform magnetic field is directed parallel to side AB and has a magnitude of 1.80 T. **(a)** Find the magnitude and direction of the magnetic force exerted on each side of the triangle. **(b)** Determine the magnitude of the net force exerted on the triangle.

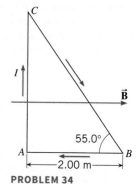
PROBLEM 34

35. **M** **CHALK** **GO** A copper rod of length 0.85 m is lying on a frictionless table (see the drawing). Each end of the rod is attached to a fixed wire by an unstretched spring that has a spring constant of $k = 75$ N/m. A magnetic field with a strength of 0.16 T is oriented perpendicular to the surface of the table. **(a)** What must be the direction of the current in the copper rod that causes the springs to stretch? **(b)** If the current is 12 A, by how much does each spring stretch?

PROBLEM 35

36. **M** **SSM** **MMH** The drawing shows a thin, uniform rod that has a length of 0.45 m and a mass of 0.094 kg. This rod lies in the plane of the paper and is attached to the floor by a hinge at point P. A uniform magnetic field of 0.36 T is directed perpendicularly into the plane of the paper. There is a current $I = 4.1$ A in the rod, which does not rotate clockwise or counterclockwise. Find the angle θ. (*Hint:* The magnetic force may be taken to act at the center of gravity.)

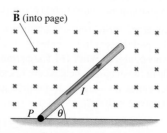
PROBLEM 36

37. **M** **CHALK** **GO** A horizontal wire is hung from the ceiling of a room by two massless strings. The wire has a length of 0.20 m and a mass of 0.080 kg. A uniform magnetic field of magnitude 0.070 T is directed from the ceiling to the floor. When a current of $I = 42$ A exists in the wire, the wire swings upward and, at equilibrium, makes an angle ϕ with respect to the vertical, as the drawing shows. Find **(a)** the angle ϕ and **(b)** the tension in each of the two strings.

PROBLEM 37

PROBLEM 43

38. H The two conducting rails in the drawing are tilted upward so they each make an angle of 30.0° with respect to the ground. The vertical magnetic field has a magnitude of 0.050 T. The 0.20-kg aluminum rod (length = 1.6 m) slides *without friction* down the rails at a constant velocity. How much current flows through the rod?

PROBLEM 38

Section 21.6 The Torque on a Current-Carrying Coil

39. E Two coils have the same number of circular turns and carry the same current. Each rotates in a magnetic field as in **Figure 21.19**. Coil 1 has a radius of 5.0 cm and rotates in a 0.18-T field. Coil 2 rotates in a 0.42-T field. Each coil experiences the same maximum torque. What is the radius (in cm) of coil 2?

40. E The 1200-turn coil in a dc motor has an area per turn of 1.1×10^{-2} m². The design for the motor specifies that the magnitude of the maximum torque is 5.8 N · m when the coil is placed in a 0.20-T magnetic field. What is the current in the coil?

41. E Two circular coils of current-carrying wire have the same magnetic moment. The first coil has a radius of 0.088 m, has 140 turns, and carries a current of 4.2 A. The second coil has 170 turns and carries a current of 9.5 A. What is the radius of the second coil?

42. E A wire has a length of 7.00×10^{-2} m and is used to make a circular coil of one turn. There is a current of 4.30 A in the wire. In the presence of a 2.50-T magnetic field, what is the maximum torque that this coil can experience?

43. E GO The coil of wire in the drawing is a right triangle and is free to rotate about an axis that is attached along side *AC*. The current in the loop is $I = 4.70$ A, and the magnetic field (parallel to the plane of the loop and side *AB*) is $B = 1.80$ T. **(a)** What is the magnetic moment of the loop, and **(b)** what is the magnitude of the net torque exerted on the loop by the magnetic field?

44. E GO You have a wire of length $L = 1.00$ m from which to make the square coil of a dc motor. The current in the coil is $I = 1.7$ A, and the magnetic field of the motor has a magnitude of $B = 0.34$ T. Find the maximum torque exerted on the coil when the wire is used to make a single-turn square coil and a two-turn square coil.

45. E SSM The rectangular loop in the drawing consists of 75 turns and carries a current of $I = 4.4$ A. A 1.8-T magnetic field is directed along the $+y$ axis. The loop is free to rotate about the z axis. **(a)** Determine the magnitude of the net torque exerted on the loop and **(b)** state whether the 35° angle will increase or decrease.

PROBLEM 45

46. M CHALK A square coil and a rectangular coil are each made from the same length of wire. Each contains a single turn. The long sides of the rectangle are twice as long as the short sides. Find the ratio $\tau_{\text{square}}/\tau_{\text{rectangle}}$ of the maximum torques that these coils experience in the same magnetic field when they contain the same current.

Section 21.7 Magnetic Fields Produced by Currents

47. E SSM Suppose in the figure that $I_1 = I_2 = 25$ A and that the separation between the wires is 0.016 m. By applying an external magnetic field (created by a source other than the wires) it is possible to cancel the mutual repulsion of the wires. This external field must point along the vertical direction. **(a)** Does the external field point up or down? Explain. **(b)** What is the magnitude of the external field?

PROBLEM 47

on this line: to the left of wire 1, between wire 1 and wire 2, and to the right of wire 2.

PROBLEM 47
(b)
Attraction

48. **E** A long solenoid has a length of 0.65 m and contains 1400 turns of wire. There is a current of 4.7 A in the wire. What is the magnitude of the magnetic field within the solenoid?

49. **E** **GO** A long solenoid has 1400 turns per meter of length, and it carries a current of 3.5 A. A small circular coil of wire is placed inside the solenoid with the normal to the coil oriented at an angle of 90.0° with respect to the axis of the solenoid. The coil consists of 50 turns, has an area of 1.2×10^{-3} m², and carries a current of 0.50 A. Find the torque exerted on the coil.

50. **E** **SSM** Two circular loops of wire, each containing a single turn, have the same radius of 4.0 cm and a common center. The planes of the loops are perpendicular. Each carries a current of 1.7 A. What is the magnitude of the net magnetic field at the common center?

51. **E** Two rigid rods are oriented parallel to each other and to the ground. The rods carry the same current in the same direction. The length of each rod is 0.85 m, and the mass of each is 0.073 kg. One rod is held in place above the ground, while the other floats beneath it at a distance of 8.2×10^{-3} m. Determine the current in the rods.

52. **E** Two long, straight wires are separated by 0.120 m. The wires carry currents of 8.0 A in opposite directions, as the drawing indicates. Find the magnitude of the net magnetic field at the points labeled (a) *A* and (b) *B*.

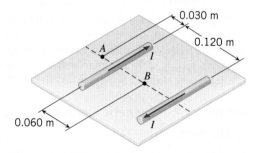

PROBLEM 52

53. **E** **GO** A long, straight wire carrying a current of 305 A is placed in a uniform magnetic field that has a magnitude of 7.00×10^{-3} T. The wire is perpendicular to the field. Find a point in space where the net magnetic field is zero. Locate this point by specifying its perpendicular distance from the wire.

54. **M** **MMH** Two circular coils are concentric and lie in the same plane. The inner coil contains 140 turns of wire, has a radius of 0.015 m, and carries a current of 7.2 A. The outer coil contains 180 turns and has a radius of 0.023 m. What must be the magnitude and direction (relative to the current in the inner coil) of the current in the outer coil, so that the net magnetic field at the common center of the two coils is zero?

55. **M** **V-HINT** Two infinitely long, straight wires are parallel and separated by a distance of one meter. They carry currents in the same direction. Wire 1 carries four times the current that wire 2 carries. On a line drawn perpendicular to both wires, locate the spot (relative to wire 1) where the net magnetic field is zero. Assume that wire 1 lies to the left of wire 2 and note that there are three regions to consider

56. **M** **GO** **CHALK** The drawing shows two perpendicular, long, straight wires, both of which lie in the plane of the paper. The current in each of the wires is $I = 5.6$ A. Find the magnitudes of the net magnetic fields at points *A* and *B*.

PROBLEM 56

57. **M** **SSM** A piece of copper wire has a resistance per unit length of 5.90×10^{-3} Ω/m. The wire is wound into a thin, flat coil of many turns that has a radius of 0.140 m. The ends of the wire are connected to a 12.0-V battery. Find the magnetic field strength at the center of the coil.

58. **H** The drawing shows two wires that both carry the same current of $I = 85.0$ A and are oriented perpendicular to the plane of the paper. The current in one wire is directed out of the paper, while the current in the other is directed into the paper. Find the magnitude and direction of the net magnetic field at point *P*.

PROBLEM 58

59. **H** The drawing shows two long, straight wires that are suspended from a ceiling. The mass per unit length of each wire is 0.050 kg/m. Each of the four strings suspending the wires has a length of 1.2 m. When the wires carry identical currents in opposite directions, the angle between the strings holding the two wires is 15°. What is the current in each wire?

PROBLEM 59

Section 21.8 Ampère's Law

60. **E** Suppose that a uniform magnetic field is everywhere perpendicular to this page. The field points directly upward toward you. A circular path is drawn on the page. Use Ampère's law to show that there can be no net current passing through the circular surface.

61. **E** **V-HINT** **CHALK** The wire in **Figure 21.37** carries a current of 12 A. Suppose that a second long, straight wire is placed right next to this wire. The current in the second wire is 28 A. Use Ampère's law to find the magnitude of the magnetic field at a distance of $r = 0.72$ m from the wires when the currents are (a) in the same direction and (b) in opposite directions.

Additional Problems

62. **E** In a certain region, the earth's magnetic field has a magnitude of 5.4×10^{-5} T and is directed north at an angle of 58° below the horizontal. An electrically charged bullet is fired north and 11° above the horizontal, with a speed of 670 m/s. The magnetic force on the bullet is 2.8×10^{-10} N, directed due east. Determine the bullet's electric charge, including its algebraic sign (+ or −).

63. **E** **SSM** An electron is moving through a magnetic field whose magnitude is 8.70×10^{-4} T. The electron experiences only a magnetic force and has an acceleration of magnitude 3.50×10^{14} m/s². At a certain instant, it has a speed of 6.80×10^{6} m/s. Determine the angle θ (less than 90°) between the electron's velocity and the magnetic field.

64. **E** **GO** A very long, straight wire carries a current of 0.12 A. This wire is tangent to a single-turn, circular wire loop that also carries a current. The directions of the currents are such that the net magnetic field at the center of the loop is zero. Both wires are insulated and have diameters that can be neglected. How much current is there in the loop?

65. **E** **SSM** The maximum torque experienced by a coil in a 0.75-T magnetic field is 8.4×10^{-4} N · m. The coil is circular and consists of only one turn. The current in the coil is 3.7 A. What is the length of the wire from which the coil is made?

66. **E** Multiple-Concept Example 7 discusses how problems like this one can be solved. A +6.00 μC charge is moving with a speed of 7.50×10^{4} m/s parallel to a very long, straight wire. The wire is 5.00 cm from the charge and carries a current of 67.0 A in a direction opposite to that of the moving charge. Find the magnitude and direction of the force on the charge.

67. **E** **V-HINT** The x, y, and z components of a magnetic field are $B_x = 0.10$ T, $B_y = 0.15$ T, and $B_z = 0.17$ T. A 25-cm wire is oriented along the z axis and carries a current of 4.3 A. What is the magnitude of the magnetic force that acts on this wire?

68. **E** In a lightning bolt, a large amount of charge flows during a time of 1.8×10^{-3} s. Assume that the bolt can be treated as a long, straight line of current. At a perpendicular distance of 27 m from the bolt, a magnetic field of 8.0×10^{-5} T is measured. How much charge has flowed during the lightning bolt? Ignore the earth's magnetic field.

69. **E** A charge is moving perpendicular to a magnetic field and experiences a force whose magnitude is 2.7×10^{-3} N. If this same charge were to move at the same speed and the angle between its velocity and the same magnetic field were 38°, what would be the magnitude of the magnetic force that the charge would experience?

70. **E** **GO** The drawing shows four insulated wires overlapping

PROBLEM 70

one another, forming a square with 0.050-m sides. All four wires are much longer than the sides of the square. The net magnetic field at the center of the square is 61 μT. Calculate the current I.

71. **M** **GO** A particle has a charge of $q = +5.60$ μC and is located at the coordinate origin. As the drawing shows, an electric field of $E_x = +245$ N/C exists along the +x axis. A magnetic field also exists, and its x and y components are $B_x = +1.80$ T and $B_y = +1.40$ T. Calculate the force (magnitude and direction) exerted on the particle by each of the three fields when it is (a) stationary, (b) moving along the +x axis at a speed of 375 m/s, and (c) moving along the +z axis at a speed of 375 m/s.

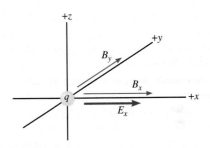

PROBLEM 71

72. **M** **GO** The figure shows two parallel, straight wires that are very long. The wires are separated by a distance of 0.065 m and carry currents of $I_1 = 15$ A and $I_2 = 7.0$ A. Find the magnitude and direction of the force that the magnetic field of wire 1 applies to a 1.5-m section of wire 2 when the currents have (a) opposite and (b) the same directions.

PROBLEM 72

Physics in Biology, Medicine, and Sports

73. E BIO SSM **21.4** In the operating room, anesthesiologists use mass spectrometers to monitor the respiratory gases of patients undergoing surgery. One gas that is often monitored is the anesthetic isoflurane (molecular mass = 3.06×10^{-25} kg). In a spectrometer, a singly ionized molecule of isoflurane (charge = $+e$) moves at a speed of 7.2×10^3 m/s on a circular path that has a radius of 0.10 m. What is the magnitude of the magnetic field that the spectrometer uses?

74. E BIO SSM **21.7** The magnetic field produced by the solenoid in a magnetic resonance imaging (MRI) system designed for measurements on whole human bodies has a field strength of 7.0 T, and the current in the solenoid is 2.0×10^2 A. What is the number of turns per meter of length of the solenoid? Note that the solenoid used to produce the magnetic field in this type of system has a length that is not very long compared to its diameter. Because of this and other design considerations, your answer will be only an approximation.

75. M BIO **21.7** The safety of living near high voltage power lines has been questioned for many years. Concerned groups believe the magnetic fields created by the large currents in the lines can lead to an increased risk of cancer and other health problems. However, multiple studies since 1995 by the American Physical Society and the National Academy of Sciences (National Research Council) have concluded that the relationship between higher rates of cancer and the proximity to high voltage power lines is not scientifically substantiated. Current research suggests that exposure to these fields does not represent a human health hazard. Consider a person standing 55 m from a high voltage line that is carrying a current of 250 A. **(a)** What is the magnitude of the magnetic field due to the current in the power line? **(b)** What is the ratio of the earth's magnetic field at its surface (0.45 gauss) to the field you obtained in part (a)?

76. M BIO **21.7** Radionics, also known as electromagnetic therapy, is an alternative medical treatment. In some cases, patients will expose themselves to magnetic fields created by electrical devices. They believe that the magnetic fields can apply forces to the iron-containing hemoglobin in the blood and increase blood flow. These claims are unproven, and no health benefits have ever been established. In fact, even a field as large as 1.0 T has no measured effect on blood hemoglobin. In an attempt to promote healing, a professional athlete inserts a broken wrist into a circular coil of wire composed of 5200 turns. If the radius of the coil is 4.5 cm, and the coil produces a 1.0-T magnetic field, what is the current in the coil?

77. M BIO **21.7** Many biological organisms, including you and your authors, are made mostly of water. Water is a *diamagnetic* substance, which is repelled by a magnetic field. This is in contrast to ferromagnetic materials described in Section 21.9, which are attracted by magnetic fields. If the magnetic fields are large enough, and the mass of the object is small, the repulsive force can cause the object to levitate (see the photo). The magnetic field at the center of the electromagnet that is levitating the frog is 16 T. These powerful magnets use a stack of thin conducting plates instead of wire windings to form a single spiraling conducting path, as shown in the photo, effectively creating a solenoid. If the magnet contains 3.2 plates/mm of length, what is the current flowing through the plates?

PROBLEM 77

78. M BIO **21.2** One technique of biological medical imaging that does not require the high magnetic fields utilized in MRI is *magneto-acoustic (MA) imaging*. The sample is placed in a uniform magnetic field, and a low frequency ac current is passed through the tissues of the body, perpendicular to the magnetic field. The current will experience a force due to the field. Since the current is switching direction, the magnetic force switches direction too. Different tissues have different conductivities, and this results in different forces at the boundaries between different types of tissue. This creates ultrasonic waves that can be detected with millimeter resolution. Consider a 0.10-mA current of sodium ions moving as an electrolyte through the extracellular fluid within the tissue. If the current is oriented perpendicular to a 0.15-T magnetic field, what is the magnitude of the force that acts on a 2.0-mm length of the current?

79. M BIO **21.7** Magnetic drug delivery is one option to target a drug onto a very localized part of the body. For example, if a blood clot, or thrombus, is detected inside a blood vessel, it should be removed in an effort to maintain normal blood flow through the vessel. Novel cancer therapies have also used magnetic drug delivery to treat tumors. Here, nanometer-sized particles (or nanoparticles) of iron oxide, a ferromagnetic material, are coated with functionalized drug components that can be precisely guided to specific areas of the body using magnetic fields. Once in position, the nanoparticles are exposed to a high frequency oscillating magnetic field that heats them, which releases the drug from their surface (see the figure). After injection into the bloodstream, a coil placed above the skin is used to guide the nanoparticles into position. If the 2.00-cm-diameter coil consists of 3750 turns of wire carrying a current of 5.00 A, what is the strength of the magnetic field at the center of the coil?

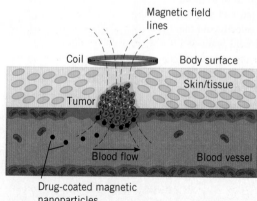

PROBLEM 79

Concepts and Calculations Problems

Online

Both magnetic and electric fields can apply forces to an electric charge. However, there are distinct differences in the way the two types of fields apply their forces. Problem 80 serves to review how the two types of fields behave. A number of forces can act on a rigid object, and some of them can produce torques, as Chapter 9 discusses. If, as a result of the forces and torques, the object has no acceleration of any kind, it is in equilibrium. Problem 81 illustrates how the magnetic force can be one of the forces keeping an object in equilibrium.

80. **M** **CHALK** **SSM** The figure shows a particle that carries a charge of $q_0 = -2.80 \times 10^{-6}$ C. It is moving along the $+y$ axis at a speed of $v = 4.8 \times 10^6$ m/s. A magnetic field $\vec{B}$ of magnitude 3.35×10^{-5} T is directed along the $+z$ axis, and an electric field $\vec{E}$ of magnitude 123 N/C points along the $-x$ axis. *Concepts:* (i) What forces make up the net force acting on the particle? (ii) How do you determine the direction of the magnetic force acting on the negative charge? (iii) How do you determine the direction of the electric force acting on the negative charge? (iv) Does the fact that the charge is moving affect the values of the magnetic and electric forces? *Calculations:* Determine the magnitude and direction of the net force that acts on the particle.

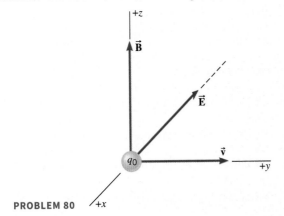

PROBLEM 80

81. **M** **CHALK** A 125-turn rectangular coil of wire is hung from one arm of a balance, as the figure shows. With the magnetic field $\vec{B}$ turned off, an object of mass M is added to the pan on the other arm to balance the weight of the coil. *Concepts:* (i) In a balanced, or equilibrium, condition the device has no angular acceleration. What does this imply about the net torque acting on the device? (ii) What is torque? (iii) In calculating the torques acting on an object in equilibrium, where do you locate the axis of rotation? *Calculations:* When a constant 0.200-T magnetic field is turned on and there is a current of 8.50 A in the coil, how much additional mass m must be added to regain the balance?

PROBLEM 81

Team Problems

Online

82. **E** **T** **WS** **A Crude Magnetometer.** You and your team have been tasked with estimating the magnetic field between the poles of a giant electromagnet. All you have to use to make this estimate is a power source that provides dc current, a piece of stiff copper wire that is 6.5 cm long, light wires to make connections, and a plastic spring that has an unstrained length of 3.5 cm and a spring constant of $k = 110$ N/m. The poles of the magnet are arranged so that the uniform field between them is horizontal. You attach one end of the spring to a support so that it hangs vertically between the poles of the magnet. You then attach the stiff copper wire to the lower end of the spring so that the wire is horizontal and perpendicular to the field. Finally, you connect the light, flexible wires to each end of the stiff wire and let them hang vertically. You run a current of 2.5 A through the wire. **(a)** With the magnet turned off, the spring stretches 0.25 cm vertically downward from its equilibrium position. What is the combined weight of the wires? **(b)** You turn on the magnet and measure the total (stretched) length of the spring to be 5.2 cm. What is the magnitude of the magnetic field?

83. **E** **T** **WS** **Building an Adjustable Electromagnet.** You and your team are tasked with designing a source to provide a uniform magnetic field for an experiment. The field inside a solenoid is uniform, provided that the length of the solenoid is much longer than its diameter. One useful characteristic of an electromagnet of this type is that the field inside can be adjusted anywhere between 0 T and some maximum value by controlling the current. In addition, the direction of the field, which is always pointed along the axis, can be reversed by reversing the current. The solenoid you are designing must be 3.60 cm in diameter and 25.0 cm long, and must generate a maximum magnetic field of magnitude $B = 0.150$ T. You intend to wind a cylinder of the given dimensions with wire that can safely pass a maximum current of 6.50 A. **(a)** How many windings should the solenoid have in order that the magnetic field at the center of the solenoid is 0.150 T when the current is 6.50 A? **(b)** What total length of wire is required? **(c)** What current should you pass through the coil to generate a smaller magnetic field of magnitude 3.50×10^{-2} T at its center, but directed

antiparallel to the field generated in part (a)? Assume the current was positive in part (a).

84. Ⓜ Ⓣ **An Electron Evaporator.** Electron beams are sometimes used to melt and evaporate metals in order to deposit thin metallic films on surfaces (similar to gold plating). One method is to put the material to be evaporated (called the "target") into a small tungsten cup (a crucible that has a very high melting point) and direct a beam of electrons at the target. Your team has been given the task of designing an electron-beam evaporator. The crucible is a cylinder, 2.0 cm in diameter and 1.5 cm in height, and

PROBLEM 84

contains a small target of pure nickel (Ni). The electrons are accelerated through a potential difference of $V = 1.20$ kV, and form a beam that originates below the crucible, exactly 3.70 cm off its center, in the $+x$ direction (see the drawing). **(a)** What is the speed of the electrons in the beam? **(b)** You must steer the electron beam with a magnetic field so that it curls over the lip of the cup and strikes the nickel target. Assuming that a uniform field exists above the cup (the field is zero below), what must be the radius of the beam's circular path? **(c)** In what direction should the field point if the beam initially approaches the cup from the $-y$ axis? **(d)** What must be the magnitude of the uniform magnetic field?

85. Ⓜ Ⓣ **An Isotope Separator.** Hydrogen has three isotopes ^{1}H ($m_1 = m_p$), ^{2}H ($m_2 \cong 2m_p$), and ^{3}H ($m_s \cong 3m_p$), where m_p is the mass of a proton (1.67×10^{-27} kg). You and your team are tasked with constructing an isotope separator that will separate a gas of mixed hydrogen isotopes. The gas first passes through a device that atomizes it (i.e., makes sure the atoms are separate, and do not form H_2 molecules), and then ionizes the atoms (strips off their only electron) so that they have a net charge of $+e$. Next, the atoms (now positive ions) are accelerated between the plates of a parallel-plate capacitor with a voltage of 2.50 kV across it, and emerge through a hole in one of the plates as a beam. **(a)** What are the speeds of the three isotopes when they emerge from the capacitor plates? **(b)** The accelerated ions enter a region of uniform magnetic field oriented perpendicular to the velocity vector of the ion beam. What should be the magnitude of the magnetic field if you want the largest diameter of the three ion paths to be 20.0 cm? **(c)** If you are collecting the atoms after completing a half-circle, one collector should be located 20.0 cm from the point where the beam enters the magnetic field. Where should the other two be located?

At the end of 2019, there were 1.5 million electric cars on the roadways in the United States, and 5.6 million worldwide. As their popularity continues to grow, so too will the unique ways to charge them. Here, an electric car is parked over a charging pad that is located on the floor. A similar pad is mounted on the bottom of the car, so that the two pads are close to each other. Embedded in each pad is a flat coil of wire with many turns. An ac current passes through the charging coil, which creates a changing magnetic field. This field passes through the coil on the car, which creates a current in that coil that is used to charge the car's battery. The transmission of electrical power between the two coils occurs wirelessly through the process of electromagnetic induction, which is the topic of this chapter.

Electromagnetic Induction

22.1 | Induced Emf and Induced Current

There are a number of ways a magnetic field can be used to generate an electric current, and **Interactive Figure 22.1** illustrates one of them. This drawing shows a bar magnet and a helical coil of wire to which an ammeter is connected. When there is no *relative* motion between the magnet and the coil, as in part *a* of the drawing, the ammeter reads zero, indicating that no current exists. However, when the magnet moves toward the coil, as in part *b*, a current *I* appears. As the magnet approaches, the magnetic field $\vec{\mathbf{B}}$ that it creates at the location of the coil becomes stronger and stronger, and it is this *changing* field that produces the current. When the magnet moves away from the coil, as in part *c*, a

LEARNING OBJECTIVES

After reading this module, you should be able to…

22.1 Predict when an induced current will flow.

22.2 Solve motional emf problems.

22.3 Calculate magnetic flux.

22.4 Solve problems using Faraday's law of induction.

22.5 Predict the direction of an induced current using Lenz's law.

22.6 Describe how sound is reproduced via induction.

22.7 Solve problems involving generators.

22.8 Define mutual induction and self-inductance.

22.9 Solve problems involving transformers.

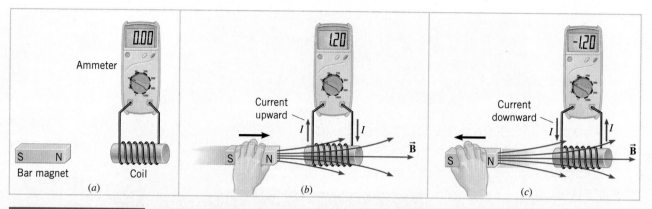

INTERACTIVE FIGURE 22.1 (a) When there is no relative motion between the coil of wire and the bar magnet, there is no current in the coil. (b) A current is created in the coil when the magnet moves toward the coil. (c) A current also exists when the magnet moves away from the coil, but the direction of the current is opposite to that in (b).

current is also produced, but with a reversed direction. Now the magnetic field at the coil becomes weaker as the magnet moves away. Once again it is the *changing* field that generates the current.

A current would also be created in **Interactive Figure 22.1** if the magnet were held stationary and the coil were moved, because the magnetic field at the coil would be changing as the coil approached or receded from the magnet. Only relative motion between the magnet and the coil is needed to generate a current; it does not matter which one moves.

The current in the coil is called an **induced current** because it is brought about (or "induced") by a changing magnetic field. Since a source of emf (electromotive force) is always needed to produce a current, the coil itself behaves as if it were a source of emf. This emf is known as an **induced emf**. Thus, a changing magnetic field induces an emf in the coil, and the emf leads to an induced current.

THE PHYSICS OF . . . an automobile cruise control. Induced emf and induced current are frequently used in the cruise controls found in many cars. **Interactive Figure 22.2** illustrates how a cruise control operates. Usually two magnets are mounted on opposite sides of the vehicle's drive shaft, with a stationary sensing coil positioned nearby. As the shaft turns, the magnets pass by the coil and cause an induced emf and current to appear in it. A microprocessor (the "brain" of a computer) counts the pulses of current and, with the aid of its internal clock and a knowledge of the shaft's radius, determines the rotational speed of the drive shaft. The rotational speed, in turn, is related to the car's speed. Thus, once the driver sets the desired cruising speed with the speed control switch (mounted near the steering wheel), the microprocessor can compare it with the measured speed. To the extent that the selected cruising speed and the measured

INTERACTIVE FIGURE 22.2 Induced emf lies at the heart of an automobile cruise control, in which an emf is induced in a sensing coil by magnets attached to the rotating drive shaft.

speed differ, a signal is sent to a servo, or control, mechanism, which causes the throttle/fuel injector to send more or less fuel to the engine. The car speeds up or slows down accordingly, until the desired cruising speed is reached.

Figure 22.3 shows another way to induce an emf and a current in a coil. An emf can be induced by *changing the area* of a coil in a constant magnetic field. Here the shape of the coil is being distorted to reduce the area. As long as the area is changing, an induced emf and current exist; they vanish when the area is no longer changing. If the distorted coil is returned to its original shape, thereby increasing the area, an oppositely directed current is generated while the area is changing.

In each of the previous examples, both an emf and a current are induced in the coil because the coil is part of a complete, or closed, circuit. If the circuit were open—perhaps because of an open switch—there would be no induced current. However, an emf would still be induced in the coil, whether the current exists or not.

Changing a magnetic field and changing the area of a coil are methods that can be used to create an induced emf. The phenomenon of producing an induced emf with the aid of a magnetic field is called **electromagnetic induction**. The next section discusses yet another method by which an induced emf can be created.

FIGURE 22.3 While the area of the coil is changing, an induced emf and current are generated.

Check Your Understanding

(The answer is given at the end of the book.)

1. Suppose that the coil and the magnet in **Interactive Figure 22.1a** were each moving with the same velocity relative to the earth. Would there be an induced current in the coil?

22.2 | Motional Emf

The Emf Induced in a Moving Conductor

When a conducting rod moves through a constant magnetic field, an emf is induced in the rod. This special case of electromagnetic induction arises as a result of the magnetic force that acts on a moving charge (see Section 21.2). Consider the metal rod of length L moving to the right in **Animated Figure 22.4a**. The velocity $\vec{v}$ of the rod is constant and is perpendicular to a uniform magnetic field $\vec{B}$. Each charge q within the rod also moves with a velocity $\vec{v}$ and experiences a magnetic force of magnitude $F = |q|vB$, according to Equation 21.1. By using RHR-1, it can be seen that the mobile, free electrons are driven to the bottom of the rod, leaving behind an equal amount of positive charge at the top. (Remember to reverse the direction of the force that RHR-1 predicts, since the electrons have a negative charge. See Section 21.2.) The positive and negative charges accumulate until the attractive electric force that they exert on each other becomes equal in magnitude to the magnetic force. When the two forces balance, equilibrium is reached and no further charge separation occurs.

The separated charges on the ends of the moving conductor give rise to an induced emf, called a **motional emf** because it originates from the motion of charges through a magnetic field. The emf exists as long as the rod moves. If the rod is brought to a halt, the magnetic force vanishes, with the result that the attractive electric force reunites the positive and negative charges and the emf disappears. The emf of the moving rod is analogous to the emf between the terminals of a battery. However, the emf of a battery is produced by chemical reactions, whereas the motional emf is created by the agent that moves the rod through the magnetic field (like the hand in **Animated Figure 22.4b**).

The fact that the electric and magnetic forces balance at equilibrium in **Animated Figure 22.4a** can be used to determine the magnitude of the motional emf $\mathscr{E}$. According to Equation 18.2, the magnitude of the electric force acting on the positive charge q at the top of the rod is Eq, where E is the magnitude of the electric field due to the separated

ANIMATED FIGURE 22.4 (*a*) When a conducting rod moves at right angles to a constant magnetic field, the magnetic force causes opposite charges to appear at the ends of the rod, giving rise to an induced emf. (*b*) The induced emf causes an induced current I to appear in the circuit.

charges. And according to Equation 19.7a (without the minus sign), the electric field magnitude is given by the voltage between the ends of the rod (the emf $\mathscr{E}$) divided by the length L of the rod. Thus, the electric force is $Eq = (\mathscr{E}/L)q$. Since we are dealing now with a positive charge, the magnetic force is $B|q|(v \sin 90°) = Bqv$, according to Equation 21.1, because the charge q moves perpendicular to the magnetic field. Since these two forces balance, it follows that $(\mathscr{E}/L)q = Bqv$. The emf, then, is

Motional emf when $\vec{v}, \vec{B}$, and L are mutually perpendicular $\mathscr{E} = vBL$ **(22.1)**

As expected, $\mathscr{E} = 0$ V when $v = 0$ m/s, because no motional emf is developed in a stationary rod. Greater speeds and stronger magnetic fields lead to greater emfs for a given length L. As with batteries, $\mathscr{E}$ is expressed in volts. In **Animated Figure 22.4b** the rod is sliding on conducting rails that form part of a closed circuit, and L is the length of the rod between the rails. Due to the emf, electrons flow in a clockwise direction around the circuit. Positive charge would flow in the direction opposite to the electron flow, so the conventional current I is drawn counterclockwise in the picture. Example 1 illustrates how to determine the electrical energy that the motional emf delivers to a device such as the light bulb in the drawing.

EXAMPLE 1 | Operating a Light Bulb with Motional Emf

Suppose that the rod in **Animated Figure 22.4b** is moving at a speed of 5.0 m/s in a direction perpendicular to a 0.80-T magnetic field. The rod has a length of 1.6 m and a negligible electrical resistance. The rails also have negligible resistance. The light bulb, however, has a resistance of 96 Ω. Find **(a)** the emf produced by the rod, **(b)** the current induced in the circuit, **(c)** the electric power delivered to the bulb, and **(d)** the energy used by the bulb in 60.0 s.

Reasoning The moving rod acts like an imaginary battery and supplies a motional emf of vBL to the circuit. The induced current can be determined from Ohm's law as the motional emf divided by the resistance of the bulb. The electric power delivered to the bulb is the product of the induced current and the potential difference across the bulb (which, in this case, is the motional emf). The energy used is the product of the power and the time.

Solution **(a)** The motional emf is given by Equation 22.1 as

$$\mathscr{E} = vBL = (5.0 \text{ m/s})(0.80 \text{ T})(1.6 \text{ m}) = \boxed{6.4 \text{ V}}$$

(b) According to Ohm's law, the induced current is equal to the motional emf divided by the resistance of the circuit:

$$I = \frac{\mathscr{E}}{R} = \frac{6.4 \text{ V}}{96 \text{ }\Omega} = \boxed{0.067 \text{ A}} \qquad \textbf{(20.2)}$$

(c) The electric power P delivered to the light bulb is the product of the current I and the potential difference across the bulb:

$$P = I\mathscr{E} = (0.067 \text{ A})(6.4 \text{ V}) = \boxed{0.43 \text{ W}} \qquad \textbf{(20.6a)}$$

(d) Since power is energy per unit time, the energy E used in 60.0 s is the product of the power and the time:

$$E = Pt = (0.43 \text{ W})(60.0 \text{ s}) = \boxed{26 \text{ J}} \qquad \textbf{(6.10b)}$$

Motional Emf and Electrical Energy

Motional emf arises because a magnetic force acts on the charges in a conductor that is moving through a magnetic field. Whenever this emf causes a current, a second magnetic force enters the picture. In **Animated Figure 22.4b**, for instance, the second force arises because the current I in the rod is perpendicular to the magnetic field. The current, and hence the rod, experiences a magnetic force $\vec{F}$ whose magnitude is given by Equation 21.3 as $F = ILB \sin 90°$. Using the values of I, L, and B given in Example 1, we see that $F = (0.067 \text{ A})(1.6 \text{ m})(0.80 \text{ T}) = 0.086$ N. The direction of $\vec{F}$ is specified by RHR-1 and is *opposite* to the velocity $\vec{v}$ of the rod, and thus points to the left (see **Figure 22.5**). By

FIGURE 22.5 A magnetic force $\vec{F}$ is exerted on the current I in the moving rod and is directed opposite to the rod's velocity $\vec{v}$. Since the force $\vec{F}_{hand}$ counterbalances the magnetic force $\vec{F}$, the rod moves to the right at a constant velocity.

itself, $\vec{\mathbf{F}}$ would *slow down* the rod, and here lies the crux of the matter. To keep the rod moving to the right with a constant velocity, a counterbalancing force must be applied to the rod by an external agent, such as the hand in the picture. This force is labeled $\vec{\mathbf{F}}_{\mathbf{hand}}$ in the drawing. The counterbalancing force must have a magnitude of 0.086 N and must be directed opposite to the magnetic force $\vec{\mathbf{F}}$. If the counterbalancing force were removed, the rod would decelerate under the influence of $\vec{\mathbf{F}}$ and eventually come to rest. During the deceleration, the motional emf would decrease and the light bulb would eventually go out.

We can now answer an important question—Who or what provides the 26 J of electrical energy that the light bulb in Example 1 uses in 60.0 seconds? The provider is the external agent that applies the 0.086-N counterbalancing force needed to keep the rod moving. This agent does work, and Example 2 shows that the work done is equal to the electrical energy used by the bulb.

Analyzing Multiple-Concept Problems

EXAMPLE 2 | The Work Needed to Keep the Light Bulb Burning

As we saw in Example 1, an induced current of 0.067 A exists in the circuit due to the moving rod. As **Figure 22.5** shows, the hand provides a force $\vec{\mathbf{F}}_{\mathbf{hand}}$ that keeps the rod moving to the right. Determine the work done by this force in a time of 60.0 s. Assume, as in Example 1, that the magnetic field has a magnitude of 0.80 T and that the rod has a length of 1.6 m and moves at a constant speed of 5.0 m/s.

Reasoning According to the discussion in Section 6.1, the work done by the hand is equal to the product of (1) the magnitude F_{hand} of the force exerted by the hand, (2) the magnitude x of the rod's displacement, and (3) the cosine of the angle θ'

between the force and the displacement. Since the rod moves to the right at a constant speed, it has no acceleration and is, therefore, in equilibrium (see Section 4.11). Thus, the force exerted by the hand must be equal in magnitude, but opposite in direction, to the magnetic force $\vec{\mathbf{F}}$ exerted on the rod, since they are the only two forces acting on the rod along the direction of the motion. We will determine the magnitude F of the magnetic force by using Equation 21.3, and this will enable us to find F_{hand}. Since the rod moves at a constant speed, the magnitude x of its displacement is the product of its speed and the time of travel.

Knowns and Unknowns The data for this problem are:

Description	Symbol	Value
Current	I	0.067 A
Length of rod	L	1.6 m
Speed of rod	v	5.0 m/s
Time during which rod moves	t	60.0 s
Magnitude of magnetic field	B	0.80 T
Unknown Variable		
Work done by hand	W	?

Modeling the Problem

STEP 1 Work The work done by the hand in **Figure 22.5** is given by $W = F_{hand}\, x \cos \theta'$ (Equation 6.1). In this equation, F_{hand} is the magnitude of the force that the hand exerts on the rod, x is the magnitude of the rod's displacement, and θ' is the angle between the force and the displacement. Since the force and displacement point in the same direction, $\theta' = 0°$, so

$$W = F_{hand}\, x \cos 0°$$

Two forces act on the rod; the force $\vec{\mathbf{F}}_{\mathbf{hand}}$, which points to the right, and the magnetic force $\vec{\mathbf{F}}$, which points to the left. Since the rod moves at a constant velocity, the magnitudes of these two forces are equal, so that $F_{hand} = F$. Substituting this relation into the expression for the work gives Equation 1 at the right. At this point, neither F nor x is known, but they will be evaluated in Steps 2 and 3, respectively.

$$W = Fx \cos 0° \qquad (1)$$

? ?

STEP 2 Magnetic Force Exerted on a Current-Carrying Rod We have seen in Section 21.5 that a rod of length L that carries a current I in a magnetic field of magnitude B experiences a magnetic force of magnitude F. The magnitude of the force is given by $F = ILB \sin \theta$ (Equation 21.3), where θ is the angle between the direction of the current and the magnetic field. In this case, the current and magnetic field are perpendicular to each other (see **Figure 22.5**), so $\theta = 90°$. Thus, the magnitude of the magnetic force is

$$\boxed{F = ILB \sin 90°}$$

The quantities, I, L, and B are known, and we substitute this expression into Equation 1 at the right.

STEP 3 Kinematics Since the rod is moving at a constant speed, the distance x it travels is the product of its speed v and the time t:

$$\boxed{x = vt}$$

The variables v and t are known. We can also substitute this relation into Equation 1, as shown in the right column.

$$W = Fx \cos 0° \quad \text{(1)}$$
$$\boxed{F = ILB \sin 90°} \quad \boxed{x = vt}$$

Solution Algebraically combining the results of the three steps, we have

$$\underset{\text{STEP 1}}{\downarrow} \quad \underset{\text{STEP 2}}{\downarrow} \qquad \underset{\text{STEP 2}}{\downarrow}$$
$$W = Fx \cos 0° = (ILB \sin 90°)x \cos 0° = (ILB \sin 90°)(vt) \cos 0°$$

The work done by the force of the hand is

$$W = (ILB \sin 90°)(vt) \cos 0°$$
$$= (0.067 \text{ A})(1.6 \text{ m})(0.80 \text{ T})(\sin 90°)(5.0 \text{ m/s})(60.0 \text{ s})(\cos 0°) = \boxed{26 \text{ J}}$$

The 26 J of work done on the rod by the hand is mechanical energy and is the same as the 26 J of energy consumed by the light bulb (see Example 1). Hence, the moving rod and the magnetic force convert mechanical energy into electrical energy, much as a battery converts chemical energy into electrical energy.

Related Homework: Problem 7

FIGURE 22.6 The current cannot be directed clockwise in this circuit, because the magnetic force $\vec{F}$ exerted on the rod would then be in the same direction as the velocity $\vec{v}$. The rod would accelerate to the right and create energy on its own, violating the principle of conservation of energy.

It is important to realize that the direction of the current in **Figure 22.5** is consistent with the principle of conservation of energy. Consider what would happen if the direction of the current were reversed, as in **Figure 22.6**. With the direction of the current reversed, the direction of the magnetic force $\vec{F}$ would also be reversed and would point in the direction of the velocity $\vec{v}$ of the rod. As a result, the force would cause the rod to accelerate rather than decelerate. The rod would accelerate without the need for an external force (like that provided by the hand in **Figure 22.5**) and would create a motional emf that supplies energy to the light bulb. Thus, this hypothetical generator would produce energy out of nothing, since there is no external agent. Such a device cannot exist because it violates the principle of conservation of energy, which states that energy cannot be created or destroyed, but can only be converted from one form to another. Therefore, the current cannot be directed clockwise around the circuit, as in **Figure 22.6**. In situations such as the one in Examples 1 and 2, when a motional emf leads to an induced current, a magnetic force always appears that opposes the motion, in accord with the principle of conservation of energy. Conceptual Example 3 deals further with the important issue of energy conservation.

CONCEPTUAL EXAMPLE 3 | Conservation of Energy

Figure 22.7a illustrates a conducting rod that is free to slide down between two vertical copper tracks. There is no kinetic friction between the rod and the tracks, although the rod maintains electrical contact with the tracks during its fall. A constant magnetic field $\vec{B}$ is directed perpendicular to the motion of the rod, as the drawing shows. Because there is no friction, the only force that acts on the rod is its weight $\vec{W}$, so the rod falls with an acceleration equal to the acceleration due to gravity, which has a magnitude of

$g = 9.8 \text{ m/s}^2$. Suppose that a resistance R is connected between the tops of the tracks, as in part b of the drawing. Is the magnitude of the acceleration with which the rod now falls **(a)** equal to g, **(b)** greater than g, or **(c)** less than g?

Reasoning As the rod falls perpendicular to the magnetic field, a motional emf is induced between its ends. This emf is induced whether or not the resistance R is attached between the tracks. However, when R is present, a complete circuit is formed, and the emf produces an induced current I that is perpendicular to the field. The direction of this current is such that the rod experiences an upward magnetic force $\vec{F}$, opposite to the weight of the rod (see part b of the drawing and use RHR-1). The net force acting on the rod is $\vec{W} + \vec{F}$, which is *less* than the weight, since $\vec{F}$ points upward and the weight $\vec{W}$ points downward. In accord with Newton's second law of motion, the downward acceleration is proportional to the net force.

Answers (a) and (b) are incorrect. Since the net downward force on the rod in **Figure 22.7b** is less than the rod's weight and since the downward acceleration is proportional to the net force, the rod cannot have an acceleration whose magnitude is equal to or greater than g.

Answer (c) is correct. Since the net downward force on the rod when R is present is less than the rod's weight and since the downward acceleration is proportional to the net force, the downward acceleration has a magnitude less than g. Thus, the rod gains speed as it falls but does so less rapidly than if R were not present. As the speed of the rod in **Figure 22.7b** increases during the descent, the magnetic force increases, until the time comes when its magnitude equals the magnitude of the rod's weight. When this occurs, the net force and the rod's acceleration will be zero. From this moment on, the rod will fall at a constant velocity. In any event, the rod always has a smaller speed than does a freely falling rod (i.e., R is absent) at the same place. The speed is smaller because only part of the gravitational potential energy (GPE) is being converted into kinetic energy (KE) as the rod falls, with part also being dissipated as heat in the resistance R. In fact, when the rod eventually attains a constant velocity, none of the GPE is converted into KE and all of it is dissipated as heat.

Related Homework: Problem 9

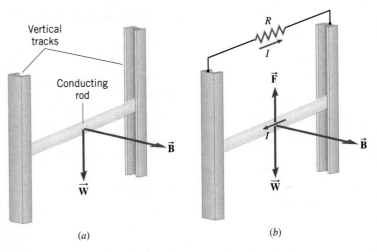

FIGURE 22.7 (a) There is no kinetic friction between the falling rod and the tracks, so the only force acting on the rod is its weight $\vec{W}$. (b) When an induced current I exists in the circuit, a magnetic force $\vec{F}$ also acts on the rod.

Check Your Understanding

(The answers are given at the end of the book.)

2. Consider the induced emf being generated in **Animated Figure 22.4**. Suppose that the length of the rod is reduced by a factor of four. For the induced emf to be the same, what should be done? **(a)** Without changing the speed of the rod, increase the strength of the magnetic field by a factor of four. **(b)** Without changing the magnetic field, increase the speed of the rod by a factor of four. **(c)** Increase both the speed of the rod and the strength of the magnetic field by a factor of two. **(d)** All of the previous three methods may be used.

3. In the discussion concerning **Figure 22.5**, we saw that a force of 0.086 N from an external agent was required to keep the rod moving at a constant velocity. Suppose that friction is absent and that the light bulb is suddenly removed from its socket while the rod is moving. How much force does the external agent then need to apply to the rod to keep it moving at a constant velocity? **(a)** 0 N **(b)** Greater than 0 N but less than 0.086 N **(c)** More than 0.086 N **(d)** 0.086 N

4. Eddy currents are electric currents that can arise in a piece of metal when it moves through a region where the magnetic field is not the same everywhere. **CYU Figure 22.1** shows, for example, a metal sheet moving to the right at a velocity $\vec{v}$ and a magnetic field $\vec{B}$ that is directed perpendicular to the sheet. At the instant represented, the field only extends over the left half of the sheet. An

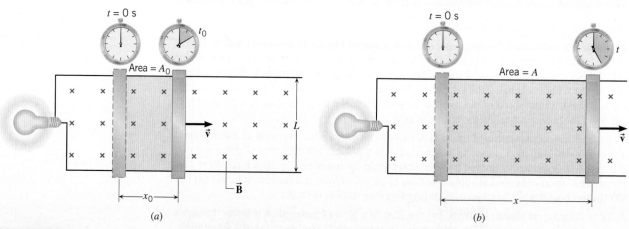

CYU FIGURE 22.1

emf is induced that leads to the eddy current indicated. Such eddy currents cause the velocity of the moving sheet to decrease and are used in various devices as a brake to damp out unwanted motion. Does the eddy current in the drawing circulate **(a)** counterclockwise or **(b)** clockwise?

22.3 | Magnetic Flux

Motional Emf and Magnetic Flux

Motional emf, as well as any other type of induced emf, can be described in terms of a concept called *magnetic flux*. Magnetic flux is analogous to electric flux, which deals with the electric field and the surface through which it passes (see Section 18.9 and Figure 18.33). Magnetic flux is defined in a similar way by bringing together the magnetic field and the surface through which it passes.

We can see how the motional emf is related to the magnetic flux with the aid of **Figure 22.8**, which shows the rod used to derive Equation 22.1 ($\mathcal{E} = vBL$). In this drawing the rod moves through a magnetic field beginning at time $t = 0$ s. In part a the rod has moved a distance x_0 to the right at time t_0, whereas in part b it has moved a greater distance x at a later time t. The speed v of the rod is the distance traveled divided by the elapsed time: $v = (x - x_0)/(t - t_0)$. Substituting this expression for v into $\mathcal{E} = vBL$ gives

$$\mathcal{E} = \left(\frac{x - x_0}{t - t_0}\right)BL = \left(\frac{xL - x_0L}{t - t_0}\right)B$$

As the drawing indicates, the term x_0L is the area A_0 swept out by the rod in moving a distance x_0, and xL is the area A swept out in moving a distance x. Thus, the emf becomes

$$\mathcal{E} = \left(\frac{A - A_0}{t - t_0}\right)B = \frac{(BA) - (BA)_0}{t - t_0}$$

FIGURE 22.8 (a) In a time t_0, the moving rod sweeps out an area $A_0 = x_0L$. (b) The area swept out in a time t is $A = xL$. In both parts of the figure the areas are shaded in color.

The product BA of the magnetic field strength and the area appears in the numerator of this expression. This product is called **magnetic flux** and is represented by the symbol Φ (Greek capital letter phi); thus $\Phi = BA$. The magnitude of the induced emf is the *change* in flux $\Delta\Phi = \Phi - \Phi_0$ divided by the time interval $\Delta t = t - t_0$ during which the change occurs:

$$\mathscr{E} = \frac{\Phi - \Phi_0}{t - t_0} = \frac{\Delta\Phi}{\Delta t}$$

In other words, the induced emf equals the time rate of change of the magnetic flux.

You will almost always see the previous equation written with a minus sign—namely, $\mathscr{E} = -\Delta\Phi/\Delta t$. The minus sign is introduced for the following reason: The direction of the current induced in the circuit is such that the magnetic force $\vec{F}$ acts on the rod to *oppose* its motion, thereby tending to slow down the rod (see **Figure 22.5**). The minus sign ensures that the polarity of the induced emf sends the induced current in the proper direction so as to give rise to this opposing magnetic force.* This issue of the polarity of the induced emf will be discussed further in Section 22.5.

The advantage of writing the induced emf as $\mathscr{E} = -\Delta\Phi/\Delta t$ is that this relation is far more general than our present discussion suggests. In Section 22.4 we will see that $\mathscr{E} = -\Delta\Phi/\Delta t$ can be applied to *all possible ways of generating induced emfs.*

A General Expression for Magnetic Flux

In **Figure 22.8** the direction of the magnetic field $\vec{B}$ is perpendicular to the surface swept out by the moving rod. In general, however, $\vec{B}$ may not be perpendicular to the surface. For instance, in **Figure 22.9** the direction perpendicular to the surface is indicated by the normal to the surface, but the magnetic field is inclined at an angle ϕ with respect to this direction. In such a case the flux is computed using only the component of the field perpendicular to the surface, $B \cos \phi$. The general expression for magnetic flux is

$$\Phi = (B \cos \phi)A = BA \cos \phi \qquad (22.2)$$

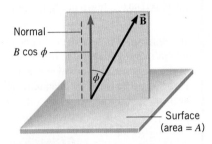

FIGURE 22.9 When computing the magnetic flux, the component of the magnetic field that is perpendicular to the surface must be used; this component is $B \cos \phi$.

Math Skills Equation 22.2 uses the component of the magnetic field perpendicular to the surface in defining the magnetic flux Φ. In order to determine this component of the field, we use the cosine function. According to the definition given in Equation 1.1, the cosine function is $\cos \phi = \dfrac{h_a}{h}$, where h_a is the length of the side of a right triangle that is adjacent to the angle ϕ and h is the length of the hypotenuse (see **Figure 22.10a**). **Figure 22.10b** shows the field $\vec{B}$ (magnitude B) and its components B_x and B_y. The field $\vec{B}$ is oriented at an angle ϕ with respect to the normal, and the component that we seek is B_y, the one that points along the direction of the normal. By comparing **Figures 22.10a** and **b**, we can see that

$$\cos \phi = \frac{h_a}{h} = \frac{B_y}{B} \quad \text{or} \quad B_y = B \cos \phi$$

FIGURE 22.10 Math Skills drawing.

*A detailed mathematical discussion of why the minus sign arises is beyond the scope of this book.

If either the magnitude B of the magnetic field or the angle ϕ is not constant over the surface (i.e., they are not the same at each point on the surface), an average value for the product $B \cos \phi$ must be used to compute the flux. Equation 22.2 shows that the unit of magnetic flux is the tesla · meter² (T · m²). This unit is called a *weber* (Wb), after the German physicist Wilhelm Weber (1804–1891): 1 Wb = 1 T · m². Example 4 illustrates how to determine the magnetic flux for three different orientations of the surface of a coil relative to the magnetic field.

EXAMPLE 4 | Magnetic Flux

A rectangular coil of wire is situated in a constant magnetic field whose magnitude is 0.50 T. The coil has an area of 2.0 m². Determine the magnetic flux for the following three orientations, $\phi = 0°$, 60.0°, and 90.0°, shown in **Figure 22.11**.

Reasoning The magnetic flux Φ is defined as $\Phi = BA \cos \phi$, where B is the magnitude of the magnetic field, A is the area of the surface through which the magnetic field passes, and ϕ is the angle between the magnetic field and the normal to the surface.

Problem-Solving Insight The magnetic flux Φ is determined by more than just the magnitude B of the magnetic field and the area A. It also depends on the angle ϕ (see **Figure 22.9** and Equation 22.2).

Solution The magnetic flux for the three cases is:

$\phi = 0°$ $\quad \Phi = (0.50 \text{ T})(2.0 \text{ m}^2) \cos 0° = \boxed{1.0 \text{ Wb}}$

$\phi = 60.0°$ $\quad \Phi = (0.50 \text{ T})(2.0 \text{ m}^2) \cos 60.0° = \boxed{0.50 \text{ Wb}}$

$\phi = 90.0°$ $\quad \Phi = (0.50 \text{ T})(2.0 \text{ m}^2) \cos 90.0° = \boxed{0 \text{ Wb}}$

FIGURE 22.11 Three orientations of a rectangular coil (drawn as an edge view) relative to the magnetic field lines. The magnetic field lines that pass through the coil are those in the regions shaded in blue.

Graphical Interpretation of Magnetic Flux

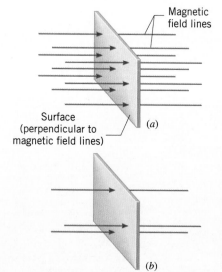

It is possible to interpret the magnetic flux graphically because the magnitude of the magnetic field $\vec{B}$ is proportional to the number of field lines per unit area that pass through a surface perpendicular to the lines (see Section 21.1). For instance, the magnitude of $\vec{B}$ in **Figure 22.12a** is three times larger than it is in part b of the drawing, since the number of field lines drawn through the identical surfaces is in the ratio of 3:1. Because Φ is directly proportional to B for a given area, the flux in part a is also three times larger than the flux in part b. Therefore, **the magnetic flux is proportional to the number of field lines that pass through a surface**.

The graphical interpretation of flux also applies when the surface is oriented at an angle with respect to $\vec{B}$. For example, as the coil in **Figure 22.11** is rotated from $\phi = 0°$ to 60° to 90°, the number of magnetic field lines passing through the surface (see the field lines in the regions shaded in blue) changes in the ratio of 8:4:0 or 2:1:0. The results of Example 4 show that the flux in the three orientations changes by the same ratio.

FIGURE 22.12 The magnitude of the magnetic field in (a) is three times as great as that in (b) because the number of magnetic field lines crossing the surfaces is in the ratio of 3:1.

Because the magnetic flux is proportional to the number of field lines passing through a surface, we often use phrases such as "the flux that passes through a surface bounded by a loop of wire."

Check Your Understanding

(The answers are given at the end of the book.)

5. A magnetic field has the same direction and the same magnitude B everywhere. A circular area A is bounded by a loop of wire. Which of the following statements is true concerning the magnitude of the magnetic flux that passes through this area? **(a)** It is zero. **(b)** It is BA. **(c)** Its maximum possible value is BA. **(d)** Its minimum possible value is BA.

6. Suppose that a magnetic field is constant everywhere on a flat 1.0-m^2 surface and that the magnetic flux through this surface is 2.0 Wb. From these data, which one of the following pieces of information can be determined about the magnetic field? **(a)** The magnitude of the field **(b)** The magnitude of the component of the field that is perpendicular to the surface **(c)** The magnitude of the component of the field that is parallel to the surface

22.4 | Faraday's Law of Electromagnetic Induction

Two scientists are given credit for the discovery of electromagnetic induction: the Englishman Michael Faraday (1791–1867) and the American Joseph Henry (1797–1878). Since Faraday investigated electromagnetic induction in more detail and published his findings first, the law that describes the phenomenon bears his name.

Faraday discovered that whenever there is a *change in magnetic flux* through a loop of wire, an emf is induced in the loop. In this context, the word "change" refers to a change as time passes. A magnetic flux that is constant in time creates no emf. Faraday's law of electromagnetic induction is expressed by bringing together the idea of magnetic flux and the time interval during which it changes. In fact, Faraday found that the magnitude of the induced emf is equal to the time rate of change of the magnetic flux. This is consistent with the relation we obtained in Section 22.3 for the specific case of motional emf: $\mathcal{E} = -\Delta\Phi/\Delta t$.

Often the magnetic flux passes through a coil of wire containing more than one loop (or turn). If the coil consists of N loops, and if the same flux passes through each loop, it is found experimentally that the total induced emf is N times that induced in a single loop. An analogous situation occurs in a flashlight when two 1.5-V batteries are stacked in series on top of one another to give a total emf of 3.0 volts. For the general case of N loops, the total induced emf is described by Faraday's law of electromagnetic induction in the following manner:

FARADAY'S LAW OF ELECTROMAGNETIC INDUCTION

The average emf $\mathcal{E}$ induced in a coil of N loops is

$$\mathcal{E} = -N\left(\frac{\Phi - \Phi_0}{t - t_0}\right) = -N\frac{\Delta\Phi}{\Delta t} \qquad (22.3)$$

where $\Delta\Phi$ is the change in magnetic flux through one loop and Δt is the time interval during which the change occurs. The term $\Delta\Phi/\Delta t$ is the average time rate of change of the flux that passes through one loop.

SI Unit of Induced Emf: volt (V)

Faraday's law states that an emf is generated if the magnetic flux changes for any reason. Since the flux is given by Equation 22.2 as $\Phi = BA \cos\phi$, it depends on the three factors, B, A, and ϕ, any of which may change. Example 5 considers a change in B.

EXAMPLE 5 | The Emf Induced by a Changing Magnetic Field

A coil of wire consists of 20 turns, or loops, each with an area of 1.5×10^{-3} m². A magnetic field is perpendicular to the surface of each loop at all times, so that $\phi = \phi_0 = 0°$. At time $t_0 = 0$ s, the magnitude of the field at the location of the coil is $B_0 = 0.050$ T. At a later time $t = 0.10$ s, the magnitude of the field at the coil has increased to $B = 0.060$ T. **(a)** Find the average emf induced in the coil during this time. **(b)** What would be the value of the average induced emf if the magnitude of the magnetic field decreased from 0.060 T to 0.050 T in 0.10 s?

Reasoning To find the induced emf, we use Faraday's law of electromagnetic induction (Equation 22.3), combining it with the definition of magnetic flux from Equation 22.2. We note that only the magnitude of the magnetic field changes in time. All other factors remain constant.

> **Problem-Solving Insight** The change in any quantity is the final value minus the initial value: e.g., the change in flux is $\Delta\Phi = \Phi - \Phi_0$ and the change in time is $\Delta t = t - t_0$.

Solution **(a)** Since $\phi = \phi_0$, the induced emf is

$$\mathscr{E} = -N\left(\frac{\Phi - \Phi_0}{t - t_0}\right) = -N\left(\frac{BA\cos\phi - B_0 A\cos\phi}{t - t_0}\right)$$

$$= -NA\cos\phi\left(\frac{B - B_0}{t - t_0}\right)$$

$$\mathscr{E} = -(20)(1.5\times 10^{-3}\text{ m}^2)(\cos 0°)\left(\frac{0.060\text{ T} - 0.050\text{ T}}{0.10\text{ s} - 0\text{ s}}\right)$$

$$= \boxed{-3.0\times 10^{-3}\text{ V}}$$

(b) The calculation here is similar to that in part (a), except the initial and final values of B are interchanged. This interchange reverses the sign of the emf, so $\boxed{\mathscr{E} = +3.0\times 10^{-3}\text{ V}}$. Because the algebraic sign or polarity of the emf is reversed, the direction of the induced current would be opposite to that in part (a).

The next example demonstrates that an emf can be created when a coil is rotated in a magnetic field.

EXAMPLE 6 | The Emf Induced in a Rotating Coil

A flat coil of wire has an area of 0.020 m² and consists of 50 turns. At $t_0 = 0$ s the coil is oriented so the normal to its surface has the same direction ($\phi_0 = 0°$) as a constant magnetic field of magnitude 0.18 T. The coil is then rotated through an angle of $\phi = 30.0°$ in a time of 0.10 s (see **Figure 22.11**). **(a)** Determine the average induced emf. **(b)** What would be the induced emf if the coil were returned to its initial orientation in the same time of 0.10 s?

Reasoning As in Example 5 we can determine the induced emf by using Faraday's law of electromagnetic induction, along with the definition of magnetic flux. In the present case, however, only ϕ (the angle between the normal to the surface of the coil and the magnetic field) changes in time. All other factors remain constant.

Solution **(a)** Faraday's law yields

$$\mathscr{E} = -N\left(\frac{\Phi - \Phi_0}{t - t_0}\right) = -N\left(\frac{BA\cos\phi - BA\cos\phi_0}{t - t_0}\right)$$

$$= -NBA\left(\frac{\cos\phi - \cos\phi_0}{t - t_0}\right)$$

$$\mathscr{E} = -(50)(0.18\text{ T})(0.020\text{ m}^2)\left(\frac{\cos 30.0° - \cos 0°}{0.10\text{ s} - 0\text{ s}}\right) = \boxed{+0.24\text{ V}}$$

(b) When the coil is rotated back to its initial orientation in a time of 0.10 s, the initial and final values of ϕ are interchanged. As a result, the induced emf has the same magnitude but opposite polarity, so $\boxed{\mathscr{E} = -0.24\text{ V}}$.

THE PHYSICS OF . . . a ground fault interrupter. One application of Faraday's law that is found in the home is a safety device called a ground fault interrupter. This device protects against electrical shock from an appliance, such as a clothes dryer. It plugs directly into a wall socket, as in **Figure 22.13** or, in new home construction, replaces the socket entirely. The interrupter consists of a circuit breaker that can be triggered to stop the current to the dryer, depending on whether an induced voltage appears across a sensing coil. This coil is wrapped around an iron ring, through which the current-carrying wires pass. In the drawing, the current going to the dryer is shown in red, and the returning current is shown in green. Each of the currents creates a magnetic field that encircles the corresponding wire, according to RHR-2 (see Section 21.7). However, the field lines have opposite directions since the currents have opposite directions. As the drawing shows, the iron ring guides the field lines through the sensing coil. Since the current is ac, the fields from the red and green current are changing, but the red and green field lines always have opposite directions and the opposing fields cancel at all times. As a result, the net flux through the coil remains zero, and no induced emf appears in the coil. Thus,

FIGURE 22.13 The clothes dryer is connected to the wall socket through a ground fault interrupter. The dryer is operating normally.

when the dryer operates normally, the circuit breaker is not triggered and does not shut down the current. The picture changes if the dryer malfunctions, as when a wire inside the unit breaks and accidentally contacts the metal case. When someone touches the case, some of the current begins to pass through the person's body and into the ground, returning to the ac generator *without using the return wire that passes through the ground fault interrupter*. Under this condition, the net magnetic field through the sensing coil is no longer zero and changes with time, since the current is ac. The changing flux causes an induced voltage to appear in the sensing coil, which triggers the circuit breaker to stop the current. Ground fault interrupters work very fast (in less than a millisecond) and turn off the current before it reaches a dangerous level.

Conceptual Example 7 discusses another application of electromagnetic induction—namely, how a stove can cook food without getting hot.

CONCEPTUAL EXAMPLE 7 | The Physics of an Induction Stove

Figure 22.14 shows two pots of water that were placed on an induction stove at the same time. There are two interesting features in this drawing. First, the stove itself is cool to the touch. Second, the water in the ferromagnetic metal pot is boiling while the water in the glass pot is not. How can such a "cool" stove boil water, and why isn't the water in the glass pot boiling?

Reasoning and Solution The key to this puzzle is related to the fact that one pot is made from a ferromagnetic metal and one from glass. We know that metals are good conductors, while glass is an insulator. Perhaps the stove causes electricity to flow directly in the metal pot. This is exactly what happens. The stove is called an *induction stove* because it operates by using electromagnetic induction. Just beneath the cooking surface is a metal coil that carries an ac current (frequency about 25 kHz). This current produces an alternating magnetic field that extends outward to the location of the metal pot. As the changing field crosses the pot's bottom surface, an emf is induced in it. Because the pot is metallic, an induced current is generated by the induced emf. The metal has a finite resistance to the induced current, however, and heats up as energy is dissipated in this resistance. The fact that the metal is ferromagnetic is important. Ferromagnetic materials contain magnetic domains (see Section 21.9), and the boundaries between them move extremely rapidly in response to the external

FIGURE 22.14 The water in the ferromagnetic metal pot is boiling—yet the water in the glass pot is not boiling, and the stove top is cool to the touch. The stove operates in this way by using electromagnetic induction.

magnetic field, thus enhancing the induction effect. A normal aluminum cooking pot, in contrast, is not ferromagnetic, so this enhancement is absent and such cookware is not used with induction stoves. An emf is also induced in the glass pot and the cooking surface of the stove. However, these materials are insulators, so very little induced current exists within them. Thus, they do not heat up very much and remain cool to the touch.

(*The answers are given at the end of the book.*)

7. In the most common form of lightning, electric charges flow between the ground and a cloud. The flow changes dramatically over short periods of time. Even without directly striking an electrical appliance in your house, a bolt of lightning that strikes nearby can produce a current in the circuits of the appliance. Note that such circuits typically contain coils or loops of wire. Why can the lightning cause the current to appear?

8. A solenoid is connected to an ac source. A copper ring and a rubber ring are placed inside the solenoid, with the normal to the plane of each ring parallel to the axis of the solenoid. An induced emf appears _____. **(a)** in the copper ring but not in the rubber ring **(b)** in the rubber ring but not in the copper ring **(c)** in both rings

9. A magnetic field of magnitude $B = 0.20$ T is reduced to zero in a time interval of $\Delta t = 0.10$ s, thereby creating an induced current in a loop of wire. Which one or more of the following choices would cause the same induced current to appear in the same loop of wire? **(a)** $B = 0.40$ T and $\Delta t = 0.20$ s **(b)** $B = 0.30$ T and $\Delta t = 0.10$ s **(c)** $B = 0.30$ T and $\Delta t = 0.30$ s **(d)** $B = 0.10$ T and $\Delta t = 0.050$ s **(e)** $B = 0.50$ T and $\Delta t = 0.40$ s

10. A coil is placed in a magnetic field, and the normal to the plane of the coil remains parallel to the field. Which one of the following options causes the magnitude of the average emf induced in the coil to be as large as possible? **(a)** The magnitude of the field is small, and its rate of change is large. **(b)** The magnitude of the field is large, and its rate of change is small. **(c)** The magnitude of the field is large, and it does not change.

22.5 | Lenz's Law

An induced emf drives current around a circuit just as the emf of a battery does. With a battery, conventional current is directed out of the positive terminal, through the attached device, and into the negative terminal. The same is true for an induced emf, although the locations of the positive and negative terminals are generally not as obvious. Therefore, a method is needed for determining the polarity or algebraic sign of the induced emf, so the terminals can be identified. As we discuss this method, it will be helpful to keep in mind that the net magnetic field penetrating a coil of wire results from two contributions. One is the original magnetic field that produces the changing flux that leads to the induced emf. The other arises because of the induced current, which, like any current, creates its own magnetic field. The field created by the induced current is called the **induced magnetic field**.

To determine the polarity of the induced emf, we will use a method based on a discovery made by the Russian physicist Heinrich Lenz (1804–1865). This discovery is known as **Lenz's law**.

> **LENZ'S LAW**
>
> **The induced emf resulting from a changing magnetic flux has a polarity that leads to an induced current whose direction is such that the induced magnetic field opposes the original flux change.**

Lenz's law is best illustrated with examples. Each will be worked out according to the following reasoning strategy:

> **REASONING STRATEGY**   **Determining the Polarity of the Induced Emf**
>
> 1. **Determine whether the magnetic flux that penetrates a coil is increasing or decreasing.**
>
> 2. **Find what the direction of the induced magnetic field must be so that it can** *oppose the change in flux* **by adding to or subtracting from the original field.**
>
> 3. **Having found the direction of the induced magnetic field, use RHR-3 (see Section 21.7) to determine the direction of the induced current. Then the polarity of the induced emf can be assigned because conventional current is directed out of the positive terminal, through the external circuit, and into the negative terminal.**

CONCEPTUAL EXAMPLE 8 | The Emf Produced by a Moving Magnet

Figure 22.15a shows a permanent magnet approaching a loop of wire. The external circuit attached to the loop consists of the resistance R, which could represent the filament in a light bulb, for instance. In **Figure 22.15a**, what is the polarity of the induced emf? In other words, **(a)** is point A positive and point B negative or **(b)** is point A negative and point B positive?

Reasoning   We will apply Lenz's law, the essence of which is that the change in magnetic flux must be opposed by the induced magnetic field. The flux through the loop is increasing, since the magnitude of the magnetic field at the loop is increasing as the magnet approaches. To oppose the increase in the flux, the direction of the induced magnetic field must be opposite to the field of the bar magnet. Thus, since the field of the bar magnet passes through the loop from left to right in part a of the drawing, the induced field must pass through the loop from right to left. An induced current creates this induced field, and from the

direction of this current we will be able to decide the polarity of the induced emf.

Answer (b) is incorrect.   If point A were negative and point B were positive, as in **Figure 22.15b**, the induced current in the loop would be as shown in that part of the drawing, because conventional current exits from the positive terminal and returns through the external circuit (the resistance R) to the negative terminal. Application of RHR-3 reveals that this induced current would lead to an induced field that passes through the loop from left to right, not right to left as needed to oppose the flux change.

Answer (a) is correct.   **Figure 22.15c** shows the situation with point A positive and point B negative and the induced current that results. An application of RHR-3 reveals that the induced field produced by this current indeed passes through the loop from right to left, as needed. We conclude, therefore, that the polarity shown in **Figure 22.15c** is correct.

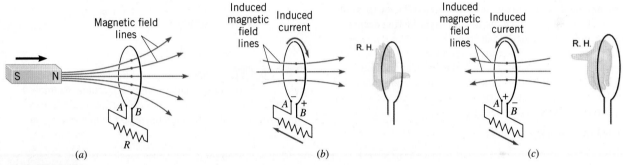

FIGURE 22.15   (a) As the magnet moves to the right, the magnetic flux through the loop increases. The external circuit attached to the loop has a resistance R. (b) One possibility and (c) another possibility for the direction of the induced current. See Conceptual Example 8.

In Conceptual Example 8 the direction of the induced magnetic field is opposite to the direction of the external field of the bar magnet.

> **Problem-Solving Insight**   The induced field is not always opposite to the external field, however, because Lenz's law requires only that it must oppose the *change* in the flux that generates the emf.

Conceptual Example 9 illustrates this point.

CONCEPTUAL EXAMPLE 9 | The Emf Produced by a Moving Copper Ring

In **Figure 22.16** there is a constant magnetic field in a rectangular region of space. This field is directed perpendicularly into the page. Outside this region there is no magnetic field. A copper ring slides through the region, from position 1 to position 5. Since the field is zero outside the rectangular region, no flux passes through the ring in positions 1 and 5, there is no change in the flux through the ring, and there is no induced emf or current in the ring. Which one of the following options correctly describes the induced current in the ring as it passes through positions 2, 3, and 4? **(a)** I_2 is clockwise, I_3 is counterclockwise, I_4 is counterclockwise. **(b)** I_2 is

counterclockwise, I_3 is clockwise, I_4 is clockwise. **(c)** I_2 is clockwise, $I_3 = 0$ A, I_4 is counterclockwise. **(d)** I_2 is counterclockwise, $I_3 = 0$ A, I_4 is clockwise.

Reasoning   Lenz's law will guide us. It requires that the induced magnetic field oppose the change in flux. Sometimes this means that the induced field is opposite to the external magnetic field, as in Example 8. However, we will see that the induced field sometimes has the same direction as the external field in order to oppose the flux change.

Answers (a) and (b) are incorrect.   Both of these answers specify that there is an induced current I_3 in the ring as it passes through position 3, contrary to fact. The external field within the rectangular region certainly produces a flux through the ring. (See **Figure 22.16**.) However, the external field is constant, so the flux through the ring does not change as the ring moves. In order for an induced emf to exist and to cause an induced current, the flux must change.

Answer (c) is incorrect.   As the ring moves out of the field region in position 4, the flux through the ring decreases, so there is an induced emf and an induced current. Lenz's law requires that the induced current must lead to an induced magnetic field that opposes this flux decrease. To oppose the decrease, the induced field must point in the same direction as the external field. To create an induced field pointing into the page, the induced current I_4 must be clockwise (use RHR-3), not counterclockwise as this answer specifies.

Answer (d) is correct.   In position 2 the flux increases and, according to Lenz's law, the induced current must create an induced magnetic field that opposes the increase. To oppose the increase the induced field must point opposite to the external field and, therefore, must point out of the page. RHR-3 indicates that the induced current must be counterclockwise, as this answer states. In position 4 the flux through the ring decreases, and the induced magnetic field must oppose the decrease by pointing in the same direction as the external field—namely, into the page. RHR-3 reveals that the induced current must be clockwise, as this answer indicates. In position 3 the flux through the loop is not changing, so there is no induced emf and current, as this answer specifies.

Related Homework: Problem 62

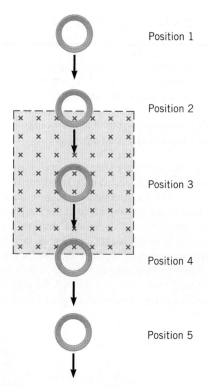

FIGURE 22.16 A constant magnetic field is directed into the page in the shaded rectangular region. Conceptual Example 9 discusses what happens to the induced emf and current in a copper ring that slides through the region from position 1 to position 5.

Lenz's law should not be thought of as an independent law, because it is a consequence of the law of conservation of energy. The connection between energy conservation and induced emf has already been discussed in Section 22.2 for the specific case of motional emf. However, the connection is valid for any type of induced emf. In fact, the polarity of the induced emf, as specified by Lenz's law, ensures that energy is conserved.

Check Your Understanding

(The answers are given at the end of the book.)

11. In **Figure 22.3** a coil of wire is being stretched. What would be the direction of the induced current if the direction of the external magnetic field in the figure were reversed? **(a)** Clockwise **(b)** Counterclockwise

12. A circular loop of wire is lying flat on a horizontal table, and you are looking down at it. An external magnetic field has a constant direction that is perpendicular to the table, and there is an induced clockwise current in the loop. Is the external field directed upward toward you or downward away from you, and is its magnitude increasing or decreasing? Note that there are two possible answers.

13. When the switch in **CYU Figure 22.2** is closed, the current in the coil increases to its equilibrium value. While the current is increasing there is an induced current in the metal ring. The ring is free to move. What happens to the ring? **(a)** It does not move. **(b)** It jumps downward. **(c)** It jumps upward.

CYU FIGURE 22.2

14. A conducting rod is free to slide along a pair of conducting rails, in a region where a uniform and constant (in time) magnetic field is directed into the plane of the paper, as **CYU Figure 22.3** illustrates. Initially the rod is at rest. There is no friction between the rails and the rod. What happens to the rod after the switch is closed? If any induced emf develops, be sure to account for its effect. **(a)** The rod accelerates to the right, its velocity increasing without limit. **(b)** The rod does not move. **(c)** The rod accelerates to the right for a while and then slows down and comes to a halt. **(d)** The rod accelerates to the right and eventually reaches a constant velocity at which it continues to move.

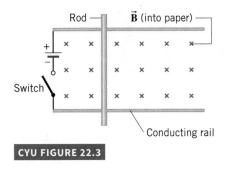

CYU FIGURE 22.3

22.6 *Applications of Electromagnetic Induction to the Reproduction of Sound

THE PHYSICS OF . . . the electric guitar pickup.   Electromagnetic induction plays an important role in the technology used for the reproduction of sound. Virtually all electric guitars, for example, use electromagnetic pickups in which an induced emf is generated in a coil of wire by a vibrating string. Each pickup is located below the strings, as **Figure 22.17** illustrates, and each is sensitive to different harmonics that the strings produce. Each string is made from a magnetizable metal, and the pickup consists of a coil of wire within which a permanent magnet is located. The magnetic field of the magnet penetrates the guitar string, causing it to become magnetized with north and south poles. When the magnetized string is plucked, it oscillates, thereby changing the magnetic flux that passes through the coil. The changing flux induces an emf in the coil, and the polarity of this emf alternates with the vibratory motion of the string. A string vibrating at 440 Hz, for example, induces a 440-Hz ac emf in the coil. This signal, after being amplified, is sent to loudspeakers, which produce a 440-Hz sound wave (concert A).

FIGURE 22.17 When the string of an electric guitar vibrates, an emf is induced in the coil of the pickup. The two ends of the coil are connected to the input of an amplifier.

THE PHYSICS OF . . . a tape-deck playback head.   The playback head of a tape deck uses a moving tape to generate an emf in a coil of wire. **Figure 22.18** shows a section of magnetized tape in which a series of "tape magnets" have been created in the magnetic layer of the tape during the recording process (see Section 21.9). The tape moves beneath the playback head, which consists of a coil of wire wrapped around an iron core. The iron core has the approximate shape of a horseshoe with a small gap between the two ends. Some of the field lines of the tape magnet under the gap are routed through the highly magnetizable iron core, and hence through the coil, as they proceed from the north pole to the south pole. Consequently, the flux through the coil changes as the tape moves past the gap. The change in flux leads to an ac emf, which is amplified and sent to the speakers, which reproduce the original sound.

FIGURE 22.18 The playback head of a tape deck. As each tape magnet goes by the gap, some magnetic field lines pass through the core and coil. The changing flux in the coil creates an induced emf. The gap width has been exaggerated.

FIGURE 22.19 A moving-coil microphone.

THE PHYSICS OF . . . microphones. There are a number of types of microphones, and **Figure 22.19** illustrates the one known as a moving-coil microphone. When a sound wave strikes the diaphragm of the microphone, the diaphragm vibrates back and forth, and the attached coil moves along with it. Nearby is a stationary magnet. As the coil alternately approaches and recedes from the magnet, the flux through the coil changes. Consequently, an ac emf is induced in the coil. This electrical signal is sent to an amplifier and then to the speakers. In a moving-magnet microphone, the magnet is attached to the diaphragm and moves relative to a stationary coil.

Example 10 describes a useful application of electromagnetic induction, called a hearing loop, that can provide hearing alternatives for those who have difficulties in places that have a lot of background noise.

EXAMPLE 10 | BIO A Hearing Loop

People that use hearing aids often encounter problems in places that have a lot of background noise, such as classrooms, theaters, and auditoriums. This occurs because, in its normal setting, a hearing aid directly detects sound waves, including those emanating from nearby sources, obscuring those from the desired source located some distance away. One solution to this problem is an audio induction loop, or a "hearing loop." The idea is that a cable is placed around a designated area, such as an auditorium, and then connected to a microphone or other electronic sound amplification source. The cable produces a magnetic field in the looped space that oscillates at the same audio frequencies as the source. Hearing aids that have an onboard telecoil or "T-coil," as shown in **Figure 22.20**, can pick up this magnetic signal via the induced emf in the coil, which can then be amplified and converted back into sound. Since none of the sound sources near the listener (e.g., whispering spectators) are picked up by the presenter's microphone, the local background noise is eliminated. Such hearing loop systems are required to produce peak magnetic field magnitudes of at least 5.00×10^{-7} T. Consider a cylindrical hearing induction coil that has 600 loops with an area of 1.50 mm². Suppose that the magnetic field is directed along the axis of the coil and ramps from 0 T to 5.00×10^{-7} T in 1.30×10^{-4} s, which approximates the rate of change of the magnetic field at a typical audio frequency. Determine the magnitude of the induced emf in the coil.

Reasoning The induced emf in the coil can be determined from Faraday's law of electromagnetic induction (Equation 22.3). Since the magnetic field is directed along the axis of the coil, $\cos \phi = 1$ in Equation 22.2 for the magnetic flux through the coil, and $\Phi = BA$. Since the area is constant, the time rate of change of the magnetic flux will be determined solely by the changing magnetic field magnitude.

Solution The time rate of change of the magnetic flux is given by $\Delta\Phi/\Delta t$, where $\Phi = BA$, since $\cos \phi = 1$ in this case. Since the

FIGURE 22.20 The telecoil, or "T-coil," inside a hearing aid.

area A does not change with time, the change in magnetic flux with time can be written as

$$\frac{\Delta\Phi}{\Delta t} = \frac{\Delta(BA)}{\Delta t} = A\left(\frac{\Delta B}{\Delta t}\right)$$

The magnitude of the emf is given by Equation 22.3. With $N = 600$, we have

$$|\mathscr{E}| = N\left|\frac{\Delta\Phi}{\Delta t}\right| = NA\left|\frac{\Delta B}{\Delta t}\right| = (600)(1.50 \times 10^{-6} \text{ m}^2)\left(\frac{5.00 \times 10^{-7} \text{ T}}{1.30 \times 10^{-4} \text{ s}}\right)$$

$$= \boxed{3.46 \times 10^{-6} \text{ V}}$$

Related homework: Problem 73

Check Your Understanding

(The answer is given at the end of the book.)

15. The string of an electric guitar vibrates in a standing wave pattern that consists of nodes and antinodes. (Section 17.5 discusses standing waves.) Where should an electromagnetic pickup be located in the standing wave pattern to produce a maximum emf? **(a)** At a node **(b)** At an antinode

---

22.7 The Electric Generator

How a Generator Produces an Emf

Electric generators, such as those in **Figure 22.21**, produce virtually all of the world's electrical energy. A generator produces electrical energy from mechanical work, which is just the opposite of what a motor does. In a motor, an *input* electric current causes a coil to rotate, thereby doing mechanical work on any object attached to the shaft of the motor. In a generator, the shaft is rotated by some mechanical means, such as an engine or a turbine, and an emf is induced in a coil. If the generator is connected to an external circuit, an electric current is the *output* of the generator.

THE PHYSICS OF . . . an electric generator. In its simplest form, an ac generator consists of a coil of wire that is rotated in a uniform magnetic field, as **Figure 22.22a** indicates. Although not shown in the picture, the wire is usually wound around an iron core. As in an electric motor, the coil/core combination is called the *armature.* Each end of the wire forming the coil is connected to the external circuit by means of a metal ring that rotates with the coil. Each ring slides against a stationary carbon brush, to which the external circuit (the lamp in the drawing) is connected.

To see how current is produced by the generator, consider the two vertical sides of the coil in **Figure 22.22b**. Since each is moving in a magnetic field $\vec{\mathbf{B}}$, the magnetic force exerted on the charges in the wire causes them

David Weintraub/Science Source

FIGURE 22.21 Electric generators such as these supply electric power by producing an induced emf according to Faraday's law of electromagnetic induction.

(a)

(b)

(c)

FIGURE 22.22 (*a*) This electric generator consists of a coil (only one loop is shown) of wire that is rotated in a magnetic field $\vec{\mathbf{B}}$ by some mechanical means. (*b*) The current I arises because of the magnetic force exerted on the charges in the moving wire. (*c*) The dimensions of the coil.

This personal energy generator (PEG) is a small device that can fit into a backpack. It uses Faraday's law of electromagnetic induction to convert some of the kinetic energy of your normal movements into electric energy, which keeps a battery in the device fully charged. The PEG serves as a back-up power source for your mobile phone or other handheld electronic equipment.

to flow, thus creating a current. With the aid of RHR-1 (fingers of extended right hand point along $\vec{\mathbf{B}}$, thumb along the velocity $\vec{\mathbf{v}}$, palm pushes in the direction of the force on a positive charge), it can be seen that the direction of the current is from bottom to top in the left side and from top to bottom in the right side. Thus, charge flows around the loop. The upper and lower segments of the loop are also moving. However, these segments can be ignored because the magnetic force on the charges within them points toward the sides of the wire and not along the length.

The magnitude of the motional emf developed in a conductor moving through a magnetic field is given by Equation 22.1. To apply this expression to the left side of the coil, whose length is L (see **Figure 22.22c**), we need to use the velocity component $v_\perp$ that is perpendicular to $\vec{\mathbf{B}}$. Letting θ be the angle between $\vec{\mathbf{v}}$ and $\vec{\mathbf{B}}$ (see **Figure 22.22b**), it follows that $v_\perp = v \sin \theta$, and, with the aid of Equation 22.1, the emf can be written as

$$\mathscr{E} = BLv_\perp = BLv \sin \theta$$

The emf induced in the right side has the same magnitude as that in the left side. Since the emfs from both sides drive current in the same direction around the loop, the emf for the complete loop is $\mathscr{E} = 2BLv \sin \theta$. If the coil consists of N loops, the net emf is N times as great as that of one loop, so

$$\mathscr{E} = N(2BLv \sin \theta)$$

It is convenient to express the variables v and θ in terms of the angular speed ω at which the coil rotates. Equation 8.2 shows that the angle θ is the product of the angular speed and the time, $\theta = \omega t$, if it is assumed that $\theta = 0$ rad when $t = 0$ s. Furthermore, any point on each vertical side moves on a circular path of radius $r = W/2$, where W is the width of the coil (see **Figure 22.22c**). Thus, the tangential speed v of each side is related to the angular speed ω via Equation 8.9 as $v = r\omega = (W/2)\omega$. Substituting these expressions for θ and v in the previous equation for $\mathscr{E}$, and recognizing that the product LW is the area A of the coil, we can write the induced emf as

Emf induced in a rotating planar coil
$$\mathscr{E} = NAB\omega \sin \omega t = \mathscr{E}_0 \sin \omega t \quad \text{where } \omega = 2\pi f \qquad (22.4)$$

In this result, the angular speed ω is in radians per second and is related to the frequency f [in cycles per second or hertz (Hz)] according to $\omega = 2\pi f$ (Equation 10.6).

Although Equation 22.4 was derived for a rectangular coil, the result is valid for any planar shape of area A (e.g., circular) and shows that the emf varies sinusoidally with time. The peak, or maximum, emf $\mathscr{E}_0$ occurs when $\sin \omega t = 1$ and has the value $\mathscr{E}_0 = NAB\omega$. **Figure 22.23** shows a plot of Equation 22.4 and reveals that the emf changes polarity as the coil rotates. This changing polarity is exactly the same as that discussed for an ac voltage in Section 20.5 and illustrated in **Figure 20.10**. If the external circuit connected to the generator is a closed circuit, an alternating current results that changes direction at the same frequency f as the emf changes polarity. Therefore, this electric generator is also called an *alternating current (ac) generator*. The next two examples show how Equation 22.4 is applied.

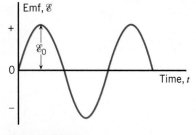

FIGURE 22.23 An ac generator produces this alternating emf $\mathscr{E}$ according to $\mathscr{E} = \mathscr{E}_0 \sin \omega t$.

EXAMPLE 11 | An Ac Generator

In **Figure 22.22** the coil of the ac generator rotates at a frequency of $f = 60.0$ Hz and develops an emf of 120 V (rms; see Section 20.5). The coil has an area of $A = 3.0 \times 10^{-3}$ m^2 and consists of $N = 500$ turns. Find the magnitude of the magnetic field in which the coil rotates.

Reasoning The magnetic field can be found from the relation $\mathscr{E}_0 = NAB\omega$. However, in using this equation we must remember that $\mathscr{E}_0$ is the peak emf, whereas the given value of 120 V is not a peak value but an rms value. The peak emf is related to the rms emf by $\mathscr{E}_0 = \sqrt{2}\,\mathscr{E}_{\text{rms}}$, according to Equation 20.13.

Problem-Solving Insight In the equation $\mathscr{E}_0 = NAB\omega$, remember that the angular frequency ω must be in rad/s and is related to the frequency f (in Hz) according to $\omega = 2\pi f$ (Equation 10.6).

Solution Solving $\mathscr{E}_0 = NAB\omega$ for B and using the facts that $\mathscr{E}_0 = \sqrt{2}\,\mathscr{E}_{\text{rms}}$ (Equation 20.13) and $\omega = 2\pi f$ (Equation 10.6), we find that the magnitude of the magnetic field is

$$B = \frac{\mathscr{E}_0}{NA\omega} = \frac{\sqrt{2}\,\mathscr{E}_{\text{rms}}}{NA2\pi f} = \frac{\sqrt{2}\,(120\text{ V})}{(500)(3.0 \times 10^{-3}\text{ m}^2)2\pi(60.0\text{ Hz})} = \boxed{0.30\text{ T}}$$

Analyzing Multiple-Concept Problems

EXAMPLE 12 | The Physics of a Bike Generator

A bicyclist is traveling at night, and a generator mounted on the bike powers a headlight. A small rubber wheel on the shaft of the generator presses against the bike tire and turns the coil of the generator at an angular speed that is 44 times as great as the angular speed of the tire itself. The tire has a radius of 0.33 m. The coil consists of 75 turns, has an area of 2.6×10^{-3} m², and rotates in a 0.10-T magnetic field. When the peak emf being generated is 6.0 V, what is the linear speed of the bike?

Reasoning Since the tires are rolling, the linear speed v of the bike is related to the angular speed ω_{tire} of its tires by

$v = r\omega_{\text{tire}}$ (see Section 8.6), where r is the radius of a tire. We are given that the angular speed ω_{coil} of the coil is 44 times as great as that of the tire. Thus, $\omega_{\text{tire}} = \frac{1}{44}\omega_{\text{coil}}$ and the linear speed of the bike can be related to the angular speed of the coil. Furthermore, according to the discussion in this section on electric generators, the angular speed of the coil is related (see Equation 22.4) to the peak emf developed by the rotating coil, the number of turns in the coil, the area of the coil, and the magnetic field, all of which are known.

Knowns and Unknowns The data for this problem are:

Description	Symbol	Value	Comment
Radius of tire	r	0.33 m	
Number of turns in coil	N	75	
Angular speed of coil	ω_{coil}	$44\omega_{\text{tire}}$	Angular speed of coil is 44 times as great as that of tire.
Area of coil	A	2.6×10^{-3} m²	
Magnitude of magnetic field	B	0.10 T	
Peak emf produced by generator	$\mathscr{E}_0$	6.0 V	
Unknown Variable			
Linear speed of bike	v	?	

Modeling the Problem

STEP 1 Rolling Motion When a tire rolls without slipping on the ground, the linear speed v of the tire (the speed at which its axle is moving forward) is related to the angular speed ω_{tire} of the tire about the axle. This relationship is given by

$$v = r\omega_{\text{tire}} \qquad (8.12)$$

where r is the radius of the tire. We are given that the angular speed ω_{coil} of the coil is 44 times as great as the angular speed of the tire, so $\omega_{\text{coil}} = 44\omega_{\text{tire}}$. Solving this equation for ω_{tire} and substituting the result into $v = r\omega_{\text{tire}}$, we obtain Equation 1 at the right. The radius of the tire is known, but the angular speed of the coil is not; we will evaluate it in Step 2.

$$v = r\left(\tfrac{1}{44}\right)\omega_{\text{coil}} \qquad (1)$$

$$\boxed{?}$$

STEP 2 Peak Emf Induced in a Rotating Planar Coil A generator produces an emf when a coil of wire rotates in a magnetic field. According to Equation 22.4, the peak emf $\mathscr{E}_0$ is given by $\mathscr{E}_0 = NAB\omega_{\text{coil}}$, where N is the number of turns in the coil, A is the area of the coil, B is the magnitude of the magnetic field, and ω_{coil} is the angular speed of the rotating coil. Solving this relation for ω_{coil} gives

$$\omega_{\text{coil}} = \frac{\mathscr{E}_0}{NAB}$$

$$v = r\left(\tfrac{1}{44}\right)\omega_{\text{coil}} \qquad (1)$$

$$\boxed{\omega_{\text{coil}} = \frac{\mathscr{E}_0}{NAB}}$$

Note from the data table that all the variables on the right side of this equation are known. We substitute this result into Equation 1, as indicated at the right.

Solution   Algebraically combining the results of the two steps, we have

$$
\overset{\text{STEP 1}}{\downarrow} \qquad \overset{\text{STEP 2}}{\downarrow}
$$

$$
v = r(\tfrac{1}{44})\omega_{\text{coil}} = r(\tfrac{1}{44})\left(\frac{\mathscr{E}}{NAB}\right)
$$

The linear speed of the bicycle is

$$
v = r(\tfrac{1}{44})\left(\frac{\mathscr{E}_0}{NAB}\right) = (0.33 \text{ m})(\tfrac{1}{44})\left[\frac{6.0 \text{ V}}{(75)(2.6 \times 10^{-3} \text{ m}^2)(0.10 \text{ T})}\right] = \boxed{2.3 \text{ m/s}}
$$

Related Homework: Problem 42

The Electrical Energy Delivered by a Generator and the Countertorque

FIGURE 22.24 The generator supplies a total current of $I = I_1 + I_2$ to the load.

Some power-generating stations burn fossil fuel (coal, gas, or oil) to heat water and produce pressurized steam for turning the blades of a turbine whose shaft is linked to the generator. Others use nuclear fuel or falling water as a source of energy. As the turbine rotates, the generator coil also rotates and mechanical work is transformed into electrical energy.

The devices to which the generator supplies electricity are known collectively as the "load," because they place a burden or load on the generator by taking electrical energy from it. If all the devices are switched off, the generator runs under a no-load condition, because there is no current in the external circuit and the generator does not supply electrical energy. Then, work needs to be done on the turbine only to overcome friction and other mechanical losses within the generator itself, and fuel consumption is at a minimum.

Figure 22.24 illustrates a situation in which a load is connected to a generator. Because there is now a current $I = I_1 + I_2$ in the coil of the generator and the coil is situated in a magnetic field, the current experiences a magnetic force $\vec{\mathbf{F}}$. **Figure 22.25** shows the magnetic force acting on the left side of the coil, with the direction of $\vec{\mathbf{F}}$ given by RHR-1. A force of equal magnitude but opposite direction also acts on the right side of the coil, although this force is not shown in the drawing. The magnetic force $\vec{\mathbf{F}}$ gives rise to a *countertorque* that opposes the rotational motion. The greater the current drawn from the generator, the greater the countertorque, and the harder it is for the turbine to turn the coil. To compensate for this countertorque and keep the coil rotating at a constant angular speed, work must be done by the turbine, which means that more fuel must be burned. This is another example of the law of conservation of energy, since the electrical energy consumed by the load must ultimately come from the energy source used to drive the turbine.

The Back Emf Generated by an Electric Motor

A generator converts mechanical work into electrical energy; in contrast, an electric motor converts electrical energy into mechanical work. Both devices are similar and consist of a coil of wire that rotates in a magnetic field. In fact, as the armature of a motor rotates, the magnetic flux passing through the coil changes and an emf is induced in the coil. Thus, when a motor is operating, two sources of emf are present: (1) the applied emf V that provides current to drive the motor (e.g., from a 120-V outlet), and (2) the emf $\mathscr{E}$ induced by the generator-like action of the rotating coil. The circuit diagram in **Figure 22.26** shows these two emfs.

Consistent with Lenz's law, the induced emf $\mathscr{E}$ acts to oppose the applied emf V and is called the **back emf** or the **counter emf** of the motor. The greater the speed of the motor, the greater is the flux change through the coil, and the greater is the back emf. Because V and $\mathscr{E}$ have opposite polarities, the net emf in the circuit is $V - \mathscr{E}$. In **Figure 22.26**, R is the

FIGURE 22.25 (a) A current I exists in the rotating coil of a generator. (b) A top view of the coil, showing magnetic force $\vec{\mathbf{F}}$ exerted on the left side of the coil.

FIGURE 22.26 The applied emf V supplies the current I to drive the motor. The circuit on the right shows V along with the electrical equivalent of the motor, including the resistance R of its coil and the back emf $\mathscr{E}$.

resistance of the wire in the coil, and the current I drawn by the motor is determined from Ohm's law as the net emf divided by the resistance:

$$I = \frac{V - \mathscr{E}}{R} \qquad (22.5)$$

The next example uses this result to illustrate that the current in a motor depends on both the applied emf V and the back emf $\mathscr{E}$.

EXAMPLE 13 | The Physics of Operating a Motor

The coil of an ac motor has a resistance of $R = 4.1\ \Omega$. The motor is plugged into an outlet where $V = 120.0$ volts (rms), and the coil develops a back emf of $\mathscr{E} = 118.0$ volts (rms) when rotating at normal constant speed. The motor is turning a wheel. Find **(a)** the current when the motor first starts up and **(b)** the current when the motor is operating at normal speed.

Reasoning Once normal operating speed is attained, the motor need only work to compensate for frictional losses. But in bringing the wheel up to speed from rest, the motor must also do work to increase the wheel's rotational kinetic energy. Thus, bringing the wheel up to speed requires more work, and hence more current, than maintaining the normal operating speed. We expect our answers to parts (a) and (b) to reflect this fact.

Problem-Solving Insight The current in an electric motor depends on both the applied emf V and any back emf $\mathscr{E}$ developed because the coil of the motor is rotating.

Solution **(a)** When the motor just starts up, the coil is not rotating, so there is no back emf induced in the coil and $\mathscr{E} = 0$ V. The start-up current drawn by the motor is

$$I = \frac{V - \mathscr{E}}{R} = \frac{120\ \text{V} - 0\ \text{V}}{4.1\ \Omega} = \boxed{29\ \text{A}} \qquad (22.5)$$

(b) At normal speed, the motor develops a back emf of $\mathscr{E} = 118.0$ volts, so the current is

$$I = \frac{V - \mathscr{E}}{R} = \frac{120.0\ \text{V} - 118.0\ \text{V}}{4.1\ \Omega} = \boxed{0.49\ \text{A}}$$

Example 13 illustrates that when a motor is just starting, there is little back emf, and, consequently, a relatively large current exists in the coil. As the motor speeds up, the back emf increases until it reaches a maximum value when the motor is rotating at normal speed. The back emf becomes almost equal to the applied emf, and the current is reduced to a relatively small value, which is sufficient to provide the torque on the coil needed to overcome frictional and other losses in the motor and to drive the load (e.g., a fan).

Check Your Understanding

(The answers are given at the end of the book.)

16. In a car, the generator-like action of the alternator occurs while the engine is running and keeps the battery fully charged. The headlights would discharge an old and failing battery quickly if it were not for the alternator. Why does the engine of a parked car run more quietly with the headlights off than with them on when the battery is in bad shape?

17. You have a fixed length of wire and need to design a generator that will produce the greatest peak emf for a given frequency and magnetic field strength. You should use **(a)** a one-turn square coil, **(b)** a two-turn square coil, **(c)** either a one- or a two-turn square coil because both give the same peak emf for a given frequency and magnetic field strength.

18. An electric motor in a hair dryer is running at its normal constant operating speed and, thus, is drawing a relatively small current, as in part (b) of Example 13. The wire in the coil of the motor has some resistance. What happens to the temperature of the coil if the shaft of the motor is prevented from turning, so that the back emf is suddenly reduced to zero? **(a)** Nothing. **(b)** The temperature decreases. **(c)** The temperature increases (the coil could even burn up).

22.8 | Mutual Inductance and Self-Inductance

Mutual Inductance

FIGURE 22.27 An alternating current I_p in the primary coil creates an alternating magnetic field. This changing field induces an emf in the secondary coil.

We have seen that an emf can be induced in a coil by keeping the coil stationary and moving a magnet nearby, or by moving the coil near a stationary magnet. **Figure 22.27** illustrates another important method of inducing an emf. Here, two coils of wire, the *primary coil* and the *secondary coil*, are placed close to each other. The primary coil is the one connected to an ac generator, which sends an alternating current I_p through it. The secondary coil is not attached to a generator, although a voltmeter is connected across it to register any induced emf.

The current-carrying primary coil is an electromagnet and creates a magnetic field in the surrounding region. If the two coils are close to each other, a significant fraction of this magnetic field penetrates the secondary coil and produces a magnetic flux. The flux is changing, since the current in the primary coil and its associated magnetic field are changing. Because of the change in flux, an emf is induced in the secondary coil.

The effect in which a changing current in one circuit induces an emf in another circuit is called **mutual induction**. According to Faraday's law of electromagnetic induction, the average emf $\mathscr{E}_s$ induced in the secondary coil is proportional to the change in flux $\Delta\Phi_s$ passing through it. However, $\Delta\Phi_s$ is produced by the change in current ΔI_p in the primary coil. Therefore, it is convenient to recast Faraday's law into a form that relates $\mathscr{E}_s$ to ΔI_p. To see how this recasting is accomplished, note that the net magnetic flux passing through the secondary coil is $N_s\Phi_s$, where N_s is the number of loops in the secondary coil and Φ_s is the flux through one loop (assumed to be the same for all loops). The net flux is proportional to the magnetic field, which, in turn, is proportional to the current I_p in the primary coil. Thus, we can write $N_s\Phi_s \propto I_p$. This proportionality can be converted into an equation in the usual manner by introducing a proportionality constant M, known as the **mutual inductance**:

$$N_s\Phi_s = MI_p \quad \text{or} \quad M = \frac{N_s\Phi_s}{I_p} \tag{22.6}$$

Substituting this equation into Faraday's law, we find that

$$\mathscr{E}_s = -N_s\frac{\Delta\Phi_s}{\Delta t} = -\frac{\Delta(N_s\Phi_s)}{\Delta t} = -\frac{\Delta(MI_p)}{\Delta t} = -M\frac{\Delta I_p}{\Delta t}$$

Emf due to mutual induction

$$\mathscr{E}_s = -M\frac{\Delta I_p}{\Delta t} \tag{22.7}$$

Writing Faraday's law in this manner makes it clear that the average emf $\mathscr{E}_s$ induced in the secondary coil is due to the change in the current ΔI_p in the primary coil.

Equation 22.7 shows that the measurement unit for the mutual inductance M is $V \cdot s/A$, which is called a henry (H) in honor of Joseph Henry: $1 \, V \cdot s/A = 1 \, H$. The mutual inductance depends on the geometry of the coils and the nature of any ferromagnetic core material that is present. Although M can be calculated for some highly symmetrical arrangements, it is usually measured experimentally. In most situations, values of M are less than 1 H and are often on the order of millihenries (1 mH = 1×10^{-3} H) or microhenries (1 μH = 1×10^{-6} H).

BIO **THE PHYSICS OF . . . transcranial magnetic stimulation (TMS).** A new technique that shows promise for the treatment of psychiatric disorders such as depression is based on mutual induction. This technique is called transcranial magnetic stimulation (TMS) and is a type of indirect and gentler electric shock therapy. In traditional electric shock therapy, electric current is delivered directly through the skull and penetrates the brain, disrupting its electrical circuitry and in the process alleviating the symptoms of the psychiatric disorder. The treatment is not gentle and requires an anesthetic, because relatively large electric currents must be used to penetrate the skull. In contrast, TMS produces its electric current by using a time-varying magnetic field. A primary coil is positioned over the part of the brain to be treated (see **Figure 22.28**), and a time-varying current is applied to this coil. The arrangement is analogous to that in **Figure 22.27**, except that the brain and the electrically conductive pathways within it take the place of the secondary coil. The magnetic field produced by the primary coil penetrates the brain and, since the field is changing in time, it induces an emf in the brain. This induced emf causes an electric current to flow in the conductive brain tissue, with therapeutic results similar to those of conventional electric shock treatment. The current delivered to the brain, however, is much smaller than the current in the conventional treatment, so that patients receive TMS treatments without anesthetic and without severe after-effects such as headaches and memory loss. TMS remains in the experimental stage, however, and the optimal protocol for applying the technique has not yet been determined.

Self-Inductance

In all the examples of induced emfs presented so far, the magnetic field has been produced by an external source, such as a permanent magnet or an electromagnet. However, the magnetic field need not arise from an external source. An emf can be induced in a current-carrying coil by a change in the magnetic field that the current *itself* produces. For instance, **Figure 22.29** shows a coil connected to an ac generator. The alternating current creates an alternating magnetic field that, in turn, creates a changing flux through the coil. The change in flux induces an emf in the coil, in accord with Faraday's law. The effect in which a changing current in a circuit induces an emf in the same circuit is referred to as **self-induction**.

When dealing with self-induction, as with mutual induction, it is customary to recast Faraday's law into a form in which the induced emf is proportional to the change in current in the coil rather than to the change in flux. If Φ is the magnetic flux that passes through one turn of the coil, then $N\Phi$ is the net flux through a coil of N turns. Since Φ is proportional to the magnetic field, and the magnetic field is proportional to the current I, it follows that $N\Phi \propto I$. By inserting a constant L, called the **self-inductance**, or simply the **inductance**, of the coil, we can convert this proportionality into Equation 22.8:

$$N\Phi = LI \quad \text{or} \quad L = \frac{N\Phi}{I} \qquad (22.8)$$

Faraday's law of induction now gives the average induced emf as

$$\mathscr{E} = -N\frac{\Delta \Phi}{\Delta t} = -\frac{\Delta(N\Phi)}{\Delta t} = -\frac{\Delta(LI)}{\Delta t} = -L\frac{\Delta I}{\Delta t}$$

Emf due to self-induction

$$\mathscr{E} = -L\frac{\Delta I}{\Delta t} \qquad (22.9)$$

Richard T. Nowitz/Science Source

FIGURE 22.28 In the technique of transcranial magnetic stimulation (TMS), a time-varying electric current is applied to a primary coil, which is positioned over a region of the brain, as this photograph illustrates. The time-varying magnetic field produced by the coil penetrates the brain and creates an induced emf within it. This induced emf leads to an induced current that disrupts the electric circuits of the brain, thereby relieving some of the symptoms of psychiatric disorders such as depression.

FIGURE 22.29 The alternating current in the coil generates an alternating magnetic field that induces an emf in the coil.

Like mutual inductance, L is measured in henries. The magnitude of L depends on the geometry of the coil and on the core material. Wrapping the coil around a ferromagnetic (iron) core substantially increases the magnetic flux—and therefore the inductance—relative to that for an air core. Because of their self-inductance, coils are known as **inductors** and are widely used in electronics. Inductors come in all sizes, typically in the range between millihenries and microhenries.

The Energy Stored in an Inductor

An inductor, like a capacitor, can store energy. This stored energy arises because a generator does work to establish a current in an inductor. Suppose that an inductor is connected to a generator whose terminal voltage can be varied continuously from zero to some final value. As the voltage is increased, the current I in the circuit rises continuously from zero to its final value. While the current is rising, an induced emf $\mathscr{E} = -L(\Delta I/\Delta t)$ appears across the inductor. Conforming to Lenz's law, the polarity of the induced emf $\mathscr{E}$ is opposite to the polarity of the generator voltage, so as to oppose the increase in the current. Thus, the generator must do work to push the charges through the inductor against this induced emf. The increment of work ΔW done by the generator in moving a small amount of charge ΔQ through the inductor is $\Delta W = -(\Delta Q)\mathscr{E} = -(\Delta Q)[-L(\Delta I/\Delta t)]$, according to Equation 19.4. Since $\Delta Q/\Delta t$ is the current I, the work done is

$$\Delta W = LI(\Delta I)$$

In this expression ΔW represents the work done by the generator to increase the current in the inductor by an amount ΔI. To determine the total work W done while the current is changed from zero to its final value, all the small increments ΔW must be added together. The result is $W = \frac{1}{2}LI^2$, where I represents the final current in the inductor. This work is stored as energy in the inductor, so that

Energy stored in an inductor
$$\text{Energy} = \tfrac{1}{2}LI^2 \qquad (22.10)$$

It is possible to regard the energy in an inductor as being stored in its magnetic field. For the special case of a long solenoid (see Problem 68), the self-inductance is $L = \mu_0 n^2 A\ell$, where n is the number of turns per unit length, A is the cross-sectional area, and ℓ is the length of the solenoid. As a result, the energy stored in a long solenoid is

$$\text{Energy} = \tfrac{1}{2}LI^2 = \tfrac{1}{2}\mu_0 n^2 A\ell I^2$$

Since $B = \mu_0 nI$ at the interior of a long solenoid (Equation 21.7), this energy can be expressed as

$$\text{Energy} = \frac{1}{2\mu_0}B^2 A\ell$$

The term $A\ell$ is the volume inside the solenoid where the magnetic field exists, so the energy per unit volume, or **energy density**, is

$$\text{Energy density} = \frac{\text{Energy}}{\text{Volume}} = \frac{1}{2\mu_0}B^2 \qquad (22.11)$$

Although this result was obtained for the special case of a long solenoid, it is quite general and is valid for any point where a magnetic field exists in air or vacuum or in a nonmagnetic material. Thus, energy can be stored in a magnetic field, just as it can in an electric field.

22.9 Transformers

One of the most important applications of mutual induction and self-induction takes place in a transformer. A **transformer** is a device that is used to increase or decrease an ac voltage. For instance, whenever a cordless device (e.g., a cell phone) is plugged into a wall receptacle to recharge the batteries, a transformer plays a role in reducing the 120-V ac voltage to a much smaller value. Typically, between 3 and 9 V are needed to energize batteries. In another example, a picture tube in a television set needs about 15 000 V to accelerate the electron beam, and a transformer is used to obtain this high voltage from the 120 V at a wall socket.

THE PHYSICS OF ... transformers. Figure 22.30 shows a drawing of a transformer. The transformer consists of an iron core on which two coils are wound: a primary coil with N_p turns and a secondary coil with N_s turns. The primary coil is the one connected to the ac generator. For the moment, suppose that the switch in the secondary circuit is open, so there is no current in this circuit. The alternating current in the primary coil establishes a changing magnetic field in the iron core. Because iron is easily magnetized, it greatly enhances the magnetic field relative to that in an air core and guides the field lines to the secondary coil. In a well-designed core, nearly all the magnetic flux Φ that passes through each turn of the primary also goes through each turn of the secondary. Since the magnetic field is changing, the flux through the primary and secondary coils is also changing, and consequently an emf is induced in both coils. In the secondary coil the induced emf $\mathcal{E}_s$ arises from mutual induction and is given by Faraday's law as

$$\mathcal{E}_s = -N_s \frac{\Delta\Phi}{\Delta t}$$

In the primary coil the induced emf $\mathcal{E}_p$ is due to self-induction and is specified by Faraday's law as

$$\mathcal{E}_p = -N_p \frac{\Delta\Phi}{\Delta t}$$

The term $\Delta\Phi/\Delta t$ is the same in both of these equations, since the same flux penetrates each turn of both coils. Dividing the two equations shows that

$$\frac{\mathcal{E}_s}{\mathcal{E}_p} = \frac{N_s}{N_p}$$

In a high-quality transformer the resistances of the coils are negligible, so the magnitudes of the emfs, $\mathcal{E}_s$ and $\mathcal{E}_p$, are nearly equal to the terminal voltages, V_s and V_p, across the coils (see Section 20.9 for a discussion of terminal voltage). The relation $\mathcal{E}_s/\mathcal{E}_p = N_s/N_p$ is called the **transformer equation** and is usually written in terms of the terminal voltages:

Transformer equation
$$\frac{V_s}{V_p} = \frac{N_s}{N_p} \qquad (22.12)$$

According to the transformer equation, if N_s is greater than N_p, the secondary (output) voltage is greater than the primary (input) voltage. In this case we have a *step-up*

FIGURE 22.30 A transformer consists of a primary coil and a secondary coil, both wound on an iron core. The changing magnetic flux produced by the current in the primary coil induces an emf in the secondary coil. At the far right is the symbol for a transformer.

Power distribution stations use high-voltage transformers similar to this one to step up or step down voltages.

transformer. On the other hand, if N_s is less than N_p, the secondary voltage is less than the primary voltage, and we have a *step-down* transformer. The ratio N_s/N_p is referred to as the *turns ratio* of the transformer. A turns ratio of 8/1 (often written as 8:1) means, for example, that the secondary coil has eight times more turns than the primary coil. Conversely, a turns ratio of 1:8 implies that the secondary has one-eighth as many turns as the primary.

A transformer operates with ac electricity and not with dc. A steady direct current in the primary coil produces a flux that does not change in time, and thus no emf is induced in the secondary coil. The ease with which transformers can change voltages from one value to another is a principal reason why ac is preferred over dc.

With the switch in the secondary circuit of **Figure 22.30** closed, a current I_s exists in the circuit and electrical energy is fed to the TV tube. This energy comes from the ac generator connected to the primary coil. Although the secondary voltage V_s may be larger or smaller than the primary voltage V_p, energy is not being created or destroyed by the transformer. Energy conservation requires that the energy delivered to the secondary coil must be the same as the energy delivered to the primary coil, provided no energy is dissipated in heating these coils or is otherwise lost. In a well-designed transformer, less than 1% of the input energy is lost in the form of heat. Noting that power is energy per unit time, and assuming 100% energy transfer, the average power $\overline{P}_p$ delivered to the primary coil is equal to the average power $\overline{P}_s$ delivered to the secondary coil: $\overline{P}_p = \overline{P}_s$. However, $\overline{P} = I_{rms}V_{rms}$ (Equation 20.15a), so $I_pV_p = I_sV_s$, or

$$\frac{I_s}{I_p} = \frac{V_p}{V_s} = \frac{N_p}{N_s} \qquad (22.13)$$

Observe that V_s/V_p is equal to the turns ratio N_s/N_p, while I_s/I_p is equal to the inverse turns ratio N_p/N_s. Note that, for clarity, we have dropped the rms subscripts in Equation 22.13, and we will continue with this truncated notation going forward with the understanding that the currents and voltages represent their rms values.

> **Problem-Solving Insight** Consequently, a transformer that steps up the voltage simultaneously steps down the current, and a transformer that steps down the voltage steps up the current.

However, the power is neither stepped up nor stepped down, since $\overline{P}_p = \overline{P}_s$. Example 14 emphasizes this fact.

EXAMPLE 14 | A Step-Down Transformer

A step-down transformer inside a stereo receiver has 330 turns in the primary coil and 25 turns in the secondary coil. The plug connects the primary coil to a 120-V wall socket, and there is a current of 0.83 A in the primary coil while the receiver is turned on. Connected to the secondary coil are the transistor circuits of the receiver. Find **(a)** the voltage across the secondary coil, **(b)** the current in the secondary coil, and **(c)** the average electric power delivered to the transistor circuits.

Reasoning The transformer equation, Equation 22.12, states that the secondary voltage V_s is equal to the product of the primary voltage V_p and the turns ratio N_s/N_p. On the other hand, Equation 22.13 indicates that the secondary current I_s is equal to the product of the primary current I_p and the inverse turns ratio N_p/N_s. The average power delivered to the transistor circuits is the product of the secondary current and the secondary voltage.

Solution **(a)** The voltage across the secondary coil can be found from the transformer equation:

$$V_s = V_p\frac{N_s}{N_p} = (120 \text{ V})\left(\frac{25}{330}\right) = \boxed{9.1 \text{ V}} \qquad (22.12)$$

(b) The current in the secondary coil is

$$I_s = I_p\frac{N_p}{N_s} = (0.83 \text{ A})\left(\frac{330}{25}\right) = \boxed{11 \text{ A}} \qquad (22.13)$$

(c) The average power P_s delivered to the secondary coil is the product of I_s and V_s:

$$\overline{P}_s = I_sV_s = (11 \text{ A})(9.1 \text{ V}) = \boxed{1.0 \times 10^2 \text{ W}} \qquad (20.15a)$$

As a check on our calculation, we verify that the power delivered to the secondary coil is the same as that sent to the primary coil from the wall receptacle:

$$\overline{P}_p = I_pV_p = (0.83 \text{ A})(120 \text{ V}) = 1.0 \times 10^2 \text{ W}$$

INTERACTIVE FIGURE 22.31   Transformers play a key role in the transmission of electric power.

Transformers play an important role in the transmission of power between electrical generating plants and the communities they serve. Whenever electricity is transmitted, there is always some loss of power in the transmission lines themselves due to resistive heating. Since the resistance of the wires is proportional to their length, the longer the wires the greater is the power loss. The resistance of the wires is also inversely proportional to their cross-sectional area, so thicker wires are used to help minimize the power loss. To reduce the loss further, power companies use transformers that step up the voltage to high levels while reducing the current. A smaller current means less power loss, since $P = I^2R$, where R is the resistance of the transmission wires (see Problem 58). **Interactive Figure 22.31** shows one possible way of transmitting power. The power plant produces a voltage of 12 000 V. This voltage is then raised to 240 000 V by a 20:1 step-up transformer. The high-voltage power is sent over the long-distance transmission line. Upon arrival at the city, the voltage is reduced to about 8000 V at a substation using a 1:30 step-down transformer. However, before any domestic use, the voltage is further reduced to 240 V (or possibly 120 V) by another step-down transformer that is often mounted on a utility pole. The power is then distributed to consumers.

Check Your Understanding

(The answers are given at the end of the book.)

19. A transformer changes the 120-V voltage at a wall socket to 12 000 V. The current delivered by the wall socket is **(a)** stepped up by a factor of 100, **(b)** stepped down by a factor of 100, **(c)** neither stepped up nor stepped down.

20. A transformer that stepped up the voltage and the current simultaneously would **(a)** produce less power at the secondary coil than was supplied at the primary coil, **(b)** produce more power at the secondary coil than was supplied at the primary coil, **(c)** produce the same amount of power at the secondary coil that was supplied at the primary coil, **(d)** violate the law of conservation of energy. Choose one or more.

EXAMPLE 15 | BIO The Physics of Wireless Charging—Implantable Medical Devices

Because of electromagnetic induction, energy can be transferred between two coils that are not in contact, as long as the changing magnetic flux created by one coil passes through the second one.

This arrangement can be used to charge batteries connected to small medical devices that are implanted inside the body for many different applications (**Figure 22.32**). The implantable device is powered by a rechargeable battery that is connected to a small, flat coil (secondary) that is positioned just below the skin. The primary coil from the external charging device is placed on the skin directly above the secondary coil and transfers energy by induction to charge the device's internal battery. Consider the example of a wireless pacemaker, where a battery is connected to a coil that produces an average power of 3.0 W. If the rms current in the charging (primary) coil is 500 mA, what is the voltage across it? Assume an energy transfer efficiency of 70%.

Reasoning The problem tells us that the average power delivered to the secondary coil by the primary occurs with an efficiency of 70%. This means that $\overline{P}_s = (70\%)\overline{P}_p = (0.70)\overline{P}_p$. We can

then use Equation 20.15a ($\overline{P} = I_{rms}V_{rms}$) to solve for the voltage across the primary coil.

Solution Substituting Equation 20.15a into the expression above, we get the following: $\overline{P}_s = (0.70)I_pV_p$, where I_p and V_p represent their rms values. We can rearrange this expression and solve for the voltage:

$$V_p = \frac{\overline{P}_s}{(0.70)I_p} = \frac{3.0\text{ W}}{(0.70)(0.50\text{ mA})} = \boxed{8.6\text{ V}}$$

Implantable devices offer the advantage of lower infection risk, since no wires have to exit the body. However, they take longer to charge, and further advances in their implementation, as well as other areas of concern, like security, will have to be improved before they are widely used.

Wireless implantable medical devices

Deep brain neurostimulators

Cochlear implants

Cardiac defibrillators/ pacemakers

Gastric stimulators

Insulin pumps

Foot drop implants

FIGURE 22.32 Examples of medical devices that could be implanted inside the body and charged inductively.

Concept Summary

22.1 Induced Emf and Induced Current Electromagnetic induction is the phenomenon in which an emf is induced in a piece of wire or a coil of wire with the aid of a magnetic field. The emf is called an induced emf, and any current that results from the emf is called an induced current.

22.2 Motional Emf An emf $\mathscr{E}$ is induced in a conducting rod of length L when the rod moves with a speed v in a magnetic field of magnitude B, according to Equation 22.1, which applies when the velocity of the rod, the length of the rod, and the magnetic field are mutually perpendicular. When the motional emf is used to operate an electrical device, such as a light bulb, the energy delivered to the device originates in the work done to move the rod, and the law of conservation of energy applies.

$$\mathscr{E} = vBL \qquad (22.1)$$

22.3 Magnetic Flux The magnetic flux Φ that passes through a surface is given by Equation 22.2, where B is the magnitude of the magnetic field, A is the area of the surface, and ϕ is the angle between the field and the normal to the surface. The magnetic flux is proportional to the number of magnetic field lines that pass through the surface.

$$\Phi = BA\cos\phi \qquad (22.2)$$

22.4 Faraday's Law of Electromagnetic Induction Faraday's law of electromagnetic induction states that the average emf $\mathscr{E}$ induced in a coil of N loops is given by Equation 22.3, where $\Delta\Phi$ is the change in magnetic flux through one loop and Δt is the time interval during which the change occurs. Motional emf is a special case of induced emf.

$$\mathscr{E} = -N\left(\frac{\Phi - \Phi_0}{t - t_0}\right) = -N\frac{\Delta\Phi}{\Delta t} \qquad (22.3)$$

22.5 Lenz's Law Lenz's law provides a way to determine the polarity of an induced emf. Lenz's law is stated as follows: The induced emf resulting from a changing magnetic flux has a polarity that leads to an induced current whose direction is such that the induced magnetic field opposes the original flux change. This statement is a consequence of the law of conservation of energy.

22.7 The Electric Generator In its simplest form, an electric generator consists of a coil of N loops that rotates in a uniform magnetic field $\vec{B}$. The emf produced by this generator is given by Equation 22.4, where A is the area of the coil, ω is the angular speed (in rad/s) of the coil, and $\mathscr{E}_0 = NAB\omega$ is the peak emf. The angular speed in rad/s is related to the frequency f in cycles/s, or Hz, according to $\omega = 2\pi f$.

$$\mathscr{E} = NAB\omega\sin\omega t = \mathscr{E}_0\sin\omega t \qquad (22.4)$$

When an electric motor is running, it exhibits a generator-like behavior by producing an induced emf, called the back emf. The current I needed to keep the motor running at a constant speed is given by Equation 22.5, where V is the emf applied to the motor by an external source, $\mathscr{E}$ is the back emf, and R is the resistance of the motor coil.

$$I = \frac{V - \mathscr{E}}{R} \quad (22.5)$$

22.8 Mutual Inductance and Self-Inductance Mutual induction is the effect in which a changing current in the primary coil induces an emf in the secondary coil. The average emf $\mathscr{E}_s$ induced in the secondary coil by a change in current ΔI_p in the primary coil is given by Equation 22.7, where Δt is the time interval during which the change occurs. The constant M is the mutual inductance between the two coils and is measured in henries (H).

$$\mathscr{E}_s = -M\frac{\Delta I_p}{\Delta t} \quad (22.7)$$

Self-induction is the effect in which a change in current ΔI in a coil induces an average emf $\mathscr{E}$ in the same coil, according to Equation 22.9. The constant L is the self-inductance, or inductance, of the coil and is measured in henries.

$$\mathscr{E} = -L\frac{\Delta I}{\Delta t} \quad (22.9)$$

To establish a current I in an inductor, work must be done by an external agent. This work is stored as energy in the inductor, the amount of energy being given by Equation 22.10. The energy stored in an inductor can be regarded as being stored in its magnetic field. At any point in air or vacuum or in a nonmagnetic material where a magnetic field $\vec{B}$ exists, the energy density, or the energy stored per unit volume, is given by Equation 22.11.

$$\text{Energy} = \tfrac{1}{2}LI^2 \quad (22.10)$$

$$\text{Energy density} = \frac{1}{2\mu_0}B^2 \quad (22.11)$$

22.9 Transformers A transformer consists of a primary coil of N_p turns and a secondary coil of N_s turns. If the resistances of the coils are negligible, the voltage V_p across the primary coil and the voltage V_s across the secondary coil are related according to the transformer equation, which is Equation 22.12, where the ratio N_s/N_p is called the turns ratio of the transformer. A transformer functions with ac electricity, not with dc. If the transformer is 100% efficient in transferring power from the primary to the secondary coil, the ratio of the secondary current I_s to the primary current I_p is given by Equation 22.13.

$$\frac{V_s}{V_p} = \frac{N_s}{N_p} \quad (22.12)$$

$$\frac{I_s}{I_p} = \frac{N_p}{N_s} \quad (22.13)$$

Focus on Concepts

Online

Additional questions are available for assignment in WileyPLUS.

Section 22.2 Motional Emf

1. You have three light bulbs; bulb A has a resistance of 240 Ω, bulb B has a resistance of 192 Ω, and bulb C has a resistance of 144 Ω, Each of these bulbs is used for the same amount of time in a setup like the one in the drawing. In each case the speed of the rod and the magnetic field strength are the same. Rank the setups in descending order, according to how much work the hand in the drawing must do (largest amount of work first). **(a)** A, B, C **(b)** A, C, B **(c)** B, C, A **(d)** B, A, C **(e)** C, B, A

QUESTION 1

Section 22.3 Magnetic Flux

2. The drawing shows a cube. The dashed lines in the drawing are perpendicular to faces 1, 2, and 3 of the cube. Magnetic fields are oriented with respect to these faces as shown, and each of the three fields $\vec{B}_1$, $\vec{B}_2$, and $\vec{B}_3$ has the same magnitude. Note that $\vec{B}_2$ is parallel to face 2 of the cube. Rank the magnetic fluxes that pass through the faces 1, 2, and 3 of the cube in decreasing order

(largest first). **(a)** Φ_1, Φ_2, Φ_3 **(b)** Φ_1, Φ_3, Φ_2 **(c)** Φ_2, Φ_1, Φ_3 **(d)** Φ_2, Φ_3, Φ_1 **(e)** Φ_3, Φ_2, Φ_1

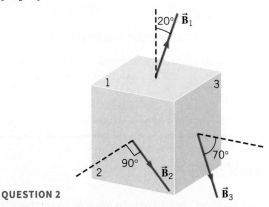

QUESTION 2

Section 22.4 Faraday's Law of Electromagnetic Induction

3. The drawing shows three flat coils, one square and two rectangular, that are each being pushed into a region where there is a uniform magnetic field directed into the page. Outside of this region the magnetic field is zero. In each case the magnetic field within the region has the same magnitude, and the coil is being pushed at the same velocity $\vec{v}$. Each coil begins with one side just at the edge of the field region. Consider the magnitude of the average emf induced as each coil is pushed from the starting position shown in the drawing until the coil is just completely within the field region. Rank the magnitudes of the average emfs in descending order (largest first). **(a)** $\mathscr{E}_A, \mathscr{E}_B, \mathscr{E}_C$ **(b)** $\mathscr{E}_A, \mathscr{E}_C, \mathscr{E}_B$ **(c)** $\mathscr{E}_B, \mathscr{E}_A$ and $\mathscr{E}_C$ (a tie) **(d)** $\mathscr{E}_C, \mathscr{E}_A$ and $\mathscr{E}_B$ (a tie)

A   B   C

QUESTION 3

4. A long, vertical, straight wire carries a current I. The wire is perpendicular to the plane of a circular metal loop and passes through the center of the loop (see the drawing). The loop is allowed to fall and maintains its orientation with respect to the straight wire while doing so. In what direction does the current induced in the loop flow? **(a)** There is no induced current. **(b)** It is flowing around the loop from A to B to C to A. **(c)** It is flowing around the loop from C to B to A to C.

QUESTION 4

Section 22.5 Lenz's Law

5. The drawing shows a top view of two circular coils of conducting wire lying on a flat surface. The centers of the coils coincide. In the larger coil there are a switch and a battery. The smaller coil contains no switch and no battery. Describe the induced current that appears in the smaller coil when the switch in the larger coil is closed. **(a)** It flows counterclockwise forever after the switch is closed. **(b)** It flows clockwise forever after the switch is closed. **(c)** It flows counterclockwise, but only for a short period just after the switch is closed. **(d)** It flows clockwise, but only for a short period just after the switch is closed.

Switch

QUESTION 5

Section 22.7 The Electric Generator

6. You have a fixed length of conducting wire. From it you can construct a single-turn flat coil that has the shape of a square, a circle, or a rectangle with the long side twice the length of the short side. Each can be used with the same magnetic field to produce a generator that operates at the same frequency. Rank the peak emfs $\mathscr{E}_0$ of the three generators in descending order (largest first). **(a)** $\mathscr{E}_{0, \text{square}}$, $\mathscr{E}_{0, \text{circle}}$, $\mathscr{E}_{0, \text{rectangle}}$ **(b)** $\mathscr{E}_{0, \text{circle}}$, $\mathscr{E}_{0, \text{square}}$, $\mathscr{E}_{0, \text{rectangle}}$ **(c)** $\mathscr{E}_{0, \text{square}}$, $\mathscr{E}_{0, \text{rectangle}}$, $\mathscr{E}_{0, \text{circle}}$ **(d)** $\mathscr{E}_{0, \text{rectangle}}$, $\mathscr{E}_{0, \text{square}}$, $\mathscr{E}_{0, \text{circle}}$ **(e)** $\mathscr{E}_{0, \text{rectangle}}$, $\mathscr{E}_{0, \text{circle}}$, $\mathscr{E}_{0, \text{square}}$

7. An electric motor is plugged into a standard wall socket and is running at normal speed. Suddenly, some dirt prevents the shaft of the motor from turning quite so rapidly. What happens to the back emf of the motor, and what happens to the current that the motor draws from the wall socket? **(a)** The back emf increases, and the current drawn from the socket decreases. **(b)** The back emf increases, and the current drawn from the socket increases. **(c)** The back emf decreases, and the current drawn from the socket decreases. **(d)** The back emf decreases, and the current drawn from the socket increases.

Section 22.8 Mutual Inductance and Self-Inductance

8. Inductor 1 stores the same amount of energy as inductor 2, although it has only one-half the inductance of inductor 2. What is the ratio I_1/I_2 of the currents in the two inductors? **(a)** 2.000 **(b)** 1.414 **(c)** 4.000 **(d)** 0.500 **(e)** 0.250

Section 22.9 Transformers

9. The primary coil of a step-up transformer is connected across the terminals of a standard wall socket, and resistor 1 with a resistance R_1 is connected across the secondary coil. The current in the resistor is then measured. Next, resistor 2 with a resistance R_2 is connected directly across the terminals of the wall socket (without the transformer). The current in this resistor is also measured and found to be the same as the current in resistor 1. How does the resistance R_2 compare to the resistance R_1? **(a)** The resistance R_2 is less than the resistance R_1. **(b)** The resistance R_2 is greater than the resistance R_1. **(c)** The resistance R_2 is the same as the resistance R_1.

Problems

Online

Additional questions are available for assignment in WileyPLUS.

SSM Student Solutions Manual	**BIO** Biomedical application	
MMH Problem-solving help	**E** Easy	
GO Guided Online Tutorial	**M** Medium	
V-HINT Video Hints	**H** Hard	
CHALK Chalkboard Videos	**WS** Worksheet	
	T Team Problem	

Section 22.2 Motional Emf

1. **E** A 0.80-m aluminum bar is held with its length parallel to the east–west direction and dropped from a bridge. Just before the bar hits the river below, its speed is 22 m/s, and the emf induced across its length is 6.5×10^{-4} V. Assuming the horizontal component of the earth's magnetic field at the location of the bar points directly north, **(a)** determine the magnitude of the horizontal component of the earth's magnetic field, and **(b)** state whether the east end or the west end of the bar is positive.

2. **E** Near San Francisco, where the vertically downward component of the earth's magnetic field is 4.8×10^{-5} T, a car is traveling forward at 25 m/s. The width of the car is 2.0 m. **(a)** Find the emf induced between the two sides of the car. **(b)** Which side of the car is positive—the driver's side or the passenger's side?

3. **E** In 1996, NASA performed an experiment called the Tethered Satellite experiment. In this experiment a 2.0×10^4-m length of wire was let out by the space shuttle *Atlantis* to generate a motional emf. The shuttle had an orbital speed of 7.6×10^3 m/s, and the magnitude of the earth's magnetic field at the location of the wire was 5.1×10^{-5} T. If the wire had moved perpendicular to the earth's magnetic field, what would have been the motional emf generated between the ends of the wire?

4. **E** **SSM** The drawing shows three identical rods (A, B, and C) moving in different planes. A constant magnetic field of magnitude 0.45 T is directed along the $+y$ axis. The length of each rod is $L = 1.3$ m, and the rods each have the same speed, $v_A = v_B = v_C = 2.7$ m/s. For each rod, find the magnitude of the motional emf, and indicate which end (1 or 2) of the rod is positive.

PROBLEM 4

5. **E** **GO** Two circuits contain an emf produced by a moving metal rod, like that shown in **Animated Figure 22.4b**. The speed of the rod is the same in each circuit, but the bulb in circuit 1 has one-half the resistance of the bulb in circuit 2. The circuits are otherwise identical. The resistance of the light bulb in circuit 1 is 55 Ω, and that in circuit 2 is 110 Ω. Determine **(a)** the ratio $\mathscr{E}_1/\mathscr{E}_2$ of the emfs and **(b)** the ratio I_1/I_2 of the currents in the circuits. **(c)** If the speed of the rod in circuit 1 were twice that in circuit 2, what would be the ratio P_1/P_2 of the powers in the circuits?

6. **M** **GO** Refer to the drawing (**CYU Figure 22.3**) that accompanies Check Your Understanding Question 14. Suppose that the voltage of the battery in the circuit is 3.0 V, the magnitude of the magnetic field (directed perpendicularly into the plane of the paper) is 0.60 T, and the length of the rod between the rails is 0.20 m. Assuming that the rails are very long and have negligible resistance, find the maximum speed attained by the rod after the switch is closed.

7. **M** **V-HINT** Multiple-Concept Example 2 discusses the concepts that are used in this problem. Suppose that the magnetic field in **Figure 22.5** has a magnitude of 1.2 T, the rod has a length of 0.90 m, and the hand keeps the rod moving to the right at a constant speed of 3.5 m/s. If the current in the circuit is 0.040 A, what is the average power being delivered to the circuit by the hand?

8. **M** **CHALK** **SSM** Suppose that the light bulb in **Animated Figure 22.4b** is a 60.0-W bulb with a resistance of 240 Ω. The magnetic field has a magnitude of 0.40 T, and the length of the rod is 0.60 m. The only resistance in the circuit is that due to the bulb. What is the shortest distance along the rails that the rod would have to slide for the bulb to remain lit for one-half second?

9. **H** Review Conceptual Example 3 and **Figure 22.7b**. A conducting rod slides down between two frictionless vertical copper tracks at a constant speed of 4.0 m/s perpendicular to a 0.50-T magnetic field. The resistance of the rod and tracks is negligible. The rod maintains electrical contact with the tracks at all times and has a length of 1.3 m. A 0.75-Ω resistor is attached between the tops of the tracks. **(a)** What is the mass of the rod? **(b)** Find the change in the gravitational potential energy that occurs in a time of 0.20 s. **(c)** Find the electrical energy dissipated in the resistor in 0.20 s.

Section 22.3 Magnetic Flux

For problems in this set, assume that the magnetic flux is a positive quantity.

10. **E** **SSM** The drawing shows two surfaces that have the same area. A uniform magnetic field $\vec{B}$ fills the space occupied by these surfaces, and it is oriented parallel to the yz plane as shown. Find the ratio Φ_{xz}/Φ_{xy} of the magnetic fluxes that pass through the surfaces.

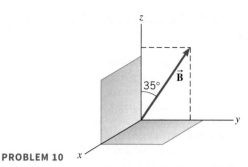

PROBLEM 10

11. **E** Two flat surfaces are exposed to a uniform, horizontal magnetic field of magnitude 0.47 T. When viewed edge-on, the first surface is tilted at an angle of 12° from the horizontal, and a net magnetic flux of 8.4×10^{-3} Wb passes through it. The same net magnetic flux passes through the second surface. **(a)** Determine the area of the first surface. **(b)** Find the smallest possible value for the area of the second surface.

12. **E** **SSM** A standard door into a house rotates about a vertical axis through one side, as defined by the door's hinges. A uniform magnetic field is parallel to the ground and perpendicular to this axis. Through what angle must the door rotate so that the magnetic flux that passes through it decreases from its maximum value to one-third of its maximum value?

13. **E** **GO** A loop of wire has the shape shown in the drawing. The top part of the wire is bent into a semicircle of radius $r = 0.20$ m. The normal to the plane of the loop is parallel to a constant magnetic field ($\phi = 0°$) of magnitude 0.75 T. What is the change $\Delta\Phi$ in the magnetic flux that passes through the loop when, starting with the position shown in the drawing, the semicircle is rotated through half a revolution?

PROBLEM 13

14. **E** A magnetic field has a magnitude of 0.078 T and is uniform over a circular surface whose radius is 0.10 m. The field is oriented at an angle of $\phi = 25°$ with respect to the normal to the surface. What is the magnetic flux through the surface?

15. **M** **GO** **CHALK** A square loop of wire consisting of a single turn is perpendicular to a uniform magnetic field. The square loop is then re-formed into a circular loop, which also consists of a single turn and is also perpendicular to the same magnetic field. The magnetic flux that passes through the square loop is 7.0×10^{-3} Wb. What is the flux that passes through the circular loop?

16. **M** **V-HINT** A five-sided object, whose dimensions are shown in the drawing, is placed in a uniform magnetic field. The magnetic field has a magnitude of 0.25 T and points along the positive y direction. Determine the magnetic flux through each of the five sides.

PROBLEM 16

Section 22.4 Faraday's Law of Electromagnetic Induction

17. **E** **GO** A magnetic field passes through a stationary wire loop, and its magnitude changes in time according to the graph in the drawing. The direction of the field remains constant, however. There are three equal time intervals indicated in the graph: 0–3.0 s, 3.0–6.0 s, and 6.0–9.0 s. The loop consists of 50 turns of wire and has an area of 0.15 m². The magnetic field is oriented parallel to the normal to the loop. For purposes of this problem, this means that $\phi = 0°$ in Equation 22.2. **(a)** For each interval, determine the induced emf. **(b)** The wire has a resistance of 0.50 Ω. Determine the induced current for the first and third intervals.

PROBLEM 17

18. **E** A rectangular loop of wire with sides 0.20 and 0.35 m lies in a plane perpendicular to a constant magnetic field (see part a of the drawing). The magnetic field has a magnitude of 0.65 T and is directed parallel to the normal of the loop's surface. In a time of 0.18 s, one-half of the loop is then folded back onto the other half, as indicated in part b of the drawing. Determine the magnitude of the average emf induced in the loop.

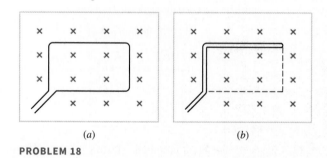

(a)   (b)

PROBLEM 18

19. **E** **SSM** A circular coil (950 turns, radius = 0.060 m) is rotating in a uniform magnetic field. At $t = 0$ s, the normal to the coil is perpendicular to the magnetic field. At $t = 0.010$ s, the normal makes an angle of $\phi = 45°$ with the field because the coil has made one-eighth of a revolution. An average emf of magnitude 0.065 V is induced in the coil. Find the magnitude of the magnetic field at the location of the coil.

20. **E** The magnetic flux that passes through one turn of a 12-turn coil of wire changes to 4.0 from 9.0 Wb in a time of 0.050 s. The average induced current in the coil is 230 A. What is the resistance of the wire?

21. **E** **MMH** A constant magnetic field passes through a single rectangular loop whose dimensions are 0.35 m × 0.55 m. The magnetic field has a magnitude of 2.1 T and is inclined at an angle of 65° with respect to the normal to the plane of the loop. **(a)** If the magnetic field decreases to zero in a time of 0.45 s, what is the magnitude of the average emf induced in the loop? **(b)** If the magnetic field remains constant at its initial value of 2.1 T, what is the magnitude of the rate $\Delta A/\Delta t$ at which the area should change so that the average emf has the same magnitude as in part (a)?

22. **E** A uniform magnetic field is perpendicular to the plane of a single-turn circular coil. The magnitude of the field is changing, so that an emf of 0.80 V and a current of 3.2 A are induced in the coil. The wire is then re-formed into a single-turn square coil, which is used in the same magnetic field (again perpendicular to the plane of

the coil and with a magnitude changing at the same rate). What emf and current are induced in the square coil?

23. **M** **MMH** A copper rod is sliding on two conducting rails that make an angle of 19° with respect to each other, as in the drawing. The rod is moving to the right with a constant speed of 0.60 m/s. A 0.38-T uniform magnetic field is perpendicular to the plane of the paper. Determine the magnitude of the average emf induced in the triangle ABC during the 6.0-s period after the rod has passed point A.

PROBLEM 23

24. **M** **CHALK** **GO** A flat coil of wire has an area A, N turns, and a resistance R. It is situated in a magnetic field, such that the normal to the coil is parallel to the magnetic field. The coil is then rotated through an angle of 90°, so that the normal becomes perpendicular to the magnetic field. The coil has an area of 1.5×10^{-3} m², 50 turns, and a resistance of 140 Ω. During the time while it is rotating, a charge of 8.5×10^{-5} C flows in the coil. What is the magnitude of the magnetic field?

25. **E** **SSM** A magnetic field is passing through a loop of wire whose area is 0.018 m². The direction of the magnetic field is parallel to the normal to the loop, and the magnitude of the field is increasing at the rate of 0.20 T/s. **(a)** Determine the magnitude of the emf induced in the loop. **(b)** Suppose that the area of the loop can be enlarged or shrunk. If the magnetic field is increasing as in part (a), at what rate (in m²/s) should the area be changed at the instant when $B = 1.8$ T if the induced emf is to be zero? Explain whether the area is to be enlarged or shrunk.

26. **M** **V-HINT** A flat circular coil with 105 turns, a radius of 4.00×10^{-2} m, and a resistance of 0.480 Ω is exposed to an external magnetic field that is directed perpendicular to the plane of the coil. The magnitude of the external magnetic field is changing at a rate of $\Delta B/\Delta t = 0.783$ T/s, thereby inducing a current in the coil. Find the magnitude of the magnetic field at the center of the coil that is produced by the induced current.

27. **M** **GO** The drawing shows a coil of copper wire that consists of two semicircles joined by straight sections of wire. In part a the coil is lying flat on a horizontal surface. The dashed line also lies in the plane of the horizontal surface. Starting from the orientation in part a, the smaller semicircle rotates at an angular frequency ω about the dashed line, until its plane becomes perpendicular to the horizontal surface, as shown in part b. A uniform magnetic field is constant in time and is directed upward, perpendicular to the horizontal surface. The field completely fills the region occupied by the coil in either part of the drawing. The magnitude of the magnetic field is 0.35 T. The resistance of the coil is 0.025 Ω, and the smaller semicircle has a radius of 0.20 m. The angular frequency at which the small semicircle rotates is 1.5 rad/s. Determine the average current I, if any, induced in the coil as the coil changes shape from that in part a of the drawing to that in part b. Be sure to include an explicit plus or minus sign along with your answer.

(a)   (b)

PROBLEM 27

28. **M** **GO** A conducting coil of 1850 turns is connected to a galvanometer, and the total resistance of the circuit is 45.0 Ω. The area of each turn is 4.70×10^{-4} m². This coil is moved from a region where the magnetic field is zero into a region where it is nonzero, the normal to the coil being kept parallel to the magnetic field. The amount of charge that is induced to flow around the circuit is measured to be 8.87×10^{-3} C. Find the magnitude of the magnetic field.

Section 22.5   Lenz's Law

29. **E** Starting from the position indicated in the drawing, the semicircular piece of wire rotates through half a revolution in the direction shown. Which end of the resistor is positive—the left or the right end? Explain your reasoning.

B (into paper)

PROBLEM 29

30. **E** **SSM** The plane of a flat, circular loop of wire is horizontal. An external magnetic field is directed perpendicular to the plane of the loop. The magnitude of the external magnetic field is increasing with time. Because of this increasing magnetic field, an induced current is flowing clockwise in the loop, as viewed from above. What is the direction of the external magnetic field? Justify your conclusion.

31. **E** **GO** The drawing shows a straight wire carrying a current I. Above the wire is a rectangular loop that contains a resistor R. If the current I is decreasing in time, what is the direction of the induced current through the resistor R—left-to-right or right-to-left?

R

I

PROBLEM 31

32. **E** **SSM** The drawing depicts a copper loop lying flat on a table (not shown) and connected to a battery via a closed switch. The current I in the loop generates the magnetic field lines shown in the drawing. The switch is then opened and the current goes to zero. There are also two smaller conducting loops A and B lying flat on the table, but not connected to batteries. Determine the direction of the induced current in **(a)** loop A and **(b)** loop B. Specify the direction of each induced current to be clockwise or counterclockwise when viewed from above the table. Provide a reason for each answer.

Magnetic field lines

Switch

Copper loop

PROBLEM 32

33. **E** The drawing shows that a uniform magnetic field is directed perpendicularly into the plane of the paper and fills the entire region to the left of the y axis. There is no magnetic field to the right of the y axis. A rigid right triangle ABC is made of copper wire. The triangle rotates counterclockwise about the origin at point C. What is the direction (clockwise or counterclockwise) of the induced current when the triangle is crossing **(a)** the $+y$ axis, **(b)** the $-x$ axis, **(c)** the $-y$ axis, and **(d)** the $+x$ axis? For each case, justify your answer.

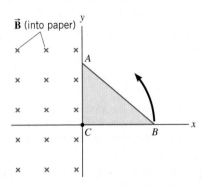

PROBLEM 33

34. **M** **MMH** A circular loop of wire rests on a table. A long, straight wire lies on this loop, directly over its center, as the drawing illustrates. The current I in the straight wire is decreasing. In what direction is the induced current, if any, in the loop? Give your reasoning.

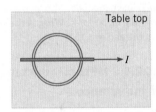

Table top

I

PROBLEM 34

35. **M** **V-HINT** The drawing shows a bar magnet falling through a metal ring. In part a the ring is solid all the way around, but in part b it has been cut through. **(a)** Explain why the motion of the magnet in part a is retarded when the magnet is above the ring and below the ring as well. Draw any induced currents that appear in the ring. **(b)** Explain why the motion of the magnet is unaffected by the ring in part b.

PROBLEM 35   (a)   (b)

Section 22.7   The Electric Generator

36. **E** A 120.0-V motor draws a current of 7.00 A when running at normal speed. The resistance of the armature wire is 0.720 Ω. **(a)** Determine the back emf generated by the motor. **(b)** What is the current at the instant when the motor is just turned on and has not begun to rotate? **(c)** What series resistance must be added to limit the starting current to 15.0 A?

37. **E** **SSM** A generator has a square coil consisting of 248 turns. The coil rotates at 79.1 rad/s in a 0.170-T magnetic field. The peak output of the generator is 75.0 V. What is the length of one side of the coil?

38. **E** You need to design a 60.0-Hz ac generator that has a maximum emf of 5500 V. The generator is to contain a 150-turn coil that has an area per turn of 0.85 m². What should be the magnitude of the magnetic field in which the coil rotates?

39. **E** **MMH** The maximum strength of the earth's magnetic field is about 6.9×10^{-5} T near the south magnetic pole. In principle, this field could be used with a rotating coil to generate 60.0-Hz ac electricity. What is the minimum number of turns (area per turn = 0.022 m²) that the coil must have to produce an rms voltage of 120 V?

40. **M** **V-HINT** A generator uses a coil that has 100 turns and a 0.50-T magnetic field. The frequency of this generator is 60.0 Hz, and its emf has an rms value of 120 V. Assuming that each turn of the coil is a square (an approximation), determine the length of the wire from which the coil is made.

41. **M** **GO** The coil of a generator has a radius of 0.14 m. When this coil is unwound, the wire from which it is made has a length of 5.7 m. The magnetic field of the generator is 0.20 T, and the coil rotates at an angular speed of 25 rad/s. What is the peak emf of this generator?

42. **M** **SSM** Consult Multiple-Concept Example 12 for background material relating to this problem. A small rubber wheel on the shaft of a bicycle generator presses against the bike tire and turns the coil of the generator at an angular speed that is 38 times as great as the angular speed of the tire itself. Each tire has a radius of 0.300 m. The coil consists of 125 turns, has an area of 3.86×10^{-3} m², and rotates in a 0.0900-T magnetic field. The bicycle starts from rest and has an acceleration of +0.550 m/s². What is the peak emf produced by the generator at the end of 5.10 s?

Section 22.8 Mutual Inductance and Self-Inductance

43. **E** **SSM** The earth's magnetic field, like any magnetic field, stores energy. The maximum strength of the earth's field is about 7.0×10^{-5} T. Find the maximum magnetic energy stored in the space above a city if the space occupies an area of 5.0×10^8 m² and has a height of 1500 m.

44. **E** The current through a 3.2-mH inductor varies with time according to the graph shown in the drawing. What is the average induced emf during the time intervals **(a)** 0–2.0 ms, **(b)** 2.0–5.0 ms, and **(c)** 5.0–9.0 ms?

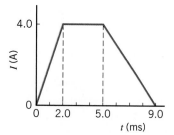

PROBLEM 44

45. **E** Two coils of wire are placed close together. Initially, a current of 2.5 A exists in one of the coils, but there is no current in the other. The current is then switched off in a time of 3.7×10^{-2} s. During this time, the average emf induced in the other coil is 1.7 V. What is the mutual inductance of the two-coil system?

46. **E** **GO** During a 72-ms interval, a change in the current in a primary coil occurs. This change leads to the appearance of a 6.0-mA current in a nearby secondary coil. The secondary coil is part of a circuit in which the resistance is 12 Ω. The mutual inductance between the two coils is 3.2 mH. What is the change in the primary current?

47. **E** Mutual induction can be used as the basis for a metal detector. A typical setup uses two large coils that are parallel to each other and

have a common axis. Because of mutual induction, the ac generator connected to the primary coil causes an emf of 0.46 V to be induced in the secondary coil. When someone without metal objects walks through the coils, the mutual inductance and, thus, the induced emf do not change much. But when a person carrying a handgun walks through, the mutual inductance increases. The change in emf can be used to trigger an alarm. If the mutual inductance increases by a factor of three, find the new value of the induced emf.

48. **E** **GO** A constant current of $I = 15$ exists in a solenoid whose inductance is $L = 3.1$ H. The current is then reduced to zero in a certain amount of time. **(a)** If the current goes from 15 to 0 A in a time of 75 ms, what is the emf induced in the solenoid? **(b)** How much electrical energy is stored in the solenoid? **(c)** At what rate must the electrical energy be removed from the solenoid when the current is reduced to 0 A in a time of 75 ms? Note that the rate at which energy is removed is the power.

49. **E** **SSM** Suppose you wish to make a solenoid whose self-inductance is 1.4 mH. The inductor is to have a cross-sectional area of 1.2×10^{-3} m² and a length of 0.052 m. How many turns of wire are needed?

50. **M** **GO** **CHALK** A long, current-carrying solenoid with an air core has 1750 turns per meter of length and a radius of 0.0180 m. A coil of 125 turns is wrapped tightly around the outside of the solenoid, so it has virtually the same radius as the solenoid. What is the mutual inductance of this system?

Section 22.9 Transformers

51. **E** The battery charger for an MP3 player contains a step-down transformer with a turns ratio of 1:32, so that the voltage of 120 V available at a wall socket can be used to charge the battery pack or operate the player. What voltage does the secondary coil of the transformer provide?

52. **E** **MMH** The secondary coil of a step-up transformer provides the voltage that operates an electrostatic air filter. The turns ratio of the transformer is 50:1. The primary coil is plugged into a standard 120-V outlet. The current in the secondary coil is 1.7×10^{-3} A. Find the power consumed by the air filter.

53. **E** **GO** The rechargeable batteries for a laptop computer need a much smaller voltage than what a wall socket provides. Therefore, a transformer is plugged into the wall socket and produces the necessary voltage for charging the batteries. The batteries are rated at 9.0 V, and a current of 225 mA is used to charge them. The wall socket provides a voltage of 120 V. **(a)** Determine the turns ratio of the transformer. **(b)** What is the current coming from the wall socket? **(c)** Find the average power delivered by the wall socket and the average power sent to the batteries.

54. **E** The resistances of the primary and secondary coils of a transformer are 56 and 14 Ω, respectively. Both coils are made from lengths of the same copper wire. The circular turns of each coil have the same diameter. Find the turns ratio N_s/N_p.

55. **E** A transformer consisting of two coils wrapped around an iron core is connected to a generator and a resistor, as shown in the drawing. There are 11 turns in the primary coil and 18 turns in the secondary coil. The peak voltage across the resistor is 67 V. What is the peak emf of the generator?

PROBLEM 55

56. E SSM A step-down transformer (turns ratio = 1:8) is used with an electric train to reduce the voltage from the wall receptacle to a value needed to operate the train. When the train is running, the current in the secondary coil is 1.6 A. What is the current in the primary coil?

57. M GO In a television set the power needed to operate the picture tube comes from the secondary of a transformer. The primary of the transformer is connected to a 120-V receptacle on a wall. The picture tube of the television set uses 91 W, and there is 5.5 mA of current in the secondary coil of the transformer to which the tube is connected. Find the turns ratio N_s/N_p of the transformer.

58. M SSM CHALK A generating station is producing 1.2×10^6 W of power that is to be sent to a small town located 7.0 km away. Each of the two wires that comprise the transmission line has a resistance per kilometer of 5.0×10^{-2} Ω/km. **(a)** Find the power used to heat the wires if the power is transmitted at 1200 V. **(b)** A 100:1 step-up transformer is used to raise the voltage before the power is transmitted. How much power is now used to heat the wires?

59. H A generator is connected across the primary coil (N_p turns) of a transformer, while a resistance R_2 is connected across the secondary coil (N_s turns). This circuit is equivalent to a circuit in which a single resistance R_1 is connected directly across the generator, without the transformer. Show that $R_1 = (N_p/N_s)^2 R_2$, by starting with Ohm's law as applied to the secondary coil.

Additional Problems

Online

60. E In each of two coils the rate of change of the magnetic flux in a single loop is the same. The emf induced in coil 1, which has 184 loops, is 2.82 V. The emf induced in coil 2 is 4.23 V. How many loops does coil 2 have?

61. E GO A planar coil of wire has a single turn. The normal to this coil is parallel to a uniform and constant (in time) magnetic field of 1.7 T. An emf that has a magnitude of 2.6 V is induced in this coil because the coil's area A is shrinking. What is the magnitude of $\Delta A/\Delta t$, which is the rate (in m^2/s) at which the area changes?

62. E CHALK SSM Review Conceptual Example 9 as an aid in understanding this problem. A long, straight wire lies on a table and carries a current I. As the drawing shows, a small circular loop of wire is pushed across the top of the table from position 1 to position 2. Determine the direction of the induced current, clockwise or counterclockwise, as the loop moves past **(a)** position 1 and **(b)** position 2. Justify your answers.

PROBLEM 62

63. E In some places, insect "zappers," with their blue lights, are a familiar sight on a summer's night. These devices use a high voltage to electrocute insects. One such device uses an ac voltage of 4320 V, which is obtained from a standard 120.0-V outlet by means of a transformer. If the primary coil has 21 turns, how many turns are in the secondary coil?

64. M GO At its normal operating speed, an electric fan motor draws only 15.0% of the current it draws when it just begins to turn the fan blade. The fan is plugged into a 120.0-V socket. What back emf does the motor generate at its normal operating speed?

65. M V-HINT Parts a and b of the drawing show the same uniform and constant (in time) magnetic field $\vec{B}$ directed perpendicularly into the paper over a rectangular region. Outside this region, there is no field. Also shown is a rectangular coil (one turn), which lies in the plane of the paper. In part a the long side of the coil (length = L) is just at the edge of the field region, while in part b the short side (width = W) is just at the edge. It is known that $L/W = 3.0$. In both parts of the drawing the coil is pushed into the field with the same velocity $\vec{v}$ until it is completely within the field region. The magnitude of the average emf induced in the coil in part a is 0.15 V. What is its magnitude in part b?

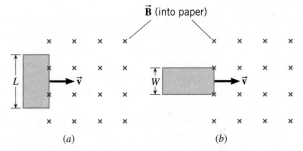

PROBLEM 65

66. M V-HINT Indicate the direction of the electric field between the plates of the parallel plate capacitor shown in the drawing if the magnetic field is decreasing in time. Give your reasoning.

PROBLEM 66

67. M SSM A piece of copper wire is formed into a single circular loop of radius 12 cm. A magnetic field is oriented parallel to the normal to the loop, and it increases from 0 to 0.60 T in a time of 0.45 s. The wire has a resistance per unit length of 3.3×10^{-2} Ω/m. What is the average electrical energy dissipated in the resistance of the wire?

68. M GO SSM A long solenoid of length 8.0×10^{-2} m and cross-sectional area 5.0×10^{-5} m^2 contains 6500 turns per meter of length. Determine the emf induced in the solenoid when the current in the solenoid changes from 0 to 1.5 A during the time interval from 0 to 0.20 s.

Physics in Biology, Medicine, and Sports

69. **E** **BIO** **22.2** The drawing shows a type of flow meter that can be used to measure the speed of blood in situations when a blood vessel is sufficiently exposed (e.g., during surgery). Blood is conductive enough that it can be treated as a moving conductor. When it flows perpendicularly with respect to a magnetic field, as in the drawing, electrodes can be used to measure the small voltage that develops across the vessel. Suppose that the speed of the blood is 0.30 m/s and the diameter of the vessel is 5.6 mm. In a 0.60-T magnetic field what is the magnitude of the voltage that is measured with the electrodes in the drawing?

PROBLEM 69

70. **E** **BIO** **22.4** Magnetic resonance imaging (MRI) is a medical technique for producing pictures of the interior of the body. The patient is placed within a strong magnetic field. One safety concern is what would happen to the positively and negatively charged particles in the body fluids if an equipment failure caused the magnetic field to be shut off suddenly. An induced emf could cause these particles to flow, producing an electric current within the body. Suppose the largest surface of the body through which flux passes has an area of 0.032 m^2 and a normal that is parallel to a magnetic field of 1.5 T. Determine the smallest time period during which the field can be allowed to vanish if the magnitude of the average induced emf is to be kept less than 0.010 V.

71. **M** **BIO** **22.4** A typical MRI machine creates a constant magnetic field of 3.0 T. However, some patients receiving a CAT scan on their brain can be exposed to changing magnetic fields, either by ramping the field up or down, or by moving their head quickly while in the field. This creates a change in magnetic flux through the head, which can cause the patient to experience *magnetophosphenes*. These are flashes of light the patient sees in their visual field. The change in magnetic flux induces currents to flow in the retina or visual cortex, which results in the illusion of light. Consider a patient in an MRI machine at a field of 3.0 T. The field is switched off and suddenly drops to zero in 4.0×10^{-4} s, creating a single loop of current enclosing an area of the brain that is 5.0×10^{-3} m^2. What is the magnitude of the induced emf in the brain? Assume the changing magnetic field passes perpendicularly through the surface of the current loop.

72. **M** **BIO** **22.4** Many athletes and cool tech lovers wear a smartwatch, like the Apple Watch, to monitor their fitness and sleep patterns, receive emails, and make phone calls. The watch has an internal battery that is charged wirelessly by induction. The watch is placed faceup onto a charging base (see the photo). There is a charging coil (the primary) in the base and a receiving coil (the secondary) just inside the back faceplate of the watch. An alternating current is sent through the base coil, which creates a changing magnetic field. This produces a change in magnetic flux through

the watch's coil, which induces an emf in that coil, thereby charging the watch's battery. Assume the secondary coil has 10 turns with an average area of 1.77 cm^2. If the emf induced in the coil is 3.80 V, what is the rate of change of the magnetic field within the charging coil?

Receiving coil at back of watch

Charging coil in base

Photo courtesy of David Young

PROBLEM 72

73. **M** **BIO** **22.4** Consider again the hearing aid equipped with the T-coil, as described in Example 10. When the loop system in the room is initially turned on, a surge of current around the loop induces an emf in the T-coil of 7.2×10^{-6} V. What is the value of the peak magnetic field in the room, if it is reached after 6.5 ms?

74. **E** **BIO** **22.9** Tesla Motors manufactures fully electric vehicles. They have gained popularity over the last several years, primarily due to the Model 3 (see the photo), and electric cars will play an essential role in the fight against global climate change. The long-range model of this car is powered by a 350-V lithium-ion battery pack with an energy of 80.5 kWh. When connected to an at-home charging station to recharge the battery pack, the transformer inside the charger only draws 80% of the current allowed by the circuit breaker that protects the circuit that powers the charger. If the homeowner is using a charger that is connected to a 220-V outlet with a 40-A circuit breaker, what is the average electrical power delivered to the Model 3 during charging?

Jteder/Pixabay

PROBLEM 74

75. [M] [BIO] **22.2** As a prelude to this problem, you should review Problem 75 in Chapter 21. Some researchers have noticed that cows and wild deer will align themselves in a north–south direction while resting or grazing (see the photo). This suggests they may sense the direction of the earth's magnetic field and align with it. One possible explanation for this ability is to coordinate movements over a large herd. For example, roe deer, when startled, tend to run north to south. However, if they are close enough to high voltage power lines, they do not exhibit this behavior. **(a)** How close would a deer have to be to a power line carrying 250 A, so that the magnetic field created by the line was equal to the earth's magnetic field (0.45 gauss)? **(b)** What would be the magnetic flux passing through the deer's body at this position? Assume the magnetic field passes perpendicularly through a 0.70-m^2 cross section of the deer.

PROBLEM 75

Olavfin, Image taken from https://commons.wikimedia.org/wiki/File:R%C3%A5dyr_1.jpg

Concepts and Calculations Problems   Online

In this chapter we have seen that there are three ways to create an induced emf in a coil: by changing the magnitude of a magnetic field, by changing the direction of the field relative to the coil, and by changing the area of the coil. Problem 76 explores the third method and, in the process, provides a review of Faraday's law of electromagnetic induction. Problem 77 explores the characteristics of the emf produced in a generator that utilizes a coil rotating in a fixed magnetic field.

76. [M] [CHALK] A circular coil of radius 0.11 m contains a single turn and is located in a constant magnetic field of magnitude 0.27 T. The magnetic field has the same direction as the normal to the plane of the coil. The radius increases to 0.30 m in a time of 0.080 s. *Concepts:* (i) Why is there an emf induced in the coil? (ii) Does the magnitude of the induced emf depend on whether the area is increasing or decreasing? Explain. (iii) What determines the amount of current induced in the coil? (iv) If the coil is cut so it is no longer one continuous piece, are there an induced emf and an induced current? Explain. *Calculations:* **(a)** Determine the magnitude of the emf induced in the coil. **(b)** The coil has a resistance of 0.70 Ω. Find the magnitude of the induced current.

77. [M] [CHALK] [SSM] The graph in the figure shows the emf produced by a generator as a function of time *t*. The coil for the generator has an area of $A = 0.15$ m^2 and consists of $N = 10$ turns. The coil rotates in a field of magnitude 0.27 T. *Concepts:* (i) Can the period of the rotating coil be determined from the graph? (ii) The emf produced by a generator depends on its angular frequency. How is the angular frequency of the coil related to its period? (iii) Starting at $t = 0$ s, how much time is required for the generator to produce its peak emf? Express the answer in terms of the period *T* of the motion (e.g., $t = (0.1) T$). (iv) How often does the polarity of the emf change in one cycle? *Calculations:* **(a)** Determine the period of the motion. **(b)** What is the angular frequency of the rotating coil? **(c)** Find the value of the emf when $t = \frac{1}{4}T$, where *T* denotes the period of the coil motion. **(d)** What is the emf when $t = 0.025$ s?

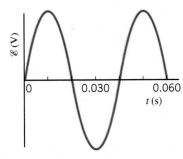

PROBLEM 77

Team Problems   Online

78. [M] [T] [WS] **A Magnetic Fan Light.** You and your team are designing a device in which a square loop of wire of side length $L = 3.30$ cm is glued to the end of a fan blade. Into one of the sides of the wire loop you insert a light-emitting diode (LED) so that whatever emf is induced in the loop appears across the LED. The idea is to orient the fan so that the end of the fan blade (and the square circuit) passes between the poles of a magnet, and therefore through a region of uniform, horizontally oriented magnetic field that has a square cross section of side length 5.00 cm. You align the fan so that the loop area is perpendicular to the field and so that the leading edge of the square loop is nearly horizontal when it enters the region of magnetic field. In order for the LED to turn on, an emf of at least 1.80 V must be induced in the loop. **(a)** If the center of the loop is 30.0 cm from the rotation axis of the fan, and the rate of rotation of the fan is 1.00×10^3 rpm, estimate the minimum magnitude of the magnetic field needed to turn on the LED. **(b)** For how long is the LED illuminated as it passes into the field region, that is, for what period of time is the required emf maintained? Note that LEDs illuminate only when the emf across them has the correct polarity (i.e., they will not illuminate with the opposite polarity). You can assume that the required emf polarity is satisfied as the loop enters the field.

79. [E] [T] [WS] **Passing by a Magnetar.** A type of neutron star called a *magnetar* is believed to have an extremely large magnetic field on the order of 10^{11} T, about a hundred million times greater than the strongest magnetic fields humans have been able to produce on earth. You and your team have been assigned to evaluate some of the risks that a probe might encounter while passing by such an object, especially regarding its electronics. Suppose that the probe passes through a region of space near a magnetar where the magnetic field is nearly uniform and has a constant magnitude of 275 T. The translational speed of the craft is 65.0 km/s relative to the star, and rotates about its axis at a rate of 1.75 rev/s. **(a)** The probe has a long, straight antenna (i.e., a conductor) that is 2.80 m in length. Neglecting the rotation of the probe, what is the maximum possible induced emf between the two ends of the antenna as a result of it moving through the magnetic field? **(b)** What is the maximum emf induced in an electrical circuit loop of area 1.33×10^{-4} m^2 due to the rotation of the craft? **(c)** There are other conducting "loops" that are a part of the structure of the craft, and not a part of the functional electronics. What is the maximum emf induced in a loop of area 1.50 m^2 due to the spinning of the probe? **(d)** Based on the results of (a) through (c), should the probe take a more distant path around the magnetar? *Note:* any induced emf that exceeds 1000 V is problematic for structures, and anything above a few volts will damage sensitive circuits.

80. Ⓜ Ⓣ **Reconfiguring a Transformer.** You and your team are exploring an abandoned science facility on the coast of western Antarctica when a large storm hits, and it is clear that you will be stranded there for a few days. You and the others search for supplies and find a generator and a tank of fuel. Having electrical power would allow you to keep your communication devices operational and make your stay more comfortable. A team member gets the generator running, but there is a complication: The output of the generator is 50 Hz at 440 VAC (rms), and your devices require 60 Hz, 110 VAC (rms). The electrical power in many European countries runs on 240 V at 50 Hz, so a few of your team members have converters. However, 440 V is still too high to use them. You search and eventually find a large transformer that, according to a worn tag on its case, is designed to step down from 5000 V to 880 V. The tag also indicates that its primary coil has 1500 turns, but you cannot read the number of turns in the secondary coil. **(a)** How many turns should its secondary coil have? **(b)** It will be a difficult job, but you can change the number of primary turns by cutting some of them out. How many turns should you leave on the primary coil so that, with the primary connected to 440 V, the secondary outputs 240 V (so that you can use the 240 V to 110 V converters)? **(c)** You find that the current at the source (i.e., that connected to the primary) is limited to a maximum of 20.0 A. What is the maximum current limit through the secondary coil? **(d)** What is the maximum average power available at the secondary coil?

81. Ⓜ Ⓣ **A Generator Bike.** You and your team are designing a generator using a stationary bike to rotate a coil in a uniform magnetic field. The gearing is set up so that the coil rotates 60 times for one complete rotation of the bike pedals. Therefore, one revolution of the pedals per second results in a 60-Hz alternating current in the coil. The circular coil has 350 turns and a diameter of 15.0 cm, and its axis of rotation is along its diameter. **(a)** If a uniform magnetic field is oriented perpendicular to the coil's axis of rotation and has a magnitude of $B = 0.225$ T, what is the peak emf produced by the generator bike? **(b)** What is the rms emf? **(c)** To what magnitude should you reduce the field if you want the rms emf to be 110 VAC? **(d)** Instead of reducing the field, you could use a step-down transformer to reduce the rms emf to 110 VAC. What should be the ratio of primary to secondary turns of the transformer coils?

Thousands of lights dot the spectacular Chicago skyline at night. Lighting control, and the delivery of electrical energy to cities in general, would not be possible without the aid of alternating current (ac) circuits. Ac circuits lie at the heart of most modern technologies, and we will study them in this chapter.

Gian Lorenzo Ferretti Photography/
E+/Getty Images

Alternating Current Circuits

23.1 Capacitors and Capacitive Reactance

Our experience with capacitors so far has been in dc circuits. As we have seen in Section 20.13, charge flows in a dc circuit only for the brief period after the battery voltage is applied across the capacitor. In other words, charge flows only while the capacitor is charging up. After the capacitor becomes fully charged, no more charge leaves the battery. However, suppose that the battery connections to the fully charged capacitor were suddenly reversed. Then charge would flow again, but in the reverse direction, until the battery recharges the capacitor according to the new connections. In an ac circuit what happens is similar. The polarity of the voltage applied to the capacitor continually switches back and forth, and, in response, charges flow first one way around the circuit and then the other way. This flow of charge, surging back and forth, constitutes an alternating current. Thus, charge flows continuously in an ac circuit containing a capacitor.

FIGURE 23.1 The resistance in a purely resistive circuit has the same value at all frequencies. The maximum emf of the generator is V_0.

To help set the stage for the present discussion, recall from Section 20.5 that the rms voltage V_{rms} across the resistor in a purely resistive ac circuit is related to the rms current I_{rms} by $V_{rms} = I_{rms}R$ (Equation 20.14). The resistance R has the same value for any frequency of the ac voltage or current. **Figure 23.1** emphasizes this fact by showing that a graph of resistance versus frequency is a horizontal straight line.

For the rms voltage across a capacitor the following expression applies, which is analogous to $V_{rms} = I_{rms}R$:

$$V_{rms} = I_{rms}X_C \tag{23.1}$$

The term X_C appears in place of the resistance R and is called the **capacitive reactance**. The capacitive reactance, like resistance, is measured in *ohms* and determines how much rms current exists in a capacitor in response to a given rms voltage across the capacitor. It is found experimentally that the capacitive reactance X_C is inversely proportional to both the frequency f and the capacitance C, according to the following equation:

$$X_C = \frac{1}{2\pi f C} \tag{23.2}$$

For a fixed value of the capacitance C, **Figure 23.2** gives a plot of X_C versus frequency, according to Equation 23.2. A comparison of this drawing with **Figure 23.1** reveals that a capacitor and a resistor behave differently. As the frequency becomes very large, **Figure 23.2** shows that X_C approaches zero, signifying that a capacitor offers only a negligibly small opposition to the alternating current. In contrast, in the limit of zero frequency (i.e., direct current), X_C becomes infinitely large, and a capacitor provides so much opposition to the motion of charges that there is no current.

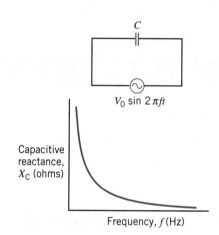

FIGURE 23.2 The capacitive reactance X_C is inversely proportional to the frequency f according to $X_C = 1/(2\pi f C)$.

Example 1 illustrates the use of Equation 23.2 and also demonstrates how frequency and capacitance determine the amount of current in an ac circuit.

EXAMPLE 1 | A Capacitor in an Ac Circuit

For the circuit in **Figure 23.2**, the capacitance of the capacitor is 1.50 μF, and the rms voltage of the generator is 25.0 V. What is the rms current in the circuit when the frequency of the generator is **(a)** 1.00×10^2 Hz and **(b)** 5.00×10^3 Hz?

Reasoning The current can be found from $I_{rms} = V_{rms}/X_C$, once the capacitive reactance X_C is determined. The values for the capacitive reactance will reflect the fact that the capacitor

provides more opposition to the current when the frequency is smaller.

Problem-Solving Insight The capacitive reactance X_C is inversely proportional to the frequency f of the voltage, so if the frequency increases by a factor of 50, as in this example, the capacitive reactance decreases by a factor of 50.

Solution   **(a)** At a frequency of 1.00×10^2 Hz, we find

$$X_C = \frac{1}{2\pi fC} = \frac{1}{2\pi(1.00 \times 10^2 \text{ Hz})(1.50 \times 10^{-6} \text{ F})} = 1060 \ \Omega \quad \textbf{(23.2)}$$

$$I_{\text{rms}} = \frac{V_{\text{rms}}}{X_C} = \frac{25.0 \text{ V}}{1060 \ \Omega} = \boxed{0.0236 \text{ A}} \qquad \textbf{(23.1)}$$

(b) When the frequency is 5.00×10^3 Hz, the calculations are similar:

$$X_C = \frac{1}{2\pi fC} = \frac{1}{2\pi(5.00 \times 10^3 \text{ Hz})(1.50 \times 10^{-6} \text{ F})} = 21.2 \ \Omega \quad \textbf{(23.2)}$$

$$I_{\text{rms}} = \frac{V_{\text{rms}}}{X_C} = \frac{25.0 \text{ V}}{21.2 \ \Omega} = \boxed{1.18 \text{ A}} \qquad \textbf{(23.1)}$$

We now consider the behavior of the instantaneous (not rms) voltage and current. For comparison, **Figure 23.3** shows graphs of voltage and current versus time in a resistive circuit. These graphs indicate that when only resistance is present, the voltage and current are proportional to each other at every moment. For example, when the voltage increases from A to B on the graph, the current follows along in step, increasing from A' to B' during the same time interval. Likewise, when the voltage decreases from B to C, the current decreases from B' to C'. For this reason, the current in a resistance R is said to be **in phase** with the voltage across the resistance.

For a capacitor, this in-phase relation between instantaneous voltage and current does *not* exist. **Interactive Figure 23.4** shows graphs of the ac voltage and current versus time for a circuit that contains only a capacitor. As the voltage increases from A to B, the charge on the capacitor increases and reaches its full value at B. The current, however, is not the same thing as the charge. The current is the rate of flow of charge and has a maximum positive value at the start of the charging process at A'. It is a maximum because there is no charge on the capacitor at the start and hence no capacitor voltage to oppose the generator voltage. When the capacitor is fully charged at B, the capacitor voltage has a magnitude equal to that of the generator and completely opposes the generator voltage. The result is that the current decreases to zero at B'. While the capacitor voltage decreases from B to C, the charges flow out of the capacitor in a direction opposite to that of the charging current, as indicated by the negative current from B' to C'. Thus, voltage and current are not in phase but are, in fact, one-quarter wave cycle out of step, or out of phase. More specifically, assuming that the voltage fluctuates as $V_0 \sin 2\pi ft$, the current varies as $I_0 \sin(2\pi ft + \pi/2) = I_0 \cos 2\pi ft$. Since $\pi/2$ radians correspond to 90° and since the current reaches its maximum value *before* the voltage does, it is said that **the current in a capacitor leads the voltage across the capacitor by a phase angle of 90°.**

The fact that the current and voltage for a capacitor are 90° out of phase has an important consequence from the point of view of electric power, since power is the product of current and voltage. For the time interval between points A and B (or A' and B') in **Interactive Figure 23.4**, both current and voltage are positive. Therefore,

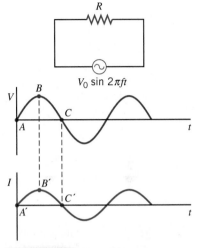

FIGURE 23.3   The instantaneous voltage V and current I in a purely resistive circuit are *in phase,* which means that they increase and decrease in step with one another.

INTERACTIVE FIGURE 23.4   In a circuit containing only a capacitor, the instantaneous voltage and current are not in phase. Instead, the current *leads* the voltage by one-quarter of a cycle or by a phase angle of 90°.

Math Skills   Why does $I_0 \sin\left(2\pi ft + \frac{\pi}{2}\right) = I_0 \cos 2\pi ft$? To see why, consider the trigonometric identity $\sin(\alpha + \beta) = \sin\alpha \cos\beta + \cos\alpha \sin\beta$ (see Appendix E.2, Other Trigonometric Identities, Equation 6). Using $\alpha = 2\pi ft$ and $\beta = \frac{\pi}{2}$ in this identity, we find that

$$I_0 \sin\left(2\pi f + \frac{\pi}{2}\right) = I_0 \sin 2\pi ft \underbrace{\cos\frac{\pi}{2}}_{0} + I_0 \cos 2\pi ft \underbrace{\sin\frac{\pi}{2}}_{1}$$

But $\frac{\pi}{2}$ radians corresponds to 90°, so that $\cos\frac{\pi}{2} = 0$ and $\sin\frac{\pi}{2} = 1$. With these two substitutions, we have

$$I_0 \sin\left(2\pi ft + \frac{\pi}{2}\right) = I_0 \cos 2\pi ft$$

FIGURE 23.5 These rotating-arrow models (or phasor models) represent the voltage and the current in ac circuits that contain (a) only a resistor and (b) only a capacitor.

the instantaneous power is also positive, meaning that the generator is delivering energy to the capacitor. However, during the period between B and C (or B' and C'), the current is negative while the voltage remains positive, and the power, which is the product of the two, is negative. During this period, the capacitor is returning energy to the generator. Thus, the power alternates between positive and negative values for equal periods of time. In other words, the capacitor alternately absorbs and releases energy.

> **Problem-Solving Insight** Consequently, the average power (and, hence, the average energy) used by a capacitor in an ac circuit is zero.

It will prove useful later on to use a model for the voltage and current when analyzing ac circuits. In this model, voltage and current are represented by rotating arrows, often called **phasors**, whose lengths correspond to the maximum voltage V_0 and maximum current I_0, as **Figure 23.5** indicates. These phasors rotate counterclockwise at a frequency f. For a resistor, the phasors are co-linear as they rotate (see part a of the drawing) because voltage and current are in phase. For a capacitor (see part b), the phasors remain perpendicular while rotating because the phase angle between the current and the voltage is 90°. Since current leads voltage for a capacitor, the current phasor is ahead of the voltage phasor in the direction of rotation.

Note from the two circuit drawings in **Figure 23.5** that the instantaneous voltage across the resistor or the capacitor is $V_0 \sin 2\pi ft$. We can find this instantaneous value of the voltage directly from the phasor diagram. Imagine that the voltage phasor V_0 in this diagram represents the hypotenuse of a right triangle. Then, the *vertical component* of the phasor would be $V_0 \sin 2\pi ft$. In a similar manner, the instantaneous current can be found as the vertical component of the current phasor.

Check Your Understanding

(The answer is given at the end of the book.)

1. One circuit contains only an ac generator and a resistor, and the rms current in this circuit is I_R. Another circuit contains only an ac generator and a capacitor, and the rms current in this circuit is I_C. The maximum, or peak, voltage of the generator is the same in both circuits and does not change. If the frequency of each generator is tripled, by what factor does the ratio I_R/I_C change? Specify whether the change is an increase or a decrease.

23.2 | Inductors and Inductive Reactance

As Section 22.8 discusses, an inductor is usually a coil of wire, and the basis of its operation is Faraday's law of electromagnetic induction. According to Faraday's law, an inductor develops a voltage that opposes a change in the current. This voltage V is given by $V = -L(\Delta I/\Delta t)$ (see Equation 22.9*), where $\Delta I/\Delta t$ is the rate at which the current changes and L is the inductance of the inductor. In an ac circuit the current is always changing, and Faraday's law can be used to show that the rms voltage across an inductor is

$$V_{rms} = I_{rms}X_L \qquad (23.3)$$

Equation 23.3 is analogous to $V_{rms} = I_{rms}R$, with the term X_L appearing in place of the resistance R and being called the **inductive reactance**. The inductive reactance is measured in ohms and determines how much rms current exists in an inductor for a given

---

*When an inductor is used in a circuit, the notation is simplified if we designate the potential difference across the inductor as the voltage V, rather than the emf $\mathscr{E}$.

rms voltage across the inductor. It is found experimentally that the inductive reactance X_L is directly proportional to the frequency f and the inductance L, as indicated by the following equation:

$$X_L = 2\pi f L \qquad \text{(23.4)}$$

This relation indicates that the larger the inductance, the larger is the inductive reactance. Note that the inductive reactance is directly proportional to the frequency ($X_L \propto f$), whereas the capacitive reactance is inversely proportional to the frequency ($X_C \propto 1/f$).

Figure 23.6 shows a graph of the inductive reactance versus frequency for a fixed value of the inductance, according to Equation 23.4. As the frequency becomes very large, X_L also becomes very large. In such a situation, an inductor provides a large opposition to the alternating current. In the limit of zero frequency (i.e., direct current), X_L becomes zero, indicating that an inductor does not oppose direct current at all. The next example demonstrates the effect of inductive reactance on the current in an ac circuit.

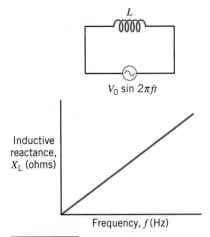

FIGURE 23.6 In an ac circuit the inductive reactance X_L is directly proportional to the frequency f, according to $X_L = 2\pi f L$.

EXAMPLE 2 | An Inductor in an Ac Circuit

The circuit in **Figure 23.6** contains a 3.60-mH inductor. The rms voltage of the generator is 25.0 V. Find the rms current in the circuit when the generator frequency is **(a)** 1.00×10^2 Hz and **(b)** 5.00×10^3 Hz.

Reasoning The current can be calculated from $I_{rms} = V_{rms}/X_L$, provided the inductive reactance is obtained first. The inductor offers more opposition to the changing current when the frequency is larger, and the values for the inductive reactance will reflect this fact.

> **Problem-Solving Insight** The inductive reactance X_L is directly proportional to the frequency f of the voltage. If the frequency increases by a factor of 50, as here, the inductive reactance also increases by a factor of 50.

Solution **(a)** At a frequency of 1.00×10^2 Hz, we find

$$X_L = 2\pi f L = 2\pi(1.00 \times 10^2 \,\text{Hz})(3.60 \times 10^{-3}\,\text{H}) = 2.26\ \Omega \qquad \text{(23.4)}$$

$$I_{rms} = \frac{V_{rms}}{X_L} = \frac{25.0\ \text{V}}{2.26\ \Omega} = \boxed{11.1\ \text{A}} \qquad \text{(23.3)}$$

(b) The calculation is similar when the frequency is 5.00×10^3 Hz:

$$X_L = 2\pi f L = 2\pi(5.00 \times 10^3 \,\text{Hz})(3.60 \times 10^{-3}\,\text{H}) = 113\ \Omega \qquad \text{(23.4)}$$

$$I_{rms} = \frac{V_{rms}}{X_L} = \frac{25.0\ \text{V}}{113\ \Omega} = \boxed{0.221\ \text{A}} \qquad \text{(23.3)}$$

Due to its inductive reactance, an inductor affects the amount of current in an ac circuit. The inductor also influences the current in another way, as **Interactive Figure 23.7** shows. This figure displays graphs of voltage and current versus time for a circuit containing only an inductor. At a maximum or minimum on the current graph, the current does not change much with time, so the voltage generated by the inductor to oppose a change in the current is zero. At the points on the current graph where the current is zero, the graph is at its steepest, and the current has the largest rate of increase or decrease. Correspondingly, the voltage generated by the inductor to oppose a change in the current has the largest positive or negative value. Thus, current and voltage are not in phase but are one-quarter of a wave cycle out of phase. If the voltage varies as $V_0 \sin 2\pi f t$, the current fluctuates as $I_0 \sin (2\pi f t - \pi/2) = -I_0 \cos 2\pi f t$. The current reaches its maximum *after* the voltage does, and it is said that *the current in an inductor lags behind the voltage by a phase angle of 90°* ($\pi/2$ radians). In a purely capacitive circuit, in contrast, the current leads the voltage by 90° (see **Interactive Figure 23.4**).

In an inductor the 90° phase difference between current and voltage leads to the same result for average power that it does in a capacitor. An inductor alternately absorbs and releases energy for equal periods of time.

INTERACTIVE FIGURE 23.7 The instantaneous voltage and current in a circuit containing only an inductor are not in phase. The current *lags behind* the voltage by one-quarter of a cycle or by a phase angle of 90°.

FIGURE 23.8 This phasor model represents the voltage and current in a circuit that contains only an inductor.

FIGURE 23.9 A series RCL circuit contains a resistor, a capacitor, and an inductor.

FIGURE 23.10 The three voltage phasors (V_R, V_C, and V_L) and the current phasor (I_0) for a series RCL circuit.

Problem-Solving Insight Thus, the average power (and, hence, the average energy) used by an inductor in an ac circuit is zero.

As an alternative to the graphs in **Interactive Figure 23.7**, **Figure 23.8** uses phasors to describe the instantaneous voltage and current in a circuit containing only an inductor. The voltage and current phasors remain perpendicular as they rotate, because there is a 90° phase angle between them. The current phasor lags behind the voltage phasor, relative to the direction of rotation, in contrast to the equivalent picture in **Figure 23.5b** for a capacitor. Once again, the instantaneous values are given by the vertical components of the phasors.

Check Your Understanding

(The answer is given at the end of the book.)

2. CYU Figure 23.1 shows three ac circuits: one contains a resistor, one a capacitor, and one an inductor. The frequency of each ac generator is reduced to one-half its initial value. Which circuit experiences **(a)** the *greatest increase* in current and **(b)** the *least change* in current?

CYU FIGURE 23.1

23.3 | Circuits Containing Resistance, Capacitance, and Inductance

Capacitors and inductors can be combined along with resistors in a single circuit. The simplest combination contains a resistor, a capacitor, and an inductor in series, as **Figure 23.9** shows. In a series RCL circuit the total opposition to the flow of charge is called the **impedance** of the circuit and comes partially from (1) the resistance R, (2) the capacitive reactance X_C, and (3) the inductive reactance X_L. It is tempting to follow the analogy of a series combination of resistors and calculate the impedance by simply adding together R, X_C, and X_L. However, such a procedure is not correct. Instead, the phasors shown in **Figure 23.10** must be used. The lengths of the voltage phasors in this drawing represent the maximum voltages V_R, V_C, and V_L across the resistor, the capacitor, and the inductor, respectively. The current is the same for each device, since the circuit is wired in series. The length of the current phasor represents the maximum current I_0. Notice that the drawing shows the current phasor to be (1) in phase with the voltage phasor for the resistor, (2) ahead of the voltage phasor for the capacitor by 90°, and (3) behind the voltage phasor for the inductor by 90°. These three facts are consistent with our earlier discussion in Sections 23.1 and 23.2.

The basis for dealing with the voltage phasors in **Figure 23.10** is Kirchhoff's loop rule. In an ac circuit this rule applies to the *instantaneous* voltages across each circuit component and the generator. Therefore, it is necessary to take into account the fact that these voltages do not have the same phase; that is, the phasors V_R, V_C, and V_L point in different directions in the drawing. Kirchhoff's loop rule indicates that the phasors add together to give the total voltage V_0 that is supplied to the circuit by the generator. The addition, however, must be like a vector addition, to take into account the different directions of the phasors. Since V_L and V_C point in opposite directions, they combine to give a resultant phasor of $V_L - V_C$, as **Figure 23.11** shows. In this drawing the resultant $V_L - V_C$ is perpendicular to V_R and may be combined with it to give the total voltage V_0. Using the Pythagorean theorem, we find

$$V_0^2 = V_R^2 + (V_L - V_C)^2$$

In this equation each of the symbols stands for a maximum voltage and when divided by $\sqrt{2}$ gives the corresponding rms voltage. Therefore, it is possible to divide both sides of the equation by $(\sqrt{2})^2$ and obtain a result for $V_{rms} = V_0/\sqrt{2}$. This result has exactly the same form as that above, but involves the rms voltages $V_{R\text{-rms}}$, $V_{C\text{-rms}}$, and $V_{L\text{-rms}}$. However, to avoid such awkward symbols, we simply interpret V_R, V_C, and V_L as rms quantities in the following expression:

$$V_{rms}^2 = V_R^2 + (V_L - V_C)^2 \tag{23.5}$$

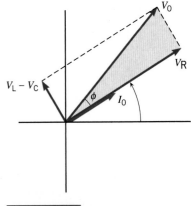

FIGURE 23.11 This simplified version of **Figure 23.10** results when the phasors V_L and V_C, which point in opposite directions, are combined to give a resultant of $V_L - V_C$.

The last step in determining the impedance of the circuit is to remember that $V_R = I_{rms}R$, $V_C = I_{rms}X_C$, and $V_L = I_{rms}X_L$. With these substitutions Equation 23.5 can be written as

$$V_{rms} = I_{rms}\sqrt{R^2 + (X_L - X_C)^2}$$

Therefore, for the entire RCL circuit, it follows that

$$V_{rms} = I_{rms}Z \tag{23.6}$$

where the impedance Z of the circuit is defined as

Series RCL combination

$$Z = \sqrt{R^2 + (X_L - X_C)^2} \tag{23.7}$$

The impedance of the circuit, like R, X_C, and X_L, is measured in ohms. In Equation 23.7, $X_L = 2\pi f L$ and $X_C = 1/(2\pi f C)$.

The phase angle between the current in and the voltage across a series RCL combination is the angle ϕ between the current phasor I_0 and the voltage phasor V_0 in **Figure 23.11**. According to **Figure 23.11**, the tangent of this angle is

$$\tan\phi = \frac{V_L - V_C}{V_R} = \frac{I_{rms}X_L - I_{rms}X_C}{I_{rms}R}$$

Series RCL combination

$$\tan\phi = \frac{X_L - X_C}{R} \tag{23.8}$$

Math Skills To see why $\tan\phi = \dfrac{V_L - V_C}{V_R}$, remember that the tangent function is defined as $\tan\phi = \dfrac{h_o}{h_a}$ (Equation 1.3). As shown in **Figure 23.12a**, h_o is the length of the side of a right triangle opposite the angle ϕ, and h_a is the length of the side adjacent to the angle ϕ. **Figure 23.11** shows the angle ϕ as the phase angle between the current phasor I_0 and the voltage phasor V_0 and is reproduced in **Figure 23.12b**. The corresponding right triangles are shaded in **Figure 23.12**, and a comparison reveals that $h_o = V_L - V_C$ and $h_a = V_R$. It follows that

$$\tan\phi = \frac{h_o}{h_a} = \frac{V_L - V_C}{V_R}$$

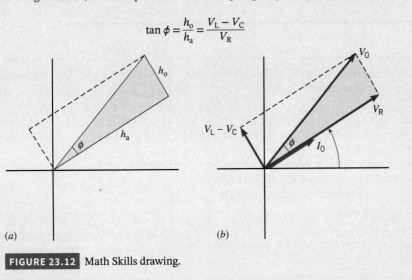

(a) (b)

FIGURE 23.12 Math Skills drawing.

The phase angle ϕ is important because it has a major effect on the average power $\overline{P}$ delivered to the circuit. Remember that, on the average, only the resistance consumes power; that is, $\overline{P} = I_{\text{rms}}^2 R$ (Equation 20.15b). According to **Figure 23.11**, $\cos \phi = V_R/V_0 = (I_{\text{rms}}R)/(I_{\text{rms}}Z) = R/Z$, so that $R = Z \cos \phi$. Therefore,

$$\overline{P} = I_{\text{rms}}^2 Z \cos \phi = I_{\text{rms}}(I_{\text{rms}}Z) \cos \phi$$
$$\overline{P} = I_{\text{rms}}V_{\text{rms}} \cos \phi \tag{23.9}$$

where $V_{\text{rms}} = I_{\text{rms}}Z$ is the rms voltage of the generator, according to Equation 23.6. The term $\cos \phi$ is called the **power factor** of the circuit. As a check on the validity of Equation 23.9, note that if no resistance is present, $R = 0\,\Omega$, and $\cos \phi = R/Z = 0$. Consequently, $\overline{P} = I_{\text{rms}}V_{\text{rms}} \cos \phi = 0$, a result that is expected since, on the average, neither a capacitor nor an inductor consumes energy. Conversely, if only resistance is present, $Z = \sqrt{R^2 + (X_L - X_C)^2} = R$, and $\cos \phi = R/Z = 1$. In this case, $\overline{P} = I_{\text{rms}}V_{\text{rms}} \cos \phi = I_{\text{rms}}V_{\text{rms}}$, which is the expression for the average power delivered to a resistor. Examples 3 and 4 deal with the current, voltages, and power for a series RCL circuit.

Analyzing Multiple-Concept Problems

EXAMPLE 3 | Current in a Series RCL Circuit

A series RCL circuit contains a 148-Ω resistor, a 1.50-μF capacitor, and a 35.7-mH inductor. The generator has a frequency of 512 Hz and an rms voltage of 35.0 V. Determine the rms current in the circuit.

Reasoning The rms current in the circuit is equal to the rms voltage of the generator divided by the impedance of the circuit, according to Equation 23.6. The impedance of the circuit can be found from the resistance of the resistor and the reactances of the capacitor and the inductor via Equation 23.7. The reactances of the capacitor and the inductor can be found with the aid of Equations 23.2 and 23.4.

Knowns and Unknowns The following table summarizes the data that we have:

Description	Symbol	Value	Comment
Resistance of resistor	R	148 Ω	
Capacitance of capacitor	C	1.50 μF	$1\,\mu\text{F} = 10^{-6}\,\text{F}$
Inductance of inductor	L	35.7 mH	$1\,\text{mH} = 10^{-3}\,\text{H}$
Frequency of generator	f	512 Hz	
Rms voltage of generator	V_{rms}	35.0 V	
Unknown Variable			
Rms current in circuit	I_{rms}	?	

Modeling the Problem

STEP 1 Current According to Equation 23.6, the rms voltage V_{rms} of the generator, the rms current I_{rms}, and the impedance Z of the circuit are related according to

$$V_{\text{rms}} = I_{\text{rms}}Z$$

Solving for the current gives Equation 1 at the right, in which V_{rms} is known. The impedance is unknown, but it can be dealt with as in Step 2.

$$I_{\text{rms}} = \frac{V_{\text{rms}}}{Z} \tag{1}$$

$$\boxed{?}$$

STEP 2 Impedance For a series RCL circuit, the impedance Z is related to the resistance R, the inductive reactance X_L, and the capacitive reactance X_C, according to

$$\boxed{Z = \sqrt{R^2 + (X_L - X_C)^2}} \tag{23.7}$$

As indicated at the right, this expression can be substituted into Equation 1. The resistance is given, and we turn to Step 3 to deal with the reactances.

$$I_{\text{rms}} = \frac{V_{\text{rms}}}{Z} \tag{1}$$

$$\boxed{Z = \sqrt{R^2 + (X_L - X_C)^2}} \tag{23.7}$$

$$\boxed{?}$$

STEP 3 Inductive and Capacitive Reactances The inductive reactance X_L and the capacitive reactance X_C are given, respectively, by Equation 23.4 and Equation 23.2 as

$$X_L = 2\pi f L \quad \text{and} \quad X_C = \frac{1}{2\pi f C}$$

where L is the inductance, C is the capacitance, and f is the frequency. Using these two expressions, we find that

$$\boxed{X_L - X_C = 2\pi f L - \frac{1}{2\pi f C}}$$

This result can now be substituted into Equation 23.7, as shown at the right.

$$I_{rms} = \frac{V_{rms}}{Z} \qquad (1)$$

$$\boxed{Z = \sqrt{R^2 + (X_L - X_C)^2}} \quad (23.7)$$

$$\boxed{X_L - X_C = 2\pi f L - \frac{1}{2\pi f C}}$$

Solution Combining the results of each step algebraically, we find that

$$
\underset{\text{STEP 1}}{\downarrow} \quad \underset{\text{STEP 2}}{\downarrow} \qquad\qquad \underset{\text{STEP 2}}{\downarrow}
$$

$$I_{rms} = \frac{V_{rms}}{Z} = \frac{V_{rms}}{\sqrt{R^2 + (X_L - X_C)^2}} = \frac{V_{rms}}{\sqrt{R^2 + \left(2\pi f L - \frac{1}{2\pi f C}\right)^2}}$$

The rms current in the circuit is

$$I_{rms} = \frac{V_{rms}}{\sqrt{R^2 + \left(2\pi f L - \frac{1}{2\pi f C}\right)^2}}$$

$$= \frac{35.0 \text{ V}}{\sqrt{(148\ \Omega)^2 + \left[2\pi(512\text{ Hz})(35.7 \times 10^{-3}\text{ H}) - \frac{1}{2\pi(512\text{ Hz})(1.50 \times 10^{-6}\text{F})}\right]^2}}$$

$$= \boxed{0.201\text{A}}$$

Related Homework: Problem 18

EXAMPLE 4 | Voltages and Power in a Series RCL Circuit

For the series RCL circuit discussed in Example 3, the resistance, capacitance, and inductance are $R = 148\ \Omega$, $C = 1.50\ \mu\text{F}$, and $L = 35.7$ mH, respectively. The generator has a frequency of 512 Hz and an rms voltage of 35.0 V. In Example 3, it is found that the rms current in the circuit is $I_{rms} = 0.201$ A. Find **(a)** the rms voltage across each circuit element and **(b)** the average electric power delivered by the generator.

Reasoning The rms voltages across each circuit element can be determined from $V_R = I_{rms}R$, $V_C = I_{rms}X_C$, and $V_L = I_{rms}X_L$. In these expressions, the rms current is known. The resistance R is given, and the capacitive reactance X_C and inductive reactance X_L can be determined from Equations 23.2 and 23.4. The average power delivered to the circuit by the generator is specified by Equation 23.9.

Solution **(a)** The individual reactances are

$$X_C = \frac{1}{2\pi f C} = \frac{1}{2\pi(512\text{ Hz})(1.50 \times 10^{-6}\text{F})} = 207\ \Omega \quad (23.2)$$

$$X_L = 2\pi f L = 2\pi(512\text{ Hz})(35.7 \times 10^{-3}\text{H}) = 115\ \Omega \quad (23.4)$$

The rms voltages across each circuit element are

$$V_R = I_{rms}\,R = (0.201\text{ A})(148\ \Omega) = \boxed{29.7\text{ V}} \quad (20.14)$$

$$V_C = I_{rms}\,X_C = (0.201\text{ A})(207\ \Omega) = \boxed{41.6\text{ V}} \quad (23.1)$$

$$V_L = I_{rms}\,X_L = (0.201\text{ A})(115\ \Omega) = \boxed{23.1\text{ V}} \quad (23.3)$$

Observe that these three voltages do not add up to give the generator's rms voltage of 35.0 V. Instead, the rms voltages satisfy Equation 23.5. It is the sum of the *instantaneous* voltages across R, C, and L that equals the generator's *instantaneous* voltage, according to Kirchhoff's loop rule. *The rms voltages do not satisfy the loop rule.*

Math Skills The sum of the three voltages calculated in Example 4 is

$$V_R + V_C + V_L = 29.7\text{ V} + 41.6\text{ V} + 23.1\text{ V} = 94.4\text{ V}$$

Since the rms voltage of the generator is 35.0 V in Example 4, the voltages V_R, V_C, and V_L (respectively across the resistor, the capacitor, and the inductor) in a series RCL circuit clearly do not add up to give the rms voltage of the generator. However, they do satisfy $V_{rms}^2 = V_R^2 + (V_L - V_C)^2$ (Equation 23.5). You can use this fact to check the correctness of your calculations in problems such as that in the example. For instance, using $V_R = 29.7$ V, $V_C = 41.6$ V, and $V_L = 23.1$ V in Equation 23.5, we find

$$V_{rms}^2 = V_R^2 + (V_L - V_C)^2 = (29.7\text{ V})^2 + (23.1\text{ V} - 41.6\text{ V})^2$$

$$V_{rms} = \sqrt{(29.7\text{ V})^2 + (23.1\text{ V} - 41.6\text{ V})^2} = 35.0\text{ V}$$

This result confirms the correctness of the calculations in Example 4, since the rms voltage of the generator is, in fact, 35.0 V.

(b) The average power delivered by the generator is $\overline{P} = I_{rms}V_{rms}\cos\phi$ (Equation 23.9). Therefore, a value for the phase angle ϕ is needed and can be obtained from Equation 23.8 as follows:

$$\tan\phi = \frac{X_L - X_C}{R} \quad \text{or} \quad \phi = \tan^{-1}\left(\frac{X_L - X_C}{R}\right)$$

$$= \tan^{-1}\left(\frac{115\ \Omega - 207\ \Omega}{148\ \Omega}\right) = -32°$$

The phase angle is negative since the circuit is more capacitive than inductive (X_C is greater than X_L), and the current leads the voltage. The average power delivered by the generator is

$$\overline{P} = I_{rms}V_{rms}\cos\phi = (0.201\ \text{A})(35.0\ \text{V})\cos(-32°) = \boxed{6.0\ \text{W}}$$

This amount of power is delivered only to the resistor, since neither the capacitor nor the inductor uses power, on average.

In addition to the series RCL circuit, there are many different ways to connect resistors, capacitors, and inductors. In analyzing these additional possibilities, it helps to keep in mind the behavior of capacitors and inductors at the extreme limits of the frequency. When the frequency approaches zero (i.e., dc conditions), the reactance of a capacitor becomes so large that no charge can flow through the capacitor. It is as if the capacitor were cut out of the circuit, leaving an open gap in the connecting wire. In the limit of zero frequency the reactance of an inductor is vanishingly small. The inductor offers no opposition to a dc current and behaves as if it were replaced with a wire of zero resistance. In the limit of very large frequency, the behaviors of a capacitor and an inductor are reversed. The capacitor has a very small reactance and offers little opposition to the current, as if it were replaced by a wire with zero resistance. In contrast, the inductor has a very large reactance when the frequency is very large. The inductor offers so much opposition to the current that it might as well be cut out of the circuit, leaving an open gap in the connecting wire. Conceptual Example 5 illustrates how to gain insight into more complicated circuits using the limiting behaviors of capacitors and inductors.

CONCEPTUAL EXAMPLE 5 | The Limiting Behavior of Capacitors and Inductors

Figure 23.13a shows two circuits. The rms voltage of the generator is the same in each case. The values of the resistance R, the capacitance C, and the inductance L in these circuits are the same. The frequency of the ac generator is very nearly zero. In which circuit does the generator supply more rms current, **(a)** circuit I or **(b)** circuit II?

Reasoning According to Equation 23.6, the rms current is given by $I_{rms} = V_{rms}/Z$. However, the impedance Z cannot be obtained from Equation 23.7, since the circuits in **Figure 23.13a** are not series RCL circuits. Since V_{rms} is the same in each case, the greater current is delivered to the circuit with the smaller impedance Z. In the limit of very small frequencies, the capacitors have very large impedances and, thus, allow very little current to flow through them. In essence, the capacitors behave as if they were cut out of the circuit, leaving gaps in the connecting wires. On the other hand, in the limit of very small frequencies, the inductors have very small impedances and behave as if they were replaced by wires with zero resistance. **Figure 23.13b** shows the circuits as they would appear according to these changes.

Answer (a) is incorrect. According to **Figure 23.13b**, circuit I behaves as if it contained only two identical resistors wired in series, with a total impedance of $Z = R + R = 2R$. In contrast, circuit II behaves as if it contained two identical resistors wired in parallel, in which case the total impedance is given by $1/Z = 1/R + 1/R = 2/R$, or $Z = R/2$. Circuit I contains the greater impedance, so the generator supplies less, not more, rms current to that circuit.

Answer (b) is correct. The rms current I_{rms} in a circuit is given by $I_{rms} = V_{rms}/Z$. Since V_{rms} is the same for both circuits and circuit II has the smaller impedance [see the explanation for why Answer (a) is incorrect], its generator supplies the greater rms current.

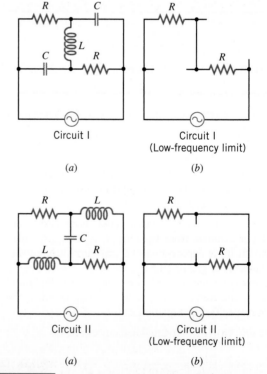

Circuit I

Circuit I
(Low-frequency limit)

(a)

(b)

Circuit II

Circuit II
(Low-frequency limit)

(a)

(b)

FIGURE 23.13 (a) These circuits are discussed in the limit of very small or low frequency in Conceptual Example 5. (b) For a frequency very near zero, the circuits in part a behave as if they were as shown here.

Related Homework: Check Your Understanding 7, 8, Problems 38, 55

BIO **THE PHYSICS OF . . . body-fat scales.** The impedance of an ac circuit contains important information about the resistance, capacitance, and inductance in the circuit. As an example of a very complex circuit, consider the human body. It contains muscle, which is a relatively good conductor of electricity due to its high water content, and also fat, which is a relatively poor conductor due to its low water content. The impedance that the body offers to ac electricity is referred to as bioelectrical impedance and is largely determined by resistance and capacitive reactance. Capacitance enters the picture because cell membranes can act like tiny capacitors. Bioelectric impedance analysis provides the basis for the determination of body-fat percentage by the body-fat scales (see **Figure 23.14**) that are widely available for home use. When you stand barefoot on such a scale, electrodes beneath your feet send a weak ac current (approximately $800\ \mu A$, 50 kHz) through your lower body in order to measure your body's impedance. The scale also measures your weight. A built-in computer combines the impedance and weight with information you provide about height, age, and sex to determine the percentage of fat in your body to an accuracy of about 5%. For men (age 20 to 39) a percentage of 8 to 19% is considered average, whereas the corresponding range of values for women of similar ages is 21 to 33%.

BIO **THE PHYSICS OF . . . transcutaneous electrical nerve stimulation (TENS).** Weak ac electricity with a much lower frequency than that used to measure bioelectrical impedance is used in a technique called transcutaneous electrical nerve stimulation (TENS). TENS is the most commonly used form of electroanalgesia in pain-management situations and, in its conventional form, uses an ac frequency between 40 and 150 Hz. Ac current is passed between two electrodes attached to the body and inhibits the transmission of pain-related nerve impulses. The technique is thought to work by affecting the "gates" in a nerve cell that control the passage of sodium ions into and out of the cell (see Section 19.6). **Figure 23.15** shows TENS being applied during assessment of pain control following suspected damage to the radial nerve in the forearm.

Electrodes for balls of feet

Electrodes for heels

FIGURE 23.14 Bathroom scales are now widely available that can provide estimates of your body-fat percentage. When you stand barefoot on the scale, electrodes beneath your feet send a small ac current through your lower body that allows the body's electrical impedance to be measured. This impedance is correlated with the percentage of fat in the body.

Martin Dohrn/Science Source

Electrodes

FIGURE 23.15 Transcutaneous electrical nerve stimulation (TENS) is shown here being applied to the forearm, in an assessment of pain control following suspected damage to the radial nerve.

Check Your Understanding

(The answers are given at the end of the book.)

3. A long wire of finite resistance is connected to an ac generator. The wire is then wound into a coil of many loops and reconnected to the generator. Is the current in the circuit with the coil greater than, less than, or the same as the current in the circuit with the uncoiled wire?

4. A light bulb and a parallel plate capacitor (containing a dielectric material between the plates) are connected in series to the 60-Hz ac voltage present at a wall outlet. When the dielectric material is removed from the space between the plates, does the brightness of the bulb increase, decrease, or remain the same?

5. An air-core inductor is connected in series with a light bulb, and this circuit is plugged into an ac electrical outlet. When a piece of iron is inserted inside the inductor, does the brightness of the bulb increase, decrease, or remain the same?

6. Consider the circuit in **Figure 23.9**. With the capacitor and the inductor present, a certain amount of current is in the circuit. Then the capacitor and the inductor are removed, so that only the resistor remains connected to the generator. Is it possible that, under a certain condition, the current in the simplified circuit has the same rms value as in the original circuit? **(a)** No **(b)** Yes, when $X_L = R$ **(c)** Yes, when $X_C = R$ **(d)** Yes, when $X_C = X_L$

7. Review Conceptual Example 5 as an aid in understanding this question. An inductor and a capacitor are connected in parallel across the terminals of an ac generator. Does the current from the generator decrease, remain the same, or increase as the frequency becomes **(a)** very large and **(b)** very small?

8. Review Conceptual Example 5 as an aid in understanding this question. For which of the two circuits discussed there does the generator deliver more current when the frequency is very large? **(a)** Circuit I **(b)** Circuit II

23.4 | Resonance in Electric Circuits

The behavior of the current and voltage in a series RCL circuit can give rise to a condition of **resonance**. Resonance occurs when the frequency of a vibrating force exactly matches a natural (resonant) frequency of the object to which the force is applied. When resonance occurs, the force can transmit a large amount of energy to the object, leading to a large-amplitude vibratory motion. We have already encountered several instances of resonance. For example, resonance can occur when a vibrating force acts on an object of mass m that is attached to a spring whose spring constant is k (Section 10.6). In this case there is one natural frequency f_0, whose value is $f_0 = [1/(2\pi)]\sqrt{k/m}$. Resonance also occurs when standing waves are set up on a string (Section 17.5) or in a tube of air (Section 17.6). The string and tube of air have many natural frequencies, one for each allowed standing wave. As we will now see, a condition of resonance can also be established in a series RCL circuit. In this case there is only one natural frequency, and the vibrating force is provided by the oscillating electric field that is related to the voltage of the generator.

Animated **Figure 23.16** helps us understand why a resonant frequency exists for an ac circuit. This drawing presents an analogy between the electrical case (ignoring resistance) and the mechanical case of an object attached to a horizontal spring (ignoring friction). Part *a* shows a fully stretched spring that has just been released, and the initial speed v of the object is zero. All the energy is stored in the form of elastic potential energy. When the object begins to move, it gradually loses potential energy and gains kinetic energy. In part *b*, the object moves with speed v_{max} and maximum kinetic energy through the position where the spring is unstretched (zero potential energy). Because of its inertia, the moving object coasts through this position and eventually comes to a halt in part *c* when the spring is fully compressed and all kinetic energy has been converted back into elastic potential energy. Part *d* of the picture is like part *b*, except that the direction of motion is reversed. The resonant frequency f_0 of the object on the spring is the natural frequency at which the object vibrates and is given as $f_0 = [1/(2\pi)]\sqrt{k/m}$ according to Equations 10.6 and 10.11. In this expression, m is the mass of the object, and k is the spring constant.

In the electrical case, Animated **Figure 23.16a** begins with a fully charged capacitor that has just been connected to an inductor. At this instant the energy is stored in the electric field between the capacitor plates. As the capacitor discharges, the electric field $\vec{E}$ between the plates decreases, while a magnetic field $\vec{B}$ builds up around the inductor because of the increasing current in the circuit. The maximum current and the maximum magnetic field exist at the instant when the capacitor is

ANIMATED FIGURE 23.16 The oscillation of an object on a spring is analogous to the oscillation of the electric and magnetic fields that occur, respectively, in a capacitor and in an inductor. (PE, potential energy; KE, kinetic energy.)

completely discharged, as in part *b* of the figure. Energy is now stored entirely in the magnetic field of the inductor. The voltage induced in the inductor keeps the charges flowing until the capacitor again becomes fully charged, but now with reversed polarity, as in part *c*. Once again, the energy is stored in the electric field between the plates, and no energy resides in the magnetic field of the inductor. Part *d* of the cycle repeats part *b*, but with reversed directions of current and magnetic field. Thus, an ac circuit can have a resonant frequency because there is a natural tendency for energy to shuttle back and forth between the electric field of the capacitor and the magnetic field of the inductor.

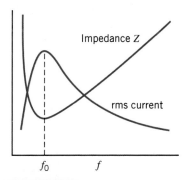

FIGURE 23.17 In a series RCL circuit the impedance is a minimum, and the current is a maximum, when the frequency *f* equals the resonant frequency f_0 of the circuit.

To determine the resonant frequency at which energy shuttles back and forth between the capacitor and the inductor, we note that the current in a series RCL circuit is $I_{rms} = V_{rms}/Z$ (Equation 23.6). In this expression *Z* is the impedance of the circuit and is given by $Z = \sqrt{R^2 + (X_L - X_C)^2}$ (Equation 23.7). As **Figure 23.17** illustrates, the rms current is a maximum when the impedance is a minimum, assuming a given generator voltage. The minimum impedance of $Z = R$ occurs when the frequency is f_0, such that $X_L = X_C$ or $2\pi f_0 L = 1/(2\pi f_0 C)$. This result can be solved for f_0, which is the resonant frequency:

$$f_0 = \frac{1}{2\pi\sqrt{LC}} \qquad (23.10)$$

The resonant frequency is determined by the inductance and the capacitance, but not the resistance.

The effect of resistance on electrical resonance is to make the "sharpness" of the circuit response less pronounced, as **Figure 23.18** indicates. When the resistance is small, the current–frequency graph falls off suddenly on either side of the maximum current. When the resistance is large, the falloff is more gradual, and there is less current at the maximum.

Example 6 demonstrates the use of Equation 23.10 to calculate the resonant frequency of an MRI magnet.

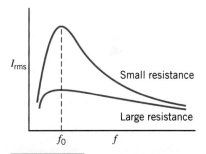

FIGURE 23.18 The effect of resistance on the current in a series RCL circuit.

EXAMPLE 6 | BIO The Physics of Nuclear Magnetic Resonance—NMR

In Chapter 21 we described the medical diagnostic technique of magnetic resonance imaging, or MRI. This technique is based on the more fundamental principle of nuclear magnetic resonance, or NMR. Here, nuclei in the substance under study will absorb and re-emit radio waves when placed in a strong magnetic field. They do this by rotating at a frequency that is typically hundreds of MHz. A pickup coil detects the AC voltage created by the rotating nuclei, and different nuclei will produce different characteristic frequencies. Consider an NMR spectrometer that is configured with a pickup coil designed to detect one particular resonant frequency. This occurs because the coil is part of an *RLC* circuit with a capacitance of 9.4×10^{-16} F. If the inductance of the pickup coil is 7.5 mH, what is the resonant frequency of the coil?

Reasoning In an *RLC* circuit, the resonance frequency depends on the value of the circuit's capacitance and inductance, as given by Equation 23.10. We can use this relationship to calculate the frequency.

Solution Applying Equation 23.10, we have:

$$f_0 = \frac{1}{2\pi\sqrt{LC}} = \frac{1}{2\pi\sqrt{(0.0075\text{ H})(9.4 \times 10^{-16}\text{ F})}} = 6.0 \times 10^7 \text{ Hz}$$

$$= \boxed{60\text{ MHz}}$$

Check Your Understanding

(*The answers are given at the end of the book.*)

9. The resistance in a series RCL circuit is doubled. **(a)** Does the resonant frequency increase, decrease, or remain the same? **(b)** Does the maximum current in the circuit increase, decrease, or remain the same?

10. In a series RCL circuit at resonance, does the current lead or lag behind the voltage across the generator, or is the current in phase with the voltage?

11. Is it possible for two series RCL circuits to have the same resonant frequencies and yet have **(a)** different *R* values and **(b)** different *C* and *L* values?

12. Suppose the generator connected to a series RCL circuit has a frequency that is greater than the resonant frequency of the circuit. Suppose, in addition, that it is necessary to match the resonant frequency of the circuit to the frequency of the generator. To accomplish this, should you add a second capacitor **(a)** in series or **(b)** in parallel with the one already present?

23.5 Semiconductor Devices

Semiconductor devices such as diodes and transistors are widely used in modern electronics, and **Figure 23.19** illustrates one application. The drawing shows an audio system in which small ac voltages (originating in a compact disc player, an FM tuner, or a cassette deck) are amplified so they can drive the speaker(s). The electric circuits that accomplish the amplification do so with the aid of a dc voltage provided by the power supply. In portable units the power supply is simply a battery. In nonportable units, however, the power supply is a separate electric circuit containing diodes, along with other elements. As we will see, the diodes convert the 60-Hz ac voltage present at a wall outlet into the dc voltage needed by the amplifier, which, in turn, performs its job of amplification with the aid of transistors.

FIGURE 23.19 In a typical audio system, diodes are used in the power supply to create a dc voltage from the ac voltage present at the wall socket. This dc voltage is necessary so the transistors in the amplifier can perform their task of enlarging the small ac voltage (in blue) originating in the compact disc player, or other device.

n-Type and p-Type Semiconductors

The materials used in diodes and transistors are semiconductors, such as silicon and germanium. However, they are not pure materials because small amounts of "impurity" atoms (about one part in a million) have been added to them to change their conductive properties. For instance, **Figure 23.20a** shows an array of atoms that symbolizes the crystal structure in pure silicon. Each silicon atom has four outer-shell* electrons, and each electron participates with electrons from neighboring atoms in forming the bonds that hold the crystal together. Since they participate in forming bonds, these electrons generally do not move throughout the crystal. Consequently, pure silicon and germanium are not good conductors of electricity. It is possible, however, to increase their conductivities by adding tiny amounts of impurity atoms, such as phosphorus or arsenic, whose atoms have five outer-shell electrons. For example, when a phosphorus atom replaces a silicon atom in the crystal, only four of the five outer-shell electrons of phosphorus fit into the crystal structure. The extra fifth electron does not fit in and is relatively free to diffuse throughout the crystal, as part b of the drawing suggests. A semiconductor containing small amounts of phosphorus can therefore be envisioned as containing immobile, positively charged phosphorus atoms and a pool of electrons that are free to wander throughout the material. These mobile electrons allow the semiconductor to conduct electricity.

FIGURE 23.20 A silicon crystal that is (a) undoped, or pure, (b) doped with phosphorus to produce an n-type material, and (c) doped with boron to produce a p-type material.

*Section 30.6 discusses the electronic structure of the atom in terms of "shells."

The process of adding impurity atoms is called *doping.* A semiconductor doped with an impurity that contributes mobile electrons is called an **n-type semiconductor**, since the mobile charge carriers have a **n**egative charge. Note that an *n*-type semiconductor is overall electrically neutral, since it contains equal numbers of positive and negative charges.

It is also possible to dope a silicon crystal with an impurity whose atoms have only three outer-shell electrons (e.g., boron or gallium). Because of the missing fourth electron, there is a "hole" in the lattice structure at the boron atom, as **Figure 23.20c** illustrates. An electron from a neighboring silicon atom can move into this hole, in which event the region around the boron atom, having acquired the electron, becomes negatively charged. Of course, when a nearby electron does move, it leaves behind a hole. This hole is positively charged, since it results from the removal of an electron from the vicinity of a neutral silicon atom. The vast majority of atoms in the lattice are silicon, so the hole is almost always next to another silicon atom. Consequently, an electron from one of these adjacent atoms can move into the hole, with the result that the hole moves to yet another location. In this fashion, a positively charged hole can wander through the crystal. This type of semiconductor can, therefore, be viewed as containing immobile, negatively charged boron atoms and an equal number of positively charged, mobile holes. Because of the mobile holes, the semiconductor can conduct electricity. In this case the charge carriers are positive. A semiconductor doped with an impurity that introduces mobile **p**ositive holes is called a *p*-type semiconductor.

The Semiconductor Diode

THE PHYSICS OF . . . a semiconductor diode. A *p-n* junction diode is a device that is formed from a *p*-type semiconductor and an *n*-type semiconductor. The *p-n* junction between the two materials is of fundamental importance to the operation of diodes and transistors. **Figure 23.21** shows separate *p*-type and *n*-type semiconductors, each electrically neutral. **Figure 23.22a** shows them joined together to form a diode. Mobile electrons from the *n*-type semiconductor and mobile holes from the *p*-type semiconductor flow across the junction and combine. This process leaves the *n*-type material with a positive charge layer and the *p*-type material with a negative charge layer, as part *b* of the drawing indicates. The positive and negative charge layers on the two sides of the junction set up an electric field $\vec{\mathbf{E}}$, much like the field in a parallel plate capacitor. This electric field tends to prevent any further movement of charge across the junction, and all charge flow quickly stops.

Suppose now that a battery is connected across the *p-n* junction, as in **Figure 23.23a**, where the negative terminal of the battery is attached to the *n*-material, and the positive terminal is attached to the *p*-material. In this situation the junction is said to be in a condition of **forward bias**, and, as a result, there is a current in the circuit. The negative terminal of the battery repels the mobile electrons in the *n*-type material, and they move toward the junction. Likewise, the positive terminal repels the positive holes in the *p*-type material, and they also move toward the junction. At the junction the electrons fill the holes. In the meantime, the negative terminal provides a fresh supply of electrons to the *n*-material, and the positive terminal pulls off electrons from the *p*-material, forming new holes in the process. Consequently, a continual flow of charge, and hence a current, is maintained.

In **Figure 23.23b** the battery polarity has been reversed, and the *p-n* junction is in a condition known as **reverse bias**. The battery forces electrons in the *n*-material and holes in the *p*-material away from the junction. As a result, the potential across the junction builds up until it opposes the battery potential, and very little current can be sustained through the diode. The diode, then, is a unidirectional device, for it allows current to pass in only one direction.

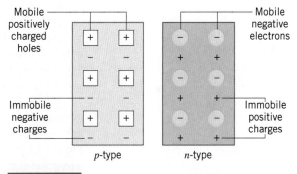

FIGURE 23.21 A *p*-type semiconductor and an *n*-type semiconductor.

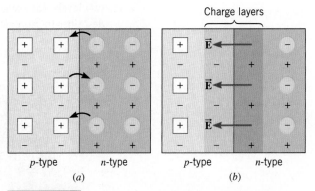

FIGURE 23.22 At the junction between *n* and *p* materials, (*a*) mobile electrons and holes combine and (*b*) create positive and negative charge layers. The electric field produced by the charge layers is $\vec{\mathbf{E}}$.

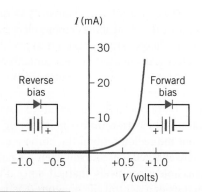

p-type   n-type

(a) Forward bias

p-type   n-type

(b) Reverse bias

FIGURE 23.23 There is an appreciable current through the diode when the diode is forward-biased. (b) Under a reverse-bias condition, there is almost no current through the diode.

FIGURE 23.24 The current–voltage characteristics of a typical *p-n* junction diode.

The graph in **Figure 23.24** shows the dependence of the current on the magnitude and polarity of the voltage applied across a *p-n* junction diode. The exact values of the current depend on the nature of the semiconductor and the extent of the doping. Also shown in the drawing is the symbol used for a diode (—▸|—). The direction of the arrowhead in the symbol indicates the direction of the conventional current in the diode under a forward-bias condition. In a forward-bias condition, the side of the symbol that contains the arrowhead has a positive potential relative to the other side.

THE PHYSICS OF . . . light-emitting diodes (LEDs). A special kind of diode is called an **LED**, which stands for light-emitting diode. You can see LEDs in the form of small bright red, green, or yellow lights that appear on most electronic devices, such as computers, TV sets, and audio systems. These diodes, like others, carry current in only one direction. Imagine a forward-biased diode, like that shown in **Figure 23.23a**, in which a current exists. An LED emits light whenever electrons and holes combine, the light coming from the *p-n* junction. Commercial LEDs are often made from gallium, suitably doped with arsenic and phosphorus atoms.

BIO **THE PHYSICS OF . . . a fetal oxygen monitor.** A fetal oxygen monitor uses LEDs to measure the level of oxygen in a fetus's blood. A sensor is inserted into the mother's uterus and positioned against the cheek of the fetus, as indicated in **Figure 23.25**. Two LEDs are located within the sensor, and each shines light of a different wavelength (or color) into the fetal tissue. The light is reflected by the oxygen-carrying red blood cells and is detected by an adjacent photodetector. Light from one of the LEDs is used to measure the level of oxyhemoglobin in the blood, and light from the other LED is used to measure the level of deoxyhemoglobin. From a comparison of these two levels, the oxygen saturation in the blood is determined. Example 3 in Chapter 24 describes in more detail the operation of an adult version of the oxygen monitor.

Sensor

FIGURE 23.25 A fetal oxygen monitor uses a sensor that contains LEDs to measure the level of oxygen in the blood of an unborn child.

THE PHYSICS OF . . . rectifier circuits. Because diodes are unidirectional devices, they are commonly used in rectifier circuits, which convert an ac voltage into a dc voltage. For instance, **Figure 23.26** shows a circuit in which charges flow through the resistance *R* only while the ac generator biases the diode in the forward direction. Since

FIGURE 23.26 A half-wave rectifier circuit, together with a capacitor and a transformer (not shown), constitutes a dc power supply because the rectifier converts an ac voltage into a dc voltage.

current occurs only during one-half of every generator voltage cycle, the circuit is called a *half-wave rectifier*. A plot of the output voltage across the resistor reveals that only the positive halves of each cycle are present. If a capacitor is added in parallel with the resistor, as indicated in **Figure 23.26**, the capacitor charges up and keeps the voltage from dropping to zero between each positive half-cycle. It is also possible to construct *full-wave rectifier circuits*, in which both halves of every cycle of the generator voltage drive current through the load resistor in the same direction (see Check Your Understanding Question 13).

When a circuit such as the one in **Figure 23.26** includes a capacitor and also a transformer to establish the desired voltage level, the circuit is called a *power supply*. In the audio system in **Figure 23.19**, the power supply receives the 60-Hz ac voltage from a wall socket and produces a dc output voltage that is used for the transistors within the amplifier. Power supplies using diodes are also found in virtually all electronic appliances, such as televisions and microwave ovens.

Solar Cells

THE PHYSICS OF . . . solar cells. Solar cells use *p-n* junctions to convert sunlight directly into electricity, as **Figure 23.27** illustrates. The solar cell in this drawing consists of a *p*-type semiconductor surrounding an *n*-type semiconductor. As discussed earlier, charge layers form at the junction between the two types of semiconductors, leading to an electric field $\vec{E}$ pointing from the *n*-type toward the *p*-type layer. The outer covering of *p*-type material is so thin that sunlight penetrates into the charge layers and ionizes some of the atoms there. In the process of ionization, the energy of the sunlight causes a negative electron to be ejected from the atom, leaving behind a positive hole. As the drawing indicates, the electric field in the charge layers causes the electron and the hole to move away from the junction. The electron moves into the *n*-type material, and the hole moves into the *p*-type material. As a result, the sunlight causes the solar cell to develop negative and positive terminals, much like the terminals of a battery. The current that a single solar cell can provide is small, so applications of solar cells often use many of them mounted to form large panels, as **Figure 23.28** illustrates.

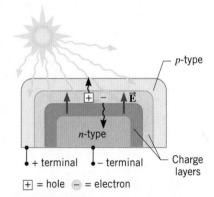

FIGURE 23.27 A solar cell formed from a *p-n* junction. When sunlight strikes it, the solar cell acts like a battery, with + and − terminals.

Transistors

THE PHYSICS OF . . . transistors. A number of different kinds of transistors are in use today. One type is the bipolar-junction transistor, which consists of two *p-n* junctions formed by three layers of doped semiconductors. As **Figure 23.29** indicates, there are *pnp* and *npn* transistors. In either case, the middle region is made very thin compared to the outer regions.

A transistor is useful because it can be used in circuits that amplify a smaller voltage into a larger one. A transistor plays the same kind of role in an amplifier circuit that a valve does when it controls the flow of water through a pipe. A small change in the valve setting produces a large change in the amount of water per second that flows through the pipe. Similarly, a small change in the voltage input to a transistor produces a large change in the output from the transistor.

CHRISTIAN CHARISIUS/Reuters/Landov LLC

FIGURE 23.28 This boat, the *PlanetSolar*, is powered by the electrical energy generated by approximately 500 m² of solar cells that cover its top surface. It was built by the shipyard of Knierim Yachtbau in Kiel, Germany.

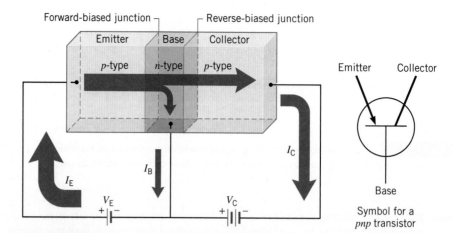

p-type *n*-type *p*-type

pnp transistor

n-type *p*-type *n*-type

npn transistor

FIGURE 23.29 There are two kinds of bipolar-junction transistors, *pnp* and *npn*.

FIGURE 23.30 A *pnp* transistor, along with its bias voltages V_E and V_C. On the symbol for the *pnp* transistor at the right, the emitter is marked with an arrow that denotes the direction of conventional current through the emitter.

Figure 23.30 shows a *pnp* transistor connected to two batteries, labeled V_E and V_C. The voltages V_E and V_C are applied in such a way that the *p-n* junction on the left has a forward bias, while the *p-n* junction on the right has a reverse bias. Moreover, the voltage V_C is usually much larger than V_E for a reason to be discussed shortly. The drawing also shows the standard symbol and nomenclature for the three sections of the *pnp* transistor—namely, the *emitter,* the *base,* and the *collector.* The arrow in the symbol points in the direction of the conventional current through the emitter.

The positive terminal of V_E pushes the mobile positive holes in the *p*-type material of the emitter toward the emitter/base junction. Since this junction has a forward bias, the holes enter the base region readily. Once in the base region, the holes come under the strong influence of V_C and are attracted to its negative terminal. Since the base is so thin (about 10^{-6} m or so), approximately 98% of the holes are drawn through the base and into the collector. The remaining 2% of the holes combine with free electrons in the base region, thereby giving rise to a small base current I_B. As the drawing shows, the moving holes in the emitter and collector constitute currents that are labeled I_E and I_C, respectively. From Kirchhoff's junction rule it follows that $I_C = I_E - I_B$.

Because the base current I_B is small, the collector current is determined primarily by current from the emitter ($I_C = I_E - I_B \approx I_E$). This means that a change in I_E will cause a change in I_C of nearly the same amount. Furthermore, a substantial change in I_E can be caused by only a small change in the forward-bias voltage V_E. To see that this is the case, look at **Figure 23.24** and notice how steep the current–voltage curve is for a *p-n* junction: small changes in the forward-bias voltage give rise to large changes in the current.

With the help of **Interactive Figure 23.31** we can now appreciate what was meant by the earlier statement that a small change in the voltage input to a transistor leads to a large change in the output. This picture shows an ac generator connected in series with the battery V_E and a resistance R connected in series with the collector. The generator voltage could originate from many sources, such as an electric guitar pickup or a compact disc player, while the resistance R could represent a loudspeaker. The generator introduces small voltage changes in the forward bias across the emitter/base junction and, thus, causes large corresponding changes in the current leaving the collector and passing through the resistance R. As a result, the output voltage across R is an enlarged or amplified version of the input voltage of the generator. The operation of an *npn* transistor is similar to that of a *pnp* transistor. The main difference is that the bias voltages and current directions are reversed.

It is important to realize that the increased power available at the output of a transistor amplifier does *not* come from the transistor itself. Rather, it comes from the power provided by the voltage source V_C. The transistor, acting like an automatic valve, merely allows the small signals from the input generator to control the power taken from the source V_C and delivered to the resistance R.

Today it is possible to combine arrays of billions of transistors, diodes, resistors, and capacitors on a tiny chip of silicon that usually measures less than a centimeter on a side. These arrays are called *integrated circuits* (ICs) and can be designed to perform almost any

Adam Hart-Davis/Science Source

FIGURE 23.32 Integrated circuit (IC) chips are manufactured on wafers of semiconductor material. Shown here is one wafer containing many chips.

INTERACTIVE FIGURE 23.31 The basic *pnp* transistor amplifier in this drawing amplifies a small generator voltage to produce an enlarged voltage across the resistance *R*.

desired electronic function. Integrated circuits composed of semiconductors, such as the type in **Figure 23.32**, have revolutionized the electronics industry and lie at the heart of computers, cellular phones, digital watches, and programmable appliances. The following example describes just one of the many unique applications of semiconductors.

CONCEPTUAL EXAMPLE 7 | **BIO** The Physics of Health-Monitoring Smart Rings

As mentioned earlier, the impurity atoms that are chemically doped into semiconductor materials alter their conductivities, and it is useful to consider the qualitative behavior of their electrical resistivity as a function of temperature. In Section 20.3, we mentioned that the resistivity of a metal decreases with decreasing temperature, whereas for semiconductors, the reverse is true. Materials for which the resistivity decreases with increasing temperature are referred to as NTCs, or *negative temperature coefficient* materials. Semiconductors become better conductors as their temperature increases. A comprehensive understanding of why this occurs requires the band theory of solids, which is beyond the scope of this book. However, it is sufficient to know that the number of mobile charge carriers increases exponentially with increasing temperature. These carriers participate in conduction, thereby lowering the resistivity.

The resistivity in many semiconductors is extremely sensitive to temperature, much more so than in metals, which makes them particularly well-suited for sensor applications. These highly temperature-sensitive resistors are also called *thermistors*. One application where thermistors are routinely utilized is in wearable health-monitoring devices. Smartwatches, smartbracelets, and smartrings use sensors to monitor in real time a person's movements, sleep patterns, and vital signs, such as heart rate, blood pressure, blood-oxygen level, and many others (see, for example, the Oura ring in **Figure 23.33**).

Thermistor sensors measure a person's temperature, and the current technology can detect temperature changes as small as 0.1 °C. Since the ring can detect an increase in a person's temperature, it can indicate the onset of a fever, which is often associated with infectious disease. In fact, during the COVID-19 pandemic in 2020, thousands of medical workers wore the smart rings to study their effectiveness at predicting the infection. Players in the NBA also

©2020 Oura, https://ouraring.com/meet-oura

FIGURE 23.33 Smart rings, like the Oura shown here, use different sensors to monitor a person's movement and vital signs. A finely calibrated thermistor measures body temperature, which, in some cases, may provide early detection of fever and possible infection.

wore the rings to monitor their wellness while finishing the 2019–2020 season at the Walt Disney World Resort in Orlando, Florida.

The Oura is powered by a 3.7-V lithium-ion battery, and the current through the thermistor is monitored by the ring's circuitry. Which of the following changes in the thermistor's current might indicate the onset of fever? **(a)** The current through the thermistor remains constant. **(b)** The current through the thermistor increases. **(c)** The current through the thermistor decreases.

Reasoning Given that the voltage output from the lithium-ion battery is a constant, a change in the thermistor's current must correspond to a change in its resistance. Ohm's law can be used to determine whether or not the resistance increases or decreases, and this change in resistance will depend on the change in temperature. An increase in temperature may indicate a fever.

Answer (a) is incorrect. If the current through the thermistor remains constant, then the thermistor's resistance would also remain constant, indicating no change in temperature.

Answer (b) is correct. An increase in the current through the thermistor corresponds to a decrease in its resistance, in accordance with Ohm's law. For an NTC thermistor, a decrease in its resistance indicates an increase in its temperature, which may be due to fever.

Answer (c) is incorrect. A decrease in the current through the thermistor corresponds to an increase in its resistance, in accordance with Ohm's law. For an NTC thermistor, an increase in its resistance indicates a decrease in its temperature, which would suggest no fever.

Check Your Understanding

(The answer is given at the end of the book.)

13. CYU Figure 23.2 shows a full-wave rectifier circuit, in which the direction of the current through the load resistor R is the same for both positive and negative halves of the generator's voltage cycle. What is the direction of the current through the resistor (left to right, or right to left) when **(a)** the top of the generator is positive and the bottom is negative and **(b)** the top of the generator is negative and the bottom is positive?

CYU FIGURE 23.2

Concept Summary

23.1 Capacitors and Capacitive Reactance In an ac circuit the rms voltage V_{rms} across a capacitor is related to the rms current I_{rms} by Equation 23.1, where X_C is the capacitive reactance. The capacitive reactance is measured in ohms (Ω) and is given by Equation 23.2, where f is the frequency and C is the capacitance. The ac current in a capacitor leads the voltage across the capacitor by a phase angle of 90° or $\pi/2$ radians. As a result, a capacitor consumes no power, on average.

$$V_{rms} = I_{rms}X_C \qquad (23.1)$$

The phasor model is useful for analyzing the voltage and current in an ac circuit. In this model, the voltage and current are represented by rotating arrows, called phasors. The length of the voltage phasor represents the maximum voltage V_0, and the length of the current phasor represents the maximum current I_0. The phasors rotate in a counterclockwise direction at a frequency f. Since the current leads the voltage by 90° in a capacitor, the current phasor is ahead of the voltage phasor by 90° in the direction of rotation. The instantaneous values of the voltage and current are equal to the vertical components of the corresponding phasors.

$$X_C = \frac{1}{2\pi fC} \qquad (23.2)$$

23.2 Inductors and Inductive Reactance In an ac circuit the rms voltage V_{rms} across an inductor is related to the rms current I_{rms} by Equation 23.3, where X_L is the inductive reactance. The inductive reactance is measured in ohms (Ω) and is given by Equation 23.4, where f is the frequency and L is the inductance. The ac current in an inductor lags behind the voltage across the inductor by a phase angle of 90° or $\pi/2$ radians. Consequently, an inductor, like a capacitor, consumes no power, on average.

$$V_{rms} = I_{rms}X_L \qquad (23.3)$$
$$X_L = 2\pi fL \qquad (23.4)$$

The voltage and current phasors in a circuit containing only an inductor also rotate in a counterclockwise direction at a frequency f. However, since the current lags the voltage by 90° in an inductor, the current phasor is behind the voltage phasor by 90° in the direction of rotation. The instantaneous values of the voltage and current are equal to the vertical components of the corresponding phasors.

23.3 Circuits Containing Resistance, Capacitance, and Inductance When a resistor, a capacitor, and an inductor are connected in series, the rms voltage across the combination is related to the rms current according to Equation 23.6, where Z is the impedance of the combination. The impedance is measured in ohms (Ω) and is given by Equation 23.7, where R is the resistance, and X_L and X_C are, respectively, the inductive and capacitive reactances.

$$V_{rms} = I_{rms}Z \qquad (23.6)$$
$$Z = \sqrt{R^2 + (X_L - X_C)^2} \qquad (23.7)$$

The tangent of the phase angle ϕ between current and voltage in a series RCL circuit is given by Equation 23.8.

$$\tan \phi = \frac{X_L - X_C}{R} \qquad (23.8)$$

Only the resistor in the RCL combination consumes power, on average. The average power $\overline{P}$ consumed in the circuit is given by Equation 23.9, where $\cos \phi$ is called the power factor of the circuit.

$$\overline{P} = I_{rms} V_{rms} \cos \phi \qquad (23.9)$$

23.4 Resonance in Electric Circuits A series RCL circuit has a resonant frequency f_0 that is given by Equation 23.10, where L is the inductance and C is the capacitance. At resonance, the impedance of the circuit has a minimum value equal to the resistance R, and the rms current has a maximum value.

$$f_0 = \frac{1}{2\pi\sqrt{LC}} \qquad (23.10)$$

23.5 Semiconductor Devices In an n-type semiconductor, mobile negative electrons carry the current. An n-type material is produced by doping a semiconductor such as silicon with a small amount of impurity atoms such as phosphorus. In a p-type semiconductor, mobile positive "holes" in the crystal structure carry the current. A p-type material is produced by doping a semiconductor with a small amount of impurity atoms such as boron. These two types of semiconductors are used in p-n junction diodes, light-emitting diodes, solar cells, and pnp and npn bipolar junction transistors.

Focus on Concepts

Online

Additional questions are available for assignment in WileyPLUS.

Section 23.1   Capacitors and Capacitive Reactance

1. A circuit contains an ac generator and a resistor. What happens to the average power dissipated in the resistor when the frequency is doubled and the rms voltage is tripled? **(a)** Nothing happens, because the average power does not depend on either the frequency or the rms voltage. **(b)** The average power doubles because it is proportional to the frequency. **(c)** The average power triples because it is proportional to the rms voltage. **(d)** The average power increases by a factor of $3^2 = 9$ because it is proportional to the square of the rms voltage. **(e)** The average power increases by a factor of $2 \times 3 = 6$ because it is proportional to the product of the frequency and the rms voltage.

Section 23.2   Inductors and Inductive Reactance

2. What happens to the capacitive reactance X_C and the inductive reactance X_L if the frequency of the ac voltage is doubled? **(a)** X_C increases by a factor of 2, and X_L decreases by a factor of 2. **(b)** X_C and X_L both increase by a factor of 2. **(c)** X_C and X_L do not change. **(d)** X_C and X_L both decrease by a factor of 2. **(e)** X_C decreases by a factor of 2, and X_L increases by a factor of 2.

3. Each of the four phasor diagrams represents a different circuit. V_0 and I_0 represent, respectively, the maximum voltage of the generator and the current in the circuit. Which circuit contains only a resistor? **(a)** A **(b)** B **(c)** C **(d)** D

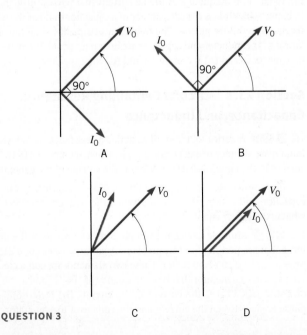

QUESTION 3

Section 23.3   Circuits Containing Resistance, Capacitance, and Inductance

4. The table shows the rms voltage V_C across the capacitor and the rms voltage V_L across the inductor for three series RCL circuits. In which circuit does the rms voltage across the entire RCL combination lead the current through the combination? **(a)** Circuit 1 **(b)** Circuit 2 **(c)** Circuit 3 **(d)** The total rms voltage across the RCL combination does not lead the current in any of the circuits.

Circuit	V_C	V_L
1	50 V	100 V
2	100 V	50 V
3	50 V	50 V

5. A capacitor and an inductor are connected to an ac generator in two ways: in series and in parallel (see the drawing). At low frequencies, which circuit has the greater current? **(a)** The series circuit, because the impedance of the circuit is small due to the small reactances of both the inductor and the capacitor. **(b)** The series circuit, because the impedance of the circuit is large due to the large reactances of both the inductor and the capacitor. **(c)** The parallel circuit, because the impedance of the circuit is small due to the large reactance of the inductor. **(d)** The parallel circuit, because the impedance of the circuit is large due to the large reactance of the inductor. **(e)** The parallel circuit, because the impedance of the circuit is small due to the small reactance of the inductor.

QUESTION 5   Series   Parallel

Section 23.4   Resonance in Electric Circuits

6. In an RCL circuit a second capacitor is added in parallel to the capacitor already present. Does the resonant frequency of the circuit increase, decrease, or remain the same? **(a)** The resonant frequency

increases, because it depends inversely on the square root of the capacitance, and the equivalent capacitance decreases when a second capacitor is added in parallel. **(b)** The resonant frequency increases, because it is directly proportional to the capacitance, and the equivalent capacitance increases when a second capacitor is added in parallel. **(c)** The resonant frequency decreases, because it is directly proportional to the capacitance, and the equivalent capacitance decreases when a second capacitor is added in parallel. **(d)** The resonant frequency decreases, because it depends inversely on the square root of the capacitance, and the equivalent capacitance increases when a second capacitor is added in parallel. **(e)** The resonant frequency remains the same.

Problems

Online

Additional questions are available for assignment in WileyPLUS.

***Note:** For problems in this set, the ac current and voltage are rms values, and the power is an average value, unless indicated otherwise.*

SSM Student Solutions Manual	**BIO** Biomedical application
MMH Problem-solving help	**E** Easy
GO Guided Online Tutorial	**M** Medium
V-HINT Video Hints	**H** Hard
CHALK Chalkboard Videos	**WS** Worksheet
	T Team Problem

Section 23.1 Capacitors and Capacitive Reactance

1. E SSM A 63.0-μF capacitor is connected to a generator operating at a low frequency. The rms voltage of the generator is 4.00 V and is constant. A fuse in series with the capacitor has negligible resistance and will burn out when the rms current reaches 15.0 A. As the generator frequency is increased, at what frequency will the fuse burn out?

2. E GO Two identical capacitors are connected in parallel to an ac generator that has a frequency of 610 Hz and produces a voltage of 24 V. The current in the circuit is 0.16 A. What is the capacitance of each capacitor?

3. E The reactance of a capacitor is 68 Ω when the ac frequency is 460 Hz. What is the reactance when the frequency is 870 Hz?

4. E GO A capacitor is connected to an ac generator that has a frequency of 3.4 kHz and produces a voltage of 2.0 V. The current in the capacitor is 35 mA. When the same capacitor is connected to a second ac generator that has a frequency of 5.0 kHz, the current in the capacitor is 85 mA. What voltage does the second generator produce?

5. E SSM A capacitor is connected across the terminals of an ac generator that has a frequency of 440 Hz and supplies a voltage of 24 V. When a second capacitor is connected in parallel with the first one, the current from the generator increases by 0.18 A. Find the capacitance of the second capacitor.

6. M CHALK GO Two parallel plate capacitors are identical, except that one of them is empty and the other contains a material with a dielectric constant of 4.2 in the space between the plates. The empty capacitor is connected between the terminals of an ac generator that has a fixed frequency and rms voltage. The generator delivers a current of 0.22 A. What current does the generator deliver after the other capacitor is connected in parallel with the first one?

7. M V-HINT A capacitor is connected across an ac generator whose frequency is 750 Hz and whose *peak* output voltage is 140 V. The rms current in the circuit is 3.0 A. **(a)** What is the capacitance of the capacitor? **(b)** What is the magnitude of the *maximum* charge on one plate of the capacitor?

8. H A capacitor (capacitance C_1) is connected across the terminals of an ac generator. Without changing the voltage or frequency of the generator, a second capacitor (capacitance C_2) is added in series with the first one. As a result, the current delivered by the generator

decreases by a factor of three. Suppose that the second capacitor had been added in parallel with the first one, instead of in series. By what factor would the current delivered by the generator have increased?

Section 23.2 Inductors and Inductive Reactance

9. E SSM An 8.2-mH inductor is connected to an ac generator (10.0 V rms, 620 Hz). Determine the *peak value* of the current supplied by the generator.

10. E An inductor has an inductance of 0.080 H. The voltage across this inductor is 55 V and has a frequency of 650 Hz. What is the current in the inductor?

11. E An inductor is to be connected to the terminals of a generator (rms voltage = 15.0 V) so that the resulting rms current will be 0.610 A. Determine the required inductive reactance.

12. E GO An ac generator has a frequency of 7.5 kHz and a voltage of 39 V. When an inductor is connected between the terminals of this generator, the current in the inductor is 42 mA. What is the inductance of the inductor?

13. E V-HINT A 40.0-μF capacitor is connected across a 60.0-Hz generator. An inductor is then connected in parallel with the capacitor. What is the value of the inductance if the rms currents in the inductor and capacitor are equal?

14. E GO An ac generator has a frequency of 2.2 kHz and a voltage of 240 V. An inductance $L_1 = 6.0$ mH is connected across its terminals. Then a second inductance $L_2 = 9.0$ mH is connected in parallel with L_1. Find the current that the generator delivers to L_1 and to the parallel combination.

15. M SSM CHALK A 30.0-mH inductor has a reactance of 2.10 kΩ. **(a)** What is the frequency of the ac current that passes through the inductor? **(b)** What is the capacitance of a capacitor that has the same reactance at this frequency? The frequency is tripled, so that the reactances of the inductor and capacitor are no longer equal. What are the new reactances of **(c)** the inductor and **(d)** the capacitor?

Section 23.3 Circuits Containing Resistance, Capacitance, and Inductance

16. E SSM A series RCL circuit includes a resistance of 275 Ω, an inductive reactance of 648 Ω, and a capacitive reactance of 415 Ω. The current in the circuit is 0.233 A. What is the voltage of the generator?

17. E A series RCL circuit contains a 47.0-Ω resistor, a 2.00-μF capacitor, and a 4.00-mH inductor. When the frequency is 2550 Hz, what is the power factor of the circuit?

18. E SSM Multiple-Concept Example 3 reviews some of the basic ideas that are pertinent to this problem. A circuit consists of a 215-Ω resistor and a 0.200-H inductor. These two elements are connected in series across a generator that has a frequency of 106 Hz and a voltage of 234 V. **(a)** What is the current in the circuit? **(b)** Determine the phase angle between the current and the voltage of the generator.

19. **E** **GO** An ac series circuit has an impedance of 192 Ω, and the phase angle between the current and the voltage of the generator is $\phi = -75°$. The circuit contains a resistor and either a capacitor or an inductor. Find the resistance R and the capacitive reactance X_C or the inductive reactance X_L, whichever is appropriate.

20. **E** When only a resistor is connected across the terminals of an ac generator (112 V) that has a fixed frequency, there is a current of 0.500 A in the resistor. When only an inductor is connected across the terminals of this same generator, there is a current of 0.400 A in the inductor. When both the resistor and the inductor are connected in series between the terminals of this generator, what are (a) the impedance of the series combination and (b) the phase angle between the current and the voltage of the generator?

21. **E** **GO** A 2700-Ω resistor and a 1.1-μF capacitor are connected in series across a generator (60.0 Hz, 120 V). Determine the power delivered to the circuit.

22. **M** **CHALK** **GO** Part a of the drawing shows a resistor and a charged capacitor wired in series. When the switch is closed, the capacitor discharges as charge moves from one plate to the other. Part b shows the amount q of charge remaining on each plate of the capacitor as a function of time. In part c of the drawing, the switch has been removed and an ac generator has been inserted into the circuit. The circuit elements in the drawing have the following values: $R = 18$ Ω, $V_{rms} = 24$ V for the generator, and $f = 380$ Hz for the generator. The time constant for the circuit in part a is $\tau = 3.0 \times 10^{-4}$ s. What is the rms current in the circuit in part c?

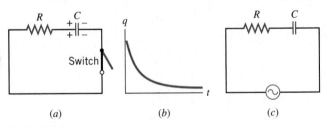

(a)   (b)   (c)

PROBLEM 22

23. **M** **SSM** A circuit consists of an 85-Ω resistor in series with a 4.0-μF capacitor, and the two are connected between the terminals of an ac generator. The voltage of the generator is fixed. At what frequency is the current in the circuit one-half the value that exists when the frequency is very large?

24. **M** **V-HINT** An 84.0-mH inductor and a 5.80-μF capacitor are connected in series with a generator whose frequency is 375 Hz. The rms voltage across the capacitor is 2.20 V. Determine the rms voltage across the inductor.

Section 23.4   Resonance in Electric Circuits

25. **E** A *tank circuit* in a radio transmitter is a series RCL circuit connected to an antenna. The antenna broadcasts radio signals at the resonant frequency of the tank circuit. Suppose that a certain tank circuit in a shortwave radio transmitter has a fixed capacitance of 1.8×10^{-11} F and a variable inductance. If the antenna is intended to broadcast

radio signals ranging in frequency from 4.0 MHz to 9.0 MHz, find the (a) minimum and (b) maximum inductance of the tank circuit.

26. **E** **SSM** A series RCL circuit has a resonant frequency of 690 kHz. If the value of the capacitance is 2.0×10^{-9} F, what is the value of the inductance?

27. **E** The power dissipated in a series RCL circuit is 65.0 W, and the current is 0.530 A. The circuit is at resonance. Determine the voltage of the generator.

28. **E** **SSM** A 10.0-Ω resistor, a 12.0-μF capacitor, and a 17.0-mH inductor are connected in series with a 155-V generator. (a) At what frequency is the current a maximum? (b) What is the maximum value of the rms current?

29. **E** **GO** The capacitance in a series RCL circuit is $C_1 = 2.60\ \mu$F, and the corresponding resonant frequency is $f_{01} = 7.30$ kHz. The generator frequency is 5.60 kHz. What is the value of the capacitance C_2 that should be added to the circuit so that the circuit will have a resonant frequency that matches the generator frequency? Note that you must decide whether C_2 is added in series or in parallel with C_1.

30. **E** A series RCL circuit is at resonance and contains a variable resistor that is set to 175 Ω. The power delivered to the circuit is 2.6 W. Assuming that the voltage remains constant, how much power is delivered when the variable resistor is set to 562 Ω?

31. **E** **GO** The resonant frequency of an RCL circuit is 1.3 kHz, and the value of the inductance is 7.0 mH. What is the resonant frequency (in kHz) when the value of the inductance is 1.5 mH?

32. **M** A series RCL circuit has a resonant frequency of 1500 Hz. When operating at a frequency other than 1500 Hz, the circuit has a capacitive reactance of 5.0 Ω and an inductive reactance of 30.0 Ω. What are the values of (a) L and (b) C?

33. **M** **GO** In a series RCL circuit the generator is set to a frequency that is *not* the resonant frequency. This nonresonant frequency is such that the ratio of the inductive reactance to the capacitive reactance of the circuit is observed to be 5.36. The resonant frequency is 225 Hz. What is the frequency of the generator?

34. **M** **CHALK** **GO** A charged capacitor and an inductor are connected as shown in the drawing (this circuit is the same as that in **Animated Figure 23.16a**). There is no resistance in the circuit. As Section 23.4 discusses, the electrical energy initially present in the charged capacitor then oscillates back and forth between the inductor and the capacitor. The initial charge on the capacitor has a magnitude of $q = 2.90\ \mu$C. The capacitance is $C = 3.60\ \mu$F, and the inductance is $L = 75.0$ mH. (a) What is the electrical energy stored initially by the charged capacitor? (b) Find the maximum current in the inductor.

PROBLEM 34

35. **M** **V-HINT** In the absence of a nearby metal object, the two inductances (L_A and L_B) in a heterodyne metal detector are the same, and the resonant frequencies of the two oscillator circuits have the same value of 630.0 kHz. When the search coil (inductor B) is brought near a buried metal object, a beat frequency of 7.30 kHz is heard. By what percentage does the buried object increase the inductance of the search coil?

Additional Problems   Online

36. **E** **GO** A circuit consists of a resistor in series with an inductor and an ac generator that supplies a voltage of 115 V. The inductive reactance is 52.0 Ω, and the current in the circuit is 1.75 A. Find the average power delivered to the circuit.

37. **E** In a series circuit, a generator (1350 Hz, 15.0 V) is connected to a 16.0-Ω resistor, a 4.10-μF capacitor, and a 5.30-mH inductor. Find the voltage across each circuit element.

38. **E** Review Conceptual Example 5 and **Figure 23.13**. Find the ratio of the current in circuit I to the current in circuit II in the *high-frequency limit* for the same generator voltage.

39. **E** **SSM** At what frequency (in Hz) are the reactances of a 52-mH inductor and a 76-μF capacitor equal?

40. **E** **GO** The resistor in a series RCL circuit has a resistance of 92 Ω, while the voltage of the generator is 3.0 V. At resonance, what is the average power delivered to the circuit?

41. **E** Two ac generators supply the same voltage. However, the first generator has a frequency of 1.5 kHz, and the second has a frequency of 6.0 kHz. When an inductor is connected across the terminals of the first generator, the current delivered is 0.30 A. How much current is delivered when this inductor is connected across the terminals of the second generator?

42. **M** **V-HINT** A series circuit contains only a resistor and an inductor. The voltage V of the generator is fixed. If $R = 16\ \Omega$ and $L = 4.0$ mH, find the frequency at which the current is one-half its value at zero frequency.

43. **M** **SSM** A series RCL circuit contains a 5.10-μF capacitor and a generator whose voltage is 11.0 V. At a resonant frequency of 1.30 kHz the power delivered to the circuit is 25.0 W. Find the values of **(a)** the inductance and **(b)** the resistance. **(c)** Calculate the power factor when the generator frequency is 2.31 kHz.

44. **M** **GO** **SSM** Part *a* of the figure shows a heterodyne metal detector being used. As part *b* of the figure illustrates, this device utilizes two capacitor/inductor oscillator circuits, A and B. Each produces its own resonant frequency, $f_{0A} = 1/[2\pi(L_A C)^{1/2}]$ and $f_{0B} = 1/[2\pi(L_B C)^{1/2}]$. Any difference between these frequencies is detected through earphones as a beat frequency $|f_{0B} - f_{0A}|$. In the absence of any nearby metal object, the inductances L_A and L_B are identical. When inductor B (the search coil) comes near a piece of metal, the inductance L_B increases, the corresponding oscillator frequency f_{0B} decreases, and a beat frequency is heard. Suppose that initially each inductor is adjusted so that $L_B = L_A$, and each oscillator has a resonant frequency of 855.5 kHz. Assuming that the inductance of search coil B increases by 1.000% due to a nearby piece of metal, determine the beat frequency heard through the earphones.

(a)

PROBLEM 44

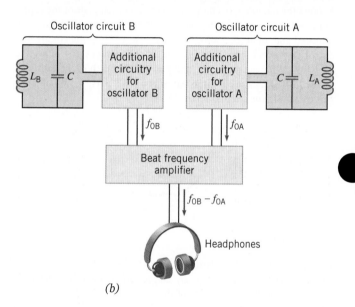

(b)

Physics in Biology, Medicine, and Sports

45. **M** **BIO** **V-HINT** **23.3** In one measurement of the body's bioelectric impedance, values of $Z = 4.50 \times 10^2\ \Omega$ and $\phi = -9.80°$ are obtained for the total impedance and the phase angle, respectively. These values assume that the body's resistance R is in series with its capacitance C and that there is no inductance L. Determine the body's resistance and capacitive reactance.

46. **M** **BIO** **23.2** A high-field superconducting magnet in an MRI machine has an inductance of 60.0 H. If this magnet were connected to a generator with an rms voltage of 2.50×10^2 V, what would be the maximum current in the magnet, if the frequency of the generator is 17.0 Hz?

47. **M** **BIO** **23.3** In Chapter 20, we analyzed a cardiac pacemaker that uses a 4.1×10^{-7}-F capacitor and a 1.8-MΩ resistor connected in series. When an rms voltage of 3.7 V is supplied to this circuit, the rms current is 1.7×10^{-6} A. What is the **(a)** impedance and **(b)** the capacitive reactance of this circuit? **(c)** What is the frequency of the rms voltage?

48. **E** **BIO** **23.2** As medical technologies advance, there is a greater need for remote monitoring of patient vital signs and the equipment that provides their care. This has led to an increase in high frequency communication signals from systems such as Bluetooth and WiFi. These high frequency signals can potentially interfere with sensitive medical equipment. One common approach to combating high frequency noise in the power input to sensitive devices is to use something called a *choke*. Chokes come in multiple types, but one of the simplest is just an inductor. Consider an inductor that is placed in series with a power supply for a medical monitor. The inductor is designed to filter out high frequency noise. An inductor with an inductance of 5.5×10^{-3} H is used to filter noise in a power line with a frequency of 150 kHz. **(a)** What is the inductive reactance? **(b)** What is the I_{rms} current in the circuit, if the power cord is plugged into a standard wall outlet with $V_{rms} = 120$ V?

49. M BIO **23.4** The game-day coordinator at NFL football games communicates wirelessly with officials on the field using a frequency of about 467 MHz. **(a)** If the detection circuit in the communication equipment uses a 2.75-pF capacitor, what inductance would be required for a resonant frequency of 467 MHz? **(b)** If during one game, the resonant frequency had to be changed to 457 MHz, and assuming the capacitance in the detection circuit could be varied, what would be the new capacitance for this lower frequency?

50. M BIO **23.4** The ionosphere is the ionized part of the upper layer of the earth's atmosphere. The air molecules there are ionized by solar radiation. This layer of the atmosphere is a fairly good conductor, and radio waves are often "bounced" off the bottom of the ionosphere back toward the earth, in a process called skip or skywave propagation. Due to these properties, the space between the surface of the earth and the bottom of the ionosphere acts like a closed waveguide that will exhibit resonance for very low frequencies. Resonance excitations in the cavity are caused by lightning strikes, which hit the earth about 50 to 100 times a second. These low atmospheric resonance frequencies are known as *Schumann resonances*, named after the physicist Winfried Otto Schumann, who

first calculated them in 1952. There are several Schumann frequencies that occur in the low frequency background, which ranges from 3 to 60 Hz. The highest intensity resonance mode (called the fundamental) occurs at 7.83 Hz. Other Schumann resonances occur at 14.3, 20.8, 27.3, and 33.8 Hz. Around 20 years ago, research showed that the majority of biological systems exhibit electrical activity in the same range as the Schumann resonances. New theoretical work suggests that primordial cells synchronized their electrical activity to the natural atmospheric resonances, and this evolutionary process is evident today across a gigantic scale—from single-celled organisms to the electrical activity in our brains. The surface of the earth and the bottom of the ionosphere act like a giant capacitor, and we can model one path that stretches between them as a circuit consisting of a resistor and capacitor connected in series. Assume the atmosphere has an average resistance of 2.00×10^2 Ω, and the capacitance is equal to 0.070 F. **(a)** What is the capacitive reactance of the circuit when it is active at the fundamental Schumann resonance frequency? **(b)** What is the impedance of this circuit? **(c)** What is the rms value of the current in the circuit, if the maximum voltage is 3.50×10^5 V?

Concepts and Calculations Problems   Online

A capacitor is one of the important elements found in ac circuits. As we have seen in this chapter, the capacitance of a capacitor influences the amount of current in a circuit. The capacitance, in turn, is determined by the geometry of the capacitor and the material that fills the space between its plates, as Section 19.5 discusses. When capacitors are connected together, the equivalent capacitance depends on the nature of the connection—for example, whether it is a series or parallel connection. Problem 51 provides a review of these issues concerning capacitors. In ac circuits that contain capacitance, inductance, and resistance, it is only the resistance that, on the average, consumes power. The average power delivered to a capacitor or an inductor is zero. However, the presence of a capacitor and/or an inductor does influence the rms current in the circuit. When the current changes for any reason, the power in the resistor also changes, as Problem 52 illustrates.

51. M CHALK Two parallel plate capacitors are filled with the same dielectric material and have the same plate area. However, the plate separation of capacitor 1 is twice that of capacitor 2. When capacitor 1 is connected across the terminals of an ac generator, the generator

delivers an rms current of 0.60 A. *Concepts:* (i) Which of the two capacitors has the greater capacitance? (ii) Is the equivalent capacitance of the parallel combination (C_P) greater or smaller than the capacitance of capacitor 1? (iii) Is the capacitive reactance of C_P greater or smaller than for C_1? (iv) When both capacitors are connected in parallel across the terminals of the generator, is the current from the generator greater or smaller than when capacitor 1 is connected alone? *Calculations:* What is the current delivered by the generator when both capacitors are connected in parallel across the terminals?

52. M CHALK SSM An ac generator has a frequency of 1200 Hz and a constant rms voltage. When a 470-Ω resistor is connected between the terminals of the generator, an average power of 0.25 W is consumed by the resistor. Then, a 0.080-H inductor is connected in series with the resistor, and the combination is connected between the generator terminals. *Concepts:* (i) In which case does the generator deliver a greater rms current? (ii) In which case is the greater average power consumed by the circuit? *Calculations:* What is the average power consumed in the inductor–resistor series circuit?

Team Problems   Online

53. E T WS **Power in Extreme Frequency Limits.** You and your team have been assigned to find a power supply for the circuit in the drawing, which can be used to supply a dc voltage at 15.0 V, or a high frequency ac signal with a root-mean-square (rms) voltage of 15.0 V. The components in the circuit have the following values: $R = 4.60$ Ω, $C = 20.0$ nF, and $L = 22.0$ mH. Your task is to estimate the peak wattage (i.e., power) required of the power supply for

PROBLEM 53

(a) the dc and **(b)** the high frequency signals. Conceptual Example 5 will provide insight into this problem.

54 M T WS **Extending the Range.** You and your team are tasked with repairing part of a communications circuit that consists of an inductor and a resistor in series, as shown in the drawing. The ac power source supplies a sinusoidal root-mean-square voltage of 30.0 V at an adjustable frequency between 0 Hz and 775 kHz. The problem was that the resistor in the circuit burned out as the ac source

PROBLEM 54

frequency was being reduced in order to produce a signal at some low frequency. The resistor had a resistance of $R = 50\ \Omega$, and an instantaneous, or peak, power rating of 0.50 W, meaning that it would burn up if forced to dissipate power above its power rating. The inductor has an inductance of $L = 3.2$ mH. **(a)** What was the minimum frequency (in Hz) that this circuit was originally designed to accommodate without burning the resistor? **(b)** Your supplies are very limited but you were able to find a few other resistors including two 25-Ω resistors (each rated for a peak power of 0.25 W), a 50-Ω (0.50 W) resistor, and two 100-Ω resistors (0.50 W). Assume all resistor values are known to two significant figures. Find a way to replace the burned resistor with a configuration of the resistors you have available that will have the same effective resistance, but will extend the low frequency limit of the circuit (i.e., so that the circuit can go to lower frequencies without burning a resistor). What is the new low frequency limit?

55. **M T A Communications Jammer.** Radio jamming is the intentional disruption or interference of radio communications by overwhelming the intended receivers of the signal with random noise. You and your team have been tasked with jamming a specific radio signal at 720 kHz. You have access to a high-powered transmitter, but the part of its circuitry that tunes the broadcast frequency, called the tank circuit, has been damaged. A tank circuit is a series RCL circuit whose resonance frequency determines the frequency broadcasted by the antenna. At your disposal are two 220-Ω resistors, one variable capacitor that ranges from 2.0 to 6.0 nF, and four inductors with the following values: $L_1 = 5.0 \times 10^{-6}$ H, $L_2 = 7.2 \times 10^{-6}$ H, $L_3 = 6.5 \times 10^{-5}$ H, and $L_4 = 5.4 \times 10^{-6}$ H. **(a)** If you set your variable capacitor at the center of its range, what must be the value of the inductance of your RCL circuit so that it resonates at 720 kHz? **(b)** How should you configure the available inductors to give you the needed equivalent inductance? (Hint: the rules for adding inductors in series and parallel are the same as for resistors.) **(c)** With the inductance set as calculated in (a), what resonant frequency range does the variable capacitor provide? **(d)** The two resistors can be configured to give different equivalent resistance values. How should you configure the resistors in the RCL circuit in order to maximize the current at the resonant frequency? (Refer to Section 23.4.)

56. **M T A Reconfigurable RCL Circuit.** A series RCL circuit is composed of a resistor ($R = 220.0\ \Omega$), two identical capacitors ($C = 3.50$ nF) connected in series, and two identical inductors ($L = 5.50 \times 10^{-5}$ H) connected in series. You and your team need to determine: **(a)** the resonant frequency of this configuration. **(b)** What are all of the other possible resonant frequencies that can be attained by reconfiguring the capacitors and inductors (while using all of the components and keeping the proper series RCL order)? **(c)** If you were to design a circuit using only one of the given inductors and one adjustable capacitor, what would the range of the variable capacitor need to be in order to cover all of the resonant frequencies found in (a) and (b)?

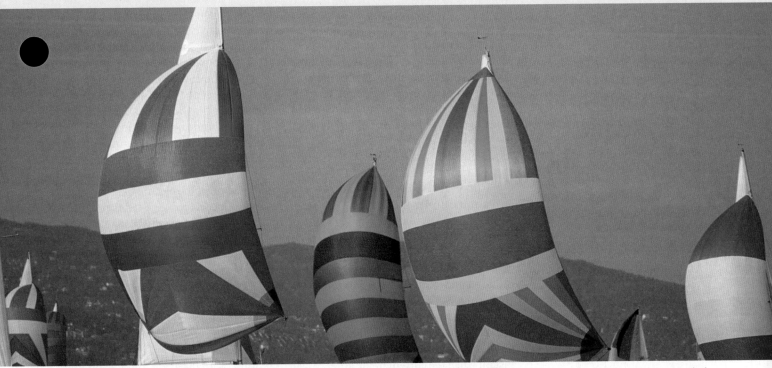

Each of the colors on the sails of these boats corresponds to a different wavelength in the visible region of the spectrum of electromagnetic waves. As we will see in this chapter, however, the visible wavelengths comprise only a small part of the total electromagnetic spectrum.

Andrew Woodley/Alamy Stock Photo

Electromagnetic Waves

24.1 The Nature of Electromagnetic Waves

In Section 13.3 we saw that energy is transported to us from the sun via a class of waves known as electromagnetic waves. This class includes the familiar visible, ultraviolet, and infrared waves. In Sections 18.6, 21.1, and 21.2 we studied the concepts of electric and magnetic fields. It was the great Scottish physicist James Clerk Maxwell (1831–1879) who showed that these two fields fluctuating together can form a propagating **electromagnetic wave**. We will now bring together our knowledge of electric and magnetic fields in order to understand this important type of wave.

Animated Figure 24.1 illustrates one way to create an electromagnetic wave. The setup consists of two straight metal wires that are connected to the terminals of an ac generator and serve as an antenna. The potential difference between the terminals changes sinusoidally with time t and has a period T. Part a shows the instant $t = 0$ s, when there is no charge at the ends of either wire. Since there is no charge, there is

(a) $t = 0$ s

(b) $t = \frac{1}{4}T$ $\vec{\mathbf{E}}$

(c) $t = \frac{2}{4}T$

(d) $t = \frac{3}{4}T$

(e) $t = T$

ANIMATED FIGURE 24.1 In each part of the drawing, the red arrow represents the electric field $\vec{\mathbf{E}}$ produced at point P by the oscillating charges on the antenna at the indicated time. The black arrows represent the electric fields created at earlier times. For simplicity, only the fields propagating to the right are shown.

FIGURE 24.2 The oscillating current I in the antenna wires creates a magnetic field $\vec{\mathbf{B}}$ at point P that is tangent to a circle centered on the wires. The field is directed as shown when the current is upward and is directed in the opposite direction when the current is downward.

no electric field at the point P just to the right of the antenna. As time passes, the top wire becomes positively charged and the bottom wire negatively charged. One-quarter of a cycle later ($t = \frac{1}{4}T$), the charges have attained their maximum values, as part b of the drawing indicates. The corresponding electric field $\vec{\mathbf{E}}$ at point P is represented by the red arrow and has increased to its maximum strength in the downward direction.* Part b also shows that the electric field created at earlier times (see the black arrow in the picture) has not disappeared but has moved to the right. Here lies the crux of the matter: At distant points, the electric field of the charges is not felt immediately. Instead, the field is created first near the wires and then, like the effect of a pebble dropped into a pond, moves outward as a wave in all directions. Only the field moving to the right is shown in the picture for the sake of clarity.

Parts c–e of **Animated Figure 24.1** show the creation of the electric field at point P (red arrow) at later times during the generator cycle. In each part, the fields produced earlier in the sequence (black arrows) continue propagating toward the right. Part d shows the charges on the wires when the polarity of the generator has reversed, so the top wire is negative and the bottom wire is positive. As a result, the electric field at P has reversed its direction and points upward. In part e of the sequence, a complete sine wave has been drawn through the tips of the electric field vectors to emphasize that the field changes sinusoidally.

Along with the electric field in **Animated Figure 24.1**, a magnetic field $\vec{\mathbf{B}}$ is also created, because the charges flowing in the antenna constitute an electric current, which produces a magnetic field. **Figure 24.2** illustrates the field direction at point P at the instant when the current in the antenna wire is upward. With the aid of Right-Hand Rule No. 2 (thumb of the right hand points along the current I, fingers curl in the direction of $\vec{\mathbf{B}}$), the magnetic field at P can be seen to point into the page. As the oscillating current changes, the magnetic field changes accordingly. The magnetic fields created at earlier times propagate outward as a wave, just as the electric fields do.

Notice that the magnetic field in **Figure 24.2** is perpendicular to the page, whereas the electric field in **Animated Figure 24.1** lies in the plane of the page. Thus, the electric and magnetic fields created by the antenna are mutually perpendicular and remain so as they move outward. Moreover, both fields are perpendicular to the direction of travel. These perpendicular electric and magnetic fields, moving together, constitute an electromagnetic wave.

The electric and magnetic fields in **Animated Figure 24.1** and **Figure 24.2** decrease to zero rapidly with increasing distance from the antenna. Therefore, they exist mainly near the antenna and together are called the **near field**. Electric and magnetic fields do form a wave at large distances from the antenna, however. These fields arise from an effect that is different from the one that produces the near field and are referred to as the **radiation field**. Faraday's law of induction provides part of the basis for the radiation field. As Section 22.4 discusses, this law describes the emf or potential difference produced by a changing magnetic field. And, as Section 19.4 explains, a potential difference can be related to an electric field. Thus, a changing magnetic field produces an electric field. Maxwell predicted that the reverse effect also occurs—namely, that a changing electric field produces a magnetic field. The radiation field arises because the changing magnetic field creates an electric field that fluctuates in time and the changing electric field creates the magnetic field.

Figure 24.3 shows the electromagnetic wave of the radiation field far from the antenna. The picture shows only the part of the wave traveling along the $+x$ axis. The parts traveling in the other directions have been omitted for clarity. It should be clear from the drawing that *an electromagnetic wave is a transverse wave* because the electric and magnetic fields are both perpendicular to the direction in which the wave travels. Moreover, an electromagnetic wave, unlike a wave on a string or a sound wave, does not require a medium in which to propagate. *Electromagnetic waves can travel through a vacuum or a material substance*, since electric and magnetic fields can exist in either one.

Electromagnetic waves can be produced in situations that do not involve a wire antenna. In general, any electric charge that is accelerating emits an electromagnetic wave, whether the charge is inside a wire or not. In an alternating current, an electron

---

*The direction of the electric field can be obtained by imagining a positive test charge at P and determining the direction in which it would be pushed because of the charges on the wires.

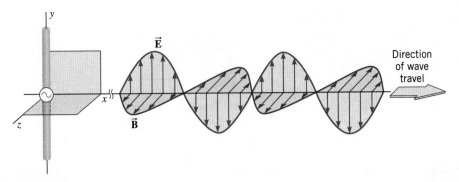

FIGURE 24.3 This picture shows the wave of the radiation field far from the antenna. Observe that $\vec{E}$ and $\vec{B}$ are perpendicular to each other, and both are perpendicular to the direction of travel.

oscillates in simple harmonic motion along the length of the wire and is one example of an accelerating charge.

All electromagnetic waves move through a vacuum at the same speed, and the symbol c is used to denote its value. This speed is called the ***speed of light in a vacuum*** and is $c = 3.00 \times 10^8$ m/s. In air, electromagnetic waves travel at nearly the same speed as they do in a vacuum, but, in general, they move through a substance such as glass at a speed that is substantially less than c.

The frequency of an electromagnetic wave is determined by the oscillation frequency of the electric charges at the source of the wave. In **Figures 24.1–24.3** the wave frequency would equal the frequency of the ac generator. Suppose, for example, that the antenna is broadcasting electromagnetic waves known as radio waves. The frequencies of AM radio waves lie between 545 and 1605 kHz, these numbers corresponding to the limits of the AM broadcast band on the radio dial. In contrast, frequencies of FM radio waves lie between 88 and 108 MHz on the dial. Television channels 2–6, on the other hand, utilize electromagnetic waves with frequencies between 54 and 88 MHz, and channels 7–13 use frequencies between 174 and 216 MHz.

THE PHYSICS OF . . . radio and television reception. Radio and television reception involves a process that is the reverse of that outlined earlier for the creation of electromagnetic waves. When broadcasted waves reach a receiving antenna, they interact with the electric charges in the antenna wires. Either the electric field or the magnetic field of the waves can be used. To take full advantage of the electric field, the wires of the receiving antenna must be parallel to the electric field, as **Figure 24.4** indicates. The electric field acts on the electrons in the wire, forcing them to oscillate back and forth along the length of the wire. Thus, an ac current exists in the antenna and the circuit connected to it. The variable-capacitor C (—||—) and the inductor L in the circuit provide one way to select the frequency of the desired electromagnetic wave. By adjusting the value of the capacitance, it is possible to adjust the corresponding resonant frequency f_0 of the circuit [$f_0 = 1/(2\pi\sqrt{LC})$, Equation 23.10] to match the frequency of the wave. Under the condition of resonance, there will be a maximum oscillating current in the inductor. Because of mutual inductance, this current creates a maximum voltage in the second coil in the drawing, and this voltage can then be amplified and processed by the remaining radio or television circuitry.

FIGURE 24.4 A radio wave can be detected with a receiving antenna wire that is parallel to the electric field of the wave. The magnetic field of the radio wave has been omitted for simplicity.

FIGURE 24.5 With a receiving antenna in the form of a loop, the magnetic field of a broadcasted radio wave can be detected. The normal to the plane of the loop should be parallel to the magnetic field for best reception. For clarity, the electric field of the radio wave has been omitted.

Peter Barritt/Alamy Stock Photo

FIGURE 24.6 This ship's mast indicates that both straight and loop antennas are being used to communicate with other vessels and on-shore stations.

To detect the magnetic field of a broadcasted radio wave, a receiving antenna in the form of a loop can be used, as **Figure 24.5** shows. For best reception, the normal to the plane of the wire loop must be oriented parallel to the magnetic field. Then, as the wave sweeps by, the magnetic field penetrates the loop, and the changing magnetic flux induces a voltage and a current in the loop, in accord with Faraday's law. Once again, the resonant frequency of a capacitor/inductor combination can be adjusted to match the frequency of the desired electromagnetic wave. Both straight wire and loop antennas can be seen on the ship's mast in **Figure 24.6**.

BIO **THE PHYSICS OF . . . cochlear implants.** Cochlear implants use the broadcasting and receiving of radio waves to provide assistance to hearing-impaired people who have auditory nerves that are at least partially intact. These implants utilize radio waves to bypass the damaged part of the hearing mechanism and access the auditory nerve directly, as **Interactive Figure 24.7** illustrates. An external microphone (often set into an ear mold) detects sound waves and sends a corresponding electrical signal to a speech processor small enough to be carried in a pocket. The speech processor encodes these signals into a radio wave, which is broadcast from an external transmitter coil placed over the site of a miniature receiver (and its receiving antenna) that has been surgically inserted beneath the skin. The receiver acts much like a radio. It detects the broadcasted wave and from the encoded audio information produces electrical signals that represent the sound wave. These signals are sent along a wire to electrodes that are implanted in the cochlea of the inner ear. The electrodes stimulate the auditory nerves that feed directly between structures within the cochlea and the brain. To the extent that the nerves are intact, a person can learn to recognize sounds.

BIO **THE PHYSICS OF . . . wireless capsule endoscopy.** The broadcasting and receiving of radio waves are also now being used in the practice of endoscopy. In this medical diagnostic technique a device called an endoscope is

INTERACTIVE FIGURE 24.7 Hearing-impaired people can sometimes recover part of their hearing with the help of a cochlear implant. Broadcasting and receiving electromagnetic waves play central roles in this device.

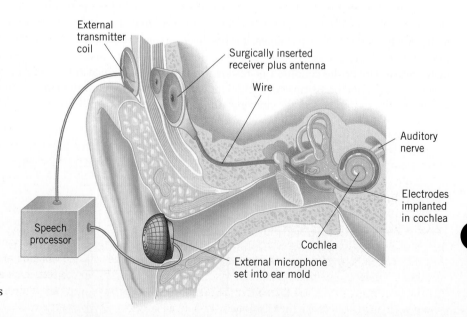

used to peer inside the body. For example, to examine the interior of the colon for signs of cancer, a conventional endoscope (known as a colonoscope) is inserted through the rectum. (See Section 26.3.) The wireless capsule endoscope shown in **Figure 24.8** bypasses this invasive procedure completely. With a size of about 11×26 mm, this capsule can be swallowed and carried through the gastrointestinal tract by the involuntary contractions of the walls of the intestines (peristalsis). The capsule is self-contained and uses no external wires. A marvel of miniaturization, it contains a radio transmitter and its associated antenna, batteries, a white-light-emitting diode (see Section 23.5) for illumination, and an optical system to capture the digital images. As the capsule moves through the intestine, the transmitter broadcasts the images to an array of small receiving antennas attached to the patient's body. These receiving antennas also are used to determine the position of the capsule within the body. The radio waves that are used lie in the ultrahigh-frequency, or UHF, band, from 3×10^8 to 3×10^9 Hz.

Radio waves are only one part of the broad spectrum of electromagnetic waves that has been discovered. The next section discusses the entire spectrum.

FIGURE 24.8 This wireless capsule endoscope is designed to be swallowed. As it passes through a patient's intestines, it broadcasts video images of the interior of the intestines.

Check Your Understanding

(The answers are given at the end of the book.)

1. Refer to **Animated Figure 24.1**. Between the times indicated in parts *c* and *d* in the drawing, what is the direction of the magnetic field at the point *P* for the electromagnetic wave being generated? Is it directed **(a)** upward along the length of the wire, **(b)** downward along the length of the wire, **(c)** into the plane of the paper, or **(d)** out of the plane of the paper?

2. A transmitting antenna is located at the origin of an *x, y, z* axis system and broadcasts an electromagnetic wave whose electric field oscillates along the *y* axis. The wave travels along the $+x$ axis. Three possible wire loops are available for use with an LC-tuned circuit to detect this wave: **(a)** a loop that lies in the *xy* plane, **(b)** a loop that lies in the *xz* plane, and **(c)** a loop that lies in the *yz* plane. Which one of the loops will detect the wave?

3. Why does the peak value of the emf induced in a loop antenna (see **Figure 24.5**) depend on the frequency of the electromagnetic wave?

24.2 | The Electromagnetic Spectrum

An electromagnetic wave, like any periodic wave, has frequency *f* and wavelength λ that are related to the speed v of the wave by $v = f\lambda$ (Equation 16.1). For electromagnetic waves traveling through a vacuum or, to a good approximation, through air, the speed is $v = c$, so $c = f\lambda$.

As **Figure 24.9** shows, electromagnetic waves exist with an enormous range of frequencies, from values less than 10^4 Hz to greater than 10^{24} Hz. Since all these waves travel through a vacuum at the same speed of $c = 3.00 \times 10^8$ m/s, Equation 16.1 can be used to find the correspondingly wide range of wavelengths that the picture also displays. The ordered series of electromagnetic wave frequencies or wavelengths in **Figure 24.9** is called the **electromagnetic spectrum**. Historically, regions of the spectrum have been given names such as radio waves and infrared waves. Although the boundary between adjacent regions is shown as a sharp line in the drawing, the boundary is not so well defined in practice, and the regions often overlap.

Beginning on the left in **Figure 24.9**, we find radio waves. Lower-frequency radio waves are generally produced by electrical oscillator circuits, while higher-frequency radio waves (called microwaves) are usually generated using electron tubes called klystrons. Infrared radiation, sometimes loosely called heat waves, originates with the vibration and rotation of molecules within a material. Visible light is emitted by hot objects, such as the sun, a burning log, or the filament of an incandescent light bulb, when the temperature is high enough to excite the electrons within an atom. Ultraviolet frequencies can be

FIGURE 24.9 The electromagnetic spectrum.

produced from the discharge of an electric arc. X-rays are produced by the sudden deceleration of high-speed electrons. And, finally, gamma rays are radiation from nuclear decay.

THE PHYSICS OF . . . astronomy and the electromagnetic spectrum. Astronomers use the different regions of the electromagnetic spectrum to gather information about distant celestial objects. **Interactive Figure 24.10**, for example, shows four views of the Crab Nebula, each in a different region of the spectrum. The Crab Nebula is located 6.0×10^{16} km away from the earth and is the remnant of a star that underwent a supernova explosion in 1054 AD.

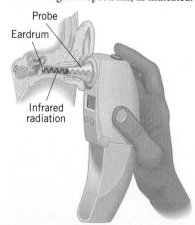

| (a) Radio wave | (b) Infrared | (c) Visible | (d) X-ray |

INTERACTIVE FIGURE 24.10 Four views of the Crab Nebula. Each view is in a different region of the electromagnetic spectrum, as indicated.

FIGURE 24.11 A pyroelectric thermometer measures body temperature by determining the amount of infrared radiation emitted by the eardrum and surrounding tissue.

BIO **THE PHYSICS OF . . . a pyroelectric ear thermometer.** The human body, like any object, radiates infrared radiation, and the amount emitted depends on the temperature of the body. Although infrared radiation cannot be seen by the human eye, it can be detected by sensors. An ear thermometer, like the pyroelectric thermometer shown in **Figure 24.11**, measures the body's temperature by determining the amount of infrared radiation that emanates from the eardrum and surrounding tissue. The ear is one of the best places for this measurement because it is close to the hypothalamus, an area at the bottom of the brain that controls body temperature. The ear is also not cooled or warmed by eating, drinking, or breathing. When the probe of the thermometer is inserted into the ear canal, infrared radiation travels down the barrel of the probe and strikes the sensor. The absorption of infrared radiation warms the sensor, and, as a result, its electrical conductivity changes. The change in electrical conductivity is measured by an electronic circuit. The output from the circuit is sent to a microprocessor, which calculates the body temperature and displays the result on a digital readout.

Of all the frequency ranges in the electromagnetic spectrum, the most familiar is that of visible light, although it is the most narrow (see **Figure 24.9**). Only waves with frequencies between about 4.0×10^{14} Hz and 7.9×10^{14} Hz are perceived by the human eye as visible light. Usually visible light is discussed in terms of wavelengths (in vacuum) rather than frequencies. As Example 1 indicates, the wavelengths of visible light are extremely small and, therefore, are normally expressed in *nanometers* (nm); 1 nm = 10^{-9} m. An obsolete (non-SI) unit still occasionally used for wavelengths is the *angstrom* (Å); 1 Å = 10^{-10} m.

EXAMPLE 1 | The Wavelengths of Visible Light

Find the range in wavelengths (in vacuum) for visible light in the frequency range between 4.0×10^{14} Hz (red light) and 7.9×10^{14} Hz (violet light). Express the answers in nanometers.

Reasoning According to Equation 16.1, the wavelength (in vacuum) λ of a light wave is equal to the speed of light c in a vacuum divided by the frequency f of the wave, $\lambda = c/f$.

Solution The wavelength corresponding to a frequency of 4.0×10^{14} Hz is

$$\lambda = \frac{c}{f} = \frac{3.00 \times 10^8 \text{ m/s}}{4.0 \times 10^{14} \text{ Hz}} = 7.5 \times 10^{-7} \text{ m}$$

Since 1 nm = 10^{-9} m, it follows that

$$\lambda = (7.5 \times 10^{-7} \text{ m})\left(\frac{1 \text{ nm}}{10^{-9} \text{ m}}\right) = \boxed{750 \text{ nm}}$$

The calculation for a frequency of 7.9×10^{14} Hz is similar:

$$\lambda = \frac{c}{f} = \frac{3.00 \times 10^8 \text{ m/s}}{7.9 \times 10^{14} \text{ Hz}} = 3.8 \times 10^{-7} \text{ m} \quad \text{or} \quad \lambda = \boxed{380 \text{ nm}}$$

The eye/brain recognizes light of different wavelengths as different colors. A wavelength of 750 nm (in vacuum) is approximately the longest wavelength of red light, whereas 380 nm (in vacuum) is approximately the shortest wavelength of violet light. Between these limits are found the other familiar colors, as **Figure 24.9** indicates.

The association between color and wavelength in the visible part of the electromagnetic spectrum is well known. The wavelength also plays a central role in governing the behavior and use of electromagnetic waves in all regions of the spectrum. For instance, Conceptual Example 2 considers the influence of the wavelength on diffraction. Example 3 describes an important medical device that simultaneously employs two wavelengths of light in order to estimate the oxygen saturation in the bloodstream.

CONCEPTUAL EXAMPLE 2 | The Physics of AM and FM Radio Reception

As we have discussed in Section 17.3, diffraction is the ability of a wave to bend around an obstacle or around the edges of an opening. Based on that discussion, which type of radio wave would you expect to bend more readily around an obstacle such as a building, **(a)** an AM radio wave or **(b)** an FM radio wave?

Reasoning Section 17.3 points out that, other things being equal, sound waves exhibit diffraction to a greater extent when the wavelength is longer than when it is shorter. Based on this information, we expect that longer-wavelength electromagnetic waves will bend more readily around obstacles than will shorter-wavelength waves.

Answer (b) is incorrect. Figure 24.9 shows that FM radio waves have considerably shorter wavelengths than do AM waves. Therefore, FM radio waves exhibit less diffraction than AM waves do and bend less readily around obstacles.

Answer (a) is correct. Since AM radio waves have greater wavelengths than FM waves do (see **Figure 24.9**), they exhibit greater diffraction and bend more readily around obstacles than FM waves do.

EXAMPLE 3 | BIO Pulse Oximetry

The amount of oxygen carried by hemoglobin in the blood is a critical parameter that is precisely regulated by the human body. *Oxygen saturation* is the fraction of the total hemoglobin that is carrying bound oxygen and should have a value between 95% and 100%. Below 90% is considered to be low and is called *hypoxemia*,

and below 80% may result in organ failure. A noninvasive device that is used to measure oxygen saturation, called a *pulse oximeter*, clips onto a fleshy semitransparent part of the body, such as a fingertip (see **Figure 24.12**). The clip uses light-emitting diodes (LEDs) to pass light of two wavelengths, usually 6.60×10^{-7} m and 9.40×10^{-7} m (in air), through the blood-carrying tissue, and monitors how much light of each wavelength gets absorbed on the way through. Since oxygenated and deoxygenated hemoglobin absorb light of those wavelengths differently, the percentage of oxygenated hemoglobin can be determined. **(a)** What are the frequencies of light emitted by the LEDs? **(b)** In what regions of the electromagnetic spectrum do the two frequencies fall? Can one see with the naked eye both types of light being emitted by the device? (Caution: Never look directly into LED devices.) Refer to **Figure 24.9** to answer this question.

Reasoning (a) According to Equation 16.1, the frequency of a light wave can be calculated from its wavelength using $v = c = f\lambda$, where c is the speed of light in vacuum. **(b)** Comparing the resulting frequencies (or wavelengths) to the electromagnetic spectrum depicted in **Figure 24.9**, we can identify the regions of the electromagnetic spectrum in which the two light frequencies fall, and can then determine whether they are visible to the naked eye.

Solution (a) From Equation 16.1, for the light of wavelength $\lambda_1 = 6.60 \times 10^{-7}$ m, we have

$$f_1 = \frac{c}{\lambda_1} = \frac{3.00 \times 10^8 \text{ m/s}}{6.60 \times 10^{-7} \text{m}} = \boxed{4.55 \times 10^{14} \text{ Hz}}$$

Similarly, for the light of wavelength $\lambda_2 = 9.40 \times 10^{-7}$ m, we have

$$f_2 = \frac{c}{\lambda_2} = \frac{3.00 \times 10^8 \text{ m/s}}{9.40 \times 10^{-7} \text{m}} = \boxed{3.19 \times 10^{14} \text{ Hz}}$$

Photo by David Young

FIGURE 24.12 The sale of fingertip pulse oximeters that are designed for at-home use, like the one shown here, soared during the COVID-19 pandemic in 2020. One of the more dangerous symptoms of COVID-19 is respiratory pneumonia, which attacks a person's lungs and inhibits the blood oxygenation process. Careful monitoring of a person's blood oxygen saturation is critical in determining the level of care that is needed.

(b) Referring to **Figure 24.9**, the light with $f_1 = 4.55 \times 10^{14}$ Hz falls into the visible (red) part of the electromagnetic spectrum, whereas the light of $f_2 = 3.19 \times 10^{14}$ Hz falls into the infrared. Thus, the light of frequency f_1 can be seen with naked eye, and that with frequency f_2 cannot.

The picture of light as a wave is supported by experiments that will be discussed in Chapter 27. However, there are also experiments indicating that light can behave as if it were composed of discrete particles rather than waves. These experiments will be discussed in Chapter 29. Wave theories and particle theories of light have been around for hundreds of years, and it is now widely accepted that light, as well as other electromagnetic radiation, exhibits a dual nature. Either wave-like or particle-like behavior can be observed, depending on the kind of experiment being performed.

24.3 | The Speed of Light

At a speed of 3.00×10^8 m/s, light travels from the earth to the moon in a little over a second, so the time required for light to travel between two places on earth is very short. Therefore, the earliest attempts at measuring the speed of light had only limited success. One of the first accurate measurements employed a rotating mirror, and **Figure 24.13** shows a simplified version of the setup. It was used first by the French scientist Jean Foucault (1819–1868) and later in a more refined version by the American physicist Albert Michelson (1852–1931). If the angular speed of the rotating eight-sided mirror in **Figure 24.13** is adjusted correctly, light reflected from one side travels to the fixed mirror, reflects, and can be detected after reflecting from another side that has rotated into place at just the right time. The minimum angular speed must be such that one side of the mirror rotates one-eighth of a revolution during the time it takes for the light to make the round trip between the mirrors. For one of his experiments, Michelson placed his fixed mirror and rotating mirror on Mt. San Antonio and Mt. Wilson in California, a distance of 35 km apart. From a value of the minimum angular speed in such experiments, he obtained the value of $c = (2.997\ 96 \pm 0.000\ 04) \times 10^8$ m/s in 1926.

Today, the speed of light has been determined with such high accuracy that it is used to define the meter. As discussed in Section 1.2, the speed of light is now *defined* to be

FIGURE 24.13 Between 1878 and 1931, Michelson used a rotating eight-sided mirror to measure the speed of light. This is a simplified version of the setup.

Speed of light in a vacuum

$$c = 299\ 792\ 458 \text{ m/s}$$

However, a value of 3.00×10^8 m/s is adequate for most calculations. The second is defined in terms of a cesium clock, and the meter is then defined as the distance light travels in a vacuum during a time of $1/(299\ 792\ 458)$ s. Although the speed of light in a vacuum is large, it is finite, so it takes a finite amount of time for light to travel from one place to another. The travel time is especially long for light traveling between astronomical objects, as Conceptual Example 4 discusses.

CONCEPTUAL EXAMPLE 4 | Looking Back in Time

A supernova is a violent explosion that occurs at the death of certain stars. For a few days after the explosion, the intensity of the emitted light can become a billion times greater than that of our own sun. After several years, however, the intensity usually returns to zero. Supernovae are relatively rare events in the universe; only six have been observed in our galaxy within the past 400 years. A supernova that occurred in a neighboring galaxy, approximately 1.66×10^{21} m away, was recorded in 1987. **Figure 24.14** shows a photograph of the sky just a few hours after the explosion. Astronomers say that viewing an event like the supernova is like looking back in time. Which one of the following statements correctly describes what we see when we view such events? **(a)** The nearer the event is to the earth, the further back in time we are looking. **(b)** The farther the event is from the earth, the further back in time we are looking.

Reasoning The light from the supernova traveled to earth at a speed of $c = 3.00 \times 10^8$ m/s. The time t required for the light to travel the distance d between the event and the earth is $t = d/c$ and is proportional to the distance.

Answer (a) is incorrect. Since the time required for the light to travel the distance between the event and the earth is proportional to the distance, the light from near-earth events reaches us sooner rather than later. Therefore, the nearer the event is to the earth, the less into the past it allows us to see, contrary to what this answer implies.

Answer (b) is correct. The travel time for light from the supernova is

$$t = \frac{d}{c} = \frac{1.66 \times 10^{21}\ \text{m}}{3.00 \times 10^8\ \text{m/s}} = 5.53 \times 10^{12}\ \text{s}$$

Celestial Image Co./Science Source

FIGURE 24.14 True color image of the 1987 supernova (bright spot at the lower right). The larger cloud-like object near the middle left is the Tarantula nebula, whose light also takes approximately 175 000 years to reach the earth.

This corresponds to 175 000 years, so when astronomers saw the explosion in 1987, they were actually seeing the light that left the supernova 175 000 years earlier. In other words, they were looking back in time. Greater values for the distance d mean greater values for the time t.

Related Homework: Problem 12

In 1865, Maxwell determined theoretically that electromagnetic waves propagate through a vacuum at a speed given by

$$c = \frac{1}{\sqrt{\varepsilon_0 \mu_0}} \tag{24.1}$$

where $\varepsilon_0 = 8.85 \times 10^{-12}\ \text{C}^2/(\text{N} \cdot \text{m}^2)$ is the (electric) permittivity of free space and $\mu_0 = 4\pi \times 10^{-7}\ \text{T} \cdot \text{m/A}$ is the (magnetic) permeability of free space. Originally ε_0 was introduced in Section 18.5 as an alternative way of writing the proportionality constant k in Coulomb's law $[k = 1/(4\pi\varepsilon_0)]$ and, hence, plays a basic role in determining the strengths of the electric fields created by point charges. The role of μ_0 is similar for magnetic fields; it was introduced in Section 21.7 as part of a proportionality constant in the expression for the magnetic field created by the current in a long, straight wire. Substituting the values for ε_0 and μ_0 into Equation 24.1 shows that

$$c = \frac{1}{\sqrt{[8.85 \times 10^{-12}\ \text{C}^2/(\text{N} \cdot \text{m}^2)](4\pi \times 10^{-7}\ \text{T} \cdot \text{m/A})}} = 3.00 \times 10^8\ \text{m/s}$$

The experimental and theoretical values for c agree. Maxwell's success in predicting c provided a basis for inferring that light behaves as a wave consisting of oscillating electric and magnetic fields.

Check Your Understanding

(The answer is given at the end of the book.)

4. The frequency of electromagnetic wave A is twice that of electromagnetic wave B. For these two waves, what is the ratio λ_A/λ_B of the wavelengths in a vacuum? **(a)** $\lambda_A/\lambda_B = 2$, because wave A has twice the speed that wave B has. **(b)** $\lambda_A/\lambda_B = 2$, because wave A has one-half the speed that wave B has. **(c)** $\lambda_A/\lambda_B = \frac{1}{2}$, because wave A has one-half the speed that wave B has. **(d)** $\lambda_A/\lambda_B = \frac{1}{2}$, because wave A has twice the speed that wave B has. **(e)** $\lambda_A/\lambda_B = \frac{1}{2}$, because both waves have the same speed.

24.4 The Energy Carried by Electromagnetic Waves

FIGURE 24.15 A microwave oven. The rotating fan blades reflect the microwaves to all parts of the oven.

THE PHYSICS OF . . . a microwave oven.   Electromagnetic waves, like water waves or sound waves, carry energy. The energy is carried by the electric and magnetic fields that comprise the wave. In a microwave oven, for example, microwaves penetrate food and deliver their energy to it, as **Figure 24.15** illustrates. The electric field of the microwaves is largely responsible for delivering the energy, and water molecules in the food absorb it. The absorption occurs because each water molecule has a permanent dipole moment; that is, one end of a molecule has a slight positive charge, and the other end has a negative charge of equal magnitude. As a result, the positive and negative ends of different molecules can form a bond. However, the electric field of the microwaves exerts forces on the positive and negative ends of a molecule, causing it to spin. Because the field is oscillating rapidly—about 2.4×10^9 times a second—the water molecules are kept spinning at a high rate. In the process, the energy of the microwaves is used to break bonds between neighboring water molecules and ultimately is converted into internal energy. As the internal energy increases, the temperature of the water increases, and the food cooks.

BIO   THE PHYSICS OF . . . the greenhouse effect.   As discussed in Section 15.7, the energy carried by electromagnetic waves in the infrared and visible regions of the spectrum plays the key role in the greenhouse effect that is a contributing factor to global warming. The infrared waves from the sun are largely prevented from reaching the earth's surface by carbon dioxide and water in the atmosphere, which reflect them back into space. The visible waves do reach the earth's surface, however, and the energy they carry heats the earth. Heat also flows to the surface from the interior of the earth. The heated surface in turn radiates infrared waves outward, which, if they could, would carry their energy into space. However, the atmospheric carbon dioxide and water reflect these infrared waves back toward the earth, just as they reflect the infrared waves from the sun. Thus, their energy is trapped, and the earth becomes warmer, like plants in a greenhouse. In a greenhouse, however, energy is trapped mainly for a different reason—namely, the lack of effective convection currents to carry warm air past the cold glass walls.

A measure of the energy stored in the electric field $\vec{E}$ of an electromagnetic wave, such as a microwave, is provided by the electric energy density. As we saw in Section 19.5, this density is the electric energy per unit volume of space in which the electric field exists:

$$\frac{\text{Electric energy}}{\text{density}} = \frac{\text{Electric energy}}{\text{Volume}} = \frac{1}{2}\kappa\varepsilon_0 E^2 = \frac{1}{2}\varepsilon_0 E^2 \qquad \textbf{(19.12)}$$

In this equation, the dielectric constant κ has been set equal to unity, since we are dealing with an electric field in a vacuum (or in air). From Section 22.8, the analogous expression for the magnetic energy density is

$$\frac{\text{Magnetic energy}}{\text{density}} = \frac{\text{Magnetic energy}}{\text{Volume}} = \frac{1}{2\mu_0}B^2 \qquad \textbf{(22.11)}$$

The **total energy density** u of an electromagnetic wave in a vacuum is the sum of these two energy densities:

$$u = \frac{\text{Total energy}}{\text{Volume}} = \frac{1}{2}\varepsilon_0 E^2 + \frac{1}{2\mu_0}B^2 \qquad \textbf{(24.2a)}$$

In an electromagnetic wave propagating through a vacuum or air, the electric field and the magnetic field carry equal amounts of energy per unit volume of space. Since $\frac{1}{2}\varepsilon_0 E^2 = \frac{1}{2}(B^2/\mu_0)$, it is possible to rewrite Equation 24.2a for the total energy density in two additional, but equivalent, forms:

$$u = \varepsilon_0 E^2 \qquad \textbf{(24.2b)}$$

$$u = \frac{1}{\mu_0}B^2 \qquad \textbf{(24.2c)}$$

The fact that the two energy densities are equal implies that the electric and magnetic fields are related. To see how, we set the electric energy density equal to the magnetic energy density and obtain

$$\frac{1}{2}\varepsilon_0 E^2 = \frac{1}{2\mu_0}B^2 \quad \text{or} \quad E^2 = \frac{1}{\varepsilon_0\mu_0}B^2$$

However, according to Equation 24.1, $c = 1/\sqrt{\varepsilon_0\mu_0}$, so it follows that $E^2 = c^2 B^2$. Taking the square root of both sides of this result shows that the relation between the magnitudes of the electric and magnetic fields in an electromagnetic wave is

$$E = cB \qquad \textbf{(24.3)}$$

In an electromagnetic wave, the electric and magnetic fields fluctuate sinusoidally in time, so Equations 24.2a–c give the energy density of the wave at any instant in time. If an average value $\bar{u}$ for the total energy density is desired, average values are needed for E^2 and B^2. In Section 20.5 we faced a similar situation for alternating currents and voltages and introduced rms (root mean square) quantities. Using an analogous procedure here, we find that the rms values for the electric and magnetic fields, E_{rms} and B_{rms}, are related to the maximum values of these fields, E_0 and B_0, by

$$E_{\text{rms}} = \frac{1}{\sqrt{2}}E_0 \quad \text{and} \quad B_{\text{rms}} = \frac{1}{\sqrt{2}}B_0$$

Equations 24.2a–c can now be interpreted as giving the average energy density $\bar{u}$, provided the symbols E and B are interpreted to mean the rms values given above. The average energy density of the sunlight reaching the earth is determined in the next example.

EXAMPLE 5 | The Average Energy Density of Sunlight

Sunlight enters the top of the earth's atmosphere with an electric field whose rms value is $E_{\text{rms}} = 720$ N/C. Find **(a)** the average total energy density of this electromagnetic wave and **(b)** the rms value of the sunlight's magnetic field.

Reasoning The average total energy density $\bar{u}$ of the sunlight can be obtained with the aid of Equation 24.2b, provided the rms value is used for the electric field. Since the magnitudes of the magnetic and electric fields are related according to Equation 24.3, the rms value of the magnetic field is $B_{\text{rms}} = E_{\text{rms}}/c$.

Solution **(a)** According to Equation 24.2b, the average total energy density is

$$\bar{u} = \varepsilon_0 E_{\text{rms}}^2 = [8.85 \times 10^{-12}\ \text{C}^2/(\text{N} \cdot \text{m}^2)](720\ \text{N/C})^2$$

$$= \boxed{4.6 \times 10^{-6}\ \text{J/m}^3}$$

(b) Using Equation 24.3, we find that the rms magnetic field is

$$B_{\text{rms}} = \frac{E_{\text{rms}}}{c} = \frac{720\ \text{N/C}}{3.0 \times 10^8\ \text{m/s}} = \boxed{2.4 \times 10^{-6}\ \text{T}}$$

As an electromagnetic wave moves through space, it carries energy from one region to another. This energy transport is characterized by the **intensity** of the wave. We have encountered the concept of intensity before, in connection with sound waves in Section 16.7. The sound intensity is the sound power that passes perpendicularly through a surface divided by the area of the surface. The intensity of an electromagnetic wave is defined similarly. For an electromagnetic wave, the intensity is the electromagnetic power divided by the area of the surface.

FIGURE 24.16 In a time t, an electromagnetic wave moves a distance ct along the x axis and passes through a surface of area A.

Using this definition of intensity, we can show that the electromagnetic intensity S is related to the energy density u. According to Equation 16.8 the intensity is the power P that passes perpendicularly through a surface divided by the area A of that surface, or $S = P/A$. Furthermore, the power is equal to the total energy passing through the surface divided by the elapsed time t (Equation 6.10b), so that $P = $ (Total energy)$/t$. Combining these two relations gives

$$S = \frac{P}{A} = \frac{\text{Total energy}}{tA}$$

> **Problem-Solving Insight** The concepts of power and intensity are similar, but they are not the same. Intensity is the power that passes perpendicularly through a surface divided by the area of the surface.

Now, consider **Figure 24.16**, which shows an electromagnetic wave traveling in a vacuum along the x axis. In a time t the wave travels the distance ct, passing through the surface of area A. Consequently, the volume of space through which the wave passes is ctA. The total (electric and magnetic) energy in this volume is

Total energy = (Total energy density) × Volume = $u(ctA)$

Using this result in the expression for the intensity, we obtain

$$S = \frac{\text{Total energy}}{tA} = \frac{uctA}{tA} = cu \qquad (24.4)$$

Thus, the intensity and the energy density are related by the speed of light, c. If the average total energy density $(\overline{u})$ is used in Equation 24.4, then the average intensity $(\overline{S})$ can be determined, as illustrated in Example 6. Substituting Equations 24.2a–c, one at a time, into Equation 24.4 shows that the intensity of an electromagnetic wave depends on the electric and magnetic fields according to the following equivalent relations:

$$S = cu = \tfrac{1}{2}c\varepsilon_0 E^2 + \frac{c}{2\mu_0}B^2 \qquad (24.5a)$$

$$S = c\varepsilon_0 E^2 \qquad (24.5b)$$

$$S = \frac{c}{\mu_0}B^2 \qquad (24.5c)$$

If the rms values for the electric and magnetic fields are used in Equations 24.5a–c, the intensity becomes an average intensity, $\overline{S}$, as Example 7 illustrates.

EXAMPLE 6 | BIO The Physics of the Laser Scalpel

A laser scalpel is a tool used for general surgery, which may include cutting or vaporizing biological tissues using the energy from coherent laser light. This surgical technique is very precise and can minimize bleeding, swelling, and general discomfort commonly associated with traditional surgical techniques. A gas laser using CO_2 (carbon dioxide) is the highest power continuous-operation laser that is currently available. It operates at wavelengths of 9.4–10.6 μm, which correspond to frequencies that are readily absorbed by water molecules, and therefore soft tissues. One such laser is being used to remove a precancerous mole from a person's arm. The average power of the laser is 7.5 W, and it creates a circular beam (spot size) with a diameter of 0.25 mm. What is the average total energy density contained in the beam? How does this value compare to the average total energy density of sunlight?

Reasoning   The average total energy density $(\bar{u})$ of the laser light in the beam is related to the average intensity $(\bar{S})$ by Equation 24.4. We can then use Equation 16.8 to write the average intensity in terms of the average power of the laser.

Solution   According to Equation 24.4, the average total energy density is directly related to the average intensity: $\bar{S} = c\bar{u}$. Using Equation 16.8, we write the average intensity in terms of the average power, $\bar{S} = \dfrac{\bar{P}}{A}$, and then substitute this into the equation above: $\dfrac{\bar{P}}{A} = c\bar{u}$. Rearranging, we solve for the average total energy density: $\bar{u} = \dfrac{\bar{P}}{cA}$. Since the beam is circular, the cross-sectional area will be equal to $A = \pi r^2 = \pi\left(\dfrac{d}{2}\right)^2 = \dfrac{\pi d^2}{4}$, where d is the diameter of the circular spot size. Substituting this expression for the area into the average total energy density equation, we get our final result:

$$\bar{u} = \frac{\bar{P}}{cA} = \frac{4\bar{P}}{c\pi d^2} = \frac{4(7.5\ \text{W})}{\pi(3.0\times10^8\ \text{m/s})(0.25\times10^{-3}\ \text{m})^2} = \boxed{0.51\ \text{J/m}3}$$

FIGURE 24.17   Surgeon using a laser scalpel to operate on the heart.

From our result in Example 5 above, we see the laser has an average total energy density that is approximately 100 000 times greater than that of sunlight!

Analyzing Multiple-Concept Problems

EXAMPLE 7 | Power and Intensity

Figure 24.18 shows a tiny source that is emitting light uniformly in all directions. At a distance of 2.50 m from the source, the rms electric field strength of the light is 19.0 N/C. Assuming that the light does not reflect from anything in the environment, determine the average power of the light emitted by the source.

Reasoning   Recall from Section 16.7 that the power crossing a surface perpendicularly is equal to the intensity at the surface times the area of the surface (see Equation 16.8). Since the source emits light uniformly in all directions, the light intensity is the same at all points on the imaginary spherical surface in

Knowns and Unknowns   The following data are available:

Description	Symbol	Value
Rms electric field strength 2.50 m from light source	E_{rms}	19.0 N/C
Distance from light source	r	2.50 m

Unknown Variable

Average power emitted by light source	$\bar{P}$	?

Figure 24.18. Moreover, the light crosses this surface perpendicularly. Equation 24.5b relates the average light intensity at the surface to its rms electric field strength (which is known), and the area of the surface can be found from a knowledge of its radius.

FIGURE 24.18   At a distance of 2.50 m from the light source, the rms electric field of the light has a value of 19.0 N/C.

Modeling the Problem

STEP 1 Average Intensity   According to the discussion in Section 16.7, the average power $\bar{P}$ that passes perpendicularly through the imaginary spherical surface is equal to the average intensity $\bar{S}$ times the area A of the surface, or $\bar{P} = \bar{S}A$. The area of a spherical surface is $A = 4\pi r^2$, where r is the radius of the sphere. Thus, the average power can be written as in Equation 1 at the right. The radius is known, but the average light intensity is not, so we turn to Step 2 to evaluate it.

$$\bar{P} = \bar{S}(4\pi r^2) \qquad (1)$$

$$\boxed{?}$$

STEP 2 Average Intensity and Electric Field The average intensity $\overline{S}$ of the light passing through the imaginary spherical surface is related to the known rms electric field strength E_{rms} at the surface by Equation 24.5b:

$$\boxed{\overline{S} = c\varepsilon_0 E_{\text{rms}}^2}$$

where c is the speed of light in a vacuum and ε_0 is the permittivity of free space. We can substitute this expression into Equation 1, as indicated at the right.

$$\overline{P} = \overline{S}(4\pi r^2) \qquad (1)$$

$$\boxed{\overline{S} = c\varepsilon_0 E_{\text{rms}}^2}$$

Solution Algebraically combining the results of each step, we have

$$\overset{\text{STEP 1}}{\downarrow} \quad \overset{\text{STEP 2}}{\downarrow}$$

$$\overline{P} = \overline{S}(4\pi r^2) = c\varepsilon_0\, E_{\text{rms}}^2(4\pi r^2)$$

The average power emitted by the light source is

$$\overline{P} = c\varepsilon_0 E_{\text{rms}}^2(4\pi r^2)$$

$$= (3.00 \times 10^8 \text{ m/s})[8.85 \times 10^{-12} \text{ C}^2/(\text{N} \cdot \text{m}^2)](19.0 \text{ N/C})^2\, 4\pi(2.50 \text{ m})^2 = \boxed{75.3 \text{ W}}$$

Related Homework: Problems 24, 25

Check Your Understanding

(The answers are given at the end of the book.)

5. If both the electric and magnetic fields of an electromagnetic wave double in magnitude, how does the intensity of the wave change? The intensity **(a)** decreases by a factor of four **(b)** decreases by a factor of two **(c)** increases by a factor of two **(d)** increases by a factor of four **(e)** increases by a factor of eight.

6. Suppose that the electric field of an electromagnetic wave decreases in magnitude. Does the magnitude of the magnetic field **(a)** increase, **(b)** decrease, or **(c)** remain the same?

24.5 The Doppler Effect and Electromagnetic Waves

Section 16.9 presents a discussion of the Doppler effect that sound waves exhibit when either the source of a sound wave, the observer of the wave, or both are moving with respect to the medium of propagation (e.g., air). This effect is one in which the observed sound frequency is greater or smaller than the frequency emitted by the source. A different Doppler effect arises when the source moves than when the observer moves.

Electromagnetic waves also can exhibit a Doppler effect, but it differs from that for sound waves for two reasons. First, sound waves require a medium such as air in which to propagate. In the Doppler effect for sound, it is the motion (of the source, the observer, and the waves themselves) relative to this medium that is important. In the Doppler effect for electromagnetic waves, motion relative to a medium plays no role, because the waves do not require a medium in which to propagate. They can travel in a vacuum. Second, in the equations for the Doppler effect in Section 16.9, the speed of sound plays an important role, and it depends on the reference frame relative to which it is measured. For example, the speed of sound with respect to moving air is different than it is with respect to stationary air. As we will see in Section 28.2, electromagnetic waves behave in a different way. The speed at which they travel has the same value, whether it is measured relative to a stationary observer or relative to one moving at a constant velocity. For these two reasons, the same Doppler effect arises for electromagnetic waves when either the source or the observer of the waves moves; only the relative motion of the source and the observer with respect to one another is important.

When electromagnetic waves and the source and the observer of the waves all travel along the same line in a vacuum (or in air, to a good degree of approximation), the single equation that specifies the Doppler effect is

$$f_o = f_s\left(1 \pm \frac{v_{rel}}{c}\right) \quad \text{if } v_{rel} \ll c \qquad \qquad \textbf{(24.6)}$$

In this expression, f_o is the observed frequency, and f_s is the frequency emitted by the source. The symbol v_{rel} stands for the speed of the source and the observer relative to one another, and c is the speed of light in a vacuum. Equation 24.6 applies only if v_{rel} is very small compared to c—that is, if $v_{rel} \ll c$. Since v_{rel} is the **relative speed** of the source and the observer, it is like any **speed** and has no algebraic sign associated with it to denote the direction. The direction of the relative motion is taken into account by choosing the plus or minus sign in Equation 24.6. *The plus sign is used when the source and the observer come together, and the minus sign is used when they move apart.*

For instance, suppose that the source and the observer are both traveling due east, the source at a speed of 28 m/s with respect to the ground and the observer at a speed of 22 m/s with respect to the ground. Neither of these speeds with respect to the ground is used for the symbol v_{rel} in Equation 24.6. Instead, the value for v_{rel} is |28 m/s − 22 m/s| = 6 m/s. If the faster source is behind the slower observer, the source and the observer come together because the source is catching up. Therefore, the plus sign is chosen in Equation 24.6. On the other hand, if the slower observer is behind the faster source, the source and the observer move apart because the source is pulling away. In this case, the minus sign is chosen in Equation 24.6.

Math Skills   The choice of the plus or minus sign in Equation 24.6 is critical. Without the right choice, the equation cannot be used successfully to solve problems. Whether the plus or the minus sign is used depends on the nature of the individual problem, and the following two lists outline the possibilities.

PLUS SIGN (SOURCE AND OBSERVER COME TOGETHER)

(1) The source is catching up with the observer.
(2) The observer is catching up with the source.
(3) The source and the observer both move toward one another.

MINUS SIGN (SOURCE AND OBSERVER MOVE APART)

(1) The source is pulling away from the observer.
(2) The observer is pulling away from the source.
(3) The source and the observer both move away from one another.

THE PHYSICS OF . . . astronomy and the Doppler effect.   The Doppler effect of electromagnetic waves provides a powerful tool for astronomers. For instance, Example 11 in Chapter 5 discusses how astronomers have identified a supermassive black hole at the center of galaxy M87 by using the Hubble Space Telescope. They focused the telescope on regions to either side of the center of the galaxy (see Figure 5.15). From the light emitted by these two regions, they were able to use the Doppler effect to determine that one side is moving away from the earth, while the other side is moving toward the earth. In other words, the galaxy is rotating. The speeds of recession and approach enabled astronomers to determine the rotational speed of the galaxy, and Example 11 in Chapter 5 shows how the value for this speed leads to the identification of the black hole. Astronomers routinely study the Doppler effect of the light that reaches the earth from distant parts of the universe. From such studies, they have determined the speeds at which distant light-emitting objects are receding from the earth.

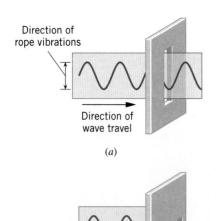

Direction of
rope vibrations

Direction of
wave travel

(a)

(b)

FIGURE 24.19 A transverse wave is linearly polarized when its vibrations always occur along one direction. (*a*) A linearly polarized wave on a rope can pass through a slit that is parallel to the direction of the rope vibrations, but (*b*) cannot pass through a slit that is perpendicular to the vibrations.

Single direction
for electric field

Direction of
wave travel

(*a*) Polarized light

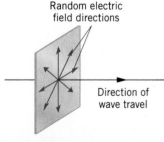

Random electric
field directions

Direction of
wave travel

(*b*) Unpolarized light

FIGURE 24.20 (*a*) In polarized light, the electric field of the electromagnetic wave fluctuates along a single direction. (*b*) Unpolarized light consists of short bursts of electromagnetic waves emitted by many different atoms. The electric field directions of these bursts are perpendicular to the direction of wave travel but are distributed randomly about it.

Check Your Understanding

(The answers are given at the end of the book.)

7. An astronomer measures the Doppler change in frequency for the light reaching the earth from a distant star. From this measurement, can the astronomer tell whether the star is moving away from the earth or the earth is moving away from the star?

8. CYU Figure 24.1 shows three situations—A, B, and C—in which an observer and a source of electromagnetic waves are moving along the same line. In each case the source emits a wave of the same frequency. The arrows in each situation denote velocity vectors relative to the ground and have magnitudes of either v or $2v$. Rank the magnitudes of the frequencies of the observed waves in descending order (largest first).

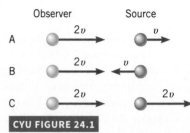

CYU FIGURE 24.1

24.6 Polarization

Polarized Electromagnetic Waves

One of the essential features of electromagnetic waves is that they are transverse waves, and because of this feature they can be polarized. **Figure 24.19** illustrates the idea of polarization by showing a transverse wave as it travels along a rope toward a slit. The wave is said to be **linearly polarized**, which means that its vibrations always occur along one direction. This direction is called the direction of polarization. In part *a* of the picture, the direction of polarization is vertical, parallel to the slit. Consequently, the wave passes through easily. However, when the slit is turned perpendicular to the direction of polarization, as in part *b*, the wave cannot pass, because the slit prevents the rope from oscillating. For longitudinal waves, such as sound waves, the notion of polarization has no meaning. In a longitudinal wave the direction of vibration is along the direction of travel, and the orientation of the slit would have no effect on the wave.

In an electromagnetic wave such as the one in **Figure 24.3**, the electric field oscillates along the *y* axis. Similarly, the magnetic field oscillates along the *z* axis. Therefore, the wave is linearly polarized, with the direction of polarization taken arbitrarily to be the direction along which the electric field oscillates. If the wave is a radio wave generated by a straight-wire antenna, the direction of polarization is determined by the orientation of the antenna. In comparison, the visible light given off by an incandescent light bulb consists of electromagnetic waves that are completely unpolarized. In this case the waves are emitted by a large number of atoms in the hot filament of the bulb. When an electron in an atom oscillates, the atom behaves as a miniature antenna that broadcasts light for brief periods of time, about 10^{-8} seconds. However, the directions of these atomic antennas change randomly as a result of collisions. Unpolarized light, then, consists of many individual waves, emitted in short bursts by many "atomic antennas," each with its own direction of polarization. **Figure 24.20** compares polarized and unpolarized light. In the unpolarized case, the arrows shown around the direction of wave travel symbolize the random directions of polarization of the individual waves that comprise the light.

Linearly polarized light can be produced from unpolarized light with the aid of certain materials. One commercially available material goes under the name of Polaroid. Such materials allow only the component of the electric field along one direction to pass through, while absorbing the field component perpendicular to this direction. As **Figure 24.21** indicates, the direction of polarization that a polarizing material allows through is called the **transmission axis**. No matter how this axis is oriented, the average intensity of the transmitted polarized light is one-half the average intensity of the incident unpolarized light. The reason for this is that the unpolarized light contains all polarization directions to an equal extent. Moreover, the electric field for each direction can be resolved into components perpendicular and parallel to the transmission axis, with the result that the average components perpendicular and parallel to the axis are equal. As a result, the polarizing material absorbs as much of the electric (and magnetic) field strength as it transmits.

Malus' Law

Once polarized light has been produced with a piece of polarizing material, it is possible to use a second piece to change the polarization direction and simultaneously adjust the intensity of the light. **Figure 24.22** shows how. As in this picture, the first piece of polarizing material is called the **polarizer** and the second piece is referred to as the **analyzer**. The transmission axis of the analyzer is oriented at an angle θ relative to the transmission axis of the polarizer. If the electric field strength of the polarized light incident on the analyzer is E, the field strength passing through is the component parallel to the transmission axis, or $E \cos \theta$. According to Equation 24.5b, the intensity is proportional to the square of the electric field strength. Consequently, the average intensity of polarized light passing through the analyzer is proportional to $\cos^2 \theta$. Thus, both the polarization direction and the intensity of the light can be adjusted by rotating the transmission axis of the analyzer relative to that of the polarizer. The average intensity $\overline{S}$ of the light leaving the analyzer, then, is

Malus' law $$\overline{S} = \overline{S}_0 \cos^2 \theta \qquad (24.7)$$

where $\overline{S}_0$ is the average intensity of the light entering the analyzer. Equation 24.7 is sometimes called **Malus' law**, for it was discovered by the French engineer Étienne Louis Malus (1775–1812). Example 8 illustrates the use of Malus' law.

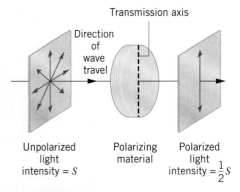

FIGURE 24.21 With the aid of a piece of polarizing material, polarized light may be produced from unpolarized light. The transmission axis of the material is the direction of polarization of the light that passes through the material.

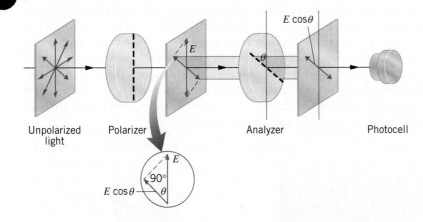

FIGURE 24.22 Two sheets of polarizing material, called the polarizer and the analyzer, may be used to adjust the polarization direction and intensity of the light reaching the photocell. This can be done by changing the angle θ between the transmission axes of the polarizer and analyzer.

EXAMPLE 8 | Using Polarizers and Analyzers

What value of θ should be used in **Figure 24.22**, so that the average intensity of the polarized light reaching the photocell will be one-tenth the average intensity of the unpolarized light?

Reasoning Both the polarizer and the analyzer reduce the intensity of the light. The polarizer reduces the intensity by a factor of one-half, as discussed earlier. Therefore, if the average intensity of the unpolarized light is $\overline{I}$, the average intensity of the polarized light leaving the polarizer and striking the analyzer is $\overline{S}_0 = \overline{I}/2$. The angle θ must now be selected so that the average intensity of the light leaving the analyzer will be $\overline{S} = \overline{I}/10$. Malus' law provides the solution.

Problem-Solving Insight Remember that when unpolarized light strikes a polarizer, only one-half of the incident light is transmitted, the other half being absorbed by the polarizer.

Solution Using $\overline{S}_0 = \overline{I}/2$ and $\overline{S} = \overline{I}/10$ in Malus' law, we find

$$\tfrac{1}{10}\overline{I} = \tfrac{1}{2}\overline{I} \cos^2 \theta$$

$$\tfrac{1}{5} = \cos^2 \theta \quad \text{or} \quad \theta = \cos^{-1}\left(\frac{1}{\sqrt{5}}\right) = \boxed{63.4°}$$

Diane Hirsch/Fundamental Photographs

INTERACTIVE FIGURE 24.23 When Polaroid sunglasses are uncrossed (left side), the transmitted light is dimmed due to the extra thickness of tinted plastic. However, when they are crossed (right side), the intensity of the transmitted light is reduced to zero because of the effects of polarization.

When $\theta = 90°$ in **Figure 24.22**, the polarizer and analyzer are said to be **crossed**, and no light is transmitted by the polarizer/analyzer combination. As an illustration of this effect, **Interactive Figure 24.23** shows two pairs of Polaroid sunglasses in uncrossed and crossed configurations.

THE PHYSICS OF . . . IMAX 3-D films. An exciting application of crossed polarizers is used in viewing IMAX 3-D movies. These movies are recorded on two separate rolls of film, using a camera that provides images from the two different perspectives that correspond to what is observed by human eyes and allow us to see in three dimensions. The camera has two apertures or openings located at roughly the spacing between our eyes. The films are projected using a projector with two lenses, as **Figure 24.24** indicates. Each lens has its own polarizer, and the two polarizers are crossed (see the drawing). In one type of theater, viewers watch the action on-screen using glasses with corresponding polarizers for the left and right eyes, as the drawing shows. Because of the crossed polarizers the left eye sees only the image from the left lens of the projector, and the right eye sees only the image from the right lens. Since the two images have the approximate perspectives that the left and right eyes would see in reality, the brain combines the images to produce a realistic 3-D effect.

FIGURE 24.24 In an IMAX 3-D film, two separate rolls of film are projected using a projector with two lenses, each with its own polarizer. The two polarizers are crossed. Viewers watch the action on-screen through glasses that have corresponding crossed polarizers for each eye. The result is a 3-D moving picture, as the text discusses.

Conceptual Example 9 illustrates an interesting result that occurs when a piece of polarizing material is inserted between a crossed polarizer and analyzer.

CONCEPTUAL EXAMPLE 9 | How Can a Crossed Polarizer and Analyzer Transmit Light?

As explained earlier, no light reaches the photocell in **Figure 24.22** when the polarizer and the analyzer are crossed. Suppose that a third piece of polarizing material is inserted between the polarizer and analyzer, as in **Figure 24.25a**. With the insert in place, will light reach the photocell when **(a)** $\theta = 0°$, **(b)** $\theta = 90°$, or **(c)** θ is between 0 and 90°?

Reasoning If any light is to pass through the analyzer, it must have an electric field component parallel to the transmission axis of the analyzer. Thus, without the insert in **Figure 24.25a**, no light reaches the photocell, because the analyzer and polarizer are crossed, which means that the electric field of the light reaching the analyzer has no component parallel to the analyzer's transmission axis. We need to consider, then, whether the presence of the insert leads to an electric field component parallel to the analyzer's transmission axis.

Answers (a) and (b) are incorrect. With $\theta = 0°$, the polarizer and the insert have parallel transmission axes, so the light leaving the polarizer passes through the insert unaffected. It reaches the analyzer with its electric field perpendicular to the analyzer's transmission axis and is, thus, prevented from reaching the photocell. With $\theta = 90°$, the polarizer and the insert are crossed, so no light leaves the insert to reach the analyzer and the photocell.

Answer (c) is correct. Parts b and c of **Figure 24.25** show that, with the insert present, the light reaching the analyzer has an electric field component that is parallel to the analyzer's transmission axis when θ is between 0 and 90°. In part b the electric field E of the light leaving the polarizer makes an angle θ with respect to the transmission axis of the insert and has a component $E \cos \theta$ with respect to that axis. This component passes through the insert. In part c the field ($E \cos \theta$) incident on the analyzer has a component parallel to the transmission axis of the analyzer—namely, ($E \cos \theta$) $\sin \theta$. This component passes through the analyzer and reaches the photocell.

Related Homework: Problem 39

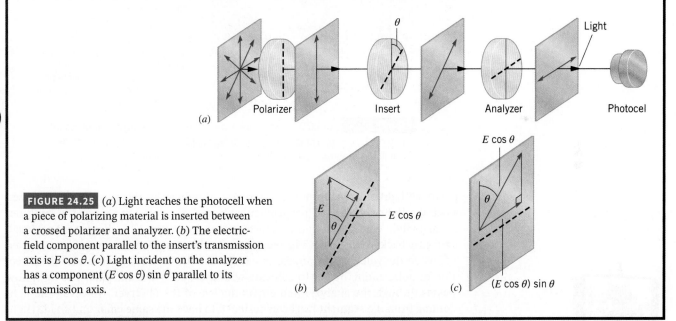

FIGURE 24.25 (a) Light reaches the photocell when a piece of polarizing material is inserted between a crossed polarizer and analyzer. (b) The electric-field component parallel to the insert's transmission axis is $E \cos \theta$. (c) Light incident on the analyzer has a component ($E \cos \theta$) $\sin \theta$ parallel to its transmission axis.

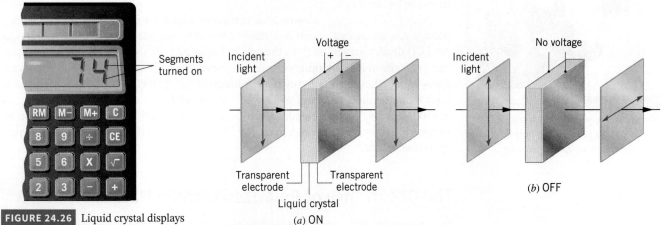

FIGURE 24.26 Liquid crystal displays (LCDs) use liquid crystal segments to form the numbers.

FIGURE 24.27 A liquid crystal in its (a) "on" state and (b) "off" state.

THE PHYSICS OF . . . a liquid crystal display (LCD). An application of a crossed polarizer/analyzer combination occurs in one kind of liquid crystal display (LCD). LCDs are widely used in pocket calculators and cell phones. The display usually consists of blackened numbers and letters set against a light gray background. As **Figure 24.26** indicates, each number or letter is formed from a combination of liquid crystal segments that have been turned on and appear black. The liquid crystal part of an LCD segment consists of the liquid crystal material sandwiched between two transparent electrodes, as in **Figure 24.27**. When a voltage is applied between the electrodes, the liquid crystal is said to be "on." Part *a* of the picture shows that linearly polarized incident light passes through the "on" material without having its direction of polarization affected. When the voltage is removed, as in part *b*, the liquid crystal is said to be "off" and now rotates the direction of polarization by 90°. A complete LCD segment also includes a crossed polarizer/analyzer combination, as **Figure 24.28** illustrates. The polarizer, analyzer, electrodes, and liquid crystal material are packaged as a single unit. The polarizer produces polarized light from incident unpolarized light. With the display segment turned on, as in **Figure 24.28**, the

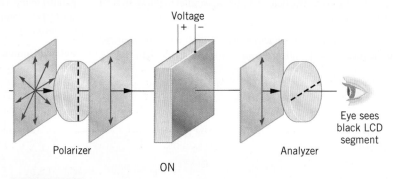

Voltage

Polarizer

ON

Analyzer

Eye sees black LCD segment

FIGURE 24.28 An LCD incorporates a crossed polarizer/analyzer combination. When the LCD segment is turned on (voltage applied), no light is transmitted through the analyzer, and the observer sees a black segment.

polarized light emerges from the liquid crystal only to be absorbed by the analyzer, since the light is polarized perpendicular to the transmission axis of the analyzer. Since no light emerges from the analyzer, an observer sees a black segment against a light gray background, as in **Figure 24.26**. On the other hand, the segment is turned off when the voltage is removed, in which case the liquid crystal rotates the direction of polarization by 90° to coincide with the axis of the analyzer. The light now passes through the analyzer and enters the eye of the observer. However, the light coming from the segment has been designed to have the same color and shade (light gray) as the background of the display, so the segment becomes indistinguishable from the background.

Color LCD display screens and computer monitors are popular because they occupy less space and weigh less than traditional cathode-ray tube (CRT) units do. An LCD display screen, such as the one in **Figure 24.29**, uses thousands of LCD segments arranged like the squares on graph paper. To produce color, three segments are grouped together to form a tiny picture element (or "pixel"). Color filters are used to enable one segment in the pixel to produce red light, one to produce green, and one to produce blue. The eye blends the colors from each pixel into a composite color. By varying the intensity of the red, green, and blue colors, the pixel can generate an entire spectrum of colors.

ermaltahiri/Pixabay

FIGURE 24.29 Want to see a color photograph of yourself? Just snap a "selfie" with a properly equipped cell phone and look at the LCD display.

The Occurrence of Polarized Light in Nature

THE PHYSICS OF . . . Polaroid sunglasses. Polaroid is a familiar material because of its widespread use in sunglasses. Such sunglasses are designed so that the transmission

Molecule

Unpolarized
light

A

B

Unpolarized
sunlight

C

Partially
polarized light

Polarized
light

FIGURE 24.30 In the process of being scattered from atmospheric molecules, unpolarized light from the sun becomes partially polarized.

axis of the Polaroid is oriented vertically when the glasses are worn in the usual fashion. Thus, the glasses prevent any light that is polarized horizontally from reaching the eye. Light from the sun is unpolarized, but a considerable amount of horizontally polarized sunlight originates by reflection from horizontal surfaces such as that of a lake. Section 26.4 discusses this effect. Polaroid sunglasses reduce glare by preventing the horizontally polarized reflected light from reaching the eyes.

Polarized sunlight also originates from the scattering of light by molecules in the atmosphere. **Figure 24.30** shows light being scattered by a single atmospheric molecule. The electric fields in the unpolarized sunlight cause the electrons in the molecule to vibrate perpendicular to the direction in which the light is traveling. The electrons, in turn, reradiate the electromagnetic waves in different directions, as the drawing illustrates. The light radiated straight ahead in direction A is unpolarized, just like the incident light; but light radiated perpendicular to the incident light in direction C is polarized. Light radiated in the intermediate direction B is partially polarized.

BIO THE PHYSICS OF . . . butterflies and polarized light. Researchers have discovered that at least one butterfly species uses polarized light to attract members of the opposite sex. The butterfly species *Heliconius* has patterns on its wings that cause light reflected from them to be polarized. This polarized light, invisible to the human eye but visible to other butterflies, is attractive to potential mates. When males were shown the female wings with their polarized light patterns, they swarmed toward the wings. When the males were shown the wings through a filter that blocked out the polarization effects, they largely ignored the wings. **Figure 24.31** shows the polarized light reflected from the wings of the *Heliconius cydno* butterfly. The left wing is shown as it normally appears. The light reflected from the white pattern is highly polarized. The right wing is shown as it appears when viewed through a polarizing filter whose transmission axis is crossed with respect to the direction in which the reflected light is polarized. The white patterns in the right wing are black when viewed through the filter, a clear indication that the light reflected from them is indeed polarized. There is also experimental evidence that some bird species use polarized light as a navigational aid.

FIGURE 24.31 A *Heliconius cydno* butterfly. The left wing is shown as it appears normally, and the right wing as it appears when viewed through a polarizing filter. The light reflected from the white patterns is polarized. These patterns in the right wing are black because the transmission axis of the filter is crossed with respect to the polarization direction of the reflected light.

Check Your Understanding

(The answers are given at the end of the book.)

9. Malus' law applies to the setup in **Figure 24.22**, which shows the analyzer rotated through an angle θ and the polarizer held fixed. Does Malus' law apply when the analyzer is held fixed and the polarizer is rotated?

10. In Example 8, we saw that when the angle between the polarizer and analyzer is 63.4°, the average intensity of the transmitted light drops to one-tenth of the average intensity of the incident unpolarized light. What happens to the light intensity that is not transmitted?

11. **CYU Figure 24.2** shows two sheets of polarizing material. The transmission axis of one is vertical, and that of the other makes an angle of 45° with the vertical. Unpolarized light shines on this arrangement first from the left and then from the right. From which direction does at least some light pass through both sheets? **(a)** From the left **(b)** From the right **(c)** From either direction **(d)** From neither direction. What is the answer when the light is horizontally polarized? What is the answer when the light is vertically polarized?

CYU FIGURE 24.2

12. You are sitting upright on the beach near a lake on a sunny day, wearing Polaroid sunglasses. When you lie down on your side, facing the lake, the sunglasses don't work as well as they do while you are sitting upright. Why not?

Concept Summary

24.1 The Nature of Electromagnetic Waves An electromagnetic wave consists of mutually perpendicular and oscillating electric and magnetic fields. The wave is a transverse wave, since the fields are perpendicular to the direction in which the wave travels. Electromagnetic waves can travel through a vacuum or a material substance. All electromagnetic waves travel through a vacuum at the same speed, which is known as the speed of light c ($c = 3.00 \times 10^8$ m/s).

24.2 The Electromagnetic Spectrum The frequency f and wavelength λ of an electromagnetic wave in a vacuum are related to its speed c through the relation $c = f\lambda$.

The series of electromagnetic waves, arranged in order of their frequencies or wavelengths, is called the electromagnetic spectrum. In increasing order of frequency (decreasing order of wavelength), the spectrum includes radio waves, infrared radiation, visible light, ultraviolet radiation, X-rays, and gamma rays. Visible light has frequencies between about 4.0×10^{14} and 7.9×10^{14} Hz. The human eye and brain perceive different frequencies or wavelengths as different colors.

24.3 The Speed of Light James Clerk Maxwell showed that the speed of light in a vacuum is given by Equation 24.1, where ε_0 is the (electric) permittivity of free space and μ_0 is the (magnetic) permeability of free space.

$$c = \frac{1}{\sqrt{\varepsilon_0 \mu_0}} \qquad (24.1)$$

24.4 The Energy Carried by Electromagnetic Waves The total energy density u of an electromagnetic wave is the total energy per unit volume of the wave and, in a vacuum, is given by Equation 24.2a, where E and B, respectively, are the magnitudes of the electric and magnetic fields of the wave. Since the electric and magnetic parts of the total energy density are equal, Equations 24.2b and 24.2c are equivalent to Equation 24.2a. In a vacuum, E and B are related according to Equation 24.3.

$$u = \frac{1}{2}\varepsilon_0 E^2 + \frac{1}{2\mu_0}B^2 \qquad (24.2a)$$

$$u = \varepsilon_0 E^2 \qquad (24.2b)$$

$$u = \frac{1}{\mu_0}B^2 \qquad (24.2c)$$

$$E = cB \qquad (24.3)$$

Equations 24.2a–c can be used to determine the average total energy density, if the rms average values E_{rms} and B_{rms} are used in place of the symbols E and B. The rms values are related to the peak values E_0 and B_0 in the usual way, as shown in Equations 1 and 2.

The intensity of an electromagnetic wave is the power that the wave carries perpendicularly through a surface divided by the area of the surface. In a vacuum, the intensity S is related to the total energy density u according to Equation 24.4.

$$E_{rms} = \frac{1}{\sqrt{2}}E_0 \quad (1) \qquad B_{rms} = \frac{1}{\sqrt{2}}B_0 \quad (2)$$

$$S = cu \qquad (24.4)$$

24.5 The Doppler Effect and Electromagnetic Waves When electromagnetic waves and the source and observer of the waves all travel along the same line in a vacuum, the Doppler effect is given by Equation 24.6, where f_o and f_s are, respectively, the observed and emitted wave frequencies and v_{rel} is the relative speed of the source and the observer. The plus sign is used when the source and the observer come together, and the minus sign is used when they move apart.

$$f_o = f_s\left(1 \pm \frac{v_{rel}}{c}\right) \quad \text{if } v_{rel} \ll c \qquad (24.6)$$

24.6 Polarization A linearly polarized electromagnetic wave is one in which the oscillation of the electric field occurs only along one direction, which is taken to be the direction of polarization. The magnetic field also oscillates along only one direction, which is perpendicular to the electric field direction. In an unpolarized wave such as the light from an incandescent bulb, the direction of polarization does not remain fixed, but fluctuates randomly in time.

Polarizing materials allow only the component of the wave's electric field along one direction (and the associated magnetic field component) to pass through them. The preferred transmission direction for the electric field is called the transmission axis of the material.

When unpolarized light is incident on a piece of polarizing material, the transmitted polarized light has an average intensity that is one-half the average intensity of the incident light.

When two pieces of polarizing material are used one after the other, the first is called the polarizer, and the second is referred to as the analyzer. If the average intensity of polarized light falling on the analyzer is $\overline{S}_0$, the average intensity $\overline{S}$ of the light leaving the analyzer is given by Malus' law, as shown in Equation 24.7, where θ is the angle between the transmission axes of the polarizer and analyzer. When $\theta = 90°$, the polarizer and the analyzer are said to be "crossed," and no light passes through the analyzer.

$$\overline{S} = \overline{S}_0 \cos^2 \theta \qquad (24.7)$$

Focus on Concepts

Online

Additional questions are available for assignment in WileyPLUS.

Section 24.1 The Nature of Electromagnetic Waves

1. The drawing shows an x, y, z coordinate system. A circular loop of wire lies in the z, x plane and, when used with an LC-tuned circuit, detects an electromagnetic wave. Which one of the following statements is correct? (**a**) The wave travels along the x axis, and its electric field oscillates along the y axis. (**b**) The wave travels along the z axis, and its electric field oscillates along the x axis. (**c**) The wave travels along the z axis, and its electric field oscillates along the y axis. (**d**) The wave travels along the y axis, and its electric field oscillates along the x axis. (**e**) The wave travels along the y axis, and its electric field oscillates along the z axis.

QUESTION 1

Section 24.2 The Electromagnetic Spectrum

2. An electromagnetic wave travels in a vacuum. The wavelength of the wave is tripled. How is this accomplished? (**a**) By tripling the frequency of the wave (**b**) By tripling the speed of the wave (**c**) By reducing the frequency of the wave by a factor of three (**d**) By reducing the speed of the wave by a factor of three (**e**) By tripling the magnitudes of the electric and magnetic fields that comprise the wave

Section 24.4 The Energy Carried by Electromagnetic Waves

3. An electromagnetic wave is traveling in a vacuum. The magnitudes of the electric and magnetic fields of the wave are _____, and the electric and magnetic energies carried by the wave are _____. (**a**) equal, proportional (but not equal) to each other (**b**) proportional (but not equal) to each other, equal (**c**) equal, equal (**d**) proportional (but not equal) to each other, unequal

Section 24.5 The Doppler Effect and Electromagnetic Waves

4. The drawing shows four situations—A, B, C, and D—in which an observer and a source of electromagnetic waves can move along the same line. In each case the source emits a wave of the same frequency, and in each case only the source or the observer is moving. The arrow in each situation denotes the velocity vector, which has the same magnitude in each situation. When there is no arrow, the observer or the source is stationary. Rank the frequencies of the observed electromagnetic waves in descending order (largest first) according to magnitude. (**a**) A and B (a tie), C and D (a tie) (**b**) C and D (a tie), A and B (a tie) (**c**) A and D (a tie), B and C (a tie) (**d**) B and D (a tie), A and C (a tie) (**e**) B and C (a tie), A and D (a tie)

QUESTION 4

Section 24.6 Polarization

5. The drawing shows two sheets of polarizing material. Polarizer 1 has its transmission axis aligned vertically, and polarizer 2 has its transmission axis aligned at an angle of 45° with respect to the vertical. Light that is completely polarized along the vertical direction is incident either from the far left or from the far right. In either case, the average intensity of the incident light is the same. Which one of the following statements is true concerning the average intensity of the light that is transmitted by the pair of sheets? (**a**) When the light is incident from either the left or the right, the transmitted intensity is one-half the incident intensity. (**b**) When the light is incident from either the left or the right, the transmitted intensity is one-fourth the incident intensity. (**c**) When the light is incident from the left, the transmitted intensity is one-half the incident intensity; when the light

is incident from the right, the transmitted intensity is zero. **(d)** When the light is incident from the left, the transmitted intensity is one-fourth the incident intensity; when the light is incident from the right, the transmitted intensity is one-half the incident intensity. **(e)** When the light is incident from the left, the transmitted intensity is one-half the incident intensity; when the light is incident from the right, the transmitted intensity is one-fourth the incident intensity.

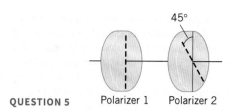

QUESTION 5   Polarizer 1   Polarizer 2

Problems

Online

Additional questions are available for assignment in WileyPLUS.

SSM Student Solutions Manual	**BIO** Biomedical application
MMH Problem-solving help	**E** Easy
GO Guided Online Tutorial	**M** Medium
V-HINT Video Hints	**H** Hard
CHALK Chalkboard Videos	**WS** Worksheet
	T Team Problem

Section 24.1   The Nature of Electromagnetic Waves

1. E The team monitoring a space probe exploring the outer solar system finds that radio transmissions from the probe take 2.53 hours to reach earth. How distant (in meters) is the probe?

2. E (a) Neil A. Armstrong was the first person to walk on the moon. The distance between the earth and the moon is 3.85×10^8 m. Find the time it took for his voice to reach the earth via radio waves. **(b)** Someday a person will walk on Mars, which is 5.6×10^{10} m from the earth at the point of closest approach. Determine the minimum time that will be required for a message from Mars to reach the earth via radio waves.

3. E SSM CHALK In astronomy, distances are often expressed in light-years. One light-year is the distance traveled by light in one year. The distance to Alpha Centauri, the closest star other than our own sun that can be seen by the naked eye, is 4.3 light-years. Express this distance in meters.

4. E GO FM radio stations use radio waves with frequencies from 88.0 to 108 MHz to broadcast their signals. Assuming that the inductance in **Figure 24.4** has a value of 6.00×10^{-7} H, determine the range of capacitance values that are needed so the antenna can pick up all the radio waves broadcasted by FM stations.

Section 24.2   The Electromagnetic Spectrum

5. E A truck driver is broadcasting at a frequency of 26.965 MHz with a CB (citizen's band) radio. Determine the wavelength of the electromagnetic wave being used. The speed of light is $c = 2.9979 \times 10^8$ m/s.

6. E In a dentist's office an X-ray of a tooth is taken using X-rays that have a frequency of 6.05×10^{18} Hz. What is the wavelength in vacuum of these X-rays?

7. E SSM In a certain UHF radio wave, the shortest distance between positions at which the electric and magnetic fields are zero is 0.34 m. Determine the frequency of this UHF radio wave.

8. E FM radio waves have frequencies between 88.0 and 108.0 MHz. Determine the range of wavelengths for these waves.

9. E GO A certain type of laser emits light that has a frequency of 5.2×10^{14} Hz. The light, however, occurs as a series of short pulses,

each lasting for a time of 2.7×10^{-11} s. **(a)** How many wavelengths are there in one pulse? **(b)** The light enters a pool of water. The frequency of the light remains the same, but the speed of the light slows down to 2.3×10^8 m/s. How many wavelengths are there now in one pulse?

10. E Two radio waves are used in the operation of a cellular telephone. To receive a call, the phone detects the wave emitted at one frequency by the transmitter station or base unit. To send your message to the base unit, your phone emits its own wave at a different frequency. The difference between these two frequencies is fixed for all channels of cell phone operation. Suppose that the wavelength of the wave emitted by the base unit is 0.34339 m and the wavelength of the wave emitted by the phone is 0.36205 m. Using a value of 2.9979×10^8 m/s for the speed of light, determine the difference between the two frequencies used in the operation of a cell phone.

11. M V-HINT CHALK A positively charged object with a mass of 0.115 kg oscillates at the end of a spring, generating ELF (extremely low frequency) radio waves that have a wavelength of 4.80×10^7 m. The frequency of these radio waves is the same as the frequency at which the object oscillates. What is the spring constant of the spring?

Section 24.3   The Speed of Light

12. E Review Conceptual Example 4 for information pertinent to this problem. When we look at the star Polaris (the North Star), we are seeing it as it was 680 years ago. How far away from us (in meters) is Polaris?

13. E GO **Figure 24.13** illustrates Michelson's setup for measuring the speed of light with the mirrors placed on Mt. San Antonio and Mt. Wilson in California, which are 35 km apart. Using a value of 3.00×10^8 m/s for the speed of light, find the minimum angular speed (in rev/s) for the rotating mirror.

14. E SSM Two astronauts are 1.5 m apart in their spaceship. One speaks to the other. The conversation is transmitted to earth via electromagnetic waves. The time it takes for sound waves to travel at 343 m/s through the air between the astronauts equals the time it takes for the electromagnetic waves to travel to the earth. How far away from the earth is the spaceship?

15. E V-HINT A laptop computer communicates with a router wirelessly, by means of radio signals. The router is connected by cable directly to the Internet. The laptop is 8.1 m from the router, and is downloading text and images from the Internet at an average rate of 260 Mbps, or 260 megabits per second. (A *bit*, or *binary digit*, is the smallest unit of digital information.) On average, how many bits are downloaded to the laptop in the time it takes the wireless signal to travel from the router to the laptop?

16. E A lidar (laser radar) gun is an alternative to the standard radar gun that uses the Doppler effect to catch speeders. A lidar gun uses an infrared laser and emits a precisely timed series of pulses of infrared electromagnetic waves. The time for each pulse to travel to the speeding vehicle and return to the gun is measured. In one situation a lidar gun in a stationary police car observes a difference of 1.27×10^{-7} s in round-trip travel times for two pulses that are emitted 0.450 s apart. Assuming that the speeding vehicle is approaching the police car essentially head-on, determine the speed of the vehicle.

17. **M** **GO** A politician holds a press conference that is televised live. The sound picked up by the microphone of a TV news network is broadcast via electromagnetic waves and heard by a television viewer. This viewer is seated 2.3 m from his television set. A reporter at the press conference is located 4.1 m from the politician, and the sound of the words travels directly from the celebrity's mouth, through the air, and into the reporter's ears. The reporter hears the words *exactly at the same instant* that the television viewer hears them. Using a value of 343 m/s for the speed of sound, determine the maximum distance between the television set and the politician. Ignore the small distance between the politician and the microphone. In addition, assume that the only delay between what the microphone picks up and the sound being emitted by the television set is that due to the travel time of the electromagnetic waves used by the network.

18. **M** **GO** **CHALK** A mirror faces a cliff located some distance away. Mounted on the cliff is a second mirror, directly opposite the first mirror and facing toward it. A gun is fired very close to the first mirror. The speed of sound is 343 m/s. How many times does the flash of the gunshot travel the round-trip distance between the mirrors before the echo of the gunshot is heard?

Section 24.4 The Energy Carried by Electromagnetic Waves

19. **E** A laser emits a narrow beam of light. The radius of the beam is 1.0×10^{-3} m, and the power is 1.2×10^{-3} W. What is the intensity of the laser beam?

20. **E** An industrial laser is used to burn a hole through a piece of metal. The average intensity of the light is $\bar{S} = 1.23 \times 10^{9} \, \text{W/m}^2$. What is the rms value of **(a)** the electric field and **(b)** the magnetic field in the electromagnetic wave emitted by the laser?

21. **E** The maximum strength of the magnetic field in an electromagnetic wave is 3.3×10^{-6} T. What is the maximum strength of the wave's electric field?

22. **E** **SSM** The microwave radiation left over from the Big Bang explosion of the universe has an average energy density of $4 \times 10^{-14} \, \text{J/m}^3$. What is the rms value of the electric field of this radiation?

23. **E** **GO** On a cloudless day, the sunlight that reaches the surface of the earth has an intensity of about $1.0 \times 10^{3} \, \text{W/m}^2$. What is the electromagnetic energy contained in 5.5 m³ of space just above the earth's surface?

24. **E** **GO** Consult Multiple-Concept Example 7 to review the concepts on which this problem depends. A light bulb emits light uniformly in all directions. The average emitted power is 150.0 W. At a distance of 5.00 m from the bulb, determine **(a)** the average intensity of the light, **(b)** the rms value of the electric field, and **(c)** the peak value of the electric field.

25. **E** Multiple-Concept Example 7 provides some pertinent background for this problem. The mean distance between earth and the sun is 1.50×10^{11} m. The average intensity of solar radiation incident on the upper atmosphere of the earth is 1390 W/m². Assuming that the sun emits radiation uniformly in all directions, determine the total power radiated by the sun.

26. **E** **GO** A stationary particle of charge $q = 2.6 \times 10^{-8}$ C is placed in a laser beam (an electromagnetic wave) whose intensity is $2.5 \times 10^{3} \, \text{W/m}^2$. Determine the magnitudes of the **(a)** electric and **(b)** magnetic forces exerted on the charge. If the charge is moving at a speed of 3.7×10^{4} m/s perpendicular to the magnetic field of the electromagnetic wave, find the magnitudes of the **(c)** electric and **(d)** magnetic forces exerted on the particle.

27. **M** The power radiated by the sun is 3.9×10^{26} W. The earth orbits the sun in a nearly circular orbit of radius 1.5×10^{11} m. The earth's axis of rotation is tilted by 27° relative to the plane of the orbit (see the drawing), so sunlight does not strike the equator perpendicularly. What power strikes a 0.75-m² patch of flat land at the equator at point Q?

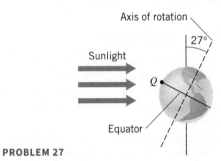

PROBLEM 27

28. **M** **V-HINT** An electromagnetic wave strikes a 1.30-cm² section of wall perpendicularly. The rms value of the wave's magnetic field is determined to be 6.80×10^{-4} T. How long does it take for the wave to deliver 1850 J of energy to the wall?

Section 24.5 The Doppler Effect and Electromagnetic Waves

29. **E** **MMH** A distant galaxy emits light that has a wavelength of 434.1 nm. On earth, the wavelength of this light is measured to be 438.6 nm. **(a)** Decide whether this galaxy is approaching or receding from the earth. Give your reasoning. **(b)** Find the speed of the galaxy relative to the earth.

30. **E** **GO** A speeder is pulling directly away and increasing his distance from a police car that is moving at 25 m/s with respect to the ground. The radar gun in the police car emits an electromagnetic wave with a frequency of 7.0×10^{9} Hz. The wave reflects from the speeder's car and returns to the police car, where its frequency is measured to be 320 Hz less than the emitted frequency. Find the speeder's speed with respect to the ground.

31. **M** **CHALK** **SSM** A distant galaxy is simultaneously rotating and receding from the earth. As the drawing shows, the galactic center is receding from the earth at a relative speed of $u_G = 1.6 \times 10^{6}$ m/s. Relative to the center, the tangential speed is $v_T = 0.4 \times 10^{6}$ m/s for locations A and B, which are equidistant from the center. When the frequencies of the light coming from regions A and B are measured on earth, they are not the same and each is different from the emitted frequency of 6.200×10^{14} Hz. Find the measured frequency for the light from **(a)** region A and **(b)** region B.

PROBLEM 31

32. **M** **GO** The drawing shows three situations— A, B, and C—in which an observer and a source of electromagnetic waves

are moving along the same line. In each case the source emits a wave that has a frequency of 4.57×10^{14} Hz. The arrows in each situation denote velocity vectors of the observer and source relative to the ground and have the magnitudes indicated (v or $2v$), where the speed v is 1.50×10^6 m/s. Calculate the observed frequency in each of the three cases.

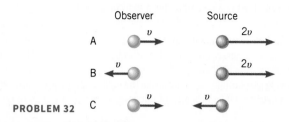

PROBLEM 32

Section 24.6 Polarization

33. **E** Unpolarized light whose intensity is 1.10 W/m² is incident on the polarizer in **Figure 24.22**. **(a)** What is the intensity of the light leaving the polarizer? **(b)** If the analyzer is set at an angle of $\theta = 75°$ with respect to the polarizer, what is the intensity of the light that reaches the photocell?

34. **E** **GO** The drawing shows three polarizer/analyzer pairs. The incident light beam for each pair is unpolarized and has the same average intensity of 48 W/m². Find the average intensity of the transmitted beam for each of the three cases (A, B, and C) shown in the drawing.

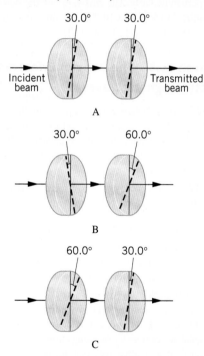

PROBLEM 34

35. **E** **SSM** The average intensity of light emerging from a polarizing sheet is 0.764 W/m², and the average intensity of the horizontally polarized light incident on the sheet is 0.883 W/m². Determine the angle that the transmission axis of the polarizing sheet makes with the horizontal.

36. **E** **GO** Light that is polarized along the vertical direction is incident on a sheet of polarizing material. Only 94% of the intensity of the light passes through the sheet and strikes a second sheet of polarizing material. No light passes through the second sheet. What angle does the transmission axis of the second sheet make with the vertical?

37. **E** **GO** The drawing shows light incident on a polarizer whose transmission axis is parallel to the z axis. The polarizer is rotated clockwise through an angle α. The average intensity of the incident light is 7.0 W/m². Determine the average intensity of the transmitted light for each of the six cases shown in the table.

Incident Light	Intensity of Transmitted Light	
	$\alpha = 0°$	$\alpha = 35°$
(a) Unpolarized		
(b) Polarized parallel to z axis		
(c) Polarized parallel to y axis		

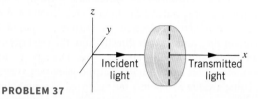

PROBLEM 37

38. **E** **MMH** For each of the three sheets of polarizing material shown in the drawing, the orientation of the transmission axis is labeled relative to the vertical. The incident beam of light is unpolarized and has an intensity of 1260.0 W/m². What is the intensity of the beam transmitted through the three sheets when $\theta_1 = 19.0°$, $\theta_2 = 55.0°$, and $\theta_3 = 100.0°$?

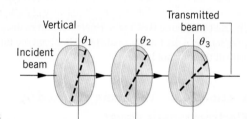

PROBLEM 38

39. **M** **V-HINT** Before attempting this problem, review Conceptual Example 9. The intensity of the light that reaches the photocell in **Figure 24.25a** is 110 W/m², when $\theta = 23°$. What would be the intensity reaching the photocell if the *analyzer* were removed from the setup, everything else remaining the same?

40. **M** **SSM** More than one analyzer can be used in a setup like the one in **Figure 24.22**, each analyzer following the previous one. Suppose that the transmission axis of the first analyzer is rotated 27° relative to the transmission axis of the polarizer, and that the transmission axis of each additional analyzer is rotated 27° relative to the transmission axis of the previous one. What is the minimum number of analyzers needed for the light reaching the photocell to have an intensity that is reduced by at least a factor of 100 relative to the intensity of the light striking the first analyzer?

41. **M** **CHALK** **GO** The drawing shows four sheets of polarizing material, each with its transmission axis oriented differently. Light that is polarized in the vertical direction is incident from the left and has an average intensity of 27 W/m². Determine the average intensity of the light that emerges on the right in the drawing when sheet A alone is removed, when sheet B alone is removed, when sheet C alone is removed, and when sheet D alone is removed.

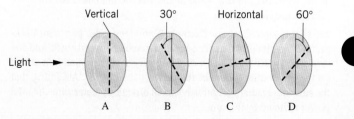

PROBLEM 41

Additional Problems

42. **E** Obtain the wavelengths in vacuum for **(a)** blue light whose frequency is 6.34×10^{14} Hz, and **(b)** orange light whose frequency is 4.95×10^{14} Hz. Express your answers in nanometers (1 nm = 10^{-9} m).

43. **E** **GO** The magnitude of the electric field of an electromagnetic wave increases from 315 to 945 N/C. **(a)** Determine the wave intensities for the two values of the electric field. **(b)** What is the magnitude of the magnetic field associated with each electric field? **(c)** Determine the wave intensity for each value of the magnetic field.

44. **E** **SSM** A future space station in orbit about the earth is being powered by an electromagnetic beam from the earth. The beam has a cross-sectional area of 135 m² and transmits an average power of 1.20×10^4 W. What are the rms values of the **(a)** electric and **(b)** magnetic fields?

45. **E** **V-HINT** Suppose that a police car is moving to the right at 27 m/s, while a speeder is coming up from behind at a speed of 39 m/s, both speeds being with respect to the ground. Assume that the electromagnetic wave emitted by the police car's radar gun has a frequency of 8.0×10^9 Hz. Find the difference between the frequency of the wave that returns to the police car after reflecting from the speeder's car and the original frequency emitted by the police car.

46. **M** **CHALK** **GO** The electromagnetic wave that delivers a cellular phone call to a car has a magnetic field with an rms value of 1.5×10^{-10} T. The wave passes perpendicularly through an open window, the area of which is 0.20 m². How much energy does this wave carry through the window during a 45-s phone call?

47. **M** **V-HINT** A beam of polarized light with an average intensity of 15 W/m² is sent through a polarizer. The transmission axis makes an angle of 25° with respect to the direction of polarization. Determine the rms value of the electric field of the transmitted beam.

48. **M** An argon–ion laser produces a cylindrical beam of light whose average power is 0.750 W. How much energy is contained in a 2.50-m length of the beam?

49. **M** **V-HINT** What fraction of the power radiated by the sun is intercepted by the planet Mercury? The radius of Mercury is 2.44×10^6 m, and its mean distance from the sun is 5.79×10^{10} m. Assume that the sun radiates uniformly in all directions.

50. **M** **SSM** **MMH** The drawing shows an edge-on view of the solar panels on a communications satellite. The dashed line specifies the normal to the panels. Sunlight strikes the panels at an angle θ with respect to the normal. If the solar power impinging on the panels is 2600 W when $\theta = 65°$, what is it when $\theta = 25°$?

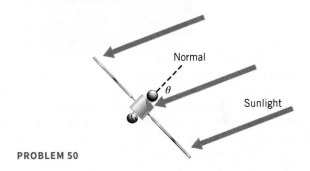

PROBLEM 50

51. **M** **GO** **SSM** Police use radar guns and the Doppler effect to catch speeders. The figure illustrates a moving car approaching a stationary police car. A radar gun emits an electromagnetic wave that reflects from the oncoming car. The reflected wave returns to the police car with a frequency (measured by on-board equipment) that is different from the emitted frequency. One such radar gun emits a wave whose frequency is 8.0×10^9 Hz. When the speed of the car is 39 m/s and the approach is essentially head-on, what is the *difference* between the frequency of the wave returning to the police car and that emitted by the radar gun?

Reflected electromagnetic wave   Outgoing electromagnetic wave

PROBLEM 51

Physics in Biology, Medicine, and Sports

52. **E** **BIO** **24.2** Magnetic resonance imaging, or MRI (see Section 21.7), and positron emission tomography, or PET scanning (see Section 32.6), are two medical diagnostic techniques. Both employ electromagnetic waves. For these waves, find the ratio of the MRI wavelength (frequency = 6.38×10^7 Hz) to the PET scanning wavelength (frequency = 1.23×10^{20} Hz).

53. **M** **BIO** **24.4** A heat lamp emits infrared radiation whose rms electric field is $E_{\text{rms}} = 2800$ N/C. **(a)** What is the average intensity of the radiation? **(b)** The radiation is focused on a person's leg over a circular

area of radius 4.0 cm. What is the average power delivered to the leg? **(c)** The portion of the leg being irradiated has a mass of 0.28 kg and a specific heat capacity of 3500 J/(kg · C°). How long does it take to raise its temperature by 2.0 C°? Assume that there is no other heat transfer into or out of the portion of the leg being heated.

54. **E** **BIO** **24.2** The human eye is most sensitive to light with a frequency of about 5.5×10^{14} Hz, which is in the yellow-green region of the electromagnetic spectrum. How many wavelengths of this light can fit across the width of your thumb, a distance of about 2.0 cm?

55. E BIO **24.2** X-rays are an important component of dental treatment and in evaluating oral health. They can help identify cavities, tooth decay, and impacted teeth. The photo shows a bitewing X-ray, in which the patient bites down on a plastic tab that holds the X-ray film or sensor in place. Here, the X-ray reveals decay inside a tooth that would otherwise go undetected. X-rays are produced in a special device called an X-ray tube (see Section 30.7). **(a)** What is the speed of the X-rays when they are produced by the tube? **(b)** If the wavelength of the X-rays is 0.0230 nm, what is their frequency?

PROBLEM 55

Twinkle Family Dentalcare Pte Ltd. https://www.twin-kledental.com.sg/xrays/

56. M BIO **24.4** *Victoria amazonica* is a giant water lily that grows some of the largest leaf pads in the world (see the photo). Some of the pads can be over 3 m in diameter and can support a weight of 100 lb. Consider the energy in the sunlight that strikes the pad. A typical hair dryer uses an average power of 1250 W. How many hair dryers could be powered by the sunlight striking the lily pad? The rms value of the electric field in sunlight at the surface of the earth is 720.0 N/C. Treat the lily pad as a circle with a diameter of 3.20 m.

PROBLEM 56

Stevebidmead/Pixabay

57. M BIO **24.5** Radar guns have many applications in sports. They are used to measure the speed of baseballs, bowling balls, and serves in tennis, as well as the running speed of athletes themselves. A Major League Baseball scout is evaluating a Minor League pitcher by standing behind home plate and measuring the speed of his pitches with a radar gun. The gun emits electromagnetic waves in the microwave part of the EM spectrum at a frequency of 24.25 GHz. Waves that reflect off of a pitched baseball return to the gun with a frequency that has increased by 7.388 kHz. What is the speed of the pitch in mph?

58. E BIO **24.2** Frequencies in the range of 30 to 300 MHz in the radio wave part of the electromagnetic spectrum correspond to the VHF (very high frequency) bands. These frequencies are heavily used by the U.S. Coast Guard (USCG), and the international distress frequency has been chosen to be channel 16, with a frequency of 156.8 MHz. The USCG, as well as most coastal stations, maintains a constant monitoring of this frequency. Cell phones are not a sufficient substitute for a good handheld VHF radio. Cell phone service is unavailable at about 1 mile offshore, whereas VHF radios can transmit and receive signals from ship to ship and ship to shore at distances of more than 20 miles. The VHF signals that are used for marine communications are also vertically polarized, so the transmitting and receiving antennas should be linear dipoles oriented vertically. The simplest form is the quarterwave antenna, in which the length of the antenna is equal to onequarter of the wavelength of the signal. What would be the length of a quarter-wave antenna for the USCG's distress frequency?

59. M BIO **24.2** The human heart creates the body's largest electromagnetic field. During normal beating, an electrocardiogram measures a voltage of 2.7 mV between two electrical contact pads on the chest that are separated by 25 cm. This creates an electromagnetic wave with a frequency of 1.2 Hz. **(a)** What is the wavelength of this wave? **(b)** What is the maximum value of the electric field in this wave? **(c)** What is the maximum value of the magnetic field? **(d)** What is the average total energy density in the wave?

Concepts and Calculations Problems   Online

One of the central ideas of this chapter is that electromagnetic waves carry energy. Two concepts are used to describe this energy—the wave's intensity and its energy density. Problem 60 reviews this important idea. We have seen how the intensities of completely polarized or completely unpolarized light beams can change when they pass through a polarizer. But what about light that is partially polarized or partially unpolarized? Can the concepts that we discussed in Section 24.6 be applied to such light? The answer is "yes," and Problem 61 illustrates how.

60. M CHALK The figure shows the popular dish antenna that receives digital TV signals from a satellite. The average intensity of the electromagnetic wave that carries a particular TV program is $\overline{S} = 7.5 \times 10^{-14} \text{ W/m}^2$, and the circular aperture of the antenna has a radius of $r = 15$ cm. *Concepts:* (i) How is the average power passing through the circular aperture of the antenna related to the average intensity of the TV signal? (ii) How much energy does the antenna receive in a time t? (iii) What is the average energy density, or average energy per unit volume, of the electromagnetic wave? *Calculations:* **(a)** Determine the electromagnetic energy

delivered to the dish during a one-hour program. **(b)** What is the average energy density of the electromagnetic wave?

PROBLEM 60

61. M CHALK SSM The light beam in the figure passes through a polarizer whose transmission axis makes an angle ϕ with the vertical. The beam is partially polarized and partially unpolarized, and the average intensity $\overline{S}_0$ of the incident light is the sum of the average intensity $\overline{S}_{0,\,\text{polar}}$ of the polarized light and the average intensity $\overline{S}_{0,\,\text{unpolar}}$ of the unpolarized light; $\overline{S}_0 = S_{0,\,\text{polar}} + S_{0,\,\text{unpolar}}$. The intensity $\overline{S}$ of the transmitted light is also the sum of two parts: $\overline{S} = \overline{S}_{\text{polar}} + \overline{S}_{\text{unpolar}}$. As the polarizer is rotated clockwise, the intensity of the transmitted light has a minimum value of $\overline{S} = 2.0$ W/m^2 when $\phi = 20.0°$ and has a maximum value of $\overline{S} = 8.0$ W/m^2 when the angle is $\phi = \phi_{\text{max}}$. *Concepts:* (i) How is $\overline{S}_{\text{unpolar}}$ related to $\overline{S}_{0,\,\text{unpolar}}$? (ii) How is $\overline{S}_{\text{polar}}$ related to $\overline{S}_{0,\,\text{polar}}$? (iii) The minimum transmitted intensity is 2.0 W/m^2. Why isn't it 0 W/m^2? *Calculations:* (a) What is the intensity $\overline{S}_{0,\,\text{unpolar}}$ of the

incident light that is unpolarized? (b) What is the intensity $\overline{S}_{0,\,\text{polar}}$ of the incident light that is polarized?

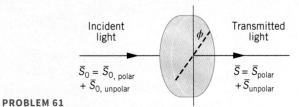

$\overline{S}_0 = \overline{S}_{0,\,\text{polar}} + \overline{S}_{0,\,\text{unpolar}}$

$\overline{S} = \overline{S}_{\text{polar}} + \overline{S}_{\text{unpolar}}$

PROBLEM 61

Team Problems

Online

62. **E T WS** **Laser Power Transmission.** You and your team are assigned to use a new power delivery system to charge the batteries of a spacecraft. The new power station consists of a large solar array that orbits the earth and harnesses solar energy to power a large laser. The laser beam has an average intensity of 2.44×10^6 W/m^2 and a circular cross section with a diameter of 20.5 cm. The power receiver on your ship that converts the light into electrical energy accepts the entire beam, but only has an efficiency of 34.0% (i.e., only 34.0% of the incoming power gets converted to energy stored in the batteries). The maximum total energy storage capacity of the batteries is 5.04×10^7 J. (a) What is the average energy density of the beam? (b) Assuming that the batteries of your craft are completely depleted, determine how long it will take to charge them.

63. **E T WS** **A Space Beacon.** You and your team are on a spacecraft far from earth (outside the solar system), and your navigation system has malfunctioned due to an encounter with a cloud of ionized gas. You have unknowingly been going off course for weeks. In addition, the ionized gas damaged your communications systems, so not only do you not know exactly where you are, but you cannot send out a distress signal for help. Fortunately, there is a high power radio beacon orbiting earth that can guide you. However, the only detection equipment on the ship that still works measures the root-mean-square electric field of incoming radio waves. The detector system is highly directional, meaning it has to be aimed directly at the source. The average output power of the beacon is 125 kW, and it radiates isotropically—that is, uniformly in all directions. If the detection device on the ship measures the magnitude of the rms electric field to be 4.84×10^{-9} N/C, how far is your ship from earth?

64. **M T** **A Turbo-shaft Helicopter.** You and your team are tasked with covertly acquiring information about a new helicopter being developed by an adversary. You are given a handheld device that has many functions, one of which is to send out a broad pulse of electromagnetic waves at a frequency of $f_0 = 9.00 \times 10^9$ Hz, and then collect the scattered light that returns from whatever it hits. The device then analyzes the scattered light and returns a graph of maximum *frequency shift* Δf as a function of collection angle θ (see the drawing). The device also has a built-in rangefinder that measures the distance to an object. As the helicopter engine powers up and the helicopter just starts to lift off the ground, you point the device at it and pull the trigger. It sends out a pulse and, a few seconds later, it gives you (i) the distance to the helicopter ($d = 82.7$ m), and (ii) a graph of the frequency shift Δf versus collection angle θ, as shown in the figure. (a) Assuming that light detected by your device has scattered from the moving rotor of the helicopter, what is the diameter of the rotor? (b) What is the speed of the tips of the rotor? (c) What is the rotational speed (in rpm) of the rotor?

PROBLEM 64

65. **M T** **A Magneto-optic Device.** The *Faraday effect* is a magneto-optic phenomenon where the polarization of linearly polarized light rotates when the light passes through a magnetized material. You and your team are tasked with analyzing the results of an experiment designed to measure how much a certain thin magnetic film rotates the polarization of the light that passes through it. As the drawing shows, an unpolarized laser beam of average intensity $\overline{S}_0 = 1.00 \times 10^3$ W/m^2 passes through a vertical polarizer. It then passes through a thin magnetic film that can be in one of three states: (i) unmagnetized, (ii) magnetized to the right (when looking from the laser), and (iii) magnetized to the left. It is assumed that the film does not absorb any of the light. The light then passes through another polarizer (called the analyzer) whose transmission axis is rotated 7.50° clockwise from the horizontal (when looking from the laser). (a) What is the intensity of the beam after it passes through the first polarizer? (b) In the case that the film is unmagnetized, the polarization will not rotate when it passes through the film. What should be the light intensity measured at the detector? (c) When the film is magnetized to the left, the intensity measured by the detector is $I_L = 12.2$ W/m^2, and when magnetized to the right the intensity is $I_R = 5.46$ W/m^2. In each case, determine the direction (clockwise or counterclockwise when looking from the laser) and the magnitude of the angle through which the magnetic film rotates the polarization of the light from the vertical.

PROBLEM 65

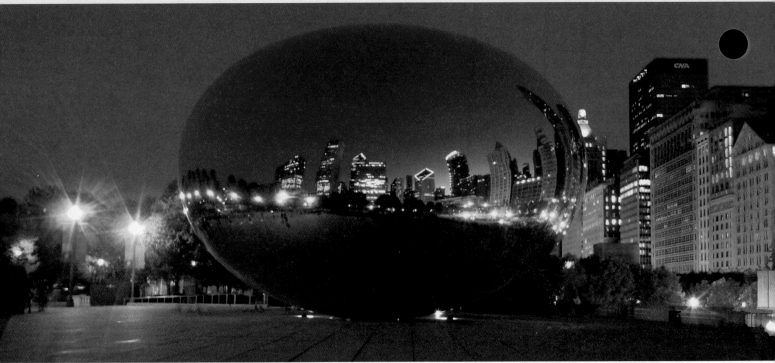

The Cloud Gate sculpture, created by artist Anish Kapoor, is affectionately known as "The Bean," due to its curved shape. Its highly polished stainless steel surface reflects the light around it, forming an image here of the buildings in the Chicago skyline. The reflection of light and the properties of the images formed by this and other kinds of mirrors will be the focus of this chapter.

rlobes/Pixabay

LEARNING OBJECTIVES

After reading this module, you should be able to...

25.1 Relate wave fronts and rays.

25.2 Apply the law of reflection to plane mirrors.

25.3 Describe image formation by a plane mirror.

25.4 Calculate the focal length of a spherical mirror.

25.5 Perform ray tracing for spherical mirrors.

25.6 Use the mirror and magnification equations to solve problems.

The Reflection of Light: Mirrors

25.1 Wave Fronts and Rays

Mirrors are usually close at hand. It is difficult, for example, to put on makeup, shave, or drive a car without them. We see images in mirrors because some of the light that strikes them is reflected into our eyes. To discuss reflection, it is necessary to introduce the concepts of a wave front and a ray of light, and we can do so by taking advantage of the familiar topic of sound waves (see Chapter 16). Both sound and light are kinds of waves. Sound is a pressure wave, whereas light is electromagnetic in nature. However, the ideas of a wave front and a ray apply to both.

Consider a small spherical object whose surface is pulsating in simple harmonic motion. A sound wave is emitted that moves spherically outward from the object at a constant speed. To represent this wave, we draw surfaces through all points of the wave that are in the same phase of motion. These surfaces of constant phase are called **wave fronts. Figure 25.1** shows a hemispherical view of the wave fronts. In this view they appear as concentric spherical shells about the vibrating object. If the wave fronts are drawn through the condensations, or crests, of the sound wave, as they are in the picture, the distance between adjacent wave fronts equals the wavelength λ. The radial lines pointing outward from the source and perpendicular to the wave fronts are called **rays**. The rays point in the direction of the velocity of the wave.

Figure 25.2a shows small sections of two adjacent spherical wave fronts. At large distances from the source, the wave fronts become less and less curved and approach the shape of flat surfaces, as in part b of the drawing. Waves whose wave fronts are flat surfaces (i.e., planes) are known as **plane waves** and are important in understanding the properties of mirrors and lenses. Since rays are perpendicular to the wave fronts, the rays for a plane wave are parallel to each other.

The concepts of wave fronts and rays can also be used to describe light waves. For light waves, the ray concept is particularly convenient when showing the path taken by the light. We will make frequent use of light rays, which can be regarded essentially as narrow beams of light much like those that lasers produce.

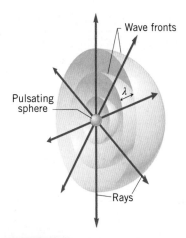

FIGURE 25.1 A hemispherical view of a sound wave emitted by a pulsating sphere. The wave fronts are drawn through the condensations of the wave, so the distance between two successive wave fronts is the wavelength λ. The rays are perpendicular to the wave fronts and point in the direction of the velocity of the wave.

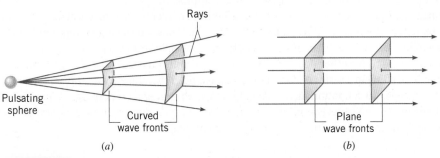

(a) *(b)*

FIGURE 25.2 (*a*) Portions of two spherical wave fronts are shown. The rays are perpendicular to the wave fronts and diverge. (*b*) For a plane wave, the wave fronts are flat surfaces, and the rays are parallel to each other.

25.2 | The Reflection of Light

Most objects reflect a certain portion of the light falling on them. Suppose that a ray of light is incident on a flat, shiny surface, such as the mirror in **Figure 25.3**. As the drawing shows, the *angle of incidence* θ_i is the angle that the incident ray makes with respect to the normal, which is a line drawn perpendicular to the surface at the point of incidence. The *angle of reflection* θ_r is the angle that the reflected ray makes with the normal. The **law of reflection** describes the behavior of the incident and reflected rays.

> **LAW OF REFLECTION**
>
> **The incident ray, the reflected ray, and the normal to the surface all lie in the same plane, and the angle of reflection θ_r equals the angle of incidence θ_i:**
>
> $$\theta_r = \theta_i$$

When parallel light rays strike a smooth, plane surface, such as the ones in **Figure 25.4a**, the reflected rays are parallel to each other. This type of reflection is one example of what is known as *specular reflection* and is important in determining the properties of mirrors. Most

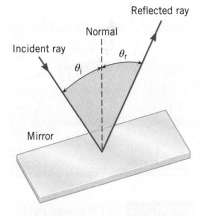

FIGURE 25.3 The angle of reflection θ_r equals the angle of incidence θ_i. These angles are measured with respect to the normal, which is a line drawn perpendicular to the surface of the mirror at the point of incidence.

FIGURE 25.4 (*a*) The drawing shows specular reflection from a polished plane surface, such as a mirror. The reflected rays are parallel to each other. (*b*) A rough surface reflects the light rays in all directions; this type of reflection is known as diffuse reflection.

(*a*) Specular reflection

(*b*) Diffuse reflection

surfaces, however, are not perfectly smooth, because they contain irregularities the sizes of which are equal to or greater than the wavelength of the light. The law of reflection applies to each ray, but the irregular surface reflects the light rays in various directions, as **Figure 25.4b** suggests. This type of reflection is known as *diffuse reflection*. Common surfaces that give rise to diffuse reflection are most papers, wood, nonpolished metals, and walls covered with a "flat" (nongloss) paint.

THE PHYSICS OF . . . digital movie projectors and micromirrors. A revolution in digital technology is occurring in the movie industry, where digital techniques are now being used to produce films. Until recently, films have been viewed primarily by using projectors that shine light directly through a strip of film containing the images. Now, however, projectors are available that allow a movie produced using digital techniques to be viewed completely without film by using a digital representation (zeros and ones) of the images. One form of digital projector depends on the law of reflection and tiny mirrors called micromirrors, each about the size of one-fourth the diameter of a human hair. Each micromirror creates a tiny portion of an individual movie frame on the screen and serves as a pixel, like one of the glowing spots that comprise the picture on a TV screen or computer monitor. This pixel action is possible because a micromirror pivots about 10° in one direction or the reverse in response to the "zero" or "one" in the digital representation of the frame. One of the directions puts a portion of the light from a powerful xenon lamp on the screen, and the other does not. The pivoting action can occur as fast as 1000 times per second, leading to a series of light pulses for each pixel that the eye and the brain combine and interpret as a continuously changing image. Present-generation digital micromirror projectors use up to several million micromirrors to reproduce each of the three primary colors (red, green, and blue) that comprise a color image.

25.3 The Formation of Images by a Plane Mirror

When you look into a plane (flat) mirror, you see an image of yourself that has three properties:

1. The image is upright.

2. The image is the same size as you are.

3. The image is located as far behind the mirror as you are in front of it.

As **Figure 25.5a** illustrates, the image of yourself in the mirror is also reversed right to left and left to right. If you wave your *right* hand, it is the *left* hand of the image that waves back. Similarly, letters and words held up to a mirror are reversed. Ambulances and other emergency vehicles are often lettered in reverse, as in **Figure 25.5b**, so that the letters will appear normal when seen in the rearview mirror of a car.

To illustrate why an image appears to originate from behind a plane mirror, **Figure 25.6a** shows a light ray leaving the top of an object. This ray reflects from the mirror (angle of reflection equals angle of incidence) and enters the eye. To the eye, it appears that the ray originates from behind the mirror, somewhere back along the dashed line. Actually, rays going in all directions leave each point on the object, but only a small bundle of such rays is intercepted by the

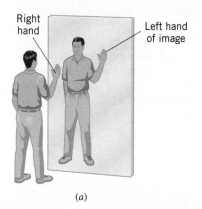

Right hand   Left hand of image

(a)

Dennis MacDonald/Age Fotostock

(b)

FIGURE 25.5 (a) The person's right hand becomes the image's left hand in a plane mirror. (b) Many emergency vehicles are reverse-lettered so the lettering appears normal when viewed in the rearview mirror of a car.

FIGURE 25.6 (a) A ray of light from the top of the chess piece reflects from the mirror. To the eye, the ray seems to come from behind the mirror. (b) The bundle of rays from the top of the object appears to originate from the image behind the mirror.

eye. Part *b* of the figure shows a bundle of two rays leaving the top of the object. All the rays that leave a given point on the object, no matter what angle θ they have when they strike the mirror, appear to originate from a corresponding point on the image behind the mirror (see the dashed lines in part *b*). For each point on the object, there is a single corresponding point on the image, and it is this fact that makes the image in a plane mirror a sharp and undistorted one.

Although rays of light *seem* to come from the image, it is evident from **Figure 25.6b** that they do not originate from behind the plane mirror where the image appears to be. Because none of the light rays actually emanate from the image, it is called a **virtual image**. In this text the parts of the light rays that appear to come from a virtual image are represented by dashed lines. *Curved* mirrors, on the other hand, can produce images from which all the light rays actually do emanate. Such images are known as **real images** and are discussed later.

With the aid of the law of reflection, it is possible to show that the image is located as far behind a plane mirror as the object is in front of it. In **Figure 25.7** the object distance is d_o and the image distance is d_i. A ray of light leaves the base of the object, strikes the mirror at an angle of incidence θ, and is reflected at the same angle. To the eye, this ray appears to come from the base of the image. For the angles β_1 and β_2 in the drawing it follows that $\theta + \beta_1 = 90°$ and $\alpha + \beta_2 = 90°$. But the angle α is equal to the angle of reflection θ, since the two are opposite angles formed by intersecting lines. Therefore, $\beta_1 = \beta_2$. As a result, triangles *ABC* and *DBC* are identical (congruent) because they share a common side *BC* and have equal angles ($\beta_1 = \beta_2$) at the top and equal angles (90°) at the base. Thus, the magnitude of the object distance d_o equals the magnitude of the image distance d_i.

By starting with a light ray from the top of the object, rather than the bottom, we can use the same line of reasoning to show also that the height of the image equals the height of the object.

Conceptual Examples 1 and 2 discuss some interesting features of plane mirrors.

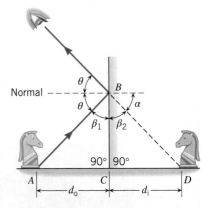

FIGURE 25.7 This drawing illustrates the geometry used with a plane mirror to show that the image distance d_i equals the object distance d_o.

CONCEPTUAL EXAMPLE 1 | Full-Length Versus Half-Length Mirrors

In **Figure 25.8** a woman is standing in front of a plane mirror. Is the minimum mirror height that is necessary for her to see her full image **(a)** equal to her height, or **(b)** equal to one-half her height?

Reasoning   The woman sees her image because light emanating from her body is reflected by the mirror (labeled *ABCD* in **Figure 25.8**) and enters her eyes. Consider, for example, a ray of light from her foot *F*. This ray strikes the mirror at *B* and enters her eyes at *E*. According to the law of reflection, the angles of incidence and reflection are both θ. This law will allow us to deduce how the height of the mirror is related to her own height.

Answer (a) is incorrect.   The mirror in **Figure 25.8** is the same height as the woman. Any light from her foot that strikes the mirror below *B* is reflected toward a point on her body that is below her eyes. Since light striking the mirror below *B* does not enter her eyes, the part of the mirror between *B* and *A* may be removed. Thus, the necessary minimum height of the mirror is not equal to the woman's height.

Answer (b) is correct.   As discussed above, the section *AB* of the mirror is not necessary in order for the woman to see her full image. The section *BC* of the mirror that produces the image is one-half the woman's height between *F* and *E*. This follows because the right triangles *FBM* and *EBM* are identical. They are identical because they share a common side *BM* and have two angles, θ and 90°, that are the same. The blowup in **Figure 25.8** illustrates a similar line of reasoning, starting with a ray from the woman's head at *H*. This ray is reflected from the mirror at *P* and enters her eyes. The top mirror section *PD* may be removed without disturbing this

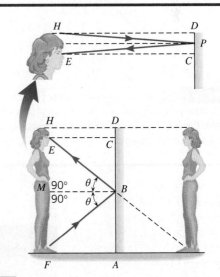

FIGURE 25.8 A woman stands in front of a plane mirror and sees her full image.

reflection. The necessary section *CP* is one-half the woman's height between her head at *H* and her eyes at *E*. We find, then, that only the sections *BC* and *CP* are needed for the woman to see her full length. The height of section *BC* plus section *CP* is exactly one-half the woman's height. The conclusions here are valid regardless of how far the person stands from the mirror. Thus, to view one's full length in a mirror, only a half-length mirror is needed.

Related Homework: Problem 34

CONCEPTUAL EXAMPLE 2 | Multiple Reflections

A person is sitting in front of two mirrors that intersect at a 90° angle. As **Figure 25.9a** illustrates, the person sees three images of herself. (The person herself is only partially visible at the bottom of the photo.) These images arise because rays of light emanate from her body, reflect from the mirrors, and enter her eyes. Consider the light that enters her eyes and appears to come from each of the three images identified in **Figure 25.9b**. The following table shows three possibilities for the number of reflections that the light undergoes before entering her eyes. Which one is correct?

| | Number of Reflections | | |
Possibility	Image 1	Image 2	Image 3
(a)	2	2	3
(b)	3	3	3
(c)	1	1	2

Reasoning Images of the woman are formed when light emanating from her body enters her eyes after being reflected by one, or both, mirrors. For each reflection, the angle of the light reflected from a mirror is equal to the angle of the light incident on the mirror (law of reflection). We will see that there are three ways that light can reach her eyes from the two mirrors.

Answers (a) and (b) are incorrect. Figure 25.9b represents a top view of the person in front of the two mirrors. It is a straightforward matter to understand two of the images that she sees. These are the images that are normally seen when one sits in front of a mirror. Sitting in front of mirror 1, she sees image 1, which is located as far behind that mirror as she is in front of it. She also sees image 2 behind mirror 2, at a distance that matches her distance in front of that mirror. Each of these images arises from light emanating from her body and reflecting only once from a single mirror. Therefore, each ray of light does not reflect two or three times before entering her eyes.

Answer (c) is correct. As discussed above, images 1 and 2 arise, respectively, from *single* reflections from mirrors 1 and 2. The third image arises when light undergoes two reflections in sequence, first from one mirror and then from the other. When such a double reflection occurs, an additional image becomes possible. **Figure 25.9b** shows two rays of light that strike mirror 1. Each one, according to the law of reflection, has an angle of reflection that equals the angle of incidence. The rays then strike mirror 2, where they again are reflected according to the law of reflection. When the outgoing rays are extended backward (see the dashed lines in the drawing), they intersect and appear to originate from image 3. Thus, the third image arises when an incident ray of light is reflected twice, once from each mirror, before entering her eyes.

(a)

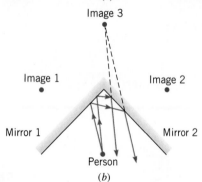

(b)

FIGURE 25.9 (a) These two perpendicular plane mirrors produce three images of the person (not completely visible) sitting in front of them. (b) A "double" reflection occurs, one from each mirror, and produces Image 3.

Check Your Understanding

(The answers are given at the end of the book.)

1. CYU Figure 25.1 shows a light ray undergoing multiple reflections from a mirrored corridor. The walls of the corridor are either parallel or perpendicular to one another. If the initial angle of incidence is 35°, what is the angle of reflection when the ray makes its last reflection?

2. A sign painted on a store window is reversed when viewed from inside the store. If a person inside the store views the reversed sign in a plane mirror, does the sign appear as it would when viewed from outside the store? (Try it by writing some letters on a transparent sheet of paper and then holding the back side of the paper up to a mirror.)

CYU FIGURE 25.1

3. If a clock is held in front of a mirror, its image is reversed left to right. From the point of view of a person looking into the mirror, does the image of the second hand rotate in the reverse (counterclockwise) direction?

25.4 | Spherical Mirrors

The most common type of curved mirror is a spherical mirror. As **Interactive Figure 25.10** shows, a spherical mirror has the shape of a section from the surface of a hollow sphere. If the inside surface of the mirror is polished, it is a **concave mirror**. If the outside surface is polished, it is a **convex mirror**. The drawing shows both types of mirrors, with a light ray reflecting from the polished surface. The law of reflection applies, just as it does for a plane mirror. For either type of spherical mirror, the normal is drawn perpendicular to the mirror at the point of incidence. For each type, the center of curvature is located at point *C*, and the radius of curvature is *R*. The **principal axis** of the mirror is a straight line drawn through *C* and the midpoint of the mirror.

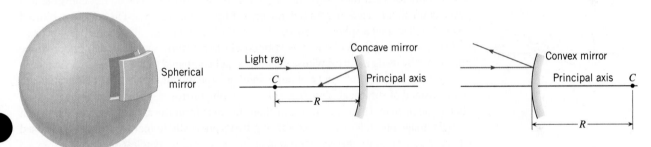

INTERACTIVE FIGURE 25.10 A spherical mirror has the shape of a segment of a spherical surface. The center of curvature is point *C* and the radius is *R*. For a concave mirror, the reflecting surface is the inner one; for a convex mirror it is the outer one.

    Figure 25.11 shows a tree in front of a concave mirror. A point on this tree lies on the principal axis of the mirror and is beyond the center of curvature *C*. Light rays emanate from this point and reflect from the mirror, consistent with the law of reflection. If the rays are near the principal axis, they cross it at a common point after reflection. This point is called the *image point*. The rays continue to diverge from the image point as if there were an object there. Since light rays actually come from the image point, the image is a real image.

    If the tree in **Figure 25.11** is infinitely far from the mirror, the rays are parallel to each other and to the principal axis as they approach the mirror. **Figure 25.12** shows rays near and parallel to the principal axis, as they reflect from the mirror and pass

FIGURE 25.11 A point on the tree lies on the principal axis of the concave mirror. Rays from this point that are near the principal axis are reflected from the mirror and cross the axis at the image point.

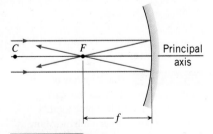

FIGURE 25.12 Light rays near and parallel to the principal axis are reflected from a concave mirror and converge at the focal point *F*. The focal length *f* is the distance between *F* and the mirror.

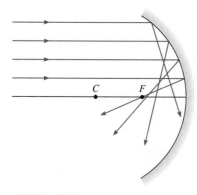

FIGURE 25.13 This drawing is used to show that the focal point *F* of a concave mirror is halfway between the center of curvature *C* and the mirror at point *B*.

through an image point. In this special case the image point is referred to as the **focal point** *F* of the mirror. Therefore, an object infinitely far away on the principal axis gives rise to an image at the focal point of the mirror. The distance between the focal point and the middle of the mirror is the **focal length** *f* of the mirror.

We can show that the focal point *F* lies halfway between the center of curvature *C* and the middle of a concave mirror. In **Figure 25.13**, a light ray parallel to the principal axis strikes the mirror at point *A*. The line *CA* is the radius of the mirror and, therefore, is the normal to the spherical surface at the point of incidence. The ray reflects from the mirror, and the angle of reflection θ equals the angle of incidence. Furthermore, the angle *ACF* is also θ because the radial line *CA* is a transversal of two parallel lines. Since two of its angles are equal, the colored triangle *CAF* is an isosceles triangle; thus, sides *CF* and *FA* are equal. However, when the incoming ray lies close to the principal axis, the angle of incidence θ is small, and the distance *FA* does not differ appreciably from the distance *FB*. Therefore, in the limit that θ is small, *CF* = *FA* = *FB*, and so the focal point *F* lies halfway between the center of curvature and the mirror. In other words, the focal length *f* is one-half of the radius *R*:

Focal length of a concave mirror

$$f = \tfrac{1}{2}R \tag{25.1}$$

Rays that lie close to the principal axis are known as **paraxial rays**,* and Equation 25.1 is valid only for such rays. Rays that are far from the principal axis do not converge to a single point after reflection from the mirror, as **Figure 25.14** shows. The result is a blurred image. The fact that a spherical mirror does not bring all rays parallel to the principal axis to a single image point is known as **spherical aberration**. Spherical aberration can be minimized by using a mirror whose height is small compared to the radius of curvature.

A sharp image point can be obtained with a large mirror, if the mirror is parabolic in shape instead of spherical. The shape of a parabolic mirror is such that all light rays parallel to the principal axis, regardless of their distance from the axis, are reflected through a single image point. However, parabolic mirrors are costly to manufacture and are used where the sharpest images are required, as in research-quality telescopes.

THE PHYSICS OF . . . capturing solar energy with mirrors. Parabolic mirrors are also used in one method of collecting solar energy for commercial purposes. **Figure 25.15** shows a long row of concave parabolic mirrors that reflect the sun's rays to the focal point. Located at the focal point and running the length of the row is an

FIGURE 25.14 Rays that are farthest from the principal axis have the greatest angle of incidence and miss the focal point *F* after reflection from the mirror.

Oil-filled pipe

FIGURE 25.15 This long row of parabolic mirrors focuses the sun's rays to heat an oil-filled pipe located at the focal point of each mirror. Many such rows are used by a solar-thermal electric plant in the Mojave Desert.

Jim West / Alamy Stock Photo

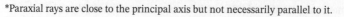

*Paraxial rays are close to the principal axis but not necessarily parallel to it.

oil-filled pipe. The focused rays of the sun heat the oil. In a solar-thermal electric plant, the heat from many such rows is used to generate steam. The steam, in turn, drives a turbine connected to an electric generator.

THE PHYSICS OF . . . automobile headlights.   Another application of parabolic mirrors is in automobile headlights. Here, however, the situation is reversed from the operation of a solar collector. In a headlight, a high-intensity light source is placed at the focal point of the mirror, and light emerges parallel to the principal axis.

A convex mirror also has a focal point, and **Figure 25.16** illustrates its meaning. In this picture, parallel rays are incident on a convex mirror. Clearly, the rays diverge after being reflected. If the incident parallel rays are paraxial, the reflected rays seem to come from a single point F behind the mirror. This point is the focal point of the convex mirror, and its distance from the midpoint of the mirror is the **focal length** f. The focal length of a convex mirror is also one-half of the radius of curvature, just as it is for a concave mirror. However, we assign the focal length of a convex mirror a negative value because it will be convenient later on:

Focal length of a convex mirror

$$f = -\tfrac{1}{2}R \qquad (25.2)$$

Spherical aberration is a problem with convex mirrors, just as it is with concave mirrors. Rays that emanate from a single point on an object but are far from the principal axis do not appear to originate from a single image point after reflection from the mirror. As with a concave mirror, the result is a blurred image.

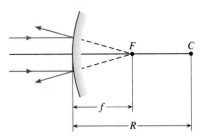

FIGURE 25.16 When paraxial light rays that are parallel to the principal axis strike a convex mirror, the reflected rays appear to originate from the focal point F. The radius of curvature is R and the focal length is f.

Check Your Understanding

(The answers are given at the end of the book.)

4. A section of the surface of a hollow sphere has a radius of curvature of 0.60 m, and both the inside and outside surfaces have a mirror-like polish. What are the focal lengths of the inside and outside surfaces?

5. **Photo 25.1** shows an experimental device at Sandia National Laboratories in New Mexico. This device is a mirror that focuses sunlight to heat sodium to a boil, which then heats helium gas in an engine. The engine does the work of driving a generator to produce electricity. The sodium unit and the engine are labeled in the photo. **(a)** What kind of mirror, concave or convex, is being used? **(b)** Where is the sodium unit located relative to the mirror? Express your answer in terms of the focal length of the mirror.

Sodium unit and engine

Courtesy Sandia National Laboratories

PHOTO 25.1

6. Refer to **Figure 25.14** and the related discussion about spherical aberration. To bring the top ray closer to the focal point F after reflection, describe how you would change the shape of the mirror. Would you open it up to produce a more gently curving shape or bring the top and bottom edges closer to the principal axis?

25.5 The Formation of Images by Spherical Mirrors

As we have seen, some of the light rays emitted from an object in front of a mirror strike the mirror, reflect from it, and form an image. We can analyze the image produced by either concave or convex mirrors by using a graphical method called **ray tracing**. This method is based on the law of reflection and the notion that a spherical mirror has a center of curvature C and a focal point F. Ray tracing enables us to find the location of the image, as well as its size, by taking advantage of the following fact: paraxial rays leave from a point on the object and intersect, or appear to intersect, at a corresponding point on the image after reflection.

Concave Mirrors

Three specific paraxial rays are especially convenient to use in the ray-tracing method. **Interactive Figure 25.17** shows an object in front of a concave mirror, and these three rays leave from a point on the top of the object. The rays are labeled 1, 2, and 3, and when tracing their paths, we use the following reasoning strategy.

> **REASONING STRATEGY**   **Ray Tracing for a Concave Mirror**
>
> *Ray 1.* **This ray is initially parallel to the principal axis and, therefore, passes through the focal point F after reflection from the mirror.**
>
> *Ray 2.* **This ray initially passes through the focal point F and is reflected parallel to the principal axis. Ray 2 is analogous to ray 1 except that the reflected, rather than the incident, ray is parallel to the principal axis.**
>
> *Ray 3.* **This ray travels along a line that passes through the center of curvature C and follows a radius of the spherical mirror; as a result, the ray strikes the mirror perpendicularly and reflects back on itself.**

If rays 1, 2, and 3 are superimposed on a scale drawing, they converge at a point on the top of the image, as can be seen in **Animated Figure 25.18a.*** Although three rays have been used here to locate the image, only two are really needed; the third ray is usually drawn to serve as a check. In a similar fashion, rays from all other points on the object locate corresponding points on the image, and the mirror forms a complete image of the object. If you were to place your eye as shown in the drawing, you would see an image that is *larger* and *inverted* relative to the object. The image is *real* because the light rays actually pass through the image.

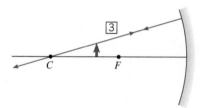

INTERACTIVE FIGURE 25.17   The rays labeled 1, 2, and 3 are useful in locating the image of an object placed in front of a concave spherical mirror. The object is represented as a vertical arrow.

ANIMATED FIGURE 25.18   (*a*) When an object is placed between the focal point F and the center of curvature C of a concave mirror, a real image is formed. The image is enlarged and inverted relative to the object. (*b*) When the object is located beyond the center of curvature C, a real image is created that is reduced in size and inverted relative to the object.

*In the drawings that follow, we assume that the rays are paraxial, although the distance between the rays and the principal axis is often exaggerated for clarity.

If the locations of the object and image in **Animated Figure 25.18a** are interchanged, the situation in part *b* of the drawing results. The three rays in part *b* are the same as those in part *a*, except that the directions are reversed. These drawings illustrate the **principle of reversibility**, which states the following:

| **Problem-Solving Insight**   **If the direction of a light ray is reversed, the light retraces its original path.**

This principle is quite general and is not restricted to reflection from mirrors. The image is *real,* and it is *smaller* and *inverted* relative to the object.

When the object is placed between the focal point *F* and a concave mirror, as in **Figure 25.19a**, three rays can again be drawn to find the image. Now, however, ray 2 does not go through the focal point on its way to the mirror, since the object is closer to the mirror than the focal point is. When projected backward, though, ray 2 appears to come from the focal point. Therefore, after reflection, ray 2 is directed parallel to the principal axis. In this case the three reflected rays diverge from each other and do not converge to a common point. However, when projected behind the mirror, the three rays appear to come from a common point; thus, a *virtual* image is formed. This virtual image is *larger* than the object and *upright.*

(a)   (b)

FIGURE 25.19 (*a*) When an object is located between a concave mirror and its focal point *F*, an enlarged, upright, and virtual image is produced. (*b*) A makeup mirror (or shaving mirror) is a concave mirror that functions in exactly this fashion, as this photograph shows.

THE PHYSICS OF . . . makeup and shaving mirrors.   Makeup and shaving mirrors are concave mirrors. When you place your face between the mirror and its focal point, you see an enlarged virtual image of yourself, as **Figure 25.19b** shows.

THE PHYSICS OF . . . a head-up display for automobiles.   Concave mirrors are also used in one method for displaying the speed of a car. The method presents a digital readout (e.g., "51 km/h") that the driver sees when looking directly through the windshield, as in **Figure 25.20a**. The advantage of the method, which is called a head-up display (HUD), is that the driver does not need to take his or her eyes off the

(a)   (b)

FIGURE 25.20 (*a*) A head-up display (HUD) presents the driver with a digital readout of the car's speed in the field of view seen through the windshield. (*b*) One version of a HUD uses a concave mirror. (See the text for explanation.)

road to monitor the speed. **Figure 25.20b** shows how one type of HUD works. Below the windshield is a readout device that displays the speed in digital form. This device is located between a concave mirror and its focal point. The arrangement is similar to the one in **Figure 25.19a** and produces a virtual, upright, and enlarged image of the speed readout (see virtual image 1 in **Figure 25.20b**). Light rays that appear to come from this image strike the windshield at a place where a so-called "combiner" is located. The purpose of the combiner is to combine the digital readout information with the field of view that the driver sees through the windshield. The combiner is virtually undetectable by the driver because it allows all colors except one to pass through it unaffected. The one exception is the color produced by the digital readout device. For this color, the combiner behaves as a plane mirror and reflects the light that appears to originate from image 1. Thus, the combiner produces image 2, which is what the driver sees. The location of image 2 is out above the front bumper. The driver can then read the speed with eyes focused just as they are to see the road.

Convex Mirrors

The ray-tracing procedure for determining the location and size of an image in a convex mirror is similar to that for a concave mirror. The same three rays are used. However, the focal point and center of curvature of a convex mirror lie behind the mirror, not in front of it. **Interactive Figure 25.21a** shows the rays. When tracing their paths, we use the following reasoning strategy, which takes into account these locations of the focal point and center of curvature.

> **REASONING STRATEGY** **Ray Tracing for a Convex Mirror**
>
> *Ray 1.* **This ray is initially parallel to the principal axis and, therefore, appears to originate from the focal point *F* after reflection from the mirror.**
>
> *Ray 2.* **This ray heads toward *F*, emerging parallel to the principal axis after reflection. Ray 2 is analogous to ray 1, except that the reflected, rather than the incident, ray is parallel to the principal axis.**
>
> *Ray 3.* **This ray travels toward the center of curvature *C*; as a result, the ray strikes the mirror perpendicularly and reflects back on itself.**

The three rays in **Interactive Figure 25.21a** appear to come from a point on a *virtual* image that is behind the mirror. The virtual image is *diminished in size* and *upright*, relative to the object. A convex mirror *always* forms a virtual image of the object, no matter where in front of the mirror the object is placed. **Interactive Figure 25.21b** shows an example of such an image.

Object   Virtual   *F*   *C*
           image

(*a*)   (*b*)

Popperfoto/Getty Images

INTERACTIVE FIGURE 25.21 (*a*) An object placed in front of a convex mirror always produces a virtual image behind the mirror; the image is reduced in size and is upright. (*b*) The skier's chromed helmet acts as a convex mirror and produces an image of the spectators who are watching the event.

THE PHYSICS OF . . . passenger-side automobile mirrors.   Because of its shape, a convex mirror gives a wider field of view than do other types of mirrors. A mirror with a wide field of view is needed to give a driver a good rear view. Thus, the outside mirror on the passenger side is often a convex mirror. Printed on such a mirror is usually the warning "OBJECTS IN MIRROR ARE CLOSER THAN THEY APPEAR." The reason for the warning is that, as in **Interactive Figure 25.21a**, the virtual image is reduced in size and therefore looks smaller, just as a distant object would appear in a plane mirror. An unwary driver, thinking that the side-view mirror is a plane mirror, might incorrectly deduce from the small size of the image that the car behind is far enough away to ignore. Because of their wide field of view, convex mirrors are also used in stores for security purposes.

Check Your Understanding

(The answers are given at the end of the book.)

7. Is it possible to use a convex mirror to produce an image that is larger than the object? Explain.

8. (a) When you look at the back side of a shiny teaspoon held at arm's length, do you see yourself upright or upside down? **(b)** When you look at the other side of the spoon, do you see yourself upright or upside down? Assume in both cases that the distance between you and the spoon is greater than the focal length of the spoon.

9. (a) Can the image formed by a concave mirror ever be projected directly onto a screen without the help of other mirrors or lenses? If so, specify where the object should be placed relative to the mirror. **(b)** Repeat part (a) assuming that the mirror is convex.

10. Suppose that you stand in front of a spherical mirror (concave or convex). Is it possible for your image to be **(a)** real and upright **(b)** virtual and inverted?

11. An object is placed between the focal point and the center of curvature of a concave mirror. The object is then moved closer to the mirror, but still remains between the focal point and the center of curvature. Do the magnitudes of **(a)** the image distance and **(b)** the image height become larger or smaller?

12. When you see the image of yourself formed by a mirror, it is because (1) light rays actually coming from a real image enter your eyes or (2) light rays appearing to come from a virtual image enter your eyes. If light rays from the image do not enter your eyes, you do not see yourself. Are there any places on the principal axis where you cannot see yourself when you are standing in front of a mirror that is **(a)** convex **(b)** concave? If so, where are these places? Assume that you have only the one mirror to use.

25.6 The Mirror Equation and the Magnification Equation

Ray diagrams drawn to scale are useful for determining the location and size of the image formed by a mirror. However, for an accurate description of the image, a more analytical technique is needed, so we will derive two equations, known as the **mirror equation** and the **magnification equation**. These equations are based on the law of reflection and provide relationships between:

f = the focal length of the mirror
d_o = the object distance, which is the distance between the object and the mirror
d_i = the image distance, which is the distance between the image and the mirror
m = the magnification of the mirror, which is the ratio of the height of the image to the height of the object.

Object

(a)

(b)

FIGURE 25.22 These diagrams are used to derive the mirror equation and the magnification equation. (a) The two colored triangles are similar triangles. (b) If the ray is close to the principal axis, the two colored regions are almost similar triangles.

Concave Mirrors

We begin our derivation of the mirror equation by referring to **Figure 25.22a**, which shows a ray leaving the top of the object and striking a concave mirror at the point where the principal axis intersects the mirror. Since the principal axis is perpendicular to the mirror, it is also the normal at this point of incidence. Therefore, the ray reflects at an equal angle and passes through the image. The two colored triangles are similar triangles because they have equal angles, so

$$\frac{h_o}{-h_i} = \frac{d_o}{d_i}$$

where h_o is the height of the object and h_i is the height of the image. The minus sign appears on the left in this equation because the image is inverted in **Figure 25.22a**. In part b another ray leaves the top of the object, this one passing through the focal point F, reflecting parallel to the principal axis, and then passing through the image. Provided the ray remains close to the axis, the two colored areas can be considered to be similar triangles, with the result that

$$\frac{h_o}{-h_i} = \frac{d_o - f}{f}$$

Setting the two equations above equal to each other yields $d_o/d_i = (d_o - f)/f$. Rearranging this result gives the **mirror equation**:

Mirror equation
$$\frac{1}{d_o} + \frac{1}{d_i} = \frac{1}{f} \qquad (25.3)$$

We have derived this equation for a real image formed in front of a concave mirror. In this case, the image distance is a positive quantity, as are the object distance and the focal length. However, we have seen in the last section that a concave mirror can also form a virtual image, if the object is located between the focal point and the mirror. Equation 25.3 can also be applied to such a situation, provided that we adopt the convention that d_i is negative for an image behind the mirror, as it is for a virtual image.

In deriving the magnification equation, we remember that the **magnification** m of a mirror is the ratio of the image height to the object height: $m = h_i/h_o$. If the image height is less than the object height, the magnitude of m is less than one, and if the image is larger than the object, the magnitude of m is greater than one. We have already shown that $h_o/(-h_i) = d_o/d_i$, so it follows that

Magnification equation
$$m = \frac{\text{Image height, } h_i}{\text{Object height, } h_o} = -\frac{d_i}{d_o} \qquad (25.4)$$

As Examples 3 and 4 show, the value of m is negative if the image is inverted and positive if the image is upright.

EXAMPLE 3 | A Real Image Formed by a Concave Mirror

A 2.0-cm-high object is placed 7.10 cm from a concave mirror whose radius of curvature is 10.20 cm. Find **(a)** the location of the image and **(b)** its size.

Reasoning For a concave mirror, Equation 25.1 gives the focal length as $f = \frac{1}{2}R$. Therefore, the focal length is $f = \frac{1}{2}(10.20 \text{ cm}) = 5.10$ cm, and the object is located between the focal point F and the center of curvature C of the mirror, as in **Animated Figure 25.18a**. Based on this figure, we expect that the image is real and that, relative to the object, it is farther away from the mirror, inverted, and larger.

Problem-Solving Insight According to the mirror equation, the image distance d_i has a reciprocal given by $d_i^{-1} = f^{-1} - d_o^{-1}$. After combining the reciprocals f^{-1} and d_o^{-1}, do not forget to take the reciprocal of the result to find d_i.

Solution **(a)** With $d_o = 7.10$ cm and $f = 5.10$ cm, the mirror equation (Equation 25.3) can be used to find the image distance:

$$\frac{1}{d_i} = \frac{1}{f} - \frac{1}{d_o} = \frac{1}{5.10 \text{ cm}} - \frac{1}{7.10 \text{ cm}} = 0.055 \text{ cm}^{-1} \quad \text{or} \quad \boxed{d_i = 18 \text{ cm}}$$

In this calculation, f and d_o are positive numbers, indicating that the focal point and the object are in front of the mirror. The positive answer for d_i means that the image is also in front of the mirror, and the reflected rays actually pass through the image, as **Animated Figure 25.18a** shows. In other words, the positive value for d_i indicates that the image is a real image.

(b) According to the magnification equation (Equation 25.4), the image height h_i is related to the object height h_o and the

magnification m by $h_i = mh_o$, where $m = -d_i/d_o$. Thus, we find that

$$h_i = -\left(\frac{d_i}{d_o}\right)h_o = -\left(\frac{18\text{ cm}}{7.10\text{ cm}}\right)(2.0\text{ cm}) = \boxed{-5.1\text{ cm}}$$

The image is larger than the object, and the negative value for h_i indicates that the image is inverted with respect to the object, as in **Animated Figure 25.18a**.

EXAMPLE 4 | A Virtual Image Formed by a Concave Mirror

An object is placed 6.00 cm in front of a concave mirror that has a 10.0-cm focal length. **(a)** Determine the location of the image. **(b)** The object is 1.2 cm high. Find the image height.

Reasoning The object is located between the focal point and the mirror, as in **Figure 25.19a**. The setup is analogous to a person using a makeup or shaving mirror. Therefore, we expect that the image is virtual and that, relative to the object, it is upright and larger.

Solution **(a)** Using the mirror equation with $d_o = 6.00$ cm and $f = 10.0$ cm, we have

$$\frac{1}{d_i} = \frac{1}{f} - \frac{1}{d_o} = \frac{1}{10.0\text{ cm}} - \frac{1}{6.00\text{ cm}}$$

$$= -0.067\text{ cm}^{-1}\quad\text{or}\quad \boxed{d_i = -15\text{ cm}}$$

The answer for d_i is negative, indicating that the image is *behind* the mirror. Thus, as expected, the image is virtual.

(b) The image height h_i can be found from the magnification equation, which indicates that $h_i = mh_o$, where h_o is the object height and $m = -d_i/d_o$. It follows, then, that

$$h_i = -\left(\frac{d_i}{d_o}\right)h_o = -\left(\frac{-15\text{ cm}}{6.00\text{ cm}}\right)(1.2\text{ cm}) = \boxed{3.0\text{ cm}}$$

The image is larger than the object, and the positive value for h_i indicates that the image is upright (see **Figure 25.19a**).

There are many applications of concave mirrors. As the following example shows, the James Webb Space Telescope will use one of the largest.

EXAMPLE 5 | BIO The Physics of Searching for Extraterrestrial Life

The James Webb Space Telescope (JWST) will be the largest and most powerful telescope ever launched into space. The successor to the Hubble Space Telescope (see Examples 10 and 11, Chapter 5), the JWST will expand on Hubble's discoveries and further improve our fundamental understanding of the universe. Both Hubble and the JWST are similar in how they operate. They each have a large primary mirror that reflects the incoming light off a smaller secondary mirror, which then directs the light back through the center of the primary mirror and focuses it at a point behind the mirror. This focal point coincides with the location of a suite of scientific instruments that analyzes the light (see **Figure 25.23a**). While the two telescopes operate in a similar fashion, the JWST is specifically designed to view the universe using longer wavelength infrared light. This makes it particularly well-suited for analyzing the atmospheres of new *exoplanets*, which are planets that exist outside the solar system. (At the time of this writing, researchers have discovered over 4395 exoplanets in 3242 star systems.) Molecules in the atmosphere of an exoplanet that might be conducive to life, such as oxygen, water vapor, and methane, have a large number of spectral features, or signatures, when viewed in the infrared, which makes the JWST the perfect instrument for this type of research. Even though the

telescope will utilize wavelengths of light outside of the visible spectrum, the law of reflection still governs how the light will reflect from the surface of its mirrors. The primary mirror of the JWST is composed of an array of 18 hexagonal mirrors that are each 1.3 m in width, making the overall diameter of the primary mirror 6.5 m. This is over 2½ times larger in diameter than Hubble's primary mirror (see **Figure 25.23b**). At a temperature of 30 K, which will be close to the operating temperature of the telescope in deep space, the radius of curvature of the primary mirror is 15.880 m. At what distance from the primary mirror should the secondary mirror be placed, if light from a faraway object is to be focused at that location?

Reasoning The primary mirror of the telescope can be treated as a concave spherical mirror. Since the radius of curvature of the mirror is known, we can calculate its focal length using Equation 25.1. Applying the mirror equation (Equation 25.3) shows that, for an object located at infinity, the image distance will correspond to the focal length.

Solution Applying Equation 25.1, we find the focal length of the primary mirror to be

$$f = \frac{R}{2} = \frac{15.880\text{ m}}{2} = 7.9400\text{ m}$$

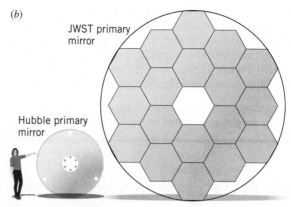

FIGURE 25.23 (*a*) The geometry of the James Webb Space Telescope (JWST). The large primary mirror consists of an array of 18 individual hexagonal mirrors coated with beryllium and gold. Light from the primary mirror reflects off a secondary mirror and back toward analysis equipment at the center of the primary. (*b*) A relative comparison of the size of the primary mirrors in the Hubble Space Telescope and the JWST.

For a faraway object, like a distant galaxy, the object distance d_o will equal infinity. Applying Equation 25.3 and solving for d_i, we find

$$\frac{1}{d_i} = \frac{1}{f} - \frac{1}{d_o} = \frac{1}{f} - \frac{1}{\infty} = \frac{1}{f} - 0 \Rightarrow d_i = f = \boxed{7.9400 \text{ m}}$$

The actual position of the secondary mirror is about 7.32 m from the primary mirror, somewhat closer than the calculated result. Thus, the focal point of the light from the primary is located behind the secondary. The secondary intercepts the light from the primary and reflects it back toward the instruments that reside behind the center of the primary mirror.

Convex Mirrors

The mirror equation and the magnification equation can also be used with convex mirrors, provided the focal length f is taken to be a *negative number*, as indicated explicitly in Equation 25.2. One way to remember this is to recall that the focal point of a convex mirror lies *behind* the mirror. Example 6 deals with a convex mirror.

EXAMPLE 6 | A Virtual Image Formed by a Convex Mirror

A convex mirror is used to reflect light from an object placed 66 cm in front of the mirror. The focal length of the mirror is $f = -46$ cm (note the minus sign). Find **(a)** the location of the image and **(b)** the magnification.

Reasoning We have seen that a convex mirror always forms a virtual image, as in **Interactive Figure 25.21a**, where the image is upright and smaller than the object. These characteristics should also be indicated by the results of our analysis here.

> **Problem-Solving Insight** When using the mirror equation, it is useful to construct a ray diagram to guide your thinking and to check your calculation.

Solution **(a)** With $d_o = 66$ cm and $f = -46$ cm, the mirror equation gives

$$\frac{1}{d_i} = \frac{1}{f} - \frac{1}{d_o} = \frac{1}{-46 \text{ cm}} - \frac{1}{66 \text{ cm}} = -0.037 \text{ cm}^{-1} \text{ or } \boxed{d_i = -27 \text{ cm}}$$

The negative sign for d_i indicates that the image is behind the mirror and, therefore, is a virtual image.

(b) According to the magnification equation, the magnification is

$$m = -\frac{d_i}{d_o} = -\frac{(-27 \text{ cm})}{66 \text{ cm}} = \boxed{0.41}$$

The image is smaller (m is less than one) and upright (m is positive) with respect to the object.

Convex mirrors, like plane (flat) mirrors, always produce virtual images behind the mirror. However, the virtual image in a convex mirror is closer to the mirror than it would be if the mirror were planar, as Example 7 illustrates.

EXAMPLE 7 | A Convex Versus a Plane Mirror

An object is placed 9.00 cm in front of a mirror. The image is 3.00 cm closer to the mirror when the mirror is convex than when it is planar (see **Figure 25.24**). Find the focal length of the convex mirror.

Reasoning For a plane mirror, the image and the object are the same distance on either side of the mirror. Thus, the image would be 9.00 cm behind a plane mirror. If the image in a convex mirror is 3.00 cm closer than this, the image must be located 6.00 cm behind the convex mirror. In other words, when the object distance is $d_o = 9.00$ cm, the image distance for the convex mirror is $d_i = -6.00$ cm (negative because the image is virtual). The mirror equation can be used to find the focal length of the mirror.

Solution According to the mirror equation, the reciprocal of the focal length is

$$\frac{1}{f} = \frac{1}{d_o} + \frac{1}{d_i} = \frac{1}{9.00 \text{ cm}} + \frac{1}{-6.00 \text{ cm}} = -0.056 \text{ cm}^{-1} \text{ or } \boxed{f = -18 \text{ cm}}$$

FIGURE 25.24 The object distance (9.00 cm) is the same for the plane mirror (top part of drawing) as for the convex mirror (bottom part of drawing). However, as discussed in Example 7, the image formed by the convex mirror is 3.00 cm closer to the mirror.

Contact lenses are worn to correct vision problems. Optometrists take advantage of the mirror equation and the magnification equation in providing lenses that fit the patient's eyes properly, as the next example illustrates.

Analyzing Multiple-Concept Problems

EXAMPLE 8 | BIO The Physics of Keratometers

A contact lens rests against the cornea of the eye. **Figure 25.25** shows an optometrist using a keratometer to measure the radius of curvature of the cornea, thereby ensuring that the prescribed lenses fit accurately. In the keratometer, light from an illuminated object reflects from the corneal surface, which acts like a convex mirror and forms an upright virtual image that is smaller than the object (see **Interactive Figure 25.21a**). With the object placed 9.0 cm in front of the cornea, the magnification of the corneal surface is measured to be 0.046. Determine the radius of the cornea.

Reasoning The radius of a convex mirror can be determined from the mirror's focal length, since the two are related. The focal length is related to the distances of the object and its image from the mirror via the mirror equation. The magnification of the mirror is also related to the object and image distances according to the magnification equation. By using the mirror equation and the magnification equation, we will be able to determine the focal length and, hence, the radius.

Knowns and Unknowns The following table summarizes the available data:

MARK THOMAS/Science Source

FIGURE 25.25 An optometrist is using a keratometer to measure the radius of curvature of the cornea of the eye, which is the surface against which a contact lens rests.

Description	Symbol	Value	Comment
Object distance	d_o	9.0 cm	Distance of object from cornea.
Magnification of corneal surface	m	0.046	Cornea acts like a convex mirror and forms a virtual image.
Unknown Variable			
Radius of cornea	R	?	

Modeling the Problem

STEP 1 Relation Between Radius and Focal Length The focal length f of a convex mirror is given by Equation 25.2 as $$f = -\frac{1}{2}R$$ where R is the radius of the spherical surface. Solving this expression for the radius gives Equation 1 at the right. In Step 2, we determine the unknown focal length.	$R = -2f$ **(1)** $\uparrow$ $\boxed{?}$
STEP 2 The Mirror Equation The focal length is related to the object distance d_o and the image distance d_i via the mirror equation, which specifies that $$\frac{1}{f} = \frac{1}{d_o} + \frac{1}{d_i} \qquad \textbf{(25.3)}$$ Solving this equation for f gives $$\boxed{f = \left(\frac{1}{d_o} + \frac{1}{d_i}\right)^{-1}}$$ which can be substituted into Equation 1 as shown in the right column. A value for d_o is given in the data table, and we turn to Step 3 to determine a value for d_i.	$R = -2f$ **(1)** $\uparrow$ $\boxed{f = \left(\frac{1}{d_o} + \frac{1}{d_i}\right)^{-1}}$ **(2)** $\uparrow$ $\boxed{?}$
STEP 3 The Magnification Equation According to the magnification equation, the magnification m is given by $$m = -\frac{d_i}{d_o} \qquad \textbf{(25.4)}$$ Solving for d_i, we obtain $$\boxed{d_i = -md_o}$$ and can substitute this result into Equation 2, as shown at the right.	$R = -2f$ **(1)** $\uparrow$ $\boxed{f = \left(\frac{1}{d_o} + \frac{1}{d_i}\right)^{-1}}$ **(2)** $\uparrow$ $\boxed{d_i = -md_o}$

Solution Combining the results of each step algebraically, we find that

$$
\begin{array}{ccccccc}
\textbf{STEP 1} & & \textbf{STEP 2} & & & \textbf{STEP 3} & \\
\downarrow & & \downarrow & & & \downarrow & \\
R & = & -2f & = & -2\left(\dfrac{1}{d_o} + \dfrac{1}{d_i}\right)^{-1} & = & -2\left[\dfrac{1}{d_o} + \dfrac{1}{(-md_o)}\right]^{-1}
\end{array}
$$

Thus, the radius is

$$R = -2\left[\frac{1}{d_o} + \frac{1}{(-md_o)}\right]^{-1} = \frac{2d_o m}{1 - m} = \frac{2(9.0\ \text{cm})(0.046)}{1 - 0.046} = \boxed{0.87\ \text{cm}}$$

Related Homework: Problem 29

Math Skills To show that the radius R is $R = \dfrac{2d_o m}{1 - m}$, we proceed in the following way. The first step is to factor out the term $\dfrac{1}{d_o}$ in the result for R:

$$R = -2\left[\frac{1}{d_o} + \frac{1}{(-md_o)}\right]^{-1} = -2\left[\frac{1}{d_o}\left(1 - \frac{1}{m}\right)\right]^{-1}$$

Rearranging the term within the brackets is the next step:

$$R = -2\left[\frac{1}{d_o}\left(1 - \frac{1}{m}\right)\right]^{-1} = -2\left[\frac{1}{d_o}\left(\frac{m}{m} - \frac{1}{m}\right)\right]^{-1} = -2\left(\frac{m-1}{d_o m}\right)^{-1}$$

Finally, taking the reciprocal of the term within the parentheses shows that

$$R = -2\left(\frac{m-1}{d_o m}\right)^{-1} = -2\left(\frac{d_o m}{m-1}\right) = \frac{2d_o m}{1 - m}$$

The following Reasoning Strategy summarizes the sign conventions that are used with the mirror equation and the magnification equation. These conventions apply to both concave and convex mirrors.

REASONING STRATEGY   **Summary of Sign Conventions for Spherical Mirrors**

Focal length

f is + for a concave mirror.

f is − for a convex mirror.

Object distance

d_o is + if the object is in front of the mirror (real object).

d_o is − if the object is behind the mirror (virtual object).*

Image distance

d_i is + if the image is in front of the mirror (real image).

d_i is − if the image is behind the mirror (virtual image).

Magnification

m is + for an image that is upright with respect to the object.

m is − for an image that is inverted with respect to the object.

Check Your Understanding

(*The answers are given at the end of the book.*)

13. An object is placed in front of a spherical mirror, and the magnification of the system is $m = -6$. What does this number tell you about the image? (Select one or more of the following choices.) **(a)** The image is larger than the object. **(b)** The image is smaller than the object. **(c)** The image is upright relative to the object. **(d)** The image is inverted relative to the object. **(e)** The image is a real image. **(f)** The image is a virtual image.

14. Plane mirrors and convex mirrors form virtual images. With a plane mirror, the image may be infinitely far behind the mirror, depending on where the object is located in front of the mirror. For an object in front of a single convex mirror, what is the greatest distance behind the mirror at which the image can be found?

EXAMPLE 9 | BIO The Physics of a Head Mirror

A head mirror is a simple medical diagnostic tool that is used to illuminate a patient's ear, throat, and nasal passages. It consists of a concave spherical mirror with a small hole drilled in the center (**Figure 25.26**). The physician places the mirror over one

FIGURE 25.26 On the left is the patient's view of a physician wearing a head mirror. The image on the right shows the bright light reflected by the mirror onto the patient under examination. Notice the lamp adjacent to the patient's head above her right shoulder.

---

*Sometimes optical systems use two (or more) mirrors, and the image formed by the first mirror serves as the object for the second mirror. Occasionally, such an object falls *behind* the second mirror. In this case the object distance is negative, and the object is said to be a virtual object.

eye and looks through the hole in the center. Light from a bright lamp that is placed near the patient's head reflects off the mirror along the line of sight of the doctor and illuminates the area on the patient that is under examination. During a routine physical, a patient sitting 0.50 m away from the physician sees the reflection of her nose in the head mirror. If the focal length of the mirror is 25 cm, **(a)** where is the image located, and **(b)** what is its magnification?

Reasoning   The head mirror is a small concave spherical mirror. The mirror equation (Equation 25.3) can be used to calculate the location of the image, and then Equation 25.4 is used to calculate the magnification.

Solution   **(a)** Rearranging the mirror equation, we have: $\frac{1}{d_i} = \frac{1}{f} - \frac{1}{d_o}$. Plugging in the values in the problem, we find:

$$\frac{1}{d_i} = \frac{1}{0.25 \text{ m}} - \frac{1}{0.50 \text{ m}} \Rightarrow d_i = \boxed{0.50 \text{ m}}$$

(b) The magnification is determined from Equation 25.4:

$$m = \frac{-d_i}{d_o} = -\frac{0.50 \text{ m}}{0.50 \text{ m}} = \boxed{-1}$$

Since the absolute value of the magnification is 1, the image is the same size as the object, and the negative sign tells us the image is real and inverted with respect to the object.

Related Homework: Problem 44

Concept Summary

25.1 Wave Fronts and Rays   Wave fronts are surfaces on which all points of a wave are in the same phase of motion. Waves whose wave fronts are flat surfaces are known as plane waves. Rays are lines that are perpendicular to the wave fronts and point in the direction of the velocity of the wave.

25.2 The Reflection of Light   When light reflects from a smooth surface, the reflected light obeys the law of reflection: The incident ray, the reflected ray, and the normal to the surface all lie in the same plane, and the angle of reflection θ_r equals the angle of incidence θ_i ($\theta_r = \theta_i$).

25.3 The Formation of Images by a Plane Mirror   A virtual image is one from which all the rays of light do not actually come, but only appear to do so. A real image is one from which all the rays of light actually do emanate.

A plane mirror forms an upright, virtual image that is located as far behind the mirror as the object is in front of it. In addition, the heights of the image and the object are equal.

25.4 Spherical Mirrors   A spherical mirror has the shape of a section from the surface of a hollow sphere. If the inside surface of the mirror is polished, it is a concave mirror. If the outside surface is polished, it is a convex mirror.

The principal axis of a mirror is a straight line drawn through the center of curvature and the middle of the mirror's surface. Rays that are close to the principal axis are known as paraxial rays. Paraxial rays are not necessarily parallel to the principal axis. The radius of curvature R of a mirror is the distance from the center of curvature to the mirror.

The focal point of a concave spherical mirror is a point on the principal axis, in front of the mirror. Incident paraxial rays that are parallel to the principal axis converge to the focal point after being reflected from the concave mirror.

The focal point of a convex spherical mirror is a point on the principal axis, behind the mirror. For a convex mirror, incident paraxial rays that are parallel to the principal axis diverge after reflecting from the mirror. These rays seem to originate from the focal point.

The fact that a spherical mirror does not bring all rays parallel to the principal axis to a single image point after reflection is known as spherical aberration.

The focal length f indicates the distance along the principal axis between the focal point and the mirror. The focal length and the radius of curvature R are related by Equations 25.1 and 25.2.

$$f = \tfrac{1}{2}R \quad \text{(Concave mirror)} \tag{25.1}$$

$$f = -\tfrac{1}{2}R \quad \text{(Convex mirror)} \tag{25.2}$$

25.5 The Formation of Images by Spherical Mirrors   The image produced by a mirror can be located by a graphical method known as ray tracing.

For a concave mirror, the following paraxial rays are useful for ray tracing (see **Interactive Figure 25.17**):

Ray 1.   This ray leaves the object traveling parallel to the principal axis. The ray reflects from the mirror and passes through the focal point.

Ray 2.   This ray leaves the object and passes through the focal point. The ray reflects from the mirror and travels parallel to the principal axis.

Ray 3.   This ray leaves the object and travels along a line that passes through the center of curvature. The ray strikes the mirror perpendicularly and reflects back on itself.

For a convex mirror, the following paraxial rays are useful for ray tracing (see **Interactive Figure 25.21a**):

Ray 1.   This ray leaves the object traveling parallel to the principal axis. After reflection from the mirror, the ray appears to originate from the focal point of the mirror.

Ray 2.   This ray leaves the object and heads toward the focal point. After reflection, the ray travels parallel to the principal axis.

Ray 3.   This ray leaves the object and travels toward the center of curvature. The ray strikes the mirror perpendicularly and reflects back on itself.

25.6 The Mirror Equation and the Magnification Equation   The mirror equation (Equation 25.3) specifies the relation between the object distance d_o, the image distance d_i, and the focal length f of the mirror. The mirror equation can be used with either concave or convex mirrors.

$$\frac{1}{d_o} + \frac{1}{d_i} = \frac{1}{f} \tag{25.3}$$

The magnification m of a mirror is the ratio of the image height h_i to the object height h_o: $m = h_i/h_o$. The magnification is also related to d_i and d_o by the magnification equation (Equation 25.4). The algebraic sign conventions for the variables appearing in these equations are summarized in the Reasoning Strategy at the end of Section 25.6.

$$m = -\frac{d_i}{d_o} \tag{25.4}$$

Focus on Concepts

Additional questions are available for assignment in WileyPLUS.

Section 25.1 Wave Fronts and Rays

1. A ray is _____. (**a**) always parallel to other rays (**b**) parallel to the velocity of the wave (**c**) perpendicular to the velocity of the wave (**d**) parallel to the wave fronts

Section 25.2 The Reflection of Light

2. The drawing shows a top view of an object located to the right of a mirror. A single ray of light is shown leaving the object. After reflection from the mirror, through which location, A, B, C, or D, does the ray pass? (**a**) A (**b**) B (**c**) C (**d**) D

QUESTION 2

Section 25.3 The Formation of Images by a Plane Mirror

3. A friend is standing 2 m in front of a plane mirror. You are standing 3 m directly behind your friend. What is the distance between you and the image of your friend? (**a**) 2 m (**b**) 3 m (**c**) 5 m (**d**) 7 m (**e**) 10 m

4. You hold the words **TOP DOG** in front of a plane mirror. What does the image of these words look like? (**a**) ꟼOT ᗡOᗡ (**b**) ᗡOᗡ ꟼOT (**c**) **TOP DOG** (**d**) **DOG TOP** (**e**) ꞮOⱢ ᗡOᗡ

Section 25.4 Spherical Mirrors

5. Rays of light coming from the sun (a very distant object) are near and parallel to the principal axis of a concave mirror. After reflecting from the mirror, where will the rays cross each other at a single point? The rays _____. (**a**) will not cross each other after reflecting from a concave mirror (**b**) will cross at the point where the principal axis intersects the mirror (**c**) will cross at the center of curvature (**d**) will cross at the focal point (**e**) will cross at a point beyond the center of curvature

Section 25.5 The Formation of Images by Spherical Mirrors

6. Which one of the following statements concerning spherical mirrors is correct? (**a**) Only a convex mirror can produce an enlarged image. (**b**) Both concave and convex mirrors can produce an enlarged image. (**c**) Only a concave mirror can produce an enlarged image, provided the object distance is less than the radius of curvature. (**d**) Only a concave mirror can produce an enlarged image, provided the object distance is greater than the radius of curvature.

7. Suppose that you hold up a small convex mirror in front of your face. Which answer describes the image of your face? (**a**) Virtual, inverted (**b**) Virtual, upright (**c**) Virtual, enlarged (**d**) Real, inverted (**e**) Real, reduced in size

Section 25.6 The Mirror Equation and the Magnification Equation

8. An object is placed at a known distance in front of a mirror whose focal length is also known. You apply the mirror equation and find that the image distance is a negative number. This result tells you that _____. (**a**) the image is larger than the object (**b**) the image is smaller than the object (**c**) the image is inverted relative to the object (**d**) the image is real (**e**) the image is virtual

9. An object is situated at a known distance in front of a convex mirror whose focal length is also known. A friend of yours does a calculation that shows that the magnification is −2. After some thought, you conclude correctly that _____. (**a**) your friend's answer is correct (**b**) the magnification should be +2 (**c**) the magnification should be $+\frac{1}{2}$ (**d**) the magnification should be $-\frac{1}{2}$

Problems

Additional questions are available for assignment in WileyPLUS.

SSM Student Solutions Manual	**BIO** Biomedical application
MMH Problem-solving help	**E** Easy
GO Guided Online Tutorial	**M** Medium
V-HINT Video Hints	**H** Hard
CHALK Chalkboard Videos	**WS** Worksheet
	T Team Problem

Section 25.2 The Reflection of Light,

Section 25.3 The Formation of Images by a Plane Mirror

1. **E** On the +y axis a laser is located at $y = +3.0$ cm. The coordinates of a small target are $x = +9.0$ cm and $y = +6.0$ cm. The +x axis represents the edge-on view of a plane mirror. At what point on the +x axis should the laser be aimed in order for the laser light to hit the target after reflection?

2. **E SSM CHALK** You are trying to photograph a bird sitting on a tree branch, but a tall hedge is blocking your view. However, as the drawing shows, a plane mirror reflects light from the bird into your camera. For what distance must you set the focus of the camera lens in order to snap a sharp picture of the bird's image?

PROBLEM 2

3. **E** **GO** Suppose that you are walking perpendicularly with a velocity of +0.90 m/s toward a stationary plane mirror. What is the velocity of your image relative to you? The direction in which you walk is the positive direction.

4. **E** **SSM** Two plane mirrors are separated by 120°, as the drawing illustrates. If a ray strikes mirror M_1 at a 65° angle of incidence, at what angle θ does it leave mirror M_2?

PROBLEM 4

5. **E** **CHALK** The drawing shows a laser beam shining on a plane mirror that is perpendicular to the floor. The beam's angle of incidence is 33.0°. The beam emerges from the laser at a point that is 1.10 m from the mirror and 1.80 m above the floor. After reflection, how far from the base of the mirror does the beam strike the floor?

PROBLEM 5

6. **M** **GO** A small mirror is attached to a vertical wall, and it hangs a distance of 1.80 m above the floor. The mirror is facing due east, and a ray of sunlight strikes the mirror early in the morning and then again later in the morning. The incident and reflected rays lie in a plane that is perpendicular to both the wall and the floor. Early in the morning, the reflected ray strikes the floor at a distance of 3.86 m from the base of the wall. Later on in the morning, the ray is observed to strike the floor at a distance of 1.26 m from the wall. The earth rotates at a rate of 15.0° per hour. How much time (in hours) has elapsed between the two observations?

7. **M** **V-HINT** In an experiment designed to measure the speed of light, a laser is aimed at a mirror that is 50.0 km due north. A detector is placed 117 m due east of the laser. The mirror is to be aligned so that light from the laser reflects into the detector. **(a)** When properly aligned, what angle should the normal to the surface of the mirror make with due south? **(b)** Suppose the mirror is misaligned, so that the actual angle between the normal to the surface and due south is too large by 0.004°. By how many meters (due east) will the reflected ray miss the detector?

8. **M** **GO** The drawing shows two plane mirrors that intersect at an angle of 50°. An incident light ray reflects from one mirror and then the other. What is the angle θ between the incident and outgoing rays?

PROBLEM 8

9. **M** Two plane mirrors are facing each other. They are parallel, 3.00 cm apart, and 17.0 cm in length, as the drawing indicates. A laser beam is directed at the top mirror from the left edge of the bottom mirror. What is the smallest angle of incidence with respect to the top mirror, such that the laser beam **(a)** hits only one of the mirrors and **(b)** hits each mirror only once?

PROBLEM 9

Section 25.4 Spherical Mirrors,

Section 25.5 The Formation of Images by Spherical Mirrors

10. **E** When an object is located very far away from a convex mirror, the image of the object is 18 cm behind the mirror. Using a ray diagram drawn to scale, determine where the image is located when the object is placed 9.0 cm in front of the mirror. Note that the mirror must be drawn to scale also. In your drawing, assume that the height of the object is 3.0 cm.

11. **E** **CHALK** The image of a very distant car is located 12 cm behind a convex mirror. **(a)** What is the radius of curvature of the mirror? **(b)** Draw a ray diagram to scale showing this situation.

12. **E** **SSM** An object is placed 11 cm in front of a concave mirror whose focal length is 18 cm. The object is 3.0 cm tall. Using a ray diagram drawn to scale, measure **(a)** the location and **(b)** the height of the image. The mirror must be drawn to scale.

13. **E** **MMH** A 2.0-cm-high object is situated 15.0 cm in front of a concave mirror that has a radius of curvature of 10.0 cm. Using a ray diagram drawn to scale, measure **(a)** the location and **(b)** the height of the image. The mirror must be drawn to scale.

14. **E** **MMH** A convex mirror has a focal length of −40.0 cm. A 12.0-cm-tall object is located 40.0 cm in front of this mirror. Using a ray diagram drawn to scale, determine the **(a)** location and **(b)** size of the image. Note that the mirror must be drawn to scale.

15. **M** **CHALK** A plane mirror and a concave mirror ($f = 8.0$ cm) are facing each other and are separated by a distance of 20.0 cm. An object is placed between the mirrors and is 10.0 cm from each mirror. Consider the light from the object that reflects first from the plane mirror and then from the concave mirror. Using a ray diagram drawn to scale, find the location of the image that this light produces in the concave mirror. Specify this distance relative to the concave mirror.

Section 25.6 The Mirror Equation and the Magnification Equation

16. **E** **SSM** The image produced by a concave mirror is located 26 cm in front of the mirror. The focal length of the mirror is 12 cm. How far in front of the mirror is the object located?

17. **E** The image behind a convex mirror (radius of curvature = 68 cm) is located 22 cm from the mirror. **(a)** Where is the object located and **(b)** what is the magnification of the mirror? Determine whether the image is **(c)** upright or inverted and **(d)** larger or smaller than the object.

18. **E** **MMH** A concave mirror ($R = 56.0$ cm) is used to project a transparent slide onto a wall. The slide is located at a distance of 31.0 cm from the mirror, and a small flashlight shines light through the slide and onto the mirror. The setup is similar to that in **Animated Figure 25.18a**. **(a)** How far from the wall should the mirror be located? **(b)** The height of the object on the slide is 0.95 cm. What is the height of the image? **(c)** How should the slide be oriented, so that the picture on the wall looks normal?

19. **E** **GO** A small statue has a height of 3.5 cm and is placed in front of a concave mirror. The image of the statue is inverted, 1.5 cm tall, and located 13 cm in front of the mirror. Find the focal length of the mirror.

20. **E** **GO** **SSM** A mirror produces an image that is located 34.0 cm behind the mirror when the object is located 7.50 cm in front of the mirror. What is the focal length of the mirror, and is the mirror concave or convex?

21. **E** **GO** A concave mirror ($f = 45$ cm) produces an image whose distance from the mirror is one-third the object distance. Determine **(a)** the object distance and **(b)** the (positive) image distance.

22. **E** The outside mirror on the passenger side of a car is convex and has a focal length of -7.0 m. Relative to this mirror, a truck traveling in the rear has an object distance of 11 m. Find **(a)** the image distance of the truck and **(b)** the magnification of the mirror.

23. **E** **GO** A convex mirror has a focal length of -27.0 cm. Find the magnification produced by the mirror when the object distance is 9.0 cm and 18.0 cm.

24. **E** **SSM** When viewed in a spherical mirror, the image of a setting sun is a virtual image. The image lies 12.0 cm behind the mirror. **(a)** Is the mirror concave or convex? Why? **(b)** What is the radius of curvature of the mirror?

25. **E** **V-HINT** A concave mirror has a focal length of 12 cm. This mirror forms an image located 36 cm in front of the mirror. What is the magnification of the mirror?

26. **M** **V-HINT** An object is located 14.0 cm in front of a convex mirror, the image being 7.00 cm behind the mirror. A second object, twice as tall as the first one, is placed in front of the mirror, but at a different location. The image of this second object has the same height as the other image. How far in front of the mirror is the second object located?

27. **M** **GO** A tall tree is growing across a river from you. You would like to know the distance between yourself and the tree, as well as its height, but are unable to make the measurements directly. However, by using a mirror to form an image of the tree and then measuring the image distance and the image height, you can calculate the distance to the tree as well as its height. Suppose that this mirror produces an image of the sun, and the image is located 0.9000 m from the mirror. The same mirror is then used to produce an image of the tree. The image of the tree is 0.9100 m from the mirror. **(a)** How far away is the tree? **(b)** The image height of the tree has a magnitude of 0.12 m. How tall is the tree?

28. **M** A spherical mirror is polished on both sides. When the concave side is used as a mirror, the magnification is $+2.0$. What is the magnification when the convex side is used as a mirror, the object remaining the same distance from the mirror?

29. **M** **GO** Consult Multiple-Concept Example 8 to see a model for solving this type of problem. A concave makeup mirror is designed so the virtual image it produces is twice the size of the object when the distance between the object and the mirror is 14 cm. What is the radius of curvature of the mirror?

30. **H** A concave mirror has a focal length of 30.0 cm. The distance between an object and its image is 45.0 cm. Find the object and image distances, assuming that **(a)** the object lies beyond the center of curvature and **(b)** the object lies between the focal point and the mirror.

Additional Problems

Online

31. **E** An object is placed in front of a convex mirror. Draw the convex mirror (radius of curvature = 15 cm) to scale, and place the object 25 cm in front of it. Make the object height 4 cm. Using a ray diagram, locate the image and measure its height. Now move the object closer to the mirror, so the object distance is 5 cm. Again, locate its image using a ray diagram. As the object moves closer to the mirror, **(a)** does the magnitude of the image distance become larger or smaller, and **(b)** does the magnitude of the image height become larger or smaller? **(c)** What is the ratio of the image height when the object distance is 5 cm to its height when the object distance is 25 cm? Give your answer to one significant figure.

32. **E** **SSM** An object that is 25 cm in front of a convex mirror has an image located 17 cm behind the mirror. How far behind the mirror is the image located when the object is 19 cm in front of the mirror?

33. **E** **GO** A concave mirror has a focal length of 42 cm. The image formed by this mirror is 97 cm in front of the mirror. What is the object distance?

34. **E** Review Conceptual Example 1 before attempting this problem. A person whose eyes are 1.70 m above the floor stands in front of a plane mirror. The top of her head is 0.12 m above her eyes. **(a)** What is the height of the shortest mirror in which she can see her entire image? **(b)** How far above the floor should the bottom edge of the mirror be placed?

35. **E** **V-HINT** A drop of water on a countertop reflects light from a flower held 3.0 cm directly above it. The flower's diameter is 2.0 cm,

and the diameter of the flower's image is 0.10 cm. What is the focal length of the water drop, assuming that it may be treated as a convex spherical mirror?

36. **M** **GO** You walk at an angle of $\theta = 50.0°$ toward a plane mirror, as in the drawing. Your walking velocity has a magnitude of 0.90 m/s. What is the velocity of your image relative to you (magnitude and direction)?

PROBLEM 36

37. **M** **CHALK** **GO** A candle is placed 15.0 cm in front of a convex mirror. When the convex mirror is replaced with a plane mirror, the image moves 7.0 cm farther away from the mirror. Find the focal length of the convex mirror.

38. **H** A man holds a double-sided spherical mirror so that he is looking directly into its convex surface, 45 cm from his face. The magnification of the image of his face is $+0.20$. What will be the image distance when he reverses the mirror (looking into its concave surface), maintaining the same distance between the mirror and his face? Be sure to include the algebraic sign ($+$ or $-$) with your answer.

Physics in Biology, Medicine, and Sports

39. **M** **BIO** **GO** **25.6** A dentist's mirror is placed 2.0 cm from a tooth. The *enlarged* image is located 5.6 cm behind the mirror. **(a)** What kind of mirror (plane, concave, or convex) is being used? **(b)** Determine the focal length of the mirror. **(c)** What is the magnification? **(d)** How is the image oriented relative to the object?

40. **E** **BIO** **25.2** At major golfing events, spectators surround the greens in an attempt to see their favorite players. A handy device that can help them see over the crowd is a periscope (see the photo). Inside the periscope there are two plane mirrors separated by the length of the periscope and oriented such that the normal to their surface makes an angle of 45° with the horizontal (see the drawing). Consider a horizontal light ray that enters the periscope. **(a)** If $\theta_1 = 45°$, what is the value of θ_2? **(b)** If it takes light 1.0 ns to travel the distance between the two mirrors, what is this distance?

41. **E** **BIO** **25.2** A laser range finder is a device that quickly and accurately determines the distance between it and an object. The devices are frequently used in hunting and in golf to measure the distance to a wild animal or to a flag on the green, respectively. The range finder works by emitting laser light toward the target. Some of that light reflects from the target and returns, where it is detected by a sensor in the range finder (see the photo). An accurate time circuit in the range finder measures the time that elapses between the emission and detection of the laser light. In the figure shown here, the range finder is used to determine the distance to mirror 2. Because of a tree line, a direct measurement of the distance to mirror 2 is not possible. Two other mirrors are used to reflect the light around the tree line. Laser light from the range finder is reflected by the three mirrors and returns to the finder in 3.0 μs. The light strikes mirror 1 at an angle of 70° with respect to the mirror's normal. **(a)** What is the angle θ_1? **(b)** What is the angle θ_2? **(c)** What is the distance between mirrors 1 and 2? **(d)** What is the distance between the range finder and mirror 2?

PROBLEM 41

PROBLEM 40

42. **M** **BIO** **25.5** Many nocturnal animals demonstrate the phenomenon of eyeshine, in which their eyes glow various colors at night when illuminated by a flashlight or the headlights of a car (see the photo). Their eyes react this way because of a thin layer of reflective tissue called the *tapetum lucidum* that is located directly behind the retina. This tissue reflects the light back through the retina, which increases the available light that can activate photoreceptors, and thus improve the animal's vision in low-light conditions. If we assume the tapetum lucidum acts like a concave spherical mirror with a radius of curvature of 0.750 cm, how far in front of the tapetum lucidum would an image form of an object located 30.0 cm away? Neglect the effects of the other structures of the eye, such as the cornea and lens.

Phil Sheldon/Popperfoto/Getty Images

Free clipart

Photo courtesy of David Young

In response to a camera flash, the eyes of a cat named Einstein glow orange-red due to the reflection of light from the tapetum lucidum located behind the retina.

PROBLEM 42

43. **H** **BIO** **25.3** Currently, nine nonhuman species of animals pass the *mirror self-recognition test* (MSR), which means they demonstrate the ability of self-recognition when they look at their reflection. Some of the animals on this list include the great apes, Asian elephants, bottlenose dolphins, and orca whales. In the figure, an Asian elephant is standing 3.5 m from a vertical wall. Given the dimensions shown in the drawing, what should be the minimum length of the mirror (*L*), such that the elephant can see the entire height of its body—from the top of its head to the bottom of its feet?

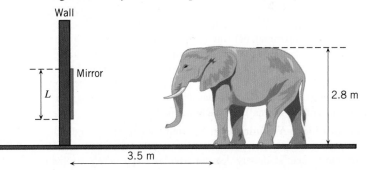

PROBLEM 43

44. **E** **BIO** **25.3** Mirror laryngoscopy is a medical diagnostic technique that traditionally uses both a head mirror (see Example 9) and a laryngeal mirror. The laryngeal mirror can come in many sizes. Size 0 corresponds to a circular plane mirror with a diameter of 10.0 mm. The length of the mirror's handle (20.3 cm) allows the physical examination of the larynx, a procedure that is often used to identify early stages of malignancy in swollen lymph nodes. A physician using a laryngeal mirror to examine a patient discovers a granularcell tumor on the patient's larynx. The approximately spherical tumor is 5.2 mm in diameter and is located 1.2 cm in front of the mirror. **(a)** What kind of image is formed by the laryngeal mirror—real or virtual? **(b)** Where is the image located? **(c)** What is the magnification of the image?

Concepts and Calculations Problems

Online

Relative to the object in front of a spherical mirror, the image can differ in a number of respects. The image can be real (in front of the mirror) or virtual (behind the mirror). It can be larger or smaller than the object, and it can be upright or inverted. As you solve problems dealing with spherical mirrors, keep these image characteristics in mind. They can help guide you to the correct answer, as Problems 45 and 46 illustrate.

45. **M** **CHALK** An object is located 7.0 cm in front of a mirror. The virtual image is located 4.5 cm away from the mirror and is smaller than the object. *Concepts:* (i) Based solely on the fact that the image is virtual, is the mirror concave or convex, or is either type possible? (ii) The image is smaller than the object, as well as virtual. Do these characteristics together indicate a concave or convex mirror, or is either type possible? (iii) Is the focal length positive or negative? Explain. *Calculation:* Find the focal length of the mirror.

46. **M** **CHALK** **SSM** The radius of curvature of a mirror is 24 cm. A diamond ring is placed in front of this mirror. The image is twice the size of the ring. *Concepts:* (i) Is the mirror concave or convex, or is either type possible? (ii) How many places are there in front of a concave mirror where the ring can be placed and produce an image twice the size of the object? (iii) What are the possible values for the magnification of the image of the ring? *Calculation:* Find the object distance of the ring.

Team Problems

Online

47. **E** **T** **WS** **A Light Pipe.** You and your team have been tasked with safely guiding a laser beam across a lab through a straight, horizontal pipe with a square cross section and mirrored internal surfaces. You are unable to aim the beam directly down the center, so it will have to reflect multiple times from the internal surfaces. You are, however, able to align the laser so that the beam is centered between the left and right sides, and reflects only from the top and bottom surfaces. The pipe is 3.50 meters in length and has a square cross section with a side length of 3.75 cm. If the laser beam enters at the lower edge of the pipe at 10.0° angle above the horizontal, **(a)** how many reflections does the beam undergo before emerging from the other side? **(b)** Does the beam emerge at an angle above or below the horizontal? **(c)** At what angle does it emerge relative to the horizontal? Assume the angle is negative if below and positive if above the horizontal. **(d)** You need to redirect the emerging beam so that it is parallel to the axis of the pipe. You do this by placing a plane mirror at the exit of the pipe to steer the emerging beam. What angle (magnitude only) should the mirror make with the horizontal? Define this angle as that between the surface plane of the mirror (i.e., not surface normal) and the horizontal.

48. E T WS A Goniometer. You and your team are designing an instrument in which a laser beam must reflect from a plane mirror into a small stationary detector. The laser and mirror are both mounted to a mechanical instrument called a *goniometer* that allows for the precise independent angular positioning of each. The laser is mounted on an arm that can rotate like the hand of a clock. As shown in the drawing, the arm is initially positioned so that the laser points vertically downward, toward the axis of rotation. In this geometry, no matter what the angular position of the arm, the laser beam will always pass though the axis of rotation. Located exactly on the axis of rotation is the top surface of a plane mirror that can rotate independently of the laser arm. The idea is that the laser reflects from the mirror and toward the detector, which is located to the right at an angle of 35.0° above an imaginary horizontal line that passes through the axis of rotation. **(a)** With the laser arm in the vertical position (i.e., the laser is aimed vertically downward), at what angle relative to the horizontal should the surface normal of the mirror be directed so that the reflected beam will strike the detector? Be sure to indicate whether the angle is clockwise or counterclockwise. **(b)** If the laser arm rotates counterclockwise at a rate of 2.00° per minute, what should the angular velocity of the mirror (in degrees/min) be so that the reflected laser beam always strikes the detector?

PROBLEM 48

49. M T An Unidentified Flying Object. You and your team spot a strange object hovering over the open ocean in broad daylight. You have nothing to reference in order to estimate its distance, height, or size, but you get the idea to use a concave mirror from a signal light. You disassemble the light, extract the mirror, and point it at the sun. You measure the distance to the sun's image from the center of the mirror to be 3.405 m. You then point the mirror at the object and find that its image forms 3.443 m from the center of the mirror. The angle between the line of sight to the object and the horizontal is 28.50°. **(a)** What is the distance between you and the object? **(b)** What is the height of the object above the ground? **(c)** If the width of the object's image is 11.50 cm, what is the width of the object?

50. M T Radius of Curvature of a Partial Sphere. You and your team need to estimate the radius of curvature of a panel that had been removed from some unknown object. You were told that the object was spherical and, by the size of the panel (about one square meter in area) and its slight curvature, you estimate that the spherical object from which it came had been quite large. You notice that the outer surface has a mirrored metallic finish (like a convex mirror) and you get an idea. You find a wrench that is 21.0 cm long, and hold it at a distance of 10.0 m from the middle of the mirrored surface. The virtual image of the wrench is 14.5 cm long. Determine the radius of the sphere from which the panel came.

When light moves from one medium into another, its direction of travel changes, and this change in direction is called refraction. The human eye is an incredible optical instrument and, as we will see, refraction plays a major role in the way it works to produce clear vision.

Design Pics/Don Hammond/
Getty Images

The Refraction of Light: Lenses and Optical Instruments

26.1 The Index of Refraction

As Section 24.3 discusses, light travels through a vacuum at a speed of $c = 3.00 \times 10^8$ m/s. It can also travel through many materials, such as air, water, and glass. Atoms in the material absorb, reemit, and scatter the light, however. Therefore, light travels through the material at a speed that is less than c, the actual speed depending on the nature of the material. In general, we will see that the change in speed as a ray of light goes from one material to another causes the ray to deviate from its incident direction. This change in direction is called **refraction**. To describe the extent to which the speed of light in a material medium differs from that in a vacuum, we use a parameter called the **index of refraction** (or **refractive index**). The index of refraction is an important parameter because it appears in Snell's law of refraction, which will be discussed in the next section. This law is the basis of all the phenomena discussed in this chapter.

> **DEFINITION OF THE INDEX OF REFRACTION**
>
> The index of refraction n of a material is the ratio of the speed c of light in a vacuum to the speed v of light in the material:
>
> $$n = \frac{\text{Speed of light in a vacuum}}{\text{Speed of light in the material}} = \frac{c}{v} \qquad (26.1)$$

LEARNING OBJECTIVES

After reading this module, you should be able to . . .

26.1 Define the index of refraction.

26.2 Use Snell's law of refraction to solve problems.

26.3 Analyze total internal reflection problems.

26.4 Define Brewster's angle.

26.5 Analyze examples involving dispersion of light.

26.6 Trace rays passing through converging and diverging lenses.

26.7 Trace rays from objects through lenses to form images.

26.8 Apply the lens and magnification equations to solve problems.

26.9 Solve problems involving lenses in combination.

26.10 Apply ray tracing and the lens and magnification equations to the human eye.

26.11 Calculate angular size and angular magnification.

26.12 Apply optical principles to the compound microscope.

26.13 Apply optical principles to the telescope.

26.14 Describe spherical and chromatic lens aberrations.

TABLE 26.1	Index of Refraction[a] for Various Substances
Substance	**Index of Refraction, n**
Solids at 20 °C	
Diamond	2.419
Glass, crown	1.523
Ice (0 °C)	1.309
Sodium chloride	1.544
Quartz	
Crystalline	1.544
Fused	1.458
Liquids at 20 °C	
Benzene	1.501
Carbon disulfide	1.632
Carbon tetrachloride	1.461
Ethyl alcohol	1.362
Water	1.333
Gases at 0 °C, 1 atm	
Air	1.000293
Carbon dioxide	1.00045
Oxygen, O_2	1.000271
Hydrogen, H_2	1.000139

[a] Measured with light whose wavelength in a vacuum is 589 nm.

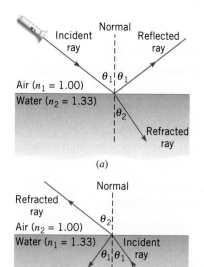

FIGURE 26.1 (a) When a ray of light is directed from air into water, part of the light is reflected at the interface and the remainder is refracted into the water. The refracted ray is bent *toward* the normal ($\theta_2 < \theta_1$). (b) When a ray of light is directed from water into air, the refracted ray in air is bent *away* from the normal ($\theta_2 > \theta_1$).

Table 26.1 lists the refractive indices for some common substances. The values of n are greater than unity because the speed of light in a material medium is less than it is in a vacuum. For example, the index of refraction for diamond is $n = 2.419$, so the speed of light in diamond is $v = c/n = (3.00 \times 10^8 \text{ m/s})/2.419 = 1.24 \times 10^8 \text{ m/s}$. In contrast, the index of refraction for air (and also for other gases) is so close to unity that $n_{\text{air}} = 1$ for most purposes. The index of refraction depends slightly on the wavelength of the light, and the values in Table 26.1 correspond to a wavelength of $\lambda = 589$ nm in a vacuum.

26.2 Snell's Law and the Refraction of Light

Snell's Law

When light strikes the interface between two transparent materials, such as air and water, the light generally divides into two parts, as **Figure 26.1a** illustrates. Part of the light is reflected, with the angle of reflection equaling the angle of incidence. The remainder is transmitted across the interface. If the incident ray does not strike the interface at normal incidence, the transmitted ray has a different direction than the incident ray. When a ray enters the second material and changes direction, it is said to be refracted and behaves in one of the following two ways:

1. When light travels from a medium where the refractive index is smaller into a medium where it is larger, the refracted ray is bent toward the normal, as in **Figure 26.1a**.

2. When light travels from a medium where the refractive index is larger into a medium where it is smaller, the refracted ray is bent away from the normal, as in **Figure 26.1b**.

These two possibilities illustrate that both the incident and refracted rays obey the principle of reversibility. Thus, the directions of the rays in part *a* of the drawing can be reversed to give the situation depicted in part *b*. In part *b* the reflected ray lies in the water rather than in the air.

In both parts of **Figure 26.1** the angles of incidence, refraction, and reflection are measured relative to the normal. Note that the index of refraction of air is labeled n_1 in part *a*, whereas it is n_2 in part *b*, because *we label all variables associated with the incident (and reflected) ray with subscript 1 and all variables associated with the refracted ray with subscript 2.*

The angle of refraction θ_2 depends on the angle of incidence θ_1 and on the indices of refraction, n_2 and n_1, of the two media. The relation between these quantities is known as **Snell's law of refraction**, after the Dutch mathematician Willebrord Snell (1591–1626), who discovered it experimentally. At the end of this section is a proof of Snell's law.

> **SNELL'S LAW OF REFRACTION**
>
> **When light travels from a material with refractive index n_1 into a material with refractive index n_2, the refracted ray, the incident ray, and the normal to the interface between the materials all lie in the same plane. The angle of refraction θ_2 is related to the angle of incidence θ_1 by**
>
> $$n_1 \sin \theta_1 = n_2 \sin \theta_2 \tag{26.2}$$

Example 1 illustrates the use of Snell's law.

EXAMPLE 1 | Determining the Angle of Refraction

A light ray strikes an air/water surface at an angle of 46° with respect to the normal. The refractive index for water is 1.33. Find the angle of refraction when the direction of the ray is **(a)** from air to water and **(b)** from water to air.

Reasoning Snell's law of refraction applies to both part (a) and part (b). However, in part (a) the incident ray is in air, whereas in part (b) it is in water. We keep track of this difference by always labeling the variables associated with the incident ray with a subscript 1 and the variables associated with the refracted ray with a subscript 2.

> **Problem-Solving Insight** The angle of incidence θ_1 and the angle of refraction θ_2 that appear in Snell's law are measured with respect to the normal to the surface, and not with respect to the surface itself.

Solution **(a)** The incident ray is in air, so $\theta_1 = 46°$ and $n_1 = 1.00$. The refracted ray is in water, so $n_2 = 1.33$. Snell's law can be used to find the angle of refraction θ_2:

$$\sin \theta_2 = \frac{n_1 \sin \theta_1}{n_2} = \frac{(1.00) \sin 46°}{1.33} = 0.54 \qquad \textbf{(26.2)}$$

$$\theta_2 = \sin^{-1}(0.54) = \boxed{33°}$$

Since θ_2 is less than θ_1, the refracted ray is bent *toward* the normal, as **Figure 26.1a** shows.

(b) With the incident ray in water, we find that

$$\sin \theta_2 = \frac{n_1 \sin \theta_1}{n_2} = \frac{(1.33) \sin 46°}{1.00} = 0.96$$

$$\theta_2 = \sin^{-1}(0.96) = \boxed{74°}$$

Since θ_2 is greater than θ_1, the refracted ray is bent *away* from the normal, as in **Figure 26.1b**.

We have seen that reflection and refraction of light waves occur simultaneously at the interface between two transparent materials. It is important to keep in mind that light waves are composed of electric and magnetic fields, which carry energy. The principle of conservation of energy (see Chapter 6) indicates that the energy reflected plus the energy refracted must add up to equal the energy carried by the incident light, provided that none of the energy is absorbed by the materials. The percentage of incident energy that appears as reflected versus refracted light depends on the angle of incidence and the refractive indices of the materials on either side of the interface. For instance, when light travels from air toward water at perpendicular incidence, most of its energy is refracted and little is reflected. But when the angle of incidence is nearly 90° and the light barely grazes the water surface, most of its energy is reflected, with only a small amount refracted into the water. On a rainy night, you probably have experienced the annoying glare that results when light from an oncoming car just grazes the wet road. Under such conditions, most of the light energy reflects into your eyes.

THE PHYSICS OF . . . rearview mirrors. The simultaneous reflection and refraction of light have applications in a number of devices. For instance, interior rearview mirrors in cars often have adjustment levers. One position of the lever is for day driving, while the other is for night driving and reduces glare from the headlights of the car behind. As **Interactive Figure 26.2a** indicates, this kind of mirror is a glass wedge with a back side that is silvered and highly reflecting. Part *b* of the picture shows the day setting. Light from the car behind follows path *ABCD* in reaching the driver's eye. At points *A* and *C*, where the light strikes the air–glass surface, there are both reflected and refracted rays. The reflected rays are drawn as thin lines, the thinness denoting that

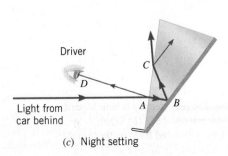

(a) (b) Day setting (c) Night setting

INTERACTIVE FIGURE 26.2 A car's interior rearview mirror with a day–night adjustment lever.

only a small percentage (about 10%) of the light during the day is reflected at A and C. The weak reflected rays at A and C do not reach the driver's eye. In contrast, almost all the light reaching the silvered back surface at B is reflected toward the driver. Since most of the light follows path $ABCD$, the driver sees a bright image of the car behind. During the night, the adjustment lever can be used to rotate the top of the mirror away from the driver (see part c of the drawing). Now, most of the light from the headlights behind follows path ABC and does not reach the driver. Only the light that is weakly reflected from the front surface along path AD is seen, and the result is significantly less glare.

Apparent Depth

One interesting consequence of refraction is that an object lying under water appears to be closer to the surface than it actually is. Example 2 sets the stage for explaining why, by showing what must be done to shine a light on such an object.

EXAMPLE 2 | Finding a Sunken Chest

A searchlight on a yacht is being used at night to illuminate a sunken chest, as in **Figure 26.3**. At what angle of incidence θ_1 should the light be aimed?

Reasoning The angle of incidence θ_1 must be such that, after refraction, the light strikes the chest. The angle of incidence can be obtained from Snell's law, once the angle of refraction θ_2 is determined. This angle can be found using the data in **Figure 26.3** and trigonometry. The light travels from a region of lower into a region of higher refractive index, so the light is bent toward the normal and we expect θ_1 to be greater than θ_2.

> **Problem-Solving Insight** Remember that the refractive indices are written as n_1 for the medium in which the incident light travels and n_2 for the medium in which the refracted light travels.

Solution From the data in the drawing it follows that $\tan \theta_2 = (2.0\text{ m})/(3.3\text{ m})$, so $\theta_2 = 31°$. With $n_1 = 1.00$ for air and $n_2 = 1.33$ for water, Snell's law gives

FIGURE 26.3 The beam from the searchlight is refracted when it enters the water.

$$\sin \theta_1 = \frac{n_2 \sin \theta_2}{n_1} = \frac{(1.33)\sin 31°}{1.00} = 0.69$$

$$\theta_1 = \sin^{-1}(0.69) = \boxed{44°}$$

As expected, θ_1 is greater than θ_2.

When the sunken chest in Example 2 is viewed from the boat (**Figure 26.4a**), light rays from the chest pass upward through the water, refract away from the normal when they enter the air, and then travel to the observer. This picture is similar to **Figure 26.3**, except that the direction of the rays is reversed and the searchlight is replaced by an observer. When the rays entering the air are extended back into the water (see the dashed lines), they indicate that the observer sees a virtual image of the chest at an *apparent depth* that is less than the actual depth. The image is virtual because light rays do not actually pass through

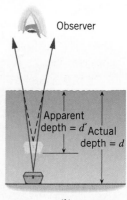

FIGURE 26.4 (a) Because light from the chest is refracted away from the normal when the light enters the air, the apparent depth of the image is less than the actual depth. (b) The observer is viewing the submerged object from directly overhead.

it. For the situation shown in **Figure 26.4a**, it is difficult to determine the apparent depth. A much simpler case is shown in part *b*, where the observer is *directly above* the submerged object, and the apparent depth d' is related to the actual depth d by

Apparent depth, observer directly above object

$$d' = d\left(\frac{n_2}{n_1}\right)$$

(26.3)

In this result, n_1 is the refractive index of the medium associated with the incident ray (the medium in which the object is located), and n_2 refers to the medium associated with the refracted ray (the medium in which the observer is situated). The proof of Equation 26.3 is the focus of Problem 23 at the end of the chapter. Example 3 illustrates that the effect of apparent depth is quite noticeable in water.

EXAMPLE 3 | The Apparent Depth of a Swimming Pool

A swimmer is treading water (with her head above the water) at the surface of a pool 3.00 m deep. She sees a coin on the bottom directly below. How deep does the coin appear to be?

Reasoning Equation 26.3 may be used to find the apparent depth, provided we remember that the light rays travel from the coin to the swimmer. Therefore, the incident ray is coming from

the coin under the water ($n_1 = 1.33$), while the refracted ray is in the air ($n_2 = 1.00$).

Solution The apparent depth d' of the coin is

$$d' = d\left(\frac{n_2}{n_1}\right) = (3.00 \text{ m})\left(\frac{1.00}{1.33}\right) = \boxed{2.26 \text{ m}}$$

(26.3)

In Example 3, a person sees a coin on the bottom of a pool at an apparent depth that is less than the actual depth. Conceptual Example 4 considers the reverse situation—namely, a person looking from under the water at a coin in the air.

CONCEPTUAL EXAMPLE 4 | On the Inside Looking Out

A swimmer is under water and looking up at the surface. Someone holds a coin in the air, directly above the swimmer's eyes. To the swimmer, the coin appears to be at a certain height above the water. Is the apparent height of the coin **(a)** greater than, **(b)** less than, or **(c)** the same as its actual height?

Reasoning **Figure 26.5** shows two rays of light leaving a point P on the coin. When the rays enter the water, they are refracted toward the normal because water has a larger index of refraction than air has. By extending the refracted rays backward (see the dashed lines in the drawing), we find that they appear to originate from a point P' on a virtual image, which is what the swimmer sees.

Answers (b) and (c) are incorrect. These answers are incorrect because the point P' in **Figure 26.5** is located at a height that is greater than, not less than or the same as, the actual height of the coin.

Answer (a) is correct. The point P' in **Figure 26.5** is on a virtual image that is located at an apparent height d' that is greater than the actual height d. Equation 26.3 [$d' = d(n_2/n_1)$] reveals the same result, because n_1 represents the medium (air) associated with the incident ray and n_2 represents the medium (water) associated with the refracted ray. Since n_2 for water is greater than n_1 for air, the ratio n_2/n_1 is greater than one and d' is larger than d. This situation is the opposite of that in **Figure 26.4b**, where an object beneath the water appears to a person above the water to be closer to the surface than it actually is.

Related Homework: Problems 19, 98

FIGURE 26.5 Rays from point P on a coin in the air above the water refract toward the normal as they enter the water. An underwater swimmer perceives the rays as originating from a point P' that is farther above the surface than the actual point P.

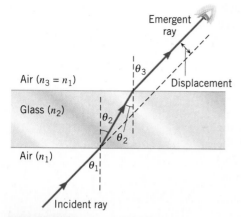

FIGURE 26.6 When a ray of light passes through a pane of glass that has parallel surfaces and is surrounded by air, the emergent ray is parallel to the incident ray ($\theta_3 = \theta_1$) but is displaced from it.

The Displacement of Light by a Transparent Slab of Material

A windowpane is an example of a transparent slab of material. It consists of a plate of glass with parallel surfaces. When a ray of light passes through the glass, the emergent ray is parallel to the incident ray but displaced from it, as **Figure 26.6** shows. This result can be verified by applying Snell's law to each of the two glass surfaces, with the result that $n_1 \sin \theta_1 = n_2 \sin \theta_2 = n_3 \sin \theta_3$. Since air surrounds the glass, $n_1 = n_3$, and it follows that $\sin \theta_1 = \sin \theta_3$. Therefore, $\theta_1 = \theta_3$, and the emergent and incident rays are parallel. However, as the drawing shows, the emergent ray is displaced laterally relative to the incident ray. The extent of the displacement depends on the angle of incidence, the thickness of the slab, and the refractive index of the slab.

Derivation of Snell's Law

Snell's law can be derived by considering what happens to the wave fronts when the light passes from one medium into another. **Figure 26.7a** shows light propagating from medium 1, where the speed is relatively large, into medium 2, where the speed is smaller; therefore, n_1 is less than n_2. The plane wave fronts in this picture are drawn perpendicular to the incident and refracted rays. Since the part of each wave front that penetrates medium 2 slows down, the wave fronts in medium 2 are rotated clockwise relative to those in medium 1. Correspondingly, the refracted ray in medium 2 is bent toward the normal, as the drawing shows.

Although the incident and refracted waves have different speeds, *they have the same frequency f.* The fact that the frequency does not change can be understood in terms of the atomic mechanism underlying the generation of the refracted wave. When the electromagnetic wave strikes the surface, the oscillating electric field forces the electrons in the molecules of medium 2 to oscillate at the same frequency as the wave. The accelerating electrons behave like atomic antennas that radiate "extra" electromagnetic waves, which combine with the original wave. The net electromagnetic wave within medium 2 is a superposition of the original wave plus the extra radiated waves, and it is this superposition that constitutes the refracted wave. Since the extra waves are radiated at the same frequency as the original wave, the refracted wave also has the same frequency as the original wave.

The distance between successive wave fronts in **Figure 26.7a** has been chosen to be the wavelength λ. Since the frequencies are the same in both media but the speeds are different, it follows from Equation 16.1 that the wavelengths are different: $\lambda_1 = v_1/f$ and $\lambda_2 = v_2/f$. Since v_1 is assumed to be larger than v_2, λ_1 is larger than λ_2, and the wave fronts are farther apart in medium 1.

Figure 26.7b shows an enlarged view of the incident and refracted wave fronts at the surface. The angles θ_1 and θ_2 within the colored right triangles are,

FIGURE 26.7 (*a*) The wave fronts are refracted as the light passes from medium 1 into medium 2. (*b*) An enlarged view of the incident and refracted wave fronts at the surface.

(*a*)

(*b*)

respectively, the angles of incidence and refraction. In addition, the triangles share the same hypotenuse h. Therefore,

$$\sin \theta_1 = \frac{\lambda_1}{h} = \frac{v_1/f}{h} = \frac{v_1}{hf}$$

and

$$\sin \theta_2 = \frac{\lambda_2}{h} = \frac{v_2/f}{h} = \frac{v_2}{hf}$$

Combining these two equations into a single equation by eliminating the common term hf gives

$$\frac{\sin \theta_1}{v_1} = \frac{\sin \theta_2}{v_2}$$

By multiplying each side of this result by c, the speed of light in a vacuum, and recognizing that the ratio c/v is the index of refraction n, we arrive at Snell's law of refraction: $n_1 \sin \theta_1 = n_2 \sin \theta_2$.

Check Your Understanding

(*The answers are given at the end of the book.*)

1. Two slabs with parallel faces are made from different types of glass. A ray of light travels through air and enters each slab at the same angle of incidence, as CYU Figure 26.1 shows. Which slab has the greater index o f refraction?

CYU FIGURE 26.1

2. CYU Figure 26.2 shows three layers of liquids, A, B, and C, each with a different index of refraction. Light begins in liquid A, passes into B, and eventually into C, as the ray of light in the drawing shows. The dashed lines denote the normals to the interfaces between the layers. Which liquid has the smallest index of refraction?

3. Light traveling through air is incident on a flat piece of glass at a 35° angle of incidence and enters the glass at an angle of refraction θ_{glass}. Suppose that a layer of water is added on top of the glass. Then the light travels through air and is incident on the water at the 35° angle of incidence. Does the light enter the glass at the same angle of refraction θ_{glass} as it did when the water was not present?

4. Two identical containers, one filled with water ($n = 1.33$) and the other with benzene ($n = 1.50$), are viewed from directly above. Which container (if either) appears to have a greater depth of fluid?

CYU FIGURE 26.2

5. When an observer peers over the edge of a deep, empty, metal bowl on a kitchen table, he does not see the entire bottom surface. Therefore, a small object lying on the bottom is hidden from view, but the object can be seen when the bowl is filled with liquid A. When the bowl is filled with liquid B, however, the object remains hidden from view. Which liquid has the greater index of refraction?

6. A man is fishing from a dock, using a bow and arrow. To strike a fish that he sees beneath the water, should he aim **(a)** somewhat above the fish, **(b)** directly at the fish, or **(c)** somewhat below the fish?

7. A man is fishing from a dock. He is using a laser gun that emits an intense beam of light. To strike a fish that he sees beneath the water, should he aim **(a)** somewhat above the fish, **(b)** directly at the fish, or **(c)** somewhat below the fish?

8. Two rays of light converge to a point on a screen. A thick plate of glass with parallel surfaces is placed in the path of this converging light, with the parallel surfaces parallel to the screen. Will the point of convergence **(a)** move away from the glass plate, **(b)** move toward the glass plate, or **(c)** remain on the screen?

26.3 | Total Internal Reflection

When light passes from a medium of larger refractive index into one of smaller refractive index—for example, from water to air—the refracted ray bends *away* from the normal, as in **Animated Figure 26.8a**. As the angle of incidence increases, the angle of refraction also increases. When the angle of incidence reaches a certain value, called the **critical angle** θ_c, the angle of refraction is 90°. Then the refracted ray points along the surface; **Animated Figure 26.8b** illustrates what happens at the critical angle. When the angle of incidence exceeds the critical angle, as in **Animated Figure 26.8c**, there is no refracted light. All the incident light is reflected back into the medium from which it came, a phenomenon called **total internal reflection**. Total internal reflection occurs only when light travels from a higher-index medium toward a lower-index medium. It does not occur when light propagates in the reverse direction—for example, from air to water.

(a)   (b)   (c)

ANIMATED FIGURE 26.8 (a) When light travels from a higher-index medium (water) into a lower-index medium (air), the refracted ray is bent away from the normal. (b) When the angle of incidence is equal to the critical angle θ_c, the angle of refraction is 90°. (c) If θ_1 is greater than θ_c, there is no refracted ray, and total internal reflection occurs.

Reinhard Dirscherl/Age Fotostock

FIGURE 26.9 This underwater photograph shows a salamander with its snout near the surface of the water. Some of the light from its body strikes the air–water interface at angles greater than the critical angle and is reflected. Thus, the interface acts like a mirror and forms the image in the upper part of the photograph.

An expression for the critical angle θ_c can be obtained from Snell's law by setting $\theta_1 = \theta_c$ and $\theta_2 = 90°$ (see **Animated Figure 26.8b**):

$$\sin \theta_c = \frac{n_2 \sin 90°}{n_1}$$

Critical angle $$\sin \theta_c = \frac{n_2}{n_1} \quad (n_1 > n_2)$$ **(26.4)**

For instance, the critical angle for light traveling from water ($n_1 = 1.33$) to air ($n_2 = 1.00$) is $\theta_c = \sin^{-1}(1.00/1.33) = 48.8°$. For incident angles greater than 48.8°, Snell's law predicts that $\sin \theta_2$ is greater than unity, a value that is not possible. Thus, light rays with incident angles exceeding 48.8° yield no refracted light, and the light is totally reflected back into the water, as **Animated Figure 26.8c** indicates. Then, the air–water interface acts like a mirror. **Figure 26.9**, for example, shows the mirror-like ability of the interface to form a reflected image of a salamander with its snout near the surface of the water. Light from the salamander's body that strikes the surface at angles exceeding the critical angle is reflected to form the image in the upper part of the photograph.

The next example illustrates how the critical angle changes when the indices of refraction change.

EXAMPLE 5 | Total Internal Reflection

A beam of light is propagating through diamond ($n_1 = 2.42$) and strikes a diamond–air interface at an angle of incidence of 28°. **(a)** Will part of the beam enter the air ($n_2 = 1.00$) or will the beam be totally reflected at the interface? **(b)** Repeat part (a), assuming that the diamond is surrounded by water ($n_2 = 1.33$) instead of air.

Reasoning Total internal reflection occurs only when the beam of light has an angle of incidence that is greater than the critical angle θ_c. The critical angle is different in parts (a) and (b), since it depends on the ratio n_2/n_1 of the refractive indices of the incident (n_1) and refracting (n_2) media.

Solution **(a)** The critical angle θ_c for total internal reflection at the diamond–air interface is given by Equation 26.4 as

$$\theta_c = \sin^{-1}\!\left(\frac{n_2}{n_1}\right) = \sin^{-1}\!\left(\frac{1.00}{2.42}\right) = 24.4°$$

Because the angle of incidence of 28° is greater than the critical angle, there is no refraction, and the light is totally reflected back into the diamond.

(b) If water, rather than air, surrounds the diamond, the critical angle for total internal reflection becomes larger:

$$\theta_c = \sin^{-1}\!\left(\frac{n_2}{n_1}\right) = \sin^{-1}\!\left(\frac{1.33}{2.42}\right) = 33.3°$$

Now a beam of light that has an angle of incidence of 28° (less than the critical angle of 33.3°) at the diamond–water interface is refracted into the water.

The critical angle plays an important role in why a diamond sparkles, as Conceptual Example 6 discusses.

CONCEPTUAL EXAMPLE 6 | The Physics of Why a Diamond Sparkles

A diamond gemstone is famous for its sparkle in air because the light coming from it glitters as the diamond is moved about. The sparkle is related to the total internal reflection of light that occurs within the diamond. What happens to the sparkle when the diamond is placed under water? **(a)** Nothing happens, for the water has no effect on total internal reflection. **(b)** The water reduces the sparkle markedly by making the total internal reflection less likely to occur.

Reasoning When a diamond is held in a certain way in air, the intensity of the light coming from within it is greatly enhanced. **Figure 26.10** helps to explain that this enhancement or sparkle is related to total internal reflection. Part *a* of the drawing shows a ray of light striking a lower facet of the diamond at an angle of incidence that exceeds the critical angle for a diamond–air interface. As a result, this ray undergoes total internal reflection back into the diamond and eventually exits the top surface. Since diamond has a relatively small critical angle in air, many of the rays striking a lower facet behave in this fashion and create the

diamond's sparkle. Part (a) of Example 5 shows that the critical angle is 24.4° and is so small because the index of refraction of diamond ($n = 2.42$) is large compared to that of air ($n = 1.00$).

Answer (a) is incorrect. The water does indeed have an effect on the total internal reflection that occurs. This is because the critical angle depends on the index of refraction of the water as well as that of the diamond (see Equation 26.4).

Answer (b) is correct. **Figure 26.10b** illustrates what happens to the same ray of light within the diamond when the diamond is surrounded by water. Because water has a larger index of refraction than air does, the critical angle for the diamond–water interface is no longer 24.4° but increases to 33.3°, as part (b) of Example 5 shows. Therefore, this particular ray is no longer totally internally reflected. As **Figure 26.10b** indicates, only some of the light is now reflected back into the diamond, while the remainder escapes into the water. Consequently, less light exits from the top of the diamond, causing it to lose much of its sparkle.

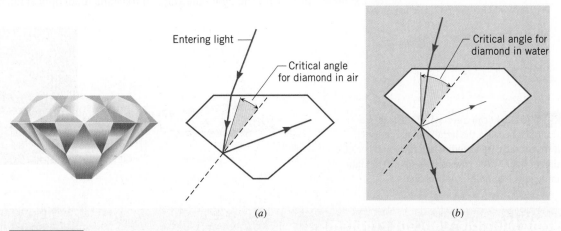

Entering light

Critical angle for diamond in air

Critical angle for diamond in water

(a)   (b)

FIGURE 26.10 (a) Near the bottom of the diamond, light is totally internally reflected, because the incident angle exceeds the critical angle for diamond and air. (b) When the diamond is in water, the same light is partially reflected and partially refracted, since the incident angle is less than the critical angle for diamond and water.

Many optical instruments, such as binoculars, periscopes, and telescopes, use glass prisms and total internal reflection to turn a beam of light through 90° or 180°. **Figure 26.11a** shows a light ray entering a 45°–45°–90° glass prism ($n_1 = 1.5$) and striking the hypotenuse of the prism at an angle of incidence of $\theta_1 = 45°$. Equation 26.4 shows that the critical angle for a glass–air interface is $\theta_c = \sin^{-1}(n_2/n_1) = \sin^{-1}(1.0/1.5) = 42°$. Since the angle of incidence is greater than the critical angle, the light is totally reflected at the hypotenuse and is directed vertically upward in the drawing, having been turned through an angle of 90°. **Figure 26.11b** shows how the same prism can turn the beam through 180° when total

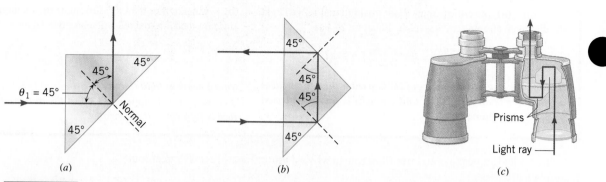

FIGURE 26.11 Total internal reflection at a glass–air interface can be used to turn a ray of light through an angle of (a) 90° or (b) 180°. (c) Two prisms, each reflecting the light twice by total internal reflection, are sometimes used in binoculars to produce a lateral displacement of a light ray.

internal reflection occurs twice. Prisms can also be used in tandem to produce a lateral displacement of a light ray, while leaving its initial direction unaltered. **Figure 26.11c** illustrates such an application in binoculars.

THE PHYSICS OF . . . fiber optics. An important application of total internal reflection occurs in fiber optics, where hair-thin threads of glass or plastic, called optical fibers, "pipe" light from one place to another. **Figure 26.12a** shows that an optical fiber consists of a cylindrical inner *core* that carries the light and an outer concentric shell, the *cladding*. The core is made from transparent glass or plastic that has a relatively high index of refraction. The cladding is also made of glass, but of a type that has a relatively low index of refraction. Light enters one end of the core, strikes the core/cladding interface at an angle of incidence greater than the critical angle, and, therefore, is reflected back into the core. Light thus travels inside the optical fiber along a zigzag path. In a well-designed fiber, little light is lost as a result of absorption by the core, so light can travel many kilometers before its intensity diminishes appreciably. Optical fibers are often bundled together to produce cables. Because the fibers themselves are so thin, the cables are relatively small and flexible and can fit into places inaccessible to larger metal wires. Example 7 deals with the light entering and traveling in an optical fiber.

FIGURE 26.12 (a) Light can travel with little loss in a curved optical fiber, because the light is totally reflected whenever it strikes the core–cladding interface and because the absorption of light by the core itself is small. (b) Light being transmitted by a bundle of optical fibers.

George Doyle/Getty Images

Analyzing Multiple-Concept Problems

EXAMPLE 7 | An Optical Fiber

Figure 26.13 shows an optical fiber that consists of a core made of flint glass ($n_{\text{flint}} = 1.667$) surrounded by a cladding made of crown glass ($n_{\text{crown}} = 1.523$). A ray of light in air enters the fiber at an angle θ_1 with respect to the normal. What is θ_1 if this light also strikes the core–cladding interface at an angle that just barely exceeds the critical angle?

Reasoning The angle of incidence θ_1 is related to the angle of refraction θ_2 (see **Figure 26.13**) by Snell's law, where θ_2 is part of the right triangle in the drawing. The critical angle θ_c for the core-cladding interface is also part of the same right triangle, so that $\theta_2 = 90° - \theta_c$. The critical angle can be determined from a knowledge of the indices of refraction of the core and the cladding.

Knowns and Unknowns   The data used in this problem are:

Description	Symbol	Value	Comment
Index of refraction of core material (flint glass)	n_{flint}	1.667	
Index of refraction of cladding material (crown glass)	n_{crown}	1.523	
Index of refraction of air	n_{air}	1.000	See Table 26.1.
Unknown Variable			
Angle of incidence of light ray entering optical fiber	θ_1	?	

FIGURE 26.13   A ray of light enters the left end of an optical fiber and strikes the core–cladding interface at an angle that just barely exceeds the critical angle θ_c.

Modeling the Problem

STEP 1 Snell's Law of Refraction   The ray of light, initially traveling in air, strikes the left end of the optical fiber at an angle of incidence labeled θ_1 in **Figure 26.13**. When the light enters the flint-glass core, its angle of refraction is θ_2. Snell's law of refraction gives the relation between these angles as

$$n_{\text{air}} \sin \theta_1 = n_{\text{flint}} \sin \theta_2 \qquad (26.2)$$

Solving this equation for θ_1 yields Equation 1 at the right. Values for n_{air} and n_{flint} are known. The angle θ_2 will be evaluated in the next step.

$$\theta_1 = \sin^{-1}\left(\frac{n_{\text{flint}} \sin \theta_2}{n_{\text{air}}}\right) \qquad (1)$$

$$\boxed{?}$$

STEP 2 The Critical Angle   We know that the light ray inside the core strikes the core–cladding interface at an angle that just barely exceeds the critical angle θ_c. When the angle of incidence exceeds the critical angle, all the light is reflected back into the core. From the right triangle in **Figure 26.13**, we see that the critical angle is related to θ_2 by $\theta_2 = 90° - \theta_c$. The critical angle depends on the indices of refraction of the core (flint glass) and cladding (crown glass) according to Equation 26.4:

$$\sin \theta_c = \frac{n_{\text{crown}}}{n_{\text{flint}}} \quad \text{or} \quad \theta_c = \sin^{-1}\left(\frac{n_{\text{crown}}}{n_{\text{flint}}}\right)$$

Substituting this expression for θ_c into $\theta_2 = 90° - \theta_c$ gives

$$\boxed{\theta_2 = 90° - \sin^{-1}\left(\frac{n_{\text{crown}}}{n_{\text{flint}}}\right)}$$

This result for θ_2 can be substituted into Equation 1, as indicated at the right.

$$\theta_1 = \sin^{-1}\left(\frac{n_{\text{flint}} \sin \theta_2}{n_{\text{air}}}\right) \qquad (1)$$

$$\theta_2 = 90° - \sin^{-1}\left(\frac{n_{\text{crown}}}{n_{\text{flint}}}\right) \qquad (2)$$

Solution   Combining the results of Steps 1 and 2 algebraically to produce a single equation gives a rather cumbersome result. Hence, we follow the simpler procedure of evaluating Equation 2 numerically and then substituting the result into Equation 1:

$$\theta_2 = 90° - \sin^{-1}\left(\frac{n_{\text{crown}}}{n_{\text{flint}}}\right) = 90° - \underbrace{\sin^{-1}\left(\frac{1.523}{1.667}\right)}_{66.01°} = 23.99° \qquad (2)$$

$$\theta_1 = \sin^{-1}\left(\frac{n_{\text{flint}} \sin \theta_2}{n_{\text{air}}}\right) = \sin^{-1}\left(\frac{1.667 \ \sin 23.99°}{1.000}\right) = \boxed{42.67°}$$

Related Homework: Problem 32

Optical fiber cables are the medium of choice for high-quality telecommunications because the cables are relatively immune to external electrical interference and because a light beam can carry information through optical fibers just as electricity carries

JAMES KING-HOLMES/Science Source

FIGURE 26.14 A doctor is using a broncho-scope to examine the lungs of a patient who has a history of asthma and allergies.

information through copper wires. The information-carrying capacity of light, however, is thousands of times greater than that of electricity. A laser beam traveling through a single optical fiber can carry tens of thousands of telephone conversations and several TV programs simultaneously.

BIO **THE PHYSICS OF . . . endoscopy.** In the field of medicine, optical fiber cables have had extraordinary impact. In the practice of endoscopy, for instance, a device called an endoscope is used to peer inside the body. **Figure 26.14** shows a bronchoscope being used, which is a kind of endoscope that is inserted through the nose or mouth, down the bronchial tubes, and into the lungs. It consists of two optical fiber cables. One provides light to illuminate interior body parts, while the other sends back an image for viewing. A bronchoscope greatly simplifies the diagnosis of pulmonary disease. Tissue samples can even be collected with some bronchoscopes. A colonoscope is another kind of endoscope, and its design is similar to that of the bronchoscope. It is inserted through the rectum and used to examine the interior of the colon (see **Figure 26.15**). The colono-scope currently offers the best hope for diagnosing colon cancer in its early stages, when it can be treated.

BIO **THE PHYSICS OF . . . arthroscopic surgery.** The use of optical fibers has also revolutionized surgical techniques. In arthroscopic surgery, a small surgi-cal instrument, several millimeters in diameter, is mounted at the end of an opti-cal fiber cable. The surgeon can insert the instrument and cable into a joint, such as the knee, with only a tiny incision and minimal damage to the surrounding tissue (see **Figure 26.16**). Consequently, recovery from the procedure is relatively rapid compared to recovery from traditional surgical techniques.

CAMAL, ISM/SCIENCE PHOTO LIBRARY

Polyp

FIGURE 26.15 A colonoscope revealed this benign (noncancerous) polyp attached to the wall of the colon (large intestine). Polyps that can turn cancerous or grow large enough to obstruct the colon are removed surgically.

Princess Margaret Rose Orthopaedic Hospital/Science Source

FIGURE 26.16 Optical fibers have made arthroscopic surgery possible, such as the repair of the damaged knee shown here.

Check Your Understanding

(The answers are given at the end of the book.)

9. CYU Figure 26.3 shows a 30°–60°–90° prism and two light rays, A and B, that both strike the prism perpendicularly. The prism is surrounded by an unknown liquid, which is the same in both parts of the drawing. When ray A reaches the hypotenuse in the drawing, it is totally internally reflected. Which one of the following statements applies to ray B when it reaches the hypotenuse?

(a) It may or may not be totally internally reflected, depending on what the surrounding liquid is. **(b)** It is not totally internally reflected, no matter what the surrounding liquid is. **(c)** It is totally internally reflected, no matter what the surrounding liquid is.

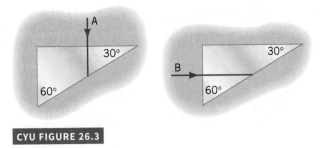

CYU FIGURE 26.3

10. A shallow swimming pool has a constant depth. A point source of light is located in the middle of the bottom of this pool and emits light in all directions. However, no light exits the surface of the water except through a relatively small circular area that is centered on and directly above the light source. Why does the light exit the water through such a limited area?

11. Refer to **Figure 26.6**. Note that the ray within the glass slab is traveling from a medium with a larger refractive index toward a medium with a smaller refractive index. Is it possible, for θ_1 less than 90°, that the ray within the glass will experience total internal reflection at the glass–air interface?

26.4 Polarization and the Reflection and Refraction of Light

For incident angles other than 0°, unpolarized light becomes partially polarized in reflecting from a nonmetallic surface, such as water. To demonstrate this fact, rotate a pair of Polaroid sunglasses in the sunlight reflected from a lake. You will see that the light intensity transmitted through the glasses is a minimum when the glasses are oriented as they are normally worn. Since the transmission axis of the glasses is aligned vertically, it follows that the light reflected from the lake is partially polarized in the horizontal direction.

There is one special angle of incidence at which the reflected light is completely polarized parallel to the surface, the refracted ray being only partially polarized. This angle is called the **Brewster angle** θ_B. **Figure 26.17** summarizes what happens when unpolarized light strikes a nonmetallic surface at the Brewster angle. The value of θ_B is given by **Brewster's law**, in which n_1 and n_2 are, respectively, the refractive indices of the materials in which the incident and refracted rays propagate:

Brewster's law
$$\tan \theta_B = \frac{n_2}{n_1} \qquad\qquad (26.5)$$

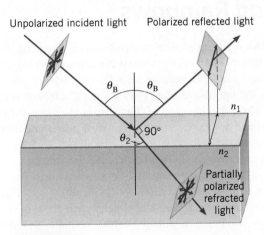

FIGURE 26.17 When unpolarized light is incident on a nonmetallic surface at the Brewster angle θ_B, the reflected light is 100% polarized in a direction parallel to the surface. The angle between the reflected and refracted rays is 90°.

This relation is named after the Scotsman David Brewster (1781–1868), who discovered it. **Figure 26.17** also indicates that the reflected and refracted rays are perpendicular to each other when light strikes the surface at the Brewster angle.

Math Skills Since the reflected and refracted rays are perpendicular in **Figure 26.17**, it follows that $\theta_B + 90° + \theta_2 = 180°$ or $\theta_B + \theta_2 = 90°$. To see that this is indeed the case, we take advantage of Snell's law (Equation 26.2). For an incident angle $\theta_1 = \theta_B$, this law is

$$\sin \theta_B = \frac{n_2 \sin \theta_2}{n_1}$$

In addition, Brewster's law states that $\tan \theta_B = \frac{n_2}{n_1}$ (Equation 26.5), and $\tan \theta_B = \frac{\sin \theta_B}{\cos \theta_B}$ (see Appendix E.2, Other Trigonometric Identities, Equation 4). Therefore, we can substitute $\frac{\sin \theta_B}{\cos \theta_B} = \frac{n_2}{n_1}$ into Snell's law and obtain

$$\sin \theta_B = \left(\frac{n_2}{n_1}\right) \sin \theta_2 = \frac{\sin \theta_B}{\cos \theta_B} \sin \theta_2 \quad \text{or} \quad \cos \theta_B = \sin \theta_2$$

This result is what we are looking for, because $\sin \theta_2 = \cos(90° - \theta_2)$ (see Appendix E.2, Other Trigonometric Identities, Equation 7). Thus, we have

$$\cos \theta_B = \sin \theta_2 = \cos(90° - \theta_2) \quad \text{or} \quad \theta_B = 90° - \theta_2 \quad \text{or} \quad \theta_B + \theta_2 = 90°$$

Check Your Understanding

(The answer is given at the end of the book.)

12. You are sitting by the shore of a lake on a sunny and windless day. When are your Polaroid sunglasses most effective in reducing the glare of the sunlight reflected from the lake surface? When the angle of incidence of the sunlight on the lake is _____. **(a)** almost 90° because the sun is low in the sky **(b)** 0° because the sun is directly overhead **(c)** somewhere between 90° and 0°

26.5 | The Dispersion of Light: Prisms and Rainbows

Figure 26.18a shows a ray of monochromatic light passing through a glass prism surrounded by air. When the light enters the prism at the left face, the refracted ray is bent toward the normal, because the refractive index of glass is greater than that of air. When the light leaves the prism at the right face, it is refracted away from the normal. Thus, the net effect of the prism is to change the direction of the ray, causing it to bend downward upon entering the prism, and downward again upon leaving. Because the refractive index of the glass depends on wavelength (see **Table 26.2**), rays corresponding to different colors are bent by different amounts by the prism and depart traveling in different directions. The greater the index of refraction for a given color, the greater the bending, and **Figure 26.18b** shows the refractions for the colors red and violet, which are at opposite ends of the visible spectrum. If a beam of sunlight, which contains all colors, is sent through the prism, the sunlight is separated into a spectrum of colors, as **Figure 26.18c** shows. The spreading of light into its color components is called **dispersion**.

In **Figure 26.18a** the ray of light is refracted twice by a glass prism surrounded by air. Conceptual Example 8 explores what happens to the light when the prism is surrounded by materials other than air.

TABLE 26.2 Indices of Refraction n of Crown Glass at Various Wavelengths

Color[a]	Vacuum Wavelength (nm)	Index of Refraction, n
Red	660	1.520
Orange	610	1.522
Yellow	580	1.523
Green	550	1.526
Blue	470	1.531
Violet	410	1.538

[a]Approximate

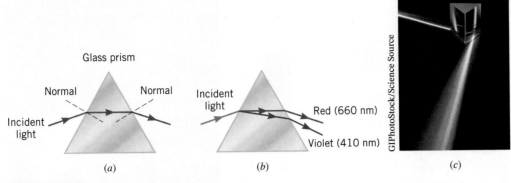

Glass prism

Normal   Normal

Incident light

Incident light   Red (660 nm)

Violet (410 nm)

GIPhotoStock/Science Source

(a)   (b)   (c)

FIGURE 26.18 (a) A ray of light is refracted as it passes through a prism. The prism is surrounded by air. (b) Two different colors are refracted by different amounts. For clarity, the amount of refraction has been exaggerated. (c) Sunlight is dispersed into its color components by this prism.

CONCEPTUAL EXAMPLE 8 | The Refraction of Light Depends on Two Refractive Indices

In **Figure 26.18a** the glass prism is surrounded by air and bends the ray of light downward. It is also possible for the prism to bend the ray upward, as in **Figure 26.19a**, or to not bend the ray at all, as in part b of the drawing. How can the situations illustrated in **Figure 26.19** arise?

Reasoning and Solution Snell's law of refraction includes the refractive indices of *both* materials on either side of an interface. With this in mind, we note that the ray bends upward, or away from the normal, as it enters the prism in **Figure 26.19a**. A ray bends away from the normal when it travels from a medium with a larger refractive index into a medium with a smaller refractive index. When the ray leaves the prism, it again bends upward, which is toward the normal at the point of exit. A ray bends toward the normal when traveling from a smaller toward a larger refractive index. Thus, *the situation in Figure 26.19a could arise if the prism were immersed in a fluid, such as carbon disulfide, that has a larger refractive index than does glass* (see **Table 26.1**).

We have seen in **Figures 26.18a** and **26.19a** that a glass prism can bend a ray of light either downward or upward, depending on whether the surrounding fluid has a smaller or larger index of refraction than the glass. It is logical to conclude, then, that *a prism will not bend a ray at all, neither up nor down, if the surrounding fluid has the same index of refraction as the glass*—a condition known as *index matching*. This is exactly what is happening in **Figure 26.19b**, where the ray proceeds straight through the prism as if the prism were not even there. If the index of refraction of the surrounding fluid equals that of the glass prism, then $n_1 = n_2$,

(a)

(b)

FIGURE 26.19 A ray of light passes through identical prisms, each surrounded by a different fluid. The ray of light is (a) refracted upward and (b) not refracted at all.

and Snell's law ($n_1 \sin \theta_1 = n_2 \sin \theta_2$) reduces to $\sin \theta_1 = \sin \theta_2$. Therefore, the angle of refraction equals the angle of incidence, and no bending of the light occurs.

Related Homework: Check Your Understanding 16

THE PHYSICS OF . . . rainbows.
Another example of dispersion occurs in rainbows, in which refraction by water droplets gives rise to the colors. You can often see a rainbow just as a storm is leaving, if you look at the departing rain with the sun at your back. When light from the sun enters a spherical raindrop, as in **Figure 26.20**, light of each color is refracted or bent by an amount that depends on the refractive index of water for that wavelength. After

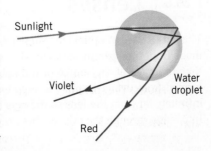

Sunlight

Violet

Water droplet

Red

FIGURE 26.20 When sunlight emerges from a water droplet, the light is dispersed into its constituent colors, of which only two are shown.

FIGURE 26.21 (*a*) The different colors seen in a rainbow originate from water droplets at different angles of elevation. (*b*) A rock climber beneath a rainbow.

reflection from the back surface of the droplet, the different colors are again refracted as they reenter the air. Although each droplet disperses the light into its full spectrum of colors, the observer in **Figure 26.21*a*** sees only one color of light coming from any given droplet, since only one color travels in the right direction to reach the observer's eyes. However, all colors are visible in a rainbow (see **Figure 26.21*b***) because each color originates from different droplets at different angles of elevation.

Check Your Understanding

(The answers are given at the end of the book.)

13. Two blocks, made from the same transparent material, are immersed in different liquids. A ray of light strikes each block at the same angle of incidence. From **CYU Figure 26.4**, determine which liquid, A or B, has the greater index of refraction.

CYU FIGURE 26.4

14. A beam of violet-colored light is propagating in crown glass. When the light reaches the boundary between the glass and the surrounding air, the beam is totally reflected back into the glass. What happens if the light is red and has the same angle of incidence θ_1 at the glass–air interface as does the violet-colored light? **(a)** Depending on the value for θ_1, red light may not be totally reflected, and some of it may be refracted into the air. **(b)** No matter what the value for θ_1, the red light behaves exactly the same as the violet-colored light. (*Hint:* Refer to **Table 26.2** and review Section 26.3.)

26.6 | Lenses

The lenses used in optical instruments, such as eyeglasses, cameras, and telescopes, are made from transparent materials that refract light. They refract the light in such a way that an image of the source of the light is formed. **Figure 26.22*a*** shows a crude lens formed from two glass prisms. Suppose that an object centered on the principal axis is infinitely far from the lens so the rays from the object are parallel to the principal axis. In passing through the prisms, these rays are bent toward the axis because of refraction.

FIGURE 26.22 (a) These two prisms cause rays of light that are parallel to the principal axis to change direction and cross the axis at different points. (b) With a converging lens, paraxial rays that are parallel to the principal axis converge to the focal point F after passing through the lens.

Unfortunately, the rays do not all cross the axis at the same place, and, therefore, such a crude lens gives rise to a blurred image of the object.

A better lens can be constructed from a single piece of transparent material with properly curved surfaces, often spherical, as in **Figure 26.22b**. With this improved lens, rays that are near the principal axis (paraxial rays) and parallel to it converge to a single point on the axis after emerging from the lens. This point is called the **focal point** F of the lens. Thus, an object located infinitely far away on the principal axis leads to an image at the focal point of the lens. The distance between the focal point and the lens is the **focal length** f. In what follows, we assume the lens is so thin compared to f that it makes no difference whether f is measured between the focal point and either surface of the lens or the center of the lens. The type of lens in **Figure 26.22b** is known as a **converging lens** because it causes incident parallel rays to converge at the focal point.

Another type of lens found in optical instruments is a **diverging lens**, which causes incident parallel rays to diverge after exiting the lens. Two prisms can also be used to form a crude diverging lens, as in **Figure 26.23a**. In a properly designed diverging lens, such as the one in part b of the picture, paraxial rays that are parallel to the principal axis appear to originate from a single point on the axis after passing through the lens. This point is the focal point F, and its distance f from the lens is the focal length. Again, we assume that the lens is thin compared to the focal length.

FIGURE 26.23 (a) These two prisms cause parallel rays to diverge. (b) With a diverging lens, paraxial rays that are parallel to the principal axis appear to originate from the focal point F after passing through the lens.

Converging and diverging lenses come in a variety of shapes, as **Figure 26.24** illustrates. Observe that converging lenses are thicker at the center than at the edges, whereas diverging lenses are thinner at the center.

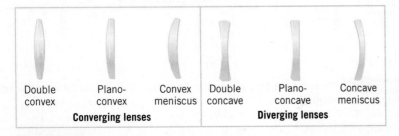

FIGURE 26.24 Converging and diverging lenses come in a variety of shapes.

Check Your Understanding

(The answers are given at the end of the book.)

15. A beacon in a lighthouse is to produce a parallel beam of light. The beacon consists of a light source and a converging lens. Should the light source be placed **(a)** between the focal point and the lens, **(b)** at the focal point of the lens, or **(c)** beyond the focal point? (*Hint:* Refer to Section 25.5 and review the principle of reversibility.)

16. Review Conceptual Example 8 as an aid in answering this question. Is it possible for a lens to behave as a converging lens when surrounded by air but to behave as a diverging lens when surrounded by another medium?

26.7 | The Formation of Images by Lenses

Ray Diagrams and Ray Tracing

Each point on an object emits light rays in all directions, and when some of these rays pass through a lens, they form an image. As with mirrors, the ray-tracing method can be used to determine the location and size of the image. Lenses differ from mirrors, however, in that light can pass through a lens from left to right or from right to left. Therefore, when constructing ray diagrams, begin by locating a focal point *F* on *each side of the lens;* each point lies on the principal axis at the same distance *f* from the lens. The lens is assumed to be thin, in that its thickness is small compared with the focal length and the distances of the object and the image from the lens. For convenience, it is also assumed that the object is located to the left of the lens and is oriented perpendicular to the principal axis. There are three paraxial rays that leave a point on the top of the object and are especially helpful in drawing ray diagrams. They are labeled 1, 2, and 3 in **Figure 26.25**. When tracing their paths, we use the following reasoning strategy.

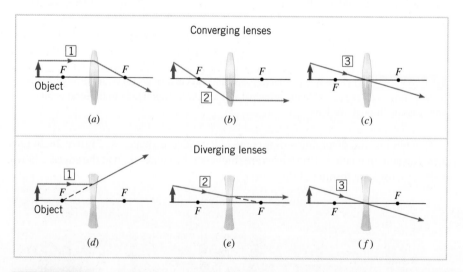

FIGURE 26.25 The rays shown here are useful in determining the nature of the images formed by converging and diverging lenses.

REASONING STRATEGY	Ray Tracing for Converging and Diverging Lenses
Converging Lens	**Diverging Lens**

Ray 1

This ray initially travels parallel to the principal axis. In passing through a converging lens, the ray is refracted toward the axis and travels through the focal point on the right side of the lens, as Figure 26.25a shows.	This ray initially travels parallel to the principal axis. In passing through a diverging lens, the ray is refracted away from the axis, and *appears* to have originated from the focal point on the left of the lens. The dashed line in Figure 26.25d represents the apparent path of the ray.

Ray 2

This ray first passes through the focal point on the left and then is refracted by the lens in such a way that it leaves traveling parallel to the axis, as in Figure 26.25b.	This ray leaves the object and moves toward the focal point on the right of the lens. Before reaching the focal point, however, the ray is refracted by the lens so as to exit parallel to the axis. See Figure 26.25e, where the dashed line indicates the ray's path in the absence of the lens.

Ray 3*

This ray travels directly through the center of the thin lens without any appreciable bending, as in Figure 26.25c.	This ray travels directly through the center of the thin lens without any appreciable bending, as in Figure 26.25f.

*Ray 3 does not bend as it proceeds through the lens because the left and right surfaces of each type of lens are nearly parallel at the center. Thus, in either case, the lens behaves as a transparent slab. As Figure 26.6 shows, the rays incident on and exiting from a slab travel in the same direction with only a lateral displacement. If the lens is sufficiently thin, the displacement is negligibly small.

Image Formation by a Converging Lens

Figure 26.26a illustrates the formation of a real image by a converging lens. Here the object is located at a distance from the lens that is greater than twice the focal length (beyond the point labeled 2F). To locate the image, any two of the three special rays can be drawn from the tip of the object, although all three are shown in the drawing. The point on the right side of the lens where these rays intersect locates the tip of the image. The ray diagram indicates that the image is real, inverted, and smaller than the object.

THE PHYSICS OF . . . a camera. This optical arrangement is similar to that used in a camera, where an image sensor* or a piece of film records the image (see Figure 26.26b).

(a)   (b)

FIGURE 26.26 (a) When the object is placed to the left of the point labeled 2F, a real, inverted, and smaller image is formed. (b) The arrangement in part a is like that used in a camera.

*One type of image sensor used in today's digital cameras utilizes a charge-coupled device (CCD). See Section 29.3 for a discussion of CCDs.

FIGURE 26.27 (*a*) When the object is placed between $2F$ and F, the image is real, inverted, and larger than the object. (*b*) This arrangement is found in projectors.

When the object is placed between $2F$ and F, as in **Figure 26.27a**, the image is still real and inverted; however, the image is now larger than the object.

THE PHYSICS OF . . . a slide or film projector. This optical system is used in a slide or film projector in which a small piece of film is the object and the enlarged image falls on a screen. However, to obtain an image that is right-side up, the film must be placed in the projector upside down.

When the object is located between the focal point and the lens, as in **Figure 26.28**, the rays diverge after leaving the lens. To a person viewing the diverging rays, they appear to come from an image behind (to the left of) the lens. Because none of the rays actually come from the image, it is a virtual image. The ray diagram shows that the virtual image is upright and enlarged.

THE PHYSICS OF . . . a magnifying glass. A magnifying glass uses this arrangement, as can be seen in part *b* of the drawing.

FIGURE 26.28 (*a*) When an object is placed between the focal point F of a converging lens and the lens, an upright, enlarged, and virtual image is created. (*b*) Such an image is seen when looking through a magnifying glass.

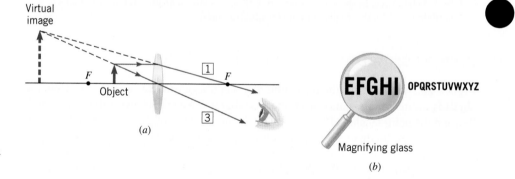

Image Formation by a Diverging Lens

We have seen that a converging lens can form a real image or a virtual image, depending on where the object is located with respect to the lens. In contrast, regardless of the position of a real object, a diverging lens always forms a virtual image that is on the same side of the lens as the object and is upright and smaller relative to the object, as **Figure 26.29** illustrates.

FIGURE 26.29 (*a*) A diverging lens always forms a virtual image of a real object. The image is upright and smaller relative to the object. (*b*) The image seen through a diverging lens.

Check Your Understanding

(The answer is given at the end of the book.)

17. A converging lens is used to produce a real image, as in **Figure 26.27a**. A piece of black tape is then placed over the upper half of the lens. Which one of the following statements is true concerning the image that results with the tape in place? **(a)** The image is of the entire object, although its brightness is reduced since fewer rays produce it. **(b)** The image is of the object's lower half only, but its brightness is not reduced. **(c)** The image is of the object's upper half only, but its brightness is not reduced.

26.8 The Thin-Lens Equation and the Magnification Equation

When an object is placed in front of a spherical mirror, we can determine the location, size, and nature of its image by using the technique of ray tracing or the mirror and magnification equations. Both options are based on the law of reflection. The mirror and magnification equations relate the distances d_o and d_i of the object and image from the mirror to the focal length f and magnification m. For an object placed in front of a lens, Snell's law of refraction leads to the technique of ray tracing and to equations that are identical to the mirror and magnification equations. Thus, mirrors work because of the reflection of light, whereas lenses work because of the refraction of light, a distinction between the two devices that is important to keep in mind.

The equations that result from applying Snell's law to lenses are referred to as the thin-lens equation and the magnification equation:

Thin-lens equation
$$\frac{1}{d_o} + \frac{1}{d_i} = \frac{1}{f} \qquad\qquad (26.6)$$

Math Skills The thin-lens equation $\left(\frac{1}{d_o} + \frac{1}{d_i} = \frac{1}{f}\right)$ is sometimes thought to imply that $d_o + d_i = f$. To emphasize that the focal length f does not equal the object distance d_o plus the image distance d_i, we can solve the thin lens equation for f. First, we multiply the left side of the thin-lens equation by 1 in the form of $\frac{d_o d_i}{d_o d_i}$:

$$\underset{1}{\left(\frac{d_o d_i}{d_o d_i}\right)}\left(\frac{1}{d_o} + \frac{1}{d_i}\right) = \frac{1}{f} \quad \text{or} \quad \left(\frac{1}{d_o d_i}\right)\left(\frac{d_o d_i}{d_o} + \frac{d_o d_i}{d_i}\right) = \frac{1}{f}$$

Simplifying this result gives

$$\left(\frac{1}{d_o d_i}\right)\left(\frac{d_o d_i}{d_o} + \frac{d_o d_i}{d_i}\right) = \frac{1}{f} \quad \text{or} \quad \frac{d_i + d_o}{d_o d_i} = \frac{1}{f}$$

Taking the reciprocal of both sides of the simplified result shows that

$$\left(\frac{d_i + d_o}{d_o d_i}\right)^{-1} = \left(\frac{1}{f}\right)^{-1} \quad \text{or} \quad \frac{d_o d_i}{d_i + d_o} = f$$

Clearly, it is not true that $d_o + d_i = f$. Do not make this mistake when solving problems.

Magnification equation
$$m = \frac{\text{Image height}}{\text{Object height}} = \frac{h_i}{h_o} = -\frac{d_i}{d_o} \qquad\qquad (26.7)$$

Figure 26.30 defines the symbols in these expressions with the aid of a thin converging lens, but the expressions also apply to a diverging lens, if it is thin. The derivations of these equations are presented at the end of this section.

FIGURE 26.30 The drawing shows the focal length f, the object distance d_o, and the image distance d_i for a converging lens. The object and image heights are, respectively, h_o and h_i.

Certain sign conventions accompany the use of the thin-lens and magnification equations, and the conventions are similar to those used with mirrors in Section 25.6. The issue of real versus virtual images, however, is slightly different with lenses than with mirrors. With a mirror, a real image is formed on the *same side* of the mirror as the object (see Figure 25.18), in which case the image distance d_i is a positive number. With a lens, a positive value for d_i also means the image is real. However, starting with an actual object, a real image is formed on the side of the lens *opposite to* the object (see **Figure 26.30**). The **sign conventions** listed in the following Reasoning Strategy apply to light rays traveling from left to right from a real object.

REASONING STRATEGY   **Summary of Sign Conventions for Lenses**

Focal length

   f is + for a converging lens.

   f is − for a diverging lens.

Object distance

   d_o is + if the object is to the left of the lens (real object), as is usual.

   d_o is − if the object is to the right of the lens (virtual object).*

Image distance

   d_i is + for an image (real) formed to the right of the lens by a real object.

   d_i is − for an image (virtual) formed to the left of the lens by a real object.

Magnification

   m is + for an image that is upright with respect to the object.

   m is − for an image that is inverted with respect to the object.

*This situation arises in systems containing more than one lens, where the image formed by the first lens becomes the object for the second lens. In such a case, the object of the second lens may lie to the right of that lens, in which event d_o is assigned a negative value and the object is called a virtual object.

Examples 9 and 10 illustrate the use of the thin-lens and magnification equations.

EXAMPLE 9 | The Real Image Formed by a Camera Lens

A person 1.70 m tall is standing 2.50 m in front of a digital camera. The camera uses a converging lens whose focal length is 0.0500 m. **(a)** Find the image distance (the distance between the lens and the image sensor) and determine whether the image is real or virtual. **(b)** Find the magnification and the height of the image on the image sensor.

Reasoning This optical arrangement is similar to that in **Figure 26.26a**, where the object distance is greater than twice the focal length of the lens. Therefore, we expect the image to be real, inverted, and smaller than the object.

Solution **(a)** To find the image distance d_i we use the thin-lens equation with $d_o = 2.50$ m and $f = 0.0500$ m:

$$\frac{1}{d_i} = \frac{1}{f} - \frac{1}{d_o} = \frac{1}{0.0500 \text{ m}} - \frac{1}{2.50 \text{ m}} = 19.6 \text{ m}^{-1} \quad \text{or} \quad \boxed{d_i = 0.0510 \text{ m}}$$

Since the image distance is a positive number, a real image is formed on the image sensor.

(b) The magnification follows from the magnification equation:

$$m = -\frac{d_i}{d_o} = -\frac{0.0510 \text{ m}}{2.50 \text{ m}} = \boxed{-0.0204}$$

The image is 0.0204 times as large as the object, and it is inverted since m is negative. Since the object height is $h_o = 1.70$ m, the image height is

$$h_i = mh_o = (-0.0204)(1.70 \text{ m}) = \boxed{-0.0347 \text{ m}}$$

EXAMPLE 10 | The Virtual Image Formed by a Diverging Lens

An object is placed 7.10 cm to the left of a diverging lens whose focal length is $f = -5.08$ cm (a diverging lens has a negative focal length). **(a)** Find the image distance and determine whether the image is real or virtual. **(b)** Obtain the magnification.

Reasoning This situation is similar to that in **Figure 26.29a**. The ray diagram shows that the image is virtual, erect, and smaller than the object.

> **Problem-Solving Insight** In the thin-lens equation, the reciprocal of the image distance d_i is given by $d_i^{-1} = f^{-1} - d_o^{-1}$, where f is the focal length and d_o is the object distance. After combining the reciprocals f^{-1} and d_o^{-1}, do not forget to take the reciprocal of the result to find d_i.

Solution **(a)** The thin-lens equation can be used to find the image distance d_i:

$$\frac{1}{d_i} = \frac{1}{f} - \frac{1}{d_o} = \frac{1}{-5.08 \text{ cm}} - \frac{1}{7.10 \text{ cm}}$$

$$= -0.338 \text{ cm}^{-1} \quad \text{or} \quad \boxed{d_i = -2.96 \text{ cm}}$$

The image distance is negative, indicating that the image is $\boxed{\text{virtual}}$ and located to the left of the lens.

(b) Since d_i and d_o are known, the magnification equation shows that

$$m = -\frac{d_i}{d_o} = -\frac{-2.96 \text{ cm}}{7.10 \text{ cm}} = \boxed{0.417}$$

The image is upright (m is positive) and smaller ($m < 1$) than the object.

The thin-lens and magnification equations can be derived by considering rays 1 and 3 in **Figure 26.31a**. Ray 1 is shown separately in part b of the drawing, where the angle θ is the same in each of the two colored triangles. Thus, $\tan \theta$ is the same for each triangle:

$$\tan \theta = \frac{h_o}{f} = \frac{-h_i}{d_i - f}$$

A minus sign has been inserted in the numerator of the ratio $h_i/(d_i - f)$ for the following reason. The angle θ in **Figure 26.31b** is assumed to be positive. Since the image is inverted relative to the object, the image height h_i is a negative number. The insertion of the minus sign ensures that the term $-h_i/(d_i - f)$, and hence $\tan \theta$, is a positive quantity.

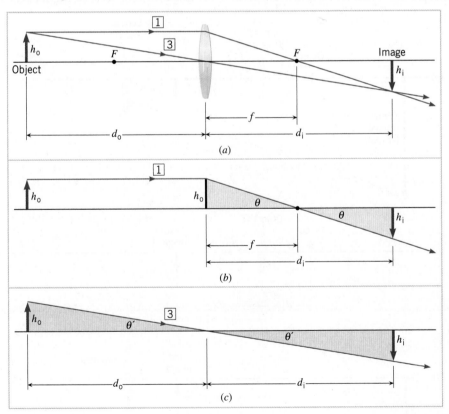

FIGURE 26.31 These ray diagrams are used for deriving the thin-lens and magnification equations.

Ray 3 is shown separately in **Figure 26.31c**, where the two angles labeled θ' are the same. Therefore,

$$\tan \theta' = \frac{h_o}{d_o} = \frac{-h_i}{d_i}$$

A minus sign has been inserted in the numerator of the term h_i/d_i for the same reason that a minus sign was inserted earlier—namely, to ensure that $\tan \theta'$ is a positive quantity. The first equation gives $h_i/h_o = -(d_i - f)/f$, while the second equation yields

$h_i/h_o = -d_i/d_o$. Equating these two expressions for h_i/h_o and rearranging the result produces the thin-lens equation, $1/d_o + 1/d_i = 1/f$. The magnification equation follows directly from the equation $h_i/h_o = -d_i/d_o$, if we recognize that h_i/h_o is the magnification m of the lens.

Check Your Understanding

(The answers are given at the end of the book.)

18. A spherical mirror and a lens are immersed in water. Compared to the way they work in air, which one do you expect will be more affected by the water?

19. An object is located at a distance d_o in front of a lens. The lens has a focal length f and produces an upright image that is twice as tall as the object. What kind of lens is it, and what is the object distance? Express your answer as a fraction or multiple of the focal length.

20. In an old movie a photographic film negative is introduced as evidence in a trial. The negative shows an image of a house that no longer exists. The verdict depends on knowing exactly how far above the ground a window ledge was (the object height h_o). The distance between the ground and the ledge on the negative (the image height h_i) can be measured. What additional information is needed to calculate h_o? **(a)** Nothing else is needed. **(b)** Just the object distance d_o, which is the distance between the house and the camera lens. **(c)** Just the focal length f of the lens. **(d)** Both d_o and f are needed.

26.9 | Lenses in Combination

Many optical instruments, such as microscopes and telescopes, use a number of lenses together to produce an image. Among other things, a multiple-lens system can produce an image that is magnified more than is possible with a single lens. For instance, **Figure 26.32a**

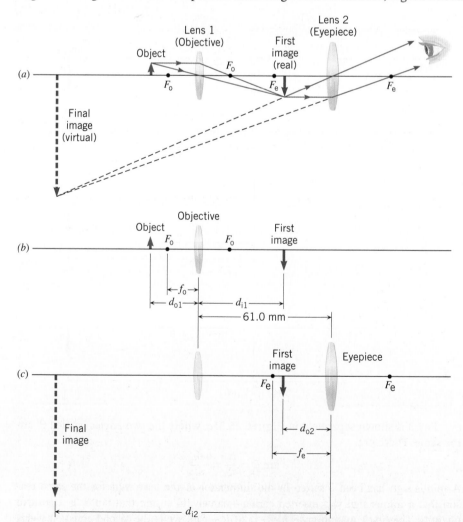

FIGURE 26.32 (a) This two-lens system can be used as a compound microscope to produce a virtual, enlarged, and inverted final image. (b) The objective forms the first image and (c) the eyepiece forms the final image.

shows a two-lens system used in a microscope. The first lens, the lens closest to the object, is referred to as the *objective*. The second lens is known as the *eyepiece* (or *ocular*). The object is placed just outside the focal point F_o of the objective. The image formed by the objective—called the "first image" in the drawing—is real, inverted, and enlarged compared to the object. This first image then serves as the object for the eyepiece. Since the first image falls between the eyepiece and its focal point F_e, the eyepiece forms an enlarged, virtual, final image, which is what the observer sees.

The location of the final image in a multiple-lens system can be determined by applying the thin-lens equation to each lens separately. The key point to remember in such situations is the following:

> **Problem-Solving Insight** The image produced by one lens serves as the object for the next lens.

The next example illustrates this point.

EXAMPLE 11 | A Microscope—Two Lenses in Combination

The objective and eyepiece of the compound microscope in **Figure 26.32** are both converging lenses and have focal lengths of $f_o = 15.0$ mm and $f_e = 25.5$ mm. A distance of 61.0 mm separates the lenses. The microscope is being used to examine an object placed $d_{o1} = 24.1$ mm in front of the objective. Find the final image distance relative to the eyepiece.

Reasoning The thin-lens equation can be used to locate the final image produced by the eyepiece. We know the focal length of the eyepiece, but to determine the final image distance from the thin-lens equation we also need to know the object distance, which is not given. To obtain this distance, we recall that the image produced by one lens (the objective) is the object for the next lens (the eyepiece). We can use the thin-lens equation to locate the image produced by the objective, since the focal length and the object distance for this lens are given. The location of this image relative to the eyepiece will tell us the object distance for the eyepiece.

Solution The final image distance relative to the eyepiece is d_{i2}, and we can determine it by using the thin-lens equation:

$$\frac{1}{d_{i2}} = \frac{1}{f_e} - \frac{1}{d_{o2}}$$

The focal length f_e of the eyepiece is known, but to obtain a value for the object distance d_{o2} we must locate the first image produced by the objective. The first image distance d_{i1} (see **Figure 26.32b**) can be determined using the thin-lens equation with $d_{o1} = 24.1$ mm and $f_o = 15.0$ mm.

$$\frac{1}{d_{i1}} = \frac{1}{f_o} - \frac{1}{d_{o1}} = \frac{1}{15.0 \text{ mm}} - \frac{1}{24.1 \text{ mm}}$$

$$= 0.0252 \text{ mm}^{-1} \quad \text{or} \quad d_{i1} = 39.7 \text{ mm}$$

The first image now becomes the object for the eyepiece, as indicated in **Figure 26.32c**. Since the distance between the lenses is 61.0 mm, the object distance for the eyepiece is $d_{o2} = 61.0$ mm $-$ $d_{i1} = 61.0$ mm $- 39.7$ mm $= 21.3$ mm. Noting that the focal length of the eyepiece is $f_e = 25.5$ mm, we can find the final image distance with the thin-lens equation:

$$\frac{1}{d_{i2}} = \frac{1}{f_e} - \frac{1}{d_{o2}} = \frac{1}{25.5 \text{ mm}} - \frac{1}{21.3 \text{ mm}}$$

$$= -0.0077 \text{ mm}^{-1} \quad \text{or} \quad \boxed{d_{i2} = -130 \text{ mm}}$$

The fact that d_{i2} is negative indicates that the final image is virtual. It lies to the left of the eyepiece, as the drawing shows.

> **Problem-Solving Insight** The overall magnification m of a two-lens system is the product of the magnifications m_1 and m_2 of the individual lenses, or $m = m_1 \times m_2$.

Suppose, for example, that the image of lens 1 is magnified by a factor of 5 relative to the original object, so that $m_1 = 5$. As we know, the image of lens 1 serves as the object for lens 2. Suppose, in addition, that lens 2 magnifies this object further by a factor of 8, so that $m_2 = 8$. The final image of the two-lens system, then, would be $5 \times 8 = 40$ times as large as the original object. In other words, the overall magnification is $m = m_1 \times m_2$, and this calculation can be extended to any number of lenses.

26.10 The Human Eye

Anatomy

Without doubt, the human eye is the most remarkable of all optical devices. **Interactive Figure 26.33** shows some of its main anatomical features. The eyeball is approximately spherical with a diameter of about 25 mm. Light enters the eye through a transparent membrane (the *cornea*). This membrane covers a clear liquid region (the *aqueous*

INTERACTIVE FIGURE 26.33 A cross-sectional view of the human eye.

humor), behind which are a diaphragm (the *iris*), the *lens*, a region filled with a jelly-like substance (the *vitreous humor*), and, finally, the *retina*. The retina is the light-sensitive part of the eye, consisting of millions of structures called *rods* and *cones*. When stimulated by light, these structures send electrical impulses via the *optic nerve* to the brain, which interprets the image detected by the retina.

The iris is the colored portion of the eye and controls the amount of light reaching the retina. The iris acts as a controller because it is a muscular diaphragm with a variable opening at its center, through which the light passes. The opening is called the *pupil*. The diameter of the pupil varies from about 2 to 7 mm, decreasing in bright light and increasing (dilating) in dim light.

Of prime importance to the operation of the eye is the fact that the lens is flexible, and its shape can be altered by the action of the *ciliary muscle*. The lens is connected to the ciliary muscle by the *suspensory ligaments* (see the drawing). We will see shortly how the shape-changing ability of the lens affects the focusing ability of the eye.

Optics

BIO **THE PHYSICS OF . . . the human eye.** Optically, the eye and the camera are similar; both have a lens system and a diaphragm with a variable opening or aperture at its center. Moreover, the retina of the eye and the image sensor in a camera serve similar functions, for both record the image formed by the lens system. In the eye, the image formed on the retina is real, inverted, and smaller than the object, just as it is in a camera. Although the image on the retina is inverted, it is interpreted by the brain as being right-side up.

For clear vision, the eye must refract the incoming light rays, so as to form a sharp image on the retina. In reaching the retina, the light travels through five different media, each with a different index of refraction n: air ($n = 1.00$), the cornea ($n = 1.38$), the aqueous humor ($n = 1.33$), the lens ($n = 1.40$, on the average), and the vitreous humor ($n = 1.34$). Each time light passes from one medium into another, it is refracted at the boundary. The greatest amount of refraction, about 70% or so, occurs at the air/cornea boundary. According to Snell's law, the large refraction at this interface occurs primarily because the refractive index of air ($n = 1.00$) is so different from that of the cornea ($n = 1.38$). The refraction at all the other boundaries is relatively small because the indices of refraction on either side of these boundaries are nearly equal. The lens itself contributes only about 20–25% of the total refraction, since the surrounding aqueous and vitreous humors have indices of refraction that are nearly the same as that of the lens.

Even though the lens contributes only a quarter of the total refraction or less, its function is an important one. The eye has a fixed image distance; that is, the distance between the lens and the retina is constant. Therefore, the only way that objects located at different distances can produce sharp images on the retina is for the focal length of the lens to be adjustable. It is the ciliary muscle that adjusts the focal length. When the eye looks at a very distant object, the ciliary muscle is not tensed. The lens has its least curvature and, consequently, its longest focal length. Under this condition the eye is said to be "fully relaxed," and the rays form a sharp image on the retina, as in **Figure 26.34a**. When the object moves closer to the eye, the ciliary muscle

FIGURE 26.34 (*a*) When fully relaxed, the lens of the eye has its longest focal length, and an image of a very distant object is formed on the retina. (*b*) When the ciliary muscle is tensed, the lens has a shorter focal length. Consequently, an image of a closer object is formed on the retina.

tenses automatically, thereby increasing the curvature of the lens, shortening the focal length, and permitting a sharp image to form again on the retina (**Figure 26.34*b***). When a sharp image of an object is formed on the retina, we say the eye is "focused" on the object. The process in which the lens changes its focal length to focus on objects at different distances is called **accommodation**.

When you hold a book too close, the print is blurred because the lens cannot adjust enough to bring the book into focus. The point nearest the eye at which an object can be placed and still produce a sharp image on the retina is called the **near point** of the eye. The ciliary muscle is fully tensed when an object is placed at the near point. For people in their early twenties with normal vision, the near point is located about 25 cm from the eye. It increases to about 50 cm at age 40 and to roughly 500 cm at age 60. Since most reading material is held at a distance of 25–45 cm from the eye, older adults typically need eyeglasses to overcome the loss of accommodation. The **far point** of the eye is the location of the farthest object on which the fully relaxed eye can focus. A person with normal eyesight can see objects very far away, such as the planets and stars, and thus has a far point located nearly at infinity.

Nearsightedness

BIO **THE PHYSICS OF . . . nearsightedness.** A person who is **nearsighted (myopic)** can focus on nearby objects but cannot clearly see objects far away. For such a person, the far point of the eye is not at infinity and may even be as close to the eye as three or four meters. When a nearsighted eye tries to focus on a distant object, the eye is fully relaxed, like a normal eye. However, the nearsighted eye has a focal length that is shorter than it should be, so rays from the distant object form a sharp image in front of the retina, as **Figure 26.35*a*** shows, and blurred vision results.

The nearsighted eye can be corrected with glasses or contacts that use *diverging* lenses, as **Figure 26.35*b*** suggests. The rays from the object diverge after leaving the eyeglass lens. Therefore, when they are subsequently refracted toward the principal axis by the eye, a sharp image is formed farther back and falls on the retina. Since the relaxed (but nearsighted) eye can focus on an object at the eye's far point—but not on objects farther away—the diverging lens is designed to transform a very distant object into an image located at the

FIGURE 26.35 (*a*) When a nearsighted person views a distant object, the image is formed in front of the retina. The result is blurred vision. (*b*) With a diverging lens in front of the eye, the image is moved onto the retina and clear vision results. (*c*) The diverging lens is designed to form a virtual image at the far point of the nearsighted eye.

far point. **Figure 26.35c** shows this transformation, and the next example illustrates how to determine the focal length of the diverging lens that accomplishes it.

EXAMPLE 12 | BIO Eyeglasses for the Nearsighted Person

A nearsighted person has a far point located only 521 cm from the eye. Assuming that eyeglasses are to be worn 2 cm in front of the eye, find the focal length needed for the diverging lenses of the glasses so the person can see distant objects.

Reasoning In **Figure 26.35c** the far point is 521 cm away from the eye. Since the glasses are worn 2 cm from the eye, the far point is 519 cm to the left of the diverging lens. The image distance, then, is −519 cm, the negative sign indicating that the image is a virtual image formed to the left of the lens. The object is assumed to be infinitely far from the diverging lens. The thin-lens equation can be used to find the focal length of the eyeglasses. We expect the focal length to be negative, since the lens is a diverging lens.

Problem-Solving Insight Eyeglasses are worn about 2 cm from the eyes. Be sure, if necessary, to take this 2 cm into account when determining the object and image distances (d_o and d_i) that are used in the thin-lens equation.

Solution With $d_i = -519$ cm and $d_o = \infty$, the focal length can be found as follows:

$$\frac{1}{f} = \frac{1}{d_o} + \frac{1}{d_i} = \frac{1}{\infty} + \frac{1}{-519 \text{ cm}} \quad \text{or} \quad \boxed{f = -519 \text{ cm}} \quad \textbf{(26.6)}$$

The value for f is negative, as expected for a diverging lens.

Farsightedness

BIO **THE PHYSICS OF . . . farsightedness.** A **farsighted (hyperopic)** person can usually see distant objects clearly, but cannot focus on those nearby. Whereas the near point of a young and normal eye is located about 25 cm from the eye, the near point of a farsighted eye may be considerably farther away than that, perhaps as far as several hundred centimeters. When a farsighted eye tries to focus on a book held closer than the near point, it accommodates and shortens its focal length as much as it can. However, even at its shortest, the focal length is longer than it should be. Therefore, the light rays from the book would form a sharp image behind the retina if they could do so, as **Figure 26.36a** suggests. In reality, no light passes through the retina, but a blurred image does form on it.

Figure 26.36b shows that farsightedness can be corrected by placing a *converging* lens in front of the eye. The lens refracts the light rays more toward the principal axis before they enter the eye. Consequently, when the rays are refracted even more by the eye, they

FIGURE 26.36 (a) When a farsighted person views an object located between the near point and the eye, a sharp image would be formed behind the retina if light could pass through it. Only a blurred image forms on the retina. (b) With a converging lens in front of the eye, the sharp image is moved onto the retina and clear vision results. (c) The converging lens is designed to form a virtual image at the near point of the farsighted eye.

converge to form an image on the retina. Part *c* of the figure illustrates what the eye sees when it looks through the converging lens. The lens is designed so that the eye perceives the light to be coming from a virtual image located at the near point. Example 13 shows how the focal length of the converging lens is determined to correct for farsightedness.

EXAMPLE 13 | BIO Contact Lenses for the Farsighted Person

A farsighted person has a near point located 210.0 cm from the eyes. Obtain the focal length of the converging lenses in a pair of contacts that can be used to read a book held 25.0 cm from the eyes.

Reasoning A contact lens is placed directly against the eye. Thus, the object distance, which is the distance from the book to the lens, is 25.0 cm. The lens forms an image of the book at the near point of the eye, so the image distance is −210.0 cm. The

minus sign indicates that the image is a virtual image formed to the left of the lens, as in **Figure 26.36c**. The focal length can be obtained from the thin-lens equation.

Solution With $d_o = 25.0$ cm and $d_i = -210.0$ cm, the focal length can be determined from the thin-lens equation as follows:

$$\frac{1}{f} = \frac{1}{d_o} + \frac{1}{d_i} = \frac{1}{25.0 \text{ cm}} + \frac{1}{-210.0 \text{ cm}} = 0.0352 \text{ cm}^{-1} \text{ or } \boxed{f = 28.4 \text{ cm}}$$

BIO **THE PHYSICS OF . . . Correcting color vision deficiency.** The most common cause of color vision deficiency in humans is an inherited condition affecting one or more of the three sets of color-sensing cells, called cones, in the *fovea centralis*, located at the back of the eye. A common problem is the inability to distinguish red and green, a condition that can sometimes be diagnosed with a simple test (see **Figure 26.37**). The three types of cones cover three overlapping wavelength ranges in the visible spectrum. The cones that are most sensitive to the red part of spectrum span a range of about 500 to 675 nm with a peak sensitivity at about 570 nm. Those most sensitive to the green portion of the spectrum sense wavelengths from about 450 to 600 nm with a peak at about 535 nm, and those that best detect blue wavelengths are most sensitive in the 400 to 500 nm range with a peak at 445 nm. These three ranges overlap, but the most significant overlap in sensitivities is between those of the red and greens cones. The ranges can vary and, in some people, the overlap between those of the red and green cones is too drastic, meaning that light that is supposed to be identified as "green" is instead is detected as "red" and vice versa, making those wavelengths indistinguishable.

One way to lessen the overlap is to remove some of the light that is entering the eye, specifically the light at the wavelengths in which the red and green cones most strongly overlap, which is typically at about 565 nm (i.e., in the yellow-orange part of the spectrum). This is done with something called a "notch" filter, which absorbs light only in a certain wavelength interval, and allows all other wavelengths to pass. Since filtering attenuates the light, it is often recommended that glasses utilizing such filters not be used at night. Notch filters can be constructed from optical materials that naturally absorb in a given wavelength range, or from multilayer thin films that exhibit interference effects, a topic that will be discussed in Chapter 27.

Example 14 explains why babies can only see nearby objects.

Ishihara's Test for Colour Deficiency: 38 Plates Edition. 2006-2018 by Colblindor. https://www.color-blindness.com/ ishiharas-test-for-colour-deficiency-38-plates-edition/#prettyPhoto

FIGURE 26.37 Someone with a red-green color vision deficiency might not be able to see the number 29 in the figure, which is called an Ishihara plate. In some cases, optical filters may provide a means to help distinguish these colors.

EXAMPLE 14 | BIO Through a Baby's Eyes—Infant Vision

Healthy newborn babies have the ability to see shapes of objects and colors, but they cannot see objects more than 8–15 inches away (**Figure 26.38**). They naturally prefer rounded objects with contrast (light and dark shades), so it is not surprising they often fixate on the faces of their parents. They also have diminished visual acuity, or the ability of their eyes to resolve objects in fine

detail. This is in large part due to the smaller size of a baby's eye compared to an adult. The cornea-to-retina distance in a baby's eye can be 6–9 mm shorter than in an adult, which results in smaller retinal images. The newborn's pupil also grows with age. Assume that an infant and an adult are both staring at a teddy bear that is located 35 cm from their eyes. If the lens-to-retina distance in the adult and infant is 25 mm and 16 mm, respectively, what is the ratio of the effective focal lengths of their eyes (f_A/f_I)?

Reasoning We can apply the thin-lens equation (Equation 26.6) twice—once for the adult, and once for the infant—to calculate the focal lengths of the eyes and then take their ratio.

Solution Applying Equation 26.6 to the adult, we have:

$$\frac{1}{f_A} = \frac{1}{d_o} + \frac{1}{d_{iA}} = \frac{1}{35\text{ cm}} + \frac{1}{2.5\text{ cm}} \Rightarrow f_A = 2.33\text{ cm.}$$

Repeating this calculation for the infant, we find:

$$\frac{1}{f_I} = \frac{1}{d_o} + \frac{1}{d_{iI}} = \frac{1}{35\text{ cm}} + \frac{1}{1.6\text{ cm}} \Rightarrow f_I = 1.53\text{ cm.}$$

We can now calculate the ratio of the two focal lengths:

$$\left(\frac{f_A}{f_I}\right) = \frac{2.33\text{ cm}}{1.53\text{ cm}} = \boxed{1.52}$$

The greater focal length in the adult eye is due in part to its larger cornea.

FIGURE 26.38 The smaller structures of an infant's eyes result in less visual acuity as compared to an adult with normal vision. However, the infant's vision improves dramatically in just a few months.

PublicDomainPictures/Pixabay

The Refractive Power of a Lens—The Diopter

The extent to which rays of light are refracted by a lens depends on its focal length. However, optometrists who prescribe correctional lenses and opticians who make the lenses do not specify the focal length directly in prescriptions. Instead, they use the concept of **refractive power** to describe the extent to which a lens refracts light:

$$\begin{array}{c}\text{Refractive power}\\ \text{of a lens (in diopters)}\end{array} = \frac{1}{f\text{ (in meters)}} \qquad (26.8)$$

The refractive power is measured in units of *diopters*. One diopter is 1 m^{-1}.

Equation 26.8 shows that a converging lens has a refractive power of 1 diopter if it focuses parallel light rays to a focal point 1 m beyond the lens. If a lens refracts parallel rays even more and converges them to a focal point only 0.25 m beyond the lens, the lens has four times more refractive power, or 4 diopters. Since a converging lens has a positive focal length and a diverging lens has a negative focal length, the refractive power of a converging lens is positive and that of a diverging lens is negative. Thus, the eyeglasses in Example 12 would be described in an optometrist's prescription in the following way: Refractive power = $1/(-5.19\text{ m}) = -0.193$ diopters. The contact lenses in Example 13 would be described in a similar fashion: Refractive power = $1/(0.284\text{ m}) = 3.52$ diopters.

Check Your Understanding

(The answers are given at the end of the book.)

21. Two people who wear glasses are camping. One is nearsighted, and the other is farsighted. Whose glasses may be useful in starting a fire by concentrating the sun's rays into a small region at the focal point of the lens used in the glasses?

22. Suppose that a person with a near point of 26 cm is standing in front of a plane mirror. How close can he stand to the mirror and still see himself in focus?

23. **BIO** To a swimmer under water, objects look blurred. However, goggles that keep the water away from the eyes allow the swimmer to see objects in sharp focus. Why?

24. When glasses use diverging lenses to correct for nearsightedness or converging lenses to correct for farsightedness, the eyes of the person wearing the glasses lie between the lenses and their focal points. When you look at the eyes of this person, they do not appear to have their normal size. Which one of the following describes what you see? **(a)** The converging lenses make the eyes appear smaller, and the diverging lenses make the eyes appear larger. **(b)** The converging lenses make the eyes appear larger, and the diverging lenses make the eyes appear smaller. **(c)** Both types of lenses make the eyes appear larger. **(d)** Both types of lenses make the eyes appear smaller.

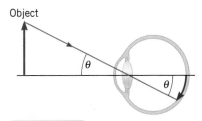

FIGURE 26.39 The angle θ is the angular size of both the image and the object.

26.11 | Angular Magnification and the Magnifying Glass

If you hold a penny at arm's length, the penny looks larger than the moon. The reason is that the penny, being so close, forms a larger image on the retina of the eye than does the more distant moon. The brain interprets the larger image of the penny as arising from a larger object. The size of the image on the retina determines how large an object appears to be. However, the size of the image on the retina is difficult to measure. Alternatively, the angle θ subtended by the image can be used as an indication of the image size. **Figure 26.39** shows this alternative, which has the advantage that θ is also the angle subtended by the object and, hence, can be measured more easily. The angle θ is called the **angular size** of both the image and the object. The larger the angular size, the larger the image on the retina, and the larger the object appears to be.

According to Equation 8.1, the angle θ (measured in radians) is the length of the circular arc that is subtended by the angle divided by the radius of the arc, as **Figure 26.40a** indicates. Part b of the drawing shows the situation for an object of height h_o viewed at a distance d_o from the eye. When θ is small, h_o is approximately equal to the arc length and d_o is nearly equal to the radius, so that

$$\theta \text{ (in radians)} = \text{Angular size} \approx \frac{h_o}{d_o}$$

This approximation is good to within one percent for angles of 9° or smaller. In the next example the angular size of a penny is compared with that of the moon.

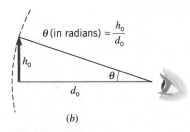

$$\theta \text{ (in radians)} = \frac{\text{Arc length}}{\text{Radius}}$$

(a)

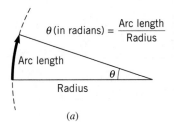

$$\theta \text{ (in radians)} \approx \frac{h_o}{d_o}$$

(b)

FIGURE 26.40 (a) The angle θ, measured in radians, is the arc length divided by the radius. (b) For small angles (less than 9°), θ in radians is approximately equal to h_o/d_o, where h_o and d_o are the object height and distance.

EXAMPLE 15 | A Penny and the Moon

Compare the angular size of a penny (diameter = h_o = 1.9 cm) held at arm's length (d_o = 71 cm) with the angular size of the moon (diameter = h_o = 3.5 × 10⁶ m, and d_o = 3.9 × 10⁸ m).

Reasoning The angular size θ of an object is given approximately by its height h_o divided by its distance d_o from the eye, $\theta \approx h_o/d_o$, provided that the angle involved is less than roughly 9°; this approximation applies here. The "heights" of the penny and the moon are their diameters.

Solution The angular sizes of the penny and moon are

Penny $$\theta \approx \frac{h_o}{d_o} = \frac{1.9 \text{ cm}}{71 \text{ cm}} = \boxed{0.027 \text{ rad } (1.5°)}$$

Moon $$\theta \approx \frac{h_o}{d_o} = \frac{3.5 \times 10^6 \text{ m}}{3.9 \times 10^8 \text{ m}} = \boxed{0.0090 \text{ rad } (0.52°)}$$

The penny thus appears to be about three times as large as the moon.

An optical instrument, such as a magnifying glass, allows us to view small or distant objects because it produces a larger image on the retina than would be possible otherwise. In other words, an optical instrument magnifies the angular size of the object. The **angular magnification** (or **magnifying power**) M is the angular size θ' of the final image produced by the instrument divided by a reference angular size θ. The reference angular size is the angular size of the object when seen without the instrument.

Angular magnification

$$M = \frac{\text{Angular size of final image produced by optical instrument}}{\text{Reference angular size of object seen without optical instrument}} = \frac{\theta'}{\theta} \qquad (26.9)$$

A magnifying glass is the simplest device that provides angular magnification. In this case, the reference angular size θ is chosen to be the angular size of the object when placed at the near point of the eye and seen without the magnifying glass. Since an object cannot be brought closer than the near point and still produce a sharp image on the retina, θ represents the largest angular size obtainable without the magnifying glass. **Figure 26.41a** indicates that the reference angular size is $\theta \approx h_0/N$, where N is the distance from the eye to the near point. To compute θ', recall from Section 26.7 and **Figure 26.28** that a magnifying glass is usually a single converging lens, with the object located between the focal point of the lens and the lens. In this situation, **Figure 26.41b** indicates that the lens produces a virtual image that is enlarged and upright with respect to the object. Assuming that the eye is next to the magnifying glass, the angular size θ' seen by the eye is $\theta' \approx h_0/d_0$, where d_0 is the object distance. The angular magnification is

$$M = \frac{\theta'}{\theta} \approx \frac{h_0/d_0}{h_0/N} = \frac{N}{d_0}$$

According to the thin-lens equation, d_0 is related to the image distance d_i and the focal length f of the lens by

$$\frac{1}{d_0} = \frac{1}{f} - \frac{1}{d_i}$$

Substituting this expression for $1/d_0$ into the previous expression for M leads to the following result:

Angular magnification of a magnifying glass

$$M = \frac{\theta'}{\theta} \approx \left(\frac{1}{f} - \frac{1}{d_i}\right)N \qquad (26.10)$$

FIGURE 26.41 (a) Without a magnifying glass, the largest angular size θ occurs when the object is placed at the near point, a distance N from the eye. (b) A magnifying glass produces an enlarged, virtual image of an object placed between the focal point F of the lens and the lens. The angular size of both the image and the object is θ'.

Two special cases of this result are of interest, depending on whether the image is located as close to the eye as possible or as far away as possible. To be seen clearly, the closest the image can be relative to the eye is at the near point, or $d_i = -N$. The minus sign indicates that the image lies to the left of the lens and is virtual. In this event, Equation 26.10 becomes $M \approx (N/f) + 1$. The farthest the image can be from the eye is at infinity ($d_i = -\infty$); this occurs when the object is placed at the focal point of the lens. When the image is at infinity, Equation 26.10 simplifies to $M \approx N/f$. Clearly, the angular magnification is greater when the image is at the near point of the eye rather than at infinity. In either case, however, the greatest magnification is achieved by using a magnifying glass with the shortest possible focal length. Example 16 illustrates how to determine the angular magnification of a magnifying glass that is used in these two ways.

EXAMPLE 16 | Examining a Diamond with a Magnifying Glass

A jeweler, whose near point is 40.0 cm from his eye and whose far point is at infinity, is using a small magnifying glass (called a loupe) to examine a diamond. The lens of the magnifying glass has a focal length of 5.00 cm, and the image of the gem is −185 cm from the lens. The image distance is negative because the image is virtual and is formed on the same side of the lens as the object.

(a) Determine the angular magnification of the magnifying glass.
(b) Where should the image be located so the jeweler's eye is fully relaxed and has the least strain? What is the angular magnification under this "least strain" condition?

Reasoning   The angular magnification of the magnifying glass can be determined from Equation 26.10. In part (a) the image distance is −185 cm. In part (b) the ciliary muscle of the jeweler's eye is fully relaxed, so the image must be located infinitely far from the eye, at its far point, as Section 26.10 discusses.

Solution   (a) With $f = 5.00$ cm, $d_i = -185$ cm, and $N = 40.0$ cm, the angular magnification is

$$M = \left(\frac{1}{f} - \frac{1}{d_i}\right) N = \left(\frac{1}{5.00 \text{ cm}} - \frac{1}{-185 \text{ cm}}\right)(40.0 \text{ cm}) = \boxed{8.22}$$

(b) With $f = 5.00$ cm, $d_i = -\infty$, and $N = 40.0$ cm, the angular magnification is

$$M = \left(\frac{1}{f} - \frac{1}{d_i}\right) N = \left(\frac{1}{5.00 \text{ cm}} - \frac{1}{-\infty}\right)(40.0 \text{ cm}) = \boxed{8.00}$$

Jewelers often prefer to minimize eyestrain when viewing objects, even though it means a slight reduction in angular magnification.

Check Your Understanding

(The answers are given at the end of the book.)

25. A bird-watcher sees the following three raptors in the air at the distances indicated: a kestrel (wing span = 0.58 m at a distance of 21 m), a bald eagle (wing span = 2.29 m at a distance of 95 m), and a red-tailed hawk (wing span = 1.27 m at a distance of 41 m). Rank the raptors in descending order (largest first) according to the angular size seen by the bird-watcher.

26. Who benefits more from using a magnifying glass, a person whose near point is located at a distance away from the eyes of **(a)** 75 cm or **(b)** 25 cm?

27. A person who has a near point of 25.0 cm is looking with unaided eyes at an object that is located at the near point. The object has an angular size of 0.012 rad. Then, holding a magnifying glass ($f = 10.0$ cm) next to her eye, she views the image of this object, the image being located at her near point. What is the angular size of the image?

26.12 | The Compound Microscope

THE PHYSICS OF . . . the compound microscope.   To increase the angular magnification beyond that possible with a magnifying glass, an additional converging lens can be included to "premagnify" the object before the magnifying glass comes into play. The result is an optical instrument known as the **compound microscope** (**Figure 26.42**). As discussed in Section 26.9, the magnifying glass is called the eyepiece, and the additional lens is called the objective.

The angular magnification of the compound microscope is $M = \theta'/\theta$ (Equation 26.9), where θ' is the angular size of the final image and θ is the reference angular size. As with the magnifying glass in **Figure 26.41**, the reference angular size is determined by the height h_o of the object when the object is located at the near point of the unaided eye: $\theta \approx h_o/N$, where N is the distance between the eye and the near point. Assuming that the object is placed just outside the focal point F_o of the objective (see **Figure 26.32a**) and that the final image is very far from the eyepiece (i.e., near infinity; see **Figure 26.32c**), it can be shown that

Angular magnification of a compound microscope $\qquad M \approx -\dfrac{(L - f_e)N}{f_o f_e} \qquad (L > f_o + f_e) \qquad$ **(26.11)**

Eyepiece

Objective

Object

FIGURE 26.42 A compound microscope.

In Equation 26.11, f_o and f_e are, respectively, the focal lengths of the objective and the eyepiece. The angular magnification is greatest when f_o and f_e are as small as possible (since they are in the denominator in Equation 26.11) and when the distance L between the lenses is as large as possible. Furthermore, L must be greater than the sum of f_o and f_e for this equation to be valid. Example 17 deals with the angular magnification of a compound microscope.

EXAMPLE 17 | The Angular Magnification of a Compound Microscope

The focal length of the objective of a compound microscope is $f_o = 0.40$ cm, and the focal length of the eyepiece is $f_e = 3.0$ cm. The two lenses are separated by a distance of $L = 20.0$ cm. A person with a near point distance of $N = 25$ cm is using the microscope. **(a)** Determine the angular magnification of the microscope. **(b)** Compare the answer in part (a) with the largest angular magnification obtainable by using the eyepiece alone as a magnifying glass.

Reasoning The angular magnification of the compound microscope can be obtained directly from Equation 26.11, since all the variables are known. When the eyepiece is used alone as a magnifying glass, as in **Figure 26.41b**, the largest angular magnification occurs when the image seen through the eyepiece is as close as possible to the eye. The image in this case is at the near point, and according to Equation 26.10, the angular magnification is $M \approx (N/f_e) + 1$.

Solution **(a)** The angular magnification of the compound microscope is

$$M \approx -\frac{(L - f_e)N}{f_o f_e} = -\frac{(20.0 \text{ cm} - 3.0 \text{ cm})(25 \text{ cm})}{(0.40 \text{ cm})(3.0 \text{ cm})} = \boxed{-350}$$

The minus sign indicates that the final image is inverted relative to the initial object.

(b) The maximum angular magnification of the eyepiece by itself is

$$M \approx \frac{N}{f_e} + 1 = \frac{25 \text{ cm}}{3.0 \text{ cm}} + 1 = \boxed{9.3}$$

The effect of the objective is to increase the angular magnification of the compound microscope by a factor of $350/9.3 = 38$ compared to the angular magnification of a magnifying glass.

26.13 | The Telescope

THE PHYSICS OF ... the telescope. A telescope is an instrument for magnifying distant objects, such as stars and planets. Like a microscope, a telescope consists of an objective and an eyepiece (also called the ocular). When the objective is a lens, as is the case in this section, the telescope is referred to as a *refracting* telescope, since lenses utilize the refraction of light.*

Usually the object being viewed is far away, so the light rays entering the telescope are nearly parallel, and the "first image" is formed just beyond the focal point F_o of the objective, as **Figure 26.43a** illustrates. The first image is real and inverted. Unlike the first image in the compound microscope, however, this image is *smaller* than the object. If, as in part *b* of the drawing, the telescope is constructed so the first image lies just inside the focal point F_e of the eyepiece, the eyepiece acts like a magnifying glass. It forms a final image that is greatly enlarged, virtual, and located near infinity. This final image can then be viewed with a fully relaxed eye.

The angular magnification M of a telescope, like that of a magnifying glass or a microscope, is the angular size θ' subtended by the final image of the telescope divided by the reference angular size θ of the object. For an astronomical object, such as a planet, it is convenient to use as a reference the angular size of the object seen in the sky with the unaided eye. Since the object is far away, the angular size seen by the unaided eye is nearly the same as the angle θ subtended at the objective of the telescope in **Figure 26.43a**. Moreover, θ is also the angle subtended by the first image, so

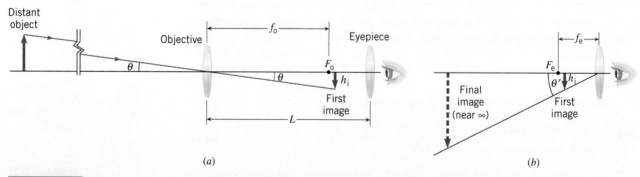

FIGURE 26.43 (a) An astronomical telescope is used to view distant objects. (Note the "break" in the principal axis, between the object and the objective.) The objective produces a real, inverted first image. (b) The eyepiece magnifies the first image to produce the final image near infinity.

*Another type of telescope utilizes a mirror instead of a lens for the objective and is called a *reflecting* telescope.

$\theta \approx -h_i/f_o$, where h_i is the height of the first image and f_o is the focal length of the objective. A minus sign has been inserted into this equation because the first image is inverted relative to the object and the image height h_i is a negative number. The insertion of the minus sign ensures that the term $-h_i/f_o$, and hence θ, is a positive quantity. To obtain an expression for θ', we refer to **Figure 26.43b** and note that the first image is located very near the focal point F_e of the eyepiece, which has a focal length f_e. Therefore, $\theta' \approx h_i/f_e$. The angular magnification of the telescope is approximately

Angular magnification of an astronomical telescope

$$M = \frac{\theta'}{\theta} \approx \frac{h_i/f_e}{-h_i/f_o} \approx -\frac{f_o}{f_e} \qquad (26.12)$$

The angular magnification is determined by the ratio of the focal length of the objective to the focal length of the eyepiece. For large angular magnifications, the objective should have a long focal length and the eyepiece a short one. Some of the design features of a telescope are the topic of the next example.

EXAMPLE 18 | The Angular Magnification of an Astronomical Telescope

A telescope similar to that in **Figure 26.44** has the following specifications: $f_o = 985$ mm and $f_e = 5.00$ mm. From these data, find **(a)** the angular magnification and **(b)** the approximate length of this telescope.

Reasoning The angular magnification of the telescope follows directly from Equation 26.12, since the focal lengths of the objective and eyepiece are known. We can find the length of the telescope by noting that it is approximately equal to the distance L between the objective and eyepiece. **Figure 26.43a** shows that the first image is located just beyond the focal point F_o of the objective. **Figure 26.43b** shows that the first image is also just to the right of the focal point F_e of the eyepiece. These two focal points are, therefore, very close together, so the distance L is approximately the sum of the two focal lengths: $L \approx f_o + f_e$.

Solution **(a)** The angular magnification is approximately

$$M \approx -\frac{f_o}{f_e} = -\frac{985 \text{ mm}}{5.00 \text{ mm}} = \boxed{-197} \qquad (26.12)$$

(b) The approximate length of the telescope is

$$L \approx f_o + f_e = 985 \text{ mm} + 5.00 \text{ mm} = \boxed{990 \text{ mm}}$$

FIGURE 26.44 An astronomical telescope typically includes a viewfinder, which is a separate small telescope with low magnification and serves as an aid in locating the object. Once the object has been found, the viewer uses the eyepiece to obtain the full magnification of the telescope.

Dorling Kindersley/Getty Images

Check Your Understanding

(The answers are given at the end of the book.)

28. In the construction of a telescope, one of two lenses is to be used as the objective and one as the eyepiece. The focal lengths of the lenses are **(a)** 3 cm and **(b)** 45 cm. Which lens should be used as the objective?

29. Two refracting telescopes have identical eyepieces, although one telescope is twice as long as the other. Which telescope has the greater angular magnification?

30. A well-designed optical instrument is composed of two converging lenses separated by 14 cm. The focal lengths of the lenses are 0.60 and 4.5 cm. Is the instrument a microscope or a telescope?

31. It is often thought that virtual images are somehow less important than real images. To show that this is not true, identify which of the following instruments normally produce final images that are virtual: **(a)** a projector, **(b)** a camera, **(c)** a magnifying glass, **(d)** eyeglasses, **(e)** a compound microscope, and **(f)** an astronomical telescope.

26.14 # Lens Aberrations

Rather than forming a sharp image, a single lens typically forms an image that is slightly out of focus. This lack of sharpness arises because the rays originating from a single point on the object are not focused to a single point on the image. As a result, each point on the image becomes a small blur. The lack of point-to-point correspondence between object and image is called an *aberration*.

One common type of aberration is **spherical aberration**, and it occurs with converging and diverging lenses made with spherical surfaces. **Figure 26.45a** shows how spherical aberration arises with a converging lens. Ideally, all rays traveling parallel to the principal axis are refracted so they cross the axis at the same point after passing through the lens. However, rays far from the principal axis are refracted more by the lens than are those closer in. Consequently, the outer rays cross the axis closer to the lens than do the inner rays, so a lens with spherical aberration does not have a unique focal point. Instead, as the drawing suggests, there is a location along the principal axis where the light converges to the smallest cross-sectional area. This area is circular and is known as the **circle of least confusion**. The circle of least confusion is where the most satisfactory image can be formed by the lens.

Spherical aberration can be reduced substantially by using a variable-aperture diaphragm to allow only those rays close to the principal axis to pass through the lens. **Figure 26.45b** indicates that a reasonably sharp focal point can be achieved by this method, although less light now passes through the lens. Lenses with parabolic surfaces are also used to reduce this type of aberration, but they are difficult and expensive to make.

Chromatic aberration also causes blurred images. It arises because the index of refraction of the material from which the lens is made varies with wavelength. Section 26.5 discusses how this variation leads to the phenomenon of dispersion, in which different colors refract by different amounts. **Figure 26.46a** shows sunlight incident on a converging lens, in which the light spreads into its color spectrum because of dispersion. For clarity, however, the picture shows only the colors at the opposite ends of the visible spectrum—red and violet. Violet is refracted more than red, so the violet ray crosses the principal axis closer to the lens than does the red ray. Thus, the focal length of the lens is shorter for violet than for red, with intermediate values of the focal length corresponding to the colors in between. As a result of chromatic aberration, an undesirable color fringe surrounds the image.

Chromatic aberration can be greatly reduced by using a compound lens, such as the combination of a converging lens and a diverging lens shown in **Figure 26.46b**. Each lens is made from a different type of glass. With this lens combination the red and violet rays almost come to a common focus and, thus, chromatic aberration is reduced. A lens combination designed to reduce chromatic aberration is called an *achromatic lens* (from the Greek "achromatos," meaning "without color"). All high-quality cameras use achromatic lenses.

FIGURE 26.45 (*a*) In a converging lens, spherical aberration prevents light rays parallel to the principal axis from converging to a common point. (*b*) Spherical aberration can be reduced by allowing only rays near the principal axis to pass through the lens. The refracted rays now converge more nearly to a single focal point *F*.

FIGURE 26.46 (*a*) Chromatic aberration arises when different colors are focused at different points along the principal axis: F_V = focal point for violet light, F_R = focal point for red light. (*b*) A converging and a diverging lens in tandem can be designed to bring different colors more nearly to the same focal point *F*.

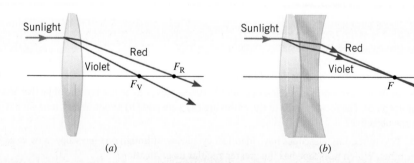

Check Your Understanding

(*The answer is given at the end of the book.*)

32. Why does chromatic aberration occur in lenses but not in mirrors?

Concept Summary

26.1 The Index of Refraction The change in speed as a ray of light goes from one material to another causes the ray to deviate from its incident direction. This change in direction is called refraction. The index of refraction n of a material is the ratio of the speed c of light in a vacuum to the speed v of light in the material, as shown in Equation 26.1. The values for n are greater than unity, because the speed of light in a material medium is less than it is in a vacuum.

$$n = \frac{c}{v} \qquad (26.1)$$

26.2 Snell's Law and the Refraction of Light The refraction that occurs at the interface between two materials obeys Snell's law of refraction. This law states that (1) the refracted ray, the incident ray, and the normal to the interface all lie in the same plane, and (2) the angle of refraction θ_2 is related to the angle of incidence θ_1 according to Equation 26.2, where n_1 and n_2 are the indices of refraction of the incident and refracting media, respectively. The angles are measured relative to the normal.

$$n_1 \sin \theta_1 = n_2 \sin \theta_2 \qquad (26.2)$$

Because of refraction, a submerged object has an apparent depth that is different from its actual depth. If the observer is directly above (or below) the object, the apparent depth (or height) d' is related to the actual depth (or height) d according to Equation 26.3, where n_1 and n_2 are the refractive indices of the materials (the media) in which the object and the observer, respectively, are located.

$$d' = d\left(\frac{n_2}{n_1}\right) \qquad (26.3)$$

26.3 Total Internal Reflection When light passes from a material with a larger refractive index n_1 into a material with a smaller refractive index n_2, the refracted ray is bent away from the normal. If the incident ray is at the critical angle θ_c, the angle of refraction is 90°. The critical angle is determined from Snell's law and is given by Equation 26.4. When the angle of incidence exceeds the critical angle, all the incident light is reflected back into the material from which it came, a phenomenon known as total internal reflection.

$$\sin \theta_c = \frac{n_2}{n_1} \quad (n_1 > n_2) \qquad (26.4)$$

26.4 Polarization and the Reflection and Refraction of Light When light is incident on a nonmetallic surface at the Brewster angle θ_B, the reflected light is completely polarized parallel to the surface. The Brewster angle is given by Equation 26.5, where n_1 and n_2 are the refractive indices of the incident and refracting media, respectively. When light is incident at the Brewster angle, the reflected and refracted rays are perpendicular to each other.

$$\tan \theta_B = \frac{n_2}{n_1} \qquad (26.5)$$

26.5 The Dispersion of Light: Prisms and Rainbows A glass prism can spread a beam of sunlight into a spectrum of colors because the index of refraction of the glass depends on the wavelength of the light. Thus, a prism bends the refracted rays corresponding to different colors by different amounts. The spreading of light into its color components is known as dispersion. The dispersion of light by water droplets in the air leads to the formation of rainbows. A prism will not bend a light ray at all, neither up nor down, if the surrounding fluid

has the same refractive index as the material from which the prism is made, a condition known as index matching.

26.6 Lenses 26.7 The Formation of Images by Lenses Converging lenses and diverging lenses depend on the phenomenon of refraction in forming an image. With a converging lens, paraxial rays that are parallel to the principal axis are focused to a point on the axis by the lens. This point is called the focal point of the lens, and its distance from the lens is the focal length f. Paraxial light rays that are parallel to the principal axis of a diverging lens appear to originate from its focal point after passing through the lens. The distance of this point from the lens is the focal length f. The image produced by a converging or a diverging lens can be located via a technique known as ray tracing, which utilizes the three rays outlined in the Reasoning Strategy given in Section 26.7.

The nature of the image formed by a converging lens depends on where the object is situated relative to the lens. When the object is located at a distance from the lens that is greater than twice the focal length, the image is real, inverted, and smaller than the object. When the object is located at a distance from the lens that is between the focal length and twice the focal length, the image is real, inverted, and larger than the object. When the object is located between the focal point and the lens, the image is virtual, upright, and larger than the object.

Regardless of the position of a real object, a diverging lens always produces an image that is virtual, upright, and smaller than the object.

26.8 The Thin-Lens Equation and the Magnification Equation The thin-lens equation (Equation 26.6) can be used with either converging or diverging lenses that are thin, and it relates the object distance d_o, the image distance d_i, and the focal length f of the lens.

$$\frac{1}{d_o} + \frac{1}{d_i} = \frac{1}{f} \qquad (26.6)$$

The magnification m of a lens is the ratio of the image height h_i to the object height h_o and is also related to d_o and d_i by the magnification equation (Equation 26.7).

$$m = \frac{h_i}{h_o} = -\frac{d_i}{d_o} \qquad (26.7)$$

The algebraic sign conventions for the variables appearing in the thin-lens and magnification equations are summarized in the Reasoning Strategy given in Section 26.8.

26.9 Lenses in Combination When two or more lenses are used in combination, the image produced by one lens serves as the object for the next lens.

26.10 The Human Eye In the human eye, a real, inverted image is formed on a light-sensitive surface, called the retina. Accommodation is the process by which the focal length of the eye is automatically adjusted, so that objects at different distances produce sharp images on the retina. The near point of the eye is the point nearest the eye at which an object can be placed and still have a sharp image produced on the retina. The far point of the eye is the location of the farthest object on which the fully relaxed eye can focus. For a young and normal eye, the near point is located 25 cm from the eye, and the far point is located at infinity.

A nearsighted (myopic) eye is one that can focus on nearby objects, but not on distant objects. Nearsightedness can be corrected with eyeglasses or contacts made from diverging lenses. A farsighted (hyperopic) eye can see distant objects clearly, but not objects close up. Farsightedness can be corrected with converging lenses.

The refractive power of a lens is measured in diopters and is given by Equation 26.8, where f is the focal length of the lens and must be expressed in meters. A converging lens has a positive refractive power, and a diverging lens has a negative refractive power.

$$\text{Refractive power (in diopters)} = \frac{1}{f \text{ (in meters)}} \qquad (26.8)$$

26.11 Angular Magnification and the Magnifying Glass The angular size of an object is the angle that it subtends at the eye of the viewer. For small angles, the angular size θ in radians is given approximately by Equation 1, where h_o is the height of the object and d_o is the object distance. The angular magnification M of an optical instrument is the angular size θ' of the final image produced by the instrument divided by the reference angular size θ of the object, which is that seen without the instrument (see Equation 26.9).

$$\theta \text{ (in radians)} \approx \frac{h_o}{d_o} \qquad (1)$$

$$M = \frac{\theta'}{\theta} \qquad (26.9)$$

A magnifying glass is usually a single converging lens that forms an enlarged, upright, and virtual image of an object placed at the focal point of the lens or between the focal point and the lens. For a magnifying glass held close to the eye, the angular magnification M is given approximately by Equation 26.10, where f is the focal length of the lens, d_i is the image distance, and N is the distance of the viewer's near point from the eye.

$$M \approx \left(\frac{1}{f} - \frac{1}{d_i}\right)N \qquad (26.10)$$

26.12 The Compound Microscope A compound microscope usually consists of two lenses, an objective and an eyepiece. The final image is enlarged, inverted, and virtual. The angular magnification M of such a microscope is given approximately by Equation 26.11, where f_o and f_e are, respectively, the focal lengths of the objective and eyepiece, L is the distance between the two lenses, and N is the distance of the viewer's near point from his or her eye.

$$M \approx -\frac{(L - f_e)N}{f_o f_e} \quad (L > f_o + f_e) \qquad (26.11)$$

26.13 The Telescope An astronomical telescope magnifies distant objects with the aid of an objective and an eyepiece, and it produces a final image that is enlarged, inverted, and virtual. The angular magnification M of a telescope is given approximately by Equation 26.12, where f_o and f_e are, respectively, the focal lengths of the objective and the eyepiece.

$$M \approx -\frac{f_o}{f_e} \qquad (26.12)$$

26.14 Lens Aberrations Lens aberrations limit the formation of perfectly focused or sharp images by optical instruments. Spherical aberration occurs because rays that pass through the outer edge of a lens with spherical surfaces are not focused at the same point as rays that pass through near the center of the lens. Chromatic aberration arises because a lens focuses different colors at slightly different points.

Focus on Concepts

Online

Additional questions are available for assignment in WileyPLUS.

Section 26.2 Snell's Law and the Refraction of Light

1. The drawings show two examples in which a ray of light is refracted at the interface between two liquids. In each example the incident ray is in liquid A and strikes the interface at the same angle of incidence. In one case the ray is refracted into liquid B, and in the other it is refracted into liquid C. The dashed lines denote the normals to the interfaces. Rank the indices of refraction of the three liquids in descending order (largest first). (a) n_A, n_B, n_C (b) n_A, n_C, n_B (c) n_C, n_A, n_B (d) n_B, n_A, n_C (e) n_C, n_B, n_A

QUESTION 1

2. A coin is resting on the bottom of an empty container. The container is then filled to the brim three times, each time with a different liquid. An observer (in air) is directly above the coin and looks down at it. With liquid A in the container, the apparent depth of the coin is 7 cm, with liquid B it is 6 cm, and with liquid C it is 5 cm. Rank the indices of refraction of the liquids in descending order (largest first). (a) n_A, n_B, n_C (b) n_A, n_C, n_B (c) n_C, n_A, n_B (d) n_C, n_B, n_A (e) n_B, n_A, n_C

Section 26.3 Total Internal Reflection

3. The refractive index of material A is greater than the refractive index of material B. A ray of light is incident on the interface between these two materials in a number of ways, as the drawings illustrate.

The dashed lines denote the normals to the interfaces. Which one of the drawings shows a situation that is *not* possible? (a) Drawing 1 (b) Drawing 2 (c) Drawing 3 (d) Drawing 4

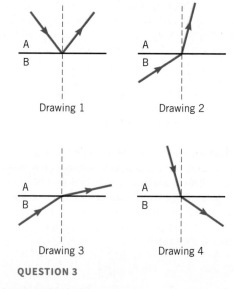

QUESTION 3

4. The drawing shows a rectangular block of glass ($n = 1.52$) surrounded by air. A ray of light starts out within the glass and travels toward point A, where some or all of it is reflected toward point B. At which points does some of the light escape the glass? (a) Only at point A (b) Only at point B (c) At both points A and B (d) At neither point A nor point B

QUESTION 4

Section 26.4 Polarization and the Reflection and Refraction of Light

5. A diamond ($n = 2.42$) is lying on a table. At what angle of incidence θ is the light that is reflected from one of the facets of the diamond completely polarized?

Section 26.5 The Dispersion of Light: Prisms and Rainbows

6. The indices of refraction for red, green, and violet light in glass are $n_{red} = 1.520$, $n_{green} = 1.526$, and $n_{violet} = 1.538$. When a ray of light passes through a transparent slab of glass, the emergent ray is parallel to the incident ray, but can be displaced relative to it. For light passing through a glass slab that is surrounded by air, which color is displaced the most? (a) All colors are displaced equally. (b) Red (c) Green (d) Violet

Section 26.7 The Formation of Images by Lenses

7. An object is situated to the left of a lens. A ray of light from the object is close to and parallel to the principal axis of the lens. The ray passes through the lens. Which one of the following statements is true? (a) The ray crosses the principal axis at a distance from the lens equal to twice the focal length, no matter whether the lens is converging or diverging. (b) The ray passes through the lens without changing direction, no matter whether the lens is converging or diverging. (c) The ray passes through a focal point of the lens, no matter whether the lens is converging or diverging. (d) The ray passes through a focal point of the lens only if the lens is a diverging lens. (e) The ray passes through a focal point of the lens only if the lens is a converging lens.

8. What type of single lens produces a virtual image that is inverted with respect to the object? (a) Both a converging and a diverging lens can produce such an image. (b) Neither a converging nor a diverging lens produces such an image. (c) A converging lens (d) A diverging lens

Section 26.9 Lenses in Combination

9. Two converging lenses have the same focal length of 5.00 cm. They have a common principal axis and are separated by 21.0 cm. An object is located 10.0 cm to the left of the left-hand lens. What is the image distance (relative to the lens on the right) of the final image produced by this two-lens system?

Section 26.10 The Human Eye

10. Here are a number of statements concerning the refractive power of lenses.

A. A positive refractive power means that a lens always creates an image that is larger than the object.
B. Two lenses with the same refractive power have the same focal lengths.
C. A lens with a positive refractive power is a converging lens, whereas a lens with a negative refractive power is a diverging lens.
D. Two lenses with different refractive powers can have the same focal length.
E. The fact that lens A has twice the refractive power of lens B means that the focal length of lens A is twice that of lens B.

Which of these statements are false? (a) A, B, C (b) C, D, E (c) A, D, E (d) B, C, E (e) B, C, D

Section 26.11 Angular Magnification and the Magnifying Glass

11. The table lists the angular sizes in radians and distances from the eye for three objects, A, B, and C. In each case the angular size is small.

Object	Angular Size (in Radians)	Distance of Object from Eye
A	θ	d_o
B	2θ	$2d_o$
C	θ	$2d_o$

Rank the heights of these objects in descending order (largest first). (a) B, C, A (b) B, A, C (c) A, B, C (d) A, C, B (e) C, A, B

Section 26.13 The Telescope

12. An astronomical telescope has an angular magnification of -125 when used properly. What would the angular magnification M be if the objective and the eyepiece were interchanged?

Section 26.14 Lens Aberrations

13. Which one of the five choices below best completes the following statement? The fact that the refractive index depends on the wavelength of light is the cause of _____. (a) dispersion (b) chromatic aberration (c) spherical aberration (d) dispersion and chromatic aberration (e) spherical aberration and chromatic aberration

Problems

Online

Additional questions are available for assignment in WileyPLUS.
Note: Unless specified otherwise, use the values given in Table 26.1 for the refractive indices.

SSM Student Solutions Manual
MMH Problem-solving help
GO Guided Online Tutorial
V-HINT Video Hints
CHALK Chalkboard Videos
BIO Biomedical application

E Easy
M Medium
H Hard
WS Worksheet
T Team Problem

Section 26.1 The Index of Refraction

1. **E** **SSM** A plate glass window ($n = 1.5$) has a thickness of 4.0×10^{-3} m. How long does it take light to pass perpendicularly through the plate?

2. **E** In an ultra-low-temperature experiment, a collection of sodium atoms enter a special state called a *Bose-Einstein condensate* in which the index of refraction is 1.57×10^7. What is the speed of light in this condensate?

3. **E** The refractive indices of materials A and B have a ratio of $n_A/n_B = 1.33$. The speed of light in material A is 1.25×10^8 m/s. What is the speed of light in material B?

4. **E** **GO** The frequency of a light wave is the same when the light travels in ethyl alcohol or in carbon disulfide. Find the ratio of the wavelength of the light in ethyl alcohol to that in carbon disulfide.

5. **E** **SSM** Light travels at a speed of 2.201×10^8 m/s in a certain substance. What substance from **Table 26.1** could this be? For the speed of light in a vacuum use 2.998×10^8 m/s; show your calculations.

6. **E** Light has a wavelength of 340.0 nm and a frequency of 5.403×10^{14} Hz when traveling through a certain substance. What substance from **Table 26.1** could this be? Show your calculations.

7. **M** **MMH** **CHALK** In a certain time, light travels 6.20 km in a vacuum. During the same time, light travels only 3.40 km in a liquid. What is the refractive index of the liquid?

8. **M** **V-HINT** A flat sheet of ice has a thickness of 2.0 cm. It is on top of a flat sheet of crystalline quartz that has a thickness of 1.1 cm. Light strikes the ice perpendicularly and travels through it and then through the quartz. In the time it takes the light to travel through the two sheets, how far (in centimeters) would it have traveled in a vacuum?

Section 26.2 Snell's Law and the Refraction of Light

9. **E** **GO** The drawing shows four different situations in which a light ray is traveling from one medium into another. In some of the cases, the refraction is not shown correctly. For cases (a), (b), and (c), the angle of incidence is 55°; for case (d), the angle of incidence is 0°. Determine the angle of refraction in each case. If the drawing shows the refraction incorrectly, explain why it is incorrect.

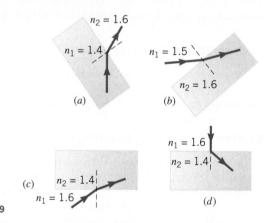

PROBLEM 9

10. **E** A layer of oil ($n = 1.45$) floats on an unknown liquid. A ray of light originates in the oil and passes into the unknown liquid. The angle of incidence is 64.0°, and the angle of refraction is 53.0°. What is the index of refraction of the unknown liquid?

11. **E** **SSM** A ray of light impinges from air onto a block of ice ($n = 1.309$) at a 60.0° angle of incidence. Assuming that this angle remains the same, find the difference $\theta_{2,\text{ice}} - \theta_{2,\text{water}}$ in the angles of refraction when the ice turns to water ($n = 1.333$).

12. **E** A narrow beam of light from a laser travels through air ($n = 1.00$) and strikes point A on the surface of the water ($n = 1.33$) in a lake. The angle of incidence is 55°. The depth of the lake is 3.0 m. On the flat lake-bottom is point B, directly below point A. (a) If refraction did not occur, how far away from point B would the laser beam strike the lake-bottom? (b) Considering refraction, how far away from point B would the laser beam strike the lake-bottom?

13. **E** **SSM** The drawing shows a coin resting on the bottom of a beaker filled with an unknown liquid. A ray of light from the coin travels to the surface of the liquid and is refracted as it enters into

the air. A person sees the ray as it skims just above the surface of the liquid. How fast is the light traveling in the liquid?

PROBLEM 13

14. **E** Amber ($n = 1.546$) is a transparent brown-yellow fossil resin. An insect, trapped and preserved within the amber, appears to be 2.5 cm beneath the surface when viewed directly from above. How far below the surface is the insect actually located?

15. **E** **SSM** A beam of light is traveling in air and strikes a material. The angles of incidence and refraction are 63.0° and 47.0°, respectively. Obtain the speed of light in the material.

16. **E** **GO** The drawing shows a ray of light traveling through three materials whose surfaces are parallel to each other. The refracted rays (but not the reflected rays) are shown as the light passes through each material. A ray of light strikes the a–b interface at a 50.0° angle of incidence. The index of refraction of material a is $n_a = 1.20$. The angles of refraction in materials b and c are, respectively, 45.0° and 56.7°. Find the indices of refraction in these two media.

PROBLEM 16

17. **E** **GO** Light in a vacuum is incident on a transparent glass slab. The angle of incidence is 35.0°. The slab is then immersed in a pool of liquid. When the angle of incidence for the light striking the slab is 20.3°, the angle of refraction for the light entering the slab is the same as when the slab was in a vacuum. What is the index of refraction of the liquid?

18. **M** **V-HINT** **CHALK** A stone held just beneath the surface of a swimming pool is released and sinks to the bottom at a constant speed of 0.48 m/s. What is the apparent speed of the stone, as viewed from directly above by an observer who is in air?

19. **M** **SSM** Review Conceptual Example 4 as background for this problem. A man in a boat is looking straight down at a fish in the water directly beneath him. The fish is looking straight up at the man. They are equidistant from the air–water interface. To the man, the fish appears to be 2.0 m beneath his eyes. To the fish, how far above its eyes does the man appear to be?

20. **M** **V-HINT** The drawing shows a rectangular block of glass ($n = 1.52$) surrounded by liquid carbon disulfide ($n = 1.63$). A ray of light is incident on the glass at point A with a 30.0° angle of incidence. At what angle of refraction does the ray leave the glass at point B?

PROBLEM 20

21. M SSM In **Figure 26.6**, suppose that the angle of incidence is $\theta_1 = 30.0°$, the thickness of the glass pane is 6.00 mm, and the refractive index of the glass is $n_2 = 1.52$. Find the amount (in mm) by which the emergent ray is displaced relative to the incident ray.

22. M GO The back wall of a home aquarium is a mirror that is a distance of 40.0 cm away from the front wall. The walls of the tank are negligibly thin. A fish, swimming midway between the front and back walls, is being viewed by a person looking through the front wall. The index of refraction of air is $n_{air} = 1.000$ and that of water is $n_{water} = 1.333$. **(a)** Calculate the apparent distance between the fish and the front wall. **(b)** Calculate the apparent distance between the image of the fish and the front wall.

23. M Refer to **Figure 26.4b** and assume the observer is nearly above the submerged object. For this situation, derive the expression for the apparent depth: $d' = d(n_2/n_1)$, Equation 26.3. (*Hint:* Use Snell's law of refraction and the fact that the angles of incidence and refraction are small, so $\tan \theta \approx \sin \theta$.)

24. H A small logo is embedded in a thick block of crown glass ($n = 1.52$), 3.20 cm beneath the top surface of the glass. The block is put under water, so there is 1.50 cm of water above the top surface of the block. The logo is viewed from directly above by an observer in air. How far beneath the top surface of the water does the logo appear to be?

Section 26.3   Total Internal Reflection

25. E For the liquids in **Table 26.1**, determine the smallest critical angle for light that originates in one of them and travels toward the air–liquid interface.

26. E SSM A glass is half-full of water, with a layer of vegetable oil ($n = 1.47$) floating on top. A ray of light traveling downward through the oil is incident on the water at an angle of 71.4°. Determine the critical angle for the oil–water interface and decide whether the ray will penetrate into the water.

27. E A point source of light is submerged 2.2 m below the surface of a lake and emits rays in all directions. On the surface of the lake, directly above the source, the area illuminated is a circle. What is the maximum radius that this circle could have?

28. E MMH A ray of light is traveling in glass and strikes a glass–liquid interface. The angle of incidence is 58.0°, and the index of refraction of glass is $n = 1.50$. **(a)** What must be the index of refraction of the liquid so that the direction of the light entering the liquid is not changed? **(b)** What is the largest index of refraction that the liquid can have, so that none of the light is transmitted into the liquid and all of it is reflected back into the glass?

29. E GO The drawing shows three layers of different materials, with air above and below the layers. The interfaces between the layers are parallel. The index of refraction of each layer is given in the drawing. Identical rays of light are sent into the layers, and light zigzags through each layer, reflecting from the top and bottom surfaces. The index of refraction for air is $n_{air} = 1.00$. For each layer, the ray of light has an angle of incidence of 75.0°. For the cases in which total internal refection is possible from either the top or bottom surface of a layer, determine the amount by which the angle of incidence exceeds the critical angle.

PROBLEM 29

30. E The drawing shows a crown glass slab with a rectangular cross section. As illustrated, a laser beam strikes the upper surface at an angle of 60.0°. After reflecting from the upper surface, the beam reflects from the side and bottom surfaces. **(a)** If the glass is surrounded by air, determine where part of the beam first exits the glass, at point A, B, or C. **(b)** Repeat part (a), assuming that the glass is surrounded by water instead of air.

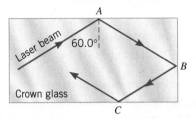

PROBLEM 30

31. E GO The drawing shows three materials, a, b, and c. A ray of light strikes the a–b interface at an angle that just barely exceeds its critical angle of 40.0°. The reflected ray then strikes the a–c interface at an angle of incidence that just barely exceeds its critical angle (which is not 40.0°). The index of refraction of material a is $n_a = 1.80$. Find the indices of refraction for the two other materials.

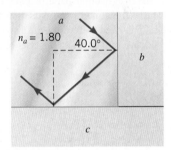

PROBLEM 31

32. M CHALK SSM MMH Multiple-Concept Example 7 provides helpful background for this problem. The drawing shows a crystalline quartz slab with a rectangular cross section. A ray of light strikes the slab at an incident angle of $\theta_1 = 34°$, enters the quartz, and travels to point P. This slab is surrounded by a fluid with a refractive index n. What is the maximum value of n for which total internal reflection occurs at point P?

PROBLEM 32

33. M V-HINT The drawing shows a ray of light traveling from point A to point B, a distance of 4.60 m in a material that has an index of refraction n_1. At point B, the light encounters a different substance whose index of refraction is $n_2 = 1.63$. The light strikes the interface at the critical angle of $\theta_c = 48.1°$. How much time does it take for the light to travel from A to B?

PROBLEM 33

34. M GO A layer of liquid B floats on liquid A. A ray of light begins in liquid A and undergoes total internal reflection at the interface between the liquids when the angle of incidence exceeds 36.5°. When liquid *B* is replaced with liquid *C*, total internal reflection occurs for angles of incidence greater than 47.0°. Find the ratio n_B/n_C of the refractive indices of liquids *B* and *C*.

Section 26.4 Polarization and the Reflection and Refraction of Light

35. **E** For light that originates within a liquid and strikes the liquid–air interface, the critical angle is 39°. What is Brewster's angle for this light?

36. **E SSM** Light is reflected from a glass coffee table. When the angle of incidence is 56.7°, the reflected light is completely polarized parallel to the surface of the glass. What is the index of refraction of the glass?

37. **E V-HINT** Light is incident from air onto the surface of a liquid. The angle of incidence is 53.0°, and the angle of refraction is 34.0°. At what angle of incidence would the reflected light be 100% polarized?

38. **E SSM CHALK** When light strikes the surface between two materials from above, the Brewster angle is 65.0°. What is the Brewster angle when the light encounters the same surface from below?

39. **E** A laser is mounted in air, at a distance of 0.476 m above the edge of a large, horizontal pane of crown glass, as shown in the drawing. The laser is aimed at the glass in such a way that the reflected beam is 100% polarized. Determine the distance *d* between the edge of the pane and the point at which the laser beam is reflected.

PROBLEM 39

Section 26.5 The Dispersion of Light: Prisms and Rainbows

40. **E** A ray of sunlight is passing from diamond into crown glass; the angle of incidence is 35.00°. The indices of refraction for the blue and red components of the ray are: blue ($n_{diamond} = 2.444$, $n_{crown\ glass} = 1.531$), and red ($n_{diamond} = 2.410$, $n_{crown\ glass} = 1.520$). Determine the angle between the refracted blue and red rays in the crown glass.

41. **E** Violet light and red light travel through air and strike a block of plastic at the same angle of incidence. The angle of refraction is 30.400° for the violet light and 31.200° for the red light. The index of refraction for violet light in plastic is greater than that for red light by 0.0400. Delaying any rounding off of calculations until the very end, find the index of refraction for violet light in plastic.

42. **E SSM** A beam of sunlight encounters a plate of crown glass at a 45.00° angle of incidence. Using the data in **Table 26.2**, find the angle between the violet ray and the red ray in the glass.

43. **E CHALK** Horizontal rays of red light ($\lambda = 660$ nm, in vacuum) and violet light ($\lambda = 410$ nm, in vacuum) are incident on the flint-glass prism shown in the drawing. The indices of refraction for the red and violet light are $n_{red} = 1.662$ and $n_{violet} = 1.698$. The prism is surrounded by air. What is the angle of refraction for each ray as it emerges from the prism?

PROBLEM 43

44. **M SSM** This problem relates to **Figure 26.18**, which illustrates the dispersion of light by a prism. The prism is made from glass, and its cross section is an equilateral triangle. The indices of refraction for the red and violet light are 1.662 and 1.698, respectively. The angle of incidence for both the red and the violet light is 60.0°. Find the angles of refraction at which the red and violet rays emerge into the air from the prism.

45. **M V-HINT** The drawing shows a horizontal ray of white light incident perpendicularly on the vertical face of a prism (crown glass). The light enters the prism, and part of it undergoes refraction at the slanted face and emerges into the surrounding material. The rest of it is totally internally reflected and exits through the horizontal base of the prism. The colors of light that emerge from the slanted face may be chosen by altering the index of refraction of the material surrounding the prism. Find the required index of refraction of the surrounding material so that **(a)** only red light and **(b)** all colors except violet emerge from the slanted face. (See **Table 26.2**.)

PROBLEM 45

Section 26.6 Lenses, Section 26.7 The Formation of Images by Lenses, Section 26.8 The Thin-Lens Equation and the Magnification Equation

(*Note:* When drawing ray diagrams, be sure that the object height h_o is much smaller than the focal length *f* of the lens or mirror.)

46. **E SSM CHALK** An object is located 9.0 cm in front of a converging lens ($f = 6.0$ cm). Using an accurately drawn ray diagram, determine where the image is located.

47. **E** The owner of a van installs a rear-window lens that has a focal length of −0.300 m. When the owner looks out through the lens at a person standing directly behind the van, the person appears to be just 0.240 m from the back of the van, and appears to be 0.34 m tall. **(a)** How far from the van is the person actually standing, and **(b)** how tall is the person?

48. **E MMH** A camera is supplied with two interchangeable lenses, whose focal lengths are 35.0 and 150.0 mm. A woman whose height is 1.60 m stands 9.00 m in front of the camera. What is the height (including sign) of her image on the image sensor, as produced by **(a)** the 35.0-mm lens and **(b)** the 150.0-mm lens?

49. **E** When a diverging lens is held 13.0 cm above a line of print, as in **Figure 26.29**, the image is 5.0 cm beneath the lens. **(a)** Is the image real or virtual? **(b)** What is the focal length of the lens?

50. **E** A slide projector has a converging lens whose focal length is 105.00 mm. **(a)** How far (in meters) from the lens must the screen be located if a slide is placed 108.00 mm from the lens? **(b)** If the slide measures 24.0 mm × 36.0 mm, what are the dimensions (in mm) of its image?

51. **E** **(a)** For a diverging lens ($f = -20.0$ cm), construct a ray diagram to scale and find the image distance for an object that is 20.0 cm from the lens. **(b)** Determine the magnification of the lens from the diagram.

52. **E** **GO** A tourist takes a picture of a mountain 14 km away using a camera that has a lens with a focal length of 50.0 mm. She then takes a second picture when she is only 5.0 km away. What is the ratio of the height of the mountain's image on the camera's image sensor for the second picture to its height on the image sensor for the first picture?

53. **E** **GO** An object is placed to the left of a lens, and a real image is formed to the right of the lens. The image is inverted relative to the object and is one-half the size of the object. The distance between the object and the image is 90.0 cm. (a) How far from the lens is the object? (b) What is the focal length of the lens?

54. **E** **MMH** A converging lens has a focal length of 88.00 cm. An object 13.00 cm tall is located 155.0 cm in front of this lens. (a) What is the image distance? (b) Is the image real or virtual? (c) What is the image height? Be sure to include the proper algebraic sign.

55. **M** **CHALK** **MMH** The distance between an object and its image formed by a diverging lens is 49.0 cm. The focal length of the lens is −233.0 cm. Find (a) the image distance and (b) the object distance.

56. **M** **SSM** The moon's diameter is 3.48×10^6 m, and its mean distance from the earth is 3.85×10^8 m. The moon is being photographed from the earth by a camera whose lens has a focal length of 50.0 mm. (a) Find the diameter of the moon's image on the slide film. (b) When the slide is projected onto a screen that is 15.0 m from the lens of the projector ($f = 110.0$ mm), what is the diameter of the moon's image on the screen?

57. **M** When a converging lens is used in a camera (as in **Figure 26.26b**), the film must be at a distance of 0.210 m from the lens to record an image of an object that is 4.00 m from the lens. The same lens and film are used in a projector (see **Figure 26.27b**), with the screen 0.500 m from the lens. How far from the projector lens should the film be placed?

58. **M** **SSM** An object is 18 cm in front of a diverging lens that has a focal length of −12 cm. How far in front of the lens should the object be placed so that the size of its image is reduced by a factor of 2.0?

59. **M** **V-HINT** An object is placed in front of a converging lens in such a position that the lens ($f = 12.0$ cm) creates a real image located 21.0 cm from the lens. Then, with the object remaining in place, the lens is replaced with another converging lens ($f = 16.0$ cm). A new, real image is formed. What is the image distance of this new image?

60. **H** A converging lens ($f = 25.0$ cm) is used to project an image of an object onto a screen. The object and the screen are 125 cm apart, and between them the lens can be placed at either of two locations. Find the two object distances.

Section 26.9   Lenses in Combination

61. **E** **GO** Two identical diverging lenses are separated by 16 cm. The focal length of each lens is −8.0 cm. An object is located 4.0 cm to the left of the lens that is on the left. Determine the final image distance relative to the lens on the right.

62. **E** **GO** Two systems are formed from a converging lens and a diverging lens, as shown in parts *a* and *b* of the drawing. (The point labeled "$F_{\text{converging}}$" is the focal point of the converging lens.) An object is placed inside the focal point of lens 1 at a distance of 10.00 cm to the left of lens 1. The focal lengths of the converging and diverging lenses are 15.00 and −20.0 cm, respectively. The distance between the lenses is 50.0 cm. Determine the final image distance for each system, measured with respect to lens 2.

(a)   (b)

PROBLEM 62

63. **E** **CHALK** Two converging lenses are separated by 24.00 cm. The focal length of each lens is 12.00 cm. An object is placed 36.00 cm to the left of the lens that is on the left. Determine the final image distance relative to the lens on the right.

64. **E** **GO** A converging lens ($f_1 = 24.0$ cm) is located 56.0 cm to the left of a diverging lens ($f_2 = -28.0$ cm). An object is placed to the left of the converging lens, and the final image produced by the two-lens combination lies 20.7 cm to the left of the diverging lens. How far is the object from the converging lens?

65. **E** **SSM** A converging lens ($f = 12.0$ cm) is located 30.0 cm to the left of a diverging lens ($f = -6.00$ cm). A postage stamp is placed 36.0 cm to the left of the converging lens. (a) Locate the final image of the stamp relative to the diverging lens. (b) Find the overall magnification. (c) Is the final image real or virtual? With respect to the original object, is the final image (d) upright or inverted, and is it (e) larger or smaller?

66. **E** **MMH** A diverging lens ($f = -10.0$ cm) is located 20.0 cm to the left of a converging lens ($f = 30.0$ cm). A 3.00-cm-tall object stands to the left of the diverging lens, exactly at its focal point. (a) Determine the distance of the final image relative to the converging lens. (b) What is the height of the final image (including the proper algebraic sign)?

67. **M** **SSM** An object is placed 20.0 cm to the left of a diverging lens ($f = -8.00$ cm). A concave mirror ($f = 12.0$ cm) is placed 30.0 cm to the right of the lens. (a) Find the final image distance, measured relative to the mirror. (b) Is the final image real or virtual? (c) Is the final image upright or inverted with respect to the original object?

68. **M** Two converging lenses ($f_1 = 9.00$ cm and $f_2 = 6.00$ cm) are separated by 18.0 cm. The lens on the left has the longer focal length. An object stands 12.0 cm to the left of the left-hand lens in the combination. (a) Locate the final image relative to the lens on the right. (b) Obtain the overall magnification. (c) Is the final image real or virtual? With respect to the original object, (d) is the final image upright or inverted and (e) is it larger or smaller?

69. **M** **V-HINT** Visitors at a science museum are invited to sit in a chair to the right of a full-length diverging lens ($f_1 = -3.00$ m) and observe a friend sitting in a second chair, 2.00 m to the left of the lens. The visitor then presses a button and a converging lens ($f_2 = +4.00$ m) rises from the floor to a position 1.60 m to the right of the diverging lens, allowing the visitor to view the friend through both lenses at once. Find (a) the magnification of the friend when viewed through the diverging lens only and (b) the overall magnification of the friend when viewed through both lenses. Be sure to include the algebraic signs (+ or −) with your answers.

Section 26.10   The Human Eye

70. **E** **BIO** A student is reading material written on a blackboard. Her contact lenses have a refractive power of 57.50 diopters; the lens-to-retina distance is 1.750 cm. (a) How far (in meters) is the blackboard from her eyes? (b) If the material written on the blackboard is 5.00 cm high, what is the size of the image on her retina?

71. **E** **BIO** A nearsighted person cannot read a sign that is more than 5.2 m from his eyes. To deal with this problem, he wears contact lenses that do not correct his vision completely, but do allow him to read signs located up to distances of 12.0 m from his eyes. What is the focal length of the contacts?

72. **E** **BIO** **GO** A woman can read the large print in a newspaper only when it is at a distance of 65 cm or more from her eyes. (a) Is she nearsighted (myopic) or farsighted (hyperopic), and what kind of lens is used in her glasses to correct her eyesight? (b) What should be the refractive power (in diopters) of her glasses (worn

2.0 cm from the eyes), so she can read the newspaper at a distance of 25 cm from her eyes?

73. **E BIO SSM** Your friend has a near point of 138 cm, and she wears contact lenses that have a focal length of 35.1 cm. How close can she hold a magazine and still read it clearly?

74. **E BIO GO** A farsighted woman breaks her current eyeglasses and is using an old pair whose refractive power is 1.660 diopters. Since these eyeglasses do not completely correct her vision, she must hold a newspaper 42.00 cm from her eyes in order to read it. She wears the eyeglasses 2.00 cm from her eyes. How far is her near point from her eyes?

75. **E BIO SSM CHALK** A person has far points of 5.0 m from the right eye and 6.5 m from the left eye. Write a prescription for the refractive power of each corrective contact lens.

76. **M BIO V-HINT** A farsighted man uses eyeglasses with a refractive power of 3.80 diopters. Wearing the glasses 0.025 m from his eyes, he is able to read books held no closer than 0.280 m from his eyes. He would like a prescription for contact lenses to serve the same purpose. What is the correct contact lens prescription, in diopters?

77. **H BIO** The far point of a nearsighted person is 6.0 m from her eyes, and she wears contacts that enable her to see distant objects clearly. A tree is 18.0 m away and 2.0 m high. **(a)** When she looks through the contacts at the tree, what is its image distance? **(b)** How high is the image formed by the contacts?

78. **H BIO** The contacts worn by a farsighted person allow her to see objects clearly that are as close as 25.0 cm, even though her uncorrected near point is 79.0 cm from her eyes. When she is looking at a poster, the contacts form an image of the poster at a distance of 217 cm from her eyes. **(a)** How far away is the poster actually located? **(b)** If the poster is 0.350 m tall, how tall is the image formed by the contacts?

Section 26.11   Angular Magnification and the Magnifying Glass

79. **E SSM** A jeweler whose near point is 72 cm from his eye uses a magnifying glass as in **Figure 26.41b** to examine a watch. The watch is held 4.0 cm from the magnifying glass. Find the angular magnification of the magnifying glass.

80. **E** A spectator, seated in the left-field stands, is watching a baseball player who is 1.9 m tall and is 75 m away. On a TV screen, located 3.0 m from a person watching the game at home, the image of this same player is 0.12 m tall. Find the angular size of the player as seen by **(a)** the spectator watching the game live and **(b)** the TV viewer. **(c)** To whom does the player appear to be larger?

81. **E** An engraver uses a magnifying glass ($f = 9.50$ cm) to examine some work, as in **Figure 26.41b**. The image he sees is located 25.0 cm from his eye, which is his near point. **(a)** What is the distance between the work and the magnifying glass? **(b)** What is the angular magnification of the magnifying glass?

82. **E GO CHALK** The near point of a naked eye is 32 cm. When an object is placed at the near point and viewed by the naked eye, it has an angular size of 0.060 rad. A magnifying glass has a focal length of 16 cm, and is held next to the eye. The enlarged image that is seen is located 64 cm from the magnifying glass. Determine the angular size of the image.

83. **E V-HINT** An object has an angular size of 0.0150 rad when placed at the near point (21.0 cm) of an eye. When the eye views this object using a magnifying glass, the largest possible angular size of the image is 0.0380 rad. What is the focal length of the magnifying glass?

84. **H SSM** A farsighted person can read printing as close as 25.0 cm when she wears contacts that have a focal length of 45.4 cm. One day, she forgets her contacts and uses a magnifying glass, as in **Figure 26.41b**. Its maximum angular magnification is 7.50 for a young person with a normal near point of 25.0 cm. What is the maximum angular magnification that the magnifying glass can provide for her?

Section 26.12   The Compound Microscope

85. **E** A forensic pathologist is viewing heart muscle cells with a microscope that has two selectable objectives with refracting powers of 100 and 300 diopters. When he uses the 100-diopter objective, the image of a cell subtends an angle of 3×10^{-3} rad with the eye. What angle is subtended when he uses the 300-diopter objective?

86. **E SSM** A compound microscope has a barrel whose length is 16.0 cm and an eyepiece whose focal length is 1.4 cm. The viewer has a near point located 25 cm from his eyes. What focal length must the objective have so that the angular magnification of the microscope will be −320?

87. **E CHALK** The distance between the lenses in a compound microscope is 18 cm. The focal length of the objective is 1.5 cm. If the microscope is to provide an angular magnification of −83 when used by a person with a normal near point (25 cm from the eye), what must be the focal length of the eyepiece?

88. **E GO** The near point of a naked eye is 25 cm. When placed at the near point and viewed by the naked eye, a tiny object would have an angular size of 5.2×10^{-5} rad. When viewed through a compound microscope, however, it has an angular size of -8.8×10^{-3} rad. (The minus sign indicates that the image produced by the microscope is inverted.) The objective of the microscope has a focal length of 2.6 cm, and the distance between the objective and the eyepiece is 16 cm. Find the focal length of the eyepiece.

89. **M** In a compound microscope, the focal length of the objective is 3.50 cm and that of the eyepiece is 6.50 cm. The distance between the lenses is 26.0 cm. **(a)** What is the angular magnification of the microscope if the person using it has a near point of 35.0 cm? **(b)** If, as usual, the first image lies just inside the focal point of the eyepiece (see **Figure 26.32**), how far is the object from the objective? **(c)** What is the magnification (not the angular magnification) of the objective?

Section 26.13   The Telescope

90. **E** An astronomical telescope has an angular magnification of −132. Its objective has a refractive power of 1.50 diopters. What is the refractive power of its eyepiece?

91. **E SSM** Mars subtends an angle of 8.0×10^{-5} rad at the unaided eye. An astronomical telescope has an eyepiece with a focal length of 0.032 m. When Mars is viewed using this telescope, it subtends an angle of 2.8×10^{-3} rad. Find the focal length of the telescope's objective lens.

92. **E GO** A telescope has an objective with a refractive power of 1.25 diopters and an eyepiece with a refractive power of 250 diopters. What is the angular magnification of the telescope?

93. **E** An amateur astronomer decides to build a telescope from a discarded pair of eyeglasses. One of the lenses has a refractive power of 11 diopters, and the other has a refractive power of 1.3 diopters. **(a)** Which lens should be the objective? **(b)** How far apart should the lenses be separated? **(c)** What is the angular magnification of the telescope?

94. **M GO CHALK** The lengths of three telescopes are $L_A = 455$ mm, $L_B = 615$ mm, and $L_C = 824$ mm. The focal length of the eyepiece for each telescope is 3.00 mm. Find the angular magnification of each telescope.

95. **H** An astronomical telescope is being used to examine a relatively close object that is only 114.00 m away from the objective of the telescope. The objective and eyepiece have focal lengths of 1.500 and 0.070 m, respectively. Noting that the expression $M \approx -f_o/f_e$ is no longer applicable because the object is so close, use the thin-lens and magnification equations to find the angular magnification of this telescope. (*Hint:* See **Figure 26.43** and note that the focal points F_o and F_e are so close together that the distance between them may be ignored.)

Additional Problems

Online

96. **E** **SSM** An object is located 30.0 cm to the left of a converging lens whose focal length is 50.0 cm. (a) Draw a ray diagram to scale and from it determine the image distance and the magnification. (b) Use the thin-lens and magnification equations to verify your answers to part (a).

97. **E** **SSM** **MMH** A glass block ($n = 1.56$) is immersed in a liquid. A ray of light within the glass hits a glass–liquid surface at a 75.0° angle of incidence. Some of the light enters the liquid. What is the smallest possible refractive index for the liquid?

98. **E** As an aid in understanding this problem, refer to Conceptual Example 4. A swimmer, who is looking up from under the water, sees a diving board directly above at an apparent height of 4.0 m above the water. What is the actual height of the diving board above the water?

99. **E** A camper is trying to start a fire by focusing sunlight onto a piece of paper. The diameter of the sun is 1.39×10^9 m, and its mean distance from the earth is 1.50×10^{11} m. The camper is using a converging lens whose focal length is 10.0 cm. (a) What is the area of the sun's image on the paper? (b) If 0.530 W of sunlight passes through the lens, what is the intensity of the sunlight at the paper?

100. **E** **GO** Red light ($n = 1.520$) and violet light ($n = 1.538$) traveling in air are incident on a slab of crown glass. Both colors enter the glass at the same angle of refraction. The red light has an angle of incidence of 30.00°. What is the angle of incidence of the violet light?

101. **E** A converging lens ($f = 12.0$ cm) is held 8.00 cm in front of a newspaper that has a print size with a height of 2.00 mm. Find (a) the image distance (in cm) and (b) the height (in mm) of the magnified print.

102. **E** **GO** To focus a camera on objects at different distances, the converging lens is moved toward or away from the image sensor, so a sharp image always falls on the sensor. A camera with a telephoto lens ($f = 200.0$ mm) is to be focused on an object located first at a distance of 3.5 m and then at 50.0 m. Over what distance must the lens be movable?

103. **E** **SSM** An office copier uses a lens to place an image of a document onto a rotating drum. The copy is made from this image. (a) What kind of lens is used, converging or diverging? If the document and its copy are to have the same size, but are inverted with respect to one another, (b) how far from the document is the lens located and (c) how far from the lens is the image located? Express your answers in terms of the focal length f of the lens.

104. **M** **V-HINT** An object is in front of a converging lens ($f = 0.30$ m). The magnification of the lens is $m = 4.0$. (a) Relative to the lens, in what direction should the object be moved so that the magnification changes to $m = -4.0$? (b) Through what distance should the object be moved?

105. **H** **SSM** The angular magnification of a telescope is 32 800 times as large when you look through the correct end of the telescope as when you look through the wrong end. What is the angular magnification of the telescope?

Physics in Biology, Medicine, and Sports

106. **M** **BIO** **SSM** **26.10** At age forty, a man requires contact lenses ($f = 65.0$ cm) to read a book held 25.0 cm from his eyes. At age forty-five, while wearing these contacts he must now hold a book 29.0 cm from his eyes. (a) By what distance has his near point *changed?* (b) What focal-length lenses does he require at age forty-five to read a book at 25.0 cm?

107. **H** **BIO** **26.10** Bill is farsighted and has a near point located 125 cm from his eyes. Anne is also farsighted, but her near point is 75.0 cm from her eyes. Both have glasses that correct their vision to a normal near point (25.0 cm from the eyes), and both wear the glasses 2.0 cm from the eyes. Relative to the eyes, what is the closest object that can be seen clearly (a) by Anne when she wears Bill's glasses and (b) by Bill when he wears Anne's glasses?

108. **M** **BIO** **26.11** Dermatologists will often use the technique of *dermatoscopy,* in which they examine skin irregularities using a device called a *dermatoscope.* The dermatoscope consists of a light source and a magnifying lens (see the photo). The light source is capable of producing both unpolarized and polarized light, which allows for the examination of surface skin lesions and abnormalities in deeper skin structures, respectively. This technique allows dermatologists to distinguish benign skin lesions from those that are malignant (or cancerous), indicating the presence of melanoma. Consider a farsighted dermatologist who has a near point located at 37.0 cm from her eyes. She is using a dermatoscope with an angular magnification of 10.0 to examine a suspicious looking mole on a patient's back. (a) What is the distance between the mole and the lens in the dermatoscope? (b) What is the focal length of the dermatoscope's lens, if the image of the mole forms at a distance of −14.2 cm from the lens?

PROBLEM 108

109. M BIO **26.10** A farsighted person has a near point of 72.0 cm. What should be the refractive power of their glasses, if they are able to see objects clearly at 25.0 cm? Assume their glasses are located 2.00 cm in front of their eyes.

110. M BIO **26.10** A person with abnormal vision dives into a swimming pool, and while underwater, can see distant objects clearly. **(a)** Is this person nearsighted or farsighted? This same person visits an optometrist who prescribes corrective contact lenses with a refractive power of −12.0 diopters. **(b)** How far from the person's eyes is their far point?

111. M BIO **26.11** A technician in a pathology laboratory is using a compound microscope to examine tissue samples. The focal length

of the objective of the microscope is 0.35 cm, and the focal length of the eyepiece is 2.5 cm. The two lenses are separated by 27 cm. One day, the technician forgot to wear his contact lenses, which have a prescription of +1.5 diopters and allow him to see objects clearly at 25 cm. When he looks into the microscope without wearing his contacts, what is the approximate angular magnification of the scope?

112. M BIO **26.10** A person with myopic vision has a right eye with a far point of 4.0 m and a left eye with a far point of 5.0 m. Write a prescription for their corrective glasses that will sit 2.7 cm in front of their eyes. **(a)** What will be the refractive power of the right lens of their glasses, and **(b)** what will be the refractive power of the left lens?

Concepts and Calculations Problems

Online

One important phenomenon discussed in this chapter is how a ray of light is refracted when it goes from one medium into another. Problem 113 reviews some of the important aspects of refraction, including Snell's law, the concept of a critical angle, and the notion of index matching. One of the most important applications of refraction is in lenses, the behaviors of which are governed by the thin-lens and magnification equations. Problem 114 reviews how these equations are applied to a two-lens system, along with the all-important sign conventions that must be followed.

113. M CHALK A ray of light is incident on a glass–water interface at the critical angle θ_c as the figure illustrates. The reflected light then passes through a liquid (immiscible with water) and into air. The indices of refraction for the four substances are given in the drawing. *Concepts:* (i) What determines the critical angle when the ray strikes the glass–water interface? (ii) When the light is incident at the glass–water interface at the critical angle, what is the angle of refraction, and how is the angle θ_1 related to the critical angle? (iii) When the reflected ray strikes the glass–liquid interface, how is the angle of refraction θ_3 related to the angle of incidence θ_2? Note that the two materials have the same indices of refraction. (iv) When the ray passes from the liquid into the air, is the ray refracted? Explain. *Calculation:* Determine the angle of refraction θ_5 for the ray as it passes into the air.

114. M CHALK SSM In the figure, a converging lens ($f_1 = +20.0$ cm) and a diverging lens ($f_2 = −15.0$ cm) are separated by a distance of 10.0 cm. An object with a height of $h_{o1} = 5.00$ mm is placed at a distance of $d_{o1} = 45.0$ cm to the left of the first (converging) lens. *Concepts:* (i) Is the image produced by the first (converging) lens real or virtual? (ii) As far as the second lens is concerned, what role does the image produced by the first lens play? (iii) Note in the figure that the image produced by the first lens is called the "first image," and it falls to the right of the second lens. This image acts as the object for the second lens. Normally, however, an object would lie to the left of the lens. How do we take into account that this object lies to the right of the diverging lens? (iv) How do we find the location of the image produced by the second lens when its object is a virtual object? *Calculation:* What are **(a)** the image distance d_{i1} and **(b)** the height h_{i1} of the image produced by the first lens? **(c)** What is the object distance for the second (diverging) lens? Find **(d)** the image distance d_{i2} and **(e)** the height h_{i2} of the image produced by the second lens.

PROBLEM 114

PROBLEM 113

Team Problems

Online

115. M T WS **Laser Ablation.** You and your team have been assigned to set up a powerful laser to be used in a process called laser ablation. Laser ablation is a process in which a powerful, focused laser beam delivers a large amount of energy to the atoms of a surface, called the target, in such a short time that the atoms are expelled from

the surface before the target material even has time to melt. Such a process can be used to shape the cornea in laser eye surgery (i.e., Lasik surgery), or to coat nearby surfaces with the ablated material to make novel coatings, such as those used in semiconductor devices. Suppose the laser emits pulses of ultraviolet light with a circular cross section of diameter 1.70 cm, each with a duration of 25.0 nanoseconds and a total energy of 650.0 mJ. **(a)** What is the energy density of each pulse? **(b)** What is the intensity of the light striking the surface during each pulse? **(c)** In order for the laser to properly ablate the target, the intensity must be 7.00×10^{11} W/m^2. What should be the magnification of a lens system that focuses the laser on the target so that this intensity is achieved? **(d)** An optical system processes the incoming beam so that the laser source can be treated as an object located in front of a converging spherical lens. If the lens has a focal length of 30.0 cm, determine the object distance so that the focused beam (i.e., the image) has an intensity of 7.00×10^{11} W/m^2. **(e)** With the object distance found in part (d), how far from the lens should the target be located, that is, what is the image distance?

116. **E** **T** **WS** **An Underwater Transparent Reflector.** You and your team have been assigned to optically engineer a transparent optical device that can reverse a laser beam under water. The idea is that the laser beam travels in water ($n_W = 1.33$) and passes through one surface of a triangular prism with an angle of incidence equal to 90.0°. It then undergoes a reflection from each of the two other internal surfaces of the prism with equal angles of incidence, and then exits through the surface from which it entered in a direction antiparallel to the incoming beam. The exiting beam has the same intensity as the incoming beam (i.e., all of the incident light is reflected) **(a)** What is the angle between the two internal reflecting surfaces of the prism? **(b)** What should be the minimum index of refraction (n_G) of the glass?

117. **M** **T** **An Optical Spectrometer.** You and your team are tasked with characterizing an equilateral, triangular prism to be used in an optical spectrometer. An optical spectrometer contains a dispersive element (in this case, the prism) that separates an incoming beam of light into its constituent wavelengths (or colors), and a photocell that measures the intensity of each color (wavelength) by measuring the intensity of the dispersed light as a function of angle. The index of refraction of the prism material you are using depends on the wavelength as

$$n(\lambda) = -(1.080 \times 10^{-4} \text{ nm}^{-1}) \lambda \text{ (nm)} + 1.586$$

PROBLEM 117

An incident beam of white light impinges on the surface (see the drawing) at an angle of 45.00 degrees below the normal. The prism is in air ($n = 1.000$). Relative to the prism base (i.e., the horizontal in the drawing), at what angles do **(a)** red light ($\lambda = 660.0$ nm) and **(b)** violet light ($\lambda = 410.0$ nm) emerge on the opposite side? **(c)** What is the angular range of the full spectrum, from red to violet?

118. **M** **T** **Emergency Replacement Glasses.** One of your team members lost her glasses in a river. She is nearsighted and cannot see long distances without them. You are on an expedition to map a remote area in southern Argentina, and her long-distance vision is crucial to her role in the group. You can request a new pair of glasses for her that will be delivered with the next airdrop of supplies, but she does not know her lens prescription. **(a)** Does she need a converging or diverging lens? **(b)** You do a simple eye test and estimate her far point to be 623 cm from her eyes. Assuming she will wear her glasses 2.00 cm in front of her eyes, what should be the focal length of her new lenses? **(c)** What should be the refractive power of her new lenses (in diopters)?

CHAPTER 27

A rainbow of colors is observed in a bubble, which is a thin film of soap surrounded by air on both sides. But unlike a real rainbow's colors, which are due to dispersion, the colors in the bubble are created by wave interference. Interference occurs when two or more waves exist simultaneously at the same place. Here, some of the light reaching the eye of the observer has reflected from the top surface of the bubble, while some of the light has reflected from the inner surface. The thickness of the bubble, which varies across its surface, affects how the light waves interfere, and thus determines the colors we see. Interference effects, which we will study in this chapter, provide direct evidence of the wave nature of light.

Squirrel_photos/Pixabay

Interference and the Wave Nature of Light

LEARNING OBJECTIVES

After reading this module, you should be able to...

27.1 Apply the principle of linear superposition to light waves.

27.2 Analyze double-slit interference.

27.3 Analyze thin-film interference.

27.4 Understand the operation of the Michelson interferometer.

27.5 Analyze single-slit diffraction.

27.6 Determine the resolving power of lenses.

27.7 Apply interference principles to the diffraction grating.

27.8 Describe X-ray diffraction in crystals.

27.1 The Principle of Linear Superposition

Chapter 17 examines what happens when several sound waves are present at the same place at the same time. The pressure disturbance that results is governed by the ***principle of linear superposition, which states that the resultant disturbance is the sum of the disturbances from the individual waves.*** Light is also a wave, an electromagnetic wave, and it too obeys the superposition principle. When two or more light waves pass through a given point, their electric fields combine

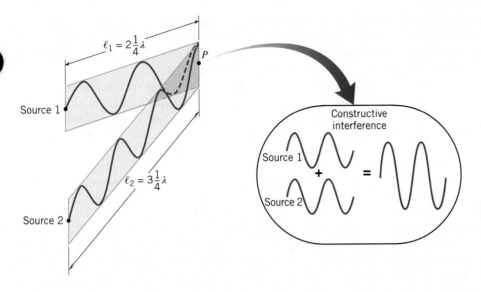

INTERACTIVE FIGURE 27.1 The waves emitted by source 1 and source 2 start out in phase and arrive at point P in phase, leading to constructive interference at that point.

according to the principle of linear superposition and produce a resultant electric field. According to Equation 24.5b, the square of the electric field strength is proportional to the intensity of the light, which, in turn, is related to its brightness. Thus, interference can and does alter the brightness of light, just as it affects the loudness of sound.

Interactive Figure 27.1 illustrates what happens when two identical waves (same wavelength λ and same amplitude) arrive at the point P in phase—that is, crest-to-crest and trough-to-trough. According to the principle of linear superposition, the waves reinforce each other and **constructive interference** occurs. The resulting total wave at P has an amplitude that is twice the amplitude of either individual wave, and, in the case of light waves, the brightness at P is greater than that due to either wave alone. The waves start out in phase and are in phase at P because the distances ℓ_1 and ℓ_2 between this spot and the sources of the waves differ by one wavelength λ. In **Interactive Figure 27.1**, these distances are $\ell_1 = 2\frac{1}{4}$ wavelengths and $\ell_2 = 3\frac{1}{4}$ wavelengths. In general, when the waves start out in phase, constructive interference will result at P whenever the distances are the same or differ by any integer number of wavelengths—in other words, assuming that ℓ_2 is the larger distance, whenever $\ell_2 - \ell_1 = m\lambda$, where $m = 0, 1, 2, 3, \ldots$.

Interactive Figure 27.2 shows what occurs when two identical waves arrive at the point P out of phase with one another, or crest-to-trough. Now the waves mutually cancel, according to the principle of linear superposition, and **destructive interference** results. With light waves this would mean that there is no brightness. The waves begin with the same phase but are out of phase at P because the distances through which they travel in reaching this spot differ by one-half of a wavelength ($\ell_1 = 2\frac{3}{4}\lambda$ and $\ell_2 = 3\frac{1}{4}\lambda$ in

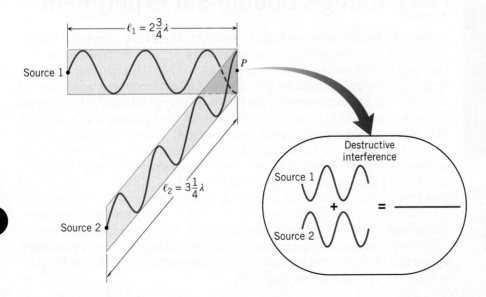

INTERACTIVE FIGURE 27.2 The waves emitted by the two sources have the same phase to begin with, but they arrive at point P out of phase. As a result, destructive interference occurs at P.

the drawing). In general, for waves that start out in phase, destructive interference will take place at P whenever the distances differ by any odd integer number of half-wavelengths—that is, whenever $\ell_2 - \ell_1 = \frac{1}{2}\lambda, \frac{3}{2}\lambda, \frac{5}{2}\lambda, \ldots$, where ℓ_2 is the larger distance. This is equivalent to $\ell_2 - \ell_1 = \left(m + \frac{1}{2}\right)\lambda$, where $m = 0, 1, 2, 3, \ldots$.

Examples illustrating the application of the principle of linear superposition to explain the interference of light waves can be found throughout this chapter. For relatively straightforward examples that deal with two sources of sound waves and the resulting constructive or destructive interference, see Examples 1 and 2 in Chapter 17.

If constructive or destructive interference is to continue occurring at a point, the sources of the waves must be **coherent sources**. Two sources are coherent if the waves they emit maintain a constant phase relation. Effectively, this means that the waves do not shift relative to one another as time passes. For instance, suppose that the wave pattern of source 1 in **Interactive Figure 27.2** shifted forward or backward by random amounts at random moments. Then, on average, neither constructive nor destructive interference would be observed at point P because there would be no stable relation between the two wave patterns. Lasers are coherent sources of light, whereas incandescent light bulbs and fluorescent lamps are incoherent sources.

Check Your Understanding

(The answers are given at the end of the book.)

1. Two separate coherent sources produce waves whose wavelengths are 0.10 m. The two waves spread out and overlap at a certain point. Does constructive or destructive interference occur at this point when **(a)** one wave travels 3.20 m and the other travels 3.00 m, **(b)** one wave travels 3.20 m and the other travels 3.05 m, **(c)** one wave travels 3.20 m and the other travels 2.95 m?

2. Suppose that a radio station broadcasts simultaneously from two transmitting antennas at *two different locations.* Compared with only one transmitting antenna, the reception with two transmitting antennas **(a)** is always better **(b)** is always worse **(c)** can be either better or worse, depending on the distance traveled by each wave.

3. Two sources of waves are in phase and produce identical waves. These sources are mounted at the corners of a square and broadcast waves uniformly in all directions. At the center of the square, will the waves always produce constructive interference no matter which two corners of the square are occupied by the sources?

FIGURE 27.3 In Young's double-slit experiment, two slits S_1 and S_2 act as coherent sources of light. Light waves from these slits interfere constructively and destructively on the screen to produce, respectively, the bright and dark fringes. The slit widths and the distance between the slits have been exaggerated for clarity.

27.2 Young's Double-Slit Experiment

In 1801 the English scientist Thomas Young (1773–1829) performed a historic experiment that demonstrated the wave nature of light by showing that two overlapping light waves can interfere with each other. His experiment was particularly important because he was also able to determine the wavelength of the light from his measurements, the first such determination of this important property. **Figure 27.3** shows one arrangement of Young's experiment, in which light of a single wavelength (monochromatic light) passes through a single narrow slit and falls on two closely spaced, narrow slits S_1 and S_2. These two slits act as coherent sources of light waves that interfere constructively and destructively at different points on the screen to produce a pattern of alternating bright and dark fringes. The purpose of the single slit is to ensure that only light from one direction falls on the double slit. Without it, light coming from different points on the light source would strike the double slit from different directions and cause the pattern on the screen to be washed out. The slits S_1 and S_2 act as coherent sources of light waves because the light from each originates from the same primary source—namely, the single slit.

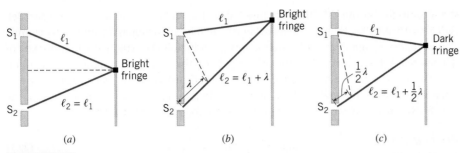

FIGURE 27.4 The waves that originate from slits S_1 and S_2 interfere constructively (parts a and b) or destructively (part c) on the screen, depending on the difference in distances between the slits and the screen. Note that the slit widths and the distance between the slits have been exaggerated for clarity.

To help explain the origin of the bright and dark fringes, **Figure 27.4** presents three top views of the double slit and the screen. Part a illustrates how a bright fringe arises directly opposite the midpoint between the two slits. In this part of the drawing the waves (identical) from each slit travel to the midpoint on the screen. At this location, the distances ℓ_1 and ℓ_2 to the slits are equal, each containing the same number of wavelengths. Therefore, constructive interference results, leading to the bright fringe. Part b indicates that constructive interference produces another bright fringe on one side of the midpoint when the distance ℓ_2 is larger than ℓ_1 by exactly one wavelength. A bright fringe also occurs symmetrically on the other side of the midpoint when the distance ℓ_1 exceeds ℓ_2 by one wavelength; for clarity, however, this additional bright fringe is not shown. Constructive interference produces other bright fringes (also not shown) on both sides of the middle wherever the difference between ℓ_1 and ℓ_2 is an integer number of wavelengths: λ, 2λ, 3λ, and so on. Part c shows how the first dark fringe arises. Here the distance ℓ_2 is larger than ℓ_1 by exactly one-half a wavelength, so the waves interfere destructively, giving rise to the dark fringe. Destructive interference creates other dark fringes on both sides of the center wherever the difference between ℓ_1 and ℓ_2 equals an odd integer number of half-wavelengths: $\frac{1}{2}\lambda$, $\frac{3}{2}\lambda$, $\frac{5}{2}\lambda$, and so on.

The brightness of the fringes in Young's experiment varies, as the fringe pattern in **Figure 27.5** shows. Below the fringe pattern is a graph to suggest the way in which the intensity varies for the fringes. The central fringe is labeled with a zero, and the other bright fringes are numbered in ascending order on either side of the center. It can be seen that the central fringe has the greatest intensity. To either side of the center, the intensities of the other fringes decrease symmetrically in a way that depends on how small the slit widths are relative to the wavelength of the light.

The position of the fringes observed on the screen in Young's experiment can be calculated with the aid of **Figure 27.6**. If the screen is located far away compared with

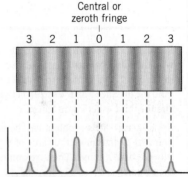

FIGURE 27.5 The results of Young's double-slit experiment, showing the pattern of the bright and dark fringes formed on the screen and a graph of the light intensity. The central or zeroth fringe is the brightest fringe (greatest intensity).

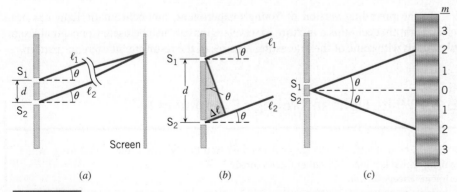

FIGURE 27.6 (a) Rays from slits S_1 and S_2, which make approximately the same angle θ with the horizontal, strike a distant screen at the same spot. (b) The difference in the path lengths of the two rays is $\Delta\ell = d\sin\theta$. (c) The angle θ is the angle at which a bright fringe ($m = 2$, here) occurs on either side of the central bright fringe ($m = 0$).

the separation d of the slits, then the lines labeled ℓ_1 and ℓ_2 in part a are nearly parallel. Being nearly parallel, these lines make approximately equal angles θ with the horizontal. The distances ℓ_1 and ℓ_2 differ by an amount $\Delta\ell$, which is the length of the short side of the colored triangle in part b of the drawing. Since the triangle is a right triangle, it follows that $\Delta\ell = d \sin \theta$. Constructive interference occurs when the distances differ by an integer number m of wavelengths λ, or $\Delta\ell = d \sin \theta = m\lambda$. Therefore, the angle θ for the interference maxima can be determined from the following expression:

Bright fringes of a double slit
$$\sin \theta = m\frac{\lambda}{d} \quad m = 0, 1, 2, 3, \ldots \tag{27.1}$$

The value of m specifies the *order* of the fringe. Thus, $m = 2$ identifies the "second-order" bright fringe. Part c of **Figure 27.6** stresses that the angle θ given by Equation 27.1 locates bright fringes on either side of the midpoint between the slits. A similar line of reasoning leads to the conclusion that the dark fringes, which lie between the bright fringes, are located according to

Dark fringes of a double slit
$$\sin \theta = \left(m + \frac{1}{2}\right)\frac{\lambda}{d} \quad m = 0, 1, 2, 3, \ldots \tag{27.2}$$

Example 1 illustrates how to determine the distance of a higher-order bright fringe from the central bright fringe with the aid of Equation 27.1.

EXAMPLE 1 | Young's Double-Slit Experiment

Red light ($\lambda = 664$ nm in vacuum) is used in Young's experiment with the slits separated by a distance $d = 1.20 \times 10^{-4}$ m. The screen in **Figure 27.7** is located at a distance of $L = 2.75$ m from the slits. Find the distance y on the screen between the central bright fringe and the third-order bright fringe.

Reasoning This problem can be solved by first using Equation 27.1 to determine the value of θ that locates the third-order ($m = 3$) bright fringe. Then trigonometry can be used to obtain the distance y.

Solution According to Equation 27.1, we find

$$\theta = \sin^{-1}\left(\frac{m\lambda}{d}\right) = \sin^{-1}\left[\frac{3(664 \times 10^{-9}\,\text{m})}{1.20 \times 10^{-4}\,\text{m}}\right] = 0.951°$$

According to **Figure 27.7**, the distance y can be calculated from $\tan \theta = y/L$:

$$y = L \tan \theta = (2.75\,\text{m}) \tan 0.951° = \boxed{0.0456\,\text{m}}$$

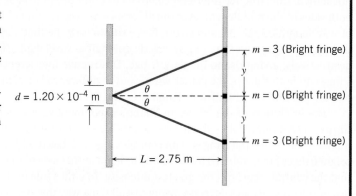

FIGURE 27.7 The third-order bright fringe ($m = 3$) is observed on the screen at a distance y from the central bright fringe ($m = 0$).

In the preceding version of Young's experiment, monochromatic light has been used. Light that contains a mixture of wavelengths can also be used. Conceptual Example 2 deals with some of the interesting features of the resulting interference pattern.

CONCEPTUAL EXAMPLE 2 | White Light and Young's Experiment

Figure 27.8 shows a photograph that illustrates the kind of interference fringes that can result when white light, which is a mixture of all colors, is used in Young's experiment. Except for the central fringe, which is white, the bright fringes are a rainbow of colors. Why does Young's experiment separate white light into its constituent colors? In any group of colored fringes, such as the two singled out in **Figure 27.8**, why is red farther out from the central

fringe than green is? And finally, why is the central fringe white rather than colored?

Reasoning and Solution To understand how the color separation arises, we need to remember that each color corresponds to a different wavelength λ and that constructive and destructive interference depend on the wavelength. According to Equation 27.1

Colored
fringes
m = 0

Andy Washnik

FIGURE 27.8 This photograph shows the results observed on the screen in one version of Young's experiment in which white light (a mixture of all colors) is used.

$(\sin \theta = m\lambda/d)$, there is a different angle that locates a bright fringe for each value of λ, and thus for each color. These different angles lead to the separation of colors on the observation screen. In fact, on either side of the central fringe, there is one group of colored fringes for $m = 1$ and another for each additional value of m.

Now, consider what it means that, within any single group of colored fringes, red is farther out from the central fringe than green is. It means that, in the equation $\sin \theta = m\lambda/d$, red light has a larger angle θ than green light does. Does this make sense? Yes, because red has the larger wavelength (see Table 26.2, where $\lambda_{red} = 660$ nm and $\lambda_{green} = 550$ nm).

In Figure 27.8, the central fringe is distinguished from all the other colored fringes by being white. In Equation 27.1, the central fringe is different from the other fringes because it is the only one for which $m = 0$. In Equation 27.1, a value of $m = 0$ means that $\sin \theta = m\lambda/d = 0$, which reveals that $\theta = 0°$, no matter what the wavelength λ is. In other words, all wavelengths have a zeroth-order bright fringe located at the same place on the screen, so that all colors strike the screen there and mix together to produce the white central fringe.

Related Homework: Problem 46

Historically, Young's experiment provided strong evidence that light has a wave-like character. If light behaved only as a stream of "tiny particles," as others believed at the time,* then the two slits would deliver the light energy into only two bright fringes located directly opposite the slits on the screen. Instead, Young's experiment shows that wave interference redistributes the energy from the two slits into many bright fringes.

Check Your Understanding

(The answers are given at the end of the book.)

4. Replace the slits S_1 and S_2 in **Figure 27.3** with identical in-phase loudspeakers and use the same ac electrical signal to drive them. The two sound waves produced will then be identical, and you will have the audio equivalent of Young's double-slit experiment. In terms of loudness and softness, what would you hear as you walk along the screen, starting from the center and going to either end? **(a)** Loud, then soft, then loud, then soft, etc., with the loud sounds *decreasing in intensity* as you walk away from the center **(b)** Loud, then soft, then loud, then soft, etc., with the loud sounds *increasing in intensity* as you walk away from the center **(c)** Soft, then loud, then soft, then loud, etc., with the loud sounds *decreasing in intensity* as you walk away from the center **(d)** Soft, then loud, then soft, then loud, etc., with the loud sounds *increasing in intensity* as you walk away from the center

5. CYU Figure 27.1 shows two double slits that have slit separations of d_1 and d_2. Light whose wavelength is either λ_1 or λ_2 passes through the slits. For comparison, the wavelengths are also illustrated in the drawing. For which combination of slit separation and wavelength would the pattern of bright and dark fringes on the observation screen be **(a)** the most spread out and **(b)** the least spread out?

CYU FIGURE 27.1

---

*It is now known that the particle, or corpuscular, theory of light, which Isaac Newton promoted, does indeed explain some experiments that the wave theory cannot explain. Today, light is regarded as having both particle and wave characteristics. Chapter 29 discusses this dual nature of light.

6. Suppose the light waves coming from *both* slits in a Young's double-slit experiment had their phases shifted by an amount equivalent to a half-wavelength. **(a)** Would the pattern be the same or would the positions of the light and dark fringes be interchanged? **(b)** Would the pattern be the same or would the positions of the light and dark fringes be interchanged if the light coming from *only one* of the slits had its phase shifted by an amount equivalent to a half-wavelength?

7. In Young's double-slit experiment, is it possible to see interference fringes when the wavelength of the light is greater than the distance between the slits?

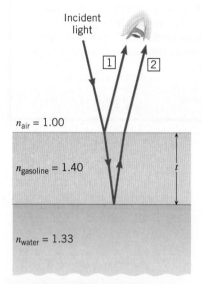

INTERACTIVE FIGURE 27.9 Because of reflection and refraction, two light waves, represented by rays 1 and 2, enter the eye when light shines on a thin film of gasoline floating on a thick layer of water.

27.3 | Thin-Film Interference

Young's double-slit experiment is one example of interference between light waves. Interference also occurs in more common circumstances. For instance, **Interactive Figure 27.9** shows a thin film such as gasoline floating on water. To begin with, let us assume that the film has a constant thickness. Consider what happens when mono-chromatic light (a single wavelength) strikes the film nearly perpendicularly. At the top surface of the film reflection occurs and produces the light wave represented by ray 1. However, refraction also occurs, and some light enters the film. Part of this light reflects from the bottom surface of the film and passes back up through the film, eventually reentering the air. Thus, a second light wave, which is represented by ray 2, also exists. Moreover, this wave, having traversed the film twice, has traveled farther than wave 1. Because of the extra travel distance, there can be interference between the two waves. If constructive interference occurs, an observer whose eyes detect the superposition of waves 1 and 2 would see a uniformly bright film. If destructive interference occurs, an observer would see a uniformly dark film.

In **Interactive Figure 27.9** the difference in path lengths between waves 1 and 2 occurs inside the thin film. Therefore, we note the following:

> **Problem-Solving Insight** The wavelength that is important for thin-film inter-ference is the wavelength within the film, not the wavelength in vacuum.

The wavelength within the film can be calculated from the wavelength in vacuum by using the index of refraction n_{film} for the film. With the aid of Equations 26.1 and 16.1, it can be shown that $n_{film} = c/v = (c/f)/(v/f) = \lambda_{vacuum}/\lambda_{film}$. In other words,

$$\lambda_{film} = \frac{\lambda_{vacuum}}{n_{film}} \qquad (27.3)$$

(a)

(b)

ANIMATED FIGURE 27.10 When a wave on a string reflects from a wall, the wave undergoes a phase change. Thus, an upward-pointing half-cycle of the wave becomes, after reflection, a downward-pointing half-cycle, and vice versa, as the numbered labels in the drawing indicate.

In explaining the interference that can occur in **Interactive Figure 27.9**, we need to add one other important part to the story. Whenever waves reflect at a bound-ary, it is possible for them to change phase. **Animated Figure 27.10**, for example, shows that a wave on a string is inverted when it reflects from the end that is tied to a wall (see also Figure 17.16). This inversion is equivalent to a half-cycle of the wave, as if the wave had traveled an additional distance of one-half of a wavelength. In contrast, a phase change does not occur when a wave on a string reflects from the end of a string that is hanging free. When light waves undergo reflection, similar phase changes occur as follows:

1. When light travels through a material with a smaller refractive index toward a ma-terial with a larger refractive index (e.g., air to gasoline), reflection at the boundary occurs along with a phase change that is equivalent to one-half of a wavelength in the film.

2. When light travels from a larger toward a smaller refractive index, there is no phase change upon reflection at the boundary.

The next example indicates how the phase change that can accompany reflection is taken into account when dealing with thin-film interference.

EXAMPLE 3 | A Colored Thin Film of Gasoline

A thin film of gasoline floats on a puddle of water. Sunlight falls almost perpendicularly on the film and reflects into your eyes. Although sunlight is white since it contains all colors, the film looks yellow because destructive interference eliminates the color of blue (λ_{vacuum} = 469 nm) from the reflected light. The refractive indices of the blue light in gasoline and in water are 1.40 and 1.33, respectively. Determine the minimum nonzero thickness t of the film.

Reasoning To solve this problem, we must express the condition for destructive interference in terms of the film thickness t and the wavelength λ_{film} in the gasoline film. We must also take into account any phase changes that occur upon reflection.

In **Interactive Figure 27.9**, the phase change for wave 1 is equivalent to one-half of a wavelength, since this light travels from a smaller refractive index (n_{air} = 1.00) toward a larger refractive index ($n_{gasoline}$ = 1.40). In contrast, there is no phase change when wave 2 reflects from the bottom surface of the film, since this light travels from a material with a larger refractive index ($n_{gasoline}$ = 1.40) toward a material with a smaller one (n_{water} = 1.33). The net phase change between waves 1 and 2 due to reflection is, thus, equivalent to one-half of a wavelength, $\frac{1}{2}\lambda_{film}$. This half-wavelength must be combined with the extra travel distance for wave 2, to determine the condition for destructive interference. For destructive interference, the combined total must be an odd integer number of half-wavelengths. Since wave 2 travels back and forth through the film and since light strikes the film nearly perpendicularly, the extra travel distance is twice the film thickness, or $2t$. Thus, the condition for destructive interference is

$$\underbrace{2t}_{\substack{\text{Extra distance}\\\text{traveled by}\\\text{wave 2}}} + \underbrace{\tfrac{1}{2}\lambda_{film}}_{\substack{\text{Half-wavelength}\\\text{net phase change}\\\text{due to reflection}}} = \underbrace{\tfrac{1}{2}\lambda_{film}, \tfrac{3}{2}\lambda_{film}, \tfrac{5}{2}\lambda_{film}, \cdots}_{\substack{\text{Condition for destructive}\\\text{interference}}}$$

or

$$2t + \tfrac{1}{2}\lambda_{film} = \left(m + \tfrac{1}{2}\right)\lambda_{film} \quad m = 0, 1, 2, 3, \ldots$$

After subtracting the term $\frac{1}{2}\lambda_{film}$ from both sides of this equation, we can solve for the thickness t of the film that yields destructive interference:

$$t = \frac{m\lambda_{film}}{2} \quad m = 0, 1, 2, 3, \ldots$$

Problem-Solving Insight When analyzing thin-film interference effects, remember to use the wavelength of the light in the film (λ_{film}) instead of the wavelength in a vacuum (λ_{vacuum}).

Solution In order to calculate t, we need to know the wavelength of the blue light in the film. Equation 27.3, with n_{film} = 1.40, gives this wavelength as

$$\lambda_{film} = \frac{\lambda_{vacuum}}{n_{film}} = \frac{469 \text{ nm}}{1.40} = 335 \text{ nm}$$

With this value for λ_{film} and $m = 1$, our result for t gives the minimum nonzero film thickness for which the blue color is missing in the reflected light as follows:

$$t = \frac{m\lambda_{film}}{2} = \frac{(1)(335 \text{ nm})}{2} = \boxed{168 \text{ nm}}$$

In Example 3 a half-wavelength net phase change occurs due to the reflections at the upper and lower surfaces of the thin film. Depending on the refractive indices of the materials above and below the film, it is also possible that these reflections yield a zero net phase change.

The thin film in Example 3 has the same yellow color everywhere. In nature, such a uniformly colored thin film would be unusual; the next example is more realistic.

CONCEPTUAL EXAMPLE 4 | Multicolored Thin Films

Under natural conditions, thin films, like gasoline on water or like the soap bubble in **Figure 27.11**, have a multicolored appearance that often changes while you are watching them. Why are such films multicolored, and what can be inferred from the fact that the colors change in time?

Reasoning and Solution In Example 3 we have seen that a thin film can appear yellow if destructive interference removes blue light from the reflected sunlight. The thickness of the film is the key. If the thickness were different, so that destructive interference removed green light from the reflected sunlight, the film would appear magenta. Constructive interference can also cause certain colors to appear brighter than others in the reflected light and thereby give the film a colored appearance. The colors that are enhanced by constructive interference, like those removed by destructive interference, depend on the thickness of the film. Thus, we conclude the following:

Problem-Solving Insight The different colors in a thin film of gasoline on water or in a soap bubble arise because the thickness is different in different places on the film. Moreover, the fact that the colors change as you watch them indicates that the thickness is changing.

A number of factors can cause the thickness to change, including air currents, temperature fluctuations, and the pull of gravity, which tends to make a vertical film sag, leading to thicker regions at the bottom than at the top.

Related Homework: Problem 17

FIGURE 27.11 This fantastic soap bubble is multicolored when viewed in sunlight because of the effects of thin-film interference.

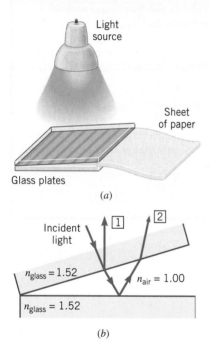

FIGURE 27.12 (*a*) The wedge of air formed between two flat glass plates causes an interference pattern of alternating dark and bright fringes to appear in reflected light. (*b*) A side view of the glass plates and the air wedge.

The colors that you see when sunlight is reflected from a thin film also depend on your viewing angle. At an oblique angle, the light corresponding to ray 2 in **Interactive Figure 27.9**, for instance, would travel a greater distance within the film than it does at nearly perpendicular incidence. The greater distance would lead to destructive interference for a different wavelength.

THE PHYSICS OF . . . nonreflecting lens coatings. Thin-film interference can be beneficial in optical instruments. For example, some cameras contain six or more lenses. Reflections from all the lens surfaces can reduce considerably the amount of light directly reaching the film. In addition, multiple reflections from the lenses often reach the film indirectly and degrade the quality of the image. To minimize such unwanted reflections, high-quality lenses are often covered with a thin nonreflective coating of magnesium fluoride ($n = 1.38$). The thickness of the coating is usually chosen to ensure that destructive interference eliminates the reflection of green light, which is in the middle of the visible spectrum. It should be pointed out that the absence of any reflected light does not mean that it has been destroyed by the nonreflective coating. Rather, the "missing" light has been transmitted into the coating and the lens.

Another interesting illustration of thin-film interference is the air wedge. As **Figure 27.12a** shows, an air wedge is formed when two flat plates of glass are separated along one side, perhaps by a thin sheet of paper. The thickness of this film of air varies between zero, where the plates touch, and the thickness of the paper. When monochromatic light reflects from this arrangement, alternate bright and dark fringes are formed by constructive and destructive interference, as the drawing indicates and Example 5 discusses.

EXAMPLE 5 | An Air Wedge

(a) Assuming that green light ($\lambda_{vacuum} = 552$ nm) strikes the glass plates nearly perpendicularly in **Figure 27.12**, determine the number of bright fringes that occur between the place where the plates touch and the edge of the sheet of paper (thickness $= 4.10 \times 10^{-5}$ m). **(b)** Explain why there is a dark fringe where the plates touch.

Reasoning A bright fringe occurs wherever there is constructive interference, as determined by any phase changes due to reflection and by the thickness of the air wedge. There is no phase change

upon reflection for wave 1, since this light travels from a larger (glass) toward a smaller (air) refractive index. In contrast, there is a half-wavelength phase change for wave 2, since the ordering of the refractive indices is reversed at the lower air/glass boundary where reflection occurs. The net phase change due to reflection for waves 1 and 2, then, is equivalent to a half-wavelength. Now we combine any extra distance traveled by ray 2 with this half-wavelength and determine the condition for the constructive interference that creates the bright fringes. Constructive interference occurs whenever the *combination* yields an integer number of wavelengths. At nearly perpendicular incidence, the extra travel distance for wave 2 is approximately twice the thickness t of the wedge at any point, so the condition for constructive interference is

$$\underbrace{2t}_{\substack{\text{Extra distance} \\ \text{traveled by} \\ \text{wave 2}}} + \underbrace{\tfrac{1}{2}\lambda_{\text{film}}}_{\substack{\text{Half-wavelength} \\ \text{net phase change} \\ \text{due to reflection}}} = \underbrace{\lambda_{\text{film}}, 2\lambda_{\text{film}}, 3\lambda_{\text{film}}, \ldots}_{\substack{\text{Condition for} \\ \text{constructive interference}}}$$

Subtracting the term $\tfrac{1}{2}\lambda_{\text{film}}$ from the left-hand side of this equation and from each term on the right-hand side yields

$$2t = \underbrace{\tfrac{1}{2}\lambda_{\text{film}}, \tfrac{3}{2}\lambda_{\text{film}}, \tfrac{5}{2}\lambda_{\text{film}}, \ldots}_{(m+\tfrac{1}{2})\lambda_{\text{film}} \quad m = 0, 1, 2, 3, \ldots}$$

Therefore,

$$t = \frac{\left(m + \tfrac{1}{2}\right)\lambda_{\text{film}}}{2} \quad m = 0, 1, 2, 3, \ldots$$

In this expression, note that the "film" is a film of air. Since the refractive index of air is nearly one, λ_{film} is virtually the same as that in a vacuum, so $\lambda_{\text{film}} = 552$ nm.

Solution   **(a)** When t equals the thickness of the paper holding the plates apart, the corresponding value of m can be obtained from the equation above:

$$m = \frac{2t}{\lambda_{\text{film}}} - \frac{1}{2} = \frac{2(4.10 \times 10^{-5}\,\text{m})}{552 \times 10^{-9}\,\text{m}} - \frac{1}{2} = 148$$

Since the first bright fringe occurs when $m = 0$, the number of bright fringes is $m + 1 = \boxed{149}$.

(b) Where the plates touch, there is a dark fringe because of destructive interference between the light waves represented by rays 1 and 2. Destructive interference occurs because the thickness of the wedge is zero here and the only difference between the rays is the half-wavelength phase change due to reflection from the lower plate.

Another type of air wedge can also be used to determine the degree to which the surface of a lens or mirror is spherical. When an accurate spherical surface is put in contact with an optically flat plate, as in **Figure 27.13a**, the circular interference fringes shown in part *b* of the figure can be observed. The circular fringes are called *Newton's rings*. They arise in the same way that the straight fringes arise in **Figure 27.12a**.

> **Problem-Solving Insight**   Reflected light will experience a phase change only if the light travels from a material with a smaller refractive index toward a material with a larger refractive index. Be sure to take such a phase change into account when analyzing thin-film interference.

As a final example of thin-film interference, we consider the phenomenon of structural coloration in the animal world in Example 6.

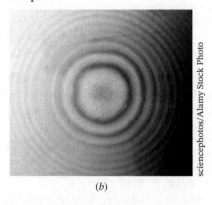

sciencephotos/Alamy Stock Photo

(a)   (b)

FIGURE 27.13   (*a*) The air wedge between a convex spherical glass surface and an optically flat plate leads to (*b*) a pattern of circular interference fringes that are known as Newton's rings.

EXAMPLE 6 | BIO The Physics of Structural Coloration

As its name suggests, the blue morpho butterfly (**Figure 27.14a**) has beautiful metallic blue wings. The color of its wings is not due to pigmentation–they are actually brown–but rather due to an interference effect. An electron microscope image of its wings reveals multiple layers of horizontal rows of shingle-like scales. Along the vertical direction of these rows are Christmas tree-like nanostructures composed of chitin–a long-chain polymer structurally similar to cellulose (**Figure 27.14b**). When light is incident on the wings, it undergoes multiple reflections from the many layers in these structures. We can model the reflections as a thin-film interference problem with the geometry shown in **Figure 27.14c**. Along the vertical direction of the tree-like structures, there are

alternating layers (or films) of chitin and air. The thickness of the chitin layers is 40.0 nm, and the thickness of the air layers is 115 nm. The thickness of the chitin layers is too small to produce constructive interference in the visible spectrum. However, the thickness of the air layers is large enough, so we treat the problem as an air film surrounded by chitin on both sides. Which wavelengths of visible light in vacuum are strongly reflected from the butterfly wings due to constructive interference? The indices of refraction of air and chitin are 1.00 and 1.56, respectively.

Reasoning   We can solve this problem by applying the thin-film interference condition for constructive interference, where we consider any phase shifts that occur upon reflection from the

2016. Published by The Company of Biologists Ltd. Giraldo. M. A., Liu, Y. C. and Stavenga. D. G. (2016). Coloration mechanisms and phylogeny of Morpho butterflies. J. Exp. Biol. 219, 3936-3944 10.1242/jeb.148726

Radislav A. Potyrailo et al., DOI: 10.1038/ncomms8959

Desertrose7/Pixabay

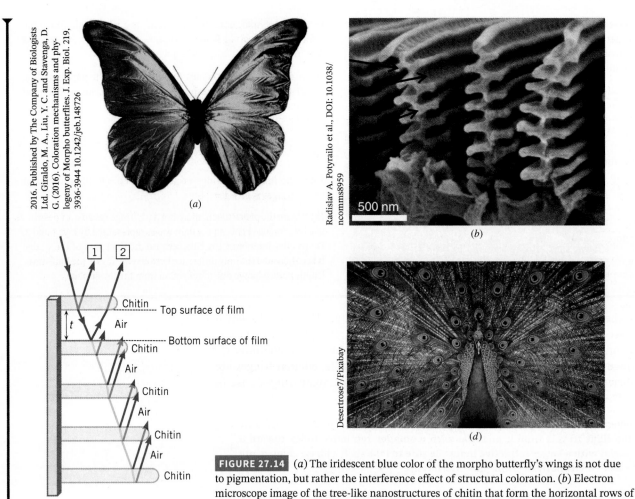

(a)

(b)

(c)

(d)

FIGURE 27.14 (a) The iridescent blue color of the morpho butterfly's wings is not due to pigmentation, but rather the interference effect of structural coloration. (b) Electron microscope image of the tree-like nanostructures of chitin that form the horizontal rows of scales in the butterfly's wings. (c) A sketch showing two waves that reflect from the top and bottom surface of the air film ($n = 1.00$) that is surrounded on both sides by chitin ($n = 1.56$). Some of the additional multiple reflections are also shown. (d) Structural coloration is also responsible for the beautiful iridescent plumage on many birds, such as the peacock.

top (wave 1) and bottom surface (wave 2) of the air film. In **Figure 27.14c**, wave 1 reflects from the top surface of the air film. It is traveling from a larger index of refraction ($n_{chitin} = 1.56$) toward a smaller index of refraction ($n_{air} = 1.00$). Therefore, wave 1 undergoes no phase shift upon reflection. Wave 2 reflects from the bottom surface of the air film. In this case, wave 2 is traveling from a smaller index of refraction ($n_{air} = 1.00$) toward a larger index ($n_{chitin} = 1.56$). Therefore, wave 2 undergoes a $\frac{1}{2}\lambda_{film}$ phase shift upon reflection. Here, the optical path length difference is just twice the film thickness ($2t$), we have one phase shift, and the condition for constructive interference is

$$\underbrace{2t}_{\substack{\text{Extra distance} \\ \text{traveled by} \\ \text{wave 2}}} + \underbrace{\tfrac{1}{2}\lambda_{film}}_{\substack{\text{Half-wavelength} \\ \text{net phase change} \\ \text{due to reflection}}} = \underbrace{\lambda_{film}, 2\lambda_{film}, 3\lambda_{film}, \ldots}_{\substack{\text{Condition for} \\ \text{constructive interference}}}$$

Subtracting the $\frac{1}{2}\lambda_{film}$ phase shift from the left-hand side of this equation and from each term on the right-hand side yields

$$2t = \left(m + \tfrac{1}{2}\right)\lambda_{film}, \quad m = 0, 1, 2, 3, \ldots$$

As in Example 5, the "film" is a film of air, so that λ_{film} is virtually the same as λ_{vacuum}. After making this substitution, we solve for λ_{vacuum} in terms of m:

$$\lambda_{vacuum} = \frac{2t}{\left(m + \frac{1}{2}\right)} =, \, m = 0, 1, 2, 3, \ldots \Rightarrow \lambda_{vacuum} = \frac{2(115 \text{ nm})}{\left(m + \frac{1}{2}\right)},$$
$$m = 0, 1, 2, 3, \ldots$$

Solution Using the first three values of m (0, 1, and 2) returns the following results for λ_{vacuum}: 460 nm, 153 nm, and 92.0 nm, respectively. The 460 nm result for $m = 0$ is the only one that resides in the visible spectrum. The other two satisfy the constructive interference condition for wavelengths of light smaller than visible light, as do all the larger values of m. As expected, 460 nm is located in the blue region of the visible spectrum. While we only considered the interference from the air film closest to the surface of the wing, light shining on the butterfly undergoes multiple reflections from the many structural components inside its wings. This, along with the size of the air gap, effectively enhances the blue light, while averaging out the other colors of visible light. When they change their viewing angle, observers will notice subtle variations in the blue color, which is referred to as *iridescence*. Examples of iridescence are also seen in shellfish, beetles, and bird feathers–most notably, the peacock (**Figure 27.14d**).

Check Your Understanding

(The answers are given at the end of the book.)

8. A camera lens is covered with a nonreflective coating that eliminates the reflection of perpendicularly incident green light. Recalling Snell's law of refraction (see Section 26.2), would you expect the reflected green light to be eliminated if it were incident on the nonreflective coating at an angle of 45° rather than perpendicularly? **(a)** No, because the distance traveled by the light in the film is less than twice the film thickness. **(b)** No, because the distance traveled by the light in the film is greater than twice the film thickness. **(c)** Yes, the green light will still be eliminated.

9. Two pieces of the same glass are covered with thin films of different materials. In reflected sunlight, however, the films have different colors. Why? **(a)** The films could have the same thickness, but different refractive indices. **(b)** The films could have different thicknesses, but the same refractive indices. **(c)** Both of the preceding answers could be correct.

10. A transparent coating is deposited on a glass plate and has a refractive index that is *larger than that of the glass.* For a certain wavelength within the coating, the thickness of the coating is a quarter-wavelength. Does the coating enhance or reduce the reflection of the light corresponding to this wavelength?

11. The drawings in **CYU Figure 27.2** show three situations—A, B, and C—in which light reflects almost perpendicularly from the top and bottom surfaces of a thin film, with the indices of refraction as shown. **(a)** For which situation(s) will there be a net phase shift (due to reflection) between waves 1 and 2 that is equivalent to either zero wavelengths or one wavelength (λ_{film}), where λ_{film} is the wavelength of the light in the film? **(b)** For which situation(s) will the film appear dark when the thickness of the film is equal to $\frac{1}{2}\lambda_{film}$?

CYU FIGURE 27.2

12. When sunlight reflects from a thin film of soapy water (air on both sides), the film appears multicolored, in part because destructive interference removes different wavelengths from the light reflected at different places, depending on the thickness of the film. What happens as the film becomes thinner and thinner? **(a)** Nothing happens, and the film remains multicolored. **(b)** The film looks brighter and brighter in reflected light, appearing totally white just before it breaks. **(c)** The film looks darker and darker in reflected light, appearing black just before it breaks.

13. Two thin films are floating on water ($n = 1.33$). The films have refractive indices of $n_1 = 1.20$ and $n_2 = 1.45$. Suppose that the thickness of each film approaches zero. In reflected light, film 1 will look _____ and film 2 will look _____. **(a)** bright, bright **(b)** bright, dark **(c)** dark, bright **(d)** dark, dark

---

27.4 | The Michelson Interferometer

THE PHYSICS OF . . . the Michelson interferometer. An interferometer is an apparatus that can be used to measure the wavelength of light by utilizing interference between two light waves. One particularly famous interferometer was developed by Albert A. Michelson (1852–1931). The Michelson interferometer uses reflection to set up conditions where two light waves interfere. **Figure 27.15** presents a schematic drawing of the instrument. Waves emitted by the monochromatic light source strike a

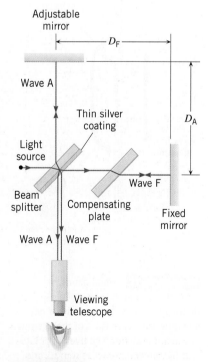

FIGURE 27.15 A schematic drawing of a Michelson interferometer.

beam splitter, so called because it splits the beam of light into two parts. The beam splitter is a glass plate, the far side of which is coated with a thin layer of silver that reflects part of the beam upward as wave A in the drawing. The coating is so thin, however, that it also allows the remainder of the beam to pass directly through as wave F. Wave A strikes an adjustable mirror and reflects back on itself. It again crosses the beam splitter and then enters the viewing telescope. Wave F strikes a fixed mirror and returns, to be partly reflected into the viewing telescope by the beam splitter. Note that wave A passes through the glass plate of the beam splitter three times in reaching the viewing scope, while wave F passes through it only once. The compensating plate in the path of wave F has the same thickness as the beam splitter plate and ensures that wave F also passes three times through the same thickness of glass on the way to the viewing scope. Thus, an observer viewing the superposition of waves A and F through the telescope sees constructive or destructive interference, depending only on the difference in path lengths D_A and D_F traveled by the two waves.

Now suppose that the mirrors are perpendicular to each other, the beam splitter makes a 45° angle with each, and the distances D_A and D_F are equal. Waves A and F travel the same distance, and the field of view in the telescope is uniformly bright due to constructive interference. However, if the adjustable mirror is moved away from the telescope by a distance of $\frac{1}{4}\lambda$, wave A travels back and forth by an amount that is twice this value, leading to an extra distance of $\frac{1}{2}\lambda$. Then, the waves are out of phase when they reach the viewing scope, destructive interference occurs, and the viewer sees a dark field. If the adjustable mirror is moved farther, full brightness returns as soon as the waves are in phase and interfere constructively. The in-phase condition occurs when wave A travels a total extra distance of λ relative to wave F. Thus, as the mirror is continuously moved, the viewer sees the field of view change from bright to dark, then back to bright, and so on. The amount by which D_A has been changed can be measured and related to the wavelength of the light, since a bright field changes into a dark field and back again each time D_A is changed by a half-wavelength. (The back-and-forth change in distance is λ.) If a sufficiently large number of wavelengths are counted in this manner, the Michelson interferometer can be used to obtain a very accurate value for the wavelength from the measured changes in D_A.

27.5 Diffraction

As we have seen in Section 17.3, **diffraction** is the bending of waves around obstacles or the edges of an opening. In **Figure 27.16**, sound waves are leaving a room through an open doorway and bend, or diffract, around the edges of the opening. Therefore, a listener hears the sound even when he is around the corner from the doorway.

Diffraction is an interference effect, and the Dutch scientist Christian Huygens (1629–1695) developed a principle that is useful in explaining why diffraction arises. **Huygens' principle** describes how a wave front that exists at one instant gives rise to the wave front that exists later on. This principle states that:

> *Every point on a wave front acts as a source of tiny wavelets that move forward with the same speed as the wave; the wave front at a later instant is the surface that is tangent to the wavelets.*

We begin by using Huygens' principle to explain the diffraction of sound waves in **Figure 27.16**. The drawing shows the top view of a plane wave front of sound approaching a doorway and identifies five points on the wave front just as it is leaving the opening. According to Huygens' principle, each of these points acts as a source of wavelets, which are shown as red circular arcs at some moment after they are emitted. The tangent to the wavelets from points 2, 3, and 4 indicates that in front of the doorway the wave front is flat and moving straight ahead. At the edges, however, points 1 and 5 are the last points that produce wavelets. Huygens' principle suggests that in conforming to the curved shape of the wavelets near the edges, the new wave front

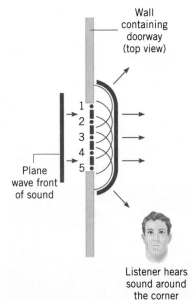

Wall containing doorway (top view)

Plane wave front of sound

Listener hears sound around the corner

FIGURE 27.16 Sound bends, or diffracts, around the edges of a doorway, so even a person who is not standing directly in front of the opening can hear the sound. The five red points within the doorway act as sources and emit the five Huygens wavelets shown in red.

(a) Smaller value for λ/W, less diffraction

(b) Larger value for λ/W, more diffraction

Courtesy Education Development Center

Courtesy Education Development Center

FIGURE 27.17 These photographs show water waves (horizontal lines) approaching an opening whose width *W* is greater in (*a*) than in (*b*). In addition, the wavelength λ of the waves is smaller in (*a*) than in (*b*). Therefore, the ratio λ/*W* increases from (*a*) to (*b*) and so does the extent of the diffraction, as the red arrows indicate.

moves into regions that it would not reach otherwise. The sound wave, then, bends or diffracts around the edges of the doorway.

Huygens' principle applies not just to sound waves, but to all kinds of waves. For instance, light has a wave-like nature and, consequently, exhibits diffraction. Therefore, you may ask, "Since I can hear around the edges of a doorway, why can't I also see around them?" As a matter of fact, light waves do bend around the edges of a doorway. However, the degree of bending is extremely small, so the diffraction of light is not enough to allow you to see around the corner.

As we will learn, the extent to which a wave bends around the edges of an opening is determined by the ratio λ/*W*, where λ is the wavelength of the wave and *W* is the width of the opening. The photographs in **Figure 27.17** illustrate the effect of this ratio on the diffraction of water waves. The degree to which the waves are diffracted or bent is indicated by the two red arrows in each photograph. In part *a*, the ratio λ/*W* is small because the wavelength (as indicated by the distance between the wave fronts) is small relative to the width of the opening. The wave fronts move through the opening with little bending or diffraction into the regions around the edges. In part *b*, the wavelength is larger and the width of the opening is smaller. As a result, the ratio λ/*W* is larger, and the wave fronts bend more into the regions around the edges of the opening.

Based on the pictures in **Figure 27.17**, we expect that light waves of wavelength λ will bend or diffract appreciably when they pass through an opening whose width *W* is small enough to make the ratio λ/*W* sufficiently large. This is indeed the case, as **Figure 27.18** illustrates. In this picture, it is assumed that parallel rays (or plane wave

(a) Without diffraction

(b) With diffraction

FIGURE 27.18 (*a*) If light were to pass through a very narrow slit *without* being diffracted, only the region on the screen directly opposite the slit would be illuminated. (*b*) Diffraction causes the light to bend around the edges of the slit into regions it would not otherwise reach, forming a pattern of alternating bright and dark fringes on the screen. The slit width has been exaggerated for clarity.

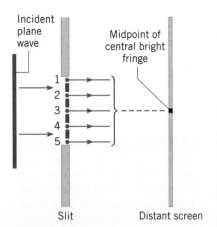

FIGURE 27.19 A plane wave front is incident on a single slit. This top view of the slit shows five sources of Huygens wavelets. The wavelets travel toward the midpoint of the central bright fringe on the screen, as the red rays indicate. The screen is very far from the slit.

fronts) of light fall on a very narrow slit and illuminate a viewing screen that is located far from the slit. Part *a* of the drawing shows what would happen if light were *not* diffracted: it would pass through the slit without bending around the edges and would produce an image of the slit on the screen. Part *b* shows what actually happens. The light diffracts around the edges of the slit and brightens regions on the screen that are not directly opposite the slit. The diffraction pattern on the screen consists of a bright central band, accompanied by a series of narrower faint fringes that are parallel to the slit itself.

To help explain how the pattern of diffraction fringes arises, **Figure 27.19** shows a top view of a plane wave front approaching the slit and singles out five sources of Huygens wavelets. Consider how the light from these five sources reaches the midpoint on the screen. To simplify things, the screen is assumed to be so far from the slit that the rays from each Huygens source are nearly parallel.* Then, all the wavelets travel virtually the same distance to the midpoint, arriving there in phase. As a result, constructive interference creates a bright central fringe on the screen, directly opposite the slit.

The wavelets emitted by the Huygens sources in the slit can also interfere destructively on the screen, as **Figure 27.20** illustrates. Part *a* shows light rays directed from each source toward the first dark fringe. The angle θ gives the position of this dark fringe relative to the line between the midpoint of the slit and the midpoint of the central bright fringe. Since the screen is very far from the slit, the rays from each Huygens source are nearly parallel and are oriented at nearly the same angle θ, as in part *b* of the drawing. The wavelet from source 1 travels the shortest distance to the screen, while the wavelet from source 5 travels the farthest. Destructive interference creates the first dark fringe when the extra distance traveled by the wavelet from source 5 is exactly one wavelength, as the colored right triangle in the drawing indicates. Under this condition, the extra distance traveled by the wavelet from source 3 at the center of the slit is exactly one-half of a wavelength. Therefore, wavelets from sources 1 and 3 in **Figure 27.20b** are exactly out of phase and interfere destructively when they reach the screen. Similarly, a wavelet that originates slightly below source 1 cancels a wavelet that originates the same distance below source 3. Thus, each wavelet from the upper half of the slit cancels a corresponding wavelet from the lower half, and no light reaches the screen. As can be seen from the

FIGURE 27.20 These drawings pertain to single-slit diffraction and show how destructive interference leads to the first dark fringe on either side of the central bright fringe. For clarity, only one of the dark fringes is shown. The screen is very far from the slit.

*When the rays are parallel, the diffraction is called Fraunhofer diffraction in tribute to the German optician Joseph von Fraunhofer (1787–1826). When the rays are not parallel, the diffraction is referred to as Fresnel diffraction, named for the French physicist Augustin Jean Fresnel (1788–1827).

colored right triangle, the angle θ locating the first dark fringe is given by $\sin \theta = \lambda/W$, where W is the width of the slit.

Math Skills To understand why the colored right triangle in **Figure 27.20b** implies that $\sin \theta = \lambda/W$, it is necessary to see why there are two angles labeled θ in the figure. To help explain these angles, we show a simplified version of **Figure 27.20b** in **Figure 27.21**, where we label the angle in the colored triangle as α. We do this so that we can show that the angle α is, in fact, the same as the angle θ between the red ray at the top of the slit and the horizontal dashed line. Since the slit is oriented vertically, we know that the line AC in **Figure 27.21** is perpendicular to the horizontal dashed line, with the result that the angles α and β form a right angle:

$$\alpha + \beta = 90° \quad \text{or} \quad \alpha = 90° - \beta$$

In addition, we know that the line AB in **Figure 27.21** has been drawn perpendicular to the red rays, which means that the angles θ and β also form a right angle:

$$\theta + \beta = 90° \quad \text{or} \quad \beta = 90° - \theta$$

Substituting this result for β into the result for α reveals that

$$\alpha = 90° - \beta = 90° - (90° - \theta) = \theta$$

FIGURE 27.21
Math Skills drawing.

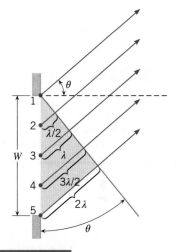

FIGURE 27.22 In a single-slit diffraction pattern, multiple dark fringes occur on either side of the central bright fringe. This drawing shows how destructive interference creates the second dark fringe on a very distant screen.

Figure 27.22 shows the condition that leads to destructive interference at the second dark fringe on either side of the midpoint on the screen. In reaching the screen, the light from source 5 now travels a distance of two wavelengths farther than the light from source 1. Under this condition, the wavelet from source 5 travels one wavelength farther than the wavelet from source 3, and the wavelet from source 3 travels one wavelength farther than the wavelet from source 1. Therefore, each half of the slit can be treated as the entire slit was in the previous paragraph; all the wavelets from the top half interfere destructively with each other, and all the wavelets from the bottom half do likewise. As a result, no light from either half reaches the screen, and another dark fringe occurs. The colored triangle in the drawing shows that this second dark fringe occurs when $\sin \theta = 2\lambda/W$. Similar arguments hold for the third- and higher-order dark fringes, with the general result being

Dark fringes for single-slit diffraction

$$\sin \theta = m\frac{\lambda}{W} \quad m = 1, 2, 3, \ldots \qquad \text{(27.4)}$$

Between each pair of dark fringes there is a bright fringe due to constructive interference. The brightness of the fringes is related to the light intensity, just as loudness is related to sound intensity. The intensity of the light at any location on the screen is the amount of light energy per second per unit area that strikes the screen there. **Figure 27.23** gives a graph of the light intensity, along with the single-slit diffraction pattern. The central bright fringe, which is approximately twice as wide as the other bright fringes, has by far the greatest intensity.

The width of the central fringe provides some indication of the extent of the diffraction, as Example 7 illustrates.

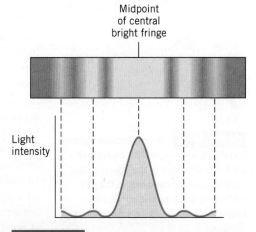

FIGURE 27.23 A single-slit diffraction pattern, with a bright and wide central fringe. The higher-order bright fringes are much less intense than the central fringe, as the graph indicates.

EXAMPLE 7 | Single-Slit Diffraction

Light passes through a slit and shines on a flat screen that is located $L = 0.40$ m away (see **Figure 27.24**). The wavelength of the light in a vacuum is $\lambda = 410$ nm. The distance between the midpoint of the central bright fringe and the first dark fringe is y. Determine the width $2y$ of the central bright fringe when the width of the slit is **(a)** $W = 5.0 \times 10^{-6}$ m and **(b)** $W = 2.5 \times 10^{-6}$ m.

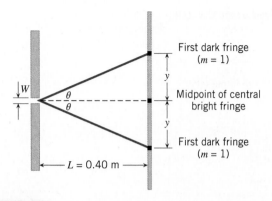

FIGURE 27.24 The distance $2y$ is the width of the central bright fringe.

Reasoning   The width of the central bright fringe is determined by two factors. One is the angle θ that locates the first dark fringe

on either side of the midpoint. The other is the distance L between the screen and the slit. Larger values for θ and L lead to a wider central bright fringe. Larger values of θ mean greater diffraction and occur when the ratio λ/W is larger. Thus, we expect the width of the central bright fringe to be greater when the slit width W is smaller.

Solution   **(a)** The angle θ in Equation 27.4 locates the first dark fringe when $m = 1$: $\sin \theta = (1)\lambda/W$. Therefore,

$$\theta = \sin^{-1}\left(\frac{\lambda}{W}\right) = \sin^{-1}\left(\frac{410 \times 10^{-9}\ \text{m}}{5.0 \times 10^{-6}\ \text{m}}\right) = 4.7°$$

According to **Figure 27.24**, $\tan \theta = y/L$, so the width $2y$ of the central bright fringe is

$$2y = 2L \tan \theta = 2(0.40\ \text{m}) \tan 4.7° = \boxed{0.066\ \text{m}}$$

(b) Repeating the same calculations as in part (a) with $W = 2.5 \times 10^{-6}$ m reveals that $\boxed{2y = 0.13\ \text{m}}$. As expected for a given wavelength, the width $2y$ of the central maximum in the diffraction pattern is greater when the width of the slit is smaller.

THE PHYSICS OF . . . producing computer chips using photolithography.
In the production of computer chips, it is important to minimize the effects of diffraction. Such chips are very small and yet contain enormous numbers of electronic components, as **Figure 23.32** illustrates. Such miniaturization is achieved using the techniques of photolithography. The patterns on the chip are created first on a "mask," which is similar to a photographic slide. Light is then directed through the mask onto silicon wafers that have been coated with a photosensitive material. The light-activated parts of the coating can be removed chemically, to leave the ultrathin lines that correspond to the miniature patterns on the chip. As the light passes through the narrow slit-like patterns on the mask, the light spreads out due to diffraction. If excessive diffraction occurs, the light spreads out so much that sharp patterns are not formed on the photosensitive material coating the silicon wafer. Ultraminiaturization of the patterns requires the absolute minimum of diffraction, and currently this is achieved by using ultraviolet light, which has a wavelength shorter than that of visible light. The shorter the wavelength λ, the smaller the ratio λ/W, and the less the diffraction. The wavelengths of X-rays are much shorter than those of ultraviolet light and, thus, will reduce diffraction even more, allowing further miniaturization.

Another example of diffraction can be seen when light from a point source falls on an opaque disk, such as a coin (**Figure 27.25**). The effects of diffraction modify the dark shadow cast by the disk in several ways. First, the light waves diffracted around the circular edge of the disk interfere constructively at the center of the shadow to produce a small bright spot. There are also circular bright fringes in the shadow area. In addition, the boundary between the circular shadow and the lighted screen is not sharply defined but consists of concentric bright and dark fringes. The various fringes are analogous to those produced by a single slit and are due to interference between Huygens wavelets that originate from different points near the edge of the disk.

FIGURE 27.25 The diffraction pattern formed by an opaque disk consists of a small bright spot in the center of the dark shadow, circular bright fringes within the shadow, and concentric bright and dark fringes surrounding the shadow.

Check Your Understanding

(The answers are given at the end of the book.)

14. A diffraction pattern is produced on a viewing screen by using a single slit with blue light. Does the pattern broaden or contract (become narrower) **(a)** when the blue light is replaced by red light **(b)** when the slit width is increased?

15. A sound wave has a much greater wavelength than does a light wave. When the two waves pass through a doorway, which one, if either, diffracts to a greater extent? **(a)** The sound wave **(b)** The light wave **(c)** Both waves diffract by the same amount.

27.6 | Resolving Power

Figure 27.26 shows three photographs of an automobile's headlights taken at progressively greater distances from the camera. In parts *a* and *b*, the two separate headlights can be seen clearly. In part *c*, however, the car is so far away that the headlights are barely distinguishable and appear almost as a single light. The **resolving power** of an optical instrument, such as a camera, is its ability to distinguish between two closely spaced objects. If a camera with a higher resolving power had taken these pictures, the photograph in part *c* would have shown two distinct and separate headlights. Any instrument used for viewing objects that are close together must have a high resolving power. This is true, for example, for a telescope used to view distant stars or for a microscope used to view tiny organisms. We will now see that diffraction occurs when light passes through the circular, or nearly circular, openings that admit light into cameras, telescopes, microscopes, and human eyes. The resulting diffraction pattern places a limit on the resolving power of these instruments.

FIGURE 27.26 These automobile headlights were photographed at various distances from the camera, closest in part *a* and farthest in part *c*. In part *c*, the headlights are so far away that they are barely distinguishable.

Truax/The Image Finders

Figure 27.27 shows the diffraction pattern created by a small circular opening when the viewing screen is far from the opening. The pattern consists of a central bright circular region, surrounded by alternating bright and dark circular fringes. These fringes are analogous to the rectangular fringes that a single slit produces. The angle θ in the picture locates the first circular dark fringe relative to the center of the central bright region and is given by

$$\sin \theta = 1.22 \frac{\lambda}{D} \qquad (27.5)$$

where λ is the wavelength of the light and D is the diameter of the opening. This expression is similar to Equation 27.4 for a slit ($\sin \theta = \lambda/W$, when $m = 1$) and is valid when the distance to the screen is much larger than the diameter D.

An optical instrument with the ability to resolve two closely spaced objects can produce images of them that can be identified separately. Think about the images on the image sensor when light from two widely separated point objects passes through the circular aperture of a camera. As **Figure 27.28** illustrates, each image is a circular diffraction pattern, but the two patterns do not overlap and are completely resolved. On the other hand, if the objects are sufficiently close together, the intensity patterns created by the diffraction overlap, as **Figure 27.29a** suggests. In fact, if the overlap is extensive, it may no longer be possible to distinguish the patterns separately. In such a case, the picture from a camera would show a single blurred object instead of two separate objects. In **Figure 27.29b** the diffraction patterns overlap, but not enough to prevent us from seeing that two objects are present. Ultimately, then, diffraction limits the ability of an optical instrument to produce distinguishable images of objects that are close together.

It is useful to have a criterion for judging whether two closely spaced objects will be resolved by an optical instrument. **Figure 27.29a** presents the **Rayleigh criterion** for resolution, first proposed by Lord Rayleigh (1842–1919):

FIGURE 27.27 When light passes through a small circular opening, a circular diffraction pattern is formed on a screen. The angle θ locates the first dark fringe relative to the central bright region. The intensities of the bright fringes and the diameter of the opening have been exaggerated for clarity.

FIGURE 27.28 When light from two point objects passes through the circular aperture of a camera, two circular diffraction patterns are formed as images on the image sensor. The images here are completely separated or resolved because the objects are widely separated.

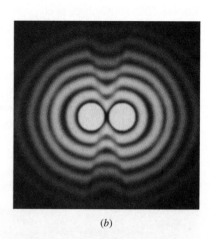

FIGURE 27.29 (*a*) According to the Rayleigh criterion, two point objects are just resolved when the first dark fringe (zero intensity) of one image falls on the central bright fringe (maximum intensity) of the other image. (*b*) This drawing shows two overlapping but still resolvable diffraction patterns.

Two point objects are just resolved when the first dark fringe in the diffraction pattern of one falls directly on the central bright fringe in the diffraction pattern of the other.

According to the Rayleigh criterion, the minimum angle θ_{min} between the two objects in **Figure 27.29a** is the angle given by Equation 27.5. If θ_{min} is small (less than about 9°) and is expressed in radians, $\sin \theta_{min} \approx \theta_{min}$. Then, Equation 27.5 becomes

$$\theta_{min} \approx 1.22 \frac{\lambda}{D} \quad (\theta_{min} \text{ in radians}) \tag{27.6}$$

Math Skills To see why $\sin \theta_{min} \approx \theta_{min}$ when θ_{min} is small and is expressed in radians, we drop the subscript (for simplicity) and consider Equation 8.1, which defines an angle θ in radians. According to this definition, θ is the length s of the circular arc that the angle subtends divided by the radius r of the arc (see **Figure 27.30a**):

$$\theta = \frac{s}{r}$$

On the other hand, Equation 1.1 defines $\sin \theta$ as

$$\sin \theta = \frac{h_o}{h}$$

where h_o is the length of the side of a right triangle that is opposite the angle θ and h is the length of the hypotenuse (see **Figure 27.30b**). As θ becomes smaller and smaller in both parts of **Figure 27.30**, we can see that h_o also becomes smaller and approaches the arc length s, whereas h approaches the radius r. Therefore, for a small angle θ expressed in radians, we conclude that

$$\sin \theta = \frac{h_o}{h} \approx \frac{s}{r} = \theta \quad \text{or} \quad \sin \theta \approx \theta$$

Note that the symbol "$\approx$" means "approximately equal to."

FIGURE 27.30 Math Skills drawing.

For a given wavelength λ and aperture diameter D, Equation 27.6 specifies the smallest angle that two point objects can subtend at the aperture and still be resolved. According to this equation, optical instruments designed to resolve closely spaced objects (small values of θ_{min}) must utilize the smallest possible wavelength and the largest possible aperture diameter. For example, when short-wavelength ultraviolet light is collected by its large 2.4-m-diameter mirror, the Hubble Space Telescope is capable of resolving two closely spaced stars that have an angular separation of about $\theta_{min} = 1 \times 10^{-7}$ rad. This angle is equivalent to resolving two objects only 1 cm apart when they are 1×10^5 m (about 62 miles) from the telescope. Example 8 deals with the resolving power of the human eye.

Analyzing Multiple-Concept Problems

EXAMPLE 8 | BIO The Physics of Comparing Human Eyes and Eagle Eyes

(a) A hang glider is flying at an altitude of 120 m. Green light (wavelength = 555 nm in vacuum) enters the pilot's eye through a pupil that has a diameter of 2.5 mm. Determine how far apart two point objects must be on the ground if the pilot is to have any hope of distinguishing between them (see **Figure 27.31**). **(b)** An eagle's eye has a pupil with a diameter of 6.2 mm. Repeat part (a) for an eagle flying at the same altitude as the glider.

Reasoning A greater distance s of separation between the objects on the ground makes it easier for the eye of the observer (the pilot or the eagle; see **Figure 27.31**) to resolve them as separate objects. This is because the angle that the two objects subtend at the pupil of the eye is greater when the separation distance is greater. This angle must be at least as large as the angle θ_{min} specified by the Rayleigh criterion for resolution. In applying this criterion, we will use the concept of the radian to express the angle as an arc length (approximately the separation distance) divided by a radius (the altitude), as discussed in Section 8.1.

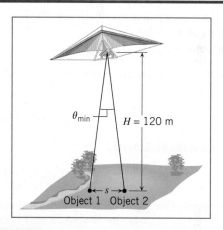

FIGURE 27.31 The Rayleigh criterion can be used to estimate the smallest distance s that can separate two objects on the ground, if a person on a hang glider is to be able to distinguish between them.

Knowns and Unknowns The following table summarizes the data that are given:

Description	Symbol	Value	Comment
Altitude	H	120 m	Same for pilot and eagle.
Wavelength of light in vacuum	λ	555 nm	1 nm = 10^{-9} m
Diameter of pupil of eye	D	2.5 mm or 6.2 mm	Smaller value is for pilot; larger value is for eagle.

Unknown Variables

Separation between objects on ground	s	?	

Modeling the Problem

STEP 1 Radian Measure The angle that the two objects subtend at the pupil of the eye must be at least as large as the angle θ_{min} specified by the Rayleigh criterion for resolution. Using radian measure as discussed in Section 8.1, we refer to **Figure 27.31** and express this angle as

$$\theta_{min} \approx \frac{s}{H}$$

This is an approximate application of Equation 8.1, which states that an angle in radians is the arc length divided by the radius. Here, the arc length is approximately the separation distance s, assuming that the altitude H is much greater than s. Solving for s gives

$$s \approx \theta_{min} H \qquad (1)$$

$\boxed{?}$

Equation 1 at the right. In this result, the altitude is known, and we proceed to Step 2 to evaluate the angle θ_{min}.

STEP 2 The Rayleigh Criterion The Rayleigh criterion specifies the angle θ_{min} in radians as

$$\boxed{\theta_{min} \approx 1.22\frac{\lambda}{D}} \qquad (27.6)$$

where λ is the wavelength of the light in vacuum* and D is the diameter of the pupil of the eye. The substitution of this expression into Equation 1 is shown at the right.

$$s \approx \theta_{min} H \qquad (1)$$

$$\boxed{\theta_{min} \approx 1.22\frac{\lambda}{D}} \qquad (27.6)$$

Problem-Solving Insight The minimum angle θ_{min} between two objects that are just resolved must be expressed in radians, not degrees, when using Equation 27.6 ($\theta_{min} \approx 1.22\,\lambda/D$).

Solution Combining the results of each step algebraically, we find that

$$\begin{array}{ccc} \text{STEP 1} & & \text{STEP 2} \\ \downarrow & & \downarrow \\ s \approx & \theta_{min} H \approx & \left(1.22\frac{\lambda}{D}\right)H \end{array}$$

The separation distance between the objects on the ground can now be obtained.

(a) For the pilot to have any hope of distinguishing between the objects, the separation distance must be at least

$$s \approx \left(1.22\frac{\lambda}{D}\right)H = 1.22\left(\frac{555 \times 10^{-9}\text{ m}}{2.5 \times 10^{-3}\text{ m}}\right)(120\text{ m}) = \boxed{0.033\text{ m}}$$

(b) For the eagle, we find that

$$s \approx \left(1.22\frac{\lambda}{D}\right)H = 1.22\left(\frac{555 \times 10^{-9}\text{ m}}{6.2 \times 10^{-3}\text{ m}}\right)(120\text{ m}) = \boxed{0.013\text{ m}}$$

Since the pupil of the eagle's eye is larger than that of a human eye, diffraction creates less of a limitation for the eagle; the two objects can be closer together and still be resolved by the eagle's eye.

Many optical instruments have a resolving power exceeding that of the human eye. The typical camera does, for instance. Conceptual Example 9 compares the abilities of the human eye and a camera to resolve two closely spaced objects.

CONCEPTUAL EXAMPLE 9 | Is What You See What You Get?

The French postimpressionist artist Georges Seurat developed a painting technique in which dots of color are placed close together on the canvas. From sufficiently far away the individual dots are not distinguishable, and the images in the picture take on a more normal appearance. **Figure 27.32** shows a person in a museum looking at one of Seurat's paintings. Suppose that the person stands close to the painting, then backs up until the dots just become indistinguishable to his eyes and takes a picture from this position. The light enters his eyes through pupils that have diameters of 2.5 mm and enters the digital camera through an aperture, or opening, with a diameter of 25 mm. He then goes home and prints an enlarged photograph of the painting. Can he see the individual dots in the photograph? **(a)** No, because if his eye cannot see the dots at the museum, the camera is also unable

---

*In applying the Rayleigh criterion, we use the given wavelength in vacuum because it is nearly identical to the wavelength in air. We use the wavelength in air or vacuum, even though the diffraction occurs within the eye, where the index of refraction is about $n_{eye} = 1.36$ and the wavelength is $\lambda_{eye} = \lambda_{vacuum}/n_{eye}$, according to Equation 27.3. The reason is that in entering the eye, the light is refracted according to Snell's law (Section 26.2), which also includes an effect due to the index of refraction. If the angle of incidence is small, the effect of n_{eye} in Snell's law cancels the effect of n_{eye} in Equation 27.3, to a good degree of approximation.

UniversalImagesGroup/Getty Images

FIGURE 27.32 This person is about to take a photograph of a famous painting by Georges Seurat, who developed the technique of using tiny dots of color to construct his images. Conceptual Example 9 discusses what the person sees when the photograph is printed.

to record the individual dots. **(b)** Yes, because the camera gathers light through a much larger aperture than does the eye. **(c)** Yes, because, unlike the eye, a photograph taken by a camera is not limited by the effects of diffraction.

Reasoning To answer this question, we turn to Equation 27.6, which expresses the Rayleigh criterion for resolving two point objects (such as the dots)—namely, $\theta_{min} \approx 1.22\lambda/D$. Here θ_{min} is the minimum angle that exists between light rays from two adjacent dots as the rays pass through the aperture (see **Figure 27.29**), λ is the wavelength of the light, and D is the diameter of the aperture. A larger value of D implies a smaller value for θ_{min}, which, in turn, means that the instrument has a greater resolving power.

Answer (a) is incorrect. Diffraction limits the ability of any instrument to see two closely spaced objects as distinct. This ability depends on the diameter of the aperture through which the light

enters the instrument. Since the eye and the camera have apertures with different diameters, the camera may record the individual dots in the painting even though the eye does not see them as distinct.

Answer (c) is incorrect. The effects of diffraction limit the resolving power of both the eye and the camera.

Answer (b) is correct. For the eye and the camera, the aperture diameters are $D_{eye} = 2.5$ mm and $D_{camera} = 25$ mm, so the diameter for the camera is ten times larger than that for the eye. Thus, at the distance at which the eye loses its ability to resolve the individual dots in the painting, the camera can still easily resolve them. As discussed in the footnote to Example 8, we can ignore the effect on the wavelength of the index of refraction of the material from which the eye is made.

Related Homework: Check Your Understanding Question 18, Problem 54

Conceptual Example 10 illustrates how diffraction affects the vision of animals that have evolved with pupils that are not circular.

CONCEPTUAL EXAMPLE 10 | BIO Predator or Prey? Pupils of Different Shapes

As discussed earlier in the chapter, pupils with different shapes will affect an animal's optical resolution along different directions. Some animals, like goats, and other large herbivores, have horizontal pupils, while others, like cats and crocodiles, have vertical pupils (**Figure 27.33**). Recent research suggests that animals that are ambush predators tend to have pupils with vertical slits, while plant-eating prey species have horizontal slits located near the sides of their heads. This is especially true for animals whose heads are close to the ground (i.e., grazing animals and small predators). Suppose a goat's eye has a horizontal, rectangular pupil that has a width-to-height ratio of 5:1. If the goat can optically resolve two-closely spaced objects separated by a distance s in the horizontal direction, how much closer would the objects have to be moved toward the goat for it to distinguish them in the vertical direction?

Reasoning We can apply Rayleigh's criterion for resolution (Equation 27.6) to the pupil in both directions.

Solution Beginning with Equation 27.6, we see that the minimum angle for resolution is inversely related to the pupil diameter:

$$\theta_{min} \propto \frac{1}{D}.$$

Using Equation 8.1, we can write the angle in terms of the separation distance (s) between the two objects and the distance from the objects (r):

$$\frac{s}{r} \propto \frac{1}{D}.$$

Rearranging this equation for r, we have: $r \propto sD$. Notice that the distance to the objects is directly proportional to the pupil diameter. Therefore, the objects would have to be moved five times closer for the goat to resolve them in the vertical direction. The goat's greater optical resolution in the horizontal direction allows it to detect predators along the ground at greater distances.

bones64/Pixabay

cocoparisienne/ Pixabay

Felix Broennimann/ Pixabay

FIGURE 27.33 The pupils of animals come in all shapes and sizes. Some are horizontal, such as in the goat, and others are vertical, such as in the domestic cat and crocodile. The shape of the pupil will affect the optical resolution of the eyes in different directions.

Check Your Understanding

(The answers are given at the end of the book.)

16. **BIO** Suppose that the pupil of your eye were elliptical instead of circular in shape, with the long axis of the ellipse oriented in the vertical direction. Would the resolving power of your eye be the same in the horizontal and vertical directions and, if not, in which direction would it be greater? The resolving power would **(a)** be the same in both directions **(b)** be greater in the horizontal direction **(c)** be greater in the vertical direction.

17. **BIO** Suppose that you were designing an eye and could select the size of the pupil and the wavelengths of the electromagnetic waves to which the eye is sensitive. As far as the limitation created by diffraction is concerned, rank the following design choices in order of decreasing resolving power (greatest first): **(a)** Large pupil and ultraviolet wavelengths **(b)** Small pupil and infrared wavelengths **(c)** Small pupil and ultraviolet wavelengths

18. **BIO** Review Conceptual Example 9 before answering this question. A person is viewing one of Seurat's paintings that consists of dots of color. She is so close to the painting that the dots are distinguishable. Without moving, she squints, thus reducing the size of the opening in her eyes. Does squinting make the painting take on a more normal appearance?

19. On many cameras one can select the *f*-number setting, or *f*-stop. The *f*-number gives the ratio of the focal length of the camera lens to the diameter of the aperture through which light enters the camera. If you want to resolve two closely spaced objects in a picture, should you use a small or a large *f*-number setting?

27.7 The Diffraction Grating

THE PHYSICS OF . . . a diffraction grating. Diffraction patterns of bright and dark fringes occur when monochromatic light passes through a single or double slit. Fringe patterns also result when light passes through more than two slits, and an arrangement consisting of a large number of parallel, closely spaced slits called a **diffraction grating** has proved very useful. Gratings with as many as 40 000 slits per centimeter can be made, depending on the production method. In one method a diamond-tipped cutting tool is used to inscribe closely spaced parallel lines on a glass plate, the spaces between the lines serving as the slits. In fact, the number of slits per centimeter is often quoted as the number of lines per centimeter.

Figure 27.34 illustrates how light travels to a distant viewing screen from each of five slits in a grating and forms the central bright fringe and the first-order bright fringes on either side. Higher-order bright fringes are also formed but are not shown in the drawing. Each bright fringe is located by an angle θ relative to the central fringe. These bright fringes are sometimes called the *principal fringes* or *principal maxima*, since they are places where the light intensity is a maximum. The term "principal" distinguishes them from other, much less bright, fringes that are referred to as *secondary fringes* or *secondary maxima*.

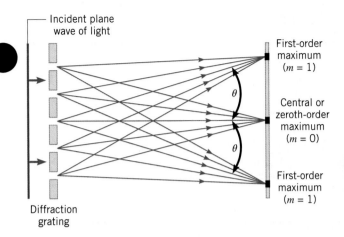

FIGURE 27.34 When light passes through a diffraction grating, a central bright fringe ($m = 0$) and higher-order bright fringes ($m = 1, 2, ...$) form when the light falls on a distant viewing screen.

Constructive interference creates the principal fringes. To show how, we assume the screen is far from the grating, so that the rays remain nearly parallel while the light travels toward the screen. In reaching the place on the screen where a first-order maximum is located, light from slit 2 travels a distance of one wavelength farther than light from slit 1, as in **Figure 27.35**. Similarly, light from slit 3 travels one wavelength

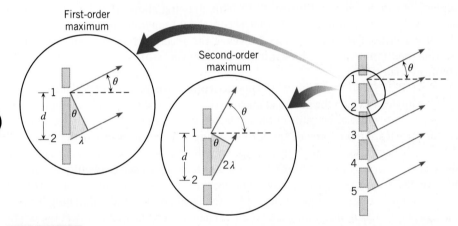

FIGURE 27.35 The conditions shown here lead to the first- and second-order intensity maxima in the diffraction pattern produced by the diffraction grating on the right.

farther than light from slit 2, and so forth, as emphasized by the four colored right triangles on the right-hand side of the drawing. For the first-order maximum, the blow-up view of slits 1 and 2 shows that constructive interference occurs when $\sin \theta = \lambda/d$, where d is the separation between slits. A second-order maximum forms when the extra distance traveled by light from adjacent slits is two wavelengths, so that $\sin \theta = 2\lambda/d$. The general result is

Principal maxima of
a diffraction grating

$$\sin \theta = m\frac{\lambda}{d} \quad m = 0, 1, 2, 3, \ldots \quad (27.7)$$

The separation d between the slits can be calculated from the number of slits per centimeter of grating; for instance, a grating with 2500 slits per centimeter has a slit separation of $d = (1/2500)$ cm $= 4.0 \times 10^{-4}$ cm. Equation 27.7 is identical to Equation 27.1 for the double slit. A grating, however, produces bright fringes that are much *narrower* or *sharper* than those from a double slit, as the intensity patterns in **Figure 27.36** reveal. Between the principal maxima of a diffraction grating there are secondary maxima with much smaller intensities. For a large number of slits, these secondary maxima are very small.

The next example illustrates the ability of a grating to separate the components in a mixture of colors.

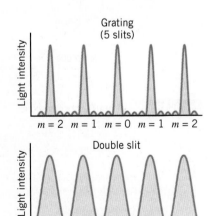

FIGURE 27.36 The bright fringes produced by a diffraction grating are much narrower than those produced by a double slit. Note the three small secondary bright fringes between the principal bright fringes of the grating.

EXAMPLE 11 | Separating Colors with a Diffraction Grating

A mixture of violet light ($\lambda = 410$ nm in vacuum) and red light ($\lambda = 660$ nm in vacuum) falls on a grating that contains 1.0×10^4 lines/cm. For each wavelength, find the angle θ that locates the first-order maxima.

Reasoning Before Equation 27.7 can be used here, a value for the separation d between the slits is needed: $d = 1/(1.0 \times 10^4 \text{ lines/cm}) = 1.0 \times 10^{-4}$ cm, or 1.0×10^{-6} m. For violet light, the angle θ_{violet} for the first-order maxima ($m = 1$) is given by $\sin \theta_{\text{violet}} = m\lambda_{\text{violet}}/d$, with an analogous equation applying for the red light.

Solution For violet light, the angle locating the first-order maxima is

$$\theta_{\text{violet}} = \sin^{-1}\frac{\lambda_{\text{violet}}}{d} = \sin^{-1}\left(\frac{410 \times 10^{-9} \text{ m}}{1.0 \times 10^{-6} \text{ m}}\right) = \boxed{24°}$$

For red light, a similar calculation with $\lambda_{\text{red}} = 660 \times 10^{-9}$ m shows that $\boxed{\theta_{\text{red}} = 41°}$. Because θ_{violet} and θ_{red} are different, separate first-order bright fringes are seen for violet and red light on a viewing screen.

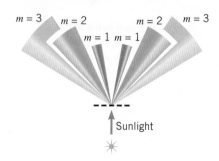

FIGURE 27.37 When sunlight falls on a diffraction grating, a rainbow of colors is produced at each principal maximum ($m = 1, 2, \ldots$). The central maximum ($m = 0$), however, is white but is not shown in the drawing.

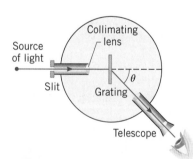

FIGURE 27.38 A grating spectroscope.

If the light in Example 11 had been sunlight, the angles for the first-order maxima would cover all values in the range between 24° and 41°, since sunlight contains all colors or wavelengths between violet and red. Consequently, a rainbow-like dispersion of the colors would be observed to either side of the central fringe on a screen, as can be seen in **Figure 27.37**. This drawing shows that the spectrum of colors associated with the $m = 2$ order is completely separate from the spectrum of the $m = 1$ order. For higher orders, however, the spectra from adjacent orders may overlap (see Problem 41). The central maximum ($m = 0$) is white because all the colors overlap there.

THE PHYSICS OF ... a grating spectroscope. An instrument designed to measure the angles at which the principal maxima of a grating occur is called a grating spectroscope. With a measured value of the angle, calculations such as those in Example 11 can be turned around to provide the corresponding value of the wavelength. As we will discuss in Chapter 30, the atoms in a hot gas emit discrete wavelengths, and determining the values of these wavelengths is one important technique used to identify the atoms. **Figure 27.38** shows the principle of a grating spectroscope. The slit that admits light from the source (e.g., a hot gas) is located at the focal point of the collimating lens, so the light rays striking the grating are parallel. The telescope is used to detect the bright fringes and, hence, to measure the angle θ.

Check Your Understanding

(The answers are given at the end of the book.)

20. CYU Figure 27.3 shows a top view of a diffraction grating and the mth-order principal maxima that are obtained with red and blue light. Red light has the longer wavelength. **(a)** Which principal maximum is associated with blue light, the one farther from or the one closer to the central maximum? **(b)** If the number of slits per centimeter in the grating were increased, would these two principal maxima move away from the central maximum, move toward the central maximum, or remain in the same place?

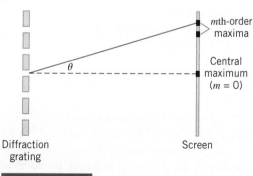

CYU FIGURE 27.3

21. What would happen to the distance between the bright fringes produced by a diffraction grating if the entire interference apparatus (light source, grating, and screen) were immersed in water?

27.8 | X-Ray Diffraction

THE PHYSICS OF . . . X-ray diffraction.   Not all diffraction gratings are commercially made. Nature also creates diffraction gratings, although these gratings do not look like an array of closely spaced slits. Instead, nature's gratings are the arrays of regularly spaced atoms that exist in crystalline solids. For example, **Figure 27.39** shows the structure of a crystal of ordinary salt (NaCl). Typically, the atoms in a crystalline solid are separated by distances of about 1.0×10^{-10} m, so we might expect a crystalline array of atoms to act like a grating with roughly this "slit" spacing for electromagnetic waves of the appropriate wavelength. Assuming that $\sin \theta = 0.5$ and that $m = 1$ in Equation 27.7, then $0.5 = \lambda/d$. A value of $d = 1.0 \times 10^{-10}$ m in this equation gives a wavelength of $\lambda = 0.5 \times 10^{-10}$ m. This wavelength is much shorter than that of visible light and falls in the X-ray region of the electromagnetic spectrum. (See Figure 24.9.)

A diffraction pattern does indeed result when X-rays are directed onto a crystalline material, as **Figure 27.40a** illustrates for a crystal of NaCl. The pattern consists of a complicated arrangement of spots because a crystal has a complex three-dimensional structure. It is from patterns such as this that the spacing between atoms and the nature of the crystal structure can be determined. X-ray diffraction has also been applied with great success toward understanding the structure of biologically important molecules, such as proteins and nucleic acids. One of the most famous results was the discovery in 1953 by James Watson and Francis Crick that the structure of the nucleic acid DNA is a double helix. X-ray diffraction patterns such as that in **Figure 27.40b** played the pivotal role in their research.

FIGURE 27.39 In this drawing of the crystalline structure of sodium chloride, the small red spheres represent positive sodium ions, and the large blue spheres represent negative chloride ions.

Courtesy Edwin Jones, University of South Carolina

Omikron/Science Source

FIGURE 27.40 The X-ray diffraction patterns of (a) crystalline NaCl and (b) DNA. The image of DNA was obtained by Rosalind Franklin in 1953.

Concept Summary

27.1 The Principle of Linear Superposition The principle of linear superposition states that when two or more waves are present simultaneously in the same region of space, the resultant disturbance is the sum of the disturbances from the individual waves.

Constructive interference occurs at a point when two waves meet there crest-to-crest and trough-to-trough, thus reinforcing each other. When two waves that start out in phase and have traveled some distance meet at a point, constructive interference occurs whenever the travel distances are the same or differ by any integer number of wavelengths: $\ell_2 - \ell_1 = m\lambda$, where ℓ_1 and ℓ_2 are the distances traveled by the waves, and $m = 0, 1, 2, 3, \ldots$.

Destructive interference occurs at a point when two waves meet there crest-to-trough, thus mutually canceling each other. When two waves that start out in phase and have traveled some distance meet at a point, destructive interference occurs whenever the travel distances differ by any odd integer number of half-wavelengths: $\ell_2 - \ell_1 = \left(m + \frac{1}{2}\right)\lambda$, where ℓ_1 and ℓ_2 are the distances traveled by the waves, and $m = 0, 1, 2, 3, \ldots$.

Two sources are coherent if the waves they emit maintain a constant phase relation. In other words, the waves do not shift relative to one another as time passes. If constructive and destructive interference are to be observed, coherent sources are necessary.

27.2 Young's Double-Slit Experiment In Young's double-slit experiment, light passes through a pair of closely spaced narrow slits and produces a pattern of alternating bright and dark fringes on a viewing screen. The fringes arise because of constructive and destructive interference. The angle θ that locates the mth-order bright fringe is given by Equation 27.1, where λ is the wavelength of the light, and d is the spacing between the slits. The angle that locates the mth dark fringe is given by Equation 27.2.

$$\sin \theta = \frac{m\lambda}{d} \qquad m = 0, 1, 2, 3, \ldots \qquad \textbf{(27.1)}$$

$$\sin \theta = \frac{\left(m + \frac{1}{2}\right)\lambda}{d} \qquad m = 0, 1, 2, 3, \ldots \qquad \textbf{(27.2)}$$

27.3 Thin-Film Interference Constructive and destructive interference of light waves can occur with thin films of transparent materials. The interference occurs between light waves that reflect from the top and bottom surfaces of the film. One important factor in thin-film interference is the thickness of a film relative to the wavelength of the light within the film. The wavelength λ_{film} within a film is given by Equation 27.3, where λ_{vacuum} is the wavelength in a vacuum, and n is the index of refraction of the film.

$$\lambda_{film} = \frac{\lambda_{vaccum}}{n} \qquad (27.3)$$

A second important factor is the phase change that can occur when light reflects at each surface of the film:

1. When light travels through a material with a smaller index of refraction toward a material with a larger index of refraction, reflection at the boundary occurs along with a phase change that is equivalent to one-half a wavelength in the film.
2. When light travels through a material with a larger index of refraction toward a material with a smaller index of refraction, there is no phase change upon reflection at the boundary.

27.4 The Michelson Interferometer An interferometer is an instrument that can be used to measure the wavelength of light by employing interference between two light waves. The Michelson interferometer splits the light into two beams. One beam travels to a fixed mirror, reflects from it, and returns. The other beam travels to a movable mirror, reflects from it, and returns. When the two returning beams are combined, interference is observed, the amount of which depends on the travel distances.

27.5 Diffraction Diffraction is the bending of waves around obstacles or around the edges of an opening. Diffraction is an interference effect that can be explained with the aid of Huygens' principle. This principle states that every point on a wave front acts as a source of tiny wavelets that move forward with the same speed as the wave; the wave front at a later instant is the surface that is tangent to the wavelets.

When light passes through a single narrow slit and falls on a viewing screen, a pattern of bright and dark fringes is formed because of the superposition of Huygens wavelets. The angle θ that specifies the mth dark fringe on either side of the central bright fringe is given by Equation 27.4, where λ is the wavelength of the light and W is the width of the slit.

$$\sin \theta = m\frac{\lambda}{W} \quad m = 1, 2, 3, \ldots \qquad (27.4)$$

27.6 Resolving Power The resolving power of an optical instrument is the ability of the instrument to distinguish between two closely spaced objects. Resolving power is limited by the diffraction that occurs when light waves enter an instrument, often through a circular opening.

The Rayleigh criterion specifies that two point objects are just resolved when the first dark fringe in the diffraction pattern of one falls directly on the central bright fringe in the diffraction pattern of the other. According to this specification, the minimum angle (in radians) that two point objects can subtend at a circular aperture of diameter D and still be resolved as separate objects is given by Equation 27.6, where λ is the wavelength of the light.

$$\theta_{min} \approx 1.22\frac{\lambda}{D} \quad (\theta_{min} \text{ in radians}) \qquad (27.6)$$

27.7 The Diffraction Grating A diffraction grating consists of a large number of parallel, closely spaced slits. When light passes through a diffraction grating and falls on a viewing screen, the light forms a pattern of bright and dark fringes. The bright fringes are referred to as principal maxima and are found at an angle θ that is specified by Equation 27.7, where λ is the wavelength of the light and d is the separation between two adjacent slits.

$$\sin \theta = m\frac{\lambda}{d} \quad m = 0, 1, 2, 3, \ldots \qquad (27.7)$$

27.8 X-Ray Diffraction A diffraction pattern forms when X-rays are directed onto a crystalline material. The pattern arises because the regularly spaced atoms in a crystal act like a diffraction grating. Because the spacing is extremely small, on the order of 1×10^{-10} m, the wavelength of the electromagnetic waves must also be very small—hence, the use of X-rays. The crystal structure of a material can be determined from its X-ray diffraction pattern.

Focus on Concepts

Online

Additional questions are available for assignment in WileyPLUS.

Section 27.1 The Principle of Linear Superposition

1. The two loudspeakers in the drawing are producing identical sound waves. The waves spread out and overlap at the point P. What is the difference $\ell_2 - \ell_1$ in the two path lengths if point P is at the third sound intensity minimum from the central sound intensity maximum? Express this difference in terms of the wavelength λ of the sound. (a) $\frac{1}{2}\lambda$ (b) λ (c) $\frac{3}{2}\lambda$ (d) 3λ (e) $\frac{5}{2}\lambda$

Section 27.2 Young's Double-Slit Experiment

2. In a certain Young's double-slit experiment, a diffraction pattern is formed on a distant screen, as the drawing shows. The angle that locates a given bright fringe is small, so that the approximation $\sin \theta \approx \theta$ is valid. Assuming that θ remains small, by what factor does it change if the wavelength λ is doubled and the slit separation d is doubled? (a) The angle does not change. (b) The angle increases by a factor of 2. (c) The angle increases by a factor of 4. (d) The angle decreases by a factor of 2. (e) The angle decreases by a factor of 4.

QUESTION 2

Section 27.3 Thin-Film Interference

3. Light of wavelength 600 nm in vacuum is incident nearly perpendicularly on a thin film whose index of refraction is 1.5. The light travels from the top surface of the film to the bottom surface, reflects from

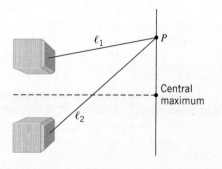

QUESTION 1

the bottom surface, and returns to the top surface, as the drawing indicates. How far has the light traveled inside the film? Express your answer in terms of the wavelength λ_{film} of the light within the film. (a) $2\lambda_{film}$ (b) $3\lambda_{film}$ (c) $4\lambda_{film}$ (d) $6\lambda_{film}$ (e) $12\lambda_{film}$

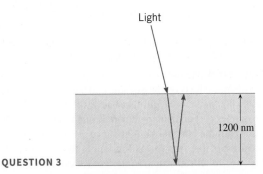

Light

1200 nm

QUESTION 3

4. Light is incident perpendicularly on four transparent films of different thickness. The thickness of each film is given in the drawings in terms of the wavelength λ_{film} of the light within the film. The index of refraction of each film is 1.5, and each is surrounded by air. Which film (or films) will appear bright due to constructive interference when viewed from the top surface, upon which the light is incident? (a) 1, 2, 3, 4 (b) 2, 3 (c) 3 (d) 3, 4 (e) 4

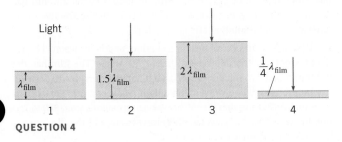

Light

λ_{film} $1.5\,\lambda_{film}$ $2\,\lambda_{film}$ $\frac{1}{4}\,\lambda_{film}$

1   2   3   4

QUESTION 4

Section 27.5   Diffraction

5. Light passes through a single slit. If the width of the slit is reduced, what happens to the width of the central bright fringe? (a) The width of the central bright fringe does not change, because it depends only on the wavelength of the light and not on the width of the slit. (b) The central bright fringe becomes wider, because the angle that locates the first dark fringe on either side of the central bright fringe becomes smaller. (c) The central bright fringe becomes wider, because the angle that locates the first dark fringe on either side of the central bright fringe becomes larger. (d) The central bright fringe becomes narrower, because the angle that locates the first dark fringe on either side of the central bright fringe becomes larger. (e) The central bright fringe becomes narrower, because the angle that locates the first dark fringe on either side of the central bright fringe becomes smaller.

6. Light of wavelength λ passes through a single slit of width W and forms a diffraction pattern on a viewing screen. If this light is then replaced by light of wavelength 2λ, the original diffraction pattern is exactly reproduced if the width of the slit _____. (a) is changed to $\frac{1}{4}W$ (b) is changed to $\frac{1}{2}W$ (c) is changed to $2W$ (d) is changed to $4W$ (e) remains the same—no change is necessary

Section 27.6   Resolving Power

7. Suppose that you are using a microscope to view two closely spaced cells. For a given lens diameter, which color of light would you use to achieve the best possible resolving power? (a) Red (b) Yellow (c) Green (d) Blue (e) All the colors give the same resolving power.

Section 27.7   The Diffraction Grating

8. A diffraction grating is illuminated with yellow light. The diffraction pattern seen on a viewing screen consists of three yellow bright fringes, one at the central maximum ($\theta = 0°$) and one on either side of it at $\theta = \pm 50°$. Then the grating is simultaneously illuminated with red light. Where a red and a yellow fringe overlap, an orange fringe is produced. The new pattern consists of _____. (a) only red fringes at 0° and ±50° (b) only yellow fringes at 0° and ±50° (c) only orange fringes at 0° and ±50° (d) an orange fringe at 0°, yellow fringes at ±50°, and red fringes farther out (e) an orange fringe at 0°, yellow fringes at ±50°, and red fringes closer in

Problems

Online

Additional questions are available for assignment in WileyPLUS.

SSM	Student Solutions Manual	BIO	Biomedical application
MMH	Problem-solving help	E	Easy
GO	Guided Online Tutorial	M	Medium
V-HINT	Video Hints	H	Hard
CHALK	Chalkboard Videos	WS	Worksheet
		T	Team Problem

Section 27.1   The Principle of Linear Superposition,

Section 27.2   Young's Double-Slit Experiment

1. **E** **SSM** In a Young's double-slit experiment, the wavelength of the light used is 520 nm (in vacuum), and the separation between the slits is 1.4×10^{-6} m. Determine the angle that locates (a) the dark fringe for which $m = 0$, (b) the bright fringe for which $m = 1$, (c) the dark fringe for which $m = 1$, and (d) the bright fringe for which $m = 2$.

2. **E** In a Young's double-slit experiment, the angle that locates the second dark fringe on either side of the central bright fringe is 5.4°. Find the ratio d/λ of the slit separation d to the wavelength λ of the light.

3. **E** **CHALK** Two in-phase sources of waves are separated by a distance of 4.00 m. These sources produce identical waves that have a wavelength of 5.00 m. On the line between them, there are two places at which the same type of interference occurs. (a) Is it constructive or destructive interference, and (b) where are the places located?

4. **E** The dark fringe for $m = 0$ in a Young's double-slit experiment is located at an angle of $\theta = 15°$. What is the angle that locates the dark fringe for $m = 1$?

5. **E** **V-HINT** In a Young's double-slit experiment, the seventh dark fringe is located 0.025 m to the side of the central bright fringe on a flat screen, which is 1.1 m away from the slits. The separation between the slits is 1.4×10^{-4} m. What is the wavelength of the light being used?

6. **E** **GO** **MMH** Two parallel slits are illuminated by light composed of two wavelengths. One wavelength is $\lambda_A = 645$ nm. The other wavelength is λ_B and is unknown. On a viewing screen, the light with wavelength $\lambda_A = 645$ nm produces its third-order bright fringe at the same place where the light with wavelength λ_B produces its fourth dark fringe. The fringes are counted relative to the central or zeroth-order bright fringe. What is the unknown wavelength?

7. **E** **GO** In a setup like that in **Figure 27.7**, a wavelength of 625 nm is used in a Young's double-slit experiment. The separation between the slits is $d = 1.4 \times 10^{-5}$ m. The total width of the screen is 0.20 m. In one version of the setup, the separation between the double slit and the screen is $L_A = 0.35$ m, whereas in another version it is $L_B = 0.50$ m. On one side of the central bright fringe, how many bright fringes lie on the screen in the two versions of the setup? Do not include the central bright fringe in your counting.

8. **M** **V-HINT** At most, how many bright fringes can be formed on either side of the central bright fringe when light of wavelength 625 nm falls on a double slit whose slit separation is 3.76×10^{-6} m?

9. **M** **CHALK** **MMH** In a Young's double-slit experiment the separation y between the second-order bright fringe and the central bright fringe on a flat screen is 0.0180 m when the light has a wavelength of 425 nm. Assume that the angles that locate the fringes on the screen are small enough so that $\sin \theta \approx \tan \theta$. Find the separation y when the light has a wavelength of 585 nm.

10. **H** In Young's experiment a mixture of orange light (611 nm) and blue light (471 nm) shines on the double slit. The centers of the first-order bright blue fringes lie at the outer edges of a screen that is located 0.500 m away from the slits. However, the first-order bright orange fringes fall off the screen. By how much and in which direction (toward or away from the slits) should the screen be moved so that the centers of the first-order bright orange fringes will just appear on the screen? It may be assumed that θ is small, so that $\sin \theta \approx \tan \theta$.

11. **H** **SSM** A sheet that is made of plastic ($n = 1.60$) covers *one slit* of a double slit (see the drawing). When the double slit is illuminated by monochromatic light ($\lambda_{\text{vacuum}} = 586$ nm), the center of the screen appears dark rather than bright. What is the minimum thickness of the plastic?

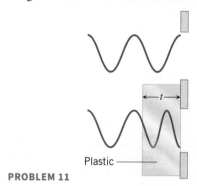

Plastic

PROBLEM 11

Section 27.3 Thin-Film Interference

12. **E** You are standing in air and are looking at a flat piece of glass ($n = 1.52$) on which there is a layer of transparent plastic ($n = 1.61$). Light whose wavelength is 589 nm in vacuum is incident nearly perpendicularly on the coated glass and reflects into your eyes. The layer of plastic looks dark. Find the two smallest possible nonzero values for the thickness of the layer.

13. **E** **SSM** A nonreflective coating of magnesium fluoride ($n = 1.38$) covers the glass ($n = 1.52$) of a camera lens. Assuming that the coating prevents reflection of yellow-green light (wavelength in vacuum = 565 nm), determine the minimum nonzero thickness that the coating can have.

14. **E** **GO** When monochromatic light shines perpendicularly on a soap film ($n = 1.33$) with air on each side, the second smallest nonzero film thickness for which destructive interference of reflected light is observed is 296 nm. What is the vacuum wavelength of the light in nm?

15. **E** **MMH** A transparent film ($n = 1.43$) is deposited on a glass plate ($n = 1.52$) to form a nonreflecting coating. The film has a thickness that is 1.07×10^{-7} m. What is the longest possible wavelength (in vacuum) of light for which this film has been designed?

16. **E** **GO** A tank of gasoline ($n = 1.40$) is open to the air ($n = 1.00$). A thin film of liquid floats on the gasoline and has a refractive index that is between 1.00 and 1.40. Light that has a wavelength of 625 nm (in vacuum) shines perpendicularly down through the air onto this film, and in this light the film looks bright due to constructive interference. The thickness of the film is 242 nm and is the minimum nonzero thickness for which constructive interference can occur. What is the refractive index of the film?

17. **E** **V-HINT** Review Conceptual Example 4 before beginning this problem. A soap film with different thicknesses at different places has an unknown refractive index n and air on both sides. In reflected light it looks multicolored. One region looks yellow because destructive interference has removed blue ($\lambda_{\text{vacuum}} = 469$ nm) from the reflected light, while another looks magenta because destructive interference has removed green ($\lambda_{\text{vacuum}} = 555$ nm). In these regions the film has the minimum nonzero thickness t required for the destructive interference to occur. Find the ratio $t_{\text{magenta}}/t_{\text{yellow}}$.

18. **M** **MMH** A film of oil lies on a shallow puddle of water. The refractive index of the oil exceeds that of the water. The film has the minimum nonzero thickness such that it appears dark due to destructive interference when viewed in red light (wavelength = 640.0 nm in vacuum). Assuming that the visible spectrum extends from 380 to 750 nm, for which visible wavelength(s) in vacuum will the film appear bright due to constructive interference?

19. **M** **CHALK** **SSM** Orange light ($\lambda_{\text{vacuum}} = 611$ nm) shines on a soap film ($n = 1.33$) that has air on either side of it. The light strikes the film perpendicularly. What is the minimum thickness of the film for which constructive interference causes it to look bright in reflected light?

Section 27.5 Diffraction

20. **E** **(a)** As Section 17.3 discusses, high-frequency sound waves exhibit less diffraction than low-frequency sound waves do. However, even high-frequency sound waves exhibit much more diffraction under normal circumstances than do light waves that pass through the same opening. The highest frequency that a healthy ear can typically hear is 2.0×10^4 Hz. Assume that a sound wave with this frequency travels at 343 m/s and passes through a doorway that has a width of 0.91 m. Determine the angle that locates the first minimum to either side of the central maximum in the diffraction pattern for the sound. This minimum is equivalent to the first dark fringe in a single-slit diffraction pattern for light. **(b)** Suppose that yellow light (wavelength = 580 nm in vacuum) passes through a doorway and that the first dark fringe in its diffraction pattern is located at the angle determined in part (a). How wide would this hypothetical doorway have to be?

21. **E** A dark fringe in the diffraction pattern of a single slit is located at an angle of $\theta_A = 34°$. With the same light, the same dark fringe formed with another single slit is at an angle of $\theta_B = 56°$. Find the ratio W_A/W_B of the widths of the two slits.

22. **E** **SSM** A diffraction pattern forms when light passes through a single slit. The wavelength of the light is 675 nm. Determine the angle that locates the first dark fringe when the width of the slit is **(a)** 1.8×10^{-4} m and **(b)** 1.8×10^{-6} m.

23. **E** **GO** A slit has a width of $W_1 = 2.3 \times 10^{-6}$ m. When light with a wavelength of $\lambda_1 = 510$ nm passes through this slit, the width of the central bright fringe on a flat observation screen has a certain value. With the screen kept in the same place, this slit is replaced with a second slit (width W_2), and a wavelength of $\lambda_2 = 740$ nm is used. The width of the central bright fringe on the screen is observed to be unchanged. Find W_2.

24. **E** **SSM** Light that has a wavelength of 668 nm passes through a slit 6.73×10^{-6} m wide and falls on a screen that is 1.85 m away. What is the distance on the screen from the center of the central bright fringe to the third dark fringe on either side?

25. **E** **GO** Light shines through a single slit whose width is 5.6×10^{-4} m. A diffraction pattern is formed on a flat screen located 4.0 m away. The distance between the middle of the central bright fringe and the first dark fringe is 3.5 mm. What is the wavelength of the light?

26. **M** **V-HINT** Light waves with two different wavelengths, 632 nm and 474 nm, pass simultaneously through a single slit whose width is 7.15×10^{-5} m and strike a screen 1.20 m from the slit. Two diffraction patterns are formed on the screen. What is the distance (in cm) between the common center of the diffraction patterns and the first occurrence of a dark fringe from one pattern falling on top of a dark fringe from the other pattern?

27. **M** **GO** **CHALK** The central bright fringe in a single-slit diffraction pattern has a width that equals the distance between the screen and the slit. Find the ratio λ/W of the wavelength λ of the light to the width W of the slit.

28. **M** **SSM** How many dark fringes will be produced on either side of the central maximum if light ($\lambda = 651$ nm) is incident on a single slit that is 5.47×10^{-6} m wide?

29. **H** In a single-slit diffraction pattern, the central fringe is 450 times as wide as the slit. The screen is 18 000 times farther from the slit than the slit is wide. What is the ratio λ/W, where λ is the wavelength of the light shining through the slit and W is the width of the slit? Assume that the angle that locates a dark fringe on the screen is small, so that $\sin\theta \approx \tan\theta$.

Section 27.6   Resolving Power

30. **E** Two stars are 3.7×10^{11} m apart and are equally distant from the earth. A telescope has an objective lens with a diameter of 1.02 m and just detects these stars as separate objects. Assume that light of wavelength 550 nm is being observed. Also assume that diffraction effects, rather than atmospheric turbulence, limit the resolving power of the telescope. Find the maximum distance that these stars could be from the earth.

31. **E** **GO** An inkjet printer uses tiny dots of red, green, and blue ink to produce an image. Assume that the dot separation on the printed page is the same for all colors. At normal viewing distances, the eye does not resolve the individual dots, regardless of color, so that the image has a normal look. The wavelengths for red, green, and blue are $\lambda_{red} = 660$ nm, $\lambda_{green} = 550$ nm, and $\lambda_{blue} = 470$ nm. The diameter of the pupil through which light enters the eye is 2.0 mm. For a viewing distance of 0.40 m, what is the maximum allowable dot separation?

32. **E** **V-HINT** Astronomers have discovered a planetary system orbiting the star Upsilon Andromedae, which is at a distance of 4.2×10^{17} m from the earth. One planet is believed to be located at a distance of 1.2×10^{11} m from the star. Using visible light with a vacuum wavelength of 550 nm, what is the minimum necessary aperture diameter that a telescope must have so that it can resolve the planet and the star?

Section 27.7   The Diffraction Grating,

33. **E** **SSM** A diffraction grating is 1.50 cm wide and contains 2400 lines. When used with light of a certain wavelength, a third-order maximum is formed at an angle of 18.0°. What is the wavelength (in nm)?

34. **E** The light shining on a diffraction grating has a wavelength of 495 nm (in vacuum). The grating produces a second-order bright fringe whose position is defined by an angle of 9.34°. How many lines per centimeter does the grating have?

35. **E** For a wavelength of 420 nm, a diffraction grating produces a bright fringe at an angle of 26°. For an unknown wavelength, the same grating produces a bright fringe at an angle of 41°. In both cases the bright fringes are of the same order m. What is the unknown wavelength?

36. **E** **GO** Two diffraction gratings, A and B, are located at the same distance from the observation screens. Light with the same wavelength λ is used for each. The separation between adjacent principal maxima for grating A is 2.7 cm, and for grating B it is 3.2 cm. Grating A has 2000 lines per meter. How many lines per meter does grating B have? (*Hint:* The diffraction angles are small enough that the approximation $\sin\theta \approx \tan\theta$ can be used.)

37. **E** **SSM** The wavelength of a laser beam is 780 nm. Suppose that a diffraction grating produces first-order maxima that are 1.2 mm apart at a distance of 3.0 mm from the grating. Estimate the spacing between the slits of the grating.

38. **E** The first-order principle maximum produced by a grating is located at an angle of $\theta = 18.0°$. What is the angle for the third-order maximum with the same light?

39. **M** **V-HINT** **CHALK** A diffraction grating has 2604 lines per centimeter, and it produces a principal maximum at $\theta = 30.0°$. The grating is used with light that contains all wavelengths between 410 and 660 nm. What is (are) the wavelength(s) of the incident light that could have produced this maximum?

40. **M** **GO** Light of wavelength 410 nm (in vacuum) is incident on a diffraction grating that has a slit separation of 1.2×10^{-5} m. The distance between the grating and the viewing screen is 0.15 m. A diffraction pattern is produced on the screen that consists of a central bright fringe and higher-order bright fringes (see the drawing). (a) Determine the distance y from the central bright fringe to the second-order bright fringe. (*Hint:* The diffraction angles are small enough that the approximation $\tan\theta \approx \sin\theta$ can be used.) (b) If the entire apparatus is submerged in water ($n_{\text{water}} = 1.33$), what is the distance y?

PROBLEM 40

41. **M** **SSM** Violet light (wavelength = 410 nm) and red light (wavelength = 660 nm) lie at opposite ends of the visible spectrum. (a) For each wavelength, find the angle θ that locates the first-order maximum produced by a grating with 3300 lines/cm. This grating converts a mixture of all colors between violet and red into a rainbow-like dispersion between the two angles. Repeat the calculation above for (b) the second-order maximum and (c) the third-order maximum. (d) From your results, decide whether there is an overlap between any of the "rainbows" and, if so, specify which orders overlap.

Additional Problems

Online

42. E GO A soap film ($n = 1.33$) is 465 nm thick and lies on a glass plate ($n = 1.52$). Sunlight, whose wavelengths (in vacuum) extend from 380 to 750 nm, travels through the air and strikes the film perpendicularly. For which wavelength(s) in this range does destructive interference cause the film to look dark in reflected light?

43. E SSM In a Young's double-slit experiment, two rays of monochromatic light emerge from the slits and meet at a point on a distant screen, as in **Figure 27.6a**. The point on the screen where these two rays meet is the eighth-order bright fringe. The difference in the distances that the two rays travel is 4.57×10^{-6} m. What is the wavelength (in nm) of the monochromatic light?

44. E GO Point A is the midpoint of one of the sides of a square. On the side opposite this spot, two in-phase loudspeakers are located at adjacent corners, as shown in the figure. Standing at point A you hear a loud sound because of constructive interference between the identical sound waves coming from the speakers. As you walk along the side of the square toward either empty corner, the loudness diminishes gradually to nothing and then increases again until you hear a maximally loud sound at the corner. If the length of each side of the square is 4.6 m, find the wavelength of the sound waves.

PROBLEM 44

45. E GO A flat screen is located 0.60 m away from a single slit. Light with a wavelength of 510 nm (in vacuum) shines through the slit and produces a diffraction pattern. The width of the central bright fringe on the screen is 0.050 m. What is the width of the slit?

46. M Review Conceptual Example 2 before attempting this problem. Two slits are 0.158 mm apart. A mixture of red light (wavelength = 665 nm) and yellow-green light (wavelength = 565 nm) falls on the slits. A flat observation screen is located 2.24 m away. What is the distance on the screen between the third-order red fringe and the third-order yellow-green fringe?

47. M GO A spotlight sends red light (wavelength = 694.3 nm) to the moon. At the surface of the moon, which is 3.77×10^8 m away, the light strikes a reflector left there by astronauts. The reflected light returns to the earth, where it is detected. When it leaves the spotlight, the circular beam of light has a diameter of about 0.20 m, and diffraction causes the beam to spread as the light travels to the moon. In effect, the first circular dark fringe in the diffraction pattern defines the size of the central bright spot on the moon. Determine the diameter (not the radius) of the central bright spot on the moon.

48. M In a single-slit diffraction pattern on a flat screen, the central bright fringe is 1.2 cm wide when the slit width is 3.2×10^{-5} m. When the slit is replaced by a second slit, the wavelength of the light and the distance to the screen remaining unchanged, the central bright fringe broadens to a width of 1.9 cm. What is the width of the second slit? It may be assumed that θ is so small that $\sin \theta \approx \tan \theta$.

49. M GO A beam of light is sent directly down onto a glass plate ($n = 1.5$) and a plastic plate ($n = 1.2$) that form a thin wedge of air (see the drawing). An observer looking down through the glass plate sees the fringe pattern shown in the lower part of the drawing, with the dark fringes at the ends A and B. The wavelength of the light is 520 nm. Using the fringe pattern shown in the drawing, determine the thickness of the air wedge at B.

PROBLEM 49

50. H There are 5620 lines per centimeter in a grating that is used with light whose wavelength is 471 nm. A flat observation screen is located at a distance of 0.750 m from the grating. What is the minimum width that the screen must have so the *centers* of all the principal maxima formed on either side of the central maximum fall on the screen?

Physics in Biology, Medicine, and Sports

51. E BIO 27.6 It is claimed that some professional baseball players can see which way the ball is spinning as it travels toward home plate. One way to judge this claim is to estimate the distance at which a batter can first hope to resolve two points on opposite sides of a baseball, which has a diameter of 0.0738 m. (a) Estimate this distance, assuming that the pupil of the eye has a diameter of 2.0 mm and the wavelength of the light is 550 nm in vacuum. (b) Considering that the distance between the pitcher's mound and home plate is 18.4 m, can you rule out the claim based on your answer to part (a)?

52. E BIO CHALK SSM 27.6 Late one night on a highway, a car speeds by you and fades into the distance. Under these conditions the pupils of your eyes have diameters of about 7.0 mm. The taillights of this car are separated by a distance of 1.2 m and emit red light (wavelength = 660 nm in vacuum). How far away from you is this car when its taillights appear to merge into a single spot of light because of the effects of diffraction?

53. E BIO GO 27.6 A hunter who is a bit of a braggart claims that from a distance of 1.6 km he can selectively shoot either of two squirrels who are sitting ten centimeters apart on the same branch of a tree. What's more, he claims that he can do this without the aid of a telescopic sight on his rifle. (a) Determine the diameter of the pupils of his eyes that would be required for him to be able to resolve the squirrels as separate objects. In this calculation use a wavelength of 498 nm (in vacuum) for the light. **(b)** State whether his claim is reasonable, and provide a reason for your answer. In evaluating his claim, consider that the human eye automatically adjusts the diameter of its pupil over a typical range of 2 to 8 mm, the larger values coming into play as the lighting becomes darker. Note also that under dark conditions, the eye is most sensitive to a wavelength of 498 nm.

54. **E** **BIO** **27.6** Review Conceptual Example 9 as background for this problem. In addition to the data given there, assume that the dots in the painting are separated by 1.5 mm and that the wavelength of the light is $\lambda_{\text{vacuum}} = 550$ nm. Find the distance at which the dots can just be resolved by **(a)** the eye and **(b)** the camera.

55. **M** **BIO** **GO** **27.6** The pupil of an eagle's eye has a diameter of 6.0 mm. Two field mice are separated by 0.010 m. From a distance of 176 m, the eagle sees them as one unresolved object and dives toward them at a speed of 17 m/s. Assume that the eagle's eye detects light that has a wavelength of 550 nm in vacuum. How much time passes until the eagle sees the mice as separate objects?

56. **E** **BIO** **27.6** A large group of football fans comes to the game with colored cards that spell out the name of their team when held up simultaneously. Most of the cards are colored blue ($\lambda_{\text{vacuum}} = 480$ nm). When displayed, the average distance between neighboring cards is 5.0 cm. If the cards are to blur together into solid blocks of color when viewed by a spectator at the other end of the stadium (160 m away), what must be the maximum diameter (in mm) of the spectator's pupils?

57. **M** **BIO** **27.6** Under dark conditions, the maximum diameter of a human pupil is 7.0 mm, where an owl's pupil may be 8.5 mm. Assume a human can optically resolve two closely spaced objects at a distance r. **(a)** By what factor could the distance between the two objects be reduced and still have the owl optically resolve them at the same distance r? **(b)** If the distance between the two objects remains fixed, by what factor could r be increased and still have the owl optically resolve the two objects? In both **(a)** and **(b)**, assume the wavelength of the light remains constant.

58. **M** **BIO** **27.7** The bold colors observed in certain bird feathers are due to diffraction. They have microstructures on their surface that act like a diffraction grating. If a first-order maximum is observed at an angle of 25° relative to the surface normal for 530-nm light, what is the spacing between the microstructures?

59. **M** **BIO** **27.7** The human eye detects the wavelengths of light in the visible spectrum, which span the range from 380 to 740 nm. White light is incident on a diffraction grating with 3500 lines per cm. **(a)** Which wavelength in the visible spectrum occurs at the greatest angle for the $m = 2$ maximum, and **(b)** what is this angle?

60. **M** **BIO** **27.6** Review Example 5 from Chapter 25. If the James Webb Space Telescope were enclosed, such that it had a circular aperture with a diameter equal to the diameter of its primary mirror, calculate its optical resolution. More specifically, what is the smallest separation distance of two nickels that are optically resolved with 520-nm light, when viewed from a distance of 4.7×10^6 m, which is the distance between New York and San Francisco?

61. **M** **BIO** **27.3** Many outdoor athletes splurge on a high-end pair of sunglasses. The lenses in these glasses are not a single piece of glass, but are actually composed of multiple layers (see the photo). The antireflective coating at the back of the lens is there to reduce back glare, which is light that strikes the back of the lens and reflects toward the eyes. The antireflective coating consists of a transparent thin film with an index of refraction equal to 1.36. The film, with a thickness of 96.0 nm, is applied to the back of the glass lens, which has an index of refraction of 1.52. Which wavelength of light in the visible spectrum is not reflected from the coating, due to destructive interference?

Anti-reflective coating

Lens

Polarizing film

Scratch resistant coating

Mirror coating

fielperson/Pixabay

PROBLEM 61

Concepts and Calculations Problems

Online

The ability to exhibit interference effects is a fundamental characteristic of any kind of wave. Our understanding of these effects depends on the principle of linear superposition, which we first encountered in Chapter 17. Only by means of this principle can we understand the constructive and destructive interference of light waves that lie at the heart of every topic in this chapter. Problem 62 serves as a review of the essence of this principle. Problem 63 deals with thin-film interference and reviews the factors that must be considered in such cases.

62. **M** **CHALK** **SSM** A square is 3.5 m on a side, and point A is the midpoint of one of the sides. On the side opposite this spot, two in-phase loud-speakers are located at adjacent corners, as shown in the figure. Standing at point A, you hear a loud sound because constructive interference occurs between the identical sound waves coming from the speakers. As you walk along the side of the square toward either empty corner, the loudness diminishes gradually but does not entirely disappear until you reach either empty corner, where you hear no sound at all. Thus, at each empty corner destructive interference occurs. *Concepts:* (i) Why does constructive interference occur at point A? (ii) What is the general condition that leads to destructive interference? (iii) The condition that leads to destructive interference entails a number of possibilities. Which of them applies at either empty corner of the square? *Calculation:* Find the wavelength of the sound waves.

PROBLEM 62

63. **M** **CHALK** A soap film ($n = 1.33$) is 375 nm thick and coats a flat piece of glass ($n = 1.52$). Thus, air is on one side of the film and glass is on the other side, as the figure illustrates. Sunlight, whose wavelengths (in vacuum) extend from 380 to 750 nm, travels through air and strikes the film nearly perpendicularly. *Concepts:* (i) What, if any, phase change occurs when light, traveling in air, reflects from the air–film interface? (ii) What, if any, phase change occurs when light, traveling in the film, reflects from the film–glass interface? (iii) Is the wavelength of the light in the film greater than, smaller than, or equal to the wavelength in a vacuum? *Calculations:* For what wavelengths in the range of 380 to 750 nm does constructive interference cause the film to look bright in reflected light?

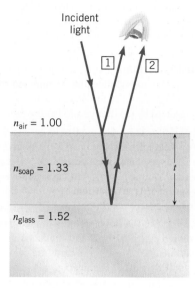

PROBLEM 63

Team Problems   Online

64. **E** **T** **WS** **Measuring small apertures.** You and your team have been tasked with reverse engineering a piece of a scientific apparatus for which you have no technical drawings, nor any other useful information. The piece in question is a square sheet of thin metal that is perforated with miniscule circular apertures arranged in a square array. There are 100 apertures in total and they are separated from each other by 0.500 cm. However, the apertures are far too small to measure by conventional means, so you get the idea to use diffraction phenomena to estimate their sizes. First, you secure the metal piece on edge at the center of a table. You then align a red laser of wavelength 632 nm so that the beam is oriented perpendicular the square piece and is directed through a single aperture. On a wall 11.6 m from the aperture, a circular diffraction pattern appears that consists of a central bright circle and concentric bright and dark rings. You measure the diameter of the smallest dark ring and then repeat the process for each aperture in the piece. You find that the 100 apertures have only three different sizes that correspond to smallest/innermost dark ring diameters of 2.75 cm, 4.10 cm, and 6.15 cm. What are the three different aperture diameters?

65. **E** **T** **WS** **Ranging With Diffraction.** You and your team need to estimate the distance between your spaceship and another spacecraft that has been damaged and is inoperable. On the front of the disabled ship are two small, bright red lights (wavelength $\lambda = 665$ nm) that you know are separated by 15.0 meters. Your own ship has bright blue lights ($\lambda = 475$ nm) on its leading edge that are separated by 12.0 m. The ships have identical cameras, each with a circular aperture of diameter 1.20 mm. Assume that you are heading directly toward the front of the damaged ship. (a) How far away from the damaged ship will your camera be able to just resolve its lights? (b) How far will you be away from the damaged ship when its camera can just resolve your lights?

66. **M** **T** **An Optical Monochromator.** You and your team are designing a device that inputs a beam of white light (i.e., a continuous spectrum of visible light spanning all wavelengths from 410 nm to 660 nm), and outputs a nearly monochromatic beam (i.e., a single color). Such a device is called an optical monochromator and is used in a wide variety of instruments and scientific experiments. In the instrument you are building, white light impinges upon the backside of a diffraction grating that has 1200 lines/cm. A movable rectangular aperture (a slit) is located on the opposite side of the grating, and can translate along a circular arc of radius 20.0 cm, the center of which is located at the grating. (a) At what angle relative to the normal of the grating should the center of the slit be located in order to pass green light ($\lambda = 550$ nm) from first order ($m = 1$) diffracted light? (b) How wide should the slit be so that the wavelengths passing through the slit fall in the range 540 nm $\leq \lambda \leq$ 560 nm?

67. **M** **T** **A Crude Thickness Monitor.** You and your team need to determine the thicknesses of two extremely thin sheets of plastic and do not have a measurement instrument capable of the job. You find a white light source with color filters (i.e., tinted glass windows that only let certain colors pass through) including red ($\lambda = 660.0$ nm), green ($\lambda = 550.0$ nm), and blue ($\lambda = 470.0$ nm). You locate two glass microscope slides, each of length 10.0 cm, stack one on top of the other, and tape their ends together on one side. On the very edge of the side opposite the tape, you sandwich a piece of plastic between the slides to form a wedge of air between the plates. (a) When you shine blue light perpendicular to the slides, 55.0 bright fringes form along the full length of the slide. How thick is the plastic? (b) When you replace the first sheet with the second sheet of plastic, the fringes that appear are too closely spaced to count. You switch to red light (a longer wavelength), and count 124 fringes. How thick is the plastic? (c) How many fringes would you expect in each case, (a) and (b), if you instead used green light?

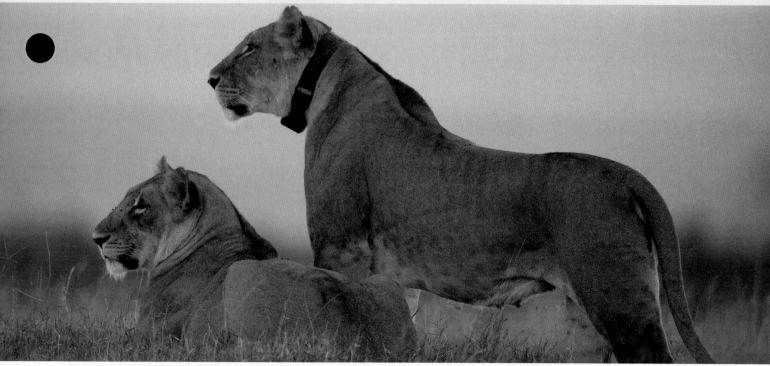

One of the lionesses is wearing a collar equipped with a Global Positioning System (GPS) tracking unit in the Ol Pejeta Conservancy in Kenya, Africa. The unit will allow researchers to track the lion's movements because GPS technology can locate objects on the earth with remarkable accuracy. The accuracy results, in part, because the system accounts for Einstein's theory of special relativity.

Ian Cumming/Axiom/Design Pics Inc/ Alamy Stock Photo

Special Relativity

28.1 Events and Inertial Reference Frames

In the theory of special relativity, an **event**, such as the launching of the space shuttle in **Figure 28.1**, is a physical "happening" that occurs at a certain place and time. In this drawing two observers are watching the lift-off, one standing on the earth and one in an airplane that is flying at a constant velocity relative to the earth. To record the event, each observer uses a **reference frame** that consists of a set of *x, y, z* axes (called a *coordinate system*) and a clock. The coordinate systems are used to establish where the event occurs, and the clocks to specify when. Each observer is at rest relative to his own reference frame. However, the earth-based observer and the airborne observer are moving relative to each other and so, also, are their respective reference frames.

The theory of special relativity deals with a "special" kind of reference frame, called an **inertial reference frame**. As Section 4.2 discusses, an inertial reference frame is one in which **Newton's** law of inertia is valid. That is, if the net force acting on a body is zero, the body either remains at rest or moves at a constant velocity. In other words, the acceleration of such a body is zero when measured in an inertial

LEARNING OBJECTIVES

After reading this module, you should be able to...

28.1 Define inertial reference frames.

28.2 List the postulates of special relativity.

28.3 Use time dilation to calculate time intervals in different frames of reference.

28.4 Use length contraction to calculate distances in different frames of reference.

28.5 Calculate the relativistic momentum of a high-speed particle.

28.6 Calculate the value of the various forms of energy a moving body possesses.

28.7 Calculate the relative velocity between relativistically moving bodies.

FIGURE 28.1 Using an earth-based reference frame, an observer standing on the earth records the location and time of an event (the space shuttle lift-off). Likewise, an observer in the airplane uses a plane-based reference frame to describe the event.

reference frame. Rotating and otherwise accelerating reference frames are not inertial reference frames. The earth-based reference frame in **Figure 28.1** is not quite an inertial frame because it is subjected to centripetal accelerations as the earth spins on its axis and revolves around the sun. In most situations, however, the effects of these accelerations are small, and we can neglect them. To the extent that the earth-based reference frame is an inertial frame, so is the plane-based reference frame, because the plane moves at a constant velocity relative to the earth. The next section discusses why inertial reference frames are important in relativity.

28.2 The Postulates of Special Relativity

Einstein built his theory of special relativity on two fundamental assumptions or postulates about the way nature behaves.

> **THE POSTULATES OF SPECIAL RELATIVITY**
>
> 1. *The Relativity Postulate.* **The laws of physics are the same in every inertial reference frame.**
> 2. *The Speed-of-Light Postulate.* **The speed of light in a vacuum, measured in any inertial reference frame, always has the same value of c, no matter how fast the source of light and the observer are moving relative to each other.**

It is not too difficult to accept the relativity postulate. For instance, in **Figure 28.1** each observer, using his own inertial reference frame, can make measurements on the motion of the space shuttle. The relativity postulate asserts that both observers find their data to be consistent with Newton's laws of motion. Similarly, both observers find that the behavior of the electronics on the space shuttle is described by the laws of electromagnetism. According to the relativity postulate, *any inertial reference frame is as good as any other for expressing the laws of physics.* As far as inertial reference frames are concerned, nature does not play favorites.

Since the laws of physics are the same in all inertial reference frames, there is no experiment that can distinguish between an inertial frame that is at rest and one that is moving at a constant velocity. When you are seated on the aircraft in **Figure 28.1**, for instance, it is just as valid to say that you are at rest and the earth is moving as it is to say the converse. It is not possible to single out one particular inertial reference frame as being at "absolute rest." Consequently, it is meaningless to talk about the "absolute velocity" of an object—that is, its velocity measured relative to a reference frame at "absolute rest." Thus, the earth moves relative to the sun, which itself moves relative to the center of our galaxy. And the galaxy moves relative to other galaxies, and so on. According to Einstein, only the relative velocity between objects, not their absolute velocities, can be measured and is physically meaningful.

FIGURE 28.2 Both the person on the truck and the observer on the earth measure the speed of the light to be c, regardless of the speed of the truck.

Whereas the relativity postulate seems plausible, the speed-of-light postulate defies common sense. For instance, **Figure 28.2** illustrates a person standing on the bed of a truck that is moving at a constant speed of 15 m/s relative to the ground. Now, suppose that you are standing on the ground and the person on the truck shines a flashlight at you. The person on the truck observes the speed of light to be c. What do you measure for the speed of light? You might guess that the speed of light would be $c + 15$ m/s. However, this guess is inconsistent with the speed-of-light postulate, which states that all observers in inertial reference frames measure the speed of light to be c—nothing more, nothing less. Therefore, you must also measure the speed of light to be c, the same as that measured by the person on the truck. According to the speed-of-light postulate, the fact that the flashlight is moving has no influence whatsoever on the speed of the light approaching you. This property of light, although surprising, has been verified many times by experiment.

Since waves, such as water waves and sound waves, require a medium through which to propagate, it was natural for scientists before Einstein to assume that light did too. This hypothetical medium was called the *luminiferous ether* and was assumed to fill all of space. Furthermore, it was believed that light traveled at the speed c only when measured with respect to the ether. According to this view, an observer moving relative to the ether would measure a speed for light that was smaller or greater than c, depending on whether the observer moved with or against the light, respectively. During the years 1883–1887, however, the American scientists A. A. Michelson and E. W. Morley carried out a series of famous experiments whose results were not consistent with the ether theory. Their results indicated that the speed of light is indeed the same in all inertial reference frames and does not depend on the motion of the observer. These experiments, and others, led eventually to the demise of the ether theory and the acceptance of the theory of special relativity.

The remainder of this chapter reexamines, from the viewpoint of special relativity, a number of fundamental concepts that have been discussed in earlier chapters from the viewpoint of classical physics. These concepts are time, length, momentum, kinetic energy, and the addition of velocities. We will see that each is modified by special relativity in a way that depends on the speed v of a moving object relative to the speed c of light in a vacuum. **Figure 28.3** illustrates that when the object moves slowly [v is much smaller than c ($v \ll c$)], the modification is negligibly small, and the classical version of each concept provides an accurate description of reality. However, when the object moves so rapidly that v is an appreciable fraction of the speed of light [v is approximately equal to c ($v \approx c$)], the effects of special relativity must be considered. The gold panel in **Figure 28.3** lists the various equations that convey the modifications imposed by special relativity. Each of these equations will be discussed in later sections of this chapter.

It is important to realize that the modifications imposed by special relativity do not imply that the classical concepts of time, length, momentum, kinetic energy, and the addition of velocities, as developed by Newton and others, are wrong. They are just limited to speeds that are very small compared to the speed of light. In contrast, the relativistic view of the concepts applies to all speeds between zero and the speed of light.

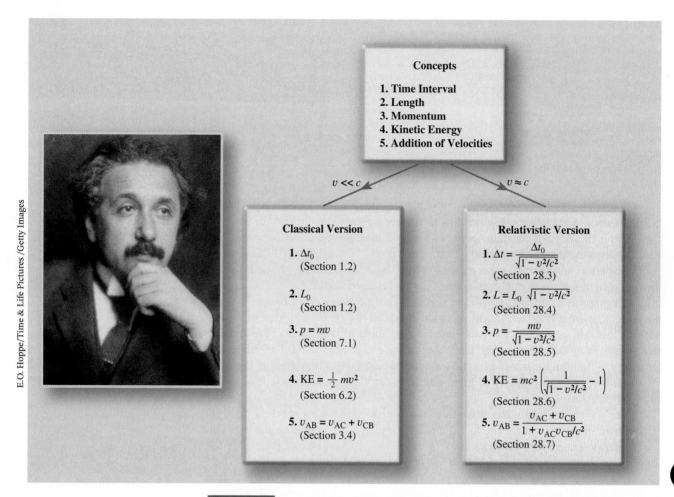

E.O. Hoppe/Time & Life Pictures /Getty Images

FIGURE 28.3   Albert Einstein (1879–1955), the author of the theory of special relativity, is one of the most famous scientists of the twentieth century. This figure emphasizes that the ratio v/c of the speed v of a moving object to the speed c of light in a vacuum is what determines how great the effects of special relativity are.

28.3 | The Relativity of Time: Time Dilation

Time Dilation

Common experience indicates that time passes just as quickly for a person standing on the ground as it does for an astronaut in a spacecraft. In contrast, special relativity reveals that the person on the ground measures time passing more slowly for the astronaut than for herself. We can see how this curious effect arises with the help of the clock illustrated in **Figure 28.4**, which uses a pulse of light to mark time. A short pulse of light is emitted by a light source, reflects from a mirror, and then strikes a detector that is situated next to the source. Each time a pulse reaches the detector, a "tick" registers on the chart recorder, another short pulse of light is emitted, and the cycle repeats. Thus, the time interval between successive "ticks" is marked by a beginning event (the firing of the light source) and an ending event (the pulse striking the detector). The source and detector are so close to each other that the two events can be considered to occur at the same location.

Suppose two identical clocks are built. One is kept on earth, and the other is placed aboard a spacecraft that travels at a constant velocity relative to the earth. The astronaut is at rest with respect to the clock on the spacecraft and, therefore, sees the light pulse move along the up/down path shown in **Animated Figure 28.5a**. According to the astronaut, the time interval Δt_0 required for the light to follow this path is the distance $2D$ divided by the speed of light c; $\Delta t_0 = 2D/c$. To the astronaut, Δt_0 is the time interval

FIGURE 28.4   A light clock.

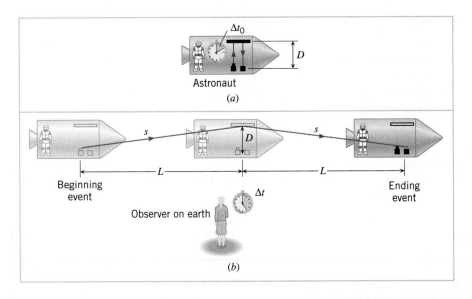

(*a*) The astronaut measures the time interval Δt_0 between successive "ticks" of his light clock. (*b*) An observer on earth watches the astronaut's clock and sees the light pulse travel a greater distance between "ticks" than it does in part *a*. Consequently, the earth-based observer measures a time interval Δt between "ticks" that is greater than Δt_0.

between the "ticks" of the spacecraft clock—that is, the time interval between the beginning and ending events of the clock. An earth-based observer, however, does *not* measure Δt_0 as the time interval between these two events. Since the spacecraft is moving, the earth-based observer sees the light pulse follow the diagonal path shown in red in part *b* of the drawing. This path is longer than the up/down path seen by the astronaut. But light travels at the *same speed c* for both observers, in accord with the speed-of-light postulate. Therefore, the earth-based observer measures a time interval Δt between the two events that is *greater* than the time interval Δt_0 measured by the astronaut. In other words, the earth-based observer, using her own earth-based clock to measure the performance of the astronaut's clock, finds that the astronaut's clock runs slowly. This result of special relativity is known as **time dilation**. (To *dilate* means to expand, and the time interval Δt is "expanded" relative to Δt_0.)

The time interval Δt that the earth-based observer measures in **Animated Figure 28.5b** can be determined as follows. While the light pulse travels from the source to the detector, the spacecraft moves a distance $2L = v\,\Delta t$ to the right, where v is the speed of the spacecraft relative to the earth. From the drawing it can be seen that the light pulse travels a total diagonal distance of $2s$ during the time interval Δt. Applying the Pythagorean theorem, we find that

$$2s = 2\sqrt{D^2 + L^2} = 2\sqrt{D^2 + \left(\frac{v\,\Delta t}{2}\right)^2}$$

But the distance $2s$ is also equal to the speed of light times the time interval Δt, so that $2s = c\,\Delta t$. Therefore,

$$c\,\Delta t = 2\sqrt{D^2 + \left(\frac{v\,\Delta t}{2}\right)^2}$$

Squaring this result and solving for Δt gives

$$\Delta t = \frac{2D}{c}\frac{1}{\sqrt{1 - \dfrac{v^2}{c^2}}}$$

However, $2D/c = \Delta t_0$, where Δt_0 is the time interval between successive "ticks" of the spacecraft's clock as measured by the astronaut. With this substitution, the equation for Δt can be expressed as

Time dilation

$$\Delta t = \frac{\Delta t_0}{\sqrt{1 - \dfrac{v^2}{c^2}}} \qquad (28.1)$$

The symbols in this formula are defined as follows:

Δt_0 = proper time interval, which is the interval between two events as measured by an observer who is at rest with respect to the events and who views them as occurring *at the same place*

Δt = dilated time interval, which is the interval measured by an observer who is in motion with respect to the events and who views them as occurring at *different places*

v = relative speed between the two observers

c = speed of light in a vacuum

For a speed v that is less than c, the term $\sqrt{1 - v^2/c^2}$ in Equation 28.1 is less than 1, and the dilated time interval Δt is greater than Δt_0. Example 1 illustrates the extent of this time dilation effect.

EXAMPLE 1 | Time Dilation

The spacecraft in **Animated Figure 28.5** is moving past the earth at a constant speed v that is 0.92 times the speed of light. Thus, $v =$ $(0.92)(3.0 \times 10^8$ m/s), which is often written as $v = 0.92c$. The astronaut measures the time interval between successive "ticks" of the spacecraft clock to be $\Delta t_0 = 1.0$ s. What is the time interval Δt that an earth observer measures between "ticks" of the astronaut's clock?

Reasoning Since the clock on the spacecraft is moving relative to the earth, the earth-based observer measures a greater time interval Δt between "ticks" than does the astronaut, who is at rest relative to the clock. The dilated time interval Δt can be determined from the time dilation relation, Equation 28.1.

Solution The dilated time interval is

$$\Delta t = \frac{\Delta t_0}{\sqrt{1 - \dfrac{v^2}{c^2}}} = \frac{1.0 \text{ s}}{\sqrt{1 - \left(\dfrac{0.92c}{c}\right)^2}} = \boxed{2.6 \text{ s}}$$

From the point of view of the earth-based observer, the astronaut is using a clock that is running slowly, because the earth-based observer measures a time between "ticks" that is longer (2.6 s) than what the astronaut measures (1.0 s).

THE PHYSICS OF . . . the Global Positioning System and special relativity. Present-day spacecrafts fly nowhere near as fast as the craft in Example 1. Yet circumstances exist in which time dilation can create appreciable errors if not accounted for. The Global Positioning System (GPS), for instance, uses highly accurate and stable atomic clocks on board each of 24 satellites orbiting the earth at speeds of about 4000 m/s. These clocks make it possible to measure the time it takes for electromagnetic waves to travel from a satellite to a ground-based GPS receiver. From the speed of light and the times measured for signals from three or more of the satellites, it is possible to locate the position of the receiver (see Section 5.5). The stability of the clocks must be better than one part in 10^{13} to ensure the positional accuracy demanded of the GPS. Using Equation 28.1 and the speed of the GPS satellites, we can calculate the difference between the dilated time interval and the proper time interval as a fraction of the proper time interval and compare the result to the stability of the GPS clocks:

$$\frac{\Delta t - \Delta t_0}{\Delta t_0} = \frac{1}{\sqrt{1 - v^2/c^2}} - 1 = \frac{1}{\sqrt{1 - (4000 \text{ m/s})^2/(3.00 \times 10^8 \text{ m/s})^2}} - 1$$

$$= \frac{1}{1.1 \times 10^{10}}$$

This result is approximately one thousand times greater than the GPS-clock stability of one part in 10^{13}. Thus, if not taken into account, time dilation would cause an error in the measured position of the earth-based GPS receiver roughly equivalent to the error caused by a thousand-fold degradation in the stability of the atomic clocks.

Proper Time Interval

In **Animated Figure 28.5** both the astronaut and the person standing on the earth are measuring the time interval between a beginning event (the firing of the light source) and an ending event (the light pulse striking the detector). For the astronaut, who is at rest with respect to the light clock, the two events occur at the same location. (Remember, we are assuming that the light source and detector are so close together that they are considered to be at the same place.) Being at rest with respect to a clock is the usual or "proper" situation, so the time interval Δt_0 measured by the astronaut is called the **proper time interval**. In general, the proper time interval Δt_0 between two events is the time interval measured by an observer who is at rest relative to the events and sees them at the *same location* in space. On the other hand, the earth-based observer does not see the two events occurring at the same location in space, since the spacecraft is in motion. The time interval Δt that the earth-based observer measures is, therefore, not a proper time interval in the sense that we have defined it.

To understand situations involving time dilation, it is essential to distinguish between Δt_0 and Δt. It is helpful if one first identifies the two events that define the time interval. These may be something other than the firing of a light source and the light pulse striking a detector. Then determine the reference frame in which the two events occur at the same place. An observer at rest in this reference frame measures the proper time interval Δt_0.

Space Travel

One of the intriguing aspects of time dilation occurs in conjunction with space travel. Since enormous distances are involved, travel to even the closest star outside our solar system would take a long time. However, as the following example shows, the travel time can be considerably less for the passengers than one might guess.

EXAMPLE 2 | **The Physics of Space Travel and Special Relativity**

Alpha Centauri, a nearby star in our galaxy, is 4.3 light-years away. This means that, as measured by a person on earth, it would take light 4.3 years to reach this star. If a rocket leaves for Alpha Centauri and travels at a speed of $v = 0.95c$ relative to the earth, by how much will the passengers have aged, according to their own clock, when they reach their destination? Assume that the earth and Alpha Centauri are stationary with respect to one another.

Reasoning The two events in this problem are the departure from earth and the arrival at Alpha Centauri. At departure, earth is just outside the spaceship. Upon arrival at the destination, Alpha Centauri is just outside. Therefore, relative to the passengers, the two events occur at the same place—namely, just outside the spaceship. Thus, the passengers measure the proper time interval Δt_0 on their clock, and it is this interval that we must find. For a person left behind on earth, the events occur at *different places*, so such a person measures the dilated time interval Δt rather than the proper time interval. To find Δt we note that the time to travel a given distance is inversely proportional to the speed. Since it takes 4.3 years to traverse the distance between earth and Alpha Centauri at the speed of light, it would take even longer at the slower speed of $v = 0.95c$. Thus, a person on earth measures the dilated

time interval to be $\Delta t = (4.3 \text{ years})/0.95 = 4.5$ years. This value can be used with the time-dilation equation to find the proper time interval Δt_0.

> **Problem-Solving Insight** In dealing with time dilation, decide which interval is the proper time interval as follows: (1) Identify the two events that define the interval. (2) Determine the reference frame in which the events occur at the same place; an observer at rest in this frame measures the proper time interval Δt_0.

Solution Using the time-dilation equation, we find that the proper time interval by which the passengers judge their own aging is

$$\Delta t_0 = \Delta t \sqrt{1 - \frac{v^2}{c^2}} = (4.5 \text{ years})\sqrt{1 - \left(\frac{0.95c}{c}\right)^2} = \boxed{1.4 \text{ years}}$$

Thus, the people aboard the rocket will have aged by only 1.4 years when they reach Alpha Centauri, and not the 4.5 years an earth-bound observer has calculated.

Verification of Time Dilation

A striking confirmation of time dilation was achieved in 1971 by an experiment carried out by J. C. Hafele and R. E. Keating.* They transported very precise cesium-beam atomic clocks around the world on commercial jets. Since the speed of a jet plane is considerably less than c, the time-dilation effect is extremely small. However, the atomic clocks were

---

*J. C. Hafele and R. E. Keating, "Around-the-World Atomic Clocks: Observed Relativistic Time Gains," *Science*, Vol. 177, July 14, 1972, p. 168.

accurate to about $\pm 10^{-9}$ s, so the effect could be measured. The clocks were in the air for 45 hours, and their times were compared to reference atomic clocks kept on earth. The experimental results revealed that, within experimental error, the readings on the clocks on the planes were different from those on earth by an amount that agreed with the prediction of relativity.

The behavior of subatomic particles called *muons* provides additional confirmation of time dilation. These particles are created high in the atmosphere, at altitudes of about 10 000 m. When at rest, muons exist only for about 2.2×10^{-6} s before disintegrating. With such a short lifetime, these particles could never make it down to the earth's surface, even traveling at nearly the speed of light. However, *a large number of muons do reach the earth.* The only way they can do so is to live longer because of time dilation, as Example 3 illustrates.

EXAMPLE 3 | The Lifetime of a Muon

The average lifetime of a muon at rest is 2.2×10^{-6} s. A muon created in the upper atmosphere, thousands of meters above sea level, travels toward the earth at a speed of $v = 0.998c$. Find, on the average, **(a)** how long a muon lives according to an observer on earth, and **(b)** how far the muon travels before disintegrating.

Reasoning The two events of interest are the generation and subsequent disintegration of the muon. When the muon is at rest, these events occur at the same place, so the muon's average (at rest) lifetime of 2.2×10^{-6} s is a proper time interval Δt_0. When the muon moves at a speed $v = 0.998c$ relative to the earth, an observer on the earth measures a dilated lifetime Δt that is given by Equation 28.1. The average distance x traveled by a muon, as measured by an earth observer, is equal to the muon's speed times the dilated time interval.

> **Problem-Solving Insight** The proper time interval Δt_0 is always shorter than the dilated time interval Δt.

Solution **(a)** The observer on earth measures a dilated lifetime. Using the time-dilation equation, we find that

$$\Delta t = \frac{\Delta t_0}{\sqrt{1 - \frac{v^2}{c^2}}} = \frac{2.2 \times 10^{-6} \text{ s}}{\sqrt{1 - \left(\frac{0.998c}{c}\right)^2}} = \boxed{35 \times 10^{-6} \text{ s}} \qquad (28.1)$$

(b) The distance traveled by the muon before it disintegrates is

$$x = v \Delta t = (0.998)(3.00 \times 10^8 \text{ m/s})(35 \times 10^{-6} \text{ s})$$
$$= \boxed{1.0 \times 10^4 \text{ m}}$$

Thus, the dilated, or extended, lifetime provides sufficient time for the muon to reach the surface of the earth. If its lifetime were only 2.2×10^{-6} s, a muon would travel only 660 m before disintegrating and could never reach the earth.

Example 4 illustrates how a person subjected to time dilation not only perceives extended time relative to another reference frame, but is subjected to the additional aging (biological processes) that would occur in such a timespan.

EXAMPLE 4 | BIO The Physics of Space Travel Revisited

Relativistic time dilation affects not only the operation of clocks, but also all physiological and biological processes, including the rates of chemical reactions. This is why a person moving on a spaceship, for example, will age slower than one that stays at home on earth. Imagine an astronaut in a spaceship that is moving away from the earth at 98% the speed of light. Given that it takes the human body only 27 days to replace the entire outer layer of the skin, how many times has the person on earth replaced their skin, if 27 days pass on the spaceship?

Reasoning We recognize this to be an example of relativistic time dilation. We can apply Equation 28.1 to calculate the dilated time interval. Since the astronaut measures the time on his clock at the same location (inside the ship), he is measuring the proper time interval Δt_0. The ship is moving relative to the reference frame of the earth, so the earth-bound observer measures the dilated time interval.

Solution Beginning with Equation 28.1, we have:

$$\Delta t = \frac{\Delta t_0}{\sqrt{1 - \frac{v^2}{c^2}}}.$$

Since we want to calculate the dilated time interval in days, we keep the proper time in those units as well. The speed of the ship is 98% the speed of light, which is $v = 0.98c$. We make these substitutions above and solve for Δt:

$$\Delta t = \frac{27 \text{ days}}{\sqrt{1 - \frac{(0.98c)^2}{c^2}}} = \frac{27 \text{ days}}{\sqrt{1 - (0.98)^2}} = \boxed{(5.0)(27 \text{ days})}$$

The length of time that passes on earth during the 27 days on the ship is five times longer. Therefore, the person on earth will have replaced their skin five times!

Check Your Understanding

(*The answers are given at the end of the book.*)

1. Which one of the following statements concerning the dilated time interval is false? **(a)** It is always greater than the proper time interval. **(b)** It depends on the relative speed between the observers who measure the proper and dilated time intervals. **(c)** It depends on the speed of light in a vacuum. **(d)** It is measured by an observer who is at rest with respect to the events that define the time interval.

2. A baseball player at home plate hits a pop fly straight up (the beginning event) that is caught by the catcher at home plate (the ending event). Which one or more of the following observers record(s) the proper time interval between the events? **(a)** A person sitting in the stands **(b)** A person watching the game on TV **(c)** The pitcher running in to cover the play

3. A playground carousel is a circular platform that can rotate about an axis perpendicular to the plane of the platform at its center. It behaves approximately as an inertial reference frame. An observer is looking down at this platform from an inertial reference frame directly above the rotational axis. Three clocks are attached to the platform. Clock A is attached directly to the axis. Clock B is attached to a point midway between the axis and the outer edge of the platform. Clock C is attached to the outer edge of the platform. Rank the clocks according to how slow the observer finds them to be running (slowest first).

28.4 | The Relativity of Length: Length Contraction

Because of time dilation, observers moving at a constant velocity relative to each other measure different time intervals between two events. For instance, Example 2 in the previous section illustrates that a trip from earth to Alpha Centauri at a speed of $v = 0.95c$ takes 4.5 years according to a clock on earth, but only 1.4 years according to a clock in the rocket. These two times differ by the factor $\sqrt{1 - v^2/c^2}$. Since the times for the trip are different, one might ask whether the observers measure different distances between earth and Alpha Centauri. The answer, according to special relativity, is yes. After all, both the earth-based observer and the rocket passenger agree that the relative speed between the rocket and earth is $v = 0.95c$. Since speed is distance divided by time and the time is different for the two observers, it follows that the distances must also be different, if the relative speed is to be the same for both individuals. Thus, the earth observer determines the distance to Alpha Centauri to be $L_0 = v\,\Delta t = (0.95c)(4.5 \text{ years}) = 4.3$ light-years. On the other hand, a passenger aboard the rocket finds the distance is only $L = v\,\Delta t_0 = (0.95c)(1.4 \text{ years}) = 1.3$ light-years. The passenger, measuring the shorter time, also measures the shorter distance. This shortening of the distance between two points is one example of a phenomenon known as **length contraction**.

The relation between the distances measured by two observers in relative motion at a constant velocity can be obtained with the aid of **Interactive Figure 28.6**. Part *a* of the drawing shows the situation from the point of view of the earth-based observer. This person measures the time of the trip to be Δt, the distance to be L_0, and the relative speed of the rocket to be $v = L_0/\Delta t$. Part *b* of the drawing presents the point of view of the passenger, for whom the rocket is at rest, and the earth and Alpha Centauri appear to move by at a speed v. The passenger determines the distance of the trip to be L, the time to be Δt_0, and the relative speed to be $v = L/\Delta t_0$. Since the relative speed computed by the passenger equals that computed by the earth-based observer, it follows that $v = L/\Delta t_0 = L_0/\Delta t$. Using this result and the time-dilation equation, Equation 28.1, we obtain the following relation between L and L_0:

Length contraction

$$L = L_0\sqrt{1 - \frac{v^2}{c^2}}$$

(28.2)

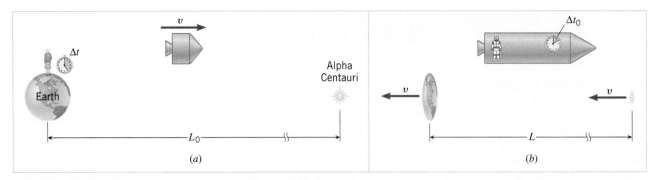

INTERACTIVE FIGURE 28.6 (a) As measured by an observer on the earth, the distance to Alpha Centauri is L_0, and the time required to make the trip is Δt. (b) According to the passenger on the spacecraft, the earth and Alpha Centauri move with speed v relative to the craft. The passenger measures the distance and time of the trip to be L and Δt_0, respectively, both quantities being less than those in part a.

The length L_0 is called the **proper length**; it is the length (or distance) between two points *as measured by an observer at rest with respect to them*. Since v is less than c, the term $\sqrt{1 - v^2/c^2}$ is less than 1, and L is less than L_0. It is important to note that this length contraction occurs only along the direction of the motion. Those dimensions that are perpendicular to the motion are not shortened, as the next example discusses.

EXAMPLE 5 | The Contraction of a Spacecraft

An astronaut, using a meter stick that is at rest relative to a cylindrical spacecraft, measures the length and diameter of the spacecraft to be 82 m and 21 m, respectively. The spacecraft moves with a constant speed of $v = 0.95c$ relative to the earth, as in **Interactive Figure 28.6**. What are the dimensions of the spacecraft, as measured by an observer on earth?

Reasoning The length of 82 m is a proper length L_0, since it is measured using a meter stick that is at rest relative to the space-craft. The length L measured by the observer on earth can be determined from the length-contraction formula, Equation 28.2. On the other hand, the diameter of the spacecraft is perpendicular to the motion, so the earth observer does not measure any change in the diameter.

Problem-Solving Insight The proper length L_0 is always larger than the contracted length L.

Solution The length L of the spacecraft, as measured by the observer on earth, is

$$L = L_0 \sqrt{1 - \frac{v^2}{c^2}} = (82 \text{ m}) \sqrt{1 - \left(\frac{0.95c}{c}\right)^2} = \boxed{26 \text{ m}}$$

Both the astronaut and the observer on earth measure the same value for the diameter of the spacecraft: Diameter = $\boxed{21 \text{ m}}$. **Interactive Figure 28.6a** shows the size of the spacecraft as measured by the earth observer, and part b shows the size as measured by the astronaut.

When dealing with relativistic effects we need to distinguish carefully between the criteria for the proper time interval and the proper length. The proper time interval Δt_0 between two events is the time interval measured by an observer who is at rest relative to the events and sees them occurring at the *same place*. All other moving inertial observers will measure a larger value for this time interval. The proper length L_0 of an object is the length measured by an observer who is *at rest* with respect to the object. All other moving inertial observers will measure a shorter value for this length. The observer who measures the proper time interval may not be the same one who measures the proper length. For instance, **Interactive Figure 28.6** shows that the astronaut measures the proper time interval Δt_0 for the trip between earth and Alpha Centauri, whereas the earth-based observer measures the proper length (or distance) L_0 for the trip.

It should be emphasized that the word "proper" in the phrases "proper time" and "proper length" does *not* mean that these quantities are the correct or preferred quantities in any absolute sense. If this were so, the observer measuring these quantities would be using a preferred reference frame for making the measurement, a situation that is prohibited by the relativity postulate. According to this postulate, there is no preferred inertial reference frame. When two observers are moving relative to each other at a constant velocity, each measures the other person's clock to run more slowly than his own, and each measures the other person's length, along that person's motion, to be contracted.

Check Your Understanding

(The answers are given at the end of the book.)

4. If the speed *c* of light in a vacuum were infinitely large instead of 3.0×10^8 m/s, would the effects of time dilation and length contraction be observable?

5. Suppose that you are standing at a railroad crossing, watching a train go by. Both you and a passenger in the train are looking at a clock on the train. Who measures the proper time interval, and who measures the proper length of a train car? **(a)** You measure the proper time interval, and the passenger measures the proper length. **(b)** You measure both the proper time interval and the proper length. **(c)** The passenger measures both the proper time interval and the proper length. **(d)** You measure the proper length, and the passenger measures the proper time interval.

6. Which one or more of the following quantities will two observers always measure to be the *same*, regardless of the relative velocity between the observers? **(a)** The time interval between two events **(b)** The length of an object **(c)** The speed of light in a vacuum **(d)** The relative speed between the observers

7. **CYU Figure 28.1** shows an object that has the shape of a square when it is at rest in inertial reference frame R. When the object moves relative to this reference frame, the object's velocity vector is in the plane of the square and is parallel to the diagonal AC. Since the speed of the motion is an appreciable fraction of the speed of light in a vacuum, noticeable length contraction occurs. Does an observer in reference frame R see the object as a square? (*Hint:* Consider what happens to each of the diagonals AC and BD.)

CYU FIGURE 28.1

28.5 Relativistic Momentum

Thus far we have discussed how time intervals and distances between two events are measured by observers moving at a constant velocity relative to each other. Special relativity also alters our ideas about momentum and energy.

Recall from Section 7.2 that when two or more objects interact, the principle of conservation of linear momentum applies if the system of objects is isolated. This principle states that the total linear momentum of an isolated system remains constant at all times. (An isolated system is one in which the sum of the external forces acting on the objects is zero.) The conservation of linear momentum is a law of physics and, in accord with the relativity postulate, is valid in all inertial reference frames. That is, when the total linear momentum is conserved in one inertial reference frame, it is conserved in all inertial reference frames.

As an example of momentum conservation, suppose that several people are watching two billiard balls collide on a frictionless pool table. One person is standing next to the pool table, and the other is moving past the table with a constant velocity. Since the two balls constitute an isolated system, the relativity postulate requires that both observers must find the total linear momentum of the two-ball system to be the same before, during, and after the collision. For this kind of situation, Section 7.1 defines the classical linear momentum $\vec{\mathbf{p}}$ of an object to be the product of its mass *m* and velocity $\vec{\mathbf{v}}$. As a result, the magnitude of the classical momentum is $p = mv$. As long as the speed of an object is considerably smaller than the speed of light, this definition is adequate. However, when the speed approaches the speed of light, an analysis of the collision shows that the total linear momentum is not conserved in all inertial reference frames if one defines linear momentum simply as the product of mass and velocity. In order to preserve the conservation of linear momentum, it is necessary to modify this definition.

The theory of special relativity reveals that the magnitude of the **relativistic momentum** must be defined as in Equation 28.3:

Magnitude of the relativistic momentum

$$p = \frac{mv}{\sqrt{1 - \dfrac{v^2}{c^2}}} \qquad\qquad (28.3)$$

The total relativistic momentum of an isolated system is conserved in all inertial reference frames.

From Equation 28.3, we can see that the magnitudes of the relativistic and nonrelativistic momenta differ by the same factor of $\sqrt{1 - v^2/c^2}$ that occurs in the time-dilation and length-contraction equations. Since this factor is always less than 1 and occurs in the denominator in Equation 28.3, the relativistic momentum is always larger than the nonrelativistic momentum. To illustrate how the two quantities differ as the speed v increases, **Figure 28.7** shows a plot of the ratio of the momentum magnitudes (relativistic/nonrelativistic) as a function of v. According to Equation 28.3, this ratio is just $1/\sqrt{1 - v^2/c^2}$. The graph shows that for speeds attained by ordinary objects, such as cars and planes, the relativistic and nonrelativistic momenta are almost equal because their ratio is nearly 1. Thus, at speeds much less than the speed of light, either the nonrelativistic momentum or the relativistic momentum can be used to describe collisions. On the other hand, when the speed of the object becomes comparable to the speed of light, the relativistic momentum becomes significantly greater than the nonrelativistic momentum and must be used. Example 6 deals with the relativistic momentum of an electron traveling close to the speed of light.

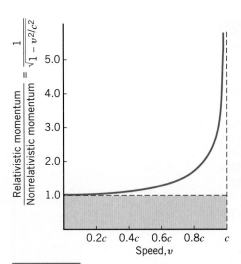

FIGURE 28.7 This graph shows how the ratio of the magnitude of the relativistic momentum to the magnitude of the nonrelativistic momentum increases as the speed of an object approaches the speed of light.

EXAMPLE 6 | The Relativistic Momentum of a High-Speed Electron

The particle accelerator at Stanford University (**Figure 28.8**) is 3 km long and accelerates electrons to a speed of $0.999\,999\,999\,7c$, which is very nearly equal to the speed of light. Find the magnitude of the relativistic momentum of an electron that emerges from the accelerator, and compare it with the nonrelativistic value.

Reasoning and Solution The magnitude of the electron's relativistic momentum can be obtained from Equation 28.3 if we recall that the mass of an electron is $m = 9.11 \times 10^{-31}$ kg:

$$p = \frac{mv}{\sqrt{1 - \dfrac{v^2}{c^2}}} = \frac{(9.11 \times 10^{-31}\ \text{kg})(0.999\,999\,999\,7c)}{\sqrt{1 - \dfrac{(0.999\,999\,999\,7c)^2}{c^2}}}$$

$$= \boxed{1 \times 10^{-17}\ \text{kg} \cdot \text{m/s}}$$

This value for the magnitude of the momentum agrees with the value measured experimentally. The relativistic momentum is greater than the nonrelativistic momentum by a factor of

$$\frac{1}{\sqrt{1 - \dfrac{v^2}{c^2}}} = \frac{1}{\sqrt{1 - \dfrac{(0.999\,999\,999\,7c)^2}{c^2}}} = \boxed{4 \times 10^4}$$

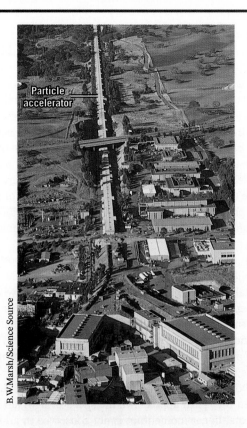

B.W.Marsh/Science Source

FIGURE 28.8 The Stanford 3-km linear accelerator accelerates electrons almost to the speed of light.

Math Skills Using your calculator to obtain the answers in Example 6 may not be possible due to the term $\sqrt{1 - \frac{v^2}{c^2}}$ in the pertinent equations. The potential difficulty is that your calculator may not let you enter a number such as 0.999 999 999 7, because it cannot accept ten places after the decimal point. Since the value for v is nearly equal to c and $\frac{v}{c} \approx 1$, one way to deal with such a situation is to note that

$$\sqrt{1 - \frac{v^2}{c^2}} = \sqrt{\left(1 + \frac{v}{c}\right)\left(1 - \frac{v}{c}\right)} \approx \sqrt{2\left(1 - \frac{v}{c}\right)}$$

Thus, we find that

$$\frac{1}{\sqrt{1 - \frac{v^2}{c^2}}} \approx \frac{1}{\sqrt{2\left(1 - \frac{v}{c}\right)}}$$

$$= \frac{1}{\sqrt{2\left(1 - \frac{0.999\ 999\ 999\ 7\ \cancel{c}}{\cancel{c}}\right)}} = \frac{1}{\sqrt{2(3 \times 10^{-10})}}$$

$$\approx \boxed{4 \times 10^4}$$

In this answer the speed v is given to ten significant figures. However, the answer is given to only one significant figure! This is because the subtraction causes a loss of significant figures as follows:

$$1 - 0.999\ 999\ 999\ 7 = 0.000\ 000\ 000\ 3 = 3 \times 10^{-10}$$

The amount of time it would take for relativistic effects to become appreciable while accelerating at $g = 9.8$ m/s^2 is explored in Example 7.

EXAMPLE 7 | BIO Artificial Gravity for Long Voyages

Weightlessness, or microgravity, can have detrimental effects on the human body. Short-term symptoms, including disorientation and nausea, can be overcome in a short time so that an astronaut can function over the course of a mission. Long-term effects, including muscle and bone atrophy, and the possible deterioration of cognitive ability, are highly undesirable, and need to be avoided in future long voyages. In Chapter 5, we encountered examples of how "artificial gravity" on a space station can be generated using rotational motion. However, in order for a spacecraft to make long voyages at high velocities relative to earth, linear accelerations are required. If the linear acceleration of a ship was $g = 9.8$ m/s^2, a person standing on the "floor" of the ship would feel as if they were standing on earth, and would not experience any of the weightlessness-induced effects. It is interesting to know how long it would take a spacecraft accelerating at g to reach relativistic speeds. Figure 28.7 gives a graphical measure of how relativistic and nonrelativistic momenta differ, and we can estimate that relativistic effects become appreciable when $v \approx 0.30c$, that is, where the relativistic and nonrelativistic momenta differ by about 5%. Use nonrelativistic kinematics for constant acceleration ($a = g = 9.8$ m/s^2) to estimate **(a)** the time it takes a ship to accelerate from rest to $v = 0.30c$, and **(b)** the distance it travels during that time.

Reasoning **(a)** Since the acceleration is constant, we can apply the kinematic equations, specifically Equation 2.4, with $a = g$ and

the initial velocity set to 0 m/s, since the ship starts from rest. **(b)** With the time calculated in part (a), and the initial velocity again set to 0 m/s, we can apply Equation 2.8 to find the total distance traveled.

Solution **(a)** Using Equation 2.4 with the initial velocity v_0 set to 0 m/s, we have

$$v = \cancel{v_0} + at \quad \Rightarrow \quad t = \frac{v}{a} = \frac{(0.30)c}{a} = \frac{(0.30)\,(3.0 \times 10^8\ \text{m/s})}{9.8\ \text{m/s}^2}$$

$$= \boxed{9.2 \times 10^6\ \text{s}}$$

This is equal to about 106 days.

(b) We know the time from part (a) and that the initial velocity is 0 m/s. Using Equation 2.8 to find the distance traveled during this time, we have

$$x = \cancel{v_0}t + \tfrac{1}{2}at^2 = \tfrac{1}{2}(9.8\ \text{m/s}^2)(9.2 \times 10^6\ \text{s})^2 = \boxed{4.1 \times 10^{14}\ \text{m}}$$

or 4.1×10^{11} km $= 0.043$ ly. For comparison, the radius of the solar system is about 4.5×10^9 km.

Although constant-acceleration (or constant-thrust) propulsion would be a quick way to travel, there are challenges that make it currently unfeasible. For instance, a large, constant thrust for an extended period of time would require a prohibitively large amount of fuel.

28.6 # The Equivalence of Mass and Energy

The Total Energy of an Object

One of the most astonishing results of special relativity is that mass and energy are equivalent, in the sense that a gain or loss of mass can be regarded equally well as a gain or loss of energy. Consider, for example, an object of mass m traveling at a speed v. Einstein showed that the **total energy** E of the moving object is related to its mass and speed by the following relation:

Total energy of an object

$$E = \frac{mc^2}{\sqrt{1 - \dfrac{v^2}{c^2}}}$$ (28.4)

To gain some understanding of Equation 28.4, consider the special case in which the object is at rest. When $v = 0$ m/s, the total energy is called the **rest energy** E_0, and Equation 28.4 reduces to Einstein's now-famous equation:

Rest energy of an object

$$E_0 = mc^2$$ (28.5)

The rest energy represents the energy equivalent of the mass of an object at rest. As Example 8 shows, even a small mass is equivalent to an enormous amount of energy.

Analyzing Multiple-Concept Problems

EXAMPLE 8 | The Energy Equivalent of a Golf Ball

A 0.046-kg golf ball is lying on the green, as **Figure 28.9** illustrates. If the rest energy of this ball were used to operate a 75-W light bulb, how long would the bulb remain lit?

Reasoning The average power delivered to the light bulb is 75 W, which means that it uses 75 J of energy per second. Therefore, the time that the bulb would remain lit is equal to the total energy used by the light bulb divided by the energy per second (i.e., the average power) delivered to it. This energy comes from the rest energy of the golf ball, which is equal to its mass times the speed of light squared.

Knowns and Unknowns The data for this problem are:

Description	Symbol	Value
Mass of golf ball	m	0.046 kg
Average power delivered to light bulb	$\overline{P}$	75 W
Unknown Variable		
Time that light bulb would remain lit	t	?

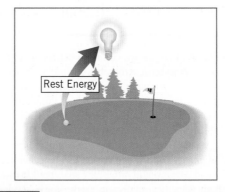

FIGURE 28.9 The rest energy of a golf ball is sufficient to keep a 75-W light bulb burning for an incredibly long time (see Example 8).

Modeling the Problem

STEP 1 Power The average power $\overline{P}$ is equal to the energy delivered to the light bulb divided by the time t (see Section 6.7 and Equation 6.10b), or $\overline{P}$ = Energy/t. In this case the energy comes from the rest energy E_0 of the golf ball, so $\overline{P} = E_0/t$. Solving for the time gives Equation 1 at the right. The average power is known, and the rest energy will be evaluated in Step 2.

$$t = \frac{E_0}{\overline{P}}$$ (1)

STEP 2 Rest Energy The rest energy E_0 is the total energy of the golf ball as it rests on the green. If the golf ball's mass is m, then its rest energy is

$$\boxed{E_0 = mc^2} \qquad (28.5)$$

where c is the speed of light in a vacuum. Both m and c are known, so we substitute this expression for the rest energy into Equation 1, as indicated at the right.

$$t = \dfrac{E_0}{P} \qquad (1)$$

$$\boxed{E_0 = mc^2} \qquad (28.5)$$

Solution Algebraically combining the results of each step, we have

$$\underset{\underset{\displaystyle t}{\downarrow}}{\text{STEP 1}} \quad = \quad \underset{\underset{\displaystyle \frac{E_0}{P}}{\downarrow}}{\text{STEP 2}} \quad = \quad \frac{mc^2}{P}$$

The time that the light bulb would remain lit is

$$t = \frac{mc^2}{P} = \frac{(0.046\ \text{kg})(3.0 \times 10^8\ \text{m/s})^2}{75\ \text{W}} = \boxed{5.5 \times 10^{13}\ \text{s}}$$

Expressed in years ($1\ \text{yr} = 3.2 \times 10^7\ \text{s}$), this time is equivalent to

$$(5.5 \times 10^{13}\ \cancel{s})\left(\frac{1\ \text{yr}}{3.2 \times 10^7\ \cancel{s}}\right) = 1.7 \times 10^6\ \text{yr or 1.7 million years!}$$

Related Homework: Problems 27, 28

When an object is accelerated from rest to a speed v, the object acquires kinetic energy in addition to its rest energy. The total energy E is the sum of the rest energy E_0 and the kinetic energy KE, or $E = E_0 + \text{KE}$. Therefore, the kinetic energy is the difference between the object's total energy and its rest energy. Using Equations 28.4 and 28.5, we can write the kinetic energy as

Kinetic energy of an object

$$\text{KE} = E - E_0 = mc^2\left(\frac{1}{\sqrt{1 - \dfrac{v^2}{c^2}}} - 1\right) \qquad (28.6)$$

This equation is the relativistically correct expression for the kinetic energy of an object of mass m moving at speed v.

Equation 28.6 looks nothing like the kinetic energy expression introduced in Section 6.2—namely, $\text{KE} = \frac{1}{2}mv^2$ (Equation 6.2). However, for speeds much less than the speed of light ($v \ll c$), the relativistic equation for the kinetic energy reduces to $\text{KE} = \frac{1}{2}mv^2$, as can be seen by using the binomial expansion* to represent the square root term in Equation 28.6:

$$\frac{1}{\sqrt{1 - \dfrac{v^2}{c^2}}} = 1 + \frac{1}{2}\left(\frac{v^2}{c^2}\right) + \frac{3}{8}\left(\frac{v^2}{c^2}\right)^2 + \cdots$$

Suppose that v is much smaller than c—say, $v = 0.01c$. The second term in the expansion has the value $\frac{1}{2}(v^2/c^2) = 5.0 \times 10^{-5}$, while the third term has the much smaller value $\frac{3}{8}(v^2/c^2)^2 = 3.8 \times 10^{-9}$. The additional terms are smaller still, so if $v \ll c$, we can neglect the third and additional terms in comparison with the first and second terms. Substituting the first two terms into Equation 28.6 gives

$$\text{KE} \approx mc^2\left(1 + \frac{1}{2}\frac{v^2}{c^2} - 1\right) = \frac{1}{2}mv^2$$

which is the familiar form for the kinetic energy. However, Equation 28.6 gives the correct kinetic energy for all speeds and must be used for speeds near the speed of light, as in Example 9.

*The binomial expansion states that $(1 - x)^n = 1 - nx + n(n-1)x^2/2 + \cdots$. In our case, $x = v^2/c^2$ and $n = -1/2$.

EXAMPLE 9 | A High-Speed Electron

An electron ($m = 9.109 \times 10^{-31}$ kg) is accelerated from rest to a speed of $v = 0.9995c$ in a particle accelerator. Determine the electron's **(a)** rest energy, **(b)** total energy, and **(c)** kinetic energy in millions of electron volts or MeV.

Reasoning and Solution **(a)** The electron's rest energy is

$$E_0 = mc^2 = (9.109 \times 10^{-31} \text{ kg})(2.998 \times 10^8 \text{ m/s})^2$$

$$= 8.187 \times 10^{-14} \text{ J} \qquad \textbf{(28.5)}$$

Since 1 eV $= 1.602 \times 10^{-19}$ J, the electron's rest energy is

$$(8.187 \times 10^{-14} \text{ J})\left(\frac{1 \text{ eV}}{1.602 \times 10^{-19} \text{ J}}\right) = \boxed{5.11 \times 10^5 \text{ eV or } 0.511 \text{ MeV}}$$

(b) The total energy of an electron traveling at a speed of $v = 0.9995c$ is

$$E = \frac{mc^2}{\sqrt{1 - \dfrac{v^2}{c^2}}} = \frac{(9.109 \times 10^{-31} \text{ kg})(2.998 \times 10^8 \text{ m/s})^2}{\sqrt{1 - \left(\dfrac{0.9995c}{c}\right)^2}} \qquad \textbf{(28.4)}$$

$$= \boxed{2.59 \times 10^{-12} \text{ J or } 16.2 \text{ MeV}}$$

(c) The kinetic energy is the difference between the total energy and the rest energy:

$$KE = E - E_0 = 2.59 \times 10^{-12} \text{ J} - 8.2 \times 10^{-14} \text{ J} \qquad \textbf{(28.6)}$$

$$= \boxed{2.51 \times 10^{-12} \text{ J or } 15.7 \text{ MeV}}$$

For comparison, if the kinetic energy of the electron had been calculated from $\frac{1}{2}mv^2$, a value of only 0.255 MeV would have been obtained.

Since mass and energy are equivalent, any change in one is accompanied by a corresponding change in the other. For instance, life on earth is dependent on electromagnetic energy (light) from the sun. Because this energy is leaving the sun (see **Figure 28.10**), there is a decrease in the sun's mass. Example 10 illustrates how to determine this decrease.

FIGURE 28.10 The sun emits electromagnetic energy over a broad portion of the electromagnetic spectrum. These photographs were obtained using that energy in the indicated regions of the spectrum.

X-ray image   Visible light image

NASA/JISAS/Science Photo Library

EXAMPLE 10 | The Sun Is Losing Mass

The sun radiates electromagnetic energy at the rate of 3.92×10^{26} W. **(a)** What is the change in the sun's mass during each second that it is radiating energy? **(b)** The mass of the sun is 1.99×10^{30} kg. What fraction of the sun's mass is lost during a human lifetime of 75 years?

Reasoning Since 1 W = 1 J/s, the amount of electromagnetic energy radiated during each second is 3.92×10^{26} J. Thus, during each second, the sun's rest energy decreases by this amount. The change ΔE_0 in the sun's rest energy is related to the change Δm in its mass by $\Delta E_0 = (\Delta m)c^2$, according to Equation 28.5.

Solution **(a)** For each second that the sun radiates energy, the change in its mass is

$$\Delta m = \frac{\Delta E_0}{c^2} = \frac{3.92 \times 10^{26} \text{ J}}{(3.00 \times 10^8 \text{ m/s})^2} = \boxed{4.36 \times 10^9 \text{ kg}}$$

Over 4 billion kilograms of mass are lost by the sun during each second.

(b) The amount of mass lost by the sun in 75 years is

$$\Delta m = (4.36 \times 10^9 \text{ kg/s})\left(\frac{3.16 \times 10^7 \text{ s}}{1 \text{ year}}\right)(75 \text{ years}) = 1.0 \times 10^{19} \text{ kg}$$

Although this is an enormous amount of mass, it represents only a tiny fraction of the sun's total mass:

$$\frac{\Delta m}{m_{\text{sun}}} = \frac{1.0 \times 10^{19} \text{ kg}}{1.99 \times 10^{30} \text{ kg}} = \boxed{5.0 \times 10^{-12}}$$

Any change ΔE_0 in the rest energy of a system causes a change in the mass of the system according to $\Delta E_0 = (\Delta m)c^2$. It does not matter whether the change in energy is due to a change in electromagnetic energy, potential energy, thermal energy, or so on. Although any change in energy gives rise to a change in mass, in most instances the change in mass is too small to be detected. For instance, when 4186 J of heat is used to raise the temperature of 1 kg of water by 1 °C, the mass changes by only $\Delta m = \Delta E_0/c^2 = (4186 \text{ J})/(3.00 \times 10^8 \text{ m/s})^2 = 4.7 \times 10^{-14}$ kg. Conceptual Example 11 illustrates further how a change in the energy of an object leads to an equivalent change in its mass.

CONCEPTUAL EXAMPLE 11 | When Is a Massless Spring Not Massless?

Interactive Figure 28.11a shows a top view of a *massless* spring on a horizontal table. Initially the spring is unstrained. Then the spring is either stretched or compressed by an amount x from its unstrained length, as Interactive Figure 28.11b illustrates. What is the mass of the spring in Interactive Figure 28.11b? **(a)** It is greater than zero by an amount that is larger when the spring is stretched. **(b)** It is greater than zero by an amount that is larger when the spring is compressed. **(c)** It is greater than zero by an amount that is the same when the spring is stretched or compressed. **(d)** It remains zero.

Reasoning When a spring is stretched or compressed, its elastic potential energy changes. As discussed in Section 10.3, the elastic potential energy of an ideal spring is equal to $\frac{1}{2}kx^2$, where k is the spring constant and x is the amount of stretch or compression. Consistent with the theory of special relativity, any change in the total energy of a system, including a change in the elastic potential energy, is equivalent to a change in the mass of the system.

Answers (a), (b), and (d) are incorrect. In being stretched or compressed by the same amount x, the spring acquires the same amount of elastic potential energy ($\frac{1}{2}kx^2$). Therefore, according to special relativity, the spring acquires the same mass regardless of whether it is stretched or compressed, so these answers must be incorrect.

Answer (c) is correct. The spring acquires elastic potential energy in being stretched or compressed. Special relativity indicates that this additional energy is equivalent to additional mass. Since the amount of stretch or compression is the same, the potential energy is the same in either case, and so is the additional mass.

INTERACTIVE FIGURE 28.11 (*a*) This spring is unstrained and assumed to have no mass. (*b*) When the spring is either stretched or compressed by an amount x, it gains elastic potential energy and, hence, mass.

It is also possible to transform matter itself into other forms of energy. For example, the positron (see Section 31.4) has the same mass as an electron but an opposite electrical charge. If these two particles of matter collide, they are completely annihilated, and a burst of high-energy electromagnetic waves is produced. Thus, matter is transformed into electromagnetic waves, the energy of the electromagnetic waves being equal to the total energies of the two colliding particles. The medical diagnostic technique known as positron emission tomography or PET scanning depends on the electromagnetic energy produced when a positron and an electron are annihilated (see Section 32.6).

The transformation of electromagnetic waves into matter also happens. In one experiment, an extremely high-energy electromagnetic wave, called a gamma ray (see Section 31.4), passes close to the nucleus of an atom. If the gamma ray has sufficient energy, it can create an electron and a positron. The gamma ray disappears, and the two particles of matter appear in its place. Except for picking up some momentum, the nearby nucleus remains unchanged. The process in which the gamma ray is transformed into the two particles is known as *pair production.*

The Relation Between Total Energy and Momentum

It is possible to derive a useful relation between the total relativistic energy E and the relativistic momentum p. We begin by rearranging Equation 28.3 for the momentum, to obtain

$$\frac{m}{\sqrt{1 - v^2/c^2}} = \frac{p}{v}$$

With this substitution, Equation 28.4 for the total energy becomes

$$E = \frac{mc^2}{\sqrt{1 - v^2/c^2}} = \frac{pc^2}{v} \quad \text{or} \quad \frac{v}{c} = \frac{pc}{E}$$

Using this expression to replace v/c in Equation 28.4 gives

$$E = \frac{mc^2}{\sqrt{1 - v^2/c^2}} = \frac{mc^2}{\sqrt{1 - p^2c^2/E^2}}$$

Math Skills To solve the equation $E = \dfrac{mc^2}{\sqrt{1 - p^2c^2/E^2}}$ for E, we begin by squaring both sides:

$$E^2 = \left(\frac{mc^2}{\sqrt{1 - p^2c^2/E^2}} \right)^2 = \frac{m^2c^4}{1 - p^2c^2/E^2} \tag{1}$$

The next step is to isolate the terms containing E^2 on the left side of Equation 1. To do this, we multiply each side of the equation by $1 - p^2c^2/E^2$:

$$E^2(1 - p^2c^2/E^2) = \frac{m^2c^4}{1 - p^2c^2/E^2}(1 - p^2c^2/E^2) \quad \text{or} \quad E^2\left(1 - \frac{p^2c^2}{E^2}\right) = m^2c^4 \tag{2}$$

Multiplying terms on the left side of Equation 2 gives

$$E^2 - p^2c^2 = m^2c^4 \tag{3}$$

Finally, to isolate E^2 on the left side of Equation 3, we add p^2c^2 to both sides of the equation:

$$E^2 - p^2c^2 + (p^2c^2) = (p^2c^2) + m^2c^4 \quad \text{or} \quad E^2 = p^2c^2 + m^2c^4$$

Solving this expression for E^2 reveals that

$$E^2 = p^2c^2 + m^2c^4 \tag{28.7}$$

The Speed of Light in a Vacuum Is the Ultimate Speed

One of the important consequences of the theory of special relativity is that objects with mass cannot reach the speed c of light in a vacuum. Thus, c represents the ultimate speed. To see that this speed limit is a consequence of special relativity, consider Equation 28.6, which gives the kinetic energy of an object moving at a speed v. As v approaches the speed of light c, the $\sqrt{1 - v^2/c^2}$ term in the denominator approaches zero. Hence, the kinetic energy becomes infinitely large. However, the work–energy theorem (Section 6.2) tells us that an infinite amount of work would have to be done to give the object an infinite kinetic energy. Since an infinite amount of work is not available, we are left with the conclusion that objects with mass cannot attain the speed of light c.

Check Your Understanding

(The answers are given at the end of the book.)

8. Consider the same cup of coffee sitting on the same table in the following four situations: **(a)** The coffee is hot (95 °C), and the table is at sea level. **(b)** The coffee is cold (60 °C), and the table is at sea level. **(c)** The coffee is hot (95 °C), and the table is on a mountain top. **(d)** The coffee is cold (60 °C), and the table is on a mountain top. In which situation does the cup of coffee have the greatest mass, and in which the smallest mass?

9. A system consists of two positive charges. Is the total mass of the system greater when the two charges are **(a)** separated by a finite distance or **(b)** infinitely far apart?

10. A parallel plate capacitor is initially uncharged. Then it is fully charged up by removing electrons from one plate and placing them on the other plate. Is the mass of the capacitor greater when it is **(a)** uncharged or **(b)** fully charged?

11. It takes work to accelerate a particle from rest to a given speed close to the speed of light in a vacuum. For which particle is less work required, **(a)** an electron or **(b)** a proton?

28.7 | The Relativistic Addition of Velocities

The velocity of an object relative to an observer plays a central role in special relativity, and to determine this velocity, it is sometimes necessary to add two or more velocities together. We first encountered relative velocity in Section 3.4, so we will begin by reviewing some of the ideas presented there. **Figure 28.12** illustrates a truck moving at a constant velocity of $v_{TG} = +15$ m/s relative to an observer standing on the ground, where the plus sign denotes a direction to the right. Suppose someone on the truck throws a baseball toward the observer at a velocity of $v_{BT} = +8.0$ m/s relative to the truck. We might conclude that the observer on the ground would see the ball approaching at a velocity of $v_{BG} = v_{BT} + v_{TG} = 8.0$ m/s $+ 15$ m/s $= +23$ m/s. These symbols are similar to those used in Section 3.4 and have the following meaning:

$$v_{\boxed{BG}} = \text{velocity of the } \boxed{\text{Baseball}} \text{ relative to the } \boxed{\text{Ground}} = +23 \text{ m/s}$$

$$v_{\boxed{BT}} = \text{velocity of the } \boxed{\text{Baseball}} \text{ relative to the } \boxed{\text{Truck}} = +8.0 \text{ m/s}$$

$$v_{\boxed{TG}} = \text{velocity of the } \boxed{\text{Truck}} \text{ relative to the } \boxed{\text{Ground}} = +15.0 \text{ m/s}$$

Although the result for the velocity of the baseball relative to the ground ($v_{BG} = +23$ m/s) seems reasonable, careful measurements would show that it is not quite right. According to special relativity, the equation $v_{BG} = v_{BT} + v_{TG}$ is not valid for the following reason. If the velocity of the truck had a magnitude sufficiently close to the speed of light in a vacuum, the equation would predict that the observer on the earth could see the baseball moving faster than the speed of light. This is not possible, since no object with a finite mass can move faster than the speed of light in a vacuum.

For the case in which the truck and ball are moving along the same straight line, the theory of special relativity reveals that the velocities are related according to

$$v_{BG} = \frac{v_{BT} + v_{TG}}{1 + \dfrac{v_{BT} v_{TG}}{c^2}}$$

The subscripts in this equation have been chosen for the specific situation shown in **Figure 28.12**. For the general situation, the relative velocities are related by the following **velocity-addition formula**:

Velocity addition
$$v_{AB} = \frac{v_{AC} + v_{CB}}{1 + \dfrac{v_{AC} v_{CB}}{c^2}} \qquad (28.8)$$

where all the velocities are assumed to be constant and the symbols have the following meanings:

$$v_{\boxed{AB}} = \text{velocity of } \boxed{\text{object A}} \text{ relative to } \boxed{\text{object B}}$$

$$v_{\boxed{AC}} = \text{velocity of } \boxed{\text{object A}} \text{ relative to } \boxed{\text{object C}}$$

$$v_{\boxed{CB}} = \text{velocity of } \boxed{\text{object C}} \text{ relative to } \boxed{\text{object B}}$$

The ordering of the subscripts in Equation 28.8 follows the discussion in Section 3.4. For motion along a straight line, the velocities can have either positive or negative values,

$v_{BT} = +8.0$ m/s

$v_{TG} = +15$ m/s

Ground-based observer

FIGURE 28.12 The truck is approaching the ground-based observer at a relative velocity of $v_{TG} = +15$ m/s. The velocity of the baseball relative to the truck is $v_{BT} = +8.0$ m/s.

depending on whether they are directed along the positive or negative direction. Furthermore, switching the order of the subscripts changes the sign of the velocity, so, for example, $v_{BA} = -v_{AB}$ (see Example 11 in Chapter 3).

Equation 28.8 differs from the nonrelativistic formula ($v_{AB} = v_{AC} + v_{CB}$) by the presence of the $v_{AC} v_{CB}/c^2$ term in the denominator. This term arises because of the effects of time dilation and length contraction that occur in special relativity. When v_{AC} and v_{CB} are small compared to c, the $v_{AC} v_{CB}/c^2$ term is small compared to 1, so the velocity-addition formula reduces to $v_{AB} \approx v_{AC} + v_{CB}$. However, when either v_{AC} or v_{CB} is comparable to c, the results can be quite different, as Example 12 illustrates.

EXAMPLE 12 | The Relativistic Addition of Velocities

Imagine a hypothetical situation in which the truck in **Figure 28.12** is moving relative to the ground with a velocity of $v_{TG} = +0.8c$. A person riding on the truck throws a baseball at a velocity relative to the truck of $v_{TG} = +0.5c$. What is the velocity v_{BG} of the baseball relative to a person standing on the ground?

Reasoning The observer standing on the ground does *not* see the baseball approaching at $v_{BG} = 0.5c + 0.8c = 1.3c$. This cannot be, because the speed of the ball would then exceed the speed of

light in a vacuum. The velocity-addition formula gives the correct velocity, which has a magnitude less than the speed of light.

Solution The ground-based observer sees the ball approaching with a velocity of

$$v_{BG} = \frac{v_{BT} + v_{TG}}{1 + \frac{v_{BT}v_{TG}}{c^2}} = \frac{0.5c + 0.8c}{1 + \frac{(0.5c)(0.8c)}{c^2}} = \boxed{0.93c} \qquad (28.8)$$

Example 12 discusses how the speed of a baseball is viewed by observers in different inertial reference frames. The next example deals with a similar situation, except that the baseball is replaced by the light of a laser beam.

CONCEPTUAL EXAMPLE 13 | The Speed of a Laser Beam

Figure 28.13 shows an intergalactic cruiser approaching a hostile spacecraft. Both vehicles move at a constant velocity. The velocity of the cruiser relative to the spacecraft is $v_{CS} = +0.7c$, the direction to the right being the positive direction. The cruiser fires a beam of laser light at the hostile renegades. The velocity of the laser beam relative to the cruiser is $v_{LC} = +c$. Which one of the following statements correctly describes the velocity v_{LS} of the laser beam relative to the renegades' spacecraft and the velocity v at which the renegades see the laser beam move away from the cruiser? **(a)** $v_{LS} = +0.7c$ and $v = +c$ **(b)** $v_{LS} = +0.3c$ and $v = +c$ **(c)** $v_{LS} = +c$ and $v = +0.7c$ **(d)** $v_{LS} = +c$ and $v = +0.3c$

Reasoning Since both vehicles move at a constant velocity, each constitutes an inertial reference frame. According to the speed-of-light postulate, *all* observers in inertial reference frames measure the speed of light in a vacuum to be c.

Answers (a) and (b) are incorrect. Since the renegades' spacecraft constitutes an inertial reference frame, the velocity of the laser beam relative to it can only have a value of $v_{LS} = +c$, according to the speed-of-light postulate.

Answer (c) is incorrect. The velocity at which the renegades see the laser beam move away from the cruiser cannot be $v = +0.7c$, because they see the cruiser moving at a velocity of $+0.7c$ and the laser beam moving at a velocity of only $+c$ (not $+1.4c$).

Answer (d) is correct. The renegades see the cruiser approach them at a relative velocity of $v_{CS} = +0.7c$ and see the laser beam approach them at a relative velocity of $v_{LS} = +c$. Both these velocities are measured relative to the *same* inertial reference frame—namely, that of their own spacecraft. Therefore, the renegades see the laser beam move away from the cruiser at a velocity that is the difference between these two velocities, or $+c - (+0.7c) = +0.3c$. The velocity-addition formula, Equation 28.8, does not apply here because both velocities are measured relative to the *same* inertial reference frame. Equation 28.8 is used only when the velocities are measured relative to *different* inertial reference frames.

Related Homework: Problem 35

FIGURE 28.13 An intergalactic cruiser, closing in on a hostile spacecraft, fires a beam of laser light.

Intergalactic cruiser   Laser beam   Hostile spacecraft

$v_{LT} = +c$

v_{TG}

Ground-based observer

FIGURE 28.14 The speed of the light emitted by the flashlight is c relative to both the truck and the observer on the ground.

It is a straightforward matter to show that the velocity-addition formula is consistent with the speed-of-light postulate. Consider **Figure 28.14**, which shows a person riding on a truck and holding a flashlight. The velocity of the light relative to the person on the truck is $v_{LT} = +c$. The velocity v_{LG} of the light relative to the observer standing on the ground is given by the velocity-addition formula as

$$v_{LG} = \frac{v_{LT} + v_{TG}}{1 + \frac{v_{LT}v_{TG}}{c^2}} = \frac{c + v_{TG}}{1 + \frac{cv_{TG}}{c^2}} = \frac{(c + v_{TG})c}{(c + v_{TG})} = c$$

Thus, the velocity-addition formula indicates that the observer on the ground and the person on the truck both measure the speed of light to be c, independent of the relative velocity v_{TG} between them. This conclusion is completely consistent with the speed-of-light postulate.

Check Your Understanding

(*The answer is given at the end of the book.*)

12. Car A and car B are both traveling due east on a straight section of an interstate highway. The speed of car A relative to the ground is 10 m/s faster than the speed of car B relative to the ground. According to special relativity, is the speed of car A relative to car B **(a)** 10 m/s, **(b)** less than 10 m/s, or **(c)** greater than 10 m/s?

Concept Summary

28.1 Events and Inertial Reference Frames An event is a physical "happening" that occurs at a certain place and time. To record the event an observer uses a reference frame that consists of a coordinate system and a clock. Different observers may use different reference frames. The theory of special relativity deals with inertial reference frames. An inertial reference frame is one in which Newton's law of inertia is valid. Accelerating reference frames are not inertial reference frames.

28.2 The Postulates of Special Relativity The theory of special relativity is based on two postulates. The relativity postulate states that the laws of physics are the same in every inertial reference frame. The speed-of-light postulate says that the speed of light in a vacuum, measured in any inertial reference frame, always has the same value of c, no matter how fast the source of the light and the observer are moving relative to each other.

28.3 The Relativity of Time: Time Dilation The proper time interval Δt_0 between two events is the time interval measured by an observer who is at rest relative to the events and views them occurring at the same place. An observer who is in motion with respect to the events and who views them as occurring at different places measures a dilated time interval Δt. The dilated time interval is greater than the proper time interval, according to the time-dilation equation (Equation 28.1). In this expression, v is the relative speed between the observer who measures Δt_0 and the observer who measures Δt.

$$\Delta t = \frac{\Delta t_0}{\sqrt{1 - \frac{v^2}{c^2}}} \qquad (28.1)$$

28.4 The Relativity of Length: Length Contraction The proper length L_0 between two points is the length measured by an observer who is at rest relative to the points. An observer moving with a relative speed v parallel to the line between the two points does not measure the proper length. Instead, such an observer measures a contracted length L given by the length-contraction formula (Equation 28.2). Length contraction occurs only along the direction of the motion. Those dimensions that are perpendicular to the motion are not shortened. The observer who measures the proper length may not be the observer who measures the proper time interval.

$$L = L_0\sqrt{1 - \frac{v^2}{c^2}} \qquad (28.2)$$

28.5 Relativistic Momentum An object of mass m, moving with speed v, has a relativistic momentum whose magnitude p is given by Equation 28.3.

$$p = \frac{mv}{\sqrt{1 - \dfrac{v^2}{c^2}}} \qquad (28.3)$$

28.6 The Equivalence of Mass and Energy Energy and mass are equivalent. The total energy E of an object of mass m, moving at speed v, is given by Equation 28.4. The rest energy E_0 is the total energy of an object at rest ($v = 0$ m/s), as given by Equation 28.5. An object's total energy is the sum of its rest energy and its kinetic energy KE, or $E = E_0 + \text{KE}$. Therefore, the kinetic energy is given by Equation 28.6.

$$E = \frac{mc^2}{\sqrt{1 - \dfrac{v^2}{c^2}}} \qquad (28.4)$$

$$E_0 = mc^2 \qquad (28.5)$$

$$\text{KE} = E - E_0 = mc^2 \left(\frac{1}{\sqrt{1 - \dfrac{v^2}{c^2}}} - 1 \right) \qquad (28.6)$$

The relativistic total energy and momentum are related according to Equation 28.7.

$$E^2 = p^2c^2 + m^2c^4 \qquad (28.7)$$

Objects with mass cannot attain the speed of light c, which is the ultimate speed for such objects.

28.7 The Relativistic Addition of Velocities According to special relativity, the velocity-addition formula specifies how the relative velocities of moving objects are related. For objects that move along the same straight line, this formula is given by Equation 28.8, where v_{AB} is the velocity of object A relative to object B, v_{AC} is the velocity of object A relative to object C, and v_{CB} is the velocity of object C relative to object B. The velocities can have positive or negative values, depending on whether they are directed along the positive or negative direction. Furthermore, switching the order of the subscripts changes the sign of the velocity, so that, for example, $v_{BA} = -v_{AB}$.

$$v_{AB} = \frac{v_{AC} + v_{CB}}{1 + \dfrac{v_{AC}\, v_{CB}}{c^2}} \qquad (28.8)$$

Focus on Concepts

Online

Additional questions are available for assignment in WileyPLUS.

Section 28.1 Events and Inertial Reference Frames

1. Consider a person along with a frame of reference in each of the following situations. In which one or more of the following situations is the frame of reference an inertial frame of reference? (a) The person is oscillating in simple harmonic motion at the end of a bungee cord. (b) The person is in a car going around a circular curve at a constant speed. (c) The person is in a plane that is landing on an aircraft carrier. (d) The person is in the space shuttle during lift-off. (e) None of the above.

Section 28.3 The Relativity of Time: Time Dilation

2. On a highway there is a flashing light to mark the start of a section of the road where work is being done. Who measures the proper time between two flashes of light? (a) A worker standing still on the road (b) A driver in a car approaching at a constant velocity (c) Both the worker and the driver (d) Neither the worker nor the driver

Section 28.4 The Relativity of Length: Length Contraction

3. Two spacecrafts A and B are moving relative to each other at a constant velocity. Observers in spacecraft A see spacecraft B. Likewise, observers in spacecraft B see spacecraft A. Who sees the proper length of either spacecraft? (a) Observers in spacecraft A see the proper length of spacecraft B. (b) Observers in spacecraft B see the proper length of spacecraft A. (c) Observers in both spacecrafts see the proper length of the other spacecraft. (d) Observers in neither spacecraft see the proper length of the other spacecraft.

4. In a baseball game the batter hits the ball into center field and takes off for first base. The catcher can only stand and watch. Assume that the batter runs at a constant velocity. Who measures the proper time it takes for the runner to reach first base, and who measures the proper length between

home plate and first base? (a) The catcher measures the proper time, and the runner measures the proper length. (b) The runner measures the proper time, and the catcher measures the proper length. (c) The catcher measures both the proper time and the proper length. (d) The runner measures both the proper time and the proper length.

5. To which one or more of the following situations do the time-dilation and length-contraction equations apply? (a) With respect to an inertial frame, two observers have different constant accelerations. (b) With respect to an inertial frame, two observers have the same constant acceleration. (c) With respect to an inertial frame, two observers are moving with different constant velocities. (d) With respect to an inertial frame, one observer has a constant velocity, and another observer has a constant acceleration. (e) All of the above.

Section 28.5 Relativistic Momentum

6. Which one of the following statements about linear momentum is true (p = magnitude of the momentum, m = mass, and v = speed)? (a) When the magnitude p of the momentum is defined as $p = \dfrac{mv}{\sqrt{1 - v^2/c^2}}$, the linear momentum of an isolated system is conserved only if the speeds of the various parts of the system are very high. (b) When the magnitude p of the momentum is defined as $p = mv$, the linear momentum of an isolated system is conserved only if the speeds of the various parts of the system are very high. (c) When the magnitude p of the momentum is defined as $p = \dfrac{mv}{\sqrt{1 - v^2/c^2}}$, the linear momentum of an isolated system is conserved no matter what the speeds of the various parts of the system are. (d) When the magnitude p of the momentum is defined as $p = mv$, the linear momentum of an isolated system is conserved no matter what the speeds of the various parts of the system are.

7. Which of the following two expressions for the magnitude p of the linear momentum can be used when the speed v of an object of mass m is very small compared to the speed of light c in a vacuum?

A. $p = mv$

B. $p = \dfrac{mv}{\sqrt{1 - \dfrac{v^2}{c^2}}}$

(a) Only A (b) Only B (c) Neither A nor B (d) Both A and B

Section 28.6   The Equivalence of Mass and Energy

8. Consider the following three possibilities for a glass of water at rest on a kitchen counter. The temperature of the water is 0 °C. Rank the mass of the water in descending order (largest first).

A. The water is half liquid and half ice.

B. The water is all liquid.

C. The water is all ice.

(a) C, A, B (b) B, A, C (c) A, C, B (d) B, C, A (e) C, B, A

9. An object has a kinetic energy KE and a potential energy PE. It also has a rest energy E_0. Which one of the following is the correct way to express the object's total energy E? (a) $E = KE + PE$ (b) $E = E_0 + KE$ (c) $E = E_0 + KE + PE$ (d) $E = E_0 + KE - PE$

10. The kinetic energy of an object of mass m is equal to its rest energy. What is the magnitude p of the object's momentum? (a) $p = \sqrt{3}\, mc$ (b) $p = 2mc$ (c) $p = 4mc$ (d) $p = \sqrt{2}mc$ (e) $p = 3mc$

Section 28.7   The Relativistic Addition of Velocities

11. Two spaceships are traveling in the same direction. With respect to an inertial frame of reference, spaceship A has a speed of 0.900c. With respect to the same inertial frame, spaceship B has a speed of 0.500c. Find the speed v_{AB} of spaceship A relative to spaceship B.

Problems

Online

Additional questions are available for assignment in WileyPLUS.

Note: *Before doing any calculations involving time dilation or length contraction, it is useful to identify which observer measures the proper time interval Δt_0 or the proper length L_0.*

SSM Student Solutions Manual	BIO Biomedical application
MMH Problem-solving help	E Easy
GO Guided Online Tutorial	M Medium
V-HINT Video Hints	H Hard
CHALK Chalkboard Videos	WS Worksheet
	T Team Problem

Section 28.3   The Relativity of Time: Time Dilation

1. E SSM A particle known as a pion lives for a short time before breaking apart into other particles. Suppose that a pion is moving at a speed of 0.990c, and an observer who is stationary in a laboratory measures the pion's lifetime to be 3.5×10^{-8} s. (a) What is the lifetime according to a hypothetical person who is riding along with the pion? (b) According to this hypothetical person, how far does the laboratory move before the pion breaks apart?

2. E A radar antenna is rotating and makes one revolution every 25 s, as measured on earth. However, instruments on a spaceship moving with respect to the earth at a speed v measure that the antenna makes one revolution every 42 s. What is the ratio v/c of the speed v to the speed c of light in a vacuum?

3. E SSM Suppose that you are planning a trip in which a spacecraft is to travel at a constant velocity for exactly six months, as measured by a clock on board the spacecraft, and then return home at the same speed. Upon your return, the people on earth will have advanced exactly one hundred years into the future. According to special relativity, how fast must you travel? Express your answer to five significant figures as a multiple of c—for example, 0.955 85c.

4. E CHALK GO Suppose that you are traveling on board a spacecraft that is moving with respect to the earth at a speed of 0.975c. You are breathing at a rate of 8.0 breaths per minute. As monitored on earth, what is your breathing rate?

5. M GO MMH A 6.00-kg object oscillates back and forth at the end of a spring whose spring constant is 76.0 N/m. An observer is traveling at a speed of 1.90×10^8 m/s relative to the fixed end of the spring. What does this observer measure for the period of oscillation?

6. M V-HINT A spaceship travels at a constant speed from earth to a planet orbiting another star. When the spacecraft arrives, 12 years have elapsed on earth, and 9.2 years have elapsed on board the ship. How far away (in meters) is the planet, according to observers on earth?

7. H As observed on earth, a certain type of bacterium is known to double in number every 24.0 hours. Two cultures of these bacteria are prepared, each consisting initially of one bacterium. One culture is left on earth and the other placed on a rocket that travels at a speed of 0.866c relative to the earth. At a time when the earthbound culture has grown to 256 bacteria, how many bacteria are in the culture on the rocket, according to an earth-based observer?

Section 28.4   The Relativity of Length: Length Contraction

8. E Suppose the straight-line distance between New York and San Francisco is 4.1×10^6 m (neglecting the curvature of the earth). A UFO is flying between these two cities at a speed of 0.70c relative to the earth. What do the voyagers aboard the UFO measure for this distance?

9. E SSM How fast must a meter stick be moving if its length is observed to shrink to one-half of a meter?

10. E V-HINT MMH The distance from earth to the center of our galaxy is about 23 000 ly (1 ly = 1 light-year = 9.47×10^{15} m), as measured by an earth-based observer. A spaceship is to make this journey at a speed of 0.9990c. According to a clock on board the spaceship, how long will it take to make the trip? Express your answer in years (1 yr = 3.16×10^7 s).

11. E A tourist is walking at a speed of 1.3 m/s along a 9.0-km path that follows an old canal. If the speed of light in a vacuum were 3.0 m/s, how long would the path be, according to the tourist?

12. E GO A Martian leaves Mars in a spaceship that is heading to Venus. On the way, the spaceship passes earth with a speed $v = 0.80c$

relative to it. Assume that the three planets do not move relative to each other during the trip. The distance between Mars and Venus is 1.20×10^{11} m, as measured by a person on earth. **(a)** What does the Martian measure for the distance between Mars and Venus? **(b)** What is the time of the trip (in seconds) as measured by the Martian?

13. **E** Two spaceships A and B are exploring a new planet. Relative to this planet, spaceship A has a speed of 0.60*c*, and spaceship B has a speed of 0.80*c*. What is the ratio D_A/D_B of the values for the planet's diameter that each spaceship measures in a direction that is parallel to its motion?

14. **E** An unstable high-energy particle is created in the laboratory, and it moves at a speed of 0.990*c*. Relative to a stationary reference frame fixed to the laboratory, the particle travels a distance of 1.05×10^{-3} m before disintegrating. What are **(a)** the proper distance and **(b)** the distance measured by a hypothetical person traveling with the particle? Determine the particle's **(c)** proper lifetime and **(d)** its dilated lifetime.

15. **M** **CHALK** As the drawing shows, a carpenter on a space station has constructed a 30.0° ramp. A rocket moves past the space station with a relative speed of 0.730*c* in a direction parallel to side x_0. What does a person aboard the rocket measure for the angle of the ramp?

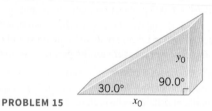

PROBLEM 15

Section 28.5 Relativistic Momentum

16. **E** At what speed is the magnitude of the relativistic momentum of a particle three times the magnitude of the nonrelativistic momentum?

17. **E** What is the magnitude of the relativistic momentum of a proton with a relativistic total energy of 2.7×10^{-10} J?

18. **E** **GO** A spacecraft has a nonrelativistic (or classical) momentum whose magnitude is 1.3×10^{13} kg · m/s. The spacecraft moves at such a speed that the pilot measures the proper time interval between two events to be one-half the dilated time interval. Find the relativistic momentum of the spacecraft.

19. **E** **SSM** A woman is 1.6 m tall and has a mass of 55 kg. She moves past an observer with the direction of the motion parallel to her height. The observer measures her relativistic momentum to have a magnitude of 2.0×10^{10} kg · m/s. What does the observer measure for her height?

20. **E** **GO** Three particles are listed in the table. The mass and speed of each particle are given as multiples of the variables *m* and *v*, which have the values $m = 1.20 \times 10^{-8}$ kg and $v = 0.200c$. The speed of light in a vacuum is $c = 3.00 \times 10^8$ m/s. Determine the momentum for each particle according to special relativity.

Particle	Mass	Speed
a	m	v
b	$\frac{1}{2}m$	$2v$
c	$\frac{1}{4}m$	$4v$

21. **M** **SSM** **CHALK** Starting from rest, two skaters push off against each other on smooth level ice, where friction is negligible. One is a woman and one is a man. The woman moves away with a velocity of +2.5 m/s

relative to the ice. The mass of the woman is 54 kg, and the mass of the man is 88 kg. Assuming that the speed of light is 3.0 m/s, so that the relativistic momentum must be used, find the recoil velocity of the man relative to the ice. (*Hint:* This problem is similar to Example 7 in Chapter 7.)

Section 28.6 The Equivalence of Mass and Energy

22. **E** Radium is a radioactive element whose nucleus emits an α particle (a helium nucleus) with a kinetic energy of about 7.8×10^{-13} J (4.9 MeV). To what amount of mass is this energy equivalent?

23. **E** **SSM** How much work must be done on an electron to accelerate it from rest to a speed of 0.990*c*?

24. **E** Suppose that one gallon of gasoline produces 1.1×10^8 J of energy, and this energy is sufficient to operate a car for twenty miles. An aspirin tablet has a mass of 325 mg. If the aspirin could be converted completely into thermal energy, how many miles could the car go on a single tablet?

25. **E** **GO** Two kilograms of water are changed **(a)** from ice at 0 °C into liquid water at 0 °C and **(b)** from liquid water at 100 °C into steam at 100 °C. For each situation, determine the change in mass of the water.

26. **E** **CHALK** **SSM** Determine the ratio of the relativistic kinetic energy to the nonrelativistic kinetic energy $(\frac{1}{2}mv^2)$ when a particle has a speed of **(a)** $1.00 \times 10^{-3}c$ and **(b)** 0.970*c*.

27. **E** Multiple-Concept Example 8 reviews the principles that play a role in this problem. A nuclear power reactor generates 3.0×10^9 W of power. In one year, what is the change in the mass of the nuclear fuel due to the energy being taken from the reactor?

28. **M** **V-HINT** Multiple-Concept Example 8 explores the approach taken in problems such as this one. Quasars are believed to be the nuclei of galaxies in the early stages of their formation. Suppose that a quasar radiates electromagnetic energy at the rate of 1.0×10^{41} W. At what rate (in kg/s) is the quasar losing mass as a result of this radiation?

29. **M** **GO** An electron is accelerated from rest through a potential difference that has a magnitude of 2.40×10^7 V. The mass of the electron is 9.11×10^{-31} kg, and the negative charge of the electron has a magnitude of 1.60×10^{-19} C. **(a)** What is the relativistic kinetic energy (in joules) of the electron? **(b)** What is the speed of the electron? Express your answer as a multiple of *c*, the speed of light in a vacuum.

Section 28.7 The Relativistic Addition of Velocities

30. **E** **GO** You are driving down a two-lane country road, and a truck in the opposite lane is traveling toward you. Suppose that the speed of light in a vacuum is $c = 65$ m/s. Determine the speed of the truck relative to you when **(a)** your speed is 25 m/s and the truck's speed is 35 m/s and **(b)** your speed is 5.0 m/s and the truck's speed is 55 m/s. The speeds given in parts (a) and (b) are relative to the ground.

31. **E** **SSM** A spacecraft approaching the earth launches an exploration vehicle. After the launch, an observer on earth sees the spacecraft approaching at a speed of 0.50*c* and the exploration vehicle approaching at a speed of 0.70*c*. What is the speed of the exploration vehicle relative to the spaceship?

32. **E** **GO** Spaceships of the future may be powered by ion-propulsion engines in which ions are ejected from the back of the ship to drive it forward. In one such engine the ions are to be ejected with a speed of 0.80*c* relative to the spaceship. The spaceship is traveling away from the earth at a speed of 0.70*c* relative to the earth. What is the velocity of the ions relative to the earth? Assume that the direction in which the spaceship is traveling is the positive direction, and be sure to assign the correct plus or minus signs to the velocities.

33. **E** The spaceship *Enterprise 1* is moving directly away from earth at a velocity that an earth-based observer measures to be +0.65c. A sister ship, *Enterprise 2*, is ahead of *Enterprise 1* and is also moving directly away from earth along the same line. The velocity of *Enterprise 2* relative to *Enterprise 1* is +0.31c. What is the velocity of *Enterprise 2*, as measured by the earth-based observer?

34. **M** **V-HINT** A person on earth notices a rocket approaching from the right at a speed of 0.75c and another rocket approaching from the left at 0.65c. What is the relative speed between the two rockets, as measured by a passenger on one of them?

35. **M** **SSM** Refer to Conceptual Example 13 as an aid in solving this problem. An intergalactic cruiser has two types of guns: a photon cannon that fires a beam of laser light and an ion gun that shoots ions at a velocity of 0.950c relative to the cruiser. The cruiser closes in on an alien spacecraft at a velocity of 0.800c relative to this spacecraft. The captain fires both types of guns. At what velocity do the aliens see **(a)** the laser light and **(b)** the ions approach them? At what velocity do the aliens see **(c)** the laser light and **(d)** the ions move away from the cruiser?

36. **M** **GO** **CHALK** Two identical spaceships are under construction. The constructed length of each spaceship is 1.50 km. After being launched, spaceship A moves away from earth at a constant velocity (speed is 0.850c) with respect to the earth. Spaceship B follows in the same direction at a different constant velocity (speed is 0.500c) with respect to the earth. Determine the length that a passenger on one spaceship measures for the other spaceship.

Additional Problems

Online

37. **E** An electron and a positron have masses of 9.11×10^{-31} kg. They collide and both vanish, with only electromagnetic radiation appearing after the collision. If each particle is moving at a speed of 0.20c relative to the laboratory before the collision, determine the energy of the electromagnetic radiation.

38. **E** **GO** The speed of an ion in a particle accelerator is doubled from 0.460c to 0.920c. The initial relativistic momentum of the ion is 5.08×10^{-17} kg · m/s. Determine **(a)** the mass and **(b)** the magnitude of the final relativistic momentum of the ion.

39. **E** **GO** A Klingon spacecraft has a speed of 0.75c with respect to the earth. The Klingons measure 37.0 h for the time interval between two events on the earth. What value for the time interval would they measure if their ship had a speed of 0.94c with respect to the earth?

40. **M** **SSM** The crew of a rocket that is moving away from the earth launches an escape pod, which they measure to be 45 m long. The pod is launched toward the earth with a speed of 0.55c relative to the rocket. After the launch, the rocket's speed relative to the earth is 0.75c. What is the length of the escape pod as determined by an observer on earth?

41. **M** **GO** The table gives the total energy and the rest energy for three objects in terms of an energy increment ε. For each object, determine the speed as a multiple of the speed c of light in a vacuum.

Object	Total Energy (E)	Rest Energy (E_0)
A	2.00ε	ε
B	3.00ε	ε
C	3.00ε	2.00ε

42. **H** **GO** Twins who are 19.0 years of age leave the earth and travel to a distant planet 12.0 light-years away. Assume that the planet and earth are at rest with respect to each other. The twins depart at the same time on different spaceships. One twin travels at a speed of 0.900c, and the other twin travels at 0.500c. **(a)** According to the theory of special relativity, what is the difference between their ages when they meet again at the earliest possible time? **(b)** Which twin is older?

Physics in Biology, Medicine, and Sports

43. **M** **BIO** **28.4** Special relativity tells us that someone sprinting will shrink, or undergo length contraction, along the direction of motion relative to someone at rest. The world's fastest sprinters, like Usain Bolt, can reach speeds near 13.0 m/s. How fast would they have to run so that their length contraction was **(a)** equal to the width of a human hair (50.0 μm) and **(b)** equal to the width of an atom (1.00 × 10^{-10} m)? Assume the width of the sprinter's body at rest is 25.0 cm.

44. **M** **BIO** **28.4** A football player returns a kickoff the full length of the field for a 100-yard touchdown. How fast would he have to run so that his return appears to be only 99 yards to him?

45. **M** **BIO** **28.3** Protons and neutrons exist together in the nuclei of atoms and form the building blocks of all living things. However, if a neutron is removed from a nucleus, so that it exists as a free neutron,

it is unstable. A free neutron at rest has a mean lifetime of 8.80×10^2 s, at which point, on average, it will decay into a proton, an electron, and another particle called an antineutrino. This process is known as β decay (see Chapter 31). If an observer measures the mean lifetime of a neutron to be twice as long (1760 s), how fast is the neutron moving relative to the observer?

46. **M** **BIO** **28.3** On earth, the greatest length of life occurs for females living in Hong Kong, whose average life expectancy is 88.2 years. If such a person lived their life on a spaceship moving at 20% the speed of light, what would be their mean lifetime, as measured by an observer at rest on earth?

47. **M** **BIO** **28.6** The African bush elephant is the largest terrestrial mammal. The tallest one ever recorded stood 4.21 m (13.8 ft) at the shoulder and weighed 18 000 lb (8160 kg). How fast would a Ferrari sports car ($m = 1660$ kg) have to be moving, such that its kinetic energy is equal to the rest mass energy of the elephant? Use $c = 3.00 \times 10^8$ m/s for the speed of light.

Concepts and Calculations Problems   **Online**

There are many astonishing consequences of special relativity, two of which are time dilation and length contraction. Problem 48 reviews these important concepts in the context of a golf game in a hypothetical world where the speed of light is only a little faster than that of a golf cart. Other important consequences of special relativity are the equivalence of mass and energy, and the dependence of kinetic energy on the total energy and on the rest energy. Problem 49 serves as a review of the roles played by mass and energy in special relativity.

48. **M** **CHALK** Imagine playing golf in a world where the speed of light is only $c = 3.40$ m/s. Golfer A drives a ball down a flat horizontal fairway for a distance that he measures as 75.0 m. Golfer B, riding in a cart, happens to pass by just as the ball is hit (see the figure). Golfer A stands at the tee and watches while golfer B moves down the fairway toward the ball at a constant speed of 2.80 m/s. *Concepts:* (i) Who measures the proper length of the drive, and who measures the

contracted length? (ii) Who measures the proper time interval, and who measures the dilated time interval? *Calculations:* **(a)** How far is the ball hit according to golfer B? **(b)** According to each golfer, how much time does it take golfer B to reach the ball?

49. **M** **CHALK** **SSM** The rest energy E_0 and the total energy E of three particles, expressed in terms of a basic amount of energy $E' = 5.98 \times 10^{-10}$ J, are listed in the table. The speeds of these particles are large, in some cases approaching the speed of light. *Concepts:* (i) Given the rest energies specified in the table, what is the ranking (largest first) of the masses of the particles? (ii) Is the kinetic energy KE given by the expression KE $= \frac{1}{2}mv^2$, and what is the ranking (largest first) of the kinetic energies of the particles? *Calculations:* For each particle, determine its **(a)** mass and **(b)** kinetic energy.

Particle	Rest Energy	Total Energy
a	E'	$2E'$
b	E'	$4E'$
c	$5E'$	$6E'$

Golfer A

Golfer B

PROBLEM 48

Team Problems   **Online**

50. **E** **T** **WS** **Space Smugglers.** You and your team are returning from a long voyage but have attracted a stalker along the way. Another ship that is transporting illegal contraband is following yours at a short distance in order that it might sneak into earth's orbit and eventually deliver its cargo. As you get closer to earth, you encounter a rotating beacon, the location of which is stationary in earth's frame of reference. It is known that, in the reference frame of earth, the beacon rotates with a frequency of 0.8030 Hz. **(a)** What is the velocity of your ship relative to the beacon, and therefore relative to earth, if the instruments on your ship measure the rotational frequency of the beacon to be 0.6080 Hz? **(b)** The smugglers' ship is traveling at exactly the same speed as yours, but is of a different type, and therefore likely to be of a different length, although you do not know whether it is longer or shorter. You decide to warn authorities on earth about the smugglers' ship, and want to give them enough information to identify it. You do this by informing them of the exact length of your ship in earth's frame of reference. Since the smugglers' ship will likely be

of a different length, the authorities will be able to distinguish it from yours. If your ship is 175.8 meters in length in your frame of reference, what is its length in earth's frame?

51. **M** **T** **WS** **Special Relativity and the Doppler Effect.** You and your team are on a spacecraft that is approaching earth at a relativistic speed, but you are low on fuel and supplies. It is crucial that you start the deceleration process at exactly the right time. If you brake too early, you will spend too much time completing the trip home at the lower speed, and will run out of food. If you brake too late, you will pass by earth and will not have enough fuel to turn the ship around. To accurately determine when to brake, you need to have an accurate measurement of your current speed relative to earth. For this purpose, there is a beacon far from earth that emits light in all directions with a wavelength of 685 nm (in its frame of reference). The beacon is stationary relative to earth and your ship is heading directly toward it. The idea is that an incoming spacecraft will measure the wavelength of the light emitted from the beacon with onboard instruments to determine the Doppler shift, from which the relative velocity between the ship and the beacon can be calculated. There is one complication, however. Since your ship is moving at a relativistic speed relative to the source, you must apply a Doppler effect formula that takes relativistic effects (i.e., time dilation and length contraction) into account. Without taking relativistic effects into account, you would apply the nonrelativistic formula given by Equation 24.6:

$$f_o = f_s\left(1 \pm \frac{v_{rel}}{c}\right) \qquad (\text{if } v_{rel} \ll c)$$

where the plus sign is used when the source (f_s) and observer (f_o) come together, and the minus sign is used when they move apart. Accounting for relativistic effects, this equation must be modified as follows:

$$f_o = f_s\sqrt{\frac{1 + \frac{v_{rel}}{c}}{1 - \frac{v_{rel}}{c}}} \qquad (\text{relativistic Doppler equation})$$

where v_{rel} is positive when the source (f_s) and observer (f_o) come together, and negative when they move apart. (a) If your ship measures the beacon wavelength to be 395 nm, what is the ship's velocity relative to the beacon? (b) What speed would you have obtained using the nonrelativistic Doppler effect equation? (c) If you had used the nonrelativistic Doppler equation, would you have stopped too early, or too late? Explain.

52. M T **A Super Thruster.** You and your team are evaluating a new thruster that will be used to propel a 6.00×10^4 kg ship. The thruster has a power output of 2.88×10^6 W, calculated to initially accelerate the ship at 9.80 m/s² (safe for humans). (a) Assuming that all of the energy of the thruster is converted into kinetic energy of the ship, what is the velocity of the ship after 1 year? (b) After 1000 years? (c) Assuming the thruster converts mass directly into energy, how much mass would be converted into energy over the 1000-year trip?

53. M T **A Long Trip.** You and your team are on a journey to a solar system in another part of our galaxy. Your destination is 4.00×10^3 ly (1 ly = 1 light-year = 9.47×10^{15} m) from earth, as measured by an earth-based observer. (a) If the ship has a speed (relative to earth) of $0.9999c$, how long will the trip take according to the clocks on board the ship? Express your answer in years. (b) You have to pass by the sun on the way out of our solar system. According to the clocks on your ship, how much time (in minutes) will it take to get to the sun from earth? The sun-earth mean distance is 1.50×10^{11} m. (c) The sun will not look spherical as you pass by. What is the ratio of the shortest to the longest dimensions (i.e., the ratio of the semi-minor to the semi-major axes) of the sun as you see it?

This photograph shows a highly magnified view of a coronavirus that was made with a scanning electron microscope (SEM). The diameter of the roughly spherical virus is approximately 120 nm. In the twentieth century, physicists were astonished when it was discovered that particles could behave like waves. In fact, we will see in this chapter that there is a wavelength associated with a moving particle such as an electron. The microscope used for the photograph takes advantage of the electron wavelength, which can be made much smaller than that of visible light. It is this small electron wavelength that is responsible for the exceptional resolution of the fine details in the photograph. The spike-like structures on the outer surface of the virus are composed of proteins that attach to the cell membrane of the host. A novel coronavirus (SARS-CoV-2) was responsible for an outbreak of coronavirus disease in 2019 (COVID-19) that infected millions of people worldwide.

PASIEKA/Getty Images

LEARNING OBJECTIVES

After reading this module, you should be able to...

29.1 Define wave–particle duality.

29.2 Explain the origin of Planck's constant from blackbody radiation.

29.3 Use photon energy to explain in detail the photoelectric effect.

29.4 Use photon momentum to explain in detail the Compton effect.

29.5 Solve problems involving the de Broglie wavelength of a particle.

29.6 Calculate quantum uncertainty using the Heisenberg uncertainty principle.

Particles and Waves

29.1 The Wave–Particle Duality

The ability to exhibit interference effects is an essential characteristic of waves. For instance, Section 27.2 discusses Young's famous experiment in which light passes through two closely spaced slits and produces a pattern of bright and dark fringes on a screen (see **Figure 27.3**). The fringe pattern is a direct indication that interference is occurring between the light waves coming from each slit.

One of the most incredible discoveries of twentieth-century physics is that particles can also behave like waves and exhibit interference effects. For instance, **Interactive Figure 29.1** shows a version of Young's experiment performed by directing *a beam of electrons* onto a double slit. In this experiment, the screen is like a television screen and glows wherever an electron strikes it. Part *a* of the drawing indicates the pattern that would be seen on the screen if each electron, behaving strictly as a particle, were to pass through one slit or the other and strike the screen. The pattern would consist of an image of each slit. Part *b* shows the pattern actually observed, which consists of bright and dark fringes, reminiscent of what is obtained when light waves pass through a double slit. The fringe pattern indicates that the electrons are exhibiting the interference effects associated with waves.

But how can electrons behave like waves in the experiment shown in **Interactive Figure 29.1b**? And what kind of waves are they? The answers to these profound questions will be discussed later in this chapter. For the moment, we intend only to emphasize that the concept of an electron as a tiny discrete particle of matter does not account for the fact that the electron can behave as a wave in some circumstances. In other words, the electron exhibits a dual nature, with both particle-like characteristics and wave-like characteristics.

Here is another interesting question: If a particle can exhibit wave-like properties, can waves exhibit particle-like behavior? As the next three sections reveal, the answer is yes. In fact, experiments that demonstrated the particle-like behavior of waves were performed near the beginning of the twentieth century, before the experiments that demonstrated the wave-like properties of the electrons. Scientists now accept the **wave–particle duality** as an essential part of nature:

Waves can exhibit particle-like characteristics, and particles can exhibit wave-like characteristics.

Section 29.2 begins the remarkable story of the wave–particle duality by discussing the electromagnetic waves that are radiated by a perfect blackbody. It is appropriate to begin with blackbody radiation, because it provided the first link in the chain of experimental evidence leading to our present understanding of the wave–particle duality.

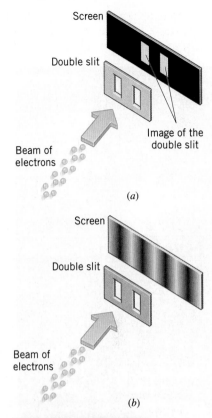

INTERACTIVE FIGURE 29.1 (*a*) If electrons behaved as discrete particles with no wave properties, they would pass through one or the other of the two slits and strike the screen, causing it to glow and produce exact images of the slits. (*b*) In reality, the screen reveals a pattern of bright and dark fringes, similar to the pattern produced when a beam of light is used and interference occurs between the light waves coming from each slit.

29.2 | Blackbody Radiation and Planck's Constant

All bodies, no matter how hot or cold, continuously radiate electromagnetic waves. For instance, we see the glow of very hot objects because they emit electromagnetic waves in the visible region of the spectrum. Our sun, which has a surface temperature of about 6000 K, appears yellow, while the cooler star Betelgeuse has a red-orange appearance due to its lower surface temperature of 2900 K. However, at relatively low temperatures, cooler objects emit visible light waves only weakly and, as a result, do not appear to be glowing. Certainly the human body, at only 310 K, does not emit enough visible light to be seen in the dark with the unaided eye. But the body does emit electromagnetic waves in the infrared region of the spectrum, and these can be detected with infrared-sensitive devices.

At a given temperature, the intensities of the electromagnetic waves emitted by an object vary from wavelength to wavelength throughout the visible, infrared, and other regions of the spectrum. **Figure 29.2** illustrates how the intensity per unit wavelength depends on wavelength for a perfect blackbody emitter. As Section 13.3 discusses, a perfect blackbody at a constant temperature absorbs and reemits all the electromagnetic radiation that falls on it. The two curves in **Figure 29.2** show that at a higher temperature the maximum emitted intensity per unit wavelength increases and shifts to shorter wavelengths, toward the visible region of the spectrum. In accounting for the shape of these curves, the German physicist Max Planck (1858–1947) took the first step toward our present understanding of the wave–particle duality.

In 1900 Planck calculated the blackbody radiation curves, using a model that represents a blackbody as a large number of atomic oscillators, each of which emits and absorbs electromagnetic waves. To obtain agreement between the theoretical and experimental curves, Planck assumed that the energy E of an atomic oscillator could have only the discrete values of $E = 0$, hf, $2hf$, $3hf$, and so on. In other words, he assumed that

$$E = nhf \quad n = 0, 1, 2, 3, \ldots \tag{29.1}$$

FIGURE 29.2 The electromagnetic radiation emitted by a perfect blackbody has an intensity per unit wavelength that varies from wavelength to wavelength, as each curve indicates. At the higher temperature, the intensity per unit wavelength is greater, and the maximum occurs at a shorter wavelength.

where n is either zero or a positive integer, f is the frequency of vibration (in hertz), and h is a constant now called **Planck's constant.*** It has been determined experimentally that Planck's constant has a value of

$$h = 6.626\,068\,96 \times 10^{-34}\ \text{J} \cdot \text{s}$$

The radical feature of Planck's assumption was that the energy of an atomic oscillator could have only discrete values (hf, $2hf$, $3hf$, etc.), with energies in between these values being forbidden. Whenever the energy of a system can have only certain definite values, and nothing in between, the energy is said to be *quantized*. This quantization of the energy was unexpected on the basis of the traditional physics of the time. However, it was soon realized that energy quantization had wide-ranging and valid implications.

Conservation of energy requires that the energy carried off by the radiated electromagnetic waves must equal the energy lost by the atomic oscillators in Planck's model. Suppose, for example, that an oscillator with an energy of $3hf$ emits an electromagnetic wave. According to Equation 29.1, the next smallest allowed value for the energy of the oscillator is $2hf$. In such a case, the energy carried off by the electromagnetic wave would have the value of hf, equaling the amount of energy lost by the oscillator. Thus, Planck's model for blackbody radiation sets the stage for the idea that electromagnetic energy occurs as a collection of discrete amounts, or packets, of energy, with the energy of a packet being equal to hf. Einstein made the proposal that light consists of such energy packets.

FIGURE 29.3 Although the spotlight beams in the photograph look like continuous beams of light, each is composed of discrete photons.

John Lee/Masterfile

INTERACTIVE FIGURE 29.4 In the photoelectric effect, light with a sufficiently high frequency ejects electrons from a metal surface. These photoelectrons, as they are called, are drawn to the positive collector, thus producing a current.

Light
Phototube (evacuated)
Collector (positive)
Metal plate (negative)
Photoelectrons
-1.20
+ |||-
V
Ammeter

29.3 Photons and the Photoelectric Effect

We have seen in Chapter 24 that light is an electromagnetic wave and that such waves are continuous patterns of electric and magnetic fields. It is not unexpected, then, that light beams, such as those in the photograph in **Figure 29.3**, or those coming from flashlights, look like continuous beams. However, we must now discuss the surprising fact that visible light and all other types of electromagnetic waves are composed of discrete particle-like entities called **photons**. As discussed in Chapters 6 and 7, the total energy E and the linear momentum $\vec{p}$ are fundamental concepts in physics that apply to moving particles such as electrons and protons. The total energy of a (nonrelativistic) particle is the sum of its kinetic energy (KE) and potential energy (PE), or $E = \text{KE} + \text{PE}$. The magnitude p of the particle's momentum is the product of its mass m and speed v, or $p = mv$. We will see that the ideas of energy and momentum also apply to photons. The defining equations for photon energy and photon momentum, however, are not the same as they are for particles such as electrons and protons.

Experimental evidence that light consists of photons comes from a phenomenon called the **photoelectric effect**, in which electrons are emitted from a metal surface when light shines on it. **Interactive Figure 29.4** illustrates the effect. The electrons are emitted if the light being used has a sufficiently high frequency. The ejected electrons move toward a positive electrode called the *collector* and cause a current to register on the ammeter. Because the electrons are ejected with the aid of light, they are called **photoelectrons**. As will be discussed shortly, a number of features of the photoelectric effect could not be explained solely with the ideas of classical physics.

In 1905 Einstein presented an explanation of the photoelectric effect that took advantage of Planck's work concerning blackbody radiation. It was primarily for his theory of the photoelectric effect that he was awarded the Nobel Prize in physics in 1921. In his photoelectric theory, Einstein proposed that light of frequency f could be regarded as a collection of discrete packets of energy (photons), each packet containing an amount of energy E given by

Energy of a photon $\qquad\qquad\qquad E = hf \qquad\qquad\qquad$ **(29.2)**

*It is now known that the energy of a harmonic oscillator is $E = \left(n + \frac{1}{2}\right)hf$; the extra term of $\frac{1}{2}$ is unimportant to the present discussion.

where h is Planck's constant. The light energy given off by a light bulb, for instance, is carried by photons. The brighter the bulb, the greater is the number of photons emitted per second. Example 1 estimates the number of photons emitted per second by a typical light bulb.

EXAMPLE 1 | Photons from a Light Bulb

In converting electrical energy into light energy, a sixty-watt incandescent light bulb operates at about 2.1% efficiency. Assuming that all the light is green light (vacuum wavelength = 555 nm), determine the number of photons per second given off by the bulb.

Reasoning The number of photons emitted per second can be found by dividing the amount of light energy emitted per second by the energy E of one photon. The energy of a single photon is $E = hf$, according to Equation 29.2. The frequency f of the photon is related to its wavelength λ by Equation 16.1 as $f = c/\lambda$.

Solution At an efficiency of 2.1%, the light energy emitted per second by a sixty-watt bulb is $(0.021)(60.0 \text{ J/s}) = 1.3 \text{ J/s}$. The energy of a single photon is

$$E = hf = \frac{hc}{\lambda} = \frac{(6.63 \times 10^{-34} \text{ J} \cdot \text{s})(3.00 \times 10^8 \text{ m/s})}{555 \times 10^{-9} \text{ m}} = 3.58 \times 10^{-19} \text{ J}$$

Therefore,

$$\text{Number of photons emitted per second} = \frac{1.3 \text{ J/s}}{3.58 \times 10^{-19} \text{ J/photon}} = \boxed{3.6 \times 10^{18} \text{ photons/s}}$$

The solution to Example 1 shows that the energy of the photons in light increases as the wavelength of the light, λ, decreases. Example 2 discusses this wavelength dependence in the context of sunlight and its ability to damage our skin.

EXAMPLE 2 | BIO The Physics of The Solar Spectrum and Sunscreen

We are all aware of the importance of protecting our skin from prolonged exposure to the sun. **Figure 29.5** shows the solar radiation, or sunlight, that impacts the earth as a function of wavelength. The pink region represents the light that strikes the earth's upper atmosphere, while the blue region shows the reduced intensity that reaches the earth's surface. The several sharp dips in intensity in the solar spectrum at the earth's surface are due to atmospheric gases absorbing the light energy at those particular wavelengths. The sun's radiation spectrum is well approximated by an ideal blackbody with a temperature of 5778 K, which is the surface temperature of the sun (see the solid green line in **Figure 29.5**). The majority of the sun's radiation that reaches the earth spans a range of wavelengths from 200 to 3000 nm, although longer wavelength infrared radiation and shorter wavelength UV, and even X-rays, are also present, but they make up a very small fraction of the sun's total power output. At earth's upper atmosphere, the composition of sunlight is approximately 10% UV, 40% visible, and 50% in the infrared region of the electromagnetic spectrum. The UV radiation can be further separated into three types by wavelength range: UV-A (315 to 400 nm), UV-B (280 to 315 nm), and UV-C (100 to 280 nm).

Around 70% of the sun's UV light is absorbed by the atmosphere. The portion that reaches the earth's surface is represented by the blue striped region shown in **Figure 29.5**, and this is the region in which we are interested. The majority of the composition of the UV light that reaches the earth is UV-A (~95%). Almost all of the UV-C and much of the UV-B are absorbed by oxygen (O_2) and ozone (O_3) in the atmosphere. This is a good thing, since UV-C, also known as hard UV, has sufficient energy to severely damage DNA molecules. In fact, without the atmospheric absorption of UV-C light, DNA-based life on earth would not be possible. UV-B makes up ~5% of the UV light reaching the earth's surface, and it has longer wavelengths than UV-C, and thus less energy, but it also damages DNA in a similar way. It is this radiation that is responsible for damaging skin cells and producing sunburn. UV-A has longer wavelengths than UV-B and even less energy, and is often referred to as soft UV.

It was believed that UV-A was far less damaging to DNA, so artificial UV-A lamps were utilized in tanning beds at salons. However, research has shown that UV-A can also destroy DNA molecules, but by more indirect routes, such as in the creation of free radicals, which are highly reactive atoms or molecules with unpaired electrons that can cause cell damage, leading to premature aging of the skin and even cancer. Sunscreen provides protection from UV-A and UV-B. Historically, both titanium oxide and zinc oxide were used as sunscreen. After application, they would leave a white layer on the skin that reflects much of the UV light away, while also absorbing some of the energy harmlessly. Present-day sunscreens use a combination of inorganic and organic compounds that contain certain chemical bonds that absorb UV light and safely dissipate its energy as heat. The chemical structure of the compounds determines which wavelengths of UV light they absorb—A, B, or both. The chemical bonds will break down over time during exposure, which is why it is so important to frequently reapply sunscreen. This breaking of chemical bonds is the same mechanism that causes the damage in DNA. If it requires 4.28 eV to break the carbon–hydrogen bond in thymine—one of the base pairs of DNA—what is the longest wavelength of UV-B light that will produce this damage?

FIGURE 29.5 The solar radiation spectrum as a function of wavelength. The overall spectrum is similar to that produced by a blackbody with a surface temperature equal to that of the sun. The UV sunlight that reaches the earth's surface is responsible for causing sunburn.

Reasoning Given the energy required to break the bond, we can set this energy equal to the energy of the photons in the light with Equation 29.2. The frequency of the light is related to the wavelength by Equation 16.1 with $v = c$.

Solution Applying Equations 29.2 and 16.1, we have

$$E = hf = \frac{hc}{\lambda} \quad \text{or} \quad \lambda = \frac{hc}{E}$$

We can now calculate the wavelength, but we first convert the energy from eV to J:

$$E = (4.28 \text{ eV})\left(\frac{1.60 \times 10^{-19} \text{ J}}{1 \text{ eV}}\right) = 6.85 \times 10^{-19} \text{ J}$$

$$\lambda = \frac{hc}{E} = \frac{(6.63 \times 10^{-34} \text{ J} \cdot \text{s})(3.00 \times 10^8 \text{ m/s})}{6.85 \times 10^{-19} \text{ J}} = 2.90 \times 10^{-7} \text{ m}$$
$$= \boxed{290 \text{ nm}}$$

This wavelength is in the middle of the UV-B band.

Related Homework: Problem 45.

According to Einstein, when light shines on a metal, a photon can give up its energy to an electron in the metal. If the photon has enough energy to do the work of removing the electron from the metal, the electron can be ejected. The work required depends on how strongly the electron is held. For the *least strongly* held electrons, the necessary work has a minimum value W_0 and is called the **work function** of the metal. If a photon has energy in excess of the work needed to remove an electron, the excess appears as kinetic energy of the ejected electron. Thus, the least strongly held electrons are ejected with the maximum kinetic energy KE_{max}. Einstein applied the conservation-of-energy principle and proposed the following relation to describe the photoelectric effect:

$$\underset{\substack{\text{Photon} \\ \text{energy}}}{hf} = \underset{\substack{\text{Maximum} \\ \text{kinetic energy} \\ \text{of ejected} \\ \text{electron}}}{KE_{max}} + \underset{\substack{\text{Minimum work} \\ \text{needed to eject} \\ \text{electron}}}{W_0} \qquad (29.3)$$

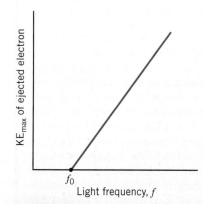

FIGURE 29.6 Photons can eject electrons from a metal when the light frequency is above a minimum value f_0. For frequencies above this value, ejected electrons have a maximum kinetic energy KE_{max} that is linearly related to the frequency, as the graph shows.

According to this equation, $KE_{max} = hf - W_0$, which is plotted in **Figure 29.6**, with KE_{max} along the y axis and f along the x axis. The graph is a straight line that crosses the x axis at $f = f_0$. At this frequency, the electron departs from the metal with no kinetic energy ($KE_{max} = 0$ J). According to Equation 29.3, when $KE_{max} = 0$ J the energy hf_0 of the incident photon is equal to the work function W_0 of the metal: $hf_0 = W_0$.

The photon concept provides an explanation for a number of features of the photoelectric experiment that are difficult to explain without photons. It is observed, for instance, that only light with a frequency above a certain minimum value f_0 will eject electrons. If the frequency is below this value, no electrons are ejected, regardless of how intense the light is. Example 3 determines the minimum frequency value for a silver surface.

EXAMPLE 3 | The Photoelectric Effect for a Silver Surface

The work function for a silver surface is $W_0 = 4.73$ eV. Find the minimum frequency that light must have to eject electrons from this surface.

Reasoning The minimum frequency f_0 is that frequency at which the photon energy equals the work function W_0 of the metal, so the electron is ejected with zero kinetic energy. Since 1 eV $= 1.60 \times 10^{-19}$ J, the work function expressed in joules is

$$W_0 = (4.73 \text{ eV})\left(\frac{1.60 \times 10^{-19} \text{ J}}{1 \text{ eV}}\right) = 7.57 \times 10^{-19} \text{ J}$$

Using Equation 29.3, we find

$$hf_0 = \underbrace{\text{KE}_{\text{max}}}_{= 0 \text{ J}} + W_0 \quad \text{or} \quad f_0 = \frac{W_0}{h}$$

Problem-Solving Insight The work function of a metal is the minimum energy needed to eject an electron from the metal. An electron that has received this minimum energy has no kinetic energy once outside the metal.

Solution The minimum frequency f_0 is

$$f_0 = \frac{W_0}{h} = \frac{7.57 \times 10^{-19} \text{ J}}{6.63 \times 10^{-34} \text{ J} \cdot \text{s}} = \boxed{1.14 \times 10^{15} \text{ Hz}}$$

Photons with frequencies less than f_0 do not have enough energy to eject electrons from a silver surface. Since $\lambda_0 = c/f_0$, the wavelength of this light is $\lambda_0 = 263$ nm, which is in the ultraviolet region of the electromagnetic spectrum.

In Example 3 the electrons are ejected with no kinetic energy, because the light shining on the silver surface has the minimum possible frequency that will eject them. When the frequency of the light exceeds this minimum value, the electrons that are ejected do have kinetic energy. The next example deals with such a situation.

Analyzing Multiple-Concept Problems

EXAMPLE 4 | The Maximum Speed of Ejected Photoelectrons

Light with a wavelength of 95 nm shines on a selenium surface, which has a work function of 5.9 eV. The ejected electrons have some kinetic energy. Determine the maximum speed with which electrons are ejected.

Reasoning The maximum speed of the ejected electrons is related to their maximum kinetic energy. Conservation of energy dictates that this maximum kinetic energy is related to the work function of the surface and the energy of the incident photons. The work function is given. The energy of the photons can be obtained from the frequency of the light, which is related to the wavelength.

Knowns and Unknowns We have the following data:

Description	Symbol	Value	Comment
Wavelength of light	λ	95 nm	1 nm $= 10^{-9}$ m
Work function of selenium surface	W_0	5.9 eV	Will be converted to joules.
Unknown Variable			
Maximum speed of photoelectrons	v_{max}	?	

Modeling the Problem

STEP 1 Kinetic Energy and Speed The maximum kinetic energy KE_{max} of the ejected electrons is $\text{KE}_{\text{max}} = \frac{1}{2}mv_{\text{max}}^2$ (Equation 6.2), where m is the mass of an electron. Solving for the maximum speed v_{max} gives Equation 1 at the right. The mass of the electron is $m = 9.11 \times 10^{-31}$ kg (see inside of front cover of the textbook). The maximum kinetic energy is unknown, but we will evaluate it in Step 2.

$$v_{\text{max}} = \sqrt{\frac{2\text{KE}_{\text{max}}}{m}} \quad (1)$$

$$\boxed{?}$$

STEP 2 Conservation of Energy According to the principle of conservation of energy, as expressed by Equation 29.3, we have

$$\underset{\substack{\text{Photon} \\ \text{energy}}}{hf} = \underset{\substack{\text{Maximum} \\ \text{kinetic energy} \\ \text{of ejected} \\ \text{electron}}}{\text{KE}_{\text{max}}} + \underset{\substack{\text{Minimum work} \\ \text{needed to eject} \\ \text{electron}}}{W_0}$$

where f is the frequency of the light. Solving for KE_{max} gives

$$\boxed{KE_{max} = hf - W_0} \qquad (2)$$

which can be substituted into Equation 1, as shown at the right. In this expression the work function W_0 is known, and we will deal with the unknown frequency f in Step 3.

$$v_{max} = \sqrt{\frac{2KE_{max}}{m}} \qquad (1)$$
$$\boxed{KE_{max} = hf - W_0} \qquad (2)$$
$$\boxed{?}$$

STEP 3 Relationship Between Frequency and Wavelength   The frequency and wavelength of the light are related to the speed of light c according to $f\lambda = c$ (Equation 16.1). Solving for the frequency gives

$$\boxed{f = \frac{c}{\lambda}}$$

which we substitute into Equation 2, as shown at the right.

$$v_{max} = \sqrt{\frac{2KE_{max}}{m}} \qquad (1)$$
$$\boxed{KE_{max} = hf - W_0} \qquad (2)$$
$$\boxed{f = \frac{c}{\lambda}}$$

Solution   Combining the results of each step algebraically, we find that

$$\overset{\text{STEP 1}}{\downarrow} \qquad \overset{\text{STEP 2}}{\downarrow} \qquad \overset{\text{STEP 3}}{\downarrow}$$
$$v_{max} = \sqrt{\frac{2KE_{max}}{m}} = \sqrt{\frac{2(hf - W_0)}{m}} = \sqrt{\frac{2(h\frac{c}{\lambda} - W_0)}{m}}$$

Thus, the maximum speed of the photoelectrons is

$$v_{max} = \sqrt{\frac{2\left(h\frac{c}{\lambda} - W_0\right)}{m}}$$

$$v_{max} = \sqrt{\frac{2\left[(6.63 \times 10^{-34}\ J \cdot s)\frac{(3.00 \times 10^8\ m/s)}{(95 \times 10^{-9}\ m)} - (5.9\ eV)\frac{(1.60 \times 10^{-19}\ J)}{(1\ eV)}\right]}{9.11 \times 10^{-31}\ kg}}$$

$$= \boxed{1.6 \times 10^6\ m/s}$$

Note that in this calculation we have converted the value of the work function from electron volts to joules.

Related Homework: Problem 8

Another significant feature of the photoelectric effect is that the maximum kinetic energy of the ejected electrons remains the same when the intensity of the light increases, provided the light frequency remains the same. As the light intensity increases, more photons per second strike the metal, and consequently more electrons per second are ejected. However, since the frequency is the same for each photon, the energy of each photon is also the same. Thus, the ejected electrons always have the same maximum kinetic energy.

Whereas the photon model of light explains the photoelectric effect satisfactorily, the electromagnetic wave model of light does not. Certainly, it is possible to imagine that the electric field of an electromagnetic wave would cause electrons in the metal to oscillate and tear free from the surface when the amplitude of oscillation becomes large enough. However, were this the case, higher-intensity light would eject electrons with a greater maximum kinetic energy, a fact that experiment does not confirm. Moreover, in the electromagnetic wave model, a relatively long time would be required with low-intensity light before the electrons would build up a sufficiently large oscillation amplitude to tear free. Instead, experiment shows that even the weakest light intensity causes electrons to be ejected almost instantaneously, provided the frequency of the light is above the minimum value f_0. The failure of the electromagnetic wave model to explain the photoelectric effect does not mean that the wave model should be abandoned. However, we must recognize that the wave model does not account for all the characteristics of light. The photon model also makes an important contribution to our understanding of the way light behaves when it interacts with matter.

Yen Hung/Shutterstock.com

Yen Hung/Shutterstock.com

FIGURE 29.7 Digital cameras like this one use an array of charge-coupled devices instead of film to capture an image.

Because a photon has energy, the photon can eject an electron from a metal surface when it interacts with the electron. However, a photon is different from a normal particle. A normal particle has a mass and can travel at speeds up to, but not equal to, the speed of light. A photon, on the other hand, travels at the speed of light in a vacuum and does not exist as an object at rest. The energy of a photon is entirely kinetic in nature, because it has no rest energy and no mass. To show that a photon has no mass, we rewrite Equation 28.4 for the total energy E as

$$E\sqrt{1 - \frac{v^2}{c^2}} = mc^2$$

The term $\sqrt{1 - (v^2/c^2)}$ is zero because a photon travels at the speed of light, $v = c$. Since the energy E of the photon is finite, the left side of the equation above is zero. Thus, the right side must also be zero, so $m = 0$ kg and the photon has no mass.

THE PHYSICS OF . . . charge-coupled devices and digital cameras. One of the most exciting and useful applications of the photoelectric effect is the charge-coupled device (CCD). An array of these devices is used instead of film in digital cameras (see **Figure 29.7**) to capture images in the form of many small groups of electrons. CCD arrays are also used in digital camcorders and electronic scanners, and they provide the method of choice with which astronomers capture those spectacular images of the planets and the stars. For use with visible light, a CCD array consists of a sandwich of semiconducting silicon, insulating silicon dioxide, and a number of electrodes, as **Figure 29.8** shows. The array is divided into many small sections, or pixels, sixteen of which are shown in the drawing. Each pixel captures a small part of a picture. Digital cameras can have up to over 100 million pixels, depending on price. The greater the number of pixels, the better is the resolution of the photograph. The blow-up in **Figure 29.8** shows a single pixel. Incident photons of visible light strike the silicon and generate electrons via the photoelectric effect. The range of energies of the visible photons is such that approximately one electron is released when a photon interacts with a silicon atom. The electrons do not escape from the silicon, but are trapped within a pixel because of a positive voltage applied to the electrodes beneath the insulating layer. Thus, the number of electrons that are released and trapped is proportional to the number of photons striking the pixel. In this fashion, each pixel in the CCD array accumulates an accurate representation of the light intensity at that point on the image. Color information is provided using red, green, or blue filters or a system of prisms to separate the colors. Astronomers use CCD arrays not only in the visible region of the electromagnetic spectrum but in other regions as well.

In addition to trapping the photoelectrons, the electrodes beneath the pixels are used to read out the electron representation of the picture. By changing the positive voltages applied to the electrodes, it is possible to cause all of the electrons trapped in one row of pixels to be transferred to the adjacent row. In this fashion, for instance, row 1 in **Figure 29.8** is transferred into row 2, row 2 into row 3, and row 3 into the bottom row, which serves a special purpose. The bottom row functions as a horizontal shift register, from which the contents of each pixel can be shifted to the right, one at a time, and read into an analog signal processor.

Incident visible photon

Semiconducting silicon

Pixel

Electron

Insulating silicon dioxide

Electrodes

Horizontal shift register

Row 1
Row 2
Row 3

Analog signal processor

Analog-to-digital converter

Digital representation of the picture

FIGURE 29.8 A CCD array can be used to capture photographic images using the photoelectric effect.

This processor senses the varying number of electrons in each pixel in the shift register as a kind of wave that has a fluctuating amplitude. After another shift in rows, the information in the next row is read out, and so forth. The output of the analog signal processor is sent to an analog-to-digital converter, which produces a digital representation of the image in terms of the zeros and ones that computers recognize.

THE PHYSICS OF . . . a safety feature of garage door openers.   Another application of the photoelectric effect depends on the fact that the moving photoelectrons in **Interactive Figure 29.4** constitute a current—a current that changes as the intensity of the light changes. All automatic garage door openers have a safety feature that prevents the door from closing when it encounters an obstruction (person, vehicle, etc.). As **Figure 29.9** illustrates, a sending unit transmits an invisible (infrared) beam across the opening of the door. The beam is detected by a receiving unit that contains a photodiode. A photodiode is a type of *p-n* junction diode (see Section 23.5). When infrared photons strike the photodiode, electrons bound to the atoms absorb the photons and become liberated. These liberated, mobile electrons cause the current in the photodiode to increase. When a person walks through the beam, the light is momentarily blocked from reaching the receiving unit, and the current in the photodiode decreases. The change in current is sensed by electronic circuitry that immediately stops the downward motion of the door and then causes it to rise up.

THE PHYSICS OF . . . photoevaporation and star formation.   **Figure 29.10a** shows a portion of the Eagle Nebula, a giant star-forming region some 7000 light-years from earth. The photo was taken by the Hubble Space Telescope and reveals clouds of molecular gas and dust, in which there is dramatic evidence of the energy carried by photons. These clouds extend more than a light-year from base to tip and are the birthplace of stars. A star begins to form within a cloud when the gravitational force pulls together sufficient gas to create a high-density "ball." When the gaseous ball becomes sufficiently dense, thermonuclear fusion (see Section 32.5) occurs at its core, and the star begins to shine. The newly born stars are buried within the cloud and cannot be seen from earth. However, the process of photoevaporation allows astronomers to see many of the high-density regions where stars are being formed. Photoevaporation is the process in which high-energy, ultraviolet (UV) photons from hot stars outside the cloud heat it up, much like microwave photons heat food in a microwave oven. **Figure 29.10a** shows streamers of gas photoevaporating from the cloud as it is illuminated by stars located beyond the photograph's upper edge. As photoevaporation proceeds, globules of gas that are denser than their surroundings are exposed. The globules are known as *evaporating gaseous globules* (EGGs), and they are slightly larger than our solar system. The drawing in part *b* of **Figure 29.10** shows that the EGGs shade the gas and dust behind them from the UV photons, creating the many finger-like protrusions seen on the surface of the cloud. Astronomers believe that some of these EGGs contain young stars within them.

FIGURE 29.9 When an obstruction prevents the infrared light beam from reaching the receiving unit, the current in the receiving unit drops. This drop in current is detected by an electronic circuit that stops the downward movement of the garage door and then causes it to rise.

FIGURE 29.10 (*a*) Photoevaporation produces finger-like projections on the surface of the gas clouds in the Eagle Nebula. At the fingertips are high-density evaporating gaseous globules (EGGs). (*b*) This drawing illustrates the photoevaporation that is occurring in the photograph in part *a*.

Check Your Understanding

(The answers are given at the end of the book.)

1. The photons emitted by a source of light do *not* all have the same energy. Is the source monochromatic? (A monochromatic light source emits light that has a single wavelength.)

2. Which colored light bulb—red, orange, yellow, green, or violet—emits photons with **(a)** the least energy and **(b)** the greatest energy? (See Example 1 in Chapter 24.)

3. Does a photon emitted by a higher-wattage red light bulb have more energy than a photon emitted by a lower-wattage red bulb?

4. Radiation of a given wavelength causes electrons to be emitted from the surface of metal 1 but not from the surface of metal 2. Why could this be? **(a)** Metal 1 has a greater work function than metal 2 has. **(b)** Metal 1 has a smaller work function than metal 2 has. **(c)** The energy of a photon striking metal 1 is greater than the energy of a photon striking metal 2.

5. In the photoelectric effect, electrons are ejected from the surface of a metal when light shines on it. Which one or more of the following would lead to an increase in the maximum kinetic energy of the ejected electrons? **(a)** Increasing the frequency of the incident light **(b)** Increasing the number of photons per second striking the surface **(c)** Using photons whose frequency f_0 is less than W_0/h, where W_0 is the work function of the metal and h is Planck's constant **(d)** Selecting a metal that has a greater work function

6. In the photoelectric effect, suppose that the intensity of the light is increased, while the frequency of the light is kept constant. The frequency is greater than the minimum frequency f_0. State whether each of the following will increase, decrease, or remain the same: **(a)** The current in the phototube **(b)** The number of electrons emitted per second from the metal surface **(c)** The maximum kinetic energy that an electron could have **(d)** The maximum momentum that an electron could have

---

29.4 | The Momentum of a Photon and the Compton Effect

Although Einstein presented his photon model for the photoelectric effect in 1905, it was not until 1923 that the photon concept began to achieve widespread acceptance. It was then that the American physicist Arthur H. Compton (1892–1962) used the photon model to explain his research on the scattering of X-rays by the electrons in graphite. X-rays are high-frequency electromagnetic waves and, like light, they are composed of photons.

Animated Figure 29.11 illustrates what happens when an X-ray photon strikes an electron in a piece of graphite. Like two billiard balls colliding on a pool table, the X-ray photon scatters in one direction after the collision, and the electron recoils in another direction. Compton observed that the scattered photon has a frequency f' that is smaller than the frequency f of the incident photon, indicating that the photon loses energy during the collision. In addition, he found that the difference between the two frequencies depends on the angle θ at which the scattered photon leaves the collision. The phenomenon in which an X-ray photon is scattered from an electron, with the scattered photon having a smaller frequency than the incident photon, is called the **Compton effect**.

In Section 7.3, collisions between two objects are analyzed using the fact that the total kinetic energy and the total linear momentum of the objects are the same before and after an elastic collision. Similar analysis can be applied to the collision between a photon and an electron. The electron is assumed to be initially at rest and essentially free—that is, not bound to the atoms of the material. According to the principle of conservation of energy,

ANIMATED FIGURE 29.11 In an experiment performed by Arthur H. Compton, an X-ray photon collides with a stationary electron. The scattered photon and the recoil electron depart from the collision in different directions.

$$\underbrace{hf}_{\substack{\text{Energy of} \\ \text{incident} \\ \text{photon}}} = \underbrace{hf'}_{\substack{\text{Energy of} \\ \text{scattered} \\ \text{photon}}} + \underbrace{\text{KE}}_{\substack{\text{Kinetic energy} \\ \text{of recoil} \\ \text{electron}}} \qquad \textbf{(29.4)}$$

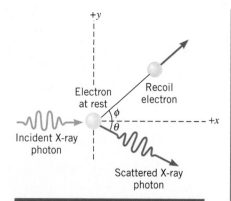

ANIMATED FIGURE 29.11 (REPEATED)
In an experiment performed by Arthur H. Compton, an X-ray photon collides with a stationary electron. The scattered photon and the recoil electron depart from the collision in different directions.

Math Skills Equation 29.5 is a relationship between the momentum $\vec{p}$ of the incident photon, the momentum $\vec{p}'$ of the scattered photon, and the momentum $\vec{p}_{electron}$ of the recoil electron in **Animated Figure 29.11**. These momenta are vector quantities. Therefore, Equation 29.5 is equivalent to two equations; one is for the x components of the vectors and one for the y components of the vectors (see Sections 1.7 and 1.8). Using components with respect to the x, y axis in **Animated Figure 29.11**, we can see that the following two equations are equivalent to Equation 29.5:

x component $\underbrace{P}_{\text{Incident photon}} = \underbrace{p'\cos\theta}_{\text{Scattered photon}} + \underbrace{p_{electron}\cos\phi}_{\text{Recoil electron}}$

y component $\underbrace{0}_{\text{Incident photon}} = \underbrace{-p'\sin\theta}_{\text{Scattered photon}} + \underbrace{p_{electron}\sin\phi}_{\text{Recoil electron}}$

In these equations, the symbols p and p' denote the vector magnitudes of the momenta. Note that the y component of the momentum is zero for the incident photon, because that photon travels along the x axis in **Animated Figure 29.11**. Note also that the y component of the momentum is negative for the scattered photon, because the direction in which that photon travels is below the x axis.

where the relation $E = hf$ has been used for the photon energies. It follows, then, that $hf' = hf - KE$, which shows that the energy and corresponding frequency f' of the scattered photon are less than the energy and frequency of the incident photon, just as Compton observed. Since $\lambda' = c/f'$ (Equation 16.1), the wavelength of the scattered X-rays is larger than that of the incident X-rays.

For an initially stationary electron, conservation of total linear momentum requires that

$$\underbrace{\vec{p}}_{\substack{\text{Momentum of} \\ \text{incident photon}}} = \underbrace{\vec{p}'}_{\substack{\text{Momentum of} \\ \text{scattered photon}}} + \underbrace{\vec{p}_{electron}}_{\substack{\text{Momentum} \\ \text{of recoil} \\ \text{electron}}} \qquad (29.5)$$

To find an expression for the magnitude p of a photon's momentum, we use Equations 28.3 and 28.4. According to these equations, the momentum p and the total energy E of any particle are

$$p = \frac{mv}{\sqrt{1 - (v^2/c^2)}} \qquad (28.3)$$

$$E = \frac{mc^2}{\sqrt{1 - (v^2/c^2)}} \qquad (28.4)$$

Rearranging Equation 28.4 to show that $\dfrac{m}{\sqrt{1 - (v^2/c^2)}} = \dfrac{E}{c^2}$ and substituting this result into Equation 28.3 gives

$$p = \frac{mv}{\sqrt{1 - (v^2/c^2)}} = \frac{Ev}{c^2}$$

A photon travels at the speed of light, so that we have $v = c$, and the momentum of a photon is

$$p = \frac{Ev}{c^2} = \frac{Ec}{c^2} = \frac{E}{c}$$

This result only applies to a photon and does not apply to a particle with mass, because such a particle cannot travel at the speed of light. We also know that the energy of a photon is related to its frequency f according to $E = hf$ (Equation 29.2) and that the speed c of a photon is related to its frequency and wavelength λ according to $c = f\lambda$ (Equation 16.1). With these substitutions, our expression for the momentum of a photon becomes

$$p = \frac{E}{c} = \frac{hf}{f\lambda} = \frac{h}{\lambda} \qquad (29.6)$$

Using Equations 29.4, 29.5, and 29.6, Compton showed that the difference between the wavelength λ' of the scattered photon and the wavelength λ of the incident photon is related to the scattering angle θ by

$$\lambda' - \lambda = \frac{h}{mc}(1 - \cos\theta) \qquad (29.7)$$

In this equation m is the mass of the electron. The quantity $h/(mc)$ is called the **Compton wavelength of the electron** and has the value $h/(mc) = 2.43 \times 10^{-12}$ m. Since $\cos\theta$ varies between $+1$ and -1, the shift $\lambda' - \lambda$ in the wavelength can vary between zero and $2h/(mc)$, depending on the value of θ, a fact observed by Compton.

The photoelectric effect and the Compton effect provide compelling evidence that light can exhibit particle-like characteristics attributable to energy packets called photons. But what about the interference phenomena discussed in Chapter 27, such as Young's double-slit experiment and single-slit diffraction, which demonstrate that light behaves as a wave? Does light have two distinct personalities, in which it behaves like a stream of particles in some experiments and like a wave in others? The answer is yes, for physicists now believe that this wave–particle duality is an inherent property of light. Light is a far more interesting (and complex) phenomenon than just a stream of particles or an electromagnetic wave.

In the Compton effect the electron recoils because it gains some of the photon's momentum. In principle, then, the momentum that photons have can be used to make other objects move. Conceptual Example 5 considers a propulsion system for an interstellar spaceship that is based on the momentum of a photon.

CONCEPTUAL EXAMPLE 5 | The Physics of Solar Sails and Spaceship Propulsion

One propulsion method that is currently being studied for interstellar travel uses a large sail. The intent is that sunlight striking the sail will create a force that pushes the ship away from the sun (**Figure 29.12**), much as the wind propels a sailboat. To get the greatest possible force, the surface of the sail facing the sun **(a)** should be shiny like a mirror, **(b)** should be black, or **(c)** could be either shiny or black, since the same force will be created for either type of surface.

Reasoning   In Conceptual Example 3 in Chapter 7, we found that hailstones striking the roof of a car exert a force on it because the collision changes their momentum. Photons also have momentum, so, like the hailstones, they can apply a force to the sail.

As in Chapter 7, we will be guided by the impulse–momentum theorem (Equation 7.4) in assessing the force. This theorem states that when a net force acts on an object, the impulse of the net force is equal to the change in momentum of the object. Greater impulses lead to greater forces for a given time interval. Thus, when a photon collides with the sail, the photon's momentum changes because of the force that the sail applies to the photon. Newton's action–reaction law (Section 4.5) indicates that the photon simultaneously applies a force of equal magnitude, but opposite direction, to the sail. It is this reaction force that propels the spaceship, and it will be greater when the momentum change experienced by the photon is greater. The surface of the sail facing the sun, then, should be such that it causes the largest possible momentum change for the impinging photons.

Answers (b) and (c) are incorrect.   In Conceptual Example 3 in Chapter 7 we examined whether hailstones or raindrops exert the greater force when they strike the roof of a car. The difference is that hailstones, being hard objects, bounce off the roof, while raindrops splatter and do not bounce very much. We concluded that hailstones, because of the bounce, experience a greater change in momentum than raindrops do, so the roof exerts a greater force on the hailstones. By Newton's action–reaction law, the car roof, then, experiences a greater force from the hailstones than from the raindrops. We saw in Section 13.3 that radiation is reflected from a shiny mirror-like surface and is absorbed by a black surface. Therefore, by analogy with raindrops that stick to the car roof, the sail experiences a smaller force when the surface facing the sun is black.

Answer (a) is correct.   The sun's radiation reflects from a shiny mirror-like surface and is absorbed by a black surface. Now, consider a photon that strikes the sail perpendicularly. When a photon reflects from a mirror-like surface, the photon's momentum changes from its value in the forward direction to a value of the same magnitude in the reverse direction. This is a greater change than the one that occurs when the photon is absorbed by a black surface. In the latter case, the momentum changes only from its value in the forward direction to a value of zero. Consequently, the surface of the sail facing the sun should be shiny in order to produce the greatest possible propulsion force. A shiny surface causes the photons to bounce like hailstones on the roof of a car and, in doing so, to apply a greater force to the sail.

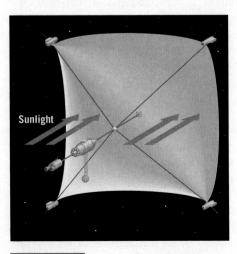

FIGURE 29.12   A solar sail provides the propulsion for this interstellar spaceship.

Sunlight

Check Your Understanding

(The answers are given at the end of the book.)

7. In the Compton effect, an incident X-ray photon of wavelength λ is scattered by an electron, and the scattered photon has a wavelength of λ'. Suppose that the incident photon is scattered by a proton instead of an electron. For a given scattering angle θ, is the change $\lambda' - \lambda$ in the wavelength of the photon scattered by the proton greater than, less than, or the same as the wavelength of the photon scattered by the electron? (*Hint:* Use Equation 29.7 for a proton instead of an electron.)

8. In a Compton scattering experiment, an incident X-ray photon is traveling along the $+x$ direction. An electron, initially at rest, is struck by the photon and is accelerated straight ahead in the same direction as the incident X-ray photon. Which way does the scattered photon move? **(a)** Along the $+y$ direction **(b)** Along the $-y$ direction **(c)** Along the $-x$ direction (*Hint:* Use the principle of conservation of momentum to guide your reasoning.)

9. The speed of a particle is much less than the speed of light. Thus, when the particle's speed doubles, the particle's momentum doubles, and its kinetic energy becomes four times greater. However, when the momentum of a photon doubles, its energy becomes **(a)** two times greater **(b)** four times greater **(c)** one-half as much **(d)** one-fourth as much.

10. Review Conceptual Example 5 as background for this question. CYU Figure 29.1 shows a device called a radiometer. The four regular panels are black on one side and shiny like a mirror on the other side. In bright light, the panel arrangement spins around in a direction from the black side of a panel toward the shiny side. Do photon collisions with both sides of the panels cause the observed spinning?

Charles D. Winters/Science Source

CYU FIGURE 29.1

29.5 The De Broglie Wavelength and the Wave Nature of Matter

As a graduate student in 1923, Louis de Broglie (1892–1987) made the astounding suggestion that since light waves could exhibit particle-like behavior, particles of matter should exhibit wave-like behavior. De Broglie proposed that all moving matter has a wavelength associated with it, just as a wave does. He made the explicit proposal that the wavelength λ of a particle is given by the same relation (Equation 29.6) that applies to a photon:

De Broglie wavelength
$$\lambda = \frac{h}{p}$$
(29.8)

where h is Planck's constant and p is the magnitude of the relativistic momentum of the particle. Today, λ is known as the **de Broglie wavelength** of the particle.

Confirmation of de Broglie's suggestion came in 1927 from the experiments of the American physicists Clinton J. Davisson (1881–1958) and Lester H. Germer (1896–1971) and, independently, those of the English physicist George P. Thomson (1892–1975). Davisson and Germer directed a beam of electrons onto a crystal of nickel and observed that the electrons exhibited a diffraction behavior, analogous to that seen when X-rays are diffracted by a crystal (see Section 27.9 for a discussion of X-ray diffraction). The wavelength of the electrons revealed by the diffraction pattern matched that predicted by de Broglie's hypothesis, $\lambda = h/p$. More recently, Young's double-slit experiment has

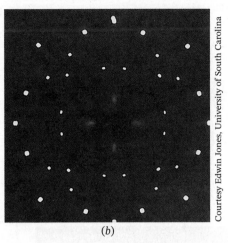

(a)   (b)

From "Phys. Rev." 73:527 (1948), by Wollan, Shull and Marney. ©1948 by the American Physical Society

Courtesy Edwin Jones, University of South Carolina

FIGURE 29.13 (*a*) The neutron diffraction pattern and (*b*) the X-ray diffraction pattern for a crystal of sodium chloride (NaCl).

been performed with electrons and reveals the effects of wave interference illustrated in **Interactive Figure 29.1**.

Particles other than electrons can also exhibit wave-like properties. For instance, neutrons are sometimes used in diffraction studies of crystal structure. **Figure 29.13** compares the neutron diffraction pattern and the X-ray diffraction pattern caused by a crystal of rock salt (NaCl).

Although all moving particles have a de Broglie wavelength, the effects of this wavelength are observable only for particles whose masses are very small, on the order of the mass of an electron or a neutron, for instance. Example 6 illustrates why.

EXAMPLE 6 | The De Broglie Wavelength of an Electron and of a Baseball

Determine the de Broglie wavelength for **(a)** an electron (mass = 9.1×10^{-31} kg) moving at a speed of 6.0×10^6 m/s and **(b)** a baseball (mass = 0.15 kg) moving at a speed of 13 m/s.

Reasoning In each case, the de Broglie wavelength is given by Equation 29.8 as Planck's constant divided by the magnitude of the momentum. Since the speeds are small compared to the speed of light, we can ignore relativistic effects and express the magnitude of the momentum as the product of the mass and the speed, as in Equation 7.2.

Solution **(a)** Since the magnitude p of the momentum is the product of the mass m of the particle and its speed v we have $p = mv$ (Equation 7.2). Using this expression in Equation 29.8 for the de Broglie wavelength, we obtain

$$\lambda = \frac{h}{p} = \frac{h}{mv} = \frac{6.63 \times 10^{-34}\,\text{J} \cdot \text{s}}{(9.1 \times 10^{-31}\,\text{kg})(6.0 \times 10^6\,\text{m/s})} = \boxed{1.2 \times 10^{-10}\,\text{m}}$$

A de Broglie wavelength of 1.2×10^{-10} m is about the size of the interatomic spacing in a solid, such as the nickel crystal used by Davisson and Germer, and, therefore, leads to the observed diffraction effects.

(b) A calculation similar to that in part (a) shows that the de Broglie wavelength of the baseball is $\boxed{\lambda = 3.3 \times 10^{-34}\,\text{m}}$. This wavelength is incredibly small, even by comparison with the size of an atom (10^{-10} m) or a nucleus (10^{-14} m). Thus, the ratio λ/W of this wavelength to the width W of an ordinary opening, such as a window, is so small that the diffraction of a baseball passing through the window cannot be observed.

The de Broglie equation for particle wavelength provides no hint as to what kind of wave is associated with a particle of matter. To gain some insight into the nature of this wave, we turn our attention to **Figure 29.14**. Part *a* shows the fringe pattern on the screen when electrons are used in a version of Young's double-slit experiment. The bright fringes occur in places where particle waves coming from each slit interfere constructively, while the dark fringes occur in places where the waves interfere destructively.

When an electron passes through the double-slit arrangement and strikes a spot on the screen, the screen glows at that spot, and parts *b*, *c*, and *d* of **Figure 29.14** illustrate how the spots accumulate in time. As more and more electrons strike the screen, the spots eventually form the fringe pattern that is evident in part *d*. Bright fringes occur where there is a high probability of electrons striking the screen, and dark fringes occur where there is a low probability. Here lies the key to understanding particle

Courtesy Akira tonomura, J. Endo, T. Matsuda and T. Kawasaki, Am. J. Phys. 57 (2): 117, February 1989.

(a)

(b) After 100 electrons

(c) After 3000 electrons

(d) After 70 000 electrons

FIGURE 29.14 In this electron version of Young's double-slit experiment, the characteristic fringe pattern becomes recognizable only after a sufficient number of electrons have struck the screen.

waves. **Particle waves are waves of probability**, waves whose magnitude at a point in space gives an indication of the probability that the particle will be found at that point. At the place where the screen is located, the pattern of probabilities conveyed by the particle waves causes the fringe pattern to emerge. The fact that no fringe pattern is apparent in part *b* of the figure does not mean that there are no probability waves present; it just means that too few electrons have struck the screen for the pattern to be recognizable.

The pattern of probabilities that leads to the fringes in **Figure 29.14** is analogous to the pattern of light intensities that is responsible for the fringes in Young's original experiment with light waves (see Figure 27.3). Section 24.4 discusses the fact that the intensity of the light is proportional to either the square of the electric field strength or the square of the magnetic field strength of the wave. In an analogous fashion in the case of particle waves, the probability is proportional to the square of the magnitude ψ (Greek letter psi) of the wave. ψ is referred to as the **wave function** of the particle.

In 1925 the Austrian physicist Erwin Schrödinger (1887–1961) and the German physicist Werner Heisenberg (1901–1976) independently developed theoretical frameworks for determining the wave function. In so doing, they established a new branch of physics called **quantum mechanics**. The word "quantum" refers to the fact that in the world of the atom, where particle waves must be considered, the particle energy is quantized, so only certain energies are allowed. To understand the structure of the atom and the phenomena related to it, quantum mechanics is essential, and the Schrödinger equation for calculating the wave function is now widely used. In the next chapter, we will explore the structure of the atom based on the ideas of quantum mechanics.

Check Your Understanding

(The answers are given at the end of the book.)

11. A stone is dropped from the top of a building. As the stone falls, does its de Broglie wavelength increase, decrease, or remain the same?

12. An electron and a neutron have different masses. Is it possible, according to Equation 29.8, that they can have the same de Broglie wavelength? **(a)** Yes, provided the magnitudes of their momenta are different. **(b)** Yes, provided their speeds are different. **(c)** No; two particles with different masses always have different de Broglie wavelengths.

13. In **Interactive Figure 29.1b**, replace the electrons with protons that have the same speed. With the aid of Equation 27.1 for the bright fringes in Young's double-slit experiment and Equation 29.8, decide whether the angular separation between the fringes would increase, decrease, or remain the same, relative to the angular separation produced by the electrons.

14. A beam of electrons passes through a single slit, and a beam of protons passes through a second, but identical, slit. The electrons and the protons have the same speed. Which one of the following correctly describes the beam that experiences the greatest amount of diffraction? **(a)** The electrons, because they have the smaller momentum and, hence, the smaller de Broglie wavelength **(b)** The electrons, because they have the smaller momentum and, hence, the larger de Broglie wavelength **(c)** The protons, because they have the smaller momentum and, hence, the smaller de Broglie wavelength **(d)** The protons, because they have the larger momentum and, hence, the smaller de Broglie wavelength **(e)** Both beams experience the same amount of diffraction, because the electrons and protons have the same de Broglie wavelength.

29.6 The Heisenberg Uncertainty Principle

As the previous section discusses, the bright fringes in **Figure 29.14** indicate the places where there is a high probability of an electron striking the screen. Since there are a number of bright fringes, there is more than one place where each electron has some probability of hitting. As a result, it is not possible to specify in advance exactly where on the screen an individual electron will hit. All we can do is speak of the probability that the electron may end up in a number of different places. No longer is it possible to say, as Newton's laws would suggest, that a single electron, fired through the double slit, will travel directly forward in a straight line and strike the screen. This simple model just does not apply when a particle as small as an electron passes through a pair of closely spaced narrow slits. Because the wave nature of particles is important in such circumstances, we lose the ability to predict with 100% certainty the path that a single particle will follow. Instead, only the average behavior of large numbers of particles is predictable, and the behavior of any individual particle is uncertain.

To see more clearly into the nature of the uncertainty, consider electrons passing through a single slit, as in **Figure 29.15**. After a sufficient number of electrons strike the screen, a diffraction pattern emerges. The electron diffraction pattern consists of alternating bright and dark fringes and is analogous to the pattern for light waves shown in **Figure 27.23**. **Figure 29.15** shows the slit and locates the first dark fringe on either side of the central bright fringe. The central fringe is bright because electrons strike the

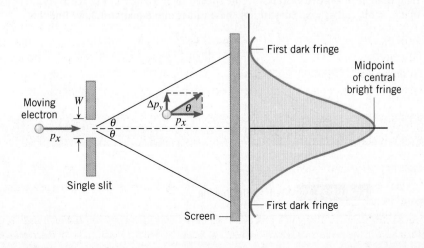

FIGURE 29.15 When a sufficient number of electrons pass through a single slit and strike the screen, a diffraction pattern of bright and dark fringes emerges. (Only the central bright fringe is shown.) This pattern is due to the wave nature of the electrons and is analogous to that produced by light waves.

screen over the entire region between the dark fringes. If the electrons striking the screen outside the central bright fringe are neglected, the extent to which the electrons are diffracted is given by the angle θ in the drawing. To reach locations within the central bright fringe, some electrons must have acquired momentum in the y direction, despite the fact that they enter the slit traveling along the x direction and have no momentum in the y direction to start with. The figure illustrates that the y component of the momentum may be as large as Δp_y. The notation Δp_y indicates the difference between the maximum value of the y component of the momentum after the electron passes through the slit and its value of zero before the electron passes through the slit. Δp_y represents the *uncertainty* in the y component of the momentum, in the sense that the y component may have any value from zero to Δp_y.

It is possible to relate Δp_y to the width W of the slit. To do this, we assume that Equation 27.4, which applies to light waves, also applies to particle waves whose de Broglie wavelength is λ. This equation, $\sin \theta = \lambda/W$, specifies the angle θ that locates the first dark fringe. If θ is small, then $\sin \theta \approx \tan \theta$. Moreover, **Figure 29.15** indicates that $\tan \theta = \Delta p_y/p_x$, where p_x is the x component of the momentum of the electron. Therefore, $\Delta p_y/p_x < \lambda/W$. However, $p_x = h/\lambda$ according to de Broglie's equation, so that

$$\frac{\Delta p_y}{p_x} = \frac{\Delta p_y}{h/\lambda} \approx \frac{\lambda}{W}$$

As a result,

$$\Delta p_y \approx \frac{h}{W} \qquad (29.9)$$

which indicates that a smaller slit width leads to a larger uncertainty in the y component of the electron's momentum.

It was Heisenberg who first suggested that the uncertainty Δp_y in the y component of the momentum is related to the uncertainty in the y position of the electron as the electron passes through the slit. To get a feel for this relationship, let's assume that the center of the slit is at $y = 0$ m. Because the width of the slit is W, the electron is somewhere within $\pm\frac{1}{2}W$ from the center of the slit. Thus, we take the uncertainty

---

Math Skills   To see why $\sin \theta \approx \tan \theta$ when θ is small, refer to **Figure 29.16a** and recall that the sine and tangent functions are defined as follows in Section 1.4:

$$\sin \theta = \frac{h_o}{h} \quad (1.1) \quad \text{and} \quad \tan \theta = \frac{h_o}{h_a} \quad (1.3)$$

where h_o is the length of the side of a right triangle opposite the angle θ, h_a is the length of the side adjacent to the angle θ, and h is the hypotenuse. When θ is small, h and h_a become approximately equal, or $h \approx h_a$, and the right-hand sides of Equations 1.1 and 1.3 become approximately the same.

To see why $\tan \theta = \Delta p_y/p_x$, refer to **Figure 29.16b**, which shows in enlarged form the portion of **Figure 29.15** that establishes the angle θ. A comparison of the shaded right triangle in **Figure 29.16b** with the shaded right triangle in **Figure 29.16a** reveals that $h_o = \Delta p_y$ and $h_a = p_x$. Substituting these values into Equation 1.3, we find that

$$\tan \theta = \frac{h_o}{h_a} = \frac{\Delta p_y}{p_x}$$

(a)

(b)

FIGURE 29.16   Math Skills drawing.

in the y position of the electron to be $\Delta y = \frac{1}{2}W$, so that $W = 2\Delta y$. Substituting this relation into Equation 29.9 shows that $\Delta p_y \approx h/(2\,\Delta y)$ or $(\Delta p_y)(\Delta y) \approx \frac{1}{2}h$. The result of Heisenberg's more complete analysis is given below in Equation 29.10 and is known as the **Heisenberg uncertainty principle**. Note that the Heisenberg uncertainty principle is a general principle with wide applicability. It does not just apply to the case of single-slit diffraction, which we have used here for the sake of convenience.

THE HEISENBERG UNCERTAINTY PRINCIPLE

Momentum and position

$$(\Delta p_y)(\Delta y) \geq \frac{h}{4\pi} \qquad (29.10)$$

Δy = uncertainty in a particle's position along the y direction
Δp_y = uncertainty in the y component of the linear momentum of the particle

Energy and time

$$(\Delta E)(\Delta t) \geq \frac{h}{4\pi} \qquad (29.11)$$

ΔE = uncertainty in the energy of a particle when the particle is in a certain state
Δt = time interval during which the particle is in the state

The Heisenberg uncertainty principle places limits on the accuracy with which the momentum and position of a particle can be specified simultaneously. These limits are not just limits due to faulty measuring techniques. They are fundamental limits imposed by nature, and there is no way to circumvent them. Equation 29.10 indicates that Δp_y and Δy cannot both be arbitrarily small at the same time. If one is small, then the other must be large, so that their product equals or exceeds ($\geq$) Planck's constant divided by 4π. For example, if the position of a particle is known exactly, so that Δy is zero, then Δp_y is an infinitely large number, and the momentum of the particle is completely uncertain. Conversely, if we assume that Δp_y is zero, then Δy is an infinitely large number, and the position of the particle is completely uncertain. In other words, the Heisenberg uncertainty principle states that it is impossible to specify precisely both the momentum and position of a particle at the same time.

There is also an uncertainty principle that deals with energy and time, as expressed by Equation 29.11. The product of the uncertainty ΔE in the energy of a particle and the time interval Δt during which the particle remains in a given energy state is greater than or equal to Planck's constant divided by 4π. Therefore, the shorter the lifetime of a particle in a given state, the greater is the uncertainty in the energy of that state.

Example 7 shows that the uncertainty principle has significant consequences for the motion of tiny particles such as electrons but has little effect on the motion of macroscopic objects, even those with as little mass as a Ping-Pong ball.

EXAMPLE 7 | The Heisenberg Uncertainty Principle

Assume that the position of an object is known so precisely that the uncertainty in the position is only $\Delta y = 1.5 \times 10^{-11}$ m. **(a)** Determine the minimum uncertainty in the momentum of the object. Find the corresponding minimum uncertainty in the speed of the object in the case when the object is **(b)** an electron (mass = 9.1×10^{-31} kg) and **(c)** a Ping-Pong ball (mass = 2.2×10^{-3} kg).

Reasoning The minimum uncertainty Δp_y in the y component of the momentum is given by the Heisenberg uncertainty principle as $\Delta p_y = h/(4\pi\Delta y)$, where Δy is the uncertainty in the position of the object along the y direction. Both the electron and the Ping-Pong ball have the same uncertainty in their momenta because they have the same uncertainty in their positions. However, these objects have very different masses. As a

result, we will find that the uncertainty in the speeds of these objects is very different.

> **Problem-Solving Insight** The Heisenberg uncertainty principle states that the product of Δp_y and Δy is greater than or equal to $h/4\pi$. For a given value of Δp_y or Δy, the minimum uncertainty in the other term occurs when the product is equal to $h/4\pi$.

Solution **(a)** From Equation 29.10, the minimum uncertainty in the y component of the momentum is given by

$$\Delta p_y = \frac{h}{4\pi\Delta y} = \frac{6.63 \times 10^{-34}\,\text{J} \cdot \text{s}}{4\pi(1.5 \times 10^{-11}\,\text{m})} = \boxed{3.5 \times 10^{-24}\,\text{kg} \cdot \text{m/s}}$$

(b) The magnitude p_y of the momentum is $p_y = mv_y$ (Equation 7.2), where m is the mass of the object and v_y is its speed. Therefore, the uncertainty Δp_y is $\Delta p_y = m\Delta v_y$, and the minimum uncertainty in the speed of the electron is

$$\Delta v_y = \frac{\Delta p_y}{m} = \frac{3.5 \times 10^{-24} \text{ kg} \cdot \text{m/s}}{9.1 \times 10^{-31} \text{ kg}} = \boxed{3.8 \times 10^6 \text{ m/s}}$$

Thus, the small uncertainty in the y position of the electron gives rise to a large uncertainty in the speed of the electron.

(c) The minimum uncertainty in the speed of the Ping-Pong ball is

$$\Delta v_y = \frac{\Delta p_y}{m} = \frac{3.5 \times 10^{-24} \text{ kg} \cdot \text{m/s}}{2.2 \times 10^{-3} \text{ kg}} = \boxed{1.6 \times 10^{-21} \text{ m/s}}$$

Because the mass of the Ping-Pong ball is relatively large, the small uncertainty in its y position gives rise to an uncertainty in its speed that is much smaller than that for the electron. Thus, in contrast to the case for the electron, we can know simultaneously where the ball is and how fast it is moving, to a very high degree of certainty.

Related Homework: Problem 36.

Example 7 emphasizes how the uncertainty principle imposes different uncertainties on the speeds of an electron (small mass) and a Ping-Pong ball (large mass). For objects like the ball, which have relatively large masses, the uncertainties in position and speed are so small that they have no effect on our ability to determine simultaneously where such objects are and how fast they are moving. The uncertainties calculated in Example 7 depend on more than just the mass, however. They also depend on Planck's constant, which is a very small number. It is interesting to speculate about what life would be like if Planck's constant were much larger than 6.63×10^{-34} J · s. Conceptual Example 8 deals with just such speculation.

CONCEPTUAL EXAMPLE 8 | What If Planck's Constant Were Large?

Suppose that you are target shooting at a stationary target. A bullet leaving the barrel of a gun is analogous to an electron passing through the single slit in **Figure 29.15**. With this analogy in mind and assuming that the magnitude of the bullet's momentum is not abnormally large, what would target shooting be like if Planck's constant had a relatively large value instead of its extremely small value of 6.63×10^{-34} J · s? **(a)** It would be more accurate because there would be less uncertainty in where the bullet hits the target. **(b)** It would be less accurate because there would be greater uncertainty in where the bullet hits the target. **(c)** There would be no difference.

Reasoning   Let's assume that the bullet is moving down the barrel of the gun in the $+x$ direction and that the target lies on the x axis. When it exits the barrel, the bullet—like the electron passing through a single slit—acquires a momentum component that is perpendicular (in the y direction) to the barrel. This happens even though inside the barrel the bullet travels only along the x direction and has no momentum component in the y direction. Analogous to the discussion relating to **Figure 29.15**, the y component of the momentum may be as large as Δp_y, where Δp_y indicates the difference between the maximum value of the y component of the momentum after the bullet leaves the barrel and its value of zero while the bullet is in the barrel. Δp_y is related to

Planck's constant h and the diameter W of the barrel opening via the relation $\Delta p_y \approx h/W$ (Equation 29.9). Since we are now postulating that Planck's constant is large, Δp_y is also large.

Answers (a) and (c) are incorrect.   Target shooting becomes more accurate if Planck's constant becomes smaller, not larger. Here's the reason: Inside the barrel the bullet is moving in the $+x$ direction. However, upon exiting the barrel, the bullet acquires a momentum component Δp_y in the y direction and begins to deviate from its original path by moving in the y direction. According to $\Delta p_y \approx h/W$ (Equation 29.9), the smaller the value of h, the smaller is Δp_y. If, in the extreme limit, Planck's constant were zero, Δp_y would also be zero, and the bullet would move only in the $+x$ direction and, thus, would hit the target.

Answer (b) is correct.   If the bullet, after leaving the barrel, had only a momentum component that was parallel to the barrel, the bullet would strike the target. However, upon leaving the barrel, the bullet also acquires a momentum component Δp_y that is perpendicular to the barrel. The relation $\Delta p_y \approx h/W$ (Equation 29.9) shows that the larger the value of Planck's constant h, the larger is the value of Δp_y. Since this momentum component is perpendicular to the barrel itself, the bullet can strike at locations other than the target. Thus, target shooting would be less accurate if Planck's constant had a relatively large value.

EXAMPLE 9 | BIO Breathing in Electrons

Our lungs contain approximately 500 million alveoli, which are the small, hollow chambers lined with a membrane that allows the exchange of oxygen and carbon dioxide in our blood. Consider the situation of an electron located in one of these sacks and **(a)** calculate the uncertainty in its momentum using the Heisenberg uncertainty principle. For the uncertainty in the electron's position, use the diameter of the alveoli, which is approximately 0.05 mm. Next,

assume that at some instant in time, the momentum of the electron is equal to the uncertainty in its momentum and **(b)** calculate the kinetic energy of the electron.

Reasoning   We simply apply the Heisenberg uncertainty principle for position and momentum (Equation 29.10) to calculate the uncertainty in the electron's momentum. We then assume this value to be equal to the electron's momentum and calculate its kinetic energy using Equation 6.2.

Solution (a) From Equation 29.10, we have:

$$\Delta p_y = \frac{h}{(4\pi)(\Delta y)} = \frac{6.63 \times 10^{-34} \text{ J} \cdot \text{s}}{4\pi(0.05 \times 10^{-3} \text{ m})} = \boxed{1.1 \times 10^{-30} \text{ kg} \cdot \text{m/s}}$$

(b) Now assume that $p = \Delta p_y = 1.1 \times 10^{-30}$ kg $\cdot$ m/s and calculate the electron's kinetic energy:

$$KE = \tfrac{1}{2}mv^2 = \frac{p^2}{2m} = \frac{(1.1 \times 10^{-30} \text{ kg} \cdot \text{m/s})^2}{2(9.11 \times 10^{-31} \text{ kg})} = \boxed{6.1 \times 10^{-31} \text{ J}}$$

While this is an incredibly small kinetic energy, the speed of the electron is still ≈ 1 m/s.

Concept Summary

29.1 The Wave–Particle Duality, 29.2 Blackbody Radiation and Planck's Constant The wave–particle duality refers to the fact that a wave can exhibit particle-like characteristics and a particle can exhibit wave-like characteristics.

At a constant temperature, a perfect blackbody absorbs and reemits all the electromagnetic radiation that falls on it. Max Planck calculated the emitted radiation intensity per unit wavelength as a function of wavelength. In his theory, Planck assumed that a black-body consists of atomic oscillators that can have only discrete, or quantized, energies. Planck's quantized energies are given by Equation 29.1, where h is Planck's constant (6.63×10^{-34} J $\cdot$ s) and f is the vibration frequency of an oscillator.

$$E = nhf \quad n = 0, 1, 2, 3, \ldots \tag{29.1}$$

29.3 Photons and the Photoelectric Effect All electromagnetic radiation consists of photons, which are packets of energy. The energy of a photon is given by Equation 29.2, where h is Planck's constant and f is the frequency of the photon. A photon has no mass and always travels at the speed of light c in a vacuum.

$$E = hf \tag{29.2}$$

The photoelectric effect is the phenomenon in which light shining on a metal surface causes electrons to be ejected from the surface. The work function W_0 of a metal is the minimum work that must be done to eject an electron from the metal. In accordance with the conservation of energy, the electrons ejected from a metal have a maximum kinetic energy KE_{max} that is related to the energy hf of the incident photon and the work function of the metal by Equation 29.3.

$$hf = KE_{max} + W_0 \tag{29.3}$$

29.4 The Momentum of a Photon and the Compton Effect The magnitude p of a photon's momentum is given by Equation 29.6, where h is Planck's constant and λ is the wavelength of the photon.

$$p = \frac{h}{\lambda} \tag{29.6}$$

The Compton effect is the scattering of a photon by an electron in a material, the scattered photon having a smaller frequency and, hence, a smaller energy than the incident photon. Part of the photon's energy and momentum are transferred to the recoiling electron. The difference between the wavelength λ' of the scattered photon and the wavelength λ of the incident photon is related by Equation 29.7 to the scattering angle θ (see **Animated Figure 29.11**), where m is the mass of the electron. The quantity $h/(mc)$ is known as the Compton wavelength of the electron.

$$\lambda' - \lambda = \frac{h}{mc}(1 - \cos \theta) \tag{29.7}$$

29.5 The De Broglie Wavelength and the Wave Nature of Matter The de Broglie wavelength λ of a particle is given by Equation 29.8, where h is Planck's constant and p is the magnitude of the relativistic momentum of the particle. Because of its wavelength, a particle can exhibit wave-like characteristics. The wave associated with a particle is a wave of probability.

$$\lambda = \frac{h}{p} \tag{29.8}$$

29.6 The Heisenberg Uncertainty Principle The Heisenberg uncertainty principle places limits on our knowledge about the behavior of a particle. The uncertainty principle is stated according to Equation 29.10, where Δy is the uncertainty in the particle's position along the y direction, and Δp_y is the uncertainty in the y component of the linear momentum of the particle.

$$(\Delta p_y)(\Delta y) \geq \frac{h}{4\pi} \tag{29.10}$$

The uncertainty principle can also be stated according to Equation 29.11, where ΔE is the uncertainty in the energy of a particle when it is in a certain state, and Δt is the time interval during which the particle is in the state.

$$(\Delta E)(\Delta t) \geq \frac{h}{4\pi} \tag{29.11}$$

Focus on Concepts

Online

Additional questions are available for assignment in WileyPLUS.

Section 29.2   Blackbody Radiation and Planck's Constant

1. An astronomer is measuring the electromagnetic radiation emitted by two stars, which are both assumed to be perfect blackbody emitters.

For each star she makes a plot of the radiation intensity per unit wavelength as a function of wavelength. She notices that the curve for star A has a maximum that occurs at a shorter wavelength than does the curve for star B. What can she conclude about the surface temperatures of the two stars? (a) Star A has the greater surface temperature. (b) Star B has the greater surface temperature. (c) Both stars, being perfect blackbody emitters, have the same surface temperature.

(d) There is not enough information to draw a conclusion about the temperatures.

Section 29.3 Photons and the Photoelectric Effect

2. Photons are generated by a microwave oven in a kitchen and by an X-ray machine at a dentist's office. Which type of photon has the greater frequency, the greater energy, and the greater wavelength?

	Greater Frequency	Greater Energy	Greater Wavelength
(a)	X-ray	X-ray	X-ray
(b)	X-ray	X-ray	Microwave
(c)	Microwave	X-ray	Microwave
(d)	Microwave	Microwave	X-ray
(e)	X-ray	Microwave	Microwave

3. A laser emits a beam of light whose photons all have the same frequency. When the beam strikes the surface of a metal, photoelectrons are ejected from the surface. What happens if the laser emits twice the number of photons per second? **(a)** The photoelectrons are ejected from the surface with twice the maximum kinetic energy. **(b)** The photoelectrons are ejected from the surface with the same maximum kinetic energy. **(c)** The number of photoelectrons ejected per second from the surface doubles. **(d)** Both b and c happen. **(e)** Both a and c happen.

4. The surface of a metal plate is illuminated with light of a certain frequency. Which of the following conditions determine whether or not photoelectrons are ejected from the metal?

1. The number of photons per second emitted by the light source
2. The length of time that the light is turned on
3. The thermal conductivity of the metal
4. The surface area of the metal illuminated by the light
5. The type of metal from which the plate is made

(a) 1 and 2 **(b)** 5 **(c)** 3 and 5 **(d)** 4 **(e)** 2 and 3

Section 29.4 The Momentum of a Photon and the Compton Effect

5. Does a photon, like a moving particle such as an electron, have a momentum? **(a)** No, because a photon is a wave, and a wave does not have a momentum. **(b)** No, because a photon has no mass, and mass is necessary in order to have a momentum. **(c)** No, because a photon, always traveling at the speed of light in a vacuum, would have an infinite momentum. **(d)** Yes, and the magnitude p of the photon's momentum is related to its wavelength λ by $p = h/\lambda$, where h is Planck's constant. **(e)** Yes, and the magnitude p of the photon's momentum is related to its wavelength λ by $p = h\lambda$, where h is Planck's constant.

Section 29.5 The De Broglie Wavelength and the Wave Nature of Matter

6. Two particles, A and B, have the same mass, but particle A has a charge of $+q$ and B has a charge of $+2q$. The particles are accelerated from rest through the same potential difference. Which one has the longer de Broglie wavelength at the end of the acceleration? **(a)** Particle A, because it has the greater momentum, and, hence, the longer de Broglie wavelength **(b)** Particle B, because it has the greater momentum, and, hence, the longer de Broglie wavelength **(c)** Particle A, because it has the smaller momentum, and, hence, the longer de Broglie wavelength **(d)** Particle B, because it has the smaller momentum, and, hence, the longer de Broglie wavelength **(e)** Both particles have the same de Broglie wavelength.

Section 29.6 The Heisenberg Uncertainty Principle

7. Suppose that the momentum of an electron is measured with complete accuracy (i.e., the uncertainty in its momentum is zero). The uncertainty in a simultaneous measurement of the electron's position _____. **(a)** is also zero **(b)** is infinitely large **(c)** is some finite value between zero and infinity **(d)** cannot be measured, because one cannot measure the position and momentum of a particle, such as an electron, simultaneously

---

Problems

Online

Additional questions are available for assignment in WileyPLUS.

Note: *In these problems, ignore relativistic effects unless instructed otherwise, and assume that wavelengths are in a vacuum unless otherwise specified.*

SSM Student Solutions Manual	**BIO** Biomedical application
MMH Problem-solving help	**E** Easy
GO Guided Online Tutorial	**M** Medium
V-HINT Video Hints	**H** Hard
CHALK Chalkboard Videos	**WS** Worksheet
	T Team Problem

Section 29.3 Photons and the Photoelectric Effect

1. **E** The dissociation energy of a molecule is the energy required to break the molecule apart into its separate atoms. The dissociation energy for the cyanogen molecule is 1.22×10^{-18} J. Suppose that this energy is provided by a single photon. Determine the **(a)** wavelength and **(b)** frequency of the photon. **(c)** In what region of the electromagnetic spectrum (see Figure 24.9) does this photon lie?

2. **E** An AM radio station broadcasts an electromagnetic wave with a frequency of 665 kHz, whereas an FM station broadcasts an electromagnetic wave with a frequency of 91.9 MHz. How many AM photons are needed to have a total energy equal to that of one FM photon?

3. **E GO SSM** Ultraviolet light with a frequency of 3.00×10^{15} Hz strikes a metal surface and ejects electrons that have a maximum kinetic energy of 6.1 eV. What is the work function (in eV) of the metal?

4. **E GO** Light is shining perpendicularly on the surface of the earth with an intensity of 680 W/m^2. Assuming that all the photons in the light have the same wavelength (in vacuum) of 730 nm, determine the number of photons per second per square meter that reach the earth.

5. **E** **SSM** Ultraviolet light is responsible for sun tanning. Find the wavelength (in nm) of an ultraviolet photon whose energy is 6.4×10^{-19} J.

6. **E** The maximum wavelength that an electromagnetic wave can have and still eject electrons from a metal surface is 485 nm. What is the work function W_0 of this metal? Express your answer in electron volts.

7. **E** Radiation of a certain wavelength causes electrons with a maximum kinetic energy of 0.68 eV to be ejected from a metal whose work function is 2.75 eV. What will be the maximum kinetic energy (in eV) with which this same radiation ejects electrons from another metal whose work function is 2.17 eV?

8. **E** **GO** Multiple-Concept Example 4 reviews the concepts necessary to solve this problem. Radiation with a wavelength of 238 nm shines on a metal surface and ejects electrons that have a maximum speed of 3.75×10^5 m/s. Which one of the following metals is it, the values in parentheses being the work functions: potassium (2.24 eV), calcium (2.71 eV), uranium (3.63 eV), aluminum (4.08 eV), or gold (4.82 eV)?

9. **M** **SSM** **MMH** An owl has good night vision because its eyes can detect a light intensity as small as 5.0×10^{-13} W/m². What is the minimum number of photons per second that an owl eye can detect if its pupil has a diameter of 8.5 mm and the light has a wavelength of 510 nm?

10. **M** **V-HINT** A proton is located at a distance of 0.420 m from a point charge of $+8.30$ μC. The repulsive electric force moves the proton until it is at a distance of 1.58 m from the charge. Suppose that the electric potential energy lost by the system were carried off by a photon. What would be its wavelength?

11. **M** **CHALK** When light with a wavelength of 221 nm is incident on a certain metal surface, electrons are ejected with a maximum kinetic energy of 3.28×10^{-19} J. Determine the wavelength (in nm) of light that should be used to double the maximum kinetic energy of the electrons ejected from this surface.

12. **M** **GO** A glass plate has a mass of 0.50 kg and a specific heat capacity of 840 J/(kg · C°). The wavelength of infrared light is 6.0×10^{-5} m, while the wavelength of blue light is 4.7×10^{-7} m. Find the number of infrared photons and the number of blue photons needed to raise the temperature of the glass plate by 2.0 C°, assuming that all the photons are absorbed by the glass.

13. **H** **SSM** A laser emits 1.30×10^{18} photons per second in a beam of light that has a diameter of 2.00 mm and a wavelength of 514.5 nm. Determine **(a)** the average electric field strength and **(b)** the average magnetic field strength for the electromagnetic wave that constitutes the beam.

Section 29.4 The Momentum of a Photon and the Compton Effect

14. **E** A light source emits a beam of photons, each of which has a momentum of 2.3×10^{-29} kg · m/s. **(a)** What is the frequency of the photons? **(b)** To what region of the electromagnetic spectrum do the photons belong? Consult Figure 24.9 if necessary.

15. **E** A photon of red light (wavelength = 720 nm) and a Ping-Pong ball (mass = 2.2×10^{-3} kg) have the same momentum. At what speed is the ball moving?

16. **E** **CHALK** **SSM** In a Compton scattering experiment, the incident X-rays have a wavelength of 0.2685 nm, and the scattered X-rays have a wavelength of 0.2703 nm. Through what angle θ in **Animated Figure 29.11** are the X-rays scattered?

17. **E** **V-HINT** A sample is bombarded by incident X-rays, and free electrons in the sample scatter some of the X-rays at an angle of $\theta = 122.0°$ with respect to the incident X-rays (see **Animated Figure 29.11**). The scattered X-rays have a momentum whose magnitude is 1.856×10^{-24} kg · m/s. Determine the wavelength (in nm) of the incident X-rays. (For accuracy, use $h = 6.626 \times 10^{-34}$ J · s, $c = 2.998 \times 10^8$ m/s, and $m = 9.109 \times 10^{-31}$ kg for the mass of an electron.)

18. **E** **MMH** An incident X-ray photon of wavelength 0.2750 nm is scattered from an electron that is initially at rest. The photon is scattered at an angle of $\theta = 180.0°$ in **Animated Figure 29.11** and has a wavelength of 0.2825 nm. Use the conservation of linear momentum to find the momentum gained by the electron.

19. **E** **GO** In the Compton effect, momentum conservation applies, so the total momentum of the photon and the electron is the same before and after the scattering occurs. Suppose that in **Animated Figure 29.11** the incident photon moves in the $+x$ direction and the scattered photon emerges at an angle of $\theta = 90.0°$, which is in the $-y$ direction. The incident photon has a wavelength of 9.00×10^{-12} m. Find the x and y components of the momentum of the scattered electron.

20. **M** **SSM** What is the maximum amount by which the wavelength of an incident photon could change when it undergoes Compton scattering from a nitrogen molecule (N_2)?

21. **M** **GO** **Animated Figure 29.11** shows the setup for measuring the Compton effect. With a fixed incident wavelength, a wavelength of λ'_1 is measured for a scattering angle of $\theta_1 = 30.0°$, whereas a wavelength of λ'_2 is measured for a scattering angle of $\theta_2 = 70.0°$. Find the difference in wavelengths, $\lambda'_2 - \lambda'_1$.

22. **M** **GO** A photon of wavelength 0.45000 nm strikes a free electron that is initially at rest. The photon is scattered straight backward. What is the speed of the recoil electron after the collision?

Section 29.5 The De Broglie Wavelength and the Wave Nature of Matter

23. **E** A bacterium (mass = 2×10^{-15} kg) in the blood is moving at 0.33 m/s. What is the de Broglie wavelength of this bacterium?

24. **E** What are **(a)** the wavelength of a 5.0-eV photon and **(b)** the de Broglie wavelength of a 5.0-eV electron?

25. **E** An electron and a proton have the same speed. Ignore relativistic effects and determine the ratio $\lambda_{electron}/\lambda_{proton}$ of their de Broglie wavelengths.

26. **E** Recall from Section 14.3 that the average kinetic energy of an atom in a monatomic ideal gas is given by $\overline{KE} = \frac{3}{2}kT$, where $k = 1.38 \times 10^{-23}$ J/K and T is the Kelvin temperature of the gas. Determine the de Broglie wavelength of a helium atom (mass = 6.65×10^{-27} kg) that has the average kinetic energy at room temperature (293 K).

27. **E** **GO** In a Young's double-slit experiment that uses electrons, the angle that locates the first-order bright fringes is $\theta_A = 1.6 \times 10^{-4}$ degrees when the magnitude of the electron momentum is $p_A = 1.2 \times 10^{-22}$ kg · m/s. With the same double slit, what momentum magnitude p_B is necessary so that an angle of $\theta_B = 4.0 \times 10^{-4}$ degrees locates the first-order bright fringes?

28. **M** **SSM** A particle has a de Broglie wavelength of 2.7×10^{-10} m. Then its kinetic energy doubles. What is the particle's new de Broglie wavelength, assuming that relativistic effects can be ignored?

29. **M** **GO** From a cliff that is 9.5 m above a lake, a young woman (mass = 41 kg) jumps from rest, straight down into the water. At the instant she strikes the water, what is her de Broglie wavelength?

30. **M** **V-HINT** The width of the central bright fringe in a diffraction pattern on a screen is identical when either electrons or red light (vacuum wavelength = 661 nm) pass through a single slit. The distance between the screen and the slit is the same in each case and is large compared to the slit width. How fast are the electrons moving?

31. **M** **GO** Particle A is at rest, and particle B collides head-on with it. The collision is completely inelastic, so the two particles stick together after the collision and move off with a common velocity. The masses of the particles are different, and no external forces act on them. The de Broglie wavelength of particle B before the collision is 2.0×10^{-34} m. What is the de Broglie wavelength of the object that moves off after the collision?

32. **M** **SSM** An electron, starting from rest, accelerates through a potential difference of 418 V. What is the final de Broglie wavelength of the electron, assuming that its final speed is much less than the speed of light?

33. **H** The kinetic energy of a particle is equal to the energy of a photon. The particle moves at 5.0% of the speed of light. Find the ratio of the photon wavelength to the de Broglie wavelength of the particle.

Section 29.6 The Heisenberg Uncertainty Principle

34. **E** **SSM** An object is moving along a straight line, and the uncertainty in its position is 2.5 m. **(a)** Find the minimum uncertainty in the momentum of the object. Find the minimum uncertainty in the object's velocity, assuming that the object is **(b)** a golf ball (mass = 0.045 kg) and **(c)** an electron.

35. **E** **CHALK** A proton is confined to a nucleus that has a diameter of 5.5×10^{-15} m. If this distance is considered to be the uncertainty in the position of the proton, what is the minimum uncertainty in its momentum?

36. **E** **SSM** For insight into this problem, review Example 9. In the lungs there are tiny sacs of air, which are called alveoli. An oxygen molecule (mass = 5.3×10^{-26} kg) is trapped within a sac, and the uncertainty in its position is 0.12 mm. What is the minimum uncertainty in the speed of this oxygen molecule?

37. **E** **GO** Particles pass through a single slit of width 0.200 mm (see **Figure 29.15**). The de Broglie wavelength of each particle is 633 nm. After the particles pass through the slit, they spread out over a range of angles. Assume that the uncertainty in the position of the particles is one-half the width of the slit, and use the Heisenberg uncertainty principle to determine the minimum range of angles.

38. **M** **GO** The minimum uncertainty Δy in the position y of a particle is equal to its de Broglie wavelength. Determine the minimum uncertainty in the speed of the particle, where this minimum uncertainty Δv_y is expressed as a percentage of the particle's speed v_y $\left(\text{Percentage} = \dfrac{\Delta v_y}{v_y} \times 100\%\right)$. Assume that relativistic effects can be ignored.

39. **M** **V-HINT** A subatomic particle created in an experiment exists in a certain state for a time of $\Delta t = 7.4 \times 10^{-20}$ s before decaying into other particles. Apply both the Heisenberg uncertainty principle and the equivalence of energy and mass (see Section 28.6) to determine the minimum uncertainty involved in measuring the mass of this short-lived particle.

Additional Problems

Online

40. **E** The interatomic spacing in a crystal of table salt is 0.282 nm. This crystal is being studied in a neutron diffraction experiment, similar to the one that produced the photograph in **Figure 29.13a**. How fast must a neutron (mass = 1.67×10^{-27} kg) be moving to have a de Broglie wavelength of 0.282 nm?

41. **E** **SSM** The de Broglie wavelength of a proton in a particle accelerator is 1.30×10^{-14} m. Determine the kinetic energy (in joules) of the proton.

42. **E** **V-HINT** Find the de Broglie wavelength of an electron with a speed of 0.88c. Take relativistic effects into account.

43. **E** **GO** The work function of a metal surface is 4.80×10^{-19} J. The maximum speed of the electrons emitted from the surface is $v_A = 7.30 \times 10^5$ m/s when the wavelength of the light is λ_A. However, a maximum speed of $v_B = 5.00 \times 10^5$ m/s is observed when the wavelength is λ_B. Find the wavelengths λ_A and λ_B.

44. **E** **GO** How fast does a proton have to be moving in order to have the same de Broglie wavelength as an electron that is moving with a speed of 4.50×10^6 m/s?

Physics in Biology, Medicine, and Sports

45. **E** **BIO** 29.3 Review Example 2 as an aid in understanding this problem. While ultraviolet light from the UV-B band has enough energy to cause sunburns and damage DNA, light from the shorter wavelength UV-C band has an even greater energy. It is so effective at damaging base pairs in DNA that it is utilized in germicidal lamps. Some of these lamps can render inactive 99.99% of bacteria and viruses, so they are used to sterilize workplaces and common objects. The popularity of the lamps surged in 2020 due to the COVID-19 pandemic. The lamps proved effective at deactivating the SARS-CoV-2 coronavirus, assuming that the exposure was

long enough. If the energy of the photons produced by the lamp is 4.88 eV, what is the wavelength of the light?

46. **E** **BIO** 29.3 Vitamin D is an essential fat-soluble secosteroid that supports many functions in the human body. It facilitates the absorption of calcium in the digestive tract, which leads to strong teeth and bones. It strengthens the immune system and plays an important role in the proper functioning of the body's major organs, such as the muscles, heart, lungs, and brain. The major source of vitamin D is a photochemical reaction between UV-B sunlight and a compound in the skin called

7-dehydrocholesterol. This compound exists in most vertebrate animals, and it reacts with UV-B light at wavelengths between 290 and 315 nm. What is the range of energies of UV-B light that reacts with 7-dehydrocholesterol? Give your answers in eV.

47. **M** **BIO** 29.3   If after reviewing Example 2 you still feel like visiting a tanning salon, the use of safety eyewear cannot be overstated. Tanning beds produce mostly UV-A and some UV-B light. Even with your eyes closed, the thin skin of your eyelids does not provide sufficient protection from the ultraviolet light. The photons in the UV-B light deposit their energy in the cornea of the eye, causing damage and an acute condition known as *photokeratitis*. This can also be caused by the flash of a welder's arc or the reflection of sunlight from snow, which can occur after a long day of downhill skiing in bright sunlight without proper eye protection, in which case the photokeratitis is referred to as snow blindness. Symptoms include intense tearing, pain, and the feeling of having sand in the eyes (see the photo). (a) Consider a person in a tanning bed whose eyes are exposed to UV-B light with an intensity of 10.0 W/m^2. If we treat their cornea as an opening with a circular cross section and radius of 5.75 mm, how much energy enters their cornea in 20 minutes? (b) If the wavelength of the UV light is 300 nm, how many photons are incident on the cornea during this time?

48. **E** **BIO** 29.3   One of the primary compounds utilized in many sunscreens is *homosalate*. Its chemical formula is $C_{16}H_{22}O_3$, and it absorbs UV light with wavelengths between 295 and 315 nm. Part of the homosalate compound is salicylic acid, which has a strong absorption at 298 nm. What is the photon energy that is absorbed by the salicylic acid?

49. **M** **BIO** 29.2   The curves in **Figure 29.2** follow a relationship known as *Wein's displacement law*. It states that the peak in the radiation intensity per unit wavelength is inversely proportional to the temperature of the blackbody, and it is given by the following:

$$\lambda_{\text{peak}} = \frac{b}{T}$$

where b is the Wein displacement constant, which has a value of $2.90 \times 10^{-3} \text{ m} \cdot \text{K}$.

A glowworm (see the photo) creates light through the process of *bioluminescence*, which is the result of a chemical reaction with a light-emitting molecule called a luciferin. Assume the light emitted by the glowworm can be treated as coming from a blackbody. (a) Use Wein's displacement law to calculate the temperature of the glowworm, assuming the intensity of the light it emits peaks at a frequency of 5.94×10^{14} Hz. (b) Based on your answer in part (a), is the blackbody approximation a good one? Explain.

PROBLEM 47

PROBLEM 49

Concepts and Calculations Problems   Online

In the photoelectric effect, electrons can be emitted from a metal surface when light shines on it. Einstein explained this phenomenon in terms of photons and the conservation of energy. Our discussion of this topic in Section 29.3 deals with a single wavelength. Problem 50, in contrast, considers a mixture of wavelengths and reviews the basic concepts that come into play in this important phenomenon. In this chapter we have seen that moving particles of matter not only possess kinetic energy and momentum but also are characterized by a wavelength that is known as the de Broglie wavelength. Problem 51 reviews kinetic energy and momentum, and focuses on the relation between these fundamental quantities and the de Broglie wavelength.

50. **M** **CHALK** Sunlight, whose visible wavelengths range from 380 to 750 nm, is incident on a sodium surface. The work function for sodium is $W_0 = 2.28$ eV. *Concepts:* (i) Will electrons with a greater value of KE_{max} be emitted when the incident photons have a relatively greater or relatively smaller amount of energy? (ii) In the range of visible wavelengths, which wavelength corresponds to incident photons that carry the greatest energy? (iii) What is the smallest value of KE_{max} with which an electron can be ejected from the sodium? *Calculations:* Find the maximum kinetic energy KE_{max} (in joules) of the photoelectrons emitted from the surface and the range of wavelengths that will cause photoelectrons to be emitted.

51. **M** **CHALK** **SSM** An electron and a proton have the same kinetic energy and are moving at speeds much less than the speed of light. *Concepts:* (i) How is the de Broglie wavelength λ related to the magnitude p of the momentum? (ii) How is the magnitude of the momentum related to the kinetic energy of a particle of mass m that is moving at a speed that is much less than the speed of light? (iii) Which has the greater de Broglie wavelength, the electron or the proton? *Calculations:* Determine the ratio of the de Broglie wavelength of the electron to that of the proton.

Team Problems

52. M T WS **A Laser Heater.** You and your team have been tasked with finding a method to heat an object located inside a vacuum chamber. The problem is that there is no way to feed electrical wires into the vessel; thus, an electrical heater is out of the question. There is, however, a window through which infrared (IR) laser light ($\lambda = 715$ nm) can pass into the chamber where the object, a 2.20-g aluminum cube, is located. The cube has a small hole that leads to an internal cavity so that all light that enters the hole is absorbed and converted into thermal energy of the cube. **(a)** If the diameter of the laser beam is $D = 1.20$ mm (smaller than the hole in the cube), what must be the intensity of the beam such that the temperature of the aluminum cube increases by 1.00 K in 5.00 s? You can neglect the effects of any heat loss mechanisms. The specific heat capacity of aluminum is 9.00×10^2 J/(kg · C°) **(b)** How many photons are delivered during this time? **(c)** Suppose you instead used ultraviolet light ($\lambda = 248$ nm) rather than infrared, but the number of photons absorbed per unit time was the same as that of the infrared light. How long would it take to increase the temperature of the cube by 1.00 K in this case?

53. M T WS **A Light Thruster.** You and your team are given the task of estimating the force provided by a hypothetical "laser thruster." The idea is to use a powerful laser to expel photons out the back of a spacecraft in order to generate a reactive force that propels the ship in the opposite direction. Suppose you intend to use an enormous green laser of wavelength 520.0 nm and a beam with circular cross section of diameter $D = 0.330$ m, which will be mounted on the back of the spaceship. **(a)** What power output of the "laser thruster" would provide a thrust of 1.00 N? **(b)** What is the intensity of the laser beam? **(c)** How much energy is carried away by the laser beam in 5.00 s?

54. M T **Identifying Unknown Metals.** You and your team are tasked with sorting a hundred metallic samples composed of pure silver (Ag), nickel (Ni), indium (In), or cobalt (Co). The problem is that they all look the same (metallic) and are not labeled. The work functions (W_0) of the materials are: Ag (4.50 eV), Ni (5.20 eV), In (4.10 eV), and Co (4.90 eV). You have a monochromatic light source that can be tuned to output wavelengths from 250 nm to 400 nm. You get the idea to use it to measure the threshold frequencies of the samples and, therefore, to identify the metals. **(a)** Calculate the threshold frequencies and wavelengths of all of the metals. **(b)** Can the light source reach the threshold frequencies of all of the metals? If not, can it still be used to sort the samples?

55. M T **Electron Diffraction.** You and your team have set up an experiment where a beam of electrons is accelerated from rest through a potential difference (V), and then passes through a crystalline material that effectively acts as a set of slits separated by 9.62×10^{-11} m. **(a)** If the first order ($m = 1$) bright fringe is located at $\theta = 12.0°$ relative to the initial direction of the electron beam, what is the de Broglie wavelength of the electrons? **(b)** What is the potential difference through which the electrons are accelerated?

Computed axial tomography, or CAT, scanning is an important noninvasive technique that utilizes X-rays to provide image "slices" of the interior of the human body. This colored 3D CAT scan of an adult jaw and part of the skull illustrates the level of detail achievable today. A computer with suitable imaging software assembles the slices into such 3D images. Surgeons can even navigate around the body using rapid animations of CAT data. The production of X-rays is related to the structure of the atom, and that structure is the main topic of this chapter.

Antoine Rosset/Science Source

The Nature of the Atom

30.1 Rutherford Scattering and the Nuclear Atom

An atom contains a small, positively charged nucleus (radius $\approx 10^{-15}$ m), which is surrounded at relatively large distances (radius $\approx 10^{-10}$ m) by a number of electrons, as **Figure 30.1** (not to scale) illustrates. In the natural state, an atom is electrically neutral because the nucleus contains a number of protons (each with a charge of $+e$) that equals the number of electrons (each with a charge of $-e$). This model of the atom is universally accepted now and is referred to as the **nuclear atom**.

The nuclear atom is a relatively recent idea. In the early part of the twentieth century a widely accepted model, developed by the English physicist Joseph J. Thomson (1856–1940), pictured the atom very differently. In Thomson's view there was no nucleus at the center of an atom. Instead, the positive charge was assumed to be spread throughout

LEARNING OBJECTIVES

After reading this module, you should be able to...

30.1 Describe the nuclear atom.

30.2 Calculate the wavelengths of hydrogen line spectra.

30.3 Use the Bohr model to predict the energy levels in hydrogen.

30.4 Explain Bohr's assumption regarding angular momentum.

30.5 Apply quantum mechanics to electron energy levels in hydrogen.

30.6 Use the Pauli exclusion principle to explain the periodic table.

30.7 Apply the Bohr model to X-ray production in atoms.

30.8 Explain how a laser operates.

30.9 Describe laser applications in medicine.

30.10 Explain how holographic images are made and viewed.

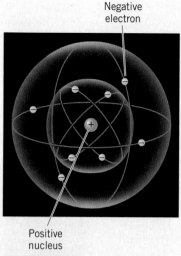

Negative electron

Positive nucleus

FIGURE 30.1 In the nuclear atom a small positively charged nucleus is surrounded at relatively large distances by a number of electrons. Drawing is not to scale.

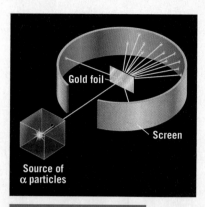

Gold foil

Screen

Source of α particles

INTERACTIVE FIGURE 30.2 A Rutherford scattering experiment in which α particles are scattered by a thin gold foil. The entire apparatus is located within a vacuum chamber (not shown).

the atom, forming a kind of paste or pudding, in which the negative electrons were suspended like plums.

The "plum-pudding" model was discredited in 1911 when the New Zealand physicist Ernest Rutherford (1871–1937) published experimental results that the model could not explain. As **Interactive Figure 30.2** indicates, Rutherford and his co-workers directed a beam of alpha particles (α particles) at a thin metal foil made of gold. Alpha particles are positively charged particles (the nuclei of helium atoms, although this was not recognized at the time) emitted by some radioactive materials. If the plum-pudding model were correct, the α particles would be expected to pass nearly straight through the foil. After all, there is nothing in this model to deflect the relatively massive α particles, since the electrons have a comparatively small mass and the positive charge is spread out in a "diluted" pudding. Using a zinc sulfide screen, which flashed briefly when struck by an α particle, Rutherford and co-workers were able to determine that not all the α particles passed straight through the foil. Instead, some were deflected at large angles, even backward. Rutherford himself said, "It was almost as incredible as if you had fired a fifteen-inch shell at a piece of tissue and it came back and hit you." Rutherford concluded that the positive charge, instead of being distributed thinly and uniformly throughout the atom, was concentrated in a small region called the nucleus.

But how could the electrons in a nuclear atom remain separated from the positively charged nucleus? If the electrons were stationary, they would be pulled inward by the attractive electric force of the nuclear charge. Therefore, the electrons must be moving around the nucleus in some fashion, like planets in orbit around the sun. In fact, the nuclear model of the atom is sometimes referred to as the "planetary" model. The dimensions of the atom, however, are such that it contains a larger fraction of empty space than our solar system does, as Conceptual Example 1 discusses.

CONCEPTUAL EXAMPLE 1 | Are Atoms Mostly Empty Space?

In the planetary model of the atom, the radius of the nucleus ($\approx 1 \times 10^{-15}$ m) is analogous to the radius of the sun ($\approx 7 \times 10^{8}$ m). The electrons orbit the nucleus at a radial distance ($\approx 1 \times 10^{-10}$ m) that is analogous to the radial distance ($\approx 1.5 \times 10^{11}$ m) at which the earth orbits the sun. Suppose that the dimensions of the sun and the earth's orbit had the same proportions as those of an atomic nucleus and an electron's orbit. What then would be true about the distance between the earth and the sun? **(a)** It would be much greater than it actually is. **(b)** It would be much smaller than it actually is. **(c)** It would be roughly the same as it actually is.

Reasoning The radius of an electron orbit is one hundred thousand times larger than the radius of the nucleus: $(1 \times 10^{-10}$ m$)/(1 \times 10^{-15}$ m$) = 10^5$. Using this factor with the radius of the sun will reveal the correct answer.

Answers (b) and (c) are incorrect. Suppose that the earth's orbital radius about the sun were indeed 10^5 times the sun's radius. The distance between the earth and the sun, then, would be $10^5 \times (7 \times 10^8$ m$) = 7 \times 10^{13}$ m, which is neither smaller than nor roughly the same as the actual distance.

Answer (a) is correct. If the earth's orbital radius about the sun were 10^5 times the sun's radius, the distance between the earth and the sun would be $10^5 \times (7 \times 10^8 \text{ m}) = 7 \times 10^{13}$ m, which is more than four hundred times greater than the actual orbital radius of 1.5×10^{11} m. In fact, the earth would be more than ten times farther from the sun than is Pluto, which has an orbital radius of about 6×10^{12} m. An atom, then, contains a much greater fraction of empty space than does our solar system.

Related Homework: Problem 3

Although the planetary model of the atom is easy to visualize, it too is fraught with difficulties. For instance, an electron moving on a curved path has a centripetal acceleration, as Section 5.2 discusses. And when an electron is accelerating, it radiates electromagnetic waves, as Section 24.1 discusses. These waves carry away energy. With their energy constantly being depleted, the electrons would spiral inward and eventually collapse into the nucleus. Since matter is stable, such a collapse does not occur. Thus, the planetary model, although providing a more realistic picture of the atom than the "plum-pudding" model, must be telling only part of the story. The full story of atomic structure is fascinating, and the next section describes another aspect of it.

30.2 | Line Spectra

We have seen in Sections 13.3 and 29.2 that all objects emit electromagnetic waves, and we will see in Section 30.3 how this radiation arises. For a solid object, such as the hot filament of a light bulb, these waves have a continuous range of wavelengths, some of which are in the visible region of the spectrum. The continuous range of wavelengths is characteristic of the entire collection of atoms that make up the solid. In contrast, individual atoms, free of the strong interactions that are present in a solid, emit only certain specific wavelengths rather than a continuous range. These wavelengths are characteristic of the atom and provide important clues about its structure. To study the behavior of individual atoms, low-pressure gases are used in which the atoms are relatively far apart.

A low-pressure gas in a sealed tube can be made to emit electromagnetic waves by applying a sufficiently large potential difference between two electrodes located within the tube. With a grating spectroscope like that in Figure 27.36, the individual wavelengths emitted by the gas can be separated and identified as a series of bright fringes. The series of fringes is called a **line spectrum** because each bright fringe appears as a thin rectangle (a "line") resulting from the large number of parallel, closely spaced slits in the grating of the spectroscope.

THE PHYSICS OF . . . neon signs and mercury vapor street lamps. Two familiar examples of low-pressure gases are the neon in neon signs and the mercury in mercury vapor street lamps. **Figure 30.3** shows the visible parts of the line spectra for these two atoms. The specific visible wavelengths that the atoms emit give neon signs and mercury vapor street lamps their characteristic colors.

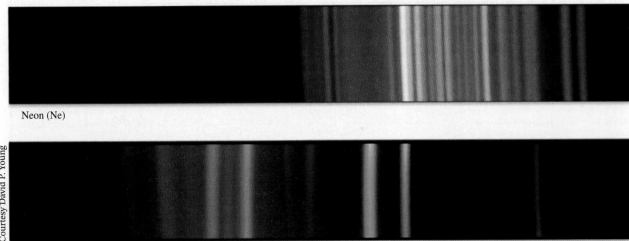

Neon (Ne)

Mercury (Hg)

Courtesy David P. Young

FIGURE 30.3 The line spectra for neon and mercury.

FIGURE 30.4 Line spectrum of atomic hydrogen. Only the Balmer series lies in the visible region of the electromagnetic spectrum.

The simplest line spectrum is that of atomic hydrogen, and much effort has been devoted to understanding the pattern of wavelengths that it contains. **Figure 30.4** illustrates in schematic form some of the groups or series of lines in the spectrum of atomic hydrogen. Only one of the groups is in the visible region of the electromagnetic spectrum; it is known as the **Balmer series**, in recognition of Johann J. Balmer (1825–1898), a Swiss schoolteacher who found an empirical equation that gave the values for the observed wavelengths. This equation is given next, along with similar equations that apply to the **Lyman series** at shorter wavelengths and the **Paschen series** at longer wavelengths, which are also shown in the drawing:

Lyman series $\qquad\qquad \frac{1}{\lambda} = R\left(\frac{1}{1^2} - \frac{1}{n^2}\right) \quad n = 2, 3, 4, \ldots$ (30.1)

Balmer series $\qquad\qquad \frac{1}{\lambda} = R\left(\frac{1}{2^2} - \frac{1}{n^2}\right) \quad n = 3, 4, 5, \ldots$ (30.2)

Paschen series $\qquad\qquad \frac{1}{\lambda} = R\left(\frac{1}{3^2} - \frac{1}{n^2}\right) \quad n = 4, 5, 6, \ldots$ (30.3)

In these equations, the constant term R has the value of $R = 1.097 \times 10^7 \text{ m}^{-1}$ and is called the *Rydberg constant*. An essential feature of each group of lines is that there are long and short wavelength limits, with the lines being increasingly crowded toward the short wavelength limit. **Figure 30.4** also gives these limits for each series, and Example 2 determines them for the Balmer series.

EXAMPLE 2 | The Balmer Series

Find **(a)** the longest and **(b)** the shortest wavelengths of the Balmer series.

Reasoning Each wavelength in the series corresponds to one value for the integer n in Equation 30.2. Longer wavelengths are associated with smaller values of n. The longest wavelength occurs when n has its smallest value of $n = 3$. The shortest wavelength arises when n has a very large value, so that $1/n^2$ is essentially zero.

Solution **(a)** With $n = 3$, Equation 30.2 reveals that for the longest wavelength

$$\frac{1}{\lambda} = R\left(\frac{1}{2^2} - \frac{1}{n^2}\right) = (1.097 \times 10^7 \text{ m}^{-1})\left(\frac{1}{2^2} - \frac{1}{3^2}\right) = 1.524 \times 10^6 \text{ m}^{-1}$$

or $\boxed{\lambda = 656 \text{ nm}}$

(b) With $1/n^2 = 0$, Equation 30.2 reveals that for the shortest wavelength

$$\frac{1}{\lambda} = (1.097 \times 10^7 \text{ m}^{-1})\left(\frac{1}{2^2} - 0\right) = 2.743 \times 10^6 \text{ m}^{-1}$$

or $\boxed{\lambda = 365 \text{ nm}}$

Equations 30.1–30.3 are useful because they reproduce the wavelengths that hydrogen atoms radiate. However, these equations are empirical and provide no insight as to *why* certain wavelengths are radiated and others are not. It was the great Danish physicist, Niels Bohr (1885–1962), who provided the first model of the atom that predicted the discrete wavelengths emitted by atomic hydrogen. Bohr's model started us on the way toward understanding how the structure of the atom restricts the radiated wavelengths to certain values. In 1922 Bohr received the Nobel Prize in Physics for his accomplishment.

30.3 | The Bohr Model of the Hydrogen Atom

In 1913 Bohr presented a model that led to equations such as Balmer's for the wavelengths that the hydrogen atom radiates. Bohr's theory begins with Rutherford's picture of an atom as a nucleus surrounded by electrons moving in circular orbits. In his theory, Bohr made a number of assumptions and combined the new quantum ideas of Planck and Einstein with the traditional description of a particle in uniform circular motion.

Adopting Planck's idea of quantized energy levels (see Section 29.2), Bohr hypothesized that in a hydrogen atom there can be only certain values of the total energy (electron kinetic energy plus potential energy). These allowed energy levels correspond to different orbits for the electron as it moves around the nucleus, the larger orbits being associated with larger total energies. **Animated Figure 30.5** illustrates two of the orbits. In addition, Bohr assumed that an electron in one of these orbits *does not* radiate electromagnetic waves. For this reason, the orbits are called **stationary orbits** or **stationary states**. Bohr recognized that radiationless orbits violated the laws of physics, as they were then known. But the assumption of such orbits was necessary, because the traditional laws indicated that an electron radiates electromagnetic waves as it accelerates around a circular path, and the loss of the energy carried by the waves would lead to the collapse of the orbit.

To incorporate Einstein's photon concept (see Section 29.3), Bohr theorized that a photon is emitted only when the electron *changes* orbits from a larger one with a higher energy to a smaller one with a lower energy, as **Animated Figure 30.5** indicates. How do electrons get into the higher-energy orbits in the first place? They get there by picking up energy when atoms collide, which happens more often when a gas is heated, or by acquiring energy when a high voltage is applied to a gas.

When an electron in an initial orbit with a larger energy E_i changes to a final orbit with a smaller energy E_f, the emitted photon has an energy of $E_i - E_f$, consistent with the law of conservation of energy. But according to Einstein, the energy of a photon is hf, where f is its frequency and h is Planck's constant. As a result, we find that

$$E_i - E_f = hf \qquad (30.4)$$

Since the frequency of an electromagnetic wave is related to the wavelength by $f = c/\lambda$, Bohr could use Equation 30.4 to determine the wavelengths radiated by a hydrogen atom. First, however, he had to derive expressions for the energies E_i and E_f.

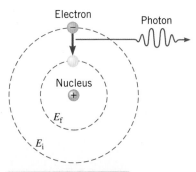

ANIMATED FIGURE 30.5 In the Bohr model, a photon is emitted when the electron drops from a larger, higher-energy orbit (energy = E_i) to a smaller, lower-energy orbit (energy = E_f).

The Energies and Radii of the Bohr Orbits

For an electron of mass m and speed v in an orbit of radius r (see **Figure 30.6**), the total energy is the kinetic energy $\left(\text{KE} = \frac{1}{2}mv^2\right)$ of the electron plus the electric potential energy EPE. The potential energy is the product of the charge $(-e)$ on the electron and the electric potential produced by the positive nuclear charge, in accord with Equation 19.3. We assume that the nucleus contains Z protons,* for a total nuclear charge of $+Ze$. The electric potential at a distance r from a point charge of $+Ze$ is given as $+kZe/r$ by Equation 19.6, where the constant k is $k = 8.988 \times 10^9 \ \text{N} \cdot \text{m}^2/\text{C}^2$. The electric potential energy is, then, EPE $= (-e)(+kZe/r)$. Consequently, the total energy E of the atom is

$$E = \text{KE} + \text{EPE}$$

$$= \frac{1}{2}mv^2 - \frac{kZe^2}{r} \qquad (30.5)$$

But a centripetal force of magnitude mv^2/r (Equation 5.3) acts on a particle in uniform circular motion. As **Figure 30.6** indicates, the centripetal force is provided by the electrostatic force of attraction $\vec{\mathbf{F}}$ that the protons in the nucleus exert on the electron.

FIGURE 30.6 In the Bohr model, the electron is in uniform circular motion around the nucleus. The centripetal force $\vec{\mathbf{F}}$ is the electrostatic force of attraction that the positive nuclear charge exerts on the electron.

*For hydrogen, $Z = 1$, but we also wish to consider situations in which Z is greater than 1.

According to Coulomb's law (Equation 18.1), the magnitude of the electrostatic force is $F = kZe^2/r^2$. Therefore, $mv^2/r = kZe^2/r^2$, or

$$mv^2 = \frac{kZe^2}{r} \tag{30.6}$$

We can use this relation to eliminate the term mv^2 from Equation 30.5, with the result that

$$E = \frac{1}{2}\left(\frac{kZe^2}{r}\right) - \frac{kZe^2}{r} = -\frac{kZe^2}{2r} \tag{30.7}$$

The total energy of the atom is negative because the negative electric potential energy is larger in magnitude than the positive kinetic energy.

A value for the radius r is needed, if Equation 30.7 is to be useful. To determine r, Bohr made an assumption about the orbital angular momentum of the electron. The magnitude L of the angular momentum is given by Equation 9.10 as $L = I\omega$, where $I = mr^2$ is the moment of inertia of the electron moving on its circular path and $\omega = v/r$ (Equation 8.9) is the angular speed of the electron in radians per second. Thus, the angular momentum is $L = (mr^2)(v/r) = mvr$. Bohr conjectured that the angular momentum can assume only certain discrete values; in other words, L is quantized. He postulated that the allowed values are integer multiples of Planck's constant divided by 2π:

$$L_n = mv_n r_n = n\frac{h}{2\pi} \quad n = 1, 2, 3, \ldots \tag{30.8}$$

Solving this equation for v_n and substituting the result into Equation 30.6 lead to the following expression for the radius r_n of the nth Bohr orbit:

$$r_n = \left(\frac{h^2}{4\pi^2 mke^2}\right)\frac{n^2}{Z} \quad n = 1, 2, 3, \ldots \tag{30.9}$$

Math Skills To obtain Equation 30.9 for the radius r_n, we begin by writing Equation 30.6 with the symbols v_n instead of v and r_n instead of r:

$$mv_n{}^2 = \frac{kZe^2}{r_n} \tag{30.6}$$

Next, we solve $mv_n r_n = n\frac{h}{2\pi}$ (Equation 30.8) for v_n, by dividing both sides by mr_n:

$$\frac{mv_n r_n}{(mr_n)} = n\frac{h}{2\pi(mr_n)} \quad \text{or} \quad v_n = \frac{nh}{2\pi mr_n} \tag{1}$$

Substituting Equation 1 into Equation 30.6, we obtain

$$m\underbrace{\left(\frac{nh}{2\pi mr_n}\right)^2}_{v_n{}^2} = \frac{kZe^2}{r_n} \quad \text{or} \quad \frac{n^2h^2}{4\pi^2 mr_n{}^2} = \frac{kZe^2}{r_n} \tag{2}$$

To isolate r_n on one side of the equals sign in Equation 2, we multiply both sides by $\frac{r_n{}^2}{kZe^2}$:

$$\frac{n^2h^2}{4\pi^2 mr_n{}^2}\left(\frac{r_n{}^2}{kZe^2}\right) = \frac{kZe^2}{r_n}\left(\frac{r_n r_n}{kZe^2}\right) \quad \text{or} \quad \frac{n^2h^2}{4\pi^2 mkZe^2} = r_n \tag{3}$$

Finally, in Equation 3, we factor out the term $\frac{n^2}{Z}$ to give

$$r_n = \left(\frac{h^2}{4\pi^2 mke^2}\right)\frac{n^2}{Z} \tag{30.9}$$

With $h = 6.626 \times 10^{-34}$ J $\cdot$ s, $m = 9.109 \times 10^{-31}$ kg, $k = 8.988 \times 10^9$ N $\cdot$ m²/C², and $e = 1.602 \times 10^{-19}$ C, this expression reveals that

Radii for Bohr orbits (in meters) $\quad r_n = (5.29 \times 10^{-11} \text{ m})\frac{n^2}{Z} \quad n = 1, 2, 3, \ldots \tag{30.10}$

Therefore, in the hydrogen atom ($Z = 1$) the smallest Bohr orbit ($n = 1$) has a radius of $r_1 = 5.29 \times 10^{-11}$ m. This particular value is called the **Bohr radius. Figure 30.7** shows the first three Bohr orbits for the hydrogen atom.

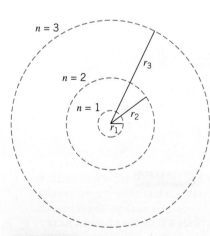

FIGURE 30.7 The first Bohr orbit in the hydrogen atom has a radius $r_1 = 5.29 \times 10^{-11}$ m. The second and third Bohr orbits have radii $r_2 = 4r_1$ and $r_3 = 9r_1$, respectively.

Equation 30.9 for the radius of a Bohr orbit can be substituted into Equation 30.7 to show that the corresponding total energy for the nth orbit is

$$E_n = -\left(\frac{2\pi^2 mk^2 e^4}{h^2}\right)\frac{Z^2}{n^2} \quad n = 1, 2, 3, \ldots \qquad \textbf{(30.11)}$$

Substituting values for h, m, k, and e into this expression yields

Bohr energy levels in joules

$$E_n = -(2.18 \times 10^{-18}\,\text{J})\frac{Z^2}{n^2} \quad n = 1, 2, 3, \ldots \qquad \textbf{(30.12)}$$

Often, atomic energies are expressed in units of electron volts rather than joules. Since $1.60 \times 10^{-19}\,\text{J} = 1\,\text{eV}$, Equation 30.12 can be rewritten as

Bohr energy levels in electron volts

$$E_n = -(13.6\,\text{eV})\frac{Z^2}{n^2} \quad n = 1, 2, 3, \ldots \qquad \textbf{(30.13)}$$

Energy Level Diagrams

It is useful to represent the energy values given by Equation 30.13 on an *energy level diagram*, as in **Figure 30.8**. In this diagram, which applies to the hydrogen atom ($Z = 1$), the highest energy level corresponds to $n = \infty$ in Equation 30.13 and has an energy of 0 eV. This is the energy of the atom when the electron is completely removed ($r = \infty$) from the nucleus and is at rest. In contrast, the lowest energy level corresponds to $n = 1$ and has a value of -13.6 eV. The lowest energy level is called the **ground state**, to distinguish it from the higher levels, which are called **excited states**. Observe how the energies of the excited states come closer and closer together as n increases.

The electron in a hydrogen atom at room temperature spends most of its time in the ground state. To raise the electron from the ground state ($n = 1$) to the highest possible excited state ($n = \infty$), 13.6 eV of energy must be supplied. Supplying this amount of energy removes the electron from the atom, producing the positive hydrogen ion H^+. This is the minimum energy needed to remove the electron and is called the **ionization energy**. Thus, the Bohr model predicts that the ionization energy of atomic hydrogen is 13.6 eV, in excellent agreement with the experimental value. In Example 3 the Bohr model is applied to doubly ionized lithium.

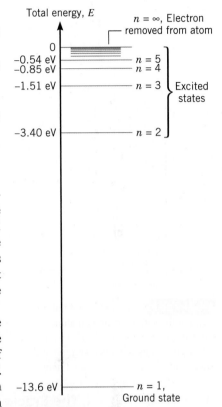

FIGURE 30.8 Energy level diagram for the hydrogen atom.

EXAMPLE 3 | The Ionization Energy of Li²⁺

The Bohr model does not apply when more than one electron orbits the nucleus because it does not account for the electrostatic force that one electron exerts on another. For instance, an electrically neutral lithium atom (Li) contains three electrons in orbit around a nucleus that includes three protons ($Z = 3$), and Bohr's analysis is not applicable. However, the Bohr model can be used for the doubly charged positive ion of lithium (Li^{2+}) that results when two electrons are removed from the neutral atom, leaving only one electron to orbit the nucleus. Obtain the ionization energy that is needed to remove the remaining electron from Li^{2+}.

Reasoning The lithium ion Li^{2+} contains three times the positive nuclear charge that the hydrogen atom contains. Therefore, the orbiting electron is attracted more strongly to the nucleus in

Li^{2+} than to the nucleus in the hydrogen atom. As a result, we expect that more energy is required to ionize Li^{2+} than the 13.6 eV required for atomic hydrogen.

Solution The Bohr energy levels for Li^{2+} are obtained from Equation 30.13 with $Z = 3$; $E_n = -(13.6\,\text{eV})(3^2/n^2)$. Therefore, the ground state ($n = 1$) energy is

$$E_1 = -(13.6\,\text{eV})\frac{3^2}{1^2} = -122\,\text{eV}$$

Removing the electron from Li^{2+} requires 122 eV of energy:

$$\boxed{\text{Ionization energy} = 122\,\text{eV}}$$

This value for the ionization energy agrees well with the experimental value of 122.4 eV and, as expected, is greater than the 13.6 eV required for atomic hydrogen.

The Line Spectra of the Hydrogen Atom

To predict the wavelengths in the line spectrum of the hydrogen atom, Bohr combined his ideas about atoms (electron orbits are stationary orbits and the angular momentum of an electron is quantized) with Einstein's idea of the photon.

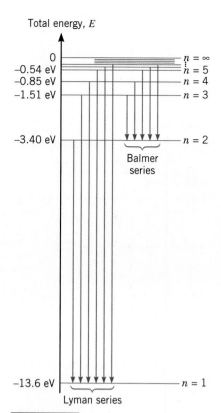

FIGURE 30.9 The Lyman and Balmer series of lines in the hydrogen atom spectrum correspond to transitions that the electron makes between higher and lower energy levels, as indicated here.

As applied by Bohr, the photon concept is inherent in Equation 30.4, $E_i - E_f = hf$, which states that the frequency f of the photon is proportional to the difference between two energy levels of the hydrogen atom. If we substitute Equation 30.11 for the total energies E_i and E_f into Equation 30.4 and recall from Equation 16.1 that $f = c/\lambda$, we obtain the following result:

$$\frac{1}{\lambda} = \frac{2\pi^2 m k^2 e^4}{h^3 c}(Z^2)\left(\frac{1}{n_f^2} - \frac{1}{n_i^2}\right) \qquad (30.14)$$

$$n_i, n_f = 1, 2, 3, \ldots \qquad \text{and} \qquad n_i > n_f$$

Using known values for h, m, k, e, and c, we find that $2\pi^2 m k^2 e^4/(h^3 c) = 1.097 \times 10^7 \text{ m}^{-1}$, in agreement with the Rydberg constant R that appears in Equations 30.1–30.3. The agreement between the theoretical and experimental values of the Rydberg constant was a major accomplishment of Bohr's theory.

With $Z = 1$ and $n_f = 1$, Equation 30.14 reproduces Equation 30.1 for the Lyman series. Thus, Bohr's model shows that the Lyman series of lines occurs when electrons make transitions from higher energy levels with $n_i = 2, 3, 4, \ldots$ to the first energy level where $n_f = 1$. **Figure 30.9** shows these transitions. Notice that when an electron makes a transition from $n_i = 2$ to $n_f = 1$, the longest wavelength photon in the Lyman series is emitted, since the energy change is the smallest possible. When an electron makes a transition from the highest level where $n_i = \infty$ to the lowest level where $n_f = 1$, the shortest wavelength is emitted, since the energy change is the largest possible. Since the higher energy levels are increasingly close together, the lines in the series become more and more crowded toward the short wavelength limit, as can be seen in **Figure 30.4**. **Figure 30.9** also shows the energy level transitions for the Balmer series, where $n_i = 3, 4, 5, \ldots$, and $n_f = 2$. In the Paschen series (see **Figure 30.4**) $n_i = 4, 5, 6, \ldots$, and $n_f = 3$. The next example deals further with the line spectrum of the hydrogen atom.

EXAMPLE 4 | The Brackett Series for Atomic Hydrogen

In the line spectrum of atomic hydrogen there is also a group of lines known as the Brackett series. These lines are produced when electrons, excited to high energy levels, make transitions to the $n = 4$ level. Determine **(a)** the longest wavelength in this series and **(b)** the wavelength that corresponds to the transition from $n_i = 6$ to $n_f = 4$. **(c)** Refer to Figure 24.9 and identify the spectral region in which these lines are found.

Reasoning The longest wavelength corresponds to the transition that has the smallest energy change, which is between the $n_i = 5$ and $n_f = 4$ levels in **Figure 30.8**. The wavelength for this transition, as well as that for the transition from $n_i = 6$ to $n_f = 4$, can be obtained from Equation 30.14.

> **Problem-Solving Insight** In the line spectrum of atomic hydrogen, all lines in a given series (e.g., the Brackett series) are identified by a single value of the quantum number n_f for the *lower energy level into which an electron falls*. Each line in a given series, however, corresponds to a different value of the quantum number n_i for the *higher energy level where an electron originates*.

Solution **(a)** Using Equation 30.14 with $Z = 1$, $n_i = 5$, and $n_f = 4$, we find that

$$\frac{1}{\lambda} = (1.097 \times 10^7 \text{ m}^{-1})(1^2)\left(\frac{1}{4^2} - \frac{1}{5^2}\right) = 2.468 \times 10^5 \text{ m}^{-1}$$

$$\text{or} \quad \boxed{\lambda = 4051 \text{ nm}}$$

(b) The calculation here is similar to that in part (a):

$$\frac{1}{\lambda} = (1.097 \times 10^7 \text{ m}^{-1})(1^2)\left(\frac{1}{4^2} - \frac{1}{6^2}\right) = 3.809 \times 10^5 \text{ m}^{-1}$$

$$\text{or} \quad \boxed{\lambda = 2625 \text{ nm}}$$

(c) According to Figure 24.9, these lines lie in the $\boxed{\text{infrared region}}$ of the spectrum.

The various lines in the hydrogen atom spectrum are produced when electrons change from higher to lower energy levels and photons are emitted. Consequently, the spectral lines are called **emission lines**. Electrons can also make transitions in the reverse direction, from lower to higher levels, in a process known as *absorption*. In this case, an atom

FIGURE 30.10 Illustration of the sun's visible spectrum. Dark lines in the sun's spectrum are absorption lines called Fraunhofer lines, in honor of their discoverer. There are thousands of these absorption lines in the sun's spectrum. The locations of three of the more prominent Fraunhofer lines are marked by arrows.

absorbs a photon that has precisely the energy needed to produce the transition. Thus, if photons with a continuous range of wavelengths pass through a gas and then are analyzed with a grating spectroscope, a series of dark **absorption lines** appears in the continuous spectrum. The dark lines indicate the wavelengths removed by the absorption process.

THE PHYSICS OF . . . absorption lines in the sun's spectrum.   Absorption lines can be seen in **Figure 30.10** in the spectrum of the sun, where they are called Fraunhofer lines, after their discoverer. They are due to atoms, located in the outer and cooler layers of the sun, that absorb radiation coming from the interior. The interior portion of the sun is too hot for individual atoms to retain their structures and, therefore, the interior emits a continuous spectrum of wavelengths.

The Bohr model provides a great deal of insight into atomic structure. However, this model is now known to be oversimplified and has been superseded by a more detailed picture provided by quantum mechanics and the Schrödinger equation (see Section 30.5).

Check Your Understanding

(The answers are given at the end of the book.)

1. Which one of the following statements is true? **(a)** An atom is less easily ionized when its outermost electron is in an excited state than when it is in the ground state. **(b)** An atom is more easily ionized when its outermost electron is in an excited state than when it is in the ground state. **(c)** The energy state (excited state or ground state) of the outermost electron in an atom has nothing to do with how easily the atom can be ionized.

2. An electron in the hydrogen atom is in the $n = 4$ energy level. When this electron makes a transition to a lower energy level, is the wavelength of the photon emitted in **(a)** the Lyman series only, **(b)** the Balmer series only, **(c)** the Paschen series only, or **(d)** could it be in the Lyman, the Balmer, or the Paschen series?

3. A tube contains atomic hydrogen, and nearly all of the electrons in the atoms are in the ground state or $n = 1$ energy level. Electromagnetic radiation with a continuous spectrum of wavelengths (including those in the Lyman, Balmer, and Paschen series) enters one end of the tube and leaves the other end. The exiting radiation is found to contain strong absorption lines. To which one or more of the series do the wavelengths of these absorption lines correspond? Assume that once an electron absorbs a photon and jumps to a higher energy level, it does not absorb yet another photon and jump to an even higher energy level.

30.4 | De Broglie's Explanation of Bohr's Assumption About Angular Momentum

Of all the assumptions Bohr made in his model of the hydrogen atom, perhaps the most puzzling is the assumption about the angular momentum of the electron $[L_n = mv_n r_n = nh/(2\pi); n = 1, 2, 3, \ldots]$. Why should the angular momentum have only those values that are integer multiples of Planck's constant divided by 2π? In 1923, ten years after Bohr's work, de Broglie pointed out that his own theory for the wavelength of a moving particle could provide an answer to this question.

In de Broglie's way of thinking, the electron in its circular Bohr orbit must be pictured as a particle wave. And like waves traveling on a string, particle waves can lead to standing waves under resonant conditions. Section 17.5 discusses these conditions for a string. Standing waves form when the total distance traveled by a wave down the string and back is one wavelength, two wavelengths, or any integer number of wavelengths. The total distance around a Bohr orbit of radius r is the circumference of the orbit or $2\pi r$. By the same reasoning, then, the condition for standing particle waves for the electron in a Bohr orbit would be

$$2\pi r = n\lambda \quad n = 1, 2, 3, \ldots$$

where n is the number of whole wavelengths that fit into the circumference of the circle. But according to Equation 29.8 the de Broglie wavelength of the electron is $\lambda = h/p$, where p is the magnitude of the electron's momentum. If the speed of the electron is much less than the speed of light, the momentum is $p = mv$, and the condition for standing particle waves becomes $2\pi r = nh/(mv)$. A rearrangement of this result gives

$$mvr = n\frac{h}{2\pi} \quad n = 1, 2, 3, \ldots$$

which is just what Bohr assumed for the angular momentum of the electron. As an example, **Figure 30.11** illustrates the standing particle wave on a Bohr orbit for which $2\pi r = 4\lambda$.

De Broglie's explanation of Bohr's assumption about angular momentum emphasizes an important fact—namely, that particle waves play a central role in the structure of the atom. Moreover, the theoretical framework of quantum mechanics includes the Schrödinger equation for determining the wave function ψ (Greek letter psi) that represents a particle wave. The next section deals with the picture that quantum mechanics gives for atomic structure, a picture that supersedes the Bohr model. In any case, the Bohr expression for the energy levels (Equation 30.11) can be applied when a single electron orbits the nucleus, whereas the theoretical framework of quantum mechanics can be applied, in principle, to atoms that contain an arbitrary number of electrons.

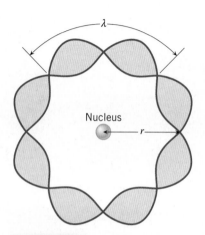

FIGURE 30.11 De Broglie suggested standing particle waves as an explanation for Bohr's angular momentum assumption. Here, a standing particle wave is illustrated on a Bohr orbit where four de Broglie wavelengths fit into the circumference of the orbit.

30.5 | The Quantum Mechanical Picture of the Hydrogen Atom

The picture of the hydrogen atom that quantum mechanics and the Schrödinger equation provide differs in a number of ways from the Bohr model. The Bohr model uses a single integer number n to identify the various electron orbits and the associated energies. Because this number can have only discrete values, rather than a continuous range of values, n is called a **quantum number**. In contrast, quantum mechanics reveals that four different quantum numbers are needed to describe each state of the hydrogen atom. These four are described as follows:

1. **The principal quantum number n.** As in the Bohr model, this number determines the total energy of the atom and can have only integer values: $n = 1, 2, 3, \ldots$. In fact, the Schrödinger equation predicts* that the energy of the hydrogen atom is identical to the energy obtained from the Bohr model: $E_n = -(13.6 \text{ eV}) Z^2/n^2$.

2. **The orbital quantum number ℓ.** This number determines the angular momentum of the electron due to its orbital motion. The values that ℓ can have depend on the value of n, and only the following integers are allowed:

$$\ell = 0, 1, 2, \ldots, (n-1)$$

*This prediction requires that small relativistic effects and small interactions within the atom be ignored, and assumes that the hydrogen atom is not located in an external magnetic field.

For instance, if $n = 1$, the orbital quantum number can have only the value $\ell = 0$, but if $n = 4$, the values $\ell = 0, 1, 2,$ and 3 are possible. The magnitude L of the angular momentum of the electron is

$$L = \sqrt{\ell(\ell + 1)}\frac{h}{2\pi} \qquad (30.15)$$

3. **The magnetic quantum number m_ℓ.** The word "magnetic" is used here because an externally applied magnetic field influences the energy of the atom, and this quantum number is used in describing the effect. Since the effect was discovered by the Dutch physicist Pieter Zeeman (1865–1943), it is known as the *Zeeman effect*. When there is no external magnetic field, m_ℓ plays no role in determining the energy. In either event, the magnetic quantum number determines the component of the angular momentum along a specific direction, which is called the z direction by convention. The values that m_ℓ can have depend on the value of ℓ, with only the following positive and negative integers being permitted:

$$m_\ell = -\ell, \ldots, -2, -1, 0, +1, +2, \ldots, +\ell$$

For example, if the orbital quantum number is $\ell = 2$, then the magnetic quantum number can have the values $m_\ell = -2, -1, 0, +1,$ and $+2$. The component L_z of the angular momentum in the z direction is

$$L_z = m_\ell \frac{h}{2\pi} \qquad (30.16)$$

4. **The spin quantum number m_s.** This number is needed because the electron has an intrinsic property called "spin angular momentum." Loosely speaking, we can view the electron as spinning while it orbits the nucleus, analogous to the way the earth spins as it orbits the sun. There are two possible values for the spin quantum number of the electron:

$$m_s = +\tfrac{1}{2} \quad \text{or} \quad m_s = -\tfrac{1}{2}$$

Sometimes the phrases "spin up" and "spin down" are used to refer to the directions of the spin angular momentum associated with the values for m_s.

Table 30.1 summarizes the four quantum numbers that are needed to describe each state of the hydrogen atom. One set of values for n, ℓ, m_ℓ, and m_s corresponds to one state. As the principal quantum number n increases, the number of possible combinations of the four quantum numbers rises rapidly, as Example 5 illustrates.

TABLE 30.1 Quantum Numbers for the Hydrogen Atom

Name	Symbol	Allowed Values
Principal quantum number	n	$1, 2, 3, \ldots$
Orbital quantum number	ℓ	$0, 1, 2, \ldots, (n-1)$
Magnetic quantum number	m_ℓ	$-\ell, \ldots, -2, -1, 0, +1, +2, \ldots, +\ell$
Spin quantum number	m_s	$-\tfrac{1}{2}$ or $+\tfrac{1}{2}$

EXAMPLE 5 | Quantum Mechanical States of the Hydrogen Atom

Determine the number of possible states for the hydrogen atom when the principal quantum number is **(a)** $n = 1$ and **(b)** $n = 2$.

Reasoning Each different combination of the four quantum numbers summarized in **Table 30.1** corresponds to a different

state. We begin with the value for n and find the allowed values for ℓ. Then, for each ℓ value we find the possibilities for m_ℓ. Finally, m_s may be $+\tfrac{1}{2}$ or $-\tfrac{1}{2}$ for each group of values for n, ℓ, and m_ℓ.

Solution **(a)** The diagram below shows the possibilities for ℓ, m_ℓ, and m_s when $n = 1$: Thus, there are two different states for the hydrogen atom. In the absence of an external magnetic field, these two states have the same energy, since they have the same value of n.

(b) When $n = 2$, there are eight possible combinations for the values of n, ℓ, m_ℓ, and m_s, as the diagram below indicates: With the same value of $n = 2$, all eight states have the same energy when there is no external magnetic field.

(a)

State			
n	ℓ	m_ℓ	m_s
1	0	0	$+\frac{1}{2}$
1	0	0	$-\frac{1}{2}$

(b)

State			
n	ℓ	m_ℓ	m_s
2	1	+1	$+\frac{1}{2}$
2	1	+1	$-\frac{1}{2}$
2	1	0	$+\frac{1}{2}$
2	1	0	$-\frac{1}{2}$
2	1	−1	$+\frac{1}{2}$
2	1	−1	$-\frac{1}{2}$
2	0	0	$+\frac{1}{2}$
2	0	0	$-\frac{1}{2}$

Quantum mechanics provides a more accurate picture of atomic structure than does the Bohr model. It is important to realize that the two pictures differ substantially, as Conceptual Example 6 illustrates.

CONCEPTUAL EXAMPLE 6 | The Bohr Model Versus Quantum Mechanics

Consider two hydrogen atoms. There are no external magnetic fields present, and the electron in each atom has the same energy. According to the Bohr model and to quantum mechanics, is it possible for the electrons in these atoms **(a)** to have zero orbital angular momentum and **(b)** to have different orbital angular momenta?

Reasoning and Solution **(a)** In both the Bohr model and quantum mechanics, the energy is proportional to $1/n^2$, according to Equation 30.13, where n is the principal quantum number. Moreover, the value of n may be $n = 1, 2, 3, \ldots$, and may not be zero. *In the Bohr model, the fact that n may not be*

zero means that it is not possible for the orbital angular momentum to be zero because the angular momentum is proportional to n, according to Equation 30.8. In the quantum mechanical picture the magnitude of the orbital angular momentum is proportional to $\sqrt{\ell(\ell + 1)}$, as given by Equation 30.15. Here, ℓ is the orbital quantum number and may take on the values $\ell = 0, 1, 2, \ldots, (n - 1)$. We note that ℓ [and therefore $\sqrt{\ell(\ell + 1)}$] may be zero, no matter what the value for n is. Consequently, *the orbital angular momentum may be zero according to quantum mechanics*, in contrast to the case for the Bohr model.

(b) If the electrons have the same energy, they have the same value for the principal quantum number *n*. ***In the Bohr model, this means that they cannot have different values for the orbital angular momentum*** L_n, since $L_n = nh/(2\pi)$, according to Equation 30.8. In quantum mechanics, the energy is also determined by *n* when external magnetic fields are absent, but the orbital angular momentum is determined by ℓ. Since $\ell = 0, 1, 2, \ldots, (n-1)$, different values of ℓ are compatible with the same value of *n*. For instance, if $n = 2$ for both electrons, one of them could have $\ell = 0$, while the other could have $\ell = 1$. ***According to quantum mechanics, then, the electrons could have different orbital angular momenta, even though they have the same energy.***

The following table summarizes the discussion from parts (a) and (b):

	Bohr Model	Quantum Mechanics
(a) For a given *n*, can the angular momentum ever be zero?	No	Yes
(b) For a given *n*, can the angular momentum have different values?	No	Yes

According to the Bohr model, the *n*th orbit is a circle of radius r_n, and every time the position of the electron in this orbit is measured, the electron is found exactly at a distance r_n away from the nucleus. This simplistic picture is now known to be incorrect, and the quantum mechanical picture of the atom has replaced it. Suppose that the electron is in a quantum mechanical state for which $n = 1$, and we imagine making a number of measurements of the electron's position with respect to the nucleus. We would find that its position is uncertain, in the sense that there is a probability of finding the electron sometimes very near the nucleus, sometimes very far from the nucleus, and sometimes at intermediate locations. The probability is determined by the wave function ψ, as Section 29.5 discusses. We can make a three-dimensional picture of our findings by marking a dot at each location where the electron is found. More dots occur at places where the probability of finding the electron is higher, and after a sufficient number of measurements, a picture of the quantum mechanical state emerges. **Figure 30.12** shows the spatial distribution for the position of an electron in a state for which $n = 1$, $\ell = 0$, and $m_\ell = 0$. This picture is constructed from so many measurements that the individual dots are no longer visible but have merged to form a kind of probability "cloud" whose density changes gradually from place to place. The dense regions indicate places where the probability of finding the electron is higher, and the less dense regions indicate places where the probability is lower. Also indicated in **Figure 30.12** is the radius where quantum mechanics predicts the greatest probability per unit radial distance of finding the electron in the $n = 1$ state. This radius matches exactly the radius of 5.29×10^{-11} m found for the first Bohr orbit.

For a principal quantum number of $n = 2$, the probability clouds are different than for $n = 1$. In fact, more than one cloud shape is possible because with $n = 2$ the orbital quantum number can be either $\ell = 0$ or $\ell = 1$. Although the value of ℓ does not affect the energy of the hydrogen atom, the value does have a significant effect on the shape of the probability clouds. **Figure 30.13a** shows the cloud for $n = 2$, $\ell = 0$, and $m_\ell = 0$. Part *b* of the drawing shows that when $n = 2$, $\ell = 1$, and $m_\ell = 0$, the cloud has a two-lobe shape with the nucleus at the center between the lobes. For larger values of *n*, the probability clouds become increasingly complex and are spread out over larger volumes of space.

The probability cloud picture of the electron in a hydrogen atom is very different from the well-defined orbit of the Bohr model. The fundamental reason for this difference is to be found in the Heisenberg uncertainty principle, as Conceptual Example 7 discusses.

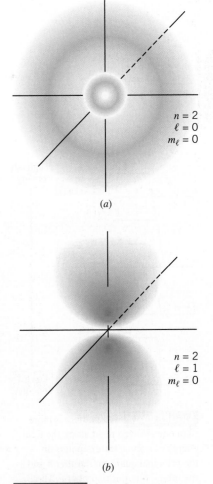

(a)

(b)

FIGURE 30.13 The electron probability clouds for the hydrogen atom when (a) $n = 2$, $\ell = 0$, $m_\ell = 0$ and (b) $n = 2$, $\ell = 1$, $m_\ell = 0$.

Most probable distance for the electron

$n = 1$
$\ell = 0$
$m_\ell = 0$

FIGURE 30.12 The electron probability cloud for the ground state ($n = 1$, $\ell = 0$, $m_\ell = 0$) of the hydrogen atom.

CONCEPTUAL EXAMPLE 7 | The Uncertainty Principle and the Hydrogen Atom

In the Bohr model of the hydrogen atom, the electron in the ground state ($n = 1$) is in an orbit that has a radius of exactly 5.29×10^{-11} m, so that the uncertainty Δy in its radial position is $\Delta y = 0$ m. According to the Heisenberg uncertainty principle, what does the fact that there is no uncertainty in the electron's radial position imply about the electron's radial speed? The uncertainty principle implies **(a)** nothing about the radial speed, **(b)** that the radial speed has only a small uncertainty, **(c)** that the radial speed has an infinitely large uncertainty.

Reasoning We need to obtain an expression for the uncertainty in the radial speed of the electron to use as a guide. As stated in Equation 29.10, the Heisenberg principle is $(\Delta p_y)(\Delta y) \geq h/(4\pi)$. In the present context, Δy is the uncertainty in the electron's radial position, so that Δp_y is the uncertainty in the electron's radial momentum. According to Equation 7.2, however, the magnitude of the momentum is $p_y = mv_y$, where m is the electron's mass and v_y is the electron's radial speed. As a result, the uncertainty in the momentum is $\Delta p_y = \Delta(mv_y) = m\Delta v_y$. With this substitution for Δp_y, the Heisenberg principle

becomes $(m\Delta v_y)(\Delta y) \geq h/(4\pi)$, which can be rearranged to show that

$$\Delta v_y \geq \frac{h}{m(\Delta y)(4\pi)}$$

Answers (a) and (b) are incorrect. Our result for Δv_y shows that the Heisenberg principle does indeed imply something about the uncertainty in the electron's radial speed. Since $\Delta y = 0$ m in the Bohr model, our result for Δv_y shows that the uncertainty in the speed is infinitely large (Δy is in the denominator on the right). Therefore, answers (a) and (b) cannot be correct.

Answer (c) is correct. Our expression for Δv_y shows that, in fact, the uncertainty in the radial speed is infinitely large ($\Delta y = 0$ m in the Bohr model and Δy is in the denominator on the right in the expression above). Such a large uncertainty in the radial speed means that the electron may be moving very rapidly in the radial direction and, therefore, would not remain in its Bohr orbit. Quantum mechanics, with its probability-cloud picture of atomic structure, correctly represents the positional and motional uncertainty that the Heisenberg principle reveals. The Bohr model does not correctly represent this aspect of reality at the atomic level.

Check Your Understanding

(The answers are given at the end of the book.)

4. In the Bohr model for the hydrogen atom, the closer the electron is to the nucleus, the smaller is the total energy of the electron. Is this also true in the quantum mechanical picture of the hydrogen atom?

5. In the quantum mechanical picture of the hydrogen atom, the orbital angular momentum of the electron may be zero in any of the possible energy states. For which energy state *must* the orbital angular momentum be zero?

6. Consider two different hydrogen atoms. The electron in each atom is in a different excited state, so that each electron has a different total energy. Is it possible for the electrons to have the same orbital angular momentum L, according to **(a)** the Bohr model and **(b)** quantum mechanics?

7. The magnitude of the orbital angular momentum of the electron in a hydrogen atom is observed to increase. According to **(a)** the Bohr model and **(b)** quantum mechanics, does this necessarily mean that the total energy of the electron also increases?

FIGURE 30.14 When there is more than one electron in an atom, the total energy of a given state depends on the principal quantum number n and the orbital quantum number ℓ. The energy increases with increasing n (with some exceptions) and, for a fixed n, with increasing ℓ. For clarity, levels for $n = 6$ and higher are not shown.

30.6 The Pauli Exclusion Principle and the Periodic Table of the Elements

Except for hydrogen, all electrically neutral atoms contain more than one electron, with the number given by the atomic number Z of the element. In addition to being attracted by the nucleus, the electrons repel each other. This repulsion contributes to the total energy of a multiple-electron atom. As a result, the one-electron energy expression for hydrogen, $E_n = -(13.6 \text{ eV})Z^2/n^2$, does not apply to other neutral atoms. However, the simplest approach for dealing with a multiple-electron atom still uses the four quantum numbers n, ℓ, m_ℓ, and m_s.

Detailed quantum mechanical calculations reveal that the energy level of each state of a multiple-electron atom depends on both the principal quantum number n and the orbital quantum number ℓ. **Figure 30.14** illustrates that the energy generally increases as

n increases, but there are exceptions, as the drawing indicates. Furthermore, for a given n, the energy also increases as ℓ increases.

In a multiple-electron atom, all electrons with the same value of n are said to be in the same **shell**. Electrons with $n = 1$ are in a single shell (sometimes called the K shell), electrons with $n = 2$ are in another shell (the L shell), those with $n = 3$ are in a third shell (the M shell), and so on. Those electrons with the same values for both n and ℓ are often referred to as being in the same **subshell**. The $n = 1$ shell consists of a single $\ell = 0$ subshell. The $n = 2$ shell has two subshells, one with $\ell = 0$ and one with $\ell = 1$. Similarly, the $n = 3$ shell has three subshells, one with $\ell = 0$, one with $\ell = 1$, and one with $\ell = 2$.

In the hydrogen atom near room temperature, the electron spends most of its time in the lowest energy level, or ground state—namely, in the $n = 1$ shell. Similarly, when an atom contains more than one electron and is near room temperature, the electrons spend most of their time in the lowest energy levels possible. The lowest energy state for an atom is called the **ground state**. However, when a multiple-electron atom is in its ground state, not every electron is in the $n = 1$ shell in general because the electrons obey a principle discovered by the Austrian physicist Wolfgang Pauli (1900–1958).

> **THE PAULI EXCLUSION PRINCIPLE**
>
> **No two electrons in an atom can have the same set of values for the four quantum numbers n, ℓ, m_ℓ, and m_s.**

Suppose two electrons in an atom have three quantum numbers that are identical: $n = 3$, $m_\ell = 1$, and $m_s = -\frac{1}{2}$. According to the exclusion principle, it is not possible for each to have $\ell = 2$, for example, since each would then have the same four quantum numbers. Each electron must have a different value for ℓ (for instance, $\ell = 1$ and $\ell = 2$) and, consequently, would be in a different subshell. With the aid of the Pauli exclusion principle, we can determine which energy levels are occupied by the electrons in an atom in its ground state, as the next example demonstrates.

EXAMPLE 8 | Ground States of Atoms

Determine which of the energy levels in **Figure 30.14** are occupied by the electrons in the ground state of hydrogen (1 electron), helium (2 electrons), lithium (3 electrons), beryllium (4 electrons), and boron (5 electrons).

Reasoning In the ground state of an atom the electrons are in the lowest available energy levels. Consistent with the Pauli exclusion principle, they fill those levels "from the bottom up"—that is, from the lowest to the highest energy.

Solution As the colored dot (●) in **Figure 30.15** indicates, the electron in the hydrogen atom (H) is in the $n = 1$, $\ell = 0$ subshell, which has the lowest possible energy. A second electron is present in the helium atom (He), and both electrons can have the quantum numbers $n = 1$, $\ell = 0$, and $m_\ell = 0$. However, in accord with the Pauli exclusion principle, each electron must have a different spin quantum number, $m_s = +\frac{1}{2}$ for one electron and $m_s = -\frac{1}{2}$ for the other. Thus, the drawing shows both electrons in the lowest energy level.

The third electron that is present in the lithium atom (Li) would violate the exclusion principle if it were also in the $n = 1$, $\ell = 0$ subshell, no matter what the value for m_s is. Thus, the $n = 1$, $\ell = 0$ subshell is filled when occupied by two electrons. With this level filled, the $n = 2$, $\ell = 0$ subshell becomes the next lowest energy level available and is where the third electron of lithium is found (see **Figure 30.15**). In the beryllium atom (Be), the fourth

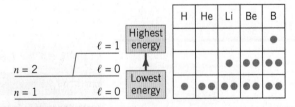

FIGURE 30.15 The electrons (●) in the ground state of an atom fill the available energy levels "from the bottom up"—that is, from the lowest to the highest energy, consistent with the Pauli exclusion principle. The ranking of the energy levels in this figure is meant to apply for a given atom only.

electron is in the $n = 2$, $\ell = 0$ subshell, along with the third electron. This is possible, since the third and fourth electrons can have different values for m_s.

With the first four electrons in place as just discussed, the fifth electron in the boron atom (B) cannot fit into the $n = 1$, $\ell = 0$ or the $n = 2$, $\ell = 0$ subshell without violating the exclusion principle. Therefore, the fifth electron is found in the $n = 2$, $\ell = 1$ subshell, which is the next available energy level with the lowest energy, as **Figure 30.15** indicates. For this electron, m_ℓ can be -1, 0, or $+1$, and m_s can be $+\frac{1}{2}$ or $-\frac{1}{2}$ in each case. However, in the absence of an external magnetic field, all six possibilities correspond to the same energy.

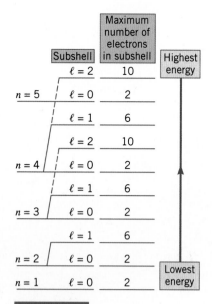

FIGURE 30.16 The maximum number of electrons that the ℓth subshell can hold is $2(2\ell + 1)$.

	The Convention of Letters Used to Refer to the Orbital Quantum Number
TABLE 30.2	

Orbital Quantum Number ℓ	Letter
0	s
1	p
2	d
3	f
4	g
5	h

Because of the Pauli exclusion principle, there is a maximum number of electrons that can fit into an energy level or subshell. Example 8 shows that the $n = 1$, $\ell = 0$ subshell can hold at most two electrons. The $n = 2$, $\ell = 1$ subshell, however, can hold six electrons because with $\ell = 1$, there are three possibilities for m_ℓ (-1, 0, and $+1$), and for each of these choices, the value of m_s can be $+\frac{1}{2}$ or $-\frac{1}{2}$. In general, m_ℓ can have the values $0, \pm1, \pm2, \ldots, \pm\ell$, for $2\ell + 1$ possibilities. Since each of these can be combined with two possibilities for m_s, the total number of different combinations for m_ℓ and m_s is $2(2\ell + 1)$. This, then, is the maximum number of electrons the ℓth subshell can hold, as **Figure 30.16** summarizes.

For historical reasons, there is a widely used convention in which each subshell of an atom is referred to by a letter rather than by the value of its orbital quantum number ℓ. For instance, an $\ell = 0$ subshell is called an s subshell. An $\ell = 1$ subshell and an $\ell = 2$ subshell are known as p and d subshells, respectively. The higher values of $\ell = 3$, 4, and so on, are referred to as f, g, and so on, in alphabetical sequence, as **Table 30.2** indicates.

This convention of letters is used in a shorthand notation that is convenient for indicating simultaneously the principal quantum number n, the orbital quantum number ℓ, and the number of electrons in the n, ℓ subshell. An example of this notation follows:

With this notation, the arrangement, or configuration, of the electrons in an atom can be specified efficiently. For instance, in Example 8, we found that the electron configuration for boron has two electrons in the $n = 1$, $\ell = 0$ subshell, two in the $n = 2$, $\ell = 0$ subshell, and one in the $n = 2$, $\ell = 1$ subshell. In shorthand notation this arrangement is expressed as $1s^2\, 2s^2\, 2p^1$. **Table 30.3** gives the ground-state electron configurations written in this fashion for elements containing up to thirteen electrons. The first five entries are those worked out in Example 8.

TABLE 30.3   **Ground-State Electronic Configurations of Atoms**

Element	Number of Element	Configuration of the Electrons
Hydrogen (H)	1	$1s^1$
Helium (He)	2	$1s^2$
Lithium (Li)	3	$1s^2\, 2s^1$
Beryllium (Be)	4	$1s^2\, 2s^2$
Boron (B)	5	$1s^2\, 2s^2\, 2p^1$
Carbon (C)	6	$1s^2\, 2s^2\, 2p^2$
Nitrogen (N)	7	$1s^2\, 2s^2\, 2p^3$
Oxygen (O)	8	$1s^2\, 2s^2\, 2p^4$
Fluorine (F)	9	$1s^2\, 2s^2\, 2p^5$
Neon (Ne)	10	$1s^2\, 2s^2\, 2p^6$
Sodium (Na)	11	$1s^2\, 2s^2\, 2p^6\, 3s^1$
Magnesium (Mg)	12	$1s^2\, 2s^2\, 2p^6\, 3s^2$
Aluminum (Al)	13	$1s^2\, 2s^2\, 2p^6\, 3s^2\, 3p^1$

Each entry in the periodic table of the elements often includes the ground-state electronic configuration, as **Interactive Figure 30.17** illustrates for argon. To save space, only the configuration of the outermost electrons and unfilled subshells is specified, using the shorthand notation just discussed. Originally the periodic table was developed by the Russian chemist Dmitri Mendeleev (1834–1907) on the basis that certain groups of elements exhibit similar chemical properties. There are eight of these groups, plus the transition elements in the middle of the table, which include the lanthanide series and the actinide series. The similar chemical properties within a group can be explained on the basis of the configurations of the outer electrons of the elements in the group. Thus, quantum mechanics and the Pauli exclusion principle offer an explanation for the chemical behavior of the atoms. The full periodic table can be found immediately before the back cover of this text.

An application that utilizes atomic emission in forensic science is discussed in Example 9.

INTERACTIVE FIGURE 30.17 The entries in the periodic table of the elements often include the ground-state configuration of the outermost electrons.

EXAMPLE 9 | BIO Atomic Emission in Forensic Medicine

Since the atomic energy levels of the elements are unique, so are their respective emission spectra. For instance, the line spectrum emitted by excited gold atoms is distinctly different than that emitted by excited lead atoms. Imagine a substance that is a mixture (or alloy) of gold and lead. Such a material, when excited in some way, would simultaneously emit the line spectra of both gold and lead. Atomic emission can therefore be used to identify the elemental compositions of materials, even in small quantities. This has great utility in medicine and forensic science. For example, arsenic has an affinity for keratin and therefore gets trapped in human hair at the time of ingestion. As the hair grows, with a known average rate of about 1.2 cm per month, the trapped arsenic remains. Therefore, not only can atomic emission detect the presence of arsenic, but it can estimate the time at which the person was exposed.

There are various methods that can be used to excite atoms in a material, including corona (high-voltage) discharge, incineration, and X-ray excitation. One commonly used method in medicine is inductively coupled plasma–atomic emission spectroscopy (ICP-AES), which consists of a device which uses electromagnetic induction to create a plasma (a highly energetic ionized gas that emits light) that includes sample atoms and their ions. The emitted light is then analyzed using an optical spectrometer.

Suppose a hair sample is suspected of containing arsenic. **(a)** What is the ground-state atomic configuration of arsenic ($Z = 33$)? **(b)** In arsenic, suppose an electron in an excited state with energy $E_i = 6.770$ eV undergoes a transition to a final state with energy $E_f = 1.353$ eV. Determine the wavelength of the light that is emitted during this transition. **(c)** If arsenic were detected only in a short hair segment, 3.5 cm above the root of the hair, estimate how much time had passed between the ingestion of the arsenic and the removal of the hair from the head.

Reasoning **(a)** Arsenic (As) has 33 electrons. We will follow the style used in Table 30.3 of Section 30.6 to fill the subshells in accordance with the Pauli exclusion principle, the order of which is depicted in Figure 30.16. **(b)** The energy of the emitted photons will be the energy difference between the levels. Knowing the energy of the emitted photons, we can use Equation 30.4 to find the frequency, and then convert to wavelength using $f = c/\lambda$. **(c)** Since the arsenic was first detected 3.5 cm above the root, and

we know the approximate rate of growth of the hair, we can estimate how much time had passed between the ingestion of the arsenic and the removal of the hair from the head.

Solution **(a)** Following the style of Table 30.3, the $\ell = 0$ subshell is denoted by s, the $\ell = 1$ subshell is denoted by p, and the $\ell = 2$ subshell is denoted by d. In Figure 30.16 we see that the 4s subshell fills before the 3d subshell, and that the 5s subshell fills before the 4d subshell. We also see in Figure 30.16 that s subshells can contain up to 2 electrons, p subshells can contain up to 6 electrons, and d subshells can contain a maximum of 10 electrons. Following these rules, in accordance with the Pauli exclusion principle, we have the following for the ground state configuration of arsenic:

$$\boxed{1s^2\, 2s^2\, 2p^6\, 3s^2\, 3p^6\, 4s^2\, 3d^{10}\, 4p^3}$$

(b) With the difference between energy levels given by $E_i - E_f$, we can apply Equation 30.4, and substitute $f = c/\lambda$ to give the following result:

$$E_i - E_f = hf = \frac{hc}{\lambda}$$

The energies were given in eV, and must first be converted to joules. Using the conversion factor 1 eV $= 1.602 \times 10^{-19}$ J, we have $E_i = 10.846 \times 10^{-19}$ J and $E_f = 2.168 \times 10^{-19}$ J. Solving for the wavelength, we have

$$\lambda = \frac{hc}{E_i - E_f} = \frac{(6.626 \times 10^{-34}\,\text{J} \cdot \text{s})(2.998 \times 10^8\,\text{m/s})}{(10.846 \times 10^{-34}\,\text{J} - 2.168 \times 10^{-34}\,\text{J})} = \boxed{228.9\,\text{nm}}$$

where more accurate values of h and c were used to accommodate the number of significant digits in the final answer. Note that this wavelength is in the ultraviolet part of the spectrum. It can therefore be detected by spectrometers, but not by the human eye.

(c) Since the arsenic was detected 3.5 cm above the root, and hair grows at an average rate of 1.2 cm/month, the hair grew for a time t after ingestion given by

$$t = \frac{3.5\,\text{cm}}{1.2\,\text{cm/month}} = \boxed{2.9\,\text{months}}$$

Thus, the person ingested the arsenic 2.9 months before the hair was removed from their head.

INTERACTIVE FIGURE 30.18 In an X-ray tube, electrons are emitted by a heated filament, accelerate through a large potential difference V, and strike a metal target. The X-rays originate when the electrons interact with the target.

Check Your Understanding

(*The answers are given at the end of the book.*)

8. Using the convention of letters to refer to the orbital quantum number, write down the ground-state configuration of the electrons in krypton ($Z = 36$).

9. Can a 5g subshell contain **(a)** 22 electrons? **(b)** 17 electrons?

10. An electronic configuration for manganese ($Z = 25$) is written as $1s^2\, 2s^2\, 2p^6\, 3s^2\, 3p^6\, 4s^2\, 3d^4\, 4p^1$. Does this configuration represent **(a)** the ground state or **(b)** an excited state?

30.7 | X-Rays

THE PHYSICS OF . . . X-rays. X-rays were discovered by the Dutch physicist Wilhelm K. Roentgen (1845–1923), who performed much of his work in Germany. X-rays can be produced when electrons, accelerated through a large potential difference, collide with a metal target made, for example, from molybdenum or platinum. The target is contained within an evacuated glass tube, as **Interactive Figure 30.18** shows. Example 10 discusses the relationship between the wavelength of the emitted X-rays and the speed of the impinging electrons.

Analyzing Multiple-Concept Problems

EXAMPLE 10 | X-Rays and Electrons

The highest-energy X-rays produced by an X-ray tube have a wavelength of 1.20×10^{-10} m. What is the speed of the electrons in **Interactive Figure 30.18** just before they strike the metal target? Ignore the effects of relativity.

Reasoning We can find the electron's speed from a knowledge of its kinetic energy, since the two are related. Moreover, it is the electron's kinetic energy that determines the energy of any photon. The energy of any photon, on the other hand, is directly proportional to its frequency, as discussed in Section 29.3. But we know from our study of waves (see Section 16.2) that the frequency of a wave is inversely proportional to its wavelength. Thus, we will be able to find the speed of an impinging electron from the given X-ray wavelength.

Knowns and Unknowns The following table summarizes what we know and what we seek:

Description	Symbol	Value
Wavelength of highest-energy X-ray photon	λ	1.20×10^{-10} m
Unknown Variable		
Speed of electron just before it strikes target	v	?

Modeling the Problem

STEP 1 Kinetic Energy The kinetic energy of an electron is the energy it has because of its motion. Since we are ignoring the effects of relativity, the kinetic energy KE is given by Equation 6.2 as $KE = \frac{1}{2} m v^2$, where m and v are the electron's mass and speed. Solving this expression for the speed gives Equation 1 at the right. The mass of the electron is known but its kinetic energy is not, so this will be evaluated in Steps 2 and 3.

$$v = \sqrt{\frac{2(KE)}{m}} \qquad (1)$$

$$\boxed{?}$$

STEP 2 Energy of a Photon   An X-ray photon is a discrete packet of electromagnetic-wave energy. The photon's energy E is given by $E = hf$, where h is Planck's constant and f is the photon's frequency (see Equation 29.2). The energy needed to produce an X-ray photon comes from the kinetic energy of an electron striking the target. We know that the photons have the highest possible energy. This means that all of an electron's kinetic energy KE goes into producing a photon, so KE = E. Substituting KE = E into $E = hf$ gives

$$\boxed{\text{KE} = hf} \tag{2}$$

This result for the energy of the photon can be substituted into Equation 1, as indicated at the right. The frequency is not known, but Step 3 discusses how it can be obtained from the given wavelength.

$$v = \sqrt{\frac{2(\text{KE})}{m}} \tag{1}$$

$$\boxed{\text{KE} = hf} \tag{2}$$

$$\uparrow$$

$$\boxed{?}$$

STEP 3 Relation Between Frequency and Wavelength   Electromagnetic waves such as X-rays travel at the speed c of light in a vacuum. According to Equation 16.1, this speed is related to the frequency f and the wavelength λ by $c = f\lambda$. Solving for the frequency gives

$$\boxed{f = \frac{c}{\lambda}}$$

All the variables on the right side of this equation are known, so we substitute it into Equation 2 for the kinetic energy, as shown in the right column.

$$v = \sqrt{\frac{2(\text{KE})}{m}} \tag{1}$$

$$\boxed{\text{KE} = hf} \tag{2}$$

$$\uparrow$$

$$\boxed{f = \frac{c}{\lambda}}$$

Solution   Algebraically combining the results of the three steps, we have

$$\text{STEP 1} \qquad \text{STEP 2} \qquad \text{STEP 3}$$

$$v = \sqrt{\frac{2(\text{KE})}{m}} \;\xrightarrow{}\; \sqrt{\frac{2(hf)}{m}} \;\xrightarrow{}\; \sqrt{\frac{2h\left(\frac{c}{\lambda}\right)}{m}} = \sqrt{\frac{2hc}{m\lambda}}$$

The speed of an electron just before it strikes the metal target is

$$v = \sqrt{\frac{2hc}{m\lambda}} = \sqrt{\frac{2(6.63 \times 10^{-34}\,\text{J} \cdot \text{s})(3.00 \times 10^{8}\,\text{m/s})}{(9.11 \times 10^{-31}\,\text{kg})(1.20 \times 10^{-10}\,\text{m})}} = \boxed{6.03 \times 10^{7}\,\text{m/s}}$$

Related Homework: Problem 41

A plot of X-ray intensity per unit wavelength versus the wavelength looks similar to **Figure 30.19** and consists of sharp peaks or lines superimposed on a broad continuous spectrum. The sharp peaks are called characteristic lines or **characteristic X-rays** because they are characteristic of the target material. The broad continuous spectrum is referred to as **Bremsstrahlung** (German for "braking radiation") and is emitted when the electrons decelerate or "brake" upon hitting the target.

In **Figure 30.19** the characteristic lines are marked K_α and K_β because they involve the $n = 1$ or K shell of a metal atom. If an electron with enough energy strikes the target, one of the K-shell electrons can be knocked out. An electron in one of the outer shells can then fall into the K shell, and an X-ray photon is emitted in the process. The K_α line arises when an electron in the $n = 2$ level falls into the vacancy that the impinging electron has created in the $n = 1$ level. Similarly, the K_β line arises when an electron in the $n = 3$ level falls to the $n = 1$ level. Example 11 shows that a large potential difference is needed to operate an X-ray tube so that the electrons impinging on the metal target will have sufficient energy to generate the characteristic X-rays. Example 12 determines an estimate for the K_α wavelength of platinum.

FIGURE 30.19 When a molybdenum target is bombarded with electrons that have been accelerated from rest through a potential difference of 45 000 V, this X-ray spectrum is produced. The vertical axis is not to scale.

EXAMPLE 11 | Operating an X-Ray Tube

Strictly speaking, the Bohr model does not apply to multiple-electron atoms, but it can be used to make estimates. Use the Bohr model to estimate the minimum energy that an incoming electron must have to knock a K-shell electron entirely out of an atom in a platinum ($Z = 78$) target in an X-ray tube.

Reasoning According to the Bohr model, the energy of a K-shell electron is given by Equation 30.13, $E_n = -(13.6\text{ eV})Z^2/n^2$, with $n = 1$. When striking a platinum target, an incoming electron must have at least enough energy to raise the K-shell electron from this low energy level up to the 0-eV level that corresponds to a very large distance from the nucleus. Only then will the incoming electron knock the K-shell electron entirely out of a target atom.

> **Problem-Solving Insight** Equation 30.13 for the Bohr energy levels [$E_n = -(13.6\text{ eV})Z^2/n^2$, $n = 1$] can be used in rough calculations of the energy levels involved in the production of K_α X-rays. In this equation, however, the atomic number Z must be reduced by one, to account approximately for the shielding of one K-shell electron by the other K-shell electron.

Solution We will use Equation 30.13 to estimate the minimum energy that an incoming electron must have. However, strictly speaking, this equation applies only to one-electron atoms, because it neglects the repulsive force between electrons in a multiple-electron atom. In the platinum K-shell, each electron exerts on the other electron a repulsive force that balances (approximately) the attractive force of one nuclear proton. In effect, one of the K-shell electrons shields the other from the force of that proton. Therefore, in our calculation we replace Z in Equation 30.13 by $Z - 1$ and find that

$$E_1 = -(13.6\text{ eV})\frac{(Z-1)^2}{n^2} = -(13.6\text{ eV})\frac{(78-1)^2}{1^2} = -8.1 \times 10^4 \text{ eV}$$

Hence, to raise the K-shell electron up to the 0-eV level, the minimum energy for an incoming electron is $\boxed{8.1 \times 10^4 \text{ eV}}$. One electron volt is the kinetic energy acquired when an electron accelerates from rest through a potential difference of one volt. Thus, a potential difference of at least 81 000 V must be applied to the X-ray tube.

EXAMPLE 12 | The K_α Characteristic X-Ray for Platinum

Use the Bohr model to estimate the wavelength of the K_α line in the X-ray spectrum of platinum ($Z = 78$).

Reasoning This example is very similar to Example 4, which deals with the emission line spectrum of the hydrogen atom. As in that example, we use Equation 30.14, this time with the initial value of n being $n_i = 2$ and the final value being $n_f = 1$. As in Example 11, a value of 77 rather than 78 is used for Z to account approximately for the shielding effect of the single K-shell electron in canceling out the attraction of one nuclear proton.

Solution Using Equation 30.14, we find that

$$\frac{1}{\lambda} = (1.097 \times 10^7 \text{ m}^{-1})(78-1)^2\left(\frac{1}{1^2} - \frac{1}{2^2}\right) = 4.9 \times 10^{10} \text{ m}^{-1}$$

or $\boxed{\lambda = 20 \times 10^{-11} \text{ m}}$

This answer is close to an experimental value of 1.9×10^{-11} m.

Another interesting feature of the X-ray spectrum in **Figure 30.19** is the sharp cutoff that occurs at a wavelength of λ_0 on the short-wavelength side of the Bremsstrahlung. This cutoff wavelength is independent of the target material but depends on the energy of the impinging electrons. An impinging electron cannot give up any more than all of its kinetic energy when decelerated by the metal target in an X-ray tube. Thus, at most, an emitted X-ray photon can have an energy equal to the kinetic energy KE of the electron and a frequency given by Equation 29.2 as $f = (\text{KE})/h$, where h is Planck's constant. But the kinetic energy acquired by an electron (charge magnitude = e) in accelerating from rest through a potential difference V is e times V, according to earlier discussions in Section 19.2; V is the potential difference applied across the X-ray tube (see **Interactive Figure 30.18**). Thus, the maximum photon frequency is $f_0 = (eV)/h$. Since $f_0 = c/\lambda_0$, a maximum frequency corresponds to a minimum wavelength, which is the cutoff wavelength λ_0:

$$\lambda_0 = \frac{hc}{eV} \tag{30.17}$$

Figure 30.19, for instance, assumes a potential difference of 45 000 V, which corresponds to a cutoff wavelength of

$$\lambda_0 = \frac{(6.63 \times 10^{-34} \text{ J} \cdot \text{s})(3.00 \times 10^8 \text{ m/s})}{(1.60 \times 10^{-9} \text{ C})(45\,000 \text{ V})} = 2.8 \times 10^{-11} \text{ m}$$

The medical profession began using X-rays for diagnostic purposes almost immediately after their discovery. When a conventional X-ray is obtained, the patient is typically positioned in front of a piece of photographic film, and a single burst of radiation is directed through the patient and onto the film. Since the dense structure of bone absorbs X-rays much more than soft tissue does, a shadow-like picture is recorded on the film. As useful as such pictures are, they have an inherent limitation. The image on the film is a superposition of all the "shadows" that result as the radiation passes through one layer of

Hank Morgan/Science Source

(a)   (b)

FIGURE 30.20 (a) In CAT scanning, a "fanned-out" array of X-ray beams is sent through the patient from different orientations. (b) A doctor and a nurse prepare a patient for a CAT scan.

body material after another. Interpreting which part of a conventional X-ray corresponds to which layer of body material is very difficult.

BIO THE PHYSICS OF . . . CAT scanning. The technique known as CAT scanning, or CT scanning, has greatly extended the ability of X-rays to provide images of specific locations within the body. The acronym CAT stands for **c**omputerized **a**xial **t**omography or **c**omputer-**a**ssisted **t**omography, and the shorter version CT stands for **c**omputerized **t**omography. In this technique a series of X-ray images are obtained as indicated in **Figure 30.20**. A number of X-ray beams form a "fanned out" array of radiation and pass simultaneously through the patient. Each of the beams is detected on the other side by a detector, which records the beam intensity. The various intensities are different, depending on the nature of the body material through which the beams have passed. The feature of CAT scanning that leads to dramatic improvements over the conventional technique is that the X-ray source can be rotated to different orientations, so that the fanned-out array of beams can be sent through the patient from various directions. **Figure 30.20a** singles out two directions for illustration. In reality many different orientations are used, and the intensity of each beam in the array is recorded as a function of orientation. The way in which the intensity of a beam changes from one orientation to another is used as input to a computer. The computer then constructs a highly resolved image of the cross-sectional slice of body material through which the fan of radiation has passed. In effect, the CAT scanning technique makes it possible to take an X-ray picture of a cross-sectional "slice" that is perpendicular to the body's long axis. In fact, the word "axial" in the phrase "computerized axial tomography" refers to the body's long axis. The chapter-opening photograph and **Figure 30.21** show three-dimensional CAT scans of parts of the human anatomy.

Living Art Enterprises/Science Source

(a)

Antoine Rosset/Science Source

(b)

FIGURE 30.21 (a) A 3D CAT scan of a human heart. The inferior vena cava and right atrium are purple. The right ventricle and pulmonary outflow tract are blue. The pulmonary veins are yellow. The left ventricle is orange, and the aorta is red. (b) A 3D CAT scan of an abdomen in frontal view. Near the top of the image the large mass is the liver (green), and beneath it the two smaller teardrop-shaped objects (green/yellow) are the kidneys.

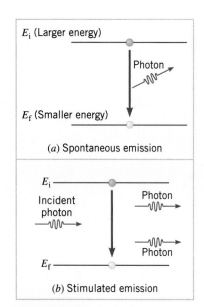

E_i (Larger energy)

Photon

E_f (Smaller energy)

(a) Spontaneous emission

E_i

Incident
photon

Photon

Photon

E_f

(b) Stimulated emission

FIGURE 30.22 (*a*) Spontaneous emission of a photon occurs when the electron (●) makes an unprovoked transition from a higher to a lower energy level, the photon departing in a random direction. (*b*) Stimulated emission of a photon occurs when an incoming photon with the correct energy induces an electron to change energy levels, the emitted photon traveling in the same direction as the incoming photon.

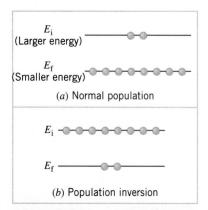

E_i
(Larger energy)

E_f
(Smaller energy)

(a) Normal population

E_i

E_f

(b) Population inversion

FIGURE 30.23 (*a*) In a normal situation at room temperature, most of the electrons in atoms are found in a lower or ground-state energy level. (*b*) If an external energy source is provided to excite electrons into a higher energy level, a population inversion can be created in which more electrons are in the higher level than in the lower level.

Check Your Understanding

(The answers are given at the end of the book.)

11. X-ray tube A and X-ray tube B use the same voltage to accelerate the electrons. However, tube A uses a copper target, whereas tube B uses a silver target. Which one of the following statements is true? **(a)** The cutoff wavelength is greater for tube A. **(b)** The cutoff wavelength is greater for tube B. **(c)** Both tubes have the same cutoff wavelengths.

12. Is it possible to adjust the electric potential V used to operate an X-ray tube so that Bremsstrahlung X-rays are created, but characteristic X-rays are not created? **(a)** Yes, if V is small enough. **(b)** Yes, if V is large enough. **(c)** No, regardless of the value of V.

13. Which one of the following statements is true? **(a)** The K_α wavelength can be smaller than the cutoff wavelength λ_0, assuming that both are produced by the same X-ray tube. **(b)** The K_α wavelength is produced when an electron undergoes a transition from the $n = 1$ energy level to the $n = 2$ energy level. **(c)** The K_β wavelength is always smaller than the K_α wavelength for a given metal target.

30.8 The Laser

THE PHYSICS OF . . . the laser. The laser is one of the most useful inventions of the twentieth century. Today, there are many types of lasers, and most of them work in a way that depends directly on the quantum mechanical structure of the atom.

When an electron makes a transition from a higher energy state to a lower energy state, a photon is emitted. The emission process can be one of two types, spontaneous or stimulated. In **spontaneous emission** (see **Figure 30.22a**), the photon is emitted spontaneously, in a random direction, without external provocation. In **stimulated emission** (see **Figure 30.22b**), an incoming photon induces, or stimulates, the electron to change energy levels. To produce stimulated emission, however, the incoming photon must have an energy that exactly matches the difference between the energies of the two levels—namely, $E_i - E_f$. Stimulated emission is similar to a resonant process, in which the incoming photon "jiggles" the electron at just the frequency to which it is particularly sensitive and causes the change between energy levels. This frequency is given by Equation 30.4 as $f = (E_i - E_f)/h$. The operation of lasers depends on stimulated emission.

Stimulated emission has three important features. First, one photon goes in and two photons come out (see **Figure 30.22b**). In this sense, the process amplifies the number of photons. In fact, this is the origin of the word "laser," which is an acronym for **l**ight **a**mplification by the **s**timulated **e**mission of **r**adiation. Second, the emitted photon travels in the same direction as the incoming photon. Third, the emitted photon is exactly in step with or has the same phase as the incoming photon. In other words, the two electromagnetic waves that these two photons represent are coherent (see Section 17.2) and are locked in step with one another. In contrast, two photons emitted by the filament of an incandescent light bulb are emitted independently. They are not coherent, since one does not stimulate the emission of the other.

Although stimulated emission plays a pivotal role in a laser, other factors are also important. For instance, an external source of energy must be available to excite electrons into higher energy levels. The energy can be provided in a number of ways, including intense flashes of ordinary light and high-voltage discharges. If sufficient energy is delivered to the atoms, more electrons will be excited to a higher energy level than remain in a lower level, a condition known as a **population inversion**. **Figure 30.23** compares a normal energy level population with a population inversion. The population inversions used in lasers involve a higher energy state that is **metastable**, in the sense that electrons remain in the metastable state for a much longer period of time than they do in an ordinary excited state (10^{-3} s versus 10^{-8} s, for example). The requirement of a metastable higher energy state is essential, so that there is more time to enhance the population inversion.

Figure 30.24 shows the widely used helium/neon laser. To sustain the necessary population inversion, a high voltage is discharged across a low-pressure mixture of 15% helium

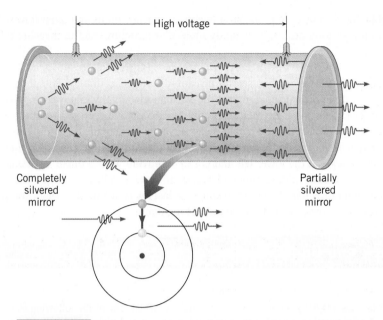

FIGURE 30.24 A schematic drawing of a helium/neon laser. The blow-up shows the stimulated emission that occurs when an electron in a neon atom is induced to change from a higher to a lower energy level.

and 85% neon contained in a glass tube. The laser process begins when an atom, via spontaneous emission, emits a photon parallel to the axis of the tube. This photon, via stimulated emission, causes another atom to emit two photons parallel to the tube axis. These two photons, in turn, stimulate two more atoms, yielding four photons. Four yield eight, and so on, in a kind of avalanche. To ensure that more and more photons are created by stimulated emission, both ends of the tube are silvered to form mirrors that reflect the photons back and forth through the helium/neon mixture. One end is only partially silvered, however, so that some of the photons can escape from the tube to form the laser beam. When the stimulated emission involves only a single pair of energy levels, the output beam has a single frequency or wavelength and is said to be monochromatic.

A laser beam is also exceptionally narrow. The width is determined by the size of the opening through which the beam exits, and very little spreading-out occurs, except that due to diffraction around the edges of the opening. A laser beam does not spread much because any photons emitted at an angle with respect to the tube axis are quickly reflected out the sides of the tube by the silvered ends (see **Figure 30.24**). These ends are carefully arranged to be perpendicular to the tube axis. Since all the power in a laser beam can be confined to a narrow region, the intensity, or power per unit area, can be quite large.

Figure 30.25 shows the pertinent energy levels for a helium/neon laser. By coincidence, helium and neon have nearly identical metastable higher energy states, respectively located 20.61 and 20.66 eV above the ground state. The high-voltage discharge across the gaseous mixture excites electrons in helium atoms to the 20.61-eV state. Then, when an excited helium atom collides inelastically with a neon atom, the 20.61 eV of energy is given to an electron in the neon atom, along with 0.05 eV of kinetic energy from the moving atoms. As a result, the electron in the neon atom is raised to the 20.66-eV state. In this fashion, a population inversion is sustained in the neon, relative to an energy level that is 18.70 eV above the ground state. In producing the laser beam, stimulated emission causes electrons in neon to drop from the 20.66-eV level to the 18.70-eV level. The energy change of 1.96 eV corresponds to a wavelength of 633 nm, which is in the red region of the visible spectrum.

The helium/neon laser is not the only kind of laser. There are many different types, including the ruby laser, the argon-ion laser, the carbon dioxide laser, the gallium arsenide solid-state laser, and chemical dye lasers. Depending on the type and whether the laser operates continuously or in pulses, the available beam power ranges from milliwatts to megawatts. Since lasers provide coherent monochromatic electromagnetic radiation that can be confined to an intense narrow beam, they are useful in a wide variety of situations. Today they are used to reproduce music in compact disc players, to weld parts of

FIGURE 30.25 These energy levels are involved in the operation of a helium/neon laser.

automobile frames together, to transmit telephone conversations and other forms of communication over long distances, to study molecular structure, and to measure distances accurately.

THE PHYSICS OF . . . a laser altimeter. Figure 30.26 shows an impressive example of how a laser can be used to measure distances accurately. The photograph in the figure is a three-dimensional map of the Martian topography that was obtained by the Mars Orbiter Laser Altimeter (MOLA) on the Mars Global Surveyor spacecraft. The map was constructed from 27 million height measurements, each made by sending laser pulses to the Martian surface and measuring their return times. The large *Hellas Planitia* impact basin (dark blue) is at the lower left and is 1800 km wide. At the upper right edge of the image is *Elysium Mons* (red, surrounded by a small band of yellow), a large volcano.

Many other uses have been found since the laser was invented in 1960, and the next section discusses some of them in the field of medicine.

Check Your Understanding

(The answers are given at the end of the book.)

14. A certain laser is designed to operate continuously. Which one of the following statements is false? **(a)** The population inversion used in this laser involves a higher energy state and a lower energy state. **(b)** The population inversion used in this laser involves a metastable higher energy state. **(c)** The laser needs an external source of energy to operate. **(d)** The external energy source that the laser uses can be disconnected once the population inversion is established.

15. Laser A produces green light. Laser B produces red light. Which laser utilizes energy levels that have a larger energy difference between them? **(a)** Laser A **(b)** Laser B **(c)** The energy difference between the levels is the same for both lasers.

FIGURE 30.26 A 3D map of the topography of Mars. The elevation is color-coded from white (highest) through red, yellow, green, blue, and purple (lowest).

NASA/Science Source

(a)   *(b)*

FIGURE 30.27 *(a)* To correct for myopia (nearsightedness) using the PRK procedure, a laser vaporizes tissue (dashed line) on the center of the cornea, thereby flattening it. *(b)* To correct for hyperopia (farsightedness), a laser vaporizes tissue on the peripheral region of the cornea, thereby steepening its contour.

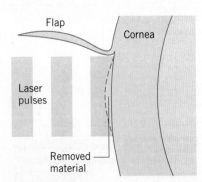

FIGURE 30.28 To correct for myopia (nearsightedness) using the LASIK technique, a laser vaporizes tissue (dashed line) on the cornea, thereby flattening it.

30.9 *Medical Applications of the Laser

One of the medical areas in which the laser has had a substantial impact is in ophthalmology, which deals with the structure, function, and diseases of the eye. Section 26.10 discusses the human eye and the use of contact lenses and eyeglasses to correct nearsightedness and farsightedness. In these conditions, the eye cannot refract light properly and produces blurred images on the retina.

BIO **THE PHYSICS OF . . . PRK eye surgery.** A laser-based procedure known as PRK (**p**hoto**r**efractive **k**eratectomy) offers an alternative treatment for nearsightedness and farsightedness that does not rely on lenses. It involves the use of a laser to remove small amounts of tissue from the cornea of the eye (see **Figure 26.33**) and thereby changes its curvature. As Section 26.10 points out, light enters the eye through the cornea, and it is at the air/cornea boundary that most of the refraction of the light occurs. Therefore, changing the curvature of that boundary can correct deficiencies in the way the eye refracts light, thus causing the image to be focused onto the retina, where it belongs. Ideally, the cornea is dome-shaped. If the dome is too steep, however, the rays of light are focused in front of the retina and nearsightedness results. As **Figure 30.27a** shows, the laser light removes tissue from the center of the cornea, thereby flattening it and increasing the eye's effective focal length. On the other hand, if the shape of the cornea is too flat, light rays would come to a focus behind the retina if they could, and farsightedness occurs. As part *b* of the drawing illustrates, the center of the cornea is now masked and the laser is used to remove peripheral tissue. This steepens the shape of the cornea, thereby shortening the eye's effective focal length and allowing rays to be focused on the retina.

BIO **THE PHYSICS OF . . . LASIK eye surgery.** The LASIK (**l**aser-**a**ssisted **i**n **s**itu **k**eratomileusis) procedure uses a motor-powered blade known as a microkeratome to partially detach a thin flap (about 0.2 mm thick) in the front of the cornea (see **Figure 30.28**). The flap is pulled back and the laser beam then remodels the corneal tissue underneath by vaporizing cells. Afterward, the flap is folded back into place, with no stitches being required. The laser light in the PRK and LASIK techniques is pulsed and comes from an ultraviolet excimer laser that produces a wavelength of 193 nm. The cornea absorbs this wavelength extremely well, so that weak pulses can be used, leading to highly precise and controllable removal of corneal tissue. Typically, 0.1 to 0.5 μm of tissue is removed by each pulse without damaging adjacent layers.

BIO **THE PHYSICS OF . . . removing port-wine stains.** Another medical application of the laser is in the treatment of congenital capillary malformations known as port-wine stains, which affect 0.3% of children at birth. These birthmarks are usually found on the head and neck. Preferred treatment for port-wine stains now utilizes a pulsed dye laser, as shown in **Figure 30.29**. The light is absorbed by oxyhemoglobin in the malformed capillaries, which are destroyed in the process without damage to adjacent normal tissue. Eventually the destroyed capillaries are replaced by normal blood vessels, which causes the port-wine stain to fade.

BIO **THE PHYSICS OF . . . photodynamic therapy for cancer.** In the treatment of cancer, the laser is being used along with light-activated drugs in photodynamic therapy. The procedure involves administering the drug intravenously, so that the tumor can absorb it from the bloodstream, the advantage being that the drug is then located right near the cancer cells. When the drug is activated by laser light, a chemical reaction ensues that disintegrates the cancer cells and the small blood vessels that feed them. In **Figure 30.30** a patient is being treated for cancer of the esophagus. An endoscope that uses optical fibers is inserted down the patient's throat to guide the red laser light to the tumor site and activate the drug. Photodynamic therapy works best with small tumors in their early stages.

FIGURE 30.29 A patient receives treatment from a pulsed dye laser. This medical application is effective at removing skin abnormalities, such as port-wine stains and rosacea.

FIGURE 30.30 Photodynamic therapy to treat cancer of the esophagus is being administered to this patient. Red laser light is routed to the tumor site with an endoscope that incorporates optical fibers.

EXAMPLE 13 | **BIO** The Physics of Laser Lithotripsy

Solid masses are sometimes produced in the kidneys, known as kidney stones, and if they are too large, they can become stuck in the ureter, which is the tube leading from the kidney to the bladder. This condition can be extremely painful, and it is often corrected with a medical treatment known as lithotripsy, which is designed to break up the stone, so that it can pass naturally through the urine. One kind of lithotripsy involves the use of a laser, where a fiber optic is inserted into the body near the stone and delivers intense laser pulses to pulverize it. The type of laser often used in this technique is a holmium YAG laser, which is a solid state laser that uses a crystal of yttrium aluminum garnet (YAG), where a small fraction of the yttrium atoms in the crystal

have been replaced by holmium atoms. The laser produces infrared light with a wavelength of 2.10 μm and operates with a frequency of 5 Hz, which means it delivers 5 pulses of light to the stone each second. If the energy delivered in each pulse is 0.5 J, how many photons of laser light strike the kidney stone in one second? Round your answer to the nearest whole number.

Reasoning Since we are given the wavelength of the laser light, we can use Equation 29.2 to calculate the energy of an individual photon in the beam. The number of photons striking the stone will be given by the total energy delivered divided by the energy per photon.

Solution **(a)** From Equation 29.2, we have:

$$E_{\text{photon}} = hf = \frac{hc}{\lambda}$$

The total energy delivered in one second is equal to the number of pulses in one second multiplied by the energy per pulse:

$$E_{\text{tot}} = (\text{number of pulses})\left(\frac{\text{energy}}{\text{pulse}}\right)$$

The number of photons N is then given by:

$$N = \frac{E_{\text{tot}}}{E_{\text{photon}}} = \frac{(\text{number of pulses})\left(\dfrac{\text{energy}}{\text{pulse}}\right)}{hc/\lambda}$$

$$= \frac{(5)(0.5\ \text{J})(2.10 \times 10^{-6}\ \text{m})}{(6.63 \times 10^{-34}\ \text{J} \cdot \text{s})(3.00 \times 10^{8}\ \text{m/s})}$$

$$= \boxed{3 \times 10^{19}\ \text{photons}}$$

30.10 *Holography

THE PHYSICS OF . . . holography. One of the most familiar applications of lasers is in holography, which is a process for producing three-dimensional images. The information used to produce a holographic image is captured on photographic film, which is referred to as a hologram. **Figure 30.31** illustrates how a hologram is made. Laser light strikes a half-silvered mirror, or beamsplitter, which reflects part and transmits part of the light. In the drawing, the reflected part is called the *object beam* because it illuminates the object (a chess piece). The transmitted part is called the *reference beam*. The object beam reflects from the chess piece at points such as A and B and, together with the reference beam, falls on the film. One of the main characteristics of laser light is that it is coherent. Thus, the light from the two beams has a stable phase relationship, like the light from the two slits in Young's double-slit experiment (see Section 27.2). Because of the stable phase relationship and because the two beams travel different distances, an interference pattern is formed on the film. This pattern is the hologram and, although much more complex, is analogous to the pattern of bright and dark fringes formed in the double-slit experiment.

Figure 30.32 shows in greater detail how a holographic interference pattern arises. This drawing considers only the reference beam and the light (wavelength $= \lambda$) coming from point A on the chess piece. As we know from Section 27.1, constructive interference between the two light waves leads to a bright fringe; it occurs when the waves, in reaching the film, travel distances that differ by an integer number m of wavelengths. In the drawing, ℓ_m is the distance between point A and the place on the film where the mth-order bright fringe occurs, and ℓ_0 is the perpendicular distance that the reference beam would travel from point A to the $m = 0$ bright fringe. In addition, r_m is the distance along the film that locates the bright fringe. In terms of these distances, we know that

$$\ell_m - \ell_0 = m\lambda \qquad \text{(condition for constructive interference)}$$
$$\ell_0{}^2 + r_m^2 = \ell_m{}^2 \qquad \text{(Pythagorean theorem)}$$

FIGURE 30.31 An arrangement used to produce a hologram.

FIGURE 30.32 This drawing helps to explain how the interference pattern arises on the film when light from point A (see **Figure 30.31**) and light from the reference beam combine there.

FIGURE 30.33 When the laser light used to produce a hologram is shone through it, both a real and a virtual image of the object are produced.

The first equation indicates that $\ell_m = m\lambda + \ell_0$, which can be substituted into the second equation. The result can be rearranged to show that

$$r_m{}^2 = m\lambda(m\lambda + 2\ell_0)$$

Since ℓ_0 is typically much larger than λ (for instance, $\ell_0 \approx 10^{-1}$ m and $\lambda \approx 10^{-6}$ m), it follows that $r_m \approx \sqrt{m\lambda 2\ell_0}$. In other words, r_m is roughly proportional to $\sqrt{m}$. Therefore, the fringes are farther apart near the top of the film than they are near the bottom. For example, for the $m = 1$ and $m = 2$ fringes, we have $r_2 - r_1 \propto \sqrt{2} - \sqrt{1} = 0.41$ whereas for the $m = 2$ and $m = 3$ fringes, we have $r_3 - r_2 \propto \sqrt{3} - \sqrt{2} = 0.32$.

In addition to the fringe pattern just discussed, the total interference pattern on the hologram includes interference effects that are related to light coming from point B and other locations on the object in **Figure 30.31**. The total pattern is very complicated. Nevertheless, the fringe pattern for point A alone is sufficient to illustrate the fact that a hologram can be used to produce both a virtual image and a real image of the object, as we will now see.

To produce the holographic images, the laser light is directed through the interference pattern on the film, as in **Figure 30.33**. The pattern can be thought of as a kind of diffraction grating, with the bright fringes analogous to the spaces between the slits of the grating. Section 27.7 discusses how light passing through a grating is split into higher-order bright fringes that are symmetrically located on either side of a central bright fringe. **Figure 30.33** shows the three rays that correspond to the central and first-order bright fringes, as they originate from a spot near the top and a spot near the bottom of the film. The angle θ, which locates the first-order fringes relative to the central fringe, is given by Equation 27.7 (with $m = 1$) as $\sin\theta = \lambda/d$, where d is the separation between the slits of the grating. When the slit separation is greater, as it is near the top of the film, the angle is smaller. When the slit separation is smaller, as it is near the bottom, the angle is larger. Thus, **Figure 30.33** has been drawn with θ_{TOP} smaller than θ_{BOTTOM}. Of the three rays emerging from the film at the top and the three at the bottom, we use the uppermost one in each case to locate the real image of point A on the chess piece. The real image is located where these two rays intersect, when extended to the right. To locate the virtual image, we use the lower ray in each of the three-ray bundles at the top and bottom of the film. When projected to the left, these rays appear to be originating from the spot where the projections intersect—that is, from the virtual image.

A holographic image differs greatly from a photographic image. The most obvious difference is that a hologram provides a three-dimensional image, whereas photographs are two-dimensional. The three-dimensional nature of the holographic image is inherent in the interference pattern formed on the film. In **Figure 30.31** part of this pattern arises because the light emitted by point A travels different distances in reaching different spots on the film than does the light in the reference beam. The same can be said about the light emitted from point B and other places as well. As a result, the total interference pattern contains information about how much farther from the film the various points on the object are, and because of this information holographic images are three-dimensional. Furthermore, it is possible to "walk around" a holographic image and view it from different angles, as you would the original object.

A vast difference exists between the methods used to produce holograms and photographs. As Section 26.7 discusses, a camera uses a converging lens to produce a photograph. The lens focuses the light rays originating from a point on the object to a corresponding point on the film. To produce a hologram, lenses are not used in this way, and a point on the object *does not correspond* to a single point on the film. In **Figure 30.32**, light from point A diverges on its way to the film, and there is no lens to make it converge to a single corresponding point. The light falls over the entire exposed region of the film and contributes everywhere to the interference pattern that is formed. Thus, a hologram may be cut into smaller pieces, and each piece will contain some information about the light originating from point A. For this reason, each smaller piece can be used to produce a three-dimensional image of the object. In contrast, it is not possible to reconstruct the entire image in a photograph from only a small piece of the original film.

The holograms discussed here are typically viewed with the aid of the laser light used to produce them. There are also other kinds of holograms. Credit cards, for example, use holograms for identification purposes, as shown in **Figure 30.34**. This kind of hologram is called a rainbow hologram and is designed to be viewed in white light that is reflected from it. Other applications of holography include head-up displays for instrument panels in high-performance fighter planes, laser scanners at checkout counters, computerized data storage and retrieval systems, and methods for high-precision biomedical measurements.

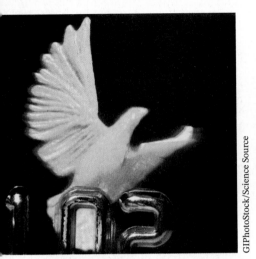

GIPhotoStock/Science Source

FIGURE 30.34 A hologram of a dove on a Visa credit card. When illuminated in white light, the hologram appears to change colors when viewed from different angles.

Concept Summary

30.1 Rutherford Scattering and the Nuclear Atom   The idea of a nuclear atom originated in 1911 as a result of experiments by Ernest Rutherford in which α particles were scattered by a thin metal foil. The phrase "nuclear atom" refers to the fact that an atom consists of a small, positively charged nucleus surrounded at relatively large distances by a number of electrons, whose total negative charge equals the positive nuclear charge when the atom is electrically neutral.

30.2 Line Spectra   A line spectrum is a series of discrete electromagnetic wavelengths emitted by the atoms of a low-pressure gas that is subjected to a sufficiently high potential difference. Certain groups of discrete wavelengths are referred to as "series." The following equations can be used to determine the wavelengths in three of the series that are found in the line spectrum of atomic hydrogen:

Lyman series $\quad \dfrac{1}{\lambda} = R\left(\dfrac{1}{1^2} - \dfrac{1}{n^2}\right) \quad n = 2, 3, 4, \ldots$ (30.1)

Balmer series $\quad \dfrac{1}{\lambda} = R\left(\dfrac{1}{2^2} - \dfrac{1}{n^2}\right) \quad n = 3, 4, 5, \ldots$ (30.2)

Paschen series $\quad \dfrac{1}{\lambda} = R\left(\dfrac{1}{3^2} - \dfrac{1}{n^2}\right) \quad n = 4, 5, 6, \ldots$ (30.3)

The constant term R is called the Rydberg constant and has the value $R = 1.097 \times 10^7 \text{ m}^{-1}$.

30.3 The Bohr Model of the Hydrogen Atom   The Bohr model applies to atoms or ions that have only a single electron orbiting a nucleus containing Z protons. This model assumes that the electron exists in circular orbits that are called stationary orbits because the electron does not radiate electromagnetic waves while in them. According to this model, a photon is emitted only when an electron changes from an orbit with a higher energy E_i to an orbit with a lower energy E_f. The orbital energies and the photon frequency f are related according to Equation 30.4, where h is Planck's constant. The model also assumes that the magnitude L_n of the orbital angular momentum of the electron can only have the discrete values indicated in Equation 30.8. With these assumptions, it can be shown that the nth Bohr orbit has a radius r_n and is associated with a total energy E_n, as given in Equations 30.10 and 30.13.

$$E_i - E_f = hf \tag{30.4}$$

$$L_n = n\frac{h}{2\pi} \quad n = 1, 2, 3, \ldots \tag{30.8}$$

$$r_n = (5.29 \times 10^{-11} \text{ m})\frac{n^2}{Z} \quad n = 1, 2, 3, \ldots \tag{30.10}$$

$$E_n = -(13.6 \text{ eV})\frac{Z^2}{n^2} \quad n = 1, 2, 3, \ldots \tag{30.13}$$

The ionization energy is the minimum energy needed to remove an electron completely from an atom. The Bohr model predicts that the wavelengths comprising the line spectrum emitted by a hydrogen atom can be calculated according to Equation 30.14.

$$\frac{1}{\lambda} = RZ^2\left(\frac{1}{n_f^2} - \frac{1}{n_i^2}\right) \tag{30.14}$$

$$n_i, n_f = 1, 2, 3, \ldots \quad \text{and} \quad n_i > n_f$$

30.4 De Broglie's Explanation of Bohr's Assumption About Angular Momentum Louis de Broglie proposed that the electron in a circular Bohr orbit should be considered as a particle wave and that standing particle waves around the orbit offer an explanation of the angular momentum assumption in the Bohr model.

30.5 The Quantum Mechanical Picture of the Hydrogen Atom Quantum mechanics describes the hydrogen atom in terms of the following four quantum numbers:

(1) The principal quantum number n, which can have the integer values $n = 1, 2, 3, \ldots$

(2) The orbital quantum number ℓ, which can have the integer values $\ell = 0, 1, 2, \ldots, (n - 1)$

(3) The magnetic quantum number m_ℓ, which can have the positive and negative values $m_\ell = -\ell, \ldots, -2, -1, 0, +1, +2, \ldots, +\ell$

(4) The spin quantum number m_s, which, for an electron, can be either $m_s = +\frac{1}{2}$ or $m_s = -\frac{1}{2}$

According to quantum mechanics, an electron does not reside in a circular orbit but, rather, has some probability of being found at various distances from the nucleus.

30.6 The Pauli Exclusion Principle and the Periodic Table of the Elements The Pauli exclusion principle states that no two electrons in an atom can have the same set of values for the four quantum numbers n, ℓ, m_ℓ, and m_s. This principle determines the way in which the electrons in multiple-electron atoms are distributed into shells (defined by the value of n) and subshells (defined by the value of ℓ).

The following notation is used to refer to the orbital quantum numbers: s denotes $\ell = 0$, p denotes $\ell = 1$, d denotes $\ell = 2$, f denotes $\ell = 3$, g denotes $\ell = 4$, h denotes $\ell = 5$, and so on.

The arrangement of the periodic table of the elements is related to the Pauli exclusion principle.

30.7 X-Rays X-rays are electromagnetic waves emitted when high-energy electrons strike a metal target contained within an evacuated glass tube. The emitted X-ray spectrum of wavelengths consists of sharp "peaks" or "lines," called characteristic X-rays, superimposed on a broad continuous range of wavelengths called Bremsstrahlung, or "braking radiation." The K_α characteristic X-ray is emitted when an electron in the $n = 2$ level (L shell) drops into a vacancy in the $n = 1$ level (K shell). Similarly, the K_β characteristic X-ray is emitted when an electron in the $n = 3$ level (M shell) drops into a vacancy in the $n = 1$ level (K shell). The minimum wavelength, or cutoff wavelength λ_0, of the Bremsstrahlung is determined by the kinetic energy of the electrons striking the target in the X-ray tube, according to Equation 30.17, where h is Planck's constant, c is the speed of light in a vacuum, e is the magnitude of the charge on an electron, and V is the potential difference applied across the X-ray tube.

$$\lambda_0 = \frac{hc}{eV} \tag{30.8}$$

30.8 The Laser/30.9 Medical Applications of the Laser A laser is a device that generates electromagnetic waves via a process known as stimulated emission. In this process, one photon stimulates the production of another photon by causing an electron in an atom to fall from a higher energy level to a lower energy level. The emitted photon travels in the same direction as the photon causing the stimulation. Because of this mechanism of photon production, the electromagnetic waves generated by a laser are coherent and may be confined to a very narrow beam. Stimulated emission contrasts with the process known as spontaneous emission, in which an electron in an atom also falls from a higher to a lower energy level, but does so spontaneously, in a random direction, without any external provocation.

Focus on Concepts

Online

Additional questions are available for assignment in WileyPLUS.

Section 30.3   The Bohr Model of the Hydrogen Atom

1. Consider applying the Bohr model to a neutral helium atom ($Z = 2$). The model takes into account a number of factors. Which one of the following does it not take into account? **(a)** The quantization of the orbital angular momentum of an electron **(b)** The centripetal acceleration of an electron **(c)** The electric potential energy of an electron **(d)** The electrostatic repulsion between electrons **(e)** The electrostatic attraction between the nucleus and an electron

2. According to the Bohr model, what determines the shortest wavelength in a given series of wavelengths emitted by the atom? **(a)** The quantum number n_i that identifies the higher energy level from which the electron falls into a lower energy level **(b)** The quantum number n_f that identifies the lower energy level into which the electron falls from a higher energy level **(c)** The ratio n_f/n_i, where n_f is the quantum number that identifies the lower energy level into which the electron falls and n_i is the quantum number that identifies the higher level from which the electron falls **(d)** The sum $n_f + n_i$ of two quantum numbers, where n_f identifies the lower energy level into which the electron falls and n_i identifies the higher level from which the electron falls **(e)** The difference $n_f - n_i$ of two quantum

numbers, where n_f identifies the lower energy level into which the electron falls and n_i identifies the higher level from which the electron falls

Section 30.5   The Quantum Mechanical Picture of the Hydrogen Atom

3. According to quantum mechanics, only one of the following combinations of the principal quantum number n and the orbital quantum number ℓ is possible for the electron in a hydrogen atom. Which combination is it? **(a)** $n = 3$, $\ell = 3$ **(b)** $n = 2$, $\ell = 3$ **(c)** $n = 1$, $\ell = 2$ **(d)** $n = 0$, $\ell = 0$ **(e)** $n = 3$, $\ell = 1$

4. Which one of the following statements is false? **(a)** The orbits in the Bohr model have precise sizes, whereas in the quantum mechanical picture of the hydrogen atom they do not. **(b)** In the absence of external magnetic fields, both the Bohr model and quantum mechanics predict the same total energy for the electron in the hydrogen atom. **(c)** The spin angular momentum of the electron plays a role in both the Bohr model and the quantum mechanical picture of the hydrogen atom. **(d)** The magnitude of the orbital angular momentum cannot be zero in the Bohr model, but it can be zero in the quantum mechanical picture of the hydrogen atom.

Section 30.6 The Pauli Exclusion Principle and the Periodic Table of the Elements

5. Each of the following answers indicates the quantum mechanical states of two electrons, A and B. Which pair of states could *not* describe two of the electrons in a multiple-electron atom?

(a)

	n	ℓ	m_ℓ	m_s
A	4	1	+1	$-\frac{1}{2}$
B	3	1	+1	$-\frac{1}{2}$

(b)

	n	ℓ	m_ℓ	m_s
A	3	2	−1	$-\frac{1}{2}$
B	3	1	−1	$+\frac{1}{2}$

(c)

	n	ℓ	m_ℓ	m_s
A	2	0	0	$-\frac{1}{2}$
B	2	1	+1	$+\frac{1}{2}$

(d)

	n	ℓ	m_ℓ	m_s
A	5	3	+1	$-\frac{1}{2}$
B	4	1	0	$+\frac{1}{2}$

(e)

	n	ℓ	m_ℓ	m_s
A	3	2	−2	$+\frac{1}{2}$
B	3	2	−2	$+\frac{1}{2}$

6. Consider the 5f and 6h subshells in a multiple-electron atom. Which of these subshells can contain 19 electrons? **(a)** Only the 6h subshell **(b)** Only the 5f subshell **(c)** Both subshells **(d)** Neither subshell

Section 30.7 X-Rays

7. Silver ($Z = 47$), copper ($Z = 29$), and platinum ($Z = 78$) can be used as the target in an X-ray tube. Rank in descending order (largest first) the energies needed for impinging electrons to knock a K-shell electron completely out of an atom in each of these targets. **(a)** Silver, copper, platinum **(b)** Silver, platinum, copper **(c)** Platinum, silver, copper **(d)** Platinum, copper, silver **(e)** Copper, silver, platinum

8. The voltage applied across an X-ray tube is doubled. What happens to the cutoff wavelength in the spectrum of wavelengths emitted by the tube's metal target? **(a)** It also doubles. **(b)** It decreases by a factor of two. **(c)** It increases by a factor of four. **(d)** It decreases by a factor of four. **(e)** Nothing happens to it.

Section 30.8 The Laser

9. Consider two energy levels that characterize the atoms of a material used in a laser. A population inversion between these two levels _____. **(a)** has the lower energy level more populated than it normally is and the higher energy level less populated than it normally is **(b)** is the same thing as a metastable state **(c)** requires no external source of energy to be sustained **(d)** has the higher energy level more populated than it normally is and the lower energy level less populated than it normally is

Problems

Online

Additional questions are available for assignment in WileyPLUS.

__Note:__ In working these problems, ignore relativistic effects.

SSM Student Solutions Manual	**BIO** Biomedical application
MMH Problem-solving help	**E** Easy
GO Guided Online Tutorial	**M** Medium
V-HINT Video Hints	**H** Hard
CHALK Chalkboard Videos	**WS** Worksheet
	T Team Problem

Section 30.1 Rutherford Scattering and the Nuclear Atom

1. **E** **SSM** The nucleus of the hydrogen atom has a radius of about 1×10^{-15} m. The electron is normally at a distance of about 5.3×10^{-11} m from the nucleus. Assuming that the hydrogen atom is a sphere with a radius of 5.3×10^{-11} m, find **(a)** the volume of the atom, **(b)** the volume of the nucleus, and **(c)** the percentage of the volume of the atom that is occupied by the nucleus.

2. **E** The nucleus of a hydrogen atom is a single proton, which has a radius of about 1.0×10^{-15} m. The single electron in a hydrogen atom normally orbits the nucleus at a distance of 5.3×10^{-11} m. What is the ratio of the density of the hydrogen nucleus to the density of the complete hydrogen atom?

3. **E** Review Conceptual Example 1 and use the information therein as an aid in working this problem. Suppose that you're building a scale model of the hydrogen atom, and the nucleus is represented by a ball that has a radius of 3.2 cm (somewhat smaller than a baseball).

How many miles away (1 mi = 1.61×10^5 cm) should the electron be placed?

4. **E** **V-HINT** In a Rutherford scattering experiment a target nucleus has a diameter of 1.4×10^{-14} m. The incoming α particle has a mass of 6.64×10^{-27} kg. What is the kinetic energy of an α particle that has a de Broglie wavelength equal to the diameter of the target nucleus? Ignore relativistic effects.

5. **M** **CHALK** **SSM** There are Z protons in the nucleus of an atom, where Z is the atomic number of the element. An α particle carries a charge of $+2e$. In a scattering experiment, an α particle, heading directly toward a nucleus in a metal foil, will come to a halt when all the particle's kinetic energy is converted to electric potential energy. In such a situation, how close will an α particle with a kinetic energy of 5.0×10^{-13} J come to a gold nucleus ($Z = 79$)?

6. **M** **GO** The nucleus of a copper atom contains 29 protons and has a radius of 4.8×10^{-15} m. How much work (in electron volts) is done by the electric force as a proton is brought from infinity, where it is at rest, to the "surface" of a copper nucleus?

Section 30.2 Line Spectra,

Section 30.3 The Bohr Model of the Hydrogen Atom

7. **E** **SSM** For a doubly ionized lithium atom Li^{2+} ($Z = 3$), what is the principal quantum number of the state in which the electron has the same total energy as a ground-state electron has in the hydrogen atom?

8. **E** A singly ionized helium atom (He^+) has only one electron in orbit about the nucleus. What is the radius of the ion when it is in the second excited state?

9. **E** Using the Bohr model, determine the ratio of the energy of the nth orbit of a triply ionized beryllium atom (Be^{3+}, $Z = 4$) to the energy of the nth orbit of a hydrogen atom (H).

10. **E** **MMH** The electron in a hydrogen atom is in the first excited state, when the electron acquires an additional 2.86 eV of energy. What is the quantum number n of the state into which the electron moves?

11. **E** **SSM** Find the energy (in joules) of the photon that is emitted when the electron in a hydrogen atom undergoes a transition from the $n = 7$ energy level to produce a line in the Paschen series.

12. **E** **GO** (a) What is the ionization energy of a hydrogen atom that is in the $n = 4$ excited state? (b) For a hydrogen atom, determine the ratio of the ionization energy for the $n = 4$ excited state to the ionization energy for the ground state.

13. **E** **CHALK** A hydrogen atom is in the ground state. It absorbs energy and makes a transition to the $n = 3$ excited state. The atom returns to the ground state by emitting two photons. What are their wavelengths?

14. **E** In the hydrogen atom, what is the total energy (in electron volts) of an electron that is in an orbit that has a radius of 4.761×10^{-10} m?

15. **M** **GO** A sodium atom ($Z = 11$) contains 11 protons in its nucleus. Strictly speaking, the Bohr model does not apply, because the neutral atom contains 11 electrons instead of a single electron. However, we can apply the model to the outermost electron as an approximation, provided that we use an effective value $Z_{effective}$ rather than 11 for the number of protons in the nucleus. (a) The ionization energy for the outermost electron in a sodium atom is 5.1 eV. Use the Bohr model with $Z = Z_{effective}$ to calculate a value for $Z_{effective}$. (b) Using $Z = 11$ and $Z = Z_{effective}$, determine the corresponding two values for the radius of the outermost Bohr orbit.

16. **M** **GO** **SSM** A wavelength of 410.2 nm is emitted by the hydrogen atoms in a high-voltage discharge tube. What are the initial and final values of the quantum number n for the energy level transition that produces this wavelength?

17. **M** **V-HINT** **MMH** A hydrogen atom emits a photon that has momentum with a magnitude of 5.452×10^{-27} kg · m/s. This photon is emitted because the electron in the atom falls from a higher energy level into the $n = 1$ level. What is the quantum number of the level from which the electron falls? Use a value of 6.626×10^{-34} J · s for Planck's constant.

18. **E** **SSM** For atomic hydrogen, the Paschen series of lines occurs when $n_f = 3$, whereas the Brackett series occurs when $n_f = 4$ in Equation 30.14. Using this equation, show that the ranges of wavelengths in these two series overlap.

19. **M** **GO** **CHALK** Doubly ionized lithium Li^{2+} ($Z = 3$) and triply ionized beryllium Be^{3+} ($Z = 4$) each emit a line spectrum. For a certain series of lines in the lithium spectrum, the shortest wavelength is 40.5 nm. For the same series of lines in the beryllium spectrum, what is the shortest wavelength?

20. **H** In the Bohr model of hydrogen, the electron moves in a circular orbit around the nucleus. Determine the angular speed of the electron, in revolutions per second, when it is in (a) the ground state and (b) the first excited state.

Section 30.5   The Quantum Mechanical Picture of the Hydrogen Atom

21. **E** A hydrogen atom is in its second excited state. Determine, according to quantum mechanics, (a) the total energy (in eV) of the atom, (b) the magnitude of the maximum angular momentum the electron can have in this state, and (c) the maximum value that the z component L_z of the angular momentum can have.

22. **E** **GO** The table lists quantum numbers for five states of the hydrogen atom. Which (if any) of them are not possible? For those that are not possible, explain why.

Process	n	ℓ	m_ℓ
(a)	3	3	0
(b)	2	1	−1
(c)	4	2	3
(d)	5	−3	2
(e)	4	0	0

23. **E** The orbital quantum number for the electron in a hydrogen atom is $\ell = 5$. What is the smallest possible value (the most negative) for the total energy of this electron? Give your answer in electron volts.

24. **E** **GO** It is known that the possible values for the magnetic quantum number m_ℓ are −4, −3, −2, −1, 0, +1, +2, +3, and +4. Determine the orbital quantum number ℓ and the smallest possible value of the principal quantum number n.

25. **E** **SSM** The maximum value for the magnetic quantum number in state A is $m_\ell = 2$, while in state B it is $m_\ell = 1$. What is the ratio L_A/L_B of the magnitudes of the orbital angular momenta of an electron in these two states?

26. **M** **V-HINT** **CHALK** The electron in a certain hydrogen atom has an angular momentum of 8.948×10^{-34} J · s. What is the largest possible magnitude for the z component of the angular momentum of this electron? For accuracy, use $h = 6.626 \times 10^{-34}$ J · s.

27. **M** **SSM** **MMH** For an electron in a hydrogen atom, the z component of the angular momentum has a *maximum* value of $L_z = 4.22 \times 10^{-34}$ J · s. Find the three smallest possible values (the most negative) for the total energy (in electron volts) that this atom could have.

28. **H** **GO** An electron is in the $n = 5$ state. What is the smallest possible value for the angle between the z component of the orbital angular momentum and the orbital angular momentum?

Section 30.6   The Pauli Exclusion Principle and the Periodic Table of the Elements

29. **E** **CHALK** Two of the three electrons in a lithium atom have quantum numbers of $n = 1$, $\ell = 0$, $m_\ell = 0$, $m_s = +\frac{1}{2}$ and $n = 1$, $\ell = 0$, $m_\ell = 0$, $m_s = -\frac{1}{2}$. What quantum numbers can the third electron have if the atom is in (a) its ground state and (b) its first excited state?

30. **E** Following the style used in **Table 30.3**, determine the electronic configuration of the ground state for yttrium Y ($Z = 39$). Refer to **Figure 30.16** to see the order in which the subshells fill.

31. **E** **Figure 30.16** was constructed using the Pauli exclusion principle and indicates that the $n = 1$ shell holds 2 electrons, the $n = 2$ shell holds 8 electrons, and the $n = 3$ shell holds 18 electrons. These numbers can be obtained by adding the numbers given in the figure for the subshells contained within a given shell. How many electrons can be put into the $n = 5$ shell, which is only partly shown in the figure?

32. **E** **GO** Which of the following subshell configurations are not allowed? For those that are not allowed, give the reason why. (a) $3s^1$ (b) $2d^2$ (c) $3s^4$ (d) $4p^8$ (e) $5f^{12}$

33. E SSM When an electron makes a transition between energy levels of an atom, there are no restrictions on the initial and final values of the principal quantum number n. According to quantum mechanics, however, there is a rule that restricts the initial and final values of the orbital quantum number ℓ. This rule is called a *selection rule* and states that $\Delta\ell = \pm 1$. In other words, when an electron makes a transition between energy levels, the value of ℓ can only increase or decrease by one. The value of ℓ may not remain the same nor may it increase or decrease by more than one. According to this rule, which of the following energy level transitions are allowed? **(a)** 2s → 1s **(b)** 2p → 1s **(c)** 4p → 2p **(d)** 4s → 2p **(e)** 3d → 3s

34. M In the ground state, the outermost shell ($n = 1$) of helium (He) is filled with electrons, as is the outermost shell ($n = 2$) of neon (Ne). The full outermost shells of these two elements distinguish them as the first two so-called "noble gases." Suppose that the spin quantum number m_s had *three* possible values, rather than two. If that were the case, which elements would be **(a)** the first and **(b)** the second noble gases? Assume that the possible values for the other three quantum numbers are unchanged, and that the Pauli exclusion principle still applies.

Section 30.7   X-Rays

35. E MMH By using the Bohr model, decide which element is likely to emit a K_α X-ray with a wavelength of 4.5×10^{-9} m.

36. E GO What is the minimum potential difference that must be applied to an X-ray tube to knock a K-shell electron completely out of an atom in a copper ($Z = 29$) target? Use the Bohr model as needed.

37. E V-HINT In the X-ray spectrum of niobium ($Z = 41$), a K_α peak is observed at a wavelength of 7.462×10^{-11} m. **(a)** Determine the magnitude of the difference between the observed wavelength of the K_α X-ray for niobium and that predicted by the Bohr model. **(b)** Express the magnitude of this difference as a percentage of the observed wavelength.

38. E CHALK SSM When a certain element is bombarded with high-energy electrons, K_α X-rays that have an energy of 9890 eV are emitted. Determine the atomic number Z of the element, and identify the element. Use the Bohr model as necessary.

39. M GO The Bohr model, although not strictly applicable, can be used to estimate the minimum energy E_{min} that an incoming electron must have in an X-ray tube in order to knock a K-shell electron entirely out of an atom in the metal target. The K_α X-ray wavelength of metal A is 2.0 times the K_α X-ray wavelength of metal B. What is the ratio of $E_{min, A}$ for metal A to $E_{min, B}$ for metal B?

40. M SSM An X-ray tube contains a silver ($Z = 47$) target. The high voltage in this tube is increased from zero. Using the Bohr model, find the value of the voltage at which the K_α X-ray just appears in the X-ray spectrum.

41. M GO Multiple-Concept Example 10 reviews the concepts that are important in this problem. An electron, traveling at a speed of 6.00×10^7 m/s, strikes the target of an X-ray tube. Upon impact, the electron decelerates to one-quarter of its original speed, an X-ray photon being emitted in the process. What is the wavelength of the photon?

Section 30.8   The Laser

42. E GO A pulsed laser emits light in a series of short pulses, each having a duration of 25.0 ms. The average power of each pulse is 5.00 mW, and the wavelength of the light is 633 nm. Find the number of photons in each pulse.

43. E GO CHALK The drawing shows three energy levels of a laser that are involved in the lasing action. These levels are analogous to the levels in the Ne atoms of a He-Ne laser. The E_2 level is a metastable level, and the E_0 level is the ground state. The difference between the energy levels of the laser is shown in the drawing. **(a)** What energy (in eV per electron) must an external source provide to start the lasing action? **(b)** What is the wavelength of the laser light? **(c)** In what region of the electromagnetic spectrum (see Figure 24.9) does the laser light lie?

PROBLEM 43

44. M V-HINT Fusion is the process by which the sun produces energy. One experimental technique for creating controlled fusion utilizes a solid-state laser that emits a wavelength of 1060 nm and can produce a power of 1.0×10^{14} W for a pulse duration of 1.1×10^{-11} s. In contrast, the helium/neon laser used in a bar-code scanner at the checkout counter emits a wavelength of 633 nm and produces a power of about 1.0×10^{-3} W. How long (in days) would the helium/neon laser have to operate to produce the same number of photons that the solid-state laser produces in 1.1×10^{-11} s?

Additional Problems

Online

45. E GO **(a)** What is the minimum energy (in electron volts) that is required to remove the electron from the ground state of a singly ionized helium atom (He$^+$, $Z = 2$)? **(b)** What is the ionization energy for He$^+$?

46. E Molybdenum has an atomic number of $Z = 42$. Using the Bohr model, estimate the wavelength of the K_α X-ray.

47. E SSM In the line spectrum of atomic hydrogen there is also a group of lines known as the Pfund series. These lines are produced when electrons, excited to high energy levels, make transitions to the $n = 5$ level. Determine **(a)** the longest wavelength and **(b)** the shortest wavelength in this series. **(c)** Refer to Figure 24.9 and identify the region of the electromagnetic spectrum in which these lines are found.

48. E GO The voltage across an X-ray tube is 35.0 kV. Suppose that the molybdenum ($Z = 42$) target in the X-ray tube is replaced by a silver ($Z = 47$) target. Determine **(a)** the tube's cutoff wavelength and **(b)** the wavelengths of the K_α X-ray photons emitted by the molybdenum and silver targets.

49. ☐M☐ ☐V-HINT☐ The energy of the $n = 2$ Bohr orbit is -30.6 eV for an unidentified ionized atom in which only one electron moves about the nucleus. What is the radius of the $n = 5$ orbit for this species?

50. ☐M☐ Consider a particle of mass m that can exist only between $x = 0$ m and $x = +L$ on the x axis. We could say that this particle is confined to a "box" of length L. In this situation, imagine the standing de Broglie waves that can fit into the box. For example, the drawing shows the first three possibilities. Note in this picture that there are either one, two, or three half-wavelengths that fit into the distance L. Use Equation 29.8 for the de Broglie wavelength of a particle and derive an expression for the allowed energies (only kinetic energy) that the particle can have. This expression involves m, L, Planck's constant, and a quantum number n that can have only the values 1, 2, 3, ...

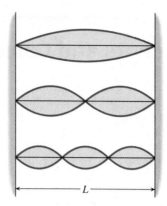

PROBLEM 50 $\longleftarrow L \longrightarrow$

Physics in Biology, Medicine, and Sports

51. ☐E☐ ☐BIO☐ ☐SSM☐ **30.8** A laser is used in eye surgery to weld a detached retina back into place. The wavelength of the laser beam is 514 nm, and the power is 1.5 W. During surgery, the laser beam is turned on for 0.050 s. During this time, how many photons are emitted by the laser?

52. ☐E☐ ☐BIO☐ ☐GO☐ **30.8** The ultraviolet excimer laser used in the PRK technique (see Section 30.9) has a wavelength of 193 nm. A carbon dioxide laser produces a wavelength of 1.06×10^{-5} m. What is the minimum number of photons that the carbon dioxide laser must produce to deliver at least as much or more energy to a target as does a single photon from the excimer laser?

53. ☐E☐ ☐BIO☐ **30.8** A laser peripheral iridotomy is a procedure for treating an eye condition known as narrow-angle glaucoma, in which pressure buildup in the eye can lead to loss of vision. A neodymium YAG laser (wavelength = 1064 nm) is used in the procedure to punch a tiny hole in the peripheral iris, thereby relieving the pressure buildup. In one application the laser delivers 4.1×10^{-3} J of energy to the iris in creating the hole. How many photons does the laser deliver?

54. ☐M☐ ☐BIO☐ **30.7** The tube in a hospital X-ray machine has an operating voltage of 90 kV. **(a)** What is the maximum energy (in eV) of a photon produced by this tube? **(b)** What is the shortest wavelength of an X-ray photon produced by the same tube?

55. ☐M☐ ☐BIO☐ **30.7** The target element in a veterinary X-ray machine is tungsten. Use the Bohr model to estimate the wavelength of **(a)** the K_α line and **(b)** the K_β line for this X-ray source.

56. ☐E☐ ☐BIO☐ **30.1** The number of atoms in a single human cell is on the order of the number of cells in the entire human body. Research suggests that a single cell consists of 100 trillion (1.0×10^{14}) atoms. Assuming the radius of these atoms is one angstrom, what would be the volume of a single cell?

57. ☐M☐ ☐BIO☐ **30.3** A pulsed dye laser that is used for the treatment (removal) of port-wine stains from the skin produces laser light at a wavelength of 486 nm. Assume this light is produced by electrons in hydrogen that undergo a transition in the Balmer series. Use the Bohr model to determine the initial energy state (n_i) of these electrons.

58. ☐M☐ ☐BIO☐ **30.1** There are several proteins that exist in nature that are *fluorescent*, which means they absorb visible light or other electromagnetic radiation and reemit that light, usually at a longer wavelength.

One of the more prevalent fluorescent proteins is GFP, or green fluorescent protein, which emits bright green fluorescent light when exposed to light in the blue to ultraviolet region of the electromagnetic spectrum (see the photo). The main absorption, or excitation, wavelength for one type of GFP is 395 nm. **(a)** What is the energy (in eV) of a photon in this absorbed light? The emission spectrum of the light from the GFP peaks near 509 nm, which is in the shorter wavelength (higher energy) portion of the green part of the visible spectrum. **(b)** What is the energy (in eV) of a photon in the emitted light? **(c)** If the emitted light corresponded to the longest wavelength in the Balmer series from a hydrogen-like atom, what would be the effective value of Z for this atom? (*Note: Z* will not be a whole number in this case.)

The mice on the left and right have been genetically modified to express GFP when illuminated with UV light. Their exposed skin cells (eyes, nose, ears, and tail) glow bright green due to fluorescence and indicate healthy cells. The mice also contain subcutaneous breast cancer tumor cells that glow red (not shown), so they can be easily distinguished from healthy cells. This aids researchers in tracking the metastization process of the cancer. The mouse in the middle has not been modified.

Ingrid Moen, Charlotte Jevne, Jian Wang, Karl-Henning Kalland, Martha Chekenya, Lars A Akslen, Linda Sleire, Per ØEnger, Rolf K Reed, Anne M Øyan and Linda EB Stuhr: *Gene expression in tumor cells and stroma in dsRed 4T1 tumors in eGFP-expressing mice with and without enhanced oxygenation.* https://commons. wikimedia.org/wiki/File:GFP_Mice_01.jpg

PROBLEM 58

Concepts and Calculations Problems

Online

The Bohr model of the hydrogen atom introduces a number of important features that characterize the quantum picture of atomic structure. Among them are the concepts of quantized energy levels and the photon emission that occurs when an electron makes a transition from a higher to a lower energy state. Problem 59 deals with these ideas. Problem 60 reviews the physics of how a K_α X-ray is produced, how its energy is related to the ionization energies of the target atom, and how to determine the minimum voltage needed to produce it in an X-ray tube.

59. M CHALK A hydrogen atom ($Z = 1$) is in the third excited state, and a photon is either emitted or absorbed. *Concepts:* (i) What is the quantum number of the third excited state? (ii) When an atom emits a photon, is the final quantum number n_f of the atom greater than or less than the initial quantum number n_i? (iii) When an atom absorbs a photon, is the final quantum number n_f of the atom greater than or less than the initial quantum number n_i? (iv) How is the wavelength of a photon related to its energy? *Calculations:* Determine the quantum number n_f of the final state and the energy of the photon when the photon is **(a)** emitted with the shortest possible wavelength, **(b)** emitted with the longest possible wavelength, and **(c)** absorbed with the longest possible wavelength.

60. M CHALK SSM The K-shell and L-shell ionization energies of a metal are 8979 eV and 951 eV, respectively. *Concepts:* (i) How is a K_α photon produced, and how much energy does it have? (ii) What must be the minimum voltage across the X-ray tube to produce a K_α photon? (iii) What is meant by the phrases "K-shell ionization energy" and "L-shell ionization energy"? (iv) What does the difference between the K-shell and L-shell ionization energies represent? *Calculations:* **(a)** Assuming that there is a vacancy in the L shell, what must be the minimum voltage across an X-ray tube with a target made from this metal to produce K_α X-ray photons? **(b)** Determine the wavelength of a K_α photon.

Team Problems

Online

61. E T WS **Detecting a Poison Gas.** You and your team are exploring an old laboratory and happen upon a sealed corridor with a sign on the door that reads "Danger: fluorine leak." You use your phone to look up fluorine (F_2) on the Internet and find that it is a halogen gas that is extremely poisonous, and one should be careful not to allow it to make contact with their skin and to not breathe it in, even in small amounts. The problem is that you have to get through the area that is sealed, and therefore must determine whether there is still any gas present in the space. You are fortunate that, in a room that you have already explored, there is a spectrometer that is capable of energizing a vial of gas by subjecting it to a high voltage, and then analyzing the atomic light emission. The spectrometer has a detectable wavelength range of 385 to 750 nm. The only atomic transitions in fluorine that you are able to find on the Internet are the following: $E_{01} = -12.697$ eV to $E_{f1} = -14.683$ eV; $E_{02} = -12.731$ eV to $E_{f2} = -14.683$ eV; $E_{03} = -12.697$ eV to $E_{f3} = -14.505$ eV; and $E_{04} = -0.05010$ eV to $E_{f4} = -13.025$ eV, where the subscripts "0" and "f" correspond to the initial and final states, respectively. **(a)** Determine the energies of the photons emitted during each transition in joules. **(b)** What are the wavelengths of the light emitted during these transitions? **(c)** What wavelengths (emission lines) will be detected by the spectrometer if fluorine gas is in the room?

62. M T WS **An Atomic Emission Device.** You and your team are designing a device that utilizes a large hydrogen-gas discharge lamp as a light source. A ruled grating is used to separate the emission lines emanating from the lamp, and the specific line that you will use is one of the Lyman lines ($n = 4 \to n = 1$). The light is collimated into a beam of rectangular cross section with length $L = 0.75$ cm and width $W = 0.55$ cm, which illuminates a flat, grounded metallic surface with an intensity of 1250 W/m². **(a)** What is the wavelength of the emitted light? **(b)** What is the energy of each emitted photon (in J)? **(c)** How many photons hit the metallic surface per second? **(d)** If 0.75% of the impinging photons result in ejected photoelectrons, what is the magnitude of the resulting electric current in the metal?

63. T M **A Simple Spectrometer.** In the Balmer series for the atomic emission of hydrogen, emission lines with n going from $3 \to 2, 4 \to 2, 5 \to 2$, and $6 \to 2$ are in the visible spectrum. **(a)** What are the wavelengths of the light emitted in these transitions? **(b)** A spectrometer is a device that disperses light into its constituent wavelengths, and provides a means of measuring these wavelengths. You and your team are designing a simple spectrometer using a transmission grating with 1.50×10^3 lines/cm. A collimated beam of light from a hydrogen discharge tube (which emits light from all possible hydrogen atomic transitions) passes through the grating. At what second-order angles ($m = 2$) do you expect each of the emission lines calculated in (a) to appear? **(c)** You place a screen with a coating that fluoresces when exposed to ultraviolet light in the light emerging from the grating, and you observe another line at $\theta = 6.84°$. Calculate the wavelength of this light. Does it come from the Balmer series?

64. T M **An X-ray Source.** You and your team are building an X-ray source for a diffraction instrument used to study the crystalline structure of materials. You decide to use the K_α radiation from a copper ($Z = 29$) target. **(a)** Estimate the minimum energy that an incoming electron must have in order to knock a K-shell electron out of the Cu atom. **(b)** Through what voltage must you accelerate the electrons? **(c)** What is the cutoff wavelength of the Bremsstrahlung radiation for the electron energy found in (a)? **(d)** What is the wavelength of the K_α line? Convert this to the energy of the K_α X-ray photons (in eV) **(e)** If a first-order peak ($m = 1$) appears at an angle of 17.0° when the K_α radiation found in (d) passes through a hypothetical diffraction grating, what would be the "line spacing" of the grating? Note: distances between atoms in solids are on the order of 10^{-10} m.

Nuclear reactors are used for research and power generation. Their operation relies on a fuel that is composed of radioactive elements. The nuclei in the atoms of these elements are unstable, and they break apart over time, releasing different types of radiation, known as decay products. The image above shows spent nuclear fuel being removed from the High Flux Isotope Reactor at Oak Ridge National Lab in Oak Ridge, Tennessee. The intense blue light, known as Cherenkov radiation, is the result of the decay products from the fuel interacting with the surrounding water that is required to keep the assembly cool.

Genevieve Martin/ORNL, U.S. Dept. of Energy

Nuclear Physics and Radioactivity

31.1 | Nuclear Structure

Atoms consist of electrons in orbit about a central nucleus. As we have seen in Chapter 30, the electron orbits are quantum mechanical in nature and have interesting characteristics. Little has been said about the nucleus, however. Since the nucleus is interesting in its own right, we now consider it in greater detail.

The nucleus of an atom consists of neutrons and protons, collectively referred to as **nucleons**. The **neutron**, discovered in 1932 by the English physicist James Chadwick (1891–1974), carries no electric charge and has a mass slightly larger than that of a proton (see **Table 31.1**).

The number of protons in the nucleus is different in different elements and is given by the **atomic number** Z. In an electrically neutral atom, the number of nuclear protons equals the number of electrons in orbit around the nucleus. The number of neutrons in the nucleus is N.

LEARNING OBJECTIVES

After reading this module, you should be able to...

31.1 Identify and explain the properties of the nucleus.

31.2 Describe the strong nuclear force.

31.3 Calculate nuclear binding energy and mass defect.

31.4 Analyze alpha, beta, and gamma decays.

31.5 Explain the role of the neutrino in weak nuclear decay.

31.6 Solve problems involving radioactive decay.

31.7 Calculate age using radioactive dating.

31.8 Analyze radioactive decay series.

31.9 Explain how radiation detectors operate.

TABLE 31.1   Properties of Select Particles

Particle	Electric Charge (C)	Mass	
		Kilograms (kg)	Atomic Mass Units (u)
Electron	-1.60×10^{-19}	$9.109\,382 \times 10^{-31}$	$5.485\,799 \times 10^{-4}$
Proton	$+1.60 \times 10^{-19}$	$1.672\,622 \times 10^{-27}$	$1.007\,276$
Neutron	0	$1.674\,927 \times 10^{-27}$	$1.008\,665$
Hydrogen atom	0	$1.673\,534 \times 10^{-27}$	$1.007\,825$

The total number of protons and neutrons is referred to as the **atomic mass number** A because the total nuclear mass is *approximately* equal to A times the mass of a single nucleon:

$$\underbrace{A}_{\substack{\text{Number of protons} \\ \text{and neutrons (atomic} \\ \text{mass number or} \\ \text{nucleon number)}}} = \underbrace{Z}_{\substack{\text{Number of protons} \\ \text{(atomic number)}}} + \underbrace{N}_{\substack{\text{Number of} \\ \text{neutrons}}} \qquad (31.1)$$

Sometimes A is also called the **nucleon number.** A shorthand notation is often used to specify Z and A along with the chemical symbol for the element. For instance, the nuclei of all naturally occurring aluminum atoms have $A = 27$, and the atomic number for aluminum is $Z = 13$. In shorthand notation, then, the aluminum nucleus is specified as $^{27}_{13}\text{Al}$. The number of neutrons in an aluminum nucleus is not given explicitly by this shorthand notation. However, it can be determined easily with the aid of Equation 31.1, which indicates that $N = A - Z = 14$. In general, for an element whose chemical symbol is X, the symbol for the nucleus is

$$^{A}_{Z}\text{X}$$

For a proton the symbol is $^{1}_{1}\text{H}$, since the proton is the nucleus of a hydrogen atom. A neutron is denoted by $^{1}_{0}\text{n}$. In the case of an electron we use $^{0}_{-1}\text{e}$, where $A = 0$ because an electron is not composed of protons or neutrons and $Z = -1$ because the electron has a negative charge.

Nuclei that contain the same number of protons, but a different number of neutrons, are known as **isotopes.** Carbon, for example, occurs in nature in two stable forms. In most carbon atoms (98.90%), the nucleus is the $^{12}_{6}\text{C}$ isotope and consists of six protons and six neutrons. A small fraction (1.10%), however, contain nuclei that have six protons and seven neutrons—namely, the $^{13}_{6}\text{C}$ isotope. The percentages given here are the natural abundances of the isotopes. The atomic masses in the periodic table are average atomic masses, taking into account the abundances of the various isotopes.

The protons and neutrons in the nucleus are clustered together to form an approximately spherical region, as **Interactive Figure 31.1** illustrates. Experiment shows that the radius r of the nucleus depends on the atomic mass number A and is given approximately in meters by

$$r \approx (1.2 \times 10^{-15}\ \text{m})A^{1/3} \qquad (31.2)$$

This equation indicates that the radius of the aluminum nucleus ($A = 27$), for example, is $r \approx (1.2 \times 10^{-15}\ \text{m})27^{1/3} = 3.6 \times 10^{-15}\ \text{m}$. Equation 31.2 leads to an important conclusion concerning the nuclear density of different atoms, as Conceptual Example 1 discusses.

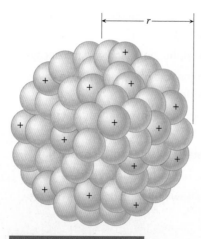

INTERACTIVE FIGURE 31.1 The nucleus in an atom is approximately spherical (radius = r) and contains protons (⊕) clustered closely together with neutrons (⬤).

CONCEPTUAL EXAMPLE 1 | Nuclear Density

It is well known that lead and oxygen contain different atoms and that the density of solid lead is much greater than the density of gaseous oxygen. Using the definition of density along with Equation 31.2, decide whether the density of the *nucleus* in a lead atom is **(a)** greater than, **(b)** approximately equal to, or **(c)** less than the density of the *nucleus* in an oxygen atom.

Reasoning The density ρ of an object, such as the nucleus, is defined as its mass M divided by its volume V: $\rho = M/V$ (Equation 11.1). The mass of a nucleus is approximately equal to the number A of nucleons in the nucleus times the mass m of a single nucleon, since the masses of a proton and a neutron are nearly the same. Thus, we have that $M \approx Am$, where A is greater for lead than for oxygen, but m is the same for both. The nucleus is approximately spherical with a radius r, so its volume V is given by $V = \frac{4}{3}\pi r^3$. The radius, however, depends on the number A of nucleons through the relation $r \approx (1.2 \times 10^{-15} \text{ m})A^{1/3}$ (Equation 31.2). Therefore, we can write the density of a nucleus as follows:

$$\rho = \frac{M}{V} \approx \frac{Am}{\frac{4}{3}\pi r^3} = \frac{Am}{\frac{4}{3}\pi[(1.2 \times 10^{-15} \text{ m})A^{1/3}]^3} \approx \frac{m}{\frac{4}{3}\pi(1.2 \times 10^{-15} \text{ m})^3}$$

Note that the nucleon number A has been eliminated algebraically from this result, as a direct consequence of Equation 31.2.

Answers (a) and (c) are incorrect. The result obtained in the Reasoning section for the nuclear density ρ depends only on numerical factors and the value of m, which is the mass of a single nucleon no matter where the nucleon is located. The nuclear density does not depend on the nuclear number A. Thus, the nuclear density of lead, which is the ratio of its mass to its volume, is neither greater than nor less than the nuclear density of oxygen.

Answer (b) is correct. The result obtained in the Reasoning section for the nuclear density ρ indicates that the density of the nucleus in a lead atom is approximately the same as it is in an oxygen atom. In general, because of Equation 31.2, the *nuclear* density has nearly the same value in all atoms. The difference in densities between solid lead and gaseous oxygen, however, arises mainly because of the difference in how closely the atoms are packed together in the solid and gaseous phases.

Check Your Understanding

(The answers are given at the end of the book.)

1. Two nuclei differ in their numbers of protons and their numbers of neutrons. Which one or more of the following statements is/are true? **(a)** They are different isotopes of the same element. **(b)** They have the same electric charge. **(c)** They could have the same radii. **(d)** They have approximately the same nuclear density.

2. A material is known to be an isotope of lead, although the particular isotope is not known. From such limited information, which of the following quantities can you specify? **(a)** Its atomic number **(b)** Its neutron number **(c)** Its atomic mass number

3. Two nuclei have different nucleon numbers A_1 and A_2. Are the two nuclei necessarily isotopes of the same element?

4. Can two nuclei have the same radius, even though they contain different numbers of protons and different numbers of neutrons?

31.2 | The Strong Nuclear Force and the Stability of the Nucleus

Two positive charges that are as close together as they are in a nucleus repel one another with a very strong electrostatic force. What, then, keeps the nucleus from flying apart? Clearly, some kind of attractive force must hold the nucleus together, since many kinds of naturally occurring atoms contain stable nuclei. The gravitational force of attraction between nucleons is too weak to counteract the repulsive electric force, so a different type of force must hold the nucleus together. This force is the **strong nuclear force** and is one of only three fundamental forces that have been discovered, fundamental in the sense that all forces in nature can be explained in terms of these three. The gravitational force is also one of these forces, as is the electroweak force (see Section 31.5).

Many features of the strong nuclear force are well known. For example, it is almost independent of electric charge. At a given separation distance, nearly the same nuclear force of attraction exists between two protons, between two neutrons, or between a proton and a neutron. The range of action of the strong nuclear force is extremely short,

FIGURE 31.2 With few exceptions, the naturally occurring stable nuclei have a number N of neutrons that equals or exceeds the number Z of protons. Each dot in this plot represents a stable nucleus.

with the force of attraction being very strong when two nucleons are as close as 10^{-15} m and essentially zero at larger distances. In contrast, the electric force between two protons decreases to zero only gradually as the separation distance increases to large values and, therefore, has a relatively long range of action.

The limited range of action of the strong nuclear force plays an important role in the stability of the nucleus. For a nucleus to be stable, the electrostatic repulsion between the protons must be balanced by the attraction between the nucleons due to the strong nuclear force. But one proton repels all other protons within the nucleus, since the electrostatic force has such a long range of action. In contrast, a proton or a neutron attracts only its nearest neighbors via the strong nuclear force. As the number Z of protons in the nucleus increases under these conditions, the number N of neutrons has to increase even more, if stability is to be maintained. **Figure 31.2** shows a plot of N versus Z for naturally occurring elements that have stable nuclei. For reference, the plot also includes the straight line that represents the condition $N = Z$. With few exceptions, the points representing stable nuclei fall above this reference line, reflecting the fact that the number of neutrons becomes greater than the number of protons as the atomic number Z increases.

As more and more protons occur in a nucleus, there comes a point when a balance of repulsive and attractive forces cannot be achieved by an increased number of neutrons. Eventually, the limited range of action of the strong nuclear force prevents extra neutrons from balancing the long-range electric repulsion of extra protons. The stable nucleus with the largest number of protons ($Z = 83$) is that of bismuth, $^{209}_{83}\text{Bi}$, which contains 126 neutrons. All nuclei with more than 83 protons (e.g., uranium, $Z = 92$) are unstable and spontaneously break apart or rearrange their internal structures as time passes. This spontaneous disintegration or rearrangement of internal structure is called **radioactivity**, first discovered in 1896 by the French physicist Antoine Becquerel (1852–1908). Section 31.4 discusses radioactivity in greater detail.

31.3 The Mass Defect of the Nucleus and Nuclear Binding Energy

Because of the strong nuclear force, the nucleons in a stable nucleus are held tightly together. Therefore, energy is required to separate a stable nucleus into its constituent protons and neutrons, as **Interactive Figure 31.3** illustrates. The more stable the nucleus is, the greater is the amount of energy needed to break it apart. The required energy is called the **binding energy** of the nucleus.

Two ideas that we have studied previously come into play as we discuss the binding energy of a nucleus. These are the rest energy of an object (Section 28.6) and mass (Section 4.2). In Einstein's theory of special relativity, energy and mass are equivalent; in fact, the rest energy E_0 and the mass m are related via $E_0 = mc^2$ (Equation 28.5), where c is the speed of light in a vacuum. Therefore, a change ΔE_0 in the rest energy of the system is equivalent to a change Δm in the mass of the system, according to $\Delta E_0 = (\Delta m)c^2$. We see, then, that the binding energy used in **Interactive Figure 31.3** to disassemble the nucleus appears as extra mass of the separated and stationary nucleons. In other words, the sum of the individual masses of the separated protons and neutrons is greater by an amount Δm than the mass of the stable nucleus. The difference in mass Δm is known as the **mass defect** of the nucleus.

As Example 2 shows, the binding energy of a nucleus can be determined from the mass defect according to Equation 31.3:

$$\text{Binding energy} = (\text{Mass defect})c^2 = (\Delta m)c^2 \tag{31.3}$$

INTERACTIVE FIGURE 31.3 Energy, called the binding energy, must be supplied to break the nucleus apart into its constituent protons and neutrons. Each of the separated nucleons is at rest and out of the range of the forces of the other nucleons.

EXAMPLE 2 | The Binding Energy of the Helium Nucleus

The most abundant isotope of helium has a ^4_2He nucleus whose mass is 6.6447×10^{-27} kg. For this nucleus, find **(a)** the mass defect and **(b)** the binding energy.

Reasoning   The symbol ^4_2He indicates that the helium nucleus contains $Z = 2$ protons and $N = 4 - 2 = 2$ neutrons. To obtain the mass defect Δm, we first determine the sum of the individual masses of the separated protons and neutrons. Then we subtract from this sum the mass of the $_2^4$He nucleus. Finally, we use Equation 31.3 to calculate the binding energy from the value for Δm.

Solution   (a) Using data from **Table 31.1**, we find that the sum of the individual masses of the nucleons is

$$\underbrace{2(1.6726 \times 10^{-27} \text{ kg})}_{\text{Two protons}} + \underbrace{2(1.6749 \times 10^{-27} \text{ kg})}_{\text{Two neutrons}} = 6.6950 \times 10^{-27} \text{ kg}$$

This value is greater than the mass of the intact $_2^4$He nucleus, and the mass defect is

$$\Delta m = 6.6950 \times 10^{-27} \text{ kg} - 6.6447 \times 10^{-27} \text{ kg} = \boxed{0.0503 \times 10^{-27} \text{ kg}}$$

(b) According to Equation 31.3, the binding energy is

$$\frac{\text{Binding}}{\text{energy}} = (\Delta m)c^2 = (0.0503 \times 10^{-27} \text{ kg})(3.00 \times 10^8 \text{ m/s})^2$$
$$= 4.53 \times 10^{-12} \text{ J}$$

Usually, binding energies are expressed in energy units of electron volts instead of joules ($1 \text{ eV} = 1.60 \times 10^{-19}$ J):

$$\frac{\text{Binding}}{\text{energy}} = (4.53 \times 10^{-12} \text{ J}) \left(\frac{1 \text{ eV}}{1.60 \times 10^{-19} \text{ J}} \right) = 2.83 \times 10^7 \text{ eV}$$
$$= \boxed{28.3 \text{ MeV}}$$

In this result, one million electron volts is denoted by the unit MeV. The value of 28.3 MeV is more than two million times greater than the energy required to remove an orbital electron from an atom.

In calculations such as that in Example 2, it is customary to use the **atomic mass unit** (u) instead of the kilogram. As introduced in Section 14.1, the atomic mass unit is one-twelfth of the mass of a $_6^{12}$C atom of carbon. In terms of this unit, the mass of a $_6^{12}$C atom is exactly 12 u. **Table 31.1** also gives the masses of the electron, the proton, and the neutron in atomic mass units. For future calculations, the energy equivalent of one atomic mass unit can be determined by observing that the mass of a proton is 1.6726×10^{-27} kg or 1.0073 u, so that

$$1 \text{ u} = (1 \text{ u}) \left(\frac{1.6726 \times 10^{-27} \text{ kg}}{1.0073 \text{ u}} \right) = 1.6605 \times 10^{-27} \text{ kg}$$

and

$$\Delta E_0 = (\Delta m)c^2 = (1.6605 \times 10^{-27} \text{ kg})(2.9979 \times 10^8 \text{ m/s})^2 = 1.4924 \times 10^{-10} \text{ J}$$

In electron volts, therefore, one atomic mass unit is equivalent to

$$1 \text{ u} = (1.4924 \times 10^{-10} \text{ J}) \left(\frac{1 \text{ eV}}{1.6022 \times 10^{-19} \text{ J}} \right) = 9.315 \times 10^8 \text{ eV} = 931.5 \text{ MeV}$$

Data tables for isotopes, such as the table in Appendix F, give masses in atomic mass units. Typically, however, the given masses are not nuclear masses. They are *atomic masses*—that is, the masses of neutral atoms, including the mass of the orbital electrons. Example 3 deals again with the $_2^4$He nucleus and shows how to take into account the effect of the orbital electrons when using such data to determine binding energies.

EXAMPLE 3 | The Binding Energy of the Helium Nucleus, Revisited

The atomic mass of helium $_2^4$He is 4.0026 u, and the atomic mass of hydrogen $_1^1$H is 1.0078 u. Using atomic mass units instead of kilograms, obtain the binding energy of the $_2^4$He nucleus.

Reasoning   To determine the binding energy, we calculate the mass defect in atomic mass units and then use the fact that one atomic mass unit is equivalent to 931.5 MeV of energy. The mass of 4.0026 u for $_2^4$He *includes the mass of the two electrons in the neutral helium atom.* To calculate the mass defect, we must subtract 4.0026 u from the sum of the individual masses of the nucleons, including the mass of the electrons. As **Figure 31.4** illustrates, the electron mass will be included if the masses of two hydrogen atoms are used in the calculation instead of the masses of two protons. The mass of a $_1^1$H hydrogen atom is given in **Table 31.1** as 1.0078 u, and the mass of a neutron as 1.0087 u.

Solution   The sum of the individual masses is

$$\underbrace{2(1.0078 \text{ u})}_{\substack{\text{Two hydrogen} \\ \text{atoms}}} + \underbrace{2(1.0087 \text{ u})}_{\text{Two neutrons}} = 4.0330 \text{ u}$$

Mass = 4.0026 u   Mass = 4.0330 u

FIGURE 31.4   Data tables usually give the mass of the neutral atom (including the orbital electrons) rather than the mass of the nucleus. When data from such tables are used to determine the mass defect of a nucleus, the mass of the orbital electrons must be taken into account, as this drawing illustrates for the $_2^4$He isotope of helium. See Example 3.

The mass defect is $\Delta m = 4.0330 \text{ u} - 4.0026 \text{ u} = 0.0304 \text{ u}$. Since 1 u is equivalent to 931.5 MeV, the binding energy is

$$\text{Binding energy} = (0.0304 \text{ u}) \left(\frac{931.5 \text{ MeV}}{1 \text{ u}} \right) = \boxed{28.3 \text{ MeV}}$$

which matches the result obtained in Example 2.

FIGURE 31.5 A plot of binding energy per nucleon versus the nucleon number A.

To see how the nuclear binding energy varies from nucleus to nucleus, it is necessary to compare the binding energy for each nucleus on a per-nucleon basis. The graph in **Figure 31.5** shows a plot in which the binding energy divided by the nucleon number A is plotted against the nucleon number itself. In the graph, the peak for the $_2^4\text{He}$ isotope of helium indicates that the $_2^4\text{He}$ nucleus is particularly stable. The binding energy per nucleon increases rapidly for nuclei with small masses and reaches a maximum of approximately 8.7 MeV/nucleon for a nucleon number of about $A = 60$. For greater nucleon numbers, the binding energy per nucleon decreases gradually. Eventually, the binding energy per nucleon decreases enough so there is insufficient binding energy to hold the nucleus together. Nuclei more massive than the $_{83}^{209}\text{Bi}$ nucleus of bismuth are unstable and hence radioactive.

Check Your Understanding

(The answers are given at the end of the book.)

5. Using **Figure 31.5**, rank the following nuclei in ascending order according to the binding energy per nucleon (smallest first): **(a)** Phosphorus $_{15}^{31}\text{P}$ **(b)** Cobalt $_{27}^{59}\text{Co}$ **(c)** Tungsten $_{74}^{84}\text{W}$ **(d)** Thorium $_{90}^{232}\text{Th}$

6. The following table gives values for the mass defect Δm for four hypothetical nuclei: A, B, C, and D. Which statement is true regarding the stability of these nuclei? **(a)** Nucleus D is the most stable, and nucleus A is the least stable. **(b)** Nucleus C is stable, whereas nuclei A, B, and D are not. **(c)** Nucleus A is the most stable, and nucleus D is not stable. **(d)** Nuclei A and B are stable, but nucleus B is more stable than nucleus A.

	A	B	C	D
Mass defect, Δm	$+6.0 \times 10^{-29}$ kg	$+2.0 \times 10^{-29}$ kg	0 kg	-6.0×10^{-29} kg

31.4 Radioactivity

When an unstable or radioactive nucleus disintegrates spontaneously, certain kinds of particles and/or high-energy photons are released. These particles and photons are collectively called "rays." Three kinds of rays are produced by naturally occurring radioactivity: α **rays**, β **rays**, and γ **rays**. They are named according to the first three

α and β rays are deflected by a magnetic field and, therefore, consist of moving charged particles. γ rays are not deflected by a magnetic field and, consequently, must be uncharged.

letters of the Greek alphabet, alpha (α), beta (β), and gamma (γ), to indicate the extent of their ability to penetrate matter. α rays are the least penetrating, being blocked by a thin (≈ 0.01 mm) sheet of lead, whereas β rays penetrate lead to a much greater distance (≈ 0.1 mm). γ rays are the most penetrating and can pass through an appreciable thickness (≈ 100 mm) of lead.

The nuclear disintegration process that produces α, β, and γ rays must obey the conservation laws of physics. These laws are called "conservation laws," because each of them deals with a property that is conserved, in the sense that it does not change during a process. The following list shows the property with which each law deals:

1. Conservation of energy/mass (Sections 6.8 and 28.6)
2. Conservation of linear momentum (Section 7.2)
3. Conservation of angular momentum (Section 9.6)
4. Conservation of electric charge (Section 18.2)
5. Conservation of nucleon number (Section 31.4)

We have studied the first four of these laws in previous chapters, and to them we now add a fifth, **the conservation of nucleon number.**

In all radioactive decay processes it has been observed that the number of nucleons (protons plus neutrons) present before the decay is equal to the number of nucleons after the decay. Therefore, the number of nucleons is conserved during a nuclear disintegration. As applied to the disintegration of a nucleus, the conservation laws require that the energy, electric charge, linear momentum, angular momentum, and nucleon number that a nucleus possesses must remain unchanged when it disintegrates into nuclear fragments and accompanying α, β, or γ rays.

The three types of radioactivity that occur naturally can be observed in a relatively simple experiment. A piece of radioactive material is placed at the bottom of a narrow hole in a lead cylinder. The cylinder is located within an evacuated chamber, as **Interactive Figure 31.6** illustrates. A magnetic field is directed perpendicular to the plane of the paper, and a photographic plate is positioned to the right of the hole. Three spots appear on the developed plate, which are associated with the radioactivity of the nuclei in the material. Since moving particles are deflected by a magnetic field only when they are electrically charged, this experiment reveals that two types of radioactivity (α and β rays, as it turns out) consist of charged particles, whereas the third type (γ rays) does not.

α Decay

When a nucleus disintegrates and produces α rays, it is said to undergo α **decay.** Experimental evidence shows that α rays consist of positively charged particles, each one being the ^4_2He nucleus of helium. Thus, an α particle has a charge of $+2e$ and a nucleon number of $A = 4$. Since the grouping of 2 protons and 2 neutrons in a ^4_2He nucleus is particularly stable, as we have seen in connection with **Figure 31.5**, it is not surprising that an α particle can be ejected as a unit from a more massive unstable nucleus.

Uranium parent nucleus

Thorium daughter nucleus

α particle (helium nucleus)

ANIMATED FIGURE 31.7 α decay occurs when an unstable parent nucleus emits an α particle and in the process is converted into a different, or daughter, nucleus.

Animated Figure 31.7 shows the disintegration process for one example of α decay:

$$\underset{\substack{\text{Parent}\\\text{nucleus}\\\text{(uranium)}}}{{}^{238}_{92}\text{U}} \longrightarrow \underset{\substack{\text{Daughter}\\\text{nucleus}\\\text{(thorium)}}}{{}^{234}_{90}\text{Th}} + \underset{\substack{\alpha \text{ particle}\\\text{(helium}\\\text{nucleus)}}}{{}^{4}_{2}\text{He}}$$

The original nucleus is referred to as the *parent nucleus* (P), and the nucleus remaining after disintegration is called the *daughter nucleus* (D). Upon emission of an α particle, the uranium ${}^{238}_{92}\text{U}$ parent is converted into the ${}^{234}_{90}\text{Th}$ daughter, which is an isotope of thorium. The parent and daughter nuclei are different, so α decay converts one element into another, a process known as **transmutation**.

Electric charge is conserved during α decay. In **Animated Figure 31.7**, for instance, 90 of the 92 protons in the uranium nucleus end up in the thorium nucleus, and the remaining 2 protons are carried off by the α particle. The total number of 92, however, is the same before and after disintegration. α decay also conserves the number of nucleons, because the number is the same before (238) and after (234 + 4) disintegration. Consistent with the conservation of electric charge and nucleon number, the general form for α decay is

α Decay

$$\underset{\substack{\text{Parent}\\\text{nucleus}}}{{}^{A}_{Z}\text{P}} \longrightarrow \underset{\substack{\text{Daughter}\\\text{nucleus}}}{{}^{A-4}_{Z-2}\text{D}} + \underset{\substack{\alpha \text{ particle (helium}\\\text{nucleus)}}}{{}^{4}_{2}\text{He}}$$

When a nucleus releases an α particle, the nucleus also releases energy. In fact, the energy released by radioactive decay is responsible, in part, for keeping the interior of the earth hot and, in some places, even molten. The following example shows how the conservation of mass/energy can be used to determine the amount of energy released in α decay.

EXAMPLE 4 | α Decay and the Release of Energy

The atomic mass of uranium ${}^{238}_{92}\text{U}$ is 238.0508 u, that of thorium ${}^{234}_{90}\text{Th}$ is 234.0436 u, and that of an α particle ${}^{4}_{2}\text{He}$ is 4.0026 u. Determine the energy released when α decay converts ${}^{238}_{92}\text{U}$ into ${}^{234}_{90}\text{Th}$.

Reasoning Since energy is released during the decay, the combined mass of the ${}^{234}_{90}\text{Th}$ daughter nucleus and the α particle is less than the mass of the ${}^{238}_{92}\text{U}$ parent nucleus. The difference in mass is equivalent to the energy released. We will determine the difference in mass in atomic mass units and then use the fact that 1 u is equivalent to 931.5 MeV.

Solution The decay and the masses are shown below:

$$\underset{238.0508\ \text{u}}{{}^{238}_{92}\text{U}} \longrightarrow \underset{234.0436\ \text{u}}{{}^{234}_{90}\text{Th}} + \underset{4.0026\ \text{u}}{{}^{4}_{2}\text{He}}$$

$$\underbrace{\phantom{234.0436\ \text{u} + 4.0026\ \text{u}}}_{238.0462\ \text{u}}$$

The decrease in mass, or mass defect for the decay process, is 238.0508 u − 238.0462 u = 0.0046 u. As usual, the masses are atomic masses and include the mass of the orbital electrons. But this causes no error here because the same total number of electrons is included for ${}^{238}_{92}\text{U}$, on the one hand, and for ${}^{234}_{90}\text{Th}$ plus ${}^{4}_{2}\text{He}$, on the other. Since 1 u is equivalent to 931.5 MeV, the released energy is

$$\text{Released energy} = (0.0046\ \cancel{\text{u}})\left(\frac{931.5\ \text{MeV}}{1\ \cancel{\text{u}}}\right) = \boxed{4.3\ \text{MeV}}$$

When α decay occurs as in Example 4, the energy released appears as kinetic energy of the recoiling ${}^{234}_{90}\text{Th}$ nucleus and the α particle, except for a small portion carried away as a γ ray. Conceptual Example 5 discusses how the ${}^{234}_{90}\text{Th}$ nucleus and the α particle share in the released energy.

CONCEPTUAL EXAMPLE 5 | How Energy Is Shared During the α Decay of $^{238}_{92}$U

In Example 4, the energy released by the α decay of $^{238}_{92}$U is found to be 4.3 MeV. Since this energy is carried away as kinetic energy of the recoiling $^{234}_{90}$Th nucleus and the α particle, it follows that $KE_{Th} + KE_{\alpha} = 4.3$ MeV. However, KE_{Th} and KE_{α} are not equal. Which particle carries away more kinetic energy, the $^{234}_{90}$Th nucleus or the α particle?

Reasoning and Solution Kinetic energy depends on the mass m and speed v of a particle, since $KE = \frac{1}{2}mv^2$ (Equation 6.2). The $^{234}_{90}$Th nucleus has a much greater mass than the α particle, and since the kinetic energy is proportional to the mass, it is tempting to conclude that the $^{234}_{90}$Th nucleus has the greater kinetic energy. This conclusion is not correct, however, since it does not take into account the fact that the $^{234}_{90}$Th nucleus and the α particle have different speeds after the decay. In fact, we expect the thorium nucleus to recoil with the smaller speed precisely *because* it has the greater mass. The decaying $^{238}_{92}$U is like a parent and their young child on ice skates, pushing off against one another. The more massive parent recoils with much less speed than the child. We can use the principle of conservation of linear momentum to verify our expectation.

As Section 7.2 discusses, the conservation principle states that the total linear momentum of an isolated system remains constant. An isolated system is one for which the vector sum of the external forces acting on the system is zero, and the decaying $^{238}_{92}$U nucleus fits this description. It is stationary initially, and since momentum is mass times velocity, its initial momentum is zero. In its final form, the system consists of the $^{234}_{90}$Th nucleus and the α particle and has a final total momentum of $m_{Th}v_{Th} + m_{\alpha}v_{\alpha}$. According to momentum conservation, the initial and final values of the total momentum of the system must be the same, so that $m_{Th}v_{Th} + m_{\alpha}v_{\alpha} = 0$. Solving this equation for the velocity of the thorium nucleus, we find that $v_{Th} = -m_{\alpha}v_{\alpha}/m_{Th}$. Since m_{Th} is much greater than m_{α}, we can see that the speed of the thorium nucleus is less than the speed of the α particle. Moreover, the kinetic energy depends on the square of the speed and only the first power of the mass. As a result of its much greater speed, *the α particle has the greater kinetic energy*.

Related Homework: Problem 27

THE PHYSICS OF . . . radioactivity and smoke detectors. One widely used application of α decay is in smoke detectors. **Figure 31.8** illustrates how a smoke detector operates. Two small and parallel metal plates are separated by a distance of about one centimeter. A tiny amount of radioactive material at the center of one of the plates emits α particles, which collide with air molecules. During the collisions, the air molecules are ionized to form positive and negative ions. The voltage from a battery causes one plate to be positive and the other negative, so that each plate attracts ions of opposite charge. As a result there is a current in the circuit attached to the plates. The presence of smoke particles between the plates reduces the current, since the ions that collide with a smoke particle are usually neutralized. The drop in current that smoke particles cause is used to trigger an alarm.

Current

α particles

Radioactive material

Battery

FIGURE 31.8 A smoke detector.

β Decay

The β rays in **Interactive Figure 31.6** are deflected by the magnetic field in a direction opposite to that of the positively charged α rays. Consequently, these β rays, which are the most common kind, consist of negatively charged particles or β⁻ particles. Experiment shows that β⁻ particles are electrons. As an illustration of β⁻ decay, consider the thorium $^{234}_{90}$Th nucleus, which decays by emitting a β⁻ particle, as in **Figure 31.9**:

$$^{234}_{90}\text{Th} \longrightarrow {}^{234}_{91}\text{Pa} + {}^{0}_{-1}\text{e}$$

Parent nucleus (thorium)   Daughter nucleus (protactinium)   β⁻ particle (electron)

β⁻ decay, like α decay, causes a transmutation of one element into another. In this case, thorium $^{234}_{90}$Th is converted into protactinium $^{234}_{91}$Pa. The law of conservation of charge is obeyed, since the net number of positive charges is the same before (90) and after $(91 - 1)$ the β⁻ emission. The law of conservation of nucleon number is obeyed, since the nucleon number remains at $A = 234$. The general form for β⁻ decay is

β⁻ decay

$$^{A}_{Z}\text{P} \longrightarrow {}^{A}_{Z+1}\text{D} + {}^{0}_{-1}\text{e}$$

Parent nucleus   Daughter nucleus   β⁻ particle (electron)

The electron emitted in β⁻ decay does *not* actually exist within the parent nucleus and is *not* one of the orbital electrons. Instead, the electron is created when a neutron decays into a proton and an electron; when this occurs, the proton number of the parent

Thorium parent nucleus   Protactinium daughter nucleus   β⁻ particle (electron)

144   90+   143   91+   ⊖

$^{234}_{90}$Th $^{234}_{91}$Pa $^{0}_{-1}$e

FIGURE 31.9 β⁻ decay occurs when a neutron in an unstable parent nucleus decays into a proton and an electron, the electron being emitted as the β⁻ particle. In the process, the parent nucleus is transformed into the daughter nucleus.

nucleus increases from Z to $Z + 1$ and the nucleon number remains unchanged. The newly created electron is usually fast-moving and escapes from the atom, leaving behind a positively charged atom.

Example 6 illustrates that energy is released during β^- decay, just as it is during α decay, and that the conservation of mass/energy applies.

EXAMPLE 6 | β^- Decay and the Release of Energy

The atomic mass of thorium $^{234}_{90}$Th is 234.043 59 u, and the atomic mass of protactinium $^{234}_{91}$Pa is 234.043 30 u. Find the energy released when β^- decay changes $^{234}_{90}$Th into $^{234}_{91}$Pa.

Reasoning To find the energy released, we follow the usual procedure of determining how much the mass has decreased because of the decay and then calculating the equivalent energy.

Problem-Solving Insight In β^- decay, be careful not to include the mass of the electron ($^{0}_{-1}$e) twice. As discussed here, the atomic mass of the daughter atom ($^{234}_{91}$Pa) already includes the mass of the emitted electron.

Solution The decay and the masses are shown below:

$$^{234}_{90}\text{Th} \longrightarrow ^{234}_{91}\text{Pa} + ^{0}_{-1}\text{e}$$

$$\underbrace{\phantom{^{234}_{90}\text{Th}}}_{234.043\ 59\ \text{u}} \qquad \underbrace{\phantom{^{234}_{91}\text{Pa}}}_{234.043\ 30\ \text{u}}$$

When the $^{234}_{90}$Th nucleus of a thorium atom is converted into a $^{234}_{91}$Pa nucleus, the number of orbital electrons remains the same, so the resulting protactinium atom is missing one orbital electron. However, the given mass includes all 91 electrons of a neutral protactinium atom. In effect, then, the value of 234.043 30 u for $^{234}_{91}$Pa already includes the mass of the β^- particle. The mass decrease that accompanies the β^- decay is

$$234.043\ 59\ \text{u} - 234.043\ 30\ \text{u} = 0.000\ 29\ \text{u}$$

Since 1 u is equivalent to 931.5 MeV, the released energy is

$$\text{Released energy} = (0.00029\ \cancel{\text{u}})\left(\frac{931.5\ \text{MeV}}{1\ \cancel{\text{u}}}\right) = \boxed{0.27\ \text{MeV}}$$

This is the maximum kinetic energy that the emitted electron can have.

A second kind of β decay sometimes occurs.* In this process the particle emitted by the nucleus is a **positron** rather than an electron. A positron, also called a β^+ particle, has the same mass as an electron but carries a charge of $+e$ instead of $-e$. The disintegration process for β^+ decay is

β^+ decay
$$\underset{\substack{\text{Parent}\\\text{nucleus}}}{^{A}_{Z}\text{P}} \longrightarrow \underset{\substack{\text{Daughter}\\\text{nucleus}}}{^{A}_{Z-1}\text{D}} + \underset{\substack{\beta^+ \text{ particle}\\\text{(positron)}}}{^{0}_{+1}\text{e}}$$

The emitted positron does *not* exist within the nucleus but, rather, is created when a nuclear proton is transformed into a neutron. In the process, the proton number of the parent nucleus decreases from Z to $Z - 1$, and the nucleon number remains the same. As with β^- decay, the laws of conservation of charge and nucleon number are obeyed, and there is a transmutation of one element into another.

γ Decay

The nucleus, like the orbital electrons, exists only in discrete energy states or levels. When a nucleus changes from an excited energy state (denoted by an asterisk*) to a lower energy state, a photon is emitted. The process is similar to the one discussed in Section 30.3 for the photon emission that leads to the hydrogen atom line spectrum. With nuclear energy levels, however, the photon has a much greater energy and is called a γ ray. The γ decay process is written as follows:

γ decay
$$\underset{\substack{\text{Excited}\\\text{energy state}}}{^{A}_{Z}\text{P}^*} \longrightarrow \underset{\substack{\text{Lower}\\\text{energy state}}}{^{A}_{Z}\text{P}} + \underset{\substack{\gamma \text{ ray}}}{\gamma}$$

γ decay does *not* cause a transmutation of one element into another.

---

*A third kind of β decay also occurs in which a nucleus pulls in, or captures, one of the orbital electrons from outside the nucleus. The process is called **electron capture**, or **K capture**, since the electron normally comes from the innermost, or K, shell.

Medical Applications of Radioactivity

BIO **THE PHYSICS OF . . . Gamma Knife radiosurgery.** Gamma Knife radiosurgery is becoming a very promising medical procedure for treating certain problems of the brain, including benign and cancerous tumors as well as blood vessel malformations. The procedure, which involves no knife at all, uses powerful, highly focused beams of γ rays aimed at the tumor or malformation. The γ rays are emitted by a radioactive cobalt-60 source. As **Figure 31.10a** illustrates, the patient wears a protective metal helmet that is perforated with many small holes. Part *b* of the figure shows that the holes focus the γ rays to a single tiny target within the brain. The target tissue thus receives a very intense dose of radiation and is destroyed, while the surrounding healthy tissue is undamaged. Gamma Knife surgery is a noninvasive, painless, and bloodless procedure that is often performed under local anesthesia. Hospital stays are 70 to 90% shorter than with conventional surgery, and patients often return to work within a few days.

(a) (b)

FIGURE 31.10 (*a*) In Gamma Knife radiosurgery, a protective metal helmet containing many small holes is placed on the patient's head. (*b*) The holes focus the beams of γ rays to a tiny target within the brain.

BIO **THE PHYSICS OF . . . an exercise thallium heart scan.** An exercise thallium heart scan is a test that uses radioactive thallium to produce images of the heart muscle. When combined with an exercise test, such as walking on a treadmill, the thallium scan helps identify regions of the heart that do not receive enough blood. The scan is especially useful in diagnosing the presence of blockages in the coronary arteries, which supply oxygen-rich blood to the heart muscle. During the test, a small amount of thallium is injected into a vein while the patient walks on a treadmill. The thallium attaches to the red blood cells and is carried throughout the body. The thallium enters the heart muscle by way of the coronary arteries and collects in heart-muscle cells that come into contact with the blood. The thallium isotope used, $^{201}_{81}\text{Tl}$, emits γ rays, which a special camera records. Since the thallium reaches those regions of the heart that have an adequate blood supply, lesser amounts show up in areas where the blood flow has been reduced due to arterial blockages (see **Figure 31.11**). A second set of images is taken several hours later, while the patient is resting. These images help differentiate between regions of the heart that temporarily do not receive enough blood (the blood flow returns to normal after the exercise) and regions that are permanently damaged due to, for example, a previous heart attack (the blood flow does not return to normal).

BIO **THE PHYSICS OF . . . brachytherapy implants.** The use of radioactive isotopes to deliver radiation to specific targets in the body is an important medical technique. In treating cancer, for example, the method of delivery should ideally apply a high dose of radiation to a malignant tumor in order to kill it, while applying only a small (non-damaging) dose to healthy surrounding tissue. Brachytherapy implants

FIGURE 31.11 An exercise thallium heart scan indicates regions of the heart that receive insufficient blood during exercise.

offer such a delivery method. In this type of treatment, radioactive isotopes are formed into small seeds and implanted directly in the tumor according to a predesigned pattern. The energy and type of radiation emitted by the isotopes can be exploited to optimize a treatment design and minimize damage to healthy tissue. Seeds containing iridium $^{192}_{77}$Ir are used to treat many cancers, and seeds containing iodine $^{125}_{53}$I and palladium $^{103}_{46}$Pd are used for prostate cancer. Research has also indicated that brachytherapy implants may have an important role to play in the treatment of atherosclerosis, in which blood vessels become blocked with plaque. Such blockages are often treated using the technique of balloon angioplasty. With the aid of a catheter inserted into an occluded coronary artery, a balloon is inflated to open the artery and place a stent (a metallic mesh that provides support for the arterial wall) at the site of the blockage. Sometimes the arterial wall is damaged in this process, and as it heals, the artery often becomes blocked again. Brachytherapy implants (using iridium $^{192}_{77}$Ir or phosphorus $^{32}_{15}$P, for instance) have been found to inhibit repeat blockages following angioplasty.

Check Your Understanding

(*The answers are given at the end of the book.*)

7. Polonium $^{216}_{84}$Po undergoes α decay to produce a daughter nucleus that itself undergoes β^- decay. Which one of the following nuclei is the one that ultimately results? **(a)** $^{211}_{82}$Pb **(b)** $^{211}_{81}$Tl **(c)** $^{212}_{81}$Tl **(d)** $^{212}_{83}$Bi **(e)** $^{213}_{82}$Pb

8. Uranium $^{238}_{92}$U decays into thorium $^{234}_{90}$Th by means of α decay, as Example 4 in the text discusses. Another possibility is that the $^{238}_{92}$U nucleus just emits a single proton instead of an α particle. This hypothetical decay scheme is shown below, along with the pertinent atomic masses:

$$^{238}_{92}\text{U} \longrightarrow {}^{237}_{91}\text{Pa} + {}^{1}_{1}\text{H}$$

Uranium	Protactinium	Proton
238.050 78 u	237.051 14 u	1.007 83 u

For a decay to be possible, it must bring the parent nucleus toward a more stable state by allowing the release of energy. Compare the total mass of the products of this hypothetical decay with the mass of $^{238}_{92}$U and decide whether the emission of a single proton is possible for $^{238}_{92}$U.

31.5 The Neutrino

When a β particle is emitted by a radioactive nucleus, energy is simultaneously released, as Example 6 illustrates. Experimentally, however, it is found that most β particles do not have enough kinetic energy to account for all the energy released. If a β particle carries away only part of the energy, where does the remainder go? The question puzzled physicists until 1930, when Wolfgang Pauli proposed that part of the energy is carried away by another particle that is emitted along with the β particle. This additional particle is called the **neutrino**, and its existence was verified experimentally in 1956. The Greek letter nu (ν) is used to symbolize the neutrino. For instance, the β^- decay of thorium $^{234}_{90}$Th (see Section 31.4) is more correctly written as

$$^{234}_{90}\text{Th} \longrightarrow {}^{234}_{91}\text{Pa} + {}^{0}_{-1}\text{e} + \bar{\nu}$$

The bar above the ν is included because the neutrino emitted in this particular decay process is an antimatter neutrino, or antineutrino. A normal neutrino (ν without the bar) is emitted when β^+ decay occurs.

The emission of neutrinos and β particles involves a force called the *weak nuclear force* because it is much weaker than the strong nuclear force. It is now known that the weak nuclear force and the electromagnetic force are two different manifestations of a single, more fundamental force, the **electroweak force**. The theory for the electroweak

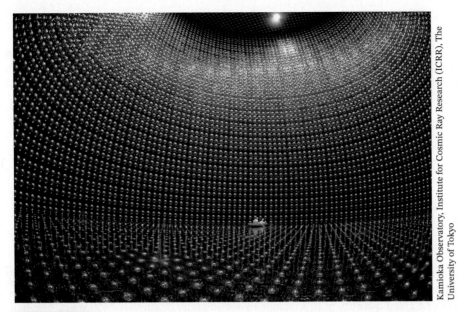

FIGURE 31.12 The Super-Kamiokande neutrino detector in Japan is an underground steel cylindrical tank with its inner wall lined with 11 000 photomultiplier tubes. It is filled with 12.5 million gallons of ultrapure water when operational. In this photograph it is partially full, and the technicians in the boat are inspecting the photomultiplier tubes.

force was developed by Sheldon Glashow (1932–), Abdus Salam (1926–1996), and Steven Weinberg (1933–), who shared a Nobel Prize for their achievement in 1979. The electroweak force, the gravitational force, and the strong nuclear force are the three fundamental forces in nature.

The neutrino has zero electric charge and is extremely difficult to detect because it interacts very weakly with matter. For example, the average neutrino can penetrate one light-year of lead (about 9.5×10^{15} m) without interacting with it. Thus, even though trillions of neutrinos pass through our bodies every second, they have no effect. One of the major scientific questions of our time is whether neutrinos have mass. The question is important because neutrinos are so plentiful in the universe. Even a very small mass could account for a significant portion of the mass in the universe and, possibly, could have an effect on the formation of galaxies.

Although difficult, it is possible to detect neutrinos. **Figure 31.12** shows the Super-Kamiokande neutrino detector in Japan. It is located 915 m underground and consists of a steel cylindrical tank, ten stories tall, whose inner wall is lined with 11 000 photomultiplier tubes (see Section 31.9). The tank is filled with 12.5 million gallons of ultrapure water. Neutrinos colliding with the water molecules produce light patterns that the photomultiplier tubes detect. In 1998 the Super-Kamiokande detector yielded the first strong, but indirect, evidence that neutrinos do indeed have a small mass. (The mass of the electron neutrino is less than 0.0004% of the mass of an electron.) This finding implies that neutrinos travel at less than the speed of light. If the neutrino's mass were zero, like that of a photon, it would travel at the speed of light.

31.6 | Radioactive Decay and Activity

The question of which radioactive nucleus in a group of nuclei disintegrates at a given instant is decided like the winning numbers in a state lottery: individual disintegrations occur randomly. As time passes, the number N of parent nuclei decreases, as **Figure 31.13** shows. This graph of N versus time indicates that the decrease occurs in a smooth fashion, with N approaching zero after enough time has passed. To help describe the graph, it is useful to define the **half-life** $T_{1/2}$ of a radioactive isotope

FIGURE 31.13 The half-life $T_{1/2}$ of a radioactive decay is the time in which one-half of the radioactive nuclei disintegrate.

TABLE 31.2 Some Half-Lives for Radioactive Decay		
Isotope		**Half-Life**
Polonium	$^{214}_{84}$Po	1.64×10^{-4} s
Krypton	$^{89}_{36}$Kr	3.16 min
Radon	$^{222}_{86}$Rn	3.83 d
Strontium	$^{90}_{38}$Sr	29.1 yr
Radium	$^{226}_{88}$Ra	1.6×10^3 yr
Carbon	$^{14}_{6}$C	5.73×10^3 yr
Uranium	$^{238}_{92}$U	4.47×10^9 yr
Indium	$^{115}_{49}$In	4.41×10^{14} yr

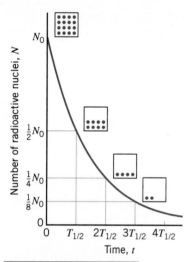

FIGURE 31.13 (REPEATED) The half-life $T_{1/2}$ of a radioactive decay is the time in which one-half of the radioactive nuclei disintegrate.

as the time required for one-half of the nuclei present to disintegrate. For example, radium $^{226}_{88}$Ra has a half-life of 1600 years, because it takes this amount of time for one-half of a given quantity of this isotope to disintegrate. In another 1600 years, one-half of the remaining radium atoms will have disintegrated, leaving only one-fourth of the original number intact. In **Figure 31.13 (Repeated)**, the number of nuclei present at time $t = 0$ s is $N = N_0$, and the number present at $t = T_{1/2}$ is $N = \frac{1}{2}N_0$. The number present at $t = 2T_{1/2}$ is $N = \frac{1}{4}N_0$, and so on. The value of the half-life depends on the nature of the radioactive nucleus. Values ranging from a fraction of a second to billions of years have been found (see **Table 31.2**).

THE PHYSICS OF . . . radioactive radon gas in houses. Radon $^{222}_{86}$Rn is a naturally occurring radioactive gas produced when radium $^{226}_{88}$Ra undergoes α decay. There is a nationwide concern about radon as a health hazard because radon in the soil is gaseous and can enter the basement of homes through cracks in the foundation. (It should be noted, however, that the mechanism of indoor radon entry is not well understood and that entry via foundation cracks is likely only part of the story.) Once inside, the concentration of radon can rise markedly, depending on the type of housing construction and the concentration of radon in the surrounding soil. Radon gas decays into daughter nuclei that are also radioactive. The radioactive nuclei can attach to dust and smoke particles that can be inhaled, and they remain in the lungs to release tissue-damaging radiation. Prolonged exposure to high levels of radon can lead to lung cancer. Since radon gas concentrations can be measured with inexpensive monitoring devices, it is recommended that all homes be tested for radon. Example 7 deals with the half-life of radon $^{222}_{86}$Rn.

EXAMPLE 7 | The Radioactive Decay of Radon Gas

Suppose that 3.0×10^7 radon atoms are trapped in a basement at the time the basement is sealed against further entry of the gas. The half-life of radon is 3.83 days. How many radon atoms remain after 31 days?

Reasoning During each half-life, the number of radon atoms is reduced by a factor of two. Thus, for each half-life in the period of 31 days, we reduce the number of radon atoms present at the beginning of that half-life by a factor of two.

Solution In a period of 31 days there are (31 days)/(3.83 days) = 8.1 half-lives. In 8 half-lives the number of radon atoms is reduced by a factor of $2^8 = 256$. Ignoring the difference between 8 and 8.1 half-lives, we find that the number of atoms remaining after 31 days is $(3.0 \times 10^7)/256 = \boxed{1.2 \times 10^5}$.

The **activity** of a radioactive sample is the number of disintegrations per second that occur. Each time a disintegration occurs, the number N of radioactive nuclei decreases. As a result, the activity can be obtained by dividing ΔN, the change in the number of nuclei, by Δt, the time interval during which the change takes place; the average activity over the time interval Δt is the magnitude of $\Delta N/\Delta t$, or $|\Delta N/\Delta t|$. Since the decay of any individual nucleus is completely random, the number of disintegrations per second that occurs in a sample is proportional to the number of radioactive nuclei present, so that

$$\frac{\Delta N}{\Delta t} = -\lambda N \tag{31.4}$$

where λ is a proportionality constant referred to as the **decay constant**. The minus sign is present in this equation because each disintegration decreases the number N of nuclei originally present.

The SI unit for activity is the *becquerel* (Bq), named after Antoine Becquerel (1852–1908). One becquerel is one disintegration per second. Activity is also measured using a unit called the *curie* (Ci), honoring Marie (1867–1934) and Pierre (1859–1906) Curie, the discoverers of radium and polonium. Historically, the curie was chosen as a unit because it is roughly the activity of one gram of pure radium. In terms of becquerels,

$$1 \text{ Ci} = 3.70 \times 10^{10} \text{ Bq}$$

The activity of the radium put into the dial of a watch to make it glow in the dark is about 4×10^4 Bq, and the activity used in radiation therapy for cancer is approximately a billion times greater, or 4×10^{13} Bq.

The mathematical expression for the graph of N versus t shown in **Figure 31.13** can be obtained from Equation 31.4 with the aid of calculus. The result for the number N of radioactive nuclei present at time t is

$$N = N_0 e^{-\lambda t} \tag{31.5}$$

assuming that the number present at $t = 0$ s is N_0. The exponential e has the value $e = 2.718\ldots$, and many calculators provide the value of e^x. We can relate the half-life $T_{1/2}$ of a radioactive nucleus to its decay constant λ in the following manner. By substituting $N = \frac{1}{2}N_0$ and $t = T_{1/2}$ into Equation 31.5, we find that $\frac{1}{2} = e^{-\lambda T_{1/2}}$. Solving this equation for $T_{1/2}$ reveals that

$$T_{1/2} = \frac{\ln 2}{\lambda} = \frac{0.693}{\lambda} \tag{31.6}$$

Examples 8 through 10 illustrate the use of Equations 31.5 and 31.6.

Math Skills   To obtain Equation 31.6, we take the natural logarithm of both sides of the equation $\frac{1}{2} = e^{-\lambda T_{1/2}}$, which gives

$$\ln\left(\frac{1}{2}\right) = \ln\left(e^{-\lambda T_{1/2}}\right)$$

According to Equation D-12 in Appendix D, the left side of this result is $\ln\left(\frac{1}{2}\right) = \ln 1 - \ln 2$. According to Equation D-9 in Appendix D, the right side is $\ln\left(e^{-\lambda T_{1/2}}\right) = -\lambda T_{1/2}$. Thus, we have

$$\ln 1 - \ln 2 = -\lambda T_{1/2}$$

However, since $\ln 1 = 0$, this result becomes

$$\ln 2 = \lambda T_{1/2} \quad \text{or} \quad T_{1/2} = \frac{\ln 2}{\lambda} \tag{31.6}$$

EXAMPLE 8 | The Activity of Radon $^{222}_{86}$Rn

As in Example 7, suppose that there are 3.0×10^7 radon atoms ($T_{1/2} = 3.83$ days or 3.31×10^5 s) trapped in a basement. **(a)** How many radon atoms remain after 31 days? Find the activity **(b)** just after the basement is sealed against further entry of radon and **(c)** 31 days later.

Reasoning   The number N of radon atoms remaining after a time t is given by $N = N_0 e^{-\lambda t}$ (Equation 31.5), where $N_0 = 3.0 \times 10^7$ is the original number of atoms when $t = 0$ s and λ is the decay constant. The decay constant is related to the half-life $T_{1/2}$ of the radon atoms by $\lambda = 0.693/T_{1/2}$. The activity can be obtained from $\Delta N/\Delta t = -\lambda N$ (Equation 31.4).

Solution   **(a)** The decay constant is

$$\lambda = \frac{0.693}{T_{1/2}} = \frac{0.693}{3.83 \text{ days}} = 0.181 \text{ days}^{-1} \qquad \textbf{(31.6)}$$

and the number N of radon atoms remaining after 31 days is

$$N = N_0 e^{-\lambda t} = (3.0 \times 10^7)e^{-(0.181 \text{ days}^{-1})(31 \text{ days})} = \boxed{1.1 \times 10^5} \qquad \textbf{(31.5)}$$

This value is slightly less than that found in Example 7 because there we ignored the difference between 8.0 and 8.1 half-lives.

(b) The activity can be obtained from Equation 31.4, provided the decay constant is expressed in reciprocal seconds:

$$\lambda = \frac{0.693}{T_{1/2}} = \frac{0.693}{3.31 \times 10^5 \text{ s}} = 2.09 \times 10^{-6} \text{ s}^{-1} \qquad \textbf{(31.6)}$$

Thus, the number of disintegrations per second is

$$\frac{\Delta N}{\Delta t} = -\lambda N = -(2.09 \times 10^{-6} \text{ s}^{-1})(3.0 \times 10^7)$$
$$= -63 \text{ disintegations/s} \qquad \textbf{(31.4)}$$

The activity is the magnitude of $\Delta N/\Delta t$, so initially $\boxed{\text{Activity} = 63 \text{ Bq}}$.

(c) From part (a), the number of radioactive nuclei remaining at the end of 31 days is $N = 1.1 \times 10^5$, and reasoning similar to that in part (b) reveals that $\boxed{\text{Activity} = 0.23 \text{ Bq}}$.

EXAMPLE 9 | BIO   The Physics of Brachytherapy Implants Revisited

As part of a treatment plan for prostate cancer, a patient undergoes brachytherapy, in which seeds containing iodine $^{125}_{53}\text{I}$ are implanted close to the prostate (see **Figure 31.14**). The iodine atoms have a half-life of 59.5 days, and they decay by electron capture into tellurium $^{125}_{52}\text{Te}^*$ that is in an excited state. The tellurium atoms immediately undergo gamma decay, which releases a gamma ray with a maximum energy of 35 keV. This provides the majority of the dose of radiation that is used to kill the cancer cells. **(a)** During a low-dose treatment, a seed of $^{125}_{53}\text{I}$ is left near the prostate for 48 hours and then removed. What fraction of the $^{125}_{53}\text{I}$ atoms remains in the seed after the treatment? **(b)** What is the correct decay process for $^{125}_{52}\text{Te}^*$?

Reasoning   **(a)** As time passes, the number of radioactive atoms of $^{125}_{53}\text{I}$ will decay. The number remaining after a time t is given by Equation 31.5. Thus, N_0 represents the original number of $^{125}_{53}\text{I}$ atoms, and N represents the final number. **(b)** The gamma decay of $^{125}_{52}\text{Te}^*$ will be described by the gamma decay process: $^A_Z\text{P}^* \rightarrow {}^A_Z\text{P} + \gamma$.

Solution   **(a)** According to Equation 31.5, the fraction of the sample that remains will be given by

$$\frac{N}{N_0} = e^{-\lambda t}$$

In order to find this ratio, we need the value of the decay constant λ, which is related to the half-life by Equation 31.6:

$$\lambda = \frac{0.693}{T_{1/2}} = \frac{0.693}{59.5 \text{ days}} = 1.16 \times 10^{-2} \text{ days}^{-1}$$

FIGURE 31.14   Photo of the seeds used in brachytherapy next to a nickel to show their size. The titanium capsules contain a radioactive element, such as $^{125}_{53}\text{I}$. The seeds are implanted directly into a tumor to kill cancer cells.

We can now calculate the fraction of $^{125}_{53}\text{I}$ atoms that remains after $t = 48.0$ hours $= 2.00$ days:

$$\frac{N}{N_0} = e^{-\lambda t} = e^{-(1.16 \times 10^{-2} \text{ days}^{-1})(2.00 \text{ days})} = \boxed{0.977}$$

Thus, 97.7% of the radioactive sample remains. **(b)** When $^{125}_{52}\text{Te}^*$ undergoes gamma decay, it will be described by the following decay process:

$$\boxed{^{125}_{52}\text{Te}^* \rightarrow {}^{125}_{52}\text{Te} + \gamma}$$

During gamma decay, no transmutation takes place. A gamma ray is produced, and the parent nucleus returns to a stable state.

EXAMPLE 10 | BIO   The Physics of ^{90}Sr in the Bones

Calcium is an essential element in the human body for maintaining proper bone density. Strontium is chemically very similar to calcium, and 99% of strontium in the body is located in the bones. This typically does not cause health problems, if the strontium in the body is a non-radioactive stable isotope. However, if a person ingests a radioactive isotope of strontium, like ^{90}Sr, for example,

then the isotope will settle in the bones and deliver harmful radiation to the nearby bone and marrow cells. This can lead to bone cancer and leukemia (**Figure 31.15**). If ^{90}Sr undergoes β^- decay with a half-life of 29.1 years, what percentage of the initial amount is still in the body after 50.0 years?

Reasoning We can use Equation 31.5 to calculate the ratio of remaining ^{90}Sr atoms N to the original number N_0. To find this ratio, we will need the decay constant λ for ^{90}Sr, which is related to its half-life $T_{1/2}$ by Equation 31.6.

Solution Beginning with Equation 31.5, we have:

$$N = N_0 e^{-\lambda t} \quad \text{or} \quad \frac{N}{N_0} = e^{-\lambda t}$$

To find the decay constant, we use Equation 31.6:

$$\lambda = \frac{0.693}{T_{1/2}} = \frac{0.693}{29.1 \text{ yr}} = 0.0238 \text{ yr}^{-1}$$

Substituting this result into the previous expression, we calculate the percentage of remaining radioactive Sr nuclei as follows:

$$\frac{N}{N_0} = e^{-(0.0238 \text{ yr}^{-1})(50.0 \text{ yr})} = 0.304 = \boxed{30.4\%}$$

The half-life of ^{90}Sr is a significant fraction of a typical lifespan. Thus, even after 50 years, there is a considerable percentage of atoms remaining.

FIGURE 31.15 A cross-section of a human femur bone showing a malignant tumor known as osteosarcoma. This is an aggressive form of cancer that can result from ^{90}Sr replacing Ca in the bone.

CNRI/Science Photo Library

Check Your Understanding

(The answers are given at the end of the book.)

9. The thallium $^{208}_{81}$Tl nucleus is radioactive, with a half-life of 3.053 min. At a given instant, the activity of a certain sample of thallium is 2400 Bq. Using the concept of a half-life, and without doing any written calculations, determine whether the activity 9 minutes later is **(a)** a little less than $\frac{1}{8}$(2400 Bq) = 300 Bq, **(b)** a little more than $\frac{1}{8}$(2400 Bq) = 300 Bq, **(c)** a little less than $\frac{1}{3}$(2400 Bq) = 800 Bq, or **(d)** a little more than $\frac{1}{3}$(2400 Bq) = 800 Bq.

10. The half-life of indium $^{115}_{49}$In is 4.41×10^{14} yr. Thus, one-half of the nuclei in a sample of this isotope will decay in this time, which is very long. Is it possible for any single nucleus in the sample to decay after only one second?

11. Could two different samples of the same radioactive element have different activities?

31.7 | Radioactive Dating

THE PHYSICS OF . . . radioactive dating. One important application of radioactivity is the determination of the age of archaeological or geological samples as in the case of the mummified remains of Queen Hatshepsut (see **Figure 31.16**). If an object contains radioactive nuclei when it is formed, then the decay of these nuclei marks the passage of time like a clock, half of the nuclei disintegrating during each half-life. If the half-life is known, a measurement of the number of nuclei present today relative to the number present initially can give the age of the sample. According to Equation 31.4, the activity of a sample is proportional to the number of radioactive nuclei, so one way to obtain the age is to compare present activity with initial activity. A more accurate way is to determine the present number of radioactive nuclei with the aid of a mass spectrometer.

The present activity of a sample can be measured, but how is it possible to know what the original activity was, perhaps thousands of years ago? Radioactive dating methods entail certain assumptions that make it possible to estimate the original activity. For instance, the radiocarbon technique utilizes the $^{14}_{6}$C isotope of carbon, which undergoes β^- decay with a half-life of 5730 yr. This isotope is currently present in the earth's atmosphere at an equilibrium concentration of about one atom for every 8.3×10^{11} atoms of normal carbon $^{12}_{6}$C. It is often assumed* that this value has remained constant over the years because $^{14}_{6}$C is created when cosmic rays interact with the earth's upper atmosphere, a production method that offsets the loss via β^- decay. Moreover, nearly all living organisms ingest the equilibrium concentration of $^{14}_{6}$C. However, once an organism dies, metabolism no longer sustains the input of $^{14}_{6}$C, and β^- decay causes half of the $^{14}_{6}$C nuclei to disintegrate every 5730 years. Example 11 illustrates how to determine the $^{14}_{6}$C activity of one gram of carbon in a living organism.

FIGURE 31.16 The mummified remains of Queen Hatshepsut, who ruled ancient Egypt from 1479 to 1458 BC. Radioactive dating is one of the techniques used to determine the age of such remains.

CRIS BOURONCLE/Getty Images

EXAMPLE 11 | $^{14}_{6}$C Activity per Gram of Carbon in a Living Organism

(a) Determine the number of carbon $^{14}_{6}$C atoms present for every gram of carbon $^{12}_{6}$C in a living organism. Find **(b)** the decay constant and **(c)** the activity of this sample.

Reasoning The total number of carbon $^{12}_{6}$C atoms in one gram of carbon $^{12}_{6}$C is equal to the corresponding number of moles times Avogadro's number (see Section 14.1). Since there is only one $^{14}_{6}$C atom for every 8.3×10^{11} atoms of $^{12}_{6}$C, the number of $^{14}_{6}$C atoms is equal to the total number of $^{12}_{6}$C atoms divided by 8.3×10^{11}. The decay constant λ for $^{14}_{6}$C is $\lambda = 0.693/T_{1/2}$, where $T_{1/2}$ is the half-life. The activity is equal to the magnitude of $\Delta N/\Delta t$, which is equal to the decay constant times the number of $^{14}_{6}$C atoms present, according to Equation 31.4.

Solution **(a)** One gram of carbon $^{12}_{6}$C (atomic mass = 12 u) is equivalent to 1.0/12 mol. Since Avogadro's number is 6.02×10^{23} atoms/mol and since there is one $^{14}_{6}$C atom for every 8.3×10^{11} atoms of $^{12}_{6}$C, the number of $^{14}_{6}$C atoms is

Number of $^{14}_{6}$C atoms for every 1.0 grams of carbon $^{12}_{6}$C
$= \left(\dfrac{1.0}{12}\text{ mol}\right)\left(6.02 \times 10^{23}\,\dfrac{\text{atoms}}{\text{mol}}\right)\left(\dfrac{1}{8.3 \times 10^{11}}\right)$

$= \boxed{6.0 \times 10^{10}\text{ atoms}}$

(b) Since the half-life of $^{14}_{6}$C is 5730 yr (1.81×10^{11} s), the decay constant is

$$\lambda = \frac{0.693}{T_{1/2}} = \frac{0.693}{1.81 \times 10^{11}\text{ s}} = \boxed{3.83 \times 10^{-12}\text{ s}^{-1}} \qquad \text{(31.6)}$$

(c) Equation 31.4 indicates that $\Delta N/\Delta t = -\lambda N$, so the magnitude of $\Delta N/\Delta t$ is λN.

Activity of $^{14}_{6}$C for every 1.0 gram of carbon $^{12}_{6}$C in a living organism
$= \lambda N = (3.83 \times 10^{-12}\text{ s}^{-1})(6.0 \times 10^{10}\text{ atoms})$

$= \boxed{0.23\text{ Bq}}$

An organism that lived thousands of years ago presumably had an activity of about 0.23 Bq per gram of carbon. When the organism died, the activity began decreasing. From a sample of the remains, the current activity per gram of carbon can be measured and compared to the value of 0.23 Bq to determine the time that has transpired since death. This procedure is illustrated in Example 12.

*The assumption that the $^{14}_{6}$C concentration has always been at its present equilibrium value has been evaluated by comparing $^{14}_{6}$C ages with ages determined by counting tree rings. More recently, ages determined using the radioactive decay of uranium $^{238}_{92}$U have been used for comparison. These comparisons indicate that the equilibrium value of the $^{14}_{6}$C concentration has indeed remained constant for the past 1000 years. However, from there back about 30 000 years, it appears that the $^{14}_{6}$C concentration in the atmosphere was larger than its present value by up to 40%. As a first approximation we ignore such discrepancies.

Analyzing Multiple-Concept Problems

EXAMPLE 12 | The Ice Man

On September 19, 1991, German tourists in the Italian Alps found a Stone-Age traveler, later dubbed the Ice Man, whose body had become trapped in a glacier. **Figure 31.17** shows the well-preserved remains, which were dated using the radiocarbon method. Material found with the body had a $^{14}_{6}C$ activity of about 0.121 Bq per gram of carbon. Find the age of the Ice Man's remains.

Reasoning In the radiocarbon method, the number of radioactive nuclei remaining at a given instant is related to the number present initially, the time that has passed since the Ice Man died, and the decay constant for $^{14}_{6}C$. Thus, to determine the age of the remains, we will need information about the number of nuclei present when the body was discovered and the number present initially, which is related to the activity of the material found with the body and the initial activity. To determine the age, we will also need the decay constant, obtainable from the half-life of $^{14}_{6}C$.

FIGURE 31.17 The frozen remains of the Ice Man or "Oetzi," as he also is called, were discovered in the ice of a glacier in the Italian Alps in 1991. Radiocarbon dating has revealed his age.

© South Tyrol Museum of Archaeology/Eurac/Samadelli/Staschitz

Knowns and Unknowns We have the following data:

Description	Symbol	Value	Comment
Explicit Data			
Activity of material found with body	A	0.121 Bq	This is the activity per gram of carbon.
Implicit Data			
Half-life of $^{14}_{6}C$	$T_{1/2}$	5730 yr	The radiocarbon dating method is specified.
Initial activity of material found with body	A_0	0.23 Bq	This activity is assumed for one gram of carbon in a living organism.
Unknown Variable			
Age of Ice Man's remains	t	?	

Modeling the Problem

STEP 1 Radioactive Decay The number N of radioactive nuclei present at a time t is

$$N = N_0 e^{-\lambda t} \qquad (31.5)$$

where N_0 is the number present initially at $t = 0$ s and λ is the decay constant for $^{14}_{6}C$. Rearranging terms gives

$$\frac{N}{N_0} = e^{-\lambda t}$$

Taking the natural logarithm of both sides of this result (see Appendix D), we find that

$$\ln\left(\frac{N}{N_0}\right) = -\lambda t$$

Solving for t shows that the age of the Ice Man's remains is given by Equation 1 at the right. To use this result, we need information about the ratio N/N_0 and λ. We deal with N/N_0 in Step 2 and with λ in Step 3.

$$t = -\left(\frac{1}{\lambda}\right) \ln\left(\frac{N}{N_0}\right) \qquad (1)$$

$$\boxed{?} \qquad \boxed{?}$$

STEP 2 Activity The activity A is the number of disintegrations per second, or $\left|\frac{\Delta N}{\Delta t}\right|$, where ΔN is the number of disintegrations that occur in the time interval Δt.

Noting that $\frac{\Delta N}{\Delta t} = -\lambda N$ according to Equation 31.4, we find for the activity A that

$$A = \left|\frac{\Delta N}{\Delta t}\right| = |-\lambda N| = \lambda N$$

Using this expression, we have that

$$\frac{N}{N_0} = \frac{\lambda N}{\lambda N_0} = \frac{A}{A_0}$$

$$t = -\left(\frac{1}{\lambda}\right) \ln\left(\frac{N}{N_0}\right) \qquad (1)$$

$$\boxed{?} \quad \boxed{\frac{N}{N_0} = \frac{A}{A_0}}$$

The substitution of this result into Equation 1 is shown at the right. We turn now to Step 3, in order to evaluate the decay constant λ.

STEP 3 Decay Constant   The decay constant is related to the half-life $T_{1/2}$ according to

$$\boxed{\lambda = \frac{0.693}{T_{1/2}}} \qquad (31.6)$$

which we substitute into Equation 1, as shown at the right.

$$t = -\left(\frac{1}{\lambda}\right) \ln\left(\frac{N}{N_0}\right) \qquad (1)$$

$$\boxed{\lambda = \frac{0.693}{T_{1/2}}} \quad \boxed{\frac{N}{N_0} = \frac{A}{A_0}}$$

Solution   Combining the results of each step algebraically, we find that

STEP 1   STEP 2   STEP 3

$$t = -\left(\frac{1}{\lambda}\right) \ln\left(\frac{N}{N_0}\right) = -\left(\frac{1}{\lambda}\right) \ln\left(\frac{A}{A_0}\right) = -\left(\frac{1}{0.693/T_{1/2}}\right) \ln\left(\frac{A}{A_0}\right)$$

This result reveals that the age of the Ice Man's remains is

$$t = -\left(\frac{T_{1/2}}{0.693}\right) \ln\left(\frac{A}{A_0}\right) = -\left(\frac{5730 \text{ yr}}{0.693}\right) \ln\left(\frac{0.121 \text{ Bq}}{0.23 \text{ Bq}}\right) = \boxed{5300 \text{ yr}}$$

Note that this solution implies for the activity that

$$A = A_0 e^{-\lambda t}$$

This can be seen by combining the result from Step 2 ($N/N_0 = A/A_0$) with Equation 31.5 ($N = N_0 e^{-\lambda t}$).

$$t = -\left(\frac{1}{\lambda}\right) \ln\left(\frac{N}{N_0}\right) \qquad (1)$$

$$\boxed{\lambda = \frac{0.693}{T_{1/2}}} \quad \boxed{\frac{N}{N_0} = \frac{A}{A_0}}$$

Related Homework: Problems 38, 40

Radiocarbon dating is not the only radioactive dating method. For example, other methods utilize uranium $^{238}_{92}U$, potassium $^{40}_{19}K$, and lead $^{210}_{82}Pb$. For such methods to be useful, the half-life of the radioactive species must be neither too short nor too long relative to the age of the sample to be dated, as Conceptual Example 13 discusses.

CONCEPTUAL EXAMPLE 13 | Dating a Bottle of Wine

A bottle of red wine is thought to have been sealed about 5 years ago. The wine contains a number of different atoms, including carbon, oxygen, and hydrogen. Each of these has a radioactive isotope. The radioactive isotope of carbon is the familiar $^{14}_{6}C$, with a half-life of 5730 yr. The radioactive isotope of oxygen is $^{15}_{8}O$ and has a half-life of 122.2 s. The radioactive isotope of hydrogen, called tritium, is $^{3}_{1}H$; its half-life is 12.33 yr. The activity of each of these isotopes is known at the time the bottle was sealed. However, only one of the isotopes is useful for determining the age of the wine accurately from a measurement of its current activity. Which is it? **(a)** $^{14}_{6}C$ **(b)** $^{15}_{8}O$ **(c)** $^{3}_{1}H$

Reasoning   To find the age of the wine, it is necessary to determine the ratio of the current activity A to the initial activity A_0 (see Example 11). If the age of the sample is very small relative to the half-life of the nuclei, relatively few of the nuclei would have decayed during the wine's life, and the measured activity would have changed little from its initial value ($A \approx A_0$). To obtain an accurate age from such a small change would require prohibitively precise measurements. On the other hand, if the age of the sample is many times greater than the half-life of the nuclei, virtually all of the nuclei would have decayed, and the current activity would be so small($A \approx 0$) that it would be virtually impossible to detect.

Answer (a) is incorrect.   The expected age of the wine is about 5 years. This period is only a tiny fraction of the 5730-yr half-life of $^{14}_{6}C$. As a result, relatively few of the $^{14}_{6}C$ nuclei would have decayed during the wine's life, and the current activity would be nearly the same as the initial activity ($A \approx A_0$), thus requiring prohibitively precise measurements.

Answer (b) is incorrect.   The $^{15}_{8}O$ isotope is not very useful either, because of its relatively short half-life of 122.2 s. During a 5-year period, so many half-lives of 122.2 s would have occurred that the current activity would be vanishingly small ($A \approx 0$) and undetectable.

Answer (c) is correct.   The only remaining option is the $^{3}_{1}H$ isotope. The expected age of 5 yr is long enough relative to the half-life of 12.33 yr that a measurable change in activity will have occurred, but not so long that the current activity will have completely vanished for all practical purposes.

Related Homework: Check Your Understanding Question 14

Check Your Understanding

(The answers are given at the end of the book.)

12. To which one or more of the following objects, each about 1000 yr old, can the radiocarbon dating technique *not* be applied? **(a)** A wooden box **(b)** A gold statue **(c)** Some well-preserved animal fur

13. Suppose there were a greater number of carbon $^{14}_{6}C$ atoms in a plant living 5000 yr ago than is currently believed. When the seeds of this plant are tested using radiocarbon dating, is the age obtained too small or too large compared to the true age?

14. Review Conceptual Example 13 as an aid in answering this question. Tritium is an isotope of hydrogen and undergoes β^{-} decay with a half-life of 12.33 yr. Like carbon $^{14}_{6}C$, tritium is produced in the atmosphere because of cosmic rays and can be used in a radioactive dating technique. Can tritium dating be used to determine a reliable date for a sample that is about 700 yr old?

31.8 Radioactive Decay Series

When an unstable parent nucleus decays, the resulting daughter nucleus is sometimes also unstable. If so, the daughter then decays and produces its own daughter, and so on, until a completely stable nucleus is produced. This sequential decay of one nucleus after another is called a **radioactive decay series.** Examples 4–6 discuss the first two steps of a series that begins with uranium $^{238}_{92}U$:

$$\text{Uranium} \qquad \text{Thorium}$$
$$^{238}_{92}U \longrightarrow \, ^{234}_{90}Th + \, ^{4}_{2}He$$
$$\longrightarrow \, ^{234}_{91}Pa + \, ^{0}_{-1}e$$
$$\text{Protactinium}$$

Furthermore, Examples 7 and 8 deal with radon $^{222}_{86}Rn$, which is formed down the line in the $^{238}_{92}U$ radioactive decay series. **Figure 31.18** shows the entire series. At several points,

FIGURE 31.18 The radioactive decay series that begins with uranium $^{238}_{92}U$ and ends with lead $^{206}_{82}Pb$. Half-lives are given in seconds (s), minutes (m), hours (h), days (d), or years (y). The inset in the upper left identifies the type of decay that each nucleus undergoes.

branches occur because more than one kind of decay is possible for an intermediate species. Ultimately, however, the series ends with lead $^{206}_{82}Pb$, which is stable.

The $^{238}_{92}U$ series and other such series are the only sources of some of the radioactive elements found in nature. Radium $^{226}_{88}Ra$, for instance, has a half-life of 1600 yr, which is short enough that all the $^{226}_{88}Ra$ created when the earth was formed billions of years ago has now disappeared. The $^{238}_{92}U$ series provides a continuing supply of $^{226}_{88}Ra$, however.

Check Your Understanding

(The answer is given at the end of the book.)

15. Because of radioactive decay, one element can be transmuted into another. Thus, a container of uranium $^{238}_{92}U$ ultimately becomes a container of lead $^{206}_{82}Pb$, as **Figure 31.18** indicates. Roughly, how long does it take for $^{238}_{92}U$ to transmute entirely into $^{206}_{82}Pb$? **(a)** Several decades **(b)** Several centuries **(c)** Thousands of years **(d)** Millions of years **(e)** Billions of years

31.9 Radiation Detectors

FIGURE 31.19 A Geiger counter.

THE PHYSICS OF . . . radiation detectors. There are a number of devices that can be used to detect the particles and photons (γ rays) emitted when a radioactive nucleus decays. Such devices detect the ionization that these particles and photons cause as they pass through matter.

The most familiar detector is the **Geiger counter***, which **Figure 31.19** illustrates. The Geiger counter consists of a gas-filled metal cylinder. The α, β, or γ rays enter the cylinder through a thin window at one end. γ rays can also penetrate directly through the metal. A wire electrode runs along the center of the tube and is kept at a high positive voltage (1000–3000 V) relative to the outer cylinder. When a high-energy particle or photon enters the cylinder, it collides with and ionizes a gas molecule. The electron produced from the gas molecule accelerates toward the positive wire, ionizing other molecules in its path. Additional electrons are formed, and an avalanche of electrons rushes toward the wire, leading to a pulse of current through the resistor R. This pulse can be counted or made to produce a "click" in a loudspeaker. The number of counts or clicks is related to the number of disintegrations that produced the particles or photons.

The **scintillation counter** is another important radiation detector. As **Figure 31.20** indicates, this device consists of a scintillator mounted on a photomultiplier tube. Often the scintillator is a crystal (e.g., cesium iodide) containing a small amount of impurity (thallium), but plastic, liquid, and gaseous scintillators are also used. In response to ionizing radiation, the scintillator emits a flash of visible light. The photons of the flash then strike the photocathode of the photomultiplier tube. The photocathode is made of a material that emits electrons because of the photoelectric effect. These photoelectrons are then attracted to a special electrode kept at a voltage of about +100 V relative to the photocathode. The electrode is coated with a substance that emits several additional electrons for every electron striking it. The additional electrons are attracted to a second similar electrode (voltage = +200 V), where they generate even more electrons. Commercial photomultiplier tubes contain as many as 15 of these special electrodes, so photoelectrons resulting from the light flash of the scintillator lead to a cascade of electrons and a pulse of current. As in a Geiger tube, the current pulses can be counted.

Ionizing radiation can also be detected with several types of **semiconductor detectors.** Such devices utilize n- and p-type materials (see Section 23.5), and their operation depends on the electrons and holes formed in the materials as a result of the radiation. One of the main advantages of semiconductor detectors is their ability to discriminate between two particles with only slightly different energies.

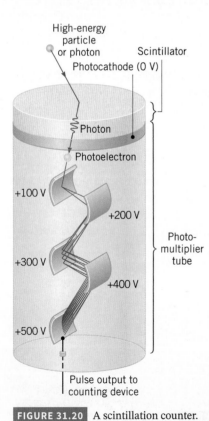

FIGURE 31.20 A scintillation counter.

*The Geiger counter is named after Hans Geiger (1882–1945), a German physicist, along with Sir Ernest Marsden (1889–1970), an English-New Zealand physicist, who co-invented the device. It was Geiger and Marsden, working under the direction of Ernest Rutherford, who performed the crucial gold foil scattering experiments that led to the discovery of the atomic nucleus (see Section 30.1).

A number of instruments provide a pictorial representation of the path that high-energy particles follow after they are emitted from unstable nuclei. In a **cloud chamber,** a gas is cooled just to the point at which it will condense into droplets, provided nucleating agents are available on which the droplets can form. When a high-energy particle, such as an α particle or a β particle, passes through the gas, the ions it leaves behind serve as nucleating agents, and droplets form along the path of the particle. A **bubble chamber** works in a similar fashion, except that it contains a liquid that is just at the point of boiling. Tiny bubbles form along the trail of a high-energy particle passing through the liquid. Paths revealed in a cloud or bubble chamber can be photographed to provide a permanent record of the event. **Figure 31.21** shows a photograph of tracks in a cloud chamber. A **photographic emulsion** also can be used directly to produce a record of the path taken by a particle of ionizing radiation. Ions formed as the particle passes through the emulsion cause silver to be deposited along the track when the emulsion is developed.

Photo Researchers / Science History Images / Alamy Stock Photo

FIGURE 31.21 A photograph showing particle tracks in a cloud chamber.

Concept Summary

31.1 Nuclear Structure   The nucleus of an atom consists of protons and neutrons, which are collectively referred to as nucleons. A neutron is an electrically neutral particle whose mass is slightly larger than that of the proton. The atomic number Z is the number of protons in the nucleus. The atomic mass number A (or nucleon number) is the total number of protons and neutrons in the nucleus and is given by Equation 31.1, where N is the number of neutrons. For an element whose chemical symbol is X, the symbol for the nucleus is ${}^{A}_{Z}$X. Nuclei that contain the same number of protons, but a different number of neutrons, are called isotopes. The approximate radius (in meters) of a nucleus is given by Equation 31.2.

$$A = Z + N \tag{31.1}$$

$$r \approx (1.2 \times 10^{-15}\ \mathrm{m})A^{1/3} \tag{31.2}$$

31.2 The Strong Nuclear Force and the Stability of the Nucleus   The strong nuclear force is the force of attraction between nucleons (protons and neutrons) and is one of the three fundamental forces of nature. This force balances the electrostatic force of repulsion between protons and holds the nucleus together. The strong nuclear force has a very short range of action and is almost independent of electric charge.

31.3 The Mass Defect of the Nucleus and Nuclear Binding Energy   The binding energy of a nucleus is the energy required to separate the nucleus into its constituent protons and neutrons. The binding energy is given by Equation 31.3, where Δm is the mass defect of the nucleus and c is the speed of light in a vacuum. The mass defect is the amount by which the sum of the individual masses of the protons and neutrons exceeds the mass of the intact nucleus.

When specifying nuclear masses, it is customary to use the atomic mass unit (u). One atomic mass unit has a mass of 1.6605×10^{-27} kg and is equivalent to an energy of 931.5 MeV.

$$\text{Binding energy} = (\Delta m)c^2 \tag{31.3}$$

31.4 Radioactivity   Unstable nuclei spontaneously decay by breaking apart or rearranging their internal structure in a process called radioactivity. Naturally occurring radioactivity produces α, β, and γ rays. α rays consist of positively charged particles, each particle being the ${}^{4}_{2}$He nucleus of helium. The general form for α decay is

$$\underbrace{{}^{A}_{Z}\mathrm{P}}_{\substack{\text{Parent} \\ \text{nucleus}}} \longrightarrow \underbrace{{}^{A-4}_{Z-2}\mathrm{D}}_{\substack{\text{Daughter} \\ \text{nucleus}}} + \underbrace{{}^{4}_{2}\mathrm{He}}_{\substack{\alpha\ \text{particle} \\ \text{(helium nucleus)}}}$$

The most common kind of β ray consists of negatively charged particles, or β^- particles, which are electrons. The general form for β^- decay is

$$\underbrace{{}^{A}_{Z}\mathrm{P}}_{\substack{\text{Parent} \\ \text{nucleus}}} \longrightarrow \underbrace{{}^{A}_{Z+1}\mathrm{D}}_{\substack{\text{Daughter} \\ \text{nucleus}}} + \underbrace{{}^{0}_{-1}\mathrm{e}}_{\substack{\beta^-\ \text{particle} \\ \text{(electron)}}}$$

β^+ decay produces another kind of β ray, which consists of positively charged particles, or β^+ particles. A β^+ particle, also called a positron, has the same mass as an electron, but carries a charge of $+e$ instead of $-e$.

If a radioactive parent nucleus disintegrates into a daughter nucleus that has a different atomic number, as occurs in α and β decay, one element has been converted into another element, the conversion being referred to as a transmutation.

γ rays are high-energy photons emitted by a radioactive nucleus. The general form for γ decay is

$$\underbrace{^A_Z P^*}_{\substack{\text{Excited} \\ \text{energy state}}} \longrightarrow \underbrace{^A_Z P}_{\substack{\text{Lower} \\ \text{energy state}}} + \underbrace{\gamma}_{\gamma \text{ ray}}$$

γ decay does not cause a transmutation of one element into another.

31.5 The Neutrino The neutrino is an electrically neutral particle that is emitted along with β particles and has a mass that is much, much smaller than the mass of an electron.

31.6 Radioactive Decay and Activity The half-life of a radioactive isotope is the time required for one-half of the nuclei present to disintegrate or decay. The activity is the number of disintegrations per second that occur. Activity is the magnitude of $\Delta N / \Delta t$, where ΔN is the change in the number N of radioactive nuclei and Δt is the time interval during which the change occurs. In other words, activity is $|\Delta N / \Delta t|$. The SI unit for activity is the becquerel (Bq), one becquerel being one disintegration per second. Activity is sometimes also measured in a unit called the curie (Ci); $1 \text{ Ci} = 3.70 \times 10^{10}$ Bq.

$$\frac{\Delta N}{\Delta t} = -\lambda N \qquad (31.4)$$

Radioactive decay obeys Equation 31.4, where λ is the decay constant. This equation can be solved by the methods of integral calculus to show that N is given by Equation 31.5, where N_0 is the original number of nuclei. The decay constant λ is related to the half-life $T_{1/2}$ according to Equation 31.6.

$$N = N_0 e^{-\lambda t} \qquad (31.5)$$

$$\lambda = \frac{0.693}{T_{1/2}} \qquad (31.6)$$

31.7 Radioactive Dating If an object contained radioactive nuclei when it was formed, then the decay of these nuclei can be used to determine the age of the object. One way to obtain the age is to relate the present activity A of an object to its initial activity A_0, according to Equation 1, where λ is the decay constant and t is the age of the object. For radiocarbon dating that uses the $^{14}_6$C isotope of carbon, the initial activity is often assumed to be $A_0 = 0.23$ Bq.

31.8 Radioactive Decay Series The sequential decay of one nucleus after another is called a radioactive decay series. A decay series starts with a radioactive nucleus and ends with a completely stable nucleus. **Figure 31.18** illustrates one such series that begins with uranium $^{238}_{92}$U and ends with lead $^{206}_{82}$Pb.

31.9 Radiation Detectors A number of devices are used to detect α and β particles as well as γ rays. These include the Geiger counter, the scintillation counter, semiconductor detectors, cloud and bubble chambers, and photographic emulsions.

Focus on Concepts

Online

Additional questions are available for assignment in WileyPLUS.

Section 31.1 Nuclear Structure

1. An indium (In) nucleus contains 49 protons and 66 neutrons. Which one of the following symbols describes this nucleus? (a) $^{49}_{115}$In (b) $^{49}_{66}$In (c) $^{115}_{66}$In (d) $^{66}_{49}$In (e) $^{115}_{49}$In

2. The notation for a particular nucleus is $^{85}_{37}$Rb. In an electrically neutral atom, how many electrons are in orbit about this nucleus? (a) $37 + 85 = 122$ (b) 85 (c) 37 (d) $85 - 37 = 48$ (e) The number of electrons cannot be determined from the notation.

Section 31.3 The Mass Defect of the Nucleus and Nuclear Binding Energy

3. Suppose that we lived in a hypothetical world in which the mass of each proton and each neutron were exactly 1 u. In this world, the atomic mass of copper $^{63}_{29}$Cu is 62.5 u. What would be the mass defect for this nucleus? (a) 63 u (b) 29 u (c) $63 - 29 = 34$ u (d) 0.5 u (e) $63 \text{ u} + 29 \text{ u} = 92$ u

Section 31.4 Radioactivity

4. Which one or more of the three decay processes (α, β^-, or γ) results in a new element? (a) α and β^- (b) Only α (c) Only β^- (d) β^- and γ (e) Only γ

5. A nucleus can undergo α, β^-, or γ decay. For each type of decay, is the radius of the daughter nucleus greater than, less than, or about the same as the radius of the parent nucleus?

	α Decay	β^- Decay	γ Decay
(a)	Greater than	Greater than	About the same as
(b)	Greater than	Less than	About the same as
(c)	Less than	About the same as	Greater than
(d)	Less than	About the same as	About the same as
(e)	About the same as	Less than	Less than

Section 31.6 Radioactive Decay and Activity

6. Two samples contain different radioactive isotopes. Is it possible for these samples to have the same activity? (a) Yes, if they have the same number of nuclei but different half-lives. (b) Yes, if they have different numbers of nuclei and different half-lives. (c) Yes, if they have different numbers of nuclei but the same half-lives. (d) No, because they can have different half-lives. (e) No, because they can have different numbers of nuclei.

7. The drawing shows the activities of three radioactive samples. Rank the samples according to half-life, largest first. (a) 2, 3, 1 (b) 1, 2, 3 (c) 3, 2, 1 (d) 1, 3, 2 (e) 3, 1, 2

QUESTION 7

Section 31.7   Radioactive Dating

8. The bones from an animal found at an archaeological dig have a $^{14}_{6}$C activity of 0.10 Bq per gram of carbon. The half-life of the radioactive isotope $^{14}_{6}$C is 5730 yr. Which one of the following best describes the age of the bones? **(a)** It is less than 2000 years. **(b)** It is between 2000 and 3000 years. **(c)** It is between 3000 and 4000 years. **(d)** It is between 4000 and 5000 years. **(e)** It is more than 5000 years.

Problems

Online

Additional questions are available for assignment in WileyPLUS.

Note: *The data given for atomic masses in these problems include the mass of the electrons orbiting the nucleus of the electrically neutral atom.*

SSM Student Solutions Manual	**BIO** Biomedical application
MMH Problem-solving help	**E** Easy
GO Guided Online Tutorial	**M** Medium
V-HINT Video Hints	**H** Hard
CHALK Chalkboard Videos	**WS** Worksheet
	T Team Problem

Section 31.1   Nuclear Structure,
Section 31.2   The Strong Nuclear Force and the Stability of the Nucleus

1. **E SSM** For $^{208}_{82}$Pb find **(a)** the net electrical charge of the nucleus, **(b)** the number of neutrons, **(c)** the number of nucleons, **(d)** the approximate radius of the nucleus, and **(e)** the nuclear density.

2. **E** A nucleus contains 18 protons and 22 neutrons. What is the radius of this nucleus?

3. **E** In each of the following cases, what element does the symbol X represent and how many neutrons are in the nucleus? Use the periodic table on the inside of the back cover of the textbook as needed. **(a)** $^{195}_{78}$X **(b)** $^{32}_{16}$X **(c)** $^{63}_{29}$X **(d)** $^{11}_{5}$X **(e)** $^{239}_{94}$X

4. **E** By what factor does the nucleon number of a nucleus have to increase in order for the nuclear radius to double?

5. **E SSM** In electrically neutral atoms, how many **(a)** protons are in the uranium $^{238}_{92}$U nucleus, **(b)** neutrons are in the mercury $^{202}_{80}$Hg nucleus, and **(c)** electrons are in orbit about the niobium $^{93}_{41}$Nb nucleus?

6. **E GO** The largest stable nucleus has a nucleon number of 209, and the smallest has a nucleon number of 1. If each nucleus is assumed to be a sphere, what is the ratio (largest/smallest) of the surface areas of these spheres?

7. **M GO CHALK** The ratio r_X/r_T of the radius of an unknown nucleus $^4_2$X to the radius of a tritium nucleus $^3_1$T is $\frac{r_X}{r_T} = 1.10$. Both nuclei contain the same number of neutrons. Identify the unknown nucleus in the form ^A_ZX. Use the periodic table immediately before the back cover of this text.

8. **M V-HINT CHALK** An unknown nucleus contains 70 neutrons and has twice the volume of the nickel $^{60}_{28}$Ni nucleus. Identify the unknown nucleus in the form ^A_ZX. Use the periodic table immediately before the back cover of this text.

Section 31.3   The Mass Defect of the Nucleus and Nuclear Binding Energy *(Note: The atomic mass for hydrogen $^1_1$H is 1.007 825 u; this includes the mass of one electron.)*

9. **E SSM** Find the binding energy (in MeV) for lithium $^7_3$Li (atomic mass = 7.016 003 u).

10. **E** The binding energy of a nucleus is 225.0 MeV. What is the mass defect of the nucleus in atomic mass units?

11. **E** Determine the mass defect (in atomic mass units) for **(a)** helium $^3_2$He, which has an atomic mass of 3.016 030 u, and **(b)** the isotope of hydrogen known as tritium $^3_1$T, which has an atomic mass of 3.016 050 u. **(c)** On the basis of your answers to parts (a) and (b), state which nucleus requires more energy to disassemble it into its separate and stationary constituent nucleons. Give your reasoning.

12. **E GO** A 245-kg boulder is dropped into a mine shaft that is 3.0×10^3 m deep. During the boulder's fall, the system consisting of the earth and the boulder loses a certain amount of gravitational potential energy. It would take an equal amount of energy to "free" the boulder from the shaft by raising it back to the top, so this can be considered the system's binding energy. **(a)** Determine the binding energy (in joules) of the earth–boulder system. **(b)** How much mass does the earth–boulder system lose when the boulder falls to the bottom of the shaft?

13. **E CHALK** For lead $^{206}_{82}$Pb (atomic mass = 205.974 440 u) obtain **(a)** the mass defect in atomic mass units, **(b)** the binding energy (in MeV), and **(c)** the binding energy per nucleon (in MeV/nucleon).

14. **M** **(a)** Energy is required to separate a nucleus into its constituent nucleons, as **Interactive Figure 31.3** indicates; this energy is the *total* binding energy of the nucleus. In a similar way one can speak of the energy that binds a single nucleon to the remainder of the nucleus. For example, separating nitrogen $^{14}_7$N into nitrogen $^{13}_7$N and a neutron takes energy equal to the binding energy of the neutron, as shown below:

$$^{14}_7\text{N} + \text{Energy} \longrightarrow {}^{13}_7\text{N} + {}^1_0\text{n}$$

Find the energy (in MeV) that binds the neutron to the $^{14}_7$N nucleus by considering the mass of $^{13}_7$N (atomic mass = 13.005 738 u) and the mass of $^1_0$n (atomic mass = 1.008 665 u), as compared to the mass of $^{14}_7$N (atomic mass = 14.003 074 u). **(b)** Similarly, one can speak of the energy that binds a single proton to the $^{14}_7$N nucleus:

$$^{14}_7\text{N} + \text{Energy} \longrightarrow {}^{13}_6\text{C} + {}^1_1\text{H}$$

Following the procedure outlined in part (a), determine the energy (in MeV) that binds the proton (atomic mass = 1.007 825 u) to the $^{14}_7$N nucleus. The atomic mass of carbon $^{13}_6$C is 13.003 355 u. **(c)** Which nucleon is more tightly bound, the neutron or the proton?

15. **M SSM** Two isotopes of a certain element have binding energies that differ by 5.03 MeV. The isotope with the larger binding energy contains one more neutron than the other isotope. Find the difference in atomic mass between the two isotopes.

16. **M GO** A copper penny has a mass of 3.0 g. Determine the energy (in MeV) that would be required to break all the copper nuclei into their constituent protons and neutrons. Ignore the energy that binds the electrons to the nucleus and the energy that binds one atom to another in the structure of the metal. For simplicity, assume that all the copper nuclei are $^{63}_{29}$Cu (atomic mass = 62.939 598 u).

Section 31.4 Radioactivity

17. **E** **SSM** Write the β^+ decay process for each of the following nuclei, being careful to include Z and A and the proper chemical symbol for each daughter nucleus: **(a)** $^{18}_{9}F$ **(b)** $^{15}_{8}O$

18. **E** Write the β^- decay process for carbon $^{14}_{6}C$, including the chemical symbols as well as the values of Z and A for the parent and daughter nuclei and the β^- particle.

19. **E** **SSM** Osmium $^{191}_{76}Os$ (atomic mass = 190.960 920 u) is converted into iridium $^{191}_{77}Ir$ (atomic mass = 190.960 584 u) via β^- decay. What is the energy (in MeV) released in this process?

20. **E** Find the energy that is released when a nucleus of lead $^{211}_{82}Pb$ (atomic mass = 210.988 735 u) undergoes β^- decay to become bismuth $^{211}_{83}Bi$ (atomic mass = 210.987 255 u).

21. **E** Find the energy (in MeV) released when α decay converts radium $^{226}_{88}Ra$ (atomic mass = 226.025 40 u) into radon $^{222}_{86}Rn$ (atomic mass = 222.017 57 u). The atomic mass of an α particle is 4.002 603 u.

22. **E** **GO** Lead $^{207}_{82}Pb$ is a stable daughter nucleus that can result from either an α decay or a β^- decay. Write the decay processes, including the chemical symbols and values for Z and A of the parent nuclei, for **(a)** the α decay and **(b)** the β^- decay.

23. **E** In the form $^A_Z X$ identify the daughter nucleus that results when **(a)** plutonium $^{242}_{94}Pu$ undergoes α decay, **(b)** sodium $^{24}_{11}Na$ undergoes β^- decay, and **(c)** nitrogen $^{13}_{7}N$ undergoes β^+ decay.

24. **E** **V-HINT** When uranium $^{235}_{92}U$ decays, it emits (among other things) a γ ray that has a wavelength of 1.14×10^{-11} m. Determine the energy (in MeV) of this γ ray.

25. **M** **CHALK** **MMH** Polonium $^{210}_{84}Po$ (atomic mass = 209.982 848 u) undergoes α decay. Assuming that all the released energy is in the form of kinetic energy of the α particle (atomic mass = 4.002 603 u) and ignoring the recoil of the daughter nucleus (lead $^{206}_{82}Pb$, 205.974 440 u), find the speed of the α particle. Ignore relativistic effects.

26. **M** **GO** Radon $^{220}_{86}Rn$ produces a daughter nucleus that is radioactive. The daughter, in turn, produces its own radioactive daughter, and so on. This process continues until lead $^{208}_{82}Pb$ is reached. What are the total number N_α of α particles and the total number N_β of β^- particles that are generated in this series of radioactive decays?

27. **M** Review Conceptual Example 5 as background for this problem. The α decay of uranium $^{238}_{92}U$ produces thorium $^{234}_{90}Th$ (atomic mass = 234.0436 u). In Example 4, the energy released in this decay is determined to be 4.3 MeV. Determine how much of this energy is carried away by the recoiling $^{234}_{90}Th$ daughter nucleus and how much by the α particle (atomic mass = 4.002 603 u). Assume that the energy of each particle is kinetic energy, and ignore the small amount of energy carried away by the γ ray that is also emitted. In addition, ignore relativistic effects.

28. **H** **MMH** An isotope of beryllium (atomic mass = 7.017 u) emits a γ ray and recoils with a speed of 2.19×10^4 m/s. Assuming that the beryllium nucleus is stationary to begin with, find the wavelength of the γ ray.

29. **H** **SSM** Find the energy (in MeV) released when β^+ decay converts sodium $^{22}_{11}Na$ (atomic mass = 21.994 434 u) into neon $^{22}_{10}Ne$ (atomic mass = 21.991 383 u). Notice that the atomic mass for $^{22}_{11}Na$ includes the mass of 11 electrons, whereas the atomic mass for $^{22}_{10}Ne$ includes the mass of only 10 electrons.

Section 31.6 Radioactive Decay and Activity

30. **E** In 9.0 days the number of radioactive nuclei decreases to one-eighth the number present initially. What is the half-life (in days) of the material?

31. **E** The $^{32}_{15}P$ isotope of phosphorus has a half-life of 14.28 days. What is its decay constant in units of s^{-1}?

32. **E** **GO** Two radioactive waste products from nuclear reactors are strontium $^{90}_{38}Sr$ ($T_{1/2} = 29.1$ yr) and cesium $^{134}_{55}Cs$ ($T_{1/2} = 2.06$ yr). These two species are present initially in a ratio of $N_{0,Sr}/N_{0,Cs} = 7.80 \times 10^{-3}$. What is the ratio N_{Sr}/N_{Cs} fifteen years later?

33. **E** **SSM** The number of radioactive nuclei present at the start of an experiment is 4.60×10^{15}. The number present twenty days later is 8.14×10^{14}. What is the half-life (in days) of the nuclei?

34. **M** **GO** A one-gram sample of radium $^{224}_{88}Ra$ (atomic mass = 224.020 186 u, $T_{1/2} = 3.66$ days) contains 2.69×10^{21} nuclei and undergoes α decay to produce radon $^{220}_{86}Rn$ (atomic mass = 220.011 368 u). The atomic mass of an α particle is 4.002 603 u. The latent heat of fusion for water is 33.5×10^4 J/kg. With the energy released in 3.66 days, how many kilograms of ice could be melted at 0 °C?

35. **M** **GO** Outside the nucleus, the neutron itself is radioactive and decays into a proton, an electron, and an antineutrino. The half-life of a neutron (mass = 1.675×10^{-27} kg) outside the nucleus is 10.4 min. On average, over what distance (in meters) would a beam of 5.00-eV neutrons travel before the number of neutrons decreased to 75.0% of its initial value?

36. **M** **SSM** Two radioactive nuclei A and B are present in equal numbers to begin with. Three days later, there are three times as many A nuclei as there are B nuclei. The half-life of species B is 1.50 days. Find the half-life of species A.

Section 31.7 Radioactive Dating

37. **E** A sample has a $^{14}_{6}C$ activity of 0.0061 Bq per gram of carbon. **(a)** Find the age of the sample, assuming that the activity per gram of carbon in a living organism has been constant at a value of 0.23 Bq. **(b)** Evidence suggests that the value of 0.23 Bq might have been as much as 40% larger. Repeat part (a), taking into account this 40% increase.

38. **E** **SSM** Review Multiple-Concept Example 11 for help in approaching this problem. An archaeological specimen containing 9.2 g of carbon has an activity of 1.6 Bq. How old (in years) is the specimen?

39. **E** **CHALK** **GO** The half-life for the α decay of uranium $^{238}_{92}U$ is 4.47×10^9 yr. Determine the age (in years) of a rock specimen that contains 60.0% of its original number of $^{238}_{92}U$ atoms.

40. **E** **GO** Multiple-Concept Example 12 reviews most of the concepts that are needed to solve this problem. Material found with a mummy in the arid highlands of southern Peru has a $^{14}_{6}C$ activity per gram of carbon that is 78.5% of the activity present initially. How long ago (in years) did this individual die?

41. **M** **SSM** When any radioactive dating method is used, experimental error in the measurement of the sample's activity leads to error in the estimated age. In an application of the radiocarbon dating technique to certain fossils, an activity of 0.100 Bq per gram of carbon is measured to within an accuracy of ±10.0%. Find the age of the fossils and the maximum error (in years) in the value obtained. Assume that there is no error in the 5730-year half-life of $^{14}_{6}C$ nor in the value of 0.23 Bq per gram of carbon in a living organism.

Additional Problems

Online

42. **E** **GO** In a nucleus, each proton experiences a repulsive electrostatic force from each of the other protons. In a nucleus of gold $^{197}_{79}$Au, what is the magnitude of the least possible electrostatic force of repulsion that one proton can exert on another?

43. **E** **V-HINT** When a sample from a meteorite is analyzed, it is determined that 93.8% of the original mass of a certain radioactive isotope is still present. Based on this finding, the age of the meteorite is calculated to be 4.51×10^9 yr. What is the half-life (in yr) of the isotope used to date the meteorite?

44. **E** **GO** The β^- decay of phosphorus $^{32}_{15}$P (atomic mass = 31.973 907 u) produces a daughter nucleus that is sulfur $^{32}_{16}$S (atomic mass = 31.972 070 u), a β^- particle, and an antineutrino. The kinetic energy of the β^- particle is 0.90 MeV. Find the maximum possible energy (in MeV) that the antineutrino could carry away.

45. **E** Complete the following decay processes by stating what the symbol X represents (X = α, β^-, β^+, or γ):

(a) $^{211}_{82}$Pb $\rightarrow$ $^{211}_{83}$Bi + X   (c) $^{231}_{90}$Th$^* \rightarrow$ $^{231}_{90}$Th + X

(b) $^{11}_{6}$C $\rightarrow$ $^{11}_{5}$B + X   (d) $^{210}_{84}$Po $\rightarrow$ $^{206}_{82}$Pb + X

46. **M** **V-HINT** A sample of ore containing radioactive strontium $^{90}_{38}$Sr has an activity of 6.0×10^5 Bq. The atomic mass of strontium is 89.908 u, and its half-life is 29.1 yr. How many grams of strontium are in the sample?

47. **M** **V-HINT** In a radioactive decay series similar to that shown in **Figure 31.18**, thorium $^{228}_{90}$Th (atomic mass = 228.028 715 u) undergoes four successive α decays, producing a daughter nucleus. (a) Determine the symbol ^A_ZX for the nucleus produced by four successive α decays of $^{228}_{90}$Th. (b) What is the total amount of energy (in MeV) released in this series of α decays? The mass of the daughter nucleus can be obtained by using the result of part (a) and consulting Appendix F at the back of the book. The mass of a single α particle is 4.002 603 u.

48. **M** **GO** **SSM** What is the wavelength (in vacuum) of the 0.186-MeV γ-ray photon emitted by $^{226}_{88}$Ra?

Physics in Biology, Medicine, and Sports

49. **E** **BIO** **31.6** Review Example 10 as background for this problem. Strontium $^{90}_{38}$Sr has a half-life of 29.1 yr. It is chemically similar to calcium, enters the body through the food chain, and collects in the bones. Consequently, $^{90}_{38}$Sr is a particularly serious health hazard. How long (in years) will it take for 99.9900% of the $^{90}_{38}$Sr released in a nuclear reactor accident to disappear?

50. **E** **BIO** **31.6** Iodine $^{131}_{53}$I is used in diagnostic and therapeutic techniques in the treatment of thyroid disorders. This isotope has a half-life of 8.04 days. What percentage of an initial sample of $^{131}_{53}$I remains after 30.0 days?

51. **M** **BIO** **31.6** A device used in radiation therapy for cancer contains 0.50 g of cobalt $^{60}_{27}$Co (59.933 819 u). The half-life of $^{60}_{27}$Co is 5.27 yr. Determine the activity of the radioactive material.

52. **M** **BIO** **V-HINT** **31.6** The isotope $^{198}_{79}$Au (atomic mass = 197.968 u) of gold has a half-life of 2.69 days and is used in cancer therapy. What mass (in grams) of this isotope is required to produce an activity of 315 Ci?

53. **M** **BIO** **31.6** Positron emission tomography, or PET, is a medical imaging technique that uses radionuclides to visualize and measure biological processes inside the body. A radioactive isotope, called a tracer, is injected into the body and decays by positron emission. Which isotope is chosen will depend on what is being imaged. Once the tracer is in its proper location, the positrons it emits will combine with electrons in the surrounding tissue to create gamma rays that are detected by a scanning ring around the body (see the photo, left). The energy and momentum of the gamma rays are measured, from which a 3-dimensional image of the tracer concentration in the body can be formed. If the tracer is designed to detect cancer, and thus collects in a tumor, the precise location and size of the tumor is revealed (see the photo, right). One of the most common tracers for cancer detection is a glucose compound called fludeoxyglucose. One of the hydroxyl groups in the compound is replaced by radioactive fluorine-18 (^{18}F). The half-life of ^{18}F is only 109.8 minutes, so it remains in the body for a short period of time. (a) If ^{18}F undergoes β^+ decay, what is the daughter nucleus? (b) If the activity of the tracer is 10 500 μCi at the beginning of the procedure, what is the activity 12.0 hours later?

Hg6996, Image taken from https://commons.wikimedia.org/wiki/File:16slicePETCT.jpg

Javier Arbizu

PROBLEM 53

54. **M** **BIO** **31.6** Cancer cells in tumors tend to accumulate more phosphate than normal cells. For this reason, a radioactive isotope of phosphorous (^{32}P) is often used in diagnostic imaging to pinpoint the location of malignant tumors. The half-life of ^{32}P is 14.3 days. After a small amount of ^{32}P is ingested, how long will it take for 99.99% of the ^{32}P to disappear?

55. **M** **BIO** **31.3** The human body is composed of 18.5% carbon. This corresponds to 5.0×10^2 moles of carbon atoms for the average human. What would be the total nuclear binding energy of all the carbon atoms in the body? For simplicity, assume all of the carbon atoms are carbon-12 (^{12}C). In comparison, the energy released by the atomic bomb dropped on Hiroshima, Japan, at the end of World War II was 60 trillion joules (6.0×10^{13} J).

56. **M** **BIO** **31.6** A radiation safety officer is using a Geiger counter to measure the activity of a radioactive material. Her initial measurement records 515 counts/min. She returns 2.00 hours later and records an activity of 125 counts/min. (a) What is the half-life of the material? (b) If the officer takes a third measurement, 2.00 hours after the second measurement, what activity will she measure then?

57. **M** **BIO** **31.6** Bananas are an excellent source of potassium. In naturally occurring potassium, a very small fraction, 0.0120%, is composed of potassium-40 (^{40}K), which is a radioactive isotope with a half-life of 1.251×10^9 years. In fact, ^{40}K is responsible for the majority of the naturally occurring radioactivity in all animals, including humans. The activity of ^{40}K in the human body is approximately 4.300×10^3 Bq.

What mass of ^{40}K is contained in the body, if its isotopic mass is 39.96 g/mol? *Note:* Your body regulates the amount of potassium you need to maintain homeostasis by excreting excess potassium through sweat and urine. Eating many bananas, or other potassium-rich foods, will not increase the amount of potassium in your body. Thus, you will not increase your annual dose of radiation from ^{40}K.

Concepts and Calculations Problems

Online

Radioactive decay obeys the conservation laws of physics, and we have studied five of these. Two of them are particularly important in understanding the types of radioactivity that occur and the nuclear changes that accompany them. These are the conservation of electric charge and the conservation of nucleon number. Problem 58 emphasizes their importance and reviews how they are applied. Problem 59 explores how radioactive decay can be a source of energy in the form of heat.

58. **M CHALK** Thorium $^{228}_{90}$Th produces a daughter nucleus that is radioactive. The daughter, in turn, produces its own radioactive daughter, and so on. This process continues until bismuth $^{212}_{83}$Bi is reached. *Concepts:* **(i)** How many of the 90 protons in the thorium nucleus are carried off by the α particles? **(ii)** How many protons are left behind when the β^- particles are emitted? **(iii)** How many of the 228 nucleons in the thorium nucleus are carried off by the α particles?

(iv) Does the departure of a β^- particle alter the number of nucleons? *Calculations:* What are the total number N_α of α particles and the total number N_β of β^- particles that are generated in this series of radioactive decays?

59. **M CHALK SSM** A one-gram sample of thorium $^{228}_{90}$Th contains 2.64×10^{21} atoms and undergoes α decay with a half-life of 1.913 yr $(1.677 \times 10^4$ h). Each disintegration releases an energy of 5.52 MeV $(8.83 \times 10^{-13}$ J). Assume that all of the energy is used to heat a 3.80-kg sample of water. *Concepts:* **(i)** How much heat Q is needed to raise the temperature of a mass m of water by ΔT degrees? **(ii)** The energy released by each disintegration is E. What is the total energy E_{total} released by a number n of disintegrations? **(iii)** What is the number n of disintegrations that occur during a time t? *Calculations:* Find the change in temperature of the 3.80-kg sample of water that occurs in one hour.

Team Problems

Online

60. **E T WS** **Californium.** Neutrons have applications in a wide variety of disciplines, including medicine, materials science, and nuclear energy. One small-scale source of neutrons is an isotope of the artificially synthesized element californium (Cf). Cf has $Z = 98$ protons in its nucleus, but has a large number of isotopes, so the number of neutrons varies. One particularly useful isotope is ^{252}Cf, which has a half-life of 2.65 years. ^{252}Cf undergoes α decay 96.9% of the time, and spontaneously fissions 3.10% of the time, a process by which it releases a variety of fragments that, on average, include about 3.77 free neutrons. This relatively large rate of spontaneous neutron emission makes ^{252}Cf a viable small-scale neutron source. You and your team have been tasked with characterizing this radioactive isotope for use in an experimental neutron source. **(a)** How many neutrons are in the ^{252}Cf nucleus? **(b)** What is the approximate radius of the ^{252}Cf nucleus? **(c)** What is the daughter nucleus in the α decay of ^{252}Cf? **(d)** How many α particles are emitted per second by 35.0 g of ^{252}Cf? **(e)** Approximately how many neutrons are emitted per second by 35.0 g of ^{252}Cf? Assume the mass per mole of ^{252}Cf is 252.08 g/mol.

61. **M T WS** **A Nuclear Clock.** You and your team are designing a device that will be used to turn on the communications systems on a space probe 350.0 years after it is launched. Americium-241 (^{241}Am) has a half-life of 432.2 years and decays through α decay, as well as through other processes such as spontaneous fission. The decay products have enough energy to ionize the atoms of a gas, creating charges that can move between the charged plates of a parallel plate capacitor. The gas pressure between the plates is set such that each disintegration of an ^{241}Am nucleus results in an average net charge flow of $1.55 \times 10^4 e$ between the plates, that is, a current flows between the plates. The idea is that when the activity of the radioactive material decays to a level such that the current drops below some predetermined value, the circuit monitoring the current will energize the communications systems of the spacecraft. **(a)** What current will flow between the plates if 1.50 kg of ^{241}Am is placed in the volume between them? **(b)** At what magnitude of the current should the communications system energize if it is to turn on 350.0 years after the launch? The molar mass of Americium-241 is 241.06 g/mol.

62. **M T** **A Radioisotope Thermoelectric Generator.** You and your team are designing a thermoelectric generator that converts the thermal energy released during the decay of a radioactive material into electrical energy. Such devices are used in deep-space probes such as Voyager 1, which was launched in 1977. Its radioisotope thermoelectric generators are expected to function until the year 2025. The isotope that your team will use is $^{238}_{94}$Pu (atomic mass = 238.049553 u), which has a half-life of 87.7 years. **(a)** If $^{238}_{94}$Pu undergoes alpha decay, what is the daughter nucleus? **(b)** Assuming the daughter nucleus found in (a) has an atomic mass of 234.0409468 u, calculate the energy released in the decay in joules (the atomic weight of an alpha particle is 4.002603 u). **(c)** How many atoms are in one gram of $^{238}_{94}$Pu? **(d)** How many atoms in one gram of $^{238}_{94}$Pu decay after one year? **(e)** How much energy is released from one gram of $^{238}_{94}$Pu during one year? **(f)** What is the average power output of one gram during one year (in watts)?

63. **M T** **Radioactive Dating of a Mystery Object.** In a science fiction movie, a strange object is discovered buried deep in the ice of western Antarctica. It appears to be a radioisotope thermoelectric device from a spacecraft, the function of which is to convert thermal energy released in the decay of its radioactive contents into electrical energy. The radioactive material is identified as americium-241 ($^{241}_{95}$Am), which has a half-life of 432 years. **(a)** If $^{241}_{95}$Am undergoes alpha decay, what is its daughter nucleus? **(b)** What is the *activity* you would expect for 1.00 g of $^{241}_{95}$Am (in Bq)? **(c)** In the movie, 1.00 g of material is extracted from the device, and the activity is measured to be 4.00×10^{10} Bq. Assuming the device was initially loaded with 100% $^{241}_{95}$Am, how old is the device? Express your answer in years.

Elementary particles are the basic building blocks for all matter. They are studied by accelerating particles such as protons to high speeds and crashing them together. The Large Hadron Collider is designed to do just such a job. It consists of a large underground ring (diameter 8.6 km or 5.3 mi) and associated facilities located beneath parts of France and Switzerland. High-speed protons travel in opposite directions in the ring, and the elementary particles that result from the collisions between them are observed using specialized detectors. This photograph shows the Compact Muon Solenoid detector for the Large Hadron Collider.

VALERIO MEZZANOTTI/The New York/Redux Pictures

Ionizing Radiation, Nuclear Energy, and Elementary Particles

32.1 Biological Effects of Ionizing Radiation

BIO THE PHYSICS OF . . . the biological effects of ionizing radiation. **Ionizing radiation** consists of photons and/or moving particles that have sufficient energy to knock an electron out of an atom or molecule, thus forming an ion. The photons usually lie in the ultraviolet, X-ray, or γ-ray regions of the electromagnetic spectrum (see Figure 24.9), whereas the moving particles can be the α and β particles emitted during radioactive decay. An energy of roughly 1 to 35 eV is needed to ionize

LEARNING OBJECTIVES

After reading this module, you should be able to...

32.1 Calculate biological effects of ionizing radiation.

32.2 Apply nuclear conservation laws to complete induced nuclear reactions.

32.3 Calculate the energy released in nuclear fission reactions.

32.4 Describe the basic components of nuclear reactors.

32.5 Calculate the energy released in nuclear fusion reactions.

32.6 Describe elementary particle theories leading to the standard model.

32.7 Use Hubble's law to explain the age and size of the universe.

32.8 Discuss recent discoveries in modern physics.

an atom or molecule, and the particles and γ rays emitted during nuclear disintegration often have energies of several million eV. Therefore, a single α particle, β particle, or γ ray can ionize thousands of molecules.

Nuclear radiation is potentially harmful to humans because the ionization it produces can significantly alter the structure of molecules within a living cell. The alterations can lead to the death of the cell and even of the organism itself. Despite the potential hazards, however, ionizing radiation is used in medicine for diagnostic and therapeutic purposes, such as locating bone fractures and treating cancer. The hazards can be minimized only if the fundamentals of radiation exposure, including dose units and the biological effects of radiation, are understood.

Exposure is a measure of the ionization produced in air by X-rays or γ rays, and it is defined in the following manner: A beam of X-rays or γ rays is sent through a mass m of dry air at standard temperature and pressure (STP: 0 °C, 1 atm pressure). In passing through the air, the beam produces positive ions whose total charge is q. Exposure is defined as the total charge per unit mass of air: exposure = q/m. The SI unit for exposure is coulombs per kilogram (C/kg). However, the first radiation unit to be defined was the *roentgen* (R), and it is still used today. With q expressed in coulombs (C) and m in kilograms (kg), the exposure in roentgens is given by

$$\text{Exposure (in roentgens)} = \left(\frac{1}{2.58 \times 10^{-4}}\right)\frac{q}{m} \tag{32.1}$$

Thus, when X-rays or γ rays produce an exposure of one roentgen, $q = 2.58 \times 10^{-4}$ C of positive charge are produced in $m = 1$ kg of dry air:

$$1\text{R} = 2.58 \times 10^{-4}\,\text{C/kg} \quad \text{(dry air, at STP)}$$

Since the concept of exposure is defined in terms of the ionizing abilities of X-rays and γ rays in air, it does not specify the effect of radiation on living tissue. For biological purposes, the **absorbed dose** is a more suitable quantity because it is the energy absorbed from the radiation per unit mass of absorbing material:

$$\text{Absorbed dose} = \frac{\text{Energy absorbed}}{\text{Mass of absorbing material}} \tag{32.2}$$

The SI unit of absorbed dose is the *gray* (Gy), which is the unit of energy divided by the unit of mass: 1 Gy = 1 J/kg. Equation 32.2 is applicable to all types of radiation and absorbing media. Another unit is often used for absorbed dose—namely, the *rad* (sometimes abbreviated rd). The word "rad" is an acronym for **r**adiation **a**bsorbed **d**ose. The rad and the gray are related by

$$1 \text{ rad} = 0.01 \text{ gray}$$

Example 1 deals with the gray and the rad as units for the absorbed dose.

Analyzing Multiple-Concept Problems

EXAMPLE 1 | Absorbed Dose of γ Rays

Figure 32.1 shows γ rays being absorbed by water. What is the absorbed dose (in rads) of γ rays that will heat the water from 20.0 to 50.0 °C?

Reasoning When γ rays are absorbed by the water, they cause it to heat up. The absorbed dose of γ rays is the energy (heat) absorbed by the water divided by its mass. According to the discussion in Section 12.7, the heat that must be absorbed by the water in order for its temperature to increase by a given amount depends on the mass and specific heat capacity of the water. We will use the concept of specific heat capacity to evaluate the absorbed dose of γ rays.

FIGURE 32.1 When the water absorbs the γ rays, its temperature rises.

Knowns and Unknowns   The following table summarizes the given information:

Description	Symbol	Value
Initial temperature of water	T_0	20.0 °C
Final temperature of water	T	50.0 °C
Unknown Variable		
Absorbed dose of γ rays (in rads)	Absorbed dose	?

Modeling the Problem

STEP 1 Absorbed Dose   The absorbed dose of γ rays is the energy (heat) Q absorbed by the water divided by its mass m (see Equation 32.2), as indicated in the right column. Neither Q nor m is known. Both variables will be dealt with in Step 2.

$$\text{Absorbed dose} = \frac{Q}{m} \qquad (32.2)$$

$$\boxed{?}$$

STEP 2 Heat Needed to Increase the Temperature of the Water   The heat Q that is needed to increase the temperature of a mass m of water by an amount ΔT is $Q = cm\Delta T$ (Equation 12.4), where c is the specific heat capacity of water. The change in temperature ΔT is equal to the higher temperature T minus the lower temperature T_0, or $\Delta T = T - T_0$. Thus, the heat can be expressed as

$$\boxed{Q = cm\Delta T = cm(T - T_0)} \qquad (12.4)$$

This expression for the heat absorbed by the water can be substituted into Equation 32.2 for the absorbed dose, as indicated in the right column. Note that the mass m of the water appears in both the numerator and denominator, so it can be eliminated algebraically.

$$\text{Absorbed dose} = \frac{Q}{m} \qquad (32.2)$$

$$\boxed{Q = cm(T - T_0)} \qquad (12.4)$$

Solution   Algebraically combining the results of the two modeling steps gives

$$\overset{\text{STEP 1}}{\downarrow} \qquad \overset{\text{STEP 2}}{\downarrow}$$

$$\text{Absorbed dose} = \frac{Q}{m} = \frac{cm(T - T_0)}{m} = c(T - T_0)$$

Taking the specific heat capacity c of water from Table 12.2, we find that the absorbed dose of γ rays [expressed in grays (Gy)] is

$$\text{Absorbed dose} = c(T - T_0) = [4186 \text{ J/(kg} \cdot \text{C}°)](50.0 \text{ °C} - 20.0 \text{ °C}) = 1.26 \times 10^5 \text{ Gy}$$

The problem asks that the absorbed dose be expressed in rads, rather than in grays. To this end, we note that 1 rad = 0.01 Gy, so

$$\text{Absorbed dose} = (1.26 \times 10^5 \text{ Gy})\left(\frac{1 \text{ rad}}{0.01 \text{ Gy}}\right) = \boxed{1.26 \times 10^7 \text{ rad}}$$

Related Homework: Problem 42

The amount of biological damage produced by ionizing radiation is different for different kinds of radiation. For instance, a 1-rad dose of neutrons is far more likely to produce eye cataracts than a 1-rad dose of X-rays. To compare the damage caused by different types of radiation, the **relative biological effectiveness** (RBE) is used.* The relative biological effectiveness of a particular type of radiation is the ratio of the dose of 200-keV X-rays needed to produce a certain biological effect to the dose of the radiation needed to produce the same biological effect:

$$\begin{array}{c}\text{Relative biological} \\ \text{effectiveness (RBE)}\end{array} = \frac{\begin{array}{c}\text{The dose of 200-keV X-rays that} \\ \text{produces a certain biological effect}\end{array}}{\begin{array}{c}\text{The dose of radiation that} \\ \text{produces the same biological effect}\end{array}} \qquad (32.3)$$

---

*The RBE is sometimes called the *quality factor* (QF).

TABLE 32.1	Relative Biological Effectiveness (RBE) for Various Types of Radiation

Type of Radiation	RBE
200-keV X-rays	1
γ rays	1
β^- particles (electrons)	1
Protons	10
α particles	10–20
Neutrons	
Slow	2
Fast	10

The RBE depends on the nature of the ionizing radiation and its energy, as well as on the type of tissue being irradiated. **Table 32.1** lists some typical RBE values for different kinds of radiation, assuming that an "average" biological tissue is being irradiated. A value of RBE = 1 for γ rays and β^- particles indicates that they produce the same biological damage as do 200-keV X-rays. The larger RBE values for protons, α particles, and neutrons indicate that they cause substantially more damage. The RBE is often used in conjunction with the absorbed dose to reflect the damage-producing character of the radiation. The product of the absorbed dose in rads (not in grays) and the RBE is the **biologically equivalent dose:**

$$\begin{array}{c}\text{Biologically equivalent dose} \\ \text{(in rems)}\end{array} = \begin{array}{c}\text{Absorbed dose} \\ \text{(in rads)}\end{array} \times \text{RBE} \qquad \textbf{(32.4)}$$

The unit for the biologically equivalent dose is the *rem,* short for **r**oentgen **e**quivalent, **m**an. Example 2 illustrates the use of the biologically equivalent dose.

EXAMPLE 2 | BIO Comparing Absorbed Doses of γ Rays and Neutrons

A biological tissue is irradiated with γ rays that have an RBE of 0.70. The absorbed dose of γ rays is 850 rad. The tissue is then exposed to neutrons whose RBE is 3.5. The biologically equivalent dose of the neutrons is the same as that of the γ rays. What is the absorbed dose of neutrons?

Reasoning The biologically equivalent doses of the neutrons and the γ rays are the same. Therefore, the tissue damage produced in each case is the same. However, the RBE of the neutrons is larger than the RBE of the γ rays by a factor of 3.5/0.70 = 5.0. Consequently, we will find that the absorbed dose of the neutrons is only one-fifth as great as that of the γ rays.

Solution According to Equation 32.4, the biologically equivalent dose is the product of the absorbed dose (in rads) and the

RBE; it is the same for the γ rays and the neutrons. Therefore, we have

$$\begin{array}{c}\text{Biologically} \\ \text{equivalent dose}\end{array} = (\text{Absorbed dose})_{\gamma\,\text{rays}}\, \text{RBE}_{\gamma\,\text{rays}}$$

$$= (\text{Absorbed dose})_{\text{neutrons}}\, \text{RBE}_{\text{neutrons}}$$

Solving for the absorbed dose of the neutrons gives

$$(\text{Absorbed dose})_{\text{neutrons}} = (\text{Absorbed dose})_{\gamma\,\text{rays}} \left(\frac{\text{RBE}_{\gamma\,\text{rays}}}{\text{RBE}_{\text{neutrons}}}\right)$$

$$= (850\,\text{rad})\left(\frac{0.70}{3.5}\right) = \boxed{170\,\text{rad}}$$

Everyone is continually exposed to background radiation from natural sources, such as cosmic rays (high-energy particles that come from outside the solar system), radioactive materials in the environment, radioactive nuclei (primarily carbon $^{14}_{6}\text{C}$ and potassium $^{40}_{19}\text{K}$) within our own bodies, and radon. **Table 32.2** lists the average biologically equivalent doses received from these sources by a citizen in the United States. According to this table, radon is a major contributor to the natural background radiation. Radon is an odorless radioactive gas and poses a health hazard because, when inhaled, it can damage the lungs and cause cancer. Radon is found in soil and rocks and enters houses via a mechanism that is not well understood. One possibility for indoor radon entry is through cracks and crevices in the foundation. The amount of radon in the soil varies greatly throughout the country, with some localities having significant amounts and others having virtually none. Accordingly, the dose that any individual receives can vary widely from the average value of 207 mrem/yr given in **Table 32.2** (1 mrem = 10^{-3} rem). In many houses, the entry of radon can be reduced significantly by sealing the foundation against entry of the gas and providing good ventilation so it does not accumulate.

To the natural background of radiation, a significant amount of human-made radiation has been added, mostly from medically related sources. Among these sources, CAT scanning (see Section 30.7) is the major contributor, as **Table 32.2** indicates.

The effects of radiation on humans can be grouped into two categories, according to the time span between initial exposure and the appearance of physiological symptoms: (1) short-term or acute effects that appear within a matter of minutes, days, or weeks, and (2) long-term or latent effects that appear years, decades, or even generations later.

Radiation sickness is the general term applied to the acute effects of radiation. Depending on the severity of the dose, a person with radiation sickness can exhibit nausea, vomiting, fever, diarrhea, and loss of hair. Ultimately, death can occur. The severity of radiation sickness is related to the dose received, and in the following discussion the biologically equivalent doses quoted are whole-body, single doses. A dose less than 50 rem causes no short-term ill effects. A dose between 50 and 300 rem brings on radiation sickness, the severity increasing with increasing dosage. A whole-body dose in the range of 400–500 rem is classified as an LD_{50} dose, meaning that it is a lethal dose (LD) for about 50% of the people so exposed; death occurs within a few months. Whole-body doses greater than 600 rem result in death for almost all individuals.

Long-term or latent effects of radiation may appear as a result of high-level brief exposure or low-level exposure over a long period of time. Some long-term effects are hair loss, eye cataracts, and various kinds of cancer. In addition, genetic defects caused by mutated genes may be passed on from one generation to the next.

Because of the hazards of radiation, the federal government has established dose limits. The permissible dose for an individual is defined as the dose, accumulated over a long period of time or resulting from a single exposure, that carries negligible probability of a severe health hazard. Federal standards (1991) state that an individual in the general population should not receive more than 500 mrem of human-made radiation each year, *exclusive* of medical sources. A person exposed to radiation in the workplace (e.g., a radiation therapist) should not receive more than 5 rem per year from work-related sources.

▐ BIO ▐ THE PHYSICS OF . . . cosmic rays and space travel.   One of the central problems in future long-term space travel is the dangerous radiation to which astronauts are exposed, even while inside their ship or space station. Cosmic rays are highly energetic particles, such as atomic nuclei and protons, which travel near the speed of light. They can originate from the sun, or from distant sources, like other galaxies. Cosmic rays carry enough energy to ionize atoms, and are therefore dangerous to humans. The surface of the earth is mostly protected from cosmic rays due to both the earth's atmosphere and its magnetic field. Astronauts in earth orbit or deep space, however, must be shielded from this dangerous radiation by materials in their spacecraft, or by other means.

One might think that a layer of a dense metal, such as lead or tungsten, built into the walls of the spaceship would suffice, but it is considerably more complicated than that. First, any added weight presents both technical issues and significant added cost (many thousands of dollars per kilogram) for launching the ship into space. More importantly, it turns out that some radiation-shielding materials, such as lead, can make the radiation inside the ship worse due to the secondary radiation that is produced when the highly energetic cosmic rays are absorbed. These secondary particles and photons can penetrate into the ship with lesser energies that can collectively cause more damage than the original cosmic ray. Thus, materials are sought that can absorb cosmic rays in a way that produces less secondary radiation than heavy-element shielding. For instance, the outer walls of the International Space Station (ISS) are composed of aluminum, a metal that is much lighter than lead or tungsten, which results in less secondary radiation and a comparatively reduced net exposure. Hydrogen-rich plastics are also a possible shielding solution for the same reason. Liquid hydrogen is an effective radiation shield that produces a smaller amount of secondary radiation, as is water, but these are both consumables on spacecraft and will diminish over time. Other hypothetical solutions include electrostatic or magnetic deflection, which are currently undeveloped, and flying close to asteroids and using them as shields! Protecting space travelers from cosmic rays and other radiation sources is a critical problem that must be solved before humans will be able to safely venture into deep space.

Example 3 discusses the biologically equivalent dose a patient receives during a CT scan.

TABLE 32.2	Average Biologically Equivalent Doses of Radiation Received by a U. S. Citizen[a]
Source of Radiation	**Biologically Equivalent Dose (mrem/yr)[b]**
Natural background radiation	
Cosmic rays	33
Radioactive earth and air	21
Internal radioactive nuclei	29
Inhaled radon	207
Human-made radiation	
Consumer products	13
CAT scanning	147
Routine medical/ dental diagnostics	33
Nuclear medicine	74

[a]National Council on Radiation Protection and Measurements, Report No. 160, "Ionizing Radiation Exposure of the Population of the United States," 2009.
[b]1 mrem = 10^{-3} rem.

EXAMPLE 3 | ▐ BIO ▐ The Energy Absorbed by a Body During a CT Scan

Consider a patient who receives a biologically equivalent dose, or BED, of 1.7 rem during a full-body CT scan. If the RBE of the X-rays in this procedure is 0.90, how much energy is absorbed by a patient with a mass of 82 kg?

Reasoning   We can use Equation 32.4 to calculate the absorbed dose and then Equation 32.2 to calculate the energy absorbed by the patient.

Solution   From Equation 32.4, we have:

$$\frac{\text{Biologically equivalent dose}}{\text{(in rems)}} = \frac{\text{Absorbed dose}}{\text{(in rads)}} \times \text{RBE}$$

Rearranging, we find:

$$\frac{\text{Absorbed}}{\text{dose (in rads)}} = \frac{\dfrac{\text{Biologically equivalent dose}}{\text{(in rems)}}}{\text{RBE}} = \frac{1.7 \text{ rem}}{0.90} = 1.9 \text{ rad}$$

In order to calculate the energy absorbed in joules, we need to convert the absorbed dose to Gy: 1.9 rad = 0.019 Gy. Using Equation 32.2, we calculate the energy absorbed:

$$\text{Energy absorbed} = (\text{Absorbed dose}) \times (\text{Mass of absorbing material})$$
$$= (0.019 \text{ Gy}) \times (82 \text{ kg}) = \boxed{1.6 \text{ J}}$$

Check Your Understanding

(The answers are given at the end of the book.)

1. Two different types of radiation have the same RBE. Is it possible for these two types of radiation to deliver different biologically equivalent doses of radiation to a given tissue sample?

2. The damage-producing character of a given type of ionizing radiation depends on **(a)** only the RBE of the radiation, **(b)** only the absorbed dose of the radiation, **(c)** both the RBE and the absorbed dose of the radiation.

3. A person faces the possibility of receiving the following absorbed doses of ionizing radiation: 20 rad of γ rays (RBE = 1), 5 rad of neutrons (RBE = 10), and 2 rad of α particles (RBE = 20). Rank the amount of biological damage that these possibilities will cause in decreasing order (greatest damage first).

32.2 Induced Nuclear Reactions

Section 31.4 discusses how a radioactive parent nucleus disintegrates spontaneously into a daughter nucleus. It is also possible to bring about, or induce, the disintegration of a stable nucleus by striking it with another nucleus, an atomic or subatomic particle, or a γ-ray photon. A **nuclear reaction** is said to occur whenever an incident nucleus, particle, or photon causes a change to occur in a target nucleus.

In 1919, Ernest Rutherford observed that when an α particle strikes a nitrogen nucleus, an oxygen nucleus and a proton are produced. This nuclear reaction is written as

$$\underbrace{{}^{4}_{2}\text{He}}_{\substack{\text{Incident} \\ \alpha \text{ particle}}} + \underbrace{{}^{14}_{7}\text{N}}_{\substack{\text{Nitrogen} \\ \text{(target)}}} \longrightarrow \underbrace{{}^{17}_{8}\text{O}}_{\text{Oxygen}} + \underbrace{{}^{1}_{1}\text{H}}_{\text{Proton, } p}$$

Because the incident α particle induces the transmutation of nitrogen into oxygen, this reaction is an example of an **induced nuclear transmutation**.

Nuclear reactions are often written in a shorthand form. For example, the reaction above is designated by ${}^{14}_{7}\text{N}\,(\alpha, p)\,{}^{17}_{8}\text{O}$. The first and last symbols represent the initial and final nuclei, respectively. The symbols within the parentheses denote the incident α particle (on the left) and the small emitted particle or proton p (on the right). Some other induced nuclear transmutations are listed below, together with the equivalent shorthand notations:

Nuclear Reaction	Notation
${}^{1}_{0}\text{n} + {}^{10}_{5}\text{B} \rightarrow {}^{7}_{3}\text{Li} + {}^{4}_{2}\text{He}$	${}^{10}_{5}\text{B}\,(n, \alpha)\,{}^{7}_{3}\text{Li}$
$\gamma + {}^{25}_{12}\text{Mg} \rightarrow {}^{24}_{11}\text{Na} + {}^{1}_{1}\text{H}$	${}^{25}_{12}\text{Mg}\,(\gamma, p)\,{}^{24}_{11}\text{Na}$
${}^{1}_{1}\text{H} + {}^{13}_{6}\text{C} \rightarrow {}^{14}_{7}\text{N} + \gamma$	${}^{13}_{6}\text{C}\,(p, \gamma)\,{}^{14}_{7}\text{N}$

Induced nuclear reactions, like the radioactive decay process discussed in Section 31.4, obey the conservation laws of physics. Each of these laws deals with a property that does not change during a process. The following list shows the property with which each law deals:

1. Conservation of energy/mass (Sections 6.8 and 28.6)

2. Conservation of linear momentum (Section 7.2)

3. Conservation of angular momentum (Section 9.6)

4. Conservation of electric charge (Section 18.2)

5. Conservation of nucleon number (Section 31.4)

In particular, items 4 and 5 on this list indicate that both the total electric charge of the nucleons and the number of nucleons in a nuclear reaction are conserved. The next example illustrates how the conservation of total electric charge and number of nucleons can be used to identify the nucleus produced in a reaction.

EXAMPLE 4 | An Induced Nuclear Transmutation

An α particle strikes an aluminum $^{27}_{13}\text{Al}$ nucleus. As a result, an unknown nucleus $^{A}_{Z}\text{X}$ and a neutron $^{1}_{0}\text{n}$ are produced:

$$^{4}_{2}\text{He} + ^{27}_{13}\text{Al} \longrightarrow ^{A}_{Z}\text{X} + ^{1}_{0}\text{n}$$

Identify the nucleus produced, including its atomic number Z (the number of protons) and its atomic mass number A (the number of nucleons).

Reasoning The total electric charge of the nucleons is conserved, so that we can set the total number of protons before the reaction equal to the total number after the reaction. The total number of nucleons is also conserved, so that we can set the total number before the reaction equal to the total number after the reaction. These two conserved quantities will allow us to identify the nucleus $^{A}_{Z}\text{X}$.

Solution The conservation of total electric charge and total number of nucleons leads to the equations listed below:

Conserved Quantity	Before Reaction		After Reaction
Total electric charge (number of protons)	$2 + 13$	$=$	$Z + 0$
Total number of nucleons	$4 + 27$	$=$	$A + 1$

Solving these equations for Z and A gives $Z = 15$ and $A = 30$. Since $Z = 15$ identifies the element as phosphorus (P), the nucleus produced is $\boxed{^{30}_{15}\text{P}}$.

Induced nuclear transmutations can be used to produce isotopes that are not found naturally. In 1934, Enrico Fermi (1901–1954) suggested a method for producing elements with a higher atomic number than uranium ($Z = 92$). These elements—neptunium ($Z = 93$), plutonium ($Z = 94$), americium ($Z = 95$), and so on—are known as *transuranium elements,* and none occurs naturally. They are created in a nuclear reaction between a suitably chosen lighter element and a small incident particle, usually a neutron or an α particle. For example, **Figure 32.2** shows a reaction that produces plutonium from uranium. A neutron is captured by a uranium $^{238}_{92}\text{U}$ nucleus, producing $^{239}_{92}\text{U}$ and a γ ray. The $^{239}_{92}\text{U}$ nucleus is radioactive and decays with a half-life of 23.5 min into neptunium $^{239}_{93}\text{Np}$. Neptunium is also radioactive and disintegrates with a half-life of 2.4 days into plutonium $^{239}_{94}\text{Pu}$. Plutonium is the final product and has a half-life of 24 100 yr.

The neutrons that participate in nuclear reactions can have kinetic energies that cover a wide range. In particular, those that have a kinetic energy of about 0.04 eV or less are called **thermal neutrons**. The name derives from the fact that such a relatively small kinetic energy is comparable to the average translational kinetic energy of a molecule in an ideal gas at room temperature.

FIGURE 32.2 An induced nuclear reaction is shown in which $^{238}_{92}\text{U}$ is transmuted into the transuranium element plutonium $^{239}_{94}\text{Pu}$.

Check Your Understanding

(The answers are given at the end of the book.)

4. Which one or more of the following nuclear reactions could possibly occur? **(a)** $^{15}_{7}\text{N} \, (\alpha, \gamma) \, ^{18}_{9}\text{F}$ **(b)** $^{14}_{6}\text{C} \, (p, n) \, ^{15}_{8}\text{O}$ **(c)** $^{14}_{7}\text{N} \, (p, \gamma) \, ^{15}_{8}\text{O}$ **(d)** $^{15}_{8}\text{O} \, (n, p) \, ^{13}_{6}\text{C}$

5. Why is each of the following reactions *not* allowed? **(a)** $^{60}_{28}\text{Ni} \, (\alpha, p) \, ^{62}_{29}\text{Cu}$ **(b)** $^{27}_{13}\text{Al} \, (n, n) \, ^{28}_{13}\text{Al}$ **(c)** $^{39}_{19}\text{K} \, (p, \alpha) \, ^{36}_{17}\text{Cl}$

32.3 | Nuclear Fission

In 1939 four German scientists, Otto Hahn, Lise Meitner, Fritz Strassmann, and Otto Frisch, made an important discovery that ushered in the atomic age. They found that a

ANIMATED FIGURE 32.3 A slowly moving neutron causes the uranium nucleus $^{235}_{92}$U to fission into barium $^{141}_{56}$Ba, krypton $^{92}_{36}$Kr, and three neutrons.

uranium nucleus, after absorbing a neutron, splits into two fragments, each with a smaller mass than the original nucleus. The splitting of a massive nucleus into two less massive fragments is known as **nuclear fission**.

Animated Figure 32.3 shows a fission reaction in which a uranium $^{235}_{92}$U nucleus is split into barium $^{141}_{56}$Ba and krypton $^{92}_{36}$Kr nuclei. The reaction begins when $^{235}_{92}$U absorbs a slowly moving neutron, creating a "compound nucleus," $^{236}_{92}$U. The compound nucleus disintegrates quickly into $^{141}_{56}$Ba, $^{92}_{36}$Kr, and three neutrons according to the following reaction:

$$\underbrace{{}^{1}_{0}n + {}^{235}_{92}U \longrightarrow {}^{236}_{92}U}_{\substack{\text{Compound}\\\text{nucleus}\\\text{(unstable)}}} \longrightarrow \underbrace{{}^{141}_{56}Ba}_{\text{Barium}} + \underbrace{{}^{92}_{36}Kr}_{\text{Krypton}} + \underbrace{3{}^{1}_{0}n}_{\text{3 neutrons}}$$

This reaction is only one of the many possible reactions that can occur when uranium fissions. For example, another reaction is

$$\underbrace{{}^{1}_{0}n + {}^{235}_{92}U \longrightarrow {}^{236}_{92}U}_{\substack{\text{Compound}\\\text{nucleus}\\\text{(unstable)}}} \longrightarrow \underbrace{{}^{140}_{54}Xe}_{\text{Xenon}} + \underbrace{{}^{94}_{38}Sr}_{\text{Strontium}} + \underbrace{2{}^{1}_{0}n}_{\text{2 neutrons}}$$

Some reactions produce as many as 5 neutrons; however, the average number produced per fission is 2.5.

When a neutron collides with and is absorbed by a uranium nucleus, the uranium nucleus begins to vibrate and becomes distorted. The vibration continues until the distortion becomes so severe that the attractive strong nuclear force can no longer balance the electrostatic repulsion between the nuclear protons. At this point, the nucleus bursts apart into fragments, which carry off energy, primarily in the form of kinetic energy. The energy carried off by the fragments is enormous and was stored in the original nucleus mainly in the form of electric potential energy. An average of roughly 200 MeV of energy is released per fission. This energy is approximately 10^8 times greater than the energy released per molecule in an ordinary chemical reaction, such as the combustion of gasoline or coal. Example 5 demonstrates how to estimate the energy released during the fission of a nucleus.

EXAMPLE 5 | The Energy Released During Nuclear Fission

Estimate the amount of energy released when a massive nucleus ($A = 240$) fissions.

Reasoning Figure 31.5 shows that the binding energy of a nucleus with $A = 240$ is about 7.6 MeV per nucleon. We assume that this nucleus fissions into two fragments, each with A $\approx$ 120. According to Figure 31.5, the binding energy of the fragments increases to about 8.5 MeV per nucleon. Consequently, when a massive nucleus fissions, there is a release of about 8.5 MeV − 7.6 MeV = 0.9 MeV of energy per nucleon.

Solution Since there are 240 nucleons involved in the fission process, the total energy released per fission is approximately (0.9 MeV/nucleon)(240 nucleons) $\approx$ $\boxed{200 \text{ MeV}}$.

Virtually all naturally occurring uranium is composed of two isotopes. These isotopes and their natural abundances are $^{238}_{92}$U (99.275%) and $^{235}_{92}$U (0.720%). Although $^{238}_{92}$U is by far the most abundant isotope, the probability that it will capture a neutron and fission is very small. For this reason, $^{238}_{92}$U is not the isotope of choice for generating nuclear energy. In contrast, the isotope $^{235}_{92}$U readily captures a neutron and fissions, *provided the neutron is a thermal neutron* (kinetic energy) ≈ 0.04 eV or less). The probability of a thermal neutron causing $^{235}_{92}$U to fission is about 500 times greater than the probability for a neutron whose energy is relatively high—say, 1 MeV. Thermal neutrons can also be used to fission other nuclei, such as plutonium $^{239}_{94}$Pu. Conceptual Example 6 deals with one of the reasons why thermal neutrons are useful for inducing nuclear fission.

CONCEPTUAL EXAMPLE 6 | Neutrons Versus Protons or Alpha Particles

A thermal neutron has a relatively small amount of kinetic energy but, nevertheless, can penetrate a nucleus. To penetrate the same nucleus, would a proton or an α particle need **(a)** the same small amount of kinetic energy as the neutron needs, **(b)** a much larger amount of kinetic energy than the neutron needs, or **(c)** much less kinetic energy than the neutron needs?

Reasoning   To penetrate a nucleus, a particle such as a neutron, a proton, or an α particle must have enough kinetic energy to do the work of overcoming any repulsive force that it encounters. A repulsive force can arise because protons in the nucleus are electrically charged. Since a neutron is electrically neutral, however, it encounters no electrostatic force of repulsion as it approaches the nuclear protons, and, hence, needs relatively little energy to reach the nucleus.

Answers (a) and (c) are incorrect.   A proton and an α particle each carry a positive charge, so that each would encounter

an electrostatic force of repulsion as it approached the nuclear protons, a force that a thermal neutron does not encounter. These answers ignore the additional kinetic energy that a proton or an α particle would need to overcome the repulsion.

Answer (b) is correct.   A proton and an α particle, each being positively charged, would each require much more kinetic energy than a neutron does, in order to overcome the electrostatic force of repulsion from the nuclear protons. It is true that each would also experience the attractive strong nuclear force from the nuclear protons and neutrons. However, this force has an extremely short range of action and, therefore, would come into play only after an impinging particle reached the target nucleus. In comparison, the electrostatic force has a long range of action and is encountered throughout the entire journey to the target.

Related Homework: Problem 38

The fact that the uranium fission reaction releases 2.5 neutrons, on the average, makes it possible for a self-sustaining series of fissions to occur. As **Figure 32.4** illustrates, each neutron released can initiate another fission event, resulting in the emission of still more neutrons, followed by more fissions, and so on. A **chain reaction** is a series of nuclear fissions whereby some of the neutrons produced by each fission cause additional fissions. During an uncontrolled chain reaction, it would not be unusual for the number of fissions to increase a thousandfold within a few millionths of a second. With an average energy of about 200 MeV being released per fission, an uncontrolled chain reaction can generate an incredible amount of energy in a very short time, as happens in an atomic bomb (which is actually a *nuclear* bomb).

By using a material that can absorb neutrons without fissioning, it is possible to limit the number of neutrons in the environment of the fissile nuclei. In this way, a condition can be established whereby each fission event contributes, on average, only *one neutron* that fissions another nucleus (see **Figure 32.5**). Thus, the chain reaction and the rate of energy production are controlled. The controlled-fission chain reaction is the principle behind nuclear reactors used in the commercial generation of electric power.

FIGURE 32.4   A chain reaction. For clarity, it is assumed that each fission generates two neutrons (2.5 neutrons are actually liberated on the average). The fission fragments are not shown.

Check Your Understanding

(The answers are given at the end of the book.)

6. When the nucleus of a certain element absorbs a thermal neutron, fission usually occurs, with the production of nuclear fragments and a number N of neutrons. However, in a collection of atoms of this element, a small fraction of the thermal neutrons absorbed by the nuclei does not lead to

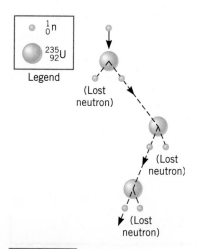

FIGURE 32.5 In a controlled chain reaction, only one neutron, on average, from each fission event causes another nucleus to fission. The "lost neutron" is absorbed by a material (not shown) that does not fission. As a result, energy is released at a steady or controlled rate.

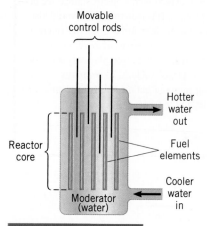

INTERACTIVE FIGURE 32.6 A nuclear reactor consists of fuel elements, control rods, and a moderator (in this case, water).

fission. For which one of the following values of N can a self-sustaining chain reaction *not* be produced using this element? **(a)** $N = 4$ **(b)** $N = 3$ **(c)** $N = 2$ **(d)** $N = 1$

7. Thermal neutrons, thermal protons, and thermal electrons all have the same kinetic energy of about 0.04 eV. Rank the speeds of these particles in descending order (greatest speed first).

8. Would a release of energy accompany the fission of a nucleus of nucleon number 25 into two fragments of about equal mass? Consult **Figure 31.5** as needed.

32.4 Nuclear Reactors

THE PHYSICS OF . . . nuclear reactors. A nuclear reactor is a type of furnace in which energy is generated by a controlled-fission chain reaction. The first nuclear reactor was built by Enrico Fermi in 1942, on the floor of a squash court under the west stands of Stagg Field at the University of Chicago. Today, there are a number of kinds and sizes of reactors, and many have the same three basic components: fuel elements, a neutron moderator, and control rods. **Interactive Figure 32.6** illustrates these components.

The **fuel elements** contain the fissile fuel and, for example, may be thin rods about 1 cm in diameter. In a large power reactor there may be thousands of fuel elements placed close together, and the entire region of fuel elements is known as the **reactor core**. Uranium $^{235}_{92}U$ is a common reactor fuel. Since the natural abundance of this isotope is only about 0.7%, there are special uranium-enrichment plants to increase the percentage. Most commercial reactors use uranium in which the amount of $^{235}_{92}U$ has been enriched to about 3%.

Whereas neutrons with energies of about 0.04 eV (or less) readily fission $^{235}_{92}U$, the neutrons released during the fission process have significantly greater energies of several MeV or so. Consequently, a nuclear reactor must contain some type of material that will decrease or moderate the speed of such energetic neutrons so they can readily fission additional $^{235}_{92}U$ nuclei. The material that slows down the neutrons is called a **moderator**. One commonly used moderator is water. When an energetic neutron leaves a fuel element, the neutron enters the surrounding water and collides with water molecules. With each collision, the neutron loses an appreciable fraction of its energy and slows down. Once slowed down to thermal energy by the moderator, a process that takes less than 10^{-3} s, the neutron is capable of initiating a fission event upon reentering a fuel element.

If the output power from a reactor is to remain constant, only one neutron from each fission event must trigger a new fission, as **Figure 32.5** illustrates. When each fission leads to one additional fission—no more or no less—the reactor is said to be *critical*. A reactor normally operates in a critical condition, because then it produces a steady output of energy. The reactor is *subcritical* when, on average, the neutrons from each fission trigger *less than one* subsequent fission. In a subcritical reactor, the chain reaction is not self-sustaining and eventually dies out. When the neutrons from each fission trigger *more than one* additional fission, the reactor is *supercritical*. During a supercritical condition, the energy released by a reactor increases. If left unchecked, the increasing energy can lead to a partial or total meltdown of the reactor core, with the possible release of radioactive material into the environment.

Clearly, a control mechanism is needed to keep the reactor in its normal, or critical, state. This control is accomplished by a number of **control rods** that can be moved into and out of the reactor core (see **Interactive Figure 32.6**). The control rods contain an element, such as boron or cadmium, that readily absorbs neutrons without fissioning. If the reactor becomes supercritical, the control rods are automatically moved farther into the core to absorb the excess neutrons causing the condition. In response, the reactor returns to its critical state. Conversely, if the reactor becomes subcritical, the control rods are partially withdrawn from the core. Fewer neutrons are absorbed, more neutrons are available for fission, and the reactor again returns to its critical state.

INTERACTIVE FIGURE 32.7 Diagram of a nuclear power plant that uses a pressurized water reactor.

Interactive Figure 32.7 illustrates a pressurized water reactor. In such a reactor, the heat generated within the fuel rods is carried away by water that surrounds the rods. To remove as much heat as possible, the water temperature is allowed to rise to a high value (about 300 °C). To prevent boiling, which occurs at 100 °C at 1 atmosphere of pressure, the water is pressurized in excess of 150 atmospheres. The hot water is pumped through a heat exchanger, where heat is transferred to water flowing in a second, closed system. The heat transferred to the second system produces steam that drives a turbine. The turbine is coupled to an electric generator, whose output electric power is delivered to consumers via high-voltage transmission lines. After exiting the turbine, the steam is condensed back into water that is returned to the heat exchanger.

32.5 | Nuclear Fusion

In Example 5 in Section 32.3, the plot of binding energy per nucleon in Figure 31.5 is used to estimate the amount of energy released in the fission process. As summarized in Figure 32.8, the massive nuclei at the right end of the curve have a binding energy of about 7.6 MeV per nucleon. The less massive fission fragments are near the center of the curve and have a binding energy of approximately 8.5 MeV per nucleon. The energy released per nucleon by fission is the difference between these two values, or about 0.9 MeV per nucleon.

FIGURE 32.8 When fission occurs, a massive nucleus divides into two fragments whose binding energy per nucleon is greater than that of the original nucleus. When fusion occurs, two low-mass nuclei combine to form a more massive nucleus whose binding energy per nucleon is greater than that of the original nuclei.

A glance at the far left end of the diagram in **Figure 32.8** suggests another means of generating energy. Two nuclei with very low mass and relatively small binding energies per nucleon could be combined or "fused" into a single, more massive nucleus that has a greater binding energy per nucleon. This process is called **nuclear fusion**. A substantial amount of energy can be released during a fusion reaction, as Example 7 shows for one possible reaction.

EXAMPLE 7 | The Energy Released During Nuclear Fusion

Two isotopes of hydrogen, ^2_1H (deuterium, D) and ^3_1H (tritium, T), fuse to form ^4_2He and a neutron according to the following reaction:

$$^2_1\text{H} + ^3_1\text{H} \longrightarrow ^4_2\text{He} + ^1_0\text{n}$$

Determine the energy that is released by this particular fusion reaction, which is illustrated in **Figure 32.9**.

Reasoning Energy is released, so the total mass of the final nuclei is less than the total mass of the initial nuclei. To determine the energy released, we find the amount (in atomic mass units u) by which the total mass has decreased. Then, we use the fact that 1 u is equivalent to 931.5 MeV of energy, as determined in Section 31.3. This approach is the same as that used in Section 31.4 for radioactive decay.

Solution The masses of the initial and final nuclei in this reaction, as well as the mass of the neutron, are

Initial Masses		Final Masses	
^2_1H	2.0141 u	^4_2He	4.0026 u
^3_1H	3.0161 u	^1_0n	1.0087 u
Total:	5.0302 u	Total:	5.0113 u

The decrease in mass, or the mass defect, is $\Delta m = 5.0302\text{ u} - 5.0113\text{ u} = 0.0189\text{ u}$. Since 1 u is equivalent to 931.5 MeV, the energy released is

$$\text{Released energy} = (0.0189\text{ u})\left(\frac{931.5\text{ MeV}}{1\text{ u}}\right) = \boxed{17.6\text{ MeV}}$$

FIGURE 32.9 Deuterium and tritium are fused together to form a helium nucleus (^4_2He). The result is the release of an enormous amount of energy, mainly carried by a single high-energy neutron (^1_0n).

The deuterium nucleus contains 2 nucleons, and the tritium nucleus contains 3. Thus, there are 5 nucleons that participate in the fusion, so the energy released per nucleon is about 3.5 MeV. This energy per nucleon is greater than the energy released in a fission process ($\approx$0.9 MeV per nucleon).

Because fusion reactions release so much energy, there is considerable interest in fusion reactors, although to date no commercial units have been constructed. The difficulties in building a fusion reactor arise mainly because the two low-mass nuclei must be brought sufficiently near each other so that the short-range strong nuclear force can pull them together, leading to fusion. However, each nucleus has a positive charge and repels the other electrically. For the nuclei to get sufficiently close in the presence of the repulsive electric force, they must have large kinetic energies, and hence large temperatures, to start with. For example, a temperature of around a hundred million °C is needed to start the deuterium–tritium reaction discussed in Example 7.

Reactions that require such extremely high temperatures are called **thermonuclear reactions**. The most important thermonuclear reactions occur in stars, such as our own sun. The energy radiated by the sun comes from deep within its core, where the temperature is high enough to initiate the fusion process. One group of reactions thought to occur in the sun is the *proton–proton* cycle, which is a series of reactions whereby six protons form a helium nucleus, two positrons, two γ rays, two protons, and two neutrinos. The energy released by the proton–proton cycle is about 25 MeV.

Human-made fusion reactions have been carried out in a fusion-type nuclear bomb—commonly called a hydrogen bomb. In a hydrogen bomb, the fusion reaction is ignited by a fission bomb using uranium or plutonium. The temperature produced by the fission bomb is sufficiently high to initiate a thermonuclear reaction where, for example, hydrogen isotopes are fused into helium, releasing even more energy. For fusion to be useful as a commercial energy source, the energy must be released in a steady, controlled manner—unlike the energy in a bomb. To date, scientists have not succeeded in constructing a fusion device that produces more energy on a continual basis than is expended in operating the device. A fusion device uses a high temperature to start a reaction, and under such a condition, all the atoms are completely ionized to form a *plasma* (a gas composed of charged particles, like $^2_1\text{H}^+$ and e^-). The problem is to confine the hot plasma for a long enough time so that collisions among the ions can lead to fusion.

THE PHYSICS OF . . . **magnetic confinement and fusion.** One ingenious method of confining the plasma is called *magnetic confinement* because it uses a magnetic field to contain and compress the charges in the plasma. Charges moving in the magnetic field are subject to magnetic forces. As the forces increase, the associated pressure builds, and the temperature rises. The gas becomes a superheated plasma, ultimately fusing when the pressure and temperature are high enough.

THE PHYSICS OF . . . **inertial confinement and fusion.** Another type of confinement scheme, known as *inertial confinement,* is also being developed. Tiny, solid pellets of fuel are dropped into a container. As each pellet reaches the center of the container, a number of high-intensity laser beams strike the pellet simultaneously. The heating causes the exterior of the pellet to vaporize almost instantaneously. However, the inertia of the vaporized atoms keeps them from expanding outward as fast as the vapor is being formed. As a result, high pressures, high densities, and high temperatures are achieved at the center of the pellet, thus causing fusion. A variant of inertial confinement fusion, called "Z pinch," is under development at Sandia National Laboratories in New Mexico. This device would also implode tiny fuel pellets, but without using lasers. Instead, scientists are using a cylindrical array of fine tungsten wires that are connected to a gigantic capacitor. When the capacitor is discharged, a huge current is sent through the wires. The heated wires vaporize almost instantly, generating a hot gas of ions, or plasma. The plasma is driven inward upon itself by the huge magnetic field produced by the current. The compressed plasma becomes superhot and generates a gigantic X-ray pulse. This pulse, it is hoped, would implode the solid fuel pellets to temperatures and pressures at which fusion would occur.

When compared to fission, fusion has some attractive features as an energy source. As we have seen in Example 7, fusion yields more energy per nucleon of fuel than fission does. Moreover, one type of fuel, ^2_1H (deuterium), is found in the waters of the oceans and is plentiful, cheap, and relatively easy to separate from the common ^1_1H isotope of hydrogen. Fissile materials like naturally occurring uranium $^{235}_{92}\text{U}$ are much less available, and supplies could be depleted within a century or two. However, the commercial use of fusion to provide cheap energy remains in the future.

Check Your Understanding

(The answers are given at the end of the book.)

9. Which one or more of the following statements correctly describe differences between fission and fusion? **(a)** Fission involves the combining of low-mass nuclei to form a more massive nucleus, whereas fusion involves the splitting of a massive nucleus into less massive fragments. **(b)** Fission involves the splitting of a massive nucleus into less massive fragments, whereas fusion involves the combining of low-mass nuclei to form a more massive nucleus. **(c)** More energy per nucleon is released when a fission event occurs than when a fusion event occurs. **(d)** Less energy per nucleon is released when a fission event occurs than when a fusion event occurs.

10. Which one or more of the following statements concerning fission and fusion are true? **(a)** Both fission and fusion reactions are characterized by a mass defect. **(b)** Both fission and fusion reactions always obey the conservation laws of physics. **(c)** Both fission and fusion take advantage of the fact that the binding energy per nucleon varies with the nucleon number of the nucleus.

11. Would the fusion of two nuclei, each with a nucleon number of 60, release energy? Consult **Figure 32.8** as needed.

32.6   Elementary Particles

Setting the Stage

By 1932 the electron, the proton, and the neutron had been discovered and were thought to be nature's three **elementary particles**, in the sense that they were the basic building blocks from which all matter is constructed. Experimental evidence obtained since then, however, shows that several hundred additional particles exist, and scientists no longer believe that the proton and the neutron are elementary particles.

Most of these new particles have masses greater than the electron's mass, and many are more massive than protons or neutrons. Most of the new particles are unstable and decay in times between about 10^{-6} and 10^{-23} s.

Often, new particles are produced by accelerating protons or electrons to high energies and letting them collide with a target nucleus. For example, **Figure 32.10** shows a collision between an energetic proton and a stationary proton. If the incoming proton has sufficient energy, the collision produces an entirely new particle, the *neutral pion* (π^0). The π^0 particle lives for only 8.4×10^{-17} s before it decays into two γ-ray photons. Since the pion did not exist before the collision, it was created from part of the incident proton's energy. Because a new particle such as the neutral pion is often created from energy, it is customary to report the mass of the particle in terms of its equivalent *rest energy* (see Equation 28.5). Often, energy units of MeV are used. For instance, detailed analyses of experiments reveal that the mass of the π^0 particle is equivalent to a rest energy of 135.0 MeV. For comparison, the more massive proton has a rest energy of 938.3 MeV. Analyses of experiments also provide the electric charge and other properties of particles created in high-energy collisions. In the limited space available here, it is not possible to describe all the new particles that have been found. However, we will highlight some of the more significant discoveries.

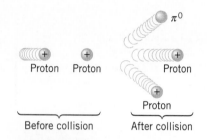

FIGURE 32.10 When an energetic proton collides with a stationary proton, a neutral pion (π^0) is produced. Part of the energy of the incident proton goes into creating the pion.

Neutrinos

In 1930, Wolfgang Pauli suggested that a particle called the **neutrino** (now known as the "electron neutrino") should accompany the β decay of a radioactive nucleus. As Section 31.5 discusses, the neutrino has no electric charge, has a very small mass (a tiny fraction of the mass of an electron), and travels at speeds approaching (but less than) the speed of light. Neutrinos were finally discovered in 1956. Today, neutrinos are created in abundance in nuclear reactors and particle accelerators and are thought to be plentiful in the universe.

Positrons and Antiparticles

The year 1932 saw the discovery of the **positron** (a contraction for "positive electron"). The positron has the same mass as the electron but carries an opposite charge of $+e$. A collision between a positron and an electron is likely to annihilate both particles, converting them into electromagnetic energy in the form of γ rays. For this reason, positrons never coexist with ordinary matter for any appreciable length of time. The mutual annihilation of a positron and an electron lies at the heart of an important medical diagnostic technique, as Conceptual Example 8 discusses.

CONCEPTUAL EXAMPLE 8 | BIO The Physics of PET Scanning

Certain radioactive isotopes decay by positron emission—for example, oxygen $^{15}_{8}$O. In the medical diagnostic technique known as PET scanning (positron emission tomography), such isotopes are injected into the body, where they collect at specific sites. A positron ($^{0}_{1}$e) emitted by the decaying isotope immediately encounters an electron ($_{-1}^{0}$e) in the body tissue, and the resulting mutual annihilation produces two γ-ray photons ($^{0}_{1}$e + $_{-1}^{0}$e → $\gamma + \gamma$), which are detected by devices mounted on a ring around the patient. As **Figure 32.11a** shows, the two photons strike oppositely positioned detectors and, in so doing, reveal the line on which the annihilation occurred. Such information leads to a computer-generated image that can be useful in diagnosing

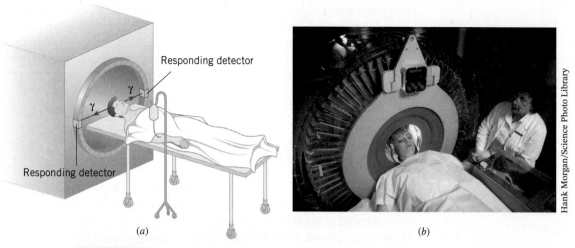

(a)

(b)

FIGURE 32.11 (a) In positron emission tomography, or PET scanning, a radioactive isotope is injected into the body. The isotope decays by emitting a positron, which annihilates an electron in the body tissue, producing two γ-ray photons. These photons strike detectors mounted on opposite sides of a ring that surrounds the patient. (b) A patient undergoing a PET scan of the brain.

FIGURE 32.12 This PET scan of the lungs reveals pulmonary edema (see the white rectangle), which is an abnormal accumulation of fluid.

abnormalities at the site where the radioactive isotope collects (see **Figure 32.12**). Which conservation principle accounts for the fact that the photons strike oppositely positioned detectors, the principle of conservation of **(a)** linear momentum or **(b)** energy?

Reasoning Momentum is a vector concept and, therefore, has a direction. The momentum-conservation principle states that the total linear momentum of an isolated system remains constant (see Section 7.2). An isolated system is one on which no net external force acts. Energy is not a vector concept and has no direction associated with it. The energy-conservation principle states that energy can neither be created nor destroyed, but can only be converted from one form to another (see Section 6.8).

Answer (b) is incorrect. Energy is not a vector and has no direction associated with it. Therefore, it cannot, by itself,

reveal the directional line along which the γ-ray photons were emitted.

Answer (a) is correct. The positron and the electron constitute an isolated system, so that momentum conservation applies. They do exert electrostatic forces on one another, since they carry electric charges. However, these forces are internal, not external, forces and cannot change the total linear momentum of the two-particle system. The total photon momentum, then, must equal the total momentum of the positron and the electron, which is nearly zero, to the extent that these particles have much less momentum than the photons do. With a total linear momentum of zero, the momentum vectors of the photons must point in opposite directions. Thus, the two photons depart from the annihilation site heading toward oppositely located detectors.

Related Homework: Problem 37

The positron is an example of an antiparticle, and after its discovery scientists came to realize that for every type of particle there is a corresponding type of antiparticle. The antiparticle is a form of matter that has the same mass as the particle but carries an opposite electric charge (e.g., the electron–positron pair) or has a magnetic moment that is oriented in an opposite direction relative to the spin (e.g., the neutrino–antineutrino pair). A few electrically neutral particles, like the photon and the neutral pion (π^0), are their own antiparticles.

Muons and Pions

In 1937, the American physicists S. H. Neddermeyer (1907–1988) and C. D. Anderson (1905–1991) discovered a new charged particle whose mass was about 207 times greater than the mass of the electron. The particle is designated by the Greek letter μ (mu) and is known as a **muon**. There are two muons that have the same mass but opposite charge: the particle μ^- and its antiparticle μ^+. The μ^- muon has the same charge as the electron, whereas the μ^+ muon has the same charge as the positron. Both muons are unstable, and have a lifetime of 2.2×10^{-6} s. The μ^- muon decays into an electron (β^-), a muon neutrino (ν_μ), and an electron antineutrino ($\bar{\nu}_e$) according to the following reaction:

$$\mu^- \longrightarrow \beta^- + \nu_\mu + \bar{\nu}_e$$

The μ^+ muon decays into a positron (β^+), a muon antineutrino ($\bar{\nu}_\mu$), and an electron neutrino (ν_e):

$$\mu^+ \longrightarrow \beta^+ + \bar{\nu}_\mu + \nu_e$$

Muons interact with protons and neutrons via the weak nuclear force (see Section 31.5).

The Japanese physicist Hideki Yukawa (1907–1981) predicted in 1935 that **pions** exist, but they were not discovered until 1947. Pions come in three varieties: one that is positively charged, the negatively charged antiparticle with the same mass, and the neutral pion, mentioned earlier, which is its own antiparticle. The symbols for these pions are, respectively, π^+, π^-, and π^0. The charged pions are unstable and have a lifetime of 2.6×10^{-8} s. The decay of a charged pion almost always produces a muon:

$$\pi^- \longrightarrow \mu^- + \bar{\nu}_\mu$$
$$\pi^+ \longrightarrow \mu^+ + \nu_\mu$$

As mentioned earlier, the neutral pion π^0 is also unstable and decays into two γ-ray photons, the lifetime being 8.4×10^{-17} s. The pions are of great interest because, unlike the muons, the pions interact with protons and neutrons via the strong nuclear force.

Classification of Particles

It is useful to group the known particles into three families—the bosons, the leptons, and the hadrons—as **Table 32.3** summarizes. The **boson family** is composed of particles (all are types of bosons) that play central roles in nature's three fundamental forces (see Section 4.6). The photon is associated with the electromagnetic force, which is one manifestation of the electroweak force. The W^-, W^+, and Z^0 particles are associated with the weak nuclear force, which is the other manifestation of the electroweak force. The gluons are associated with the strong nuclear force, whereas the graviton is thought to be associated with the gravitational force.

The **lepton family** consists of particles that interact by means of the *weak nuclear force*. Leptons can also exert gravitational and (if the leptons are charged) electromagnetic forces on other particles. The four better-known leptons are the electron, the muon, the electron neutrino ν_e, and the muon neutrino ν_μ. **Table 32.3** lists these particles together with their antiparticles. Two other leptons have also been discovered, the tau particle τ and its neutrino ν_τ, bringing the number of particles in the lepton family to six.

The **hadron family** contains the particles that interact by means of *both the strong nuclear force and the weak nuclear force.* Hadrons can also interact by gravitational and electromagnetic forces, but at short distances ($\leq 10^{-15}$ m) the strong nuclear force dominates. Among the hadrons are the proton, the neutron, and the pions. As **Table 32.3** indicates, most hadrons are short-lived. The hadrons are subdivided into two groups, the **mesons** and the **baryons**, for a reason that will be discussed in connection with the idea of quarks.

TABLE 32.3   Some Particles and Their Properties

Family	Particle	Particle Symbol	Antiparticle Symbol	Rest Energy (MeV)	Lifetime (s)
Boson[a]	Photon	γ	Self[b]	0	Stable
	W$^\pm$	W$^-$	W$^+$	8.04×10^4	3×10^{-25}
	Z^0	Z^0	Self[b]	9.12×10^4	3×10^{-25}
	Gluons[c]	g	—	0	—
	Graviton[c]	—	—	0	—
Lepton	Electron	e^- or β^-	e^+ or β^+	0.511	Stable
	Muon	μ^-	μ^+	105.7	2.2×10^{-6}
	Tau	τ^-	τ^+	1777	2.9×10^{-13}
	Electron neutrino	ν_e	$\bar{\nu}_e$	$<2 \times 10^{-6}$	Stable
	Muon neutrino	ν_μ	$\bar{\nu}_\mu$	<0.19	Stable
	Tau neutrino	ν_τ	$\bar{\nu}_\tau$	<18.2	Stable
Hadron					
Mesons					
	Pion	π^+	π^-	139.6	2.6×10^{-8}
		π^0	Self[b]	135.0	8.4×10^{-17}
	Kaon	K^+	K^-	493.7	1.2×10^{-8}
		K^0	$\bar{K}^0$	497.6	—[d]
	Eta	η^0	Self[b]	547.3	$<10^{-18}$
Baryons					
	Proton	p	$\bar{p}$	938.3	Stable
	Neutron	n	$\bar{n}$	939.6	886
	Lambda	Λ^0	$\bar{\Lambda}^0$	1116	2.6×10^{-10}
	Sigma	Σ^+	$\bar{\Sigma}^-$	1189	8.0×10^{-11}
		Σ^0	$\bar{\Sigma}^0$	1193	7.4×10^{-20}
		Σ^-	$\bar{\Sigma}^+$	1197	1.5×10^{-10}
	Omega	Ω^-	Ω^+	1672	8.2×10^{-11}

[a]The particles in this family are types of bosons that are associated with (or mediate) nature's fundamental forces.
[b]The particle is its own antiparticle.
[c]Free gluons and the graviton have not been observed experimentally.
[d]This particle and its antiparticle do not have definite lifetimes.

TABLE 32.4 **Quarks and Antiquarks**

| Name | Quarks | | Antiquarks | |
	Symbol	Charge	Symbol	Charge
Up	u	$+\frac{2}{3}e$	$\bar{u}$	$-\frac{2}{3}e$
Down	d	$-\frac{1}{3}e$	$\bar{d}$	$+\frac{1}{3}e$
Strange	s	$-\frac{1}{3}e$	$\bar{s}$	$+\frac{1}{3}e$
Charmed	c	$+\frac{2}{3}e$	$\bar{c}$	$-\frac{2}{3}e$
Top	t	$+\frac{2}{3}e$	$\bar{t}$	$-\frac{2}{3}e$
Bottom	b	$-\frac{1}{3}e$	$\bar{b}$	$+\frac{1}{3}e$

Quarks

As more and more hadrons were discovered, it became clear that they were not all elementary particles. The suggestion was made that the hadrons are made up of smaller, more elementary particles called **quarks**. In 1963, a quark theory was advanced independently by M. Gell-Mann (1929–2019) and G. Zweig (1937–   ). The theory proposed that there are three quarks and three corresponding antiquarks, and that hadrons are constructed from combinations of these. Thus, the quarks are elevated to the status of elementary particles for the hadron family. The particles in the lepton family are considered to be elementary, and as such they are not composed of quarks.

The three quarks were named *up* (*u*), *down* (*d*), and *strange* (*s*), and were assumed to have, respectively, fractional charges of $+\frac{2}{3}e$, $-\frac{1}{3}e$, and $-\frac{1}{3}e$. In other words, a quark possesses a charge magnitude smaller than that of an electron, which has a charge of $-e$. **Table 32.4** lists the symbols and electric charges of these quarks and the corresponding antiquarks. Experimentally, quarks should be recognizable by their fractional charges, but in spite of an extensive search for them, free quarks have never been found.

According to the original quark theory, the mesons are different from the baryons, because each meson consists of only two quarks—a quark and an antiquark—whereas a baryon contains three quarks. For instance, the π^- pion (a meson) is composed of a d quark and a $\bar{u}$ antiquark, $\pi^- = d + \bar{u}$, as **Figure 32.13** shows. These two quarks combine to give the π^- pion a net charge of $-e$. Similarly, the π^+ pion is a combination of the $\bar{d}$ and u quarks, $\pi^+ = \bar{d} + u$. In contrast, protons and neutrons, being baryons, consist of three quarks. A proton contains the combination $d + u + u$, and a neutron contains the combination $d + d + u$ (see **Figure 32.13**). These groups of three quarks give the correct charges for the proton and neutron.

The original quark model was extremely successful in predicting not only the correct charges for the hadrons, but other properties as well. However, in 1974 a new particle, the J/ψ meson, was discovered. This meson has a rest energy of 3097 MeV, much larger than the rest energies of other known mesons. The existence of the J/ψ meson could be explained only if a new quark–antiquark pair existed; this new quark was called *charmed* (*c*). With the discovery of more and more particles, it has been necessary to postulate a fifth and a sixth quark; their names are *top* (*t*) and *bottom* (*b*), although some scientists prefer to call these quarks *truth* and *beauty*. Today, there is firm evidence for all six quarks, each with its corresponding antiquark. All of the hundreds of known hadrons can be accounted for in terms of these six quarks and their antiquarks.

In addition to electric charge, quarks also have other properties. For example, each quark possesses a characteristic called **color**, for which there are three possibilities: blue, green, or red. The corresponding possibilities for the antiquarks are antiblue, antigreen, and antired. The use of the term "color" and the specific choices of blue, green, and red are arbitrary, for the visible colors of the electromagnetic spectrum have nothing to do with quark properties. The quark property of color, however, is important, because it

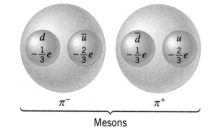

FIGURE 32.13 According to the original quark model of hadrons, all mesons consist of a quark and an antiquark, whereas baryons contain three quarks.

brings the quark model into agreement with the Pauli exclusion principle and enables the model to account for experimental observations that are otherwise difficult to explain.

The Standard Model

The various elementary particles that have been discovered can interact via one or more of the following four forces: the gravitational force, the strong nuclear force, the weak nuclear force, and the electromagnetic force. In particle physics, the phrase "the standard model" refers to the currently accepted explanation for the strong nuclear force, the weak nuclear force, and the electromagnetic force. In this model, the strong nuclear force between quarks is described in terms of the concept of color, and the theory is referred to as quantum chromodynamics. According to the standard model, the weak nuclear force and the electromagnetic force are separate manifestations of a single, even more fundamental, force, referred to as the electroweak force, as we have seen in Section 31.5.

In the standard model, our understanding of the building blocks of matter follows the hierarchical pattern illustrated in **Interactive Figure 32.14**. Molecules, such as water (H_2O) and glucose ($C_6H_{12}O_6$), are composed of atoms. Each atom consists of a nucleus that is surrounded by a cloud of electrons. The nucleus, in turn, is made up of protons and neutrons, which are composed of quarks.

10^{-9} m 10^{-10} m $10^{-15} - 10^{-14}$ m 10^{-15} m   Less than 10^{-18} m

Molecule   Atom   Nucleus   Neutron (or proton)   Quark

INTERACTIVE FIGURE 32.14 The current view of how matter is composed of basic units, starting with a molecule and ending with a quark. The approximate sizes of each unit are also listed.

Check Your Understanding

(The answers are given at the end of the book.)

12. Of the following particles, which ones are *not* composed of quarks and antiquarks? **(a)** A proton **(b)** An electron **(c)** A neutron **(d)** A neutrino

13. The sigma-minus particle Σ^- has a charge of $-e$. Which one of the following quark combinations corresponds to this particle? **(a)** *dds* **(b)** *ssu* **(c)** *uus*

32.7 Cosmology

Cosmology is the study of the structure and evolution of the universe. In this study, both the very large and the very small aspects of the universe are important. Astronomers, for example, study stars located at enormous distances from the earth, up to billions of light-years away. In contrast, particle physicists focus their efforts on the very small elementary particles (10^{-18} m or smaller) that comprise matter. The synergy between the work of astronomers and that of particle physicists has led to significant advances in our understanding of the universe. Central to that understanding is the belief that the universe is expanding, and we begin by discussing the evidence that justifies this belief.

The Expanding Universe and the Big Bang

THE PHYSICS OF . . . an expanding universe. The idea that the universe is expanding originated with the astronomer Edwin P. Hubble (1889–1953). He found that light reaching the earth from distant galaxies is Doppler-shifted toward greater wavelengths—that is, toward the red end of the visible spectrum. As Section 24.5 discusses, this type of Doppler shift results when the observer and the source of the light are moving away from each other. The speed at which a galaxy is receding from the earth can be determined from the measured Doppler shift in wavelength. Hubble found that a galaxy located at a distance d from the earth recedes from the earth at a speed v given by

Hubble's law $$v = Hd \tag{32.5}$$

where H is a constant known as the Hubble parameter. In other words, the recession speed is proportional to the distance d, so that more distant galaxies are moving away from the earth at greater speeds. Equation 32.5 is referred to as Hubble's law.

Hubble's picture of an expanding universe does not mean that the earth is at the center of the expansion. In fact, there is no literal center. Imagine a loaf of raisin bread expanding as it bakes. Each raisin moves away from every other raisin, without any single one acting as a center for the expansion. Galaxies in the universe behave in a similar fashion. Observers in other galaxies would see distant galaxies moving away, just as we do.

THE PHYSICS OF . . . "dark energy." Not only is the universe expanding, it is doing so at an accelerated rate, according to recent astronomical measurements of the brightness of supernovas, or exploding stars. To account for the accelerated rate, astronomers have postulated that "dark energy" pervades the universe. The normal gravitational force between galaxies slows the rate at which they are moving away from each other. The dark energy gives rise to a force that counteracts gravity and pushes galaxies apart. As yet, little is known about dark energy.

Experimental measurements by astronomers indicate that an approximate value for the Hubble parameter is

$$H = 0.022 \frac{\text{m}}{\text{s} \cdot \text{light-year}}$$

The value for the Hubble parameter is believed to be accurate within 10%. Scientists are very interested in obtaining an accurate value for H because it can be related to an age for the universe, as the next example illustrates.

EXAMPLE 9 | An Age for the Universe

Determine an estimate of the age of the universe using Hubble's law.

Reasoning Consider a galaxy currently located at a distance d from the earth. According to Hubble's law, this galaxy is moving away from us at a speed of $v = Hd$. At an earlier time, therefore, this galaxy must have been closer. We can imagine, in fact, that in the remote past the separation distance was relatively small and that the universe originated at such a time. To estimate the age of the universe, we calculate the time it has taken the galaxy to recede to its present position. For this purpose, time is simply distance divided by speed, or $t = d/v$.

Solution Using Hubble's law and the fact that a distance of 1 light-year is 9.46×10^{15} m, we estimate the age of the universe to be

$$t = \frac{d}{v} = \frac{d}{Hd} = \frac{1}{H}$$

$$t = \frac{1}{0.022 \dfrac{\text{m}}{\text{s} \cdot \text{light-year}}} = \frac{1}{\left(0.022 \dfrac{\text{m}}{\text{s} \cdot \text{light-year}}\right)\left(\dfrac{1 \text{ light-year}}{9.46 \times 10^{15} \text{ m}}\right)}$$

$$= 4.3 \times 10^{17} \text{ s} \quad \text{or} \quad \boxed{1.4 \times 10^{10} \text{ yr}}$$

The idea presented in Example 9, that our galaxy and other galaxies in the universe were very close together at some earlier instant in time, lies at the heart of the **Big Bang theory**. This theory postulates that the universe had a definite beginning in a cataclysmic event, sometimes called the primeval fireball. Dramatic evidence supporting the theory was discovered in 1965 by Arno A. Penzias (1933–) and Robert W. Wilson (1936–). Using a radio telescope, they discovered that the earth is being bathed in weak electro-magnetic waves in the microwave region of the spectrum (wavelength = 7.35 cm, see **Figure 24.9**). They observed that the intensity of these waves is the same, no matter where in the sky they pointed their telescope, and concluded that the waves originated outside of our galaxy. This microwave background radiation, as it is called, represents radiation left over from the Big Bang and is a kind of cosmic afterglow. Subsequent measurements have confirmed the research of Penzias and Wilson and have shown that the microwave radiation is consistent with a perfect blackbody (see Sections 13.3 and 29.2) radiating at a temperature of 2.7 K, in agreement with theoretical analysis of the Big Bang. In 1978, Penzias and Wilson received the Nobel Prize in Physics for their discovery.

The Standard Model for the Evolution of the Universe

Based on the recent experimental and theoretical research in particle physics, scientists have proposed an evolutionary sequence of events following the Big Bang. This sequence is known as the **standard cosmological model** and is illustrated in **Figure 32.15**.

Immediately following the Big Bang, the temperature of the universe was incredibly high, about 10^{32} K. During this initial period, the three fundamental forces (the gravitational force, the strong nuclear force, and the electroweak force) all behaved as a single unified force. Very quickly, in about 10^{-43} s, the gravitational force took on a separate identity all its own, as **Figure 32.15** indicates. Meanwhile, the strong nuclear force and the electroweak force continued to act as a single force, which is sometimes referred to as the GUT force. GUT stands for the **g**rand **u**nified **t**heory that presumably would explain such a force. Slightly later, at about 10^{-35} s after the Big Bang, the GUT force separated into the strong nuclear force and the electroweak force, the universe expanding and cooling somewhat to a temperature of roughly 10^{28} K (see **Figure 32.15**). From this point on, the strong nuclear force behaved as we know it today, while the electroweak force maintained its identity. In this scenario, note that the weak nuclear force and the electromagnetic force have not yet manifested themselves as separate entities. The disappearance of the electroweak force and the appearance of the weak nuclear force and the electromagnetic force eventually occurred at approximately 10^{-10} s after the Big Bang, when the temperature of the expanding universe had cooled to about 10^{15} K.

From the Big Bang up until the strong nuclear force separated from the GUT force at a time of 10^{-35} s, all particles of matter were similar, and there was no distinction between quarks and leptons. After this time, quarks and leptons became distinguishable. Eventually the quarks and antiquarks formed hadrons, such as protons and neutrons and their

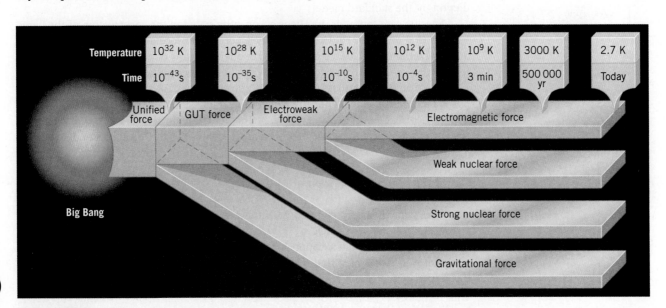

| Temperature | 10^{32} K | 10^{28} K | 10^{15} K | 10^{12} K | 10^9 K | 3000 K | 2.7 K |
| Time | 10^{-43}s | 10^{-35}s | 10^{-10}s | 10^{-4}s | 3 min | 500 000 yr | Today |

Unified force
GUT force
Electroweak force
Electromagnetic force
Weak nuclear force
Big Bang
Strong nuclear force
Gravitational force

FIGURE 32.15 According to the standard cosmological model, the universe has evolved as illustrated here. In this model, the universe is presumed to have originated with a cataclysmic event known as the Big Bang. The times shown are those following this event.

antiparticles. By a time of 10^{-4} s after the Big Bang, however, the temperature had cooled to approximately 10^{12} K, and the hadrons had mostly disappeared. Protons and neutrons survived only as a very small fraction of the total number of particles, the majority of which were leptons such as electrons, positrons, and neutrinos. Like most of the hadrons before them, most of the electrons and positrons eventually disappeared. However, they did leave behind a relatively small number of electrons to join the small number of protons and neutrons at a time of about 3 min following the Big Bang. At this time the temperature of the expanding universe had decreased to about 10^9 K, and small nuclei such as that of helium began forming. Later, when the universe was about 500 000 years old and the temperature had dropped to near 3000 K, hydrogen and helium atoms began forming. As the temperature decreased further, stars and galaxies formed, and today we find a temperature of 2.7 K characterizing the cosmic background radiation of the universe.

32.8   Recent Discoveries in Modern Physics

The advancement of science is a continuing process that trudges along slowly at times, but sometimes leaps forward with entirely new and unexpected discoveries, or observations that verify or disprove the predictions of longstanding theories. In the past decade, two Nobel Prizes in Physics have been awarded as the result of two important experiments: the discovery of the **Higgs boson** and the observation of **gravitational waves**.

The Higgs Boson

In the **standard model** (see Section 32.6), a type of field is required in order to explain certain symmetry properties of particles, and to give the predicted particles their correct masses. This field, which is assumed to exist in all space, is called the Higgs field, after Peter Ware Higgs (1929–   ), the British physicist who contributed to its theoretical development and predictions. The Higgs boson, sometimes referred to as the God particle, is a manifestation of the Higgs field, and therefore its detection would constitute strong evidence of the existence of the theoretically predicted field, and confirm the predictions of the standard model.

The 40-year search for the Higgs boson culminated in the construction of the Large Hadron Collider (LHC) at CERN, a particle accelerator located on the border of France and Switzerland. The device produces high-energy beams of particles (e.g., protons and lead nuclei) that travel close to the speed of light and then collide, breaking up into a cornucopia of different particles. Early in 2013, scientists at CERN confirmed that the Higgs boson had been detected. The 2013 Nobel Prize in Physics was awarded to Peter W. Higgs and François Englert (1932–   ) for their theoretical work that led to the discovery of the Higgs boson, and therefore verified a prediction consistent with the standard model.

Gravitational Waves

Einstein's general theory of relativity describes, among other things, the curvature of spacetime near massive objects. This warping of spacetime amounts to a geometric description of gravity that accounts for the bending of light around stars, which was precisely the one experiment that was used to verify this prediction of the theory. In 1919, the results of the Eddington experiment, organized by the British astronomers Frank Watson Dyson and Arthur Stanley Eddington, were reported, confirming the predicted bending of starlight around the sun.

Another prediction of the general theory of relativity is the creation of disturbances in spacetime by accelerating masses in a way that is analogous to the production of electromagnetic waves by accelerating charges. These ripples in spacetime are called

Courtesy Caltech/MIT/LIGO Laboratory

FIGURE 32.16 The Laser Interferometer Gravitational-Wave Observatory (LIGO) at Livingston, Louisiana.

gravitational waves, and they travel at the speed of light. They can only be detected using very sensitive instrumentation, such as the enormous laser interferometer at the Laser Interferometer Gravitational-Wave Observatory (LIGO) at Livingston, Louisiana (see Figure 32.16), and an identical one 3000 miles away at Hanover, Washington. With two perpendicular arms, each 4 kilometers in length, the LIGO interferometers can detect very slight distortions in their lengths, on the order of 1/10 000 the width of a proton, a miniscule effect that would occur as the result of the spacetime distortions produced by a gravitational wave as it passes through. Even though the LIGO detectors are incredibly sensitive, the intensity of gravitational waves is so weak that it requires collisions in the universe that are, well, on an astronomical scale. LIGO can detect gravity waves from events such as the merger of two neutron stars or the collision of two black holes. In order for a gravitational-wave detection to be considered legitimate, it would have to be detected by both facilities.

Albert Einstein predicted the existence of gravitational waves in 1916. Nearly a century later, on September 14, 2015, the LIGO collaboration detected the first gravitational waves due to the merging of two back holes 1.3 billion light-years from earth. Not only did this observation confirm the prediction of Einstein's general theory of relativity, but it provided a new observational tool for astronomers, and it will continue to offer valuable information to astrophysicists and cosmologists in the future. Rainer Weiss (1932–), Kip Thorne (1940–), and Barry Barish (1936–) were awarded the 2017 Nobel Prize in Physics for their role in the direct detection of gravitational waves.

Concept Summary

32.1 Biological Effects of Ionizing Radiation Ionizing radiation consists of photons and/or moving particles that have enough energy to ionize an atom or molecule. Exposure is a measure of the ionization produced in air by X-rays or γ rays. When a beam of X-rays or γ rays is sent through a mass m of dry air (0 °C, 1 atm pressure) and produces positive ions whose total charge is q, the exposure in coulombs per kilogram (C/kg) is

$$\text{Exposure (in coulombs per kilogram)} = \frac{q}{m}$$

With q in coulombs (C) and m in kilograms (kg), the exposure in roentgens is given by Equation 32.1.

$$\begin{array}{c}\text{Exposure}\\\text{(in roentgens)}\end{array} = \left(\frac{1}{2.58 \times 10^{-4}}\right)\frac{q}{m} \qquad (32.1)$$

The absorbed dose is the amount of energy absorbed from the radiation per unit mass of absorbing material, as specified by Equation 32.2.

The SI unit of absorbed dose is the gray (Gy); 1 Gy = 1 J/kg. However, the rad is another unit that is often used: 1 rad = 0.01 Gy.

$$\text{Absorbed dose} = \frac{\text{Energy absorbed}}{\text{Mass of absorbing material}} \qquad (32.2)$$

The amount of biological damage produced by ionizing radiation is different for different types of radiation. The relative biological effectiveness (RBE) is the absorbed dose of 200-keV X-rays required to produce a certain biological effect divided by the dose of a given type of radiation that produces the same biological effect, as given by Equation 32.3.

$$\text{RBE} = \frac{\begin{array}{c}\text{The dose of 200-keV X-rays that}\\\text{produces a certain biological effect}\end{array}}{\begin{array}{c}\text{The dose of radiation that}\\\text{produces the same biological effect}\end{array}} \qquad (32.3)$$

The biologically equivalent dose (in rems) is the product of the absorbed dose (in rads) and the RBE, as shown in Equation 32.4.

$$\begin{array}{c} \text{Biologically equivalent dose} \\ \text{(in rems)} \end{array} = \begin{array}{c} \text{Absorbed dose} \\ \text{(in rads)} \end{array} \times \text{RBE} \quad \textbf{(32.4)}$$

32.2 Induced Nuclear Reactions An induced nuclear reaction occurs whenever a target nucleus is struck by an incident nucleus, an atomic or subatomic particle, or a γ-ray photon and undergoes a change as a result. An induced nuclear transmutation is a reaction in which the target nucleus is changed into a nucleus of a new element.

All nuclear reactions (induced or spontaneous) obey the conservation laws of physics as they relate to mass/energy, electric charge, linear momentum, angular momentum, and nucleon number.

Nuclear reactions are often written in a shorthand form, such as $^{14}_{7}\text{N}\,(\alpha, p)\,^{17}_{8}\text{O}$. The first and last symbols $^{14}_{7}\text{N}$ and $^{17}_{8}\text{O}$ denote, respectively, the initial and final nuclei. The symbols within the parentheses denote the incident α particle (on the left) and the small emitted particle or proton p (on the right).

A thermal neutron is one that has a kinetic energy of about 0.04 eV.

32.3 Nuclear Fission Nuclear fission occurs when a massive nucleus splits into two less massive fragments. Fission can be induced by the absorption of a thermal neutron. When a massive nucleus fissions, energy is released because the binding energy per nucleon is greater for the fragments than for the original nucleus. Neutrons are also released during nuclear fission. These neutrons can, in turn, induce other nuclei to fission and lead to a process known as a chain reaction. A chain reaction is said to be controlled if each fission event contributes, on average, only one neutron that fissions another nucleus.

32.4 Nuclear Reactors A nuclear reactor is a device that generates energy by a controlled chain reaction. Many reactors in use today have the same three basic components: fuel elements, a neutron moderator, and control rods. The fuel elements contain the fissile fuel, and the entire region of fuel elements is known as the reactor core. The neutron moderator is a material (water, for example) that slows down the neutrons released in a fission event to thermal energies so they can initiate additional fission events. Control rods contain material that readily absorbs neutrons without fissioning. They are used to keep the reactor in its normal, or critical, state, in which each fission event leads to one additional fission, no more, no less. The reactor is subcritical when, on average, the neutrons from each fission trigger less than one subsequent fission. The reactor is supercritical when, on average, the neutrons from each fission trigger more than one additional fission.

32.5 Nuclear Fusion In a fusion process, two nuclei with smaller masses combine to form a single nucleus with a larger mass. Energy is released by fusion when the binding energy per nucleon is greater for the larger nucleus than for the smaller nuclei. Fusion reactions are said to be thermonuclear because they require extremely high temperatures to proceed. Current studies of nuclear fusion utilize either magnetic confinement or inertial confinement to contain the fusing nuclei at the high temperatures that are necessary.

32.6 Elementary Particles Subatomic particles are divided into three families: the boson family (which includes the photon), the lepton family (which includes the electron), and the hadron family (which includes the proton and the neutron). Elementary particles are the basic building blocks of matter. All members of the boson and lepton families are elementary particles.

The quark theory proposes that the hadrons are not elementary particles but are composed of elementary particles called quarks. Currently, the hundreds of hadrons can be accounted for in terms of six quarks (up, down, strange, charmed, top, and bottom) and their antiquarks.

The standard model consists of two parts: (1) the currently accepted explanation for the strong nuclear force in terms of the quark concept of "color" and (2) the theory of the electroweak interaction.

32.7 Cosmology Cosmology is the study of the structure and evolution of the universe. Our universe is expanding. The speed v at which a distant galaxy recedes from the earth is given by Hubble's law (Equation 32.5), where $H = 0.022 \text{ m/(s} \cdot \text{light-year)}$ is called the Hubble parameter and d is the distance of the galaxy from the earth.

$$v = Hd \quad \textbf{(32.5)}$$

The Big Bang theory postulates that the universe had a definite beginning in a cataclysmic event, sometimes called the primeval fireball. The radiation left over from this event is in the microwave region of the electromagnetic spectrum, and it is consistent with a perfect blackbody radiating at a temperature of 2.7 K, in agreement with theoretical analysis of the Big Bang.

The standard cosmological model for the evolution of the universe is summarized in **Figure 32.15**.

32.8 Recent Discoveries in Modern Physics In the last decade, we have witnessed profound discoveries in elementary particle physics. The 2013 Nobel Prize in physics was awarded to Peter W. Higgs and François Englert for their theoretical work that led to the discovery of the Higgs boson, or "God particle" that verified a prediction consistent with the standard model of particle physics that explains the building blocks of matter.

Rainer Weiss, Kip Thorne, and Barry Barish were awarded the 2017 Nobel Prize in physics for their role in the direct detection of gravitational waves, ripples in spacetime that move at the speed of light and were first predicted by Albert Einstein in 1916. These disturbances are created by the motion of matter, such as merging neutron stars, and their detection not only confirmed Einstein's general theory of relativity but demonstrated a new tool to be used by astronomers and astrophysicists in the future.

Focus on Concepts

Online

Additional questions are available for assignment in WileyPLUS.

Section 32.1 Biological Effects of Ionizing Radiation

1. Biologically equivalent doses are specified in units called_____.
(a) rads (b) grays (c) rems (d) J/kg (e) roentgens

Section 32.2 Induced Nuclear Reactions

2. Determine the unknown nuclear species $^{A}_{Z}\text{X}$ in the following nuclear reaction:

$$^{A}_{Z}\text{X} + {}^{14}_{7}\text{N} \longrightarrow {}^{14}_{6}\text{C} + {}^{1}_{1}\text{H}$$

(a) $^{2}_{1}\text{H}$ (b) $^{1}_{0}\text{n}$ (c) γ ray (d) $^{4}_{2}\text{He}$

Section 32.3   Nuclear Fission

3. The fission of $^{235}_{92}U$ can occur via many different reactions. In general, they can be written as follows:

$$^{1}_{0}n + ^{235}_{92}U \longrightarrow ^{A_X}_{Z_X}X + ^{A_Y}_{Z_Y}Y + \eta\,^{1}_{0}n$$

where X and Y refer to the identities of the fission fragments and η is the number of neutrons produced. Which one or more of the following statements are true?

A. The compound nucleus that is formed to initiate the fission process is the same, no matter what X and Y refer to.

B. The greater the number η of neutrons produced by the reaction, the smaller is the sum of the nucleon numbers A_X and A_Y.

C. The sum of the atomic numbers Z_X and Z_Y is the same, no matter what X and Y refer to.

(a) A (b) A and B (c) A and C (d) B and C (e) A, B, and C

Section 32.5   Nuclear Fusion

4. In each of the following three nuclear fusion reactions, the masses of the nuclei are given beneath each nucleus. Rank the energy produced by each reaction in descending order (greatest first).

Reaction I $^{2}_{1}H$ + $^{2}_{1}H$ ⟶ $^{3}_{2}He$ + $^{1}_{0}n$
                 2.0141 u   2.0141 u   3.0160 u   1.0087 u

Reaction II $^{3}_{2}He$ + $^{3}_{2}He$ ⟶ $^{4}_{2}He$ + $^{1}_{1}H$ + $^{1}_{1}H$
               3.0160 u   3.0160 u   4.0026 u   1.0078 u   1.0078 u

Reaction III $^{15}_{7}n$ + $^{1}_{1}H$ ⟶ $^{12}_{6}C$ + $^{4}_{2}He$
                15.0001 u   1.0078 u   12.0000 u   4.0026 u

(a) I, II, III (b) I, III, II (c) II, III, I (d) II, I, III (e) III, II, I

Section 32.6   Elementary Particles

5. The drawings show four possibilities for hadrons in the quark theory. In each of the possibilities, the symbols for the quarks are shown together with the corresponding electric charges. Note that e stands for the magnitude of the charge on an electron. Which one shows the quark structure for an antiproton? (a) A (b) B (c) C (d) D

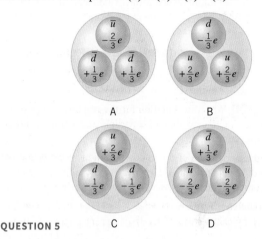

QUESTION 5

Section 32.7   Cosmology

6. Which one of the following statements is *not* accepted as a part of the current picture of the universe? (a) The universe is expanding. (b) Early in the history of the universe, there was only one fundamental force. (c) The universe began with a cataclysmic event known as the Big Bang. (d) The weak electromagnetic radiation in the microwave region of the spectrum that bathes the earth provides evidence of the Big Bang theory. (e) Hubble's law indicates that a galaxy located at a distance d from the earth is moving away from the earth at a speed that is inversely proportional to d.

Problems   Online

Additional questions are available for assignment in WileyPLUS.

SSM	Student Solutions Manual	BIO	Biomedical application
MMH	Problem-solving help	E	Easy
GO	Guided Online Tutorial	M	Medium
V-HINT	Video Hints	H	Hard
CHALK	Chalkboard Videos	WS	Worksheet
		T	Team Problem

Section 32.1   Biological Effects of Ionizing Radiation

1. E SSM Neutrons (RBE = 2.0) and α particles have the same biologically equivalent dose. However, the absorbed dose of the neutrons is six times the absorbed dose of the α particles. What is the RBE for the α particles?

2. E What absorbed dose (in rads) of α particles (RBE = 15) causes as much biological damage as a 60-rad dose of protons (RBE = 10)?

3. E BIO Over a full course of treatment, two different tumors are to receive the same absorbed dose of therapeutic radiation. The smaller of the tumors (mass = 0.12 kg) absorbs a total of 1.7 J of energy. (a) Determine the absorbed dose, in Gy. (b) What is the total energy absorbed by the larger of the tumors (mass = 0.15 kg)?

4. E Over a year's time, a person receives a biologically equivalent dose of 24 mrem (millirems) from cosmic rays, which consist primarily of high-energy protons bombarding earth's atmosphere from space. The relative biological effectiveness of protons is 10. (a) What is the person's absorbed dose in rads? (b) The person absorbs 1.9×10^{-3} J of energy from cosmic rays in a year. What is the person's mass?

5. E BIO GO SSM A beam of particles is directed at a 0.015-kg tumor. There are 1.6×10^{10} particles per second reaching the tumor, and the energy of each particle is 4.0 MeV. The RBE for the radiation is 14. Find the biologically equivalent dose given to the tumor in 25 s.

6. E GO A 75-kg person is exposed to 45 mrem of α particles (RBE = 12). How much energy (in joules) has this person absorbed?

7. E GO The biologically equivalent dose for a typical chest X-ray is 2.5×10^{-2} rem. The mass of the exposed tissue is 21 kg, and it absorbs 6.2×10^{-3} J of energy. What is the relative biological effectiveness (RBE) for the radiation on this particular type of tissue?

8. M BIO A 2.0-kg tumor is being irradiated by a radioactive source. The tumor receives an absorbed dose of 12 Gy in a time of 850 s. Each

disintegration of the radioactive source produces a particle that enters the tumor and delivers an energy of 0.40 MeV. What is the activity $\Delta N/\Delta t$ (see Section 31.6) of the radioactive source?

9. **M CHALK SSM** A beam of nuclei is used for cancer therapy. Each nucleus has an energy of 130 MeV, and the relative biological effectiveness (RBE) of this type of radiation is 16. The beam is directed onto a 0.17-kg tumor, which receives a biologically equivalent dose of 180 rem. How many nuclei are in the beam?

Section 32.2 Induced Nuclear Reactions

10. **E** What is the atomic number Z, the atomic mass number A, and the element X in the reaction $^{10}_{5}B\,(\alpha, p)\,^{A}_{Z}X$?

11. **E SSM** Write the equation for the reaction $^{17}_{8}O\,(\gamma, \alpha n)\,^{12}_{6}C$. The notation "$an$" means that an α particle and a neutron are produced by the reaction.

12. **E GO** For each of the nuclear reactions listed below, determine the unknown particle $^{A}_{Z}X$. Use the periodic table on the inside of the back cover of the text as needed.

    (a) $^{A}_{Z}X + ^{14}_{7}N \longrightarrow ^{1}_{1}H + ^{17}_{8}O$ (c) $^{1}_{1}H + ^{27}_{13}Al \longrightarrow ^{A}_{Z}X + ^{1}_{0}n$

    (b) $^{15}_{7}N + ^{A}_{Z}X \longrightarrow ^{12}_{6}C + ^{4}_{2}He$ (d) $^{7}_{3}Li + ^{1}_{1}H \longrightarrow ^{4}_{2}He + ^{A}_{Z}X$

13. **E** A neutron causes $^{232}_{90}Th$ to change according to the reaction

$$^{1}_{0}n + ^{232}_{90}Th \longrightarrow ^{A}_{Z}X + \gamma$$

(a) Identify the unknown nucleus $^{A}_{Z}X$, giving its atomic mass number A, its atomic number Z, and the symbol X for the element. **(b)** The $^{A}_{Z}X$ nucleus subsequently undergoes β^- decay, and its daughter does too. Identify the final nucleus, giving its atomic mass number, atomic number, and name.

14. **E** Write the reactions below in the shorthand form discussed in the text.

    (a) $^{4}_{2}He + ^{27}_{13}Al \longrightarrow ^{30}_{15}P + ^{1}_{0}n$

    (b) $^{1}_{1}H + ^{9}_{4}Be \longrightarrow ^{6}_{3}Li + ^{4}_{2}He$

    (c) $^{1}_{0}n + ^{55}_{25}Mn \longrightarrow ^{56}_{25}Mn + \gamma$

15. **E SSM** Complete the following nuclear reactions, assuming that the unknown quantity signified by the question mark is a single entity: **(a)** $^{34}_{18}Ar\,(n, \alpha)?$ **(b)** $^{82}_{34}Se\,(?, n)\,^{82}_{35}Br$ **(c)** $^{58}_{28}Ni\,(^{40}_{18}Ar, ?)\,^{57}_{27}Co$ **(d)** $?(\gamma, \alpha)\,^{16}_{8}O$

16. **M CHALK** During a nuclear reaction, an unknown particle is absorbed by a copper $^{63}_{29}Cu$ nucleus, and the reaction products are $^{62}_{29}Cu$, a neutron, and a proton. What are the name, atomic number, and nucleon number of the nucleus formed temporarily when the copper $^{63}_{29}Cu$ nucleus absorbs the unknown particle?

Section 32.3 Nuclear Fission,
Section 32.4 Nuclear Reactors

17. **E** Determine the atomic number Z, the atomic mass number A, and the element X for the unknown species $^{A}_{Z}X$ in the following reaction for the fission of uranium $^{235}_{92}U$:

$$^{1}_{0}n + ^{235}_{92}U \longrightarrow ^{133}_{51}Sb + ^{A}_{Z}X + 4^{1}_{0}n$$

Consult the periodic table on the inside of the back cover of the text as needed.

18. **E SSM** When a $^{235}_{92}U$ (235.043 924 u) nucleus fissions, about 200 MeV of energy is released. What is the ratio of this energy to the rest energy of the uranium nucleus?

19. **E CHALK** How many neutrons are produced when $^{235}_{92}U$ fissions in the following way? $^{1}_{0}n + ^{235}_{92}U \rightarrow ^{132}_{50}Sn + ^{101}_{42}Mo + $ neutrons

20. **E** Neutrons released by a fission reaction must be slowed by collisions with the moderator nuclei before the neutrons can cause further fissions. Suppose a 1.5-MeV neutron leaves each collision with 65% of its incident energy. How many collisions are required to reduce the neutron's energy to at least 0.040 eV, which is the energy of a thermal neutron?

21. **E V-HINT** The energy released by each fission within the core of a nuclear reactor is 2.0×10^2 MeV. The number of fissions occurring each second is 2.0×10^{19}. Determine the power (in watts) that the reactor generates.

22. **E** Uranium $^{235}_{92}U$ fissions into two fragments plus three neutrons: $^{1}_{0}n + ^{235}_{92}U \rightarrow$ (2 fragments) $+ 3^{1}_{0}n$. The mass of a neutron is 1.008 665 u and the mass of $^{235}_{92}U$ is 235.043 924 u. If 225.0 MeV of energy is released, what is the total mass of the two fragments?

23. **M GO CHALK** When 1.0 kg of coal is burned, approximately 3.0×10^7 J of energy is released. If the energy released during each $^{235}_{92}U$ fission is 2.0×10^2 MeV, how many kilograms of coal must be burned to produce the same energy as 1.0 kg of $^{235}_{92}U$?

24. **M GO** The water that cools a reactor core enters the reactor at 216 °C and leaves at 287 °C. (The water is pressurized, so it does not turn to steam.) The core is generating 5.6×10^9 W of power. Assume that the specific heat capacity of water is 4420 J/(kg · C°) over the temperature range stated above, and find the mass of water that passes through the core each second.

25. **H SSM** A nuclear power plant is 25% efficient, meaning that only 25% of the power it generates goes into producing electricity. The remaining 75% is wasted as heat. The plant generates 8.0×10^8 watts of electric power. If each fission releases 2.0×10^2 MeV of energy, how many kilograms of $^{235}_{92}U$ are fissioned per year?

Section 32.5 Nuclear Fusion

26. **E** Two deuterium atoms ($^{2}_{1}H$) react to produce tritium ($^{3}_{1}H$) and hydrogen ($^{1}_{1}H$) according to the following fusion reaction:

$$\underset{2.014\,102\,\text{u}}{^{2}_{1}H} \quad + \quad \underset{2.014\,102\,\text{u}}{^{2}_{1}H} \quad \longrightarrow \quad \underset{3.016\,050\,\text{u}}{^{3}_{1}H} \quad + \quad \underset{1.007\,825\,\text{u}}{^{1}_{1}H}$$

What is the energy (in MeV) released by this deuterium–deuterium reaction?

27. **E GO** In one type of fusion reaction a proton fuses with a neutron to form a deuterium nucleus:

$$^{1}_{1}H + ^{1}_{0}n \longrightarrow ^{2}_{1}H + \gamma$$

The masses are $^{1}_{1}H$ (1.0078 u), $^{1}_{0}n$ (1.0087 u), and $^{2}_{1}H$ (2.0141 u). The γ-ray photon is massless. How much energy (in MeV) is released by this reaction?

28. **E MMH** The fusion of two deuterium nuclei ($^{2}_{1}H$, mass $=$ 2.0141 u) can yield a helium nucleus ($^{3}_{2}He$, mass $=$ 3.0160 u) and a neutron ($^{1}_{0}n$, mass $=$ 1.0087 u). What is the energy (in MeV) released in this reaction?

29. **E V-HINT** Tritium ($^{3}_{1}H$) is a rare isotope of hydrogen that can be produced by the following fusion reaction:

$$\underset{1.0087\,\text{u}}{^{A}_{Z}X} \quad + \quad \underset{2.0141\,\text{u}}{^{4}_{1}Y} \longrightarrow \underset{3.0161\,\text{u}}{^{3}_{1}H} \quad + \quad \gamma$$

(a) Determine the atomic mass number A, the atomic number Z, and the names X and Y of the unknown particles. **(b)** Using the masses given in the reaction, determine how much energy (in MeV) is released by this reaction.

30. **M CHALK SSM** One proposed fusion reaction combines lithium $^{6}_{3}Li$ (6.015 u) with deuterium $^{2}_{1}H$ (2.014 u) to give helium $^{4}_{2}He$ (4.003 u):

$^2_1H + ^6_3Li \rightarrow 2^4_2He$. How many kilograms of lithium would be needed to supply the energy needs of one household for a year, estimated to be 3.8×10^{10} J?

31. **M** **GO** In Example 7 it was determined that 17.6 MeV of energy is released when the following fusion reaction occurs:

$$\underbrace{^2_1H}_{2.0141\,u} + \underbrace{^3_1H}_{3.0161\,u} \longrightarrow \underbrace{^4_2He}_{4.0026\,u} + \underbrace{^1_0n}_{1.0087\,u}$$

Ignore relativistic effects and determine the kinetic energies of the neutron and the α particle.

32. **M** Deuterium (^2_1H) is an attractive fuel for fusion reactions because it is abundant in the oceans, where about 0.015% of the hydrogen atoms in the water (H_2O) are deuterium atoms. **(a)** How many deuterium atoms are there in one kilogram of water? **(b)** If each deuterium nucleus produces about 7.2 MeV in a fusion reaction, how many kilograms of water would be needed to supply the energy needs of the United States for one year, estimated to be 1.1×10^{20} J?

Section 32.6 Elementary Particles

33. **E** The main decay mode for the negative pion is $\pi^- \rightarrow \mu^- + \bar{\nu}_\mu$. Find the energy (in MeV) released in this decay. Consult **Table 32.3** for rest energies and assume that the rest energy for $\bar{\nu}_\mu$ is ≈ 0 MeV.

34. **E** **V-HINT** A neutral pion π^0 (rest energy = 135.0 MeV) produced in a high-energy particle experiment moves at a speed of 0.780 c. After a very short time, it decays into two γ-ray photons. One of the γ-ray photons has an energy of 192 MeV. What is the energy (in MeV) of the second γ-ray photon? Take relativistic effects into account.

35. **E** **GO** In addition to its rest energy, a moving proton (p') has kinetic energy. This proton collides with a stationary proton (p), and the reaction forms a stationary neutron (n), a stationary proton (p), and a stationary pion (π^+), according to the following reaction: $p' + p \rightarrow n + p + \pi^+$. The rest energy of each proton is 938.3 MeV, and the rest energy of the neutron is 939.6 MeV. The rest energy of the pion is 139.6 MeV. What is the kinetic energy (in MeV) of the moving proton?

36. **E** **SSM** Suppose a neutrino is created and has an energy of 35 MeV. **(a)** Assuming the neutrino, like the photon, has no mass and travels at the speed of light, find the momentum of the neutrino. **(b)** Determine the de Broglie wavelength of the neutrino.

37. **M** **GO** Review Conceptual Example 8 as background for this problem. An electron and its antiparticle annihilate each other, producing two γ-ray photons. The kinetic energies of the particles are negligible. Determine the magnitude of the momentum of each photon.

38. **M** Review Conceptual Example 6 as background for this problem. An energetic proton is fired at a stationary proton. For the reaction to produce new particles, the two protons must approach each other to within a distance of about 8.0×10^{-15} m. The moving proton must have a sufficient speed to overcome the repulsive Coulomb force. What must be the minimum initial kinetic energy (in MeV) of the proton?

Additional Problems

Online

39. **E** **GO** Someone stands near a radioactive source and receives doses of the following types of radiation: γ rays (20 mrad, RBE = 1), electrons (30 mrad, RBE = 1), protons (5 mrad, RBE = 10), and slow neutrons (5 mrad, RBE = 2). Rank the types of radiation, highest first, as to which produces the largest biologically equivalent dose.

40. **E** Identify the unknown species $^A_Z X$ in the following nuclear reaction: $^{22}_{11}Na\,(d, \alpha)\,^A_Z X$. Here, d stands for the deuterium isotope ^2_1H of hydrogen.

41. **E** **GO** **MMH** What energy (in MeV) is released by the following fission reaction?

$$\underbrace{^1_0n}_{1.009\,u} + \underbrace{^{235}_{92}U}_{235.044\,u} \longrightarrow \underbrace{^{140}_{54}Xe}_{139.922\,u} + \underbrace{^{94}_{38}Sr}_{93.915\,u} + \underbrace{2^1_0n}_{2(1.009\,u)}$$

42. **M** **V-HINT** Multiple-Concept Example 1 discusses some of the physics principles that are used to solve this problem. What absorbed dose (in rads) of γ rays is required to change a block of ice at 0.0 °C into steam at 100.0 °C?

43. **M** **SSM** Imagine that your car is powered by a fusion engine in which the following reaction occurs: $3^2_1H \rightarrow ^4_2He + ^1_1H + ^1_0n$. The masses are ^2_1H (2.0141 u), ^4_2He (4.0026 u), ^1_1H (1.0078 u), and ^1_0n (1.0087 u). The engine uses 6.1×10^{-6} kg of deuterium ^2_1H fuel. If one gallon of gasoline produces 2.1×10^9 J of energy, how many gallons of gasoline would have to be burned to equal the energy released by all the deuterium fuel?

44. **M** **GO** The energy consumed in one year in the United States is about 1.1×10^{20} J. With each $^{235}_{92}U$ fission, about 2.0×10^2 MeV of energy is released. How many kilograms of $^{235}_{92}U$ would be needed to generate this energy if all the nuclei fissioned?

45. **M** **(a)** If each fission reaction of a $^{235}_{92}U$ nucleus releases about 2.0×10^2 MeV of energy, determine the energy (in joules) released by the complete fissioning of 1.0 gram of $^{235}_{92}U$. **(b)** How many grams of $^{235}_{92}U$ would be consumed in one year to supply the energy needs of a household that uses 30.0 kWh of energy per day, on the average?

46. **H** One kilogram of dry air at STP conditions is exposed to 1.0 R of X-rays. One roentgen is defined by Equation 32.1. An equivalent definition can be based on the fact that an exposure of one roentgen deposits 8.3×10^{-3} J of energy per kilogram of dry air. Using the two definitions and assuming that all ions produced are singly charged, determine the average energy (in eV) needed to produce one ion in air.

Physics in Biology, Medicine, and Sports

47. **M** **BIO** **32.1** In March 2011, the Fukushima Daiichi nuclear power station, which is located about 130 miles northeast of Tokyo, Japan, experienced a powerful 9.0 earthquake and subsequent tsunami that caused three of its reactor cores to melt down. The damaged reactors leaked radiation into the atmosphere and Pacific Ocean, forcing over 150 000 residents within 20 km of the plant to be evacuated. Elevated levels of radiation will exist at the plant site for decades.

The torch relay for the 2020 Summer Olympics in Tokyo was to begin at the J-Village Sports Facility in Fukushima in December 2019. However, background radiation measurements at the facility in October of that year measured a value of 7.1 mrem/hr at one location. Businesses in and around Fukushima often display current radiation levels (see the photo). **(a)** If people receive, on average, 630 mrem of radiation per year from natural sources, how many days at the sports facility would it take to receive an annual dose of radiation? **(b)** If a 95-kg person stays at the facility for the number of days in part (a), how much energy does their body absorb from the radiation during that time? Assume the RBE of the radiation is 1. *Note:* The Olympic Torch Relay in 2019, and the entire 2020 Summer Games, were rescheduled due to the COVID-19 pandemic.

A hotel in Naraha, near Fukushima, Japan, displays the radiation dose rate in microsieverts (μSv) per hour, five years after the Fukushima disaster. The sievert (Sv) is the SI unit of dose. 1 μSv is equal to 0.1 mrem.

PROBLEM 47

48. **M** **BIO** **32.1** A now-retired Russian particle physicist named Anatoli Bugorski survived accidental exposure to a very high dose of radiation in 1978 while working at the U-70 synchrotron in Protvino, Russia. At the time, the U-70 synchrotron was one of the world's largest and most powerful particle accelerators. It is similar to the present-day Large Hadron Collider (LHC) near Geneva, in that both are proton synchrotrons. In order to repair a piece of malfunctioning equipment, it was necessary for Bugorski to enter the proton beam tunnel. A breakdown in several safety protocols allowed him to enter the tunnel while the beam was still on. The beam of protons is invisible, so once inside, there was no indication that he was in any danger. He leaned over to perform a visual inspection of the malfunctioning equipment and placed his head in the proton beam. Relativistic protons moving near the speed of light with an energy of 76 GeV cut a 2- to 3-mm-wide path through his face, head, and brain in a fraction of a second (see the photo). He recalls seeing a flash of light, "brighter than a thousand suns," but said he felt no pain. He knew immediately what had happened. Surprisingly, he finished the repairs and noted the accident in his lab notebook.

The damage to Bugorski's cells was extensive. Over the next few weeks, his face would swell up, his skin would die and peel off, he would suffer nerve damage to the left side of his face that led to facial paralysis, he would lose hearing in his left ear, and he began to suffer seizures. Although the prognosis from his medical doctors was grim, he remarkably survived and began to recover. The overall damage to his brain was small, but so extensive along the path of the beam that cancer cells were unable to form. He would eventually recover nearly all of his intellectual capacity, finish his PhD, get married, and have children. Sources disagree on the exact absorbed dose of radiation he received, but the value is near 3.0 × 10⁵ rad. **(a)** Assuming this value for the absorbed dose, how much energy was

deposited in his head? Take the mass of his head to be 3.7 kg. **(b)** If the RBE of the protons was 10.0, what biologically equivalent dose of radiation did he receive?

(Left, top image) A picture of Anatoli Bugorski several days after the accident. Notice the large amount of swelling on the left side of his face. The red line shows the path of the proton beam through Bugorski's head. The beam entered through the back of his head above his left ear. Notice the small patch of hair that is missing at the entrance point. (Left, bottom image) A sketch of the cross section of his head, showing how the beam bored through nonessential parts of his brain. The beam exited his face near the left side of his nose. (Right) A more recent picture of Bugorski. Note the slight droop in the left side of his face and a lack of wrinkles on that side due to the paralysis.

PROBLEM 48

49. **E** **BIO** **32.7** Over the past few decades, scientists have discovered over 4000 exoplanets, which are planets that exist outside of our solar system. This is quite remarkable, when one considers that only 30 years ago, we only knew of 9 planets. Current estimates suggest there could be billions of exoplanets in just the Milky Way Galaxy alone. One of these planets is KOI-456.04, which orbits the star Kepler-160 within the constellation Lyra. This exoplanet is being heavily studied, since it is similar to earth and orbits a star like our sun. Measurements suggest that KOI-456.04 is less than twice the size of earth and that it orbits its star at roughly the same distance as the earth orbits the sun. This places the planet in the so-called habitable zone or Goldilocks zone, which means it's neither too hot nor too cold, but just right to potentially support liquid water and possibly life. The Kepler-160 system is located 3.14 × 10³ ly from earth. Apply Hubble's law to estimate the speed at which this system is moving away from earth. *Note:* Hubble's law should, technically, only be applied to very distant galaxies. The motion of galaxies that are relatively close to earth is dominated by local forces, such as gravitational attraction, and is not representative of the overall expansion of the universe.

50. **M** **BIO** **32.1** A patient with a brain tumor is undergoing a gamma knife treatment (see Section 31.4). Multiple sources of ⁶⁰Co emit two gamma rays with a total energy of 2.50 MeV. If the activity of all the sources combined is 5.00 mCi, what is the biologically equivalent dose delivered to the brain tumor in one hour? The mass of the tumor is 0.150 kg. Since the half-life of ⁶⁰Co is 5.27 yr, we can assume the activity remains constant during the procedure.

51. **M** **BIO** **32.1** A scientist working at a neutron reactor facility accidentally exposes his hand to slow neutrons and receives an absorbed dose of radiation of 2.25 rad. **(a)** How much energy is absorbed by his hand, which has a mass of 0.570 kg? **(b)** If we assume his hand is completely composed of water, what would be the rise in temperature of his hand due to the exposure?

Concepts and Calculations Problems

Online

When considering the biological effects of ionizing radiation, the concept of biologically equivalent dose is especially important. Its importance lies in the fact that the biologically equivalent dose incorporates both the amount of energy per unit mass that is absorbed and the effectiveness of a particular type of radiation in producing a certain biological effect. Problem 52 examines this concept and also reviews the notions of power (Section 6.7) and intensity (Section 16.7) of a wave. Problem 53 illustrates the decay of a particle into two photons, and provides a review of the principles of conservation of energy and conservation of momentum.

52. **M** **CHALK** **SSM** A patient is being given a chest X-ray. The X-ray beam is turned on for 0.20 s, and its intensity is 0.40 W/m^2. The area of the chest being exposed is 0.072 m^2, and the radiation is absorbed by 3.6 kg of tissue. The relative biological effectiveness (RBE) of the X-ray for this tissue is 1.1. *Concepts*: (i) How is the power of the beam related to the beam intensity? (ii) How is the energy absorbed by the tissue related to the power of the beam? (iii) What is the absorbed dose? (iv) How is the biologically equivalent dose related to the absorbed dose? *Calculation*: Calculate the biologically equivalent dose received by the patient.

53. **M** **CHALK** The π^0 meson is a particle that has a rest energy of 135.0 MeV (see **Table 32.3**). It lives for a very short time and then decays into two γ-ray photons: $\pi^0 \rightarrow \gamma + \gamma$. Suppose that one of the γ-ray photons travels along the $+x$ axis. *Concepts*: (i) How is the energy E of each γ-ray photon related to the rest energy E_0 of the π_0 particle? (ii) How can the frequency and wavelength of a photon be determined from its energy? (iii) How is the total linear momentum of the photons related to the momentum of the π^0 particle, and what is the momentum of each particle? *Calculations*: If the π^0 is at rest when it decays, find **(a)** the energy (in MeV), **(b)** the frequency and wavelength, and **(c)** the momentum of each γ-ray photon.

Team Problems

Online

54. **E** **T** **WS** **The Safety of a Neutron Source.** Californium-252 is a strong neutron source. You and your team have been assigned to assess the risk associated with storing 1.50 grams of californium-252 that is to be used in an experiment that requires neutrons. Through spontaneous fission processes, the 1.50-g ^{252}Cf source emits 3.47×10^{12} neutrons per second. The material is contained in a shielded container, and your task is to estimate the biologically equivalent dose (in rems) delivered to an unsuspecting person who has opened the shielded vessel without taking proper precautions. The neutrons have an average energy of 2.30 MeV and have an RBE = 10.0. Since the source emits neutrons uniformly in all directions, the geometry of the container is such that a person opening its hatch will only receive a fraction (7.50%) of the total emitted neutrons. The affected area on the body would be the head and chest, and the estimated mass in which the neutrons would be absorbed is 21.0 kg. If a person opens the hatch and is exposed for 5.00 s, **(a)** what is the absorbed dose in rad? **(b)** What is the biologically equivalent dose in rem? **(c)** How long would it take for a person standing in front of the open container to get a dose of neutrons that would result in radiation sickness (i.e., greater than 50 rem)? Note that this result pertains only to the emitted neutrons. Californium-252 also emits other ionizing radiation including α particles, fission fragments, and gamma rays.

55. **E** **T** **WS** **A New Heavy Element.** Scientists can induce nuclear transmutations through collisions, as in the case of an α particle striking a nitrogen nucleus to produce an oxygen nucleus (see Section 32.2). In the case of $^{48}_{20}$Ca nuclei striking $^{243}_{95}$Am nuclei, some of the collisions result in the formation of one large nucleus plus two free neutrons. You and your team have been assigned to identify the unknown nucleus and its decay products. **(a)** Assuming the number of protons and total number of nucleons are conserved in the collision, what is the unknown nucleus? **(b)** The unknown nucleus has been observed to undergo α decay. What is the daughter nucleus that results from this process? *Note:* both the parent and daughter nuclei have only recently been discovered. They were artificially synthesized and are not found in nature since they are extremely unstable and decay quickly.

56. **M** **T** **A Rough Measure of Exposure.** You and your team are designing a crude device to estimate radiation exposure. The device consists of a set of parallel plates with a large voltage across them. The circular plates have a radius of 3.50 cm and are separated by distance of 1.00 cm. The region between the plates is filled with dry air at standard temperature and pressure (STP: $T = 0°$, $P = 1$). A cubic meter of dry air at STP has a mass of 1.29 kg. When radiation ionizes an air molecule, the stripped electron is accelerated by the electric field between the plates and is registered as a current. When the device is located near a particularly strong source, it registers a current of 1.30×10^{-9} A. What is the exposure between the plates after 5.00 s (in roentgens)?

57. **M** **T** **A Hypothetical Fusion Reactor.** You and your team are tasked with evaluating the following deuterium-deuterium fusion reaction for use in a future fusion power reactor:

$$\underset{2.014\ 102\ u}{{}^{2}_{1}\text{H}} + \underset{2.014\ 102\ u}{{}^{2}_{1}\text{H}} \longrightarrow \underset{3.016\ 050\ u}{{}^{3}_{1}\text{H}} + \underset{1.007\ 825\ u}{{}^{1}_{1}\text{H}}$$

(a) What is the energy released in this reaction (in joules)? **(b)** How many reactions per second would be required to run a 5000 MW reactor? **(c)** What mass of deuterium (in kg) would be needed to run the reactor for a year with a power output of 5000 MW?

Powers of Ten and Scientific Notation

In science, very large and very small decimal numbers are conveniently expressed in terms of powers of ten, some of which are listed below:

$$10^3 = 10 \times 10 \times 10 = 1000 \qquad 10^{-3} = \frac{1}{10 \times 10 \times 10}$$
$$= 0.001$$

$$10^2 = 10 \times 10 = 100 \qquad 10^{-2} = \frac{1}{10 \times 10} = 0.01$$

$$10^1 = 10 \qquad 10^{-1} = \frac{1}{10} = 0.1$$

$$10^0 = 1$$

Using powers of ten, we can write the radius of the earth in the following way, for example:

$$\text{Earth radius} = 6\ 380\ 000 \text{ m} = 6.38 \times 10^6 \text{ m}$$

The factor of ten raised to the sixth power is ten multiplied by itself six times, or one million, so the earth's radius is 6.38 million meters. Alternatively, the factor of ten raised to the sixth power indicates that the decimal point in the term 6.38 is to be moved six places *to the right* to obtain the radius as a number without powers of ten.

For numbers less than one, negative powers of ten are used. For instance, the Bohr radius of the hydrogen atom is

$$\text{Bohr radius} = 0.000\ 000\ 000\ 0529 \text{ m} = 5.29 \times 10^{-11} \text{ m}$$

The factor of ten raised to the minus eleventh power indicates that the decimal point in the term 5.29 is to be moved eleven places *to the left* to obtain the radius as a number without powers of ten. Numbers expressed with the aid of powers of ten are said to be in ***scientific notation.***

Calculations that involve the multiplication and division of powers of ten are carried out as in the following examples:

$$(2.0 \times 10^6)(3.5 \times 10^3) = (2.0 \times 3.5) \times 10^{6+3} = 7.0 \times 10^9$$

$$\frac{9.0 \times 10^7}{2.0 \times 10^4} = \left(\frac{9.0}{2.0}\right) \times 10^7 \times 10^{-4}$$

$$= \left(\frac{9.0}{2.0}\right) \times 10^{7-4} = 4.5 \times 10^3$$

The general rules for such calculations are

$$\frac{1}{10^n} = 10^{-n} \qquad \text{(A-1)}$$

$$10^n \times 10^m = 10^{n+m} \qquad \text{(Exponents added)} \qquad \text{(A-2)}$$

$$\frac{10^n}{10^m} = 10^{n-m} \qquad \text{(Exponents subtracted)} \qquad \text{(A-3)}$$

where n and m are any positive or negative number.

Scientific notation is convenient because of the ease with which it can be used in calculations. Moreover, scientific notation provides a convenient way to express the significant figures in a number, as Appendix B discusses.

Significant Figures

The number of *significant figures* in a number is the number of digits whose values are known with certainty. For instance, a person's height is measured to be 1.78 m, with the measurement error being in the third decimal place. All three digits are known with certainty, so that the number contains three significant figures. If a zero is given as the last digit to the right of the decimal point, the zero is presumed to be significant. Thus, the number 1.780 m contains four significant figures. As another example, consider a distance of 1500 m. This number contains only two significant figures, the one and the five. The zeros immediately to the left of the unexpressed decimal point are not counted as significant figures. However, zeros located between significant figures are significant, so a distance of 1502 m contains four significant figures.

Scientific notation is particularly convenient from the point of view of significant figures. Suppose it is known that a certain distance is fifteen hundred meters, to four significant figures. Writing the number as 1500 m presents a problem because it implies that only two significant figures are known. In contrast, the scientific notation of 1.500×10^3 m has the advantage of indicating that the distance is known to four significant figures.

When two or more numbers are used in a calculation, the number of significant figures in the answer is limited by the number of significant figures in the original data. For instance, a rectangular garden with sides of 9.8 m and 17.1 m has an area of (9.8 m)(17.1 m). A calculator gives 167.58 m^2 for this product. However, one of the original lengths is known only to two significant figures, so the final answer is limited to only two

significant figures and should be rounded off to 170 m^2. In general, *when numbers are multiplied or divided, the number of significant figures in the final answer equals the smallest number of significant figures in any of the original factors.*

The number of significant figures in the answer to an addition or a subtraction is also limited by the original data. Consider the total distance along a biker's trail that consists of three segments with the distances shown as follows:

	2.5 km
	11 km
	5.26 km
Total	18.76 km

The distance of 11 km contains no significant figures to the right of the decimal point. Therefore, neither does the sum of the three distances, and the total distance should not be reported as 18.76 km. Instead, the answer is rounded off to 19 km. In general, *when numbers are added or subtracted, the last significant figure in the answer occurs in the last column (counting from left to right) containing a number that results from a combination of digits that are all significant.* In the answer of 18.76 km, the eight is the sum of 2 + 1 + 5, each digit being significant. However, the seven is the sum of 5 + 0 + 2, and the zero is not significant, since it comes from the 11-km distance, which contains no significant figures to the right of the decimal point.

Algebra

C.1 Proportions and Equations

Physics deals with physical variables and the relations between them. Typically, variables are represented by the letters of the English and Greek alphabets. Sometimes, the relation between variables is expressed as a proportion or inverse proportion. Other times, however, it is more convenient or necessary to express the relation by means of an equation, which is governed by the rules of algebra.

If two variables are **directly proportional** and one of them doubles, then the other variable also doubles. Similarly, if one variable is reduced to one-half its original value, then the other is also reduced to one-half its original value. In general, if x is directly proportional to y, then increasing or decreasing one variable by a given factor causes the other variable to change in the same way by the same factor. This kind of relation is expressed as $x \propto y$, where the symbol $\propto$ means "is proportional to."

Since the proportional variables x and y always increase and decrease by the same factor, the ratio of x to y must have a constant value, or $x/y = k$, where k is a constant, independent of the values for x and y. Consequently, a proportionality such as $x \propto y$ can also be expressed in the form of an equation: $x = ky$. The constant k is referred to as a **proportionality constant.**

If two variables are **inversely proportional** and one of them increases by a given factor, then the other decreases by the same factor. An inverse proportion is written as $x \propto 1/y$. This kind of proportionality is equivalent to the following equation: $xy = k$, where k is a proportionality constant, independent of x and y.

C.2 Solving Equations

Some of the variables in an equation typically have known values, and some do not. It is often necessary to solve the equation so that a variable whose value is unknown is expressed in terms of the known quantities. **In the process of solving an equation, it is permissible to manipulate the equation in any way, as long as a change made on one side of the equals sign is also made on the other side.** For example,

consider the equation $v = v_0 + at$. Suppose values for v, v_0, and a are available, and the value of t is required. To solve the equation for t, we begin by subtracting v_0 from *both* sides:

$$\begin{array}{rcl} v &=& v_0 + at \\ -v_0 &=& -v_0 \\ \hline v - v_0 &=& at \end{array}$$

Next, we divide both sides of $v - v_0 = at$ by the quantity a:

$$\frac{v - v_0}{a} = \frac{at}{a} = (1)t$$

On the right side, the a in the numerator divided by the a in the denominator equals one, so that

$$t = \frac{v - v_0}{a}$$

It is always possible to check the correctness of the algebraic manipulations performed in solving an equation by substituting the answer back into the original equation. In the previous example, we substitute the answer for t into $v = v_0 + at$:

$$v = v_0 + a\left(\frac{v - v_0}{a}\right) = v_0 + (v - v_0) = v$$

The result $v = v$ implies that our algebraic manipulations were done correctly.

Algebraic manipulations other than addition, subtraction, multiplication, and division may play a role in solving an equation. The same basic rule applies, however: Whatever is done to the left side of an equation must also be done to the right side. As another example, suppose it is necessary to express v_0 in terms of v, a, and x, where $v^2 = v_0^2 + 2ax$. By subtracting $2ax$ from both sides, we isolate v_0^2 on the right:

$$\begin{array}{rcl} v^2 &=& v_0^2 + 2ax \\ -2ax &=& -2ax \\ \hline v^2 - 2ax &=& v_0^2 \end{array}$$

To solve for v_0, we take the positive and negative square root of *both* sides of $v^2 - 2ax = v_0^2$:

$$v_0 = \pm\sqrt{v^2 - 2ax}$$

C.3 Simultaneous Equations

When more than one variable in a single equation is unknown, additional equations are needed if solutions are to be found for all of the unknown quantities. Thus, the equation $3x + 2y = 7$ cannot be solved by itself to give unique values for both x and y. However, if x and y also (i.e., simultaneously) obey the equation $x - 3y = 6$, then both unknowns can be found.

There are a number of methods by which such simultaneous equations can be solved. One method is to solve one equation for x in terms of y and substitute the result into the other equation to obtain an expression containing only the single unknown variable y. The equation $x - 3y = 6$, for instance, can be solved for x by adding $3y$ to each side, with the result that $x = 6 + 3y$. The substitution of this expression for x into the equation $3x + 2y = 7$ is shown below:

$$3x + 2y = 7$$
$$3(6 + 3y) + 2y = 7$$
$$18 + 9y + 2y = 7$$

We find, then, that $18 + 11y = 7$, a result that can be solved for y:

$$\begin{array}{r} 18 + 11y = 7 \\ -18 -18 \\ \hline 11y = -11 \end{array}$$

Dividing both sides of this result by 11 shows that $y = -1$. The value of $y = -1$ can be substituted in either of the original equations to obtain a value for x:

$$\begin{array}{rcl} x - 3y & = & 6 \\ x - 3(-1) & = & 6 \\ x + 3 & = & 6 \\ -3 & & -3 \\ \hline x & = & 3 \end{array}$$

C.4 The Quadratic Formula

Equations occur in physics that include the square of a variable. Such equations are said to be *quadratic* in that variable, and often can be put into the following form:

$$ax^2 + bx + c = 0 \qquad \text{(C-1)}$$

where a, b, and c are constants independent of x. This equation can be solved to give the **quadratic formula,** which is

$$x = \frac{-b \pm \sqrt{b^2 - 4ac}}{2a} \qquad \text{(C-2)}$$

The $\pm$ in the quadratic formula indicates that there are two solutions. For instance, if $2x^2 - 5x + 3 = 0$, then $a = 2$, $b = -5$, and $c = 3$. The quadratic formula gives the two solutions as follows:

Solution 1:
Plus sign

$$x = \frac{-b + \sqrt{b^2 - 4ac}}{2a}$$

$$= \frac{-(-5) + \sqrt{(-5)^2 - 4(2)(3)}}{2(2)}$$

$$= \frac{+5 + \sqrt{1}}{4} = \frac{3}{2}$$

Solution 2:
Minus sign

$$x = \frac{-b - \sqrt{b^2 - 4ac}}{2a}$$

$$= \frac{-(-5) - \sqrt{(-5)^2 - 4(2)(3)}}{2(2)}$$

$$= \frac{+5 - \sqrt{1}}{4} = 1$$

Exponents and Logarithms

Appendix A discusses powers of ten, such as 10^3, which means ten multiplied by itself three times, or $10 \times 10 \times 10$. The three is referred to as an **exponent.** The use of exponents extends beyond powers of ten. In general, the term y^n means the factor y is multiplied by itself n times. For example, y^2, or y squared, is familiar and means $y \times y$. Similarly, y^5 means $y \times y \times y \times y \times y$.

The rules that govern algebraic manipulations of exponents are the same as those given in Appendix A (see Equations A-1, A-2, and A-3) for powers of ten:

$$\frac{1}{y^n} = y^{-n} \qquad \text{(D-1)}$$

$$y^n y^m = y^{n+m} \quad \text{(Exponents added)} \qquad \text{(D-2)}$$

$$\frac{y^n}{y^m} = y^{n-m} \quad \text{(Exponents subtracted)} \qquad \text{(D-3)}$$

To the three rules above we add two more that are useful. One of these is

$$y^n z^n = (yz)^n \qquad \text{(D-4)}$$

The following example helps to clarify the reasoning behind this rule:

$$3^2 5^2 = (3 \times 3)(5 \times 5) = (3 \times 5)(3 \times 5) = (3 \times 5)^2$$

The other additional rule is

$$(y^n)^m = y^{nm} \quad \text{(Exponents multiplied)} \qquad \text{(D-5)}$$

To see why this rule applies, consider the following example:

$$(5^2)^3 = (5^2)(5^2)(5^2) = 5^{2+2+2} = 5^{2 \times 3}$$

Roots, such as a square root or a cube root, can be represented with fractional exponents. For instance,

$$\sqrt{y} = y^{1/2} \qquad \text{and} \qquad \sqrt[3]{y} = y^{1/3}$$

In general, the nth root of y is given by

$$\sqrt[n]{y} = y^{1/n} \qquad \text{(D-6)}$$

The rationale for Equation D-6 can be explained using the fact that $(y^n)^m = y^{nm}$. For instance, the fifth root of y is the number that, when multiplied by itself five times, gives back y. As shown below, the term $y^{1/5}$ satisfies this definition:

$$(y^{1/5})(y^{1/5})(y^{1/5})(y^{1/5})(y^{1/5}) = (y^{1/5})^5 = (y^{1/5})^{\times 5} = y$$

Logarithms are closely related to exponents. To see the connection between the two, note that it is possible to express any number y as another number B raised to the exponent x. In other words,

$$y = B^x \qquad \text{(D-7)}$$

The exponent x is called the **logarithm** of the number y. The number B is called the **base number.** One of two choices for the base number is usually used. If $B = 10$, the logarithm is known as the *common logarithm,* for which the notation "log" applies:

Common logarithm $\qquad y = 10^x \quad \text{or} \quad x = \log y \qquad$ **(D-8)**

If $B = e = 2.718 \ldots$, the logarithm is referred to as the *natural logarithm,* and the notation "ln" is used:

Natural logarithm $\qquad y = e^z \quad \text{or} \quad z = \ln y \qquad$ **(D-9)**

The two kinds of logarithms are related by

$$\ln y = 2.3026 \log y \qquad \text{(D-10)}$$

Both kinds of logarithms are often given on calculators.

The logarithm of the product or quotient of two numbers A and C can be obtained from the logarithms of the individual numbers according to the rules below. These rules are illustrated here for natural logarithms, but they are the same for any kind of logarithm.

$$\ln(AC) = \ln A + \ln C \qquad \text{(D-11)}$$

$$\ln\left(\frac{A}{C}\right) = \ln A - \ln C \qquad \text{(D-12)}$$

Thus, the logarithm of the product of two numbers is the sum of the individual logarithms, and the logarithm of the quotient of two numbers is the difference between the individual logarithms. Another useful rule concerns the logarithm of a number A raised to an exponent n:

$$\ln A^n = n \ln A \qquad \text{(D-13)}$$

Rules D-11, D-12, and D-13 can be derived from the definition of the logarithm and the rules governing exponents.

Geometry and Trigonometry

E.1 | Geometry

Angles

Two angles are equal if

1. They are vertical angles (see **Figure E1**).
2. Their sides are parallel (see **Figure E2**).

FIGURE E1   **FIGURE E2**

3. Their sides are mutually perpendicular (see **Figure E3**).

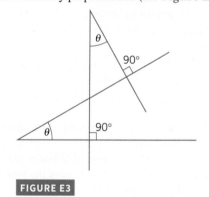

FIGURE E3

Triangles

1. The **sum of the angles** of any triangle is 180° (see **Figure E4**).

$$\alpha + \beta + \gamma = 180°$$

FIGURE E4

2. A **right triangle** has one angle that is 90°.
3. An **isosceles triangle** has two sides that are equal.
4. An **equilateral triangle** has three sides that are equal. Each angle of an equilateral triangle is 60°.

5. Two triangles are **similar** if two of their angles are equal (see **Figure E5**). The corresponding sides of similar triangles are proportional to each other:

$$\frac{a_1}{a_2} = \frac{b_1}{b_2} = \frac{c_1}{c_2}$$

FIGURE E5

6. Two similar triangles are **congruent** if they can be placed on top of one another to make an exact fit.

Circumferences, Areas, and Volumes of Some Common Shapes

1. Triangle of base b and altitude h (see **Figure E6**):

$$\text{Area} = \frac{1}{2}bh$$

FIGURE E6

2. Circle of radius r:

$$\text{Circumference} = 2\pi r$$

$$\text{Area} = \pi r^2$$

3. Sphere of radius r:

$$\text{Surface area} = 4\pi r^2$$

$$\text{Volume} = \frac{4}{3}\pi r^3$$

4. Right circular cylinder of radius r and height h (see **Figure E7**):

$$\text{Surface area} = 2\pi r^2 + 2\pi rh$$
$$\text{Volume} = \pi r^2 h$$

FIGURE E7

E.2 Trigonometry

Basic Trigonometric Functions

1. For a right triangle, the sine, cosine, and tangent of an angle θ are defined as follows (see **Figure E8**):

$$\sin \theta = \frac{\text{Side opposite } \theta}{\text{Hypotenuse}} = \frac{h_o}{h}$$

$$\cos \theta = \frac{\text{Side adjacent } \theta}{\text{Hypotenuse}} = \frac{h_a}{h}$$

$$\tan \theta = \frac{\text{Side opposite } \theta}{\text{Side adjacent to } \theta} = \frac{h_o}{h_a}$$

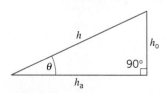

FIGURE E8

2. The secant ($\sec \theta$), cosecant ($\csc \theta$), and cotangent ($\cot \theta$) of an angle θ are defined as follows:

$$\sec \theta = \frac{1}{\cos \theta} \qquad \csc \theta = \frac{1}{\sin \theta} \qquad \cot \theta = \frac{1}{\tan \theta}$$

Triangles and Trigonometry

1. The **Pythagorean theorem** states that the square of the hypotenuse of a right triangle is equal to the sum of the squares of the other two sides (see **Figure E8**):

$$h^2 = h_o{}^2 + h_a{}^2$$

2. The **law of cosines** and the **law of sines** apply to any triangle, not just a right triangle, and they relate the angles and the lengths of the sides (see **Figure E9**):

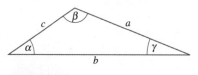

FIGURE E9

Law of cosines	$c^2 = a^2 + b^2 - 2ab \cos \gamma$
Law of sines	$\dfrac{a}{\sin \alpha} = \dfrac{b}{\sin \beta} = \dfrac{c}{\sin \gamma}$

Other Trigonometric Identities

1. $\sin(-\theta) = -\sin \theta$

2. $\cos(-\theta) = \cos \theta$

3. $\tan(-\theta) = -\tan \theta$

4. $(\sin \theta)/(\cos \theta) = \tan \theta$

5. $\sin^2 \theta + \cos^2 \theta = 1$

6. $\sin(\alpha \pm \beta) = \sin \alpha \cos \beta \pm \cos \alpha \sin \beta$

   If $\alpha = 90°$, $\sin(90° \pm \beta) = \cos \beta$

   If $\alpha = \beta$, $\sin 2\beta = 2 \sin \beta \cos \beta$

7. $\cos(\alpha \pm \beta) = \cos \alpha \cos \beta \mp \sin \alpha \sin \beta$

   If $\alpha = 90°$, $\cos(90° \pm \beta) = \mp\sin \beta$

   If $\alpha = \beta$, $\cos 2\beta = \cos^2 \beta - \sin^2 \beta = 1 - 2 \sin^2 \beta$

Selected Isotopes[a]

Atomic No. Z	Element	Symbol	Atomic Mass No. A	Atomic Mass (u)	% Abundance, or Decay Mode if Radioactive	Half-Life (if Radioactive)
0	(Neutron)	n	1	1.008 665	β^-	10.37 min
1	Hydrogen	H	1	1.007 825	99.985	
	Deuterium	D	2	2.014 102	0.015	
	Tritium	T	3	3.016 050	β^-	12.33 yr
2	Helium	He	3	3.016 030	0.000 138	
			4	4.002 603	≈ 100	
3	Lithium	Li	6	6.015 121	7.5	
			7	7.016 003	92.5	
4	Beryllium	Be	7	7.016 928	EC, γ	53.29 days
			9	9.012 182	100	
5	Boron	B	10	10.012 937	19.9	
			11	11.009 305	80.1	
6	Carbon	C	11	11.011 432	β^+, EC	20.39 min
			12	12.000 000	98.90	
			13	13.003 355	1.10	
			14	14.003 241	β^-	5730 yr
7	Nitrogen	N	13	13.005 738	β^+, EC	9.965 min
			14	14.003 074	99.634	
			15	15.000 108	0.366	
8	Oxygen	O	15	15.003 065	β^+, EC	122.2 s
			16	15.994 915	99.762	
			18	17.999 160	0.200	
9	Fluorine	F	18	18.000 937	EC, β^+	1.8295 h
			19	18.998 403	100	
10	Neon	Ne	20	19.992 435	90.51	
			22	21.991 383	9.22	
11	Sodium	Na	22	21.994 434	β^+, EC, γ	2.602 yr
			23	22.989 767	100	
			24	23.990 961	β^-, γ	14.659 h
12	Magnesium	Mg	24	23.985 042	78.99	
13	Aluminum	Al	27	26.981 539	100	
14	Silicon	Si	28	27.976 927	92.23	
			31	30.975 362	β^-, γ	2.622 h
15	Phosphorus	P	31	30.973 762	100	
			32	31.973 907	β^-	14.282 days
16	Sulfur	S	32	31.972 070	95.02	
			35	34.969 031	β^-	87.51 days
17	Chlorine	Cl	35	34.968 852	75.77	
			37	36.965 903	24.23	
18	Argon	Ar	40	39.962 384	99.600	
19	Potassium	K	39	38.963 707	93.2581	
			40	39.963 999	β^-, EC, γ	1.277×10^9 yr

[a]Data for atomic masses are taken from *Handbook of Chemistry and Physics,* 66th ed., CRC Press, Boca Raton, FL. The masses are those for the neutral atom, including the Z electrons. Data for percent abundance, decay mode, and half-life are taken from E. Browne and R. Firestone, *Table of Radioactive Isotopes,* V. Shirley, Ed., Wiley, New York, 1986. α = alpha particle emission, β^- = negative beta emission, β^+ = positron emission, γ = γ-ray emission, EC = electron capture.

Appendix F Selected Isotopes (*continued*)

Atomic No. Z	Element	Symbol	Atomic Mass No. A	Atomic Mass (u)	% Abundance, or Decay Mode if Radioactive	Half-Life (if Radioactive)
20	Calcium	Ca	40	39.962 591	96.941	
21	Scandium	Sc	45	44.955 910	100	
22	Titanium	Ti	48	47.947 947	73.8	
23	Vanadium	V	51	50.943 962	99.750	
24	Chromium	Cr	52	51.940 509	83.789	
25	Manganese	Mn	55	54.938 047	100	
26	Iron	Fe	56	55.934 939	91.72	
27	Cobalt	Co	59	58.933 198	100	
			60	59.933 819	β^-, γ	5.271 yr
28	Nickel	Ni	58	57.935 346	68.27	
			60	59.930 788	26.10	
29	Copper	Cu	63	62.939 598	69.17	
			65	64.927 793	30.83	
30	Zinc	Zn	64	63.929 145	48.6	
			66	65.926 034	27.9	
31	Gallium	Ga	69	68.925 580	60.1	
32	Germanium	Ge	72	71.922 079	27.4	
			74	73.921 177	36.5	
33	Arsenic	As	75	74.921 594	100	
34	Selenium	Se	80	79.916 520	49.7	
35	Bromine	Br	79	78.918 336	50.69	
36	Krypton	Kr	84	83.911 507	57.0	
			89	88.917 640	β^-, γ	3.16 min
			92	91.926 270	β^-, γ	1.840 s
37	Rubidium	Rb	85	84.911 794	72.165	
38	Strontium	Sr	86	85.909 267	9.86	
			88	87.905 619	82.58	
			90	89.907 738	β^-	29.1 yr
			94	93.915 367	β^-, γ	1.235 s
39	Yttrium	Y	89	88.905 849	100	
40	Zirconium	Zr	90	89.904 703	51.45	
41	Niobium	Nb	93	92.906 377	100	
42	Molybdenum	Mo	98	97.905 406	24.13	
43	Technetium	Tc	98	97.907 215	β^-, γ	4.2×10^6 yr
44	Ruthenium	Ru	102	101.904 348	31.6	
45	Rhodium	Rh	103	102.905 500	100	
46	Palladium	Pd	106	105.903 478	27.33	
47	Silver	Ag	107	106.905 092	51.839	
			109	108.904 757	48.161	
48	Cadmium	Cd	114	113.903 357	28.73	
49	Indium	In	115	114.903 880	95.7; β^-	4.41×10^{14} yr
50	Tin	Sn	120	119.902 200	32.59	
51	Antimony	Sb	121	120.903 821	57.3	
52	Tellurium	Te	130	129.906 229	38.8; β^-	2.5×10^{21} yr
53	Iodine	I	127	126.904 473	100	
			131	130.906 114	β^-, γ	8.040 days

Appendix F Selected Isotopes (*continued*)

Atomic No. Z	Element	Symbol	Atomic Mass No. A	Atomic Mass (u)	% Abundance, or Decay Mode if Radioactive	Half-Life (if Radioactive)
54	Xenon	Xe	132	131.904 144	26.9	
			136	135.907 214	8.9	
			140	139.921 620	β^-, γ	13.6 s
55	Cesium	Cs	133	132.905 429	100	
			134	133.906 696	β^-, EC, γ	2.062 yr
56	Barium	Ba	137	136.905 812	11.23	
			138	137.905 232	71.70	
			141	140.914 363	β^-, γ	18.27 min
57	Lanthanum	La	139	138.906 346	99.91	
58	Cerium	Ce	140	139.905 433	88.48	
59	Praseodymium	Pr	141	140.907 647	100	
60	Neodymium	Nd	142	141.907 719	27.13	
61	Promethium	Pm	145	144.912 743	EC, α, γ	17.7 yr
62	Samarium	Sm	152	151.919 729	26.7	
63	Europium	Eu	153	152.921 225	52.2	
64	Gadolinium	Gd	158	157.924 099	24.84	
65	Terbium	Tb	159	158.925 342	100	
66	Dysprosium	Dy	164	163.929 171	28.2	
67	Holmium	Ho	165	164.930 319	100	
68	Erbium	Er	166	165.930 290	33.6	
69	Thulium	Tm	169	168.934 212	100	
70	Ytterbium	Yb	174	173.938 859	31.8	
71	Lutetium	Lu	175	174.940 770	97.41	
72	Hafnium	Hf	180	179.946 545	35.100	
73	Tantalum	Ta	181	180.947 992	99.988	
74	Tungsten (wolfram)	W	184	183.950 928	30.67	
75	Rhenium	Re	187	186.955 744	62.60; β^-	4.6×10^{10} yr
76	Osmium	Os	191	190.960 920	β^-, γ	15.4 days
			192	191.961 467	41.0	
77	Iridium	Ir	191	190.960 584	37.3	
			193	192.962 917	62.7	
78	Platinum	Pt	195	194.964 766	33.8	
79	Gold	Au	197	196.966 543	100	
			198	197.968 217	β^-, γ	2.6935 days
80	Mercury	Hg	202	201.970 617	29.80	
81	Thallium	Tl	205	204.974 401	70.476	
			208	207.981 988	β^-, γ	3.053 min
82	Lead	Pb	206	205.974 440	24.1	
			207	206.975 872	22.1	
			208	207.976 627	52.4	
			210	209.984 163	α, β^-, γ	22.3 yr
			211	210.988 735	β^-, γ	36.1 min
			212	211.991 871	β^-, γ	10.64 h
			214	213.999 798	β^-, γ	26.8 min
83	Bismuth	Bi	209	208.980 374	100	
			211	210.987 255	α, β^-, γ	2.14 min
			212	211.991 255	β^-, α, γ	1.0092 h

Appendix F Selected Isotopes (*continued*)

Atomic No. Z	Element	Symbol	Atomic Mass No. A	Atomic Mass (u)	% Abundance, or Decay Mode if Radioactive	Half-Life (if Radioactive)
84	Polonium	Po	210	209.982 848	α, γ	138.376 days
			212	211.988 842	α, γ	45.1 s
			214	213.995 176	α, γ	163.69 μs
			216	216.001 889	α, γ	150 ms
85	Astatine	At	218	218.008 684	α, β^-	1.6 s
86	Radon	Rn	220	220.011 368	α, γ	55.6 s
			222	222.017 570	α, γ	3.825 days
87	Francium	Fr	223	223.019 733	α, β^-, γ	21.8 min
88	Radium	Ra	224	224.020 186	α, γ	3.66 days
			226	226.025 402	α, γ	1.6×10^3 yr
			228	228.031 064	β^-, γ	5.75 yr
89	Actinium	Ac	227	227.027 750	α, β^-, γ	21.77 yr
			228	228.031 015	β^-, γ	6.13 h
90	Thorium	Th	228	228.028 715	α, γ	1.913 yr
			231	231.036 298	β^-, γ	1.0633 days
			232	232.038 054	100; α, γ	1.405×10^{10} yr
			234	234.043 593	β^-, γ	24.10 days
91	Protactinium	Pa	231	231.035 880	α, γ	3.276×10^4 yr
			234	234.043 303	β^-, γ	6.70 h
			237	237.051 140	β^-, γ	8.7 min
92	Uranium	U	232	232.037 130	α, γ	68.9 yr
			233	233.039 628	α, γ	1.592×10^5 yr
			235	235.043 924	0.7200; α, γ	7.037×10^8 yr
			236	236.045 562	α, γ	2.342×10^7 yr
			238	238.050 784	99.2745; α, γ	4.468×10^9 yr
			239	239.054 289	β^-, γ	23.47 min
93	Neptunium	Np	239	239.052 933	β^-, γ	2.355 days
94	Plutonium	Pu	239	239.052 157	α, γ	2.411×10^4 yr
			242	242.058 737	α, γ	3.763×10^5 yr
95	Americium	Am	243	243.061 375	α, γ	7.380×10^3 yr
96	Curium	Cm	245	245.065 483	α, γ	8.5×10^3 yr
97	Berkelium	Bk	247	247.070 300	α, γ	1.38×10^3 yr
98	Californium	Cf	249	249.074 844	α, γ	350.6 yr
99	Einsteinium	Es	254	254.088 019	α, γ, β^-	275.7 days
100	Fermium	Fm	253	253.085 173	EC, α, γ	3.00 days
101	Mendelevium	Md	255	255.091 081	EC, α	27 min
102	Nobelium	No	255	255.093 260	EC, α	3.1 min
103	Lawrencium	Lr	257	257.099 480	α, EC	646 ms
104	Rutherfordium	Rf	261	261.108 690	α	1.08 min
105	Dubnium	Db	262	262.113 760	α	34 s

Answers to Check Your Understanding

Chapter 1

CYU 1: (a) Yes. (b) No.

CYU 2: No.

CYU 3: a, b, c, f

CYU 4: No.

CYU 5: b, d

CYU 6: (a) 11 m (b) 5 m

CYU 7: No.

CYU 8: Yes.

CYU 9: (a) The magnitude of $\vec{B}$ is equal to the magnitude of $\vec{A}$.
(b) The direction of $\vec{B}$ is opposite to the direction of $\vec{A}$.

CYU 10: Vector $\vec{A}$ is perpendicular to vector $\vec{B}$.

CYU 11: Vector $\vec{A}$ points in the same direction as vector $\vec{B}$.

CYU 12: $\vec{A}$ and $\vec{D}$

CYU 13: (a) A_x is − and A_y is +
(b) B_x is + and B_y is −
(c) R_x is + and R_y is +

CYU 14: No.

CYU 15: Yes.

CYU 16: (a) $A_x = 0$ units and $A_y = +12$ units
(b) $A_x = -12$ units and $A_y = 0$ units
(c) $A_x = 0$ units and $A_y = -12$ units
(d) $A_x = +12$ units and $A_y = 0$ units

CYU 17: No.

CYU 18: a

Chapter 2

CYU 1: 0 m

CYU 2: scalar quantity

CYU 3: No.

CYU 4: a

CYU 5: average velocity = 2.7 m/s due east, average speed = 8.0 m/s

CYU 6: Yes.

CYU 7: No.

CYU 8: c

CYU 9: No.

CYU 10: No.

CYU 11: the rifle with the short barrel

CYU 12: 1.73 v

CYU 13: a

CYU 14: b

CYU 15: b

CYU 16: b

Chapter 3

CYU 1: b

CYU 2: c

CYU 3: a and c

CYU 4: (a) Yes; when the object is at its highest point.
(b) No.

CYU 5: No.

CYU 6: b

CYU 7: (a) when the ball is at its highest point in the trajectory
(b) at the initial and final positions of the motion

CYU 8: Yes.

CYU 9: Both bullets reach the ground at the same time.

CYU 10: (a) The displacement is greater for the stone thrown horizontally.
(b) The impact speed is greater for the stone thrown horizontally.
(c) The time of flight is the same for both stones.

CYU 11: No.

CYU 12: Ball A has the greater launch speed.

CYU 13: (a) +70 m/s (b) +30 m/s
(c) +40 m/s (d) −60 m/s

CYU 14: No.

CYU 15: The two times are the same.

CYU 16: (a) The range toward the front is the same as the range toward the rear.
(b) The range toward the front is greater than the range toward the rear.

CYU 17: swimmer A

Chapter 4

CYU 1: b

CYU 2: c

CYU 3: d

CYU 4: No, because two or more forces can cancel each other, leading to a net force of zero.

CYU 5: c

CYU 6: a and d

CYU 7: b

CYU 8: Yes, because the ratio of the two weights depends only on the masses of the objects, which are the same on the earth and on Mars.

CYU 9: a

CYU 10: d

CYU 11: No.

CYU 12: a

CYU 13: b

CYU 14: c

CYU 15: To pull, because the upward component of the pulling force reduces the normal force and, therefore, also reduces the force of kinetic friction acting on the sled.

CYU 16: 43°

CYU 17: c, a, b

CYU 18: a

CYU 19: a

CYU 20: b

CYU 21: d

CYU 22: No, because there must always be a vertical (upward) component of the tension force in the rope to balance the weight of the crate.

CYU 23: c

CYU 24: No, because the transfer described does not change the total mass being pulled by the engine.

CYU 25: a

Chapter 5

CYU 1: (a) The velocity is due south and the acceleration is due west.
(b) The velocity is due west and the acceleration is due north.

CYU 2: Yes, if you are going around a curve.

CYU 3: the person at the equator

CYU 4: a and b

CYU 5: *AB* or *DE*, *CD*, *BC*

CYU 6: (a) 4*r*
(b) 4*r*

CYU 7: No.

CYU 8: the same

CYU 9: edge of the turntable

CYU 10: car B

CYU 11: less than

CYU 12: (a) less than
(b) equal to

CYU 13: (a) Yes.
(b) Yes.
(c) Yes.
(d) Yes.

CYU 14: vertical

Chapter 6

CYU 1: b

CYU 2: d

CYU 3: d

CYU 4: a

CYU 5: No.

CYU 6: c

CYU 7: false

CYU 8: c

CYU 9: a

CYU 10: a, b, and c

CYU 11: b

CYU 12: d

CYU 13: b and d

CYU 14: e

CYU 15: c

CYU 16: a

CYU 17: No.

Chapter 7

CYU 1: No.

CYU 2: The total linear momentum is approximately zero because of the random directions and random speeds of the moving people.

CYU 3: (a) Yes.
(b) No.

CYU 4: (a) No.
(b) Yes.

CYU 5: b

CYU 6: (a) No.
(b) The impulse of the thrust is equal in magnitude and opposite in direction to the impulse of the force due to air resistance.

CYU 7: (a) No.
(b) No.

CYU 8: equal to

CYU 9: Yes.

CYU 10: a

CYU 11: decrease

CYU 12: (a) No.
(b) decrease

CYU 13: b, c, d

CYU 14: the cannonball

CYU 15: d

CYU 16: No. It is the total kinetic energy of the *system* that is the same before and after the collision.

CYU 17: c

CYU 18: nearer the heavier end

CYU 19: (a) zero
(b) Yes, opposite to the motion of the sunbather.

CYU 20: a

Chapter 8

CYU 1: Both axes lie in the plane of the paper. One passes through point *A* and is parallel to the line *BC*. The other passes through point *A* and the midpoint of the line *BC*.

CYU 2: B, C, A

CYU 3: No. The instantaneous angular speed of each blade is the same, but the blades are rotating in opposite directions.

CYU 4: c

CYU 5: 1.0 rev/s

CYU 6: b

CYU 7: Case A

CYU 8: a

CYU 9: at the north pole or at the south pole

CYU 10: c

CYU 11: 0.30 m

CYU 12: d

CYU 13: c

CYU 14: b

CYU 15: 8.0 m/s^2

CYU 16: a

CYU 17: Among the many possible answers are the motions of a Frisbee through the air, the earth in its orbit, a twirling baton that has been thrown into the air, the blades on a moving lawn mower cutting the grass, and an ice skater performing a quadruple jump.

Chapter 9

CYU 1: 0°, 45°, 90°

CYU 2: greater torque

CYU 3: (a) Yes, if the lever arm is very small.
(b) Yes, if the lever arm is very large.

CYU 4: the box at the far right

CYU 5: (a) C
(b) A
(c) B

CYU 6: Additional forces are necessary.

CYU 7: a

CYU 8: b

CYU 9: Bob

CYU 10: A, B, C

CYU 11: axis B

CYU 12: (a) remains the same
(b) remains the same

CYU 13: (a) remains the same
(b) increases

CYU 14: (a) Both have the same translational speed.
(b) Both have the same translational speed.

CYU 15: axis B

CYU 16: solid sphere, solid cylinder, spherical shell, hoop

CYU 17: (a) decreases
(b) remains the same

CYU 18: decrease

CYU 19: greater than

CYU 20: No.

Chapter 10

CYU 1: No, because the force of gravity acting on the ball is constant, unlike the restoring force of simple harmonic motion.

CYU 2: Both boxes experience the same net force due to the springs.

CYU 3: 180 N/m

CYU 4: The spring stretches more when attached to the wall.

CYU 5: object II

CYU 6: at the position $x = 0$ m

CYU 7: The particle can cover the greater distance in the same time because at larger amplitudes the maximum speed is greater.

CYU 8: The same amount of energy is stored in both cases, since the elastic potential energy is proportional to the square of the displacement x.

CYU 9: b, c, a

CYU 10: The amplitude is unchanged. The frequency and maximum speed each decrease by a factor of $\sqrt{2}$.

CYU 11: a

CYU 12: the simple-pendulum clock, because its period depends on the acceleration due to gravity

CYU 13: Use a shoe and the shoe laces to make a simple pendulum whose period is related to the magnitude g of the acceleration due to gravity (see Equations 10.5 and 10.16). Measure the period of your pendulum and calculate g.

CYU 14: Yes, because for small angles the period of each person's motion is the same.

CYU 15: Yes, because the frequency depends on the mass of the car and its occupants (see Equations 10.6 and 10.11).

CYU 16: $v = \frac{d}{2\pi}\sqrt{\frac{k}{m}}$

CYU 17: b

CYU 18: The rod with the square cross section is longer.

CYU 19: No, because the value of B given in Table 10.3 applies to solid aluminum, not to a can that is mostly empty space.

CYU 20: No, because pressure involves a force that acts *perpendicular* to an area. In Equation 10.18 for shear deformation, the force acts parallel, not perpendicular, to the area A (see Figure 10.29).

CYU 21: Face B experiences the largest stress, and face C experiences the smallest stress.

Chapter 11

CYU 1: (a) outward
(b) inward

CYU 2: b

CYU 3: a

CYU 4: (a) increase
(b) decrease
(c) remain constant

CYU 5: A noticeable amount of water will remain in the tank.

CYU 6: b

CYU 7: Yes; see Equation 11.4, in which P_2 is the pressure at his wrist and P_1 is the pressure above the water.

CYU 8: c

CYU 9: Both beams experience the same buoyant force.

CYU 10: (a) The readings are the same.
(b) The final reading is greater than the initial reading.

CYU 11: No. You float because the weight of the water you displace equals your weight. Each weight is proportional to g, so its value makes no difference.

CYU 12: No. F_B depends only on the weight of the water she displaces, which doesn't change.

CYU 13: b

CYU 14: No.

CYU 15: d

CYU 16: c

CYU 17: e

CYU 18: c

CYU 19: c

CYU 20: b

CYU 21: a

CYU 22: c

Chapter 12

CYU 1: 178 °X

CYU 2: (a) No. (b) Yes. (c) No.

CYU 3: It decreases (see Equations 10.5 and 10.16).

CYU 4: With equal values for α, concrete and steel expand (contract) by the same amount as the temperature increases (decreases), thus minimizing problems with thermal stress.

CYU 5: The bottom is bowed outward, because it acts like a bimetallic strip.

CYU 6: b and d

CYU 7: cooled

CYU 8: No. When the temperature changes, the change in volume of the cavity within the glass would exactly compensate for the change in volume of the mercury, which would never rise or fall in the capillary tube of the thermometer.

CYU 9: Less than. The buoyant force is equal to the weight of the displaced water (see Section 11.6, Archimedes' principle), which is proportional to the water's density. Here, warmer water has a smaller density than cooler water does (see Figure 12.20).

CYU 10: a and b

CYU 11: the object with the smaller mass

CYU 12: c, b, d, a

CYU 13: Because heat is released when the water freezes at 0 °C (consistent with the latent heat of fusion of water), and this heat warms the blossoms.

CYU 14: c, a, b

CYU 15: No, because at sea level water boils at a higher temperature and the stove may not generate enough heat.

CYU 16: Because water in an open pot boils at 100 °C, thus preventing the temperature from rising further, whereas under the elevated pressure in the autoclave water has a boiling point above 100 °C.

CYU 17: Boiling water has a vapor pressure of one atmosphere, and the cool water in the sealed jar has a lower vapor pressure. The excess external pressure creates a net force pushing on the lid, making it hard to unscrew.

CYU 18: Under pressure in the sealed bottle, the soda has a freezing point lower than normal (see Figure 12.35b). The outside temperature is not cold enough to freeze it. When the bottle is opened, the pressure on the liquid decreases to one atmosphere, and the freezing point rises to its normal value. The liquid is now cold enough to freeze.

CYU 19: As the water vapor is removed, more forms in an attempt to reestablish equilibrium between liquid and vapor. When the pumping is rapid, the required latent heat is supplied mostly by the remaining liquid, which cools and eventually freezes.

CYU 20: 100%

CYU 21: Yes. The dew points on the two nights could be different, Tuesday's being higher than Monday's due to a greater partial pressure of water vapor in the air on Tuesday than on Monday.

CYU 22: The air above the swimming pool probably has a greater partial pressure of water vapor (due to inefficient humidity control) and, therefore, a higher dew point than that in the other room. Evidently, the temperature at the inner window-surfaces is below the dew point of the room with the swimming pool but above the dew point in the other room.

Chapter 13

CYU 1: a

CYU 2: b

CYU 3: the house with the snow on the roof

CYU 4: b

CYU 5: c

CYU 6: c

CYU 7: hollow, air-filled strands

CYU 8: c

CYU 9: b

CYU 10: forced convection

CYU 11: strip B

CYU 12: a

CYU 13: d

CYU 14: b

CYU 15: e

Chapter 14

CYU 1: Both have the same number of molecules, but oxygen has the greater mass.

CYU 2: In general, the number of molecules would be different. But they could be the same, if the molecular masses of the two types of molecules happen to be the same.

CYU 3: 66.4%

CYU 4: The ideal gas law gives the pressure as $P = nRT/V$, where T and V are constant. The fan reduces n in the house and increases it in the attic, so pressure decreases in the house and increases in the attic. The fan has a harder job pushing air out against the higher attic pressure.

CYU 5: The ideal gas law gives the gas pressure as $P = nRT/V$, where V and n are constant. As T increases, the pressure increases and could cause the can to burst.

CYU 6: The ideal gas law gives the gas pressure as $P = nRT/V$, where V and n are constant. As T increases, the pressure increases.

CYU 7: The ideal gas law gives the gas pressure as $P = nRT/V$, where T and n are constant. As V decreases due to the incoming tide, the pressure increases, and your ears pop inward, as if you were climbing down a mountain.

CYU 8: The ideal gas law gives the gas volume as $V = nRT/P$, where T and n are constant. As the pressure P decreases during the balloon's ascent, the volume increases. The balloon would overinflate if not underinflated to start with.

CYU 9: Boyle's law gives the final pressure in the bottle after the cork is pressed in: $P_f = P_i(V_i/V_f)$, where V_i/V_f is the volume of air above the wine before the cork is pressed in divided by the volume after the cork is pressed in. This ratio is much larger for the full bottle than for the half-full bottle, creating a pressure large enough to push the cork out.

CYU 10: Xenon has the greatest and argon the smallest temperature.

CYU 11: less than, which follows directly from the impulse–momentum theorem

CYU 12: No. The average kinetic energy is proportional to the Kelvin, not the Celsius, temperature.

CYU 13: It remains unchanged.

CYU 14: argon

CYU 15: $v_{\text{rms, new}}/v_{\text{rms, initial}} = 0.707$

CYU 16: L must be small and there must be many alveoli so that the total effective area A is large.

CYU 17: c, a, b

Chapter 15

CYU 1: d

CYU 2: c

CYU 3: b

CYU 4: $A \rightarrow B$: $Q = +$ and $W = +$
$B \rightarrow C$: $\Delta U = +$ and $W = 0$

CYU 5: a

CYU 6: b

CYU 7: a

CYU 8: c

CYU 9: c

CYU 10: c

CYU 11: b

CYU 12: No, because Carnot's principle only states that a reversible engine operating between two temperatures is more efficient than an irreversible engine operating between the *same temperatures.*

CYU 13: d

CYU 14: d

CYU 15: b

CYU 16: c

CYU 17: a

CYU 18: c

CYU 19: b

CYU 20: a

CYU 21: d

CYU 22: c and d

CYU 23: the popcorn that results from the kernels; a salad after it has been tossed; a messy apartment

CYU 24: b

CYU 25: c

Chapter 16

CYU 1: c

CYU 2: No. The coil moves back and forth in simple harmonic motion.

CYU 3: The wavelength increases.

CYU 4: The person pulling on string B should pull harder to increase the tension in the string.

CYU 5: In Equation 16.2, the speed would be infinitely large if m were zero, so it would take no time at all.

CYU 6: decrease

CYU 7: No, because the particles exhibit simple harmonic motion, in which the acceleration is not always zero.

CYU 8: increase

CYU 9: a

CYU 10: No, because each particle executes simple harmonic motion as the wave passes by.

CYU 11: hot day

CYU 12: CO and N_2

CYU 13: increase

CYU 14: Large outer ears intercept and direct more sound power into the auditory system than smaller ones do.

CYU 15: No, because not all points on the surface are at the same distance from the source.

CYU 16: No, because it is the intensities I_1 and I_2 that add to give a total intensity I_{total}. The intensity levels β_1 and β_2 do not add to give a total intensity level β_{total}.

CYU 17: (a) 1/4
  (b) 2

CYU 18: (a) f_o is smaller than f_s, and f_o decreases during the fall.
  (b) f_o is greater than f_s, and f_o increases during the fall.

CYU 19: No, because the observed frequency is less than the source frequency, so the car is moving away from him.

CYU 20: (a) greater in air
  (b) greater under water

CYU 21: No, because there is no relative motion of the cars.

CYU 22: (a) minus sign in both places
  (b) the truck driver

Chapter 17

CYU 1: (a) −3 cm
  (b) −2 cm

CYU 2: No, because if the two sound waves have the same amplitude and frequency, they might cancel in a way analogous to that illustrated in Figure 17.2*b* and no sound will be heard.

CYU 3: b

CYU 4: c

CYU 5: a

CYU 6: d

CYU 7: a

CYU 8: c

CYU 9: b

CYU 10: d

CYU 11: (a) 4
  (b) 3
  (c) node
  (d) 110 Hz

CYU 12: b

CYU 13: d

CYU 14: b

CYU 15: (a) antinode
  (b) node
  (c) $\frac{1}{4}\lambda$
  (d) lowered

CYU 16: b

CYU 17: c ($\frac{1}{4}\lambda$ is the distance between an antinode and an adjacent node.)

CYU 18: a

Chapter 18

CYU 1: c

CYU 2: $+3.2 \times 10^{-13}$ C on object A and -3.2×10^{-13} C on object B

CYU 3: $+1.6 \times 10^{-13}$ C on object A and -3.2×10^{-13} C on object B

CYU 4: b and e

CYU 5: Yes, because the charge on the balloon will induce a slight charge of opposite polarity in the surface of the ceiling, analogous to that in Figure 18.8.

CYU 6: a

CYU 7: b

CYU 8: C, A, B

CYU 9: the electron, because, being less massive, it has the greater acceleration

CYU 10: No, because the force of the spring changes direction when the spring is stretched compared to when it is compressed, while the electrostatic force does not have this characteristic.

CYU 11: d

CYU 12: 0 N/C

CYU 13: (a) corner C
  (b) negative
  (c) greater

CYU 14: a

CYU 15: (a) No.
  (b) No.

CYU 16: For rod A, the field points perpendicularly away from the rod. For rod B, it points parallel to the rod and is directed from the positive toward the negative half.

CYU 17: (a) false
  (b) false
  (c) true
  (d) false
  (e) false

CYU 18: d

CYU 19: The flux does not change, as long as the charge remains within the Gaussian surface.

CYU 20: The same flux passes through each, since each encloses the same net charge.

CYU 21: (a) q_1 and q_2
  (b) q_1, q_2, and q_3

Chapter 19

CYU 1: (a) Yes.
  (b) No.
  (c) Yes.
  (d) Yes.

CYU 2: The work is the same in all three cases (see Equation 19.4).

CYU 3: The electron arrives at a plate first.

CYU 4: a

CYU 5: b

CYU 6: c

CYU 7: a

CYU 8: d

CYU 9: (a) remains the same
  (b) decreases

CYU 10: the electron

CYU 11: (a) +2.0 V
  (b) 0 V
  (c) +2.0 V

CYU 12: The electric field is zero.

CYU 13: b

CYU 14: a

CYU 15: c

CYU 16: (a) bottom of a valley
  (b) top of a mountain

CYU 17: (a) decreases
       (b) increases
       (c) increases
       (d) increases

CYU 18: (a) decreases
       (b) increases
       (c) remains the same
       (d) increases

Chapter 20

CYU 1: d

CYU 2: 0.50 A

CYU 3: b

CYU 4: a

CYU 5: b, d, and e

CYU 6: c (A value for the current is also needed.)

CYU 7: a

CYU 8: c

CYU 9: b and d

CYU 10: The 75-W bulb. See Equation 20.15c.

CYU 11: d

CYU 12: e

CYU 13: in parallel

CYU 14: b

CYU 15: c

CYU 16: a, b, d, and e

CYU 17: There are two ways. One is to form two groups of two parallel resistors and then connect the groups in series. The other is to form two groups of two series resistors and then connect the groups in parallel.

CYU 18: Junction rule: $I_1 + I_3 = I_2$
Loop rule, loop $ABCD$:
$3.0\,\text{V} + 7.0\,\text{V} + I_3R_3 = I_1R_1$
Loop rule, loop $BEFC$:
$5.0\,\text{V} = I_3R_3 + 7.0\,\text{V} + I_2R_2$

CYU 19: c

CYU 20: b

CYU 21: ohm × farad
$= (\text{volt/ampere})(\text{coulomb/volt})$
$= \text{coulomb/ampere}$
$= \text{coulomb/(coulomb/second)}$
$= \text{second}$

CYU 22: e

Chapter 21

CYU 1: c

CYU 2: d

CYU 3: (a) Yes.
       (b) No, because the particle could move either parallel or anti-parallel to the magnetic field.

CYU 4: b

CYU 5: d

CYU 6: particle 3

CYU 7: b

CYU 8: c

CYU 9: b

CYU 10: c

CYU 11: (a) The direction of the magnetic force reverses.
       (b) The direction of the magnetic force does not change.

CYU 12: B and D (a tie), A, C

CYU 13: a

CYU 14: (a) repelled
       (b) repelled

CYU 15: (a) attracted
       (b) repelled

CYU 16: c

CYU 17: a

CYU 18: Part a: There is a point to the right of both wires where the total magnetic field is zero. Part b: There is a point between the wires where the total magnetic field is zero. This point is closer to the wire carrying the current I_2.

CYU 19: A, D, C, B

CYU 20: d

CYU 21: No, because aluminum is a non-ferromagnetic material.

CYU 22: b

Chapter 22

CYU 1: No. With both the magnet and coil moving at the same velocity with respect to the earth, there is no relative motion between the magnet and the coil, which is needed for there to be an induced current in the coil.

CYU 2: d

CYU 3: a

CYU 4: b

CYU 5: c

CYU 6: b

CYU 7: A lightning bolt is a large electric current that changes in time and, thus, produces a magnetic field that also changes in time. When this changing field passes through a coil or loop of wire in an appliance, it can, via Faraday's law, create an induced emf, which can lead to an induced current.

CYU 8: c

CYU 9: a and d

CYU 10: a

CYU 11: b

CYU 12: Answer 1: downward and decreasing
Answer 2: upward and increasing

CYU 13: c

CYU 14: d

CYU 15: b

CYU 16: With the headlights off, the engine does not need to do the work of keeping the battery charged.

CYU 17: a

CYU 18: c

CYU 19: b

CYU 20: b and d

Chapter 23

CYU 1: The ratio decreases by a factor of 3.

CYU 2: (a) the circuit containing the inductor
       (b) the circuit containing the resistor

CYU 3: less than

CYU 4: decreases

CYU 5: decreases

CYU 6: d

CYU 7: (a) increases
       (b) increases

CYU 8: a

CYU 9: (a) remains the same
       (b) decreases

CYU 10: in phase (see Equation 23.8, in which $X_L = X_C$)

CYU 11: (a) Yes.
       (b) Yes.

CYU 12: a

CYU 13: (a) left to right
       (b) left to right

Chapter 24

CYU 1: d

CYU 2: a

CYU 3: because, according to Faraday's law of electromagnetic induction, the emf depends on how rapidly the magnetic field of the wave is changing and this is determined by the frequency of the wave

CYU 4: e

CYU 5: d

CYU 6: b

CYU 7: No. The same Doppler change results when the star moves away from the earth and when the earth moves away from the star. Only the relative motion between the star and the earth can be detected.

CYU 8: B, A, C

CYU 9: Yes.

CYU 10: The light intensity that is not transmitted is absorbed by the polarizer and the analyzer. The polarizer absorbs one-half of the incident intensity, and the analyzer absorbs four-tenths of the incident intensity.

CYU 11: unpolarized: c
horizontally polarized: b
vertically polarized: c

CYU 12: because the transmission axis of the Polaroid material is nearly horizontal, in the same direction as the polarized light reflected from the lake

Chapter 25

CYU 1: 55°

CYU 2: Yes.

CYU 3: Yes.

CYU 4: $f_{inside} = +0.30 \text{ m}, f_{outside} = -0.30 \text{ m}$

CYU 5: (a) concave
(b) The sodium unit and engine are located at the focal point of the mirror.

CYU 6: Open the surface up to produce a more gently curving shape.

CYU 7: No.

CYU 8: (a) upright
(b) upside down

CYU 9: (a) Yes, provided the object distance is greater than the focal length of the mirror.
(b) It is not possible for a convex mirror to project an image directly onto a screen.

CYU 10: (a) No.
(b) No.

CYU 11: (a) The magnitude of the image distance becomes larger.
(b) The magnitude of the image height becomes larger.

CYU 12: (a) No. You can see yourself anywhere on the principal axis.
(b) You cannot see yourself when you are between the center of curvature and the focal point of the mirror because your image is behind you.

CYU 13: A, D, and E

CYU 14: The image will never be located beyond the focal point (behind the mirror).

Chapter 26

CYU 1: slab B

CYU 2: liquid B

CYU 3: Yes. To see why, apply Snell's law at the air–water interface and at the water–glass interface.

CYU 4: the one filled with water

CYU 5: liquid A

CYU 6: c

CYU 7: b

CYU 8: a

CYU 9: c

CYU 10: The critical angle for a water–air interface is 48.8° (see Equation 26.4). Any light emitted at an angle greater than 48.8° with respect to the vertical is incident on the surface at an angle exceeding the critical angle. It is totally internally reflected and doesn't exit the water.

CYU 11: No. To see why, apply Snell's law at both surfaces of the glass slab and use Equation 26.4.

CYU 12: c (They are most effective when the angle of incidence is the Brewster angle and the reflected light is 100% polarized.)

CYU 13: liquid A

CYU 14: a (Since $n = 1.520$ for red light and $n = 1.538$ for violet-colored light, the critical angle for total internal reflection is greater for red than for violet-colored light.)

CYU 15: b

CYU 16: Yes.

CYU 17: a

CYU 18: the lens

CYU 19: converging lens, $d_o = \frac{1}{2}f$

CYU 20: d

CYU 21: the glasses of the farsighted person, since they use converging lenses

CYU 22: 13 cm

CYU 23: Light normally passes from air ($n = 1.00$) into the cornea ($n = 1.38$), at which time most of the eye's refraction of the light occurs. If water ($n = 1.33$) replaces air, the similarity of the index of refraction of water to that of the cornea reduces the eye's normal refraction and causes blurred vision. Goggles preserve the air–cornea boundary.

CYU 24: b

CYU 25: hawk, kestrel, eagle

CYU 26: a

CYU 27: 0.042 rad

CYU 28: b

CYU 29: the longer telescope

CYU 30: microscope

CYU 31: c, d, e, f

CYU 32: because chromatic aberration is related to the refraction of light and not to the reflection of light

Chapter 27

CYU 1: (a) constructive
(b) destructive
(c) destructive

CYU 2: c

CYU 3: Yes.

CYU 4: a

CYU 5: (a) d_1 and λ_2
(b) d_2 and λ_1

CYU 6: (a) The pattern would be the same.
(b) The positions of the light and dark fringes would be interchanged.

CYU 7: No, because θ in Equations 27.1 and 27.2 approaches 90° as λ becomes larger and larger.

CYU 8: b

CYU 9: c

CYU 10: enhances

CYU 11: (a) A and C
(b) B

CYU 12: c

CYU 13: b

CYU 14: (a) broadens
(b) contracts

CYU 15: a

CYU 16: c

CYU 17: a, c, b

CYU 18: Yes.

CYU 19: small f-number setting

CYU 20: (a) the maximum that is closer to the central maximum
(b) away from the central maximum

CYU 21: The distance between the bright fringes would decrease.

Chapter 28

CYU 1: d

CYU 2: a, b

CYU 3: C, B, A

CYU 4: No, because the term v^2/c^2 in Equations 28.1 and 28.2 would then be zero.

CYU 5: c

CYU 6: c, d

CYU 7: No, because the two diagonals are perpendicular, so that diagonal AC is contracted, whereas diagonal BD is not contracted.

CYU 8: greatest mass: c, smallest mass: b

CYU 9: a, because then they have more electric potential energy (see Example 8 in Chapter 19)

CYU 10: b, because the fully charged capacitor stores electric potential energy (see Section 19.5)

CYU 11: a. The work is the change in kinetic energy, which is proportional to the mass (see the work–energy theorem in Section 6.2). The electron has the smaller mass.

CYU 12: c

Chapter 29

CYU 1: No.

CYU 2: (a) red
(b) violet

CYU 3: No.

CYU 4: b

CYU 5: a

CYU 6: (a) increases
(b) increases
(c) remains the same
(d) remains the same

CYU 7: less than

CYU 8: c

CYU 9: a

CYU 10: No. Photon collisions would cause spinning in a direction from the shiny side of a panel toward the black side.

CYU 11: decreases

CYU 12: b

CYU 13: decreases

CYU 14: b

Chapter 30

CYU 1: b

CYU 2: d

CYU 3: The absorption lines belong only to the Lyman series, since very few electrons are present with $n = 2$ or $n = 3$.

CYU 4: No, because the location of the electron in a given quantum mechanical energy state is uncertain.

CYU 5: when the electron is in the $n = 1$ state, because then the only possible value for the orbital quantum number is $\ell = 0$

CYU 6: (a) No, because the Bohr model uses the same quantum number n for the total energy and the orbital angular momentum (see Equations 30.13 and 30.8).
(b) Yes, because quantum mechanics uses the quantum number n for the total energy but the quantum number ℓ for the orbital angular momentum.

CYU 7: (a) Yes, because the Bohr model uses the same quantum number n for the orbital angular momentum and the total energy (see Equations 30.8 and 30.13).
(b) No, because quantum mechanics uses the quantum number ℓ for the orbital angular momentum but the quantum number n for the total energy.

CYU 8: $1s^2\ 2s^2\ 2p^6\ 3s^2\ 3p^6\ 4s^2\ 3d^{10}\ 4p^6$

CYU 9: (a) No.
(b) Yes.

CYU 10: b

CYU 11: c

CYU 12: a

CYU 13: c

CYU 14: d

CYU 15: a

Chapter 31

CYU 1: c, d

CYU 2: a

CYU 3: No, because they could have different numbers of protons (different atomic numbers).

CYU 4: Yes, because the total number A of nucleons could be the same, and it is the value of A that determines the radius.

CYU 5: d, c, a, b

CYU 6: c

CYU 7: d

CYU 8: It is not possible, because the total mass of the decay products is greater than the mass of the parent nucleus, $^{238}_{92}\text{U}$, indicating that energy would not be released.

CYU 9: b

CYU 10: Yes, because the decay of any single nucleus occurs randomly and can happen at any moment.

CYU 11: Yes.

CYU 12: b, because the gold statue does not contain carbon atoms

CYU 13: too small

CYU 14: No, because in 700 years the activity of a sample would have decreased to an immeasurably small fraction of its initial value.

CYU 15: e

Chapter 32

CYU 1: Yes, if the absorbed dose of the radiation is different for each type of radiation (see Equation 32.4).

CYU 2: c

CYU 3: neutrons, α particles, γ rays

CYU 4: c

CYU 5: (a) because it violates the conservation of nucleon number
(b) because it violates the conservation of nucleon number
(c) because it violates the conservation of electric charge

CYU 6: d

CYU 7: electrons, protons, neutrons

CYU 8: No, because the binding energy per nucleon is greater for the original nucleus than for the two fragments, as indicated in Figure 31.5.

CYU 9: b and d

CYU 10: a, b, and c

CYU 11: No, because the binding energy per nucleon is greater for the original nuclei than for the nucleus resulting from the fusion, as indicated in Figure 32.8.

CYU 12: b and d

CYU 13: a

Chapter 1

1. 124 m²
3. 10 159 m
5. 0.75 m²/s
7. 2.0 magnums
9. 6.65×10^5 m
11. $[M]/[T]^2$
13. 25.9° south of west, 80.1 km
15. 0.25 m
17. 54.1 m
19. (a) 4500 m
    (b) 2.8 mi
21. 32°
23. due east, 200 N
    due west, 600 N
25. (a) 89 units
    (b) 89 units
27. 25 cm
29. (a) 8.6 units
    (b) 34.9° north of west
    (c) 8.6 units
    (d) 34.9° south of west
31. (a) 48° east of south
    (b) 48° west of south
33. (a) 465 N
    (b) –338 N
35. (a) 4.3 m, due east
    (b) 7.4 m, due south
37. 150 m/s
39. (a) 2.0×10^2 N
    (b) 41°
41. (a) 25° above the $+x$ axis
    (b) 34.8 N
43. 3.00 m, 42.8° above the negative x axis
45. 0.532 m
47. (a) 2.7 km
    (b) 6.0×10^1 degrees, north of east
49. 6.88 km, 26.9°
51. (a) 5550 m
    (b) 6160 m
53. (a) 226 km
    (b) 51.3° north of east
55. 127 km, 74°
57. 18.7 N, 76°

59. 29.6 mL
61. 55 people
63. 9.72 cm
65. 71°
67. 20.4 m, 59.9°
69. $\theta_2 = 55.3°$, $\theta_1 = 95.5°$
71. (a) 2780 m
    (b) 37.0° south of west
    (c) 6.60°

Chapter 2

1. 9.1
3. 5×10^4 yr
5. (a) 7.07 km
    (b) 2.12 km 45.0° (north of east)
7. 64 s
9. 0.81 km
11. +16 m/s
13. (a) −0.40 m/s² (accelerating)
    (b) +0.80 m/s² (decelerating)
15. (a) -3.94×10^4 m/s
    (b) -5.79×10^4 m/s
17. (a) 2.8 m/s (b) 2.8 m/s
19. (a) due south
    (b) 9.4 m/s, due north
21. +30.0 m/s
23. 4.5 m
25. (a) 11.4 s (b) 44.7 m/s²
27. 3.1 m/s², directed southward
29. −1.0 m/s²
31. 39.2 m
33. 52.8 m
35. 0.87 m/s² (in the same
    direction as the car's velocity)
37. 1.7 s
39. 44.1 m/s
41. +39 m/s
43. −1.5 m/s²
45. 12 m
47. 5.33 m
49. 22 m
51. 3.43 s
53. 2.46 m
55. 0.40 s

57. −11 m/s
59. +2.7 m/s

61. A: -2.0×10^1 km/h B: $+1.0 \times 10^1$ km/h
    C: $+4.0 \times 10^1$ km/h
63. (a)

(b)

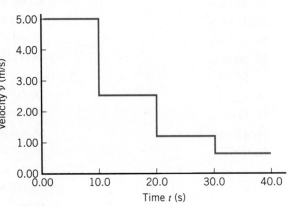

65. 109 m
67. 2.3 min
69. 5.63 s
71. (a) 2.2 m/s (b) 4.4 m/s (c) 0.021 m/s²

73. +62 m

75. 1.6×10^{-2} s

77. 283 m/s

79. 1.20 m/s^2

81. 7250 m

83. (a) 80 m/s [at $t = 2.0$ s]; 160 m/s
 [at $t = 4.0$ s]

 (b) 80 m/s [at $t = 2.0$ s]; 320 m/s
 [at $t = 4.0$ s]

85. (a) 11.5 m/s^2 (not safe) (b) 0.32 in (additional distance)

87. 3840 m, 50.6° west of south

Chapter 3

1. 2.8 m

3. $v_x = 11$ m/s, $v_y = 13$ m/s

5. 8.8×10^2 m

7. $x = 75.3$ km, $y = 143$ km

9. (a) 2.47 m/s^2

 (b) 2.24 m/s^2

11. (a) 2.99×10^4 m/s

 (b) 2.69×10^4 m/s

13. 8.6 m/s

15. (a) +1.6 m

 (b) +2.0 m

17. 9.4 mm

19. (a) 4.37 s

 (b) 93.5 m

21. 18 m/s

23. (a) 239 m/s, 57.1°

 (b) 239 m/s at an angle of 57.1° with
 respect to the horizontal

25. $a_x = 4.79$ m/s^2, $a_y = 7.59$ m/s^2

27. 30.0 m

29. 45.5 m/s

31. 62.8 m/s

33. (a) 85 m

 (b) 610 m

35. 4

37. 53 m

39. 4.52 m/s, 59.4°

41. 0.96 s

43. 3.4 m/s

45. 4

47. 27.2°

49. 190 s

51. 5.1 m/s, 51° north of west

53. 6.3 m/s, 18° north of east

55. 8.92°, south of west

57. 5.2 m/s, 52° west of south

59. 1130 m/s, 37.7°

61. 1.7 s

63. (a) 9.02 m

 (b) 1.05 s

65. 12 m

67. 14.9 m

69. 5.79 m/s

71. (a) 4.58 m

 (b) 18.3 m

73. (a) 51.1 s

 (b) 1.68 m/s

75. 23 cm

77. 34 m/s

79. (a) 41.8°

 (b) 36.4 m

 (c) 4.07 m

81. (a) 86.2 m

 (b) 37.0°

Chapter 4

1. 93 N

3. (a) +6.0 N (b) −24 N (c) −9.0 N

5. 32 N

7. 3560 N

9. (a) 3.6 N (b) 0.40 N

11. 30.9 m/s^2, 27.2° above the +x axis

13. 1.20 m/s^2, directed straight upward

15. 33 s

17. 18.4 N, 68° north of east

19. (a) 3.75 m/s^2 (b) 2.4×10^2 N

21. (a) 5.1×10^{-6} N (b) 5.1×10^{-6} N

23. 9.6×10^{-9} N

25 0.223 m/s^2

27. (a) 5.67×10^{-5} N (to the right)

 (b) 3.49×10^{-5} N (to the right)

 (c) 9.16×10^{-5} N (to the left)

29. 3.47×10^8 m

31. 178

33. (a) 980 N (b) 640 N

35. 7.3×10^2 N

37. 645 N

39. (a) 267 N, +x direction (b) 360 N

41. 2.9 m/s^2

43. 16.3 N

45. 68°

47. 1.00×10^2 N, 53.1° south of east

49. 929 N

51. 436 N, 5.7° above the horizontal

53. 406 N

55. +0.62 m/s^2

57. (a) 45 N (b) 37 N

59. 1320 N

61. (a) 914 N (b) 822 N

63. 5.03 m/s^2

65. 5.9×10^{-3} m/s^2

67. 0.14 m/s^2

69. (a) 1.1×10^4 N (b) 5.4×10^2 N

71. 4.0×10^2 N

73. (a) Case A: 2.9 N; Case B: 3.9 N

 (b) Case A: 0.97 m/s^2; Case B: 0.64 m/s^2

75. 8.17 s

77. 1.2 s

79. (a) 1.10×10^3 N (b) 931 N (c) 808 N

81. 39 N

83. 1730 N, directed due west

85. (a) 1.04×10^3 N (b) 1.04×10^3 N

 (c) 2.45 m/s^2 (d) 1.74×10^{-22} m/s^2

87. 0.265 m

89. (a) 0.60 m/s^2 (b) 104 N (left string);
 230 N (right string)

91. 72 m/s

93. (a) 29.4 N (b) 44.1 N

95. 3090 N

97. −470 N

99. (a) 1100 N (b) 650 N

101. (a) 0.980 m/s^2, opposite to the direction
 of motion (b) 29.5 m

103. 42°

105. +2.3 m/s^2

107. 925 N (208 lb)

109. (a) $\mu_s = 0.22$ and $\theta = 10°$ (b) 12.4°

 (c) Yes. It is safe.

Chapter 5

1. 0.79 m/s^2

3. 0.71

5. 6.9 m/s^2

7. 332 m

9. (a) 464 m/s, 3.37×10^{-2} m/s^2

 (b) 402 m/s, 2.92×10^{-2} m/s^2

11. 1.5 m

13. 73 kg

15. 12 m/s

17. 0.187

19. (a) 0.91

 (b) 38 m/s

21. (a) 19 m/s

 (b) 23 m/s

23. 170 m

25. 184 m

27. 45 s

29. 3070 m/s

31. 5.92×10^3 m/s

33. 2.45×10^4 N

35. (a) 1.70×10^3 N

 (b) 1.66×10^3 N

37. 17 m/s

39. 8.48 m/s

41. 606 N

43. (a) 0.189 N

 (b) 4

45. 14.0 m/s

47. 28°

49. 816 m

51. 10 600 rev/min

53. 33 m/s

55. 0.38 m

57. 0.224 kg

59. 25 m/s, 18 m/s

61. 110 m

63. (a)

 (b) $r_{Mars} = 115$ m, $r_{moon} = 49.7$ m,
 $r_E = 300$ m

 (c) 34.8 s

Chapter 6

1. 2.2×10^3 J

3. -2.6×10^6 J

5. 42.8°

7. (a) More net work is done during the dive.

 (b) 6.8×10^7 J

9. 25°

11. 256 N

13. 2.5×10^7 J

15. 1450 kg

17. 3.2×10^3 J

19. 0.13

21. (a) 8.3×10^2 N

 (b) 9.1×10^3 J

 (c) -8.5×10^3 J (d) 4 m/s

23. (a) 10.0 m (b) 3.64 s

25. 5.24×10^5 J

27. (a) 27 J (b) 36 J (c) 8.8 J

 (d) The change in gravitational potential
 energy is -27 J $= -W$, where W is the
 work done by the weight.

29. (a) -3.0×10^4 J (b) The resistive force is
 not a conservative force.

31. (a) -1086 J (b) The skater is 2.01 m
 below the starting point.

33. 6.6 m/s

35. (a) 28.3 m/s (b) 28.3 m/s (c) 28.3 m/s

37. (a) 52.2 J (b) 48.8 m/s

39. 0.60 m

41. 43°

43. (a) 2.50 m (b) 1.98 m

45. 48°

47. -3.8 J

49. 8.6 m/s

51. 2.4 m/s

53. 4.17 m/s

55. (a) 3.3×10^4 W

 (b) 5.1×10^4 W

57. (a) 1.0×10^4 W (b) 13 hp

59. 49.6 J

61. 7.07 m/s

63. at $h = 20.0$ m: KE $= 0$ J;
 PE $= 392$ J; E $= 392$ J
 at $h = 10.0$ m: KE $= 196$ J;
 PE $= 196$ J; E $= 392$ J
 at $h = 0.0$ m: KE $= 392$ J;
 PE $= 0$ J; E $= 392$ J

65. 45 N

67. (a) -270 J (b) 140 N

69. 5.82×10^4 W (78 hp)

71. (a) 49.7 m/s (b) -4.89×10^4 J

73. 1.0×10^2 N

75. (a) 7110 J (b) 467 J

77. 61 J

79. $+1470$ J

81. (a) 14.4 m/s (b) 1.58 hp

83. (a) 479 m; It therefore has enough energy to lift the elevator and passenger
 (without friction).

 (b) $E_{grav} = 5.51 \times 10^5$ J and $E_{net} = 5.32 \times 10^5$ J. Since $E_{net} < E_{grav}$, the elevator
 will NOT make it to the top when the
 frictional force is included.

Chapter 7

1. $+69$ N, opposite to the velocity

3. -1.8 N, downward

5. -8.7 kg · m/s

7. (a) $+2.2 \times 10^3$ N

 (b) $+4.4 \times 10^3$ N

9. 960 N

11. 3.7 N · s

13. 1.4 m

15. (a) Bonzo has the larger mass, since he
 has the smaller recoil velocity.

 (b) 1.7

17. $+7.5$ m/s

19. 9.5%

21. -4.6 m/s

23. 0.34 s

25. 1.5 m

27. (a) 4.89 m/s

 (b) 1.22 m

29. $+9.3$ m/s

31. (a) -0.400 m/s, $+1.60$ m/s

 (b) $+0.800$ m/s

33. 7.4%

35. $+9.09$ m/s

37. 0.062 kg

39. (a) 73.0°

 (b) 4.28 m/s

41. 2.175×10^{-3} N

43. (a) 5.56 m/s

 (b) -2.83 m/s, $+2.73$ m/s

 (c) 0.409 m, 0.380 m

45. 1.6 m toward the origin

47. (a) 0 nm

 (b) $+0.0358$ nm

49. $-11\,600$ N

51. 0.707

53. (a) $+2.4 \times 10^5$ N

 (b) $+4.8 \times 10^3$ N

 (c) Stiff-legged: $F_{Ground} = +2.4 \times 10^5$ N,
 Knees-bent: $F_{Ground} = +5.5 \times 10^3$ N

55. $x_{cm} = 0.11$ m, $y_{cm} = 0.20$ m

57. (a) 0.84 m/s

 (b) 29 N

59. 2500 N

61. (a) 1160 J, 580 kg · m/s

 (b) 1160 J, 140 kg · m/s

63. $x_{cm} = 11.2$ cm, $y_{cm} = 2.30$ cm

65. (a) 2.49 m/s

 (b) 3.3×10^{-3} m/s

 (c) 3.00 m/s

Chapter 8

1. 21 rad

3. (a) $+7.3 \times 10^{-5}$ rad/s (b) $+2.0 \times 10^{-7}$ rad/s

5. (a) 18 rad/s (b) 6.0 rad/s (c) 6.0 rad/s
   (d) 18 rad/s

7. 163 m

9. 59 m

11. (a) 111 rad/s (b) 134 rad/s (c) 118 rad/s

13. 825 m

15. 1.3 s

17. 1.43×10^{-1} m/s

19. 0.125 s

21. (a) 117 rad/s (b) 140 s

23. (a) 10.0 s (b) −2.00 rad/s

25. 28 rad/s

27. 24 rad

29. 12.5 s

31. (a) 0.332 rad/s (b) 128 m

33. 9.1 rad/s

35. 4.63 m/s

37. 3.61 rad/s (b) 6.53 rad/s²

39. 0.87

41. (a) 177 m/s² (b) 14.0°

43. (a) 2.49 m/s² (b) 17.7°

45. 8.71 rad/s²

47. 693 rad

49. (a) −1.4 rad/s² (b) +33 rad

51. 21.6 rad

53. 11.8 rad

55. 157.3 rad/s

57. (a) 125 m/s (b) 7.98 rev/s

59. (a) 8 (b) 5

61. −485 rad

63. (a) 1500 rad/s (b) 4.2×10^{-3} s

65. (a) 36 rad/s (b) 9.0 rev

67. 31.8 μm

69. 33.3 rotations

71. (a) 8.2×10^{-2} rad/s (b) 2.6 m/s²
    (c) 3.2 m/s²

73. (a) -6.13×10^{-3} rad/s² (b) 12.8 s

75. (a) Clockwise (b) 1.12 rad/s; 0.26 m/s
    (c) The largest tangential speed occurs
    when the wheel with the largest di-
    ameter is in the drive-wheel position:
    $v = 0.59$ m/s

Chapter 9

1. 843 N

3. 2.1×10^2 N

5. (a) 13 500 N · m
   (b) 132 000 N · m

7. (a) FL
   (b) FL
   (c) FL

9. 43.7°

11. (a) 2590 N
    (b) 2010 N

13. 6.0 N pointing toward corner B

15. (a) 1.60×10^5 N
    (b) 4.20×10^5 N

17. 37.6°

19. (a) 2260 N
    (b) $S_x = 1450$ N, $S_y = 1450$ N

21. 430 N

23. 8.0×10^{-4} N · m

25. 0.027 kg · m²

27. Hoop: 0.20 N · m, Disk: 0.10 N · m

29. (a) 5.94 rad/s²
    (b) 44.0 N

31. (a) System A: 229 kg · m²,
        System B: 321 kg · m²
    (b) System A: −1270 N · m,
        System B: 0 N · m
    (c) System A: −27.7 rad/s, System B: 0
        rad/s

33. 0.78 N

35. 1700 N · m

37. (a) 12.0 m/s, 9.00 m/s, 18.0 m/s
    (b) 1.08×10^3 J
    (c) 60.0 kg · m³
    (d) 1.08×10^3 J

39. 6.1×10^5 rev/min

41. 2/7

43. (a) 3.7 m
    (b) 7.0 m/s

45. 2.0 m

47. 0.075 rev/d

49. 0.26 rad/s

51. 4.5 rad/s

53. 0.037 rad/s

55. 0.70 m

57. 180 rad/s², 5.0 N

59. (a) Rolling wheel: 54 J, Sliding wheel: 36 J
    (b) Rolling wheel: 2.8 m, Sliding wheel:
        1.8 m

61. 6.3 rad/s², 5.00×10^3 N

63. $F_{\text{Each Hand}} = 196$ N, $F_{\text{Each Foot}} = 96$ N

65. $M = 1.21 \times 10^3$ N, $F = 1.01 \times 10^3$ N,
    downward

67. 460 N

69. 1400 N

71. 4700 N

73. (a) 2.5 rotations
    (b) 0.129 J

75. 290 N

77. Choose solid cylinders over hallow ones.

79. (a) Roll the barrels down an inclined
        plane and time how long it takes,
        starting from rest, to roll a given
        distance. Due to their differing
        moments of inertia, they should take
        different times, and we can therefore
        identify which are solid and which
        are hollow cylinders.
    (b) $a_{\text{hallow}} = 2.45$ m/s², $a_{\text{solid}} = 3.27$ m/s²
    (c) $t_{\text{hallow}} = 2.86$ s, $t_{\text{solid}} = 2.47$ s

Chapter 10

1. 237 N

3. 650 N/m

5. (a) 8.5×10^4 N/m (b) 290 N

7. 1.4 kg

9. 0.240 m

11. 2.29×10^{-3} m

13. (a) 0.407 m (b) 397 N

15. (a) 50.0×10^{-6} s (b) 2.00×10^4 Hz
    (c) 1.26×10^5 rad/s

17. (a) 0.080 m (b) 1.6 rad (c) 2.0 N/m
    (d) $v = 0$ m/s (e) 0.20 m/s²

19. 696 N/m

21. 0.285 s

23. 0.806 s

25. (a) -1.84×10^2 J (b) $+1.84 \times 10^2$ J (c) 0 J

27. 4.4 J

29. 6.55 m/s

31. 303 N/m

33. 1.95 m

35. (a) 3.3×10^{-2} m (b) 7.6×10^{-2} m

37. 180 N/m

39. 2.37×10^3 N/m

41. 0.40 s

43. (a) 3.5 rad/s (b) 2.0×10^{-2} J (c) 0.41 m/s

45. (a) 1.64 s (b) 1.64 s

47. 5.2×10^{-4} m

49. 260 m

51. 4.2×10^4 N

53. 0.091°

55. 2.7×10^{-4} m

57. 1.2×10^4 N

59. (a) 710 N/m² (b) 3.6×10^{-8}
    (c) 3.6×10^{-10} m

61. -4.4×10^{-5}

63. (a) 0.450 m (b) 3.31 rad/s (c) 1.49 m/s

65. +0.50 m

67. (a) 0.037 m (b) 0.074 m

69. 8.00

71. 2.4×10^{-2} m

73. 1.6×10^{-5} m

75. 0.56 m

77. 2.3×10^{-6} m

79. 5.00×10^{7} m

81. 2.1×10^{6} N/m

83. (a) 6.20×10^{-3} m (b) 88.4 J

85. (a) 13 m/s (b) 11.7 m/s

87. (a) 0.67 Hz (b) 0.55 m

(c) $x = L$; i.e., the equilibrium displacement x of the spring is equal to the length L of the pendulum.

89. (a) For each full oscillation of the spring/mass, the beam will pass through the slot twice.

(b) 1.75 Hz (c) 0.60 kg

(d) $L = 8.27$ cm; $w = 3.00$ cm

Chapter 11

1. 8.3×10^{3} lb

3. 317 m^2

5. 7.0×10^{-2} m

7. 1.9 gal

9. 4240 s

11. 1.1×10^{3} N

13. 24

15. 32 N

17. 2400 Pa

19. (a) 3.5×10^{6} N

(b) 1.2×10^{6} N

21. (a) 9.95×10^{4} Pa

(b) 1.001×10^{5} Pa

23. (a) 2.45×10^{5} Pa

(b) 1.73×10^{5} Pa

25. 0.74 m

27. 2.3×10^{8} N

29. 3.8×10^{5} N

31. (a) 93.0 N

(b) 94.9 N

33. 8.50×10^{5} N · m

35. 108 N

37. 59 N

39. 250 kg/m^3

41. 390 kg/m^3

43. At least 20 logs are needed.

45. 0.20 m

47. 7.6×10^{-2} m

49. 2.52 m/s

51. 9.3×10^{-5} m^2

53. 46 Pa, (Enters at B and exits at A.)

55. 1.92×10^{5} N

57. (a) 17.1 m/s

(b) 2.22×10^{-2} m^3/s

59. 5.0×10^{-2} m^3/s

61. 1.7×10^{5} Pa

63. 9600 N

65. 21 m/s

67. 6.4×10^{5} Pa

69. 2.9×10^{4} Pa

71. 290 gal/min

73. (a) 1.6×10^{-4} m^3/s

(b) 2.0×10^{1} m/s

75. 7.9×10^{-4} m^3

77. 55 Pa

79. 46.2 mm Hg

81. 1.4×10^{4} Pa

83. (a) 4.2×10^{-2} m/s

(b) 0.38 m/s

85. 96 Pa

87. 0.34 m

89. 3.5 m

91. 1.55 g/cm^3

93. 44 400 gal

95. 2.0×10^{1} m/s

97. 7.72 cm

99. (a) $N_{barrels} = 6$

(b) $N_{barrels} = 8$

Chapter 12

1. (a) 56 F° (b) 31 K

3. (a) 102 °C (day); −173 °C (night)

(b) 215 °F (day); −2.80 × 10^2 °F (night)

5. 44.0 °C

7. 4.9×10^{-2} m

9. 3.2×10^{2} m

11. 110 C°

13. 8.6×10^{-6} m/s

15. 4.1×10^{-4}

17. 7.7×10^{10} N/m^2

19. 91 C°

21. Arrangement A: 31.7 °C; Arrangement B: 34.1 °C; Arrangement C: 30.4 °C

23. Gasoline

25. 2.26×10^{-4} m^3

27. 1 penny

29. 1.8×10^{-4} m^3

31. 2.1×10^{-4} (C°)$^{-1}$

33. 9.0 mm

35. 13 200 kg/m^3

37. 0.765 m

39. 36.2 °C

41. 2500 J/(kg · C°)

43. 940 °C

45. 4.50 m

47. 0.016 C°

49. 3.9×10^{5} J

51. (a) 2.4×10^{6} J (b) 5.3 C°

53. 9.49×10^{-3} kg

55. 2.46×10^{3} m/s

57. 5.2×10^{7} J

59. 4.38×10^{-3} kg

61. (a) 1.49×10^{14} J (b) 993 homes

63. 1.51×10^{5} J/kg

65. 5.5×10^{6} Pa

67. 33%

69. 2.8×10^{5} J

71. 0.027 kg

73. 230 C°

75. 1.4×10^{-3} m

77. 1.2×10^{-2} kg

79. 32.1 °C

81. 9.89×10^{-2} m^3

83. 21.03 °C

85. 4.4×10^{3} N

87. 87%

89. 8.47 hours

91. 187 Calories

93. 174 Calories

95. (Block A) 18 C°; (Block B) 6.0 C°; (Block C) 2.8 C°

97. (a) The rate of rotation decreases

(b) 44.5 rad.s

99. (a) 1750.00 rpm (b) 1749.50 rpm

(c) Set rotation rate of aluminum drive wheel to 35.010 rpm

Chapter 13

1. 14 h

3. 1/4

5. Case A: 4.0×10^{3} J, Case B: 9.0×10^{3} J, Case C: 3.6×10^{4} J

7. 287 °C

9. 85 J

11. (a) 2200 J

(b) 0.26 C°

13. 103.3 °C

15. 0.74

17. 5800 K

19. (a) 6.3 J/s

(b) 4.8 J/s

21. 275 W

23. 380 K

25. 732 K

27. 0.012 m

29. 386 K

31. (a) 1.27

(b) 424 K

33. 3.1×10^{-5} kg/s

35. 1.5 C°

37. 1.2×10^4 s

39. (a) 67 W

(b) 58 Calories

41. Ratio = 23.7

43. 139 J/s

45. 220 s

47. 718 °C

49. (a) 1072 J/s

(b) 123 J/s

(c) 3216 J/s

Chapter 14

1. 1.07×10^{-22} kg

3. (a) 196.967 u (b) 3.2706×10^{-25} kg

(c) 1.45 mol

5. (a) 2.3×10^3 mol (b) 1.4×10^{27}

7. 42.4 mol

9. 2.4×10^4 Pa

11. $T_A = 120$ K; $T_B = 180$ K; $T_C = 120$ K; $T_D = 180$ K

13. 12

15. 882 K

17. 8.1 g

19. (a) 8.17×10^{-4} m³ (b) 2.3 g

21. 3.70×10^2 K

23. 0.93 mol/m³

25. 6.19×10^5 Pa

27. 1.6×10^{-15} kg

29. 3.9×10^5 J

31. 2.098

33. 2820 m

35. 2.10×10^{26} atoms/s

37. 3.0×10^{-3} kg/m³

39. (a) 5.00×10^{-13} kg/s (b) 5.8×10^{-3} kg/m³

41. 0.32 m

43. 1.4

45. 11.7

47. 1.98×10^3 N/m

49. (a) 1.29×10^{21} (b) 5.84×10^{20}

51. 11

53. 0.14 kg/m³

55. (a) 1.66×10^{-4} moles

(b) 9.99×10^{19} molecules

57. Argon

59. 1.1×10^{-5} μg

61. (a) 1.5×10^{-16} N (b) 3.8×10^{-17} Pa

(c) 2.7 K

63. 87.4 kg (nitrogen); 29.2 kg (oxygen)

65. 1.84×10^6 Pa

Chapter 15

1. (a) $+1.6 \times 10^4$ J

(b) -4.2×10^4 J

(c) -2.6×10^4 J

3. (a) −87 J

(b) +87 J

5. 436 K

7. 1.2×10^7 Pa

9. 3.0×10^5 Pa

11. (a) $+3.0 \times 10^3$ J

(b) work is done by the system

13. 0.24 m

15. 3.1×10^5 Pa

17. 4.99×10^{-6}

19. (a) $+5.0 \times 10^3$ J

(b) -5.0×10^3 J

21. −4700 J

23. 1.81

25. 19.3

27. Step A to B: 4990 J, 3320 J, 8310 J

Step B to C: −4990 J, 0 J, −4990 J

Step C to D: −2490 J, −1660 J, −4150 J

Step D to A: 2490 J, 0 J, 2490 J

29. (a) -8.00×10^4 J

(b) Heat flows out of the gas.

31. (a) $T_{1f} = 477$ K

(b) $T_{2f} = 323$ K

33. (a) 1.1×10^4 J

(b) 1.8×10^4 J

35. 45 K

37. (a) 1.40×10^2 K

(b) 5.24×10^3 J

(c) 2.33×10^3 Pa

39. 0.264 m

41. (a) 8600 J

(b) 3100 J

43. 65 J

45. (a) 5/9

(b) 1/3

47. 0.21

49. 1090 K

51. lowering

53. 1.23

55. 21

57. 13

59. 284 K

61. 275 K

63. (a) 2.0×10^1

(b) 1.5×10^4 J

65. 1.4×10^4 J

67. 11.6 J/K

69. 1.4×10^{26} J/K

71. 0.66

73. 4.5×10^{-3} m³

75. (a) reversible

(b) −125 J/K

77. 5.86×10^5 J

79. (a) 0 J

(b) $+2.1 \times 10^3$ J

81. $e = e_1 + e_2 - e_1 e_2$

83. (a) $+1.2 \times 10^5$ J

(b) 5.9×10^5 J

85. (a) 4.6×10^5 J

(b) 4.1×10^5 J

87. 0.170 kg

89. (a) 21.7%

(b) 7.93×10^4 J

91. 2.87×10^6 J

93. 5450 s

95. (a) 881 K

(b) 916 K

(c) It is likely that the engine with the higher compression ratio is more efficient.

Chapter 16

1. 5.50×10^{14} Hz

3. (a) 10.0 s (b) 0.100 Hz (c) 32 m

(d) 3.2 m/s

(e) Not possible to determine amplitude

5. 0.25 m

7. 0.20 m/s

9. 5.0×10^1 m/s

11. 0.68 m

13. 340 m/s

15. 0.17 s

17. 21.2 m/s

19. 9.1 cm

21. 3.26×10^{-3} s

23. $y = (0.37 \text{ m}) \sin [(8.2 \text{ rad/s})t + (0.68 \text{ m}^{-1})x]$

25. $y = (0.35 \text{ m}) \sin [(88 \text{ rad/s})t - (17 \text{ m}^{-1})x]$

27. (a) 4.2 m/s (b) 0.35 m

(c) $y = (3.6 \times 10^{-2} \text{ m}) \sin [(75 \text{ rad/s})t + (18 \text{ m}^{-1})x]$

29. 110 m

31. 28.8 K

33. 1730 m/s

35. 690 rad/s

37. 61 m

39. (a) First in metal, second in water, third in air

(b) Second sound arrives 0.059 s later. Third sound arrives 0.339 s later.

41. 650 m

43. Tungsten

45. 8.0×10^5 m

47. 65 W

49. 2.4×10^{-5} W/m²

51. (a) 3.54×10^{-4} W/m²

(b) 4.39×10^{-4} W/m²

53. 82.0 m and 246 m

55. 6300

57. −6 dB

59. 0.316 W/m²

61. 76.8 dB

63. 3.1%

65. 13 Hz

67. 17 m/s

69. (a) Away (b) 8.7 m/s

71. 1.4 m/s²

73. (a) 0.140 m; 2450 Hz

(b) 0.129 m; 2660 Hz

(c) 0.129 m; 2770 Hz

75. 3.02×10^{-6} W/m²

77. 64 N

79. 1000

81. 78 cm

83. 3.4 m/s

85. 1.0×10^2

87. 2.6

89. (a) $v = (yg)^{1/2}$

(b) 2.2 m/s at $y = 0.50$ m; 4.4 m/s at $y = 2.0$ m

91. 113 dB

93. 6.7×10^{-9} W

95. 6.0

97. 5.1×10^{-10} W

99. (a) 0.700 mm (b) 14.0 cm (c) 1.8×10^{-4} s

101. 5.9×10^{-4} J

103. (a) 2310 Hz (b) 1970 Hz

105. Highest: 46.7 m/s; Lowest: 18.3 m/s

107. (a) 2004 m (b) 23.7 m/s (c) 83.2 s

Chapter 17

1. 8.42 m

3.

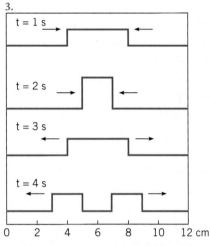

5. (a) +13.3 mm

(b) +48.8 mm

7. 3.89 m

9. 28 Hz and 42 Hz

11. 3.90 m, 1.55 m from speaker A, 6.25 m from speaker A

13. 16.0°

15. 44°

17. 1.5×10^4 Hz

19. 0.25 s

21. (a) 50 kHz

(b) 90 kHz

23. 14 Hz

25. 1.7 Hz

27. (a) 1.00 m

(b) 85.0 m/s

(c) 17.0 Hz

29. Wire A: 790 m/s, Wire B: 530 m/s

31. 2

33. 130 Hz

35. 11 N

37. 13 Hz

39. (a) 261.6 Hz

(b) 0.656 m

41. 20 Hz: 8.6 m, 20 kHz: 8.6×10^{-3} m

43. 0.57 m

45. 6.1 m

47. 0.28 m

49.

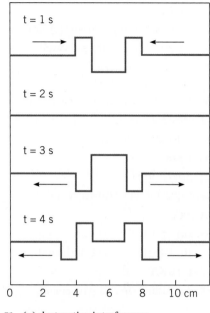

51. (a) destructive interference

(b) constructive interference

53. 0.669 m

55. 2.92

57. 20.83 minutes

59. 1.47 m

61. Air: 8.0°, Water: 37°

63. $(156.6)(m + \frac{1}{2})$ Hz

65. (a) $f = 95.50$ kHz: $L = 2.65$ cm, $f = 102.550$ kHz: $L = 2.47$ cm

(b) $f = 95.50$ kHz: $L = 2.70$ cm, $f = 102.550$ kHz: $L = 2.52$ cm

Chapter 18

1. $+3.04 \times 10^{-18}$ C

3. (a) -1.6μC (b) 1.0×10^{13} electrons

5. (a) $+1.5q$ (b) $+4q$ (c) $+4q$

7. (a) 3.35×10^{26} electrons (b) -5.36×10^7 C

9. 8 electrons

11. (a) 0.83 N (a) attractive

13. 1.8×10^{-5} C

15. (a) Like signs (i.e., either both positive or both negative)

(b) 1.7×10^{-16} C

17. (a) 4.56×10^{-8} C (b) 3.25×10^{-6} kg

19. 7.19×10^{23} m/s²

21. (a) 0.166 N directed along the $+y$ axis

(b) 111 m/s² directed along the $+y$ axis

23. 1.96×10^{-17} J

25. -3.3×10^{-6} C

27. 1.37

29. 1.8 N due east

31. 54 N/C

33. (a) 7700 N/C (b) 1300 N/C (c) 5500 N/C

35. 6.5×10^3 N/C directed downward

37. 1.81×10^2 N/C in Fig. 18.20a and 3.11×10^2 N/C in Fig. 18.20b

39. 2.2×10^5 N/C directed along the $-x$ axis

41. 2900 N/C directed 47° above the $+x$ axis

43. (a) $+2.0\,\mu C$ (b) $-6.0\,\mu C$

45. 6.5 m

47. 3.25×10^{-8} C

49. (a) 350 N · m²/C (b) 460 N · m²/C

51. 58°

53. (a) 7.9×10^5 N/C, radially outward

   (b) 1.4×10^6 N/C, radially inward

   (c) 0 N/C

55. (a) Positive (b) 2.53×10^7 protons

57. 0.80 m

59. $F_A = 430$ N; $F_B = 250$ N; $F_C = 250$ N

61. (a) 8.2×10^{-8} C (b) 8.2×10^{-3} N

63. 8.0×10^5 N/C (b) 1.3×10^{-13} N

65. 2.6×10^{-9} N

67. 2.7×10^{-14} C

69. 1340 m

71. (a) 7.56×10^4 N/C (along the $+y$ axis)
   (b) 5.04×10^2 m/s² (along the $+y$ axis)

73. (a) 2.9×10^{-9} C (b) 2.1×10^{-4} kg

75. (a) 1.23×10^5 N/C (b) -1.94×10^5 N/C

Chapter 19

1. -2.1×10^{-11} J

3. (a) 4.5×10^{-3} N directed from A toward B

   (b) 3.0×10^3 N/C directed from A toward B

5. 19 m/s

7. $+4.80 \times 10^{-18}$ C

9. -4.05×10^4 V

11. 2.4

13. -9.4×10^3 V

15. $+7.8 \times 10^6$ V

17. -3.1×10^{-6} C

19. -0.746 J

21. 1.53×10^{-14} m

23. 0.0342 m

25. 1.41×10^{-2} m

27. 1.6×10^{-8} C

29. 3.5×10^4 V

31. (a) 179 V (b) 143 V

   (c) 155 V

33. (a) 0 V/m

   (b) 1.0×10^1 V/m

   (c) 5.0 V/m

35. (a) 0 V

   (b) +290 V

   (c) −290 V

37. 36 V

39. 6.1×10^{-5} C

41. 5.66 V

43. 52 V

45. 18 V

47. (a) -5.6×10^{-5} J

   (b) 0 J

49. 3.82

51. 0.213 J

53. 1.1×10^{-20} J

55. 1.1×10^3 V

57. 2×10^{-8} F

59. (a) decreases

   (b) 36

61. (a) 3.75×10^9 J

   (b) 1.45×10^3 kg

63. 0.068 J

65. (a) 2.06×10^3 V

   (b) 1.37×10^{-4} W

67. (a) 3657 capacitors

   (b) 398 capacitors

Chapter 20

1. 174 C

3. 8.6 A

5. 16 Ω

7. (a) 0.600 Ω (for a); 0.0375 Ω (for b); 0.150 Ω (for c)

   (b) 5.00 A (for a); 80.0 A (for b); 20.0 A (for c)

9. Aluminum

11. 58 m

13. 9.3%

15. 1.8 V

17. 6.0×10^2 Ω

19. $\$5.9 \times 10^6$

21. 8.9 hours

23. 3.1×10^{-4} m

25. 250 °C

27. (a) 786 W (b) 1572 W

29. 150 W

31. 21 V

33. (a) 50.0 Hz (b) 2.40×10^2 Ω (c) 60.0 W

35. 4.0×10^1 Ω

37. $P(2.0\,\Omega) = 8.0$ W; $P(4.0\,\Omega) = 16$ W; $P(6.0\,\Omega) = 24$ W

39. 85.9 Ω and 242 Ω

41. 140 W

43. 1800 W

45. (a) $I_1 = I_2 = I_3 = 0.282$ A; $V_1 = 14.1$ V; $V_2 = 7.05$ V; $V_3 = 2.82$ V

   (b) $I_1 = 0.480$ A; $I_2 = 0.960$ A; $I_3 = 2.40$ A; $V_1 = V_2 = V_3 = 24.0$ V

47. 2.3 Ω

49. (a) 130 Ω (b) 7.2 W

51. 2.00 Ω and 4.00 Ω

53. 4

55. 9.2 A

57. 4.6 Ω

59. 42 Ω

61. 6.00 Ω, 0.545 Ω, 3.67 Ω, 2.75 Ω, 2.20 Ω, 1.50 Ω, 1.33 Ω, and 0.833 Ω

63. 25 Ω

65. 12.0 V

67. 24.0 V

69. 4.84 A

71. 2.0 A

73. 0.75 V, left end at higher potential

75. (a) 15.0 A (b) 6.5 Ω (c) 3.5 Ω

77. 2.49×10^4 Ω

79. 820 Ω and 8.00×10^{-3} A

81. 9.0 V

83. 18 μF

85. 2.0 μF

87. Demonstration of the equality is proof

89. 11 V

91. 3.9 s

93. 1.61

95. (a) 24.0 W (for a); 12.0 W (for b); 96.0 W (for c); 48.0 W (for d)

   (b) 2.00 A (for a); 1.00 A (for b); 4.00 A (for c); 2.00 A (for d)

97. −34.6 °C

99. 21 V

101. (a) 1.2 Ω (b) 110 V

103. 0.049 m

105. (a) 1.5×10^{-11} A (b) 4.7×10^7 ions

107. 4.1×10^{-7} F

109. (a) 860 W (b) 5700 cells (c) 5700

111. 1.92×10^{-15} moles

113. (a) 9.6 W (b) 2.4 W (c) 9.6 W

115. (a) 6.65 A (b) 1.37×10^4 W

117. (a) 12 Ω

   (b) Four 47-Ω resistors in parallel have a resistance of 11.8 Ω, which is within 10% of the 12-Ω requirement.

Chapter 21

1. 5.9×10^{12} m/s^2
3. 4.1×10^{-3} m/s
5. $58°$
7. 1.1×10^{-2} N
9. 1.3×10^{-10} m, directed out of the page
11. (a) due south
    (b) 2.55×10^{14} m/s^2
13. 1.5×10^{-8} s
15. 2.7×10^{-4} T
17. 0.0904 m
19. 19 cm
21. 1.41
23. 5.25×10^3 m/s
25. 0.71 m
27. 6.8×10^5 C/kg
29. 5.1×10^{-5} T
31. 3.4×10^{-3} T
33. 2.7 m
35. (a) left-to-right
    (b) 1.1×10^{-2} m
37. (a) $37°$
    (b) 0.49 N
39. 3.3 cm
41. 0.053 m
43. (a) 13.4 A · m^2
    (b) 24.1 N · m
45. (a) 170 N · m
    (b) $35°$ angle increases
47. 3.1×10^{-4} T
49. 1.8×10^{-4} N·m
51. 190 A
53. 8.71×10^{-3} m
55. 0.800 m
57. 1.04×10^{-2} T
59. 320 A
61. (a) 1.1×10^{-5} T
    (b) 4.4×10^{-6} T
63. $19.7°$
65. 0.062 m
67. 0.19 N
69. 1.7×10^{-3} N
71. (a) $F_E = +1.37 \times 10^{-3}$ N, along $+ x$ axis, $F_{Bx} = 0$ N, $F_{By} = 0$ N
    (b) $F_E = +1.37 \times 10^{-3}$ N, along the $+ x$ axis, $F_{Bx} = 0$ N, $F_{By} = 2.94 \times 10^{-3}$ N, along $+ z$ axis
    (c) $F_E = +1.37 \times 10^{-3}$ N, along $+ x$ axis, $F_{Bx} = 3.78 \times 10^{-3}$ N, along $+ y$ axis, $F_{By} = 2.94 \times 10^{-3}$ N, along $-x$ axis
73. 0.14 T

75. (a) 9.1×10^{-7} T
    (b) 49
77. 4.0×10^3 A
79. 1.18 T
81. 0.325 kg
83. (a) 4590 turns
    (b) 519 m
    (c) -1.52 A
85. (a) ^{1}H: 6.9×10^5 m/s, ^{2}H: 4.9×10^5 m/s, ^{3}H: 4.0×10^5 m/s
    (b) 0.125 T
    (c) ^{1}H: 11.6 cm, ^{2}H: 16.4 cm

Chapter 22

1. (a) 3.7×10^{-5} T (b) East end is positive
3. 7800 V
5. (a) 1 (b) 2 (c) 8
7. 0.15 W
9. (a) 0.23 kg (b) -1.8 J (c) 1.8 J
11. (a) 0.086 m^2 (b) 0.018 m^2
13. -0.094 Wb
15. 8.9×10^{-3} Wb
17. (a) -1.0 V $(0 - 3.0$ s); 0 V $(3.0 - 6.0$ s); $+0.50$ V $(6.0 - 9.0$ s)
    (b) -2.0 A $(0 - 3.0$ s); $+1.0$ A $(6.0 - 9.0$ s)
19. 8.6×10^{-5} T
21. (a) 0.38 V (b) 0.43 m^2/s
23. 0.14 V
25. (a) 3.6×10^{-3} V (b) The area of the loop must shrink at a rate of 2.0×10^{-3} m^2/s
27. -0.84 A
29. The right end is positive
31. Left-to-right
33. (a) counterclockwise
    (b) No induced current
    (c) clockwise (d) no induced current
35. (a) Lenz's law explains the behavior
    (b) No induced current can flow in the cut ring
37. 0.150 m
39. 3.0×10^5
41. 2.0 V
43. 1.5×10^9 J
45. 2.5×10^{-2} H
47. 1.4 V
49. 220 turns
51. 3.8 V
53. (a) 1:13
    (b) 1.7×10^{-2} A
    (c) 2.0 W delivered by wall socket
    (d) 2.0 W delivered to the battery

55. 41 V
57. 140
59. $R_1 = (N_p/N_s)R_2$
61. 1.5 m^2/s
63. 756
65. 0.050 V
67. 6.6×10^{-2} J
69. 1.0×10^{-3} V
71. 38 V
73. 5.2×10^{-5} T
75. (a) 1.1 m (b) 3.2×10^{-5} Wb
77. (a) 0.040 s (b) 160 rad/s (c) 65 V
    (d) -49 V
79. (a) 5.01×10^4 V (b) 0.402 V
    (c) 4.54×10^3 V
    (d) Yes, the probe should take a more distant path.
81. (a) 534 V (b) 378 V (c) 0.066 T
    (d) $N_p : N_s = 3.44{:}1$

Chapter 23

1. 9470 Hz
3. 36 Ω
5. 2.7×10^{-6} F
7. (a) 6.4×10^{-6} F
    (b) 9.0×10^{-4} C
9. 0.44 A
11. 24.6 Ω
13. 176 mH
15. (a) 1.11×10^4 Hz
    (b) 6.83×10^{-9} F
    (c) 6.30×10^3 Ω
    (d) 7.00×10^2 Ω
17. 0.819
19. $R = 49.7$ Ω, $X_C = 185$ Ω
21. 3.0 W
23. 270 Hz
25. (a) 1.7×10^{-5} H
    (b) 8.8×10^{-5} H
27. 123 V
29. 1.8 μF
31. 2.8 kHz
33. 521 Hz
35. 2.4%
37. $V_R = 10.5$ V, $V_C = 19.0$ V, $V_L = 29.6$ V
39. 8.0×10^1 Hz
41. 0.075 A
43. (a) 2.94×10^{-3} H
    (b) 4.84 Ω
    (c) 0.163

45. $R = 443\ \Omega$, $X_C = 76.6\ \Omega$

47. (a) $2.2 \times 10^5\ \Omega$

   (b) $1.3 \times 10^5\ \Omega$

   (c) 0.30 Hz

49. (a) 4.20×10^{-8} H

   (b) 2.89×10^{-12} F

51. 1.8 A

53. (a) 24.5 W

   (b) 138 W

55. (a) 1.22×10^{-5} H

   (b) Two possible configurations: (i) L_1 and L_2 in series. (ii) L_1 in series with $L_3 \| L_4$.

   (c) 588 kHz $\leq f_o \leq$ 1019 kHz

   (d) Lowest resistance is obtained with the resistors wired in parallel.

Chapter 24

1. 2.73×10^{12} m

3. 4.1×10^{16} m

5. 11.118 m

7. 4.4×10^8 Hz

9. (a) 1.4×10^4 (b) 1.4×10^4

11. 177 N/m

13. 540 rev/s

15. 7.0 bits

17. 1.6×10^6 m

19. 3.8×10^2 W/m²

21. 990 N/C

23. 1.8×10^{-5} J

25. 3.93×10^{26} W

27. 1.7×10^{11} W

29. (a) Receding

   (b) 3.1×10^6 m/s

31. (a) 6.175×10^{14} Hz (b) 6.159×10^{14} Hz

33. (a) 0.55 W/m² (b) 3.7×10^{-2} W/m²

35. 21.5°

37. For $\alpha = 0°$ and 35°, respectively:
   (a) 3.5 W/m², 3.5 W/m²
   (b) 7.0 W/m², 4.7 W/m²
   (c) 0 W/m², 2.3 W/m²

39. 720 W/m²

41. Sheet A removed: 3.8 W/m²; Sheet B removed: 0 W/m²; Sheet C removed: 0 W/m²; Sheet D removed: 5.1 W/m²

43. For $E = 315$ N/C and 945 N/C, respectively:
   (a) 263 W/m², 2370 W/m²
   (b) 1.05×10^{-6} T, 3.15×10^{-6} T
   (c) 263 W/m², 2370 W/m²

45. 640 Hz

47. 68 N/C

49. 4.44×10^{-10}

51. 2.1×10^3 Hz

53. (a) 2.1×10^4 W/m²

   (b) 1.1×10^2 W (c) 18 s

55. (a) 3.0×10^8 m/s (b) 1.30×10^{19} Hz

57. 102.2 mph

59. (a) 2.5×10^8 m (b) 0.011 V/m

   (c) 3.7×10^{-11} T (d) 5.4×10^{-16} J/m³

61. (a) 4.0 W/m² (b) 6.0 W/m²

63. 4.00×10^{11} m

65. (a) 500 W/m² (b) 8.52 W/m²

   (c) −1.5° (clockwise); 1.5° (counterclockwise)

Chapter 25

1. $x = +3.0$ m

3. −1.80 m/s

5. 1.67 m

7. (a) 0.0670°

   (b) 7 m

9. (a) 70.6°

   (b) 62.1°

11. (a) 24 cm

   (b) Ray diagram similar to Figure 25.16.

13. (a) 7.5 cm in front of the mirror

   (b) Image height is 1.0 cm

15. Image located at 10.9 cm. Ray diagram below.

17. (a) $d_o = +62$ cm

   (b) $m = +0.35$

   (c) upright

   (d) smaller

19. $f = 9.1$ cm

21. (a) 180 cm

   (b) 6.0×10^1 cm

23. Smaller object distance: $m = 0.750$, Greater object distance: $m = 0.600$

25. −2.0

27. (a) $d_o = 82$ m

   (b) $h_o = 11$ m

29. 56 cm

31. (a) As object moves closer to the mirror, the magnitude of the image distance becomes smaller.

   (b) As object moves closer to the mirror, the magnitude of the image height becomes larger.

33. 74 cm

35. −0.16 cm

37. −17 cm

39. (a) concave (b) +3.1 cm (c) +2.8

   (d) upright

41. (a) 70° (b) 20° (c) 225 m (d) 420 m

43. 1.4 m

45. −13 cm

47. (a) 16

   (b) above the horizontal

   (c) +10.0°

   (d) $\beta = 5.0°$

49. (a) $d_o = 308.5$ m

   (b) $h = 147.2$ m

   (c) $w_o = 10.3$ m

Chapter 26

1. 2.0×10^{-11} s

3. 1.66×10^8 m/s

5. Ethyl alcohol

7. 1.82

9. (a) 46°

   (b) 50° (Drawing wrongly shows an angle of refraction greater than 55°)

   (c) 69°

   (d) 0° (Drawing wrongly shows an angle of refraction greater than 0°)

11. 0.9°

13. 1.92×10^8 m/s

15. 2.46×10^8 m/s

17. 1.65

19. 2.7 m

21. 1.19 mm

23. The answer is a derivation

25. 37.79°

27. 2.5 m

29. Layer a top surface: 24.7°; Layer b top surface: 14.9°; Layer b bottom surface: 6.0°; Layer c bottom surface: 29.4°

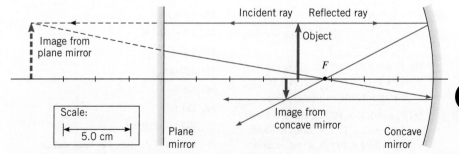

DIAGRAM FOR CHAPTER 25, PROBLEM 15.

31. $n_b = 1.16$; $n_c = 1.38$

33. 3.36×10^{-8} s

35. 32°

37. 55.0°

39. 0.725 m

41. 1.73

43. 44.6° (red), 45.9° (violet)

45. (a) 1.075 (b) 1.083

47. (a) 1.2 m (b) 1.7 m

49. (a) virtual (b) −8.1

51. (a) −10.0 cm (b) +0.500

53. (a) 60.0 cm (b) 20.0 cm

55. (a) −85.1 cm (b) 134.1 cm

57. 0.333 m

59. 37.3 cm

61. −5.6 cm

63. −12 cm

65. (a) 4.00 cm to the left of the diverging lens
(b) −0.167 (c) virtual (d) inverted
(e) smaller

67. (a) 18.1 cm (b) real (c) inverted

69. (a) +0.600 (b) +2.00

71. −9.2 m

73. 28.0 cm

75. −2.0 diopters (right eye); −0.15 diopters (left eye)

77. (a) −4.5 m (b) 0.50 m

79. 18

81. (a) 6.88 cm (b) 3.63

83. 13.7 cm

85. 9×10^{-3} rad

87. 3.0 cm

89. (a) −30.0 (b) 4.27 cm (c) −4.57

91. 1.1 m

93. (a) 1.3-dopter lens (b) 0.86 m (c) −8.5

95. −31

97. 1.51

99. (a) 6.74×10^{-7} m² (a) 7.86×10^{5} W/m²

101. (a) −24 cm (b) 6.0 mm

103. (a) converging (b) 2f (c) 2f

105. −181

107. (a) 22.4 cm (b) 28.4 cm

109. +2.92 diopters

111. −1100

113. 47.5°

115. (a) 382 J/m³ (b) 1.15×10^{11} W/m²
(c) −0.405 (d) 1.04 m (e) 0.422 m

117. (a) 23.8° (b) 26.4° (c) 2.60°

Chapter 27

1. (a) 11°
(b) 22°
(c) 34°
(d) 48°

3. (a) destructive interference is occurring
(b) 3.25 m and 0.75 m from one of the sources

5. 4.9×10^{-7} m

7. Version A: 6 bright fringes, Version B: 4 bright fringes

9. 0.0248 m

11. 487 nm

13. 102 nm

15. 6.12×10^{-7} m

17. 1.18

19. 115 nm

21. 1.5

23. 3.3×10^{-6} m

25. 4.9×10^{-7} m

27. 0.447

29. 0.013

31. 1.1×10^{-4} m

33. 644 nm

35. 630 nm

37. 4.0×10^{-6} m

39. 640 nm and 480 nm

41. (a) 7.9° for violet light, 13° for red light
(b) For $m = 2$: 16° for violet light and 26° for red light
(c) For $m = 3$: 24° for violet light and 41° for red light
(d) The second and third orders overlap.

43. 571 nm

45. 1.2×10^{-5} m

47. 3.2×10^{3} m

49. 1560 nm

51. (a) 220 nm
(b) No

53. (a) 9.7×10^{-3} m = 9.7 mm
(b) The hunter's claim is not reasonable.

55. 5.1 s

57. (a) 0.82
(b) 1.2

59. (a) 740 nm
(b) 31°

61. 522 nm

63. 499 nm

65. (a) 22.2×10^{3} m
(b) 24.8×10^{3} m

67. (a) 12.8 μm
(b) 40.7 μm
(c) $N = 47$ for sample in (a). $N = 149$ for sample in (b).

Chapter 28

1. (a) 4.9×10^{-9} s (b) 1.5 m

3. 0.999 95c

5. 2.28 s

7. 16

9. 2.6×10^{8} m/s

11. 8.1 km

13. 1.3

15. 40.2°

17. 7.5×10^{-19} kg · m/s

19. 1.0 m

21. −2.0 m/s

23. 5.0×10^{-13} J

25. (a) 7.44×10^{-12} kg (b) 5.02×10^{-11} kg

27. 1.1 kg

29. (a) 3.84×10^{-12} J (b) 0.999 781c

31. 0.31c

33. +0.80c

35. Assuming the direction of the cruiser, ions, and laser light is in the positive direction; (a) +c (b) +0.994c (c) +0.200c (d) +0.194c

37. 1.7×10^{-13} J

39. 72 h

41. 0.866c (A); 0.943c (B); 0.745c (C)

43. (a) 0.020c (6.0×10^{6} m/s)
(b) 2.83×10^{-5}c (8490 m/s)

45. 2.60×10^{8} m/s

47. 2.96×10^{8} m/s

49. (a) $m_a = 6.64 \times 10^{-27}$ kg; $m_b = 6.64 \times 10^{-27}$ kg; $m_c = 33.2 \times 10^{-27}$ kg
(b) $KE_a = 5.98 \times 10^{-10}$ J; $KE_b = 17.9 \times 10^{-10}$ J; $KE_c = 5.98 \times 10^{-10}$ J

51. (a) 0.501c (b) 0.733c (c) too early

53. (a) 56.6 yr (b) 0.12 min = 7.1 s (c) 0.014:1

Chapter 29

1. (a) 1.63×10^{-7} m
(b) 1.84×10^{15} Hz
(c) UV region of the electromagnetic spectrum

3. 6.3 eV

5. 310 nm

7. 1.26 eV

9. 73 photons/s

11. 162 nm

13. (a) 7760 N/C

    (b) 2.59×10^{-5} T

15. 4.2×10^{-25} m/s

17. 0.3533 nm

19. $p_x = 7.37 \times 10^{-23}$ kg·m/s, $p_y = 5.82 \times 10^{-23}$ kg · m/s

21. 1.27×10^{-12} m

23. 1×10^{-18} m

25. 1830

27. 4.8×10^{-23} kg · m/s

29. 1.2×10^{-36} m

31. $\lambda_f = 2.0 \times 10^{-34}$ m

33. 4.0×10^1

35. 9.6×10^{-21} kg · m/s

37. $-0.0289°$ to $+0.0289°$

39. 7.9×10^{-33} kg

41. 7.77×10^{-13} J

43. $\lambda_A = 2.75 \times 10^{-7}$ m, $\lambda_B = 3.35 \times 10^{-7}$ m

45. 254 nm

47. (a) 1.25 J

    (b) 1.89×10^{18} photons

49. (a) 5740 K

    (b) No. The answer in part (a) is unreasonably hot.

51. 42.8

53. (a) 3.00×10^8 W

    (b) 3.51×10^9 W/m²

    (c) 1.50×10^9 J

55. (a) 2.00×10^{-11} m

    (b) 3.80 kV

Chapter 30

1. (a) 6.2×10^{-31} m³ (b) 4×10^{-45} m³

    (c) 7×10^{-13} %

3. 2 mi

5. 7.3×10^{-14} m

7. $n = 3$

9. 16

11. 1.98×10^{-19} J

13. 6.56×10^{-7} m and 1.22×10^{-7} m

15. (a) 1.8

    (b) 4.3×10^{-11} m ($Z = 11$) and 2.6×10^{-10} m ($Z = 1.8$)

17. 2

19. 22.8 nm

21. (a) -1.51 eV (b) 2.58×10^{-34} J · s

    (c) 2.11×10^{-34} J · s

23. -0.378 eV

25. 1.732

27. -0.278 eV, -0.378 eV, -0.544 eV

29. (a)

n	ℓ	m_ℓ	m_s
2	0	0	1/2
2	0	0	−1/2

    (b)

n	ℓ	m_ℓ	m_s
2	1	1	1/2
2	1	1	−1/2
2	1	0	1/2
2	1	0	−1/2
2	1	−1	1/2
2	1	−1	−1/2

31. 50

33. (a) Not allowed (b) Allowed

    (c) Not allowed (d) Allowed

    (e) Not allowed

35. Carbon ($Z = 6$)

37. (a) 1.34×10^{-12} m (b) 1.80 %

39. 0.50

41. 1.29×10^{-10} m

43. (a) 0.289 eV (b) 1.00×10^{-5} m

    (c) infrared

45. (a) 40.8 eV (b) 54.4 eV

47. (a) 7458 nm (b) 2279 nm

    (c) Infrared region

49. 4.41×10^{-10} m

51. 1.9×10^{17} photons

53. 2.2×10^{16} photons

55. (a) 2.28×10^{-11} m (b) 1.92×10^{-11} m

57. 4

59. (a) $n_f = 1$, 12.8 eV (b) $n_f = 3$, 0.661 eV

    (c) $n_f = 5$, 0.306 eV

61. (a) $E_1 = 3.182 \times 10^{-19}$ J, $E_2 = 3.127 \times 10^{-19}$ J, $E_3 = 2.897 \times 10^{-19}$ J, $E_4 = 2.079 \times 10^{-18}$ J

    (b) $\lambda_1 = 624.3$ nm, $\lambda_2 = 635.3$ nm, $\lambda_3 = 685.7$ nm, $\lambda_4 = 95.55$ nm

    (c) 624.3 nm, 635.3 nm, and 685.7 nm

63. (a) $\lambda_1(3\rightarrow2) = 656.3$ nm (red); $\lambda_2(4\rightarrow2) = 486.2$ nm (aqua); $\lambda_3(5\rightarrow2) = 434.1$ nm (blue); $\lambda_4(6\rightarrow2) = 410.2$ nm (violet)

    (b) $\theta_1 = 11.3°$; $\theta_2 = 8.38°$; $\theta_3 = 7.48°$; $\theta_4 = 7.07°$

    (c) 3.97 nm, this is the 7→2 transition of the Balmer series

Chapter 31

1. (a) $+1.31 \times 10^{-17}$ C

    (b) 126

    (c) 208

    (d) 7.1×10^{-15} m

    (e) 2.3×10^{17} kg/m³

3. (a) 117 neutrons (platinum, Pt)

    (b) 16 neutrons (sulfur, S)

    (c) 34 neutrons (copper, Cu)

    (d) 6 neutrons (boron, B)

    (e) 145 neutrons (plutonium, Pu)

5. (a) 92

    (b) 122

    (c) 41

7. $^4_2$He

9. 39.25 MeV

11. (a) 0.008 285 u

    (b) 0.009 105 u

    (c) More energy must be supplied to tritium ($^3_1$T) than to helium ($^3_2$He).

13. (a) 1.741 670 u

    (b) 1622 MeV

    (c) 7.87 MeV/nucleon

15. 1.003 27 u

17. (a) $^{18}_9$F $\rightarrow$ $^{18}_8$O $+$ $^0_{+1}$e

    (b) $^{15}_8$O $\rightarrow$ $^{15}_7$N $+$ $^0_{+1}$e

19. 0.313 MeV

21. 4.87 MeV

23. (a) $^{238}_{92}$U

    (b) $^{24}_{12}$Mg

    (c) $^{13}_6$C

25. 1.61×10^7 m/s

27. $KE_{Th} = 0.072$ MeV, $KE_\alpha = 4.2$ MeV

29. 1.82 MeV

31. 5.62×10^{-7} s⁻¹

33. 8.00 days

35. 8.01×10^6 m

37. (a) 3.0×10^4 yr

    (b) 3.3×10^4 yr

39. 3.29×10^9 yr

41. Age of the fossils is 6100 yr. Maximum error is 900 yr.

43. 4.88×10^{10} yr

45. (a) X represents a β^- particle (electron).

(b) X represents a β^+ particle (positron).

(c) X represents a γ ray.

(d) X represents an α particle (helium nucleus).

47. (a) $^{212}_{82}\text{Pb}$

(b) 24.62 MeV

49. 387 yr

51. 2.1×10^{13} Bq

53. (a) oxygen

(b) 112 μCi

55. 2.78×10^{28} MeV

57. 16.24 mg

59. 6.1 C°

61. (a) 4.74×10^{-13} A

(b) 2.73×10^{-13} A

63. (a) $^{237}_{93}\text{Np}$

(b) 1.27×10^{11} Bq

(c) 720.6 yr

Chapter 32

1. 12

3. (a) 14 Gy (b) 2.1 J

5. 2.4×10^4 rem

7. 0.85

9. 9.2×10^8 nuclei

11. $\gamma + {}^{17}_{8}\text{O} \rightarrow {}^{12}_{6}\text{C} + {}^{4}_{2}\text{He} + {}^{1}_{0}\text{n}$

13. (a) $A = 233$, $Z = 90$, X = Th (Thorium)

(b) $A = 233$, $Z = 92$, X = U (Uranium)

15. (a) Sulfur ${}^{31}_{16}\text{S}$ (b) proton ${}^{1}_{1}\text{H}$

(c) potassium ${}^{41}_{19}\text{K}$ (d) neon ${}^{20}_{10}\text{Ne}$

17. ${}^{A}_{Z}\text{X} = {}^{99}_{41}\text{Nb}$

19. 3

21. 6.4×10^8 W

23. 2.7×10^6 kg

25. 1200 kg

27. 2.2 MeV

29. (a) $A = 2$, $Z = 0$, X = neutron, Y = deuteron nucleus

(b) 6.2 MeV

31. $\text{KE}_n = 14.1$ MeV, $\text{KE}_\alpha = 3.5$ MeV

33. 33.9 MeV

35. 140.9 MeV

37. 2.73×10^{-22} kg · m/s

39. Protons, electrons, gamma rays, slow neutrons

41. 184 MeV

43 1.0 gal

45. (a) 8.2×10^{10} J (b) 0.48 g

47. (a) 89 h = 3.7 days (b) 0.60 J

49. 69 m/s

51. (a) 0.0128 J (b) 5.36×10^{-6} C°

53. (a) 67.5 MeV

(b) 1.63×10^{22} Hz; 1.84×10^{-14} m

(c) $+3.60 \times 10^{-20}$ kg · m/s, -3.60×10^{-20} kg · m/s

55. (a) $^{289}_{115}\text{Mc}$ (moscovium)

(b) $^{285}_{113}\text{Nh}$ (nihonium)

57. (a) 6.46×10^{-13} J

(b) 7.73×10^{21} reactions/s

(c) 1.63×10^3 kg

Index

Fundamental Constants

Quantity	Symbol	Value*
Avogadro's number	N_A	$6.022\ 140\ 76 \times 10^{23}\ \text{mol}^{-1}$
Boltzmann's constant	k	$1.380\ 649 \times 10^{-23}\ \text{J/K}$
Electron charge magnitude	e	$1.602\ 176\ 634 \times 10^{-19}\ \text{C}$
Permeability of free space	μ_0	$4\pi \times 10^{-7}\ \text{T} \cdot \text{m/A}$
Permittivity of free space	ε_0	$8.854\ 187\ 8128 \times 10^{-12}\ \text{C}^2/(\text{N} \cdot \text{m}^2)$
Planck's constant	h	$6.626\ 070\ 15 \times 10^{-34}\ \text{J} \cdot \text{s}$
Mass of electron	m_e	$9.109\ 383\ 7015 \times 10^{-31}\ \text{kg}$
Mass of neutron	m_n	$1.674\ 927\ 471 \times 10^{-27}\ \text{kg}$
Mass of proton	m_p	$1.672\ 621\ 923\ 69 \times 10^{-27}\ \text{kg}$
Speed of light in vacuum	c	$2.997\ 924\ 58 \times 10^{8}\ \text{m/s}$
Universal gravitational constant	G	$6.674\ 30 \times 10^{-11}\ \text{N} \cdot \text{m}^2/\text{kg}^2$
Universal gas constant	R	$8.314\ 462\ 618\ \text{J/(mol} \cdot \text{K)}$

*2018 CODATA recommended values.

Useful Physical Data

Acceleration due to earth's gravity	$9.80\ \text{m /s}^2 = 32.2\ \text{ft /s}^2$
Atmospheric pressure at sea level	$1.013 \times 10^5\ \text{Pa} = 14.70\ \text{lb/in.}^2$
Density of air (0 °C, 1 atm pressure)	$1.29\ \text{kg /m}^3$
Speed of sound in air (20 °C)	$343\ \text{m/s}$
Water	
Density (4 °C)	$1.000 \times 10^3\ \text{kg /m}^3$
Latent heat of fusion	$3.35 \times 10^5\ \text{J/ kg}$
Latent heat of vaporization	$2.26 \times 10^6\ \text{J/kg}$
Specific heat capacity	$4186\ \text{J/(kg} \cdot \text{C°)}$
Earth	
Mass	$5.98 \times 10^{24}\ \text{kg}$
Radius (equatorial)	$6.38 \times 10^6\ \text{m}$
Mean distance from sun	$1.50 \times 10^{11}\ \text{m}$
Moon	
Mass	$7.35 \times 10^{22}\ \text{kg}$
Radius (mean)	$1.74 \times 10^6\ \text{m}$
Mean distance from earth	$3.85 \times 10^8\ \text{m}$
Sun	
Mass	$1.99 \times 10^{30}\ \text{kg}$
Radius (mean)	$6.96 \times 10^8\ \text{m}$

Frequently Used Mathematical Symbols

Symbol	Meaning
=	is equal to
≠	is not equal to
∝	is proportional to
>	is greater than
<	is less than
≈	is approximately equal to
$\lvert x \rvert$	absolute value of x (always treated as a positive quantity)
Δ	the difference between two variables (e.g., ΔT is the final temperature minus the initial temperature)
Σ	the sum of two or more variables (e.g., $\sum_{i=1}^{3} x_i = x_1 + x_2 + x_3$)

Conversion Factors

Length
1 in. = 2.54 cm

1 ft = 0.3048 m

1 mi = 5280 ft = 1.609 km

1 m = 3.281 ft

1 km = 0.6214 mi

1 angstrom (Å) = 10^{-10} m

Mass
1 slug = 14.59 kg

1 kg = 1000 grams = 6.852×10^{-2} slug

1 atomic mass unit (u) = 1.6605×10^{-27} kg

(1 kg has a weight of 2.205 lb where the acceleration due to gravity is 32.174 ft/s^2)

Time
1 d = 24 h = 1.44×10^3 min = 8.64×10^4 s

1 yr = 365.24 days = 3.156×10^7 s

Speed
1 mi/h = 1.609 km/h = 1.467 ft/s = 0.4470 m/s

1 km/h = 0.6214 mi/h = 0.2778 m/s = 0.9113 ft/s

Force
1 lb = 4.448 N

1 N = 10^5 dynes = 0.2248 lb

Work and Energy
1 J = 0.7376 ft · lb = 10^7 ergs

1 kcal = 4186 J

1 Btu = 1055 J

1 kWh = 3.600×10^6 J

1 eV = 1.602×10^{-19} J

Power
1 hp = 550 ft · lb/s = 745.7 W

1 W = 0.7376 ft · lb/s

Pressure
1 Pa = 1 N/m^2 = 1.450×10^{-4} lb/in.^{2}

1 lb/in.^{2} = 6.895×10^3 Pa

1 atm = 1.013×10^5 Pa = 1.013 bar = 14.70 lb/in.^{2} = 760 torr

Volume
1 liter = 10^{-3} m^3 = 1000 cm^3 = 0.03531 ft^3

1 ft^3 = 0.02832 m^3 = 7.481 U.S. gallons

1 U.S. gallon = 3.785×10^{-3} m^3 = 0.1337 ft^3

Angle
1 radian = 57.30°

1° = 0.01745 radian

Standard Prefixes Used to Denote Multiples of Ten

Prefix	Symbol	Factor
Tera	T	10^{12}
Giga	G	10^{9}
Mega	M	10^{6}
Kilo	k	10^{3}
Hecto	h	10^{2}
Deka	da	10^{1}
Deci	d	10^{-1}
Centi	c	10^{-2}
Milli	m	10^{-3}
Micro	μ	10^{-6}
Nano	n	10^{-9}
Pico	p	10^{-12}
Femto	f	10^{-15}

Basic Mathematical Formulas

Area of a circle = πr^2

Circumference of a circle = $2\pi r$

Surface area of a sphere = $4\pi r^2$

Volume of a sphere = $\frac{4}{3}\pi r^3$

Pythagorean theorem: $h^2 = h_o^2 + h_a^2$

Sine of an angle: $\sin \theta = h_o/h$

Cosine of an angle: $\cos \theta = h_a/h$

Tangent of an angle: $\tan \theta = h_o/h_a$

Law of cosines: $c^2 = a^2 + b^2 - 2ab \cos \gamma$

Law of sines: $a/\sin \alpha = b/\sin \beta = c/\sin \gamma$

Quadratic formula:

If $ax^2 + bx + c = 0$, then, $x = (-b \pm \sqrt{b^2 - 4ac})/(2a)$

SI Units

Quantity	Name of Unit	Symbol	Expression in Terms of Other SI Units
Length	meter	m	Base unit
Mass	kilogram	kg	Base unit
Time	second	s	Base unit
Electric current	ampere	A	Base unit
Temperature	kelvin	K	Base unit
Amount of substance	mole	mol	Base unit
Velocity	—	—	m/s
Acceleration	—	—	m/s^2
Force	newton	N	$kg \cdot m/s^2$
Work, energy	joule	J	$N \cdot m$
Power	watt	W	J/s
Impulse, momentum	—	—	$kg \cdot m/s$
Plane angle	radian	rad	m/m
Angular velocity	—	—	rad/s
Angular acceleration	—	—	rad/s^2
Torque	—	—	$N \cdot m$
Frequency	hertz	Hz	s^{-1}
Density	—	—	kg/m^3

Quantity	Name of Unit	Symbol	Expression in Terms of Other SI Units
Pressure, stress	pascal	Pa	N/m^2
Viscosity	—	—	$Pa \cdot s$
Electric charge	coulomb	C	$A \cdot s$
Electric field	—	—	N/C
Electric potential	volt	V	J/C
Resistance	ohm	Ω	V/A
Capacitance	farad	F	C/V
Inductance	henry	H	$V \cdot s/A$
Magnetic field	tesla	T	$N \cdot s/(C \cdot m)$
Magnetic flux	weber	Wb	$T \cdot m^2$
Specific heat capacity	—	—	$J/(kg \cdot K)$ or $J/(kg \cdot C°)$
Thermal conductivity	—	—	$J/(s \cdot m \cdot K)$ or $J/(s \cdot m \cdot C°)$
Entropy	—	—	J/K
Radioactive activity	becquerel	Bq	s^{-1}
Absorbed dose	gray	Gy	J/kg
Exposure	—	—	C/kg

The Greek Alphabet

Alpha	A	α	Iota	I	ι	Rho	P	ρ
Beta	B	β	Kappa	K	κ	Sigma	Σ	σ
Gamma	Γ	γ	Lambda	Λ	λ	Tau	T	τ
Delta	Δ	δ	Mu	M	μ	Upsilon	Υ	υ
Epsilon	E	ε	Nu	N	ν	Phi	Φ	ϕ
Zeta	Z	ζ	Xi	Ξ	ξ	Chi	X	χ
Eta	H	η	Omicron	O	o	Psi	ψ	ψ
Theta	Θ	θ	Pi	Π	π	Omega	Ω	ω

Periodic Table of the Elements

Legend: Symbol — **Cl** 17 (Atomic number); Atomic mass* — 35.453; Electron configuration — $3p^5$

Group I	Group II	Transition elements										Group III	Group IV	Group V	Group VI	Group VII	Group 0
H 1 / 1.00794 / $1s^1$																	**He** 2 / 4.00260 / $1s^2$
Li 3 / 6.941 / $2s^1$	**Be** 4 / 9.01218 / $2s^2$											**B** 5 / 10.81 / $2p^1$	**C** 6 / 12.011 / $2p^2$	**N** 7 / 14.0067 / $2p^3$	**O** 8 / 15.9994 / $2p^4$	**F** 9 / 18.9984 / $2p^5$	**Ne** 10 / 20.180 / $2p^6$
Na 11 / 22.9898 / $3s^1$	**Mg** 12 / 24.305 / $3s^2$											**Al** 13 / 26.9815 / $3p^1$	**Si** 14 / 28.0855 / $3p^2$	**P** 15 / 30.9738 / $3p^3$	**S** 16 / 32.07 / $3p^4$	**Cl** 17 / 35.453 / $3p^5$	**Ar** 18 / 39.948 / $3p^6$
K 19 / 39.0983 / $4s^1$	**Ca** 20 / 40.08 / $4s^2$	**Sc** 21 / 44.9559 / $3d^14s^2$	**Ti** 22 / 47.87 / $3d^24s^2$	**V** 23 / 50.9415 / $3d^34s^2$	**Cr** 24 / 51.996 / $3d^54s^1$	**Mn** 25 / 54.9380 / $3d^54s^2$	**Fe** 26 / 55.845 / $3d^64s^2$	**Co** 27 / 58.9332 / $3d^74s^2$	**Ni** 28 / 58.69 / $3d^84s^2$	**Cu** 29 / 63.546 / $3d^{10}4s^1$	**Zn** 30 / 65.41 / $3d^{10}4s^2$	**Ga** 31 / 69.72 / $4p^1$	**Ge** 32 / 72.64 / $4p^2$	**As** 33 / 74.9216 / $4p^3$	**Se** 34 / 78.96 / $4p^4$	**Br** 35 / 79.904 / $4p^5$	**Kr** 36 / 83.80 / $4p^6$
Rb 37 / 85.4678 / $5s^1$	**Sr** 38 / 87.62 / $5s^2$	**Y** 39 / 88.9059 / $4d^15s^2$	**Zr** 40 / 91.224 / $4d^25s^2$	**Nb** 41 / 92.9064 / $4d^45s^1$	**Mo** 42 / 95.94 / $4d^55s^1$	**Tc** 43 / (98) / $4d^55s^2$	**Ru** 44 / 101.07 / $4d^75s^1$	**Rh** 45 / 102.906 / $4d^85s^1$	**Pd** 46 / 106.42 / $4d^{10}5s^0$	**Ag** 47 / 107.868 / $4d^{10}5s^1$	**Cd** 48 / 112.41 / $4d^{10}5s^2$	**In** 49 / 114.82 / $5p^1$	**Sn** 50 / 118.71 / $5p^2$	**Sb** 51 / 121.76 / $5p^3$	**Te** 52 / 127.60 / $5p^4$	**I** 53 / 126.904 / $5p^5$	**Xe** 54 / 131.29 / $5p^6$
Cs 55 / 132.905 / $6s^1$	**Ba** 56 / 137.33 / $6s^2$	**La** 57 / 138.906 / $5d^16s^2$	**Hf** 72 / 178.49 / $5d^26s^2$	**Ta** 73 / 180.948 / $5d^36s^2$	**W** 74 / 183.84 / $5d^46s^2$	**Re** 75 / 186.207 / $5d^56s^2$	**Os** 76 / 190.2 / $5d^66s^2$	**Ir** 77 / 192.22 / $5d^76s^2$	**Pt** 78 / 195.08 / $5d^96s^1$	**Au** 79 / 196.967 / $5d^{10}6s^1$	**Hg** 80 / 200.59 / $5d^{10}6s^2$	**Tl** 81 / 204.383 / $6p^1$	**Pb** 82 / 207.2 / $6p^2$	**Bi** 83 / 208.980 / $6p^3$	**Po** 84 / (209) / $6p^4$	**At** 85 / (210) / $6p^5$	**Rn** 86 / (222) / $6p^6$
Fr 87 / (223) / $7s^1$	**Ra** 88 / (226) / $7s^2$	**Ac** 89 / (227) / $6d^17s^2$	**Rf** 104 / (261) / $6d^27s^2$	**Db** 105 / (262) / $6d^37s^2$	**Sg** 106 / (266) / $6d^47s^2$	**Bh** 107 / (264) / $6d^57s^2$	**Hs** 108 / (277) / $6d^67s^2$	**Mt** 109 / (268) / $6d^77s^2$	**Ds** 110 / (281)	**Rg** 111 / (282)	**Cn** 112 / (285)	**Nh** 113 / (286)	**Fl** 114 / (289)	**Mc** 115 / (289)	**Lv** 116 / (293)	**Ts** 117 / (294)	**Og** 118 / (294)

Lanthanide series (57–71):

Ce 58	Pr 59	Nd 60	Pm 61	Sm 62	Eu 63	Gd 64	Tb 65	Dy 66	Ho 67	Er 68	Tm 69	Yb 70	Lu 71
140.12 $4f^26s^2$	140.908 $4f^36s^2$	144.24 $4f^46s^2$	(145) $4f^56s^2$	150.36 $4f^66s^2$	151.96 $4f^76s^2$	157.25 $5d^14f^76s^2$	158.925 $4f^96s^2$	162.50 $4f^{10}6s^2$	164.930 $4f^{11}6s^2$	167.26 $4f^{12}6s^2$	168.934 $4f^{13}6s^2$	173.04 $4f^{14}6s^2$	174.967 $5d^14f^{14}6s^2$

Actinide series (89–103):

Th 90	Pa 91	U 92	Np 93	Pu 94	Am 95	Cm 96	Bk 97	Cf 98	Es 99	Fm 100	Md 101	No 102	Lr 103
232.038 $6d^27s^2$	231.036 $5f^26d^17s^2$	238.029 $5f^36d^17s^2$	(237) $5f^46d^17s^2$	(244) $5f^66d^07s^2$	(243) $5f^76d^07s^2$	(247) $5f^76d^17s^2$	(247) $5f^96d^07s^2$	(251) $5f^{10}6d^07s^2$	(252) $5f^{11}6d^07s^2$	(257) $5f^{12}6d^07s^2$	(258) $5f^{13}6d^07s^2$	(259) $6d^17s^2$	(262) $6d^17s^2$

* Atomic mass values are averaged over isotopes according to the percentages that occur on the earth's surface. For unstable elements, the mass number of the most stable known isotope is given in parentheses. *Source: IUPAC Commission on Atomic Weights and Isotopic Abundances, 2001.*